Advanced
Engineering Mathematics

Fourth Edition

R.K. Jain □ S.R.K. Iyengar

Advanced Engineering Mathematics

Fourth Edition

Alpha Science International Ltd.
Oxford, U.K.

R.K. Jain *(Retd.)*
Professor of Mathematics
Indian Institute of Technology, Delhi
New Delhi, India

S.R.K. Iyengar *(Retd.)*
Professor & Head
Department of Mathematics
Indian Institute of Technology, Delhi
New Delhi, India

Copyright © 2002, 2003, 2007, 2014
First Edition 2002
Second Edition 2003
Third Edition 2007
Fourth Edition 2014

ALPHA SCIENCE INTERNATIONAL LTD.
7200 The Quorum, Oxford Business Park North
Garsington Road, Oxford OX4 2JZ, U.K.

www.alphasci.com

ISBN 978-1-84265-846-8

Printed in India

To Our Parents

Bhagat Ram Jain and Sampati Devi Jain
&
S.T.V. Raghavacharya and Rajya Lakshmi

Whose memories had always been an inspiration

Preface to the Fourth Edition

We sincerely thank the faculty members and the students of various Institutes and Engineering Colleges for their suggestions to improve the book.

Based on these suggestions, we have included the following new material.

(i) Condition number of a matrix and Singular Value Decomposition (Chapter 3).

(ii) Application of Z-transforms to find the sum of series (Chapter 17).

(iii) Cubic splines, B-splines, Romberg integration, Gauss quadrature rules and Two-point boundary value problems (Chapter 18)

We hope that the book in the present form includes most of the topics covered in the core courses for Engineering students.

We look forward to get more suggestions from the faculty members and the students to improve the book further.

R.K. Jain
S.R.K. lyengar

Preface to the First Edition

This book is based on the experience and the lecture notes of the authors while teaching mathematics courses to engineering students at the Indian Institute of Technology, Delhi for more than three decades. A number of available textbooks have been a source of inspiration for introduction of concepts and formulation of problems. We are thankful to the authors of these books for their indirect help.

This comprehensive textbook covers syllabus for two courses in Mathematics for engineering students in various Institutes, Universities and Engineering Colleges. The emphasis is on the presentation of the fundamentals and theoretical concepts in an intelligible and easy to understand manner.

Each chapter in the book has been carefully planned to make it an effective tool to arouse interest in the study and application of mathematics to solve engineering and scientific problems. Simple and illustrative examples are used to explain each theoretical concept. Graded sets of examples and exercises are given in each chapter, which will help the students to understand every important concept. The book contains 682 solved examples and 2984 problems in the exercises. Answers to every problem and hints for difficult problems are given at the end of each chapter which will motivate the students for self-learning. While some problems emphasize the theoretical concepts, others provide enough practice and generate confidence to use these concepts in problem solving. This textbook offers a logical and lucid presentation of both the theory and problem solving techniques so that the student is not lost in unnecessary details.

We hope that this textbook will meet the requirement and the expectations of the engineering students.

We will gratefully receive and acknowledge every comment, suggestion for inclusion/exclusion of topics and errors in the book, both from the faculty and the students.

We are grateful to our former teachers, colleagues and well wishers for their encouragement and valuable suggestions. We are also thankful to our students for their feed back. We are grateful to the authorities of IIT Delhi for providing us their support.

We extend our thanks to the editorial and the production staff of M/s Narosa Publishing House, in particular Mr. Mohinder Singh Sejwal for their care and enthusiasm in the preparation of this book.

Last, but not the least, we owe a lot to our family members, in particular, our wives Vinod Jain and Seetha Lakshmi whose encouragement and support had always been inspiring and rejuvenating. We appreciate their patience during our long hours of work day and night.

New Delhi
October 2001

R.K. Jain
S.R.K. Iyengar

Contents

10.   Functions of a Complex Variable: Analytic Functions   10.1

11.   Integration of Complex Functions   11.1

Chapter 1

Functions of a Real Variable

1.1 Introduction

In a first course in Mathematics, you have studied limits, continuity, differentiability and integration of functions of one variable $y = f(x)$. We will now discuss the application of derivatives and integration to solve various engineering problems.

1.2 Application of Derivatives

We now discuss some applications of derivatives like finding approximate value of a function, mean value theorems, increasing and decreasing functions, maximum and minimum values of a function.

1.2.1 Differentials and Approximations

Let $y = f(x)$ be a real valued differentiable function and x_0 be a point in its domain. Let $x_0 + \Delta x$ be a point in the neighborhood of x_0. Then, Δx may be considered as an increment in x. The corresponding increment in $f(x)$ is given by

$$\Delta f_0 = \Delta f(x_0) = f(x_0 + \Delta x) - f(x_0).$$

From the definition of derivative, we have

$$f'(x_0) = \lim_{\Delta x \to 0} \frac{f(x_0 + \Delta x) - f(x_0)}{\Delta x} = \lim_{\Delta x \to 0} \frac{\Delta f_0}{\Delta x}. \tag{1.1}$$

Since $f'(x_0)$ exists, we can write from Eq. (1.1) that

$$\frac{\Delta f_0}{\Delta x} = f'(x_0) + \alpha \quad \text{or} \quad \Delta f_0 = f'(x_0)\,\Delta x + \alpha\,\Delta x \tag{1.2}$$

where α is an infinitesimal quantity dependent on Δx and tends to zero as $\Delta x \to 0$. Thus, the increment Δf_0 consists of the following two parts.

(i) Principal part $f'(x_0)\,\Delta x$, which is called the *differential* of f.

(ii) Residual part $\alpha\,\Delta x$ which tends to zero as $\Delta x \to 0$.

In the limit, the differential is also written as

$$df(x_0) = dy_0 = f'(x_0)\, dx. \tag{1.3}$$

Hence, an approximation to $f(x_0 + \Delta x)$ can be written as

$$f(x_0 + \Delta x) = f(x_0) + f'(x_0)\, dx. \tag{1.4}$$

Differentials have application in calculating errors in functions due to small errors in the independent variable. We define $|dy|$ as the *absolute error*; dy/y as the *relative error* and $(dy/y) \times 100$ as the *percentage error in* computations.

Example 1.1 Find an approximate value of

$$y = 3(4.02)^2 - 2(4.02)^{3/2} + 8/\sqrt{4.02}.$$

Solution Let a function be defined as

$$y = f(x) = 3x^2 - 2x^{3/2} + 8/\sqrt{x}.$$

Let $x_0 = 4$ and $\Delta x = 0.02$. Then, we need an approximation to $f(x_0 + \Delta x) = f(4.02)$. The approximate value is given by (see Eq. 1.4)).

$$f(4.02) \approx f(4) + (0.02)\, f'(4)$$

We have

$$f(4) = 48 - 2(8) + 8/2 = 36,$$

$$f'(x) = 6x - 3x^{1/2} - 4x^{-3/2} \text{ and } f'(4) = 24 - 6 - 4/8 = 35/2.$$

Therefore, the required approximation is

$$f(4.02) \approx 36 + 0.02\,(35/2) = 36.35.$$

Example 1.2 If there is a possible error of 0.02 cm in the measurement of the diameter of a sphere, then find the possible percentage error in its volume, when the radius is 10 cm.

Solution Let the radius of the sphere be r cm. Volume of the sphere $= V = 4\pi r^3/3$ and $dr = \pm 0.01$ when $r = 10$ cm.

Differentiating V, we obtain $dV = 4\pi r^2 dr$.

When $r = 10$, we get from Eq. (1.3), $dV = 4\pi (10)^2 (\pm 0.01) = \pm 4\pi$.

Hence, the percentage error in volume is

$$\left(\frac{dV}{V}\right) \times 100 = 100 \left[\frac{\pm 12\pi}{4\pi(10)^3}\right] = \pm 0.3 \text{ cubic cm.}$$

1.2.2 Mean Value Theorems

We now prove the three basic mean value theorems of the functions of one variable.

Theorem 1.1 (Rolle's theorem) Let a real valued function $f(x)$ be continuous on a closed interval $[a, b]$ and differentiable in the open interval (a, b). If $f(a) = f(b)$, then there exists at least one value c, $a < c < b$ such that $f'(c) = 0$.

Proof Since the function $f(x)$ is continuous on the closed interval $[a, b]$, it is bounded and attains its maximum value M and minimum value m at some points in $[a, b]$. Let $f(x)$ attain respectively its minimum and maximum values at the points c and $d \in [a, b]$, that is

$$f(c) = m \quad \text{and} \quad f(d) = M.$$

If $m = M$, then the function $f(x)$ is constant over $[a, b]$ and therefore, its derivative $f'(x)$ is zero for all x in $[a, b]$.

If $m \neq M$, then both of these cannot be equal to the same quantity $f(a)$ or $f(b)$. We note that $f(a) = f(b)$. Thus, atleast one of these, say m, is different from $f(a)$ and $f(b)$. Hence,

$$f(c) = m \neq f(a), \text{ implies } c \neq a,$$

$$f(c) = m \neq f(b), \text{ implies } c \neq b.$$

Therefore, $c \in (a, b)$. We shall now show that at this point c, $f'(c) = 0$.

If $f'(c) < 0$, then for every x in the interval $(c, c + \varepsilon_1)$, $\varepsilon_1 > 0$,

$$f(x) < f(c) = m$$

which contradicts the assumption that m is the minimum value of $f(x)$.

If $f'(c) > 0$, then for every x in the interval $(c - \varepsilon_2, c)$, $\varepsilon_2 > 0$,

$$f(x) < f(c) = m$$

which is again a contradiction. Hence, $f'(c) = 0$.

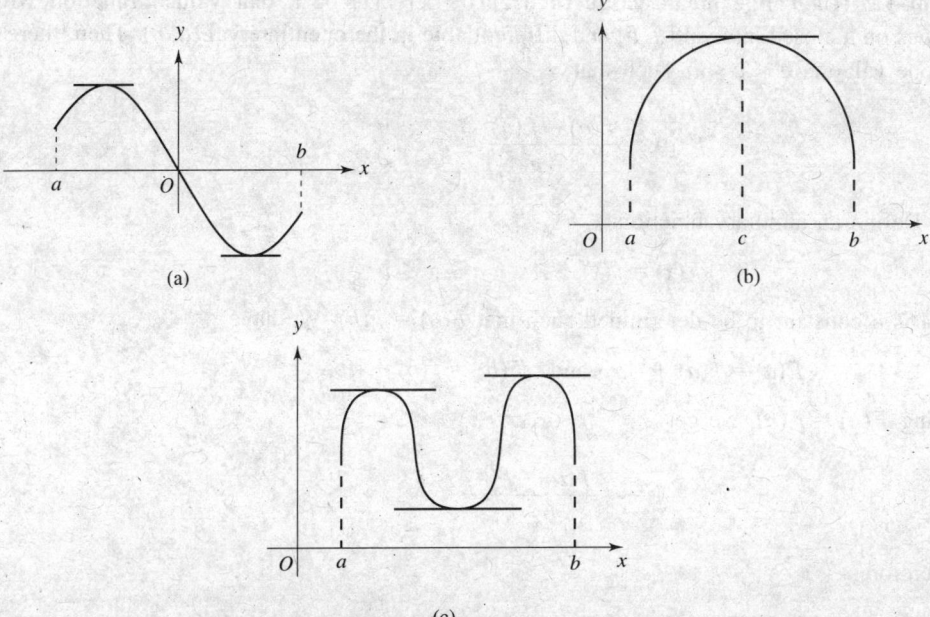

(a)

(b)

(c)

Fig. 1.1. Rolle's theorem.

Remark 1

(a) Differentiability of $f(x)$ in an open interval (a, b) is a necessary condition for the applicability of the Rolle's theorem.

For example, consider the function $f(x) = |x|, -1 \leq x \leq 1$. Now, $f(x)$ is continuous on $[-1, 1]$ and is differentiable at all points in the interval $(-1, 1)$ except at the point $x = 0$. Now,

$$f'(x) = \begin{cases} 1, & x > 0 \\ -1, & x < 0 \end{cases}$$

does not vanish at any point in the interval $(-1, 1)$. This shows that the Rolle's theorem cannot be applied as the function $f(x)$ is not differentiable in $(-1, 1)$.

(b) Rolle's theorem gives sufficient conditions for the existence of a value c such that $f'(c) = 0$. For example, the function

$$f(x) = \begin{cases} 0, & 1 \leq x \leq 2 \\ 2, & 2 < x \leq 3 \end{cases}$$

is not continuous on $[1, 3]$, but $f'(c) = 0$ for all c in $[1, 3]$.

(c) Geometrically, the theorem states that if a function satisfies the conditions of Rolle's theorem and has the same value at the end points of an interval $[a, b]$, then there exists at least one point $(c, f(c))$, $a < c < b$ where the tangent to the curve $y = f(x)$, $a \leq x \leq b$ is parallel to the x-axis.

Theorem 1.2 (Lagrange mean value theorem)   Let $f(x)$ be a real valued function which is continuous on a closed interval $[a, b]$ and differentiable in the open interval (a, b). Then, there exists atleast one value c, $a < c < b$, such that

$$f'(c) = \frac{f(b) - f(a)}{b - a}. \tag{1.5}$$

Proof   Define an auxiliary function

$$F(x) = f(x) + Ax, \quad a \leq x \leq b$$

where A is a constant to be determined such that $F(a) = F(b)$. We have

$$F(a) = f(a) + Aa \quad \text{and} \quad F(b) = f(b) + Ab.$$

Using $F(a) = F(b)$, we get

$$A = -\frac{f(b) - f(a)}{b - a}.$$

Therefore,

$$F(x) = f(x) - \left[\frac{f(b) - f(a)}{b - a} \right] x.$$

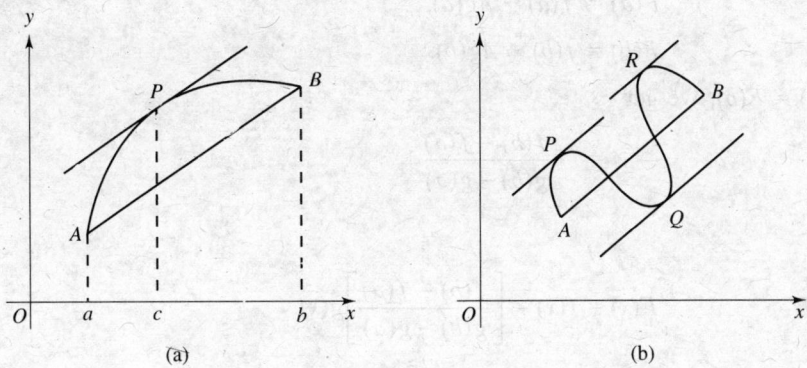

Fig. 1.2   Lagrange mean value theorem.

Now, $F(x)$ is continuous on the closed interval $[a, b]$ and differentiable in the open interval (a, b) and $F(a) = F(b)$. Since, the function $F(x)$ satisfies all conditions of the Rolle's theorem, there exists a point $(c, F(c))$, $a < c < b$ such that

$$F'(c) = 0, \quad \text{or} \quad f'(c) = \frac{f(b) - f(a)}{b - a}.$$

Remark 2

(a) If $f(a) = f(b)$, then Lagrange mean value theorem reduces to the Rolle's theorem.

(b) Geometrically, Lagrange mean value theorem states that there exists a point $(c, f(c))$, $a < c < b$ on the curve $C: y = f(x)$, $a \le x \le b$, such that the tangent to the curve C at this point is parallel to the chord joining the points $(a, f(a))$ and $(b, f(b))$ on the curve.

(c) Using Eq. (1.5), we can write

$$\min_{a \le x \le b} f'(x) \le \frac{f(b) - f(a)}{b - a} \le \max_{a \le x \le b} f'(x). \tag{1.6}$$

Theorem 1.3 (Cauchy mean value theorem)   Let $f(x)$ and $g(x)$ be two real valued functions defined on a closed interval $[a, b]$ such that (i) they are continuous on $[a, b]$, (ii) they are differentiable in (a, b) and (iii) $g'(x) \ne 0$ for every x in (a, b). Then, there exists at least one value c, $a < c < b$ such that

$$\frac{f(b) - f(a)}{g(b) - g(a)} = \frac{f'(c)}{g'(c)}, \quad a < c < b. \tag{1.7}$$

Proof   Define an auxiliary function

$$F(x) = f(x) + Ag(x), \quad a \le x \le b$$

where A is a constant to be determined such that $F(a) = F(b)$. We have

$$F(a) = f(a) + Ag(a),$$

$$F(b) = f(b) + Ag(b).$$

Using $F(a) = F(b)$, we get

$$A = -\frac{f(b) - f(a)}{g(b) - g(a)}$$

Therefore,

$$F(x) = f(x) - \left[\frac{f(b) - f(a)}{g(b) - g(a)}\right] g(x).$$

Now, $F(x)$ is continuous on $[a, b]$, differentiable in (a, b) and $F(a) = F(b)$. Since, the function $F(x)$ satisfies all conditions of the Rolle's theorem, there exists a point $(c, F(c))$, $a < c < b$, such that

$$F'(c) = 0,$$

or

$$f'(c) - \left[\frac{f(b) - f(a)}{g(b) - g(a)}\right] g'(c) = 0,$$

or

$$\frac{f(b) - f(a)}{g(b) - g(a)} = \frac{f'(c)}{g'(c)}, \quad a < c < b.$$

Remark 3

(a) For $g(x) = x$, Cauchy mean value theorem reduces to Lagrange mean value theorem.

(b) Let a curve C be represented parametrically as $x = f(t)$, $y = g(t)$, $a \le t \le b$. Then, Cauchy mean value theorem states that there exists a point $(f(c), g(c))$, $c \in (a, b)$ on the curve such that the slope $g'(c)/f'(c)$ of the tangent to the curve at this point is equal to the slope of the chord joining the end points of the curve. Hence, Cauchy mean value theorem has the same geometrical interpretation as the Lagrange mean value theorem.

(c) Cauchy mean value theorem cannot be proved by applying the Lagrange mean value theorem separately to the numerator and denominator on the left side of Eq. (1.7). If we apply the Lagrange mean value theorem to the numerator and the denominator separately, we obtain

$$\frac{f(b) - f(a)}{g(b) - g(a)} = \frac{f'(c_1)}{g'(c_2)}, \quad a < c_1 < b, \quad a < c_2 < b, \quad c_1 \ne c_2.$$

Example 1.3 A twice differentiable function f is such that $f(a) = f(b) = 0$ and $f(c) > 0$ for $a < c < b$. Prove that there is atleast one value ξ, $a < \xi < b$ for which $f''(\xi) < 0$.

Solution Consider the function $f(x)$ defined on $[a, b]$. Since $f''(x)$ exists, both f and f' exist and are continuous on $[a, b]$. Let $a < c < b$. Applying the Lagrange mean value theorem to $f(x)$ on $[a, c]$ and $[c, b]$ separately, we get

$$\frac{f(c)-f(a)}{c-a}=f'(\xi_1), \quad a<\xi_1<c, \quad \text{and} \quad \frac{f(b)-f(c)}{b-c}=f'(\xi_2), \quad c<\xi_2<b.$$

Using $f(a) = f(b) = 0$, we obtain from the above equations

$$f'(\xi_1) = \frac{f(c)}{c-a} \quad \text{and} \quad f'(\xi_2) = -\frac{f(c)}{b-c}.$$

Now, $f'(x)$ is continuous and differentiable on $[\xi_1, \xi_2]$. Using the Lagrange mean value theorem again, we obtain

$$\frac{f'(\xi_2)-f'(\xi_1)}{\xi_2-\xi_1}=f''(\xi), \quad \xi_1<\xi<\xi_2.$$

Substituting the values of $f'(\xi_1)$ and $f'(\xi_2)$, we get

$$f''(\xi) = -\frac{f(c)}{\xi_2-\xi_1}\left[\frac{1}{b-c}+\frac{1}{c-a}\right] = -\frac{(b-a)\,f(c)}{(b-c)\,(c-a)\,(\xi_2-\xi_1)} < 0.$$

Example 1.4   Using the Lagrange mean value theorem, show that

$$|\cos b - \cos a| \le |b-a|.$$

Solution   Let $f(x) = \cos x$, $a \le x \le b$. Using the Lagrange mean value theorem to $f(x)$, we obtain

$$\frac{\cos b - \cos a}{b-a} = f'(c) = -\sin c, \quad \text{or} \quad \left|\frac{\cos b - \cos a}{b-a}\right| = |-\sin c| \le 1.$$

Hence, the result.

Example 1.5   Let $f'(x) = 1/(3-x^2)$ and $f(0) = 1$. Find an interval in which $f(1)$ lies.

Solution   Using Eq. (1.6), we obtain for $a = 0$ and $b = 1$

$$\min_{0\le x\le 1} f'(x) \le \frac{f(1)-f(0)}{1-0} \le \max_{0\le x\le 1} f'(x)$$

or

$$\min_{0\le x\le 1}\left[\frac{1}{3-x^2}\right] \le f(1) - 1 \le \max_{0\le x\le 1}\left[\frac{1}{3-x^2}\right]$$

or

$$\frac{1}{3} \le f(1) - 1 \le \frac{1}{2}, \quad \text{or} \quad \frac{4}{3} \le f(1) \le \frac{3}{2}.$$

Example 1.6   Let C be a curve defined parametrically as $x = a\cos^3\theta$, $y = a\sin^3\theta$, $0 \le \theta \le \pi/2$. Determine a point P on C, where the tangent to C is parallel to the chord joining the points $(a, 0)$ and $(0, a)$.

Solution   We have $x = f(\theta) = a \cos^3 \theta$ and $y = g(\theta) = a \sin^3 \theta$. Using the Cauchy mean value theorem, we have for some θ, $0 \leq \theta \leq \pi/2$, slope of the tangent to C = slope of the chord joining the points $(a, 0)$ and $(0, a)$

or
$$\frac{g'(\theta)}{f'(\theta)} = \frac{3a \sin^2 \theta \cos \theta}{-3a \sin^2 \theta \cos \theta} = \frac{g(\pi/2) - g(0)}{f(\pi/2) - g(0)} = \frac{a - 0}{0 - a}$$

or
$$- \tan \theta = - 1, \quad \text{or} \quad \theta = \pi/4.$$

Therefore, the required point is $(a/2\sqrt{2}, a/2\sqrt{2})$.

1.2.3   Indeterminate Forms

Consider the ratio $f(x)/g(x)$ of two functions $f(x)$ and $g(x)$. If at any point $x = a$, $f(a) = g(a) = 0$, then the ratio $f(x)/g(x)$ takes the form $0/0$ and it is called an *indeterminate form*. The problem is to determine $\lim\limits_{x \to a} [f(x)/g(x)]$, if it exists. Since $f(a) = g(a) = 0$, we can write

$$\lim_{x \to a} \frac{f(x)}{g(x)} = \lim_{x \to a} \frac{f(x) - f(a)}{g(x) - g(a)} = \lim_{x \to a} \frac{[f(x) - f(a)]/(x - a)}{[g(x) - g(a)]/(x - a)} = \lim_{x \to a} \frac{f'(x)}{g'(x)}$$

provided the limit on the right hand side exists. This result is known as *L' Hospital's rule*.

L' Hospital's rule   Suppose that the real valued functions f and g are differentiable in some open interval containing the point $x = a$ and $f(a) = 0 = g(a)$. Then,

$$\lim_{x \to a} \frac{f(x)}{g(x)} = \lim_{x \to a} \frac{f'(x)}{g'(x)} = \frac{f'(a)}{g'(a)}, \quad g'(a) \neq 0. \tag{1.8}$$

Suppose now that $f'(a) = 0 = g'(a)$. Then, we repeat the application of L' Hospital's rule on $f'(x)/g'(x)$ and obtain

$$\lim_{x \to a} \frac{f(x)}{g(x)} = \lim_{x \to a} \frac{f'(a)}{g'(a)} = \lim_{x \to a} \frac{f''(x)}{g''(x)} = \frac{f''(a)}{g''(a)}$$

provided the limits exist. This application of the rule can be continued as long as the indeterminate form is obtained.

When both $f(a) = \pm \infty$ and $g(a) = \pm \infty$, we get another indeterminate form. In this case also, L' Hospital's rule can be applied. We write

$$\lim_{x \to a} \frac{f(a)}{g(a)} = \lim_{x \to a} \frac{[1/g(x)]}{[1/f(x)]}$$

which is of $0/0$ form.

Remark 4

(a) L' Hospital rule can be used only when the ratio is of indeterminate form, that is, either it is of form $0/0$ or ∞/∞.

(b) The other indeterminate forms are $0 \cdot \infty$, 0^0, ∞^0, 1^∞ and $\infty - \infty$. In each of these cases, we can reduce the ratio function to the form $0/0$ or ∞/∞ and use this rule. For the indeterminate forms 0^0, ∞^0 and 1^∞, we take logarithm of the given function and then take the limits.

(c) When the function is of the form 0^∞, $\infty \cdot \infty$, $\infty + \infty$, ∞^∞ or $\infty^{-\infty}$, it is not of indeterminate form and we cannot apply L' Hospital's rule. We note that $0^\infty = 0$, $\infty \cdot \infty = \infty$, $\infty + \infty = \infty$, $\infty^\infty = \infty$ and $\infty^{-\infty} = 0$.

(d) L' Hospital's rule can also be applied to find the limits as $x \to \pm \infty$.

Example 1.7 Evaluate the following limits

(i) $\lim\limits_{x \to 0} \left[\dfrac{\ln(1+x)}{\sin x} \right]$, (ii) $\lim\limits_{x \to 0} [x^n (\ln\ x)]$, (iii) $\lim\limits_{x \to \infty} \left[\dfrac{e^x}{x} \right]$.

Solution Using L' Hospital's rule, we get

(i) $\lim\limits_{x \to 0} \left[\dfrac{\ln(1+x)}{\sin x} \right] = \lim\limits_{x \to 0} \dfrac{1/(1+x)}{\cos x} = 1.$

(ii) $\lim\limits_{x \to 0} [x^n (\ln\ x)] = \lim\limits_{x \to 0} \dfrac{[\ln x]}{[1/x^n]} = \lim\limits_{x \to 0} \dfrac{[1/x]}{[-n/x^{n+1}]} = \lim\limits_{x \to 0} \dfrac{-x^n}{n} = 0.$

(iii) $\lim\limits_{x \to \infty} \left[\dfrac{e^x}{x} \right] = \lim\limits_{x \to \infty} \left[\dfrac{e^x}{1} \right] = \infty.$

Example 1.8 Evaluate $\lim\limits_{x \to 0} x^x$.

Solution The given limit is of the form 0^0 which is an indeterminate form. Let $y = x^x$. Then, $\ln y = x \ln x$. Now,

$$\lim\limits_{x \to 0} [\ln y] = \lim\limits_{x \to 0} [x \ln x] = \lim\limits_{x \to 0} \left[\dfrac{\ln x}{1/x} \right] = \lim\limits_{x \to 0} \dfrac{[1/x]}{[-1/x^2]} = - \lim\limits_{x \to 0} x = 0.$$

Therefore, $\lim\limits_{x \to 0} y = e^0 = 1.$

Example 1.9 Evaluate $\lim\limits_{x \to \infty} x \tan (1/x)$.

Solution As $x \to \infty$, the function takes the form $\infty \cdot 0$. We first write it as $\lim\limits_{x \to \infty} \dfrac{x}{\cot (1/x)}$ which is of the form ∞/∞. Applying the L'Hospital's rule, we obtain

$$\lim_{x\to\infty} x \tan(1/x) = \lim_{x\to\infty} \frac{x}{\cot(1/x)} = \lim_{x\to\infty} \frac{1}{(1/x^2)\operatorname{cosec}^2(1/x)}$$

$$= \lim_{x\to\infty} \frac{\sin^2(1/x)}{(1/x)^2} = \lim_{y\to 0} \frac{\sin^2 y}{y^2} = \lim_{y\to 0} \left(\frac{\sin y}{y}\right)^2 = 1.$$

1.2.4 Increasing and Decreasing Functions

Let $y = f(x)$ be a function defined on an interval I contained in the domain of the function $f(x)$. Let x_1, x_2 be any two points in I, where x_1, x_2 are not the end points of the interval. On the interval I, the function $f(x)$ is said to be

(i) an *increasing* function, if $f(x_1) \le f(x_2)$ whenever $x_1 \le x_2$.

(ii) a *strictly* increasing function, if $f(x_1) < f(x_2)$ whenever $x_1 < x_2$.

(iii) a *decreasing* function, if $f(x_1) \ge f(x_2)$ whenever $x_1 < x_2$.

(iv) a *strictly* decreasing function, if $f(x_1) > f(x_2)$ whenever $x_1 < x_2$.

A function which is either increasing or decreasing in the entire interval I is called a *monotonic* function.

Let a real valued function f defined on an interval I, have a derivative at every point x in I. Then, using the Lagrange mean value theorem, we have

$$\frac{f(x_2) - f(x_1)}{x_2 - x_1} = f'(c), \quad x_1 < c < x_2.$$

Therefore, we conclude that

(i) f increases in I if $f'(x) > 0$ for all x in I.

(ii) f decreases in I if $f'(x) < 0$ for all x in I.

Thus, a differentiable function increases when its graph has positive slopes and decreases when its graph has negative slopes. Now, if $f'(x)$ is continuous, then $f'(x)$ can go from positive to negative values or from negative to positive values only by going through the value 0. The values of x for which $f'(x) = 0$ are called the *turning points* or the *critical points*. At a turning point, the tangent to the curve is parallel to the x-axis. On the left and right of a turning point, tangents to the curve have different directions.

Example 1.10 Find the intervals in which the function $f(x) = \sin 3x$, $0 \le x \le \pi/2$ is increasing or decreasing.

Solution We have $f'(x) = 3\cos 3x$. Now, $f'(x) = 0$ when $3x = \pi/2, 3\pi/2, \ldots$ for positive x. Hence $x = \pi/6$ is the only turning point in $(0, \pi/2)$. We consider the intervals $(0, \pi/6)$ and $(\pi/6, \pi/2)$. We have in

$0 < x < \pi/6$: $f'(x) = 3 \cos 3x > 0$, $f(x)$ is an increasing function,

$\pi/6 < x < \pi/2$: $f'(x) = 3 \cos 3x < 0$, $f(x)$ is a decreasing function.

Example 1.11   Show that for all $x > 0$

$$1 - x < e^{-x} < 1 - x + \frac{x^2}{2}.$$

Solution   Let $f(x) = e^{-x} + x - 1$. Now,

$$f'(x) = 1 - e^{-x} > 0 \text{ for all } x > 0.$$

Hence, $f(x)$ is an increasing function for all $x > 0$. Therefore,

$$f(x) > f(0) = 0, \quad \text{or} \quad e^{-x} + x - 1 > 0 \quad \text{or} \quad e^{-x} > 1 - x.$$

Now, consider $g(x) = e^{-x} - 1 + x - \dfrac{x^2}{2}$.

We have $\qquad\qquad\qquad g'(x) = 1 - x - e^{-x} < 0$ for all $x > 0$.

Hence, $g(x)$ is a decreasing function for all $x > 0$. Therefore,

$$g(x) < g(0) = 0, \quad \text{or} \quad e^{-x} < 1 - x + \frac{x^2}{2}.$$

Combining the above two results, we obtain

$$1 - x < e^{-x} < 1 - x + \frac{x^2}{2}, \quad x > 0.$$

1.2.5   Maximum and Minimum Values of a Function

Let a real valued function $f(x)$ be continuous on a closed interval $[a, b]$. Since a continuous function in a closed interval is bounded and attains these bounds at least once in the interval, we wish to determine the points where $f(x)$ attains these bounds. Let x_0 be a point in (a, b) and $I = (x_0 - h, x_0 + h)$ be an infinitesimal interval around x_0. Then, the function $f(x)$ is said to have a

local maximum (or a *relative maximum*) at the point x_0, if $f(x_0) \geq f(x)$, for all x in I.

local minimum (or a *relative minimum*) at the point x_0, if $f(x_0) \leq f(x)$ for all x in I.

The points of local maximum and local minimum are called the *critical points* or the *stationary points*. The values of the function at these points are called the *extreme values*.

The following theorem gives the necessary condition for the existence of a local maximum or a local minimum.

Theorem 1.4   (First derivative test) Let $f(x)$ be differentiable at $x_0 \in (a, b)$. Then, a necessary condition for the function $f(x)$ to have a local maximum or a local minimum at x_0 is that $f'(x_0) = 0$.

At a critical point, $f'(x)$ changes direction. Thus, to find the local maximum/minimum values of the function in an interval I, we find the critical points in I by solving $f'(x) = 0$. By studying the sign of $f'(x)$ as it passes through the critical point, we decide whether it is a point of a local maximum ($f'(x)$ changes sign from positive to negative) or a point of local minimum ($f'(x)$ changes sign from negative to positive).

Example 1.12   Examine the function

(i) $f(x) = x^3 - 3x + 3$, $x \in R$,   (ii) $f(x) = \sin^2 x$, $0 < x < \pi$

for maximum and minimum values.

Solution   We have

(i) $f'(x) = 3x^2 - 3$. Now, $f'(x) = 0$ gives $x = 1, -1$.
   For $x < 1$, $f'(x) < 0$ and for $x > 1$, $f'(x) > 0$. Since $f'(x)$ changes sign from negative to positive as it passes through the critical point $x = 1$, the function has a local minimum value $f(1) = 1$ at $x = 1$. For $x < -1$, $f'(x) > 0$ and for $x > -1$, $f'(x) < 0$. Since $f'(x)$ changes sign from positive to negative as it passes through the critical point $x = -1$, the function has a local maximum value $f(-1) = 5$ at $x = -1$.

(ii) $f'(x) = 2 \sin x \cos x = \sin 2x = 0$ at $x = \pi/2$.
   For $x < \pi/2$, $f'(x) > 0$ and for $x > \pi/2$, $f'(x) > 0$. Since $f'(x)$ changes sign from positive to negative as it passes through the critical point $x = \pi/2$, the function has a local maximum value $f(\pi/2) = 1$ at $x = \pi/2$.

Theorem 1.5   (Second derivative test) Let $f(x)$ be differentiable at x_0, $a \leq x_0 \leq b$ and let $f'(x_0) = 0$. If $f''(x)$ exists and is continuous in a neighborhood of x_0, then

$$f''(x_0) = \lim_{h \to 0} \frac{f'(x_0 + h) - f'(x_0)}{h} = \lim_{h \to 0} \frac{f'(x_0 - h) - f'(x_0)}{-h}, \ h > 0.$$

Therefore,

(i) $f(x)$ has a local maximum value at $x = x_0$, when $f''(x_0) < 0$,

(ii) $f(x)$ has a local minimum value at $x = x_0$, when $f''(x_0) > 0$.

When $f''(x_0) = 0$, further investigation is needed to decide whether $x = x_0$ is a point of local maximum or local minimum. In this case, we have the following result.

Theorem 1.6 Let $f^{(n)}(x)$ exist for x in (a, b) and be continuous there. Let

$$f'(x_0) = f''(x_0) = \ldots = f^{(n-1)}(x_0) = 0 \text{ and } f^{(n)}(x_0) \neq 0.$$

Then,

(i) when n is even, $f(x)$ has a maximum if $f^{(n)}(x_0) < 0$ and a minimum if $f^{(n)}(x_0) < 0$

(ii) when n is odd, $f(x)$ has neither a maximum, nor a minimum.

Absolute maximum/minimum values of a function $f(x)$ in an interval $[a, b]$ are defined as follows:

Absolute maximum value = max $\{f(a), f(b), \text{all local maximum values}\}$.

Absolute minimum value = min $\{f(a), f(b), \text{all local minimum values}\}$.

Example 1.13   Find the absolute maximum/minimum values of the function

$$f(x) = \sin x(1 + \cos x), \ \ 0 \le x \le 2\pi.$$

Solution   We have

$$f(x) = \sin x(1 + \cos x) = \sin x + \frac{1}{2} \sin 2x, \ \ f'(x) = \cos x + \cos 2x.$$

Setting $f'(x) = 0$, we get

$$\cos x + \cos 2x = 0, \ \ \text{or} \ \ \cos x + 2 \cos^2 x - 1 = 0, \ \ \text{or} \ \ \cos x = -1, \ 1/2$$

Therefore, the critical points are $x = \pi/3, \ \pi$ and $5\pi/3$.

Now, $\hspace{3cm} f''(x) = -\sin x - 2\sin 2x.$

At $x = \pi/3$, $f''(\pi/3) = -3\sqrt{3}/2 < 0$. Hence, $f(x)$ has a local maximum at $x = \pi/3$ and the local maximum value is $f(\pi/3) = 3\sqrt{3}/4$.

At $x = \pi$, $f''(\pi) = 0$. We find that

$$f'''(x) = -\cos x - 4 \cos 2x \ \ \text{and} \ \ f'''(\pi) = -3 \ne 0.$$

Since, $f^{(n)}(\pi) \ne 0$ and $n = 3$ is odd, the function has neither a maximum nor a minimum at $x = \pi$.

At $x = 5\pi/3$, $f''(5\pi/3) = 3\sqrt{3}/2 > 0$. Hence, $f(x)$ has a local minimum at $x = 5\pi/3$. The local minimum value is $f(5\pi/3) = -3\sqrt{3}/4$.

We also have $f(0) = f(2\pi) = 0$. Therefore,

absolute maximum value of $f(x) = \max \ \{f(0), \ f(2\pi), \text{local maximum value at } x = \pi/3\}$

$$= \max\{0, \ 0, \ 3\sqrt{3}/4 \ \} = 3\sqrt{3}/4.$$

absolute minimum value of $f(x) = \min\{f(0), \ f(2\pi), \text{local minimum value at } x = 5\pi/3\}$

$$= \min\{0, \ 0, \ -3\sqrt{3}/4\} = -3\sqrt{3}/4.$$

Example 1.14   Find a right angled triangle of maximum area with hypotenuse h.

Solution   Let x be the base of the right angled triangle. The area of the right angled triangle is

$$A(x) = \frac{1}{2} x \sqrt{h^2 - x^2}, \ \ 0 < x < h.$$

Now, $\hspace{2cm} A'(x) = \frac{1}{2}\left[\sqrt{h^2 - x^2} - \frac{x^2}{\sqrt{h^2 - x^2}} \right] = \frac{h^2 - 2x^2}{2\sqrt{h^2 - x^2}}.$

Setting $A'(x) = 0$, we obtain the critical point as $x = h/\sqrt{2}$.

Now, $A'(x) > 0$ for $x < h/\sqrt{2}$ and $A'(x) < 0$ for $x > h/\sqrt{2}$.

Therefore, $A(x)$ is maximum when $x = h/\sqrt{2}$ and the maximum area is $A(h/\sqrt{2}) = h^2/4$.

Leibniz formula   Let f and g be two differentiable functions. Then, the nth order derivative of the product fg is given by the Leibniz formula as

$$(f \cdot g)^{(n)} = {}^nC_0 f^{(n)}(x) g(x) + {}^nC_1 f^{(n-1)}(x) g'(x) + \ldots + {}^nC_r f^{(n-r)}(x) g^{(r)}(x) + \ldots + {}^nC_n f(x) g^{(n)}(x) \quad (1.9)$$

This formula can be proved by induction.

Example 1.15   Find the fourth order derivative of $e^{ax} \sin bx$ at the point $x = 0$.

Solution   Let $f(x) = e^{ax}$, $g(x) = \sin bx$ and $F(x) = f(x) g(x)$. Using the Leibniz formula, we obtain

$$F^{(4)}(x) = \frac{d^4}{dx^4} (e^{ax} \sin bx) = {}^4C_0 (e^{ax})^{(4)} \sin bx + {}^4C_1 (e^{ax})^{(3)} (\sin bx)'$$

$$+ {}^4C_2 (e^{ax})'' (\sin bx)'' + {}^4C_3 (e^{ax})' (\sin bx)^{(3)} + {}^4C_4 e^{ax} (\sin bx)^{(4)}$$

$$= e^{ax} [a^4 \sin bx + 4a^3 b \cos bx - 6a^2 b^2 \sin bx - 4ab^3 \cos bx + b^4 \sin bx]$$

Hence, $F^{(4)}(0) = 4a^3 b - 4ab^3 = 4ab(a^2 - b^2)$.

1.2.6   Taylor's Theorem and Taylor's Series

A very useful technique in the analysis of real valued functions is the approximation of continuous functions by polynomials. Taylor's theorem (Taylor's formula) is an important tool which provides such an approximation by polynomials. Taylor's theorem can be regarded as an extension of the mean value theorems to higher order derivatives. Mean value theorems relate the value of the function and its first order derivative, whereas the Taylor's theorem relates the value of the function and its higher order derivatives.

Theorem 1.7   (Taylor's theorem with remainder)   Let $f(x)$ be defined and have continuous derivatives upto $(n + 1)$th order in some interval I, containing a point a. Then, Taylor's expansion of the function $f(x)$ about the point $x = a$ is given by

$$f(x) = f(a) + \frac{(x-a)}{1!} f'(a) + \frac{(x-a)^2}{2!} f''(a) + \ldots + \frac{(x-a)^n}{n!} f^{(n)}(a) + R_n(x) \quad (1.10)$$

where, $$R_n(x) = \frac{(x-a)^{n+1}}{(n+1)!} f^{(n+1)}(c), \quad a < c < x \quad (1.11)$$

is the *remainder* or the error term of the expansion.

Proof   We first find a polynomial $P_n(x)$, of degree n, which satisfies the conditions

$$P_n(a) = f(a), \quad P_n^{(k)}(a), = f^{(k)}(a), \quad k = 1, 2, \ldots, n.$$

In a certain sense, $P_n(x)$ is a polynomial approximation to $f(x)$. Write the required polynomial as

$$P_n(x) = c_0 + c_1(x - a) + c_2(x - a)^2 + \ldots + c_n(x - a)^n.$$

Using the given conditions, we obtain

$$P_n(a) = f(a) = c_0, \ P_n'(a) = f'(a) = c_1, \ P_n''(a) = f''(a) = 2\,c_2, \ldots,$$

$$P_n^{(n)}(a) = f^{(n)}(a) = (n!)\,c_n.$$

Hence, we have $$c_k = \frac{1}{k!} f^{(k)}(a), \quad k = 0, 1, 2, \ldots, n.$$

Therefore, $f(x) \approx P_n(x) = f(a) + \dfrac{(x-a)}{1!} f'(a) + \dfrac{(x-a)^2}{2!} f''(a) + \ldots + \dfrac{(x-a)^n}{n!} f^{(n)}(a).$

The error of approximation is given by $R_n(x) = f(x) - P_n(x)$. Therefore,

$$f(x) = P_n(x) + R_n(x) = \sum_{k=0}^{n} \frac{(x-a)^k}{k!} f^{(k)}(a) + R_n(x).$$

Now, we derive a form of $R_n(x)$. Write $R_n(x)$ as

$$R_n(x) = \frac{(x-a)^{n+1}}{(n+1)!} h(x)$$

where $h(x)$ is to be determined.

Consider the auxiliary function

$$F(t) = f(x) - \left[f(t) + (x-t)f'(t) + \ldots + \frac{(x-t)^n}{n!} f^{(n)}(t) + \frac{(x-t)^{n+1}}{(n+1)!} h(x) \right], \ a < t < x.$$

We have t as a variable and x is fixed. The function $F(t)$ has the following properties:

(i) $F(t)$ is continuous in $a \le t \le x$ and differentiable in $a < t < x$,

(ii) $F(x) = 0$,

(iii) $F(a) = f(x) - \left[f(a) + (x-a)f'(a) + \ldots + \dfrac{(x-a)^n}{n!} f^{(n)}(a) + \dfrac{(x-a)^{n+1}}{(n+1)!} h(x) \right]$

$$= f(x) - f(x) = 0.$$

Hence, $F(t)$ satisfies the hypothesis of the Rolle's theorem on $[a, x]$. Therefore, there exists a point c, $a < c < x$ such that $F'(c) = 0$. Now,

$$F'(t) = 0 - \left[f'(t) - f'(t) + (x-t)f''(t) - \frac{2(x-t)}{2!} f''(t) + \cdots \right.$$

$$\left. + \frac{(x-t)^n}{n!} f^{(n+1)}(t) - \frac{(n+1)(x-t)^n}{(n+1)!} h(x) \right] = \frac{(x-t)^n}{n!} \left[h(x) - f^{(n+1)}(t) \right]$$

and $\qquad F'(c) = 0 = \dfrac{(x-c)^n}{n!} \, [h(x) - f^{(n+1)}(c)].$

We obtain $h(x) = f^{(n+1)}(c)$. Therefore,

$$R_n(x) = \frac{(x-a)^{n+1}}{(n+1)!} \, f^{(n+1)}(c), \quad a < c < x.$$

The error term can also be written as

$$R_n(x) = \frac{(x-a)^{n+1}}{(n+1)!} \, f^{(n+1)} \, (a + \theta(x-a)), \quad 0 < \theta < 1 \tag{1.12}$$

which is called the *Lagrange form of the remainder*.

If $a = 0$, we get

$$f(x) = f(0) = \frac{x}{1!} f'(0) + \frac{x^2}{2!} f''(0) + \cdots + \frac{x^n}{n!} f^{(n)}(0) + \frac{x^{n+1}}{(n+1)!} f^{(n+1)}(c), \quad 0 < c < x \tag{1.13}$$

which is called the *Maclaurin's theorem* with remainder.

Writing $x = a + h$ in Eq. (1.10), we obtain

$$f(a + h) \approx f(a) + \frac{h}{1!} f'(a) + \frac{h^2}{2!} f''(a) + \cdots + \frac{h^n}{n!} f^{(n)}(a). \tag{1.14}$$

The error of approximation simplifies as

$$R_n(x) = \frac{h^{n+1}}{(n+1)!} f^{(n+1)}(c), \quad a < c < a + h. \tag{1.15}$$

If we neglect the error term in Eq. (1.10), we obtain

$$f(x) \approx P_n(x) = \sum_{m=0}^{n} \frac{(x-a)^m}{m!} f^{(m)}(a) \tag{1.16}$$

which is called the nth degree Taylor's polynomial approximation to $f(x)$.

Since c or θ in the remainder term (see Eqs.(1.11), (1.12)) is not known, we cannot evaluate $R_n(x)$ exactly for a given x in the interval I. However, a bound on the error can be obtained as

$$|R_n(x)| = \left| \frac{(x-a)^{n+1}}{(n+1)!} f^{(n+1)}(c) \right| \le \max_{x \in I} \frac{|x-a|^{n+1}}{(n+1)!} \left[\max_{x \in I} |f^{(n+1)}(x)| \right]. \tag{1.17}$$

For a given error bound ε, we can use Eq. (1.17) to determine

  (i) n for a given x and a,
  (ii) $x = x^*$ for a given n and a such that $|R_n(x^*)| < \varepsilon$

Cauchy form of remainder

Consider a function $\phi(x)$ defined on $[a, a + h]$ as

$$\phi(x) = f(x) + (a + h - x) f'(x) + \ldots + \frac{(a + h - x)^n}{n!} f^{(n)}(x) + A(a + h - x)$$

where A is a constant to be determined such that $\phi(a + h) = \phi(a)$.

The function $\phi(x)$ satisfies all conditions of the Rolle's theorem. Therefore,

$$\phi'(a + \theta h) = 0, \quad 0 < \theta < 1.$$

Now,

$$\phi'(x) = \frac{1}{n!} (a + h - x)^n f^{(n+1)}(x) - A$$

and

$$\phi'(a + \theta h) = \frac{h^n}{n!} (1 - \theta)^n f^{(n+1)}(a + \theta h) - A = 0.$$

Hence,

$$A = \frac{h^n}{n!} (1 - \theta)^n f^{(n+1)} (a + \theta h).$$

From $\phi(a + h) = \phi(a)$, we get

$$f(a + h) = f(a) + h f'(a) + \ldots + \frac{h^n}{n!} f^{(n)}(a) + hA.$$

Therefore, the remainder in the Taylor's theorem is

$$R_n(x) = hA = \frac{h^{n+1}}{n!} (1 - \theta)^n f^{(n+1)}(a + \theta h), \quad 0 < \theta < 1. \tag{1.18}$$

Integral form of remainder

Consider the result

$$\int_{x_n}^{x_{n+1}} f'(x)\, dx = f(x_{n+1}) - f(x_n),$$

where $x_{n+1} = x_n + h$. Write the transformation $x = x_{n+1} - t$. Then, we have

$$\int_{x_n}^{x_{n+1}} f'(x)\, dx = -\int_{h}^{0} f'(x_{n+1} - t)\, dt = \int_{0}^{h} f'(x_{n+1} - t)\, dt.$$

Integrating by parts, we get

$$\int_0^h f'(x_{n+1} - t) \, dt = \left[t f'(x_{n+1} - t) \right]_0^h + \int_0^h t f''(x_{n+1} - t) \, dt$$

$$= h f'(x_n) + \int_0^h t f''(x_{n+1} - t) \, dt.$$

Integrating by parts again, we get

$$\int_0^h f'(x_{n+1} - t) \, dt = h f'(x_n) + \left[\frac{t^2}{2} f''(x_{n+1} - t) \right]_0^h + \int_0^h \frac{t^2}{2} f'''(x_{n+1} - t) \, dt$$

$$= h f'(x_n) + \frac{h^2}{2!} f''(x_n) + \frac{1}{2!} \int_0^h t^2 f'''(x_{n+1} - t) \, dt.$$

Integrating by parts repeatedly, we get

$$\int_0^h f'(x_{n+1} - t) \, dt = h f'(x_n) + \frac{h^2}{2!} f''(x_n) + \ldots + \frac{h^n}{n!} f^{(n)}(x_n) + \frac{1}{n!} \int_0^h t^n f^{(n+1)}(x_{n+1} - t) \, dt$$

$$= f(x_{n+1}) - f(x_n).$$

Therefore,

$$f(x_{n+1}) = f(x_n) + \sum_{k=1}^n \frac{h^k}{k!} f^{(k)}(x_n) + \frac{1}{n!} \int_{x_n}^{x_{n+1}} (x_{n+1} - s)^n f^{(n+1)}(s) \, ds$$

where $s = x_{n+1} - t$.

Hence, the remainder in the Taylor's theorem is

$$R_n = \frac{1}{n!} \int_{x_n}^{x_{n+1}} (x_{n+1} - s)^n f^{(n+1)}(s) \, ds.$$

Setting $x_n = a$, we may also write

$$R_n = \frac{1}{n!} \int_a^{a+h} (a + h - s)^n f^{(n+1)}(s) \, ds. \tag{1.19}$$

Example 1.16 The function $f(x) = \sin x$ is approximated by Taylor's polynomial of degree three about the point $x = 0$. Find c such that the error satisfies $|R_3(x)| \le 0.001$ for all x in the interval $[0, c]$.

Solution We have

$$f(x) = f(0) + x f'(0) + \frac{x^2}{2!} f''(0) + \frac{x^3}{3!} f'''(0).$$

For $f(x) = \sin x$, we obtain

$$f'(x) = \cos x, \; f''(x) = -\sin x, \; f'''(x) = -\cos x \quad \text{and} \quad f^{(4)}(x) = \sin x.$$

Hence, $\qquad f'(0) = 1, \; f''(0) = 0, \; f'''(0) = -1 \quad \text{and} \quad f^{(4)}(\xi) = \sin \xi.$

The required approximation is $f(x) = \sin x \approx x - \dfrac{x^3}{6}.$

The maximum error in the interval $[0, c]$ is given by

$$|R_3(x)| = \left| \frac{x^4}{4!} \sin \xi \right| \leq \max_{0 \leq x \leq c} \left[\frac{x^4}{24} \right] \max_{0 \leq x \leq c} |\sin x| \leq \frac{c^4}{24}.$$

Now, c is to be determined such that

$$\frac{c^4}{24} \leq 0.001 \quad \text{or} \quad c^4 \leq 0.024.$$

We obtain $c \approx 0.3936$. Hence, for all x in the interval $[0, 0.3936]$, this error criterion is satisfied.

Taylor's Series

In the Taylor's formula with remainder (Eqs. (1.10), (1.11)), if the remainder $R_n(x) \to 0$ as $n \to \infty$, then we obtain

$$f(x) = f(a) + \frac{(x-1)}{1!} f'(a) + \frac{(x-a)^2}{2!} f''(a) + \ldots + \frac{(x-a)^n}{n!} f^{(n)}(a) + \ldots \qquad (1.20)$$

which is called the *Taylor's series*. When $a = 0$, we obtain the *Maclaurin's series*

$$f(x) = f(0) + \frac{x}{1!} f'(0) + \frac{x^2}{2!} f''(0) + \ldots + \frac{x^n}{n!} f^{(n)}(0) + \ldots \qquad (1.21)$$

Since it is assumed that $f(x)$ has continuous derivatives upto $(n + 1)$th order, $f^{(n+1)}(x)$ is bounded in the interval (a, x). Hence, to establish that $\lim_{n \to \infty} |R_n(x)| = 0$, it is sufficient to show that

$$\lim_{n \to \infty} \frac{|x - a|^{n+1}}{(n+1)!} = 0$$ for any fixed numbers x and a. Now, for any fixed numbers x and a, we can always

find a finite positive integer N such that $|x - a| < N$. Denote $q = |x - a|/N$. Then,

$$\left| \frac{(x-a)^{n+1}}{(n+1)!} \right| = \left| \frac{x-a}{1} \right| \left| \frac{x-a}{2} \right| \cdots \left| \frac{x-a}{N-1} \right| \left| \frac{x-a}{N} \right| \cdots \left| \frac{x-a}{n+1} \right|$$

$$< \left| \frac{(x-a)^{N-1}}{(N-1)!} \right| q \cdot q \cdots q = \left| \frac{(x-a)^{N-1}}{(N-1)!} \right| q^{n-N+2}.$$

Now, $\left|\dfrac{(x-a)^{N-1}}{(N-1)!}\right|$ is a finite quantity and is independent of n. Also $q < 1$. Hence,

$$\lim_{n \to \infty}\left|\frac{(x-a)^{n+1}}{(n+1)!}\right| = 0 \text{ for any fixed } x \text{ and } a, \text{ and } \lim_{n \to \infty} |R_n(x)| = 0.$$

Example 1.17   Obtain the Taylor's polynomial expansion of the function $f(x) = \sin x$ about the point $x = \pi/4$. Show that the error term tends to zero as $n \to \infty$ for any real x. Hence, write the Taylor's series expansion of $f(x)$.

Solution   For $f(x) = \sin x$, we have

$$f^{(2n)}(x) = (-1)^n \sin x \quad \text{and} \quad f^{(2n+1)}(x) = (-1)^n \cos x$$

for any integer n. Therefore,

$$f^{(2n)}(\pi/4) = (-1)^n/\sqrt{2} \quad \text{and} \quad f^{(2n+1)}(\pi/4) = (-1)^n/\sqrt{2}.$$

Hence, the Taylor's expansion of $f(x) = \sin x$ about $x = \pi/4$ is given by

$$f(x) = f\left(\frac{\pi}{4}\right) + \left(x - \frac{\pi}{4}\right)f'\left(\frac{\pi}{4}\right) + \cdots + \frac{1}{n!}\left(x - \frac{\pi}{4}\right)^n f^{(n)}\left(\frac{\pi}{4}\right) + R_n(x).$$

Now,

$$|R_{2n}(x)| = \left|\frac{1}{(2n+1)!}\left(x - \frac{\pi}{4}\right)^{2n+1} f^{(2n+1)}(\xi)\right| \le \frac{1}{(2n+1)!}\left(x - \frac{\pi}{4}\right)^{2n+1}$$

since $f^{(2n+1)}(c) = |(-1)^n \cos c| < 1$. Hence, $R_{2n}(x) \to 0$ as $n \to \infty$.

Similarly, we find that $R_{2n+1}(x) \to 0$ as $n \to \infty$.

Therefore, the required Taylor's series expansion is given by

$$\sin x = \frac{1}{\sqrt{2}} + \frac{1}{\sqrt{2}}\left(x - \frac{\pi}{4}\right) - \frac{1}{(2!)\sqrt{2}}\left(x - \frac{\pi}{4}\right)^2 - \frac{1}{(3!)\sqrt{2}}\left(x - \frac{\pi}{4}\right)^3 + \cdots$$

$$= \frac{1}{\sqrt{2}}\left[1 + \left(x - \frac{\pi}{4}\right) - \frac{1}{2!}\left(x - \frac{\pi}{4}\right)^2 - \frac{1}{3!}\left(x - \frac{\pi}{4}\right)^3 + \cdots\right].$$

Example 1.18   Obtain the Maclaurin's series expansion of $f(x) = \sin(m \sin^{-1} x)$, where m is a constant.

Solution   The Maclaurin's series is given by

$$f(x) = f(0) + xf'(0) + \frac{x^2}{2!} f''(0) + \frac{x^3}{3!} f'''(0) + \cdots \qquad (1.22)$$

For $y(x) = f(x) = \sin(m \sin^{-1} x)$, we find

$$y'(x) = f'(x) = \frac{m}{\sqrt{1-x^2}} \cos(m \sin^{-1} x). \tag{1.23}$$

We proceed as follows:

Let $y_r(x) = y^{(r)}(x) = f^{(r)}(x)$. From (1.23), we get

$$(1 - x^2)\, y_1^2 = m^2 \cos^2(m \sin^{-1} x) = m^2(1 - \sin^2(m \sin^{-1} x)) = m^2(1 - y^2).$$

Differentiating, we get

$$2(1 - x^2)\, y_1 y_2 - 2x\, y_1^2 = -2m^2 y y_1$$

or

$$(1 - x^2)\, y_2 - xy_1 + m^2 y = 0, \quad y_1 \neq 0. \tag{1.24}$$

Differentiating (1.24) n times using Leibniz theorem, we obtain

$$(1 - x^2)\, y_{n+2} + n(-2x)\, y_{n+1} + \frac{n(n-1)}{2}(-2)\, y_n - (x\, y_{n+1} + ny_n) + m^2 y_n = 0$$

or

$$(1 - x^2)\, y_{n+2} - (2n + 1)\, xy_{n+1} - (n^2 - m^2)\, y_n = 0.$$

For $x = 0$, we get

$$y_{n+2}^{(0)} = (n^2 - m^2)\, y_n(0), \quad n = 0, 1, 2, \ldots$$

Therefore, we find

$$f(0) = y_0(0) = 0,\ f'(0) = y_1(0) = m,\ f''(0) = y_2(0) = 0,$$

$$f'''(0) = y_3(0) = m(1^2 - m^2),\ f^{iv}(0) = y_4(0) = 0,$$

$$f^{v}(0) = y_5(0) = m(1^2 - m^2)(3^2 - m^2), \ldots$$

Substituting in (1.22), we obtain

$$f(x) = mx + \frac{m(1^2 - m^2)}{3!}x^3 + \frac{m(1^2 - m^2)(3^2 - m^3)}{5!}x^5 + \cdots$$

Example 1.19   Using Taylor's series, obtain the value of cos 31° correct to four decimal places.

Solution   Let $f(x) = \cos x$. We have

$f'(x) = -\sin x,\ f''(x) = -\cos x,\ f'''(x) = \sin x$ and so on. Using Taylor's series, we obtain

$$\cos(x + h) = f(x + h) = f(x) + hf'(x) + \frac{h^2}{2!}f''(x) + \frac{h^3}{3!}f'''(x) + \cdots$$

$$= \cos x - h \sin x - \frac{h^2}{2}\cos x + \frac{h^3}{6}\sin x + \cdots$$

Substituting $x = 30° = \pi/6$ and $h = 1° = \pi/180 \approx 0.01745$, we get

$$\cos 31° = \cos(\pi/6) - (\pi/180)\sin(\pi/6) - \frac{1}{2}(\pi/180)^2 \cos(\pi/6) + \frac{1}{6}(\pi/180)^3 \sin(\pi/6) + \ldots$$

$$= \frac{\sqrt{3}}{2} - (0.01745)\left(\frac{1}{2}\right) - \frac{1}{2}(0.01745)^2 \left(\frac{\sqrt{3}}{2}\right) + \frac{1}{6}(0.01745)^3 \left(\frac{1}{2}\right) + \ldots$$

$$= 0.86602 - 0.00872 - 0.00013 + 0.00000 + \ldots$$

$$= 0.85717.$$

Hence, we obtain $\cos 31° = 0.8572$ correct to four decimal places.

1.2.7   Exponential, Logarithmic and Binomial Series

Exponential series

Consider the Taylor's polynomial approximation of degree $\leq n$ about the point $x = 0$ for the function $f(x) = e^x$. The Taylor's polynomial approximation is given by

$$f(x) = f(0) + \frac{x}{1!}f'(0) + \frac{x^2}{2!}f''(0) + \cdots + \frac{x^n}{n!}f^{(n)}(0).$$

For $f(x) = e^x$, we obtain

$$f^{(r)}(x) = e^x, \ f^{(r)}(0) = 1, \ r = 0, \ 1, \ \ldots, \ n \ \text{ and } \ f^{(n+1)}(x) = e^x.$$

Hence,
$$f(x) = e^x \approx 1 + x + \frac{x^2}{2!} + \cdots + \frac{x^n}{n!}.$$

Using the Lagrange form of the remainder, we get

$$R_n(x) = \frac{x^{n+1}}{(n+1)!}f^{(n+1)}(c) = \frac{x^{n+1}}{(n+1)!}e^c$$

or as
$$R_n(x) = \frac{x^{n+1}}{(n+1)!}e^{\theta x}, \ 0 < \theta < 1.$$

Therefore,
$$\lim_{n \to \infty} |R_n(x)| = \lim_{n \to \infty} \left|\frac{x^{n+1}}{(n+1)!}e^{\theta x}\right| = \lim_{n \to \infty}\left[\frac{|x|^{n+1}}{(n+1)!}\right]e^{\theta x} = 0$$

for all x, since $e^{\theta x}$ is bounded for a given x.

Hence, we obtain the *exponential series*

$$e^x = 1 + x + \frac{x^2}{2!} + \cdots + \frac{x^n}{n!} + \ldots \tag{1.25}$$

Example 1.20 For the Taylor's polynomial approximation of degree $\leq n$ about the point $x = 0$ for the function $f(x) = e^x$, determine the value of n such that the error satisfies $|R_n(x)| \leq 0.005$, when $-1 \leq x \leq 1$.

Solution We have the Taylor's polynomial approximation of e^x as

$$f(x) = e^x \approx 1 + x + \frac{x^2}{2!} + \cdots + \frac{x^n}{n!}.$$

The maximum error in the interval $[-1, 1]$ is given by

$$|R_n(x)| = \left| \frac{x^{n+1}}{(n+1)!} f^{(n+1)}(c) \right| \leq \max_{-1 \leq x \leq 1} \left[\frac{|x|^{n+1}}{(n+1)!} \right] \max_{-1 \leq x \leq 1} \left[|e^x| \right] \leq \frac{e}{(n+1)!}.$$

Now, n is to be determined such that

$$\frac{e}{(n+1)!} \leq 0.005 \quad \text{or} \quad (n+1)! \geq 200\, e.$$

We find that $n \geq 5$. Hence, we will require at least 6 terms in the Taylor's polynomial approximation to achieve the given accuracy.

Example 1.21 Obtain the fourth degree Taylor's polynomial approximation to $f(x) = e^{2x}$ about $x = 0$. Find the maximum error when $0 \leq x \leq 0.5$.

Solution We have

$$f(x) \approx f(0) + x f'(0) + \frac{x^2}{2!} f''(0) + \frac{x^3}{3!} f'''(0) + \frac{x^4}{4!} f^{(4)}(0).$$

For $f(x) = e^{2x}$, we obtain $f^{(r)}(x) = 2^r e^{2x}$, $f^{(r)}(0) = 2^r$, $r = 0, 1, 2, \ldots$ and $f^{(5)}(c) = 32\, e^{2c}$.

Therefore, $f(x) = e^{2x} \approx 1 + 2x + \frac{4x^2}{2!} + \frac{8x^3}{3!} + \frac{16x^4}{4!} = 1 + 2x + 2x^2 + \frac{4}{3}x^3 + \frac{2}{3}x^4$.

The error term is given by

$$R_4(x) = \frac{x^5}{5!} f^{(5)}(c) = \frac{32}{5!} x^5 e^{2c}, \quad 0 < c < x.$$

and

$$|R_4(x)| \leq \frac{32}{120} \left[\max_{0 \leq x \leq 0.5} x^5 \right] \left[\max_{0 \leq x \leq 0.5} e^{2x} \right] \leq \frac{e}{120}.$$

Logarithmic series

Consider the Taylor's polynomial approximation of degree $\leq n$ about the point $x = 0$ for the function $f(x) = \ln(1 + x)$. The Taylor's polynomial approximation is given by

$$f(x) = f(0) + xf'(0) + \frac{x^2}{2!} f''(0) + \ldots + \frac{x^n}{n!} f^{(n)}(0).$$

For $f(x) = \ln(1 + x)$, we obtain

$$f(x) = \ln(1 + x), \; f(0) = 0; \; f'(x) = \frac{1}{1+x}, \; f'(0) = 1;$$

$$f^{(r)}(x) = \frac{(-1)^{r-1}(r-1)!}{(1+x)^r}, \; f^{(r)}(0) = (-1)^{r-1}(r-1)!, \; r = 2, 3, \ldots, n;$$

$$f^{(n+1)}(x) = \frac{(-1)^n n!}{(1+x)^{n+1}}.$$

Hence, $$f(x) = \ln(1+x) \approx x + \frac{(-1)(1!)}{2!} x^2 + \frac{(-1)^2(2!)}{3!} x^3 + \cdots + \frac{(-1)^{n-1}(n-1)!}{n!} x^n$$

$$= x - \frac{x^2}{2} + \frac{x^3}{3} - \cdots + (-1)^{n-1} \frac{x^n}{n}.$$ (1.26)

We note that $f(x)$ and all its derivatives exist and are continuous for $-1 < x \le 1$. Using the Lagrange form of the remainder, we get

$$R_n(x) = \frac{x^{n+1}}{(n+1)!} f^{(n+1)}(\theta x) = \frac{x^{n+1}}{(n+1)!} \left[\frac{(-1)^n n!}{(1+\theta x)^{n+1}} \right]$$

$$= \frac{(-1)^n}{(n+1)} \left[\frac{x}{1+\theta x} \right]^{n+1}, \; 0 < \theta < 1.$$

We consider the following two cases:

Case 1   Let $0 \le x \le 1$. Since $0 < \theta < 1$, we have

$$0 < \theta x < x \le 1 \quad \text{and} \quad \frac{x}{1+\theta x} < 1.$$

Therefore, we obtain

$$\lim_{n \to \infty} |R_n(x)| = \lim_{n \to \infty} \frac{1}{(n+1)} \left| \frac{x}{1+\theta x} \right|^{n+1} = 0.$$

Case 2   Let $-1 < x < 0$. Since $0 < \theta < 1$, $|x/(1 + \theta x)|$ may or may not be less than 1. Hence, we cannot use the Lagrange form of the remainder to find $\lim_{n \to \infty} |R_n(x)|$.

Now, consider the Cauchy's form of the remainder. We have

$$R_n(x) = \frac{x^{n+1}}{n!}(1-\theta)^n f^{(n+1)}(\theta x) = \frac{(-1)^n(1-\theta)^n}{(1+\theta x)^{n+1}} x^{n+1}$$

$$= \frac{(-1)^n}{(1+\theta x)}\left(\frac{1-\theta}{1+\theta x}\right)^n x^{n+1}, \quad 0 < \theta < 1.$$

Since,
$$\left|\frac{1-\theta}{1+\theta x}\right| < 1 \quad \text{and} \quad \left|\frac{1}{1+\theta x}\right| < \frac{1}{1-|x|},$$

we have
$$\lim_{n\to\infty} |R_n(x)| < \lim_{n\to\infty} \left[\frac{|x|^{n+1}}{(1-|x|)}\left|\frac{1-\theta}{1+\theta x}\right|^n\right] = 0.$$

Hence, we obtain

$$\ln(1 + x) = x - \frac{x^2}{2} + \frac{x^3}{3} - \ldots + (-1)^{n-1}\frac{x^n}{n} + \ldots, \quad -1 < x \le 1. \tag{1.27}$$

Note that for $|x| > 1$, $\lim_{n\to\infty} |R_n(x)| = \infty$.

Writing $1 + x = y$ or $x = y - 1$ in Eq. (1.27), we get

$$\ln y = (y - 1) - \frac{1}{2}(y - 1)^2 + \ldots + \frac{(-1)^{n-1}}{n}(y - 1)^n + \ldots, \quad 0 < y \le 2. \tag{1.28}$$

The series given in Eqs. (1.27) and (1.28) are called the logarithmic series.

Binomial series

Consider the expansion of the function $(x + y)^m$. We can write
$$(x + y)^m = x^m[1 + (y/x)]^m = x^m(1 + z)^m, \text{ where } z = y/x.$$

Therefore, it is sufficient to obtain the expansion of the function $f(z) = (1 + z)^m$ or $f(x) = (1 + x)^m$. Consider the Taylor's polynomial approximation for the function $f(x) = (1 + x)^m$ about the point $x = 0$. We consider the following two cases.

Case 1   When m is a positive integer, $f(x) = (1 + x)^m$ possesses continuous derivatives of all orders and $f^{(r)}(x) = 0$, $r \ge m + 1$ for all x. We have

$$f(x) = (1 + x)^m, \quad f(0) = 1; \quad f'(x) = m(1 + x)^{m-1}, \quad f'(0) = m;$$

$$f''(x) = m(m - 1)(1 + x)^{m-2}, \quad f''(0) = m(m - 1); \ldots$$

$$f^{(m)}(x) = m(m - 1)\ldots 2.1, \quad f^{(m)}(0) = m! \quad \text{and} \quad f^{(r)}(x) = 0, \, r > m.$$

Therefore, we obtain

$$f(x) = (1 + x)^m = f(0) + xf'(0) + \frac{x^2}{2!} f''(0) + \ldots + \frac{x^m}{m!} f^{(m)}(0)$$

$$= 1 + mx + \frac{m(m-1)}{2!} x^2 + \ldots + x^m$$

$$= \binom{m}{0} + \binom{m}{1} x + \binom{m}{2} x^2 + \cdots + \binom{m}{m} x^m \qquad (1.29)$$

We also have

$$R_n(x) = \frac{x^{n+1}}{(n+1)!} f^{(n+1)} (c) = 0, \quad n \geq m.$$

Case 2   When m is not a positive integer, $f(x) = (1 + x)^m$ possesses continuous derivatives of all orders provided $x \neq -1$.

Let $-1 < x < 1$. We have

$$f^{(n+1)}(x) = m(m - 1) \ldots (m - n) (1 + x)^{m-n-1}$$

Using the Cauchy's form of the remainder, we get

$$R_n(x) = \frac{x^{n+1}}{n!} (1 - \theta)^n f^{(n+1)}(\theta x)$$

$$= \frac{x^{n+1}}{n!} (1 - \theta)^n m(m - 1) \ldots (m - n) (1 + \theta x)^{m-n-1}$$

$$= \frac{m(m-1)\ldots(m-n)}{n!} \left(\frac{1-\theta}{1+\theta x}\right)^n (1 + \theta x)^{m-1} x^{n+1}, \quad 0 < \theta < 1. \qquad (1.30)$$

Now, for $|x| < 1$, $\quad 0 < \dfrac{1-\theta}{1+\theta x} < 1$ and

$$\lim_{n \to \infty} \left(\frac{1-\theta}{1+\theta x}\right)^n = 0, \quad \lim_{n \to \infty} |x|^{n+1} = 0, \quad \lim_{n \to \infty} \left|\frac{m(m-1)\ldots(m-n)}{n!}\right| = a$$

where a is a finite quantity. Since $(1 + \theta x)^{m-1}$ is independent of n and bounded, we obtain from Eq. (1.30)

$$\lim_{n \to \infty} |R_n(x)| = 0.$$

Therefore, when m is not a positive integer and $|x| < 1$, we obtain

$$f(x) = 1 + mx + \frac{m(m-1)}{2!}x^2 + \cdots + \frac{m(m-1)\ldots(m-n)}{n!}x^n + \ldots \tag{1.31}$$

Alternative proof

Consider the series $\Sigma\, a_n$ where $a_n = \dfrac{m(m-1)\ldots(m-n)}{n!}x^n$.

Using the ratio test, we get

$$\lim_{n\to\infty}\left|\frac{a_{n+1}}{a_n}\right| = \lim_{n\to\infty}\left|\frac{(m-n-1)}{(n+1)}x\right| = |x|.$$

Hence, the series (1.31) converges for $-1 < x < 1$.

Further, it can be shown that the binomial series (1.31) converges at $x = 1$ when $m > -1$.

For example, consider the series

$$(1+x)^{-1} = 1 - x + x^2 - x^3 + \ldots$$

For $x = 1$, the series on the right hand side has two limit points 0 and 1 and hence the series is not convergent.

We have the following binomial series

$$(1+x)^{-2} = 1 - 2x + 3x^2 - 4x^3 + \ldots, \quad -1 < x < 1.$$

$$(1+x)^{-1/2} = 1 - \frac{1}{2}x + \frac{1\cdot3}{2\cdot4}x^2 - \frac{1\cdot3\cdot5}{2\cdot4\cdot6}x^3 + \ldots, \quad -1 < x \le 1.$$

$$(1+x)^{1/3} = 1 + \frac{1}{3}x - \frac{2}{3\cdot6}x^2 + \frac{2\cdot5}{3\cdot6\cdot9}x^3 - \ldots, \quad -1 < x \le 1.$$

Exercise 1.1

Find the approximate values of the following quantities using differentials.

1. $(1005)^{1/3}$.
2. $(999)^{1/3}$
3. $(1.001)^3 + 2(1.001)^{4/3} + 5$.
4. $\sin 60° \, 10'$.
5. $\tan 45° \, 5' \, 30''$.

6. State why Rolle's theorem cannot be applied to the following functions.
   (i) $f(x) = \tan x$ in the interval $[0, \pi]$,   (ii) $f(x) = \lfloor x \rfloor$ in the interval $[-1/2, 3/2]$,

   (iii) $f(x) = \begin{cases} x, & 0 \le x \le 1 \\ 2 - x, & 1 \le x \le 2. \end{cases}$

7. It is given that the Rolle's theorem holds for the function $f(x) = x^3 + bx^2 + cx, \; 1 \le x \le 2$ at the point $x = 4/3$. Find the values of b and c.

8. The functions $f(x)$ and $g(x)$ are continuous on $[a, b]$ and differentiable in (a, b) such that $f(a) = 4$, $f(b) = 10, g(a) = 1$ and $g(b) = 3$. Then, show that $f'(c) = 3g'(c), a < c < b$.

9. Prove that between any two real roots of $e^x \sin x = 1$, there exists atleast one root of $e^x \cos x + 1 = 0$.

10. Let $f'(x)$, $g'(x)$ be continuous and differentiable functions on $[a, b]$. Then, show that for $a < c < b$

$$\frac{f(b) - f(a) - (b - a)f'(a)}{g(b) - g(a) - (b - a)g'(a)} = \frac{f''(c)}{g''(c)}, \quad g''(c) \neq 0.$$

11. Let $f(x)$ be continuous on $[a - 1, a + 1]$ and differentiable in $(a - 1, a + 1)$. Show that there exists a θ, $0 < \theta < 1$ such that $f(a - 1) - 2 f(a) + f(a + 1) = f'(a + \theta) - f'(a - \theta)$.

12. Using the Lagrange mean value theorem, show that

    (i) $1 + x < e^x < 1 + xe^x$;

    (ii) $\ln(1 + x) < x, \quad x > 0$;

    (iii) $x < \sin^{-1} x < x/\sqrt{1 - x^2}, \quad 0 < x < 1$;

    (iv) $\dfrac{\pi}{6} + \dfrac{1}{5\sqrt{3}} < \sin^{-1} x < \dfrac{\pi}{6} + \dfrac{1}{8}$.

13. Suppose that $f(x)$ is differentiable for all values of x such that $f(a) = a$, $f(-a) = -a$ and $|f'(x)| \leq 1$ for all x. Show that $f(0) = 0$.

14. Let $F(x)$ and $G(x)$ be two functions defined on $[a, b]$ satisfying the hypothesis of the mean value theorem with $G(x) \neq 0$ for any x in $[a, b]$. Show that there exists a point c in (a, b) such that

$$\frac{F(b) - F(a)}{G(b) - G(a)} = \frac{F'(c)}{G'(c)} \left[\frac{G^2(c)}{G(a)G(b)} \right].$$

Evaluate the limits in problems **15** to **28**.

15. $\lim\limits_{x \to 1} \dfrac{x - 1}{x^n - 1}$

16. $\lim\limits_{x \to 0} \dfrac{e^x - 2\cos x + e^{-x}}{x \sin x}$

17. $\lim\limits_{x \to \pi/2} \dfrac{\ln(\sin x)}{(\pi - 2x)^2}$

18. $\lim\limits_{x \to 0} \dfrac{\sin^2 x - x^2}{x^2 \sin^2 x}$

19. $\lim\limits_{x \to 1} (1 - x) \tan(\pi x/2)$.

20. $\lim\limits_{x \to 2} \left[\dfrac{x - 1}{x - 2} - \dfrac{1}{\ln(x - 1)} \right]$

21. $\lim\limits_{x \to 1} x^{1/(x-1)}$.

22. $\lim\limits_{x \to \pi/2} (\sin x)^{\tan x}$.

23. $\lim\limits_{x \to 0} \dfrac{e^{f(x)} - 1}{f(x)}, \quad f(0) = 0$.

24. $\lim\limits_{x \to \infty} \dfrac{x^2}{e^x}$.

25. $\lim\limits_{x \to \infty} [1 + f(x)]^{1/f(x)}, \quad \lim\limits_{x \to \infty} f(x) = 0$.

26. $\lim\limits_{x \to \infty} \left(1 + \dfrac{1}{x} \right)^x$.

27. $\lim\limits_{x \to \infty} \sqrt{\dfrac{x + \sin x}{x - \cos^2 x}}$.

28. $\lim\limits_{x \to \infty} \left(\dfrac{x + 4}{x + 2} \right)^{x+3}$.

In problems **29** to **36**, find the intervals in which $f(x)$ is increasing or decreasing.

29. $\ln(2 + x) - 2x/(2 + x), \quad x \in \mathbb{R}$

30. $x|x|, \quad x \in \mathbb{R}$.

31. $\tan^{-1} x + x, \quad x \in \mathbb{R}$.

32. $\sin x + |\sin x|, \quad 0 < x \leq 2\pi$.

33. $\ln(\sin x), \quad 0 < x < \pi$

34. $(\ln x)/x, \quad x > 0$.

35. $\sin x(1 + \cos x), \quad 0 < x < \pi/2$.

36. $x^x, \quad x > 0$.

37. Let $a > b > 0$ and n be a positive integer satisfying $n \geq 2$. Prove that $a^{1/n} - b^{1/n} < (a - b)^{1/n}$.

In problems **38** to **43**, find the extreme values of the given function $f(x)$.

38. $(x-1)^2(x+1)^3$.

39. $\sin x + \cos x$.

40. $x^{1/x}$.

41. $(\sin x)^{\sin x}$.

42. $2\sin x + \cos 2x$, $\quad 0 \le x \le 2\pi$.

43. $\sin^2 x \sin 2x + \cos^2 x \cos 2x$, $\quad 0 < x < \pi$.

44. Show that the function $f(x) = (ax+b)/(cx+d)$ has no extreme value regardless of the values of a, b, c, d.

45. Let $f(x) = \begin{cases} -x^3 + [(b^3 - b^2 + b - 1)/(b^2 + 3b + 2)], & 0 \le x < 1 \\ 2x - 3 & , \quad 1 \le x \le 3. \end{cases}$

Find all possible real values of b such that $f(x)$ has minimum value at $x = 1$.

46. If $y = x^3 e^{2x}$, then find $d^n y/dx^n$ at $x = 0$.

47. Find the nth order derivative of $f(x) = \sqrt{ax+b}$.

48. Find the nth order derivative of $f(x) = e^{ax} \sin(bx + c)$.

49. If $y = \cos^{-1} x$, $-(\pi/2) \le x \le (\pi/2)$, then find $d^n y/dx^n$ at $x = 0$.

50. If $y = e^{a \sin^{-1} x}$, then find $d^n y/dx^n$ at $x = 0$.

In problems **51** to **55**, obtain the Taylor's polynomial approximation of degree n to the function $f(x)$ about the point $x = a$. Estimate the error in the given interval.

51. $f(x) = \sqrt{x}$, $n = 3$, $a = 1$, $1 \le x \le 1.5$.

52. $f(x) = e^{-x^2}$, $n = 3$, $a = 0$, $-1 \le x \le 1$.

53. $f(x) = x \sin x$, $n = 4$, $a = 0$, $-1 \le x \le 1$.

54. $f(x) = x^2 e^{-x}$, $n = 4$, $a = 1$, $0.5 \le x \le 1.5$.

55. $f(x) = 1/(1-x)$, $n = 3$, $a = 0$, $0 \le x \le 0.25$.

In problems **56** to **59**, obtain the Taylor's polynomial approximation of degree n to the function $f(x)$ about the point $x = a$. Find the error term and show that it tends to zero as $n \to \infty$. Hence, write its Taylor's series.

56. $f(x) = \sin 3x$, $\quad a = 0$.

57. $f(x) = \sin^2 x$, $\quad a = 0$.

58. $f(x) = x^2 \ln x$, $\quad a = 1$.

59. $f(x) = 2^x$, $\quad a = 1$.

60. Show that the number θ which occurs in the Taylor's formula with Lagrange form of remainder (given in Eq. 1.12) after n terms approaches the limit $1/(n+1)$ as $h \to 0$ provided $f^{(n+1)}(x)$ is continuous and not zero at $x = a$.

61. Find the number of terms that must be retained in the Taylor's polynomial approximation about the point $x = 0$ for the function $\cosh x$ in the interval $[0, 1]$ such that $|\text{Error}| < 0.001$.

62. Find the number of terms that must be retained in the Taylor's polynomial approximation about the point $x = 0$ for the function $\sin x \cos x$ in the interval $[0, 1]$ such that $|\text{Error}| < 0.0005$.

63. The function $\ln(1 - x^2)$ is approximated about $x = 0$ by an nth degree Taylor's polynomial. Find n such that $|\text{Error}| < 0.1$ on $0 \le x \le 0.5$.

64. The function $\sin^2 x$ is approximated by the first two non-zero terms in the Taylor's polynomial expansion about the point $x = 0$. Find c such that $|\text{Error}| < 0.005$, when $0 < x < c$.

65. The function $\tan^{-1} x$ is approximated by the first two non-zero terms in the Taylor's polynomial expansion about the point $x = 0$. Find c such that $|\text{Error}| < 0.005$, when $0 < x < c$.

66. Obtain the Maclaurin's series expansion of $y(x) = e^{m \sin^{-1} x}$, where m is a constant.

67. Using Taylor's series, find the approximate value of (i) $\sqrt{15}$, (ii) $\sin 29°$.

Obtain the Taylor's series expansions as given in problems **68** to **75**.

68. $a^x = 1 + x \ln a + \dfrac{(x \ln a)^2}{2!} + \dfrac{(x \ln a)^3}{3!} + \ldots$, $\quad -\infty < x < \infty$.

69. $\ln(1-x) = -x - \dfrac{x^2}{2} - \dfrac{x^3}{3} - \ldots, \quad -1 \le x < 1.$

70. $\ln\left(\dfrac{1+x}{1-x}\right) = 2\left[x + \dfrac{x^3}{3} + \dfrac{x^5}{5} + \cdots\right], \quad -1 < x < 1.$

71. $\ln x = 2\left[\left(\dfrac{x-1}{x+1}\right) + \dfrac{1}{3}\left(\dfrac{x-1}{x+1}\right)^3 + \dfrac{1}{5}\left(\dfrac{x-1}{x+1}\right)^5 + \cdots\right], \quad x > 0.$

72. $\ln x = \left(\dfrac{x-1}{x}\right) + \dfrac{1}{2}\left(\dfrac{x-1}{x}\right)^2 + \dfrac{1}{3}\left(\dfrac{x-1}{x}\right)^3 + \ldots, \quad x \ge 1/2.$

73. $\tan^{-1}\left(\dfrac{2x}{1-x^2}\right) = 2\left[x - \dfrac{x^3}{3} + \dfrac{x^5}{5} - \cdots\right], \quad -1 < x < 1.$

74. $\tan^{-1}\left(\dfrac{\sqrt{1+x^2} - \sqrt{1-x^2}}{\sqrt{1+x^2} + \sqrt{1-x^2}}\right) = \left[\dfrac{x^2}{2} + \dfrac{x^6}{12} + \cdots\right].$

75. $\sin^{-1}\left(\dfrac{2x^2}{1+x^4}\right) = 2\left[x^2 - \dfrac{x^6}{3} + \dfrac{x^{10}}{5} - \cdots\right].$

1.3   Integration and Its Applications

Let $f(x)$ be defined and be continuous on a closed interval $[a, b]$. Let there exist a function $F(x)$ such that $F'(x) = f(x)$, $a \le x \le b$. Then, the function $F(x)$ is called the *anti-derivative* of $f(x)$. We observe that if $F(x)$ is an anti-derivative of $f(x)$, then $F(x) + c$, where c is an arbitrary constant, is also an anti-derivative of $f(x)$. We write

$$\int f(x)\, dx = F(x) + c \qquad (1.32)$$

We note that, not every function is integrable, for example the function $f(x)$ defined on $[0, 1]$ as

$$f(x) = \begin{cases} 0, & x \text{ is rational} \\ 1, & x \text{ is irrational} \end{cases}$$

does not have an antiderivative and hence is not integrable. We note the following:

(a) Every function which is continuous on a closed and bounded interval is integrable.

(b) For integrability, the condition that $f(x)$ is continuous on $[a, b]$ can be relaxed. The function $f(x)$ may only be piecewise continuous on $[a, b]$.

(c) Let m and M be the minimum and maximum values of $f(x)$ on $[a, b]$. Then,

$$m(b-a) \le \int_a^b f(x)\, dx \le M(b-a). \qquad (1.33)$$

(d) $$\int_a^b f(x)\,dx = (b-a)\,f(\xi), \quad a < \xi, b \tag{1.34}$$

(*mean value theorem of integrals*).

(e) If $f(x)$ is bounded and integrable on $[a, b]$, then $|f(x)|$ is also bounded and integrable on $[a, b]$, and

$$\left| \int_a^b f(x)\,dx \right| \le \int_a^b |f(x)|\,dx. \tag{1.35}$$

(f) The *average value* of an integrable function $f(x)$ defined on $[a, b]$ is given by $\dfrac{1}{(b-a)} \displaystyle\int_a^b f(x)\,dx$.

(g) Let $f(x)$ and $g(x)$ be integrable functions on $[a, b]$ and let $f(x) \le g(x)$, $a \le x \le b$. Then,

$$\int_a^b f(x)\,dx \le \int_a^b g(x)\,dx. \tag{1.36}$$

(h) **(First fundamental theorem of integrals)** Let $f(x)$ be a continuous function on a closed and bounded interval $[a, b]$. Then the function

$$F(x) = \int_a^x f(t)\,dt$$

is continuous on $[a, b]$, differentiable in (a, b) and $F'(x) = f(x)$.

(i) **(Second fundamental theorem of integrals, Newton-Leibniz formula)** Let $F(x)$ be an anti-derivative of a continuous function $f(x)$ on $[a, b]$. Then,

$$\int_a^b f(x) = F(b) - F(a). \tag{1.37}$$

Definite integrals have many applications. In particular, they can be used to find the (i) areas of bounded regions, (ii) lengths of curves, (iii) volumes of solids, (iv) areas of surfaces of revolution etc.

1.3.1   Areas of Bounded Regions

C_1 The area of the region bounded by the curve $y = f(x)$, the x-axis and the lines $x = a$, $x = b$ is given by (Fig. 1.3)

$$\text{Area} = \int_a^b y\,dx = \int_a^b f(x)\,dx. \tag{1.38}$$

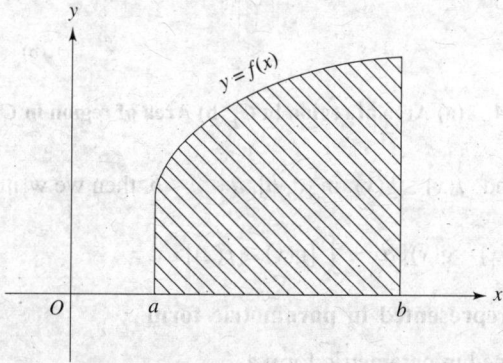

Fig. 1.3.   Area of region in C_1.

We note the following:

(i) If the curve $y = f(x)$, $a \leq x \leq b$ is above the x-axis, then the value of the integral given in Eq. (1.38) is positive. If the curve $y = f(x)$, $a \leq x \leq b$ is below the x-axis, then the value of the integral given in Eq. (1.38) is negative. Since area is a positive quantity, we take the magnitude of this value.

(ii) If the curve $y = f(x)$ is above the x-axis in the interval $a \leq x \leq c$ and below the x-axis in the interval $c \leq x \leq b$, then we write

$$\text{Area} = \int_a^b f(x)\, dx + \left| \int_c^b f(x)dx \right|.$$

C_2 The area of the region bounded by the curve $x = \phi(y)$, the y-axis and the lines $y = c$, $y = d$ is given by (Fig. 4a)

$$\text{Area} = \int_a^b x\, dy = \int_a^b \phi(y)\, dy. \tag{1.39}$$

C_3 The area of the region enclosed between the curves $y = f(x)$, $y = g(x)$ and the lines $x = a$, $x = b$ is given by (Fig. 4b)

$$\text{Area} = \int_a^b [f(x) - g(x)]\, dx, \quad \text{where} \quad f(x) \geq g(x) \text{ in } [a, b]. \tag{1.40}$$

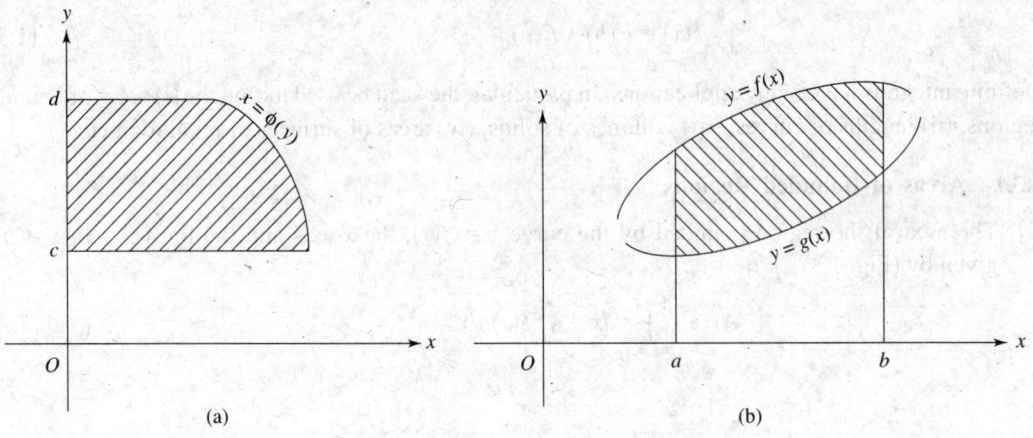

Fig. 1.4.   (a) Area of region in C_2 (b) Area of region in C_3.

C_4 If $f(x) \geq g(x)$ in $[a, c]$ and $f(x) \leq g(x)$ in $[c, b]$, $a < c < b$, then we write the area as (Fig. 1.5)

$$\text{Area} = \int_a^c [f(x) - g(x)]\, dx + \int_c^b [g(x) - f(x)]\, dx. \tag{1.41}$$

Area bounded by a curve represented in parametric form

Let the curve $y = f(x)$ be defined in parametric form as

$$x = \phi(t), \quad y = \psi(t), \quad a \leq t \leq b$$

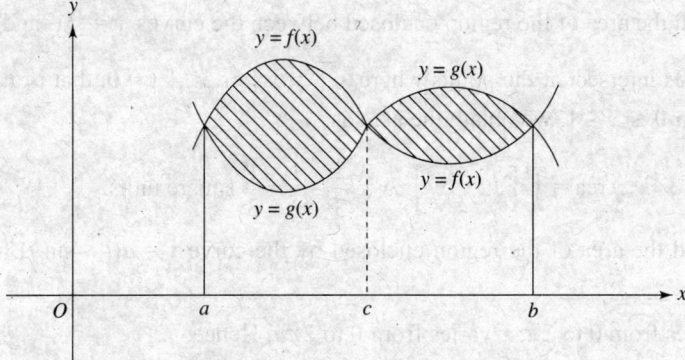

Fig. 1.5. Area of region in C_4.

where $\phi(t)$ and $\psi(t)$ are continuous functions of t in the interval $[a, b]$. Let $x_0 = \phi(a)$ and $x_1 = \phi(b)$. Then, from Eq. (1.38), the area is given by

$$\text{Area} = \int_{x_0}^{x_1} y\,dx = \int_a^b \psi(t)\,\phi'(t)\,dt. \tag{1.42}$$

Area of a sector

Let the curve be defined in polar form as

$$r = f(\theta), \quad \alpha \leq \theta \leq \beta \tag{1.43}$$

where $f(\theta)$ is a continuous function in $[\alpha, \beta]$. Let A be the area of the sector bounded by the curve and the radial lines $\theta = \alpha$ and $\theta = \beta$ (Fig. 1.6). Let the angle AOB be divided into n parts.

In an element area, we approximate area of the sector OPQ, by the area of the triangle OPN, with base $PN = r \sin \Delta\theta_i \approx r\Delta\theta_i$ and height $ON = OP = r$. (PN is perpendicular to OQ). Then, $dA_i = \dfrac{1}{2}\,r^2\,\Delta\theta_i$ and area of sector $AOB = \displaystyle\sum_{i=1}^{n} dA_i = \sum_{i=1}^{n}\frac{1}{2}r^2\Delta\theta_i$. Let $n \to \infty$ such that $\max(\Delta x_i) \to 0$. In the limit, we obtain

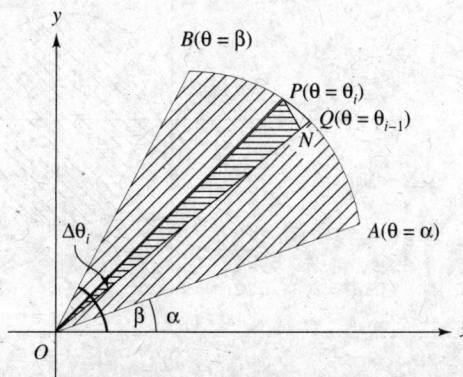

Fig. 1.6. Area of a sector.

$$\text{Area} = A = \frac{1}{2}\int_\alpha^\beta r^2 d\theta. \tag{1.44}$$

Example 1.22   Find the area of the region enclosed between the curves $y = \sqrt{x}$ and $y = x^2$.

Solution   The curves intersect at the points where $\sqrt{x} = x^2$, or $x^4 - x = 0$, that is at $x = 0$ and $x = 1$. Since $\sqrt{x} \geq x^2$ when $0 \leq x \leq 1$, we obtain the area as

$$\text{Area} = \int_0^1 \left[\sqrt{x} - x^2 \right] dx = \frac{2}{3} - \frac{1}{3} = \frac{1}{3} \text{ square units.}$$

Example 1.23   Find the area of the region enclosed by the curve $x = a(t - \sin t)$, $y = a(1.- \cos t)$, $0 \leq t \leq 2\pi$.

Solution   As t varies from 0 to 2π, x varies from 0 to $2\pi a$. Hence,

$$\text{Area} = \int_0^{2\pi} [a(1 - \cos t)] [a(1 - \cos t)] \, dt$$

$$= a^2 \int_0^{2\pi} (1 - 2\cos t + \cos^2 t) \, dt = a^2 \int_0^{2\pi} \left[1 - 2\cos t + \frac{1}{2}(1 + \cos 2t) \right] dt$$

$$= \frac{a^2}{2} \int_0^{2\pi} (3 - 4\cos t + \cos 2t) \, dt = \frac{a^2}{2} \left[3t - 4\sin t + \frac{1}{2}\sin 2t \right]_0^{2\pi}$$

$$= 3\pi a^2 \text{ square units.}$$

Example 1.24   Find the area of the region that lies inside the circle $r = a \cos \theta$ and outside the cardioid $r = a(1 - \cos \theta)$.

Solution   The region is given in Fig. 1.7. The curves intersect at $\theta = \pm\ \pi/3$. Let $r_1 = a \cos \theta$ and $r_2 = a(1 - \cos \theta)$. Therefore, the required area is given by

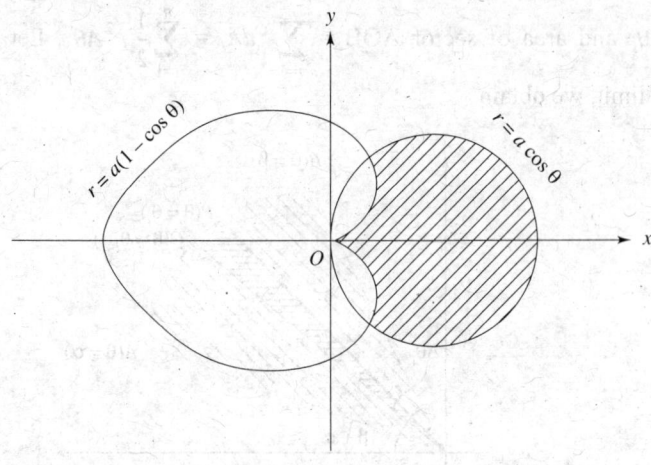

Fig. 1.7.   Example 1.24.

$$\text{Area} = \frac{1}{2} \int_{-\pi/3}^{\pi/3} (r_1^2 - r_2^2)\, d\theta = \frac{1}{2}\, a^2 \int_{-\pi/3}^{\pi/3} [\cos^2 \theta - (1 - \cos \theta)^2]\, d\theta$$

$$= a^2 \int_0^{\pi/3} (2 \cos \theta - 1)\, d\theta = a^2\, [2 \sin \theta - \theta]_0^{\pi/3} = a^2 \left[2\left(\frac{\sqrt{3}}{2}\right) - \frac{\pi}{3} \right]$$

$$= \frac{a^2}{3} [3\sqrt{3} - \pi] \text{ square units.}$$

1.3.2   Arc Length of a Plane Curve

Consider a portion of the curve $y = f(x)$ between $A(a, f(a))$, $B(b, f(b))$. Let $f(x)$ be continuous and differentiable in $[a, b]$. Divide the arc AB into n parts and join the successive points of division by straight lines. Let the end points of a typical straight line PQ be $P(x, y)$ and $Q(x + \Delta x_k, y + \Delta y_k)$, (see Fig. 1.8). Approximate the length of the arc PQ by the length of the line PQ.

Length of arc PQ $\approx$ length of line PQ $= \sqrt{\Delta x_k^2 + \Delta y_k^2}$.

Total length of the curve $= s \approx$ sum of lengths of n straight lines $= \sum_{k=1}^{n} \sqrt{\Delta x_k^2 + \Delta y_k^2}$.

By Lagrange mean value theorem, there exists a point $R(c_k, d_k)$ between P and Q on the curve where the tangent to the curve is parallel to the chord PQ. That is, $f'(c_k) = \Delta y_k / \Delta x_k$.

Now,

$$\sum_{k=1}^{n} \sqrt{\Delta x_k^2 + \Delta y_k^2} = \sum_{k=1}^{n} \Delta x_k \sqrt{1 + (\Delta y_k / \Delta x_k)^2} = \sum_{k=1}^{n} \sqrt{1 + [f'(c_k)]^2}\, \Delta x_k$$

Taking the limit as $n \to \infty$, such that (max $\Delta x_k) \to 0$, we obtain the length of the arc of the curve between $x = a$ and $x = b$ as

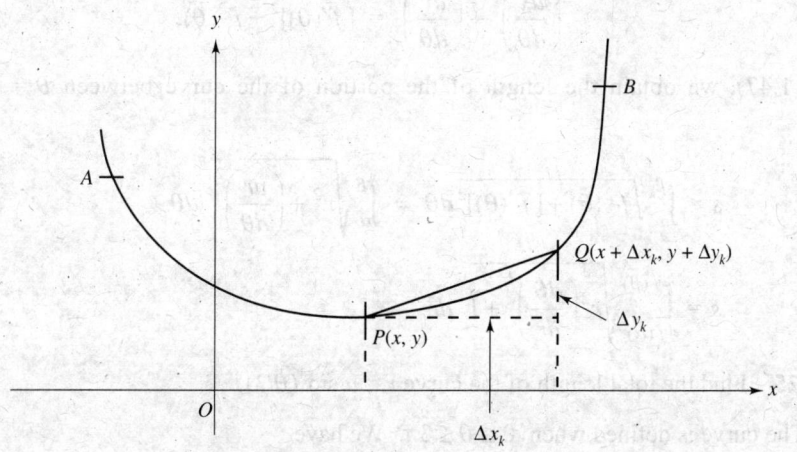

Fig. 1.8.   Length of a curve.

$$s = \int_a^b \sqrt{1 + \left(\frac{dy}{dx}\right)^2}\, dx. \tag{1.45}$$

If the curve is defined by $x = \phi(y),\ c \le y \le d,$ then the length of the arc is given by

$$s = \int_c^d \sqrt{1 + \left(\frac{dx}{dy}\right)^2}\, dy. \tag{1.46}$$

Arc length of a curve represented in parametric form

Let the parametric form of the curve be given by

$$x = \phi(t),\quad y = \psi(t),\quad t_0 \le t \le t_1$$

where $\phi(t)$ and $\psi(t)$ are continuously differentiable functions on $[t_0, t_1]$. If $\phi(t_0) = a$ and $\phi(t_1) = b$, then from Eq. (1.45), the arc length is given by

$$s = \int_a^b \sqrt{1 + \left(\frac{dy}{dx}\right)^2}\, dx = \int_{t_0}^{t_1} \sqrt{1 + \left(\frac{dy/dt}{dx/dt}\right)^2}\, \frac{dx}{dt}\, dt = \int_{t_0}^{t_1} \sqrt{\left(\frac{dx}{dt}\right)^2 + \left(\frac{dy}{dt}\right)^2}\, dt. \tag{1.47}$$

Arc length of a curve represented in polar form

Consider the portion of the curve defined by $r = f(\theta),\ \alpha \le \theta \le \beta,$ where $f(\theta)$ is continuous and differentiable on $[\alpha, \beta]$. The curve can be represented in parametric form as

$$x = r \cos\theta = f(\theta) \cos\theta,\quad y = r \sin\theta = f(\theta) \sin\theta,\quad \alpha \le \theta \le \beta.$$

Therefore,

$$\frac{dx}{d\theta} = f'(\theta) \cos\theta - f(\theta) \sin\theta,\quad \frac{dy}{d\theta} = f'(\theta) \sin\theta + f(\theta) \cos\theta,$$

and

$$\left(\frac{dx}{d\theta}\right)^2 + \left(\frac{dy}{d\theta}\right)^2 = [f'(\theta)]^2 + f^2(\theta).$$

Using Eq. (1.47), we obtain the length of the portion of the curve between $\theta = \alpha$ and $\theta = \beta$ as

$$s = \int_\alpha^\beta \sqrt{f^2(\theta) + [f'(\theta)]^2}\, d\theta = \int_\alpha^\beta \sqrt{r^2 + \left(\frac{dr}{d\theta}\right)^2}\, d\theta \tag{1.48}$$

or

$$s = \int_{f(\alpha)}^{f(\beta)} \sqrt{r^2 \left(\frac{d\theta}{dr}\right)^2 + 1}\; dr. \tag{1.49}$$

Example 1.25 Find the total length of the curve $r = a \sin^3(\theta/3)$.

Solution The curve is defined when $0 \le \theta \le 3\pi$. We have

$$f(\theta) = a \sin^3\left(\frac{\theta}{3}\right), \quad f'(\theta) = a \sin^2\left(\frac{\theta}{3}\right) \cos\left(\frac{\theta}{3}\right) \quad \text{and} \quad f^2(\theta) + [f'(\theta)]^2 = a^2 \sin^4\left(\frac{\theta}{3}\right).$$

Therefore,

$$s = \int_0^{3\pi} \sqrt{f^2(\theta) + [f'(\theta)]^2}\, d\theta = a \int_0^{3\pi} \sin^2\left(\frac{\theta}{3}\right) d\theta = 3a \int_0^{\pi} \sin^2 \phi\, d\phi$$

where $\theta = 3\phi$. Integrating, we obtain

$$s = \frac{3a}{2} \int_0^{\pi} (1 - \cos 2\phi)\, d\phi = \frac{3a}{2}\left[\phi - \frac{1}{2} \sin 2\phi\right]_0^{\pi} = \frac{3a\pi}{2}.$$

1.3.3 Volume of Solids

In this section we discuss methods for finding the volume of solids.

Method of slicing

Let a solid be bounded by two parallel planes $x = a$ and $x = b$ (Fig. 1.9). Divide the interval $[a, b]$ for x into n subintervals $[x_0, x_1], [x_1, x_2], \ldots, [x_{n-1}, x_n]$, where $a = x_0 < x_1 < x_2 \ldots < x_n = b$.

Let $\Delta x_k = x_k - x_{k-1}, k = 1, 2, \ldots, n$. Draw the planes $x = x_0, x = x_1, \ldots, x = x_n$. This will cut the solid into slices of thickness Δx_k. We now approximate the volume of the sliced solid part S_k by the volume of a cylinder with base as a cross section of the sliced solid S_k and the height as Δx_k. Therefore, an approximation to the volume of the sliced solid between $x = x_{k-1}$ and $x = x_k$ is given by

$$V_k = \text{Area of base} \times \text{height} = A(\xi_k)\, \Delta x_k, \quad x_{k-1} < \xi_k \leq x_k$$

where $A(\xi_k)$ is the cross-sectional area of the sliced solid. Now, consider the sum of all the approximate volumes of the sliced solids. We obtain

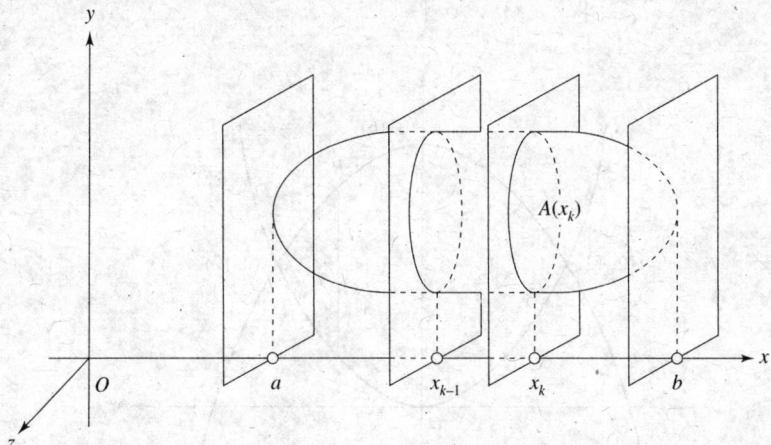

Fig. 1.9.   Method of slicing.

$$V_n = \sum_{k=1}^{n} V_k = \sum_{k=1}^{n} A(\xi_k)\, \Delta x_k.$$

Let $n \to \infty$ such that max $\Delta x_k \to 0$. In the limit $V_n \to V$, volume of the solid and the summation reduces to an integral. Therefore, volume of the solid is given by

$$V = \int_a^b A(x)\, dx. \tag{1.50}$$

If the solid is bounded by the planes $y = c$ and $y = d$, then volume of the solid can be written as

$$V = \int_c^d A(y)\, dy$$

where $A(y)$ is the cross-sectional area.

Example 1.26   The cross sections of a certain solid made by planes perpendicular to the x-axis are circles with diameters extending from the curve $y = 3x^2$ to the curve $y = 16 - x^2$. Find the volume of the solid which lies between the points of intersection of these curves.

Solution   At the points of intersection of the curves, we have $3x^2 = 16 - x^2$, or $x^2 = 4$, that is $x = \pm 2$. Therefore, the points of intersection of the curves are $(-2, 12)$ and $(2, 12)$ (Fig. 1.10).
Any point on the curve $y = 16 - x^2$ is $R(x, 16 - x^2)$.
Any point on the curve $y = 3x^2$ is $S(x, 3x^2)$.
Diameter of the circle $= RS = 16 - 4x^2$.

$$\text{Area of the circle} = A(x) = \frac{\pi}{4}\,(RS)^2 = 4\pi(4 - x^2)^2.$$

Since the solid is symmetric about the y-axis, the required volume is obtained as

$$V = 2\int_0^2 A(x)\, dx = 8\pi \int_0^2 (4 - x^2)^2\, dx = 8\pi \left[16x - \frac{8}{3}x^3 + \frac{x^5}{5} \right]_0^2$$

$$= 8\pi \left[32 - \frac{64}{3} + \frac{32}{5} \right] = \frac{2048}{15}\,\pi \text{ cubic units.}$$

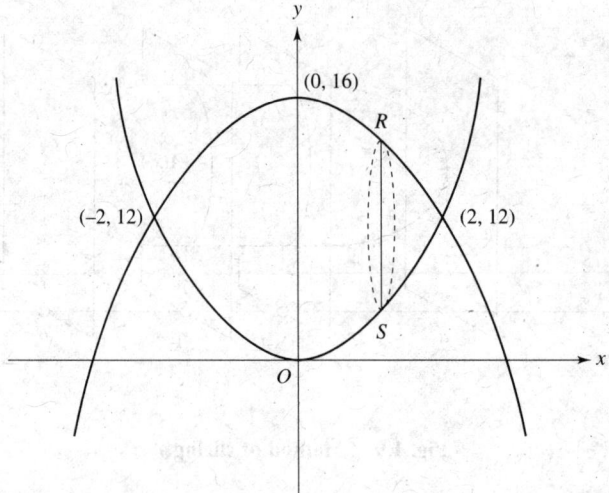

Fig. 1.10.   Problem 1.26.

Volume of a solid of revolution

Let AB be the portion of a curve $y = f(x)$, $f(x) > 0$, between $x = a$ and $x = b$. Consider the area bounded by the arc AB of the curve $y = f(x)$, the x-axis, and the lines $x = a$ and $x = b$. A solid is generated by revolving this area about the x-axis (Fig. 1.11).

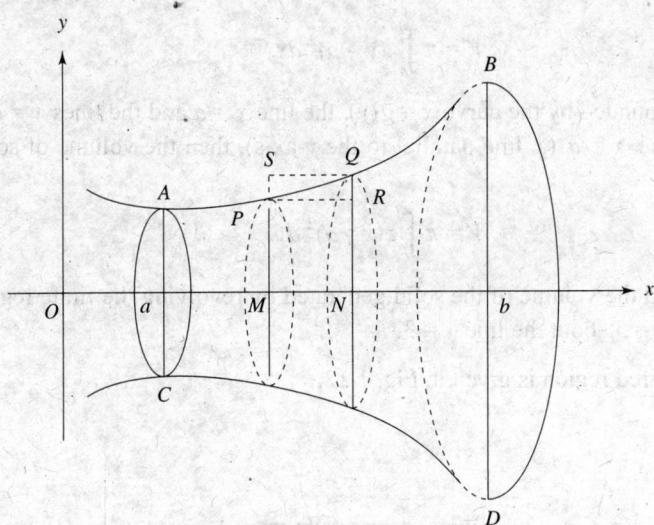

Fig. 1.11. Solid of revolution.

Divide the arc AB into n parts by considering the subintervals $[x_0, x_1]$, $[x_1, x_2]$, ..., $[x_{n-1}, x_n]$, where $a = x_0 < x_1 < x_2 ... < x_n = b$. Let $\Delta x_k = x_k - x_{k-1}$, $k = 1, 2, ..., n$. Consider a typical subinterval $[x_{k-1}, x_k]$ of length $\Delta x_k = MN$. A solid is generated by rotating the area $MNQP$ about the x-axis. The volume V_k of this solid lies in magnitude between the volumes generated by revolving the areas $MNRP$ and $MNQS$ about the x-axis. Now,

$$MP = y_{k-1} = f(x_{k-1}) \quad \text{and} \quad NQ = y_k = f(x_k).$$

Hence, the volume V_k of the typical solid is bounded as

$$\pi y_{k-1}^2 \, \Delta x_k \leq V_k \leq \pi y_k^2 \, \Delta x_k.$$

Adding the inequalities corresponding to all the subintervals, we get

$$\pi \sum_{k=1}^{n} y_{k-1}^2 \, \Delta x_k \leq \sum_{k=1}^{n} V_k \leq \pi \sum_{k=1}^{n} y_k^2 \, \Delta x_k.$$

Let $n \to \infty$ such that max $\Delta x_k \to 0$. In the limit, we obtain the volume of the solid of revolution as

$$V = \int_a^b \pi y^2 \, dx. \tag{1.51}$$

Smilarly, if the area bounded by the arc AB of the curve $x = \phi(y)$, the y-axis, and the lines $y = c$ and $y = d$ is revolved about the y-axis, then the volume of the solid of revolution can be written as

$$V = \int_c^d \pi x^2 \, dy. \tag{1.52}$$

Remark 5

(a) If the area bounded by the curve $y = f(x)$, the line $y = p$ and the lines $x = a$, $x = b$ is revolved about the line $y = p$ (a line parallel to the x-axis), then the volume of the solid of revolution is given by

$$V = \pi \int_a^b (y - p)^2 \, dx. \tag{1.53}$$

(b) If the area bounded by the curve $x = g(y)$, the line $x = q$ and the lines $y = c$, $y = d$ is revolved about the line $x = q$ (a line parallel to the y-axis), then the volume of solid of revolution is given by

$$V = \pi \int_c^d (x - q)^2 \, dy. \tag{1.54}$$

Example 1.27   Find the volume of the solid generated by revolving the finite region bounded by the curves $y = x^2 + 1$, $y = 5$ about the line $x = 3$.

Solution   The required region is given in Fig. 1.12.

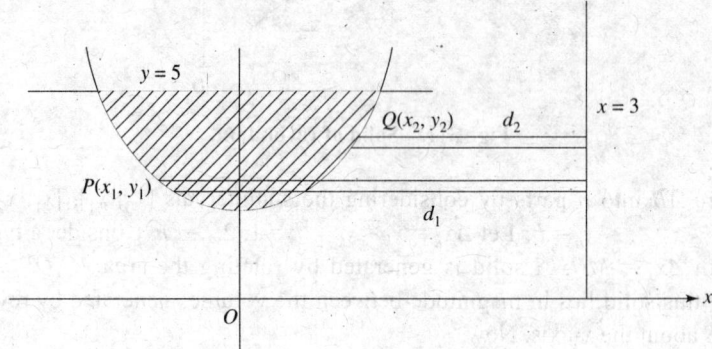

Fig. 1.12.   Region of revolution in Example 1.27.

The volume is given by

$$V = \pi \int_1^5 (d_1^2 - d_2^2) \, dy = \pi \int_1^5 [(3 + \sqrt{y - 1})^2 - (3 - \sqrt{y - 1})^2] \, dy$$

$$= 12\,\pi \int_1^5 \sqrt{y - 1} \, dy = 12\pi \left(\frac{2}{3}\right) \left[(y - 1)^{3/2}\right]_1^5 = 8\pi(8) = 64\,\pi \text{ cubic units.}$$

Example 1.28   Find the volume of the solid generated by revolving the region bounded by the curves $y = 3 - x^2$ and $y = -1$ about the line $y = -1$.

Solution   The required region is given in Fig. 1.13. The region PAQ is revolved about the line $y = -1$. Since the region is symmetric about the y-axis, the volume is obtained as

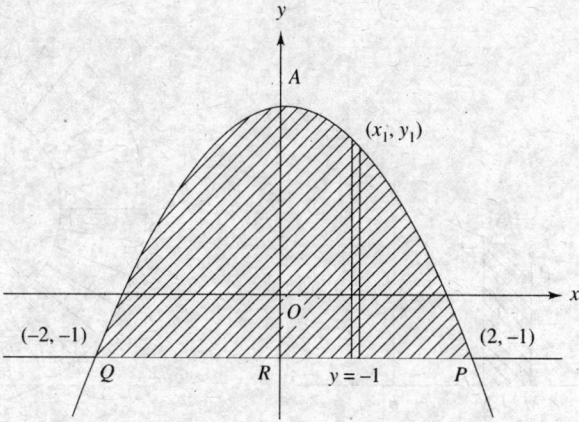

Fig. 1.13.   Example 1.28.

$V = 2$[volume of the solid obtained by revolving the region *PAR* about the line $y = -1$]

$$= 2\pi \int_0^2 (1 + y)^2 \, dx = 2\pi \int_0^2 (1 + 3 - x^2)^2 \, dx$$

$$= 2\pi \int_0^2 (16 - 8x^2 + x^4) \, dx = 2\pi \left[16x - \frac{8x^3}{3} + \frac{x^5}{5} \right]_0^2 = \frac{512}{15} \pi \text{ cubic units.}$$

Example 12.9   Find the volume of the solid generated by revolving an arch of the cycloid $x = a(t - \sin t)$, $y = a(1 - \cos t)$ and x-axis about the x-axis.

Solution   Setting $y = 0$, we obtain $\cos t = 1$, or $t = 0$, and $t = 2\pi$. Hence, one arch of the cycloid intersects the x-axis at the points $(0, 0)$ and $(2\pi a, 0)$. Therefore, the required volume is given by

$$V = \pi \int_0^{2\pi a} y^2 dx = \pi \int_0^{2\pi} a^2 (1 - \cos t)^2 \left[a \, (1 - \cos t) \right] dt$$

$$= \pi a^3 \int_0^{2\pi} 8 \sin^6 \left(\frac{t}{2} \right) dt = 16\pi a^3 \int_0^{\pi} \sin^6 T \, dT$$

$$= 32 \, \pi a^3 \int_0^{\pi/2} \sin^6 T \, dT = 32 \, \pi a^3 \left[\frac{5}{6} \cdot \frac{3}{4} \cdot \frac{1}{2} \cdot \frac{\pi}{2} \right] = 5\pi^2 \, a^3 \text{ cubic units.}$$

Volume of solid of revolution by the method of cylindrical shells

Suppose that a region in the x-y plane bounded by the curve $y = f(x)$, the x-axis and lines $x = a$, $x = b$ is revolved about the y-axis. Divide the interval $[a, b]$ into n subintervals $[x_0, x_1]$, $[x_1, x_2]$, ..., $[x_{n-1}, x_n]$, where $a = x_0 < x_1 < x_2 ... < x_n = b$. Let $\Delta x_k = x_k - x_{k-1}$, $k = 1, 2, ..., n$. Consider a typical region *MNQP* (Fig. 1.14a), where $MN = \Delta x_k$ and the coordinates of *M*, *N* are $M(x_{k-1}, 0)$ and $N(x_k, 0)$. When we revolve this strip about the y-axis, it generates a hollow thin walled shell of inner radius x_{k-1} and outer radius x_k and volume ΔV_k. The base of this shell is a ring bounded by the concentric circles with inner radius x_{k-1} and outer radius $x_k = x_{k-1} + \Delta x_k$. A cross-section of this solid is given in Fig. 1.14b.

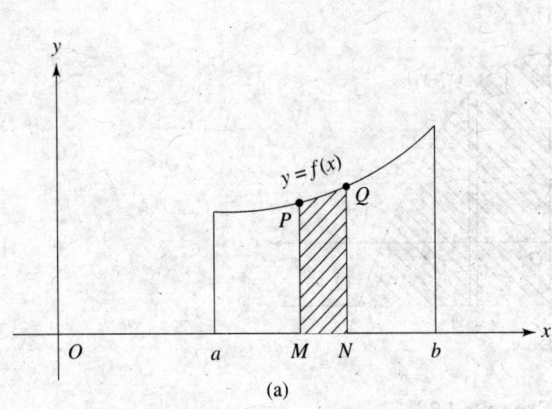

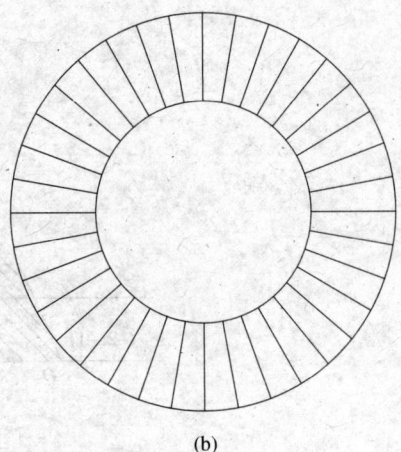

Fig. 1.14.   (a) Region of revolution, (b) Cross-section of solid of revolution.

The area of this ring is given by

$$\Delta A_k = \pi x_k^2 - \pi x_{k-1}^2 = \pi(x_k + x_{k-1})(x_k - x_{k-1}) = 2\pi \xi_k \, \Delta x_k$$

where $\xi_k = (x_k + x_{k-1})/2$ is the radius of the circle midway between the inner and outer boundaries of the ring and $2\pi \xi_k$ is its circumference. Now, if we take a cylindrical shell of constant height y standing on this base, we obtain volume as $\Delta V_k = (\Delta A_k)y$. Since $f(x)$ is continuous, y can take any value between the minimum and maximum values of $f(x)$ on $[x_{k-1}, x_k]$. If we take $y = f(\eta_k)$, $x_{k-1} \le \eta_k \le x_k$, then we can write approximately the volume of the shell as

$$\Delta V_k = (2\pi \xi_k) f(\eta_k) \, \Delta x_k, \quad k = 1, 2, \ldots, n.$$

Adding the volumes corresponding to all the subintervals, we obtain

$$V_n = \sum_{k=1}^{n} \Delta V_k = \sum_{k=1}^{n} (2\pi \xi_k) f(\eta_k) \, \Delta x_k.$$

Let $n \to \infty$ such that max $\Delta x_k \to 0$. In the limit, we obtain the volume of the solid as

$$V = \int_a^b 2\pi x f(x) \, dx. \tag{1.55}$$

If the region given in Fig. 1.15 is revolved about the x-axis, the volume of the solid of revolution is obtained as

$$V = \int_c^d 2\pi y g(y) \, dy \tag{1.56}$$

where $x = g(y)$ is the equation of the bounding curve $APQB$.

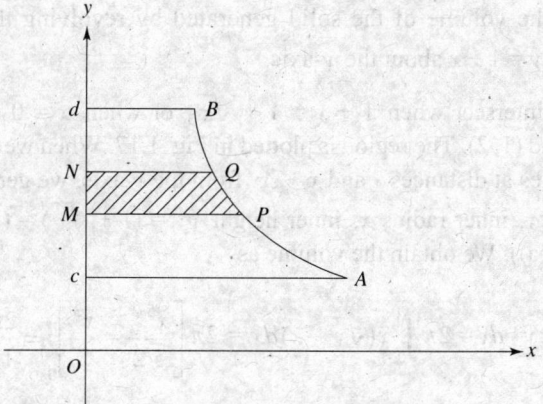

Fig. 1.15. Region of revolution.

Example 1.30 Find the volume of the solid generated by revolving the region bounded by $y = \sqrt{x}$, $y = 0$ and $x = 9$ about the y-axis.

Solution The region is plotted in Fig. 1.16. When we revolve the vertical strip of the area between the lines at distances x and $x + \Delta x$ from the y-axis, we generate a cylindrical shell of inner circumference $2\pi x$, inner radius x, inner height y and wall thickness Δx (Fig. 1.16(a)). We obtain the volume as

$$V = \int_0^9 2\pi xy \, dx = 2\pi \int_0^9 x\sqrt{x} \, dx = 2\pi \left(\frac{2}{5}\right) \left[x^{5/2}\right]_0^9 = \frac{972\pi}{5} \text{ cubic units.}$$

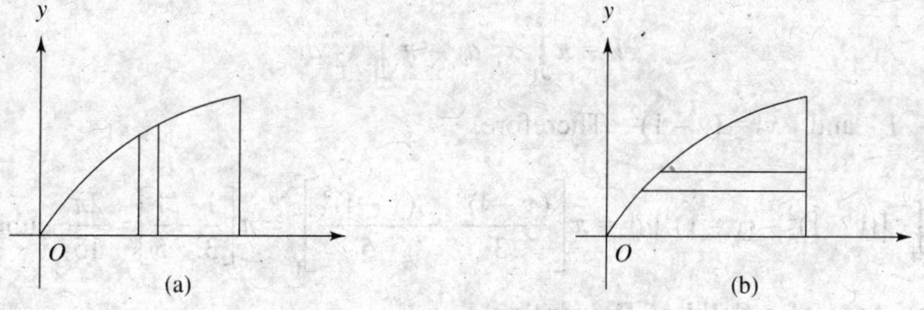

Fig. 1.16. Example 1.30.

Alternative If we revolve the horizontal strip about the y-axis (Fig. 1.16(b)), we obtain the volume as

$$V = \int_0^3 \pi x_2^2 \, dy - \int_0^3 \pi x_1^2 \, dy$$

where $x_2 = 9$, $x_1 = y^2$. Therefore,

$$V = \pi \int_0^3 81 dy - \pi \int_0^3 y^4 dy = \pi \left[243 - \frac{243}{5}\right] = \frac{972\pi}{5} \text{ cubic units.}$$

Example 1.31   Find the volume of the solid generated by revolving the region bounded by the curves $y = 1 + \sqrt{x}$ and $y = 1 + x$ about the y-axis.

Solution   The curves intersect when $1 + x = 1 + \sqrt{x}$, or when $x = 0$ and $x = 1$. The points of intersection are $(0, 1)$ and $(1, 2)$. The region is plotted in Fig. 1.17. When we revolve the vertical strip of the area between the lines at distances x and $x + \Delta x$ from the y-axis, we generate a cylindrical shell of inner circumference $2\pi x$, inner radius x, inner height $y^* = (1 + \sqrt{x}) - (1 + x) = \sqrt{x} - x$ and wall thickness Δx (Fig. 1.17(a)). We obtain the volume as

$$V = \int_0^1 2\pi x y^* \, dx = 2\pi \int_0^1 x(\sqrt{x} - x) \, dx = 2\pi \left[\frac{x^{5/2}}{5/2} - \frac{x^3}{3} \right]_0^1 = \frac{2\pi}{15} \text{ cubic units.}$$

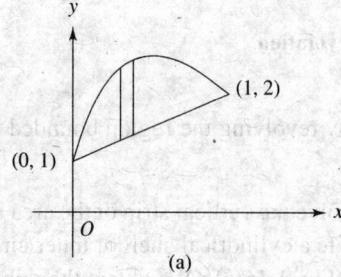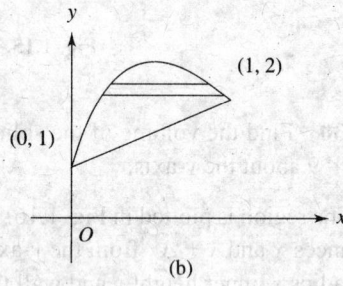

Fig. 1.17.   Example 1.31.

Alternative   If we revolve the horizontal strip about the y-axis (Fig. 1.17(b)), we obtain the volume as

$$V = \pi \int_1^2 x_1^2 \, dy - \pi \int_1^2 x_2^2 \, dy$$

where $x_1 = y - 1$ and $x_2 = (y - 1)^2$. Therefore,

$$V = \pi \int_1^2 [(y - 1)^2 - (y - 1)^4] dy = \pi \left[\frac{(y - 1)^3}{3} - \frac{(y - 1)^5}{5} \right]_1^2 = \pi \left[\frac{1}{3} - \frac{1}{5} \right] = \frac{2\pi}{15} \text{ cubic units.}$$

1.3.4   Surface Area of a Solid of Revolution

Let $y = f(x)$, $f(x) \geq 0$ between $x = a$ and $x = b$ define a curve. Let this curve be revolved about the x-axis to generate a surface S (Fig. 1.18). Divide the interval $[a, b]$ into n subintervals $[x_0, x_1]$, $[x_1, x_2]$, ..., $[x_{n-1}, x_n]$, where, $a = x_0 < x_1 < x_2 \ldots < x_n = b$. Let $\Delta x_k = x_k - x_{k-1}$ and $\Delta y_k = y_k - y_{k-1}$ $= f(x_k) - f(x_{k-1})$, $k = 1, 2, \ldots, n$. Consider the portion of the curve PQ in the interval $[x_{k-1}, x_k]$. Let S_k be the surface generated by revolving this portion of the curve about the x-axis. In this interval $[x_{k-1}, x_k]$, we approximate arc $(PQ) \approx$ chord (PQ). If we revolve the chord PQ about the x-axis, we obtain a frustum of a cone (Fig. 1.19). Now, the area of the surface S_k is approximated by the area of the surface of the frustum of the cone. We have

$$PM = y_{k-1}, \quad QN = y_k, \quad PQ = l = \sqrt{(\Delta x_k)^2 + (\Delta y_k)^2}$$

and

$$S_k \approx \pi (y_{k-1} + y_k) \, l = \pi (y_{k-1} + y_k) \sqrt{(\Delta x_k)^2 + (\Delta y_k)^2}$$

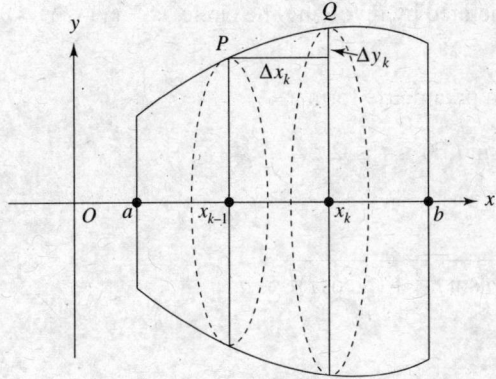

Fig. 1.18.   **Surface of revolution.**

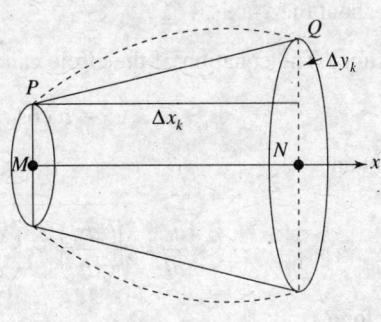

Fig. 1.19.   **Surface of revolution.**

$$= \pi(y_{k-1} + y_k) \sqrt{1 + \left(\frac{\Delta y_k}{\Delta x_k}\right)^2} \; \Delta x_k.$$

Adding the approximations corresponding to each of the subintervals $[x_{k-1}, x_k]$, $k = 1, 2, \ldots, n$, we obtain

$$S \approx \sum_{k=1}^{n} S_k = \sum_{k=1}^{n} \pi(y_{k-1} + y_k) \sqrt{1 + \left(\frac{\Delta y_k}{\Delta x_k}\right)^2} \; \Delta x_k.$$

Let $n \to \infty$ such that max $\Delta x_k \to 0$. In the limit, we obtain the surface area of the solid of revolution as

$$S = \int_a^b 2\pi y \sqrt{1 + \left(\frac{dy}{dx}\right)^2} \; dx = \int_a^b 2\pi y \; ds \tag{1.57}$$

where, $ds = \sqrt{1 + (dy/dx)^2} \; dx$.

If the given region is revolved about the y-axis, then we obtain the surface area as

$$S = \int_c^d 2\pi x \sqrt{1 + \left(\frac{dx}{dy}\right)^2} \; dy = \int_c^d 2\pi x \, ds \tag{1.58}$$

where, $ds = \sqrt{1 + (dx/dy)^2} \; dy$

If the curve is given in parametric form as $x = \phi(t)$, $y = \psi(t)$, $t_0 \le t \le t_1$, then we have

$$ds = \sqrt{\left(\frac{dx}{dt}\right)^2 + \left(\frac{dy}{dt}\right)^2} \; dt. \tag{1.59}$$

If the curve is given in polar form $r = f(\theta)$, $\theta_0 \le \theta \le \theta_1$, then we have

$$ds = \sqrt{r^2 + \left(\frac{dr}{d\theta}\right)^2} \; d\theta. \tag{1.60}$$

Example 1.32   Find the surface area of the solid generated by revolving the circle $x^2 + (y - b)^2 = a^2$, $b \geq a$ about the x-axis.

Solution   The equation of the circle can be written in parametric form as

$$x = a \cos t, \ y = b + a \sin t, \ 0 \leq t \leq 2\pi.$$

We obtain

$$\frac{ds}{dt} = \sqrt{\left(\frac{dx}{dt}\right)^2 + \left(\frac{dy}{dt}\right)^2} = \sqrt{(-a \sin t)^2 + (a \cos t)^2} = a.$$

Therefore,

$$S = \int_0^{2\pi} 2\pi (b + a \sin t) \, a \, dt = 2\pi a \left[bt - a \cos t\right]_0^{2\pi} = 4\pi^2 ab \text{ square units.}$$

Example 1.33   The part of the lemniscate $r^2 = 2a^2 \cos 2\theta, \ 0 \leq \theta \leq \pi/4$, is revolved about the x-axis. Find the surface area of the solid generated.

Solution   We have $x = r \cos \theta$, and $y = r \sin \theta$. We find that

$$ds^2 = r^2 + \left(\frac{dr}{d\theta}\right)^2 = r^2 + \frac{4a^4 \sin^2 2\theta}{r^2} = \frac{1}{r^2}[4a^4 \cos^2 2\theta + 4a^4 \sin^2 2\theta] = \frac{4a^4}{r^2}.$$

Therefore,

$$s = \int_0^{\pi/4} 2\pi y \, ds = \int_0^{\pi/4} 2\pi r \sin \theta \left(\frac{2a^2}{r}\right) d\theta$$

$$= 4\pi a^2 \left[-\cos \theta\right]_0^{\pi/4} = 4\pi a^2 \left[1 - \frac{1}{\sqrt{2}}\right] = 2\pi a^2 (2 - \sqrt{2}) \text{ square units.}$$

Example 1.34   The line segment $x = \sin^2 t, \ y = \cos^2 t, 0 \leq t \leq \pi/2$ is revolved about the y-axis. Find the surface area of the solid generated.

Solution   The surface area is given by

$$S = \int_c^d 2\pi x \sqrt{1 + \left(\frac{dx}{dy}\right)^2} \, dy.$$

We have

$$\frac{dx}{dt} = 2 \sin t \cos t, \quad \frac{dy}{dt} = -2 \cos t \sin t \quad \text{and} \quad \frac{dx}{dy} = -1.$$

As t varies from 0 to $\pi/2$, y varies from 1 to 0. Taking the anti-clockwise direction as the positive direction, we get

$$S = \int_0^1 2\sqrt{2}\,\pi x\,dy = -\int_0^{\pi/2} 2\sqrt{2}\,\pi\,\sin^2 t\,(-2\sin t\cos t)\,dt$$

$$= 4\sqrt{2}\,\pi\int_0^{\pi/2}\sin^3 t\cos t\,dt = 4\sqrt{2}\,\pi\left[\frac{1}{4}\sin^4 t\right]_0^{\pi/2} = \pi\sqrt{2}\ \text{square units.}$$

Exercise 1.2

In problems **1** to **9**, find the area of the region bounded by the given curves.

1. $y = x^2 - 5x + 6$, the x-axis and the lines $x = 0$, $x = 3$.

2. $y = x$, $y = \sqrt{x}$ and the lines $x = 0$, $x = 1$.   **3.** $y^2 = x + 1$ and $y = x + 1$.

4. $\dfrac{x^2}{a^2} + \dfrac{y^2}{b^2} = 1$ and $\dfrac{x^2}{b^2} + \dfrac{y^2}{a^2} = 1, 0, < a < b$.   **5.** $\sqrt{x} + \sqrt{y} = 1$ and the coordinate axes.

6. $y = ex\ln x$ and $y = \ln x/(ex)$.

7. $x = 2t + 1$, $y = 4t^2 - 1$, $-1/2 \le t \le 1/2$ and the x-axis.

8. $x = a\cos^3 t$, $y = b\sin^3 t$, $0 \le t \le 2\pi$.   **9.** $r = a(2 - \cos 2\theta)$, $0 \le \theta \le 2\pi$.

10. Find the area of the region enclosed between the curve $y = 2x^4 - x^2$, the x-axis and the ordinates of the points where the curve has local minimum.

11. Find the area of a loop of the curve $x(x^2 + y^2) = a(x^2 - y^2)$.

12. Find the area of the region inside the curve $r^2 = 2a^2\cos 2\theta$ and outside the circle $r = a$.

13. Find the area that is inside the circle $r = a$ and outside the cardioid $r = a(1 - \cos\theta)$.

In problems **14** to **26**, find the length of the indicated portion of the curve.

14. $9x^2 = y^3$, from $x = 0$ to $x = 9$.

15. $x^{2/3} + y^{2/3} = a^{2/3}$, from $x = 0$ to $x = a$ in the first quardrant.

16. $x = \dfrac{1}{4}y^4 + \dfrac{1}{8}y^{-2}$, from $y = 1$ to $y = 2$.

17. $x^2 + y^2 - 2ax = 0$ and above the line $y = a/2$, $a > 0$.

18. $y = \int_0^x \sqrt{\cos t}\,dt$, from $x = 0$ to $x = \pi/2$.   **19.** $y = \ln[(e^x + 1)/(e^x - 1)]$, from $x = 1$ to $x = 2$.

20. $x = 3at^2$, $y = a(t - 3t^3)$, from $t = 0$ to $t = 1$.

21. $x = a(t - \sin t)$, $y = a(1 - \cos t)$, from $t = 0$ to $t = 2\pi$.

22. $x = e^{2t}\cos t$, $y = e^{2t}\sin t$, from $t = 0$ to $t = 1$.

23. $x = [\ln(a^2 + t^2)]/2$, $y = \tan^{-1}(t/a)$, from $t = 0$ to $t = a$.

24. $x = 2\cos t + \cos 2t + 1$, $y = 2\sin t + \sin 2t$, from $t = 0$ to $t = \pi$.

25. $r = a\theta$, from $r = r_1$ to $r = r_2$.

26. $r = ae^{b\theta}$, from $r = r_1$ to $r = r_2$.

27. Find the volume of the ellipsoid $\dfrac{x^2}{a^2} + \dfrac{y^2}{b^2} + \dfrac{z^2}{c^2} = 1$.

28. The base of a certain solid is the circle $x^2 + y^2 = a^2$. Each cross-section of the solid cut out by a plane perpendicular to the x-axis is a square with one side of the square in the base of the solid. Find the volume of the solid.

29. The base of a certain solid is the circle $x^2 + y^2 = a^2$. Each cross-section of the solid cut out by a plane perpendicular to the x-axis is an isosceles right triangle with one of the equal sides in the base of the solid. Find its volume.

30. The base of a certain solid is the region between the x-axis and the curve $y = \cos x$ between $x = 0$ and $x = \pi/2$. Each cross-section of the solid cut out by a plane perpendicular to the x-axis is an equilateral triangle with one side in the plane of the solid. Find its volume.

In problems **31** to **36**, find the volume of the solid of revolution generated by revolving the specified region about the given axis.

31. Region bounded by $y = \cos x$, $y = 0$ from $x = 0$ to $x = \pi/2$ about the x-axis.

32. Region bounded by $y = \sqrt{x}$, $y = 0$ from $x = 0$ to $x = 4$ about the x-axis.

33. Region bounded by $y = \sqrt{x}$, $y = 0$ from $x = 0$ to $x = 4$ about the line $y = 2$.

34. Region bounded by $y = x^2 + 1$ and $y = 3 - x$ about the x-axis.

35. Region bounded by $x = a \sin^3 t$, $y = a \cos^3 t$, $0 \le t \le \pi/2$, $x = 0$, $y = 0$ about the x-axis.

36. Region bounded by $x = 2t + 1$, $y = 4t^2 - 1$, $-1/2 \le t \le 0$, $y = 0$ about the line $x = 1$.

In problems **37** to **41**, use the method of cylindrical shells to find the volume of the solid generated by revolving the specified region about the given axis.

37. Region bounded by $y = x$, $y = 2$ and $x = 0$ about the y-axis.

38. Region bounded by $y = 2x - x^2$ and $y = x$ about the y-axis.

39. Region bounded by $y = x^2$ and $y = x$ about the y-axis.

40. Region inside the triangle with vertices at $(0, 0)$, $(a, 0)$ and $(0, b)$ about the y-axis.

41. Region inside the circle $x^2 + y^2 = a^2$ about the line $y = b$, $b > a > 0$.

In problems **42** to **50**, find the surface area of the solid generated by revolving the curve C about the given line.

42. $(x - b)^2 + y^2 = a^2$, $b \ge a$ about the y-axis.   43. $\dfrac{x^2}{a^2} + \dfrac{y^2}{b^2} = 1$, $a \ge b$, $y \ge 0$ about the x-axis.

44. $\dfrac{x^2}{a^2} + \dfrac{y^2}{b^2} = 1$, $a \ge b$, $x \ge 0$ about the y-axis.   45. $y = \dfrac{x^4}{4} + \dfrac{1}{8x^2}$, $1 \le x \le 2$ about the line $y = -1$.

46. $x = \dfrac{y^3}{3} + \dfrac{1}{4y}$, $1 \le y \le 2$ about the line $x = -1$.

47. $x = a(1t - \sin t)$, $y = a(1 - \cos t)$, $0 \le t \le 2\pi$ about the x-axis.

48. $x = a \cos^3 t$, $y = a \sin^3 t$, $0 \le t \le \pi/2$ about the x-axis.

49. $x = e^t \cos t$, $y = e^t \sin t$, $0 \le t \le \pi/2$ about the y-axis.

50. $r = a(1 + \cos \theta)$, $0 \le \theta \le \pi$ about the initial line.

1.4   Improper Integrals

While defining the definite integral $\displaystyle\int_a^b f(x)\, dx$, we had assumed that

   (i) a and b are finite constants.

   (ii) $f(x)$ is bounded for all x in $[a, b]$.

If in the above integral, (i) a or b or both a and b are infinite, or (ii) a, b are finite but $f(x)$ becomes infinite at $x = a$ or $x = b$ or at one or more points within the interval (a, b), then the definite integral is respectively called

(i) *improper integral of the first kind.*

(ii) *improper integral of the second kind.*

To define the improper integrals, we assume the following:

(i) The integrand $f(x)$ is of the same sign within its range of integration. Without any loss of generality, we assume that $f(x) \geq 0$ (when $f(x) \leq 0$, we can write $g(x) = -f(x)$ so that $g(x) \geq 0$). We shall discuss later, the case when $f(x)$ changes sign within its range of integration.

(ii) $f(x)$ is continuous over each finite subinterval $[\alpha, \beta]$ contained in the range of integration. Hence, there exists a positive constant K independent of α and β such that

$$\int_{\alpha}^{\beta} f(x)\, dx < K.$$

The improper integrals are evaluated by a limiting process.

1.4.1 Improper Integrals of the First Kind (Range of Integration is Infinite)

We shall now discuss methods to evaluate improper integrals of the form

$$\text{(i)} \int_{a}^{\infty} f(x)\, dx, \quad \text{(ii)} \int_{-\infty}^{b} f(x)\, dx, \quad \text{and} \quad \text{(iii)} \int_{-\infty}^{\infty} f(x)\, dx$$

if they exist. We define these improper integrals as follows:

(i)
$$\int_{a}^{\infty} f(x)\, dx = \lim_{p \to \infty} \int_{a}^{p} f(x)\, dx. \tag{1.61}$$

If the limit exists and is finite, say equal to l_1, then the improper integral converges and has the value l_1. Otherwise, the improper integral diverges.

(ii)
$$\int_{-\infty}^{b} f(x)\, dx = \lim_{p \to -\infty} \int_{p}^{b} f(x)\, dx. \tag{1.62}$$

If the limit exists and is finite, say equal to l_2, then the improper integral converges and has the value l_2. Otherwise, the improper integral diverges.

(iii)
$$\int_{-\infty}^{\infty} f(x)\, dx = \lim_{a \to -\infty} \int_{a}^{c} f(x)\, dx + \lim_{b \to \infty} \int_{c}^{b} f(x)\, dx \tag{1.63}$$

where c is any finite constant including zero. If both the limits on the right hand side exist separately and are finite, say equal to l_3 and l_4 respectively, then the improper integral converges and has the value $l_3 + l_4$. If one or both the limits do not exist or are infinite, then the improper integral diverges.

Example 1.35 Evaluate the following improper integrals, if they exist.

$$\text{(i)} \int_{0}^{\infty} \frac{dx}{a^2 + x^2}, \quad a > 0, \qquad \text{(ii)} \int_{-\infty}^{0} e^x \, dx,$$

(iii) $\displaystyle\int_0^\infty x \sin x \, dx,$ 　　　　　　　(iv) $\displaystyle\int_0^\infty e^{-ax} \cos px \, dx, \quad a > 0, p \text{ constant.}$

Solution

(i) $\displaystyle\int_0^\infty \frac{dx}{a^2 + x^2} = \lim_{b \to \infty} \int_0^b \frac{dx}{a^2 + x^2} = \lim_{b \to \infty} \left[\frac{1}{a} \tan^{-1}\left(\frac{b}{a} \right) \right] = \frac{\pi}{2a}.$

Therefore, the improper integral converges to $\pi/(2a)$.

(ii) $\displaystyle\int_{-\infty}^0 e^x dx = \int_0^\infty e^{-x} \, dx = \lim_{b \to \infty} \int_0^b e^{-x} \, dx = \lim_{b \to \infty} (1 - e^{-b}) = 1.$

Therefore, the improper integral converges to 1.

(iii) $\displaystyle\int_0^\infty x \sin x \, dx = \lim_{b \to \infty} \int_0^b x \sin x \, dx = \lim_{b \to \infty} (\sin b - b \cos b).$

Since this limit does not exist, the improper integral diverges.

(iv) Using the result

$$\int e^{-ax} \cos px \, dx = \frac{e^{-ax}}{a^2 + p^2} (p \sin px - a \cos px),$$

we obtain 　　$\displaystyle\int_0^b e^{-ax} \cos px \, dx = \left[\frac{e^{-ax}}{a^2 + p^2} (p \sin px - a \cos px) \right]_0^b$

$$= \frac{1}{a^2 + p^2} [e^{-ab} (p \sin bp - a \cos bp) + a].$$

Now, $\sin bp$ and $\cos bp$ have finite values and $\displaystyle\lim_{b \to \infty} e^{-ab} = 0$. Hence,

$$\int_0^\infty e^{-ax} \cos px \, dx = \lim_{b \to \infty} \int_0^b e^{-ax} \cos px \, dx = \frac{a}{a^2 + p^2}.$$

Therefore, the improper integral converges to $a/(a^2 + p^2)$.

Example 1.36 Discuss the convergence of the improper integral $\displaystyle\int_1^\infty \frac{dx}{x^p}$.

Solution　We have

$$\int_1^b \frac{dx}{x^p} = \frac{1}{1-p} \left[x^{1-p} \right]_1^b = \frac{1}{1-p} [b^{1-p} - 1]$$

Now, 　　　　　　$\displaystyle\lim_{b \to \infty} [b^{1-p}] = \begin{cases} \infty, & \text{if } p < 1 \\ 0, & \text{if } p > 1. \end{cases}$

Therefore, the improper integral converges if $p > 1$ and diverges if $p < 1$.

For $p = 1$, we have

$$\int_1^\infty \frac{dx}{x} = \lim_{b \to \infty} \int_1^b \frac{dx}{x} = \lim_{b \to \infty} \left[\ln x \right]_1^b = \lim_{b \to \infty} \ln b.$$

Since the limit does not exist, the improper integral diverges. Hence, the given improper integral converges to $1/(p-1)$ for $p > 1$ and diverges for $p \le 1$.

Example 1.37   Discuss the convergence of the integral $\int_{-\infty}^\infty xe^{-x^2}\, dx$.

Solution   We write

$$I = \int_{-\infty}^\infty xe^{-x^2}\, dx = \int_{-\infty}^c xe^{-x^2}\, dx + \int_c^\infty xe^{-x^2}\, dx$$

where c is any finite constant. We have

$$I = \lim_{a \to -\infty} \int_a^c xe^{-x^2}\, dx + \lim_{b \to \infty} \int_c^b xe^{-x^2}\, dx$$

$$= \lim_{a \to -\infty} \left[\frac{1}{2}\left(e^{-a^2} - e^{-c^2} \right) \right] + \lim_{b \to \infty} \left[\frac{1}{2}\left(e^{-c^2} - e^{-b^2} \right) \right]$$

$$= \frac{1}{2}\left(-e^{-c^2} + e^{c^2} \right) = 0.$$

Therefore, the given improper integral converges to 0.

It is not always possible to study the convergence/divergence of an improper integral by evaluating it as was done in the previous examples. A simple example is the integral $\int_0^\infty e^{-x^2}\, dx$ which cannot be evaluated directly. We now present some results which can be used to discuss the convergence or divergence of improper integrals. In this case, we cannot find the value of the improper integral, that is the value to which it converges. However, we may be able to find a bound of the integral.

Theorem 1.8   (Comparison Test 1)   If $0 \le f(x) \le g(x)$ for all x, then

(i) $\int_a^\infty f(x)\, dx$ converges if $\int_a^\infty g(x)\, dx$ converges.

(ii) $\int_a^\infty g(x)\, dx$ diverges if $\int_a^\infty f(x)\, dx$ diverges.

Theorem 1.9   (Comparison Test 2)   Suppose that $f(x)$ and $g(x)$ are positive functions and let

$$\lim_{x \to \infty} \left[\frac{f(x)}{g(x)} \right] = L, \quad 0 < L < \infty. \tag{1.64}$$

Then, the improper integrals $\int_a^\infty f(x)\, dx$ and $\int_a^\infty g(x)\, dx$ converge or diverge together.

Example 1.38 Discuss the convergence of the following improper integrals

(i) $\int_1^\infty e^{-x^2}\, dx$.

(ii) $\int_1^\infty \dfrac{dx}{(e^{-x}+1)x^2}$,

(iii) $\int_2^\infty \dfrac{dx}{\ln x}$,

(iv) $\int_2^\infty \dfrac{dx}{x(\ln x)^p}$,

(v) $\int_1^\infty \dfrac{x\tan^{-1}x}{\sqrt{4+x^3}}\, dx$.

Solution

(i) We have $e^{-x^2} < e^{-x}$ for all $x \geq 1$. Consider the improper integral $\int_1^\infty e^{-x}\, dx$.

We have $\quad\quad \int_1^\infty e^{-x}\, dx = \lim_{b\to\infty} \int_1^b e^{-x}\, dx = \lim_{b\to\infty}\, [1 - e^{-b}] = 1.$

Therefore, the integral $\int_1^\infty e^{-x}\, dx$ is convergent. By Comparison Test 1 (i), the given integral is also convergent. Further, its value is less than 1.

(ii) Let $f(x) = \dfrac{1}{(e^{-x}+1)x^2}$ and $g(x) = \dfrac{1}{x^2}$.

Now $\quad\quad \lim_{x\to\infty}\left[\dfrac{f(x)}{g(x)}\right] = \lim_{x\to\infty}\left[\dfrac{1}{(e^{-x}+1)x^2}\right]\left[\dfrac{x^2}{1}\right] = \lim_{x\to\infty}\dfrac{1}{e^{-x}+1} = 1.$

Also, $\int_1^\infty g(x)dx = \int_1^\infty \dfrac{dx}{x^2}$ converges to 1 (see Example 1.36). Therefore, by Comparison Test 2, the given improper integral is also convergent. Its value is less than 1.

Alternative We have $\dfrac{1}{(e^{-x}+1)x^2} < \dfrac{1}{x^2}$ for all $x \geq 1$. The improper integral $\int_1^\infty \dfrac{dx}{x^2}$ is convergent. Therefore, by Comparison Test 1 (i), the given improper integral converges.

(iii) We have $\ln x < x$ for all $x > 0$. Hence,

$$\dfrac{1}{\ln x} > \dfrac{1}{x} \quad \text{and} \quad \int_2^\infty \dfrac{dx}{\ln x} > \int_2^\infty \dfrac{dx}{x}.$$

Let $g(x) = 1/(\ln x)$ and $f(x) = 1/x$. We have $g(x) > f(x)$. Now, the integral

$$\int_2^\infty f(x)\, dx = \int_2^\infty \dfrac{dx}{x} \quad \text{is divergent (see Example 1.36).}$$

Therefore, by Comparison Test 1 (ii), the integral $\int_2^\infty g(x)\, dx = \int_2^\infty \dfrac{dx}{\ln x}$ is also divergent.

(iv) Substitute $\ln x = t$. We get

$$I = \int_2^\infty \dfrac{dx}{x(\ln x)^p} = \int_{\ln 2}^\infty \dfrac{dt}{t^p}$$

which is convergent for $p > 1$ and divergent for $p \leq 1$ (see Example 1.36).

(v) Let $\qquad f(x) = \dfrac{x \tan^{-1} x}{\sqrt{4 + x^3}} = \dfrac{\tan^{-1} x}{\sqrt{x}\sqrt{1 + 4x^{-3}}}$ and $g(x) = \dfrac{1}{\sqrt{x}}$.

We find that $\qquad \lim_{x \to \infty} \dfrac{f(x)}{g(x)} = \lim_{x \to \infty} \dfrac{\tan^{-1} x}{\sqrt{1 + 4x^{-3}}} = \dfrac{\pi}{2}$.

Hence, by Comparison Test 2, the integrals $\int_1^\infty f(x)dx$ and $\int_1^\infty g(x)dx$ converge or diverge

together. Now, $\int_1^\infty g(x)\,dx$ is divergent. Therefore, $\int_1^\infty f(x)\,dx$ is also divergent.

1.4.2   Improper Integral of the Second Kind

Now consider an improper integral of the form $\int_a^b f(x)\,dx$, where a, b are finite constants, $f(x)$ is continuous in (a, b) and has infinite discontinuity (becomes infinite) at (i) $x = a$, or (ii) $x = b$, or (iii) $x = a$ and $x = b$, or (iv) $f(x)$ is continuous in (a, b) except at $x = c$, $a < c < b$, where $f(x)$ has an infinite discontinuity.

If $f(x)$ has a finite number of points of discontinuity, $c_1, c_2, \ldots, c_m$, $a \leq c_1 < c_2 \ldots < c_m \leq b$, then we write the integral as

$$\int_a^b f(x)dx = \int_a^{c_1} f(x)dx + \int_{c_1}^{c_2} f(x)dx + \ldots + \int_{c_m}^b f(x)dx \qquad (1.65)$$

and consider each integral on the right hand side separately.

Infinite discontinuity at $x = a$ Since the function $f(x)$ is continuous at all points except at $x = a$, the integral $\int_{a+\varepsilon}^b f(x)dx$ is a proper integral and exists for every ε, $0 < \varepsilon < b - a$.

We evaluate the improper integral as

$$\int_a^b f(x)dx = \lim_{\varepsilon \to 0} \int_{a+\varepsilon}^b f(x)dx.$$

If this limit exists and is finite, say equal to l_1, then the improper integral converges to l_1. Otherwise, it diverges.

Infinite discontinuity at $x = b$ Since the function $f(x)$ is continuous at all points except at $x = b$, the integral $\int_a^{b-\varepsilon} f(x)dx$ is a proper integral and exists for every ε, $0 < \varepsilon < b - a$.

We evaluate the improper integral as

$$\int_a^b f(x)dx = \lim_{\varepsilon \to 0} \int_a^{b-\varepsilon} f(x)dx.$$

If this limit exists and in finite, say equal to l_2, then the improper integral converges to l_2, otherwise it diverges.

Infinite discontinuity at $x = a$ **and** $x = b$. We write the improper integral as

$$\int_a^b f(x)\,dx = \int_a^\alpha f(x)\,dx + \int_\alpha^b f(x)\,dx$$

where α is any finite constant between a and b at which f is defined. We evaluate the improper integral as

$$\int_a^b f(x)\,dx = \lim_{\varepsilon \to 0} \int_{a+\varepsilon}^\alpha f(x)\,dx + \lim_{\xi \to 0} \int_\alpha^{b-\xi} f(x)\,dx.$$

If both the limits exist and are finite, then the improper integral converges. Otherwise, it diverges.

Infinite discontinuity at $x = c$, $a < c < b$. We write the improper integral as

$$\int_a^b f(x)\,dx = \int_a^c f(x)\,dx + \int_c^b f(x)\,dx = \lim_{\varepsilon \to 0} \int_a^{c-\varepsilon} f(x)\,dx + \lim_{\xi \to 0} \int_{c+\xi}^b f(x)\,dx.$$

The given improper integral converges, if both the integrals on the right hand side converge. If one or both the integrals on the right hand side diverge, then the given improper integral diverges.

The following tests can be used to discuss the convergence or divergence of the above improper integrals. In this case, we cannot find the value of the improper integral, that is the value to which it converges. However, we may be able to find a bound of the integral.

Theorem 1.10 (Comparison Test 3)   If $0 \le f(x) \le g(x)$ for all x in $[a, b]$, then

(i) $\displaystyle \int_a^b f(x)\,dx$ converges if $\displaystyle \int_a^b g(x)\,dx$ converges.

(ii) $\displaystyle \int_a^b g(x)\,dx$ diverges if $\displaystyle \int_a^b f(x)\,dx$ diverges.

Theorem 1.11 (Comparison Test 4)   If $f(x)$ and $g(x)$ are two positive functions and

(i) a is a point of infinite discontinuity such that

$$\lim_{x \to a^+} \frac{f(x)}{g(x)} = \lim_{h \to 0} \frac{f(a+h)}{g(a+h)} = l_1, \quad 0 < l_1 < \infty,$$

or (ii) b is a point of infinite discontinuity such that

$$\lim_{x \to b^-} \frac{f(x)}{g(x)} = \lim_{h \to 0} \frac{f(b-h)}{g(b-h)} = l_2, \quad 0 < l_2 < \infty,$$

then, the improper integrals $\displaystyle \int_a^b f(x)\,dx$ and $\displaystyle \int_a^b g(x)\,dx$ converge or diverge together.

Example 1.39   Evaluate the following improper integrals, if they exist:

(i) $\displaystyle \int_0^4 \frac{dx}{\sqrt{x}}$,

(ii) $\displaystyle \int_0^2 \frac{dx}{\sqrt{4 - x^2}}$,

(iii) $\int_{-1}^{1} \sqrt{\dfrac{1+x}{1-x}}\, dx,$

(iv) $\int_{0}^{3} \dfrac{dx}{3x - x^2},$

(v) $\int_{-1}^{1} \dfrac{dx}{x^2},$

(vi) $\int_{0}^{3} \dfrac{dx}{x^2 - 3x + 2},$

(vii) $\int_{1}^{\infty} \dfrac{dx}{x\sqrt{1 - x^2}}.$

Solution

(i) $\int_{0}^{4} \dfrac{dx}{\sqrt{x}} = \lim_{\varepsilon \to 0} \int_{\varepsilon}^{4} \dfrac{dx}{\sqrt{x}} = 2 \lim_{\varepsilon \to 0} [2 - \sqrt{\varepsilon}] = 4.$

Therefore, the improper integral converges to 4.

(ii) $\int_{0}^{2} \dfrac{dx}{\sqrt{4 - x^2}} = \lim_{\varepsilon \to 0} \int_{0}^{2-\varepsilon} \dfrac{dx}{\sqrt{4 - x^2}} = \lim_{\varepsilon \to 0} \sin^{-1}\left(1 - \dfrac{\varepsilon}{2}\right) = \sin^{-1} 1 = \dfrac{\pi}{2}.$

Therefore, the improper integral converges to $\pi/2$.

(iii) $\int_{-1}^{1} \sqrt{\dfrac{1+x}{1-x}}\, dx = \lim_{\varepsilon \to 0} \int_{-1}^{1-\varepsilon} \sqrt{\dfrac{1+x}{1-x}}\, dx = \lim_{\varepsilon \to 0} \left[\int_{-1}^{1-\varepsilon} \dfrac{dx}{\sqrt{1 - x^2}} - \dfrac{1}{2} \int_{-1}^{1-\varepsilon} \dfrac{-2x}{\sqrt{1 - x^2}}\, dx \right]$

$= \lim_{\varepsilon \to 0} \left[\left\{ \sin^{-1}(1 - \varepsilon) - \sin^{-1}(-1) \right\} - \left\{ \sqrt{1 - (1 - \varepsilon)^2} - \sqrt{1 - 1} \right\} \right]$

$= \sin^{-1}(1) - \sin^{-1}(-1) = 2 \sin^{-1}(1) = \pi.$

Therefore, the improper integral converges to π.

(iv) Here, the integrand $f(x)$ has infinite discontinuity, at both the end points $x = 0$ and $x = 3$. We take any point, say $x = c$, inside the interval of integration, at which $f(x)$ is defined. We write

$\int_{0}^{3} \dfrac{dx}{3x - x^2} = \int_{0}^{c} \dfrac{dx}{3x - x^2} + \int_{c}^{3} \dfrac{dx}{3x - x^2} = \lim_{\varepsilon \to 0} \int_{\varepsilon}^{c} \dfrac{dx}{x(3 - x)} + \lim_{\xi \to 0} \int_{c}^{3-\xi} \dfrac{dx}{x(3 - x)}$

$= \dfrac{1}{3} \lim_{\varepsilon \to 0} \left[\ln\left(\dfrac{x}{3 - x}\right) \right]_{\varepsilon}^{c} + \dfrac{1}{3} \lim_{\xi \to 0} \left[\ln\left(\dfrac{x}{3 - x}\right) \right]_{c}^{3-\xi}$

$= \dfrac{1}{3} \lim_{\varepsilon \to 0} \left[\ln\left(\dfrac{c}{3 - c}\right) - \ln\left(\dfrac{\varepsilon}{3 - \varepsilon}\right) \right] + \dfrac{1}{3} \lim_{\xi \to 0} \left[\ln\left(\dfrac{3 - \xi}{\xi}\right) - \ln\left(\dfrac{c}{3 - c}\right) \right].$

Since the limits do not exist, the improper integral diverges.

(v) The integrand has infinite discontinuity at $x = 0$ which lies inside the interval of integration. We write

$$\int_{-1}^{1} \frac{dx}{x^2} = \int_{-1}^{0} \frac{dx}{x^2} + \int_{0}^{1} \frac{dx}{x^2} = \lim_{\varepsilon \to 0} \int_{-1}^{-\varepsilon} \frac{dx}{x^2} + \lim_{\xi \to 0} \int_{\xi}^{1} \frac{dx}{x^2}$$

$$= \lim_{\varepsilon \to 0} \left[\frac{1}{\varepsilon} - 1 \right] + \lim_{\xi \to 0} \left[\frac{1}{\xi} - 1 \right] \to \infty.$$

Therefore, the improper integral diverges.

(vi) The integrand has infinite discontinuities at $x = 1$ and $x = 2$, both of which lie inside the interval of integration. We write

$$\int_{0}^{3} \frac{dx}{x^2 - 3x + 2} = \int_{0}^{1} \frac{dx}{(x-1)(x-2)} + \int_{1}^{2} \frac{dx}{(x-1)(x-2)} + \int_{2}^{3} \frac{dx}{(x-1)(x-2)}.$$

$$= I_1 + I_2 + I_3.$$

We find that

(a) the integrand in I_1 has infinite discontinuity at $x = 1$,

(b) the integrand $f(x)$ in I_2 has infinite discontinuity at both the end points $x = 1$ and $x = 2$. We take any point, say $x = c$ inside the limits of integration, at which $f(x)$ is defined. We also find that $f(x) < 0$ when $1 < x < 2$. We write $g(x) = -f(x)$ so that $g(x) > 0$ when $1 < x < 2$. Therefore, we can write

$$I_2 = -\int_{1}^{c} \frac{dx}{(x-1)(2-x)} - \int_{c}^{2} \frac{dx}{(x-1)(2-x)},$$

(c) the integrand in I_3 has infinite discontinuity at $x = 2$.

Hence, we can write

$$\int_{0}^{3} \frac{dx}{x^2 - 3x + 2} = \lim_{\varepsilon_1 \to 0} \int_{0}^{1-\varepsilon_1} \frac{dx}{(x-1)(x-2)} - \lim_{\varepsilon_2 \to 0} \int_{1+\varepsilon_2}^{c} \frac{dx}{(x+1)(2-x)}$$

$$- \lim_{\varepsilon_3 \to 0} \int_{c}^{2-\varepsilon_3} \frac{dx}{(x-1)(2-x)} + \lim_{\varepsilon_4 \to 0} \int_{2+\varepsilon_4}^{3} \frac{dx}{(x-1)(x-2)}$$

$$= \lim_{\varepsilon_1 \to 0} \left[\ln \left(\frac{\varepsilon_1 + 1}{\varepsilon_1} \right) - \ln 2 \right] - \lim_{\varepsilon_2 \to 0} \left[\ln \left(\frac{c-1}{2-c} \right) - \ln \left(\frac{\varepsilon_2}{1-\varepsilon_2} \right) \right]$$

$$- \lim_{\varepsilon_3 \to 0} \left[\ln \left(\frac{1-\varepsilon_3}{\varepsilon_3} \right) - \ln \left(\frac{c-1}{2-c} \right) \right] + \lim_{\varepsilon_4 \to 0} \left[\ln \left(\frac{1}{2} \right) - \ln \left(\frac{\varepsilon_4}{\varepsilon_4 + 1} \right) \right].$$

Since the limits do not exist, the improper integral diverges.

Note that the improper integral I_1 diverges. We could have concluded that the improper integral diverges without discussing the convergence/divergence of I_2 and I_3.

(vii) $\int_{1}^{\infty} \frac{dx}{x\sqrt{x^2 - 1}} = \int_{1}^{c} \frac{dx}{x\sqrt{x^2 - 1}} + \int_{c}^{\infty} \frac{dx}{x\sqrt{x^2 - 1}}.$

$$= \lim_{\varepsilon \to 0} \int_{1+\varepsilon}^{c} \frac{dx}{x\sqrt{x^2-1}} + \lim_{b \to \infty} \int_{c}^{b} \frac{dx}{x\sqrt{x^2-1}} = \lim_{\varepsilon \to 0} \left[\sec^{-1} x\right]_{1+\varepsilon}^{c} + \lim_{b \to \infty} \left[\sec^{-1} x\right]_{c}^{b}$$

$$= \lim_{\varepsilon \to 0} \left[\sec^{-1} c - \sec^{-1}(1+\varepsilon)\right] + \lim_{b \to \infty} \left[\sec^{-1} b - \sec^{-1} c\right]$$

$$= \sec^{-1} c - \sec^{-1} 1 + \frac{\pi}{2} - \sec^{-1} c = \frac{\pi}{2}.$$

Therefore, the improper integral converges to $\pi/2$.

Example 1.40 Discuss the convergence of the improper integral $\displaystyle\int_a^b \frac{dx}{(x-a)^p}$, $p > 0$.

Solution The integrand has infinite discontinuity at $x = a$. We write

$$\int_a^b \frac{dx}{(x-a)^p} = \lim_{\varepsilon \to 0} \int_{a+\varepsilon}^b \frac{dx}{(x-a)^p} = \frac{1}{1-p} \lim_{\varepsilon \to 0} \left[\frac{1}{(b-a)^{p-1}} - \frac{1}{\varepsilon^{p-1}}\right]$$

$$= \begin{cases} 1/\left[(1-p)(b-a)^{p-1}\right] & \text{if } p < 1 \\ \infty, & \text{if } p > 1. \end{cases}$$

For $p = 1$, we get

$$\int_c^b \frac{dx}{x-a} = \lim_{\varepsilon \to 0} \int_{a+\varepsilon}^b \frac{dx}{x-a} = \lim_{\varepsilon \to 0} \ln\left[\frac{b-a}{\varepsilon}\right] = \infty.$$

Therefore, the improper integral converges for $p < 1$ and diverges for $p \geq 1$.

Example 1.41 Show that the improper integral $\displaystyle\int_{-\pi/2}^{\pi/2} \tan x \, dx$ is divergent.

Solution The integrand has infinite discontinuity at $x = \pm\,\pi/2$. We write

$$\int_{-\pi/2}^{\pi/2} \tan x \, dx = \lim_{\varepsilon \to 0} \int_{-(\pi/2)+\varepsilon}^{c} \tan x \, dx + \lim_{\xi \to 0} \int_{c}^{(\pi/2)-\xi} \tan x \, dx$$

$$= \lim_{\varepsilon \to 0} \left[-\ln(\cos x)\right]_{-(\pi/2)+\varepsilon}^{c} + \lim_{\xi \to 0} \left[-\ln(\cos x)\right]_{c}^{(\pi/2)-\xi}$$

$$= \lim_{\varepsilon \to 0} \left\{\ln\left[\cos\left(-\frac{\pi}{2}+\varepsilon\right)\right] - \ln\left[\cos(c)\right]\right\}$$

$$- \lim_{\xi \to 0} \left\{\ln\left[\cos\left(\frac{\pi}{2}-\xi\right)\right] - \ln\left[\cos(c)\right]\right\}.$$

Since the limits do not exist, the improper integral diverges.

Note that if we write

$$\int_{-\pi/2}^{\pi/2} \tan x \, dx = \lim_{\varepsilon \to 0} \int_{-(\pi/2)+\varepsilon}^{(\pi/2)-\varepsilon} \tan x \, dx$$

we get $\displaystyle\int_{-\pi/2}^{\pi/2} \tan x \, dx = 0$, which is not the correct solution.

Example 1.42   Discuss the convergence of the following improper integrals

(i) $\int_1^2 \dfrac{\sqrt{x}}{\ln x}\, dx,$ (ii) $\int_0^{\pi/2} \dfrac{\sin x}{x\sqrt{x}}\, dx.$

Solution

(i)   We have $f(x) = (\sqrt{x}/\ln x) \geq 0,\ 1 < x \leq 2$. The point $x = 1$ is the only point of infinite discontinuity.

Let $g(x) = 1/(x \ln x)$. Then, we have

$$\lim_{x \to 1^+} \frac{f(x)}{g(x)} = \lim_{h \to 0} \frac{f(1+h)}{g(1+h)} = \lim_{h \to 0} \left[\frac{\sqrt{1+h}}{\ln(1+h)} \right] \left[(1+h) \ln(1+h) \right]$$

$$= \lim_{h \to 0} (1+h)^{3/2} = 1.$$

Therefore, both the integrals $\int_1^2 f(x)\, dx$ and $\int_1^2 g(x)\, dx$ converge or diverge together.

Now, $\int_1^2 g(x)\, dx = \int_1^2 \dfrac{dx}{x \ln x} = \lim_{\varepsilon \to 0} \int_{1+\varepsilon}^2 \dfrac{dx}{x \ln x} = \lim_{\varepsilon \to 0} \left[\ln(\ln x) \right]_{1+\varepsilon}^2$

$$= \lim_{\varepsilon \to 0} \left[\ln(\ln 2) - \ln(\ln(1 + \varepsilon)) \right] \to \infty.$$

Since $\int_1^2 g(x)\, dx$ is divergent, the given integral $\int_1^2 f(x)\, dx$ is also divergent by Comparison Test 4.

(ii)   We have $f(x) = \dfrac{\sin x}{x\sqrt{x}} = \left(\dfrac{\sin x}{x} \right)\left(\dfrac{1}{\sqrt{x}} \right) \leq \dfrac{1}{\sqrt{x}}$, since $\sin x / x$ is bounded and $(\sin x/x) \leq$

$1,\ 0 \leq x \leq \pi/2$. Let $g(x) = 1/\sqrt{x}$. We have $f(x) \leq g(x),\ 0 < x < \pi/2$.

Now, $g(x)$ has a point of discontinuity at $x = 0$. Hence

$$\int_0^{\pi/2} g(x)\, dx = \int_0^{\pi/2} \frac{dx}{\sqrt{x}} = \lim_{\varepsilon \to 0} \int_{\varepsilon}^{\pi/2} \frac{dx}{\sqrt{x}} = \lim_{\varepsilon \to 0} \left[\sqrt{2\pi} - 2\sqrt{\varepsilon} \right] = \sqrt{2\pi}.$$

Since $\int_0^{\pi/2} g(x)\, dx$ is convergent, the given integral $\int_0^{\pi/2} f(x)\, dx$ is also convergent by Comparison Test 3 (i).

Example 1.43   Show that the improper integral $\int_0^{\pi/2} \dfrac{\cos^m x}{x^n}\, dx$ converges when $n < 1$.

Solution   We have $f(x) = \dfrac{\cos^m x}{x^n} < \dfrac{1}{x^n},\ 0 < x < \pi/2.\ x = 0$ is the point of infinite discontinuity

of $f(x)$. Let $g(x) = 1/x^n$. Then $f(x) < g(x)$.

Since the integral $\int_0^{\pi/2} g(x)\,dx = \int_0^{\pi/2} \dfrac{dx}{x^n}$ is convergent for $n < 1$ (see Example 1.40), the given integral is also convergent for $n < 1$ by Comparison Test 3(i).

1.4.3   Absolute Convergence of Improper Integrals

In the previous sections, we had assumed that $f(x)$ is of the same sign throughout the interval of integration. Now, assume that $f(x)$ changes sign within the interval of integration. In this case, we consider absolute convergence of the improper integral.

Absolute convergence   The improper integral $\int_a^b f(x)\,dx$ is said to be *absolutely convergent* if $\int_a^b |f(x)|\,dx$ is convergent.

Theorem 1.12   An absolutely convergent improper integral is convergent, that is if $\int_a^b |f(x)|\,dx$ converges, then $\int_a^b f(x)\,dx$ converges.

Since, $|f|$ is always positive within the interval of integration, all the comparison tests can be used to discuss the absolute convergence of the given improper integral.

Example 1.44   Show that the improper integral $\int_0^1 \dfrac{\sin(1/x)}{x^p}\,dx$ converges absolutely for $p < 1$.

Solution   The integrand changes sign within the interval of integration. Hence, we consider the absolute convergence of the given integral. The function $f(x) = \sin(1/x)/x^p$ has a point of infinite discontinuity at $x = 0$. We have

$$|f(x)| = \left| \frac{\sin(1/x)}{x^p} \right| \le \frac{1}{x^p}.$$

Since $\int_0^1 \dfrac{1}{x^p}\,dx$ converges for $p < 1$, the given improper integral converges absolutely for $p < 1$.

Example 1.45   Show that the improper integral $\int_{-\infty}^{\infty} \dfrac{\sin x}{1 + x^2}\,dx$ converges.

Solution   The integrand $f(x)$ changes sign within the interval of integration. Hence, we consider the absolute convergence of the given integral. We have

$$|I| = \left| \int_{-\infty}^{\infty} \frac{\sin x}{1+x^2}\,dx \right| \le \int_{-\infty}^{\infty} \left| \frac{\sin x}{1+x^2} \right| dx$$

$$= \lim_{a \to -\infty} \int_a^c \left| \frac{\sin x}{1+x^2} \right| dx + \lim_{b \to \infty} \int_c^b \left| \frac{\sin x}{1+x^2} \right| dx = I_1 + I_2.$$

Now,

$$I_1 = \lim_{a \to -\infty} \int_a^c \left| \frac{\sin x}{1+x^2} \right| dx \le \lim_{a \to -\infty} \int_a^c \frac{dx}{1+x^2} = \lim_{a \to -\infty} [\tan^{-1}c - \tan^{-1}a] = \tan^{-1}c + \frac{\pi}{2}.$$

$$I_2 = \lim_{b \to \infty} \int_c^b \left| \frac{\sin x}{1+x^2} \right| dx \le \lim_{b \to \infty} \int_c^b \frac{dx}{1+x^2} = \lim_{b \to \infty} [\tan^{-1} b - \tan^{-1} c] = \frac{\pi}{2} - \tan^{-1} c.$$

Hence, $|I| \le I_1 + I_2 \le \pi$. Therefore, the given improper integral converges.

1.4.4   Beta and Gamma Functions

Beta and Gamma functions are improper integrals which are commonly encountered in many science and engineering applications. These functions are used in evaluating many definite integrals.

Gamma function

Consider the improper integral $\qquad I(\alpha) = \int_0^\infty x^{\alpha-1} e^{-x} dx, \quad \alpha > 0.$ $\qquad\qquad$ (1.66)

We write the integral as $\qquad I(\alpha) = \int_0^c x^{\alpha-1} e^{-x} dx + \int_c^\infty x^{\alpha-1} e^{-x} dx = I_1 + I_2, \quad 0 < c < \infty.$

The integral I_1 is an improper integral of the second kind as the integrand has a point of discontinuity at $x = 0$, whenever $0 < \alpha < 1$. For $\alpha \ge 1$, it is a proper integral. The integral I_2 is an improper integral of the first kind as its upper limit is infinite. We consider the two integrals separately.

Convergence at $x = 0$, $0 < \alpha < 1$, of the first integral I_1

In the integral I_1, let $f(x) = x^{\alpha-1} e^{-x}$ and $g(x) = x^{\alpha-1}$. Now, $\displaystyle\lim_{x \to 0^+} \frac{f(x)}{g(x)} = \lim_{x \to 0^+} \frac{x^{\alpha-1} e^{-x}}{x^{\alpha-1}} = 1.$

Since $\displaystyle\int_0^c g(x) dx = \int_0^c \frac{dx}{x^{1-\alpha}}$ converges when $1 - \alpha < 1$, or $\alpha > 0$, the improper integral I_1 is convergent for all $\alpha > 0$.

Convergence at ∞, of the second integral I_2

Without loss of generality, let $c \ge 1$. Otherwise, the integral can be written as the sum of two integrals with the intervals $(c, 1)$, $(1, \infty)$. The first integral is a proper integral.

Let n be a positive integer such that $n > \alpha - 1$, $\alpha > 0$. Then,

$$\alpha - 1 < n, \quad x^{\alpha-1} < x^n \quad \text{and} \quad x^{\alpha-1} e^{-x} < x^n e^{-x}, \quad 1 < x < \infty.$$

Therefore, $\qquad \displaystyle\int_c^\infty x^{\alpha-1} e^{-x} dx < \int_c^\infty x^n e^{-x} dx = \lim_{b \to \infty} \int_c^b x^n e^{-x} dx$

$$= \lim_{b \to \infty} [e^{-x} \{\text{polynomial of degree } n \text{ in } x, P_n(x)\}]_c^b$$

$$= \lim_{b \to \infty} [e^{-b} P_n(b) - e^{-c} P_n(c)] = -e^{-c} P_n(c)$$

since $\displaystyle\lim_{b \to \infty} [b^k / e^b] = 0$ for fixed k.

The limit exists and the integral I_2 converges for $\alpha > 0$.

Hence the given improper integral (Eq. (1.66)) converges when $\alpha > 0$.

This improper integral is called the *Gamma function* and is denoted by $\Gamma(\alpha)$. Therefore,

$$\Gamma(\alpha) = \int_0^\infty x^{\alpha-1} e^{-x} dx, \quad \alpha > 0. \tag{1.67}$$

Some identities of Gamma functions

1. $\Gamma(1) = \int_0^\infty e^{-x} dx = 1.$ $\hspace{4cm}$ (1.68)

2. $\Gamma(\alpha + 1) = \alpha\Gamma(\alpha).$ $\hspace{4.5cm}$ (1.69)

   Integrating Eq. (1.66) by parts, we get

   $$\Gamma(\alpha + 1) = \int_0^\infty x^\alpha e^{-x} dx = -\left[x^a e^{-x}\right]_0^\infty + \alpha \int_0^\infty x^{\alpha-1} e^{-x} dx = \alpha\Gamma(\alpha).$$

   If α is negative and not an integer, then we write $\Gamma(\alpha) = \dfrac{1}{\alpha} \Gamma(\alpha + 1).$

3. $\Gamma(m + 1) = m!$, for any position integer m. $\hspace{2.5cm}$ (1.70)

   We have $\Gamma(m + 1) = m\Gamma(m) = m(m - 1) \Gamma(m - 1) = \ldots = m(m - 1) \ldots 1\Gamma(1) = m!.$

4. $\Gamma(1/2) = \sqrt{\pi}.$ $\hspace{5cm}$ (1.71)

   We have $\Gamma(1/2) = \int_0^\infty x^{-1/2} e^{-x} dx = 2 \int_0^\infty e^{-u^2} du.$ (set $x = u^2$).

   We write

   $$\left[\Gamma\left(\frac{1}{2}\right)\right]^2 = \left[2 \int_0^\infty e^{-u^2} du\right]\left[2 \int_0^\infty e^{-v^2} dv\right] = 4 \int_0^\infty \int_0^\infty e^{-(u^2+v^2)} du\, dv.$$

   Changing to polar coordinates $u = r \cos \theta$, $v = r \sin \theta$, we obtain $du\,dv = r\, dr\, d\theta$ and

   $$\left[\Gamma\left(\frac{1}{2}\right)\right]^2 = 4 \int_{\theta=0}^{\pi/2} \int_{r=0}^\infty re^{-r^2} dr\, d\theta = 2\pi \int_0^\infty re^{-r^2} dr = -\pi \left[e^{-r^2}\right]_0^\infty = \pi.$$

   Hence, $\hspace{4cm}$ $\Gamma(1/2) = \sqrt{\pi}.$

   (In Chapter 2, we shall discuss evaluation of double integrals and change of variables.)

5. $\Gamma(-1/2) = -2\sqrt{\pi}.$ $\hspace{5cm}$ (1.72)

   We have $\Gamma(\alpha) = [\Gamma(\alpha + 1)]/\alpha.$ Substituting $\alpha = -1/2$, we get

   $$\Gamma\left(-\frac{1}{2}\right) = \frac{\Gamma(1/2)}{(-1/2)} = -2\sqrt{\pi}.$$

Beta function

Consider the improper integral

$$I = \int_0^1 x^{m-1} (1 - x)^{n-1} dx, \quad 0 < m < 1, \quad 0 < n < 1. \tag{1.73}$$

Note that I is a proper integral for $m \geq 1$ and $n \geq 1$. The improper integral has points of infinite discontinuity at (i) $x = 0$, when $m < 1$ and (ii) $x = 1$, when $n < 1$. When $m < 1$ and $n < 1$, we take a number, say c between 0 and 1 and write the improper integral as

$$I = \int_0^c x^{m-1}(1-x)^{n-1} \, dx + \int_c^1 x^{m-1}(1-x)^{n-1} \, dx = I_1 + I_2$$

where $\qquad I_1 = \int_0^c x^{m-1}(1-x)^{n-1} \, dx \quad$ and $\quad I_2 = \int_c^1 x^{m-1}(1-x)^{n-1} \, dx.$

I_1 is an improper integral, since $x = 0$ is a point of infinite discontinuity, while I_2 is an improper integral, since $x = 1$ is a point of infinite discontinuity. We consider these two integrals separately.

Convergence at $x = 0$, $0 < m < 1$, of the integral I_1

In the integral I_1, let $f(x) = x^{m-1}(1-x)^{n-1}$ and $g(x) = x^{m-1}$.

Now, $\qquad\qquad \lim_{x \to 0^+} \dfrac{f(x)}{g(x)} = \lim_{x \to 0^+} \dfrac{x^{m-1}(1-x)^{n-1}}{x^{m-1}} = 1$

and $\displaystyle \int_0^c g(x)dx = \int_0^c \frac{dx}{x^{1-m}}$ is convergent only when $1 - m < 1$, or $m > 0$.

Therefore, the improper integral I_1 converges when $m > 0$.

Convergence at $x = 1$, $0 < n < 1$, of the integral I_2

In the integral I_2, let $f(x) = x^{m-1}(1-x)^{n-1}$ and $g(x) = (1-x)^{n-1}$.

Now, $\qquad\qquad \lim_{x \to 1} \dfrac{f(x)}{g(x)} = \lim_{x \to 1} \dfrac{x^{m-1}(1-x)^{n-1}}{(1-x)^{n-1}} = 1$

and $\qquad\qquad \displaystyle \int_c^1 g(x)dx = \int_c^1 \frac{dx}{(1-x)^{1-n}}$ converges when $1 - n < 1$, or $n > 0$.

Therefore, the improper integral I_2 converges when $n > 0$. Combining the two results, we deduce that the given improper integral (Eq. (1.73)) converges when $m > 0$ and $n > 0$. This improper integral is called the *Beta function* and is denoted by $\beta(m, n)$. Therefore,

$$\beta(m, n) = \int_0^1 x^{m-1}(1-x)^{n-1} \, dx, \quad m > 0, n > 0. \tag{1.74}$$

Some identities of Beta functions

    1. $\beta(m, n) = \beta(n, m)$ (1.75)

        (substitute $x = 1 - t$ in Eq. (1.74) and simplify).

    2. $\beta(m, n) = 2 \displaystyle\int_0^{\pi/2} \sin^{2m-1}(\theta) \cos^{2n-1}(\theta) \, d\theta = 2 \int_0^{\pi/2} \sin^{2n-1}(\theta) \cos^{2m-1}(\theta) \, d\theta.$ (1.76)

        (substitute $x = \sin^2 \theta$ in Eq. (1.74) and simplify).

3. $\beta(m, n) = \int_0^\infty \dfrac{x^{m-1}}{(1+x)^{m+n}} \, dx$ (1.77)

(substitute $x = t/(1 + t)$ in Eq. (1.74) and simplify).

4. $\beta(m, n) = \dfrac{\Gamma(m)\,\Gamma(n)}{\Gamma(m+n)}.$ (1.78)

We can prove this result using double integrals and change of variables. We have

$$\Gamma(m) = \int_0^\infty x^{m-1}\, e^{-x}\, dx = 2 \int_0^\infty u^{2m-1}\, e^{-u^2}\, du, \quad (\text{set } x = u^2)$$

$$\Gamma(n) = \int_0^\infty x^{n-1}\, e^{-x}\, dx = 2 \int_0^\infty v^{2n-1}\, e^{-v^2}\, dv, \quad (\text{set } x = v^2)$$

$$\Gamma(m)\,\Gamma(n) = 4 \int_0^\infty \int_0^\infty u^{2m-1}\, v^{2n-1}\, e^{-(u^2+v^2)}\, du\, dv.$$

Changing to polar coordinates, $u = r \cos\theta$, $v = r \sin\theta$, we get

$$\Gamma(m)\,\Gamma(n) = 4 \int_{\theta=0}^{\pi/2} \int_0^\infty \cos^{2m-1}(\theta)\, \sin^{2n-1}(\theta)\, r^{2m+2n-1}\, e^{-r^2}\, dr\, d\theta$$

$$= 4 \left[\int_0^\infty r^{2m+2n-1}\, e^{-r^2}\, dr\right]\left[\int_0^{\pi/2} \cos^{2m-1}(\theta)\, \sin^{2n-1}(\theta)\, d\theta\right]$$

$$= 2\beta(m, n) \int_0^\infty r^{2m+2n-1}\, e^{-r^2}\, dr, \quad (\text{using Eq. (1.76)}).$$

We also have

$$\Gamma(m + n) = \int_0^\infty x^{m+n-1}\, e^{-x}\, dx = 2 \int_0^\infty r^{2m+2n-1}\, e^{-r^2}\, dr, \quad (\text{set } x = r^2).$$

Combining the two results, we obtain

$$\Gamma(m)\, \Gamma(n) = \beta(m, n)\, \Gamma(m + n), \quad \text{or} \quad \beta(m, n) = \dfrac{\Gamma(m)\Gamma(n)}{\Gamma(m+n)}.$$

5. $\beta(m, n) = \beta(m + 1, n) + \beta(m, n + 1).$

We have

$$\beta(m + 1, n) = 2 \int_0^{\pi/2} \sin^{2m+1}(\theta)\, \cos^{2n-1}(\theta)\, d\theta = 2 \int_0^{\pi/2} \sin^{2m-1}(\theta)\, \sin^2\theta\, \cos^{2n-1}(\theta)\, d\theta$$

$$= 2 \int_0^{\pi/2} \sin^{2m-1}(\theta)\, \cos^{2n-1}(\theta)\, (1 - \cos^2\theta)\, d\theta$$

$$= 2 \int_0^{\pi/2} \sin^{2m-1}(\theta)\, \cos^{2n-1}(\theta)\, d\theta - 2 \int_0^{\pi/2} \sin^{2m-1}(\theta)\, \cos^{2n+1}(\theta)\, d\theta$$

$$= \beta(m, n) - \beta(m, n + 1).$$

Therefore, $\beta(m, n) = \beta(m + 1, n) + \beta(m, n + 1).$

Example 1.46 Given that $\int_0^\infty \dfrac{x^{p-1}}{1+x} = \dfrac{\pi}{\sin p\pi}$, show that $\Gamma(p)\,\Gamma(1-p) = \dfrac{\pi}{\sin p\pi}$.

Solution Let $\dfrac{x}{1+x} = y$. Solving for x, we get $x = \dfrac{y}{1-y}$ and $dx = \dfrac{1}{(1-y)^2}\,dy.$

Then, $I = \displaystyle\int_0^\infty \dfrac{x^{p-1}}{1+x}\,dx = \int_0^1 y^{p-1}(1-y)^{-p}\,dy = \beta(p, 1-p) = \dfrac{\Gamma(p)\Gamma(1-p)}{\Gamma(1)} = \Gamma(p)\,\Gamma(1-p).$

Hence, the result.

Example 1.47 Evaluate the following improper integrals

(i) $\displaystyle\int_0^\infty \sqrt{x}\, e^{-x^2}\, dx,$

(ii) $\displaystyle\int_0^\infty e^{-x^3}\, dx$

in terms of Gamma functions.

Solution

(i) Substitute $x = \sqrt{t}$. We get $dx = dt/(2\sqrt{t})$ and

$$I = \int_0^\infty \sqrt{x}\, e^{-x^2}\, dx = \frac{1}{2}\int_0^\infty t^{-1/4}\, e^{-t}\, dt = \frac{1}{2}\int_0^\infty t^{(3/4)-1}\, e^{-t}\, dt = \frac{1}{2}\Gamma\left(\frac{3}{4}\right).$$

(ii) Substitute $x = t^{1/3}$. We get $dx = \dfrac{1}{3}\, t^{-2/3}\, dt$ and

$$I = \int_0^\infty e^{-x^3}\, dx = \frac{1}{3}\int_0^\infty t^{-2/3}\, e^{-t}\, dt = \frac{1}{3}\int_0^\infty t^{(1/3)-1}\, e^{-t}\, dt = \frac{1}{3}\Gamma\left(\frac{1}{3}\right).$$

Example 1.48 Using Beta and Gamma functions, evaluate the integral

$$I = \int_{-1}^1 (1 - x^2)^n\, dx, \text{ where } n \text{ is a positive integer.}$$

Solution We have $I = \displaystyle\int_{-1}^1 (1 + x)^n (1 - x)^n\, dx.$

Let $1 + x = 2t$. Then, $dx = 2dt$ and $1 - x = 2(1 - t)$. We obtain

$$I = 2^{2n+1} \int_0^1 t^n (1 - t)^n\, dt = 2^{2n+1}\, \beta(n + 1, n + 1)$$

$$= 2^{2n+1}\, \frac{\Gamma(n+1)\Gamma(n+1)}{\Gamma(2n+2)} = \frac{2^{2n+1}(n!)^2}{(2n+1)!}.$$

Example 1.49 Express $\int_0^1 x^m(1-x^p)^n \, dx$ in terms of Beta function and hence evaluate the integral $\int_0^1 x^{3/2}(1-\sqrt{x})^{1/2} \, dx$.

Solution Let $x^p = y$. Then $px^{p-1} \, dx = dy$. We obtain

$$I = \int_0^1 x^m(1-x^p)^n \, dx = \frac{1}{p} \int_0^1 y^{(m-p+1)/p}(1-y)^n \, dy$$

$$= \frac{1}{p} \int_0^1 y^{[(m+1)/p-1]}(1-y)^n \, dy = \frac{1}{p} B\left(\frac{m+1}{p}, n+1\right)$$

Now, comparing the integral $\int_0^1 x^{3/2}(1-\sqrt{x})^{1/2} \, dx$ with the given integral, we find that $m = 3/2$, $p = 1/2$ and $n = 1/2$. Therefore,

$$\int_0^1 x^{3/2}(1-\sqrt{x})^{1/2} \, dx = 2B\left(5, \frac{3}{2}\right) = \frac{2\Gamma(5)\Gamma(3/2)}{\Gamma(13/2)}.$$

Now,

$$\Gamma(5) = 4! = 24, \quad \Gamma\left(\frac{13}{2}\right) = \frac{11}{2} \cdot \frac{9}{2} \cdot \frac{7}{2} \cdot \frac{5}{2} \cdot \frac{3}{2} \Gamma\left(\frac{3}{2}\right) = \frac{10395}{32} \Gamma\left(\frac{3}{2}\right).$$

Hence,

$$I = \frac{2(24)(32)\,\Gamma(3/2)}{10395\,\Gamma(3/2)} = \frac{1536}{10395} = \frac{512}{3465}.$$

Example 1.50 Using Beta and Gamma functions, show that for any positive integer m

(i) $\displaystyle \int_0^{\pi/2} \sin^{2m-1}(\theta) \, d\theta = \frac{(2m-2)(2m-4)\ldots 2}{(2m-1)(2m-3)\ldots 3}.$

(ii) $\displaystyle \int_0^{\pi/2} \sin^{2m}(\theta) \, d\theta = \frac{(2m-1)(2m-3)\ldots 1}{(2m)(2m-2)\ldots 2} \frac{\pi}{2}.$

Solution From Eq. (1.76), we obtain

$$B\left(m, \frac{1}{2}\right) = 2\int_0^{\pi/2} \sin^{2m-1}(\theta) \, d\theta \quad \text{and} \quad B\left(m+\frac{1}{2}, \frac{1}{2}\right) = 2\int_0^{\pi/2} \sin^{2m}(\theta) \, d\theta.$$

(i) $I = \displaystyle \int_0^{\pi/2} \sin^{(2m-1)}(\theta) \, d\theta = \frac{1}{2} B\left(m, \frac{1}{2}\right) = \frac{\Gamma(m)\Gamma(1/2)}{2\Gamma(m+1/2)}.$

We have $\Gamma(m) = (m-1)!$, and

$$\Gamma\left(m+\frac{1}{2}\right) = \left(m-\frac{1}{2}\right)\left(m-\frac{3}{2}\right)\ldots\left(\frac{1}{2}\right)\Gamma\left(\frac{1}{2}\right)$$

$$= \frac{1}{2^m}[(2m-1)(2m-3)\ldots 3.1]\,\Gamma\left(\frac{1}{2}\right).$$

Therefore,

$$I = \frac{(m-1)! \, 2^m \, \Gamma(1/2)}{2(2m-1)(2m-3)\ldots 3 \cdot 1 \cdot \Gamma(1/2)} = \frac{2^{m-1}\left[(m-1)(m-2)\ldots 2 \cdot 1\right]}{(2m-1)(2m-3)\ldots 3 \cdot 1}$$

$$= \frac{(2m-2)(2m-4)\ldots 4 \cdot 2}{(2m-1)(2m-3)\ldots 3 \cdot 1}.$$

(ii) $I = \int_0^{\pi/2} \sin^{2m}(\theta) \, d\theta = \frac{1}{2} \beta\left(m + \frac{1}{2}, \frac{1}{2}\right) = \frac{\Gamma(m+1/2)\Gamma(1/2)}{2\Gamma(m+1)}$

$$= \frac{1}{2(m!)} \left[\frac{(2m-1)(2m-3)\ldots 3 \cdot 1}{2^m}\right] (\sqrt{\pi})^2 = \frac{(2m-1)(2m-3)\ldots 3 \cdot 1}{2^{m+1}\left[m(m-1)\ldots 2 \cdot 1\right]} (\pi)$$

$$= \frac{(2m-1)(2m-3)\ldots 3 \cdot 1}{(2m)(2m-2)\ldots 4 \cdot 2} \frac{\pi}{2}.$$

Example 1.51   Evaluate $\int_0^\infty 2^{-9x^2} \, dx$ using the Gamma function.

Solution   We write

$$I = \int_0^\infty 2^{-9x^2} \, dx = \int_0^\infty e^{-9x^2 \ln 2} \, dx$$

Substitute $9x^2 \ln 2 = y$. Then, $x = \dfrac{\sqrt{y}}{3\sqrt{\ln 2}}$ and $dx = \dfrac{y^{-1/2} \, dy}{6\sqrt{\ln 2}}$.

Therefore, $I = \dfrac{1}{6\sqrt{\ln 2}} \int_0^\infty y^{-1/2} \, e^{-y} \, dy = \dfrac{1}{6\sqrt{\ln 2}} \cdot \int_0^\infty y^{(1/2)-1} \, e^{-y} \, dy$

$$= \frac{\Gamma(1/2)}{6\sqrt{\ln 2}} = \frac{1}{6} \sqrt{\frac{\pi}{\ln 2}}.$$

Example 1.52   Show that

$$\Gamma(2n) = \frac{2^{2n-1}}{\sqrt{\pi}} \Gamma\left(n + \frac{1}{2}\right) \Gamma(n) \tag{1.79}$$

and

$$\Gamma\left(\frac{1}{4}\right) \Gamma\left(\frac{3}{4}\right) = \pi\sqrt{2}. \tag{1.80}$$

Solution   From Eq. (1.78), we have

$$\frac{\Gamma(m) \, \Gamma(n)}{\Gamma(m+n)} = \beta(m, n) = 2 \int_0^{\pi/2} \sin^{2m-1}(\theta) \cos^{2n-1}(\theta) \, d\theta. \tag{1.81}$$

Setting $m = n$, we get

$$\frac{[\Gamma(n)]^2}{\Gamma(2n)} = \beta(n, n) = 2 \int_0^{\pi/2} \sin^{2n-1}(\theta) \cos^{2n-1}(\theta) \, d\theta$$

$$= \frac{1}{2^{2n-2}} \int_0^{\pi/2} \sin^{2n-1}(2\theta) \, d\theta.$$

Substituting, $2\theta = \frac{\pi}{2} - \phi$, we get $d\theta = -\frac{1}{2} d\phi$. Hence, we obtain

$$\frac{[\Gamma(n)]^2}{\Gamma(2n)} = \frac{-1}{2^{2n-1}} \int_{\pi/2}^{-\pi/2} \cos^{2n-1}(\phi) \, d\phi$$

$$= \frac{1}{2^{2n-1}} \int_{-\pi/2}^{\pi/2} \cos^{2n-1} \phi \, d\phi = \frac{2}{2^{2n-1}} \int_0^{\pi/2} \cos^{2n-1}(\theta) \, d\theta \qquad (1.82)$$

since $\cos \theta$ is an even function.

Setting $m = 1/2$ in Eq. (1.81), we obtain

$$\frac{\Gamma(n)\Gamma(1/2)}{\Gamma(n+1/2)} = 2 \int_0^{\pi/2} \cos^{2n-1}(\theta) \, d\theta. \qquad (1.83)$$

Comparing Eqs. (1.82) and (1.83), we have

$$\frac{[\Gamma(n)]^2}{\Gamma(2n)} = \frac{1}{2^{2n-1}} \left[\frac{\Gamma(n)\,\Gamma(1/2)}{\Gamma(n+1/2)} \right]$$

or

$$\Gamma(2n) = \frac{2^{2n-1}}{\sqrt{\pi}} \Gamma(n)\,\Gamma(n+1/2), \quad \left(\text{since } \Gamma\left(\frac{1}{2}\right) = \sqrt{\pi} \right)$$

which is the required result.

Setting $n = 1/4$ in Eq. (1.79), we obtain

$$\Gamma\left(\frac{1}{2}\right) = \frac{2^{-1/2}}{\sqrt{\pi}} \Gamma\left(\frac{1}{4}\right) \Gamma\left(\frac{3}{4}\right)$$

$$\Gamma\left(\frac{1}{4}\right) \Gamma\left(\frac{3}{4}\right) = \pi\sqrt{2}\,.$$

Example 1.53   Show that $\int_0^{\pi/2} \sqrt{\tan x} \, dx = \frac{\pi}{\sqrt{2}}$.

Solution   We have

$$I = \int_0^{\pi/2} \sqrt{\tan x} \, dx = \int_0^{\pi/2} \sin^{1/2} x \cos^{-1/2} x \, dx = \frac{1}{2} \beta\left(\frac{3}{4}, \frac{1}{4}\right)$$

$$= \frac{1}{2} \frac{\Gamma(1/4)\Gamma(3/4)}{\Gamma(1)} = \frac{\pi\sqrt{2}}{2} = \frac{\pi}{\sqrt{2}} \quad \text{(using Eq. (1.80))}.$$

1.4.5   Improper Integrals Involving a Parameter

Often, we come across integrals of the form

$$\phi(\alpha) = \int_{a(\alpha)}^{b(\alpha)} f(x, \alpha)\, dx \qquad (1.84)$$

where α is a parameter and the integrand f is such that the integral cannot be evaluated by standard methods. We can evaluate some of these integrals by differentiating the integral with respect to the parameter, that is first obtain $\phi'(\alpha)$, evaluate the integral (that is integrate with respect to x) and then integrate $\phi'(\alpha)$ with respect to α. Note that f is a function of two variables x and α. When we differentiate f with respect to α, we treat x as a constant and denote the derivative as $\partial f / \partial \alpha$ (partial derivative of f with respect to α, Chapter 2 discusses partial derivatives in detail). We assume that f, $\partial f / \partial \alpha$, $a(\alpha)$ and $b(\alpha)$ are continuous functions of α.

We now present the formula which gives the derivative of $\phi(\alpha)$.

Theorem 1.16 (Leibniz formula)   If $a(\alpha)$, $b(\alpha)$, $f(x, \alpha)$ and $\partial f / \partial \alpha$ are continuous functions of α, then

$$\frac{d\phi}{d\alpha} = \int_{a(\alpha)}^{b(\alpha)} \frac{\partial f}{\partial \alpha}(x, \alpha)\, dx + f(b, \alpha)\frac{db}{d\alpha} - f(a, \alpha)\frac{da}{d\alpha}. \qquad (1.85)$$

Proof   Let $\Delta\alpha$ be an increment in α and Δa, Δb be the corresponding increments in a and b. We have

$$\Delta\phi = \phi(\alpha + \Delta\alpha) - \phi(\alpha) = \int_{a+\Delta a}^{b+\Delta b} f(x, \alpha + \Delta a)\, dx - \int_a^b f(x, \alpha)\, dx$$

$$= \int_{a+\Delta a}^{a} f(x, \alpha + \Delta\alpha)\, dx + \int_a^b f(x, \alpha + \Delta\alpha)\, dx + \int_b^{b+\Delta b} f(x, \alpha + \Delta\alpha)\, dx - \int_a^b f(x, \alpha)\, dx$$

or
$$\frac{\Delta\phi}{\Delta\alpha} = \int_{a+\Delta a}^{a} \frac{1}{\Delta\alpha} f(x, \alpha + \Delta\alpha)\, dx + \int_a^b \frac{1}{\Delta\alpha}[f(x, \alpha + \Delta\alpha) - f(x, \alpha)]\, dx$$

$$+ \int_b^{b+\Delta b} \frac{1}{\Delta\alpha} f(x, \alpha + \Delta\alpha)\, dx. \qquad (1.86)$$

Using the mean value theorem of integrals

$$\int_{x_0}^{x_1} f(x)\, dx = (x_1 - x_0)\, f(\xi), \quad x_0 < \xi < x_1$$

we get
$$\int_{a+\Delta a}^{a} f(x, \alpha + \Delta\alpha)\, dx = -\Delta a\, f(\xi_1, \alpha + \Delta\alpha), \quad a < \xi_1 < a + \Delta a \qquad (1.87)$$

and
$$\int_b^{b+\Delta b} f(x, \alpha + \Delta\alpha)\, dx = \Delta b\, f(\xi_2, \alpha + \Delta\alpha), \quad b < \xi_2 < b + \Delta b. \qquad (1.88)$$

Using the Lagrange mean value theorem, we get

$$f(x, \alpha + \Delta\alpha) - f(x, \alpha) = \Delta\alpha\, \frac{\partial f}{\partial \alpha}(x, \xi_3), \quad \alpha < \xi_3 < \alpha + \Delta\alpha. \qquad (1.89)$$

We note that

$$\lim_{\Delta\alpha\to 0}\xi_1 = a, \ \lim_{\Delta\alpha\to 0}\xi_2 = b \ \text{ and } \ \lim_{\Delta\alpha\to 0}\xi_3 = \alpha. \tag{1.90}$$

Taking limits as $\Delta\alpha \to 0$ on both sides of Eq. (1.86) and using the results in Eqs. (1.87) to (1.90), we obtain

$$\frac{d\phi}{d\alpha} = \int_a^b \frac{\partial f}{\partial \alpha}(x, \alpha)\, dx + f(b, \alpha)\frac{db}{d\alpha} - f(a, \alpha)\frac{da}{d\alpha}.$$

Remark 6

(a) If the limits $a(\alpha)$ and $b(\alpha)$ are constants, then we obtain from Eq. (1.85)

$$\frac{d\phi}{d\alpha} = \int_a^b \frac{\partial f}{\partial \alpha}(x, \alpha)\, dx. \tag{1.90}$$

(b) If the integrand f is independent of α, then we obtain from Eq. (1.85)

$$\frac{d\phi}{d\alpha} = f(b)\frac{db}{d\alpha} - f(a)\frac{da}{d\alpha}. \tag{1.92}$$

Example 1.54   Evaluate the integral $\int_0^\infty \dfrac{e^{-\alpha x}\sin x}{x}\, dx,\ \alpha > 0$ and deduce that

(i) $\displaystyle\int_0^\infty \frac{\sin x}{x}\, dx = \frac{\pi}{2},$

(ii) $\displaystyle\int_0^\infty \frac{\sin ax}{x}\, dx = \frac{\pi}{2},\ a > 0.$

Solution   Let $\phi(\alpha) = \displaystyle\int_0^\infty \frac{e^{-\alpha x}\sin x}{x}\, dx.$ \hfill (1.93)

The limits of integration are independent of the parameter α. we obtain

$$\frac{d\phi}{d\alpha} = \int_0^\infty \frac{\partial}{\partial \alpha}\left[\frac{e^{-\alpha x}\sin x}{x}\right] dx = -\int_0^\infty \frac{x e^{-\alpha x}\sin x}{x}\, dx = -\int_0^\infty e^{-\alpha x}\sin x\, dx.$$

Using the result $\displaystyle\int e^{-\alpha x}\sin x\, dx = \frac{e^{-\alpha x}}{1+\alpha^2}(\alpha\sin x + \cos x)$, we obtain

$$\frac{d\phi}{d\alpha} = \left[\frac{e^{-\alpha x}}{1+\alpha^2}(\alpha\sin x + \cos x)\right]_0^\infty = -\frac{1}{1+\alpha^2}.$$

Integrating with respect to α, we get

$$\phi(\alpha) = -\tan^{-1}\alpha + c, \text{ where } c \text{ is the constant of integration.}$$

From Eq. (1.93), we get the condition $\phi(\infty) = 0$. Hence,

$$\phi(\infty) = 0 = -\tan^{-1}\infty + c, \quad \text{or} \quad c = \pi/2.$$

Therefore, $$\phi(\alpha) = \int_0^\infty \frac{e^{-\alpha x} \sin x}{x}\, dx = \frac{\pi}{2} - \tan^{-1}\alpha.$$

(i) Setting $\alpha = 0$, we obtain $\int_0^\infty \frac{\sin x}{x}\, dx = \frac{\pi}{2}.$ (1.94)

(ii) Substituting $x = ay$ on the left hand side of Eq. (1.94), we obtain

$$\int_0^\infty \frac{\sin x}{x} = \int_0^\infty \frac{\sin ay}{y}\, dy = \frac{\pi}{2}.$$

Example 1.55 Using the result $\int_0^\infty e^{-x^2}\, dx = \frac{\sqrt{\pi}}{2}$, evaluate the integral $\int_0^\infty e^{-x^2} \cos(2\alpha x)\, dx.$

Solution Let $\phi(\alpha) = \int_0^\infty e^{-x^2} \cos(2\alpha x)\, dx.$ (1.95)

The limits of integration are independent of the parameter α. Hence,

$$\frac{d\phi}{d\alpha} = \int_0^\infty \frac{\partial}{\partial \alpha}[e^{-x^2}\cos(2\alpha x)]\, dx = \int_0^\infty (-2x)\, e^{-x^2} \sin(2\alpha x)\, dx$$

$$= [e^{-x^2}\sin(2\alpha x)]_0^\infty - 2\alpha \int_0^\infty e^{-x^2}\cos(2\alpha x)\, dx = -2\alpha\phi.$$

Integrating the differential equation $\dfrac{d\phi}{d\alpha} + 2\alpha\phi = 0$, we obtain $\phi(\alpha) = ce^{-\alpha^2}.$

From Eq. (1.95), we get the condition $\phi(0) = \int_0^\infty e^{-x^2}\, dx = \frac{\sqrt{\pi}}{2}.$

Using this condition, we obtain $\phi(0) = \dfrac{\sqrt{\pi}}{2} = c.$

Therefore, $\phi(\alpha) = \int_0^\infty e^{-x^2}\cos(2\alpha x)\, dx = \dfrac{\sqrt{\pi}}{2}e^{-\alpha^2}.$

Example 1.56 Evaluate the integral $\int_0^\infty \dfrac{\tan^{-1}(ax)}{x(1+x^2)}\, dx,\ a > 0$ and $a \neq 1.$

Solution Let $\phi(a) = \int_0^\infty \dfrac{\tan^{-1}(ax)}{x(1+x^2)}\, dx.$ (1.96)

We have

$$\frac{d\phi}{da} = \int_0^\infty \frac{\partial}{\partial a}\left[\frac{\tan^{-1}(ax)}{x(1+x^2)}\right]\, dx = \int_0^\infty \frac{dx}{(1+x^2)(1+a^2x^2)}$$

$$= \frac{1}{a^2 - 1}\int_0^\infty \left[\frac{a^2}{a^2x^2 + 1} - \frac{1}{1+x^2}\right]\, dx$$

$$= \frac{1}{a^2 - 1} \left[\left\{ a \tan^{-1}(ax) \right\}_0^\infty - \left\{ \tan^{-1}(x) \right\}_0^\infty \right] = \frac{\pi}{2} \left[\frac{a-1}{a^2 - 1} \right] = \frac{\pi}{2(a+1)}.$$

Integrating with respect to a, we obtain

$$\phi(a) = \frac{\pi}{2} \ln(a+1) + c.$$

From Eq. (1.96), we get the condition $\phi(0) = 0$. Using this condition, we obtain $\phi(0) = 0 = c$.

Therefore, $\phi(a) = \displaystyle\int_0^\infty \frac{\tan^{-1}(ax)}{x(1+x^2)} dx = \frac{\pi}{2} \ln(a+1).$

1.4.6 Error Functions

Error functions arise in the theory of probability and solution of certain types of partial differential equations (see section 8.7).

Let us first consider the following function that arises in defining the normal probability distribution (the case when mean $= \mu = 0$ and variance $= \sigma^2 = 1$)

$$f(x) = \frac{1}{\sqrt{2\pi}} e^{-x^2/2}. \tag{1.97}$$

This function is also called the *Gaussian function*. The bell shaped *normal curve* defined by Eq. (1.97) is given in Fig. 1.20. The area under the curve and above the x-axis is given by

$$I = \int_{-\infty}^\infty f(x)\, dx = \frac{1}{\sqrt{2\pi}} \int_{-\infty}^\infty e^{-x^2/2}\, dx. \tag{1.98}$$

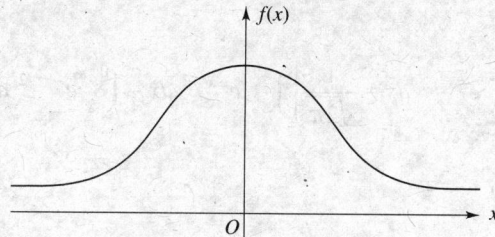

Fig. 1.20. Normal curve.

Setting $u = x/\sqrt{2}$, we get

$$I = \frac{1}{\sqrt{\pi}} \int_{-\infty}^\infty e^{-u^2}\, du.$$

It was shown in equation (1.71), that

$$\int_0^\infty e^{-u^2}\, du = \frac{1}{2} \Gamma\left(\frac{1}{2}\right) = \frac{\sqrt{\pi}}{2}.$$

Hence,
$$I = \int_{-\infty}^{\infty} f(x)\, dx = \frac{1}{\sqrt{\pi}} \int_{-\infty}^{\infty} e^{-u^2}\, du = \frac{1}{\sqrt{\pi}} (2)\left(\frac{\sqrt{\pi}}{2}\right) = 1,$$

that is, the total area under the normal curve is 1. Since the area is symmetric about the y-axis, we get

$$\int_{-\infty}^{0} f(x)\, dx = \int_{0}^{\infty} f(x)\, dx = \frac{1}{2}.$$

The area under the curve, from $-\infty$ to any point z, is given by

$$\phi(z) = \frac{1}{\sqrt{2\pi}} \int_{-\infty}^{z} e^{-x^2/2}\, dx. \tag{1.99}$$

Hence, by definition $\phi(0) = 1/2$. The function $\phi(z)$ is called the distribution function of the normal distribution with mean 0 and variance 1. Setting $x = -y$ in Eq. (1.99), we get $dx = -dy$ and

$$\phi(z) = -\frac{1}{\sqrt{2\pi}} \int_{\infty}^{-z} e^{-y^2/2}\, dy = \frac{1}{\sqrt{2\pi}} \int_{-z}^{\infty} e^{-y^2/2}\, dy$$

$$= \frac{1}{\sqrt{2\pi}} \int_{-\infty}^{\infty} e^{-y^2/2}\, dy - \frac{1}{\sqrt{2\pi}} \int_{-\infty}^{-z} e^{-y^2/2}\, dy$$

$$= 1 - \phi(-z)$$

or
$$\phi(-z) = 1 - \phi(z). \tag{1.100}$$

Values of the distribution function are tabulated for various values of z. Further, the area under the curve from $x = 0$ to $x = z$ is given by

$$I(z) = \frac{1}{\sqrt{2\pi}} \int_{0}^{z} e^{-x^2/2}\, dx = \frac{1}{\sqrt{2\pi}} \left[\int_{-\infty}^{z} e^{-x^2/2}\, dx - \int_{-\infty}^{0} e^{-x^2/2}\, dx \right] = \phi(z) - \frac{1}{2} \tag{1.101}$$

or
$$\phi(z) = \frac{1}{2} + I(z).$$

Error function $erf(x)$

The *error function* is also called the *error integral function*. It is defined by

$$erf(x) = \frac{2}{\sqrt{\pi}} \int_{0}^{x} e^{-t^2}\, dt. \tag{1.102}$$

Let $t^2 = u$. Then, $dt = \frac{1}{2t}\, du = \frac{1}{2\sqrt{u}}\, du$, and

$$erf(x) = \frac{1}{\sqrt{\pi}} \int_{0}^{x^2} u^{-1/2}\, e^{-u}\, du. \tag{1.103}$$

This is another form of the error function. Using this definition, we obtain

$$erf(\infty) = \frac{1}{\sqrt{\pi}} \int_0^{\infty} u^{-1/2} e^{-u} \, du = \frac{1}{\sqrt{\pi}} \Gamma\left(\frac{1}{2}\right) = \frac{1}{\sqrt{\pi}} (\sqrt{\pi}) = 1. \tag{1.104}$$

Let $t^2 = u^2/2$ in Eq. (1.102). Then,

$$2t \, dt = u \, du, \quad dt = \frac{u}{2t} \, du = \frac{du}{\sqrt{2}}, \quad \text{and}$$

$$erf(x) = \frac{2}{\sqrt{\pi}} \int_0^{\sqrt{2}x} e^{-u^2/2} \frac{du}{\sqrt{2}} = \frac{\sqrt{2}}{\sqrt{\pi}} \int_0^{\sqrt{2}x} e^{-u^2/2} \, du. \tag{1.105}$$

Using Eq. (1.101), we can write

$$I(\sqrt{2}\,x) = \frac{1}{\sqrt{2\pi}} \int_0^{\sqrt{2}x} e^{-x^2/2} \, dx = \phi(\sqrt{2}\,x) - \frac{1}{2}.$$

Therefore, $\qquad\qquad erf(x) = 2I(\sqrt{2}\,x) = 2\phi(\sqrt{2}\,x) - 1. \tag{1.106}$

Hence, the error function can be evaluated using this relation.

Complementary error function $erfc(x)$

Using the definition of the error function given in Eqs. (1.103) and (1.104), we write

$$erf(x) = \frac{1}{\sqrt{\pi}} \int_0^{x^2} u^{-1/2} e^{-u} \, du = \frac{1}{\sqrt{\pi}} \int_0^{\infty} u^{-1/2} e^{-u} \, du - \frac{1}{\sqrt{\pi}} \int_{x^2}^{\infty} u^{-1/2} e^{-u} \, du$$

$$= 1 - \frac{1}{\sqrt{\pi}} \int_{x^2}^{\infty} u^{-1/2} e^{-u} \, du = 1 - erfc(x) \tag{1.107}$$

where we define

$$erfc(x) = \frac{1}{\sqrt{\pi}} \int_{x^2}^{\infty} u^{-1/2} e^{-u} \, du. \tag{1.108}$$

The function $erfc(x)$ is called the complementary error function.

Using Eqs. (1.107), (1.102) and (1.104), we can write

$$erfc(x) = 1 - erf(x) = 1 - \frac{2}{\sqrt{\pi}} \int_0^x e^{-t^2} \, dt$$

$$= \frac{2}{\sqrt{\pi}} \int_0^{\infty} e^{-t^2} \, dt - \frac{2}{\sqrt{\pi}} \int_0^x e^{-t^2} \, dt = \frac{2}{\sqrt{\pi}} \int_x^{\infty} e^{-t^2} \, dt. \tag{1.109}$$

Eqs. (1.102) and (1.109) are the commonly used definitions of error function and complementary error function respectively. The graphs of $erf(x)$ and $erfc(x)$ are given in Fig. 1.21.

Some properties of error functions

   1. $erf(-x) = -erf(x)$.   (1.110)

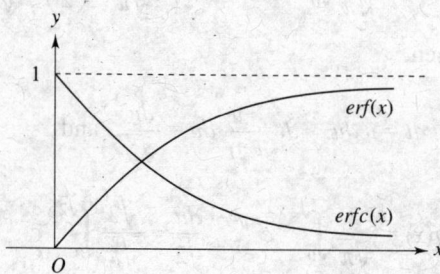

Fig. 1.21.   Error function and complementary error function.

Using the definition given in Eq. (1.102), we get

$$erf(-x) = \frac{2}{\sqrt{\pi}} \int_0^{-x} e^{-t^2} dt = \frac{2}{\sqrt{\pi}} \int_0^x e^{-u^2} (-du) \quad (\text{setting } t = -u)$$

$$= -erf(x).$$ (1.111)

2. $erfc(-x) = 1 + erf(x) = 2 - erfc(x)$.

Using Eq. (1.107), we get

$$erfc(-x) = 1 - erf(-x) = 1 + erf(x)$$
$$= 1 + [1 - erfc(x)] = 2 - erfc(x).$$

3. *Derivative of error function:* We have

$$\frac{d}{dx}[erf(\alpha x)] = \frac{2\alpha}{\sqrt{\pi}} e^{-\alpha^2 x^2}.$$ (1.112)

From the definition, we have

$$erf(\alpha x) = \frac{2}{\sqrt{\pi}} \int_0^{\alpha x} e^{-t^2} dt.$$ (1.113)

Consider x as a parameter. Comparing Eq. (1.113) with Eq. (1.84)

$$\phi(a) = \int_{a(\alpha)}^{b(\alpha)} f(x, \alpha) \, d\alpha, \quad \text{where } \alpha \text{ is a parameter}$$ (1.114)

we get $f(t, x) = \frac{2}{\sqrt{\pi}} e^{-t^2}$, $b(x) = \alpha x$, $a(x) = 0$, $\phi(x) = erf(\alpha x)$.

Using Eq. (1.85)

$$\frac{d\phi}{d\alpha} = \int_{a(\alpha)}^{b(\alpha)} \frac{\partial f}{\partial \alpha}(x, \alpha) \, dx + f(b, \alpha) \frac{db}{d\alpha} - f(a, \alpha) \frac{da}{d\alpha}$$ (1.115)

we obtain

$$\frac{d}{dx}[erf(\alpha x)] = \frac{2}{\sqrt{\pi}} \int_0^{\alpha x} \frac{\partial}{\partial x}(e^{-t^2})dt + f(\alpha x, x)\frac{d}{dx}(\alpha x) - 0$$

$$= \frac{2\alpha}{\sqrt{\pi}} e^{-\alpha^2 x^2}.$$

4. *Integral of error function:* We have

$$\int_0^t erf(\alpha x)dx = t\, erf(\alpha t) + \frac{1}{\alpha\sqrt{\pi}}[e^{-\alpha^2 t^2} - 1]. \tag{1.116}$$

Integrating the left hand side by parts, we obtain

$$\int_0^t 1 \cdot erf(\alpha x)\, dx = [x\, erf(\alpha x)]_0^t - \int_0^t x\frac{d}{dx}[erf(\alpha x)]dx$$

$$= t\, erf(\alpha t) - \frac{2\alpha}{\sqrt{\pi}}\int_0^t x e^{-\alpha^2 x^2}\, dx$$

using Eq. (1.112). Let $\alpha^2 x^2 = u$. Then, $2\alpha^2 x\, dx = du$ or $x\, dx = du/(2\alpha^2)$. Hence,

$$\int_0^t erf(\alpha x)\, dx = t\, erf(\alpha t) - \left(\frac{2\alpha}{\sqrt{\pi}}\right)\left(\frac{1}{2\alpha^2}\right)\int_0^{\alpha^2 t^2} e^{-u}\, du$$

$$= t\, erf(\alpha t) + \frac{1}{\alpha\sqrt{\pi}}\left[e^{-u}\right]_0^{\alpha^2 t^2}$$

$$= t\, erf(\alpha t) + \frac{1}{\alpha\sqrt{\pi}}\left[e^{-\alpha^2 t^2} - 1\right].$$

Exercise 1.3

In problems **1** to **25**, discus the convergence or divergence of the given improper integral. Find its value if it exists.

1. $\displaystyle\int_0^\infty \frac{dx}{4+x}.$

2. $\displaystyle\int_2^\infty \frac{\ln x}{x}\, dx.$

3. $\displaystyle\int_3^\infty \frac{dx}{x^2 + 2x}.$

4. $\displaystyle\int_0^\infty \frac{x\, dx}{x^4 + 1}.$

5. $\displaystyle\int_0^\infty x^2 e^{-ax}\, dx,\ a > 0.$

6. $\displaystyle\int_0^\infty e^{-ax}\sin bx\, dx,\ a > 0.$

7. $\displaystyle\int_1^\infty x e^{-x^2}\, dx.$

8. $\displaystyle\int_{-\infty}^\infty \frac{dx}{x^2 + 2x + 2}.$

9. $\displaystyle\int_{-\infty}^\infty \frac{dx}{e^x + e^{-x}}.$

10. $\displaystyle\int_0^\infty e^{-x}\, dx.$

11. $\displaystyle\int_0^\infty \frac{x^{p-1}}{1+x}\, dx,\ 0 < p < 1.$

12. $\displaystyle\int_1^\infty \frac{x+4}{x^{3/2}}\, dx.$

13. $\displaystyle\int_0^\infty \frac{dx}{x^3 + 1}.$

14. $\displaystyle\int_1^3 \frac{dx}{x\ln x}.$

15. $\displaystyle\int_0^4 \frac{dx}{x^2 - 2x - 8}.$

16. $\displaystyle\int_0^{\pi/2} \frac{dx}{\cos x}.$

17. $\displaystyle\int_0^2 \ln x \, dx.$

18. $\displaystyle\int_0^1 \frac{x}{\sqrt{1-x^2}} \, dx.$

19. $\displaystyle\int_{-1}^1 \frac{dx}{x^4}.$

20. $\displaystyle\int_0^1 \frac{dx}{\sqrt{x}+x^3}.$

21. $\displaystyle\int_1^3 \frac{\sqrt{x}}{\ln x} \, dx.$

22. $\displaystyle\int_0^{\pi/2} \frac{\sin^n x}{x^m} \, dx.$

23. $\displaystyle\int_2^\infty \frac{\sin x}{x(\ln x)^2} \, dx.$

24. $\displaystyle\int_0^\pi \frac{\cos x}{\sqrt{x}} \, dx.$

25. $\displaystyle\int_0^\infty \frac{x^p}{1+x^q} \, dx,$ (i) $q \geq 0$, (ii) $q < 0$.

In problems **27** to **40**, evaluate the integrals using the Beta and Gamma functions.

26. $\displaystyle\int_0^{\pi/2} \frac{dx}{\sqrt{\sin x}}.$

27. $\displaystyle\int_0^{\pi/2} \sin^2\theta \cos^4\theta \, d\theta.$

28. $\displaystyle\int_0^{\pi/2} \sin^3\theta \cos^5\theta \, d\theta.$

29. $\displaystyle\int_0^{\pi/2} \cos^m\theta \, d\theta,$ m integer.

30. $\displaystyle\int_0^a x\sqrt{a^3-x^3} \, dx.$

31. $\displaystyle\int_0^1 \frac{dx}{\sqrt[3]{1-x^3}}.$

32. $\displaystyle\int_0^1 x^n(\ln x)^m \, dx.$

33. $\displaystyle\int_0^a \frac{x^{3/2}}{\sqrt{a^2-x^2}} \, dx.$

34. $\displaystyle\int_0^1 \frac{dx}{\sqrt{-\ln x}}.$

35. $\displaystyle\int_0^1 x^k(1-x)^{n-k} \, dx, k>0.$

36. $\displaystyle\int_0^\infty \frac{dx}{1+x^4}.$

37. $\displaystyle\int_0^\infty \frac{x^a}{a^x} \, dx,\ a>1.$

38. $\displaystyle\int_0^\infty t^k e^{-st} \, dt, s>0, k>0.$

39. $\displaystyle\int_0^\infty t^4 e^{-2t^2} \, dt.$

40. $\displaystyle\int_0^\infty x^{1/3} e^{-x^2} \, dx.$

Establish the following results.

41. $\displaystyle\int_0^p x^m(p^q - x^q)^n \, dx = \left(\frac{p^{m+nq+1}}{q}\right) \beta\left(n+1, \frac{m+1}{q}\right),$ m, n, p, q are positive constants.

42. $\displaystyle\int_a^b (x-a)^{m-1}(b-x)^{n-1} \, dx = (b-a)^{m+n-1} \beta(m,n),$ m, n, a, b are positive constants.

43. $\displaystyle\int_{-\infty}^\infty \frac{e^{mx}}{ae^{nx}+b} \, dx = \frac{\pi}{n}\left(\frac{b}{a}\right)^{m/n}\left[\frac{1}{b\sin(m\pi/n)}\right],$ a, b, m, n are positive constants.

44. $\displaystyle\int_{-1}^1 (1-x^2)^n \, dx = \frac{2^{2n+1}(n!)^2}{(2n+1)!},$ n is positive integer.

45. $\displaystyle\int_0^m x^n\left(1-\frac{x}{m}\right)^{m-1} \, dx = m^{n+1} \beta(m, n+1),$ $m, n,$ are positive constants.

46. $\displaystyle\int_0^\infty x^m e^{-\alpha x^n} \, dx = \frac{1}{n\alpha^{(m+1)/n}} \Gamma\left(\frac{m+1}{n}\right),$ m, n, α are positive constants.

47. $\int_0^\infty e^{-mx}(1 - e^{-x})^n \, dx = \beta \,(m, \, n+1), \quad m, \, n,$ are positive constants.

48. $\int_0^1 x^m (\ln x)^n \, dx = \dfrac{(-1)^n \, n!}{(m+1)^{n+1}}, \quad m > 1$ and n is a positive integer.

49. $\int_0^\infty \dfrac{\sin x}{x^p} \, dx = \dfrac{p}{2\Gamma(p) \sin (p\pi/2)}, \quad 0 < p < 1, \quad \left(\text{given that } \int_0^\infty \dfrac{x^{p-1}}{1+x} \, dx = \dfrac{\pi}{\sin p\pi} \right).$

50. For large n, $n! \approx \sqrt{2\pi n} \; n^n \, e^{-n}$ (*Stirling's formula*).

51. $\int_0^1 \dfrac{x^{m-1}(1-x)^{n-1}}{(a+bx)^{m+n}} \, dx = \dfrac{1}{a^n (a+b)^m} \, \beta(m, n).$

Using the concept of differentiation of integrals (assuming that the differentiation is valid) evaluate the following integrals:

52. $\int_0^\infty \dfrac{e^{-ax} - e^{-bx}}{x} \, dx, \quad a > 0, \, b > 0.$

53. $\int_0^1 \dfrac{x^a - x^b}{\log x} \, dx, \quad a > b > -1.$

54. $\int_0^1 x^n (\log x)^k \, dx, \quad n$ any integer $> -1.$

55. $\int_0^{\pi/2\alpha} \alpha \sin \alpha x \, dx.$

56. $\int_{-\infty}^\infty x^2 \, e^{-x^2} \, dx, \quad$ where $\int_{-\infty}^\infty e^{-x^2} \, dx = \sqrt{\pi}$

57. $\int_0^{\alpha^3} \cot^{-1}(x/\alpha^3) \, dx.$

58. $\int_0^\pi \dfrac{\cos x}{(a+b\cos x)^3} \, dx, \quad$ given that $\int_0^\pi \dfrac{dx}{a+b\cos x} = \dfrac{\pi}{\sqrt{a^2 - b^2}}, \quad a > b > 0,$

59. $\int_0^\infty e^{-x^2 - (a^2/x^2)} \, dx, \, a > 0, \quad$ given that $\int_0^\infty e^{-x^2} \, dx = \dfrac{\sqrt{\pi}}{2},$

60. $\int_0^{\pi/2} \log(1 - \alpha^2 \sin^2 x) \, dx, \quad |x| < 1.$

61. $\int_0^\infty \dfrac{dx}{(x^2+1)^{n+1}}, \quad n$ any positive integer.

62. Show that $\dfrac{d}{dx} \, [erfc \,(\alpha x)] = -\dfrac{2\alpha}{\sqrt{\pi}} \, e^{-\alpha^2 x^2}.$

63. Show that $erf(x) = \dfrac{2}{\sqrt{\pi}} \left[x - \dfrac{x^3}{3(1!)} + \dfrac{x^5}{5(2!)} - \dfrac{x^7}{7(3!)} + \ldots \right].$

64. Show that $\int_0^t erfc \,(\alpha x) \, dx = t \, erfc \,(\alpha t) - \dfrac{1}{\alpha\sqrt{\pi}} \, [e^{-\alpha^2 t^2} - 1].$

65. Show that $\int_0^\infty e^{-t^2 - 2\alpha t} \, dt = \dfrac{\sqrt{\pi}}{2} \, e^{\alpha^2} \, [1 - erf(\alpha)].$

66. The relation $\int_0^\infty e^{-x^2} \cos (2bx) \, dx = \dfrac{\sqrt{\pi}}{2} \, e^{-b^2}$ is given (see Example 13.40). Deduce the result for

$\int_0^\infty e^{-\alpha^2 x^2} \cos (px) \, dx.$ Integrate this result with respect to p, taken as a parameter, from $p = 0$ to $p = s$

and show that $\int_0^\infty e^{-\alpha^2 x^2} \left(\dfrac{\sin(sx)}{x} \right) dx = \dfrac{\pi}{2} \, erf \left(\dfrac{s}{2\alpha} \right).$

1.5 Some Properties of Curves and Curve Sketching

We now discuss some properties of curves defined by $y = f(x)$ or $f(x, y) = 0$ and sketch the graphs of these curves.

1.5.1   Convexity and Concavity of a Curve

Let $f(x)$ have continuous first and second order derivatives in an open interval (a, b). Then, we define the convexity and concavity of a curve as follows.

Concave upward. A curve defined by $y = f(x)$, $a \le x \le b$, is said to be concave upward if and only if the derivative $f'(x)$ is an increasing function on (a, b). In this case, all points on the curve in the interval (a, b) lie above the tangent to the curve at any point in this interval. This means that, as x increases, the tangent at the point $(x, f(x))$ turns in the anti-clockwise direction (Fig. 1.22). In terms of the second order derivative, we define that a curve is concave upward if $f''(x) > 0$ on (a, b). Such a function $f(x)$ is also called a *convex function*.

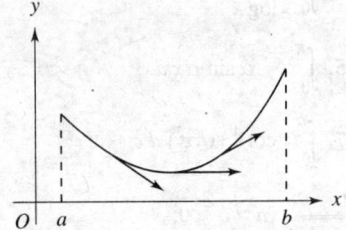

Fig. 1.22.   Concave upward curve.

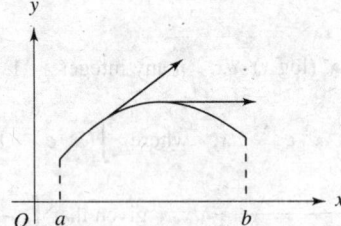

Fig. 1.23.   Concave downward curve.

Concave downward.   A curve defined by $y = f(x)$, $a \le x \le b$, is said to be concave downward if and only if the derivative $f'(x)$ is a decreasing function on (a, b). In this case, all points on the curve in the interval (a, b) lie below the tangent to the curve at any point in the interval. This means that, as x increases, the tangent at the point $(x, f(x))$ turns in the clockwise direction (Fig. 1.23). In terms of the second order derivative, we define that a curve is concave downward if $f''(x) < 0$ on (a, b). Such a function $f(x)$ is also called a *concave function*.

It is possible that a given curve may be concave upward in part of an interval and concave downward in the other part of the interval. In such cases we define the following:

Point of inflection. A point $P(c, f(c))$, $a < c < b$, on the curve $y = f(x)$ at which the curve changes its concavity from concave upward to concave downward or from concave downward to concave upward is called a point of inflection. Since concavity changes at the point of inflection $P(c, f(c))$, $f''(c - h)$ and $f''(c + h)$ must be of opposite signs, when $h > 0$ is sufficiently small. Hence, if $f''(x)$ is continuous, then we obtain the condition that at the point of inflection, $f''(c) = 0$. Then tangent line at the point of inflection, having on one side a part of the curve which is concave downward (or upward) and having on the other side a part of the curve which is concave upward (or downward) must cross the curve at the point of inflection (Fig. 1.24).

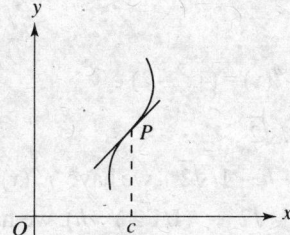

Fig. 1.24. Point of inflection.

A point at which $f''(x)$ does not exist, but $f''(x+h)$ and $f''(x-h)$ are of opposite signs for sufficiently small $h > 0$, is also called a point of inflection.

Therefore, to find the points of inflection, we first find the points at which $f''(x) = 0$. Then, we examine whether the sign of $f''(x)$ changes as x crosses these points.

It may be pointed out that $f''(x)$ may also vanish at a point which is not a point of inflection. For example, the curve $y = f(x) = x^4$ is always concave upwards as $y' = f'(x) = 4x^3$ is always an increasing function. Now, $f''(x) = 12x^2 = 0$ gives $x = 0$ as a possible point of inflection. But $f''(0 + h)$ and $f''(0 - h)$ have the same sign. Hence, $x = 0$ is not a point of inflection.

Example 1.57 Find the intervals in which the following curves

(i) $y = 3x^4 + 4x^3 - 6x^2 + 12x + 12$,   (ii) $y = e^{-x^2}$,   (iii) $y = (x + 1)^{1/3}$

are concave upward or concave downward. Also find the points of inflection.

Solution

(i) We have

$$f(x) = 3x^4 + 4x^3 - 6x^2 + 12x + 12, \quad f'(x) = 12x^3 + 12x^2 - 12x + 12,$$
$$f''(x) = 36x^2 + 24x - 12 = 12(3x^2 + 2x - 1) = 12(3x - 1)(x + 1).$$

Setting $f''(x) = 0$, we get $x = 1/3$ and $x = -1$.

Now, in the interval $(1/3 - h, 1/3)$ we have $f''(x) < 0$. The curve is concave downward in this interval.

In the interval $((1/3), (1/3) + h)$ we have $f''(x) > 0$. The curve is concave upward in this interval. We find that $f(1/3) = 419/27$. Therefore, the point $(1/3, 419/27)$ is a point of inflection.

In the interval $(-1 - h, -1)$, we have $f''(x) > 0$. The curve is concave upward is this interval. In the interval $(-1, -1 + h)$, we have $f''(x) < 0$. The curve is concave downward is this interval. We find that $f(-1) = -7$. Therefore, the point $(-1, -7)$ is a point of inflection.

We find that the curve is concave upward is the intervals $(-\infty, -1)$ and $(1/3, \infty)$ and concave downward in the interval $(-1, 1/3)$ (Fig. 1.25i)

	Concave		Concave		Concave	
$-\infty$	upward	-1	downward	$1/3$	upward	$-\infty$

Fig. 125 i. Intervals of concavity for Example 1.57 i.

(ii) We have

$$f(x) = e^{-x^2}, \ f'(x) = -2xe^{-x^2}, \ f''(x) = (4x^2 - 2)\,e^{-x^2}.$$

Setting $f''(x) = 0$, we get $x = \pm\, 1/\sqrt{2}$.

Now, in the interval $(-(1/\sqrt{2}) - h, -1/\sqrt{2})$, we have $f''(x) > 0$. The curve is concave upward in this interval. In the interval $(-1/\sqrt{2}, -(1/\sqrt{2}) + h)$, we have $f''(x) < 0$. The curve is concave downward in this interval. We find that $f(-1/\sqrt{2}) = 1/\sqrt{e}$. Therefore, the point $(-1/\sqrt{2}, 1/\sqrt{e})$ is a point of inflection.

In the interval $((1/\sqrt{2}) - h, 1/\sqrt{2})$, we have $f''(x) < 0$. The curve is concave downward in this interval. In the interval $(1/\sqrt{2}, (1/\sqrt{2}) + h)$, we have $f''(x) > 0$. The curve is concave upward in this interval. We find that $f(1/\sqrt{2}) = 1/\sqrt{e}$. Therefore, the point $(1/\sqrt{2}, 1/\sqrt{e})$ is a point of inflection.

We find that the curve is concave upward in the intervals $(-\infty, -1/\sqrt{2})$ and $(1/\sqrt{2}, \infty)$ and concave downward in the interval $(-1/\sqrt{2}, 1/\sqrt{2})$ (Fig. 1.25ii).

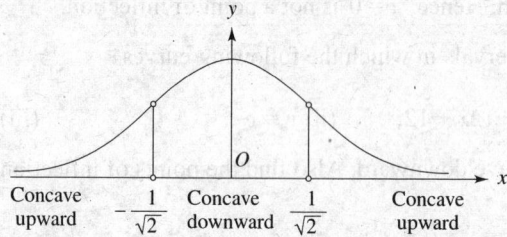

Fig. 1.25 ii.   Intervals of concavity for Example 1.57 ii.

(iii) We have

$$f(x) = (x+1)^{1/3}, \ f'(x) = \frac{1}{3}\,(x+1)^{-2/3}, \ f''(x) = -\frac{2}{9}\,(x+1)^{-5/3}.$$

We find that $f''(x)$ is not zero for any finite x. However, $f''(x)$ is not defined at $x = -1$. Now, $f''(x) > 0$ for $x < -1$ and $f''(x) < 0$ for $x > -1$. Since $f''(x)$ changes sign at $x = -1$, it follows that the point $(-1, 0)$ is a point of inflection. The tangent at this point is parallel to the y-axis, since $f'(-1) = \infty$, (Fig. 1.25iii). We find that the curve is concave upward in the interval $(-\infty, -1)$ and concave downward in the interval $(-1, \infty)$.

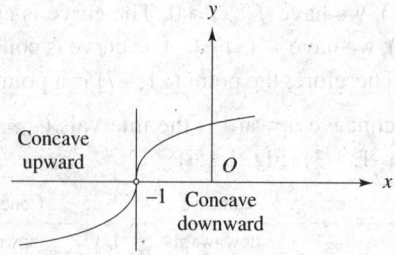

Fig. 1.25 iii.   Intervals of concavity for Example 1.57 iii.

1.5.2 Curvature, Circle of Curvature and Radius of Curvature

An important property of a curve is its curvature. Curvature measures the degree of sharpness of the bending of a curve at that point on the curve. Curvature plays an important role in laying curved railway tracks etc. The rate of change of the direction of the tangent line, at a point on the curve, with respect to the arc length s along the curve is called the curvature of the curve.

Let A be a fixed point on the curve from which the arc length is measured. Let the arc lengths of the points P and Q on the curve be s and $s + \Delta s$ respectively. Then, Δs is the length of arc PQ (Fig. 1.26). Let the tangents at the points P and Q on the curve make angles α and $\alpha + \Delta \alpha$ respectively with the positive direction of the x-axis. Then, $\Delta \alpha$ is the angle between the tangents at the points P and Q on the curve. The curvature of the curve at the point P is defined as

$$curvature = \kappa = \lim_{\Delta s \to 0} \frac{\Delta \alpha}{\Delta s} = \frac{d\alpha}{ds}. \tag{1.117}$$

If the angle α is measured in radians, then the units of the curvature κ is in radians per unit length.

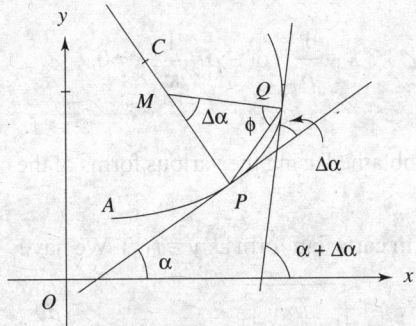

Fig. 1.26. Curvature of a curve.

Radius of curvature

Let a curve be given (Fig. 1.26). As discussed above, the angle between the tangents at the points P and Q on the curve is given by $\Delta \alpha$. Let the normals at the points P and Q intersect at M. Then, the angle between the normals PM and QM is $\Delta \alpha$, that is $\angle PMQ = \Delta \alpha$. Join the points P and Q. As $Q \to P$, the point of intersection M of the normals moves and takes a fixed position C in the limit. This point C is called the *centre of curvature*. The length PC is called the *radius of curvature*. The radius of curvature is usually denoted by ρ. The circle with centre at C and radius ρ is called the *circle of curvature*. It touches the curve at the point P. Every chord of this circle, which passes through the point of contact P is called the *chord of curvature*.

Denote chord $PQ = \overline{PQ}$, $\Delta s = $ arc $PQ = \widehat{PQ}$ and $\angle PQM = \phi$.

In the triangle PMQ, we have $\dfrac{PM}{\sin \phi} = \dfrac{\overline{PQ}}{\sin \Delta \alpha}$.

Therefore,
$$PM = \overline{PQ}\left(\frac{\sin\phi}{\sin\Delta\alpha}\right) = \frac{\overline{PQ}}{\widehat{PQ}} \cdot \frac{\widehat{PQ}}{\Delta\alpha} \cdot \frac{\Delta\alpha}{\sin\Delta\alpha}\sin\phi$$

$$= \frac{\overline{PQ}}{\widehat{PQ}} \cdot \frac{\Delta s}{\Delta\alpha} \cdot \frac{\Delta\alpha}{\sin\Delta\alpha}\sin\phi. \tag{1.118}$$

Now, as $Q \to P$, $\Delta\alpha \to 0$, $\phi \to \pi/2$,

$$\lim_{Q\to P}\frac{\overline{PQ}}{\widehat{PQ}} = 1, \quad \lim_{\Delta\alpha\to 0}\frac{\Delta s}{\Delta\alpha} = \frac{ds}{d\alpha} \quad \text{and} \quad \lim_{\Delta\alpha\to 0}\frac{\Delta\alpha}{\sin\Delta\alpha} = 1.$$

Hence, in the limit as $Q \to P$, we get from Eq. (1.118),

$$PC = \rho = \frac{ds}{d\alpha}. \tag{1.119}$$

Therefore, from Eq. (1.117), we obtain the curvature as

$$\kappa = \frac{1}{\rho} \quad \text{or} \quad \rho = \frac{1}{\kappa}, \ \kappa \neq 0. \tag{1.120}$$

If $\kappa = 0$, then we do not define ρ.

The expression for κ or ρ can be obtained using the various forms of the curve. We consider the following cases.

Case 1   Let the curve be given in cartesian form as $y = f(x)$. We have

$$\frac{ds}{dx} = \sqrt{1 + \left(\frac{dy}{dx}\right)^2} \quad \text{and} \quad \tan\alpha = \frac{dy}{dx}.$$

Differentiating the second expression, with respect to x, we get

$$\sec^2\alpha\frac{d\alpha}{dx} = \frac{d^2y}{dx^2}, \quad \text{or} \quad \frac{d\alpha}{dx} = \frac{y''(x)}{1 + \tan^2\alpha} = \frac{y''(x)}{1 + (y'(x))^2}.$$

Hence, we obtain

$$\kappa = \frac{d\alpha}{ds} = \frac{d\alpha/dx}{ds/dx} = \frac{y''}{[1 + (y')^2]^{3/2}} \tag{1.121}$$

and

$$\rho = \frac{1}{\kappa} = \frac{[1 + (y')^2]^{3/2}}{y''}. \tag{1.122a}$$

If the curve is given is the form $x = g(y)$, then

$$\kappa = \frac{x''}{[1 + (x')^2]^{3/2}}, \quad \text{where} \quad x' = \frac{dx}{dy}, \ x'' = \frac{d^2x}{dy^2}. \tag{1.22b}$$

Case 2   Let the curve be given in parametric form as $x = \phi(t)$, $y = \psi(t)$. Then, we have

$$\frac{ds}{dt} = \sqrt{\left(\frac{dx}{dt}\right)^2 + \left(\frac{dy}{dt}\right)^2} = \sqrt{(x'(t))^2 + (y'(t))^2}$$

and

$$\tan \alpha = \frac{dy}{dx} = \frac{dy/dt}{dx/dt} = \frac{y'(t)}{x'(t)}. \qquad (1.123)$$

Differentiating (1.123), with respect to t, we obtain

$$\sec^2 \alpha \, \frac{d\alpha}{dt} = \frac{1}{[x'(t)]^2} \, [x'(t) y''(t) - y'(t) x''(t)].$$

Now,

$$\sec^2 \alpha = 1 + \tan^2 \alpha = 1 + \left[\frac{y'(t)}{x'(t)}\right]^2 = \frac{(x'(t))^2 + (y'(t))^2}{(x'(t))^2}.$$

Hence,

$$\frac{d\alpha}{dt} = \frac{x'(t) y''(t) - y'(t) x''(t)}{(x'(t))^2 + (y'(t))^2}.$$

Therefore,

$$\kappa = \frac{d\alpha}{ds} = \frac{d\alpha/dt}{ds/dt} = \frac{x'(t) y''(t) - y'(t) x''(t)}{[(x'(t))^2 + (y'(t))^2]^{3/2}} \qquad (1.124)$$

and

$$\rho = 1/\kappa.$$

Case 3   Let the curve be given in polar form as $r = f(\theta)$. We have $x = r \cos \theta$, $y = r \sin \theta$. Hence,

$$\frac{dx}{d\theta} = r' \cos \theta - r \sin \theta, \quad \frac{dy}{d\theta} = r' \sin \theta + r \cos \theta,$$

where $r' = dr/d\theta$, and

$$\frac{dy}{dx} = \frac{dy/d\theta}{dx/d\theta} = \frac{r' \sin \theta + r \cos \theta}{r' \cos \theta - r \sin \theta}.$$

Differentiating both sides with respect to x and simplifying, we get

$$\frac{d^2 y}{dx^2} = \frac{1}{(r' \cos \theta - r \sin \theta)^2} \Big[(r' \cos \theta - r \sin \theta)(r'' \sin \theta + 2r' \cos \theta - r \sin \theta)$$

$$- (r' \sin \theta + r \cos \theta)(r'' \cos \theta - 2r' \sin \theta - r \cos \theta) \Big] \left(\frac{d\theta}{dx}\right)$$

$$= \frac{r^2 + 2(r')^2 - rr''}{(r' \cos\theta - r \sin\theta)^2} \left(\frac{1}{dx/d\theta}\right) = \frac{r^2 + 2(r')^2 - rr''}{(r' \cos\theta - r \sin\theta)^3}.$$

Now,

$$1 + \left(\frac{dy}{dx}\right)^2 = 1 + \frac{(r' \sin\theta + r \cos\theta)^2}{(r' \cos\theta - r \sin\theta)^2}$$

$$= \frac{(r' \cos\theta - r \sin\theta)^2 + (r' \sin\theta + r \cos\theta)^2}{(r' \cos\theta - r \sin\theta)^2} = \frac{r^2 + (r')^2}{(r' \cos\theta - r \sin\theta)^2}.$$

Substituting for dy/dx and d^2y/dx^2 in Eq. (1.121), we get

$$\kappa = \frac{r^2 + 2(r')^2 - rr''}{(r'\cos\theta - r\sin\theta)^3} \cdot \frac{(r'\cos\theta - r\sin\theta)^3}{[r^2 + (r')^2]^{3/2}} = \frac{r^2 + 2(r')^2 - rr''}{[r^2 + (r')^2]^{3/2}}. \qquad (1.125)$$

and $\qquad\qquad \rho = 1/\kappa.$

Curvature may be positive or negative. Since $y'' > 0$ when the curve is concave upward and $y'' < 0$ when the curve is concave downward, the sign associated with the curvature is same as the sign associated with the concavity of the curve. Generally, we take $|\kappa|$ as the measure of curvature.

Remark 7

A straight line is the only curve whose curvature is zero at every point, since $y''(x) = 0$ for all x. Hence, we do not define radius of curvature for a straight line.

Example 1.58 Find the curvature and radius of curvature of the following curves at the indicated points.

(i) $x^2 + y^2 = a^2$ at (x, y).   (ii) $x = a(t - \sin t)$, $y = a(1 - \cos t)$ at $t = \pi$.

(iii) $x = a\cos^3 t$, $y = a\sin^3 t$ at $t = \pi/4$.   (iv) $r = a\sin 2\theta$ at $\theta = \pi/4$.

(v) $r = a\sin\theta + b\cos\theta$ at any θ.

Solution

(i) Differentiating the given equation two times, we get

$$2x + 2yy' = 0, \quad \text{or} \quad x + yy' = 0 \quad \text{and} \quad 1 + yy'' + (y')^2 = 0.$$

Hence, $\qquad\qquad y' = -\dfrac{x}{y} \quad \text{and} \quad 1 + yy'' + \dfrac{x^2}{y^2} = 0 \quad \text{or} \quad \dfrac{(x^2 + y^2)}{y^2} + yy'' = 0.$

Substituting $x^2 + y^2 = a^2$, we get

$$\frac{a^2}{y^2} + yy'' = 0 \quad \text{or} \quad y'' = -\frac{a^2}{y^3}.$$

Using Eqs. (1.121) and (1.122), we get

$$\kappa = \frac{y''}{[1 + (y')^2]^{3/2}} = -\frac{(a^2/y^3)}{[1 + (x^2/y^2)]^{3/2}} = -\frac{a^2}{y^3} \cdot \frac{y^3}{a^3} = -\frac{1}{a}.$$

Hence, curvature $= |\kappa| = 1/a$ and radius of curvature $= \rho = a$.

(ii) We have

$$\frac{dx}{dt} = a(1 - \cos t), \quad \frac{dy}{dt} = a\sin t, \quad \left(\frac{dx}{dt}\right)^2 + \left(\frac{dy}{dt}\right)^2 = 2a^2(1 - \cos t),$$

$$\frac{d^2x}{dt^2} = a\sin t, \quad \frac{d^2y}{dt^2} = a\cos t.$$

At $t = \pi$, we get

$$\frac{dx}{dt} = 2a, \quad \frac{dy}{dt} = 0, \quad \left(\frac{dx}{dt}\right)^2 + \left(\frac{dy}{dt}\right)^2 = 4a^2, \quad \frac{d^2x}{dt^2} = 0, \quad \frac{d^2y}{dt^2} = -a.$$

Using Eq. (1.124), we get

$$\kappa = \frac{x' y'' - y' x''}{[(x')^2 + (y')^2]^{3/2}} = \frac{2a(-a)}{(4a^2)^{3/2}} = -\frac{2a^2}{8a^3} = -\frac{1}{4a}.$$

Hence, curvature $= |\kappa| = 1/(4a)$ and radius of curvature $= \rho = 4a$.

(iii) We have

$$x'(t) = -3a \cos^2 t \sin t, \quad y'(t) = 3a \sin^2 t \cos t,$$

$$x''(t) = -3a[\cos^3 t - 2\cos t \sin^2 t], \quad y''(t) = 3a[2 \sin t \cos^2 t - \sin^3 t].$$

At $t = \pi/4$, we get

$$x' = -\frac{3a}{2\sqrt{2}}, \quad y' = \frac{3a}{2\sqrt{2}}, \quad (x')^2 + (y')^2 = \frac{9a^2}{4}, \quad x'' = \frac{3a}{2\sqrt{2}}, \quad y'' = \frac{3a}{2\sqrt{2}}.$$

Using Eq. (1.124), we get

$$\kappa = \frac{x' y'' - y' x''}{[(x')^2 + (y')^2]^{3/2}} = \frac{-[3a/(2/\sqrt{2})]^2 - [3a/(2\sqrt{2})]^2}{(9a^2/4)^{3/2}}.$$

$$= -\left(\frac{9a^2}{4}\right)\left(\frac{8}{27a^3}\right) = -\frac{2}{3a}.$$

Hence, curvature $= |\kappa| = 2/(3a)$ and radius of curvature $= \rho = (3a)/2$.

(iv) We have

$$r = a \sin(2\theta), \quad r' = 2a \cos(2\theta), \quad r'' = -4a \sin(2\theta).$$

At $\theta = \pi/4$, we get $r = a$, $r' = 0$, $r'' = -4a$.

Using Eq. (1.125), we get

$$\kappa = \frac{r^2 + (2r')^2 - r\, r''}{[r^2 + (r')^2]^{3/2}} = \frac{a^2 + 4a^2}{(a^2)^{3/2}} = \frac{5a^2}{a^3} = \frac{5}{a}.$$

Hence, curvature $= |\kappa| = 5/a$ and radius of curvature $= \rho = a/5$.

(v) We have

$$r = a \sin \theta + b \cos \theta, \quad r' = a \cos \theta - b \sin \theta, \quad r'' = -(a \sin \theta + b \cos \theta).$$

Using Eq. (1.125), we get

$$\kappa = \frac{r^2 + (2r')^2 - rr''}{[r^2 + (r')^2]^{3/2}} = \frac{2(a\cos\theta - b\sin\theta)^2 + 2(a\sin\theta + b\cos\theta)^2}{[(a\sin\theta + b\cos\theta)^2 + (a\cos\theta - b\sin\theta)^2]^{3/2}}$$

$$= \frac{2(a^2 + b^2)}{(a^2 + b^2)^{3/2}} = \frac{2}{(a^2 + b^2)^{1/2}}.$$

Hence, curvature $= |\kappa| = \dfrac{2}{(a^2 + b^2)^{1/2}}$ and radius of curvature $= \rho = \dfrac{(a^2 + b^2)^{1/2}}{2}$.

Radius of curvature at origin for a curve represented by an algebraic equation $f(x, y) = 0$ (Newton's method)

Let the curve pass through the origin (the equation of the curve does not have constant term). Set the lowest degree term (linear part) to zero. This gives the equation of the tangent at the origin.

Let x-axis be the tangent at origin to the curve. Then,

$$y' = 0, \quad \text{and} \quad \rho(0, 0) = \frac{1}{y''(0, 0)}.$$

Now,
$$\lim_{x \to 0, y \to 0} \frac{x^2}{2y} = \lim_{x \to 0, y \to 0} \frac{2x}{2y'} = \frac{1}{y''(0, 0)} = \rho(0, 0).$$

Let y-axis be the tangent at origin to the curve. Then, $(dx/dy) = 0$. Radius of curvature is given by

$$\rho = \frac{[1 + (x')^2]^{3/2}}{x''}, \quad \text{where } x' = \frac{dx}{dy}, \quad \text{and} \quad x'' = \frac{d^2x}{dy^2}. \quad \rho(0, 0) = \frac{1}{x''(0, 0)}.$$

Now,
$$\lim_{x \to 0, y \to 0} \frac{y^2}{2x} = \lim_{x \to 0, y \to 0} \frac{2y}{2x'} = \frac{1}{x''(0, 0)} = \rho(0, 0).$$

Example 1.59   Find the equation of the tangent and the radius of curvature at the origin to the following curves.

   (i) $2x^4 + y^4 + x^3 + y^3 + 3x^2 - y^2 + x = 0$.

  (ii) $3x^2y + 3xy^2 + y^3 + x^2 - 2y^2 - 5y = 0$.

Solution

   (i) Setting the linear part to zero, we get $x = 0$, that is y-axis is the tangent to the curve. Divide the equation by x. We obtain

$$2x^3 + 2y^2\{y^2/(2x)\} + x^2 + 2y\{y^2/(2x)\} + 3x - 2\{y^2/(2x)\} + 1 = 0.$$

      Taking the limit $x \to 0$, $y \to 0$, we obtain $-2\rho + 1 = 0$,   or $\rho = 1/2$.

  (ii) Setting the linear part to zero, we get $5y = 0$, that is x-axis is the tangent to the curve. Divide the equation by y. We obtain

$$3x^2 + 3xy + y^2 + 2\{x^2/(2y)\} - 2y - 5 = 0.$$

      Taking the limit $x \to 0$, $y \to 0$, we obtain $2\rho - 5 = 0$,   or $\rho = 5/2$.

Centre of Circle of Curvature

Consider the circle of curvature with centre at $C(\alpha, \beta)$, (Fig. 1.27). The normal at $P(x, y)$ passes through the centre C. Draw PN perpendicular to AC. Let the tangent at P make an angle θ with the positive direction of x-axis. We have $CP = \rho$ and $\angle PCN = \theta$. From Fig. 1.27, we get

$$CN = CP \cos \theta = \rho \cos \theta, \; PN = CP \sin \theta = \rho \sin \theta,$$

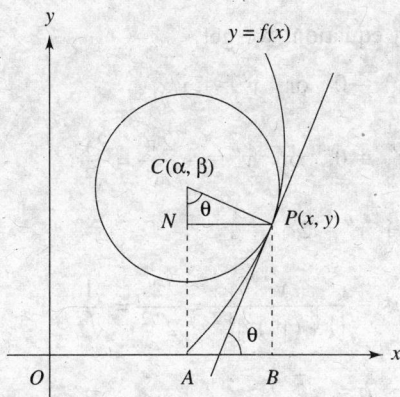

Fig. 1.27. Centre of curvature.

$$\alpha = OA = OB - AB = x - PN = x - \rho \sin \theta,$$

and
$$\beta = AC = AN + CN = BP + CN = y + \rho \cos \theta. \qquad (1.126a)$$

Since $\dfrac{dy}{dx} = y' = \tan \theta$, we get

$$\sin \theta = \frac{y'}{\sqrt{1 + (y')^2}} \quad \text{and} \quad \cos \theta = \frac{1}{\sqrt{1 + (y')^2}}.$$

Hence, from Eq. (1.126a), the coordinates of the centre of circle of curvature are given by

$$\alpha = x - \rho \sin \theta = x - \frac{[1 + (y')^2]^{3/2}}{y''} \cdot \frac{y'}{[1 + (y')^2]^{1/2}}$$

$$= x - \left(\frac{y'}{y''}\right) [1 + (y')^2], \qquad (1.126b)$$

$$\beta = y + \rho \cos \theta = y + \frac{[1 + (y')^2]^{3/2}}{y''} \cdot \frac{1}{[1 + (y')^2]^{1/2}}$$

$$= y + \left(\frac{1}{y''}\right) [1 + (y')^2]. \qquad (1.126c)$$

Example 1.60 Find the coordinates of the centre and radius of the circle of curvature at the indicated points. Also, obtain the equation of the circle of curvature for the following curves.

(i) $xy = 1$ at $(1, 1)$,   (ii) $y = e^x$ at $(0, 1)$,   (iii) $y = \tan x$ at $(\pi/4, 1)$.

Solution

(i) Differentiating the given equation, we get

$$y + xy' = 0 \quad \text{or} \quad y' = -y/x.$$

$$2y' + xy'' = 0, \quad \text{or} \quad y'' = -\frac{2y'}{x} = \frac{2y}{x^2}.$$

At $(1, 1)$, we get $y' = -1$, $y'' = 2$.

Therefore,
$$\kappa = \frac{y''}{[1+(y')^2]^{3/2}} = \frac{2}{2^{3/2}} = \frac{1}{\sqrt{2}}$$

and radius of curvature $= \rho = \sqrt{2}$.

The coordinates of the centre are

$$\alpha = x - \left(\frac{y'}{y''}\right)[1+(y')^2] = 1 + \frac{1}{2}[2] = 2,$$

$$\beta = y + \left(\frac{1}{y''}\right)[1+(y')^2] = 1 + \frac{1}{2}[2] = 2.$$

The circle of curvature is given by

$$(x-\alpha)^2 + (y-\beta)^2 = \rho^2 \quad \text{or} \quad (x-2)^2 + (y-2)^2 = 2.$$

(ii) From the given equation $y = e^x$, we get $y' = e^x$, $y'' = e^x$. At the point $(0, 1)$, we get $y' = 1 = y''$. Hence,

$$\kappa = \frac{y''}{[1+(y')^2]^{3/2}} = \frac{1}{2\sqrt{2}} \quad \text{and} \quad \rho = \frac{1}{\kappa} = 2\sqrt{2}.$$

The coordinates of the centre are

$$\alpha = x - \left(\frac{y'}{y''}\right)[1+(y')^2] = 0 - 2 = -2,$$

$$\beta = y + \left(\frac{1}{y''}\right)[1+(y')^2] = 1 + 2 = 3.$$

The circle of curvature is given by

$$(x-\alpha)^2 + (y-\beta)^2 = \rho^2 \quad \text{or} \quad (x+2)^2 + (y-3)^2 = 8.$$

(iii) From the given equation $y = \tan x$, we get $y' = \sec^2 x$ and $y'' = 2\sec^2 x \tan x$.

At $(\pi/4, 1)$, we get $y' = 2$, $y'' = 4$. Hence,

$$\kappa = \frac{y''}{[1+(y')^2]^{3/2}} = \frac{4}{5^{3/2}} = \frac{4}{5\sqrt{5}} \quad \text{and} \quad \rho = \frac{1}{\kappa} = \frac{5\sqrt{5}}{4}.$$

The coordinates of the centre are

$$\alpha = x - \left(\frac{y'}{y''}\right)[1+(y')^2] = \frac{\pi}{4} - \frac{2}{4}(5) = \frac{\pi - 10}{4},$$

$$\beta = y + \left(\frac{1}{y''}\right)[1+(y')^2] = 1 + \frac{1}{4}(5) = \frac{9}{4}.$$

The circle of curvature is given by

$$(x - \alpha)^2 + (y - \beta^2) = \rho^2 \quad \text{or} \quad \left[x - \frac{\pi - 10}{4}\right]^2 + \left[y - \frac{9}{4}\right]^2 = \frac{125}{16}.$$

Example 1.61 The centre of curvature of the ellipse $(x^2/a^2) + (y^2/b^2) = 1$, at one end of the minor axis is at the other end of the minor axis. Then, show that the eccentricity of the ellipse is $e = 1/\sqrt{2}$.

Solution Let one end of the minor axis be $P(0, b)$. The other end is $Q(0, -b)$. Differentiating the given equation, we get

$$\frac{2x}{a^2} + \frac{2yy'}{b^2} = 0, \quad \text{or} \quad y' = -\frac{b^2 x}{a^2 y}$$

Differentiating again, we obtain

$$\frac{1}{a^2} + \frac{1}{b^2}[yy'' + (y')^2] = 0, \quad \text{or} \quad y'' = -\frac{1}{y}\left[\frac{b^2}{a^2} + (y')^2\right].$$

$$y'' = -\frac{1}{y}\left[\frac{b^2}{a^2} + \frac{b^4 x^2}{a^4 y^2}\right] = -\frac{b^2}{a^2 y}\left[\frac{a^2 y^2 + b^2 x^2}{a^2 y^2}\right] = -\frac{b^2}{a^2 y}\left[\frac{a^2 b^2}{a^2 y^2}\right]\left[\frac{y^2}{b^2} + \frac{x^2}{a^2}\right] = -\frac{b^4}{a^2 y^3}.$$

$$\alpha = x - \left(\frac{y'}{y''}\right)[1+(y')^2] = x - \left(-\frac{b^2 x}{a^2 y}\right)\left(-\frac{a^2 y^3}{b^4}\right)\left[\frac{a^4 y^2 + b^4 x^2}{a^4 y^2}\right]$$

$$= x - \frac{x}{a^4 b^2}[a^4 y^2 + b^4 x^2] = 0, \quad \text{at} \quad P(0, b).$$

$$\beta = y + \left(\frac{1}{y''}\right)[1+(y')^2] = y + \left(-\frac{a^2 y^3}{b^4}\right)\left[\frac{a^4 y^2 + b^4 x^2}{a^4 y^2}\right]$$

$$= y - \frac{y}{a^2 b^4}[a^4 y^2 + b^4 x^2] = b - \frac{b}{a^2 b^4}[a^4 b^2] = b - \frac{a^2}{b}, \quad \text{at} \quad P(0, b).$$

It is given that $(\alpha, \beta) = (0, -b)$, Hence

$$b - \frac{a^2}{b} = -b, \quad \text{or} \quad a^2 = 2b^2.$$

The eccentricity of the ellipse is given by

$$b^2 = a^2(1 - e^2) = 2b^2(1 - e^2), \quad \text{or} \quad 1 - e^2 = \frac{1}{2}, \quad \text{or} \quad e = \frac{1}{\sqrt{2}}.$$

Example 1.62 Show that for the cardioid $r = a(1 + \cos \theta)$, $(\rho^2/r) = $ constant, where ρ is the radius of curvature, Hence, show that if ρ_1, ρ_2 are the radii of curvature of the cardioid at the extremities of a chord through the pole, then $\rho_1^2 + \rho_2^2 = (16a^2/9)$.

Solution From the equation of the curve $r = a(1 + \cos \theta)$, we get

$$r' = -a \sin \theta, \quad r'' = -a \cos \theta$$

$$\rho = \frac{[r^2 + (r')^2]^{3/2}}{r^2 + 2(r')^2 - rr''} = \frac{[a^2(1 + \cos \theta)^2 + a^2 \sin^2 \theta]^{3/2}}{a^2(1 + \cos \theta)^2 + 2a^2 \sin^2 \theta + a^2 \cos \theta (1 + \cos \theta)}$$

$$= \frac{[a^2(1 + \cos \theta)^2 + a^2(1 - \cos^2 \theta)]^{3/2}}{a^2(1 + \cos \theta)^2 + 2a^2(1 - \cos^2 \theta) + a^2 \cos \theta (1 + \cos \theta)}$$

$$= \frac{a[(1 + \cos \theta)(1 + \cos \theta + 1 - \cos \theta)]^{3/2}}{(1 + \cos \theta)((1 + \cos \theta) + 2 - 2 \cos \theta + \cos \theta)}$$

$$= \frac{a(1 + \cos \theta)^{3/2}(2)^{3/2}}{3(1 + \cos \theta)}$$

$$= \frac{2\sqrt{2}}{3} a(1 + \cos \theta)^{1/2} = \frac{2\sqrt{2}}{3} a \left(\frac{r}{a}\right)^{1/2} = \frac{2}{3}\sqrt{2ar}.$$

Therefore, $\dfrac{\rho^2}{r} = \dfrac{8a}{9}$, which is a constant.

The extremities of a chord through the pole are $A(r_1, \alpha)$ and $B(r_2, \alpha + \pi)$.

We obtain

$$\rho_1^2 + \rho_2^2 = \frac{8a}{9}(r_1 + r_2) = \frac{8a}{9}[a\{(1 + \cos \alpha) + 1 + \cos(\alpha + \pi)\}] = \frac{16}{9}a^2.$$

1.5.3   Evolute and Involute of a Curve

Evolute   Let $y = f(x)$ be a given curve, $P(x, y)$ be a point on it and $C(\alpha, \beta)$ be the centre of circle of curvature at the point P, (Fig. 1.27). As P moves along the curve, we obtain different circles of curvature and the centre $C(\alpha, \beta)$ traces another curve. The locus of the centre of curvature $C(\alpha, \beta)$ is called the *evolute* of the curve. We can regard equations (1.126b) and (1.126c) as the parametric equations of the evolute, where x is the parameter.

Remark 8

If the given curve is a circle $(x - p)^2 + (y - q)^2 = a^2$, then $C(p, q)$ is the centre of all the circles of curvature, since normal at any point on the circle passes through the centre of the circle. Therefore, the evolute of a circle is a single point.

Involute   The *involute* of a curve is a curve for which the given curve is an evolute. Therefore, the given curve is the involute.

Example 1.63   Find the evolutes of the following curves.

(i) $y^2 = 4ax$,   (ii) $\dfrac{x^2}{a^2} + \dfrac{y^2}{b^2} = 1$,   (iii) $x = a\cos^3\theta$, $y = a\sin^3\theta$.

Solution   Let (α, β) be the centre of curvature. Then,

$$\alpha = x - \left(\frac{y'}{y''}\right)[1 + (y')^2], \quad \beta = y + \left(\frac{1}{y''}\right)[1 + (y')^2].$$

(i) We have $2yy' = 4a$,   or $y' = \dfrac{2a}{y}$ and $y'' = -\dfrac{2a}{y^2}\,y' = -\dfrac{4a^2}{y^3}$.

Hence, $\alpha = x + \left(\dfrac{y^2}{2a}\right)\left[1 + \dfrac{4a^2}{y^2}\right] = x + \dfrac{1}{2a}\,[y^2 + 4a^2].$

$$= x + \frac{1}{2a}\,[4ax + 4a^2] = 3x + 2a,$$

$$\beta = y - \left(\frac{y^3}{4a^2}\right)\left[\frac{y^2 + 4a^2}{y^2}\right] = y - \frac{y}{4a^2}\,[4ax + 4a^2]$$

$$= \frac{y}{4a^2}\,[4a^2 - 4ax - 4a^2] = -\frac{xy}{a} = \pm\frac{2x\sqrt{ax}}{a} = \pm\frac{2x^{3/2}}{\sqrt{a}}$$

and $\beta^2 = \dfrac{4x^3}{a}.$

Eliminating x, we obtain the locus of (α, β) as

$$Y^2 = \frac{4}{a}\left[\frac{1}{3}(X - 2a)\right]^3 = \frac{4}{27a}(X - 2a)^3.$$

(ii) Differentiating the given equation $b^2x^2 + a^2y^2 = a^2b^2$, we get

$$2b^2x + 2a^2yy' = 0, \quad \text{or} \quad y' = -b^2x/(a^2y),$$

$$2b^2 + 2a^2[yy'' + (y')^2] = 0$$

or

$$y'' = -\frac{1}{a^2y}\left[b^2 + a^2(y')^2\right] = -\frac{1}{a^2y}\left[b^2 + \frac{b^4x^2}{a^2y^2}\right]$$

$$= -\frac{b^2}{a^4y^3}\left[a^2y^2 + b^2x^2\right] = -\frac{b^4}{a^2y^3}.$$

Hence,

$$\alpha = x - \left(\frac{b^2x}{a^2y}\right)\left(\frac{a^2y^3}{b^4}\right)\left[1 + \frac{b^4x^2}{a^4y^2}\right] = x - \frac{x}{a^4b^2}\left[a^4y^2 + b^4x^2\right], \qquad (1.27a)$$

$$\beta = y - \frac{a^2y^3}{b^4}\left[1 + \frac{b^4x^2}{a^4y^2}\right] = y - \frac{y}{a^2b^4}\left[a^4y^2 + b^4x^2\right], \qquad (1.127b)$$

Substitute $a^2y^2 = (a^2b^2 - b^2x^2)$ in Eq. (1.127a) and $b^2x^2 = (a^2b^2 - a^2y^2)$ in Eq. (1.127b) and simplify. We get

$$\alpha = x - \frac{x}{a^4b^2}\left[a^2(a^2b^2 - b^2x^2) + b^4x^2\right]$$

$$= \frac{x}{a^4}\left[a^4 - a^2(a^2 - x^2) - b^2x^2\right] = \frac{(a^2 - b^2)}{a^4}x^3$$

or

$$x = \left[\frac{a^4\alpha}{a^2 - b^2}\right]^{1/3},$$

$$\beta = y - \frac{y}{a^2b^4}\left[a^4y^2 + b^2(a^2b^2 - a^2y^2)\right]$$

$$= \frac{y}{b^4}\left[b^4 - a^2y^2 - b^2(b^2 - y^2)\right] = \frac{(b^2 - a^2)}{b^4}y^3$$

or

$$y = \left[\frac{b^4\beta}{b^2 - a^2}\right]^{1/3} = -\left[\frac{b^4\beta}{a^2 - b^2}\right]^{1/3}.$$

Substituting for x, y in the given equation, we get

$$\frac{1}{a^2}\left[\frac{a^4\alpha}{a^2-b^2}\right]^{2/3} + \frac{1}{b^2}\left[\frac{b^4\beta}{a^2-b^2}\right]^{2/3} = 1$$

or $\qquad a^{2/3}\,\alpha^{2/3} + b^{2/3}\,\beta^{2/3} = (a^2-b^2)^{2/3}.$

Hence, the evolute of the given curve is

$$a^{2/3}\,X^{2/3} + b^{2/3}\,Y^{2/3} = (a^2-b^2)^{2/3}.$$

(iii) From $x = a\cos^3\theta$, $y = a\sin^3\theta$, we get

$$\frac{dx}{d\theta} = -3a\cos^2\theta\sin\theta, \quad \frac{dy}{d\theta} = 3a\sin^2\theta\cos\theta$$

$$y' = \frac{dy}{dx} = \frac{dy/d\theta}{dx/d\theta} = -\frac{3a\sin^2\theta\cos\theta}{3a\cos^2\theta\sin\theta} = -\tan\theta.$$

$$y'' = -\sec^2\theta\,\frac{d\theta}{dx} = \frac{1}{3a\sin\theta\cos^4\theta}$$

Hence, $\qquad \alpha = x - \dfrac{y'}{y''}\,[1+(y')^2] = a\cos\theta(\cos^2\theta + 3\sin^2\theta),$

$$\beta = y + \frac{1}{y''}\,[1+y'^2] = a\sin\theta(\sin^2\theta + 3\cos^2\theta).$$

Now, $\quad \alpha + \beta = a\,(\cos^3\theta + 3\sin^2\theta\cos\theta + 3\cos^2\theta\sin\theta + \sin^3\theta)$

$$= a\,(\cos\theta + \sin\theta)^3$$

$$\alpha - \beta = a\,(\cos^3\theta - 3\cos^2\theta\sin\theta + 3\cos\theta\sin^2\theta - \sin^3\theta)$$

$$= a\,(\cos\theta - \sin\theta)^3.$$

Hence, $\quad (\alpha+\beta)^{2/3} + (\alpha-\beta)^{2/3} = a^{2/3}[(\cos\theta + \sin\theta)^2 + (\cos\theta - \sin\theta)^2]$

$$= 2a^{2/3}\,(\cos^2\theta + \sin^2\theta) = 2a^{2/3}.$$

Therefore, the locus of (α,β) is

$$(X+Y)^{2/3} + (X-Y)^{2/3} = 2a^{2/3}.$$

Example 1.64 Show that the normal to a given curve is a tangent to its evolute.

Solution The coordinates of the centre of curvature are

$$\alpha = x - \frac{y'}{y''}\,[1+(y')^2], \quad \beta = y + \left(\frac{1}{y''}\right)[1+(y')^2].$$

These equations are the parametric equations of the evolute, where x is a parameter. We note that CP (Fig. 1.27) is normal to the given curve at the point P. Differentiating α, β with respect to x, we get

$$\frac{d\alpha}{dx} = 1 - \frac{1}{(y'')^2} \; [y''\{y'' + 3(y')^2 y''\} - \{y' + (y')^3\}y''']$$

$$= \frac{1}{(y'')^2} \; [(y'')^2 - y''\{y'' + 3(y')^2 y''\} + \{y' + (y')^3\}y''']$$

$$= - \frac{y'}{(y'')^2} \; [3y'(y'')^2 - \{1 + (y')^2\} \; y'''],$$

$$\frac{d\beta}{dx} = y' + \frac{1}{(y'')^2} \; [y''(2y' \, y'') - \{1 + (y')^2\} \; y''']$$

$$= \frac{1}{(y'')^2} \; [y'(y'')^2 + 2y'(y'')^2 - \{1 + (y')^2\}y''']$$

$$= \frac{1}{(y'')^2} \; [3y'(y'')^2 - \{1 + (y')^2\} \; y''']$$

Hence, $$\frac{d\beta}{d\alpha} = \frac{d\beta/dx}{d\alpha/dx} = -\frac{1}{y'} \quad \text{or} \quad \left(\frac{d\beta}{d\alpha}\right) y' = -1.$$

Now, $d\beta/d\alpha$ is the slope of the tangent to the evolute and y' is the slope of the tangent to the given curve at P. Therefore, the tangent to the evolute and the tangent to the given curve are mutually perpendicular, that is, the normal to a given curve is a tangent to its evolute.

1.5.4   Envelope of a Family of Curves

Consider the equation of a family of curves in the form

$$y = g(x, p), \quad \text{or} \quad \text{as,} \; f(x, y, p) = 0 \tag{1.128}$$

where p is a parameter. We say that Eq. (1.128) represents a one parameter family of curves. For different values of p, we obtain different members of the family of curves. If each member of an infinite family (one parameter family) of curves is tangent to a certain curve C, and if at each point of the curve C, at least one member of the family is tangent, then the curve C is called the envelope of the given one parameter family of curves.

Now, to every value p_1 to p in Eq. (1.128), there corresponds a point on the envelope where $f(x, y, p_1) = 0$ touches the envelope. Hence, the coordinates of the points on the envelope are functions of p. Therefore, the equation of the envelope can be represented in parametric form as

$$x = \phi(p), \quad y = \psi(p). \tag{1.129}$$

The slope m of the tangent of the envelope is given by

$$m = \frac{dy}{dx} = \frac{dy/dp}{dx/dp}. \tag{1.130}$$

But, the slope of the envelope at a point of contact with a curve of the family is the same as the slope of the tangent to this curve.

Differentiating Eq. (1.128) partially with respect to x, we get

$$\frac{\partial f}{\partial x} + \frac{\partial f}{\partial y}\frac{dy}{dx} = 0 \quad \text{or} \quad \frac{dy}{dx} = -\frac{\partial f/\partial x}{\partial f/\partial y}, \quad \text{where } x = \phi(p),\ y = \psi(p).$$

Hence,
$$m = \frac{dy/dp}{dx/dp} = -\frac{\partial f/\partial x}{\partial f/\partial y},$$

or
$$\left[\frac{\partial f}{\partial x}\frac{dx}{dp} + \frac{\partial f}{\partial y}\frac{dy}{dp}\right] = 0, \quad \text{where } x = \phi(p),\ y = \psi(p). \tag{1.131}$$

Now, the point $(x, y) = (\phi(p),\ \psi(p))$ lies on the curve (1.128).

Therefore, we have

$$f(x, y, p) = 0$$

and
$$\frac{df}{dp} = 0, \quad \text{or} \quad \frac{\partial f}{\partial x}\frac{dx}{dp} + \frac{\partial f}{\partial y}\frac{dy}{dp} + \frac{\partial f}{\partial p} = 0 \tag{1.132}$$

where $x = \phi(p)$, $y = \psi(p)$.

Using Eqs. (1.131) and (1.132), we get

$$\frac{\partial f}{\partial p} = 0. \tag{1.133}$$

Hence, the envelope of the family of curves (1.128) is the curve obtained after eliminating p from Eqs. (1.128) and (1.133).

Example 1.65   Find the equation of the envelope of the family of straight lines $y = cx + c^2$, where c is a parameter.

Solution   We have

$$f(x, y, c) = y - cx - c^2 = 0, \quad \frac{\partial f}{\partial c}(x, y, c) = -x - 2c = 0.$$

The second equation gives $c = -x/2$. Eliminating c from $f(x, y, c) = 0$ we obtain

$$y = -\frac{x^2}{2} + \frac{x^2}{4} \quad \text{or} \quad y = -\frac{x^2}{4} \quad \text{or} \quad x^2 + 4y = 0$$

which is the equation of the envelope. The envelope is a parabola. (see also sections 4.3, 4.8.3 and Fig. 4.1).

Example 1.66   Find the equation of the envelope of the family of curves $x \cos \alpha + y \sin \alpha = p$, where p is a constant and α is a parameter.

Solution   We have

$$f(x, y, \alpha) = x \cos \alpha + y \sin \alpha - p = 0, \quad \frac{\partial f}{\partial \alpha} = -x \sin \alpha + y \cos \alpha = 0.$$

Solving the equations for x, y, we get $x = p \cos \alpha$, $y = p \sin \alpha$.
Eliminating α from $f(x, y, \alpha) = 0$ we get

$$\frac{x^2}{p^2} + \frac{y^2}{p^2} = 1 \quad \text{or} \quad x^2 + y^2 = p^2$$

which is the equation of the envelope. The envelope is a circle with centre at $(0, 0)$ and radius p.

Example 1.67   Find the equation of the envelope of the family of curves $x^2 + y^2 - 2px + p^2/2 = 0$ where p is a   parameter.

Solution   We have

$$f(x, y, p) = x^2 + y^2 - 2px + \frac{p^2}{2} = 0, \quad \frac{\partial f}{\partial p} = -2x + p = 0.$$

Solving the second equation, we get $p = 2x$. Eliminating p from $f(x, y, p) = 0$, we get

$$x^2 + y^2 - 4x^2 + 2x^2 = 0 \quad \text{or} \quad x^2 = y^2 \quad \text{or} \quad y = \pm x.$$

1.5.5   Asymptotes to a Curve

Let $P(x, y)$ be a variable point on a curve, which is given implicitly as $g(x, y) = 0$. If the distance of this point $P(x, y)$ from a straight line L tends to zero as x or y or both $x, y \to \infty$, then the line L is called an asymptote of the curve.

Asymptotes are usually classified as the following:

(i) Vertical asymptotes (ii) Horizontal asymptotes, and (iii) Inclined or oblique asymptotes.

Now, we define these asymptotes.

Vertical asymptotes

Assume that the given equation $g(x, y) = 0$ can be solved for y as

$$y = f(x) = \frac{p(x)}{q(x)}. \tag{1.134}$$

Let there exist a number a such that

$$\lim_{x \to a} f(x) = \pm \infty. \tag{1.135}$$

Then, the line $x = a$ is called a *vertical asymptote* to the curve (Fig. 1.28). This implies that $x = a$ is a root of $q(x) = 0$. If this root is of odd order, then the curve approaches the opposite ends of the asymptote from the two sides of the asymptote. If the root is of even order, then the curve approaches the same end of the asymptote from the both sides of the asymptote. For example, consider the curve

$$y = f(x) = \frac{1}{(x-1)^3}. \tag{1.136}$$

Now, as $x \to 1^+$, $f(x) \to +\infty$, and as $x \to 1^-$, $f(x), \to -\infty$. Therefore, the curve approaches the opposite ends of the vertical asymptote $x = 1$ (Fig. 1.29).

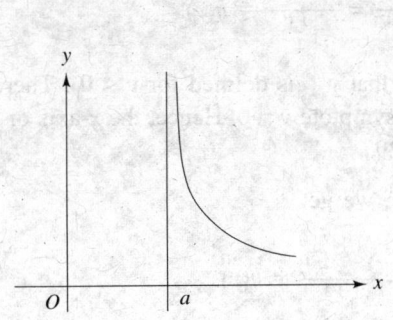

Fig. 1.28.   Vertical asymptote.

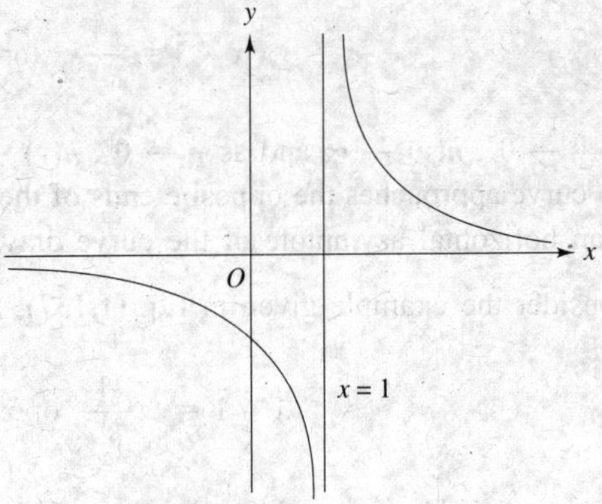

Fig. 1.29.   Vertical asymptote of $y = 1/(x-1)^3$.

Consider now the curve $y = f(x) = \dfrac{1}{(x-1)^2}$ $\qquad\qquad$ (1.137)

Now, as $x \to 1^+$, $f(x) \to +\infty$ and as $x \to 1^-$, $f(x) \to +\infty$.

Therefore, the curve approaches the same end of the asymptote $x = 1$. (Fig. 1.30)

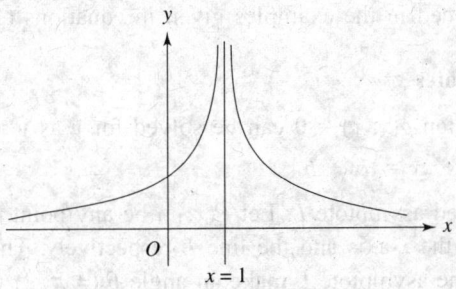

Fig. 1.30.   Vertical asymptote of $y = 1/(x-1)^2$.

Horizontal asymptotes

Assume that the given equation $g(x, y) = 0$ can be solved for x as

$$x = h(y) = \frac{p(y)}{q(y)}.$$ (1.138)

Let there exist a number b such that

$$\lim_{y \to b} h(y) = \pm \infty.$$ (1.139)

Then, the line $y = b$ is called a *horizontal asymptote* to the curve. This implies that $y = b$ is a root of $q(y) = 0$.

Consider the example given in Eq. (1.136). Solving for x, we get

$$x - 1 = \frac{1}{y^{1/3}} \quad \text{or} \quad x = 1 + \frac{1}{y^{1/3}} = \frac{y^{1/3} + 1}{y^{1/3}} = h(y).$$

As $y \to 0^+$, $h(y) \to \infty$ and as $y \to 0^-$, $h(y) \to -\infty$ (note that $y^{1/3}$ is defined for $y < 0$). Therefore, the curve approaches the opposite ends of the horizontal asymptote $y = 0$. Hence, the x-axis or $y = 0$ is an horizontal asymptote of the curve drawn in Fig. 1.29.

Consider the example given in Eq. (1.137). Solving for x, we get

$$x - 1 = \pm \frac{1}{y^{1/2}} \quad \text{or} \quad x = 1 \pm \frac{1}{y^{1/2}} = \frac{y^{1/2} \pm 1}{y^{1/2}} = h(y).$$

We note that $h(y)$ is not defined for $y < 0$. As $y \to 0^+$, $h(y) \to +\infty$. Therefore, the curve approaches the opposite ends of horizontal asymptote $y = 0$. Hence, the x-axis or $y = 0$ is an horizontal asymptote of the curve drawn in Fig. 1.30.

We can give an alternate method for finding an horizontal asymptote. Let the curve be given as $y = f(x) = p(x)/q(x)$. Let there exist a number b such that

$$\lim_{x \to \pm \infty} f(x) = b.$$ (1.140)

Then, the line $y = b$ is an horizontal asymptote of the curve. We can verify that $y = 0$ is an horizontal asymptote of the curves defined in the examples given in equations (1.136) and (1.137).

Inclined or oblique asymptotes

Assume that the given equation $g(x, y) = 0$ can be solved for y as $y = f(x)$. Let

$$y = ax + b$$ (1.141)

be the equation of the inclined asymptote L. Let $P(x, y)$ be any point on the curve. Draw the lines PS and PR perpendicular to the x-axis and the line L respectively. Then, the coordinates of Q are (x, y_1) (Figs. 1.31a, b). Let the asymptote L make an angle θ ($\neq \pi/2$) with the positive direction of x-axis. From the triangles TQS and QPR, we obtain

$$\angle QTS = \angle QPR = \theta.$$ (1.142)

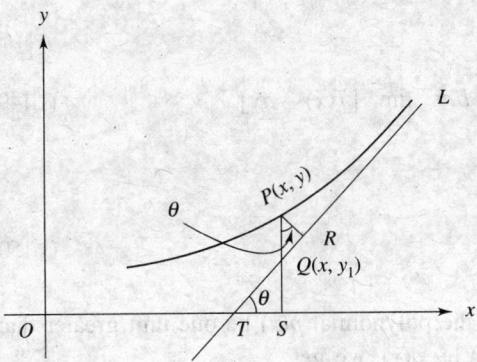

Fig. 1.31(a).   Inclined asymptote.   **Fig. 1.31(b).   Inclined asymptote.**

Using the definition, we have that the line L will be an asymptote if

$$\lim_{x \to +\infty} PR = 0. \tag{1.143}$$

Now, $PR = PQ \cos \theta$. Since θ is a known angle, we obtain

$$\lim_{x \to +\infty} PR = \lim_{x \to +\infty} PQ \cos \theta = 0 \quad \text{or} \quad \lim_{x \to +\infty} PQ = 0. \tag{1.144}$$

Now, $$PQ = |PS - QS| = |y - y_1| = |f(x) - (ax + b)|$$

since P is a point on the curve $y = f(x)$ and y_1 is a point on the line L, $y = ax + b$. Therefore, Eq. (1.144) gives

$$\lim_{x \to +\infty} PQ = \lim_{x \to +\infty} |f(x) - (ax + b)| = 0 \tag{1.145}$$

or $$\lim_{x \to +\infty} x \left| \frac{f(x)}{x} - a - \frac{b}{x} \right| = 0. \tag{1.146}$$

Now, as $x \to \infty$, $(b/x) = 0$. If the limit on the left hand side of (1.146) exists and equals 0, then we must have

$$\lim_{x \to +\infty} \left| \frac{f(x)}{x} - a \right| = 0, \quad \text{or} \quad a = \lim_{x \to +\infty} \left| \frac{f(x)}{x} \right|. \tag{1.147}$$

Now, using Eq. (1.145), we obtain

$$b = \lim_{x \to +\infty} [f(x) - ax]. \tag{1.148}$$

If one or both the limits in Eqs. (1.147), (1.148) do not exist, then there is no inclined asymptote. If the limit in Eq. (1.147) is zero, that is $a = 0$, the L is an horizontal asymptote $y = b$.

Similarly, we can define that the line L will be an asymptote, if

$$\lim_{x \to -\infty} |f(x) - (ax + b)| = 0.$$

In this case, we have again

$$a = \lim_{x \to -\infty} \left| \frac{f(x)}{x} \right| \quad \text{and} \quad b = \lim_{x \to -\infty} [f(x) - ax]. \tag{1.149}$$

Alternative   Consider the case, when

$$y = f(x) = \frac{p(x)}{q(x)}, \tag{1.150}$$

and $p(x)$, $q(x)$ are polynomials in x. Let the degree of the polynomial $p(x)$ be one unit greater than the degree of the polynomial $q(x)$. Then, dividing $p(x)$ by $q(x)$, we get

$$f(x) = (ax + b) + \frac{r(x)}{q(x)} \tag{1.151}$$

where the degree of x in $r(x)$ is less than the degree of x in $q(x)$. As $x \to \pm \infty$, $[r(x)/q(x)] \to 0$ and the distance between the curve $y = f(x)$ and the line $y = ax + b$, goes to zero. Hence, $y = ax + b$ is an inclined asymptote.

For example, consider the curve

$$y = \frac{3x^2 + x + 5}{2x}.$$

Now, as $x \to 0^+$, $y \to + \infty$ and as $x \to 0^-$, $y \to - \infty$.
Therefore, $x = 0$ or y-axis is a vertical asymptote. Now, write y as

$$y = \frac{1}{2} \left[3x + 1 + \frac{5}{x} \right].$$

Now, $(5/x) \to 0$ as $x \to \pm \infty$. Hence, $y = (3x + 1)/2$ is an inclined asymptote of the curve.

Remark 9

1. The graph of a given curve may intersect an horizontal or an inclined asymptote (Fig. 1.33).
2. If in Eq. (1.150), the degree of the polynomial $p(x)$ is two or more units higher than the degree of the polynomial $q(x)$, then there are no horizontal or inclined asymptotes.

Example 1.68   Find the asymptotes of the curve

$$(2x + 3) y = (x - 1)^2.$$

Solution   Write the given equation as

$$y = \frac{(x - 1)^2}{2x + 3} = \frac{(x - 1)^2}{2[x + (3/2)]}.$$

As $x \to (-3/2)^+$, $y \to + \infty$ and as $x \to (-3/2)^-$, $y \to - \infty$.
Therefore, $x = - 3/2$ is a vertical asymptote.

Now, write the given equation as

$$y = \frac{x^2 - 2x + 1}{2x + 3} = \frac{1}{2}\left[\frac{x - 2 + (1/x)}{x + (3/x)}\right].$$

As $x \to \infty$, $y \to \infty$. Therefore, there are no horizontal asymptotes.
Now, write the given equation as

$$y = \frac{x^2 - 2x + 1}{2x + 3} = \frac{1}{2}\left[\left(x - \frac{7}{2}\right) + \frac{(25/4)}{x + (3/2)}\right].$$

As $x \to \pm \infty$, the second term on the right hand side goes to 0.

Therefore, $y = \frac{1}{2}\left(x - \frac{7}{2}\right)$ is an inclined asymptote.

Alternative   We have from Eqs. (1.147), (1.148)

$$a = \lim_{x \to +\infty}\left[\frac{f(x)}{x}\right] = \lim_{x \to +\infty}\left(\frac{x^2 - 2x + 1}{2x^2 + 3x}\right) = \frac{1}{2},$$

$$b = \lim_{x \to +\infty}[f(x) - ax] = \lim_{x \to +\infty}\left[\frac{x^2 - 2x + 1}{2x + 3} - \frac{x}{2}\right]$$

$$= \lim_{x \to +\infty}\left[\frac{-7x + 2}{2(2x + 3)}\right] = -\frac{7}{4}.$$

Therefore, $y = ax + b = \frac{x}{2} - \frac{7}{4}$ is an inclined asymptote. If $x \to -\infty$, then also we get the same line
as asymptote.

Example 1.169   Find the asymptotes of the curve $y = \frac{x^2}{\sqrt{x^2 - 4}}$.

Solution   We find that

$$\lim_{x \to 2}\frac{x^2}{\sqrt{x^2 - 4}} = \infty \quad \text{and} \quad \lim_{x \to -2}\frac{x^2}{\sqrt{x^2 - 4}} = \infty.$$

Hence, $x = 2$ and $x = -2$ are the vertical asymptotes.
We have from Eqs. (1.147), (1.148)

$$a = \lim_{x \to +\infty}\left[\frac{f(x)}{x}\right] = \lim_{x \to +\infty}\left[\frac{x}{\sqrt{x^2 - 4}}\right] = \lim_{x \to +\infty}\left[\frac{1}{\sqrt{1 - (4/x^2)}}\right] = 1.$$

$$b = \lim_{x \to +\infty} [f(x) - ax] = \lim_{x \to +\infty} \left[\frac{x^2}{\sqrt{x^2 - 4}} - x \right]$$

$$= \lim_{x \to +\infty} \left[\frac{x^2 - x\sqrt{x^2 - 4}}{\sqrt{x^2 - 4}} \right] = \lim_{x \to +\infty} \left[\frac{4x^2}{\sqrt{x^2 - 4}\{x^2 + x\sqrt{x^2 - 4}\}} \right] = 0.$$

Therefore, $y = x$ is an inclined asymptote.

Similarly,

$$a = \lim_{x \to -\infty} \left[\frac{f(x)}{x} \right] = \lim_{x \to -\infty} \left[\frac{x}{\sqrt{x^2 - 4}} \right] = \lim_{t \to +\infty} \left[\frac{-t}{\sqrt{t^2 - 4}} \right] = -1,$$

$$b = \lim_{x \to -\infty} [f(x) - ax] = \lim_{x \to -\infty} \left[\frac{x^2}{\sqrt{x^2 - 4}} + x \right]$$

$$= \lim_{x \to -\infty} \left[\frac{x^2 + x\sqrt{x^2 - 4}}{\sqrt{x^2 - 4}} \right] = \lim_{x \to -\infty} \left[\frac{4x^2}{\sqrt{x^2 - 4}\{x^2 - x\sqrt{x^2 - 4}\}} \right] = 0.$$

Therefore, $y = -x$ is also an inclined asymptote.

There are no horizontal asymptotes.

Asymptotes of curves defined in parametric form

Let the given curve be represented in the parametric form as

$$x = \phi(t), \, y = \psi(t). \tag{1.152}$$

We first find the values of t (if any) for which one of the functions $\phi(t)$ or $\psi(t)$ becomes infinite, while the other remains finite.

When $\phi(t_0) = \pm \infty$ and $\psi(t_0) = c$, then the curve has a horizontal asymptote $y = c$.

When $\phi(t_0) = a$ and $\psi(t_0) = \pm \infty$, then the curve has a vertical asymptote $x = a$.

When $\phi(t_0) = \pm \infty$ and $\psi(t_0) = \pm \infty$, then the curve has an inclined asymptote $y = ax + b$, where

$$a = \lim_{t \to t_0} \frac{\psi(t)}{\phi(t)} \quad \text{and} \quad b = \lim_{t \to t_0} [\psi(t) - a\phi(t)]. \tag{1.153}$$

Asymptotes of curves defined in polar form

Let the given curve be represented in the polar form as $r = f(\theta)$. Then, we write

$$x = r \cos \theta = f(\theta) \cos \theta = \phi(\theta), \quad y = r \sin \theta = f(\theta) \sin \theta = \psi(\theta). \tag{1.54}$$

We now use the above procedure to obtain the asymptotes.

Example 1.70   Find the asymptotes of the curve (*Folium of Descartes*)

$$x = \frac{3at}{1+t^3}, \quad y = \frac{3at^2}{1+t^3}, \quad a > 0.$$

Solution   The curve has no horizontal or vertical asymptotes. We now find the inclined asymptote of the form $y = mx + b$.

As $t \to -1^+$, $x(t) \to -\infty$ and $y \to +\infty$. As $t \to -1^-$, $x(t) \to +\infty$ and $y \to -\infty$.

Using Eq. (1.153), We get

$$m = \lim_{t \to -1} \left(\frac{3at^2}{1+t^3} \right) \left(\frac{1+t^3}{3at} \right) = \lim_{t \to -1} (t) = -1,$$

$$b = \lim_{t \to -1} \left[\frac{3at^2}{1+t^3} + \frac{3at}{1+t^3} \right] = \lim_{t \to -1} \left[\frac{3at(1+t)}{1+t^3} \right] = \lim_{t \to -1} \left[\frac{3at}{t^2 - t + 1} \right] = -a.$$

Hence, $y = -(x + a)$ is the inclined asymptote. The cartesian form of the curve is $x^3 + y^3 = 3axy$. The plot of the curve and its asymptote is given in Fig. 1.32.

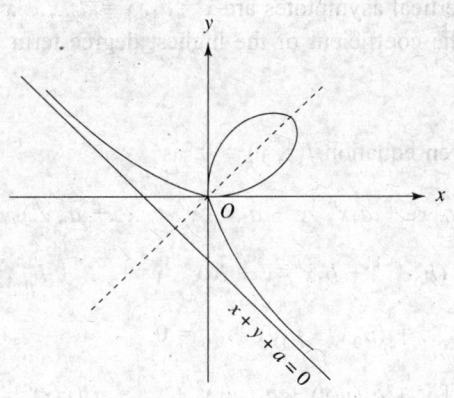

Fig. 1.32   Asymptote of the folium of Descartes.

Asymptotes of a curve given implicitly as an algebraic equation $f(x, y) = 0$ of degree n.

1. Horizontal asymptotes

To find the horizontal asymptotes or asymptotes parallel to *x*-axis, write the equation in descending powers of *x* in the form

$$x^n g_0(y) + x^{n-1} g_1(y) + x^{n-2} g_2(y) + \dots + g_n(y) = 0. \tag{1.155}$$

Dividing the equation (1.155) by x^n, we obtain

$$g_0(y) + \frac{1}{x} g_1(y) + \frac{1}{x^2} g_2(y) + \dots + \frac{1}{x^n} g_n(y) = 0. \tag{1.156}$$

Taking the limit as $x \to \infty$, we obtain the equation $g_0(y) = 0$. If $g_0(y)$ is a constant or real roots do not exist to this equation, then there are no horizontal asymptotes. If the real roots are given by $y = c_1, y = c_2, \dots, y = c_r$, then the horizontal asymptotes are $y = c_1, y = c_2, \dots, y = c_r$. Thus, to obtain horizontal asymptotes, equate to zero the coefficient of the highest degree term in x.

2. Vertical asymptotes

To find the vertical asymptotes or asymptotes parallel to y-axis, write the equation in descending power of y in the form

$$y^n g_0(x) + y^{n-1} g_1(x) + y^{n-2} g_2(x) + \dots + g_n(x) = 0. \tag{1.157}$$

Dividing the equation (1.157) by y^n, we obtain

$$g_0(x) + \frac{1}{y} g_1(x) + \frac{1}{y^2} g_2(x) + \dots + \frac{1}{y^n} g_n(x) = 0. \tag{1.158}$$

Taking the limit as $y \to \infty$, we obtain the equation $g_0(x) = 0$. If $g_0(x)$ is a constant or real roots do not exist to this equation, then there are no vertical asymptotes. If the real roots are given by $x = d_1$, $x = d_2, \dots, x = d_r$, then the vertical asymptotes are $x = d_1, x = d_2, \dots, x = d_r$. Thus, to obtain vertical asymptotes, equate to zero the coefficient of the highest degree term in y.

3. Inclined asymptotes

Arrange the terms in the given equation $f(x, y) = 0$, as

$$f(x, y) = (a_0 x^n + a_1 x^{n-1} y + a_2 x^{n-2} y^2 + \dots + a_n y^n)$$

$$+ (b_0 x^{n-1} + b_1 x^{n-2} y + b_2 x^{n-3} y^2 + \dots + b_{n-1} y^{n-1})$$

$$+ \dots + (p_0 x + p_1 y) + q_0 = 0,$$

or $$f(x, y) = x^n[a_0 + a_1(y/x) + a_2(y/x)^2 + \dots + a_n(y/x)^n]$$

$$+ x^{n-1}[b_0 + b_1(y/x) + b_2(y/x)^2 + \dots + b_{n-1}(y/x)^{n-1})$$

$$+ \dots + x[p_0 + p_1(y/x)] + q_0 = 0,$$

or $$f(x, y) = x^n \, \phi_n(y/x) + x^{n-1} \, \phi_{n-1}(y/x) + \dots + x \, \phi_1(y/x) + q_0 = 0, \tag{1.159}$$

where $\phi_m(y/x)$, $m = 1, 2, \dots, n$ are polynomials of degree m in (y/x).

Let $y = ax + b$ be an inclined asymptote.

Dividing the equation (1.159) by x^n, we obtain

$$\phi_n(y/x) + \frac{1}{x}\phi_{n-1}(y/x) + \frac{1}{x^2}\phi_{n-2}(y/x) + \ldots + \frac{1}{x^n}q_0 = 0. \qquad (1.160)$$

From the equation of the asymptote, we obtain $(y/x) = a + (b/x)$.

Taking limit as $x \to \infty$, we obtain, $\lim(y/x) = a$.

Taking limit as $x \to \infty$, of the equation (1.160), we obtain $\phi_n(a) = 0$.

Solving $\phi_n(a) = 0$, we obtain all possible values of a. If no real solutions exist, then there are no inclined asymptotes.

Substituting $(y/x) = a + (b/x)$ in (1.159), we obtain

$$x^n\,\phi_n\!\left(a + \frac{b}{x}\right) + x^{n-1}\phi_{n-1}\!\left(a + \frac{b}{x}\right) + \cdots x\phi_1\!\left(a + \frac{b}{x}\right) + q_0 = 0.$$

Now, expand each term in Taylor series about a. We obtain

$$x^n\!\left[\phi_n(a) + \left(\frac{b}{x}\right)\phi_n'(a) + \frac{1}{2!}\left(\frac{b}{x}\right)^2\phi_n''(a) + \cdots\right] + x^{n-1}\!\left[\phi_{n-1}(a) + \left(\frac{b}{x}\right)\phi_{n-1}'(a) + \frac{1}{2!}\left(\frac{b}{x}\right)^2\phi_{n-1}''(a) + \cdots\right]$$

$$+ x^{n-2}\!\left[\phi_{n-2}(a) + \left(\frac{b}{x}\right)\phi_{n-2}'(a) + \cdots\right] + \cdots + x\!\left[\phi_1(a) + \left(\frac{b}{x}\right)\phi_1'(a) + \cdots\right] + q_0 = 0.$$

Setting $\phi_n(a) = 0$, writing the equation in descending powers of x and dividing by x^{n-1}, we get

$$[b\phi_n'(a) + \phi_{n-1}(a)] + \frac{1}{x}\left[\left(\frac{b^2}{2!}\right)\phi_n''(a) + b\phi_{n-1}'(a) + \phi_{n-2}(a)\right]$$

$$+ \frac{1}{x^2}\left[\left(\frac{b^3}{3!}\right)\phi_n'''(a) + \left(\frac{b^2}{2!}\right)\phi_{n-1}''(a) + b\phi_{n-2}'(a) + \phi_{n-3}(a)\right] + \ldots = 0. \qquad (1.161)$$

Taking limit as $x \to \infty$, in the equation (1.161), we obtain

$$b\,\phi_n'(a) + \phi_{n-1}(a) = 0, \quad \phi_n'(a) \neq 0. \qquad (1.162)$$

Hence, $b = -\,\phi_{n-1}(a)/\phi_n'(a)$. For each value of a computed earlier, we obtain the corresponding value of b.

If $\phi_n'(a) = 0$, and $\phi_{n-1}(a) \neq 0$, then b is not defined. That is, there is no inclined asymptote corresponding to this value of a.

If $\phi_n'(a) = 0$, and $\phi_{n-1}(a) = 0$, then (1.162) is an identity. That is, the first term in (1.161) vanishes. Multiplying the equation by x and taking limit as $x \to \infty$, we get the equation

$$\left(\frac{b^2}{2!}\right) \phi_n''(a) + b\phi_{n-1}'(a) + \phi_{n-2}(a) = 0. \tag{1.163}$$

If the equation has two real roots b_1, b_2, then we obtain two inclined asymptotes $y = ax + b_1$ and $y = ax + b_2$.

If $\phi_n''(a) = 0$, $\phi_{n-1}'(a) = 0$, and $\phi_{n-2}(a) = 0$, then the first two terms in (1.161) vanish and we consider the next term and so on.

A way to write $\phi_m(a)$, for any m, is to set $x = 1$ and $y = a$ in the mth degree terms.

Alternative method to find inclined asymptotes

Let $y = mx + b$ be an asymptote. Then, the points of intersection of the curve and the asymptote satisfy the equation $f(x, mx + b) = 0$. We now, equate to zero, the coefficients of x^n and x^{n-1}. This gives two equations in two unknowns m and b. Solving, we obtain the values of m and b. The first equation gives all possible values of m and the second equation gives the corresponding values of b. If for a particular value of m, the coefficient of x^{n-1} becomes zero, we equate to zero the coefficient of x^{n-2}. This will give two values for b for the same m. In this case we get two parallel asymptotes.

For example, consider the Folium of Descartes (see Example 1.70) given by

$$f(x, y) = x^3 + y^3 - 3axy = 0.$$

Substituting $y = mx + b$, we obtain

$$f(x, y) = g(x) = x^3 + (mx + b)^3 - 3ax(mx + b) = 0$$

or

$$x^3 + (m^3x^3 + 3m^2bx^2 + 3mb^2x + b^3) - (3amx^2 + 3abx) = 0$$

or

$$(1 + m^3)\, x^3 + 3m(bm - a)\, x^2 + 3b(mb - a)\, x + b^3 = 0.$$

Setting the coefficients of x^3 and x^2 to zero, we get $1 + m^3 = 0$, or $m^3 = -1$, or $m = -1$

and $bm - a = 0$, or $b = -a$. Therefore, the equation of the inclined asymptote is

$$y = mx + b = -x - a, \quad \text{or} \quad x + y + a = 0.$$

Remark 10

An algebraic equation of degree n has a maximum of n asymptotes.

Example 1.71   Find all the asymptotes of the curves

(i) $y^2(x - 2a) = x^3 - a^3$,   (ii) $y^2(a^2 - x^2) = a^3x$,

(iii) $x^3 + 3x^2y - 4y^3 - x + y + 3 = 0$.

Solution

(i) Equating to zero the coefficient of y^2 (highest power of y), we get the vertical asymptote $x - 2a = 0$. Since the coefficient of x^3 (highest power of x) is a constant, there is no horizontal asymptote.

Let $y = mx + b$ be the inclined asymptote. Writing the given equation as

$$(x^3 - xy^2) + 2a\, y^2 - a^3 = 0$$

we obtain $\qquad \phi_3(m) = 1 - m^2, \quad \phi_2(m) = 2am^2.$

From $\qquad\qquad \phi_3(m) = 0, \quad$ we get $m = \pm 1.$

Now, $\qquad\qquad b = -\dfrac{\phi_2(m)}{\phi_3'(m)} = -\dfrac{2am^2}{-2m} = am.$

For $m = 1,$ we get $b = a$ and for $m = -1,$ we get $b = -a.$

Hence, the inclined asymptotes are $y = x + a$ and $y = -x - a.$

(ii) Equating to zero the coefficient of y^2, we get $a^2 - x^2 = 0$ or $x = \pm a.$
Therefore, $x = a$ and $x = -a$ are the vertical asymptotes. Equating to zero the coefficient of x^2, we get $y = 0.$ Therefore, $y = 0$ is the horizontal asymptote.
Let $y = mx + b$ be the inclined asymptote. Writing the given equation as

$$x^2y^2 - a^2y^2 - a^3x = 0,$$

we obtain $\qquad\qquad \phi_4(m) = m^2, \quad \phi_3(m) = 0, \quad \phi_2(m) = -a^2m^2.$

From $\qquad\qquad \phi_4(m) = 0, \quad$ we get $m = 0, 0.$

Since $\phi_4'(0) = 0$ and $\phi_3(0) = 0,$ we determine b using the equation

$$\frac{b^2}{2}\phi_4''(0) + b\phi_3'(0) + \phi_2(0) = 0.$$

Now, $\qquad\qquad \phi_4''(0) = 2, \quad \phi_3'(0) = 0, \quad \phi_2(0) = 0.$

We get $b^2 = 0$ or $b = 0.$ Therefore, we obtain $y = 0$ which is the horizontal asymptote. Thus the curve has two vertical asymptotes $x = \pm a,$ one horizontal asymptote $y = 0$ and no inclined asymptote.

(iii) Since the coefficients of x^3 and y^3 (the highest degree terms) are constants, there are no horizontal and vertical asymptotes.
Let $y = mx + b$ be the inclined asymptote. We write the given equation as

$$(x^3 - 4y^3 + 3x^2y) - x + y + 3 = 0$$

and get $\qquad\qquad \phi_3(m) = 1 + 3m - 4m^3 = (1 - m)(1 + 2m)^2,$

$$\phi_2(m) = 0, \quad \phi_1(m) = -1 + m, \quad \phi_0(m) = 3.$$

From $\qquad\qquad \phi_3(m) = 0, \quad$ we get $m = 1, -\dfrac{1}{2}, -\dfrac{1}{2}$

For $\qquad\qquad m = 1, \quad$ we obtain $b = -\dfrac{\phi_2(1)}{\phi_3'(1)} = 0.$

Therefore, $y = x$ is the inclined asymptote

For $m = -\dfrac{1}{2}$,   we find that $\phi'_3\left(-\dfrac{1}{2}\right) = \phi_2\left(-\dfrac{1}{2}\right) = 0$. We determine b from

$$\frac{b^2}{2}\phi''_3\left(-\frac{1}{2}\right) + b\phi'_2\left(-\frac{1}{2}\right) + \phi_1\left(-\frac{1}{2}\right) = 0$$

Now, $\phi''_3(m) = -24\,m$. We get $\phi''_3(-1/2) = 12$, $\phi'_2(-1/2) = 0$ and $\phi_1(-1/2) = \dfrac{3}{2}$.

Therefore, $6b^2 - \dfrac{3}{2} = 0$ or $b = \pm\dfrac{1}{2}$.

Hence, $y = -\dfrac{x}{2} \pm \dfrac{1}{2}$ are inclined asymptotes.

1.5.6. Curve Sketching

Using the knowledge of the differential calculus and the properties of a curve, we can analyse the shape of the graph of a given curve of the form $y = f(x)$ or $f(x, y) = 0$. Such an analysis enables us to sketch or trace the graph of the given curve. We generally use the following information to sketch the curve.

(i) Find the natural domain and the range of the curve. For example

   (a) the curve $y = \sqrt{1 - x^2}$ is not defined for $|x| > 1$,

   (b) the curve $y^2(a^2 + x^2) = x^2(a^2 - x^2)$ is not defined for $|x| > a$, since the right hand side becomes negative for $|x| > a$.

(ii) Find the points of intersection of the curve with the coordinate axes and find the intercepts made by the curve on the coordinate axes. Also, find whether the curve passes through the origin.

(iii) Find whether the curve is symmetric.

   (a) If $f(x, y) = f(x, -y)$ then the curve is symmetric about the x-axis. The curve has the same shape above and below the x-axis. We may sketch the curve above the x-axis, that is for $y > 0$ and draw its reflection below the x-axis.

   (b) If $f(x, y) = f(-x, y)$, then the curve is symmetric about the y-axis. The curve has the same shape on the right and left of the y-axis. We may sketch the curve on the right of the y-axis, that is for $x > 0$ and draw its reflection on the left of the y-axis.

   (c) If $f(x, y) = f(-x, -y)$, then the curve is symmetric about the origin. In this case, the curve has the same shape in the first and the third quadrants.

   (d) If $f(x, y) = f(y, x)$, then the curve is symmetric about the line $y = x$.

   (e) If $f(x, y) = f(-y, -x)$, then the curve is symmetric about the line $y = -x$.

(iv) Find the periodicity of the curve, if any. If $f(x) = f(x + T)$, then the curve is periodic with period T. We sketch the curve in any interval $[a, a + T]$ and then extend the curve on either side due to periodicity.

(v) Find the points of local maximum and the local minimum of the curve (critical points) and obtain the local maximum and local minimum values.

(vi) Using the critical points, find the intervals in which the curve is increasing or the curve is decreasing.

(vii) Find the intervals in which the curve is concave upward or concave downward.

(viii) Find the points of inflection of the curve. At these points, the concavity of the curve changes.

(ix) Find the asymptotes to the curve.

Example 1.72   Sketch the graph of the curve $y = \dfrac{(x-1)(x-3)}{x^2}$.

Solution   We have the following information about the curve:

(i) The curve is defined for all x except at $x = 0$.

(ii) The curve intersects the x-axis at $x = 1$ and $x = 3$. For $1 < x < 3$, $y < 0$ and $y > 0$ otherwise.

(iii) We have $y = 1 - \dfrac{4}{x} + \dfrac{3}{x^2}$. Setting $\dfrac{dy}{dx} = \dfrac{4}{x^2} - \dfrac{6}{x^3} = 0$, we get $x = 3/2$ as the critical point.

Since $y'' = -\dfrac{8}{x^3} + \dfrac{18}{x^4} > 0$ at $x = 3/2$, the curve has local minimum at $x = 3/2$ and this value is $-1/3$.

(iv) We have $y'(x) = 2(2x - 3)/x^3$. For $x \in (0, 3/2)$, $y'(x) < 0$ and for $x \in (-\infty, 0)$ or $x \in (3/2, \infty)$, $y'(x) > 0$. Hence, the curve is increasing in the interval $(-\infty, 0)$, decreasing in the interval $(0, 3/2)$ and again increasing in the interval $(3/2, \infty)$.

(v) We have $y''_{(x)} = [2(9 - 4x)/x^4] = 0$ when $x = 9/4$. For $x < 9/4$, $y'' > 0$ and for $x > 9/4$, $y'' < 0$. Hence, the point $(9/4, -5/27)$ is a point of inflection. The curve is concave upward in the interval $(0, 9/4)$ and concave downward in the interval $(9/4, \infty)$.

(vi) We have $\lim\limits_{x \to 0} y(x) = \lim\limits_{x \to 0} \dfrac{(x-1)(x-3)}{x^2} = \infty$.

Therefore, $x = 0$ is a vertical asymptote.

Also, $\lim\limits_{x \to \pm\infty} y(x) = \lim\limits_{x \to \pm\infty} \dfrac{(x-1)(x-3)}{x^2} = 1$.

Therefore, $y = 1$ is a horizontal asymptote.

The graph of the curve is given in Fig. 1.33.

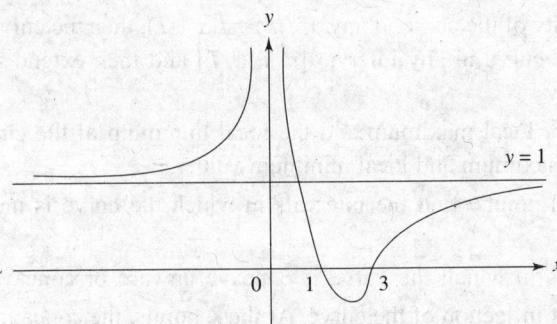

Fig. 1.33.   Graph of the curve $y = (x - 1)(x - 3)/x^2$.

Example 1.73   Sketch the graph of the curve $y = x + \dfrac{1}{x}$.

Solution   We have the following information about the curve:

(i)   The function is defined for all x, except at $x = 0$.

(ii)   For $x > 0$, $y > x > 0$. Hence, the curve is above the line $y = x$ in the first quadrant.

(iii)   For $x < 0$, $y < x < 0$. Hence, the curve is below the line $y = x$ in the third quadrant.

(iv)   Since $f(-x, -y) = f(x, y)$, the curve is symmetric about the origin.

(v)   We have $y'(x) = 1 - (1/x^2)$. Setting $y'(x) = 0$, we get $x = \pm 1$ as the critical points. Now, $y''(x) = 2/x^3$. Also, $y''(x) < 0$ when $x = -1$ and $y''(x) > 0$ when $x = 1$. Therefore, the curve has a local maximum at $x = -1$ and this value is -2. The curve has a local minimum at $x = 1$ and this value is 2.

(vi)   We have $y'(x) = (x - 1)(x + 1)/x^2$. Now, $y'(x) > 0$ when $x \in (1, \infty)$ or $x \in (-\infty, -1)$ and $y'(x) < 0$ when $x \in (-1, 0)$ or $x \in (0, 1)$. Therefore, the curve is increasing in the interval $(-\infty, -1)$, decreasing in the intervals $(-1, 0)$, $(0, 1)$ and again increasing in the interval $(1, \infty)$.

(vii)   We have $y'' = 2/x^3$. Now, $y''(x) > 0$, when $x > 0$ and $y''(x) < 0$, when $x < 0$. Hence, the curve is concave upward when $x > 0$ and concave downward when $x < 0$.

(viii)   We have $\lim\limits_{x \to 0} y(x) = \pm \infty$. Therefore, $x = 0$ is a vertical asymptote and the curve extends to both ends of the asymptote. Also, $y = x$ is an inclined asymptote.

The graph of the curve is given in Fig. 1.34.

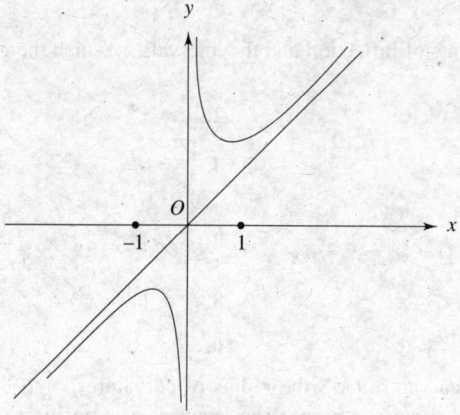

Fig. 1.34.   Graph of the curve $y = x + (1/x)$.

Example 1.74   Sketch the graph of the curve $y^2 = \dfrac{x-3}{x^2 - 6x - 7}$.

Solution   We write the equation of the curve as

$$y = \pm \sqrt{\frac{x-3}{(x+1)(x-7)}}$$

We have the following information about the curve.

(i) The curve is not defined when $3 < x < 7$.

(ii) The curve intersects the x-axis at $(3, 0)$ and the y-axis at $(0, \pm \sqrt{3/7}\,)$.

(iii) The curve is symmetric about the x-axis. Thus, we sketch the curve for $y > 0$ and extend it for $y < 0$.

(iv) We have $\lim\limits_{x \to -1} y(x) = \infty$ and $\lim\limits_{x \to 7} y(x) = \infty$. Therefore, $x = -1$ and $x = 7$ are the vertical asymptotes. Also, $\lim\limits_{x \to \pm\infty} y(x) = 0$. Hence, $y = 0$ is a horizontal asymptote.

The graph of the curve is given in Fig. 1.35.

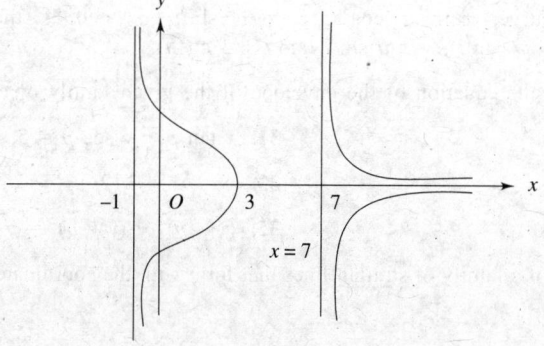

Fig. 1.35.   Graph of the curve $y^2 = (x-3)/(x^2 - 6x - 7)$.

Exercise 1.4

In problems **1** to **10**, find the points of inflection and the intervals in which the given curves are concave upward or concave downward.

1. $y = x^3 - 3x^2 + 6x + 5$.

2. $y = x^4 - x^3$.

3. $y = x^4 + 2x^2 + 4$.

4. $y = \sqrt{x} - x^2$, $x > 0$.

5. $y = \dfrac{1}{x-3}$.

6. $y = \dfrac{x^3}{x^2 + 4}$.

7. $y = x - \sin(x)$, $-\pi < x < \pi$.

8. $y = (1 + x^2)\, e^{-x}$.

9. $y = x^3 \ln(x)$, $x > 0$.

10. $y = \cot^{-1} x + x$.

In problems **11** to **20**, find the curvature $|\kappa|$, the radius of curvature ρ and the centre (α, β) of the circle of curvature for the given curve at the given point. The constant a is positive.

11. $y = x^2 - 6x + 10$ at $(3, 1)$.

12. $ay^2 = x^3$ at (a, a).

13. $x^3 + y^3 - 2axy = 0$ at (a, a).

14. $y = a \cosh(x/a)$ at $(0, a)$.

15. $x^2 = 4ay$ at $(2a, a)$.

16. $y = x^2 + \ln(x + \sqrt{1 + x^2})$ at $(0, 0)$.

17. $y = 4/(x^2 + 2)$ at $(0, 2)$.

18. $y = \sin(x)$ at $(\pi/2, 1)$.

19. $x = a(\cos t + t \sin t)$, $y = a(\sin t - t \cos t)$ at $t = \pi/4$.

20. $x = a \ln(\sec t + \tan t)$, $y = a \sec t$ at $t = 0$.

In problems **21** to **22**, find the radius of curvature at the origin using Newton's method.

21. $4x^3 + 3y^2 - 2x^2 + 6y = 0$.

22. $2x^4 - 3y^4 + x^3 y + xy - y^2 + 2x = 0$.

In problems **23** to **27**, find the evolute of the given curves. The constant a is positive.

23. $x^2 = 4ay$.

24. $xy = 1$.

25. $\dfrac{x^2}{a^2} - \dfrac{y^2}{b^2} = 1$, $a > b$.

26. $x^{2/3} + y^{2/3} = a^{2/3}$.

27. $x = a(\theta - \sin\theta)$, $y = a(1 - \cos\theta)$.

28. Show that the curve $x = a(\cos t + t \sin t)$, $y = (a \sin t - t \cos t)$, $a > 0$ is the involute of the curve $x = a \cos t$, $y = a \sin t$.

29. Show that the curve, $x = a \cos^3 t$, $y = a \sin^3 t$, $a > 0$ is the involute of the curve $x = a \cos t (\cos^2 t + 3 \sin^2 t)$, $y = a \sin t (\sin^2 t + 3 \cos^2 t)$.

In problems **30** to **35**, find the equation of the envelope of the given family of curves (p is a parameter).

30. $py + p^2 x - 10 = 0$.

31. $x \tan p + y \sec p = 5$.

32. $y = (x - p)^2$.

33. $y = px + 3/(2p)$.

34. $y = 3px - p^3$.

35. $(x - p)^2 + (y - p)^2 = p^2$.

36. Find the envelope of a family of straight lines that form with the coordinate axes a triangle of constant area A.

37. Find the envelope of all ellipses $\dfrac{x^2}{a^2} + \dfrac{y^2}{b^2} = 1$, which have a constant area $A = \pi ab$.

In problems **38** to **50** find all the asymptotes to the given curve

38. $(y - 2)(x^2 - 1) = 5.$

39. $x^2 y^2 = a^2(y^2 - x^2)$.

40. $y = \dfrac{x - 4}{x^2 + 4x + 3}$.

41. $y = \dfrac{(x - 1)^3}{x^2(x + 1)}$

42. $y = \dfrac{x + 1}{\sqrt{x^2 - 4}}$.

43. $y = e^{2/x} - 1$.

44. $\dfrac{x^2}{a^2} - \dfrac{y^2}{b^2} = 1$.

45. $x^5 + y^5 = 5\, ax^2 y^2$.

46. $x = t,\; y = t + 2\cot(t),\; 0 < t < 2\pi$.

47. The *hyperbolic spiral* $r = a/\theta$.

48. $y^3 + 2xy^2 - x^2 y - 2x^3 + 2xy + y^2 + x + 1 = 0$.

49. $y^3 - xy^2 - x^2 y + x^3 + x^2 - y^2 = 0$.

50. $(x + y)^2(x + y + 2) = x + 9y - 2$.

Sketch the graph of the curves given in problems **51** to **55**

51. $y = x/(1 + x^2)$.

52. $x^2 y^2 = a^2(y^2 - x^2)$.

53. $y^2 x = a^2(a - y)$.

54. $x(x^2 + y^2) = a(x^2 - y^2)$.

55. $x^5 + y^5 = 5\, ax^2 y^2$.

1.6   Answers and Hints

Exercise 1.1

1. $f(x) = x^{1/3}$, $x_0 = 1000$ and $dx = 5$; 10.0167.

2. $f(x) = x^{1/3}$, $x_0 = 1000$ and $dx = -1$; 9.9967.

3. $f(x) = x^3 + 2x^{4/3} + 5$, $x_0 = 1$ and $dx = 0.001$; 8.0057.

4. $f(x) = \sin x$, $x_0 = 60°$ and $dx = 10' \approx 0.0029$ radians; 0.8675.

5. $f(x) = \tan x$, $x_0 = 45°$ and $dx = 5'30'' \approx 0.0016$ radians; 1.0032.

6. (i) $f(x)$ is not continuous at $\pi/2$, (ii) $f(-1/2) \neq f(3/2)$. Also, $f(x)$ is not continuous at $x = 0$ and $x = 1$; (iii) $f(x)$ is not differentiable at $x = 1$.

7. $b = -5$, $c = 8$.

8. Define $\phi(x) = f(x) - 3g(x)$ and apply Rolle's theorem on $\phi(x)$.

9. Let a and b be the roots of $f(x) = e^x \sin x - 1 = 0$. Therefore, $f(a) = f(b) = 0$.

   Define $\phi(x) = e^{-x} - \sin x$, $a \le x \le b$. Now, apply Rolle's theorem on $\phi(x)$.

10. Define $\phi(x) = f(x) + (b - x)f'(x) + A[g(x) + (b - x)g'(x)]$. Find A such that $\phi(a) = \phi(b)$. Now, apply Rolle's theorem on $\phi(x)$.

11. Define $\phi(t) = f(a + t) + f(a - t)$. Apply Lagrange mean value theorem on $\phi(t)$ on the interval $[0, 1]$.

12. (i) On $[0, x]$, use Lagrange mean value theorem to $f(x) = e^x - x - 1$ to prove the left inequality and to $f(x) = e^x - 1 - xe^x$ to prove the right inequality.

(ii) Consider $f(x) = \ln(1 + x) - x$ on the interval $[0, x]$. Use the Lagrange mean value theorem on $f(x)$ and note that $[1/(1 + c) - 1] < 0$ for all c in $(0, x)$.

(iii) Consider $f(x) = x - \sin^{-1}x$ on the interval $[0, 1]$. Use the Lagrange mean value theorem to obtain $x < \sin^{-1}x$, (since $[1 - 1/\sqrt{1 - c^2}] < 0$ for all c in $(0, 1)$). Now, consider $g(x) = \sin^{-1} x - (x/\sqrt{1 - x^2})$ on the interval $[0, 1]$ and use the Lagrange mean value theorem.

(iv) On $[a, b]$, choose $f(x) = \sin^{-1}x$ and show that on $[a, b]$

$$[(b - a)/\sqrt{1 - a^2}] < \sin^{-1} b - \sin^{-1}a < [(b - a)/\sqrt{1 - b^2}].$$ Now, choose $b = 3/5$ and $a = 1/2$.

13. Using the Lagrange mean value theorem on $[-a, a]$, we obtain $f'(x) = 1$, where x is any point in the interval $[-a, a]$. Integrating, we get $f(x) = x + c$. Using $f(a) = a$, we obtain $c = 0$. Therefore, $f(x) = x$ and $f(0) = 0$.

14. Take $f(x) = F(x)$, $g(x) = 1/G(x)$ and apply Cauchy mean value theorem.

To evaluate limits in problems **15** to **28**, use the L'Hosptial's rule.

15. $1/n$.   **16.** 2.   **17.** $-1/8$.   **18.** $-1/3$

19. $2/\pi$.   **20.** $1/2$ **21.** e.   **22.** 1.

23. 1.   **24.** 0.   **25.** e.   **26.** e.

27. 1   **28.** e^2.

29. Decreasing for $-\infty < x < 2$ and increasing for $2 < x < \infty$.

30. Increasing for all x.   **31.** Increasing for all x.

32. Increasing for $0 < x < \pi/2$, decreasing for $\pi/2 < x \leq \pi$, neither increasing nor decreasing for $\pi < x < 2\pi$.

33. Increasing for $0 < x < \pi/2$ and decreasing for $\pi/2 < x < \pi$.

34. Increasing for $0 < x < e$ and decreasing for $e < x < \infty$.

35. Increasing for $0 < x < \pi/3$ and decreasing for $\pi/3 < x < \pi/2$.

36. Increasing for $x > 1/e$ and decreasing for $0 < x < 1/e$.

37. Let $f(x) = x^{1/n} - (x - 1)^{1/n}$. Now, $f(x)$ is a decreasing function for $x > 1$. We find that $f(1) = 1$. Hence, $f(a/b) = [a^{1/n} - (a - b)^{1/n}]/b^{1/n} < 1$. The result follows.

38. (Minimum at $x = 1$) $= 0$, (maximum at $x = 1/5$) $= 3456/3125$. At $x = -1$, there is neither a maximum nor a minimum.

39. Critical points are $x = n\pi + \pi/4$. When $n = 0$ or even, $f(x)$ has a maximum and the maximum value is $\sqrt{2}$. When n is odd, $f(x)$ has a minimum and the minimum value is $-\sqrt{2}$.

40. Maximum at $x = e$.

41. Minimum at $x = \sin^{-1}(1/e)$ and maximum at $x = (4n + 1)\,\pi/2$.

42. Maximum at $x = \pi/6, 5\pi/6$, maximum value $= 3/2$; minimum at $x = \pi/2, 3\pi/2$, (minimum value at $x = \pi/2) = 1$, (minimum value at $x = 3\pi/2) = -3$.

43. Critical points are at $\pi/4$, $\pi/3$; $2\pi/3$; (minimum value at $x = \pi/4$) = 1/2; (maximum value at $x = \pi/3$) = $(3\sqrt{3} - 1)/8$; (minimum value at $x = 2\pi/3$) = $-(3\sqrt{3} + 1)/8$.

45. For the function to be continuous at $x = 1$, we get $b = 1$. Since $f(x)$ is decreasing in [0, 1] and increasing in [1, 3], the function has minimum value $f(1) = -1$.

46. Use Leibniz formula; $y^{(n)}(0) = n(n-1)(n-2)2^{n-3}$.

47. $\dfrac{(-1)^{n-1}}{2^n} a^n[1 \cdot 3 \cdot 5 \ldots (2n-3)] (ax+b)^{-(2n-1)/2}$.

48. $\dfrac{dy}{dx} = re^{ax} \sin(bx + c + \theta)$, $r = \sqrt{a^2 + b^2}$, $\tan \theta = b/a$; $\dfrac{d^n y}{dx^n} = r^n e^{ax} \sin(bx + c + n\theta)$.

49. $\sqrt{1-x^2}\, y' + 1 = 0$, $y'(0) = -1$; $(1-x^2) y'' - xy' = 0$, $y''(0) = 0$; Use Leibniz formula.

$$y^{(n)}(0) = \begin{cases} 0, & n \text{ even} \\ -(n-2)^2(n-4)^2 \ldots 1^2, & n \text{ odd} \end{cases}.$$

50. $\sqrt{1-x^2}\, y' - ay = 0$, $y(0) = 1$, $y'(0) = a$; $(1-x^2) y'' - xy' - a^2 y = 0$, $y''(0) = a^2$. Use Leibniz formula.

$$y^{(n)}(0) = \begin{cases} [(n-2)^2 + a^2][(n-4)^2 + a^2] \ldots [1^2 + a^2]a, & n \text{ odd} \\ [(n-2)^2 + a^2][(n-4)^2 + a^2] \ldots [0^2 + a^2], & n \text{ even} \end{cases}.$$

51. $1 + \dfrac{1}{2}(x-1) - \dfrac{1}{8}(x-1)^2 + \dfrac{1}{16}(x-1)^3$, 0.00244.

52. $1 - x^2$, 19/6.   **53.** $x^2 - x^4/6$, 0.05.

54. $e^{-1}\left[1 + (x-1) - \dfrac{1}{2!}(x-1)^2 - \dfrac{1}{3!}(x-1)^3 + \dfrac{5}{4!}(x-1^4)\right]$, $(61e^{-0.5})/15360$.

55. $1 + x + x^2 + x^3$, 4/243.   **56.** $\displaystyle\sum_{m=0}^{\infty} \dfrac{(-1)^m (3x)^{2m+1}}{(2m+1)!}$ **57.** $\dfrac{1}{2}\displaystyle\sum_{m=1}^{\infty} \dfrac{(-1)^{m-1}(2x)^{2m}}{(2m)!}$

58. $(x-1) + \dfrac{3}{2!}(x-1)^2 + 2\displaystyle\sum_{m=3}^{\infty} \dfrac{(m-3)!(-1)^{m-1}}{m!}(x-1)^m$.   **59.** $2\displaystyle\sum_{m=0}^{\infty} \dfrac{(\ln 2)^m}{m!}(x-1)^m$.

60. Write the Taylor's formula for $f(a + h)$ with Lagrange form of remainder after n and $(n + 1)$ terms and subtract to obtain

$$\frac{h^n}{n!} f^{(n)}(a + \theta h) - \frac{h^n}{n!} f^{(n)}(a) - \frac{h^{n+1}}{(n+1)!} f^{(n+1)}(a + \theta_1 h) = 0$$

or $f^{(n)}(a + \theta h) - f^{(n)}(a) = \dfrac{h}{n+1} f^{(n+1)}(a + \theta_1 h)$, $0 < \theta < 1$, $0 < \theta_1 < 1$.

Using the Lagrange mean value theorem on the left hand side, we get

$$\theta h f^{(n+1)}(a + \theta_2 h) = \frac{h}{n+1} f^{(n+1)}(a + \theta_1 h), \quad 0 < \theta_2 < 1$$

or

$$\theta = \frac{1}{n+1}\left[\frac{f^{(n+1)}(a+\theta_1 h)}{f^{(n+1)}(a+\theta_2 h)}\right].$$

Taking limits as $h \to 0$, we obtain the result.

61. $|R_n(x)| \le \dfrac{e+1}{(2)(n+1)!} < \dfrac{2}{(n+1)!}$.

Now, $|R_n(x)| \le 0.001$ gives $n \ge 6$, that is, a minimum of seven terms are to be retained.

62. $|R_n(x)| \le \dfrac{2^{2n}}{(2n)!} < 0.0005$ gives $n \ge 5$, that is, a minimum of six terms are to be retained.

63. $|R_n(x)| \le \max\limits_{0 \le x \le 0.5}\left|\dfrac{x^{n+1}}{(n+1)!}\right|\left[\max\limits_{0 \le x \le 0.5}\left\{\left|\dfrac{n!}{(1-x)^{n+1}}\right| + \left|\dfrac{n!}{(1+x)^{n+1}}\right|\right\}\right] \le \dfrac{1}{n+1}\left[1 + \dfrac{1}{2^{n+1}}\right] < 0.1.$

We get $n \ge 10$, that is, a minimum of eleven terms are to be retained.

64. $|R_5(x)| \le \dfrac{2}{15}c^5 \sin 2c < \dfrac{2}{15}c^5 < 0.005$ gives $c \approx 0.51$.

65. $|R_4(x)| \le c^4(c + c^3) < 0.005$ gives $c \approx 0.33$.

66. Differentiating two times, we get $(1-x^2)y_2 - xy_1 = m^2 y$. Using the Leibniz theorem and setting $x = 0$, we get $y_{n+2}(0) = (n^2 + m^2)y_n(0)$, $(y_1 \ne 0)$.

67. (i) Let $f(x) = \sqrt{x}$ or $f^2(x) = x$. Differentiate and take $x = 16$, $h = -1$; 3.8730.

(ii) Let $f(x) = \sin x$. Differentiate and take $x = 30° = \pi/6$ and $h = -1° = -\pi/180$; 0.48481.

68. Write $a^x = a^{x \ln a}$ and use Eq. (1.25).

69. Replace x by $-x$ in Eq. (1.26). The series is valid in $-1 < (-x) \le 1$ or $-1 \le x < 1$.

70. Write $\ln\left(\dfrac{1+x}{1-x}\right) = \ln(1+x) - \ln(1-x)$. The series is valid in $-1 < x < 1$, since the function is not defined at $x = 1$.

71. Write $y = [(x-1)/(x+1)]$. Then, $x = [(1+y)/(1-y)]$. The series in valid in $-1 < y < 1$, that is when $-1 < [(x-1)/(x+1)] < 1$. We obtain $x > 0$.

72. Write $\dfrac{1}{x} = 1 - \dfrac{x-1}{x}$. We get

$\ln x = -\ln\dfrac{1}{x} = -\ln\left(1 - \dfrac{x-1}{x}\right)$. The series is valid when $-1 < [-(x-1)/x] \le 1$ or when $x \ge 1/2$.

73. Let $x = \tan\theta$. $(1+x^2)y' = 2$, $(1+x^2)y'' + 2xy' = 0$. Using the Leibniz theorem and setting $x = 0$, we get $y_{n+2}(0) = -n(n+1)y_n(0)$.

74. Let $x^2 = \cos 2\theta$. Then, $y = (\pi/4 - \theta) = (\pi/4) - (\cos^{-1} x^2)/2$. Differentiate and set $x = 0$.

75. Let $x^2 = \tan\theta$. Then, $y = 2\theta = 2\tan^{-1} x^2$. Differentiate and set $x = 0$.

Exercise 1.2

1. 29/6

2. 1/6.

3. 1/6.

4. $4ab \tan^{-1}(a/b)$.

5. 1/6.

6. $(e^2 - 5)/(4e)$.

7. 4/3.

8. $3\pi ab/8$.

9. $9\pi a^2/2$.

10. 7/120.

11. $a^2(4 - \pi)/2$.

12. $a^2[3\sqrt{3} - \pi]/3$.

13. $a^2(8 - \pi)/4$.

14. $(13\sqrt{13} - 8)/3$.

15. $3a/2$.

16. 123/32.

17. $2a\pi/3$.

18. 2.

19. $\ln(e + e^{-1})$.

20. $4a$.

21. $8a$.

22. $\sqrt{5}(e^2 - 1)/2$.

23. $\ln(1 + \sqrt{2})$.

24. 8.

25. $[f(r_2) - f(r_1)]/a$, where $f(r) = \dfrac{r}{2}\sqrt{a^2 + r^2} + \dfrac{a^2}{2}\ln[r + \sqrt{a^2 + r^2}]$. **26.** $(r_2 - r_1)\sqrt{1 + b^2}/b$.

27. For any given x, the cross-section is an ellipse with semi-manor and semiminor axes as $(b/a)\sqrt{a^2 - x^2}$ and $(c/a)\sqrt{a^2 - x^2}$. Area of cross-section is $\pi bc(a^2 - x^2)/a^2$; volume $= 4\pi abc/3$.

28. Area of cross-section is $4(a^2 - x^2)$; volume $= 16a^3/3$.

29. Area of cross-section is $2(a^2 - x^2)$; volume $= 8a^3/3$.

30. Area of cross-section is $\sqrt{3}\cos^2 x/4$; volume $= \pi\sqrt{3}/16$.

31. $\pi^2/4$.

32. 8π.

33. $40\pi/3$.

34. $117\pi/5$.

35. $16\pi a^3/105$.

36. $\pi/2$.

37. $8\pi/3$.

38. $\pi/6$.

39. $\pi/6$.

40. $\pi ba^2/3$.

41. $2\pi^2 a^2 b$.

42. $4\pi^2 ab$.

43. $2\pi b^2 + \dfrac{2\pi ba^2}{\sqrt{a^2 - b^2}}\sin^{-1}\left[\dfrac{\sqrt{a^2 - b^2}}{a}\right]$. **44.** $2\pi a^2 + \dfrac{2\pi ab^2}{\sqrt{a^2 - b^2}}\ln\left[\dfrac{a + \sqrt{a^2 - b^2}}{b}\right]$.

45. $24783\pi/1024$. **46.** $2489\pi/192$. **47.** $64\pi a^2/3$. **48.** $6\pi a^2/5$.

49. $2\sqrt{2}\,\pi(e^\pi - 2)/5$. **50.** $32\pi a^2/5$.

Exercise 1.3

1. Diverges.

2. Diverges.

3. Converges, $[\ln(5/3)]/2$.

4. Converges, $\pi/4$.

5. Converges, $2/a^3$.

6. Converges, $b/(a^2 + b^2)$.

7. Converges, $1/(2e)$.

8. Converges, π.

9. Converges, $\pi/2$.

10. Converges, 1.

11. Converges, $0 < p < 1$.

12. Diverges.

13. Converges.

14. Diverges.

15. Diverges.

16. Diverges.

17. Converges, $2(\ln 2 - 1)$.

18. Converges, 1.

19. Diverges.

20. Converges.

21. Let $g(x) = 1/(x - 1)$, diverges.

22. Let $g(x) = 1/x^{m-n}$, converges for $m < n + 1$.

23. Absolutely convergent.

24. Absolutely convergent.

25. (i) For arbitrarily large x, the integrand behaves like x^{p-q}, the integral will converge if $p - q + 1 < 0$. In the neighborhood of $x = 0$, the integrand behaves like x^p, the integral will converge if $p + 1 > 0$. Therefore, for convergence $- 1 < p < q - 1, q > 0$.

(ii) At ∞, the integrand behaves like x^p, the integral will converge if $p + 1 < 0$. At 0, the integrand behaves like x^{p-q}. The integral will converge if $p - q + 1 > 0$. Therefore, for convergence $q - 1 < p < 1, q < 0$.

26. $\dfrac{1}{2} \beta \left(\dfrac{1}{4}, \dfrac{1}{2} \right).$ **27.** $\dfrac{\pi}{32}.$ **28.** $\dfrac{1}{24}.$ **29.** $\dfrac{1}{2} \beta \left(\dfrac{1}{2}, \dfrac{m+1}{2} \right).$

30. Let $x = a \sin^{2/3} \theta$, $\dfrac{1}{3} a^{7/2} \beta \left(\dfrac{2}{3}, \dfrac{3}{2} \right).$ **31.** $x = \sin^{2/3} \theta$, $\dfrac{1}{3} \beta \left(\dfrac{1}{3}, \dfrac{2}{3} \right).$

32. Let $x = e^{-t}$, $\dfrac{(-1)^m}{(n+1)^{m+1}} \Gamma(m + 1).$ **33.** Let $x = a \sin \theta$, $\dfrac{1}{2} a^{3/2} \beta \left(\dfrac{5}{4}, \dfrac{1}{2} \right).$

34. $x = e^{-u}$, $\sqrt{\pi}$.   **35.** Let $x = \sin^2 \theta$, $\beta (k + 1, n - k + 1).$

36. Let $x^2 = \tan \theta$, $\dfrac{1}{4} \beta \left(\dfrac{1}{4}, \dfrac{3}{4} \right) = \dfrac{1}{4} \Gamma \left(\dfrac{1}{4} \right) \Gamma \left(\dfrac{3}{4} \right) = \dfrac{\pi}{2\sqrt{2}}.$

37. Write $a^x = e^{x \ln a}$ and let $x \ln a = t$, $\dfrac{\Gamma(a + 1)}{(\ln a)^{n+1}}.$

38. Let $st = T$, $\dfrac{\Gamma(k + 1)}{s^{k+1}}.$ **39.** Let $2t^2 = T$, $3\sqrt{2\pi}/64.$ **40.** Let $x = \sqrt{t}$, $\dfrac{1}{2} \Gamma \left(\dfrac{2}{3} \right)$

41. Let $x^q = p^q y.$ **42.** Let $x = b \sin^2 \theta + a \cos^2 \theta.$ **43.** Let $ae^{nx} = bt.$

44. Let $x = \sin \theta$ and use Eq. (1.79).   **45.** Let $x = my.$

46. Let $\alpha x^n = t.$ **47.** Let $1 - e^{-x} = t.$ **48.** Let $x = e^{-y}$, then let $(m + 1) y = t.$

49. We have $\dfrac{1}{x^p} = \dfrac{1}{\Gamma(p)} \displaystyle\int_0^\infty y^{p-1} e^{-xy} dy$. Substitute for $1/x^p$ and integrate $\displaystyle\int_0^\infty \sin x \, e^{-xy} dx$. Then, let $y = \sqrt{t}$.

50. We have $\Gamma(n + 1) = \displaystyle\int_0^\infty x^n e^{-x} dx = \int_0^\infty e^{n \ln x - x} dx$. The function $n \ln x - x$ has maximum at $x = n$. Write $x = n + y$. We obtain

$$\Gamma(n + 1) = n^n e^{-n} \int_{-n}^\infty e^{n \ln [1 + (y/n)] - y} dy$$

Expand $\ln [1 + (y/n)]$, $|y/n| < 1$, approximate to first term and let $y = \sqrt{2n} \, u.$

51. Let $x/(a + bx) = z/(a + b).$

In problems **52** to **61**, let the given integral be denoted by ϕ, unless mentioned otherwise.

52. Find $d\phi/da$, integrate with respect to a, use $\phi(b) = 0$. We get $\phi(a) = \ln (b/a).$

53. Find $d\phi/da$, integrate with respect to a, use $\phi(b) = 0$. We get $\phi(a) = \ln [(a + 1)/(b + 1)].$

54. Use $\int_0^1 x^n\, dx = 1/(n+1)$ and differentiate it k times with respect to the parameter n.

We get $\phi = [(-1)^k k!]/(n+1)^{k+1}$.

55. Use the Leibniz rule to get $\phi'(\alpha) = 0$. Integrate and use $\phi(0) = 0$. We obtain $\phi(\alpha) = 0$.

56. In $\int_{-\infty}^{\infty} e^{-x^2}\, dx = \sqrt{\pi}$, let $x = \alpha y$ and define $\phi(\alpha) = \int_{-\infty}^{\infty} e^{-\alpha^2 y^2}\, dy = \dfrac{\sqrt{\pi}}{\alpha}$. Differentiate with respect

to α and set $\alpha = 1$. We get $\phi = \sqrt{\pi}/2$.

57. Find $d\phi/d\alpha$, integrate with respect to α, use $\phi(1) = (\pi/4) + (\ln 2)/2$. We get $\phi(\alpha) = [(\pi + 2\ln 2)\,\alpha^3]/4$.

58. Differentiating $\int_0^{\pi} \dfrac{dx}{a + b\cos x} = \dfrac{\pi}{\sqrt{a^2 - b^2}}$ with respect to b and readjusting the terms, we get

$$\int_0^{\pi} \frac{dx}{(a + b\cos x)^2} = \frac{\pi a}{(a^2 - b^2)^{3/2}}.$$

Differentiate again with respect to b. We get $\phi = -3\pi\, ab/[2(a^2 - b^2)^{5/2}]$.

59. Find $\dfrac{d\phi}{da}$ and substitute $x = \dfrac{a}{y}$. We get $\dfrac{d\phi}{da} = -2\phi$. Use $\phi(0) = \int_0^{\infty} e^{-x^2}\, dx = \dfrac{\sqrt{\pi}}{2}$. We obtain

$\phi(a) = \dfrac{\sqrt{\pi}}{2}\, e^{-2a}$.

60. Find $\dfrac{d\phi}{d\alpha}$. Integrate and use $\phi(0) = 0$. We get $\phi(\alpha) = \pi \ln \left[\{1 + \sqrt{1 - \alpha^2}\,\}/2\right]$.

61. Consider $\phi(\alpha) = \int_0^{\infty} \dfrac{dx}{x^2 + \alpha^2} = \dfrac{\pi}{2\sqrt{\alpha}}$. Differentiate n times with respect to α and set $\alpha = 1$. We get

$$\phi(\alpha) = \frac{\pi}{2}\left[\frac{(2n)!}{2^{2n}(n!)^2}\right].$$

62. Use Eqs. (1.114) and (1.115). With $f(t, x) = (2/\sqrt{\pi})e^{-t^2}$ we get

$$\frac{d}{dx}[erfc(\alpha x)] = 0 + 0 - f(\alpha x, x)\frac{d}{dx}(\alpha x) = -\frac{2\alpha}{\sqrt{\pi}}\, e^{-\alpha^2 x^2}.$$

63. $erf(x) = \dfrac{2}{\sqrt{\pi}}\int_0^x e^{-t^2}\, dt = \dfrac{2}{\sqrt{\pi}}\int_0^x \left[1 - \dfrac{t^2}{1!} + \dfrac{t^4}{2!} - \dfrac{t^6}{3!} + \cdots\right] dt = \dfrac{2}{\sqrt{\pi}}\left[x - \dfrac{x^3}{3(1!)} + \dfrac{x^5}{5(2!)} - \dfrac{x^7}{7(3!)} + \cdots\right]$

64. Integrate by parts and use the result of problem 61.

65. $\int_0^{\infty} e^{-t^2 - 2\alpha t}\, dt = \int_0^{\infty} e^{-(t+\alpha)^2 + \alpha^2}\, dt = e^{\alpha^2}\int_0^{\infty} e^{-(t+\alpha)^2}\, dt$

$$= e^{\alpha^2}\int_{\alpha}^{\infty} e^{-u^2}\, du = \frac{\sqrt{\pi}}{2}\, e^{\alpha^2}\, erfc(\alpha) = \frac{\sqrt{\pi}}{2}\, e^{\alpha^2}\, [1 - erf(\alpha)].$$

66. Let $x = \alpha u$.

$$I = \int_0^\infty e^{-\alpha^2 u^2} \cos(2b\alpha\, u)\,(\alpha\, du) = \frac{\sqrt{\pi}}{2}\, e^{-b^2}$$

or

$$\int_0^\infty e^{-\alpha^2 u^2} \cos(2b\alpha\, u)\, du = \frac{\sqrt{\pi}}{2\alpha}\, e^{-b^2}.$$

Integrating with respect to p, from $p = 0$ to $p = s$, we get

$$\int_0^\infty e^{-\alpha^2 u^2} \left[\frac{\sin(pu)}{u}\right]_0^s du = \frac{\sqrt{\pi}}{2\alpha} \int_0^s e^{-p^2/(4\alpha^2)}\, dp$$

or $\displaystyle \int_0^\infty e^{-\alpha^2 u^2} \left(\frac{\sin(su)}{u}\right) du = \frac{\sqrt{\pi}}{2\alpha} \int_0^{s/(2\alpha)} e^{-u^2}\,(2\alpha\, du) = (\sqrt{\pi})\left(\frac{\sqrt{\pi}}{2}\right) erf\left(\frac{s}{2\alpha}\right) = \frac{\pi}{2}\, erf\left(\frac{s}{2\alpha}\right)$

Replace the dummy variable u by x.

Exercise 1.4

1. Point of inflection: $(1, 9)$; concave upward in the interval $(1, \infty)$; concave downward in the interval $(-\infty, 1)$.

2. Points of inflection: $(0, 0)$ and $(1/2, -1/6)$; concave upward in the intervals $(-\infty, 0)$ and $(1/2, \infty)$; concave downward in the interval $(0, 1/2)$.

3. Concave upward in $(-\infty, \infty)$.　　　　　**4.** Concave downward in $(0, \infty)$.

5. Concave upward in the interval $(3, \infty)$; concave downward in the interval $(-\infty, 3)$.

6. Points of inflection: $(-2\sqrt{3}, -3\sqrt{3}/2)$, $(0, 0)$ and $(2\sqrt{3}, 3\sqrt{3}/2)$; concave upward in the intervals $(-\infty, -2\sqrt{3})$ and $(0, 2\sqrt{3})$; concave downward in the intervals $(-2\sqrt{3}, 0)$ and $(2\sqrt{3}, \infty)$.

7. Point of inflection: $(0, 0)$; concave upward in the interval $(0, \pi)$; concave downward in the interval $(-\pi, 0)$.

8. Points of inflection: $(1, 2e^{-1})$, $(3, 10e^{-3})$; concave upward in the intervals $(-\infty, 1)$ and $(3, \infty)$; concave downward in the interval $(1, 3)$.

9. Point of inflection: $(e^{-5/6}, -(5/6)\, e^{-5/2})$; concave upward in the interval $(e^{-5/6}, \infty)$; concave downward in the interval $(0, e^{-5/6})$.

10. Point of inflection: $(0, 0)$; concave upward in the interval $(0, \infty)$; concave downward in the interval $(-\infty, 0)$.

11. $2, 1/2, (3, 3/2)$.　　　　　**12.** $6/(13\sqrt{13}\, a)$, $(13\sqrt{13}\, a/6)$, $(-11a/2, 16a/3)$,

13. $4\sqrt{2}/a$, $a/4\sqrt{2}$, $(7a/8, 7a/8)$.　　　　　**14.** $1/a$, a. $(0, 2a)$.

15. $\sqrt{2}/(8a)$, $8a/\sqrt{2}$, $(-2a, 5a)$.　　　　　**16.** $1/\sqrt{2}$, $\sqrt{2}$, $(-1, 1)$.

17. $2, 1/2, (0, 3/2)$.　　　　　**18.** $1, 1, (\pi/2, 0)$.

19. $4/(a\pi)$, $a\pi/4$, $(a/\sqrt{2}, a/\sqrt{2})$.　　　　　**20.** $1/a$, a, $(0, 2a)$.

21. $\rho = 3/2$.   **22.** $\rho = 1$

23. $4(Y - 2a)^3 = 27a\, X^2$.

24. $X = (3b^4 + 1)/(2b^3)$, $Y = (3 + b^4)/(2b)$, where b is the abscissa of any point on the given curve.

25. $\alpha = \left[\dfrac{a^2 + b^2}{a^4}\right] x^3$, $\beta = -\left[\dfrac{a^2 + b^2}{b^4}\right] y^3$. Therefore, $x = \left[\dfrac{a^4 \alpha}{a^2 + b^2}\right]^{1/3}$, $y = -\left[\dfrac{b^4 \beta}{a^2 + b^2}\right]^{1/3}$.

Substituting in the equation of the given curve, we obtain the evolute of the given curve as

$$a^{2/3}X^{2/3} - b^{2/3}Y^{2/3} = (a^2 + b^2)^{2/3}.$$

26. $\alpha = x + 3x^{1/3}y^{2/3}$, $\beta = y + 3x^{2/3}y^{1/3}$. We obtain

$$\alpha + \beta = (x^{1/3} + y^{1/3})^3, \quad \alpha - \beta = (x^{1/3} - y^{1/3})^3.$$

We get $x^{1/3} = [(\alpha + \beta)^{1/3} + (\alpha - \beta)^{1/3}]/2$, $y^{1/3} = [(\alpha + \beta)^{1/3} - (\alpha - \beta)^{1/3}]/2$.

Substituting in the equation of the given curve, we obtain the evolute as

$$(X + Y)^{2/3} + (X - Y)^{2/3} = 2a^{2/3}.$$

27. $X = a(\theta + \sin\theta)$, $Y = -a(1 - \cos\theta)$, where θ is any point on the given curve.

30. $y^2 + 40x = 0$.   **31.** $y^2 - x^2 = 25$.

32. $y = 0$.   **33.** $y^2 = 6x$.

34. $y^2 = 4x^3$.   **35.** $x = 0$ and $y = 0$.

36. The area of the triangle is $A = pq/2$, where $O(0, 0)$, $P(p, 0)$ and $Q(0, q)$ are the vertices of the triangle. The equation of the line is $(x/p) + (y/q) = 1$, where $q = 2A/p$. The envelope is $2xy = \pm A$.

37. Use $b = A/(\pi a)$. Envelope is $2\pi xy = \pm A$.

38. $x = \pm 1$, $y = 2$.   **39.** $x = \pm a$.

40. $x = -1$, $x = -3$, $y = 0$.   **41.** $x = 0$, $x = -1$, $y = 1$.

42. $x = \pm 2$, $y = \pm 1$.   **43.** $x = 0$, $y = 0$.

44. $y = \pm (bx/a)$.

45. The curve has no vertical or horizontal asymptotes. Substituting $y = mx + n$ in the given equation of the curve, we get

$$(1 + m^5)\, x^5 + 5m^2(m^2n - a)\, x^4 + 10mn(m^2n - a)\, x^3 + 5n^2(2m^2n - a)x^2 + 5\, mn^4x + n^5 = 0$$

Equating the coefficients of x^5 and x^4 to zero, we get $m^5 + 1 = 0$ and $m^2n - a = 0$. Therefore, $m = -1$ and $n = a$. We obtain the asymptote as $x + y = a$.

46. We have $y(\pi) = \infty$ and $x(\pi) = \pi$. The curve has a vertical asymptote $x = \pi$, ($t = 0$ is not part of the domain).

47. We have $x = r \cos\theta = a \cos\theta/\theta = \phi(\theta)$ and $y = r \sin\theta = a \sin\theta/\theta = \psi(\theta)$. As $\theta \to 0$, $x \to \infty$ and $y \to a$. Hence, the curve has an horizontal asymptote $y = a$.

48. $y = -2x$, $y = x - (1/2)$, $y = -x - (1/2)$.

49. $y = x$, $y = -x$, $y = x + 1$.

50. $x + y + 1 = 0$, $x + y - 1 = 0$, $x + y + 2 = 0$.

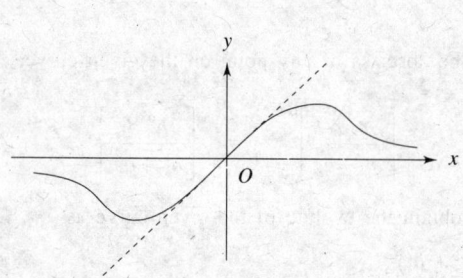

Fig. 1.35   Problem 51.

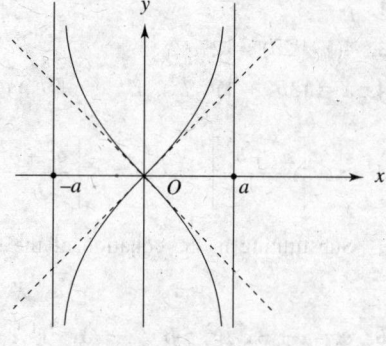

Fig. 1.36   Problem 52.

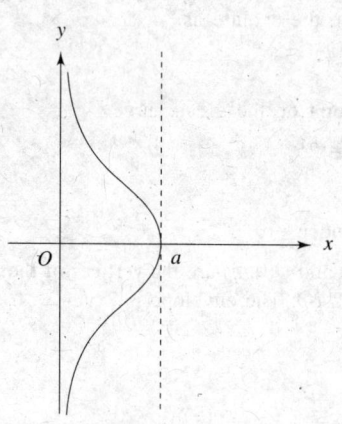

Fig. 1.37   Problem 53.

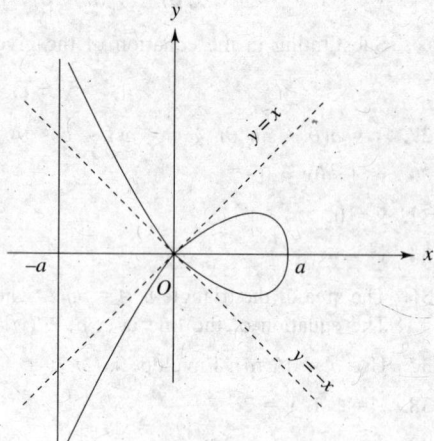

Fig. 1.38   Problem 54.

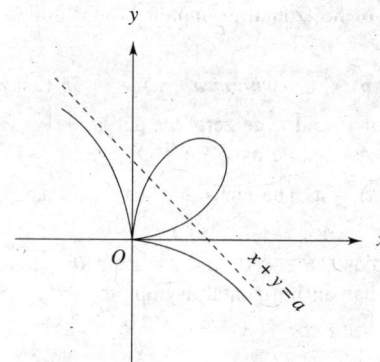

Fig. 1.39   Problem 55.

Chapter 2

Functions of Several Real Variables

2.1 Introduction

In Chapter 1 we studied the calculus of functions of a single real variable defined by $y = f(x)$. In this chapter we shall extend the concepts of functions of one variable to functions of two or more variables.

If to each point (x, y) of a certain part of the x-y plane, $x \in \mathbb{R}$, $y \in \mathbb{R}$ or $(x, y) \in \mathbb{R} \times \mathbb{R} = \mathbb{R}^2$, there corresponds a real value z according to some rule $f(x, y)$, then $f(x, y)$ is called a *real valued function of two variables* x and y and is written as

$$z = f(x, y), \ x \in \mathbb{R}, \ y \in \mathbb{R}, \text{ or } (x, y) \in \mathbb{R}^2, z \in \mathbb{R}. \tag{2.1}$$

We call x, y as the independent variables and z as the dependent variable.

In general, we define a real valued function of n variables as

$$z = f(x_1, x_2, \ldots, x_n), \ (x_1, x_2, \ldots, x_n) \in \mathbb{R}^n, z \in \mathbb{R} \tag{2.2}$$

where $x_1, x_2, \ldots, x_n$ are the n independent variables and z is the dependent variable. The point $(x_1, x_2, \ldots, x_n)$ is called an *n-tuple* and lies in an n-dimensional space. In this case, the function f maps $\mathbb{R}^n$ into $\mathbb{R}$.

The function as defined by Eq. (2.2) is called an *explicit* function, whereas a function defined by $\phi(z, x_1, x_2, \ldots, x_n) = 0$ is called an *implicit* function.

We shall discuss the calculus of the functions of two variables in detail and then generalize to the case of several variables.

2.2 Functions of Two Variables

Consider the function of two variables

$$z = f(x, y). \tag{2.3}$$

The set of points (x, y) in the x-y plane for which $f(x, y)$ is defined is called the *domain* of definition of the function and is denoted by D. This domain may be the entire x-y plane or a part of the x-y plane. The collection of the corresponding values of z is called the *range* of the function. The following are some examples

$z = \sqrt{1 - x^2 - y^2}$: z is real. Therefore, we have $1 - x^2 - y^2 \geq 0$, or $x^2 + y^2 \leq 1$, that is, the domain is the region $x^2 + y^2 \leq 1$. The range is the set of all real, positive numbers.

$z = 1/(x^2 - y^2)$:   The domain is the set of all points (x, y) such that $x^2 - y^2 \neq 0$, that is $y \neq \pm x$. The range is $\mathbb{R}$.

$z = \log(x + y)$:   The domain is the set of all points (x, y) such that $x + y > 0$. The range is $\mathbb{R}$.

The domain of a function and its *natural domain* can be different. For example, we have

$$f(x, y) = \text{area of a triangle} = x\,y/2$$

where x and y are respectively the base and the altitude of the triangle. The domain is $x > 0$, $y > 0$, whereas the natural domain of the function is the entire x-y plane.

Consider the rectangular coordinate system $Oxyz$ (Fig. 2.1).

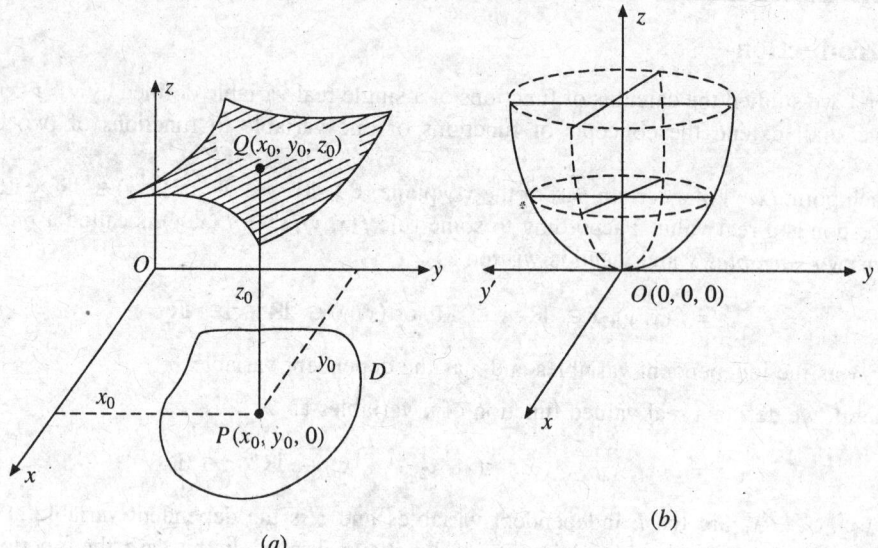

(a)

(b)

Fig. 2.1.   Function of two variables.

At each point $P(x_0, y_0, 0)$ in the x-y plane, construct a perpendicular to the x-y plane. Take a point Q on it such that $PQ = z_0 = f(x_0, y_0)$. This gives a point $Q(x_0, y_0, z_0)$, or $Q(x_0, y_0, f(x_0, y_0))$ in space. The locus of all such points (x, y, z) satisfying $z = f(x, y)$ is called a surface. For example, the graph of the function $z = x^2 + y^2$, $(x, y) \in \mathbb{R}^2$ is the paraboloid of revolution as given in Fig. 2.1b. Each perpendicular to the x-y plane intersects the surface $z = f(x, y)$ at exactly one point if $(x, y) \in D$ and at no point if $(x, y) \notin D$.

The graph of $z = f(x, y) = c$, where c is a real constant is called a *level curve*. For example, for the paraboloid of revolution $z = x^2 + y^2$, the level curves are the circles $x^2 + y^2 = c$, $c > 0$.

We define the following:

Distance between two points   Let $P(x_0, y_0)$ and $Q(x_1, y_1)$ be any two points in $\mathbb{R}^2$. Then

$$d(P, Q) = |PQ| = \sqrt{(x_1 - x_0)^2 + (y_1 - y_0)^2} \tag{2.4}$$

is called the distance between the points P and Q.

Neighborhood of a point Let $P(x_0, y_0)$ be a point in $\mathbb{R}^2$. Then the δ-neighborhood of the point $P(x_0, y_0)$ is the set of all points (x, y) which lie inside a circle of radius δ with centre at the point (x_0, y_0), (Fig. 2.2). We usually denote this neighborhood by $N_\delta(P)$ or by $N(P, \delta)$.

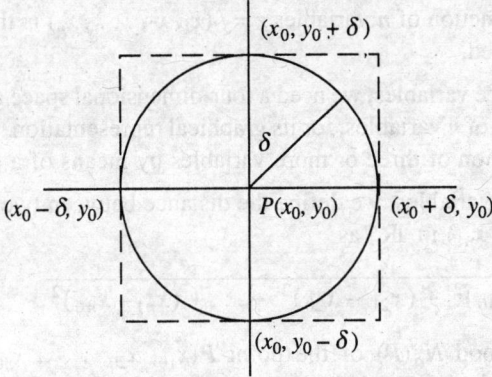

Fig. 2.2.   Neighborhood of a point $P(x_0, y_0)$.

Therefore,

$$N_\delta(P) = \left\{(x, y): \sqrt{(x - x_0)^2 + (y - y_0)^2} < \delta\right\}. \tag{2.5}$$

Since $\ |x - x_0| \le \sqrt{(x - x_0)^2 + (y - y_0)^2}\ $ and $\ |y - y_0| \le \sqrt{(x - x_0)^2 + (y - y_0)^2}$,

the neighborhood of the point $P(x_0, y_0)$ can also be defined as

$$N_\delta(P) = \{(x, y): |x - x_0| < \delta \ \text{ and } \ |y - y_0| < \delta\}. \tag{2.6}$$

that is, the set of all points which lie inside a square of side 2δ with centre at (x_0, y_0) and sides parallel to the coordinate axes (Fig. 2.2).

If the point $P(x_0, y_0)$ is not included in the set, then it is called the *deleted δ-neighborhood* of the point, that is, the set of points which satisfy

$$0 < \sqrt{(x - x_0)^2 + (y - y_0)^2} < \delta \tag{2.7}$$

is called the deleted neighborhood of $P(x_0, y_0)$.

Open domain A domain D is open, if for every point P in D, there exists a $\delta > 0$ such that all points in the δ-neighborhood of P are in D.

Connected domain A domain D is connected, if any two points $P, Q \in D$ can be joined by finitely many number of line segments all of which lie entirely in D.

Bounded domain A domain D is bounded, if there exists a real finite positive number M (no matter how large) such that D can be enclosed within a circle with radius M and centre at the origin. That is, the distance of any point P in D from the origin is less than M, $|OP| < M$.

Closed region A closed region is a bounded domain together with its boundary.

Bounded function A function $f(x, y)$ defined in some domain D in $\mathbb{R}^2$ is bounded, if there exists a real finite positive number M such that $|f(x, y)| \leq M$ for all $(x, y) \in D$.

Remark 1

(a) The domain of a function of n variables $z = f(x_1, x_2, \ldots, x_n)$ is the set of all n-tuples in $\mathbb{R}^n$ for which f is defined.

(b) For functions of three variables, we need a four-dimensional space and an $(n + 1)$ - dimensional space for a function of n variables, for its graphical representation. Therefore, it is not possible to represent a function of three or more variables by means of a graph in space.

(c) For a function of n variables, we define the distance between two points $P(x_{10}, x_{20}, \ldots, x_{n0})$ and $Q(x_{11}, x_{21}, \ldots, x_{n1})$ in $\mathbb{R}^n$ as

$$|PQ| = \sqrt{(x_{11} - x_{10})^2 + (x_{21} - x_{20})^2 + \ldots + (x_{n1} - x_{n0})^2}$$

and the neighborhood $N_\delta(P)$ of the point $P(x_{10}, x_{20}, \ldots, x_{n0})$ is the set of all points $(x_1, x_2, \ldots, x_n)$ inside an open ball

$$\sqrt{(x_1 - x_{10})^2 + (x_2 - x_{20})^2 + \ldots + (x_n - x_{n0})^2} < \delta.$$

2.2.1 Limits

Let $z = f(x, y)$ be a function of two variables defined in a domain D. Let $P(x_0, y_0)$ be a point in D. If for a given real number $\varepsilon > 0$, however small, we can find a real number $\delta > 0$ such that for every point (x, y) in the δ-neighborhood of $P(x_0, y_0)$

$$|f(x, y) - L| < \varepsilon, \quad \text{whenever} \quad \sqrt{(x - x_0)^2 + (y - y_0)^2} < \delta \tag{2.8}$$

then the real, finite number L is called the limit of the function $f(x, y)$ as $(x, y) \to (x_0, y_0)$. Symbolically, we write it as

$$\lim_{(x, y) \to (x_0, y_0)} f(x, y) = L.$$

Note that for the limit to exist, the function $f(x, y)$ may or may not be defined at (x_0, y_0). If $f(x, y)$ is not defined at $P(x_0, y_0)$, then we write

$$|f(x, y) - L| < \varepsilon, \quad \text{whenever} \quad 0 < \sqrt{(x - x_0)^2 + (y - y_0)^2} < \delta.$$

This definition is called the δ-ε approach to study the existence of limits.

Remark 2

(a) $\lim\limits_{(x, y) \to (x_0, y_0)} f(x, y)$, if it exists is unique.

(b) Let $x = r \cos\theta$, $y = r \sin\theta$ so that $x^2 + y^2 = r^2$ and $\theta = \tan^{-1}(y/x)$. Then, we can define the limit given in Eq. (2.8) as

$$\lim_{r \to 0} |f(r \cos\theta, r \sin\theta) - L| < \varepsilon, \quad \text{whenever} \quad r < \delta, \text{ independent of } \theta.$$

(c) Since $(x, y) \to (x_0, y_0)$ in the two-dimensional plane, there are infinite number of paths joining (x, y) to (x_0, y_0). Since the limit is unique, the limit is same along all the paths, that is the limit is independent of the path. Thus, the limit of a function cannot be obtained by approaching the point P along a particular path and finding the limit of $f(x, y)$. If the limit is dependent on a path, then the limit does not exist.

Let $u = f(x, y)$ and $v = g(x, y)$ be two real valued functions defined in a domain D. Let

$$\lim_{(x, y) \to (x_0, y_0)} f(x, y) = L_1 \quad \text{and} \quad \lim_{(x, y) \to (x_0, y_0)} g(x, y) = L_2.$$

Then, the following results can be easily established.

(i) $\displaystyle \lim_{(x, y) \to (x_0, y_0)} [k f(x, y)] = k L_1$ for any real constant k.

(ii) $\displaystyle \lim_{(x, y) \to (x_0, y_0)} [f(x, y) \pm g(x, y)] = L_1 \pm L_2$.

(iii) $\displaystyle \lim_{(x, y) \to (x_0, y_0)} [f(x, y) g(x, y)] = L_1 L_2$.

(iv) $\displaystyle \lim_{(x, y) \to (x_0, y_0)} [f(x, y)/g(x, y)] = L_1/L_2, \ L_2 \neq 0$.

Remark 3

Let $z = f(x_1, x_2, \ldots, x_n)$ be a function of n variables defined in some domain D in $\mathbb{R}^n$. Then, for any fixed point $P_0(x_{10}, x_{20}, \ldots, x_{n0})$ in D

$$\lim_{P \to P_0} f(x_1, x_2, \ldots, x_n) = L.$$

if $|f(x_1, x_2, \ldots, x_n) - L| < \varepsilon$, whenever $\sqrt{(x_1 - x_{10})^2 + (x_2 - x_{20})^2 + \ldots + (x_n - x_{n0})^2} < \delta$ where $P(x_1, x_2, \ldots, x_n)$ is a point in the neighborhood or the deleted neighborhood of P_0.

Example 2.1 Using the δ-ε approach, show that

(i) $\displaystyle \lim_{(x, y) \to (2, 1)} (3x + 4y) = 10$,

(ii) $\displaystyle \lim_{(x, y) \to (1, 1)} (x^2 + 2y) = 3$.

Solution

(i) Here $f(x, y) = 3x + 4y$ is defined at $(2, 1)$. We have

$$|f(x, y) - 10| = |3x + 4y - 10| = |3(x - 2) + 4(y - 1)| \leq 3|x - 2| + 4|y - 1|.$$

If we take $|x - 2| < \delta$ and $|y - 1| < \delta$, we get $|f(x, y) - 10| < 7\delta < \varepsilon$, which is satisfied when $\delta < \varepsilon/7$.

Hence, $\displaystyle \lim_{(x, y) \to (2, 1)} f(x, y) = 10$.

Note that the value of δ is not unique.

(ii) Here $f(x, y) = x^2 + 2y$ is defined at $(1, 1)$. We have

$$|f(x, y) - 3| = |x^2 + 2y - 3| = |(x - 1 + 1)^2 + 2(y - 1 + 1) - 3|$$

$$= |(x - 1)^2 + 2(x - 1) + 2(y - 1)| \leq |x - 1|^2 + 2|x - 1| + 2|y - 1|$$

If we take $|x - 1| < \delta$ and $|y - 1| < \delta$, we get $|f(x, y) - 3| < \delta^2 + 4\delta < \varepsilon$ which is satisfied when

$$(\delta + 2)^2 < \varepsilon + 4 \text{ or } \delta < \sqrt{\varepsilon + 4} - 2.$$

Hence, $\lim\limits_{(x,y)\to(1,1)} f(x,y) = 3$.

We can also write $|f(x,y) - 3| < \delta^2 + 4\delta < 5\delta < \varepsilon$

which is satisfied when $\delta < \varepsilon/5$.

Example 2.2 Using δ-ε approach, show that

(i) $\lim\limits_{(x,y)\to(0,0)}\left(\dfrac{xy}{\sqrt{x^2+y^2}}\right) = 0,$ (ii) $\lim\limits_{(x,y,z)\to(0,0,0)}\left(\dfrac{xy+xz+yz}{\sqrt{x^2+y^2+z^2}}\right) = 0.$

Solution

(i) Here $f(x,y) = xy/(\sqrt{x^2+y^2})$ is not defined at $(0,0)$. We have

$$\left|\frac{xy}{\sqrt{x^2+y^2}} - 0\right| = \left|\frac{xy}{\sqrt{x^2+y^2}}\right| \le \frac{1}{2}\frac{(x^2+y^2)}{\sqrt{x^2+y^2}} = \frac{1}{2}\sqrt{x^2+y^2} < \varepsilon, (x,y) \ne (0,0)$$

since $|xy| \le (x^2+y^2)/2$. If we choose $\delta < 2\varepsilon$, then we get

$$\left|\frac{xy}{\sqrt{x^2+y^2}} - 0\right| < \varepsilon, \text{ whenever } 0 < \sqrt{x^2+y^2} < \delta.$$

Hence, $\lim\limits_{(x,y)\to(0,0)} \dfrac{xy}{\sqrt{x^2+y^2}} = 0$.

Alternative Writing $x = r\cos\theta$, $y = r\sin\theta$, we obtain

$$\lim\limits_{(x,y)\to(0,0)}\left|\frac{xy}{\sqrt{x^2+y^2}}\right| = \lim\limits_{r\to0}\left|\frac{r^2\sin\theta\cos\theta}{r}\right| = 0$$

which is independent of θ.

(ii) Here $f(x,y,z) = (xy+xz+yz)/\sqrt{x^2+y^2+z^2}$ is not defined at $(0,0,0)$.

Since $|xy| \le (x^2+y^2)/2$, $|xz| \le (x^2+z^2)/2$, $|yz| < (y^2+z^2)/2$, we get

$$\left|\frac{xy+xz+yz}{\sqrt{x^2+y^2+z^2}} - 0\right| \le \frac{1}{2}\left[\frac{x^2+y^2+x^2+z^2+y^2+z^2}{\sqrt{x^2+y^2+z^2}}\right] = \left|\sqrt{x^2+y^2+z^2}\right| < \varepsilon.$$

If we choose $\delta < \varepsilon$, we obtain

$$\left|\frac{xy+xz+yz}{\sqrt{x^2+y^2+z^2}} - 0\right| < \varepsilon \text{ whenever } 0 < \sqrt{x^2+y^2+z^2} < \delta.$$

Hence, $\lim\limits_{(x,y,z)\to(0,0,0)}\left[\dfrac{xy+xz+yz}{\sqrt{x^2+y^2+z^2}}\right] = 0$.

Example 2.3   Show that the following limits

(i) $\displaystyle\lim_{(x,\,y)\to(0,\,0)} \frac{xy}{x^2+y^2}$,   (ii) $\displaystyle\lim_{(x,\,y)\to(0,\,0)} \frac{x+\sqrt{y}}{x^2+y^2}$,

(iii) $\displaystyle\lim_{(x,\,y)\to(0,\,0)} \frac{x^3y}{x^6+y^2}$.   (iv) $\displaystyle\lim_{(x,\,y)\to(0,\,1)} \tan^{-1}\left(\frac{y}{x}\right)$.

do not exist.

Solution   The limit does not exist if it is not finite, or if it depends on a particular path.

(i) Consider the path $y = mx$. As $(x, y) \to (0, 0)$, we get $x \to 0$. Therefore

$$\lim_{(x,\,y)\to(0,\,0)} \frac{xy}{x^2+y^2} = \lim_{x\to0} \frac{mx^2}{(1+m^2)x^2} = \frac{m}{1+m^2}.$$

which depends on m. For different values of m, we obtain different limits. Hence, the limit does not exist.

Alternative Setting $x = r\cos\theta$, $y = r\sin\theta$, we obtain

$$\lim_{(x,\,y)\to(0,\,0)} \frac{xy}{x^2+y^2} = \lim_{r\to0} \frac{r^2\sin\theta\cos\theta}{r^2} = \sin\theta\cos\theta$$

which depends on θ. Hence, the limit is dependent on different radial paths θ = constant. Hence, the limit does not exist.

(ii) Choose the path $y = mx^2$. As $(x, y) \to (0, 0)$, we get $x \to 0$. Therefore,

$$\lim_{(x,\,y)\to(0,\,0)} \frac{x+\sqrt{y}}{x^2+y} = \lim_{x\to0} \frac{1+\sqrt{m}}{(1+m)x} = \infty.$$

Since the limit is not finite, the limit does not exist.

(iii) Choose the path $y = mx^3$. As $(x, y) \to (0, 0)$, we get $x \to 0$. Therefore

$$\lim_{(x,\,y)\to(0,\,0)} \frac{x^3y}{x^6+y^2} = \lim_{x\to0} \frac{mx^6}{(1+m^2)x^6} = \frac{m}{1+m^2}$$

which depends on m. For different values of m, we obtain different limits. Hence, the limit does not exist.

(iv) We have

$$\lim_{(x,\,y)\to(0,\,1)} \tan^{-1}\frac{y}{x} = \tan^{-1}(\pm\infty) = \pm\frac{\pi}{2}$$

depending on whether the point $(0, 1)$ is approached from left or from right along the line $y = 1$. If we approach from left, we obtain the limit as $-\pi/2$ and if we approach from right, we obtain the limit as $\pi/2$. Since the limit is not unique, the limit does not exist as $(x, y) \to (0, 1)$.

2.2.2   Continuity

A function $z = f(x, y)$ is said to be *continuous* at a point (x_0, y_0). if

(i) $f(x, y)$ is defined at the point (x_0, y_0),   (ii) $\displaystyle\lim_{(x,\,y)\to(x_0,\,y_0)} f(x, y)$ exists, and

(iii) $\displaystyle\lim_{(x,\,y)\to(x_0,\,y_0)} f(x, y) = f(x_0, y_0)$.

If any one of the above conditions is not satisfied, then the function is said to be discontinuous at the point (x_0, y_0).

Therefore, a function $f(x, y)$ is continuous at (x_0, y_0) if

$$|f(x,y) - f(x_0, y_0)| < \varepsilon, \quad \text{whenever } \sqrt{(x - x_0)^2 + (y - y_0)^2} < \delta.$$

If $f(x_0, y_0)$ is defined and $\lim\limits_{(x,y) \to (x_0, y_0)} f(x, y) = L$ exists, but $f(x_0, y_0) \neq L$, then the point (x_0, y_0) is called a point of *removable discontinuity*. We can redefine the function at the point (x_0, y_0) as $f(x_0, y_0) = L$ so that the new function becomes continuous at the point (x_0, y_0).

If the function $f(x, y)$ is continuous at every point in a domain D, then it is said to be continuous in D.

In the definition of continuity, $\lim\limits_{(x,y) \to (x_0, y_0)} f(x,y) = f(x_0, y_0)$ holds for all paths going to the point (x_0, y_0).

Hence, if the continuity of a function is to be proved, we cannot choose a path and find the limit. However, to show that a function is discontinuous, it is sufficient to choose a path and show that the limit does not exist.

A continuous function has the following properties:

P1 A continuous function in a closed and bounded domain D attains atleast once its maximum value M and its minimum value m at some point inside or on the boundary of D.

P2 For any number μ that satisfies $m < \mu < M$, there exists a point (x_0, y_0) in D such that

$$f(x_0, y_0) = \mu.$$

P3 A continuous function, in a closed and bounded domain D. that attains both positive and negative values will have the value zero at some point in D.

Example 2.4 Show that the following functions are continuous at the point $(0, 0)$.

(i) $f(x,y) = \begin{cases} \dfrac{2x^4 + 3y^4}{x^2 + y^2}, & (x, y) \neq (0, 0) \\ 0, & (x, y) = (0, 0) \end{cases}$ (ii) $f(x,y) = \begin{cases} \dfrac{2x(x^2 - y^2)}{x^2 + y^2}, & (x, y) \neq (0, 0) \\ 0, & (x, y) = (0, 0), \end{cases}$

(iii) $f(x,y) = \begin{cases} \dfrac{\sin^{-1}(x + 2y)}{\tan^{-1}(2x + 4y)}, & (x, y) \neq (0, 0) \\ 1/2, & (x, y) = (0, 0). \end{cases}$

Solution

(i) Let $x = r \cos \theta, y = r \sin \theta$. Then $r = \sqrt{x^2 + y^2} \neq 0$. We have

$$|f(x,y) - f(0,0)| = \left| \frac{2x^4 + 3y^4}{x^2 + y^2} \right| = \left| \frac{r^4 (2 \cos^4 \theta + 3 \sin^4 \theta)}{r^2 (\cos^2 \theta + 2 \sin^2 \theta)} \right|$$

$$< r^2 [2 | \cos^4 \theta | + 3 | \sin^4 \theta |] < 5r^2 < \varepsilon$$

or
$$r = \sqrt{x^2 + y^2} < \sqrt{\varepsilon/5}.$$

If we choose $\delta < \sqrt{\varepsilon/5}$, we find that $|f(x, y) - f(0, 0)| < \varepsilon$, whenever $0 < \sqrt{x^2 + y^2} < \delta$. Therefore, $\lim\limits_{(x, y) \to (0, 0)} f(x, y) = f(0, 0) = 0$. Hence, $f(x, y)$ is continuous at $(0, 0)$.

(ii) Let $x = r \cos\theta$, $y = r \sin\theta$. Then, $r = \sqrt{x^2 + y^2} \neq 0$. We have

$$|f(x, y) - f(0, 0)| = \left| \frac{2x(x^2 - y^2)}{x^2 + y^2} \right| = \left| \frac{2r^3 (\cos^2\theta - \sin^2\theta) \cos\theta}{r^2(\cos^2\theta + \sin^2\theta)} \right|$$

$$= |2r \cos 2\theta \cos\theta| \le 2r < \varepsilon$$

or
$$r = \sqrt{x^2 + y^2} < \varepsilon/2.$$

If we choose $\delta < \varepsilon/2$, we find that

$$|f(x, y) - f(0, 0)| < \varepsilon, \quad \text{whenever} \quad 0 < \sqrt{x^2 + y^2} < \delta.$$

Therefore, $\lim\limits_{(x, y) \to (0, 0)} f(x, y) = f(0, 0) = 0$. Hence, $f(x, y)$ is continuous at $(0, 0)$.

(iii) Let $x + 2y = t$. Therefore, $t \to 0$ as $(x, y) \to (0, 0)$.

We can now write

$$\lim_{(x, y) \to (0, 0)} f(x, y) = \lim_{t \to 0} \frac{\sin^{-1} t}{\tan^{-1} 2t} = \lim_{t \to 0} \left[\frac{(\sin^{-1} t)/t}{(\tan^{-1}(2t))/(2t)} \right] \left[\frac{t}{2t} \right] = \frac{1}{2}.$$

Since $\lim\limits_{(x, y) \to (0, 0)} f(x, y) = f(0, 0) = \frac{1}{2}$, the given function is continuous at $(x, y) = (0, 0)$.

Example 2.5 Show that the following functions are discontinuous at the given points

(i) $f(x, y) = \begin{cases} \dfrac{x - y}{x + y}, & (x, y) \neq (0, 0) \\ 0, & (x, y) = (0, 0) \end{cases}$

at the point $(0, 0)$.

(ii) $f(x, y) = \begin{cases} \dfrac{x^2 - x\sqrt{y}}{x^2 + y}, & (x, y) \neq (0, 0) \\ 0, & (x, y) = (0, 0) \end{cases}$

at the point $(0, 0)$.

(iii) $f(x, y) = \begin{cases} \dfrac{x^2 + xy + x + y}{x + y}, & (x, y) \neq (2, 2) \\ 4, & (x, y) = (2, 2) \end{cases}$

at the point $(2, 2)$.

Solution

(i) Choose the path $y = mx$. As $(x, y) \to (0, 0)$, we get $x \to 0$. Therefore,

$$\lim_{(x, y) \to (0, 0)} \frac{x - y}{x + y} = \lim_{x \to 0} \frac{(1 - m)x}{(1 + m)x} = \frac{1 - m}{1 + m}$$

which depends on m. Since, the limit does not exist, the function is not continuous at $(0, 0)$.

(ii) Choose the path $y = m^2x^2$. As $(x, y) \to (0, 0)$, we get $x \to 0$. Therefore,

$$\lim_{(x, y)\to(0, 0)} \frac{x^2 - x\sqrt{y}}{x^2 + y} = \lim_{x\to 0} \frac{(1 - m)x^2}{(1 + m^2)x^2} = \frac{1 - m}{1 + m^2}$$

which depends on m. Since the limit does not exist, the function is not continuous at $(0, 0)$.

(iii) $\displaystyle\lim_{(x, y)\to(2, 2)} f(x, y) = \lim_{(x, y)\to(2, 2)} \frac{(x + y)(x + 1)}{(x + y)} = \lim_{(x, y)\to(2, 2)} (x + 1) = 3.$

Since $\displaystyle\lim_{(x, y)\to(2, 2)} f(x, y) \ne f(2, 2)$, the function is not continuous at $(2, 2)$.

Note that the point $(2, 2)$ is a point of removable discontinuity.

Example 2.6 Let $f(x, y) = \begin{cases} \dfrac{x^4y - 3x^2y^3 + y^5}{(x^2 + y^2)^2}, & (x, y) \ne (0, 0) \\ 0, & (x, y) = (0, 0). \end{cases}$

Find a $\delta > 0$ such that $|f(x, y) - f(0, 0)| < 0.01$, whenever $\sqrt{x^2 + y^2} < \delta$.

Solution We have

$$|f(x, y) - f(0, 0)| = \left| \frac{x^4y - 3x^2y^3 + y^5}{(x^2 + y^2)^2} \right|.$$

Substituting $x = r\cos\theta$, $y = r\sin\theta$, we obtain

$$|f(x, y) - f(0, 0)| = \left| \frac{r^5(\cos^4\theta\sin\theta - 3\cos^2\theta\sin^3\theta + \sin^5\theta)}{r^4(\cos^2\theta + \sin^2\theta)^2} \right|$$

$$= |r(\cos^4\theta\sin\theta - 3\cos^2\theta\sin^3\theta + \sin^5\theta)|$$

$$\le r(1 + 3 + 1) = 5r = 5\sqrt{x^2 + y^2} < 0.01.$$

Therefore, $\sqrt{x^2 + y^2} \le 0.01/5 = 0.002$. Hence, $\delta < 0.002$.

Exercise 2.1

Using the δ-ε approach, establish the following limits.

1. $\displaystyle\lim_{(x, y)\to(1, 1)} (x^2 + y^2 - 1) = 1.$

2. $\displaystyle\lim_{(x, y)\to(2, 1)} (x^2 + 2x - y^2) = 7.$

3. $\displaystyle\lim_{(x, y)\to(0, 0)} \frac{x + y}{x^2 + y^2 + 1} = 0.$

4. $\displaystyle\lim_{(x, y)\to(0, 0)} \frac{x^3 + y^3}{x^2 + y^2} = 0.$

5. $\displaystyle\lim_{(x, y)\to(0, 0)} \left[y + x\cos\left(\frac{1}{y}\right) \right] = 0.$

6. $\displaystyle\lim_{(x, y)\to(0, 0)} (x^2 + y^2)\sin\frac{1}{xy} = 0.$

Determine the following limits if they exist.

7. $\displaystyle \lim_{(x,y)\to(0,0)} \frac{x}{\sqrt{x^2+y^2}}$.

8. $\displaystyle \lim_{(x,y)\to(1,-1)} \frac{x^3-y^3}{x-y}$.

9. $\displaystyle \lim_{(x,y)\to(\alpha,0)} \left(1+\frac{x}{y}\right)^y$.

10. $\displaystyle \lim_{(x,y)\to(0,0)} \cot^{-1}\left(\frac{1}{\sqrt{x^2+y^2}}\right)$.

11. $\displaystyle \lim_{(x,y)\to(0,1)} \frac{(y-1)\tan^2 x}{x^2(y^2-1)}$.

12. $\displaystyle \lim_{(x,y)\to(1,0)} \frac{(x-1)\sin y}{y\ln x}$.

13. $\displaystyle \lim_{(x,y)\to(0,0)} \frac{1-x-y}{x^2+y^2}$.

14. $\displaystyle \lim_{(x,y)\to(0,0)} \frac{x}{x^2+y^2}$.

15. $\displaystyle \lim_{(x,y)\to(0,0)} \frac{x^2}{x^3+y^3}$.

16. $\displaystyle \lim_{(x,y)\to(0,0)} \frac{x^4 y^2}{(x^4+y^2)^2}$.

17. $\displaystyle \lim_{(x,y,z)\to(0,0,0)} \log\left(\frac{z}{xy}\right)$.

18. $\displaystyle \lim_{(x,y,z)\to(0,0,0)} \frac{xy+z}{x+y+z^2}$.

19. $\displaystyle \lim_{(x,y,z)\to(0,0,0)} \frac{x\,y^2 z^2}{x^4+y^4+z^8}$.

20. $\displaystyle \lim_{(x,y,z)\to(0,0,0)} \frac{x(x+y+z)}{x^2+y^2+z^2}$.

Discuss the continuity of the following functions at the given points.

21. $f(x,y) = \begin{cases} \dfrac{(x-y)^2}{x^2+y^2}, & (x,y)\neq(0,0) \\ 0, & (x,y)=(0,0) \end{cases}$

at $(0,0)$.

22. $f(x,y) = \begin{cases} \dfrac{1}{1+e^{1/x}}+y^2, & (x,y)\neq(0,0) \\ 0, & (x,y)=(0,0) \end{cases}$

at $(0,0)$.

23. $f(x,y) = \begin{cases} \dfrac{e^{xy}}{x^2+1}, & (x,y)\neq(0,0) \\ 0, & (x,y)=(0,0) \end{cases}$

at $(0,0)$.

24. $f(x,y) = \begin{cases} \dfrac{x^2+y^2}{xy}, & (x,y)\neq(0,0) \\ 0, & (x,y)=(0,0) \end{cases}$

at $(0,0)$.

25. $f(x,y) = \begin{cases} \dfrac{x^2+y^2}{\tan xy}, & (x,y)\neq(0,0) \\ 0, & (x,y)=(0,0) \end{cases}$

at $(0,0)$.

26. $f(x,y) = \begin{cases} \dfrac{x^2-2xy+y^2}{x-y}, & (x,y)\neq(1,-1) \\ 0, & (x,y)=(1,-1) \end{cases}$

at $(1,-1)$.

27. $f(x,y) = \begin{cases} \dfrac{xy(x-y)}{x^2+y^2}, & (x,y)\neq(0,0) \\ 0, & (x,y)=(0,0) \end{cases}$

at $(0,0)$.

28. $f(x,y) = \begin{cases} \dfrac{x^4 y^4}{(x^2+y^4)^3}, & (x,y)\neq(0,0) \\ 0, & (x,y)=(0,0) \end{cases}$

at $(0,0)$.

29. $f(x,y) = \begin{cases} \dfrac{\sin\sqrt{|xy|}-\sqrt{|xy|}}{\sqrt{x^2+y^2}}, & (x,y)\neq(0,0) \\ 0, & (x,y)=(0,0) \end{cases}$

at $(0,0)$.

30. $f(x,y) = \begin{cases} \dfrac{2x^2+y^2}{3+\sin x}, & (x,y)\neq(0,0) \\ 0, & (x,y)=(0,0) \end{cases}$

at $(0,0)$.

31. $f(x, y) = \begin{cases} \dfrac{x^2 y^2}{x^3 + y^3}, & (x, y) \neq (0, 0) \\ \quad 0 \quad , & (x, y) = (0, 0) \end{cases}$

at $(0, 0)$.

32. $f(x, y) = \begin{cases} \dfrac{x^5 - y^5}{x^2 + y^2}, & (x, y) \neq (0, 0) \\ \quad 0 \quad , & (x, y) = (0, 0) \end{cases}$

at $(0, 0)$.

33. $f(x, y) = \begin{cases} \dfrac{x^2 y}{1 + x}, & x \neq -1 \\ \quad y \quad , & (x, y) = (-1, \alpha) \end{cases}$

at $(-1, \alpha)$.

34. $f(x, y, z) = \begin{cases} \dfrac{xyz}{x^2 + y^2 + z^2}, & (x, y, z) \neq (0, 0, 0) \\ \quad 0 \quad , & (x, y, z) = (0, 0, 0) \end{cases}$

at $(0, 0, 0)$.

35. $f(x, y, z) = \begin{cases} \dfrac{2xy}{x^2 - 3z^2}, & (x, y, z) \neq (0, 0, 0) \\ \quad 0 \quad , & (x, y, z) = (0, 0, 0) \end{cases}$

at $(0, 0, 0)$.

2.3 Partial Derivatives

The derivative of a function of several variables with respect to one of the independent variables keeping all the other independent variables as constant is called the *partial derivative* of the function with respect to that variable.

Consider the function of two variables $z = f(x, y)$ defined in some domain D of the x-y plane. Let y be held constant, say $y = y_0$. Then, the function $f(x, y_0)$ depends on x alone and is defined in an interval about x, that is $f(x, y_0)$ is a function of one variable x. Let the points (x, y_0) and $(x + \Delta x, y_0)$ be in D, where Δx is an increment in the independent variable x. Then

$$\Delta_x z = f(x + \Delta x, y_0) - f(x, y_0) \qquad (2.10)$$

is called the *partial increment* in z with respect to x and is a function of x and Δx.

Similarly, if x is held constant, say $x = x_0$, then the function $f(x_0, y)$ depends only on y and is defined in some interval about y, that is $f(x_0, y)$ is a function of one variable y. Let the points (x_0, y) and $(x_0, y + \Delta y)$ be in D, where Δy is an increment in the independent variable y. Then

$$\Delta_y z = f(x_0, y + \Delta y) - f(x_0, y) \qquad (2.11)$$

is called the partial increment in z with respect to y and is a function of y and Δy.

When both x and y are given increments Δx and Δy respectively, then the increment Δz in z is given by

$$\Delta z = f(x + \Delta x, y + \Delta y) - f(x, y). \qquad (2.12)$$

This increment is called the *total increment* in z and is a function of x, y, Δx and Δy.

In general, $\Delta z \neq \Delta_x z + \Delta_y z$. For example, consider the function $z = f(x, y) = xy$ and a point (x_0, y_0). We have

$$\Delta_x z = (x_0 + \Delta x) y_0 - x_0 y_0 = y_0 \Delta x$$

$$\Delta_y z = x_0(y_0 + \Delta y) - x_0 y_0 = x_0 \Delta y$$

$$\Delta z = (x_0 + \Delta x)(y_0 + \Delta y) - x_0 y_0 = x_0 \Delta y + y_0 \Delta y + \Delta x \Delta y \ne \Delta_x z + \Delta_y z.$$

Now, consider the limit

$$\lim_{\Delta x \to 0} \frac{\Delta_x z}{\Delta x} = \lim_{\Delta x \to 0} \frac{f(x_0 + \Delta x, y_0) - f(x_0, y_0)}{\Delta x}. \tag{2.13}$$

If this limit exists, then this limit is called the first order partial derivative of z or $f(x, y)$ with respect to x at the point (x_0, y_0) and is denoted by $z_x(x_0, y_0)$ or $f_x(x_0, y_0)$ or $(\partial f / \partial x)(x_0, y_0)$ or $(\partial z / \partial x)(x_0, y_0)$.

Similarly, if the limit

$$\lim_{\Delta y \to 0} \frac{\Delta_y z}{\Delta y} = \lim_{\Delta y \to 0} \frac{f(x_0, y_0 + \Delta y) - f(x_0, y_0)}{\Delta y} \tag{2.14}$$

exists, then this limit is called the first order partial derivative of z or $f(x, y)$ with respect to y at the point (x_0, y_0) and is denoted by $z_y(x_0, y_0)$ or $f_y(x_0, y_0)$ or $(\partial z / \partial y)(x_0, y_0)$ or $(\partial f / \partial y)(x_0, y_0)$.

Remark 4

Let $z = f(x_1, x_2, \ldots, x_n)$ be a function of n variables defined in some domain D in $\mathbb{R}^n$. Let $P_0(x_1, x_2, \ldots, x_n)$ be a point in D. If the limit

$$\lim_{\Delta x_i \to 0} \frac{\Delta_{x_i} z}{\Delta x_i} = \lim_{\Delta x_i \to 0} \frac{f(x_1, x_2, \ldots, x_i + \Delta x_i, \ldots, x_n) - f(x_1, x_2, \ldots, x_i, \ldots x_n)}{\Delta x_i}$$

exists, then it is called the partial derivative of f at the point P_0 and is denoted by $(\partial f / \partial x_i)(P_0)$.

Remark 5

The definition of continuity, $\lim\limits_{(x, y) \to (x_0, y_0)} f(x, y) = f(x_0, y_0)$ can be written in alternate forms. Set

$x = x_0 + \Delta x, y = y_0 + \Delta y$. Define $\Delta \rho = \sqrt{(\Delta x)^2 + (\Delta y)^2}$. Then, $\Delta x \to 0, \Delta y \to 0$ implies that $\Delta \rho \to 0$. We note that $|\Delta x| < \Delta \rho$ and $|\Delta y| < \Delta \rho$.

The above definition of continuity is equivalent to the following forms:

(i) $\lim\limits_{\Delta x \to 0, \Delta y \to 0} [f(x_0 + \Delta x, y_0 + \Delta y) - f(x_0, y_0)] = 0$.

(ii) $\lim\limits_{\Delta \rho \to 0} [f(x_0 + \Delta x, y_0 + \Delta y) - f(x_0, y_0)] = 0$.

(iii) $\lim\limits_{\Delta \rho \to 0} \Delta z = 0$.

Example 2.7 Find the first order partial derivatives of the following functions

(i) $f(x, y) = x^2 + y^2 + x$, (ii) $f(x, y) = y e^{-x}$, (iii) $f(x, y) = \sin(2x + 3y)$

at the point (x, y) from the first principles.

Solution we have

(i) $\dfrac{\partial f}{\partial x} = \lim\limits_{\Delta x \to 0} \dfrac{f(x + \Delta x, y) - f(x, y)}{\Delta x} = \lim\limits_{\Delta x \to 0} \dfrac{[(x + \Delta x)^2 + y^2 + (x + \Delta x)] - [x^2 + y^2 + x]}{\Delta x}$

$= \lim\limits_{\Delta x \to 0} \dfrac{(2x + 1)\Delta x + (\Delta x)^2}{\Delta x} = \lim\limits_{\Delta x \to 0} [2x + 1 + \Delta x] = 2x + 1$.

$$\frac{\partial f}{\partial y} = \lim_{\Delta y \to 0} \frac{f(x, y + \Delta y) - f(x, y)}{\Delta y} = \lim_{\Delta y \to 0} \frac{[x^2 + (y + \Delta y)^2 + x] - [x^2 + y^2 + x]}{\Delta y}$$

$$= \lim_{\Delta y \to 0} \frac{2y \Delta y + (\Delta y)^2}{\Delta y} = \lim_{\Delta y \to 0} [2y + \Delta y] = 2y.$$

(ii) $\dfrac{\partial f}{\partial x} = \lim_{\Delta x \to 0} \dfrac{ye^{-(x + \Delta x)} - ye^{-x}}{\Delta x} = \lim_{\Delta x \to 0} \dfrac{-ye^{-x}(1 - e^{-\Delta x})}{\Delta x} = -ye^{-x} \lim_{\Delta x \to 0} \dfrac{1 - e^{-\Delta x}}{\Delta x} = -ye^{-x}$

$$\frac{\partial f}{\partial y} = \lim_{\Delta y \to 0} \frac{(y + \Delta y)e^{-x} - ye^{-x}}{\Delta y} = e^{-x}.$$

(iii) $\dfrac{\partial f}{\partial x} = \lim_{\Delta x \to 0} \dfrac{\sin(2(x + \Delta x) + 3y) - \sin(2x + 3y)}{\Delta x} = \lim_{\Delta x \to 0} \dfrac{2\cos(2x + 3y + \Delta x)\sin \Delta x}{\Delta x}$

$$= 2 \cos(2x + 3y).$$

$$\frac{\partial f}{\partial y} = \lim_{\Delta y \to 0} \frac{\sin(2x + 3(y + \Delta y)) - \sin(2x + 3y)}{\Delta y} = \lim_{\Delta y \to 0} \frac{2\cos(2x + 3y + 3\Delta y/2)\sin(3\Delta y/2)}{\Delta y}$$

$$= \lim_{\Delta y \to 0} [3 \cos(2x + 3y + 3\Delta y/2)] \frac{\sin(3\Delta y/2)}{(3\Delta y/2)} = 3 \cos(2x + 3y).$$

Example 2.8 Show that the function

$$f(x, y) = \begin{cases} (x + y) \sin\left(\dfrac{1}{x + y}\right), & x + y \neq 0 \\ 0, & x + y = 0 \end{cases}$$

is continuous at $(0, 0)$ but its partial derivatives f_x and f_y do not exist at $(0, 0)$.

Solution We have

$$|f(x, y) - f(0, 0)| = \left| (x + y) \sin\left(\frac{1}{x + y}\right) \right| \leq |x + y| \leq |x| + |y| \leq 2\sqrt{x^2 + y^2} < \varepsilon.$$

If we choose $\delta < \varepsilon/2$, then

$$|f(x, y) - 0| < \varepsilon, \quad \text{whenever} \quad 0 < \sqrt{x^2 + y^2} < \delta.$$

Therefore, $\lim\limits_{(x, y) \to (0,0)} f(x, y) = 0 = f(0, 0).$

Hence, the given function is continuous at $(0, 0)$.

Now, at $(0, 0)$, the limit

$$\lim_{\Delta x \to 0} \frac{\Delta_x z}{\Delta x} = \lim_{\Delta x \to 0} \frac{f(\Delta x, 0) - f(0, 0)}{\Delta x} = \lim_{\Delta x \to 0} \frac{\Delta x \sin(1/\Delta x)}{\Delta x} = \lim_{\Delta x \to 0} \sin\left(\frac{1}{\Delta x}\right)$$

does not exist. Therefore, the partial derivative f_x does not exist at $(0, 0)$.

Similarly at $(0, 0)$, the limit

$$\lim_{\Delta y \to 0} \frac{\Delta_y z}{\Delta y} = \lim_{\Delta y \to 0} \frac{f(0, \Delta y) - f(0, 0)}{\Delta y} = \lim_{\Delta y \to 0} \frac{\Delta y \sin(1/\Delta y)}{\Delta y} = \lim_{\Delta y \to 0} \sin\left(\frac{1}{\Delta y}\right)$$

does not exist. Therefore, the partial derivative f_y does not exist at $(0, 0)$.

Example 2.9 Show that the function

$$f(x, y) = \begin{cases} \dfrac{x^2 + y^2}{|x| + |y|}, & (x, y) \neq (0, 0) \\ 0, & (x, y) = (0, 0) \end{cases}.$$

is continuous at $(0, 0)$ but its partial derivatives f_x and f_y do not exist at $(0, 0)$.

Solution   We have

$$|f(x, y) - f(0, 0)| = \left| \dfrac{x^2 + y^2}{|x| + |y|} \right| \leq \dfrac{[\,|x| + |y|\,]^2}{|x| + |y|} = |x| + |y| \leq 2\sqrt{x^2 + y^2} < \varepsilon.$$

Taking $\delta < \varepsilon / 2$, we find that

$$|f(x, y) - 0| < \varepsilon, \quad \text{whenever} \quad 0 < \sqrt{x^2 + y^2} < \delta.$$

Therefore, $\displaystyle \lim_{(x, y) \to (0,0)} f(x, y) = 0 = f(0, 0).$

Hence, the given function is continuous at $(0, 0)$.

Now, at $(0, 0)$ we have

$$\lim_{\Delta x \to 0} \dfrac{\Delta_x f}{\Delta x} = \lim_{\Delta x \to 0} \dfrac{f(\Delta x, 0) - f(0, 0)}{\Delta x} = \lim_{\Delta x \to 0} \dfrac{\Delta x}{|\Delta x|} = \begin{cases} 1, & \text{when } \Delta x > 0 \\ -1, & \text{when } \Delta x < 0. \end{cases}$$

Hence, the limit does not exist. Therefore, f_x does not exist at $(0, 0)$.

Also at $(0, 0)$, the limit

$$\lim_{\Delta y \to 0} \dfrac{\Delta_y f}{\Delta y} = \lim_{\Delta y \to 0} \dfrac{f(0, \Delta y) - f(0, 0)}{\Delta y} = \lim_{\Delta y \to 0} \dfrac{\Delta y}{|\Delta y|} = \begin{cases} 1, & \text{when } \Delta y > 0 \\ -1, & \text{when } \Delta y < 0 \end{cases}$$

does not exist. Therefore, f_y does not exist at $(0, 0)$.

Example 2.10 Show that the function

$$f(x, y) = \begin{cases} \dfrac{xy}{x^2 + 2y^2}, & (x, y) \neq (0, 0) \\ 0, & (x, y) = (0, 0) \end{cases}$$

is not continuous at $(0, 0)$ but its partial derivatives f_x and f_y exist at $(0, 0)$.

Solution   Choose the path $y = mx$. Since the limit

$$\lim_{(x, y) \to (0,0)} f(x, y) = \lim_{x \to 0} \dfrac{mx^2}{(1 + 2m^2)x^2} = \dfrac{m}{1 + 2m^2}$$

depends on m, the function is not continuous at $(0, 0)$. We now have

$$f_x(0, 0) = \lim_{\Delta x \to 0} \dfrac{f(\Delta x, 0) - f(0, 0)}{\Delta x} = \lim_{\Delta x \to 0} \dfrac{0 - 0}{\Delta x} = 0$$

$$f_y(0, 0) = \lim_{\Delta y \to 0} \frac{f(0, \Delta y) - f(0, 0)}{\Delta y} = \lim_{\Delta y \to 0} \frac{0 - 0}{\Delta y} = 0.$$

Therefore, the partial derivatives f_x and f_y exist at $(0, 0)$.

Theorem 2.1 (Sufficient condition for continuity) A sufficient condition for a function $f(x, y)$ to be continuous at a point (x_0, y_0) is that one of its first order partial derivatives exists and is bounded in the neighborhood of (x_0, y_0) and that the other exists at (x_0, y_0).

Proof Let the partial derivative f_x exist and be bounded in the neighborhood of the point (x_0, y_0) and f_y exist at (x_0, y_0). Since f_y exists at (x_0, y_0), we have

$$\lim_{\Delta y \to 0} \frac{f(x_0, y_0 + \Delta y) - f(x_0, y_0)}{\Delta y} = f_y(x_0, y_0).$$

Therefore, we can write

$$f(x_0, y_0 + \Delta y) - f(x_0, y_0) = \Delta y f_y(x_0, y_0) + \varepsilon_1 \Delta y \tag{2.15}$$

where ε_1 depends on Δy and tends to zero as $\Delta y \to 0$. Since f_x exists in the neighborhood of (x_0, y_0), we can write using the Lagrange mean value theorem

$$f(x_0 + \Delta x, y_0 + \Delta y) - f(x_0, y_0 + \Delta y) = \Delta x f_x(x_0 + \theta \Delta x, y_0 + \Delta y), 0 < \theta < 1. \tag{2.16}$$

Now, using Eqs. (2.15) and (2.16), we obtain

$$f(x_0 + \Delta x, y_0 + \Delta y) - f(x_0, y_0) = [f(x_0 + \Delta x, y_0 + \Delta y) - f(x_0, y_0 + \Delta y)] + [f(x_0, y_0 + \Delta y) - f(x_0, y_0)]$$

$$= \Delta x f_x(x_0 + \theta \Delta x, y_0 + \Delta y) + \Delta y f_y(x_0, y_0) + \varepsilon_1 \Delta y. \tag{2.17}$$

Since f_x is bounded in the neighborhood of the point (x_0, y_0), we obtain from Eq. (2.17)

$$\lim_{\Delta x \to 0, \Delta y \to 0} f(x_0 + \Delta x, y_0 + \Delta y) = f(x_0, y_0).$$

Hence, the function $f(x, y)$ is continuous at the point (x_0, y_0).

Geometrical interpretation of partial derivatives

Let $z = f(x, y)$ represent a surface as shown in Fig. 2.3. Let the plane $x = x_0 = $ constant intersect the surface $z = f(x, y)$ along the curve $z = f(x_0, y)$. Let $P(x_0, y, 0)$ be a particular point in the x-y plane and $R(x_0, y, z)$ be the corresponding point on the surface, where $z = f(x_0, y)$. Let $Q(x_0, y + \Delta y, 0)$ be a point in the x-y plane in the neighborhood of P and $S(x_0, y + \Delta y, z + \Delta_y z)$ be the corresponding point on the surface $z = f(x, y)$. From Fig. 2.3, we find that $\Delta y = PQ = RS'$ and the function z is increased by $SS' = (z + \Delta_y z) - z = \Delta_y z$. Now, let θ^* be the angle which the chord RS makes with the positive y-axis. Then, from $\triangle RSS'$, we have

$$\tan \theta^* = \frac{SS'}{RS'} = \frac{\Delta_y z}{\Delta y}.$$

Let $\Delta y \to 0$. Then, $\Delta_y z \to 0$. Hence,

$$\lim_{\Delta y \to 0} \frac{\Delta_y z}{\Delta y} = \frac{\partial z}{\partial y} = \tan \theta$$

where in the limit, θ is the angle made by the tangent to the curve $z = f(x_0, y)$ at the point $R(x_0, y, z)$ on the surface $z = f(x, y)$ with the positive y-axis.

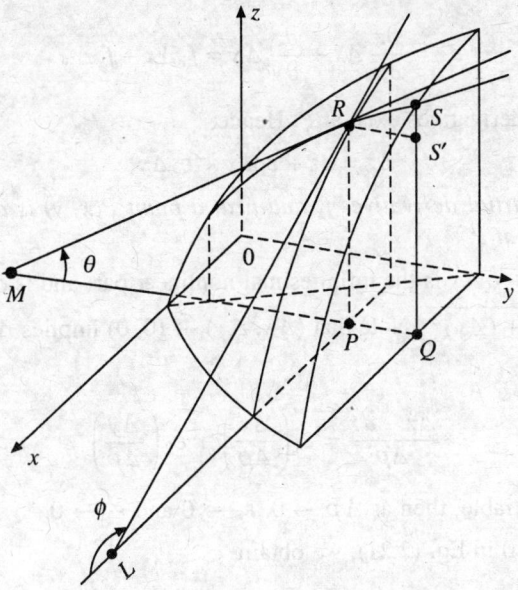

Fig. 2.3.   Geometrical representation of partial derivatives.

Now, consider the intersection of the plane $y = y_0 = $ constant with the surface $z = f(x, y)$. Following the similar procedure, we obtain $\partial z / \partial x = \tan \phi$, where ϕ is the angle made by the tangent to the curve $z = f(x, y_0)$ at the point (x, y_0, z) on the surface $z = f(x, y)$ with the positive x-axis.

It can be observed that this representation of partial derivatives is a direct extension of the one dimensional case.

2.3.1 Total Differential and Differentiability

Let a function of two variables $z = f(x, y)$ be defined in some domain D in the x-y plane. Let $P(x, y)$ be any point in D and $(x + \Delta x, y + \Delta y)$ be a point in the neighborhood of (x, y), in D. Then,

$$\Delta z = f(x + \Delta x, y + \Delta y) - f(x, y)$$

is called the *total increment* in z corresponding to the increments Δx in x and Δy in y.

The function $z = f(x, y)$ is said to be *differentiable* at the point (x, y), if at this point Δz can be written as

$$\Delta z = (a \, \Delta x + b \, \Delta y) + (\varepsilon_1 \Delta x + \varepsilon_2 \Delta y) \tag{2.18}$$

where a, b are independent of Δx, Δy and $\varepsilon_1 = \varepsilon_1(\Delta x, \Delta y)$, $\varepsilon_2 = \varepsilon_2(\Delta x, \Delta y)$ are infinitesimals and functions of Δx, Δy such that $\varepsilon_1 \to 0$, $\varepsilon_2 \to 0$ as $(\Delta x, \Delta y) \to (0, 0)$.

The first part $a \, \Delta x + b \, \Delta y$ in Eq. (2.18) which is linear in Δx and Δy is called the *total differential* or simply the differential of z at the point (x, y) and is denoted by dz or df. That is

$$dz = a \, \Delta x + b \Delta y \quad \text{or} \quad dz = a \, dx + b \, dy.$$

Let $\Delta y = 0$ in Eq. (2.18). Then, $\Delta z = a \, \Delta x + \varepsilon_1 \Delta x$. Dividing by Δx and taking limits as $\Delta x \to 0$, we obtain $a = \partial z/\partial x$. Similarly, letting $\Delta x = 0$ in Eq. (2.18), dividing by Δy and taking limits as $\Delta y \to 0$, we obtain $b = \partial z/\partial y$. Therefore,

$$dz = \frac{\partial z}{\partial x} \Delta x + \frac{\partial z}{\partial y} \Delta y = f_x \Delta x + f_y \Delta_y \tag{2.19}$$

assuming that the partial derivatives exist at P. Hence,

$$\Delta z = dz + \varepsilon_1 \Delta x + \varepsilon_2 \Delta y. \tag{2.20}$$

Therefore, *existence of partial derivatives f_x and f_y at a point $P(x, y)$ is a necessary condition for differentiability of $f(x, y)$ at P.*

The second part $\varepsilon_1 \Delta x + \varepsilon_2 \Delta y$ is the infinitesimal nonlinear part and is of higher order relative to Δx, Δy or $\Delta \rho = \sqrt{(\Delta x)^2 + (\Delta y)^2}$. Note that $(\Delta x, \Delta y) \to (0, 0)$ implies $\Delta \rho \to 0$. Eq. (2.20) can be written as

$$\frac{\Delta z - d\dot{z}}{\Delta \rho} = \varepsilon_1 \left(\frac{\Delta x}{\Delta \rho} \right) + \varepsilon_2 \left(\frac{\Delta y}{\Delta \rho} \right) \tag{2.21}$$

Now, if $f(x, y)$ is differentiable, then as $\Delta \rho \to 0$, $\varepsilon_1 \to 0$ and $\varepsilon_2 \to 0$.

Taking the limit as $\Delta \rho \to 0$ in Eq. (2.21), we obtain

$$\lim_{\Delta \rho \to 0} \frac{\Delta z - dz}{\Delta \rho} = \lim_{\Delta \rho \to 0} \left[\varepsilon_1 \left(\frac{\Delta x}{\Delta \rho} \right) + \varepsilon_2 \left(\frac{\Delta y}{\Delta \rho} \right) \right] = 0 \tag{2.22}$$

since $| \Delta x/\Delta \rho | \leq 1$ and $| \Delta y/\Delta \rho | \leq 1$.

Therefore, to test differentiability at a point $P(x, y)$, we can use either of the following two approaches.

(i)　Show that $\displaystyle\lim_{\Delta \rho \to 0} \frac{\Delta z - dz}{\Delta \rho} = 0$ $\tag{2.23}$

(ii) Find the expressions for $\varepsilon_1(\Delta x, \Delta y)$, $\varepsilon_2(\Delta x, \Delta y)$ from Eq. (2.20) and then show that $\lim \varepsilon_1 \to 0$ and $\lim \varepsilon_2 \to 0$ as $(\Delta x, \Delta y) \to (0, 0)$ or $\Delta \rho \to 0$.

Note that the function $f(x, y)$ may not be differentiable at a point $P(x, y)$, even if the partial derivatives f_x, f_y exist at P (see Example 2.12). However, if the first order partial derivatives are continuous at the point P, then the function is differentiable at P.
We present this result in the following theorem.

Theorem 2.2 (Sufficient condition for differentiability) If the function $z = f(x, y)$ has continuous first order partial derivatives at a point $P(x, y)$ in D, then $f(x, y)$ is differentiable at P.

Proof Let $P(x, y)$ be a fixed point in D. By the Lagrange mean value theorem, we have

$$f(x + \Delta x, y) - f(x, y) = \Delta x \, f_x(x + \theta_1 \Delta x, y), \, 0 < \theta_1 < 1$$

and $\quad f(x + \Delta x, y + \Delta y) - f(x + \Delta x, y) = \Delta y f_y(x + \Delta x, y + \theta_2 \Delta y), \, 0 < \theta_2 < 1.$

Since f_x and f_y are continuous at (x, y), we can write

$$f_x(x + \theta_1 \Delta x, y) = f_x(x, y) + \varepsilon_1$$

and $\quad\quad\quad\quad f_y(x + \Delta x, y + \theta_2 \Delta y) = f_y(x, y) + \varepsilon_2$

where ε_1, ε_2 are infinitesimals, are functions of Δx, Δy and tend to zero as $\Delta x \to 0$, $\Delta y \to 0$, that is, as $\Delta \rho = \sqrt{(\Delta x)^2 + (\Delta y)^2} \to 0$. Therefore, we have

$$f(x + \Delta x, y) - f(x, y) = \Delta x f_x(x, y) + \varepsilon_1 \Delta x \tag{2.24}$$

and

$$f(x + \Delta x, y + \Delta y) - f(x + \Delta x, y) = \Delta y f_y(x, y) + \varepsilon_2 \Delta y \tag{2.25}$$

Now, the total increment is given by

$$\Delta z = f(x + \Delta x, y + \Delta y) - f(x, y)$$

$$= [f(x + \Delta x, y) - f(x, y)] + [f(x + \Delta x, y + \Delta y) - f(x + \Delta x, y)].$$

Using Eqs. (2.24) and (2.25), we obtain

$$\Delta z = f_x \Delta x + f_y \Delta y + \varepsilon_1 \Delta x + \varepsilon_2 \Delta y \tag{2.26}$$

where the partial derivatives are evaluated at the point $P(x, y)$. Hence, $f(x, y)$ is differentiable at P.

Remark 6

(a) For a function of n variables $z = f(x_1, x_2, \ldots, x_n)$, we write the total differential as

$$dz = f_{x_1} dx_1 + f_{x_2} dx_2 + \ldots + f_{x_n} dx_n. \tag{2.27}$$

(b) Note that continuity of the first partial derivatives f_x and f_y at a point P is a sufficient condition for differentiability at P, that is, a function may be differentiable even if f_x and f_y are not continuous (Problem 5, Exercise 2.2).

(c) The conditions of Theorem 2.2 can be relaxed. It is sufficient that one of the first order partial derivatives is continuous at (x_0, y_0) and the other exists at (x_0, y_0).

Example 2.11 Find the total differential of the following functions

(i) $z = \tan^{-1}(x/y)$, $(x, y) \neq (0, 0)$, (ii) $u = \left(xz + \dfrac{x}{z}\right)^y$, $z \neq 0$.

Solution

(i)
$$f(x, y) = \tan^{-1}\left(\frac{x}{y}\right), \quad f_x = \frac{1}{1 + (x/y)^2}\left(\frac{1}{y}\right) = \frac{y}{x^2 + y^2}$$

and
$$f_y = \frac{1}{1 + (x/y)^2}\left(-\frac{x}{y^2}\right) = -\frac{x}{x^2 + y^2}.$$

Therefore, we obtain the total differential as

$$dz = f_x \, dx + f_y \, dy = \frac{1}{x^2 + y^2}(y \, dx - x \, dy).$$

(ii)
$$f(x, y, z) = \left(xz + \frac{x}{z}\right)^y, \quad f_x = y\left(xz + \frac{x}{z}\right)^{y-1}\left(z + \frac{1}{z}\right)$$

$$f_y = \left(xz + \frac{x}{z}\right)^y \ln\left(xz + \frac{x}{z}\right), \quad f_z = y\left(xz + \frac{x}{z}\right)^{y-1}\left(x - \frac{x}{z^2}\right).$$

Therefore, we obtain the total differential as

$$du = \left(xz + \frac{x}{z}\right)^{y-1}\left[y\left(z + \frac{1}{z}\right)dx + xy\left(1 - \frac{1}{z^2}\right)dz\right] + \left[\left(xz + \frac{x}{z}\right)^y \ln\left(xz + \frac{x}{z}\right)\right]dy.$$

Example 2.12   Show that the function

$$f(x, y) = \begin{cases} \dfrac{x^3 + 2y^3}{x^2 + y^2}, & (x, y) \neq (0, 0) \\ 0, & (x, y) = (0, 0) \end{cases}$$

(i)   is continuous at (0, 0),

(ii) possesses partial derivatives $f_x(0, 0)$ and $f_y(0, 0)$,

(iii) is not differentiable at (0, 0).

Solution

(i)   Let $x = r \cos \theta$ and $y = r \sin \theta$. We have

$$|f(x, y) - f(0, 0)| = \left| \frac{r^3 (\cos^3\theta + 2 \sin^3\theta)}{r^2} \right| \leq r [\, |\cos^3\theta| + 2\, |\sin^3\theta| \,]$$

$$\leq 3r = 3\sqrt{x^2 + y^2} < \varepsilon.$$

Taking $\delta < \varepsilon/3$, we find that

$$|f(x, y) - 0| < \varepsilon, \quad \text{whenever} \quad 0 < \sqrt{x^2 + y^2} < \delta.$$

Therefore, $\displaystyle \lim_{(x, y) \to f(0,0)} f(x, y) = 0 = f(0, 0)$.

Hence, $f(x, y)$ is continuous at (0, 0).

(ii) $f_x(0, 0) = \displaystyle \lim_{\Delta x \to 0} \frac{f(\Delta x, 0) - f(0, 0)}{\Delta x} = \lim_{\Delta x \to 0} \frac{\Delta x - 0}{\Delta x} = 1$

$f_y(0, 0) = \displaystyle \lim_{\Delta y \to 0} \frac{f(0, \Delta y) - f(0, 0)}{\Delta y} = \lim_{\Delta y \to 0} \frac{2\Delta y - 0}{\Delta y} = 2.$

Therefore, the partial derivatives $f_x(0, 0)$ and $f_y(0, 0)$ exist.

(iii) We have $dz = \Delta x + 2\Delta y$. Using Eq. (2.20), we get

$$\Delta z = \Delta x + 2\Delta y + \varepsilon_1 \Delta x + \varepsilon_2 \Delta y$$

Let $\Delta\rho = \sqrt{(\Delta x)^2 + (\Delta y)^2}$. Now,

$$\Delta z = f(\Delta x, \Delta y) - f(0, 0) = \frac{(\Delta x)^3 + 2(\Delta y)^3}{(\Delta x)^2 + (\Delta y)^2}$$

Hence

$$\lim_{\Delta\rho \to 0} \frac{\Delta z - dz}{\Delta\rho} = \lim_{\Delta\rho \to 0} \frac{1}{\Delta\rho} \left[\frac{(\Delta x)^3 + 2(\Delta y)^3}{(\Delta x)^2 + (\Delta y)^2} - (\Delta x + 2\Delta y) \right]$$

$$= \lim_{\Delta\rho \to 0} - \left[\frac{\Delta x \Delta y (\Delta y + 2\Delta x)}{\{(\Delta x)^2 + (\Delta y)^2\}^{3/2}} \right]$$

Let $\Delta x = r \cos \theta$ and $\Delta y = r \sin \theta$. As $(\Delta x, \Delta y) \to (0, 0)$, $\Delta \rho = r \to 0$ for arbitrary θ. Therefore,

$$\lim_{\Delta \rho \to 0} \frac{\Delta z - dz}{\Delta \rho} = - \lim_{r \to 0} [\cos \theta \sin \theta (\sin \theta + 2 \cos \theta)]$$

$$= - [\cos \theta \sin \theta (\sin \theta + 2 \cos \theta)].$$

The limit depends on θ and does not tend to zero for arbitrary θ. Hence, the given function is not differentiable. Alternately, we can write

$$\frac{\Delta z - dz}{\Delta \rho} = - \frac{1}{\Delta \rho} \left[\frac{\Delta x (\Delta y)^2 + 2(\Delta x)^2 \Delta y}{(\Delta x)^2 + (\Delta y)^2} \right] = \varepsilon_1 \left(\frac{\Delta x}{\Delta \rho} \right) + \varepsilon_2 \left(\frac{\Delta y}{\Delta \rho} \right)$$

where $\qquad \varepsilon_1 = - \dfrac{(\Delta y)^2}{(\Delta x)^2 + (\Delta y)^2} \quad$ and $\quad \varepsilon_2 = - \dfrac{2(\Delta x)^2}{(\Delta x)^2 + (\Delta y)^2}.$

Substituting $\Delta x = r \cos \theta$, $\Delta y = r \sin \theta$, we find that ε_1 and ε_2 depend on θ and do not tend to zero for arbitrary θ, in the limit as $r \to 0$.

Example 2.13 Show that the function

$$f(x, y) = \begin{cases} \dfrac{x^2 - y^2}{x - y}, & (x, y) \neq (1, -1) \\ 0, & (x, y) = (1, -1) \end{cases}$$

is continuous and differentiable at $(1, -1)$.

Solution We have

$$\lim_{(x, y) \to (1, -1)} \frac{x^2 - y^2}{x - y} = \lim_{(x, y) \to (1, -1)} (x + y) = 0 = f(1, -1).$$

Therefore, the function is continuous at $(1, -1)$.

The partial derivatives are given by

$$f_x(1, -1) = \lim_{\Delta x \to 0} \frac{f(1 + \Delta x, -1) - f(1, -1)}{\Delta x} = \lim_{\Delta x \to 0} \frac{1}{\Delta x} \left[\frac{(1 + \Delta x)^2 - 1}{(1 + \Delta x) + 1} - 0 \right] = \lim_{\Delta x \to 0} \frac{2 + \Delta x}{2 + \Delta x} = 1.$$

$$f_y(1, -1) = \lim_{\Delta y \to 0} \frac{f(1, -1 + \Delta y) - f(1, -1)}{\Delta y} = \lim_{\Delta y \to 0} \frac{1}{\Delta y} \left[\frac{1 - (-1 + \Delta y)^2}{1 - (-1 + \Delta y)} - 0 \right] = \lim_{\Delta y \to 0} \frac{2 - \Delta y}{2 - \Delta y} = 1.$$

Therefore, the first order partial derivatives exist at $(1, -1)$.

Now, we have

$$f_x(x, y) = \frac{(x - y)(2x) - (x^2 - y^2)(1)}{(x - y)^2} = \frac{x^2 - 2xy + y^2}{(x - y)^2} = \frac{(x - y)^2}{(x - y)^2}, (x, y) \neq (1, -1)$$

and $\qquad\qquad\qquad f_x(x, y) = 1, (x, y) = (1, -1).$

Since $\qquad\qquad \lim_{(x, y) \to (1, -1)} f_x(x, y) = \lim_{(x, y) \to (1, -1)} \frac{(x - y)^2}{(x - y)^2} = 1 = f_x(1, -1)$

the partial derivative f_x is continuous at $(1, -1)$. Also $f_y(1, -1)$ exists. Hence, $f(x, y)$ is differentiable at $(1, -1)$.

Alternately, we can show that $\lim\limits_{\Delta\rho \to 0} [(\Delta z - dz)/\Delta\rho] = 0$.

2.3.2 Approximation by Total Differentials

From Theorem 2.2, we have for a function $f(x, y)$ of two variables

$$f(x + \Delta x, y + \Delta y) - f(x, y) \approx f_x \Delta x + f_y \Delta y$$

or $$f(x + \Delta x, y + \Delta y) \approx f(x, y) + f_x \Delta x + f_y \Delta y \qquad (2.28)$$

where the partial derivatives are evaluated at the given point (x, y). This result has applications in estimating errors in calculations.

Consider now a function of n variables $x_1, x_2, \ldots, x_n$. Let the function $z = f(x_1, x_2, \ldots, x_n)$ be differentiable at the point $P(x_1, x_2, \ldots, x_n)$. Let there be errors $\Delta x_1, \Delta x_2, \ldots, \Delta x_n$ in measuring the values of $x_1, x_2, \ldots, x_n$ respectively. Then, the computed value of z using the inexact values of the arguments will be obtained with an error

$$\Delta z = f(x_1 + \Delta x_1, x_2 + \Delta x_2, \ldots, x_n + \Delta x_n) - f(x_1, x_2, \ldots, x_n). \qquad (2.29)$$

When the errors $\Delta x_1, \Delta x_2, \ldots, \Delta x_n$ are small in magnitude, we obtain (using the Remark 6 (a), Eq. (2.27))

$$f(x_1 + \Delta x_1, x_2 + \Delta x_2, \ldots, x_n + \Delta x_n) \approx f(x_1, x_2, \ldots, x_n) + f_{x_1} \Delta x_1 + f_{x_2} \Delta x_2 + \ldots + f_{x_n} \Delta x_n \quad (2.30)$$

where the partial derivatives are evaluated at the point $(x_1, x_2, \ldots, x_n)$. This is the generalization of the result for functions of two variables given in Eq. (2.28).

Since the partial derivatives and errors in arguments can be both positive and negative, we define the *absolute error* as (using Eq. (2.29))

$$|\Delta z| \approx |dz| = |df| = |f_{x_1} \Delta x_1 + f_{x_2} \Delta x_2 + \ldots + f_{x_n} \Delta x_n|.$$

Then, $$|df| \leq |f_{x_1}||\Delta x_1| + |f_{x_2}||\Delta x_2| + \ldots + |f_{x_n}||\Delta x_n| \qquad (2.31)$$

gives the *maximum absolute error* in z. If max $|\Delta x_i| \leq \Delta x$, then we can write

$$|df| \leq \Delta x \,[\,|f_{x_1}| + |f_{x_2}| + \ldots + |f_{x_n}|\,].$$

The expression $|df|/|f|$ is called the *maximum relative error* and $[\,|df|/|f|\,] \times 100$ is called the *percentage error*.

The maximum relative error can also be written as

$$\frac{|df|}{|f|} \leq \left|\frac{\partial f/\partial x_1}{f}\right| |\Delta x_1| + \left|\frac{\partial f/\partial x_2}{f}\right| |\Delta x_2| + \ldots + \left|\frac{\partial f/\partial x_n}{f}\right| |\Delta x_n|$$

$$\leq \left|\frac{\partial}{\partial x_1}[\ln|f|]\right| |\Delta x_1| + \left|\frac{\partial}{\partial x_2}[\ln|f|]\right| |\Delta x_2| + \ldots + \left|\frac{\partial}{\partial x_n}[\ln|f|]\right| |\Delta x_n|.$$

Example 2.14 Find the total increment and the total differential of the function $z = x + y + xy$ at the point $(1, 2)$ for $\Delta x = 0.1$ and $\Delta y = -0.2$. Find the maximum absolute error and the maximum relative error.

Solution   We are given that $f(x, y) = x + y + xy$, $(x, y) = (1, 2)$.

Therefore, $f(1, 2) = 5$, $f_x(1, 2) = 3$, $f_y(1, 2) = 2$. We have

total increment $= f(x + \Delta x, y + \Delta y) - f(x, y)$

$$= [(x + \Delta x) + (y + \Delta y) + (x + \Delta x)(y + \Delta y)] - [x + y + xy]$$

$$= \Delta x + \Delta y + x\, \Delta y + y\, \Delta x + \Delta x\, \Delta y.$$

At the point $(1, 2)$ with $\Delta x = 0.1$ and $\Delta y = -0.2$, we obtain

total increment $= 0.1 - 0.2 + 1(-0.2) + 2(0.1) + (0.1)(-0.2) = -0.12$

total differential $= f_x(1, 2)\, \Delta x + f_y(1, 2)\, \Delta y = 3(0.1) + (2)(-0.2) = -0.1$

maximum absolute error $= |df| = \left|\dfrac{\partial f}{\partial x}\right| |\Delta x| + \left|\dfrac{\partial f}{\partial y}\right| |\Delta y| = 3(0.1) + 2(0.2) = 0.7$

maximum relative error $= \dfrac{|df|}{|f|} = \dfrac{0.7}{5} = 0.14$.

Example 2.15   Using differentials, find an approximate value of

(i) $f(4.1, 4.9)$, where $f(x, y) = \sqrt{x^3 + x^2 y}$,

(ii) $f(2.1, 3.2)$, where $f(x, y) = x^y$.

Solution

(i) Let $(x, y) = (4, 5)$, $\Delta x = 0.1$, $\Delta y = -0.1$. We have

$$f(x, y) = \sqrt{x^3 + x^2 y},\ f(4, 5) = 12,\ f_x(x, y) = \frac{3x^2 + 2xy}{2\sqrt{x^3 + x^2 y}},\ f_x(4, 5) = \frac{11}{3},$$

$$f_y(x, y) = \frac{x^2}{2\sqrt{x^3 + x^2 y}},\ f_y(4, 5) = \frac{2}{3}.$$

Therefore,

$$f(4.1, 4.9) \approx f(4, 5) + f_x(4, 5)\, \Delta x + f_y(4, 5)\, \Delta y$$

$$= 12 + (11/3)(0.1) + (2/3)(-0.1) = 12.3.$$

The exact value is $f(4.1, 4.9) = 12.3$

(ii) Let $(x, y) = (2, 3)$, $\Delta x = 0.1$, $\Delta y = 0.2$. We have

$$f(x, y) = x^y,\ f(2, 3) = 8,\ f_x(x, y) = yx^{y-1},\ f_x(2, 3) = 12,$$

$$f_y(x, y) = x^y \ln x,\ f_y(2, 3) = 8 \ln 2 = 2.54518.$$

Therefore, $f(2.1, 3.2) \approx f(2, 3) + f_x(2, 3)\, \Delta x + f_y(2, 3)\, \Delta y$

$$= 8 + 12(0.1) + (0.2)(5.54518) = 10.3090.$$

The exact value is $f(2.1, 3.2) = 10.7424$.

Therefore, $f(2.1, 3.2) \approx f(2, 3) + f_x(2, 3)\,\Delta x + f_y(2, 3)\,\Delta y$

$$= 8 + 12(0.1) + (2.408)(0.2) = 9.6816.$$

The exact value is $f(2.1, 3.2) = 10.7424$.

Example 2.16 Find the percentage error in the computed area of an ellipse when an error of 2% is made in measuring the semi major and semi minor axes.

Solution Let the major and minor axes of the ellipse be $2a$ and $2b$ respectively. The errors Δa and Δb in computing the lengths of the semi major and minor axes are

$$\Delta a = a(0.02) = 0.02\,a \quad \text{and} \quad \Delta b = b(0.02) = 0.02\,b.$$

The area of the ellipse is given by $A = \pi\,ab$. Therefore, we have the following:
Maximum absolute error in computing the area of ellipse is

$$|dA| = \left|\frac{\partial A}{\partial a}\right||\Delta a| + \left|\frac{\partial A}{\partial b}\right||\Delta b| = \pi b(0.02a) + \pi a(0.02b) = 0.04\,\pi ab.$$

Maximum relative error is

$$\left|\frac{dA}{A}\right| = (0.04\,\pi ab)\left(\frac{1}{\pi ab}\right) = 0.04.$$

Percentage error $= \left|\dfrac{dA}{A}\right| \times 100 = 4\%$.

2.3.3 Derivatives of Composite and Implicit Functions (*Chain Rule*)

Let $z = f(x, y)$ be a function of two independent variables x and y. Suppose that x and y are themselves functions of some independent variable t, say $x = \phi(t)$, $y = \psi(t)$. Then, $z = f[\phi(t), \psi(t)]$ is a composite function of the independent variable t. Now, assume that the partial derivatives f_x, f_y are continuous functions of x, y and $\phi(t)$, $\psi(t)$ are differentiable functions of t.

Let Δx, Δy and Δz be the increments respectively in x, y and z corresponding to the increment Δt in t. Then we have

$$\Delta z = \frac{\partial f}{\partial x}\,\Delta x + \frac{\partial f}{\partial y}\,\Delta y + \varepsilon_1\,\Delta x + \varepsilon_2\,\Delta y.$$

Dividing both sides by Δt, we get

$$\frac{\Delta z}{\Delta t} = \frac{\partial f}{\partial x}\frac{\Delta x}{\Delta t} + \frac{\partial f}{\partial y}\frac{\Delta y}{\Delta t} + \varepsilon_1\frac{\Delta x}{\Delta t} + \varepsilon_2\frac{\Delta y}{\Delta t}. \tag{2.32}$$

Now as $\Delta t \to 0$; $\Delta x \to 0$, $\Delta y \to 0$ and $\varepsilon_1\left(\dfrac{\Delta x}{\Delta t}\right) \to 0$, $\varepsilon_2\left(\dfrac{\Delta y}{\Delta t}\right) \to 0$. Therefore, taking limits on both sides in Eq. (2.32) as $\Delta t \to 0$, we obtain

$$\frac{dz}{dt} = \frac{\partial f}{\partial x}\frac{dx}{dt} + \frac{\partial f}{\partial y}\frac{dy}{dt}. \tag{2.33}$$

Now, let x and y be functions of two independent variables u and v, say $x = \phi(u, v)$, $y = \psi(u, v)$. Then, $z = f[\phi(u, v), \psi(u, v)]$ is a composite function of two independent variables u and v. Assume that the functions $f(x, y)$, $\phi(u, v)$, $\psi(u, v)$ have continuous partial derivatives with respect to their arguments. Now, consider v as a constant and give an increment Δu to u. Let $\Delta_u x$ and $\Delta_u y$ be the corresponding increments in x and y. Then, the increment Δz in z is given by (using Eq. (2.20))

$$\Delta z = \frac{\partial f}{\partial x} \Delta_u x + \frac{\partial f}{\partial y} \Delta_u y + \varepsilon_1 \Delta_u x + \varepsilon_2 \Delta_u y$$

where $\varepsilon_1, \varepsilon_2 \to 0$ as $\Delta u \to 0$.

Dividing both sides by Δu, we get

$$\frac{\Delta z}{\Delta u} = \frac{\partial f}{\partial x} \frac{\Delta_u x}{\Delta u} + \frac{\partial f}{\partial y} \frac{\Delta_u y}{\Delta u} + \varepsilon_1 \frac{\Delta_u x}{\Delta u} + \varepsilon_2 \frac{\Delta_u y}{\Delta u}. \tag{2.34}$$

Taking limits on both sides in Eq. (2.34) as $\Delta u \to 0$, we obtain

$$\frac{\partial z}{\partial u} = \frac{\partial f}{\partial x} \frac{\partial x}{\partial u} + \frac{\partial f}{\partial y} \frac{\partial y}{\partial u}. \tag{2.35}$$

Similarly, keeping u as constant and varying v, we obtain

$$\frac{\partial z}{\partial v} = \frac{\partial f}{\partial x} \frac{\partial x}{\partial v} + \frac{\partial f}{\partial y} \frac{\partial y}{\partial v}. \tag{2.36}$$

The rules given in Eqs. (2.35) and (2.36) are called the *chain rules*. These rules can be easily extended to a function of n variables $z = f(x_1, x_2, \ldots, x_n)$. If the partial derivatives of f with respect to all its arguments are continuous and $x_1, x_2, \ldots, x_n$ are differentiable functions of some independent variable t, then

$$\frac{dz}{dt} = \frac{\partial f}{\partial x_1} \frac{dx_1}{dt} + \frac{\partial f}{\partial x_2} \frac{dx_2}{dt} + \ldots + \frac{\partial f}{\partial x_n} \frac{dx_n}{dt}. \tag{2.37}$$

Example 2.17 Find df/dt at $t = 0$, where

(i) $f(x, y) = x \cos y + e^x \sin y$, $x = t^2 + 1$, $y = t^3 + t$.

(ii) $f(x, y, z) = x^3 + x z^2 + y^3 + xyz$, $x = e^t$, $y = \cos t$, $z = t^3$.

Solution

(i) When $t = 0$, we get $x = 1$, $y = 0$. Using the chain rule, we obtain

$$\frac{df}{dt} = \frac{\partial f}{\partial x} \frac{dx}{dt} + \frac{\partial f}{\partial y} \frac{dy}{dt} = (\cos y + e^x \sin y)(2t) + (-x \sin y + e^x \cos y)(3t^2 + 1).$$

Substituting $t = 0$, $x = 1$ and $y = 0$, we obtain $(df/dt) = e$.

(ii) When $t = 0$, we get $x = 1$, $y = 1$, $z = 0$. Using the chain rule, we obtain

$$\frac{df}{dt} = \frac{\partial f}{\partial x} \frac{dx}{dt} + \frac{\partial f}{\partial y} \frac{dy}{dt} + \frac{\partial f}{\partial z} \frac{dz}{dt}$$

$$= (3x^2 + z^2 + yz)(e^t) + (3y^2 + xz)(-\sin t) + (2xz + xy)(3t^2).$$

Substituting $t = 0$, $x = 1$, $y = 1$, $z = 0$, we obtain $(df/dt) = 3$.

Example 2.18 If $z = f(x, y)$, $x = e^{2u} + e^{-2v}$, $y = e^{-2u} + e^{2v}$, then show that

$$\frac{\partial f}{\partial u} - \frac{\partial f}{\partial v} = 2\left[x \frac{\partial f}{\partial x} - y \frac{\partial f}{\partial y} \right].$$

Solution   Using the chain rule, we obtain

$$\frac{\partial f}{\partial u} = \frac{\partial f}{\partial x} \frac{\partial x}{\partial u} + \frac{\partial f}{\partial y} \frac{\partial y}{\partial u} = 2e^{2u} \frac{\partial f}{\partial x} - 2e^{-2u} \frac{\partial f}{\partial y}$$

$$\frac{\partial f}{\partial v} = \frac{\partial f}{\partial x} \frac{\partial x}{\partial v} + \frac{\partial f}{\partial y} \frac{\partial y}{\partial v} = -2e^{-2v} \frac{\partial f}{\partial x} + 2e^{2v} \frac{\partial f}{\partial y}.$$

Therefore,

$$\frac{\partial f}{\partial u} - \frac{\partial f}{\partial v} = 2(e^{2u} + e^{-2v}) \frac{\partial f}{\partial x} - 2(e^{-2u} + e^{2v}) \frac{\partial f}{\partial y}$$

$$= 2x \frac{\partial f}{\partial x} - 2y \frac{\partial f}{\partial y}.$$

Change of variables

Suppose that $f(x, y)$ is a function of two independent variables x, y and x, y are functions of two new independent variables u, v given by $x = \phi(u, v)$, $y = \psi(u, v)$. By chain rule, we have

$$\frac{\partial f}{\partial u} = \frac{\partial f}{\partial x} \frac{\partial x}{\partial u} + \frac{\partial f}{\partial y} \frac{\partial y}{\partial u} \quad \text{and} \quad \frac{\partial f}{\partial v} = \frac{\partial f}{\partial x} \frac{\partial x}{\partial v} + \frac{\partial f}{\partial y} \frac{\partial y}{\partial v}.$$

We want to determine $\partial f/\partial x$, $\partial f/\partial y$ in terms of $\partial f/\partial u$ and $\partial f/\partial v$. Solving the above system of equations by Cramer's rule, we get

$$\frac{\partial f/\partial x}{\dfrac{\partial f}{\partial u} \dfrac{\partial y}{\partial v} - \dfrac{\partial f}{\partial v} \dfrac{\partial y}{\partial u}} = \frac{\partial f/\partial y}{\dfrac{\partial f}{\partial v} \dfrac{\partial x}{\partial u} - \dfrac{\partial f}{\partial u} \dfrac{\partial x}{\partial v}} = \frac{1}{\dfrac{\partial x}{\partial u} \dfrac{\partial y}{\partial v} - \dfrac{\partial x}{\partial v} \dfrac{\partial y}{\partial u}}.$$

The determinant

$$J = \begin{vmatrix} \partial x/\partial u & \partial x/\partial v \\ \partial y/\partial u & \partial y/\partial v \end{vmatrix} = \frac{\partial(x, y)}{\partial(u, v)}$$

is called the *Jacobian* of the variables of transformation. Similarly, we write

$$\frac{\partial f}{\partial u} \frac{\partial y}{\partial v} - \frac{\partial f}{\partial v} \frac{\partial y}{\partial u} = \frac{\partial(f, y)}{\partial(u, v)} = \begin{vmatrix} \partial f/\partial u & \partial f/\partial v \\ \partial y/\partial u & \partial y/\partial v \end{vmatrix}$$

and

$$\frac{\partial f}{\partial v} \frac{\partial x}{\partial u} - \frac{\partial f}{\partial u} \frac{\partial x}{\partial v} = \frac{\partial(x, f)}{\partial(u, v)} = \begin{vmatrix} \partial x/\partial u & \partial x/\partial v \\ \partial f/\partial u & \partial f/\partial v \end{vmatrix} = -\frac{\partial(f, x)}{\partial(u, v)}.$$

Hence, we obtain

$$\frac{\partial f}{\partial x} = \frac{1}{J}\left[\frac{\partial(f, y)}{\partial(u, v)} \right] \quad \text{and} \quad \frac{\partial f}{\partial y} = -\frac{1}{J}\left[\frac{\partial(f, x)}{\partial(u, v)} \right]. \tag{2.38}$$

Similarly, if $f(x, y, z)$ is a function of three independent variables x, y, z and x, y, z are functions of three new independent variables u, v, w given by $x = F(u, v, w)$, $y = G(u, v, w)$, $z = H(u, v, w)$, then by chain rule, we have

$$\frac{\partial f}{\partial u} = \frac{\partial f}{\partial x}\frac{\partial x}{\partial u} + \frac{\partial f}{\partial y}\frac{\partial y}{\partial u} + \frac{\partial f}{\partial z}\frac{\partial z}{\partial u}$$

$$\frac{\partial f}{\partial v} = \frac{\partial f}{\partial x}\frac{\partial x}{\partial v} + \frac{\partial f}{\partial y}\frac{\partial y}{\partial v} + \frac{\partial f}{\partial z}\frac{\partial z}{\partial v}$$

$$\frac{\partial f}{\partial w} = \frac{\partial f}{\partial x}\frac{\partial x}{\partial w} + \frac{\partial f}{\partial y}\frac{\partial y}{\partial w} + \frac{\partial f}{\partial z}\frac{\partial z}{\partial w}.$$

Solving the above system of equations by Cramer's rule, we get

$$\frac{\partial f}{dx} = \frac{1}{J}\left[\frac{\partial(f,y,z)}{\partial(u,v,w)}\right] = \frac{1}{J}\begin{vmatrix} \partial f/\partial u & \partial f/\partial v & \partial f/\partial w \\ \partial y/\partial u & \partial y/\partial v & \partial y/\partial w \\ \partial z/\partial u & \partial z/\partial v & \partial z/\partial w \end{vmatrix}$$

$$\frac{\partial f}{dy} = \frac{1}{J}\left[\frac{\partial(x,f,z)}{\partial(u,v,w)}\right] = -\frac{1}{J}\left[\frac{\partial(f,x,z)}{\partial(u,v,w)}\right] = -\frac{1}{J}\begin{vmatrix} \partial f/\partial u & \partial f/\partial v & \partial f/\partial w \\ \partial x/\partial u & \partial x/\partial v & \partial x/\partial w \\ \partial z/\partial u & \partial z/\partial v & \partial z/\partial w \end{vmatrix}$$

$$\frac{\partial f}{dz} = \frac{1}{J}\left[\frac{\partial(x,y,f)}{\partial(u,v,w)}\right] = \frac{1}{J}\left[\frac{\partial(f,x,y)}{\partial(u,v,w)}\right] = \frac{1}{J}\begin{vmatrix} \partial f/\partial u & \partial f/\partial v & \partial f/\partial w \\ \partial x/\partial u & \partial x/\partial v & \partial x/\partial w \\ \partial y/\partial u & \partial y/\partial v & \partial y/\partial w \end{vmatrix} \qquad (2.39)$$

where

$$J = \begin{vmatrix} \partial x/\partial u & \partial x/\partial v & \partial x/\partial w \\ \partial y/\partial u & \partial y/\partial v & \partial y/\partial w \\ \partial z/\partial u & \partial z/\partial v & \partial z/\partial w \end{vmatrix}$$

is the Jacobian of the variables of transformation.

Example 2.19   If $z = f(x, y)$, $x = r\cos\theta$, $y = r\sin\theta$, then show that

$$\left(\frac{\partial f}{\partial x}\right)^2 + \left(\frac{\partial f}{\partial y}\right)^2 = \left(\frac{\partial f}{\partial r}\right)^2 + \frac{1}{r^2}\left(\frac{\partial f}{\partial \theta}\right)^2$$

Solution   The variables of transformation are r and θ. We have

$$J = \frac{\partial(x,y)}{\partial(r,\theta)} = \begin{vmatrix} \partial x/\partial r & \partial x/\partial \theta \\ \partial y/\partial r & \partial y/\partial \theta \end{vmatrix} = \begin{vmatrix} \cos\theta & -r\sin\theta \\ \sin\theta & r\cos\theta \end{vmatrix} = r$$

$$\frac{\partial(f,y)}{\partial(r,\theta)} = \begin{vmatrix} \partial f/\partial r & \partial f/\partial \theta \\ \partial y/\partial r & \partial y/\partial \theta \end{vmatrix} = \begin{vmatrix} \partial f/\partial r & \partial f/\partial \theta \\ \sin\theta & r\cos\theta \end{vmatrix} = r\cos\theta\frac{\partial f}{\partial r} - \sin\theta\frac{\partial f}{\partial \theta}$$

$$\frac{\partial(f,x)}{\partial(r,\theta)} = \begin{vmatrix} \partial f/\partial r & \partial f/\partial \theta \\ \partial x/\partial r & \partial x/\partial \theta \end{vmatrix} = \begin{vmatrix} \partial f/\partial r & \partial f/\partial \theta \\ \cos\theta & -r\sin\theta \end{vmatrix} = -r\sin\theta\frac{\partial f}{\partial r} - \cos\theta\frac{\partial f}{\partial \theta}.$$

Hence, using Eq. (2.38), we obtain

$$\frac{\partial f}{\partial x} = \frac{1}{J}\left[\frac{\partial(f,y)}{\partial(r,\theta)}\right] = \cos\theta\frac{\partial f}{\partial r} - \frac{\sin\theta}{r}\frac{\partial f}{\partial \theta}$$

$$\frac{\partial f}{\partial y} = -\frac{1}{J}\left[\frac{\partial(f,x)}{\partial(r,\theta)}\right] = \sin\theta\frac{\partial f}{\partial r} + \frac{\cos\theta}{r}\frac{\partial f}{\partial \theta}.$$

Squaring and adding, we obtain the required result.

Example 2.20(a)   If $u = f(x,y,z)$ and $x = r\sin\theta\cos\phi, y = r\sin\theta\sin\phi, z = r\cos\theta$, then show that

$$\left(\frac{\partial f}{\partial x}\right)^2 + \left(\frac{\partial f}{\partial y}\right)^2 + \left(\frac{\partial f}{\partial z}\right)^2 = \left(\frac{\partial f}{\partial r}\right)^2 + \frac{1}{r^2}\left(\frac{\partial f}{\partial \theta}\right)^2 + \frac{1}{r^2\sin^2\theta}\left(\frac{\partial f}{\partial \phi}\right)^2.$$

Solution   The variables of transformation are r, θ and ϕ. We have

$$J = \frac{\partial(x,y,z)}{\partial(r,\theta,\phi)} = \begin{vmatrix} \sin\theta\cos\phi & r\cos\theta\cos\phi & -r\sin\theta\sin\phi \\ \sin\theta\sin\phi & r\cos\theta\sin\phi & r\sin\theta\cos\phi \\ \cos\theta & -r\sin\theta & 0 \end{vmatrix} = r^2\sin\theta$$

$$\frac{\partial(f,x,z)}{\partial(r,\theta,\phi)} = \begin{vmatrix} \partial f/\partial r & \partial f/\partial \theta & \partial f/\partial \phi \\ \sin\theta\sin\phi & r\cos\theta\sin\phi & r\sin\theta\cos\phi \\ \cos\theta & -r\sin\theta & 0 \end{vmatrix}$$

$$= r^2\sin^2\theta\cos\phi\frac{\partial f}{\partial r} + r\sin\theta\cos\theta\cos\phi\frac{\partial f}{\partial \theta} - r\sin\phi\frac{\partial f}{\partial \phi}$$

$$\frac{\partial(f,x,z)}{\partial(r,\theta,\phi)} = \begin{vmatrix} \partial f/\partial r & \partial f/\partial \theta & \partial f/\partial \phi \\ \sin\theta\cos\phi & r\cos\theta\cos\phi & -r\sin\theta\sin\phi \\ \cos\theta & -r\sin\theta & 0 \end{vmatrix}$$

$$= -r^2 \sin^2\theta \sin\phi \, \frac{\partial f}{\partial r} - r\sin\theta\cos\theta\sin\phi \, \frac{\partial f}{\partial\theta} - r\cos\phi \, \frac{\partial f}{\partial\phi}.$$

$$\frac{\partial(f,x,y)}{\partial(r,\theta,\phi)} = \begin{vmatrix} \partial f/\partial r & \partial f/\partial\theta & \partial f/\partial\phi \\ \sin\theta\cos\phi & r\cos\theta\cos\phi & -r\sin\theta\sin\phi \\ \sin\theta\sin\phi & r\cos\theta\sin\theta & r\sin\theta\cos\phi \end{vmatrix}$$

$$= r^2 \sin\theta\cos\theta \, \frac{\partial f}{\partial r} - r\sin^2\theta \, \frac{\partial f}{\partial\theta}.$$

Using Eq. (2.39), we obtain

$$\frac{\partial f}{\partial x} = \frac{1}{J}\left[\frac{\partial(f,y,z)}{\partial(r,\theta,\phi)} \right] = \sin\theta\cos\phi \, \frac{\partial f}{\partial r} + \frac{\cos\theta\cos\phi}{r} \, \frac{\partial f}{\partial\theta} - \frac{\sin\phi}{r\sin\theta} \, \frac{\partial f}{\partial\theta}$$

$$\frac{\partial f}{\partial y} = -\frac{1}{J}\left[\frac{\partial(f,x,z)}{\partial(r,\theta,\phi)} \right] = \sin\theta\sin\phi \, \frac{\partial f}{\partial r} + \frac{\cos\theta\sin\phi}{r} \, \frac{\partial f}{\partial\theta} + \frac{\cos\phi}{r\sin\theta} \, \frac{\partial f}{\partial\phi}$$

$$\frac{\partial f}{\partial z} = \frac{1}{J}\left[\frac{\partial(f,x,y)}{\partial(r,\theta,\phi)} \right] = \cos\theta \, \frac{\partial f}{\partial r} - \frac{\sin\theta}{r} \, \frac{\partial f}{\partial\theta}.$$

Squaring and adding, we obtain the required result.

Remark 7

The variables of transformation $u = f(x, y, z)$, $v = g(x, y, z)$, $w = h(x, y, z)$ are functionally related if

$$\frac{\partial(u, v, w)}{\partial(x, y, z)} = 0,$$

that is, there exists a relationship between the variables u, v, w and the transformation is not independent.

Example 2.20(b)   Show that the variables $u = x - y + z$, $v = x + y - z$, $w = x^2 + xz - xy$, are functionally related. Find the relationship between them.

Solution   The Jacobian of transformation is given by

$$J = \begin{vmatrix} \partial u/\partial x & \partial u/\partial y & \partial u/\partial z \\ \partial v/\partial x & \partial v/\partial y & \partial v/\partial z \\ \partial w/\partial x & \partial w/\partial y & \partial w/\partial z \end{vmatrix} = \begin{vmatrix} 1 & -1 & 1 \\ 1 & 1 & -1 \\ 2x+z-y & -x & x \end{vmatrix} = \begin{vmatrix} 2 & 0 & 0 \\ 1 & 1 & -1 \\ 2x+z-y & -x & x \end{vmatrix} = 0.$$

Hence, the variables are related.

Now, $w = x(x - y + z) = xu$,   and $u + v = 2x$. Therefore, $2w = u(u + v)$.

$$\frac{\partial f}{\partial y} = \frac{2y}{x^2 + y^2} + \frac{1}{1 + y^2/x^2}\left(\frac{1}{x}\right) = \frac{2y}{x^2 + y^2} + \frac{x}{x^2 + y^2} = \frac{2y + x}{x^2 + y^2}.$$

Therefore, $\qquad \dfrac{dy}{dx} = -\dfrac{\partial f/\partial x}{\partial f/\partial y} = -\dfrac{2x - y}{2y + x} = \dfrac{y - 2x}{2y + x}, \quad y \neq -\dfrac{x}{2}.$

Exercises 2.2

1. Show that the function

$$f(x, y) = \begin{cases} \dfrac{xy}{\sqrt{x^2 + y^2}}, & (x, y) \neq (0, 0) \\ 0, & (x, y) = (0, 0) \end{cases}$$

has partial derivatives $f_x(0, 0), f_y(0, 0)$, but the partial derivatives are not continuous at $(0, 0)$.

2. Show that the function

$$f(x, y) = \begin{cases} \dfrac{x^2 + y^2}{x - y}, & (x, y) \neq (0, 0) \\ 0, & (x, y) = (0, 0) \end{cases}$$

possesses partial derivatives at $(0, 0)$, though it is not continuous at $(0, 0)$.

3. For the function

$$f(x, y) = \begin{cases} \dfrac{y(x^2 - y^2)}{x^2 + y^2}, & (x, y) \neq (0, 0) \\ 0, & (x, y) = (0, 0) \end{cases}$$

compute $f_x(0, y), f_y(x, 0), f_x(0, 0)$ and $f_y(0, 0)$, if they exist.

4. Show that the function $f(x, y) = \sqrt{x^2 + y^2}$ is not differentiable at $(0, 0)$.

5. Show that the function

$$f(x, y) = \begin{cases} (x^2 + y^2) \cos\left[\dfrac{1}{\sqrt{x^2 + y^2}}\right], & (x, y) \neq (0, 0) \\ 0, & (x, y) = (0, 0) \end{cases}$$

is differentiable at $(0, 0)$ and that f_x, f_y are not continuous at $(0, 0)$. Does this result contradict Theorem 2.2?

Find the first order partial derivatives for the following functions at the specified point:

6. $f(x, y) = x^4 - x^2y^2 + y^4$ at $(-1, 1)$.

7. $f(x, y) = \ln(x/y)$ at $(2, 3)$.

8. $f(x, y) = x^2 e^{y/x}$ at $(4, 2)$.

9. $f(x, y) = x/\sqrt{x^2 + y^2}$ at $(6, 7)$.

10. $f(x, y) = \cot^{-1}(x + y)$ at $(1, 2)$.

11. $f(x, y) = \ln\left[\dfrac{\sqrt{x^2 + y^2} - x}{\sqrt{x^2 + y^2} + x}\right]$ at $(3, 4)$.

12. $f(x, y, z) = (x^2 + y^2 + z^2)^{-1/2}$ at $(2, 1, 2)$.

13. $f(x, y, z) = e^{x/y} + e^{z/y}$ at $(1, 1, 1)$.

14. $f(x, y, z) = (xy)^{\sin z}$ at $(3, 5, \pi/2)$.

15. $f(x, y, z) = \ln(x + \sqrt{y^2 + z^2})$ at $(2, 3, 4)$.

Find dw/dt in following problems.

16. $w = x^2 + y^2$, $x = (t^2 - 1)/t$, $y = t/(t^2 + 1)$ at $t = 1$.

17. $w = x^2 + y^2 + z^2$, $x = \cos t$, $y = \ln(t + 1)$, $z = e^t$ at $t = 0$.

18. $w = e^x \sin(y + 2z)$, $x = t$, $y = 1/t$, $z = t^2$.   **19.** $w = xy + yz + zx$, $x = t^2$, $y = te^t$, $z = te^{-t}$.

20. $w = z \ln y + y \ln z + xyz$, $x = \sin t$, $y = t^2 + 1$, $z = \cos^{-1} t$ at $t = 0$.

Verify the given results in the following problems:

21. If $z = f(ax + by)$, then $b\dfrac{\partial z}{\partial x} - a\dfrac{\partial z}{\partial y} = 0$.

22. If $z = \log[(x^2 - y^2)/(x^2 + y^2)]$, then $x\dfrac{\partial z}{\partial x} + y\dfrac{\partial z}{\partial y} = 0$.

23. If $u = f(x - y, y - z, z - x)$, then $\dfrac{\partial u}{\partial x} + \dfrac{\partial u}{\partial y} + \dfrac{\partial u}{\partial z} = 0$.

24. If $z = f(x, y)$, $x = r \cosh \theta$, $y = r \sinh \theta$, then

$$\left(\frac{\partial z}{\partial x}\right)^2 - \left(\frac{\partial z}{\partial y}\right)^2 = \left(\frac{\partial z}{\partial r}\right)^2 - \frac{1}{r^2}\left(\frac{\partial z}{\partial \theta}\right)^2.$$

25. If $z = y + f(u)$, $u = \dfrac{x}{y}$, then $u\dfrac{\partial z}{\partial x} + \dfrac{\partial z}{\partial y} = 1$.

26. If $w = f(u, \upsilon)$, $u = \sqrt{x^2 + y^2}$, $\upsilon = \cot^{-1}(y/x)$, then

$$\left(\frac{\partial f}{\partial x}\right)^2 + \left(\frac{\partial f}{\partial y}\right)^2 = \frac{1}{x^2 + y^2}\left[(x^2 + y^2)\left(\frac{\partial f}{\partial u}\right)^2 + \left(\frac{\partial f}{\partial \upsilon}\right)^2\right].$$

27. If $z = f(x, y)$, $x = u \cos \alpha - \upsilon \sin \alpha$, $y = u \sin \alpha + \upsilon \cos \alpha$, where α is a constant, then

$$\left(\frac{\partial f}{\partial u}\right)^2 + \left(\frac{\partial f}{\partial \upsilon}\right)^2 = \left(\frac{\partial f}{\partial x}\right)^2 + \left(\frac{\partial f}{\partial y}\right)^2.$$

28. If $z = \ln(u^2 + \upsilon)$, $u = e^{x+y^2}$, $\upsilon = x + y^2$, then $2y\dfrac{\partial z}{\partial x} - \dfrac{\partial z}{\partial y} = 0$.

29. If $w = \sqrt{x^2 + y^2 + z^2}$, $x = u \cos \upsilon$, $y = u \sin \upsilon$, $z = u\upsilon$, then

$$u\frac{\partial w}{\partial u} - \upsilon\frac{\partial w}{\partial \upsilon} = \frac{u}{\sqrt{1 + \upsilon^2}}$$

30. If $w = \sin^{-1} u$, $u = (x^2 + y^2 + z^2)/(x + y + z)$, then

$$x\frac{\partial w}{\partial x} + y\frac{\partial w}{\partial y} + z\frac{\partial w}{\partial z} = \tan w.$$

Check whether the variables in the following transformations are functionally related. If so, find the relationship between them.

31. $u = x^2 - y^2 - z^2$, $v = x^2 - y^2 + z^2$, $w = x^4 + y^4 + z^4 - 2x^2 y^2$.

32. $u = x + 3z$, $v = x - y - z$, $w = y^2 + 16z^2 + 8yz$.

33. $u = x + y + z$, $v = x^2 + y^2 + z^2 - 2xy - 2yz - 2zx$, $w = x^3 + y^3 + z^3 - 3xyz$.

34. $u = (x + y)/(1 - xy)$, $v = \tan^{-1} x + \tan^{-1} y$, $x > 0$, $y > 0$, $xy < 1$.

35. $u = x\sqrt{1 - y^2} + y\sqrt{1 - x^2}$, $v = \sin^{-1} x + \sin^{-1} y$, $x \geq 0$, $y \geq 0$, $x^2 + y^2 \leq 1$.

Using implicit differentiation, obtain the following:

36. $\dfrac{dy}{dx}$, when $x^y + y^x = \alpha$, α any constant, $x > 0$, $y > 0$.

37. $\dfrac{dy}{dx}$, when $\cot^{-1}(x/y) + y^3 + 1 = 0$, $x > 0$, $y > 0$.

38. $\left(\dfrac{\partial z}{\partial x}\right)_y$ and $\left(\dfrac{\partial z}{\partial y}\right)_x$, when $\cos xy + \cos yz + \cos zx = 1$.

39. $\left(\dfrac{\partial z}{\partial x}\right)_y$ and $\left(\dfrac{\partial z}{\partial y}\right)_x$, when $x^3 + 3xy - 2y^2 + 3xz + z^2 = 0$.

40. $y\left(\dfrac{\partial x}{\partial y}\right)_y + z\left(\dfrac{\partial x}{\partial z}\right)_y$, when $f\left(\dfrac{z}{y}, \dfrac{x}{y}\right) = 0$.

Using differentials, obtain the approximate value of the following quantities:

41. $\sqrt{(298)^2 + (401)^2}$.

42. $(4.05)^{1/2}(7.97)^{1/3}$.

43. $\cos 44° \sin 32°$.

44. $\dfrac{1}{\sqrt{1.05}} + \dfrac{1}{\sqrt{3.97}} + \dfrac{1}{\sqrt{9.01}}$.

45. $\sin 26° \cos 57° \tan 48°$.

46. A certain function $z = f(x, y)$ has values $f(2, 3) = 5$, $f_x(2, 3) = 3$ and $f_y(2, 3) = 7$. Find an approximate value of $f(1.98, 3.01)$.

47. The radius r and the height h of a conical tank increases at the rate of $(dr/dt) = 0.2''$/hr and $(dh/dt) = 0.1''$/hr. Find the rate of increase dV/dt in volume V when the radius is 5 feet and the height is 20 feet.

48. The dimensions of a rectangular block of wood are $60''$, $80''$ and $100''$ with possible absolute error of $3''$ in each measurement. Find the maximum absolute error and the percentage error in the surface area.

49. Two sides of a triangle are measured as 5 cm and 3 cm and the included angle as $30°$. If the possible absolute errors are 0.2 cm in measuring the sides and $1°$ in the angle, then find the percentage error in the computed area of the triangle.

50. The sides of a rectangular box are found to be a feet, b feet and c feet with a possible error of 1% in magnitude in each of the measurements. Find the percentage eror in the volume of the box caused by the errors in individual measurements.

51. The diameter and the altitude of a can in the shape of a right circular cylinder are measured as 6 cm and 8 cm respectively. The maximum absolute error in each measurement is 0.2 cm. Find the maximum absolute error and the percentage error in the computed value of the volume.

52. The power consumed in an electric resistor is given by $P = E^2/R$ (in watts). If $E = 80$ volts and $R = 5$ Ohms, by how much the power consumption will change if E is increased by 3 volts and R is decreased by 0.1 Ohms.

53. If two resistors with resistences R_1 and R_2 in Ohms are connected in parallel, then the resistence of the resulting circuit is $R = [(1/R_1) + (1/R_2)]^{-1}$. Find an approximate value of the percentage change in resistence that results by changing R_1 from 2 to 1.9 Ohms and R_2 from 6 to 6.2 Ohms.

54. Suppose that $u = xze^y$ and x, y, z can be measured with maximum absolute errors 0.1, 0.2 and 0.3 respectively. Find the percentage error in the computed value of u from the measured values $x = 3$, $y = \ln 2$ and $z = 5$.

55. If the radius r and the altitude h of a cone are measured with an absolute error of 1% in each measurement, then find the approximate percentage change in the lateral area of the cone if the measured values are $r = 3$ feet and $h = 4$ feet.

2.4 Higher Order Partial Derivatives

Let $z = f(x, y)$ be a function of two variables and let its first order partial derivatives exist at all the points in the domain of definition D of the function f. Then, the first order partial derivatives are also functions of x and y. We define the second order partial derivatives as

$$\frac{\partial^2 f}{\partial x^2} = \frac{\partial}{\partial x}\left[\frac{\partial f}{\partial x}\right] = f_{xx}(x, y) = \lim_{\Delta x \to 0}\left[\frac{f_x(x + \Delta x, y) - f_x(x, y)}{\Delta x}\right]$$

$$\frac{\partial^2 f}{\partial y\,\partial x} = \frac{\partial}{\partial y}\left[\frac{\partial f}{\partial x}\right] = f_{yx}(x, y) = \lim_{\Delta y \to 0}\left[\frac{f_x(x, y + \Delta y) - f_x(x, y)}{\Delta y}\right]$$

(differentiate partially first with respect to x and then with respect to y)

$$\frac{\partial^2 f}{\partial x\,\partial y} = \frac{\partial}{\partial x}\left[\frac{\partial f}{\partial y}\right] = f_{xy}(x, y) = \lim_{\Delta x \to 0}\left[\frac{f_y(x + \Delta x, y) - f_y(x, y)}{\Delta x}\right]$$

(differentiate partially first with respect to y and then with respect to x)

$$\frac{\partial^2 f}{\partial y^2} = \frac{\partial}{\partial y}\left[\frac{\partial f}{\partial y}\right] = f_{yy}(x, y) = \lim_{\Delta y \to 0}\left[\frac{f_y(x, y + \Delta y) - f_y(x, y)}{\Delta y}\right]$$

if the limits exist. The derivatives f_{xy} and f_{yx} are called *mixed derivatives*. If f_{xy} and f_{yx} are continuous at a point $P(x, y)$, then at this point $f_{xy} = f_{yx}$. That is, the order of differentiation is immaterial in this case. There are four partial derivatives of second order for $f(x, y)$. If all the second order partial derivatives exist at all points in D, then these derivatives are also functions of x and y and can be further differentiated.

Example 2.22 Find all the second order partial derivatives of the function
$$f(x, y) = \ln(x^2 + y^2) + \tan^{-1}(y/x), \ (x, y) \neq (0, 0).$$

Solution We have

$$f_x(x, y) = \frac{2x}{x^2 + y^2} + \frac{1}{1 + (y/x)^2}\left(-\frac{y}{x^2}\right) = \frac{2x - y}{x^2 + y^2}$$

$$f_y(x, y) = \frac{2y}{x^2 + y^2} + \frac{1}{1 + (y/x)^2}\left(\frac{1}{x}\right) = \frac{2y + x}{x^2 + y^2}$$

$$f_{yx}(x, y) = \frac{\partial}{\partial y}(f_x) = \frac{\partial}{\partial y}\left(\frac{2x - y}{x^2 + y^2}\right) = \frac{(x^2 + y^2)(-1) - (2x - y)(2y)}{(x^2 + y^2)^2} = \frac{y^2 - x^2 - 4xy}{(x^2 + y^2)^2}$$

$$f_{xy}(x, y) = \frac{\partial}{\partial x}(f_y) = \frac{\partial}{\partial x}\left(\frac{2y + x}{x^2 + y^2}\right) = \frac{(x^2 + y^2)(1) - (2y + x)(2x)}{(x^2 + y^2)^2} = \frac{y^2 - x^2 - 4xy}{(x^2 + y^2)^2}$$

$$f_{xx}(x, y) = \frac{\partial}{\partial x}(f_x) = \frac{\partial}{\partial x}\left(\frac{2x - y}{x^2 + y^2}\right) = \frac{(x^2 + y^2)(2) - (2x - y)(2x)}{(x^2 + y^2)^2} = \frac{2y^2 - 2x^2 + 2xy}{(x^2 + y^2)^2}$$

$$f_{yy}(x, y) = \frac{\partial}{\partial y}(f_y) = \frac{\partial}{\partial y}\left(\frac{2y + x}{x^2 + y^2}\right) = \frac{(x^2 + y^2)(2) - (2y + x)(2y)}{(x^2 + y^2)^2} = \frac{2x^2 - 2y^2 - 2xy}{(x^2 + y^2)^2}$$

We note that $f_{xy} = f_{yx}$.

Example 2.23 For the function

$$f(x, y) = \begin{cases} \dfrac{xy(2x^2 - 3y^2)}{x^2 + y^2}, & (x, y) \neq (0, 0) \\ 0, & (x, y) = (0, 0) \end{cases}$$

show that $f_{xy}(0, 0) \neq f_{yx}(0, 0)$.

Solution We obtain the required derivatives as

$$f_x(0,0) = \lim_{\Delta x \to 0} \frac{f(\Delta x, 0) - f(0,0)}{\Delta x} = 0, \ f_y(0,0) = \lim_{\Delta y \to 0} \frac{f(0, \Delta y) - f(0,0)}{\Delta y} = 0$$

$$f_x(0, y) = \lim_{\Delta x \to 0} \frac{f(\Delta x, y) - f(0, y)}{\Delta x} = \lim_{\Delta x \to 0} \frac{y[2(\Delta x)^2 - 3y^2]\Delta x}{[(\Delta x)^2 + y^2]\,\Delta x} = -3y$$

$$f_y(x, 0) = \lim_{\Delta y \to 0} \frac{f(x, \Delta y) - f(x, 0)}{\Delta y} = \lim_{\Delta y \to 0} \frac{x[2x^2 - 3(\Delta y)^2]\Delta y}{[x^2 + (\Delta y)^2]\,\Delta y} = 2x.$$

Now,

$$f_{xy}(0,0) = \frac{\partial}{\partial x}\left(\frac{\partial f}{\partial y}\right)_{(0,0)} = \lim_{\Delta x \to 0} \frac{f_y(\Delta x, 0) - f_y(0,0)}{\Delta x} = \lim_{\Delta x \to 0} \frac{2\Delta x - 0}{\Delta x} = 2$$

$$f_{yx}(0,0) = \frac{\partial}{\partial y}\left(\frac{\partial f}{\partial x}\right)_{(0,0)} = \lim_{\Delta y \to 0} \frac{f_x(0, \Delta y) - f_x(0,0)}{\Delta y} = \lim_{\Delta y \to 0} \frac{-3\Delta y - 0}{\Delta y} = -3.$$

Hence, $f_{xy}(0,0) \neq f_{yx}(0,0)$.

Example 2.24 Compute $f_{xy}(0,0)$ and $f_{yx}(0,0)$ for the function

$$f(x, y) = \begin{cases} \dfrac{xy^3}{x + y^2}, & (x, y) \neq (0, 0) \\ 0, & (x, y) = (0, 0). \end{cases}$$

Also discuss the continuity of f_{xy} and f_{yx} at $(0, 0)$.

Solution We have

$$f_x(0,0) = \lim_{\Delta x \to 0} \frac{f(\Delta x, 0) - f(0,0)}{\Delta x} = 0, \ f_y(0,0) = \lim_{\Delta y \to 0} \frac{f(0, \Delta y) - f(0,0)}{\Delta y} = 0$$

$$f_x(0, y) = \lim_{\Delta x \to 0} \frac{f(\Delta x, y) - f(0, y)}{\Delta x} = \lim_{\Delta x \to 0} \frac{y^3 \Delta x}{[\Delta x + y^2]\,\Delta x} = y$$

$$f_y(x, 0) = \lim_{\Delta y \to 0} \frac{f(x, \Delta y) - f(x, 0)}{\Delta y} = \lim_{\Delta y \to 0} \frac{x(\Delta y)^3}{[x + (\Delta y)^2]\,\Delta y} = 0$$

$$f_{xy}(0,0) = \lim_{\Delta x \to 0} \frac{f_y(\Delta x, 0) - f_y(0,0)}{\Delta x} = 0$$

$$f_{yx}(0,0) = \lim_{\Delta y \to 0} \frac{f_x(0, \Delta y) - f_x(0,0)}{\Delta y} = \lim_{\Delta y \to 0} \frac{\Delta y}{\Delta y} = 1.$$

Since $f_{xy}(0,0) \neq f_{yx}(0,0)$, f_{xy} and f_{yx} are not continuous at $(0, 0)$.

Alternative We find that for $(x, y) \neq (0, 0)$

$$f_{yx}(x, y) = \frac{y^6 + 5xy^4}{(x + y^2)^3} = f_{xy}(x, y).$$

Along the path $x = my^2$, we obtain

$$\lim_{(x,y) \to (0,0)} f_{yx}(x, y) = \lim_{y \to 0} \frac{y^6(1 + 5m)}{y^6(1 + m)^3} = \frac{1 + 5m}{(1 + m)^3}.$$

Since the limit does not exist, f_{yx} is not continuous at $(0, 0)$.

Example 2.25   For the implicit function $f(x, y) = 0$ of one independent variable x, obtain $y'' = d^2y/dx^2$. Assume that $f_{xy} = f_{yx}$.

Solution   Taking the differential of $f(x, y) = 0$, we obtain

$$y' = \frac{dy}{dx} = -\left(\frac{f_x}{f_y}\right).$$

Therefore,

$$\frac{d^2y}{dx^2} = \frac{d}{dx}\left[\frac{dy}{dx}\right] = -\frac{d}{dx}\left[\frac{f_x}{f_y}\right] = -\frac{f_y\dfrac{d}{dx}(f_x) - f_x\dfrac{d}{dx}(f_y)}{f_y^2}$$

$$= -\frac{f_y[f_{xx} + (f_{yx})y'] - f_x[f_{xy} + (f_{yy})y']}{f_y^2}$$

$$= -\frac{(f_y f_{xx} - f_x f_{xy}) + (f_y f_{yx} - f_x f_{yy})y'}{f_y^2}.$$

Substituting $y' = -f_x/f_y$, we obtain

$$\frac{d^2y}{dx^2} = -\frac{f_y^2 f_{xx} - 2f_x f_y f_{xy} + f_x^2 f_{yy}}{f_y^3}, \text{ since } f_{yx} = f_{xy}.$$

2.4.1   Homogeneous Functions

A function $f(x, y)$ is said to be *homogeneous* of degree n in x and y, if it can be written in any one of the following forms

(i) $f(\lambda x, \lambda y) = \lambda^n f(x, y)$. (2.46)

(ii) $f(x, y) = x^n g(y/x)$. (2.47)

(iii) $f(x, y) = y^n g(x/y)$. (2.48)

Similarly, a function $f(x, y, z)$ of three variables is said to be homogeneous, of degree n, if it can be

written as $f(\lambda x, \lambda y, \lambda z) = \lambda^n f(x, y, z)$, or $f(x, y, z) = x^n g\left(\dfrac{y}{x}, \dfrac{z}{x}\right)$ etc.

Some examples of homogeneous functions are the following:

f	degree of homogeneity
$x^2 + xy$	2
$\tan^{-1}(y/x)$	0
$1/(x + y)$	-1
$1/(x^4 + y^4 + z^4)$	-4
$xyz/(x^4 + y^4 + z^4)$	-1
$\sqrt{x}/\sqrt{x^2 + y^2 + z^2}$	$-1/2$

The function $f(x, y) = (x^2 + y)/(x + y^2)$ is not homogeneous.

An important result concerning homogeneous functions is the following.

Theorem 2.4 (Euler's theorem) If $f(x, y)$ is a homogeneous function of degree n in x and y and has continuous first and second order partial derivatives, then

(i)
$$x \frac{\partial f}{\partial x} + y \frac{\partial f}{\partial y} = nf.$$
(2.49)

(ii)
$$x^2 \frac{\partial^2 f}{\partial x^2} + 2xy \frac{\partial^2 f}{\partial x \, \partial y} + y^2 \frac{\partial^2 f}{\partial y^2} = n(n-1)f.$$
(2.50)

Proof Since $f(x, y)$ is a homogeneous function of degree n in x and y, we can write $f(x, y) = x^n g(y/x)$.

Differentiating partially with respect to x and y, we get

$$\frac{\partial f}{\partial x} = nx^{n-1} g\left(\frac{y}{x}\right) + x^n g'\left(\frac{y}{x}\right)\left(-\frac{y}{x^2}\right) = nx^{n-1} g\left(\frac{y}{x}\right) - yx^{n-2} g'\left(\frac{y}{x}\right).$$

$$\frac{\partial f}{\partial y} = x^n g'\left(\frac{y}{x}\right)\left(\frac{1}{x}\right) = x^{n-1} g'\left(\frac{y}{x}\right).$$

Hence, we obtain

$$x \frac{\partial f}{\partial x} + y \frac{\partial f}{\partial y} = nx^n g\left(\frac{y}{x}\right) - yx^{n-1} g'\left(\frac{y}{x}\right) + yx^{n-1} g'\left(\frac{y}{x}\right) = nx^n g\left(\frac{y}{x}\right) = nf.$$

Differentiating Eq. (2.49) partially with respect to x and y, we get

$$x \frac{\partial^2 f}{\partial x^2} + \frac{\partial f}{\partial x} + y \frac{\partial^2 f}{\partial x \, \partial y} = n \frac{\partial f}{\partial x}$$
(2.51)

and
$$x \frac{\partial^2 f}{\partial y \, \partial x} + \frac{\partial f}{\partial y} + y \frac{\partial^2 f}{\partial y^2} = n \frac{\partial f}{\partial y}.$$
(2.52)

Multiplying Eq. (2.51) by x and Eq. (2.52) by y and adding, we obtain

$$x^2 \frac{\partial^2 f}{\partial x^2} + \left(x \frac{\partial f}{\partial x} + y \frac{\partial f}{\partial y}\right) + xy \left(\frac{\partial^2 f}{\partial x \, \partial y} + \frac{\partial^2 f}{\partial y \, \partial x}\right) + y^2 \frac{\partial^2 f}{\partial y^2} = n \left(x \frac{\partial f}{\partial x} + y \frac{\partial f}{\partial y}\right)$$

or
$$x^2 \frac{\partial^2 f}{\partial x^2} + 2xy \frac{\partial^2 f}{\partial x \, \partial y} + y^2 \frac{\partial^2 f}{\partial y^2} = n(n-1)f.$$

Example 2.26 If $u(x, y) = \cos^{-1}\left(\frac{x + y}{\sqrt{x} + \sqrt{y}}\right)$, $0 < x, y < 1$, then prove that

$$x \frac{\partial u}{\partial x} + y \frac{\partial u}{\partial y} = -\frac{1}{2} \cot u.$$

Solution For all x, y, $0 < x, y < 1$, $(x + y)/[\sqrt{x} + \sqrt{y}] < 1$, so that $u(x, y)$ is defined. The given function can be written as

$$\cos u = \frac{x+y}{\sqrt{x}+\sqrt{y}} = \frac{x[1+y/x]}{\sqrt{x}\,[1+\sqrt{y/x}]} = \sqrt{x}\left[\frac{1+(y/x)}{1+\sqrt{y/x}}\right]$$

Therefore, $\cos u$ is a homogeneous function of degree 1/2. Using the Euler's theorem for $f = \cos u$ and $n = 1/2$, we obtain

$$x\frac{\partial}{\partial x}(\cos u) + y\frac{\partial}{\partial y}(\cos u) = \frac{1}{2}\cos u$$

or $\qquad -x(\sin u)\dfrac{\partial u}{\partial x} - y(\sin u)\dfrac{\partial u}{\partial y} = \dfrac{1}{2}\cos u$, or $x\dfrac{\partial u}{\partial x} + y\dfrac{\partial u}{\partial y} = -\dfrac{1}{2}\cot u.$

Example 2.27 If $u(x, y) = x^2 \tan^{-1}(y/x) - y^2 \tan^{-1}(x/y)$, $x > 0$, $y > 0$, then evaluate

$$x^2\frac{\partial^2 u}{\partial x^2} + 2xy\frac{\partial^2 u}{\partial x\,\partial y} + y^2\frac{\partial^2 u}{\partial y^2}.$$

Solution We have $u(\lambda x, \lambda y) = \lambda^2 u(x, y)$. Therefore, $u(x, y)$ is a homogeneous function of degree 2. Using Theorem 2.4 (ii) for $f = u$ and $n = 2$, we obtain

$$x^2\frac{\partial^2 u}{\partial x^2} + 2xy\frac{\partial^2 u}{\partial x\,\partial y} + y^2\frac{\partial^2 u}{\partial y^2} = 2(2-1)u = 2u.$$

Example 2.28 Let $u(x, y) = [x^3 + y^3]/[x + y]$, $(x, y) \neq (0, 0)$. Then evaluate

$$x\frac{\partial^2 u}{\partial x^2} + y\frac{\partial^2 u}{\partial x\,\partial y} - \frac{\partial u}{\partial x}.$$

Solution We have $u(x, y) = \dfrac{x^2[1 + (y/x)^3]}{[1 + (y/x)]}$. Therefore, $u(x, y)$ is a homogeneous function of degree 2. Using Euler's theorem, we get

$$x\frac{\partial u}{\partial x} + y\frac{\partial u}{\partial y} = 2u.$$

Differentiating partially with respect to x, we obtain

$$\frac{\partial u}{\partial x} + x\frac{\partial^2 u}{\partial x^2} + y\frac{\partial^2 u}{\partial x\,\partial y} = 2\frac{\partial u}{\partial x}, \quad \text{or} \quad x\frac{\partial^2 u}{\partial x^2} + y\frac{\partial^2 u}{\partial x\,\partial y} - \frac{\partial u}{\partial x} = 0.$$

Example 2.29 Let $f(x, y)$ and $g(x, y)$ be two homogeneous functions of degree m and n respectively where $m \neq 0$. Let $h = f + g$. If $x\dfrac{\partial h}{\partial x} + y\dfrac{\partial h}{\partial y} = 0$, then show that $f = \alpha g$ for some scalar α.

Solution Since f and g are homogeneous functions of degrees m and n respectively, we obtain on using Euler's theorem

$$x\frac{\partial f}{\partial x} + y\frac{\partial f}{\partial y} = mf \quad \text{and} \quad x\frac{\partial g}{\partial x} + y\frac{\partial g}{\partial y} = ng.$$

Adding the two results, we get

$$x\left(\frac{\partial f}{\partial x} + \frac{\partial g}{\partial x}\right) + y\left(\frac{\partial f}{\partial y} + \frac{\partial g}{\partial y}\right) = mf + ng$$

or
$$x \frac{\partial h}{\partial x} + y \frac{\partial h}{\partial y} = mf + ng = 0, \text{ where } h = f + g.$$

Therefore, $f = -\frac{n}{m} g = \alpha g$, where $\alpha = -\frac{n}{m}$ is a scalar.

2.4.2 Taylor's Theorem

In section 1.3.6 we have derived the Taylor's theorem in one variable. If $f(x)$ has continuous derivatives upto $(n + 1)$th order in some interval containing $x = a$, then

$$f(x) = f(a) + (x - a) f'(a) + \ldots + \frac{(x - a)^n}{n!} f^{(n)}(a) + R_n(x) \qquad (2.53)$$

where $R_n(x)$ is the remainder term given by

$$R_n(x) = \frac{(x - a)^{n+1}}{(n + 1)!} f^{(n+1)}(\xi) = \frac{(x - a)^{n+1}}{(n + 1)!} f^{(n+1)}[a + \theta(x - a)], a < \xi < x, 0 < \theta < 1. \quad (2.54)$$

We now extend this theorem to functions of two variables.

Theorem 2.5 (**Taylor's theorem**) Let a function $f(x, y)$ defined in some domain D in $\mathbb{R}^2$ have continuous partial derivatives upto $(n + 1)$th order in some neighborhood of a point $P(x_0, y_0)$ in D. Then, for some point $(x_0 + h, y_0 + k)$ in this neighborhood, we have

$$f(x_0 + h, y_0 + k) = f(x_0, y_0) + \left(h \frac{\partial}{\partial x} + k \frac{\partial}{\partial y} \right) f(x_0, y_0) + \frac{1}{2!} \left(h \frac{\partial}{\partial x} + k \frac{\partial}{\partial y} \right)^2 f(x_0, y_0)$$

$$+ \ldots + \frac{1}{n!} \left(h \frac{\partial}{\partial x} + k \frac{\partial}{\partial y} \right)^n f(x_0, y_0) + R_n \qquad (2.55)$$

where R_n is the remainder term given by

$$R_n = \frac{1}{(n + 1)!} \left(h \frac{\partial}{\partial x} + k \frac{\partial}{\partial y} \right)^{n+1} f(x_0 + \theta h, y_0 + \theta k), 0 < \theta < 1. \qquad (2.56)$$

Proof Let $x = x_0 + th$, $y = y_0 + tk$, where the parameter t takes values in the interval $[0, 1]$. Define a function $\phi(t)$ as $\phi(t) = f(x, y) = f(x_0 + th, y_0 + tk)$.

Using the chain rule, we get

$$\phi'(t) = \frac{\partial f}{\partial x} \frac{dx}{dt} + \frac{\partial f}{\partial y} \frac{dy}{dt} = h \frac{\partial f}{\partial x} + k \frac{\partial f}{\partial y} = \left(h \frac{\partial}{\partial x} + k \frac{\partial}{\partial y} \right) f$$

$$\phi''(t) = \left(h \frac{\partial}{\partial x} + k \frac{\partial f}{\partial y} \right)^2 f, \ldots, \quad \phi^{(n+1)}(t) = \left(h \frac{\partial}{\partial x} + k \frac{\partial}{\partial y} \right)^{n+1} f.$$

Using the Taylor's theorem for a function of one variable (see Eq. (2.53)) with $t = 1$ and $a = 0$, we obtain

$$\phi(1) = \phi(0) + \phi'(0) + \frac{1}{2!} \phi''(0) + \ldots + \frac{1}{n!} \phi^{(n)}(0) + \frac{1}{(n + 1)!} \phi^{(n+1)}(\theta) \qquad (2.57)$$

where
$$\phi(0) = f(x_0, y_0)$$

$$\phi(1) = f(x_0 + h, y_0 + k)$$

$$\phi^{(i)}(0) = \left(h\frac{\partial}{\partial x} + k\frac{\partial}{\partial y} \right)^i f(x_0, y_0), i = 1, 2, \ldots, n$$

$$\phi^{(n+1)}(\theta) = \left(h\frac{\partial}{\partial x} + k\frac{\partial}{\partial y} \right)^{n+1} f(x_0 + \theta h, y_0 + \theta k), 0 < \theta < 1.$$

Substituting the expressions for $\phi(1)$, $\phi(0)$, $\phi'(0)$, ..., $\phi^{(n)}(0)$ and $\phi^{(n+1)}(\theta)$ in Eq. (2.57), we obtain the Taylor's theorem for functions of two variables as given in Eqs. (2.55) and (2.56).

Substituting $x = x_0 + h$, $y = y_0 + k$ in Eq. (2.55), we can also write the Taylor's theorem as

$$f(x, y) = f(x_0, y_0) + \left[(x - x_0)\frac{\partial}{\partial x} + (y - y_0)\frac{\partial}{\partial y} \right] f(x_0, y_0)$$

$$+ \frac{1}{2!} \left[(x - x_0)\frac{\partial}{\partial x} + (y - y_0)\frac{\partial}{\partial y} \right]^2 f(x_0, y_0)$$

$$+ \cdots + \frac{1}{n!} \left[(x - x_0)\frac{\partial}{\partial x} + (y - y_0)\frac{\partial}{\partial y} \right]^n f(x_0, y_0) + R_n \qquad (2.58)$$

where,
$$R_n = \frac{1}{(n+1)!} \left[(x - x_0)\frac{\partial}{\partial x} + (y - y_0)\frac{\partial}{\partial y} \right]^{n+1} f(\xi, \eta) \qquad (2.59)$$

and
$$\xi = (1 - \theta)x_0 + \theta x, \eta = (1 - \theta)y_0 + \theta y, 0 < \theta < 1.$$

For $n = 1$, we get the *linear polynomial approximation* to $f(x, y)$ as

$$f(x, y) \approx f(x_0, y_0) + (x - x_0)f_x + (y - y_0)f_y \qquad (2.60)$$

where the partial derivatives are evaluated at (x_0, y_0). This equation is same as the equation (2.28) which was obtained using differentials.

For $n = 2$, we get the *second degree (quadratic) polynomial approximation* to $f(x, y)$ as

$$f(x, y) \approx f(x_0, y_0) + (x - x_0)f_x + (y - y_0)f_y$$
$$+ \frac{1}{2}\left[(x - x_0)^2 f_{xx} + 2(x - x_0)(y - y_0)f_{xy} + (y - y_0)^2 f_{yy} \right] \qquad (2.61)$$

where the partial derivatives are evaluated at (x_0, y_0).

Remark 8

(a) If we set $(x_0, y_0) = (0, 0)$ in Eq. (2.55), we obtain the *Maclaurin's theorem* for functions of two variables as

$$f(x, y) = f(0, 0) + \left(x\frac{\partial}{\partial x} + y\frac{\partial}{\partial y} \right) f(0, 0) + \frac{1}{2!}\left(x\frac{\partial}{\partial x} + y\frac{\partial}{\partial y} \right)^2 f(0, 0)$$

$$+ \cdots + \frac{1}{n!}\left(x\frac{\partial}{\partial x} + y\frac{\partial}{\partial y} \right)^n f(0, 0) + R_n \qquad (2.62)$$

where $R_n = \dfrac{1}{(n+1)!}\left(x\dfrac{\partial}{\partial x} + y\dfrac{\partial}{\partial y}\right)^{n+1} f(\theta x, \theta y), 0 < \theta < 1.$

(b) When $\lim\limits_{n\to\infty} R_n = 0$, we obtain the *Taylor's series* expansion of the function $f(x, y)$ about the point (x_0, y_0).

(c) Taylor's theorem can be easily extended to functions of m variables $f(x_1, x_2, \ldots, x_m)$.

Error estimate

Since the point (ξ, η) or the value of θ in the error term given in Eq. (2.59) is not known, we cannot evaluate the error term exactly. However, it is possible to find a bound of the error term in a given rectangular region $R: |x - x_0| < \delta_1, |y - y_0| < \delta_2$. We assume that all the partial derivatives of the required order are continuous throughout this region.

For $n = 1$ (linear approximation), the error term is given by

$$R_1 = \frac{1}{2!}\,[(x-x_0)^2 f_{xx} + 2(x-x_0)(y-y_0) f_{xy} + (y-y_0)^2 f_{yy}] \tag{2.63}$$

where the partial dirivatives are evaluated at the point $(\xi, \eta) = [x_0 + \theta(x - x_0), y_0 + \theta(y - y_0)]$, $0 < \theta < 1$. Hence, we get

$$|R_1| \le \frac{1}{2}\,[\,|x-x_0|^2 |f_{xx}| + 2\,|x-x_0||y-y_0||f_{xy}| + |y-y_0|^2 |f_{yy}|\,].$$

If we assume that

$$B = \max\,[\,|f_{xx}|,\,|f_{xy}|,\,|f_{yy}|\,]\ \text{for all}\ (x, y)\ \text{in}\ R,\ \text{then we obtain}$$

$$|R_1| \le \frac{B}{2}\,[\,|x-x_0|^2 + 2\,|x-x_0||y-y_0| + |y-y_0|^2]$$

$$= \frac{B}{2}\,[\,|x-x_0| + |y-y_0|\,]^2 \le \frac{B}{2}\,[\delta_1 + \delta_2]^2. \tag{2.64}$$

This value of $|R_1|$ is called the *maximum absolute error* in the linear approximation of $f(x, y)$ about the point (x_0, y_0).

For $n = 2$ (quadratic approximation), the error term is given by

$$R_2 = \frac{1}{3!}\,[(x-x_0)^3 f_{xxx} + 3(x-x_0)^2 (y-y_0) f_{xxy} + 3(x-x_0)(y-y_0)^2 f_{xyy} + (y-y_0)^3 f_{yyy}] \tag{2.65}$$

where the partial derivatives are evaluated at the point

$$(\xi, \eta) = [x_0 + \theta(x - x_0), y_0 + \theta(y - y_0)], 0 < \theta < 1.$$

From Eq. (2.65), we get

$$|R_2| \le \frac{1}{6}\,[\,|x-x_0|^3 |f_{xxx}| + 3\,|x-x_0|^2 |y-y_0||f_{xxy}| + 3\,|x-x_0||y-y_0|^2 |f_{xyy}|$$

$$+ |y-y_0|^3 |f_{yyy}|\,]$$

$$\le \frac{B}{6}\,[\,|x-x_0|^3 + 3\,|x-x_0|^2 |y-y_0| + 3\,|x-x_0||y-y_0|^2 + |y-y_0|^3]$$

$$= \frac{B}{6} [\, | \, x - x_0 | + | \, y - y_0 | \,]^3 \le \frac{B}{6} (\delta_1 + \delta_2)^3 \qquad (2.66)$$

where $B = \max [\, |f_{xxx}|, |f_{xxy}|, |f_{xyy}|, |f_{yyy}| \,]$ for all points (x, y) in R.

Remark 9

In a similar manner, we can obtain error estimates for approximations of functions of three or more variables. For example, if $f(x, y, z)$ is to be approximated by a first degree polynomial (linear approximation) about the point (x_0, y_0, z_0), then we have

$$f(x, y, z) \approx P_1(x, y, z) = f(x_0, y_0, z_0) + [(x - x_0) f_x + (y - y_0) f_y + (z - z_0) f_z]$$

where the partial derivatives are evaluated at (x_0, y_0, z_0). The error associated with this approximation is given by

$$R_1 = \frac{1}{2!} [(x - x_0)^2 f_{xx} + (y - y_0)^2 f_{yy} + (z - z_0)^2 f_{zz} + 2(x - x_0)(y - y_0) f_{xy}$$

$$+ 2(x - x_0)(z - z_0) f_{xz} + 2(y - y_0)(z - z_0) f_{yz} \,].$$

If we consider the region $R: |x - x_0| \le \delta_1, |y - y_0| < \delta_2, |z - z_0| < \delta_3$

and assume that $B = \max [\, |f_{xx}|, |f_{yy}|, |f_{zz}|, |f_{xy}|, |f_{xz}|, |f_{yz}| \,]$

for all points (x, y, z) in this region, we can write

$$|R_1| \le \frac{B}{2} [\, | \, x - x_0 | + | \, y - y_0 | + | \, z - z_0 | \,]^2 \le \frac{B}{2} (\delta_1 + \delta_2 + \delta_3)^2.$$

Example 2.30   Find the linear and the quadratic Taylor series polynomial approximations to the function $f(x, y) = 2x^3 + 3y^3 - 4x^2 y$ about the point $(1, 2)$. Obtain the maximum absolute error in the region $|x - 1| < 0.01$ and $|y - 2| < 0.1$.

Solution   We have

$$f(x, y) = 2x^3 + 3y^3 - 4x^2y \quad ; \quad f(1, 2) = 18$$

$$f_x(x, y) = 6x^2 - 8xy \quad ; \quad f_x(1, 2) = -10$$

$$f_y(x, y) = 9y^2 - 4x^2 \quad ; \quad f_y(1, 2) = 32$$

$$f_{xx}(x, y) = 12x - 8y \quad ; \quad f_{xx}(1, 2) = -4$$

$$f_{xy}(x, y) = -8x \quad ; \quad f_{xy}(1, 2) = -8$$

$$f_{yy}(x, y) = 18y \quad ; \quad f_{yy}(1, 2) = 36$$

$$f_{xxx}(x, y) = 12, \, f_{xxy}(x, y) = -8, \quad f_{xyy}(x, y) = 0, \, f_{yyy}(x, y) = 18.$$

The linear approximation is given by

$$f(x, y) \approx f(1, 2) + [(x - 1)f_x(1, 2) + (y - 2) f_y(1, 2)]$$

$$= 18 + (x - 1)(-10) + (y - 2)(32) = 18 - 10(x - 1) + 32(y - 2).$$

Maximum absolute error in the linear approximation is given by

$$|R_1| \le \frac{B}{2} [\, | \, x - 1 | + | \, y - 2 | \,]^2 \le \frac{B}{2} [(0.01) + (0.1)]^2 = 0.00605 \, B$$

where B = max $[\,|f_{xx}|,|f_{xy}|,|f_{yy}|\,]$ in the given region $|x-1|<0.01$, $|y-2|<0.1$.

Now,   max $|f_{xx}|$ = max $|\,12\,x-8y\,|$ = max $|\,12(x-1)-8(y-2)-4\,|$

$$\leq \text{max }[\,12\,|x-1|+8\,|y-2|+4]=4.92$$

max $|f_{xy}|$ = max $|-8x\,|$ = max $|\,8(x-1)+8\,| \leq$ max $[8\,|x-1|+8]=8.08$

max $|f_{yy}|$ = max $|\,18y\,|$ = max $[\,18(y-2)+36] \leq$ max $[18\,|y-2|+36\,]=37.8$.

Hence, $|B|=37.8$ and $|R_1|\leq 0.00605(37.8)\approx 0.23$.

The quadratic approximation is given by

$$f(x,y)\approx f(1,2)+[(x-1)\,f_x\,(1,2)+(y-2)\,f_y\,(1,2)]$$

$$+\frac{1}{2}\,[(x-1)^2\,f_{xx}\,(1,2)+2(x-1)\,(y-2)\,f_{xy}\,(1,2)+(y-2)^2\,f_{yy}\,(1,2)]$$

$$=18-10(x-1)+32(y-2)+\frac{1}{2}\,[-4\,(x-1)^2-16(x-1)\,(y-2)+36\,(y-2)^2]$$

$$=18-10(x-1)+32(y-2)-2\,[(x-1)^2+4(x-1)\,(y-2)-9(y-2)^2].$$

Using Eq. (2.66), the maximum absolute error in the quadratic approximation is given by

$$|R_2|\leq \frac{B}{6}\,[\,|x-1|+|y-2|\,]^3 \leq \frac{B}{6}\,(0.11)^3=\frac{B}{6}\,(0.001331)$$

where B = max $[\,|f_{xxx}|,|f_{xxy}|,|f_{xyy}|,|f_{yyy}|\,]$ = max $[\,12,8,0,18\,]=18$.

Hence, we obtain

$$|R_2|\leq \frac{18}{6}\,(0.001331)\approx 0.004.$$

Example 2.31 Expand $f(x,y)=21+x-20y+4x^2+xy+6y^2$ in Taylor series of maximum order about the point $(-1,2)$.

Solution Since all the third order partial derivatives of $f(x,y)$ are zero, the maximum order of the Taylor series expansion of $f(x,y)$ about the point $(-1,2)$ is two. We obtain

$$f(x,y)=f(-1,2)+\left[(x+1)\frac{\partial}{\partial x}+(y-2)\frac{\partial}{\partial y}\right]f(-1,2)+\frac{1}{2!}\left[(x+1)\frac{\partial}{\partial x}+(y-2)\frac{\partial}{\partial y}\right]^2 f(-1,2).$$

We have

$$f(-1,2)=6,\ f_x(x,y)=1+8x+y,\ f_x(-1,2)=-5,$$

$$f_y(x,y)=-20+x+12y,\ f_y(-1,2)=3,$$

$$f_{xx}(x,y)=8, f_{xy}(x,y)=1,\ f_{yy}(x,y)=12.$$

Therefore,

$$f(x,y)=6-5(x+1)+3(y-2)+4(x+1)^2+(x+1)(y-2)+6(y-2)^2.$$

This is an rearrangement of the terms in the given function.

Example 2.32   The function $f(x, y) = x^2 - xy + y^2$ is approximated by a first degree Taylor's polynomial about the point (2, 3). Find a square $|x - 2| < \delta, |y - 3| < \delta$ with centre at (2, 3) such that the error of approximation is less than or equal to 0.1 in magnitude for all points within this square.

Solution   We have $f_x = 2x - y$, $f_y = 2y - x$, $f_{xx} = 2$, $f_{xy} = -1$, $f_{yy} = 2$.
The maximum absolute error in the first degree approximation is given by

$$|R_1| \le \frac{B}{2} [\,|x - 2| + |y - 3|\,]^2$$

where $B = \max [\,[\,|f_{xx}|, |f_{xy}|, |f_{yy}|\,]\,] = \max [2, 1, 2] = 2$.

We also have $|x - 2| < \delta, |y - 3| < \delta$. Therefore, we want to determine δ such that

$$|R_1| \le \frac{2}{2} [\delta + \delta]^2 < 0.1, \text{ or } 4\delta^2 < 0.1, \text{ or } \delta < \sqrt{0.025} \approx 0.1581.$$

Example 2.33   If $f(x, y) = \tan^{-1}(xy)$, find an approximate value of $f(1.1, 0.8)$ using the Taylor's series (i) linear approximation and (ii) quadratic approximation.

Solution   Let $(x_0, y_0) = (1.0, 1.0)$, $h = 0.1$, $k = -0.2$. Then $f(1.1, 0.8) = f(1 + 0.1, 1 - 0.2)$.

  (i) Using the Taylor series linear approximation, we have

$$f(1.1, 0.8) \approx f(1, 1) + \left(h\frac{\partial}{\partial x} + k\frac{\partial}{\partial y} \right) f(1, 1).$$

From $f(x, y) = \tan^{-1}(xy)$, we get

$$f(1, 1) = \tan^{-1}(1) = \pi/4 \approx 0.7854$$

$$f_x(x, y) = \frac{y}{1 + x^2 y^2}, \quad f_x(1, 1) = \frac{1}{2}, \quad f_y(x, y) = \frac{x}{1 + x^2 y^2}, \quad f_y(1, 1) = \frac{1}{2}.$$

Therefore,

$$f(1.1, 0.8) \approx 0.7854 + \left\{ \frac{1}{2}(0.1) + \frac{1}{2}(-0.2) \right\} = 0.7354.$$

  (ii) Using the Taylor series quadratic approximation, we have

$$f(1.1, 0.8) \approx f(1, 1) + (h f_x + k f_y)_{(1,1)} + \frac{1}{2} [h^2 f_{xx} + 2hk f_{xy} + k^2 f_{yy}]_{(1,1)}.$$

We have

$$f_{xx}(x, y) = -\frac{2xy^3}{(1 + x^2 y^2)^2}, \quad f_{xx}(1, 1) = -\frac{1}{2}; \quad f_{yy}(x, y) = -\frac{2x^3 y}{(1 + x^2 y^2)^2}, \quad f_{yy}(1, 1) = -\frac{1}{2}$$

$$f_{xy}(x, y) = \frac{(1 + x^2 y^2) - y(2x^2 y)}{(1 + x^2 y^2)^2} = \frac{1 - x^2 y^2}{(1 + x^2 y^2)^2}, \quad f_{xy}(1, 1) = 0.$$

Therefore, using the result of (i), we obtain

$$f(1.1, 0.8) \approx 0.7354 + \frac{1}{2} \left\{ (0.01) \left(-\frac{1}{2} \right) + 2(0.1)(-0.2)(0) + (0.04) \left(-\frac{1}{2} \right) \right\}$$

$$= 0.7354 - 0.0125 = 0.7229.$$

The exact value of $f(1.1, 0.8)$ to four decimal places is 0.7217. Thus, the accuracy increases as the order of approximation increases.

Exercises 2.3

Find all the partial derivatives of the specified order for the following functions at the given point:

1. $f(x, y) = [x - y]/[x + y]$, second order at (1, 1).

2. $f(x, y) = x \ln y$, third order at (2, 3).

3. $f(x, y) = \ln [(1/x) - (1/y)]$, second order at (1, 2).

4. $f(x, y) = e^x \ln y + (\cos y) \ln x$, third order at $(1, \pi/2)$.

5. $f(x, y) = e^{\sin(x/y)}$, second order at $(\pi/2, 1)$.

6. $f(x, y, z) = [x + y]/[x + z]$, second order at $(1, -1, 1)$.

7. $f(x, y, z) = e^{x^2 + y^2 + z^2}$, second order at $(-1, 1, -1)$.

8. $f(x, y, z) = \sin xy + \sin xz + \sin yz$, second order at $(1, \pi/2, \pi/2)$.

9. $f(x, y, z) = \dfrac{x}{y} + \dfrac{y}{z} + \dfrac{z}{x}$, second order at (1, 2, 3).

10. $f(x, y, z) = x^x y^y z^z$, $\dfrac{\partial^2 f}{\partial x \, \partial y}$ at any point $(x, y, z) \neq (0, 0, 0)$.

11. For the function
$$f(x, y) = \begin{cases} \dfrac{x^2 y(x - y)}{x^2 + y^2}, & (x, y) \neq (0, 0) \\ 0, & (x, y) = (0, 0) \end{cases}$$
show that $f_{xy} \neq f_{yx}$ at (0, 0).

12. Show that $f_{xy} = f_{yx}$ for all $(x, y) \neq (0, 0)$, when $f(x, y) = x^y$.

13. Show that $f_{xy} = f_{yx}$ for all $(x, y) \neq (0, 0)$, when $f(x, y) = \log [x + \sqrt{y^2 + x^2}]$.

14. Show that $f_{xyz} = f_{yzx}$ for all (x, y, z), when $f(x, y, z) = e^{xy} \sin z$.

15. Show that $f_{xyyz} = f_{yyxz}$ for all (x, y, z), when $f(x, y, z) = z^2 e^{x + y^2}$.

16. If $z = e^x \sin y + e^y \cos x$, where x and y are implicit functions of t defined by the equations $x^3 + x + e^t + t^2 + t - 1 = 0$ and $y t^3 + y^3 t + t + y = 0$, then find dz/dt at $t = 0$.

17. If x and y are defined as functions of u, v by the implicit equations $x^2 - y^2 + 2u^2 + 3v^2 - 1 = 0$ and $2x^2 - y^2 - u^2 + 4v^2 - 2 = 0$, then find $\partial x/\partial u$, $\partial y/\partial u$, $\partial^2 x/\partial u^2$ and $\partial^2 y/\partial u^2$.

18. If u and v are defined as functions of x and y by the implicit equations $4x^2 + 3y^2 - z^2 - u^2 + v^2 = 6$, $3x^2 - 2y^2 + z^2 + u^2 + 2v^2 = 14$, then find $(\partial u/\partial x)_{y,z}$ and $(\partial v/\partial y)_{x,z}$ at $x = 1$, $y = -1$, $z = 2$. Assume that $u > 0$, $v > 0$.

19. If $x \sqrt{1 - y^2} + y \sqrt{1 - x^2} = c$, c any constant, $|x| < 1$, $|y| < 1$, then find dy/dx and d^2y/dx^2.

20. Find dy/dx and d^2y/dx^2 at the point $(x, y) = (1, 1)$, for $e^y - e^x + xy = 1$.

21. If $z = u^v$, $u = (x/y)$, $v = xy$, then find $\partial^2 z/\partial x^2$.

22. If $u = \ln(1/r)$, $r = \sqrt{(x-a)^2 + (y-b)^2}$, then show that $u_{xx} + u_{yy} = 0$.

23. If $F = f(u, v)$, $u = y + ax$, $v = -y - ax$, a any constant, then show that $F_{xx} = a^2 F_{yy}$.

24. If $f(x, y) = x \log(y/x)$, $(x, y) \neq (0, 0)$, then show that $x^2 f_{xx} + 2xy f_{xy} + y^2 f_{yy} = 0$.

25. If $f(x, y) = y/(x^2 + y^2)$, $(x, y) \neq (0, 0)$, then show that $f_{xx} + f_{yy} = 0$.

26. Find α and β such that $u(x, y) = e^{\alpha x + \beta y}$ satisfies the equation $u_{xx} - 7u_{xy} + 12u_{yy} = 0$.

27. If $z = f(u, v)$, $u = x/(x^2 + y^2)$, $v = y/(x^2 + y^2)$, $(x, y) \neq (0, 0)$, then show that

$$z_{uu} + z_{vv} = (x^2 + y^2)^2 (z_{xx} + z_{yy}).$$

28. If $x = r \cos \theta$, $v = r \sin \theta$, then show that

(i) $\dfrac{\partial^2 \theta}{\partial x \, \partial y} = -\dfrac{\cos 2\theta}{r^2}$, (ii) $\dfrac{\partial^2 r}{\partial x \, \partial y} = -\dfrac{\sin 2\theta}{2r}$.

Using Euler's theorem, establish the following results.

29. If $u = \sin^{-1}\left(\dfrac{x^2 + y^2}{x + y}\right)$, then $x \dfrac{\partial u}{\partial x} + y \dfrac{\partial u}{\partial y} = \tan u$.

30. If $u = \log\left[\dfrac{\sqrt{x^2 + y^2}}{x}\right]$, then $x \dfrac{\partial u}{\partial x} + y \dfrac{\partial u}{\partial y} = 0$.

31. If $u = \sqrt{y^2 - x^2} \, \sin^{-1}\left(\dfrac{x}{y}\right)$, then $x \dfrac{\partial u}{\partial x} + y \dfrac{\partial u}{\partial y} = u$.

32. If $u = \dfrac{y^3 - x^3}{y^2 + x^2}$, then $x \dfrac{\partial u}{\partial x} + y \dfrac{\partial u}{\partial y} = u$ and $x^2 \dfrac{\partial^2 u}{\partial x^2} + 2xy \dfrac{\partial^2 u}{\partial x \, \partial y} + y^2 \dfrac{\partial^2 u}{\partial y^2} = 0$.

33. If $\tan u = \dfrac{x^3 + y^3}{x - y}$, then $x \dfrac{\partial u}{\partial x} + y \dfrac{\partial u}{\partial y} = \sin 2u$ and

$$x^2 \dfrac{\partial^2 u}{\partial x^2} + 2xy \dfrac{\partial^2 u}{\partial x \partial y} + y^2 \dfrac{\partial^2 u}{\partial y^2} = (1 - 4\sin^2 u)\sin 2u.$$

34. Obtain the Taylor's series expansion of the maximum order for the function

$f(x, y) = x^2 + 3y^2 - 9x - 9y + 26$ about the point $(2, 2)$.

35. Obtain the Taylor's linear approximation to the function $f(x, y) = 2x^2 - xy + y^2 + 3x - 4y + 1$ about the point $(-1, 1)$. Find the maximum error in the region $|x + 1| < 0.1$, $|y - 1| < 0.1$.

36. Obtain the first degree Taylor's series approximation to the function $f(x, y) = e^y \ln(x + y)$ about the point $(1, 0)$. Estimate the maximum absolute error over the rectangle $|x - 1| < 0.1$, $|y| < 0.1$.

37. Obtain the second order Taylor's series approximation to the function $f(x, y) = xy^2 + y \cos(x - y)$ about the point $(1, 1)$. Find the maximum absolute error in the region $|x - 1| < 0.05$, $|y - 1| < 0.1$.

38. Expand $f(x, y) = \sqrt{x + y}$ in Taylor's series upto second order terms about the point $(1, 3)$. Estimate the maximum absolute error in the region $|x - 1| < 0.2$, $|y - 3| < 0.1$.

39. Obtain the Taylor's series expansion, upto third degree terms, of the function $f(x, y) = e^{2x+y}$ about the point $(0, 0)$. Obtain the maximum error in the region $|x| < 0.1$, $|y| < 0.2$.

40. Expand $f(x, y) = \sin(x + 2y)$ in Taylor's series upto third order terms about the point $(0, 0)$. Find the maximum error over the rectangle $|x| < 0.1$, $|y| < 0.1$.

41. Expand $f(x, y) = \sin x \sin y$ in Taylor's series upto second order terms about the point $(\pi/4, \pi/4)$. Find the maximum error in the region $|x - \pi/4| < 0.1$, $|y - \pi/4| < 0.1$.

42. Expand $f(x, y, z) = \sqrt{x^2 + y^2 + z^2}$ in Taylor series upto first order terms about the point $(2, 2, 1)$. Obtain the maximum error in the region $|x - 2| < 0.1$, $|y - 2| < 0.1$, $|z - 1| < 0.1$.

43. Expand $f(x, y, z) = \sqrt{xy + yz + xz}$ in Taylor's series upto first order terms about the point $(1, 3, 3/2)$. Obtain the maximum error in the region $|x - 1| < 0.1$, $|y - 3| < 0.1$, $|z - 3/2| < 0.1$.

44. Expand $f(x, y, z) = e^z \sin(x + y)$ in Taylor's series upto second order terms about the point $(0, 0, 0)$. Obtain the maximum error in the region $|x| < 0.1$, $|y| < 0.1$, $|z| < 0.1$.

45. Expand $f(x, y, z) = e^x \sin(yz)$ in Taylor's series upto second order terms about the point $(0, 1, \pi/2)$. Obtain the maximum error in the region $|x| < 0.1$, $|y - 1| < 0.1$, $|z - \pi/2| < 0.1$.

2.5 Maximum and Minimum Values of a Function

Let a function $f(x, y)$ be defined and continuous in some closed and bounded region R. Let (a, b) be an interior point of R and $(a + h, b + k)$ be a point in its neighborhood and lies inside R. We define the following.

(i) The point (a, b) is called a point of *relative* (or *local*) *minimum*, if

$$f(a + h, b + k) \geq f(a, b) \tag{2.67a}$$

for all h, k. Then, $f(a, b)$ is called the *relative* (or *local*) *minimum* value.

(ii) The point (a, b) is called a point of *relative* (or *local*) *maximum*, if

$$f(a + h, b + k) \leq f(a, b) \tag{2.67b}$$

for all h, k. Then $f(a, b)$ is called the *relative* (or *local*) *maximum* value.

A function $f(x, y)$ may also attain its minimum or maximum values on the boundary of the region. The smallest and the largest values attained by a function over the entire region including the boundary are called the *absolute* (or *global*) *minimum* and *absolute* (or *global*) *maximum* values respectively.

The points at which minimum / maximum values of the function occur are also called *points of extrema* or the *stationary points* and the minimum and the maximum values taken together are called the *extreme values* of the function.

We now present the necessary conditions for the existence of an extremum of a functon.

Theorem 2.6 (Necessary conditions for a function to have an extremum) Let the function $f(x, y)$ be continuous and possess first order partial derivatives at a point $P(a, b)$. Then, the necessary conditions for the existence of an extreme value of f at the point P are $f_x(a, b) = 0$ and $f_y(a, b) = 0$.

Proof Let $(a + h, b + k)$ be a point in the neighborhood of the point $P(a, b)$. Then, P will be a point of maximum, if

$$\Delta f = f(a + h, b + k) - f(a, b) \leq 0 \quad \text{for all} \quad h, k \tag{2.68}$$

and a point of minimum, if

$$\Delta f = f(a + h, b + k) - f(a, b) \geq 0 \quad \text{for all} \quad h, k. \tag{2.69}$$

Using the Taylor's series expansion about the point (a, b), we obtain

$$f(a + h, b + k) = f(a, b) + (hf_x + kf_y)_{(a,b)} + \frac{1}{2}[h^2f_{xx} + 2hkf_{xy} + k^2f_{yy}]_{(a,b)} + \dots \quad (2.70)$$

Neglecting the second and higher order terms, we get

$$\Delta f \approx hf_x(a, b) + kf_y(a, b). \quad (2.71)$$

The sign of Δf in Eq. (2.71) depends on the sign of $hf_x(a, b) + kf_y(a, b)$ which is a function of h and k. Letting $h \to 0$, we find that Δf changes sign with k. Therefore, the function cannot have an extremum unless $f_y = 0$. Similarly, letting $k \to 0$, we find that the function f cannot have an extremum unless $f_x = 0$.

Therefore, the necessary conditions for the existence of an extremum at the point (a, b) is that

$$f_x(a, b) = 0 \quad \text{and} \quad f_y(a, b) = 0. \quad (2.72)$$

A point $P(a, b)$, where $f_x(a, b) = 0$ and $f_y(a, b) = 0$ is called a *critical point* or a *stationary point*. A point P is also called a critical point when one or both of the first order partial derivatives do not exist at this point.

Remark 10

To find the minimum/maximum values of a function f, we first find all the critical points. We then examine each critical point to decide whether at this point the function has a minimum value or a maximum value using the sufficient conditions.

Theorem 2.7 (Sufficient conditions for a function to have a minimum/maximum) Let a function $f(x, y)$ be continuous and possess first and second order partial derivatives at a point $P(a, b)$. If $P(a, b)$ is a critical point, then the point P is a point of

$$\text{relative minimum if } rt - s^2 > 0 \text{ and } r > 0 \quad (2.73a)$$

$$\text{relative maximum if } rt - s^2 > 0 \text{ and } r < 0 \quad (2.73b)$$

where $r = f_{xx}(a, b)$, $s = f_{xy}(a, b)$ and $t = f_{yy}(a, b)$.

No conclusion about an extremum can be drawn if $rt - s^2 = 0$ and further investigation is needed. If $rt - s^2 < 0$, then the function f has no minimum or maximum at this point. In this case, the point P is called a *saddle point*.

Proof Let $(a + h, b + k)$ be a point in the neighborhood of the point $P(a, b)$. Since P is a critical point, we have $f_x(a, b) = 0$, and $f_y(a, b) = 0$. Neglecting the third and higher order terms in the Taylor's series expansion of $f(a + h, b + k)$ about the point (a, b), we get

$$\Delta f = f(a + h, b + k) - f(a, b) \approx \frac{1}{2}[h^2f_{xx}(a, b) + 2hkf_{xy}(a, b) + k^2f_{yy}(a, b)]$$

$$= \frac{1}{2}[h^2r + 2hks + k^2t] = \frac{1}{2r}[h^2r^2 + 2hkrs + k^2rt]$$

$$= \frac{1}{2r}[(hr + ks)^2 + k^2(rt - s^2)]. \quad (2.74)$$

Since $(hr + ks)^2 > 0$, the sufficient condition for the expression $(hr + ks)^2 + k^2(rt - s^2)$ to be positive is that $rt - s^2 > 0$.

Hence, if $rt - s^2 > 0$, then

$$\Delta f > 0 \text{ if } r > 0 \quad \text{and} \quad \Delta f < 0 \text{ if } r < 0.$$

Therefore, a sufficient condition for the critical point $P(a, b)$ to be a point of relative minimum is $rt - s^2 > 0$ and $r > 0$

point of relative maximum is $rt - s^2 > 0$ and $r < 0$.

If $rt - s^2 < 0$, then the sign of Δf in Eq. (2.74) depends on h and k. Hence, no maximum/minimum of f can occur at $P(a, b)$ in this case.

If $rt - s^2 = 0$ or $r = t = s = 0$, no conclusion can be drawn and the terms involving higher order partial derivatives must be considered.

Remark 11

(a) We can also write Eq. (2.74) as

$$\Delta f = \frac{1}{2t} \, [k^2 t^2 + 2hkst + h^2 rt] = \frac{1}{2t} \, [(kt + hs)^2 + (rt - s^2)h^2].$$

Hence, a sufficient condition for a critical point $P(a, b)$ to be a

point of relative minimum is $rt - s^2 > 0$ and $t > 0$

point of relative maximum is $rt - s^2 > 0$ and $t < 0$.

From these conditions and Eqs. (2.73a, 2.73b), we find that when an extremum exists, then $rt - s^2 > 0$, and both r and t have the same sign either positive or negative.

(b) Alternate statement of Theorem 2.7

A real symmetric matrix $\mathbf{A} = (a_{ij})$ is called a positive definite matrix, if

$$\mathbf{x}^T \mathbf{A} \, \mathbf{x} > 0 \text{ for all real vectors } \mathbf{x} \neq \mathbf{0}$$

or $\qquad \sum\limits_{i=1}^{n} \sum\limits_{j=1}^{n} a_{ij} \, x_i \, x_j > 0$ for all x_i, x_j (see section 3.5.3).

A sufficient condition for the matrix $\mathbf{A}$ to be positive definite is that the minors of all its leading submatrices are positive. Now we state the result. Let

$$\mathbf{A} = \begin{bmatrix} r & s \\ s & t \end{bmatrix}$$

where $r = f_{xx}(a, b)$, $s = f_{xy}(a, b) = f_{yx}(a, b)$ and $t = f_{yy}(a, b)$. Then, the function $f(x, y)$ has a relative minimum at a critical point $P(a, b)$, if the matrix $\mathbf{A}$ is positive definite. Since all the leading minors of $\mathbf{A}$ are positive, we obtain the conditions $r > 0$ and $rt - s^2 > 0$.

The function $f(x, y)$ has a relative maximum at $P(a, b)$, if the matrix $\mathbf{B} = -\mathbf{A} = \begin{bmatrix} -r & -s \\ -s & -t \end{bmatrix}$ is

positive definite. Since all the leading minors of $\mathbf{B}$ are positive, we obtain the conditions $-r > 0$ and $rt - s^2 > 0$, that is $r < 0$ and $rt - s^2 > 0$.

This alternative statement of the Theorem 2.7 is useful when we consider the extreme values of the functions of three or more variables. For example, for the function $f(x, y, z)$ of three variables, we have

$$\mathbf{A} = \begin{bmatrix} f_{xx} & f_{xy} & f_{xz} \\ f_{yx} & f_{yy} & f_{yz} \\ f_{zx} & f_{zy} & f_{zz} \end{bmatrix}$$

where $f_{yx} = f_{xy}, f_{zx} = f_{xz}, f_{zy} = f_{yz}$. The matrix $\mathbf{A}$ or the matrix $\mathbf{B} = -\mathbf{A}$ can be tested whether it is positive definite, to find the points of minimum/maximum. Therefore, a critical point (a point at which $f_x = 0 = f_y = f_z$)

   (i) is a point of relative minimum if $\mathbf{A}$ is positive definite and f_{xx}, f_{yy}, f_{zz} are all positive.

   (ii) is a point of relative maximum if $\mathbf{B} = -\mathbf{A}$ is positive definite (that is, the leading minors of $\mathbf{A}$ are alternately negative and positive) and f_{xx}, f_{yy}, f_{zz} are all negative.

Example 2.34 Find the relative maximum and minimum values of the function

$$f(x, y) = 2(x^2 - y^2) - x^4 + y^4.$$

Solution   We have

$$f_x = 4x - 4x^3 = 0, \text{ or } x = 0, \pm 1$$

$$f_y = -4y + 4y^3 = 0, \text{ or } y = 0, \pm 1.$$

Hence, $(0, 0), (0, \pm 1), (\pm 1, 0), (\pm 1, \pm 1)$ are the critical points. We find that

$$r = f_{xx} = 4 - 12x^2, \ s = f_{xy} = 0, \ t = f_{yy} = -4 + 12y^2$$

and $\qquad\qquad rt - s^2 = -16(1 - 3x^2)(1 - 3y^2).$

At the points $(0, 1)$ and $(0, -1)$, we have $rt - s^2 = 32 > 0$ and $r = 4 > 0$. Therefore, the points $(0, 1)$ and $(0, -1)$ are points of relative minimum and the minimum value at each point is -1.

At the points $(-1, 0)$ and $(1, 0)$, we have $rt - s^2 = 32 > 0$ and $r = -8 < 0$. The points $(-1, 0)$, $(1, 0)$ are points of relative maximum and the maximum value at each point is 1.

At $(0, 0)$, we have $rt - s^2 = -16 < 0$. At $(\pm 1, \pm 1)$, we have $rt - s^2 = -64 < 0$. Hence, the points $(0, 0)$ $(\pm 1, \pm 1)$ are neither the points of maximum nor minimum.

Example 2.35 Find the absolute maximum and minimum values of

$$f(x, y) = 4x^2 + 9y^2 - 8x - 12y + 4$$

over the rectangle in the first quadrant bounded by the lines $x = 2, y = 3$ and the coordinate axes.

Solution   The function f can attain maximum/minimum values at the critical points or on the boundary of the rectangle $OABC$ (Fig. 2.4).

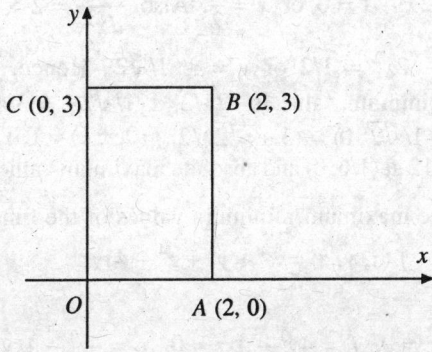

Fig. 2.4. Region in Example 2.35.

We have $f_x = 8x - 8 = 0$, $f_y = 18y - 12 = 0$. The critical point is $(x, y) = (1, 2/3)$. Now,

$$r = f_{xx} = 8, \quad s = f_{xy} = 0, \quad t = f_{yy} = 18, \quad rt - s^2 = 144.$$

Since $rt - s^2 > 0$ and $r > 0$, the point $(1, 2/3)$ is a point of relative minimum. The minimum value is $f(1, 2/3) = -4$.

On the boundary line OA, we have $y = 0$ and $f(x, y) = f(x, 0) = g(x) = 4x^2 - 8x + 4$, which is a function of one variable. Setting $dg/dx = 0$, we get $8x - 8 = 0$ or $x = 1$. Now $d^2g/dx^2 = 8 > 0$. Therefore, at $x = 1$, the function has a minimum. The minimum value is $g(1) = 0$. Also, at the corners $(0, 0)$, $(2, 0)$, we have $f(0, 0) = g(0) = 4$, $f(2, 0) = g(2) = 4$.

Similarly, along the other boundary lines, we have the following results:

$x = 2$: $h(y) = 9y^2 - 12y + 4$; $dh/dy = 18y - 12 = 0$ gives $y = 2/3$; $d^2h/dy^2 = 18 > 0$. Therefore, $y = 2/3$ is a point of minimum. The minimum value is $f(2, 2/3) = 0$. At the corner $(2, 3)$, we have $f(2, 3) = 49$.

$y = 3$: $g(x) = 4x^2 - 8x + 49$; $dg/dx = 8x - 8 = 0$ gives $x = 1$; $d^2g/dx^2 = 8 > 0$. Therefore, $x = 1$ is a point of minimum. The minimum value is $f(1, 3) = 45$. At the corner point $(0, 3)$, we have $f(0, 3) = 49$.

$x = 0$: $h(y) = 9y^2 - 12y + 4$, which is the same case as for $x = 2$.

Therefore, the absolute minimum value is -4 which occurs at $(1, 2/3)$ and the absolute maximum value is 49 which occurs at the points $(2, 3)$ and $(0, 3)$.

Example 2.36   Find the absolute maximum and minimum values of the function

$$f(x, y) = 3x^2 + y^2 - x \text{ over the region } 2x^2 + y^2 \le 1.$$

Solution   We have $f_x = 6x - 1 = 0$ and $f_y = 2y = 0$. Therefore, the critical point is $(x, y) = (1/6, 0)$.

Now, $\quad r = f_{xx} = 6, \quad s = f_{xy} = 0, \quad t = f_{yy} = 2, \quad rt - s^2 = 12 > 0.$

Therefore, $(1/6, 0)$ is a point of minimum. The minimum value at this point is $f(1/6, 0) = -1/12$.

On the boundary, we have $y^2 = 1 - 2x^2$, $-1/\sqrt{2} \le x \le 1/\sqrt{2}$. Substituting in $f(x, y)$, we obtain

$$f(x, y) = 3x^2 + (1 - 2x^2) - x = 1 - x + x^2 = g(x)$$

which is a function of one variable. Setting $dg/dx = 0$, we get

$$\frac{dg}{dx} = 2x - 1 = 0, \text{ or } x = \frac{1}{2}. \text{ Also } \frac{d^2g}{dx^2} = 2 > 0.$$

For $x = 1/2$, we get $y^2 = 1 - 2x^2 = 1/2$ or $y = \pm 1/\sqrt{2}$. Hence, the points $(1/2, \pm 1/\sqrt{2})$ are points of minimum. The minimum value is $f(1/2, \pm 1/\sqrt{2}) = 3/4$. At the vertices, we have $f(1/\sqrt{2}, 0) = (3 - \sqrt{2})/2$, $f(-1/\sqrt{2}, 0) = (3 + \sqrt{2})/2$, $f(0, \pm 1) = 1$. Therefore, the given function has absolute minimum value $-1/12$ at $(1/6, 0)$ and absolute maximum value $(3 + \sqrt{2})/2$ at $(-1/\sqrt{2}, 0)$.

Example 2.37   Find the relative maximum/minimum values of the function

$$f(x, y, z) = x^4 + y^4 + z^4 - 4xyz.$$

Solution   We have

$$f_x = 4x^3 - 4yz = 0, \quad f_y = 4y^3 - 4xz = 0, \quad f_z = 4z^3 - 4xy = 0.$$

Therefore, $\quad x^3 = yz, \quad y^3 = xz, \quad z^3 = xy \quad$ or $\quad x^3y^3z^3 = x^2y^2z^2 \quad$ or $\quad x^2y^2z^2 (xyz - 1) = 0.$

Therefore, all points which satisfy $xyz = 0$ or $xyz = 1$ are critical points. The solutions of these equations are $(0, 0, 0)$, $(1, 1, 1)$, $(\pm1, \pm1, 1)$, $(1, \pm1, \pm1)$, $(\pm1, 1, \pm1)$ with the same sign taken for the two coordinates. Now,

$$f_{xx} = 12x^2, \ f_{yy} = 12y^2, \ f_{zz} = 12z^2, \ f_{xy} = -4z, \ f_{xz} = -4y, \ f_{yz} = -4x.$$

At $(0, 0, 0)$, all the second order partial derivatives are zero. Therefore, no conclusion can be drawn.

We have
$$\mathbf{A} = \begin{bmatrix} f_{xx} & f_{xy} & f_{xz} \\ f_{yx} & f_{yy} & f_{yz} \\ f_{zx} & f_{zy} & f_{zz} \end{bmatrix} = \begin{bmatrix} 12x^2 & -4z & -4y \\ -4z & 12y^2 & -4x \\ -4y & -4x & 12z^2 \end{bmatrix}$$

Depending on whether $\mathbf{A}$ or $\mathbf{B} = -\mathbf{A}$ is positive definite, we can decide the points of minimum or maximum. The leading minors are

$$M_1 = 12x^2, \ M_2 = \begin{vmatrix} 12x^2 & -4z \\ -4z & 12y^2 \end{vmatrix} = 16\,(9x^2y^2 - z^2)$$

and
$$M_3 = |\mathbf{A}| = 192x^2\,(9y^2z^2 - x^2) - 192z^4 - 64xyz - 64xyz - 192\,y^4$$
$$= 192\,[9x^2y^2z^2 - (x^4 + y^4 + z^4)] - 128xyz.$$

At all points $(1, 1, 1)$, $(\pm1, \pm1, 1)$, $(\pm1, 1, \pm1)$, $(1, \pm1, \pm1)$ with the same sign taken for two coordinates, we find that $M_1 > 0$, $M_2 > 0$ and $M_3 > 0$. Hence, $\mathbf{A}$ is a positive definite matrix and the given function has relative minimum at all these points, since $f_{xx} > 0$, $f_{yy} > 0$, and $f_{zz} > 0$. The relative minimum value at all these points is same and is given by $f(1, 1, 1) = -1$.

Conditional maximum/minimum

In many practical problems, we need to find the maximum/minimum value of a function $f(x_1, x_2, \ldots, x_n)$ when the variables are not independent but are connected by one or more constraints of the form

$$\phi_i(x_1, x_2, \ldots, x_n) = 0, \ i = 1, 2, \ldots, k$$

where generally $n > k$. We present the Lagrange method of multipliers to find the solution of such problems.

2.5.1 Lagrange Method of Multipliers

We want to find the extremum of the function $f(x_1, x_2, \ldots, x_n)$ under the conditions
$$\phi_i(x_1, x_2, \ldots, x_n) = 0, \ i = 1, 2, \ldots, k. \tag{2.75}$$
We construct an auxiliary function of the form

$$F(x_1, x_2, \ldots, x_n, \lambda_1, \lambda_2, \ldots, \lambda_k) = f(x_1, x_2, \ldots, x_n) + \sum_{i=1}^{k} \lambda_i\, \phi_i(x_1, x_2, \ldots, x_n) \tag{2.76}$$

where λ_i's are undetermined parameters and are known as *Lagrange multipliers*. Then, to determine the stationary points of F, we have the necessary conditions

$$\frac{\partial F}{\partial x_1} = 0 = \frac{\partial F}{\partial x_2} = \ldots = \frac{\partial F}{\partial x_n}$$

which give the equations

$$\frac{\partial f}{\partial x_j} + \sum_{i=1}^{k} \lambda_i \frac{\partial \phi_i}{\partial x_j} = 0, \ j = 1, 2, \ldots, n. \tag{2.77}$$

From Eqs. (2.75) and (2.77), we obtain $(n + k)$ equations in $(n + k)$ unknowns $x_1, x_2, \ldots, x_n$, $\lambda_1, \lambda_2, \ldots, \lambda_k$. Solving these equations, we obtain the required stationary points $(x_1, x_2, \ldots, x_n)$ at which the function f has an extremum. Further investigation is needed to determine the exact nature of these points.

Example 2.38 Find the minimum value of $x^2 + y^2 + z^2$ subject to the condition $xyz = a^3$.

Solution Consider the auxiliary function

$$F(x, y, z, \lambda) = x^2 + y^2 + z^2 + \lambda (xyz - a^3) .$$

We obtain the necessary conditions for extremum as

$$\frac{\partial F}{\partial x} = 2x + \lambda yz = 0, \quad \frac{\partial F}{\partial y} = 2y + \lambda xz = 0, \quad \frac{\partial F}{\partial z} = 2z + \lambda xy = 0.$$

From these equations, we obtain

$$\lambda yz = -2x \text{ or } \lambda xyz = -2x^2$$

$$\lambda xz = -2y \text{ or } \lambda xyz = -2y^2$$

$$\lambda xy = -2z \text{ or } \lambda xyz = -2z^2.$$

Therefore, $x^2 = y^2 = z^2$. Using the condition $xyz = a^3$, we obtain the solutions as (a, a, a), $(a, -a, -a)$, $(-a, a, -a)$ and $(-a, -a, a)$. At each of these points, the value of the given function is $x^2 + y^2 + z^2 = 3a^2$.

Now,   the arithmetic mean of x^2, y^2, z^2 is $AM = (x^2 + y^2 + z^2)/3$

the geometric mean of x^2, y^2, z^2 is $GM = (x^2 y^2 z^2)^{1/3} = a^2$.

Since, $AM \geq GM$, we obtain $x^2 + y^2 + z^2 \geq 3a^2$.

Hence, all the above points are the points of constrained minimum and the minimum value of $x^2 + y^2 + z^2$ is $3a^2$.

Example 2.39 Find the extreme values of $f(x, y, z) = 2x + 3y + z$ such that $x^2 + y^2 = 5$ and $x + z = 1$.

Solution Consider the auxiliary function

$$F(x, y, z, \lambda_1, \lambda_2) = 2x + 3y + z + \lambda_1 (x^2 + y^2 - 5) + \lambda_2 (x + z - 1).$$

For the extremum, we have the necessary conditions

$$\frac{\partial F}{\partial x} = 2 + 2\lambda_1 x + \lambda_2 = 0; \quad \frac{\partial F}{\partial y} = 3 + 2\lambda_1 y = 0, \quad \frac{\partial F}{\partial z} = 1 + \lambda_2 = 0.$$

From these equations, we get

$$\lambda_2 = -1, \ 3 + 2\lambda_1 y = 0 \text{ and } 1 + 2\lambda_1 x = 0$$

or $$x = -1/(2\lambda_1) \text{ and } y = -3/(2\lambda_1).$$

Substituting in the constraint $x^2 + y^2 = 5$, we get

$$\frac{1}{4\lambda_1^2} + \frac{9}{4\lambda_1^2} = 5 \quad \text{or} \quad \lambda_1^2 = \frac{1}{2} \quad \text{or} \quad \lambda_1 = \pm \frac{1}{\sqrt{2}}.$$

For $\lambda_1 = 1/\sqrt{2}$, we get $x = -\sqrt{2}/2, \ y = -3\sqrt{2}/2, \ z = 1 - x = (2 + \sqrt{2})/2$

and $f(x, y, z) = -\sqrt{2} - \dfrac{9\sqrt{2}}{2} + \dfrac{2 + \sqrt{2}}{2} = \dfrac{2 - 10\sqrt{2}}{2} = 1 - 5\sqrt{2}.$

For $\lambda_1 = -1/\sqrt{2}$, we get $x = \sqrt{2}/2, \ y = 3\sqrt{2}/2, \ z = 1 - x = (2 - \sqrt{2})/2$

and $f(x, y, z) = \sqrt{2} + \dfrac{9\sqrt{2}}{2} + \dfrac{2 - \sqrt{2}}{2} = \dfrac{2 + 10\sqrt{2}}{2} = 1 + 5\sqrt{2}.$

Example 2.40 Find the shortest distance between the line $y = 10 - 2x$ and the ellipse $(x^2/4) + (y^2/9) = 1$.

Solution Let (x, y) be a point on the ellipse and (u, v) be a point on the line. Then, the shortest distance between the line and the ellipse is the square root of the minimum value of

$$f(x, y, u, v) = (x - u)^2 + (y - v)^2$$

subject to the constraints

$$\phi_1(x, y) = \frac{x^2}{4} + \frac{y^2}{9} - 1 = 0 \quad \text{and} \quad \phi_2(u, v) = 2u + v - 10 = 0.$$

We define the auxiliary function as

$$F(x, y, u, v, \lambda_1, \lambda_2) = (x - u)^2 + (y - v)^2 + \lambda_1 \left(\frac{x^2}{4} + \frac{y^2}{9} - 1 \right) + \lambda_2 (2u + v - 10).$$

For extremum, we have the necessary conditions

$$\frac{\partial F}{\partial x} = 2(x - u) + \frac{x}{2}\lambda_1 = 0, \quad \text{or} \quad \lambda_1 x = 4(u - x)$$

$$\frac{\partial F}{\partial y} = 2(y - v) + \frac{2y}{9}\lambda_1 = 0, \quad \text{or} \quad \lambda_1 y = 9(v - y)$$

$$\frac{\partial F}{\partial u} = -2(x - u) + 2\lambda_2 = 0, \quad \text{or} \quad \lambda_2 = x - u$$

$$\frac{\partial F}{\partial v} = -2(y - v) + \lambda_2 = 0, \quad \text{or} \quad \lambda_2 = 2(y - v).$$

Eliminating λ_1 and λ_2 from the above equations, we get

$$4(u - x)y = 9(v - y)x \quad \text{and} \quad x - u = 2(y - v).$$

Dividing the two equations, we obtain $8y = 9x$. Substituting in the equation of the ellipse, we get

$$\frac{x^2}{4} + \frac{9x^2}{64} = 1, \quad \text{or} \quad x^2 = \frac{64}{25}.$$

Therefore, $x = \pm 8/5$ and $y = \pm 9/5$. Corresponding to $x = 8/5, \ y = 9/5$, we get

$$\frac{8}{5} - u = 2\left(\frac{9}{5} - v\right), \quad \text{or} \quad 2v - u = 2, \quad \text{or} \quad u = 2v - 2.$$

Substituting in the equation of the line $2u + v - 10 = 0$, we get $u = 18/5$ and $v = 14/5$.

Hence, an extremum is obtained when $(x, y) = (8/5, 9/5)$ and $(u, v) = (18/5, 14/5)$. The distance between the two points is $\sqrt{5}$.

Corresponding to $x = -8/5$, $y = -9/5$, we get $u - 2v = 2$. Substituting in the equation $2u + v - 10 = 0$, we obtain $u = 22/5$, $v = 6/5$. Hence, another extremum is obtained when $(x, y) = (-8/5, -9/5)$ and $(u, v) = (22/5, 6/5)$. The distance between these two points is $3\sqrt{5}$.

Hence, the shortest distance between the line and the ellipse is $\sqrt{5}$.

Exercise 2.4

Test the following functions for relative maximum and minimum.

1. $xy + (9/x) + (3/y)$.
2. $\sqrt{a^2 - x^2 - y^2}\ a > 0$.
3. $x^2 + 2bxy + y^2$.
4. $x^2 + xy + y^2 + (1/x) + (1/y)$.
5. $x^2 + 2/(x^2y) + y^2$.
6. $\cos 2x + \cos y + \cos (2x + y)$, $0 < x, y < \pi$.
7. $4x^2 + 4y^2 - z^2 + 12xy - 6y + z$.
8. $18xz - 6xy - 9x^2 - 2y^2 - 54z^2$.
9. $x^4 + y^4 + z^4 + 4xyz$.
10. $2 \ln(x + y + z) - (x^2 + y^2 + z^2)$, $x + y + z > 0$.

Find the relative and absolute maximum and minimum values for the following functions in the given closed region R in problemes **11** to **20**.

11. $x^2 - y^2 - 2y$, $R: x^2 + y^2 \le 1$.
12. xy, $R: x^2 + y^2 \le 1$.
13. $x + y$, $R: 4x^2 + 9y^2 \le 36$.
14. $4x^2 + y^2 - 2x + 1$, $R: 2x^2 + y^2 \le 1$.
15. $x^2 + y^2 - x - y + 1$, R: rectangular region; $0 \le x \le 2, 0 \le y \le 2$.
16. $2x^2 + y^2 - 2x - 2y - 4$, R: triangular region bounded by the lines $x = 0$, $y = 0$ and $2x + y = 1$.
17. $x^3 + y^3 - xy$, R: triangular region bounded by the lines $x = 1$, $y = 0$ and $y = 2x$.
18. $4x^2 + 2y^2 + 4xy - 10x - 2y - 3$, R: rectangular region; $0 \le x \le 3, -4 \le y \le 2$.
19. $\cos x + \cos y + \cos (x + y)$, R: rectangular region; $0 \le x \le \pi, 0 \le y \le \pi$.
20. $\cos x \cos y \cos (x + y)$, R: rectangular region; $0 \le x \le \pi, 0 \le y \le \pi$.
21. Show that the necessary condition for the existence of an extreme value of $f(x, y)$ such that $\phi(x, y) = 0$ is that x, y satisfy the equation $f_x\ \phi_y - f_y\ \phi_x = 0$.
22. Find the smallest and the largest value of xy on the line segment $x + 2y = 2$, $x \ge 0$, $y \ge 0$.
23. Find the smallest and the largest value of $x + 2y$ on the circle $x^2 + y^2 = 1$.
24. Find the smallest and the largest value of $2x - y$ on the curve $x - \sin y = 0$, $0 \le y \le 2\pi$.
25. Find the extreme value of $x^2 + y^2$ when $x^4 + y^4 = 1$.
26. Find the points on the curve $x^2 + xy + y^2 = 16$, which are nearest and farthest from the origin.
27. Find the rectangle of constant perimeter whose diagonal is maximum.
28. Find the triangle whose perimeter is constant and has largest area.
29. Find a point on the plane $Ax + By + cz = D$ which is nearest to origin.
30. Find the extreme value of xyz, when $x + y + z = a$, $a > 0$.

31. Find the extreme value of $a^3x^2 + b^3y^2 + c^3z^2$ such that $x^{-1} + y^{-1} + z^{-1} = 1$, where $a > 0$, $b > 0$, $c > 0$.

32. Find the extreme value of $x^p + y^p + z^p$ on the surface $x^q + y^q + z^q = 1$, where $0 < p < q$, $x > 0$, $y > 0$, $z > 0$.

33. Find the extreme value of $x^3 + 8y^3 + 64z^3$, when $xyz = 1$.

34. Find the dimensions of a rectangular parallelopiped of maximum volume with edges parallel to the coordinate axes that can be inscribed in the ellipsoid $(x^2/a^2) + (y^2/b^2) + (z^2/c^2) = 1$.

35. Divide a number into three parts such that the product of the first, square of the second and cube of the third is maximum.

36. Find the dimensions of a rectangular parallelopiped of fixed total edge length with maximum surface area.

37. Find the dimensions of a rectangular parallelopiped of greatest volume having constant surface area S.

38. A rectangular box without top is to have a given volume. How should the box be made so as to use the least material.

39. Find the dimensions of a right circular cone of fixed lateral area with minimum volume.

40. A tent is to be made in the form of a right circular cylinder surmounted by a cone. Find the ratios of the height H of the cylinder and the height h of the conical part to the radius r of the base, if the volume V of the tent is maximum for a given surface area S of the tent.

41. Find the maximum value of xyz under the constraints $x^2 + z^2 = 1$ and $y - x = 0$.

42. Find the extreme value of $x^2 + 2xy + z^2$ under the constraints $2x + y = 0$ and $x + y + z = 1$.

43. Find the extreme value of $x^2 + y^2 + z^2 + xy + xz + yz$ under the constraints $x + y + z = 1$ and $x + 2y + 3z = 3$.

44. Find the points on the ellipse obtained by the intersection of the plane $x + z = 1$ and the ellipsoid $x^2 + y^2 + 2z^2 = 1$ which are nearest and farthest from the origin.

45. Find the smallest and the largest distance between the points P and Q such that P lies on the plane $x + y + z = 2a$ and Q lies on the sphere $x^2 + y^2 + z^2 = a^2$, where a is any constant.

2.6 Multiple Integrals

In the previous chapter, we studied methods for evaluating the definite integral $\int_a^b f(x)dx$, where the integrand $f(x)$ is piecewise continuous on the interval $[a, b]$. In this section, we shall discuss methods for evaluating the double and triple integrals, that is integrals of the forms

$$\iint_R f(x, y)dx\, dy \quad \text{and} \quad \iiint_T f(x, y, z)dx\, dy\, dz.$$

We assume that the integrand f is continuous at all points inside and on the boundary of the region R or T. These integrals are called *multiple integrals*. The multiple integral over $\mathbb{R}^n$ is written as

$$\iint_R \cdots \int f(x_1, x_2, \ldots, x_n)dx_1 dx_2 \ldots dx_n.$$

2.6.1 Double Integrals

Let $f(x, y)$ be a continuous function in a simply connected, closed and bounded region R in a two dimensional space $\mathbb{R}^2$, bounded by a simple closed curve C (Fig. 2.5).

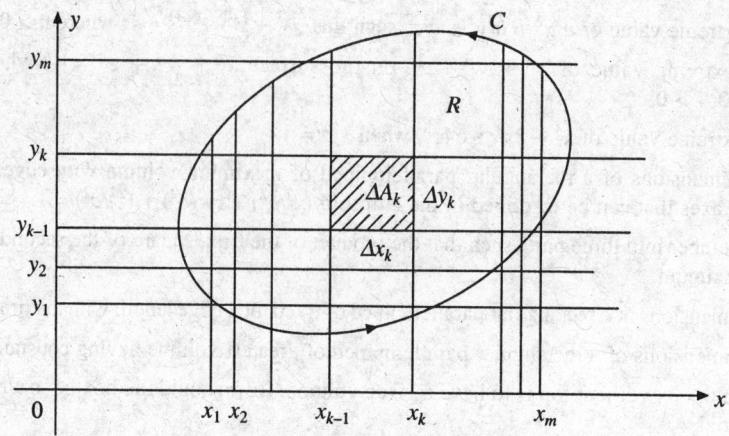

Fig. 2.5.   Region R for double integral.

Subdivide the region R by drawing lines $x = x_k$, $y = y_k$, $k = 1, 2, \ldots, m$, parallel to the coordinate axes. Number the rectangles which are inside R from 1 to n. In each such rectangle, take an arbitrary point, say (ξ_k, η_k) in the kth rectangle and form the sum

$$J_n = \sum_{k=1}^{n} f(\xi_k, \eta_k) \Delta A_k$$

where $\Delta A_k = \Delta x_k \, \Delta y_k$ is the area of the kth rectangle and $d_k = \sqrt{(\Delta x_k)^2 + (\Delta y_x)^2}$ is the length of the diagonal of this rectangle. The maximum length of the diagonal, that is max d_k of the subdivisions is also called the *norm* of the subdivision. For different values of n, say $n_1, n_2, \ldots, n_m, \ldots$, we obtain a sequence of sums $J_{n_1}, J_{n_2}, \ldots, J_{n_m}, \ldots$. Let $n \to \infty$, such that the length of the largest diagonal $d_k \to 0$. If $\lim\limits_{n \to \infty} J_n$ exists, independent of the choice of the subdivision and the point (ξ_k, η_k), then we say that $f(x, y)$ is integrable over R. This limit is called the *double integral* of $f(x, y)$ over R and is denoted by

$$J = \iint\limits_{R} f(x, y) dx \, dy. \tag{2.78}$$

Evaluation of double integrals by two successive integrations

A double integral can be evaluated by two successive integrations. We evaluate it with respect to one variable (treating the other variable as constant) and reduce it to an integral of one variable. Thus, there are two possible ways to evaluate a double integral, which are the following:

$$J = \iint\limits_{R} f(x, y) dy \, dx = \iint\limits_{R} [f(x, y) dy] \, dx: \quad \text{first integrate with respect to } y \text{ and then integrate with respect to } x.$$

or $\quad J = \iint\limits_{R} f(x, y) dx \, dy = \iint\limits_{R} [f(x, y) dx] \, dy: \quad \text{first integrate with respect to } x \text{ and then integrate with respect to } y.$

Let f be a continuous function over R. We consider the following cases.

Case 1 Let the region R be expressed in the form

$$R = \{(x, y) : \phi(x) \leq y \leq \psi(x), a \leq x \leq b\} \tag{2.79}$$

where $\phi(x)$ and $\psi(x)$ are integrable functions, such that $\phi(x) \leq \psi(x)$ for all x in $[a, b]$. We write (Fig. 2.6)

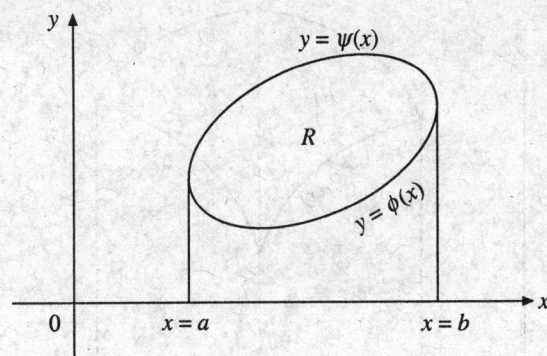

Fig. 2.6. Region of integration.

$$J = \int_{x=a}^{b} \left[\int_{y=\phi(x)}^{\psi(x)} f(x, y) dy \right] dx. \tag{2.80}$$

While evaluating the inner integral, x is treated as constant.

Case 2 Let the region R be expressed in the form

$$R = \{(x, y) : g(y) \leq x \leq h(y), c \leq y \leq d\} \tag{2.81}$$

where $g(y)$ and $h(y)$ are integrable functions, such that $g(y) \leq h(y)$ for all y in $[c, d]$. We write (Fig. 2.7)

$$J = \int_{y=c}^{d} \left[\int_{x=g(y)}^{h(y)} f(x, y) dx \right] dy. \tag{2.82}$$

While evaluating the inner integral, y is treated as constant.

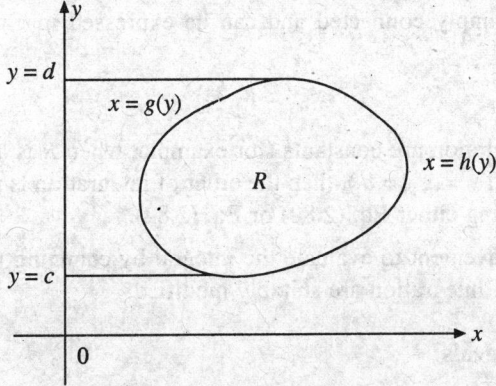

Fig. 2.7. Region of integration.

Often, the region R may be such that it cannot be represented in either of the forms given in Eqs. (2.79) or (2.81). In such cases, the region R can be subdivided such that each of these can be expressed in either of the forms given in Eqs. (2.79) or (2.81). For example, R may be expressed as shown in Fig. 2.8 and we write $R = R_1 \cup R_2$ where R_1, R_2 have no common interior points.

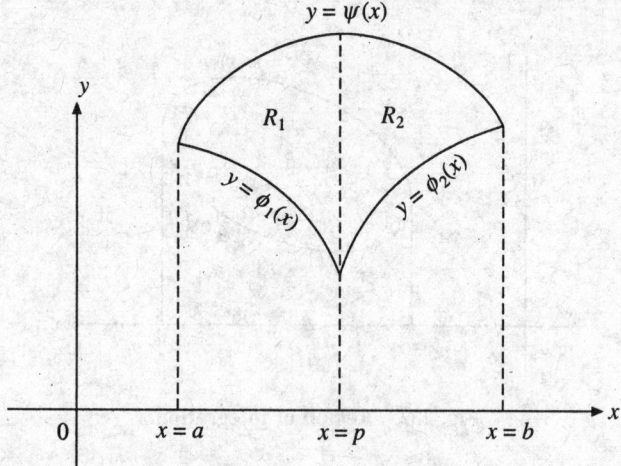

Fig. 2.8.   Region of integration.

Then, we have

$$\iint\limits_{R} f(x, y)dy\,dx = \iint\limits_{R_1} f(x, y)dy\,dx + \iint\limits_{R_2} f(x, y)dy\,dx$$

$$= \int_a^p \left[\int_{\phi_1(x)}^{\psi(x)} f(x, y)\,dy \right] dx + \int_p^b \left[\int_{\phi_2(x)}^{\psi(x)} f(x, y)dy \right] dx. \tag{2.83}$$

In the general case, the region R may be subdivided into a number of parts so that

$$\iint\limits_{R} f(x, y)dy\,dx = \sum_{i=1}^{m} \left[\iint\limits_{R_i} f(x, y)\,dy\,dx \right] \tag{2.84}$$

where each region R_i is simply connected and can be expressed in either of the forms given in Eqs. (2.79) or (2.81).

Remark 12

(a) If the limits of integration are constants (for example, when R is a rectangle bounded by the lines $x = a$, $x = b$ and $y = c$, $y = d$), then the order of integration is not important. The integral can be evaluated using either Eq. (2.80) or Eq. (2.82).

(b) Sometimes, it is convenient to evaluate the integral by changing the order of integration. In such cases, limits of integration are suitably modified.

Properties of double integrals

1. If $f(x, y)$ and $g(x, y)$ are integrable functions, then

$$\iint\limits_{R} [f(x, y) \pm g(x, y)]dx\,dy = \iint\limits_{R} f(x, y)dx\,dy \pm \iint\limits_{R} g(x, y)dx\,dy.$$

2. $\iint\limits_{R} kf(x, y)dx\,dy = k\iint\limits_{R} f(x, y)dx\,dy$, where k is any real constant.

3. When $f(x, y)$ is integrable, then $|f(x, y)|$ is also integrable, and

$$\left| \iint\limits_{R} f(x, y)dx\,dy \right| \le \iint\limits_{R} |f(x, y)|\,dx\,dy. \tag{2.85}$$

4.
$$\iint\limits_{R} f(x, y)dx\,dy = f(\xi, \eta)\,A \tag{2.86}$$

where A is the area of the region R and (ξ, η) is any arbitrary point in R. This result is called the *mean value theorem* of the double integrals.

If $m \le f(x, y) \le M$ for all (x, y) in R, then

$$mA \le \iint\limits_{R} f(x, y)dx\,dy \le MA. \tag{2.87}$$

5. If $0 < f(x, y) \le g(x, y)$ for all (x, y) in R, then

$$\iint\limits_{R} f(x, y)dx\,dy \le \iint\limits_{R} g(x, y)dx\,dy. \tag{2.88}$$

6. If $f(x, y) \ge 0$ for all (x, y) in R, then

$$\iint\limits_{R} f(x, y)dx\,dy \ge 0. \tag{2.89}$$

Application of double integrals

Double integrals have large number of applications. We state some of them.

1. If $f(x, y) = 1$, then $\iint\limits_{R} dx\,dy$ gives the *area A* of the region R.

For example, if R is the rectangle bounded by the lines $x = a$, $x = b$, $y = c$ and $y = d$, then

$$A = \int_{c}^{d} \int_{a}^{b} dx\,dy = \int_{c}^{d} \left[\int_{a}^{b} dx \right] dy = (b - a) \int_{c}^{d} dy = (b - a)(d - c)$$

gives the area of the rectangle.

2. If $z = f(x, y)$ is a surface, then

$$\iint\limits_{R} z\,dx\,dy \quad \text{or} \quad \iint\limits_{R} f(x, y)dx\,dy$$

gives the *volume* of the region beneath the surface $z = f(x, y)$ and above the x-y plane.

For example, if $z = \sqrt{a^2 - x^2 - y^2}$ and $R : x^2 + y^2 \leq a^2$, then

$$V = \iint_R \sqrt{a^2 - x^2 - y^2} \, dx \, dy$$

gives the volume of the hemisphere $x^2 + y^2 + z^2 = a^2$, $z \geq 0$.

3. Let $f(x, y) = \rho(x, y)$ be a density function (mass per unit area) of a distribution of mass in the x-y plane. Then

$$M = \iint_R f(x, y) dx \, dy \qquad (2.90)$$

give the total *mass* of R.

4. Let $f(x, y) = \rho(x, y)$ be a density function. Then

$$\bar{x} = \frac{1}{M} \iint_R x f(x, y) dx \, dy, \quad \bar{y} = \frac{1}{M} \iint_R y f(x, y) dx \, dy \qquad (2.91)$$

give the coordinates of the *centre of gravity* $(\bar{x}, \bar{y})$ of the mass M in R.

5. Let $f(x, y) = \rho(x, y)$ be a density function. Then

$$I_x = \iint_R y^2 f(x, y) dx \, dy \quad \text{and} \quad I_y = \iint_R x^2 f(x, y) dx \, dy \qquad (2.92)$$

give the *moments of inertia* of the mass in R about the x-axis and the y-axis respectively, whereas $I_0 = I_x + I_y$ is called the moment of inertia of the mass in R about the origin. Similarly,

$$I_y = \iint_R (x - a)^2 f(x, y) dx \, dy \quad \text{and} \quad I_x = \iint_R (y - b)^2 f(x, y) dx \, dy \qquad (2.93)$$

give the moment of inertia of the mass in R about the lines $x = a$ and $y = b$ respectively.

6. $\dfrac{1}{A} \displaystyle\iint_R f(x, y) dx \, dy$ gives the *average value* of $f(x, y)$ over R, where A is the area of the region R.

Example 2.41   Evaluate the double integral $\displaystyle\iint_R xy \, dx \, dy$, where R is the region bounded by the x-axis, the line $y = 2x$ and the parabola $y = x^2/(4a)$.

Solution   The points of intersection of the curves $y = 2x$ and $y = x^2/(4a)$ are $(0, 0)$ and $(8a, 16a)$. The region

$$R = \{(x, y): (x^2/4a) \leq y \leq 2x, 0 \leq x \leq 8a\}$$

is given in Fig. 2.9.

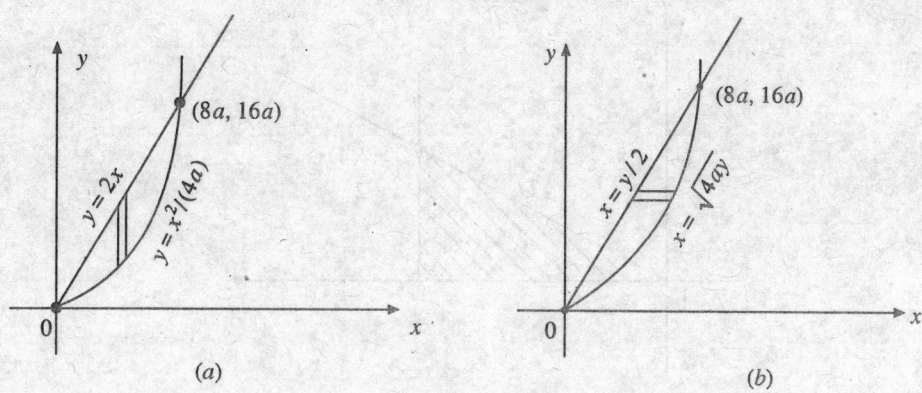

Fig. 2.9.   Region in Example 2.41.

We evaluate the double integral as

$$I = \iint\limits_{R} xy \, dx \, dy = \int_0^{8a} \left[\int_{x^2/(4a)}^{2x} xy \, dy \right] dx = \int_0^{8a} \left[\frac{xy^2}{2} \right]_{x^2/(4a)}^{2x} dx$$

$$= \int_0^{8a} \frac{x}{2} \left(4x^2 - \frac{x^4}{16a^2} \right) dx = \left[\frac{x^4}{2} - \frac{x^6}{192a^2} \right]_0^{8a} = 4096 \left[\frac{1}{2} - \frac{64}{192} \right] a^4 = \frac{2048}{3} a^4.$$

Alternative   We can evaluate the integral as

$$I = \iint\limits_{R} xy \, dx \, dy = \int_0^{16a} \left[\int_{y/2}^{\sqrt{4ay}} xy \, dx \right] dy = \int_0^{16a} \left[\frac{1}{2} yx^2 \right]_{y/2}^{\sqrt{4ay}} dy$$

$$= \frac{1}{2} \int_0^{16a} y \left(4ay - \frac{y^2}{4} \right) dy = \frac{1}{2} \left[\frac{4ay^3}{3} - \frac{y^4}{16} \right]_0^{16a} = \frac{4096 \, a^3}{2} \left[\frac{4a}{3} - \frac{16a}{16} \right] = \frac{2048}{3} a^4.$$

Example 2.42   Evaluate the double integral $\iint\limits_{R} e^{x^2} dx \, dy$, where the region R is given by

$R : 2y \leq x \leq 2$ and $0 \leq y \leq 1$.

Solution   The integral cannot be evaluated by integrating first with respect to x. We try to evaluate it by integrating it first with respect to y. The region of integration is given in Fig. 2.10. We have

$$I = \int_0^2 \left[\int_0^{x/2} e^{x^2} dy \right] dx = \int_0^2 \left[y \, e^{x^2} \right]_0^{x/2} dx$$

$$= \frac{1}{2} \int_0^2 x \, e^{x^2} dx = \left[\frac{1}{4} e^{x^2} \right]_0^2 = \frac{1}{4} (e^4 - 1).$$

Example 2.43   Evaluate the integral $\displaystyle\int_0^2 \int_0^{y^2/2} \frac{y}{\sqrt{x^2 + y^2 + 1}} \, dx \, dy$.

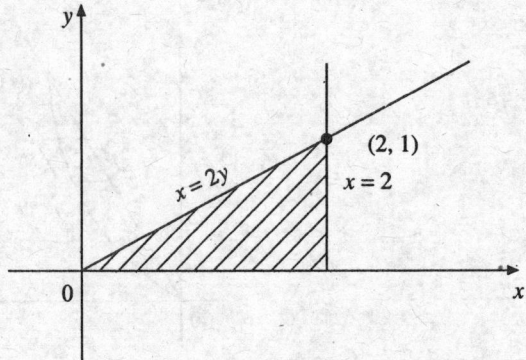

Fig. 2.10.   Region in Example 2.42.

Solution   Because of the form of the integrand, it would be easier to integrate it first with respect to y. The point of intersection of the line $y = 2$ and the curve $y^2 = 2x$ is $(2, 2)$. The region of integration is given in Fig. 2.11.

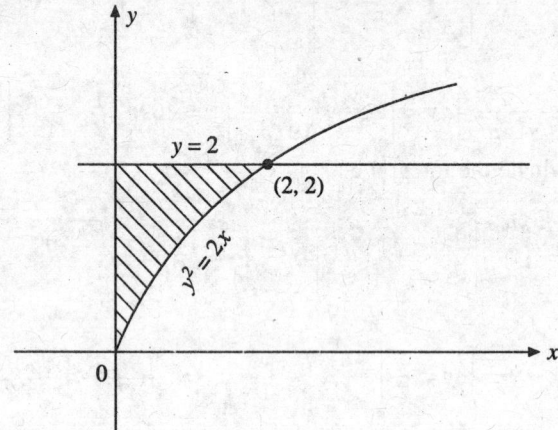

Fig. 2.11.   Region in Example 2.43.

The given region of integration $0 \leq y \leq 2$ and $0 \leq x \leq y^2/2$ can also be written as $0 \leq x \leq 2$ and $\sqrt{2x} \leq y \leq 2$. Hence, we obtain

$$I = \int_0^2 \left[\int_{\sqrt{2x}}^2 \frac{y}{\sqrt{x^2 + y^2 + 1}} \, dy \right] dx = \int_0^2 \left[\sqrt{x^2 + y^2 + 1} \right]_{\sqrt{2x}}^2 dx = \int_0^2 \left[\sqrt{x^2 + 5} - (x + 1) \right] dx$$

$$= \left[\frac{x\sqrt{x^2 + 5}}{2} + \frac{5}{2} \ln (x + \sqrt{x^2 + 5}) - \frac{1}{2} (x + 1)^2 \right]_0^2$$

$$= 3 + \frac{5}{2} (\ln 5 - \ln \sqrt{5}) - \frac{1}{2} (9 - 1) = \frac{5}{4} \ln 5 - 1.$$

Example 2.44   The cylinder $x^2 + z^2 = 1$ is cut by the planes $y = 0$, $z = 0$ and $x = y$. Find the volume of the region in the first octant.

Solution In the first octant we have $z = \sqrt{1 - x^2}$. The projection of the surface in the x-y plane is bounded by $x = 0$, $x = 1$, $y = 0$ and $y = x$. Therefore,

$$V = \iint_R z \, dx \, dy = \int_0^1 \left[\int_0^x \sqrt{1 - x^2} \, dy \right] dx = \int_0^1 \sqrt{1 - x^2} \, [y]_0^x \, dx$$

$$= \int_0^1 x\sqrt{1 - x^2} \, dx = -\frac{1}{3} \left[(1 - x^2)^{3/2} \right]_0^1 = \frac{1}{3} \text{ cubic units.}$$

Example 2.45 Find the volume of the ellipsoid $\dfrac{x^2}{a^2} + \dfrac{y^2}{b^2} + \dfrac{z^2}{c^2} = 1$.

Solution We have volume = 8 (volume in the first octant). The projection of the surface $z = c\sqrt{1 - \dfrac{x^2}{a^2} - \dfrac{y^2}{b^2}}$ in the x-y plane is the region in the first quadrant of the ellipse $\dfrac{x^2}{a^2} + \dfrac{y^2}{b^2} = 1$. Therefore,

$$V = 8 \int_0^a \left[\int_0^{b\sqrt{1 - x^2/a^2}} c\sqrt{1 - \frac{x^2}{a^2} - \frac{y^2}{b^2}} \, dy \right] dx = 8c \int_0^a \left[\int_0^{bk} \sqrt{k^2 - \frac{y^2}{b^2}} \, dy \right] dx$$

where $k^2 = 1 - (x^2/a^2)$. Setting $y = b \, k \sin \theta$, we obtain

$$V = 8c \int_0^a \left[\int_0^{\pi/2} \sqrt{k^2 - k^2 \sin^2 \theta} \, (bk \cos \theta) \, d\theta \right] dx = 8bc \int_0^a \left[\int_0^{\pi/2} k^2 \cos^2 \theta \, d\theta \right] dx$$

$$= 4bc \left(\frac{\pi}{2} \right) \int_0^a \left(1 - \frac{x^2}{a^2} \right) dx = \frac{2\pi bc}{a^2} \int_0^a (a^2 - x^2) \, dx$$

$$= \frac{2\pi bc}{a^2} \left[a^2 x - \frac{x^3}{3} \right]_0^a = \frac{4\pi abc}{3} \text{ cubic units.}$$

Example 2.46 Find the centre of gravity of a plate whose density $\rho(x, y)$ is constant and is bounded by the curves $y = x^2$ and $y = x + 2$. Also, find the moments of inertia about the axes.

Solution The mass of the plate is given by (see Eq. 2.90)

$$M = \iint_R \rho(x, y) \, dx \, dy = k \iint_R dx \, dy \qquad (\rho(x, y) = k \text{ constant}).$$

The boundary of the plate is given in Fig. 2.12. The line $y = x + 2$ intersects the parabola $y = x^2$ at the points $(-1, 1)$ and $(2, 4)$. The limits of integration can be written as $-1 \leq x \leq 2$, $x^2 \leq y \leq x + 2$. Therefore,

$$M = k \int_{-1}^2 \left[\int_{x^2}^{x+2} dy \right] dx = k \int_{-1}^2 (x + 2 - x^2) \, dx$$

$$= k \left[-\frac{x^3}{3} + \frac{x^2}{2} + 2x \right]_{-1}^{2} = k \left(-\frac{9}{3} + \frac{3}{2} + 6 \right) = \frac{9}{2} k.$$

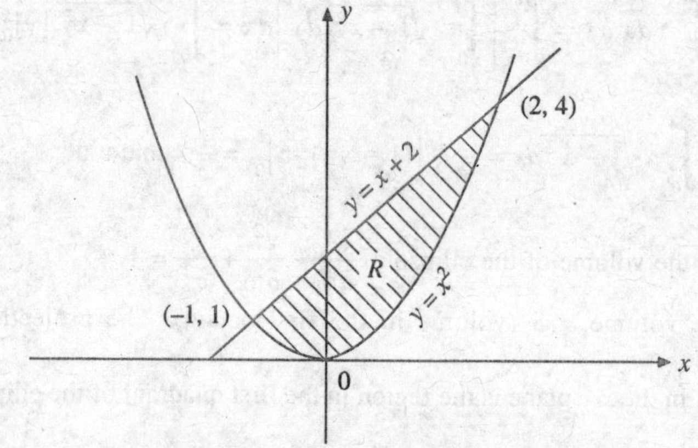

Fig. 2.12.   Region in Example 2.46.

The centre of gravity $(\bar{x}, \bar{y})$ is given by (see Eq. 2.91)

$$\bar{x} = \frac{1}{M} \iint_R x \, \rho(x, y) dx \, dy = \frac{2}{9} \int_{-1}^{2} \left[\int_{x^2}^{x+2} dy \right] x \, dx$$

$$= \frac{2}{9} \int_{-1}^{2} x(x + 2 - x^2) dx = \frac{2}{9} \left[\frac{x^3}{3} + x^2 - \frac{x^4}{4} \right]_{-1}^{2} = \frac{1}{2}.$$

$$\bar{y} = \frac{1}{M} \iint_R y \, \rho(x, y) dx \, dy = \frac{2}{9} \int_{-1}^{2} \left[\int_{x^2}^{x+2} y \, dy \right] dx = \frac{2}{9} \int_{-1}^{2} \left[\frac{y^2}{2} \right]_{x^2}^{x+2} dx$$

$$= \frac{1}{9} \int_{-1}^{2} [(x + 2)^2 - x^4] dx = \frac{1}{9} \left[\frac{(x+2)^3}{3} - \frac{x^5}{5} \right]_{-1}^{2}$$

$$= \frac{1}{9} \left[\frac{1}{3} (64 - 1) - \frac{1}{5} (32 + 1) \right] = \frac{1}{9} \left[21 - \frac{33}{5} \right] = \frac{8}{5}.$$

Therefore, the centre of gravity is located at (1/2, 8/5).

Moment of inertia about the x-axis is given by (see Eq. 2.92)

$$I_x = \iint_R y^2 \rho(x, y) dx \, dy = k \int_{-1}^{2} \left[\int_{x^2}^{x+2} y^2 dy \right] dx = k \int_{-1}^{2} \left[\frac{y^3}{3} \right]_{x^2}^{x+2} dx$$

$$= \frac{k}{3} \int_{-1}^{2} [(x + 2)^3 - x^6] dx = \frac{k}{3} \left[\frac{(x+2)^4}{4} - \frac{x^7}{7} \right]_{-1}^{2}$$

$$= \frac{k}{3} \left(\frac{255}{4} - \frac{129}{7} \right) = \frac{423}{28} k.$$

Moment of inertia about the y-axis is given by (see Eq. 2.92)

$$I_y = \iint\limits_R x^2 \rho(x, y)\, dx\, dy = k \int_{-1}^{2} \left[\int_{x^2}^{x+2} dy \right] x^2 dx = k \int_{-1}^{2} x^2 (x + 2 - x^2)\, dx$$

$$= k \left[\frac{x^4}{4} + \frac{2x^3}{3} - \frac{x^5}{5} \right]_{-1}^{2} = k \left[\frac{15}{4} + 6 - \frac{33}{5} \right] = \frac{63}{20}\, k.$$

2.6.2 Triple Integrals

Let $f(x, y, z)$ be a continuous function defined over a closed and bounded region T in $\mathbb{R}^3$. Divide the region T into a number of parallelopipeds by drawing planes parallel to the coordinate planes. Number the parallelopipeds inside T from 1 to n and form the sum

$$J_n = \sum_{k=1}^{n} f(x_k, y_k, z_k)\, \Delta V_k$$

where (x_k, y_k, z_k) is an arbitrary point in the kth parallelopiped and ΔV_k is its volume. For different values of n, say $n_1, n_2, \ldots, n_m, \ldots$, we obtain a sequence of sums $J_{n_1}, J_{n_2}, \ldots, J_{n_m}, \ldots$ The length of the diagonal of the kth parallelopiped is $d_k = \sqrt{(\Delta x_k)^2 + (\Delta y_k)^2 + (\Delta z_k)^2}$. Let $n \to \infty$ such that $\max d_k \to 0$. If $\lim\limits_{n \to \infty} J_n$ exists, independent of the choice of the subdivision and the point (x_k, y_k, z_k), then we say that $f(x, y, z)$ is integrable over T. This limit is called the *triple integral* of $f(x, y, z)$ over T and is denoted by

$$J = \iiint\limits_T f(x, y, z)\, dx\, dy\, dz. \tag{2.94}$$

Triple integrals satisfy properties similar to double integrals.

Application of triple integrals

1. If $f(x, y, z) = 1$, then the triple integral

$$V = \iiint\limits_T dx\, dy\, dz \tag{2.95}$$

    gives the volume of the region T.

2. If $f(x, y, z) = \rho(x, y, z)$ is the density of a mass, then the triple integral

$$M = \iiint\limits_T f(x, y, z)\, dx\, dy\, dz \tag{2.96}$$

    gives the *mass* of the solid.

3. $$\bar{x} = \frac{1}{M} \iiint\limits_T x f(x, y, z)\, dx\, dy\, dz, \quad \bar{y} = \frac{1}{M} \iiint\limits_T y f(x, y, z)\, dx\, dy\, dz,$$

$$\bar{z} = \frac{1}{M} \iiint\limits_T z f(x, y, z)\, dx\, dy\, dz \tag{2.97}$$

give the coordinates of the *centre of mass* (or the *centre of gravity*) of the solid of mass M in T, where $f(x, y, z) = \rho(x, y, z)$ is the density function.

4.
$$I_x = \iiint\limits_T (y^2 + z^2) f(x, y, z) dx\, dy\, dz, \quad I_y = \iiint\limits_T (x^2 + z^2) f(x, y, z) dx\, dy\, dz,$$

$$I_z = \iiint\limits_T (x^2 + y^2) f(x, y, z) dx\, dy\, dz \tag{2.98}$$

give the *moments of inertia* of the mass in T about the x-axis, y-axis and z-axis respectively where $f(x, y, z) = \rho(x, y, z)$ is the density function.

Evaluation of triple integrals

We evaluate the triple integral by three successive integrations. If the region T can be described by

$$x_1 \leq x \leq x_2, \quad y_1(x) \leq y \leq y_2(x), \quad z_1(x, y) \leq z \leq z_2(x, y)$$

then we evaluate the triple integral as

$$\int_{x_1}^{x_2} \int_{y_1(x)}^{y_2(x)} \int_{z_1(x,y)}^{z_2(x,y)} f(x, y, z)\, dz\, dy\, dx = \int_{x_1}^{x_2} \left[\int_{y_1(x)}^{y_2(x)} \left[\int_{z_1(x,y)}^{z_2(x,y)} f(x, y, z) dz \right] dy \right] dx \tag{2.99}$$

We note that there are six possible ways in which a triple integral can be evaluated (order of variables of integration). We choose the one which is simple to use.

Example 2.47 Evaluate the triple integral $\displaystyle\iiint\limits_T y\, dx\, dy\, dz$, where T is the region bounded by the surfaces $x = y^2$, $x = y + 2$, $4z = x^2 + y^2$ and $z = y + 3$.

Solution The variable z varies from $(x^2 + y^2)/4$ to $y + 3$. The projection of T on the x-y plane is the region bounded by the curves $x = y^2$ and $x = y + 2$. These curves intersect at the points $(1, -1)$ and $(4, 2)$. Also, $y^2 \leq y + 2$ for $-1 \leq y \leq 2$. Hence, the required region can be written as

$$-1 \leq y \leq 2, \quad y^2 \leq x < y + 2 \quad \text{and} \quad [(x^2 + y^2)/4] \leq z \leq y + 3.$$

Therefore, we can evaluate the triple integral as

$$J = \int_{-1}^{2} \left[\int_{y^2}^{y+2} \left[\int_{(x^2+y^2)/4}^{(y+3)} y\, dz \right] dx \right] dy = \int_{-1}^{2} \left[\int_{y^2}^{y+2} y \left\{ y + 3 - \frac{x^2 + y^2}{4} \right\} dx \right] dy$$

$$= \int_{-1}^{2} \left[\left(y^2 + 3y - \frac{y^3}{4} \right) x - \frac{x^3 y}{12} \right]_{y^2}^{y+2} dy$$

$$= \int_{-1}^{2} \left[\left(y^2 + 3y - \frac{y^3}{4} \right) (y + 2 - y^2) - \frac{1}{12} y \left\{ (y + 2)^3 - y^6 \right\} \right] dy$$

$$= \int_{-1}^{2} \left[\frac{y^7}{12} + \frac{y^5}{4} - \frac{4y^4}{3} - 3y^3 + 4y^2 + \frac{16y}{3} \right] dy$$

$$= \left[\frac{y^8}{96} + \frac{y^6}{24} - \frac{4y^5}{15} - \frac{3y^4}{4} + \frac{4y^3}{3} + \frac{8y^2}{3}\right]_{-1}^{2} = \frac{837}{160}.$$

Example 2.48   Evaluate the integral $\iiint\limits_{T} z\, dx\, dy\, dz$, where T is the region bounded by the cone

$z^2 = x^2 \tan^2 \alpha + y^2 \tan^2 \beta$ and the planes $z = 0$ to $z = h$ in the first octant.

Solution   The required region can be written as

$$0 \le z \le \sqrt{x^2 \tan^2\alpha + y^2 \tan^2\beta}, \ 0 \le y \le (\sqrt{h^2 - x^2 \tan^2\alpha})\cot \beta, \ 0 \le x \le h \cot \alpha$$

Therefore,

$$J = \int_0^{h\cot\alpha}\left[\int_0^{(\sqrt{h^2 - x^2\tan^2\alpha})\cot\beta} \frac{1}{2}(x^2 \tan^2\alpha + y^2 \tan^2\beta)dy\right]dx$$

$$= \frac{1}{2}\int_0^{h\cot\alpha}\left[x^2(h^2 - x^2\tan^2\alpha)^{1/2}\tan^2\alpha + \frac{1}{3}(h^2 - x^2\tan^2\alpha)^{3/2}\right]\cot\beta\, dx.$$

Substituting $x \tan \alpha = h \sin \theta$, we obtain

$$J = \frac{\cot\beta}{2}\int_0^{\pi/2}\left[h^2\sin^2\theta\,(h\cos\theta) + \frac{1}{3}(h^3\cos^3\theta)\right]h\cot\alpha\cos\theta\, d\theta$$

$$= \frac{1}{2}h^4\cot\beta\cot\alpha\left[\int_0^{\pi/2}(\sin^2\theta\cos^2\theta + \frac{1}{3}\cos^4\theta)d\theta\right]$$

$$= \frac{1}{2}h^4\cot\beta\cot\alpha\left[\int_0^{\pi/2}(\sin^2\theta - \sin^4\theta + \frac{1}{3}\cos^4\theta)d\theta\right]$$

$$= \frac{1}{2}h^4\cot\beta\cot\alpha\left[\frac{\pi}{4} - \frac{3\pi}{16} + \frac{\pi}{16}\right] = \frac{h^4\pi}{16}\cot\alpha\cot\beta.$$

Example 2.49   Find the volume of the solid in the first octant bounded by the paraboloid $z = 36 - 4x^2 - 9y^2$.

Solution   We have

$$V = \iiint\limits_{T} dz\, dy\, dx.$$

The projection of the paraboloid (in the first octant) in the x-y plane is the region in the first quadrant of the ellipse $4x^2 + 9y^2 = 36$.

Therefore, the region T is given by

$$0 \le z \le 36 - 4x^2 - 9y^2, \ \ 0 \le y \le \frac{1}{3}\sqrt{36 - 4x^2}, \ \ 0 \le x \le 3.$$

Hence,

$$V = \int_0^3 \left[\int_0^{(2\sqrt{9-x^2}/3)} (36 - 4x^2 - 9y^2)dy \right] dx$$

$$= \int_0^3 [4(9-x^2)y - 3y^3]_0^{(2\sqrt{9-x^2}/3)} \, dx$$

$$= \int_0^3 \left[\frac{8}{3}(9-x^2)^{3/2} - \frac{8}{9}(9-x^2)^{3/2} \right] dx = \frac{16}{9} \int_0^3 (9-x^2)^{3/2} \, dx.$$

Substituting $x = 3 \sin \theta$, we obtain

$$V = \frac{16}{9} \int_0^{\pi/2} (27 \cos^3 \theta)(3 \cos \theta)d\theta = 144 \int_0^{\pi/2} \cos^4 \theta \, d\theta$$

$$= 144 \left(\frac{3}{4} \cdot \frac{1}{2} \cdot \frac{\pi}{2} \right) = 27\pi \text{ cubic units.}$$

Example 2.50   Find the volume of the solid enclosed between the surfaces $x^2 + y^2 = a^2$ and $x^2 + z^2 = a^2$.

Solution   We have the region as

$$-\sqrt{a^2 - x^2} \le z \le \sqrt{a^2 - x^2}, \quad -\sqrt{a^2 - x^2} \le y \le \sqrt{a^2 - x^2}, \quad -a \le x \le a.$$

Therefore,

$$V = \int_{-a}^a \int_{-\sqrt{a^2-x^2}}^{\sqrt{a^2-x^2}} \int_{-\sqrt{a^2-x^2}}^{\sqrt{a^2-x^2}} dz \, dy \, dx = 8 \int_0^a \int_0^{\sqrt{a^2-x^2}} \sqrt{a^2 - x^2} \, dy \, dx$$

$$= 8 \int_0^a (a^2 - x^2)dx = 8 \left(a^2 x - \frac{x^3}{3} \right)_0^a = \frac{16a^3}{3} \text{ cubic units.}$$

2.6.3   Change of Variables in Integrals

In the case of definite integrals $\int_a^b f(x)dx$ of one variable, we have seen that the evaluation of the integral is often simplified by using some substitution and thus changing the variable of integration. Similarly, the double and triple integrals can be evaluated by using some substitutions and changing the variables of integration.

Double integrals

Let the variables x, y defined in a region R of the x-y plane be transformed as

$$x = x(u, v), \, y = y(u, v). \tag{2.100}$$

We assume that the functions $x(u, v), y(u, v)$ are defined and have continuous partial derivatives in the region R^* of interest in the u-v plane. We also assume that the inverse functions $u = u(x, y)$, $v = v(x, y)$ are defined and are continuous in the region of interest in the x-y plane, so that the mapping is one-to-one. Since the function $f(x, y)$ is continuous in R, the function $f[x(u, v), y(u, v)]$ is also continuous in R^*. Then, the double integral transforms as

$$\iint_R f(x, y)dx \, dy = \iint_{R^*} f[x(u, v), y(u, v)] \, |J| \, du \, dv = \iint_{R^*} F(u, v)du \, dv \qquad (2.101)$$

where

$$J = \frac{\partial(x, y)}{\partial(u, v)} = \begin{vmatrix} \partial x/\partial u & \partial x/\partial v \\ \partial y/\partial u & \partial y/\partial v \end{vmatrix}$$

is the *Jacobian* of the variables of transformation.

For example, if we change the cartesian coordinates to *polar* coordinates, we have

$$x = r \cos \theta, \quad y = r \sin \theta, \quad r_1 \leq r \leq r_2, \quad \theta_1 \leq \theta \leq \theta_2$$

$$J = \begin{vmatrix} \partial x/\partial r & \partial x/\partial \theta \\ \partial y/\partial r & \partial y/\partial \theta \end{vmatrix} = \begin{vmatrix} \cos \theta & -r \sin \theta \\ \sin \theta & r \cos \theta \end{vmatrix} = r \qquad (2.102)$$

Therefore,

$$\iint_R f(x, y)dx \, dy = \iint_{R^*} f(r \cos \theta, r \sin \theta) \, r \, dr \, d\theta = \iint_{R^*} F(r, \theta) \, r \, dr \, d\theta$$

where R^* is the region corresponding to R is the r-θ plane.

Triple integrals

Analogous to double integrals, we define x, y, z as functions of three new variables

$$x = x(u, v, w), \quad y = y(u, v, w), \quad z = z(u, v, w). \qquad (2.103)$$

Then,

$$\iiint_T f(x, y, z)dx \, dy \, dz = \iiint_{T^*} f[x(u, v, w), y(u, v, w), z(u, v, w)] \, |J| \, du \, dv \, dw \qquad (2.104)$$

where

$$J = \frac{\partial(x, y, z)}{\partial(u, v, w)} = \begin{vmatrix} \partial x/\partial u & \partial x/\partial v & \partial x/\partial w \\ \partial y/\partial u & \partial y/\partial v & \partial y/\partial w \\ \partial z/\partial u & \partial z/\partial v & \partial z/\partial w \end{vmatrix}$$

is the Jacobian of the variables of transformation.

For example, if we change the cartesian coordinates to *cylindrical* coordinates, we have

$$x = r \cos \theta, \quad y = r \sin \theta, \quad z = z$$

$$J = \begin{vmatrix} \partial x/\partial r & \partial x/\partial \theta & \partial x/\partial z \\ \partial y/\partial r & \partial y/\partial \theta & \partial y/\partial z \\ \partial z/\partial r & \partial z/\partial \theta & \partial z/\partial z \end{vmatrix} = \begin{vmatrix} \cos \theta & -r \sin \theta & 0 \\ \sin \theta & r \cos \theta & 0 \\ 0 & 0 & 1 \end{vmatrix} = r \qquad (2.105)$$

and

$$\iiint_T f(x, y, z)dx \, dy \, dz = \iiint_{T^*} f(r \cos \theta, r \sin \theta, z) \, r \, dr \, d\theta \, dz$$

If we change the cartesian coordinates to *spherical* coordinates, we have (Fig. 2.13)

$$x = r \sin \phi \cos \theta, \ \ y = r \sin \phi \sin \theta, \ \ z = r \cos \phi, \ \ 0 \le \theta \le 2\pi, \ \ 0 \le \phi \le \pi$$

$$J = \begin{vmatrix} \partial x/\partial r & \partial x/\partial \phi & \partial x/\partial \theta \\ \partial y/\partial r & \partial y/\partial \phi & \partial y/\partial \theta \\ \partial z/\partial r & \partial z/\partial \phi & \partial z/\partial \theta \end{vmatrix} = \begin{vmatrix} \sin \phi \cos \theta & r \cos \phi \cos \theta & -r \sin \phi \sin \theta \\ \sin \phi \sin \theta & r \cos \phi \sin \theta & r \sin \phi \cos \theta \\ \cos \phi & -r \sin \phi & 0 \end{vmatrix}$$

$$= \cos \phi [r^2 \sin \phi \cos \phi \cos^2 \theta + r^2 \sin \phi \cos \phi \sin^2 \theta] + r \sin \phi [r \sin^2 \phi \cos^2 \theta + r \sin^2 \phi \sin^2 \theta]$$

$$= r^2 [\sin \phi \cos^2 \phi + \sin^3 \phi] = r^2 \sin \phi \tag{2.106}$$

and

$$\iiint_T f(x, y, z) dx\, dy\, dz = \iiint_{T^*} F(r, \theta, \phi)\, r^2 \sin \phi\, dr\, d\theta\, d\phi.$$

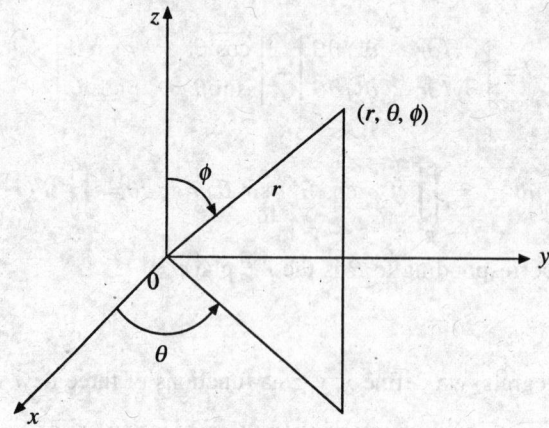

Fig. 2.13.   Spherical coordinates.

Example 2.51   Evaluate the integral $\iint\limits_R (a^2 - x^2 - y^2)\, dx\, dy$, where R is the region $x^2 + y^2 \le a^2$.

Solution   We can evaluate the integral directly by writing it as

$$I = \int_{-a}^{a} \left[\int_{-\sqrt{a^2 - x^2}}^{\sqrt{a^2 - x^2}} (a^2 - x^2 - y^2) dy \right] dx.$$

However, it is easier to evaluate, if we change to polar coordinates. Transforming cartesian coordinates to polar coordinates, we have (see Eq. 2.102)

$$x = r \cos \theta, \ \ y = r \sin \theta, \ \ J = r.$$

Therefore,

$$I = \int_0^a \int_0^{2\pi} (a^2 - r^2)\, r\, dr\, d\theta = \int_0^a \left[\int_0^{2\pi} d\theta \right] (a^2 r - r^3) dr$$

$$= 2\pi \int_0^a (a^2 r - r^3)\, dr = 2\pi \left(\frac{a^2 r^2}{2} - \frac{r^4}{4} \right)_0^a = \frac{\pi a^4}{2}.$$

Example 2.52 Evaluate the integral $\displaystyle\iint_R (x - y)^2 \cos^2(x + y)\, dx\, dy$, where R is the rhombus with successive vertices at $(\pi, 0)$, $(2\pi, \pi)$, $(\pi, 2\pi)$ and $(0, \pi)$.

Solution The region R is given in Fig. 2.14. The equations of the sides AB, BC, CD and DA are respectively

$$x - y = \pi, \quad x + y = 3\pi, \quad x - y = -\pi \quad \text{and} \quad x + y = \pi.$$

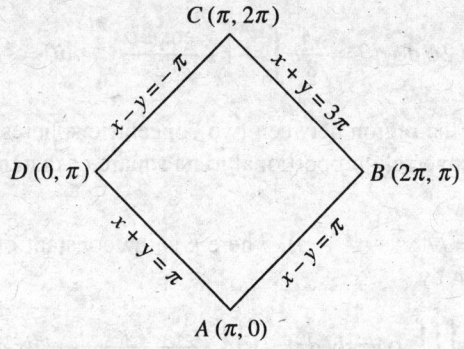

Fig. 2.14. Region in Example 2.52.

Substitute $y - x = u$ and $y + x = v$. Then, $-\pi \le u \le \pi$ and $\pi \le v \le 3\pi$. We obtain

$$x = (v - u)/2, \quad y = (v + u)/2$$

and
$$J = \frac{\partial(x, y)}{\partial(u, v)} = \begin{vmatrix} \partial x/\partial u & \partial x/\partial v \\ \partial y/\partial u & \partial y/\partial v \end{vmatrix} = \begin{vmatrix} -1/2 & 1/2 \\ 1/2 & 1/2 \end{vmatrix} = -\frac{1}{2}, \quad |J| = \frac{1}{2}.$$

Therefore,

$$I = \iint_R (x - y)^2 \cos^2(x + y)\, dx\, dy = \frac{1}{2} \int_\pi^{3\pi} \int_{-\pi}^\pi u^2 \cos^2 v\, du\, dv$$

$$= \frac{\pi^3}{3} \int_\pi^{3\pi} \cos^2 v\, dv = \frac{\pi^3}{6} \int_\pi^{3\pi} (1 + \cos 2v)\, dv = \frac{\pi^4}{3}.$$

Example 2.53 Evaluate the integral $\displaystyle\iint_R \sqrt{x^2 + y^2}\, dx\, dy$ by changing to polar coordinates, where R is the region in the x-y plane bounded by the circles $x^2 + y^2 = 4$ and $x^2 + y^2 = 9$.

Solution Using $x = r \cos\theta$, $y = r \sin\theta$, we get $dx\, dy = r\, dr\, d\theta$, and

$$I = \int_0^{2\pi} \int_2^3 r(r\, dr\, d\theta) = \int_0^{2\pi} \left[\frac{r^3}{3} \right]_2^3 d\theta = \frac{19}{3} \int_0^{2\pi} d\theta = \frac{38\pi}{3}.$$

Example 2.54   Evaluate the integral $\iiint\limits_{T} z \, dx \, dy \, dz$, where T is the hemisphere of radius a, $x^2 + y^2 + z^2 = a^2$, $z \geq 0$.

Solution   Changing to spherical coordinates

$$x = r \sin \phi \cos \theta, \quad y = r \sin \phi \sin \theta, \quad z = r \cos \phi, \quad 0 \leq \theta \leq 2\pi, \quad 0 \leq \phi \leq \pi/2,$$

we obtain $dx \, dy \, dz = r^2 \sin \phi \, dr \, d\phi \, d\theta$ (see Eq. 2.106). Therefore,

$$I = \int_0^{2\pi} \int_0^{\pi/2} \int_0^a (r \cos \phi) \, r^2 \sin \phi \, dr \, d\phi \, d\theta = \frac{a^4}{4} \int_0^{2\pi} \int_0^{\pi/2} \sin \phi \cos \phi \, d\phi \, d\theta$$

$$= \frac{a^4}{8} \int_0^{2\pi} \int_0^{\pi/2} \sin 2\phi \, d\phi \, d\theta = \frac{a^4}{8} \int_0^{2\pi} \left[-\frac{\cos 2\phi}{2} \right]_0^{\pi/2} d\theta = \frac{a^4}{8} \int_0^{2\pi} d\theta = \frac{\pi a^4}{4}.$$

Example 2.55   A solid fills the region between two concentric spheres of radii a and b, $0 < a < b$. The density at each point is inversely proportional to its square of distance from the origin. Find the total mass.

Solution   The density is $\rho = k/(x^2 + y^2 + z^2)$, where k is the constant of proportionality. Therefore, the mass of the solid is given by

$$M = \iiint\limits_{T} \rho \, dx \, dy \, dz = \iiint\limits_{T} \frac{k}{x^2 + y^2 + z^2} \, dx \, dy \, dz$$

where $a^2 < x^2 + y^2 + z^2 < b^2$. Changing to spherical coordinates, we obtain

$$x = r \sin \phi \cos \theta, \quad y = r \sin \phi \sin \theta, \quad z = r \cos \phi, \quad x^2 + y^2 + z^2 = r^2, \quad a \leq r \leq b,$$

$$dx \, dy \, dz = r^2 \sin \phi \, dr \, d\theta \, d\phi, \quad 0 \leq \theta \leq 2\pi, \quad 0 \leq \phi \leq \pi.$$

Therefore,

$$M = k \int_0^{2\pi} \int_0^{\pi} \int_a^b \frac{r^2 \sin \phi}{r^2} \, dr \, d\phi \, d\theta = k(b - a) \int_0^{2\pi} \int_0^{\pi} \sin \phi \, d\phi \, d\theta$$

$$= k(b - a) \int_0^{2\pi} [-\cos \phi]_0^{\pi} \, d\theta = 2k(b - a) \int_0^{2\pi} d\theta = 4\pi k(b - a).$$

2.6.4   Dirichlet Integrals

Let T be a closed region in the first octant in $\mathbb{R}^3$, bounded by the surface $(x/a)^p + (y/b)^q + (z/c)^r = 1$ and the coordinate planes. Then, an integral of the form

$$I = \iiint\limits_{T} x^{\alpha-1} y^{\beta-1} z^{\gamma-1} \, dx \, dy \, dz \tag{2.107}$$

is called a Dirichlet integral, where all the constants α, β, γ, a, b, c and p, q, r are assumed to be positive.

We now show that

$$I = \iiint\limits_{T} x^{\alpha-1} y^{\beta-1} z^{\gamma-1} dx\, dy\, dz = \frac{a^\alpha b^\beta c^\gamma}{pqr} \frac{\Gamma(\alpha/p)\Gamma(\beta/q)\Gamma(\gamma/r)}{\Gamma\left(\dfrac{\alpha}{p} + \dfrac{\beta}{q} + \dfrac{\gamma}{r} + 1\right)}. \tag{2.108}$$

Let $\quad \left(\dfrac{x}{a}\right)^p = u, \left(\dfrac{y}{b}\right)^q = v, \left(\dfrac{z}{c}\right)^r = w,$ or $x = au^{1/p}, y = bv^{1/q}, z = cw^{1/r}.$

The Jacobian of the transformation is given by

$$J = \begin{vmatrix} \partial x/\partial u & \partial x/\partial v & \partial x/\partial w \\ \partial y/\partial u & \partial y/\partial v & \partial y/\partial w \\ \partial z/\partial u & \partial z/\partial v & \partial z/\partial w \end{vmatrix} = \begin{vmatrix} (a/p)u^{(1/p)-1} & 0 & 0 \\ 0 & (b/q)v^{(1/q)-1} & 0 \\ 0 & 0 & (c/r)w^{(1/r)-1} \end{vmatrix}$$

$$= \frac{abc}{pqr} u^{(1/p)-1} v^{(1/q)-1} w^{(1/r)-1}$$

and $\; dx\,dy\,dz = |J|\, du\,dv\,dw = \dfrac{abc}{pqr} u^{(1/p)-1} v^{(1/q)-1} w^{(1/r)-1}\, du\,dv\,dw.$

Now, $x \geq 0, y \geq 0, z \geq 0 \;$ gives $\; u \geq 0, v \geq 0, w \geq 0$ respectively.

Hence, we obtain

$$I = \iiint\limits_{R} \left[au^{(1/p)}\right]^{\alpha-1} \left[bv^{(1/q)}\right]^{\beta-1} \left[cw^{(1/r)-1}\right] \frac{abc}{pqr} u^{(1/p)-1} v^{(1/q)-1} w^{(1/r)-1}\, du\,dv\,dw$$

$$= \frac{a^\alpha b^\beta c^\gamma}{pqr} \iiint\limits_{R} u^{(\alpha/p)-1} v^{(\beta/q)-1} w^{(\gamma/r)-1}\, du\,dv\,dw$$

where R is the region in the uvw-space bounded by the plane $u + v + w = 1$ and the uv, vw and uw coordinate planes, (Fig. 2.15), that is, R is defined by

$$0 \leq w \leq 1 - u - v, \quad 0 \leq v \leq 1 - u, \quad 0 \leq u \leq 1.$$

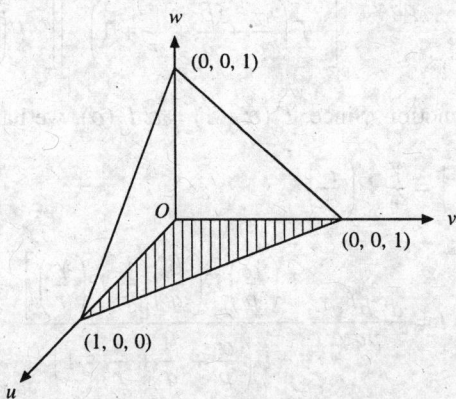

Fig. 2.15. Dirichlet integral.

Therefore, we get

$$I = \frac{a^\alpha b^\beta c^\gamma}{pqr} \int_{u=0}^{1} \int_{v=0}^{1-u} \int_{w=0}^{1-u-v} u^{(\alpha/p)-1} v^{(\beta/q)-1} w^{(\gamma/r)-1} du\,dv\,dw$$

$$= \frac{a^\alpha b^\beta c^\gamma}{pqr} \int_{u=0}^{1} \int_{v=0}^{1-u} u^{(\alpha/p)-1} v^{(\beta/q)-1} \left[\frac{w^{(\gamma/r)}}{(\gamma/r)} \right]_0^{1-u-v} du\,dv$$

$$= \frac{a^\alpha b^\beta c^\gamma}{pq\gamma} \int_{u=0}^{1} \int_{v=0}^{1-u} u^{(\alpha/p)-1} v^{(\beta/q)-1} (1-u-v)^{(\gamma/r)} du\,dv$$

Substituting $v = (1-u)\,t$, $dv = (1-u)dt$, we obtain

$$I = \frac{a^\alpha b^\beta c^\gamma}{pq\gamma} \int_{u=0}^{1} \int_{t=0}^{1} u^{(\alpha/p)-1} (1-u)^{[(\beta/q)+(\gamma/r)]} t^{(\beta/q)-1} (1-t)^{(\gamma/r)} du\,dt\,.$$

Since the limits are constants, we can write

$$I = \frac{a^\alpha b^\beta c^\gamma}{pq\gamma} \left[\int_0^1 u^{(\alpha/p)-1} (1-u)^{[(\beta/q)+(\gamma/r)]} du \right] \left[\int_0^1 t^{(\beta/q)-1} (1-t)^{(\gamma/r)} dt \right]$$

Using the definition of Beta function

$$\int_0^1 x^{m-1} (1-x)^{n-1} dx = \beta(m,n)$$

we obtain

$$I = \frac{a^\alpha b^\beta c^\gamma}{pq\gamma} \beta\left(\frac{\alpha}{p}, \frac{\beta}{q} + \frac{\gamma}{r} + 1 \right) \beta\left(\frac{\beta}{q}, \frac{\gamma}{r} + 1 \right)$$

$$= \frac{a^\alpha b^\beta c^\gamma}{pq\gamma} \left[\frac{\Gamma\left(\frac{\alpha}{p}\right) \Gamma\left(\frac{\beta}{q} + \frac{\gamma}{r} + 1\right)}{\Gamma\left(\frac{\alpha}{p} + \frac{\beta}{q} + \frac{\gamma}{r} + 1\right)} \right] \left[\frac{\Gamma\left(\frac{\beta}{q}\right) \Gamma\left(\frac{\gamma}{r} + 1\right)}{\Gamma\left(\frac{\beta}{q} + \frac{\gamma}{r} + 1\right)} \right]$$

where $\Gamma(x)$ is the Gamma function. Since, $\Gamma(\alpha + 1) = \alpha\,\Gamma(\alpha)$, we have

$$\Gamma\left(\frac{\gamma}{r} + 1 \right) = \frac{\gamma}{r} \Gamma\left(\frac{\gamma}{r} \right)$$

Hence,

$$I = \frac{a^\alpha b^\beta c^\gamma}{pq\gamma} \frac{\Gamma\left(\frac{\alpha}{p}\right) \Gamma\left(\frac{\beta}{q}\right) \left[\frac{\gamma}{r} \Gamma\left(\frac{\gamma}{r}\right) \right]}{\Gamma\left(\frac{\alpha}{p} + \frac{\beta}{q} + \frac{\gamma}{r} + 1 \right)}$$

$$= \frac{a^\alpha b^\beta c^\gamma}{pqr} \frac{\Gamma(\alpha/p)\Gamma(\beta/q)\Gamma(\gamma/r)}{\Gamma\left(\dfrac{\alpha}{p} + \dfrac{\beta}{q} + \dfrac{\gamma}{r} + 1\right)}$$

which is the required result.

Example 2.56   Evaluate the Dirichlet integral

$$I = \iiint_T x^3 y^3 z^3 \, dx \, dy \, dz$$

where T is the region in the first octant bounded by the sphere $x^2 + y^2 + z^2 = 1$ and the coordinate planes.

Solution   Comparing the given integral with Eq. (2.107), we get

$$\alpha = \beta = \gamma = 4, \, p = q = r = 2, \, a = b = c = 1.$$

Substituting in Eq. (2.108), we obtain

$$I = \frac{1}{8} \frac{[\Gamma(2)]^3}{\Gamma(7)} = \frac{1}{8(6!)} = \frac{1}{5760}$$

since $\Gamma(n + 1) = n!$, when n is an integer.

Example 2.57   Evaluate the Dirichlet integral

$$I = \iiint_T x^{1/2} y^{1/2} z^{1/2} \, dx \, dy \, dz$$

where T is the region in the first octant bounded by the plane $x + y + z = 1$ and the coordinate planes.

Solution   Comparing the given integral with Eq. (2.107), we get

$$\alpha = \beta = \gamma = 3/2, \, p = q = r = 1, \, a = b = c = 1.$$

Substituting in Eq. (2.108), we obtain

$$I = \frac{[\Gamma(3/2)]^3}{\Gamma(11/2)}.$$

Using the results, $\Gamma(\alpha + 1) = \alpha \, \Gamma(\alpha)$ and $\Gamma(1/2) = \sqrt{\pi}$, we obtain

$$I = \frac{[(1/2)\Gamma(1/2)]^3}{(9/2)(7/2)(5/2)(3/2)(1/2)\Gamma(1/2)} = \frac{4\pi}{945}.$$

Exercises 2.5

1. Find the area bounded by the curves $y = x^2$, $y = 4 - x^2$.
2. Find the area bounded by the curves $x = y^2$, $x + y - 2 = 0$.
3. Find the area bounded by the curves $y^2 = 4 - 2x$, $x \geq 0$, $y \geq 0$.

4. Find the area bounded by the curves $x^2 = y^3$, $x = y$.

5. By changing to polar coordinates, find the area bounded by the curves $x^2 + y^2 = 2y$, $x^2 + y^2 = 4y$, $x = y$, and $x = 0$.

Change the order of integration and evaluate the following double integrals.

6. $\displaystyle\int_{y=0}^{1} \int_{x=y}^{\sqrt{2-y^2}} \frac{y\,dx\,dy}{\sqrt{x^2+y^2}}$.

7. $\displaystyle\int_{y=0}^{1} \int_{x=0}^{y+4} \frac{2y+1}{x+1}\,dx\,dy$.

8. $\displaystyle\int_{y=0}^{1} \int_{x=y}^{y^{1/3}} e^{x^2}\,dx\,dy$.

9. $\displaystyle\int_{x=0}^{2} \int_{y=0}^{x^2/2} \frac{x}{\sqrt{x^2+y^2+1}}\,dy\,dx$.

10. $\displaystyle\int_{x=0}^{1} \int_{y=0}^{1-x} e^{y/(x+y)}\,dy\,dx$ (use the substitution $x + y = u$ and $y = u\,v$).

11. Find the volume of the solid which is below the plane $z = 2x + 3$ and above the x-y plane and bounded by $y^2 = x$, $x = 0$ and $x = 2$.

12. Find the volume of the solid which is below the plane $z = x + 3y$ and above the ellipse $25x^2 + 16y^2 = 400$, $x \geq 0$, $y \geq 0$.

13. Find the volume of the solid which is bounded by the cylinder $x^2 + y^2 = 1$ and the planes $y + z = 1$ and $z = 0$.

14. Find the volume of the solid which is bounded by the paraboloid $z = 9 - x^2 - 4y^2$ and the coordinate planes $x \geq 0$, $y \geq 0$, $z \geq 0$.

15. Find the volume of the solid which is enclosed between the cylinders $x^2 + y^2 = 2ay$ and $z^2 = 2ay$.

16. Find the volume of the solid which is bounded by the surfaces $2z = x^2 + y^2$ and $z = x$.

17. Find the volume of the solid which is bounded by the surfaces $z = 0$, $3z = x^2 + y^2$ and the cylinder $x^2 + y^2 = 9$.

18. Find the volume of the solid which is in the first octant bounded by the cylinders $x^2 + y^2 = a^2$ and $y^2 + z^2 = a^2$.

19. Find the volume of the solid which is bounded by the paraboloid $4z = x^2 + y^2$, the cone $z^2 = x^2 + y^2$ and the cylinder $x^2 + y^2 = 2x$.

20. Find the volume of the solid which is common to the right circular cylinders $x^2 + z^2 = 1$, $y^2 + z^2 = 1$ and $x^2 + y^2 = 1$.

21. Find the volume of the solid which is above the cone $z^2 = x^2 + y^2$ and inside the sphere $x^2 + y^2 + (z - a)^2 = a^2$.

22. Find the volume of the solid which is below the surface $z = 4x^2 + 9y^2$ and above the square with vertices at $(0, 0)$, $(2, 0)$, $(2, 2)$ and $(0, 2)$.

23. Find the volume of the solid which is bounded by the paraboloids $z = x^2 + y^2$ and $z = 4 - 3(x^2 + y^2)$.

24. Find the volume of the solid which is bounded by $\sqrt{x} + \sqrt{y} + \sqrt{z} = \sqrt{a}$ and the coordinate planes.

25. Find the volume of the solid which is contained between the cone $z^2 = 2(x^2 + y^2)$ and the hyperboloid $z^2 = x^2 + y^2 + a^2$.

26. Find the volume of the region under the cone $z = 3r$ and over the rose petal with boundary $r = \sin 4\theta$, $0 \leq \theta \leq \pi/4$.

27. Find the volume of the portion of the unit sphere which lies inside the right circular cone having its vertex at the origin and making an angle α with the positive z-axis.

28. Find the volume of the region under the plane $z = 1 + 3x + 2y$, $z \geq 0$ and above the region bounded by $x = 1$, $x = 2$, $y = x^2$, and $y = 2x^2$.

29. Find the volume of the portion of the sphere $x^2 + y^2 + z^2 \leq 2ay$ between the planes $y = 0$ and $y = a$.

30. Find the moment of inertia about the axes, of the circular lamina $x^2 + y^2 \leq a^2$, when the density function is $\rho = \sqrt{x^2 + y^2}$

31. Find the total mass and the centre of gravity of the region bounded by $x^{2/3} + y^{2/3} = a^{2/3}$, $x \geq 0$, $y \geq 0$, when the density is constant k.

32. Show that $I = \displaystyle\iint\limits_{R} \frac{dx\,dy}{(x^2 + y^2)^p}$, p integer, $R: x^2 + y^2 \geq 1$ converges for $p > 1$.

    Hence, evaluate the integral.

Evaluate the following integrals (change the variables if necessary) in the given region.

33. $\displaystyle\iint\limits_{R} (x^2 + y^2)\,dx\,dy$, boundary of R: triangle with vertices $(0, 0)$, $(1, 0)$, $(1, 1)$.

34. $\displaystyle\iint\limits_{R} x^2\,dx\,dy$, boundary of $R : y = x^2$, $y = x + 2$.

35. $\displaystyle\iint\limits_{R} (x^2 + y^2)\,dx\,dy$, $R: 0 \leq y \leq \sqrt{1 - x^2}$, $0 \leq x \leq 1$.

36. $\displaystyle\iint\limits_{R} \sqrt{1 - \frac{x^2}{a^2} - \frac{y^2}{b^2}}\, dx\,dy$, boundary of $R : \dfrac{x^2}{a^2} + \dfrac{y^2}{b^2} = 1$.

37. $\displaystyle\iint\limits_{R} e^{2(x^2 + y^2)}\, dx\,dy$, $R : x^2 + y^2 \geq 4$, $x^2 + y^2 \leq 25$, $y = x$, $x \geq 0$, $y \geq 0$.

38. $\displaystyle\iint\limits_{R} x^3 y^3\,dx\,dy$, $R : x^2 + y^2 \leq 1$, $x \geq 0$, $y \geq 0$.

39. $\displaystyle\iint\limits_{R} xy\,dx\,dy$, $R : \sqrt{x} + \sqrt{y} = \sqrt{a}$, $x \geq 0$, $y \geq 0$.

40. $\displaystyle\iint\limits_{R} (1 - x^2 - y^2)\,dx\,dy$, boundary of R : the square with vertices $(\pm 1, 0)$, $(0, \pm 1)$

    (change coordinates : $x - y = u$, $x + y = v$).

41. $\displaystyle\iint\limits_{R} (x + y)^2\, dx\,dy$, boundary of R : parallelogram with sides $x + y = 1$, $x + y = 4$, $x - 2y = -2$,

    $x - 2y = 1$, (change coordinates: $x + y = u$, $x - 2y = v$).

42. $\displaystyle\iint\limits_{R} (4 - 3x^2 - y^2)\,dx\,dy$, boundary of $R : x = 0$, $y = 0$, $x + y - 2 = 0$.

43. $\iint\limits_{R} xy\, dx\, dy$, region (in polar coordinates) $R : r = \sin 2\theta,\ 0 \le \theta \le \pi/2$.

44. $\iiint\limits_{T} x^2 y^2 z\, dx\, dy\, dz,\ T : x^2 + y^2 \le 1,\ 0 \le z \le 1$.

45. $\iiint\limits_{T} \dfrac{dx\, dy\, dz}{(x+y+z+1)^3}$, boundary of $T : x = 0,\ y = 0,\ z = 0,\ x + y + z = 1$.

46. $\iiint\limits_{T} (x + 3y - 2z)dx\, dy\, dz,\ T : 0 \le y \le x^2,\ 0 \le z \le x + y,\ 0 \le x \le 1$.

47. $\iiint\limits_{T} x\, dx\, dy\, dz$, boundary of $T : y = x^2,\ y = x + 2,\ 4z = x^2 + y^2,\ z = x + 3$.

48. $\iiint\limits_{T} (2x - y - z)dx\, dy\, dz,\ T : 0 \le x \le 1,\ 0 \le y \le x^2,\ 0 \le z \le x + y$.

49. $\iiint\limits_{T} \dfrac{dx\, dy\, dz}{(x^2 + y^2 + z^2)^{3/2}}$, boundary of $T : x^2 + y^2 + z^2 = a^2,\ x^2 + y^2 + z^2 = b^2,\ a > b$.

50. $\iiint\limits_{T} z\, dx\, dy\, dz$, boundary of $T : z^2 = x^2 + y^2,\ x^2 + y^2 + z^2 = 1$.

51. $\iiint\limits_{T} \sqrt{1 - \dfrac{x^2}{a^2} - \dfrac{y^2}{b^2} - \dfrac{z^2}{c^2}}\ dx\, dy\, dz$, boundary of $T : \dfrac{x^2}{a^2} + \dfrac{y^2}{b^2} + \dfrac{z^2}{c^2} = 1$.

52. $\iiint\limits_{T} \sqrt{x^2 + y^2 + z^2}\, dx\, dy\, dz,\ T : x^2 + y^2 + z^2 \le y$.

53. $\iiint\limits_{T} (x^2 + y^2)dx\, dy\, dz$, boundary of $T :$ the plane $\dfrac{x}{a} + \dfrac{y}{b} + \dfrac{z}{c} = 1$ and the coordinate planes.

54. $\iiint\limits_{T} (y^2 + z^2)dx\, dy\, dz$, boundary of $T : y^2 + z^2 \le a^2,\ 0 \le x \le h$.

55. $\iiint\limits_{T} x^2 y\, dx\, dy\, dz,\ T : x^2 + y^2 \le 1,\ 0 \le z \le 1$.

Evaluate the following Dirichlet integrals.

56. $\iiint\limits_{T} xyz\, dx\, dy\, dz,\ T :$ Region bounded by $x + y + z = 2$ and the coordinate planes.

57. $\iiint\limits_{T} xy^2z^3 \, dx \, dy \, dz$, T: Region bounded by $x + y + z = 1$ and the coordinate planes.

58. $\iiint\limits_{T} \sqrt{xyz} \, dx \, dy \, dz$, T: Region bounded by $x^3 + y^3 + z^3 = 8$ and the coordinate planes.

59. $\iiint\limits_{T} xy^{1/2}z \, dx \, dy \, dz$, T: Region bounded by $x + y^3 + z^4 = 1$.

60. $\iiint\limits_{T} x^2y \, dx \, dy \, dz$, T: Region bounded by $\dfrac{x^2}{1} + \dfrac{y^2}{4} + \dfrac{z^2}{9} = 1$.

2.7 Answers and Hints

Exercise 2.1

1. $|\,f(x, y) - 1\,| = |\,(x - 1)^2 + (y - 1)^2 + 2(x - 1) + 2(y - 1)\,|$

$$< |\,x - 1\,|^2 + |\,y - 1\,|^2 + 2\,|\,x - 1\,| + 2\,|\,y - 1\,| < \varepsilon$$

(i) if $|\,x - 1\,| < \delta,\ |\,y - 1\,| < \delta$ is used, we get $2\delta^2 + 4\delta < \varepsilon$ or $\delta < [\sqrt{(\varepsilon + 2)/2} - 1]$

(ii) if $\delta^2 < \delta$ is used, we get $\delta < \varepsilon/6$

(iii) if $(x - 1)^2 + (y - 1)^2 < \delta^2$ and $|\,x - 1\,| < \delta,\ |\,y - 1\,| < \delta$ is used, we get $\delta < \sqrt{\varepsilon + 4} - 2$.

2. $|f(x, y) - 7| = |\,(x - 2)^2 - (y - 1)^2 + 6(x - 2) - 2(y - 1)\,|$

$$< |\,x - 2\,|^2 + |\,y - 1\,|^2 + 6\,|\,x - 2\,| + 2\,|\,y - 1\,| < \varepsilon.$$

(i) if $|\,x - 2\,| < \delta,\ |\,y - 1\,| < \delta$ is used, we get $2\delta^2 + 8\delta < \varepsilon$, or $\delta < \sqrt{(\varepsilon + 8)/2} - 2$.

(ii) if $\delta^2 < \delta$ is used, we get $\delta < \varepsilon/10$.

(iii) if $(x - 2)^2 + (y - 1)^2 < \delta^2$ and $|\,x - 2\,| < \delta,\ |\,y - 1\,| < \delta$ is used, we get $\delta < \sqrt{\varepsilon + 16} - 4$.

3. $\left|\dfrac{x + y}{x^2 + y^2 + 1}\right| < |\,x + y\,| < |\,x\,| + |\,y\,| < 2\sqrt{x^2 + y^2} < \varepsilon$. Take $\delta < \varepsilon/2$.

4. Let $x = r \cos\theta,\ y = r \sin\theta$. Therefore

$\left|\dfrac{x^3 + y^3}{x^2 + y^2}\right| < |\,r\,(\cos^3\theta + \sin^3\theta)\,| < 2r < \varepsilon$. Take $\delta < \varepsilon/2$.

5. $|f(x, y) - 0| < |\,x\,| + |\,y\,| < 2\sqrt{x^2 + y^2} < \varepsilon$. Take $\delta < \varepsilon/2$.

6. $|f(x, y) - 0| < x^2 + y^2 < \varepsilon$. Take $\delta < \sqrt{\varepsilon}$.

7. Choose the path $y = mx$. Limit does not exist.

8. Factorize and cancel $x - y$; 1.

9. $[1 + (x/y)]^y = [[1 + (x/y)]^{y/x}]^x$; e^α.

10. 0.   **11.** 1/2.   **12.** 1.

13. Limit does not exist.   **14.** Limit does not exist.

15. Let $x = r \cos \theta$, $y = r \sin \theta$; $\dfrac{1}{r} \left(\dfrac{\cos^2\theta}{\cos^3\theta + \sin^3\theta} \right) \to \infty$ as $r \to 0$. Limit does not exist.

16. Choose the path $y = mx^2$. Limit does not exist.

17. Choose the path $z = x^2$, $y = mx$. Limit does not exist.

18. Choose the path $y = mx$, $z = mx$. Limit does not exist.

19. Choose the path $z = \sqrt{x}$, $y = mx$. Limit does not exist.

20. Choose the path $z = 0$, $y = mx$. Limit does not exist.

21. Choose the path $y = mx$. Discontinuous.

22. Limit is 0 for $x > 0$ and 1 for $x < 0$. Discontinuous.

23. Discontinuous.   24. Choose the path $y = mx$. Discontinuous.

25. Choose the path $y = mx$. Discontinuous.   26. Cancel $(x - y)$. Discontinuous.

27. Let $x = r \cos \theta$, $y = r \sin \theta$. Continuous.   28. Choose the path $y^2 = mx$. Discontinuous.

29. Since $x^2 + y^2 \geq 2|x||y|$, we have $\dfrac{1}{\sqrt{x^2 + y^2}} \leq \dfrac{1}{\sqrt{2|xy|}}$. Therefore, $|f(x, y)| \leq \dfrac{\left| \sin \sqrt{|xy|} - \sqrt{|xy|} \right|}{\sqrt{2} \sqrt{|xy|}}$.

Continuous.

30. Since $2 \leq 3 + \sin x \leq 4$, we have $[1/(3 + \sin x)] \leq 1/2$. Therefore, $|f(x, y)| \leq [(2x^2 + y^2)/2] \leq x^2 + y^2$.
Continuous.

31. The function is not defined along the path $y = -x$. Discontinuous.

32. $\left| \dfrac{x^5 - y^5}{x^2 + y^2} \right| \leq \dfrac{|x|^5 + |y|^5}{x^2 + y^2} \leq \dfrac{(x^2 + y^2)^{5/2} + (x^2 + y^2)^{5/2}}{x^2 + y^2}$. Continuous.

33. Function is unbounded in any neighborhood of $x = -1$. Discontinuous.

34. Since $|x|, |y|, |z|$ are all $\leq \sqrt{x^2 + y^2 + z^2}$, $|f| \leq \sqrt{x^2 + y^2 + z^2}$. Continuous.

35. The function is unbounded along $x = \sqrt{3}z$. Discontinuous.

Exercise 2.2

1. $f_x(0, 0) = 0, f_y(0, 0) = 0$. For $(x, y) \neq (0, 0)$, find f_x, f_y and choose the path $y = mx$. The limits do not exist as $(x, y) \to (0, 0)$.

2. $f(x, y)$ is unbounded as $(x, y) \to (0, 0)$, for example along $x = y$; $f_x(0, 0) = 1, f_y(0, 0) = -1$.

3. $f_x(0, 0) = 0, f_y(0, 0) = -1, f_x(0, y) = 0, f_y(x, 0) = 1$.

4. $f_x(0, 0) = 1, f_y(0, 0) = 1, dz = \Delta x + \Delta y$, $\lim\limits_{\Delta\rho \to 0} [(\Delta z - dz)/\Delta\rho]$ does not exist.

5. $f_x(0, 0) = 0 = f_y(0, 0), dz = 0$. $\lim\limits_{\Delta\rho \to 0} [(\Delta z - dz)/\Delta\rho] = 0$.

No contradiction since continuity of f_x, f_y is only a sufficient condition.

In problems **6** to **15**, f_x, f_y and f_z are given in that order at the given point.

6. $-2, 2$.   7. $1/2, -1/3$.

8. $6e^{1/2}, 4e^{1/2}$.   9. $49/(85)^{3/2}, -42/(85)^{3/2}$.

10. $-1/10, -1/10$.

11. $f(x, y) = 2 \ln [\sqrt{x^2 + y^2} - x] - 2 \ln y, -2/5, 3/10.$

12. $-2/27, -1/27, -2/27.$ **13.** $e, -2e, e.$

14. $5, 3, 0.$ **15.** $1/7, 3/35, 4/35.$

16. $0.$ **17.** $2.$

18. $e^x[\sin(y + 2z) + \{(4t^3 - 1)/t^2\} \cos(y + 2z)].$ **19.** $2(y + z)t + (x + z)(t + 1) e^t + (x + y)(1 - t) e^{-t}.$

20. $(\pi/2) - (2/\pi).$ **23.** Set $s = x - y, \upsilon = y - z, w = z - x.$

31. $u^2 + v^2 = 2w.$ **32.** $w = (u - v)^2.$

33. $4w = u(u^2 + 3v).$ **34.** $u = \tan v.$

35. $u = \sin v,$ or $\upsilon = \sin^{-1} u.$

36. $-[yx^{y-1} + y^x \ln y]/[xy^{y-1} + x^y \ln x].$ **37.** $y/[x + 3y^2(x^2 + y^2)].$

38. $\left(\dfrac{\partial z}{\partial x}\right)_y = -\left(\dfrac{\partial f/\partial x}{\partial f/\partial z}\right) = -\dfrac{y(\sin xy) + z(\sin xz)}{y(\sin yz) + x(\sin xz)}, \left(\dfrac{\partial z}{\partial y}\right)_x = -\left(\dfrac{\partial f/\partial y}{\partial f/\partial z}\right) = -\dfrac{x(\sin xy) + z(\sin yz)}{y(\sin yz) + x(\sin xz)}$

39. $\left(\dfrac{\partial z}{\partial x}\right)_y = -\left(\dfrac{\partial f/\partial x}{\partial f/\partial z}\right) = -\dfrac{3x^2 + 3y + 3z}{3x + 2z}, \left(\dfrac{\partial z}{\partial y}\right)_x = -\left(\dfrac{\partial f/\partial y}{\partial f/\partial z}\right) = -\dfrac{3x - 4y}{3x + 2z}$

40. Let $u = z/y, \upsilon = x/y;$ then $f(u, \upsilon) = 0; x.$ **41.** $499.6.$

42. $4.02.$ **43.** $\dfrac{1}{2\sqrt{2}}\left[1 + \dfrac{\pi}{180}(2\sqrt{3} + 1)\right].$

44. $1.81.$ **45.** $\dfrac{1}{720}[180 + \pi(6 - \sqrt{3})] = 0.2686.$

46. $5.01.$ **47.** $V = \pi r^2 h/3, dV/dt = 85\pi/72 \approx 3.71 \text{ ft}^3/\text{hr}.$

48. $S = 2(xy + xz + yz)$, max. absolute error $= 2880 \text{ in}^2$, max. relative error $= 0.0766$ in, percentage error $\approx 7.66\%.$

49. $A = \dfrac{1}{2} xy \sin \alpha$, percentage error $\approx 13.7\%.$

50. $V = abc$, percentage error $= 3\%.$ **51.** $V = \pi r^2 h$, percentage error $\approx 9.2\%.$

52. 121.6 watts.   **53.** $2.92\%.$

54. $29.33\%.$

55. Lateral length $l = \sqrt{r^2 + h^2}$, lateral area $= \pi rl, dr = r/100, dh = h/100, dl = \sqrt{(dr)^2 + (dh)^2} = 1/20,$ percentage error $= 2\%.$

Exercise 2.3

1. At $(1, 1)$: $f_{xx} = -1/2, f_{xy} = 0, f_{yy} = 1/2.$

2. At $(2, 3)$: $f_{xxx} = 0, f_{xxy} = 0, f_{xyy} = -1/9, f_{yyy} = 4.27.$

3. At $(1, 2)$: $f_{xx} = 0, f_{xy} = 1, f_{yy} = -3/4.$

4. At $(1, \pi/2)$: $f_{xxx} = e \ln(\pi/2), f_{xxy} = (2e/\pi) + 1, f_{xyy} = -4e/\pi^2, f_{yyy} = 16e/\pi^3.$

5. At $(\pi/2, 1)$: $f_{xx} = -e, f_{xy} = \pi e/2, f_{yy} = -\pi^2 e/4.$

6. At $(1, -1, 1)$: $f_{xx} = -1/2, f_{xy} = -1/4, f_{xz} = -1/4, f_{yy} = 0, f_{yz} = -1/4, f_{zz} = 0.$

7. At $(-1, 1, -1)$: $f_{xx} = 6e^3, f_{xy} = -4e^3, f_{xz} = 4e^3, f_{yy} = 6e^3, f_{yz} = -4e^3, f_{zz} = 6e^3.$

8. At $(1, \pi/2, \pi/2)$: $f_{xx} = -\pi^2/2, f_{xy} = -\pi/2, f_{xz} = -\pi/2, f_{yy} = -[1 + (\pi^2 S/4)], f_{yz} = -[(\pi^2 S/4) - c],$
$f_{zz} = -[1 + (\pi^2 S/4)], S = \sin(\pi^2/4), c = \cos(\pi^2/4).$

9. At $(1, 2, 3)$: $f_{xx} = 6, f_{xy} = -1/4, f_{xz} = -1, f_{yy} = 1/4, f_{yz} = -1/9, f_{zz} = 4/27.$

10. $f_{xy} = f \ln(ex) \ln(ey).$

11. $f_x(0, 0) = 0, f_y(0, 0) = 0, f_x(0, y) = 0, f_y(x, 0) = x, f_{xy}(0, 0) = 1, f_{yx}(0, 0) = 0.$

12. $f_{xy}(x, y) = f_{yx}(x, y) = x^{y-1}(1 + y \ln x).$ **13.** $f_{xy}(x, y) = f_{yx}(x, y) = -y/(x^2 + y^2)^{3/2}.$

14. $(1 + xy)(\cos z)e^{xy}.$ **15.** $4(1 + 2y^2) z\ e^{x+y^2}.$

16. For $t = 0$, we get $x = 0$, $y = 0$, $dz/dt = -2.$

17. $\partial x/\partial u = 3u/x,\ \partial y/\partial u = 5u/y;\ \partial^2 x/\partial u^2 = 3(x^2 - 3u^2)/x^3,\ \partial^2 y/\partial u^2 = 5(y^2 - 5u^2)/y^3.$

18. For $x = 1$, $y = -1$, $z = 2$, we get $u = 1, v = 2;\ (\partial u/\partial x)_{y,z} = 5/3;\ (\partial v/\partial y)_{x,z} = 1/6.$

19. $\dfrac{dy}{dx} = -\sqrt{\dfrac{1-y^2}{1-x^2}},\ \dfrac{d^2 y}{dx^2} = -\dfrac{c}{(1-x^2)^{3/2}}.$

20. $dy/dx = (e-1)/(e+1),\ d^2y/dx^2 = 2(e^2+1)/(e+1)^3.$

21. $\dfrac{\partial z}{\partial x} = u^v (v/u)^{1/2} \ln(eu),\ \dfrac{\partial^2 z}{\partial x^2} = u^{v-1}[1 + v(\ln eu)^2].$

26. $\alpha = 3\beta$ or $\alpha = 4\beta$ and $\beta \neq 0$ arbitrary.

27. Note that $u_x^2 + u_y^2 = v_x^2 + v_y^2 = 1/(x^2 + y^2)^2$, $u_{xx} + u_{yy} = 0 = v_{xx} + v_{yy}$. We have

$$z_{xx} + z_{yy} = f_u(u_{xx} + u_{yy}) + f_v(v_{xx} + v_{yy}) + f_{uu}(u_x^2 + u_y^2) + f_{vv}(v_x^2 + v_y^2).$$

28. Use $x^2 + y^2 = r^2$, $\theta = \tan^{-1}(y/x)$ and differentiate.

29. $\sin u = (x^2 + y^2)/(x + y)$ is a homogeneous function of degree 1.

30. $e^u = [\sqrt{x^2 - y^2}/x]$ is a homogeneous function of degree 0.

31. u is a homogeneous function of degree 1. **32.** u is a homogeneous function of degree 1.

33. $w = \tan u$ is a homogeneous function of degree 2.

34. $f(x, y) = 6 - 5(x - 2) + 3(y - 2) + (x - 2)^2 + 3(y - 2)^2.$

35. $f(x, y) \approx -2 - 2(x - 1) - (y - 1); B = 4; |E| \leq 0.08.$

36. $f(x, y) \approx (x - 1) + y; B = 4.6912; |E| \leq 0.0938.$

37. $f(x, y) \approx 2 + [(x - 1) + 3(y - 1)] + \dfrac{1}{2}[-(x - 1)^2 + 6(x - 1)(y - 1) + (y - 1)^2]; B = 5.1; |E| \leq 0.0029.$

38. $f(x, y) \approx 2 + \dfrac{1}{4}[(x - 1) + (y - 3)] - \dfrac{1}{64}[(x - 1)^2 + 2(x - 1)(y - 3) + (y - 3)^2]; B = 0.0142,$
$|E| \leq 0.64 \times 10^{-4}.$

39. $f(x, y) \approx 1 + (2x + y) + \dfrac{1}{2}(2x + y)^2 + \dfrac{1}{6}(2x + y)^3; B \doteq 23.87; |E| \leq 0.008.$

40. $f(x, y) \approx (x + 2y) - \dfrac{1}{6}(x + 2y)^3; B = 16[\sin(0.3)] = 4.7283; |E| \leq 0.315 \times 10^{-3}.$

41. $f(x, y) \approx \dfrac{1}{2} + \dfrac{1}{2}\left[\left(x - \dfrac{\pi}{4}\right) + \left(y - \dfrac{\pi}{4}\right)\right] - \dfrac{1}{4}\left[\left(x - \dfrac{\pi}{4}\right)^2 - 2\left(x - \dfrac{\pi}{4}\right)\left(y - \dfrac{\pi}{4}\right) + \left(y - \dfrac{\pi}{4}\right)^2\right]; B = 1;$
$|E| \leq 0.0013.$

42. $f(x, y, z) \approx 3 + \dfrac{2}{3}[(x - 2) + (y - 2) + (z - 1)]; B = 0.3872; |E| \leq 0.017.$

43. $f(x, y, z) \approx 3 + \dfrac{3}{4}(x - 1) + \dfrac{5}{12}(y - 3) + \dfrac{2}{3}\left(z - \dfrac{3}{2}\right); B = 0.3985; |E| \leq 0.0179.$

44. $f(x, y, z) \approx x + y + xz + yz$; $B = 1.11$; $|E| \leq 0.005$.

45. $f(x, y, z) \approx 1 + x + \dfrac{1}{2}\left[x^2 - \dfrac{\pi^2}{4}(y-1)^2 - \left(z - \dfrac{\pi}{2}\right)^2 - \pi(y-1)\left(z - \dfrac{\pi}{2}\right)\right]$; $|B| = 7.0817$;

$|E| \leq 0.0319$.

Exercises 2.4

1. minimum value 9 at $(3, 1)$.

2. maximum value a at $(0, 0)$.

3. minimum value 0 at $(0, 0)$ if $|b| < 1$.

4. minimum value $(3)^{4/3}$ at $(3^{-1/3}, 3^{-1/3})$.

5. minimum value $5(2)^{-2/5}$ at $(\pm 2^{3/10}, 2^{-1/5})$.

6. minimum value $-3/2$ at $(\pi/3, 2\pi/3)$.

7. The matrix $\mathbf{A}$ or the matrix $\mathbf{B} = -\mathbf{A}$ is not positive definite. The function has no relative minimum or maximum.

8. The matrix $\mathbf{B} = -\mathbf{A}$ is positive definite and $f_{xx}, f_{yy}, f_{zz} < 0$ at $(0, 0, 0)$. Maximum value is 0.

9. $\mathbf{A}$ is positive definite and $f_{xx}, f_{yy}, f_{zz} > 0$ at $(-1, -1, -1)$, $(-1, 1, 1)$, $(1, -1, 1)$, $(1, 1, -1)$. Minimum value is -1 at all these points.

10. $\mathbf{B} = -\mathbf{A}$ is positive definite and $f_{xx}, f_{yy}, f_{zz} < 0$ at $(1/\sqrt{3}, 1/\sqrt{3}, 1/\sqrt{3})$. Maximum value is $(\log 3) - 1$.

11. No relative maximum and minimum. Absolute minimum value -3 at $(0, 1)$. Absolute maximum value $3/2$ at $(\pm\sqrt{3}/2, -1/2)$.

12. No relative maximum and minimum. Absolute maximum value $1/2$ at $(1/\sqrt{2}, 1/\sqrt{2})$ and $(-1/\sqrt{2}, -1/\sqrt{2})$. Absolute minimum $-1/2$ at $(-1/\sqrt{2}, 1/\sqrt{2})$ and $(1/\sqrt{2}, -1/\sqrt{2})$.

13. No relative maximum and minimum. Absolute maximum value $\sqrt{13}$ at $(9/\sqrt{13}, 4/\sqrt{13})$. Absolute minimum value $-\sqrt{13}$ at $(-9/\sqrt{13}, -4/\sqrt{13})$.

14. Relative minimum value $3/4$ at $(1/4, 0)$. Minimum value $3/2$ on the boundary at $(1/2, \pm 1/\sqrt{2})$. Absolute minimum value $3/4$ at $(1/4, 0)$.

15. Absolute minimum value $1/2$ at $(1/2, 1/2)$. Absolute maximum value 5 at $(2, 2)$.

16. Absolute minimum value $-93/18$ at $(1/6, 2/3)$. Absolute maximum value -4 at $(0, 0)$.

17. Absolute minimum value $-1/27$ at $(1/3, 1/3)$. Absolute maximum value 7 at $(1, 2)$.

18. Absolute minimum value $-23/2$ at $(2, -3/2)$. Absolute maximum value 37 at $(0, -4)$.

19. Absolute minimum value $-3/2$ at $(2\pi/3, 2\pi/3)$. Absolute maximum value 3 at $(0, 0)$.

20. Absolute maximum value 1 at $(0, 0)$, $(0, \pi)$, $(\pi, 0)$ and (π, π). Absolute minimum value $-1/8$ at $(\pi/3, \pi/3)$, $(2\pi/3, 2\pi/3)$.

21. $F = f(x, y) + \lambda\phi(x, y) \Rightarrow f_x + \lambda\phi_x = 0$ and $f_y + \lambda\phi_y = 0$. Eliminate λ.

22. $\lambda = -1/2$, $(x, y) = (1, 1/2)$; maximum value is $1/2$; minimum value is 0.

23. $\lambda = \sqrt{5}/2$, $(x, y) = (-1/\sqrt{5}, -2/\sqrt{5})$, minimum value is $-\sqrt{5}$;

$\lambda = -\sqrt{5}/2$, $(x, y) = (1/\sqrt{5}, 2/\sqrt{5})$, maximum value is $\sqrt{5}$.

24. Maximum value $(3\sqrt{3} - \pi)/3$ at $(\sqrt{3}/2, \pi/3)$. Minimum value $-(3\sqrt{3} + 5\pi)/3$ at $(-\sqrt{3}/2, 5\pi/3)$.

25. Extreme value is $\sqrt{2}$.

26. The points $(4, -4)$, $(-4, 4)$ are farthest, $d^2 = 32$. The points $(4/\sqrt{3}, 4/\sqrt{3})$, $(-4/\sqrt{3}, -4/\sqrt{3})$ are nearest, $d^2 = 32/3$.

27. Rectangle must be a square.

28. Triangle must be an equilateral triangle.

29. The point is $(AD/t, BD/t, CD/t)$, $t = A^2 + B^2 + C^2$.

30. Extreme value is $a^3/27$ at $(a/3, a/3, a/3)$.

31. Extreme value is $(a + b + c)^3$ at $(t/a, t/b, t/c)$, $t = a + b + c$.

32. Extreme value is $3^{(q-p)/q}$ at (t, t, t), $t = 3^{-1/q}$.

33. Extreme value is 24 at $(2, 1, 1/2)$.

34. Maximise $V = 8xyz$ such that $(x^2/a^2) + (y^2/b^2) + (z^2/c^2) = 1$. We get $(x, y, z) = (2a/\sqrt{3}, 2b/\sqrt{3}, 2c/\sqrt{3})$.

35. Maximise xy^2z^3 such that $x + y + z = a$, a constant. We get $x = a/6$, $y = a/3$, $z = a/2$.

36. Maximise $2(xy + xz + yz)$ such that $4(x + y + z) = a$, a constant. We get $x = y = z = a/12$, that is the parallelopiped is a cube.

37. Maximise $V = xyz$ such that $xy + xz + \hat{y}z = S/2$, we get $x = y = z = \sqrt{S/6}$.

38. Minimise $S = xy + 2xz + 2yz$ such that $xyz = a$. We get $x = y = (2a)^{1/3}$ and $z = x/2$.

39. Maximise $V = \pi r^2 h/3$ such that $\pi rl = a$ where $l = \sqrt{r^2 + h^2}$. We get $h = \sqrt{2}r$.

40. Maximise $V = \pi r^2 H + (\pi r^2 h)/3$ such that $2\pi rH + \pi rl = S$, $l = \sqrt{r^2 + h^2}$. We get $h/r = 2/\sqrt{5}$ and $H/r = 1/\sqrt{5}$.

41. Maximum value is $2/(3\sqrt{3})$ at $(\pm 2/\sqrt{3}, \pm 2/\sqrt{3}, 1/\sqrt{3})$.

42. Extreme value is $3/2$ at $(1/2, -1, 3/2)$.   43. Extreme value is $11/12$ at $(-1/6, 1/3, 5/6)$.

44. Farthest point $(1, 0, 0)$, $d = 1$; nearest point $(1/3, 0, 2/3)$ $d = \sqrt{5}/3$.

45. The coordinates of the points P and Q are obtained as $(2a/3, 2a/3, 2a/3)$ and $(\pm a/\sqrt{3}, \pm a/\sqrt{3}, \pm a/\sqrt{3})$. Shortest distance : $d^2 = a^2(7 - 4\sqrt{3})/3$; Largest distance : $d^2 = a^2(7 + 4\sqrt{3})/3$.

Exercise 2.5

1. Curves intersect at $x = \pm\sqrt{2}$, $y = 2$; $16\sqrt{2}/3$.

2. Curves intersect at $(1, 1)$ and $(4, -2)$; $9/2$.   3. $8/3$.

4. Curves intersect at $(0, 0)$ and $(1, 1)$; $1/10$.   5. $R: \pi/4 \le \theta \le \pi/2$, $2\sin \theta \le r \le 4 \sin \theta$; $3(\pi + 2)/4$.

6. $I = \int_{x=0}^{1} \int_{y=0}^{x} f(x, y)dy\, dx + \int_{x=1}^{\sqrt{2}} \int_{0}^{\sqrt{2-x^2}} f(x, y)dy\, dx$, where $f(x, y) = \dfrac{y}{\sqrt{x^2 + y^2}}$; $(2 - \sqrt{2})/2$.

7. $I = \int_{x=0}^{4} \int_{y=0}^{1} f(x, y)dy\, dx + \int_{x=4}^{5} \int_{y=x-4}^{1} f(x, y)dy\, dx$, where $f(x, y) = \dfrac{2y + 1}{x + 1}$;
   $20 \ln (5) - 18 \ln (6) + (7/2)$.

8. $I = \int_{x=0}^{1} \int_{y=x^3}^{x} e^{x^2} dy\, dx = (e - 2)/2$.   9. $I = \int_{y=0}^{2} \int_{x=\sqrt{2y}}^{2} \dfrac{x}{\sqrt{x^2 + y^2 + 1}} dx\, dy = \dfrac{1}{4}(5 \ln 5 - 4)$.

10. $I = \int_{0}^{1} \int_{0}^{1} ue^v du\, dv = \dfrac{1}{2}(e - 1)$.   11. $14\sqrt{2}/5$.

12. 380/3.

13. π.

14. $81\pi/16$.

15. $128\,a^3/15$.

16. $\pi/4$.

17. $27\pi/2$.

18. $2a^3/3$.

19. $(256 - 27\pi)/72$.

20. $8(2 - \sqrt{2})$.

21. πa^3.

22. 208/3.

23. 2π.

24. $a^3/90$.

25. $4\pi a^3(\sqrt{2}-1)/3$.

26. 1/3.

27. $2\pi(1 - \cos\alpha)/3$.

28. 1931/60.

29. $2\pi a^3/3$.

30. $I_y = a^5\pi/5 = I_x$.

31. $M = 3\pi\,k\,a^2/32,\ \bar{x} = \bar{y} = 8ka^3/(105\,M)$.

32. Evaluate the integral over $1 \le x^2 + y^2 \le a^2$ and take the limit as $a \to \infty$. $I = \pi/(p - 1)$.

33. 1/3.

34. 63/20.

35. $\pi/8$.

36. $2\pi ab/3$.

37. $(e^{50} - e^8)\,\pi/16$.

38. 1/96.

39. $a^4/280$.

40. 4/3.

41. 21.

42. 8/3.

43. 1/15.

44. $\pi/48$.

45. $(8 \ln 2 - 5)/16$.

46. 11/42.

47. 837/160.

48. 8/35.

49. $4\pi \ln(a/b)$.

50. $\pi/8$.

51. $\pi^2 abc/4$.

52. $\pi/10$.

53. $abc(a^2 + b^2)/60$.

54. $\pi h a^4/2$.

55. 0.

In problems **56** to **60** compare the given integral with Eq. (2.107).

56. $\alpha = \beta = \gamma = 2,\ a = b = c = 2,\ p = q = r = 1;\ I = 4/45$.

57. $\alpha = 2,\ \beta = 3,\ \gamma = 4,\ a = b = c = 1,\ p = q = r = 1,\ I = 12/9!$.

58. $\alpha = \beta = \gamma = 3/2,\ a = b = c = 1,\ p = q = r = 3,\ I = 64\sqrt{2}\ \pi/81$.

59. $\alpha = 2,\ \beta = 3/2,\ \gamma = 2,\ a = b = c = 1,\ p = 1,\ q = 3,\ r = 4,\ I = \pi/288$.

60. $\alpha = 3,\ \beta = 2,\ \gamma = 1,\ a = 1,\ b = 2,\ c = 3,\ p = q = r = 2,\ I = \pi/8$.

Chapter 3

Matrices and Eigenvalue Problems

3.1 Introduction

In modern mathematics, matrix theory occupies an important place and has applications in almost all branches of engineering and physical sciences. Matrices of order $m \times n$ form a vector space and they define linear transformations which map vector spaces consisting of vectors in $\mathbb{R}^n$ or $\mathbb{C}^n$ into another vector space consisting of vectors in $\mathbb{R}^n$ or $\mathbb{C}^m$ under a given set of rules of vector addition and scalar multiplication. A matrix does not denote a number and no value can be assigned to it. The usual rules of arithmetic operations do not hold for matrices. The rules defining the operations on matrices are usually called its algebra. In this chapter, we shall discuss the matrix algebra and its use in solving linear system of algebraic equations $\mathbf{Ax} = \mathbf{b}$ and solving the eigenvalue problem $\mathbf{Ax} = \lambda\mathbf{x}$.

3.2 Matrices

An $m \times n$ matrix is an arrangement of mn objects (not necessarily distinct) in m *rows* and n *columns* in the form

$$\mathbf{A} = \begin{bmatrix} a_{11} & a_{12} & \cdots & a_{1n} \\ a_{21} & a_{22} & \cdots & a_{2n} \\ \vdots & & & \\ a_{m1} & a_{m2} & \cdots & a_{mn} \end{bmatrix}. \tag{3.1}$$

We say that the matrix is of *order* $m \times n$ (m by n). The objects $a_{11}, a_{12}, \ldots, a_{mn}$ are called the *elements* of the matrix. Each element of the matrix can be a real or a complex number or a function of one or more variables or any other object. The element a_{ij} which is common to the ith row and the jth column is called its *general element*. The matrices are usually denoted by boldface uppercase letters $\mathbf{A}, \mathbf{B}, \mathbf{C}, \ldots$ etc. When the order of the matrix is understood, we can simply write $\mathbf{A} = (a_{ij})$. If all the elements of a matrix are real, it is called a *real matrix*, whereas if one or more elements of a matrix are complex it is called a *complex matrix*. We define the following particular types of matrices.

Row Vector   A matrix of order $1 \times n$, that is, it has one row and n columns is called a *row vector* or a row matrix of order n and is written as

$$[a_{11}\ a_{12}\ \dots\ a_{1n}], \text{ or } [a_1\ a_2\ \dots\ a_n]$$

in which a_{1j} (or a_j) is the jth element.

Column vector   A matrix of order $m \times 1$, that is, it has m rows and one column is called a *column vector* or a *column matrix* of order m and is written as

$$\begin{bmatrix} b_{11} \\ b_{21} \\ \vdots \\ b_{m1} \end{bmatrix}, \text{ or } \begin{bmatrix} b_1 \\ b_2 \\ \vdots \\ b_m \end{bmatrix}$$

in which b_{j1} (or b_j) is the jth element.

The number of elements in a row/column vector is called its *order*. The vectors are usually denoted by boldface lower case letters **a**, **b**, **c**, ... etc. If a vector has n elements and all its elements are real numbers, then it is called an *ordered n-tuple* in $\mathbb{R}^n$, whereas if one or more elements are complex numbers, then it is called an ordered *n-tuple* in $\mathbb{C}^n$.

Rectangular matrix   A matrix **A** of order $m \times n$, $m \neq n$ is called a *rectangular matrix*.

Square matrices   A matrix **A** of order $m \times n$ in which $m = n$, that is number of rows is equal to the number of columns is called a square matrix of order n. The elements a_{ii}, that is the elements $a_{11}, a_{22}, \dots, a_{nn}$ are called the *diagonal elements* and the line on which these elements lie is called the *principal diagonal* or the *main diagonal* of the matrix. The elements a_{ij}, when $i \neq j$ are called the *off-diagonal elements*. The sum of the diagonal elements of a square matrix is called the *trace* of the matrix.

Null matrix   A matrix **A** of order $m \times n$ in which all the elements are zero is called a *null matrix* or a *zero matrix* and is denoted by **0**.

Diagonal matrix   A square matrix **A** in which all the off-diagonal elements a_{ij}, $i \neq j$ are zero is called a diagonal matrix. For example

$$\mathbf{A} = \begin{bmatrix} a_{11} & & & \mathbf{0} \\ & a_{22} & & \\ & & \ddots & \\ \mathbf{0} & & & a_{nn} \end{bmatrix} \text{ is a } \textit{diagonal matrix of order } n.$$

A diagonal matrix is denoted by **D**. It is also written as diag $[a_{11}\ a_{22}\ \dots\ a_{nn}]$.

If all the elements of a diagonal matrix of order n are equal, that is $a_{ii} = \alpha$ for all i, then the matrix is called a *scalar matrix* of order n.

If all the elements of a diagonal matrix of order n are 1, then the matrix

$$A = \begin{bmatrix} 1 & & & \mathbf{0} \\ & 1 & & \\ & & \ddots & \\ \mathbf{0} & & & 1 \end{bmatrix} \quad \text{is called an } \textit{unit matrix} \text{ or an } \textit{identity matrix} \text{ of order } n.$$

An identity matrix is denoted by $\mathbf{I}$.

Equal matrices Two matrices $\mathbf{A} = (a_{ij})_{m \times n}$ and $\mathbf{B} = (b_{ij})_{p \times q}$ are said to be equal, when

  (i) they are of the same order, that is $m = p$, $n = q$ and

  (ii) their corresponding elements are equal, that is $a_{ij} = b_{ij}$ for all i, j.

Submatrix A matrix obtained by omitting some rows and or columns from a given matrix $\mathbf{A}$ is called a *submartix* of $\mathbf{A}$. As a convention, the given matrix $\mathbf{A}$ is also taken as the submatrix of $\mathbf{A}$.

3.2.1 Matrix Algebra

The basic operations allowed on matrices are

  (i) multiplication of a matrix by a scalar,

  (ii) addition/subtraction of two matrices,

  (iii) multiplication of two matrices.

Note that there is no concept of dividing a matrix by a matrix. Therefore, the operation $\mathbf{A}/\mathbf{B}$ where $\mathbf{A}$ and $\mathbf{B}$ are matrices is not defined.

Multiplication of a matrix by a scalar

Let α be a scalar (real or complex) and $\mathbf{A} = (a_{ij})$ be a given matrix of order $m \times n$. Then

$$\mathbf{B} = \alpha \mathbf{A} = \alpha(a_{ij}) = (\alpha a_{ij}) \quad \text{for all } i \text{ and } j. \tag{3.2}$$

The order of the new matrix $\mathbf{B}$ is same as that of the matrix $\mathbf{A}$.

Addition/subtraction of two matrices

Let $\mathbf{A} = (a_{ij})$ and $\mathbf{B} = (b_{ij})$ be two matrices of the same order. Then

$$\mathbf{C} = (c_{ij}) = \mathbf{A} + \mathbf{B} = (a_{ij}) + (b_{ij}) = (a_{ij} + b_{ij}), \quad \text{for all } i \text{ and } j \tag{3.3a}$$

and $\qquad\qquad \mathbf{D} = (d_{ij}) = \mathbf{A} - \mathbf{B} = (a_{ij}) - (b_{ij}) = (a_{ij} - b_{ij}), \quad \text{for all } i \text{ and } j. \tag{3.3b}$

The order of the new matrix $\mathbf{C}$ or $\mathbf{D}$ is the same as that of the matrices $\mathbf{A}$ and $\mathbf{B}$. Matrices of the same order are said to be *conformable* for addition/subtraction.

If $\mathbf{A}_1, \mathbf{A}_2, ..., \mathbf{A}_p$ are p matrices which are conformable for addition and $\alpha_1, \alpha_2, ..., \alpha_p$ are any scalars, then

$$\mathbf{C} = \alpha_1 \mathbf{A}_1 + \alpha_2 \mathbf{A}_2 + ... + \alpha_p \mathbf{A}_p \tag{3.4}$$

is called a linear combination of the matrices $\mathbf{A}_1, \mathbf{A}_2, ..., \mathbf{A}_p$. The order of the matrix $\mathbf{C}$ is same as that of $\mathbf{A}_i$, $i = 1, 2, ..., p$.

Properties of the matrix addition and scalar multiplication

Let **A**, **B**, **C** be the matrices which are conformable for addition and α, β be scalars. Then

1. $\mathbf{A} + \mathbf{B} = \mathbf{B} + \mathbf{A}$.   (commutative law)
2. $(\mathbf{A} + \mathbf{B}) + \mathbf{C} = \mathbf{A} + (\mathbf{B} + \mathbf{C})$ (associative law).
3. $\mathbf{A} + \mathbf{0} = \mathbf{A}$ (**0** is the null matrix of the same order as **A**).
4. $\mathbf{A} + (-\mathbf{A}) = \mathbf{0}$.   5. $\alpha\,(\mathbf{A} + \mathbf{B}) = \alpha\mathbf{A} + \alpha\mathbf{B}$.
6. $(\alpha + \beta)\mathbf{A} = \alpha\mathbf{A} + \beta\mathbf{A}$.   7. $\alpha\,(\beta\mathbf{A}) = \alpha\beta\mathbf{A}$.
8. $1 \times \mathbf{A} = \mathbf{A}$ and $0 \times \mathbf{A} = \mathbf{0}$.

Multiplication of two matrices

The product **AB** of two matrices **A** and **B** is defined only when the number of columns in **A** is equal to the number of rows in **B**. Such matrices are said to be *conformable* for multiplication. Let $\mathbf{A} = (a_{ij})$ be an $m \times n$ matrix and $\mathbf{B} = (b_{ij})$ be an $n \times p$ matrix. Then the product matrix

$$\mathbf{C} = (c_{ij}) = \mathbf{AB} = \begin{bmatrix} a_{11} & a_{12} & \cdots & a_{1n} \\ a_{21} & a_{22} & \cdots & a_{2n} \\ \vdots & & & \\ a_{i1} & a_{i2} & \cdots & a_{in} \\ \vdots & & & \\ a_{m1} & a_{m2} & \cdots & a_{mn} \end{bmatrix} \begin{bmatrix} b_{11} & b_{12} & \cdots & b_{1j} & \cdots & b_{1p} \\ b_{21} & b_{22} & \cdots & b_{2j} & \cdots & b_{2p} \\ \vdots & & & & & \\ b_{n1} & b_{n2} & \cdots & b_{nj} & \cdots & b_{np} \end{bmatrix}$$

$$\qquad\qquad m \times n \qquad\qquad\qquad\qquad n \times p$$

is a matrix of order $m \times p$. The general element of the product matrix **C** is given by

$$c_{ij} = a_{i1}b_{1j} + a_{i2}b_{2j} + \ldots + a_{in}b_{nj} = \sum_{k=1}^{n} a_{ik}b_{kj}. \qquad (3.5)$$

In the product **AB**, **B** is said to be pre-multiplied by **A** or **A** is said to be post-multiplied by **B**. If **A** is a row matrix of order $1 \times n$ and **B** is a column matrix of order $n \times 1$, then **AB** is a matrix of order 1×1, that is a single element, and **BA** is a matrix of order $n \times n$.

Remark 1

(a) It is possible that for two given matrices **A** and **B**, the product matrix **AB** is defined but the product matrix **BA** may not be defined. For example, if **A** is a 2×3 matrix and **B** is a 3×4 matrix, then the product matrix **AB** is defined and is a matrix of order 2×4, whereas the product matrix **BA** is not defined.

(b) If both the product matrices **AB** and **BA** are defined, then both the matrices **AB** and **BA** are square matrices. In general $\mathbf{AB} \neq \mathbf{BA}$. Thus, the matrix product is not commutative.

If $\mathbf{AB} = \mathbf{BA}$, then the matrices **A** and **B** are said to *commute* with each other.

(c) If $\mathbf{AB} = \mathbf{0}$, then it does not always imply that either $\mathbf{A} = \mathbf{0}$ or $\mathbf{B} = \mathbf{0}$. For example, let

$$\mathbf{A} = \begin{bmatrix} x & 0 \\ y & 0 \end{bmatrix} \quad \text{and} \quad \mathbf{B} = \begin{bmatrix} 0 & 0 \\ a & b \end{bmatrix}$$

then
$$\mathbf{AB} = \begin{bmatrix} 0 & 0 \\ 0 & 0 \end{bmatrix} \quad \text{and} \quad \mathbf{BA} = \begin{bmatrix} 0 & 0 \\ ax + by & 0 \end{bmatrix} \neq \mathbf{AB}.$$

(d) If $\mathbf{AB} = \mathbf{AC}$, it does not always imply that $\mathbf{B} = \mathbf{C}$.

(e) Define $\mathbf{A}^k = \mathbf{A} \times \mathbf{A} \ldots \times \mathbf{A}$ (k times). Then, a matrix $\mathbf{A}$ such that $\mathbf{A}^k = \mathbf{0}$ for some positive integer k is said to be *nilpotent*. The smallest value of k for which $\mathbf{A}^k = \mathbf{0}$ is called the *index of nilpotency* of the matrix $\mathbf{A}$.

(f) If $\mathbf{A}^2 = \mathbf{A}$, then $\mathbf{A}$ is called an *idempotent matrix*.

Properties of matrix multiplication

1. If $\mathbf{A}$, $\mathbf{B}$, $\mathbf{C}$ are matrices of order $m \times n$, $n \times p$ and $p \times q$ respectively, then

   $$(\mathbf{AB})\mathbf{C} = \mathbf{A}(\mathbf{BC}) \qquad \text{(associative law)}$$

   is a matrix of order $m \times q$.

2. If $\mathbf{A}$ is a matrix of order $m \times n$ and $\mathbf{B}$, $\mathbf{C}$ are matrices of order $n \times p$, then

   $$\mathbf{A}(\mathbf{B} + \mathbf{C}) = \mathbf{AB} + \mathbf{AC} \qquad \text{(left distributive law)}.$$

3. If $\mathbf{A}$, $\mathbf{B}$ are matrices of order $m \times n$ and $\mathbf{C}$ is a matrix of order $n \times p$, then

   $$(\mathbf{A} + \mathbf{B})\mathbf{C} = \mathbf{AC} + \mathbf{BC} \qquad \text{(right distributive law)}.$$

4. If $\mathbf{A}$ is a matrix of order $m \times n$ and $\mathbf{B}$ is a matrix of order $n \times p$, then

   $$\alpha(\mathbf{AB}) = \mathbf{A}(\alpha\mathbf{B}) = (\alpha\mathbf{A})\mathbf{B}$$

   for any scalar α.

3.2.2 Some Special Matrices

We now define some special matrices.

Transpose of a matrix   The matrix obtained by interchanging the corresponding rows and columns of a given matrix $\mathbf{A}$ is called the *transpose matrix* of $\mathbf{A}$ and is denoted by $\mathbf{A}^T$ or $\mathbf{A}'$, that is, if

$$\mathbf{A} = \begin{bmatrix} a_{11} & a_{12} & \cdots & a_{1n} \\ a_{21} & a_{22} & \cdots & a_{2n} \\ \vdots & & & \vdots \\ a_{m1} & a_{m2} & \cdots & a_{mn} \end{bmatrix}, \quad \text{then} \quad \mathbf{A}^T = \begin{bmatrix} a_{11} & a_{21} & \cdots & a_{m1} \\ a_{12} & a_{22} & \cdots & a_{m2} \\ \vdots & \vdots & & \vdots \\ a_{1n} & a_{2n} & & a_{mn} \end{bmatrix}.$$

If $\mathbf{A}$ is an $m \times n$ matrix, then $\mathbf{A}^T$ is an $n \times m$ matrix. Also, both the product matrices $\mathbf{A}^T\mathbf{A}$ and $\mathbf{AA}^T$ are defined, and

$$\mathbf{A}^T\mathbf{A} = (n \times m)(m \times n) \text{ is an } n \times n \text{ square matrix}$$

and
$$\mathbf{A}\mathbf{A}^T = (m \times n)(n \times m) \text{ is an } m \times m \text{ square matrix.}$$

A column vector $\mathbf{b}$ can also be written as $[b_1 \; b_2 \; \ldots \; b_n]^T$.

The following results can be easily verified

1. The transpose of a row matrix is a column matrix and the transpose of a column matrix is a row matrix.
2. $(\mathbf{A}^T)^T = \mathbf{A}$.
3. $(\mathbf{A} + \mathbf{B})^T = \mathbf{A}^T + \mathbf{B}^T$, when the matrices $\mathbf{A}$ and $\mathbf{B}$ are conformable for addition.
4. $(\mathbf{A}\mathbf{B})^T = \mathbf{B}^T\mathbf{A}^T$, when the matrices $\mathbf{A}$ and $\mathbf{B}$ are conformable for multiplication.

   If the product $\mathbf{A}_1 \mathbf{A}_2 \ldots \mathbf{A}_p$ is defined, then
   $$[\mathbf{A}_1 \mathbf{A}_2 \ldots \mathbf{A}_p]^T = \mathbf{A}_p^T \; \mathbf{A}_{p-1}^T \; \ldots \; \mathbf{A}_1^T.$$

Remark 2

The product of a row vector $\mathbf{a}_i = (a_{i1} \; a_{i2} \; \ldots \; a_{in})$ of order $1 \times n$ and a column vector $\mathbf{b}_j = (b_{1j} \; b_{2j} \; \ldots \; b_{nj})^T$ of order $n \times 1$ is called the *dot product* or the *inner product* of the vectors $\mathbf{a}_i$ and $\mathbf{b}_j$, that is

$$c_{ij} = \mathbf{a}_i \cdot \mathbf{b}_j = \sum_{k=1}^{n} a_{ik} b_{kj}$$

which is a scalar. In terms of the inner products, the product matrix $\mathbf{C}$ in Eq. (3.5) can be written as

$$\mathbf{C} = \mathbf{A}\mathbf{B} = \begin{bmatrix} \mathbf{a}_1 \cdot \mathbf{b}_1 & \mathbf{a}_1 \cdot \mathbf{b}_2 & \cdots & \mathbf{a}_1 \cdot \mathbf{b}_p \\ \mathbf{a}_2 \cdot \mathbf{b}_1 & \mathbf{a}_2 \cdot \mathbf{b}_2 & \cdots & \mathbf{a}_2 \cdot \mathbf{b}_p \\ \cdots & \cdots & \cdots & \\ \mathbf{a}_m \cdot \mathbf{b}_1 & \mathbf{a}_m \cdot \mathbf{b}_2 & \cdots & \mathbf{a}_m \cdot \mathbf{b}_p \end{bmatrix}. \tag{3.6}$$

Symmetric and skew-symmetric matrices   A real square matrix $\mathbf{A} = (a_{ij})$ is said to be

   symmetric,   if $a_{ij} = a_{ji}$ for all i and j, that is $\mathbf{A} = \mathbf{A}^T$

skew-symmetric,   if $a_{ij} = -a_{ji}$ for all i and j, that is $\mathbf{A} = -\mathbf{A}^T$.

Remark 3

(a) In a skew-symmetric matrix $\mathbf{A} = (a_{ij})$, all its diagonal elements are zero.

(b) The matrix which is both symmetric and skew-symmetric must be a null matrix.

(c) For any real square matrix $\mathbf{A}$, the matrix $\mathbf{A} + \mathbf{A}^T$ is always symmetric and the matrix $\mathbf{A} - \mathbf{A}^T$ is always skew-symmetric. Therefore, a real square matrix $\mathbf{A}$ can be written as the sum of a symmetric matrix and a skew-symmetric matrix. That is

$$\mathbf{A} = \frac{1}{2}(\mathbf{A} + \mathbf{A}^T) + \frac{1}{2}(\mathbf{A} - \mathbf{A}^T).$$

Triangular matrices   A square matrix $\mathbf{A} = (a_{ij})$ is called a *lower triangular matrix* if $a_{ij} = 0$, whenever $i < j$, that is all elements above the principal diagonal are zero and an *upper triangular matrix* if $a_{ij} = 0$, whenever $i > j$, that is all the elements below the principal diagonal are zero.

Conjugate matrix   Let $\mathbf{A} = (a_{ij})$ be a complex matrix. Let $\bar{a}_{ij}$ denote the complex conjugate of a_{ij}. Then, the matrix $\overline{\mathbf{A}} = (\bar{a}_{ij})$ is called the *conjugate matrix* of A.

Hermitian and skew-Hermitian matrices   A complex matrix $\mathbf{A}$ is called an *Hermitian matrix* if $\overline{\mathbf{A}} = \mathbf{A}^T$ or $\mathbf{A} = (\overline{\mathbf{A}})^T$ and a *skew-Hermitian* matrix if $\overline{\mathbf{A}} = -\mathbf{A}^T$ or $\mathbf{A} = -(\overline{\mathbf{A}})^T$. Sometimes, a Hermitian matrix is denoted by $\mathbf{A}^H$ or $\mathbf{A}^*$.

Remark 4

(a) If $\mathbf{A}$ is a real matrix, then an Hermitian matrix is same as a symmetric matrix and a skew-Hermitian matrix is same as a skew-symmetric matrix.

(b) In an Hermitian matrix, all the diagonal elements are real (let $a_{jj} = x_j + iy_j$; then $a_{jj} = \bar{a}_{jj}$ gives $x_j + iy_j = x_j - iy_j$ or $y_j = 0$ for all j).

(c) In a skew-Hermitian matrix, all the diagonal elements are either 0 or pure imaginary (let $a_{jj} = x_j + iy_j$; then $a_{jj} = -\bar{a}_{jj}$ gives $x_j + iy_j = -(x_j - iy_j)$ or $x_j = 0$ for all j).

(d) For any complex square matrix $\mathbf{A}$, the matrix $\mathbf{A} + \overline{\mathbf{A}}^T$ is always an Hermitian matrix and the matrix $\mathbf{A} - \overline{\mathbf{A}}^T$ is always a skew-Hermitian matrix. Therefore, a complex square matrix $\mathbf{A}$ can be written as the sum of an Hermitian matrix and a skew-Hermitian matrix, that is

$$\mathbf{A} = \frac{1}{2}(\mathbf{A} + \overline{\mathbf{A}}^T) + \frac{1}{2}(\mathbf{A} - \overline{\mathbf{A}}^T).$$

Example 3.1   Let $\mathbf{A}$ and $\mathbf{B}$ be two symmetric matrices of the same order. Show that the matrix $\mathbf{AB}$ is symmetric if and only if $\mathbf{AB} = \mathbf{BA}$, that is the matrices $\mathbf{A}$ and $\mathbf{B}$ commute.

Solution   Since the matrices $\mathbf{A}$ and $\mathbf{B}$ are symmetric, we have

$$\mathbf{A}^T = \mathbf{A} \quad \text{and} \quad \mathbf{B}^T = \mathbf{B}.$$

Let $\mathbf{AB}$ be symmetric. Then

$$(\mathbf{AB})^T = \mathbf{AB}, \quad \text{or} \quad \mathbf{B}^T\mathbf{A}^T = \mathbf{AB}, \quad \text{or} \quad \mathbf{BA} = \mathbf{AB}.$$

Now, let $\mathbf{AB} = \mathbf{BA}$. Taking transpose on both sides, we get

$$(\mathbf{AB})^T = (\mathbf{BA})^T = \mathbf{A}^T\mathbf{B}^T = \mathbf{AB}.$$

Hence, the result.

3.2.3   Determinants

With every square matrix $\mathbf{A}$ of order n, we associate a determinant of order n which is denoted by $det(\mathbf{A})$ or $|\mathbf{A}|$. The determinant has a value and this value is real if the matrix $\mathbf{A}$ is real and may be real or complex, if the matrix is complex. A determinant of order n is defined as

$$det(\mathbf{A}) = |\mathbf{A}| = \begin{vmatrix} a_{11} & a_{12} & \cdots & a_{1n} \\ a_{21} & a_{22} & \cdots & a_{2n} \\ \vdots & & & \\ a_{n1} & a_{n2} & \cdots & a_{nn} \end{vmatrix}$$

$$= \sum_{j=1}^{n} (-1)^{i+j} \, a_{ij} \, M_{ij} = \sum_{j=1}^{n} a_{ij} \, A_{ij}$$

$$= \sum_{i=1}^{n} (-1)^{i+j} \, a_{ij} \, M_{ij} = \sum_{i=1}^{n} a_{ij} \, A_{ij} \qquad (3.7)$$

where M_{ij} and A_{ij} are the *minors* and *cofactors* of a_{ij} respectively.

We give now some important properties of determinants.

1. If all the elements of a row (or column) are zero then the value of the determinant is zero.
2. $|\mathbf{A}| = |\mathbf{A}^T|$.
3. If any two rows (or columns) are interchanged, then the value of the determinant is multiplied by (-1).
4. If the corresponding elements of two rows (or columns) are proportional to each other, then the value of the determinant is zero.
5. If each element of a row (or column) is multiplied by a scalar α then the value of the determinant is multiplied by the scalar α. Therefore, if β is a factor of each element of a row (or column), then this factor β can be taken out of the determinant.

   Note that when we multiply a matrix by a scalar α, then every element of the matrix is multiplied by α. Therefore, $|\alpha \mathbf{A}| = \alpha^n |\mathbf{A}|$ where $\mathbf{A}$ is a matrix of order n.
6. If a non-zero constant multiple of the elements of some row (or column) is added to the corresponding elements of some other row (or column), then the value of the determinant remains unchanged.
7. $|\mathbf{A} + \mathbf{B}| \neq |\mathbf{A}| + |\mathbf{B}|$, in general.

Remark 5

When the elements of the jth row are multiplied by a non-zero constant k and added to the corresponding elements of the ith row, we denote this operation as $R_i \leftarrow R_i + kR_j$, where R_i is the ith row of $|\mathbf{A}|$. The elements of the jth row remain unchanged whereas the elements of the ith row get changed. This operation is called an *elementary row operation*. Similarly, the operation $C_i \leftarrow C_i + kC_j$, where C_i is the ith column of $|\mathbf{A}|$, is called the *elementary column operation*. Therefore, under elementary row (or column) operations, the value of a determinant is unchanged.

Product of two determinants

If $\mathbf{A}$ and $\mathbf{B}$ are two square matrices of the same order, then

$$|\mathbf{AB}| = |\mathbf{A}| \, |\mathbf{B}|.$$

Since $|\mathbf{A}| = |\mathbf{A}^T|$, we can multiply two determinants in any one of the following ways

(i) row by row, (ii) column by column,

(iii) row by column, (iv) column by row.

The value of the determinant is same in each case.

Rank of a matrix

The *rank* of a matrix **A**, denoted by r or $r(\mathbf{A})$ is the order of the largest non-zero minor of $|\mathbf{A}|$. Therefore, the rank of a matrix is the largest value of r, for which there exists at least one $r \times r$ submartix of **A** whose determinant is not zero. Thus, for an $m \times n$ matrix $r \leq \min(m, n)$. For a square matrix **A** of order n, the rank $r = n$ if $|\mathbf{A}| \neq 0$, otherwise $r < n$. The rank of a null matrix is zero and if the rank of matrix is 0, then it must be a null matrix.

Example 3.2   Find all values of μ for which rank of the matrix

$$\mathbf{A} = \begin{bmatrix} \mu & -1 & 0 & 0 \\ 0 & \mu & -1 & 0 \\ 0 & 0 & \mu & -1 \\ -6 & 11 & -6 & 1 \end{bmatrix}$$

is equal to 3.

Solution   Since the matrix **A** is of order 4, $r(\mathbf{A}) \leq 4$. Now, $r(\mathbf{A}) = 3$, if $|\mathbf{A}| = 0$ and there is at least one submatrix of order 3 whose determinant is not zero. Expanding the determinant through the elements of first row, we get

$$|\mathbf{A}| = \mu \begin{vmatrix} \mu & -1 & 0 \\ 0 & \mu & -1 \\ 11 & -6 & 1 \end{vmatrix} + \begin{vmatrix} 0 & -1 & 0 \\ 0 & \mu & -1 \\ -6 & -6 & 1 \end{vmatrix} = \mu \left[\mu(\mu - 6) + 11 \right] - 6$$

$$= \mu^3 - 6\mu^2 + 11\mu - 6 = (\mu - 1)(\mu - 2)(\mu - 3).$$

Setting $|\mathbf{A}| = 0$, we obtain $\mu = 1, 2, 3$. For $\mu = 1, 2, 3$, the determinant of the leading third order submatrix

$$|\mathbf{A}_1| = \begin{vmatrix} \mu & -1 & 0 \\ 0 & \mu & -1 \\ 0 & 0 & \mu \end{vmatrix} = \mu^3 \neq 0.$$

Hence, $r(\mathbf{A}) = 3$, when $\mu = 1$ or 2 or 3. For other values of μ, $r(\mathbf{A}) = 4$.

3.2.4   Inverse of a Square Matrix

Let $\mathbf{A} = (a_{ij})$ be a square matrix of order n. Then, **A** is called a

   (i) *singular matrix* if $|\mathbf{A}| = 0$,

   (ii) *non-singular matrix* if $|\mathbf{A}| \neq 0$.

In other words, a square matrix of order n is singular if its rank $r(\mathbf{A}) < n$ and non-singular if its rank $r(\mathbf{A}) = n$. A square non-singular matrix **A** of order n is said to be *invertible*, if there exists a non-singular square matrix **B** of order n such that

$$\mathbf{AB} = \mathbf{BA} = \mathbf{I} \tag{3.8}$$

where **I** is an identity matrix of order n. The matrix **B** is called the *inverse matrix* of **A** and we write $\mathbf{B} = \mathbf{A}^{-1}$ or $\mathbf{A} = \mathbf{B}^{-1}$. Hence, we say that $\mathbf{A}^{-1}$ is the inverse of the matrix **A**, if

$$\mathbf{A}^{-1}\mathbf{A} = \mathbf{A}\mathbf{A}^{-1} = \mathbf{I}. \tag{3.9}$$

The inverse, $\mathbf{A}^{-1}$ of the matrix **A** is given by

$$\mathbf{A}^{-1} = \frac{1}{|\mathbf{A}|} \, adj(\mathbf{A}) \tag{3.10}$$

where $adj(\mathbf{A})$ = adjoint matrix of **A**

$\qquad\qquad$ = transpose of the matrix of cofactors of **A**.

Remark 6

(a) $\qquad\qquad\qquad\qquad (\mathbf{AB})^{-1} = \mathbf{B}^{-1}\mathbf{A}^{-1}.$

We have

$$(\mathbf{AB})(\mathbf{AB})^{-1} = \mathbf{I}.$$

Pre-multiplying both sides first by $\mathbf{A}^{-1}$ and then by $\mathbf{B}^{-1}$ we obtain

$$\mathbf{B}^{-1}\mathbf{A}^{-1}(\mathbf{AB})(\mathbf{AB})^{-1} = \mathbf{B}^{-1}(\mathbf{A}^{-1}\mathbf{A})\,\mathbf{B}(\mathbf{AB})^{-1} = \mathbf{B}^{-1}\mathbf{A}^{-1} \text{ or } (\mathbf{AB})^{-1} = \mathbf{B}^{-1}\mathbf{A}^{-1}.$$

In general, we have $(\mathbf{A}_1\mathbf{A}_2 \ldots \mathbf{A}_p)^{-1} = \mathbf{A}_p^{-1} \mathbf{A}_{p-1}^{-1} \ldots \mathbf{A}_1^{-1}.$

(b) If **A** and **B** are non-singular matrices, then **AB** is also a non-singular matrix.

(c) If $\mathbf{AB} = 0$ and **A** is a non-singular matrix, then **B** must be null matrix, since $\mathbf{AB} = 0$ can be pre-multiplied by $\mathbf{A}^{-1}$. If **B** is non-singular matrix, then **A** must be a null matrix, since $\mathbf{AB} = 0$ can be post-multiplied by $\mathbf{B}^{-1}$.

(d) If $\mathbf{AB} = \mathbf{AC}$ and **A** is a non-singular matrix, then $\mathbf{B} = \mathbf{C}$ (see Remark 1(d)).

(e) $(\mathbf{A} + \mathbf{B})^{-1} \neq \mathbf{A}^{-1} + \mathbf{B}^{-1}$, in general.

Properties of inverse martices

1. If $\mathbf{A}^{-1}$ exists, then it is unique.

2. $(\mathbf{A}^{-1})^{-1} = \mathbf{A}.$

3. $(\mathbf{A}^T)^{-1} = (\mathbf{A}^{-1})^T.$ (From $(\mathbf{AA}^{-1})^T = \mathbf{I}^T = \mathbf{I}$, we get $(\mathbf{A}^{-1})^T\mathbf{A}^T = \mathbf{I}$. Hence, the result).

4. Let $\mathbf{D} = \text{diag}\,(d_{11}, d_{22}, \ldots, d_{nn}),\ d_{ii} \neq 0$. Then, $\mathbf{D}^{-1} = \text{diag}\,(1/d_{11}, 1/d_{22}, \ldots 1/d_{nn}).$

5. The inverse of a non-singular upper or lower triangular matrix is respectively an upper or a lower triangular matrix.

6. The inverse of a non-singular symmetric matrix is a symmetric matrix.

7. $(\mathbf{A}^{-1})^n = \mathbf{A}^{-n}$ for any positive integer n.

Example 3.3 Show that the matrix $\mathbf{A} = \begin{bmatrix} 2 & 0 & -1 \\ 5 & 1 & 0 \\ 0 & 1 & 3 \end{bmatrix}$ satisfies the matrix equation $\mathbf{A}^3 - 6\mathbf{A}^2 +$

$11\mathbf{A} - \mathbf{I} = \mathbf{0}$ where $\mathbf{I}$ is an identity matrix of order 3. Hence, find the matrix (i) $\mathbf{A}^{-1}$ and (ii) $\mathbf{A}^{-2}$.

Solution   We have

$$\mathbf{A}^2 = \begin{bmatrix} 2 & 0 & -1 \\ 5 & 1 & 0 \\ 0 & 1 & 3 \end{bmatrix}\begin{bmatrix} 2 & 0 & -1 \\ 5 & 1 & 0 \\ 0 & 1 & 3 \end{bmatrix} = \begin{bmatrix} 4 & -1 & -5 \\ 15 & 1 & -5 \\ 5 & 4 & 9 \end{bmatrix}.$$

$$\mathbf{A}^3 = \mathbf{A}^2\mathbf{A} = \begin{bmatrix} 4 & -1 & -5 \\ 15 & 1 & -5 \\ 5 & 4 & 9 \end{bmatrix}\begin{bmatrix} 2 & 0 & -1 \\ 5 & 1 & 0 \\ 0 & 1 & 3 \end{bmatrix} = \begin{bmatrix} 3 & -6 & -19 \\ 35 & -4 & -30 \\ 30 & 13 & 22 \end{bmatrix}.$$

Substituting in $\mathbf{B} = \mathbf{A}^3 - 6\mathbf{A}^2 + 11\mathbf{A} - \mathbf{I}$, we get

$$\mathbf{B} = \begin{bmatrix} 3 & -6 & -19 \\ 35 & -4 & -30 \\ 30 & 13 & 22 \end{bmatrix} - \begin{bmatrix} 24 & -6 & -30 \\ 90 & 6 & -30 \\ 30 & 24 & 54 \end{bmatrix} + \begin{bmatrix} 22 & 0 & -11 \\ 55 & 11 & 0 \\ 0 & 11 & 33 \end{bmatrix} - \begin{bmatrix} 1 & 0 & 0 \\ 0 & 1 & 0 \\ 0 & 0 & 1 \end{bmatrix}$$

$$= \begin{bmatrix} 0 & 0 & 0 \\ 0 & 0 & 0 \\ 0 & 0 & 0 \end{bmatrix} = \mathbf{0}.$$

(i) Premultiplying $\mathbf{A}^3 - 6\mathbf{A}^2 + 11\mathbf{A} - \mathbf{I} = \mathbf{0}$ by $\mathbf{A}^{-1}$, we get

$$\mathbf{A}^{-1}\mathbf{A}^3 - 6\mathbf{A}^{-1}\mathbf{A}^2 + 11\mathbf{A}^{-1}\mathbf{A} - \mathbf{A}^{-1} = \mathbf{0}$$

or $\qquad \mathbf{A}^{-1} = \mathbf{A}^2 - 6\mathbf{A} + 11\,\mathbf{I}$

$$= \begin{bmatrix} 4 & -1 & -5 \\ 15 & 1 & -5 \\ 5 & 4 & 9 \end{bmatrix} - \begin{bmatrix} 12 & 0 & -6 \\ 30 & 6 & 0 \\ 0 & 6 & 18 \end{bmatrix} + \begin{bmatrix} 11 & 0 & 0 \\ 0 & 11 & 0 \\ 0 & 0 & 11 \end{bmatrix} = \begin{bmatrix} 3 & -1 & 1 \\ -15 & 6 & -5 \\ 5 & -2 & 2 \end{bmatrix}.$$

(ii) $\mathbf{A}^{-2} = (\mathbf{A}^{-1})^2 = \begin{bmatrix} 3 & -1 & 1 \\ -15 & 6 & -5 \\ 5 & -2 & 2 \end{bmatrix}\begin{bmatrix} 3 & -1 & 1 \\ -15 & 6 & -5 \\ 5 & -2 & 2 \end{bmatrix} = \begin{bmatrix} 29 & -11 & 10 \\ -160 & 61 & -55 \\ 55 & -21 & 19 \end{bmatrix}.$

We can also write

$$\mathbf{A}^{-2} = (\mathbf{A}^{-1})\,(\mathbf{A}^{-1}) = \mathbf{A} - 6\,\mathbf{I} + 11(\mathbf{A}^{-1}).$$

3.2.5 Solution of $n \times n$ Linear System of Equations

Consider the system of n equations in n unknowns

$$a_{11}x_1 + a_{12}x_2 + \ldots + a_{1n}x_n = b_1$$

$$a_{21}x_1 + a_{22}x_2 + \ldots + a_{2n}x_n = b_2$$

$$\ldots\ldots\ldots\ldots\ldots\ldots\ldots\ldots\ldots\ldots\ldots\ldots$$

$$a_{n1}x_1 + a_{n2}x_2 + \ldots + a_{nn}x_n = b_n. \qquad (3.11)$$

In matrix form, we can write the system of equations (3.11) as

$$\mathbf{A}\mathbf{x} = \mathbf{b} \qquad (3.12)$$

where
$$\mathbf{A} = \begin{bmatrix} a_{11} & a_{12} & \ldots & a_{1n} \\ a_{21} & a_{22} & \ldots & a_{2n} \\ \vdots & & & \\ a_{m1} & a_{n2} & \ldots & a_{nn} \end{bmatrix}, \quad \mathbf{b} = \begin{bmatrix} b_1 \\ b_2 \\ \vdots \\ b_n \end{bmatrix}, \quad \mathbf{x} = \begin{bmatrix} x_1 \\ x_2 \\ \vdots \\ x_n \end{bmatrix}$$

and $\mathbf{A}$, $\mathbf{b}$, $\mathbf{x}$ are respectively called the *coefficient matrix*, the right hand side column vector and the solution vector. If $\mathbf{b} \neq \mathbf{0}$, that is, at least one of the elements $b_1, b_2, \ldots, b_n$ is not zero, then the system of equations is called *non-homogeneous*. If $\mathbf{b} = \mathbf{0}$, then the system of equations is called *homogeneous*. The system of equations is called *consistent* if it has at least one solution and *inconsistent* if it has no solution.

Non-homogeneous system of equations

The non-homogeneous system of equations $\mathbf{A}\mathbf{x} = \mathbf{b}$ can be solved by the following methods.

Matrix method

Let $\mathbf{A}$ be non-singular. Pre-multiplying $\mathbf{A}\mathbf{x} = \mathbf{b}$ by $\mathbf{A}^{-1}$, we obtain

$$\mathbf{x} = \mathbf{A}^{-1}\mathbf{b}. \qquad (3.13)$$

The system of equations is consistent and has a unique solution. If $\mathbf{b} = \mathbf{0}$, then $\mathbf{x} = \mathbf{0}$ (trivial solution) is the only solution.

Cramer's rule

Let $\mathbf{A}$ be non-singular. The Cramer's rule for the solution of $\mathbf{A}\mathbf{x} = \mathbf{b}$ is given by

$$x_i = \frac{|\mathbf{A}_i|}{|\mathbf{A}|}, \quad i = 1, 2, \ldots, n \qquad (3.14)$$

where $|\mathbf{A}_i|$ is the determinant of the matrix $\mathbf{A}_i$ obtained by replacing the ith column of $\mathbf{A}$ by the right hand side column vector $\mathbf{b}$.

We discuss the following cases.

Case 1   When $|A| \neq 0$, the system of equations is consistent and the unique solution is obtained by using Eq. (3.14).

Case 2   When $|A| = 0$ and one or more of $|A_i|$, $i = 1, 2, ..., n$, are not zero, then the system of equations has no solution, that is the system is inconsistent.

Case 3   When $|A| = 0$ and all $|A_i| = 0$, $i = 1, 2, ..., n$, then the system of equations is consistent and has infinite number of solutions. The system of equations has at least a one-parameter family of solutions.

Homogeneous system of equations

Consider the homogeneous system of equations

$$Ax = 0. \tag{3.15}$$

Trivial solution $x = 0$ is always a solution of this system.

If A is non-singular, then again $x = A^{-1} 0 = 0$ is the solution.

Therefore, a homogeneous system of equations is always consistent. We conclude that non-trivial solutions for $Ax = 0$ exist if and only if A is singular. In this case, the homogeneous system of equations has infinite number of solutions.

Example 3.4   Show that the system of equations

$$\begin{bmatrix} 1 & -1 & 1 \\ 2 & 1 & -3 \\ 1 & 1 & 1 \end{bmatrix} \begin{bmatrix} x \\ y \\ z \end{bmatrix} = \begin{bmatrix} 4 \\ 0 \\ 2 \end{bmatrix}.$$

has a unique solution. Solve this system using (i) matrix method, (ii) Cramer's rule.

Solution   We find that

$$|A| = \begin{vmatrix} 1 & -1 & 1 \\ 2 & 1 & -3 \\ 1 & 1 & 1 \end{vmatrix} = 1(1 + 3) - 2(-1 - 1) + 1 (3 - 1) = 10 \neq 0.$$

Therefore, the coefficient matrix A is non-singular and the given system of equations has a unique solution. Let $x = [x, y, z]^T$.

(i) We obtain

$$A^{-1} = \frac{1}{10} \begin{bmatrix} 4 & 2 & 2 \\ -5 & 0 & 5 \\ 1 & -2 & 3 \end{bmatrix} \text{ and } b = \begin{bmatrix} 4 \\ 0 \\ 2 \end{bmatrix}.$$

Therefore,

$$x = A^{-1}b = \frac{1}{10} \begin{bmatrix} 4 & 2 & 2 \\ -5 & 0 & 5 \\ 1 & -2 & 3 \end{bmatrix} \begin{bmatrix} 4 \\ 0 \\ 2 \end{bmatrix} = \begin{bmatrix} 2 \\ -1 \\ 1 \end{bmatrix}.$$

Hence, $x = 2$, $y = -1$ and $z = 1$.

(ii) We have

$$|A_1| = \begin{vmatrix} 4 & -1 & 1 \\ 0 & 1 & -3 \\ 2 & 1 & 1 \end{vmatrix} = 4(1 + 3) - 0 + 2(3 - 1) = 20.$$

$$|A_2| = \begin{vmatrix} 1 & 4 & 1 \\ 2 & 0 & -3 \\ 1 & 2 & 1 \end{vmatrix} = 1(0 + 6) - 2(4 - 2) + 1(-12 - 0) = -10.$$

$$|A_3| = \begin{vmatrix} 1 & -1 & 4 \\ 2 & 1 & 0 \\ 1 & 1 & 2 \end{vmatrix} = 1(2 - 0) - 2(-2 - 4) + 1(0 - 4) = 10.$$

Therefore, $x = \dfrac{|A_1|}{|A|} = 2,\ y = \dfrac{|A_2|}{|A|} = -1,\ z = \dfrac{|A_3|}{|A|} = 1.$

Example 3.5   Show that the system of equations

$$\begin{bmatrix} 1 & -1 & 3 \\ 2 & 3 & 1 \\ 3 & 2 & 4 \end{bmatrix} \begin{bmatrix} x_1 \\ x_2 \\ x_3 \end{bmatrix} = \begin{bmatrix} 3 \\ 2 \\ 5 \end{bmatrix}.$$

has infinite number of solutions. Hence, find the solutions.

Solutions   We find that

$$|A| = \begin{vmatrix} 1 & -1 & 3 \\ 2 & 3 & 1 \\ 3 & 2 & 4 \end{vmatrix} = 0,\quad |A_1| = \begin{vmatrix} 3 & -1 & 3 \\ 2 & 3 & 1 \\ 5 & 2 & 4 \end{vmatrix} = 0,$$

$$|A_2| = \begin{vmatrix} 1 & 3 & 3 \\ 2 & 2 & 1 \\ 3 & 5 & 4 \end{vmatrix} = 0,\quad |A_3| = \begin{vmatrix} 1 & -1 & 3 \\ 2 & 3 & 2 \\ 3 & 2 & 5 \end{vmatrix} = 0.$$

Therefore, the system of equations has infinite number of solutions. Using the first two equations

$$x_1 - x_2 = 3 - 3x_3$$
$$2x_1 + 3x_2 = 2 - x_3$$

and solving, we obtain $x_1 = (11 - 10\,x_3)/5$ and $x_2 = (5x_3 - 4)/5$ where x_3 is arbitrary. This solution satisfies the third equation.

Example 3.6 Show that the system of equations

$$\begin{bmatrix} 4 & 9 & 3 \\ 2 & 3 & 1 \\ 2 & 6 & 2 \end{bmatrix} \begin{bmatrix} x \\ y \\ z \end{bmatrix} = \begin{bmatrix} 6 \\ 2 \\ 7 \end{bmatrix}$$

is inconsistent.

Solution We find that

$$|\mathbf{A}| = \begin{vmatrix} 4 & 9 & 3 \\ 2 & 3 & 1 \\ 2 & 6 & 2 \end{vmatrix} = 0, \quad |\mathbf{A}_1| = \begin{vmatrix} 6 & 9 & 3 \\ 2 & 3 & 1 \\ 7 & 6 & 2 \end{vmatrix} = 0, \quad |\mathbf{A}_2| = \begin{vmatrix} 4 & 6 & 3 \\ 2 & 7 & 1 \\ 2 & 7 & 2 \end{vmatrix} = 6.$$

Since $|\mathbf{A}| = 0$ and $|\mathbf{A}_2| \neq 0$, the system of equations is inconsistent.

Example 3.7 Solve the homogeneous system of equations

$$\begin{bmatrix} 1 & 2 & 3 \\ 2 & 3 & -2 \\ 4 & 7 & 4 \end{bmatrix} \begin{bmatrix} x \\ y \\ z \end{bmatrix} = \begin{bmatrix} 0 \\ 0 \\ 0 \end{bmatrix}.$$

Solution We find that $|\mathbf{A}| = 0$. Hence, the given system has infinite number of solutions. Solving the first two equations

$$\begin{bmatrix} 1 & 2 \\ 2 & 3 \end{bmatrix} \begin{bmatrix} x \\ y \end{bmatrix} = \begin{bmatrix} -3z \\ 2z \end{bmatrix}$$

we obtain $x = 13z$, $y = -8z$ where z is arbitrary. This solution satisfies the third equation.

Exercise 3.1

1. Given the matrices $\mathbf{A} = \begin{bmatrix} 1 & 2 & 1 \\ -1 & 1 & 1 \\ 2 & 1 & 3 \end{bmatrix}$, $\mathbf{B} = \begin{bmatrix} 2 & 2 & 1 \\ 3 & 0 & -1 \\ 1 & 1 & -1 \end{bmatrix}$, verify that

    (i) $|\mathbf{AB}| = |\mathbf{A}| |\mathbf{B}|$, (ii) $|\mathbf{A} + \mathbf{B}| \neq |\mathbf{A}| + |\mathbf{B}|$.

2. If $\mathbf{A}^T = [1, -5, 7]$, $\mathbf{B} = [3, 1, 2]$, verify that $(\mathbf{AB})^T = \mathbf{B}^T \mathbf{A}^T$.

3. Show that the matrix $\mathbf{A} = \begin{bmatrix} 1 & 2 & 2 \\ 2 & 1 & 2 \\ 2 & 2 & 1 \end{bmatrix}$ satisfies the matrix equation $\mathbf{A}^2 - 4\mathbf{A} - 5\mathbf{I} = \mathbf{0}$. Hence, find $\mathbf{A}^{-1}$.

4. Show that the matrix $\mathbf{A} = \begin{bmatrix} 1 & 1 & 1 \\ 1 & 2 & -3 \\ 2 & -1 & 3 \end{bmatrix}$ satisfies the matrix equation $\mathbf{A}^3 - 6\mathbf{A}^2 + 5\mathbf{A} + 11\mathbf{I} = \mathbf{0}$. Hence, find $\mathbf{A}^{-1}$.

5. For the matrix $\mathbf{A} = \begin{bmatrix} 0 & 1 & -1 \\ 4 & -3 & 4 \\ 3 & -3 & 4 \end{bmatrix}$, verify that

    (i) $[adj\,(\mathbf{A})]^T = adj\,(\mathbf{A}^T)$,   (ii) $[adj\,(\mathbf{A})]^{-1} = adj\,(\mathbf{A}^{-1})$.

6. For the matrix $\mathbf{A} = \begin{bmatrix} 1 & -2 & 1 \\ -2 & 3 & 1 \\ 1 & 1 & 5 \end{bmatrix}$, verify that

    (i) $(\mathbf{A}^{-1})^T = (\mathbf{A}^T)^{-1}$,   (ii) $(\mathbf{A}^{-1})^{-1} = \mathbf{A}$.

7. For the matrices $\mathbf{A} = \begin{bmatrix} 2 & -1 & 1 \\ -1 & 2 & -1 \\ 1 & -1 & 2 \end{bmatrix}$, $\mathbf{B} = \begin{bmatrix} 1 & 2 & 1 \\ 1 & 3 & 4 \\ 2 & 0 & 9 \end{bmatrix}$, verify that

    (i) $adj\,(\mathbf{AB}) = adj\,(\mathbf{A})\,adj\,(\mathbf{B})$,   (ii) $(\mathbf{A} + \mathbf{B})^{-1} \neq \mathbf{A}^{-1} + \mathbf{B}^{-1}$.

8. For any non-singular matrix $\mathbf{A} = (a_{ij})$ of order n, show that
    (i) $|adj\,(\mathbf{A})| = |\mathbf{A}|^{n-1}$,   (ii) $adj\,(adj(\mathbf{A})) = |\mathbf{A}|^{n-2}\,\mathbf{A}$.

9. For any non-singular matrix $\mathbf{A}$, show that $|\mathbf{A}^{-1}| = 1/|\mathbf{A}|$.

10. For any symmetric matrix $\mathbf{A}$, show that $\mathbf{BAB}^T$ is symmetric, where $\mathbf{B}$ is any matrix for which the product matrix $\mathbf{BAB}^T$ is defined.

11. If $\mathbf{A}$ is a symmetric matrix, prove that $(\mathbf{BA}^{-1})^T (\mathbf{A}^{-1}\mathbf{B}^T)^{-1} = \mathbf{I}$ where $\mathbf{B}$ is any matrix for which the product matrices are defined.

12. If $\mathbf{A}$ and $\mathbf{B}$ are symmetric matrices, then prove that
    (i) $\mathbf{A} + \mathbf{B}$ is symmetric,   (ii) $\mathbf{AA}^T$ and $\mathbf{A}^T\mathbf{A}$ are both symmetric,

    (iii) $\mathbf{AB} - \mathbf{BA}$ is skew-symmetric.

13. If $\mathbf{A}$ and $\mathbf{B}$ are non-singular commutative and symmetric matrices, then prove that
    (i) $\mathbf{AB}^{-1}$,   (ii) $\mathbf{A}^{-1}\mathbf{B}$,   (iii) $\mathbf{A}^{-1}\mathbf{B}^{-1}$

are symmetric.

14. Let $\mathbf{A}$ be a non-singular matrix. Show that
    (i) if $\mathbf{I} + \mathbf{A} + \mathbf{A}^2 + \ldots + \mathbf{A}^n = \mathbf{0}$, then $\mathbf{A}^{-1} = \mathbf{A}^n$,

    (ii) if $\mathbf{I} - \mathbf{A} + \mathbf{A}^2 - \ldots + (-1)^n \mathbf{A}^n = \mathbf{0}$, then $\mathbf{A}^{-1} = (-1)^{n-1}\mathbf{A}^n$

15. Let $\mathbf{P}$, $\mathbf{Q}$ and $\mathbf{A}$ be non-singular square matrices of order n and $\mathbf{PAQ} = \mathbf{I}$, then show that $\mathbf{A}^{-1} = \mathbf{QP}$.

16. If $\mathbf{I} - \mathbf{A}$ is a non-singular matrix, then show that
$$(\mathbf{I} - \mathbf{A})^{-1} = \mathbf{I} + \mathbf{A} + \mathbf{A}^2 + \ldots.$$
assuming that the series on the right hand side converges.

17. For any three non-singular matrices $\mathbf{A}$, $\mathbf{B}$, $\mathbf{C}$, each of order n, show that $(\mathbf{ABC})^{-1} = \mathbf{C}^{-1}\mathbf{B}^{-1}\mathbf{A}^{-1}$.

Solve the following system of equations:

18. $x - y + z = 2$, $2x + 3y - z = 5$, $x + y - z = 0$.

19. $x + 2y + 3z = 6$, $2x + 4y + z = 7$, $3x + 2y + 9z = 14$.

20. $-x + y + 2z = 2$, $3x - y + z = 3$, $-x + 3y + 4z = 6$.

21. $2x - z = 1$, $5x + y = 7$, $y + 3z = 5$.

22. Determine the values of k for which the system of equations
$$x - ky + z = 0, \quad kx + 3y - kz = 0, \quad 3x + y - z = 0$$
has (i) only trivial solution, (ii) non-trivial solution.

23. Find the value of θ for which the system of equations
$$2(\sin \theta) x + y - 2z = 0, \quad 3x + 2(\cos 2\theta)y + 3z = 0, \quad 5x + 3y - z = 0$$
has a non-trivial solution.

24. If the system of equations $x + ay + az = 0$, $bx + y + bz = 0$, $cx + cy + z = 0$, where a, b, c are non-zero and non-unity, has a non-trivial solution, then show that
$$\frac{a}{1-a} \frac{b}{1-b} + \frac{c}{1-c} = -1.$$

25. Find the values of λ and μ for which the system of equations
$$x + 2y + z = 6, \quad x + 4y + 3z = 10, \quad x + 4y + \lambda z = \mu$$
has (i) a unique solution, (ii) infinite number of solutions, (iii) no solution.

Find the rank of the matrix A, where A is given by

26. $\begin{bmatrix} 2 & 3 & 4 \\ 3 & 5 & 7 \end{bmatrix}$.

27. $\begin{bmatrix} 1 & 3 & -4 \\ -1 & -3 & 4 \\ 2 & 6 & -8 \end{bmatrix}$.

28. $\begin{bmatrix} 1 & 2 & -3 \\ 2 & 1 & -1 \\ 1 & -1 & 2 \\ 5 & 4 & -5 \end{bmatrix}$.

29. $\begin{bmatrix} 1 & 1 & 1 \\ p & q & r \\ p^3 & q^3 & r^3 \end{bmatrix}$.

30. (a) $\begin{bmatrix} 2 & 1 & 5 & -1 \\ -1 & 2 & 5 & 3 \\ 3 & 2 & 9 & -1 \end{bmatrix}$.   (b) $\begin{bmatrix} 0 & c_1 & -b_1 & a_2 \\ -c_1 & 0 & a_1 & b_2 \\ b_1 & -a_1 & 0 & c_2 \\ -a_2 & -b_2 & -c_2 & 0 \end{bmatrix}$, $a_i, b_i, c_i \neq 0, i = 1, 2$.

31. Prove that if A is an Hermitian matrix, then iA is a skew-Hermitian matrix and if A is a skew-Hermitian matrix, then iA is a Hermitian matrix.

32. Prove that if A is a real matrix and $A^n \to 0$ as $n \to \infty$, then $I + A$ is invertible.

33. Let A, B be $n \times n$ real matrices Then, show that

(i) Trace $(\alpha A + \beta B) = \alpha$ Trace $(A) + \beta$ Trace (B) for any scalars α and β.

(ii) Trace $(AB) =$ Trace (BA),   (iii) $AB - BA = I$ is never true.

34. If B, C are $n \times n$ matrices, $A = B + C$, $BC = CB$ and $C^2 = 0$, then show that
$$A^{p+1} = B^p[B + (p + 1) C] \text{ for any positive integer } p.$$

35. Let $A = (a_{ij})$ be a square matrix of order n, such that $a_{ij} = d, i \neq j$ and $a_{ij} = c, i = j$. Then, show that
$$|A| = (c - d)^{n-1}[c + (n - 1)d].$$

Identity the following matrices as symmetric, skew-symmetric, Hermitian, skew-Hermitian or none of these;

36. $\begin{bmatrix} 1 & 2 & 3 \\ -2 & 5 & 4 \\ -3 & -4 & 6 \end{bmatrix}$.

37. $\begin{bmatrix} a & b & c \\ b & d & e \\ c & e & f \end{bmatrix}$.

38. $\begin{bmatrix} 0 & b & c \\ -b & 0 & e \\ -c & -e & 0 \end{bmatrix}$.

39. $\begin{bmatrix} 1 & 2+4i & 1-i \\ 2-4i & -5 & 3-5i \\ 1+i & 3+5i & 6 \end{bmatrix}$.

40. $\begin{bmatrix} 1 & 2+4i & 1-i \\ -2+4i & -5 & 3-5i \\ -1-i & 3-5i & 6 \end{bmatrix}$.

41. $\begin{bmatrix} 0 & 2+4i & 1-i \\ -2+4i & 0 & 3-5i \\ -1-i & -3-5i & 0 \end{bmatrix}$.

42. $\begin{bmatrix} 0 & i & i \\ i & 0 & i \\ i & i & 0 \end{bmatrix}$

43. $\begin{bmatrix} 0 & -i & 1+i \\ -i & -2i & 0 \\ -1+i & 0 & i \end{bmatrix}$.

44. $\begin{bmatrix} 1 & -1 & i \\ -1 & 0 & 1-i \\ -i & 1+i & 2 \end{bmatrix}$.

45. $\begin{bmatrix} 1 & 2i & -i \\ -2i & i & 1 \\ i & 1 & 2 \end{bmatrix}$.

3.3 Vector Spaces

Let V be a non-empty set of certain objects, which may be vectors, matrices, functions or some other objects. Each object is an element of V and is called a vector. The elements of V are denoted by **a, b, c, u, v,** etc. Assume that the two algebraic operations

(i) vector addition and (ii) scalar multiplication

are defined on elements of V.

If the vector addition is defined as the usual addition of vectors, then

$$\mathbf{a} + \mathbf{b} = (a_1, a_2, ..., a_n) + (b_1, b_2, ..., b_n) = (a_1 + b_1, a_2 + b_2, ..., a_n + b_n).$$

If the scalar multiplication is defined as the usual scalar multiplication of a vector by the scaler α, then

$$\alpha\mathbf{a} = \alpha(a_1, a_2, ..., a_n) = (\alpha a_1, \alpha a_2, ..., \alpha a_n).$$

The set V defines a vector space if for any elements **a, b, c** in V and any scalars α, β the following properties (axioms) are satisfied.

Properties (axioms) with respect to vector addition

  1. **a + b** is in V.

  2. **a + b = b + a.**   (commutative law)

  3. **(a + b) + c = a + (b + c).**   (associative law)

  4. **a + 0 = 0 + a = a.**   (existence of a unique zero element in V)

  5. **a + (− a) = 0.**   (existence of additive inverse or negative vector in V)

Properties (axioms) with respect to scalar multiplication

6. $\alpha\mathbf{a}$ is in V.
7. $(\alpha + \beta)\,\mathbf{a} = \alpha\mathbf{a} + \beta\mathbf{a}$.   (left distributive law)
8. $(\alpha\beta)\mathbf{a} = \alpha(\beta\mathbf{a})$.
9. $\alpha(\mathbf{a} + \mathbf{b}) = \alpha\mathbf{a} + \alpha\mathbf{b}$.   (right distributive law)
10. $1\mathbf{a} = \mathbf{a}$.   (existence of multiplicative identity)

The properties defined in **1** and **6** are called the *closure* properties. When these two properties are satisfied, we say that the vector space is closed under the vector addition and scalar multiplication. The vector addition and scalar multiplication defined above need not always be the usual addition and multiplication operators. Thus, *the vector space depends not only on the set V of vectors, but also on the definition of vector addition and scalar multiplication on V.*

If the elements of V are real, then it is called a *real vector space* when the scalars α, β are real numbers, whereas V is called a *complex vector space,* if the elements of V are complex and the scalars α, β may be real or complex numbers or if the elements of V are real and the scalars α, β are complex numbers.

Remark 7

(a) If even one of the above properties is not satisfied, then V is not a vector space. We usually check the closure properties first before checking the other properties.

(b) The concepts of length, dot product, vector product etc. are not part of the properties to be satisfied.

(c) The set of real numbers and complex numbers are called *fields* of scalars. We shall consider vector space only on the fields of scalars. In an advanced course on linear algebra, vector spaces over arbitrary fields are considered.

(d) The vector space $V = \{\mathbf{0}\}$ is called a trivial vector space.

The following are some examples of vector spaces under the usual operations of vector addition and scalar multiplication.

1. The set V of real or complex numbers.
2. The set of real valued continuous functions f on any closed interval $[a, b]$. The **0** vector defined in property **4** is the zero function.
3. The set of polynomials P_n of degree less than or equal to n.
4. The set V of n-tuples in $\mathbb{R}^n$ or $\mathbb{C}^n$.
5. The set V of all $m \times n$ matrices. The element **0** defined in property **4** is the null matrix of order $m \times n$.

The following are some examples which are not vector spaces. Assume that usual operations of vector addition and scalar multiplication are being used.

1. The set V of all polynomials of degree n. Let P_n and Q_n be two polynomials of degree n in V. Then, $\alpha P_n + \beta Q_n$ need not be a polynomial of degree n and thus may not be in V. For example, if $P_n = x^n + a_1 x^{n-1} + \ldots + a_n$ and $Q_n = -x^n + b_1 x^{n-1} + \ldots + b_n$, then $P_n + Q_n$ is a polynomial of degree $(n - 1)$.

2. The set V of all real-valued functions of one variable x, defined and continuous on the closed interval $[a, b]$ such that the value of the function at b is some non-zero constant p. For example, let $f(x)$ and $g(x)$ be two elements in V. Now, $f(b) = g(b) = p$. Since $f(b) + g(b) = 2p$, $f(x) + g(x)$ is not in V. Note that if $p = 0$, then V forms a vector space.

Example 3.8 Let V be the set of all polynomials, with real coefficients, of degree n, where addition is defined by $\mathbf{a} + \mathbf{b} = \mathbf{ab}$ and under usual scalar multiplication. Show that V is not a vector space.

Solution Let P_n and Q_n be two elements in V. Now, $P_n + Q_n = (P_n)(Q_n)$ is a polynomial of degree $2n$, which is not in V. Therefore, V does not define a vector space.

Example 3.9 Let V be the set of all ordered pairs (x, y), where x, y are real numbers. Let $\mathbf{a} = (x_1, y_1)$ and $\mathbf{b} = (x_2, y_2)$ be two elements in V. Define the addition as

$$\mathbf{a} + \mathbf{b} = (x_1, y_1) + (x_2, y_2) = (2x_1 - 3x_2, y_1 - y_2)$$

and the scalar multiplication as

$$\alpha(x_1, y_1) = (\alpha x_1/3, \ \alpha y_1/3).$$

Show that V is not a vector space.

Solution We illustrate the properties that are not satisfied.

(i) $(x_2, y_2) + (x_1, y_1) = (2x_2 - 3x_1, y_2 - y_1) \neq (x_1, y_1) + (x_2, y_2)$.

Therefore, property **2** (commutative law) does not hold.

(ii) $((x_1, y_1) + (x_2, y_2)) + (x_3, y_3) = (2x_1 - 3x_2, y_1 - y_2) + (x_3, y_3)$

$$= (4x_1 - 6x_2 - 3x_3, y_1 - y_2 - y_3)$$

$(x_1, y_1) + ((x_2, y_2) + (x_3, y_3)) = (x_1, y_1) + (2x_2 - 3x_3, y_2 - y_3)$

$$= (2x_1 - 6x_2 + 9x_3, y_1 - y_2 + y_3).$$

Therefore, property **3** (associative law) is not satisfied.

Hence, V is not a vector space.

Example 3.10 Let V be the set of all ordered pairs (x, y), where x, y are real numbers. Let $\mathbf{a} = (x_1, y_1)$ and $\mathbf{b} = (x_2, y_2)$ be two elements in V. Define the addition as

$$\mathbf{a} + \mathbf{b} = (x_1, y_1) + (x_2, y_2) = (x_1 x_2, y_1 y_2)$$

and the scalar multiplication as

$$\alpha(x_1, y_1) = (\alpha x_1, \alpha y_1).$$

Show that V is not a vector space.

Solution Note that $(1, 1)$ is an element of V. From the given definition of vector addition, we find that

$$(x_1, y_1) + (1, 1) = (x_1, y_1).$$

This is true only for the element $(1, 1)$. Therefore, the element $(1, 1)$ plays the role of $\mathbf{0}$ element as defined in property **4**.

Now, there exists the element $(1/x_1, 1/y_1)$ such that $(x_1, y_1) + (1/x_1 + 1/y_1) = (1, 1)$. The element $(1/x_1, 1/y_1)$ plays the role of additive inverse.

Therefore, property **5** is satisfied.

Now, let $\alpha = 1$, $\beta = 2$ be any two scalars. We have

$$(\alpha + \beta)(x_1, y_1) = 3(x_1, y_1) = (3x_1, 3y_1)$$

and $\qquad \alpha(x_1, y_1) + \beta(x_1, y_1) = 1(x_1, y_1) + 2(x_1, y_1) = (x_1, y_1) + (2x_1, 2y_1) = (2x_1^2, 2y_1^2)$.

Therefore, $(\alpha + \beta)(x_1, y_1) \neq \alpha(x_1, y_1) + \beta(x_1, y_1)$ and property **7** is not satisfied.

Similarly, it can be shown that property **9** is not satisfied. Hence, V is not a vector space.

3.3.1 Subspaces

Let V be an arbitrary vector space defined under a given vector addition and scalar multiplication. A non-empty subset W of V, such that W is also a vector space under the same two operations of vector addition and scalar multiplication, is called a *subspace* of V. Thus, W is also closed under the two given algebraic operations on V. As a convention, the vector space V is also taken as a subspace of V.

Remark 8

To show that W is a subspace of a vector space V, it is not necessary to verify all the 10 properties as given in section 3.3. If it is shown that W is closed under the given definition of vector addition and scalar multiplication, then the properties **2, 3, 7, 8, 9** and **10** are automatically satisfied because these properties are valid for all elements in V and hence are also valid for all elements in W. Thus, we need to verify the remaining properties, that is, the existence of the zero element and the additive inverse in W.

Consider the following examples:

1. Let V be the set of n-tuples $(x_1, x_2, \ldots x_n)$ in $\mathbb{R}^n$ with usual addition and scalar multiplication. Then

    (i) W consisting of n-tuples $(x_1, x_2, \ldots, x_n)$ with $x_1 = 0$ is a subspace of V.

    (ii) W consisting of n-tuples $(x_1, x_2, \ldots, x_n)$ with $x_1 \geq 0$ is not a subspace of V, since W is not closed under scalar multiplication ($\alpha\mathbf{x}$, when α is a negative real number, is not in W).

    (iii) W consisting of n-tuples $(x_1, x_2, \ldots, x_n)$ with $x_2 = x_1 + 1$ is not a subspace of V, since W is not closed under addition.

    (Let $\mathbf{x} = (x_1, x_2, \ldots, x_n)$ with $x_2 = x_1 + 1$ and $\mathbf{y} = (y_1, y_2, \ldots, y_n)$ with $y_2 = y_1 + 1$ be two elements in W. Then

    $$\mathbf{x} + \mathbf{y} = (x_1 + y_1, x_2 + y_2, \ldots, x_n + y_n)$$

    is not in W as $x_2 + y_2 = x_1 + y_1 + 2 \neq x_1 + y_1 + 1$).

2. Let V be the set of all real polynomials P of degree $\leq m$ with usual addition and scalar multiplication. Then

    (i) W consisting of all real polynomials of degree $\leq m$ with $P(0) = 0$ is a subspace of V.

(ii) W consisting of all real polynomials of degree $\leq m$ with $P(0) = 1$ is not a subspace of V, since W is not closed under addition (if P and $Q \in W$, then $P + Q \notin W$).

(iii) W consisting of all polynomials of degree $\leq m$ with real positive coefficients is not a subspace of V since W is not closed under scalar multiplication (if P is an element of W, then $-P \notin W$).

3. Let V be the set of all $n \times n$ real square matrices with usual matrix addition and scalar multiplication. Then

(i) W consisting of all symmetric/skew-symmetric matrices of order n is a subspace of V.

(ii) W consisting of all upper/lower triangular matrices of order n is a subspace of V.

(iii) W consisting of all $n \times n$ matrices having real positive elements is not a subspace of V since W is not closed under scalar multiplication (if $\mathbf{A}$ is an element of W, then $-\mathbf{A} \notin W$).

4. Let V be the set of all $n \times n$ complex matrices with usual matrix addition and scalar multiplication. Then

(i) W consisting of all Hermitian matrices of order n forms a vector space when scalars are real numbers and does not form a vector space when scalars are complex numbers (W is not closed under scalar multiplication).

Let
$$\mathbf{A} = \begin{pmatrix} a & x+iy \\ x-iy & b \end{pmatrix} \in W.$$

Let $\alpha = i$. We get $\alpha\mathbf{A} = i\mathbf{A} = \begin{pmatrix} ai & xi-y \\ xi+y & bi \end{pmatrix} \notin W.$

(ii) W consisting of all skew-Hermitian matrices of order n forms a vector space when scalars are real numbers and does not form a vector space when scalars are complex numbers.

Let
$$\mathbf{A} = \begin{pmatrix} i & x+iy \\ -x+iy & 2i \end{pmatrix} \in W.$$

Let $\alpha = i$. We get $i\mathbf{A} = \begin{pmatrix} -1 & ix-y \\ -ix-y & -2 \end{pmatrix} \notin W.$

Example 3.11 Let F and G be subspaces of a vector space V such that $F \cap G = \{0\}$. The sum of F and G is written as $F + G$ and is defined by

$$F + G = \{\mathbf{f} + \mathbf{g}: \mathbf{f} \in F, \mathbf{g} \in G\}.$$

Show that $F + G$ is a subspace of V assuming the usual definition of vector addition and scalar multiplication.

Solution Let $W = F + G$ and $\mathbf{f} \in F$, $\mathbf{g} \in G$. Since $\mathbf{0} \in F$, and $\mathbf{0} \in G$ we have $\mathbf{0} + \mathbf{0} = \mathbf{0} \in W$. Let $\mathbf{f}_1 + \mathbf{g}_1$ and $\mathbf{f}_2 + \mathbf{g}_2$ belong to W where $\mathbf{f}_1, \mathbf{f}_2 \in F$ and $\mathbf{g}_1, \mathbf{g}_2 \in G$. Then

$$(\mathbf{f}_1 + \mathbf{g}_1) + (\mathbf{f}_2 + \mathbf{g}_2) = (\mathbf{f}_1 + \mathbf{f}_2) + (\mathbf{g}_1 + \mathbf{g}_2) \in F + G = W.$$

Also, for any scalar α, $\alpha\,(\mathbf{f} + \mathbf{g}) = \alpha\,\mathbf{f} + \alpha\,\mathbf{g} \in F + G = W.$
Therefore, $W = F + G$ is a subspace of V.

We now state an important result on subspaces.

Theorem 3.1   Let $\mathbf{v}_1, \mathbf{v}_2, \ldots, \mathbf{v}_r$ be any r elements of a vector space V under usual vector addition and scalar multiplication. Then, the set of all linear combinations of these elements, that is the set of all elements of the form

$$\alpha_1 \mathbf{v}_1 + \alpha_2 \mathbf{v}_2 + \ldots + \alpha_r \mathbf{v}_r \qquad (3.16)$$

is a subspace of V, where $\alpha_1, \alpha_2, \ldots, \alpha_r$ are scalars.

Spanning set   Let S be a subset of a vector space V and suppose that every element in V can be obtained as a linear combination of the elements taken from S. Then S is said to be the *spanning set* for V. We also say that S spans V.

Example 3.12   Let V be the vector space of all 2×2 real matrices. Show that the sets

(i)
$$S = \left\{ \begin{pmatrix} 1 & 0 \\ 0 & 0 \end{pmatrix}, \begin{pmatrix} 0 & 1 \\ 0 & 0 \end{pmatrix}, \begin{pmatrix} 0 & 0 \\ 1 & 0 \end{pmatrix}, \begin{pmatrix} 0 & 0 \\ 0 & 1 \end{pmatrix} \right\}$$

(ii)
$$S = \left\{ \begin{pmatrix} 1 & 0 \\ 0 & 0 \end{pmatrix}, \begin{pmatrix} 1 & 1 \\ 0 & 0 \end{pmatrix}, \begin{pmatrix} 1 & 1 \\ 1 & 0 \end{pmatrix}, \begin{pmatrix} 1 & 1 \\ 1 & 1 \end{pmatrix} \right\}$$

span V.

Solution   Let $\mathbf{x} = \begin{bmatrix} a & b \\ c & d \end{bmatrix}$ be an arbitrary element of V.

(i) We write

$$\begin{bmatrix} a & b \\ c & d \end{bmatrix} = a \begin{bmatrix} 1 & 0 \\ 0 & 0 \end{bmatrix} + b \begin{bmatrix} 0 & 1 \\ 0 & 0 \end{bmatrix} + c \begin{bmatrix} 0 & 0 \\ 1 & 0 \end{bmatrix} + d \begin{bmatrix} 0 & 0 \\ 0 & 1 \end{bmatrix}.$$

Since every element of V can be written as a linear combination of the elements of S, the set S spans the vector space V.

(ii) We need to determine the scalars $\alpha_1, \alpha_2, \alpha_3, \alpha_4$ so that

$$\begin{bmatrix} a & b \\ c & d \end{bmatrix} = \alpha_1 \begin{bmatrix} 1 & 0 \\ 0 & 0 \end{bmatrix} + \alpha_2 \begin{bmatrix} 1 & 1 \\ 0 & 0 \end{bmatrix} + \alpha_3 \begin{bmatrix} 1 & 1 \\ 1 & 0 \end{bmatrix} + \alpha_4 \begin{bmatrix} 1 & 1 \\ 1 & 1 \end{bmatrix}.$$

Equating the corresponding elements, we obtain the system of equations

$$\alpha_1 + \alpha_2 + \alpha_3 + \alpha_4 = a, \quad \alpha_2 + \alpha_3 + \alpha_4 = b,$$

$$\alpha_3 + \alpha_4 = c, \quad \alpha_4 = d.$$

The solution of this system of equations is

$$\alpha_4 = d, \quad \alpha_3 = c - d, \quad \alpha_2 = b - c, \quad \alpha_1 = a - b.$$

Therefore, we can write

$$\begin{bmatrix} a & b \\ c & d \end{bmatrix} = (a - b)\begin{bmatrix} 1 & 0 \\ 0 & 0 \end{bmatrix} + (b - c)\begin{bmatrix} 1 & 1 \\ 0 & 0 \end{bmatrix} + (c - d)\begin{bmatrix} 1 & 1 \\ 1 & 0 \end{bmatrix} + d\begin{bmatrix} 1 & 1 \\ 1 & 1 \end{bmatrix}.$$

Since every element of V can be written as a linear combination of the elements of S, the set S spans the vector space V.

Example 3.13   Let V be the vector space of all polynomials of degree ≤ 3. Determine whether or not the set

$$S = \{t^3, t^2 + t, t^3 + t + 1\}$$

spans V ?

Solution   Let $P(t) = \alpha t^3 + \beta t^2 + \gamma t + \delta$ be an arbitrary element in V. We need to find whether or not there exist scalars a_1, a_2, a_3 such that

$$\alpha t^3 + \beta t^2 + \gamma t + \delta = a_1 t^3 + a_2(t^2 + t) + a_3(t^3 + t + 1)$$

$$\alpha t^3 + \beta t^2 + \gamma t + \delta = (a_1 + a_3)t^3 + a_2 t^2 + (a_2 + a_3)\, t + a_3.$$

Comparing the coefficients of various powers of t, we get

$$a_1 + a_3 = \alpha, \quad a_2 = \beta, \quad a_2 + a_3 = \gamma, \quad a_3 = \delta.$$

The solution of the first three equations is given by

$$a_1 = \alpha + \beta - \gamma, \quad a_2 = \beta, \quad a_3 = \gamma - \beta.$$

Substituting in the last equation, we obtain $\gamma - \beta = \delta$, which may not be true for all elements in V. For example, the polynomial $t^3 + 2t^2 + t + 3$ does not satisfy this condition and therefore, it cannot be written as a linear combination of the elements of S. Therefore, S does not span the vector space V.

3.3.2   Linear Independence of Vectors

Let V be a vector space. A finite set $\{v_1, v_2, \ldots, v_n\}$ of the elements of V is said to be *linearly dependent* if there exist scalars $\alpha_1, \alpha_2, \ldots, \alpha_n$, not all zero, such that

$$\alpha_1 v_1 + \alpha_2 v_2 + \ldots + \alpha_n v_n = 0. \qquad (3.17)$$

If Eq. (3.17) is satisfied only for $\alpha_1 = \alpha_2 = \ldots = \alpha_n = 0$, then the set of vectors is said to be *linearly independent*.

The above definition of linear dependence of $v_1, v_2, \ldots v_n$ can be written alternately as follows.

Theorem 3.2   The set of vectors $\{v_1, v_2, \ldots, v_n\}$ is linearly dependent if and only if at least one element of the set is a linear combination of the remaining elements.

Remark 9

Eq. (3.17) gives a homogeneous system of algebraic equations. Non-trivial solutions exist if det(coefficient matrix) = 0, that is the vectors are linearly dependent in this case. If the det(coefficient matrix) $\neq 0$, then by Cramer's rule, $\alpha_1 = \alpha_2 = \ldots = \alpha_n = 0$ and the vectors are linearly independent.

Example 3.14   Let $v_1 = (1, -1, 0)$, $v_2 = (0, 1, -1)$ and $v_3 = (0, 0, 1)$ be elements of $\mathbb{R}^3$. Show that the set of vectors $\{v_1, v_2, v_3\}$ is linearly independent.

Solution   We consider the vector equation

$$\alpha_1 v_1 + \alpha_2 v_2 + \alpha_3 v_3 = 0.$$

Substituting for v_1, v_2, v_3, we obtain

$$\alpha_1(1, -1, 0) + \alpha_2(0, 1, -1) + \alpha_3(0, 0, 1) = 0$$

or
$$(\alpha_1, -\alpha_1 + \alpha_2, -\alpha_2 + \alpha_3) = 0.$$

Comparing, we obtain $\alpha_1 = 0$, $-\alpha_1 + \alpha_2 = 0$ and $-\alpha_2 + \alpha_3 = 0$. The solution of these equations is $\alpha_1 = \alpha_2 = \alpha_3 = 0$. Therefore, the given set of vectors is linearly independent.

Alternative

$$det(v_1, v_2, v_3) = \begin{vmatrix} 1 & 0 & 0 \\ -1 & 1 & 0 \\ 0 & -1 & 1 \end{vmatrix} = 1 \neq 0.$$

Therefore, the given vectors are linearly independent.

Example 3.15   Let $v_1 = (1, -1, 0)$, $v_2 = (0, 1, -1)$, $v_3 = (0, 2, 1)$ and $v_4 = (1, 0, 3)$ be elements of $\mathbb{R}^3$. Show that the set of vectors $\{v_1 \ v_2 \ v_3 \ v_4\}$ is linearly dependent.

Solution   The given set of elements will be linearly dependent if there exist scalars $\alpha_1, \alpha_2, \alpha_3, \alpha_4$, not all zero, such that

$$\alpha_1 v_1 + \alpha_2 v_2 + \alpha_3 v_3 + \alpha_4 v_4 = 0. \tag{3.18}$$

Substituting for v_1, v_2, v_3, v_4 and comparing, we obtain

$$\alpha_1 + \alpha_4 = 0, \quad -\alpha_1 + \alpha_2 + 2\alpha_3 = 0, \quad -\alpha_2 + \alpha_3 + 3\alpha_4 = 0.$$

The solution of this system of equations is

$$\alpha_1 = -\alpha_4, \quad \alpha_2 = 5\alpha_4/3, \quad \alpha_3 = -4\alpha_4/3, \quad \alpha_4 \text{ arbitrary.}$$

Substituting in Eq. (3.18) and cancelling α_4, we obtain

$$-v_1 + \frac{5}{3} v_2 - \frac{4}{3} v_3 + v_4 = 0.$$

Hence, there exist scalars not all zero, such that Eq. (3.18) is satisfied. Therefore, the set of vectors is linearly dependent.

3.3.3 Dimension and Basis

Let V be a vector space. If for some positive integer n, there exists a set S of n linearly independent elements of V and if every set of $n + 1$ or more elements in V is linearly dependent, then V is said to have *dimension n*. Then, we write $dim\ (V) = n$. Thus, the maximum number of linearly independent elements of V is the dimension of V. The set S of n linearly independent vectors is called the *basis* of V. Note that a vector space whose only element is zero has dimension zero.

Theorem 3.3 Let V be a vector space of dimension n. Let $v_1, v_2, ..., v_n$ be the linearly independent elements of V. Then, every other element of V can be written as a linear combination of these elements. Further, this representation is unique.

Proof Let v be an element of V. Then, the set $\{v, v_1, ..., v_n\}$ is linearly dependent as it has $n + 1$ elements. Therefore, there exist scalars $\alpha_0, \alpha_1, ..., \alpha_n$, not all zero, such that

$$\alpha_0 v + \alpha_1 v_1 + ... + \alpha_n v_n = 0. \tag{3.19}$$

Now, $\alpha_0 \neq 0$. Because, if $\alpha_0 = 0$, we get $\alpha_1 v_1 + ... + \alpha_n v_n = 0$ and since $v_1, v_2, ..., v_n$ are linearly independent, we get $\alpha_1 = \alpha_2 = ... = \alpha_n = 0$. This implies that the set of $n + 1$ elements $v, v_1, ..., v_n$ is linearly independent, which is not possible as the dimension of V is n.

Therefore, we obtain from Eq. (3.19)

$$v = \sum_{i=1}^{n} (-\alpha_i / \alpha_0) v_i. \tag{3.20}$$

Hence, v is a linear combination of n linearly independent vectors of V.

Now, let there be two representations of v given by

$$v = a_1 v_1 + a_2 v_2 + ... + a_n v_n \quad \text{and} \quad v = b_1 v_1 + b_2 v_2 + ... + b_n v_n$$

where $b_i \neq a_i$ for at least one i. Subtracting these two equations, we get

$$0 = (a_1 - b_1) v_1 + (a_2 - b_2) v_2 + ... + (a_n - b_n) v_n.$$

Since $v_1, v_2, ... v_n$ are linearly independent, we get

$$a_i - b_i = 0 \quad \text{or} \quad a_i = b_i, \ i = 1, 2, ..., n.$$

Therefore, both the representations of v are same and the representation of v given by Eq. (3.20) is unique.

Remark 10

(a) A set of $(n + 1)$ vectors in $\mathbb{R}^n$ is linearly dependent.

(b) A set of vectors containing 0 as one of its elements is linearly dependent as 0 is the linear combination of any set of vectors.

Theorem 3.4 Let V be an n-dimensional vector space. If $v_1, v_2, ..., v_k$, $k < n$ are linearly independent elements of V, then there exist elements $v_{k+1}, v_{k+2}, ..., v_n$ such that $\{v_1, v_2, ..., v_n\}$ is a basis of V.

Proof There exists an element $\mathbf{v}_{k+1}$ such that $\mathbf{v}_1, \mathbf{v}_2, ..., \mathbf{v}_k, \mathbf{v}_{k+1}$ are linearly independent. Otherwise, every element of V can be written as a linear combination of the vectors $\mathbf{v}_1, \mathbf{v}_2, ..., \mathbf{v}_k$ and therefore V has dimension $k < n$. This argument can be continued. If $n > k + 1$, we keep adding elements $\mathbf{v}_{k+1}, \mathbf{v}_{k+2}, ..., \mathbf{v}_n$ such that $\{\mathbf{v}_1, \mathbf{v}_2, ..., \mathbf{v}_n\}$ is a basis of V.

Since all the elements of a vector space V of dimension n can be represented as linear combinations of the n elements in the basis of V, the basis of V spans V. However, there can be many basis for the same vector space. For example, consider the vector space $\mathbb{R}^3$. Each of the following set of vectors

(i) $[1, -1, 0], [0, 1, -1], [0, 0, 1]$

(ii) $[1, -1, 0], [0, 0, 1], [1, 2, 3]$

(iii) $[1, 0, 0], [0, 1, 0], [0, 0, 1]$

are linearly independent and therefore forms a basis in $\mathbb{R}^3$. Some of the standard basis are the following.

1. If V consists of n-tuples in $\mathbb{R}^n$, then

$$\mathbf{e}_1 = (1, 0, 0, ..., 0), \ \mathbf{e}_2 = (0, 1, ..., 0), \ ..., \ \mathbf{e}_n = (0, 0, ..., 0, 1)$$

is called a standard basis in $\mathbb{R}^n$.

2. If V consists of all $m \times n$ matrices, then

$$\mathbf{E}_{rs} = \begin{bmatrix} 0 & ... & 0 & ... & 0 \\ \vdots & & & & \\ 0 & ... & 1 & ... & 0 \\ \vdots & & & & \\ 0 & ... & 0 & ... & 0 \end{bmatrix}, r = 1, 2, ..., m \text{ and } s = 1, 2, ..., n$$

where 1 is located in the (r, s) location, that is the rth row and the sth column, is called its standard basis.

For example, if V consists of all 2×3 matrices, then any matrix $\begin{bmatrix} a & b & c \\ x & y & z \end{bmatrix}$ in V can be written as

$$\begin{bmatrix} a & b & c \\ x & y & z \end{bmatrix} = a\mathbf{E}_{11} + b\mathbf{E}_{12} + c\mathbf{E}_{13} + x\mathbf{E}_{21} + y\mathbf{E}_{22} + z\mathbf{E}_{23}$$

where $\mathbf{E}_{11} = \begin{bmatrix} 1 & 0 & 0 \\ 0 & 0 & 0 \end{bmatrix}, \mathbf{E}_{12} = \begin{bmatrix} 0 & 1 & 0 \\ 0 & 0 & 0 \end{bmatrix}$ etc.

3. If V consists of all polynomials $P(t)$ of degree $\leq n$, then $\{1, t, t^2, ..., t^n\}$ is taken as its standard basis.

Example 3.16 Determine whether the following set of vectors $\{\mathbf{u}, \mathbf{v}, \mathbf{w}\}$ forms a basis in $\mathbb{R}^3$, where

(i) $\mathbf{u} = (2, 2, 0), \mathbf{v} = (3, 0, 2), \mathbf{w} = (2, -2, 2)$

(ii) $\mathbf{u} = (0, 1, -1), \mathbf{v} = (-1, 0, -1), \mathbf{w} = (3, 1, 3)$.

Solution If the set $\{\mathbf{u}, \mathbf{v}, \mathbf{w}\}$ forms a basis in $\mathbb{R}^3$, then $\mathbf{u}, \mathbf{v}, \mathbf{w}$ must be linearly independent. Let $\alpha_1, \alpha_2, \alpha_3$ be scalars. Then, the only solution of the equation

$$\alpha_1 \mathbf{u} + \alpha_2 \mathbf{v} + \alpha_3 \mathbf{w} = \mathbf{0} \qquad (3.21)$$

must be $\alpha_1 = \alpha_2 = \alpha_3 = 0$.

(i) Using Eq. (3.21), we obtain the system of equations

$$2\alpha_1 + 3\alpha_2 + 2\alpha_3 = 0, \quad 2\alpha_1 - 2\alpha_3 = 0 \quad \text{and} \quad 2\alpha_2 + 2\alpha_3 = 0.$$

The solution of this system of equations is $\alpha_1 = \alpha_2 = \alpha_3 = 0$. Therefore, $\mathbf{u}, \mathbf{v}, \mathbf{w}$ are linearly independent and they form a basis in $\mathbb{R}^3$.

(ii) Using Eq. (3.21), we obtain the system of equations

$$-\alpha_2 + 3\alpha_3 = 0, \quad \alpha_1 + \alpha_3 = 0, \quad \text{and} \quad -\alpha_1 - \alpha_2 + 3\alpha_3 = 0.$$

The solution of this system of equations is $\alpha_1 = \alpha_2 = \alpha_3 = 0$. Therefore, $\mathbf{u}, \mathbf{v}, \mathbf{w}$ are linearly independent and they form a basis in $\mathbb{R}^3$.

Example 3.17 Find the dimension of the subspace of $\mathbb{R}^4$ spanned by the set $\{(1\ 0\ 0\ 0), (0\ 1\ 0\ 0), (1\ 2\ 0\ 1), (0\ 0\ 0\ 1)\}$. Hence find its basis.

Solution The dimension of the subspace is ≤ 4. If it is 4, then the only solution of the vector equation

$$\alpha_1 (1\ 0\ 0\ 0) + \alpha_2 (0\ 1\ 0\ 0) + \alpha_3 (1\ 2\ 0\ 1) + \alpha_4 (0\ 0\ 0\ 1) = \mathbf{0} \qquad (3.22)$$

should be $\alpha_1 = \alpha_2 = \alpha_3 = \alpha_4 = 0$. Comparing, we obtain the system of equations

$$\alpha_1 + \alpha_3 = 0, \quad \alpha_2 + 2\alpha_3 = 0, \quad \alpha_3 + \alpha_4 = 0.$$

The solution of this system of equations is given by

$$\alpha_1 = \alpha_4, \quad \alpha_2 = 2\alpha_4, \quad \alpha_3 = -\alpha_4, \quad \text{where} \quad \alpha_4 \text{ is arbitrary.}$$

Hence, the vector equation (3.22) is satisfied for non-zero values of $\alpha_1, \alpha_2, \alpha_3,$ and α_4. Therefore, the dimension of the set is less than 4.

Now, consider any three elements of the set, say $(1\ 0\ 0\ 0), (0\ 1\ 0\ 0)$ and $(1\ 2\ 0\ 1)$. Consider the vector equation

$$\alpha_1 (1\ 0\ 0\ 0) + \alpha_2 (0\ 1\ 0\ 0) + \alpha_3 (1\ 2\ 0\ 1) = \mathbf{0}. \qquad (3.23)$$

Comparing, we obtain the system of equations

$$\alpha_1 + \alpha_3 = 0, \quad \alpha_2 + 2\alpha_3 = 0 \quad \text{and} \quad \alpha_3 = 0.$$

The solution of this system of equations is $\alpha_1 = \alpha_2 = \alpha_3 = 0$. Hence, these three elements are linearly independent. Therefore, the dimension of the given subspace is 3 and the basis is the set of vectors $\{(1\ 0\ 0\ 0), (0\ 1\ 0\ 0), (1\ 2\ 0\ 1)\}$. We find that the fourth vector can be written as

$$(0\ 0\ 0\ 1) = (1\ 0\ 0\ 0) - 2(0\ 1\ 0\ 0) + 1(1\ 2\ 0\ 1).$$

Example 3.18 Let $\mathbf{u} = \{(a, b, c, d)$, such that $a + c + d = 0$, $b + d = 0\}$ be a subspace of IR^4. Find the dimension and the basis of the subspace.

Solution $\mathbf{u}$ satisfies the closure properties. From the given equations, we have

$$a + c + d = 0 \text{ and } b + d = 0 \quad \text{or} \quad a = -c - d \text{ and } b = -d.$$

We have two free parameters, say, c and d. Therefore, the dimension of the given subspace is 2. Choosing $c = 0$, $d = 1$ and $c = 1$, $d = 0$, we may write a basis as $\{(-1 \ -1 \ 0 \ 1), (-1 \ 0 \ 1 \ 0)\}$.

3.3.4 Linear Tranformations

Let A and B be two arbitrary sets. A rule that assigns to elements of A exactly one element of B is called a *function* or a *mapping* or a *transformation*. Thus, a transformation maps the elements of A into the elements of B. The set A is called the *domain* of the transformation. We use capital letters T, S etc. to denote a transformation. If T is a transformation from A into B, we write

$$T : A \to B. \tag{3.24}$$

For each element $\mathbf{a} \in A$, we get a unique element $\mathbf{b} \in B$. We write $\mathbf{b} = T(\mathbf{a})$ or $\mathbf{b} = T\mathbf{a}$ and $\mathbf{b}$ is called the image of $\mathbf{a}$ under the mapping T. The collection of all such images in B is called the *range* or the image set of the transformation T.

In this section, we shall discuss mapping from a vector space into a vector space. Let V and W be two vector spaces, both real or complex, over the same field F of scalars. Let T be a mapping from V into W. The mapping T is said to be a *linear transformation* or a *linear mapping*, if it satisfies the following two properties:

(i) For every scalar α and every element $\mathbf{v}$ in V

$$T(\alpha \mathbf{v}) = \alpha T(\mathbf{v}). \tag{3.25}$$

(ii) For any two elements $\mathbf{v}_1$, $\mathbf{v}_2$ in V

$$T(\mathbf{v}_1 + \mathbf{v}_2) = T(\mathbf{v}_1) + T(\mathbf{v}_2). \tag{3.26}$$

Since V is a vector space, the product $\alpha \mathbf{v}$ and the sum $\mathbf{v}_1 + \mathbf{v}_2$ are defined and are elements in V. Then, T defines a mapping from V into W. Since $T(\mathbf{v}_1)$ and $T(\mathbf{v}_2)$ are in W, the product $\alpha T(\mathbf{v})$ and the sum $T(\mathbf{v}_1) + T(\mathbf{v}_2)$ are in W. The conditions given in Eqs. (3.25) and (3.26) are equivalent to

$$T(\alpha \mathbf{v}_1 + \beta \mathbf{v}_2) = T(\alpha \mathbf{v}_1) + T(\alpha \mathbf{v}_2) = \alpha T(\mathbf{v}_1) + \beta T(\mathbf{v}_2)$$

for $\mathbf{v}_1$ and $\mathbf{v}_2$ in V and any scalars α, β.

Let V be a vector space of dimension n and let the set $\{\mathbf{v}_1, \mathbf{v}_2, ..., \mathbf{v}_n\}$ be its basis. Then, any element $\mathbf{v}$ in V can be written as a linear combination of the elements $\mathbf{v}_1, \mathbf{v}_2, ..., \mathbf{v}_n$.

Remark 11

A linear transformation is completely determined by its action on basis vectors of a vector space.

Letting $\alpha = 0$ in Eq. (3.25), we find that for every element $\mathbf{v}$ in V

$$T(0\mathbf{v}) = T(\mathbf{0}) = 0T(\mathbf{v}) = \mathbf{0}.$$

Therefore, the zero element in V is mapped into zero element in W by the linear transformation T. The collection of all elements $w = T(\mathbf{v})$ is called the *range* of T and is written as $ran(T)$. The set of all elements of V that are mapped into the zero element by the linear transformation T is called the *kernel* or the *null-space* of T and is denoted by $ker(T)$. Therefore, we have

$$ker(T) = \{\mathbf{v} \mid T(\mathbf{v}) = \mathbf{0}\} \quad \text{and} \quad ran(T) = \{T(\mathbf{v}) \mid \mathbf{v} \in V\}.$$

Thus, the null space of T is a subspace of V and the range of T is a subspace of W.

The dimension of $ran(T)$ is called the rank(T) and the dimension of $ker(T)$ is called the *nullity* of T. We have the following result.

Theorem 3.5   If T has rank r and the dimension of V is n, then the nullity of T is $n - r$, that is

$$\text{rank } (T) + \text{nullity} = n = dim\ (V).$$

We shall discuss the linear transformation only in the context of matrices.

Let $\mathbf{A}$ be an $m \times n$ real (or complex) matrix. Let the rows of $\mathbf{A}$ represent the elements in IR^n (or $\mathbb{C}^n$) and the columns of $\mathbf{A}$ represent the elements in IR^m (or $\mathbb{C}^m$). If $\mathbf{x}$ is in IR^n, then $\mathbf{Ax}$ is in IR^m. Thus, an $m \times n$ matrix maps the elements in IR^n into the elements in IR^m. We write

$$T = \mathbf{A} \ : \ IR^n \rightarrow IR^m, \quad \text{and} \quad T\mathbf{x} = \mathbf{Ax}.$$

The mapping $\mathbf{A}$ is a linear transformation. The range of T is a linear subspace of IR^m and the kernel of T is a linear subspace of IR^n.

Remark 12

Let T_1 and T_2 be linear transformations from V into W. We define the sum $T_1 + T_2$ to be the transformation S such that

$$S\mathbf{v} = T_1\mathbf{v} + T_2\mathbf{v}, \quad \mathbf{v} \in V.$$

Then, $T_1 + T_2$ is a linear transformation and $T_1 + T_2 = T_2 + T_1$.

Example 3.19   Let T be a a linear transformation defined by

$$T\left[\begin{pmatrix} 1 & 1 \\ 1 & 1 \end{pmatrix}\right] = \begin{pmatrix} 1 \\ 2 \\ 3 \end{pmatrix}, \quad T\left[\begin{pmatrix} 0 & 1 \\ 1 & 1 \end{pmatrix}\right] = \begin{pmatrix} 1 \\ -2 \\ 3 \end{pmatrix}, \quad T\left[\begin{pmatrix} 0 & 0 \\ 1 & 1 \end{pmatrix}\right] = \begin{pmatrix} 1 \\ -2 \\ -3 \end{pmatrix}, \quad T\left[\begin{pmatrix} 0 & 0 \\ 0 & 1 \end{pmatrix}\right] = \begin{pmatrix} -1 \\ 2 \\ 3 \end{pmatrix}.$$

Find $T\left[\begin{pmatrix} 4 & 5 \\ 3 & 8 \end{pmatrix}\right]$.

Solution   The matrices $\begin{bmatrix} 1 & 1 \\ 1 & 1 \end{bmatrix}, \begin{bmatrix} 0 & 1 \\ 1 & 1 \end{bmatrix}, \begin{bmatrix} 0 & 0 \\ 1 & 1 \end{bmatrix}, \begin{bmatrix} 0 & 0 \\ 0 & 1 \end{bmatrix}$ are linearly independent and hence form a basis in the space of 2×2 matrices. We write for any scalars $\alpha_1, \alpha_2, \alpha_3, \alpha_4$, not all zero

$$\begin{pmatrix} 4 & 5 \\ 3 & 8 \end{pmatrix} = \alpha_1 \begin{pmatrix} 1 & 1 \\ 1 & 1 \end{pmatrix} + \alpha_2 \begin{pmatrix} 0 & 1 \\ 1 & 1 \end{pmatrix} + \alpha_3 \begin{pmatrix} 0 & 0 \\ 1 & 1 \end{pmatrix} + a_4 \begin{pmatrix} 0 & 0 \\ 0 & 1 \end{pmatrix}$$

$$= \begin{bmatrix} \alpha_1 & \alpha_1 + \alpha_2 \\ \alpha_1 + \alpha_2 + \alpha_3 & \alpha_1 + \alpha_2 + \alpha_3 + \alpha_4 \end{bmatrix}.$$

Comparing the elements and solving the resulting system of equations, we get $\alpha_1 = 4$, $\alpha_2 = 1$, $\alpha_3 = -2$, $\alpha_4 = 5$. Since T is a linear transformation, we get

$$T \left[\begin{pmatrix} 4 & 5 \\ 3 & 8 \end{pmatrix} \right] = \alpha_1 T \left[\begin{pmatrix} 1 & 1 \\ 1 & 1 \end{pmatrix} \right] + \alpha_2 T \left[\begin{pmatrix} 0 & 1 \\ 1 & 1 \end{pmatrix} \right] + \alpha_3 T \left[\begin{pmatrix} 0 & 0 \\ 1 & 1 \end{pmatrix} \right] + \alpha_4 T \left[\begin{pmatrix} 0 & 0 \\ 0 & 1 \end{pmatrix} \right]$$

$$= 4 \begin{pmatrix} 1 \\ 2 \\ 3 \end{pmatrix} + 1 \begin{pmatrix} 1 \\ -2 \\ 3 \end{pmatrix} - 2 \begin{pmatrix} 1 \\ -2 \\ -3 \end{pmatrix} + 5 \begin{pmatrix} -1 \\ 2 \\ 3 \end{pmatrix} = \begin{pmatrix} -2 \\ 20 \\ 36 \end{pmatrix}.$$

Example 3.20 For the set of vectors $\{x_1, x_2\}$, where $x_1 = (1, 3)^T$, $x_2 = (4, 6)^T$, are in $\mathbb{R}^2$, find the matrix of linear transformation $T : \mathbb{R}^2 \to \mathbb{R}^3$, such that

$$T x_1 = (-2 \quad 2 \quad -7)^T \quad \text{and} \quad T x_2 = (-2 \quad -4 \quad -10)^T.$$

Solution The transformation T maps column vector in $\mathbb{R}^2$ into column vectors in $\mathbb{R}^3$. Therefore, T must be a matrix A of order 3×2. Let

$$A = \begin{bmatrix} a_1 & b_1 \\ a_2 & b_2 \\ a_3 & b_3 \end{bmatrix}.$$

Therefore, we have

$$\begin{bmatrix} a_1 & b_1 \\ a_2 & b_2 \\ a_3 & b_3 \end{bmatrix} \begin{bmatrix} 1 \\ 3 \end{bmatrix} = \begin{bmatrix} -2 \\ 2 \\ -7 \end{bmatrix} \quad \text{and} \quad \begin{bmatrix} a_1 & b_1 \\ a_2 & b_2 \\ a_3 & b_3 \end{bmatrix} \begin{bmatrix} 4 \\ 6 \end{bmatrix} = \begin{bmatrix} -2 \\ -4 \\ -10 \end{bmatrix}.$$

Multiplying and comparing the corresponding elements, we get

$$\begin{aligned} a_1 + 3b_1 &= -2, & 4a_1 + 6b_1 &= -2, \\ a_2 + 3b_2 &= 2, & 4a_2 + 6b_2 &= -4, \\ a_3 + 3b_3 &= -7, & 4a_3 + 6b_3 &= -10. \end{aligned}$$

Solving these equations, we obtain

$$A = \begin{bmatrix} 1 & -1 \\ -4 & 2 \\ 2 & -3 \end{bmatrix}.$$

Example 3.21 Let T be a linear transformation from $\mathbb{R}^3$ into $\mathbb{R}^2$, where

$$T\mathbf{x} = \mathbf{A}\mathbf{x}, \ \mathbf{A} = \begin{bmatrix} 1 & 1 & 0 \\ -1 & 0 & 1 \end{bmatrix}, \ \mathbf{x} = (x \ y \ z)^T. \text{ Find } ker(T), ran(T) \text{ and their dimensions.}$$

Solution To find $ker(T)$, we need to determine all $\mathbf{v} = (v_1 \ v_2 \ v_3)^T$ such that $T\mathbf{v} = \mathbf{0}$. Now, $T\mathbf{v} = \mathbf{A}\mathbf{v} = \mathbf{0}$ gives the equations

$$v_1 + v_2 = 0, \ -v_1 + v_3 = 0$$

whose solution is $v_1 = -v_2 = v_3$. Therefore $\mathbf{v} = v_1[1 \ -1 \ 1]^T$.
Hence, dimension of $ker(T)$ is 1.
Now, $ran(T)$ is defined as $\{T(\mathbf{v}) \mid \mathbf{v} \in V\}$. We have

$$T(\mathbf{v}) = \mathbf{A}\mathbf{v} = \begin{bmatrix} 1 & 1 & 0 \\ -1 & 0 & 1 \end{bmatrix}\begin{bmatrix} v_1 \\ v_2 \\ v_3 \end{bmatrix}\begin{bmatrix} v_1 + v_2 \\ -v_1 + v_3 \end{bmatrix}$$

$$= v_1\begin{pmatrix} 1 \\ -1 \end{pmatrix} + v_2\begin{pmatrix} 1 \\ 0 \end{pmatrix} + v_3\begin{pmatrix} 0 \\ 1 \end{pmatrix}.$$

Since $\begin{pmatrix} 1 \\ -1 \end{pmatrix} = \begin{pmatrix} 1 \\ 0 \end{pmatrix} - \begin{pmatrix} 0 \\ 1 \end{pmatrix}$, the dimension of $ran(T)$ is 2.

Example 3.22 Find the matrix of a linear transformation T from $\mathbb{R}^3$ into $\mathbb{R}^3$ such that

$$T\begin{pmatrix} 1 \\ 1 \\ 1 \end{pmatrix} = \begin{pmatrix} 6 \\ 2 \\ 4 \end{pmatrix}, \ T\begin{pmatrix} 1 \\ -1 \\ 1 \end{pmatrix} = \begin{pmatrix} 2 \\ -4 \\ 2 \end{pmatrix}, \ T\begin{pmatrix} 1 \\ -2 \\ 3 \end{pmatrix} = \begin{pmatrix} 6 \\ 6 \\ 5 \end{pmatrix},$$

Solution The transformation T maps elements in $\mathbb{R}^3$ into $\mathbb{R}^3$. Therefore, the transformation is a matrix of order 3×3. Let this matrix be written as

$$T = \mathbf{A} = \begin{bmatrix} \alpha_1 & \alpha_2 & \alpha_3 \\ \beta_1 & \beta_2 & \beta_3 \\ \gamma_1 & \gamma_2 & \gamma_3 \end{bmatrix}.$$

We determine the elements of the matrix $\mathbf{A}$ such that

$$\begin{bmatrix} \alpha_1 & \alpha_2 & \alpha_3 \\ \beta_1 & \beta_2 & \beta_3 \\ \gamma_1 & \gamma_2 & \gamma_3 \end{bmatrix}\begin{bmatrix} 1 \\ 1 \\ 1 \end{bmatrix} = \begin{bmatrix} 6 \\ 2 \\ 4 \end{bmatrix}, \ \begin{bmatrix} \alpha_1 & \alpha_2 & \alpha_3 \\ \beta_1 & \beta_2 & \beta_3 \\ \gamma_1 & \gamma_2 & \gamma_3 \end{bmatrix}\begin{bmatrix} 1 \\ -1 \\ 1 \end{bmatrix} = \begin{bmatrix} 2 \\ -4 \\ 2 \end{bmatrix}, \ \begin{bmatrix} \alpha_1 & \alpha_2 & \alpha_3 \\ \beta_1 & \beta_2 & \beta_3 \\ \gamma_1 & \gamma_2 & \gamma_3 \end{bmatrix}\begin{bmatrix} 1 \\ -2 \\ 3 \end{bmatrix} = \begin{bmatrix} 6 \\ 6 \\ 5 \end{bmatrix}.$$

Equating the elements and solving the resulting equations, we obtain

$$A = \begin{bmatrix} 1 & 2 & 3 \\ -15/2 & 3 & 13/2 \\ 1 & 1 & 2 \end{bmatrix}.$$

Example 3.23   Let T be a transformation from IR^3 into IR^1 defined by

$$T(x_1, x_2, x_3) = x_1^2 + x_2^2 + x_3^2.$$

Show that T is not a linear transformation.

Solution   Let $\mathbf{x} = (x_1, x_2, x_3)$ and $\mathbf{y} = (y_1, y_2, y_3)$ be any two elements in IR^3. Then

$$\mathbf{x} + \mathbf{y} = (x_1 + y_1, x_2 + y_2, x_3 + y_3).$$

We have

$$T(\mathbf{x}) = x_1^2 + x_2^2 + x_3^2, \quad T(\mathbf{y}) = y_1^2 + y_2^2 + y_3^2$$

$$T(\mathbf{x} + \mathbf{y}) = (x_1 + y_1)^2 + (x_2 + y_2)^2 + (x_3 + y_3)^2 \neq T(\mathbf{x}) + T(\mathbf{y}).$$

Therefore, T is not a linear transformation.

Matrix representation of a linear transformation

We observe from the earlier discussion that a matrix $\mathbf{A}$ of order $m \times n$ is a linear transformation which maps the elements in IR^n into the elements in IR^m. Now, let T be a linear transformation from a finite dimensional vector space into another finite dimensional vector space over the same field F. We shall now show that with this linear transformation, we may associate a matrix $\mathbf{A}$.

Let V and W be respectively, n-dimensional and m-dimensional vector spaces over the same field F. Let T be a linear transformation such that $T : V \rightarrow W$. Let

$$\mathbf{x} = \{\mathbf{v}_1, \mathbf{v}_2, ..., \mathbf{v}_n\}, \ \mathbf{y} = \{\mathbf{w}_1, \mathbf{w}_2, ..., \mathbf{w}_m\}$$

be the ordered basis of V and W respectively. Let $\mathbf{v}$ be an arbitrary element in V and $\mathbf{w}$ be an arbitrary element in W. Then, there exist scalars, $\alpha_1, \alpha_2, ..., \alpha_n$ and $\beta_1, \beta_2, ..., \beta_m$, not all zero, such that

$$\mathbf{v} = \alpha_1 \mathbf{v}_1 + \alpha_2 \mathbf{v}_2 + ... + \alpha_n \mathbf{v}_n \tag{3.27i}$$

$$\mathbf{w} = \beta_1 \mathbf{w}_1 + \beta_2 \mathbf{w}_2 + ... + \beta_m \mathbf{w}_m \tag{3.27 ii}$$

and

$$\mathbf{w} = T\mathbf{v} = T(\alpha_1 \mathbf{v}_1 + \alpha_2 \mathbf{v}_2 + ... + \alpha_n \mathbf{v}_n)$$

$$= \alpha_1 T\mathbf{v}_1 + \alpha_2 T\mathbf{v}_2 + ... + \alpha_n T\mathbf{v}_n \tag{3.27 iii}$$

Since every element $T\mathbf{v}_i$, $i = 1, 2, ..., n$ is in W, it can be written as a linear combination of the basis vectors $\mathbf{w}_1, \mathbf{w}_2, ..., \mathbf{w}_m$ in W. That is, there exist scalars a_{ij}, $i = 1, 2, ..., n$, $j = 1, 2, ..., m$ not all zero, such that

$$T\mathbf{v}_i = a_{1i}\mathbf{w}_1 + a_{2i}\mathbf{w}_2 + ... + a_{mi}\mathbf{w}_m$$

$$= [\mathbf{w}_1, \mathbf{w}_2, ..., \mathbf{w}_m] [a_{1i}, a_{2i}, ..., a_{mi}]^T, \ i = 1, 2, ..., n. \tag{3.27 iv}$$

Hence, we can write

$$T [\mathbf{v}_1, \mathbf{v}_2, ..., \mathbf{v}_n] = [\mathbf{w}_1, \mathbf{w}_2, ..., \mathbf{w}_m] \begin{bmatrix} a_{11} & a_{12} & ... & a_{1n} \\ a_{21} & a_{22} & ... & a_{2n} \\ \vdots & & & \\ a_{m1} & a_{m2} & & a_{mn} \end{bmatrix} \qquad (3.27 \text{ v})$$

or $\qquad\qquad\qquad\qquad T\mathbf{x} = \mathbf{y}A$

where A is the $m \times n$ matrix

$$\mathbf{A} = \begin{bmatrix} a_{11} & a_{12} & ... & a_{1n} \\ a_{21} & a_{22} & ... & a_{2n} \\ \vdots & & & \\ a_{m1} & a_{m2} & & a_{mn} \end{bmatrix}. \qquad (3.27 \text{ vi})$$

The $m \times n$ matrix $\mathbf{A}$ is called the matrix representation of T or the matrix of T with respect to the ordered basis $\mathbf{x}$ and $\mathbf{y}$. It may be observed that $\mathbf{x}$ is a basis of the vector space V, on which T acts and $\mathbf{y}$ is the basis of the vector space W that contains the range of T. Therefore, the matrix representation of T depends not only on T but also on the basis $\mathbf{x}$ and $\mathbf{y}$. For a given linear transformation T, the elements a_{ij} of the matrix $\mathbf{A} = (a_{ij})$ are determined from (3.27 v), using the given basis vectors in $\mathbf{x}$ and $\mathbf{y}$. From (3. 27 iii), we have (using 3.27iv)

$$\mathbf{w} = \alpha_1(a_{11}\mathbf{w}_1 + a_{21}\mathbf{w}_2 + ... + a_{m1}\mathbf{w}_m) + \alpha_2(a_{12}\mathbf{w}_1 + a_{22}\mathbf{w}_2 + ... + a_{m2}\mathbf{w}_m)$$

$$+ ... + \alpha_n(a_{1n}\mathbf{w}_1 + a_{2n}\mathbf{w}_2 + ... + a_{mn}\mathbf{w}_m)$$

$$= (\alpha_1 a_{11} + \alpha_2 a_{12} + ... + \alpha_n a_{1n}) \mathbf{w}_1 + (\alpha_1 a_{21} + \alpha_2 a_{22} + ... + \alpha_n a_{2n})\mathbf{w}_2$$

$$+ ... + (\alpha_1 a_{m1} + \alpha_2 a_{m2} + ... + \alpha_n a_{mn})\mathbf{w}_m$$

$$= \beta_1 \mathbf{w}_1 + \beta_2 \mathbf{w}_2 + ... + \beta_m \mathbf{w}_m$$

where $\qquad\qquad \beta_i = \alpha_1 a_{i1} + \alpha_2 a_{i2} + ... + \alpha_n a_{in}, \quad i = 1, 2, ..., m.$

Hence, $\qquad \begin{bmatrix} \beta_1 \\ \beta_2 \\ \vdots \\ \beta_m \end{bmatrix} = \begin{bmatrix} a_{11} & a_{12} & ... & a_{1n} \\ a_{21} & a_{22} & ... & a_{2n} \\ \vdots & & & \\ a_{m1} & a_{m2} & & a_{mn} \end{bmatrix} \begin{bmatrix} \alpha_1 \\ \alpha_2 \\ \vdots \\ \alpha_n \end{bmatrix}$

or $\qquad\qquad\qquad\qquad \boldsymbol{\beta} = \mathbf{A}\boldsymbol{\alpha}$

where the matrix $\mathbf{A}$ is as defined in (3.27 vi) and

$$\boldsymbol{\beta} = [\beta_1, \beta_2, ..., \beta_m]^{\mathbf{T}}, \quad \boldsymbol{\alpha} = [\alpha_1, \alpha_2, ..., \alpha_n]^{\mathbf{T}}.$$

For a given ordered basis vectors $\mathbf{x}$ and $\mathbf{y}$ of vector spaces V and W respectively, and a linear transformation $T : V \to W$, the matrix $\mathbf{A}$ obtained from (3.27 v) is unique. We prove this result as follows:

Let $A = (a_{ij})$ and $B = (b_{ij})$ be two matrices each of order $m \times n$ such that

$$T\mathbf{x} = \mathbf{y}A \quad \text{and} \quad T\mathbf{x} = \mathbf{y}B.$$

Therefore, we have

$$\mathbf{y}A = \mathbf{y}B$$

or

$$\sum_{i=1}^{m} \mathbf{w}_i a_{ij} = \sum_{i=1}^{m} \mathbf{w}_i b_{ij}, \quad j = 1, 2, ..., n.$$

Since $Y = \{\mathbf{w}_1, \mathbf{w}_2, ..., \mathbf{w}_m\}$ is a given basis, we obtain $a_{ij} = b_{ij}$ for all i and j and hence $A \equiv B$.

Example 3.24 Let $T : \text{IR}^3 \to \text{IR}^2$ be a linear transformation defined by

$$T\begin{pmatrix} x \\ y \\ z \end{pmatrix} = \begin{pmatrix} y + z \\ y - z \end{pmatrix}.$$

Determine the matrix of the linear transformation T, with respect to the ordered basis

(i) $\quad \mathbf{x} = \left\{ \begin{pmatrix} 1 \\ 0 \\ 0 \end{pmatrix}, \begin{pmatrix} 0 \\ 1 \\ 0 \end{pmatrix}, \begin{pmatrix} 0 \\ 0 \\ 1 \end{pmatrix} \right\}$ in IR^3 and $\mathbf{y} = \left\{ \begin{pmatrix} 1 \\ 0 \end{pmatrix}, \begin{pmatrix} 0 \\ 1 \end{pmatrix} \right\}$ in IR^2

(standard basis e_1, e_2, e_3 in IR^3 and e_1, e_2 in IR^2).

(ii) $\quad \mathbf{x} = \left\{ \begin{pmatrix} 0 \\ 1 \\ 1 \end{pmatrix}, \begin{pmatrix} 1 \\ 0 \\ 1 \end{pmatrix}, \begin{pmatrix} 1 \\ 1 \\ 0 \end{pmatrix} \right\}$ in IR^3 and $\mathbf{y} = \left\{ \begin{pmatrix} 1 \\ 1 \end{pmatrix}, \begin{pmatrix} 1 \\ -1 \end{pmatrix} \right\}$ in IR^2.

Solution Let $V = \text{IR}^3$, $W = \text{IR}^2$. Let $\mathbf{x} = \{\mathbf{v}_1, \mathbf{v}_2, \mathbf{v}_3\}$, $\mathbf{y} = \{\mathbf{w}_1, \mathbf{w}_2\}$.

(i) We have $\quad \mathbf{v}_1 = \begin{pmatrix} 1 \\ 0 \\ 0 \end{pmatrix}, \quad \mathbf{v}_2 = \begin{pmatrix} 0 \\ 1 \\ 0 \end{pmatrix}, \quad \mathbf{v}_3 = \begin{pmatrix} 0 \\ 0 \\ 1 \end{pmatrix}, \quad \mathbf{w}_1 = \begin{pmatrix} 1 \\ 0 \end{pmatrix}, \quad \mathbf{w}_2 = \begin{pmatrix} 0 \\ 1 \end{pmatrix}.$

We obtain $\quad T\begin{pmatrix} 1 \\ 0 \\ 0 \end{pmatrix} = \begin{pmatrix} 0 \\ 0 \end{pmatrix} = \begin{pmatrix} 1 \\ 0 \end{pmatrix}(0) + \begin{pmatrix} 0 \\ 1 \end{pmatrix}(0), \quad T\begin{pmatrix} 0 \\ 1 \\ 0 \end{pmatrix} = \begin{pmatrix} 1 \\ 1 \end{pmatrix} = \begin{pmatrix} 1 \\ 0 \end{pmatrix}(1) + \begin{pmatrix} 0 \\ 1 \end{pmatrix}(1),$

$$T\begin{pmatrix} 0 \\ 0 \\ 1 \end{pmatrix} = \begin{pmatrix} 1 \\ -1 \end{pmatrix} = \begin{pmatrix} 1 \\ 0 \end{pmatrix}(1) + \begin{pmatrix} 0 \\ 1 \end{pmatrix}(-1)$$

Using the notation given in (3.27 v), that is $T\mathbf{x} = \mathbf{y}A$, we write

$$T[\mathbf{v}_1, \mathbf{v}_2, \mathbf{v}_3] = [\mathbf{w}_1, \mathbf{w}_2] \begin{bmatrix} 0 & 1 & 1 \\ 0 & 1 & -1 \end{bmatrix}$$

or
$$T\left[\begin{pmatrix} 1 \\ 0 \\ 0 \end{pmatrix}, \begin{pmatrix} 0 \\ 1 \\ 0 \end{pmatrix}, \begin{pmatrix} 0 \\ 0 \\ 1 \end{pmatrix}\right] = \left[\begin{pmatrix} 1 \\ 0 \end{pmatrix}, \begin{pmatrix} 0 \\ 1 \end{pmatrix}\right] \begin{bmatrix} 0 & 1 & 1 \\ 0 & 1 & -1 \end{bmatrix}.$$

Therefore, the matrix of the linear transformation T with respect to the given basis vectors is given by

$$A = \begin{bmatrix} 0 & 1 & 1 \\ 0 & 1 & -1 \end{bmatrix}$$

(ii) We have
$$\mathbf{v}_1 = \begin{pmatrix} 0 \\ 1 \\ 1 \end{pmatrix}, \quad \mathbf{v}_2 = \begin{pmatrix} 1 \\ 0 \\ 1 \end{pmatrix}, \quad \mathbf{v}_3 = \begin{pmatrix} 1 \\ 1 \\ 0 \end{pmatrix}, \quad \mathbf{w}_1 = \begin{pmatrix} 1 \\ 1 \end{pmatrix}, \quad \mathbf{w}_2 = \begin{pmatrix} 1 \\ -1 \end{pmatrix}.$$

We obtain
$$T\begin{pmatrix} 0 \\ 1 \\ 1 \end{pmatrix} = \begin{pmatrix} 2 \\ 0 \end{pmatrix} = \begin{pmatrix} 1 \\ 1 \end{pmatrix}(1) + \begin{pmatrix} 1 \\ -1 \end{pmatrix}(1), \quad T\begin{pmatrix} 1 \\ 0 \\ 1 \end{pmatrix} = \begin{pmatrix} 1 \\ -1 \end{pmatrix} = \begin{pmatrix} 1 \\ 1 \end{pmatrix}(0) + \begin{pmatrix} 1 \\ -1 \end{pmatrix}(1)$$

$$T\begin{pmatrix} 1 \\ 1 \\ 0 \end{pmatrix} = \begin{pmatrix} 1 \\ 1 \end{pmatrix} = \begin{pmatrix} 1 \\ 1 \end{pmatrix}(1) + \begin{pmatrix} 1 \\ -1 \end{pmatrix}(0).$$

Using (3.27 v), that is $T\mathbf{x} = \mathbf{y}A$, we write

$$T\left[\begin{pmatrix} 0 \\ 1 \\ 1 \end{pmatrix}, \begin{pmatrix} 1 \\ 0 \\ 1 \end{pmatrix}, \begin{pmatrix} 1 \\ 1 \\ 0 \end{pmatrix}\right] = \left[\begin{pmatrix} 1 \\ 1 \end{pmatrix}, \begin{pmatrix} 1 \\ -1 \end{pmatrix}\right] \begin{bmatrix} 1 & 0 & 1 \\ 1 & 1 & 0 \end{bmatrix}$$

Therefore, the matrix of the linear transformation T with respect to the given basis vectors is given by

$$A = \begin{bmatrix} 1 & 0 & 1 \\ 1 & 1 & 0 \end{bmatrix}.$$

Exercise 3.2

Discuss whether V defined in problems **1** to **10** is a vector space. If V is not a vector space, state which of the properties are not satisfied.

1. Let V be the set of the real polynomials of degree $\leq m$ and having 2 as a root with the usual addition and scalar multiplication.

2. Let V be the set of all real polynomials of degree 4 or 6 with the usual addition and scalar multiplication.

3. Let V be the set of all real polynomials of degree ≥ 4 with the usual addition and scalar multiplication.

4. Let V be the set of all rational numbers with the usual addition and scalar multiplication.

5. Let V be the set of all positive real numbers with addition defined as $x + y = xy$ and usual scalar multiplication.

6. Let V be the set of all ordered pairs (x, y) in $\dot{\mathrm{IR}}^2$ with vector addition defined as $(x, y) + (u, v) = (x + u, y + v)$ and scalar multiplication defined as $\alpha(x, y) = (3\alpha x, y)$.

7. Let V be the set of all ordered triplets (x, y, z), $x, y, z \in \mathrm{IR}$, with vector addition defined as

$$(x, y, z) + (u, v, w) = (3x + 4u, y - 2v, z + w)$$

   and scalar multiplication defined as

$$\alpha(x, y, z) = (\alpha x, \alpha y, \alpha z/3).$$

8. Let V be the set of all positive real numbers with addition defined as $x + y = xy$ and scalar multiplication defined as $\alpha x = x^{\alpha}$.

9. Let V be the set of all positive real valued continuous functions f on $[a, b]$ such that

   (i) $\int_a^b f(x)\,dx = 0$ and (ii) $\int_a^b f(x)\,dx = 2$ with usual addition and scalar multiplication.

10. Let V be the set of all solutions of the

   (i) homogeneous linear differential equation $y'' - 3y' + 2y = 0$.
   (ii) non-homogeneous linear differential equation $y'' - 3y' + 2y = x$.

   under the usual addition and scalar multiplication.

Is W a subspace of V in problems **11** to **15**? If not, state why?

11. Let V be the set of all 3×1 real matrices with usual matrix addition and scalar multiplication and W consisting of all 3×1 real matrices of the from

   (i) $\begin{bmatrix} a \\ b \\ a+b \end{bmatrix}$,   (ii) $\begin{bmatrix} a \\ a \\ a^2 \end{bmatrix}$,   (iii) $\begin{bmatrix} a \\ b \\ 2 \end{bmatrix}$,   (iv) $\begin{bmatrix} a \\ b \\ 0 \end{bmatrix}$.

12 . Let V be the set of all 3×3 real matrices with the usual matrix addition and scalar multiplication and W consisting of all 3×3 matrices **A** which

   (i) have positive elements,   (ii) are non-singular,
   (iii) are symmetric,   (iv) $\mathbf{A}^2 = \mathbf{A}$.

13. Let V be the set of all 2×2 complex matrices with the usual matrix addition and scalar multiplication and W consisting of all matrices with the usual addition and scalar multiplication and W consisting

   of all matrices of the form $\begin{bmatrix} z & x+iy \\ x-iy & u \end{bmatrix}$, where x, y, z, u are real numbers and (i) scalars are real numbers, (ii) scalars are complex numbers.

14. Let V consist of all real polynomials of degree ≤ 4 with the usual polynomial addition and scalar multiplication and W consisting of polynomials of degree ≤ 4 having

    (i) constant term 1,   (ii) coefficient of t^2 as 0,

    (iii) coefficient of t^3 as 1,   (iv) only real roots.

15. Let V be the vector space of all triplets of the form (x_1, x_2, x_3) in IR^3 with the usual addition and scalar multiplication and W be the set of triplets of the form (x_1, x_2, x_3) such that

    (i) $x_1 = 2x_2 = 3x_3$,   (ii) $x_1 = x_2 = x_3 + 1$,

    (iii) $x_1 \geq 0, x_2, x_3$ arbitrary,   (iv) $x_1^2 + x_2^2 + x_3^2 \leq 4$.   (v) x_3 is an integer.

16. Let $\mathbf{u} = (1, 2, -1)$, $\mathbf{v} = (2, 3, 4)$ and $\mathbf{w} = (1, 5, -3)$. Determine whether or not $\mathbf{x}$ is a linear combination of $\mathbf{u}, \mathbf{v}, \mathbf{w}$, where $\mathbf{x}$ is given by

    (i) $(4, 3, 10)$,   (ii) $(3, 2, 5)$ (iii) $(-2, 1, -5)$.

17. Let $\mathbf{u} = (1, -2, 1, 3)$, $\mathbf{v} = (1, 2, -1, 1)$ and $\mathbf{w} = (2, 3, 1, -1)$. Determine whether or not $\mathbf{x}$ is a linear combination of $\mathbf{u}, \mathbf{v}, \mathbf{w}$, where $\mathbf{x}$ is given by

    (i) $(3, 0, 5, -1)$,   (ii) $(2, -7, 1, 11)$,   (iii) $(4, 3, 0, 3)$.

18. Let $P_1(t) = t^2 - 4t - 6$, $P_2(t) = 2t^2 - 7t - 8$, $P_3(t) = 2t - 3$, Write $P(t)$ as a linear combination of $P_1(t)$, $P_2(t)$, $P_3(t)$, when

    (i) $P(t) = -t^2 + 1$,   (ii) $P(t) = 2t^2 - 3t - 25$.

19. Let V be the set of all 3×1 real matrices. Show that the set

$$S = \left\{ \begin{pmatrix} 1 \\ 1 \\ 0 \end{pmatrix}, \begin{pmatrix} 1 \\ -1 \\ 0 \end{pmatrix}, \begin{pmatrix} 0 \\ 0 \\ 1 \end{pmatrix} \right\} \text{ spans } V,$$

20. Let V be the set of all 2×2 real matrices. Show that the set

$$S = \left\{ \begin{pmatrix} 2 & 1 \\ 1 & -2 \end{pmatrix}, \begin{pmatrix} 1 & 1 \\ 1 & 0 \end{pmatrix}, \begin{pmatrix} 0 & 0 \\ 1 & 1 \end{pmatrix}, \begin{pmatrix} 0 & 2 \\ 0 & -1 \end{pmatrix} \right\} \text{ spans } V.$$

21. Examine whether the following vectors in $\text{IR}^3/\mathbb{C}^3$ are linearly independent.

    (i) $(2, 2, 1), (1, -1, 1), (1, 0, 1)$,   (ii) $(1, 2, 3), (3, 4, 5), (6, 7, 8)$,

    (iii) $(0, 0, 0), (1, 2, 3), (3, 4, 5)$,   (iv) $(2, i, -1), (1, -3, i), (2i, -1, 5)$,

    (v) $(1, 3, 4), (1, 1, 0), (1, 4, 2), (1, -2, 1)$.

22. Examine whether the following vectors in IR^4 are linearly independent.

    (i) $(4, 1, 2, -6), (1, 1, 0, 3), (1, -1, 0, 2), (-2, 1, 0, 3)$,

    (ii) $(1, 2, 3, 1), (2, 1, -1, 1), (4, 5, 5, 3), (5, 4, 1, 3)$,

    (iii) $(1, 2, 3, 4), (2, 0, 1, -2), (3, 2, 4, 2)$,

    (iv) $(1, 1, 0, 1), (1, 1, 1, 1), (-1, -1, 1, 1), (1, 0, 0, 1)$,

    (v) $(1, 2, 3, -1), (0, 1, -1, 2), (1, 5, 1, 8), (-1, 7, 8, 3)$.

23. If $\mathbf{x}, \mathbf{y}, \mathbf{z}$ are linearly independent vectors in IR^3, then show that

    (i) $\mathbf{x} + \mathbf{y}, \mathbf{y} + \mathbf{z}, \mathbf{z} + \mathbf{x}$;   (ii) $\mathbf{x}, \mathbf{x} + \mathbf{y}, \mathbf{x} + \mathbf{y} + \mathbf{z}$

    are also linearly independent in IR^3.

24. Write $(-4, 7, 9)$ as a linear combination of the elements of the set S: $\{(1, 2, 3), (-1, 3, 4), (3, 1, 2)\}$. Show that S is not a spanning set in IR^3.

25. Write $t^2 + t + 1$ as a linear combination of the elements of the set S: $\{3t, t^2 - 1, t^2 + 2t + 2\}$. Show that S is the spanning set for all polynomials of degree 2 and can be taken as its basis.

26. Let V be the set of all vectors in IR^4 and S be a subset of V consisting of all vectors of the form

(i) $(x, y, -y, -x)$,   (ii) (x, y, z, w) such that $x + y + z - w = 0$,

(iii) $(x, 0, z, w)$,   (iv) (x, x, x, x).

Find the dimension and the basis of S.

27. For what values of k do the following set of vectors form a basis in IR^3 ?

(i) $\{(k, 1 - k, k), (0, 3k - 1, 2), (-k, 1, 0)\}$,

(ii) $\{(k, 1, 1), (0, 1, 1), (k, 0, k)\}$,

(iii) $\{(k, k, k), (0, k, k), (k, 0, k)\}$,

(iv) $\{(1, k, 5), (1, -3, 2), (2, -1, 1)\}$.

28. Find the dimension and the basis for the vector space V, when V is the set of all 2×2 (i) real matrices (ii) symmetric matrices, (iii) skew-symmetric matrices, (iv) skew-Hermitian matrices, (v) real matrices $\mathbf{A} = (a_{ij})$ with $a_{11} + a_{22} = 0$, (vi) real matrices $\mathbf{A} = (a_{ij})$ with $a_{11} + a_{12} = 0$.

29. Find the dimension and the basis for the vector space V, when V is the set of all 3×3 (i) diagonal matrices (ii) upper triangular matrices, (iii) lower triangular matrices.

30. Find the dimension of the vector space V, when V is the set of all $n \times n$ (i) real matrices, (ii) diagonal matrices, (iii) symmetric matrices (iv) skew-symmetric matrices.

Examine whether the transformation T given in problems **31** to **35** is linear or not. If not linear, state why?

31. $T : \text{IR}^2 \to \text{IR}^1$; $T\begin{pmatrix} x \\ y \end{pmatrix} = x + y + a$, $a \neq 0$, a real constant.

32. $T : \text{IR}^3 \to \text{IR}^2$; $T\begin{pmatrix} x \\ y \\ z \end{pmatrix} = \begin{pmatrix} y \\ x + z \end{pmatrix}$.

33. $T : \text{IR}^1 \to \text{IR}^2$; $T(x) = \begin{pmatrix} x^2 \\ 3x \end{pmatrix}$.

34. $T : \text{IR}^2 \to \text{IR}^1$; $T\begin{pmatrix} x \\ y \end{pmatrix} = \begin{cases} 0 & x \neq 0, y \neq 0 \\ 2y, & x = 0 \\ 3x, & y = 0. \end{cases}$

35. $T : \text{IR}^3 \to \text{IR}^1$; $T\begin{pmatrix} x \\ y \\ z \end{pmatrix} = xy + x + z$.

Find $ker(T)$ and $ran(T)$ and their dimensions in problems **36** to **42**.

36. $T : \text{IR}^3 \to \text{IR}^3$; $T\begin{pmatrix} x \\ y \\ z \end{pmatrix} = \begin{pmatrix} x + y \\ z \\ x - y \end{pmatrix}$.

37. $T : \text{IR}^2 \to \text{IR}^3$; $T\begin{pmatrix} x \\ y \end{pmatrix} = \begin{pmatrix} 2x + y \\ y - x \\ 3x + 4y \end{pmatrix}$.

38. $T : \text{IR}^4 \to \text{IR}^3$; $T\begin{pmatrix} x \\ y \\ z \\ w \end{pmatrix} = \begin{pmatrix} x + y + w \\ z \\ y + 2w \end{pmatrix}$.

39. $T : \text{IR}^2 \to \text{IR}^1$; $T\begin{pmatrix} x \\ y \end{pmatrix} = x + 3y$.

40. $T : \text{IR}^3 \to \text{IR}^1$; $T\begin{pmatrix} x \\ y \\ z \end{pmatrix} = x + 3y$.

41. $T : \text{IR}^2 \to \text{IR}^2$; $T\begin{pmatrix} x \\ y \end{pmatrix} = \begin{pmatrix} x - y \\ x - y \end{pmatrix}$.

42. $T : \mathbb{R}^3 \to \mathbb{R}^2;\ T \begin{pmatrix} x \\ y \\ z \end{pmatrix} = \begin{pmatrix} 2x - y \\ 3x + z \end{pmatrix}.$

43. Let $T : \mathbb{R}^3 \to \mathbb{R}^2$ be a linear transformation defined by $T \begin{pmatrix} x \\ y \\ z \end{pmatrix} = \begin{pmatrix} x + y \\ x - z \end{pmatrix}.$

Find the matrix representation of T with respect to the ordered basis

$$\mathbf{x} = \left\{ \begin{pmatrix} 1 \\ 0 \\ 1 \end{pmatrix}, \begin{pmatrix} 1 \\ 1 \\ 0 \end{pmatrix}, \begin{pmatrix} 0 \\ 1 \\ 1 \end{pmatrix} \right\} \text{ in } \mathbb{R}^3 \text{ and } \mathbf{y} = \left\{ \begin{pmatrix} 1 \\ 3 \end{pmatrix}, \begin{pmatrix} 2 \\ 5 \end{pmatrix} \right\} \text{ in } \mathbb{R}^2.$$

44. Let V and W be two vector spaces in $\mathbb{R}^3$. Let $T : V \to W$ be a linear transformation defined by

$$T \begin{pmatrix} x \\ y \\ z \end{pmatrix} = \begin{pmatrix} 0 \\ x + y \\ x + y + z \end{pmatrix}.$$

Find the matrix representation of T with respect to the ordered basis

$$\mathbf{x} = \left\{ \begin{pmatrix} 1 \\ 0 \\ 0 \end{pmatrix}, \begin{pmatrix} 0 \\ 1 \\ 0 \end{pmatrix}, \begin{pmatrix} 0 \\ 0 \\ 1 \end{pmatrix} \right\} \text{ in } V \text{ and } \mathbf{y} = \left\{ \begin{pmatrix} 1 \\ 0 \\ 1 \end{pmatrix}, \begin{pmatrix} 1 \\ 1 \\ 0 \end{pmatrix}, \begin{pmatrix} 0 \\ 1 \\ 1 \end{pmatrix} \right\} \text{ in } W.$$

45. Let V and W be two vector spaces in $\mathbb{R}^3$. Let $T : V \to W$ be a linear transformation defined by

$$T \begin{pmatrix} x \\ y \\ x \end{pmatrix} = \begin{pmatrix} x + z \\ x + y \\ x + y + z \end{pmatrix}.$$

Find the matrix representation of T with respect to the ordered basis

$$\mathbf{x} = \left\{ \begin{pmatrix} -1 \\ 1 \\ 1 \end{pmatrix}, \begin{pmatrix} 1 \\ -1 \\ 1 \end{pmatrix}, \begin{pmatrix} 1 \\ 1 \\ -1 \end{pmatrix} \right\} \text{ in } V \text{ and } \mathbf{y} = \left\{ \begin{pmatrix} 1 \\ -1 \\ -1 \end{pmatrix}, \begin{pmatrix} -1 \\ 1 \\ -1 \end{pmatrix}, \begin{pmatrix} -1 \\ -1 \\ 1 \end{pmatrix} \right\} \text{ in } W$$

46. Let $T : \mathbb{R}^3 \to \mathbb{R}^4$ be a linear transformation defined by $T \begin{pmatrix} x \\ y \\ z \end{pmatrix} = \begin{pmatrix} x + y \\ y + z \\ x + z \\ x + y + z \end{pmatrix}.$

Find the matrix representation of T with respect to the ordered basis

$$\mathbf{x} = \left\{ \begin{pmatrix} 1 \\ 0 \\ 1 \end{pmatrix}, \begin{pmatrix} 1 \\ 1 \\ 0 \end{pmatrix}, \begin{pmatrix} 0 \\ 1 \\ 1 \end{pmatrix} \right\} \text{ in } \mathbb{R}^3 \text{ and } \mathbf{y} = \left\{ \begin{pmatrix} 0 \\ 1 \\ 1 \\ 1 \end{pmatrix}, \begin{pmatrix} 1 \\ 0 \\ 1 \\ 1 \end{pmatrix}, \begin{pmatrix} 1 \\ 1 \\ 0 \\ 1 \end{pmatrix}, \begin{pmatrix} 1 \\ 1 \\ 1 \\ 0 \end{pmatrix} \right\} \text{ in } \mathbb{R}^4$$

47. Let $T : \text{IR}^2 \to \text{IR}^3$ be a linear transformation. Let $\mathbf{A} = \begin{bmatrix} 1 & 2 \\ 2 & 3 \\ 3 & 4 \end{bmatrix}$ be the matrix representation of the linear

    transformation T with respect to the ordered basis vectors $\mathbf{v}_1 = [1, 2]^T$, $\mathbf{v}_2 = [3, 4]^T$ in IR^2 and $\mathbf{w}_1 = [-1, 1, 1]^T$, $\mathbf{w}_2 = [1, -1, 1]^T$, $\mathbf{w}_3 = [1, 1, -1]^T$ in IR^3. Then, determine the linear transformation T.

48. Let $T : \text{IR}^3 \to \text{IR}^2$ be a linear transformation. Let $\mathbf{A} = \begin{bmatrix} 1 & 2 & 1 \\ 2 & -3 & -4 \end{bmatrix}$ be the matrix representation of

    the linear transformation with respect to the ordered basis vectors $\mathbf{v}_1 = [1, -1, 1]^T$, $\mathbf{v}_2 = [2, 3, -1]^T$, $\mathbf{v}_3 = [1, 1, -1]^T$ in IR^3 and $\mathbf{w}_1 = [1, 1]^T$, $\mathbf{w}_2 = [2, 3]^T$ in IR^2. Then, determine the linear transformation T.

49. Let $T : P_1(t) \to P_2(t)$ be a linear transformation. Let $\mathbf{A} = \begin{bmatrix} 1 & 2 \\ 2 & 3 \\ -1 & 1 \end{bmatrix}$ be the matrix representation of the

    linear transformation with respect to the ordered basis $[1 + t, t]$ in $P_1(t)$ and $[1 - t, 2t, 2 + 3t - t^2]$ in $P_2(t)$. Then, determine the linear transformation T.

50. Let V be the set of all vectors of the form (x_1, x_2, x_3) in IR^3 satisfying (i) $x_1 - 3x_2 + 2x_3 = 0$; (ii) $3x_1 - 2x_2 + x_3 = 0$ and $4x_1 + 5x_2 = 0$. Find the dimension and basis for V.

3.4 Solution of General linear System of Equations

In section 3.2.5, we have discussed the matrix method and the Cramer's rule for solving a system of n equations in n unknowns, $\mathbf{A}\mathbf{x} = \mathbf{b}$. We assumed that the coefficient matrix $\mathbf{A}$ is non-singular, that is $|\mathbf{A}| \neq 0$, or the rank of the matrix $\mathbf{A}$ is n. The matrix method requires evaluation of n^2 determinants each of order $(n - 1)$, to generate the cofactor matrix, and one determinant of order n, whereas the Cramer's rule requires evaluation of $(n + 1)$ determinants each of order n. Since the evaluation of high order determinants is very time consuming, these methods are not used for large values of n, say $n > 4$. In this section, we discuss a method for solving a general system of m equations in n unknowns, given by

$$\mathbf{A}\mathbf{x} = \mathbf{b} \tag{3.28}$$

where $\qquad \mathbf{A} = \begin{bmatrix} a_{11} & a_{12} & \dots & a_{1n} \\ a_{21} & a_{22} & \dots & a_{2n} \\ \vdots & & & \\ a_{m1} & a_{m2} & \dots & a_{mn} \end{bmatrix}, \; \mathbf{b} = \begin{bmatrix} b_1 \\ b_2 \\ \vdots \\ b_m \end{bmatrix}, \; \mathbf{x} = \begin{bmatrix} x_1 \\ x_2 \\ \vdots \\ x_n \end{bmatrix}$

are respectively called the *coefficient matrix*, *right hand side column vector* and the *solution vector*. The order of the matrices $\mathbf{A}$, $\mathbf{b}$, $\mathbf{x}$ are respectively $m \times n$, $m \times 1$ and $n \times 1$.

The matrix

$$(\mathbf{A} \mid \mathbf{b}) = \begin{bmatrix} a_{11} & a_{12} & \dots & a_{1n} & b_1 \\ a_{21} & a_{22} & \dots & a_{2n} & b_2 \\ \vdots & & & & \vdots \\ a_{m1} & a_{m2} & \dots & a_{mn} & b_m \end{bmatrix} \tag{3.29}$$

is called the *augmented matrix* and has m rows and $(n + 1)$ columns. The augmented matrix describes completely the system of equations. The solution vector of the system of equations (3.28) is an n-tuple $(x_1, x_2, ..., x_n)$ that satisfies all the equations. There are three possibilities:

(i) the system has a unique solution,

(ii) the system has no solution,

(iii) the system has infinite number of solutions.

The system of equations is said to be *consistent*, if it has atleast one solution and *inconsistent*, if it has no solution. Using the concepts of ranks and vector spaces, we now obtain the necessary and sufficient conditions for the existence and uniqueness of the solution of the linear system of equations.

3.4.1   Existence and Uniqueness of the Solution

Let V_n be a vector space consisting of n-tuples in IR^n (or $\mathbb{C}^n$). The row vectors $R_1, R_2, ..., R_m$ of the $m \times n$ matrix $\mathbf{A}$ are n-tuples which belong to V_n. Let S be the subspace of V_n generated by the rows of $\mathbf{A}$. Then, S is called the *row-space* of the matrix $\mathbf{A}$ and its dimension is called the *row-rank* of $\mathbf{A}$ and is denoted by $rr(\mathbf{A})$. Therefore,

$$\text{row-rank of } \mathbf{A} = rr(\mathbf{A}) = dim(S). \tag{3.30}$$

Similarly, we define the *column-space* of $\mathbf{A}$ and the *column-rank* of $\mathbf{A}$ denoted by $cr(\mathbf{A})$.

Since the row-space of $m \times n$ matrix $\mathbf{A}$ is generated by m row vectors of $\mathbf{A}$, we have $\dim(S) \leq m$ and since S is a subspace of V_n, we have $dim(S) \leq n$. Therefore, we have

$$rr(\mathbf{A}) \leq \min(m, n) \quad \text{and} \quad \text{similarly} \quad cr(\mathbf{A}) \leq \min(m, n). \tag{3.31}$$

Theorem 3.6   Let $\mathbf{A} = (a_{ij})$ be an $m \times n$ matrix. Then the row-rank and column-rank of $\mathbf{A}$ are same.

Now, we state an important result which is known as the *fundamental theorem of linear algebra*.

Theorem 3.7   The non-homogeneous system of equations $\mathbf{A}\mathbf{x} = \mathbf{b}$, where $\mathbf{A}$ is an $m \times n$ matrix, has a solution if and only if the matrix $\mathbf{A}$ and the augmented matrix $(\mathbf{A} \mid \mathbf{b})$ have the same rank.

In section 3.2.3, we defined the rank of $m \times n$ matrix $\mathbf{A}$ in terms of the determinants of the submatrices of $\mathbf{A}$. An $m \times n$ matrix has rank r if it has at least one square submatrix of order r which is non-singular and all square submatrices of order greater than r are singular. This approach is very time consuming when n is large. Now, we discuss an alternative procedure to obtain the rank of a matrix.

3.4.2   Elementary Row and Column Operation

The following three operations on a matrix $\mathbf{A}$ are called the *elementary row operations*:

(i) Interchange of any two rows (written as $R_i \sim R_j$).

(ii) Multiplication/division of any row by a non-zero scalar (written as αR_i)

(iii) Adding/subtracting a scalar multiple of any row to another row (written as $R_i \leftarrow R_i + \alpha R_j$, that is α multiples of the elements of the jth row are added to the corresponding elements of the ith row. The elements of the jth row remain unchanged, whereas, the elements of the ith row get changed).

These operations change the form of **A** but do not change the row-rank of **A** as they do not change the row-space of **A**. A matrix **B** is said to be *row equivalent* to a matrix **A**, if the matrix **B** can be obtained from the matrix **A** by a finite sequence of elementary row operations. Then, we usually write **B** ≈ **A**. We observe that

(i) every matrix is row equivalent to itself.

(ii) if **A** is row equivalent to **B**, then **B** is row equivalent to **A**.

(iii) if **A** is row equivalent to **B** and **B** is row equivalent to **C**, then **A** is row equivalent to **C**.

The above operations performed on columns (that is column in place of row) are called *elementary column operations*.

3.4.3 Echelon Form of a Matrix

An $m \times n$ matrix is called a *row echelon* matrix or in *row echelon* form if the number of zeros preceeding the first non-zero entry of a row increases row by row until a row having all zero entries (or no other elimination is possible) is obtained. Therefore, a matrix is in row echelon form if the following are satisfied.

(i) If the ith row contains all zeros, it is true for all subsequent rows.

(ii) If a column contains a non-zero entry of any row, then every subsequent entry in this column is zero, that is, if the ith and $(i + 1)$th rows are both non-zero rows, then the initial non-zero entry of the $(i + 1)$th row appears in a later column than that of the ith row.

(iii) Rows containing all zeros occur only after all non-zero rows.

For example, the following matrices are in row echelon form.

$$\begin{bmatrix} 1 & 3 & 5 & 7 \\ 0 & 5 & 4 & 1 \\ 0 & 0 & 0 & 9 \end{bmatrix}, \begin{bmatrix} 1 & -1 & 2 & 3 \\ 0 & 0 & 3 & 5 \\ 0 & 0 & 0 & 0 \end{bmatrix}, \begin{bmatrix} 1 & 2 & 3 & 4 \\ 0 & 0 & 1 & 2 \\ 0 & 0 & 0 & 0 \\ 0 & 0 & 0 & 0 \end{bmatrix}.$$

Let **A** = (a_{ij}) be a given $m \times n$ matrix. Assume that $a_{11} \neq 0$. If $a_{11} = 0$, we interchange the first row with some other row to make the element in the $(1, 1)$ position as non-zero. Using elementary row operations, we reduce the matrix **A** to its row echelon form (elements of first column below a_{11} are made zero, then elements in the second column below a_{22} are made zero and so on). Similarly, we define the column echelon form of a matrix.

Rank of A The number of non-zero rows in the row echelon form of a matrix **A** gives the rank of the matrix **A** (that is, the dimension of the row-space of the matrix **A**) and the set of the non-zero rows in the row echelon form gives the basis of the row-space.

Similar results hold for column echelon matrices.

Remark 13

(i) If **A** is a square matrix, then the row-echelon form is an upper triangular matrix and the column echelon form is a lower triangular matrix.

(ii) This approach can be used to examine whether a given set of vectors are linearly independent or not. We form the matrix with each vector as its row (or column) and reduce it to the row (column) echelon form. The given vectors are linearly independent, if the row echelon form has no row with all its elements as zeros. The number of non-zero rows is the dimension of the given set of vectors and the set of vectors consisting of the non-zero rows is the basis.

Example 3.25 Reduce the following matrices to row echelon form and find their ranks.

(i) $\begin{bmatrix} 1 & 3 & 5 \\ 2 & -1 & 4 \\ -2 & 8 & 2 \end{bmatrix}$,

(ii) $\begin{bmatrix} 1 & 2 & 3 & 4 \\ 2 & 1 & 4 & 5 \\ 1 & 5 & 5 & 7 \\ 8 & 1 & 14 & 17 \end{bmatrix}$.

Solution Let the given matrix be denoted by **A**. We have

(i) $\mathbf{A} = \begin{bmatrix} 1 & 3 & 5 \\ 2 & -1 & 4 \\ -2 & 8 & 2 \end{bmatrix} \begin{array}{l} R_2 - 2R_1 \\ R_3 + 2R_1 \end{array} \approx \begin{bmatrix} 1 & 3 & 5 \\ 0 & -7 & -6 \\ 0 & 14 & 12 \end{bmatrix} R_3 + 2R_2 \approx \begin{bmatrix} 1 & 3 & 5 \\ 0 & -7 & -6 \\ 0 & 0 & 0 \end{bmatrix}$.

This is the row echelon form of **A**. Since the number of non-zero rows in the row echelon form is 2, we get rank (**A**) = 2.

(ii) $\mathbf{A} = \begin{bmatrix} 1 & 2 & 3 & 4 \\ 2 & 1 & 4 & 5 \\ 1 & 5 & 5 & 7 \\ 8 & 1 & 14 & 17 \end{bmatrix} \begin{array}{l} R_2 - 2R_1 \\ R_3 - R_1 \\ R_4 - 8R_1 \end{array} \approx \begin{bmatrix} 1 & 2 & 3 & 4 \\ 0 & -3 & -2 & -3 \\ 0 & 3 & 2 & 3 \\ 0 & -15 & -10 & -15 \end{bmatrix} \begin{array}{l} R_3 + R_2 \\ R_4 - 5R_2 \end{array} \approx \begin{bmatrix} 1 & 2 & 3 & 4 \\ 0 & -3 & -2 & -3 \\ 0 & 0 & 0 & 0 \\ 0 & 0 & 0 & 0 \end{bmatrix}$.

Since the number of non-zero rows in the echelon form of **A** is 2, we get rank (**A**) = 2.

Example 3.26 Reduce the following matrices to column echelon form and find their ranks.

(i) $\begin{bmatrix} 3 & 1 & 7 \\ 1 & 2 & 4 \\ 4 & -1 & 7 \\ 4 & -1 & 5 \end{bmatrix}$,

(ii) $\begin{bmatrix} 1 & 1 & -1 & 1 \\ -1 & 1 & -3 & -3 \\ 1 & 0 & 1 & 2 \\ 1 & -1 & 3 & 3 \end{bmatrix}$.

Solution Let the given matrix be denoted by **A**. We have

(i) $\mathbf{A} = \begin{bmatrix} 3 & 1 & 7 \\ 1 & 2 & 4 \\ 4 & -1 & 7 \\ 2 & 1 & 5 \end{bmatrix} \begin{array}{l} C_2 - C_1/3 \\ C_3 - 7C_1/3 \end{array} \approx \begin{bmatrix} 3 & 0 & 0 \\ 1 & 5/3 & 5/3 \\ 4 & -7/3 & -7/3 \\ 2 & 1/3 & 1/3 \end{bmatrix} C_3 - C_2 \approx \begin{bmatrix} 3 & 0 & 0 \\ 1 & 5/3 & 0 \\ 4 & -7/3 & 0 \\ 2 & 1/3 & 0 \end{bmatrix}$.

Since the column echelon form of **A** has two non-zero columns, rank (**A**) = 2.

$$
\text{(ii) } \mathbf{A} = \begin{bmatrix} 1 & 1 & -1 & 1 \\ -1 & 1 & -3 & -3 \\ 1 & 0 & 1 & 2 \\ 1 & -1 & 3 & 3 \end{bmatrix} \begin{matrix} C_2 - C_1 \\ C_3 + C_1 \\ C_4 - C_1 \end{matrix} \approx \begin{bmatrix} 1 & 0 & 0 & 0 \\ -1 & 2 & -4 & -2 \\ 1 & -1 & 2 & 1 \\ 1 & -2 & 4 & 2 \end{bmatrix} \begin{matrix} C_3 + 2C_2 \\ C_4 + C_2 \end{matrix} \approx \begin{bmatrix} 1 & 0 & 0 & 0 \\ -1 & 2 & 0 & 0 \\ 1 & -1 & 0 & 0 \\ 1 & 2 & 0 & 0 \end{bmatrix}.
$$

Since the column echelon form of **A** has 2 non-zero columns, rank (**A**) = 2.

Example 3.27 Examine whether the following set of vectors is linearly independent. Find the dimension and the basis of the given set of vectors.

(i) (1, 2, 3, 4), (2, 0, 1, – 2), (3, 2, 4, 2),

(ii) (1, 1, 0, 1), (1, 1, 1, 1), (–1, 1, 1, 1), (1, 0, 0, 1),

(iii) (2, 3, 6, –3, 4), (4, 2, 12, –3, 6), (4, 10, 12, –9, 10).

Solution Let each given vector represent a row of a matrix **A**. We reduce **A** to row echelon form. If all the rows of the echelon form have some non-zero elements, then the given set of vectors are linearly independent.

$$
\text{(i) } \mathbf{A} = \begin{bmatrix} 1 & 2 & 3 & 4 \\ 2 & 0 & 1 & -2 \\ 3 & 2 & 4 & 2 \end{bmatrix} \begin{matrix} R_2 - 2R_1 \\ R_3 - 3R_1 \end{matrix} \approx \begin{bmatrix} 1 & 2 & 3 & 4 \\ 0 & -4 & -5 & -10 \\ 0 & -4 & -5 & -10 \end{bmatrix} R_3 - R_2 \approx \begin{bmatrix} 1 & 2 & 3 & 4 \\ 0 & -4 & -5 & -10 \\ 0 & 0 & 0 & 0 \end{bmatrix}.
$$

Since all the rows in the row echelon form of **A** are not non-zero, the given set of vectors are linearly dependent. Since the number of non-zero rows is 2, the dimension of the given set of vectors is 2. The basis can be taken as the set of vectors {(1 2 3 4), (0, –4, –5, –10)}.

$$
\text{(ii) } \mathbf{A} = \begin{bmatrix} 1 & 1 & 0 & 1 \\ 1 & 1 & 1 & 1 \\ -1 & 1 & 1 & 1 \\ 1 & 0 & 0 & 1 \end{bmatrix} \begin{matrix} R_2 - R_1 \\ R_3 + R_1 \\ R_4 - R_1 \end{matrix} \approx \begin{bmatrix} 1 & 1 & 0 & 1 \\ 0 & 0 & 1 & 0 \\ 0 & 2 & 1 & 2 \\ 0 & -1 & 0 & 0 \end{bmatrix} R_2 \sim R_3 \approx \begin{bmatrix} 1 & 1 & 0 & 1 \\ 0 & 2 & 1 & 2 \\ 0 & 0 & 1 & 0 \\ 0 & -1 & 0 & 0 \end{bmatrix} R_4 + R_2/2
$$

$$
\approx \begin{bmatrix} 1 & 1 & 0 & 1 \\ 0 & 2 & 1 & 2 \\ 0 & 0 & 1 & 0 \\ 0 & 0 & 1/2 & 1 \end{bmatrix} R_4 - R_3/2 \approx \begin{bmatrix} 1 & 1 & 0 & 1 \\ 0 & 2 & 1 & 2 \\ 0 & 0 & 1 & 0 \\ 0 & 0 & 0 & 1 \end{bmatrix}.
$$

Since all the rows in the row echelon form of **A** are non-zero, the given set of vectors are linearly independent and the dimension of the given set of vectors is 4. The set of vectors {(1, 1, 0, 1), (0, 2, 1, 2), (0, 0, 1, 0), (0, 0, 0, 1)} or the given set itself forms the basis.

(iii) $\mathbf{A} = \begin{bmatrix} 2 & 3 & 6 & -3 & 4 \\ 4 & 2 & 12 & -3 & 6 \\ 4 & 10 & 12 & -9 & 10 \end{bmatrix} \begin{matrix} R_2 - 2R_1 \\ R_3 - 2R_1 \end{matrix} \approx \begin{bmatrix} 2 & 3 & 6 & -3 & 4 \\ 0 & -4 & 0 & 3 & -2 \\ 0 & 4 & 0 & -3 & 2 \end{bmatrix} R_3 + R_4 \approx \begin{bmatrix} 2 & 3 & 6 & -3 & 4 \\ 0 & -4 & 0 & 3 & -2 \\ 0 & 0 & 0 & 0 & 0 \end{bmatrix}.$

Since all the rows in the echelon form of $\mathbf{A}$ are not non-zero, the given set of vectors are linearly dependent. Since the number of non-zero rows is 2, the dimension of the given set of vectors is 2 and its basis can be taken as the set $\{(2, 3, 6, -3, 4), (0, -4, 0, 3, -2)\}$.

3.4.4 Gauss Elimination Method for Non-homogeneous Systems

Consider a non-homogeneous system of m equations in n unknowns

$$\mathbf{Ax} = \mathbf{b} \tag{3.32}$$

where
$$\mathbf{A} = \begin{bmatrix} a_{11} & a_{12} & \cdots & a_{1n} \\ a_{21} & a_{22} & \cdots & a_{2n} \\ \vdots & & & \\ a_{m1} & a_{m2} & \cdots & a_{mn} \end{bmatrix}, \; \mathbf{b} = \begin{bmatrix} b_1 \\ b_2 \\ \vdots \\ b_m \end{bmatrix}, \; \mathbf{x} = \begin{bmatrix} x_1 \\ x_2 \\ \vdots \\ x_n \end{bmatrix}.$$

We assume that at least one element of $\mathbf{b}$ is not zero. We write the augmented matrix of order $m \times (n + 1)$ as

$$(\mathbf{A} \mid \mathbf{b}) = \begin{bmatrix} a_{11} & a_{12} & \cdots & a_{1n} & b_1 \\ a_{21} & a_{22} & \cdots & a_{2n} & b_2 \\ \vdots & & & & \vdots \\ a_{m1} & a_{m2} & \cdots & a_{mn} & b_m \end{bmatrix}$$

and reduce it to the row echelon form by using elementary row operations. We need a maximum of $(m - 1)$ stages of eliminations to reduce the given augmented matrix to the equivalent row echelon form. This process may terminate at an earlier stage. We then have an equivalent system of the form

$$(\mathbf{A} \mid \mathbf{b}) = \begin{bmatrix} a_{11} & a_{12} & \cdots & a_{1r} & \cdots & a_{1n} & b_1 \\ 0 & \bar{a}_{22} & \cdots & \bar{a}_{2r} & \cdots & \bar{a}_{2n} & \bar{b}_2 \\ \vdots & & & & & & \vdots \\ 0 & 0 & \cdots & a_{rr}^* & \cdots & a_{rn}^* & b_r^* \\ 0 & 0 & \cdots & 0 & \cdots & 0 & b_{r+1}^* \\ \vdots & & & & & & \vdots \\ 0 & 0 & \cdots & 0 & \cdots & 0 & b_m^* \end{bmatrix}. \tag{3.33}$$

where $r \le m$ and $a_{11} \ne 0$, $\bar{a}_{22} \ne 0$, $\ldots$, $a_{rr}^* \ne 0$ are called *pivots*. We have the following cases:

(a) Let $r < m$ and one or more of the elements b^*_{r+1}, b^*_{r+2}, ..., b^*_m are not zero. Then, rank $(\mathbf{A}) \neq$ rank $(\mathbf{A} \mid \mathbf{b})$ and the system of equations has no solution.

(b) Let $m \geq n$ and $r = n$ (the number of columns in $\mathbf{A}$) and b^*_{r+1}, b^*_{r+2}, ..., b^*_m are all zeros. In this case, rank $(\mathbf{A}) =$ rank $(\mathbf{A} \mid \mathbf{b}) = n$ and the system of equations has a unique solution. We solve the nth equation for x_n, the $(n-1)$th equation for x_{n-1} and so on. This procedure is called the *back substitution method*.

For example, if we have 10 equations in 5 variables, then the augmented matrix is of order 10×6. When rank $(\mathbf{A}) =$ rank $(\mathbf{A} \mid \mathbf{b}) = 5$, the system has a unique solution.

(c) Let $r < n$ and b^*_{r+1}, b^*_{r+2}, ..., b^*_m are all zeros. In this case, r unknowns, x_1, x_2, ..., x_r can be determined in terms of the remaining $(n - r)$ unknowns x_{r+1}, x_{r+2}, ..., x_n by solving the rth equation for x_r, $(r-1)$th equation for x_{r-1} and so on. In this case, we obtain an $(n - r)$ parameter family of solutions, that is infinitely many solutions.

Remark 14

(a) We do not, normally use column elementary operations in solving the linear system of equations. When we interchange two columns, the order of the unknowns in the given system of equations is also changed. Keeping track of the order of unknowns is quite difficult.

(b) Gauss elimination method may be written as

$$(\mathbf{A} \mid \mathbf{b}) \xrightarrow[\text{row operations}]{\text{Elementary}} (\mathbf{B} \mid \mathbf{c}).$$

The matrix $\mathbf{B}$ is the row echelon form of the matrix $\mathbf{A}$ and $\mathbf{c}$ is the new right hand side column vector. We obtain the solution vector (if it exists) using the back substitution method.

(c) If $\mathbf{A}$ is a square matrix of order n, then $\mathbf{B}$ is an upper triangular matrix of order n.

(d) Gauss elimination method can be used to solve p systems of the form $\mathbf{A}\mathbf{x} = \mathbf{b}_1$, $\mathbf{A}\mathbf{x} = \mathbf{b}_2$, ..., $\mathbf{A}\mathbf{x} = \mathbf{b}_p$ which have the same coefficient matrix but different right hand side column vectors. We form the augmented matrix as $(\mathbf{A} \mid \mathbf{b}_1, \mathbf{b}_2, ..., \mathbf{b}_p)$, which has m rows and $(n + p)$ columns. Using the elementary row operations, we obtain the row equivalent system $(\mathbf{B} \mid \mathbf{c}_1, \mathbf{c}_2, ..., \mathbf{c}_p)$, where $\mathbf{B}$ is the row echelon form of $\mathbf{A}$. Now, we solve the systems $\mathbf{B}\mathbf{x} = \mathbf{c}_1$, $\mathbf{B}\mathbf{x} = \mathbf{c}_2$, ..., $\mathbf{B}\mathbf{x} = \mathbf{c}_p$, using the back substitution method.

Remark 15

(a) If at any stage of elimination, the pivot element becomes zero, then we interchange this row with any other row below it such that we obtain a non-zero pivot element. We normally choose the row such that the pivot element becomes largest in magnitude.

(b) For an $n \times n$ system, we require $(n - 1)$ stages of elimination. It is possible to compute the total number of additions, subtractions, multiplications and divisions. This number is called the *operation count* of the method. The operation count of the Gauss elimination method for solving an $n \times n$ system is $n(n^2 + 3n - 1)/3$. For large n, the operation count is approximately $n^3/3$.

Example 3.28 Solve the following systems of equations (if possible) using Gauss elimination method.

(i) $\begin{bmatrix} 2 & 1 & -1 \\ 1 & -1 & 2 \\ -1 & 2 & -1 \end{bmatrix} \begin{bmatrix} x \\ y \\ z \end{bmatrix} = \begin{bmatrix} 4 \\ -2 \\ 2 \end{bmatrix}$, (ii) $\begin{bmatrix} 2 & 0 & 1 \\ 1 & -1 & 1 \\ 4 & -2 & 3 \end{bmatrix} \begin{bmatrix} x \\ y \\ z \end{bmatrix} = \begin{bmatrix} 3 \\ 1 \\ 3 \end{bmatrix}$,

(iii) $\begin{bmatrix} 1 & -1 & 1 \\ 2 & 1 & -1 \\ 5 & -2 & 2 \end{bmatrix} \begin{bmatrix} x \\ y \\ z \end{bmatrix} = \begin{bmatrix} 1 \\ 2 \\ 5 \end{bmatrix}$.

Solution We write the augmented matrix and reduce it to row echelon form by applying elementary row operations.

(i) $(\mathbf{A} \mid \mathbf{b}) = \begin{bmatrix} 2 & 1 & -1 & | & 4 \\ 1 & -1 & 2 & | & -2 \\ -1 & 2 & -1 & | & 2 \end{bmatrix} \begin{matrix} \\ R_2 - R_1/2 \\ R_3 + R_1/2 \end{matrix} \approx \begin{bmatrix} 2 & 1 & -1 & | & 4 \\ 0 & -3/2 & 5/2 & | & -4 \\ 0 & 5/2 & -3/2 & | & 4 \end{bmatrix} R_3 + 5R_2/3$

$\approx \begin{bmatrix} 2 & 1 & -1 & | & 4 \\ 0 & -3/2 & 5/2 & | & -4 \\ 0 & 0 & 8/3 & | & -8/3 \end{bmatrix}$.

Using the back substitution method, we obtain the solution as

$$\frac{8}{3}z = -\frac{8}{3}, \quad \text{or} \quad z = -1,$$

$$-\frac{3}{2}y + \frac{5}{2}z = -4, \quad \text{or} \quad y = 1,$$

$$2x + y - z = 4, \quad \text{or} \quad x = 1.$$

Therefore, the system of equations has the unique solution $x = 1$, $y = 1$, $z = -1$.

(ii) $(\mathbf{A} \mid \mathbf{b}) = \begin{bmatrix} 2 & 0 & 1 & | & 3 \\ 1 & -1 & 1 & | & 1 \\ 4 & -2 & 3 & | & 3 \end{bmatrix} \begin{matrix} \\ R_2 - R_1/2 \\ R_3 - 2R_1 \end{matrix} \approx \begin{bmatrix} 2 & 0 & 1 & | & 3 \\ 0 & -1 & 1/2 & | & -1/2 \\ 0 & -2 & 1 & | & -3 \end{bmatrix} R_3 - 2R_2 \approx \begin{bmatrix} 2 & 0 & 1 & | & 3 \\ 0 & -1 & 1/2 & | & -1/2 \\ 0 & 0 & 0 & | & -2 \end{bmatrix}$.

We find that rank $(\mathbf{A}) = 2$ and rank $(\mathbf{A} \mid \mathbf{b}) = 3$. Therefore, the system of equations has no solution.

(iii) $(\mathbf{A} \mid \mathbf{b}) = \begin{bmatrix} 1 & -1 & 1 & | & 1 \\ 2 & 1 & -1 & | & 2 \\ 5 & -2 & 2 & | & 5 \end{bmatrix} \begin{matrix} \\ R_2 - 2R_1 \\ R_3 - 5R_1 \end{matrix} \approx \begin{bmatrix} 1 & -1 & 1 & | & 1 \\ 0 & 3 & -3 & | & 0 \\ 0 & 3 & -3 & | & 0 \end{bmatrix} R_3 - R_2 \approx \begin{bmatrix} 1 & -1 & 1 & | & 1 \\ 0 & 3 & -3 & | & 0 \\ 0 & 0 & 0 & | & 0 \end{bmatrix}$.

The system is consistent and has infinite number of solutions. We find that the last equation is satisfied for all values of x, y, z. From the second equation, we get $3y - 3z = 0$, or $y = z$. From the first equation, we get $x - y + z = 1$, or $x = 1$. Therefore, we obtain the solution $x = 1$, $y = z$ and z is arbitrary.

Example 3.29 Solve the following system of equations using Gauss elimination method.

(i) $\begin{aligned} 4x - 3y - 9z + 6w &= 0 \\ 2x + 3y + 3z + 6w &= 6 \\ 4x - 21y - 39z - 6w &= -24, \end{aligned}$

(ii) $\begin{aligned} x + 2y - 2z &= 1 \\ 2x - 3y + z &= 0 \\ 5x + y - 5z &= 1 \\ 3x + 14y - 12z &= 5. \end{aligned}$

Solution   We have

(i) $(\mathbf{A} \mid \mathbf{b}) = \begin{bmatrix} 4 & -3 & -9 & 6 & | & 0 \\ 2 & 3 & 3 & 6 & | & 6 \\ 4 & -21 & -39 & -6 & | & -24 \end{bmatrix} \begin{matrix} R_2 - R_1/2 \\ R_3 - R_1 \end{matrix} \approx \begin{bmatrix} 4 & -3 & -9 & 6 & | & 0 \\ 0 & 9/2 & 15/2 & 3 & | & 6 \\ 0 & -18 & -30 & -12 & | & -24 \end{bmatrix} R_3 + 4R_2$

$= \begin{bmatrix} 4 & -3 & -9 & 6 & | & 0 \\ 0 & 9/2 & 15/2 & 3 & | & 6 \\ 0 & 0 & 0 & 0 & | & 0 \end{bmatrix}.$

The system of equations is consistent and has infinite number of solutions. Choose w as arbitrary. From the second equation, we obtain

$$\frac{9}{2}y + \frac{15}{2}z = 6 - 3w, \quad \text{or} \quad y = \frac{2}{9}\left(6 - 3w - \frac{15}{2}z\right) = \frac{1}{3}(4 - 5z - 2w).$$

From the first equation, we obtain

$$4x = 3y + 9z - 6w = 4 - 5z - 2w + 9z - 6w = 4 + 4z - 8w$$

or $\qquad\qquad x = 1 + z - 2w.$

Thus, we obtain a two parameter family of solutions

$$x = 1 + z - 2w \quad \text{and} \quad y = (4 - 5z - 2w)/3$$

where z and w are arbitrary.

(ii) $(\mathbf{A} \mid \mathbf{b}) = \begin{bmatrix} 1 & 2 & -2 & | & 1 \\ 2 & -3 & 1 & | & 0 \\ 5 & 1 & -5 & | & 1 \\ 3 & 14 & -12 & | & 5 \end{bmatrix} \begin{matrix} R_2 - 2R_1 \\ R_3 - 5R_1 \\ R_4 - 3R_1 \end{matrix} \approx \begin{bmatrix} 1 & 2 & -2 & | & 1 \\ 0 & -7 & 5 & | & -2 \\ 0 & -9 & 5 & | & -4 \\ 0 & 8 & -6 & | & 2 \end{bmatrix} \begin{matrix} R_3 - 9R_2/7 \\ R_4 + 8R_2/7 \end{matrix}$

$= \begin{bmatrix} 1 & 2 & -2 & | & 1 \\ 0 & -7 & 5 & | & -2 \\ 0 & 0 & -10/7 & | & -10/7 \\ 0 & 0 & -2/7 & | & -2/7 \end{bmatrix} R_4 - R_3/5 \approx \begin{bmatrix} 1 & 2 & -2 & | & 1 \\ 0 & -7 & 5 & | & -2 \\ 0 & 0 & -10/7 & | & -10/7 \\ 0 & 0 & 0 & | & 0 \end{bmatrix}.$

The last equation is satisfied for all values of x, y, z. From the third equation, we obtain $z = 1$. Back substitution gives $y = 1$, $x = 1$. Hence, the system of equations has a unique solution $x = 1$, $y = 1$ and $z = 1$. Since $R_4 = (24R_1 - 7R_2 + R_3)/5$, the last equation is redundant.

3.4.5 Gauss-Jordan Method

In this method, we perform elementary row transformations on the augmented matrix $[\mathbf{A} \mid \mathbf{b}]$ and reduce it to the form

$$[\mathbf{A} \mid \mathbf{b}] \xrightarrow[\text{row operations}]{\text{Elementary}} [\mathbf{I} \mid \mathbf{c}]$$

where $\mathbf{I}$ is the identity matrix and $\mathbf{c}$ is the solution vector. This reduction is equivalent to finding the solution as $\mathbf{x} = \mathbf{A}^{-1}\mathbf{b}$. The first step is same as in the Gauss elimination method. From second step onwards, we make elements below and above the pivot as zeros, using elementary row transformations. Finally, we divide each row by its pivot to obtain the form $[\mathbf{I} \mid \mathbf{c}]$. Alternately, at every step, the pivot can be made as 1 before elimination. Then, $\mathbf{c}$ is the solution vector.

This method is more expensive (larger operation count) than the Gauss elimination. Hence, we do not normally use the Gauss-Jordan method for finding the solution of a system. However, this method is very useful for finding the inverse ($\mathbf{A}^{-1}$) of a matrix $\mathbf{A}$. We consider the augmented matrix $[\mathbf{A} \mid \mathbf{I}]$ and reduce it to the form

$$[\mathbf{A} \mid \mathbf{I}] \xrightarrow[\text{row operations}]{\text{Elementary}} [\mathbf{I} \mid \mathbf{A}^{-1}]$$

using elementary row transformations. If we are solving the system of equations (3.28), then we have $\mathbf{x} = \mathbf{A}^{-1}\mathbf{b}$, and the matrix multiplication on the right hand side gives the solution vector.

Remark 16

If any pivot element at any stage of elimination becomes zero, then we interchange rows as in the Gauss elimination method.

Example 3.30   Using the Gauss-Jordan method, solve the system of equations $\mathbf{A}\,\mathbf{x} = \mathbf{b}$, where

$$\mathbf{A} = \begin{bmatrix} 1 & -1 & 1 \\ 2 & 1 & -3 \\ 1 & 1 & 1 \end{bmatrix}, \quad \text{and} \quad \mathbf{b} = \begin{bmatrix} 0 \\ 4 \\ 1 \end{bmatrix}.$$

Solution   We perform elementary row transformations on the augmented matrix and reduce it the form $[\mathbf{I} \mid \mathbf{c}]$. We get

$$[\mathbf{A} \mid \mathbf{b}] = \begin{bmatrix} 1 & -1 & 1 & | & 0 \\ 2 & 1 & -3 & | & 4 \\ 1 & 1 & 1 & | & 1 \end{bmatrix} \begin{matrix} \\ R_2 - 2R_1 \\ R_3 - R_1 \end{matrix} \approx \begin{bmatrix} 1 & -1 & 1 & | & 0 \\ 0 & 3 & -5 & | & 4 \\ 0 & 2 & 0 & | & 1 \end{bmatrix} R_2/3$$

$$\approx \begin{bmatrix} 1 & -1 & 1 & 0 \\ 0 & 1 & -5/3 & 4/3 \\ 0 & 2 & 0 & 1 \end{bmatrix} \begin{matrix} R_1 + R_2 \\ \\ R_3 - 2R_2 \end{matrix} \approx \begin{bmatrix} 1 & 0 & -2/3 & 4/3 \\ 0 & 1 & -5/3 & 4/3 \\ 0 & 0 & 10/3 & -5/3 \end{bmatrix} R_3/(10/3)$$

$$\approx \begin{bmatrix} 1 & 0 & -2/3 & 4/3 \\ 0 & 1 & -5/3 & 4/3 \\ 0 & 0 & 1 & -1/2 \end{bmatrix} \begin{matrix} R_1 + 2R_3/3 \\ \\ R_2 + 5R_3/3 \end{matrix} \approx \begin{bmatrix} 1 & 0 & 0 & 1 \\ 0 & 1 & 0 & 1/2 \\ 0 & 0 & 1 & -1/2 \end{bmatrix}.$$

Hence, the solution vector is

$$\mathbf{x} = \begin{bmatrix} 1 & 1/2 & -1/2 \end{bmatrix}^T.$$

Example 3.31   Using Gauss-Jordan method, find the inverse of the matix $\mathbf{A} = \begin{bmatrix} -1 & 1 & 2 \\ 3 & -1 & 1 \\ -1 & 3 & 4 \end{bmatrix}$

Solution   We have

$$(\mathbf{A} \mid \mathbf{I}) = \begin{bmatrix} -1 & 1 & 2 & 1 & 0 & 0 \\ 3 & -1 & 1 & 0 & 1 & 0 \\ -1 & 3 & 4 & 0 & 0 & 1 \end{bmatrix}.$$

The pivot element a_{11} is -1. We make it 1 by multiplying the first row by -1. Therefore,

$$(\mathbf{A} \mid \mathbf{I}) \approx \begin{bmatrix} 1 & -1 & -2 & -1 & 0 & 0 \\ 3 & -1 & 1 & 0 & 1 & 0 \\ -1 & 3 & 4 & 0 & 0 & 1 \end{bmatrix} \begin{matrix} R_2 - 3R_1 \\ \\ R_3 + R_1 \end{matrix} \approx \begin{bmatrix} 1 & -1 & -2 & -1 & 0 & 0 \\ 0 & 2 & 7 & 3 & 1 & 0 \\ 0 & 2 & 2 & -1 & 0 & 1 \end{bmatrix} R_2/2$$

$$\approx \begin{bmatrix} 1 & -1 & -2 & -1 & 0 & 0 \\ 0 & 1 & 7/2 & 3/2 & 1/2 & 0 \\ 0 & 2 & 2 & -1 & 0 & 1 \end{bmatrix} \begin{matrix} R_1 + R_2 \\ \\ R_3 - 2R_2 \end{matrix} \approx \begin{bmatrix} 1 & 0 & 3/2 & 1/2 & 1/2 & 0 \\ 0 & 1 & 7/2 & 3/2 & 1/2 & 0 \\ 0 & 0 & -5 & -4 & -1 & 1 \end{bmatrix} (-R_3)/5$$

$$\approx \begin{bmatrix} 1 & 0 & 3/2 & 1/2 & 1/2 & 0 \\ 0 & 1 & 7/2 & 3/2 & 3/2 & 0 \\ 0 & 0 & 1 & 4/5 & 1/5 & -1/5 \end{bmatrix} \begin{matrix} R_1 - 3R_3/2 \\ \\ R_3 - 7R_3/2 \end{matrix} \approx \begin{bmatrix} 1 & 0 & 0 & -7/10 & 2/10 & 3/10 \\ 0 & 1 & 0 & -13/10 & -2/10 & 7/10 \\ 0 & 0 & 1 & 4/5 & 1/5 & -1/5 \end{bmatrix}.$$

Hence,

$$\mathbf{A}^{-1} = \frac{1}{10} \begin{bmatrix} -7 & 2 & 3 \\ -13 & -2 & 7 \\ 8 & 2 & -2 \end{bmatrix}.$$

3.4.6 Homogeneous System of Linear Equations

Consider the homogeneous system of equations

$$\mathbf{A}\mathbf{x} = \mathbf{0} \tag{3.34}$$

where $\mathbf{A}$ is an $m \times n$ matrix. The homogenous system is always consistent since $\mathbf{x} = \mathbf{0}$ (trivial solution) is always a solution. In this case, rank $(\mathbf{A})$ = rank $(\mathbf{A} \mid \mathbf{0})$. Therefore, for the homogeneous system to have a non-trivial solution, we require that rank $(\mathbf{A}) < n$. If rank $(\mathbf{A}) = r < n$, we obtain an $(n - r)$ parameter family of solutions which form a vector space of dimension $(n - r)$ as $(n - r)$ parameters can be chosen arbitrarily.

The solution space of the homogeneous system is called the *null space* and its dimension is called the *nullity* of $\mathbf{A}$. Therefore, we obtain the result

$$\text{rank } (\mathbf{A}) + \text{nullity } (\mathbf{A}) = n \ (\text{see Theorem 3.5}).$$

Remark 17

(a) If $\mathbf{x}_1$ and $\mathbf{x}_2$ are two solutions of a linear homogeneous system, then $\alpha\mathbf{x}_1 + \beta\mathbf{x}_2$ is also a solution of the homogenous system for any scalars α, β. This result does not hold for non-homogenous systems.

(b) A homogeneous system of m equations in n unknowns and $m < n$, always possesses a non-trivial solution.

Theorem 3.8 If a non-homogeneous system of linear equations $\mathbf{A}\mathbf{x} = \mathbf{b}$ has solutions, then all these solutions are of the form $\mathbf{x} = \mathbf{x}_0 + \mathbf{x}_h$ where $\mathbf{x}_0$ is any fixed solution of $\mathbf{A}\mathbf{x} = \mathbf{b}$ and $\mathbf{x}_h$ is any solution of the corresponding homogeneous system.

Proof Let $\mathbf{x}$ be any solution and $\mathbf{x}_0$ be any fixed solution of $\mathbf{A}\mathbf{x} = \mathbf{b}$. Therefore, we have

$$\mathbf{A}\mathbf{x} = \mathbf{b} \quad \text{and} \quad \mathbf{A}\mathbf{x}_0 = \mathbf{b}.$$

Subtracting, we get

$$\mathbf{A}\mathbf{x} - \mathbf{A}\mathbf{x}_0 = \mathbf{0}, \quad \text{or} \quad \mathbf{A}(\mathbf{x} - \mathbf{x}_0) = \mathbf{0}.$$

Thus, the difference $\mathbf{x} - \mathbf{x}_0$ between any solution $\mathbf{x}$ of $\mathbf{A}\mathbf{x} = \mathbf{b}$ and any fixed solution $\mathbf{x}_0$ of $\mathbf{A}\mathbf{x} = \mathbf{b}$ is a solution of the homogeneous system $\mathbf{A}\mathbf{x} = \mathbf{0}$, say $\mathbf{x}_h$. Hence, the result.

Remark 18

If the non-homogeneous system $\mathbf{A}\mathbf{x} = \mathbf{b}$ where $\mathbf{A}$ is an $m \times n$ matrix $(m \geq n)$ has a unique solution, that is rank $(\mathbf{A}) = n$, then the corresponding homogeneous system $\mathbf{A}\mathbf{x} = \mathbf{0}$ has only the trivial solution, that is $\mathbf{x}_h = \mathbf{0}$.

Example 3.32 Solve the following homogeneous system of equation $\mathbf{A}\mathbf{x} = \mathbf{0}$, where $\mathbf{A}$ is given by

$$\text{(i) } \begin{bmatrix} 2 & 1 \\ 1 & -1 \\ 3 & 2 \end{bmatrix}, \qquad \text{(ii) } \begin{bmatrix} 1 & 2 & -3 \\ 1 & 1 & -1 \\ 1 & -1 & 1 \end{bmatrix}, \qquad \text{(iii) } \begin{bmatrix} 1 & 1 & -1 & 1 \\ 2 & 3 & 1 & 4 \\ 3 & 2 & -6 & 1 \end{bmatrix}.$$

Find the rank $(\mathbf{A})$ and nullity $(\mathbf{A})$.

Solution   We write the augmented matrix $(\mathbf{A} \mid \mathbf{0})$ and reduce it to row echelon form.

(i) $(\mathbf{A} \mid \mathbf{0}) = \begin{bmatrix} 2 & 1 & 0 \\ 1 & -1 & 0 \\ 3 & 2 & 0 \end{bmatrix} \begin{matrix} R_2 - R_1/2 \\ R_3 - 3R_1/2 \end{matrix} \approx \begin{bmatrix} 2 & 1 & 0 \\ 0 & -3/2 & 0 \\ 0 & 1/2 & 0 \end{bmatrix} R_3 + R_2/3 \approx \begin{bmatrix} 2 & 1 & 0 \\ 0 & -3/2 & 0 \\ 0 & 0 & 0 \end{bmatrix}.$

Since, rank $(\mathbf{A}) = 2 =$ number of unknowns, the system has only a trivial solution. Hence, nullity $(\mathbf{A}) = 0$.

(ii) $(\mathbf{A} \mid \mathbf{0}) = \begin{bmatrix} 1 & 2 & -3 & 0 \\ 1 & 1 & -1 & 0 \\ 1 & -1 & 1 & 0 \end{bmatrix} \begin{matrix} R_2 - R_1 \\ R_3 - R_1 \end{matrix} \approx \begin{bmatrix} 1 & 2 & -3 & 0 \\ 0 & -1 & 2 & 0 \\ 0 & -3 & 4 & 0 \end{bmatrix} R_3 - 3R_2 \approx \begin{bmatrix} 1 & 2 & -3 & 0 \\ 0 & -1 & 2 & 0 \\ 0 & 0 & -2 & 0 \end{bmatrix}.$

Since rank $(\mathbf{A}) = 3 =$ number of unknowns, the homogeneous system has only a trivial solution. Therefore, nullity $(\mathbf{A}) = 0$.

(iii) $(\mathbf{A} \mid \mathbf{0}) = \begin{bmatrix} 1 & 1 & -1 & 1 & 0 \\ 2 & 3 & 1 & 4 & 0 \\ 3 & 2 & -6 & 1 & 0 \end{bmatrix} \begin{matrix} R_2 - 2R_1 \\ R_3 - 3R_1 \end{matrix} \approx \begin{bmatrix} 1 & 1 & -1 & 1 & 0 \\ 0 & 1 & 3 & 2 & 0 \\ 0 & -1 & -3 & -2 & 0 \end{bmatrix} R_3 + R_2 \approx \begin{bmatrix} 1 & 1 & -1 & 1 & 0 \\ 0 & 1 & 3 & 2 & 0 \\ 0 & 0 & 0 & 0 & 0 \end{bmatrix}.$

Therefore, rank $(\mathbf{A}) = 2$ and the number of unknowns is 4. Hence, we obtain a two parameter family of solutions as $x_2 = -3x_3 - 2x_4$, $x_1 = -x_2 + x_3 - x_4 = 4x_3 + x_4$, where x_3 and x_4 are arbitrary. Therefore, nullity $(\mathbf{A}) = 2$.

Exercise 3.3

Using the elementary row operations, determine the ranks of the following matrices.

1. $\begin{bmatrix} 2 & 0 & -1 \\ 5 & 1 & 0 \\ 0 & 1 & 3 \end{bmatrix}.$
2. $\begin{bmatrix} 1 & -2 & 3 \\ 2 & 1 & 2 \\ 5 & -5 & 11 \end{bmatrix}.$
3. $\begin{bmatrix} 2 & 1 & -2 \\ -1 & -1 & 1 \\ 3 & 1 & -2 \end{bmatrix}.$

4. $\begin{bmatrix} 1 & 2 & -1 & 1 \\ 2 & 3 & 4 & 5 \\ 1 & 4 & -13 & -5 \end{bmatrix}.$
5. $\begin{bmatrix} 1 & -2 & 1 & -1 \\ 1 & 1 & -2 & 3 \\ 4 & 1 & -5 & 8 \end{bmatrix}.$
6. $\begin{bmatrix} 1 & 2 & 1 \\ 2 & 3 & 4 \\ 1 & 3 & -1 \\ 8 & 13 & 14 \end{bmatrix}.$

7. $\begin{bmatrix} 1 & 2 & 3 & 4 \\ 2 & 3 & 4 & 5 \\ 7 & 11 & 15 & 19 \\ 9 & 15 & 21 & 27 \end{bmatrix}.$
8. $\begin{bmatrix} 1 & 1 & 1 & 1 \\ -1 & 1 & 1 & -1 \\ 1 & 0 & 1 & 1 \\ 1 & 1 & 0 & 1 \end{bmatrix}.$
9. $\begin{bmatrix} 2 & 0 & -1 & 0 \\ 4 & 1 & 0 & 5 \\ 0 & 1 & 3 & 6 \\ 6 & 1 & -2 & 6 \end{bmatrix}.$

10. $\begin{bmatrix} 2 & 3 & 1 & 0 & 4 \\ 3 & 1 & 2 & -1 & 1 \\ 4 & -1 & 3 & -2 & -2 \\ 5 & 4 & 3 & -1 & 5 \end{bmatrix}.$

Using the elementary column operations, determine the rank of the following matrices.

11. $\begin{bmatrix} 1 & 1 & -1 \\ 1 & 0 & 1 \\ 1 & -1 & 3 \end{bmatrix}.$

12. $\begin{bmatrix} -1 & 1 & 2 \\ 3 & -1 & 1 \\ -1 & 3 & 4 \end{bmatrix}.$

13. $\begin{bmatrix} 1 & 2 & 3 & 2 \\ -1 & 1 & 3 & -5 \\ 2 & 3 & 4 & 5 \end{bmatrix}.$

14. $\begin{bmatrix} 1 & 2 & -3 \\ 2 & 1 & -1 \\ 1 & -1 & 2 \\ 5 & 4 & -5 \end{bmatrix}.$

15. $\begin{bmatrix} 2 & 3 & 1 & 0 & 4 \\ 3 & 1 & 2 & -1 & 1 \\ 4 & -1 & 3 & -2 & -2 \\ 5 & 4 & 3 & -1 & 5 \end{bmatrix}.$

Determine whether the following set of vectors is linearly independent. Find also its dimension.

16. $\{(3, 2, 4), (1, 0, 2), (1, -1, -1)\}$.

17. $\{(2, 2, 1), (1, -1, 1), (1,0,1)\}$.

18. $\{(2, 1, 0), (1, -1, 1), (4, 1, 2), (2, -3, 3)\}$.

19. $\{(2, 2, 1), (2, i, -1), (1 + i, - i, 1)\}$.

20. $\{(1, 1, 1), (i, i, i), (1 + i, -1 -i, i)\}$

21. $\{(1, 1, 1, 1), (-1, 1, 1, -1), (1, 0, 1, 1), (1, 1, 0,1)\}$.

22. $\{(1, 2, 3, 1), (2, 1, -1, 1), (4, 5, 5, 3), (5, 4, 1, 3)\}$.

23. $\{(1, 2, 3, 4), (0, 1, -1, 2), (1, 4, 1, 8), (3, 7, 8, 14)\}$.

24. $\{(1, 1, 0, 1), (1, 1, 1, 1), (4, 4, 1, 1), (1, 0, 0, 1)\}$.

25. $\{(2, 2, 0, 2), (4, 1, 4, 1), (3, 0, 4, 0)\}$.

Determine which of the following systems are consistent and find all the solutions for the consistent systems.

26. $\begin{bmatrix} 2 & -3 & 1 \\ 1 & -1 & 2 \\ 2 & 1 & -3 \end{bmatrix} \begin{bmatrix} x \\ y \\ z \end{bmatrix} = \begin{bmatrix} -2 \\ 3 \\ -2 \end{bmatrix}.$

27. $\begin{bmatrix} 1 & 4 & 7 \\ 2 & 5 & 8 \\ 1 & 2 & 3 \end{bmatrix} \begin{bmatrix} x \\ y \\ z \end{bmatrix} = \begin{bmatrix} 1 \\ 2 \\ 1 \end{bmatrix}.$

28. $\begin{bmatrix} 1 & -4 & 7 \\ 3 & 8 & -2 \\ 7 & -8 & 26 \end{bmatrix} \begin{bmatrix} x \\ y \\ z \end{bmatrix} = \begin{bmatrix} 8 \\ 6 \\ 3 \end{bmatrix}.$

29. $\begin{bmatrix} 1 & 1 & 1 \\ 3 & -9 & 2 \\ 5 & -3 & 4 \end{bmatrix} \begin{bmatrix} x \\ y \\ z \end{bmatrix} = \begin{bmatrix} 3 \\ -4 \\ 6 \end{bmatrix}.$

30. $\begin{bmatrix} 1 & 1 & 1 \\ 1 & 2 & 3 \\ 1 & 3 & 4 \end{bmatrix} \begin{bmatrix} x \\ y \\ z \end{bmatrix} = \begin{bmatrix} 7 \\ 16 \\ 22 \end{bmatrix}.$

31. $\begin{bmatrix} 2 & 0 & -3 \\ 0 & 2 & -3 \\ 1 & -1 & 1 \end{bmatrix} \begin{bmatrix} x \\ y \\ z \end{bmatrix} = \begin{bmatrix} 0 \\ 0 \\ 1 \end{bmatrix}.$

32. $\begin{bmatrix} 5 & 3 & 14 \\ 0 & 1 & 2 \\ 1 & 1 & 2 \\ 2 & 1 & 6 \end{bmatrix} \begin{bmatrix} x \\ y \\ z \end{bmatrix} = \begin{bmatrix} 4 \\ 1 \\ 0 \\ 2 \end{bmatrix}.$

33. $\begin{bmatrix} 1 & -2 & 1 & 2 \\ 1 & 1 & -1 & 1 \\ 1 & 7 & -5 & -1 \end{bmatrix} \begin{bmatrix} x \\ y \\ z \\ w \end{bmatrix} = \begin{bmatrix} 1 \\ 2 \\ 4 \end{bmatrix}.$

34. $\begin{bmatrix} 1 & 1 & 1 & 1 \\ 1 & 1 & 1 & -1 \\ 1 & -1 & 1 & -1 \end{bmatrix} \begin{bmatrix} x \\ y \\ z \\ w \end{bmatrix} = \begin{bmatrix} 4 \\ 2 \\ 0 \end{bmatrix}.$

35. $\begin{bmatrix} -1 & 1 & 1 & 1 \\ 1 & -1 & 1 & 1 \\ 1 & 1 & -1 & 1 \\ 1 & 1 & 1 & -1 \end{bmatrix} \begin{bmatrix} x \\ y \\ z \\ w \end{bmatrix} = \begin{bmatrix} 1 \\ 0 \\ 0 \\ 0 \end{bmatrix}.$

Find all the solutions of the following homogeneous systems $\mathbf{Ax} = \mathbf{0}$, where $\mathbf{A}$ is given as the following.

36. $\begin{bmatrix} 3 & 1 & 2 \\ 1 & -2 & 3 \\ 1 & 5 & -4 \end{bmatrix}.$

37. $\begin{bmatrix} 1 & 1 & 2 \\ 3 & 4 & -7 \\ -1 & -2 & 11 \end{bmatrix}.$

38. $\begin{bmatrix} 3 & -11 & 5 \\ 4 & 1 & -10 \\ 4 & 9 & -6 \end{bmatrix}$

39. $\begin{bmatrix} 1 & 2 & 3 & 4 \\ 1 & 1 & 1 & 1 \\ 1 & 2 & 6 & 12 \end{bmatrix}.$

40. $\begin{bmatrix} 2 & -1 & -3 & 1 \\ 1 & 1 & 1 & 1 \\ 2 & -7 & -13 & -1 \\ -1 & 5 & 9 & 1 \end{bmatrix}.$

41. $\begin{bmatrix} 1 & 1 & 1 & 1 \\ -1 & 1 & 1 & -1 \\ -1 & -1 & 1 & 1 \\ 1 & 1 & -1 & 1 \end{bmatrix}.$

42. $\begin{bmatrix} 3 & 1 & 1 & 4 \\ 0 & 4 & 10 & 1 \\ 1 & 7 & 17 & 3 \\ 2 & 2 & 4 & 3 \end{bmatrix}$

43. $\begin{bmatrix} 1 & 1 & -3 & 2 \\ 2 & -1 & -2 & -3 \\ 3 & 0 & -5 & -1 \\ 5 & -1 & -7 & -4 \end{bmatrix}.$

44. $\begin{bmatrix} 1 & -2 & 1 & -1 \\ 1 & 1 & -2 & 3 \\ 4 & 1 & -5 & 8 \\ 5 & -7 & 2 & -1 \end{bmatrix}.$

45. $\begin{bmatrix} 1 & 1 & -2 & -1 \\ 2 & 1 & 1 & -2 \\ 3 & 2 & -1 & -3 \\ 4 & 2 & 2 & -4 \end{bmatrix}.$

Using the Gauss-Jordan method, find the inverses of the following matrices.

46. $\begin{bmatrix} 1 & 1 & 1 \\ 1 & 2 & 3 \\ 1 & 3 & 4 \end{bmatrix}.$

47. $\begin{bmatrix} 2 & 3 & 1 \\ 1 & 3 & 3 \\ 0 & 1 & 2 \end{bmatrix}.$

48. $\begin{bmatrix} -1 & 1 & 1 & 1 \\ 1 & -1 & 1 & 1 \\ 1 & 1 & -1 & 1 \\ 1 & 1 & 1 & -1 \end{bmatrix}.$

49. $\begin{bmatrix} 1 & 1 & 1 & 1 \\ -1 & 1 & 1 & -1 \\ -1 & -1 & 1 & 1 \\ 1 & 1 & -1 & 1 \end{bmatrix}.$

50. $\begin{bmatrix} 1 & 1 & 0 & 1 \\ 1 & 1 & 1 & 1 \\ 4 & 4 & 1 & 1 \\ 1 & 0 & 0 & 1 \end{bmatrix}.$

3.5   Eigenvalue Problems

Let $\mathbf{A} = (a_{ij})$ be a square matrix of order n. The matrix $\mathbf{A}$ may be singular or non-singular. Consider the homogeneous system of equations

$$\mathbf{Ax} = \lambda \mathbf{x} \quad \text{or} \quad (\mathbf{A} - \lambda \mathbf{I})\, \mathbf{x} = \mathbf{0} \tag{3.35}$$

where λ is a scalar and $\mathbf{I}$ is an identity matrix of order n. The homogeneous system of equations (3.35) always has a trivial solution. We need to find values of λ for which the homogeneous system (3.35) has non-trivial solutions. The values of λ for which non-trivial solutions of the homogeneous system (3.35) exist, are called the *eigenvalues* or the *characteristic values* of $\mathbf{A}$ and the corresponding non-trivial solution vectors $\mathbf{x}$ are called the *eigenvectors* or the *characteristic vectors* of $\mathbf{A}$. If $\mathbf{x}$ is a non-trivial solution of the homogeneous system (3.35), then $\alpha \, \mathbf{x}$, where α is any constant is also a solution of the homogeneous system. Hence, an eigenvector is unique only upto a constant multiple. The problem of determining the eigenvalues and the corresponding eigenvectors of a square matrix $\mathbf{A}$ is called an *eigenvalues problem*.

3.5.1   Eigenvalues and Eigenvectors

If the homogeneous system (3.35) has a non-trivial solution, then the rank of the coefficient matrix $(\mathbf{A} - \lambda \mathbf{I})$ is less than n, that is the coefficient matrix must be singular. Therefore,

$$det \, (\mathbf{A} - \lambda \mathbf{I}) = \begin{vmatrix} a_{11} - \lambda & a_{12} & \cdots & a_{1n} \\ a_{21} & a_{22} - \lambda & \cdots & a_{2n} \\ \vdots & & \cdots & \\ a_{n1} & a_{n2} & \cdots & a_{nn} - \lambda \end{vmatrix} = 0.. \tag{3.36}$$

Expanding the determinant given in Eq. (3.36), we obtain a polynomial of degree n in λ, which is of the form

$$P_n(\lambda) = |\mathbf{A} - \lambda \mathbf{I}| = (-1)^n [\lambda^n - c_1 \lambda^{n-1} + c_2 \lambda^{n-2} - \ldots + (-1)^n c_n] = 0$$

or
$$\lambda^n - c_1 \lambda^{n-1} + c_2 \lambda^{n-2} - \ldots + (-1)^n c_n = 0. \tag{3.37}$$

where $c_1, c_2, \ldots, c_n$ can be expressed in terms of the elements a_{ij} of the matrix $\mathbf{A}$. This equation is called the *characteristic equation* of the matrix $\mathbf{A}$. The polynomial equation $P_n(\lambda) = 0$ has n roots which can be real or complex, simple or repeated. The roots $\lambda_1, \lambda_2, \ldots, \lambda_n$ of the polynomial equation $P_n(\lambda) = 0$ are called the *eigenvalues*. By using the relation between the roots and the coefficients, we can write

$$\lambda_1 + \lambda_2 + \ldots + \lambda_n = c_1 = a_{11} + a_{22} + \ldots + a_{nn}$$

$$\lambda_1 \lambda_2 + \lambda_1 \lambda_3 + \ldots + \lambda_{n-1} \lambda_n = c_2$$

$$\vdots$$

$$\lambda_1 \lambda_2 \ldots \lambda_n = c_n. \tag{3.38}$$

If we set $\lambda = 0$ in Eq. (3.36), then we get

$$|\mathbf{A}| = (-1)^{2n} c_n = c_n = \lambda_1 \lambda_2 \ldots \lambda_n.$$

Therefore, we get

sum of eigenvalues = trace $(\mathbf{A})$,   and   product of eigenvalues = $|\mathbf{A}|$.

The set of the eigenvalues is called the *spectrum* of $\mathbf{A}$ and the largest eigenvalue in magnitude is called the *spectral radius* of $\mathbf{A}$ and is denoted by $\rho(\mathbf{A})$. If $|\mathbf{A}| = 0$, that is the martix is singular, then from Eq. (3.38), we find that atleast one of the eigenvalues must be zero. Conversely, if one of the eigenvalues is zero, then $|\mathbf{A}| = 0$. Note that if $\mathbf{A}$ is a diagonal or an upper triangular or a lower triangular matrix, then the diagonal elements of the matrix $\mathbf{A}$ are the eigenvalues of $\mathbf{A}$.

After determining the eigenvalues λ_i's, we solve the homogeneous system $(\mathbf{A} - \lambda_i \mathbf{I})\mathbf{x} = \mathbf{0}$ for each λ_i, $i = 1, 2, \ldots, n$ to obtain the corresponding eigenvectors.

Properties of eigenvalues and eigenvectors

Let λ be an eigenvalue of $\mathbf{A}$ and $\mathbf{x}$ be its corresponding eigenvector. Then, we have the following results.

1. $\alpha \mathbf{A}$ has eigenvalue $\alpha \lambda$ and the corresponding eigenvector is $\mathbf{x}$.

$$\mathbf{A}\mathbf{x} = \lambda \mathbf{x} \Rightarrow \alpha \mathbf{A}\mathbf{x} = (\alpha \lambda)\mathbf{x}.$$

2. $\mathbf{A}^m$ has eigenvalue λ^m and the corresponding eigenvector is $\mathbf{x}$ for any positive interger m. Pre-multiplying both sides of $\mathbf{A}\mathbf{x} = \lambda \mathbf{x}$ by $\mathbf{A}$, we get

$$\mathbf{A}\mathbf{A}\mathbf{x} = \mathbf{A}\lambda \mathbf{x} = \lambda \mathbf{A}\mathbf{x} = \lambda(\lambda \mathbf{x}) \text{ or } \mathbf{A}^2\mathbf{x} = \lambda^2 \mathbf{x}.$$

Therefore, $\mathbf{A}^2$ has the eigenvalue λ^2 and the corresponding eigenvector is $\mathbf{x}$. Pre-multiplying successively m times, we obtain the result.

3. $\mathbf{A} - k\mathbf{I}$ has the eigenvalue $\lambda - k$, for any scalar k and the corresponding eigenvector is $\mathbf{x}$.

$$\mathbf{A}\mathbf{x} = \lambda \mathbf{x} \Rightarrow \mathbf{A}\mathbf{x} - k\mathbf{I}\mathbf{x} = \lambda \mathbf{x} - k\mathbf{x}$$

or $\qquad\qquad (\mathbf{A} - k\mathbf{I})\mathbf{x} = (\lambda - k)\mathbf{x}.$

4. $\mathbf{A}^{-1}$(if it exists) has the eigenvalue $1/\lambda$ and the corresponding eigenvector is $\mathbf{x}$. Pre-multiplying both sides of $\mathbf{A}\mathbf{x} = \lambda \mathbf{x}$ by $\mathbf{A}^{-1}$, we get

$$\mathbf{A}^{-1}\mathbf{A}\mathbf{x} = \lambda \mathbf{A}^{-1}\mathbf{x} \text{ or } \mathbf{A}^{-1}\mathbf{x} = (1/\lambda)\mathbf{x}.$$

5. $(\mathbf{A} - k\mathbf{I})^{-1}$ has the eigenvalue $1/(\lambda - k)$ and the corresponding eigenvector is $\mathbf{x}$ for any scalar k.

6. $\mathbf{A}$ and $\mathbf{A}^T$ have the same eigenvalues (since a determinant can be expanded by rows or by columns) but different eigenvectors, (see Example 3.41).

7. For a real matrix $\mathbf{A}$, if $\alpha + i\beta$ is an eigenvalue, then its conjugate $\alpha - i\beta$ is also an eigenvalue (since the characteristic equation has real coefficients). When the matrix $\mathbf{A}$ is complex, this property does not hold.

We now present an important result which gives the relationship of a matrix $\mathbf{A}$ and its characteristic equation.

Theorem 3.9 (Cayley-Hamilton theorem) Every square matrix $\mathbf{A}$ satisfies its own characteristic equation

$$\mathbf{A}^n - c_1\mathbf{A}^{n-1} + \ldots + (-1)^{n-1} c_{n-1} \mathbf{A} + (-1)^n c_n \mathbf{I} = \mathbf{0}. \qquad (3.39)$$

Proof The cofactors of the elements of the determinant $|\mathbf{A} - \lambda\mathbf{I}|$ are polynomials in λ of degree $(n - 1)$ or less. Therefore, the elements of the adjoint matrix (transpose of the cofactor matrix) are also polynomials in λ of degree $(n - 1)$ or less. Hence, we can express the adjoint matrix as a polynomial in λ whose coefficients $\mathbf{B}_1, \mathbf{B}_2, ..., \mathbf{B}_n$ are square matrices of order n having elements as functions of the elements of the matrix $\mathbf{A}$. Thus, we can write

$$adj(\mathbf{A} - \lambda\mathbf{I}) = \mathbf{B}_1\lambda^{n-1} + \mathbf{B}_2\lambda^{n-2} + ... + \mathbf{B}_{n-1}\lambda + \mathbf{B}_n.$$

We also have

$$(\mathbf{A} - \lambda\mathbf{I})\, adj\, (\mathbf{A} - \lambda\mathbf{I}) = |\mathbf{A} - \lambda\mathbf{I}|\, \mathbf{I}.$$

Therefore, we can write for any λ

$$(\mathbf{A} - \lambda\mathbf{I})(\mathbf{B}_1\lambda^{n-1} + \mathbf{B}_2\lambda^{n-2} + ... + \mathbf{B}_{n-1}\lambda + \mathbf{B}_n)$$
$$= \lambda^n\mathbf{I} - c_1\lambda^{n-1}\mathbf{I} + ... + (-1)^{n-1} c_{n-1}\, \lambda\mathbf{I} + (-1)^n c_n\mathbf{I}$$

Comparing the coefficients of various powers of λ, we obtain

$$-\mathbf{B}_1 = \mathbf{I}$$
$$\mathbf{A}\mathbf{B}_1 - \mathbf{B}_2 = c_1\mathbf{I}$$
$$\mathbf{A}\mathbf{B}_2 - \mathbf{B}_3 = c_2\mathbf{I}$$
$$...$$
$$\mathbf{A}\mathbf{B}_{n-1} - \mathbf{B}_n = (-1)^{n-1} c_{n-1}\mathbf{I}$$
$$\mathbf{A}\mathbf{B}_n = (-1)^n c_n\mathbf{I}.$$

Pre-multiplying these equations by $\mathbf{A}^n, \mathbf{A}^{n-1}, ..., \mathbf{A}, \mathbf{I}$ respectively and adding, we get

$$\mathbf{A}^n - c_1\mathbf{A}^{n-1} + ... + (-1)^{n-1}c_{n-1}\,\mathbf{A} + (-1)^n c_n\,\mathbf{I} = \mathbf{0}$$

which proves the theorem.

Remark 19

(a) We can use Eq. (3.39) to find $\mathbf{A}^{-1}$ (if it exists) in terms of the powers of the matrix $\mathbf{A}$.

Pre-multiplying both sides in Eq. (3.39) by $\mathbf{A}^{-1}$, we get

$$\mathbf{A}^{-1}\mathbf{A}^n - c_1\mathbf{A}^{-1}\mathbf{A}^{n-1} + ... + (-1)^{n-1}c_{n-1}\mathbf{A}^{-1}\mathbf{A} + (-1)^n c_n\mathbf{A}^{-1}\mathbf{I} = \mathbf{A}^{-1}\mathbf{0} = \mathbf{0}$$

or
$$\mathbf{A}^{-1} = -\frac{(-1)^n}{c_n}\,[\mathbf{A}^{n-1} - c_1\mathbf{A}^{n-2} + ... + (-1)^{n-1}\,c_{n-1}\mathbf{I}] \qquad (3.40)$$

(b) We can use Eq.(3.39) to obtain $\mathbf{A}^n$ in terms of lower powers of $\mathbf{A}$ as

$$\mathbf{A}^n = c_1\mathbf{A}^{n-1} - c_2\mathbf{A}^{n-2} + ... + (-1)^{n-1}c_n\mathbf{I}. \qquad (3.41)$$

Example 3.33 Verify Cayley-Hamilton theorem for the martrix

$$\mathbf{A} = \begin{bmatrix} 1 & 2 & 0 \\ -1 & 1 & 2 \\ 1 & 2 & 1 \end{bmatrix}.$$

Also, (i) obtain $\mathbf{A}^{-1}$ and $\mathbf{A}^3$, (ii) find eigenvalues of $\mathbf{A}$, $\mathbf{A}^2$ and verify that eigenvalues of $\mathbf{A}^2$ are squares of those of $\mathbf{A}$, (iii) find the spectral radius of $\mathbf{A}$.

Solution The characteristic equation of $\mathbf{A}$ is given by

$$|\mathbf{A} - \lambda \mathbf{I}| = \begin{vmatrix} 1-\lambda & 2 & 0 \\ -1 & 1-\lambda & 2 \\ 1 & 2 & 1-\lambda \end{vmatrix} = (1-\lambda)\{(1-\lambda)^2 - 4\} - 2\{-(1-\lambda) - 2\}$$

$$= (1-\lambda)(\lambda^2 - 2\lambda - 3) - 2(\lambda - 3) = -\lambda^3 + 3\lambda^2 - \lambda + 3 = 0.$$

Now,

$$\mathbf{A}^2 = \begin{bmatrix} 1 & 2 & 0 \\ -1 & 1 & 2 \\ 1 & 2 & 1 \end{bmatrix}\begin{bmatrix} 1 & 2 & 0 \\ -1 & 1 & 2 \\ 1 & 2 & 1 \end{bmatrix} = \begin{bmatrix} -1 & 4 & 4 \\ 0 & 3 & 4 \\ 0 & 6 & 5 \end{bmatrix}.$$

$$\mathbf{A}^3 = \mathbf{A}^2\mathbf{A} = \begin{bmatrix} -1 & 4 & 4 \\ 0 & 3 & 4 \\ 0 & 6 & 5 \end{bmatrix}\begin{bmatrix} 1 & 2 & 0 \\ -1 & 1 & 2 \\ 1 & 2 & 1 \end{bmatrix} = \begin{bmatrix} -1 & 10 & 12 \\ 1 & 11 & 10 \\ -1 & 16 & 17 \end{bmatrix}.$$

We have

$$-\mathbf{A}^3 + 3\mathbf{A}^2 - \mathbf{A} + 3\mathbf{I} = -\begin{bmatrix} -1 & 10 & 12 \\ 1 & 11 & 10 \\ -1 & 16 & 17 \end{bmatrix} + 3\begin{bmatrix} -1 & 4 & 4 \\ 0 & 3 & 4 \\ 0 & 6 & 5 \end{bmatrix} - \begin{bmatrix} 1 & 2 & 0 \\ -1 & 1 & 2 \\ 1 & 2 & 1 \end{bmatrix} + 3\begin{bmatrix} 1 & 0 & 0 \\ 0 & 1 & 0 \\ 0 & 0 & 1 \end{bmatrix}$$

$$= \begin{bmatrix} 0 & 0 & 0 \\ 0 & 0 & 0 \\ 0 & 0 & 0 \end{bmatrix} = \mathbf{0}. \tag{3.42}$$

Hence, $\mathbf{A}$ satisfies the characteristic equation $-\lambda^3 + 3\lambda^2 - \lambda + 3 = 0$.

(i) From Eq. (3.42), we get

$$\mathbf{A}^{-1} = \frac{1}{3}[\mathbf{A}^2 - 3\mathbf{A} + \mathbf{I}] = \frac{1}{3}\left[\begin{pmatrix} -1 & 4 & 4 \\ 0 & 3 & 4 \\ 0 & 6 & 5 \end{pmatrix} - \begin{pmatrix} 3 & 6 & 0 \\ -3 & 3 & 6 \\ 3 & 6 & 3 \end{pmatrix} + \begin{pmatrix} 1 & 0 & 0 \\ 0 & 1 & 0 \\ 0 & 0 & 1 \end{pmatrix}\right] = \frac{1}{3}\begin{bmatrix} -3 & -2 & 4 \\ 3 & 1 & -2 \\ -3 & 0 & 3 \end{bmatrix}.$$

From Eq. (3.42), we get

$$\mathbf{A}^3 = 3\mathbf{A}^2 - \mathbf{A} + 3\mathbf{I} = \begin{bmatrix} -3 & 12 & 12 \\ 0 & 9 & 12 \\ 0 & 18 & 15 \end{bmatrix} - \begin{bmatrix} 1 & 2 & 0 \\ -1 & 1 & 2 \\ 1 & 2 & 1 \end{bmatrix} + \begin{bmatrix} 3 & 0 & 0 \\ 0 & 3 & 0 \\ 0 & 0 & 3 \end{bmatrix} = \begin{bmatrix} -1 & 10 & 12 \\ 1 & 11 & 10 \\ -1 & 16 & 17 \end{bmatrix}.$$

(ii)   Eigenvalues of **A** are the roots of

$$\lambda^3 - 3\lambda^2 + \lambda - 3 = 0 \text{ or } (\lambda - 3)(\lambda^2 + 1) = 0 \text{ or } \lambda = 3, i, -i.$$

The characteristic equation of $\mathbf{A}^2$ is given by

$$\begin{vmatrix} -1-\lambda & 4 & 4 \\ 0 & 3-\lambda & 4 \\ 0 & 6 & 5-\lambda \end{vmatrix} = (-1-\lambda)\left[(3-\lambda)(5-\lambda) - 24\right] = 0$$

or $(\lambda + 1)(\lambda^2 - 8\lambda - 9) = 0$ or $(\lambda + 1)(\lambda - 9)(\lambda + 1) = 0.$

The eigenvalues of $\mathbf{A}^2$ are $9, -1, -1$ which are the squares of the eigenvalues of **A**.

(iii)   The spectral radius of **A** is given by

$$\rho\,(\mathbf{A}) = \text{largest eigenvalue in magnitude} = \max_i |\lambda_i| = 3.$$

Example 3.34   If $\mathbf{A} = \begin{bmatrix} 1 & 0 & 0 \\ 1 & 0 & 1 \\ 0 & 1 & 0 \end{bmatrix}$, then show that $\mathbf{A}^n = \mathbf{A}^{n-2} + \mathbf{A}^2 - \mathbf{I}$ for $n \geq 3$. Hence, find $\mathbf{A}^{50}$.

Solution   The characteristic equation of **A** is given by

$$|\mathbf{A} - \lambda\mathbf{I}| = \begin{vmatrix} 1-\lambda & 0 & 0 \\ 1 & -\lambda & 1 \\ 0 & 1 & -\lambda \end{vmatrix} = (1-\lambda)(\lambda^2 - 1) = 0, \quad \text{or} \quad \lambda^3 - \lambda^2 - \lambda + 1 = 0.$$

Using Cayley-Hamilton theorem, we get

$$\mathbf{A}^3 - \mathbf{A}^2 - \mathbf{A} + \mathbf{I} = \mathbf{0}, \quad \text{or} \quad \mathbf{A}^3 - \mathbf{A}^2 = \mathbf{A} - \mathbf{I}.$$

Pre-multiplying both sides successively by **A**, we obtain

$$\mathbf{A}^3 - \mathbf{A}^2 = \mathbf{A} - \mathbf{I}$$

$$\mathbf{A}^4 - \mathbf{A}^3 = \mathbf{A}^2 - \mathbf{A}$$

$$\cdots$$

$$\mathbf{A}^{n-1} - \mathbf{A}^{n-2} = \mathbf{A}^{n-3} - \mathbf{A}^{n-4}$$

$$\mathbf{A}^n - \mathbf{A}^{n-1} = \mathbf{A}^{n-2} - \mathbf{A}^{n-3}.$$

Adding these equations, we get

$$\mathbf{A}^n - \mathbf{A}^2 = \mathbf{A}^{n-2} - \mathbf{I}, \quad \text{or} \quad \mathbf{A}^n = \mathbf{A}^{n-2} + \mathbf{A}^2 - \mathbf{I}, \quad n \geq 3.$$

Using this equation recursively, we get

$$A^n = (A^{n-4} + A^2 - I) + A^2 - I = A^{n-4} + 2(A^2 - I)$$
$$= (A^{n-6} + A^2 - I) + 2(A^2 - I) = A^{n-6} + 3(A^2 - I)$$

$$\cdots$$

$$= A^{n-(n-2)} + \frac{1}{2}(n-2)(A^2 - I) = \frac{n}{2}A^2 - \frac{1}{2}(n-2)I.$$

Substituting $n = 50$, we get

$$A^{50} = 25A^2 - 24I = 25 \begin{bmatrix} 1 & 0 & 0 \\ 1 & 1 & 0 \\ 1 & 0 & 1 \end{bmatrix} - 24 \begin{bmatrix} 1 & 0 & 0 \\ 0 & 1 & 0 \\ 0 & 0 & 1 \end{bmatrix} = \begin{bmatrix} 1 & 0 & 0 \\ 25 & 1 & 0 \\ 25 & 0 & 1 \end{bmatrix}.$$

Example 3. 35 Find the eigenvalues and the corresponding eigenvectors of the following matrices.

(i) $A = \begin{bmatrix} 1 & 4 \\ 3 & 2 \end{bmatrix}$,
(ii) $A = \begin{bmatrix} 1 & 1 \\ -1 & 1 \end{bmatrix}$,
(iii) $A = \begin{bmatrix} 1 & 0 & 0 \\ 0 & 2 & 1 \\ 2 & 0 & 3 \end{bmatrix}$.

Solution

(i) The characteristic equation of A is given by

$$|A - \lambda I| = \begin{vmatrix} 1-\lambda & 4 \\ 3 & 2-\lambda \end{vmatrix} = 0 \quad \text{or} \quad \lambda^2 - 3\lambda - 10 = 0, \quad \text{or} \quad \lambda = -2, 5.$$

Corresponding to the eigenvalue $\lambda = -2$, we have

$$(A + 2I)\,x = \begin{pmatrix} 3 & 4 \\ 3 & 4 \end{pmatrix} \begin{pmatrix} x_1 \\ x_2 \end{pmatrix} = \begin{pmatrix} 0 \\ 0 \end{pmatrix} \quad \text{or} \quad 3x_1 + 4x_2 = 0 \text{ or } x_1 = -\frac{4}{3}x_2.$$

Hence, the eigenvector x is given by

$$x = \begin{bmatrix} x_1 \\ x_2 \end{bmatrix} = \begin{bmatrix} -4x_2/3 \\ x_2 \end{bmatrix} = x_2 \begin{bmatrix} -4/3 \\ 1 \end{bmatrix}.$$

Since an eigenvector is unique upto a constant multiple, we can take the eigenvector as $[-4, 3]^T$.

Corresponding to the eigenvalue $\lambda = 5$, we have

$$(A - 5I)\,x = \begin{pmatrix} -4 & 4 \\ 3 & -3 \end{pmatrix} \begin{pmatrix} x_1 \\ x_2 \end{pmatrix} = \begin{pmatrix} 0 \\ 0 \end{pmatrix} \quad \text{or } x_1 - x_2 = 0, \text{ or } x_1 = x_2.$$

Therefore, the eigenvalue is given by $x = (x_1, x_2)^T = x_1(1, 1)^T$ or $(1, 1)^T$.

(ii) The characteristic equation of A is given by

$$|A - \lambda I| = \begin{vmatrix} 1-\lambda & 1 \\ -1 & 1-\lambda \end{vmatrix} = 0, \quad \text{or} \quad \lambda^2 - 2\lambda + 2 = 0, \quad \text{or} \quad \lambda = 1 \pm i.$$

Corresponding to the eigenvalue $\lambda = 1 + i$, we have

$$[\mathbf{A} - (1 + i)\,\mathbf{I}]\,\mathbf{x} = \begin{bmatrix} -i & 1 \\ -1 & -i \end{bmatrix} \begin{bmatrix} x_1 \\ x_2 \end{bmatrix} = \begin{bmatrix} 0 \\ 0 \end{bmatrix}$$

or $\qquad\qquad -ix_1 + x_2 = 0$ and $-x_1 - ix_2 = 0.$

Both the equations reduce to $-x_1 - ix_2 = 0$. Choosing $x_2 = 1$, we get $x_1 = -i$. Therefore, the eigenvector is $\mathbf{x} = [-i,\ 1]^T$.

Corresponding to the eigenvalue $\lambda = 1 - i$, we have

$$[\mathbf{A} - (1 - i)\mathbf{I}]\,\mathbf{x} = \begin{bmatrix} i & 1 \\ -1 & i \end{bmatrix} \begin{bmatrix} x_1 \\ x_2 \end{bmatrix} = \begin{bmatrix} 0 \\ 0 \end{bmatrix}$$

or $\qquad\qquad ix_1 + x_2 = 0$ and $-x_1 + ix_2 = 0.$

Both the equations reduce to $-x_1 + ix_2 = 0$. Choosing $x_2 = 1$, we get $x_1 = i$. Therefore, the eigenvector is $x = [i,\ 1]^T$.

Remark 20

For a real matrix $\mathbf{A}$, the eigenvalues and the corresponding eigenvectors can be complex.

(iii)   The characteristic equation of $\mathbf{A}$ is given by

$$|\mathbf{A} - \lambda\,\mathbf{I}| = \begin{vmatrix} 1 - \lambda & 0 & 0 \\ 0 & 2 - \lambda & 1 \\ 2 & 0 & 3 - \lambda \end{vmatrix} = 0 \ \text{ or } \ (1 - \lambda)\,(2 - \lambda)\,(3 - \lambda) = 0 \ \text{ or } \ \lambda = 1, 2, 3.$$

Corresponding to the eigenvalue $\lambda = 1$, we have

$$(\mathbf{A} - \mathbf{I})\mathbf{x} = \begin{bmatrix} 0 & 0 & 0 \\ 0 & 1 & 1 \\ 2 & 0 & 2 \end{bmatrix} \begin{bmatrix} x_1 \\ x_2 \\ x_3 \end{bmatrix} \begin{bmatrix} 0 \\ 0 \\ 0 \end{bmatrix} \ \text{ or } \ \begin{cases} x_2 + x_3 = 0 \\ x_1 + x_3 = 0 \end{cases}$$

We obtain two equations in three unknowns. One of the variables x_1, x_2, x_3 can be chosen arbitrarily. Taking $x_3 = 1$, we obtain the eigenvector as $[-1,\ -1,\ 1]^T$.

Corresponding to the eigenvalue $\lambda = 2$, we have

$$(\mathbf{A} - 2\mathbf{I})\mathbf{x} = \begin{bmatrix} -1 & 0 & 0 \\ 0 & 0 & 1 \\ 2 & 0 & 1 \end{bmatrix} \begin{bmatrix} x_1 \\ x_2 \\ x_3 \end{bmatrix} = \begin{bmatrix} 0 \\ 0 \\ 0 \end{bmatrix}$$

or $x_1 = 0$, $x_3 = 0$ and x_2 arbitrary. Taking $x_2 = 1$, we obtain the eigenvector as $[0,\ 1,\ 0]^T$.

Corresponding to the eigenvalue $\lambda = 3$, we have

$$(A - 3I)x = \begin{bmatrix} -2 & 0 & 0 \\ 0 & 1 & 1 \\ 2 & 0 & 1 \end{bmatrix} \begin{bmatrix} x_1 \\ x_2 \\ x_3 \end{bmatrix} \begin{bmatrix} 0 \\ 0 \\ 0 \end{bmatrix} \quad \text{or} \quad \begin{cases} x_1 = 0 \\ x_2 + x_3 = 0 \end{cases}$$

Choosing $x_3 = 1$, we obtain the eigenvector as $[\,0, -1, 1\,]^T$.

Example 3.36 Find the eigenvalues and the corresponding eigenvectors of the following matrices.

$$(i)\ A = \begin{bmatrix} 1 & 1 & 0 \\ 0 & 1 & 1 \\ 0 & 0 & 1 \end{bmatrix}, \qquad (ii)\ A = \begin{bmatrix} 1 & 1 & 0 \\ 0 & 1 & 0 \\ 0 & 0 & 1 \end{bmatrix}, \qquad (iii)\ A = \begin{bmatrix} 1 & 0 & 0 \\ 0 & 1 & 0 \\ 0 & 0 & 1 \end{bmatrix}.$$

Solution In each of the above problems, we obtain the characteristic equation as $(1 - \lambda)^3 = 0$. Therefore, the eigenvalues are $\lambda = 1, 1, 1$, a repeated value. Since a 3×3 matrix has 3 eigenvalues, it is important to know, whether the given matrix has 3 linearly independent eigenvectors, or it has lesser number of linearly independent eigenvectors.

Corresponding to the eigenvalue $\lambda = 1$, we obtain the following eigenvectors.

(i)
$$(A - I)x = \begin{pmatrix} 0 & 1 & 0 \\ 0 & 0 & 1 \\ 0 & 0 & 0 \end{pmatrix} x = \begin{pmatrix} 0 \\ 0 \\ 0 \end{pmatrix} \quad \text{or} \quad \begin{cases} x_2 = 0 \\ x_3 = 0 \\ x_1 \text{ arbitrary.} \end{cases}$$

Choosing $x_1 = 1$, we obtain the solution as $[1, 0, 0]^T$.

Hence, **A** has only one independent eigenvector.

(ii)
$$(A - I)x = \begin{bmatrix} 0 & 1 & 0 \\ 0 & 0 & 0 \\ 0 & 0 & 0 \end{bmatrix} \begin{bmatrix} x_1 \\ x_2 \\ x_3 \end{bmatrix} = \begin{bmatrix} 0 \\ 0 \\ 0 \end{bmatrix}, \quad \text{or} \quad \begin{cases} x_2 = 0 \\ x_1, x_3 \text{ arbitrary.} \end{cases}$$

Taking $x_1 = 0$, $x_3 = 1$ and $x_1 = 1$, $x_3 = 0$, we obtain two linearly independent solutions

$$x_1 = [0, 0, 1]^T, \quad x_2 = [1, 0, 0]^T.$$

In this case **A** has two linearly independent eigenvectors.

(iii)
$$(A - I)x = \begin{bmatrix} 0 & 0 & 0 \\ 0 & 0 & 0 \\ 0 & 0 & 0 \end{bmatrix} \begin{bmatrix} x_1 \\ x_2 \\ x_3 \end{bmatrix} = \begin{bmatrix} 0 \\ 0 \\ 0 \end{bmatrix}.$$

This system is satisfied for arbitrary values of all the three variables. Hence, we obtain three linearly independent eigenvectors, which can be taken as

$$\mathbf{x}_1 = [1, 0, 0]^T, \quad \mathbf{x}_2 = [0, 1, 0]^T, \quad \mathbf{x}_3 = [0, 0, 1]^T.$$

We now state some important results regarding the relationship between the eigenvalues of a matrix and the corresponding linearly independent eigenvectors.

1. Eigenvectors corresponding to distinct eigenvalues are linearly independent.
2. If λ is an eigenvalue of multiplicity m of a square matix $\mathbf{A}$ of order n, then the number of linearly independent eigenvectors associated with λ is given by

$$p = n - r, \quad \text{where} \quad r = \text{rank } (\mathbf{A} - \lambda \mathbf{I}), \ 1 \le p \le m.$$

Remark 21

In Example 3.35, all the eigenvalues are distinct and therefore, the corresponding eigenvectors are linearly indenpendent. In Example 3.36, the eigenvalue $\lambda = 1$ is of multiplicity 3. We find that in

(i) Example 3.36(i), the rank of the matrix $\mathbf{A} - \mathbf{I}$ is 2 and we obtain one linearly independent eigenvector.

(ii) Example 3.36(ii), the rank of the matrix $\mathbf{A} - \mathbf{I}$ is 1 and we obtain two linearly independent eigenvectors.

(iii) Example 3.36(iii), the rank of the matrix $\mathbf{A} - \mathbf{I}$ is 0 and we obtain three linearly independent eigenvectors.

3.5.2   Similar and Diagonalizable Matrices

Similar matrices

Let $\mathbf{A}$ and $\mathbf{B}$ be square matrices of the same order. The matrix $\mathbf{A}$ is said to be similar to the matrix $\mathbf{B}$ if there exists an invertible matrix $\mathbf{P}$ such that

$$\mathbf{A} = \mathbf{P}^{-1}\mathbf{B}\mathbf{P} \quad \text{or} \quad \mathbf{P}\mathbf{A} = \mathbf{B}\mathbf{P}. \tag{3.43}$$

Post-multiplying both sides in Eq. (3.43) by $\mathbf{P}^{-1}$, we get

$$\mathbf{P}\mathbf{A}\mathbf{P}^{-1} = \mathbf{B}.$$

Therefore, $\mathbf{A}$ is similar to $\mathbf{B}$ if and only if $\mathbf{B}$ is similar to $\mathbf{A}$. The matrix $\mathbf{P}$ is called the *similarity matrix*. The transformation in Eq. (3.43) is called a *similarity transformation*. We now prove a result regarding eigenvalues of similar matrices.

Theorem 3.10   Similar matrices have the same characteristic equation (and hence the same eigenvalues). Further, if $\mathbf{x}$ is an eigenvector of $\mathbf{A}$ corresponding to the eigenvalue λ, then $\mathbf{P}^{-1}\mathbf{x}$ is an eigenvector of $\mathbf{B}$ corresponding to the eigenvalue λ, where $\mathbf{P}$ is the similarity matrix.

Proof   Let λ be an eigenvalue and $\mathbf{x}$ be the corresponding eigenvector of $\mathbf{A}$. That is

$$\mathbf{A}\mathbf{x} = \lambda\mathbf{x}.$$

Pre-multiplying both sides by an invertible matrix $\mathbf{P}^{-1}$, we obtain

$$\mathbf{P}^{-1}\mathbf{A}\mathbf{x} = \lambda\mathbf{P}^{-1}\mathbf{x}.$$

Set $\mathbf{x} = \mathbf{P}\mathbf{y}$. We get

$$P^{-1}APy = \lambda P^{-1}Py, \quad \text{or} \quad (P^{-1}AP)y = \lambda y \quad \text{or} \quad By = \lambda y.$$

where $B = P^{-1}AP$. Therefore, B has the same eigenvalues as A, that is, the characterstic equation of B is same as the characteristic equation of A. Now, A and B are similar matrices. Therefore, similar matrices have the same characteristic equation (and hence the same eigenvalues). Also, $x = Py$, that is eigenvectors of A and B are related by $x = Py$ or $y = P^{-1}x$.

Remark 22

(a) Theorem 3.10 states that if two matrices are similar, then they have the same characterstic equation and hence the same eigenvalues. However, the converse of this theorem is not true. Two matrices which have the same characteristic equation need not always be similar.

(b) If A is similar to B and B is similar to C, then A is similar to C.

Let there be two invertible matrices P and Q such that

$$A = P^{-1}BP \quad \text{and} \quad B = Q^{-1}CQ.$$

Then $\qquad A = P^{-1}Q^{-1}CQP = R^{-1}CR, \quad \text{where} \quad R = QP.$

Example 3.37 Examine whether A is similar to B, where

(i) $A = \begin{bmatrix} 5 & 5 \\ -2 & 0 \end{bmatrix}$ and $B = \begin{bmatrix} 1 & 2 \\ -3 & 4 \end{bmatrix}$, (ii) $A = \begin{bmatrix} 1 & 0 \\ 0 & 1 \end{bmatrix}$ and $B = \begin{bmatrix} 1 & 1 \\ 0 & 1 \end{bmatrix}$.

Solution The given matrices are similar if there exists an invertible matrix P such that

$$A = P^{-1}BP \quad \text{or} \quad PA = BP.$$

Let $P = \begin{bmatrix} a & b \\ c & d \end{bmatrix}$. We shall determine a, b, c and d such that $PA = BP$ and then check whether P is non-singular.

(i) $\begin{bmatrix} a & b \\ c & d \end{bmatrix}\begin{bmatrix} 5 & 5 \\ -2 & 0 \end{bmatrix} = \begin{bmatrix} 1 & 2 \\ -3 & 4 \end{bmatrix}\begin{bmatrix} a & b \\ c & d \end{bmatrix}$, or $\begin{bmatrix} 5a - 2b & 5a \\ 5c - 2d & 5c \end{bmatrix} = \begin{bmatrix} a + 2c & b + 2d \\ -3a + 4c & -3b + 4d \end{bmatrix}$.

Equating the corresponding elements, we obtain the system of equations

$$5a - 2b = a + 2c, \qquad \text{or} \qquad 4a - 2b - 2c = 0$$
$$5a = b + 2d, \qquad \text{or} \qquad 5a - b - 2d = 0$$
$$5c - 2d = -3a + 4c, \qquad \text{or} \qquad 3a + c - 2d = 0$$
$$5c = -3b + 4d, \qquad \text{or} \qquad 3b + 5c - 4d = 0.$$

A solution to this system of equations is $a = 1$, $b = 1$, $c = 1$, $d = 2$.

Therefore, we get $P = \begin{bmatrix} 1 & 1 \\ 1 & 2 \end{bmatrix}$, which is a non-singular matrix. Hence, the matrices A and B are similar.

(ii) $\begin{bmatrix} a & b \\ c & d \end{bmatrix}\begin{bmatrix} 1 & 0 \\ 0 & 1 \end{bmatrix} = \begin{bmatrix} 1 & 1 \\ 0 & 1 \end{bmatrix}\begin{bmatrix} a & b \\ c & d \end{bmatrix}$, or $\begin{bmatrix} a & b \\ c & d \end{bmatrix} = \begin{bmatrix} a+c & b+d \\ c & d \end{bmatrix}$.

Equating the corresponding elements, we get

$$a = a + c, \quad b = b + d \quad \text{or} \quad c = d = 0.$$

Therefore, we get $P = \begin{bmatrix} a & b \\ 0 & 0 \end{bmatrix}$, which is a singular matrix.

Since an invertible matrix P does not exist, the matrices A and B are not similar.

It can be verified that the eigenvalues of A are 1, 1 whereas the eigenvalues of B are 0, 2.

In practice, it is usually difficult to obtain a non-singular matrix P which satisfies the equation $A = P^{-1}BP$ for any two matrices A and B. However, it is possible to obtain the matrix P when A or B is a diagonal matrix. Thus, our interest is to find a similarity matrix P such that for a given matrix A, we have

$$D = P^{-1}AP \quad \text{or} \quad PDP^{-1} = A$$

where D is a diagonal matrix. If such a matrix exists, then we say that the matrix A is *diagonalizable*.

Diagonalizable matrices

A matrix A is diagonalizable, if it is similar to a diagonal matrix, that is there exists an invertible matrix P such that $P^{-1}AP = D$, where D is a diagonal matrix. Since, similar matrices have the same eigenvalues, the diagonal elements of D are the eigenvalues of A. A necessary and sufficient condition for the existence of P is given in the following theorem.

Therorem 3.11 A square matrix A of order n is diagonalizable if and only if it has n linearly independent eigenvectors.

Proof We shall prove the case that if A has n linearly independent eigenvectors, then A is diagonalizable. Let $x_1, x_2, ..., x_n$ be n linearly independent eigenvectors corresponding to the eigenvalues $\lambda_1, \lambda_2, ..., \lambda_n$ (not necessarily distinct) of the matrix A in the same order, that is the eigenvector x_j corresponds to the eigenvalue λ_j, $j = 1, 2, ..., n$. Let

$$P = [x_1, x_2, ..., x_n] \quad \text{and} \quad D = \text{diag}(\lambda_1, \lambda_2, ..., \lambda_n)$$

be the diagonal matrix with eigenvalues of A as its diagonal elements. The matrix P is called the *modal matrix* of A and D is called the *spectral matrix* of A. We have

$$AP = A[x_1, x_2, ..., x_n] = (Ax_1, Ax_2, ..., Ax_n)$$

$$= (\lambda_1 x_1, \lambda_2 x_2, ..., \lambda_n x_n) = (x_1, x_2, ..., x_n)D = PD. \tag{3.44}$$

Since the columns of P are linearly independent, the rank of P is n and therefore the matrix P is invertible. Pre-multiplying both sides in Eq. (3.44) by P^{-1}, we obtain

$$P^{-1}AP = P^{-1}PD = D \tag{3.45}$$

which implies that **A** is similar to **D**. Therefore, the matrix of eigenvectors **P** reduces a matrix **A** to its diagonal form.

Post-multiplying both sides in Eq. (3.44) by $\mathbf{P}^{-1}$, we obtain

$$\mathbf{A} = \mathbf{PDP}^{-1}. \tag{3.46}$$

Remark 23

(a) A square matrix **A** of order n has always n linearly independent eigenvectors when its eigenvalues are distinct. The matrix may also have n linearly independent eigenvectors even when some eigenvalues are repeated (see Example 3.36 (iii)). Therefore, there is no restriction imposed on the eigenvalues of the matrix **A** in Theorem 3.11.

(b) From Eq. (3.46), we obtain

$$\mathbf{A}^2 = \mathbf{AA} = (\mathbf{PDP}^{-1})\,(\mathbf{PDP}^{-1}) = \mathbf{PD}^2\mathbf{P}^{-1}.$$

Repeating the pre-multiplication (post-multiplication) m times, we get

$$\mathbf{A}^m = \mathbf{PD}^m\mathbf{P}^{-1} \text{ for any positive integer } m.$$

Therefore, if **A** is diagonalizable, so is $\mathbf{A}^m$.

(c) If **D** is a diagonal matrix of order n, and

$$\mathbf{D} = \begin{bmatrix} \lambda_1 & & & \mathbf{0} \\ & \lambda_2 & & \\ & & \ddots & \\ \mathbf{0} & & & \lambda_n \end{bmatrix}, \text{ then } \mathbf{D}^m = \begin{bmatrix} \lambda_1^m & & & \\ & \lambda_2^m & & \mathbf{0} \\ & & \ddots & \\ \mathbf{0} & & & \lambda_n^m \end{bmatrix}.$$

for any positive integer m. If $Q(\mathbf{D})$ is a polynomial in **D**, then we get

$$Q(\mathbf{D}) = \begin{bmatrix} Q(\lambda_1) & & & \mathbf{0} \\ & Q(\lambda_2) & & \\ & & \ddots & \\ \mathbf{0} & & & Q(\lambda_n) \end{bmatrix}.$$

Now, let a matrix **A** be diagonalizable. Then, we have

$$\mathbf{A} = \mathbf{PDP}^{-1} \quad \text{and} \quad \mathbf{A}^m = \mathbf{PD}^m\,\mathbf{P}^{-1}$$

for any positive integer m. Hence, we obtain

$$Q(\mathbf{A}) = \mathbf{P}Q(\mathbf{D})\mathbf{P}^{-1}$$

for any matrix polynomial $Q(\mathbf{A})$.

Example 3.38 Show that the matrix

$$\mathbf{A} = \begin{bmatrix} 3 & 1 & -1 \\ -2 & 1 & 2 \\ 0 & 1 & 2 \end{bmatrix}$$

is diagonalizable. Hence, find **P** such that $\mathbf{P}^{-1}\mathbf{A}\mathbf{P}$ is a diagonal matrix. Then, obtain the matrix
$$\mathbf{B} = \mathbf{A}^2 + 5\mathbf{A} + 3\mathbf{I}.$$

Solution   The characteristic equation of **A** is given by

$$|\mathbf{A} - \lambda\mathbf{I}| = \begin{vmatrix} 3-\lambda & 1 & -1 \\ -2 & 1-\lambda & 2 \\ 0 & 1 & 2-\lambda \end{vmatrix} = \lambda^3 - 6\lambda^2 + 11\lambda - 6 = 0, \quad \text{or} \quad \lambda = 1, 2, 3.$$

Since the matrix **A** has three distinct eigenvalues, it has three linearly independent eigenvectors and hence it is diagonalizable.

The eigenvector corresponding to the eigenvalue $\lambda = 1$ is the solution of the system

$$(\mathbf{A} - \mathbf{I})\mathbf{x} = \begin{bmatrix} 2 & 1 & -1 \\ -2 & 0 & 2 \\ 0 & 1 & 1 \end{bmatrix} \begin{bmatrix} x_1 \\ x_2 \\ x_3 \end{bmatrix} = \begin{bmatrix} 0 \\ 0 \\ 0 \end{bmatrix}. \text{ The solution is } \mathbf{x}_1 = \begin{bmatrix} 1 \\ -1 \\ 1 \end{bmatrix}.$$

The eigenvector corresponding to the eigenvalue $\lambda = 2$ is the solution of the system

$$(\mathbf{A} - 2\mathbf{I})\mathbf{x} = \begin{bmatrix} 1 & 1 & -1 \\ -2 & -1 & 2 \\ 0 & 1 & 0 \end{bmatrix} \begin{bmatrix} x_1 \\ x_2 \\ x_3 \end{bmatrix} = \begin{bmatrix} 0 \\ 0 \\ 0 \end{bmatrix}. \text{ The solution is } \mathbf{x}_2 = \begin{bmatrix} 1 \\ 0 \\ 1 \end{bmatrix}.$$

The eigenvector corresponding to the eigenvalue $\lambda = 3$ is the solution of the system

$$(\mathbf{A} - 3\mathbf{I})\mathbf{x} = \begin{bmatrix} 0 & 1 & -1 \\ -2 & -2 & 2 \\ 0 & 1 & -1 \end{bmatrix} \begin{bmatrix} x_1 \\ x_2 \\ x_3 \end{bmatrix} = \begin{bmatrix} 0 \\ 0 \\ 0 \end{bmatrix}. \text{ The solution is } \mathbf{x}_3 = \begin{bmatrix} 0 \\ 1 \\ 1 \end{bmatrix}.$$

Hence, the modal matrix is given by

$$\mathbf{P} = [\mathbf{x}_1, \mathbf{x}_2, \mathbf{x}_3] = \begin{bmatrix} 1 & 1 & 0 \\ -1 & 0 & 1 \\ 1 & 1 & 1 \end{bmatrix} \quad \text{and} \quad \mathbf{P}^{-1} = \begin{bmatrix} -1 & -1 & 1 \\ 2 & 1 & -1 \\ -1 & 0 & 1 \end{bmatrix}.$$

It can be verified that $\mathbf{P}^{-1}\mathbf{A}\mathbf{P} = \text{diag}(1, 2, 3)$.

We have $\mathbf{D} = \text{diag}(1, 2, 3)$, $\mathbf{D}^2 = \text{diag}(1, 4, 9)$.

Therefore, $\mathbf{A}^2 + 5\mathbf{A} + 3\mathbf{I} = \mathbf{P}(\mathbf{D}^2 + 5\mathbf{D} + 3\mathbf{I})\mathbf{P}^{-1}.$

Now, $\mathbf{D}^2 + 5\mathbf{D} + 3\mathbf{I} = \begin{bmatrix} 1 & 0 & 0 \\ 0 & 4 & 0 \\ 0 & 0 & 9 \end{bmatrix} + \begin{bmatrix} 5 & 0 & 0 \\ 0 & 10 & 0 \\ 0 & 0 & 15 \end{bmatrix} + \begin{bmatrix} 3 & 0 & 0 \\ 0 & 3 & 0 \\ 0 & 0 & 3 \end{bmatrix} = \begin{bmatrix} 9 & 0 & 0 \\ 0 & 17 & 0 \\ 0 & 0 & 27 \end{bmatrix}.$

Hence, we obtain

$$A^2 + 5A + 3I = \begin{bmatrix} 1 & 1 & 0 \\ -1 & 0 & 1 \\ 1 & 1 & 1 \end{bmatrix} \begin{bmatrix} 9 & 0 & 0 \\ 0 & 17 & 0 \\ 0 & 0 & 27 \end{bmatrix} \begin{bmatrix} -1 & -1 & 1 \\ 2 & 1 & -1 \\ -1 & 0 & 1 \end{bmatrix} = \begin{bmatrix} 25 & 8 & -8 \\ -18 & 9 & 18 \\ -2 & 8 & 19 \end{bmatrix}.$$

Example 3.39 Examine whether the matrix **A**, where **A** is given by

$$(i)\ A = \begin{bmatrix} 1 & 2 & 2 \\ 0 & 2 & 1 \\ -1 & 2 & 2 \end{bmatrix}, \qquad (ii)\ A = \begin{bmatrix} -2 & 2 & -3 \\ 2 & 1 & -6 \\ -1 & -2 & 0 \end{bmatrix},$$

is diagonalizable. If so, obtain the matrix **P** such that $P^{-1}AP$ is a diagonal matrix.

Solution

(i) The characteristic equation of the matrix **A** is given by

$$|A - \lambda I| = \begin{vmatrix} 1-\lambda & 2 & 2 \\ 0 & 2-\lambda & 1 \\ -1 & 2 & 2-\lambda \end{vmatrix}$$

$$= (1 - \lambda)[(2 - \lambda)(2 - \lambda) - 2] - [2 - 2(2 - \lambda)] = (1 - \lambda)(2 - \lambda)(2 - \lambda) = 0,$$

or $\lambda = 1, 2, 2$. We first find the eigenvectors corresponding to the repeated eigenvalue $\lambda = 2$. We have the system

$$(A - 2I)x = \begin{bmatrix} -1 & 2 & 2 \\ 0 & 0 & 1 \\ -1 & 2 & 0 \end{bmatrix} \begin{bmatrix} x_1 \\ x_2 \\ x_3 \end{bmatrix} = \begin{bmatrix} 0 \\ 0 \\ 0 \end{bmatrix}.$$

Since the rank of the coefficient matrix is 2, it has one linearly independent eigenvector. We obtain another linearly independent eigenvector corresponding to the eigenvalue $\lambda = 1$. Since the matrix **A** has only two linearly independedent eigenvectors, the matrix is not diagonalizable.

(ii) The characteristic equation of the matrix **A** is given by

$$|A - \lambda I| = \begin{vmatrix} -2-\lambda & 2 & -3 \\ 2 & 1-\lambda & -6 \\ -1 & -2 & -\lambda \end{vmatrix} = 0 \text{ or } \lambda^3 + \lambda^2 - 21\lambda - 45 = 0, \text{ or } \lambda = 5, -3, -3.$$

Eigenvector corresponding to the eigenvalue $\lambda = 5$ is the solution of the system

$$(A - 5I)x = \begin{bmatrix} -7 & 2 & -3 \\ 2 & -4 & -6 \\ -1 & -2 & -5 \end{bmatrix} \begin{bmatrix} x_1 \\ x_2 \\ x_3 \end{bmatrix} = \begin{bmatrix} 0 \\ 0 \\ 0 \end{bmatrix}.$$

A solution of this system is $[1, 2, -1]^T$.

Eigenvectors corresponding to $\lambda = -3$ are the solutions of the system

$$(\mathbf{A} + 3\mathbf{I})\mathbf{x} = \begin{bmatrix} 1 & 2 & -3 \\ 2 & 4 & -6 \\ -1 & -2 & 3 \end{bmatrix} \begin{bmatrix} x_1 \\ x_2 \\ x_3 \end{bmatrix} = \begin{bmatrix} 0 \\ 0 \\ 0 \end{bmatrix} \text{ or } x_1 + 2x_2 - 3x_3 = 0.$$

The rank of the coefficient matrix is 1. Therefore, the system has two linearly independent eigenvectors. We use the equation $x_1 + 2x_2 - 3x_3 = 0$ to find two linearly independent eigenvectors. Taking $x_3 = 0$, $x_2 = 1$, we obtain the eigenvector $[-2, 1, 0]^T$ and taking $x_2 = 0$, $x_3 = 1$, we obtain the eigenvector $[3, 0, 1]^T$. The given 3×3 matrix has three linearly independent eigenvectors. Therefore, the matrix $\mathbf{A}$ is diagonalizable. The modal matrix $\mathbf{P}$ is given by

$$\mathbf{P} = \begin{bmatrix} 1 & -2 & 3 \\ 2 & 1 & 0 \\ -1 & 0 & 1 \end{bmatrix} \text{ and } \mathbf{P}^{-1} = \frac{1}{8} \begin{bmatrix} 1 & 2 & -3 \\ -2 & 4 & 6 \\ 1 & 2 & 5 \end{bmatrix}.$$

It can be verified that $\mathbf{P}^{-1}\mathbf{A}\mathbf{P} = \text{diag}(5, -3, -3)$.

Example 3.40 The eigenvectors of a 3×3 matrix $\mathbf{A}$ corresponding to the eigenvalues 1, 1, 3 are $[1, 0, -1]^T$, $[0, 1, -1]^T$ and $[1, 1, 0]^T$ respectively. Find the matrix $\mathbf{A}$.

Solution We have

$$\text{modal matrix } \mathbf{P} = \begin{bmatrix} 1 & 0 & 1 \\ 0 & 1 & 1 \\ -1 & -1 & 0 \end{bmatrix} \text{ and the spectral matrix } \mathbf{D} = \begin{bmatrix} 1 & 0 & 0 \\ 0 & 1 & 0 \\ 0 & 0 & 3 \end{bmatrix}.$$

We find that

$$\mathbf{P}^{-1} = \frac{1}{2} \begin{bmatrix} 1 & -1 & -1 \\ -1 & 1 & -1 \\ 1 & 1 & 1 \end{bmatrix}.$$

Therefore,

$$\mathbf{A} = \mathbf{P}\,\mathbf{D}\,\mathbf{P}^{-1} = \frac{1}{2} \begin{bmatrix} 1 & 0 & 1 \\ 0 & 1 & 1 \\ -1 & -1 & 0 \end{bmatrix} \begin{bmatrix} 1 & 0 & 0 \\ 0 & 1 & 0 \\ 0 & 0 & 3 \end{bmatrix} \begin{bmatrix} 1 & -1 & -1 \\ -1 & 1 & -1 \\ 1 & 1 & 1 \end{bmatrix}$$

$$= \frac{1}{2} \begin{bmatrix} 1 & 0 & 1 \\ 0 & 1 & 1 \\ -1 & -1 & 0 \end{bmatrix} \begin{bmatrix} 1 & -1 & -1 \\ -1 & 1 & -1 \\ 3 & 3 & 3 \end{bmatrix} = \begin{bmatrix} 2 & 1 & 1 \\ 1 & 2 & 1 \\ 0 & 0 & 1 \end{bmatrix}.$$

3.5.3 Special Matrices

In this section, we define some special matrices and study the properties of the eigenvalues and eigenvectors of these matrices. These matrices have applications in many areas. We first give some definitions.

Let $\mathbf{x} = (x_1, x_2, ..., x_n)^T$ and $\mathbf{y} = (y_1, y_2, ..., y_n)^T$ be two vectors of dimension n in $\mathbb{R}^n$ or $\mathbb{C}^n$. Then we define the following:

Inner Product (dot product) of vectors   Let $\mathbf{x}$ and $\mathbf{y}$ be two vectors in $\mathbb{R}^n$. Then

$$\mathbf{x} \cdot \mathbf{y} = x^T y = \sum_{i=1}^{n} x_i y_i. \tag{3.47}$$

is called the *inner product* of the vectors $\mathbf{x}$ and $\mathbf{y}$ and is a scalar. The inner product is also denoted by $< \mathbf{x}, \mathbf{y}>$. In this case $\mathbf{x} \cdot \mathbf{y} = \mathbf{y} \cdot \mathbf{x}$. Note that $\mathbf{x} \cdot \mathbf{x} \geq 0$ and $\mathbf{x} \cdot \mathbf{x} = 0$ if and only if $\mathbf{x} = \mathbf{0}$.

If $\mathbf{x}$ and $\mathbf{y}$ are in $\mathbb{C}^n$, then the inner product of these vectors is defined as

$$\mathbf{x} \cdot \mathbf{y} = x^T \, \overline{y} = \sum_{i=1}^{n} x_i \overline{y}_i \quad \text{and} \quad \mathbf{y} \cdot \mathbf{x} = y^T \overline{x} = \sum_{i=1}^{n} y_i \, \overline{x}_i$$

where $\overline{\mathbf{x}}$ and $\overline{\mathbf{y}}$ are complex conjugate vectors of $\mathbf{x}$ and $\mathbf{y}$ respectively. Note that $\mathbf{x} \cdot \mathbf{y} = \overline{\mathbf{y} \cdot \mathbf{x}}$. It can be easily verified that

$$(\alpha\mathbf{x} + \beta\mathbf{y}) \cdot \mathbf{z} = \alpha(\mathbf{x} \cdot \mathbf{z}) + \beta(\mathbf{y} \cdot \mathbf{z})$$

for any vectors $\mathbf{x}, \mathbf{y}, \mathbf{z}$ and scalars α, β.

Length (norm of a vector)   Let $\mathbf{x}$ be a vector in $\mathbb{R}^n$ or $\mathbb{C}^n$. Then

$$\|\mathbf{x}\| = \sqrt{\mathbf{x} \cdot \mathbf{x}} = \sqrt{x_1^2 + x_2^2 + ... + x_n^2}$$

is called the *length* or the *norm* of the vector $\mathbf{x}$.

Unit vector   The vector $\mathbf{x}$ is called a *unit vector* if $\|\mathbf{x}\| = 1$. If $\mathbf{x} \neq \mathbf{0}$, then vector $\mathbf{x}/\|\mathbf{x}\|$ is always a unit vector.

Orthogonal vectors   The vectors $\mathbf{x}$ and $\mathbf{y}$ for which $\mathbf{x} \cdot \mathbf{y} = 0$ are said to be *orthogonal vectors*.

Orthonormal vectors   The vectors $\mathbf{x}$ and $\mathbf{y}$ for which

$$\mathbf{x} \cdot \mathbf{y} = 0 \quad \text{and} \quad \|\mathbf{x}\| = 1, \|\mathbf{y}\| = 1$$

are called orthonormal vectors. If $\mathbf{x}, \mathbf{y}$ are any vectors and $\mathbf{x} \cdot \mathbf{y} = 0$, then $\mathbf{x}/\|\mathbf{x}\|$, $\mathbf{y}/\|\mathbf{y}\|$ are orthonormal. For example, the set of vectors

(i) $\begin{pmatrix} 1 \\ 0 \\ 0 \end{pmatrix}, \begin{pmatrix} 0 \\ 1 \\ 0 \end{pmatrix}, \begin{pmatrix} 0 \\ 0 \\ 1 \end{pmatrix}$ form an orthonormal set in $\mathbb{R}^3$.

(ii) $\begin{pmatrix} 3i \\ 4i \\ 0 \end{pmatrix}, \begin{pmatrix} -4i \\ 3i \\ 0 \end{pmatrix}, \begin{pmatrix} 0 \\ 0 \\ 1+i \end{pmatrix}$ form an orthogonal set in $\mathbb{C}^3$ and $\begin{pmatrix} 3i/5 \\ 4i/5 \\ 0 \end{pmatrix}, \begin{pmatrix} -4i/5 \\ 3i/5 \\ 0 \end{pmatrix}, \begin{pmatrix} 0 \\ 0 \\ (1+i)/\sqrt{2} \end{pmatrix}$

form an orthonormal set in $\mathbb{C}^3$.

Orthonormal and unitary system of vectors Let $x_1, x_2, \ldots, x_n$ be n vectors in $\mathbb{R}^n$. Then, this set of vectors forms an *orthonormal system* of vectors, if

$$x_i \cdot x_j = x_i^T x_j = \begin{cases} 0, & i \neq j \\ 1, & i = j. \end{cases}$$

Let $x_1, x_2, \ldots, x_n$ be n vectors in $\mathbb{C}^n$. Then, this set of vectors forms an *unitary system* of vectors, if

$$x_i \cdot x_j = x_i^T \bar{x}_j = \begin{cases} 0, & i \neq j \\ 1, & i = j. \end{cases}$$

In section 3.2.2, we have defined symmetric, skew-symmetric, Hermitian and skew-Hermitian matrices. We now define a few more special matrices.

Orthogonal matrices A real matrix A is *orthogonal* if $A^{-1} = A^T$. A simple example is

$$A = \begin{bmatrix} \cos\theta & -\sin\theta \\ \sin\theta & \cos\theta \end{bmatrix}.$$

A linear transformation in which the matrix of transformation is an orthogonal matrix is called an *orthogonal transformation*.

Unitary matrices A complex matrix A is *unitary* if $A^{-1} = (\bar{A})^T$, or $(\bar{A})^{-1} = A^T$. If A is real, then unitary matrix is same as orthogonal matrix.

A linear transformation in which the matrix of transformation is a unitary matrix is called a *unitary transformation*.

We note the following:

1. If A and B are Hermitian matrices, then $\alpha A + \beta B$ is also Hermitian for any real scalars α, β, since

$$(\overline{\alpha A + \beta B})^T = (\alpha \bar{A} + \beta \bar{B})^T = \alpha \bar{A}^T + \beta \bar{B}^T = \alpha A + \beta B.$$

2. Eigenvalues and eigenvectors of $\bar{A}$ are the conjugates of the eigenvalues and eigenvectors of A, since

$$A x = \lambda x \text{ gives } \bar{A}\,\bar{x} = \bar{\lambda}\,\bar{x}.$$

3. The inverse of a unitary (orthogonal) matrix is unitary (orthogonal). We have $A^{-1} = \bar{A}^T$. Let $B = A^{-1}$. Then

$$B^{-1} = A = (\bar{A}^T)^{-1} = [(\bar{A})^{-1}]^T = [\overline{(A^{-1})}]^T = \bar{B}^T.$$

Diagonally dominant matrix   A matrix $\mathbf{A} = (a_{ij})$ is said to be diagonally dominant, if

$$|a_{ii}| \geq \sum_{j=1, i \neq j}^{n} |a_{ij}|, \text{ for all } i.$$

The system of equations $\mathbf{Ax} = \mathbf{b}$, is called a *diagonally dominant system*, if the above conditions are satisfied and the strict inequality is satisfied for at least one i. If the strict inequality is satisfied for all i, then it is called a *strictly diagonally dominant system*.

Permutation matrix   A matrix $\mathbf{P}$ is called a *permutation matrix* if it has exactly one 1 in each row and column and all other elements are 0.

Property *A* of a matrix   Let $\mathbf{B}$ be a sparse matrix. Then, the matrix $\mathbf{B}$ is said to satisfy the *property A*, if and only if there exists a permutation matrix $\mathbf{P}$ such that

$$\mathbf{PBP}^T = \begin{bmatrix} \mathbf{A}_{11} & \mathbf{A}_{12} \\ \mathbf{A}_{21} & \mathbf{A}_{22} \end{bmatrix},$$

where $\mathbf{A}_{11}$ and $\mathbf{A}_{22}$ are diagonal matrices. The similarity transformation performs row interchanges followed by corresponding column interchanges in $\mathbf{B}$ such that $\mathbf{A}_{11}$ and $\mathbf{A}_{22}$ become diagonal matrices. The following procedure is a simple way of testing whether $\mathbf{B}$ can be reduced to the required form. It finds the locations of the non-zero elements and tests whether the interchanges of rows and corresponding interchanges of columns are possible to bring $\mathbf{B}$ to the required form. Let n be the order of the matrix $\mathbf{B}$ and $b_{ii} \neq 0$. Denote the set $U = \{1, 2, 3, ..., n\}$. Let there exist disjoint subsets U_1 and U_2 such that $U = U_1 \cup U_2$, where the suffixes of the non-zero off diagonal elements $b_{ik} \neq 0$, $i \neq k$, can be grouped as either $(i \in U_1, k \in U_2)$ or $(i \in U_2, k \in U_1)$. Then, the matrix $\mathbf{B}$ satisfies *property A*.

Consider, for example the matrix $\mathbf{B} = \begin{bmatrix} -2 & 1 & 0 \\ 1 & -2 & 1 \\ 0 & 1 & -2 \end{bmatrix}$.

Let the permutation matrix be taken as $\mathbf{P} = \begin{bmatrix} 1 & 0 & 0 \\ 0 & 0 & 1 \\ 0 & 1 & 0 \end{bmatrix}$.

Then, $\mathbf{PBP}^T = \begin{bmatrix} 1 & 0 & 0 \\ 0 & 0 & 1 \\ 0 & 1 & 0 \end{bmatrix} \begin{bmatrix} -2 & 1 & 0 \\ 1 & -2 & 1 \\ 0 & 1 & -2 \end{bmatrix} \begin{bmatrix} 1 & 0 & 0 \\ 0 & 0 & 1 \\ 0 & 1 & 0 \end{bmatrix}$

$$= \begin{bmatrix} 1 & 0 & 0 \\ 0 & 0 & 1 \\ 0 & 1 & 0 \end{bmatrix} \begin{bmatrix} -2 & 0 & 1 \\ 1 & 1 & -2 \\ 0 & -2 & 1 \end{bmatrix} = \begin{bmatrix} -2 & 0 & 1 \\ 0 & -2 & 1 \\ 1 & 1 & -2 \end{bmatrix} = \begin{bmatrix} \mathbf{A}_{11} & \mathbf{A}_{12} \\ \mathbf{A}_{21} & \mathbf{A}_{22} \end{bmatrix}$$

where A_{11} and A_{22} are diagonal matrices. Hence, **B** has *property A*. Note that the above similarity transformation is equivalent to interchanging rows 2 and 3, followed by an interchange of columns 2 and 3.

Now, $a_{ii} \neq 0$, $i = 1, 2, 3$. $a_{12} \neq 0$, $1 \in U_1$, $2 \in U_2$; $a_{21} \neq 0$, $2 \in U_2$, $1 \in U_1$; $a_{23} \neq 0$, $2 \in U_2$, $3 \in U_1$; $a_{32} \neq 0$, $3 \in U_1$, $2 \in U_2$. Subsets $U_1 = \{1, 3\}$, $U_2 = \{2\}$ exist such that $U = \{1, 2, 3\} = U_1 \cup U_2$. Hence, matrix **B** has *property A*.

We now establish some important results.

Therorem 3.12   An orthogonal set of vectors is linearly independent.

Proof   Let $x_1, x_2, ..., x_m$ be an orthogonal set of vectors, that is $x_i \cdot x_j = 0$, $i \neq j$. Consider the vector equation

$$x = \alpha_1 x_1 + \alpha_2 x_2 + ... + \alpha_m x_m = 0 \qquad (3.48)$$

where $\alpha_1, \alpha_2, ..., \alpha_m$ are scalars. Taking the inner product of the vector **x** in Eq. (3.48) with x_1, we get

$$x \cdot x_1 = (\alpha_1 x_1 + \alpha_2 x_2 + ... + \alpha_m x_m) \cdot x_1 = 0 \cdot x_1 = 0$$

or

$$\alpha_1 (x_1 \cdot x_1) = 0 \quad \text{or} \quad \alpha_1 \|x_1\|^2 = 0.$$

Since $\|x_1\|^2 \neq 0$, we get $\alpha_1 = 0$. Similarly, taking the inner products of **x** with $x_2, x_3, ..., x_m$ successively, we find that $\alpha_2 = \alpha_3 = ... = \alpha_m = 0$. Therefore, the set of orthogonal vectors $x_1, x_2, ..., x_m$ is linearly independent.

Theorem 3.13   The eigenvalues of

(i) an Hermitian matrix are real.

(ii) a skew-Hermitian matrix are zero or pure imaginary.

(iii) an unitary matrix are of magnitude 1.

Proof   Let λ be an eigenvalue and **x** be the corresponding eigenvector of the matrix **A**. We have $Ax = \lambda x$. Pre-multiplying both sides by $\bar{x}^T$, we get

$$\bar{x}^T A x = \lambda \bar{x}^T x \quad \text{or} \quad \lambda = \frac{\bar{x}^T A x}{\bar{x}^T x}. \qquad (3.49)$$

Note that $\bar{x}^T A x$ and $\bar{x}^T x$ are scalars. Also, the denominator $\bar{x}^T x$ is always real and positive. Therefore, the behavior of λ is governed by the scalar $\bar{x}^T A x$.

(i) Let **A** be an Hermitian matrix, that is $\bar{A} = A^T$. Now,

$$\overline{(\bar{x}^T A x)} = x^T \bar{A} \bar{x} = x^T A^T \bar{x} = (x^T A^T \bar{x})^T = \bar{x}^T A x$$

since $x^T A^T \bar{x}$ is a scalar. Therefore, $\bar{x}^T A x$ is real. From Eq. (3.49), we conclude that λ is real.

(ii) Let **A** be a skew-Hermitian matrix, that is $A^T = -\bar{A}$. Now,

$$\overline{(\bar{x}^T A x)} = x^T \bar{A} \bar{x} = -x^T A^T \bar{x} = -(x^T A^T \bar{x})^T = -\bar{x}^T A x$$

since $\mathbf{x}^T\mathbf{A}^T\bar{\mathbf{x}}$ is a scalar. Therefore, $\bar{\mathbf{x}}^T\mathbf{A}\mathbf{x}$ is zero or pure imaginary. From Eq. (3.49), we conclude that λ is zero or pure imaginary.

(iii) Let $\mathbf{A}$ be an unitary matrix, that is $\mathbf{A}^{-1} = (\bar{\mathbf{A}})^T$. Now, from

$$\mathbf{A}\mathbf{x} = \lambda\mathbf{x} \quad \text{or} \quad \bar{\mathbf{A}}\bar{\mathbf{x}} = \bar{\lambda}\,\bar{\mathbf{x}} \tag{3.50}$$

we get

$$(\bar{\mathbf{A}}\,\bar{\mathbf{x}})^T = \left(\bar{\lambda}\,\bar{\mathbf{x}}^T\right)^T \quad \text{or} \quad \bar{\mathbf{x}}^T\,\bar{\mathbf{A}}^T = \bar{\lambda}\,\bar{\mathbf{x}}^T.$$

or

$$\bar{\mathbf{x}}^T\mathbf{A}^{-1} = \bar{\lambda}\,\bar{\mathbf{x}}^T. \tag{3.51}$$

Using Eqs. (3.50) and (3.51), we can write

$$(\bar{\mathbf{x}}^T\,\mathbf{A}^{-1})(\mathbf{A}\mathbf{x}) = (\bar{\lambda}\,\bar{\mathbf{x}}^T)(\lambda\mathbf{x}) = |\lambda|^2\,\bar{\mathbf{x}}^T\mathbf{x}$$

or

$$\bar{\mathbf{x}}^T\mathbf{x} = |\lambda|^2\,\bar{\mathbf{x}}^T\mathbf{x}.$$

Since $\mathbf{x} \neq \mathbf{0}$, we have $\bar{\mathbf{x}}^T\mathbf{x} \neq 0$. Therefore, $|\lambda|^2 = 1$, or $|\lambda| = 1$. Hence, the result.

Remark 24

From Theorem 3.13, we conclude that the eigenvalues of

(i) a symmetric matrix are real.

(ii) a skew-symmetric matrix are zero or pure imaginary.

(iii) an orthogonal matrix are of magnitude 1 and are real or complex conjugate pairs.

Theorem 3.14 The column vectors (and also row vectors) of an unitary matrix form an unitary system of vectors.

Proof Let $\mathbf{A}$ be an unitary matrix of order n, with column vectors $\mathbf{x}_1, \mathbf{x}_2, ..., \mathbf{x}_n$. Then

$$\mathbf{A}^{-1}\mathbf{A} = \bar{\mathbf{A}}^T\mathbf{A} = \begin{bmatrix} \bar{\mathbf{x}}_1^T \\ \bar{\mathbf{x}}_2^T \\ \vdots \\ \bar{\mathbf{x}}_n^T \end{bmatrix} [\mathbf{x}_1, \mathbf{x}_2, ..., \mathbf{x}_n] = \begin{bmatrix} \bar{\mathbf{x}}_1^T\mathbf{x}_1 & \bar{\mathbf{x}}_1^T\mathbf{x}_2 & ... & \bar{\mathbf{x}}_1^T\mathbf{x}_n \\ \bar{\mathbf{x}}_2^T\mathbf{x}_1 & \bar{\mathbf{x}}_2^T\mathbf{x}_2 & ... & \bar{\mathbf{x}}_2^T\mathbf{x}_n \\ \vdots & & & \\ \bar{\mathbf{x}}_n^T\mathbf{x}_1 & \bar{\mathbf{x}}_n^T\mathbf{x}_2 & ... & \bar{\mathbf{x}}_n^T\mathbf{x}_n \end{bmatrix} = \mathbf{I}$$

Therefore,

$$\bar{\mathbf{x}}_i^T\mathbf{x}_j = \begin{cases} 0, & i \neq j \\ 1, & i = j \end{cases}.$$

Hence, the column vectors of $\mathbf{A}$ form an unitary system. Since the inverse of an unitary matrix is also an unitary matrix and the columns of $\mathbf{A}^{-1}$ are the conjugate of the rows of $\mathbf{A}$, we conclude that the row vectors of $\mathbf{A}$ also form an unitary system.

Remark 25

(a) From Theorem 3.14, we conclude that the column vectors (and also the row vectors) of an orthogonal matrix form an orthonormal system of vectors.

(b) A symmetric matrix of order n has n linearly independent eigenvectors and hence is diagonlizable.

Example 3.41 Show that the matrices $\mathbf{A}$ and $\mathbf{A}^T$ have the same eigenvalues and for distinct eigenvalues the eigenvectors corresponding to $\mathbf{A}$ and $\mathbf{A}^T$ are mutually orthogonal.

Solution We have

$$|\mathbf{A} - \lambda\mathbf{I}| = |(\mathbf{A}^T)^T - \lambda\mathbf{I}^T| = |[\mathbf{A}^T - \lambda\mathbf{I}]^T| = |\mathbf{A}^T - \lambda\mathbf{I}|.$$

Since $\mathbf{A}$ and $\mathbf{A}^T$ have the same characteristic equation, they have the same eigenvalues.

Let λ and μ be two distinct eigenvalues of $\mathbf{A}$. Let $\mathbf{x}$ be the eigenvector corresponding to the eigenvalue λ for $\mathbf{A}$ and $\mathbf{y}$ be the eigenvector corresponding to the eigenvalue μ for $\mathbf{A}^T$. We have $\mathbf{A}\mathbf{x} = \lambda\mathbf{x}$. Pre-multiplying by $\mathbf{y}^T$, we get

$$\mathbf{y}^T\mathbf{A}\mathbf{x} = \lambda\mathbf{y}^T\mathbf{x}. \tag{3.52}$$

We also have $\mathbf{A}^T\mathbf{y} = \mu\mathbf{y}$,   or $(\mathbf{A}^T\mathbf{y})^T = (\mu\mathbf{y})^T$ or $\mathbf{y}^T\mathbf{A} = \mu\mathbf{y}^T$.
Post-multiplying by $\mathbf{x}$, we get

$$\mathbf{y}^T\mathbf{A}\mathbf{x} = \mu\mathbf{y}^T\mathbf{x} \tag{3.53}$$

Subtracting Eqs. (3.52) and (3.53), we obtain

$$(\lambda - \mu)\mathbf{y}^T\mathbf{x} = 0.$$

Since $\lambda \neq \mu$, we obtain $\mathbf{y}^T\mathbf{x} = 0$. Therefore, the vectors $\mathbf{x}$ and $\mathbf{y}$ are mutually orthogonal.

3.6 Quadratic Forms

Let $\mathbf{x} = (x_1, x_2, \ldots, x_n)^T$ be an arbitrary vector in IR^n. A real *quadratic form* is an homogeneous expression of the form

$$Q = \sum_{i=1}^{n} \sum_{j=1}^{n} a_{ij}\, x_i\, x_j \tag{3.54}$$

in which the total power in each term is 2. Expanding, we can write

$$
\begin{aligned}
Q &= a_{11}x^2 + (a_{12} + a_{21})\, x_1 x_2 + \ldots + (a_{1n} + a_{n1})\, x_1 x_n \\
&\quad + a_{22}\, x_2^2 + (a_{23} + a_{32})\, x_2 x_3 + \ldots + (a_{2n} + a_{n2})\, x_2 x_n \\
&\quad + \ldots + a_{nn}\, x_n^2 \\
&= \mathbf{x}^T\mathbf{A}\mathbf{x} \tag{3.55}
\end{aligned}
$$

using the definition of matrix multiplication. Now, set $b_{ij} = (a_{ij} + a_{ji})/2$. The matrix $\mathbf{B} = (b_{ij})$ is symmetric since $b_{ij} = b_{ji}$. Further, $b_{ij} + b_{ji} = a_{ij} + a_{ji}$. Hence, Eq. (3.55) can be wtitten as

$$Q = \mathbf{x}^T\mathbf{B}\mathbf{x}$$

where $\mathbf{B}$ is a symmetric matrix and $b_{ij} = (a_{ij} + a_{ji})/2$.

For example, for $n = 2$, we have

$$b_{11} = a_{11}, \quad b_{12} = b_{21} = (a_{12} + a_{21})/2 \quad \text{and} \quad b_{22} = a_{22}.$$

Example 3.42 Obtain the symmetric matrix **B** for the quadratic form

(i) $Q = 2x_1^2 + 3x_1x_2 + x_2^2.$

(ii) $Q = x_1^2 + 2x_1x_2 - 4x_1x_3 + 6x_2x_3 - 5x_2^2 + 4x_3^2.$

Solution

(i) $\qquad\qquad a_{11} = 2, \quad a_{12} + a_{21} = 3 \text{ and } a_{22} = 1. \text{ Therefore,}$

$$b_{11} = a_{11} = 2, \quad b_{12} = b_{21} = \frac{1}{2}(a_{12} + a_{21}) = \frac{3}{2} \quad \text{and} \quad b_{22} = a_{22} = 1.$$

Therefore, $\qquad \mathbf{B} = \begin{bmatrix} 2 & 3/2 \\ 3/2 & 1 \end{bmatrix}.$

(ii) $a_{11} = 1, \ a_{12} + a_{21} = 2, \ a_{13} + a_{31} = -4, \ a_{23} + a_{32} = 6, \ a_{22} = -5, \ a_{33} = 4.$ Therefore,

$$b_{11} = a_{11} = 1, \ b_{12} = b_{21} = \frac{1}{2}(a_{12} + a_{21}) = 1, \ b_{13} = b_{31} = \frac{1}{2}(a_{13} + a_{31}) = -2,$$

$$b_{23} = b_{32} = \frac{1}{2}(a_{23} + a_{32}) = 3, \ b_{22} = a_{22} = -5, \ b_{33} = a_{33} = 4.$$

Therefore, $\qquad \mathbf{B} = \begin{bmatrix} 1 & 1 & -2 \\ 1 & -5 & 3 \\ -2 & 3 & 4 \end{bmatrix}.$

If **A** is a complex matrix, then the quadratic form is defined as

$$Q = \sum_{i=1}^{n} \sum_{j=1}^{n} a_{ij} \bar{x}_i x_j = \bar{\mathbf{x}}^T \mathbf{A} \mathbf{x} \tag{3.56}$$

where $\mathbf{x} = (x_1, x_2, ..., x_n)$ is an arbitrary vector in $\mathbb{C}^n$. However, this quadratic form is usually defined for an Hermitian matrix **A**. Then, it is called a *Hermitian form* and is always real.

For example, consider the Hermitian matrix $\mathbf{A} = \begin{bmatrix} 1 & 1+i \\ 1-i & 2 \end{bmatrix}$. The quadratic form becomes

$$Q = \bar{\mathbf{x}}^T \mathbf{A} \mathbf{x} = \begin{bmatrix} \bar{x}_1, \bar{x}_2 \end{bmatrix} \begin{bmatrix} 1 & 1+i \\ 1-i & 2 \end{bmatrix} \begin{bmatrix} x_1 \\ x_2 \end{bmatrix}$$

$$= |x_1|^2 + (1+i)\,\bar{x}_1 x_2 + (1-i)x_1\bar{x}_2 + 2\,|x_2|^2.$$

$$= |x_1|^2 + (\bar{x}_1 x_2 + x_1 \bar{x}_2) + i(\bar{x}_1 x_2 - x_1 \bar{x}_2) + 2|x_2|^2.$$

Now, $\bar{x}_1 x_2 + x_1 \bar{x}_2$ is real and $\bar{x}_1 x_2 - x_1 \bar{x}_2$ is imaginary. For example if $x_1 = p_1 + iq_1$, $x_2 = p_2 + iq_2$, we obtain

$$\bar{x}_1 x_2 + x_1 \bar{x}_2 = 2(p_1 p_2 + q_1 q_2) \text{ and } \bar{x}_1 x_2 - x_1 \bar{x}_2 = 2i(p_1 q_2 - p_2 q_1).$$

We can also write

$$(\bar{x}_1 x_2 + \bar{x}_2 x_1) + i(\bar{x}_1 x_2 - x_1 \bar{x}_2) = 2[(p_1 p_2 + q_1 q_2) - (p_1 q_2 - p_2 q_1)] = 2 \operatorname{Re}[(1 + i)\bar{x}_1 x_2].$$

Therefore, $\qquad Q = |x_1|^2 + 2\operatorname{Re}[(1 + i)\bar{x}_1 x_2] + |x_2|^2.$

Positive definite matrices

Let $\mathbf{A} = (a_{ij})$ be a square matrix. Then, the matrix $\mathbf{A}$ is said to be *positive definite* if

$$Q = \bar{\mathbf{x}}^T \mathbf{A} \mathbf{x} > 0 \text{ for any vector } \mathbf{x} \neq \mathbf{0} \text{ and } \bar{\mathbf{x}}^T \mathbf{A} \mathbf{x} = 0, \text{ if and only if } \mathbf{x} = \mathbf{0}.$$

If $\mathbf{A}$ is real, then $\mathbf{x}$ can be taken as real.

Positive definite matrices have the following properties.

1. The eigenvalues of a positive definite matrix are all real and positive.

   This is easily proved when $\mathbf{A}$ is a real matrix. From Eq. (3.49), we have

   $$\lambda = (\mathbf{x}^T \mathbf{A} \mathbf{x})/(\mathbf{x}^T \mathbf{x}).$$

   Since $\mathbf{x}^T \mathbf{x} > 0$ and $\mathbf{x}^T \mathbf{A} \mathbf{x} > 0$, we obtain $\lambda > 0$. If $\mathbf{A}$ is Hermetian, then $\bar{\mathbf{x}}^T \mathbf{A} \mathbf{x}$ is real and λ is real (see Theorem 3.13). Therefore, if the Hermitian form $Q > 0$, then the eigenvalues are real and positive.

2. All the leading minors of $\mathbf{A}$ are positive.

Remark 26

(a) If $\mathbf{A}$ is Hermitian and strictly diagonally dominant with positive real elements on the diagonal, then $\mathbf{A}$ is positive definite.

(b) If $\bar{\mathbf{x}}^T \mathbf{A} \mathbf{x} \geq 0$, then the matrix $\mathbf{A}$ is called *semi-positive definite*.

(c) A matrix $\mathbf{A}$ is called *negative definite* if $(-\mathbf{A})$ is positive definite. All the eigenvalues of a negative definite matrix are real and negative.

Example 3.43   Examine which of the following matrices are positive definite.

$$\text{(a) } \mathbf{A} = \begin{bmatrix} 3 & 1 \\ 2 & 4 \end{bmatrix}, \qquad \text{(b) } \mathbf{A} = \begin{bmatrix} 3 & -2i \\ 2i & 4 \end{bmatrix}, \qquad \text{(c) } \mathbf{A} = \begin{bmatrix} 1 & 0 & i \\ 0 & 1 & 0 \\ -i & 0 & 3 \end{bmatrix}.$$

Solution

(a) (i) $Q = \mathbf{x}^T \mathbf{A} \mathbf{x} = [x_1, x_2] \begin{bmatrix} 3 & 1 \\ 2 & 4 \end{bmatrix} \begin{bmatrix} x_1 \\ x_2 \end{bmatrix} = 3x_1^2 + 3x_1 x_2 + 4x_2^2$

$$= 3\left(x_1 + \frac{1}{2}x_2\right)^2 + \frac{13}{4}x_2^2 > 0 \text{ for all } \mathbf{x} \neq \mathbf{0}.$$

(ii) Eigenvalues of **A** are 2 and 5 which are both positive.

(iii) Leading minors $|3| = 3$, $\begin{vmatrix} 3 & 1 \\ 2 & 4 \end{vmatrix} = 10$ are both positive.

Hence, the matrix **A** is positive definite (it is not necessary to show all the three parts).

(b) $Q = \bar{\mathbf{x}}^T \mathbf{A} \mathbf{x} = [\bar{x}_1, \bar{x}_2] \begin{bmatrix} 3 & -2i \\ 2i & 4 \end{bmatrix} \begin{bmatrix} x_1 \\ x_2 \end{bmatrix} = [\bar{x}_1, \bar{x}_2] \begin{bmatrix} 3x_1 - 2ix_2 \\ 2ix_1 + 4x_2 \end{bmatrix}$

$$= 3x_1\bar{x}_1 - 2i\bar{x}_1 x_2 + 2ix_1\bar{x}_2 + 4x_2\bar{x}_2.$$

Taking $x_1 = p_1 + iq_1$ and $x_2 = p_2 + iq_2$ and simplifying, we get

$$Q = 3(p_1^2 + q_1^2) + 4(p_2^2 + q_2^2) + 4(p_1 q_2 - p_2 q_1)$$
$$= p_1^2 + q_1^2 + 2p_2^2 + 2q_2^2 + 2(p_2 - q_1)^2 + 2(p_1 + q_2)^2 > 0.$$

Therefore, the given matrix is positive definite.

Note that **A** is Hermitian, strictly diagonally dominant ($3 > |-2i|$, $4 > |2i|$) with positive real diagonal entries. Therefore, **A** is positive definite (see Remark 26(a).)

(c) $Q = \bar{\mathbf{x}}^T \mathbf{A} \mathbf{x} = [\bar{x}_1, \bar{x}_2, \bar{x}_3] \begin{bmatrix} 1 & 0 & i \\ 0 & 1 & 0 \\ -i & 0 & 3 \end{bmatrix} \begin{bmatrix} x_1 \\ x_2 \\ x_3 \end{bmatrix} = [\bar{x}_1, \bar{x}_2, \bar{x}_3] \begin{bmatrix} x_1 + ix_3 \\ x_2 \\ -ix_1 + 3x_3 \end{bmatrix}$

$$= x_1\bar{x}_1 + i\bar{x}_1 x_3 + x_2\bar{x}_2 - ix_1\bar{x}_3 + 3x_3\bar{x}_3$$

$$= |x_1|^2 + |x_2|^2 + 3|x_3|^2 + i(\bar{x}_1 x_3 - x_1\bar{x}_3)$$

Taking $x_1 = p_1 + iq_1$, $x_2 = p_2 + iq_2$, $x_3 = p_3 + iq_3$ and simplifying, we obtain

$$Q = (p_1^2 + q_1^2) + (p_2^2 + q_2^2) + 3(p_3^2 + q_3^2) - 2(p_1 q_3 - p_3 q_1)$$
$$= (p_1 - q_3)^2 + (p_3 + q_1)^2 + (p_2^2 + q_2^2) + 2(p_3^2 + q_3^2) > 0.$$

Therefore, the matrix **A** is positive definite. It can be verified that the eigenvalues of **A** are 1, 2, 2 which are all positive.

Example 3.44 Let **A** be a real square matrix. Show that the matrix $\mathbf{A}^T\mathbf{A}$ has real and positive eigenvalues.

Solution Since $\left(\mathbf{A}^T \mathbf{A}\right)^T = \mathbf{A}^T\mathbf{A}$, the matrix $\mathbf{A}^T\mathbf{A}$ is symmetric. Therefore, the eigenvalues of $\mathbf{A}^T\mathbf{A}$ are all real. Now,

$$\mathbf{x}^T\mathbf{A}^T\mathbf{A}\mathbf{x} = (\mathbf{A}\mathbf{x})^T(\mathbf{A}\mathbf{x}) = \mathbf{y}^T\mathbf{y}, \text{ where } \mathbf{A}\mathbf{x} = \mathbf{y}.$$

Since $\mathbf{y}^T\mathbf{y} > 0$ for any vector $\mathbf{y} \neq \mathbf{0}$, the matrix $\mathbf{A}^T\mathbf{A}$ is positive definite and hence all the eigenvalues of $\mathbf{A}^T\mathbf{A}$ are positive. Therefore, all the eigenvalues of $\mathbf{A}^T\mathbf{A}$ are real and positive.

Nature, rank, index and signature of a quadratic form

Let $\mathbf{A}$ be the matrix of the quadratic form $Q = \mathbf{x}^T\mathbf{A}\mathbf{x}$, where $\mathbf{A}$ is a symmetric matrix. The quadratic form Q is said to be

positive definite if all the eigenvalues of $\mathbf{A}$ are real and positive,
negative definite if all the eigenvalues of $\mathbf{A}$ are real and negative,
semi positive definite if all the eigenvalues of $\mathbf{A}$ are real and non-negative,
semi negative definite if all the eigenvalues of $\mathbf{A}$ are real and non-positive,
indefinite if some eigenvalues of $\mathbf{A}$ are positive and some are negative.

This defines the nature of the quadratic form.

Rank of the quadratic form The rank r of $\mathbf{A}$ is called the rank of the quadratic form, that is, the number of non-zero eigenvalues.

Index of a quadratic form The number of positive eigenvalues is called the index of the quadratic form and is denoted by k.

Signature of a quadratic form We define

Signature = (Number of positive eigenvalues) – (Number of negative eigenvalues)

$$= k - (r - k) = 2k - r.$$

Signature can be a negative integer.

We now, prove the invariance of a quadratic form under non-singular linear transformations.

Theorem 3.15 Under a non-singular linear transformation, a quadratic form $\mathbf{x}^T\mathbf{A}\mathbf{x}$, remains a quadratic form.

Proof Let $\mathbf{x}^T\mathbf{A}\mathbf{x}$ be a quadratic form, where $\mathbf{A}$ is symmetric and $\mathbf{x} = (x_1, x_2, \ldots, x_n)^T$. Let $\mathbf{x} = \mathbf{P}\mathbf{y}$ be a non-singular linear transformation, which transforms the quadratic form from the variables $x_1, x_2, \ldots, x_n$ to $y_1, y_2, \ldots, y_n$. Then,

$$\mathbf{x}^T\mathbf{A}\mathbf{x} = (\mathbf{P}\mathbf{y})^T\mathbf{A}(\mathbf{P}\mathbf{y}) = \mathbf{y}^T\,\mathbf{P}^T\,\mathbf{A}\mathbf{P}\mathbf{y} = \mathbf{y}^T\,\mathbf{B}\mathbf{y}$$

where $\mathbf{B} = \mathbf{P}^T\mathbf{A}\mathbf{P}$. Now,

$$\mathbf{B}^T = (\mathbf{P}^T\,\mathbf{A}\mathbf{P})^T = \mathbf{P}^T\mathbf{A}^T\mathbf{P} = \mathbf{P}^T\mathbf{A}\mathbf{P} = \mathbf{B}.$$

Hence, $\mathbf{B}$ is symmetric and $\mathbf{y}^T\,\mathbf{B}\mathbf{y}$ is also a quadratic form.

This proves the invariance of a quadratic form under a non-singular linear transformation.

3.6.1 Canonical Form of a Quadratic Form

A quadratic form $Q = \mathbf{x}^T\mathbf{A}\mathbf{x}$ is said to be in *canonical form* if all the mixed terms such as x_1x_2, x_1x_3, ... are absent, that is, $a_{ij} = 0$, $i \neq j$. We may also say that a canonical form is a *sum of squares form*. The canonical form is, therefore, given by

$$Q = a_1 y_1^2 + a_2 y_2^2 + \ldots + a_r y_r^2 \tag{3.57}$$

if rank $(\mathbf{A}) = r < n$

and

$$Q = a_1 y_1^2 + a_2 y_2^2 + \ldots + a_n y_n^2 \tag{3.58}$$

if rank $(\mathbf{A}) = n$, where $a_1, a_2, \ldots, a_n$ are any real numbers.

For example, $Q = 6x_1^2 + 5x_2^2$, $Q = 3x_1^2 - 4x_2^2$ are canonical forms.

Remark 27

Since the matrix $\mathbf{A}$ is symmetric, it is diagonalizable. Hence, every quadratic form $Q = \mathbf{x}^T \mathbf{A} \mathbf{x}$ can be reduced to a sum of squares form. The number of square terms is equal to the rank r.

Theorem 3.16 An orthogonal transformation $\mathbf{x} = \mathbf{P}\,\mathbf{y}$, where $\mathbf{P}$ is an orthogonal matrix, transforms a quadratic form $Q = \mathbf{x}^T \mathbf{A} \mathbf{x}$ to the sum of squares form $Q = \mathbf{y}^T \mathbf{D}\,\mathbf{y}$ where $\mathbf{D}$ is the diagonal matrix, $\mathbf{D} = \text{diag}\,(\lambda_1, \lambda_2, \ldots, \lambda_r)$.

Proof Let the rank of the quadratic form $Q = \mathbf{x}^T \mathbf{A} \mathbf{x}$ be r. Now, $\mathbf{A}$ is a symmetric matrix. Let $\mathbf{P}$ be the normalised modal matrix of $\mathbf{A}$. Therefore,

$$\mathbf{P}^{-1}\mathbf{A}\,\mathbf{P} = \mathbf{P}^T \mathbf{A} \mathbf{P} = \begin{bmatrix} \mathbf{D} & \mathbf{0} \\ \mathbf{0} & \mathbf{0} \end{bmatrix}, \text{ where } \mathbf{D} = \text{diag}\,(\lambda_1, \lambda_2, \ldots, \lambda_r).$$

If $r = n$, then $\mathbf{P}^T \mathbf{A} \mathbf{P} = \mathbf{D}$, where $\mathbf{D} = \text{diag}\,(\lambda_1, \lambda_2, \ldots, \lambda_n)$.

Under the orthogonal transformation $\mathbf{x} = \mathbf{P}\,\mathbf{y}$, we obtain

$$\mathbf{x}^T \mathbf{A} \mathbf{x} = (\mathbf{P}\,\mathbf{y})^T \mathbf{A}(\mathbf{P}\,\mathbf{y}) = \mathbf{y}^T \mathbf{A} \mathbf{P}\,\mathbf{y} = \mathbf{y}^T \begin{bmatrix} \mathbf{D} & \mathbf{0} \\ \mathbf{0} & \mathbf{0} \end{bmatrix} \mathbf{y}$$

$$= [y_1, y_2, \ldots, y_n] \begin{bmatrix} \lambda_1 & & & & & 0 \\ & \lambda_2 & & & & \\ & & \ddots & & & \\ & & & \lambda_r & & \\ 0 & & & & 0 & \\ & & & & & \ddots \\ & & & & & & 0 \end{bmatrix} \begin{bmatrix} y_1 \\ y_2 \\ \vdots \\ y_n \end{bmatrix} = \lambda_1 y_1^2 + \lambda_2 y_2^2 + \ldots + \lambda_r y_r^2. \tag{3.59}$$

If rank $= r = n$, then we get $\lambda_1 y_1^2 + \lambda_2 y_2^2 + \ldots + \lambda_n y_n^2$.

Let k coeffiecients in the sum of squares in Eq. (3.57) be positive and $r - k$ coefficients be negative. Arrange the terms in Eq. (3.57) such that the terms with positive coefficients appear first and then the terms with negative coefficients. That is, Eq. (3.57) is arranged as

$$Q = a_1 y_1^2 + a_2 y_2^2 + \ldots + a_k y_k^2 - a_{k+1} y_{k+1}^2 - \ldots - a_r y_r^2. \tag{3.60}$$

where $a_i > 0$.

Index and signature of a quadratic form The number of positive terms, k, is called the *index*. The difference between the number of positive and negative terms, that is, $2k - r$, is called the signature of the quadratic form.

Sylvester's law of inertia The rank r and index k of a real quadratic form Q are invariants under all real, non-singular transformations.

Reduction of a quadratic form to a canonical form

We give below two methods for reducing a quadratic form to a canonical form.

1. *Lagrange reduction* Let the quadratic form contain the variables x_1, x_2, x_3. We write the non singular transformation as

$$\left.\begin{aligned} y_1 &= x_1 + px_2 + qx_3, \\ y_2 &= x_2 + rx_3, \\ y_3 &= x_3, \end{aligned}\right\} \text{ or } \mathbf{y} = \begin{bmatrix} 1 & p & q \\ 0 & 1 & r \\ 0 & 0 & 1 \end{bmatrix} \mathbf{x}, \text{ or } \mathbf{y} = \mathbf{Px}. \tag{3.61}$$

Let the sum of squares form be $ay_1^2 + by_2^2 + cy_3^2$. Substitute (3.61) in $ay_1^2 + by_2^2 + cy_3^2$, simplify and compare the terms with the terms in the given quadratic form. Solve for a, p, q, b, r and c.

2. *Orthogonalisation method* Find the eigenvalues and eigenvectors of $\mathbf{A}$. Obtain the normalized modal matrix $\mathbf{P}$. Under the transformation $\mathbf{x} = \mathbf{Py}$, the quadratic form reduces to $\lambda_1 y_1^2 + \lambda_2 y_2^2 + \ldots + \lambda_r y_r^2$, or $\lambda_1 y_1^2 + \lambda_1 y_2^2 + \ldots + \lambda_n y_n^2$, depending on the rank of $\mathbf{A}$, where λ_i's are the eigenvalues of $\mathbf{A}$.

Example 3.45 Reduce the quadratic form $2x_1^2 + 2x_2^2 + 2x_3^2 - 2x_1x_2 - 2x_2x_3 - 2x_3x_1$ to canonical form through an orthogonal transformation. Find the index and signature.

Solution The matrix of the quadratic form is

$$\mathbf{A} = \begin{bmatrix} 2 & -1 & -1 \\ -1 & 2 & -1 \\ -1 & -1 & 2 \end{bmatrix}.$$

Eigenvalues of $\mathbf{A}$ are 0, 3, 3.

Eigenvector corresponding to $\lambda = 0$, is $\mathbf{v}_1 = [1\ 1\ 1]^T$.

Corresponding to the repeated eigenvalue $\lambda = 3$, we have the equation for a finding the eigevector as $x_1 + x_2 + x_3 = 0$. One eigenvector can be taken as $\mathbf{v}_2 = [1\ 0\ -1]^T$. Since the model matrix should be orthogonal, the third eigenvector must satisfy the above equation and also be orthogonal to both $\mathbf{v}_1$ and $\mathbf{v}_2$. Assuming $\mathbf{v}_3 = [a\ b\ c]^T$ and orthogonalising with $\mathbf{v}_1$ and $\mathbf{v}_2$, we obtain $\mathbf{v}_3 = [1\ -2\ 1]^T$. The normalized modal matrix is given by

$$\mathbf{P} = \begin{bmatrix} 1/\sqrt{3} & 1/\sqrt{2} & 1/\sqrt{6} \\ 1/\sqrt{3} & 0 & -2/\sqrt{6} \\ 1/\sqrt{3} & -1/\sqrt{2} & 1/\sqrt{6} \end{bmatrix}.$$

The orthogonal transformation is $\mathbf{x} = \mathbf{P}\mathbf{y}$, and

canonical form $= \mathbf{x}^T\mathbf{A}\mathbf{x} = \mathbf{y}^T\mathbf{P}^T\mathbf{A}\mathbf{P}\mathbf{y} = \mathbf{y}^T\mathbf{D}\mathbf{y} = (0)\, y_1^2 + 3y_2^2 + 3y_3^2 = 3y_2^2 + 3y_3^2$,

index $= 2$, signature $= 2$.

3.7 Condition Number of a Matrix

Norm of a matix   Let $\mathbf{A}$ be a real or a complex matrix. Then, the norm of a matrix denoted by $\|\mathbf{A}\|$, is defined as follows:

(i) *Euclidean norm:* $\|\mathbf{A}\| = \sqrt{\sum_{i,j} |a_{ij}|^2}$.

(ii) *Spectral norm or Hilbert norm:* Compute $\mathbf{A}^*\mathbf{A} = (\overline{\mathbf{A}})^T\, \mathbf{A}$. Define

$\qquad\qquad \lambda =$ spectral radius (largest eigenvalue in magnitude) of $\mathbf{A}^*\mathbf{A}$.

Then, $\qquad\qquad \|\mathbf{A}\| = \sqrt{\lambda}$.

If $\mathbf{A}$ is an Hermitian matrix $(\mathbf{A}^* = \mathbf{A})$ or $\mathbf{A}$ is a real symmetric matrix, then $\lambda =$ spectral radius of $\mathbf{A}^*\mathbf{A} =$ spectral radius of $\mathbf{A}^2 =$ (spectral radius of $\mathbf{A}$)^2.

Therefore, $\|\mathbf{A}\| = \sqrt{\lambda} =$ spectral radius of $\mathbf{A}$.

For most engineering applications, we use the spectral norm.

Condition number of a matrix

Condition number of a matrix is an important concept in the theory of solution of linear algebraic equations. In engineering applications, we often require to solve a large system of linear algebraic equations. Since the system is large, we solve it by iterative methods. Naturally, iterative methods produce round off errors. The round off errors should not magnify during iteration. Condition number of the coefficient matrix $\mathbf{A}$ of the system of equations $\mathbf{A}\mathbf{x} = \mathbf{b}$, gives a measure of the sensitivity of the system to round off errors.

Using the spectral norm, we define the *condition number* of a matrix $\mathbf{A}$ as

$$\text{cond}(\mathbf{A}) = \kappa(\mathbf{A}) = \|\mathbf{A}\|\, \|\mathbf{A}^{-1}\| = \sqrt{\frac{\lambda}{\mu}} \qquad (3.62)$$

where $\lambda =$ largest eigenvalue in magnitude of $\mathbf{A}^*\mathbf{A}$,
and $\mu =$ smallest eigenvalue in magnitude of $\mathbf{A}^*\mathbf{A}$.

If $\mathbf{A}$ is an Hermitian matrix or $\mathbf{A}$ is a real and symmetric matrix, then (3.62) simplifies to

$$\text{cond}(\mathbf{A}) = \kappa(\mathbf{A}) = \|\mathbf{A}\|\, \|\mathbf{A}^{-1}\| = \frac{\lambda_1}{\mu_1} \qquad (3.63)$$

where $\lambda_1 =$ largest eigenvalue in magnitude of $\mathbf{A}$,
and $\mu_1 =$ smallest eigenvalue in magnitude of $\mathbf{A}$.

The larger the value of the condition number more is the sensitivity of the system to round off errors. For example, when the condition number is large, the solution obtained using say, four decimal places arithmetic may differ completely from the solution obtained by say, six decimal places arithmetic.

We can use other norms also to compute the condition number.

Example 3.46   Find the condition numbers of the following matrices.

$$\text{(i)} \begin{bmatrix} 1 & 4 \\ 3 & 2 \end{bmatrix}, \qquad \text{(ii)} \begin{bmatrix} 5 & -2 & 0 \\ -2 & 6 & 2 \\ 0 & 2 & 7 \end{bmatrix}. \qquad \text{(iii)} \begin{bmatrix} 2 & 3+4i \\ 3-4i & 2 \end{bmatrix}.$$

Solution

(i) **A** is real. We have

$$\mathbf{A}^T \mathbf{A} = \begin{bmatrix} 1 & 3 \\ 4 & 2 \end{bmatrix} \begin{bmatrix} 1 & 4 \\ 3 & 2 \end{bmatrix} = \begin{bmatrix} 10 & 10 \\ 10 & 20 \end{bmatrix}.$$

The characteristic equation of $\mathbf{A}^T \mathbf{A}$ is given by

$$|\mathbf{A}^T \mathbf{A} - \lambda \mathbf{I}| = \begin{vmatrix} 10-\lambda & 10 \\ 10 & 20-\lambda \end{vmatrix} = \lambda^2 - 30\lambda + 100 = 0.$$

The eigenvalues are given by $\lambda = [15 + 5\sqrt{5}], [15 - 5\sqrt{5}]$.

Largest eigenvalue in magnitude of $\mathbf{A}^T \mathbf{A} = \lambda_1 = [15 + 5\sqrt{5}]$.

Smallest eigenvalue in magnitude of $\mathbf{A}^T \mathbf{A} = \mu_1 = [15 - 5\sqrt{5}]$.

Hence, $\qquad \text{cond}(\mathbf{A}) = \sqrt{\dfrac{\lambda_1}{\mu_1}} = \sqrt{\dfrac{15+5\sqrt{5}}{15-5\sqrt{5}}} \approx 2.618.$

(ii) **A** is real and symmetric. The characteristic equation of **A** is given by

$$|\mathbf{A} - \lambda \mathbf{I}| = \begin{vmatrix} 5-\lambda & -2 & 0 \\ -2 & 6-\lambda & 2 \\ 0 & 2 & 7-\lambda \end{vmatrix} = \lambda^3 - 18\lambda^2 + 99\lambda - 162 = 0.$$

The eigenvalues are given by $\lambda = 3, 6, 9$.

Hence, $\qquad \text{cond}(\mathbf{A}) = \dfrac{\text{Largest eigenvalue in magnitude of } \mathbf{A}}{\text{Smallest eigenvalue in magnitude of } \mathbf{A}} = \dfrac{9}{3} = 3.$

(iii) **A** is a complex (Hermitian) matrix. We have

$$\mathbf{A}^*\mathbf{A} = (\overline{\mathbf{A}})^T \mathbf{A} = \begin{bmatrix} 2 & 3+4i \\ 3-4i & 2 \end{bmatrix}\begin{bmatrix} 2 & 3+4i \\ 3-4i & 2 \end{bmatrix} = \begin{bmatrix} 29 & 12+16i \\ 12-16i & 29 \end{bmatrix}.$$

The characteristic equation of $\mathbf{A}^*\mathbf{A}$ is given by

$$|\mathbf{A}^*\mathbf{A} - \lambda\mathbf{I}| = \begin{bmatrix} 29-\lambda & 12+16i \\ 12-16i & 29-\lambda \end{bmatrix} = (29-\lambda)^2 - 400 = 0.$$

The eigenvalues are given by $\lambda = 49, 9$.

Largest eigenvalue in magnitude of $\mathbf{A}^*\mathbf{A} = \lambda_1 = 49$.
Smallest eigenvalue in magnitude of $\mathbf{A}^*\mathbf{A} = \mu_1 = 9$.

Hence,
$$\text{cond }(\mathbf{A}) = \sqrt{\frac{\lambda_1}{\mu_1}} = \sqrt{\frac{49}{9}} = \frac{7}{3}.$$

Note that since **A** is a Hermitian matrix, we could have used the formula

$$\text{cond}(\mathbf{A}) = \frac{\text{Largest eigenvalue in magnitude of A}}{\text{Smallest eigenvalue in magnitude of A}} = \frac{7}{3}.$$

3.8   Singular Value Decomposition

Singular value decomposition is an important concept which has applications in many areas of engineering, computer science etc. In an earlier section, we discussed the diagonalization of a square matrix. We now discuss diagonalization of a rectangular matrix **A**. **A** may be a real or a complex matrix. We shall discuss in detail the case when **A** is a real matrix.

Theorem 3.17   An arbitrary $m \times n$ real matrix **A** can be decomposed as $\mathbf{A} = \mathbf{PDQ}$, where **D** is a generalized diagonal matrix of order $m \times n$ and **P** and **Q** are orthogonal matrices of orders $m \times m$ and $n \times n$ respectively.

Proof   Consider the $n \times n$ matrix $\mathbf{B} = \mathbf{A}^T\mathbf{A}$. Now, **B** is a symmetric and positive semi-definite matrix, since

$$\mathbf{B}^T = (\mathbf{A}^T\mathbf{A})^T = \mathbf{A}^T\mathbf{A} = \mathbf{B} \tag{3.64}$$

and
$$\mathbf{x}^T\mathbf{B}\mathbf{x} = \mathbf{x}^T\mathbf{A}^T\mathbf{A}\mathbf{x} = (\mathbf{A}\mathbf{x})^T(\mathbf{A}\mathbf{x}) = \|\mathbf{A}\mathbf{x}\|^2 \geq 0, \tag{3.65}$$

where **x** is an $n \times 1$ arbitrary vector. Therefore, the eigenvalues of **B** are either positive or zero. Denote the eigenvalues of **B** as $\lambda_1 = \mu_1^2, \lambda_2 = \mu_2^2, ..., \lambda_n = \mu_n^2$. If the eigen values are repeated, we count its multiplicity also. If some of the eigenvalues are zero, we order the eigenvalues such that the non-zero eigenvalues are taken first and then the zero eigenvalues. If rank $(\mathbf{B}) = r$, then we take the r positive eigenvalues as $\mu_1^2, \mu_2^2, ..., \mu_r^2$ and $n - r$ zero eigenvalues as $\mu_{r+1}^2, \mu_{r+2}^2, ..., \mu_n^2$. Define the generalized diagonal matrix **D** of order $m \times n$, with $\mu_1, \mu_2, ..., \mu_r$ on its diagonal and zeros elsewhere.

Find the normalised eigenvectors of $\mathbf{B}$ and denote them as $\mathbf{u}_1, \mathbf{u}_2, \dots, \mathbf{u}_n$. The normalised eigenvectors form an orthonormal system. That is, $\mathbf{u}_1, \mathbf{u}_2, \dots, \mathbf{u}_n$ are solutions of

$$\mathbf{B}\mathbf{u}_i = \mathbf{A}^T\mathbf{A}\mathbf{u}_i = \mu_i^2\,\mathbf{u}_i. \tag{3.66}$$

and
$$\mathbf{u}_i^T\,\mathbf{u}_j = 0, \quad i \neq j,$$
$$= 1, \quad i = j.$$

Now, form the orthogonal matrix $\mathbf{Q}$ with $\mathbf{u}_1^T, \mathbf{u}_2^T, \dots, \mathbf{u}_n^T$ as its rows.

Now, $\|\mathbf{A}\mathbf{u}_i\|^2 = (\mathbf{A}\mathbf{u}_i)^T\,(\mathbf{A}\mathbf{u}_i) = \mathbf{u}_i^T\,\mathbf{A}^T\,\mathbf{A}\mathbf{u}_i = \mu_i^2\,(\mathbf{u}_i^T\,\mathbf{u}_i) = \mu_i^2,$

Therefore, $\|\mathbf{A}\mathbf{u}_i\|^2 > 0$, for $i = 1, 2, \dots, r$, and $\|\mathbf{A}\mathbf{u}_i\|^2 = 0$, for $i = r+1, r+2, \dots, n.$

This implies that $\mathbf{A}\mathbf{u}_i = \mathbf{0}$, for $i = r+1, r+2, \dots, n.$ $\tag{3.67}$

Define the vectors $\mathbf{v}_1, \mathbf{v}_2, \dots, \mathbf{v}_r$ such that

$$\mathbf{v}_i = \frac{1}{\mu_i}\mathbf{A}\mathbf{u}_i, \quad i = 1, 2, \dots, r. \tag{3.68}$$

Now,
$$\mathbf{v}_i^T\,\mathbf{v}_j = \frac{1}{\mu_i\,\mu_j}(\mathbf{A}\mathbf{u}_i)^T\,(\mathbf{A}\mathbf{u}_j) = \frac{1}{\mu_i\,\mu_j}\,\mathbf{u}_i^T(\mathbf{A}^T\,\mathbf{A}\mathbf{u}_j) = \frac{\mu_j^2}{\mu_i\,\mu_j}\,\mathbf{u}_i^T\,\mathbf{u}_j.$$

That is, $\mathbf{v}_i^T\,\mathbf{v}_j = 1$, for $i = j = 1, 2, \dots, r$ and $\mathbf{v}_j^T\,\mathbf{v}_j = 0$, $i \neq j.$ $\tag{3.69}$

Therefore, $\mathbf{v}_1, \mathbf{v}_2, \dots, \mathbf{v}_r$ form an orthonormal system. Since $r \leq m$, we select the vectors $\mathbf{v}_{r+1}, \mathbf{v}_{r+2}, \dots, \mathbf{v}_m$ such that they form an orthonormal system with the vectors $\mathbf{v}_1, \mathbf{v}_2, \dots, \mathbf{v}_r$. This choice of $\mathbf{v}_{r+1}, \mathbf{v}_{r+2}, \dots, \mathbf{v}_m$ is arbitrary. Define the $m \times m$ matrix $\mathbf{P}$, such that $\mathbf{v}_1, \mathbf{v}_2, \dots, \mathbf{v}_m$ form its columns. That is, $\mathbf{P} = [\mathbf{v}_1, \mathbf{v}_2, \dots, \mathbf{v}_m]$. Note that $\mathbf{P}$ is an orthogonal matrix.

Now,
$$\mathbf{P}^T\mathbf{A}\mathbf{Q}^T = \begin{bmatrix} \mathbf{v}_1^T \\ \mathbf{v}_2^T \\ \dots \\ \mathbf{v}_m^T \end{bmatrix} \mathbf{A}[\mathbf{u}_1, \mathbf{u}_2, \dots, \mathbf{u}_n] = \begin{bmatrix} \mathbf{v}_1^T \\ \mathbf{v}_2^T \\ \dots \\ \mathbf{v}_m^T \end{bmatrix} [\mathbf{A}\mathbf{u}_1, \mathbf{A}\mathbf{u}_2, \dots, \mathbf{A}\mathbf{u}_n] = (c_{ij}).$$

Therefore, $\mathbf{P}^T\mathbf{A}\mathbf{Q}^T$ defines an $m \times n$ matrix whose elements are $c_{ij} = \mathbf{v}_i^T\,\mathbf{A}\mathbf{u}_j.$

From (3.67), $\mathbf{A}\mathbf{u}_j = \mathbf{0}$, for $j = r+1, r+2, \dots, n$. Therefore, $c_{ij} = 0$, for $j = r+1, r+2, \dots, n.$

From (3.68) and (3.69), $c_{ij} = \mathbf{v}_j^T\,\mathbf{A}\mathbf{u}_j = \mu_j\,\mathbf{v}_i^T\,\mathbf{v}_j = \mu_j$, for $i = j = 1, 2, \dots, r$; and 0 for $i \neq j.$

Therefore, $\mathbf{P}^T\mathbf{A}\mathbf{Q}^T$ defines the $m \times n$ generalized diagonal matrix $\mathbf{D}$, that is $\mathbf{P}^T\mathbf{A}\mathbf{Q}^T = \mathbf{D}.$

Pre-multiplying by $\mathbf{P}$ and post multiplying by $\mathbf{Q}$, we obtain

$$\mathbf{P}(\mathbf{P}^T\mathbf{A}\mathbf{Q}^T)\,\mathbf{Q} = \mathbf{P}\mathbf{D}\mathbf{Q}, \quad \text{or} \quad \mathbf{A} = \mathbf{P}\mathbf{D}\mathbf{Q}, \tag{3.70}$$

since $\mathbf{P}$ and $\mathbf{Q}$ are orthogonal matrices.

The decomposition $\mathbf{A} = \mathbf{PDQ}$, is called a *singular value decomposition*, and the numbers μ_1, μ_2, ..., μ_r, which are the positive square roots of μ_1^2, μ_2^2, ..., μ_r^2 are called the *singular values* of $\mathbf{A}$. If rank $(\mathbf{B}) = n$, the singular values are μ_1, μ_2, ..., μ_n. Since, the ordering of the non-zero eigenvalues is arbitrary and the choice of $\mathbf{v}_{r+1}$, $\mathbf{v}_{r+2}$, ..., $\mathbf{v}_m$ is arbitrary, singular value decomposition of a matrix is not unique.

For example, we have the following decompositions:

$$\mathbf{A}_{4\times3} = \mathbf{P}_{4\times4}\, \mathbf{D}_{4\times3}\, \mathbf{Q}_{3\times3}, \quad \mathbf{A}_{3\times3} = \mathbf{P}_{3\times3}\, \mathbf{D}_{3\times3}\, \mathbf{Q}_{3\times3}, \quad \mathbf{A}_{2\times3} = \mathbf{P}_{2\times2}\, \mathbf{D}_{2\times3}\, \mathbf{Q}_{3\times3}$$

where $\mathbf{D}$ is suitably defined.

Remark 28

We can determine $\mathbf{v}_i$ in an alternate way. From (3.68), we have $\mathbf{Au}_i = \mu_i\, \mathbf{v}_i$.

But from (3.66), we get

$$\mathbf{A}^T \mathbf{Au}_i = \mu_i^2\, \mathbf{u}_i, \quad \text{or} \quad \mathbf{A}^T(\mu_i\, \mathbf{v}_i) = \mu_i^2\, \mathbf{u}_i,$$

or $$\mathbf{A}(\mathbf{A}^T \mathbf{v}_i) = \mu_i(\mathbf{Au}_i), \quad \text{or} \quad (\mathbf{AA}^T)\mathbf{v}_i = \mu_i^2\, \mathbf{v}_i. \tag{3.71}$$

Therefore, μ_i^2 are also eigenvalues of $\mathbf{AA}^T$ and $\mathbf{v}_i$ are the normalised eigenvectors of $\mathbf{AA}^T$. That is, we can determine $\mathbf{v}_i$ by solving (3.71). If rank $(\mathbf{B}) = r$, we determine $\mathbf{v}_{r+1}$, $\mathbf{v}_{r+2}$, ..., $\mathbf{v}_m$ as described above.

Remark 29

If $\mathbf{A}$ is a $n \times n$ real symmetric matrix, we have $\mathbf{B} = \mathbf{A}^T \mathbf{A} = \mathbf{A}^2$. The eigenvalues of $\mathbf{B}$ are squares of eigenvalues of $\mathbf{A}$. The eigenvectors $\mathbf{u}_1$, $\mathbf{u}_2$, ..., $\mathbf{u}_n$; and $\mathbf{v}_1$, $\mathbf{v}_2$, ..., $\mathbf{v}_n$ are identical. Therefore, $\mathbf{P}^T = \mathbf{Q}$, or $\mathbf{P} = \mathbf{Q}^T$. In this case, singular value decomposition is given by $\mathbf{A} = \mathbf{PDP}^T$.

Remark 30

In the case of a complex matrix, we have the following result:

An arbitrary $m \times n$ complex matrix $\mathbf{A}$ can be decomposed as $\mathbf{A} = \mathbf{PDQ}$, where $\mathbf{D}$ is a generalized diagonal matrix of order $m \times n$ and $\mathbf{P}$ and $\mathbf{Q}$ are unitary matrices of orders $m \times m$ and $n \times n$ respectively.

Example 3.47 Find singular value decompositions of the following matrices.

(i) $\begin{bmatrix} 1 & 2 \\ 1 & 3 \end{bmatrix}$.   (ii) $\begin{bmatrix} 1 & 3 \\ 3 & 9 \end{bmatrix}$.   (iii) $\begin{bmatrix} 1 & 1 \\ 1 & 1 \\ 1 & 1 \end{bmatrix}$.

Solution

(i) We have $\mathbf{B} = \mathbf{A}^T\mathbf{A} = \begin{bmatrix} 1 & 1 \\ 2 & 3 \end{bmatrix}\begin{bmatrix} 1 & 2 \\ 1 & 3 \end{bmatrix} = \begin{bmatrix} 2 & 5 \\ 5 & 13 \end{bmatrix}$.

Eigenvalues of $\mathbf{B}$ are given by $\lambda^2 - 15\lambda + 1 = 0$.
We obtain $\lambda_1 = \mu_1^2 = 14.93303$, $\lambda_2 = \mu_2^2 = 0.066966$, $\mu_1 = 3.86433$, $\mu_2 = 0.25878$.

The diagonal matrix $\mathbf{D}$ is given by

$$\mathbf{D} = \begin{bmatrix} 3.86433 & 0 \\ 0 & 0.25878 \end{bmatrix}.$$

The eigenvectors are obtained as the following:

$\lambda_1 = 14.93303$: $\mathbf{x}_1 = [0.38661, 1]^T$; $\mathbf{u}_1 = \dfrac{\mathbf{x}_1}{\|\mathbf{x}_1\|} = [0.36060, 0.93272]^T$.

$\lambda_2 = 0.066966$: $\mathbf{x}_2 = [1, -0.386607]^T$; $\mathbf{u}_2 = \dfrac{\mathbf{x}_2}{\|\mathbf{x}_2\|} = [0.93272, -0.36060]^T$.

The matrix $\mathbf{Q}$ is given by

$$\mathbf{Q} = \begin{bmatrix} \mathbf{u}_1^T \\ \mathbf{u}_2^T \end{bmatrix} = \begin{bmatrix} 0.36060 & 0.93272 \\ 0.93272 & -0.36060 \end{bmatrix}.$$

Note that $\mathbf{Q}$ is an orthogonal matrix.
Define the vectors $\mathbf{v}_1$, $\mathbf{v}_2$ as the following.

$$\mathbf{v}_1 = \frac{1}{\mu_1} \mathbf{A}\mathbf{u}_1 = \frac{1}{3.86433} \begin{bmatrix} 1 & 2 \\ 1 & 3 \end{bmatrix} \begin{bmatrix} 0.36060 \\ 0.93272 \end{bmatrix} = \begin{bmatrix} 0.57605 \\ 0.81741 \end{bmatrix}.$$

$$\mathbf{v}_2 = \frac{1}{\mu_2} \mathbf{A}\mathbf{u}_2 = \frac{1}{0.25878} \begin{bmatrix} 1 & 2 \\ 1 & 3 \end{bmatrix} \begin{bmatrix} 0.93272 \\ -0.36060 \end{bmatrix} = \begin{bmatrix} 0.81737 \\ -0.57609 \end{bmatrix}.$$

The matrix $\mathbf{P}$ is given by

$$\mathbf{P} = [\mathbf{v}_1 \ \mathbf{v}_2] = \begin{bmatrix} 0.57605 & 0.81737 \\ 0.81741 & -0.57609 \end{bmatrix}.$$

Note that $\mathbf{P}$ is also an orthogonal matrix.

Alternate way to find $\mathbf{v}_i$

The vectors $\mathbf{v}_i$ are solutions of $(\mathbf{A}\mathbf{A}^T)\,\mathbf{v}_i = \mu_i^2\,\mathbf{v}_i$. We have

$$\mathbf{A}\mathbf{A}^T = \begin{bmatrix} 1 & 2 \\ 1 & 3 \end{bmatrix} \begin{bmatrix} 1 & 1 \\ 2 & 3 \end{bmatrix} = \begin{bmatrix} 5 & 7 \\ 7 & 10 \end{bmatrix}.$$

The eigenvectors are obtained as the following.

$\lambda_1 = \mu_1^2 = 14.93303$: $\mathbf{x}_1 = [1, 1.41900]^T$; $\mathbf{v}_1 = \dfrac{\mathbf{x}_1}{\|\mathbf{x}_1\|} = [0.57605, 0.81741]^T$.

$\lambda_2 = \mu_2^2 = 0.066966$: $\mathbf{x}_2 = [1, -0.70472]^T$; $\mathbf{v}_2 = \dfrac{\mathbf{x}_2}{\|\mathbf{x}_2\|} = [0.81741, -0.57605]^T$.

The singular value decomposition is $\mathbf{A} = \mathbf{PDQ}$.

(ii) The given matrix is real and symmetric. The eigenvalues of $\mathbf{B} = \mathbf{A}^T\mathbf{A}$ are squares of eigenvalues of $\mathbf{A}$. The singular value decomposition is given by $\mathbf{A} = \mathbf{PDP}^T = \mathbf{Q}^T\mathbf{DQ}$. The eigenvalues of $\mathbf{A}$ are 10, 0. Eigenvalues of $\mathbf{B}$ are 100, 0.

The diagonal matrix $\mathbf{D}$ is given by

$$\mathbf{D} = \begin{bmatrix} 10 & 0 \\ 0 & 0 \end{bmatrix}.$$

Now,
$$\mathbf{B} = \mathbf{A}^T\mathbf{A} = \begin{bmatrix} 1 & 3 \\ 3 & 9 \end{bmatrix}\begin{bmatrix} 1 & 3 \\ 3 & 9 \end{bmatrix} = \begin{bmatrix} 10 & 30 \\ 30 & 90 \end{bmatrix}.$$

The eigenvectors are obtained as the following:

$$\lambda_1 = 100: \mathbf{x}_1 = [1, 3]^T; \mathbf{u}_1 = \frac{\mathbf{x}_1}{\|\mathbf{x}_1\|} = [1/\sqrt{10}, 3/\sqrt{10}]^T.$$

$$\lambda_2 = 0: \mathbf{x}_2 = [-3, 1]^T; \mathbf{u}_2 = \frac{\mathbf{x}_2}{\|\mathbf{x}_2\|} = [-3/\sqrt{10}, 1/\sqrt{10}]^T.$$

The matrix $\mathbf{Q}$ is given by

$$\mathbf{Q} = \begin{bmatrix} \mathbf{u}_1^T \\ \mathbf{u}_2^T \end{bmatrix} = \begin{bmatrix} 1/\sqrt{10} & 3/\sqrt{10} \\ -3/\sqrt{10} & 1/\sqrt{10} \end{bmatrix}.$$

Note that $\mathbf{Q}$ is an orthogonal matrix. We have

$$\mathbf{P} = \mathbf{Q}^T = \begin{bmatrix} 1/\sqrt{10} & -3/\sqrt{10} \\ 3/\sqrt{10} & 1/\sqrt{10} \end{bmatrix}.$$

The singular value decomposition is $\mathbf{A} = \mathbf{PDP}^T$.

(iii) We have the decomposition as $\mathbf{A}_{3\times2} = \mathbf{P}_{3\times3}\,\mathbf{D}_{3\times2}\,\mathbf{Q}_{2\times2}$.

We have
$$\mathbf{B} = \mathbf{A}^T\mathbf{A} = \begin{bmatrix} 1 & 1 & 1 \\ 1 & 1 & 1 \end{bmatrix}\begin{bmatrix} 1 & 1 \\ 1 & 1 \\ 1 & 1 \end{bmatrix} = \begin{bmatrix} 3 & 3 \\ 3 & 3 \end{bmatrix}.$$

Eigenvalues of $\mathbf{B}$ are given by

$$\begin{vmatrix} 3-\lambda & 3 \\ 3 & 3-\lambda \end{vmatrix} = \lambda^2 - 6\lambda = 0. \quad \lambda_1 = \mu_1^2 = 6, \lambda_2 = \mu_2^2 = 0, \mu_1 = \sqrt{6}, \mu_2 = 0.$$

The matrix $\mathbf{D}$ is given by

$$\mathbf{D} = \begin{bmatrix} \sqrt{6} & 0 \\ 0 & 0 \\ 0 & 0 \end{bmatrix}.$$

The eigenvectors are obtained as the following:

$$\lambda_1 = 6: \ \mathbf{x}_1 = [1, 1]^T; \ \mathbf{u}_1 = \frac{\mathbf{x}_1}{\|\mathbf{x}_1\|} = [1/\sqrt{2}, \ 1/\sqrt{2}]^T.$$

$$\lambda_2 = 0: \ \mathbf{x}_2 = [1, -1]^T; \ \mathbf{u}_2 = \frac{\mathbf{x}_2}{\|\mathbf{x}_2\|} = [1/\sqrt{2}, \ -1/\sqrt{2}]^T.$$

The matrix $\mathbf{Q}$ is given by

$$\mathbf{Q} = \begin{bmatrix} \mathbf{u}_1^T \\ \mathbf{u}_2^T \end{bmatrix} = \frac{1}{\sqrt{2}} \begin{bmatrix} 1 & 1 \\ 1 & -1 \end{bmatrix}.$$

Note that $\mathbf{Q}$ is an orthogonal matrix.

Define the vector $\mathbf{v}_1$ as the following.

$$\mathbf{v}_1 = \frac{1}{\mu_1} \mathbf{A} \mathbf{u}_1 = \frac{1}{\sqrt{6}} \begin{bmatrix} 1 & 1 \\ 1 & 1 \\ 1 & 1 \end{bmatrix} \frac{1}{\sqrt{2}} \begin{bmatrix} 1 \\ 1 \end{bmatrix} = \frac{1}{\sqrt{3}} \begin{bmatrix} 1 \\ 1 \\ 1 \end{bmatrix}.$$

The vectors $\mathbf{v}_2, \mathbf{v}_3$ are arbitrary, but $\mathbf{v}_1, \mathbf{v}_2, \mathbf{v}_3$, should form an orthonormal system.

Choose $\mathbf{v}_2$, and $\mathbf{x}_3$ as
$$\mathbf{v}_2 = \frac{1}{\sqrt{2}} \begin{bmatrix} 1 \\ -1 \\ 0 \end{bmatrix}, \ \mathbf{x}_3 = \begin{bmatrix} a \\ b \\ c \end{bmatrix}.$$

Making $\mathbf{v}_1, \mathbf{v}_2, \mathbf{x}_3$ orthogonal, we get the equations $a + b + c = 0$, $a - b = 0$. The solution is $a = b$, $c = -2b$. We obtain

$$\mathbf{x}_3 = \begin{bmatrix} b \\ b \\ -2b \end{bmatrix}, \ \mathbf{v}_3 = \frac{1}{\sqrt{6}} \begin{bmatrix} 1 \\ 1 \\ -2 \end{bmatrix}.$$

The matrix $\mathbf{P}$ is given by

$$\mathbf{P} = [\mathbf{v}_1 \ \mathbf{v}_2 \ \mathbf{v}_3] = \begin{bmatrix} 1/\sqrt{3} & 1/\sqrt{2} & 1/\sqrt{6} \\ 1/\sqrt{3} & -1/\sqrt{2} & 1/\sqrt{6} \\ 1/\sqrt{3} & 0 & -2/\sqrt{6} \end{bmatrix}.$$

Note that $\mathbf{P}$ is an orthogonal matrix.

The singular value decomposition is $\mathbf{A} = \mathbf{PDQ}$.

Exercise 3.4

Verify the Cayley-Hamilton theorem for the matrix **A**. Find **A**$^{-1}$, if it exists, where **A** is as given in Problems **1** to **6.**

1. $\begin{bmatrix} 1 & 0 & -4 \\ 0 & 5 & 4 \\ -4 & 4 & 3 \end{bmatrix}.$

2. $\begin{bmatrix} 3 & 2 & 1 \\ 0 & 2 & 0 \\ 1 & 2 & 3 \end{bmatrix}$

3. $\begin{bmatrix} 1 & -2 & 1 \\ 2 & 3 & -2 \\ 3 & 1 & -1 \end{bmatrix}.$

4. $\begin{bmatrix} 1 & 2 & -2 \\ 1 & 1 & 1 \\ 1 & 3 & -1 \end{bmatrix}.$

5. $\begin{bmatrix} 1 & 1 & 2 \\ 1 & 3 & -1 \\ -2 & -1 & 1 \end{bmatrix}.$

6. $\begin{bmatrix} 1 & i & i \\ i & 1 & i \\ i & i & 1 \end{bmatrix}.$

Find all the eigenvalues and the corresponding eigenvectors of the matrices given in Problems **7** to **18.** Which of the matrices are diagonalizable?

7. $\begin{bmatrix} 1 & 2 & 2 \\ 0 & 2 & 1 \\ -1 & 2 & 2 \end{bmatrix}.$

8. $\begin{bmatrix} 1 & -1 & -1 \\ 1 & -1 & 0 \\ 1 & 0 & -1 \end{bmatrix}.$

9. $\begin{bmatrix} 1 & 2 & 3 \\ 3 & 1 & 0 \\ -2 & 0 & 1 \end{bmatrix}.$

10. $\begin{bmatrix} 1 & 3 & 3 \\ 1 & 4 & 3 \\ -1 & 3 & 4 \end{bmatrix}.$

11. $\begin{bmatrix} 1 & 1 & i \\ 1 & 0 & i \\ -i & -i & 1 \end{bmatrix}.$

12. $\begin{bmatrix} 0 & i & i \\ i & 0 & i \\ i & i & 0 \end{bmatrix}.$

13. $\begin{bmatrix} 0 & 1 & 0 & 0 \\ 0 & 0 & 1 & 0 \\ 0 & 0 & 0 & 1 \\ 0 & 0 & 0 & 0 \end{bmatrix}.$

14. $\begin{bmatrix} 0 & 2 & -2 & 0 \\ 1 & 1 & 0 & -1 \\ -1 & 1 & -2 & 1 \\ -1 & 1 & -2 & 1 \end{bmatrix}.$

15. $\begin{bmatrix} 0 & 1 & 1 & 1 \\ 1 & 0 & 1 & 1 \\ 1 & 1 & 0 & 1 \\ 1 & 1 & 1 & 0 \end{bmatrix}.$

16. $\begin{bmatrix} 1 & 2 & 3 & 4 \\ 2 & 1 & 4 & 3 \\ 3 & 4 & 2 & 1 \\ 4 & 3 & 1 & 2 \end{bmatrix}.$

17. $\begin{bmatrix} 0 & 0 & 0 & 0 & 1 \\ 0 & 0 & 0 & 1 & 0 \\ 0 & 0 & 1 & 0 & 0 \\ 0 & 1 & 0 & 0 & 0 \\ 1 & 0 & 0 & 0 & 0 \end{bmatrix}.$

18. $\begin{bmatrix} 0 & 1 & 0 & 0 & 0 \\ 0 & 0 & 1 & 0 & 0 \\ 0 & 0 & 0 & 1 & 0 \\ 0 & 0 & 0 & 0 & 1 \\ 1 & 0 & 0 & 0 & 0 \end{bmatrix}.$

Show that the matrices given in Problems **19** to **24** are diagonalizable. Find the matrix **P** such that **P**$^{-1}$**AP** is a diagonal matrix.

19. $\begin{bmatrix} 3 & 2 & 1 \\ 0 & 2 & 0 \\ 1 & 2 & 3 \end{bmatrix}.$

20. $\begin{bmatrix} -3 & -2 & 1 \\ -2 & 0 & 4 \\ -6 & -3 & 5 \end{bmatrix}.$

21. $\begin{bmatrix} 0 & 2 & 1 \\ 2 & 0 & 3 \\ 1 & -3 & 0 \end{bmatrix}.$

22. $\begin{bmatrix} 1 & 1 & 1 \\ 1 & 1 & 1 \\ 1 & -1 & 1 \end{bmatrix}.$

23. $\begin{bmatrix} 5 & -6 & -6 \\ -1 & 4 & 2 \\ 3 & -6 & -4 \end{bmatrix}.$

24. $\begin{bmatrix} 1 & 2 & -2 \\ 1 & 1 & 1 \\ 1 & 3 & -1 \end{bmatrix}.$

Find the matrix **A** whose eigenvalues and the corresponding eigenvectors are as given in Problems **25** to **30**.

25. Eigenvalues: 2, 2, 4; Eigenvectors: $(-2, 1, 0)^T$, $(-1, 0, 1)^T$, $(1, 0, 1)^T$.

26. Eigenvalues: 1, −1, 2; Eigenvectors: $(1, 1, 0)^T$, $(1, 0, 1)^T$, $(3, 1, 1)^T$.

27. Eigenvalues: 1, 2, 3; Eigenvectors: $(1, 2, 1)^T$, $(2, 3, 4)^T$, $(1, 4, 9)^T$.

28. Eigenvalues: 1, 1, 1; Eigenvectors: $(-1, 1, 1)^T$, $(1, -1, 1)^T$, $(1, 1, -1)^T$.

29. Eigenvalues: 0, −1, 1; Eigenvectors: $(-1, 1, 0)^T$, $(1, 0, -1)^T$, $(1, 1, 1)^T$.

30. Eigenvalues: 0, 0, 3; Eigenvectors: $(1, 2, -1)^T$, $(-2, 1, 0)^T$, $(3, 0, 1)^T$.

31. Let a 4×4 matrix **A** have eigenvalues 1, −1, 2, −2, Find the value of the determinant of the matrix $\mathbf{B} = 2\mathbf{A} + \mathbf{A}^{-1} - \mathbf{I}$.

32. Let a 3×3 matrix **A** have eigenvalues 1, 2, −1. Find the trace of the matrix $\mathbf{B} = \mathbf{A} - \mathbf{A}^{-1} + \mathbf{A}^2$.

33. Show that the matrices **A** and $\mathbf{P}^{-1}\mathbf{AP}$ have the same eigenvalues.

34. Let **A** and **B** be square matrices of the same order. Then, show that **AB** and **BA** have the same eigenvalues but different eigenvectors.

35. Show that the matrices $\mathbf{A}^{-1}\mathbf{B}$ and $\mathbf{BA}^{-1}$ have the same eigenvlaues but different eigenvectors.

36. An $n \times n$ matrix **A** is *nilpotent* if for some positive integer k, $\mathbf{A}^k = \mathbf{0}$. Show that all the eigenvalues of a nilpotent matrix are zero.

37. If **A** is an $n \times n$ diagonalizable matrix and $\mathbf{A}^2 = \mathbf{A}$, then show that each eigenvalue of **A** is 0 or 1.

38. Show that the matrix $\mathbf{A} = \begin{pmatrix} a & h \\ h & b \end{pmatrix}$, $a \neq b$, is transformed to a diagonal matrix $\mathbf{D} = \mathbf{P}^{-1}\mathbf{AP}$, where

    P is of the form $\mathbf{P} = \begin{pmatrix} \cos\theta & -\sin\theta \\ \sin\theta & \cos\theta \end{pmatrix}$ and $\tan 2\theta = \dfrac{2h}{a-b}$.

39. Let **A** be similar to **B**. Then show that (i) $\mathbf{A}^{-1}$ is similar to $\mathbf{B}^{-1}$, (ii) $\mathbf{A}^m$ is similar to $\mathbf{B}^m$ for any positive integer m, (iii) $|\mathbf{A}| = |\mathbf{B}|$.

40. Let **A** and **B** be symmetric matrices of the same order. Then, show that **AB** is symmetric if and only if **AB** = **BA**.

41. For any square matrix **A**, show that $\mathbf{A}^T\mathbf{A}$ is symmetric.

42. Let **A** be a non-singular matrix. show that $\mathbf{A}^T\mathbf{A}^{-1}$ is symmetric if and only if $\mathbf{A}^2 = (\mathbf{A}^T)^2$.

43. If **A** is a symmetric matrix and $\mathbf{P}^{-1}\mathbf{AP} = \mathbf{D}$, then show that **P** is an orthogonal matrix.

44. Show that the product of two orthogonal matrices of the same order is also an orthogonal matrix.

45. Find the conditions that a matrix $\mathbf{A} = \begin{bmatrix} l_1 & m_1 & n_1 \\ l_2 & m_2 & n_2 \\ l_3 & m_3 & n_3 \end{bmatrix}$ is orthogonal.

46. If **A** is an orthogonal matrix, show that $|\mathbf{A}| = \pm 1$.

47. Prove that the eigenvectors of a symmetric matrix corresponding to distinct eigenvalues are orthogonal.

48. A matrix **A** is called a *normal matrix* if $\mathbf{A}\overline{\mathbf{A}}^T = \overline{\mathbf{A}}^T\mathbf{A}$. Show that the Hermitian, skew-Hermitian and unitary matrices are normal.

49. If a matrix **A** can be diagonalized using an orthogonal matrix, then show that **A** is symmetric.

50. Suppose that a matrix **A** is both unitary and Hermitian. Then, show that $\mathbf{A} = \mathbf{A}^{-1}$.

51. If A is a symmetric matrix and $x^T A x > 0$ for every real vector $x \neq 0$, then show that $\bar{z}^T A z$ is real and positive for any complex vector $z \neq 0$.

52. Show that an unitary transformation $y = A x$, where A is an unitary matrix preserves the value of inner product.

53. Do the following matrices satisfy *property A*?

$$\text{(i)} \quad \begin{bmatrix} 2 & 1 & 1 \\ 1 & 2 & 1 \\ 1 & 1 & 3 \end{bmatrix}, \qquad \text{(ii)} \quad \begin{bmatrix} -4 & 1 & 0 & 1 \\ 1 & -4 & 1 & 0 \\ 0 & 1 & -4 & 1 \\ 1 & 0 & 1 & -4 \end{bmatrix}.$$

54. Prove that a real 2×2 symmetric matrix $\begin{bmatrix} a & b \\ b & c \end{bmatrix}$ is positive definite if and only if $a > 0$ (1×1 leading minor) and $ac - b^2 > 0$ (2×2 leading minor).

55. Show that the matrix $\begin{bmatrix} 2 & 1 & 3 \\ -3 & 4 & -1 \\ -1 & 1 & 2 \end{bmatrix}$ is positive definite.

56. Show that the matrix $\begin{bmatrix} -3 & -2 & 1 \\ -2 & 0 & 4 \\ -6 & -3 & 5 \end{bmatrix}$ is not positive definite.

Find the symmetric or the Hermitian matrix A for the quadratic forms given in Problems **57** to **61**.

57. $x_1^2 - 2x_1 x_2 + 4x_2 x_3 - x_2^2 + x_3^2$.

58. $3x_1^2 + 2x_1 x_2 - 4x_1 x_3 + 8x_2 x_3 + x_2^2$.

59. $x_1^2 + 2ix_1 x_2 - 8x_1 x_3 + 4ix_2 x_3 + 4x_3^2$.

60. $x_1^2 - (2 + 4i)\, x_1 x_2 - (4 - 6i)x_2 x_3 + x_2^2$.

61. $2x_1^2 - 3x_2^2 + (6 + 8i)x_1 x_2 + (4 - 2i)x_2 x_3$.

Reduce the quadratic form in Problems 62 and 63 to canonical form using Lagrange reduction.

62. $x_1^2 + 7x_2^2 + 7x_3^2 + 4x_1 x_2 - 18x_2 x_3 - 6x_3 x_1$.

63. $x_1^2 + 7x_2^2 + 26x_3^2 + 4x_1 x_2 - 22x_2 x_3 - 2x_3 x_1$.

Reduce the quadratic form in Problems 64 and 65 to canonical form using orthogonal reduction.

64. $x_1^2 + 3x_2^2 + 3x_3^2 - 2x_2 x_3$.

65. $6x_1^2 + 3x_2^2 + 3x_3^2 - 4x_1 x_2 - 2x_2 x_3 + 4x_3 x_1$.

'Find the condition numbers of the following matrices.

66. $\begin{bmatrix} 2 & 4 \\ 3 & 3 \end{bmatrix}$.

67. $\begin{bmatrix} 3 & 5 \\ 5 & 4 \end{bmatrix}$.

68. $\begin{bmatrix} 0 & 1 & 1 \\ 1 & 0 & 1 \\ 1 & 1 & 0 \end{bmatrix}$.

69. $\begin{bmatrix} 3 & 5-i \\ 5+i & 4 \end{bmatrix}$.

Find the singular value decomposition of the following matrices.

70. $\begin{bmatrix} 1 & 1 \\ 6 & 2 \end{bmatrix}$.

71. $\begin{bmatrix} 1 & 2 \\ -1 & 3 \end{bmatrix}$

72. $\begin{bmatrix} 1 & 1 & 1 \\ 1 & 1 & 1 \end{bmatrix}$.

73. $\begin{bmatrix} 5 & -2 & 0 \\ -2 & 6 & 2 \\ 0 & 2 & 7 \end{bmatrix}$.

74. $\begin{bmatrix} 1 & 0 & 0 \\ 0 & 3 & -1 \\ 0 & -1 & 3 \end{bmatrix}$.

75. $\begin{bmatrix} 2 & 1 & 1 \\ -1 & 2 & 1 \end{bmatrix}$.

3.9 Answers and Hints

Exercise 3.1

3. $\mathbf{A}^{-1} = \dfrac{1}{5}\begin{bmatrix} -3 & 2 & 2 \\ 2 & -3 & 2 \\ 2 & 2 & -3 \end{bmatrix}$.

4. $\mathbf{A}^{-1} = \dfrac{1}{11}\begin{bmatrix} -3 & 4 & 5 \\ 9 & -1 & -4 \\ 5 & -3 & -1 \end{bmatrix}$.

8. (i) $|\mathbf{A}\, adj\,(\mathbf{A})| = \text{diag}\,(|\mathbf{A}|, |\mathbf{A}|, \ldots |\mathbf{A}|) = |\mathbf{A}|^n$. Therefore, $|Adj\,(\mathbf{A})| = |\mathbf{A}|^{n-1}$.

(ii) Let $\mathbf{B} = adj\,(\mathbf{A})$. Since $\mathbf{B}^{-1} = adj(\mathbf{B})/|\mathbf{B}|$, we have $\mathbf{B}\,adj(\mathbf{B}) = |\mathbf{B}|\mathbf{I}$. Therefore, $adj(\mathbf{A})\, adj(adj(\mathbf{A})) = |adj(\mathbf{A})|\mathbf{I} = |\mathbf{A}|^{n-1}\mathbf{I}$.

Pre-multiplying by $\mathbf{A}$ and using $adj(\mathbf{A}) = \mathbf{A}^{-1}|\mathbf{A}|$, we get

$\mathbf{A}[\mathbf{A}^{-1}|\mathbf{A}|]\, adj(adj(\mathbf{A})) = |\mathbf{A}|^{n-1}\, \mathbf{AI}$ or $adj(adj(\mathbf{A})) = |\mathbf{A}|^{n-2}\mathbf{A}$.

9. $|\mathbf{A}\mathbf{A}^{-1}| = |\mathbf{A}||\mathbf{A}^{-1}| = |\mathbf{I}|$ or $|\mathbf{A}^{-1}| = 1/|\mathbf{A}|$.

10. $(\mathbf{B}\mathbf{A}\mathbf{B}^T)^T = \mathbf{B}\mathbf{A}^T\mathbf{B}^T = \mathbf{B}\mathbf{A}\mathbf{B}^T$.

13. $\mathbf{A}\mathbf{B} = \mathbf{B}\mathbf{A} \Rightarrow \mathbf{B}^{-1}\mathbf{A}\mathbf{B} = \mathbf{A} \Rightarrow \mathbf{B}^{-1}\mathbf{A} = \mathbf{A}\mathbf{B}^{-1}$. Similarly, $\mathbf{A}^{-1}\mathbf{B} = \mathbf{B}\mathbf{A}^{-1}$.

(i) $(\mathbf{A}\mathbf{B}^{-1})^T = (\mathbf{B}^{-1})^T\mathbf{A}^T = (\mathbf{B}^T)^{-1}\mathbf{A}^T = \mathbf{B}^{-1}\mathbf{A} = \mathbf{A}\mathbf{B}^{-1}$.

(ii) $(\mathbf{A}^{-1}\mathbf{B})^T = \mathbf{B}^T(\mathbf{A}^{-1})^T = \mathbf{B}^T(\mathbf{A}^T)^{-1} = \mathbf{B}\mathbf{A}^{-1} = \mathbf{A}^{-1}\mathbf{B}$.

(iii) $(\mathbf{A}^{-1}\mathbf{B}^{-1})^T = [(\mathbf{B}\mathbf{A})^{-1}]^T = [(\mathbf{A}\mathbf{B})^{-1}]^T = (\mathbf{A}^T)^{-1}(\mathbf{B}^T)^{-1} = \mathbf{A}^{-1}\mathbf{B}^{-1}$.

14. Pre-multiply both sides by (i) $\mathbf{I} - \mathbf{A}$, (ii) $\mathbf{I} + \mathbf{A}$.

15. $(\mathbf{P}\mathbf{A}\mathbf{Q})^{-1} = \mathbf{Q}^{-1}\mathbf{A}^{-1}\mathbf{P}^{-1} = \mathbf{I} \Rightarrow \mathbf{A}^{-1}\mathbf{P}^{-1} = \mathbf{Q} \Rightarrow \mathbf{A}^{-1} = \mathbf{Q}\mathbf{P}$.

16. Use $(\mathbf{I} - \mathbf{A})(\mathbf{I} + \mathbf{A} + \mathbf{A}^2 + \ldots) = \mathbf{I}$.

17. $(\mathbf{A}\mathbf{B}\mathbf{C})(\mathbf{A}\mathbf{B}\mathbf{C})^{-1} = \mathbf{I}$. Pre-multiply successively by $\mathbf{A}^{-1}$, $\mathbf{B}^{-1}$ and $\mathbf{C}^{-1}$.

18. 1, 2, 3.   **19.** 1, 1, 1.   **20.** 1, 1, 1.

21. 1, 2, 1.   **22.** (i) $k \neq 2$ and $k \neq -3$, (ii) $k = 2$, or $k = -3$.

23. $\theta = \pi/6$, or $\theta = \sin^{-1}[(9 - \sqrt{161})/4]$.

25. (i) $\lambda \neq 3$, μ arbitrary, (ii) $\lambda = 3$, $\mu = 10$, (iii) $\lambda = 3$, $\mu \neq 10$.

26. 2.   **27.** 1.   **28.** 2.

29. $|\mathbf{A}| = (p - q)(q - r)(r - p)(p + q + r)$; rank $(\mathbf{A})$ is

(i) 3, if $p \neq q \neq r$ and $p + q + r \neq 0$;

(ii) 2, if $p \neq q \neq r$ and $p + q + r = 0$,

(iii) 2, if exactly two of p, q and r are identical;

(iv) 1, if $p = q = r$.

30. (a) 2; (b) $|\mathbf{A}| = (a_1a_2 + b_1b_2 + c_1c_2)^2$, rank $(\mathbf{A})$ is

(i) 4, if $a_1a_2 + b_1b_2 + c_1c_2 \neq 0$;

(ii) 2, if $a_1a_2 + b_1b_2 + c_1c_2 = 0$, since all determinants of third order have the value zero.

32. Consider $(\mathbf{I} + \mathbf{A})(\mathbf{I} - \mathbf{A} + \mathbf{A}^2 - \ldots + (-1)^{n-1}\mathbf{A}^{n-1}) = \mathbf{I} + (-1)^{n-1}\mathbf{A}^n$. In the limit $n \to \infty$, $\mathbf{A}^n \to 0$. Therefore, $(\mathbf{I} + \mathbf{A})(\mathbf{I} - \mathbf{A} + \mathbf{A}^2 - \ldots) = \mathbf{I}$.

33. (i) $Trace\ (\alpha\mathbf{A} + \beta\mathbf{B}) = \alpha\sum_{i=1}^{n} a_{ii} + \beta\sum_{i=1}^{n} b_{ii} = \alpha\ Trace\ (\mathbf{A}) + \beta\ Trace\ (\mathbf{B})$,

(ii) $Trace\ (\mathbf{AB}) = \sum_{i=1}^{n}\sum_{k=1}^{n} a_{ik}b_{ki} = \sum_{i=1}^{n}\sum_{m=1}^{n} b_{im}a_{mi} = Trace\ (\mathbf{BA})$,

(iii) If the result is true, then Trace $(\mathbf{AB} - \mathbf{BA}) = $ Trace $(\mathbf{I})$ which gives $0 = n$ which is not possible.

34. Result is true for $p = 0$ and 1. Let it be true for $p = k$ and show that it is true for $p = k + 1$. Note that when $\mathbf{BC} = \mathbf{CB}$ and $\mathbf{C}^2 = 0$, we have $\mathbf{CB}^{k+1} = \mathbf{B}^{k+1}\mathbf{C}$ and $\mathbf{CB}^k\mathbf{C} = 0$.

35. Apply the operation $C_1 \leftarrow C_1 + C_2 + \ldots + C_n$ and then the operation $R_i \leftarrow R_i - R_1$, $i = 2, 3, \ldots, n$.

36. None.　　　　37. Symmetric.　　　　38. Skew-symmetric.

39. Hermitian.　　　40. None.　　　　41. Skew-Hermitian.

42. None.　　　　43. Skew–Hermitian.　　44. Hermitian.

45. None.

Exercise 3.2

1. Yes.　　　　　2. No, 1, 4, 5, 6.　　　3. No, 1, 4, 5, 6.

4. No, when the scalar α is irrational, Property 6 is not satisfied. If the field of scalars is taken only as rationals, then it defines a veactor space.

5. Yes, since $1 + x = 1x = x = \mathbf{x}$ and $x + 1 = x1 = x = \mathbf{x}$, the zero vector $\mathbf{0}$ is $1 = 1$. Define $-\mathbf{x} = 1/x$. Then, $\mathbf{x} + (-\mathbf{x}) = x(1/x) = 1 = 1 = \mathbf{0}$. Therefore, negative vector is its reciprocal.

6. No, 8, 10.　　　　　　　　　　　　7. No, 2, 3, 8, 10.

8. Yes (same arguments as in Problem 5). $(\alpha + \beta)x = x^{\alpha+\beta} = x^{\alpha}x^{\beta} = x^{\alpha} + x^{\beta} = \alpha x + \beta x$.

9. (i)　　Yes,　　(ii)　　No, 1, 6.

10. (i) Yes,　　　　　　　　　　　　(ii) No, 1, 4, 6.

11. (i) Yes,　　　　　　　　　　　　(ii) No, when $\mathbf{x}, \mathbf{y} \in W$, $\mathbf{x} + \mathbf{y} \notin W$,

(iii) No, when $\mathbf{x}, \mathbf{y} \in W$, $\mathbf{x} + \mathbf{y} \notin W$,　　(iv) Yes.

12. (i) No, when $\mathbf{A} \in W$, $\alpha\mathbf{A} \notin W$ for α negative,

(ii) No, sum of two non-singular matrices need not be non-singular,

(iii) Yes,

(iv) No, $\alpha\mathbf{A}$ and $\mathbf{A} + \mathbf{B}$ need not belong to W, $(\mathbf{A} = \mathbf{I}, \mathbf{A}^2 = \mathbf{I} = \mathbf{A}$ but $2\mathbf{A} \neq (2\mathbf{A})^2)$.

13. (i) Yes,　　　　　　　　　　　　(ii) No; let $\alpha = i$, Then, $\alpha\mathbf{A} = i\mathbf{A} \notin W$.

14. (i) No; for $P, Q \in W, P + Q \notin W$,   (ii) Yes.

  (iii) No; for $P, Q \in W, \alpha P \notin W$ and also $P + Q \notin W$,

  (iv) No, for $P, Q \in W$ having real roots, $P + Q$ need not have real roots. For example, take
    $P = 2t^2 - 1, Q = -t^2 + 3$.

15. (i) Yes,

  (ii) No, $\mathbf{x}, \mathbf{y} \in W, \mathbf{x} + \mathbf{y} \notin W$. For example, if $\mathbf{x} = (x_1, x_1, x_1 - 1), \mathbf{y} = (y_1, y_1, y_1 - 1)$;
    $\mathbf{x} + \mathbf{y} = (x_1 + y_1, x_1 + y_1, x_1 + y_1 - 2) \notin W$,

  (iii) No, $\mathbf{x} \in W, \alpha \mathbf{x} \notin W$, for α negative,

  (iv) No, $\mathbf{x} \in W, \alpha \mathbf{x} \notin W$,   (v) No, $\mathbf{x} \in W, \alpha \mathbf{x} \notin W$, (for α a rational number).

16. (i) $\mathbf{u} + 2\mathbf{v} - \mathbf{w}$,   (ii) $2\mathbf{u} + \mathbf{v} - \mathbf{w}$,

  (iii) $(-33\mathbf{u} - 11\mathbf{v} + 23\mathbf{w})/16$.

17. (i) $\mathbf{u} - 2\mathbf{v} + 2\mathbf{w}$,   (ii) $3\mathbf{u} + \mathbf{v} - \mathbf{w}$,   (iii) not possible.

18. (i) $3P_1(t) - 2P_2(t) - P_3(t)$,   (ii) $4P_1(t) - P_2(t) + 3P_3(t)$.

19. Let $S = \{\mathbf{u}, \mathbf{v}, \mathbf{w}\}$. Then, $\mathbf{x} = (a, b, c)^T = \alpha\mathbf{u} + \beta\mathbf{v} + \gamma\mathbf{w}$, where $\alpha = (a + b)/2, \beta = (a - b)/2$ and $\gamma = c$.

20. Let $S = \{\mathbf{A}, \mathbf{B}, \mathbf{C}, \mathbf{D}\}$. Then, $\mathbf{E} = \begin{pmatrix} a & b \\ c & d \end{pmatrix} = \alpha\mathbf{A} + \beta\mathbf{B} + \gamma\mathbf{C} + \delta\mathbf{D}$, where $\alpha = (-a - b + 2c - 2d)/3$,

  $\beta = (5a + 2b - 4c + 4d)/3, \gamma = (-4a - b + 5c - 2d)/3$ and $\delta = (-2a + b + c - d)/3$.

21. (i) independent,   (ii) dependent,   (iii) dependent,
  (iv) independent,   (v) dependent,

22. (i) independent,   (ii) dependent,   (iii) dependent,
  (iv) independent,   (v) independent,

24. $(-4, 7, 9) = (1, 2, 3) + 2(-1, 3, 4) - (3, 1, 2)$. The vectors in S are linearly dependent.

25. $t^2 + t + 1 = [-t + (t^2 - 1) + 2(t^2 + 2t + 2)]/3$. The elements in S are linearly independent.

26. (i) dimension: 2, a basis : $\{(1, 0, 0, -1), (0, 1, -1, 0)\}$,

  (ii) dimension: 3, a basis: $\{(1, 0, 0, 1), (0, 1, 0, 1), (0, 0, 1, 1)\}$,

  (iii) dimension: 3, a basis: $\{(1, 0, 0, 0), (0, 0, 1, 0), (0, 0, 0, 1)\}$,

  (iv) dimension: 1, a basis: $\{(1, 1, 1, 1)\}$.

27. The given vectors must be linearly independent.
    (i) $k \neq 0, 1 - 4/3$,   (ii) $k \neq 0$.   (iii) $k \neq 0$,   (iv) $k \neq -8$.

28. (i) dimension: 4, basis: $\{\mathbf{E}_{11}, \mathbf{E}_{12}, \mathbf{E}_{21}, \mathbf{E}_{22}\}$ where $\mathbf{E}_{rs}$ is the standard basis of order 2,

  (ii) dimension: 3, basis: $\left\{ \begin{pmatrix} 1 & 0 \\ 0 & 0 \end{pmatrix}, \begin{pmatrix} 0 & 1 \\ 1 & 0 \end{pmatrix}, \begin{pmatrix} 0 & 0 \\ 0 & 1 \end{pmatrix} \right\}$,

  (iii) dimension: 1, basis: $\left\{ \begin{pmatrix} 0 & 1 \\ -1 & 0 \end{pmatrix} \right\}$,

(iv) a 2×2 skew-Hermitian matrix (diagonal elements are 0 or pure imaginary) is given by

$$\mathbf{A} = \begin{pmatrix} ia_1 & b_1 + ib_2 \\ -b_1 + ib_2 & ia_2 \end{pmatrix} = \begin{pmatrix} 0 & b_1 \\ -b_1 & 0 \end{pmatrix} + i \begin{pmatrix} a_1 & b_2 \\ b_2 & a_2 \end{pmatrix} = \mathbf{B} + i\mathbf{C}$$

where $\mathbf{B}$ is a skew-symmetric and $\mathbf{C}$ is a symmetric matrix,

dimension: 4, basis: $\left\{ \begin{pmatrix} 0 & 1 \\ -1 & 0 \end{pmatrix}, \begin{pmatrix} 1 & 0 \\ 0 & 0 \end{pmatrix}, \begin{pmatrix} 0 & 1 \\ 1 & 0 \end{pmatrix}, \begin{pmatrix} 0 & 0 \\ 0 & 1 \end{pmatrix} \right\}$,

(v) dimension: 3, basis: $\left\{ \begin{pmatrix} 0 & 1 \\ 0 & 0 \end{pmatrix}, \begin{pmatrix} 0 & 0 \\ 1 & 0 \end{pmatrix}, \begin{pmatrix} 1 & 0 \\ 0 & -1 \end{pmatrix} \right\}$.

(vi) dimension: 3, basis: $\left\{ \begin{pmatrix} 1 & -1 \\ 0 & 0 \end{pmatrix}, \begin{pmatrix} 0 & 0 \\ 1 & 0 \end{pmatrix}, \begin{pmatrix} 0 & 0 \\ 0 & 1 \end{pmatrix} \right\}$.

29. (i) dimension: 3, basis: $\{\mathbf{E}_{11}, \mathbf{E}_{22}, \mathbf{E}_{33}\}$,

(ii) dimension: 6, basis: $\{\mathbf{E}_{11}, \mathbf{E}_{12}, \mathbf{E}_{13}\ \mathbf{E}_{22}, \mathbf{E}_{23}, \mathbf{E}_{33}\}$,

(iii) dimension: 6, basis: $\{\mathbf{E}_{11}, \mathbf{E}_{21}, \mathbf{E}_{22}\ \mathbf{E}_{31}, \mathbf{E}_{32}, \mathbf{E}_{33}\}$,

where $\mathbf{E}_{rs}$ is the standard basis of order 3.

30. (i) n^2, 　　　　(ii) n, 　　　　(iii) $n(n+1)/2$, 　　　　(iv) $n(n-1)/2$.

31. Not linear, $T(\mathbf{x}) + T(\mathbf{y}) \neq T(\mathbf{x} + \mathbf{y})$. 　　　　**32.** Linear.

33. Not linear, $T(\mathbf{x}) + T(\mathbf{y}) \neq T(\mathbf{x} + \mathbf{y})$.

34. Not linear, $T(1, 0) = 3$, $T(0, 1) = 2$, $T(1, 1) = 0 \neq T(1, 0) + T(0, 1)$.

35. Not linear, $T(\mathbf{x}) + T(\mathbf{y}) \neq T(\mathbf{x} + \mathbf{y})$.

36. $ker(T) = (0, 0, 0)^T$, $ran(T) = x(1, 0, 1)^T + y(1, 0, -1)^T + z(0, 1, 0)^T$.

$dim(ker(T)) = 0$, $dim(ran(T)) = 3$.

37. $ker(T) = (0, 0)^T$, $ran(T) = x(2, -1, 3)^T + y(1, 1, 4)^T$. $dim(ker(T)) = 0$, $dim(ran(T)) = 2$.

38. $ker(T) = w(1, -2, 0, 1)^T$,

$ran(T) = x(1, 0, 0)^T + y(1, 0, 1)^T + z(0, 1, 0)^T + w(1, 0, 2)^T$

$= r(1, 0, 0)^T + s(1, 0, 1)^T + z(0, 1, 0)^T$,

where $r = x - w$, $s = y + 2w$. $dim(ker(T)) = 1$, $dim(ran(T)) = 3$.

39. $ker(T) = x(-3, 1)^T$, $ran(T) = $ real number. $dim(ker(T)) = 1$, $dim(ran(T)) = 1$.

40. $ker(T) = x(1, -3, 0)^T + z(0, 0, 1)^T$, $ran(T) = $ real number. $dim(ker(T)) = 2$,

$dim(ran(T)) = 1$.

41. $ker(T) = x(1, 1)^T$, $ran(T) = x(1, 1)^T - y(1, 1)^T = r(1, 1)^T$, where $r = x - y$.

$dim(ker(T)) = 1$, $dim(ran(T)) = 1$.

42. $ker(T) = x(1, 2, -3)^T$, $ran(T) = x(2, 3)^T + y(-1, 0)^T + z(0, 1)^T$ or $ran(T) = r(-1, 0)^T + s(0, 1)^T$, where $r = y + 2x$, $s = z + 3x$, $dim(ker(T)) = 1$, $dim(ran(T)) = 2$.

43. $A = \begin{bmatrix} -5 & -8 & -7 \\ 3 & 5 & 4 \end{bmatrix}$.

44. $A = \begin{bmatrix} 0 & 0 & 1/2 \\ 0 & 0 & -1/2 \\ 1 & 1 & 1/2 \end{bmatrix}$.

45. $A = \begin{bmatrix} -1/2 & -1/2 & -3/2 \\ -1/2 & -3/2 & -1/2 \\ 0 & -1 & -1 \end{bmatrix}$.

46. $\begin{bmatrix} 1 & 0 & 1 \\ 1 & 1 & 0 \\ 0 & 1 & 1 \\ 0 & 0 & 0 \end{bmatrix}$.

47. We have $T[\mathbf{v}_1, \mathbf{v}_2] = [\mathbf{w}_1, \mathbf{w}_2, \mathbf{w}_3]\,A = \begin{bmatrix} -1 & 1 & 1 \\ 1 & -1 & 1 \\ 1 & 1 & -1 \end{bmatrix}\begin{bmatrix} 1 & 2 \\ 2 & 3 \\ 3 & 4 \end{bmatrix} = \begin{bmatrix} 4 & 5 \\ 2 & 3 \\ 0 & 1 \end{bmatrix}$.

Now, any vector $\mathbf{x} = (x_1, x_2)^T$ in IR^2 with respect to the given basis can be written as

$$\begin{bmatrix} x_1 \\ x_2 \end{bmatrix} = \alpha \begin{bmatrix} 1 \\ 2 \end{bmatrix} + \beta \begin{bmatrix} 3 \\ 4 \end{bmatrix}.$$

We obtain $\alpha = (-4x_1 + 3x_2)/2$, $\beta = (2x_1 - x_2)/2$. Hence, we have

$$T\mathbf{x} = \alpha\,T\mathbf{v}_1 + \beta T\mathbf{v}_2 = \alpha \begin{bmatrix} 4 \\ 2 \\ 0 \end{bmatrix} + \beta \begin{bmatrix} 5 \\ 3 \\ 1 \end{bmatrix} = \begin{bmatrix} 4\alpha + 5\beta \\ 2\alpha + 3\beta \\ \beta \end{bmatrix} = \frac{1}{2}\begin{bmatrix} -6x_1 + 7x_2 \\ -2x_1 + 3x_2 \\ 2x_1 - x_2 \end{bmatrix}.$$

48. $T\mathbf{x} = \begin{bmatrix} -x_1 + 2x_2 + 8x_3 \\ -2x_1 + 3x_2 + 12x_3 \end{bmatrix}$.

49. $TP_1(t) = (4x_2 - 5x_1) + 7(x_2 - x_1)t + (2x_1 - x_2)t^2$.

50. (i) Two degrees of freedom, dimension is 2, a basis is $\{[3, 1, 0], [-2, 0, 1]\}$.

    (ii) One degree of freedom, dimension is 1, a basis is $\{(-5, 4, 23)\}$.

Exercise 3.3

1. 3.   2. 2.   3. 3.   4. 2.

5. 2.   6. 2.   7. 2.   8 3.

9. 4.   10. 2.   11. 2.   12. 3.

13. 2.   14. 2.   15. 2.

16. Independent, 3.   17. Independent, 3.   18. Dependent, 3.

19. Independent, 3.   20. Dependent, 2.   21. Dependent, 3.

22. Dependent, 2.   23. Dependent, 2.   24. Independent, 4.

25. Dependent, 2.   26. $[1, 2, 2]$.

27. $[1 + \alpha, -2\alpha, \alpha]$, α arbitrary.   28. Inconsistent.

29. $[1, 1, 1]$.   30. $[1, 3, 3]$.   31. $[3/2, 3/2, 1]$.   32. $[-1, -1/2, 3/4]$.

33. $[(5 + \alpha - 4\beta)/3, (1 + 2\alpha + \beta)/3, \alpha, \beta]$, α, β arbitrary.

34. $[2 - \alpha, 1, \alpha, 1]$, α arbitrary.

35. $[-1/4, 1/4, 1/4, 1/4]$.

36. $[-\alpha, \alpha, \alpha]$, α arbitrary.

37. $[-15\alpha, 13\alpha, \alpha]$, α arbitrary.

38. $[0, 0, 0]$.

39. $[-2\alpha/3, 7\alpha/3, -8\alpha/3, \alpha]$, α arbitrary.

40. $[2(\beta - \alpha)/3, -(5\beta + \alpha)/3, \beta, \alpha]$, α, β arbitary.

41. $[0, 0, 0, 0]$.

42. $[(2\beta - 5\alpha)/4, -(10\beta + \alpha)/4, \beta, \alpha]$, α, β arbitrary.

43. $[(\alpha + 5\beta)/3, (4\beta - 7\alpha)/3, \beta, \alpha]$, α, β arbitrary.

44. $[(3\beta - 5\alpha)/3, (3\beta - 4\alpha)/3, \beta, \alpha]$, α, β arbitrary.

45. $[\alpha - 3\beta, 5\beta, \beta, \alpha,]$ α, β arbitrary.

46. $\begin{bmatrix} 1 & 1 & -1 \\ 1 & -3 & 2 \\ -1 & 2 & -1 \end{bmatrix}$.

47. $\begin{bmatrix} 3 & -5 & 6 \\ -2 & 4 & -5 \\ 1 & -2 & 3 \end{bmatrix}$.

48. $\dfrac{1}{4}\begin{bmatrix} -1 & 1 & 1 & 1 \\ 1 & -1 & 1 & 1 \\ 1 & 1 & -1 & 1 \\ 1 & 1 & 1 & -1 \end{bmatrix}$.

49. $\dfrac{1}{2}\begin{bmatrix} 1 & -1 & -1 & -1 \\ 0 & 1 & 0 & 1 \\ 1 & 0 & 0 & -1 \\ 0 & 0 & 1 & 1 \end{bmatrix}$.

50. $\begin{bmatrix} -1 & -1/3 & 1/3 & 1 \\ 1 & 0 & 0 & -1 \\ -1 & 1 & 0 & 0 \\ 1 & 1/3 & -1/3 & 0 \end{bmatrix}$.

Exercise 3.4

1. $P(\lambda) = \lambda^3 - 9\lambda^2 - 9\lambda + 81 = 0$; $\mathbf{A}^{-1} = \dfrac{1}{81}\begin{bmatrix} 1 & 16 & -20 \\ 16 & 13 & 4 \\ -20 & 4 & -5 \end{bmatrix}$.

2. $P(\lambda) = \lambda^3 - 8\lambda^2 + 20\lambda - 16 = 0$; $\mathbf{A}^{-1} = \dfrac{1}{16}\begin{bmatrix} 6 & -4 & -2 \\ 0 & 8 & 0 \\ -2 & -4 & 6 \end{bmatrix}$.

3. $P(\lambda) = \lambda^3 - 3\lambda^2 + 2\lambda = 0$; Inverse does not exist.

4. $P(\lambda) = \lambda^3 - \lambda^2 - 4\lambda + 4 = 0$; $\mathbf{A}^{-1} = \dfrac{1}{4}\begin{bmatrix} 4 & 4 & -4 \\ -2 & -1 & 3 \\ -2 & 1 & 1 \end{bmatrix}$.

5. $P(\lambda) = \lambda^3 - 5\lambda^2 + 9\lambda - 13 = 0$; $\mathbf{A}^{-1} = \dfrac{1}{13}\begin{bmatrix} 2 & -3 & -7 \\ 1 & 5 & 3 \\ 5 & -1 & 2 \end{bmatrix}$.

6. $P(\lambda) = \lambda^3 - 3\lambda^2 + 6\lambda - 4 + 2i = 0$; $\mathbf{A}^{-1} = -\dfrac{1+3i}{10}\begin{bmatrix} i-1 & 1 & 1 \\ 1 & i-1 & 1 \\ 1 & 1 & i-1 \end{bmatrix}$.

7. $\lambda = 1$: $(1, 1, -1)^T$; $\lambda = 2, 2$: $(2, 1, 0)^T$; not diagonalizable.

8. $\lambda = -1$: $(0, -1, 1)^T$; $\lambda = i$: $(1 + i, 1, 1)^T$; $\lambda = -i$: $(1 - i, 1, 1)^T$; diagonlizable.

9. $\lambda = 1, 1, 1$: $[0, 3, -2]^T$; not diagonlizable.

10. $\lambda = 1, 1$: $[0, 1, -1]^T$; $\lambda = 7$: $(6, 7, 5)^T$; not diagonalizable.

11. $\lambda = 0$: $[i, 0, -1]^T$; $\lambda = 1 + \sqrt{3}$: $[1, \sqrt{3} - 1, -i]^T$;
$\lambda = 1 - \sqrt{3}$: $[1, -(\sqrt{3} + 1), -i]^T$; diagonalizable.

12. $\lambda = -i, -i$: $[1, 0, -1]^T$, $[1, -1, 0]^T$; $\lambda = 2i$: $[1, 1, 1]^T$; diagonalizable.

13. $\lambda = 0, 0, 0, 0$: $[1, 0, 0, 0]^T$; not diagonlizable.

14. $\lambda = 0, 0$: $[1, 0, 0, 1]^T$, $[1, -1, -1, 0]^T$; $\lambda = 2$: $[1, 1, 0, 0]^T$;
$\lambda = -2$: $[1, 0, 1, 1]^T$; diagonalizable.

15. $\lambda = -1, -1, -1$: $[1, -1, 0, 0]^T$, $[1, 0, -1, 0]^T$, $[1, 0, 0, -1]^T$;
$\lambda = 3$: $[1, 1, 1, 1]^T$; diagonalizable.

16. $\lambda = -4$: $[1, 1, -1, -1]^T$; $\lambda = 10$: $[1, 1, 1, 1]^T$; $\lambda = \sqrt{2}$: $[\sqrt{2} - 1, 1 - \sqrt{2}, -1, 1]^T$,
$\lambda = -\sqrt{2}$: $[-(1 + \sqrt{2}), 1 + \sqrt{2}, -1, 1]^T$; diagonalizable.

17. $\lambda = -1, -1$: $[1, 0, 0, 0, -1]^T$, $[0, 1, 0, -1, 0]^T$; $\lambda = 1, 1, 1$: $[1, 0, 0, 0, 1]^T$, $[0, 1, 0, 1, 0]^T$,
$[0, 0, 1, 0, 0]^T$; diagonlizable.

18. $\lambda = 1, w, w^2, w^3, w^4$, w is fifth root of unity. Let $\xi_j = w^j$, $j = 0, 1, 2, 3, 4$.
$\lambda = \xi_j$: $[1, \xi_j, \xi_j^2, \xi_j^3, \xi_j^4]^T$, $j = 0, 1, 2, 3, 4$; diagonalizable.

19. $\lambda = 2, 2$: $[1, 0, -1]^T$, $[-2, 1, 0]^T$; $\lambda = 4$: $[1, 0, 1]^T$.

$$\mathbf{P} = \begin{bmatrix} 1 & -2 & 1 \\ 0 & 1 & 0 \\ -1 & 0 & 1 \end{bmatrix}; \mathbf{P}^{-1} = \frac{1}{2}\begin{bmatrix} 1 & 2 & -1 \\ 0 & 2 & 0 \\ 1 & 2 & 1 \end{bmatrix}.$$

20. $\lambda = 1$: $[1, -2, 0]^T$; $\lambda = -1$: $[3, -2, 2]^T$; $\lambda = 2$: $[-1, 3, 1]^T$.

$$\mathbf{P} = \begin{bmatrix} 1 & 3 & -1 \\ -2 & -2 & 3 \\ 0 & 2 & 1 \end{bmatrix}; \mathbf{P}^{-1} = \frac{1}{2}\begin{bmatrix} -8 & -5 & 7 \\ 2 & 1 & -1 \\ -4 & -2 & 4 \end{bmatrix}.$$

21. $\lambda = 0$: $[3, 1, -2]^T$; $\lambda = 2i$: $[3 + i, 1 + 3i, -4]^T$; $\lambda = -2i$: $[3 - i, 1 - 3i, -4]^T$.

$$\mathbf{P} = \begin{bmatrix} 3 & 3+i & 3-i \\ 1 & 1+3i & 1-3i \\ -2 & -4 & -4 \end{bmatrix}; \mathbf{P}^{-1} = \frac{1}{32}\begin{bmatrix} 24 & -8 & 16 \\ 2i-6 & 2-6i & -8 \\ -2i-6 & 2+6i & -8 \end{bmatrix}.$$

22. $\lambda = 0$: $[1, 0, -1]^T$; $\lambda = 1$: $[-1, -1, 1]^T$; $\lambda = 2$: $[1, 1, 0]^T$.

$$\mathbf{P} = \begin{bmatrix} 1 & -1 & 1 \\ 0 & -1 & 1 \\ -1 & 1 & 0 \end{bmatrix}; \mathbf{P}^{-1} = \begin{bmatrix} 1 & -1 & 0 \\ 1 & -1 & 1 \\ 1 & 0 & 1 \end{bmatrix}.$$

23. $\lambda = 1$: $[3, -1, 3]^T$; $\lambda = 2, 2$: $[2, 0, 1]^T$, $[2, 1, 0]^T$.

$$\mathbf{P} = \begin{bmatrix} 3 & 2 & 2 \\ -1 & 0 & 1 \\ 3 & 1 & 0 \end{bmatrix}; \mathbf{P}^{-1} = \begin{bmatrix} -1 & 2 & 2 \\ 3 & -6 & -5 \\ -1 & 3 & 2 \end{bmatrix}.$$

24. $\lambda = 1$: $[1, -1, -1]^T$; $\lambda = 2$: $[0, 1, 1]^T$; $\lambda = -2$: $[8 - 5, 7]^T$.

$$\mathbf{P} = \begin{bmatrix} 1 & 0 & 8 \\ -1 & 1 & -5 \\ -1 & 1 & 7 \end{bmatrix}; \mathbf{P}^{-1} = \frac{1}{12}\begin{bmatrix} 12 & 8 & -8 \\ 12 & 15 & -3 \\ 0 & -1 & 1 \end{bmatrix}.$$

25. $$\mathbf{P} = \begin{bmatrix} -2 & -1 & 1 \\ 1 & 0 & 0 \\ 0 & 1 & 1 \end{bmatrix}; \mathbf{P}^{-1} = \frac{1}{2}\begin{bmatrix} 0 & 2 & 0 \\ -1 & -2 & 1 \\ 1 & 2 & 1 \end{bmatrix}; \mathbf{A} = \mathbf{PDP}^{-1} = \begin{bmatrix} 3 & 2 & 1 \\ 0 & 2 & 0 \\ 1 & 2 & 3 \end{bmatrix}.$$

26. $$\mathbf{P} = \begin{bmatrix} 1 & 1 & 3 \\ 1 & 0 & 1 \\ 0 & 1 & 1 \end{bmatrix}; \mathbf{P}^{-1} = \begin{bmatrix} -1 & 2 & 1 \\ -1 & 1 & 2 \\ 1 & -1 & -1 \end{bmatrix}; \mathbf{A} = \mathbf{PDP}^{-1} = \begin{bmatrix} 6 & -5 & -7 \\ 1 & 0 & -1 \\ 3 & -3 & -4 \end{bmatrix}.$$

27. $$\mathbf{P} = \begin{bmatrix} 1 & 2 & 1 \\ 2 & 3 & 4 \\ 1 & 4 & 9 \end{bmatrix}; \mathbf{P}^{-1} = \frac{1}{12}\begin{bmatrix} -11 & 14 & -5 \\ 14 & -8 & 2 \\ -5 & 2 & 1 \end{bmatrix}; \mathbf{A} = \mathbf{PDP}^{-1} = \frac{1}{12}\begin{bmatrix} 30 & -12 & 6 \\ 2 & 4 & 14 \\ -34 & 4 & 38 \end{bmatrix}.$$

28. $$\mathbf{P} = \begin{bmatrix} -1 & 1 & 1 \\ 1 & -1 & 1 \\ 1 & 1 & -1 \end{bmatrix}; \mathbf{P}^{-1} = \frac{1}{4}\begin{bmatrix} 0 & 2 & 2 \\ 2 & 0 & 2 \\ 2 & 2 & 0 \end{bmatrix}; \mathbf{A} = \mathbf{PDP}^{-1} = \begin{bmatrix} 1 & 0 & 0 \\ 0 & 1 & 0 \\ 0 & 0 & 1 \end{bmatrix}.$$

29. $$\mathbf{P} = \begin{bmatrix} -1 & 1 & 1 \\ 1 & 0 & 1 \\ 0 & -1 & 1 \end{bmatrix}; \mathbf{P}^{-1} = \frac{1}{3}\begin{bmatrix} -1 & 2 & -1 \\ 1 & 1 & -2 \\ 1 & 1 & 1 \end{bmatrix}; \mathbf{A} = \mathbf{PDP}^{-1} = \frac{1}{3}\begin{bmatrix} 0 & 0 & 3 \\ 1 & 1 & 1 \\ 2 & 2 & -1 \end{bmatrix}.$$

30. $$\mathbf{P} = \begin{bmatrix} 1 & -2 & 3 \\ 2 & 1 & 0 \\ -1 & 0 & 1 \end{bmatrix}; \mathbf{P}^{-1} = \frac{1}{8}\begin{bmatrix} 1 & 2 & -3 \\ -2 & 4 & 6 \\ 1 & 2 & 5 \end{bmatrix}; \mathbf{A} = \mathbf{PDP}^{-1} = \frac{1}{8}\begin{bmatrix} 9 & 18 & 45 \\ 0 & 0 & 0 \\ 3 & 6 & 15 \end{bmatrix}.$$

31. Eigenvalues of $\mathbf{B}$ are $2\lambda_j + (1/\lambda_j) - 1$, $j = 1, 2, 3, 4$ or $2, -4, 7/2, -11/2$. $|\mathbf{B}|$ = product of eigenvalues of $\mathbf{B}$ = 154.

32. Eigenvalues of $\mathbf{B}$ are $\lambda_j + \lambda_j^2 - (1/\lambda_j)$, $j = 1, 2, 3$, or $1, 11/2, 1$. Trace of $\mathbf{B}$ = sum of eigenvalues of $\mathbf{B}$ = 15/2.

33. Premultiply $\mathbf{Ax} = \lambda\mathbf{x}$ by $\mathbf{P}^{-1}$ and substitute $\mathbf{x} = \mathbf{Py}$.

34. Let λ be an eigenvalue and $\mathbf{x}$ be the corresponding eigenvector of $\mathbf{AB}$, that is $\mathbf{ABx} = \lambda\mathbf{x}$. Pre-multiply by $\mathbf{A}^{-1}$ and substitute $\mathbf{x} = \mathbf{Ay}$. We get $\mathbf{BAy} = \lambda\mathbf{y}$. Therefore, λ is also an eigenvalue of $\mathbf{BA}$ and eigenvectors are related by $\mathbf{x} = \mathbf{Ay}$.

35. Let λ be an eigenvalue and $\mathbf{x}$ be the corresponding eigenvector of $\mathbf{A}^{-1}\mathbf{B}$, that is $\mathbf{A}^{-1}\mathbf{Bx} = \lambda\mathbf{x}$. Premultiply by $\mathbf{A}$ and set $\mathbf{x} = \mathbf{A}^{-1}\mathbf{y}$. We obtain $\mathbf{BA}^{-1}\mathbf{y} = \lambda\mathbf{y}$. Therefore, λ is also an eigenvalue of $\mathbf{BA}^{-1}$ with the corresponding eigenvector $\mathbf{y} = \mathbf{Ax}$.

36. From $\mathbf{Ax} = \lambda\mathbf{x}$, we obtain $\mathbf{A}^k\mathbf{x} = \lambda^k\mathbf{x} = \mathbf{0}$. Therefore, $\lambda^k = 0$ or $\lambda = 0$, since $\mathbf{x} \neq \mathbf{0}$.

37. Since $\mathbf{A}$ is a diagonalizable matrix, there exists a non-singular matrix $\mathbf{P}$ such that $\mathbf{P}^{-1}\mathbf{AP} = \mathbf{D}$ and the eigenvalues of $\mathbf{A}$ and $\mathbf{D}$ are same. We have $\mathbf{P}^{-1}\mathbf{A}^2\mathbf{P} = \mathbf{D}^2$. Since $\mathbf{A}^2 = \mathbf{A}$, we get $\mathbf{P}^{-1}\mathbf{AP} = \mathbf{D}^2$. Therefore, we obtain $\mathbf{D}^2 - \mathbf{D} = \mathbf{0}$. Thus $\mathbf{D} = \mathbf{0}$ or $\mathbf{D} = \mathbf{I}$. Hence, the eigenvalues of $\mathbf{A}$ are 0 or 1.

38. Simplify the right hand side and set the off-diagonal element to zero.

39. Since $\mathbf{A}$ and $\mathbf{B}$ are similar, we have $\mathbf{A} = \mathbf{P}^{-1}\mathbf{BP}$. From this equation, show that $\mathbf{A}^{-1} = \mathbf{P}^{-1}\mathbf{B}^{-1}\mathbf{P}$ and $\mathbf{A}^m = \mathbf{P}^{-1}\mathbf{B}^m\mathbf{P}$. Also $|\mathbf{A}| = |\mathbf{P}^{-1}||\mathbf{B}||\mathbf{P}| = |\mathbf{B}|$.

40. We have $\mathbf{A} = \mathbf{A}^T$ and $\mathbf{B} = \mathbf{B}^T$. Therefore, $(\mathbf{AB})^T = \mathbf{B}^T\mathbf{A}^T = \mathbf{BA}$.

41. $(\mathbf{A}^T\mathbf{A})^T = \mathbf{A}^T\mathbf{A}$.

42. Let $\mathbf{A}^T\mathbf{A}^{-1}$ be a symmetric matrix. We have $(\mathbf{A}^T\mathbf{A}^{-1})^T = (\mathbf{A}^{-1})^T\mathbf{A} = \mathbf{A}^T\mathbf{A}^{-1}$, or $(\mathbf{A}^{-1})^T\mathbf{A}^2 = \mathbf{A}^T$ or $\mathbf{A}^2 = (\mathbf{A}^T)^2$. Now, let $\mathbf{A}^2 = (\mathbf{A}^T)^2$. We have $\mathbf{AA} = \mathbf{A}^T\mathbf{A}^T \Rightarrow \mathbf{A} = \mathbf{A}^{-1}\mathbf{A}^T\mathbf{A}^T \Rightarrow \mathbf{A}(\mathbf{A}^T)^{-1} = \mathbf{A}^{-1}\mathbf{A}^T$, or $\mathbf{A}(\mathbf{A}^{-1})^T = (\mathbf{A}^{-1}\mathbf{A}^T)^T = \mathbf{A}^{-1}\mathbf{A}^T$. Therefore, $\mathbf{A}^T\mathbf{A}^{-1}$ is symmetric.

43. Since $\mathbf{A}$ is symmetric, we have

    $\mathbf{I} = \mathbf{A}^{-1}\mathbf{A} = \mathbf{A}^{-1}\mathbf{A}^T = (\mathbf{PDP}^{-1})^{-1}(\mathbf{PDP}^{-1})^T = (\mathbf{PD}^{-1}\mathbf{P}^{-1})[(\mathbf{P}^{-1})^T\mathbf{DP}^T\mathbf{j}$, since $\mathbf{D}^T = \mathbf{D}$. This result is true only when $\mathbf{P}^{-1}(\mathbf{P}^{-1})^T = \mathbf{I}$, or $\mathbf{P}^{-1} = \mathbf{P}^T$.

44. Let $\mathbf{A}$ and $\mathbf{B}$ be the orthogonal matrices, that is $\mathbf{A}^{-1} = \mathbf{A}^T$ and $\mathbf{B}^{-1} = \mathbf{B}^T$. Then $(\mathbf{AB})^T = \mathbf{B}^T\mathbf{A}^T = \mathbf{B}^{-1}\mathbf{A}^{-1} = (\mathbf{AB})^{-1}$.

45. $\mathbf{A}^{-1} = \mathbf{A}^T$ gives $\mathbf{AA}^T = \mathbf{I}$. We obtain conditions as $l_i^2 + m_i^2 + n_i^2 = 1$, $i = 1, 2, 3$ and $l_1l_2 + m_1m_2 + n_1n_2 = 0$, $l_1l_3 + m_1m_3 + n_1n_3 = 0$, $l_2l_3 + m_2m_3 + n_2n_3 = 0$.

46. Since $\mathbf{A}$ is an orthogonal matrix, we have $\mathbf{A}^{-1} = \mathbf{A}^T$. Hence, $|\mathbf{A}^{-1}| = |\mathbf{A}^T| = |\mathbf{A}|$ or $1/|\mathbf{A}| = |\mathbf{A}| \Rightarrow |\mathbf{A}|^2 = 1$ or $|\mathbf{A}| = \pm 1$.

47. Let λ and μ be two distinct eigenvalues and $\mathbf{x}$, $\mathbf{y}$ be the corresponding eigenvectors. We have $\mathbf{Ax} = \lambda\mathbf{x}$ and $\mathbf{Ay} = \mu\mathbf{y}$. From the first equation, we get $\mathbf{x}^T\mathbf{A}^T = \lambda\mathbf{x}^T$ or $\mathbf{x}^T\mathbf{A} = \lambda\mathbf{x}^T$. Postmultiplying by $\mathbf{y}$, we obtain $\mathbf{x}^T\mathbf{Ay} = \lambda\mathbf{x}^T\mathbf{y}$. From the second equation, we get $\mathbf{x}^T\mathbf{Ay} = \mu\mathbf{x}^T\mathbf{y}$. Subtracting the two results, we obtain $(\lambda - \mu)\mathbf{x}^T\mathbf{y} = \mathbf{0}$, which gives $\mathbf{x}^T\mathbf{y} = 0$ since $\lambda \neq \mu$.

49. There exists an orthogonal matrix $\mathbf{P}$ such that $\mathbf{P}^{-1}\mathbf{AP} = \mathbf{D}$. Now, $\mathbf{A} = \mathbf{PDP}^{-1} = \mathbf{PDP}^T$, since $\mathbf{P}$ is orthogonal. We have $\mathbf{A}^T = (\mathbf{PDP}^T)^T = \mathbf{PD}^T\mathbf{P}^T = \mathbf{PDP}^T = \mathbf{A}$, since a diagonal matrix is always symmetric.

51. Let $\mathbf{z} = \mathbf{U} + i\mathbf{V}$, where $\mathbf{U} \neq \mathbf{0}$, $\mathbf{V} \neq \mathbf{0}$ be real vectors. Then

    $\bar{\mathbf{z}}^T\mathbf{Az} = (\mathbf{U}^T\mathbf{AU} + \mathbf{V}^T\mathbf{AV}) + i(\mathbf{U}^T\mathbf{AV} - \mathbf{V}^T\mathbf{AU}) = \mathbf{U}^T\mathbf{AU} + \mathbf{V}^T\mathbf{AV} > 0$

    since $\mathbf{U}^T\mathbf{AV} = (\mathbf{U}^T\mathbf{AV})^T = \mathbf{V}^T\mathbf{A}^T\mathbf{U} = \mathbf{V}^T\mathbf{AU}$.

52. Let the vectors $\mathbf{a}$, $\mathbf{b}$ be transformed to vectors $\mathbf{u}$, $\mathbf{v}$ respectively. Then

    $\langle\mathbf{u}, \mathbf{v}\rangle = \mathbf{u} \cdot \mathbf{v} = \bar{\mathbf{u}}^T \cdot \mathbf{v} = (\overline{\mathbf{Aa}})^T(\mathbf{Ab}) = \bar{\mathbf{a}}^T\bar{\mathbf{A}}^T\mathbf{Ab} = \bar{\mathbf{a}}^T\mathbf{b} = \mathbf{a} \cdot \mathbf{b}$.

53. (i) No. (ii) Yes. (interchange rows 2 and 3 followed by interchange of columns 2 and 3).

$U_1 = \{1, 3\}.$, $U_2 = \{2, 4\}.$

54. $x^T A x = [x_1, x_2] \begin{bmatrix} a & b \\ b & c \end{bmatrix} \begin{bmatrix} x_1 \\ x_2 \end{bmatrix} = ax_1^2 + 2bx_1x_2 + cx_2^2$

$$= a[(x_1 + bx_2/a)^2 + x_2^2 \, (ac - b^2)/a^2] > 0, \text{ for all } x_1, x_2.$$

Therefore, $a > 0$, $ac - b^2 > 0$.

55. $x^T A x = [x_1, x_2, x_3] \begin{bmatrix} 2 & 1 & 3 \\ -3 & 4 & -1 \\ -1 & 1 & 2 \end{bmatrix} \begin{bmatrix} x_1 \\ x_2 \\ x_3 \end{bmatrix}$

$$= 2x_1^2 - 2x_1x_2 + 2x_1x_3 + 4x_2^2 + 2x_3^2 = (x_1 - x_2)^2 + (x_1 + x_3)^2 + 3x_2^2 + x_3^2 > 0.$$

56. All the leading minors are not positive. It can also be verified that all the eigenvalues are not positive.

57. $\begin{bmatrix} 1 & -1 & 0 \\ -1 & -1 & 2 \\ 0 & 2 & 1 \end{bmatrix}$

58. $\begin{bmatrix} 3 & 1 & -2 \\ 1 & 1 & 4 \\ -2 & 4 & 0 \end{bmatrix}.$

59. $\begin{bmatrix} 1 & i & -4 \\ -i & 0 & 2i \\ -4 & -2i & 4 \end{bmatrix}.$

60. $\begin{bmatrix} 1 & -1-2i & 0 \\ -1+2i & 1 & -2+3i \\ 0 & -2-3i & 0 \end{bmatrix}.$

61. $\begin{bmatrix} 2 & 3+4i & 0 \\ 3-4i & -3 & 2-i \\ 0 & 2+i & 0 \end{bmatrix}$

62. $y_1^2 + 3y_2^2 - 5y_3^2.$

63. $y_1^2 + 3y_2^2 - 2y_3^2.$

64. $y_1^2 + 2y_2^2 + 4y_3^2.$

65. $8y_1^2 + 2y_2^2 + 2y_3^2.$

66. 6.1713.

67. 5.59.

68. 2.

69. 5.312.

70. $P = \begin{bmatrix} 0.19795 & -0.98021 \\ 0.98022 & 0.19795 \end{bmatrix}$, $D = \begin{bmatrix} 6.45101 & 0 \\ 0 & 0.62006 \end{bmatrix}$, $Q = \begin{bmatrix} 0.94237 & 0.33458 \\ 0.33458 & -0.94237 \end{bmatrix}.$

71. $P = \begin{bmatrix} 0.52573 & 0.87065 \\ -0.87065 & -0.52573 \end{bmatrix}$, $D = \begin{bmatrix} 3.61803 & 0 \\ 0 & 1.38197 \end{bmatrix}$, $Q = \begin{bmatrix} -0.08981 & 0.99596 \\ 0.99596 & 0.08981 \end{bmatrix}.$

72. $P = \dfrac{1}{\sqrt{2}} \begin{bmatrix} 1 & 1 \\ 1 & -1 \end{bmatrix}$, $D = \begin{bmatrix} \sqrt{6} & 0 & 0 \\ 0 & 0 & 0 \end{bmatrix}$, $Q = \begin{bmatrix} 1/\sqrt{3} & 1/\sqrt{3} & 1/\sqrt{3} \\ -1/\sqrt{2} & 0 & 1/\sqrt{2} \\ 1/\sqrt{6} & -2/\sqrt{6} & 1/\sqrt{6} \end{bmatrix}.$

73. $P = Q^T$, $D = \begin{bmatrix} 3 & 0 & 0 \\ 0 & 6 & 0 \\ 0 & 0 & 9 \end{bmatrix}$, $Q = \begin{bmatrix} -2/3 & -2/3 & 1/3 \\ 2/3 & -1/3 & 2/3 \\ -1/3 & 2/3 & 2/3 \end{bmatrix}.$

74. $\mathbf{P} = \mathbf{Q}^T$, $\mathbf{D} = \begin{bmatrix} 1 & 0 & 0 \\ 0 & 2 & 0 \\ 0 & 0 & 4 \end{bmatrix}$, $\mathbf{Q} = \begin{bmatrix} 1 & 0 & 0 \\ 0 & 1/\sqrt{2} & 1/\sqrt{2} \\ 0 & 1/\sqrt{2} & -1/\sqrt{2} \end{bmatrix}$.

75. $\mathbf{P} = \dfrac{1}{\sqrt{2}}\begin{bmatrix} 1 & -1 \\ 1 & 1 \end{bmatrix}$, $\mathbf{D} = \begin{bmatrix} \sqrt{7} & 0 & 0 \\ 0 & \sqrt{5} & 0 \end{bmatrix}$, $\mathbf{Q} = \begin{bmatrix} 1/\sqrt{14} & 3/\sqrt{14} & 2/\sqrt{14} \\ -3/\sqrt{10} & 1/\sqrt{10} & 0 \\ 1/\sqrt{35} & 3/\sqrt{35} & -5/\sqrt{35} \end{bmatrix}$.

Chapter 4

Ordinary Differential Equations of First Order

4.1 Introduction

Many practical problems in science and engineering are formulated by finding how one quantity is related to, or depends upon, one or more (other) quantities defined in the problem. Often, it is easier to model a relation between the rates of changes in the variables rather than between the variables themselves. The study of this relationship gives rise to differential equations. Derivatives can always be interpreted as rates. For example, if x is a function of t then dx/dt is the rate of change of x with respect to t. If x denotes the displacement of a particle, then dx/dt represents the velocity of the particle. If x represents the electric charge (q), then dx/dt or dq/dt represents the rate of flow of charge, that is the current. Derivatives of higher orders represent rate of rates. If x denotes the displacement of a particle, then d^2x/dt^2 represents the acceleration. A *differential equation* can be defined as an equation containing derivatives of various orders and the variables. Differential equations which involve one independent variable are called *ordinary differential equations*. If the differential equation involves more than one independent variable and partial derivatives of the dependent variable with respect to them, then it is called a *partial differential equation*.

Differential equations are used to model the physical systems. The systems which change with time are called *dynamic systems*. For example, models which describe heat transfer, chemical engineering processes which depend on time, electronic circuits with time dependent currents and voltages are dynamic systems. Therefore, differential equations provide a tool for studying the phenomena which varies continuously.

Let y be the dependent variable and x be the independent variable. We denote the derivatives as

$$\frac{dy}{dx} = y', \quad \frac{d^2y}{dx^2} = y'', \quad \frac{d^3y}{dx^3} = y''', \text{ etc.}$$

Some examples of ordinary differential equations are given below.

(i) $y' = 6x^2$, (ii) $y'' + 16y = 2x$, (iii) $x^2y'' - xy' + 6y = \log x$,

(iv) $y'y'' + y^2 = x^2$, (v) $[1 + (y')^2]^{1/2} = 5y$, (vi) $y' = \sqrt{x} + \sqrt{y}$,

(vii) $(y'')^2 = 1 + (y')^3$, (viii) $y''' + 3y'' + 3y' + y = 0$.

We define the following:

Order The order of a differential equation is the order of the highest order derivative occuring in the equation.

In the above examples, the orders of the differential equations are

order 1: examples (i), (v), (vi),
order 2: examples (ii), (iii), (iv), (vii),
order 3: example (viii).

Degree The degree of a differential equation is the degree (or power) of the highest order derivative occuring in the equation, after the equation has been made free of radicals and fractions in its derivatives.

In the above examples, the differential equations (v) and (vii) are of degree 2, the others are of degree 1.

Linear A differential equation is linear, when the dependent variable and its derivatives occur only in the first degree and no products of the dependent variable and its derivatives or of various order derivatives occur. The form of a linear ordinary differential equation, in general, is

$$y^{(n)} + c_1 y^{(n-1)} + c_2 y^{(n-2)} + \ldots + c_{n-1} y' + c_n y = r(x)$$

where $c_i = c_i(x)$ are some functions of x or are constants.

Non-linear A differential equation which is not linear, is called a non-linear differential equation. In the above examples, (i), (ii), (iii), (viii) are linear, and (iv), (v),(vi), (vii) are non-linear.

4.2 Formation of Differential Equations

Let y and x be the dependent and the independent variables respectively. The equation

$$f(x, y, c) = 0 \qquad (4.1)$$

containing one arbitrary constant c, represents a family of curves. For example, the equation $x^2 + y^2 = r^2$ where r is arbitrary, represents circles with centre at the origin and radius r. The equation

$$g(x, y, c, d) = 0 \qquad (4.2)$$

containing two arbitrary constants c and d also represents a family of curves. We often say that it represents a two parameter family of curves. For example, the equation $y = mx + k$, where m and k are arbitrary constants, represents a two parameter family of straight lines having slope m and passing through the point $(0, k)$.

To eliminate the arbitrary constant c in Eq. (4.1), we need two equations. One equation is given by Eq. (4.1) itself and the second equation is obtained by differentiating Eq. (4.1) with respect to x. On eliminating c from the two equations, we obtain an equation containing x, y and y' which is a first order differential equation.

For example, consider $y = cx^2$. Differentiating, we get $y' = 2cx$. Eliminating c, we get

$$y' = 2x \left(\frac{y}{x^2} \right) = \frac{2y}{x}, \quad \text{or} \quad xy' - 2y = 0, \, x \neq 0.$$

Hence, $y = cx^2$ satisfies the differential equation $xy' - 2y = 0$.

Similarly, to eliminate the arbitrary constants c and d in Eq. (4.2), we need three equations. One equation is given by Eq. (4.2), and the remaining two equations are obtained by differentiating Eq. (4.2) with respect to x two times. On eliminating c and d from these three equations, we obtain a second order differential equation.

Example 4.1 Let $y = A \cos 3x + B \sin 3x$.
This equation has two arbitrary constants. Differentiating successively two times, we get

$$y' = -3A \sin 3x + 3B \cos 3x$$

$$y'' = -9A \cos 3x - 9B \sin 3x = -9 (A \cos 3x + B \sin 3x) = -9y$$

or $\qquad y'' + 9y = 0$

which is a linear, second order differential equation. Therefore, the differential equation that governs $y = A \cos 3x + B \sin 3x$ is $y'' + 9y = 0$.

Example 4.2 Let $y = cx + (1/c)$, $c \neq 0$.
This equation has one arbitrary constant. Differentiating once, we get $y' = c$. Eliminating c, we obtain

$$y = xy' + \frac{1}{y'}, \quad \text{or} \quad x(y')^2 - yy' + 1 = 0$$

which is a first order, second degree, non-linear differential equation.

4.3 Solution of a Differential Equation

A function $y = f(x)$ is called a solution of a differential equation on $-\infty < x < \infty$ or on a finite interval $a < x < b$, if $f(x)$ is continuous and differentiable (required number of times) throughout the interval and if substitution of $y(x)$ and its derivatives into the differential equation reduces it to an identity.

For example, the function $y = e^{3x}$ is a solution of the differential equation $y' = 3y$, in the interval $-\infty < x < \infty$.

The function $y = \sin 2x$ is a solution of the differential equation $y' = 2 \cos 2x$, in the interval $-\infty < x < \infty$.

The functions $y = (a^2 - 2x^2)^{1/2}$ and $y = -(a^2 - 2x^2)^{1/2}$ are both solutions of the differential equation $yy' + 2x = 0$ in the interval $-(a/\sqrt{2}) < x < (a/\sqrt{2})$. The solutions in this case can be combined and written as $y^2 = a^2 - 2x^2$, $-(a/\sqrt{2}) < x < (a/\sqrt{2})$.

The solutions in the first two examples are of the form $y = f(x)$, and are called *explicit forms of solutions*. An equation $g(x, y) = 0$ is a solution, in *implicit form*, of a differential equation if $y(x)$ satisfies the differential equation and $g(x, y(x)) = 0$. In the third example, the solution $y^2 = a^2 - 2x^2$, is in implict form.

Solution of a differential equation is also called an *integral* of the differential equation. The curves representing the solution are called *integral curves*. For example, the ellipses $4x^2 + y^2 = a^2$ are integral curves of the differential equation $yy' + 4x = 0$.

To obtain the solution of an ordinary differential equation, we integrate it as many times as the order of the differential equation, since each integration reduces the order of the differential equation by one. Also, each integration introduces one arbitrary constant in the solution. Therefore, the *general solution* of a differential equation of order n contains n arbitrary constants. For example, the general solution of a first order ordinary differential equation contains one arbitrary constant, the general solution of a second order ordinary differential equation contains two arbitrary constants etc. The general solution is also called the *complete integral* or *complete primitive* of the differential equation. Any solution obtained from the general solution, by giving particular values to the arbitrary constants is called a *particular solution*.

Remark 1 ·

(a) It is important to note that the general solution of a differential equation can be expressed in different (but equivalent) forms. For example,

$$\log x - \log (y + 2) = k \tag{4.3}$$

where k is an arbitrary constant is the general solution of the differential equation $xy' = y + 2$. The solution given by Eq. (4.3) can also be re-written as

$$\log \left(\frac{x}{y + 2} \right) = k, \quad \text{or} \quad \frac{x}{y + 2} = e^k = c_1 \tag{4.4}$$

or
$$x = c_1 (y + 2) \tag{4.5}$$

where $c_1 = e^k$ is another arbitrary constant. The solution (4.5) can also be written as

$$y + 2 = c_2 x,$$

where $c_2 = 1/c_1$ is another arbitrary constant.

(b) Not all differential equations that we come across have unique solutions or a family of solutions. For example, the differential equation

$$\left| \frac{dy}{dx} \right| + |y| = 0 \tag{4.6}$$

has only the trivial solution, that is $y \equiv 0$.
The differential equation

$$\left| \frac{dy}{dx} \right| + |y| + c = 0, \quad c > 0 \tag{4.7}$$

has no solution.

We shall discuss later the conditions that should be satisfied in order that a differential equation has a unique solution or a family of solutions.

Singular solution of a differential equation

In the case of most of the differential equations, every solution can be obtained from the general solution by assigning suitable values to the arbitrary constants. However, in some cases there exists a solution which cannot be obtained from the general solution. Such a solution is called a *singular solution*. Usually, singular solutions are of interest only under special topics. For example, it can be verified that $y = cx + c^2$ is the general solution of the differential equation

$$(y')^2 + xy' = y. \tag{4.8}$$

The general solution is a one parameter family of straight lines, one straight line for each value of c (positive or negative). However, we find that (by substitution) $4y + x^2 = 0$ is also a solution of the differential equation, which cannot be obtained from the general solution. Therefore, the parabola $4y + x^2 = 0$ is the singular solution of the differential equation (4.8). It may be noted that this singular solution is the envelope of the family of the straight lines represented by the general solution (Fig. 4.1).

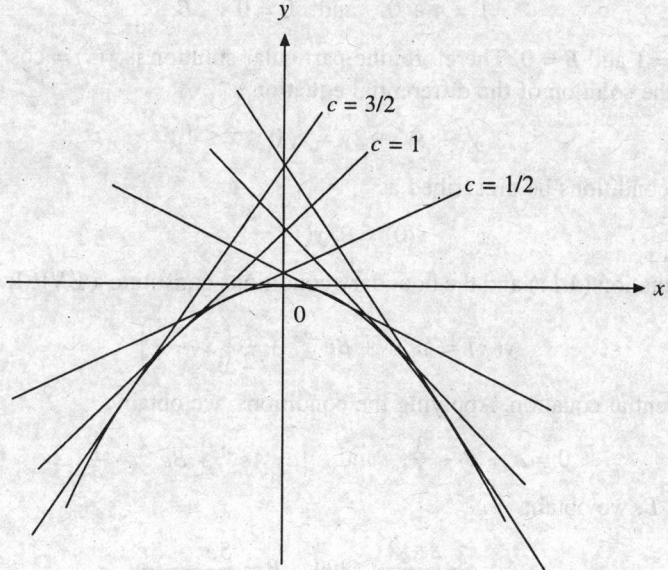

Fig. 4.1.   Singular solution of Eq. (4.8).

4.4   Initial and Boundary Value Problems

Consider a general first order differential equation $F(x, y, y') = 0$ expressed in canonical form as

$$y' = f(x, y). \tag{4.9}$$

The general solution of this differential equation contains one arbitrary constant. If a particular solution is to be obtained, then one condition is to be prescribed. Since the general solution of an mth order differential equation

$$\frac{d^m y}{dx^m} + a_1(x)\frac{d^{m-1}y}{dx^{m-1}} + a_2(x)\frac{d^{m-2}y}{dx^{m-2}} + \ldots + a_m(x)y = r(x), \, x > x_0 \tag{4.10}$$

contains m arbitrary constants, m conditions are needed to obtain a particular solution. If all the m conditions are prescribed at a single point, say $x = x_0$, then the differential equation together with the conditions is called an *initial value problem* (IVP) and the conditions are called the *initial conditions*. The point x_0 is called the initial point. In this case, the solution of the IVP gives the history of the solutions for $x > x_0$. If the independent variable is t(time), then it gives the time history of the solution for $t > t_0$.

If the required conditions are prescribed at two points, say at $x = a$ and $x = b$, that is the differential equation is valid in $a \le x \le b$, then the differential equation together with these conditions is called a *boundary value problem* (BVP) and the conditions are called the *boundary conditions*. Out of the m conditions, k conditions may be prescribed at $x = a$ and the remaining $m - k$ conditions may be prescribed at $x = b$.

For example, $y = A \cos 3x + B \sin 3x$ satisfies the linear differential equation $y'' + 9y = 0$. Since $y = A \cos 3x + B \sin 3x$ contains two arbitrary constants, it is the general solution of the differential equation. Two conditions are to be prescribed to determine the constants A, B, that is to find a particular solution. For example, these conditions may be taken as $y = 1$ and $y' = 0$ at $x = 0$. These conditions are the initial conditions. Applying these conditions, we obtain

$$1 = A + 0, \quad \text{and} \quad 0 = 0 + 3B.$$

The solution is $A = 1$ and $B = 0$. Therefore, the particular solution is $y(x) = \cos 3x$, $x > 0$.

Consider now the solution of the differential equation

$$y'' + 3y' + 2y = x, \ 0 \le x \le 1. \tag{4.11}$$

Let the boundary conditions be prescribed as

$$y(0) = 0, \ y(1) = 1.$$

The differential equation (4.11) and the boundary conditions constitute a BVP. It can be verified that

$$y(x) = Ae^{-x} + Be^{-2x} + \frac{1}{2}\left(x - \frac{3}{2}\right)$$

satisfies the differential equation. Applying the conditions, we obtain

$$0 = A + B - \frac{3}{4}, \quad \text{and} \quad 1 = Ae^{-1} + Be^{-2} - \frac{1}{4}.$$

Solving for A and B, we obtain

$$A = \frac{3 - 5e^2}{4(1 - e)} \quad \text{and} \quad B = \frac{5e^2 - 3e}{4(1 - e)}.$$

Exercise 4.1

Find the order and the degree of the following differential equations. State also whether they are linear or non-linear.

1. $y'' + 3y' + 4y = 0.$

2. $x^2 y'' + xy' + 3y = 5x.$

3. $(y')^2 + 3xy' + y = 0.$

4. $\sqrt{1 + 2x^2}\ dx + \sqrt{1 + 2y^2}\ dy = 0.$

5. $[1 + (y')^2]^{1/2} = x^2 + y.$

6. $yy'' + t^2 y' + 4y = \cos t.$

7. $(y'')^2 + 3y' + x = 0.$

8. $y'y'' + y' + 5y = \sin x.$

9. $(1 + y')^{1/2} = y''.$

10. $y' = \sin y.$

Eliminate the arbitrary constants and obtain the differential equation satisfied by it.

11. $y = ce^{qx}$, q: fixed constant.

12. $y = a \cos \theta x + b \sin \theta x$, θ: fixed constant.

13. $y = c \cos (pt - a)$, p: fixed constant.

14. $y = ae^x + be^{2x}.$

15. $y = e^{-2x}(a \cos 2x + b \sin 2x).$

16. $y = (a/x^2) + bx.$

17. $x^2 + y^2 = a^2.$

18. $x^2 + y^2 - 2ay = 0.$

19. $y = c \cosh (x/c)$, $c \ne 0.$

20. $y = c \sin x.$

21. $(x - p)^2 + (y - q)^2 = a^2$, a: fixed constant.

22. $x^2 - y^2 = a(x^2 + y^2)^2.$

23. $y = ae^{2x} + be^{-2x} + c.$

24. $y = ae^{-x} + be^{-2x} + ce^{-3x}.$

25. $y = 2cx - c^2.$

Verify that the given function satisfies the differential equation.

26. $y = ce^{-x^2}$, $y' + 2xy = 0.$

27. $y = x \log x - x$, $y' = \log x.$

28. $y = \sin^{-1} x$, $y'' = x/[1 - x^2]^{3/2}.$

29. $y = \sec x + \tan x$, $(1 - \sin x)^2 y'' = \cos x.$

30. $y = 2 \tan (x/2) - x$, $(1 + \cos x)y' = 1 - \cos x$.

Find all values of m for which $y = e^{mx}$ is a solution of the following differential equations.

31. $y'' + 3y' + 2y = 0$.

32. $y''' - 6y'' + 11y' - 6y = 0$.

33. $y''' - 2y'' - y' + 2y = 0$.

34. $y'' - 4y' + y = 0$.

35. $y'' - 2y' + 4y = 0$.

4.5   Solution of Equations in Separable Form

Consider the solution of the first order ordinary differential equation of the form

$$y' = f(x, y), \quad \text{or} \quad F(x, y, y') = 0 \tag{4.12}$$

or an initial value problem associated with it. It is not always possible to express the solution of a given differential equation explicitly in terms of known functions. However, in the following we shall discuss those cases in which the equation admits a solution in terms of known functions. Consider the special case when $f(x, y) = g(x)$, a function of x alone, in Eq. (4.12). Then, from integral calculus, we get

$$\int \frac{dy}{dx}\, dx = \int g(x) + c, \quad \text{or} \quad y = F(x) + c.$$

Suppose now that the function $f(x, y)$ in Eq. (4.12) can be written in the separable form as $f(x, y) = g(x)\, h(y)$. Then, Eq. (4.12) can be written as

$$\frac{dy}{dx} = g(x)\, h(y) \tag{4.13}$$

or, in terms of differentials, we write it as

$$\frac{dy}{h(y)} = g(x)\, dx. \tag{4.14}$$

The variables are now separated. Integrating both sides, we get

$$\int \frac{dy}{h(y)} = \int g(x)dx + c.$$

Using the indefinite integration in evaluating the integrals on both sides, the solution can be expressed as $A(y) = B(x) + c$.

Example 4.3 Find the general solution of the differential equation $y' - 2y + a = 0$, where a is fixed constant.

Solution   We write the equation as

$$\frac{dy}{dx} = 2y - a, \quad \text{or} \quad \frac{dy}{2y - a} = dx, \text{ for } y \neq a/2. \tag{4.15}$$

For $y = a/2$, we find that $dy/dx = 0$ and the differential equation is satisfied. Hence, $y = a/2$ is a solution. Since it does not contain an arbitrary constant, it is not a general solution. Integrating Eq. (4.15) on both sides, we get

$$\frac{1}{2} \ln |2y - a| = x + c_1, \quad \text{or} \quad |2y - a| = e^{2x+2c_1} = e^{2x} e^{2c_1}$$

or $2y - a = ke^{2x}$, where $k = \pm e^{2c_1}$ is an arbitrary constant. We can also write this solution as

$$y = ce^{2x} + \frac{a}{2}, \quad \text{where} \quad c = \frac{k}{2}$$

Note that $c > 0$ when $2y - a > 0$ and $c < 0$ when $2y - a < 0$.

Example 4.4   Solve the initial value problem

$$L\frac{dI}{dt} + RI = 0, \quad I(0) = I_0$$

where I, R, L are, respectively, the current, resistance and inductance in electrical circuits (R and L being constants).

Solution   For $I \neq 0$, $L \neq 0$, we write the given differential equation as

$$\frac{dI}{I} = -\frac{R}{L} dt.$$

Integrating, we get

$$\log |I| = -\frac{R}{L} t + k$$

or

$$|I| = e^{-(R/L)t + k} = e^{-(R/L)t} e^k, \quad \text{or} \quad I = ce^{-(R/L)t}$$

where $c = e^k$, $(I > 0)$. Applying the initial condition, we obtain

$$I(0) = I_0 = c.$$

Therefore, $I = I_0 e^{-(R/L)t}$.

Example 4.5   Find the curve through the point $(1, 0)$ and having at each of its points the slope $-x/y$.

Solution   The problem can be reframed as the initial value problem

$$\text{slope} = \frac{dy}{dx} = -\frac{x}{y}, \quad y(1) = 0.$$

Separating the variables, we get $y\,dy = -x\,dx$. Integrating, we obtain $y^2 = c - x^2$.
Applying the initial condition, we get

$$0 = c - 1, \quad \text{or} \quad c = 1.$$

Therefore, the solution curve is the circle $x^2 + y^2 = 1$.

Example 4.6   Reduce the differential equation $y'' + e^{2y}(y')^3 = 0$ to a lower order equation and hence find the solution of the differential equation.

Solution   Set $y' = u$. Then

$$\frac{d^2y}{dx^2} = \frac{du}{dx} = \frac{du}{dy} \cdot \frac{dy}{dx} = u\frac{du}{dy}.$$

The given equation simplifies to $u\dfrac{du}{dy} + e^{2y}u^3 = 0$. Separating the variables, we get $\dfrac{du}{u^2} = -e^{2y}dy$, $u \neq 0$.
Integrating, we obtain

$$\frac{1}{u} = \frac{1}{2}e^{2y} + c_1, \quad \text{or} \quad \frac{dx}{dy} = \frac{1}{2}e^{2y} + c_1$$

or
$$dx = \left(\frac{1}{2} e^{2y} + c_1 \right) dy.$$

Integrating again, we obtain $\qquad x = \frac{1}{4} e^{2y} + c_1 y + c_2,$

where c_1 and c_2 are arbitrary constants. For $y' = u = 0$, we obtain $y = c$ which is also a solution.

Exercise 4.2

Find the general solution of the following differential equations.

1. $(x \ln x)y' = y, x > 0.$
2. $y' = y \tanh x.$
3. $axy' = by, a \neq 0.$
4. $yy' = \cos^2 wx.$
5. $y^2 y' + x^2 = 0.$
6. $y' + ay + b = 0, a \neq 0.$
7. $y' + 2y \tan 2x = 0.$
8. $(x^2 + x + 1)y' + (6x + 3)y = 0.$
9. $y' = e^{x+y} + x^2 e^y.$
10. $(1 + x)y - (1 + y)xy' = 0, x > 0, y > 0.$
11. $yy' = xe^{-x} \sqrt{1 - y^2}.$
12. $y' = 1 + 2x + y + 2xy.$
13. $x(e^{4y} - 1)y' + (x^2 - 1)e^{2y} = 0, x > 0.$
14. $ydx - 3xdy = 3xydx.$

Reduce the following to first order equations and solve. (Hint: set $y' = u$)

15. $y'' + 16y' = 0.$
16. $xy'' + y' = 0.$
17. $y'' + (y')^2 = 0.$
18. $y'' + e^y(y')^3 = 0.$
19. $y'' + (1 + y^{-1})(y')^2 = 0.$
20. $y'' - 3y' = 0.$

Solve the following initial value problems.

21. $y' = 4x^3 e^{-y}, y(1) = 0.$
22. $(x \ln x)y' = 2y, y(2) = (\ln 2)^2.$
23. $(y + 2)y' = \sin x, y(0) = 0.$
24. $y^2 y' = \cos^2 x, y(0) = 0.$
25. $y' = y^2 \sin x, y(2\pi) = 1.$
26. $y' = y \tan 2x, y(0) = 2.$
27. $dr/d\theta = 2r \sin^2\theta, r(\pi/2) = 1.$
28. $(ds/dt) + 2s = st^2, s(0) = 1.$
29. $(\cot x)y' + y + 3 = 0, y(0) = 1.$
30. $e^x[(dx/dt) + 1] = 1, x(0) = 1.$
31. $y' = \frac{x}{y} - \frac{x}{1+y}, y(0) = 2.$

Using the indicated substitution, reduce to separable form and solve the differential equation.

32. $(xy' - y) \cos (y/x) + x = 0, y/x = t.$
33. $(x + 1)(y' - 1) = 2(y - x), y - x = v.$
34. $y' = 1 + (y - x) \cot x, y - x = v.$
35. $y' = (y - x)/(y - x + 2), y - x = v.$
36. $xy' = e^{-xy} - y, xy = v.$
37. $xy' = y + x^2 \tan (y/x), y/x = t.$
38. $y' = (y - x)^2, y - x = v.$
39. $2x^2 yy' = \tan (x^2 y^2) - 2xy^2, x^2 y^2 = u.$
40. $xy' = (y - x)^3 + y, y - x = v.$

41. Observations show that the rate of change of the atmospheric pressure p with altitude h is proportional to pressure. Assuming that the pressure at 6000 meters is half of its value p_0 at sea level, find the formula for the pressure at any height.
42. The growth rate of a bacteria population is proportional to its size. Initially the population is 10,000, while after 10 days its size is 25,000. What will be the population after 20 days?

43. A drop of liquid evaporates at a rate proportional to its area of surface. If the radius initially is 4 mm and 5 minutes later, the radius is reduced to 2 mm, find the radius of the drop as a function of time.

44. A radio active substance disintegrates at a rate proportional to the amount of the substance present. 50% of the amount disintegrates in 1000 years (half life of the substance is 1000 years). Approximately, what percentage of the substance will disintegrate in 50 years?

45. The initial value problem governing the current i flowing in an RL circuit when a step voltage of magnitude E is applied to the circuit at $t = 0$ is given by (R, L, E are constants)

$$iR + L\frac{di}{dt} = E, \quad t > 0, \quad i(0) = 0.$$

Find the solution $i(t)$. Determine the limiting value of i as $t \to \infty$.

46. Find all curves in the x-y plane such that the tangents to the curves pass through the origin.

47. Find all curves in the x-y plane such that the tangents at each point (x, y) of the curves intersects the x-axis at $(x - 1, 0)$.

48. Find the curve $y = f(x)$ through the origin for which $y'' = y'$ and the tangent at the origin is $y = x$.

49. Find the curve in the x-y plane which passes through the point $(1, 1)$, intersects the line $y = x$ at right angle at that point and satisfies $xy'' + 2y' = 0$.

50. A particle moves on a straight line so that its acceleration is equal to four times its velocity. At time $t = 0$, its displacement from the origin is 1 feet and its velocity is 1.5 ft/sec. Find an approximation to the time when the displacement is 10 feet.

4.5.1 Equations Reducible to Separable Form

The following forms of the differential equations can be reduced to separable form by substitution.

Equations of the form $\dfrac{dy}{dx} = f(ax + by + c)$

Substituting $ax + by + c = t$, we get

$$a + b\frac{dy}{dx} = \frac{dt}{dx}, \quad \text{or} \quad \frac{dy}{dx} = \frac{1}{b}\left(\frac{dt}{dx} - a\right).$$

The diffential equations simplifies to

$$\frac{1}{b}\left(\frac{dt}{dx} - a\right) = f(t), \quad \text{or} \quad \frac{dt}{dx} = a + bf(t), \quad \text{or} \quad \frac{dt}{a + bf(t)} = dx.$$

Integrating, we obtain $\displaystyle\int \frac{dt}{a + bf(t)} = x + c.$

Example 4.7 Solve the differential equation $y' = \tan(6x + 3y + 5) - 2$.

Solution Substitute $6x + 3y + 5 = t$. Then, $6 + 3y' = t'$. We obtain

$$\frac{t' - 6}{3} = \tan t - 2, \quad \text{or} \quad t' = 3\tan t, \quad \text{or} \quad (\cot t)\,dt = 3\,dx.$$

Integrating, we get $\ln|\sin t| = 3x + c$,   or $\sin t = Ae^{3x}$, where $A = e^c$.

Therefore, $\sin(6x + 3y + 5) = Ae^{3x}$.

Homogeneous first order differential equation

A first order differential equation $y' = f(x, y)$ is said to be a homogenous equation, if $f(x, y)$ is a homogeneous function of degree 0, that is, the equation can be written in either of the forms

$$y' = g(y/x), \quad \text{or} \quad y' = h(x/y).$$

For example, the differential equations $y' = \dfrac{y}{x} \tan\left(\dfrac{x - y}{x + y}\right), y' = \dfrac{x^3 + y^3 + x^2 y}{x^3 + y^3}$

or the equation in the differential form $(y^2 - x^2 + xy)dx - (x^2 + y^2)dy = 0$
are homogeneous.

If a first order differential equation is homogeneous, then the substitution $y = vx$, or $x = uy$ reduces the equation to a separable form.

Substituting $y = vx$ in $y' = g(y/x)$, we get

$$v + xv' = g(v), \quad \text{or} \quad xv' = g(v) - v$$

or

$$\frac{dv}{g(v) - v} = \frac{dx}{x}$$

which is in separable form. Integrating, we obtain

$$\int \frac{dv}{g(v) - v} = \ln|x| + \ln|c| = \ln|cx|.$$

After integrating, we replace v by y/x to obtain the general solution of the given differential equation. To solve the homogeneous equation $y' = h(x/y)$, we use the substitution $x = uy$.

Example 4.8   Solve the differential equation

$$(x^2 + 4y^2 + xy)\, dx - x^2 dy = 0.$$

Solution   The given equation

$$\frac{dy}{dx} = \frac{x^2 + 4y^2 + xy}{x^2}$$

is homogenous. Substituting $y = vx$, we obtain

$$v + xv' = \frac{x^2(1 + 4v^2 + v)}{x^2} = 1 + 4v^2 + v, \quad \text{or} \quad xv' = 1 + 4v^2.$$

Separating the variables, we obtain

$$\frac{dv}{1+4v^2} = \frac{dx}{x}.$$

Integrating, we get $\frac{1}{2} \tan^{-1}(2v) = \ln|x| + \ln|c| = \ln|cx|$

or $2v = \tan[2\ln|cx|]$,   or $2y = x\tan[2\ln|cx|]$.

Example 4.9   Solve the initial value problem

$$(3xy + y^2)\,dx + (x^2 + xy)\,dy = 0,\; y(1) = 1.$$

Solution   The given equation

$$\frac{dy}{dx} = -\frac{3xy + y^2}{x^2 + xy}$$

is homogeneous. Substituting $y = vx$, we get

$$v + xv' = -\frac{3v + v^2}{1+v}$$

or $xv' = -\left[v + \dfrac{3v + v^2}{1+v}\right] = -\left[\dfrac{2v^2 + 4v}{1+v}\right] = -\dfrac{2v(2+v)}{1+v}.$

Separating the variables, we obtain

$$\frac{(1+v)}{v(2+v)}\,dv = -\frac{2dx}{x}, \quad \text{or} \quad \frac{1}{2}\left[\frac{1}{v} + \frac{1}{2+v}\right]dv = -\frac{2}{x}\,dx.$$

Integrating, we obtain

$$\frac{1}{2}\left[\ln|v| + \ln|2+v|\right] = -2\ln|x| + \ln|c|$$

or $\ln|v(2+v)| + 4\ln|x| = 2\ln|c|$,   or $v(2+v)x^4 = c^2.$

Substituting $v = y/x$, we get $x^2y\,(2x+y) = c^2 = k.$

Using the initial condition $y(1) = 1$, we obtain $k = 3$. The particular solution is $x^2y\,(2x+y) = 3$.

Equations of the form $dy/dx = (ax + by + c)/(lx + my + n)$

Consider the differential equation in the form

$$\frac{dy}{dx} = \frac{ax + by + c}{lx + my + n} \tag{4.17}$$

which is not homogeneous.

(a) If $\dfrac{a}{l} = \dfrac{b}{m} = s$, then Eq. (4.17) can be written as

$$\frac{dy}{dx} = \frac{s(lx + my) + c}{lx + my + n}.$$ (4.18)

Substituting $lx + my = v$, we obtain $l + my' = v'$. Then, Eq. (4.18) becomes

$$\frac{dy}{dx} = \frac{1}{m}(v' - l) = \frac{sv + c}{v + n}$$

or

$$v' = l + \frac{m(sv + c)}{v + n} = \frac{l(v + n) + m(sv + c)}{v + n}$$

or

$$\frac{(v + n)\, dv}{l(v + n) + m(sv + c)} = dx$$

which is in separable form. Integrating and replacing v by $lx + my$, we obtain the general solution.

(b) If $\dfrac{a}{l} \neq \dfrac{b}{m}$, then we substitute $x = X + h$, $y = Y + k$ in the differential equation to get

$$\frac{dy}{dx} = \frac{dY}{dX} = \frac{a(X + h) + b(Y + k) + c}{l(X + h) + m(Y + k) + n} = \frac{aX + bY + (ah + bk + c)}{lX + mY + (lh + mk + n)}.$$ (4.19)

Choose h and k such that $ah + bk + c = 0$ and $lh + mk + n = 0$.

Then, Eq. (4.19) simplifies to

$$\frac{dY}{dX} = \frac{aX + bY}{lX + mY}$$

which is a homogeneous equation in the variables, X, Y. We solve this equation and substitute $X = x - h$ and $Y = y - k$ to obtain the general solution.

Example 4.10   Find the solution of the differential equation

$$(x - 2y + 1)\, dy - (3x - 6y + 2)\, dx = 0.$$

Solution   We have $\dfrac{dy}{dx} = \dfrac{3(x - 2y) + 2}{(x - 2y) + 1}.$

Substituting $x - 2y = v$, we get $1 - 2y' = v'$. Hence,

$$y' = \frac{1}{2}(1 - v') = \frac{3v + 2}{v + 1}, \quad \text{or} \quad v' = 1 - \frac{2(3v + 2)}{v + 1} = -\frac{5v + 3}{v + 1}.$$

Separating the variables, we obtain

$$\frac{(v + 1)dv}{5v + 3} = -dx, \quad \text{or} \quad \frac{1}{5}\left[1 + \frac{(2/5)}{v + (3/5)}\right] dv = -dx.$$

Integrating, we obtain

$$\frac{1}{5}v + \frac{2}{25}\ln\left|v + \frac{3}{5}\right| = -x + c$$

or

$$\frac{1}{5}(x - 2y) + \frac{2}{25}\ln\left|x - 2y + \frac{3}{5}\right| + x = c.$$

Example 4.11   Find the solution of the differential equation

$$(y - x + 1)dy - (y + x + 2)\,dx = 0.$$

Solution   The given differential equation

$$\frac{dy}{dx} = \frac{y + x + 2}{y - x + 1}, \quad y - x + 1 \neq 0$$

is of the form given by Eq. (4.19). Set $x = X + h$, $y = Y + k$. We get

$$\frac{dY}{dX} = \frac{Y + X + (k + h + 2)}{Y - X + (k - h + 1)}.$$

Choose h, k such that $h + k + 2 = 0$, $k - h + 1 = 0$. Solving, we obtain $h = -1/2$ and $k = -3/2$. For this choice of h, k, we get

$$\frac{dY}{dX} = \frac{Y + X}{Y - X}, \quad Y \neq X.$$

Substituting $Y = vX$, we obtain

$$Xv' + v = \frac{v + 1}{v - 1}, \quad \text{or} \quad Xv' = \frac{v + 1}{v - 1} - v = \frac{2v + 1 - v^2}{v - 1}, \quad v \neq 1.$$

Separating the variables, we get

$$\frac{(v - 1)dv}{v^2 - 2v - 1} = -\frac{dX}{X}, \quad \text{or} \quad \frac{(v - 1)dv}{(v - 1)^2 - 2} = -\frac{dX}{X}.$$

Integrating, we get

$$\frac{1}{2}\ln\left|(v - 1)^2 - 2\right| = -\ln|X| - \ln|c| = -\ln|cX|$$

or

$$[(v - 1)^2 - 2]c^2X^2 = 1.$$

Substituting

$$v = \frac{Y}{X} = \frac{y - k}{x - h} = \frac{y + (3/2)}{x + (1/2)} = \frac{2y + 3}{2x + 1}$$

and simplifying, we obtain the solution as

$$\left[\left(\frac{2y + 3}{2x + 1} - 1\right)^2 - 2\right]c^2\left(\frac{2x + 1}{2}\right)^2 = 1$$

or
$$[4(y - x + 1)^2 - 2(2x + 1)^2]\, c^2 = 4$$

or
$$2(y - x + 1)^2 - (2x + 1)^2 = \frac{2}{c^2} = k.$$

Exercise 4.3

Find the general solution of the following differential equations.

1. $y' = (4x + y)^2.$
2. $y' = (2x - y + 1)^2.$
3. $y' = (2y - y^2)^{1/2}.$
4. $xy' = xe^{-y/x} + y.$
5. $(x + y)(xy' - y)y = x^3.$
6. $x^2y' + xy = x^2 + y^2.$
7. $xy' = y + x \sec (y/x).$
8. $x^2y' - xy = x^2 + y^2.$
9. $(y + x)y' = y - x.$
10. $(2xy + x^2)y' = 3y^2 + 2xy.$

Solve the following initial value problems.

11. $2x (x + y)y' = 3y^2 + 4xy,\ y(1) = 1.$
12. $[x(x^2 - y^2)^{-1/2} + e^{y/x}]xy' = x + [x(x^2 - y^2)^{-1/2} + e^{y/x}]y,\ y(1) = 1.$
13. $3xy' - 3y + (x^2 - y^2)^{1/2} = 0,\ y(1) = 1.$ 14. $(x^2 + y^2)y' = xy,\ y(1) = 2.$

Find the general solution of the following differential equations.

15. $(2x + y - 1)\, dy + (4x + 2y - 3)\, dx = 0.$ 16. $(2x + 6y + 1)\, dy - (x + 3y - 2)\, dx = 0.$
17. $(x + y + 3)\, dx - (2x + 2y - 1)\, dy = 0.$ 18. $(x - 2y)\, dy - (2x - 4y - 3)\, dx = 0.$
19. $(y + x - 2)\, dy - (y - x + 1)\, dx = 0.$ 20. $(x + y + 2)\, dy = (y + 3)\, dx.$
21. $y\, dx - [x - y \cos (x/y)]\, dy = 0.$ 22. $x^2\, dy - xy\, dx + y^2\, e^{x^2/y^2}\, dy = 0.$

23. Let $M(x, y)\, dx - N(x, y)\, dy = 0$ be a homogeneous differential equation. Show that the substitution in terms of the polar coordinates r and θ, $x = r \cos \theta$, $y = r \sin \theta$ reduces the equation to separable form.

Find the general solution of the following differential equations after reducing the given equation into polar form.

24. $(x - 2y)\, dy - (2x + y)\, dx = 0.$ 25. $(3x - y)\, dy - (x + 3y)\, dx = 0.$

4.6 Exact First Order Differential Equations

Let a differentiable function $f(x, y)$ be defined over $(x, y) \in S \in \mathbb{R}^2$. Then $df(x, y) = 0$ defines an exact differential equation. Its general solution is $f(x, y) = c$, where c is an arbitrary constant.

For a function $f(x, y)$ of two variables, we know that if $f(x, y)$ has continuous first order partial derivatives, then the *total* or *exact differential* of $f(x, y)$ is expressed as

$$df = \frac{\partial f}{\partial x}\, dx + \frac{\partial f}{\partial y}\, dy. \tag{4.20a}$$

Consider now, a first order differential equation in the differential form as

$$M(x, y)\, dx + N(x, y)\, dy = 0. \tag{4.20b}$$

Comparing Eqs. (4.20a) and (4.20b), we get

$$\frac{\partial f}{\partial x} = M(x, y) \quad \text{and} \quad \frac{\partial f}{\partial y} = N(x, y). \tag{4.21}$$

We assume that $M(x, y)$, $N(x, y)$ are defined and have continuous partial derivatives of order 2 in the region in which the given differential equation is valid. From Eq. (4.21), we obtain

$$\frac{\partial^2 f}{\partial y \partial x} = \frac{\partial M}{\partial y} \quad \text{and} \quad \frac{\partial^2 f}{\partial x \partial y} = \frac{\partial N}{\partial x}.$$

Continuity of second order partial derivatives gives that

$$\frac{\partial M}{\partial y} = \frac{\partial N}{\partial x}. \tag{4.22}$$

Therefore, the differential equation $M(x, y)\, dx + N(x, y)\, dy = 0$ *is an exact differential equation if there exists a function $f(x, y)$ such that*

$$\frac{\partial f}{\partial x} = M(x, y), \quad \text{and} \quad \frac{\partial f}{\partial y} = N(x, y) \tag{4.23}$$

and

$$\frac{\partial M}{\partial y} = \frac{\partial N}{\partial x} \tag{4.24}$$

in the region in which the given equation is valid.

If the given differential equation is exact, then $f(x, y)$ can be determined using the equations (4.23). Integrating the first equation, we get

$$f(x, y) = \int M(x, y)\, dx + g(y). \tag{4.25}$$

Note that f is a function of two variables and $\partial f / \partial x$ is the partial derivative of f with respect to x. Therefore, the integration on the right hand side of Eq. (4.25) is partial integration, that is, y is taken as a constant in $M(x, y)$. The second term on the right hand side, $g(y)$, plays the role of an arbitrary constant of integration. Differentiating Eq. (4.25) partially with respect to x, we recover the equation $\partial f / \partial x = M(x, y)$. From Eq. (4.25), we get

$$f(x, y) = k(x, y) + g(y), \quad \text{where} \quad k(x, y) = \int M(x, y)\, dx. \tag{4.26}$$

Now, $f(x, y)$ must also satisfy the second equation in (4.23).

Substituting $f(x, y)$ from Eq. (4.26) in the second equation of Eq. (4.23), we obtain

$$\frac{\partial f}{\partial y} = \frac{\partial k}{\partial y} + \frac{\partial g}{\partial y} = N(x, y) \quad \text{or} \quad \frac{dg}{dy} = N(x, y) - \frac{\partial k}{\partial y}.$$

Note that the right hand side is a function of y alone. Integrating this equation, we obtain $g(y)$. Substituting $g(y)$ in Eq. (4.26), we obtain $f(x, y)$. The solution of the exact differential equation is now given by $f(x, y) = c$. Note that, it is not necessary to introduce an arbitrary constant while integrating for $g(y)$ as the final solution $f(x, y) = c$ contains an arbitrary constant. We may also integrate the second equation in (4.23) partially with respect to y first and proceed in a similar way.

Example 4.12   Check the equation

$$(3x^2 + 2e^y)\, dx + (2xe^y + 3y^2)\, dy = 0$$

for exactness. If it is exact, find the solution.

Solution   We have

$$M(x, y) = 3x^2 + 2e^y, \ N(x, y) = 2xe^y + 3y^2.$$

$$\frac{\partial M}{\partial y} = 2e^y, \quad \frac{\partial N}{\partial x} = 2e^y.$$

Since $\partial M/\partial y = \partial N/\partial x$, the given equation is exact. Therefore, there exists a function $f(x, y)$ such that

$$\frac{\partial f}{\partial x} = M = 3x^2 + 2e^y \quad \text{and} \quad \frac{\partial f}{\partial y} = N = 2xe^y + 3y^2.$$

Integrating the first equation, we obtain

$$f(x, y) = \int (3x^2 + 2e^y) \, dx + g(y) = x^3 + 2xe^y + g(y).$$

Now, $g(y)$ is determined such that $f(x, y)$ satisfies the second equation. Substituting, we obtain

$$\frac{\partial f}{\partial y} = N = 2xe^y + 3y^2 = 2xe^y + \frac{dg}{dy}, \quad \text{or} \quad \frac{dg}{dy} = 3y^2.$$

Integrating, we get $g(y) = y^3$. Hence, $f(x, y) = x^3 + 2x \, e^y + y^3$. The solution of the given exact differential equation $df(x, y) = 0$ is given by

$$f(x, y) = c, \quad \text{or} \quad x^3 + 2xe^y + y^3 = c.$$

Example 4.13   Solve the initial value problem

$$e^x (\cos y \, dx - \sin y \, dy) = 0, \quad y(0) = 0.$$

Solution   We have $M(x, y) = e^x \cos y, \ N(x, y) = -e^x \sin y.$

$$\frac{\partial M}{\partial y} = -e^x \sin y, \quad \frac{\partial N}{\partial x} = -e^x \sin y.$$

Since $\partial M/\partial y = \partial N/\partial x$, the equation is exact. Therefore, we have

$$\frac{\partial f}{\partial x} = M = e^x \cos y, \quad \text{and} \quad \frac{\partial f}{\partial y} = N = -e^x \sin y.$$

Integrating the first equation, we obtain

$$f(x, y) = e^x \cos y + g(y).$$

Substituting in the second equation, we get

$$\frac{\partial f}{\partial y} = -e^x \sin y + \frac{dg}{dy} = -e^x \sin y, \quad \text{or} \quad \frac{dg}{dy} = 0$$

giving $g(y) = c_1$. Hence, the solution is

$$f(x, y) = e^x \cos y + c_1 = c_2, \quad \text{or} \quad e^x \cos y = c, \quad \text{where} \quad c = c_2 - c_1.$$

Applying the initial condition $y(0) = 0$, we obtain $c = 1$. Hence, the solution of the IVP is $e^x \cos y = 1$.

Example 4.14   Determine for what values of a and b, the following differential equation is exact and obtain the general solution of the exact equation

$$(y + x^3)\, dx + (ax + by^3)\, dy = 0.$$

Solution   We have $M(x, y) = y + x^3$, $N(x, y) = ax + by^3$

$$\frac{\partial M}{\partial y} = 1, \quad \frac{\partial N}{\partial x} = a.$$

Hence, if $a = 1$, the equation is exact, that is the equation is exact for $a = 1$, irrespective of the value of b.

We have $\qquad \dfrac{\partial f}{\partial x} = M = y + x^3 \quad$ and $\quad \dfrac{\partial f}{\partial y} = N = x + by^3.$

Integrating the first equation, we get

$$f(x, y) = \int (y + x^3)\, dx + g(y) = xy + \frac{x^4}{4} + g(y).$$

Substituting in the second equation, we obtain

$$\frac{\partial f}{\partial y} = x + \frac{dg}{dy} = x + by^3 \quad \text{or} \quad \frac{dg}{dy} = by^3.$$

Integrating, we get $g(y) = by^4/4$. Hence, the solution is

$$xy + \frac{x^4}{4} + \frac{by^4}{4} = c$$

for all b and c is the arbitrary constant.

4.6.1   Integrating Factors

If the given first order equation is not exact, then sometimes it can be made exact by multiplying it by an *integrating factor*. For example, consider the equation $3y dx - 2x dy = 0$. Here

$$M(x, y) = 3y, N(x, y) = -2x, \frac{\partial M}{\partial y} = 3 \quad \text{and} \quad \frac{\partial N}{\partial x} = -2.$$

Therefore, the given equation is not exact. If we multiply the equation by $1/(xy)$, we get

$$\frac{3}{x}\, dx - \frac{2}{y}\, dy = 0$$

which is exact. The equation is integrable and we obtain the solution as

$$3 \ln |x| - 2 \ln |y| = \ln |c|, \quad \text{or} \quad x^3 = cy^2.$$

We call $1/(xy)$, an integrating factor (I.F.) of the differential equation.

Therefore, if the first order differential equation

$$M(x, y)\, dx + N(x, y)\, dy = 0 \tag{4.27}$$

is not exact, then we multiply it by an appropriate factor $F(x, y)$ such that the new equation

$$M(x, y)\, F(x, y)\, dx + N(x, y)\, F(x, y)\, dy = 0 \tag{4.28}$$

is exact. $F(x, y)$ is called an integrating factor. Determination of an integrating factor for an arbitrary first order differential equation is very difficult. However, we can find integrating factors in some special cases. Let the Eq. (4.27) be not exact. Then, we have the following results.

Theorem 4.1   If
$$\left[\left(\frac{\partial M}{\partial y} - \frac{\partial N}{\partial x}\right)\Big/ N\right] = f(x)$$

a function of x alone, then $e^{\int f(x)\,dx}$ is an integrating factor of $M(x, y)\,dx + N(x, y)\,dy = 0$.

Proof   Let for some function $f(x)$, $e^{\int f(x)\,dx}$ be an integrating factor. Then

$$e^{\int f(x)\,dx} M(x, y)\,dx + e^{\int f(x)\,dx} N(x, y)\,dy = 0 \tag{4.29}$$

is exact. This equation is exact if

$$\frac{\partial}{\partial y}\left[e^{\int f(x)\,dx} M(x, y)\right] = \frac{\partial}{\partial x}\left[e^{\int f(x)\,dx} N(x, y)\right]$$

or
$$e^{\int f(x)\,dx}\frac{\partial M}{\partial y} = e^{\int f(x)\,dx}\frac{\partial N}{\partial x} + f(x)\,e^{\int f(x)\,dx} N.$$

Cancelling $e^{\int f(x)\,dx}$, we get

$$\frac{\partial M}{\partial y} = \frac{\partial N}{\partial x} + f(x)N \quad \text{or} \quad f(x) = \left(\frac{\partial M}{\partial y} - \frac{\partial N}{\partial x}\right)\Big/ N$$

showing that $\left(\dfrac{\partial M}{\partial y} - \dfrac{\partial N}{\partial x}\right)\Big/ N$ is a function of x only. Since Eq. (4.29) is exact, $e^{\int f(x)\,dx}$ is an integrating factor of $M(x, y)\,dx + N(x, y)\,dy = 0$.

Theorem 4.2   If $\left[\left(\dfrac{\partial M}{\partial y} - \dfrac{\partial N}{\partial x}\right)\Big/ M\right] = -g(y)$, a function of y alone, then $e^{\int g(y)\,dy}$ is an integrating factor of $M(x, y)\,dx + N(x, y)\,dy = 0$.

Proof   Let for some function $g(y)$, $e^{\int g(y)\,dy}$ be an integrating factor. Then

$$e^{\int g(y)\,dy} M(x, y)\,dx + e^{\int g(y)\,dy} N(x, y)\,dy = 0 \tag{4.30}$$

is exact. This equation is exact if

$$\frac{\partial}{\partial y}\left[e^{\int g(y)\,dy} M\right] = \frac{\partial}{\partial x}\left[e^{\int g(y)\,dy} N\right]$$

or
$$e^{\int g(y)\,dy}\frac{\partial M}{\partial y} + g(y)\,e^{\int g(y)\,dy} M = e^{\int g(y)\,dy}\frac{\partial N}{\partial x}.$$

Cancelling $e^{\int g(y)\,dy}$, we get

$$\frac{\partial M}{\partial y} + g(y)\,M = \frac{\partial N}{\partial x} \quad \text{or} \quad g(y) = -\left(\frac{\partial M}{\partial y} - \frac{\partial N}{\partial x}\right)\Big/ M.$$

Since Eq. (4.30) is exact, $e^{\int g(y)dy}$ is an integrating factor of $M(x, y)\, dx + N(x, y)\, dy = 0$.

Theorem 4.3   If the functions $M(x, y)$ and $N(x, y)$ in the equation $M(x, y)\, dx + N(x, y)\, dy = 0$ are homogeneous functions of degree n and $Mx + Ny \neq 0$, then $1/(Mx + Ny)$ is an integrating factor. If $Mx + Ny = 0$, then $1/(xy)$, or $1/x^2$, or $1/y^2$ is an integrating factor.

Proof   If $1/(Mx + Ny)$ is an integrating factor, then

$$\frac{M\, dx}{Mx + Ny} + \frac{N\, dy}{Mx + Ny} = 0$$

is an exact equation. Therefore,

$$\frac{\partial}{\partial y}\left[\frac{M}{Mx + Ny}\right] = \frac{\partial}{\partial x}\left[\frac{N}{Mx + Ny}\right]$$

or

$$\frac{(Mx + Ny)(\partial M/\partial y) - M\left[x(\partial M/\partial y) + N + y(\partial N/\partial y)\right]}{(Mx + Ny)^2}$$

$$= \frac{(Mx + Ny)(\partial N/\partial x) - N\left[x(\partial M/\partial x) + M + y(\partial N/\partial x)\right]}{(Mx + Ny)^2}$$

or

$$Ny\frac{\partial M}{\partial y} - My\frac{\partial N}{\partial y} = Mx\frac{\partial N}{\partial x} - Nx\frac{\partial M}{\partial x}$$

or

$$M\left(x\frac{\partial N}{\partial x} + y\frac{\partial N}{\partial y}\right) - N\left(x\frac{\partial M}{\partial x} + y\frac{\partial M}{\partial y}\right) = 0. \tag{4.31}$$

Since $M(x, y)$ and $N(x, y)$ are homogeneous functions of degree n, we have from the Euler's theorem

$$x\frac{\partial N}{\partial x} + y\frac{\partial N}{\partial y} = nN, \quad \text{and} \quad x\frac{\partial M}{\partial x} + y\frac{\partial M}{\partial y} = nM.$$

Substituting in Eq. (4.31) we find that the equation is satisfied. Hence, the result.

If $Mx + Ny = 0$, for all x and y, then $M/N = -y/x$. Then, the differential equation reduces to

$$\frac{M}{N}\, dx + dy = 0, \quad \text{or} \quad -\frac{y}{x}\, dx + dy = 0, \quad \text{or} \quad x\, dy - y\, dx = 0.$$

Therefore, $1/(xy)$, or $1/x^2$, or $1/y^2$ is an integrating factor.

Very often, by regrouping the terms of the equation, we can obtain the integrating factor by inspection.

1. If the differential equation contains the group of terms $x\, dy - y\, dx$, then $1/x^2$, or $1/y^2$, or $1/(xy)$, or $1/(x^2 + y^2)$ or some function of these expressions can be tried as an integrating factor. We have

$$\frac{x\, dy - y\, dx}{x^2} = d\left(\frac{y}{x}\right), \quad \frac{x\, dy - y\, dx}{y^2} = -d\left(\frac{x}{y}\right),$$

$$\frac{x\, dy - y\, dx}{xy} = \frac{dy}{y} - \frac{dx}{x} = d\left(\ln\left|\frac{y}{x}\right|\right)$$

$$\frac{x\,dy - y\,dx}{x^2 + y^2} = \frac{(x\,dy - y\,dx)/x^2}{1 + (y/x)^2} = d\left[\tan^{-1}\left(\frac{y}{x}\right)\right]$$

$$\frac{x\,dy - y\,dx}{x^2 + y^2} = \frac{(x\,dy - y\,dx)/y^2}{1 + (x/y)^2} = -d\left[\tan^{-1}\left(\frac{x}{y}\right)\right].$$

2. If the differential equation contains the group of terms $x\,dx + y\,dy$, then $1/(x^2 + y^2)$, or $1/(xy)$ or some function of these expressions can be tried as an integrating factor. We have

$$\frac{x\,dx + y\,dy}{x^2 + y^2} = \frac{1}{2}\,d[\ln(x^2 + y^2)]$$

Example 4.15 By inspection, obtain an integrating factor and solve the differential equation

$$x\,dx + y\,dy + 2(x^2 + y^2)\,dx = 0.$$

Solution Since the differential equation contains the term $x\,dx + y\,dy$ and $x^2 + y^2$ is the coefficient of the third term, $1/(x^2 + y^2)$ is an integrating factor. Multiplying with the integrating factor, we have

$$\frac{x\,dx + y\,dy}{x^2 + y^2} + 2\,dx = 0.$$

Integrating, we get $\qquad \dfrac{1}{2}\ln(x^2 + y^2) + 2x = k$

or $\qquad\qquad \ln(x^2 + y^2) = 2k - 4x, \quad \text{or} \quad x^2 + y^2 = ce^{-4x}$

where $c = e^{2k}$ is an arbitrary constant.

Example 4.16 Solve the differential equation

$$y\,dx - x\,dy + e^{1/x}\,dx = 0$$

by finding an integrating factor by inspection.

Solution Since the differential equation contains the term $y\,dx - x\,dy$ and $e^{1/x}$ is the coefficient of the third term, $1/x^2$ may be tried as an integrating factor. Multiplying throughout by the integrating factor, we obtain

$$\frac{y\,dx - x\,dy}{x^2} + \frac{e^{1/x}}{x^2}\,dx = 0$$

or $\qquad\qquad -d\left(\dfrac{y}{x}\right) - d(e^{1/x}) = 0.$

Integrating, we obtain $\qquad \dfrac{y}{x} + e^{1/x} = c, \quad \text{or} \quad y + xe^{1/x} = cx.$

Example 4.17 Solve the differential equation

$$x(1 + y^2)\,dy + y(1 + x^2)\,dx = 0$$

by finding an integrating factor by inspection.

Solution We rewrite the differential equation as

$$(x\,dy + y\,dx) + xy(y\,dy + x\,dx) = 0.$$

Since the equation contains the term $x\,dy + y\,dx$ and the second term has the factor xy, we may try an integrating factor in the form $1/(xy)$. Multiplying by the integrating factor, we obtain

$$\frac{x\,dy + y\,dx}{xy} + y\,dy + x\,dx = 0.$$

Integrating, we get

$$\ln|xy| + \frac{y^2}{2} + \frac{x^2}{2} = k, \quad \text{or} \quad 2\ln|xy| + x^2 + y^2 = 2k$$

or $\qquad \ln(x^2y^2) + x^2 + y^2 = c,$ where $c = 2k$ is an arbitrary constant.

Example 4.18　Solve the differential equation

$$(5x^3 + 12x^2 + 6y^2)\,dx + 6xy\,dy = 0.$$

Solution　We have $M = 5x^3 + 12x^2 + 6y^2, \quad N = 6xy$

$$\frac{\partial M}{\partial y} = 12y, \quad \frac{\partial N}{\partial x} = 6y.$$

The equation is not exact.

Since $\qquad \dfrac{(\partial M/\partial y) - (\partial N/\partial x)}{N} = \dfrac{12y - 6y}{6xy} = \dfrac{1}{x} = f(x)$

a function of x alone, $e^{\int f(x)\,dx} = e^{\int (1/x)\,dx}$ is an integrating factor. We have

$$\text{I.F.} = e^{\int (1/x)\,dx} = e^{\ln x} = x.$$

Multiplying the differential equation throughout by the integrating factor x, we get

$$(5x^4 + 12x^3 + 6xy^2)\,dx + 6x^2y\,dy = 0, \quad \text{or} \quad (5x^4 + 12x^3)\,dx + 6(xy^2\,dx + x^2y\,dy) = 0,$$

or $\qquad (5x^4 + 12x^3)\,dx + 3d(x^2y^2) = 0.$

Integrating, we get the solution as $\qquad x^5 + 3x^4 + 3x^2y^2 = c.$

Example 4.19　Solve the differential equation

$$(3x^2y^3e^y + y^3 + y^2)\,dx + (x^3y^3e^y - xy)\,dy = 0.$$

Solution　We have $M = 3x^2y^3e^y + y^3 + y^2, \quad N = x^3y^3e^y - xy$

$$\frac{\partial M}{\partial y} = 9x^2y^2e^y + 3x^2y^3e^y + 3y^2 + 2y, \quad \frac{\partial N}{\partial x} = 3x^2y^3e^y - y.$$

The equation is not exact. We have

$$\frac{\partial M}{\partial y} - \frac{\partial N}{\partial x} = 9x^2y^2e^y + 3y^2 + 3y = 3(3x^2y^2e^y + y^2 + y)$$

and $\qquad \dfrac{1}{M}\left[\dfrac{\partial M}{\partial y} - \dfrac{\partial N}{\partial x}\right] = \dfrac{3(3x^2y^2e^y + y^2 + y)}{y(3x^2y^2e^y + y^2 + y)} = \dfrac{3}{y} = -g(y)$

which is a function of y alone. The integrating factor is

$$\text{I.F.} = e^{\int g(y)dy} = e^{-\int \frac{3}{y}dy} = e^{-3\ln y} = \frac{1}{y^3}.$$

Multiplying the given differential equation throughout by the integrating factor $1/y^3$, we get

$$\left(3x^2e^y + 1 + \frac{1}{y}\right)dx + \left(x^3e^y - \frac{x}{y^2}\right)dy = 0 \tag{4.32}$$

or

$$(3x^2e^y\,dx + x^3e^y\,dy) + dx + \left(\frac{1}{y}\,dx - \frac{x}{y^2}\,dy\right) = 0,$$

or

$$d(x^3\,e^y) + dx + d\left(\frac{x}{y}\right) = 0.$$

Integrating, we get the solution as

$$x^3e^y + x + \frac{x}{y} = c, \quad \text{or} \quad y(x^3e^y + x) + x = cy.$$

Alternative Since Eq. (4.32) is exact, it is of the form

$$\frac{\partial f}{\partial x}\,dx + \frac{\partial f}{\partial y}\,dy = 0.$$

Comparing with the Eq. (4.32), we get

$$\frac{\partial f}{\partial x} = 3x^2e^y + 1 + \frac{1}{y} \quad \text{and} \quad \frac{\partial f}{\partial y} = x^3e^y - \frac{x}{y^2}.$$

Integrating the first equation with respect to x (keeping y as constant), we get

$$f(x, y) = \int \left(3x^2e^y + 1 + \frac{1}{y}\right)dx + g(y) = x^3e^y + x + \frac{x}{y} + g(y).$$

Substituting in the second equation, we obtain

$$\frac{\partial f}{\partial y} = x^3e^y - \frac{x}{y^2} + \frac{dg}{dy} = x^3e^y - \frac{x}{y^2}, \quad \text{or} \quad \frac{dg}{dy} = 0, \quad \text{or} \quad g(y) = k, \quad \text{a constant.}$$

The solution is $f(x, y) = c_1$, or $x^3e^y + x + \dfrac{x}{y} = c$

where $c = c_1 - k$ is an arbitrary constant.

Example 4.20 Solve the differential equation

$$(2xy + x^2)y' = 3y^2 + 2xy.$$

(see Problem 10, Exercise 4.3)

Solution We write the equation as

$$(3y^2 + 2xy)\,dx - (2xy + x^2)\,dy = 0 = M(x, y)\,dx + N(x, y)\,dy.$$

The equation is not exact and $M(x, y)$, $N(x, y)$ are homogeneous functions of degree 2. Hence,

$$\frac{1}{Mx + Ny} = \frac{1}{(3y^2 + 2xy)x - (2xy + x^2)y} = \frac{1}{xy^2 + x^2y} = \frac{1}{xy(x + y)}$$

is an integrating factor. Multiplying the given equation throughout by the integrating factor, we obtain

$$\frac{y(3y + 2x)}{xy(x + y)} dx - \frac{x(2y + x)}{xy(x + y)} dy = 0.$$

This equation is exact. We have, therefore

$$\frac{\partial f}{\partial x} = \frac{3y + 2x}{x(x + y)} = \frac{3}{x} - \frac{1}{x + y} \quad \text{and} \quad \frac{\partial f}{\partial y} = -\frac{2y + x}{y(x + y)} = -\left[\frac{1}{y} + \frac{1}{x + y}\right].$$

Integrating the first equation with respect to x (keeping y as a constant), we get

$$f(x, y) = \int \left(\frac{3}{x} - \frac{1}{x + y}\right) dx + g(y) = 3 \ln |x| - \ln |x + y| + g(y).$$

Substituting in the second equation, we obtain

$$\frac{\partial f}{\partial y} = -\frac{1}{x + y} + g'(y) = -\frac{1}{y} - \frac{1}{x + y} \quad \text{or} \quad g'(y) = -\frac{1}{y}.$$

Integrating, we obtain $g(y) = -\ln |y|$.

Hence, $$f(x, y) = 3 \ln |x| - \ln |x + y| - \ln |y| = \ln c$$

or $$\frac{x^3}{y(x + y)} = c \quad \text{or} \quad x^3 = cy(x + y)$$

is the required solution, where c is an arbitrary constant.

Exercise 4.4

For the following differential equations, check whether the equation is exact and obtain its general solution.

1. $(1 + e^x) dx + y dy = 0.$
2. $y dx + x(1 + y) dy = 0.$
3. $2 \cosh x dx + \sinh x dy = 0.$
4. $\sinh x \cos y dx - \cosh x \sin y dy = 0.$
5. $(3x^2y + (y/x)) dx + (x^3 + \ln x) dy = 0.$
6. $(xe^{xy} + 2y) dy + ye^{xy} dx = 0.$
7. $x dy + 2y dx = xy dy.$
8. $x dy - y dx = e^y(x^2 + y^2) dy.$
9. $x dx + y dy = 2y(x^2 + y^2) dy.$
10. $x dy - y dx + y^2 dx = 0.$
11. $y(1 + 6xy) dx + (4y - x) dy = 0.$
12. $(2x + e^y) dx + xe^y dy = 0.$
13. $(1 + x^2) dy + 2xy dx = 0.$
14. $2xy dx + (x^2 + 1) dy = 0.$
15. $(e^{2y} + 1) \cos x dx + 2e^{2y} \sin x dy = 0.$

Under what conditions, the following differential equations are exact?

16. $xy^3 dx + ax^2y^2 dy = 0.$
17. $[f(x) + g(y)] dx + [h(x) + k(y)] dy = 0.$
18. $(ax + y) dx + (kx + by) dy = 0.$
19. $(a \sinh x \cos y + b \cosh x \sin y) dx + (c \sinh x \cos y + d \cosh x \sin y) dy = 0.$

Find the integrating factor and hence solve the following differential equations

20. $(y - 1) dx - x dy = 0.$
21. $dx + e^{(y-x)} dy = 0.$
22. $(x^3 + y^3 + 1) dx + xy^2 dy = 0.$
23. $(4y + x^3) dx + x dy = 0.$

24. $(2y^3xe^y + y^2 + y)\, dx + (y^3x^2e^y - xy - 2x)\, dy = 0.$
25. $y(1 + 3x^3 + 12x^2)\, dx + (x + 4)\, dy = 0.$ 26. $y(1 + xy^2)\, dx + 2(x^2y^2 + x + y^4)\, dy = 0.$
27. $(12y + 3y^4 + 4x^3)\, dx + 6x(1 + y^3)\, dy = 0.$ 28. $(x^2 + y^2)\, dx - (2xy)\, dy = 0.$
29. $(2x + y)\, dy - (x + 2y)\, dx = 0.$ 30. $y^2\, dx + x(x - y)\, dy = 0.$

Solve the following initial value problems.

31. $3x^2y^4dx + 4x^3y^3dy = 0, \quad y(1) = 2.$ 32. $(1 + y)\, dy - (1 - x)\, dx = 0, \quad y(1) = 0.$
33. $3y\, dx + 2x\, dy = 0, \quad y(1) = 1.$ 34. $2xy\, dx + (x^2 + \pi \cos \pi y)\, dy = 0, \quad y(1) = 1.$
35. $(\cos x + y \sin x)\, dx = (\cos x)\, dy, \quad y(\pi) = 0.$

36. $xe^{x^2+y^2}\, dx + y(1 + e^{x^2+y^2})\, dy = 0, \quad y(0) = 0.$
37. $xy\, dx - (x^2 + y^2)\, dy = 0, \quad y(0) = 1.$
38. $\left(4x^3y^3 + \dfrac{1}{x}\right) dx + \left(3x^4y^2 - \dfrac{1}{y}\right) dy = 0, \quad y(1) = 1.$
39. $(x - y \cos x)\, dx - \sin x\, dy = 0, \quad y(\pi/2) = 1.$
40. $(ye^{xy} + 4y^3)\, dx + (xe^{xy} + 12xy^2 - 2y)dy = 0, \quad y(0) = 2.$
41. $(2xy + e^y)\, dx + (x^2 + xe^y)\, dy = 0, \quad y(1) = 1.$
42. $(x^2 + y^2 + x)\, dx + y\, dy = 0, \quad y(1) = 1.$ 43. $xy\, dx + (x^2 + 2y^2 + 2)\, dy = 0, \quad y(0) = 1.$
44. Prove that if M and N in $M(x, y)\, dx + N(x, y)\, dy = 0$ satisfy the equation

$$\frac{\partial M}{\partial y} = \frac{\partial N}{\partial x} + \frac{k}{x} N$$

    then, $F = x^k$ is an integrating factor. Hence, solve $4y\, dx + x\, dy = 0.$

45. Show that $F(x, y)$ is an integrating factor of $M(x, y)\, dx + N(x, y)\, dy = 0$, if and only if

$$\left(M \frac{\partial F}{\partial y} - N \frac{\partial F}{\partial x}\right) + \left(\frac{\partial M}{\partial y} - \frac{\partial N}{\partial x}\right) F = 0.$$

4.7   Linear First Order Equations

Consider the linear first order equation

$$f(x) \frac{dy}{dx} + g(x)y = h(x).$$

Dividing the equation by $f(x)$, we can write it in canonical form as

$$\frac{dy}{dx} + p(x)y = r(x) \tag{4.33}$$

where $$p(x) = g(x)/f(x) \quad \text{and} \quad r(x) = h(x)/f(x).$$

If $r(x) = 0$, then the equation is easily solved, as it is in the separable form. In this case, the equation $y' + p(x)y = 0$ is also called a *homogeneous first order equation*. For the homogeneous equation, we have

$$\frac{dy}{dx} = -p(x)y, \quad \text{or} \quad \frac{dy}{y} = -p(x)dx.$$

Integrating, we obtain

$$\ln |y| = - \int p(x)\, dx + k, \quad \text{or} \quad y = ce^{-\int p(x)\,dx}$$

where $c = \pm e^{k}$ when $y \gtrless 0$. Since, the trivial solution is also a solution of the homogeneous equation, we may also take $c = 0$.

For the general case, that is, for equation (4.33), we have the following result.

Theorem 4.4   The integrating factor of the non-homogeneous first order linear equation $y' + p(x)y = r(x)$, is given by $e^{\int p(x)\,dx}$

Proof   We can rewrite the given equation as

$$[p(x)y - r(x)]\, dx + dy = 0.$$

Comparing this equation with the equation $M(x, y)\, dx + N(x, y)\, dy = 0$, we obtain

$$M(x, y) = p(x)y - r(x), \quad N(x, y) = 1.$$

We have

$$\frac{\partial M}{\partial y} = p(x), \quad \frac{\partial N}{\partial x} = 0$$

and

$$\frac{(\partial M/\partial y) - (\partial N/\partial x)}{N} = \frac{p(x)}{1} = p(x).$$

Hence, by Theorem 4.1, $e^{\int p(x)\,dx}$ is an integrating factor. Multiplying the given equation throughout by the integrating factor, we obtain

$$e^{\int p(x)\,dx} \frac{dy}{dx} + p(x)\, e^{\int p(x)\,dx}\, y = r(x)\, e^{\int p(x)\,dx}$$

or

$$\frac{d}{dx} [ye^{\int p(x)\,dx}] = r(x)\, e^{\int p(x)\,dx}.$$

Integrating, we obtain the solution as

$$ye^{\int p(x)\,dx} = \int [r(x)e^{\int p(x)\,dx}]\, dx + c. \tag{4.34}$$

The solution can also be written in the form

$$y = e^{-\int p(x)\,dx} \left[\int r(x)e^{\int p(x)\,dx}\, dx \right] + ce^{-\int p(x)\,dx}. \tag{4.35}$$

(Note that if the arbitrary constant was not introduced in Eq. (4.34) before simplification, then an important component of the solution could be lost.)

Example 4.21   Solve the differential equation

$$x \frac{dy}{dx} = 2y + x^4 + 6x^2 + 2x, \quad x \neq 0.$$

Solution   Write the equation as

$$\frac{dy}{dx} - \frac{2}{x} y = x^3 + 6x + 2$$

An integrating factor is

$$\text{I.F.} = e^{-\int (2/x)dx} = e^{-2\ln x} = e^{\ln(1/x^2)} = \frac{1}{x^2}.$$

Multiplying the given equation throughout by the integrating factor, we get

$$\frac{1}{x^2} \frac{dy}{dx} - \frac{2}{x^3} y = x + \frac{6}{x} + \frac{2}{x^2}, \quad \text{or} \quad \frac{d}{dx}\left(\frac{1}{x^2} y\right) = x + \frac{6}{x} + \frac{2}{x^2}.$$

Integrating, we obtain

$$\frac{1}{x^2} y = \frac{x^2}{2} + 6 \ln |x| - \frac{2}{x} + c \quad \text{or} \quad y = \frac{x^4}{2} + 6x^2 \ln |x| - 2x + cx^2.$$

Example 4.22 Solve the differential equation

$$(x - a) \frac{dy}{dx} + 3y = 12(x - a)^3, \quad x > a > 0.$$

Solution Write the equation as

$$\frac{dy}{dx} + \frac{3}{x - a} y = 12(x - a)^2.$$

An integrating factor is

$$\text{I.F.} = e^{\int [3/(x-a)]dx} = e^{3\ln(x-a)} = (x - a)^3.$$

Multiplying the given equation throughout by $(x - a)^3$ and integrating, we get

$$(x - a)^3 y = \int 12(x - a)^5 + c = 2(x - a)^6 + c$$

or

$$y = 2(x - a)^3 + \frac{c}{(x - a)^3}.$$

Example 4.23 Solve the differential equation

$$\cot 3x \frac{dy}{dx} - 3y = \cos 3x + \sin 3x, \quad 0 < x < \pi/2.$$

Solution Write the equation as

$$\frac{dy}{dx} - 3(\tan 3x)y = (\tan 3x)(\cos 3x + \sin 3x) = \sin 3x + \frac{\sin^2 3x}{\cos 3x}$$

An integrating factor is

$$\text{I.F} = e^{-3\int \tan 3x\, dx} = e^{\log \cos 3x} = \cos 3x.$$

Multiplying the given equation throughout by $\cos 3x$ and integrating, we obtain

$$y \cos 3x = \int (\sin 3x \cos 3x + \sin^2 3x) \, dx + c$$

$$= \frac{1}{2} \int (\sin 6x + 1 - \cos 6x) \, dx + c$$

$$= \frac{1}{2}\left[-\frac{\cos 6x}{6} + x - \frac{\sin 6x}{6}\right] + c = \frac{1}{12} [6x - \cos 6x - \sin 6x] + c.$$

Example 4.24   The initial value problem governing the current i flowing in a series RL circuit when a voltage $v(t) = t$ is applied, is given by

$$iR + L\frac{di}{dt} = t, \quad t \geq 0, \quad i(0) = 0$$

where R and L are constants. Find the current $i(t)$ at time t.

Solution   Write the differential equation as

$$\frac{di}{dt} + \frac{R}{L}i = \frac{1}{L}t.$$

An integrating factor is

$$\text{I.F.} = e^{\int (R/L)\,dt} = e^{(Rt)/L}.$$

Multiplying the given equation by $e^{(Rt)/L}$ and integrating, we obtain

$$i\,e^{(Rt)/L} = \int \frac{1}{L}e^{(Rt)/L}\,t\,dt + c = \frac{1}{L}\left[\frac{e^{(Rt)/L}}{(R/L)}t - \frac{1}{(R/L)}\int e^{(Rt)/L}\,dt\right] + c$$

$$= \frac{1}{L(R/L)}\left[e^{(Rt)/L}\,t - \frac{e^{(Rt)/L}}{(R/L)}\right] + c$$

Therefore,

$$i(t) = \frac{1}{R}\left[t - \frac{L}{R}\right] + ce^{-(Rt)/L}.$$

Applying the initial condition $i(0) = 0$, we get

$$0 = -\frac{L}{R^2} + c, \quad \text{or} \quad c = \frac{L}{R^2}.$$

The current $i(t)$ at time t is given by

$$i(t) = \frac{t}{R} + \frac{L}{R^2}[e^{-(Rt)/L} - 1].$$

Exercise 4.5

Find the general solution of the following first order linear differential equations.

1. $y' + y = \sin x$.
2. $xy' + y = \sin x$.
3. $y' + xy = 2x$.
4. $xy' = y + (x + 1)^2$.
5. $y' - 2y = \cos 3x$.
6. $\frac{dx}{dt} - 2x = t^2 e^{2t}$.
7. $\frac{ds}{du} + s = ue^{-u} + 1$.
8. $y' - 3y = \sin 2x$.
9. $y' + 4y = 2x + 4x^2$.
10. $y' = 2x(y - x^2 + 1)$.
11. $xy' + y = x^3 + x$.
12. $y' + 3y = e^{2x} + 6$.
13. $y' = (y + 1)\tan x$.
14. $xy' + 2(1 + x^2)y = 6$.
15. $y' = \cos^3 x + y \cot x$.
16. $(1 + x^2)y' + 2xy = x \sin x$.
17. $y' + 2xy = xe^{-x^2}$.
18. $x^2 y' + xy = 2x^2 e^{x^2}$.
19. $(x^2 - 2y)\,dx = x\,dy$.
20. $xy' + (1 + 2x)y = 1 + xe^{-2x}$.

Solve the following initial value problems.

21. $y' - y = e^x$, $y(0) = 1$. $\qquad$ **22.** $y' - (3/x)y = x^3$, $y(1) = 4$.

23. $y' + y \tan x = \sin 2x$, $y(0) = 1$. $\qquad$ **24.** $y' - \left(1 + \dfrac{3}{x}\right)y = x + 2$, $y(1) = e - 1$.

25. $y' + y = (x + 1)^2$, $y(0) = 0$.

26. $xy' - 3y = x^4(e^x + \cos x) - 2x^2$, $y(\pi) = \pi^3 e^\pi + 2\pi^2$.

27. $xy' + y = 2x + x^3 + x^5$, $y(1) = 0$. $\qquad$ **28.** $(1 + x^2)y' - 2xy = 2x(1 + x^2)$, $y(0) = 1$.

29. $y' + (y/x) = \ln x$, $y(1) = 1$. $\qquad$ **30.** $y' = \sec^2 x + y \tan x$, $y(0) = 0$.

Taking x as the dependent variable and y as the independent variable, solve the following differential equations.

31. $y' = 1/[e^{-y} - x]$. $\qquad$ **32.** $(2x + y^4)y' = y$.

33. $y' = 1/[e^y + x]$. $\qquad$ **34.** $(x + 1)y' = y^2$.

35. The initial value problem governing the current i, flowing in a series RL circuit when a sinusoidal voltage $v(t) = \sin \omega t$ is applied, is given by (R, ω and L are constants)

$$iR + L \frac{di}{dt} = \sin \omega t, \quad t \geq 0, \quad i(0) = 0.$$

Find the current $i(t)$, $t > 0$.

4.8 Some Special First Order Differential Equations

In this section, we shall consider the solution of the Bernoulli equation, Riccati equation and Clairaut's equation.

4.8.1 Bernoulli Equation

Some non-linear first order differential equations can be reduced to the linear form by change of the dependent variable. The *Bernoulli equation*

$$y' + p(x)y = q(x)y^n \tag{4.36}$$

where n is any real number, is one such equation. If $n = 0$ or 1, then the equation is linear. For all other values of n, the equation is non-linear. Note that if $n > 0$, $y = 0$ is a solution. To find the non-trivial solutions, consider the following transformation of the dependent variable

$$v(x) = [y(x)]^{1-n}. \tag{4.37}$$

We obtain

$$\frac{dv}{dx} = (1 - n)y^{-n} \frac{dy}{dx} = (1 - n)y^{-n}[q(x)y^n - p(x)y]$$

$$= (1 - n)[q(x) - p(x)y^{1-n}] = (1 - n)[q(x) - p(x)v]$$

using Eq. (4.37). Hence, Eq. (4.36) reduces to the linear equation

$$\frac{dv}{dx} + (1 - n)p(x)v = (1 - n)q(x). \tag{4.38}$$

When the solution of this linear equation is obtained and $v(x)$ is replaced by y^{1-n}, we get the general solution of the Bernoulli equation.

Example 4.25 Solve the differential equation

$$y' + 4xy + xy^3 = 0.$$

Solution Write the given equation as

$$y' + 4xy = -xy^3$$

which is a Bernoulli equation with $n = 3$. Consider the transformation $v(x) = y^{1-3} = y^{-2}$. We obtain

$$\frac{dv}{dx} = -2y^{-3}\frac{dy}{dx} = -2y^{-3}[-xy^3 - 4xy]$$

$$= 2[x + 4xy^{-2}] = 2[x + 4xv]$$

or

$$\frac{dv}{dx} - 8xv = 2x.$$

An integrating factor of this equation is

$$\text{I.F.} = e^{-\int 8x\,dx} = e^{-4x^2}.$$

The general solution of the transformed equation is obtained as

$$ve^{-4x^2} = \int e^{-4x^2} 2x\,dx + c = -\frac{1}{4}e^{-4x^2} + c$$

or

$$v = -\frac{1}{4} + ce^{4x^2}, \quad \text{or} \quad \frac{1}{y^2} = -\frac{1}{4} + ce^{4x^2}$$

or

$$y = \left(ce^{4x^2} - \frac{1}{4}\right)^{-1/2}.$$

Note that $y = 0$ is a singular solution.

Example 4.26 Solve the differential equation

$$\frac{dy}{dx} - y = y^2(\sin x + \cos x).$$

Solution The given equation is a Bernoulli equation with $n = 2$. Set

$$v(x) = y^{1-2} = y^{-1}.$$

Therefore, $\quad \dfrac{dv}{dx} = -\dfrac{1}{y^2}\dfrac{dy}{dx} = -\dfrac{1}{y^2}[y + y^2(\sin x + \cos x)] = -\dfrac{1}{y} - (\sin x + \cos x)$

or

$$\frac{dv}{dx} + v = -(\sin x + \cos x).$$

An integrating factor is

$$\text{I.F.} = e^{\int dx} = e^x.$$

The solution is

$$ve^x = - \int e^x (\sin x + \cos x)\, dx + \dot{c} = - [e^x \sin x] + c$$

or
$$v = \frac{1}{y} = - \sin x + ce^{-x}, \quad \text{or} \quad y = \frac{1}{ce^{-x} - \sin x}.$$

4.8.2 Riccati Equation

One of the important first order non-linear differential equations is the *Riccati equation*

$$y' = p(x)y^2 + q(x)y + r(x). \tag{4.39}$$

If $r(x) = 0$, then it reduces to the Bernoulli's equation with $n = 2$. The solution in this case can be obtained by substituting $v = 1/y$ and solving the resulting first order linear equation. Trivial solution $y = 0$ is also a solution is this case.

When $r(x) \neq 0$, there does not exist any simple method for solving the equation. However, the Riccati equation (4.39) can be reduced to a Bernoulli's equation if one solution, say $y = v(x)$ of the Riccati equation is known. This Bernoulli's equation can then be reduced to a linear first order equation by the substitution $y = 1/z$, in the new dependent variable z. In other words, the substitution $y = v(x) + (1/z)$ reduces Eq. (4.39) to a first order linear equation when one solution $v(x)$ is known. Then, we have

$$y' = v' - \frac{1}{z^2} z'.$$

Substituting in Eq. (4.39), we get

$$v' - \frac{1}{z^2} z' = p \left[v + \frac{1}{z} \right]^2 + q \left[v + \frac{1}{z} \right] + r$$

or
$$- \frac{1}{z^2} z' = p \left[\frac{2v}{z} + \frac{1}{z^2} \right] + \frac{q}{z}$$

since $v(x)$ is a solution of Eq. (4.39), that is, $v' = pv^2 + qv + r$. Simplifying, we get

$$z' = - p\, [2vz + 1] - qz, \quad \text{or} \quad z' + (2pv + q)\, z = - p$$

which is a first order linear equation.

Example 4.27   Find the general solution of the differential equation
$$y' = y^2 - (2x - 1)y + x^2 - x + 1$$

if $y = x$ is a solution of the differential equation.

Solution   Substitute $y = x + (1/z)$. We obtain $y' = 1 - (z'/z^2)$. Substituting in the given equation, we get

$$1 - \frac{1}{z^2} z' = \left(x + \frac{1}{z} \right)^2 - (2x - 1)\left(x + \frac{1}{z} \right) + x^2 - x + 1$$

$$= x^2 + \frac{2x}{z} + \frac{1}{z^2} - \left(2x^2 + \frac{2x}{z} - x - \frac{1}{z} \right) + x^2 - x + 1$$

or
$$z' = - z^2 \left(\frac{1}{z^2} + \frac{1}{z} \right), \quad \text{or} \quad z' + z = - 1.$$

The integrating factor is e^x. Multiplying by the integrating factor and integrating, we get

$$ze^x = -e^x + c, \quad \text{or} \quad z = -1 + ce^{-x}.$$

The general solution is

$$y = x + \frac{1}{z} = x + \frac{1}{ce^{-x} - 1}.$$

Example 4.28   Find the general solution of the equation

$$y' = 2xy^2 + (1 - 4x)y + 2x - 1$$

if $y = 1$ is a solution of the equation.

Solution   Substitute $y = 1 + (1/z)$. We obtain

$$-\frac{1}{z^2} z' = 2x\left(1 + \frac{2}{z} + \frac{1}{z^2}\right) + (1 - 4x)\left(1 + \frac{1}{z}\right) + 2x - 1 = \frac{2x}{z^2} + \frac{1}{z}$$

or $z' = -(2x + z), \quad \text{or} \quad z' + z = -2x.$

The integrating factor is e^x. Multiplying by the integrating factor and integrating, we get

$$ze^x = -2\int xe^x dx + c = -2(x - 1)e^x + c$$

or $$z = -2(x - 1) + ce^{-x}.$$

The general solution is

$$y = 1 + \frac{1}{z} = 1 + \frac{1}{ce^{-x} - 2(x - 1)}.$$

4.8.3   Clairaut's Equation

A non-linear differential equation of the form

$$y = xy' + f(y') \tag{4.40}$$

is called the *Clairaut's equation*. This equation is an interesting equation as it always has a singular solution. Solving Eq. (4.40) for x, we obtain

$$x = \frac{1}{y'}[y - f(y')] = \frac{1}{p}[y - f(p)]$$

where $p = y'$. Differentiating with respect to y, we obtain

$$\frac{dx}{dy} = \frac{1}{p}\left[1 - f'(p)\frac{dp}{dy}\right] + [y - f(p)]\left[-\frac{1}{p^2}\frac{dp}{dy}\right]$$

or $$\frac{1}{p} = \frac{1}{p} - \frac{1}{p^2}[y - f(p) + pf'(p)]\frac{dp}{dy}$$

or $$\frac{1}{p^2}[y - f(p) + pf'(p)]\frac{dp}{dy} = 0. \tag{4.41}$$

If $\dfrac{dp}{dy} = 0$, we get $p = y' = c$.

Hence, substituting in Eq. (4.40), we get the general solution as

$$y = cx + f(c). \tag{4.42}$$

If the first term in Eq. (4.41) is set to zero, we obtain

$$y = f(p) - pf'(p).$$

Substituting in Eq. (4.40), we obtain

$$f(p) - pf'(p) = xp + f(p), \quad (\because y' = p)$$

or

$$x = -f'(p).$$

Hence, the parametric equations

$$x = -f'(t), \quad y = f(t) - tf'(t) \tag{4.43}$$

define another solution (not involving any constant), which is the singular solution.

The general solution (4.42) defines a one parameter family of straight lines. These straight lines are all tangential to the curve defined by the parametric equations (4.43), that is to the singular solution. The singular solution curve is the *envelope* of the family of straight lines defined by the general solution $y = cx + f(c)$ (see section 4.3 and Fig. 4.1).

Example 4.29   Obtain the general solution and the singular solution of the non-linear equation

$$y = xy' + (y')^2. \tag{4.44}$$

Solution   Let us retrace the general procedure of obtaining the solution outlined earlier. Solving for x, we obtain

$$x = \frac{1}{p}(y - p^2), \quad \text{where} \quad p = y'.$$

Differentiating with respect to y, we get

$$\frac{1}{p} = \frac{1}{p} - \frac{y}{p^2}\frac{dp}{dy} - \frac{dp}{dy}, \quad \text{or} \quad \left[\frac{y}{p^2} + 1\right]\frac{dp}{dy} = 0.$$

If we set $dp/dy = 0$, we get $p = c$.

Therefore, the general solution is

$$y = cx + c^2. \tag{4.45}$$

Setting the first term to zero, we obtain $y = -p^2$. Substituting in Eq. (4.44), we get

$$-p^2 = xp + p^2, \quad \text{or} \quad x = -2p.$$

Hence, the singular solution is defined by the curve whose parametric equations are

$$x = -2t, \ y = -t^2, \quad \text{or} \quad x^2 = -4y, \quad \text{or} \quad x^2 + 4y = 0. \tag{4.46}$$

From Eq. (4.44), we find that two distinct directions are obtained if the discriminant $x^2 + 4y > 0$. For example, consider a point $(1, 2)$ satisfying $x^2 + 4y > 0$. At this point, the two directions are obtained by solving $p^2 + p - 2 = 0$, or $p = 1, -2$. The two values of c at this point are obtained by solving $c^2 + c - 2 = 0$, or $c = 1, -2$. Hence, the two lines passing through this point are $y = x + 1$ and $y = 4 - 2x$. Consider now any point $(x_1, -x_1^2/4)$ on the curve $x^2 + 4y = 0$. At this point, the values of c are obtained from the equation

$$c^2 + cx_1 + \frac{x_1^2}{4} = 0, \quad \text{or} \quad (2c + x_1)^2 = 0.$$

Therefore, there is only one value of $c = -x_1/2$. That is, only one straight line passes through any point on the curve $x^2 + 4y = 0$. This straight line is $y = -x(x_1/2) + (x_1^2/4)$. Obviously, this straight line is tangential to the singular solution curve. The parabola $x^2 + 4y = 0$ is the envelope of the family of straight lines $y = cx + c^2$.

Exercise 4.6

Find the solution of the following Bernoulli equations.

1. $yy' + xy^2 = x$.

2. $y' + y = y^2$.

3. $2xy' = 10x^3y^5 + y$.

4. $xy' = (y^2 - 1)/y$.

5. $xy' + y = y^2$.

6. $3y^2y' + xy^3 = x$.

7. $yy' = 2x - y^2$.

8. $y' + y = xy^{5/3}$.

9. $xy' + y = x^2y^2 \ln x$.

10. $xy' + 3y = x^3y^2$.

Find the general solution of the following Riccati equations.

11. $y' = 4xy^2 + (1 - 8x)y + 4x - 1$, $y = 1$ is a particular solution.

12. $y' = 3y^2 - (1 + 6x)y + 3x^2 + x + 1$, $y = x$ is a particular solution.

13. $y' = 4x^2(y - x)^2 + (y/x)$, $y = x$ is a particular solution.

14. $y' = 2e^{-x}y^2 + 3y - 4e^x$, $y = e^x$ is a particular solution.

Find the general solution and the singular solution of the following Clairaut's equations.

15. $y = xy' - (y')^3$.

16. $y = xy' - (1/y')$.

17. $y = xy' - (y')^2/2$.

18. $y = xy' - e^{2y'}$.

19. $y = xy' + e^{-y'}$.

20. If $f(t)$ in the Clairaut's equation is linear, then show that the parametric equations $x = -f'(t)$, $y = f(t) - tf'(t)$ satisfy a linear equation of the form $px + qy + r = 0$, where p, q, r are constants.

4.9 Orthogonal Trajectories of a Given Family of Curves

The general solution of a first order differential equation contains one arbitrary constant. For each value of the arbitrary constant, we obtain an integral curve. The totality of all these curves forms a one parameter family of curves. Thus, the general solution of a first order differential equation represents a one parameter family of curves. We now, formulate the problem of finding another one parameter family of curves which is orthogonal to the given one parameter family of curves (using the first order differential equation). We know that two curves intersect *orthogonally* if the angle of intersection between the curves is a right angle, that is, the tangents to the curves at the point of intersection are perpendicular.

For example, the concentric circles $x^2 + y^2 = r^2$, with centre at the origin and radius r are orthogonal to the radial lines $y = mx$ (Fig. 4.2). In polar form, these lines are $\theta = \alpha$, where α is a constant.

Therefore, the one parameter family of integral curves $x^2 + y^2 = r^2$ are the orthogonal trajectories

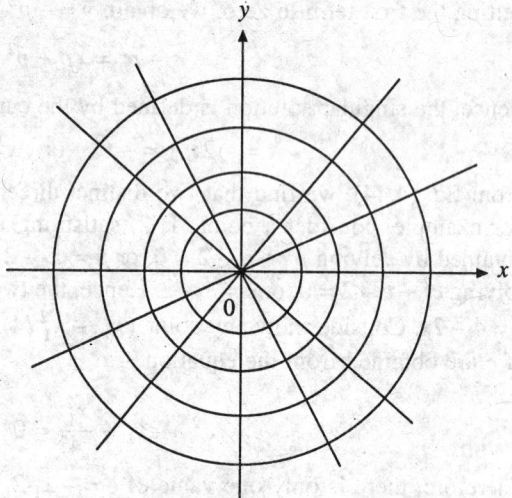

Fig. 4.2. Orthogonal familes.

of the one parameter family of straight lines $y = mx$. Let us now verify this statement using the definition that the tangents to the curves at the point of intersection are perpendicular. Differentiating $x^2 + y^2 = r^2$ we obtain the slope at any point (x, y), $y > 0$ on the curve as $m_1 = y' = -x/y$. The slopes of the straight lines are $m_2 = m = y/x$. Since $m_1 m_2 = -1$, the circles and the straight lines are orthogonal to each other.

Formulation of the Problem

Let

$$f(x, y, c) = 0 \tag{4.47}$$

be a one parameter family of curves. Differentiating and eliminating c, we can obtain the first order differential equation for which $f(x, y, c) = 0$ are the integral curves. Let the differential equation be

$$F(x, y, y') = 0. \tag{4.48}$$

Since the slope of an integral curve given by Eq. (4.47) is y', the slope of the curve which is orthogonal to this curve is $-1/y'$. Hence, the differential equation governing the family of orthogonal curves is given by

$$F\left(x, y, -\frac{1}{y'}\right) = 0. \tag{4.49}$$

Solving this equation, we obtain the one parameter family of orthogonal curves.

Example 4.30   Find the orthogonal trajectories of the hyperbolas $x^2 - y^2 = c.$

Solution   Differentiating, we obtain

$$2x - 2yy' = 0$$

which is the differential equation governing the family of hyperbolas. The differential equation governing the orthogonal trajectories is

$$2x - 2y\left(-\frac{1}{y'}\right) = 0, \quad \text{or} \quad xy' + y = 0$$

$$\text{or} \quad x\,dy + y\,dx = 0, \quad \text{or} \quad d(xy) = 0.$$

The solution of this equation is $xy = c$. Therefore, the rectangular hyperbolas $xy = c$ are the orthogonal trajectories of the hyperbolas $x^2 - y^2 = c$ (Fig. 4.3).

Example 4.31   Find the orthogonal trajectories of the family of circles passing through the points $(0, 2)$ and $(0, -2)$.

Solution   Since the circles pass through the points $(0, -2)$, $(0, 2)$ which are on the y-axis and are symmetrically placed with respect to the x-axis, the centres of the circles lie on the x-axis. Let the centres of the circles be $(h, 0)$, where h is arbitrary. The one parameter family of circles are given by

$$(x - h)^2 + (y - 0)^2 = h^2 + 4, \quad \text{or} \quad x^2 - 2xh + y^2 = 4. \tag{4.50}$$

Differentiating, we obtain $\qquad 2x - 2h + 2yy' = 0.$

Eliminating h, we get

$$2x - \frac{1}{x}(x^2 + y^2 - 4) + 2yy' = 0, \quad \text{or} \quad x^2 - y^2 + 4 + 2xyy' = 0,$$

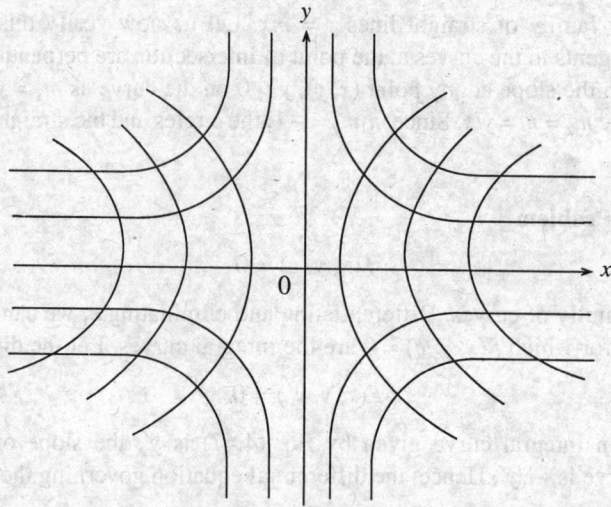

Fig. 4.3.   Example 4.30.

which is the differential equation governing the given family of curves. Therefore, the orthogonal trajectories are the solutions of the differential equation

$$x^2 - y^2 + 4 + 2xy\left(-\frac{1}{y'}\right) = 0, \quad \text{or} \quad y' = \frac{2xy}{x^2 - y^2 + 4}$$

or

$$2xy\,dx - (x^2 - y^2 + 4)\,dy = 0.$$

Comparing with $M(x, y)\,dx + N(x, y)\,dy = 0$, we have

$$M(x, y) = 2xy, N(x, y) = y^2 - x^2 - 4 \text{ and } \frac{1}{M}\left[\frac{\partial M}{\partial y} - \frac{\partial N}{\partial x}\right] = \frac{1}{2xy}[2x + 2x] = \frac{2}{y}.$$

The integrating factor is (see Theorem 4.2)

$$\text{I.F.} = e^{-\int (2/y)\,dy} = 1/y^2.$$

Multiplying the equation by $1/y^2$, we obtain

$$\frac{2x}{y}\,dx - \frac{1}{y^2}(x^2 - y^2 + 4)\,dy = 0, \quad \text{or} \quad dy + \frac{(2xy\,dx - x^2\,dy)}{y^2} - \frac{4}{y^2}\,dy = 0$$

or

$$dy + d\left(\frac{x^2}{y}\right) + d\left(\frac{4}{y}\right) = 0.$$

Integrating, we obtain $y + \dfrac{x^2}{y} + \dfrac{4}{y} = 4\,c, \quad \text{or} \quad x^2 + y^2 + 4 = 4cy.$

or

$$x^2 + (y - 2c)^2 = 4(c^2 - 1), \quad |c| > 1.$$

The orthogonal trajectories are the circles with centres at $(0, 2c)$ and radius $2\sqrt{c^2 - 1}, |c| > 1$ (Fig. 4.4).

Example 4.32   Show that the one parameter family of curves $y^2 = 4c(c + x)$ are self orthogonal (a family of curves is *self-orthogonal* if it is its own orthogonal family).

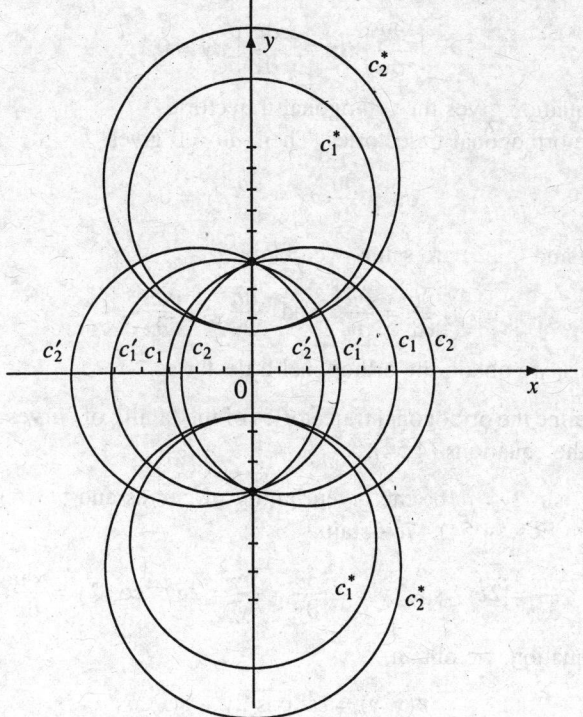

Fig. 4.4.　Example 4.31.

Solution　Differentiating, we obtain $2yy' = 4c$. Eliminating c, we obtain

$$y^2 = 2yy'\left(\frac{1}{2}yy' + x\right) = yy'(yy' + 2x) \tag{4.51}$$

which is the differential equation governing the given family. Replacing y' by $-1/y'$, we obtain the differential equation governing the orthogonal family as

$$y^2 = -\frac{y}{y'}\left(-\frac{y}{y'} + 2x\right) = \frac{y^2 - 2x\,yy'}{(y')^2}$$

or
$$y^2 = y^2(y')^2 + 2x\,yy' = yy'(yy' + 2x)$$

which is same as Eq. (4.51). Therefore, the orthogonal trajectories are the given curves themselves.

Alternate method for finding orthogonal trajectories

Let $u(x, y) = c$ be a given family of curves.

Then,
$$du = \frac{\partial u}{\partial x}\,dx + \frac{\partial u}{\partial y}\,dy = 0, \quad \text{or} \quad \frac{dy}{dx} = y' = -\frac{(\partial u/\partial x)}{(\partial u/\partial y)}$$

is the differential equation governing the given family. The orthogonal trajectories are governed by the equation

$$-\frac{1}{y'} = -\frac{(\partial u/\partial x)}{(\partial u/\partial y)}, \quad \text{or} \quad y' = \frac{(\partial u/\partial y)}{(\partial u/\partial x)} \tag{4.52a}$$

or
$$\left(\frac{\partial u}{\partial y}\right) dx - \left(\frac{\partial u}{\partial x}\right) dy = 0. \tag{4.52b}$$

The solution of this equation gives the orthogonal trajectories.

Let $v(x, y) = k$ be the orthogonal trajectories. Then, $dv = 0$ gives

$$dv = \frac{\partial v}{\partial x} dx + \frac{\partial v}{\partial y} dy = 0. \tag{4.53}$$

Since the Eqs. (4.52b) and (4.53) are same, we obtain

$$\frac{\partial v}{\partial x} = \frac{\partial u}{\partial y}, \text{ and } \frac{\partial v}{\partial y} = -\frac{\partial u}{\partial x}. \tag{4.54}$$

Solving these equations, we obtain the orthogonal trajectories.

Example 4.33 Determine the orthogonal trajectories of the family of curves $e^{2x} \sin 2y = c$, using the method described by the equations (4.54).

Solution Let $u = e^{2x} \sin 2y = c$ be the given family of curves and $v(x, y) = k$ be the orthogonal trajectories. Then, from Eqs. (4.54), we obtain

$$\frac{\partial u}{\partial x} = 2e^{2x} \sin 2y = -\frac{\partial v}{\partial y}, \quad \frac{\partial u}{\partial y} = 2e^{2x} \cos 2y = \frac{\partial v}{\partial x}.$$

Integrating the first equation, we obtain

$$v(x, y) = e^{2x} \cos 2y + g(x).$$

Substituting in the second equation, we get

$$\frac{\partial v}{\partial x} = 2e^{2x} \cos 2y + \frac{dg}{dx} = 2e^x \cos 2y, \text{ or } \frac{dg}{dx} = 0$$

or
$$g(x) = \text{constant } c_1.$$

Therefore, the orthogonal trajectories are $v(x, y) = e^{2x} \cos 2y + c_1 = k$, or $e^{2x} \cos 2y = c^*$, where c^* is an arbitrary constant.

Remark 2 Let $f(r, \theta, c) = 0$ represent a family of curves in polar coordinates form. Differentiating and eliminating c, we obtain the differential equation governing this family as $F(r, \theta, r') = 0$, where $r' = dr/d\theta$. The family of orthogonal trajectories is governed by the differential equation $F(r, \theta, (-r^2/r')) = 0$. Solving this differential equation, we obtain the orthogonal trajectories of the given family of curves.

Example 4.34 Find the orthogonal trajectories of the family of curves

    (i) $r^2 = c \sin(2\theta)$,   (ii) $r = c(\sec \theta + \tan \theta)$.

Solution

(i) Differentiating $r^2 = c \sin(2\theta)$ with respect to θ, we get

$$2r r' = 2c \cos(2\theta).$$

Eliminating c, we obtain $r' = r \cot(2\theta),$

which is the differential equation governing the given family of curves. Replacing r' by $(-r^2/r')$, we obtain the differential equation governing the orthogonal trajectories as

$$-\frac{r^2}{r'} = r \cot(2\theta), \quad \text{or} \quad r' = -r \tan(2\theta).$$

Separating the variables, we get

$$\frac{dr}{r} = -\tan(2\theta)\, d\theta$$

Integrating, we obtain

$$\ln r = \frac{1}{2} \ln \cos(2\theta) + k, \quad \text{or} \quad r^2 = c^* \cos(2\theta)$$

where $c^* = e^{2k}$ is an arbitrary constant. Hence, the orthogonal trajectories of the given family of curves are given by $r^2 = c^* \cos(2\theta)$.

(ii) Differentiating $r = c(\sec\theta + \tan\theta)$ with respect to θ, we get

$$r' = c \sec\theta (\sec\theta + \tan\theta)$$

Eliminating c, we obtain $r' = r \sec\theta$, which is the differential equation governing the given family of curves. Replacing r' by $(-r^2/r')$, we obtain the differential equation governing the orthogonal trajectories as

$$r' = -r \cos\theta.$$

Separating the variables, we get

$$\frac{dr}{r} = -\cos\theta\, d\theta.$$

Integrating, we obtain

$$\ln r = -\sin\theta + k, \quad \text{or} \quad r = c^* e^{-\sin\theta}$$

where $c^* = e^k$ is an arbitrary constant. Hence, the orthogonal trajectories of the given family of curves are given by $r = c^* e^{-\sin\theta}$.

Exercise 4.7

Find the orthogonal trajectories of the following curves.

1. $y = x + c$.
2. $x^2 + y^2 = c$.
3. $y^2 = x + c$.
4. $y = cx^2$.
5. $y = ce^x$.
6. $y = x + ce^{-x}$.
7. $y^2 = cx^3$.
8. $(x - c)^2 + y^2 = 1$.
9. $16x^2 + y^2 = c$.
10. $y = \tan x + c$.

11. Find the equation of the family of all orthogonal trajectories of the family of circles which pass through the points $(2, 0)$, $(-2, 0)$.

12. Find the equation of the family of all orthogonal trajectories of the family of circles which pass through $(0, 0)$ and have centres on the y-axis.

13. The family of curves intersecting a given family at a fixed angle α is called the *isogonal* family (also called α-trajectories of the family). Show that if the differential equation of the family is $g(x, y, y') = 0$, then the differential equation governing the isogonal family is

$$g\left(x, y, \frac{y'-m}{1+my'}\right) = 0, \quad \text{where } m = \tan \alpha.$$

14. Using the method given in problem 13, find the curves which cut the family of curves $xy = c$ at an angle of 45°.

15. Show that the family of curves $\dfrac{x^2}{c} + \dfrac{y^2}{c+2} + 1 = 0$, is self-orthogonal.

Find the orthogonal trajectories of the following family of curves.

16. $r = ce^{\theta}$.

17. $r = c\theta^2$.

18. $r = c (1 + \cos \theta)$.

19. $r = c (\operatorname{cosec} \theta + \cot \theta)$.

20. $r = 2c/(1 - \cos \theta)$.

4.10   Existence and Uniqueness of Solutions

In section 4.3, we made a remark that not all differential equations that we come across have unique solutions. The examples that we considered were

$$\left|\frac{dy}{dx}\right| + |y| = 0$$

which has only a trivial solution $y = 0$ and the differential equation

$$\left|\frac{dy}{dx}\right| + |y| + c = 0, \quad c > 0$$

which has no solution.

Consider now the solution of the initial value problem (IVP)

$$xy' = 3y, \quad y(0) = 1.$$

The differential equation has the general solution $y = cx^3$, but the IVP has no solution. However, if we modify the initial condition as $y(0) = 0$, then the new IVP admits a one parameter family of solutions $y = cx^3$, that is there exists infinite number of solutions. Therefore, in general, an initial value problem may have a unique solution, more than one solution or no solution.

Consider now, the general initial value problem for the first order equation as

$$y' = f(x, y), \quad y(x_0) = y_0. \tag{4.55}$$

Often, the independent variable x is the time denoted by t and x_0 (or t_0) is an initial point. An initial value problem then gives the time history of solutions for $t > t_0$.

Often, in mathematical modelling of a physical system, many assumptions are made to simplify the problem. Eventhough, the physical system is well defined and behaves uniquely, the mathematical model may or may not have a solution. Therefore, it is necessary to examine the existence and uniqueness of solutions of the mathematical model. With respect to the IVP given by Eq. (4.55), we would like to find answers for the following questions.

(a) **Existence**   (i) Under what conditions does the IVP given by Eq. (4.55) admits a solution? (ii) If a solution exists, then for what values of x (bounded or unbounded domain) it is defined?

(b) **Uniqueness**   If there is a solution to the IVP given by Eq. (4.55), is it unique?

We present below some results on the existence and uniqueness of the IVP given by Eq. (4.55).

Theorem 4.5 (Existence Theorem) (*Sufficient conditions*) Let $f(x, y)$ be continuous at all points in the closed rectangular region R: $|x - x_0| \le a, |y - y_0| \le b$ (Fig. 4.5) and $|f(x, y)| \le M$. Then, the initial value problem $y' = f(x, y)$, $y(x_0) = y_0$ has at least one solution. This solution is defined at least for all x in the interval

$$I = |x - x_0| < h, \quad \text{where} \quad h = \min\{a, b/M\}$$

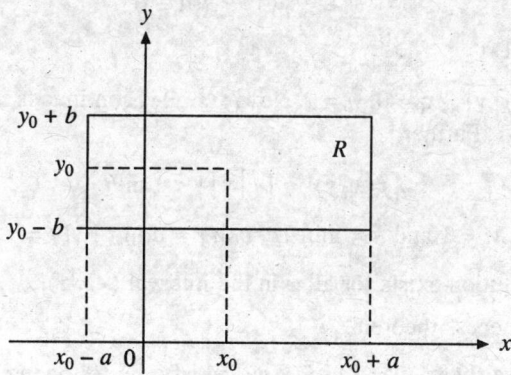

Fig. 4.5. Region where $f(x, y)$ is continuous.

The proof is omitted.

Since $y' = f(x, y)$ and $|f(x, y)| \le M$ implies $|y'| \le M$, the slope of every solution curve lies in the interval $[-M, M]$. If $h = \min\{a, b/M\} = a$, then the solution exists at least in the interval $[x_0 - a, x_0 + a]$. If $h = \min\{a, b/M\} = b/M$, then the solution exists at least in the smaller interval $[x_0 - (b/M), x_0 + (b/M)]$. For values of x outside these ranges, the curve leaves the region R and hence nothing can be said about the existence of solution at these values.

It may be pointed out that the conditions given in the theorem are sufficient and not necessary, that is, even if the conditions are violated, the problem may have a solution.

Example 4.34 Investigate the existence of solutions of the initial value problem $y' = 2x^2 + 3y^2$, $y(0) = 1$ over the rectangle $|x| \le 1$, $|y - 1| \le 1$.

Solution Here $f(x, y) = 2x^2 + 3y^2$, $x_0 = 0$ and $y_0 = 1$. The given rectangle is $|x - 0| \le 1$ and $|y - 1| \le 1$, or $-1 \le x \le 1$ and $0 \le y \le 2$. Now, $f(x, y)$ is continuous everywhere in the rectangle. Further,

$$|f(x, y)| = |2x^2 + 3y^2| \le 14, \quad -1 \le x \le 1, \ 0 \le y \le 2.$$

Therefore, at least one solution of the IVP exists. Now, $M = 14$ and

$$h = \min\{a, b/M\} = \min\{1, 1/14\} = 1/14.$$

Solution exists for all x at least in the interval $[-1/14, 1/14]$.

Example 4.35 Study the existence of solutions of the initial value problem

$$xy' = 3y, \quad y(1) = 1$$

over the rectangle $|x - 1| \le 2$, $|y - 1| \le 4$.

Solution Here, $f(x, y) = 3y/x$, $x_0 = 1$ and $y_0 = 1$. The given rectangle is $-1 \le x \le 3, -3 \le y \le 5$. Now,

$f(x, y)$ is continuous everywhere in the rectangle except at $x = 0$. Hence, Theorem 4.5 cannot be applied, that is, the theorem does not provide any conclusion about the existence of solutions in the region considered. However, solving the differential equation we find that $y = x^3$ is the unique solution of the initial value problem, showing that the conditions in the theorem are sufficient, but not necessary.

Example 4.36 Study the existence of solutions of the initial value problem

$$y' = \sqrt{|y|}, \quad y(0) = 0$$

over the rectangle $|x| < 1, |y| < 1$.

Solution Here $f(x, y) = \sqrt{|y|}$, $x_0 = 0$, $y_0 = 0$. Now, $f(x, y)$ is continuous everywhere in the rectangle $R: -1 < x < 1, -1 < y < 1$. Further,

$$|f(x, y)| = |\sqrt{|y|}| < 1 \text{ in } R$$

We have $\qquad M = 1$ and $h = \min \{a, b/M\} = \min \{1, 1\} = 1$.

Therefore, at least one solution exists for all x in the interval $(-1, 1)$.

We now present the uniqueness theorem.

Theorem 4.6 (Uniqueness theorem) Let $f(x, y)$ satisfy the following conditions

(i) $f(x, y)$ is continuous at all points in the closed rectangular region $R: |x - x_0| \le a$, $|y - y_0| \le b$ and $|f(x, y) \le M$.

(ii) $f(x, y)$ satisfies the *Lipschitz condition*

$$|f(x, y_2) - f(x, y_1)| \le L|y_2 - y_1| \qquad (4.56)$$

where L is the *Lipschitz constant* and $(x, y_1), (x, y_2) \in R$. Then, the initial value problem has a unique solution. This solution is defined at least for all x in the interval

$$I: |x - x_0| < h, \quad \text{where} \quad h = \min \{a, b/M\}.$$

The proof in omitted.

Therefore, the solution of the IVP, $y' = f(x, y)$, $y(x_0) = y_0$ exists and is unique if $f(x, y)$ is continuous and satisfies the Lipschitz condition in R.

Sometimes, it is easy to apply the following stronger condition than the condition given in Eq. (4.56). Let $\partial f/\partial y$ be bounded, that is $|\partial f/\partial y| \le N$ in R. Then the Lipschitz condition is satsified and Eq. (4.56) can be replaced by the condition $|\partial f/\partial y| \le N$. It may be noted that this is a sufficient condition and not a necessary condition. This result is also known as the Picard's theorem, which is stated below.

Theorem 4.7 (Picard's theorem) Let $f(x, y)$ and $\partial f/\partial y$ be continuous at all points in a closed rectangular region R. If (x_0, y_0) is any point in the interior of R, then there exists a positive number h such that the initial value problem (4.55) has a unique solution for all x in the interval $[x_0 - h, x_0 + h]$.

The proof in omitted.

Example 4.37 Test the existence and uniqueness of the solutions of the initial value problem $y' = \sqrt{y}$, $y(1) = 0$, in a suitable rectangle R. If more than one solution exists, then find all solutions.

Solution Here $f(x, y) = \sqrt{y}$, $x_0 = 1$, $y_0 = 0$. Now, $f(x, y)$ is continuous and bounded in R. Hence, by Theorem 4.5, at least one solution exists in some rectangle containing $(1, 0)$. Let us now test the

Lipschitz condition. We have

$$|f(x, y_2) - f(x, y_1)| = |\sqrt{y_2} - \sqrt{y_1}| = \left| \frac{(y_2 - y_1)(\sqrt{y_2} - \sqrt{y_1})}{y_2 - y_1} \right| = \left| \frac{y_2 - y_1}{\sqrt{y_2} + \sqrt{y_1}} \right|$$

or

$$\frac{|f(x, y_2) - f(x, y_1)|}{|y_2 - y_1|} = \frac{1}{\sqrt{y_2} + \sqrt{y_1}}.$$

This quantity can be made as large as possible by choosing y_1 and y_2 sufficiently small, that is, a finite value for the Lipschitz constant L cannot be determined.

We have

$$\sqrt{y_1} + \sqrt{y_2} < 2\sqrt{y}, \quad \text{if} \quad y = \max \{y_1, y_2\}$$

and

$$\left| \frac{1}{\sqrt{y_1} + \sqrt{y_2}} \right| > \frac{1}{2\sqrt{y}} > M, \quad \text{if} \quad \sqrt{y} < \frac{1}{2M}.$$

In the neighbourhood of $y = 0$, this criterion is satisfied for every $M > 0$.

Therefore, the initial value problem does not possess a unique solution. Infact, there are two solutions of the IVP, which are

$$y = (x - 1)^2/4, \quad \text{and} \quad y \equiv 0.$$

as can be easily verified. For $y \neq 0$, we have from the differential equation

$$\frac{dy}{\sqrt{y}} = dx.$$

Integrating, we have $2\sqrt{y} = x + c$, or $y = \left(\dfrac{x + c}{2} \right)^2$.

The condition $y(1) = 0$ is satisfied both by (i) $y \equiv 0$, and (ii) $[(c + 1)^2/4] = 0$, or $c = -1$.

Hence, the solutions are $y = 0$, and $y = (x - 1)^2/4$.

Example 4.38 Test for uniqueness of solutions of the initial value problem

$$(x - 2y + 1) \, dy - (3x - 6y + 2) \, dx = 0, \; y(0) = 0$$

over the rectangle R: $|x| \leq 1/4$, $|y| \leq 1/4$. Determine the interval for x, over which the solution is guaranteed.

Solution We have

$$f(x, y) = \frac{3x - 6y + 2}{x - 2y + 1}, \quad x_0 = 0, \, y_0 = 0$$

$$|f(x, y)| = \frac{|3x - 6y + 2|}{|x - 2y + 1|} < \frac{3|x| + 6|y| + 2}{1 - |x| - 2|y|}$$

$$< \frac{3(1/4) + 6(1/4) + 2}{1 - (1/4) - 2(1/4)} = 17 = M$$

$$|f_y| = \left| \frac{(x - 2y + 1)(-6) - (3x - 6y + 2)(-2)}{(x - 2y + 1)^2} \right| = \left| \frac{-2}{(x - 2y + 1)^2} \right| < \frac{2}{(1/4)^2} = 32 = N.$$

Also
$$h = \min \left\{ a, \frac{b}{M} \right\} = \min \left\{ \frac{1}{4}, \frac{1}{68} \right\} = \frac{1}{68}.$$

Conditions of Theorem 4.7 are satisfied. Therefore, the IVP has a unique solution atleast for all x in the interval $|x| \leq 1/68$. From Example 4.10, the solution of the problem is

$$\frac{1}{5}(x - 2y) + \frac{2}{25} \ln \left| x - 2y + \frac{3}{5} \right| + x = \frac{2}{25} \ln \left(\frac{3}{5} \right).$$

It can be seen that the solution is defined in a much larger interval for x than predicted.

4.10.1   Picard's Iteration Method of Solution

Picard's theorem (Theorem 4.7) gives a sufficient condition for the existence and uniqueness of the solution of the initial value problem given by Eq. (4.55). Integrating the differential equation $y' = f(x, y)$, we obtain

$$\int_{x_0}^{x} y' \, dx = \int_{x_0}^{x} f(x, y) \, dx, \quad \text{or} \quad y(x) = y(x_0) + \int_{x_0}^{x} f(x, y) \, dx. \tag{4.57}$$

The solution of the initial value problem defined in Eq. (4.55) is given by Eq. (4.57). Conversely, if $y(x)$ satisfies Eq. (4.57), then it satisfies the initial value problem given by Eq. (4.55). The integral equation (4.57) is solved iteratively assuming the first approximation to $y(x)$ as $y_0(x) = y(x_0) = y_0$. This iterate is called $y_1(x)$ and is given by

$$y_1(x) = y(x_0) + \int_{x_0}^{x} f[x, y_0(x)] \, dx. \tag{4.58}$$

The second approximation $y_2(x)$ is given by

$$y_2(x) = y(x_0) + \int_{x_0}^{x} f[x, y_1(x)] \, dx. \tag{4.59}$$

The sequence of approximations $y_0(x), y_1(x), y_2(x), \ldots$ given by

$$y_{n+1}(x) = y(x_0) + \int_{x_0}^{x} f[x, y_n(x)] \, dx \tag{4.60}$$

converges to $y(x)$, for x sufficiently close to x_0. The solution $y(x)$, thus obtained, is the unique solution of the initial value problem defined by Eq. (4.55).

Example 4.39   Find the solution of the initial value problem $y' = 2y - x$, $y(0) = 1$, using the Picard's iteration method. Compare with the exact solution.

Solution   We have

$$y(x) = y(x_0) + \int_{x_0}^{x} f(x, y) \, dx = 1 + \int_{0}^{x} (2y - x) \, dx.$$

The iteration is defined by

$$y_{n+1}(x) = 1 + \int_{0}^{x} [2y_n - x] \, dx, \quad y_0 = 1.$$

Setting $n = 0, 1, 2, \ldots$ we get

$$y_1(x) = 1 + \int_0^x [2y_0 - x] \, dx = 1 + \int_0^x (2 - x) \, dx = 1 + 2x - \frac{x^2}{2}$$

$$y_2(x) = 1 + \int_0^x [2 + 4x - x^2 - x] \, dx = 1 + 2x + \frac{3}{2}x^2 - \frac{x^3}{3}$$

$$y_3(x) = 1 + \int_0^x [2 + 4x + 3x^2 - \frac{2}{3}x^3 - x] \, dx$$

$$= 1 + 2x + \frac{3}{2}x^2 + x^3 - \frac{1}{6}x^4$$

$$y_4(x) = 1 + \int_0^x \left[2 + 4x + 3x^2 + 2x^3 - \frac{1}{3}x^4 - x \right] dx$$

$$= 1 + 2x + \frac{3}{2}x^2 + x^3 + \frac{1}{2}x^4 - \frac{1}{15}x^5.$$

The given equation is a linear differential equation, whose integrating factor is e^{-2x}. The solution of the differential equation is

$$y(x) = \frac{1}{4}(2x + 1) + ce^{2x}.$$

Applying the initial condition, we obtain $c = 3/4$. The solution of the IVP is

$$y(x) = \frac{1}{4}[(2x + 1) + 3e^{2x}] = \frac{1}{4}\left[(2x + 1) + 3\left(1 + \frac{2x}{1!} + \frac{4x^2}{2!} + \frac{8x^3}{3!} + \ldots \right) \right]$$

$$= \frac{1}{4}\left[4 + 8x + 6x^2 + 4x^3 + 2x^4 + \frac{4}{5}x^5 + \ldots \right]$$

$$= 1 + 2x + \frac{3}{2}x^2 + x^3 + \frac{1}{2}x^4 + \frac{1}{5}x^5 + \ldots$$

whose first five terms agree with the first five terms of $y_4(x)$.

Example 4.40 Using Picard's approximation method find two iterates of the initial value problem

$$y' = y + y^2, \quad y(0) = 1.$$

Compare with the exact solution.

Solution We have

$$y(x) = y(x_0) + \int_0^x (y + y^2) \, dx.$$

The iteration is defined by $\quad y_{n+1}(x) = 1 + \int_0^x (y_n + y_n^2) \, dx, \quad y_0 = 1.$

Setting $n = 0, 1, 2, \ldots$ we get

$$y_1(x) = 1 + \int_0^x (y_0 + y_0^2)\, dx = 1 + 2x,$$

$$y_2(x) = 1 + \int_0^x (y_1 + y_1^2)\, dx = 1 + \int_0^x [(1 + 2x) + (1 + 2x)^2]\, dx$$

$$= 1 + \int_0^x (2 + 6x + 4x^2)\, dx = 1 + 2x + 3x^2 + \frac{4}{3}x^3.$$

The given equation is a Bernoulli equation with $n = 2$. Set $v = y^{-1}$. We get

$$\frac{dv}{dx} = -y^{-2}\frac{dy}{dx} = -y^{-2}[y + y^2] = -\left[\frac{1}{y} + 1\right] = -[v + 1]$$

or

$$\frac{dv}{dx} + v = -1.$$

The integrating factor is e^x. Therefore,

$$ve^x = -\int e^x\, dx + c = -e^x + c, \quad \text{or} \quad v = ce^{-x} - 1$$

or

$$y(x) = \frac{1}{ce^{-x} - 1}$$

Using the initial condition, we obtain $c = 2$. Hence,

$$y(x) = \frac{1}{2e^{-x} - 1} = \frac{1}{1 - [2x - x^2 + (x^3/3) - \ldots]} = [1 - \{2x - x^2 + (x^3/3) - \ldots\}]^{-1}$$

$$= 1 + \{2x - x^2 + \frac{x^3}{3} - \ldots\} + \{\ldots\}^2 + \{\ldots\}^3 + \ldots = 1 + 2x + 3x^2 + \frac{13}{3}x^3 + \ldots$$

whose first three terms agree with the first three terms of $y_2(x)$.

Exercise 4.8

Study the existence and uniqueness of solutions of the following initial value problems.

1. $y' = y$, $y(0) = 1$, $R: |x| \le 1$, $|y - 1| \le 1$.
2. $y' = (1 + 2x + 3y)/(2 + x^2 + y^2)$, $y(0) = 0$, $R: |x| \le 2, |y| \le 1$.
3. $y' = (1 - x + y)/(2 + x^2 + y^2)$, $y(1) = 1$, $R: |x - 1| \le 1, |y - 1| \le 1$.
4. $y' = 2xy/(4 + x^2 + y^2)$, $y(-1) = 1$, $R: |x + 1| \le 1$, $|y - 1| \le 2$.
5. $y' = 3xy/(2 + \cos xy)$, $y(0) = 0$, $R: |x| \le 1$, $|y| \le 2$.
6. Discuss the applicability of the existence theorem to $y' = 4y/x$, $y(0) = 0$. Find the solution directly. What do you conclude?
7. Discuss the applicability of the existence theorem to $x\, dx - y\, dy = 0$, $y(0) = 0$.
8. Discuss the existence and uniqueness of solution of the initial value problem $y' = |y|^{3/4}$, $y(0) = 0$.

9. Show that $f(x, y)$ in $y' = |\sin y|$ satisfies the Lipschitz condition in the whole x-y plane. Does $\partial f/\partial y$ exist everywhere in the x-y plane?

10. Find the largest possible interval for x, for which the theorem of existence guarantees at least one solution for the initial value problem $y' = 16 + y^2$, $y(0) = 0$. Find the interval for x in which the solution actually exists.

11. Let $y' + p(x)y = q(x)$, be a linear first order differential equation where $p(x)$, $q(x)$ are continuous in $|x - x_0| \le a$. Show that this problem has always a unique solution. Find the Lipschitz constant in this case.

12. Show that $\cos^2 x$ and $(1 + \cos 2x)/2$ are solutions of the initial value problem $y' + \sin 2x = 0$, $y(0) = 1$. Use the existence and uniqueness theorem to show that $\cos^2 x = (1 + \cos 2x)/2$.

Using Picard's iteration method find the first four approximations to the solution of the following initial value problems. Compare with the exact solution.

13. $y' = x + y$, $y(0) = 1$.

14. $y' = x - y$, $y(0) = 1$.

15. $y' = x + y$, $y(1) = 1$.

16. $y' = xy$, $y(1) = 2$.

17. $y' = x - y$, $y(0) = 0$.

18. $y' = x - 2y$, $y(-1) = 1/4$.

Using Picard's iteration method find the first two approximations to the solution of the following initial value problems.

19. $y' = x^2 + y^2$, $y(0) = 1$.

20. $y' = x^2/(4 + y^2)$, $y(0) = 0$.

21. $y' = 1 + y^2$, $y(0) = 0$.

22. $y' + y = y^2$, $y(1) = 2$.

4.11   Answers and Hints

Exercise 4.1

1. two, one, linear.
3. one, two, non-linear.

2. two, one, linear.
4. one, one, non-linear.

5. one, two, non-linear.
7. two, two, non-linear.

6. two, one, non-linear.
8. two, one, non-linear.

9. two, two, non-linear.
11. $y' = qy$.

10. one, one, non-linear.
12. $y'' + \theta^2 y = 0$.

13. $y'' + p^2 y = 0$.
15. $y'' + 4y' + 8y = 0$.
17. $yy' + x = 0$.

14. $y'' - 3y' + 2y = 0$.
16. $x^2 y'' + 2xy' - 2y = 0$.
18. $(x^2 - y^2)y' = 2xy$.

19. $y \log [\{1 + (y')^2\}^{1/2} + y'] = x \{1 + (y')^2\}^{1/2}$.

20. $(\tan x)y' - y = 0$.

21. $a^2(y'')^2 = [1 + (y')^2]^3$.

22. $(3x^2 - y^2)yy' = x(3y^2 - x^2)$.

23. $y''' = 4y'$.

24. $y''' + 6y'' + 11y' + 6y = 0$.

25. $y = xy' - \dfrac{1}{4}(y')^2$.

31. $m = -1, -2$.

32. $m = 1, 2, 3$.

33. $m = -1, 1, 2$.

34. $m = 2 \pm \sqrt{3}$.

35. $m = 1 \pm \sqrt{3}\, i$.

Exercise 4.2

1. $y = c \ln |x|$.

2. $y = c \cosh x$.

3. $y = cx^{b/a}$.

4. $y^2 = x + (\sin 2wx)/(2w) + c$.

5. $x^3 + y^3 = c$.

6. $ay = ce^{-ax} - b$.

7. $y = c \cos 2x$.

8. $y = c/(1 + x + x^2)^3$.

9. $e^{-y} + (x^3/3) + e^x = c$.

10. $x - y + \log |(x/y)| = c$.

11. $\sqrt{1 - y^2} = (x + 1)e^{-x} + c$.

12. $y = ce^{x+x^2} - 1$.

13. $\cosh (2y) = \log |x| - (x^2/2) + c$.

14. $cy^3 = xe^{-3x}$.

15. $y = c_1 e^{-16x} + c_2$.

16. $y = c_1 \ln |x| + c_2$.

17. $y = \ln |c_1 x + c_2|$.

18. $x = e^y + c_1 y + c_2$.

19. $(y - 1)e^y = c_1 x + c_2$,

20. $y = c_1 e^{3x} + c_2$.

21. $e^y = x^4$,

22. $y = (\ln |x|)^2$.

23. $y^2 + 4y = 2(1 - \cos x)$,

24. $4y^3 = 3(2x + \sin 2x)$.

25. $y \cos x = 1$,

26. $y^2 = 4 \sec 2x$.

27. $\ln r = \theta - (\sin 2\theta)/2 - \pi/2$,

28. $s = e^{(t^3/3) - 2t}$.

29. $y = 4 \cos x - 3$,

30. $e^x = 1 - (1 - e)e^{-t}$.

31. $3y^2 + 2y^3 = 3x^2 + 28$,

32. $\sin (y/x) + \ln |x| = c$.

33. $y - x = c(x + 1)^2$,

34. $y = x + c \sin x$.

35. $(y - x)^2 = c - 4y$,

36. $e^{xy} = x + c$.

37. $\sin (y/x) = ce^x$,

38. $y = x + (1 + ce^{2x})/(1 - ce^{2x})$.

39. $\sin (x^2 y^2) = ce^x$,

40. $y = x + x(c - x^2)^{-1/2}$.

41. $dp/dh = kp$ along with the conditions $p(0) = p_0$, $p(6000) = p_0/2$, $p(h) = p_0 e^{kh}$, where $k = -\ln 2/6000$.

42. $dp/dt = kp$, along with the conditions $p(0) = 10,000$, $p(10) = 25,000$, $p(t) = 10,000 \, e^{kt}$, where $k = \ln (2.5)/10$, $p(20) = 62,500$.

43. $dV/dt = -kS$, where $V = 4\pi r^3/3$ and $S = 4\pi r^2$. We get $dr/dt = -k$ or $r(t) = -kt + c$. Applying the conditions, we get $r(t) = 4 - 0.4 \, t$.

44. The IVP is $y' = -ky$, $y(0) = y_0$. The solution is $y(t) = y_0 \, e^{-kt}$. The amount of substance present after 1000 years is $y_0 - (y_0/2) = y_0/2$. Therefore, $k = \ln 2/1000$. In 50 years, the amount present is $y(50) = y_0 e^{-50k} \approx (1 - 50k)y_0$. The amount that disintegrates is $50 \, ky_0$ and the percentage is $5000 \, k \approx 5 \ln 2 \approx 3.5\%$.

45. $i = E[1 - e^{-(Rt)/L}]/R$. As $t \to \infty$, $i \to E/R$.

46. $xy' = y$, $y = cx$.

47. $y' = y$, $y = ce^x$.

48. $y = e^x - 1$.

49. $y(1) = 1$, $y'(1) = -1$; $xy = 1$.

50. $y = (3e^{4t} + 5)/8$, 0.8 sec.

Exercise 4.3

1. Substitute $4x + y = u$; $\tan^{-1}[(4x + y)/2] = 2(c + x)$.

2. Substitute $2x - y + 1 = u$; $2x - y + 1 - \sqrt{2} = c(2x - y + 1 + \sqrt{2})e^{-2\sqrt{2}x}$.

3. Substitute $y - 1 = u$; $\sin^{-1}(y - 1) = x + c$.

4. $e^{y/x} = \ln|cx|$.

5. $3xy^2 + 2y^3 = 6x^3 \ln|cx|$.

6. $(y - x)\ln|cx| + x = 0$.

7. $\sin(y/x) = \ln|cx|$.

8. $y = x \tan(\ln|cx|)$.

9. $2\tan^{-1}(y/x) + \ln(x^2 + y^2) = c$.

10. $y(y + x) = cx^3$.

11. $y^2 + 2xy = 3x^3$.

12. $e^{y/x} + \sin^{-1}(y/x) = \ln|x| + e + \pi/2$.

13. $3\cos^{-1}(y/x) - \ln|x| = 0$.

14. $2y^2 \ln(cy) = x^2$, $c = e^{1/8}/2$.

15. $4x^2 + y^2 + 4xy - 6x - 2y = c$.

16. $2y + \ln|x + 3y - 1| = x + c$.

17. $6x + 6y - 7\ln|3x + 3y + 2| = 9x + c$.

18. $2x - y + \ln|x - 2y - 2| = c$.

19. $2\tan^{-1}v + \ln(1 + v^2) = -\ln X^2 + c$. $2\tan^{-1}\left(\dfrac{2y - 1}{2x - 3}\right) + \ln\left[\dfrac{1}{4}(2x - 3)^2 + \dfrac{1}{4}(2y - 1)^2\right] = c$.

20. $\ln|x - 1| - \dfrac{x - 1}{y + 3} + \ln\left|\dfrac{y + 3}{x - 1}\right| = c$.

21. $y[\sec(x/y) + \tan(x/y)] = c$.

22. $\ln y^2 + e^{-x^2/y^2} = c$.

23. $\dfrac{dy}{dx} = \dfrac{M(x, y)}{N(x, y)} = f(y/x) = f(\tan \theta)$

$\dfrac{dy}{dx} = \dfrac{dy/d\theta}{dx/d\theta} = \dfrac{\sin\theta\,(dr/d\theta) + r\cos\theta}{\cos\theta\,(dr/d\theta) - r\sin\theta}$, $\dfrac{dr}{d\theta} = -\dfrac{r[\cos\theta + \sin\theta f(\tan\theta)]}{[\sin\theta - \cos\theta f(\tan\theta)]}$.

24. $2\dfrac{dr}{d\theta} = r$, $\ln|cr^2| = \theta$, or $\tan^{-1}(y/x) = \ln|c(x^2 + y^2)|$.

25. $\dfrac{dr}{d\theta} = 3r; 3\theta = \ln|cr|$ or $3\tan^{-1}(y/x) = \ln|c(x^2 + y^2)^{1/2}|$.

Exercise 4.4

1. $x + e^x + (y^2/2) = c$.

2. I.F. $= 1/(xy)$, $xye^y = c$.

3. I.F. $= 1/\sinh x$, $\sinh^2 x = ce^{-y}$.

4. Exact, $\cosh x \cos y = c$.

5. $d(x^3y) + d(y \ln x) = 0$, $x^3y + y \ln x = c$.

6. $d(e^{xy}) + d(y^2) = 0$, $e^{xy} + y^2 = c$.

7. I.F. $= 1/(xy)$, $x^2y = ce^y$.

8. I.F. $= 1/(x^2 + y^2)$, $\tan^{-1}(y/x) = e^y + c$.

9. $x^2 + y^2 = ce^{2y^2}$

10. I.F. $= 1/y^2$, $x(y - 1) = cy$.

11. I.F. $= 1/y^2$, $(x/y) + 3x^2 + 4\ln|y| = c$.

12. Exact, $x(x + e^y) = c$.

13. Exact, $y(1 + xy) = c$.

14. Exact, $y(1 + x^2) = c$.

15. Exact, $(e^{2y} + 1)\sin x = c$.

16. $a = 3/2$.

17. $g'(y) = h'(x)$.

18. $k = 1$.

19. $a = -d$, $b = c$.

20. $\left[\left(\dfrac{\partial M}{\partial y} - \dfrac{\partial N}{\partial x}\right)\Big/N\right] = -\dfrac{2}{x}$, I.F. $= \dfrac{1}{x^2}$, $1 - y = cx$.

21. I.F. $= e^x$, $e^x + e^y = c$.

22. $\left[\left(\dfrac{\partial M}{\partial y} - \dfrac{\partial N}{\partial x}\right)\Big/N\right] = \dfrac{2}{x}$, I.F. $= x^2$, $x^6 + 2x^3 + 2x^3y^3 = c$.

23. $\left[\left(\dfrac{\partial M}{\partial y} - \dfrac{\partial N}{\partial x}\right)\Big/ N\right] = \dfrac{3}{x}$, I.F. $= x^3$, $x^4 y + (x^7/7) = c$.

24. $\left[\left(\dfrac{\partial M}{\partial y} - \dfrac{\partial N}{\partial x}\right)\Big/ M\right] = \dfrac{3}{y}$, I.F. $= \dfrac{1}{y^3}$ · Integrate $\dfrac{\partial f}{\partial x} = M$

$f(x, y) = x^2 e^y + (x/y) + (x/y^2) + g(y)$; $g(y) = k$; $x^2 e^y + (x/y) + (x/y^2) = c$.

25. $\left[\left(\dfrac{\partial M}{\partial y} - \dfrac{\partial N}{\partial x}\right)\Big/ N\right] = 3x^2$; I.F. $= e^{x^3}$; $y(x + 4)e^{x^3} = c$.

26. $\left[\left(\dfrac{\partial M}{\partial y} - \dfrac{\partial N}{\partial x}\right)\Big/ M\right] = -\dfrac{1}{y}$; I.F. $= y$; $d(y^2 x) + \dfrac{1}{2} d(x^2 y^4) + 2y^5 dy = 0$

$6xy^2 + 3x^2 y^4 + 2y^6 = c$.

27. $\left[\left(\dfrac{\partial M}{\partial y} - \dfrac{\partial N}{\partial x}\right)\Big/ N\right] = \dfrac{1}{x}$, I.F. $= x$; $d(6x^2 y) + \dfrac{3}{2} d(x^2 y^4) + 4x^4 dx = 0$,

$6 x^2 y + \dfrac{3}{2} x^2 y^4 + \dfrac{4}{5} x^5 = c$.

28. $\dfrac{1}{Mx + Ny} = \dfrac{1}{x(x^2 - y^2)}$ is the I.F. Integrating $\dfrac{\partial f}{\partial y}$, we get $f = \ln |x^2 - y^2| + g(x)$. We get

$g'(x) = -1/x$ or $g(x) = -\ln |x| + k$; $x^2 - y^2 = cx$.

29. $\dfrac{1}{Mx + Ny} = \dfrac{1}{y^2 - x^2}$ is the I.F. Integrating $\dfrac{\partial f}{\partial y}$. We get

$f(x, y) = \ln |y - x| - \ln |y + x| + \dfrac{1}{2} \ln |y^2 - x^2| + g(x)$. We get

$g'(x) = 0$, or $g(x) = c$; $(y - x)^3 = c(y + x)$.

30. $\dfrac{1}{Mx + Ny} = \dfrac{1}{x^2 y}$ is the I.F. $y = ce^{y/x}$.

31. Exact, $x^3 y^4 = 16$. **32.** Exact, $(x - 1)^2 + (y + 1)^2 = 1$.

33. I.F. $= \sqrt{x}$, $yx^{3/2} = 1$. **34.** Exact, $x^2 y + \sin \pi y = 1$.

35. Exact, $y \cos x = \sin x$. **36.** Exact, $y^2 + e^{x^2 + y^2} = 1$.

37. I.F. $= 1/y^3$; $y^2 = e^{x^2/y^2}$. **38.** Exact, $x^4 y^3 + \ln |x/y| = 1$.

39. Exact, $x^2 - 2y \sin x = (\pi^2/4) - 2$. **40.** $e^{xy} + 4xy^3 - y^2 + 3 = 0$.

41. Exact, $x^2 y + xe^y = e + 1$. **42.** I.F. $= 1/(x^2 + y^2)$, $x^2 + y^2 = 2e^{2 - 2x}$.

43. I.F. $= y$, $y^2(x^2 + y^2 + 2) = 3$.

44. The condition of exactness of $Mx^k\, dx + Nx^k\, dy = 0$ gives the relation:

$4 = (k + 1)$, or $k = 3$; $x^4 y = c$.

45. The condition of exactness of $MF\, dx + NF\, dy = 0$ gives the equation.

Exercise 4.5

1. I.F. $= e^x$, $y = \dfrac{1}{2} (\sin x - \cos x) + ce^{-x}$. **2.** I.F. $= x$, $xy = c - \cos x$.

3. I.F. $= e^{x^2/2}$, $y = 2 + ce^{-x^2/2}$.
4. I.F. $= 1/x$, $y = x^2 + 2x \ln |x| - 1 + cx$.

5. I.F. $= e^{-2x}$, $y = \dfrac{1}{13} (3 \sin 3x - 2 \cos 3x) + ce^{2x}$.

6. I.F. $= e^{-2t}$, $xe^{-2t} = (t^3/3) + c$.
7. I.F. $= e^u$, $s = 1 + \left(c + \dfrac{u^2}{2} \right) e^{-u}$.

8. I.F. $= e^{-3x}$, $y = -\dfrac{1}{13} (3 \sin 2x + 2 \cos 2x) + ce^{3x}$.

9. I.F. $= e^{4x}$, $y = x^2 + ce^{-4x}$.
10. I.F. $= e^{-x^2}$, $y = x^2 + ce^{x^2}$.

11. I.F. $= x$, $y = \dfrac{x}{2} + \dfrac{x^3}{4} + \dfrac{c}{x}$.
12. I.F. $= e^{3x}$, $y = \dfrac{1}{5} e^{2x} + 2 + ce^{-3x}$.

13. I.F. $= \cos x$, $y = c \sec x - 1$.
14. I.F. $= x^2 e^{x^2}$, $x^2 y = 3 + ce^{-x^2}$.

15. I.F. $= \operatorname{cosec} x$, $y = \sin x \left(\log |\sin x| - \dfrac{1}{2} \sin^2 x + c \right)$.

16. $y = [c + \sin x - x \cos x]/(1 + x^2)$.
17. I.F. $= e^{x^2}$, $y = [(x^2/2) + c] e^{-x^2}$.

18. $y = [c + e^{x^2}]/x$.
19. $y = [c + (x^4/4)]/x^2$.

20. $2xe^{2x}y = x^2 + e^{2x} + c$.
21. $y = (1 + x) e^x$.

22. $y = (3 + x) x^3$.
23. $y = (3 - 2 \cos x) \cos x$.

24. $y = x^3 e^x - x$.
25. $y = 1 + x^2 - e^{-x}$.

26. $y = 2x^2 + (e^x + \sin x) x^3$.
27. $xy = x^2 + (x^4/4) + (x^6/6) - (17/12)$.

28. $y = (1 + x^2) [1 + \ln (1 + x^2)]$.
29. $y = (5 - x^2 + 2x^2 \ln x)/(4x)$.

30. $y = \sec x \ln (\sec x + \tan x)$.
31. $x = (c + y)e^{-y}$.

32. $x = y^2(c + y^2)/2$.
33. $x = (c + y) e^y$.

34. $x = ce^{-1/y} - 1$.
35. $i = [R \sin \omega t + \omega L (e^{-(Rt)/L} - \cos \omega t)]/(R^2 + \omega^2 L^2)$.

Exercise 4.6

1. $v = y^2$, $y^2 = 1 + ce^{-x^2}$.
2. $v = 1/y$, $y = 1/(1 + ce^x)$.

3. $v = y^{-4}$, $y^4 = x^2/(c - 4x^5)$.
4. $y = (1 + cx^2)^{1/2}$.

5. $v = 1/y$, $y = 1/(cx + 1)$.
6. $v = y^3$, $y = [1 + ce^{-x^2/2}]^{1/3}$.

7. $v = y^2$, $y^2 = (2x - 1) + ce^{-2x}$.
8. $v = y^{-2/3}$, $y^{-2/3} = \dfrac{1}{2} (2x + 3) + ce^{(2x)/3}$.

9. $v = 1/y$, $y = 1/[x(c + x - x \ln |x|)]$.
10. $v = 1/y$, $y = 1/[x^3(c - \ln |x|)]$.

11. $y = 1 + (1/z)$, where $z = ce^{-x} - 4(x - 1)$.

12. $y = x + (1/z)$, where $z = 3 + ce^x$.
13. $y = x + (1/z)$, where $z = (c - x^4)/x$.

14. $y = e^x + (1/z)$, where $z = (ce^{-7x} - e^{-x})/3$.

15. $y = cx - c^3$, $y^2 = 4x^3/27$.
16. $y = cx - (1/c)$, $y^2 + 4x = 0$.

17. $y = cx - (c^2/2)$, $2y = x^2$.
18. $y = cx - e^{2c}$, $y = [\ln (x/2) - 1](x/2)$, $x > 0$.

19. $y = cx + e^{-c}$, $y = (1 - \ln x) x$, $x > 0$.

20. $f(t) = at + b$, $x = -a$, $y = b$. Therefore, $px + qy + r = 0$ where p, q, r are suitable constants.

Exercise 4.7

1. $y + x = k$.

2. $y = kx$.

3. $y = ke^{-2x}$.

4. $x^2 + 2y^2 = k$.

5. $2x + y^2 = k$.

6. $x = y - 2 + ke^{-y}$.

7. $2x^2 + 3y^2 = k$.

8. $\log\left[\dfrac{1 - \sqrt{1 - y^2}}{1 + \sqrt{1 - y^2}}\right] + 2\sqrt{1 - y^2} = 2x + k$.

9. $y^{16} = kx$.

10. $2x + 4y + \sin 2x = k$.

11. Centres are on y-axis. Circles are $x^2 + (y - c)^2 = 4 + c^2$. Orthogonal trajectories are governed by a Bernoulli equation. Orthogonal trajectories are

$$(x - (k/2))^2 + y^2 = (k^2 - 16)/4, \quad k^2 > 16.$$

12. $(x - (k/2))^2 + y^2 = k^2/4$.

13. At P both curves have the same x, y values. (Fig. 4.6)

$$y' = \tan\phi = \tan(\psi - \alpha) = \frac{\tan\psi - m}{1 + m\tan\psi} = \frac{(y^*)' - m}{1 + m(y^*)'}$$

where $m = \tan\alpha$, slope at $P(x^*, y^*)$ for $C_1 = (y^*)'$ and slope at $P(x, y)$ for $C_2 = y'$. At P,

$$g(x, y, y') = g\left(x, y, \frac{(y^*)' - m}{1 + m(y^*)'}\right) = 0, \quad \text{or} \quad g\left(x, y, \frac{y' - m}{1 + my'}\right) = 0.$$

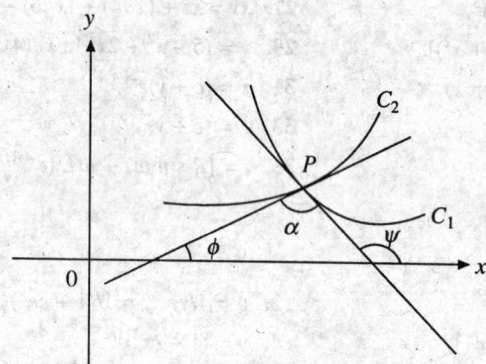

Fig. 4.6.   Problem 13, Exercise 4.7.

14. Replacing y' by $(y' - m)/(1 + my')$, where $m = \tan 45° = 1$, we obtain the equation governing the required trajectories as $y' = (x - y)/(x + y)$. The required trajectories are $y^2 - x^2 + 2xy = c$.

15. The differential equation of the family is $(x + yy')(y - xy') + 2y' = 0$. Replacing y' by $-1/y'$, we obtain the same equation.

16. $r = ke^{-\theta}$.

17. $r = ke^{-\theta^2/4}$.

18. $r = k\sin^2(\theta/2)$.

19. $r = ke^{-\cos\theta}$.

20. $r = k/\cos^2(\theta/2)$, or as $r = k*/(1 + \cos\theta)$.

Exercise 4.8

1. Since R is $-1 \le x \le 1$, $0 \le y \le 2$, $f(x, y) = y$, $|f| \le 2, |f_y| = 1 = N$. Solution exists and is unique in at least $|x| < 1/2$.

2. $|f| \leq 4 = M, |f_y| = 31/4 = N.$ Solution exists and is unique in at least $|x| \leq 1/4.$

3. $|f| \leq 3/2 = M, \quad |f_y| \leq 11/2 = N.$ Solution exists and is unique in at least $|x - 1| \leq 2/3$

4. $|f| \leq 3 = M, \quad |f_y| \leq 17/4 = N.$ Solution exists and is unique in at least $|x + 1| \leq 2/3.$

5. $|f| \leq 6 = M, \quad |f_y| \leq 15 = N.$ Solution exists and is unique in at least $|x| \leq 1/3.$

6. $f = 4y/x$ is not defined for $x = 0$. Theorem cannot be applied. Infinite number of solutions given by $y = cx^4.$

7. $f = x/y$ is not defined for $x = 0, y = 0$. The existence theorem is not applicable. Solution is $x^2 - y^2 = 0.$

8. f is continuous, $|f| \leq M$ over any rectangle $|x| \leq a, |y| \leq b$. At least one solution exists.
$$\left| \frac{f(x, y_2) - f(x, 0)}{y_2 - 0} \right| = \frac{1}{|y_2|^{1/4}}$$ is not bounded. Lipschitz condition is not satisfied. The problem does not have a non-trivial solution. $y = 0$ is the only solution.

9. $\partial f/\partial y$ does not exist when $y = 0.$

10. $f(x, y)$ is continuous and $|f(x, y)| < 16 + b^2 = M, \quad |f_y| < 2b = N.$ Solution exists and is unique. $h = \min \{a, b/(16 + b^2)\}$ gives $h = 1/8$ where h is the largest value of $b/(16 + b^2)$. Hence, the required interval is $|x| < 1/8$. The solution is $y = 4 \tan 4x, |x| < \pi/8.$

11. $f(x, y)$ is continuous and bounded. Lipschitz condition gives $L = \max |p(x)|$. Hence, the equation has a unique solution in $|x - x_0| \leq a.$

12. Substitution shows that both the given functions are solutions. Now $f(x, y)$ is continuous and bounded in $|x| \leq a, |y - 1| \leq b$ and f_y is bounded. Hence, the solution exists and is unique.

13. $y_1 = 1 + x + (x^2/2), y_2 = 1 + x + x^2 + (x^3/6), \quad y_3 = 1 + x + x^2 + (x^3/3) + (x^4/24),$

$y_4 = 1 + x + x^2 + (x^3/3) + (x^4/12) + (x^5/120).$ All terms of y_4 agree with the exact solution $y(x) = 2e^x - x - 1.$

14. $y_1 = 1 - x + (x^2/2), \quad y_2 = 1 - x + x^2 - (x^3/6), \quad y_3 = 1 - x + x^2 - (x^3/3) + (x^4/4!),$

$y_4 = 1 - x + x^2 - (x^3/3) + (x^4/12) - (x^5/120).$ First five terms of y_4 agree with the first five terms of the exact solution $y(x) = x - 1 + 2e^{-x}.$

15. $y_1 = 1 + 2(x - 1) + (1/2)(x - 1)^2, \quad y_2 = 1 + 2(x - 1) + (3/2)(x - 1)^2 + (1/6)(x - 1)^3$

$y_3 = 1 + 2(x - 1) + (3/2)(x - 1)^2 + (1/2)(x - 1)^3 + (1/24)(x - 1)^4$

$y_4 = 1 + 2(x - 1) + (3/2)(x - 1)^2 + (1/2)(x - 1)^3 + (1/8)(x - 1)^4 + (1/120)(x - 1)^5.$

First five terms of y_4 agree with the first five terms of the exact solution $y(x) = -2 - (x - 1) + 3e^{(x-1)}.$

16. $y_1 = 2 + (x^2 - 1), \quad y_2 = 2 + (x^2 - 1) + \frac{1}{4}(x^2 - 1)^2.$

$y_3 = 2 + (x^2 - 1) + \frac{1}{4}(x^2 - 1)^2 + \frac{1}{24}(x^2 - 1)^3.$

$y_4 = 2 + (x^2 - 1) + \frac{1}{4}(x^2 - 1)^2 + \frac{1}{24}(x^2 - 1)^3 + \frac{1}{192}(x^2 - 1)^4.$

All the terms of y_4 agree with the exact solution $y = 2e^{(x^2 - 1)/2}.$

17. $y_1 = \frac{x^2}{2}, \quad y_2 = \frac{x^2}{2!} - \frac{x^3}{3!}, \quad y_3 = \frac{x^2}{2!} - \frac{x^3}{3!} + \frac{x^4}{4!}, \quad y_4 = \frac{x^2}{2!} - \frac{x^3}{3!} + \frac{x^4}{4!} - \frac{x^5}{5!}.$

All the terms of y_4 agree with the exact solution $y = (x - 1) + e^{-x}.$

18. $y_1 = \frac{1}{4} - \frac{3}{2}(x + 1) + \frac{1}{2}(x + 1)^2, \quad y_2 = \frac{1}{4} - \frac{3}{2}(x + 1) + 2(x + 1)^2 - \frac{1}{3}(x + 1)^3,$

$$y_3 = \frac{1}{4} - \frac{3}{2}(x+1) + 2(x+1)^2 - \frac{4}{3}(x+1)^3 + \frac{1}{6}(x+1)^4,$$

$$y_4 = \frac{1}{4} - \frac{3}{2}(x+1) + 2(x+1)^2 - \frac{4}{3}(x+1)^3 + \frac{2}{3}(x+1)^4 - \frac{1}{15}(x+1)^5.$$

First five terms of y_4 agree with the first five terms of the exact solution $y(x) = \frac{1}{4}(2x-1) + e^{-2(x+1)}$.

19. $y_2 = 1 + x + x^2 + \frac{2}{3}x^3 + \frac{1}{6}x^4 + \frac{2}{15}x^5 + \frac{x^7}{63}$.

20. $y_2 = 2 \tan^{-1}(x^3/24)$. **21.** $y_2 = (3x + x^3)/3$.

22. $y_2 = (4x^3 - 3x^2 + 5)/3$.

Chapter 5

Linear Differential Equations

5.1 Introduction

Linear differential equations occur in the study of many practical problems in science and engineering. Constant coefficient equations arise in the theory of electric circuits, vibrations etc. Variable coefficient equations arise in many areas of physics, electric circuits, mathematical modelling of physical problems etc. Some of the important variable coefficient differential equations are Bessel equation, Legendre equation, Chebyshev equation etc. The solution of constant coefficient equations can be obtained in terms of known standard functions. However, no such solution procedure exists for variable coefficient equations. Often, we attempt their solution in the form of an infinite series. These solutions may sometimes reduce to known standard functions.

A linear ordinary differential equation of order n, is written as

$$a_0(x)\frac{d^n y}{dx^n} + a_1(x)\frac{d^{n-1}y}{dx^{n-1}} + \ldots + a_{n-1}(x)\frac{dy}{dx} + a_n(x)y = r(x)$$

or
$$a_0(x)y^{(n)}(x) + a_1(x)y^{(n-1)}(x) + \ldots + a_{n-1}(x)y'(x) + a_n(x)y(x) = r(x) \qquad (5.1)$$

where y is the dependent variable and x is the independent variable and $a_0(x) \neq 0$. If $r(x) = 0$, then it is called a *homogeneous equation*, otherwise it is called a *non-homogeneous equation*. For example, a second order homogeneous equation is of the form

$$a_0(x)y'' + a_1(x)y' + a_2(x)y = 0, \quad a_0(x) \neq 0 \qquad (5.2)$$

and a non-homogeneous second order equation is of the form

$$a_0(x)y'' + a_1(x)y' + a_2(x)y = r(x), \quad a_0(x) \neq 0. \qquad (5.3)$$

If $a_i(x)$, $i = 0, 1, 2$ are constants then the equations are linear second order constant coefficient equations. A few examples of linear second order equations are

$$y'' + 4y' + 3y = x^2 e^x, \qquad (5.4)$$

$$y'' + 2y' + y = \sin x, \qquad (5.5)$$

$$x^2 y'' + xy' + (x^2 - 4)y = 0, \qquad (5.6)$$

$$(1 - x^2)y'' - 2xy' + 20y = 0. \qquad (5.7)$$

Equations (5.4), (5.5) are constant coefficient second order equations and Eqs. (5.6), (5.7) are variable coefficient second order equations.

5.2 Solutions of Linear Differential Equations

We assume that x in Eq. (5.1) varies on some interval I, where the interval may be open, closed, semi-open or infinite. For example, the differential equation may be valid for all $x \in (0, \infty)$ or $x \in (-\infty, \infty)$. If $y_1(x)$ is a solution of the Eq. (5.1), then it must identically satisfy the equation. Hence, $y_1(x)$ must be continuously differentiable $n - 1$ times and $y_1^{(n)}(x)$ must be continuous on I.

We now state an important result regarding the uniqueness of solutions.

Theorem 5.1 If the functions $a_0(x), a_1(x), \ldots, a_n(x)$ and $r(x)$ are continuous over I and $a_0(x) \neq 0$ on I, then there exists a unique solution to the initial value problem

$$a_0 y^{(n)} + a_1 y^{(n-1)} + \ldots + a_{n-1} y' + a_n y = r(x) \tag{5.8}$$

$$y(x_0) = c_1, \, y'(x_0) = c_2, \ldots, \, y^{(n-1)}(x_0) = c_n \tag{5.9}$$

where $x_0 \in I$, and $c_1, c_2, \ldots, c_n$ are n known constants.

This theorem does not give us a procedure to find the solutions but guarantees that there exists a unique solution if the conditions stated in the theorem are satisfied.

If the conditions of the Theorem 5.1 are satisfied, then the differential Equation (5.8) is said to be *normal* on I (these conditions are both necessary and sufficient for the differential equation to be normal).

A point $x_0 \in I$, for which $a_0(x) \neq 0$, is called an *ordinary point* or a *regular point* of the differential equation.

Example 5.1 Find the intervals on which the following differential equations are normal.

(a) $(1 - x^2)y'' - 2xy' + n(n + 1)y = 0$, n an interger.
(b) $x^2 y'' + xy' + (n^2 - x^2)y = 0$, n real.
(c) $\sqrt{x}\, y'' + 6xy' + 15y = \ln(x^4 - 256)$.

Solution

(a) Here, $a_0(x) = (1 - x^2)$, $a_1(x) = -2x$, and $a_2(x) = n(n + 1)$. Now, a_0, a_1 and a_2 are continuous everywhere in $(-\infty, \infty)$. Also, $a_0(x) = 1 - x^2 \neq 0$ for all $x \in (-\infty, \infty)$, except at the points $x = -1, 1$. Hence, the differential equation is normal on every subinterval I of the open intervals $(-\infty, -1)$, $(-1, 1)$ and $(1, \infty)$.

(b) Here, $a_0(x) = x^2$, $a_1(x) = x$, $a_2(x) = n^2 - x^2$. We find that a_0, a_1 and a_2 are continuous everywhere in $(-\infty, \infty)$. Also, $a_0(x) = x^2 \neq 0$ for all $x \in (-\infty, \infty)$ except at $x = 0$. Hence, the differential equation is normal on every subinterval I of the open intervals $(-\infty, 0)$, $(0, \infty)$.

(c) Here, $a_0(x) = \sqrt{x}$, $a_1(x) = 6x$, $a_2(x) = 15$, and $r(x) = \ln(x^4 - 256)$. Now, a_0, a_1, a_2 and $r(x)$ are continuous for all x satisfying $x > 4$. Hence, the differential equation is normal on every subinterval I of the open interval $(4, \infty)$.

Remark 1

If the functions $a_0(x), a_1(x), \ldots, a_n(x)$ are continuous over I and $a_0(x) \neq 0$ on I, then the only solution of the homogeneous initial value problem

$$a_0 y^{(n)} + a_1 y^{(n-1)} + \ldots + a_{n-1} y' + a_n y = 0 \tag{5.10}$$

$$y(x_0) = 0, \ y'(x_0) = 0, \ldots, y^{(n-1)}(x_0) = 0 \tag{5.11}$$

where $x_0 \in I$, is the trivial solution $y \equiv 0$ on I.

Linear combination of functions   Let $f_1(x), f_2(x), \ldots, f_n(x)$ be n functions. Then

$$c_1 f_1(x) + c_2 f_2(x) + \ldots + c_n f_n(x), \text{ where } c_1, c_2, \ldots, c_n \text{ are constants}$$

is called a linear combination of the given functions.

We had earlier defined that a function $y_1(x)$ is a solution of a non-homogeneous or a homogeneous equation, if the equation reduces to an identity when $y_1(x)$ is substituted into it. Let $y_1(x), y_2(x), \ldots, y_m(x)$ be m solutions of the linear homogeneous equation (5.10). Then, we show in the following that the *superposition principle* or *linearity principle* holds.

Theorem 5.2   If $y_1(x), y_2(x), \ldots, y_m(x)$ are m solutions of the linear homogeneous equation (5.10) on I, then a linear combination of the solutions $c_1 y_1(x) + c_2 y_2(x) + \ldots + c_m y_m(x)$, where $c_1, c_2, \ldots, c_m$ are constants is also a solution of Eq. (5.10) on I.

Proof   Substituting the linear combination $y = c_1 y_1(x) + c_2 y_2(x) + \ldots + c_m y_m(x)$ into Eq. (5.10), we get

$$a_0(x) \frac{d^n}{dx^n} [c_1 y_1(x) + c_2 y_2(x) + \ldots + c_m y_m(x)]$$

$$+ a_1(x) \frac{d^{n-1}}{dx^{n-1}} [c_1 y_1(x) + c_2 y_2(x) + \ldots + c_m y_m(x)]$$

$$+ \ldots + a_{n-1}(x) \frac{d}{dx} [c_1 y_1(x) + c_2 y_2(x) + \ldots + c_m y_m(x)]$$

$$+ a_n(x)[c_1 y_1(x) + c_2 y_2(x) + \ldots + c_m y_m(x)]$$

$$= c_1 \left[a_0(x) \frac{d^n y_1}{dx^n} + a_1(x) \frac{d^{n-1} y_1}{dx^{n-1}} + \ldots + a_{n-1}(x) \frac{dy_1}{dx} + a_n(x)y_1 \right]$$

$$+ c_2 \left[a_0(x) \frac{d^n y_2}{dx^n} + a_1(x) \frac{d^{n-1} y_2}{dx^{n-1}} + \ldots + a_{n-1}(x) \frac{dy_2}{dx} + a_n(x)y_2 \right]$$

$$+ \ldots + c_m \left[a_0(x) \frac{d^n y_m}{dx^n} + a_1(x) \frac{d^{n-1} y_m}{dx^{n-1}} + \ldots + a_{n-1}(x) \frac{dy_m}{dx} + a_n(x)y_m \right]$$

$$= c_1[0] + c_2[0] + \ldots + c_m[0] = 0$$

since $y_1(x), y_2(x), \ldots, y_m(x)$ are solutions of the linear homogeneous equation.

Remark 2

Superposition principle does not hold for a non-homogeneous equation or a nonlinear equation.

Example 5.2   Show that e^{-x}, e^x and their linear combination $c_1 e^{-x} + c_2 e^x$ are solutions of the homogeneous equation $y'' - y = 0$.

Solution   For $y_1 = e^{-x}$, we have $y_1' = -e^{-x}, y_1'' = e^{-x}, y_1'' - y_1 = 0$.

For $y_2 = e^x$, we have $y_2' = e^x, y_2'' = e^x, y_2'' - y_2 = 0$.

Hence, e^{-x} and e^x are solutions of $y'' - y = 0$.

Substituting $y = c_1 e^{-x} + c_2 e^x = c_1 y_1 + c_2 y_2$, we obtain

$$y'' - y = (c_1 y_1 + c_2 y_2)'' - (c_1 y_1 + c_2 y_2) = c_1(y_1'' - y_1) + c_2(y_2'' - y_2) = c_1(0) + c_2(0) = 0.$$

5.2.1 Linear Independence and Dependence

Let $f_1(x), f_2(x), \ldots, f_n(x)$ be n functions. Then, these functions are said to be *linearly independent* on some interval I (where they are defined), if the equation

$$c_1 f_1(x) + c_2 f_2(x) + \ldots + c_n f_n(x) = 0 \tag{5.12}$$

implies $c_1 = 0 = c_2 = \ldots = c_n$.

These functions are said to be *linearly dependent* on I, if Eq. (5.12) holds for $c_1, c_2, \ldots, c_n$ not all zero. In this case, one or more functions can be expressed as a linear combination of the remaining functions. For example, if $c_1 \neq 0$, then

$$f_1(x) = -\frac{1}{c_1}[c_2 f_2(x) + \ldots + c_n f_n(x)].$$

Conversely, if any function $f_i(x)$ can be expressed as a linear combination of the functions $f_1, f_2, \ldots$ $f_{i-1}, f_{i+1} \ldots, f_n$ then the given set of functions are linearly dependent.

Example 5.3 Show that the functions $f_1(x) = x^2, f_2(x) = x^3, f_3(x) = 6x^2 - x^3$ are linearly dependent on any interval I.

Solution We have $f_3(x) = 6x^2 - x^3 = 6f_1(x) - f_2(x)$. Hence, the given functions are linearly dependent on any interval I.

Example 5.4 Show that the functions $x^2 - 1$, $3x^2$ and $2 - 5x^2$ are linearly dependent.

Solution The given functions are linearly dependent if the equation

$$c_1(x^2 - 1) + c_2(3x^2) + c_3(2 - 5x^2) = 0 \tag{5.13}$$

holds for c_1, c_2, c_3 not all zero. We have from Eq. (5.13)

$$(c_1 + 3c_2 - 5c_3)x^2 + (2c_3 - c_1) = 0, \quad \text{for all } x.$$

We have, $c_1 + 3c_2 - 5c_3 = 0$, and $2c_3 - c_1 = 0$. The solution of these equations is $c_1 = 2c_3, c_2 = c_3$ where c_3 is arbitrary. For example, if $c_3 = 1$, then $c_1 = 2$, $c_2 = 1$ and $f_3(x) = -2f_1(x) - f_2(x)$. The given functions are linearly dependent.

Example 5.5 Show that the functions x, x^2, x^3 are linearly independent on any interval I.

Solution We have $f_1(x) = x$, $f_2(x) = x^2$, $f_3(x) = x^3$. Substituting in equation (5.12), we get $c_1 x + c_2 x^2 + c_3 x^3 = 0$. For finding the values of the three constants, take three distinct arbitrary points $x_0, x_1, x_2 (\neq 0)$ on I. Hence

$$c_1 x_0 + c_2 x_0^2 + c_3 x_0^3 = 0, \, c_1 x_1 + c_2 x_1^2 + c_3 x_1^3 = 0, \, c_1 x_2 + c_2 x_2^2 + c_3 x_2^3 = 0.$$

This system of homogeneous algebraic equations has a non-trivial solution if the determinant of the coefficient matrix vanishes, that is

$$\det = \begin{vmatrix} x_0 & x_0^2 & x_0^3 \\ x_1 & x_1^2 & x_1^3 \\ x_2 & x_2^2 & x_2^3 \end{vmatrix} = 0.$$

Evaluating the determinant, we have

$$\det = (x_1 - x_0)(x_2 - x_0)(x_2 - x_1)x_0 x_1 x_2.$$

Since x_0, x_1, x_2 are distinct, $\det \neq 0$. Therefore, the only solution is $c_1 = 0$, $c_2 = 0$, $c_3 = 0$. Hence, the given functions are linearly independent.

The procedure used in Example 5.5 is lengthy and a difficult one. It is not always possible to examine the linear dependence or independence in this way. A very elegant procedure to test the linear independence or dependence of a given set of functions is the application of *Wronskians*. Let $f_1(x)$, $f_2(x)$, ... , $f_n(x)$ be n functions. The Wronskian of these functions is denoted by $W(f_1, f_2, \ldots, f_n)$ and is defined by

$$W(f_1, f_2, \ldots, f_n) = \begin{vmatrix} f_1 & f_2 & \cdots & f_n \\ f_1' & f_2' & \cdots & f_n' \\ \cdots & \cdots & \cdots & \cdots \\ f_1^{(n-1)} & f_2^{(n-1)} & \cdots & f_n^{(n-1)} \end{vmatrix} = W(x). \qquad (5.14)$$

The Wronskian of the n functions exists if all the functions $f_1, f_2, \ldots, f_n$ are differentiable $n-1$ times on the interval I. If any one or more functions are not differentiable then the Wronskian does not exist.

We have the following result for testing the linear dependence or independence of the solutions of the linear homogeneous differential equation (5.10).

Theorem 5.3 If the coefficients $a_0(x)$, $a_1(x)$, ... , $a_n(x)$ in the linear homogeneous equation

$$a_0 y^{(n)} + a_1 y^{(n-1)} + \ldots + a_{n-1} y' + a_n y = 0, \quad a_0 \neq 0 \qquad (5.15)$$

are continuous on I and $y_1(x)$, $y_2(x)$, ... , $y_n(x)$ are n solutions of this equation, then

(i) $W(x) = W(y_1, y_2, \ldots, y_n) \neq 0$ for all $x \in I \Leftrightarrow y_1(x), y_2(x), \ldots, y_n(x)$ are linearly independent on I,

(ii) $W(x_0) = 0$ where $x_0 \in I$ is any fixed point, implies $W(x) = 0$ for all x in I and the functions $y_1(x)$, $y_2(x)$, ... , $y_n(x)$ are linearly dependent.

Proof Let $y_1(x)$, $y_2(x)$, ... , $y_n(x)$ be linearly dependent on I. By definition, there exist constants c_1, c_2, ... , c_n not all zero, such that

$$c_1 y_1(x) + c_2 y_2(x) + \ldots + c_n y_n(x) = 0, \text{ for all } x \in I. \qquad (5.16)$$

Differentiating Eq. (5.16) successively, $n - 1$ times, we get

$$c_1 y_1'(x) + c_2 y_2'(x) + \ldots + c_n y_n'(x) = 0$$

$$c_1 y_1''(x) + c_2 y_2''(x) + \ldots + c_n y_n''(x) = 0$$

$$\cdots$$

$$c_1 y_1^{(n-1)}(x) + c_2 y_2^{(n-1)}(x) + \ldots + c_n y_n^{(n-1)}(x) = 0. \qquad (5.17)$$

Eqs. (5.16), (5.17) form a homogeneous, linear system of algebraic equations. Non-trivial solutions of the system exist if and only if the determinant of the coefficient matrix is zero for all $x \in I$. But this determinant is the Wronskian $W(x)$ of the solutions. Hence, if the solutions are dependent then $W(x) = 0$ for all $x \in I$.

Let now $W(x_0) = 0$ for some fixed point $x_0 \in I$. Then, the system of equations given by (5.16), (5.17) has a nontrivial solution $c_1 = c_1^*, c_2 = c_2^*, \ldots, c_n = c_n^*$ not all zero. Hence, $y^*(x) = c_1^* y_1(x) + c_2^* y_2(x) + \ldots + c_n^* y_n(x)$ is a solution of the linear homogeneous Equation (5.15). Using Eqs. (5.16) and (5.17), we find that $y^*(x)$ also satisfies the initial conditions $y^*(x_0) = 0$, $(y^*)'(x_0) = 0$, $\ldots, (y^*)^{(n-1)}(x_0) = 0$. Now, the differential equation (5.15) and these conditions form a homogeneous initial value problem. Hence, $y^*(x) \equiv 0$ is the solution of the initial value problem. Since the solution of the initial value problem is unique, we obtain $y^*(x) = y(x) = 0$, or, for all x, $c_1^* y_1(x) + c_2^* y_2(x) + \ldots + c_n^* y_n(x) = 0$, where not all c_i are zero. Hence, the solutions are dependent. Since x_0 is arbitrary, $W(x_0) = 0$ for some $x_0 \in I$ implies $W(x) = 0$ for all $x \in I$.

We now define the general solution of the homogeneous Eq. (5.15).

Theorem 5.4 If the coefficients $a_0(x), a_1(x), \ldots, a_n(x), a_0(x) \neq 0$, in the linear homogeneous equation (5.15) are continuous on I, then the equation (5.15) has n linearly independent solutions. If $y_1(x)$, $y_2(x), \ldots, y_n(x)$ are n linearly independent solutions, then the general solution is given by $y(x) = c_1 y_1(x) + c_2 y_2(x) + \ldots + c_n y_n(x)$, that is, their linear combination.

The n linearly independent solutions $y_1(x), y_2(x), \ldots, y_n(x)$ are also called the *fundamental solutions* of Eq. (5.15) on I. This set of fundamental solutions forms a *basis* of the nth order linear homogeneous equation.

Example 5.6 Show that the functions x, x^2, x^3 are linearly independent on any interval I, not containing zero (see Example 5.5).

Solution The Wronskian of the functions is

$$W(x) = \begin{vmatrix} x & x^2 & x^3 \\ 1 & 2x & 3x^2 \\ 0 & 2 & 6x \end{vmatrix} = x(12x^2 - 6x^2) - (6x^3 - 2x^3) = 2x^3.$$

Therefore, $W(x) \neq 0$ on any interval not containing zero. Hence, the functions are linearly independent in $(-\infty, 0), (0, \infty)$.

Example 5.7 Show that the functions $1, \sin x, \cos x$ are linearly independent.

Solution The Wronskian of the functions is

$$W(x) = \begin{vmatrix} 1 & \sin x & \cos x \\ 0 & \cos x & -\sin x \\ 0 & -\sin x & -\cos x \end{vmatrix} = -1.$$

Hence, the given functions are linearly independent on any interval I.

Example 5.8 Show that e^x, e^{2x}, e^{3x} are the fundamental solutions of $y''' - 6y'' + 11y' - 6y = 0$, on any interval I.

Solution Substituting $y = e^x, e^{2x}, e^{3x}$, we find that they satisfy the differential equation

$$y''' - 6y'' + 11y' - 6y = 0.$$

The Wronskian of these functions is

$$W(x) = \begin{vmatrix} e^x & e^{2x} & e^{3x} \\ e^x & 2e^{2x} & 3e^{3x} \\ e^x & 4e^{2x} & 9e^{3x} \end{vmatrix} = e^{6x} \begin{vmatrix} 1 & 1 & 1 \\ 1 & 2 & 3 \\ 1 & 4 & 9 \end{vmatrix} = 2e^{6x} \neq 0.$$

Therefore, the solutions are linearly independent and they form a set of fundamental solutions on any interval I.

Example 5.9 Show that the set of functions $\{x, 1/x\}$ forms a basis of the equation $x^2 y'' + x y' - y = 0$. Obtain a particular solution when $y(1) = 1$, $y'(1) = 2$.

Solution   We have

$$y_1(x) = x,\ y_1' = 1,\ y_1'' = 0, \quad \text{and} \quad x^2 y_1'' + x y_1' - y_1 = x - x = 0$$

$$y_2(x) = 1/x,\ y_2' = -1/x^2,\ y_2'' = 2/x^3$$

and

$$x^2 y_2'' + x y_2' - y_2 = x^2 \left(\frac{2}{x^3} \right) + x \left(-\frac{1}{x^2} \right) - \left(\frac{1}{x} \right) = 0.$$

Hence $y_1(x)$ and $y_2(x)$ are solutions of the given equation. The Wronskian is given by

$$W(y_1, y_2) = \begin{vmatrix} x & 1/x \\ 1 & -1/x^2 \end{vmatrix} = -\frac{2}{x} \neq 0, \quad \text{for } x \geq 1.$$

Therefore, the set $\{y_1(x), y_2(x)\}$ forms a basis of the equation. The general solution is

$$y(x) = c_1 y_1(x) + c_2 y_2(x) = c_1 x + \frac{c_2}{x}.$$

Substituting in the given conditions, we get

$$y(1) = 1 = c_1 + c_2,\ y'(1) = 2 = c_1 - c_2.$$

Solving, we obtain $c_1 = 3/2$, $c_2 = -1/2$.

The particular solution is $y(x) = \frac{1}{2} \left(3x - \frac{1}{x} \right)$.

Exercise 5.1

From the following linear differential equations, find the constant coefficient and variable coefficient equations.

1. $y'' - a^2 y = 0$.
2. $y' = y/x$.
3. $y''' + 3y'' + 6y' + 12y = x^2$.
4. $x^3 y''' + 9x^2 y'' + 18xy' + 6y = 0$.
5. $(1 - x)y'' + xy' - y = 0$.
6. $y'' - (1 + x^2)y = 0$.

Find the intervals on which the following differential equations are normal.

7. $y' = 3y/x$.
8. $(1 + x^2)y'' + 2xy' + y = 0$.
9. $x^2 y'' - 4xy' + 6y = x$.
10. $y'' + 3y' + \sqrt{x}\, y = \sin x$.
11. $y''' + 9y' + y = \log(x^2 - 9)$.
12. $y'' + |x|\, y' + y = x \ln x$.

13. $x(1 - x)y'' - 3xy' - y = 0$.   14. $y'' + xy' + 6y = \ln \sin (\pi x/4)$.

15. Verify that $y = x^2$ is a solution of $x^2 y'' + xy' - 4y = 0$, $x \in (0, \infty)$ and satisfies the conditions $y(0) = 0$, $y'(0) = 0$. Does Theorem 5.1 guarantee the existence and uniqueness of such a solution? Is the Remark 1 applicable in this case?

16. By inspection find a solution of $x^2 y'' + xy' - y = 0$, $x \in (-\infty, \infty)$ which satisfies the conditions $y(0) = 0$, $y'(0) = 2$. Does Theorem 5.1 guarantee the existence and uniqueness of such a solution?

17. Show that

$$y_1(x) = x^3 - x^2, -3 \le x \le 3, \quad \text{and} \quad y_2(x) = \begin{cases} x^2 - x^3, & -3 \le x \le 0, \\ x^3 - x^2, & 0 \le x \le 3 \end{cases}$$

both satisfy the differential equation $x^2 y'' - 4xy' + 6y = 0$ and the conditions $y(2) = 4$, $y'(2) = 8$. But $y_1(x)$, $y_2(x)$ are different. Does this contradict Theorem 5.1?

Verify that the given functions are solutions of the associated differential equation. Verify also that a linear combination of these functions is also a solution.

18. $1, x, e^x; y''' - y'' = 0$.   19. $e^x, e^{-2x}; y'' + y' - 2y = 0$.

20. $e^{-x} \cos 2x, e^{-x} \sin 2x; y'' + 2y' + 5y = 0$.

Examine whether the following functions are linearly independent for $x \in (0, \infty)$.

21. $2x, 6x + 3, 3x + 2$.   22. $x^2 - x, 3x^2 + x + 1, 9x^2 - x + 2$.

23. $x^2 - 2x, 3x^2 + x + 2, 4x^2 - x + 1$.   24. $\sin x, \sin 2x, \sin 3x$.

25. $1, \cos x, \sin x$.   26. $e^x, \sinh x, \cosh x$.

27. $x^2, 1/x^2$.   28. $\ln x, \ln x^2, \ln x^3$.

29. $x - 1, x + 1, (x - 1)^2$.   30. $e^{-x}, \sinh x, \cosh x$.

31. Find the intervals on which the three functions $1, \cos x, \sec x, x > 0$ are linearly independent.

32. Determine how many of the given functions are linearly independent on $[0, 1]$.

   (i) $1, 1 + x, x^2, x(1 - x), x$;   (ii) $1 + x, 1 - x, 1, x^2, 1 + x^2$.

33. Show that $y_1(x) = \sin x$, and $y_2(x) = 4 \sin x - 2 \cos x$ are linearly independent solutions of $y'' + y = 0$. Write the solution $y_3(x) = \cos x$ as a linear combination of y_1 and y_2.

34. Let $a_0(x)y'' + a_1(x)y' + a_2(x)y = 0$ be a second order differential equation. Let $a_0(x)$, $a_1(x)$, $a_2(x)$ be continuous and $a_0(x) \ne 0$ on I and $y_1(x)$, $y_2(x)$ be two linearly independent solutions. Show that the Wronskian of $y_1(x)$, $y_2(x)$ satisfies the differential equation $a_0(x)W'(x) + a_1(x)W(x) = 0$. Also, show that the Wronskian is given by

$$W(x) = c e^{-\int [a_1(x)/a_0(x)]dx}.$$

   (This is called the *Abel's formula*).

35. Show that $\cos at$, $\sin at$ are solutions of the equation $y'' + a^2 y = 0$, $a \ne 0$ on any interval. Show that they are independent. Use the result (Abel's formula) given in Problem 34 and find the Wronskian. Are the two Wronskians same?

36. Show that e^{2x} and xe^{2x} are solutions of the equation $y'' - 4y' + 4y = 0$ on any interval. Show that they are independent. Use the result given in problem 34 and find the Wronskian. Are the two Wronskians same?

Show that in the following problems, $\{y_i(x)\}$ forms a set of fundamental solutions (basis) to the corresponding differential equation.

37. $x^{1/4}, x^{5/4}; 16x^2 y'' - 8xy' + 5y = 0$, $x > 0$.

38. $e^{2x} \cos 3x$, $e^{2x} \sin 3x$; $2y'' - 8y' + 26y = 0$.

39. 1, x^2; $x^2 y'' - xy' = 0$, $x > 0$.

40. e^x, e^{2x}, e^{-3x}; $y''' - 7y' + 6y = 0$.

41. e^x, $e^x \cos x$, $e^x \sin x$; $y''' - 3y'' + 4y' - 2y = 0$.

42. e^{2x}, $e^{-x} \cos(\sqrt{3}x)$, $e^{-x} \sin(\sqrt{3}x)$; $y''' - 8y = 0$.

43. $\sin(\ln x^2)$, $\cos(\ln x^2)$; $x^2 y'' + xy' + 4y = 0$, $x > 0$.

44. Let the coefficients $a_0(x)$, $a_1(x)$, $a_2(x)$ in the equation $a_0(x)y'' + a_1(x)y' + a_2(x)y = 0$ be continuous and $a_0(x) \neq 0$ on I. Let $\{y_1(x), y_2(x)\}$ be the basis (set of fundamental solutions) of the equation. Show that the set $\{u(x), v(x)\}$ such that $u = ay_1(x) + by_2(x)$, $v = cy_1(x) + dy_2(x)$, is also a basis of the equation if $ad - bc \neq 0$. If $y_1(x) = \cosh kx$, $y_2 = \sinh kx$, obtain a simple form of u and v.

45. Let $y_1(x)$, $y_2(x)$ be the linearly independent solutions of the equation $y'' + a(x)y' + b(x)y = 0$ on I. Show that there is no point $x_0 \in I$ at which (i) both $y_1(x)$, $y_2(x)$ vanish, (ii) both $y_1(x)$, $y_2(x)$ take extreme values.

46. Let $\{y_1(x), y_2(x)\}$ be the basis of the equation $y'' + a(x)y' + b(x)y = 0$. Show that the equation can be written as the Wronskian $W(y, y_1, y_2) = 0$.

47. Let $y_1(x)$ be a solution of the homogeneous equation $y'' + a(x)y' + b(x)y = 0$, on the interval $I : \alpha \leq x \leq \beta$. The coefficients $a(x)$ and $b(x)$ are continuous on I. If the curve $y = y_1(x)$ is tangential to the x-axis at a point x_1 in I, then prove that $y_1(x) \equiv 0$.

Using the problem 46, find a differential equation of the form $y'' + a(x)y' + b(x)y = 0$ for which the following functions are solutions.

48. e^{3x}, e^{-2x}.

49. $e^{-(\alpha + i\omega)x}$, $e^{-(\alpha - i\omega)x}$.

50. e^{5x}, xe^{5x}.

5.3 Methods for Solution of Linear Equations

In this section, we shall discuss various methods of finding solution of linear equations. We first define the differential operator D.

5.3.1 Differential Operator D

Sometimes, it is convenient to write the given linear differential equation in a simple form using the differential operator $D = d/dx$. We define an operator T as a transformation $T : V \to W$ that transforms a function f in V into another function $T(V)$ in W. Let the operator D be defined, over the set V_1 of all differentiable functions f on I, by $D = d/dx$.

Then, we write
$$Df(x) = Df = (Df)(x) = \frac{df}{dx} = f'. \tag{5.18}$$

We have, for example $D(x^n) = \frac{d}{dx}(x^n) = nx^{n-1}$, n constant; $D(\cos x) = -\sin x$, etc.

Let $f(x)$ and $g(x)$ be differentiable functions. Since D is a linear operator, we have
$$D(af + bg) = aDf + bDg, \quad a, b \text{ constants.}$$

We also have for $f \in V_2$, the set of functions having a second derivative on I
$$D(Df) = D(f') = \frac{d}{dx}(f') = f''.$$

We simply write $D(Df) = D(D)f = D^2 f$ so that

$$D^3(f) = D(D^2f) = D(f'') = f''', \ldots, D^k(f) = f^{(k)}.$$

where f is sufficiently differentiable.

We define $D^0 \equiv 1$, so that if 1 is the operator defined by $1(f) = f$, we have $D^0(f) = 1(f) = f$. We, now define the operator L by

$$L = a_0(x)\frac{d^n}{dx^n} + a_1(x)\frac{d^{n-1}}{dx^{n-1}} + \ldots + a_{n-1}(x)\frac{d}{dx} + a_n(x)$$

$$= a_0(x)D^n + a_1(x)D^{n-1} + \ldots + a_{n-1}(x)D + a_n(x) = P(D) \qquad (5.19)$$

which is a polynomial in D, so that

$$Ly = a_0(x)\frac{d^n y}{dx^n} + a_1(x)\frac{d^{n-1}y}{dx^{n-1}} + \ldots + a_{n-1}(x)\frac{dy}{dx} + a_n(x)y$$

$$= a_0(x)D^n y + a_1(x)D^{n-1}y + \ldots + a_{n-1}(x)Dy + a_n(x)y$$

$$= [a_0(x)D^n + a_1(x)D^{n-1} + \ldots + a_{n-1}(x)D + a_n(x)]y = P(D)y. \qquad (5.20)$$

For example, the differential equation

$$\frac{d^2 y}{dx^2} + 5\frac{dy}{dx} + 6y = 0$$

can be written as

$$Ly = (D^2 + 5D + 6)y = 0 \qquad (5.21)$$

where the operator L is given by $L = P(D) = D^2 + 5D + 6$.

Similarly, the equation

$$\frac{d^2 y}{dx^2} + 2\frac{dy}{dx} + 2y = x^2$$

can be written as

$$Ly = (D^2 + 2D + 2)y = x^2 \qquad (5.22)$$

where the operator L is defined by $L = P(D) = D^2 + 2D + 2$.

Suppose $y = e^{mx}$. Then, $D(y) = D(e^{mx}) = me^{mx}$ and $D^2(y) = m^2 e^{mx}$. Substituting in Eq. (5.21), we obtain

$$Ly = (D^2 + 5D + 6)y = (m^2 + 5m + 6)y = P(m)y$$

using Eq. (5.19). Therefore,

$$P(D)y = (D^2 + 5D + 6)e^{mx} = P(m)y.$$

In general, substituting $y = e^{mx}$ in the equation (5.20), we get

$$P(D)y = (a_0 D^n + a_1 D^{n-1} + \ldots + a_{n-1}D + a_n)e^{mx}$$

$$= (a_0 m^n + a_1 m^{n-1} + \ldots + a_{n-1}m + a_n)e^{mx} = P(m)e^{mx}. \qquad (5.23)$$

When a_i, $i = 0, 1, \ldots, n$ are constants, the operator $L = P(D)$ can often be factorised. For example, we have

(i) $D^2 + 5D + 6 = (D + 2)(D + 3)$

(ii) $D^3 - 6D^2 + 11D - 6 = (D - 1)(D - 2)(D - 3)$.

When a_i are functions of x, factorisation is often not possible. For example, $x^2 Dy \neq D(x^2 y)$, or in general $a(x)Dy \neq D[a(x)y]$, since the right hand side is $D[a(x)y] = a(x)y' + a'(x)y$.

5.3.2   Solution of Second Order Linear Homogeneous Equations with Constant Coefficients

Consider the linear homogeneous second order equation

$$ay'' + by' + cy = 0, \quad a, b, c \text{ are constants.} \tag{5.24}$$

In the operator notation, we write the equation as

$$Ly = P(D)y = aD^2 y + bDy + cy = (aD^2 + bD + c)y = 0. \tag{5.25}$$

In the previous chapter, we have shown that the solution of the first order equation $y' + my = 0$ is $y = e^{-mx} + c$; and the solution of the equation $y' - my = 0$ is $y = e^{mx} + c$. Therefore, it is natural to try for a particular solution of the form $y = e^{mx}$, for Eq. (5.25), where m is an unknown constant to be determined. Since $y' = me^{mx}$, $y'' = m^2 e^{mx}$, we obtain from Eq. (5.24)

$$(am^2 + bm + c)e^{mx} = 0.$$

Since $e^{mx} \neq 0$, we obtain

$$am^2 + bm + c = 0. \tag{5.26}$$

This is an algebraic equation in m. It is called the *characteristic equation* or the *auxiliary equation* of the linear homogeneous equation (5.24) (we can write the characteristic equation by replacing y'' by m^2, y' by m and y by 1 in Eq. (5.24) implicitly noting that solutions of the form e^{mx} are being determined). The roots of this equation are called the *characteristic roots*. The quadratic equation (5.26) has the roots

$$m = [-b \pm \sqrt{b^2 - 4ac}]/2a.$$

We have the following three cases.

(i) The roots are real and distinct, say $m = m_1, m_2$; $m_1 \neq m_2$ if $b^2 - 4ac > 0$.

(ii) The roots are real and equal, say $m = m_1, m_1$ if $b^2 - 4ac = 0$.

(iii) The roots are complex if $b^2 - 4ac < 0$.

To find the complete solution in the above three cases, we proceed as follows.

Real and distinct roots

Let the distinct roots be $m = m_1$ and $m = m_2$. Then, we obtain two solutions of the equation (5.24) as $e^{m_1 x}$ and $e^{m_2 x}$. The two solutions are linearly independent on any interval I, since the Wronskian,

$$W(y_1, y_2) = y_1 y_2' - y_2 y_1' = m_2 e^{m_1 x} e^{m_2 x} - m_1 e^{m_1 x} e^{m_2 x}$$

$$= (m_2 - m_1)e^{(m_1 + m_2)x} \neq 0.$$

Hence, the general solution of Eq. (5.24) is

$$y(x) = c_1 e^{m_1 x} + c_2 e^{m_2 x}. \tag{5.27}$$

Example 5.10   Find the solution of the differential equation $y'' - y' - 6y = 0$.

Solution   Substituting $y = e^{mx}$, we obtain the characteristic equation as

$$m^2 - m - 6 = 0, \quad \text{or} \quad (m-3)(m+2) = 0, \quad \text{or} \quad m = -2, 3.$$

The two linearly independent solutions are e^{3x} and e^{-2x}. The general solution is

$$y(x) = Ae^{3x} + Be^{-2x}.$$

Example 5.11   Solve the initial value problem

$$4y'' - 8y' + 3y = 0, \; y(0) = 1, \; y'(0) = 3.$$

Solution   Substituting $y = e^{mx}$, we obtain the characteristic equation as

$$4m^2 - 8m + 3 = 0, \quad \text{or} \quad m = 1/2, 3/2.$$

Hence, the linearly independent solutions are $e^{x/2}$ and $e^{(3x)/2}$. The general solution is

$$y(x) = Ae^{(3x)/2} + Be^{x/2}.$$

Substituting the initial conditions, we get

$$y(0) = 1 = A + B, \quad y'(0) = 3 = \frac{3A}{2} + \frac{B}{2}.$$

Solving the above equations, we get $A = 5/2$ and $B = -3/2$. The solution of the initial value problem is
$y(x) = [5e^{(3x)/2} - 3e^{x/2}]/2$.

Real and equal roots

Real and equal roots are obtained for the characteristic equation (5.26) when $b^2 - 4ac = 0$. In this case the repeated root is $m = -b/(2a)$. This value gives one solution as $y_1(x) = e^{mx} = e^{-(bx)/(2a)}$. We need to determine another linearly independent solution $y_2(x)$, so that $\{y_1(x), y_2(x)\}$ forms a basis for the equation. The second solution $y_2(x)$ can be determined in a number of ways. We shall show that if m is a repeated root then e^{mx} and xe^{mx} are the two linearly independent solutions.
For $y_2(x) = xe^{mx}$, $m = -b/(2a)$, we have

$$y_2 = xe^{mx}, \; y_2' = (1 + mx)e^{mx}, \; y_2'' = (2 + mx)me^{mx}.$$

Substituting in Eq. (5.24), we get

$$[ma(2 + mx) + b(1 + mx) + cx]e^{mx} = 0$$

or

$$[(2ma + b) + (am^2 + bm + c)x]e^{mx} = 0.$$

Since $2ma + b = -b + b = 0$ and $am^2 + bm + c = 0$, this equation is automatically satisfied. Therefore, xe^{mx} is also a solution. Since e^{mx} and xe^{mx} are linearly independent, they form a set of the two fundamental solutions. Hence, the general solution is

$$y(x) = Ae^{mx} + Bxe^{mx} = (A + Bx)e^{mx}, \quad m = -b/(2a). \tag{5.28}$$

Alternative   We can use the following method (*reduction of order*) to find the second linearly independent solution.

Let

$$y_2(x) = u(x)\, y_1(x)$$

where $y_1(x) = e^{mx}$, $m = -b/(2a)$ be a solution of Eq. (5.24). We have

$$y_2' = uy_1' + u'y_1, \quad y_2'' = uy_1'' + 2u'y_1' + u''y_1.$$

Substituting in the differential equation, we obtain

$$a(uy_1'' + 2u'y_1' + u''y_1) + b(uy_1' + u'y_1) + cuy_1$$

$$= ay_1u'' + (2ay_1' + by_1)u' + (ay_1'' + by_1' + cy_1)u = 0. \qquad (5.29)$$

Since $y_1(x)$ is a solution, we have $ay_1'' + by_1' + cy_1 = 0$.

Also, $\qquad 2ay_1' + by_1 = 2a\left(-\dfrac{b}{2a}\right)e^{-(bx)/(2a)} + be^{-(bx)/(2a)} = 0.$

Hence, Eq. (5.29) reduces to $ay_1u'' = 0$. Since $a \neq 0$, $y_1 \neq 0$, we get $u'' = 0$, whose solution is $u = c_1x + c_2$. Therefore, $y_2(x) = (c_1x + c_2)y_1(x) = c_1xy_1(x) + c_2 y_1(x)$. Since $y_1(x)$ is a solution, the second linearly independent solution is $xy_1(x)$, (note that a linear combination of the two linearly independent solutions is also a solution). The general solution is

$$y(x) = Ay_1(x) + Bxy_1(x) = (A + Bx)e^{mx}, \quad m = -b/(2a)$$

which is same as Eq. (5.28).

Alternative The second linearly independent solution can be determined by factorising the differential operator and reducing the given second order equation to a first order equation. We have

$$(aD^2 + bD + c)y = a\left[D^2 + \frac{b}{a}D + \frac{c}{a}\right]y = 0, \quad a \neq 0.$$

Since $b^2 - 4ac = 0$ and $m = m_1 = -b/(2a)$ is a repeated root, the operator is factorisable so that we can write the equation as

$$(D - m_1)(D - m_1)y = 0. \qquad (5.30)$$

Set $(D - m_1)y = u$. Then, Eq. (5.30) reduces to $(D - m_1)u = 0$ or $u' - m_1u = 0$ whose solution is $u = c_1e^{m_1x}$. Substituting in the equation $(D - m_1)y = u$, we obtain

$$(D - m_1)y = y' - m_1y = u = c_1e^{m_1x}.$$

The integrating factor of this equation is e^{-m_1x}. Therefore, the solution of this equation is

$$ye^{-m_1x} = \int c_1e^{m_1x}e^{-m_1x}dx + c_2 = c_1x + c_2$$

or $\qquad\qquad y = (c_1x + c_2)c^{m_1x}$

which is same as $y_2(x)$ obtained in the previous case.

Example 5.12 Find the solution of the differential equation $4y'' + 4y' + y = 0$.

Solution Assume a solution of the form $y = e^{mx}$. The characteristic equation is given by

$$4m^2 + 4m + 1 = 0, \text{ or } (2m + 1)^2 = 0, \text{ or } m = -1/2, -1/2,$$

which is a repeated root. Hence, the general solution is $y(x) = (A + Bx)e^{-x/2}$.

Example 5.13 Solve the initial value problem

$$y'' + 6y' + 9y = 0, \ y(0) = 2, \ y'(0) = 3.$$

Solution Assume a solution of the form $y = e^{mx}$. The characteristic equation is given by

$$m^2 + 6m + 9 = 0, \text{ or } (m + 3)^2 = 0, \text{ or } m = -3, -3,$$

which is a repeated root. The general solution is $y(x) = (A + Bx)e^{-3x}$. Substituting in the initial conditions, we get

$$y(0) = 2 = A, \ y' = Be^{-3x} - 3(A + Bx)e^{-3x}, \ y'(0) = 3 = B - 3A.$$

The solution is $A = 2$, $B = 9$. The solution of the given initial value problem is $y(x) = (2 + 9x)e^{-3x}$.

Example 5.14 Factorising the differential operator and reducing it into first order equations, solve the differential equation $y'' - 4y' - 5y = 0$.

Solution In the operator notation, the differential equation can be written as

$$(D^2 - 4D - 5)y = 0, \quad \text{or} \quad (D - 5)(D + 1)y = 0. \tag{5.31}$$

Set $(D + 1)y = u$. Then, we obtain from Eq. (5.31), $(D - 5)u = 0$. This is a first order equation whose solution is $u = Ae^{5x}$. Hence,

$$(D + 1)y = Ae^{5x}.$$

This is a first order linear equation, whose integrating factor is e^x. Hence, we have the solution as

$$e^x y = \int Ae^{6x} dx + B = \frac{A}{6} e^{6x} + B,$$

or

$$y = \frac{A}{6} e^{5x} + Be^{-x} = Ce^{5x} + Be^{-x}$$

where $C = A/6$ is an arbitrary constant.
We could have written Eq. (5.31) as $(D + 1)(D - 5)y = 0$ and obtain the same answer.

Example 5.15 Factorising the differential operator and reducing it to first order equations, solve the differential equation $4y'' + 12y' + 9y = 0$

Solution In the operator notation, the differential equation can be written as

$$(4D^2 + 12D + 9)y = (2D + 3)^2 y = 0. \tag{5.32}$$

Set $(2D + 3)y = u$. Then, we obtain from Eq. (5.32),

$$(2D + 3)u = 0.$$

The solution of this equation is $u = Ae^{-(3x)/2}$. Therefore,

$$(2D + 3)y = Ae^{-(3x)/2}, \quad \text{or} \quad \left(D + \frac{3}{2}\right)y = \frac{A}{2} e^{-(3x)/2}.$$

This is a linear first order equation whose integrating factor is $e^{(3x)/2}$. The solution is given by

$$ye^{(3x)/2} = \int \frac{A}{2} dx + B = \frac{Ax}{2} + B, \quad \text{or} \quad y = (Cx + B)e^{-(3x)/2},$$

where $C = A/2$.

Complex roots

When $b^2 - 4ac < 0$, then the roots of the characteristic equation (5.26) are complex. We have

$$m = \frac{-b \pm \sqrt{b^2 - 4ac}}{2a} = \frac{-b \pm i\sqrt{4ac - b^2}}{2a} = p \pm iq$$

where $p = -b/(2a)$ and $q = \sqrt{4ac - b^2}/(2a)$. Since the characteristic equation (5.26) has real coefficients, the complex roots occur in conjugate pairs and are of the form $p \pm iq$. Then, the solution of the equation can be written as

$$y(x) = Ae^{(p+iq)x} + Be^{(p-iq)x} = Ae^{px}e^{iqx} + Be^{px}e^{-iqx} = (Ae^{iqx} + Be^{-iqx})e^{px}$$

$$= [A(\cos qx + i \sin qx) + B(\cos qx - i \sin qx)]e^{px}$$

by the Euler formula. Simplifying, we obtain

$$y(x) = [c_1 \cos qx + c_2 \sin qx]e^{px} \tag{5.33}$$

where $c_1 = A + B$ and $c_2 = i(A - B)$. Therefore, the two linearly independent solutions are $y_1 = e^{px} \cos qx$ and $y_2 = e^{px} \sin qx$. The Wronskian is given by

$$W(y_1, y_2) = \begin{vmatrix} e^{px} \cos qx & e^{px} \sin qx \\ e^{px}(p \cos qx - q \sin qx) & e^{px}(p \sin qx + q \cos qx) \end{vmatrix} = qe^{2px} \neq 0$$

showing that $y_1(x)$ and $y_2(x)$ are linearly independent.

Example 5.16 Find the solution of the differential equation $y'' + 2y' + 2y = 0$.

Solution Assume a solution of the form $y = e^{mx}$. The characteristic equation is given by

$$m^2 + 2m + 2 = 0, \quad \text{or} \quad m = \frac{-2 \pm \sqrt{-4}}{2} = -1 \pm i = p \pm iq.$$

The general solution is

$$y(x) = (A \cos qx + B \sin qx)e^{px} = (A \cos x + B \sin x)e^{-x}.$$

Example 5.17 Find the solution of the initial value problem

$$y'' + 4y' + 13y = 0, \, y(0) = 0, \, y'(0) = 1.$$

Solution Assume a solution of the form $y = e^{mx}$. The characteristic equation is given by

$$m^2 + 4m + 13 = 0, \quad \text{or} \quad m = \frac{-4 \pm \sqrt{16 - 52}}{2} = -2 \pm 3i = p \pm iq.$$

The general solution is given by

$$y(x) = [A \cos qx + B \sin qx]e^{px} = [A \cos 3x + B \sin 3x]e^{-2x}.$$

Substituting in the initial conditions, we obtain

$$y(0) = 0 = A,$$
$$y'(x) = Be^{-2x}(3 \cos 3x - 2 \sin 3x), \quad y'(0) = 1 = 3B, \quad \text{or} \quad B = 1/3.$$

The solution of the initial value problem is

$$y(x) = (e^{-2x} \sin 3x)/3.$$

Example 5.18 Find all the non-trivial solutions, if any, of the boundary value problem

$$y'' + \omega^2 y = 0, \ y(0) = 0, \ y(l) = 0.$$

Solution Assume a solution of the form $y = e^{mx}$. The characteristic equation is given by

$$m^2 + \omega^2 = 0, \quad \text{or} \quad m = \pm i\omega.$$

The general solution is

$$y(x) = A \cos \omega x + B \sin \omega x. \tag{5.34}$$

Substituting in the boundary conditions, we obtain

$$y(0) = 0 = A, \ y(l) = 0 = B \sin (\omega l).$$

If $B = 0$, then we obtain the trivial solution $y = 0$.

For $B \neq 0$, we get $\sin \omega l = 0 = \sin n\pi, n = 1, 2, \ldots$
Therefore, $\omega = n\pi/l$. The general solution is

$$y_n(x) = B_n \sin [(n\pi x)/l], \ n = 1, 2, \ldots$$

where B_i's are arbitrary. There are infinite number of solutions. Since the boundary value problem is homogenous, by the superposition principle, the sum of these solutions is also a solution. Therefore, the general solution is given by

$$y(x) = \sum_{n=1}^{\infty} B_n \sin \left(\frac{n\pi x}{l} \right).$$

(The convergence of such an infinite series called the Fourier series, is discussed in chapter 9.)

5.3.3 Method of Reduction of Order for Variable Coefficient Linear Homogeneous Second Order Equations

Suppose that we know one of the solutions of the second order equation

$$a_0(x)y'' + a_1(x)y' + a_2(x)y = 0, \quad a_0(x) \neq 0 \text{ on } I. \tag{5.35}$$

Then, we can obtain the second linearly independent solution by the method of reduction of order. Let $y = y_1(x)$ be a non-trivial solution of Eq. (5.35), that is

$$a_0(x)y_1'' + a_1(x)y_1' + a_2(x)y_1 = 0. \tag{5.36}$$

Then, we write the second solution as $y_2(x) = u(x)y_1(x)$. Since $u(x) = y_2(x)/y_1(x)$ is not a constant, y_1 and y_2 are two linearly independent solutions of Eq. (5.35). Now,

$$y_2' = u'y_1 + uy_1', \quad \text{and} \quad y_2'' = u''y_1 + 2u'y_1' + uy_1''.$$

Substituting in Eq. (5.35) and collecting the terms, we get

$$a_0(x)y_1 u'' + [2a_0(x) y_1' + a_1(x)y_1]u' + [a_0(x) y_1'' + a_1(x) y_1' + a_2(x)y_1]u = 0.$$

Using Eq. (5.36), we obtain

$$a_0(x)y_1 u'' + [2a_0(x)y_1' + a_1(x)y_1]u' = 0.$$

Now, let $v = u'$. Then, we have

$$a_0(x)y_1 v' + [2a_0(x)y_1' + a_1(x)y_1]v = 0 \tag{5.37}$$

which is a first order equation in v.

Separating the variables, we obtain

$$\frac{v'}{v} = -\frac{(2a_0 y_1' + a_1 y_1)}{a_0 y_1} = -\left[\frac{2y_1'}{y_1} + \frac{a_1}{a_0}\right].$$

Integrating, we obtain

$$W(y_1, y_2) = \begin{vmatrix} y_1 & y_2 \\ y_1' & y_2' \end{vmatrix} = \begin{vmatrix} y_1 & uy_1 \\ y_1' & uy_1' + y_1 u' \end{vmatrix} = \begin{vmatrix} y_1 & uy_1 \\ y_1' & uy_1' + \dfrac{1}{y_1} e^{-\int pdx} \end{vmatrix} = e^{-\int pdx}.$$

where $p(x) = a_1(x)/a_0(x)$. Integrating $u' = v$, we obtain $u = \displaystyle\int v(x)\,dx$. The second linearly independent solution is given by $y_2(x) = u(x)y_1(x)$. It can be verified that the Wronskian of y_1, y_2 is equal to

$$W(y_1, y_2) = e^{-\int p(x)dx} \neq 0$$

showing that $y_1(x)$ and $y_2(x)$ are linearly independent.

Example 5.19 It is known that $1/x$ is a solution of the differential equation $x^2 y'' + 4xy' + 2y = 0$. Find the second linearly independent solution and write the general solution.

Solution Write $y_2(x) = u(x)y_1(x) = u(x)/x$. Here, $p(x) = a_1(x)/a_0(x) = 4/x$. Hence,

$$v(x) = \frac{1}{y_1^2} e^{-\int p(x)dx} = x^2 e^{-\int (4/x)dx} = x^2 \left(\frac{1}{x^4}\right) = \frac{1}{x^2}.$$

$$u(x) = \int v(x)\,dx = \int \frac{dx}{x^2} = -\frac{1}{x}, \quad \text{and} \quad y_2(x) = u(x)y_1(x) = -\frac{1}{x^2}.$$

The general solution is $y(x) = Ay_1(x) + By_2(x) = \dfrac{A}{x} + \dfrac{B}{x^2}$.

Exercise 5.2

Show that the given set of functions $\{y_1(x), y_2(x)\}$ forms a basis of the equation and hence solve the initial value problem.

1. $e^x, e^{4x}, \quad y'' - 5y' + 4y = 0, \quad y(0) = 2, \quad y'(0) = 1.$
2. $e^{2x}, e^{-2x}, \quad y'' - 4y = 0, \quad y(0) = 1, \quad y'(0) = 4.$
3. $e^{-3x}, xe^{-3x}, \quad y'' + 6y' + 9y = 0, \quad y(0) = 1, \quad y'(0) = 2.$
4. $x^2, 1/x^2, \quad x^2 y'' + xy' - 4y = 0, \quad y(1) = 2, \quad y'(1) = 6.$
5. $x, x \ln x, \quad x^2 y'' - xy' + y = 0, \quad y(1) = 3, \quad y'(1) = 4.$

Find a general solution of the following differential equations.

6. $y'' - 4y = 0.$ 7. $y'' - y' - 2y = 0.$
8. $y'' + y' - 2y = 0.$ 9. $y'' - 4y' - 12y = 0.$

10. $y'' + 4y' + y = 0.$

11. $4y'' - 9y' + 2y = 0.$

12. $4y'' + 8y' - 5y = 0.$

13. $y'' + 2y' + y = 0.$

14. $y'' + 2\pi y' + \pi^2 y = 0.$

15. $9y'' - 12y' + 4y = 0.$

16. $4y'' + 4y' + y = 0.$

17. $25y'' - 20y' + 4y = 0.$

18. $y'' + 25y = 0.$

19. $y'' + 4y' + 5y = 0.$

20. $y'' - 2y' + 2y = 0.$

21. $(4D^2 - 4D + 17)y = 0.$

22. $(D^2 - 6D + 18)y = 0.$

23. $(D^2 + 9D)y = 0.$

24. $[D^2 - 2aD + (a^2 + b^2)]y = 0.$

Find a differential equation of the form $ay'' + by' + cy = 0$, for which the following functions are solutions.

25. $e^{3x}, e^{-2x}.$

26. $e^{x/4}, e^{-(3x)/4}.$

27. $1, e^{-2x}.$

28. $e^{2x}, xe^{2x}.$

29. $e^{-x}, xe^{-x}.$

30. $e^{-3ix}, e^{3ix}.$

31. $e^{-(a+ib)x}, e^{-(a-ib)x}.$

32. $e^{(5+3i)x}, e^{(5-3i)x}.$

Solve the following initial value problems.

33. $y'' - y = 0, y(0) = 0, y'(0) = 2.$

34. $y'' - y' - 12y = 0, y(0) = 4, y'(0) = -5.$

35. $y'' + y' - 2y = 0, y(0) = 0, y'(0) = 3.$

36. $\dfrac{d^2\theta}{dt^2} + g\theta = 0$, g constant, $\theta(0) = a$, constant, $\dfrac{d\theta}{dt}(0) = 0.$

37. $y'' - 4y' + 5y = 0, y(0) = 2, y'(0) = -1.$

38. $25y'' - 10y' + 2y = 0, y(0) = 1, y'(0) = 0.$

39. $4y'' + 12y' + 9y = 0, y(0) = -1, y'(0) = 2.$

40. $9y'' + 6y' + y = 0, y(0) = 0, y'(0) = 1.$

Solve the following boundary value problems.

41. $y'' + 25y = 0, y(0) = 1, y(\pi) = -1.$

42. $y'' - 36y = 0, y(0) = 2, y(1/6) = 1/e.$

43. $y'' + 2y' + 2y = 0, y(0) = 1, y(\pi/2) = e^{-\pi/2}.$

44. $9y'' - 6y' + y = 0, y(1) = e^{1/3}, y(2) = 1.$

45. $y'' - 4y' + 3y = 0, y(0) = 1, y(1) = 0.$

46. Verify that $(D - 2)(D + 3) \sin x = (D + 3)(D - 2) \sin x = (D^2 + D - 6) \sin x.$

47. Show that $x^2 Dy \neq D(x^2 y).$

48. Find the conditions under which the following equations hold.

(i) $(D + a)[D + b(x)]f(x) = [D + b(x)][D + a]f(x)$, a constant.

(ii) $[D + a(x)][D + b(x)]f(x) = [D + b(x)][D + a(x)]f(x).$

Factorize the operator and find the solution of the following differential equations using the method of reduction of order or by the direct method.

49. $(D^2 + 5D + 4)y = 0.$

50. $(4D^2 + 8D + 3)y = 0.$

51. $(4D^2 + 12D + 9)y = 0$.

52. $(D^2 + 6D + 9)y = 0$.

53. $(D^2 - 4)y = 0$.

54. $(9D^2 + 6D + 1)y = 0$.

55. The displacement $x(t)$ of a particle is governed by the differential equation $\ddot{x} + \dot{x} + bx = c\dot{x}, b > 0$. For what values of b and c is the motion of the particle oscillatory?

56. Find all non-trivial solutions of the boundary value problem

$$y'' + \omega^2 y = 0, \ y(0) = 0, \ y(\pi) = 0.$$

57. Find all the non-trivial solutions of the boundary value problem

$$y'' + \omega^2 y = 0, \ y'(0) = 0, \ y'(\pi) = 0.$$

58. Find all non-trivial solutions of the boundary value problem

$$y'' + \omega^2 y = 0, \ y(0) = 0, \ y'(\pi) = 0.$$

59. If $a^2 > 4b$, then show that the solution of the differential equation $y'' + ay' + by = 0$ can be expressed as

$y(x) = e^{px} (A \cosh qx + B \sinh qx)$ where $p = -a/2$ and $q = \sqrt{a^2 - 4b}/2$.

60. The motion of a damped mechanical system is governed by the linear differential equation $m\ddot{y} + c\dot{y} + ky = 0$ in which m (mass), k (spring modulus), c (damping factor) are positive constants and dot denotes derivative with respect to time t. Discuss the behaviour of the general solution when $t \to \infty$ in the following three cases: (i) $c^2 > 4mk$ (over damping), (ii) $c^2 < 4mk$ (under damping), (iii) $c^2 = 4mk$ (critical damping). In each case, obtain the solution subject to the initial conditions $y(0) = 0$, $\dot{y}(0) = v_0$.

Find the solution of the following differential equations, if one of its solutions is known.

61. $y'' - y' - 6y = 0, \ y_1 = e^{-2x}$.

62. $y'' + 3y' - 4y = 0, \ y_1 = e^x$.

63. $(x^2 - 1)y'' - 2xy' + 2y = 0, \ y_1 = x, \ x \neq \pm 1$.

64. $x^2 y'' + xy' + (x^2 - 1/4)y = 0, \ x > 0, \ y_1 = x^{-1/2} \sin x$.

65. $(x - 2)y'' - xy' + 2y = 0, \ x \neq 2, \ y_1 = e^x$.

5.3.4 Solution of Higher Order Homogeneous Linear Equations with Constant Coefficients

In this section, we shall extend the methods discussed in section 5.3.2, for the solution of higher order linear homogeneous equations with constant coefficients.

Consider the nth order homogeneous linear equation with constant coefficients

$$a_0 y^{(n)} + a_1 y^{(n-1)} + \ldots + a_{n-1} y' + a_n y = 0. \tag{5.38}$$

We attempt to find a solution of the form $y = e^{mx}$, as in the case of second order equations. Substituting $y = e^{mx}, y^{(k)} = m^k e^{mx}, k = 1, 2, \ldots, n$ in Eq. (5.38) and cancelling e^{mx}, we obtain the characteristic equation as

$$a_0 m^n + a_1 m^{n-1} + \ldots + a_{n-1} m + a_n = 0. \tag{5.39}$$

The degree of this algebraic equation is same as the order of the differential equation. This equation has n roots. All the roots may be real and distinct, all or some of the roots may be equal, all or some of the roots may be complex. Consider the following cases.

Real and distinct roots

Let the polynomial equation (5.39) have all real and distinct roots as $m_1, m_2, \ldots m_n$. Then the n solutions

$$y_1(x) = e^{m_1 x}, y_2(x) = e^{m_2 x}, \ldots, y_n(x) = e^{m_n x} \tag{5.40}$$

are the linearly independent solutions of the differential equation (5.38). Since $m_1 \neq m_2 \neq \ldots \neq m_n$, it can be easily shown that the Wronskian of the solutions $y_1, y_2, \ldots, y_n$ given in Eq. (5.40) does not vanish and therefore they are linearly independent solutions.

Hence, the set of the solutions forms a basis and the general solution is given by

$$y(x) = c_1 e^{m_1 x} + c_2 e^{m_2 x} + \ldots + c_n e^{m_n x}. \tag{5.41}$$

Example 5.20 Find the general solution of the differential equation

$$y''' - 2y'' - 5y' + 6y = 0.$$

Solution Substituting $y = e^{mx}$, we obtain the characteristic equation as

$$m^3 - 2m^2 - 5m + 6 = 0.$$

The roots of this equation are $m = 1, -2, 3$. Since the roots are real and distinct, the general solution of the equation is given by

$$y(x) = Ae^x + Be^{-2x} + Ce^{3x}.$$

Example 5.21 Solve the differential equation $y''' - y'' - 4y' + 4y = 0$.

Solution Substituting $y = e^{mx}$, we obtain the characteristic equation as

$$m^3 - m^2 - 4m + 4 = 0 \text{ or } (m - 1)(m^2 - 4) = 0.$$

The roots of this equation are $m = 1, -2, 2$ which are real and distinct. The general solution of the equation is given by

$$y(x) = Ae^x + Be^{-2x} + Ce^{2x}.$$

Example 5.22 Solve the differential equation $y^{iv} - 5y'' + 4y = 0$.

Solution Substituting $y = e^{mx}$, we obtain the characteristic equation as.

$$m^4 - 5m^2 + 4 = 0 \quad \text{or} \quad (m^2 - 4)(m^2 - 1) = 0.$$

The roots of this equation are $m = -1, 1, -2, 2$. The general solution is

$$y(x) = Ae^{-x} + Be^x + Ce^{-2x} + De^{2x}.$$

Example 5.23 Solve the differential equation $4y^{iv} - 12y''' - y'' + 27y' - 18y = 0$.

Solution Substituting $y = e^{mx}$, we obtain the characteristic equation as

$$4m^4 - 12m^3 - m^2 + 27m - 18 = 0.$$

We find that $m = 1$ is a root. We write the equation as

$$(m - 1)(4m^3 - 8m^2 - 9m + 18) = 0, (m - 1)(m - 2)(4m^2 - 9) = 0.$$

The roots of the characteristic equation are $m = 1, 2, 3/2, -3/2$. The general solution is

$$y(x) = Ae^x + Be^{2x} + Ce^{-3x/2} + De^{3x/2}.$$

Example 5.24 Solve the initial value problem

$$y''' - 6y'' + 11y' - 6y = 0, \; y(0) = 0, \; y'(0) = -4, \; y''(0) = -18.$$

Solution Substituting $y = e^{mx}$, we obtain the characteristic equation as

$$m^3 - 6m^2 + 11m - 6 = 0, \quad \text{or} \quad (m-1)(m-2)(m-3) = 0.$$

The roots of this equation are $m = 1, 2, 3$ and the general solution is

$$y(x) = Ae^x + Be^{2x} + Ce^{3x}.$$

Substituting the initial conditions, we get

$$y(0) = 0 = A + B + C, \; y'(0) = -4 = A + 2B + 3C, \; y''(0) = -18 = A + 4B + 9C.$$

Solving, we obtain $A = 1$, $B = 2$ and $C = -3$. Hence, the particular solution is $y(x) = e^x + 2e^{2x} - 3e^{3x}$.

Real multiple roots

The characteristic equation (5.39) may have some multiple roots. Let r be the multiplicity of the root m_1, that is the root $m = m_1$ is repeated r times. Let the remaining $n - r$ roots be real and distinct. Substituting $m = m_1$ we obtain $y_1(x) = e^{m_1 x}$ as one of the solutions. We shall now show that the remaining $r - 1$ linearly independent solutions corresponding to the multiple root $m = m_1$ are given by

$$x\,y_1, \, x^2 y_1, \, \ldots, \, x^{r-1} y_1.$$

That is, the linearly independent solutions in this case are

$$e^{m_1 x}, x e^{m_1 x}, x^2 e^{m_1 x}, \ldots, x^{r-1} e^{m_1 x} \tag{5.42}$$

since the Wronskian of these solutions $W \neq 0$.

If

$$L[y] = a_0 y^{(n)} + a_1 y^{(n-1)} + \ldots + a_{n-1} y' + a_n y$$

then, substituting $y = e^{mx}$ in this equation, we get

$$L[e^{mx}] = [a_0 m^n + a_1 m^{n-1} + \ldots + a_n] e^{mx}$$

$$= (m - m_1)^r g(m) e^{mx}, \; g(m_1) \neq 0 \tag{5.43}$$

since $m = m_1$ is a multiple root of multiplicity r. Consider now m as a parameter. Differentiating Eq. (5.43) with respect to m, we get

$$\frac{d}{dm} L[e^{mx}] = r(m - m_1)^{r-1} g(m) e^{mx} + (m - m_1)^r \frac{d}{dm}[g(m) e^{mx}].$$

Now, L is a linear differentiable operator with respect to the independent variable x. Since m and x are independent, we obtain

$$\frac{d}{dm} L[e^{mx}] = L\left[\frac{d}{dm} e^{mx}\right] = L[x e^{mx}]$$

$$= r(m - m_1)^{r-1} g(m) e^{mx} + (m - m_1)^r \frac{d}{dm}[g(m) e^{mx}]. \tag{5.44}$$

Since the right hand side of Eq. (5.44) vanishes at $m = m_1$, $x e^{m_1 x}$ is also a solution of the differential equation. Differentiating Eq. (5.44) with respect to m, we get

$$\frac{d}{dm} L[xe^{mx}] = L\left[\frac{d}{dm}(xe^{mx})\right] = L[x^2 e^{mx}]$$

$$= r(r-1)(m-m_1)^{r-2} g(m)e^{mx} + 2r(m-m_1)^{r-1} \frac{d}{dm}[g(m)e^{mx}]$$

$$+ (m-m_1)^r \frac{d^2}{dm^2}[g(m)e^{mx}]. \tag{5.45}$$

The right hand side of Eq. (5.45) vanishes at $m = m_1$ again. Hence, $x^2 e^{m_1 x}$ is also a solution. After $r-1$ differentiations, the first term on the right hand side is obtained as $r!(m-m_1)g(m)e^{mx}$ which vanishes for $m = m_1$. The other terms also vanish for $m = m_1$. Therefore, $x^{r-1}e^{m_1 x}$ is also a solution. If we differentiate one more time, that is r times, the first term on the right hand side becomes $r!g(m)e^{mx}$ which does not vanish at $m = m_1$, showing that $x^r e^{m_1 x}$ is not a solution. Hence, we find that $e^{m_1 x}, xe^{m_1 x}, x^2 e^{m_1 x}, \ldots, x^{r-1}e^{m_1 x}$ are the linearly independent solutions corresponding to the multiple root $m = m_1$. For example, if $m = m_1$ is a multiple root of order 3, then $e^{m_1 x}, xe^{m_1 x}$ and $x^2 e^{m_1 x}$ are the linearly independent solutions.

Example 5.25   Solve the differential equation $y''' - 3y' - 2y = 0$.

Solution   Substituting $y = e^{mx}$, we obtain the characteristic equation as

$$m^3 - 3m - 2 = 0, \quad \text{or} \quad (m+1)(m^2 - m - 2) = 0$$

or
$$(m+1)^2(m-2) = 0, \quad \text{or} \quad m = -1, -1, 2.$$

Corresponding to the double root $m = -1$, the linearly independent solutions are e^{-x} and xe^{-x}. Hence, the general solution is

$$y(x) = Ae^{2x} + (Bx + C)e^{-x}.$$

Example 5.26   Solve the differentiable equation $8y''' - 12y'' + 6y' - y = 0$.

Solution   Substituting $y = e^{mx}$, we obtain the characteristic equation as

$$8m^3 - 12m^2 + 6m - 1 = 0, \quad \text{or} \quad (2m-1)^3 = 0, \quad \text{or} \quad m = 1/2, 1/2, 1/2.$$

The general solution is
$$y(x) = (A + Bx + Cx^2)e^{x/2}.$$

Example 5.27   Solve the initial value problem

$$y''' + 3y'' - 4y = 0, \, y(0) = 1, \, y'(0) = 0, \, y''(0) = 1/2.$$

Solution   Substituting $y = e^{mx}$, we obtain the characteristic equation as

$$m^3 + 3m^2 - 4 = 0, \quad \text{or} \quad (m-1)((m^2 + 4m + 4) = 0, \quad \text{or} \quad (m-1)(m+2)^2 = 0.$$

The roots of this equation are $m = 1, -2, -2$. The general solution is

$$y(x) = Ae^x + (Bx + C)e^{-2x}.$$

Substituting in the initial conditions, we get

$$y(0) = 1 = A + C,$$

$$y'(x) = Ae^x + Be^{-2x} - 2(Bx + C)e^{-2x}, \, y'(0) = 0 = A + B - 2C,$$

$$y''(x) = Ae^x - 4Be^{-2x} + 4(Bx + C)e^{-2x}, \, y''(0) = \frac{1}{2} = A - 4B + 4C.$$

The solution of the system is $A = 1/2$, $B = 1/2$ and $C = 1/2$. The particular solution is

$$y(x) = [e^x + (x + 1)e^{-2x}]/2.$$

Simple complex roots

Since the coefficients in the characteristic equation (5.39) are real, complex roots occur in conjugate pairs. That is, if $p + iq$ is a root, then $p - iq$ is also a root. In this case, the linearly independent solutions are given by $e^{px} \cos qx$ and $e^{px} \sin qx$. If the characteristic equation (5.39) has r complex conjugate pairs of roots $p_k \pm i q_k$, $k = 1, 2, \ldots r$, then the corresponding linearly independent solutions are $e^{p_1 x} \cos q_1 x$, $e^{p_1 x} \sin q_1 x$, $e^{p_2 x} \cos q_2 x$, $e^{p_2 x} \sin q_2 x, \ldots, e^{p_r x} \cos q_r x$ and $e^{p_r x} \sin q_r x$.

Example 5.28   Solve the differential equation $y^{iv} + 5y'' + 4y = 0$.

Solution   Substituting $y = e^{mx}$, we obtain the characteristic equation as

$$m^4 + 5m^2 + 4 = 0, \quad \text{or} \quad (m^2 + 4)(m^2 + 1) = 0.$$

The roots are $m = \pm i, \pm 2i$. The general solution is

$$y(x) = A \cos x + B \sin x + C \cos 2x + D \sin 2x.$$

Example 5.29   Solve the initial value problem

$$y^{iv} + 2y''' + 11y'' + 18y' + 18 = 0, \ y(0) = 2, \ y'(0) = 3, \ y''(0) = -11, \ y'''(0) = -23.$$

Solution   Substituting $y = e^{mx}$, we obtain the characteristic equation as

$$m^4 + 2m^3 + 11m^2 + 18m + 18 = 0 \quad \text{or} \quad (m^2 + 9)(m^2 + 2m + 2) = 0.$$

The roots are $m = \pm 3i, -1 \pm i$. The general solution is

$$y(x) = A \cos 3x + B \sin 3x + e^{-x} (C \cos x + D \sin x).$$

Substituting in the initial conditions, we get

$$y(0) = 2 = A + C,$$
$$y'(x) = -3A \sin 3x + 3B \cos 3x + e^{-x} (-C \sin x + D \cos x - C \cos x - D \sin x),$$
$$y'(0) = 3 = 3B + D - C,$$
$$y''(x) = -9A \cos 3x - 9B \sin 3x$$
$$\qquad + e^{-x}[-(C + D) \cos x + (C - D) \sin x + (C + D) \sin x + (C - D) \cos x]$$
$$\qquad = -9A \cos 3x - 9B \sin 3x + 2e^{-x} [C \sin x - D \cos x],$$
$$y''(0) = -11 = -9A - 2D,$$
$$y'''(x) = 27A \sin 3x - 27B \cos 3x + 2e^{-x} [C \cos x + D \sin x - C \sin x + D \cos x],$$
$$y'''(0) = -23 = -27B + 2C + 2D.$$

Therefore, we have the system of equations

$$A + C = 2, \quad 3B - C + D = 3,$$
$$-9A - 2D = -11, \quad -27B + 2C + 2D = -23.$$

The solution of this system is $A = 1$, $B = 1$, $C = 1$, $D = 1$. The particular solution is

$$y(x) = \cos 3x + \sin 3x + e^{-x} (\cos x + \sin x).$$

Example 5.30   Find the non trivial solutions of the boundary value problem

$$y^{iv} - \omega^4 y = 0,\ y(0) = 0,\ y''(0) = 0,\ y(l) = 0,\ y''(l) = 0.$$

Solution   Assume the solution to be of the form $y = e^{mx}$. The characteristic equation is given by

$$m^4 - \omega^4 = 0,\quad \text{or}\quad m^2 = \pm\,\omega^2,\quad \text{or}\quad m = \pm\,\omega, \pm\, i\omega.$$

The general solution is given by

$$y(x) = A_1 e^{\omega x} + B_1 e^{-\omega x} + C \cos \omega x + D \sin \omega x$$

$$= A \cosh \omega x + B \sinh \omega x + C \cos \omega x + D \sin \omega x$$

Substituting in the initial conditions, we get

$$y(0) = A + C = 0.$$

$$y'' = \omega^2\,[A \cosh \omega x + B \sinh \omega x - C \cos \omega x - D \sin \omega x];$$

$$y''(0) = \omega^2(A - C) = 0,\quad \text{or}\quad A - C = 0.$$

Solving the two equations, we get $A = 0$, $C = 0$. We also have

$$y(l) = 0 = B \sinh \omega l + D \sin \omega l,\ y''(l) = 0 = B \sinh \omega l - D \sin \omega l.$$

Adding, we obtain $2B \sinh \omega l = 0$, or $B = 0$. Therefore, we obtain $D \sin \omega l = 0$. Since, we require non-trivial solutions, we have $D \neq 0$. Hence, $\sin \omega l = 0 = \sin n\pi,\ n = 1, 2, \ldots$.

Therefore, $\omega = n\pi/l,\ n = 1, 2, \ldots$

The solution of the boundary value problem is

$$y_n(x) = D_n \sin (n\pi x/l),\ n = 1, 2, \ldots,$$

By superposition principle, the solution can be written as

$$y(x) = \sum_{n=1}^{\infty} D_n \sin (n\pi x/l).$$

Multiple complex roots

This case is a combination of the two earlier cases of real multiple roots and simple complex roots. Now, if $p + iq$ is a multiple root of order m, then $p - iq$ is also a multiple root of order m. For example, if $p_1 + iq_1$ is a double root, then $p_1 - iq_1$ is also a double root. The corresponding linearly independent solutions are

$$e^{p_1 x} \cos q_1 x,\ e^{p_1 x} \sin q_1 x,\ x e^{p_1 x} \cos q_1 x,\ x e^{p_1 x} \sin q_1 x.$$

Example 5.31   Solve the differential equation $y^{iv} + 32y'' + 256y = 0$.

Solution   Substituting $y = e^{mx}$, we obtain the characteristic equation as

$$m^4 + 32m^2 + 256 = 0,\quad \text{or}\quad (m^2 + 16)^2 = 0.$$

The roots of this equation are the double roots $m = \pm\,4i$. Therefore, the general solution is

$$y(x) = (Ax + B) \cos 4x + (Cx + D) \sin 4x.$$

Exercise 5.3

Find the general solution of the following differential equations.

1. $y''' - 9y' = 0$.

2. $2y''' + y'' - 13y' + 6y = 0$.

3. $3y''' - 2y'' - 3y' + 2y = 0$.

4. $y^{iv} - 13y'' + 36y = 0$.

5. $4y^{iv} - 12y''' + 7y'' + 3y' - 2y = 0$.

6. $y^{iv} + y''' - 4y'' - 4y' = 0$.

7. $8y^{iv} - 6y''' - 7y'' + 6y' - y = 0$.

8. $144y^{iv} - 25y'' + y = 0$.

9. $y''' - 2y'' + y' = 0$.

10. $y''' + 4y'' + 5y' + 2y = 0$.

11. $y''' - 2y'' - 4y' + 8y = 0$.

12. $27y''' - 27y'' + 9y' - y = 0$.

13. $y^{iv} - 11y''' + 35y'' - 25y' = 0$.

14. $y^{iv} - 3y''' + 3y'' - y' = 0$.

15. $4y^{iv} + 4y''' - 3y'' - 2y' + y = 0$.

16. $9y^{iv} - 66y''' + 157y'' - 132y' + 36y = 0$.

17. $y''' + y' = 0$.

18. $y''' - 2y'' + 4y' - 8y = 0$.

19. $y''' + 5y'' + 8y' + 6y = 0$.

20. $y''' - 7y'' + 19y' - 13y = 0$.

21. $y^{iv} + 8y'' - 9y = 0$.

22. $y^{iv} + y''' + 14y'' + 16y' - 32y = 0$.

23. $4y^{iv} + 101y'' + 25y = 0$.

24. $y^{iv} + 2y''' - 9y'' - 10y' + 50y = 0$.

25. $y^{iv} + 50y'' + 625y = 0$.

26. $y^{iv} + 2y'' + y = 0$.

Find a homogeneous linear differential equation with real constant coefficients of lowest order which has the following particular solution.

27. $5 + e^x + 2e^{3x}$.

28. $e^{-x} + \cos 5x + 3 \sin 5x$.

29. $xe^{-x} + e^{2x}$.

30. $1 + x + e^x - 3e^{3x}$.

31. $x^2 e^{2x} + 2e^{-2x}$.

32. $3 \cos 2x + 5 \sinh 3x$.

Solve the following initial value problems.

33. $y''' - 2y'' - 5y' + 6y = 0$, $y(0) = 0$, $y'(0) = 0$, $y''(0) = 1$.

34. $4y''' - 4y'' - 9y' + 9y = 0$, $y(0) = 1$, $y'(0) = 0$, $y''(0) = 0$.

35. $y''' - 5y'' + 7y' - 3y = 0$, $y(0) = 1$, $y'(0) = 0$, $y''(0) = -5$.

36. $y^{iv} - 2y''' - 3y'' + 4y' + 4y = 0$, $y(0) = 3$, $y'(0) = 3$, $y''(0) = 3$, $y'''(0) = 6$.

37. $y^{iv} + y'' = 0$, $y(0) = 1$, $y'(0) = 2$, $y''(0) = -1$, $y'''(0) = -1$.

38. $y''' - y'' + 4y' - 4y = 0$, $y(0) = 0$, $y'(0) = 3$, $y''(0) = -5$.

39. $y''' + y'' - 2y = 0$, $y(0) = 2$, $y'(0) = 2$, $y''(0) = -3$.

40. $y^{iv} - 3y''' = 0$, $y(0) = 2$, $y'(0) = 5$, $y''(0) = 15$, $y'''(0) = 27$.

Find the solution of the following differential equations satisfying the given conditions.

41. $y''' + \pi^2 y' = 0$, $y(0) = 0$, $y(1) = 0$, $y'(0) + y'(1) = 0$.

42. $y''' - 36y' = 0$, $y(0) = 2$, $y'(0) = 12$, $y'(1) = 6 \sinh(6) + 12 \cosh(6)$.

43. $y^{iv} + 13y'' + 36y = 0$, $y(0) = 0$, $y''(0) = 0$, $y(\pi/2) = -1$, $y'(\pi/2) = -4$.

44. $y^{iv} - \omega^4 y = 0$, $\omega \neq 0$, $y(0) = 0$, $y''(0) = 0$, $y(\pi) = 0$, $y''(\pi) = 0$.

45. $y^{iv} + 10y'' + 9y = 0$, $y'(0) = 0$, $y'''(0) = 0$, $y'(\pi/2) = 5$, $y'''(\pi/2) = -53$.

5.4   Solution of Non-Homogeneous Linear Equations

In the previous section, we have discussed methods for finding the general and particular solutions of homogeneous linear equations. In this section, we shall discuss methods for finding the general solution of a non-homogeneous linear equation (see Eq. (5.1)) of the form

$$L[y] = a_0(x)y^{(n)}(x) + a_1(x)y^{(n-1)}(x) + \ldots + a_{n-1}(x)y' + a_n(x)y = r(x), \, a_0(x) \neq 0, \quad (5.46)$$

when the general solution of the corresponding homogeneous linear equation $L[y] = 0$ is known. We present the following theorem.

Theorem 5.5 If $\{y_1(x), y_2(x), \ldots, y_n(x)\}$ is a basis and $c_1 y_1(x) + c_2 y_2(x) + \ldots + c_n y_n(x)$ is the general solution of the corresponding homogeneous linear equation $L[y] = 0$ and if $y_p(x)$ is any particular solution (a solution not containing any arbitrary constants) of the non-homogeneous equation (5.46), then the general solution of equation (5.46) is given by

$$y(x) = c_1 y_1(x) + c_2 y_2(x) + \ldots + c_n y_n(x) + y_p(x). \quad (5.47)$$

Proof Since $y_p(x)$ is a particular solution, we have

$$L[y_p(x)] = a_0 y_p^{(n)} + a_1 y_p^{(n-1)} + \ldots + a_{n-1} y_p' + a_n y_p = r(x). \quad (5.48)$$

Subtracting Eq. (5.48) from (5.46), we obtain

$$a_0 (y^{(n)} - y_p^{(n)}) + a_1 (y^{(n-1)} - y_p^{(n-1)}) + \ldots + a_{n-1}(y' - y_p') + a_n (y - y_p) = 0. \quad (5.49)$$

Denote $y - y_p = z$. Then, from Eq. (5.49) we obtain

$$a_0 z^{(n)} + a_1 z^{(n-1)} + \ldots + a_{n-1} z' + a_n z = 0. \quad (5.50)$$

But, this equation is the corresponding homogeneous equation of Eq. (5.46), whose basis is $\{y_1(x), y_2(x), \ldots, y_n(x)\}$. Hence, the general solution of Eq. (5.50) is given by

$$z = c_1 y_1(x) + c_2 y_2(x) + \ldots + c_n y_n(x).$$

Replacing $z = y - y_p$, and taking y_p to the right hand side, we obtain

$$y(x) = c_1 y_1(x) + c_2 y_2(x) + \ldots + c_n y_n(x) + y_p(x). \quad (5.51)$$

Since, this solution contains n arbitrary constants, it is the general solution of the Eq. (5.46).

From the above theorem, we conclude that the solution of a non-homogeneous equation consists of the sum of the following two parts.

(i) The general solution of the corresponding homogeneous equation. This solution is called the *complementary function* and is denoted by $y_c(x)$.

(ii) A particular solution of the non-homogeneous equation. This solution is also called a *particular integral* of the non-homogeneous equation and is denoted by $y_p(x)$.

The general solution of the non-homogeneous equation is then written as

$$y(x) = y_c(x) + y_p(x).$$

Now, suppose that the right hand side $r(x)$ is the sum of a number of functions

$$r(x) = r_1(x) + r_2(x) + \ldots + r_m(x). \quad (5.52)$$

Let $y_{p_i}(x), i = 1, 2, \ldots, m$ be any particular solutions, not containing any arbitrary constants, of the equations

$$a_0 y^{(n)} + a_1 y^{(n-1)} + \ldots + a_{n-1} y' + a_n y = r_i(x), i = 1, 2, \ldots, m. \quad (5.53)$$

Then, $y_{p_1} + y_{p_2} + \ldots + y_{p_m}$ is the particular integral of the equation

$$a_0 y^{(n)} + a_1 y^{(n-1)} + \ldots + a_{n-1} y' + a_n y = r_1(x) + r_2(x) + \ldots + r_m(x) = r(x)$$

and hence of the given non-homogeneous linear equation. This can be proved by summing Eq. (5.53) over i. In other words, if the right hand side of Eq. (5.46) consists of sum of a number of functions, then particular integrals of the Eq. (5.53) can be obtained with respect to each of the functions and the particular integral of Eq. (5.46) is then given by the sum of these particular integrals.

The methods for finding $y_c(x)$ have been discussed in the previous section. In the remaining part of this section, we shall derive methods for finding the particular integral $y_p(x)$ of the non-homogeneous equation.

5.4.1 Method of Variation of Parameters

Consider the second order non-homogeneous linear equation

$$a_0(x)y'' + a_1(x)y' + a_2(x)y = r(x), \quad a_0(x) \neq 0. \tag{5.54}$$

We shall discuss a general method of solution, called the method of *variation of parameters*, which can always be used to find a particular integral whenever the complementary function of the equation is known. Consider first, the solution of the corresponding homogeneous equation

$$a_0(x)y'' + a_1(x)y' + a_2(x)y = 0, \quad a_0(x) \neq 0. \tag{5.55}$$

Using the methods given in the previous section, we can find two linearly independent solutions $y_1(x)$ and $y_2(x)$ of the equation (5.55). The complementary function is given by

$$y_c(x) = A y_1(x) + B y_2(x) \tag{5.56}$$

where A and B are arbitrary constants. The idea behind the method of variation of parameters is to vary the parameters A and B. That is, we assume A and B to be functions of x and determine $A(x), B(x)$ such that

$$y(x) = A(x)y_1(x) + B(x)y_2(x) \tag{5.57}$$

is the general solution of Eq. (5.54). Now, $y(x)$ contains two functions $A(x)$ and $B(x)$ which are to be determined. Therefore, we need two equations to determine them. One equation is obtained by substituting $y(x)$ from Eq. (5.57) in Eq. (5.54). The determination of the second equation is at our disposal. This equation is chosen such that the determination of $A(x)$ and $B(x)$ is simple. Differentiating Eq. (5.57), we obtain

$$y'(x) = A'y_1 + Ay_1' + By_2' + B'y_2 = (A'y_1 + B'y_2) + (Ay_1' + By_2'). \tag{5.58}$$

If we differentiate this equation again, then the equation would contain the second derivatives A'' and B'' of the unknown functions. In order that these derivatives are not used, we set in Eq. (5.58)

$$A'y_1 + B'y_2 = 0. \tag{5.59}$$

which gives us the second equation to determine $A(x)$ and $B(x)$. Now, differentiating $y'(x) = Ay_1' + By_2'$, we obtain

$$y''(x) = Ay_1'' + A'y_1' + By_2'' + B'y_2'. \tag{5.60}$$

Substituting the expressions for $y(x)$, $y'(x)$ and $y''(x)$ in Eq. (5.54), we obtain

$$a_0(x)[Ay_1'' + A'y_1' + By_2'' + B'y_2'] + a_1(x)[Ay_1' + By_2'] + a_2(x)[Ay_1 + By_2] = r(x)$$

or
$$a_0(x)[A'y_1' + B'y_2'] + A[a_0(x)y_1'' + a_1(x)y_1' + a_2(x)y_1]$$
$$+ B[a_0(x)y_2'' + a_1(x)y_2' + a_2(x)y_2] = r(x).$$

Since, $y_1(x)$ and $y_2(x)$ are the solutions of the homogeneous equation (5.55), we obtain

$$a_0(x)[A'y_1' + B'y_2'] = r(x), \quad \text{or} \quad A'y_1' + B'y_2' = \frac{r(x)}{a_0(x)} = g(x). \tag{5.61}$$

Since $a_0(x) \neq 0$ on the given interval I, $g(x)$ is continuous on I. Solving the equations

$$A'y_1 + B'y_2 = 0$$

$$A'y_1' + B'y_2' = g(x),$$

we obtain

$$A' = -\frac{g(x)y_2}{y_1y_2' - y_2y_1'}, \quad B' = \frac{g(x)y_1}{y_1y_2' - y_2y_1'}. \tag{5.62}$$

We note that the Wronskian $W(y_1, y_2)$ is

$$W = \begin{vmatrix} y_1 & y_2 \\ y_1' & y_2' \end{vmatrix} = y_1y_2' - y_2y_1' \neq 0 \cdot$$

since y_1, y_2 are the linearly independent solutions of the homogeneous equation. Hence, we can write Eqs. (5.62) as

$$A' = -\frac{g(x)y_2}{W(x)}, \quad \text{and} \quad B' = \frac{g(x)y_1}{W(x)}. \tag{5.63}$$

Integrating, we obtain

$$A(x) = -\int \frac{g(x)y_2(x)}{W(x)} \, dx + c_1 \quad \text{and} \quad B(x) = \int \frac{g(x)y_1(x)}{W(x)} \, dx + c_2. \tag{5.64}$$

Substituting in Eq. (5.57), we obtain the general solution which contains two arbitrary constants. If we do not add the arbitrary constants while carrying out integrations of Eqs. (5.63), then we obtain the particular solution as $y_p(x) = A(x)y_1(x) + B(x)y_2(x)$, which does not contain any arbitrary constants. The general solution is then given by $y(x) = y_c(x) + y_p(x)$.

The method is applicable both for constant coefficient and variable coefficient problems. The method can also be easily extended to equations of any order. At each differentiation step, we set the part containing the derivatives of the unknown functions to zero, until we arrive at the final substitution step. For example, consider the third order equation

$$a_0(x)y''' + a_1(x)y'' + a_2(x)y' + a_3(x)y = r(x), \quad a_0(x) \neq 0. \tag{5.65}$$

The complementary function is

$$y(x) = Ay_1(x) + By_2(x) + Cy_3(x)$$

where y_1, y_2, y_3 are the linearly independent solutions of the corresponding homogeneous equation and A, B, C are arbitrary constants. We assume the solution as

$$y(x) = A(x)y_1(x) + B(x)y_2(x) + C(x)y_3(x). \tag{5.66}$$

Following the procedure discussed earlier, we obtain the required equations for determining $A(x)$, $B(x)$ and $C(x)$ as

$$A'(x)\,y_1 + B'(x)y_2 + C'(x)y_3 = 0$$

$$A'(x)y_1' + B'(x)y_2' + C'(x)y_3' = 0$$

and

$$A'(x)y_1'' + B'(x)y_2'' + C'(x)y_3'' = \frac{r(x)}{a_0(x)} = g(x). \tag{5.67}$$

The determinant of the coefficient matrix is the Wronskian $W(y_1, y_2, y_3) \neq 0$. We determine $A(x)$, $B(x)$. $C(x)$ and substitute in Eq. (5.66) to obtain the general solution.

Example 5.32 Find the general solution of the equation $y'' + 3y' + 2y = 2e^x$, using the method of variation of parameters.

Solution The corresponding homogeneous equation is $y'' + 3y' + 2y = 0$. The characteristic equation is $m^2 + 3m + 2 = 0$ and its roots are $m = -1, -2$. Hence, the complementary function is

$$y_c(x) = Ay_1(x) + By_2(x) = A e^{-x} + B e^{-2x}$$

where $y_1(x) = e^{-x}$ and $y_2(x) = e^{-2x}$ are two linearly independent solutions of the homogeneous equation. Assume the general solution as

$$y(x) = A(x)e^{-x} + B(x)e^{-2x}$$

We have $g(x) = r(x)/a_0(x) = 2e^x$.

The Wronskian of $y_1(x)$, $y_2(x)$ is given by

$$W(x) = \begin{vmatrix} y_1 & y_2 \\ y_1' & y_2' \end{vmatrix} = \begin{vmatrix} e^{-x} & e^{-2x} \\ -e^{-x} & -2e^{-2x} \end{vmatrix} = -e^{-3x}.$$

Using Eq. (5.64), we obtain the solutions for $A(x)$ and $B(x)$ as

$$A(x) = -\int \frac{g(x)y_2(x)}{W}\,dx + c_1 = -\int \frac{2e^x e^{-2x}}{-e^{-3x}}\,dx + c_1 = e^{2x} + c_1$$

$$B(x) = \int \frac{g(x)y_1(x)}{W}\,dx + c_2 = \int \frac{2e^x e^{-x}}{-e^{-3x}}\,dx + c_2 = -\frac{2}{3}e^{3x} + c_2.$$

The general solution is

$$y(x) = A(x)e^{-x} + B(x)e^{-2x}$$

$$= (e^{2x} + c_1)e^{-x} + \left(-\frac{2}{3}e^{3x} + c_2\right)e^{-2x} = c_1 e^{-x} + c_2 e^{-2x} + \frac{1}{3}e^x.$$

Example 5.33 Find the general solution of the equation $y'' + 16y = 32 \sec 2x$, using the method of variation of parameters.

Solution The characteristic equation of the corresponding homogeneous equation is $m^2 + 16 = 0$. The characteristic roots are $m = \pm 4i$. The complementary function is given by

$$y_c(x) = Ay_1(x) + By_2(x) = A \cos 4x + B \sin 4x$$

where $y_1(x) = \cos 4x$ and $y_2(x) = \sin 4x$ are two linearly independent solutions of the homogeneous equation. By the method of the variation of parameters, we write the general solution as

$$y(x) = A(x) \cos 4x + B(x) \sin 4x.$$

We have $g(x) = r(x)/a_0(x) = 32 \sec 2x$. The Wronskian of y_1, y_2 is given by

$$W(x) = \begin{vmatrix} y_1 & y_2 \\ y_1' & y_2' \end{vmatrix} = \begin{vmatrix} \cos 4x & \sin 4x \\ -4 \sin 4x & 4 \cos 4x \end{vmatrix} = 4.$$

Therefore, from Eq. (5.64), we obtain

$$A(x) = -\int \frac{g(x) y_2(x)}{W} \, dx + c_1 = -\frac{1}{4} \int 32 \sec 2x \sin 4x \, dx + c_1$$

$$= -16 \int \sin 2x \, dx + c_1 = 8 \cos 2x + c_1.$$

$$B(x) = \int \frac{g(x) y_1(x)}{W} \, dx + c_2 = \frac{1}{4} \int 32 \sec 2x \cos 4x \, dx + c_2$$

$$= 8 \int \frac{2 \cos^2 2x - 1}{\cos 2x} \, dx + c_2 = 8 \int (2 \cos 2x - \sec 2x) \, dx + c_2$$

$$= 8 \sin 2x - 4 \ln | \sec 2x + \tan 2x | + c_2.$$

The general solution is

$$y(x) = A(x) \cos 4x + B(x) \sin 4x$$

$$= c_1 \cos 4x + c_2 \sin 4x + 8 \cos 2x \cos 4x + 8 \sin 2x \sin 4x$$

$$- 4 \sin 4x \ln | \sec 2x + \tan 2x |$$

$$= c_1 \cos 4x + c_2 \sin 4x + 8 \cos 2x - 4 \sin 4x \ln | \sec 2x + \tan 2x |.$$

Example 5.34 Find the general solution of the equation $y''' - 6y'' + 11y' - 6y = e^{-x}$.

Solution   The characteristic equation of the corresponding homogeneous equation is $m^3 - 6m^2 + 11m - 6 = 0$ and its roots are $m = 1, 2, 3$. The complementary function is given by

$$y_c(x) = A e^x + B e^{2x} + C e^{3x}.$$

By the method of variation of parameters, we assume the solution as

$$y(x) = A(x) e^x + B(x) e^{2x} + C(x) e^{3x}.$$

We have
$$g(x) = r(x)/a_0(x) = e^{-x}.$$

From Eqs. (5.67), the equations for determining $A(x)$, $B(x)$ and $C(x)$ are

$$A' e^x + B' e^{2x} + C' e^{3x} = 0$$

$$A' e^x + 2B' e^{2x} + 3C' e^{3x} = 0$$

$$A' e^x + 4B' e^{2x} + 9C' e^{3x} = e^{-x}.$$

The Wronskian of $y_1 = e^x$, $y_2 = e^{2x}$, $y_3 = e^{3x}$ is given by

$$W(x) = \begin{vmatrix} e^x & e^{2x} & e^{3x} \\ e^x & 2e^{2x} & 3e^{3x} \\ e^x & 4e^{2x} & 9e^{3x} \end{vmatrix} = e^{6x} \begin{vmatrix} 1 & 1 & 1 \\ 1 & 2 & 3 \\ 1 & 4 & 9 \end{vmatrix} = 2e^{6x}.$$

By the Cramer's rule, we obtain

$$WA' = \begin{vmatrix} 0 & e^{2x} & e^{3x} \\ 0 & 2e^{2x} & 3e^{3x} \\ e^{-x} & 4e^{2x} & 9e^{3x} \end{vmatrix} = e^{4x}, \text{ or } A' = \frac{e^{4x}}{2e^{6x}} = \frac{1}{2}e^{-2x}.$$

Integrating, we get $A = -\frac{1}{4}e^{-2x} + c_1$.

Similarly, we have

$$WB' = \begin{vmatrix} e^x & 0 & e^{3x} \\ e^x & 0 & 3e^{3x} \\ e^x & e^{-x} & 9e^{3x} \end{vmatrix} = -2e^{3x}, \text{ or } B' = -\frac{2e^{3x}}{2e^{6x}} = -e^{-3x}.$$

$$WC' = \begin{vmatrix} e^x & e^{2x} & 0 \\ e^x & 2e^{2x} & 0 \\ e^x & 4e^{2x} & e^{-x} \end{vmatrix} = e^{2x}, \text{ or } C' = \frac{e^{2x}}{2e^{6x}} = \frac{1}{2}e^{-4x}.$$

Integrating, we obtain $B(x) = \frac{1}{3}e^{-3x} + c_2$ and $C(x) = -\frac{1}{8}e^{-4x} + c_3$. The general solution is

$$y(x) = A(x)e^x + B(x)e^{2x} + C(x)e^{3x}$$

$$= \left(-\frac{1}{4}e^{-2x} + c_1 \right)e^x + \left(\frac{1}{3}e^{-3x} + c_2 \right)e^{2x} + \left(-\frac{1}{8}e^{-4x} + c_3 \right)e^{3x}$$

$$= c_1 e^x + c_2 e^{2x} + c_3 e^{3x} - \frac{1}{24}e^{-x}.$$

Example 5.35 It is given that $y_1 = x$ and $y_2 = 1/x$ are two linearly independent solutions of the associated homogeneous equation of $x^2 y'' + xy' - y = x$, $x \neq 0$. Find a particular integral and the general solution of the equation.

Solution By the method of variation of parameters, we write

$$y(x) = A(x)x + B(x)\left(\frac{1}{x} \right).$$

The Wronskian of $y_1(x) = x$ and $y_2(x) = 1/x$ is given by

$$W = \begin{vmatrix} x & 1/x \\ 1 & -1/x^2 \end{vmatrix} = -\frac{2}{x}, \quad x \neq 0.$$

We have $g(x) = r(x)/a_0(x) = 1/x$. Using Eq. (5.64), we obtain

$$A(x) = -\int \frac{g(x)y_2(x)}{W} dx = -\int \frac{1}{x^2}\left(-\frac{x}{2}\right) dx = \frac{1}{2}\ln|x| + c_1.$$

$$B(x) = \int \frac{g(x)y_1(x)}{W} dx = \int \frac{1}{x}\left(-\frac{x^2}{2}\right) dx = -\frac{1}{4}x^2 + c_2.$$

The particular integral is

$$y_p(x) = A(x)x + B(x)\left(\frac{1}{x}\right) = \frac{x}{2}\ln|x| - \frac{x}{4}.$$

The general solution is

$$y(x) = y_c(x) + y_p(x) = c_1 x + \frac{1}{x}c_2 + \frac{x}{2}\ln|x| - \frac{x}{4}$$

or

$$y(x) = c_1^* x + \frac{1}{x}c_2 + \frac{x}{2}\ln|x|, \quad \text{where} \quad c_1^* = c_1 - \frac{1}{4}.$$

Exercise 5.4

Find the general solution of the following differential equations, using the method of variation of parameters.

1. $y'' - 2y' - 3y = e^x$.
2. $y'' - 4y' + 4y = e^{-2x}$.
3. $y'' + 4y = \cos x$.
4. $y'' + y = \sec x$.
5. $y'' + y = \operatorname{cosec} x$.
6. $y'' + y = \tan x$.
7. $y'' - 4y' + 3y = e^x$.
8. $y'' + 4y = \sec 2x$.
9. $y'' + 4y = \cos 2x$.
10. $y'' + 4y' + 4y = e^{-2x}\sin x$.
11. $y'' + 6y' + 9y = e^{-3x}/x$.
12. $y'' + 2y' + 2y = e^{-x}\cos x$.

In the following problems, using the method of variation of parameters and the given linearly independent solutions, find a particular integral and the general solution.

13. $x^2 y'' + xy' - y = x^3$, $y_1 = x$, $y_2 = 1/x$. 14. $x^2 y'' + xy' - 4y = x^2\ln|x|$, $y_1 = x^2$, $y_2 = 1/x^2$.

15. $x^2 y'' - xy' + y = 1/x^4$, $y_1 = x$, $y_2 = x\ln|x|$.

16. $x^2 y'' - 2xy' + 2y = x^3 + x$, $y_1 = x$, $y_2 = x^2$.

17. $y'' + 4y' + 8y = 16 e^{-2x}\operatorname{cosec}^2 2x$, $y_1 = e^{-2x}\cos 2x$, $y_2 = e^{-2x}\sin 2x$.

18. $y''' + 4y' = \sec 2x$, $y_1 = 1$, $y_2 = \cos 2x$, $y_3 = \sin 2x$.

19. $y''' - 6y'' + 12y' - 8y = e^{2x}/x$, $y_1 = e^{2x}$, $y_2 = xe^{2x}$, $y_3 = x^2 e^{2x}$.

20. Show that the general solution of the equation $y'' + k^2 y = g(x)$, where $k \neq 0$ and $g(x)$ is continuous on I, can always be written as

$$y(x) = A\cos kx + B\sin kx + \frac{1}{k}\int_0^x \sin k(x - t)g(t)\,dt.$$

5.4.2 Method of Undetermined Coefficients

In the previous section, we have discussed the method of variation of parameters for finding the solution of the differential equation

$$a_0 y^{(n)} + a_1 y^{(n-1)} + \ldots + a_{n-1} y' + a_n y = r(x)$$

where $a_0, a_1, \ldots, a_n$ are constants. In the cases when the right hand side $r(x)$ is of a special form containing exponentials, polynomials, cosine and sine functions, sums or products of these functions, then the particular integral can be easily obtained by the method of undetermined coefficients. The basic idea behind this approach is as follows.

If $r(x)$ is of exponential form e^{mx}, then its derivatives also contain exponentials e^{mx} only, that is, if $r(x) = pe^{mx}$, p constant, then we can choose the particular integral as $y_p(x) = ce^{mx}$, c constant and determine c by substituting $y_p(x)$ in the given equation and comparing both sides of the equation. That is, the equation is identically satisfied.

If $r(x)$ is a cosine or a sine function, $\cos mx$ or $\sin mx$, then their derivatives contain the terms $\cos mx$ and $\sin mx$. In other words, if $r(x) = p \cos mx$ or $p \sin mx$, p constant, then we can choose the particular integral as $y_p(x) = c_1 \cos mx + c_2 \sin mx$. The constants c_1, c_2 are determined by substituting $y_p(x)$ in the given equation and comparing both sides of the equation.

If $r(x)$ is of the form x^m, then its derivatives contain the terms $x^m, x^{m-1}, \ldots, x, 1$. Hence, when $r(x) = px^m$, p constant then we can choose the particular integral as

$$y_p(x) = c_0 x^m + c_1 x^{m-1} + \ldots + c_{m-1} x + c_m \tag{5.68}$$

where $c_0, c_1, \ldots, c_m$ are constants.

If $r(x)$ is of the forms $e^{ax} \cos bx$ or $e^{ax} \sin bx$ then their derivatives contain the terms $e^{ax} \cos bx$ and $e^{ax} \sin bx$. Hence, when $r(x) = e^{ax} \cos bx$ or $e^{ax} \sin bx$, then we can choose the particular integral as

$$y_p(x) = e^{ax}(c_1 \cos bx + c_2 \sin bx). \tag{5.69}$$

However, if any term in the choice of the particular integral is also a solution of the corresponding homogeneous equation, that is, a term in the complementary function, then we multiply this term by x or by x^m (if the term in the complementary function corresponds to a multiple root of multiplicity m). If $r(x)$ is the sum of a number of functions, then the contribution with respect to each of the terms is included in the choice of the particular integral.

Example 5.36   Using the method of undetermined coefficients find the general solution of the differential equation $y'' + y = 32x^3$.

Solution   The characteristic equation of the homogeneous equation is $m^2 + 1 = 0$ and its roots are $m = \pm i$. The complementary function is $y_c(x) = A \cos x + B \sin x$.
Since $r(x) = 32x^3$, we choose the particular integral as

$$y_p(x) = c_1 x^3 + c_2 x^2 + c_3 x + c_4.$$

Substituting in the given equation, we get

$$(6c_1 x + 2c_2) + (c_1 x^3 + c_2 x^2 + c_3 x + c_4) = 32x^3.$$

Comparing the coefficients of various powers of x, we get

$$c_1 = 32, \ c_2 = 0, \ 6c_1 + c_3 = 0, \ 2c_2 + c_4 = 0.$$

The solution of the system is $c_1 = 32, \ c_2 = 0, \ c_3 = -192, \ c_4 = 0$. Therefore, $y_p(x) = 32x^3 - 192x$. The general solution is

$$y(x) = A \cos x + B \sin x + 32x(x^2 - 6).$$

Example 5.37   Find the general solution of the differential equation $y'' - 2y' - 3y = 6e^{-x} - 8e^x$.

Solution   The characteristic equation of the homogeneous equation is $m^2 - 2m - 3 = 0$ and its roots are $m = -1, 3$.

The complementary function is $y_c(x) = Ae^{-x} + Be^{3x}$.

We note that e^{-x} appears both as a term in $y_c(x)$ (due to the simple root $m = -1$) and the right hand side $r(x)$. The term e^x appears only in $r(x)$. Hence, we choose the particular integral as

$$y_p(x) = c_1 x e^{-x} + c_2 e^x.$$

We have $y_p'(x) = c_1(1 - x)e^{-x} + c_2 e^x, y_p''(x) = -c_1(2 - x)e^{-x} + c_2 e^x.$

Substituting in the given equation, we get

$$c_1[-(2 - x) - 2(1 - x) - 3x]e^{-x} + c_2[1 - 2 - 3]e^x = 6x^{-x} - 8e^x$$

or $-4c_1 e^{-x} - 4c_2 e^x = 6e^{-x} - 8e^x.$

Comparing the coefficients of e^{-x} and e^x, we get $c_1 = -3/2$, $c_2 = 2$. The general solution is

$$y(x) = Ae^{-x} + Be^{3x} - \frac{3}{2}xe^{-x} + 2e^x.$$

Example 5.38   Find the general solution of the equation $y'' + 9y = \cos 3x$.

Solution   The characteristic equation of the homogeneous equation is $m^2 + 9 = 0$ and its roots are $m = \pm 3i$. The complementary function is

$$y_c(x) = A \cos 3x + B \sin 3x.$$

We note that $\cos 3x$ appears as a term in $y_c(x)$ and the right hand side $r(x)$. Hence, we choose the particular integral as

$$y_p(x) = x(c_1 \cos 3x + c_2 \sin 3x).$$

We have $y_p'(x) = c_1 \cos 3x + c_2 \sin 3x + 3x(-c_1 \sin 3x + c_2 \cos 3x)$

$$y_p''(x) = 6(-c_1 \sin 3x + c_2 \cos 3x) + 9x(-c_1 \cos 3x - c_2 \sin 3x).$$

Substituting in the given equation, we get

$$y_p'' + 9y_p = \sin 3x\,[-6c_1 - 9xc_2 + 9xc_2] + \cos 3x\,[6c_2 - 9xc_1 + 9xc_1] = \cos 3x$$

or $-6c_1 \sin 3x + 6c_2 \cos 3x = \cos 3x.$

Comparing both sides, we get $c_1 = 0$ and $c_2 = 1/6$. The particular integral is $y_p(x) = (x \sin 3x)/6$. The general solution is

$$y(x) = A \cos 3x + B \sin 3x + \frac{1}{6}x \sin 3x.$$

Example 5.39   Find the general solution of the equation $y'' + 4y' + 4y = 12e^{-2x}$.

Solution   The characteristic equation of the homogeneous equation is $m^2 + 4m + 4 = (m + 2)^2 = 0$ and its roots are $m = -2, -2$.

The complementary function is $y_c(x) = (Ax + B)e^{-2x}$.

We note that e^{-2x} and xe^{-2x} are present in the complementary function (due to the double root

$m = -2$) and e^{-2x} is also a term on the right hand side $r(x)$. Therefore, we choose the particular integral as

$$y_p(x) = c_1 x^2 e^{-2x}.$$

We have $\qquad y_p'(x) = c_1[2x - 2x^2]e^{-2x}, \; y_p''(x) = c_1[2 - 8x + 4x^2]e^{-2x}.$

Substituting in the given equation, we get

$$y_p'' + 4y_p' + 4y_p = c_1[(2 - 8x + 4x^2) + 4(2x - 2x^2) + 4x^2]e^{-2x} = 12e^{-2x}$$

or $\qquad 2c_1 e^{-2x} = 12e^{-2x}.$

Comparing both sides, we get $c_1 = 6$. Therefore, the particular integral is $y_p(x) = 6x^2 e^{-2x}$. The general solution is

$$y(x) = (Ax + B)e^{-2x} + 6x^2 e^{-2x} = (Ax + B + 6x^2)e^{-2x}.$$

Example 5.40 Find the general solution of the equation $y'' - 4y' + 13y = 12e^{2x} \sin 3x$.

Solution The characteristic equation of the homogeneous equation is $m^2 - 4m + 13 = 0$. The roots of this equation are

$$m = \frac{4 \pm \sqrt{16 - 52}}{2} = 2 \pm 3i.$$

The complementary function is $y_c(x) = e^{2x}(A \cos 3x + B \sin 3x)$.

We note that $e^{2x} \sin 3x$ appears both in the complementary function and the right hand side $r(x)$. Therefore, we choose

$$y_p(x) = xe^{2x}(c_1 \cos 3x + c_2 \sin 3x).$$

We have

$$y_p'(x) = (1 + 2x)e^{2x}(c_1 \cos 3x + c_2 \sin 3x) + 3xe^{2x}(-c_1 \sin 3x + c_2 \cos 3x)$$

$$y_p''(x) = (4 + 4x)e^{2x}(c_1 \cos 3x + c_2 \sin 3x)$$

$$+ 6(1 + 2x)e^{2x}(-c_1 \sin 3x + c_2 \cos 3x) + 9xe^{2x}(-c_1 \cos 3x - c_2 \sin 3x).$$

Substituting in the given equation, we get

$$y_p'' - 4y_p' + 13y_p = e^{2x} \cos 3x \, [c_1(4 + 4x) + 6c_2(1 + 2x) - 9c_1 x - 4c_1(1 + 2x)$$

$$- 12xc_2 + 13c_1 x] + e^{2x} \sin 3x \, [c_2(4 + 4x) - 6c_1(1 + 2x) - 9c_2 x$$

$$- 4c_2(1 + 2x) + 12c_1 x + 13xc_2] = 12e^{2x} \sin 3x$$

or $\qquad 6c_2 e^{2x} \cos 3x - 6c_1 e^{2x} \sin 3x = 12 \, e^{2x} \sin 3x.$

Comparing both sides, we get $c_1 = -2$ and $c_2 = 0$. Therefore, the particular integral is $y_p(x) = -2xe^{2x} \cos 3x$. The general solution is

$$y(x) = e^{2x} [A \cos 3x + B \sin 3x - 2x \cos 3x].$$

Example 5.41 Find the general solution of the differential equation $y''' - 2y'' - 5y' + 6y = 18e^x$.

Solution The characteristic equation of the homogeneous equation is

$$m^3 - 2m^2 - 5m + 6 = (m-1)(m+2)(m-3) = 0, \text{ or } m = 1, -2, 3.$$

The complementary function is $y_c(x) = Ae^x + Be^{-2x} + Ce^{3x}$.

Choose the particular integral as $y_p(x) = c_1 x e^x$.

We have $\qquad y_p' = c_1(1+x)e^x, y_p'' = c_1(2+x)e^x, y_p''' = c_1(3+x)e^x.$

Substituting in the given equation, we get

$$y_p''' - 2y_p'' - 5y_p' + 6y_p = c_1 e^x[(3+x) - 2(2+x) - 5(1+x) + 6x]$$

$$= -6c_1 e^x = 18e^x.$$

Comparing both sides, we get $c_1 = -3$. Hence, the particular integral is $y_p = -3xe^x$. The general solution is

$$y(x) = Ae^x + Be^{-2x} + Ce^{3x} - 3xe^x.$$

Example 5.42 Find the general solution of the differential equation

$$y''' - 6y'' + 12y' - 8y = 12e^{2x} + 27e^{-x}.$$

Solution The characteristic equation of the homogeneous equation is

$$m^3 - 6m^2 + 12m - 8 = (m-2)^3 = 0, \text{ or } m = 2, 2, 2.$$

The complementary function is $y_c(x) = (Ax^2 + Bx + C)e^{2x}$. Note that $m = 2$ is a triple root and e^{2x} is contained in a term in $r(x)$. Therefore, we choose the particular integral as

$$y_p(x) = c_1 x^3 e^{2x} + c_2 e^{-x}.$$

We have $\qquad y_p' = c_1(3x^2 + 2x^3)e^{2x} - c_2 e^{-x}, y_p'' = c_1(6x + 12x^2 + 4x^3)e^{2x} + c_2 e^{-x},$

$$y_p''' = c_1(6 + 36x + 36x^2 + 8x^3)e^{2x} - c_2 e^{-x}.$$

Substituting in the given equation, we get

$$y_p''' - 6y_p'' + 12y_p' - 8y_p = c_1 e^{2x}[(6 + 36x + 36x^2 + 8x^3) - 6(6x + 12x^2 + 4x^3)$$

$$+ 12(3x^2 + 2x^3) - 8x^3] + c_2 e^{-x}[-1 - 6 - 12 - 8]$$

$$= 6c_1 e^{2x} - 27c_2 e^{-x} = 12e^{2x} + 27e^{-x}.$$

Comparing both sides, we get $c_1 = 2$ and $c_2 = -1$. Therefore, the particular integral is $y_p(x) = 2x^3 e^{2x} - e^{-x}$. The general solution is

$$y(x) = (Ax^2 + Bx + C)e^{2x} + 2x^3 e^{2x} - e^{-x}.$$

Exercise 5.5

Find the general solution of the following differential equations by the method of undetermined coefficients.

   1. $y'' - 3y' - 10y = 1 + x^2.$ **2.** $2y'' - y' - 3y = x^3 + x + 1.$

3. $4y'' - y = e^x + e^{3x}$.

4. $3y'' + 2y' - y = e^{-2x} + x$.

5. $y'' + 6y' + 8y = e^{-3x} + e^x$.

6. $y'' + 4y' + 3y = 6e^{-x}$.

7. $2y'' + 3y' - 2y = 5e^{-2x} + e^x$.

8. $y'' - y' - 6y = 5e^{-2x} + 10e^{3x}$.

9. $3y'' + 5y' - 2y = 14e^{x/3}$.

10. $y'' + 3y' + 2y = \cos x + \sin x$.

11. $y'' + y' - 6y = 39 \cos 3x$.

12. $y'' + 4y' - 5y = 34 \cos 2x - 2 \sin 2x$.

13. $y'' + 25y = 50 \cos 5x + 30 \sin 5x$.

14. $y'' + 16y = 16 \sin 4x$.

15. $y'' - 4y' + 4y = 8e^{2x} + e^{3x}$.

16. $4y'' - 4y' + y = 6e^{x/2}$.

17. $y'' + 6y' + 9y = 26e^{-3x} + 5e^{2x}$.

18. $y'' + y = e^x \sin x$.

19. $y'' + 2y' + 10y = e^{-x} \sin 3x$.

20. $y'' - 4y' + 5y = 16e^{2x} \cos x$.

21. $y'' - 6y' + 13y = 6e^{3x} \sin x \cos x$.

22. $y'' + 4y' + 4y = 6e^{-2x} \cos^2 x$.

23. $y'' + 3y' + 2y = 12e^{-x} \sin^3 x$.

24. $y'' - 4y' + 3y = 4 \cosh 3x$.

25. $y''' + 4y'' - y' - 4y = 18e^{-x}$.

26. $y''' + 3y'' - 4y = 12e^{-2x} + 9e^x$.

27. $y''' - 9y'' + 27y' - 27y = 36e^{3x}$.

28. $y''' - y'' + y' - y = 6 \cos 2x$.

29. $y''' - 2y'' + 4y' - 8y = 8(x^2 + \cos 2x)$.

30. $y^{iv} - 256y = 128 \cos 4x$.

31. $y^{iv} - y = x^4 + 1$.

32. $y^{iv} + 3y''' + 3y'' + y' = 2x + 4$.

33. $y^{iv} - 3y'' - 4y = 60e^{2x}$.

34. $y^{iv} + 6y''' + 12y'' + 8y' = 60e^{-2x}$.

35. $y^{iv} - 16y'' = 8x + 16$.

5.4.3 Solution of Euler-Cauchy Equation

In the previous sections, we have discussed methods for finding the solution of the constant coefficient differential equations. Closed form solutions do not exist, in general, for the variable coefficient linear equations. However, for the *Euler-Cauchy equation*

$$a_0 x^n y^{(n)} + a_1 x^{n-1} y^{(n-1)} + \ldots + a_{n-1} x y' + a_n y = r(x), \quad x \neq 0 \tag{5.70}$$

where $a_0, a_1, \ldots, a_n$ are constants, closed form solutions can be obtained by using one of the following two procedures.

We shall illustrate these procedures using the second order equation

$$a_0 x^2 y'' + a_1 x y' + a_2 y = r(x), \quad a_0 \neq 0, x \neq 0. \tag{5.71}$$

Consider first, the corresponding homogeneous equation

$$a_0 x^2 y'' + a_1 x y' + a_2 y = 0. \tag{5.72}$$

We attempt to find a solution of the form $y = x^m$. We have $y' = mx^{m-1}$ and $y'' = m(m-1)x^{m-2}$. Substituting in Eq. (5.72), we get

$$[a_0 m(m-1) + a_1 m + a_2] x^m = 0. \tag{5.73}$$

Cancelling x^m, we get

$$a_0 m(m-1) + a_1 m + a_2 = a_0 m^2 + (a_1 - a_0)m + a_2 = 0 \tag{5.74}$$

which is called the *auxiliary* equation corresponding to the Eq. (5.72). Equation (5.74) has two roots $m = m_1, m_2$, which may be real and distinct, real and equal or complex conjugates. In these cases, we obtain the following solutions.

Real and distinct roots

If the roots m_1 and m_2 are real and distinct, then the two linearly independent solutions are

$$y_1(x) = x^{m_1} \quad \text{and} \quad y_2(x) = x^{m_2}. \tag{5.75}$$

The general solution is given by

$$y(x) = Ax^{m_1} + Bx^{m_2} \tag{5.76}$$

where A, B are arbitrary constants.

Example 5.43 Find the solution of the differential equation $x^2y'' + 2xy' - 2y = 0$.

Solution Here, $a_0 = 1$, $a_1 = 2$ and $a_2 = -2$. The auxiliary equation is

$$a_0m(m-1) + a_1m + a_2 = m^2 + m - 2 = 0, \quad \text{or} \quad (m+2)(m-1) = 0.$$

_he roots of this equation are $m = 1, -2$. Hence, the two linearly independent solutions are

$$y_1(x) = x, \quad \text{and} \quad y_2(x) = x^{-2}.$$

The general solution is $y(x) = Ax + (B/x^2)$.

Example 5.44 Find the solution of the differential equation $2x^2y'' + xy' - 6y = 0$.

Solution Here, $a_0 = 2$, $a_1 = 1$, and $a_2 = -6$. The auxiliary equation is

$$a_0m(m-1) + a_1m + a_2 = 2m^2 - m - 6 = 0, \quad \text{or} \quad (m-2)(2m+3) = 0.$$

The roots of this equation are $m = 2, -3/2$. The two linearly independent solutions are

$$y_1(x) = x^2, \quad \text{and} \quad y_2(x) = x^{-3/2}.$$

The general solution is $y(x) = Ax^2 + \dfrac{B}{x\sqrt{x}}$.

Real and equal roots

Let the roots of the auxiliary equation be real and equal, that is, $m = m_1$ is a double root. Then $m_1 = (a_0 - a_1)/(2a_0)$. Since, the discriminant of Eq. (5.74) vanishes in this case, we can also write $m_1^2 = a_2/a_0$ (product of roots). Then, $y_1(x) = x^{m_1}$ is one of the linearly independent solutions. The second linearly independent solution can now be obtained by the method of reduction of order (see section 5.3.3). Write $y_2(x) = u(x)y_1(x)$. We have

$$y_2' = uy_1' + u'y_1, \; y_2'' = uy_1'' + 2u_1'y_1' + u''y_1.$$

Substituting in Eq. (5.72) and simplifying, we get

$$a_0x^2(uy_1'' + 2u'y_1' + u''y_1) + a_1x(uy_1' + u'y_1) + a_2uy_1 = 0$$

or

$$a_0y_1x^2u'' + xu'(2a_0xy_1' + a_1y_1) + u(a_0x^2y_1'' + a_1xy_1' + a_2y_1) = 0. \tag{5.77}$$

Since $y_1(x)$ is a solution of Eq. (5.72), the third term in Eq. (5.77) vanishes. Further, since $y_1(x) = x^{m_1}$ where $m_1 = (a_0 - a_1)/(2a_0)$, we obtain

$$2a_0 x y_1' + a_1 y_1 = (2a_0 m_1 + a_1)x^{m1} = a_0 x^{m1} = a_0 y_1.$$

Therefore, Eq. (5.77) simplifies to

$$a_0 y_1 x^2 u'' + a_0 x u' y_1 = (x u'' + u')a_0 x y_1 = 0$$

Since $x \neq 0$, $y_1 \neq 0$, $a_0 \neq 0$, we get $x u'' + u' = 0$. Separating the variables, we get

$$\frac{u''}{u'} = -\frac{1}{x}$$

Integrating, we get for $x > 0$

$$\ln|u'| = -\ln x, \quad \text{or} \quad u' = \frac{1}{x}.$$

Integrating again, we get $u = \ln x$.

Therefore, $y_2 = u y_1 = y_1 \ln x$. Since $y_2/y_1 = \ln x$, is not a constant, the two solutions y_1, y_2 are linearly independent. The general solution in this case is

$$y(x) = A y_1 + B y_2 = (A + B \ln x)y_1 = (A + B \ln x)x^{m1} \tag{5.78}$$

where $m_1 = (a_0 - a_1)/(2a_0)$.

Example 5.45   Find the solution of the differential equation $4x^2 y'' + y = 0$.

Solution   Here, $a_0 = 4$, $a_1 = 0$, $a_2 = 1$. The auxiliary equation is

$$a_0 m(m-1) + a_1 m + a_2 = 4m^2 - 4m + 1 = 0, \quad \text{or} \quad (2m-1)^2 = 0.$$

The equation has the double root $m = 1/2$. The general solution is (from Eq. (5.78))

$$y(x) = (A + B \ln x)\, x^{1/2}, \quad x > 0.$$

Complex roots

Let the roots of the auxiliary equation (5.74) be a complex conjugate pair, $m = p \pm iq$. Then the solutions are given by

$$x^m = x^{p \pm iq} = x^p\, x^{\pm iq} = x^p (e^{\ln x})^{\pm iq}$$

$$= x^p e^{\pm iq \ln x} = x^p [\cos(q \ln x) \pm i \sin(q \ln x)], \quad x > 0.$$

Therefore, we can take the two linearly independent solutions as

$$y_1(x) = x^p \cos(q \ln x), \quad \text{and} \quad y_2(x) = x^p \sin(q \ln x). \tag{5.79}$$

Example 5.46   Find the general solution of the equation $4x^2 y'' + 8x y' + 17y = 0$.

Solution   Here, $a_0 = 4$, $a_1 = 8$ and $a_2 = 17$. The auxiliary equation is

$$a_0 m(m-1) + a_1 m + a_2 = 4m^2 + 4m + 17 = 0.$$

The roots of this equation are $m = \dfrac{-4 \pm \sqrt{16 - 272}}{8} = \dfrac{-4 \pm 16i}{8} = -\dfrac{1}{2} \pm 2i = p \pm iq.$

The general solution is (from Eq. (5.79))

$$y(x) = A x^{-1/2} \cos(2 \ln x) + B x^{-1/2} \sin(2 \ln x).$$

The method considered here is easily applicable for the homogeneous equations. However, for non-homogeneous equations, finding a particular solution is difficult.

We now discuss a method which can be applied for the solution of general Euler-Cauchy equation given by Eq. (5.71). For $x > 0$, we change the independent variable to

$$x = e^t, \quad \text{or} \quad t = \ln x, \quad x > 0. \tag{5.80}$$

The case $x < 0$ can also be considered by writing the transformation as

$$|x| = e^t, \quad \text{or} \quad t = \ln |x|. \tag{5.81}$$

This transformation always reduces the Euler-Cauchy equation into a linear equation with constant coefficients. The solution of this equation can tnen be obtained using the methods discussed in the previous sections. Finally, the solution of the given equation, in terms of the original variable x, is obtained by replacing t by $\ln x$.

When $x = e^t$, $t = \ln x$, we have

$$\frac{d}{dx} = \frac{d}{dt} \cdot \frac{dt}{dx} = \frac{1}{x} \frac{d}{dt}, \quad \text{or} \quad x \frac{d}{dx} = \frac{d}{dt} \tag{5.82}$$

$$\frac{d^2}{dx^2} = -\frac{1}{x^2} \frac{d}{dt} + \frac{1}{x} \frac{d}{dt} \left(\frac{d}{dt} \right) \frac{dt}{dx} = -\frac{1}{x^2} \frac{d}{dt} + \frac{1}{x^2} \frac{d^2}{dt^2}$$

or

$$x^2 \frac{d^2}{dx^2} = \frac{d^2}{dt^2} - \frac{d}{dt} = \frac{d}{dt} \left(\frac{d}{dt} - 1 \right). \tag{5.83}$$

In operator notation, set $D = d/dx$, $D^2 = d^2/dx^2$, $\theta = d/dt$, $\theta^2 = d^2/dt^2$ etc. Then, Eqs. (5.82), (5.83) can be written as

$$xD = \theta, \quad x^2 D^2 = \theta^2 - \theta = \theta(\theta - 1) \tag{5.84}$$

or

$$x Dy = \theta y, \quad x^2 D^2 y = \theta(\theta - 1)y. \tag{5.85}$$

By induction, we can prove that

$$x^n D^n y = \theta(\theta - 1) \ldots [\theta - (n-1)]y. \tag{5.86}$$

Substituting in the non-homogeneous second order linear equation (5.71), we obtain the reduced equation as

$$a_0 \theta(\theta - 1) y + a_1 \theta y + a_2 y = a_0 \theta^2 y + (a_1 - a_0) \theta y + a_2 y = r(e^t). \tag{5.87}$$

This is a linear equation with constant coefficients. The methods described in the previous sections can be applied to find its solution.

Example 5.47 Find the general solution of the equation $2x^2 y'' + 3xy' - 3y = x^3$.

Solution Using the transformation $x = e^t$, we get (using Eqs. (5.82) and 5.83))

$$2 \left(\frac{d^2 y}{dt^2} - \frac{dy}{dt} \right) + 3 \frac{dy}{dt} - 3y = e^{3t}, \quad \text{or} \quad 2 \frac{d^2 y}{dt^2} + \frac{dy}{dt} - 3y = e^{3t}. \tag{5.88}$$

This is a linear, constant coefficient equation. Substituting, $y = e^{mt}$, the characteristic equation of the corresponding homogeneous equation is obtained as

$$2m^2 + m - 3 = 0, \quad \text{or} \quad (m - 1)(2m + 3) = 0, \quad \text{or} \quad m = 1, -3/2.$$

The complementary function is $y_c(t) = Ae^t + Be^{-3t/2}$.

Let the particular integral be written as $y_p = ce^{3t}$. Substituting in Eq. (5.88), we obtain

$$(18 + 3 - 3)ce^{3t} = e^{3t}, \quad \text{or} \quad c = 1/18.$$

The particular integral is $y_p = e^{3t}/18$.

The general solution is $y(t) = Ae^t + Be^{-3t/2} + \dfrac{1}{18} e^{3t}$.

Substituting $e^t = x$, we get the general solution as

$$y(x) = Ax + \frac{B}{x\sqrt{x}} + \frac{x^3}{18}.$$

Example 5.48 Find the general solution of the equation $x^2 y'' + 5xy' + 3y = \ln x, \quad x > 0$.

Solution Using the transformation $x = e^t$, we obtain

$$\left(\frac{d^2 y}{dt^2} - \frac{dy}{dt}\right) + 5\frac{dy}{dt} + 3y = \ln(e^t), \quad \text{or} \quad \frac{d^2 y}{dt^2} + 4\frac{dy}{dt} + 3y = t. \tag{5.89}$$

The characteristic equation of the corresponding homogeneous equation is

$$m^2 + 4m + 3 = 0, \quad \text{or} \quad (m + 1)(m + 3) = 0, \quad \text{or} \quad m = -1, -3.$$

The complementary function is $y_c(t) = Ae^{-t} + Be^{-3t}$.

Let the particular integral be written as $y_p = c_1 t + c_2$. Substituting in Eq. (5.89), we get

$$4c_1 + 3(c_1 t + c_2) = t.$$

Comparing the coefficients of t and the constant terms on both sides, we obtain $3c_1 = 1$ and $4c_1 + 3c_2 = 0$. The solution is $c_1 = 1/3, c_2 = -4/9$.

The particular integral is $y_p = \dfrac{t}{3} - \dfrac{4}{9}$.

The general solution of the given equation is

$$y(t) = Ae^{-t} + Be^{-3t} + \frac{t}{3} - \frac{4}{9}.$$

Substituting $e^t = x$, we get the general solution as

$$y(x) = \frac{A}{x} + \frac{B}{x^3} + \frac{1}{3}\ln x - \frac{4}{9}.$$

Example 5.49 Find the general solution of the equation $x^2 y'' - 5xy' + 13y = 30x^2$.

Solution Using the transformation $x = e^t$, we obtain

$$\frac{d^2 y}{dt^2} - \frac{dy}{dt} - 5\frac{dy}{dt} + 13y = 30e^{2t}, \quad \text{or} \quad \frac{d^2 y}{dt^2} - 6\frac{dy}{dt} + 13y = 30e^{2t}. \tag{5.90}$$

The characteristic equation of the corresponding homogeneous equation is

$$m^2 - 6m + 13 = 0, \quad \text{or} \quad m = \frac{6 \pm \sqrt{36 - 52}}{2} = \frac{6 \pm 4i}{2} = 3 \pm 2i.$$

The complementary function is $y_c(t) = e^{3t}(A\cos 2t + B\sin 2t)$.

Let the particular integral be written as $y_p = ce^{2t}$. Substituting in equation (5.90), we obtain

$$(4 - 12 + 13)ce^{2t} = 30e^{2t}, \quad \text{or} \quad c = 6.$$

The particular integral is $y_p = 6e^{2t}$.

The general solution is $y(t) = e^{3t}(A \cos 2t + B \sin 2t) + 6e^{2t}$.

Substituting $e^t = x$, we get

$$y(x) = x^3 [A \cos (2 \ln x) + B \sin (2 \ln x)] + 6x^2.$$

Example 5.50   Find the general solution of the equation

$$x^3 y''' + 5x^2 y'' + 5xy' + y = x^2 + \ln x, \quad x > 0.$$

Solution   Using the transformation $x = e^t$, we get (in operator notation)

$$[\theta(\theta - 1)(\theta - 2) + 5\theta(\theta - 1) + 5\theta + 1] \, y = e^{2t} + t$$

or

$$[\theta^3 - 3\theta^2 + 2\theta + 5\theta^2 - 5\theta + 5\theta + 1] \, y = e^{2t} + t$$

or

$$[\theta^3 + 2\theta^2 + 2\theta + 1] \, y = e^{2t} + t \qquad (5.91)$$

where $\theta = d/dt$.

The characteristic equation of the corresponding homogeneous equation is

$$m^3 + 2m^2 + 2m + 1 = 0, \quad \text{or} \quad (m + 1)(m^2 + m + 1) = 0.$$

Its roots are $m = -1, \dfrac{-1 \pm i\sqrt{3}}{2}$.

The complementary function is

$$y_c(t) = Ae^{-t} + [B \cos (\sqrt{3}t/2) + C \sin (\sqrt{3}t/2)]e^{-t/2}.$$

Let the particular integral be written as $y_p = c_1 e^{2t} + c_2 t + c_3$.

Then,

$$y_p' = 2c_1 e^{2t} + c_2, \quad y_p'' = 4c_1 e^{2t}, \quad y_p''' = 8c_1 e^{2t}.$$

Substituting in Eq. (5.91), we obtain

$$(8 + 8 + 4 + 1)c_1 e^{2t} + c_2 t + 2c_2 + c_3 = e^{2t} + t, \quad \text{or} \quad 21c_1 e^{2t} + c_2 t + 2c_2 + c_3 = e^{2t} + t,$$

Comparing both sides, we get $c_1 = 1/21$, $c_2 = 1$, $c_3 = -2$.

The particular integral is $y_p = \dfrac{1}{21} e^{2t} + t - 2$.

The general solution is

$$y(t) = Ae^{-t} + [B \cos (\sqrt{3}t/2) + C \sin (\sqrt{3}t/2)]e^{-t/2} + \frac{1}{21} e^{2t} + t - 2.$$

Substituting $e^t = x$, we get

$$y(x) = \frac{A}{x} + \frac{1}{\sqrt{x}} [B \cos (\sqrt{3} \ln x/2) + C \sin (\sqrt{3} \ln x/2)] + \left(\frac{x^2}{21} + \ln x - 2 \right).$$

Example 5.51   Find the general solution of the equation

$$x^3 y''' - 3xy' + 3y = 16x + 9x^2 \ln x, \quad x > 0.$$

Solution   Using the transformation $x = e^t$, we get (in operator notation)

$$[\theta(\theta - 1)(\theta - 2) - 3\theta + 3]y = 16e^t + 9te^{2t}$$

or

$$(\theta^3 - 3\theta^2 - \theta + 3)y = 16e^t + 9te^{2t} \qquad (5.92)$$

where $\theta = d/dt$. The characteristic equation of the corresponding homogeneous equation is

$$m^3 - 3m^2 - m + 3 = 0, \quad \text{or} \quad (m-1)(m+1)(m-3) = 0, \quad \text{or} \quad m = \pm 1, 3.$$

The complementary function is given by $y_c(t) = Ae^t + Be^{-t} + Ce^{3t}$. Note that e^t, which is one of the linearly independent solutions, also appears as a term on the right hand side of Eq. (5.92). Hence, by the method of undetermined parameters, we write the particular solution as

$$y_p = (c_1 t + c_2)e^{2t} + c_3 t e^t$$

We have
$$y_p' = (c_1 + 2c_1 t + 2c_2)e^{2t} + (1+t)c_3 e^t$$

$$y_p'' = (4c_1 + 4c_1 t + 4c_2)e^{2t} + (2+t)c_3 e^t$$

$$y_p''' = (12c_1 + 8c_1 t + 8c_2)e^{2t} + (3+t)c_3 e^t.$$

Substituting in Eq. (5.92), we obtain

$$[(12c_1 + 8c_1 t + 8c_2) - 3(4c_1 + 4c_1 t + 4c_2) - (c_1 + 2c_1 t + 2c_2) + 3(c_1 t + c_2)]e^{2t}$$

$$+ [(3+t)c_3 - 3(2+t)c_3 - (1+t)c_3 + 3c_3 t]e^t = 16e^t + 9te^{2t}$$

or
$$- (c_1 + 3c_1 t + 3c_2)e^{2t} - 4c_3 e^t = 16e^t + 9te^{2t}.$$

Comparing both sides, we obtain $c_1 + 3c_2 = 0, -3c_1 = 9, -4c_3 = 16$. The solution is $c_1 = -3, c_2 = 1$ and $c_3 = -4$.

The particular integral is $y_p(t) = (1 - 3t)e^{2t} - 4t e^t$.

The general solution is $y(t) = A e^t + B e^{-t} + C e^{3t} + (1 - 3t)e^{2t} - 4t e^t$.

Substituting $x = e^t$, we obtain the general solution as

$$y(x) = Ax + \frac{B}{x} + Cx^3 + (1 - 3\ln x)x^2 - 4x \ln x.$$

Exercise 5.6

Find the general solution of the following homogeneous differential equations (Assume $x > 0$ in Problems 1 to 20).

1. $x^2 y'' + xy' - 4y = 0.$
2. $x^2 y'' + 4xy' + 2y = 0.$
3. $x^2 y'' + xy' - y = 0.$
4. $9x^2 y'' + 15xy' + y = 0.$
5. $4x^2 y'' + 16xy' + 9y = 0.$
6. $2x^2 y'' + 2xy' + y = 0.$
7. $x^2 y'' + 3xy' + y = 0.$
8. $x^2 y'' - xy' + 5y = 0.$
9. $x^2 y'' + 3xy' + 10y = 0.$
10. $9x^2 y'' + 3xy' + 10y = 0.$
11. $x^3 y''' + 2x^2 y'' = 0.$
12. $x^3 y''' + xy' - y = 0.$
13. $x^3 y''' + 4x^2 y'' + 2xy' - 2y = 0.$
14. $x^3 y''' + 9x^2 y'' + 18xy' + 6y = 0.$
15. $x^3 y''' - 2xy' + 4y = 0.$
16. $x^3 y''' + 3x^2 y'' + 14xy' + 34y = 0.$
17. $x^4 y^{iv} + 3x^3 y''' = 0.$
18. $x^4 y^{iv} + 6x^3 y''' + 4x^2 y'' - 2xy' - 4y = 0.$
19. $4x^4 y^{iv} + 24x^3 y''' + 43x^2 y'' + 19xy' - 4y = 0.$
20. $x^4 y^{iv} + 6x^3 y''' + 5x^2 y'' - xy' + y = 0.$

Find the general solution of the following differential equations (Assume $x > 0$ in Problems 21 to 40).

21. $x^2 y'' - 2y = 2x + 6.$
22. $x^2 y'' - 3xy' + 3y = 2 + 3 \ln x,$

23. $x^2y'' + 2xy' - 2y = 6x - 14$.

24. $x^2y'' + 2xy' - 6y = 15x^2$.

25. $x^2y'' + 2xy' = \cos(\ln x)$.

26. $x^2y'' + 5xy' - 5y = 24x \ln x$.

27. $4x^2y'' + y = 25 \sin(\ln x)$.

28. $x^2y'' - 3xy' + 4y = x^3$.

29. $4x^2y'' + 16xy' + 9y = 19 \cos(\ln x) + 22 \sin(\ln x)$.

30. $x^2y'' + 2xy' - 2y = x \sin(\ln x)$.

31. $x^2y'' - 2xy' - 4y = 6x^2 + 4 \ln x$.

32. $x^3y''' + 8x^2y'' + 5xy' - 5y = 42x^2$.

33. $x^3y''' + 6x^2y'' - 12y = 12/x^2$.

34. $x^3y''' - 3x^2y'' + 7xy' - 8y = 3x^3 + 8x$.

35. $4x^3y''' + 12x^2y'' + xy' + y = 50 \sin(\ln x)$.

36. $(3x + 1)^2y'' + (3x + 1)y' + y = 6x$.

37. $(x + 2)^3y''' + (x + 2)^2y'' + (x + 2)y' - y = 24x^2$.

38. $x^4y^{iv} + 6x^3y''' + 2x^2y'' - 4xy' + 4y = 10/x^3$.

39. $4x^4y^{iv} + 16x^3y''' - x^2y'' + 9xy' - 9y = 14x^2 + 1$.

40. $x^4y^{iv} + 6x^3y''' + 12x^2y'' + 6xy' + 4y = 2/x^2$.

Find the solutions of the following differential equations, which satisfy the given conditions.

41. $2x^2y'' + 3xy' - y = x$, $y(1) = 1$, $y(4) = 41/16$.

42. $4x^2y'' + y = \ln x$, $x > 0$, $y(1) = 0$, $y(e) = 5$.

43. $x^2y'' - 3xy' + 3y = 5x^2 - x$, $y(1) = 1$, $y'(1) = 3/2$.

44. $x^2y'' - xy' + 2y = 6$, $y(1) = 1$, $y'(1) = 2$.

45. $x^2y'' + 3xy' + 10y = 9x^2$, $y(1) = 5/2$, $y'(1) = 8$.

5.5 Operator Methods for Finding Particular Integrals

In section 5.3.1, we have introduced the differential operator D, where $D = d/dx$. For example, we can write

$$L(y) = a\frac{d^2y}{dx^2} + b\frac{dy}{dx} + cy = (aD^2 + bD + c)y = F(D)y.$$

Since D is a differential operator, its inverse D^{-1} defines the integral operator, such that $D^{-1}Df(x) = f(x)$.

In this section, we develop symbolic short cut methods for finding a particular integral of a linear non-homogeneous equation with constant coefficients.

Consider the linear non-homogeneous equation with constant coefficients

$$L(y) = a_0\frac{d^ny}{dx^n} + a_1\frac{d^{n-1}y}{dx^{n-1}} + \ldots + a_{n-1}\frac{dy}{dx} + a_ny = r(x) \tag{5.93}$$

or $$L(y) = (a_0D^n + a_1D^{n-1} + \ldots + a_{n-1}D + a_n)y = F(D)y = r(x) \tag{5.94}$$

where $F(D) = a_0D^n + a_1D^{n-1} + \ldots + a_{n-1}D + a_n$, and $a_0, a_1, \ldots, a_n$ are constants. From Eq. (5.94), we write the particular integral as

$$y_p(x) = [F(D)]^{-1}r(x). \tag{5.95}$$

In the following, we develop methods for finding $[F(D)]^{-1}r(x)$ for particular cases of $r(x)$.

5.5.1   Case $r(x) = e^{\alpha x}$.

When $y = e^{\alpha x}$, we have

$$F(D)y = (a_0 D^n + a_1 D^{n-1} + \ldots + a_{n-1} D + a_n)e^{\alpha x}$$

$$= (a_0 \alpha^n + a_1 \alpha^{n-1} + \ldots + a_{n-1}\alpha + a_n)e^{\alpha x} = F(\alpha)e^{\alpha x}.$$

Case $F(\alpha) \neq 0$

We may now symbolically write this equation as

$$y = [F(D)]^{-1}F(\alpha)e^{\alpha x} = F(\alpha)[F(D)]^{-1}e^{\alpha x}$$

since $F(\alpha)$ is a constant. We can further write

$$\frac{1}{F(\alpha)} y = [F(D)]^{-1} e^{\alpha x}, \quad \text{or} \quad [F(D)]^{-1} e^{\alpha x} = \frac{1}{F(\alpha)} e^{\alpha x}.$$

Hence, if $r(x) = e^{\alpha x}$, we obtain

$$y_p(x) = [F(D)]^{-1} e^{\alpha x} = \frac{1}{F(\alpha)} e^{\alpha x}, \quad F(\alpha) \neq 0. \tag{5.96}$$

We can verify that this result is true. Operating with $F(D)$ on both sides, we get

$$F(D)y_p(x) = F(D) \cdot \frac{1}{F(\alpha)} e^{\alpha x} = \frac{1}{F(\alpha)} F(D)e^{\alpha x}$$

$$= \frac{1}{F(\alpha)} F(\alpha)e^{\alpha x} = e^{\alpha x}.$$

Example 5.52   Find the general solution of the differential equation $y'' - 2y' - 3y = 3e^{2x}$.

Solution   In operator notation, the given equation is $(D^2 - 2D - 3)y = 3e^{2x}$. The characteristic equation of the corresponding homogeneous equation is

$(m^2 - 2m - 3) = (m - 3)(m + 1) = 0$. Its roots are $m = -1, 3$.

The complementary function is given by $y_c(x) = Ae^{-x} + Be^{3x}$.

We have $F(D) = D^2 - 2D - 3$. The particular integral is

$$y_p(x) = [F(D)]^{-1} r(x) = [D^2 - 2D - 3]^{-1}(3e^{2x}) = \frac{3e^{2x}}{F(2)} = \frac{3}{-3} e^{2x} = -e^{2x}.$$

The general solution is

$$y(x) = y_c(x) + y_p(x) = Ae^{-x} + Be^{3x} - e^{2x}.$$

Example 5.53   Find the general solution of the equation $y''' - 2y'' - 5y' + 6y = 4e^{-x} - e^{2x}$.

Solution   The given equation in operator notation is

$$F(D)y = (D^3 - 2D^2 - 5D + 6)y = 4e^{-x} - e^{2x}, \quad \text{where} \quad F(D) = D^3 - 2D^2 - 5D + 6.$$

The characteristic equation of the corresponding homogeneous equation is

$$m^3 - 2m^2 - 5m + 6 = 0, \quad \text{or} \quad (m - 1)(m + 2)(m - 3) = 0.$$

The roots of this equation are $m = 1, -2, 3$. The complementary function is

$$y_c(x) = Ae^x + Be^{-2x} + Ce^{3x}.$$

The particular integral is

$$y_p(x) = [F(D)]^{-1}(4e^{-x} - e^{2x})$$

$$= [F(D)]^{-1}(4e^{-x}) - [F(D)]^{-1}e^{2x}$$

$$= \frac{4}{F(-1)} e^{-x} - \frac{1}{F(2)} e^{2x} = \frac{e^{-x}}{2} + \frac{e^{2x}}{4}.$$

The general solution is

$$y(x) = y_c(x) + y_p(x) = Ae^x + Be^{-2x} + Ce^{3x} + \frac{e^{-x}}{2} + \frac{e^{2x}}{4}.$$

Before we discuss the case when $F(\alpha) = 0$, let us derive the following result

$$F(D)[g(x)e^{\alpha x}] = e^{\alpha x} F(D + \alpha) g(x). \tag{5.97}$$

By Leibniz theorem, we have

$$D^n[e^{\alpha x}g(x)] = (D^n e^{\alpha x})g + {}^nC_1(D^{n-1}e^{\alpha x})(Dg) + \ldots + e^{\alpha x}(D^n g)$$

$$= \alpha^n e^{\alpha x}g + {}^nC_1\alpha^{n-1}e^{\alpha x}Dg + {}^nC_2\alpha^{n-2}e^{\alpha x}(D^2 g) + \ldots + e^{\alpha x}(D^n g)$$

$$= e^{\alpha x}[D^n g + {}^nC_1(D^{n-1}g)\alpha + {}^nC_2(D^{n-2}g)\alpha^2 + \ldots + \alpha^n g]$$

(by reversing the order of terms and using the result ${}^nC_r = {}^nC_{n-r}$)

$$= e^{\alpha x}[D^n + {}^nC_1\alpha D^{n-1} + {}^nC_2\alpha^2 D^{n-2} + \ldots + \alpha^n]g$$

$$= e^{\alpha x}[D + \alpha]^n g.$$

Substituting the expressions for $n = 1, 2, \ldots$ on the left hand side of Eq. (5.97), we obtain

$$F(D)[e^{\alpha x}g(x)] = [a_0 D^n + a_1 D^{n-1} + a_2 D^{n-2} + \ldots + a_n](e^{\alpha x}g)$$

$$= a_0 D^n(e^{\alpha x}g) + a_1 D^{n-1}(e^{\alpha x}g) + \ldots + a_n(e^{\alpha x}g)$$

$$= a_0 e^{\alpha x}(D + \alpha)^n g + a_1 e^{\alpha x}(D + \alpha)^{n-1}g + \ldots + a_n e^{\alpha x}g$$

$$= e^{\alpha x}[a_0(D + \alpha)^n + a_1(D + \alpha)^{n-1} + \ldots + a_n]g$$

$$= e^{\alpha x}F(D + \alpha)g.$$

Case $F(\alpha) = 0$.

Let us now consider the case $F(\alpha) = 0$. From the theory of polynomial equations, $(D - \alpha)$ is a factor of $F(D)$. If $F'(\alpha)$ also vanishes, then $(D - \alpha)^2$ is a factor of $F(D)$. If $F(\alpha) = 0 = F'(\alpha) = \ldots = F^{(r-1)}(\alpha)$, $F^r(\alpha) \neq 0$, then $(D - \alpha)^r$ is a factor of $F(D)$. Then, we can write

$$F(D) = (D - \alpha)^r G(D), \quad G(\alpha) \neq 0. \tag{5.98}$$

Let us now write the particular integral of $F(D)y = e^{\alpha x}$ as

$$y_p(x) = [F(D)]^{-1}e^{\alpha x} = [(D - \alpha)^r G(D)]^{-1} e^{\alpha x} = [(D - \alpha)^r]^{-1}[G(D)]^{-1}e^{\alpha x}$$

$$= [(D - \alpha)^r]^{-1}[G(\alpha)]^{-1}e^{\alpha x} \qquad \text{(since } G(\alpha) \neq 0 \text{ and using Eq. (5.96))}$$

$$= \frac{1}{G(\alpha)}[(D - \alpha)^r]^{-1}[e^{\alpha x} \cdot 1] = \frac{1}{G(\alpha)} e^{\alpha x}[(D + \alpha - \alpha)^r]^{-1}[1]$$

$$\text{(using Eq. (5.97))}$$

$$= \frac{1}{G(\alpha)} e^{\alpha x}[D^r]^{-1}[1] = \frac{1}{G(\alpha)} e^{\alpha x}[D^{-r}][1]$$

$$= \frac{1}{G(\alpha)} e^{\alpha x} \frac{x^r}{r!} = \frac{x^r}{r!} \cdot \frac{e^{\alpha x}}{G(\alpha)} \qquad (5.99)$$

since D^{-r} represents integration r times.

Therefore, when $F(\alpha) = 0$, $F'(\alpha) \neq 0$, we have $F(D) = (D - \alpha)G(D)$ and the particular integral of $F(D)y = e^{\alpha x}$ is given by

$$y_p(x) = \frac{x}{1!} \frac{e^{\alpha x}}{G(\alpha)}. \qquad (5.100)$$

Generalization to the case $r(x) = e^{\alpha x}h(x)$.

Irrespective of whether $F(\alpha)$ vanishes or does not vanish, the above result can be extended to the case $r(x) = e^{\alpha x}h(x)$. We have the particular integral in this case as

$$y_p(x) = [F(D)]^{-1}[e^{\alpha x}h(x)] = e^{\alpha x}[F(D + \alpha)]^{-1}h(x) \qquad (5.101)$$

using Eq. (5.97). Now, $[F(D + \alpha)]^{-1}h(x)$ can be evaluated when $h(x)$ is of some particular forms.

Example 5.54 Find the general solution of the equation $y'' + y' - 6y = 5e^{-3x}$.

Solution The equation in operator notation is $(D^2 + D - 6)y = 5e^{-3x}$, where

$$F(D) = D^2 + D - 6 = (D + 3)(D - 2).$$

The characteristic equation of the corresponding homogeneous equation $(D^2 + D - 6)y = 0$ is

$$m^2 + m - 6 = 0, \quad \text{or} \quad (m + 3)(m - 2) = 0, \quad \text{or} \quad m = 2, -3.$$

The complementary function is $y_c(x) = Ae^{2x} + Be^{-3x}$.

Now, $F(m) = m^2 + m - 6$, $F(-3) = 0$ and $F'(-3) = -5 \neq 0$. Therefore,

$$y_p(x) = [(D + 3)(D - 2)]^{-1}(5e^{-3x}) = 5(D + 3)^{-1}[(D - 2)^{-1}e^{-3x}]$$

$$= 5(D + 3)^{-1}(-5)^{-1}e^{-3x} = -(D + 3)^{-1}[e^{-3x} \cdot 1] \qquad \text{(using Eq. (5.96))}$$

$$= -e^{-3x}(D - 3 + 3)^{-1} \cdot 1 = -e^{-3x}D^{-1}(1) = -x e^{-3x}. \qquad \text{(using Eq. (5.99))}$$

We might have also used the formula (5.100) directly where $G(D) = D - 2$. The general solution is

$$y(x) = y_c(x) + y_p(x) = Ae^{2x} + Be^{-3x} - x e^{-3x}.$$

Example 5.55 Find the general solution of the equation $4y'' - 4y' + y = e^{x/2}$.

Solution The characteristic equation of the corresponding homogeneous equation is

$$4m^2 - 4m + 1 = 0, \quad \text{or} \quad (2m - 1)^2 = 0. \text{ Its roots are } m = 1/2, 1/2.$$

The complementary function is $y_c(x) = (A + Bx) e^{x/2}$. We have

$$F(D) = 4D^2 - 4D + 1 = (2D - 1)^2, \quad \text{where } F(1/2) = 0, \text{ and } F'(1/2) = 0.$$

The particular integral is

$$y_p(x) = (2D - 1)^{-2} (e^{x/2} \cdot 1) = e^{x/2} \left[2\left(D + \frac{1}{2}\right) - 1 \right]^{-2} (1)$$

$$= \frac{1}{4} e^{x/2} D^{-2}(1) = \frac{x^2}{8} e^{x/2}.$$

The general solution is $y(x) = (A + Bx)e^{x/2} + (x^2 e^{x/2})/8$.

Example 5.56   Find the general solution of the equation $9y''' + 3y'' - 5y' + y = 42e^x + 64e^{x/3}$

Solution   The characteristic equation of the corresponding homogeneous equation $9y''' + 3y'' - 5y' + y = 0$ is

$$9m^3 + 3m^2 - 5m + 1 = 0, \quad \text{or} \quad (m + 1)(3m - 1)^2 = 0.$$

The roots of this equation are $m = -1, 1/3, 1/3$. The complementary function is

$$y_c(x) = Ae^{-x} + (Bx + C)e^{x/3}$$

We have $F(D) = 9D^3 + 3D^2 - 5D + 1 = (D + 1)(3D - 1)^2$ and $F(1/3) = 0, F'(1/3) = 0$.

The particular integral is

$$y_p(x) = [(D + 1)(3D - 1)^2]^{-1}(42e^x + 64e^{x/3})$$

$$= [(D + 1)(3D - 1)^2]^{-1}(42e^x) + [(D + 1)(3D - 1)^2]^{-1}(64e^{x/3}).$$

Since $F(1) \neq 0$ and $F(1/3) = 0$, we obtain

$$y_p(x) = [(1 + 1)(3 - 1)^2]^{-1}(42e^x) + (3D - 1)^{-2}[(D + 1)^{-1}(64e^{x/3})]$$

$$= \frac{21}{4} e^x + (3D - 1)^{-2}\left[\frac{64}{(4/3)} e^{x/3}\right]$$

$$= \frac{21}{4} e^x + 48e^{x/3}\left[3\left(D + \frac{1}{3}\right) - 1\right]^{-2} (1)$$

$$= \frac{21}{4} e^x + \frac{48}{9} e^{x/3}D^{-2} (1) = \frac{21}{4} e^x + \frac{8}{3} x^2 e^{x/3} .$$

The general solution is

$$y(x) = Ae^{-x} + (Bx + C)e^{x/3} + \frac{21}{4} e^x + \frac{8}{3} x^2 e^{x/3} .$$

Example 5.57   Find the general solution of the equation $16y'' + 8y' + y = 48xe^{-x/4}$.

Solution   The characteristic equation of the corresponding homogeneous equation is

$$16m^2 + 8m + 1 = 0, \quad \text{or} \quad (4m + 1)^2 = 0. \text{ Its roots are } m = -1/4, -1/4.$$

The complementary function is $y_c(x) = (Ax + B)e^{-x/4}$.

We have $F(D) = 16D^2 + 8D + 1 = (4D + 1)^2$ where $F(-1/4) = 0$, and $F'(-1/4) = 0$.

The particular integral is

$$y_p(x) = (4D + 1)^{-2} (48xe^{-x/4}) = 48e^{-x/4}\left[4\left(D - \frac{1}{4}\right) + 1\right]^{-2} x$$

$$= 48e^{-x/4} (4D)^{-2} (x) = 3e^{-x/4} D^{-2} (x) = \frac{1}{2} x^3 e^{-x/4} .$$

The general solution is $y(x) = (Ax + B)e^{-x/4} + \frac{1}{2} x^3 e^{-x/4}$.

5.5.2   Case $r(x) = \cos \alpha x$ or $\sin \alpha x$.

Consider first, the case when $F(D)$ contains even powers of D.
When $f(x) = \cos \alpha x$, we have

$$D^2 f = -\alpha^2 \cos \alpha x, \ D^4 f = (-\alpha^2)^2 \cos \alpha x, \ D^6 f = (-\alpha^2)^3 \cos \alpha x, \ldots$$

Let
$$F(D^2)y = [a_0(D^2)^n + a_1(D^2)^{n-1} + a_2(D^2)^{n-2} + \ldots + a_n]y.$$

Now, let $y = \cos \alpha x$, then

$$F(D^2) \cos \alpha x = [a_0(D^2)^n + a_1(D^2)^{n-1} + \ldots + a_n] \cos \alpha x$$

$$= [a_0(-\alpha^2)^n + a_1(-\alpha^2)^{n-1} + \ldots + a_n] \cos \alpha x = F(-\alpha^2) \cos \alpha x \quad (5.102)$$

Case $F(-\alpha^2) \neq 0$

From Eq. (5.102), we symbolically write

$$\cos \alpha x = [F(D^2)]^{-1}[F(-\alpha^2) \cos \alpha x] = F(-\alpha^2)[F(D^2)]^{-1} \cos \alpha x$$

or
$$[F(D^2)]^{-1} \cos \alpha x = \frac{\cos \alpha x}{F(-\alpha^2)}.$$

Therefore, the particular integral of the equation $F(D^2)y = \cos \alpha x$ is given by

$$y_p(x) = [F(D^2)]^{-1} \cos \alpha x = \frac{\cos \alpha x}{F(-\alpha^2)}, \quad F(-\alpha^2) \neq 0. \quad (5.103)$$

It is easy to show that similar formula holds when $r(x) = \sin \alpha x$. That is, if $F(D^2)y = \sin \alpha x$, then

$$y_p(x) = [F(D^2)]^{-1} \sin \alpha x = \frac{\sin \alpha x}{F(-\alpha^2)}. \quad (5.104)$$

When odd powers of D also exist in $F(D)$, we can follow the same procedure to obtain $y_p(x)$. Let $F(D) = F_1(D^2) + F_2(D)$, where $F_2(D)$ contains odd powers of D. Then

$$[F_1(D^2) + F_2(D)] \cos \alpha x = [F_1(-\alpha^2) + F_2(D)] \cos \alpha x$$

Since $F(D)$ has constant coefficients, we obtain

$$[F_1(D^2) + F_2(D)]^{-1} \cos \alpha x = [F_1(-\alpha^2) + F_2(D)]^{-1} \cos \alpha x, \quad (5.105)$$

We now simplify $F_1(-\alpha^2) + F_2(D)$ and multiply it by $F_3(D)[F_3(D)]^{-1}$, where $F_3(D)$ contains odd powers of D, such that $[F_1(-\alpha^2) + F_2(D)]F_3(D)$ contains only even powers of D. Formula (5.105) is applied and the procedure is repeated to obtain $y_p(x)$. We illustrate this technique through examples.

Example 5.58   Find the general solution of the equation $y'' + 4y = 6 \cos x$.

Solution   It is easy to verify that the complementary function is

$$y_c(x) = A \cos 2x + B \sin 2x.$$

We have $F(D^2) = D^2 + 4$ and $r(x) = 6 \cos x$, that is $\alpha = 1$. Since $F(-\alpha^2) = -\alpha^2 + 4$ and $F(-1) = -1 + 4 = 3 \neq 0$, we have

$$y_p(x) = [(D^2 + 4)]^{-1}(6 \cos x) = \frac{6 \cos x}{F(-\alpha^2)} = \frac{6 \cos x}{F(-1)} = 2 \cos x.$$

The general solution is $y(x) = A \cos 2x + B \sin 2x + 2 \cos x$.

Example 5.59 Find the general solution of the equation $2y'' + y' - y = 16 \cos 2x$.

Solution The characteristic equation of the corresponding homogeneous equation is

$$2m^2 + m - 1 = 0, \quad \text{or} \quad (m+1)(2m-1) = 0.$$

Its roots are $m = -1, 1/2$. The complementary function is

$$y_c(x) = Ae^{-x} + Be^{x/2}.$$

We have $F(D) = 2D^2 + D - 1$, $r(x) = 16 \cos 2x$. Therefore, $\alpha = 2$. Using Eq. (5.105), we get

$$y_p(x) = [(2D^2 + D - 1)]^{-1}(16 \cos 2x) = 16[2(-4) + D - 1]^{-1} \cos 2x$$

$$= 16(D - 9)]^{-1} \cos 2x = 16(D + 9)[(D + 9)(D - 9)]^{-1} \cos 2x$$

$$= 16(D + 9)[(D^2 - 81)]^{-1} \cos 2x = -\frac{16}{85}(D + 9) \cos 2x$$

$$= -\frac{16}{85}(9 \cos 2x - 2 \sin 2x)$$

The general solution is

$$y(x) = Ae^{-x} + Be^{x/2} - \frac{16}{85}(9 \cos 2x - 2 \sin 2x).$$

Example 5.60 Find the general solution of the equation $y'' - 5y' + 4y = 65 \sin 2x$.

Solution The characteristic equation of the corresponding homogeneous equation is

$$m^2 - 5m + 4 = 0, \quad \text{or} \quad (m-1)(m-4) = 0. \text{ Its roots are } m = 1, 4.$$

The complementary function is $y_c(x) = Ae^x + Be^{4x}$.

We have $F(D) = D^2 - 5D + 4$, $r(x) = 65 \sin 2x$. Therefore $\alpha = 2$. Using Eq. (5.105) we get the particular integral as

$$y_p(x) = (D^2 - 5D + 4)^{-1}(65 \sin 2x) = 65[-4 - 5D + 4]^{-1}(\sin 2x)$$

$$= -\frac{65}{5} D^{-1}(\sin 2x) = \frac{13}{2} \cos 2x.$$

since integral of $\sin 2x$ is $(-\cos 2x)/2$. The general solution is

$$y(x) = Ae^x + Be^{4x} + \frac{13}{2} \cos 2x.$$

Example 5.61 Find the general solution of the equation $y''' - y'' + 4y' - 4y = \sin 3x$.

Solution The characteristic equation of the homogeneous equation is

$$m^3 - m^2 + 4m - 4 = 0, \quad \text{or} \quad (m-1)(m^2 + 4) = 0. \text{ Its roots are } m = 1, \pm 2i.$$

The complementary function is $y_c(x) = Ae^x + B \cos 2x + C \sin 2x$.

We have $F(D) = D^3 - D^2 + 4D - 4 = (D - 1)(D^2 + 4)$, $r(x) = \sin 3x$, $\alpha = 3$.

Using Eq. (5.105) we get the particular integral as

$$y_p(x) = [(D - 1)(D^2 + 4)]^{-1}(\sin 3x) = [(D - 1)(-9 + 4)]^{-1} \sin 3x$$

$$= -\frac{1}{5}(D+1)(D+1)^{-1}(D-1)^{-1}\sin 3x = -\frac{1}{5}(D+1)(D^2-1)^{-1}\sin 3x$$

$$= -\frac{1}{5}(D+1)(-9-1)^{-1}\sin 3x = \frac{1}{50}(D+1)\sin 3x = \frac{1}{50}(\sin 3x + 3\cos 3x).$$

The general solution is

$$y(x) = A'e^x + B\cos 2x + C\sin 2x + (3\cos 3x + \sin 3x)/50.$$

Remark 3

Note that the above results hold good when $r(x)$ is also of the form $\cos(\alpha x + a)$ or $\sin(\alpha x + b)$.

Case $F(-\alpha^2) = 0$.

When $F(-\alpha^2) = 0$, we write $\cos \alpha x = \text{Re}(e^{i\alpha x})$ and $\sin \alpha x = \text{Im}(e^{i\alpha x})$ and apply the formula (5.97). We shall illustrate this technique through the following examples.

Example 5.62 Find the general solution of the equation $y'' + y = 6\sin x$.

Solution The complementary function is $y_c(x) = A\cos x + B\sin x$.

We have $F(D^2) = D^2 + 1$, $r(x) = 6\sin x$. Therefore $\alpha = 1$ and $F(-\alpha^2) = F(-1) = 0$.

We write the particular integral as

$$y_p(x) = (D^2 + 1)^{-1}(6\sin x) = \text{Im}\,(D^2 + 1)^{-1}(6e^{ix})$$

$$= 6\,\text{Im}\,\{e^{ix}[(D+i)^2 + 1]^{-1}(1)\} = 6\,\text{Im}\,\{e^{ix}[D^2 + 2iD]^{-1}(1)\}$$

$$= 6\,\text{Im}\,\{e^{ix}D^{-1}[(D+2i)^{-1}](1)\} = 6\,\text{Im}\,\{e^{ix}D^{-1}(0+2i)^{-1}(1)\} \quad (\because 1 = e^{0x})$$

$$= 3\,\text{Im}\,\left\{\frac{1}{i}e^{ix}x\right\} = 3x\,\text{Im}\,\{-i(\cos x + i\sin x)\} = -3x\cos x.$$

The general solution is $y(x) = A\cos x + B\sin x - 3x\cos x$.

Example 5.63 Find the general solution of the equation $y'' - 4y' + 13y = 18e^{2x}\sin 3x$.

Solution The characteristic equation of the homogeneous equation is

$$m^2 - 4m + 13 = 0. \text{ Its roots are } m = \frac{4 \pm \sqrt{16 - 52}}{2} = 2 \pm 3i.$$

The complementary function is $y_c(x) = e^{2x}(A\cos 3x + B\sin 3x)$.

We have $F(D) = D^2 - 4D + 13$ and $r(x) = 18e^{2x}\sin 3x$.

We write the particular integral as

$$y_p(x) = 18[D^2 - 4D + 13]^{-1}(e^{2x}\sin 3x)$$

$$= 18e^{2x}[(D+2)^2 - 4(D+2) + 13]^{-1}(\sin 3x) \qquad \text{(using Eq. (5.97))}$$

$$= 18e^{2x}[D^2 + 9]^{-1}(\sin 3x) = 18e^{2x}\{\text{Im}\,(D^2 + 9)^{-1}(e^{3ix})\}$$

$$= 18e^{2x}\{\text{Im}\,e^{3ix}[(D+3i)^2 + 9]^{-1}(1)\}$$

$$= 18e^{2x}\{\text{Im}\,e^{3ix}[D^2 + 6iD]^{-1}(1)\}$$

$$= 18e^{2x}\{\text{Im } e^{3ix}D^{-1}(D + 6i)^{-1}(1)\}$$

$$= 18e^{2x}\{\text{Im } e^{3ix}D^{-1}(0 + 6i)^{-1}(1)\} = 18e^{2x}\left\{\text{Im } \frac{1}{6i}xe^{3ix}\right\}$$

$$= 3x\,e^{2x}\text{ Im }\{-i(\cos 3x + i\sin 3x)\} = -3xe^{2x}\cos 3x.$$

The general solution is

$$y(x) = e^{2x}(A\cos 3x + B\sin 3x) - 3xe^{2x}\cos 3x.$$

5.5.3 Case $r(x) = x^{\alpha}$, $\alpha > 0$ and integer.

The particular integral of $F(D)y = x^{\alpha}$, is

$$y_p(x) = [F(D)]^{-1}x^{\alpha}$$

Symbolically, we expand the operator $[F(D)]^{-1}$ as an infinite series in ascending powers of D and operate on x^{α}.

Example 5.64 Find the general solution of the equation $y'' + 16y = 64x^2$.

Solution The complementary function is $y_c(x) = A\cos 4x + B\sin 4x$.
The particular integral is

$$y_p(x) = (D^2 + 16)^{-1}(64x^2) = \frac{64}{16}\left[1 + \frac{D^2}{16}\right]^{-1}(x^2)$$

$$= 4\left[1 - \frac{D^2}{16} + \frac{D^4}{256} - \cdots\right]x^2 = 4\left[x^2 - \frac{1}{8}\right]$$

The general solution is $y(x) = A\cos 4x + B\sin 4x + 4x^2 - (1/2)$.

Example 5.65 Find the general solution of the equation $y'' + 4y' + 3y = x\sin 2x$.

Solution The characteristic equation of the corresponding homogeneous equation is

$$m^2 + 4m + 3 = 0, \quad \text{or} \quad (m + 1)(m + 3) = 0.\text{ Its roots are } m = -1, -3.$$

The complementary function is $y_c(x) = Ae^{-x} + Be^{-3x}$.

The particular integral is

$$y_p(x) = [D^2 + 4D + 3]^{-1}(\text{Im } xe^{2ix}) = \text{Im }\{e^{2ix}[D + 2i)^2 + 4(D + 2i) + 3]^{-1}(x)\}$$

$$= \text{Im }\{e^{2ix}[D^2 + 4(1 + i)D + (8i - 1)]^{-1}(x)\}$$

$$= \text{Im }\left\{\frac{e^{2ix}}{8i - 1}\left[1 + \frac{4(1 + i)D}{8i - 1} + \frac{D^2}{8i - 1}\right]^{-1}(x)\right\}$$

$$= \text{Im }\left\{\frac{e^{2ix}}{8i - 1}\left[1 - \frac{4(1 + i)D}{8i - 1} + \cdots\right](x)\right\}$$

$$= \text{Im }\left\{\frac{(8i + 1)}{(-65)}e^{2ix}\left[x - \frac{4(8i + 1)(1 + i)}{(-65)}\right]\right\}$$

$$= \text{Im} \left\{ -\frac{1}{65} (8i + 1)(\cos 2x + i \sin 2x) \left[x + \frac{4}{65} (9i - 7) \right] \right\}$$

$$= \text{Im} \left\{ -\frac{1}{65} [(\cos 2x - 8 \sin 2x) + i(\sin 2x + 8 \cos 2x)] \left[\left(x - \frac{28}{65} \right) + \frac{36}{65} i \right] \right\}$$

$$= -\frac{1}{4225} \left[65x(8 \cos 2x + \sin 2x) - 28(8 \cos 2x + \sin 2x) + 36(\cos 2x - 8 \sin 2x) \right]$$

$$= -\frac{1}{4225} \left[65x(8 \cos 2x + \sin 2x) - 188 \cos 2x - 316 \sin 2x \right].$$

The general solution is

$$y(x) = Ae^{-x} + Be^{-3x} - \frac{1}{4225} \left[65x(8 \cos 2x + \sin 2x) - 188 \cos 2x - 316 \sin 2x \right].$$

Example 5.66 Find the general solution of the equation $y^{iv} + 3y'' = 108x^2$.

Solution The characteristic equation of the homogeneous equation is

$$m^4 + 3m^2 = 0, \quad \text{or} \quad m^2(m^2 + 3) = 0. \text{ Its roots are } m = 0, 0, \pm \sqrt{3}i.$$

The complementary function is $y_c(x) = A + Bx + (C \cos \sqrt{3}x + D \sin \sqrt{3}x)$.
We have $F(D) = D^4 + 3D^2 = D^2(D^2 + 3)$. The particular integral is given by

$$y_p(x) = 108[D^2(D^2 + 3)]^{-1}(x^2) = 108[D^{-2}] \frac{1}{3} \left[1 + \frac{D^2}{3} \right]^{-1} (x^2)$$

$$= 36[D^{-2}] \left[1 - \frac{D^2}{3} + \frac{D^4}{9} - \dots \right](x^2) = 36D^{-2} \left[x^2 - \frac{2}{3} \right]$$

$$= 36 \left[\frac{x^4}{12} - \frac{x^2}{3} \right] = 3x^4 - 12x^2.$$

The general solution is $y(x) = A + Bx + (C \cos \sqrt{3}x + D \sin \sqrt{3}x) + 3x^4 - 12x^2$.

Exercise 5.7

Find the general solution of the following differential equations.

1. $(D^2 + 5D + 4)y = 18e^{2x}$.
2. $(D^2 - 1)y = 8e^{3x}$.
3. $(D^2 - 3D - 4)y = e^x + 6e^{5x}$.
4. $(D^2 + D + 2)y = e^{x/2}$.
5. $(D^2 + 3D + 3)y = 7e^x$.
6. $(D^2 - 2D + 1)y = 5e^{4x} + 4e^{2x}$.
7. $(9D^2 - 6D + 1)y = 4e^{-x}$.
8. $(D^2 - 6D + 9)y = 14e^{3x}$.
9. $(D^2 + D - 6)y = e^{2x}$.
10. $(2D^2 - 3D - 2)y = xe^{-x/2}$.
11. $(D^2 - 1)y = 6xe^x$.
12. $(4D^2 + 9D + 2)y = xe^{-2x}$.
13. $(9D^2 + 6D + 1)y = e^{-x/3}$.
14. $(2D^2 + 7D - 4)y = xe^{-4x}$.
15. $(D^3 + 2D^2 - 5D - 6)y = 4e^x$.
16. $(2D^3 + 3D^2 - 3D - 2)y = 10e^{2x}$.
17. $(D^3 - 2D^2 - D + 2)y = e^{3x}$.
18. $(D^3 - 6D^2 + 12D - 8)y = 18e^{2x}$.

19. $(2D^3 - 3D^2 + 1)y = 16e^x$.

20. $(D^3 + 3D^2 - 4D - 12)y = 12xe^{-2x}$.

21. $(D^2 + 16)y = \cos 2x$.

22. $(2D^2 - 5D + 3)y = \sin x$.

23. $(3D^2 - 7D + 2)y = \sin x + \cos x$.

24. $(2D^2 - 7D + 3)y = \sin 2x$.

25. $(D^2 + D + 1)y = 16 \cos x$.

26. $(8D^2 - 12D + 5)y = 16 \sin x$.

27. $(D^2 + 9)y = \sin 3x$.

28. $(D^2 + 3)y = \cos \sqrt{3}x$.

29. $(D^2 + 2D + 5)y = e^{-x} \cos 2x$.

30. $(D^2 - 4D + 5)y = 24e^{2x} \sin x$.

31. $(D^2 - 6D + 13)y = 28e^{3x} \sin 2x$.

32. $(D^2 - 2D + 10)y = 16e^x \cos 3x + 24e^x \sin 3x$.

33. $(D^3 - 3D^2 + D - 3)y = 6 \cos x$.

34. $(D^3 - D^2 + 9D - 9)y = 30 \cos 3x$.

35. $(D^3 - 4D^2 + 9D - 10)y = 24e^x \sin 2x$.

36. $(4D^3 - 12D^2 + 13D - 10)y = 16e^{x/2} \cos x$.

37. $(D^4 + 5D^2 + 4)y = 16 \sin x + 64 \cos 2x$.

38. $(D^2 + 25)y = 9x^3 + 4x^2$.

39. $(D^2 + 6D + 9)y = 4x^2 - 1$.

40. $(D^2 - 2D - 3)y = 2x^2 + 6x$.

41. $(D^2 - 5D + 6)y = x \cos 2x$.

42. $(D^2 + D - 2)y = x^2 \sin x$.

43. $(D^2 - D - 6)y = xe^{-2x}$.

44. $(D^2 + 7D + 12)y = e^x \sin 2x$.

45. $(D^2 + 4D + 3)y = e^{2x} \cos x$.

46. $(D^2 + 3D + 4)y = e^x \cos (\sqrt{7}x/2)$.

47. $(D^2 + 3D + 2)y = xe^x \sin x$.

48. $(D^2 + 9)y = xe^{2x} \cos x$.

49. $(4D^2 + 8D + 3)y = xe^{-x/2} \cos x$.

50. $(D^4 + 3D^2 + 2)y = 16x^2 \cos x$.

51. If $(2D - 1)y = e^{3x}$, then prove that $(D - 3)(2D - 1)y = 0$. Find the general solution of the second equation and substituting in the first equation obtain the general solution of the first order equation.

52. If $F(D)y = (D - m)y = r(x)$, then show that the particular integral can be written as

$$y_p(x) = e^{mx} \int e^{-mx} r(x) dx.$$

53. Show that $y = \dfrac{1}{n} \displaystyle\int_a^x r(t) \sin n(x - t) dt$ is the solution of the equation $y'' + n^2 y = r(x)$.

54. If u is a function of x, then show that

$$F(D)xu = xF(D)u + F'(D)u$$

where $F(D) = a_0 D^n + a_1 D^{n-1} + \ldots + a_n$, and a_i are constants.

55. Let a given differential equation be of the form $F(D)y = r(x) = xu(x)$.
Then, using the result in problem 54 prove that the particular integral $y(x)$ can be written as

$$y(x) = [F(D)]^{-1}xu(x) = x[F(D)]^{-1}u(x) - [F'(D)\{F(D)\}^{-2}]u(x).$$

56. The particular integral of the equation $F(D)y = e^{mx}$ is

$$y_p(x) = \frac{x}{1!} \frac{e^{mx}}{G(m)}, \quad \text{where} \quad F(D) = (D - m)G(D), \; G(m) \neq 0,$$

$$y_p(x) = \frac{x^2}{2!} \frac{e^{mx}}{G(m)}, \quad \text{where} \quad F(D) = (D - m)^2 G(D), \; G(m) \neq 0, \text{ etc.}$$

Show that these particular integrals can be written as

$$[F(D)]^{-1} e^{mx} = x \left[\frac{1}{F'(m)} \right] e^{mx}, \quad F(m) = 0, F'(m) \neq 0$$

$$[F(D)]^{-1} e^{mx} = x^2 \left[\frac{1}{F''(m)} \right] e^{mx}, \quad F(m) = 0, F'(m) = 0, F''(m) \neq 0, \text{ etc.}$$

Use these formulas to evaluate the particular integral in Problems 8, 9, 13 and 19 of this exercise.

57. If $F(D)$ can be factorised into n distinct factors $F(D) = (D - m_1)(D - m_2) \dots (D - m_n)$, then show that the particular integral of $F(D)y = r(x)$, can be written as

$$y_p(x) = A_1 e^{m_1 x} \int e^{-m_1 x} r(x) \, dx + A_2 e^{m_2 x} \int e^{-m_2 x} r(x) \, dx + \dots + A_n e^{m_n x} \int e^{-m_n x} r(x) \, dx$$

Use this formula to evaluate the particular integral in problem 40 of this exercise.

58. The forced oscillations of a mechanical system with periodic input are governed by the non-homogeneous equation

$$m\ddot{y} + c\dot{y} + ky = F_0 \cos \omega t,$$

where $m > 0$, $c > 0$ and $k > 0$. Obtain its general solution when (i) $c \neq 0$ (forced damped oscillations), (ii) $c = 0$ (forced undamped oscillations).

5.6   Simultaneous Linear Equations

In the previous sections, we have discussed the solution of a single linear differential equation, in which y is the dependent variable and x is the independent variable. In this section, we consider the solution of a system of two linear first order equations in two dependent variables y_1 and y_2 and one independent variable t. We shall restrict ourselves to the solution of constant coefficient equations. For example, the equations

(i) $6 \dfrac{dy_1}{dt} + 5 \dfrac{dy_2}{dt} + 3y_1 + y_2 = 0, \quad \dfrac{dy_2}{dt} - 5y_1 + 3y_2 = e^t$

(ii) $3 \dfrac{dy_1}{dt} + 2y_1 + y_2 = e^{-t}, \quad \dfrac{dy_1}{dt} + \dfrac{dy_2}{dt} - 2y_1 + 3y_2 = t$

are two systems of linear, constant coefficient first order equations. These two systems can respectively be written in operator form as

(i) $(6D + 3)y_1 + (5D + 1)y_2 = 0,$ and   (ii) $(3D + 2)y_1 + y_2 = e^{-t},$

$\quad -5y_1 + (D + 3)y_2 = e^t,$ $\qquad\qquad\qquad (D - 2)y_1 + (D + 3)y_2 = t$

where $D = d/dt$.

The solution of such systems can be obtained by eliminating one of the variables and solving the resulting linear, second order equation for the second variable. Sometimes, elimination of one of the variables may also produce a first order equation for the second variable.

We illustrate the method of obtaining the solution through the following examples.

Example 5.67   Find the solution of the system of equations

$$\frac{dy_1}{dt} + 2\frac{dy_2}{dt} - 2y_1 - y_2 = e^{2t} \tag{5.106}$$

$$\frac{dy_2}{dt} + y_1 - 2y_2 = 0. \tag{5.107}$$

Solution

Method 1

Eliminate one of the dependent variables directly. Differentiating Eq. (5.107) with respect to t, we get

$$y_2'' + y_1' - 2y_2' = 0$$

where dash denotes differentiation with respect to t. Substituting for y_1' from Eq. (5.106), we obtain

$$y_2'' - 2y_2' + 2y_1 + y_2 + e^{2t} - 2y_2' = 0, \quad \text{or} \quad y_2'' - 4y_2' + 2y_1 + y_2 = -e^{2t}.$$

Substituting for y_1 from Eq. (5.107), we get

$$y_2'' - 4y_2' + 4y_2 - 2y_2' + y_2 = -e^{2t}, \quad \text{or} \quad y_2'' - 6y_2' + 5y_2 = -e^{2t} \tag{5.108}$$

which is a second order equation in the variable y_2.

The complementary function is $(y_2)_c = Ae^t + Be^{5t}$. The particular integral is

$$(y_2)_p = (D^2 - 6D + 5)^{-1}(-e^{2t}) = \frac{1}{3}e^{2t}.$$

The general solution is $y_2(t) = Ae^t + Be^{5t} + \frac{1}{3}e^{2t}$.

From Eq. (5.107), we obtain

$$y_1 = 2y_2 - y_2' = 2\left(Ae^t + Be^{5t} + \frac{1}{3}e^{2t} \right) - \left(Ae^t + 5Be^{5t} + \frac{2}{3}e^{2t} \right)$$

$$= Ae^t - 3Be^{5t}.$$

This procedure can be very cumbersome in general.

Method 2

We eliminate one of the dependent variables after writing the equations in operator notation. We have

$$(D - 2)y_1 + (2D - 1)y_2 = e^{2t} \tag{5.109}$$

$$y_1 + (D - 2)y_2 = 0. \tag{5.110}$$

Operating with $(D - 2)$ on equation (5.110), we get $(D - 2)y_1 + (D - 2)^2 y_2 = 0$.

Subtracting Eq. (5.109) from this equation, we get

$$[(D - 2)^2 - (2D - 1)]y_2 = -e^{2t}, \quad \text{or} \quad (D^2 - 6D + 5)y = -e^{2t}$$

which is the same as Eq. (5.108). The remaining solution procedure is same as in method 1.

Method 3

In this method, we find the equations governing y_1 and y_2 using the determinants, by considering (symbolically) the given equations as algebraic equations. Solving the equations (5.109) and (5.110) by Cramer's rule we obtain

$$\begin{vmatrix} D-2 & 2D-1 \\ 1 & D-2 \end{vmatrix} y_1 = \begin{vmatrix} e^{2t} & 2D-1 \\ 0 & D-2 \end{vmatrix} \quad \text{and} \quad \begin{vmatrix} D-2 & 2D-1 \\ 1 & D-2 \end{vmatrix} y_2 = \begin{vmatrix} D-2 & e^{2t} \\ 1 & 0 \end{vmatrix}$$

or $\quad [(D-2)^2 - (2D-1)]y_1 = (D-2)e^{2t}, \quad$ or $\quad (D^2 - 6D + 5)y_1 = 2e^{2t} - 2e^{2t} = 0 \quad$ (5.111)

and $$(D^2 - 6D + 5)y_2 = -e^{2t}. \tag{5.112}$$

(Note the order of evaluation of the determinants on the right hand side. Otherwise, the method does not make any sense). We would have obtained the first equation if we had eliminated y_2 in method 1 or method 2. Care must be taken to properly choose the arbitrary constants. If we solve Eqs. (5.111) and (5.112), we obtain

$$y_1(t) = Ae^t + Be^{5t} \quad \text{and} \quad y_2(t) = C*e^t + D*e^{5t} + \frac{1}{3}e^{2t}.$$

These solutions should satisfy the given equation. Substituting in either of the equations, say (5.110), we get

$$(Ae^t + Be^{5t}) + \left(C*e^t + 5D*e^{5t} + \frac{2}{3}e^{2t} - 2C*e^t - 2D*e^{5t} - \frac{2}{3}e^{2t}\right) = 0$$

or $$(A - C*)e^t + (B + 3D*)e^{5t} = 0.$$

Since this equation is to be identically satisfied, we get

$$A - C* = 0, \quad \text{and} \quad B + 3D* = 0, \quad \text{or} \quad A = C* \quad \text{and} \quad B = -3D*.$$

We obtain the general solution as

$$y_1(t) = C*e^t - 3D*e^{5t}, y_2(t) = C*e^t + D*e^{5t} + \frac{1}{3}e^{2t}$$

which is same as the solution obtained earlier.

Example 5.68   Solve the system of equations

$$(2D - 4)y_1 + (3D + 5)y_2 = 3t + 2$$

$$(D - 2)y_1 + (D + 1)y_2 = t.$$

Solution   Multiply the second equation by 2 and subtract from the first equation. We obtain

$$(D + 3)y_2 = t + 2$$

which is a linear first order equation in y_2.

The integrating factor is e^{3t}. The solution is

$$e^{3t}y_2 = \int (t+2)e^{3t}\, dt + A = \frac{1}{3}(t+2)e^{3t} - \frac{e^{3t}}{9} + A$$

or $$y_2 = Ae^{-3t} + \frac{1}{9}(3t + 5).$$

Substituting in the second equation, we get

$$(D-2)y_1 + \left[-3Ae^{-3t} + \frac{1}{3} + Ae^{-3t} + \frac{1}{9}(3t+5)\right] = t$$

$$(D-2)y_1 = 2Ae^{-3t} + t - \frac{1}{9}(3t+8) = 2Ae^{-3t} + \frac{1}{9}(6t-8)$$

which is a linear first order equation in y_1.

The integrating factor of this equation is e^{-2t}. The solution is

$$e^{-2t}y_1 = \int \left[2Ae^{-5t} + \frac{1}{9}(6t-8)e^{-2t}\right]dt + B$$

$$= -\frac{2}{5}Ae^{-5t} + \frac{1}{9}\left[(6t-8)\frac{e^{-2t}}{(-2)} + \frac{1}{2}\frac{(6e^{-2t})}{(-2)}\right] + B$$

$$= -\frac{2}{5}Ae^{-5t} - \frac{1}{18}(6t-5)e^{-2t} + B$$

or
$$y_1 = Be^{2t} - \frac{2}{5}Ae^{-3t} - \frac{1}{18}(6t-5).$$

Example 5.69 Find the solution of the system of equations

$$(3D+1)y_1 + 3Dy_2 = 3t + 1$$

$$(D-3)y_1 + Dy_2 = 2t.$$

Solution Multiply the second equation by 3 and substract from the first equation. We obtain

$$10y_1 = (3t+1) - 6t = 1 - 3t, \quad \text{or} \quad y_1 = (1-3t)/10.$$

Substituting in the second equation, we get

$$\frac{1}{10}[-3-3+9t] + Dy_2 = 2t, \quad \text{or} \quad Dy_2 = 2t - \frac{1}{10}(9t-6) = \frac{1}{10}(11t+6)$$

which is a linear first order equation in y_2.

Integrating, we obtain
$$y_2 = \frac{11}{20}t^2 + \frac{6}{10}t + A.$$

Note that the system has only one arbitrary constant as the eliminant is a first order equation. This can also be verified by writing the determinant of the coefficient matrix, which is

$$\begin{vmatrix} 3D+1 & 3D \\ D-3 & D \end{vmatrix} = 10D$$

which is of first order only.

5.6.1 Solution of First Order Systems by Matrix Method

The method presented in this section uses some concepts of matrix theory. We discuss the application of this method for solving a 2×2 system which can be generalized to an $n \times n$ system.

Homogeneous systems

Consider a linear homogeneous constant coefficient 2×2 system of the form

$$y_1' = a_{11} y_1 + a_{12} y_2 \tag{5.113}$$

$$y_2' = a_{21} y_1 + a_{22} y_2 \tag{5.114}$$

where y_1, y_2 are the dependent variables, t is the independent variable and a_{11}, a_{12}, a_{21}, a_{22} are constants. Denote

$$\mathbf{y} = \begin{bmatrix} y_1 \\ y_2 \end{bmatrix}, \mathbf{A} = \begin{bmatrix} a_{11} & a_{12} \\ a_{21} & a_{22} \end{bmatrix}.$$

In matrix notation, we can write Eqs. (5.113), (5.114) as

$$\mathbf{y}' = \mathbf{A}\mathbf{y}. \tag{5.115}$$

Note that a higher order, constant coefficient homogeneous equation can be reduced to this form. For example, consider the second order equation $y'' + ay' + by = 0$. Denote $y = y_1$ and write

$$y_1' = y_2, \quad (y_2 = y')$$

$$y_2' = -ay' - by = -ay_2 - by_1, \quad (y_2' = y'')$$

or
$$\begin{bmatrix} y_1' \\ y_2' \end{bmatrix} = \begin{bmatrix} 0 & 1 \\ -b & -a \end{bmatrix} \begin{bmatrix} y_1 \\ y_2 \end{bmatrix}, \quad \text{or} \quad \mathbf{y}' = \mathbf{A}\mathbf{y}. \tag{5.116}$$

Therefore, the equation $y'' + ay' + by = 0$ is equivalent to the first order system given by Eq. (5.116).

We know that the scalar equation $y' = my$ has the solution $y = ce^{mt}$. Therefore, for the solution of Eq. (5.115), we examine a solution of the form

$$\mathbf{y} = e^{\lambda t}\mathbf{x} \tag{5.117}$$

where $\mathbf{x} = [x_1, x_2]^T$, or equivalently $y_1 = e^{\lambda t}x_1$, and $y_2 = e^{\lambda t}x_2$.

Substituting Eq. (5.117) into Eq. (5.115), we obtain

$$\lambda e^{\lambda t}\mathbf{x} = \mathbf{A}e^{\lambda t}\mathbf{x} = e^{\lambda t}\mathbf{A}\mathbf{x}.$$

Cancelling $e^{\lambda t}$, we get

$$\mathbf{A}\mathbf{x} = \lambda\mathbf{x} \tag{5.118}$$

which is an algebraic eigenvalue problem. The eigenvalues are the roots of the characteristic equation

$$|\mathbf{A} - \lambda\mathbf{I}| = 0, \quad \text{or} \quad \begin{vmatrix} a_{11} - \lambda & a_{12} \\ a_{21} & a_{22} - \lambda \end{vmatrix} = 0$$

or
$$\lambda^2 - (a_{11} + a_{22})\lambda + (a_{11}a_{22} - a_{12}a_{21}) = 0.$$

Note that the coefficient of λ is equal to $-$ (trace of $\mathbf{A}$) and the constant term is $|\mathbf{A}|$. The roots of this equation, that is $\lambda = \lambda_1, \lambda_2$ are called the eigenvalues of $\mathbf{A}$. The eigenvalues may be real and distinct, real and equal or a complex conjugate pair. We assume that the system has the complete set of eigenvectors, that is, in the present case the system has two linearly independent eigenvectors. Let the eigenvectors be denoted by $\mathbf{x}^{(1)}$ and $\mathbf{x}^{(2)}$. (If $\lambda_1 \neq \lambda_2$, then linear independence of $\mathbf{x}^{(1)}$, $\mathbf{x}^{(2)}$ is guaranteed). Therefore, we obtain the two linearly independent solutions as

$$\mathbf{y}_1^* = e^{\lambda_1 t}\mathbf{x}^{(1)}, \quad \mathbf{y}_2^* = e^{\lambda_2 t}\mathbf{x}^{(2)}. \tag{5.119}$$

The general solution is

$$\mathbf{y}(t) = A_1 \mathbf{y}_1^* + B_1 \mathbf{y}_2^* = A_1 e^{\lambda_1 t} \mathbf{x}^{(1)} + B_1 e^{\lambda_2 t} \mathbf{x}^{(2)}. \tag{5.120}$$

Let $\mathbf{x}^{(1)} = [x_{11}, x_{12}]^T$ and $\mathbf{x}^{(2)} = [x_{21}, x_{22}]^T$. Componentwise, we can write the solution $\mathbf{y}(t)$ as

$$y_1(t) = A_1 e^{\lambda_1 t} x_{11} + B_1 e^{\lambda_2 t} x_{21}, \quad y_2(t) = A_1 e^{\lambda_1 t} x_{12} + B_1 e^{\lambda_2 t} x_{22}. \tag{5.121}$$

The method can be extended to an $n \times n$ system of linear (constant coefficient) first order equations.

Example 5.70 Find the general solution of the homogeneous linear system

$$y_1' = -2y_1 + y_2, \quad y_2' = y_1 - 2y_2.$$

Solution In matrix notation, the given system can be written as

$$\mathbf{y}' = \begin{bmatrix} -2 & 1 \\ 1 & -2 \end{bmatrix} \mathbf{y} = \mathbf{A}\mathbf{y}.$$

where $\mathbf{y} = [y_1, y_2]^T$. Substituting $\mathbf{y} = e^{\lambda t} \mathbf{x}$ and cancelling $e^{\lambda t}$, we obtain the eigenvalue problem $\mathbf{A}\mathbf{x} = \lambda \mathbf{x}$, that is

$$\begin{bmatrix} -2 & 1 \\ 1 & -2 \end{bmatrix} \begin{bmatrix} x_1 \\ x_2 \end{bmatrix} = \lambda \begin{bmatrix} x_1 \\ x_2 \end{bmatrix}. \tag{5.122}$$

The characteristic equation of $\mathbf{A}$ is given by

$$\begin{vmatrix} -2-\lambda & 1 \\ 1 & -2-\lambda \end{vmatrix} = 0, \quad \text{or} \quad (2+\lambda)^2 - 1 = 0, \quad \text{or} \quad \lambda^2 + 4\lambda + 3 = 0.$$

The roots of this equation are $\lambda = -1, -3$.

For $\lambda = -1$, we get from Eq. (5.122)

$$-x_1 + x_2 = 0 \quad \text{and} \quad x_1 - x_2 = 0.$$

The solution is $x_1 = x_2$, so that we can take $\mathbf{x}^{(1)} = [1 \ \ 1]^T$

For $\lambda = -3$, we get from Eq. (5.122), $x_1 + x_2 = 0$, so that we can take $\mathbf{x}^{(2)} = [1 \ -1]^T$.

These two vectors, $\mathbf{x}^{(1)}$, $\mathbf{x}^{(2)}$ are linearly independent. The general solution of the system is

$$\mathbf{y} = A_1 e^{-t} \mathbf{x}^{(1)} + B_1 e^{-3t} \mathbf{x}^{(2)}$$

or

$$\begin{bmatrix} y_1 \\ y_2 \end{bmatrix} = A_1 e^{-t} \begin{bmatrix} 1 \\ 1 \end{bmatrix} + B_1 e^{-3t} \begin{bmatrix} 1 \\ -1 \end{bmatrix}.$$

Componentwise we can write the solution as

$$y_1 = A_1 e^{-t} + B_1 e^{-3t}, \quad y_2 = A_1 e^{-t} - B_1 e^{-3t}.$$

Example 5.71 Find the general solution of the linear homogeneous system

$$y_1' = -ay_1 + ay_2, \quad y_2' = -ay_1 - ay_2, \quad a \neq 0.$$

Solution In matrix notation, the given system can be written as

$$\mathbf{y}' = \begin{bmatrix} -a & a \\ -a & -a \end{bmatrix} \begin{bmatrix} y_1 \\ y_2 \end{bmatrix} = \mathbf{A}\mathbf{y}$$

Substituting $\mathbf{y} = e^{\lambda t}\mathbf{x}$ and cancelling $e^{\lambda t}$, we obtain the eigenvalue problem $\mathbf{Ax} = \lambda\mathbf{x}$, that is

$$\begin{bmatrix} -a & a \\ -a & -a \end{bmatrix}\begin{bmatrix} x_1 \\ x_2 \end{bmatrix} = \lambda\begin{bmatrix} x_1 \\ x_2 \end{bmatrix}. \qquad (5.123)$$

The characteristic equation of $\mathbf{A}$ is

$$\begin{vmatrix} -a - \lambda & a \\ -a & -a - \lambda \end{vmatrix} = 0, \text{ or } (a + \lambda)^2 + a^2 = 0 \text{ and its roots are } \lambda = -a \pm ia = -a(1 \mp i).$$

For $\lambda = -a(1 + i)$, we get from Eq. (5.123)

$$aix_1 + ax_2 = 0, \quad -ax_1 + aix_2 = 0.$$

The solution is $x_2 = -ix_1$. We can take $\mathbf{x}^{(1)}$ as $\mathbf{x}^{(1)} = [1 \quad -i]^T$.

For $\lambda = -a(1 - i)$, we get from Eq. (5.123)

$$-aix_1 + ax_2 = 0, \quad -ax_1 - aix_2 = 0.$$

The solution is $x_2 = ix_1$. We can take $\mathbf{x}^{(2)}$ as $\mathbf{x}^{(2)} = [1 \quad i]^T$.

Therefore, the general solution is

$$\mathbf{y} = A_1 e^{-a(1+i)t}\mathbf{x}^{(1)} + B_1 e^{-a(1-i)t}\mathbf{x}^{(2)}$$

$$= A_1 e^{-at}(\cos at - i\sin at)\begin{bmatrix} 1 \\ -i \end{bmatrix} + B_1 e^{-at}(\cos at + i\sin at)\begin{bmatrix} 1 \\ i \end{bmatrix}$$

$$= A_1 e^{-at}\begin{bmatrix} \cos at - i\sin at \\ -\sin at - i\cos at \end{bmatrix} + B_1 e^{-at}\begin{bmatrix} \cos at + i\sin at \\ -\sin at + i\cos at \end{bmatrix}$$

$$= (A_1 + B_1)e^{-at}\begin{bmatrix} \cos at \\ -\sin at \end{bmatrix} + (B_1 - A_1)ie^{-at}\begin{bmatrix} \sin at \\ \cos at \end{bmatrix} = C^* e^{-at}\begin{bmatrix} \cos at \\ -\sin at \end{bmatrix} + D^* e^{-at}\begin{bmatrix} \sin at \\ \cos at \end{bmatrix}$$

where $C^* = A_1 + B_1$, and $D^* = (B_1 - A_1)i$. Componentwise, the solution is

$$y_1 = e^{-at}(C^*\cos at + D^*\sin at), \quad y_2 = e^{-at}(-C^*\sin at + D^*\cos at).$$

Non-homogeneous systems

Consider now a non-homogeneous, linear constant coefficient system of equations of the form

$$y_1' = a_{11}y_1 + a_{12}\,y_2 + h_1(t) \qquad (5.124)$$

$$y_2' = a_{21}y_1 + a_{22}\,y_2 + h_2(t) \qquad (5.125)$$

or $\qquad\qquad\qquad \mathbf{y}' = \mathbf{Ay} + \mathbf{h}$ $\qquad\qquad\qquad\qquad\qquad\qquad (5.126)$

where $\mathbf{A}, \mathbf{y}$ are as defined earlier and $\mathbf{h} = [h_1(t) \quad h_2(t)]^T$. The solution of the system is $\mathbf{y} = \mathbf{y}_c + \mathbf{y}_p$ where $\mathbf{y}_c$ is the complementary function and $\mathbf{y}_p$ is the particular integral. The complementary function $\mathbf{y}_c$ is the solution of the homogeneous equation which can be obtained by the methods described above. The particular integral $\mathbf{y}_p$ can be obtained by the method of undetermined coefficients, if the components of $\mathbf{h}(t)$ are simple functions like a polynomial, exponential, sine or cosine function. However, in general, we can use the *diagonalisation method* to find $\mathbf{y}_p$. We now illustrate both these methods.

5.6.2 Method of Undetermined Coefficients to Find the Particular Integral

Since the method is straightforward we illustrate it through examples.

Examples 5.72 Find the general solution of the linear system of equations $y' = Ay + h$, where

$$A = \begin{bmatrix} 1 & 2 \\ 4 & 3 \end{bmatrix}, h = \begin{bmatrix} 6t^2 + 12t + 9 \\ 4t^2 + 3t + 6 \end{bmatrix}, \quad \text{and} \quad y = \begin{bmatrix} y_1 \\ y_2 \end{bmatrix}.$$

Solution Consider the corresponding homogeneous equation $y' = Ay$. Substituting $y = e^{\lambda t}x$ and cancelling $e^{\lambda t}$, we obtain the eigenvalue problem $Ax = \lambda x$; that is

$$\begin{bmatrix} 1 & 2 \\ 4 & 3 \end{bmatrix} \begin{bmatrix} x_1 \\ x_2 \end{bmatrix} = \lambda \begin{bmatrix} x_1 \\ x_2 \end{bmatrix}.$$

The characteristic equation of A is

$$\begin{vmatrix} 1 - \lambda & 2 \\ 4 & 3 - \lambda \end{vmatrix} = 0, \quad \text{or} \quad \lambda^2 - 4\lambda - 5 = 0, \text{ and its roots are } \lambda = -1, 5.$$

For $\lambda = -1$, $(A - \lambda I)x = 0$ gives the equations

$$2x_1 + 2x_2 = 0, \quad \text{and} \quad 4x_1 + 4x_2 = 0,$$

whose solution is $x_1 = -x_2$. The eigenvector can be taken as $x^{(1)} = [1 \quad -1]^T$.

For $\lambda = 5$, $(A - \lambda I)x = 0$ gives the equations

$$-4x_1 + 2x_2 = 0, \quad \text{and} \quad 4x_1 - 2x_2 = 0$$

whose solution is $x_2 = 2x_1$. The eigenvector can be taken as $x^{(2)} = [1 \quad 2]^T$.

The complementary function is given by

$$y_c = c_1 e^{-t} x^{(1)} + c_2 e^{5t} x^{(2)}.$$

Since the elements of h are polynomials, we write the particular integral as

$$y_p = d t^2 + e t + f.$$

Differentiating, we have $y_p' = 2dt + e$.

Substituting in the differential equation, we obtain

$$2dt + e = A(d t^2 + e t + f) + h.$$

Comparing the coefficients of t^2, t and the constant term, we obtain

$$Ad + \begin{bmatrix} 6 \\ 4 \end{bmatrix} = 0, \quad Ae + \begin{bmatrix} 12 \\ 3 \end{bmatrix} = 2d, \quad \text{and} \quad Af + \begin{bmatrix} 9 \\ 6 \end{bmatrix} = \begin{bmatrix} e_1 \\ e_2 \end{bmatrix}.$$

Consider now the first system. We have the equations

$$\begin{bmatrix} 1 & 2 \\ 4 & 3 \end{bmatrix} \begin{bmatrix} d_1 \\ d_2 \end{bmatrix} + \begin{bmatrix} 6 \\ 4 \end{bmatrix} = \begin{bmatrix} 0 \\ 0 \end{bmatrix}$$

or $\qquad d_1 + 2d_2 = -6, \quad 4d_1 + 3d_2 = -4$, whose solution is $d_1 = 2$, and $d_2 = -4$.

The second system gives the equations

$$\begin{bmatrix} 1 & 2 \\ 4 & 3 \end{bmatrix} \begin{bmatrix} e_1 \\ e_2 \end{bmatrix} + \begin{bmatrix} 12 \\ 3 \end{bmatrix} = 2 \begin{bmatrix} d_1 \\ d_2 \end{bmatrix} = 2 \begin{bmatrix} 2 \\ -4 \end{bmatrix}$$

or $\qquad e_1 + 2e_2 = -8, \; 4e_1 + 3e_2 = -11$, whose solution is $e_1 = 2/5$, and $e_2 = -21/5$.

The third system gives the equations

$$\begin{bmatrix} 1 & 2 \\ 4 & 3 \end{bmatrix} \begin{bmatrix} f_1 \\ f_2 \end{bmatrix} + \begin{bmatrix} 9 \\ 6 \end{bmatrix} = \begin{bmatrix} e_1 \\ e_2 \end{bmatrix} = \begin{bmatrix} 2/5 \\ -21/5 \end{bmatrix}$$

or $\qquad f_1 + 2f_2 = -43/5, \; 4f_1 + 3f_2 = -51/5$, whose solution is $f_1 = 27/25$, and $f_2 = -121/25$.

The general solution of the given system is

$$\mathbf{y} = \mathbf{y}_c + \mathbf{y}_p = c_1 e^{-t} \mathbf{x}^{(1)} + c_2 e^{5t} \mathbf{x}^{(2)} + \mathbf{d}t^2 + \mathbf{e}t + \mathbf{f}$$

$$= c_1 e^{-t} \begin{bmatrix} 1 \\ -1 \end{bmatrix} + c_2 e^{5t} \begin{bmatrix} 1 \\ 2 \end{bmatrix} + t^2 \begin{bmatrix} 2 \\ -4 \end{bmatrix} + \frac{t}{5} \begin{bmatrix} 2 \\ -21 \end{bmatrix} + \frac{1}{25} \begin{bmatrix} 27 \\ -121 \end{bmatrix}.$$

Componentwise, the solution is

$$y_1 = c_1 e^{-t} + c_2 e^{5t} + 2t^2 + (2t/5) + (27/25),$$

$$y_2 = -c_1 e^{-t} + 2c_2 e^{5t} - 4t^2 - (21t/5) - (121/25).$$

Example 5.73 Find the general solution of the linear system of equations

$$\mathbf{y}' = \mathbf{Ay} + \mathbf{h} = \begin{bmatrix} 5 & -7 \\ 2 & -4 \end{bmatrix} \mathbf{y} - \begin{bmatrix} 2 \\ 4 \end{bmatrix} e^t.$$

Solution The eigenvalue problem corresponding to the homogeneous equation is $\mathbf{Ax} = \lambda \mathbf{x}$, that is

$$\begin{bmatrix} 5 & -7 \\ 2 & -4 \end{bmatrix} \mathbf{x} = \lambda \mathbf{x}.$$

The characteristic equation is given by

$$\begin{vmatrix} 5 - \lambda & -7 \\ 2 & -4 - \lambda \end{vmatrix} = 0, \quad \text{or} \quad \lambda^2 - \lambda - 6 = 0, \text{ and its roots are } \lambda = 3, -2.$$

For $\lambda = 3$, $(\mathbf{A} - \lambda\mathbf{I})\mathbf{x} = \mathbf{0}$ gives the equations

$$2x_1 - 7x_2 = 0, \quad \text{and} \quad 2x_1 - 7x_2 = 0$$

whose solution is $x_2 = 2x_1/7$. The eigenvector can be taken as $\mathbf{x}^{(1)} = [7 \quad 2]^T$.

For $\lambda = -2$, $(\mathbf{A} - \lambda\mathbf{I})\mathbf{x} = \mathbf{0}$ gives the equations

$$7x_1 - 7x_2 = 0, \quad 2x_1 - 2x_2 = 0$$

whose solution is $x_1 = x_2$. The eigenvector can be taken as $\mathbf{x}^{(2)} = [1 \quad 1]^T$.

The complementary function is given by

$$\mathbf{y}_c = c_1 e^{3t} \mathbf{x}^{(1)} + c_2 e^{-2t} \mathbf{x}^{(2)}.$$

As in the scalar case, we write the particular integral as $\mathbf{y}_p = \mathbf{d}\,e^t$. We have, $\mathbf{y}_p' = \mathbf{d}\,e^t$. Substituting in the given equation, we have

$$\mathbf{d}\,e^t = \mathbf{A}\,\mathbf{d}\,e^t - \begin{bmatrix} 2 \\ 4 \end{bmatrix} e^t.$$

Cancelling e^t, we obtain

$$\begin{bmatrix} 5 & -7 \\ 2 & -4 \end{bmatrix} \begin{bmatrix} d_1 \\ d_2 \end{bmatrix} - \begin{bmatrix} d_1 \\ d_2 \end{bmatrix} = \begin{bmatrix} 2 \\ 4 \end{bmatrix}$$

or
$$4d_1 - 7d_2 = 2, \quad \text{and} \quad 2d_1 - 5d_2 = 4.$$

The solution of these equations is $d_1 = -3$ and $d_2 = -2$, or $\mathbf{d} = [-3 \quad -2]^T$.

The general solution of the system is

$$\mathbf{y} = \mathbf{y}_c + \mathbf{y}_p = c_1 e^{3t} \mathbf{x}^{(1)} + c_2 e^{-2t} \mathbf{x}^{(2)} + \mathbf{d}\,e^t$$

$$= c_1 e^{3t} \begin{bmatrix} 7 \\ 2 \end{bmatrix} + c_2 e^{-2t} \begin{bmatrix} 1 \\ 1 \end{bmatrix} - \begin{bmatrix} 3 \\ 2 \end{bmatrix} e^t.$$

Example 5.74   Find the general solution of the linear system of equations

$$\mathbf{y}' = \mathbf{A}\mathbf{y} + \mathbf{h} = \begin{bmatrix} -1 & 3 \\ 2 & -2 \end{bmatrix} \mathbf{y} + \begin{bmatrix} 4 \\ 1 \end{bmatrix} e^{-4t}.$$

Solution   The eigenvalue problem corresponding to the homogeneous equation is $\mathbf{A}\mathbf{x} = \lambda\mathbf{x}$, that is

$$\begin{bmatrix} -1 & 3 \\ 2 & -2 \end{bmatrix} \mathbf{x} = \lambda\,\mathbf{x}.$$

The characteristic equation is

$$\begin{vmatrix} -1-\lambda & 3 \\ 2 & -2-\lambda \end{vmatrix} = 0, \quad \text{or} \quad \lambda^2 + 3\lambda - 4 = 0, \quad \text{or} \quad \lambda = 1, -4.$$

For $\lambda = 1$, $(\mathbf{A} - \lambda\mathbf{I})\mathbf{x} = \mathbf{0}$ gives the equations

$$-2x_1 + 3x_2 = 0, \quad \text{and} \quad 2x_1 - 3x_2 = 0$$

whose solution is $2x_1 = 3x_2$. The eigenvector can be taken as $\mathbf{x}^{(1)} = [3 \quad 2]^T$.

For $\lambda = -4$, $(\mathbf{A} - \lambda\mathbf{I})\mathbf{x} = \mathbf{0}$ gives the equations

$$3x_1 + 3x_2 = 0, \quad 2x_1 + 2x_2 = 0$$

whose solution is $x_1 = -x_2$. The eigenvector can be taken as $\mathbf{x}^{(2)} = [1 \quad -1]^T$.

The complementary function is given by

$$\mathbf{y}_c = c_1 e^t \mathbf{x}^{(1)} + c_2 e^{-4t} \mathbf{x}^{(2)}.$$

We note that e^{-4t} occurs both in $\mathbf{y}_c$ and on the right hand side. Hence, as in the scalar case, we write the particular integral as

$$\mathbf{y}_p = (\mathbf{d}t + \mathbf{e})e^{-4t}$$

We have

$$\mathbf{y}_p' = (\mathbf{d} - 4\mathbf{d}t - 4\mathbf{e})e^{-4t}.$$

Substituting in the given equation, we obtain

$$(-4\mathbf{d}t + \mathbf{d} - 4\mathbf{e})e^{-4t} = \mathbf{A}(\mathbf{d}t + \mathbf{e})e^{-4t} + \begin{bmatrix} 4 \\ 1 \end{bmatrix} e^{-4t}.$$

Comparing the coefficients of te^{-4t}, we get

$$-4\mathbf{d} = \mathbf{A}\mathbf{d}.$$

Hence, $\mathbf{d}$ is the eigenvector corresponding to the eigenvalue $\lambda = -4$. Hence, $\mathbf{d} = p\mathbf{x}^{(2)}$, where p is a constant. Comparing the coefficients of e^{-4t}, we obtain

$$\mathbf{d} - 4\mathbf{e} = \mathbf{A}\mathbf{e} + \begin{bmatrix} 4 \\ 1 \end{bmatrix}$$

or

$$\begin{bmatrix} -1 & 3 \\ 2 & -2 \end{bmatrix} \begin{bmatrix} e_1 \\ e_2 \end{bmatrix} + 4 \begin{bmatrix} e_1 \\ e_2 \end{bmatrix} = p \begin{bmatrix} 1 \\ -1 \end{bmatrix} - \begin{bmatrix} 4 \\ 1 \end{bmatrix}$$

or

$$3e_1 + 3e_2 = p - 4, \quad 2e_1 + 2e_2 = -(p+1).$$

Therefore, $2p - 8 = -3(p+1)$, or $p = 1$. Hence, $\mathbf{d} = \mathbf{x}^{(2)}$. With this value, we get $e_1 + e_2 = -1$. We can choose $e_1 = -1$, $e_2 = 0$, so that $\mathbf{e} = [-1 \quad 0]^T$. The general solution of the given system is

$$\mathbf{y} = \mathbf{y}_c + \mathbf{y}_p = c_1 e^t \mathbf{x}^{(1)} + c_2 e^{-4t} \mathbf{x}^{(2)} + (\mathbf{d}t + \mathbf{e})e^{-4t}$$

$$= c_1 e^t \mathbf{x}^{(1)} + (c_2 \mathbf{x}^{(2)} + \mathbf{e})e^{-4t} + \mathbf{d}t e^{-4t}$$

$$= c_1 e^t \begin{bmatrix} 3 \\ 2 \end{bmatrix} + \left\{ c_2 \begin{bmatrix} 1 \\ -1 \end{bmatrix} + \begin{bmatrix} -1 \\ 0 \end{bmatrix} \right\} e^{-4t} + t e^{-4t} \begin{bmatrix} 1 \\ -1 \end{bmatrix}.$$

5.6.3 Method of Diagonalisation to Find the Particular Integral

Let the matrix $\mathbf{A}$ of the non-homogeneous system $\mathbf{y}' = \mathbf{A}\mathbf{y} + \mathbf{h}$, have a complete system of eigenvectors. That is, if $\mathbf{A}$ is an $n \times n$ matrix, then there exist n linearly independent eigenvectors $\mathbf{x}^{(1)}, \mathbf{x}^{(2)}, \ldots \mathbf{x}^{(n)}$. If the eigenvalues λ_i of $\mathbf{A}$ are distinct, then linear independence of the eigenvectors is guaranteed. Let $\mathbf{x}$ denote the matrix of eigenvectors

$$\mathbf{x} = [\mathbf{x}^{(1)}, \mathbf{x}^{(2)} \ldots, \mathbf{x}^{(n)}]. \tag{5.127}$$

Since $\mathbf{x}^{(i)}$, $i = 1, 2, \ldots, n$ are linearly independent, $\mathbf{x}$ is non-singular and $\mathbf{x}^{-1}$ exists. Premultiplying $\mathbf{y}' = \mathbf{A}\mathbf{y} + \mathbf{h}$ by $\mathbf{x}^{-1}$, we get

$$\mathbf{x}^{-1}\mathbf{y}' = \mathbf{x}^{-1}\mathbf{A}\mathbf{y} + \mathbf{x}^{-1}\mathbf{h}. \tag{5.128}$$

Let $\mathbf{y} = \mathbf{x}\mathbf{u}$. We have, $\mathbf{y}' = \mathbf{x}\mathbf{u}'$. Substituting in Eq. (5.128), we get

$$x^{-1}xu' = x^{-1}Axu + x^{-1}h. \tag{5.129}$$

We know from matrix theory that the matrix x (of eigenvectors) diagonalises the matrix A, that is $x^{-1}Ax = D$, where D is a diagonal matrix with its diagonal entries as the eigenvalues of A. Therefore, we obtain from Eq. (5.129)

$$u' = Du + g, \quad \text{where} \quad g = x^{-1}h. \tag{5.130}$$

Now, Eq. (5.130) is of the form

$$\begin{bmatrix} u_1' \\ u_2' \\ \vdots \\ u_n' \end{bmatrix} = \begin{bmatrix} \lambda_1 & 0 & 0 & \cdots & 0 \\ 0 & \lambda_2 & 0 & \cdots & 0 \\ \vdots & \vdots & \vdots & \ddots & \vdots \\ 0 & 0 & 0 & & \lambda_n \end{bmatrix} \begin{bmatrix} u_1 \\ u_2 \\ \vdots \\ u_n \end{bmatrix} + \begin{bmatrix} g_1 \\ g_2 \\ \vdots \\ g_n \end{bmatrix}. \tag{5.131}$$

Therefore, Eq. (5.130) degenerates into n independent scalar equations

$$u_j' = \lambda_j u_j + g_j, \quad j = 1, 2, \ldots, n. \tag{5.132}$$

The method of finding the solution of these first order equations was discussed in the previous chapter. The solution of these equations can be written as

$$u_j = c_j e^{\lambda_j t} + e^{\lambda_j t}\left[\int g_j e^{-\lambda_j t}\, dt\right]. \tag{5.133}$$

The solution of the given non-homogeneous system is then obtained from the equation $y = xu$.

Example 5.75 Find the general solution of the linear system of equations

$$y' = Ay + h = \begin{bmatrix} 5 & -7 \\ 2 & -4 \end{bmatrix} y - \begin{bmatrix} 2 \\ 4 \end{bmatrix} e^t$$

(see Example 5.73)

Solution From Example 5.73, we have

$$x = [x^{(1)} \ x^{(2)}] = \begin{bmatrix} 7 & 1 \\ 2 & 1 \end{bmatrix}, \quad \text{and} \quad x^{-1} = \frac{1}{5}\begin{bmatrix} 1 & -1 \\ -2 & 7 \end{bmatrix}.$$

Now,

$$g = x^{-1}h = -\frac{1}{5}\begin{bmatrix} 1 & -1 \\ -2 & 7 \end{bmatrix}\begin{bmatrix} 2 \\ 4 \end{bmatrix} e^t = -\frac{1}{5}\begin{bmatrix} -2 \\ 24 \end{bmatrix} e^t.$$

The eigenvalues of A are $\lambda_1 = 3$ and $\lambda_2 = -2$. Eqs. (5.132) become

$$u_1' = \lambda_1 u_1 + g_1 = 3u_1 + \frac{2}{5} e^t$$

$$u_2' = \lambda_2 u_2 + g_2 = -2u_2 - \frac{24}{5} e^t.$$

The solutions of these equations are $u_1 = c_1 e^{3t} - \frac{1}{5} e^t, u_2 = c_2 e^{-2t} - \frac{8}{5} e^t$, respectively. Hence

$$\mathbf{y} = \mathbf{xu} = \begin{bmatrix} 7 & 1 \\ 2 & 1 \end{bmatrix} \begin{bmatrix} c_1 e^{3t} - (1/5)e^t \\ c_2 e^{-2t} - (8/5)e^t \end{bmatrix} = \begin{bmatrix} 7c_1 e^{3t} + c_2 e^{-2t} - 3e^t \\ 2c_1 e^{3t} + c_2 e^{-2t} - 2e^t \end{bmatrix}$$

which is same as the solution obtained in Example 5.73.

Exercise 5.8

Find the solution of the following systems of equations using the elimination method.

1. $y_1' = 2y_1 + y_2,\ y_2' = y_1 + 2y_2.$ 2. $y_1' = y_1 + y_2,\ y_2' = 9y_1 + y_2.$

3. $y_1' = y_2,\ y_2' = -9y_1.$ 4. $y_1' = 2y_1 + y_2,\ y_2' = -18y_1 - 7y_2.$

5. $y_1' + y_2 = 4 \sin t,\ y_2' + y_1 = 8 \cos t.$ 6. $y_1' + y_1 + 3y_2 = 4e^{-t},\ y_2' + 4y_1 - 3y_2 = 8t.$

7. $y_1' + 3y_1 + y_2 = 6e^t,\ y_2' - 5y_1 - 3y_2 = 3e^{-t}.$

8. $y_1' + 4y_1 - 5y_2 = 16 \sin t,\ y_2' + 5y_1 - 4y_2 = e^t.$

9. $y_1' + 3y_1 - 5y_2 = 64 \sin 4t,\ y_2' + 5y_1 - 3y_2 = 12 \cos 2t.$

10. $y_1' + y_1 - 3y_2 = 6e^{-t},\ y_2' + 2y_1 - 4y_2 = 12e^t.$

Find the solution of the following systems of equations.

11. $(D - 2)y_1 + (D + 3)y_2 = t + 4,\ (3D + 5)y_1 + (2D + 6)y_2 = 5t + 3.$

12. $(2D + 3)y_1 + (D - 1)y_2 = e^t,\ (4D + 6)y_1 + 3Dy_2 = e^{-t}.$

13. $(D - 1)y_1 + 2Dy_2 = t^2 + 1,\ (3D + 5)y_1 + 6Dy_2 = t + 3.$

14. $Dy_1 + (3D - 1)y_2 = e^{-t},\ 3Dy_1 + (11D - 1)y_2 = 2(e^t + e^{-t}).$

15. $(D + 3)y_1 + (3D + 23)y_2 = e^{-2t},\ (D + 2)y_1 + (4D + 14)y_2 = e^{2t}.$

16. $(2D + 3)y_1 + (D + 5)y_2 = t^2,\ (8D + 14)y_1 + (11D + 28)y_2 = t + 3.$

17. $(2D + 1)y_1 + (D + 1)y_2 = t,\ (D + 2)y_1 + (3D + 2)y_2 = 2t + 1.$

18. $(D - 1)y_1 - (D + 1)y_2 = t,\ (D + 1)y_1 + (2D + 1)y_2 = e^t.$

Using the matrix method, find the solution of the systems of equations $\mathbf{y}' = \mathbf{Ay}$, where $\mathbf{y} = [y_1\ \ y_2]^T$ and

19. $\mathbf{A} = \begin{bmatrix} 1 & -2 \\ 1 & 4 \end{bmatrix}.$ 20. $\mathbf{A} = \begin{bmatrix} 5 & -3 \\ 2 & -2 \end{bmatrix}.$

21. $\mathbf{A} = \begin{bmatrix} 3 & -2 \\ 9 & -3 \end{bmatrix}.$ 22. $\mathbf{A} = \begin{bmatrix} 2 & -4 \\ 2 & -2 \end{bmatrix}.$

23. $\mathbf{A} = \begin{bmatrix} 1 & -1 \\ 1 & 1 \end{bmatrix}.$ 24. $\mathbf{A} = \begin{bmatrix} 1/2 & -1 \\ 1 & 1/2 \end{bmatrix}.$

Using the matrix method and the method of undetermined parameters, find the solution of the non-homogeneous system of equations $\mathbf{y}' = \mathbf{Ay} + \mathbf{h}$, where $\mathbf{y} = [y_1\ \ y_2]^T$, and

25. $\mathbf{A} = \begin{bmatrix} 1 & 1 \\ 4 & -2 \end{bmatrix},\ \mathbf{h} = \begin{bmatrix} 3t + 1 \\ 2t + 5 \end{bmatrix}.$ 26. $\mathbf{A} = \begin{bmatrix} 1 & -3/2 \\ 1/2 & -1 \end{bmatrix},\ \mathbf{h} = \begin{bmatrix} 2 \\ 6 \end{bmatrix} e^{2t}.$

27. $\mathbf{A} = \begin{bmatrix} 3 & -6 \\ 1 & -4 \end{bmatrix},\ \mathbf{h} = \begin{bmatrix} 2 \\ 3 \end{bmatrix} e^{2t}.$ 28. $\mathbf{A} = \begin{bmatrix} 5 & -4 \\ 2 & -1 \end{bmatrix},\ \mathbf{h} = \begin{bmatrix} 24 \\ 18 \end{bmatrix} e^t.$

29. $A = \begin{bmatrix} 3 & 2 \\ 4 & 1 \end{bmatrix}$, $h = \begin{bmatrix} 25 \\ 13 \end{bmatrix} e^{5t}$.

30. In Problems 25 and 26, use the method of diagonalisation to find the solution of the systems.

5.7 Answers and Hints

Exercise 5.1

1. Constant coeff.   **2.** Variable coeff.   **3.** Constant coeff.

4. Variable coeff.   **5.** Variable coeff.   **6.** Variable coeff.

7. Any subinterval on $(-\infty, 0)$, $(0, \infty)$.   **8.** Any subinterval on $(-\infty, \infty)$.

9. Any subinterval on $(-\infty, 0)$, $(0, \infty)$.   **10.** Any subinterval on $[0, \infty)$.

11. Any subinterval on $(3, \infty)$.   **12.** Any subinterval on $(0, \infty)$.

13. Any subinterval on $(-\infty, 0)$, $(0, 1)$, $(1, \infty)$.

14. $4m < x < 4(m + 1)$, $m = 0, 2, 4, \ldots$.

15. No, because the equation is not normal on any interval containing $x = 0$, Remark 1 is also not applicable.

16. $2x$. No, because the equation is not normal on any interval containing $x = 0$.

17. No, because $x = 0$ at which the equation is not normal is included in the interval $[-3, 3]$, even though the conditions are specified at $x = 2$.

21. $6x + 3 = (3/4)(2x) + (3/2)(3x + 2)$, linearly dependent.

22. Dependent, $9x^2 - x + 2 = 3(x^2 - x) + 2(3x^2 + x + 1)$.

23. Independent, no linear combination can be found, alternately $W = 14$.

24. $W = -16 \sin^6 x$, linearly indpendent.   **25.** $W = 1$, linearly independent.

26. Dependent, $W = 0$, $x \in I$. Alternately, $\cosh x = e^x - \sinh x$.

27. Linearly independent, $W = -4/x$.   **28.** Dependent, $W = 0$.

29. Linearly independent, $W = -4$.   **30.** Dependent, $\sinh x = \cosh x - e^{-x}$.

31. $W = -2 \tan^3 x$, linearly independent on $(0, \pi/2)$, $\left((2n - 1)\dfrac{\pi}{2}, (2n + 1)\dfrac{\pi}{2} \right)$, $n = 1, 2, \ldots$.

32. (i) Three, (ii) Three.   **33.** $W(y_1, y_2) = 2$, $y_3 = 2y_1 - y_2/2$.

34. $y_i'' = -(a_1/a_0)y_i' - (a_2/a_0)y_i$, $W(x) = y_1 y_2' - y_2 y_1'$. Differentiating $W(x)$ and substituting for y_i'', we obtain $a_0 W'(x) + a_1 W(x) = 0$. Finding the integrating factor we obtain the solution as given. The value of c depends on y_1, y_2.

35. Substitution shows that $\cos at$, $\sin at$ are solutions. $W = a \neq 0$. y_1, y_2 are linearly independent on any interval I. Using the Abel's formula we get $W = c$, where c can be taken as a. Yes.

36. Substitution shows that e^{2x} and xe^{2x} are solutions of the equation. $W = e^{4x} \neq 0$, y_1, y_2 are linearly independent on any interval I. Using Abel's formula we get $W = ce^{4x}$ which is same as the earlier value when $c = 1$.

37. Normal in $(0, \infty)$, $W = x^{1/2}$. $\{y_1, y_2\}$ forms a basis.

38. Normal in any I, $W = 3e^{4x}$. $\{y_1, y_2\}$ forms a basis.

39. Normal in $(0, \infty)$, $W = 2x$. $\{y_1, y_2\}$ forms a basis.

40. Normal in $(-\infty, \infty)$, $W = 20$. $\{y_1, y_2, y_3\}$ forms a basis.

41. Normal in $(-\infty, \infty)$, $W = e^{3x}$. $\{y_1, y_2, y_3\}$ forms a basis.

42. Normal in $(-\infty, \infty)$, $W = 12\sqrt{3}$. $\{y_1, y_2, y_3\}$ forms a basis.

43. Normal in $(0, \infty)$, $W = -2/x$. $\{y_1, y_2\}$ forms a basis.

44. $W(u, v) = (ad - bc)(y_1 y_2' - y_2 y_1')$. Since $y_1 y_2' - y_2 y_1' \neq 0$, $W(u, v) \neq 0$ if $ad - bc \neq 0$, (the determinant of the coefficient matrix of the transformation). Take $a = 1$, $b = 1$, $c = 1$, $d = -1$, $ad - bc = -2$, $u = e^{kx}$, $v = e^{-kx}$.

45. $W(y_1, y_2) \neq 0$. If for $x_0 \in I$, either $y_1(x_0)$, $y_2(x_0)$ vanish or $y_1'(x_0)$, $y_2'(x_0)$ vanish, then $W(y_1, y_2) = 0$.

46. Simplify $W(y, y_1, y_2)$ and substitute $y_i'' = -(a y_i' + b y_i)$, $i = 1, 2$. We obtain
$$W(y, y_1, y_2) = (y'' + a y' + by)(y_1 y_2' - y_2 y_1') = 0.$$

47. At the given point $y_1(x_1) = y'(x_1) = 0$. Therefore, $y_1 \equiv 0$.

48. The differential equation is $W(y, y_1, y_2) = 0$, where $y_1 = e^{3x}$, $y_2 = e^{-2x}$, $y'' - y' - 6y = 0$.

49. $y'' + 2\alpha y' + (\alpha^2 + \omega^2)y = 0$.

50. $y'' - 10y' + 25y = 0$.

Exercise 5.2

1. $(7e^x - e^{4x})/3$.

2. $(3e^{2x} - e^{-2x})/2$.

3. $(1 + 5x)e^{-3x}$.

4. $\dfrac{1}{2}(5x^2 - (1/x^2))$.

5. $(3 + \ln x)x$.

6. $Ae^{2x} + Be^{-2x}$.

7. $Ae^{2x} + Be^{-x}$.

8. $Ae^x + Be^{-2x}$.

9. $Ae^{6x} + Be^{-2x}$.

10. $Ae^{m_1 x} + Be^{m_2 x}$, $m_1 = -2 + \sqrt{3}$, $m_2 = -2 - \sqrt{3}$.

11. $Ae^{2x} + Be^{x/4}$.

12. $Ae^{x/2} + Be^{-(5x)/2}$.

13. $(A + Bx)e^{-x}$.

14. $(A + Bx)e^{-\pi x}$.

15. $(A + Bx)e^{(2x)/3}$.

16. $(A + Bx)e^{-x/2}$.

17. $(A + Bx)e^{(2x)/5}$.

18. $A \cos 5x + B \sin 5x$.

19. $(A \cos x + B \sin x)e^{-2x}$.

20. $e^x(A \cos x + B \sin x)$.

21. $e^{x/2}(A \cos 2x + B \sin 2x)$.

22. $e^{3x}(A \cos 3x + B \sin 3x)$.

23. $A + Be^{-9x}$.

24. $e^{ax}(A \cos bx + B \sin bx)$.

25. $m = 3, -2$, ch. equation is $m^2 - m - 6 = 0$, diff. equation is $y'' - y' - 6y = 0$.

26. $m = 1/4, -3/4$, ch. equation is $16m^2 + 8m - 3 = 0$, diff. equation is $16y'' + 8y' - 3y = 0$.

27. $m = 0, -2$, ch. equation is $m(m + 2) = 0$, diff. equation is $y'' + 2y' = 0$.

28. $m = 2, 2$, ch. equation is $(m - 2)^2 = 0$, diff. equation is $y'' - 4y' + 4y = 0$.

29. $m = -1, -1$, ch. equation is $(m + 1)^2 = 0$, diff. equation is $y'' + 2y' + y = 0$.

30. $y'' + 9y = 0$.

31. $y'' + 2ay' + (a^2 + b^2)y = 0$.

32. $y'' - 10y' + 34y = 0$.

33. $e^x - e^{-x}$.

34. $e^{4x} + 3e^{-3x}$.

35. $e^x - e^{-2x}$.

36. $a \cos \sqrt{g}\, t$.

37. $e^{2x}(2 \cos x - 5 \sin x)$.

38. $e^{x/5}[\cos (x/5) - \sin (x/5)]$.

39. $((x/2) - 1) e^{-(3x)/2}$.

40. $xe^{-x/3}$.

41. $\cos 5x + B \sin 5x$, B arbitrary.

42. $[(2e^2 - 1)e^{-6x} - e^{6x}]/(e^2 - 1)$.

43. $e^{-x}(\cos x + \sin x)$.

44. $(Ax + B)e^{x/3}$, $A = e^{-2/3} - 1$, $B = 2 - e^{-2/3}$.

45. $(e^{x+2} - e^{3x})/(e^2 - 1)$.

48. (i) $b = $ constant, (ii) $a(x) = b(x)$.

49. $(D + 4)(D + 1)y = 0$, set $(D + 1)y = v$ and $(D + 4)v = 0$; $v = A_1 e^{-4x}$, $y = Ae^{-4x} + Be^{-x}$.

50. $(2D + 1)(2D + 3)y = 0$, set $(2D + 3)y = v$ and $(2D + 1)v = 0$, $v = A_1 e^{-x/2}$, $y = Ae^{-x/2} + Be^{(-3x)/2}$.

51. $(2D + 3)(2D + 3)y = 0$, set $(2D + 3)y = v$, $(2D + 3)v = 0$, $v = A_1 e^{-(3x)/2}$, $y = (Ax + B)e^{-(3x)/2}$.

52. $(D + 3)(D + 3)y = 0$, set $(D + 3)y = v$, $(D + 3)v = 0$, $v = A_1 e^{-3x}$, $y = (Ax + B)e^{-3x}$.

53. $(D + 2)(D - 2)y = 0$, set $(D - 2)y = v$, $(D + 2)v = 0$, $v = A_1 e^{-2x}$, $y = Ae^{-2x} + Be^{2x}$.

54. $(3D + 1)(3D + 1)y = 0$, set $(3D + 1)y = v$, $(3D + 1)v = 0$, $v = A_1 e^{-x/3}$, $y = (Ax + B)e^{-x/3}$.

55. For oscillatory solutions, the discriminant of the characteristic equation should be less than zero. $|1 - c| < 2\sqrt{b}$, $1 - 2\sqrt{b} < c < 1 + 2\sqrt{b}$.

56. $\omega = n$, $y(x) = B_n \sin nx$, B_n arbitrary.

57. $y_n(x) = A_n \cos nx$, A_n arbitrary $y(x) = \sum\limits_{n=1}^{\infty} y_n(x)$.

58. $y_n(x) = B_n \sin [(2n + 1)x/2]$, B_n arbitrary $y(x) = \sum\limits_{n=1}^{\infty} y_n(x)$.

59. $y(x) = e^{px}(A'e^{qx} + B'e^{-qx}) = e^{px}[A \cosh qx + B \sinh qx]$.

60. (i) For $c^2 > 4mk$, both the characteristic roots $-p \pm q$ where $p = c/(2m)$ and $q = \sqrt{c^2 - 4mk}/(2m)$, are negative and $q < p$. Therefore, the solution $y(t) = e^{-pt}(Ae^{qt} + Be^{-qt}) \to 0$ as $t \to \infty$, that is, there exists a t_0 such that for $t > t_0$ the system is in equilibrium. $y = [av_0 e^{-pt} \sinh qt]/q$.

(ii) For $c^2 < 4mk$, the characteristic roots are $-p \pm iq$, where $p = c/(2m)$ and $q = \sqrt{4mk - c^2}/(2m)$ are complex. The solutions are oscillatory in this case. The solution is $y(t) = e^{-pt}(A \cos qt + B \sin qt)$. The oscillations are damped and they decay as $t \to \infty$. $y = (e^{-pt}v_0 \sin qt)/q$.

(iii) For $c^2 = 4mk$, the characteristic roots are repeated roots $-p$. The solution is $y(t) = (A + Bt)e^{-pt}$. $y = v_0 te^{-pt}$.

61. $Ae^{3x} + Be^{-2x}$.

62. $Ae^x + Be^{-4x}$.

63. $u = x + 1/x$, $y_2 = 1 + x^2$, $Ax + B(1 + x^2)$.

64. $u = -\cot x$, $y_2 = -x^{-1/2} \cos x$, $x^{-1/2}(A \cos x + B \sin x)$.

65. $u = -e^{-x}(x^2 - 2x + 2)$, $y_2 = -(x^2 - 2x + 2)$, $Ae^x + B(x^2 - 2x + 2)$.

Exercise 5.3

1. $A + Be^{3x} + Ce^{-3x}$.

2. $Ae^{x/2} + Be^{2x} + Ce^{-3x}$.

3. $Ae^x + Be^{-x} + Ce^{2x/3}$.

4. $Ae^{2x} + Be^{-2x} + Ce^{3x} + De^{-3x}$.

5. $Ae^x + Be^{2x} + Ce^{-x/2} + De^{x/2}$.

6. $A + Be^{2x} + Ce^{-2x} + De^{-x}$.

7. $Ae^{x/4} + Be^{x/2} + Ce^x + De^{-x}$.

8. $Ae^{x/3} + Be^{-x/3} + Ce^{x/4} + De^{-x/4}$.

9. $A + (Bx + C)e^x$.

10. $Ae^{-2x} + (Bx + C)e^{-x}$.

11. $Ae^{-2x} + (Bx + C)e^{2x}$.

12. $(A + Bx + Cx^2)e^{x/3}$.

13. $A + Be^x + (Cx + D)e^{5x}$.

14. $A + (Bx^2 + Cx + D)e^x$.

15. $(Ax + B)e^{-x} + (Cx + D)e^{x/2}$.

16. $(Ax + B)e^{3x} + (Cx + D)e^{2x/3}$.

17. $A + B \cos x + C \sin x$.

18. $Ae^{2x} + B \cos 2x + C \sin 2x$.

19. $Ae^{-3x} + e^{-x}(B \cos x + C \sin x)$.

20. $Ae^x + e^{3x}(B \cos 2x + C \sin 2x)$.

21. $Ae^x + Be^{-x} + C \cos 3x + D \sin 3x$.

22. $Ae^x + Be^{-2x} + C \cos 4x + D \sin 4x$.

23. $A \cos 5x + B \sin 5x + C \cos (x/2) + D \sin (x/2)$.

24. $e^{2x}(A \cos x + B \sin x) + e^{-3x}(C \cos x + D \sin x)$.

25. $(A + Bx) \cos 5x + (C + Dx) \sin 5x$.

26. $(A + Bx) \cos x + (C + Dx) \sin x$.

27. $m = 0, 1, 3, \ y''' - 4y'' + 3y' = 0$.

28. $m = -1, \pm 5i, \ y''' + y'' + 25y' + 25y = 0$.

29. $m = -1, -1, 2, \ y''' - 3y' - 2y = 0$.

30. $m = 0, 0, 1, 3, \ y^{iv} - 4y''' + 3y'' = 0$.

31. $m = 2, 2, 2, -2, \ y^{iv} - 4y''' + 16y' - 16y = 0$.

32. $m = \pm 3, \pm 2i, \ y^{iv} - 5y'' - 36y = 0$.

33. $(3e^{3x} + 2e^{-2x} - 5e^x)/30$.

34. $(9e^x - 5e^{3x/2} + e^{-3x/2})/5$.

35. $(2 + x)e^x - e^{3x}$.

36. $(1 + x)e^{-x} + (2 - x)e^{2x}$.

37. $x + \cos x + \sin x$.

38. $\cos 2x + 2 \sin 2x - e^x$.

39. $e^x + e^{-x}(\cos x + 2 \sin x)$.

40. $1 + 2x + 3x^2 + e^{3x}$.

41. $A \sin \pi x$, A arbitrary.

42. $1 + 2 \sinh 6x + \cosh 6x$.

43. $2 \sin 2x + \sin 3x$.

44. $D_n \sin nx, \Sigma D_n \sin nx$.

45. $2 \cos 3x + \cos x$.

Exercise 5.4

1. $A(x) = -e^{2x}/8$, $B(x) = -e^{-2x}/8$, $y = c_1 e^{-x} + c_2 e^{3x} - (e^x/4)$.

2. $A(x) = -e^{-4x}/4$, $B(x) = (4x + 1)e^{-4x}/16$, $y = (c_1 x + c_2) e^{2x} + e^{-2x}/16$.

3. $A(x) = \cos^3 x/3$, $B(x) = (\sin 3x + 3 \sin x)/12$, $y_p = (\cos x)/3$, $y = c_1 \cos 2x + c_2 \sin 2x + y_p$.

4. $A(x) = \ln | \cos x |$, $B(x) = x$, $y_p = \cos x \ln | \cos x | + x \sin x$, $y = c_1 \cos x + c_2 \sin x + y_p$.

5. $A(x) = -x$, $B(x) = \ln | \sin x |$, $y_p = \sin x \ln | \sin x | - x \cos x$, $y = c_1 \cos x + c_2 \sin x + y_p$.

6. $A(x) = \sin x - \ln | \sec x + \tan x |$, $B(x) = -\cos x$, $y_p = -\cos x \ln | \sec x + \tan x |$,
   $y = c_1 \cos x + c_2 \sin x + y_p$.

7. $A(x) = -x/2$, $B(x) = -e^{-2x}/4$, $y(x) = c_1 e^x + c_2 e^{3x} - (xe^x)/2$.

8. $A(x) = \frac{1}{4} \ln | \cos 2x |$, $B(x) = x/2$, $y_p = \frac{1}{4} \cos 2x \ln | \cos 2x | + \frac{1}{2} x \sin 2x$.
   $y(x) = c_1 \cos 2x + c_2 \sin 2x + y_p$.

9. $A(x) = (\cos 4x)/16$, $B(x) = (4x + \sin 4x)/16$, $y_p = (\cos 2x + 4x \sin 2x)/16$.
   $y(x) = c_1 \cos 2x + c_2 \sin 2x + (x \sin 2x)/4$.

10. $A(x) = \sin x + x \cos x$, $B(x) = -\cos x$, $y_p = -e^{-2x} \sin x$, $y(x) = (c_1 x + c_2)e^{-2x} + y_p$.

11. $A(x) = -x$, $B(x) = \ln | x |$, $y_p = x [\ln | x | - 1] e^{-3x}$, $y(x) = (c_1 x + c_2) e^{-3x} + y_p$.

12. $A(x) = (\cos 2x)/4$, $B(x) = (2x + \sin 2x)/4$, $y(x) = c_1 e^{-x} \cos x + c_2 e^{-x} \sin x + (xe^{-x} \sin x)/2$.

13. $g(x) = x$, $A(x) = x^2/4$, $B(x) = -x^4/8$, $y_p = x^3/8$, $y(x) = c_1 x + (c_2/x) + y_p$.

14. $g(x) = \ln|x|$, $A(x) = [\ln|x|]^2/8$, $B(x) = -x^4[4 \ln|x| - 1]/64$.

$y_p = x^2[8(\ln|x|)^2 - 4 \ln|x| + 1]/64$, $y(x) = c_1 x^2 + c_2/x^2 + y_p$.

15. $g(x) = 1/x^6$, $A(x) = [1 + 5 \ln|x|]/(25x^5)$, $B(x) = -1/(5x^5)$,

$y_p = 1/(25x^4)$, $y(x) = c_1 x + c_2 x \ln|x| + y_p$.

16. $g(x) = x + (1/x)$, $A(x) = -[(x^2/2) + \ln|x|]$, $B(x) = x - (1/x)$,

$y_p = (x^3/2) - x(1 + \ln|x|)$, $y(x) = c_1 x + c_2 x^2 + y_p$.

17. $g(x) = 16e^{-2x} \operatorname{cosec}^2 2x$, $A(x) = 4 \ln|\operatorname{cosec} 2x + \cot 2x|$, $B(x) = -4/\sin 2x$.

$y_p = 4e^{-2x} \cos 2x \ln|\operatorname{cosec} 2x + \cot 2x| - 4e^{-2x}$, $y(x) = e^{-2x}(c_1 \cos 2x + c_2 \sin 2x) + y_p$.

18. $A(x) = (\ln|\sec 2x + \tan 2x|)/8$, $B(x) = -x/4$, $C(x) = (\ln|\cos 2x|)/8$,

$y(x) = c_1 + c_2 \cos 2x + c_3 \sin 2x - (x \cos 2x)/4 + (\sin 2x \ln|\cos 2x|)/8 + (\ln|\sec 2x + \tan 2x|)/8$.

19. $A(x) = x^2/4$, $B(x) = -x$, $C(x) = (\ln|x|)/2$,

$y(x) = (c_1 + c_2 x + c_3 x^2)e^{2x} + (x^2 \ln|x| e^{2x})/2$.

20. $y_p = \dfrac{1}{k} \displaystyle\int_0^x g(t)[\sin kx \cos kt - \cos kx \sin kt]\,dt = \dfrac{1}{k} \displaystyle\int_0^x g(t) \sin[k(x-t)]\,dt$.

Exercise 5.5

1. $y_p = -(50x^2 - 30x + 69)/500$, $y_c = Ae^{-2x} + Be^{5x}$.

2. $y_p = (20 - 51x + 9x^2 - 9x^3)/27$, $y_c = Ae^{-x} + Be^{3x/2}$.

3. $y_p = (35e^x + 3e^{3x})/105$, $y_c = Ae^{x/2} + Be^{-x/2}$.

4. $y_p = (e^{-2x} - 7x - 14)/7$, $y_c = Ae^{-x} + Be^{x/3}$.

5. $y_p = -e^{-3x} + e^x/15$, $y_c = Ae^{-2x} + Be^{-4x}$.

6. $y_p = 3xe^{-x}$, $y_c = Ae^{-x} + Be^{-3x}$.

7. $y_p = -xe^{-2x} + e^x/3$, $y_c = Ae^{-2x} + Be^{x/2}$.

8. $y_p = 2xe^{3x} - xe^{-2x}$, $y_c = Ae^{-2x} + Be^{3x}$.

9. $y_p = 2xe^{x/3}$, $y_c = Ae^{-2x} + Be^{x/3}$.

10. $y_p = (2 \sin x - \cos x)/5$, $y_c = Ae^{-x} + Be^{-2x}$.

11. $y_p = (\sin 3x - 5 \cos 3x)/2$, $y_c = Ae^{2x} + Be^{-3x}$.

12. $y_p = 2(\sin 2x - \cos 2x)$, $y_c = Ae^x + Be^{-5x}$.

13. $y_p = x(-3 \cos 5x + 5 \sin 5x)$, $y_c = A \cos 5x + B \sin 5x$.

14. $y_p = -2x \cos 4x$, $y_c = A \cos 4x + B \sin 4x$.

15. $y_p = 4x^2 e^{2x} + e^{3x}$, $y_c = (Ax + B)e^{2x}$.

16. $y_p = 3x^2 e^{(x/2)}/4$, $y_c = (Ax + B)e^{x/2}$.

17. $y_p = 13x^2 e^{-3x} + e^{2x}/5$, $y_c = (Ax + B)e^{-3x}$.

18. $y_p = e^x(\sin x - 2 \cos x)/5$, $y_c = A \cos x + B \sin x$.

19. $y_p = -(xe^{-x} \cos 3x)/6$, $y_c = e^{-x}(A \cos 3x + B \sin 3x)$.

20. $y_p = 8xe^{2x} \sin x$, $y_c = e^{2x}(A \cos x + B \sin x)$.

21. $y_p = -3xe^{3x}\cos 2x/4$, $y_c = e^{3x}(A\cos 2x + B\sin 2x)$.

22. $r(x) = 3e^{-2x}(1+\cos 2x)$, $y_p = e^{-2x}(c_1 x^2 + c_2\cos 2x + c_3\sin 2x) = [3e^{-2x}(2x^2 - \cos 2x)]/4$, $y_c = (Ax+B)e^{-2x}$.

23. $r(x) = 3e^{-x}(3\sin x - \sin 3x)$, $y_p = e^{-x}[-45(\cos x + \sin x) + (\cos 3x + 3\sin 3x)]/10$, $y_c = Ae^{-x} + Be^{-2x}$.

24. $r(x) = 2(e^{3x} + e^{-3x})$, $y_p = (e^{-3x} + 12xe^{3x})/12$, $y_c = Ae^x + Be^{3x}$.

25. $y_p = -3xe^{-x}$, $y_c = Ae^x + Be^{-x} + Ce^{-4x}$.

26. $y_p = xe^x - 2x^2 e^{-2x}$, $y_c = (Ax+B)e^{-2x} + Ce^x$.

27. $y_p = 6x^3 e^{3x}$, $y_c = (Ax^2 + Bx + C)e^{3x}$.

28. $y_p = 2(\cos 2x - 2\sin 2x)/5$, $y_c = Ae^x + B\cos x + C\sin x$.

29. $y_p = -[2(x^2 + x) + x(\cos 2x + \sin 2x)]/2$, $y_c = Ae^{2x} + B\cos 2x + C\sin 2x$.

30. $y_p = -x\sin 4x/2$, $y_c = Ae^{4x} + Be^{-4x} + C\cos 4x + D\sin 4x$.

31. $y_p = -(x^4 + 25)$, $y_c = Ae^x + Be^{-x} + C\cos x + D\sin x$.

32. $y_p = x^2 - 2x$, $y_c = A + (Bx^2 + Cx + D)e^{-x}$.

33. $y_p = 3xe^{2x}$, $y_c = Ae^{2x} + Be^{-2x} + C\cos x + D\sin x$.

34. $y_p = -5x^3 e^{-2x}$, $y_c = A + (Bx^2 + Cx + D)e^{-2x}$.

35. $y_p = -(x^3 + 6x^2)/12$, $y_c = Ax + B + Ce^{4x} + De^{-4x}$.

Exercise 5.6

1. $y = Ax^2 + B/x^2$.
2. $y = (A/x) + (B/x^2)$.

3. $y = Ax + B/x$.
4. $y = (A + B\ln x)x^{-1/3}$.

5. $y = (A + B\ln x)x^{-3/2}$.
6. $y = A\cos(\ln x/\sqrt{2}) + B\sin(\ln x/\sqrt{2})$.

7. $y = (A + B\ln x)/x$.
8. $y = x[A\cos(2\ln x) + B\sin(2\ln x)]$.

9. $y = x^{-1}[A\cos(3\ln x) + B\sin(3\ln x)]$.

10. $y = x^{1/3}[A\cos(\ln x) + B\sin(\ln x)]$.
11. $y = A + Bx + C\ln x$.

12. $y = [A + B\ln x + C\ln^2 x]x$.
13. $y = Ax + x^{-1}[B\cos(\ln x) + C\sin(\ln x)]$.

14. $y = (A/x) + (B/x^2) + (C/x^3)$.
15. $y = (A/x) + (B + C\ln x)x^2$.

16. $y = (A/x^2) + x[B\cos(4\ln x) + C\sin(4\ln x)]$.

17. $y = A + Bx + Cx^2 + D\ln x$.
18. $y = Ax^2 + (B/x^2) + C\cos(\ln x) + D\sin(\ln x)$.

19. $y = A\sqrt{x} + (B/\sqrt{x}) + C\cos(2\ln x) + D\sin(2\ln x)$.

20. $y = (A + B\ln x)x + (C + D\ln x)/x$.

21. $y = Ax^2 + (B/x) - x - 3$.
22. $y = Ax + Bx^3 + \ln x + 2$.

23. $y = Ax + (B/x^2) + 2x\ln x + 7$.
24. $y = Ax^2 + (B/x^3) + 3x^2\ln x$.

25. $y = A + (B/x) + [\sin(\ln x) - \cos(\ln x)]/2$.

26. $y = Ax + (B/x^5) + 2x(3\ln^2 x - \ln x)/3$.

27. $y = (A + B\ln x)x^{1/2} + 4\cos(\ln x) - 3\sin(\ln x)$.

28. $y = (A + B\ln x)x^2 + x^3$.
29. $y = (A + B\ln x)x^{-3/2} + 2\sin(\ln x) - \cos(\ln x)$.

30. $y = Ax + (B/x^2) - x[3\cos(\ln x) + \sin(\ln x)]/10$.

31. $y = (A/x) + Bx^4 - x^2 - \ln x + 3/4$.
32. $y = Ax + (B/x) + (C/x^5) + 2x^2$.

33. $y = Ax^2 + (B/x^2) + (C/x^3) - (3 \ln x)/x^2$.

34. $y = (A + B \ln x + C \ln^2 x)x^2 + 3x^3 - 8x$.

35. $y = (A + B \ln x)x^{1/2} + (C/x) + \sin(\ln x) + 7 \cos(\ln x)$.

36. Set $3x + 1 = z$, $y = [A + B \ln(3x + 1)](3x + 1)^{1/3} + \dfrac{3}{2}(x - 1)$.

37. Set $x + 2 = z$, $y = A(x + 2) + (x + 2)^{1/2}[B \cos t + C \sin t] + 8(x + 2)^2 - 96(x + 2) \ln(x + 2) - 96$,
where $t = \sqrt{3} \ln(x + 2)/2$.

38. $y = Ax + (B/x) + Cx^2 + (D/x^2) + 1/(4x^3)$.

39. $y = Ax^{3/2} + Bx^{-3/2} + (C + D \ln x)x + 2x^2 - 1/9$.

40. $y = A \cos(\ln x) + B \sin(\ln x) + C \cos(2 \ln x) + D \sin(2 \ln x) + 1/(20x^2)$.

41. $y = \dfrac{1}{4}\left(\sqrt{x} + \dfrac{1}{x}\right) + \dfrac{x}{2}$.

42. $y = 4(\ln x - 1)\sqrt{x} + \ln x + 4$.

43. $y = [7x - 10x^2 + 5x^3 + x \ln x]/2$.

44. $y = x[4 \sin(\ln x) - 2 \cos(\ln x)] + 3$.

45. $y = \dfrac{1}{x}\left[2 \cos(3 \ln x) + 3 \sin(3 \ln x) + \dfrac{x^2}{2}\right]$.

Exercise 5.7

1. $Ae^{-x} + Be^{-4x} + e^{2x}$.

2. $Ae^{x} + Be^{-x} + e^{3x}$.

3. $Ae^{-x} + Be^{4x} + e^{5x} - (e^x)/6$.

4. $e^{-x/2}[A \cos(\sqrt{7}x/2) + B \sin(\sqrt{7}x/2)] + \dfrac{4}{11}e^{x/2}$.

5. $e^{-3x/2}[A \cos(\sqrt{3}x/2) + B \sin(\sqrt{3}x/2)] + e^x$.

6. $(A + Bx)e^x + 4e^{2x} + (5e^{4x})/9$.

7. $(A + Bx)e^{x/3} + (e^{-x})/4$.

8. $(A + Bx)e^{3x} + 7x^2e^{3x}$.

9. $Ae^{2x} + Be^{-3x} + (xe^{2x})/5$.

10. $Ae^{2x} + Be^{-x/2} - e^{-x/2}(4x + 5x^2)/50$.

11. $Ae^{x} + Be^{-x} + [3e^x(x^2 - x)]/2$.

12. $Ae^{-2x} + Be^{-x/4} - \dfrac{1}{98}(7x^2 + 8x)e^{-2x}$.

13. $(A + Bx)e^{-x/3} + (x^2e^{-x/3})/18$.

14. $Ae^{x/2} + Be^{-4x} - e^{-4x}(9x^2 + 4x)/162$.

15. $Ae^{-x} + Be^{2x} + Ce^{-3x} - (e^x)/2$.

16. $Ae^{x} + Be^{-2x} + Ce^{-x/2} + (e^{2x})/2$.

17. $Ae^{x} + Be^{-x} + Ce^{2x} + (e^{3x})/8$.

18. $(A + Bx + Cx^2)e^{2x} + 3x^3e^{2x}$.

19. $(A + Bx)e^x + Ce^{-x/2} + (8x^2e^x)/3$.

20. $Ae^{2x} + Be^{-2x} + Ce^{-3x} - 3e^{-2x}(2x^2 - 3x)/4$.

21. $A \cos 4x + B \sin 4x + (\cos 2x)/12$.

22. $Ae^{x} + Be^{3x/2} + (\sin x + 5 \cos x)/26$.

23. $Ae^{2x} + Be^{x/3} + (3 \cos x - 4 \sin x)/25$.

24. $Ae^{3x} + Be^{x/2} + (14 \cos 2x - 5 \sin 2x)/221$.

25. $e^{-x/2}[A \cos(\sqrt{3}x/2) + B \sin(\sqrt{3}x/2)] + 16 \sin x$.

26. $e^{3x/4}[A \cos(x/4) + B \sin(x/4)] + 16(4 \cos x - \sin x)/51$.

27. $A \cos 3x + B \sin 3x - (x \cos 3x)/6$.

28. $A \cos(\sqrt{3}x) + B \sin(\sqrt{3}x) + (x \sin \sqrt{3}x)/(2\sqrt{3})$.

29. $e^{-x}(A \cos 2x + B \sin 2x) + (xe^{-x} \sin 2x)/4$.

30. $e^{2x}(A \cos x + B \sin x) - 12 x \cos x e^{2x}$.

31. $e^{3x}(A \cos 2x + B \sin 2x) - 7x \cos 2x e^{3x}$.

32. $e^{x}[A \cos 3x + B \sin 3x + x (8 \sin 3x - 12 \cos 3x)/3]$.

33. $A e^{3x} + B \cos x + C \sin x - 3x(\cos x + 3 \sin x)/10$.

34. $A e^{x} + B \cos 3x + C \sin 3x - x(3 \cos 3x + \sin 3x)/2$.

35. $A e^{2x} + e^{x}(B \cos 2x + C \sin 2x) - 6xe^{x}(2 \sin 2x - \cos 2x)/5$.

36. $A e^{2x} + e^{x/2}(B \cos x + C \sin x) - 4xe^{x/2}(2 \cos x + 3 \sin x)/13$.

37. $A \cos x + B \sin x + C^* \cos 2x + D^* \sin 2x - 8x(\cos x + 2 \sin 2x)/3$.

38. $A \cos 5x + B \sin 5x + (225x^3 + 100x^2 - 54x - 8)/625$.

39. $(A + Bx)e^{-3x} + (12x^2 - 16x + 5)/27$.

40. $A e^{-x} + B e^{3x} - (18x^2 + 30x - 8)/27$.

41. $A e^{2x} + B e^{3x} + [(52x + 25)(\cos 2x - 5 \sin 2x) - 21(5 \cos 2x + \sin 2x)]/2704$.

42. $A e^{x} + B e^{-2x} - [(25x^2 + 5x - 9)(3 \sin x + \cos x) + (35x + 12)(3 \cos x - \sin x)]/250$.

43. $A e^{3x} + B e^{-2x} - e^{-2x}(5x^2 + 2x)/50$.

44. $A e^{-3x} + B e^{-4x} + e^{x}(8 \sin 2x - 9 \cos 2x)/290$.

45. $A e^{-x} + B e^{-3x} + e^{2x}(7 \cos x + 4 \sin x)/130$.

46. $e^{-3x/2}[A \cos p + B \sin p] + 4e^{x}(25 \cos p + 10\sqrt{7} \sin p)/1325$, $p = \sqrt{7}x/2$.

47. Write $xe^{x} \sin x = \text{Im}\,[xe^{(1+i)x}]$, $A e^{-x} + B e^{-2x} + e^{x}[5(1 - x) \cos x + (5x - 2) \sin x]/50$.

48. Write $xe^{2x} \cos x = \text{Re}\,[xe^{(2+i)x}]$, $A \cos 3x + B \sin 3x + e^{2x}[(30x - 11) \cos x + (10x - 2) \sin x]/400$.

49. $A e^{-x/2} + B e^{-3x/2} - e^{-x/2}[(x - 2) \cos x - (x + 1) \sin x]/8$.

50. $A \cos x + B \sin x + C^* \cos \sqrt{2}x + D^* \sin \sqrt{2}x - 4[9x^2 \cos x - (2x^3 - 51x) \sin x]/3$.

51. $y = A e^{x/2} + B e^{3x}$, $B = 1/5$.

52. $\displaystyle \int e^{-mx} r(x)\, dx = \int e^{-mx} (D - m)y\, dx = e^{-mx}y$, or $\displaystyle y = e^{mx} \int e^{-mx} r(x)\, dx$.

53. Use the result

$$\frac{d}{dx} \int_a^b f(x, t)\, dt = f(x, b)\frac{db}{dx} - f(x, a)\frac{da}{dx} + \int_a^b \frac{\partial f}{\partial x}\, dt$$

$$\frac{dy}{dx} = \int_a^x r(t) \cos n(x - t)\, dt, \quad \frac{d^2 y}{dx^2} = r(x) - n \int_a^b r(t) \sin n(x - t)\, dt = r(x) - n^2 y.$$

54. $D^m(x\,u) = xD^m u + mD^{m-1}u = xD^m u + \left[\dfrac{d}{dD}D^m\right]u$ $m = 1, 2, \ldots$

$$F(D)(x\,u) = x[a_0 D^n + a_1 D^{n-1} + \ldots + a_n]u + \frac{d}{dD}[a_0 D^n + a_1 D^{n-1} + \ldots + a_n]u$$

$$= xF(D)u + F'(D)u.$$

55. $F(D)(xv) = x F(D)v + F'(D)v$. Let $F(D)v = u$.

$$F(D)[x\{F(D)\}^{-1}u] = xF(D)[F(D)]^{-1}u + F'(D)[F(D)]^{-1}u = xu + F'(D)[F(D)]^{-1}u$$

$$xu = F(D)[x\{F(D)\}^{-1}u] - F'(D)[F(D)]^{-1}u$$

$$[F(D)]^{-1}(xu) = x[F(D)]^{-1}u - F'(D)[F(D)]^{-2}u.$$

56. When $F(m) = 0$, $F'(m) \neq 0$, $F(D) = (D - m)G(D)$ and $F'(m) = G(m)$.

$$[F(D)]^{-1}e^{mx} = \frac{xe^{mx}}{G(m)} = x\left(\frac{1}{F'(m)}\right)e^{mx},$$

When $F(m) = 0$, $F'(m) = 0$, $F''(m) \neq 0$, $F(D) = (D - m)^2 G(D)$ and $F''(m) = 2G(m)$.

$$[F(D)]^{-1}e^{mx} = \frac{x^2}{2!}\frac{e^{mx}}{G(m)} = x^2\left(\frac{1}{F''(m)}\right)e^{mx}.$$

Problem 8: $F(3) = 0$, $F'(3) = 0$, $F''(3) = 2$, $y_p = 7x^2e^{3x}$.

Problem 9: $F(2) = 0$, $F'(2) = 5$, $y_p = (xe^{2x})/5$.

Problem 13: $F(-1/3) = 0$, $F'(-1/3) = 0$, $F''(-1/3) = 18$, $y_p = (x^2e^{-x/3})/18$.

Problem 19: $F(1) = 0$, $F'(1) = 0$, $F''(1) = 6$, $y_p = (8x^2e^x)/3$.

57. Note that $[F(D)]^{-1}$ can be written as

$$[F(D)]^{-1} = A_1(D - m_1)^{-1} + A_2(D - m_2)^{-1} + \ldots + A_n(D - m_n)^{-1}$$

$$\left(\text{equivalent to writing in partial fractions of } \frac{1}{F(m)} \text{ as } \frac{1}{F(m)} = \frac{A_1}{m - m_1} + \frac{A_2}{m - m_2} + \ldots + \frac{A_n}{m - m_n}\right)$$

Now apply the solution of Problem 52.

Problem 40: $y_p = \frac{1}{4}e^{3x}\int e^{-3x}(2x^2 + 6x)\,dx - \frac{1}{4}e^{-x}\int e^x(2x^2 + 6x)\,dx$

$$= -\frac{1}{27}(18x^2 + 30x - 8).$$

58. For forced damped oscillations, $c^2 < 4mk$, $y_c = e^{-ct/(2m)}.[A\cos dt + B\sin dt]$

$$d = \sqrt{4mk - c^2}/(2m).$$

For forced undamped oscillations, $y_c = A\cos(\sqrt{k/m}\,t) + B\sin(\sqrt{k/m}\,t)$.

$c \neq 0$, $y_p = F_0[(k - m\omega^2)\cos\omega t + c\omega\sin\omega t]/[(k - m\omega^2)^2 + c^2\omega^2]$

$c = 0$, $y_p = F_0\cos\omega t/(k - m\omega^2)$.

Exercise 5.8

1. $y_1'' - 4y_1' + 3y_1 = 0$, $y_1 = Ae^t + Be^{3t}$, $y_2 = Be^{3t} - Ae^t$.

2. $y_1'' - 2y_1' - 8y_1 = 0$, $y_1 = Ae^{-2t} + Be^{4t}$, $y_2 = 3Be^{4t} - 3Ae^{-2t}$.

3. $y_1 = A\cos 3t + B\sin 3t$, $y_2 = 3(B\cos 3t - A\sin 3t)$.

4. $y_1 = Ae^{-t} + Be^{-4t}$, $y_2 = -(3Ae^{-t} + 6Be^{-4t})$.

5. $y_1 = Ae^t + Be^{-t} + 2\cos t$, $y_2 = Be^{-t} - Ae^t + 6\sin t$.

6. $y_1 = Ae^{5t} + Be^{-3t} + 4e^{-t}/3 + 8(15t - 2)/75$.

$3y_2 = 4e^{-t} - 6Ae^{5t} + 2Be^{-3t} - 8(15t + 13)/75$.

7. $y_1 = Ae^{2t} + Be^{-2t} + 4e^t + e^{-t}$, $y_2 = -(5Ae^{2t} + Be^{-2t} + 10e^t + 2e^{-t})$.

8. $y_1 = A \cos 3t + B \sin 3t + (1/2)e^t - 8 \sin t + 2 \cos t.$

$5y_2 = (4B - 3A) \sin 3t + (3B + 4A) \cos 3t + (5/2)e^t - 50 \sin t.$

9. $y_1 = A \cos 4t + B \sin 4t + 5 \cos 2t + 32t \sin 4t + 24t \cos 4t.$

$5y_2 = (3B - 4A) \sin 4t + (4B + 3A) \cos 4t - 10 \sin 2t + 15 \cos 2t - 32 \sin 4t + (24 + 200t) \cos 4t.$

10. $y_1 = Ae^t + Be^{2t} - 5e^{-t} - 36te^t, \ 3y_2 = 2Ae^t + 3Be^{2t} - 6e^{-t} - 36(1 + 2t)e^t.$

11. $y_1 = Ae^{-9t} + (9t - 16)/27, \ y_2 = -(11/6) Ae^{-9t} + (1/81) (52 + 45t) + Be^{-3t}.$

12. $y_1 = (1/5)e^t + 2e^{-t} - 3Ae^{-2t} + Be^{-3t/2}, \ y_2 = Ae^{-2t} + e^{-t} - (2/3)e^t.$

13. $y_1 = (t - 3t^2)/8, \ y_2 = A + (10t^3 + 21t^2 + 42t)/96.$

14. $y_1 = [(4t - 1 - 8A)e^{-t} - 2e^t]/2, \ y_2 = [(2A - t)e^{-t} + e^t]/2.$

15. $y_1 = (29/6)e^{2t} - 5Ae^{4t} - 10Be^{-t} + e^{-2t}, \ y_2 = Ae^{4t} + Be^{-t} - (5/6)e^{2t}.$

16. $y_1 = -(9/4)Ae^{-t/2} + 3Be^{-2t} + (1413 - 738t + 168t^2)/84.$

$y_2 = Ae^{-t/2} + Be^{-2t} - (207 - 114t + 28t^2)/28.$

17. $y_1 = -(1 + A) + (1/3)Be^{-4t/5}, \ y_2 = A + Be^{-4t/5} + t.$

18. $y_1 = [e^t - 2A - Be^{-t/3} - 6 - 2t + t^2]/2, \ y_2 = A + Be^{-t/3} + 2t - (t^2/2).$

19. $\mathbf{y} = Ce^{2t}\mathbf{x}^{(1)} + De^{3t}\mathbf{x}^{(2)}, \ \mathbf{x}^{(1)} = [-2, 1]^T, \ \mathbf{x}^{(2)} = [-1, 1]^T.$

20. $\mathbf{y} = Ce^{-t}\mathbf{x}^{(1)} + De^{4t}\mathbf{x}^{(2)}, \ \mathbf{x}^{(1)} = [1, 2]^T, \ \mathbf{x}^{(2)} = [3, 1]^T.$

21. $\mathbf{y} = Ce^{3it}\mathbf{x}^{(1)} + De^{-3it}\mathbf{x}^{(2)}, \ \mathbf{x}^{(1)} = [1 + i, 3]^T, \ \mathbf{x}^{(2)} = [1 - i, 3]^T,$ or as

$\mathbf{y} = C\mathbf{z}^{(1)} + D\mathbf{z}^{(2)}, \ \mathbf{z}^{(1)} = [\cos 3t - \sin 3t, \ 3 \cos 3t]^T, \ \mathbf{z}^{(2)} = [\cos 3t + \sin 3t, \ 3 \sin 3t].$

22. $\mathbf{y} = A_1e^{2it}\mathbf{x}^{(1)} + B_1e^{-2it}\mathbf{x}^{(2)}, \ \mathbf{x}^{(1)} = [2, 1 - i]^T, \ \mathbf{x}^{(2)} = [2, 1 + i]^T,$ or as

$\mathbf{y} = C\mathbf{z}^{(1)} + D\mathbf{z}^{(2)}, \ \mathbf{z}^{(1)} = [2 \cos 2t, \ \cos 2t + \sin 2t]^T, \ \mathbf{z}^{(2)} = [2 \sin 2t, \ \sin 2t - \cos 2t]^T.$

23. $\mathbf{y} = Ce^t\mathbf{z}^{(1)} + De^t\mathbf{z}^{(2)}, \ \mathbf{z}^{(1)} = [\cos t, \ \sin t]^T, \ \mathbf{z}^{(2)} = [\sin t, \ -\cos t]^T.$

24. $\mathbf{y} = e^{t/2} (C\mathbf{z}^{(1)} + D\mathbf{z}^{(2)}), \ \mathbf{z}^{(1)}, \ \mathbf{z}^{(2)}$ as in problem 23.

25. $\mathbf{y} = c_1e^{2t}\mathbf{x}^{(1)} + c_2 e^{-3t}\mathbf{x}^{(2)} - \mathbf{d}t - \mathbf{e}, \ \mathbf{x}^{(1)} = [1, 1]^T, \ \mathbf{x}^{(2)} = [1, -4]^T, \ \mathbf{d} = [4/3, 5/3]^T, \ \mathbf{e} = [17/9, 4/9]^T.$

26. $\mathbf{y} = c_1e^{t/2}\mathbf{x}^{(1)} + c_2 e^{-t/2}\mathbf{x}^{(2)} + \mathbf{d}e^{2t}, \ \mathbf{x}^{(1)} = [3, 1]^T, \ \mathbf{x}^{(2)} = [1, 1]^T, \ \mathbf{d} = [-12/15, 28/15].$

27. $\mathbf{y} = c_1e^{2t}\mathbf{x}^{(1)} + c_2e^{-3t}\mathbf{x}^{(2)} + (\mathbf{d}t + \mathbf{e})e^{2t}, \ \mathbf{x}^{(1)} = [6, 1]^T, \ \mathbf{x}^{(2)} = [1, 1]^T, \ \mathbf{d} = [-6/5, -1/5]^T, \ \mathbf{e} = [-16/5, 0]^T.$

28. $\mathbf{y} = c_1e^t\mathbf{x}^{(1)} + c_2e^{3t}\mathbf{x}^{(2)} + (\mathbf{d}t + \mathbf{e})e^t, \ \mathbf{x}^{(1)} = [1, 1]^T, \ \mathbf{x}^{(2)} = [2, 1]^T, \ \mathbf{d} = [12, 12]^T, \ \mathbf{e} = [-3, 0]^T.$

29. $\mathbf{y} = c_1e^{-t}\mathbf{x}^{(1)} + c_2e^{5t}\mathbf{x}^{(2)} + (\mathbf{d}t + \mathbf{e})e^{5t}, \ \mathbf{x}^{(1)} = [1, -2]^T, \ \mathbf{x}^{(2)} = [1, 1]^T, \ \mathbf{d} = [21, 21]^T, \ \mathbf{e} = [2, 0]^T.$

30. *Problem 25:* $\mathbf{x} = \begin{bmatrix} 1 & 1 \\ 1 & -4 \end{bmatrix}, \mathbf{g} = \mathbf{x}^{-1}\mathbf{h} = \frac{1}{5}\begin{bmatrix} 14t + 9 \\ t - 4 \end{bmatrix}, \lambda_1 = 2, \lambda_2 = -3,$

$u_j' = \lambda_j u_j + g_j, \ u_1 = c_1e^{2t} - (7t + 8)/5, \ u_2 = c_2e^{-3t} + (3t - 13)/45,$

$\mathbf{y} = \mathbf{xu} = \begin{bmatrix} c_1e^{2t} + c_2e^{-3t} - (12t + 17)/9 \\ c_1e^{2t} - 4c_2 e^{-3t} - (15t + 4)/9 \end{bmatrix}$

Problem 26: $\mathbf{x} = \begin{bmatrix} 3 & 1 \\ 1 & 1 \end{bmatrix}, \mathbf{g} = \mathbf{x}^{-1}\mathbf{h} = \begin{bmatrix} -2 \\ 8 \end{bmatrix}e^{2t}, \lambda_1 = 1/2, \lambda_2 = -1/2.$

$u_j' = \lambda_j u_j + g_j, \ u_1 = c_1e^{t/2} - (4/3)e^{2t}, \ u_2 = c_2 e^{-t/2} + (16/5)e^{2t}.$

$\mathbf{y} = \mathbf{xu} = \begin{bmatrix} 3c_1e^{t/2} + c_2e^{-t/2} - (4/5)e^{2t} \\ c_1e^{t/2} + c_2e^{-t/2} + (28/15)e^{2t} \end{bmatrix}.$

Chapter 6

Series Solutions of Differential Equations

6.1 Introduction

In the previous two chapters, we have studied the methods of solving general linear first order differential equations, constant coefficient second and higher order differential equations and a special case of a variable coefficient differential equation (Euler-Cauchy type equation). The solutions of these equations were all closed form solutions in terms of standard functions. However, it is often not possible to express the solutions of variable coefficient equations in closed form using the standard functions. In such cases, we seek the solution as an infinite series in terms of the independent variable. Many of the important physical problems can be described by second order variable coefficient equations. Solutions of such equations can be obtained in terms of infinite series. The series solution methods can be classified into two categories: power series method and general series solution method (*Frobenius method*). In the following sections we discuss the application of these methods.

6.2 Ordinary and Singular Points of an Equation

Consider the variable coefficient second order, linear homogeneous equation

$$a_0(x)y'' + a_1(x)y' + a_2(x)y = 0, \; a_0(x) \neq 0. \tag{6.1}$$

Dividing the equation by $a_0(x)$, we can write it in the standard form (*normal form* or *canonical form*) as

$$y'' + p(x) \, y' + q(x) \, y = 0 \tag{6.2}$$

where $p(x) = a_1(x)/a_0(x)$ and $q(x) = a_2(x)/a_0(x)$. Let x_0 be a point in the interval I and $a_1(x)$, $a_2(x)$ be analytic at x_0 (that is $a_1(x)$, $a_2(x)$ are differentiable at x_0 and at every point in its neighborhood). We define the following.

Ordinary point   A point $x_0 \in I$ is said to be an *ordinary point* of Eq. (6.1) if $a_0(x_0) \neq 0$. Ordinary point is also called a regular point of the equation.

Singular point   A point $x_0 \in I$ is said to be a *singular point* of Eq. (6.1) if $a_0(x_0) = 0$.

Example 6.1 Find the regular and singular points of the differential equations

(i) $(1 - x^2)\, y'' - 2xy' + n(n + 1)\, y = 0$,

(ii) $x^2 y'' + axy' + by = 0$, a, b are constants.

Solution

(i) Setting $a_0(x) = 1 - x^2 = 0$, we get $x = \pm 1$. Hence, $x = \pm 1$ are the singular points of the equation, while all others points are regular points. Note that $a_1(x) = -2x$ and $a_2(x) = n(n + 1)$ are analytic at $x = \pm 1$.

(ii) Setting $a_0(x) = x^2 = 0$, we get $x = 0$, Hence, $x = 0$ is the singular point of the equation, while all other points are regular points.

Let $x_0 \in I$ be a singular point of Eq. (6.1), where $a_1(x)$ and $a_2(x)$ are analytic at x_0. Consider now the standard form of Eq. (6.1), which is given by

$$y'' + p(x)\, y' + q(x)\, y = 0$$

where
$$p(x) = a_1(x)/a_0(x) \quad \text{and} \quad q(x) = a_2(x)/a_0(x).$$

Now, write this equation as

$$y'' + \frac{p_1(x)}{x - x_0}\, y' + \frac{q_1(x)}{(x - x_0)^2}\, y = 0 \tag{6.3}$$

where $p_1(x) = (x - x_0)\, p(x) = \dfrac{(x - x_0) a_1(x)}{a_0(x)}$, and $q_1(x) = (x - x_0)^2\, q(x) = \dfrac{(x - x_0)^2 a_2(x)}{a_0(x)}$.

We define the following.

Regular singular point A singular point x_0 of Eq. (6.1) is said to be a *regular singular point* if and only if the functions $p_1(x)$ and $q_1(x)$ defined in Eq. (6.3) have removable discontinuities at x_0 and become analytic when these discontinuities are removed. In other words, after the discontinuity at x_0 is removed, the functions $p_1(x)$ and $q_1(x)$ have Taylor series expansions about the point $x = x_0$. What is intended in the above is that the definitions of $p_1(x)$, $q_1(x)$ given in Eq. (6.3) may be continuously extended to include the point x_0 and these continuous extensions will lead to analytic functions at the point x_0.

Irregular singular point A singular point x_0 of Eq. (6.1) is said to be an *irregular singular point* if and only if x_0 is not a regular singular point. In other words, if either $p_1(x)$ or $q_1(x)$ or both $p_1(x)$ and $q_1(x)$ do not have Taylor series expansions about the point $x = x_0$, then x_0 is an irregular singular point.

Example 6.2 Classify the singular points of the following equations

(i) $x^2 y'' + axy' + by = 0$, a, b constants,

(ii) $x^2 y'' + xy' + (x^2 - n^2)\, y = 0$, n constant,

(iii) $(1 - x^2)\, y'' - 2xy' + n(n + 1)\, y = 0$, n constant,

(iv) $x^3(x - 2)\, y'' + x^3 y' + 6y = 0$.

Solution

(i) Setting $a_0(x) = x^2 = 0$, we get the singular point of the equation as $x = 0$. Dividing the equation by x^2, we get

$$y'' + \frac{ax}{x^2}y' + \frac{b}{x^2}y = 0, \quad \text{or} \quad y'' + \frac{a}{x}y' + \frac{b}{x^2}y = 0, \quad x \neq 0. \tag{6.4}$$

Comparing with Eq. (6.3), we get $p_1(x) = a$ and $q_1(x) = b$. (Note that $x = 0$ is a removable discontinuity of $p_1(x) = (ax^2)/x^2$, and $q_1(x) = (bx^2)/x^2$.) Since a and b are constants, the expressions of $p_1(x)$ and $q_1(x)$ are the Taylor series expansions. Hence, $x = 0$ is a regular singular point of the given equation.

(ii)   The singular point of the equation is $x = 0$. Dividing by x^2, we get

$$y'' + \frac{x}{x^2}y' + \frac{1}{x^2}(x^2 - n^2)y = 0, \quad \text{or} \quad y'' + \frac{1}{x}y' + \frac{1}{x^2}(x^2 - n^2)y = 0, x \neq 0.$$

Comparing with Eq. (6.3), we get $p_1(x) = 1$ and $q_1(x) = x^2 - n^2$. (Note that $x = 0$ is a removable discontinuity of $p_1(x) = x^2/x^2$, and $q_1(x) = x^2(x^2 - n^2)/x^2$). The expressions of $p_1(x)$ and $q_1(x)$ are the Taylor series expansions about $x = 0$. Hence, $x = 0$ is a regular singular point of the given equation.

(iii)   Setting $a_0(x) = 1 - x^2 = 0$, we get the singular points of the equation as $x = \pm 1$. Dividing the equation by $(1 - x^2)$ and writing it in the required form, we get

$$y'' - \frac{2x}{1 - x^2}y' + \frac{n(n + 1)}{1 - x^2}y = 0, \quad x \neq \pm 1 \tag{6.5}$$

or $\qquad y'' - \frac{1}{(1 - x)}\left[\frac{2x}{1 + x}\right]y' + \frac{1}{(1 - x)^2}\left[\frac{(1 - x)n(n + 1)}{1 + x}\right]y = 0$

where the singularity at $x = 1$ is being considered. Now,

$$p_1(x) = -\frac{2x}{1 + x}, \quad \text{and} \quad q_1(x) = \frac{(1 - x)n(n + 1)}{1 + x}$$

are both analytic at $x = 1$ and hence Taylor series expansions of these functions about $x = 1$ exist. For example,

$$p_1(x) = -\frac{2(x - 1 + 1)}{(x - 1 + 2)} = -[1 + (x - 1)]\left[1 + \frac{(x - 1)}{2}\right]^{-1}$$

$$= -[1 + (x - 1)]\left[1 - \frac{(x - 1)}{2} + \frac{(x - 1)^2}{4} - \ldots\right] = -\left[1 + \frac{(x - 1)}{2} - \frac{(x - 1)^2}{4} + \ldots\right].$$

Similarly, Taylor series expansion for $q_1(x)$ can be written.
Write now, the Eq. (6.5) as

$$y'' - \frac{1}{1 + x}\left[\frac{2x}{1 - x}\right] + \frac{1}{(1 + x)^2}\left[\frac{(1 + x)n(n + 1)}{1 - x}\right]y = 0$$

where the singularly at $x = -1$ is being considered. Now

$$p_1(x) = -\frac{2x}{1 - x}, \quad \text{and} \quad q_1(x) = \frac{(1 + x)n(n + 1)}{1 - x}$$

are both analytic at $x = -1$ and hence have Taylor series expansions about $x = -1$. Therefore, $x = \pm 1$ are regular singular points of the given equation.

(iv) Setting $a_0(x) = x^3(x - 2) = 0$, we get the singular points as $x = 0, 2$. Dividing the equation throughout by $x^3(x - 2)$, we get

$$y'' + \frac{x^3}{x^3(x - 2)} y' + \frac{6}{x^3(x - 2)} y = 0, \quad x \neq 0, 2. \tag{6.6}$$

Consider now, the singularity at $x = 2$.

Comparing Eq. (6.6) with

$$y'' + \frac{p_1(x)}{x - 2} y' + \frac{q_1(x)}{(x - 2)^2} y = 0$$

we get $p_1(x) = 1$, and $q_1(x) = 6(x - 2)/x^3$. Since $p_1(x)$ and $q_1(x)$ are both analytic at $x = 2$, their Taylor series expansions about $x = 2$ exist. Hence, $x = 2$ is a regular singular point of the given equation. Consider now, the singularity at $x = 0$.

Comparing Eq. (6.6) with

$$y'' + \frac{p_1(x)}{x} y' + \frac{q_1(x)}{x^2} y = 0,$$

we get $p_1(x) = x/(x - 2)$, and $q_1(x) = 6/[x(x - 2)]$. Now, $q_1(x)$ is not analytic at $x = 0$, and Taylor series expansion of $q_1(x)$ about $x = 0$ does not exist. Hence, $x = 0$ is an irregular singular point of the equation.

Exercise 6.1

Find the singular points of the following differential equations and classify them.

1. $x^2 y'' + (x + x^2) y' - y = 0$.
2. $x^2 y'' + 2xy' + (x^2 - n^2) y = 0$, n constant.
3. $x^2 y'' - 5y' + 3x^2 y = 0$.
4. $xy'' + y' + xy = 0$.
5. $x^2 y'' + (\sin x) y' + (\cos x) y = 0$.
6. $x^3(x^2 - 1) y'' - x(x + 1) y' - (x - 1) y = 0$.
7. $(x^2 + x - 2)^2 y'' + 3(x + 2) y' + (x - 1) y = 0$.
8. $x^2 y'' + 4xy' + (x^2 + 2) y = 0$.
9. $x^4 y'' + 4x^3 y' + y = 0$.
10. $x^3 y'' + 3xy' + 6y = 0$.

6.3   Power Series Solution

We now present the results regarding the existence of a power series solution of a differential equation about an ordinary point $x = x_0$.

Theorem 6.1 Let $x = x_0$ be an ordinary point (regular point) of the equation $a_0(x) y'' + a_1(x) y' + a_2(x) y = 0$. Then, every solution of the equation is analytic at $x = x_0$ and hence has a power series expansion about the point $x = x_0$, of the form

$$y(x) = c_0 + c_1(x - x_0) + c_2(x - x_0)^2 + \ldots \tag{6.7}$$

where $c_0, c_1, \ldots$ are constants.

The proof is obvious. Since $a_0(x_0) \neq 0$, we can write the given equation as $y'' + p(x)y' + q(x) y = 0$, where $p(x) = a_1(x)/a_0(x)$, and $q(x) = a_2(x)/a_0(x)$ are analytic at $x = x_0$. Hence, $y''(x_0), y'''(x_0), \ldots$ exist and the Taylor expansion of $y(x)$, that is , power series solution about $x = x_0$ exists. We note that every function which is analytic in the region $|x - x_0| < R$ admits a converging power series representation $\sum\limits_{m=0}^{\infty} c_m (x - x_0)^m$ in the region.

Example 6.3 Find the power series solution about $x = 0$, of the differential equation $y' - 2y = 0$.

Solution Let the power series solution about the point $x = 0$ be written as

$$y(x) = c_0 + c_1 x + c_2 x^2 + c_3 x^3 + \ldots$$

Differentiating, we obtain

$$y'(x) = c_1 + 2c_2 x + 3c_3 x^2 + \ldots.$$

Substituting for $y(x)$ and $y'(x)$ in the differential equation, we obtain

$$(c_1 + 2c_2 x + 3c_3 x^2 + \ldots) - 2(c_0 + c_1 x + c_2 x^2 + \ldots) = 0$$

or $\qquad (c_1 - 2c_0) + 2(c_2 - c_1)x + \ldots + [(m + 1) c_{m+1} - 2c_m] x^m + \ldots = 0.$

Equating the coefficients of various powers of x, we get

$$c_1 - 2c_0 = 0, \quad 2(c_2 - c_1) = 0, \ldots, (m + 1) c_{m+1} - 2c_m = 0, \ldots.$$

Solving, we get $c_1 = 2c_0$, $c_2 = c_1 = 2c_0$, $c_3 = 2c_2/3 = 4c_0/3, \ldots.$

Therefore, the solution becomes

$$y(x) = c_0 + c_1 x + c_2 x^2 + c_3 x^3 + \ldots$$

$$= c_0 + 2c_0 x + 2c_0 x^2 + \frac{4}{3} c_0 x^3 + \ldots$$

$$= c_0 \left[1 + 2x + \frac{(2x)^2}{2!} + \frac{(2x)^3}{3!} + \ldots \right] = c_0 e^{2x}$$

where c_0 is an arbitrary constant. The solution is expressible in terms of a standard function.

If an initial value problem is given, an alternative method can be used to find the power series solution. We determine $y^{(n)}(x_0)$, $n = 1, 2, \ldots$ using the differential equation and substitute in the Taylor series expansion of $y(x)$ about $x = x_0$.

Example 6.4 Find the power series solution about $x = 2$, of the initial value problem

$$4y'' - 4y' + y = 0, \quad y(2) = 0, \quad y'(2) = 1/e.$$

Express the solution in the closed form.

Solution We shall use the Taylor series to determine the solution. We have the solution as

$$y(x) = y(2) + (x - 2) y'(2) + \frac{(x - 2)^2}{2!} y''(2) + \ldots.$$

From the given equation, we have $\quad y'' = \frac{1}{4}(4y' - y).$

Differentiating $(m - 2)$ times we get $\quad y^{(m)} = \frac{1}{4}[4y^{(m-1)} - y^{(m-2)}].$

Substituting $x = 2$, we obtain $y^{(m)}(2) = \frac{1}{4}[4y^{(m-1)}(2) - y^{(m-2)}(2)]$, $m = 2, 3, \ldots.$

Using the values $y(2) = 0$ and $y'(2) = 1/e$, we obtain

$$y''(2) = \frac{1}{4}[4y'(2) - y(2)] = \frac{1}{4}\left[\frac{4}{e} - 0\right] = \frac{1}{e},$$

$$y'''(2) = \frac{1}{4}\left[4y''(2) - y'(2)\right] = \frac{1}{4}\left[\frac{4}{e} - \frac{1}{e}\right] = \frac{3}{4e},$$

$$y^{iv}(2) = \frac{1}{2e}, \quad y^{v}(2) = \frac{5}{16e}, \dots$$

The solution is given by

$$y(x) = 0 + (x - 2)\left(\frac{1}{e}\right) + \frac{(x-2)^2}{2!}\left(\frac{1}{e}\right) + \frac{(x-2)^3}{3!}\left(\frac{3}{4e}\right) + \frac{(x-2)^4}{4!}\left(\frac{1}{2e}\right) + \dots$$

$$= \frac{(x-2)}{e}\left[1 + \frac{\{(x-2)/2\}}{1!} + \frac{\{(x-2)/2\}^2}{2!} + \frac{\{(x-2)/2\}^3}{3!} + \dots\right]$$

$$= \frac{1}{e}(x-2)e^{(x-2)/2}.$$

This solution can be verified. The solution of the differential equation is $y(x) = (Ax + B)e^{x/2}$. Applying the initial conditions, we get

$$y(2) = 0 = (2A + B)\,e, \quad \text{or} \quad 2A + B = 0$$

$$y'(2) = \frac{1}{e} = \left(2A + \frac{B}{2}\right)e, \quad \text{or} \quad 2A + \frac{B}{2} = \frac{1}{e^2}.$$

The solution is $A = 1/e^2$ and $B = -2/e^2$. Hence, we have

$$y(x) = \frac{1}{e^2}(x - 2)\,e^{x/2} = \frac{1}{e}(x - 2)\,e^{(x-2)/2}.$$

Example 6.5 Find the power series solution about $x = 0$, of the differential equation $y'' - 4y = 0$.

Solution Let the power series solution be

$$y(x) = c_0 + c_1 x + c_2 x^2 + c_3 x^3 + \dots$$

We have $\quad y'(x) = c_1 + 2c_2 x + 3c_3 x^2 + \dots, \quad y''(x) = 2c_2 + 6c_3 x + 12c_4 x^2 + \dots.$

Substituting for $y(x)$, $y'(x)$ and $y''(x)$ in the given equation, we get

$$(2c_2 + 6c_3 x + 12c_4 x^2 + \dots) - 4(c_0 + c_1 x + c_2 x^2 + \dots) = 0$$

or $\qquad (2c_2 - 4c_0) + (6c_3 - 4c_1)\,x + (12c_4 - 4c_2)\,x^2 + \dots = 0.$

Equating the coefficients of various powers of x to zero, we obtain

$$c_2 = 2c_0, c_3 = \frac{2}{3}c_1, c_4 = \frac{1}{3}c_2 = \frac{2}{3}c_0, c_5 = \frac{1}{5}c_3 = \frac{2}{15}c_1, \dots$$

Therefore, the solution is

$$y(x) = c_0 + c_1 x + 2c_0 x^2 + \frac{2}{3}c_1 x^3 + \frac{2}{3}c_0 x^4 + \frac{2}{15}c_1 x^5 + \dots$$

$$= c_0\left[1 + 2x^2 + \frac{2}{3}x^4 + \dots\right] + c_1\left[x + \frac{2}{3}x^3 + \frac{2}{15}x^5 + \dots\right].$$

In terms of standard functions, we can write $y(x)$ as

$$y(x) = \frac{c_0}{2}\left[e^{2x} + e^{-2x}\right] + \frac{1}{4}c_1[e^{2x} - e^{-2x}] = Ae^{2x} + Be^{-2x}$$

where $\qquad A = (2c_0 + c_1)/4$, and $B = (2c_0 - c_1)/4$ are arbitrary constants.

Example 6.6 Find the power series solution about $x = 0$, of the differential equation

$$(1 - x^2)\, y'' - 2xy' + 2y = 0.$$

Solution Substituting for $y(x) = c_0 + c_1 x + c_2 x^2 + \ldots$ and its derivatives $y'(x)$, $y''(x)$ in the given equation we get

$$(1 - x^2)(2c_2 + 6c_3 x + 12c_4 x^2 + \ldots) - 2x(c_1 + 2c_2 x + 3c_3 x^2 + \ldots)$$

$$+ 2(c_0 + c_1 x + c_2 x^2 + \ldots) = 0$$

or $\qquad 2(c_2 + c_0) + 6c_3 x + 4(3c_4 - c_2)\, x^2 + 10\, (2c_5 - c_3)\, x^3 + \ldots = 0.$

Equating the coefficients of various powers of x to zero, we obtain

$$c_2 = -c_0, \quad c_3 = 0, \quad c_4 = c_2/3 = -c_0/3, \quad c_5 = (c_3/2) = 0, \ldots$$

We have c_0, c_1 arbitrary and $c_3 = 0 = c_5 = c_7 = \ldots$. The solution becomes

$$y(x) = \left(c_0 - c_0 x^2 - \frac{1}{3} c_0 x^4 - \ldots \right) + c_1 x = c_0 \left[1 - x^2 - \frac{x^4}{3} - \ldots \right] + c_1 x.$$

In the above example, we know that $x = 0$ is an ordinary point of the equation and therefore the power series solution exists. However, $x = \pm 1$ are the singular points of the equation. In other words, the power series solution cannot exist in any interval I which contains $+1$ or -1. Therefore, the interval of largest length, in which the power series solution of Example 6.6 holds is $(-1, 1)$. We call this interval as the *interval of convergence*. The distance R, between the point x_0 about which the power series solution is sought and the nearest singularity of the equation is called the *radius of convergence* of the series. In Example 6.6, $R = 1$.

We have the following result for the convergence of power series.

Theorem 6.2 The power series representation

$$y(x) = \sum_{m=0}^{\infty} c_m (x - x_0)^m \tag{6.8}$$

about an ordinary point $x = x_0$, of the equation $a_0(x)y'' + a_1(x)y' + a_2(x)\, y = 0$ always converges. The maximum possible radius of convergence R is the distance from x_0 to the nearest singularity of the equation. The interval I of convergence is $|\, x - x_0\,| < R$.

The interval of convergence may sometimes be $(-\infty, \infty)$, that is, the power series converges for all x. We then write, $R = \infty$. This result is always true for a constant coefficient equation, as the equation does not have any singularity.

Remark 1

Various tests for convergence are available for testing the convergence and finding the interval of convergence of the power series. Two of the most commonly used tests are the *ratio test* and the *root test*.

We now state these results.

Ratio test The series $y(x) = \sum_{m=0}^{\infty} c_m (x - x_0)^m = \sum_{m=0}^{\infty} u_m$ converges, if

$$\lim_{m \to \infty} \left| \frac{u_{m+1}}{u_m} \right| < 1. \tag{6.9}$$

The radius of convergence of the power series (Eq. (6.8)) is given by

$$R = \lim_{m \to \infty} \left| \frac{c_m}{c_{m+1}} \right|. \tag{6.10}$$

The series converges in the interval $|x - x_0| < R$. If the limit is ∞, then the series converges for all x.

Root test The series $y(x) = \sum_{m=0}^{\infty} c_m (x - x_0)^m = \sum_{m=0}^{\infty} u_m$ converges, if

$$\lim_{m \to \infty} \sqrt[m]{|u_m|} < 1. \tag{6.11}$$

The radius of convergence of the power series (Eq. (6.8)) is given by

$$R = 1 \Big/ \left[\lim_{m \to \infty} \sqrt[m]{|c_m|} \right] \tag{6.12}$$

provided the limit exists. If the limit is 0, then $R = \infty$.

Example 6.7 Find the radius of convergence of the series

$$1 + \frac{x + 1}{6} + \frac{(x + 1)^2}{6^2} + \dots.$$

Solution We have $c_m = 1/6^m$. Now,

$$\lim_{m \to \infty} \sqrt[m]{|c_m|} = \lim_{m \to \infty} \sqrt[m]{1/6^m} = \frac{1}{6}.$$

Therefore, $R = 6$. The series converges in the interval $|x + 1| < 6$.

Example 6.8 Find the radius of convergence of the series

$$1 + \frac{(x - 1)^2}{2} + \frac{(x - 1)^4}{2^2} + \frac{(x - 1)^6}{2^3} + \dots.$$

Solution Let $(x - 1)^2 = u$. Then, the given series is $1 + \frac{u}{2} + \frac{u^2}{2^2} + \frac{u^3}{2^3} + \dots.$ Here, $c_m = 1/2^m$. The radius of convergence of the series in u is

$$R = \lim_{m \to \infty} \left| \frac{c_m}{c_{m+1}} \right| = \lim_{m \to \infty} \left| \frac{2^{m+1}}{2^m} \right| = 2.$$

Now, $|u| < R = 2$, gives $|(x - 1)^2| = |x - 1|^2 < 2$, or $|x - 1| < \sqrt{2}$.

The following operations on power series are allowed.

1. A power series may be differentiated (integrated) term by term. If the given power series is convergent on the interval $|x - x_0| < R$, then the differentiated (integrated) series is also convergent on the same interval $|x - x_0| < R$.

2. We can add two power series term by term. If

$$s_1 = \sum_{m=0}^{\infty} b_m (x - x_0)^m, \quad s_2 = \sum_{m=0}^{\infty} c_m (x - x_0)^m \tag{6.13}$$

then
$$s = s_1 + s_2 = \sum_{m=0}^{\infty} (b_m + c_m)(x - x_0)^m.$$

Further, the series s is convergent for all x in each of the intervals of convergence of s_1 and s_2, that is, the intersection of the intervals of convergence of s_1 and s_2 is the interval of convergence of s. Therefore, if s_1 converges to y_1 in the interval I_1 and s_2 converges to y_2 in the interval I_2, then the sum $s_1 + s_2$ converges to $y_1 + y_2$ in the interval $I_1 \cap I_2$.

3. We can multiply two power series term by term. If the power series are as given in Eq. (6.13), then

$$s_1 s_2 = [b_0 + b_1(x - x_0) + b_2(x - x_0)^2 + \ldots] [c_0 + c_1(x - x_0) + c_2(x - x_0)^2 + \ldots]$$

$$= \sum_{i=0}^{\infty} (b_0 c_i + b_1 c_{i-1} + \ldots + b_{i-1} c_1 + b_i c_0)(x - x_0)^i.$$

If the series s_1 converges to y_1 in the interval I_1 and s_2 converges to y_2 in the interval I_2, then the product $s_1 s_2$ converges to $y_1 y_2$ in the interval $I_1 \cap I_2$.

Example 6.9 Write the power series expansion of $\cos \alpha x$, differentiate term by term and verify the derivative formula

$$\frac{d}{dx}[\cos(\alpha x)] = -\alpha \sin(\alpha x).$$

Solution The power series expansion of $\cos \alpha x$ is

$$\cos \alpha x = 1 - \frac{\alpha^2 x^2}{2!} + \frac{\alpha^4 x^4}{4!} - \ldots .$$

Differentiating the right hand side term by term, we obtain

$$\frac{d}{dx}\left[1 - \frac{\alpha^2 x^2}{2!} + \frac{\alpha^4 x^4}{4!} - \ldots \right] = -\alpha^2 x + \frac{\alpha^4 x^3}{3!} - \frac{\alpha^6 x^5}{5!} + \ldots$$

$$= -\alpha\left[\alpha x - \frac{\alpha^3 x^3}{3!} + \frac{\alpha^5 x^5}{5!} - \ldots \right] = -\alpha \sin(\alpha x).$$

Example 6.10 Expand $1/(1 - x^2)^{1/2}$ as a power series in terms of x, then integrate term by term to find a power series representation of $\sin^{-1} x$.

Solution The power series expansion of $1/(1 - x^2)^{1/2}$ can be written as (binomial expansion)

$$(1 - x^2)^{-1/2} = 1 + \frac{1}{2}(x^2) + \frac{(-1/2)(-3/2)}{2!} x^4 - \frac{(-1/2)(-3/2)(-5/2)}{3!} x^6 - \ldots$$

$$= 1 + \frac{1}{2}(x^2) + \frac{3}{8}(x^4) + \frac{15}{48}(x^6) + \ldots .$$

Alternately, the Maclaurin expansion can be written. Integrating term by term, we obtain

$$\int \frac{dx}{(1 - x^2)^{1/2}} = \int \left[1 + \frac{1}{2}x^2 + \frac{3}{8}x^4 + \frac{15}{48}x^6 + \ldots \right]dx$$

$$= x + \frac{x^3}{6} + \frac{3}{40}x^5 + \frac{5}{112}x^7 + \ldots = \sin^{-1} x.$$

The general term of the power series is $\dfrac{1 \cdot 3 \cdot 5 \cdot \ldots (2n - 1)}{n! \, 2^n} \, x^{2n}$. The general term of the integrated series is

$$\frac{1 \cdot 3 \cdot 5 \cdot \ldots (2n - 1)}{n! \, 2^n \, (2n + 1)} \, x^{2n+1}.$$

The radius of convergence of the integrated series is

$$R = \lim_{m \to \infty} \left| \frac{c_m}{c_{m+1}} \right| = \lim_{m \to \infty} \left| \frac{1 \cdot 3 \cdot 5 \ldots (2m - 1)}{m! \, 2^m \, (2m + 1)} \cdot \frac{(m + 1)! \, 2^{m+1} (2m + 3)}{1 \cdot 3 \cdot 5 \ldots (2m + 1)} \right|$$

$$= \lim_{m \to \infty} \left| \frac{2(m + 1)(2m + 3)}{(2m + 1)^2} \right| = 1.$$

The given function $(1 - x^2)^{-1/2}$ is valid in $I = (-1, 1)$. We find that the integrated series is also valid in I.

In the beginning of the section, we have presented a method for finding the power series solution. However, the coefficients in the power series can be obtained more elegantly by writing a recurrence relation between them. We shall illustrate this technique through examples.

Write the power series as

$$y(x) = \sum_{m=0}^{\infty} c_m (x - x_0)^m. \tag{6.14}$$

Then, we have

$$y'(x) = \sum_{m=1}^{\infty} m c_m (x - x_0)^{m-1}, \text{ and } y''(x) = \sum_{m=2}^{\infty} m(m - 1) c_m (x - x_0)^{m-2}. \tag{6.15}$$

Example 6.11 Find the power series solution about the origin of the equation

$$(1 - x^2) \, y'' - 4xy' + 2y = 0.$$

Solution Substituting the expressions for y, y', y'' from Eqs. (6.14) and (6.15) with $x_0 = 0$, we get

$$(1 - x^2) \sum_{m=2}^{\infty} m(m - 1) c_m x^{m-2} - 4x \sum_{m=1}^{\infty} m c_m x^{m-1} + 2 \sum_{m=0}^{\infty} c_m x^m = 0$$

or $\quad \displaystyle\sum_{m=2}^{\infty} m(m - 1) c_m x^{m-2} - \sum_{m=2}^{\infty} m(m - 1) c_m x^m - 4 \sum_{m=1}^{\infty} m c_m x^m + 2 \sum_{m=0}^{\infty} c_m x^m = 0.$

We note that the first term in each sum contains x^0, x^2, x and x^0 respectively. In order that these four terms can be combined into a single term, we require that the starting power of x in each sum should be the same. Therefore, we first remove some terms from the sums such that all of them have the same starting power of x. We write

$$2c_2 + 6c_3 x + \sum_{m=4}^{\infty} m(m - 1) c_m x^{m-2} - \sum_{m=2}^{\infty} m(m - 1) c_m x^m - 4c_1 x$$

$$- 4 \sum_{m=2}^{\infty} m c_m x^m + 2(c_0 + c_1 x) + 2 \sum_{m=2}^{\infty} c_m x^m = 0.$$

In the third term, set $m - 2 = t$, or $m = t + 2$. We have

$$2(c_2 + c_0) + 2(3c_3 - c_1)x + \sum_{t=2}^{\infty}(t+2)(t+1)c_{t+2}\, x^t$$

$$- \sum_{m=2}^{\infty} m(m-1)c_m x^m - 4 \sum_{m=2}^{\infty} mc_m x^m + 2 \sum_{m=2}^{\infty} c_m x^m = 0.$$

Since t is a dummy variable, we can write

$$2(c_2 + c_0) + 2(3c_3 - c_1)x + \sum_{m=2}^{\infty} [(m+2)(m+1)c_{m+2} - m(m-1)c_m - 4mc_m + 2c_m]x^m = 0$$

or $2(c_2 + c_0) + 2(3c_3 - c_1)x + \sum_{m=2}^{\infty} [(m+2)(m+1)c_{m+2} - (m^2 + 3m - 2)c_m]x^m = 0.$

This procedure is called the *shift of the summation index.* Setting the coefficients of successive powers of x to zero, we obtain

$$c_2 + c_0 = 0, \quad 3c_3 - c_1 = 0, \quad (m+2)(m+1)c_{m+2} = (m^2 + 3m - 2)c_m, \quad m \geq 2.$$

The solution is

$$c_2 = -c_0, c_3 = \frac{1}{3}c_1, \ c_{m+2} = \frac{(m^2 + 3m - 2)}{(m+2)(m+1)} c_m, \quad m \geq 2$$

where c_0 and c_1 are arbitrary constants. We find that

$$c_4 = \frac{2}{3}c_2 = -\frac{2}{3}c_0, \quad c_5 = \frac{4}{5}c_3 = \frac{4}{15}c_1, \dots.$$

The power series solution is

$$y(x) = c_0 \left[1 - x^2 - \frac{2}{3}x^4 - \dots\right] + c_1 \left[x + \frac{1}{3}x^3 + \frac{4}{15}x^5 + \dots\right].$$

Example 6.12   Find a fourth degree polynomial approximation (a power series about $x = 0$) to the solution of the initial value problem

$$y'' - y = 0, \quad y(0) = 2, \quad y'(0) = 0.$$

Solution   Using the power series expansions (6.14) and (6.15) with $x_0 = 0$, we get

$$\sum_{m=2}^{\infty} m(m-1) c_m x^{m-2} - \sum_{m=0}^{\infty} c_m x^m = 0.$$

In the first term set $m - 2 = t$, or $m = 2 + t$. Then, we have

$$\sum_{t=0}^{\infty} (t+2)(t+1)c_{t+2}\, x^t - \sum_{m=0}^{\infty} c_m x^m = 0.$$

Since t is a dummy variable, we can write

$$\sum_{m=0}^{\infty} [(m+2)(m+1)c_{m+2} - c_m]x^m = 0.$$

Setting the coefficients of successive powers of x to zero, we obtain

$$(m + 2)(m + 1)\, c_{m+2} = c_m, \quad \text{or} \quad c_{m+2} = \frac{c_m}{(m + 2)(m + 1)}, \quad m = 0, 1, 2. \ldots$$

where c_0 and c_1 are arbitrary constants. We have

$$c_2 = \frac{c_0}{2}, \quad c_3 = \frac{c_1}{6}, \quad c_4 = \frac{c_2}{12} = \frac{c_0}{24}, \quad c_5 = \frac{c_3}{20} = \frac{c_1}{120}, \ldots$$

The power series solution is

$$y(x) = c_0 \left[1 + \frac{x^2}{2!} + \frac{x^4}{4!} + \ldots \right] + c_1 \left[x + \frac{x^3}{3!} + \frac{x^5}{5!} + \ldots \right] = c_0 \cosh x + c_1 \sinh x.$$

The initial condition $y(0) = 2$, gives $c_0 = 2$. The initial condition $y'(0) = 0$, gives $c_1 = 0$. Therefore, the required fourth degree polynomial approximation to the solution is

$$y(x) \simeq 2 \left[1 + \frac{x^2}{2!} + \frac{x^4}{4!} \right].$$

Example 6.13 Find the first five non-vanishing terms in the power series solution of the initial value problem

$$(1 - x^2)\, y'' + 2xy' + y = 0, \; y(0) = 1, \; y'(0) = 1.$$

Solution Substituting the power series expansions (6.14) and (6.15) with $x_0 = 0$, in the given equation, we get

$$\sum_{m=2}^{\infty} m(m - 1)\, c_m x^{m-2} - \sum_{m=2}^{\infty} m(m - 1)\, c_m x^m + 2 \sum_{m=1}^{\infty} m c_m x^m + \sum_{m=0}^{\infty} c_m x^m = 0$$

or $\qquad 2\,c_2 + 6\,c_3 x + \sum_{m=4}^{\infty} m(m - 1)\, c_m x^{m-2} - \sum_{m=2}^{\infty} m(m - 1) c_m x^m + 2\,c_1 x$

$$+ 2 \sum_{m=2}^{\infty} m c_m x^m + c_0 + c_1 x + \sum_{m=2}^{\infty} c_m x^m = 0.$$

In the third term set $m - 2 = t$, or $m = 2 + t$. Since, t is a dummy variable, we obtain

$$(2\,c_2 + c_0) + 3(2c_3 + c_1)\, x + \sum_{m=2}^{\infty} [(m + 2)(m + 1)c_{m+2} - m(m - 1)c_m + 2mc_m + c_m]\, x^m = 0$$

or $\quad (2\,c_2 + c_0) + 3(2\,c_3 + c_1)x + \sum_{m=2}^{\infty} [(m + 2)(m + 1)c_{m+2} - (m^2 - 3m - 1)c_m]\, x^m = 0.$

Setting the coefficients of successive powers of x to zero, we obtain

$$2c_2 + c_0 = 0, \; 2c_3 + c_1 = 0, \; (m + 2)\, (m + 1)\, c_{m+2} = (m^2 - 3m - 1)\, c_m, \quad m \geq 2.$$

We have

$$c_2 = -\frac{1}{2}\, c_0, \; c_3 = -\frac{1}{2}\, c_1, \; c_{m+2} = \frac{(m^2 - 3m - 1)}{(m + 2)(m + 1)} c_m, \quad m \geq 2.$$

where c_0 and c_1 are arbitrary constants. We have $c_4 = -\frac{3}{12}\, c_2 = \frac{1}{8}\, c_0$. The power series solution is

$$y(x) = c_0 \left[1 - \frac{1}{2}\, x^2 + \frac{1}{8}\, x^4 - \ldots \right] + c_1 \left[x - \frac{1}{2}\, x^3 + \ldots \right].$$

The initial condition $y(0) = 1$ gives $c_0 = 1$ and $y'(0) = 1$ gives $c_1 = 1$. The first five non-vanishing terms in the power series solution are

$$y(x) = 1 + x - \frac{1}{2}x^2 - \frac{1}{2}x^3 + \frac{1}{8}x^4.$$

Example 6.14 Find the power series solution about the point $x_0 = 2$ of the equation

$$y'' + (x - 1)\,y' + y = 0.$$

Solution Write the power series as $y(x) = \sum\limits_{m=0}^{\infty} c_m (x - 2)^m$. Substituting in the given equation, we get

$$\sum_{m=2}^{\infty} m(m - 1)c_m (x - 2)^{m-2} + [(x - 2) + 1]\sum_{m=1}^{\infty} mc_m (x - 2)^{m-1} + \sum_{m=0}^{\infty} c_m (x - 2)^m = 0$$

or $\displaystyle \sum_{n=2}^{\infty} m(m-1)c_m (x - 2)^{m-2} + \sum_{m=1}^{\infty} mc_m (x - 2)^m + \sum_{m=1}^{\infty} mc_m (x - 2)^{m-1} + \sum_{m=0}^{\infty} c_m (x - 2)^m = 0$

or $\displaystyle 2c_2 + \sum_{m=3}^{\infty} m(m - 1)c_m (x - 2)^{m-2} + \sum_{m=1}^{\infty} mc_m (x - 2)^m$

$$+ c_1 + \sum_{m=2}^{\infty} mc_m (x - 2)^{m-1} + c_0 + \sum_{m=1}^{\infty} c_m (x - 2)^m = 0.$$

Set $m - 2 = t$ in the second term and $m - 1 = u$ in the fourth term. We obtain

$$(2c_2 + c_1 + c_0) + \sum_{t=1}^{\infty} (t + 2)(t + 1)c_{t+2} (x - 2)^t + \sum_{u=1}^{\infty} (u + 1)\,c_{u+1} (x - 2)^u$$

$$+ \sum_{m=1}^{\infty} (m + 1)\,c_m (x - 2)^m = 0.$$

Since t and u are dummy variables, we obtain

$$(2c_2 + c_1 + c_0) + \sum_{m=1}^{\infty} [(m + 2)(m + 1)c_{m+2} + (m + 1)c_{m+1} + (m + 1)\,c_m](x - 2)^m = 0.$$

Setting the coefficients of successive powers of x to zero, we obtain

$$2c_2 + c_1 + c_0 = 0, \quad c_{m+2} = -\frac{(m + 1)(c_{m+1} + c_m)}{(m + 2)(m + 1)} = -\frac{(c_{m+1} + c_m)}{m + 2}, \quad m \geq 1$$

where c_0 and c_1 are arbitrary constants. We obtain

$$c_2 = -\frac{1}{2}(c_0 + c_1),\; c_3 = -\frac{1}{3}(c_2 + c_1) = -\frac{1}{3}\left[-\frac{1}{2}(c_0 + c_1) + c_1\right] = -\frac{1}{6}[-c_0 + c_1], \ldots.$$

The power series solution is

$$y(x) = c_0 + c_1(x - 2) + c_2 (x - 2)^2 + c_3 (x - 2)^3 + \ldots$$

$$= c_0 + c_1 (x - 2) - \frac{1}{2}(c_0 + c_1)(x - 2)^2 + \frac{1}{6}(c_0 - c_1)(x - 2)^3 + \ldots$$

$$= c_0\left[1 - \frac{1}{2}(x - 2)^2 + \frac{1}{6}(x - 2)^3 - \ldots\right] + c_1\left[(x - 2) - \frac{1}{2}(x - 2)^2 - \frac{1}{6}(x - 2)^3 + \ldots\right].$$

Remark 2

Let $x = x_0$ be a regular singular point of Eq. (6.1). Then, a power series solution of Eq. (6.1) about $x = x_0$, does not exist. In other words, a power series is inadequate to represent a solution at a regular singular point. This implies that a solution may also contain negative powers and/or fractional powers of $x - x_0$. We shall illustrate this remark using the following equations.

Consider the Euler-Cauchy equation $x^2 y'' + 5xy' + 3y = 0$ whose solution is $y(x) = (A/x) + (B/x^3)$. We find that $x = 0$ is a regular singular point of the equation. The solution is not a power series about $x = 0$, as it contains negative powers of x.

Consider now the Euler-Cauchy equation $4x^2 y'' + 4xy' - y = 0$ whose solution is $y(x) = A\sqrt{x} + (B/\sqrt{x})$. Again, $x = 0$ is a regular singular point of the equation. The solution now contains positive and negative fractional powers of x and is not a power series about $x = 0$.

We shall discuss the method of finding series solutions about regular singular points in the next section.

Note that no series solution exists about an irregular singular point.

Exercise 6.2

Find the power series solutions about the origin of the following first order equations.

1. $y' + 3y = 0$.

2. $y' - 4y = 0$.

3. $y' = xy$.

4. $(1 - x^2) y' = 2xy$.

5. $(x - 1) y' = xy$.

6. $(1 + x) y' + xy = 0$.

7. $x(x + 1) y' - (2x + 1) y = 0$.

8. $xy' - (x + 2) y = 0$.

Find the power series solutions about the origin of the following second order equations.

9. $y'' = y$.

10. $y'' + 4y = 0$.

11. $y'' - xy' + y = 0$.

12. $y'' - 3x^2 y' = 0$.

13. $y'' + 2xy' + y = 0$.

14. $(1 - x^2) y'' - 2xy' + 6y = 0$.

15. $(1 + x^2) y'' - 9y = 0$.

16. $(4 + x^2)y'' - 6xy' + 8y = 0$.

17. $y'' + xy' - y = 0$.

18. $y'' + xy = 0$.

Find the power series solutions of the following equations about the given point.

19. $y' = 2y, x_0 = 1$.

20. $y'' - y = 0, x_0 = 1$.

21. $y'' + xy' + y = 0, x_0 = 2$.

22. $(x + 1)y' - (x + 2) y = 0, x_0 = -2$.

23. $xy' - y = 0, x_0 = 1$.

Solve the following initial value problems using the power series method (by expanding about the given initial point).

24. $y'' + 4y = 0, y(2) = 2, y'(2) = 2$.

25. $y'' - xy = 0, y(1) = 2, y'(1) = 0$.

26. $(2 + x^2) y'' - 2xy' + 3y = 0, y(1) = 1, y'(1) = -1$.

Write the power series expansions (about $x = 0$) of the given functions, differentiate term by term and verify the given differentiation formulas.

27. $[\sin(mx)]' = m \cos(mx)$.

28. $[1/(1 + x)]' = -1/(1 + x)^2$.

Expand the given function $f(x)$ as a power series in terms of $x - x_0$, then integrate term by term to find a power series representation of the second function $g(x)$ and find the radius of convergence.

29. $f(x) = 1/(a^2 + x^2)$, $x_0 = 0$, $g(x) = \dfrac{1}{a} \tan^{-1}\left(\dfrac{x}{a}\right)$.

30. $f(x) = 1/x$, $x_0 = 1$, $g(x) = \ln x$.

Find the general solution of the following equations using the power series method about the indicated point. Find the recurrence relations between the coefficients.

31. $y'' + \lambda^2 y = 0$, λ constant, $x_0 = 0$. **32.** $y'' - x^2 y = 0$, $x_0 = 0$.

33. $y'' + 2xy' + y = 0$, $x_0 = 0$. **34.** $(1 + x^2)\, y'' - 8y = 0$, $x_0 = 0$.

35. $(1 - x^2)y'' - 2xy' + 2y = 0$, $x_0 = 0$. **36.** $(1 + x^2)y'' + 6xy = 0$, $x_0 = 0$.

37. $(1 + x)y'' - 2y' + y = 0$, $x_0 = 0$. **38.** $y'' + xy' + y = 0$, $x_0 = 0$.

39. $y'' + xy = 0$, $x_0 = 1$. **40.** $y'' + xy' - y = 0$, $x_0 = 1$.

41. $xy'' = 2y$, $x_0 = 2$. **42.** $xy' - y = 0$, $x_0 = 3$

Using the power series method, solve the following initial value problems (write the expansion about the initial point). Find also the recurrence relation between the coefficients.

43. $y'' + 4y = 0$, $x_0 = 0$, $y(0) = 3$, $y'(0) = 4$. **44.** $y'' + 3xy' + 2y = 0$, $x_0 = 0$, $y(0) = 1$, $y'(0) = 0$.

45. $(1 + 2x)y'' + xy' + y = 0$, $x_0 = 1$, $y(1) = 1$, $y'(1) = -1$.

46. $(1 + x)y'' - xy' - y = 0$, $x_0 = 2$, $y(2) = 1$, $y'(2) = 0$.

Find a polynomial approximation of the fourth degree to the solution of the following initial value problems.

47. $(1 + x)y'' + y' + 3y = 0$, $y(0) = 1$, $y'(0) = 2$.

48. $y'' + xy' + (1 + x)y = 0$, $y(0) = -1$, $y'(0) = 0$.

49. $(1 - x)y'' + xy' + 2y = 0$, $y(0) = 2$, $y'(0) = 1$.

50. $(1 + 2x)y'' - y' + y = 0$, $y(0) = 0$, $y'(0) = 1$.

6.4 Series Solution about a Regular Singular Point: Frobenius Method

In the previous sections, we have discussed methods for finding series solution about an ordinary point, that is power series solution of the equation. We have also pointed out in section 6.3, that a power series is inadequate to represent a solution about a regular singular point $x = x_0$. In other words, the solution may also contain negative powers and/or fractional powers of $(x - x_0)$. In this section, we shall study the Frobenius method for obtaining a series solution about a regular singular point of the equation

$$A_0(x)y'' + A_1(x)y' + A_2(x)\, y = 0. \tag{6.16}$$

Let $x = x_0$ be a regular singular point of the equation, that is, $A_0(x_0) = 0$. Since x_0 is a regular singular point, we can write equation (6.16) as (see Eq. (6.3))

$$y'' + \frac{p_1(x)}{(x - x_0)}\, y' + \frac{q_1(x)}{(x - x_0)^2}\, y = 0 \tag{6.17}$$

where $\qquad p_1(x) = \dfrac{A_1(x)}{A_0(x)}\, (x - x_0)$ and $q_1(x) = \dfrac{A_2(x)}{A_0(x)}\, (x - x_0)^2$,

or as $\qquad (x - x_0)^2 y'' + (x - x_0)\, p_1(x)y' + q_1(x)y = 0 \tag{6.18}$

where $p_1(x)$ and $q_1(x)$ are analytic at $x = x_0$. We write the series solution of Eq. (6.18) in the form

$$y(x) = (x - x_0)^r [c_0 + c_1(x - x_0) + c_2(x - x_0)^2 + \ldots]$$

$$= (x - x_0)^r \sum_{m=0}^{\infty} c_m (x - x_0)^m = \sum_{m=0}^{\infty} c_m (x - x_0)^{m+r}, \quad c_0 \neq 0 \tag{6.19}$$

where r is any real number.

Since $p_1(x)$ and $q_1(x)$ are analytic at $x = x_0$, we first write the power series expansions of these functions as

$$p_1(x) = a_0 + a_1(x - x_0) + a_2(x - x_0)^2 + \ldots \tag{6.20}$$

$$q_1(x) = b_0 + b_1(x - x_0) + b_2(x - x_0)^2 + \ldots.$$

Now, differentiate the series expansion of $y(x)$, to obtain

$$y'(x) = rc_0(x - x_0)^{r-1} + (r+1)c_1(x - x_0)^r + \ldots = \sum_{m=0}^{\infty} (m+r)c_m(x - x_0)^{m+r-1}$$

$$y''(x) = r(r-1)c_0(x - x_0)^{r-2} + (r+1)rc_1(x - x_0)^{r-1} + \ldots$$

$$= \sum_{m=0}^{\infty} (m+r)(m+r-1)c_m(x - x_0)^{m+r-2}. \tag{6.21}$$

Substituting the expressions from Eqs. (6.20) and (6.21) in Eq. (6.18), we obtain

$$\sum_{m=0}^{\infty} (m+r)(m+r-1)c_m(x - x_0)^{m+r}$$

$$+ [a_0 + a_1(x - x_0) + a_2(x - x_0)^2 + \ldots] \sum_{m=0}^{\infty} (m+r)c_m(x - x_0)^{m+r}$$

$$+ [b_0 + b_1(x - x_0) + b_2(x - x_0)^2 + \ldots] \sum_{m=0}^{\infty} c_m(x - x_0)^{m+r} = 0.$$

We now collect the coefficients of various powers of $(x - x_0)$. We note that the lowest power of $(x - x_0)$ is r. Since the equation is an identity, the coefficients of various powers of $(x - x_0)$ should vanish. Equating the coefficient of the lowest power of $(x - x_0)$, that is the coefficient of $(x - x_0)^r$ to zero, we obtain

$$[r(r-1) + ra_0 + b_0]c_0 = 0.$$

Since $c_0 \neq 0$, we obtain

$$r(r-1) + a_0 r + b_0 = 0. \tag{6.22}$$

This equation is called the *indicial equation* of the differential equation. This quadratic equation has two roots $r = r_1$ and $r = r_2$. These roots are called the *indicial roots*. One of these roots gives a solution in the form (6.19). Let this solution be denoted by $y_1(x)$. The form of the second independent solution $y_2(x)$ depends on the value of the second indicial root. Equating the coefficients of the remaining powers of $(x - x_0)$ to zero, we obtain a recurrence relation relating the coefficients c_m. Usually, this relation is a two or three term relation. Substitution of the indicial roots r_1 and r_2 gives the values of the coefficients c_m. The general solution of the differential equation can now be written as

$$y(x) = Ay_1(x) + By_2(x). \tag{6.23}$$

where A and B are arbitrary constants. Depending upon the nature of the indicial roots, three different types of solution procedures are followed. The three cases are given by the following.

Case 1 $r_1 - r_2 \neq n$, n an integer. That is, the roots are distinct and do not differ by an integer. For example, we may have $r_1 = 1/2$, and $r_2 = 1$. In this case, the leading terms of the solution for $r = r_1$ and $r = r_2$ are $c_0(x - x_0)^{r_1}$ and $c_0(x - x_0)^{r_2}$ respectively. Therefore, the two solutions obtained by setting $r = r_1$ and $r = r_2$ respectively are linearly independent. The complete solution is given by Eq. (6.23), where $y_1(x)$, $y_2(x)$ are obtained from Eq. (6.19) by substituting $r = r_1$ and $r = r_2$ respectively.

Case 2 $r_1 - r_2 = 0$, that is $r = r_1$ is a double root. Now, substituting $r = r_1$ in Eq. (6.19), we obtain one of the solutions $y_1(x)$. The second solution can be obtained by setting $y_2(x) = u(x) y_1(x)$ and using the method of variation of parameters. However, there is an alternate, simple method to obtain the second linearly independent solution which we shall discuss below.

Case 3 $r_1 - r_2 = n$, n an integer. As in Case 2, we can use the method of variation of parameters for finding the second solution. We shall also discuss an alternate method for obtaining the second linearly independent solution.

We now discuss the above three cases in detail.

Case 1. Indicial roots are distinct and do not differ by an integer

Let r_1 and r_2 be the roots of the indicial equation, such that $r_1 - r_2 \neq n$, n an integer. Substituting the values of $r = r_1$ and r_2 in the recurrence relation for the coefficients, we obtain the values of the coefficients. Let these coefficients be denoted by $c_0, d_1, d_2, \ldots$ and $c_0^*, d_1^*, d_2^*, \ldots$ respectively. Note that $d_1, d_2, \ldots$ are expressed in terms of c_0 and $d_1^*, d_2^*, \ldots$ are expressed in terms of c_0^*. That is

$$y_1(x) = (x - x_0)^{r_1}[c_0 + d_1(x - x_0) + d_2(x - x_0)^2 + \ldots] \tag{6.24}$$

and
$$y_2(x) = (x - x_0)^{r_2}[c_0^* + d_1^*(x - x_0) + d_2^*(x - x_0)^2 + \ldots]. \tag{6.25}$$

Each of the solutions contains one arbitrary constant. The general solution is given by

$$y(x) = A y_1(x) + B y_2(x)$$

which contains two arbitrary constants.

Example 6.15 Find two linearly independent solutions of the equation

$$2x^2 y'' + xy' - (x^2 + 1)y = 0$$

using the Frobenius method. Determine the radius of convergence of each of the series (solutions).

Solution The point $x = 0$ is a regular singular point of the equation. Let the solution be written as

$$y(x) = \sum_{m=0}^{\infty} c_m x^{m+r}. \tag{6.26}$$

Differentiating, we obtain

$$y'(x) = \sum_{m=0}^{\infty} (m + r)c_m x^{m+r-1}, \quad y''(x) = \sum_{m=0}^{\infty} (m + r)(m + r - 1)c_m x^{m+r-2}. \tag{6.27}$$

Substituting in the differential equation and simplifying, we obtain

$$2 \sum_{m=0}^{\infty} (m+r)(m+r-1)c_m x^{m+r} + \sum_{m=0}^{\infty} (m+r)c_m x^{m+r} - \sum_{m=0}^{\infty} c_m x^{m+r+2} - \sum_{m=0}^{\infty} c_m x^{m+r} = 0.$$

The lowest degree term is the term containing x^r. Setting the coefficient of x^r to zero, we obtain

$$c_0[2r(r-1)+r-1] = 0, \quad \text{or} \quad c_0[2r^2 - r - 1] = 0, \quad c_0 \neq 0.$$

The indicial roots are obtained from the equation $2r^2 - r - 1 = 0$. The roots of this equation are $r = 1, -1/2$. The leading term in the third sum contains x^{r+2}. Therefore, collecting the coefficients of x^{r+1} from the remaining three sums and equating it to zero, we obtain

$$c_1[2(1+r)r+(1+r)-1] = 0, \quad \text{or} \quad c_1[2r^2 + 3r] = 0.$$

For $r = 1, -1/2$, we obtain $c_1 = 0$. For the remaining terms, we get

$$\sum_{m=2}^{\infty} [2(m+r)(m+r-1)+(m+r)-1] c_m x^{m+r} - \sum_{m=0}^{\infty} c_m x^{m+r+2} = 0.$$

Setting $m + 2 = t$ in the second term, we obtain

$$\sum_{m=2}^{\infty} [(m+r)(2m+2r-1)-1]c_m x^{m+r} - \sum_{t=2}^{\infty} c_{t-2} x^{r+t} = 0.$$

Since t is a dummy variable, we obtain

$$\sum_{m=2}^{\infty} \{[(m+r)(2m+2r-1)-1]c_m - c_{m-2}\}x^{m+r} = 0.$$

Setting the coefficient of x^{m+r} to zero, we obtain the recurrence relation

$$[(m+r)(2m+2r-1)-1] c_m - c_{m-2} = 0$$

or
$$c_m = \frac{c_{m-2}}{[(m+r)(2m+2r-1)-1]}, \quad m \geq 2.$$

Since $c_1 = 0$, we obtain $c_3 = 0 = c_5 = \ldots$

For $r = 1$, we get
$$c_m = \frac{c_{m-2}}{[(m+1)(2m+1)-1]},$$

$$c_2 = \frac{c_0}{14}, \quad c_4 = \frac{c_2}{44} = \frac{c_0}{616}, \ldots$$

For $r = -\frac{1}{2}$, we get
$$c_m = \frac{c_{m-2}}{[(m-1/2)(2m-2)-1]}$$

$$c_2 = \frac{c_0}{2}, \quad c_4 = \frac{c_2}{20} = \frac{c_0}{40}, \ldots$$

The two linearly independent solutions are

$$y_1(x) = c_0 x \left[1 + \frac{x^2}{14} + \frac{x^4}{616} + \cdots \right] = c_0 u(x)$$

and

$$y_2(x) = c_0 x^{-1/2} \left[1 + \frac{x^2}{2} + \frac{x^4}{40} + \cdots \right] = c_0 v(x).$$

The general solution is $y(x) = A_1 y_1(x) + B_1 y_2(x) = A u(x) + B v(x)$, where $A = A_1 c_0$ and $B = B_1 c_0$ are the arbitrary constants.

For convergence, we require that the ratio of successive terms satisfy the condition (ratio test)

$$\lim_{m \to \infty} \left| \frac{x^2}{(m+r)(2m+2r-1)-1} \right| < 1.$$

This condition is satisfied for all x since the limit is 0. Therefore, both the series are convergent in $(-\infty, \infty)$.

We can also obtain the radius of convergence of the series solutions by studying the behaviour of the coefficients in the differential equation. Write the equation as

$$y'' + \frac{1}{x}\left(\frac{1}{2}\right) y' - \frac{1}{x^2}\left(\frac{1+x^2}{2}\right) y = 0.$$

We find that $p_1(x) = 1/2$ and $q_1(x) = -(1+x^2)/2$ have only finite number of terms when looked as a power series and hence convergence is trivially true. Therefore, the Frobenius series solution is convergent for all x.

Case 2. Indicial roots are equal

In this case, we have $r_1 = r_2$. One of the linearly independent solutions is obtained by substituting $r = r_1$ in the series solution. The second solution can be obtained by setting $y_2(x) = u(x) y_1(x)$ and determining $u(x)$ by the method of variation of parameters. We shall illustrate an alternate simple procedure using the following example.

Example 6.16 Find the series solution, about $x = 0$, of the equation $xy'' + y' - xy = 0$, by the Frobenius method.

Solution We find that $x = 0$ is a regular singular point of the equation. Substituting the expressions for y, y' and y'' from Eqs. (6.26) and (6.27) in the given differential equation, we get

$$\sum_{m=0}^{\infty} (m+r)(m+r-1)c_m x^{m+r-1} + \sum_{m=0}^{\infty} (m+r)c_m x^{m+r-1} - \sum_{m=0}^{\infty} c_m x^{m+r+1} = 0$$

or

$$\sum_{m=0}^{\infty} (m+r)^2 c_m x^{m+r-1} - \sum_{m=0}^{\infty} c_m x^{m+r+1} = 0.$$

The lowest degree term is the term containing x^{r-1}. Setting the coefficient of x^{r-1} to zero, we obtain

$$c_0 r^2 = 0, \quad \text{or} \quad r = 0, 0, \text{ since } c_0 \neq 0.$$

The indicial root is a double root. The leading term in the second sum contains x^{r+1}. Therefore, setting the coefficient of x^r in the first sum to zero, we obtain

$$(r+1)^2 c_1 = 0, \quad \text{or} \quad c_1 = 0, \quad \text{since} \quad r = 0.$$

For the remaining terms, we obtain

$$\sum_{m=2}^{\infty} (m + r)^2 c_m x^{m+r-1} - \sum_{m=0}^{\infty} c_m x^{m+r+1} = 0.$$

Setting $m - 2 = t$, in the first sum we get

$$\sum_{t=0}^{\infty} (t + r + 2)^2 c_{t+2} x^{t+r+1} - \sum_{m=0}^{\infty} c_m x^{m+r+1} = 0.$$

Since t is a dummy variable, we get

$$\sum_{m=0}^{\infty} [(m + r + 2)^2 c_{m+2} - c_m]x^{m+r+1} = 0.$$

Setting the coefficient of x^{m+r+1} to zero, we obtain

$$c_{m+2} = \frac{c_m}{(m + r + 2)^2}, \quad m \geq 0.$$

Setting $m = 1, 3, 5, \ldots$ we obtain $c_3 = c_5 = \ldots = 0$. For $m = 2, 4, \ldots$, we get

$$c_2 = \frac{c_0}{(r + 2)^2}, \quad c_4 = \frac{c_2}{(r + 4)^2} = \frac{c_0}{(r + 2)^2 (r + 4)^2},$$

$$c_6 = \frac{c_4}{(r + 6)^2} = \frac{c_0}{(r + 2)^2 (r + 4)^2 (r + 6)^2}, \ldots$$

Therefore,

$$y(x) = c_0 x^r \left[1 + \frac{x^2}{(r + 2)^2} + \frac{x^4}{(r + 2)^2 (r + 4)^2} + \ldots \right]. \tag{6.28}$$

For $r = 0$, we obtain one of the linearly independent solutions as

$$y_1(x) = c_0 \left[1 + \frac{x^2}{2^2} + \frac{x^4}{2^2 \cdot 4^2} + \frac{x^6}{2^2 \cdot 4^2 \cdot 6^2} + \ldots \right] = c_0 v(x).$$

Let us now substitute $y(x)$ given by Eq. (6.28) in the differential equation. We obtain

$$xy'' + y' - xy = c_0 r^2 x^{r-1} \tag{6.29}$$

(where the right hand side is simply the indicial equation). Consider now r as a parameter (variable). Differentiating Eq. (6.29) partially with respect to r, we obtain

$$\frac{\partial}{\partial r} [xy'' + y' - xy] = \frac{\partial}{\partial r} [c_0 r^2 x^{r-1}]$$

or

$$\left(x \frac{d^2}{dx} + \frac{d}{dx} - x \right) \left(\frac{\partial y}{\partial r} \right) = c_0 [2rx^{r-1} + r^2 x^{r-1} \ln x] \tag{6.30}$$

where we assume that the interchange of differential operators is allowed. As the right hand side of Eq. (6.30) vanishes at $r = 0$, we conclude that $(\partial y/\partial r)_{r=0}$ is also a solution of the differential equation. Differentiating Eq. (6.28) partially with respect to r, we obtain

$$\frac{\partial y}{\partial r} = c_0 x^r (\ln x) \left[1 + \frac{x^2}{(r + 2)^2} + \frac{x^4}{(r + 2)^2 (r + 4)^2} + \ldots \right]$$

$$+ c_0 x^r \left[-\frac{2x^2}{(r + 2)^3} - x^4 \left\{ \frac{2}{(r + 2)^3 (r + 4)^2} + \frac{2}{(r + 2)^2 (r + 4)^3} \right\} - \ldots \right]$$

$$= (\ln x) y(x) + c_0 x^r \left[-\frac{2x^2}{(r + 2)^3} - 2x^4 \left\{ \frac{1}{(r + 2)^3 (r + 4)^2} + \frac{1}{(r + 2)^2 (r + 4)^3} \right\} - \cdots \right].$$

Setting $r = 0$, the second linearly independent solution is obtained as

$$y_2(x) = \left(\frac{\partial y}{\partial r} \right)_{r=0} = (\ln x) y_1(x) - c_0 \left[\frac{x^2}{4} + \frac{3}{128} x^4 + \cdots \right]$$

$$= c_0 \left[v(x) \ln x - \left\{ \frac{x^2}{4} + \frac{3}{128} x^4 + \cdots \right\} \right] = c_0 u(x).$$

The general solution is $y(x) = A_1 y_1(x) + B_1 y_2(x) = A v(x) + B u(x)$, where $A = A_1 c_0$ and $B = B_1 c_0$ are the arbitrary constants.

Remark 3

When the indicial equation has a double root $r_1 = r_2 = p$, then the two linearly independent solutions of Eq. (6.18) are given by $y_1(x) = [y(x)]_{r=p}$ and $y_2(x) = [\partial y/\partial r]_{r=p}$. The second solution is always of the form

$$y_2(x) = \left(\frac{\partial y}{\partial r} \right)_{r=p} = y_1(x) \ln (x - x_0) + (x - x_0)^p [d_0 + d_1 (x - x_0) + d_2 (x - x_0)^2 + \ldots], x > x_0.$$

$$(6.31)$$

where x_0 is the regular singular point about which the series expansion is being written.

Case 3. Indicial roots differ by an integer

Let the roots of the indicial equation be r_1, r_2 such that $r_1 - r_2 = n$, an integer. In this case, two types of problems are encountered. In one type, the two linearly independent solutions are obtained by substituting $r = r_1$ and $r = r_2$ in $y(x)$ as in Case 1. Let us first illustrate an example of this type.

Example 6.17   Find a series solution of the equation $x^2 y'' + x^3 y' + (x^2 - 2) y = 0$, about the point $x = 0$.

Solution   We find that $x = 0$ is a regular singular point of the equation. Therefore, Frobenius series solution can be obtained. Substituting the expressions for y, y' and y'' from Eqs. (6.26) and (6.27) in the given differential equation, we get

$$\sum_{m=0}^{\infty} (m + r)(m + r - 1) c_m x^{m+r} + \sum_{m=0}^{\infty} (m + r) c_m x^{m+r+2} + \sum_{m=0}^{\infty} c_m x^{m+r+2} - 2 \sum_{m=0}^{\infty} c_m x^{m+r} = 0$$

or

$$\sum_{m=0}^{\infty} [(m + r)(m + r - 1) - 2] c_m x^{m+r} + \sum_{m=0}^{\infty} (m + r + 1) c_m x^{m+r+2} = 0.$$

The lowest degree term is the term containing x^r. Setting the coefficient of x^r to zero, we obtain

$$c_0 [r(r - 1) - 2] = 0, \quad \text{or} \quad r^2 - r - 2 = (r - 2)(r + 1) = 0, \quad \text{or} \quad r = 2, -1,$$

since $c_0 \neq 0$. The indicial roots differ by an integer. The leading term in the second sum contains x^{r+2}. Therefore, setting the coefficient of x^{r+1} in the first sum to zero, we get

$$[(r + 1) r - 2] c_1 = 0, \quad \text{or} \quad c_1 = 0 \text{ for } r = 2, -1.$$

The remaining terms are given by

$$\sum_{m=2}^{\infty} [(m + r)(m + r - 1) - 2]c_m x^{m+r} + \sum_{m=0}^{\infty} (m + r + 1)c_m x^{m+r+2} = 0.$$

Setting $m - 2 = t$ in the first sum, we get

$$\sum_{t=0}^{\infty} [(t + r + 2)(t + r + 1) - 2]c_{t+2} x^{r+t+2} + \sum_{m=0}^{\infty} (m + r + 1)c_m x^{m+r+2} = 0.$$

Since t is a dummy variable, we obtain

$$\sum_{m=0}^{\infty} [\{(m + r + 2)(m + r + 1) - 2\}c_{m+2} + (m + r + 1)c_m]x^{m+r+2} = 0.$$

Setting the coefficient of x^{m+r+2} to zero, we get $\quad c_{m+2} = -\dfrac{(m + r + 1)c_m}{(m + r + 1)(m + r + 2) - 2}, \; m \geq 0,$

We have

$$c_2 = -\frac{(r + 1)c_0}{(r + 1)(r + 2) - 2}, \; c_3 = -\frac{(r + 2)c_1}{(r + 2)(r + 3) - 2} = 0, \; c_4 = -\frac{(r + 3)c_2}{(r + 3)(r + 4) - 2}, \ldots$$

For $r = 2$, we get $c_2 = -\dfrac{3}{10} c_0, \; c_4 = -\dfrac{5}{28} c_2 = \dfrac{3}{56} c_0, \ldots$

One of the linearly independent solutions is

$$y_1(x) = c_0 x^2 \left[1 - \frac{3}{10} x^2 + \frac{3}{56} x^4 - \ldots\right] = c_0 v(x).$$

For $r = -1$, we get $c_2 = 0, c_3 = 0 = c_4 = \ldots$

The second linearly independent solution is $y_2(x) = \dfrac{c_0}{x} = c_0 u(x)$. The general solution is

$$y(x) = A_1 y_1(x) + B_1 y_2(x) = Av(x) + Bu(x) = A\left[x^2 - \frac{3}{10} x^4 + \frac{3}{56} x^4 - \ldots\right] + \frac{B}{x}$$

where $A = A_1 c_0, B = B_1 c_0$ are arbitrary constants.

In the second type of problems, two cases arise. *In the first case, one of the coefficients becomes infinite (undefined) for one of the indicial roots. The other coefficients connected to this coefficient through the recurrence relation also become infinite.* This type of problem can be tackled in two ways. Let us illustrate through an example.

Example 6.18 Find the series solution about $x = 0$, of the differential equation

$$x(1 + x) y'' + 3xy' + y = 0.$$

Solutions We find that $x = 0$ is a regular singular point of the equation. Therefore, Frobenius series solution can be obtained. Substituting the expressions for y, y' and y'' from Eqs. (6.26) and (6.27) in the given differential equation, we get

$$\sum_{m=0}^{\infty} (m + r)(m + r - 1)c_m x^{m+r-1} + \sum_{m=0}^{\infty} (m + r)(m + r - 1)c_m x^{m+r}$$

$$+ 3 \sum_{m=0}^{\infty} (m + r)c_m x^{m+r} + \sum_{m=0}^{\infty} c_m x^{m+r} = 0.$$

The lowest degree term is the term containing x^{r-1}. Setting the coefficient of x^{r-1} to zero, we obtain

$$c_0 r(r-1) = 0, \quad \text{or} \quad r = 0, 1, \text{ since } c_0 \neq 0. \tag{6.32}$$

Combining the remaining terms, we obtain

$$\sum_{m=1}^{\infty} (m+r)(m+r-1)c_m x^{m+r-1} + \sum_{m=0}^{\infty} [(m+r)(m+r-1) + 3(m+r) + 1]\, c_m x^{m+r} = 0.$$

Letting $m - 1 = t$ in the first sum, and changing the dummy variable t to m, we get

$$\sum_{m=0}^{\infty} [(m+r+1)(m+r)c_{m+1} + \{(m+r)(m+r+2) + 1\}c_m]x^{m+r} = 0.$$

Setting the coefficient of x^{m+r} to zero, we obtain

$$c_{m+1} = -\frac{(m+r)(m+r+2) + 1}{(m+r)(m+r+1)}\, c_m = -\frac{(m+r+1)^2}{(m+r)(m+r+1)}\, c_m = -\frac{(m+r+1)}{(m+r)}\, c_m, \quad m \geq 0.$$

We have

$$c_1 = -\frac{r+1}{r}\, c_0, \ c_2 = -\frac{r+2}{r+1}\, c_1 = \frac{r+2}{r}\, c_0, \ c_3 = -\frac{r+3}{r+2}\, c_2 = -\frac{r+3}{r}\, c_0, \dots.$$

The solution can be written as

$$y(x) = c_0 x^r \left[1 - \frac{r+1}{r}x + \frac{r+2}{r}x^2 - \frac{r+3}{r}x^3 + \dots \right].$$

We note that c_1 becomes infinite when $r = 0$, and hence all the remaining coefficients also become infinite. We now modify the above procedure to obtain the solution. The series solution was written as

$$y(x) = x^r [c_0 + c_1 x + c_2 x^2 + \dots], \quad c_0 \neq 0. \tag{6.33}$$

We modify the series (Eq. (6.33)) by assuming $c_0 = a_0 r$, $a_0 \neq 0$. With this assumption, we get

$$c_1 = -\frac{r+1}{r}\, c_0 = -\frac{r+1}{r}\, a_0 r = -(r+1)a_0, \quad c_2 = -\frac{r+2}{r+1}\, c_1 = (r+2)a_0, \dots$$

and all the coefficients are now well defined. Therefore, the series solution is now assumed as

$$y(x) = x^r [a_0 r + c_1 x + c_2 x^2 + \dots], \quad a_0 \neq 0, \tag{6.34}$$

In general, if the coefficient c_1 becomes infinite at $r = r_0$, then we replace c_0 by $c_0 = a_0(r - r_0)$, $a_0 \neq 0$ so that all the coefficients are well defined.

The indicial equation (Eq. (6.32)) now becomes $a_0 r^2 (r-1) = 0$ and the solution is given by

$$y(x) = a_0 x^r [r - (r+1)x + (r+2)x^2 - (r+3)x^3 + \dots] \tag{6.35}$$

If we substitute this solution in the differential equation, we obtain

$$x(1+x)y'' + 3xy' + y = a_0 r^2 (r-1). \tag{6.36}$$

The right hand side is the indicial equation which vanishes when $r = 0$. Therefore, $[y(x)]_{r=0}$ given by Eq. (6.35), is a solution of the differential equation. Setting $r = 0$ in Eq. (6.35), we obtain

$$y_1(x) = a_0[-x + 2x^2 - 3\,x^3 + \ldots] = -a_0 x[1 - 2x + 3x^2 - \ldots] = -a_0 x(1+x)^{-2}.$$

For $r = 1$, we get from Eq. (6.35)

$$y(x) = a_0 x[1 - 2x + 3x^2 - \ldots] = a_0 x\,(1+x)^{-2} = -y_1(x).$$

Often, the solution corresponding to the second indicial root becomes a part of one of the linearly independent solutions.

Differentiating Eq. (6.36) partially with respect to r, we get

$$\frac{\partial}{\partial r}\left[x(1+x)\frac{d^2 y}{dx^2} + 3x\frac{dy}{dx} + y\right] = \frac{\partial}{\partial r}\,[a_0 r^2\,(r-1)] = a_0[2r(r-1) + r^2]$$

or
$$\left[x(1+x)\frac{d^2}{dx^2} + 3x\frac{d}{dx} + 1\right]\left(\frac{\partial y}{\partial r}\right) = a_0\,(3r^2 - 2r).$$

The right hand side vanishes when $r = 0$. Therefore, $[\partial y/\partial r]_{r=0}$ is also a solution of the differential equation.

Now, $\quad \dfrac{\partial y}{\partial r} = a_0 x^r\,(\ln x)\,[r - (r+1)x + (r+2)x^2 - \ldots] + a_0 x^r[1 - x + x^2 - x^3 + \ldots].$

Putting $r = 0$ in this equation, we obtain the second linearly independent solution as

$$y_2(x) = \left(\frac{\partial y}{\partial r}\right)_{r=0} = (\ln x)y_1(x) + a_0[1 - x + x^2 - x^3 + \ldots] = (\ln x)y_1(x) + a_0(1+x)^{-1}$$

The general solution is given by

$$y(x) = A_1 y_1(x) + B_1 y_2(x) = -A x(1+x)^{-2} + B[-x(1+x)^{-2}\ln x + (1+x)^{-1}]$$

where $A = A_1 a_0$ and $B = B_1 a_0$ are arbitrary constants.

Therefore, in problems of the above type, one of the coefficients becomes infinite for one of the indicial roots, say $r = r_0$. Then, we set $c_0 = a_0(r - r_0)$, $a_0 \neq 0$. One of the linearly independent solutions is given by $y_1(x) = [y(x)]_{r=r_0}$ and the second linearly independent solution is given by $y_2(x) = (\partial y/\partial r)_{r=r_0}$. This solution always contains a logarithmic term. The solution obtained by setting the second indicial root $r = r_1$ in $y(x)$ produces a linearly dependent solution. It is usually a multiple of $y_1(x)$ or a part of $y_2(x)$.

In the second case, one of the coefficients becomes indeterminate for one of the indicial roots, say $r = r_0$. This root produces the complete solution as it contains two arbitrary constants. The solution obtained by setting the second indicial root $r = r_1$ in $y(x)$ produces a linearly dependent solution.

We illustrate this case through the following example.

Example 6.19 Find the series solution about the origin, of the differential equation

$$x^2 y'' + 6xy' + (6 + x^2)\,y = 0.$$

Solution We find that $x = 0$ is a regular singular point of the equation. Therefore, Frobenius series solution can be obtained. Substituting the expressions for y, y' and y'' from Eqs. (6.26) and (6.27) in the given differential equation, we get

$$\sum_{m=0}^{\infty} (m+r)(m+r-1)c_m x^{m+r} + 6 \sum_{m=0}^{\infty} (m+r)c_m x^{m+r} + 6 \sum_{m=0}^{\infty} c_m x^{m+r} + \sum_{m=0}^{\infty} c_m x^{m+r+2} = 0.$$

or
$$\sum_{m=0}^{\infty} [(m+r)(m+r+5)+6]c_m x^{m+r} + \sum_{m=0}^{\infty} c_m x^{m+r+2} = 0.$$

The lowest degree term is the term containing x^r. Setting the coefficient of x^r to zero, we obtain

$$[r(r+5)+6]\, c_0 = 0, \quad \text{or} \quad (r+2)(r+3) = 0, \, c_0 \neq 0.$$

The indicial roots are $r = -2, -3$. Setting the coefficient of x^{r+1} to zero, we get

$$[(r+1)(r+6)+6]\, c_1 = 0.$$

For $r = -2$, $c_1 = 0$ and for $r = -3$, c_1 is arbitrary. Therefore, the indicial root $r = -3$ gives the complete solution as the corresponding solution contains two arbitrary constants. The remaining terms are

$$\sum_{m=2}^{\infty} [(m+r)(m+r+5)+6]c_m x^{m+r} + \sum_{m=0}^{\infty} c_m x^{m+r+2} = 0.$$

Setting $m - 2 = t$ in the first sum and changing the dummy variable t to m, we get

$$\sum_{m=0}^{\infty} \left\{ [(m+r+2)(m+r+7)+6]c_{m+2} + c_m \right\} x^{m+r+2} = 0.$$

Setting the coefficient of x^{m+r+2} to zero, we get

$$c_{m+2} = -\frac{c_m}{(m+r+2)(m+r+7)+6}, \quad m \geq 0.$$

We have

$$c_2 = -\frac{c_0}{(r+2)(r+7)+6}, \quad c_3 = -\frac{c_1}{(r+3)(r+8)+6}, \quad c_4 = -\frac{c_2}{(r+4)(r+9)+6}, \cdots$$

For $r = -3$, we get

$$c_2 = -\frac{c_0}{2}, \quad c_3 = -\frac{c_1}{6}, \quad c_4 = -\frac{c_2}{12} = \frac{c_0}{24}, \quad c_5 = -\frac{c_3}{20} = \frac{c_1}{120}, \cdots$$

The solution is given by

$$y(x) = x^{-3} \left[c_0 \left(1 - \frac{x^2}{2!} + \frac{x^4}{4!} - \cdots \right) + c_1 \left(x - \frac{x^3}{3!} + \frac{x^5}{5!} - \cdots \right) \right]$$

$$= c_0 \left(\frac{\cos x}{x^3} \right) + c_1 \left(\frac{\sin x}{x^3} \right) = c_0 y_1(x) + c_1 y_2(x).$$

For $r = -2$, we get

$$c_1 = 0, \quad c_2 = -\frac{c_0}{6}, \quad c_3 = -\frac{c_1}{12} = 0, \quad c_4 = -\frac{c_2}{20} = \frac{c_0}{120}, \cdots$$

Therefore, the solution is

$$y^*(x) = c_0 x^{-2} \left[1 - \frac{x^2}{3!} + \frac{x^4}{5!} - \cdots \right] = c_0 x^{-3} \left[x - \frac{x^3}{3!} + \frac{x^5}{5!} - \cdots \right] = c_0 y_2(x).$$

We find that the indicial root $r = -2$ produced a linearly dependent solution.

Remark 4

We know that if a series solution is written about an ordinary point, then we obtain a power series solution. Now, what happens if we attempt to find a series solution of Frobenius type about an ordinary point. The index r has to be a non-negative integer since Frobenius solution should be same as the power series solution. In this case, one of the indicial roots, say $r = r_0$ produces the complete solution as it contains two arbitrary constants. The solution obtained by setting the second indicial root $r = r_1$ in $y(x)$ produces a linearly dependent solution. Let us illustrate this case by an example.

Example 6.20 Find the Frobenius series solution about $x = 0$, of the equation

$$(1 - x^2)y'' - 2xy' + 6y = 0.$$

Solution The point $x = 0$ is a regular point of the differential equation. Substituting

$$y(x) = \sum_{m=0}^{\infty} c_m x^{m+r}, \; y'(x) = \sum_{m=0}^{\infty} (m + r) c_m x^{m+r-1},$$

$$y''(x) = \sum_{m=0}^{\infty} (m + r)(m + r - 1) c_m x^{m+r-2}$$

in the given equation, we obtain

$$\sum_{m=0}^{\infty} (m + r)(m + r - 1) c_m x^{m+r-2} - \sum_{m=0}^{\infty} (m + r)(m + r - 1) c_m x^{m+r}$$

$$-2 \sum_{m=0}^{\infty} (m + r) c_m x^{m+r} + 6 \sum_{m=0}^{\infty} c_m x^{m+r} = 0.$$

The lowest degree term is the term containing x^{r-2}. Setting the coefficient of x^{r-2} to zero, we get

$$c_0 r(r - 1) = 0, \; c_0 \neq 0, \; \text{giving} \; r = 0, 1.$$

Setting the coefficient of x^{r-1} to zero, we obtain

$$c_1 r(r + 1) = 0.$$

For $r = 1$, $c_1 = 0$ and for $r = 0$, c_1 is arbitrary. We shall show that $r = 0$ gives the complete solution. Combining the remaining terms, we get

$$\sum_{m=2}^{\infty} (m + r)(m + r - 1) c_m x^{m+r-2} - \sum_{m=0}^{\infty} [(m + r)(m + r - 1) + 2(m + r) - 6] c_m x^{m+r} = 0.$$

Letting $m - 2 = t$ in the first sum and changing the dummy variable t to m, we get

$$\sum_{m=0}^{\infty} [(m + r + 2)(m + r + 1) c_{m+2} - \{(m + r)(m + r + 1) - 6\} c_m] x^{m+r} = 0.$$

Setting the coefficient of x^{m+r} to zero, we obtain

$$c_{m+2} = \frac{(m + r)(m + r + 1) - 6}{(m + r + 1)(m + r + 2)} c_m, \quad m \geq 0.$$

We have for $r = 0$, $\qquad c_{m+2} = \dfrac{m(m + 1) - 6}{(m + 1)(m + 2)} c_m, \; m \geq 0.$

Therefore,

$$c_2 = -3c_0, \ c_3 = -\frac{2}{3}c_1, \ c_4 = 0, \ c_5 = \frac{3}{10}c_3 = -\frac{1}{5}c_1, \ c_6 = 0 = c_8 = \ldots.$$

The solution is given by

$$y(x) = c_0(1 - 3x^2) + c_1\left(x - \frac{2}{3}x^3 - \frac{1}{5}x^5 - \ldots\right). \tag{6.37}$$

For $r = 1$, we have

$$c_1 = 0 \quad \text{and} \quad c_{m+2} = \frac{(m+1)(m+2) - 6}{(m+2)(m+3)} c_m, \quad m \geq 0.$$

We have $c_2 = -\frac{2}{3}c_0, c_4 = \frac{3}{10} c_2 = -\frac{1}{5} c_0, \ldots, c_3 = 0 = c_5 = \ldots$ Therefore, we have

$$y_2(x) = c_0 x\left[1 - \frac{2}{3}x^2 - \frac{1}{5}x^4 - \ldots\right] = c_0\left[x - \frac{2}{3}x^3 - \frac{1}{5}x^5 - \ldots\right]$$

But, this solution is a constant multiple of the second solution in Eq. (6.37). The singular points of the equation are $x = \pm 1$, and the series expansion is written about $x = 0$. Therefore, the radius of convergence is $R = 1$.

Exercise 6.3

Find the series solutions of the following differential equations by the Frobenius method. Wherever possible, identify the series as expansions of known functions.

1. $x^2 y'' + xy' + (x^2 - n^2) y = 0$, n a fraction.
2. $2x(1 + x)y'' + (1 + x) y' - 3y = 0$.
3. $9x(1 + x) y'' - 6y' + 2y = 0$.
4. $4x^2 y'' - 8xy' + 5y = 0$.
5. $3x(1 - x)y'' + 2(1 - x)y' + 4y = 0$.
6. $2x(1 - x)y'' + (1 - x) y' + 3y = 0$.
7. $2x^2(1 + x^2)y'' - 3x(1 + x^2)y' + 2y = 0$.
8. $xy'' + (1 - 2x)y' + (x - 1) y = 0$.
9. $xy'' + (1 - x)y' + 3y = 0$.
10. $xy'' + y' + xy = 0$.
11. $(x + x^2)y'' + (1 + x)y' - y = 0$.
12. $xy'' + (1 + x)y' - 2y = 0$.
13. $4(x^2 + x^4)y'' + 16x^3 y' + y = 0$.
14. $xy'' - y = 0$.
15. $2x^2 y'' + xy' - 3y = 0,.$
16. $16x^2 y'' + 3y = 0$.
17. $x^2 y'' + 4xy' + (x^2 + 2) y = 0$.
18. $y'' + x^2 y = 0$.
19. $x^2 y'' + xy' + (x^2 - 1)y = 0$.
20. $x^2 y''. + 6xy + (6 - 4x^2)y = 0$.

21. Does the equation $x^4 y'' + 2x^3 y' - 4y = 0$ have a Frobenius series solution about the origin? Transform the equation using the substitution $x = 1/(2z)$. Does a series solution about $z = 0$ exist for the transformed equation?

22. Show that the equation $x^2 y'' - (1 - 2x)y' + y = 0$ has no solution that is regular in ascending powers of x, since the series that is obtainable diverges for all x. Why it does not have a solution?

Find the series solutions about the indicated point of the following differential equations by Frobenius method.

23. $2(1 - x)y'' - xy' + y = 0$, $x = 1$.
24. $9x(1 + x)y'' - 6y' + 2y = 0$, $x = -1$.
25. $x(x - 2)y'' + 4y' + 3y = 0$, $x = 2$.

6.5   Answers and Hints

Exercise 6.1

1. $x = 0$, $p_1(x) = 1 + x$, $q_1(x) = -1$, regular singular point.

2. $x = 0$, regular singular point.

3. $x = 0$, $p_1(x) = -5/x$, $q_1(x) = 3x^2$, irregular singular point.

4. $x = 0$, regular singular point.   **5.** $x = 0$, regular singular point.

6. $x = 0$, irregular singular point, $x = \pm 1$, regular singular points.

7. $x = 1$, irregular singular point, $x = -2$, regular singular point.

8. $x = 0$, regular singular point.   **9.** $x = 0$, irregular singular point.

10. $x = 0$, irregular singular point.

Exercise 6.2

1. $mc_m + 3\,c_{m-1} = 0$, $y = c_0\left(1 - 3x + \dfrac{3^2}{2!}x^2 - \dfrac{3^3}{3!}x^3 - + \ldots\right) = c_0 e^{-3x}$.

2. $mc_m - 4c_{m-1} = 0$, $y = c_0\left(1 + 4x + \dfrac{4^2}{2!}x^2 + \dfrac{4^3}{3!}x^3 + \ldots\right) = c_0 e^{4x}$.

3. $c_1 = c_3 = \ldots = 0$, $mc_m - c_{m-2} = 0$, $m = 2, 4, 6, \ldots$,

$$y(x) = c_0\left[1 + (x^2/2) + \frac{(x^2/2)^2}{2!} + \frac{(x^2/2)^3}{3!} + \ldots\right] = c_0 e^{x^2/2}.$$

4. $c_1 = c_3 = \ldots 0$, $c_0 = c_2 = c_4, \ldots$, $y(x) = c_0(1 + x^2 + x^4 + \ldots) = c_0/(1 - x^2)$.

5. $c_1 = 0$, $y(x) = c_0\left[1 - \dfrac{x^2}{2} - \dfrac{x^3}{3} - \dfrac{x^4}{8} - \ldots\right] = c_0\,(1 - x)\,e^x$.

6. $c_1 = 0$, $y(x) = c_0\left[1 - \dfrac{x^2}{2} + \dfrac{x^3}{3} - \dfrac{x^4}{8} + \ldots\right] = c_0\,(1 + x)e^{-x}$.

7. $c_0 = c_3 = c_5 = \ldots = 0$, $c_2 = c_1$, $y(x) = c_1(x + x^2)$.

8. $c_0 = c_1 = 0$, $c_3 = c_2$, $c_4 = c_2/2$, $c_5 = c_2/6$, $y(x) = c_2\left(x^2 + x^3 + \dfrac{x^4}{2!} + \dfrac{x^5}{3!} + \ldots\right) = c_2 x^2 e^x$.

9. $y(x) = c_0\left(1 + \dfrac{x^2}{2!} + \dfrac{x^4}{4!} + \ldots\right) + c_1\left(x + \dfrac{x^3}{3!} + \dfrac{x^5}{5!} + \ldots\right) = Ae^x + Be^{-x}$, where

$A = (c_0 + c_1)/2$, $B = (c_0 - c_1)/2$.

10. $y(x) = c_0\left(1 - 2x^2 + \dfrac{2}{3}x^4 + \ldots\right) + c_1\left(x - \dfrac{2}{3}x^3 + \dfrac{2}{15}x^5 + \ldots\right)$

$$= c_0\left[1 - \frac{(2x)^2}{2!} + \frac{(2x)^4}{4!} + \ldots\right] + \frac{1}{2}c_1\left[(2x) - \frac{(2x)^3}{3!} + \frac{(2x)^5}{5!} + \ldots\right] = c_0\cos 2x + (c_1/2)\sin 2x.$$

11. $y(x) = c_0\left[1 - \dfrac{x^2}{2} - \dfrac{x^4}{24} - \dfrac{x^6}{240} - \ldots\right] + c_1 x$.

12. $y(x) = c_0 + c_1\left[x + \dfrac{x^4}{4} + \dfrac{x^7}{14} + \ldots\right].$

13. $y(x) = c_0\left[1 - \dfrac{x^2}{2} + \dfrac{5x^4}{24} - \ldots\right] + c_1\left[x - \dfrac{x^3}{2} + \dfrac{7x^5}{40} - \ldots\right].$

14. $y(x) = c_0(1 - 3x^2) + c_1\left(x - \dfrac{2x^3}{3} - \dfrac{1}{5}x^5 - \ldots\right).$

15. $y(x) = c_0\left[1 + \dfrac{9}{2}x^2 + \dfrac{21}{8}x^4 + \ldots\right] + c_1\left[x + \dfrac{3}{2}x^3 + \dfrac{9}{40}x^5 + \ldots\right].$

16. $y(x) = c_0\left[1 - x^2 - \dfrac{1}{24}x^4 - \ldots\right] + c_1\left[x - \dfrac{1}{12}x^3 - \dfrac{1}{240}x^5 - \ldots\right].$

17. $y(x) = c_0\left[1 + \dfrac{1}{2}x^2 - \dfrac{1}{24}x^4 + \dfrac{1}{240}x^6 + \ldots\right] + c_1 x.$

18. $y(x) = c_0\left[1 - \dfrac{1}{6}x^3 + \dfrac{1}{180}x^6 - \ldots\right] + c_1\left[x - \dfrac{1}{12}x^4 + \ldots\right].$

19. $y(x) = c_0\left[1 + 2(x - 1) + 2(x - 1)^2 + \dfrac{4}{3}(x - 1)^3 + \ldots\right].$

20. $y(x) = c_0\left[1 + \dfrac{1}{2}(x - 1)^2 + \dfrac{1}{24}(x - 1)^4 + \ldots\right] + c_1\left[(x - 1) + \dfrac{1}{6}(x - 1)^3 + \dfrac{1}{120}(x - 1)^5 + \ldots\right].$

21. $y(x) = c_0\left[1 - \dfrac{1}{2}(x - 2)^2 + \dfrac{1}{3}(x - 2)^3 + \ldots\right] + c_1\left[(x - 2) - (x - 2)^2 + \dfrac{1}{3}(x - 2)^3 - \ldots\right].$

22. $y(x) = c_0\left[1 - \dfrac{1}{2}(x + 2)^2 - \dfrac{1}{3}(x + 2)^3 + \ldots\right].$

23. $y(x) = c_0[1 + (x - 1)].$

24. $y(x) = 2 + 2(x - 2) - 4(x - 2)^2 - \dfrac{4}{3}(x - 2)^3 - \ldots.$

25. $y(x) = 2 + (x - 1)^2 + \dfrac{1}{3}(x - 1)^3 + \dfrac{1}{12}(x - 1)^4 + \ldots.$

26. $y(x) = 1 - (x - 1) - \dfrac{5}{6}(x - 1)^2 + \dfrac{1}{18}(x - 1)^3 - \ldots.$

29. $\dfrac{1}{a}\left[\left(\dfrac{x}{a}\right) - \dfrac{1}{3}\left(\dfrac{x}{a}\right)^3 + \dfrac{1}{5}\left(\dfrac{x}{a}\right)^5 - \ldots\right], \quad |x| < a.$

30. $\left[(x - 1) - \dfrac{1}{2}(x - 1)^2 + \dfrac{1}{3}(x - 1)^3 - \ldots\right], \quad |x - 1| < 1.$

31. $c_{m+2} = -\dfrac{\lambda^2 c_m}{(m + 2)(m + 1)}, \quad m \geq 0.$

$y(x) = c_0\left[1 - \dfrac{\lambda^2 x^2}{2!} + \dfrac{\lambda^4 x^4}{4!} - \ldots\right] + c_1^*\left[\lambda x - \dfrac{\lambda^3 x^3}{3!} + \ldots\right], \quad c_1^* = c_1/\lambda.$

32. $c_{m+4} = c_m/[(m + 4)(m + 3)], \ m \geq 0, \ c_2 = 0 = c_3.$

$y(x) = c_0\left[1 + \dfrac{x^4}{4 \cdot 3} + \dfrac{x^8}{8 \cdot 7 \cdot 4 \cdot 3} + \ldots\right] + c_1\left[x + \dfrac{x^5}{5 \cdot 4} + \dfrac{x^9}{9 \cdot 8 \cdot 5 \cdot 4} + \ldots\right].$

33. $c_2 = -c_0/2, \ c_{m+2} = -(2m + 1)\,c_m/[(m + 2))(m + 1)], \quad m \geq 1,$

$$y(x) = c_0 \left[1 - \frac{x^2}{2!} + \frac{5x^4}{4!} - \ldots \right] + c_1 \left[x - \frac{3x^3}{3!} + \frac{7 \cdot 3x^5}{5!} - \ldots \right].$$

34. $c_2 = 4c_0$, $c_3 = 4c_1/3$, $c_{m+2} = [8 - m(m-1)]c_m/[(m+2)(m+1)]$, $m \geq 2$,

$$y(x) = c_0 [1 + 4x^2 + 2x^4 + \ldots] + c_1 \left[x + \frac{4}{3}x^3 + \frac{2}{15}x^5 + \ldots \right].$$

35. $c_2 = -c_0$, $c_{m+2} = (m^2 + m - 2) c_m /[(m+2)(m+1)]$, $m \geq 2$, $c_3 = 0 = c_5 \ldots$.

$$y(x) = c_0 \left[1 - x^2 - \frac{x^4}{3} - \ldots \right] + c_1 x.$$

36. $c_2 = 0$, $c_3 = -c_0$, $c_{m+2} = -[m(m-1)c_m + 6c_{m-1}]/[(m+2)(m+1)]$, $m \geq 2$,

$$y(x) = c_0 \left[1 - x^3 + \frac{3}{10} x^5 - \ldots \right] + c_1 \left[x - \frac{1}{2} x^4 + \ldots \right].$$

37. $c_2 = \frac{1}{2}(2c_1 - c_0)$, $c_{m+2} = -\dfrac{(m+1)(m-2)c_{m+1} + c_m}{(m+2)(m+1)}$, $m \geq 1$.

$$y(x) = c_0 \left[1 - \frac{x^2}{2!} - \frac{x^3}{3!} + \frac{x^4}{4!} + \ldots \right] + c_1 \left[x + \frac{2x^2}{2!} + \frac{x^3}{3!} - \frac{2x^4}{4!} + \ldots \right].$$

38. $c_2 = -c_0/2$, $c_{m+2} = -c_m/(m+2)$, $m \geq 1$,

$$y(x) = c_0 \left[1 - \frac{x^2}{2} + \frac{x^4}{8} - \ldots \right] + c_1 \left[x - \frac{x^3}{3} + \frac{x^5}{15} + \ldots \right].$$

39. $c_2 = -c_0/2$, $c_{m+2} = -(c_m + c_{m-1})/[(m+2)(m+1)]$, $m \geq 1$.

$$y(x) = c_0 \left[1 - \frac{1}{2!}(x-1)^2 - \frac{1}{3!}(x-1)^3 + \ldots \right] + c_1 \left[(x-1) - \frac{1}{3!}(x-1)^3 - \frac{2}{4!}(x-1)^4 + \ldots \right].$$

40. $c_2 = \frac{1}{2}(c_0 - c_1)$, $c_{m+2} = -\dfrac{(m+1)c_{m+1} + (m-1)c_m}{(m+1)(m+2)}$, $m \geq 1$,

$$y(x) = c_0 \left[1 + \frac{1}{2}(x-1)^2 - \frac{1}{6}(x-1)^3 + \frac{1}{60}(x-1)^5 - \ldots \right]$$
$$+ c_1 \left[(x-1) - \frac{1}{2}(x-1)^2 + \frac{1}{6}(x-1)^3 - \ldots \right].$$

41. $c_2 = \frac{1}{2}c_0$, $c_{m+2} = -\dfrac{m(m+1)c_{m+1} - 2c_m}{2(m+2)(m+1)}$, $m \geq 1$

$$y(x) = c_0 \left[1 + \frac{1}{2}(x-2)^2 - \frac{1}{12}(x-2)^3 + \frac{1}{16}(x-2)^4 - \ldots \right]$$
$$+ c_1 \left[(x-2) + \frac{1}{6}(x-2)^3 - \frac{1}{24}(x-2)^4 + \ldots \right].$$

42. $c_1 = \frac{1}{3}c_0$, $c_{m+1} = -\dfrac{(m-1)}{3(m+1)} c_m$, $m \geq 1$, $c_2 = 0 = c_3 = \ldots$, $y(x) = c_0 \left[1 + \frac{1}{3}(x-3) \right]$.

43. $c_{m+2} = -\dfrac{4c_m}{(m+2)(m+1)}$, $m \geq 0$, $y(x) = 3 \left[1 - 2x^2 + \frac{2}{3}x^4 - \ldots \right] + 4 \left[x - \frac{2}{3}x^3 + \frac{2}{15} x^5 - \right]$.

44. $c_2 = -c_0$, $c_{m+2} = -\dfrac{(3m+2)c_m}{(m+2)(m+1)}$, $m \geq 1$, $y(x) = 1 - x^2 + \frac{2}{3}x^4 - \ldots$.

45. $c_2 = -\frac{1}{6}(c_0 + c_1)$, $c_{m+2} = -\dfrac{(m+1)(2m+1)c_{m+1} + (m+1)c_m}{3(m+2)(m+1)}$, $m \geq 1$,

$$y(x) = \left[1 - \frac{1}{6}(x-1)^2 + \frac{1}{18}(x-1)^3 - \frac{1}{108}(x-1)^4 - \ldots\right]$$

$$- \left[(x-1) - \frac{1}{6}(x-1)^2 - \frac{1}{18}(x-1)^3 + \frac{1}{27}(x-1)^4 - \ldots\right].$$

46. $c_2 = \frac{1}{6}(c_0 + 2c_1)$, $c_{m+2} = -\dfrac{(m+1)(m-2)c_{m+1} - (m+1)c_m}{3(m+2)(m+1)}$, $\quad m \geq 1$,

$$y(x) = 1 + \frac{1}{6}(x-2)^2 + \frac{1}{54}(x-2)^3 + \ldots.$$

47. $y(x) \approx 1 + 2x - \dfrac{5}{2}x^2 + \dfrac{2}{3}x^3 + \dfrac{1}{8}x^4.$
    48. $y(x) \approx -1 + \dfrac{1}{2}x^2 + \dfrac{1}{6}x^3 - \dfrac{1}{8}x^4.$

49. $y(x) \approx 2 + x - 2x^2 - \dfrac{7}{6}x^3 + \dfrac{1}{12}x^4.$
    50. $y(x) \approx x + \dfrac{1}{2}x^2 - \dfrac{1}{3}x^3 + \dfrac{5}{24}x^4.$

Exercise 6.3

1. $r = \pm n.$ $c_1 = c_3 = \ldots = 0$, $c_{m+2} = -\dfrac{c_m}{(m+r+2+n)(m+r+2-n)}$, $\quad m \geq 0$

$$y(x) = x^r\left[1 - \frac{x^2}{(r+2+n)(r+2-n)} + \frac{x^4}{(r+4+n)(r+4-n)(r+2+n)(r+2-n)} - \ldots\right]$$

$y_1(x) = y(x)$ at $r = n$ and $y_2(x) = y(x)$ at $r = -n$.

2. $r = 0, 1/2$, $c_{m+1} = -\dfrac{[(m+r)(2m+2r-1)-3]}{(m+r+1)(2m+2r+1)}c_m$, $\quad m \geq 0$,

$$y_1(x) = \left[1 + 3x + x^2 - \frac{1}{5}x^3 + \ldots\right]; y_2(x) = x^{1/2}[1+x].$$

3. $r = 0, 5/3$, $c_{m+1} = -\dfrac{[9(m+r)(m+r-1)+2]}{3(m+r+1)(3m+3r-2)}c_m$, $\quad m \geq 0$,

$$y_1(x) = 1 + \frac{x}{3} - \frac{1}{9}x^2 + \frac{5}{81}x^3 - \ldots = (1+x)^{1/3}; \quad y_2(x) = x^{5/3}\left[1 - \frac{1}{2}x + \frac{7}{22}x^2 - \ldots\right].$$

4. $r = 1/2, 5/2$, $c_1 = c_3 = \ldots = 0$, $y = Ax^{1/2} + Bx^{5/2}.$

5. $r = 0, 1/3$, $c_{m+1} = \dfrac{\{(m+r)(3m+3r-1)-4\}}{(m+r+1)(3m+3r+2)}c_m$, $\quad m \geq 0$,

$$y_1(x) = 1 - 2x + \frac{2}{5}x^2 + \frac{1}{10}x^3 + \ldots; \quad y_2(x) = x^{1/3}(1-x).$$

6. $r = 0, 1/2$, $c_{m+1} = \dfrac{(m+r)(2m+2r-1)-3}{(m+r+1)(2m+2r+1)}c_m$, $\quad m \geq 0$,

$$y_1(x) = 1 - 3x + x^2 + \frac{1}{5}x^3 + \ldots; \quad y_2(x) = x^{1/2}(1-x).$$

7. $r = 2, 1/2$, $c_{m+2} = -\dfrac{(m+r)(2m+2r-5)c_m}{(m+r+2)(2m+2r-1)+2}$, $\quad m \geq 0$,

$$y_1(x) = x^2\left[1 + \frac{1}{7}x^2 - \frac{3}{77}x^4 + \ldots\right]; \quad y_2(x) = x^{1/2}[1+x^2].$$

8. $r = 0, 0$, $c_1 = \dfrac{(2r+1)}{(r+1)^2}c_0$, $c_{m+2} = \dfrac{(2m+2r+3)c_{m+1} - c_m}{(m+r+2)^2}$, $\quad m \geq 0$,

$$y_1(x) = \left[1 + x + \frac{2}{1^2 \cdot 2^2} x^2 + \frac{6}{1^2 \cdot 2^2 \cdot 3^2} x^3 + \dots\right]; y_2(x) = (\ln x) y_1(x).$$

9. $r = 0, 0, \quad c_{m+1} = \frac{(m + r - 3)}{(m + r + 1)^2} c_m, \quad m \geq 0,$

$$y_1(x) = \left[1 - \frac{3}{1^2} x + \frac{2 \cdot 3}{1^2 \cdot 2^2} x^2 - \frac{1 \cdot 2 \cdot 3}{1^2 \cdot 2^2 \cdot 3^2} x^3 + \dots\right]; y_2(x) = (\ln x) y_1(x) + \left[7x - \frac{23}{4} x^2 + \dots\right].$$

10. $r = 0, 0, \quad c_{m+2} = -\frac{c_m}{(m + r + 2)^2}, m \geq 0, \quad c_1 = c_3 = \dots = 0,$

$$y_1(x) = \left[1 - \frac{x^2}{2^2} + \frac{x^4}{2^2 \cdot 4^2} - \frac{x^6}{2^2 \cdot 4^2 \cdot 6^2} + \dots\right],$$

$$y_2(x) = (\ln x) y_1(x) + \left[\frac{2}{2^3} x^2 - \frac{3}{2 \cdot 4^3} x^4 + \frac{11}{4^3 \cdot 6^3} x^6 - \dots\right].$$

11. $r = 0, 0, \quad c_{m+1} = -\frac{(m + r - 1)}{(m + r + 1)} c_m, \quad m \geq 0,$

$$y_1(x) = 1 + x; \quad y_2(x) = (\ln x) y_1(x) + \left[-2x - \frac{x^2}{2} + \frac{x^3}{6} + \dots\right].$$

12. $r = 0, 0, \quad c_{m+1} = -\frac{(m + r - 2)}{(m + r + 1)^2} c_m, \quad m \geq 0,$

$$y_1(x) = 1 + 2x + \frac{x^2}{2}; \quad y_2(x) = (\ln x) y_1(x) - \left[5x + \frac{9}{4} x^2 + \frac{1}{18} x^3 - \dots\right].$$

13. $r = 1/2, 1/2, \quad c_{m+2} = -\frac{4(m + r)(m + r + 3)}{[2(m + r) + 3]^2} c_m, \quad m \geq 0,$

$$y_1(x) = x^{1/2}\left[1 - \frac{7}{4^2} x^2 + \frac{5 \cdot 7 \cdot 11}{4^2 \cdot 8^2} x^4 - \dots\right]; \quad y_2(x) = (\ln x) y_1(x) + x^{1/2}\left[-\frac{9}{16} x^2 + \frac{1053}{2048} x^4 - \dots\right].$$

14. $r = 0, 1, \quad c_{m+1} = \frac{c_m}{(m + r)(m + r + 1)}, \quad m \geq 0, \quad \text{set} \quad c_0 = a_0 r, \quad a_0 \neq 0,$

$$y_1(x) = x + \frac{x^2}{1^2 \cdot 2} + \frac{x^3}{1^2 \cdot 2^2 \cdot 3} + \dots$$

$$y_2(x) = y_1(x) \ln x +$$

$$\left[1 - \frac{x}{1^2} - \left(\frac{2}{1^3 \cdot 2} + \frac{1}{1^2 \cdot 2^2}\right) x^2 - \left(\frac{2}{1^3 \cdot 2^2 \cdot 3} + \frac{2}{1^2 \cdot 2^3 \cdot 3} + \frac{1}{1^2 \cdot 2^2 \cdot 3^2}\right) x^3 - \dots\right].$$

15. $r = -1, 3/2, \quad y_1(x) = 1/x; y_2(x) = x^{3/2}.$

16. $r = 1/4, 3/4, \quad y_1(x) = x^{1/4}; y_2(x) = x^{3/4}.$

17. $r = -1, -2, \quad r = -2$ gives complete solution, $c_{m+2} = -\frac{c_m}{(m + r + 2)(m + r + 5) + 2}, \quad m \geq 0$

$y_1(x) = \cos x / x^2, \quad y_2(x) = \sin x / x^2.$

18. $r = 0, 1. \quad r = 0$ gives complete solution, $c_{m+4} = -\frac{c_m}{(m + r + 3)(m + r + 4)}, \quad m \geq 0$

$$y_1(x) = 1 - \frac{x^4}{3 \cdot 4} + \frac{x^8}{3 \cdot 4 \cdot 7 \cdot 8} - \dots; \quad y_2(x) = x - \frac{x^5}{4 \cdot 5} + \frac{x^9}{4 \cdot 5 \cdot 8 \cdot 9} - \dots.$$

19. $r = \pm 1$, $c_{m+2} = -\dfrac{c_m}{(m+r+1)(m+r+3)}$, $m \geq 0$,

Set $c_0 = a_0(r+1)$, $a_0 \neq 0$, $y_1(x) = -\dfrac{1}{2}\left[x - \dfrac{x^3}{2 \cdot 4} + \ldots\right]$,

$$y_2(x) = \left[(\ln x)y_1(x) + \dfrac{1}{x}\left\{1 + \dfrac{x^2}{2^2} - \left(\dfrac{2}{2^3 \cdot 4} + \dfrac{1}{2^2 \cdot 4^2}\right)x^4 + \ldots\right\}\right].$$

20. $r = -2, -3$, $c_{m+2} = \dfrac{4c_m}{(m+r+2)(m+r+7)+6}$, $r = -3$ gives complete solution.

$$y(x) = c_0 x^{-3}\left(1 + 2x^2 + \dfrac{2}{3}x^4 + \ldots\right) + c_1 x^{-3}\left(x + \dfrac{2}{3}x^3 + \dfrac{2}{15}x^5 + \ldots\right).$$

21. No. It has no solution in ascending powers of x. The fundamental solutions are $e^{2/x}$ and $e^{-2/x}$ which can only be written in descending powers of x. The transformed equation is $\dfrac{d^2 y}{dz^2} - 16 y = 0$ whose solution is e^{4z} and e^{-4z}. This equation has power series solution about the origin $z = 0$.

22. $r = 0$ is the only value. For $r = 0$, $\lim\limits_{m \to \infty}\left|\dfrac{c_{m+1}}{c_m}\right| \to \infty$ and the series diverges.

23. $r = 0, 1/2$, $c_{m+1} = -\dfrac{(m+r-1)c_m}{(m+r+1)(2m+2r+1)}$, $m \geq 0$,

$$y_1(x) = [1 + (x-1)]; \quad y_2(x) = (x-1)^{1/2}\left[1 + \dfrac{(x-1)}{6} - \dfrac{(x-1)^2}{120} + \dfrac{(x-1)^3}{1680} - \ldots\right].$$

24. $r = 0, 1/3$, $c_{m+1} = \dfrac{9(m+r)(m+r-1)+2}{(m+r+1)(9m+9r+6)} c_m$, $m \geq 0$,

$$y_1(x) = 1 + \dfrac{1}{3}(x+1) + \dfrac{1}{45}(x+1)^2 + \dfrac{1}{162}(x+1)^3 + \ldots; \quad y_2(x) = (x+1)^{1/3}.$$

25. $r = 0, -1$, $c_{m+1} = -\dfrac{(m+r)(m+r-1)+3}{2(m+r+1)(m+r+2)} c_m$, $m \geq 0$, Set $c_0 = a_0(r+1)$, $a_0 \neq 0$,

$$y_1(x) = 1 - \dfrac{3}{4}(x-2) + \dfrac{3}{16}(x-2)^2 - \ldots.$$

$$y_2(x) = (\ln(x-2))y_1(x) + (x-2)^{-1}\left[1 + 4(x-2) - \dfrac{103}{16}(x-2)^2 - \ldots\right].$$

Legendre Polynomials, Chebyshev Polynomials, Bessel Functions and Sturm-Liouville Problem

7.1 Introduction

In many problems of mathematical physics and engineering, we come across second order linear differential equations whose solutions give rise to *special functions*, which are so called as they are different from the standard functions like sine, cosine, exponential, logarithmic functions etc.

In this chapter, we obtain the solutions of this important class of linear differential equations and define the special functions.

7.2 Legendre Differential Equation and Legendre Polynomials

One of the important differential equations which gives rise to special functions is the *Legendre differential equation*

$$(1 - x^2)\, y'' - 2xy' + n(n + 1)y = 0 \tag{7.1}$$

where n is a real constant. In most applications, n takes positive integral values. We can also write this equation as

$$[(1 - x^2)y']' + n(n + 1)y = 0. \tag{7.2}$$

The singularities of this equation are $x = \pm 1$. If we now seek a power series solution about $x = 0$, of the form $y(x) = \sum\limits_{m=0}^{\infty} c_m x^m$, then this series solution is convergent in $|x| < 1$, since the distance between $x = 0$ and the nearest singularity is 1. Substituting

$$y(x) = \sum_{m=0}^{\infty} c_m x^m, \quad y'(x) = \sum_{m=1}^{\infty} m c_m x^{m-1} \quad \text{and} \quad y''(x) = \sum_{m=2}^{\infty} m(m - 1)c_m x^{m-2} \tag{7.3}$$

in Eq. (7.1), we obtain

$$\sum_{m=2}^{\infty} m(m-1)c_m x^{m-2} - \sum_{m=2}^{\infty} m(m-1)c_m x^m - 2\sum_{m=1}^{\infty} mc_m x^m + n(n+1)\sum_{m=0}^{\infty} c_m x^m = 0$$

or
$$2c_2 + 6c_3 x + \sum_{m=4}^{\infty} m(m-1)c_m x^{m-2} - \sum_{m=2}^{\infty} m(m-1)c_m x^m - 2c_1 x$$

$$- 2\sum_{m=2}^{\infty} mc_m x^m + n(n+1)(c_0 + c_1 x) + n(n+1)\sum_{m=2}^{\infty} c_m x^m = 0.$$

Substituting $m - 2 = t$ in the first sum, we get

$$[2c_2 + n(n+1)c_0] + [6c_3 - 2c_1 + n(n+1)c_1]x + \sum_{t=2}^{\infty} (t+2)(t+1)c_{t+2} x^t$$

$$- \sum_{m=2}^{\infty} [m(m-1) + 2m - n(n+1)]c_m x^m = 0.$$

Since t is a dummy variable, we can combine the third and fourth terms of this equation. Equating the coefficients of various powers of x to zero, we get

$$2c_2 + n(n+1)c_0 = 0, \quad 6c_3 - 2c_1 + n(n+1)c_1 = 0,$$

and
$$(m+2)(m+1)c_{m+2} - (m^2 + m - n^2 - n)c_m = 0, \quad m = 2, 3, \ldots$$

Therefore, $\quad c_2 = -\dfrac{1}{2!}n(n+1)c_0, \quad c_3 = \dfrac{1}{6}[2 - n^2 - n]c_1 = -\dfrac{1}{3!}(n-1)(n+2)c_1$

$$c_{m+2} = -\frac{(n-m)(m+n+1)}{(m+2)(m+1)} c_m, \quad m \geq 2. \tag{7.4}$$

We have

$$c_4 = -\frac{(n-2)(n+3)}{4 \cdot 3} c_2 = \frac{(n-2)n(n+1)(n+3)}{4 \cdot 3 \cdot 2!} c_0 = \frac{1}{4!}(n-2)n(n+1)(n+3)c_0$$

$$c_5 = -\frac{(n-3)(n+4)}{5 \cdot 4} c_3 = \frac{1}{5!}(n-3)(n-1)(n+2)(n+4)c_1, \ldots.$$

Substituting in the power series solution, we obtain

$$y(x) = c_0\left[1 - \frac{1}{2!}n(n+1)x^2 + \frac{1}{4!}(n-2)n(n+1)(n+3)x^4 - \ldots\right]$$

$$+ c_1\left[x - \frac{1}{3!}(n-1)(n+2)x^3 + \frac{1}{5!}(n-3)(n-1)(n+2)(n+4)x^5 - \ldots\right]$$

$$= c_0 y_0(x) + c_1 y_1(x). \tag{7.5}$$

where $\quad y_0(x) = 1 - \dfrac{1}{2!}n(n+1)x^2 + \dfrac{1}{4!}(n-2)n(n+1)(n+3)x^4 - \ldots$

$$y_1(x) = x - \frac{1}{3!}(n-1)(n+2)x^3 + \frac{1}{5!}(n-3)(n-1)(n+2)(n+4)x^5 - \ldots. \tag{7.6}$$

The series for $y_0(x)$, $y_1(x)$ converge for $|x| < 1$. We note that $y_0(x)$ contains even powers of x only and $y_1(x)$ contains odd powers of x only. Hence, the two solutions $y_0(x)$, $y_1(x)$ are the linearly independent solutions of the Legendre's equation.

As n takes the value zero and even positive integral values, we obtain for

$$n = 0: y_0(x) = 1, \quad n = 2: y_0(x) = 1 - 3x^2, \quad n = 4: y_0(x) = 1 - 10x^2 + \frac{35}{3}x^4, \ldots \quad (7.6a)$$

This implies that $y_0(x)$ reduces to an even polynomial (polynomial of even powers) as n takes even positive integral values, whereas $y_1(x)$ remains an infinite series.

As n takes odd positive integral values, we obtain for

$$n = 1: y_1(x) = x, \quad n = 3: y_1(x) = x - \frac{5}{3}x^3, \quad n = 5: y_1(x) = x - \frac{14}{3}x^3 + \frac{21}{5}x^5, \ldots \quad (7.6b)$$

Therefore, $y_1(x)$ reduces to a polynomial of odd powers (odd polynomial) as n takes odd positive integral values, whereas $y_0(x)$ remains an infinite series. These polynomials multiplied by suitable constants are called *Legendre polynomials*. The Legendre polynomials are denoted by $P_n(x)$ where n denotes the order of the polynomial. Therefore, when n takes integral values, one of the linearly independent solutions of Eq. (7.1) is a Legendre polynomial and the second independent solution is an infinite series. The second solution is denoted by $Q_n(x)$. In order to explicitly write the expressions for the Legendre polynomials, we need to evaluate the multiplicative constants. The values of the multiplicative constants are obtained by setting $P_n(1) = 1$. It is too cumbersome to use the recurrence relation given in Eq. (7.4) or the polynomials given in Eqs. (7.6a), (7.6b) to determine the multiplicative constants. It is easy to use the *Rodrigue's* formula (see next section) to find the expressions for the Legendre polynomials.

7.2.1 Rodrigue's Formula

We shall now show that by applying the binomial theorem to $(x^2 - 1)^n$ and differentiating n times, the resulting expression, which is a polynomial of degree n, is a solution of the Legendre differential equation. We can then determine the expression for $P_n(x)$ by requiring that $P_n(1) = 1$.

Let $u = (x^2 - 1)^n$. Differentiating, we get

$$u_1 = 2nx(x^2 - 1)^{n-1} = \frac{2nxu}{x^2 - 1}$$

or
$$(1 - x^2)u_1 + 2nxu = 0, \quad \text{where} \quad u_1 = u'.$$

Differentiating this equation $(n + 1)$ times by using the Leibniz theorem, we obtain

$$(1 - x^2)u_{n+2} + (n + 1)(-2x)u_{n+1} + \frac{1}{2}(n + 1)n(-2)u_n + 2n[xu_{n+1} + (n + 1)u_n] = 0$$

or
$$(1 - x^2)u_{n+2} + [-2(n + 1)x + 2nx]u_{n+1} + [2n(n + 1) - n(n + 1)]u_n = 0$$

or
$$(1 - x^2)(u_n)'' - 2x(u_n)' + n(n + 1)(u_n) = 0$$

which is the Legendre differential equation for $y = u_n$.

However, $P_n(x)$ is the finite series solution of the Legendre's equation. Therefore,

$$P_n(x) = a\, u_n, \quad a \text{ is a constant}$$

$$= a\,\frac{d^n}{dx^n}(x^2 - 1)^n. \quad (7.7)$$

The constant a is determined by setting $P_n(1) = 1$. Now, denoting $D = d/dx$, we have

$$P_n(1) = 1 = a \left[\frac{d^n}{dx^n} (x-1)^n (x+1)^n \right]_{x=1}$$

$$= a \left[\sum_{j=0}^{n} {}^nC_j \, D^j (x-1)^n \, D^{n-j} (x+1)^n \right]_{x=1}$$

$$= a[n! \, (x+1)^n + \text{terms containing product of various powers of } (x-1) \text{ and } (x+1)]_{x=1}$$

$$= a[n! \, 2^n]$$

since all the terms containing positive powers of $(x-1)$ vanish at $x = 1$. Therefore,

$$a = \frac{1}{n! \, 2^n}, \quad \text{and} \quad P_n(x) = \frac{1}{n! \, 2^n} \frac{d^n}{dx^n} (x^2 - 1)^n. \tag{7.8}$$

This equality is called the *Rodrigue's formula*. Using the Binomial theorem, we get

$$(x^2 - 1)^n = \sum_{r=0}^{n} {}^nC_r (x^2)^{n-r} (-1)^r = \sum_{r=0}^{n} \frac{(-1)^r \, n! \, x^{2n-2r}}{r! \, (n-r)!}.$$

Differentiating n times, we get from Eq. (7.8)

$$P_n(x) = \frac{1}{n! \, 2^n} \sum_{r=0}^{n} \frac{(-1)^r \, n!}{r! \, (n-r)!} \frac{d^n}{dx^n} (x^{2n-2r})$$

$$= \frac{1}{2^n} \sum_{r=0}^{N} \frac{(-1)^r (2n-2r)! \, x^{n-2r}}{r! \, (n-r)! \, (n-2r)!} \tag{7.9}$$

where $N = n/2$ or $(n-1)/2$ whichever is an integer. In notation, we usually write as $N = \lfloor n/2 \rfloor$. Substituting $n = 0, 1, 2, \ldots$ in Eq. (7.9), we get the Legendre polynomials as

$$P_0(x) = 1, \quad P_1(x) = x, \quad P_2(x) = \frac{1}{2}(3x^2 - 1)$$

$$P_3(x) = \frac{1}{2}(5x^3 - 3x), \quad P_4(x) = \frac{1}{8}(35x^4 - 30x^2 + 3), \ldots. \tag{7.10}$$

The graphs of $P_0(x)$, $P_1(x)$, $P_2(x)$, $P_3(x)$ and $P_4(x)$ are given in Fig. 7.1. We note that $| P_n(x) | \leq 1$, $x \in [-1, 1]$.

The Legendre polynomials belong to an important class of orthogonal polynomials, which we shall discuss in a later section.

Example 7.1 Express $P(x) = 3P_3(x) + 2P_2(x) + 4P_1(x) + 5P_0(x)$ as a polynomial in x, where $P_m(x)$ is the Legendre polynomial of order m.

Solution Substituting the expressions for P_0, P_1, P_2 and P_3, we obtain

$$P(x) = \frac{3}{2}(5x^3 - 3x) + (3x^2 - 1) + 4x + 5$$

$$= \frac{1}{2}(15x^3 - 9x + 6x^2 - 2 + 8x + 10) = \frac{1}{2}(15x^3 + 6x^2 - x + 8).$$

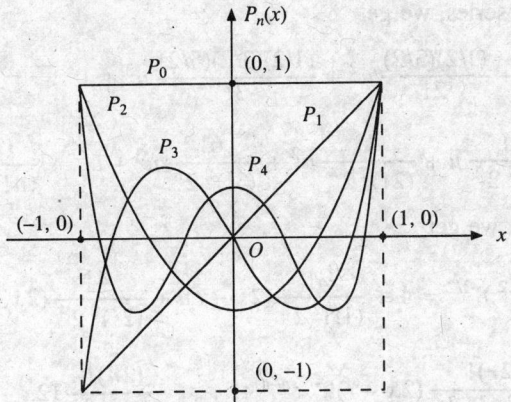

Fig. 7.1. Legendre polynomials.

Example 7.2 Express $f(x) = x^4 + 2x^3 - 6x^2 + 5x - 3$ in terms of Legendre polynomials.

Solution Writing various powers of x in terms of Legendre polynomials, we get

$$1 = P_0(x), \quad x = P_1(x), \quad x^2 = \frac{1}{3}(2P_2 + 1) = \frac{1}{3}(2P_2 + P_0),$$

$$x^3 = \frac{1}{5}(2P_3 + 3x) = \frac{1}{5}(2P_3 + 3P_1),$$

$$x^4 = \frac{1}{35}(8P_4 + 30x^2 - 3) = \frac{1}{35}[8P_4 + 10(2P_2 + P_0) - 3P_0]$$

$$= \frac{1}{35}(8P_4 + 20P_2 + 7P_0).$$

Therefore, $f(x) = \frac{1}{35}(8P_4 + 20P_2 + 7P_0) + \frac{2}{5}(2P_3 + 3P_1) - 2(2P_2 + P_0) + 5P_1 - 3P_0$

$$= \frac{1}{35}(8P_4 + 28P_3 - 120P_2 + 217P_1 - 168P_0).$$

7.2.2 Generating Function for Legendre Polynomials

Legendre's polynomials can also be obtained by using a generating function. We shall prove that

$$\frac{1}{\sqrt{1 - 2xt + t^2}} = \sum_{n=0}^{\infty} t^n P_n(x), \quad t \neq 1. \tag{7.11}$$

The function on the left hand side is called the *generating function* of the Legendre polynomials. We have

$$(1 - 2xt + t^2)^{-1/2} = [1 - (2x - t)t]^{-1/2} = [1 - u]^{-1/2}$$

where $u = (2x - t)t$.

Expanding in a binomial series, we get

$$(1 - u)^{-1/2} = 1 + \frac{1}{2} u + \frac{(1/2)(3/2)}{2!} u^2 + \frac{(1/2)(3/2)(5/2)}{3!} u^3 + \dots$$

$$= 1 + \frac{2!}{(1!)^2 2^2} u + \frac{4!}{(2!)^2 2^4} u^2 + \frac{6!}{(3!)^2 2^6} u^3 + \dots + \frac{(2n)!}{(n!)^2 2^{2n}} u^n + \dots.$$

Substituting $u = (2x - t)t$, we get

$$(1 - 2xt + t^2)^{-1/2} = 1 + \frac{2!}{(1!)^2 2^2} (2x - t)t + \frac{4!}{(2!)^2 2^4} (2x - t)^2 t^2 + \dots$$

$$+ \frac{(2n - 2r)!}{[(n - r)!]^2 2^{2n-2r}} (2x - t)^{n-r} t^{n-r} + \dots + \frac{(2n)!}{(n!)^2 2^{2n}} (2x - t)^n t^n + \dots.$$

Now, the term in t^n, in the product $(2x - t)^{n-r} t^{n-r}$ is given by

$$t^{n-r} [^{n-r} C_r (-t)^r (2x)^{n-2r}] = \frac{(n - r)! \, 2^{n-2r}}{r!(n - 2r)!} (-1)^r t^n x^{n-2r}.$$

Therefore, the contribution of this term to t^n, in the series is

$$\frac{(2n - 2r)!}{[(n - r)!]^2 2^{2n-2r}} \cdot \frac{(n - r)! \, 2^{n-2r}}{r!(n - 2r)!} (-1)^r t^n x^{n-2r} = \frac{(-1)^r (2n - 2r)!}{2^n (n - r)! \, r! \, (n - 2r)!} t^n x^{n-2r}.$$

Summing for $r = 0, \ 1, \dots, N$, where $N = n/2$ or $(n - 1)/2$, whichever is an integer, we get on using Eq. (7.9)

$$\sum_{r=0}^{N} \frac{(-1)^r (2n - 2r)! \, x^{n-2r}}{2^n \, r!(n - r)!(n - 2r)!} t^n = P_n(x) t^n.$$

Therefore,

$$(1 - 2xt + t^2)^{-1/2} = \sum_{n=0}^{\infty} P_n(x) t^n, \quad t \neq 1.$$

This proves that $P_n(x)$ can be obtained by the generating function.

Example 7.3 Show that (a) $P_n(1) = 1$, (b) $P_n(-x) = (-1)^n P_n(x)$. Hence find $P_n(-1)$.

Solution

(a) Substituting $x = 1$ in the generating function, we get

$$\frac{1}{\sqrt{1 - 2t + t^2}} = \sum_{n=0}^{\infty} P_n(1) t^n$$

or

$$\sum_{n=0}^{\infty} P_n(1) t^n = \frac{1}{1 - t} = (1 - t)^{-1} = 1 + t + t^2 + \dots + t^n + \dots.$$

Comparing the coefficient of t^n on both sides, we get $P_n(1) = 1$, $n = 0, 1, \dots$

(b) Replacing x and t by $-x$ and $-t$ respectively in the generating function given in Eq. (7.11), we get

$$\frac{1}{\sqrt{1 - 2xt + t^2}} = \sum_{n=0}^{\infty} P_n(-x)(-t)^n = \sum_{n=0}^{\infty} (-1)^n P_n(-x) t^n.$$

Since the left hand side is $\sum\limits_{n=0}^{\infty} P_n(x)t^n$, we get

$$\sum_{n=0}^{\infty} P_n(x)t^n = \sum_{n=0}^{\infty} (-1)^n P_n(-x)t^n.$$

Comparing the coefficients of t^n on both sides, we get

$$P_n(x) = (-1)^n P_n(-x) \quad \text{or} \quad P_n(-x) = (-1)^n P_n(x).$$

Setting $x = 1$, we get $P_n(-1) = (-1)^n P_n(1) = (-1)^n$.

7.2.3 Recurrence Relations for Legendre Polynomials

In many applications, use of recurrence relations between Legendre polynomials simplifies the solution of a given problem. We prove the following recurrence relations.

(a)
$$(n+1)P_{n+1}(x) = (2n+1)xP_n(x) - nP_{n-1}(x). \tag{7.12}$$

(This formula is also called the *Bonnet*'s recurrence relation.)

(b)
$$nP_n(x) = xP_n'(x) - P_{n-1}'(x). \tag{7.13}$$

(c)
$$P_{n+1}'(x) - xP_n'(x) = (n+1)P_n(x). \tag{7.14}$$

Differentiating the generating function

$$(1 - 2xt + t^2)^{-1/2} = \sum_{n=0}^{\infty} P_n(x)t^n \tag{7.15}$$

partially with respect to t, we get

$$-\frac{1}{2}(1 - 2xt + t^2)^{-3/2}(-2x + 2t) = \sum_{n=0}^{\infty} nP_n(x)t^{n-1} \tag{7.16}$$

or
$$(x - t)(1 - 2xt + t^2)^{-1/2} = (1 - 2xt + t^2)\sum_{n=0}^{\infty} nP_n(x)t^{n-1}$$

or
$$(x - t)\sum_{n=0}^{\infty} P_n(x)t^n = (1 - 2xt + t^2)\sum_{n=0}^{\infty} nP_n(x)t^{n-1}.$$

Comparing the coefficient of t^n on both sides, we obtain

$$xP_n(x) - P_{n-1}(x) = (n+1)P_{n+1}(x) - 2nxP_n(x) + (n-1)P_{n-1}(x)$$

or
$$(n+1)P_{n+1}(x) = (2n+1)xP_n(x) - nP_{n-1}(x).$$

This proves (a).

Differentiating Eq. (7.11) partially with respect to x, we obtain

$$-\frac{1}{2}(1 - 2xt + t^2)^{-3/2}(-2t) = \sum_{n=0}^{\infty} P_n'(x)t^n. \tag{7.17}$$

Dividing the expressions in Eqs. (7.16) and (7.17), we get

$$\frac{x - t}{t} = \frac{\sum n P_n(x) t^{n-1}}{\sum P_n'(x) t^n}$$

or $\qquad (x - t) \sum_{n=0}^{\infty} P_n'(x) t^n = \sum_{n=0}^{\infty} n P_n(x) t^n.$

Comparing the coefficient of t^n on both sides, we get

$$x P_n'(x) - P_{n-1}'(x) = n P_n(x).$$

This proves (b). Similarly, (c) is proved (see Problem 19 in Exercise 7.1).

Example 7.4   Using the recurrence relation (7.12), generate the Legendre polynomials P_2, P_3, P_4 given that $P_0(x) = 1$ and $P_1(x) = x$.

Solution   The required recurrence relation is (see Eq. (7.12))

$$(n + 1) P_{n+1}(x) = (2n + 1) x P_n(x) - n P_{n-1}(x)$$

We are given that $P_0(x) = 1$, $\quad P_1(x) = x$.
For $n = 1$, we get

$$2 P_2(x) = 3x P_1(x) - P_0(x) = 3x(x) - 1,$$

or $\qquad P_2(x) = \frac{1}{2}(3x^2 - 1).$

For $n = 2$, we get

$$3 P_3(x) = 5x P_2(x) - 2 P_1(x)$$

$$= \frac{5}{2} x (3x^2 - 1) - 2x = \frac{1}{2}(15x^3 - 9x)$$

or $\qquad P_3(x) = \frac{1}{2}(5x^3 - 3x).$

For $n = 3$, we get

$$4 P_4(x) = 7x P_3(x) - 3 P_2(x)$$

$$= \frac{7}{2} x (5x^3 - 3x) - \frac{3}{2}(3x^2 - 1) = \frac{1}{2}(35x^4 - 30x^2 + 3)$$

or $\qquad P_4(x) = \frac{1}{8}(35x^4 - 30x^2 + 3).$

Example 7.5   Using the recurrence relation

$$(n + 1) P_{n+1}(x) = (2n + 1) \, x P_n(x) - n P_{n-1}(x).$$

(recursively) evaluate $P_2(1.5)$ and $P_3(2.1)$.

Solution   Setting $n = 1$, we get

$$2 P_2(x) = 3x P_1(x) - P_0(x).$$

Since, $P_0(x) = 1$ for all x and $P_1(x) = x$, we get $P_1(1.5) = 1.5$. Therefore,

$$P_2(1.5) = \frac{1}{2}[3(1.5)(1.5) - 1] = \frac{1}{2}(5.75) = 2.875.$$

Setting $n = 2$, we get

$$3P_3(x) = 5xP_2(x) - 2P_1(x), \quad \text{and} \quad 2P_2(x) = 3xP_1(x) - P_0(x).$$

Now, $\qquad\qquad\qquad P_0(2.1) = 1, \quad P_1(2.1) = 2.1.$

Therefore $\qquad\qquad P_2(2.1) = \frac{1}{2}[3(2.1)(2.1) - 1] = 6.115.$

and $\qquad\qquad P_3(2.1) = \frac{1}{3}[5(2.1)(6.115) - 2(2.1)] = 20.0025.$

Example 7.6 Show that

$$\int P_n(x)\,dx = \frac{1}{2n+1}[P_{n+1}(x) - P_{n-1}(x)].$$

Solution Adding the two recurrence relations given in Eqs. (7.13) and (7.14), we get

$$P'_{n+1}(x) - P'_{n-1}(x) = (2n+1)P_n(x).$$

Integrating, we obtain the result.

7.2.4 Orthogonal and Orthonormal Functions

A set of functions $\{\phi_i(x)\}$ is said to be *orthogonal* on an interval $[a, b]$, with respect to a weight function $W(x)$, $W(x) > 0$ in $[a, b]$, if

$$\int_a^b W(x)\phi_i(x)\phi_j(x)\,dx = 0, \quad i \neq j. \tag{7.18}$$

If $W(x) = 1$, we usually say that the functions $\phi_1(x)$, $\phi_2(x)$, ... are orthogonal over $[a, b]$.

If a discrete data is given, then we have the following definition:

A set of functions $\{\phi_i(x)\}$ is said to be orthogonal over a set of $N + 1$ points $\{x_j\}$, with respect to a weight function $W(x) > 0$, if

$$\sum_{i=0}^N W(x_i)\phi_j(x_i)\phi_k(x_i) = 0, \quad j \neq k. \tag{7.19}$$

The *norm* $\| \phi_i(x) \|$ of $\phi_i(x)$ is defined as

$$\|\phi_i(x)\| = \left[\int_a^b W(x)\phi_i^2(x)\,dx\right]^{1/2}. \tag{7.20}$$

It is obvious that

$$\int_a^b W(x)\phi_i^2(x)\,dx > 0.$$

For discrete data, we define $\qquad \|\phi_j(x)\| = \left[\sum_{i=0}^N W(x_i)\phi_j^2(x_i)\right]^{1/2}.$

The functions $\phi_1(x), \phi_2(x), \ldots$ are said to be *orthonormal* if they are orthogonal on the interval $[a, b]$ and all functions have norm 1, that is, $\| \phi_i(x) \| = 1$ for all i. It is easy to construct orthonormal functions from the corresponding orthogonal functions. Let the functions $\{\phi_i(x)\}$ be orthogonal. Define

$$f_i(x) = \phi_i(x)/\|\phi_i(x)\|. \tag{7.21}$$

Then, the functions $\{f_i(x)\}$ are orthonormal. We have

$$\int_a^b W(x) f_i(x) f_j(x)\, dx = \int_a^b W(x) \frac{\phi_i(x)}{\|\phi_i(x)\|} \cdot \frac{\phi_j(x)}{\|\phi_j(x)\|}\, dx$$

$$= \frac{1}{\|\phi_i(x)\|\, \|\phi_j(x)\|} \int_a^b W(x)\phi_i(x)\phi_j(x)\, dx = 0$$

and

$$\int_a^b W(x) f_i^2(x)\, dx = \int_a^b W(x) \cdot \frac{\phi_i^2(x)}{\|\phi_i(x)\|^2}\, dx$$

$$= \frac{1}{\|\phi_i(x)\|^2} \int_a^b W(x)\phi_i^2(x)\, dx = \frac{\|\phi_i(x)\|^2}{\|\phi_i(x)\|^2} = 1.$$

Example 7.7   Show that with $W(x) = 1$, the functions $\phi_i(x) = \cos ix$, $i = 0, 1, 2, \ldots$ are orthogonal over the interval $[-\pi, \pi]$. Construct the corresponding orthonormal set of functions.

Solution   We have for $i \neq j$

$$\int_{-\pi}^{\pi} \phi_i(x)\phi_j(x)\, dx = \int_{-\pi}^{\pi} \cos ix \cos jx\, dx = \frac{1}{2} \int_{-\pi}^{\pi} [\cos(i+j)x + \cos(i-j)x]\, dx$$

$$= \frac{1}{2} \left[\frac{\sin(i+j)x}{i+j} + \frac{\sin(i-j)x}{i-j} \right]_{-\pi}^{\pi} = 0.$$

Also

$$\|\phi_i(x)\|^2 = \int_{-\pi}^{\pi} \phi_i^2(x)\, dx = \int_{-\pi}^{\pi} \cos^2 ix\, dx = \frac{1}{2} \int_{-\pi}^{\pi} (\cos 2ix + 1)\, dx$$

$$= \frac{1}{2} \left[\frac{\sin 2ix}{2i} + x \right]_{-\pi}^{\pi} = \pi,\ i \neq 0.$$

For $i = 0$, we have

$$\|\phi_0(x)\|^2 = \int_{-\pi}^{\pi} dx = 2\pi.$$

Hence, the functions

$$\frac{1}{\sqrt{2\pi}}, \frac{\cos x}{\sqrt{\pi}}, \frac{\cos 2x}{\sqrt{\pi}}, \ldots$$

are orthonormal over the interval $[-\pi, \pi]$.

An important application of the orthogonal or orthonormal functions is the series expansion of (suitable) functions in terms of the orthogonal functions. We first define a complete set of orthogonal functions.

Completeness Let $\{y_i(x)\}$ be a set of orthogonal functions defined on $[a, b]$ and $f(x)$ be any function defined on (a, b). If the relation $\int_a^b f(x) y_i(x) dx = 0$ can hold for all values of i, only if $f(x)$ is a null function on (a, b), then the orthogonal set $\{y_i(x)\}$ is said to be complete.

Let $\{\phi_j(x)\}$ be a complete set of orthogonal functions, orthogonal with respect to a weight function $W(x)$ over an interval $[a, b]$. Under suitable conditions, a function $f(x)$ can be expressed as a convergent series

$$f(x) = c_0\phi_0(x) + c_1\phi_1(x) + c_2\phi_2(x) + \dots . \tag{7.22}$$

This series is called an orthogonal expansion of the function $f(x)$ over $[a, b]$. Sometimes, it is also called a *generalised Fourier series*. If $f(x)$ can be expressed as the series (7.22), then the coefficients are easily obtained. Multiplying the equation (7.22) by $W(x)\phi_j(x)$ and integrating over $[a, b]$, we obtain on using the orthogonal property given in Eq. (7.18)

$$\int_a^b W(x) f(x)\phi_j(x) dx = \sum_{m=0}^{\infty} c_m \int_a^b W(x)\phi_m(x)\phi_j(x) dx = c_j \int_a^b W(x)\phi_j^2(x) dx.$$

Hence, $\qquad c_j = \left[\int_a^b W(x) f(x)\phi_j(x) dx\right] \Big/ \left[\int_a^b W(x)\phi_j^2(x) dx\right]$

or $\qquad c_j = \left[\int_a^b W(x) f(x)\phi_j(x) dx\right] \Big/ \|\phi_j(x)\|^2, \quad j = 0, 1, 2, \dots . \tag{7.23}$

The *Fourier series* belongs to this class of series, where the given function can be expanded in terms of 1, $\cos x$, $\sin x$, $\cos 2x$, $\sin 2x$, ... over the interval $[-\pi, \pi]$, (with $W(x) = 1$) as

$$f(x) = \sum_{n=0}^{\infty} (a_n \cos nx + b_n \sin nx). \tag{7.24}$$

Using the orthogonal property, we obtain the coefficients as

$$a_0 = \frac{1}{2\pi} \int_{-\pi}^{\pi} f(x) dx, \quad a_n = \frac{1}{\pi} \int_{-\pi}^{\pi} f(x) \cos nx \, dx, \quad n = 1, 2, \dots$$

$$b_n = \frac{1}{\pi} \int_{-\pi}^{\pi} f(x) \sin nx \, dx, \quad n = 1, 2, \dots \tag{7.25}$$

7.2.5 Orthogonal Property of Legendre Polynomials

The Legendre polynomials $P_n(x)$ satisfy the following orthogonal property ($W(x) = 1$)

$$\int_{-1}^{1} P_m(x) P_n(x) dx = \begin{cases} 0, & m \neq n & (7.26) \\ \dfrac{2}{2n+1}, & m = n & (7.27) \end{cases}$$

We shall prove property (7.26) by using the differential equation. The Legendre polynomials $P_m(x)$, $P_n(x)$ satisfy the differential equation

$$(1 - x^2)y''(x) - 2xy'(x) + n(n+1)y(x) = 0$$

that is,
$$(1 - x^2)P_m'' - 2xP_m' + m(m + 1)P_m = 0 \tag{7.28}$$

and
$$(1 - x^2)P_n'' - 2xP_n' + n(n + 1)P_n = 0. \tag{7.29}$$

Multiply Eq. (7.28) by $P_n(x)$ and (7.29) by $P_m(x)$, and subtract. We obtain

$$(1 - x^2)[P_m''P_n - P_n''P_m] - 2x[P_m'P_n - P_mP_n'] + [m(m + 1) - n(n + 1)]P_mP_n = 0.$$

Combining the first two terms, we get

$$\frac{d}{dx}[(1 - x^2)\{P_m'P_n - P_mP_n'\}] + (m - n)(m + n + 1)P_mP_n = 0.$$

Integrating over $[-1, 1]$, we obtain

$$\left[(1 - x^2)\{P_m'P_n - P_mP_n'\}\right]_{-1}^{1} + (m - n)(m + n + 1)\int_{-1}^{1} P_m(x)P_n(x)\, dx = 0.'$$

The first term vanishes at $x = \pm 1$. Since $m \neq n$, we get

$$\int_{-1}^{1} P_m(x)P_n(x)\, dx = 0, \quad m \neq n.$$

When $m = n$, it is easy to prove the result given in Eq. (7.27) by using the generating function

$$(1 - 2xt + t^2)^{-1/2} = \sum_{n=0}^{\infty} P_n(x)t^n.$$

Squaring both sides of the equation and integrating with respect to x over $[-1, 1]$, we obtain

$$\int_{-1}^{1} \frac{dx}{(1 - 2xt + t^2)} = \int_{-1}^{1} \left[\sum_{n=0}^{\infty} P_n(x)t^n\right]^2 dx.$$

From the left hand side, we get

$$\int_{-1}^{1} \frac{dx}{1 - 2xt + t^2} = \left[\frac{\ln(1 - 2xt + t^2)}{-2t}\right]_{-1}^{1}$$

$$= -\frac{1}{2t}[\ln(1 - 2t + t^2) - \ln(1 + 2t + t^2)]$$

$$= \frac{1}{2t}[\ln(1 + t)^2 - \ln(1 - t)^2] = \frac{1}{t}[\ln(1 + t) - \ln(1 - t)]$$

$$= \frac{1}{t}\left[\left(t - \frac{t^2}{2} + \frac{t^3}{3} - \cdots\right) - \left(-t - \frac{t^2}{2} - \frac{t^3}{3} - \cdots\right)\right]$$

$$= 2\left[1 + \frac{t^2}{3} + \frac{t^4}{5} + \cdots + \frac{t^{2n}}{2n + 1} + \cdots\right]. \tag{7.30}$$

Using the orthogonal property given in Eq. (7.26), we obtain from the right hand side

$$\int_{-1}^{1} \left[\sum_{n=0}^{\infty} P_n(x)t^n\right]^2 dx = \sum_{n=0}^{\infty} \int_{-1}^{1} P_n^2(x)t^{2n}\, dx = \sum_{n=0}^{\infty} t^{2n} \int_{-1}^{1} P_n^2(x)\, dx. \tag{7.31}$$

Comparing the coefficients of t^{2n}, in Eqs. (7.30) and (7.31), we obtain

$$\int_{-1}^{1} P_n^2(x)\,dx = \frac{2}{2n+1}.$$

7.2.6 Fourier-Legendre Series

Let $f(x)$ be a continuous function, having continuous derivatives over the interval $[-1, 1]$. Then, $f(x)$ can be written as the infinite series

$$f(x) = c_0 P_0(x) + c_1 P_1(x) + c_2 P_2(x) + \dots. \tag{7.32}$$

The series converges uniformly in $[-1, 1]$ and is unique. Multiplying both sides by $P_m(x)$ and integrating over $[-1, 1]$, we get on using the orthogonal property of $P_n(x)$

$$\int_{-1}^{1} f(x)P_m(x)\,dx = \sum_{n=0}^{\infty} c_n \int_{-1}^{1} P_m(x)P_n(x)\,dx = \frac{2c_m}{2m+1}.$$

Hence,

$$c_m = \frac{2m+1}{2} \int_{-1}^{1} f(x)P_m(x)\,dx, \quad m = 0, 1, 2, \dots. \tag{7.33}$$

If $f(x)$ is a polynomial, then it is easy to evaluate c_m using the Rodrigue's formula (7.8). In this case, we have

$$c_m = \frac{2m+1}{2} \int_{-1}^{1} \frac{1}{2^m\, m!} \frac{d^m}{dx^m}[(x^2-1)^m]\,f(x)\,dx$$

$$= \frac{(2m+1)}{2^{m+1}\, m!} \int_{-1}^{1} \frac{d^m}{dx^m}[(x^2-1)^m]\,f(x)\,dx.$$

Integrating by parts, we obtain

$$c_m = \frac{(2m+1)(-1)^m}{2^{m+1}\, m!} \int_{-1}^{1} (x^2-1)^m\, f^{(m)}(x)\,dx$$

since

$$\left\{\frac{d^n}{dx^n}[(x^2-1)^m]\right\}_{x=\pm 1} = 0, \quad \text{for } n < m.$$

If $f^{(m)}(x)$ is an odd function, then $c_m = 0$.

If $f^{(m)}(x)$ is an even function, then $\quad c_m = \dfrac{(-1)^m(2m+1)}{2^m\, m!} \displaystyle\int_0^1 (x^2-1)^m\, f^{(m)}(x)\,dx.$

If $f(x)$ is a polynomial, then at some stage $f^{(n)}(x) = $ constant, and $c_i = 0, \quad i > n$.

Example 7.8 Expand $f(x) = x^3 + x, -1 \le x \le 1$, as a Fourier-Legendre series.

Solution We have $f(x) = c_0 P_0(x) + c_1 P_1(x) + c_2 P_2(x) + \dots$, where

$$c_m = \frac{(2m+1)}{2} \int_{-1}^{1} f(x)P_m(x)\,dx = \frac{(-1)^m(2m+1)}{2^{m+1}\, m!} \int_{-1}^{1} (x^2-1)^m\, f^{(m)}(x)\,dx.$$

We have for $f(x) = x^3 + x$, $\quad f^{(4)}(x) = 0 = f^{(5)}(x) = \dots$. We get for $m = 0$: $\qquad c_0 = \dfrac{1}{2} \displaystyle\int_{-1}^{1} f(x)\, dx = \dfrac{1}{2} \int_{-1}^{1} (x^3 + x)\, dx = 0$

for $m = 1$: $\qquad c_1 = -\dfrac{3}{4} \displaystyle\int_{-1}^{1} (x^2 - 1) f'(x)\, dx = -\dfrac{3}{4} \int_{-1}^{1} (x^2 - 1)(3x^2 + 1)\, dx$

$$= -\dfrac{3}{2} \int_{0}^{1} (3x^4 - 2x^2 - 1)\, dx = -\dfrac{3}{2}\left(\dfrac{3}{5} - \dfrac{2}{3} - 1\right) = \dfrac{8}{5}$$

for $m = 2$: $\qquad c_2 = \dfrac{5}{16} \displaystyle\int_{-1}^{1} (x^2 - 1)^2 f''(x)\, dx = \dfrac{5}{16} \int_{-1}^{1} (x^2 - 1)^2 (6x)\, dx = 0$

for $m = 3$: $\qquad c_3 = -\dfrac{7}{96} \displaystyle\int_{-1}^{1} (x^2 - 1)^3 f'''(x)\, dx = -\dfrac{7}{96} \int_{-1}^{1} 6(x^2 - 1)^3\, dx$

$$= -\dfrac{7}{8} \int_{0}^{1} (x^6 - 3x^4 + 3x^2 - 1)\, dx = -\dfrac{7}{8}\left(\dfrac{1}{7} - \dfrac{3}{5} + \dfrac{3}{3} - 1\right) = \dfrac{2}{5}.$$

Hence, $f(x) = x^3 + x = [8P_1(x) + 2P_3(x)]/5$.

Example 7.9 If $f(x) = c_1\phi_1(x) + c_2\phi_2(x) + \ldots + c_n\phi_n(x)$, where c_k are constants and $\{\phi_k(x)\}$ is an orthonormal set on $[a, b]$, then show that

$$\int_{a}^{b} f^2(x)\, dx = c_1^2 + c_2^2 + \ldots + c_n^2.$$

Solution We have $\qquad \displaystyle\int_{a}^{b} f^2(x)\, dx = \int_{a}^{b} (c_1\phi_1 + c_2\phi_2 + \ldots + c_n\phi_n)^2\, dx$

$$= \int_{a}^{b} \Big[c_1^2\phi_1^2 + c_2^2\phi_2^2 + \ldots + c_n^2\phi_n^2)$$

$$+ 2(c_1 c_2 \phi_1 \phi_2 + \ldots + c_i c_j \phi_i \phi_j + \ldots + c_{n-1} c_n \phi_{n-1} \phi_n) \Big]\, dx$$

$$= c_1^2 + c_2^2 + \ldots + c_n^2$$

since $\{\phi_k(x)\}$ is an orthonomal set on $[a, b]$.

Exercise 7.1

1. Show that when $n = 0, 1$ one of the solutions $y_0(x)$, $y_1(x)$, (see Eqs. (7.6) of the Legendre's equation reduces to a polynomial and the other solution becomes a polynomial multiplied by $\ln[(1 + x)/(1 - x)]$.

2. Using the Rodrigue's formula directly, find the expressions for the Legendre polynomials, $P_1(x)$, $P_2(x)$, $P_3(x)$ and $P_4(x)$.

Express the following sums of Legendre polynomials in terms of powers of x.

   3. $6P_3(x) - 2P_1(x) + P_0(x)$. $\qquad\qquad$ 4. $8P_4(x) + 2P_2(x) + P_0(x)$.

   5. $4P_3(x) + 6P_2(x) - 3P_1(x) - 2P_0(x)$. $\qquad$ 6. $5P_4(x) + 10P_3(x) + 2P_2(x) + P_1(x)$.

Express the following polynomials in terms of Legendre polynomials.

7. $3x^2 + 5x - 6$.

8. $x^3 + x + 1$.

9. $4x^3 + 3x^2 + 2x - 6$.

10. $7x^4 + 6x^3 + 3x^2 + x - 6$.

11. $5x^4 + 3x^3 - 6x^2 - 2x + 3$.

Using the Rodrigue's formula prove the following.

12. $\displaystyle\int_{-1}^{1} P_n(x)\,dx = 0$.

13. $\displaystyle\int_{-1}^{1} x^m P_n(x)\,dx = 0, \quad m < n$.

14. $\displaystyle\int_{-1}^{1} f(x)P_n(x)\,dx = \frac{(-1)^n}{2^n n!}\int_{-1}^{1}(x^2-1)^n f^{(n)}(x)\,dx$.

15. Using the definition of Legendre polynomials, show that $P_n'(-x) = (-1)^{n+1} P_n'(x)$.

16. Using the generating function of Legendre polynomials prove that

$$P_{2n+1}(0) = 0, \quad P_{2n}(0) = (-1)^n \frac{1 \cdot 3 \cdot 5 \ldots (2n-1)}{2 \cdot 4 \cdot 6 \ldots 2n}.$$

17. Show that $\displaystyle\int_{-1}^{1} x P_n(x)P_{n-1}(x)\,dx = \frac{2n}{4n^2-1}, \quad n = 1, 2, \ldots$.

Prove the following recurrence formulas for the Legendre polynomials using the generating function.

18. $(2n+1)P_n(x) = P_{n+1}'(x) - P_{n-1}'(x)$.

19. $(n+1)P_n(x) = P_{n+1}'(x) - xP_n'(x)$.

20. $P_n'(x) = xP_{n-1}'(x) + nP_{n-1}(x)$.

21. $(1-x^2)P_n'(x) = n[P_{n-1}(x) - xP_n(x)]$.

22. $(1-x^2)P_n'(x) = (n+1)[xP_n(x) - P_{n+1}(x)]$.

Prove the following

23. $P_n'(1) = n(n+1)/2$.

24. $P_n'(-1) = (-1)^{n-1}\, n(n+1)/2$.

25. $P_{n+1}'(x) + P_n'(x) = P_0(x) + 3P_1(x) + 5P_2(x) + \ldots + (2n+1)P_n(x)$.

26. Using the recurrence relation given in Eq. (7.12), find $P_2(1.1)$, $P_3(2.5)$ and $P_4(1.2)$.

27. Using the recurrence relation given in problem 21, find $P_3'(1.6)$ and $P_4'(2.1)$.

28. In the r-θ plane, let P, Q be two distinct points such that $OP = r_1$ and $OQ = r_2$. Let $PQ = r$ (Fig. 7.2). From trigonometry, cosine formula gives $r^2 = r_1^2 + r_2^2 - 2r_1 r_2 \cos\theta$. Using the generating function of the Legendre polynomials, prove that

$$\frac{1}{r} = \frac{1}{[r_1^2 + r_2^2 - 2r_1 r_2 \cos\theta]^{1/2}} = \frac{1}{r_2}\sum_{m=0}^{\infty} P_m(\cos\theta)\left(\frac{r_1}{r_2}\right)^m, \quad r_2 \neq 0.$$

or as $\displaystyle\frac{1}{r} = \frac{1}{r_1}\sum_{m=0}^{\infty} P_m(\cos\theta)\left(\frac{r_2}{r_1}\right)^m, \quad r_1 \neq 0.$

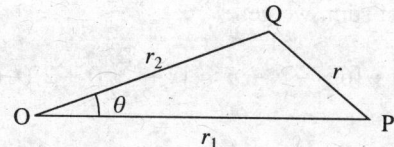

Fig. 7.2.   Problem 28.

29. Show that the functions $\phi_m(x) = \sin mx$, $m = 1, 2, \ldots$ are orthogonal over the interval $[-\pi, \pi]$. Construct the corresponding orthonormal set of functions.

30. Show that the functions 1, $\cos \pi x$, $\sin \pi x$, $\cos 2\pi x$, $\sin 2\pi x$, $\ldots$ are orthogonal over the interval $[-1, 1]$. Construct the corresponding orthonormal set of functions.

31. Show that the functions 1, $\cos x$, $\sin x$, $\cos 2x$, $\sin 2x$, $\ldots$ are orthogonal over the interval $[-\pi, \pi]$. Construct the corresponding orthonormal set of functions.

32. Using the Rodrigue's formula show that $\displaystyle\int_{-1}^{1} P_n^2(x)\,dx = \frac{2}{2n+1}$.

33. Prove that $\displaystyle\int_{-1}^{1} x P_n(x) P_m'(x) = 0$, or 2, or $(2n)/(2n+1)$.

34. Prove that $\displaystyle\int_{-1}^{1} (1 - x^2) P_n'(x) P_m'(x) = 0$, $m \neq n$.

35. Find the Fourier-Legendre series of the following functions.

(a) $f(x) = x^4 + x + 1$.   (b) $f(x) = 3x^3 + 2x^2 + x + 3$.

(c) $f(x) = 1 - 3x + 3x^2$.   (d) $f(x) = x^3 - 3x + 6$.

7.3   Chebyshev Differential Equation and Chebyshev Polynomials

One of the important differential equations which gives rise to special functions is the *Chebyshev differential equation*

$$(1 - x^2)\, y'' - xy' + n^2 y = 0 \tag{7.34}$$

where n is a real positive integer. The singularities of this equation are $x = \pm 1$. If we seek a power series solution of this equation about $x = 0$, of the form $y(x) = \displaystyle\sum_{m=0}^{\infty} c_m x^m$, then this series solution is convergent in $|x| < 1$, since the distance between $x = 0$ and the nearest singularity is 1. Substituting

$$y(x) = \sum_{m=0}^{\infty} c_m x^m, \quad y'(x) = \sum_{m=1}^{\infty} m c_m x^{m-1} \quad \text{and} \quad y''(x) = \sum_{m=2}^{\infty} m(m-1) c_m x^{m-2}$$

in Eq. (7.34), we obtain

$$\sum_{m=2}^{\infty} m(m-1) c_m x^{m-2} - \sum_{m=2}^{\infty} m(m-1) c_m x^m - \sum_{m=1}^{\infty} m c_m x^m + n^2 \sum_{m=0}^{\infty} c_m x^m = 0$$

or $2c_2 + 6c_3 x + \displaystyle\sum_{m=4}^{\infty} m(m-1) c_m x^{m-2} - \sum_{m=2}^{\infty} m(m-1) c_m x^m - c_1 x$

$$- \sum_{m=2}^{\infty} m c_m x^m + n^2 (c_0 + c_1 x) + n^2 \sum_{m=2}^{\infty} c_m x^m = 0.$$

Substituting $m - 2 = t$ in the first sum, we get

$$[2c_2 + n^2 c_0] + [6c_3 - c_1 + n^2 c_1] x + \sum_{t=2}^{\infty} (t+2)(t+1) c_{t+2} x^{t+2}$$

$$- \sum_{m=2}^{\infty} [m(m-1) + m - n^2] c_m x^m = 0.$$

Since t is a dummy variable, we can combine the third and fourth terms of this equation. Equating the coefficients of various powers of x to zero, we obtain

$$2c_2 + n^2 c_0 = 0, \quad 6c_3 - c_1 + n^2 c_1 = 0$$

and

$$(m + 2)(m + 1) c_{m+2} - (m^2 - n^2) c_m = 0, \, m = 2, 3, \ldots$$

Therefore,

$$c_2 = -\frac{n^2}{2!} c_0, \quad c_3 = \frac{1 - n^2}{3!} c_1 = -\frac{(n^2 - 1)}{3!} c_1,$$

$$c_{m+2} = -\frac{(n^2 - m^2)}{(m + 2)(m + 1)} c_m, \, m \geq 2. \tag{7.35}$$

We have,

$$c_4 = -\frac{(n^2 - 2^2)}{4 \cdot 3} c_2 = \frac{1}{4!} n^2 (n^2 - 2^2) c_0,$$

$$c_5 = -\frac{(n^2 - 3^2)}{5 \cdot 4} c_3 = \frac{1}{5!} (n^2 - 1^2)(n^2 - 3^2) c_1, \, \ldots\ldots$$

Substituting in the power series solution, we obtain

$$y(x) = c_0 \left[1 - \frac{1}{2!} n^2 x^2 + \frac{1}{4!} (n^2 - 2^2) n^2 x^4 + \ldots \right]$$

$$+ c_1 \left[x - \frac{1}{3!} (n^2 - 1^2) x^3 + \frac{1}{5!} (n^2 - 3^2)(n^2 - 1^2) x^5 + \ldots \right]$$

$$= c_0 y_0(x) + c_1 y_1(x) \tag{7.36}$$

where

$$y_0(x) = 1 - \frac{1}{2!} n^2 x^2 + \frac{1}{4!} (n^2 - 2^2) n^2 x^4 + \ldots$$

$$y_1(x) = x - \frac{1}{3!} (n^2 - 1^2) x^3 + \frac{1}{5!} (n^2 - 3^2)(n^2 - 1^2) x^5 + \ldots$$

The series $y_0(x)$, $y_1(x)$ converge for $| x | < 1$. Now, $y_0(x)$ contains even powers of x only and $y_1(x)$ contains odd powers of x only. Hence, the two solutions $y_0(x)$, $y_1(x)$ are linearly independent solutions of the Chebyshev differential equation.

As n takes the value zero and even positive integral values, we obtain for

$$n = 0: y_0(x) = 1; \quad n = 2: y_0(x) = 1 - 2x^2; \quad n = 4: y_0(x) = 1 - 8x^2 + 8x^4; \, \ldots$$

Hence, $y_0(x)$ reduces to an even polynomial as n takes even positive integral values, whereas $y_1(x)$ remains an infinite series. As n takes odd positive integral values, we obtain for

$$n = 1: y_1(x) = x; \quad n = 3: y_1(x) = \frac{1}{3} (3x - 4x^3); \quad n = 5: y_1(x) = \frac{1}{5} (5x - 20x^3 + 16x^5); \, \ldots$$

Hence, $y_1(x)$ reduces to an odd polynomial as n takes odd positive integral values, whereas $y_0(x)$ remains an infinite series. These polynomials give rise to an important class of polynomials called *Chebyshev polynomials*.

Now, substitute $x = \cos\theta$ in the Chebyshev differential equation (7.34). We have

$$\frac{dy}{dx} = \frac{dy}{d\theta}\frac{d\theta}{dx} = \left(-\frac{1}{\sin\theta}\right)\frac{dy}{d\theta},$$

$$\frac{d^2y}{dx^2} = \frac{d}{d\theta}\left[-\frac{1}{\sin\theta}\frac{dy}{d\theta}\right]\frac{d\theta}{dx} = \left(-\frac{1}{\sin\theta}\right)\left[-\frac{1}{\sin\theta}\frac{d^2y}{d\theta^2} + \frac{\cos\theta}{\sin^2\theta}\frac{dy}{d\theta}\right]$$

$$= \frac{1}{\sin^2\theta}\frac{d^2y}{d\theta^2} - \frac{\cos\theta}{\sin^3\theta}\frac{dy}{d\theta}.$$

Substituting in the differential equation, we obtain

$$\sin^2\theta\left[\frac{1}{\sin^2\theta}\frac{d^2y}{d\theta^2} - \frac{\cos\theta}{\sin^3\theta}\frac{dy}{d\theta}\right] - \cos\theta\left[-\frac{1}{\sin\theta}\frac{dy}{d\theta}\right] + n^2y = 0$$

or $\qquad\qquad \dfrac{d^2y}{d\theta^2} + n^2y = 0.$

The general solution of this differential equation is given by

$$y(\theta) = A\cos(n\theta) + B\sin(n\theta).$$

Therefore, the solution of Eq. (7.34) can be written as

$$y(x) = A\cos[n\cos^{-1}x] + B\sin[n\cos^{-1}x]. \tag{7.37}$$

Hence, $\cos(n\cos^{-1}x)$ and $\sin(n\cos^{-1}x)$ are two linearly independent solutions of Eq. (7.34). We denote the first solution by $T_n(x)$, that is,

$$T_n(x) = \cos(n\cos^{-1}x), \quad -1 \le x \le 1. \tag{7.38}$$

This defines the *Chebyshev polynomial of first kind*. We denote the second linearly independent solution as

$$U_n(x) = \sin(n\cos^{-1}x), \quad -1 \le x \le 1. \tag{7.39}$$

However, we shall show in the next section that $U_n(x)$ is not a polynomial.

7.3.1   Chebyshev Polynomials of First Kind

The Chebyshev polynomials of first kind are given by

$$T_n(x) = \cos(n\theta) = \cos(n\cos^{-1}x), \quad -1 \le x \le 1. \tag{7.40}$$

We note that $|T_n(x)| \le 1$.

These polynomials can be generated through a recurrence relation.

Recurrence relation for Chebyshev polynomials $T_n(x)$

Chebyshev polynomials $T_n(x)$ satisfy the following recurrence relation

$$T_{n+1}(x) - 2x\,T_n(x) + T_{n-1}(x) = 0. \tag{7.41}$$

We have

$$T_{n+1}(x) = \cos[(n+1)\cos^{-1}x] = \cos[(n+1)\theta],$$

$$T_{n-1}(x) = \cos[(n-1)\cos^{-1}x] = \cos[(n-1)\theta]$$

where $\cos^{-1}x = \theta$. Hence,

$$T_{n+1}(x) = \cos(n\theta + \theta) = \cos(n\theta)\cos\theta - \sin(n\theta)\sin\theta$$

and

$$T_{n-1}(x) = \cos(n\theta - \theta) = \cos(n\theta)\cos\theta + \sin(n\theta)\sin\theta.$$

Adding, we obtain

$$T_{n+1}(x) + T_{n-1}(x) = 2\cos(n\theta)\cos\theta = 2x\,T_n(x)$$

or

$$T_{n+1}(x) - 2x\,T_n(x) + T_{n-1}(x) = 0.$$

From Eq. (7.40), we have

$$T_0(x) = \cos(0) = 1, \quad T_1(x) = \cos(\cos^{-1}x) = x.$$

Using the recurrence relation given in Eq. (7.41), we get

$$T_2(x) = 2x\,T_1(x) - T_0(x) = 2x^2 - 1,$$

$$T_3(x) = 2x\,T_2(x) - T_1(x) = 2x\,(2x^2 - 1) - x = 4x^3 - 3x,$$

$$T_4(x) = 2x\,T_3(x) - T_2(x) = 2x\,(4x^3 - 3x) - (2x^2 - 1) = 8x^4 - 8x^2 + 1,$$

$$T_5(x) = 2x\,T_4(x) - T_3(x) = 2x\,(8x^4 - 8x^2 + 1) - (4x^3 - 3x) = 16x^5 - 20x^3 + 5x. \tag{7.42}$$

Therefore, $T_n(x)$ is a polynomial of degree n. If n is even, $T_n(x)$ is an even polynomial and if n is odd, $T_n(x)$ is an odd polynomial.

The first few Chebyshev polynomials are plotted in Fig. 7.3.

In many applications, we need to express x^n in terms of the Chebyshev polynomials $T_n(x)$. From Eq. (7.42), we obtain

$$1 = T_0, \quad x = T_1, \quad x^2 = \frac{1}{2}(T_0 + T_2), \quad x^3 = \frac{1}{4}(3T_1 + T_3)$$

$$x^4 = \frac{1}{8}\left[T_4 + 8\left\{\frac{1}{2}(T_0 + T_2)\right\} - T_0\right] = \frac{1}{8}[3T_0 + 4T_2 + T_4]$$

$$x^5 = \frac{1}{16}\left[T_5 + 20\left\{\frac{1}{4}(3T_1 + T_3)\right\} - 5T_1\right] = \frac{1}{16}[10T_1 + 5T_3 + T_5] \tag{7.43}$$

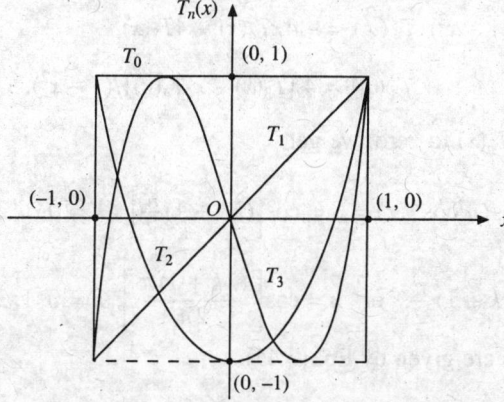

Fig. 7.3.   Chebyshev polynomials.

The relationship between $T_n'(x)$ and $T_n(x)$ is given by the recurrence relation

$$(1 - x^2) T_n'(x) = -nx\, T_n(x) + n T_{n-1}(x), \tag{7.44}$$

Set $x = \cos\theta$. Then, we get

$$(1 - x^2) T_n'(x) = (1 - \cos^2\theta) \frac{d}{d\theta} [\cos(n\theta)] \frac{d\theta}{dx} = \sin^2\theta[-n\sin(n\theta)] \left[-\frac{1}{\sin\theta}\right]$$

$$= n \sin\theta \sin(n\theta)$$

and

$$nT_{n-1}(x) - nx\, T_n(x) = n \cos[(n-1)\,\theta] - n \cos\theta \cos(n\theta)$$

$$= n [\cos(n\theta) \cos\theta + \sin(n\theta) \sin\theta] - n \cos\theta \cos(n\theta)$$

$$= n \sin\theta \sin(n\theta).$$

Hence, the relation given in Eq. (7.44) is proved.

Example 7.10 Write the following polynomials in terms of the Chebyshev polynomials of first kind.

(i) $x^3 + 3x^2 - 5x + 2$,   (ii) $5x^4 - x^3 + 3$.

Solution   Using the expressions given in Eq. (7.43), we have the following results.

(i)
$$x^3 + 3x^2 - 5x + 2 = \frac{1}{4}(3T_1 + T_3) + \frac{3}{2}(T_0 + T_2) - 5T_1 + 2T_0$$

$$= \frac{1}{4} T_3 + \frac{3}{2} T_2 - \frac{17}{4} T_1 + \frac{7}{2} T_0.$$

(ii)
$$5x^4 - x^3 + 3 = \frac{5}{8}(3T_0 + 4T_2 + T_4) - \frac{1}{4}(3T_1 + T_3) + 3T_1$$

$$= \frac{5}{8} T_4 - \frac{1}{4} T_3 + \frac{5}{2} T_2 + \frac{9}{4} T_1 + \frac{15}{8} T_0.$$

Example 7.11   Find the expression for $T_4'(x)$ in terms of $T_4(x)$ and $T_3(x)$.

Solution   We use the recurrence formula

$$(1 - x^2)\ T_n'(x) = -nx\, T_n(x) + n T_{n-1}(x)$$

For $n = 4$, we get

$$(1 - x^2)\ T_4'(x) = -4x T_4(x) + 4 T_3(x)$$

or
$$T_4'(x) = 4\,[T_3(x) - x\, T_4(x)]/(1 - x^2).$$

Zeros of $T_n(x)$ Equating $T_n(x)$ to zero, we get

$$T_n(x) = \cos(n \cos^{-1}x) = 0 = \cos\left[(2k + 1)\frac{\pi}{2}\right], \quad k = 0, 1, 2, \ldots, n - 1$$

or
$$n\cos^{-1}x = (2k + 1)\frac{\pi}{2}, \ \text{ or }\ x = \cos\left[\frac{(2k + 1)\pi}{2n}\right], \ k = 0, 1, 2, \ldots, n - 1. \tag{7.45}$$

The n simple zeros of $T_n(x)$ are given by Eq. (7.45).

Turning points (extreme points) of $T_n(x)$ Setting $T_n'(x) = 0$, we get

$$T_n'(x) = \frac{n \sin (n \cos^{-1} x)}{\sqrt{1 - x^2}} = 0,$$

or $\qquad \sin (n \cos^{-1} x) = 0 = \sin (k\pi)$

or $\qquad n \cos^{-1} x = k\pi, \quad \text{or } x = \cos \left(\dfrac{k\pi}{n} \right), \quad k = 1, 2, \ldots, n - 1.$ $\qquad$ (7.46)

Therefore, $T_n(x)$ attains its relative maximum/minimum at $n - 1$ points given by Eq. (7.46). At these points, we have

$$T_n(x_k) = \cos (n \cos^{-1} x_k) = \cos (k\pi) = (-1)^k. \qquad (7.47)$$

Also, at the end points on the interval $[-1, 1]$, we have

$$T_n(-1) = \cos (n\pi) = (-1)^n \quad \text{and} \quad T_n(1) = \cos (0) = 1.$$

Therefore, $T_n(x)$ attains its maximum and minimum values at $n + 1$ points

$$x_k = \cos (k\pi/n), \quad k = 0, 1, 2, \ldots, n.$$

Leading coefficient of $T_n(x)$ $\quad$ We have

$$
\begin{aligned}
T_n(x) &= \cos n\theta = \text{Re} \, (e^{in\theta}) = \text{Re} \, [\cos \theta + i \sin \theta]^n \\
&= \text{Re} \, [\cos^n \theta + {}^nC_1 \cos^{n-1} \theta \, (i \sin \theta) + {}^nC_2 \cos^{n-2} \theta \, (i \sin \theta)^2 + \ldots] \\
&= \cos^n \theta - {}^nC_2 (\cos^{n-2} \theta) (\sin^2 \theta) + {}^nC_4 (\cos^{n-4} \theta) (\sin^4 \theta) + \ldots \\
&= x^n - {}^nC_2 x^{n-2} (1 - x^2) + {}^nC_4 x^{n-4} (1 - x^2)^2 + \ldots \\
&= [1 + {}^nC_2 + {}^nC_4 + \ldots] \, x^n + \text{terms of lower degree} \\
&= 2^{n-1} x^n + \ldots
\end{aligned}
$$
$\qquad$ (7.48)

where Re is the real part and since $1 + {}^nC_2 + {}^nC_4 + \ldots = 2^{n-1}$.

This shows that $T_n(x)$ is a polynomial of degree n and its leading coefficient is 2^{n-1}.

Generating function of Chebyshev polynomials $T_n(x)$

Chebyshev polynomials $T_n(x)$ can also be obtained by using a generating function. We shall prove that

$$\frac{1 - xt}{1 - 2xt + t^2} = \sum_{n=0}^{\infty} T_n(x) t^n. \qquad (7.49)$$

The function on the left hand side is called the *generating function* of the Chebyshev polynomials $T_n(x)$. We have

$$\sum_{n=0}^{\infty} T_n(x) t^n = \sum_{n=0}^{\infty} \cos(n\theta) t^n = \text{Re} \left[\sum_{n=0}^{\infty} e^{in\theta} t^n \right] \qquad (7.50)$$

where Re is the real part. Now, since $x = \cos \theta$, we have

$$\sum_{n=0}^{\infty} e^{in\theta} t^n = 1 + (e^{i\theta} t) + (e^{i\theta} t)^2 + \ldots = \frac{1}{1 - e^{i\theta} t}$$

$$= \frac{1}{(1 - t\cos\theta) - it\sin\theta} = \frac{1}{(1 - tx) - it(1 - x^2)^{1/2}}$$

$$= \frac{(1 - tx) + it(1 - x^2)^{1/2}}{(1 - tx)^2 + t^2(1 - x^2)} = \frac{(1 - tx) + it(1 - x)^{1/2}}{1 - 2tx + t^2}$$

Substituting in Eq. (7.50), we obtain

$$\sum_{n=0}^{\infty} T_n(x)t^n = \frac{1 - xt}{1 - 2xt + t^2}$$

Hence, the result is proved.

Expanding the left hand side in Eq. (7.49), we can obtain an alternate (equivalent) expansion of $T_n(x)$. We write

$$[1 - 2xt + t^2]^{-1} = [1 - (2x - t)t]^{-1}$$

$$= 1 + (2x - t)t + (2x - t)^2t^2 + \ldots + (2x - t)^{n-r}t^{n-r} + \ldots$$

Now, the term involving t^n, in the product $(2x - t)^{n-r}t^{n-r}$ is given by

$$t^{n-r}[{}^{(n-r)}C_r(-t)^r(2x)^{n-2r}] = (-1)^r[{}^{(n-r)}C_r(2x)^{n-2r}t^n].$$

Therefore, the contribution of all terms to t^n is given by

$$b_n = \sum_{r=0} (-1)^r[{}^{(n-r)}C_r(2x)^{n-2r}].$$

Now, the term involving t^{n-1}, in the product $(2x - t)^{n-r}t^{n-r}$ is given by

$$t^{n-r}[{}^{(n-r)}C_{r-1}(-t)^{r-1}(2x)^{n-2r+1}] = (-1)^{r-1}[{}^{(n-r)}C_{r-1}(2x)^{n-2r+1}t^{n-1}].$$

Therefore, the contribution of all terms to t^{n-1} is given by

$$b_{n-1} = \sum_{r=1} (-1)^{r-1}[{}^{(n-r)}C_{r-1}(2x)^{n-2r+1}].$$

Now, the left hand side of Eq. (7.49) gives

$$(1 - xt)[1 + \ldots + b_{n-1}t^{n-1} + b_n t^n + \ldots] = 1 + \ldots + (b_n - xb_{n-1})t^n + \ldots$$

Hence, the coefficient of t^n is obtained as

$$b_n - xb_{n-1} = \sum_{r=0} (-1)^r[{}^{(n-r)}C_r(2x)^{n-2r}] - \sum_{r=1} (-1)^{r-1}[{}^{(n-r)}C_{r-1}2^{n-2r+1}x^{n-2r+2}].$$

Set $r - 1 = p$ in the second sum and change the dummy variable p to r. We obtain

$$b_n - xb_{n-1} = \sum_{r=0} (-1)^r[{}^{(n-r)}C_r(2x)^{n-2r}] - \frac{1}{2}\sum_{r=0} (-1)^r[{}^{(n-r-1)}C_r(2x)^{n-2r}]$$

$$= \frac{1}{2}\Big[\{2[{}^nC_0] - {}^{n-1}C_0\}(2x)^n - \{2[{}^{(n-1)}C_1] - {}^{(n-2)}C_1\}(2x)^{n-2} + \ldots$$

$$+ \{2[{}^{(n-r)}C_r] - {}^{(n-r-1)}C_r\}(2x)^{n-2r} + \ldots\Big]$$

Hence,
$$T_n(x) = \frac{1}{2}\Big[(2x)^n - \big\{2[^{(n-1)}C_1] - {}^{(n-2)}C_1\big\}(2x)^{n-2}$$

$$+ \big\{2[^{(n-2)}C_2] - {}^{(n-3)}C_2\big\}(2x)^{n-4} - \ldots\Big] \tag{7.51}$$

The corresponding expression for x^n in terms of the Chebyshev polynomials can also be obtained. We write

$$x^n = (\cos\theta)^n = \frac{1}{2^n}(e^{i\theta} + e^{-i\theta})^n$$

$$= \frac{1}{2^n}\Big[e^{in\theta} + {}^nC_1 e^{i(n-1)\theta}e^{-i\theta} + {}^nC_2 e^{i(n-2)\theta}e^{-2i\theta} + \ldots + {}^nC_{n-1}e^{i(n-(n-1))\theta}e^{-(n-1)\theta} + e^{-in\theta}\Big]$$

On the right hand side, we collect the first and last terms, second and nth terms, ... and use the relation $\cos(r\theta) = (e^{ir\theta} + e^{-ir\theta})/2$. We obtain

$$x^n = \frac{1}{2^{n-1}}\big[\cos(n\theta) + {}^nC_1\cos\{(n-2)\theta\} + {}^nC_2\cos\{(n-4)\theta\} + \ldots\big]$$

$$= \frac{1}{2^{n-1}}\big[T_n(x) + {}^nC_1 T_{n-2}(x) + {}^nC_2 T_{n-4}(x) + \ldots\big] \tag{7.52}$$

where, for even n, we divide the coefficient of $T_0(x)$ by 2.

Integration of $T_n(x)$

Since $x = \cos\theta$, we have

$$\int T_n(x)\,dx = -\int \cos(n\theta)\sin\theta\,d\theta$$

$$= -\frac{1}{2}\int [\sin\{(n+1)\theta\} - \sin\{(n-1)\theta\}]\,d\theta$$

$$= \frac{1}{2}\left[\frac{\cos\{(n+1)\theta\}}{n+1} - \frac{\cos\{(n-1)\theta\}}{n-1}\right]$$

$$= \frac{1}{2}\left[\frac{1}{n+1}T_{n+1}(x) - \frac{1}{n-1}T_{n-1}(x)\right], n \geq 2. \tag{7.53}$$

For the cases $n = 0$ and $n = 1$, we get

$$\int T_0(x)\,dx = x = T_1, \quad \int T_1(x)\,dx = \frac{x^2}{2} = \frac{1}{4}[T_0 + T_2]. \tag{7.54}$$

Orthogonality of Chebyshev Polynomials $T_n(x)$

The Chebyshev polynomials $T_n(x)$ satisfy the orthogonality relations

$$\int_{-1}^{1} \frac{T_m(x)T_n(x)}{\sqrt{1-x^2}}\,dx = 0, \quad \text{if } m \neq n$$

$$= \pi, \quad \text{if } m = n = 0$$

$$= \pi/2, \quad \text{if } m = n \neq 0. \tag{7.55}$$

If $m = n = 0$, we get

$$\int_{-1}^{1} \frac{dx}{\sqrt{1 - x^2}} = 2\left[\sin^{-1} x\right]_0^1 = 2\left[\frac{\pi}{2}\right] = \pi$$

since $T_0(x) = 1$.

If $m = n \neq 0$, we get

$$I = \int_{-1}^{1} \frac{T_n^2}{\sqrt{1 - x^2}}\, dx = \int_{-1}^{1} \frac{\cos^2(n \cos^{-1} x)}{\sqrt{1 - x^2}}\, dx.$$

Substitute $\cos^{-1} x = t$ or $x = \cos t$. We have $dx = -\sin t\, dt$ and

$$I = -\int_{\pi}^{0} \frac{\cos^2(nt)}{\sin t}\, \sin t\, dt = \frac{1}{2} \int_{0}^{\pi} [1 + \cos(2nt)]\, dt$$

$$= \frac{1}{2}\left[t + \frac{\sin(2nt)}{2n}\right]_0^{\pi} = \frac{\pi}{2}.$$

If $m \neq n$, we get

$$I = \int_{-1}^{1} \frac{\cos(m \cos^{-1} x) \cos(n \cos^{-1} x)}{\sqrt{1 - x^2}}\, dx.$$

Again, substitute $\cos^{-1} x = t$. We get

$$I = -\int_{\pi}^{0} \frac{\cos(mt) \cos(nt)}{\sin t}\, (\sin t)\, dt$$

$$= \frac{1}{2} \int_{0}^{\pi} [\cos(m + n)t + \cos(m - n)t]\, dt$$

$$= \frac{1}{2}\left[\frac{\sin(m + n)t}{m + n} + \frac{\sin(m - n)t}{m - n}\right]_0^{\pi} = 0.$$

We say that the Chebyshev polynomials $T_n(x)$ are orthogonal with respect to the weight function $W(x) = 1/\sqrt{1 - x^2}$ on $[-1, 1]$.

Chebyshev series

Let $f(x)$ be a continuous function, having continuous derivatives over the interval $[-1, 1]$. Then, $f(x)$ can be written as the infinite series

$$f(x) = c_0 T_0(x) + c_1 T_1(x) + c_2 T_2(x) + \dots \tag{7.56}$$

The series converges uniformly in $[-1, 1]$ and is unique. Multiplying both sides by $T_m(x)/\sqrt{1 - x^2}$ and integrating over $[-1, 1]$, we get on using the orthogonal property of the Chebyshev polynomials

$$\int_{-1}^{1} \frac{f(x) T_m(x)}{\sqrt{1 - x^2}}\, dx = \sum_{n=0}^{\infty} c_n \int_{-1}^{1} \frac{T_m(x) T_n(x)}{\sqrt{1 - x^2}}\, dx = \pi c_0, \quad \text{for} \quad m = n = 0$$

$$= \frac{\pi}{2} c_m, \quad \text{for} \quad m = n \neq 0 \tag{7.57}$$

Hence, $$c_0 = \frac{1}{\pi} \int_{-1}^{1} \frac{f(x)}{\sqrt{1-x^2}} \, dx, \quad c_m = \frac{2}{\pi} \int_{-1}^{1} \frac{f(x) T_m(x)}{\sqrt{1-x^2}} \, dx, \quad m = 1, 2, \ldots \tag{7.58}$$

If we write the series given in Eq. (7.56) as

$$f(x) = \frac{1}{2} c_0 T_0(x) + c_1 T_1(x) + c_2 T_2(x) + \ldots \tag{7.59}$$

then, from Eq. (7.58), we can write

$$c_m = \frac{2}{\pi} \int_{-1}^{1} \frac{f(x) T_m(x)}{\sqrt{1-x^2}} \, dx, \quad m = 0, 1, 2, \ldots \tag{7.60}$$

This series is also written as

$$f(x) = \sum_{n=0}^{\infty}{}' c_n T_n(x)$$

where $\sum'$ means that the coefficient of the first term $T_0(x)$ is divided by 2.

Example 7.12 Expand $f(x) = x^3 + x$, $-1 \le x \le 1$, in a Chebyshev series.

Solution We have $f(x) = x^3 + x$.

Write $$f(x) = c_0 T_0(x) + c_1 T_1(x) + c_2 T_2(x) + \ldots$$

We obtain

$$c_0 = \frac{1}{\pi} \int_{-1}^{1} \frac{f(x)}{\sqrt{1-x^2}} \, dx = \frac{1}{\pi} \int_{-1}^{1} \frac{x^3 + x}{\sqrt{1-x^2}} \, dx = 0, \text{ (odd function)}.$$

$$c_m = \frac{2}{\pi} \int_{-1}^{1} \frac{(x^3 + x) T_m(x)}{\sqrt{1-x^2}} \, dx = 0, \text{ for } m > 3.$$

$$c_1 = \frac{2}{\pi} \int_{-1}^{1} \frac{(x^3 + x) T_1(x)}{\sqrt{1-x^2}} \, dx = \frac{2}{\pi} \int_{-1}^{1} \frac{x^4 + x^2}{\sqrt{1-x^2}} \, dx$$

$$= \frac{4}{\pi} \int_{0}^{1} \frac{x^4 + x^2}{\sqrt{1-x^2}} \, dx = \frac{4}{\pi} \int_{0}^{\pi/2} \frac{(\cos^4\theta + \cos^2\theta)}{\sin\theta} (\sin\theta) \, d\theta$$

$$= \frac{4}{\pi} \left[\frac{3}{4} \cdot \frac{1}{2} \cdot \frac{\pi}{2} + \frac{1}{2} \cdot \frac{\pi}{2} \right] = \frac{7}{4}.$$

$$c_2 = \frac{2}{\pi} \int_{-1}^{1} \frac{(x^3 + x)(2x^2 - 1)}{\sqrt{1-x^2}} \, dx = 0, \text{ (odd function)}$$

$$c_3 = \frac{2}{\pi} \int_{-1}^{1} \frac{(x^3 + x)(4x^3 - 3x)}{\sqrt{1-x^2}} \, dx = \frac{4}{\pi} \int_{0}^{\pi/2} (\cos^3\theta + \cos\theta)(4\cos^3\theta - 3\cos\theta) \, d\theta$$

$$= \frac{4}{\pi} \int_0^{\pi/2} (4\cos^6\theta + \cos^4\theta - 3\cos^2\theta)\, d\theta$$

$$= \frac{4}{\pi} \left[4 \left(\frac{5}{6} \cdot \frac{3}{4} \cdot \frac{1}{2} \cdot \frac{\pi}{2} \right) + \frac{3}{4} \cdot \frac{1}{2} \cdot \frac{\pi}{2} - 3 \left(\frac{1}{2} \cdot \frac{\pi}{2} \right) \right] = \frac{1}{4}.$$

Hence, $f(x) = \frac{1}{4}(T_3 + 7T_1)$.

This result can also be directly verified by writing x and x^3 in terms of Chebyshev polynomials.

Error of approximation

Let $f(x)$ be a continuous function and $P_n(x)$ be a polynomial approximation, of degree n, to $f(x)$. Then , the error of approximation is

$$e_n(x) = f(x) - P_n(x).$$

We wish to find that polynomial $P_n(x)$ of degree n, such that the maximum value of the absolute error, that is, $\max |e_n(x)|$ is minimum on $-1 \le x \le 1$. That is, we wish to find $P_n(x)$ such that

$$\{ \max |e_n(x)| \} = \text{minimum}, \quad -1 \le x \le 1.$$

This is also called the *minimax criterion*.

Consider the case when $f(x) \equiv 0$. Then, we need to find a polynomial of degree n, $P_n(x)$, with leading coefficient unity (called *monic polynomial*), for which $\max |e_n(x)|$ is a minimum. If this condition is to be satisfied, then we require that $P_n(x)$ must have alternate maximum and minimum values of equal magnitude at $n + 1$ points $-1 = x_0 < x_1 < \ldots < x_n = 1$. Now, $|T_n(x)| \le 1$ and $T_n(x)$ has alternate maximum and minimum values $+1$ and -1. Since the leading coefficient of $P_n(x)$ is unity, we have the required polynomial as

$$P_n(x) = \frac{1}{2^{n-1}} T_n(x).$$

Then, the maximum and minimum values in magnitude are 2^{1-n}. Now, let $p_n(x)$ be any polynomial of degree n with leading coefficient unity and $\tilde{T}_n(x) = T_n(x)/2^{n-1}$ be the monic Chebyshev polynomial. Then,

$$\max_{-1 \le x \le 1} |\tilde{T}_n(x)| \le \max_{-1 \le x \le 1} |p_n(x)|. \tag{7.61}$$

7.3.2 Chebyshev Polynomials of Second Kind

We have shown earlier that $u_n(x) = \sin(n \cos^{-1}x)$ is a linearly independent solution of the Chebyshev differential equation. The second linearly independent solution $T_n(x) = \cos(n \cos^{-1} x)$ is a polynomial of degree n and defines the Chebyshev polynomial of first kind. In Eq. (7.48), it was shown that the leading coefficient of $T_n(x)$ is 2^{n-1} and the expression for $T_n(x)$ can be obtained as the real part of $(\cos\theta + i\sin\theta)^n$. Now, $\sin(n\theta)$ which is the imaginary part of $(\cos\theta + i\sin\theta)^n$ is not a polynomial, since $\sin\theta = \sqrt{1 - \cos^2\theta} = \sqrt{1 - x^2}$ is a common factor of all the terms.

We define,
$$U_n(x) = \frac{\sin[(n+1)\theta]}{\sin\theta} = \frac{\sin[(n+1)\cos^{-1}x]}{\sin(\cos^{-1}x)}$$

$$= \frac{\sin[(n+1)\cos^{-1}x]}{\sqrt{1 - x^2}}, \quad -1 \le x \le 1 \tag{7.62}$$

as the Chebyshev polynomial of second kind. Since, $U_n(x)$ contains division by $\sin\theta$, the imaginary part of $(\cos\theta + i\sin\theta)^{n+1}$ defines a polynomal of degree n.

For $x = 1$, we get

$$U_n(1) = \lim_{x \to 1} \frac{\sin[(n+1)\cos^{-1}x]}{\sqrt{1-x^2}} = \lim_{\theta \to 0} \frac{\sin[(n+1)\theta]}{\sin\theta}$$

$$= \lim_{\theta \to 0} \frac{(n+1)\cos[(n+1)\theta]}{\cos\theta} = n+1. \qquad (7.63)$$

Similarly, the limiting value for $x = -1$ can be obtained (see problem 9, Exercise 7.2).

Recurrence relation

Chebyshev polynomials of second kind $U_n(x)$ satisfies the recurrence ralation

$$U_{n+1}(x) = 2x\, U_n(x) - U_{n-1}(x). \qquad (7.64)$$

We have

$$\sin[(n+2)\,\theta] + \sin(n\theta) = 2\sin\{(n+1)\theta\}\cos\theta.$$

Therefore,

$$\frac{\sin[(n+2)\theta] + \sin(n\theta)}{\sin\theta} = 2\cos\theta\left[\frac{\sin\{(n+1)\theta\}}{\sin\theta}\right]$$

or $\qquad\qquad U_{n+1}(x) + U_{n-1}(x) = 2x\, U_n(x).$

Expression for $U_n(x)$ We have

$$U_n(x) = \frac{\sin[(n+1)\theta]}{\sin\theta} = \frac{1}{\sin\theta}\,\mathrm{Im}[e^{i(n+1)\theta}]$$

where Im, denotes the imaginary part. We get

$$U_n(x) = \frac{1}{\sin\theta}\,\mathrm{Im}\,[\cos\theta + i\sin\theta]^{n+1}$$

$$= \frac{1}{\sin\theta}\,\mathrm{Im}\,[\cos^{n+1}\theta + {}^{(n+1)}C_1\cos^n\theta(i\sin\theta) + {}^{(n+1)}C_2\cos^{n-1}\theta(i\sin\theta)^2$$

$$+ {}^{(n+1)}C_3\cos^{n-2}\theta(i\sin\theta)^3 + \dots]$$

$$= \frac{1}{\sin\theta}\,[{}^{(n+1)}C_1\cos^n\theta\sin\theta - {}^{(n+1)}C_3\cos^{n-2}\theta\sin^3\theta + \dots]$$

$$= {}^{(n+1)}C_1\cos^n\theta - {}^{(n+1)}C_3\cos^{n-2}\theta\sin^2\theta + {}^{(n+1)}C_5\cos^{n-4}\theta\sin^4\theta - \dots$$

$$= {}^{(n+1)}C_1 x^n - {}^{(n+1)}C_3 x^{n-2}(1-x^2) + {}^{(n+1)}C_5 x^{n-4}(1-x^2)^2 - \dots$$

$$= 2^n x^n + \dots. \qquad (7.65)$$

which is a polynomial of degree n. The first few polynomials are obtained as

$$U_0(x) = 1, \; U_1(x) = \frac{\sin(2\theta)}{\sin\theta} = 2\cos\theta = 2x,$$

$$U_2(x) = 2x\, U_1(x) - U_0(x) = 4x^2 - 1,$$

$$U_3(x) = 2x\, U_2(x) - U_1(x) = 2x\,(4x^2 - 1) - 2x = 8x^3 - 4x,$$

$$U_4(x) = 2x\, U_3(x) - U_2(x) = 2x\,(8x^3 - 4x) - (4x^2 - 1) = 16x^4 - 12x^2 + 1.$$

Hence, we can write

$$1 = U_0, \; x = \frac{1}{2}U_1, \; x^2 = \frac{1}{4}[1 + U_2] = \frac{1}{2^2}[U_0 + U_2],$$

$$x^3 = \frac{1}{8}[4x + U_3] = \frac{1}{2^3}[2U_1 + U_3],$$

$$x^4 = \frac{1}{16}[U_4 + 12x^2 - 1] = \frac{1}{16}[U_4 + 3(U_0 + U_2) - U_0] = \frac{1}{2^4}[2U_0 + 3U_2 + U_4].$$

Generating Function for $U_n(x)$

Chebyshev polynomials of second kind, $U_n(x)$, can also be obtained by using a generating function. We shall prove that

$$\frac{1}{1 - 2xt + t^2} = \sum_{n=0}^{\infty} U_n(x)t^n. \tag{7.66}$$

The function on the left hand side is called the *generating function* of $U_n(x)$. We have

$$\sum_{n=0}^{\infty} U_n(x)t^n = \sum_{n=0}^{\infty} \frac{\sin[(n+1)\theta]}{\sin\theta} t^n = \frac{1}{\sin\theta} \, \text{Im}\left[\sum_{n=0}^{\infty} e^{i(n+1)\theta} t^n\right] \tag{7.67}$$

where Im is the imaginary part. Now, since $x = \cos\theta$, we have

$$\sum_{n=0}^{\infty} e^{i(n+1)\theta} t^n = e^{i\theta}[1 + (e^{i\theta}t) + (e^{i\theta}t)^2 + \ldots] = \frac{e^{i\theta}}{1 - e^{i\theta}t}$$

$$= \frac{e^{i\theta}}{(1 - t\cos\theta) - it\sin\theta} = \frac{e^{i\theta}[(1 - t\cos\theta) + it\sin\theta]}{(1 - t\cos\theta)^2 + t^2\sin^2\theta}$$

Substituting in Eq. (7.67), we obtain

$$\sum_{n=0}^{\infty} U_n(x)t^n = \frac{1}{\sin\theta}\left[\frac{t\cos\theta\sin\theta + \sin\theta\,(1 - t\cos\theta)}{(1 - t\cos\theta)^2 + t^2\sin^2\theta}\right]$$

$$= \frac{1}{(1 - tx)^2 + t^2(1 - x^2)} = \frac{1}{1 - 2xt + t^2}$$

Hence, the result is proved.

Orthogonality of Chebyshev Polynomials of Second Kind $U_n(x)$

The Chebyshev polynomials of second kind $U_n(x)$ satisfy the orthogonality relations

$$\int_{-1}^{1} \sqrt{1-x^2}\, U_m(x) U_n(x)\, dx = 0, \quad \text{for} \quad m \neq n,$$

$$= \frac{\pi}{2}, \quad \text{for} \quad m = n. \tag{7.68}$$

Let $m \neq n$. Substituting $x = \cos\theta$, we get

$$\int_{-1}^{1} \sqrt{1-x^2}\, U_m(x) U_n(x)\, dx = -\int_{\pi}^{0} \sin\theta \cdot \frac{\sin[(m+1)\theta]}{\sin\theta} \cdot \frac{\sin[(n+1)\theta]}{\sin\theta} \sin\theta\, d\theta$$

$$= \int_{0}^{\pi} \sin[(m+1)\theta]\, \sin[(n+1)\theta]\, d\theta$$

$$= \frac{1}{2} \int_{0}^{\pi} \left\{ \cos[(m-n)\theta] - \cos[(m+n+2)\theta] \right\} d\theta$$

$$= \frac{1}{2} \left\{ \frac{\sin[(m-n)\theta]}{m-n} - \frac{\sin[(m+n+2)]\theta}{m+n+2} \right\}_{0}^{\pi} = 0.$$

For $m = n$, we get

$$\int_{-1}^{1} \sqrt{1-x^2}\, U_n^2(x)\, dx = -\int_{\pi}^{0} \sin\theta \left\{ \frac{\sin[(n+1)\theta]}{\sin\theta} \right\}^2 \sin\theta\, d\theta$$

$$= \frac{1}{2} \int_{0}^{\pi} \left\{ 1 - \cos[2(n+1)\theta] \right\} d\theta = \frac{\pi}{2},$$

We say that the Chebyshev polynomials of second kind $U_n(x)$ are orthogonal with respect to the weight function $W(x) = \sqrt{1-x^2}$ on $[-1, 1]$.

Orthogonal series in terms of $U_n(x)$

Let $f(x)$ be a continuous function, having continuous derivatives over the interval $[-1, 1]$. Then, $f(x)$ can be written as the infinite series

$$f(x) = c_0 U_0(x) + c_1 U_1(x) + c_2 U_2(x) + \dots \tag{7.69}$$

Multiplying both sides by $\sqrt{1-x^2}\, U_m(x)$ and integrating over $[-1, 1]$, we get on using the orthogonal property of $U_n(x)$

$$\int_{-1}^{1} \sqrt{1-x^2}\, f(x) U_m(x)\, dx = \sum_{n=0}^{\infty} c_n \int_{-1}^{1} \sqrt{1-x^2}\, U_m(x) U_n(x)\, dx$$

$$= \frac{\pi}{2} c_m.$$

Therefore,
$$c_m = \frac{2}{\pi} \int_{-1}^{1} \sqrt{1-x^2}\, f(x)\, U_m(x)\, dx. \tag{7.70}$$

Example 7.13 Expand $f(x) = x^3 + x, -1 \le x \le 1$ in terms of Chebyshev polynomials of second kind $U_n(x)$.

Solution   we have

$$c_m = \frac{2}{\pi} \int_{-1}^{1} \sqrt{1 - x^2} \, (x^3 + x) U_m(x) dx.$$

Hence,

$$c_0 = \frac{2}{\pi} \int_{-1}^{1} \sqrt{1 - x^2} \, (x^3 + x) dx = \frac{2}{\pi} \int_0^{\pi} \sin\theta \, (\cos^3\theta + \cos\theta) \sin\theta \, d\theta$$

$$= \frac{2}{\pi} \int_0^{\pi} [(\cos^3\theta + \cos\theta)(1 - \cos^2\theta)] d\theta = 0. \quad [\text{since } f(\pi - \theta) = -f(\theta)].$$

$$c_1 = \frac{2}{\pi} \int_{-1}^{1} \sqrt{1 - x^2} \, (x^3 + x)(2x) dx = \frac{2}{\pi} \int_0^{\pi} 2 \sin^2\theta \cos\theta \, (\cos^3\theta + \cos\theta) \, d\theta$$

$$= \frac{8}{\pi} \int_0^{\pi/2} (\cos^2\theta - \cos^6\theta) d\theta = \frac{8}{\pi} \left[\frac{1}{2} \cdot \frac{\pi}{2} - \frac{5}{6} \cdot \frac{3}{4} \cdot \frac{1}{2} \cdot \frac{\pi}{2} \right] = \frac{3}{4}.$$

$$c_2 = \frac{2}{\pi} \int_{-1}^{1} \sqrt{1 - x^2} \, (x^3 + x)(4x^2 - 1) dx = \frac{2}{\pi} \int_0^{\pi} \sin^2\theta (\cos^3\theta + \cos\theta)(4\cos^2\theta - 1) d\theta$$

$$= 0, \quad (\text{since } f(\pi - \theta) = -f(\theta)).$$

$$c_3 = \frac{2}{\pi} \int_{-1}^{1} \sqrt{1 - x^2} \, (x^3 + x)(8x^3 - 4x) dx = \frac{8}{\pi} \int_0^{\pi} \sin^2\theta (\cos^3\theta + \cos\theta)(2\cos^3\theta - \cos\theta) d\theta$$

$$= \frac{16}{\pi} \int_0^{\pi/2} [-2 \cos^8\theta + \cos^6\theta + 2\cos^4\theta - \cos^2\theta] d\theta$$

$$= \frac{16}{\pi} \left[-2 \left(\frac{7}{8} \cdot \frac{5}{6} \cdot \frac{3}{4} \cdot \frac{\pi}{4} \right) + \left(\frac{5}{6} \cdot \frac{3}{4} \cdot \frac{\pi}{4} \right) + 2 \left(\frac{3}{4} \cdot \frac{\pi}{4} \right) - \frac{\pi}{4} \right] = \frac{1}{8}.$$

$c_m = 0, m > 3.$

Hence, $f(x) = \frac{1}{8} [6U_1(x) + U_3(x)].$

Relationship between $T_n(x)$ and $U_n(x)$

The Chebyshev polynomials of the first and second kinds $T_n(x)$, $U_n(x)$ are related as

$$T_n(x) = U_n(x) - x U_{n-1}(x) \tag{7.71}$$

We have

$$U_n(x) - x U_{n-1}(x) = \frac{\sin[(n+1)\theta]}{\sin\theta} - \frac{\cos\theta \sin(n\theta)}{\sin\theta}$$

$$= \frac{1}{\sin\theta} [\sin(n\theta) \cos\theta + \cos(n\theta) \sin\theta - \cos\theta \sin(n\theta)]$$

$$= \cos(n\theta) = T_n(x).$$

The solution of the Chebyshev differential equation (7.34) can now be written as

$$y(x) = A \cos{(n \cos^{-1}x)} + B \sin{(n \cos^{-1}x)}$$

$$= A\,T_n(x) + B\,\sqrt{1 - x^2}\,U_{n-1}(x), \quad \text{for} \quad n = 1, 2, 3, \ldots \tag{7.72}$$

and
$$y(x) = A + B \sin^{-1}x, \quad \text{for} \quad n = 0. \tag{7.73}$$

Exercise 7.2

Prove the following

1. $T_n(1) = 1$.

2. $T_n(-1) = (-1)^n$.

3. $T_n(-x) = (-1)^n\,T_n(x)$.

4. $T_{2n}(0) = (-1)^n$.

5. $T_{2n+1}(0) = 0$.

6. $\displaystyle\int_{-1}^{1} \frac{T_n(x)}{\sqrt{1 - x^2}}\,dx = 0$.

7. $\displaystyle\int_{-1}^{1} \frac{x^m\,T_n(x)}{\sqrt{1 - x^2}}\,dx = 0,\ m < n$.

8. $U_n(-x) = (-1)^n\,U_n(x)$.

9. $U_n(-1) = (-1)^n\,(n + 1)$.

10. $U_{2n}(0) = (-1)^n$.

11. $U_{2n+1}(0) = 0$.

12. Show that $y = U_n(x)$ satisfies the differential equation

$$(1 - x^2)\,y'' - 3xy' + n(n + 2)\,y = 0.$$

13. Show that $T_n'(x) = n\,U_{n-1}(x)$.

14. Show that $\displaystyle\int U_n(x)\,dx = \frac{1}{n + 1}\,T_{n+1}(x)$.

15. Show that $(1 - x^2)\,U_n'(x) = x\,U_n(x) - (n + 1)\,T_{n+1}(x)$.

16. Show that $\displaystyle\frac{1 - t^2}{1 - 2xt + t^2} = T_0(x) + 2 \sum_{n=1}^{\infty} T_n(x)\,t^n$.

(The function on the left hand side can be considered as a *generating function* for the Chebyshev polynomials $T_n(x)$).

17. Show that the Chebyshev polynomials of first kind $T_n(x)$ can be defined by the formula

$$T_\dot{n}(x) = c_n\,(1 - x^2)^{1/2}\,\frac{d^n}{dx^n}\,[(1 - x^2)^{n-1/2}]$$

where c_n is a constant to be determined from some normalisation criterion.

18. Show that the Chebyshev polynomials of second kind $U_n(x)$ can be defined by the formula

$$U_n(x) = c_n\,(1 - x^2)^{-1/2}\,\frac{d^n}{dx^n}\,[(1 - x^2)^{n+1/2}]$$

where c_n is a constant to be determined from some normalisation criterion.

19. The finite Chebyshev series $p(x) = \displaystyle\sum_{r=0}^{n}{}' c_r T_r(x)$ is given ($\sum'$ means that the first term is divided by 2). The derivative $p'(x)$ is a polynomial of degree $n - 1$. Express $p'(x)$ as

$$p'(x) = \sum_{r=0}^{n-1}{}' a_r T_r(x)$$

and evaluate the coefficients a_r in terms of c_r.

20. (a) Express $p(x) = x^4 + 3x^3 + 2x^2 + 5x + 1$ as a Chebyshev series in terms of $T_n(x)$.
(b) Evaluate $p'(x)$ in terms of $T_n(x)$ using problem 19. Verfy the result using direct differentiation.

7.4   Bessel's Differential Equation and Bessel Functions

One of the important differential equations that is commonly encountered in science and engineering is the Bessel's equation

$$x^2 y'' + xy' + (x^2 - v^2)y = 0. \tag{7.74}$$

of order v, where v is a non-negative real number. The point $x = 0$ is a regular singular point of the equation and Frobenius series solution exists for the equation. Substituting

$$y(x) = \sum_{m=0}^{\infty} c_m x^{m+r}$$

in Eq. (7.74), we obtain

$$\sum_{m=0}^{\infty} (m+r)(m+r-1)c_m x^{m+r} + \sum_{m=0}^{\infty} (m+r)c_m x^{m+r} + \sum_{m=0}^{\infty} c_m x^{m+r+2} - v^2 \sum_{m=0}^{\infty} c_m x^{m+r} = 0$$

or

$$\sum_{m=0}^{\infty} [(m+r)^2 - v^2]c_m x^{m+r} + \sum_{m=0}^{\infty} c_m x^{m+r+2} = 0.$$

Setting the coefficient of x^r to zero, we get the indicial equation as $r^2 - v^2 = 0$, giving the values $r = v, -v$. Equating the coefficient of x^{r+1} to zero, we obtain

$$[(r+1)^2 - v^2]c_1 = 0, \quad \text{or} \quad c_1 = 0, \quad \text{for} \quad r = \pm v.$$

The remaining terms of the summations are

$$\sum_{m=2}^{\infty} [(m+r)^2 - v^2]c_m x^{m+r} + \sum_{m=0}^{\infty} c_m x^{m+r+2} = 0.$$

Setting $m - 2 = t$ in the first summation and changing the dummy variable t to m, we get

$$\sum_{m=0}^{\infty} \{[(m+r+2)^2 - v^2]c_{m+2} + c_m\}x^{m+r+2} = 0.$$

Setting the coefficient of x^{m+r+2} to zero, we get

$$c_{m+2} = -\frac{c_m}{(m+r+2-v)(m+r+2+v)}, \quad m \geq 0.$$

Since $c_1 = 0$, we obtain

$$c_3 = 0 = c_5 = \ldots.$$

For $r = v$, we get

$$c_{m+2} = -\frac{c_m}{(m+2)(m+2v+2)}, \quad m \geq 0$$

$$c_2 = -\frac{c_0}{2^2(1+v)}, \quad c_4 = -\frac{c_2}{2^3(2+v)} = \frac{c_0}{2^4(2!)(1+v)(2+v)}$$

$$c_6 = -\frac{c_4}{2^2(3)(3+v)} = -\frac{c_0}{2^6(3!)(1+v)(2+v)(3+v)}, \ldots$$

or

$$c_{2m} = \frac{(-1)^m c_0}{2^{2m}(m!)(1+v)(2+v)\ldots(m+v)}, \quad m = 1, 2, \ldots. \tag{7.75}$$

Therefore, one of the linearly independent solutions of the Bessel's equation is

$$y_1(x) = c_0 x^v \left[1 - \frac{x^2}{2^2(1+v)} + \frac{x^4}{2^4(2!)(1+v)(2+v)} - \cdots \right] \tag{7.76}$$

If we make a suitable choice of a value for c_0, then the expression on the right side of Eq. (7.76) defines a very important function called the Bessel's function $J_v(x)$, which is a solution of the Bessel's differential equation.

Let the value of c_0 be chosen as

$$c_0 = 1/[2^v \Gamma(v+1)] \tag{7.77}$$

where Γ is the *Gamma function* defined by

$$\Gamma(v) = \int_0^\infty e^{-t} t^{v-1} \, dt \tag{7.78}$$

We know that $\Gamma(v+1) = v\,\Gamma(v)$. Substituting the value of c_0 from Eq. (7.77) in Eq. (7.75), we obtain

$$c_{2m} = \frac{(-1)^m}{2^{2m+v}(m!)\Gamma(v+1)[(1+v)(2+v)\ldots(m+v)]} = \frac{(-1)^m}{2^{2m+v}(m!)\Gamma(m+v+1)}. \tag{7.79}$$

Then, $y_1(x)$ defines the *Bessel's function of the first kind* of order v. The Bessel's function of the first kind is denoted by $J_v(x)$.

Therefore, from Eqs. (7.76) and (7.79) we obtain

$$J_v(x) = \sum_{m=0}^\infty \frac{(-1)^m}{m!\,\Gamma(m+v+1)} \left(\frac{x}{2} \right)^{2m+v} \tag{7.80}$$

$$= \left(\frac{x}{2} \right)^v \left[\frac{1}{\Gamma(v+1)} - \frac{1}{1!\,\Gamma(v+2)} \left(\frac{x}{2} \right)^2 + \frac{1}{2!\,\Gamma(v+3)} \left(\frac{x}{2} \right)^4 - \cdots \right]. \tag{7.81}$$

The series converges very fast.

$J_v(x)$ is one of the linearly independent solutions of the Bessel's equation. To find the complete (general) solution of the Bessel's equation, we need to find the second linearly independent solution. We have earlier shown that the indicial roots of the Bessel's equation are $r = \pm v$. We can check whether $J_{-v}(x)$ gives us the second linearly independent solution . We have

$$J_{-v}(x) = \sum_{m=0}^\infty \frac{(-1)^m}{m!\,\Gamma(m-v+1)} \left(\frac{x}{2} \right)^{2m-v}$$

$$= \left(\frac{x}{2} \right)^{-v} \left[\frac{1}{\Gamma(1-v)} - \frac{1}{1!\,\Gamma(2-v)} \left(\frac{x}{2} \right)^2 + \cdots \right].$$

The leading terms of $J_v(x)$ and $J_{-v}(x)$ contain x^v and x^{-v} respectively. Therefore, if v is not an integer, then $J_v(x)$ and $J_{-v}(x)$ are linearly independent. Hence, when v is not an integer, we have the general solution of the Bessel's equation as

$$y(x) = A J_v(x) + B J_{-v}(x). \tag{7.82}$$

Example 7.14   Find the solution of the differential equation

$$x^2 y'' + xy' + \left(x^2 - \frac{1}{16} \right) y = 0.$$

Soluton   The given equation is the Bessel's differential equation with $v = 1/4$. Therefore, the complete solution is

$$y(x) = A J_v(x) + B J_{-v}(x) = A J_{1/4}(x) + B J_{-1/4}(x).$$

Example 7.15   Transform the equation

$$4x^2 y'' + 4xy' + (64x^2 - 9)y = 0$$

using the substitution $4x = z$ and hence find the general solution of the equation.

Solution   When $4x = z$, we have

$$\frac{dy}{dx} = \frac{dy}{dz}\frac{dz}{dx} = 4\frac{dy}{dz}, \quad \text{and} \quad \frac{d^2 y}{dx^2} = 4\frac{d^2 y}{dz^2}\frac{dz}{dx} = 16\frac{d^2 y}{dz^2}.$$

The given equation is transformed to

$$4\left(\frac{z^2}{16} \right)(16)\frac{d^2 y}{dz^2} + 4\left(\frac{z}{4} \right)(4)\frac{dy}{dz} + (4z^2 - 9)y = 0$$

or

$$\frac{d^2 y}{dz^2} + z\frac{dy}{dz} + \left(z^2 - \frac{9}{4} \right) y = 0.$$

This equation is the Bessel's differential equation with $v = 3/2$. Hence, the general solution is

$$y(z) = A J_{3/2}(z) + B J_{-3/2}(z), \text{ or } y(x) = A J_{3/2}(4x) + B J_{-3/2}(4x).$$

Case when v is an integer

Let $v = n$ be an integer. Then the expression for c_{2m} given in Eq. (7.79) can be simplified. Since $\Gamma(v + 1) = v\Gamma(v)$, we have, when $v = n$

$$\Gamma(1) = 1, \quad \Gamma(2) = \Gamma(1) = 1!, \quad \Gamma(3) = 2\Gamma(2) = 2!, \dots, \Gamma(n + 1) = n\Gamma(n) = n!.$$

Therefore, from Eq. (7.77) we obtain $c_0 = 1/(2^n n!)$ and from Eq. (7.79),

$$c_{2m} = \frac{(-1)^m}{2^{2m+n}(m!)(m + n)!}. \tag{7.83}$$

The Bessel function $J_n(x)$ is now defined as

$$J_n(x) = \sum_{m=0}^{\infty} \frac{(-1)^m}{m!(m + n)!}\left(\frac{x}{2} \right)^{2m+n} \tag{7.84}$$

The series converges very rapidly, for all x, since the denominators of the terms become larger and larger very fast. The Bessel's functions of orders zero and one are given below.

Bessel's function of order zero

Setting $n = 0$ in Eq. (7.84), we get

$$J_0(x) = \sum_{m=0}^{\infty} \frac{(-1)^m}{(m!)^2}\left(\frac{x}{2}\right)^{2m} = 1 - \frac{x^2}{2^2(1!)^2} + \frac{x^4}{2^4(2!)^2} - \ldots \tag{7.85}$$

We find that $J_0(0) = 1$ so that the graph of $y = J_0(x)$ cuts the y-axis at $(0, 1)$. Further, $|J_0(x)| \le 1$. The presence of positive and negative terms show that the graph of $y = J_0(x)$ oscillates and decays fast as $x \to \infty$. The graph of $J_0(x)$ is given in Fig. 7.4.

An approximation to the first zero of $J_0(x)$, (the first point where the graph cuts the x-axis) can be obtained by setting the first two terms of the series to zero. We obtain the approximation as

$$1 - \frac{x^2}{4} = 0, \quad \text{or} \quad x = 2. \tag{7.86}$$

A better approximation is obtained by adding one more term to the approximation. The next approximation to the zero is obtained, when

$$1 - \frac{x^2}{4} + \frac{x^4}{64} = 0, \quad \text{or} \quad x^4 - 16x^2 + 64 = 0,$$

or
$$(x^2 - 8)^2 = 0, \quad \text{or} \quad x = \sqrt{8} \approx 2.828. \tag{7.87}$$

Bessel function of order 1

Setting $n = 1$ in Eq. (7.84), we get

$$J_1(x) = \sum_{m=0}^{\infty} \frac{(-1)^m}{m!(m+1)!}\left(\frac{x}{2}\right)^{2m+1} = \frac{x}{2} - \frac{x^3}{(2^3)(1!)(2!)} + \frac{x^5}{(2^5)(2!)(3!)} - \ldots$$

We find that $J_1(0) = 0$ so that the graph of $y = J_1(x)$ passes through the origin. The curve $y = J_1(x)$ oscillates and decays rapidly as $x \to \infty$. The graph of $J_1(x)$ is given in Fig. 7.4. The first five zeros of $J_0(x)$ and $J_1(x)$ are given below:

$$J_0(x): 2.4048, \ 5.5201, \ 8.6537, \ 11.7915, \ 14.9309,$$

$$J_1(x): 3.8317, \ 7.0156, \ 10.1735, \ 13.3237, \ 16.4706.$$

It is interesting to note that the successive large roots differ approximately by $\pi = 3.1415\ldots$

When $v = n$, an integer, we can show that $J_v(x)$ and $J_{-v}(x)$ are linearly dependent. We have

$$J_{-n}(x) = \sum_{m=0}^{\infty} \frac{(-1)^m}{m!\Gamma(m-n+1)!}\left(\frac{x}{2}\right)^{2m-n} = \sum_{m=n}^{\infty} \frac{(-1)^m}{m!(m-n)!}\left(\frac{x}{2}\right)^{2m-n}$$

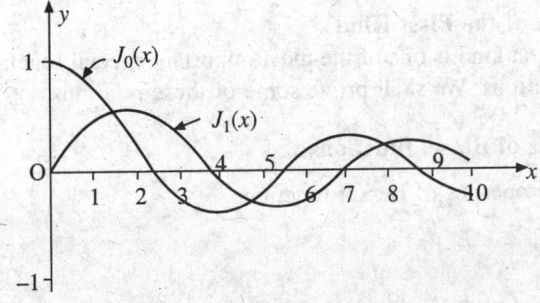

Fig. 7.4. Graphs of $J_0(x)$ and $J_1(x)$.

since $\Gamma(m - n + 1) = (m - n)!$, and $\Gamma(m - n + 1)$ becomes infinite for $m = 0, 1, 2, \ldots, n - 1$. Now, let $m - n = k$. Then

$$J_{-n}(x) = \sum_{k=0}^{\infty} \frac{(-1)^{n+k}}{(n+k)!\, k!}\left(\frac{x}{2}\right)^{2k+n} = (-1)^n \sum_{k=0}^{\infty} \frac{(-1)^k}{k!(n+k)!}\left(\frac{x}{2}\right)^{2k+n} = (-1)^n J_n(x). \quad (7.88)$$

Therefore, $J_n(x)$ and $J_{-n}(x)$ are linearly dependent.

We will discuss the method of obtaining the second linearly independent solution in section 7.4.2.

Exercise 7.3

1. Show that $J_n(x)$ is an even function for n even and an odd function for n odd where n is an integer.

2. For small $|x|$, $J_0(x)$ can be approximated as

   $J_0(x) \simeq 1 - 0.25x^2$, or $J_0(x) \simeq 1 - 0.25x^2 + 0.015625x^4$.

   Compute $J_0(x)$ for $x = 0.1, 0.2, 0.3, 0.4$ and 0.5 and compare with the exact values from the table of Bessel functions.

3. Show that the smallest positive root of $J_0(x) = 0$ lies in the interval $(2, \sqrt{8})$.

4. Using the value $\Gamma(1/2) = \sqrt{\pi}$, show that $J_{1/2}(x)$ and $J_{-1/2}(x)$ can be expressed as

   $$J_{1/2}(x) = \sqrt{\frac{2}{\pi x}} \sin x, \quad J_{-1/2}(x) = \sqrt{\frac{2}{\pi x}} \cos x.$$

5. Using the series expansions for $J_0(x)$ and $J_1(x)$ show that $J_0'(x) = -J_1(x)$.

6. Show that between every successive pair of zeros of $J_0(x)$, there exists a zero of $J_1(x)$.

Find the general solution of the following differential equations.

7. $x^2 y'' + xy' + (x^2 - 1/4) y = 0$,   8. $36x^2 y'' + 36xy' + (36x^2 - 4) y = 0$.

Using the indicated substitution, transform the given differential equation and hence find the general solution in terms of Bessel's functions.

9. $4x^2 y'' + 4xy' + (x^2 - 1)y = 0$, $x = 2z$.   10. $9x^2 y'' + 9xy' + (81x^2 - 1)y = 0$. $z = 3x$.

11. $x^2 y'' + xy' + (\lambda^2 x^2 - v^2)y = 0$, $z = \lambda x$, λ is a real number and v a fraction.

12. $xy'' + (1 + 2\lambda)y' + xy = 0$, $y = x^{-\lambda}u$, λ is a real fraction.

13. $x^2 y'' + xy' + 4(x^4 - v^2)y = 0$, $x^2 = z$, v is a real fraction.

14. $y'' + xy = 0$, $y = x^{1/2}u$, $2x^{3/2} = 3z$.

15. $x^2 y'' + (1 - 2v)xy' + v^2(x^{2v} + 1 - v^2)y = 0$, $y = x^v u$, $x^v = z$, v is a fraction.

7.4.1   Bessel's Function of the First Kind

Bessel's function of the first kind is one of the most important special functions and satisfies a large number of recurrence relations. We shall prove some of these relations.

Derivatives and integrals of Bessel functions

We prove the following properties of Bessel functions.

P_1:   (a) $[x^v J_v(x)]' = x^v J_{v-1}(x)$.

                                                       (7.89)

   (b) $\displaystyle\int x^v J_{v-1}(x)\, dx = x^v J_v(x) + c.$

                                                       (7.90)

P_2:   (a) $[x^{-v}J_v(x)]' = -x^{-v}J_{v+1}(x).$ (7.91)

      (b) $\displaystyle\int x^{-v}J_{v+1}(x)\,dx = -x^{-v}J_v(x) + c.$ (7.92)

Multiplying the Bessel's function $J_v(x)$, given in Eq. (7.80), by x^v we get

$$x^v J_v(x) = \sum_{m=0}^{\infty} \frac{(-1)^m x^{2m+2v}}{m!\,\Gamma(m+v+1)\,2^{2m+v}}.$$

Differentiating this equation with respect to x, and using the property,

$$\Gamma(m+v+1) = (m+v)\,\Gamma(m+v)$$

we get $\displaystyle [x^v J_v(x)]' = \sum_{m=0}^{\infty} \frac{(-1)^m (2m+2v)x^{2m+2v-1}}{m!\,2^{2m+v}\,\Gamma(m+v+1)}$

$$= x^v \sum_{m=0}^{\infty} \frac{(-1)^m}{m!\,\Gamma(m+v)}\left(\frac{x}{2}\right)^{2m+v-1} = x^v J_{v-1}(x).$$

Therefore, $[x^v J_v(x)]' = x^v J_{v-1}(x)$.

Setting $v = 0$, we get $J_0'(x) = J_{-1}(x) = -J_1(x)$, since $J_{-n}(x) = (-1)^n J_n(x)$.

Integrating Eq. (7.89), we get

$$\int x^v J_{v-1}(x)\,dx = \int [x^v J_v(x)]'\,dx + c = x^v J_v(x) + c.$$

Multiplying the Bessel's function $J_v(x)$ by x^{-v} and differentiating, we get

$$[x^{-v}J_v(x)]' = \sum_{m=0}^{\infty} \frac{(-1)^m (2m)x^{2m-1}}{m!\,2^{2m+v}\,\Gamma(m+v+1)}$$

$$= \sum_{m=1}^{\infty} \frac{(-1)^m x^{2m-1}}{(m-1)!\,2^{2m+v-1}\,\Gamma(m+v+1)}.$$

Setting $m - 1 = k$, we obtain

$$[x^{-v}J_v(x)]' = \sum_{k=0}^{\infty} \frac{(-1)^{k+1} x^{2k+1}}{k!\,2^{2k+v+1}\,\Gamma(k+v+2)}$$

$$= x^{-v} \sum_{k=0}^{\infty} \frac{(-1)^{k+1}}{k!\,\Gamma(k+v+2)}\left(\frac{x}{2}\right)^{2k+v+1} = -x^{-v}J_{v+1}(x).$$

Therefore, $[x^{-v}J_v(x)]' = -x^{-v}J_{v+1}(x).$

Integrating, we get

$$\int x^{-v}J_{v+1}(x)\,dx = -\int [x^{-v}J_v(x)]'\,dx + c = -x^{-v}J_v(x) + c.$$

Example 7.16   Evaluate the following integrals in terms of the Bessel's functions.

   (a) $\displaystyle\int \frac{1}{x} J_2(x)\,dx,$ (b) $\displaystyle\int x^2 J_1(x)\,dx,$ (c) $\displaystyle\int x^3 J_0(x)\,dx.$

Solution

(a) Using Eq. (7.92), we get $\int x^{-1} J_2(x)\, dx = -x^{-1} J_1(x) + c.$

(b) Using Eq. (7.90), we get $\int x^2 J_1(x)\, dx = x^2 J_2(x) + c.$

(c) Using Eq. (7.90), we write the given integral as

$$\int x^2 (xJ_0)\, dx = x^2(xJ_1) - \int (xJ_1)(2x)\, dx$$

$$= x^3 J_1 - 2 \int x^2 J_1\, dx = x^3 J_1 - 2x^2 J_2 + c.$$

Example 7.17   Show that

$$\int x J_0^2(x)\, dx = \frac{1}{2} x^2 [J_0^2(x) + J_1^2(x)].$$

Solution   Integrating by parts, we get

$$\int x J_0^2(x)\, dx = \frac{x^2}{2} J_0^2(x) - \int \frac{x^2}{2} (2 J_0 J_0')\, dx$$

$$= \frac{x^2}{2} J_0^2(x) + \int x^2 J_0 J_1\, dx \qquad (\because J_0' = -J_1)$$

$$= \frac{x^2}{2} J_0^2(x) + \int (xJ_1)\frac{d}{dx}(xJ_1)\, dx.$$

since from Eq. (7.89), we have $(xJ_1)' = xJ_0$. Therefore,

$$\int x J_0^2(x)\, dx = \frac{x^2}{2} J_0^2(x) + \frac{1}{2}(xJ_1(x))^2 = \frac{1}{2} x^2 [J_0^2(x) + J_1^2(x)].$$

Recurrence relations

Bessel's function of the first kind satisfies the following recurrence relations.

P_3: (a) $x J_\nu'(x) = x J_{\nu-1}(x) - \nu J_\nu(x).$ (7.93)

   (b) $x J_\nu'(x) = \nu J_\nu(x) - x J_{\nu+1}(x).$ (7.94)

   (c) $2 J_\nu'(x) = J_{\nu-1}(x) - J_{\nu+1}(x).$ (7.95)

   (d) $2\nu J_\nu(x) = x(J_{\nu-1}(x) + J_{\nu+1}(x)).$ (7.96)

Simplifying Eq. (7.89), we get

$$x^\nu J_\nu'(x) + \nu x^{\nu-1} J_\nu(x) = x^\nu J_{\nu-1}(x).$$

Cancelling $x^{\nu-1}$, we get

$$x J_\nu'(x) + \nu J_\nu(x) = x J_{\nu-1}(x)$$

or
$$x J_v'(x) = x J_{v-1}(x) - v J_v(x).$$

When, v is an integer n, we write

$$x J_n'(x) = x J_{n-1}(x) - n J_n(x).$$

Similarly, simplifying Eq. (7.91), we get

$$x^{-v} J_v'(x) - v x^{-v-1} J_v(x) = - x^{-v} J_{v+1}(x).$$

Cancelling x^{-v-1}, we get

$$x J_v'(x) - v J_v(x) = - x J_{v+1}(x),$$

or
$$x J_v'(x) = v J_v(x) - x J_{v+1}(x).$$

Again, when v is an integer n, we write

$$x J_n'(x) = n J_n(x) - x J_{n+1}(x).$$

Adding Eqs. (7.93) and (7.94), we get

$$2 x J_v'(x) = x[J_{v-1}(x) - J_{v+1}(x)], \quad \text{or} \quad 2 J_v'(x) = J_{v-1}(x) - J_{v+1}(x).$$

Subtracting Eq. (7.93) from Eq. (7.94), we get

$$2 v J_v(x) - x J_{v+1}(x) - x J_{v-1}(x) = 0$$

or
$$2 v J_v(x) = x[J_{v-1}(x) + J_{v+1}(x)].$$

Example 7.18   Show that

$$J_3(x) = \left(\frac{8}{x^2} - 1 \right) J_1(x) - \frac{4}{x} J_0(x).$$

Solution   Using the recurrence relation, given by Eq. (7.96), we get

$$J_{n+1}(x) = \frac{2n}{x} J_n(x) - J_{n-1}(x).$$

Setting $n = 1$, we get
$$J_2(x) = \frac{2}{x} J_1(x) - J_0(x).$$

Setting $n = 2$, we get $J_3(x) = \dfrac{4}{x} J_2(x) - J_1(x) = \dfrac{4}{x} \left[\dfrac{2}{x} J_1(x) - J_0(x) \right] - J_1(x)$

$$= \left(\frac{8}{x^2} - 1 \right) J_1(x) - \frac{4}{x} J_0(x).$$

Example 7.19   Show that

(a) $J_{5/2}(x) = \sqrt{\dfrac{2}{\pi x}} \left[\dfrac{1}{x^2}(3 - x^2) \sin x - \dfrac{3}{x} \cos x \right].$

(b) $4 J_0''' + 3 J_0' + J_3 = 0.$

Solution
(a) Using the recurrence relation given by Eq. (7.96), we get

$$J_{v+1}(x) = \frac{2v}{x} J_v(x) - J_{v-1}(x).$$

Setting $v = \frac{3}{2}$, we get $J_{5/2}(x) = \frac{3}{x} J_{3/2}(x) - J_{1/2}(x).$

Setting $v = \frac{1}{2}$, we get $J_{3/2}(x) = \frac{1}{x} J_{1/2}(x) - J_{-1/2}(x).$

Therefore, $J_{5/2}(x) = \frac{3}{x} \left[\frac{1}{x} J_{1/2}(x) - J_{-1/2}(x) \right] - J_{1/2}(x) = \left(\frac{3}{x^2} - 1 \right) J_{1/2}(x) - \frac{3}{x} J_{-1/2}(x).$

But $\qquad\qquad J_{1/2}(x) = \sqrt{\frac{2}{\pi x}} \sin x$ and $J_{-1/2}(x) = \sqrt{\frac{2}{\pi x}} \cos x$

(see problem 4, Exercise 7.3).

Therefore, $\qquad\qquad J_{5/2}(x) = \sqrt{\frac{2}{\pi x}} \left[\frac{1}{x^2} (3 - x^2) \sin x - \frac{3}{x} \cos x \right].$

(b) We have $J_0' = -J_1$. Differentiating, we get

$$J_0'' = -J_1' = -\frac{1}{2} [J_0 - J_2], \qquad \text{(using Eq. (7.95))}.$$

Differentiating again, we get

$$J_0''' = -\frac{1}{2} [J_0' - J_2'] = -\frac{1}{2} \left[J_0' - \frac{1}{2} (J_1 - J_3) \right] \quad \text{(using Eq. (7.95))}$$

or $\qquad\qquad\qquad 4J_0''' + 2J_0' - J_1 + J_3 = 0$

or $\qquad\qquad\qquad 4J_0''' + 3J_0' + J_3 = 0, \quad \text{since } J_1 = -J_0'.$

Example 7.20   Show that $J_0^2 + 2(J_1^2 + J_2^2 + J_3^2 + \ldots) = 1.$

Solution   Consider first the following derivative

$$\frac{d}{dx} \left[J_n^2 + J_{n+1}^2 \right] = 2J_n J_n' + 2J_{n+1} J_{n+1}'. \qquad (7.97)$$

From Eq. (7.94), we get $J_n' = \frac{n}{x} J_n - J_{n+1}.$ $\qquad\qquad\qquad\qquad (7.98)$

From Eq. (7.93), we get $J_n' = J_{n-1} - \frac{n}{x} J_n.$

Setting $n = n + 1$, we get $J_{n+1}' = J_n - \frac{(n + 1)}{x} J_{n+1}.$ $\qquad\qquad (7.99)$

Substituting Eqs. (7.98) and (7.99) in Eq. (7.97), we get

$$[J_n^2 + J_{n+1}^2]' = 2J_n \left[\frac{n}{x} J_n - J_{n+1} \right] + 2J_{n+1} \left[J_n - \frac{(n + 1)}{x} J_{n+1} \right]$$

$$= 2 \left[\frac{n}{x} J_n^2 - \frac{(n + 1)}{x} J_{n+1}^2 \right].$$

Setting $n = 0, 1, 2, \ldots$ we get

$$[J_0^2 + J_1^2]' = 2\left[0 - \frac{1}{x} J_1^2\right]$$

$$[J_1^2 + J_2^2]' = 2\left[\frac{1}{x} J_1^2 - \frac{2}{x} J_2^2\right]$$

$$[J_2^2 + J_3^2]' = 2\left[\frac{2}{x} J_2^2 - \frac{3}{x} J_3^2\right]$$

$$\cdots \quad \cdots \quad \cdots \quad \cdots$$

$$[J_n^2 + J_{n+1}^2]' = 2\left[\frac{n}{x} J_n^2 - \frac{(n+1)}{x} J_{n+1}^2\right], \ldots$$

Adding, we obtain

$$[J_0^2 + 2(J_1^2 + J_2^2 + J_3^2 + \ldots)]' = 0$$

since $J_n \to 0$ as $n \to \infty$. Integrating, we obtain

$$J_0^2 + 2(J_1^2 + J_2^2 + J_3^2 + \ldots) = c.$$

Let $x = 0$. Since $J_0(0) = 1$ and $J_n(0) = 0$, $n > 0$, we obtain $c = 1$. Therefore, we have

$$J_0^2 + 2(J_1^2 + J_2^2 + J_3^2 + \ldots) = 1.$$

Exercise 7.4

Express the following Bessel functions in terms of $J_0(x)$ and $J_1(x)$.

   1. $J_2(x)$.   **2.** $J_3(x)$.   **3.** $J_4(x)$.

Prove the following.

   4. $J_1'(x) = J_0(x) - \dfrac{1}{x} J_1(x)$.   **5.** $J_2'(x) = \left(1 - \dfrac{4}{x^2}\right) J_1(x) + \dfrac{2}{x} J_0(x)$.

   6. $J_3'(x) = \left(\dfrac{12}{x^2} - 1\right) J_0(x) - \left(\dfrac{24}{x^3} - \dfrac{5}{x}\right) J_1(x)$.

   7. $J_1''(x) = \dfrac{1}{x} J_2(x) - J_1(x)$.   **8.** $J_n''(x) = \dfrac{1}{4}(J_{n-2}(x) - 2 J_n(x) + J_{n+2}(x))$.

   9. $[J_\nu^2(x)]' = \dfrac{x}{4\nu}(J_{\nu-1}^2(x) - J_{\nu+1}^2(x))$.   **10.** $\left(\dfrac{1}{x} \dfrac{d}{dx}\right)^m (x^{-\nu} J_\nu(x)) = (-1)^m x^{-\nu-m} J_{m+\nu}(x)$.

   11. $[x J_n(x) J_{n+1}(x)]' = x[J_n^2(x) - J_{n+1}^2(x)]$.

   12. $J_{-5/2}(x) = \sqrt{\dfrac{2}{\pi x}} \left[\dfrac{1}{x^2}(3 - x^2) \cos x + \dfrac{3}{x} \sin x\right]$.

   13. $J_{1/2}^2(x) + J_{-1/2}^2(x) = 2/(\pi x)$.   **14.** $x^2 J_n''(x) = (n^2 - n - x^2) J_n(x) + x J_{n+1}(x)$.

Prove the following.

   15. $\displaystyle\int J_{n+1}(x)\, dx = \int J_{n-1}(x)\, dx - 2 J_n(x) + c.$

16. $\int J_0(x)J_1(x)\,dx = -\frac{1}{2}J_0^2(x) + c.$ **17.** $\int xJ_0^2(x)\,dx = \frac{x^2}{2}[J_0^2 + J_1^2] + c.$

18. $\int J_2(x)\,dx = -2J_1(x) + \int J_0(x)\,dx + c.$

19. $\int J_3(x)\,dx = -J_2(x) - \frac{2}{x}J_1(x) + c.$ **20.** $\int J_5(x)\,dx = -J_4(x) - \frac{4}{x}J_3(x) - \frac{8}{x^2}J_2(x) + c.$

21. $\int J_5(x)\,dx = -[2J_4(x) + 2J_2(x) + J_0(x)] + c.$

22. $\int x^2 J_0(x)\,dx = x^2 J_1(x) + x J_0(x) - \int J_0(x)\,dx + c.$

23. $\int x^3 J_0(x)\,dx = x(x^2 - 4)J_1(x) + 2x^2 J_0(x) + c.$

24. $\int x^{-1} J_4(x)\,dx = -x^{-1}J_3(x) - 2x^{-2}J_2(x) + c.$

25. $\int x^5 J_0(x)\,dx = x^5 J_1(x) - 4x^4 J_2(x) + 8x^3 J_3(x) + c.$

26. $x\int_0^1 J_n(xt)t^{n+1}\,dt = J_{n+1}(x).$ **27.** $\int_0^a xJ_0(rx)\,dx = \frac{a}{r}J_1(ar),$ a, r are constants.

28. $\int_0^1 J_1(\alpha x)\,dx = \frac{1}{\alpha},$ where α is a root of $J_0(x) = 0$, that is $J_0(\alpha) = 0.$

29. $\int_0^1 (x - x^3)J_0(x)\,dx = 4J_1(1) - 2J_0(1).$

30. $\int_0^1 x^2 J_0(x)\,dx = J_1(1) + J_0(1) - \int_0^1 J_0(x)\,dx.$

31. $\int J_0(x)\cos x\,dx = xJ_0(x)\cos x + xJ_1(x)\sin x + c.$

32. $\int J_1(x)\sin x\,dx = xJ_1(x)\sin x + J_0(x)(x\cos x - \sin x) + c.$

Let $u = J_0(\alpha x)$ and $v = J_0(\beta x)$, Then show the following.

33. $xu'' + u' + \alpha^2 xu = 0,$ $xv'' + v' + \beta^2 xv = 0.$

34. $[x(u'v - uv')]' = (\beta^2 - \alpha^2)xuv.$

35. $(\beta^2 - \alpha^2)\int xJ_0(\alpha x)J_0(\beta x)\,dx = x[\alpha J_0'(\alpha x)J_0(\beta x) - \beta J_0(\alpha x)J_0'(\beta x)].$

7.4.2 Bessel's Function of the Second Kind

We have shown in section 7.4 that if v is not an integer then $J_v(x)$ and $J_{-v}(x)$ are linearly independent. Hence, when v is not an integer, the complete solution of the Bessel's equation is given by

$$y(x) = A J_\nu(x) + B J_{-\nu}(x).$$

However, when $\nu = n$ is an integer, $J_n(x)$ and $J_{-n}(x)$ are linearly dependent since $J_{-n}(x) = (-1)^n J_n(x)$. Therefore, we have only one independent solution. We have to obtain the second independent solution to find the general solution. The second solution can be obtained by using the methods described in section 6.4 for finding the series solution. Let us illustrate this method for the case $n = 0$, that is, Bessel's equation of zeroth order

$$xy'' + y' + xy = 0. \tag{7.100}$$

If we substitute the series solution $y(x) = \sum\limits_{m=0}^{\infty} c_m x^{m+r}$ in Eq. (7.100), we obtain the indicial roots as the double root $r = 0$. Then, the two linearly independent solutions are given by $[y(x)]_{r=0}$, and $[\partial y/\partial r]_{r=0}$ (see section 6.4, case 2). If $y_1(x) = [y(x)]_{r=0}$, then the second solution is of the form

$$y_2(x) = y_1(x) \ln x + [d_1 x + d_2 x^2 + \ldots], \quad x > 0$$

$$= J_0(x) \ln x + [d_1 x + d_2 x^2 + \ldots]$$

since $y_1(x) = J_0(x)$. We have

$$y_2'(x) = J_0' \ln x + \frac{1}{x} J_0 + [d_1 + 2d_2 x + 3d_3 x^2 + \ldots]$$

$$y_2''(x) = J_0'' \ln x + \frac{2}{x} J_0' - \frac{1}{x^2} J_0 + [2d_2 + 6d_3 x + \ldots].$$

Substituting in Eq. (7.100), we obtain

$$[xJ_0'' + J_0' + xJ_0] \ln x + 2J_0' + [d_1 + 4d_2 x + (d_1 + 9d_3)x^2 + \ldots$$

$$+ (d_{m-1} + (m + 1)^2 d_{m+1})x^m + \ldots] = 0.$$

The first term vanishes as $J_0(x)$ is a solution of Eq. (7.100). Substituting for J_0'

$$J_0' = \sum_{m=1}^{\infty} \frac{(-1)^m (2m) x^{2m-1}}{(m!)^2 2^{2m}} = \sum_{m=1}^{\infty} \frac{(-1)^m x^{2m-1}}{m!(m-1)! 2^{2m-1}} = -\frac{x}{2} + \frac{x^3}{(2!)(2^3)} - \frac{x^5}{(3!)(2!)(2^5)} + \ldots$$

we obtain

$$\left[-x + \frac{x^3}{(2!)(2^2)} - \frac{x^5}{(3!)(2!)(2^4)} + \ldots \right] + [d_1 + 4d_2 x + (d_1 + 9d_3)x^2 + (d_2 + 16d_4)x^3 + \ldots] = 0.$$

Equating the coefficients of even powers of x to zero, we obtain

$$d_1 = 0, \quad d_1 + 9d_3 = 0, \quad \text{or} \quad d_3 = 0, \ldots, \quad d_{2m-1} + (2m + 1)^2 d_{2m+1} = 0, \quad m = 1, 2, \ldots$$

Therefore, $d_1 = 0 = d_3 = d_5 = \ldots$
Equating the coefficient of x to zero, we obtain

$$-1 + 4d_2 = 0, \quad \text{or} \quad d_2 = 1/4 = 1/2^2.$$

Equating the coefficient of x^{2m+1} to zero, we obtain

$$\frac{(-1)^{m+1}}{m!(m+1)! 2^{2m}} + d_{2m} + (2m + 2)^2 d_{m+2} = 0, \quad m = 1, 2, \ldots$$

For $m = 1$, we get

$$\frac{1}{(2!)(2^2)} + d_2 + 16d_4 = 0, \quad \text{or} \quad d_4 = -\frac{1}{16}\left[\frac{1}{(2!)(2^2)} + \frac{1}{2^2}\right] = -\frac{1}{2^2 \cdot 4^2}\left[1 + \frac{1}{2}\right].$$

For $m = 2$, we get

$$-\frac{1}{(3!)(2!)(2^4)} + d_4 + 36d_6 = 0,$$

or $\quad d_6 = \frac{1}{6^2}\left[\frac{1}{(3!)(2!)(2^4)} + \frac{1}{2^2 \cdot 4^2}\left(1 + \frac{1}{2}\right)\right] = \frac{1}{2^2 \cdot 4^2 \cdot 6^2}\left[1 + \frac{1}{2} + \frac{1}{3}\right], \ldots$

Therefore,

$$y_2(x) = J_0(x) \ln x + \left[\frac{1}{2^2} x^2 - \frac{1}{2^2 \cdot 4^2}\left(1 + \frac{1}{2}\right)x^4 + \frac{1}{2^2 \cdot 4^2 \cdot 6^2}\left(1 + \frac{1}{2} + \frac{1}{3}\right)x^6 - \ldots\right] \quad (7.101)$$

is the second linearly independent solution of Eq. (7.100). However, any other linear combination of $y_1(x)$ and $y_2(x)$ can also be taken as the second linearly independent solution. In particular

$$y_2^*(x) = \frac{2}{\pi}[y_2(x) + (\gamma - \ln 2)J_0(x)] \quad (7.102)$$

where γ is the *Euler constant* ($\gamma = 0.577215 \ldots$) is usually taken as the second linearly independent solution. This solution is called the *Bessel's function of the second kind* of order zero. It is denoted by $Y_0(x)$. It is also called the *Neumann function* of second kind. Hence

$$Y_0(x) = \frac{2}{\pi}\left[J_0(x)\left\{\ln\left(\frac{x}{2}\right) + \gamma\right\} + \left\{\frac{1}{2^2} x^2 - \frac{1}{2^2 \cdot 4^2}\left(1 + \frac{1}{2}\right)x^4 + \ldots\right\}\right] \quad (7.103)$$

Since $Y_0(x)$ contains the $\ln(x)$ term, we have $Y_0(x) \to -\infty$ as $x \to 0$. For small x, $Y_0(x)$ behaves like $\ln(x)$ (Fig. 7.5).

In the general case, when $v = n \neq 0$, we can obtain the solution in a similar manner. The second linearly independent solution always contains a logarithmic term. This solution can be written as

$$Y_n(x) = J_n(x) \ln x - \frac{1}{2}\sum_{j=0}^{n-1}\frac{(n-j-1)!}{j!}\left(\frac{x}{2}\right)^{-n+2j}$$

$$-\frac{1}{2}\sum_{j=0}^{\infty}\frac{(-1)^j}{j!(n+j)!}\left(\frac{x}{2}\right)^{n+2j}[\phi(j) + \phi(n+j)] \quad (7.104)$$

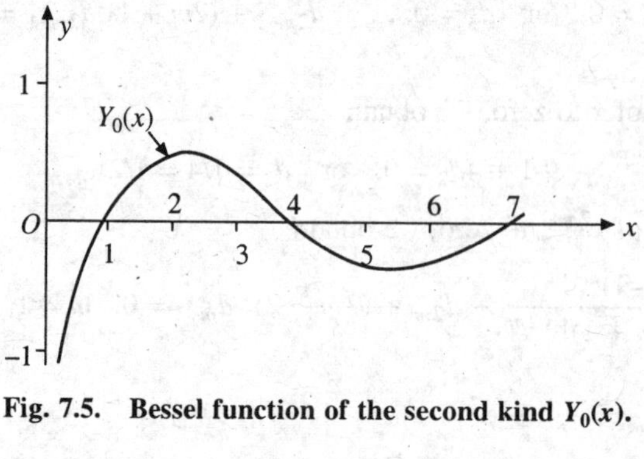

Fig. 7.5. **Bessel function of the second kind $Y_0(x)$.**

where $\phi(j) = 1 + \dfrac{1}{2} + \ldots + \dfrac{1}{j}$ and $\phi(0) = 0$.

Without proof, we state the generalisation which is valid for all orders. We have the Bessel's function of the second kind defined as

$$Y_v(x) = \frac{1}{\sin v\pi}[J_v(x)\cos v\pi - J_{-v}(x)] \tag{7.105}$$

and

$$Y_n(x) = \lim_{v \to n} Y_v(x). \tag{7.106}$$

The roles of $\cos v\pi$ and $\sin v\pi$ in Eq. (7.105) can be explained. As $v \to n$, an integer, the right hand side is an indeterminate form $0/0$ and the limit in Eq. (7.105) can be evaluated. Therefore, the general solution of the Bessel's differential equation for all v, is

$$Y(x) = AJ_v(x) + BY_v(x).$$

where $Y_v(x)$ is defined by Eq. (7.105).

Example 7.21   Find the solution of the differential equation

$$x^2y'' + xy' + (x^2 - 9)y = 0.$$

Solution   The given equation is the Bessel equation of order three, $v = 3$. Therefore, the solution is

$$y(x) = AJ_3(x) + BY_3(x).$$

Example 7.22   Find the solution of the differential equation

$$xy'' + y' + y = 0$$

using the substitution $z = 2\sqrt{x}$.

Solution

When $z = 2\sqrt{x}$, we have $\dfrac{dy}{dx} = \dfrac{1}{\sqrt{x}}\dfrac{dy}{dz}$ and $\dfrac{d^2y}{dx^2} = \dfrac{1}{x}\dfrac{d^2y}{dz^2} - \dfrac{1}{2x\sqrt{x}}\dfrac{dy}{dz}$.

Substituting in the given differential equation and using $\sqrt{x} = z/2$, we get

$$x\left(\frac{1}{x}\frac{d^2y}{dz^2} - \frac{1}{2x\sqrt{x}}\frac{dy}{dz}\right) + \frac{1}{\sqrt{x}}\frac{dy}{dz} + y = 0$$

or

$$\frac{d^2y}{dz^2} - \frac{1}{z}\frac{dy}{dz} + \frac{2}{z}\frac{dy}{dz} + y = 0, \quad \text{or} \quad z\frac{d^2y}{dz^2} + \frac{dy}{dz} + zy = 0.$$

This equation is the Bessel equation of order zero. Therefore, the solution is

$$y(z) = AJ_0(z) + BY_0(z), \quad \text{or} \quad y(x) = AJ_0(2\sqrt{x}) + BY_0(2\sqrt{x}).$$

Generating function and integral representation for Bessel function $J_n(x)$

Bessel function of the first kind, $J_n(x)$ can also be obtained by using a *generating function*. We shall prove that

$$e^{\frac{1}{2}x\left(t-\frac{1}{t}\right)} = \sum_{-\infty}^{\infty} J_n(x)t^n. \tag{7.107}$$

The function on the left hand side is called the generating function of the Bessel's function $J_n(x)$. We have

$$e^{\frac{1}{2}x\left(t-\frac{1}{t}\right)} = e^{(xt)/2} \cdot e^{-x/(2t)} = \left[1 + \frac{xt}{2} + \frac{1}{2!}\left(\frac{xt}{2}\right)^2 + \ldots\right]\left[1 - \frac{x}{2t} + \frac{1}{2!}\left(\frac{x}{2t}\right)^2 - \ldots\right].$$

Multiply the two series on the right hand side and collect the coefficients of t^n, $n = 0, 1, 2, \ldots$
   The terms that are independent of t, that is t^0, are given as

$$1 - \left(\frac{x}{2}\right)^2 + \frac{1}{(2!)^2}\left(\frac{x}{2}\right)^4 - \frac{1}{(3!)^2}\left(\frac{x}{2}\right)^6 + \ldots = \sum_{m=0}^{\infty} \frac{(-1)^m}{(m!)^2}\left(\frac{x}{2}\right)^{2m} = J_0(x).$$

Consider now, the terms containing t^n. We have these terms as

$$\frac{1}{n!}\left(\frac{x}{2}\right)^n - \frac{1}{(n+1)!}\left(\frac{x}{2}\right)^{n+2} + \frac{1}{(n+2)!2!}\left(\frac{x}{2}\right)^{n+4} - \ldots$$

$$= \sum_{m=0}^{\infty} \frac{(-1)^m}{(n+m)!(m!)}\left(\frac{x}{2}\right)^{n+2m} = J_n(x). \tag{7.108}$$

Now, the coefficient of $(-t)^{-n}$ is same as in Eq. (7.108), that is, $J_n(x)$. Therefore, the coefficient of t^{-n} is $(-1)^n J_n(x) = J_{-n}(x)$. Hence, we obtain the result given in Eq. (7.107).

From Eq. (7.107), we have

$$e^{\frac{1}{2}x\left(t-\frac{1}{t}\right)} = J_0(x) + J_1(x)t + J_2(x)t^2 + \ldots + J_n(x)t^n + \ldots$$

$$- J_1(x)t^{-1} + J_2(x)t^{-2} + \ldots + (-1)^n J_n(x)t^{-n} + \ldots$$

$$= J_0(x) + \left(t - \frac{1}{t}\right)J_1(x) + \left(t^2 + \frac{1}{t^2}\right)J_2(x) + \ldots. \tag{7.109}$$

Let now, $t = \cos\theta + i\sin\theta$. Then $(1/t) = \cos\theta - i\sin\theta$. Therefore,

$$e^{\frac{1}{2}x\left(t-\frac{1}{t}\right)} = e^{ix\sin\theta} = \cos(x\sin\theta) + i\sin(x\sin\theta). \tag{7.110}$$

The right hand side of Eq. (7.109) becomes

$$J_0(x) + 2iJ_1(x)\sin\theta + 2J_2(x)\cos 2\theta + 2iJ_3(x)\sin 3\theta + \ldots$$

$$= [J_0(x) + 2\{J_2(x)\cos 2\theta + J_4(x)\cos 4\theta + \ldots\}] + 2i[J_1(x)\sin\theta + J_3(x)\sin 3\theta + \ldots]. \tag{7.111}$$

Comparing the real and imaginary parts of Eqs. (7.110) and (7.111), we obtain

$$J_0(x) + 2\{J_2(x)\cos 2\theta + J_4(x)\cos 4\theta + \ldots\} = \cos(x\sin\theta) \tag{7.112}$$

and

$$2[J_1(x)\sin\theta + J_3(x)\sin 3\theta + \ldots\} = \sin(x\sin\theta). \tag{7.113}$$

These series are usually called the *Jacobi series*.

Multiplying both sides of Eq. (7.112) by cos $n\theta$, and integrating over the interval $[0, \pi]$, we obtain

$$\frac{1}{\pi} \int_0^{\pi} \cos{(x \sin{\theta})} \cos{n\theta} d\theta = J_n(x), \quad n \text{ is even}$$

$$= 0, \quad n \text{ is odd.} \tag{7.114}$$

Similarly, multiplying both sides of Eq. (7.113) by sin $n\theta$, and integrating over the interval $[0, \pi]$, we obtain

$$\frac{1}{\pi} \int_0^{\pi} \sin{(x \sin{\theta})} \sin{n\theta} \, d\theta = J_n(x), \quad n \text{ is odd}$$

$$= 0, \quad n \text{ is even.} \tag{7.115}$$

Adding Eq. (7.114) and (7.115), we obtain

$$\frac{1}{\pi} \int_0^{\pi} [\cos{n\theta} \cos{(x \sin{\theta})} + \sin{n\theta} \sin{(x \sin{\theta})}] \, d\theta = J_n(x)$$

or

$$\frac{1}{\pi} \int_0^{\pi} \cos{(n\theta - x \sin{\theta})} \, d\theta = J_n(x) \tag{7.116}$$

for all integral values of n.

This equation provides the integral representation of the Bessel function of the first kind $J_n(x)$ and is called the *Bessel integral*. For $n = 0$, Eq. (7.116) gives the integral representation of $J_0(x)$ as

$$J_0(x) = \frac{1}{\pi} \int_0^{\pi} \cos{(x \sin{\theta})} \, d\theta. \tag{7.117}$$

Example 7.23 Show that

$$\cos{(x \cos{\theta})} = J_0 - 2J_2 \cos{2\theta} + 2J_4 \cos{4\theta} - \ldots$$

and

$$\sin{(x \cos{\theta})} = 2[J_1 \cos{\theta} - J_3 \cos{3\theta} + \ldots].$$

Solution In Eq. (7.109), substitute $t = i (\cos{\theta} + i \sin{\theta}) = ie^{i\theta}$.

Then,

$$\frac{1}{t} = - ie^{-i\theta}, \quad t - \frac{1}{t} = 2i \cos{\theta}, \quad t^2 + \frac{1}{t^2} = - 2 \cos{2\theta}, \ldots.$$

Therefore,

$$e^{\frac{1}{2}x\left(t - \frac{1}{t}\right)} = e^{ix \cos{\theta}} = \cos{(x \cos{\theta})} + i \sin{(x \sin{\theta})}. \tag{7.118}$$

The right hand side of Eq. (7.109) becomes

$$J_0 + 2i J_1 \cos{\theta} - 2J_2 \cos{2\theta} - 2i J_3 \cos{3\theta} + \ldots$$

$$= [J_0 - 2J_2 \cos{2\theta} + 2J_4 \cos{4\theta} - \ldots] + 2i [J_1 \cos{\theta} - J_3 \cos{3\theta} + \ldots] \tag{7.119}$$

Comparing Eqs. (7.118) and (7.119), we obtain

$$\cos{(x \cos{\theta})} = J_0 - 2J_2 \cos{2\theta} + 2J_4 \cos{4\theta} - \ldots \tag{7.120}$$

$$\sin{(x \cos{\theta})} = 2[J_1 \cos{\theta} - J_3 \cos{3\theta} + \ldots]. \tag{7.121}$$

Example 7.24 Show that $J_0(x) = \frac{1}{\pi} \int_0^\pi \cos(x \cos \theta) \, d\theta$.

Solution We shall use the results of Example 7.23. Integrating Eq. (7.120) over the interval $[0, \pi]$, we obtain

$$\int_0^\pi \cos(x \cos \theta) \, d\theta = J_0(x) \int_0^\pi d\theta = \pi J_0(x).$$

since the other terms vanish. Hence the result.

Exercise 7.5

Reduce the given differential equation to Bessel's equation, using the given substitution. Hence, find the general solution of the given differential equation.

1. $x^2 y'' + xy' + (x^2 - 25)y = 0$, $z = x$.   2. $x^2 y'' + xy' + (9x^2 - 4)y = 0$, $z = 3x$.

3. $x^2 y'' + xy' + (4x^4 - 16)y = 0$, $z = x^2$.   4. $x^2 y'' + xy' + \frac{1}{4}(x - 1)y = 0$, $z = \sqrt{x}$.

5. $xy'' + 5y' + xy = 0$, $y = z/x^2$.

6. Show that the Bessel's equation reduces to

$$z'' + \left[1 + \frac{1}{4x^2}(1 - 4n^2)\right]z = 0$$

   under the transformation $y = z/\sqrt{x}$.

7. The differential equation

$$x^2 y'' + (1 - 2a)xy' + [b^2 c^2 x^{2c} + (a^2 - n^2 c^2)]y = 0$$

   is given. Show that the change of variables $y = ux^a$, $t = x^c$ and $z = bt$ reduces it to the Bessel equation

$$z^2 \frac{d^2 u}{dz^2} + z \frac{du}{dz} + (z^2 - n^2)u = 0.$$

8. Using the substitutions in problem 7, reduce the equation $x^2 y'' + xy' + \frac{1}{4}(x - n^2)y = 0$ to the Bessel equation and hence find the solution.

9. Find the solution of the differential equation $xy'' + y = 0$ in terms of Bessel's functions.

10. Find the solution of the differential equation $xy'' + y' + (y/4) = 0$ in terms of Bessel functions.

7.5 Sturm-Liouville Problem

Any differential equation of the form

$$\frac{d}{dx}\left[r(x)\frac{dy}{dx}\right] + [q(x) + \lambda p(x)]y = 0 \qquad (7.122)$$

where λ is a real constant, is called a *Sturm-Liouville equation*. Consider the following examples.

(i) The differential equation $y'' + \lambda y = 0$ can be written as

$$\frac{d}{dx}\left[\frac{dy}{dx}\right] + \lambda y = 0 \tag{7.123}$$

which is of the form (7.122) with $r(x) = 1$, $q(x) = 0$ and $p(x) = 1$.

(ii) Legendre differential equation

$$(1 - x^2)y'' - 2xy' + n(n + 1)y = 0 \tag{7.124}$$

can be written as

$$[(1 - x^2)y']' + \lambda y = 0$$

where $\lambda = n(n + 1)$. Therefore, Legendre's differential equation is a Sturm-Liouville equation where $r(x) = 1 - x^2$, $q(x) = 0$ and $p(x) = 1$.

(iii) In the Bessel differential equation

$$x^2 y'' + xy' + (x^2 - n^2)y = 0, \quad x \neq 0 \tag{7.125}$$

set $x = kz$. Then, $\dfrac{dy}{dx} = \dfrac{1}{k}\dfrac{dy}{dz}$ and $\dfrac{d^2 y}{dx^2} = \dfrac{1}{k^2}\dfrac{d^2 y}{dz^2}$.

Substituting in the Bessel's equation, we get

$$z^2 \frac{d^2 y}{dz^2} + z \frac{dy}{dz} + (k^2 z^2 - n^2)y = 0$$

or

$$z \frac{d^2 y}{dz^2} + \frac{dy}{dz} + \left[-\frac{n^2}{z} + k^2 z\right]y = 0$$

or

$$(zy')' + \left[-\frac{n^2}{z} + k^2 z\right]y = 0.$$

Note that $y = y(kz)$. Replacing the variable z by x, we get

$$(xy')' + \left[-\frac{n^2}{x} + k^2 x\right]y = 0 \tag{7.126}$$

which is a Sturm-Liouville equation with $r(x) = x$, $q(x) = -n^2/x$, $p(x) = x$ and $\lambda = k^2$. Note that $y = y(kx)$.

Consider a general homogeneous, linear differential equation expressed in the canonical form as

$$y'' + R(x)y'(x) + [Q(x) + \lambda P(x)]y = 0. \tag{7.127}$$

Now, $r(x) = e^{\int R(x)\,dx}$ is an integrating factor of the first two terms, since

$$e^{\int R(x)\,dx}[y'' + R(x)y'] = [e^{\int R(x)\,dx} y']'. \tag{7.128}$$

Multiplying Eq. (7.127) by the integrating factor, we obtain (using Eq. (7.128))

$$[r(x)y']' + [q(x) + \lambda p(x)]y = 0$$

where $q(x) = Q(x)e^{\int R(x)\,dx}$ and $p(x) = P(x)e^{\int R(x)\,dx}$.

Therefore, it is possible to reduce a general homogeneous, linear differential equation into the Sturm-Liouville form. Consider the following examples.

(i) Laguerre differential equation

$$xy'' + (1 - x)y' + ny = 0, \quad x \neq 0 \qquad (7.129)$$

or

$$y'' + \left(\frac{1}{x}\right)(1 - x)y' + \frac{n}{x}y = 0. \qquad (7.130)$$

or

$$y'' + R(x)\,y' + nP(x)y = 0$$

where $R(x) = (1 - x)/x$ and $P(x) = 1/x$.

Now

$$\int R(x) = \int \left(\frac{1}{x} - 1\right)dx = \ln x - x,$$

and integrating factor $= r(x) = e^{\ln x - x} = xe^{-x}$.

Multiplying Eq. (7.130) with the integrating factor, we get

$$xe^{-x}y'' + (1 - x)e^{-x}y' + ne^{-x}y = 0$$

or

$$[xe^{-x}y']' + (ne^{-x})y = 0. \qquad (7.131)$$

This is a Sturm-Liouville equation with $r(x) = xe^{-x}$, $q(x) = 0$, $p(x) = e^{-x}$ and $\lambda = n$.

(ii) Chebyshev differential equation

$$(1 - x^2)y'' - xy' + n^2 y = 0, \, -1 < x < 1,$$

or

$$y'' - \left(\frac{x}{1 - x^2}\right)y' + \frac{n^2}{1 - x^2}\,y = 0. \qquad (7.132)$$

We have

$$R(x) = -\frac{x}{1 - x^2}, \quad \int R(x)\,dx = \frac{1}{2}\ln(1 - x^2) = \ln(1 - x^2)^{1/2},$$

and

$$r(x) = e^{\ln(1-x^2)^{1/2}} = (1 - x^2)^{1/2}.$$

Multiplying Eq. (7.132) by $r(x)$, we obtain

$$[(1 - x^2)^{1/2}\,y']' + \left[\frac{n^2}{(1 - x^2)^{1/2}}\right]y = 0$$

which is a Sturm-Liouville equation with $r(x) = (1 - x^2)^{1/2}$, $q(x) = 0$, $p(x) = (1 - x^2)^{-1/2}$ and $\lambda = n^2$.

(iii) Hermite differential equation

$$y'' - 2xy' + 2ny = 0. \qquad (7.133)$$

We have $R(x) = -2x$, $\displaystyle\int R(x)\,dx = -x^2$ and $r(x) = e^{-x^2}$.

Multiplying the Eq. (7.133) by $r(x)$, we obtain

$$[e^{-x^2} y']' + [2ne^{-x^2}]y = 0$$

which is a Sturm-Liouville equation with $r(x) = e^{-x^2}$, $q(x) = 0$, $p(x) = e^{-x^2}$ and $\lambda = 2n$.

Sometimes, $y'' - xy' + ny = 0$ is taken as the Hermite equation. In this case, $r(x) = e^{-x^2/2}$, $q(x) = 0$, $p(x) = e^{-x^2/2}$ and $\lambda = n$.

Sturm-Liouville boundary value problem

Eq. (7.122) is usually considered on some given interval $[a, b]$. We assume that $p(x)$, $q(x)$, $r(x)$ and $r'(x)$ are real valued continuous functions on $[a, b]$. We also assume that $p(x) > 0$ on $[a, b]$ and the boundary conditions are of the form

$$a_1 y(a) + a_2 y'(a) = 0 \tag{7.134a}$$

$$b_1 y(b) + b_2 y'(b) = 0 \tag{7.134b}$$

where a_1, a_2, b_1, b_2 are real constants. We assume that both a_1, a_2 are not zero, and both b_1, b_2 are not zero. The differential equation (7.122) along with the boundary conditions (Eqs. (7.134a), (7.134b)) is called the *Sturm-Liouville boundary value problem.*

   Now, $y \equiv 0$ (trivial solution) is always a solution of the homogeneous boundary value problem (*bvp*). We are interested in finding the non-trivial solutions. The values of λ for which non-trivial solutions exist are called the *eigenvalues* of the *bvp* and the corresponding solutions $y(x)$ are called the *eigen functions*. It is obvious that if $y(x)$ is an eigenfunction then $ky(x)$, k real constant, is also a solution. The set of eigenvalues of the *bvp* is called the *spectrum* of the problem. It may be noted that the Sturm-Liouville equation is *self-adjoint.*

Consider the second order linear differential operator

$$L = a(x) \frac{d^2}{dx^2} + b(x) \frac{d}{dx} + c(x). \tag{7.135}$$

Then, the formal adjoint operator is given by

$$L^* = a(x) \frac{d^2}{dx^2} + [2a'(x) - b(x)] \frac{d}{dx} + [a''(x) - b'(x) + c(x)].$$

If $L = L^*$, then L is called formally self adjoint or simply a self adjoint operator. Therefore, L is self adjoint if

$$2a' - b = b, \quad \text{and} \quad a'' - b' + c = c \quad \text{or} \quad a' = b.$$

Then, L can be written as

$$L = a(x) \frac{d^2}{dx^2} + a'(x) \frac{d}{dx} + c(x) = \frac{d}{dx} \left[a(x) \frac{d}{dx} \right] + c(x). \tag{7.136}$$

In the case of second order, linear differential operators, even if L is not self adjoint it can be made self adjoint by multiplying it by a factor $r(x)$, (see Eqs. (7.127), (7.128)).

If $r(a) = r(b)$, then the boundary conditions (Eqs. (7.134)) are replaced by the periodic boundary conditions

$$y(a) = y(b), \quad y'(a) = y'(b), \tag{7.137}$$

It can be shown that, often, infinitely many eigenvalues exist for the Sturm-Liouville problem and the eigenvalues are all real.

Example 7.25 Find the eigenvalues and eigenfunctions of the boundary value problem

$$y'' + \lambda y = 0, \quad y(0) = 0, \quad y(l) = 0$$

(vibration of an elastic string of length l, where y(x) is the displacement).

Solution It is easy to show that $y \equiv 0$ is the only solution for $\lambda = -k^2$ and $\lambda = 0$.
When $\lambda = -k^2$, the general solution is

$$y(x) = Ae^{kx} + Be^{-kx}.$$

Applying the boundary conditions, we obtain

$$y(0) = 0 = A + B, \quad \text{and} \quad y(l) = 0 = Ae^{kl} + Be^{-kl}.$$

The solutions is $A = 0, B = 0$ or $y(x) = 0$.

When $\lambda = 0$, the general solution is $y(x) = Ax + B$. Applying the boundary conditions, we obtain $A = 0, B = 0$ or $y(x) = 0$.

Therefore, for obtaining non trivial solutions, we take $\lambda = k^2, k \in R$. The general solution is

$$y(x) = A \cos kx + B \sin kx.$$

The boundary condition $y(0) = 0$ gives $A = 0$. The boundary condition $y(l) = 0$ gives $B \sin kl = 0$. In order that non-trivial solutions exist, we need $B \neq 0$. Therefore,

$$\sin kl = 0 = \sin n\pi, \quad n = 1, 2, 3, \ldots$$

or $$kl = n\pi, \quad \text{or} \quad k = (n\pi)/l, n = 1, 2, 3, \ldots$$

The eigenvalues of the *bvp* are $\lambda = k^2 = (n^2\pi^2)/l^2, n = 1, 2, \ldots$ The corresponding eigenfunctions are

$$y_n(x) = \sin [(n\pi x)/l], \quad n = 1, 2, 3, \ldots$$

We now prove an important result of the Sturm-Liouville *bvp*, which proves the orthogonality of the eigenfunctions.

Theorem 7.1 (**Sturm-Liouville theorem**) Let in the Sturm-Liouville equation (7.122), $p(x), q(x), r(x), r'(x)$ be real valued continuous functions and let $p(x) > 0$ on $[a, b]$. If $y_m(x), y_n(x)$ are the eigenfunctions corresponding to distinct eigenvalues λ_m, λ_n, then $y_m(x), y_n(x)$ are orthogonal to each other on the interval $[a, b]$ with respect to the weight function $p(x)$, that is

$$\int_a^b p(x)y_m(x)y_n(x)\, dx = 0, \quad m \neq n. \tag{7.138}$$

If $r(a) = 0$, then the first boundary condition (7.134a) can be neglected. If $r(b) = 0$, then the second boundary condition (7.134b) can be neglected. If $r(a) = r(b)$, then the problem is solved under the periodic boundary conditions (Eq. (7.137)).

(If $r(a)$ or $r(b)$ vanishes then the regularity condition is violated. The *bvp* is then called a *singular bvp*. If $r(a) = r(b)$, then the problem is called a *periodic bvp*.)

Proof Since λ_m, λ_n are distinct eigenvalues and $y_m(x)$, $y_n(x)$ are the corresponding eigenfunctions, we have

$$(ry_n')' + (q + \lambda_n p)y_n = 0 \quad \text{and} \quad (ry_m')' + (q + \lambda_m p)y_m = 0.$$

Multiply the first equation by y_m, the second equation by y_n and subtract. We obtain

$$y_m(ry_n')' - y_n(ry_m')' + (\lambda_n - \lambda_m)py_m y_n = 0$$

or

$$[y_m(ry_n') - y_n(ry_m')]' + (\lambda_n - \lambda_m)py_m y_n = 0.$$

By the assumptions on p and r, this equation is continuous. Integrating over $[a, b]$, we obtain

$$(\lambda_n - \lambda_m) \int_a^b py_m y_n dx = \int_a^b [y_m(ry_n') - y_n(ry_m')]' \, dx = \big[y_m(ry_n') - y_n(ry_m') \big]_a^b$$

$$= r(b)[y_m(b)y_n'(b) - y_n(b)y_m'(b)] - r(a)[y_m(a)y_n'(a) - y_n(a)y_m'(a)]. \qquad (7.139)$$

Let $\Delta(x)$ denote the determinant

$$\Delta(x) = \begin{vmatrix} y_n(x) & y_n'(x) \\ y_m(x) & y_m'(x) \end{vmatrix} = y_n(x)y_m'(x) - y_m(x)y_n'(x). \qquad (7.140)$$

Then, Eq. (7.139) can be written as

$$(\lambda_n - \lambda_m) \int_a^b p(x) y_m(x) y_n(x) \, dx = r(a)\Delta(a) - r(b)\,\Delta(b) \qquad (7.141)$$

where $\lambda_n \neq \lambda_m$. Since, the eigenfunctions $y_n(x)$ and $y_m(x)$ satisfy the boundary conditions (7.134), we have

$$a_1 y_n(a) + a_2 y_n'(a) = 0 \qquad (7.142a)$$

$$a_1 y_m(a) + a_2 y_m'(a) = 0. \qquad (7.142b)$$

Since both a_1, a_2 are not zero, non-trivial solution exists for these homogeneous equations. The necessary condition for the existence of the non-trivial solutions is

$$\begin{vmatrix} y_n(a) & y_n'(a) \\ y_m(a) & y_m'(a) \end{vmatrix} = 0, \quad \text{or} \quad \Delta(a) = 0. \qquad (7.143)$$

Since both b_1, b_2 are not zero, we obtain from the boundary conditions (7.134) that

$$\begin{vmatrix} y_n(b) & y_n'(b) \\ y_m(b) & y_m'(b) \end{vmatrix} = 0 \quad \text{or} \quad \Delta(b) = 0. \qquad (7.144)$$

Substituting in Eq. (7.141), we obtain

$$(\lambda_n - \lambda_m) \int_a^b p(x) y_m(x) y_n(x)\, dx = 0.$$

Since $\lambda_n \neq \lambda_m$, we obtain the orthogonality condition

$$\int_a^b p(x) y_m(x) y_n(x)\, dx = 0. \tag{7.145}$$

If $r(a) = 0$ and $r(b) = 0$, then from Eq. (7.141), we immediately obtain the result (7.145). Hence, we do not need any boundary conditions in this case.

If $r(a) = 0$, and $r(b) \neq 0$, then the right hand side of Eq. (7.141) is $[-r(b)\Delta(b)]$, even if $\Delta(a) \neq 0$. Since both b_1, b_2 are not zero, we obtain from Eq. (7.144) that $\Delta(b) = 0$ and hence the orthogonality result. Now, if $\Delta(a) \neq 0$, unique solution exists for the system (7.142) which is the trivial solution $a_1 = 0$, $a_2 = 0$. Therefore, the first boundary condition (7.134a) can be dropped. Similarly, if $r(b) = 0$, but $r(a) \neq 0$, we find that the second boundary condition (7.134b) can be dropped.

If $r(a) = r(b)$, then we use the periodic boundary conditions. In this case, the right hand side of Eq. (7.141) is

$$-r(a)[\Delta(b) - \Delta(a)] = -r(a)[\{y_n(b)y_m'(b) - y_m(b)y_n'(b)\} - \{y_n(a)y_m'(a) - y_m(a)y_n'(a)\}]$$

$$= -r(a)[\{y_n(b)y_m'(b) - y_n(a)y_m'(a)\} + \{y_m(a)y_n'(a) - y_m(b)y_n'(b)\}] = 0$$

since $y(a) = y(b)$ and $y'(a) = y'(b)$. The theorem is proved.

Consider the Sturm-Liouville *bvp* given in Example 7.25. We have $r(x) = 1$, $q(x) = 0$ and $p(x) = 1$. Therefore, by Theorem 7.1 the eigenfunctions $\sin[(n\pi x)/l]$, $n = 1, 2, 3, \ldots$ are orthogonal over the interval $[0, l]$, with respect to the weight function $p(x) = 1$. That is,

$$\int_0^l \sin\left(\frac{n\pi x}{l}\right) \sin\left(\frac{m\pi x}{l}\right) dx = 0, \quad m \neq n. \tag{7.146}$$

Example 7.26　Considering the eigenfunctions of the periodic Sturm-Liouville *bvp*

$$y''(x) + \lambda y(x) = 0, \quad y(-\pi) = y(\pi), \quad y'(-\pi) = y'(\pi)$$

show that the functions $1, \cos x, \sin x, \cos 2x, \sin 2x, \ldots$ are orthogonal.

Solution　For $\lambda = k^2$, we obtain the general solution as

$$y(x) = A \cos kx + B \sin kx.$$

Applying the boundary conditions, we obtain

$$A \cos k\pi - B \sin k\pi = A \cos k\pi + B \sin k\pi$$

and

$$Ak \sin k\pi + Bk \cos k\pi = -Ak \sin k\pi + Bk \cos k\pi$$

We obtain $\sin k\pi = 0 = \sin n\pi$, or $k = n$, $n = 0, 1, 2, \ldots$
The eigenvalues are $\lambda = k^2 = n^2$, $n = 0, 1, 2, \ldots$

Hence, the eigenfunctions $\cos kx$, $\sin kx$ are 1, $\cos x$, $\sin x$, $\cos 2x$, $\sin 2x$, ... Theorem 7.1 states that these eigenfunctions are orthogonal (with respect to the weight function $p(x) = 1$) over the interval $[-\pi, \pi]$ for distinct eigenvalues λ_i, λ_j.

It can be verified that for the eigenvalue λ_i

$$\int_{-\pi}^{\pi} \cos ix \sin ix \, dx = \frac{1}{2} \int_{-\pi}^{\pi} \sin 2ix \, dx = 0.$$

Hence, the given set of eigenfunctions are orthogonal over $[-\pi, \pi]$.

Example 7.27 Prove the orthogonality of Legendre polynomials using Theorem 7.1.

Solution We have shown in Eq. (7.124) that the Legendre's equation is a Sturm-Liouville equation

$$[(1 - x^2)y']' + \lambda y = 0, \quad -1 \le x \le 1 \tag{7.147}$$

with $\qquad r(x) = 1 - x^2$, $q(x) = 0$, $p(x) = 1$ and $\lambda = n(n + 1)$.

Since $r(-1) = 0 = r(1)$, we do not need any boundary conditions and the problem is a singular *bvp*. The Legendre polynomials $P_n(x)$ are the solutions of Eq. (7.147). Hence, by Theorem 7.1, the Legendre polynomials $P_n(x)$ are orthogonal over $[-1, 1]$, with respect to the weight function $p(x) = 1$, that is

$$\int_{-1}^{1} P_m(x)P_n(x) \, dx = 0.$$

7.5.1 Orthogonality of Bessel Functions

Consider the Bessel function $J_n(x)$. Using Eq. (7.126), we can write the Bessel equation in the Sturm-Liouville form as

$$[xJ_n'(kx)]' + \left[-\frac{n^2}{x} + k^2 x\right] J_n(kx) = 0 \tag{7.148}$$

with $r(x) = x$, $q(x) = -n^2/x$, $p(x) = x$ and $\lambda = k^2$. Since $r(0) = 0$, the first boundary condition, Eq. (7.134a) can be dropped. If we choose $b_1 \ne 0$ and $b_2 = 0$ in Eq. (7.134b), we obtain $y(R) = 0$, where $b = R$. That is, for fixed n

$$y(R) = J_n(kR) = 0.$$

This equation has infinite number of real zeros, say, a_{1n}, a_{2n}, ... Hence

$$kR = a_{mn}, \quad \text{or} \quad k = k_{mn} = a_{mn}/R, m = 1, 2, 3, \ldots. \tag{7.149}$$

Therefore, Theorem 7.1 implies orthogonality of the Bessel function $J_n(kx)$ for those values of k for which $J_n(kR) = 0$, over the interval $0 \le x \le R$, with respect to the weight function $p(x) = x$. Hence, we can state the orthogonality result for Bessel functions as follows:

For every fixed $n \ge 0$, the Bessel's functions $J_n(k_{1n}x)$, $J_n(k_{2n}x)$, $J_n(k_{3n}x)$, ... where $k_{mn} = a_{mn}/R$, are orthogonal over the interval $0 \le x \le R$, with respect to the weight function $p(x) = x$, that is

$$\int_{0}^{R} xJ_n(k_{in}x)J_n(k_{jn}x) \, dx = 0, \quad i \ne j. \tag{7.150}$$

Therefore, orthogonality is with respect to the distinct zeros of Bessel's function $J_n(kR) = 0$. For $R = 1$, these are zeros of $J_n(k) = 0$. This result implies that we can obtain infinitely many orthogonal sets (corresponding to $n = 0, 1, 2, \ldots$) involving the Bessel's functions.

The corresponding orthonormal functions are

$$J_n(k_{in}x)/\| J_n(k_{in}x) \|.$$

The value of

$$\| J_n(k_{in}x) \|^2 = \int_0^R xJ_n^2(k_{in}x) \, dx$$

can be obtained directly using the Bessel's equation. The Bessel functions $J_n(kx)$ are the solutions of the equation (see Eq. (7.126))

$$x^2y'' + xy' + (k^2x^2 - n^2) \, y = 0$$

where $y = y(kx)$. Therefore, $J_n(k_{in}x) = y(k_{in}x) = u$ and $J_n(k_{jn}x) = y(k_{jn}x) = v$, respectively satisfy the equations

$$x^2u'' + xu' + (k_{in}^2 x^2 - n^2)u = 0 \qquad (7.151)$$

and

$$x^2v'' + xv' + (k_{jn}^2x^2 - n^2)v = 0. \qquad (7.152)$$

Multiplying Eq. (7.151) by v/x and Eq. (7.152) by u/x, and subtracting, we get (compare with the proof of Theorem 7.1)

$$x(u''v - v''u) + (u'v - v'u) + (k_{in}^2 - k_{jn}^2)xuv = 0$$

or

$$[x(u'v - uv')]' + (k_{in}^2 - k_{jn}^2)xuv = 0.$$

Integrating over $[0, R]$, we obtain

$$(k_{jn}^2 - k_{in}^2) \int_0^R xuv \, dx = [x(u'v - uv')]_0^R$$

$$= [x\{k_{in}J_n'(k_{in}x)J_n(k_{jn}x) - k_{jn}J_n(k_{in}x)J_n'(k_{jn}x)\}]_0^R \qquad (7.153)$$

since $u' = \dfrac{du}{dx} = \dfrac{d}{dx}J_n(k_{in}x) = k_{in}\dfrac{d}{d(k_{in}x)}J_n(k_{in}x) = k_{in}J_n'(k_{in}x)$, and similarly for v'.

Since $J_n(0) = 0$, we obtain

$$(k_{jn}^2 - k_{in}^2) \int_0^R xJ_n(k_{in}x)J_n(k_{jn}x) \, dx = R[k_{in}J_n'(k_{in}R)J_n(k_{jn}R) - k_{jn}J_n(k_{in}R)J_n'(k_{jn}R)]$$

or $\displaystyle\int_0^R x J_n(k_{in}x)J_n(k_{jn}x) \, dx = \dfrac{R}{k_{jn}^2 - k_{in}^2}[k_{in}J_n'(k_{in}R)J_n(k_{jn}R) - k_{jn}J_n(k_{in}R)J_n'(k_{jn}R)].$ (7.154)

If $k_{in}R, k_{jn}R$ are the distinct zeros of $J_n(kR) = 0$, then the right hand side vanishes and we have the orthogonality relation

$$\int_0^R x J_n(k_{in}x) J_n(k_{jn}x)\, dx = 0, \quad i \neq j. \tag{7.155}$$

Now, suppose that $k_{in}R$ is a zero of $J_n(kR) = 0$ and $k_{jn} \to k_{in}$ so that in the limit

$$x J_n(k_{in}x) J_n(k_{jn}x) \to x J_n^2(k_{in}x).$$

Then, from Eq. (7.154), we obtain

$$\lim_{k_{jn}\to k_{in}} \int_0^R x J_n(k_{in}x) J_n(k_{jn}x)\, dx = R \lim_{k_{jn}\to k_{in}} \left[\frac{k_{in} J_n'(k_{in}R) J_n(k_{jn}R)}{k_{jn}^2 - k_{in}^2} \right].$$

In the limit, the right hand side is of the form 0/0. Hence, we obtain in the limit

$$\int_0^R x J_n^2(k_{in}x)\, dx = R k_{in} J_n'(k_{in}R) \lim_{k_{jn}\to k_{in}} \left[\frac{R J_n'(k_{jn}R)}{2 k_{jn}} \right] = \frac{1}{2} R^2 [J_n'(k_{in}R)]^2.$$

Since

$$x J_n'(x) = n J_n(x) - x J_{n+1}(x), \quad \text{and} \quad J_n(k_{in}R) = 0,$$

we obtain

$$\| J_n(k_{in}x) \|^2 = \int_0^R x J_n^2(k_{in}x)\, dx = \frac{1}{2} R^2 J_{n+1}^2(k_{in}R). \tag{7.156}$$

It can be shown that the orthonormal sets, $\{y_i\}$, derived above (Legendre orthonormal polynomials defined on $[-1, 1]$, Bessel orthonormal polynomials defined on $[0, R]$ etc.) are *complete* over a set of functions S defined on their respective intervals. That is, we can approximate the function $f(x) \in S$, arbitrarily closely by $c_0 y_0 + c_1 y_1 + \ldots + c_m y_m$. Mathematically, we can state that given an $\varepsilon > 0$, we can find the constants $c_0, c_1, \ldots, c_m$ such that

$$\| f(x) - (c_0 y_0 + c_1 y_1 + \ldots + c_m y_m) \| < \varepsilon. \tag{7.157}$$

7.5.2 Fourier-Bessel Series

In the section 7.2.6, we have derived the Fourier-Legendre series (expansion of a function $f(x)$, $x \in (-1, 1)$ in terms of Legendre polynomials $P_n(x)$). In this section, we present the expansion of a function $f(x)$, $x \in (0, R)$ in terms of the Bessel function of a given order $J_n(x)$. We know from Eq. (7.150) that for a fixed n, the Bessel functions $J_n(k_{1n}x)$, $J_n(k_{2n}x)$, $\ldots$, where k_{in}, defined by Eq. (7.149), are orthogonal over $[0, R]$ with respect to the weight function $p(x) = x$. Hence, we write the Fourier-Bessel series as

$$f(x) = c_1 J_n(k_{1n}x) + c_2 J_n(k_{2n}x) + c_3 J_n(k_{3n}x) + \ldots = \sum_{i=1}^{\infty} c_i J_n(k_{in}x). \tag{7.158}$$

Multiplying both sides of this equation by $x J_n(k_{jn}x)$ and integrating on $[0, R]$, we obtain (using the orthogonal property given in Eq. (7.150))

$$\int_0^R x f(x) J_n(k_{jn}x)\, dx = \sum c_i \int_0^R x J_n(k_{in}x) J_n(k_{jn}x)\, dx$$

$$= c_j \int_0^R x J_n^2(k_{jn}x)\, dx = c_j \| J_n(k_{jn}x) \|^2$$

Therefore,
$$c_j = \frac{1}{\|J_n(k_{jn}x)\|^2} \int_0^R xf(x)J_n(k_{jn}x)\,dx$$

$$= \frac{2}{R^2 J_{n+1}^2(k_{jn}R)} \int_0^R xf(x)J_n(k_{jn}x)\,dx. \tag{7.159}$$

If $f(x)$ is defined on $(0, 1)$, then c_j simplifies as

$$c_j = \frac{2}{J_{n+1}^2(k_{jn})} \int_0^1 xf(x)J_n(k_{jn}x)\,dx \tag{7.160}$$

where k_{jn} are the zeros of $J_n(k) = 0$ or $J_n(x) = 0$.

Example 7.28   Expanding $f(x) = x$, $0 < x < 1$, in terms of the Bessel functions $J_1(k_{i1}x)$, $i = 1, 2, \ldots$ where k_{i1} are the zeros of $J_1(x) = 0$, establish the identity

$$\sum_{j=1}^\infty \frac{J_1(k_{j1}x)}{k_{j1}J_2(k_{j1})} = \frac{x}{2}, \quad 0 < x < 1.$$

Solution   We write the Fourier-Bessel series as

$$f(x) = x = c_1 J_1(k_{11}x) + c_2 J_1(k_{21}x) + \ldots.$$

Multiplying both sides of the equation by $xJ_1(k_{j1}x)$ and integrating on $(0, 1)$, we obtain

$$c_j = \frac{2}{J_2^2(k_{j1})} \int_0^1 x^2 J_1(k_{j1}x)\,dx = \frac{2}{J_2^2(k_{j1})} \int_0^{k_{j1}} \frac{1}{k_{j1}^3} t^2 J_1(t)\,dt \quad (\text{set } k_{j1}x = t)$$

$$= \frac{2}{k_{j1}^3 J_2^2(k_{j1})} \left[t^2 J_2(t)\right]_0^{k_{j1}} = \frac{2}{k_{j1}J_2(k_{j1})}$$

where k_{j1} are the zeros of $J_1(x) = 0$. Substituting in the Fourier-Bessel series, we obtain

$$\sum_{j=1}^\infty \frac{J_1(k_{j1}x)}{k_{j1}J_2(k_{j1})} = \frac{x}{2}, \quad 0 < x < 1.$$

Example 7.29   Represent the function $f(x) = R^2 - x^2$, $0 < x < R$ in Fourier-Bessel series of the form

$$f(x) = c_1 J_0(k_{10}x) + c_2 J_0(k_{20}x) + c_3 J_0(k_{30}x) + \ldots.$$

Solution   The coefficients c_j are given by (from Eq. (7.159))

$$c_j = \frac{2}{R^2 J_1^2(k_{j0}R)} \int_0^R x(R^2 - x^2)J_0(k_{j0}x)\,dx$$

$$= \frac{2}{R^2 J_1^2(k_{j0}R)} \left[R^2 \int_0^R xJ_0(k_{j0}x)\,dx - \int_0^R x^3 J_0(k_{j0}x)\,dx\right]$$

$$= \frac{2}{R^2 J_1^2(k_{j0}R)} \left[\frac{R^2}{k_{j0}^2} \int_0^{Rk_{j0}} tJ_0(t)\,dt - \frac{1}{k_{j0}^4} \int_0^{Rk_{j0}} t^3 J_0(t)\,dt\right].$$

Since $\int t J_0 \, dt = t J_1$, and $\int t^3 J_0 dt = \int t^2 (t J_0) \, dt = t^3 J_1 - 2 \int t^2 J_1 dt = t^3 J_1 - 2 t^2 J_2$

we obtain $\quad c_j = \dfrac{2}{R^2 J_1^2(k_{j0}R)} \left[\dfrac{R^3}{k_{j0}} J_1(k_{j0}R) - \dfrac{R^2}{k_{j0}^2} \{ k_{j0} R J_1(k_{j0}R) - 2 J_2(k_{j0}R) \} \right]$

$$= \dfrac{2}{k_{j0}^2 J_1^2(k_{j0}R)} [k_{j0} R J_1(k_{j0}R) - k_{j0} R J_1(k_{j0}R) + 2 J_2(k_{j0}R)]$$

$$= \dfrac{4 J_2(k_{j0}R)}{k_{j0}^2 J_1^2(k_{j0}R)}. \tag{7.161}$$

Hence, $$R^2 - x^2 = \sum_{j=1}^{\infty} c_j J_0(k_{j0}x), \quad 0 < x < R$$

where c_j are given in Eq. (7.161).

Exercise 7.6

Find the eigenvalues and eigenfunctions of the following Sturm-Liouville *bvp*. Verify orthogonality by direct calculations.

1. $y'' + \lambda y = 0$, $y(0) = 0$, $y(\pi) = 0$.   2. $y'' + \lambda y = 0$, $y'(0) = 0$, $y'(l) = 0$.

3. $y'' + \lambda y = 0$, $y(0) = 0$, $y'(l) = 0$.   4. $y'' + \lambda y = 0$, $y'(0) = 0$, $y(\pi) = 0$.

5. $y'' + \lambda y = 0$, $y(0) = y(2\pi)$, $y'(0) = y'(2\pi)$.

6. $(xy')' + (\lambda/x)y = 0$, $1 < x < l$, $y(1) = 0$, $y(l) = 0$.

7. $(xy')' + (\lambda/x)y = 0$, $1 < x < a$, $y'(1) = 0$, $y(a) = 0$.

8. Verify the orthogonality of Bessel functions, using the Bessel equation directly.

9. If the functions $y_0(x)$, $y_1(x)$, $y_2(x)$, $\ldots$, $c \le x \le d$, form an orthogonal set, show that a linear transformation $x = at + b$, retains the orthogonality of the functions in the new interval $(c - b)/a \le t \le (d - b)/a$.

10. Using the eigenfunctions obtained in Example 7.22, find the eigenfunctions of Problem 5, by using a linear transformation. Compare the two sets of eigenfunctions.

11. The functions $T_n(x) = \cos(n\theta)$, $x = \cos\theta$, $-1 \le x \le 1$ are called the *Chebyshev polynomials* of the first kind. $T_n(x)$ can be written as

    $T_n(x) = \cos(n \cos^{-1} x) = x^n - {}^nC_2 \, x^{n-2}(1 - x^2) + {}^nC_4 x^{n-4}(1 - x^2)^2 - \ldots$.

    (a) Show that the first few polynomials are

    $$T_0(x) = 1, \quad T_1(x) = x, \quad T_2(x) = 2x^2 - 1, \quad T_3(x) = 4x^3 - 3x, \quad T_4(x) = 8x^4 - 8x^2 + 1.$$

    (b) Show that $T_n(x)$ satisfies the Chebyshev differential equation (Eq. (7.132))

    $$(1 - x^2)y'' - xy' + n^2 y = 0.$$

    (c) Show that the generating function for $T_n(x)$ is

    $$\frac{1 - xt}{1 - 2xt + t^2} = \sum_{n=0}^{\infty} T_n(x) t^n, \quad t \ne 0.$$

(d) Show that $T_n(x)$ satisfies the recurrence relation

$$T_{n+1}(x) - 2xT_n(x) + T_{n-1}(x) = 0.$$

(e) Prove the orthogonality of $T_n(x)$

$$\int_{-1}^{1} \frac{T_m(x)T_n(x)}{\sqrt{1-x^2}}\, dx = 0, \quad m \neq n,$$

$$= \pi, \quad \text{if} \quad m = n = 0$$

$$= \pi/2, \quad \text{if} \quad m = n \neq 0.$$

(f) Show that a function $f(x)$, $-1 < x < 1$, can be expanded in a Chebyshev series as

$$f(x) = \frac{1}{2}c_0 T_0(x) + c_1 T_1(x) + c_2 T_2(x) + \ldots$$

where

$$c_j = \frac{2}{\pi}\int_{-1}^{1} \frac{f(x)T_j(x)}{\sqrt{1-x^2}}\, dx.$$

12. The functions

$$L_n(x) = e^x \frac{d^n}{dx^n}(x^n e^{-x}), \quad 0 \le x < \infty$$

are called the *Laguerre polynomials*.

(a) Show that $L_n(x)$ satisfies the Laguerre differential equation (Eq. (7.129))

$$xy'' + (1 - x)y' + ny = 0.$$

(b) Show that the first few polynomials are

$$L_0(x) = 1, \quad L_1(x) = -x + 1, \quad L_2(x) = x^2 - 4x + 2, \quad L_3(x) = -x^3 + 9x^2 - 18x + 6.$$

(c) Show that $L_n(x)$ satisfies the recurrence relation

$$L_{n+1}(x) - (2n + 1 - x)L_n(x) + n^2 L_{n-1}(x) = 0.$$

(d) Prove the orthogonality of the Laguerre polynomials

$$\int_{0}^{\infty} e^{-x} L_m(x) L_n(x)\, dx = 0, \, m \neq n.$$

Prove also that

$$\int_{0}^{\infty} e^{-x} L_n^2(x)\, dx = (n!)^2.$$

(e) Show that a function $f(x)$, $0 \le x < \infty$ can be expanded in a Laguerre series as

$$f(x) = c_0 L_0(x) + c_1 L_1(x) + c_2 L_2(x) + \ldots$$

where

$$c_j = \frac{1}{(j!)^2}\int_{0}^{\infty} e^{-x} f(x) L_j(x)\, dx.$$

13. The functions

$$H_n(x) = (-1)^n e^{x^2} \frac{d^n}{dx^n}(e^{-x^2}), \quad -\infty < x < \infty$$

are called the *Hermite polynomials*.

(a) Show that $H_n(x)$ satisfies the Hermite differential equation (Eq. (7.133))

$$y'' - 2xy' + 2ny = 0.$$

(b) Show that the first few Hermite polynomials are

$$H_0(x) = 1, \quad H_1(x) = 2x, \quad H_2(x) = 4x^2 - 2, \quad H_3(x) = 8x^3 - 12x.$$

(c) Show that $H_n(x)$ satisfies the recurrence relation

$$H_{n+1}(x) = 2x\,H_n(x) - 2nH_{n-1}(x).$$

(d) Prove the orthogonality of the Hermite polynomials

$$\int_{-\infty}^{\infty} e^{-x^2} H_m(x)\, H_n(x)\, dx = 0, \ m \neq n$$

Prove also that $\quad\displaystyle\int_{-\infty}^{\infty} e^{-x^2} H_n^2(x)\, dx = 2^n(n!)\sqrt{\pi}.$

(e) Show that a function $f(x)$, $-\infty < x < \infty$ can be expanded in a Hermite series of the form

$$f(x) = c_0 H_0(x) + c_1 H_1(x) + c_2 H_2(x) + \ldots$$

where $\quad\displaystyle c_j = \frac{1}{2^j(j!)\sqrt{\pi}} \int_{-\infty}^{\infty} e^{-x^2} f(x) H_j(x)\, dx.$

For the following functions $f(x)$ develop a Fourier-Bessel series of the form

$$f(x) = c_1 J_0(k_{10}x) + c_2 J_0(k_{20}x) + c_3 J_0(k_{30}x) + \ldots .$$

14. $f(x) = 1, \quad 0 < x < R.$ **15.** $f(x) = c, \ 0 < x \leq a$ and $f(x) = 0, \ a < x < R.$

16. $f(x) = 0, \quad 0 < x \leq 1/2$ and $f(x) = c, \ 1/2 < x < 1.$

17. $f(x) = cx^2, \quad 0 < x < 1.$ **18.** $f(x) = x^4, \quad 0 < x < R.$

For the following functions $f(x)$ develop a Fourier-Bessel series of the form

$$f(x) = c_1 J_1(k_{11}x) + c_2 J_1(k_{21}x) + c_3 J_1(k_{31}x) + \ldots .$$

19. $f(x) = x, \quad 0 < x < R.$ **20.** $f(x) = x, \quad 0 < x \leq a$ and $f(x) = 0, \quad a < x < R.$

21. $f(x) = x - x^3, \quad 0 < x < 1.$ **22.** $f(x) = x^5, \quad 0 < x < 1.$

For the following functions $f(x)$ develop a Fourier-Bessel series of the form

$$f(x) = c_1 J_2(k_{12}x) + c_2 J_2(k_{22}x) + c_3 J_2(k_{32}x) + \ldots .$$

23. $f(x) = x^2, \quad 0 < x < R.$ **24.** $f(x) = 0, \quad 0 < x < a$ and $f(x) = x^2, \quad a < x < R.$

25. $f(x) = x^2, \quad 0 < x < 1/2$ and $f(x) = 0, \ 1/2 < x < 1.$

7.6 Answers and Hints

Exercise 7.1

1. $n = 0$: $y_0 = 1, \quad y_1 = \dfrac{1}{2}\ln[(1+x)/(1-x)], \quad n = 1$: $y_0 = 1 - \dfrac{x}{2}\ln[(1+x)/(1-x)], y_1(x) = x.$

2. $P_1(x) = x, \quad P_2(x) = (3x^2 - 1)/2, \quad P_3(x) = (5x^3 - 3x)/2, \quad P_4(x) = (35x^4 - 30x^2 + 3)/8.$

3. $15x^3 - 11x + 1.$ **4.** $35x^4 - 27x^2 + 3.$

5. $10x^3 + 9x^2 - 9x - 5.$ **6.** $(175x^4 + 200x^3 - 126x^2 - 112x + 7)/8.$

7. $2P_2 + 5P_1 - 5P_0.$ **8.** $(2P_3 + 8P_1 + 5P_0)/5.$

9. $(8P_3 + 10P_2 + 22P_1 - 25P_0)/5.$ **10.** $(8P_4 + 12P_3 + 30P_2 + 23P_1 - 18P_0)/5.$

11. $(40P_4 + 42P_3 - 40P_2 - 7P_1 + 70P_0)/35.$

12. Result is true since $\dfrac{d^m}{dx^m}[(x^2 - 1)^n] = 0$ at $x = \pm 1$ for $m < n$.

13. Integrate by parts m times.   **14.** Integrate by parts n times.

16. Set $x = 0$ in the equation for generating function. $P_{2n+1}(0) = 0$, $n = 0, 1, 2, \ldots$,

$$P_{2n}(0) = \frac{(-1)^n 1 \cdot 3 \ldots (2n-1)}{2^n (n)!} = \frac{(-1)^n 1 \cdot 3 \ldots (2n-1)}{2 \cdot 4 \cdot 6 \ldots 2n}.$$

17. Multiply both sides of the recurrence relation given in Eq. (7.12) with $P_{n-1}(x)$ and integrate on $[-1, 1]$. Use the orthogonal properties.

18. Differentiate Eq. (7.12) with respect to x, eliminate $xP_n'(x)$ using Eq. (7.13) and cancel $(n + 1)$ to get the result.

19. Subtract Eq. (7.13) from the result of Problem 18.

20. Replace n by $n - 1$ in Problem 19.

21. Multiply Eq. (7.13) by x and subtract from the result of Problem 20.

22. In the result of problem 21, eliminate $P_{n-1}(x)$ using Eq. (7.12).

23. Set $x = 1$ in the recurrence relation $P_n'(x) = xP_{n-1}'(x) + nP_{n-1}(x)$; $P_n'(1) = P_{n-1}'(1) + n$. Putting $n = 1, 2, \ldots$, we get $P_n'(1) = n(n + 1)/2$.

24. Set $x = -1$ in the solution of problem 23. We obtain $P_n'(-1) = (-1)^{n-1} n(n + 1)/2$.

25. Set $n = 1, 2, \ldots, n$, in $(2n + 1)P_n = P_{n+1}' - P_{n-1}'$. We obtain

$$P_{n+1}' + P_n' - P_0' - P_1' = 3P_1 + 5P_2 + \ldots + (2n + 1)P_n$$

Since $P_0' = 0$, $P_1' = 1 = P_0$, we obtain the result.

26. $P_{n+1}(x) = [(2n + 1)xP_n(x) - nP_{n-1}(x)]/(n + 1)$,

$P_2(1.1) = 1.315$, $P_3(2.5) = 35.3125$, $P_4(1.2) = 4.047$.

27. $P_3'(1.6) = 17.7$, $P_4'(2.1) = 146.3175$.

28. In the equation of the generating function, put $t = r_1/r_2$, $x = \cos \theta$.

29. $\sin(mx)/\sqrt{\pi}$, $m = 1, 2, \ldots$.

30. $1/\sqrt{2}$, $\cos \pi x$, $\sin \pi x$, $\cos 2\pi x$, $\sin 2\pi x, \ldots$.

31. $1/\sqrt{2\pi}$, $\cos x/\sqrt{\pi}$, $\sin x/\sqrt{\pi}$, $\cos 2x/\sqrt{\pi}$, $\sin 2x/\sqrt{\pi}, \ldots$.

32. Write as $\displaystyle\int_{-1}^{1} P_n^2(x)\, dx = \frac{1}{(n!2^n)^2} \int_{-1}^{1} D^n(x^2 - 1)^n D^n(x^2 - 1)^n\, dx$

$$= \frac{(-1)^n}{(n!2^n)^2} \int_{-1}^{1} (2n)!\,(x^2 - 1)^n\, dx \quad \text{(integrate by parts)}$$

$$= \frac{2(2n)!}{(n!2^n)^2} \int_{0}^{1} (1 - x^2)^n\, dx = \frac{2(2n)!}{(n!2^n)^2} \int_{0}^{\pi/2} \cos^{2n+1}\theta\, d\theta = \frac{2}{2n + 1}.$$

33. For $m = n$, $I = \frac{1}{2} \int_{-1}^{1} x(P_n^2)' dx = \frac{2n}{2n + 1}$ (integrate by parts). For $m \neq n$, I is either equal to 0 or 2.

34. Integrate by parts and use the differential equation.

35. (a) $c_0 = 6/5$, $c_1 = 1$, $c_2 = 4/7$, $c_3 = 0$ $c_4 = 8/35$.

    (b) $c_0 = 11/3$, $c_1 = 14/5$, $c_2 = 4/3$, $c_3 = 6/5$.

    (c) $c_0 = 2$, $c_1 = -3$, $c_2 = 2$.

    (d) $c_0 = 6$, $c_1 = -12/5$, $c_2 = 0$, $c_3 = 2/5$.

Exercise 7.2

1. $T_n(1) = \cos(0) = 1$.   **2.** $T_n(-1) = \cos(n\pi) = (-1)^n$.

3. $T_n(-x) = \cos[n(\pi - \theta)] = \cos(n\pi)\cos(n\theta) = (-1)^n T_n(x)$.

4. $T_{2n}(0) = \cos\left[(2n)\frac{\pi}{2}\right] = (-1)^n$.   **5.** $T_{2n+1}(0) = \cos\left[(2n+1)\frac{\pi}{2}\right] = 0$.

6. Set $x = \cos\theta$.

7. x^m can be expressed in terms of $T_m, T_{m-1}, \ldots, T_1$ or T_0. Using the orthogonal properties, we find that the integral vanishes, since $m < n$.

8. $U_n(-x) = \{\sin[(n+1)(\pi - \theta)]\}/\sin\theta = (-1)^n U_n(x)$.

9. $U_n(-1) = \lim_{x \to -1} \frac{\sin[(n+1)\cos^{-1}x]}{\sqrt{1 - x^2}} = \lim_{\theta \to \pi} \frac{\sin[(n+1)\theta]}{\sin\theta}$

$$= \lim_{\theta \to \pi} \frac{(n+1)\cos[(n+1)\theta]}{\cos\theta} = (-1)^n(n+1).$$

10. $U_{2n}(0) = \sin\left[(2n+1)\frac{\pi}{2}\right] = (-1)^n$.   **11.** $U_{2n+1}(0) = \sin\left[(2n+2)\frac{\pi}{2}\right] = 0$.

12. We have $\sqrt{1 - x^2}\, U_n = \sin[(n+1)\cos^{-1}x]$. Differentiating and simplifying, we get

$$(1 - x^2)U_n' - xU_n = -(n+1)\cos[(n+1)]\cos^{-1}x].$$

Differentiating again and simplifying, we get

$$(1 - x^2)U_n'' - 3xU_n' + n(n+2)U_n = 0.$$

13. $T_n'(x) = \frac{n\sin(n\theta)}{\sin\theta} = nU_{n-1}(x)$.

14. $T_{n+1}'(x) = (n+1)U_n(x)$. Integrating, we get $\int U_n(x)\, dx = \frac{1}{n+1} T_{n+1}(x)$.

15. Using the result from problem 12, we have
$$(1 - x^2)U_n' - xU_n = -(n+1)T_{n+1}.$$

16. We have $2x = 2\cos\theta = e^{i\theta} + e^{-i\theta}$.

$$\frac{1 - t^2}{1 - 2tx + t^2} = \frac{1 - t^2}{1 - t(e^{i\theta} + e^{-i\theta}) + t^2} = \frac{1 - t^2}{(1 - e^{i\theta}t)(1 - e^{-i\theta}t)}$$

$$= (1 - t^2)\sum_{m,l=0}^{\infty} e^{i(m-l)\theta}t^{m+l} = \sum_{m,l=0}^{\infty} e^{i(m-l)\theta}t^{m+l} - \sum_{m,l=0}^{\infty} e^{i(m-l)\theta}t^{m+l+2}$$

$$= \sum_{n=0}^{\infty} c_n t^n.$$

Now, coefficient of $t^0 = c_0 = 1 = T_0(x)$

coefficient of $t^1 = c_1 = (e^{i\theta} + e^{-i\theta}) = 2\cos\theta = 2x = 2T_1(x)$.

coefficient of $t^n = c_n =$ (coefficient of t^{m+l} when $m + l = n$)

$\qquad$ + (coefficient of t^{m+l+2}, when $m + l + 2 = n$)

$$= \sum_{m=0}^{\infty} e^{i[m-(n-m)]\theta} - \sum_{m=0}^{n-2} e^{i[m-(n-m-2)]\theta}$$

$$= e^{-in\theta} \sum_{m=0}^{n} e^{2im\theta} - e^{-(n-2)\theta} \sum_{m=0}^{n-2} e^{2im\theta}$$

$$= \frac{1}{1-e^{2i\theta}} [e^{-in\theta}\{1-(e^{2i\theta})^{n+1}\} - e^{-(n-2)\theta}\{1-(e^{2i\theta})^{n-1}\}]$$

$$= \frac{1}{1-e^{2i\theta}} [e^{-in\theta} - e^{i(n+2)\theta} - e^{-i(n-2)\theta} + e^{in\theta}]$$

$$= \frac{1}{1-e^{2i\theta}} [(e^{in\theta} + e^{-in\theta})(1-e^{2i\theta})] = e^{in\theta} + e^{-in\theta} = 2\cos(n\theta) = 2T_n(x).$$

17. Let $u = (1-x^2)^{n-1/2}$. Denote $u_n = d^n u/dx^n$. Then

$$(1-x^2)u_1 + (2n-1)xu = 0.$$

Using Leibniz rule, we get

$$(1-x^2)u_{n+2} - 3xu_{n+1} + (n^2-1)u_n = 0. \qquad (7.162)$$

Now, let $v = c_n(1-x^2)^{1/2}u_n$. Then

$$(1-x^2)^{1/2}v' = c_n[(1-x^2)u_{n+1} - xu_n], \text{ and}$$

$$(1-x^2)v'' - xv' = c_n(1-x^2)^{1/2}[(1-x^2)u_{n+2} - 3xu_{n+1} - u_n] = -n^2v.$$

(using (7.162)). Hence,

$$(1-x^2)v'' - xv' + n^2v = 0, \text{ and } v(x) = T_n(x).$$

18. Let $u = (1-x^2)^{n+1/2}$. Denote $u_n = d^n u/dx^n$. Then,

$$(1-x^2)u_1 + (2n+1)xu = 0.$$

Using Leibniz rule, we get

$$(1-x^2)u_{n+2} - xu_{n+1} + (n+1)^2 u_n = 0. \qquad (7.163)$$

Now, let $v = c_n(1-x^2)^{-1/2}u_n$. Then, we get

$$(1-x^2)v' - xv = c_n(1-x^2)^{1/2}u_{n+1}, \text{ and}$$

$$(1-x^2)v'' - 3xv' - v = -(n+1)^2 v, \text{ (using (7.163))}.$$

Hence, $(1-x^2)v'' - 3xv' + n(n+2)v = 0$, and $v(x) = U_n(x)$.

19. $p'(x) = \dfrac{a_0}{2}T_0 + a_1 T_1 + \ldots + a_r T_r + \ldots + a_{n-1}T_{n-1}$.

Integrate and use Eq. (7.53).

$$p(x) = \frac{a_0}{2}T_1 + \frac{a_1}{4}(T_0 + T_2) + \sum_{r=2}^{n-1} \frac{1}{2}a_r\left[\frac{1}{r+1}T_{r+1} - \frac{1}{r-1}T_{r-1}\right] = \sum_{r=0}^{n}{}' c_r T_r(x).$$

Comparing, we solve for the coefficients backwards. We get

$$c_n = \frac{1}{2n} a_{n-1}, c_{n-1} = \frac{1}{2(n-1)} a_{n-2},$$

$$c_r = \frac{1}{2r}(a_{r-1} - a_{r+1}), r = n-2, n-3, \ldots$$

Hence, $a_{n-1} = 2nc_n$, $a_{n-2} = 2(n-1)c_{n-1}$, $a_{n-3} = 2(n-2)c_{n-2} + 2nc_n, \ldots,$

$a_1 = 4c_2 + 8c_4 + 12c_6 + \ldots$, $a_0 = 2c_1 + 6c_3 + 10c_5 + \ldots$, ending with a_n or a_{n-1}.

20. (a) $p(x) = \frac{1}{8}T_4 + \frac{3}{4}T_3 + \frac{3}{2}T_2 + \frac{29}{4}T_1 + \frac{19}{8}T_0 = \Sigma' c_r T_r(x)$

(b) $c_0 = 19/4$, $c_1 = 29/4$, $c_2 = 3/2$, $c_3 = 3/4$, $c_4 = 1/8$.

$a_3 = 8c_4 = 1$, $a_2 = 6c_3 = 9/2$, $a_1 = 4c_2 + 8c_4 = 7$, $a_0 = 2c_1 + 6c_3 = 19$.

Hence, $p'(x) = [19 T_0 + 14T_1 + 9T_2 + 2T_3]/2$.

Exercise 7.3

2. First approximation: 0.9975, 0.99, 0.9775, 0.96, 0.9375;

Second approximation: 0.9975, 0.99, 0.9776, 0.9604, 0.9385.

There is no error in the second approximation (taken to four places).

3. Approximation $J_0(x) \approx 1 - (x^2/4)$ gives $x > 2$.

Approximation $J_0 \approx 1 - (x^2/4) + (x^2/64)$ gives $x < \sqrt{8}$.

4. $J_{1/2}(x) = \left(\frac{x}{2}\right)^{1/2}\left[\frac{1}{\Gamma(3/2)} - \frac{1}{\Gamma(5/2)}\left(\frac{x}{2}\right)^2 + \frac{1}{2\Gamma(7/2)}\left(\frac{x}{2}\right)^4 - \ldots\right] = \sqrt{\frac{2}{\pi x}}\sin x,$

$J_{-1/2}(x) = \left(\frac{x}{2}\right)^{-1/2}\left[\frac{1}{\Gamma(1/2)} - \frac{1}{\Gamma(3/2)}\left(\frac{x}{2}\right)^2 + \frac{1}{2\Gamma(5/2)}\left(\frac{x}{2}\right)^4 - \ldots\right] = \sqrt{\frac{2}{\pi x}}\cos x.$

6. $J_0(a) = 0 = J_0(b)$, where a and b are two successive zeros of $J_0(x)$. By Rolle's theorem, we get $J_0' = -J_1 = 0$ for some $c \in (a, b)$, that is, $J_1(x) = 0$ has a root $c \in (a, b)$.

7. $AJ_{1/2}(x) + BJ_{-1/2}(x) = \sqrt{\frac{2}{\pi x}}(A \sin x + B \cos x)$.

8. $AJ_{1/3}(x) + BJ_{-1/3}(x)$.

9. Bessel's equation with $\nu = 1/2$, $y(x) = AJ_{1/2}(x/2) + BJ_{-1/2}(x/2)$.

10. Bessel's equation with $\nu = 1/3$, $y(x) = AJ_{1/3}(3x) + BJ_{-1/3}(3x)$.

11. $AJ_\nu(\lambda x) + BJ_{-\nu}(\lambda x)$.
12. $x^{-\lambda}[AJ_\lambda(x) + BJ_{-\lambda}(x)]$.

13. $AJ_\nu(x^2) + BJ_{-\nu}(x^2)$.
14. $x^{1/2}\left[AJ_{1/3}\left(\frac{2}{3}x^{3/2}\right) + BJ_{-1/3}\left(\frac{2}{3}x^{3/2}\right)\right]$.

15. $x^\nu[AJ_\nu(x^\nu) + BJ_{-\nu}(x^\nu)]$.

Exercise 7.4

1. $\frac{2}{x} J_1 - J_0$.

2. $\left(\frac{8}{x^2} - 1 \right) J_1 - \frac{4}{x} J_0$.

3. $\left(\frac{6}{x^2} - 1 \right)\left(\frac{8}{x} \right) J_1 - \left(\frac{24}{x^2} - 1 \right) J_0$.

4. Set $v = 1$ in Eqs. (7.95), (7.96) and simplify.

5. Set $v = 2$ in Eq. (7.93) and $v = 1$ in Eq. (7.96) and simplify.

6. Set $v = 3$ in Eq. (7.93) and use solutions of problems 1 and 2.

7. Set $v = 1$ in Eq. (7.95), differentiate and use Eq. (7.93) to simplify.

8. Take $v = n$. Differentiate Eq. (7.95) and use Eq. (7.95) again.

9. $2 J_v J_v' = J_v (J_{v-1} - J_{v+1})$. Using Eqs. (7.93) and (7.94), we get

   $$2 J_v J_v' = \frac{x}{v} (J_{v-1}^2 - J_{v+1}^2) - \frac{x}{v} J_v' (J_{v-1} + J_{v+1}).$$

   Using Eq. (7.95), we get $J_v J_v' = \frac{x}{4v} (J_{v-1}^2 - J_{v+1}^2)$.

10. Using Eq. (7.91), we obtain $\frac{1}{x} [x^{-v} J_v]' = - x^{-(v+1)} J_{v+1}$.

    Repeatedly applying Eq. (7.91), we obtain

    $$\left(\frac{1}{x} \frac{d}{dx} \right)^2 (x^{-v} J_v) = (-1)^2 x^{-(v+2)} J_{v+2}, \ldots, \left(\frac{1}{x} \frac{d}{dx} \right)^m (x^{-v} J_v) = (-1)^m x^{-(v+m)} J_{v+m}.$$

11. Write left hand side as $[(x^{-n} J_n)(x^{n+1} J_{n+1})]'$, differentiate and use Eqs. (7.89) and (7.91).

12. Use Eq. (7.96) with $v = - 3/2$ and $v = - 1/2$. Eliminate $J_{-3/2}(x)$ and use the expressions for $J_{1/2}(x)$ and $J_{-1/2}(x)$.

13. Use the expressions for $J_{1/2}(x)$ and $J_{-1/2}(x)$.

14. Multiply Eq. (7.94) by x and differentiate to obtain $(v = n)$

    $$x^2 J_n'' = n J_n + (n - 2) x J_n' - 2 x J_{n+1} - x^2 J_{n+1}'.$$

    Use Eq. (7.94) to eliminate $x J_{n+1}$, Eq. (7.93) to eliminate $x J_{n+1}'$ and again Eq. (7.94) to eliminate $x J_n'$.

15. Use Eq. (7.95), with $v = n$.

16. Use the identity $(J_0^2)' = - 2 J_0 J_1$.

17. $\int x J_0^2 \, dx = \frac{x^2}{2} J_0^2 + \int x^2 J_0 J_1 dx = \frac{x^2}{2} J_0^2 + \int (x J_0)(x J_1) \, dx$

    $$= \frac{x^2}{2} J_0^2 + (x J_1)^2 - \int (x J_1)(x J_1)' \, dx = \frac{x^2}{2} (J_0^2 + J_1^2) + c.$$

18. $J_2 = J_0 - 2 J_1'$. Integrate.

19. Write $J_3 = x^2 (x^{-2} J_3)$ and integrate. Use Eq. (7.92).

20. Write $J_5 = x^4 (x^{-4} J_5)$ and integrate. Use Eq. (7.92) repeatedly.

21. Use Eq. (7.95) repeatedly.

    $$J_5 = J_3 - 2 J_4' = J_1 - 2 J_2' - 2 J_4' = - (J_0' + 2 J_2' + 2 J_4').$$

22. Write $x^2 J_0 = x (x J_0)$ and use Eq. (7.90). Then write $J_0' = - J_1$ and integrate.

23. Write $x^3 J_0 = x^2 (x J_0)$ and use Eq. (7.90). Then use Eq. (7.96).

24. Use Eq. (7.92).

$$\int x^{-1} J_4 \, dx = -x^{-1} J_3 + 2 \int x^{-2} J_3 \, dx = -x^{-1} J_3 - 2x^{-2} J_2 + c.$$

25. $\int x^5 J_0 \, dx = x^5 J_1 - 4 \int x^4 J_1 \, dx = x^5 J_1 - 4x^4 J_2 + 8 \int x^3 J_2 \, dx = x^5 J_1 - 4x^4 J_2 + 8x^3 J_2 + c.$

26. Set $xt = y$ and use Eq. (7.90).   **27.**   Let $rx = t$ and use Eq. (7.90).

28. Set $\alpha x = t$ and integrate using $J_0'(x) = -J_1(x)$.

29. Using Eq. (7.90), we obtain $\int_0^1 (x - x^3) J_0 \, dx = 2J_2(1) = 2[2J_1(1) - J_0(1)]$,   using Eq. (7.96).

30. Write $x^2 J_0 = x(x J_0)$ and use Eq. (7.90). Use $J_0' = -J_1$ and integrate.

31. Integrate by parts taking 1 as the first function

$$\int J_0 \cos x \, dx = x J_0 \cos x + \int x(J_1 \cos x + J_0 \sin x) \, dx$$

$$= x J_0 \cos x + x J_1 \sin x - \int \sin x (x J_1' + J_1 - x J_0) \, dx = x J_0 \cos x + x J_1 \sin x + c,   \text{using Eq. (7.93).}$$

32. Integrate by parts, use Eq. (7.93) and again integrate by parts.

33. In the Bessel equation of order zero, set $x = \alpha t$ and change the dummy variable t to x.

34. Multiply the two equations in Problem 33, by v and u respectively, subtract and simplfy.

35. Integrate the equation in Problem 34.

Exercise 7.5

1. $AJ_5(x) + BY_5(x)$.

2. $AJ_2(3x) + BY_2(3x)$.

3. $AJ_2(x^2) + BY_2(x^2)$.

4. $AJ_1(\sqrt{x}) + BY_1(\sqrt{x})$.

5. $[AJ_2(x) + BY_2(x)]/x^2$.

8. Comparing with Problem 7, we get $a = 0$, $c = 1/2$, $b = 1$, or

$y = u$, $t = \sqrt{x}$, $z = t$. $y = AJ_n(\sqrt{x}) + BJ_{-n}(\sqrt{x})$ or $AJ_n(\sqrt{x}) + BY_n(\sqrt{x})$ depending on n being a non-integer or an integer.

9. Comparing with Problem 7, we get $a = 1/2$, $c = 1/2$, $b = 2$, $n = 1$ or

$y = u\sqrt{x}$, $t = \sqrt{x}$, $z = 2t$. $y = \sqrt{x}[AJ_1(2\sqrt{x}) + BY_1(2\sqrt{x})]$.

10. Comparing with Problem 7, we get $y = u$, $t = \sqrt{x}$, $z = t$, $y = AJ_0(\sqrt{x}) + BY_0(\sqrt{x})$.

Exercise 7.6

1. $\lambda = n^2$, $n = 1, 2, \ldots$ $y(x) = \sin nx$.

2. $\lambda = (n\pi/l)^2$, $y(x) = \cos(n\pi x/l)$, $n = 0, 1, 2, \ldots$.

3. $\lambda = [(2n + 1)\pi/(2l)]^2$, $y(x) = \sin[(2n + 1)\pi x/(2l)]$, $n = 0, 1, 2, \ldots$.

4. $\lambda = [(2n + 1)/2]^2$, $y(x) = \cos[(2n + 1)x/2]$, $n = 0, 1, 2, \ldots$.

5. $\lambda = n^2$, $y(x) = \cos nx$, $\sin nx$, $n = 0, 1, 2, \ldots$, that is $1, \cos x, \sin x, \cos 2x, \sin 2x, \ldots$.

6. Set $x = e^t$. Then $0 < t < \ln l$, $\lambda = a^2$, $y(x) = \sin (a \ln x)$ where $a = (n\pi/\ln l)$, $n = 1, 2, \ldots$.

7. Set $x = e^t$. Then $0 < t < \ln a$, $\lambda = k^2$, $y(x) = \cos (k \ln x)$, where $k = (2n + 1)\pi/(2 \ln a)$, $n = 1, 2, \ldots$.

8. Use Eq. (7.113) and the fact $k_i \neq k_j$ and $J_n(k_{in}R) = 0 = J_n(k_{Jn}R)$.

14. $c_j = 2/[Rk_{j0}J_1(k_{j0}R)]$.

15. $c_j = [2caJ_1(k_{j0}a)]/[R^2 k_{j0} J_1^2 (k_{j0}R)]$.

16. $c_j = c[2J_1(k_{j0}) - J_1(k_{j0}/2)]/[k_{j0} J_1^2 (k_{j0})]$.

17. $c_j = 2c[k_{j0}J_1(k_{j0}) - 2J_2(k_{j0})]/[k_{j0}^2 J_1^2 (k_{j0})]$.

18. $c_j = 2R[k_{j0}^2 R^2 J_1(k_{j0}R) - 4k_{j0}RJ_2(k_{j0}R) + 8J_3(k_{j0}R)]/[k_{j0}^3 J_1^2 (k_{j0}R)]$.

19. $c_j = 2/[k_{j1}J_2(k_{j1}R)]$.

20. $c_j = 2a^2 J_2(k_{j1}a)/[R^2 k_{j1} J_2^2 (k_{j1}R)]$.

21. $c_j = 4J_3(k_{j1})/[k_{j1}^2 J_2^2 (k_{j1})]$.

22. $c_j = 2[k_{j1}^2 J_2 (k_{j1}) - 4k_{j1}J_3(k_{j1}) + 8J_4(k_{j1})]/[k_{j1}^3 J_2^2 (k_{j1})]$.

23. $c_j = 2R/k_{j2}J_3 (k_{j2}R)$.

24. $c_j = 2[R^3 J_3(k_{j2}R) - a^3 J_3(k_{j2}a)]/[R^2 k_{j2} J_3^2 (k_{j2}R)]$.

25. $c_j = J_3(k_{j2}/2)/[4k_{j2} J_3^2 (k_{j2})]$.

Chapter 8

Laplace Transformation

8.1 Introduction

Laplace transformation is one of the important tools for solving linear, constant coefficient ordinary or partial differential equations under suitable initial and boundary conditions. Laplace transformation when applied to the given initial value problem consisting of a single or a system of linear, ordinary differential equations, converts it into a single or a system of linear, algebraic equations in terms of the Laplace transform of the dependent variable. This equation is called the *subsidiary equation*. The initial conditions are automatically absorbed during the derivation of this algebraic equation. The solution of this algebraic equation gives the expression for the Laplace transform of the dependent variable. Taking the inverse Laplace transformation, we find the solution of the original initial value problem. Thus, we note the following.

 (i) Laplace transformation converts a given initial value problem to an algebraic equation.
 (ii) The solution of the initial value problem is directly found without first finding the general solution.

In the case of partial differential equations in terms of two independent variables, the Laplace transformation is applied with respect to one of the variables, usually the variable t (time). The resulting ordinary differential equation in terms of the second variable is solved by the usual methods of solving ordinary differential equations. The inverse Laplace transform of this solution gives the solution of the given partial differential equation.

One of the important applications of Laplace transformation is the solution of the mathematical models of physical systems in which the right hand side of the differential equation, representing the driving force is discontinuous or acts for a short time only or is a periodic function (which is not necessarily a sine or a cosine function).

8.2 Basic Theory of Laplace Transforms

Let $f(t)$ be a function defined for $t \geq 0$. Then, the integral

$$F(s) = \int_0^\infty e^{-st} f(t)\, dt \qquad (8.1)$$

is called the Laplace transform of $f(t)$, provided the integral exists. The Laplace transform of $f(t)$ is usually denoted by $\mathcal{L}[f(t)]$, where $\mathcal{L}$ is called the Laplace transform operator. That is,

$$\mathcal{L}[f(t)] = F(s). \tag{8.2}$$

The original function $f(t)$ is called the inverse Laplace transform and we write

$$\mathcal{L}^{-1}[F(s)] = f(t). \tag{8.3}$$

Theorem 8.1   (Linearity of Laplace transformation) Let $f(t)$ and $g(t)$ be any two functions whose Laplace transforms exist. Then, for any two constants α and β, we have

$$\mathcal{L}[\alpha f(t) + \beta g(t)] = \alpha \mathcal{L}[f(t)] + \beta \mathcal{L}[g(t)]. \tag{8.4}$$

Proof   Using the definition, we have

$$\mathcal{L}[\alpha f(t) + \beta g(t)] = \int_0^\infty e^{-st} [\alpha f(t) + \beta g(t)] dt$$

$$= \alpha \int_0^\infty e^{-st} f(t) dt + \beta \int_0^\infty e^{-st} g(t) dt = \alpha \mathcal{L}[f(t)] + \beta \mathcal{L}[g(t)].$$

Example 8.1   Find the Laplace transform of $f(t) = 1$, $t \geq 0$.

Solution   From the definition, we have

$$\mathcal{L}[1] = \int_0^\infty e^{-st} \cdot 1 \, dt = \left[-\frac{e^{-st}}{s} \right]_0^\infty = \frac{1}{s}, \ s > 0.$$

Example 8.2   Find the Laplace transform of $f(t) = t$, $t \geq 0$.

Solution   We have

$$\mathcal{L}[t] = \int_0^\infty e^{-st} \cdot t \, dt = \left[-\frac{e^{-st} t}{s} - \frac{e^{-st}}{s^2} \right]_0^\infty = \frac{1}{s^2}, \ s > 0.$$

Example 8.3   Find the Laplace transform of $f(t) = e^{at}$, $t \geq 0$.

Solution   We have

$$\mathcal{L}[e^{at}] = \int_0^\infty e^{-st} e^{at} \, dt = \int_0^\infty e^{-(s-a)t} \, dt = \left[-\frac{e^{-(s-a)t}}{s-a} \right]_0^\infty = \frac{1}{s-a}, \ s > a.$$

Example 8.4   Find the Laplace transform of $f(t) = \sinh \omega t$, $t \geq 0$.

Solution   We write $\sinh \omega t = (e^{\omega t} - e^{-\omega t})/2$. Using the linearity principle, we obtain

$$\mathcal{L}[\sinh \omega t] = \frac{1}{2} \int_0^\infty e^{-st} (e^{\omega t} - e^{-\omega t}) \, dt = \frac{1}{2} \int_0^\infty [e^{-(s-\omega)t} - e^{-(s+\omega)t}] dt$$

$$= \frac{1}{2} \left[-\frac{e^{-(s-\omega)t}}{s-\omega} + \frac{e^{-(s+\omega)t}}{s+\omega} \right]_0^\infty = \frac{1}{2} \left[\frac{1}{s-\omega} - \frac{1}{s+\omega} \right]$$

$$= \frac{\omega}{s^2 - \omega^2}, \ s > \omega.$$

Example 8.5 Find the Laplace transform of $f(t) = \cos at,\ t \geq 0$.

Solution We write $\cos at = \text{Re}\ (e^{iat})$ and consider $\mathscr{L}[e^{iat}]$.
We have

$$\mathscr{L}[e^{iat}] = \mathscr{L}[\cos at + i \sin at] = \mathscr{L}[\cos at] + i\ \mathscr{L}[\sin at].$$

Now, from Example 8.3

$$\mathscr{L}[e^{iat}] = \frac{1}{s - ia} = \frac{s + ia}{s^2 + a^2}.$$

Therefore, comparing the real parts on both sides, we obtain

$$\mathscr{L}[\cos at] = \frac{s}{s^2 + a^2}.$$

We present below in the table, the Laplace transforms of some basic functions.

Table : Laplace transform of some basic functions.

$f(t)$	$\mathscr{L}[f(t)]$	$f(t)$	$\mathscr{L}[f(t)]$
1	$\dfrac{1}{s}$	$\sin at$	$\dfrac{a}{s^2 + a^2}$
t^n	$\dfrac{(n!)}{s^{n+1}},\ n = 1, 2, \ldots$	$\cos at$	$\dfrac{s}{s^2 + a^2}$
t^α	$\dfrac{\Gamma(\alpha + 1)}{s^{\alpha+1}},\ \alpha$ positive	$\sinh \omega t$	$\dfrac{\omega}{s^2 - \omega^2}$
e^{at}	$\dfrac{1}{s - a}$	$\cosh \omega t$	$\dfrac{s}{s^2 - \omega^2}$

Using the definition, we write

$$\mathscr{L}^{-1}\left[\frac{1}{s}\right] = 1,\ \mathscr{L}^{-1}\left[\frac{n!}{s^{n+1}}\right] = t^n,\ \mathscr{L}^{-1}\left[\frac{1}{s-a}\right] = e^{at},$$

$$\mathscr{L}^{-1}\left[\frac{a}{s^2 + a^2}\right] = \sin at,\ \mathscr{L}^{-1}\left[\frac{s}{s^2 - \omega^2}\right] = \cosh \omega t,\ \text{etc.}$$

Sufficient conditions for the existence of Laplace transform

Let $f(t)$ be a piecewise continuous function on $0 \leq t < \infty$. Then, in any interval $0 \leq a \leq t \leq b$, there are at most a finite number of points $t_k,\ k = 1, 2, \ldots, m,\ t_{k-1} < t_k,$ at which the function $f(t)$ has finite jumps (discontinuities) and is continuous on each subinterval $t_{k-1} < t < t_k$.

A function $f(t)$ is said to be of *exponential order* α, if there exist constants α and $M > 0$ such that

$$|f(t)| \leq M e^{\alpha t},\quad t \geq 0. \tag{8.5}$$

Geometrically, this condition implies that the graph of $f(t),\ t > 0$ does not grow faster than the graph of the exponential function $g(t) = M e^{\alpha t},\ \alpha > 0$.

For example, since

$$|t| \le e^t, \quad |e^{-2t}| \le e^t, \quad |\cos t| \le e^t, \quad t \ge 0$$

the functions $f_1(t) = t$, $f_2(t) = e^{-2t}$ and $f_3(t) = \cos t$ are all of exponential order. For a function of the form $f(t) = e^{ct^2}$, $c > 0$, it is not possible to determine constants α and M such that $e^{ct^2} \le Me^{\alpha t}$. Therefore, e^{ct^2} is not a function of exponential order.

Theorem 8.2 (**Sufficient conditions for existence of Laplace transform**) If $f(t)$ is a piecewise continuous function on the interval $[0, \infty)$ and is of exponential order α for $t \ge 0$, then $\mathscr{L}[f(t)]$ exists for $s > \alpha$.

Proof If $f(t)$ is piecewise continuous on $[0, T]$ for $T > 0$, then $e^{-st} f(t)$ is also piecewise continuous on $[0, T]$. Hence, $\displaystyle\int_0^T e^{-st} f(t)dt$ exists. Therefore, the existence of the Laplace transform of $f(t)$ depends on whether this integral converges or has a finite limit as $T \to \infty$. To prove the convergence, we use the following results from the theory of improper integrals.

If for all t $(t \ge t_0)$ the inequality $0 \le f(t) \le g(t)$ is satisfied and if $\displaystyle\int_{t_0}^\infty g(t)dt$ converges, then $\displaystyle\int_{t_0}^\infty f(t)dt$ also converges. If $\displaystyle\int_{t_0}^\infty |f(t)|\, dt$ converges, then the integral $\displaystyle\int_{t_0}^\infty f(t)dt$ also converges. In this case, the later integral is called an absolutely convergent integral. To use these results, we can determine a function $g(t)$ such that $\displaystyle\int_0^\infty g(t)dt$ converges and

$$|e^{-st} f(t)| \le g(t), \quad t \ge 0. \tag{8.6}$$

Now, choose some numbers a and M such that

$$|f(t)| \le Me^{at}.$$

Then, from Eq. (8.6), we obtain

$$|e^{-st} f(t)| \le Me^{(a-s)t} = g(t).$$

Now,

$$\int_0^\infty g(t)dt = \int_0^\infty Me^{(a-s)t}\, dt$$

converges if $s > a$. Therefore, if we choose $g(t) = Me^{(a-s)t}$, $s > a$, the integral $\displaystyle\int_0^\infty e^{-st} f(t)dt$ is absolutely convergent and hence convergent. Hence, $\mathscr{L}[f(t)]$ exists.

Alternatively, we have

$$|\mathscr{L}(f)| = \left| \int_0^\infty e^{-st} f(t)dt \right| \le \int_0^\infty e^{-st} |f(t)|\, dt \le M \int_0^\infty e^{-st} e^{at}\, dt$$

$$= M \int_0^\infty e^{-(s-a)t}\, dt = \frac{M}{s-a}, \quad s > a.$$

proving the existence of the Laplace transform.

It is important to note that the above theorem gives only the sufficient conditions for the existence of the Laplace transform. That is, a function may have Laplace transform even if it violates the existence conditions.

For example, consider the function $f(t) = t^{-1/2}$. This function is not continuous on any interval $[0, T]$ because it has limit ∞ as $t \to 0$ from the right. However, $\int_0^T e^{-st} t^{-1/2} dt$ exists for all $T > 0$. The Laplace transform of $t^{-1/2}$ can be computed as

$$\mathcal{L}[t^{-1/2}] = \int_0^\infty e^{-st} t^{-1/2} dt = 2 \int_0^\infty e^{-su^2} du \quad \text{(setting } t = u^2)$$

$$= \frac{2}{\sqrt{s}} \int_0^\infty e^{-z^2} dz \quad \text{(by setting } z = u\sqrt{s})$$

$$= \sqrt{\pi/s}, \quad s > 0.$$

Remark 1
Let $f(t)$ satisfy the conditions of the existence theorem and $\mathcal{L}[f(t)] = F(s)$. Then

(i) $\lim_{s \to \infty} F(s) = 0,$　　　　　　　　　　　(ii) $\lim_{s \to \infty} [sF(s)]$ is bounded.

These two results indicate that not all functions of s are Laplace transforms of some funtion $f(t)$.

Example 8.6 Find the Laplace transform of the piecewise continuous function

$$f(t) = \begin{cases} 0, & 0 \le t < 2 \\ k, & t \ge 2, \ k \text{ constant.} \end{cases}$$

Solution We have

$$\mathcal{L}[f(t)] = \int_0^\infty e^{-st} f(t) dt = \int_2^\infty k e^{-st} dt = -k \left[\frac{e^{-st}}{s} \right]_2^\infty = \frac{k e^{-2s}}{s}, s > 0.$$

Example 8.7 Find the Laplace transform of $\cos^2 (at)$.

Solution We have $\cos^2 at = (1 + \cos 2at)/2$. Therefore,

$$\mathcal{L}[\cos^2 at] = \mathcal{L}\left[\frac{1}{2}\right] + \mathcal{L}\left[\frac{\cos 2at}{2}\right] = \frac{1}{2s} + \frac{1}{2} \cdot \frac{s}{s^2 + 4a^2} = \frac{s^2 + 2a^2}{s(s^2 + 4a^2)}.$$

Example 8.8 Find $\mathcal{L}^{-1}\left[\dfrac{2s + 5}{s^2 + 25}\right]$

Solution We have by the linearity principle

$$\mathcal{L}^{-1}\left[\frac{2s + 5}{s^2 + 25}\right] = \mathcal{L}^{-1}\left[\frac{2s}{s^2 + 25}\right] + \mathcal{L}^{-1}\left[\frac{5}{s^2 + 25}\right] = 2 \cos 5t + \sin 5t.$$

Example 8.9 Find $\mathcal{L}^{-1}\left[\dfrac{5s^2 + 3s - 16}{(s - 1)(s - 2)(s + 3)}\right]$

Solution We use partial fractions to write the given function $F(s)$ as the sum of three factors.

Write
$$\frac{5s^2 + 3s - 16}{(s-1)(s-2)(s+3)} = \frac{A}{s-1} + \frac{B}{s-2} + \frac{C}{s+3}.$$

Simplifying and equating the numerators, we obtain

$$5s^2 + 3s - 16 = A\ (s-2)\ (s+3) + B(s-1)\ (s+3) + C(s-1)\ (s-2).$$

Setting $s = 1, 2, -3$, we obtain

$$-8 = -4A \quad \text{or} \quad A = 2; \quad 10 = 5B \quad \text{or} \quad B = 2; \quad 20 = 20C \quad \text{or} \quad C = 1.$$

Therefore,

$$\mathcal{L}^{-1}\left[\frac{5s^2 + 3s - 16}{(s-1)(s-2)(s+3)}\right] = \mathcal{L}^{-1}\left[\frac{2}{s-1}\right] + \mathcal{L}^{-1}\left[\frac{2}{s-2}\right] + \mathcal{L}^{-1}\left[\frac{1}{s+3}\right]$$

$$= 2e^t + 2e^{2t} + e^{-3t}.$$

Exercise 8.1

Using the definition, find the Laplace transform of the following functions.

1. $2t - 5$.
2. $at^2 + bt + c$.
3. $(2 + 3t)^2$.

4. $\sin(3t + 2)$.
5. $\cos(at + b)$.
6. $\sin^2(\omega t)$.

7. $\sinh^2 t$.
8. $\cosh^2 at$.
9. $\cos t - \sin t$.

10. $3t^2 - \cos 2t$.
11. te^{2t}.
12. e^{-2t+5}.

13. $t^2 e^t$.
14. $t \sin t$.
15. $t \cos 2t$.

16. $e^t \cos t$.
17. $e^{-t} \sin t$.
18. $(\cos t + \sin t)^2$.

19. $f(t) = \begin{cases} 1, & 0 \le t < 1, \\ -1, & t \ge 1, \end{cases}$

20. $f(t) = \begin{cases} 2, & 0 \le t < 3, \\ 0, & t \ge 3. \end{cases}$

21. $f(t) = \begin{cases} \cos t, & 0 \le t < \pi, \\ 0, & t \ge \pi. \end{cases}$

22. $f(t) = \begin{cases} 0, & 0 \le t < \pi \\ \sin t, & t \ge \pi. \end{cases}$

Find the Laplace transform of the following functions.

23. $(t + 3)^2$.
24. $6 - e^{3t}$.

25. $3t^2 - 5e^{-2t} + 6$.
26. $\sin 5t + \cos 4t$.

27. $\sinh 2t + \cosh 2t$.
28. $e^{-t} \sinh t$.

29. $e^t \cosh t$.
30. $e^t \sin t$.

In the following problems, the Laplace transform $F(s) = \mathcal{L}[f(t)]$ is given. Find the inverse Laplace transform $f(t)$.

31. $\dfrac{3}{s+5}$.
32. $\dfrac{\pi}{s^2 + \pi^2}$.

33. $\dfrac{s}{s^2 - 4}$.
34. $\dfrac{6}{s^4}$.

35. $\dfrac{s+3}{(s-1)(s+2)}$.
36. $\dfrac{2}{s^3} + \dfrac{6}{s^2} - \dfrac{5}{s}$.

37. $\dfrac{3}{s^2 + 2s}$.

38. $\dfrac{13}{s^2 - 9}$.

39. $\dfrac{s^2 + 2s + 5}{(s-1)(s-2)(s-3)}$.

40. $\dfrac{s}{(s+1)(s+2)(s-3)}$.

41. Show that $\mathcal{L}[t^a] = \dfrac{\Gamma(1+a)}{s^{1+a}}, \ a > -1$.

42. Find the Laplace transform of the function $f(t) = t^{5/2}$ given that $\Gamma(1/2) = \sqrt{\pi}$.

43. If $\mathcal{L}[f(t)] = F(s)$, show that $\mathcal{L}[f(at)] = (1/a)F(s/a)$.

44. Show that $\displaystyle\int_{nT}^{(n+1)T} e^{-st} f(t)dt = e^{-nsT} \int_0^T e^{-st} f(t)dt$, n is an integer, when $f(t)$ is a periodic function of

period T.

45. Find $\mathcal{L}\{e^{(a+ib)t}\}$. Hence, show that

$$\mathcal{L}[e^{at} \cos bt] = \frac{s-a}{(s-a)^2 + b^2}, \ \mathcal{L}[e^{at} \sin bt] = \frac{b}{(s-a)^2 + b^2}.$$

8.3   Laplace Transform Solution of Initial Value Problems

Laplace transformation is one of the important tools of solving the initial value problems. We first prove the following results on which the method of solution depends.

8.3.1   Laplace Transform of Derivatives

Theorem 8.3   Let $f(t)$, $t \geq 0$ be a continuous function and be of exponential order on $[0, \infty)$. Let $f'(t)$ be also of exponential order and at least piecewise continuous on $[0, \infty)$. Then

$$\mathcal{L}[f'(t)] = s \, \mathcal{L}[f(t)] - f(0). \tag{8.7}$$

Proof   First, consider the case when $f'(t)$ is continuous on $[0, \infty)$. Then

$$\mathcal{L}[f'(t)] = \int_0^\infty e^{-st} f'(t)dt = \left[e^{-st} f(t)\right]_0^\infty + s \int_0^\infty e^{-st} f(t)dt$$

$$= -f(0) + s \, \mathcal{L}[f(t)], \ s > \alpha \quad \text{(see Theorem 8.2)}.$$

Hence, the result,

Suppose now, that $f'(t)$ is piecewise continuous. Assume that at $t = T, f'(t)$ has a finite jump. Then, we write

$$\mathcal{L}[f'(t)] = \int_0^\infty e^{-st} f'(t)dt = \lim_{\varepsilon \to 0} \left[\int_0^{T-\varepsilon} e^{-st} f'(t)dt + \int_{T+\varepsilon}^\infty e^{-st} f'(t)dt\right]$$

$$= \lim_{\varepsilon \to 0} \left[\left\{e^{-st} f(t)\right\}_0^{T-\varepsilon} + s \int_0^{T-\varepsilon} e^{-st} f(t)dt + \left\{e^{-st} f(t)\right\}_{T+\varepsilon}^\infty + s \int_{T+\varepsilon}^\infty e^{-st} f(t)dt\right]$$

$$\tag{8.8}$$

In the limit, the two integrals on the right hand side of Eq. (8.8) combine to give

$$s \int_0^\infty e^{-st} f(t) dt = s \mathcal{L}[f(t)].$$

In the limit, the first term on the right hand side of Eq. (8.8) gives $e^{-st} f(T^-) - f(0^+)$ and the third term gives $-e^{-sT} f(T^+)$. Since, $f(t)$ is continuous on $[0, \infty)$, we have $f(0^+) = f(0)$ and $f(T^+) = f(T^-)$. Hence, the first and third terms give $-f(0)$. Substituting in Eq. (8.8), we obtain

$$\mathcal{L}[f'(t)] = s \mathcal{L}[f(t)] - f(0).$$

Remark 2

Let $f(t)$ satisfy the conditions given in Theorem 8.3 and $f(0) = 0$. Then

$$\mathcal{L}[f'(t)] = s F(s), \quad \text{or} \quad \mathcal{L}^{-1}[sF(s)] = f'(t) = \frac{d}{dt} \mathcal{L}^{-1}[F(s)] \tag{8.9}$$

where $\mathcal{L}[f(t)] = F(s)$. This result is useful in finding the inverse Laplace transform when the given function $\phi(s)$ can be written as $\phi(s) = s F(s)$.

The result given in Eq. (8.7) can be extended to find the Laplace transform of higher order derivatives. We have

$$\mathcal{L}[f''(t)] = s \mathcal{L}[f'(t)] - f'(0) = s\{s \mathcal{L}[f(t)] - f(0)\} - f'(0)$$

$$= s^2 \mathcal{L}[f(t)] - sf(0) - f'(0). \tag{8.10}$$

$$\mathcal{L}[f'''(t)] = s \mathcal{L}[f''(t)] - f''(0) = s\{s^2 \mathcal{L}[f(t)] - sf(0) - f'(0)\} - f''(0)$$

$$= s^3 \mathcal{L}[f(t)] - s^2 f(0) - sf'(0) - f''(0). \tag{8.11}$$

By induction, we can prove the following result.

Theorem 8.4 Let $f(t), f'(t), \ldots, f^{(n-1)}(t)$ be continuous on $[0, \infty)$ and be of exponential order. Let $f^{(n)}(t)$ be at least piecewise continuous on $[0, \infty)$. Then

$$\mathcal{L}[f^{(n)}(t)] = s^n \mathcal{L}[f(t)] - s^{n-1} f(0) - s^{n-2} f'(0) - \ldots - f^{(n-1)}(0). \tag{8.12}$$

The above results, (Eqs. (8.9) to (8.12)) allow us to compute the Laplace transform of many functions which satisfy the given conditions.

Remark 3

Let $f(t)$ satisfy the conditions given in Theorem 8.3 and $f(0) = 0$ and $f'(0) = 0$. Then, from Eq. (8.10), we have

$$\mathcal{L}[f''(t)] = s^2 \mathcal{L}[f(t)], \quad \text{or} \quad \mathcal{L}^{-1}[s^2 F(s)] = f''(t) = \frac{d^2}{dt^2} \mathcal{L}^{-1}[F(s)]. \tag{8.13}$$

If $f(0) = f'(0) = f''(0) = \ldots = f^{(n-1)}(0) = 0$, then from Eq. (8.12), we have

$$\mathcal{L}[f^{(n)}(t)] = s^n \mathcal{L}[f(t)], \text{ or } \mathcal{L}^{-1}[s^n F(s)] = f^{(n)}(t). \tag{8.14}$$

Example 8.10 Find the Laplace transform of $f(t) = \sin^2 t$ using the differentiation formula given in Eq. (8.7).

Solution We have $f(t) = \sin^2 t$ and $f(0) = 0$. Differentiating, we obtain $f'(t) = 2 \sin t \cos t = \sin 2t$. Using the formula given in Eq. (8.7), we obtain

$$\mathcal{L}[f'(t)] = \mathcal{L}[\sin 2t] = s \, \mathcal{L}[f(t)] - f(0)$$

or
$$\frac{2}{s^2 + 4} = s\mathcal{L}[f(t)], \quad \text{or} \quad \mathcal{L}[f(t)] = \frac{2}{s(s^2 + 4)}.$$

Example 8.11   Find the Laplace transform of $f(t) = t^3$, using the formula given in Eq. (8.11).

Solution   We have $f(t) = t^3, f'(t) = 3t^2, f''(t) = 6t, f'''(t) = 6$ and $f(0) = f'(0) = f''(0) = 0$. Hence,
$$\mathcal{L}[f'''(t)] = \mathcal{L}[6] = s^3 \mathcal{L}[f(t)] - s^2 f(0) - s f'(0) - f''(0).$$

or
$$\frac{6}{s} = s^3 \mathcal{L}[f(t)], \quad \text{or} \quad \mathcal{L}[f(t)] = \frac{6}{s^4}.$$

Example 8.12   Find $\mathcal{L}^{-1}[s/(s^2 + 9)]$ using the formula given in Eq. (8.9).

Solution   Let $F(s) = 1/(s^2 + 9)$. Then $f(t) = \mathcal{L}^{-1}[F(s)] = \mathcal{L}^{-1}\left[\dfrac{1}{s^2 + 9}\right] = \dfrac{1}{3} \sin 3t$ and $f(0) = 0$.

Therefore, using Eq. (8.9), we obtain

$$\mathcal{L}^{-1}[sF(s)] = f'(t) = \frac{d}{dt}\left(\frac{1}{3} \cdot \sin 3t\right) = \cos 3t.$$

We are now in a position to outline the procedure for solving the initial value problem

$$ay'' + by' + cy = f(t), \, a, \, b, \, c \text{ constants}$$

$$y(0) = y_0, \, y'(0) = y_0'. \tag{8.15}$$

Taking the Laplace transform on both sides and denoting

$$\mathcal{L}[y(t)] = Y(s), \quad \text{and} \quad \mathcal{L}[f(t)] = F(s), \text{ we obtain}$$

$$a \, \mathcal{L}[y''] + b \, \mathcal{L}[y'] + c \, \mathcal{L}[y] = \mathcal{L}[f(t)]$$

or
$$a[s^2 Y(s) - sy(0) - y'(0)] + b[sY(s) - y(0)] + cY(s) = F(s)$$

or
$$(as^2 + bs + c) \, Y(s) = F(s) + a[sy(0) + y'(0)] + by(0) \tag{8.16}$$

or
$$Y(s) = \frac{F(s) + a[sy(0) + y'(0)] + by(0)}{as^2 + bs + c} \tag{8.17}$$

which is an algebraic equation for $Y(s)$. Eq. (8.16) is called the *subsidiary equation*. Taking the inverse Laplace transform, we obtain

$$y(t) = \mathcal{L}^{-1}[Y(s)] = \mathcal{L}^{-1}\left[\frac{F(s) + a\{sy(0) + y'(0)\} + by(0)}{as^2 + bs + c}\right]. \tag{8.18}$$

We need more results to obtain the inverse Laplace transform, given in Eq. (8.18), in the general case. However, in this section, we shall solve some simple problems.

Example 8.13   Solve the initial value problem

$$y'' + 4y = 0, \quad y(0) = 1, \quad y'(0) = 6.$$

Solution   Taking Laplace transform of the given differential equation, we obtain

$$[s^2Y(s) - sy(0) - y'(0)] + 4Y(s) = 0$$

where $\mathcal{L}[y(t)] = Y(s)$. Substituting the values of $y(0)$ and $y'(0)$, we obtain

$$(s^2 + 4)\, Y(s) = s + 6, \quad \text{or} \quad Y(s) = \frac{s+6}{s^2+4} = \frac{s}{s^2+4} + \frac{6}{s^2+4}.$$

Therefore, $y(t) = \mathcal{L}^{-1}[Y(s)] = \mathcal{L}^{-1}\left[\dfrac{s}{s^2+4}\right] + \mathcal{L}^{-1}\left[\dfrac{6}{s^2+4}\right] = \cos 2t + 3 \sin 2t.$

Example 8.14   Solve the initial value problem

$$y'' + 2y' - 3y = 3, \quad y(0) = 4,\ y'(0) = -7.$$

Solution   Taking Laplace transform of the differential equation, we obtain

$$[s^2Y(s) - sy(0) - y'(0)] + 2\,[sY(s) - y(0)] - 3Y(s) = \mathcal{L}[3]$$

or $\qquad (s^2 + 2s - 3)Y(s) = \dfrac{3}{s} + 4s - 7 + 8 = \dfrac{3}{s} + 4s + 1$

where $\mathcal{L}[y(t)] = Y(s)$. Therefore,

$$Y(s) = \frac{3}{s(s-1)(s+3)} + \frac{4s+1}{(s-1)(s+3)}.$$

Using the partial fractions, we obtain

$$Y(s) = \left[-\frac{1}{s} + \frac{3}{4}\cdot\frac{1}{s-1} + \frac{1}{4}\cdot\frac{1}{s+3}\right] + \frac{1}{4}\left[\frac{5}{s-1} + \frac{11}{s+3}\right]$$

$$= -\frac{1}{s} + \frac{2}{s-1} + \frac{3}{s+3}.$$

Taking the inverse Laplace transform, we get

$$y(t) = \mathcal{L}^{-1}[Y(s)] = \mathcal{L}^{-1}\left(-\frac{1}{s}\right) + \mathcal{L}^{-1}\left(\frac{2}{s-1}\right) + \mathcal{L}^{-1}\left(\frac{3}{s+3}\right) = -1 + 2e^t + 3e^{-3t}.$$

Example 8.15   Solve the initial value problem

$$y'' - 5y' + 4y = e^{2t}, \quad y(0) = 19/12,\ y'(0) = 8/3.$$

Solution   Taking Laplace transform of the differential equation, we obtain

$$[s^2Y(s) - sy(0) - y'(0)] - 5\,[sY(s) - y(0)] + 4y(s) = \mathcal{L}[e^{2t}]$$

or $\qquad (s^2 - 5s + 4)Y(s) = \dfrac{1}{s-2} + \dfrac{19}{12}s + \dfrac{8}{3} - 5\left(\dfrac{19}{12}\right) = \dfrac{1}{s-2} + \dfrac{19s}{12} - \dfrac{63}{12}$

where $\mathcal{L}[y(t)] = Y(s)$. Therefore,

$$Y(s) = \frac{1}{(s-2)(s-1)(s-4)} + \frac{19s - 63}{12(s-1)(s-4)}.$$

Using partial fractions, we obtain

$$Y(s) = -\frac{1}{2(s-2)} + \frac{1}{3(s-1)} + \frac{1}{6(s-4)} + \frac{11}{9(s-1)} + \frac{13}{36(s-4)}$$

$$= -\frac{1}{2(s-2)} + \frac{14}{9(s-1)} + \frac{19}{36(s-4)}$$

Therefore, $\quad y(t) = -\dfrac{1}{2}e^{2t} + \dfrac{14}{9}e^t + \dfrac{19}{36}e^{4t}$

8.3.2 Laplace Transform of Integrals

In many problems of electrical engineering, we encounter integro-differential equations. Consider a series electric circuit (Fig. 8.1). Using the *Kirchoff's second law,* we obtain that the flow of current i satisfies the integro-differential equation

$$L\frac{di}{dt} + Ri + \frac{1}{C}\int_0^t i\, d\tau = E_0 \cos \omega t$$

Many other integro-differential equations arise in the theory of electrical circuits. If Laplace transform method is to be applied, we need the formula for the Laplace transform of an integral. Such a formula is presented in the following theorem.

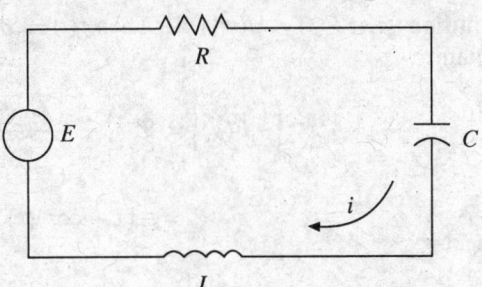

Fig. 8.1. **Series electric circuit.** *C*: capacitance, *E*: impressed voltage, *L*: inductance, *R*: resistance.

Theorem 8.5 Let $f(t)$, $t \geq 0$ be a piecewise continuous function and be of exponential order on $[0, \infty)$. Then

$$\mathscr{L}\left[\int_0^t f(\tau)d\tau\right] = \frac{1}{s}\mathscr{L}[f(t)], s > 0, \ s > \alpha. \tag{8.19}$$

Proof It is given that $f(t)$ is piecewise continuous on $[0, \infty)$ and is of exponential order. Therefore,

$\phi(t) = \displaystyle\int_0^t f(\tau)d\tau$ is at least piecewise continuous on $[0, \infty)$ and is of exponential order. Further $\phi'(t) = f(t)$ except for points at which $f(t)$ is discontinuous. That is, $\phi'(t)$ is piecewise continuous on each finite sub-interval. Hence, Theorem 8.3 can be applied to the function $\phi'(t)$. We have

$$\mathscr{L}[f(t)] = \mathscr{L}[\phi'(t)] = s\,\mathscr{L}[\phi(t)] - \phi(0), \quad s > \alpha$$

$$= s\,\mathscr{L}[\phi(t)]$$

since $\phi(0) = 0$. Therefore, $\qquad \mathscr{L}\left[\displaystyle\int_0^t f(\tau)d\tau\right] = \dfrac{1}{s}\,\mathscr{L}[f(t)]$

or $$\mathscr{L}^{-1}\left[\frac{1}{s}F(s)\right] = \int_0^t f(\tau)d\tau \tag{8.20}$$

where $F(s) = \mathscr{L}[f(t)]$.

Example 8.16 Find $\mathscr{L}^{-1}[1/\{s(s^2 + 9)\}]$.

Solution Let $F(s) = 1/(s^2 + 9)$. Then $f(t) = \mathscr{L}^{-1}[F(s)] = (\sin 3t)/3$. Therefore, using Eq. (8.20) we obtain

$$\mathcal{L}^{-1}\left[\frac{1}{s}F(s)\right] = \int_0^t \frac{1}{3}\sin 3\tau\, d\tau = -\left[\frac{\cos 3\tau}{9}\right]_0^t = \frac{1}{9}(1 - \cos 3t).$$

Example 8.17   Find $\mathcal{L}^{-1}\left[\dfrac{1}{s^2(s^2+4)}\right]$

Solution   Let $F(s) = 1/(s^2+4)$. Then $f(t) = \mathcal{L}^{-1}[F(s)] = (\sin 2t)/2$. Therefore, using Eq. (8.20) we obtain

$$\mathcal{L}^{-1}\left[\frac{1}{s}F(s)\right] = \mathcal{L}^{-1}\left[\frac{1}{s(s^2+4)}\right] = \int_0^t \frac{1}{2}\sin 2\tau\, d\tau = -\frac{1}{4}\left[\cos 2\tau\right]_0^t$$

$$= \frac{1}{4}[1 - \cos 2t]$$

and

$$\mathcal{L}^{-1}\left[\frac{1}{s^2}F(s)\right] = \mathcal{L}^{-1}\left[\frac{1}{s^2(s^2+4)}\right] = \int_0^t \frac{1}{4}(1 - \cos 2\tau)\, d\tau$$

$$= \frac{1}{4}\left[\tau - \frac{\sin 2\tau}{2}\right]_0^t = \frac{1}{8}[2t - \sin 2t].$$

Alternately, we can write

$$\frac{1}{s^2(s^2+4)} = \frac{1}{4}\left[\frac{1}{s^2} - \frac{1}{s^2+4}\right]$$

and

$$\mathcal{L}^{-1}\left[\frac{1}{s^2(s^2+4)}\right] = \frac{1}{4}\left[\mathcal{L}^{-1}\left(\frac{1}{s^2}\right) - \mathcal{L}^{-1}\left(\frac{1}{s^2+4}\right)\right] = \frac{1}{4}\left[t - \frac{1}{2}\sin 2t\right].$$

Example 8.18   Find the solution of the initial value problem

$$y' + 3y + 2\int_0^t y(\tau)\, d\tau = t, \quad y(0) = 0.$$

Solution   Taking the Laplace transform on both sides of the differential equation, we obtain

$$sY(s) - y(0) + 3Y(s) + \frac{2}{s}Y(s) = \frac{1}{s^2}$$

where $\mathcal{L}[y(t)] = Y(s)$. Simplifying, we obtain

$$(s^2 + 3s + 2)Y(s) = \frac{1}{s}, \quad \text{or} \quad Y(s) = \frac{1}{s(s+1)(s+2)} = \frac{1}{s}F(s)$$

where

$$F(s) = \frac{1}{(s+1)(s+2)} = \frac{1}{s+1} - \frac{1}{s+2}.$$

Therefore, $f(t) = e^{-t} - e^{-2t}.$

Hence,

$$y(t) = \int_0^t (e^{-\tau} - e^{-2\tau})\, d\tau = \left[\frac{1}{2}e^{-2\tau} - e^{-\tau}\right]_0^t = \frac{1}{2}e^{-2t} - e^{-t} + \frac{1}{2}.$$

Exercise 8.2

Using the formulas of Laplace transforms of derivatives find the Laplace transforms of the following functions.

   1. $\cos^2 2t$.

   2. te^{at}

   3. $t \sin at$.

   4. $t \cos at$.

   5. $t \sinh at$.

Use Theorem 8.3 to find the following Laplace transforms.

   6. Given that $\mathscr{L}[\cos at] = s/(s^2 + a^2)$, find $\mathscr{L}[\sin at]$.

   7. Given that $\mathscr{L}[\sin at] = a/(s^2 + a^2)$, find $\mathscr{L}[\cos at]$.

Using the result in Remark 2, find the following inverse Laplace transforms.

   8. Given that $\mathscr{L}^{-1}[1/(s^2 + a^2)] = (\sin at)/a$, find $\mathscr{L}^{-1}[s/(s^2 + a^2)]$.

   9. Given that $\mathscr{L}^{-1}[1/\{(s + 1)(s + 2)\}] = e^{-t} - e^{-2t}$, find $\mathscr{L}^{-1}[s/\{(s + 1)(s + 2)\}]$.

   10. Find $\mathscr{L}^{-1}[1/(4s^2 + 1)]$. Hence, find $\mathscr{L}^{-1}[2s/(4s^2 + 1)]$.

   11. Find $\mathscr{L}^{-1}[1/\{(s^2 + 1)(s^2 + 4)\}]$. Hence, find $\mathscr{L}^{-1}[s/\{(s^2 + 1)(s^2 + 4)\}]$.

   12. Find $\mathscr{L}^{-1}[1/\{(s - 1)(s^2 + 1)\}]$. Hence, find $\mathscr{L}^{-1}[s/\{(s - 1)(s^2 + 1)\}]$.

Using Theorem 8.5, find the inverse Laplace transform of the following functions.

   13. $\dfrac{1}{s^2 + 5s}$.

   14. $\dfrac{16}{s^3 + 9s}$.

   15. $\dfrac{s - 2}{s(s + 3)}$.

   16. $\dfrac{1}{s^3 + s^2}$.

   17. $\dfrac{\omega}{s^2(s^2 + \omega^2)}$.

   18. $\dfrac{s - a}{s^2(s^2 + a^2)}$.

   19. $\dfrac{3}{s^2(s^2 + 4)(s^2 + 1)}$.

   20. $\dfrac{1}{s^4 + 3s^3}$.

   21. Assume that the function $f(t)$ satisfies the conditions of Theorem 8.3 except that $f(t)$ has a jump discontinuity at $t = T$. Show that

   $$\mathscr{L}[f'(t)] = sF(s) - f(0) - e^{-sT}[f(T^+) - f(T^-)]$$

   where $f(T^+) = \lim_{t \to T^+} f(t)$, $\quad f(T^-) = \lim_{t \to T^-} f(t)$ and $\mathscr{L}[f(t)] = F(s)$.

   22. Assume that the function $f(t)$ satisfies the conditions of Theorem 8.5. Then, show that

   $$\mathscr{L}\left[\int_a^t f(\tau)d\tau\right] = \frac{1}{s} F(s) - \frac{1}{s} \int_0^a f(\tau)d\tau, \quad a > 0, \text{ where } \mathscr{L}[f(t)] = F(s).$$

Solve the following initial value problems.

   23. $y' + 3y = 1, y(0) = 1$.

   24. $y' - 4y = t, y(0) = -1$.

   25. $y' + 2y = \sin t, y(0) = 1$.

   26. $y' + 3y = \cos t, y(0) = 0$.

   27. $y' - 2y = 1 + t, y(0) = 2$.

   28. $y'' + y = t, y(0) = 1, y'(0) = 0$.

   29. $y'' + 3y' + 2y = 3, y(0) = 1, y'(0) = 1$.

   30. $y'' + 5y' + 4y = e^{3t}, y(0) = 0, y'(0) = 3$.

   31. $y'' - 6y' + 5y = e^{2t}, y(0) = 1, y'(0) = -1$.

   32. $4y'' - 8y' + 3y = \sin t, y(0) = 0, y'(0) = 2$.

   33. $2y'' - y' - y = \cos t, y(0) = 1, y'(0) = 0$.

   34. $y' - 4y + 3\displaystyle\int_0^t y(\tau)d\tau = t, y(0) = 1$.

35. $y' + 6y + 5 \int_0^t y(\tau)d\tau = 1 + t,\ y(0) = 1.$　　**36.** $y' - y - 6 \int_0^t y(\tau)d\tau = \sin t,\ y(0) = 2.$

Find the solution of the following initial value problems.

37. $y_1' - y_2 = 1,\ y_1(0) = -1.$

　　$y_2' + 4y_1 = t,\ y_2(0) = 1.$

38. $y_1' - 3y_2 = 3 \sin 3t,\ y_1(0) = -3.$

　　$3y_1 + y_2' = 3(1 - \cos 3t),\ y_2(0) = 1.$

39. $y_1' + y_2 = 2 - \sin t,\ y_1(0) = 3.$

　　$y_1 - y_2' = t + \cos t,\ y_2(0) = 0.$

40. $y_1' + y_2 = 3t,\ y_1(0) = 0.$

　　$y_1 - y_2' = t^2 - 1,\ y_2(0) = 1.$

8.4　Translation Theorems (Shifting Theorems)

In the previous section we have solved some simple initial value problems. In the general case (see Eq. 8.18), to find the inverse Laplace transform we require some more formulas and also some results which simplify the computation of the inverse.

Theorem 8.6　(**Translation on the *s*-axis or shifting on the *s*-axis: First shifting theorem**)

Let $\mathcal{L}[f(t)] = F(s),\ s > \alpha \geq 0$ and a be a real number. Then,

$$\mathcal{L}[e^{at} f(t)] = F(s - a),\ s > a + \alpha \tag{8.21}$$

Proof　By definition, we have

$$\mathcal{L}[e^{at} f(t)] = \int_0^\infty e^{-st} e^{at} f(t)dt = \int_0^\infty e^{-(s-a)t} f(t)dt$$

$$= F(s - a),\ s > a + \alpha.$$

The translation on the *s*-axis is given in Fig. 8.2.

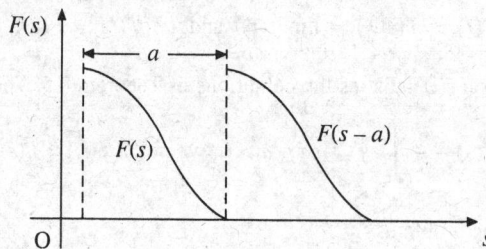

Fig. 8.2.　Translation on the *s*-axis (first shifting theorem).

Remark 4

Using Eq. (8.21), we can write

$$f(t) = e^{-at} \mathcal{L}^{-1}[F(s - a)]\quad \text{or}\quad \mathcal{L}^{-1}[F(s)] = e^{-at} \mathcal{L}^{-1}[F(s - a)]. \tag{8.22}$$

Example 8.19　Using the first shifting theorem, find the Laplace transforms of the following functions.

(a) $e^{at} \cos bt$,　(b) $e^{-\alpha t} \sin \beta t$.

Solution

(a) We have $\mathcal{L}[\cos bt] = s/(s^2 + b^2) = F(s)$. Hence, by the translation theorem, we obtain

$$\mathcal{L}[e^{at} \cos bt] = F(s - a) = \frac{s - a}{(s - a)^2 + b^2}.$$

(b) We have $\mathcal{L}[\sin \beta t] = \beta/(s^2 + \beta^2) = F(s)$. Hence, by the translation theorem, we obtain

$$\mathcal{L}[e^{-\alpha t} \sin \beta t] = \frac{\beta}{(s + \alpha)^2 + \beta^2}.$$

Example 8.20   Find the Laplace transforms of the following functions.

(a) $e^{\alpha t} t^3$,   (b) $e^{-\omega t} t^2$.

Solution

(a) $\mathcal{L}[t^3] = 3!/s^4$. Hence, we obtain

$$\mathcal{L}[e^{\alpha t} t^3] = 3!/(s - \alpha)^4.$$

(b) $\mathcal{L}[t^2] = 2/s^3$. Hence, we obtain

$$\mathcal{L}[e^{-\omega t} t^2] = 2/(s + \omega)^3.$$

Example 8.21   Find the inverse Laplace transforms of the following functions.

(a) $\dfrac{1}{s^2 - 4s + 8}$,   (b) $\dfrac{4}{s^2 - s + 2}$.

Solution

(a) We write $\dfrac{1}{s^2 - 4s + 8} = \dfrac{1}{(s - 2)^2 + 2^2}$.

Hence, $\mathcal{L}^{-1}\left[\dfrac{1}{(s - 2)^2 + 2^2}\right] = \dfrac{1}{2} e^{2t} \sin 2t$.

(b) We write $\dfrac{1}{s^2 - s + 2} = \dfrac{1}{(s - 1/2)^2 + (\sqrt{7}/2)^2}$

Hence, $\mathcal{L}^{-1}\left[\dfrac{4}{(s - 1/2)^2 + (\sqrt{7}/2)^2}\right] = \dfrac{4}{(\sqrt{7}/2)} e^{t/2} \sin\left(\dfrac{\sqrt{7}}{2}t\right) = \dfrac{8}{\sqrt{7}} e^{t/2} \sin(\sqrt{7}t/2)$.

Example 8.22   Find the inverse Laplace transforms of the following functions.

(a) $\dfrac{3s - 1}{(s - 2)^2}$,   (b) $\dfrac{6 + s}{s^2 + 6s + 13}$.

Solution

(a) We write $\dfrac{3s - 1}{(s - 2)^2} = \dfrac{3(s - 2) + 5}{(s - 2)^2} = \dfrac{3}{(s - 2)} + \dfrac{5}{(s - 2)^2}$.

Since $\mathcal{L}[t] = 1/s^2$, we obtain

$$\mathcal{L}^{-1}\left[\frac{3}{s-2} + \frac{5}{(s-2)^2}\right] = \mathcal{L}^{-1}\left[\frac{3}{s-2}\right] + 5\mathcal{L}^{-1}\left[\frac{1}{(s-2)^2}\right] = 3e^{2t} + 5te^{2t}.$$

(b) We write $\dfrac{6+s}{s^2+6s+13} = \dfrac{6+s}{(s+3)^2+4} = \dfrac{(s+3)+3}{(s+3)^2+4}$

$$= \frac{s+3}{(s+3)^2+2^2} + \frac{3}{(s+3)^2+2^2}.$$

Therefore, $\mathcal{L}^{-1}\left[\dfrac{s+3}{(s+3)^2+2^2} + \dfrac{3}{(s+3)^2+2^2}\right] = e^{-3t}\cos 2t + \dfrac{3}{2}e^{-3t}\sin 2t$.

Example 8.23   Find the solution of the initial value problem

$$y'' + 4y' + 4y = 12t^2e^{-2t}, \; y(0) = 2, \; y'(0) = 1.$$

Solution   Taking the Laplace transform of the differential equation, we obtain

$$[s^2Y(s) - s(2) - 1] + 4[sY(s) - 2] + 4Y(s) = \frac{24}{(s+2)^3}$$

since $\mathcal{L}[t^2] = 2/s^3$ and $\mathcal{L}[e^{-2t}t^2] = 2/(s+2)^3$. We have

$$(s^2 + 4s + 4)Y(s) = \frac{24}{(s+2)^3} + 2s + 9$$

or $\quad Y(s) = \dfrac{24}{(s+2)^5} + \dfrac{2s}{(s+2)^2} + \dfrac{9}{(s+2)^2} = \dfrac{24}{(s+2)^5} + \dfrac{2(s+2) - 4}{(s+2)^2} + \dfrac{9}{(s+2)^2}$

$$= \frac{24}{(s+2)^5} + \frac{2}{(s+2)} + \frac{5}{(s+2)^2}.$$

Therefore, $y(t) = t^4e^{-2t} + 2e^{-2t} + 5te^{-2t} = (2 + 5t + t^4)\,e^{-2t}$.

Example 8.24   Find the solution of the initial value problem

$$y'' + 4y' + 13y = e^{-t}, \quad y(0) = 0, \quad y'(0) = 2.$$

Solution   Taking the Laplace transform of the differential equation, we obtain

$$[s^2Y(s) - s(0) - 2] + 4[sY(s) - 0] + 13\,Y(s) = \frac{1}{s+1}$$

or $\qquad (s^2 + 4s + 13)\,Y(s) = \dfrac{1}{s+1} + 2 = \dfrac{2s+3}{s+1}$

or $\qquad Y(s) = \dfrac{2s+3}{(s+1)(s^2+4s+13)}.$

Taking the partial fractions, we obtain

$$Y(s) = \frac{1}{10}\left[\frac{1}{s+1} - \frac{s-17}{s^2+4s+13}\right] = \frac{1}{10}\left[\frac{1}{s+1} - \frac{s+2}{(s+2)^2+3^2} + \frac{19}{(s+2)^2+3^2}\right].$$

Therefore, $$y(t) = \frac{1}{10}\left[e^{-t} - e^{-2t}\cos 3t + \frac{19}{3}e^{-2t}\sin 3t \right]$$

We now present methods for solving certain initial value problems with discontinuous forcing functions. For example, an external force acting on a mechanical system or a voltage applied to an electrical circuit can be turned off after a certain period of time. Therefore, we shall discuss the case of functions having jump discontinuities. A function $f(t)$ has a jump discontinuity at $t = a$, if the left and right hand limits, as $t \to a$, both exist but are not equal (Fig. 8.3). We define such a function and find its Laplace transform.

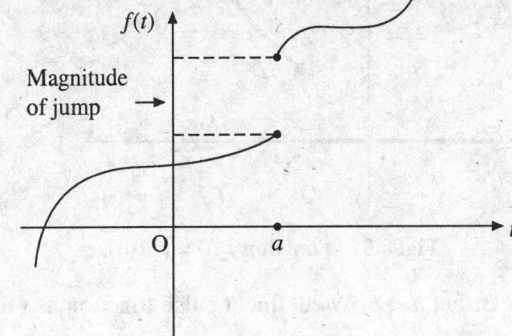

Fig. 8.3.   Jump discontinuity at $t = a$.

8.4.1   Heaviside Function or Unit Step Function

The *Heaviside function* is the function $H(t)$ defined by

$$H(t) = \begin{cases} 0, & \text{if } t < 0 \\ 1, & \text{if } t \geq 0. \end{cases} \tag{8.23a}$$

The Heaviside function is also called a *unit step function* and is denoted by $u(t)$. If the jump discontinuity is at the point $t = a$, then we define

$$H(t-a) = u(t-a) = \begin{cases} 0, & \text{if } t < a, \\ 1, & \text{if } t \geq a. \end{cases} \tag{8.23b}$$

The jump is of magnitude 1. The unit step function $u(t-a)$ is also denoted by $u_a(t)$, that is, we write $H(t-a) = u(t-a) = u_a(t)$ (Figs. 8.4a, b).

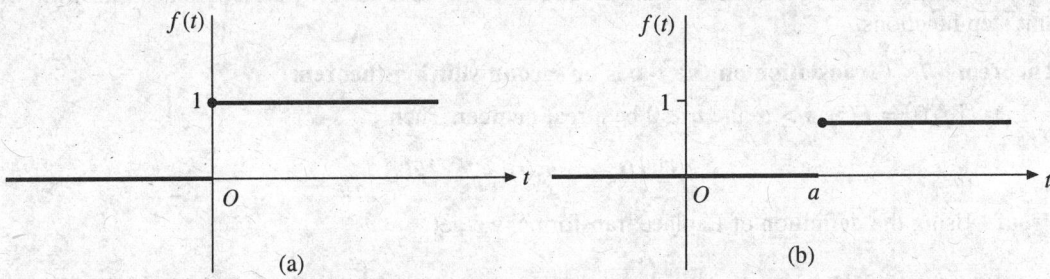

Fig. 8.4. (a) Heaviside function $H(t)$ or unit step function $u_0(t)$. (b) Heaviside function $H(t-a)$ or unit step function $u_a(t)$.

The following are some examples of functions represented in terms of unit step functions.

1. $f_1(t) = \sin(t)\, u_\pi(t) = \begin{cases} 0, & \text{if } t < \pi, \\ \sin(t), & \text{if } t \geq \pi. \end{cases}$

2. $f_2(t) = t^2 u_2(t) = \begin{cases} 0, & \text{if} \quad t < 2, \\ t^2, & \text{if} \quad t \geq 2 \end{cases}$ (Fig. 8.5).

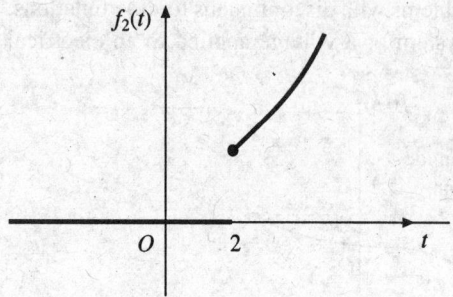

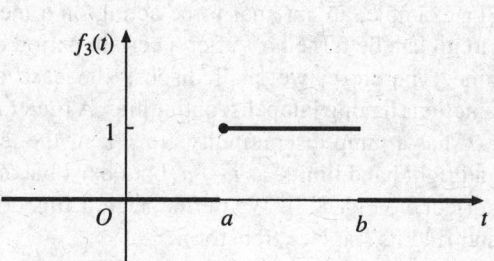

Fig. 8.5. Function $f_2(t) = t^2 u_2(t)$.   **Fig. 8.6.** Function $f_3(t) = u_a(t) - u_b(t)$.

3. Let $a < b$. We define a pulse function as (Fig. 8.6)

$$f_3(t) = u_a(t) - u_b(t) = \begin{cases} 0, & \text{if} \quad t < a, \\ 1, & \text{if} \quad a \leq t < b, \\ 0, & \text{if} \quad t \geq b. \end{cases}$$

Example 8.25 Express the following piecewise function in terms of the unit step functions

$$f(t) = \begin{cases} 5t, & 0 \leq t < 2 \\ 0, & t \geq 2. \end{cases}$$

Solution We can express $f(t)$ is terms of the unit step function $u_2(t)$. We have

$$f(t) = 5t - 5t u_2(t) = 5t[1 - u_2(t)] = 5t [u_0(t) - u_2(t)].$$

The second translation theorem or the second shift theorem deals with the Laplace transform of unit step functions.

Theorem 8.7 (**Translation on the t-axis or second shifting theorem**)

Let $\mathcal{L}[f(t)] = F(s)$, $s > \alpha$ and $a \geq 0$ be a real number. Then

$$\mathcal{L}[f(t - a) u_a(t)] = e^{-as} F(s). \tag{8.24}$$

Proof Using the definition of Laplace transforms, we get

$$\mathcal{L}[f(t - a)u_a(t)] = \int_0^\infty e^{-st} f(t - a)u_a(t)dt = \int_a^\infty e^{-st} f(t - a)dt.$$

Let $\tau = t - a$. Then $dt = d\tau$ and

$$\int_a^\infty e^{-st} f(t - a)dt = \int_0^\infty e^{-s(\tau+a)} f(\tau)d\tau$$

$$= e^{-as} \int_0^\infty e^{-s\tau} f(\tau)d\tau = e^{-as} \mathcal{L}[f(t)] = e^{-as} F(s).$$

Using the result given in Eq. (8.24), we have

(i) $\mathcal{L}^{-1}[e^{-as}F(s)] = f(t - a)\, u_a(t)$. $\hspace{4cm}$ (8.25)

(ii) $\mathcal{L}[u_a(t)] = \mathcal{L}\{1.\, u_a(t)\} = \dfrac{e^{-as}}{s}$. $\hspace{3cm}$ (8.26)

Very often, we may encounter the problem of finding the Laplace transform of a function of the form $f(t)\, u_a(t)$, where the function $f(t)$ lacks the shifted form $f(t - a)$. To find the Laplace transform of such a function, we use the following procedure. We write

$$f(t)\, u_a(t) = f(t - a + a)\, u_a(t) = F(t - a)\, u_a(t)$$

where $F(t - a) = f(t)$ or $F(t) = f(t + a)$. Using the Theorem 8.7, we get

$$\mathcal{L}[F(t - a)\, u_a(t)] = e^{-as}\mathcal{L}[F(t)] = e^{-as}\,\mathcal{L}[f(t + a)]$$

or $\hspace{3cm}$ $\mathcal{L}[f(t)\, u_a(t)] = e^{-as}\mathcal{L}[f(t + a)]$. $\hspace{3cm}$ (8.27)

Example 8.26 Find the Laplace transform of the function

$$f(t) = \begin{cases} 0, & 0 \le t < 3, \\ (t - 3)^2, & t \ge 3. \end{cases}$$

Solution We can write $f(t)$ as $f(t) = (t - 3)^2 u_3(t)$. Therefore,

$$\mathcal{L}[f(t)] = \mathcal{L}[(t - 3)^2\, u_3(t)] = \frac{2e^{-3s}}{s^3}$$

since $\mathcal{L}[t^2] = 2/s^3$.

Example 8.27 Find the Laplace transform of the function $f(t) = t^2 u_3(t)$.

Solution Using the result given in Eq. (8.27), we obtain

$$\mathcal{L}[t^2 u_3(t)] = e^{-3s}\,\mathcal{L}[(t + 3)^2] = e^{-3s}\,\mathcal{L}[t^2 + 6t + 9]$$

$$= e^{-3s}\left[\frac{2}{s^3} + \frac{6}{s^2} + \frac{9}{s}\right].$$

Example 8.28 Find the inverse Laplace transform of the following functions.

(a) $\dfrac{e^{-3s}}{s + 2}$, $\hspace{2cm}$ (b) $\dfrac{4e^{-(s\pi/2)}}{s^2 + 16}$, $\hspace{2cm}$ (c) $\dfrac{e^{-2s}}{s^2}$.

Solution

(a) Comparing with Eq. (8.25), we have $a = 3$ and $F(s) = 1/(s + 2)$. Therefore,

$$\mathcal{L}^{-1}\left[\frac{e^{-3s}}{s + 2}\right] = e^{-2(t-3)} \cdot u_3(t)$$

since $\mathcal{L}^{-1}[F(s)] = \mathcal{L}^{-1}[1/(s + 2)] = e^{-2t}$.

(b) We have $a = \pi/2$ and $F(s) = 4/(s^2 + 16)$. Hence, $f(t) = \sin 4t$ and

$$\mathcal{L}^{-1}\left[\frac{4e^{-(s\pi/2)}}{s^2 + 16}\right] = \sin 4\left(t - \frac{\pi}{2}\right)u_{\pi/2}(t) = \sin(4t - 2\pi)u_{\pi/2}(t) = (\sin 4t)u_{\pi/2}(t).$$

(c) We have $a = 2$ and $F(s) = 1/s^2$. Hence, $f(t) = t$ and

$$\mathcal{L}^{-1}\left[\frac{e^{-2s}}{s^2}\right] = (t - 2)u_2(t).$$

Example 8.29 Find the solution of the initial value problem

$$y' + y = f(t), \quad y(0) = 2 \quad \text{where} \quad f(t) = \begin{cases} 0, & 0 \le t < \pi/2, \\ \cos t, & t \ge \pi/2. \end{cases}$$

Solution The graph of $f(t)$ is given in Fig. 8.7.

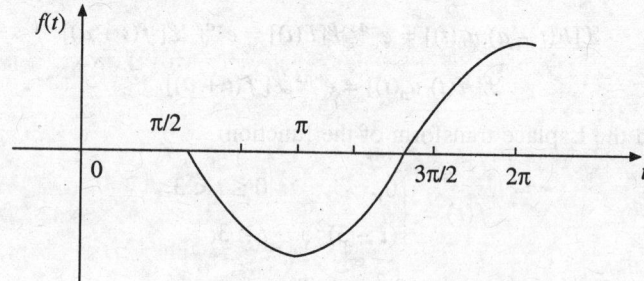

Fig. 8.7. The function $f(t)$, Example 8.29.

The function $f(t)$ can be represented in terms of unit step functions. We have

$$f(t) = (\cos t)\, u_{\pi/2}(t).$$

Taking Laplace transform of the differential equation, we obtain

$$sY(s) - 2 + Y(s) = \mathcal{L}[f(t)] = \mathcal{L}[(\cos t)\, u_{\pi/2}(t)]$$

where $\mathcal{L}[y(t)] = Y(s)$. Now, using Eq. (8.27), we obtain

$$\mathcal{L}[(\cos t)\, u_{\pi/2}(t)] = e^{-\pi s/2}\mathcal{L}[\cos(t + \pi/2)] = -e^{-\pi s/2}\,\mathcal{L}[\sin t] = \frac{-e^{-\pi s/2}}{s^2 + 1}.$$

Hence, we get

$$(s + 1)Y(s) = 2 - \frac{e^{-\pi s/2}}{s^2 + 1}, \quad \text{or} \quad Y(s) = \frac{2}{s + 1} - \frac{e^{-\pi s/2}}{(s + 1)(s^2 + 1)}$$

or

$$Y(s) = \frac{2}{s + 1} - \frac{1}{2}\left[\frac{1}{s + 1} - \frac{s}{s^2 + 1} + \frac{1}{s^2 + 1}\right]e^{-\pi s/2}.$$

Therefore,

$$y(t) = 2e^{-t} - \frac{1}{2}\left[e^{-(t - \pi/2)} - \cos\left(t - \frac{\pi}{2}\right) + \sin\left(t - \frac{\pi}{2}\right)\right]u_{\pi/2}(t)$$

$$= 2e^{-t} - \frac{1}{2}[e^{-(t - \pi/2)} - \sin t - \cos t]\,u_{\pi/2}(t).$$

The solution can also be written as

$$y(t) = \begin{cases} 2e^{-t}, & 0 \le t < \pi/2, \\ 2e^{-t} - \frac{1}{2}\left[e^{-(t-\pi/2)} - \sin t - \cos t\right], & t \ge \pi/2. \end{cases}$$

Example 8.30 Find the solution of the integro-differential equation

$$y' + 5y + 4\int_0^t y(\tau)d\tau = f(t)$$

under the condition $y(0) = 2$ and $f(t)$ is a rectangular pulse as given in Fig. 8.8.

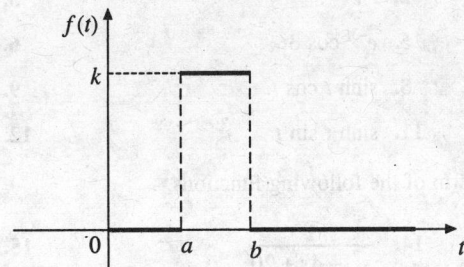

Fig. 8.8.   Example 8.30.

Solution   In terms of the unit step functions, $f(t)$ can be written as

$$f(t) = k[u_a(t) - u_b(t)].$$

Taking Laplace transform of the given equation, we obtain

$$sY(s) - 2 + 5Y(s) + \frac{4}{s}Y(s) = \mathcal{L}[f(t)].$$

Now, $\mathcal{L}[f(t)] = k[\mathcal{L}\{u_a(t)\} - \mathcal{L}\{u_b(t)\}] = \frac{k}{s}(e^{-as} - e^{-bs})$.

Hence, we have

$$(s^2 + 5s + 4)\,Y(s) = 2s + k(e^{-as} - e^{-bs})$$

or

$$Y(s) = \frac{2s}{(s+4)(s+1)} + \frac{k(e^{-as} - e^{-bs})}{(s+1)(s+4)}$$

$$= \frac{2}{3}\left[\frac{4}{s+4} - \frac{1}{s+1}\right] + \frac{k}{3}\left[\frac{1}{s+1} - \frac{1}{s+4}\right](e^{-as} - e^{-bs}).$$

Using the result given in Eq. (8.25), we obtain

$$y(t) = \frac{2}{3}(4e^{-4t} - e^{-t}) + \frac{k}{3}\left[\{e^{-(t-a)} - e^{-4(t-a)}\}u_a(t) - \{e^{-(t-b)} - e^{-4(t-b)}\}u_b(t)\right].$$

This solution can also be written as

$$y(t) = \begin{cases} \frac{2}{3}(4e^{-4t} - e^{-t}), & 0 \le t < a \\ \frac{2}{3}(4e^{-4t} - e^{-t}) + \frac{k}{3}\left[e^{-(t-a)} - e^{-4(t-a)}\right], & a \le t < b \\ \frac{2}{3}(4e^{-4t} - e^{-t}) + \frac{k}{3}[\{e^{-(t-a)} - e^{-4(t-a)}\} - \{e^{-(t-b)} - e^{-4(t-b)}\}], & t \ge b. \end{cases}$$

Exercise 8.3

Find the Laplace transform of the following functions.

1. $(t^2 - 2t - 3)e^{2t}$.   **2.** $t^5 e^{-4t}$.   **3.** $(t-2)^2 e^{3t}$.

4. $e^t \sin 5t$.   **5.** $e^{-3t} \cos 3t$.   **6.** $e^{-t}(\cos t - \sin t)$.

7. $t^3 \cosh 3t$ **8.** $\sinh t \cos t$.   **9.** $t^2 \sinh t$.

10. $\cosh \omega t \cos \omega t$.   **11.** $\sinh t \sin t$.   **12.** $\cosh \omega t \sin \omega t$.

Find the inverse Laplace transform of the following functions.

13. $\dfrac{1}{s^2 + 6s + 15}$.   **14.** $\dfrac{1}{s^2 - 4s + 20}$.   **15.** $\dfrac{s}{s^2 + 4s + 8}$.

16. $\dfrac{5s + 6}{(s-1)^2}$.   **17.** $\dfrac{s}{(s-2)^3}$.   **18.** $\dfrac{16 + 3s}{s^2 - 8s + 20}$.

19. $\dfrac{3s + 2}{(s+3)^3}$.   **20.** $\dfrac{(s+1)^2}{(s-2)^4}$.   **21.** $\dfrac{3s + 5}{s^2 + 6s + 12}$.

Use the Laplace transforms to solve the following initial value problems.

22. $y' + y = 1 + te^t$, $y(0) = 1$.   **23.** $y'' + 6y' + 9y = 8te^{2t}$, $y(0) = 0$, $y'(0) = -1$.

24. $y'' + 8y' + 16y = te^{-4t}$, $y(0) = 1$, $y'(0) = 2$.   **25.** $y'' + 6y' + 13y = e^{-t}$, $y(0) = 0$, $y'(0) = 4$.

26. $y'' + y = e^t \sin t$, $y(0) = 0$, $y'(0) = 0$.   **27.** $y'' + 2y' + 5y = 1 + t$, $y(0) = 4$, $y'(0) = -3$.

28. $12y'' - 24y' + 9y = 2t$, $y(0) = 0$, $y'(0) = 3$.

Write the following functions, whose graphs are given, in terms of unit step functions and find their Laplace transforms.

29. Rectangular pulse (Fig. 8.9).   **30.** Undamped single square pulse (Fig. 8.10).

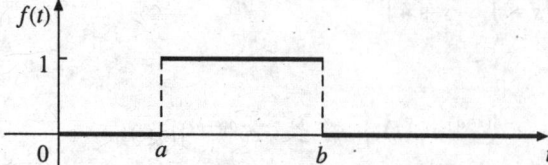

Fig. 8.9.   Problem 29.

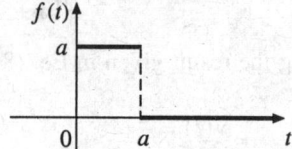

Fig. 8.10.   Problem 30.

31.

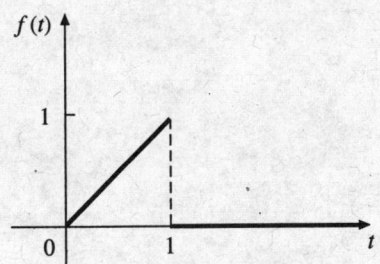

Fig. 8.11. Problem 31.

32. Periodic function of period 2*a*.

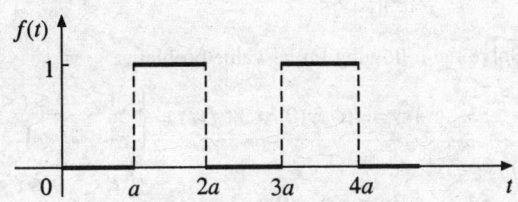

Fig. 8.12. Problem 32.

33. Staircase function.

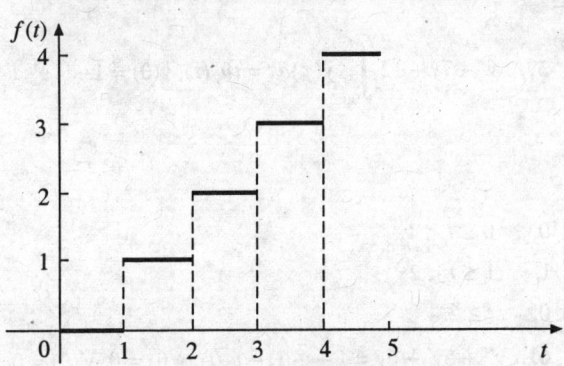

Fig. 8.13. Problem 33.

34.

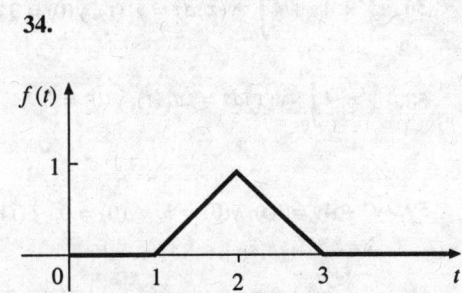

Fig. 8.14. Problem 34.

In each of the following problems, find the Laplace transform of the given functions.

35. $(t^2 - 1)\, u_2(t)$.

36. $(t^2 + 1)u_1(t)$.

37. $(\cos t)\, u_1(t)$.

38. $(\sin t)\, u_\pi(t)$.

39. $\cos (t - 3)\, u_3(t)$.

40. $e^{3-t} u_3(t)$.

41. $f(t) = \begin{cases} 5, & 0 \le t < 2 \\ -5, & t \ge 2. \end{cases}$

42. $f(t) = \begin{cases} t, & 0 \le t < 3 \\ 0, & t \ge 3. \end{cases}$

43. $f(t) = \begin{cases} 0, & 0 \le t < \pi/2 \\ \sin t, & t \ge \pi/2. \end{cases}$

44. $f(t) = \begin{cases} k, & 0 \le t < 2 \\ 0, & 2 \le t < 4 \\ k, & t \ge 4. \end{cases}$

45. $f(t) = \begin{cases} t^2, & 0 \le t < 1 \\ 0, & t \ge 1. \end{cases}$

46. $f(t) = \begin{cases} 0, & 0 \le t < 1 \\ t - 1, & 1 \le t < 2 \\ 0, & t \ge 2. \end{cases}$

In each of the following problems, find the inverse Laplace transform of the given functions.

47. $\dfrac{e^{-s}}{s^3}$.

48. $\dfrac{e^{-s\pi/2}}{s^2 + 1}$.

49. $\dfrac{se^{-s\pi}}{s^2 + 9}$.

50. $\dfrac{(1 + e^{-s\pi})^2}{s + 5}$.

51. $\dfrac{e^{-s\pi} - e^{-2s\pi}}{s^2 - 8s + 25}$.

52. $\dfrac{s(1 + e^{-s\pi/2})}{s^2 + 4}$.

Solve the following initial value problems.

53. $y' + y = f(t),\ y(0) = 3,\ f(t) = \begin{cases} 1, & 0 \le t < 1 \\ -1, & t \ge 1. \end{cases}$

54. $y' + 3y = f(t),\ y(0) = 2,\ f(t) = \begin{cases} t, & 0 \le t < 1 \\ 0, & t \ge 1. \end{cases}$

55. $y' + 2y = f(t),\ y(0) = 1,\ f(t) = \begin{cases} 1, & 0 \le t < 2 \\ 0, & t \ge 2. \end{cases}$

56. $y' + 4y + 4\displaystyle\int_0^t y(\tau)d\tau = u_1(t),\ y(0) = 3.$

57. $y' + 7y + 12\displaystyle\int_0^t y(\tau)d\tau = tu_2(t),\ y(0) = 1.$

58. $y' + 4\displaystyle\int_0^t y(\tau)d\tau = tu_1(t),\ y(0) = 2.$

59. $y'' + 4y = f(t),\ y(0) = 1,\ y'(0) = 0,\ f(t) = \begin{cases} 0, & 0 \le t < 1 \\ 1, & 1 \le t < 2 \\ 0, & t \ge 2. \end{cases}$

60. $y'' - 3y' + 2y = u_1(t),\ y(0) = 1,\ y'(0) = 1.$

61. $y'' + 5y' + 6y = 1 - u_3(t) - u_5(t),\ y(0) = 0,\ y'(0) = 0.$

62. $y'' + 4y' + 3y = f(t),\ y(0) = 0,\ y'(0) = 1,\ f(t) = \begin{cases} -1, & 0 \le t < 3 \\ 0, & t \ge 3. \end{cases}$

63. The integro-differential equation governing the flow of current $i(t)$ in an *RC* - circuit (Fig. 8.15) is given by (where the resistance R and capacitance C are constants)

$$Ri(t) + \frac{1}{C}\int_0^t i(\tau)d\tau = E(t).$$

If initially at $t = 0$ there is no current and $E(t) = v[u_1(t) - u_2(t)]$, v constant, find the current $i(t)$.

64. The differential equation governing the flow of current $i(t)$ in an *LR* series circuit (Fig. 8.16) is given by (where the inductance L and resistance R are constants)

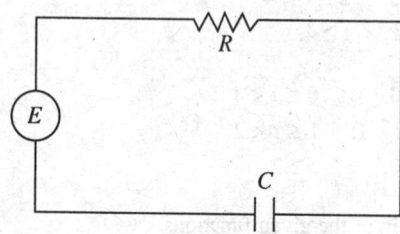

Fig. 8.15.   RC series circuit.

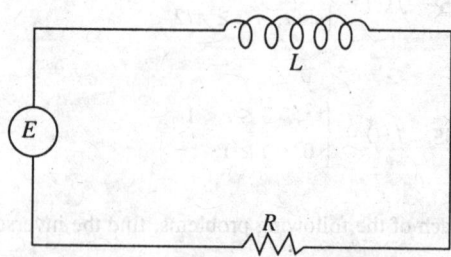

Fig. 8.16.   LR series circuit.

$$L\frac{di}{dt} + Ri = E(t).$$

Find the current $i(t)$ if the current is initially zero and $E(t) = \begin{cases} 0, & 0 \le t < 2, \\ 5, & t \ge 2. \end{cases}$

65. Solve for the current in the circuit of problem 64, if the current is initially zero and

$$E(t) = \begin{cases} \sin t, & 0 \le t < \pi/2, \\ 0, & t \ge \pi/2. \end{cases}$$

8.5 Laplace Transform of Dirac–Delta Function and More Properties of Laplace Transforms

8.5.1 Laplace Transform of Dirac-delta Function

Many problems in Electrical engineering, Physics and Mechanical engineering involve the concept of *impulse*. Impulse may be interpreted as a force of very large magnitude applied for just an instant. This impulse function is called the *Dirac-delta function.* It is not a function in the usual sense but is called a generalised function. Consider first the following examples.

Press a coin of given thickness, with unit force, over the edge of a metal plate ($-\infty < x < \infty$). We are interested in determining the pressure distribution $w(x)$ introduced along the plate. Exact shape of $w(x)$ and stress field is determined as part of the problem (Fig. 8.17a, b)

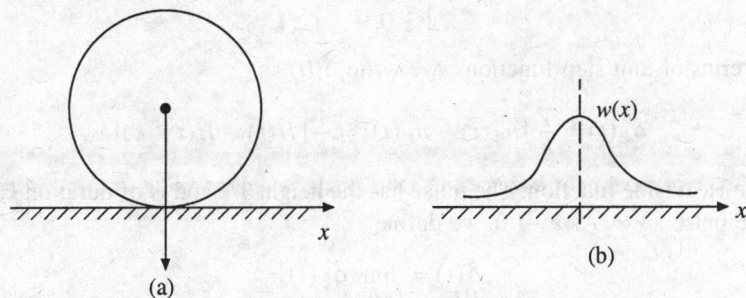

(a)

(b)

Fig. 8.17.   Pressure distribution and shape of curve $w(x)$.

Since a unit force is applied, we have $\displaystyle\int_{-\infty}^{\infty} w(x)dx = 1$. For example, in a general case of application of a force, consider some specific forms of $w(x)$ as a sequence $\{w_k(x)\}$ of functions, called δ-sequences,

(i) $w_k(x) = \begin{cases} k/2, & |x| < 1/k, \\ 0, & |x| \ge 1/k. \end{cases}$ (Fig. 8.18).

We have $\displaystyle\int_{-\infty}^{\infty} w_k(x)\,dx = \int_{-1/k}^{1/k} \frac{k}{2}\,dx = 1.$

(ii) $w_k(x) = \dfrac{k}{\pi(1 + k^2 x^2)},\ k > 0.$ (Fig. 8.19).

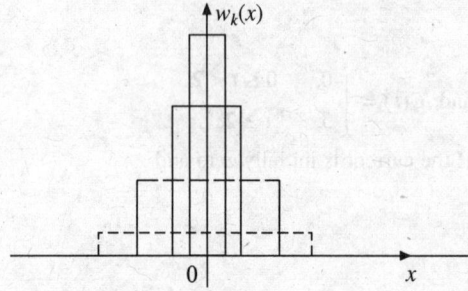

Fig. 8.18.   Shape of $w_k(x)$.

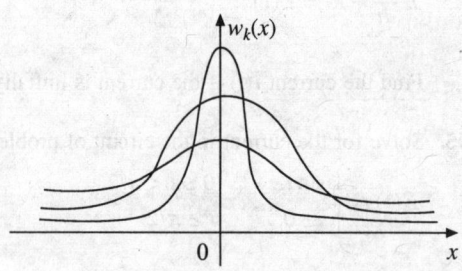

Fig. 8.19.   Shape of $w_k(x)$.

We have $\displaystyle \int_{-\infty}^{\infty} w_k(x) = \frac{k}{\pi} \int_{-\infty}^{\infty} \frac{dx}{1 + k^2 x^2} = 1.$

As $k \to \infty$, we get a "point" force, situated at $x = 0$, of unit strength. Denoting this force by $\delta(x)$, we define

$$\delta(x) = \lim_{k \to \infty} w_k(x). \tag{8.28}$$

Consider now, another example using the unit step function. Define

$$\delta_k(t) = \begin{cases} 0, & t < 0 \\ 1/k, & 0 \le t < k \\ 0, & t \ge k. \end{cases}$$

(Fig. 8.20). In terms of unit step functions, we write $\delta_k(t)$ as

$$\delta_k(t) = \frac{1}{k}[u_0(t) - u_k(t)] = \frac{1}{k}[H(t) - H(t - k)] \tag{8.29}$$

where $H(t)$ is the Heaviside function. The pulse has the height $1/k$ and is of duration k. As $k \to 0$, the amplitude of the pulse $\to \infty$. As $k \to 0$, we define

$$\delta(t) = \lim_{k \to 0^+} \delta_k(t). \tag{8.30}$$

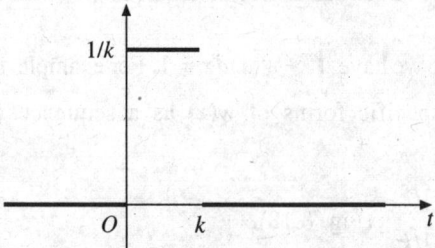

Fig. 8.20.   Graph of $\delta_k(t)$.

Therefore, we may interpret that $\delta(t) = 0$ for $t \ne 0$ and at $t = 0$ it is infinite. The delta function can be made to act at any other point. The delta function $\delta(t - a)$ acts at $t = a$. Again, we define $\delta(t - a) = 0$ for $t \ne a$ and at $t = a$ it is infinite. Taking the limit as $k \to 0$ in Eq. (8.29), we obtain

$$H'(t) = \delta(t). \tag{8.31}$$

The ordinary derivative of $H(t)$ does not exist as it is discontinuous at $t = 0$. It is to be understood as a generalised function. Similarly, we can define $H'(t - a) = \delta(t - a)$. We now prove an important property of delta function.

Theorem 8.8 (**Filtering property of Dirac-delta function**) Let $f(t)$ be continuous and integrable in $[0, \infty)$. Then

$$\int_0^\infty f(t)\delta(t - a)\,dt = f(a). \tag{8.32}$$

Proof   We prove this result by using the definition of the Dirac-delta function. We have

$$\int_0^\infty f(t)\delta_k(t - a)\,dt = \frac{1}{k}\int_a^{a+k} f(t)\,dt. \tag{8.33}$$

Using the mean value theorem of integral calculus, we obtain

$$\frac{1}{k}\int_a^{a+k} f(t)\,dt = \frac{1}{k}f(t_0)\int_a^{a+k} dt = f(t_0), \quad a < t_0 < a + k.$$

Substituting in Eq. (8.33), we get

$$\int_0^\infty f(t)\delta_k(t - a)\,dt = f(t_0).$$

Taking the limit $k \to 0$, we obtain

$$\int_0^\infty f(t)\delta(t - a)\,dt = f(a)$$

since $f(t_0) \to f(a)$.

Laplace transform of Dirac-delta function   The delta function does not satisfy the conditions of the existence theorem. However, we can obtain the Laplace transform of the delta function using the definition of the delta function. We have

$$\delta_k(t - a) = \frac{1}{k}[H(t - a) - H(t - a - k)] = \frac{1}{k}[u_a(t) - u_{a+k}(t)].$$

Taking Laplace transform, we get

$$\mathcal{L}[\delta_k(t - a)] = \frac{1}{k}\left[\frac{e^{-as}}{s} - \frac{e^{-(a+k)s}}{s}\right] = \frac{e^{-as}}{ks}(1 - e^{-ks}).$$

Taking the limit $k \to 0^+$, we obtain

$$\lim_{k \to 0^+} \frac{1 - e^{-ks}}{k} = s.$$

Hence,

$$\lim_{k \to 0^+} \mathcal{L}[\delta_k(t - a)] = \mathcal{L}[\delta(t - a)] = \frac{e^{-as}}{s}(s) = e^{-as}. \tag{8.34}$$

When $a = 0$, we have $\qquad \mathcal{L}[\delta(t)] = 1$ and $\mathcal{L}^{-1}[1] = \delta(t)$. $\qquad\qquad$ (8.35)

This result could also have been obtained by using the filtering property for the function $f(t) = e^{-st}$. We have from Eq. (8.32)

$$\int_0^\infty e^{-st}\delta(t-a)\,dt = f(a) = e^{-as}.$$

Remark 5

$\mathcal{L}[H'(t)] = \mathcal{L}[\delta(t)] = 1$.

Example 8.31 Find the solution of the initial value problem

$$y'' + 2y' + 5y = \delta(t-2), \; y(0) = 0, \; y'(0) = 0.$$

Solution Taking the Laplace transform of the differential equation, we obtain

$$s^2 Y(s) + 2sY(s) + 5Y(s) = e^{-2s}, \quad \text{or } Y(s) = \frac{e^{-2s}}{(s+1)^2 + 4}$$

where $\mathcal{L}[y(t)] = Y(s)$. Therefore, using Eq. (8.25) and $\mathcal{L}^{-1}\left[\dfrac{1}{(s+1)^2 + 4}\right] = \dfrac{1}{2} e^{-t} \sin(2t)$, we

obtain

$$y(t) = \frac{1}{2} e^{-(t-2)} \sin 2(t-2) u_2(t)$$

$$= \begin{cases} 0, & 0 \le t < 2 \\ \dfrac{1}{2} e^{-(t-2)} \sin 2(t-2), & t \ge 2 \end{cases}$$

The solution $y(t)$ is continuous and differentiable for $t > 0$. However, $y'(t)$ is discontinuous and has a jump discontinuity of magnitude 1 at $t = 2$. This magnitude is equal to the coefficient of the delta function $\delta(t-2)$ in the differential equation.

Electrical engineers often use the Dirac-delta function to study the behaviour of circuits. When a circuit is switched on, sometimes it is subjected to transients, that is, high input voltages occur. Such transients are modelled by the delta function and the behaviour of the circuit is studied.

8.5.2 Differentiation of Laplace Transform

In the earlier sections, we have discussed the solution of constant coefficient differential equations along with its initial conditions. The properties of Laplace transforms discussed earlier are not sufficient to solve the differential equations with polynomial coefficients in terms of the independent variable. The concept of differentiation of Laplace transform helps us in the solution of a class of such problems.

Theorem 8.9 (Differentiation of Laplace transform) Let $f(t)$ be piecewise continuous on $[0, \infty)$ and be of exponential order. Then

$$\mathcal{L}[t f(t)] = -F'(s) \quad \text{and} \quad \mathcal{L}[t^n f(t)] = (-1)^n \, F^{(n)}(s), \, s > \alpha \qquad\qquad (8.36)$$

where $F(s) = \mathcal{L}[f(t)]$ and $F^{(n)}(s) = d^n F(s)/ds^n$.

Proof Assuming that interchange of differentiation and integration is possible, we obtain

$$\frac{dF}{ds} = \frac{d}{ds}\int_0^\infty e^{-st}f(t)dt = \int_0^\infty \frac{d}{ds}(e^{-st})f(t)dt = -\int_0^\infty e^{-st}[tf(t)]dt = -\mathcal{L}[tf(t)].$$

Hence, $\mathcal{L}[tf(t)] = -F'(s)$.

Now,
$$\mathcal{L}[t^2 f(t)] = \mathcal{L}[t\{tf(t)\}] = -\frac{d}{ds}[\mathcal{L}(tf(t))]$$

$$= -\frac{d}{ds}\left[-\frac{d}{ds}F(s)\right] = (-1)^2 F''(s).$$

By induction, we can prove that $\mathcal{L}[t^n f(t)] = (-1)^n F^{(n)}(s).$

Remark 6

From Eq. (8.36), we obtain

$$\mathcal{L}^{-1}[F'(s)] = -tf(t) \quad \text{and} \quad \mathcal{L}^{-1}[F^{(n)}(s)] = (-1)^n t^n f(t). \tag{8.37}$$

Example 8.32 Find the Laplace transforms of the following functions.

(a) $t \sin 4t$,   (b) $t^2 \cos 3t$,   (c) te^{5t},   (d) $t^2 e^{-2t}$.

Solution

(a) Since $\mathcal{L}(\sin 4t) = 4/(s^2 + 16)$, we obtain

$$\mathcal{L}[t \sin 4t] = -\frac{d}{ds}\left[\frac{4}{s^2 + 16}\right] = \frac{8s}{(s^2 + 16)^2}.$$

(b) Since $\mathcal{L}(\cos 3t) = s/(s^2 + 9)$, we obtain

$$\mathcal{L}[t \cos 3t] = -\frac{d}{ds}\left[\frac{s}{s^2 + 9}\right] = \frac{s^2 - 9}{(s^2 + 9)^2}$$

$$\mathcal{L}[t^2 \cos 3t] = -\frac{d}{ds}\left[\frac{s^2 - 9}{(s^2 + 9)^2}\right] = -\frac{(s^2 + 9)^2(2s) - (s^2 - 9)2(s^2 + 9)(2s)}{(s^2 + 9)^4}$$

$$= -\frac{2s(s^2 + 9) - 4s(s^2 - 9)}{(s^2 + 9)^3} = \frac{2s(s^2 - 27)}{(s^2 + 9)^3}.$$

(c) We have $\mathcal{L}[e^{5t}] = 1/(s - 5)$. Hence

$$\mathcal{L}[te^{5t}] = -\frac{d}{ds}\left[\frac{1}{s - 5}\right] = \frac{1}{(s - 5)^2}.$$

(d) We have $\mathcal{L}[e^{-2t}] = 1/(s + 2)$. Hence

$$\mathcal{L}[t^2 e^{-2t}] = (-1)^2 \frac{d^2}{ds^2}\left[\frac{1}{s + 2}\right] = \frac{2}{(s + 2)^3}.$$

Example 8.33 Find the inverse Laplace transform of the following functions.

(a) $\dfrac{2(s + 1)}{(s^2 + 2s + 2)^2}$,   (b) $\dfrac{1}{(s + 5)^4}$.   (c) $\dfrac{1}{(s^2 + 9)^2}$.

Solution

(a) We have $\dfrac{2(s+1)}{(s^2+2s+2)^2} = \dfrac{2(s+1)}{[(s+1)^2+1]^2} = -\dfrac{d}{ds}\left[\dfrac{1}{(s+1)^2+1}\right] = -F'(s).$

Hence, $\mathscr{L}^{-1}[-F'(s)] = tf(t)$ and

$$f(t) = \mathscr{L}^{-1}[F(s)] = \mathscr{L}^{-1}\left[\dfrac{1}{(s+1)^2+1}\right] = e^{-t}\sin t.$$

Therefore, $\mathscr{L}^{-1}\left[\dfrac{2(s+1)}{(s^2+2s+2)^2}\right] = te^{-t}\sin t.$

(b) We have $\dfrac{1}{(s+5)^4} = -\dfrac{1}{6}\dfrac{d^3}{ds^3}\left[\dfrac{1}{s+5}\right]$

and $\qquad\qquad \mathscr{L}^{-1}\left[\dfrac{d^3}{ds^3}\left(\dfrac{1}{s+5}\right)\right] = (-1)^3 t^3 e^{-5t}$ \hfill (using Eq. 8.37)

Therefore, $\mathscr{L}^{-1}\left[\dfrac{1}{(s+5)^4}\right] = \dfrac{1}{6}t^3 e^{-5t}.$

(c) We write $\dfrac{1}{(s^2+9)^2} = \dfrac{(s^2+9)+(9-s^2)}{18(s^2+9)^2} = \dfrac{1}{18}\left[\dfrac{1}{s^2+9} - \dfrac{s^2-9}{(s^2+9)^2}\right]$

Therefore, $\mathscr{L}^{-1}\left[\dfrac{1}{(s^2+9)^2}\right] = \dfrac{1}{18}\left[\mathscr{L}^{-1}\left(\dfrac{1}{s^2+9}\right) - \mathscr{L}^{-1}\left(\dfrac{s^2-9}{(s^2+9)^2}\right)\right]$

$$= \dfrac{1}{18}\left[\dfrac{1}{3}\sin 3t - t\cos 3t\right].$$

(see Example 8.32 (b)).

Example 8.34 Find the solution of the initial value problem

$$y'' + t\,y' - 2y = 6 - t, \quad y(0) = 0,\ y'(0) = 1$$

given that $\mathscr{L}[y(t)]$ exists.

Solution Taking the Laplace transform of the differential equation, we get

$$\mathscr{L}[y''] + \mathscr{L}[ty'] - 2\,\mathscr{L}[y] = \mathscr{L}[6-t]$$

or $\qquad s^2 Y(s) - 1 + \left[-\dfrac{d}{ds}\{sY(s) - 0\}\right] - 2Y(s) = \dfrac{6}{s} - \dfrac{1}{s^2}$

or $\qquad s^2 Y(s) - [sY'(s) + Y(s)] - 2Y(s) = 1 + \dfrac{6}{s} - \dfrac{1}{s^2}$

or
$$Y'(s) + \left(\frac{3}{s} - s\right)Y(s) = -\left(\frac{1}{s} + \frac{6}{s^2} - \frac{1}{s^3}\right).$$

Therefore, application of Laplace transform to a differential equation with polynomial coefficients has resulted into a linear constant coefficient differential equation for $Y(s)$. This equation can be solved for $Y(s)$. The integrating factor of the differential equation is

$$\text{I. F} = e^{\int [(3/s) - s]ds} = s^3 e^{-s^2/2}.$$

Hence, we obtain

$$s^3 e^{-s^2/2} Y(s) = \int e^{-s^2/2} ds - 6 \int s e^{-s^2/2}\, ds - \int s^2 e^{-s^2/2} ds + c$$

$$= \int e^{-s^2/2} ds + 6 e^{-s^2/2} - \left[-s e^{-s^2/2} + \int e^{-s^2/2} ds \right] + c$$

$$= (6 + s) e^{-s^2/2} + c$$

Therefore, $Y(s) = \dfrac{6}{s^3} + \dfrac{1}{s^2} + \dfrac{c}{s^3} e^{s^2/2}.$

Since $y(t)$ satisfies the conditions of the existence theorem, we require that (see Remark 1) $\lim\limits_{s \to \infty} Y(s) = 0$. This requirement gives $c = 0$. Therefore,

$$Y(s) = \frac{6}{s^3} + \frac{1}{s^2} \quad \text{and} \quad y(t) = 3t^2 + t$$

which is the solution of the given initial value problem.

Example 8.35 Find the solution of the initial value problem

$$t y'' + 2t y' + 2y = 2, \; y(0) = 1, \; y'(0) \text{ is arbitrary.}$$

Solution Taking Laplace transform of the differential equation, we get

$$-\frac{d}{ds}[s^2 Y(s) - s - y'(0)] + 2\left[-\frac{d}{ds}\{s Y(s) - 1\} \right] + 2Y(s) = \frac{2}{s} \tag{8.38}$$

or
$$-[s^2 Y'(s) + 2s Y(s) - 1] - 2[s Y'(s) + Y(s)] + 2Y(s) = \frac{2}{s}$$

or
$$-s(s + 2)Y'(s) - 2s Y(s) = \frac{2}{s} - 1 = \frac{2 - s}{s}$$

or
$$Y'(s) + \left(\frac{2}{s + 2}\right)Y(s) = \frac{s - 2}{s^2(s + 2)}.$$

The integrating factor of the equation is

$$\text{I} \cdot \text{F} = e^{\int [2/(s+2)]ds} = (s + 2)^2.$$

The solution of the equation is

$$(s + 2)^2 Y(s) = \int \frac{(s - 2)(s + 2)}{s^2} \, ds + c = s + \frac{4}{s} + c$$

or

$$Y(s) = \frac{4 + s^2 + cs}{s(s + 2)^2} = \frac{1}{s} - \frac{4}{(s + 2)^2} + \frac{c}{(s + 2)^2}.$$

Taking inverse Laplace transform, we obtain $y(t) = 1 - (4 - c) \, te^{-2t}$ where c is an arbitrary constant. This problem does not have a unique solution. Note that $y'(0) = c - 4$. Hence, the solution is valid for arbitrary values of $y'(0)$. This fact can be observed from Eq. (8.38) as the contribution of $y'(0)$ vanishes due to the derivative with respect to s.

8.5.3 Integration of Laplace Transform

We now present a useful result involving the integration of Laplace transforms.

Theorem 8.10 (Integration of Laplace transform) Let $f(t)$ be piecewise continuous on $[0, \infty)$ and be of exponential order. Further, let $\lim\limits_{t \to 0^+} [f(t)/t]$ exist. Then

$$\mathcal{L}\left[\frac{f(t)}{t}\right] = \int_s^\infty F(s^*) ds^*, \quad s > \alpha \tag{8.39}$$

where $\mathcal{L}[f(t)] = F(s)$.

Proof From the definition, we have

$$F(s) = \int_0^\infty e^{-st} f(t) dt.$$

Integrating from s to ∞, we obtain

$$\int_s^\infty F(s^*) ds^* = \int_s^\infty \left[\int_0^\infty e^{-s^* t} f(t) dt \right] ds^*.$$

Now, $f(t)$ is piecewise continuous, is of exponential order and limit of $f(t)/t$ exists as t approaches zero from the right. Under these assumptions, the order of integration can be reversed and we can write

$$\int_s^\infty F(s^*) ds^* = \int_0^\infty \left[\int_s^\infty e^{-s^* t} ds^* \right] f(t) dt = \int_0^\infty \left[\frac{e^{-s^* t}}{-t} \right]_s^\infty f(t) dt$$

$$= \int_0^\infty e^{-st} \left[\frac{f(t)}{t} \right] dt = \mathcal{L}\left[\frac{f(t)}{t} \right], s > \alpha.$$

Remark 7.

From Eq. (8.39), we obtain

$$\mathcal{L}^{-1}\left[\int_s^\infty F(s^*) ds^* \right] = \frac{f(t)}{t}, \quad \text{or} \quad f(t) = t \, \mathcal{L}^{-1}\left[\int_s^\infty F(s^*) ds^* \right]. \tag{8.40}$$

This result is useful when the integral of a transform is simpler than the transform itself.

Example 8.36   Find the Laplace transform of $(\sin \omega t)/t$.

Solution   Using Theorem 8.10, we obtain

$$\mathscr{L}\left[\frac{\sin \omega t}{t}\right] = \int_s^\infty \mathscr{L}[\sin \omega t]\,ds^* = \int_s^\infty \frac{\omega}{s^{*2} + \omega^2}\,ds^*$$

$$= \left[\tan^{-1}\left(\frac{s^*}{\omega}\right)\right]_s^\infty = \frac{\pi}{2} - \tan^{-1}\left(\frac{s}{\omega}\right) = \cot^{-1}\left(\frac{s}{\omega}\right).$$

Example 8.37   Find the inverse Laplace transform of the following functions.

(a) $\ln \dfrac{s+c}{s+d}$; c, d constants,   (b) $\dfrac{1}{s(s+3)^2}$,   (c) $\dfrac{s}{(s^2-9)^2}$.

Solution

(a) Denote $G(s) = \ln [(s+c)/(s+d)]$. Let $F(s) = -d[G(s)]/ds$.

Therefore, $F(s) = -\dfrac{d}{ds}[\ln (s+c) - \ln (s+d)] = \dfrac{1}{s+d} - \dfrac{1}{s+c}.$

We note that

$$\int_s^\infty F(s^*)\,ds^* = -\int_s^\infty \frac{d}{ds}G(s^*)\,ds^* = \int_s^\infty \left[\frac{1}{s^*+d} - \frac{1}{s^*+c}\right]ds^* = \left[\ln \frac{s^*+d}{s^*+c}\right]_s^\infty = \ln \frac{s+c}{s+d}.$$

Now, $f(t) = \mathscr{L}^{-1}[F(s)] = e^{-dt} - e^{-ct}$. Therefore,

$$\mathscr{L}^{-1}\left[\ln\left(\frac{s+c}{s+d}\right)\right] = \mathscr{L}^{-1}\left[\int_s^\infty F(s^*)\,ds^*\right] = \frac{f(t)}{t} = \frac{1}{t}(e^{-dt} - e^{-ct}).$$

(b) Let $G(s) = \dfrac{1}{s(s+3)^2}$ and $F(s) = -\dfrac{d}{ds}G(s) = -\dfrac{d}{ds}\left[\dfrac{1}{9}\left(\dfrac{1}{s} - \dfrac{1}{s+3} - \dfrac{3}{(s+3)^2}\right)\right]$

$$= \frac{1}{9}\left[\frac{1}{s^2} - \frac{1}{(s+3)^2} - \frac{6}{(s+3)^3}\right].$$

Now, $\displaystyle\int_s^\infty F(s^*)\,ds^* = -\int_s^\infty \left[\frac{d}{ds}G(s^*)\right]ds^* = -\left[\frac{1}{s^*(s^*+3)^2}\right]_s^\infty = \frac{1}{s(s+3)^2}.$

We have $f(t) = \mathscr{L}^{-1}[F(s)] = \mathscr{L}^{-1}\left\{\dfrac{1}{9}\left[\dfrac{1}{s^2} - \dfrac{1}{(s+3)^2} - \dfrac{6}{(s+3)^3}\right]\right\} = \dfrac{1}{9}[t - te^{-3t} - 3t^2e^{-3t}]$

Therefore,

$$\mathscr{L}^{-1}\left[\frac{1}{s(s+3)^2}\right] = \frac{1}{9t}(t - te^{-3t} - 3t^2e^{-3t}) = \frac{1}{9}(1 - e^{-3t} - 3te^{-3t}).$$

This result could also have been obtained directly.

(c) We shall use Remark 7 to find the inverse. We have

$$f(t) = t\mathcal{L}^{-1}\left[\int_s^\infty \frac{s^*}{(s^{*2}-9)^2}\,ds^*\right] = t\mathcal{L}^{-1}\left[\left\{\frac{-1}{2(s^{*2}-9)}\right\}_s^\infty\right] = \frac{t}{2}\,\mathcal{L}^{-1}\left[\frac{1}{s^2-9}\right]$$

$$= \frac{t}{12}\,\mathcal{L}^{-1}\left[\frac{1}{s-3} - \frac{1}{s+3}\right] = \frac{t}{12}[e^{3t} - e^{-3t}] = \frac{1}{6}\,t\sinh 3t.$$

8.5.4 Convolution Theorem

Consider, for example, a function

$$H(s) = \frac{1}{s(s^2+16)} = F(s)G(s)$$

where $F(s) = 1/s$ and $G(s) = 1/(s^2+16)$. The inverse Laplace transformation of $H(s)$ can be determined by first writing its partial fractions. However, we know the inverse Laplace transforms of $F(s)$ and $G(s)$. This information could not be used as the Laplace transform of product of functions $f(t)$ and $g(t)$ is not the product of Laplace transforms of $f(t)$ and $g(t)$. We now present a result which gives the Laplace transform of a product of functions.

Convolution Let $f(t)$ and $g(t)$ be defined in $[0, \infty)$. Then, the convolution of $f(t)$ and $g(t)$, denoted by $(f * g)(t)$, is defined by

$$(f * g)(t) = \int_0^t f(\tau)\,g(t-\tau)\,d\tau,\ t \geq 0. \tag{8.41}$$

The following properties of convolution are easily proved.

   (i) $f * g = g * f$,
                     (ii) $f * (g_1 + g_2) = f * g_1 + f * g_2$,

   (iii) $(f * g) * h = f * (g * h)$,
         (iv) $f * 0 = 0 * f = 0$.

That is, the commutative, distributive and associative laws are satisfied. But, $f * 1 \neq 1 * f \neq f$ in general. For example, for $f(t) = t$, we have

$$(1 * f)(t) = \int_0^t 1 \cdot (t-\tau)\,d\tau = \frac{t^2}{2} \neq f(t).$$

If we use $f(t) = e^{-st}$ in Theorem 8.8 (filtering property of delta function), we obtain

$$\int_0^\infty e^{-st}\,\delta(t-a)dt = e^{-as}$$

which is same as the Laplace transform of $\delta(t-a)$. If we write Theorem 8.8 as

$$\int_0^\infty f(\tau)\,\delta(\tau-t)d\tau = f(t)$$

then we recognise it as the convolution $f * \delta = f$.

We now prove the convolution theorem.

Theorem 8.11 (Convolution theorem) Let $f(t)$, $g(t)$ be piecewise continuous functions on $[0, \infty)$ and be of exponential orders. Then

$$\mathcal{L}[(f * g)(t)] = \mathcal{L}[f(t)]\,\mathcal{L}[g(t)] = F(s)\,G(s). \tag{8.42}$$

Proof   We have

$$F(s)\,G(s)=\left[\int_0^\infty e^{-s\tau}\,f(\tau)d\tau\right]\left[\int_0^\infty e^{-su}\,g(u)du\right]$$

$$=\int_0^\infty\int_0^\infty e^{-s(\tau+u)}f(\tau)g(u)\,d\tau\,du=\int_0^\infty f(\tau)d\tau\int_0^\infty e^{-s(\tau+u)}g(u)\,du.$$

Let $\tau+u=t$, where τ is fixed. Then, $u=t-\tau$ and the limits of integration for t are from τ to ∞. Therefore, we have

$$F(s)\,G(s)=\int_0^\infty f(\tau)d\tau\int_\tau^\infty e^{-st}g(t-\tau)dt.$$

The region of integration, $0\le\tau<\infty$, $\tau\le t<\infty$, is given in Fig. 8.21. Changing the order of integration, we obtain the region of integration as $0\le t<\infty$, $0\le\tau\le t$. Therefore,

$$F(s)\,G(s)=\int_0^\infty e^{-st}dt\int_0^t f(\tau)g(t-\tau)d\tau$$

$$=\int_0^\infty e^{-st}\left[\int_0^t f(\tau)g(t-\tau)d\tau\right]dt=\int_0^\infty e^{-st}[(f*g)(t)]dt$$

$$=\mathscr{L}[(f*g)\,(t)].$$

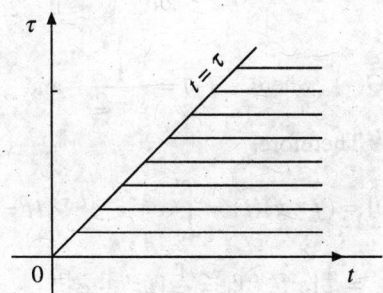

Fig. 8.21.   Region of integration.

Remark 8
From the Theorem 8.11, we have

$$\mathscr{L}^{-1}[F(s)\,G(s)]=(f*g)\,(t). \tag{8.43}$$

Example 8.38   Find the convolution $\sin\omega t*\cos\omega t$.

Solution   We have

$$\sin\omega t*\cos\omega t=\int_0^t\sin\omega\tau\cos\omega(t-\tau)d\tau=\frac{1}{2}\int_0^t[\sin\omega t+\sin\omega(2\tau-t)]d\tau$$

$$=\frac{1}{2}\left[\tau\sin\omega t-\frac{1}{2\omega}\cos\omega(2\tau-t)\right]_0^t=\frac{1}{2}\left[t\sin\omega t-\frac{1}{2\omega}\cos\omega t+\frac{1}{2\omega}\cos\omega t\right]$$

$$=\frac{1}{2}\,t\sin\omega t.$$

Example 8.39 Find the inverse Laplace transforms of the following functions using convolution.

(a) $\dfrac{1}{(s^2 + \omega^2)^2}$, (b) $\dfrac{1}{(s-2)(s+3)}$.

Solution

(a) Write

$$\frac{1}{(s^2 + \omega^2)^2} = \frac{1}{(s^2 + \omega^2)(s^2 + \omega^2)} = F(s)G(s)$$

where $F(s) = G(s) = 1/(s^2 + \omega^2)$. We have $f(t) = g(t) = (\sin \omega t)/\omega$.
Therefore,

$$\mathcal{L}^{-1}[F(s)\,G(s)] = (f * g)(t) = \int_0^t f(\tau)g(t - \tau)d\tau$$

$$= \frac{1}{\omega^2} \int_0^t \sin \omega\tau \sin \omega(t - \tau)d\tau$$

$$= \frac{1}{2\omega^2} \int_0^t [\cos \omega(2\tau - t) - \cos \omega t]d\tau = \frac{1}{2\omega^2}\left[\frac{1}{2\omega}\sin \omega(2\tau - t) - \tau \cos \omega t\right]_0^t$$

$$= \frac{1}{2\omega^2}\left[\frac{1}{2\omega}\sin \omega t - t\cos \omega t + \frac{1}{2\omega}\sin \omega t\right] = \frac{1}{2\omega^3}[\sin \omega t - \omega t \cos \omega t].$$

(b) Write $\dfrac{1}{(s-2)(s+3)} = F(s)G(s)$ where $F(s) = \dfrac{1}{s-2}$ and $G(s) = \dfrac{1}{s+3}$.

We have $f(t) = e^{2t}$ and $g(t) = e^{-3t}$. Therefore,

$$\mathcal{L}^{-1}[F(s)G(s)] = (f * g)(t) = \int_0^t e^{2\tau}e^{-3(t-\tau)}d\tau = \int_0^t e^{5\tau - 3t}d\tau$$

$$= \frac{1}{5}[e^{5\tau - 3t}]_0^t = \frac{1}{5}[e^{2t} - e^{-3t}].$$

Example 8.40 Find $f(t)$ as the solution of the integral equation

$$f(t) = t + e^{-2t} + \int_0^t f(\tau)e^{2(t-\tau)}d\tau.$$

Solution We identify the integral on the right hand side as the convolution

$$(f * g)(t) = \int_0^t f(\tau)g(t - \tau)d\tau, \quad \text{where } g(t) = e^{2t}.$$

Taking Laplace transform of the given equation, we obtain

$$F(s) = \frac{1}{s^2} + \frac{1}{s+2} + F(s)\left(\frac{1}{s-2}\right)$$

or

$$\left[1 - \frac{1}{s-2}\right]F(s) = \frac{1}{s^2} + \frac{1}{s+2} \quad \text{or} \quad F(s) = \frac{s-2}{s-3}\left[\frac{s^2 + s + 2}{s^2(s+2)}\right].$$

Writing the partial fractions, we get

$$F(s) = \frac{1}{45}\left[\frac{14}{s-3} - \frac{5}{s} + \frac{30}{s^2} + \frac{36}{s+2}\right].$$

Taking the inverse transform, we obtain

$$f(t) = \frac{1}{45}[14e^{3t} - 5 + 30t + 36e^{-2t}].$$

Example 8.41 Using convolution, solve the initial value problem

$$y'' + 9y = \sin 3t, \quad y(0) = 0, \quad y'(0) = 0.$$

Solution Taking Laplace transform of the differential equation, we obtain

$$s^2 Y(s) + 9Y(s) = \frac{3}{s^2 + 9}, \quad \text{or} \quad Y(s) = \frac{3}{(s^2 + 9)^2}.$$

Write $Y(s)$ as
$$Y(s) = \frac{1}{3}\left(\frac{3}{s^2 + 9}\right)\left(\frac{3}{s^2 + 9}\right) = \frac{1}{3}F(s)G(s)$$

where $F(s) = G(s) = 3/(s^2 + 9)$. We have $f(t) = g(t) = \sin 3t$. Using the convolution theorem, we obtain

$$y(t) = \mathcal{L}^{-1}\left[\frac{1}{3}F(s)G(s)\right] = \frac{1}{3}\int_0^t \sin 3\tau \, \sin 3(t - \tau)\,d\tau$$

$$= \frac{1}{6}\int_0^t [\cos 3(2\tau - t) - \cos 3t]\,d\tau = \frac{1}{6}\left[\frac{1}{6}\sin 3(2\tau - t) - \tau\cos 3t\right]_0^t$$

$$= \frac{1}{6}\left[\frac{1}{6}(\sin 3t + \sin 3t) - t\cos 3t\right] = \frac{1}{18}[\sin 3t - 3t\cos 3t].$$

Exercise 8.4

Evaluate the following integrals.

1. $\displaystyle\int_0^3 t^3 \delta(t - 4)\,dt.$

2. $\displaystyle\int_0^\infty \cos t \, \delta\!\left(t - \frac{\pi}{4}\right)dt.$

3. $\displaystyle\int_0^\infty f(t)\delta(t-1)\,dt$ where $f(t) = t^2, 0 \le t < 1$, $f(t) = 2$ at $t = 1$ and $f(t) = t$ for $t > 1$.

Solve the following initial value problems.

4. $y'' + 4y' + 5y = \delta(t - 3), \quad y(0) = 0, \quad y'(0) = 0.$

5. $y'' + 16y = \delta(t - 2), \quad y(0) = 0, \quad y'(0) = 0.$

6. $y'' + 2y' + 10y = 6\delta(t - 2) - 3\delta(t - 3), \quad y(0) = 0, y'(0) = 0.$

7. $y'' + 9y = 4\delta(t), \quad y(0) = 0, \quad y'(0) = 0.$

8. $y'' + 4y' + 8y = 16\,\delta(t - 1) + 8\delta(t - 2), \quad y(0) = 0, \quad y'(0) = 0.$

9. $y'' + 4y = 8\delta(t - \pi/6), \quad y(0) = 2, \quad y'(0) = 0.$

Find the Laplace transform of the following functions.

10. $t \sin 2t$.

11. $t^2 e^{3t}$.

12. $t^2 \sinh 3t$.

13. $t e^{-4t} \sin 3t$.

14. $t \int_0^t e^{-2\tau} \cos 3\tau d\tau$.

15. $t \int_0^t e^{-\tau} \sin 2\tau d\tau$.

16. $t^2 \int_0^t e^{-5\tau} d\tau$.

Find the inverse Laplace transform of the following functions.

17. $\dfrac{1}{(s+a)^2}$.

18. $\dfrac{1}{(s-a)^3}$.

19. $\dfrac{8(s+2)}{(s^2+4s+8)^2}$.

20. $\dfrac{1}{(s^2+16)^2}$.

21. $\dfrac{6s}{(s^2-16)^2}$.

22. $\dfrac{6s+5}{(s^2+2s+2)^2}$.

Find the solution of the following differential equations/initial value problems using Laplace transforms. In problems **29** to **31**, $y'(0)$ is arbitrary.

23. $ty' - 2y = 6$.

24. $ty' - 3y = 2t$.

25. $y'' - ty' + 4y = 3$, $y(0) = 0$, $y'(0) = 0$.

26. $y'' + 4ty' - 12y = 0$, $y(0), = 0$, $y'(0) = -2$.

27. $y'' + 6ty' - 12y = 1$, $y(0) = 2$, $y'(0) = 0$.

28. $y'' + 6ty' - 30y = 0$, $y(0) = 0$, $y'(0) = 2$.

29. $ty'' + 4ty' + 4y = 8$, $y(0) = 2$.

30. $ty'' + (6t - 2)y' - 6y = 0$, $y(0) = 1$.

31. $ty'' + (8t - 2)y' - 8y = 0$, $y(0) = 2$.

Find the Laplace transform of the following functions.

32. $(\sinh t)/t$.

33. $(1 - \cos bt)/t$.

34. $(e^{-2t} \sin 3t)/t$.

Find the inverse Laplace transform of the following functions.

35. $\dfrac{s}{(s+4)^3}$.

36. $\dfrac{18}{(s^2+9)^2}$.

37. $\ln\left(\dfrac{s^2+1}{s^2}\right)$.

38. $\ln\left(\dfrac{s-1}{s+1}\right)$.

39. $\ln\left(\dfrac{s^2+1}{s(s+1)}\right)$.

40. $\cot^{-1}s$.

41. $\dfrac{1}{s}\tan^{-1}\left(\dfrac{1}{s}\right)$.

42. $\tanh^{-1}\left(\dfrac{1}{s}\right)$.

43. Give an example to show that the convolution $f * f$ may not always be non-negative.

44. Show that $\mathcal{L}[(f * g * h)(t)] = \mathcal{L}[f(t)]\,\mathcal{L}[g(t)]\,\mathcal{L}[h(t)]$.

45. Show that $1 * 1 = t$, $1 * 1 * 1 = t^2/2, \ldots, 1 * 1 * 1 * (n\ \text{times}) = t^n/n!$

Find the following convolutions.

46. $1 * e^{-2t}$.

47. $t * e^{at}$.

48. $e^{at} * e^{bt}$ $(a \neq b)$.

49. $\sin \omega t * \sin \omega t$.

In the following problems use the convolution theorem to find the inverse Laplace transform.

50. $\dfrac{1}{(s-a)(s-b)}$.

51. $\dfrac{1}{s^2(s^2+16)}$.

52. $\dfrac{1}{(s^2+9)^2}$.

53. $\dfrac{s}{(s^2 + 4)^2}$. **54.** $\dfrac{s}{(s^2 + 4)(s^2 + 9)}$. **55.** $\dfrac{6}{(s + 1)^2 (s + 2)}$.

56. $\dfrac{8s}{(s^2 + 16)(s^2 + 1)^2}$. **57.** $\dfrac{1}{(s + a)^2 (s + b)^2}$, $a \neq b$.

Using convolution, solve the following initial value problems.

58. $y'' + 16y = \cos 4t$, $y(0) = 0$, $y'(0) = 0$. **59.** $y'' + 3y' + 2y = e^{-t}$, $y(0) = 0$, $y'(0) = -1$.

60. $y'' + 4y' + 4y = te^{-t}$, $y(0) = 0$, $y'(0) = 2$.

61. $y'' - \omega^2 y = \cosh \omega t$, $y(0) = 1$, $y'(0) = 2$. **62.** $y'' - 5y' - 6y = e^{-t}$, $y(0) = 1$, $y'(0) = 1$.

63. $y' - y = te^t \cos t$, $y(0) = 0$. **64.** $y' + y = t \cos t$, $y(0) = 0$.

Solve the following integral equations using convolution theorem.

65. $f(t) = t + 6 \displaystyle\int_0^t f(\tau)e^{(t-\tau)}\,d\tau$. **66.** $f(t) = e^t + \displaystyle\int_0^t \tau f(t - \tau)\,d\tau$.

67. $f(t) = 1 + t + 2 \displaystyle\int_0^t \sin \tau f(t - \tau)\,d\tau$. **68.** $f(t) + \displaystyle\int_0^t f(\tau) \cos (t - \tau)\,d\tau = e^{-t}$.

69. $f(t) = \cos t + e^{-t} \displaystyle\int_0^t f(\tau)e^{\tau}\,d\tau$. **70.** $4f(t) = t - \displaystyle\int_0^t (e^{\tau} + e^{-\tau})f(t - \tau)\,d\tau$.

8.6 Laplace Transform of Periodic Functions

Periodic functions appear in many applications in science and engineering. Let a function $f(t)$ be periodic with period T, that is

$$f(t + T) = f(t), \; t > 0. \tag{8.44}$$

Geometrically, this implies that the graph of the function $y = f(t)$ repeats itself after every interval of length T. The following are some examples of periodic functions.

(i) *Triangular wave*

$$f(t) = \begin{cases} t/a, & 0 \leq t < a \\ (2a - t)/a, & a \leq t \leq 2a. \end{cases}$$

$$f(t + T) = f(t + 2a) = f(t). \text{ (Fig. 8.22)}.$$

(ii) *Square wave*

$$f(t) = \begin{cases} k, & 0 \leq t < a \\ -k, & a \leq t < 2a. \end{cases}$$

$$f(t + T) = f(t + 2a) = f(t) \quad \text{(Fig. 8.23)}.$$

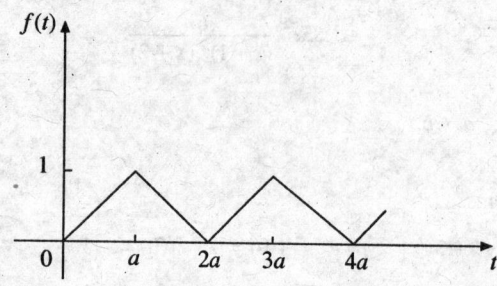

Fig. 8.22. Triangular wave.

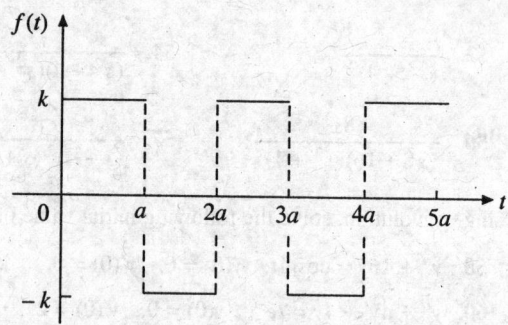

Fig. 8.23. Square wave.

(iii) *Square wave*

$$f(t) = \begin{cases} k, & 0 \le t < a \\ 0, & a \le t < 2a. \end{cases}$$

$$f(t + T) = f(t + 2a) = f(t). \text{ (Fig. 8.24).}$$

(iv) *Sawtooth wave*

$$f(t) = t, \quad 0 \le t < a.$$

$$f(t + T) = f(t + a) = f(t). \text{ (Fig. 8.25).}$$

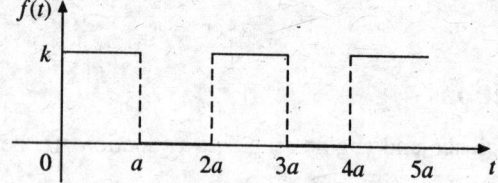

Fig. 8.24.   Square wave.

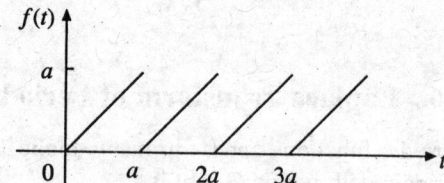

Fig. 8.25.   Sawtooth wave.

In the case of $f(t)$ defined in (i), it is continuous while in the cases (ii), (iii) and (iv), $f(t)$ is piecewise continuous. The Laplace transform of a periodic function can be obtained by integrating over one period. The result is presented in the following theorem.

Theorem 8.12 Let $f(t)$ be piecewise continuous on $[0, \infty)$, be of exponential order and periodic with period T. Then

$$\mathscr{L}[f(t)] = \frac{1}{1 - e^{-sT}} \int_0^T e^{-st} f(t)\,dt, \quad s > 0. \tag{8.45}$$

Proof We write the Laplace transform as the sum of the following integrals.

$$\mathscr{L}[f(t)] = \int_0^\infty e^{-st} f(t)\,dt = \int_0^T e^{-st} f(t)\,dt + \int_T^{2T} e^{-st} f(t)\,dt + \int_{2T}^{3T} e^{-st} f(t)\,dt + \ldots.$$

Substitute $t = u + T$ in the second integral, $t = u + 2T$ in the third integral, . . . on the right hand side of this equation. We have

$$\mathcal{L}[f(t)] = \int_0^T e^{-st} f(t)dt + \int_0^T e^{-s(u+T)} f(u+T)\,du + \int_0^T e^{-s(u+2T)} f(u+2T)\,du + \ldots.$$

Since $f(t)$ is periodic with period T, we obtain

$$\mathcal{L}[f(t)] = \int_0^T e^{-su} f(u)du + e^{-sT} \int_0^T e^{-su} f(u)\,du + e^{-2sT} \int_0^T f(u)\,du + \ldots$$

$$= (1 + e^{-sT} + e^{-2sT} + \ldots) \int_0^T e^{-st} f(t)dt = \frac{1}{1-e^{-sT}} \int_0^T e^{-st} f(t)dt.$$

Example 8.42 Find the Laplace transform of the periodic function defined by the triangular wave

$$f(t) = \begin{cases} t/a, & 0 \le t \le a, \\ (2a-t)/a, & a \le t \le 2a. \end{cases}$$

and $f(t+2a) = f(t)$.

Solution We have $T = 2a$. Therefore,

$$\mathcal{L}[f(t)] = \frac{1}{1-e^{-2as}} \left[\int_0^a \frac{t}{a} e^{-st} dt + \int_a^{2a} \frac{(2a-t)}{a} e^{-st} dt \right]$$

$$= \frac{1}{1-e^{-2as}} \left[-\frac{1}{a}\left\{ \left(\frac{t}{s} + \frac{1}{s^2} \right) e^{-st} \right\}_0^a + \left\{ \left(-\frac{2}{s} + \frac{1}{a}\left(\frac{t}{s} + \frac{1}{s^2} \right) \right) e^{-st} \right\}_a^{2a} \right]$$

$$= \frac{1}{1-e^{-2as}} \left[\frac{1}{as^2}(1 - 2e^{-as} + e^{-2as}) \right] = \frac{(1-e^{-as})^2}{as^2(1-e^{-2as})}$$

$$= \frac{1-e^{-as}}{as^2(1+e^{-as})} = \frac{e^{as/2} - e^{-as/2}}{as^2(e^{as/2} + e^{-as/2})} = \frac{1}{as^2} \tanh\left(\frac{as}{2} \right), \quad s > 0.$$

Example 8.43 Find the Laplace transform of the periodic function defined by the sawtooth wave

$$f(t) = t, \quad 0 \le t \le a, \quad f(t+a) = f(t).$$

Solution We have $T = a$. Therefore,

$$\mathcal{L}[f(t)] = \frac{1}{1-e^{-as}} \int_0^a te^{-st} dt = \frac{1}{1-e^{-as}} \left[-\left(\frac{t}{s} + \frac{1}{s^2} \right) e^{-st} \right]_0^a$$

$$= \frac{1}{1-e^{-as}} \left[-\left(\frac{a}{s} + \frac{1}{s^2} \right) e^{-as} + \frac{1}{s^2} \right] = \frac{1}{1-e^{-as}} \left[-\frac{a}{s} e^{-as} + \frac{1}{s^2}(1-e^{-as}) \right]$$

$$= \frac{1}{s^2} - \frac{ae^{-as}}{s(1-e^{-as})}, \quad s > 0.$$

Example 8.44   Using the Laplace transforms of periodic functions, verify that

$$\mathcal{L}[\sin \omega t] = \frac{\omega}{\omega^2 + s^2}.$$

Solution   The period of $\sin \omega t$ is $T = 2\pi/\omega$. Therefore,
$\mathcal{L}[\sin \omega t]$

$$= \frac{1}{1 - e^{-2\pi s/\omega}} \int_0^{2\pi/\omega} \sin \omega t \, e^{-st} dt = \frac{-1}{1 - e^{-2\pi s/\omega}} \left[\frac{s^2}{s^2 + \omega^2} \left(\frac{\sin \omega t}{s} + \frac{\omega \cos \omega t}{s^2} \right) e^{-st} \right]_0^{2\pi/\omega}$$

$$= - \frac{1}{1 - e^{-2\pi s/\omega}} \left[\frac{s^2}{s^2 + \omega^2} \left\{ \frac{\omega}{s^2} e^{-2\pi s/\omega} - \frac{\omega}{s^2} \right\} \right] = \frac{\omega}{s^2 + \omega^2}.$$

Example 8.45   Find the solution of the initial value problem

$$y'' + 8y' + 17y = f(t), \quad y(0) = 0, \quad y'(0) = 0$$

where $f(t)$ is the periodic function $f(t) = \begin{cases} 1, & 0 < t < \pi \\ 0, & \pi < t < 2\pi, \end{cases}$ and $f(t + 2\pi) = f(t)$.

Solution   Let $\mathcal{L}[y(t)] = Y(s)$ and $\mathcal{L}[f(t)] = F(s)$. Taking the Laplace transform of the differential equation, we get

$$[s^2 Y(s) - sy(0) - y'(0)] + 8 [sY(s) - y(0)] + 17 Y(s) = F(s), \quad s > 0.$$

where $F(s) = \dfrac{1}{1 - e^{-2\pi s}} \displaystyle\int_0^{2\pi} e^{-st} f(t) dt = \dfrac{1}{1 - e^{-2\pi s}} \displaystyle\int_0^{\pi} e^{-st} dt = \dfrac{1 - e^{-s\pi}}{s(1 - e^{-2\pi s})} = \dfrac{1}{s(1 + e^{-\pi s})}.$

Substituting the initial values, we obtain

$$Y(s) = \frac{1}{s(s^2 + 8s + 17)(1 + e^{-\pi s})} = \frac{1}{s(s^2 + 8s + 17)} (1 - e^{-\pi s} + e^{-2\pi s} - e^{-3\pi s} + \ldots).$$

Now,

$$\mathcal{L}^{-1}\left[\frac{1}{s^2 + 8s + 17} \right] = \mathcal{L}^{-1}\left[\frac{1}{(s + 4)^2 + 1} \right] = e^{-4t} \sin t$$

and

$$\mathcal{L}^{-1}\left[\frac{1}{s\{(s + 4)^2 + 1\}} \right] = \int_0^t e^{-4\tau} \sin \tau d\tau = \frac{1}{17} - \frac{1}{17} e^{-4t} (4 \sin t + \cos t) = \frac{1}{17} - g(t)$$

where $g(t) = e^{-4t}(4 \sin t + \cos t)/17$. Therefore,

$$y(t) = \left[\frac{1}{17} - g(t) \right] - \left[\frac{1}{17} - g(t - \pi) \right] u_\pi(t) + \left[\frac{1}{17} - g(t - 2\pi) \right] u_{2\pi}(t) - \ldots$$

Now,

$$g(t - \pi) = \frac{1}{17} e^{-4(t-\pi)} [4 \sin (t - \pi) + \cos (t - \pi)] = - e^{4\pi} g(t)$$

$$g(t - 2\pi) = \frac{1}{17} e^{-4(t-2\pi)} [4 \sin (t - 2\pi) + \cos (t - 2\pi)] = e^{8\pi} g(t), \ldots$$

Hence,

$$y(t) = \left[\frac{1}{17} - g(t) \right] u_0(t) - \left[\frac{1}{17} + e^{4\pi} g(t) \right] u_\pi(t) + \left[\frac{1}{17} - e^{8\pi} g(t) \right] u_{2\pi}(t) - \ldots$$

$$= \frac{1}{17} - g(t), \quad 0 < t < \pi,$$

$$= \frac{1}{17} - g(t) - \frac{1}{17} - e^{4\pi} g(t) = -(1 + e^{4\pi})g(t), \quad \pi < t < 2\pi,$$

$$= \left[\frac{1}{17} - g(t)\right] - \left[\frac{1}{17} + e^{4\pi} g(t)\right] + \left[\frac{1}{17} - e^{8\pi} g(t)\right] = \frac{1}{17} - (1 + e^{4\pi} + e^{8\pi})g(t), 2\pi < t < 3\pi,$$

....

$$= \frac{1}{34}[(-1)^n + 1] - (1 + A + A^2 + \ldots + A^n)g(t), \quad n\pi < t < (n+1)\pi, \quad A = e^{4\pi}.$$

Therefore, $$y(t) = \frac{1}{34}[(-1)^n + 1] - \frac{A^{n+1} - 1}{A - 1} g(t), \quad n = 0, 1, 2, \ldots \text{ and } A = e^{4\pi}.$$

Exercise 8.5

1. Using the Laplace transforms of periodic functions verify that
   $$\mathscr{L}[\cos \omega t] = s/(s^2 + \omega^2).$$

Find the Laplace transforms of the following functions which are defined graphically.

2. Square wave with period 2a, as given in Fig. 8.26.

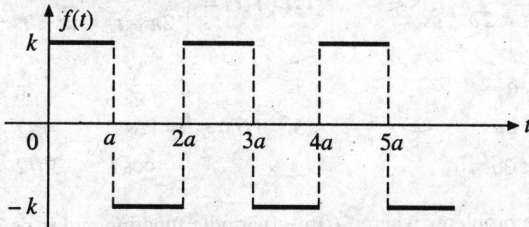

Fig. 8.26.   Problem 2.

3. Square wave with period 2a, as given in Fig. 8.27.

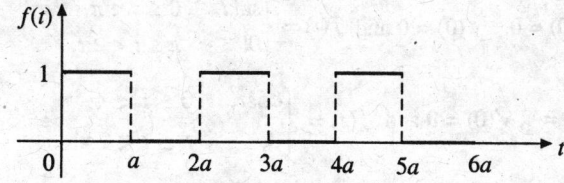

Fig. 8.27.   Problem 3.

4. Half wave rectification of sin t, as given in Fig. 8.28.

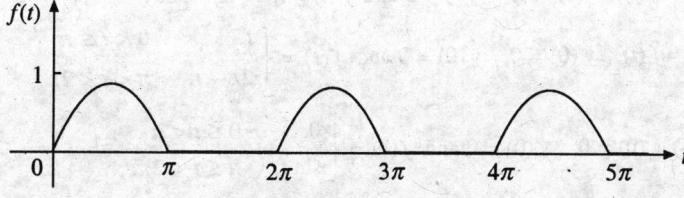

Fig. 8.28.   Problem 4.

5. Full wave rectification of $|\sin \omega t|$, as given in Fig. 8.29.

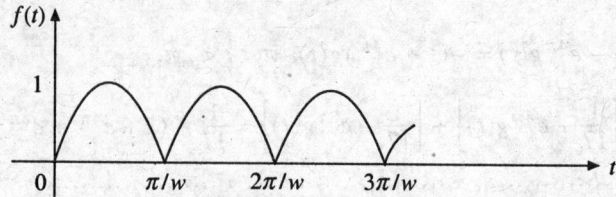

Fig. 8.29.   Problem 5.

Find the Laplace transforms of the periodic functions which are defined in one period as the following.

6. $f(t) = kt,\ \ 0 < t < 2\pi,\ \ k$ a constant.   **7.** $f(t) = t^2,\ \ 0 < t < 2\pi.$

8. $f(t) = \begin{cases} t, & 0 < t < a \\ 0, & a < t < 2a. \end{cases}$ **9.** $f(t) = \begin{cases} 1, & 0 < t < \pi \\ 0, & \pi < t < 2\pi \\ -1, & 2\pi < t < 3\pi \\ 0, & 3\pi < t < 4\pi. \end{cases}$

10. $f(t) = \begin{cases} 0, & 0 < t < a \\ t - a, & a < t < 2a. \end{cases}$ **11.** $f(t) = \begin{cases} t, & 0 < t \le \pi \\ 2\pi - t, & \pi < t < 2\pi. \end{cases}$

12. $f(t) = \begin{cases} 0, & 0 < t < a \\ 1, & a < t < 2a \\ 2, & 2a < t < 3a. \end{cases}$ **13.** $f(t) = \begin{cases} \cos t, & 0 < t < \pi/2 \\ -\cos t, & \pi/2 < t < 3\pi/2 \\ \cos t, & 3\pi/2 < t < 2\pi. \end{cases}$

Solve the following initial value problems where $f(t)$ is a periodic function and is defined in one period.

14. $y'' + 4y' + 5y = f(t),\ \ y(0) = 0,\ \ y'(0) = 0,$ and $f(t) = \begin{cases} 1, & 0 \le t < \pi \\ -1, & \pi \le t < 2\pi. \end{cases}$

15. $y'' + 4y = f(t),\ \ y(0) = 0,\ \ y'(0) = 0$ and $f(t) = \begin{cases} \sin t, & 0 \le t < \pi \\ 0, & \pi \le t < 2\pi. \end{cases}$

16. $y'' + 9y = f(t),\ y(0) = 1,\ y'(0) = 0$ and $f(t) = \begin{cases} \cos t, & 0 \le t < \pi \\ 0, & \pi \le t < 2\pi. \end{cases}$

17. $y'' + 6y' + 10y = f(t),\ \ y(0) = 0,\ \ y'(0) = 1$ and $f(t) = \begin{cases} 1, & 0 \le t < \pi \\ 0, & \pi \le t < 2\pi. \end{cases}$

18. $y'' + 3y' + 2y = f(t),\ \ y(0) = 0,\ \ y'(0) = 1$ and $f(t) = t,\ \ 0 \le t < a.$

19. $y'' + 4y' + 3y = f(t),\ \ y(0) = 0,\ \ y'(0) = 0$ and $f(t) = \begin{cases} t, & 0 < t \le \pi \\ 2\pi - t, & \pi < t < 2\pi. \end{cases}$

20. $y'' + y = f(t),\ \ y(0) = 0,\ \ y'(0) = 0$ and $f(t) = \begin{cases} 0, & 0 \le t < \pi \\ \sin t, & \pi \le t < 2\pi. \end{cases}$

8.7   Laplace Transform Methods for the Solution of Some Partial Differential Equations

In the previous sections, the inverse Laplace transform of $F(s)$ was obtained by one of the following methods.

(i) Writing partial fractions of $F(s)$ and for each fraction writing down the inverse Laplace transform.

(ii) Using the shift theorems, differentiation or integration of Laplace transform to find the inverse.

If we apply the Laplace transform method for solving an initial value problem in ordinary differential equations, we obtain an algebraic equation for the Laplace transform of the dependent variable. This equation is solved for the Laplace transform and its inverse gives the solution of the initial value problem.

Suppose now, that a partial differential equation along with its boundary and initial conditions is given. Laplace transform is applied with respect to one of the independent variables, say t. The partial derivatives with respect to this variable t are transformed using Laplace transforms. For partial derivatives with respect to other independent variables, we assume that operations of differentiation and integration of Laplace transforms can be interchanged. That is,

$$\mathcal{L}\left[\frac{\partial u(x, t)}{\partial x}\right] = \int_0^\infty \frac{\partial u}{\partial x} e^{-st}\, dt = \frac{\partial}{\partial x}\int_0^\infty e^{-st} u(x; t)\, dt = \frac{d}{dx}\left[\mathcal{L}\{u(x, t\}\right] \qquad (8.46)$$

which is a total derivative with respect to x, since $U(x, s) = \mathcal{L}[u(x, t)]$ is not a function of t. This results into an ordinary differential equation for $U(x, s)$ in which s is a parameter. This ordinary differential equation for $U(x, s)$ is solved using the analytic techniques. The arbitrary constants are functions of s. The inverse of $U(x, s)$ gives the solution of the given partial differential equation along with its conditions. However, due to complexity of the function $U(x, s)$, its inverse cannot usually be obtained by the methods described in the previous sections. In such cases, we use some special techniques or the *inversion theorem*. Application of the inversion theorem requires the knowledge of functions of a complex variable and the residue theorem.

First, we present an example in which the inverse can be obtained by the methods given in the earlier sections. Then, we give some examples which require the use of some special techniques. Finally, we present the inversion theorem and its application.

Example 8.46   Using Laplace transforms, find the solution of the initial value problem

$$x\frac{\partial u}{\partial t} + \frac{\partial u}{\partial x} = xt, \quad u(x, 0) = 0, \quad u(0, t) = t.$$

Solution   Denote $\mathcal{L}[u(x, t)] = U(x, s)$. Taking Laplace transforms (with respect to t) of the differential equation and the boundary condition $u(0, t) = t$, we obtain

$$x[sU(x, s) - u(x, 0)] + \frac{d}{dx} U(x, s) = \frac{x}{s^2}, \quad U(0, s) = \frac{1}{s^2}$$

or

$$\frac{dU}{dx} + sx\, U = \frac{x}{s^2}.$$

The integrating factor of this ordinary differential equation is $e^{sx^2/2}$. Integrating, we obtain

$$e^{sx^2/2}U = \int \frac{x}{s^2} e^{sx^2/2} dx + a(s) = \frac{1}{s^3} e^{sx^2/2} + a(s)$$

or

$$U(x, s) = \frac{1}{s^3} + a(s)e^{-sx^2/2}$$

where $a(s)$ is an arbitrary function of s.

Using the condition $U(0, s) = 1/s^2$, we obtain

$$U(0, s) = \frac{1}{s^2} = \frac{1}{s^3} + a(s), \quad \text{or} \quad a(s) = \frac{1}{s^2} - \frac{1}{s^3}.$$

Hence,

$$U(x, s) = \frac{1}{s^3} + \left(\frac{1}{s^2} - \frac{1}{s^3} \right) e^{-sx^2/2}.$$

Therefore,

$$u(x, t) = \frac{t^2}{2!} + \mathcal{L}^{-1} \left[\left(\frac{1}{s^2} - \frac{1}{s^3} \right) e^{-sx^2/2} \right].$$

Using the shift theorem, we obtain

$$u(x, t) = \frac{t^2}{2!} + \left[\left(t - \frac{x^2}{2} \right) - \frac{1}{2} \left(t - \frac{x^2}{2} \right)^2 \right] u_{x^2/2}(t)$$

$$= \begin{cases} \dfrac{t^2}{2}, & t < \dfrac{x^2}{2} \\[3mm] \dfrac{t^2}{2} + \left(t - \dfrac{x^2}{2} \right) - \dfrac{1}{2} \left(t - \dfrac{x^2}{2} \right)^2, & t \geq \dfrac{x^2}{2}. \end{cases}$$

Example 8.47 Consider the integral

$$f(\lambda) = \int_0^\infty \frac{e^{-z} e^{-\lambda/z}}{\sqrt{z}} dz.$$

and show that $f(\lambda)$ satisfies the differential equation $\sqrt{\lambda} f'(\lambda) + f(\lambda) = 0$. Hence, show that $f(\lambda) = \sqrt{\pi} e^{-2\sqrt{\lambda}}$. Using this result, prove the following

(i) $\mathcal{L}^{-1} \left[\dfrac{e^{-a\sqrt{s}}}{\sqrt{s}} \right] = \dfrac{1}{\sqrt{\pi t}} e^{-a^2/(4t)},$ (8.47)

(ii) $\mathcal{L}^{-1} \left[e^{-a\sqrt{s}} \right] = \dfrac{a}{2\sqrt{\pi t^3}} e^{-a^2/(4t)},$ (8.48)

(iii) $\mathcal{L}^{-1} \left[\dfrac{e^{-a\sqrt{s}}}{s} \right] = \dfrac{2}{\sqrt{\pi}} \int_{a/(2\sqrt{t})}^\infty e^{-u^2} du = \text{erfc} \left(\dfrac{a}{2\sqrt{t}} \right) = 1 - \text{erf} \left(\dfrac{a}{2\sqrt{t}} \right).$ (8.49)

Solution   We have

$$f'(\lambda) = \int_0^\infty \frac{e^{-z}}{\sqrt{z}}\left(-\frac{1}{z}e^{-\lambda/z}\right)dz = -\int_0^\infty \frac{e^{-z}e^{-\lambda/z}}{z^{3/2}}dz. \tag{8.50}$$

Set $v = \lambda/z$ in $f(\lambda)$. Then, we obtain

$$f(\lambda) = \int_\infty^0 \frac{e^{-\lambda/v}e^{-v}}{\sqrt{\lambda/v}}\left(-\frac{\lambda}{v^2}\right)dv = \sqrt{\lambda}\int_0^\infty \frac{e^{-v}e^{-\lambda/v}}{v^{3/2}}\,dv. \tag{8.51}$$

Using Eqs. (8.50) and (8.51), we obtain

$$f'(\lambda) = -\frac{f(\lambda)}{\sqrt{\lambda}}, \quad \text{or} \quad \sqrt{\lambda}\,f'(\lambda) + f(\lambda) = 0.$$

Write the equation as $\dfrac{f'(\lambda)}{f(\lambda)} = -\dfrac{1}{\sqrt{\lambda}}$.

Integrating, we obtain $\ln|f(\lambda)| = -2\sqrt{\lambda} + c_1$, or $f(\lambda) = ce^{-2\sqrt{\lambda}}$.

Using the condition that

$$f(0) = \int_0^\infty e^{-z}z^{-1/2}dz = \Gamma(1/2) = \sqrt{\pi}$$

we obtain $\quad f(0) = c = \sqrt{\pi}$. Therefore, $f(\lambda) = \sqrt{\pi}\,e^{-2\sqrt{\lambda}}$. $\tag{8.52}$

(i) Set $(\lambda/z) = a^2/(4t)$ in $f(\lambda)$. We obtain

$$f(\lambda) = \int_0^\infty \frac{e^{-4\lambda t/a^2}e^{-a^2/(4t)}}{\sqrt{4\lambda t/a^2}}\left(\frac{4\lambda}{a^2}\right)dt = \frac{2\sqrt{\lambda}}{a}\int_0^\infty \frac{e^{-4\lambda t/a^2}e^{-a^2/(4t)}}{\sqrt{t}}\,dt.$$

Let $s = 4\lambda/a^2$. We have

$$f(\lambda) = \sqrt{s}\int_0^\infty e^{-st}\left(\frac{e^{-a^2/(4t)}}{\sqrt{t}}\right)dt = \sqrt{s}\,\mathscr{L}\left[\frac{e^{-a^2/(4t)}}{\sqrt{t}}\right].$$

But, from Eq. (8.52), we have $f(\lambda) = \sqrt{\pi}\,e^{-2\sqrt{\lambda}} = \sqrt{\pi}\,e^{-a\sqrt{s}}$.

Hence, $\quad \mathscr{L}\left[\dfrac{e^{-a^2/(4t)}}{\sqrt{t}}\right] = \dfrac{\sqrt{\pi}}{\sqrt{s}}e^{-a\sqrt{s}}, \quad$ or $\quad \mathscr{L}^{-1}\left[\dfrac{e^{-a\sqrt{s}}}{\sqrt{s}}\right] = \dfrac{e^{-a^2/(4t)}}{\sqrt{\pi t}}.$

(ii) Denote $F(s) = e^{-a\sqrt{s}}/\sqrt{s}$. Then

$$\mathscr{L}^{-1}[F(s)] = \frac{e^{-a^2/(4t)}}{\sqrt{\pi t}} = f(t).$$

Now,
$$\int_s^\infty F(s*)ds* = \int_s^\infty \frac{e^{-a\sqrt{s*}}}{\sqrt{s*}}ds* = \left[-\frac{2}{a}e^{-a\sqrt{s*}}\right]_s^\infty = \frac{2}{a}e^{-a\sqrt{s}}.$$

Therefore,
$$\mathcal{L}\left[\frac{f(t)}{t}\right] = \int_s^\infty F(s*)ds* = \frac{2}{a}e^{-a\sqrt{s}}.$$

Hence,
$$\mathcal{L}^{-1}[e^{-a\sqrt{s}}] = \frac{a}{2\sqrt{\pi t^3}}e^{-a^2/(4t)}.$$

(iii) Denote $F(s) = e^{-a\sqrt{s}}$ and $f(t) = \mathcal{L}^{-1}[F(s)]$.

Using the result $\dfrac{1}{s}\mathcal{L}[f(t)] = \mathcal{L}\left[\displaystyle\int_0^t f(\tau)d\tau\right]$,   we obtain

$$\mathcal{L}^{-1}\left[\frac{F(s)}{s}\right] = \int_0^t \frac{a}{2\sqrt{\pi\tau^3}}e^{-a^2/(4\tau)}d\tau.$$

Let $z = \dfrac{a}{2\sqrt{\tau}}$ or $z^2 = \dfrac{a^2}{4\tau}$. Then, $d\tau = -\dfrac{a^2}{2z^3}dz.$

Therefore,
$$\mathcal{L}^{-1}\left[\frac{F(s)}{s}\right] = \frac{a}{2\sqrt{\pi}}\int_\infty^{a/(2\sqrt{t})} \frac{e^{-z^2}}{(a^3/8z^3)}\left(\frac{-a^2}{2z^3}\right)dz$$

$$= \frac{2}{\sqrt{\pi}}\int_{a/(2\sqrt{t})}^\infty e^{-z^2}dz = erfc\left(\frac{a}{2\sqrt{t}}\right)$$

where *erfc* is the complimentary error function. We can also write

$$erfc\left(\frac{a}{2\sqrt{t}}\right) = \frac{2}{\sqrt{\pi}}\int_0^\infty e^{-z^2}dz - \frac{2}{\sqrt{\pi}}\int_0^{a/(2\sqrt{t})} e^{-z^2}dz = 1 - erf\left(\frac{a}{2\sqrt{t}}\right)$$

where, we have used the result $\displaystyle\int_0^\infty e^{-z^2}dz = \frac{\sqrt{\pi}}{2}$ and the definition of the error function

$$erf\left(\frac{a}{2\sqrt{t}}\right) = \frac{2}{\sqrt{\pi}}\int_0^{a/(2\sqrt{t})} e^{-z^2}dz.$$

Hence,
$$\mathcal{L}^{-1}\left[\frac{e^{-a\sqrt{s}}}{s}\right] = erfc\left(\frac{a}{2\sqrt{t}}\right) = 1 - erf\left(\frac{a}{2\sqrt{t}}\right).$$

Example 8.48   The temperature distribution in a semi-infinite, thin, insulated rod is governed by the one-dimensional heat equation

$$\frac{\partial u}{\partial t} = c^2 \frac{\partial^2 u}{\partial x^2}.$$

Assume that the left end of the rod is maintained at an arbitrary time dependent temperature $u(0, t) = f(t)$. Initially, the rod was at zero temperature $u(x, 0) = 0$. Find the distribution of the heat flow, if the temperature is bounded as $x \to \infty$.

Solution   The initial condition is $u(x, 0) = 0$, $0 \le x < \infty$, and the boundary condition is $u(0, t) = f(t)$, $t > 0$ and $u(x, t)$ is finite as $x \to \infty$.

Taking Laplace transform of the differential equation, with respect to t, we get

$$\mathcal{L}\left[\frac{\partial u}{\partial t}\right] = \mathcal{L}\left[c^2 \frac{\partial^2 u}{\partial x^2}\right].$$

Denote $\mathcal{L}[u(x, t)] = U(x, s)$. Then

$$sU(x, s) - u(x, 0) = c^2 \frac{d^2}{dx^2} U(x, s)$$

or

$$c^2 \frac{d^2 U(x, s)}{dx^2} - sU(x, s) = 0.$$

The solution of this differential equation is

$$U(x, s) = A(s)\, e^{(\sqrt{s}/c)x} + B(s) e^{-(\sqrt{s}/c)x}.$$

Applying Laplace transform to the condition that $u(x, t)$ is finite as $x \to \infty$, we obtain that $U(x, s)$ is finite as $x \to \infty$. Hence, $A(s) = 0$ and

$$U(x, s) = B(s)\, e^{-(\sqrt{s}/c)x}. \qquad (8.53)$$

Applying Laplace transforms to the boundary condition $u(0, t) = f(t)$, we obtain

$$U(0, s) = \mathcal{L}[f(t)] = F(s).$$

Substituting in Eq. (8.53), we obtain

$$U(0, s) = B(s) = F(s).$$

Hence, $U(x, s) = F(s)e^{-(\sqrt{s}/c)x} = F(s)G(s). \qquad (8.54)$

where $G(s) = e^{-(\sqrt{s}/c)x} = \mathcal{L}[g(t)].$

Setting $a = x/c$ and using the convolution theorem, we obtain (using the result given in Eq. (8.48))

$$u(x, t) = \int_0^t g(\tau) f(t - \tau) d\tau = \int_0^t \frac{a}{2\sqrt{\pi \tau^3}} e^{-a^2/(4\tau)} f(t - \tau) d\tau$$

$$= \frac{x}{2c\sqrt{\pi}} \int_0^t \frac{f(t - \tau)}{\tau^{3/2}} e^{-x^2/(4c^2\tau)} d\tau. \tag{8.55}$$

We could also have written $u(x, t)$ as

$$u(x, t) = \int_0^t f(\tau) g(t - \tau) d\tau = \int_0^t f(\tau) \frac{a}{2\sqrt{\pi}(t - \tau)^{3/2}} e^{-a^2/[4(t-\tau)]} d\tau$$

$$= \frac{x}{2c\sqrt{\pi}} \int_0^t \frac{f(\tau)}{(t - \tau)^{3/2}} e^{-x^2/[4c^2(t-\tau)]} d\tau.$$

Example 8.49 If in Example 8.48 $f(t)$ is the unit step temperature, that is, $f(t) = u_\tau(t)$, then find the temperature distribution in the rod.

Solution We have $f(t) = u_\tau(t) = \begin{cases} 0, & \text{if} \quad t < \tau \\ 1, & \text{if} \quad t \geq \tau \end{cases}$

that is, $f(t) = 1$, if $\tau \leq t$. Substituting in Eq. (8.55), we obtain

$$u(x, t) = \frac{x}{2c\sqrt{\pi}} \int_0^t \frac{1}{\tau^{3/2}} e^{-x^2/(4c^2\tau)} d\tau$$

Substitute $\quad \dfrac{x^2}{4c^2\tau} = T^2$. Then $\dfrac{x^2}{4c^2\tau^2} d\tau = -2T \, dT$.

We obtain

$$u(x, t) = \frac{x}{2c\sqrt{\pi}} \int_\infty^{x/(2c\sqrt{t})} \frac{e^{-T^2}}{\tau^{3/2}} \left(-\frac{8c^2\tau^2 T}{x^2} \right) dT$$

$$= \frac{4c}{x\sqrt{\pi}} \int_{x/(2c\sqrt{t})}^\infty e^{-T^2} \tau^{1/2} T \, dT = \frac{2}{\sqrt{\pi}} \int_{x/(2c\sqrt{t})}^\infty e^{-T^2} dT$$

$$= erfc\left(\frac{x}{2c\sqrt{t}} \right) = 1 - erf\left(\frac{x}{2c\sqrt{t}} \right).$$

We now state the *inversion theorem* and give its applications.

Theorem 8.13 (**Inversion theorem**) Let $f(t)$ have a continuous derivative and $|f(t)| < k e^{\alpha t}$, where k and α are positive real constants. Let $F(s)$ be the Laplace transform of $f(t)$, that is

$$F(s) = \mathscr{L}[f(t)] = \int_0^\infty e^{-st} f(t) dt, \quad Re(s) > \alpha. \tag{8.56}$$

Then, the inverse Laplace transform is given by

$$f(t) = \mathcal{L}^{-1}[F(s)] = \frac{1}{2\pi i} \lim_{\omega \to \infty} \int_{\gamma - i\omega}^{\gamma + i\omega} e^{st} F(s)\, ds, \quad \gamma > \alpha. \tag{8.57}$$

In the limit, we obtain the improper integral $\displaystyle\int_{\gamma - i\infty}^{\gamma + i\infty} e^{st} F(s)\, ds$.

The line integral in Eq. (8.57), is evaluated in the complex s-plane by converting it into an integral over a closed contour and then applying the residue theorem. This transformation is possible, when $F(s)$ satisfies certain conditions. We state this result in the following theorem.

Theorem 8.14   Let $F(s)$ satisfy the condition

$$|F(s)| < MR^{-k}, \quad \text{when} \quad s = Re^{i\theta}, -\pi \le \theta \le \pi$$

where $R > R_0$; R_0, M and $k > 0$ are constants. Let C be the contour taken first along the straight line path AB: Re $(s) = \gamma > 0$ and then along the curved path BPQRA: $s = Re^{i\theta}$. Then, $\displaystyle\int e^{st} F(s)\, ds, t > 0$ taken along the curves BPQ and ARQ tend to zero as $R \to \infty$ (Fig. 8.30).

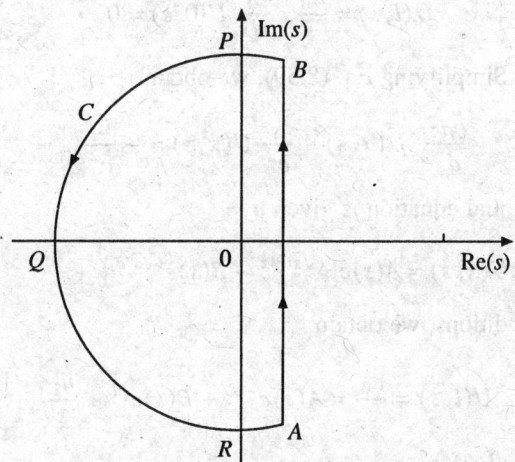

Fig. 8.30.   Theorem 8.14.

In all applications, $F(s)$ satisfies the conditions of the theorem and it is analytic except at a finite number of poles all of which lie to the left of the line Re $(s) = \gamma$. Therefore, the line integral from $\gamma - i\infty$ to $\gamma + i\infty$ of the Inversion theorem is replaced by the contour integral given in Fig. 8.30, when $R \to \infty$. The integrand of this contour integral is analytic inside C except at a finite number of poles inside C. The contour integral is then evaluated using the residue theorem. Therefore, we have

$$f(t) = \frac{1}{2\pi i} \lim_{\omega \to \infty} \int_{r - i\omega}^{r + i\omega} e^{st} F(s)\, ds = \frac{1}{2\pi i} \lim_{R \to \infty} \int_C e^{st} f(s)\, ds$$

$$= \Sigma \, [\text{residues of } e^{st} F(s) \text{ at poles of } F(s) \text{ inside } C]. \tag{8.58}$$

Example 8.50　Consider heat flow in a thin rod of length l, $0 \le x \le l$. The end $x = l$ is maintained at a constant temperature u_0. The other end $x = 0$ is insulated. The rod was initially at the temperature u_1. Find the temperature distribution $u(x, t)$ in the rod, where $u(x, t)$ is governed by the partial differential equation

$$\frac{\partial u}{\partial t} = c^2 \frac{\partial^2 u}{\partial x^2}, \quad 0 < x < l, \ t > 0.$$

Solution　The conditions in the problem are the following:

Initial condition:
$$u(x, 0) = u_1, \quad 0 < x \le l.$$

Boundary conditions:
$$u(l, t) = u_0, \quad t > 0; \quad \frac{\partial u}{\partial x}(0, t) = 0, \quad t > 0.$$

Taking Laplace transform of the differential equation and the boundary conditions with respect to t, we obtain

$$s U(x, s) - u(x, 0) = c^2 \frac{d^2}{dx^2} U(x, s) \tag{8.59}$$

$$U(l, s) = \frac{u_0}{s}, \quad \frac{d}{dx} U(0, s) = 0$$

where $\mathcal{L}[u(x, t)] = U(x, s)$. Simplifying Eq. (8.59), we obtain

$$\frac{d^2}{dx^2} U(x, s) - \frac{s}{c^2} U(x, s) = -\frac{u_1}{c^2}.$$

The solution of this differential equation is given by

$$U(x, s) = A(s) e^{-(\sqrt{s}/c)x} + B(s) e^{(\sqrt{s}/c)x} + \frac{u_1}{s}. \tag{8.60}$$

Applying the boundary conditions, we obtain

$$U(l, s) = \frac{u_0}{s} = A(s) e^{-pl} + B(s) e^{pl} + \frac{u_1}{s}$$

$$\frac{dU(0, s)}{dx} = 0 = A(s)(-p) + B(s)(p)$$

where $p = \sqrt{s}/c$. The second equation gives $A(s) = B(s)$. Substituting in the first equation, we obtain

$$A(s)[2 \cosh(pl)] = \frac{1}{s}(u_0 - u_1) \quad \text{or} \quad A(s) = \frac{u_0 - u_1}{2s \cosh(pl)}.$$

Substituting in Eq. (8.60), we obtain the solution for $U(x, s)$ as

$$U(x, s) = \frac{(u_0 - u_1) \cosh px}{s \cosh(pl)} + \frac{u_1}{s}. \tag{8.61}$$

By inversion theorem, we have the solution as

$$u(x, t) = u_1 + \frac{u_0 - u_1}{2\pi i} \int_{\gamma - i\infty}^{\gamma + i\infty} e^{st} F(s) ds = u_1 + \frac{u_0 - u_1}{2\pi i} \int_C e^{st} F(s) ds \tag{8.62}$$

where

$$F(s) = [\cosh (\sqrt{s}/c)x]/[s \cosh (\sqrt{s}/c)l]$$

and C is the closed contour as given in Fig. 8.30. The integrand has singularities at $s = 0$ and when

$$\cosh (\sqrt{s}/c)l = 0 \quad \text{or} \quad e^{2(\sqrt{s}/c)l} = -1 \quad \text{or} \quad 2(\sqrt{s}/c) l = (2n + 1)\pi i$$

or

$$s = s_n = -\frac{(2n + 1)^2 \pi^2 c^2}{4l^2}, \quad n = 0, 1, 2, \dots \tag{8.63}$$

All the singularites are simple poles and lie to the left of the line $s = \gamma_0 > 0$. It can be shown that as $R \to \infty$, the integral over BQA (see Fig. 8.30) tends to zero. Hence, from Eq. (8.62), we obtain

$$u(x, t) = u_1 + (u_0 - u_1) [\Sigma \text{ residues of } e^{st}F(s) \text{ at poles } s = 0 \text{ and those given in Eq. (8.63)}]$$

Now,

$$[\text{residue at } s = 0] = \lim_{s \to 0} \left[\frac{e^{st} \cosh (\sqrt{s}/c)x}{\cosh (\sqrt{s}/c)l} \right] = 1.$$

Let $e^{st} F(s) = p(s)/q(s)$. Then, the residue at a simple pole $s = s_n$ is given by

$$[\text{residue at } s = s_n] = \frac{p(s_n)}{q'(s_n)}$$

where

$$p(s_n) = e^{s_n t} \cosh [(2n + 1) \pi i x/(2l)] = e^{-(2n+1)^2 \pi^2 c^2 t/(4l^2)} \cos \left[\frac{(2n + 1)\pi x}{2l} \right]$$

$$q'(s_n) = [\cosh (\sqrt{s_n}/c)l] + \frac{\sqrt{s_n} \, l}{2c} [\sinh (\sqrt{s_n}/c)l]$$

$$= \cosh \left[\frac{(2n + 1)\pi i}{2} \right] + \frac{(2n + 1)\pi i}{4} \sinh \left[\frac{(2n + 1)\pi i}{2} \right]$$

$$= \cos \left[\frac{(2n + 1)\pi}{2} \right] - \frac{(2n + 1)\pi}{4} \sin \left[\frac{(2n + 1)\pi}{2} \right] = \frac{(-1)^{n+1} (2n + 1)\pi}{4}.$$

Therefore,

$$[\text{residue at } s = s_n] = \frac{(-1)^{n+1} 4}{(2n + 1) \pi} e^{-(2n+1)^2 \pi^2 c^2 t/(4l^2)} \cos \left[\frac{(2n + 1)\pi x}{2l} \right], \quad n = 0, 1, 2, \dots$$

The solution of the given problem is

$$u(x, t) = u_0 + \frac{4(u_0 - u_1)}{\pi} \sum_{n=0}^{\infty} \frac{(-1)^{n+1}}{(2n + 1)} e^{-(2n+1)^2 \pi^2 c^2 t/(4l^2)} \cos \left[\frac{(2n + 1)\pi x}{2l} \right].$$

Alternative

We have from Eq. (8.61)

$$U(x, s) = \frac{u_1}{s} + \frac{(u_0 - u_1) \cosh px}{s \cosh pl} = \frac{u_1}{s} + \frac{(u_0 - u_1)}{s} \frac{e^{px} + e^{-px}}{e^{pl} + e^{-pl}}$$

$$= \frac{u_1}{s} + \frac{(u_0 - u_1)}{s} \frac{(e^{px} + e^{-px})}{e^{pl}} [1 + e^{-2pl}]^{-1}$$

$$= \frac{u_1}{s} + \frac{(u_0 - u_1)}{s} [e^{-p(l-x)} + e^{-p(l+x)}] \sum_{n=0}^{\infty} (-1)^n e^{-2npl}$$

$$= \frac{u_1}{s} + (u_0 - u_1) \sum_{n=0}^{\infty} \frac{(-1)^n}{s} [e^{-p(l-x+2nl)} + e^{-p(l+x+2nl)}]$$

$$= \frac{u_1}{s} + (u_0 - u_1) \sum_{n=0}^{\infty} \frac{(-1)^n}{s} [e^{-(l-x+2nl)\sqrt{s}/c} + e^{-(l+x+2nl)\sqrt{s}/c}]$$

Using Eq. (8.49), we can write the inverse as

$$u(x, t) = u_1 + (u_0 - u_1) \sum_{n=0}^{\infty} (-1)^n \left[erfc \left(\frac{l - x + 2nl}{2c\sqrt{t}} \right) + erfc \left(\frac{l + x + 2nl}{2c\sqrt{t}} \right) \right].$$

Example 8.51 Using Laplace transforms solve the initial boundary value problem

$$u_{tt} = u_{xx}, \quad 0 < x < l, \quad t > 0$$

$$u(x, 0) = 0, \, u_t(x, 0) = \sin(\pi x/l), \quad 0 < x < l.$$

$$u(0, t) = 0, \quad u(l, t) = 0, \quad t > 0.$$

Solution Taking Laplace transform (with respect to t) of the differential equation and the boundary conditions, we obtain

$$s^2 U(x, s) - su(x, 0) - u_t(x, 0) = \frac{d^2}{dx^2} U(x, s)$$

$$U(0, s) = 0, \quad U(l, s) = 0$$

where $U(x, s) = \mathcal{L}[u(x, t)]$. Substituting the initial conditions, we obtain

$$\frac{d^2 U}{dx^2} - s^2 U(x, s) = - \sin \left(\frac{\pi x}{l} \right). \tag{8.64}$$

The complimentary function is $U_c(x, s) = A(s) \cosh sx + B(s) \sinh sx$.

The particular integral is $\qquad U_p(x, s) = \dfrac{\sin px}{p^2 + s^2}, \quad p = \dfrac{\pi}{l}.$

Hence, the complete solution of Eq. (8.64) is given by

$$U(x, s) = A(s) \cosh sx + B(s) \sinh sx + \frac{\sin px}{p^2 + s^2}.$$

Applying the boundary conditions, we get

$$U(0, s) = 0 = A(s), \quad \text{and} \quad U(l, s) = 0 = B(s).$$

Therefore,
$$U(x, s) = \frac{\sin px}{p^2 + s^2}.$$

Inverse Laplace transform gives

$$u(x, t) = \frac{1}{p} \sin px \sin pt = \frac{l}{\pi} \sin\left(\frac{\pi x}{l}\right) \sin\left(\frac{\pi t}{l}\right).$$

Example 8.52 A stretched elastic string of length l has its ends fixed at $x = 0$ and $x = l$. Initially, the string is pulled to a distance h, at a point P which is at a distance of b from the origin and released. Find the displacement $u(x, t)$. It is given that $u(x, t)$ is governed by the partial differential equation

$$u_{tt} = c^2 u_{xx}, \quad 0 < x < l, \quad t > 0.$$

Solution The conditions on the problem are as follows.

Initial conditions:
$$u(x, 0) = f(x) = hx/b, \quad 0 \le x \le b$$
$$= h(l - x)/(l - b), \quad b \le x \le l \text{ (Fig. 8.31)}.$$

$$\frac{\partial u}{\partial t}(x, 0) = 0. \tag{8.65}$$

Boundary conditions:
$$u(0, t) = 0 = u(l, t), \quad t > 0. \tag{8.66}$$

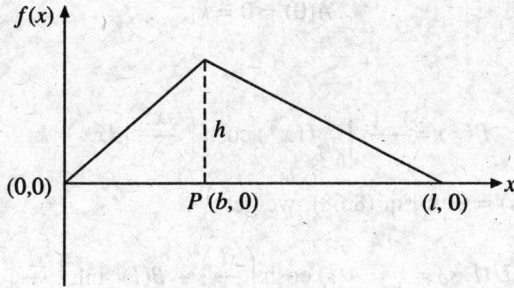

Fig. 8.31. Example 8.52.

Taking the Laplace transforms of the differential equation and the boundary conditions, we obtain

$$s^2 U(x, s) - su(x, 0) - u_t(x, 0) = c^2 \frac{d^2}{dx^2} U(x, s)$$

$$U(0, s) = 0 = U(l, s)$$

where $U(x, s) = \mathcal{L}[u(x, t)]$. Substituting the initial conditions, we obtain

$$\frac{d^2 U}{dx^2} - \frac{s^2}{c^2} U = -\frac{s}{c^2} f(x). \tag{8.67}$$

The solution of this ordinary differential equation can be obtained by the method of variation of parameters. The solution of the homogeneous equation is

$$U(x, s) = A \cosh (sx/c) + B \sinh (sx/c).$$

Let the solution of the non-homogeneous equation be

$$U(x, s) = A(x) \cosh (sx/c) + B(x) \sinh (sx/c). \tag{8.68}$$

Using the variation of parameters, we obtain the equations

$$A' \cosh (sx/c) + B' \sinh (sx/c) = 0$$

$$A'(x) \sinh \left(\frac{sx}{c}\right) + B'(x) \cosh \left(\frac{sx}{c}\right) = - \frac{s}{c^2} f(x)$$

or $\qquad\qquad A' \sinh \left(\frac{sx}{c}\right) + B' \cosh \left(\frac{sx}{c}\right) = - \frac{1}{c} f(x).$

Solving these equations, we obtain

$$A'(x) = \frac{1}{c} f(x) \sinh \left(\frac{sx}{c}\right) \quad \text{and} \quad B'(x) = - \frac{1}{c} f(x) \cosh \left(\frac{sx}{c}\right).$$

Integrating $A'(x)$, we get

$$A(x) = \frac{1}{c} \int_0^x f(x^*) \sinh \left(\frac{sx^*}{c}\right) dx^* + k_1. \tag{8.69}$$

Using the condition $U(0, s) = 0$ in Eq. (8.68), we get $A(0) = 0$. Hence, from Eq. (8.69), we get

$$A(0) = 0 = k_1.$$

Integrating $B'(x)$, we get

$$B(x) = - \frac{1}{c} \int_0^x f(x^*) \cosh \left(\frac{sx^*}{c}\right) dx^* + k_2. \tag{8.70}$$

Using the condition $U(l, s) = 0$ in Eq. (8.68), we get

$$U(l, s) = 0 = A(l) \cosh \left(\frac{sl}{c}\right) + B(l) \sinh \left(\frac{sl}{c}\right).$$

Solving for $B(l)$, we obtain

$$B(l) = - A(l) \coth \left(\frac{sl}{c}\right).$$

Hence, from Eq. (8.70), we get

$$B(l) = - A(l) \coth \left(\frac{sl}{c}\right) = - \frac{1}{c} \int_0^l f(x^*) \cosh \left(\frac{sx^*}{c}\right) dx^* + k_2.$$

Substituting for $A(l)$, we obtain

$$k_2 = \frac{1}{c} \int_0^l f(x^*) \cosh \left(\frac{sx^*}{c}\right) dx^* - \frac{1}{c} \coth \left(\frac{sl}{c}\right) \int_0^l f(x^*) \sinh \left(\frac{sx^*}{c}\right) dx^*.$$

Let $p = s/c$. From Eq. (8.70), the solution for $B(x)$ is

$$B(x) = -\frac{1}{c} \int_0^x f(x^*) \cosh (px^*)\, dx^* + \frac{1}{c} \int_0^l f(x^*) \cosh (px^*)\, dx^*$$

$$-\frac{1}{c} \coth (pl) \int_0^l f(x^*) \sinh (px^*)\, dx^*$$

$$= \frac{1}{c} \int_x^l f(x^*) \cosh (px^*)\, dx^* - \frac{1}{c} \coth (pl) \int_0^l f(x^*) \sinh (px^*)\, dx^*.$$

Therefore, the solution of the ordinary differential equation (8.67) is given by

$$U(x, s) = \frac{1}{c} \cosh (px) \int_0^x f(x^*) \sinh (px^*)\, dx^* + \frac{1}{c} \sinh (px) \int_x^l f(x^*) \cosh (px^*)\, dx^*$$

$$-\frac{1}{c} \sinh (px) \coth (pl) \int_0^l f(x^*) \sinh (px^*)\, dx^*$$

or $c \sinh (pl)\, U(x, s) = \sinh (pl) \cosh (px) \int_0^x f(x^*) \sinh (px^*)\, dx^* + \sinh (pl) \sinh (px) \times$

$$\int_x^l f(x^*) \cosh (px^*)\, dx^* - \sinh (px) \cosh (pl) \int_0^l f(x^*) \sinh (px^*)\, dx^*.$$

Writing the integral

$$\int_0^l \quad \text{as} \quad \int_0^x + \int_x^l$$

and combining the terms, we obtain

$$c \sinh (pl)\, U(x, s) = [\sinh (pl) \cosh (px) - \sinh (px) \cosh (pl)] \int_0^x f(x^*) \sinh (px^*)\, dx^*$$

$$+ \sinh (px) \int_x^l [\sinh (pl) \cosh (px^*) - \cosh (pl) \sinh (px^*)] f(x^*)\, dx^*$$

$$= \sinh p(l - x) \int_0^x f(x^*) \sinh (px^*)\, dx^* + \sinh (px) \int_x^l \sinh p(l - x^*) f(x^*)\, dx^*$$

$$= I_1 + I_2.$$

We write

$$I_1 = \int_0^x f(x^*) \sinh p(l - x) \sinh (px^*)\, dx^* = \frac{1}{2} \int_0^x f(x^*)(\cosh \theta_1 - \cosh \theta_2)\, dx^*$$

where $\theta_1 = p(l - x + x^*)$ and $\theta_2 = p(x^* - l + x)$. Similarly, write

$$I_2 = \int_x^l f(x^*) \sinh(px) \sinh p(l - x^*) dx^* = \frac{1}{2} \int_x^l f(x^*)(\cosh \theta_1^* - \cosh \theta_2) dx^*$$

where $\theta_1^* = p(x + l - x^*)$.

Substituting for $f(x^*)$ from Eq. (8.65) and integrating, we obtain

$$U(x, s) = \frac{hx}{bs} - \frac{hlc}{b(l - b)s^2} \left[\frac{\sinh(px) \sinh p(l - b)}{\sinh(pl)} \right], \quad 0 \le x \le b \qquad (8.71(a))$$

$$= \frac{h(l - x)}{s(l - b)} - \frac{hlc}{b(l - b)s^2} \left[\frac{\sinh(pb) \sinh p(l - x)}{\sinh(pl)} \right], \quad b \le x \le l \qquad (8.71(b))$$

(see Problem 16, Exercise 8.6). Note that by replacing x by $l - x$, and b by $l - b$ in the expression on the right hand side of Eq. (8.71a), we obtain the expression on the right hand side of Eq. (8.71b). Consider now, inversion of $U(x, s)$ in the interval $0 \le x \le b$. The first term (hx/bs) on the right hand side of Eq. (8.71a) gives (hx/b).

For finding the inverse of $[\sinh(px) \sinh p(l - b)]/[s^2 \sinh(pl)]$, we use the inversion theorem and consider

$$\frac{1}{2\pi i} \int_{\gamma - i\infty}^{\gamma + i\infty} \frac{e^{st} \sinh(px) \sinh p(l - b)}{\sinh(pl) s^2} ds = \frac{1}{2\pi i} \int_C \frac{e^{st} \sinh(px) \sinh p(l - b)}{\sinh(pl) s^2} ds \qquad (8.72)$$

where C is the contour given in Fig. 8.30. It can be shown that, as $n \to \infty$, the integral over the arc BQA tends to zero. Then,

  Value of the integral in Eq. (8.72) = Sum of residues of the integrand at poles inside C.

The integrand has poles at $s = 0$ and when

$$\sinh pl = 0 \quad \text{or} \quad pl = \frac{sl}{c} = n\pi i, \quad \text{or} \quad s = s_n = \frac{n\pi ci}{l}, \quad n = 0, \pm 1, \pm 2, \ldots.$$

Now, $s = 0$ is a simple pole (the degree of s in the leading term of the expansion of the numerator of the integrand is 2 and the degree of s in the denominator is 3). The remaining poles s_n, $n = \pm 1$, $\pm 2, \ldots$ are also simple poles. We have

$$[\text{Residue at } s = 0] = \lim_{s \to 0} \left[\frac{\sinh[(x/c)s] \sinh[(l - b)s/c]}{s \sinh[sl/c]} \right]$$

$$= \lim_{s \to 0} \left[\frac{\left\{ \frac{xs}{c} + O(s^3) \right\}\left\{ \frac{(l - b)s}{c} + O(s^3) \right\}}{s\left\{ \frac{ls}{c} + O(s^3) \right\}} \right] = \left(\frac{x}{c} \right)\left(\frac{l - b}{c} \right)\left(\frac{c}{l} \right) = \frac{x(l - b)}{cl}$$

$$[\text{Residue at } s = s_n, \quad n = 1, 2, \ldots] = \frac{p(s_n)}{q'(s_n)}$$

where $p(s)$ and $q(s)$ are the numerator and denominator respectively. Now

$$p(s_n) = \sinh\left[\frac{x}{c} \cdot \frac{n\pi c}{l} i \right] \sinh\left[\frac{l - b}{c} \cdot \frac{n\pi c}{l} i \right] e^{s_n t}$$

$$= \left[i \sin \left(\frac{n\pi x}{l} \right) \right] \left[i \sin \frac{n\pi(l-b)}{l} \right] e^{(n\pi cti)/l}, \quad n = 1, 2, 3, \ldots$$

$$q'(s_n) = s_n^2 \cosh \left(\frac{l}{c} s_n \right) \left(\frac{l}{c} \right) + 2 s_n \sinh \left(\frac{l}{c} s_n \right)$$

$$= - \left(\frac{c^2}{l^2} \right) n^2 \pi^2 \left(\frac{l}{c} \right) \cosh \left[\frac{l}{c} \cdot \frac{n\pi ci}{l} \right] = - \frac{n^2 \pi^2 c}{l} (-1)^n, \quad n = 1, 2, 3, \ldots.$$

Substituting $-n$ for n, we obtain expressions for $p(s_n)$ and $q'(s_n)$, $n = -1, -2, \ldots$. Therefore,

$$\text{Sum of residues} = \frac{x(l-b)}{cl} + \sum_{n=1}^{\infty} \frac{2 \sin(n\pi x/l) \sin(n\pi\{(l-b)/l\}) \cos(n\pi ct/l)}{n^2 \pi^2 (c/l)(-1)^n}$$

where we have used the identity $2 \cos(n\pi ct/l) = e^{(n\pi cti)/l} + e^{-(n\pi cti/l)}$.
Hence,

$$u(x, t) = \frac{hx}{b} - \frac{hlc}{b(l-b)} \left[\frac{x(l-b)}{cl} + \sum_{n=1}^{\infty} \frac{2l(-1)^n}{n^2 \pi^2 c} \sin \left(\frac{n\pi x}{l} \right) \sin \left(\frac{n\pi(l-b)}{l} \right) \cos \left(\frac{n\pi ct}{l} \right) \right]$$

$$= \frac{2hl^2}{b(l-b)\pi^2} \sum_{n=1}^{\infty} \frac{1}{n^2} \sin \left(\frac{n\pi x}{l} \right) \sin \left(\frac{n\pi b}{l} \right) \cos \left(\frac{n\pi ct}{l} \right) \tag{8.73}$$

since $\sin \left(\frac{n\pi(l-b)}{l} \right) = \sin \left(n\pi - \frac{n\pi b}{l} \right) = -\cos(n\pi) \sin \left(\frac{n\pi b}{l} \right) = -(-1)^n \sin \left(\frac{n\pi b}{l} \right)$.

To find the solution in the interval $b \leq x \leq l$, replace x by $l - x$, b by $l - b$ in Eq. (8.73). We obtain, for $b \leq x \leq l$

$$u(x, t) = \frac{2hl^2}{b(l-b)\pi^2} \sum_{n=1}^{\infty} \frac{1}{n^2} \sin \left(\frac{n\pi(l-x)}{l} \right) \sin \left(\frac{n\pi(l-b)}{l} \right) \cos \left(\frac{n\pi ct}{l} \right)$$

Since $\sin \left(\frac{n\pi(l-x)}{l} \right) = \sin \left(n\pi - \frac{n\pi x}{l} \right) = -\cos(n\pi) \sin \left(\frac{n\pi x}{l} \right) = -(-1)^n \sin \left(\frac{n\pi x}{l} \right)$

and $$\sin \left(\frac{n\pi(l-b)}{l} \right) = -(-1)^n \sin \left(\frac{n\pi b}{l} \right)$$

we obtain the same expression as in Eq. (8.73). Hence, the solution $u(x, t)$ given by Eq. (8.72) is valid in the whole interval $[0, l]$.

Exercise 8.6

Solve by the Laplace transforms method, the following first order initial-boundary value problems.

1. $\dfrac{\partial u}{\partial x} + x \dfrac{\partial u}{\partial t} = 0$, $\quad u(x, 0) = 1$, $\quad u(0, t) = t$.

2. $\dfrac{\partial u}{\partial x} + cx\dfrac{\partial u}{\partial t} = 0$, $u(x, 0) = 0$, $u(0, t) = pt^2$, c and p are constants.

3. $\dfrac{\partial u}{\partial x} + x\dfrac{\partial u}{\partial t} = xt^2$, $u(x, 0) = 0$, $u(0, t) = t$.

4. $\dfrac{\partial u}{\partial x} + x\dfrac{\partial u}{\partial t} = 4xt$, $u(x, 0) = 0$, $u(0, t) = 0$.

5. $\dfrac{\partial u}{\partial x} + 6x\dfrac{\partial u}{\partial t} = 12x$, $u(x, 0) = 2$, $u(0, t) = 2$.

Solve the following initial-boundary value problems.

6. $u_t = Ku_{xx}$, $x > 0$, $t > 0$.

   $u(x, 0) = 0$, $u(0, t) = 1$, $u(x, t)$ is finite as $x \to \infty$.

7. $u_t = Ku_{xx}$, $x > 0$, $t > 0$.

   $u(x, 0) = A$, $u(0, t) = B$, A and B constants. $u(x, t)$ is finite as $x \to \infty$.

8. $u_t = Ku_{xx}$, $x > 0$, $t > 0$. $u(x, t)$ is bounded as $x \to \infty$.

   $u(x, 0) = u_0, u(0, t) = \begin{cases} u_1, & 0 < t < T \\ 0, & t \geq T. \end{cases}$

9. $u_t = Ku_{xx}$, $x > 0$, $t > 0$. $u(x, t)$ is bounded as $x \to \infty$.
   $u(0, t) = t$, $u(x, 0) = 0$.

10. $u_t = Ku_{xx} + g$, $x > 0$, $t > 0$, g is a constant. $u(x, t)$ is bounded as $x \to \infty$

    $u(0, t) = 0$, $u(x, 0) = 0$.

11. $u_t = Ku_{xx}$, $0 < x < 1$, $t > 0$.

    $u(x, 0) = u_0, 0 < x < 1$; $u(1, t) = u_1$, $u_x(0, t) = 0$.

12. $u_t = Ku_{xx}$, $0 < x < 1$, $t > 0$.

    $u(x, 0) = 0$, $0 < x < 1$; $u(0, t) = 1$, $u(1, t) = 0$.

Solve the following initial-boundary value problems.

13. $u_{tt} = u_{xx}$, $x > 0$, $t > 0$

    $u(x, 0) = 0$, $u_t(x, 0) = e^{-x}$, $u(0, t) = \sin t$, $u(x, t)$ is bounded as $x \to \infty$.

14. $u_{tt} = u_{xx}$, $0 < x < 1$, $t > 0$

    $u(x, 0) = 0$, $u_t(x, 0) = \sin 3\pi x$, $0 < x < 1$, $u(0, t) = 0$, $u(1, t) = 0$.

15. $u_{tt} = u_{xx}$, $x > 0$, $t > 0$

    $u(x, 0) = 0$, $u_t(x, 0) = 0$,

    $u(0, t) = \begin{cases} \sin t, 0 \leq t \leq \pi, \\ 0, t > \pi. \end{cases}$ $u(x, t)$ is bounded as $x \to \infty$.

16. Derive the Eqs. (8.71a) and (8.71b) in Example 8.52.

17. $u_{tt} = u_{xx}$, $x > 0$, $t > 0$

    $u(x, 0) = e^{-x}$, $u_t(x, 0) = 0$, $u(0, t) = 0$, $u(x, t)$ is bounded as $x \to \infty$.

18. $u_{tt} = c^2u_{xx}$, $0 < x < l$, $t > 0$,

    $u(0, t) = 0$, $u(l, t) = 0$,

$$u_t(x, 0) = 0, u(x, 0) = f(x) = \begin{cases} \dfrac{2hx}{l}, 0 \le x \le \dfrac{l}{2} \\ \dfrac{2h}{l}(l-x), \dfrac{l}{2} \le x \le l. \end{cases}$$

19. $u_{tt} = c^2 u_{xx} - g$, $x > 0$, $t > 0$, g is acceleration due to gravity

  $u(x, 0) = 0$, $u_t(x, 0) = 0$; $u(0, t) = 0$, $u_x \to 0$ as $x \to \infty$.

20. $u_{tt} = c^2 u_{xx} + g$, $0 < x < l$, $t > 0$, g is acceleration due to gravity

  $u(x, 0) = 0$, $u_t(x, 0) = 0$; $u(0, t) = 0$, $u_x(l, t) = 0$.

8.8 Answers and Hints

Exercise 8.1

1. $\dfrac{2}{s^2} - \dfrac{5}{s}$.

2. $\dfrac{2a}{s^3} + \dfrac{b}{s^2} + \dfrac{c}{s}$.

3. $\dfrac{4}{s} + \dfrac{12}{s^2} + \dfrac{18}{s^3}$.

4. $(s \sin 2 + 3 \cos 2)/(s^2 + 9)$.

5. $(s \cos b - a \sin b)/(s^2 + a^2)$.

6. $f(t) = (1 - \cos 2\, \omega t)/2$; $2\omega^2/[s(s^2 + 4\omega^2)]$.

7. $f(t) = (e^{2t} - 2 + e^{-2t})/4$; $2/[s(s^2 - 4)]$, $s > 2$.

8. $f(t) = (e^{2t} + 2 + e^{-2t})/4$; $(s^2 - 2)/[s(s^2 - 4)]$, $s > 2$.

9. $(s - 1)/(s^2 + 1)$.

10. $(6s^2 + 24 - s^4)/[s^3(s^2 + 4)]$.

11. Integrate by parts; $1/(s - 2)^2$, $s > 2$.

12. $e^5/(s + 2)$, $s > -2$.

13. Integrate by parts; $2/(s - 2)^3$, $s > 2$.

14. $\sin t = \text{Im}\,(e^{it})$; integrate by parts; $2s/(s^2 + 1)^2$.

15. $\cos 2t = \text{Re}\,(e^{2it})$; integrate by parts; $(s^2 - 4)/(s^2 + 4)^2$.

16. $\cos t = \text{Re}\,(e^{it})$; $(s - 1)/[(s - 1)^2 + 1]$.

17. $\sin t = \text{Im}\,(e^{it})$; $1/[(s + 1)^2 + 1]$.

18. $(1/s) + [2/(s^2 + 4)]$.

19. $(1 - 2e^{-s})/s$.

20. $2(1 - e^{-3s})/s$.

21. $s(1 + e^{-s\pi})/(s^2 + 1)$.

22. $-e^{-s\pi}/(s^2 + 1)$.

23. $(2/s^3) + (6/s^2) + (9/s)$.

24. $(5s - 18)/[s(s - 3)]$.

25. $(6/s^3) - 5/(s + 2) + (6/s)$.

26. $[5/(s^2 + 25)] + [s/(s^2 + 16)]$.

27. $1/(s - 2)$, $s > 2$.

28. $f(t) = (1 - e^{-2t})/2$; $1/[s(s + 2)]$.

29. $f(t) = (1 + e^{2t})/2$; $(s - 1)/[s(s - 2)]$.

30. $\sin t = \text{Im}\,(e^{it})$; $1/[(s - 1)^2 + 1]$.

31. $3e^{-5t}$.

32. $\sin(\pi t)$.

33. $\cosh(2t)$.

34. t^3.

35. Use partial fractions; $(4e^t - e^{-2t})/3$.

36. $t^2 + 6t - 5$.

37. $3(1 - e^{-2t})/2$.

38. $13 \sinh(3t)/3$.

39. Use partial fractions; $4e^t - 13e^{2t} + 10e^{3t}$.

40. Use partial fractions; $(1/4)\, e^{-t} - (2/5)e^{-2t} + (3/20)\, e^{3t}$.

41. Substitute $st = u$.

42. Substitute $a = 5/2$ in Problem 41.

43. Substitute $at = u$.

44. Substitute $u = t - nT$ and use $f(u + nT) = f(u)$.

45. $F(s) = 1/[s - (a + ib)] = [(s - a) + ib]/[(s - a)^2 + b^2]$. Compare the real and imaginary parts.

Exercise 8.2

1. $f(t) = \cos^2 2t$, $f(0) = 1$, $f'(t) = -2 \sin 4t$; $F(s) = (s^2 + 8)/[s(s^2 + 16)]$.

2. $f(t) = te^{at}$, $f(0) = 0$, $f'(t) = af(t) + e^{at}$; $F(s) = 1/(s - a)^2$.

3. $f(t) = t \sin at$, $f(0) = 0$, $f'(0) = 0$, $f'' = -a^2 f(t) + 2a \cos at$; $F(s) = 2as/(s^2 + a^2)^2$.

4. $f(t) = t \cos at$, $f(0) = 0$, $f'(0) = 1$, $f'' = -a^2 f(t) - 2a \sin at$; $F(s) = (s^2 - a^2)/(s^2 + a^2)^2$.

5. $f(t) = t \sinh at$, $f(0) = 0$, $f'(0) = 0$, $f'' = a^2 f(t) + 2a \cosh at$; $F(s) = 2as/(s^2 - a^2)^2$.

6. $f(t) = \cos at$, $f(0) = 1$, $f' = -a \sin at$.   **7.** $f(t) = \sin at$, $f(0) = 0$, $f' = a \cos at$.

8. $f'(t) = [(\sin at)/a]' = \cos at$.   **9.** $f'(t) = (e^{-t} - e^{-2t})' = 2e^{-2t} - e^{-t}$.

10. $\mathcal{L}^{-1}[1/(4s^2 + 1)] = [\sin (t/2)]/2 = f(t)$; $f'(t) = [\cos (t/2)]/4$.

11. $\mathcal{L}^{-1}\left[\dfrac{1}{(s^2 + 1)(s^2 + 4)}\right] = \dfrac{1}{3}\left[\sin t - \dfrac{1}{2} \sin 2t\right] = f(t)$; $f'(t) = \dfrac{1}{3}[\cos t - \cos 2t]$.

12. $\mathcal{L}^{-1}\left[\dfrac{1}{(s - 1)(s^2 + 1)}\right] = \dfrac{1}{2}[e^t - \cos t - \sin t]$; $f'(t) = \dfrac{1}{2}[e^t + \sin t - \cos t]$.

13. Write as $F(s)/s$ where $F(s) = 1/(s + 5)$; $(1 - e^{-5t})/5$.

14. Write as $F(s)/s$ where $F(s) = 16/(s^2 + 9)$; $16(1 - \cos 3t)/9$.

15. Write as $\dfrac{1}{s} - \dfrac{5}{s(s + 1)} = \dfrac{1}{s} - \dfrac{F(s)}{s}$ where $F(s) = \dfrac{5}{s + 1}$; $-\dfrac{1}{3}(2 - 5e^{-3t})$.

16. Write as $\dfrac{1}{s}\left[\dfrac{1}{s} F(s)\right]$ where $F(s) = 1/(s + 1)$; $t + e^{-t} - 1$.

17. Write as $\dfrac{1}{s}\left[\dfrac{1}{s} F(s)\right]$ where $F(s) = \dfrac{\omega}{s^2 + \omega^2}$; $\dfrac{1}{\omega^2}(\omega t - \sin \omega t)$.

18. Write as $\dfrac{1}{s}\left[\dfrac{1}{s}\left\{\dfrac{s}{s^2 + a^2} - \dfrac{a}{s^2 + a^2}\right\}\right]$; $\dfrac{1}{a^2}[\sin at - \cos at + 1] - \dfrac{t}{a}$.

19. Write as $\dfrac{1}{s}\left[\dfrac{1}{s}\left\{\dfrac{1}{s^2 + 1} - \dfrac{1}{s^2 + 4}\right\}\right]$; $\dfrac{3}{4} t - \sin t + \dfrac{1}{8} \sin 2t$.

20. Write as $\dfrac{1}{s}\left[\dfrac{1}{s}\left\{\dfrac{1}{s} \cdot \dfrac{1}{s + 3}\right\}\right] = \dfrac{1}{18}(3t^2 - 2t) + \dfrac{1}{27}(1 - e^{-3t})$.

21. Deduce from the discussion after Eq. (8.8).

22. Write $\displaystyle\int_a^t f(\tau)d\tau = \int_0^t f(\tau)d\tau - \int_0^a f(\tau)d\tau$. The second integral is a constant.

23. $Y(s) = \dfrac{1}{3s} + \dfrac{2}{3(s + 3)}$, $y(t) = \dfrac{1}{3} + \dfrac{2}{3}e^{-3t}$.

24. $Y(s) = -\dfrac{1}{16s} - \dfrac{1}{4s^2} - \dfrac{15}{16(s - 4)}$, $y(t) = -\dfrac{1}{16} - \dfrac{1}{4}t - \dfrac{15}{16} e^{4t}$.

25. $Y(s) = -\dfrac{s}{5(s^2+1)} + \dfrac{2}{5(s^2+1)} + \dfrac{6}{5(s+2)}$, $y(t) = -\dfrac{1}{5}\cos t + \dfrac{2}{5}\sin t + \dfrac{6}{5}e^{-2t}$.

26. $Y(s) = -\dfrac{3}{10(s+3)} + \dfrac{3s}{10(s^2+1)} + \dfrac{1}{10(s^2+1)}$, $y(t) = -\dfrac{3}{10}e^{-3t} + \dfrac{3}{10}\cos t + \dfrac{1}{10}\sin t$.

27. $Y(s) = \dfrac{11}{4(s-2)} - \dfrac{3}{4s} - \dfrac{1}{2s^2}$, $y(t) = \dfrac{11}{4}e^{2t} - \dfrac{3}{4} - \dfrac{t}{2}$.

28. $Y(s) = \dfrac{s}{s^2+1} + \dfrac{1}{s^2} - \dfrac{1}{s^2+1}$, $y(t) = \cos t + t - \sin t$.

29. $Y(s) = \dfrac{3}{2s} - \dfrac{1}{2(s+2)}$, $y(t) = \dfrac{3}{2} - \dfrac{1}{2}e^{-2t}$.

30. $Y(s) = \dfrac{11}{12(s+1)} - \dfrac{20}{21(s+4)} + \dfrac{1}{28(s-3)}$, $y(t) = \dfrac{11}{12}e^{-t} - \dfrac{20}{21}e^{-4t} + \dfrac{1}{28}e^{3t}$.

31. $Y(s) = -\dfrac{1}{3(s-2)} - \dfrac{5}{12(s-5)} + \dfrac{7}{4(s-1)}$, $y(t) = \dfrac{7}{4}e^{t} - \dfrac{5}{12}e^{5t} - \dfrac{1}{3}e^{2t}$.

32. $Y(s) = \dfrac{8s}{65(s^2+1)} - \dfrac{1}{65(s^2+1)} - \dfrac{11}{5(s-1/2)} + \dfrac{27}{13(s-3/2)}$

$y(t) = \dfrac{8}{65}\cos t - \dfrac{1}{65}\sin t - \dfrac{11}{5}e^{t/2} + \dfrac{27}{13}e^{3t/2}$.

33. $Y(s) = -\dfrac{3s}{10(s^2+1)} - \dfrac{1}{10(s^2+1)} + \dfrac{1}{2(s-1)} + \dfrac{8}{5(2s+1)}$,

$y(t) = -\dfrac{3}{10}\cos t - \dfrac{1}{10}\sin t + \dfrac{1}{2}e^{t} + \dfrac{4}{5}e^{-t/2}$.

34. $Y(s) = \dfrac{1}{3s} + \dfrac{5}{3(s-3)} - \dfrac{1}{s-1}$, $y(t) = \dfrac{1}{3} - e^{t} + \dfrac{5}{3}e^{3t}$.

35. $Y(s) = \dfrac{1}{5s} - \dfrac{1}{4(s+1)} + \dfrac{21}{20(s+5)}$, $y(t) = \dfrac{1}{5} - \dfrac{1}{4}e^{-t} + \dfrac{21}{20}e^{-5t}$.

36. $Y(s) = -\dfrac{7s}{50(s^2+1)} - \dfrac{1}{50(s^2+1)} + \dfrac{63}{50(s-3)} + \dfrac{22}{25(s+2)}$,

$y(t) = -\dfrac{7}{50}\cos t - \dfrac{1}{50}\sin t + \dfrac{63}{50}e^{3t} + \dfrac{22}{25}e^{-2t}$.

37. $Y_1(s) = \dfrac{1}{4s^2} - \dfrac{s}{s^2+4} + \dfrac{7}{4(s^2+4)}$, $y_1(t) = \dfrac{t}{4} - \cos 2t + \dfrac{7}{8}\sin 2t$,

$Y_2(s) = -\dfrac{3}{4s} + \dfrac{7s}{4(s^2+4)} + \dfrac{4}{s^2+4}$, $y_2(t) = -\dfrac{3}{4} + \dfrac{7}{4}\cos 2t + 2\sin 2t$.

38. $Y_1(s) = \dfrac{3}{s^2+9} - \dfrac{3s}{s^2+9} + \dfrac{9}{s(s^2+9)}$, $y_1(t) = 1 + \sin 3t - 4\cos 3t$,

$Y_2(s) = \dfrac{s}{s^2+9} + \dfrac{9}{s^2+9}$, $y_2(t) = \cos 3t + 3\sin 3t$.

39. $Y_2(s) = \dfrac{s}{s^2+9} + \dfrac{9}{s^2+9}$, $y_2(t) = \cos 3t + 3\sin 3t$.

$$Y_2(s) = \frac{2}{s^2+1} - \frac{s}{s^2+1} + \frac{1}{s}, \quad y_2(t) = 2\sin t - \cos t + 1.$$

40. $Y_1(s) = \dfrac{2}{s^3(1+s^2)} + \dfrac{2}{s(1+s^2)} - \dfrac{1}{1+s^2}.$

$y_1(t) = 2(t^2/2 + \cos t - 1) + 2(1 - \cos t) - \sin t = t^2 - \sin t.$

$Y_2(s) = \dfrac{1}{s^2(1+s^2)} + \dfrac{s}{1+s^2} + \dfrac{1}{1+s^2}, \quad y_2(t) = (t - \sin t) \overset{?}{+} \cos t + \sin t = t + \cos t.$

Exercise 8.3

1. $\dfrac{2}{(s-2)^3} - \dfrac{2}{(s-2)^2} - \dfrac{3}{s-2}.$

2. $\dfrac{120}{(s+4)^6}.$

3. $\dfrac{2}{(s-3)^3} - \dfrac{4}{(s-3)^2} + \dfrac{4}{(s-3)}.$

4. $\dfrac{5}{s^2 - 2s + 26}.$

5. $\dfrac{s+3}{s^2 + 6s + 18}.$

6. $\dfrac{s}{s^2 + 2s + 2}.$

In Problems 7 to 12 write sinh *at*, cosh *at* in terms of exponential functions.

7. $\dfrac{6(s^4 + 54s^2 + 81)}{(s^2 - 9)^4}.$

8. $\dfrac{s^2 - 2}{s^4 + 4}.$

9. $\dfrac{2(3s^2 + 1)}{(s^2 - 1)^3}.$

10. $\dfrac{s^3}{s^4 + 4\omega^4}.$

11. $\dfrac{2s}{s^4 + 4}.$

12. $\dfrac{\omega(s^2 + 2\omega^2)}{s^4 + 4\omega^4}.$

13. $(e^{-3t} \sin \sqrt{6}t)/\sqrt{6}.$

14. $(e^{2t} \sin 4t)/4.$

15. $(\cos 2t - \sin 2t) e^{-2t}.$

16. $(5 + 11t)e^t.$

17. $(t + t^2)e^{2t}.$

18. $(3 \cos 2t + 14 \sin 2t) e^{4t}.$

19. $(6t - 7t^2) e^{-3t}/2.$

20. $(2t + 6t^2 + 3t^3) e^{2t}/2.$

21. $(3\sqrt{3} \cos \sqrt{3}t - 4 \sin \sqrt{3}t)e^{-3t}/\sqrt{3}.$

22. $1 + [e^{-t} - e^t + 2te^t]/4.$

23. $[(16 - 85t) e^{-3t} + (40t - 16) e^{2t}]/125.$

24. $e^{-4t}(6 + 36t + t^3)/6.$

25. $[e^{-t} + (15 \sin 2t - \cos 2t) e^{-3t}]/8.$

26. $[(2 \cos t + \sin t) + e^t (\sin t - 2 \cos t)]/5.$

27. $[6 + 10t + (17 \sin 2t + 194 \cos 2t) e^{-t}]/50.$

28. $[16 + 6t + \{182 \sinh (t/2) - 16 \cosh (t/2)\}e^t]/27.$

29. $f(t) = u_a(t) - u_b(t), \; (e^{-as} - e^{-bs})/s.$

30. $f(t) = a[u_0(t) - u_a(t)], \quad a(1 - e^{-as})/s.$

31. $f(t) = t[u_0(t) - u_1(t)], \quad [1 - e^{-s}(s+1)]/s^2.$

32. $f(t) = u_a(t) - u_{2a}(t) + u_{3a}(t) - u_{4a}(t) + \dots, \quad e^{-as/2}/[2s \cosh (as/2)].$

33. $e^{-s/2}/[(2s) \sinh (s/2)].$

34. $e^{-s}(1 - e^{-s})^2/s^2.$

35. $e^{-2s}(2 + 4s + 3s^2)/s^3.$

36. $2e^{-s}(1 + s + s^2)/s^3.$

37. $e^{-s}(s \cos 1 - \sin 1)/(s^2 + 1).$

38. $-e^{-\pi s}/(s^2 + 1).$

39. $se^{-3s}/(s^2 + 1)$.

40. $e^{-3s}/(s + 1)$.

41. $5(1 - 2e^{-2s})/s$.

42. $[1 - e^{-3s}(1 + 3s)]/s^2$.

43. $se^{-\pi s/2}/(s^2 + 1)$.

44. $k[1 - e^{-2s} + e^{-4s}]/s$.

45. $[2 - e^{-s}(2 + 2s + s^2)]/s^3$.

46. $[e^{-s} - e^{-2s}(1 + s)]/s^2$.

47. $u_1(t)\,(t - 1)^2/2$.

48. $-\cos t\, u_{\pi/2}(t)$.

49. $-\cos 3t\, u_\pi(t)$.

50. $e^{-5t}[u_0(t) + 2e^{5\pi}u_\pi(t) + e^{10\pi}u_{2\pi}(t)]$.

51. $-e^{4t}\sin 3t\,[e^{-4\pi}u_\pi(t) + e^{-8\pi}u_{2\pi}(t)]/3$.

52. $\cos 2t\,[u_0(t) - u_{\pi/2}(t)]$.

53. $(1 + 2e^{-t})u_0(t) - 2[1 - e^{-(t-1)}]u_1(t)$.

54. $[(3t - 1 + 19e^{-3t})\,u_0(t) + (1 - 3t + 2e^{-3(t-1)})u_1(t)]/9$.

55. $[(1 + e^{-2t})\,u_0(t) - (1 - e^{-2(t-2)})\,u_2(t)]/2$.

56. $(3e^{-2t} - 6te^{-2t})\,u_0(t) + (t - 1)\,e^{-2(t-1)}u_1(t)$.

57. $(4e^{-4t} - 3e^{-3t})u_0(t) + u_2(t)\,(1 + 20\,e^{-3(t-2)} - 21\,e^{-4(t-2)})/12$.

58. $2\cos 2t\, u_0(t) + [1 - \cos 2(t - 1) + 2\sin 2\,(t - 1)]\,u_1(t)/4$.

59. $u_0(t)\cos 2t + [\{1 - \cos 2(t - 1)\}\,u_1(t) - \{1 - \cos 2(t - 2)\}\,u_2(t)]/4$.

60. $e^t u_0(t) + [1 - 2e^{(t-1)} + e^{2(t-1)}]u_1(t)/2$.

61. $(1/6)\,[1 + 2e^{-3t} - 3e^{-2t}]u_0(t) - (1/6)\,[1 + 2e^{-3(t-3)} - 3e^{-2(t-3)}]\,u_3(t) - (1/6)\,[1 + 2e^{-3(t-5)} - 3e^{-2(t-5)}]\,u_5(t)$.

62. $(1/3)\,(3e^{-t} - 2e^{-3t} - 1)\,u_0(t) + (1/6)[2 + e^{-3(t-3)} - 3e^{-(t-3)}]\,u_3(t)$.

63. $(v/R)\,[e^{-p(t-1)}u_1(t) - e^{-p(t-2)}\,u_2(t)], \quad p = 1/(RC)$.

64. $(5/R)\,[1 - e^{-p(t-2)}]\,u_2(t), \quad p = R/L$.

65. $k[e^{-pt} - \cos t + p\sin t]\,u_0(t) - k[p\sin t - \cos t - pe^{-p(t-\pi/2)}]u_{\pi/2}(t)$.

$\quad p = R/L$ and $k = 1/[L(1 + p^2)]$.

Exercise 8.4

1. 0.   **2.** $1/\sqrt{2}$.   **3.** 2.   **4.** $e^{-2(t-3)}\sin(t - 3)\,u_3(t)$.

5. $[\sin 4\,(t - 2)\,u_2(t)]/4$.

6. $2e^{-(t-2)}\sin 3(t - 2)\,u_2(t) - e^{-(t-3)}\sin 3(t - 3)\,u_3(t)$.

7. $[4\sin 3t]/3$.

8. $8e^{-2(t-1)}\sin 2(t - 1)\,u_1(t) + 4e^{-2(t-2)}\sin 2(t - 2)\,u_2(t)$.

9. $2\cos 2t + 4\sin 2\,(t - \pi/6)\,u_{\pi/6}(t)$.

10. $4s/(s^2 + 4)^2$.

11. $2/(s - 3)^3$.

12. $18(s^2 + 3)/(s^2 - 9)^3$.

13. $6(s + 4)/(s^2 + 8s + 25)^2$.

14. $2(s^3 + 5s^2 + 8s + 13)/[s^2(s^2 + 4s + 13)^2]$.

15. $2(3s^2 + 4s + 5)/[s^2(s^2 + 2s + 5)^2]$.

16. $2(3s^2 + 15s + 25)/[s^3(s + 5)^3]$.

17. te^{-at}.

18. $t^2 e^{at}/2$.

19. $2te^{-2t}\sin 2t$.

20. $(\sin 4t - 4t\cos 4t)/128$.

21. $(3t\sinh 4t)/4$.

22. $3te^{-t} \sin t - (e^{-t} \sin t - te^{-t} \cos t)/2$. Write the given expression as

$$\frac{6(s+1)}{[(s+1)^2+1]^2} - \frac{1}{2}\left[\frac{1}{(s+1)^2+1} - \frac{(s+1)^2-1}{\{(s+1)^2+1\}^2}\right].$$

23. $-3 + (ct^2/2)$, c arbitrary constant.

24. $-t + (ct^3/6)$, c arbitrary constant.

25. $(6t^2 - t^4)/4$.

26. $-2(3t + 4t^3)/3$.

27. $(4 + 25\ t^2)/2$.

28. $2(5t + 20t^3 + 12t^5)/5$.

29. $2 + (16 - c)\ te^{-4t}$, c arbitrary constant.

30. $(1 + 3t)\ e^{-6t} + c\ [-1 + 3t + (1 + 3t)\ e^{-6t}]/108$, c arbitrary constant.

31. $2(1 + 4t)\ e^{-8t} + c\ [-1 + 4t + (1 + 4t)\ e^{-8t}]/256$, c arbitrary constant.

32. $\dfrac{1}{2} \ln\left(\dfrac{s+1}{s-1}\right).$

33. $\ln\left(\dfrac{\sqrt{s^2+b^2}}{s}\right).$

34. $\cot^{-1}[(s + 2)/3]$.

35. $t(1 - 2t)e^{-4t}$.

36. $(\sin 3t - 3t \cos 3t)/3$.

37. $2(1 - \cos t)/t$.

38. $-(2 \sinh t)/t$.

39. $(1 + e^{-t} - 2 \cos t)/t$.

40. $(\sin t\ )/t$.

41. $\displaystyle\int_0^t \frac{\sin \tau}{\tau}\, d\tau$

42. $(\sinh t)/t$.

43. $f(t) = \begin{cases} \cos t, & (\pi/2) \le t \le \pi \\ 0, & \text{otherwise.} \end{cases}$

46. $(1 - e^{-2t})/2$.

47. $(e^{at} - 1 - at)/a^2$.

48. $(e^{at} - e^{bt})/(a - b)$.

49. $[\sin \omega t - \omega t \cos \omega t]/(2\omega)$.

50. $(e^{at} - e^{bt})/(a - b)$.

51. $(4t - \sin 4t)/64$.

52. $(\sin 3t - 3t \cos 3t)/54$.

53. $(t \sin 2t)/4$.

54. $(\cos 2t - \cos 3t)/5$.

55. $6\ [(t - 1)e^{-t} + e^{-2t}]$.

56. $(60\ t \sin t - 8 \cos t + 8 \cos 4t)/225$.

57. $[\{2 + (a - b)t\}\ e^{-at} - \{2 - (a - b)\ t\}\ e^{-bt}]/(a - b)^3$.

In Problems 58 to 70, use convolution theorem for one or both terms of $Y(s)$.

58. $(t \sin 4t)/8$.

59. $(t - 2)e^{-t} + 2e^{-2t}$.

60. $(3t + 2)e^{-2t} + (t - 2)e^{-t}$.

61. $[2\omega \cosh \omega t + 4 \sinh \omega t + t \sinh \omega t]/(2\omega)$.

62. $[(34 - 7t)\ e^{-t} + 15\ e^{6t}]/49$.

63. $[t \sin t + \cos t - 1]\ e^t$.

64. $[t \cos t - (1 - t) \sin t]/2$.

65. $[-6 + 7t + 6e^{7t}]/49$.

66. $[3e^t + 2te^t + e^{-t}]/4$.

67. $[2e^t - t - 1]$.

68. $2e^{-t} - e^{-t/2} [\cos (\sqrt{3}t/2) + (1/\sqrt{3}) \sin (\sqrt{3}t/2)]$.

69. $\cos t + \sin t$.

70. $\dfrac{1}{8}\left[1 + 2t - \left\{\cosh\left(\dfrac{\sqrt{17}}{4}t\right) + \dfrac{1}{\sqrt{17}} \sinh\left(\dfrac{\sqrt{17}}{4}t\right)\right\}e^{-t/4}\right].$

Exercise 8.5

1. $T = 2\pi/\omega$.

2. $f(t) = \begin{cases} k, 0 < t < a \\ -k, a < t < 2a \end{cases}$, $T = 2a$, $F(s) = \frac{k}{s} \tanh\left(\frac{as}{2}\right)$.

3. $f(t) = \begin{cases} 1, 0 < t < a \\ 0, a < t < 2a \end{cases}$, $T = 2a$, $F(s) = \frac{1}{s(1 + e^{-as})}$.

4. $f(t) = \begin{cases} \sin t, 0 < t < \pi \\ 0, \pi < t < 2\pi \end{cases}$, $T = 2\pi$, $F(s) = \frac{1}{(s^2+1)(1-e^{-\pi s})}$.

5. $f(t) = \begin{cases} \sin \omega t, 0 < t < \pi/\omega \\ -\sin \omega t, \pi/\omega < t < 2\pi/\omega \end{cases}$, $T = \frac{2\pi}{\omega}$, $F(s) = \frac{\omega \coth[\pi s/(2\omega)]}{s^2 + \omega^2}$.

6. $\dfrac{k}{s^2} - \dfrac{2k\pi e^{-2\pi s}}{s(1 - e^{-2\pi s})}$,

7. $\dfrac{2}{s^3} - \dfrac{4\pi(\pi s + 1)e^{-2\pi s}}{s^2(1 - e^{-2\pi s})}$.

8. $\dfrac{(1 - e^{-as}) - as\, e^{-as}}{s^2(1 - e^{-2as})}$,

9. $\dfrac{1 - e^{-\pi s} - e^{-2\pi s} + e^{-3\pi s}}{s(1 - e^{-4\pi s})} = \dfrac{1 - e^{-\pi s}}{s(1 + e^{-2\pi s})}$.

10. $\dfrac{e^{-as} - (as + 1)e^{-2as}}{s^2(1 - e^{-2as})}$.

11. $\dfrac{1 - e^{-\pi s}}{s^2(1 + e^{-\pi s})}$.

12. $\dfrac{e^{-as} + e^{-2as} - 2e^{-3as}}{s(1 - e^{-3as})} = \dfrac{e^{-as}(1 + 2e^{-as})}{s(1 + e^{-as} + e^{-2as})}$.

13. $\left[\dfrac{1}{s} + \dfrac{2e^{-\pi s/2}}{s^2(1 - e^{-\pi s})}\right]\left(\dfrac{s^2}{1 + s^2}\right)$.

14. $Y(s) = \dfrac{1}{s[(s+2)^2 + 1]}\left[\dfrac{2}{1 + e^{-\pi s}} - 1\right] = \dfrac{1}{s[(s+2)^2 + 1]}[1 - 2e^{-\pi s} + 2e^{-2\pi s} - \ldots]$

$\mathcal{L}^{-1}\left[\dfrac{1}{s\{(s+2)^2 + 1\}}\right] = \int_0^t e^{-2t} \sin t\, dt = \frac{1}{5}[1 - e^{-2t}(2\sin t + \cos t)] = \frac{1}{5}[1 - g(t)]$

$y(t) = \frac{1}{5}[1 - g(t)]u_0(t) - \frac{2}{5}[1 - g(t - \pi)]u_\pi(t) + \frac{2}{5}[1 - g(t - 2\pi)]u_{2\pi}(t) - \ldots$

15. $Y(s) = \dfrac{1}{(s^2+1)(s^2+4)}[1 + e^{-\pi s} + e^{-2\pi s} + \ldots]$

$y(t) = g(t)\, u_0(t) + g(t - \pi)\, u_\pi(t) + g(t - 2\pi)\, u_{2\pi}(t) + \ldots$, where $g(t) = \frac{1}{6}(2\sin t - \sin 2t)$.

16. $Y(s) = \dfrac{s}{s^2+9} + \dfrac{s}{(s^2+9)(s^2+1)(1 - e^{-\pi s})} = \dfrac{s}{s^2+9} + \dfrac{s}{(s^2+9)(s^2+1)}(1 + e^{-\pi s} + e^{-2\pi s} + \ldots)$

$y(t) = (\cos 3t)\, u_0(t) + g(t)u_0(t) + g(t - \pi)\, u_\pi(t) + g(t - 2\pi)\, u_{2\pi}(t) + \ldots$

where $g(t) = (\cos t - \cos 3t)/8$. Since $g(t - \pi) = -g(t)$, $g(t - 2\pi) = g(t)$, etc.

we have $y(t) = \cos 3t\, u_0(t) + g(t)[u_0(t) - u_\pi(t) + u_{2\pi}(t) - u_{3\pi}(t) + \ldots]$.

17. $Y(s) = \dfrac{1}{s^2 + 6s + 10} + \dfrac{1}{s(s^2 + 6s + 10)(1 + e^{-\pi s})}$

$= \dfrac{1}{s^2 + 6s + 10} + \dfrac{1}{s(s^2 + 6s + 10)}(1 - e^{-\pi s} + e^{-2\pi s} - \ldots)$

$$y(t) = e^{-3t} \sin t \, u_0(t) + \frac{1}{10} \left[u_0(t) - u_\pi(t) + u_{2\pi}(t) - \ldots \right] + g(t)[u_0(t) + e^{3\pi} u_\pi(t) + e^{6\pi} u_{2\pi}(t) + \ldots]$$

where $g(t) = -e^{-3t}(3 \sin t + \cos t)/10$.

18. $Y(s) = \dfrac{1}{(s+1)(s+2)} \left[\dfrac{1}{s^2} - \dfrac{ae^{-as}}{s(1-e^{-as})} \right]$

$$y(t) = g_1(t)u_0(t) - a \left[\left\{ \frac{1}{2} + g_2(t-a) \right\} u_a(t) + \left\{ \frac{1}{2} + g_2(t-2a) \right\} u_{2a}(t) + \ldots \right]$$

where $g_1(t) = -\dfrac{3}{4} + \dfrac{t}{2} + 2e^{-t} - \dfrac{5}{4} e^{-2t}$, $g_2(t) = \dfrac{1}{2} e^{-2t} - e^{-t}$.

19. $Y(s) = \dfrac{1}{s^2(s+1)(s+3)} \left(\dfrac{1-e^{-\pi s}}{1+e^{-\pi s}} \right)$

$$y(t) = \left[-\frac{4}{9} + g(t) \right] u_0(t) - 2 \left[\left\{ -\frac{4}{9} + g(t-\pi) \right\} u_\pi(t) - \left\{ -\frac{4}{9} + g(t-2\pi) \right\} u_{2\pi}(t) + \ldots \right]$$

$$g(t) = \frac{1}{3} t + \frac{1}{2} e^{-t} - \frac{1}{18} e^{-3t}.$$

20. $Y(s) = -\dfrac{e^{-\pi s}}{(s^2+1)^2(1-e^{-\pi s})}$

$$y(t) = g(t-\pi) \, u_\pi(t) + g(t-2\pi) \, u_{2\pi}(t) + \ldots \, ; \quad g(t) = \frac{1}{2} (t \cos t - \sin t).$$

Exercise 8.6

Let $\mathscr{L}[u(x,t)] = U(x,s)$.

1. $U(x,s) = \dfrac{1}{s} + \left(\dfrac{1}{s^2} - \dfrac{1}{s} \right) e^{-sx^2/2}$, $u(x,t) = 1 + [(t-\alpha) - 1] u_\alpha(t)$, $\alpha = x^2/2$.

2. $U(x,s) = (2p/s^3) e^{-scx^2/2}$, $u(x,t) = p(t-\alpha)^2 u_\alpha(t)$, $\alpha = cx^2/2$.

3. $U(x,s) = \dfrac{2}{s^4} + \left[\dfrac{1}{s^2} - \dfrac{2}{s^4} \right] e^{-(sx^2/2)}$, $u(x,t) = \dfrac{t^3}{3} + \left[(t-\alpha) - \dfrac{1}{3}(t-\alpha)^3 \right] u_\alpha(t)$, $\alpha = x^2/2$.

4. $U(x,s) = \dfrac{4}{s^3} - \dfrac{4}{s^3} e^{-sx^2/2}$, $u(x,t) = 2t^2 - 2(t-\alpha)^2 u_\alpha(t)$, $\alpha = x^2/2$.

5. $U(x,s) = \dfrac{2}{s^2} + \dfrac{2}{s} - \dfrac{2}{s^2} e^{-3x^2 s}$, $u(x,t) = 2(t+1) - 2(t-\alpha)u_\alpha(t)$, $\alpha = 3x^2$.

6. $U(x,s) = e^{-p\sqrt{s}}/s$, $u(x,t) = erfc(x^*)$, $x^* = x/(2\sqrt{tk})$, $p = x/\sqrt{k}$.

7. $U(x,s) = (A/s) + [(B-A)/s]e^{-p\sqrt{s}}$, $u(x,t) = A + (B-A) \, erfc(x^*)$,

$x^* = x/(2\sqrt{tk})$, $p = x/\sqrt{k}$.

8. $U(x,s) = \dfrac{u_0}{s} + \dfrac{(u_1 - u_0)}{s} e^{-p\sqrt{s}} - \dfrac{u_1}{s} e^{-sT} e^{-p\sqrt{s}}$, $p = x/\sqrt{k}$.

$$u(x,t) = u_0 + (u_1 - u_0) \, erfc(x^*) - u_1 \, erfc \left(\frac{x}{2\sqrt{k(t-T)}} \right) u_T(t),$$

$x^* = x/(2\sqrt{t\,k}\,)$ and $u_T(t)$ is the unit step function.

9. $U(x, s) = e^{-p\sqrt{s}}/s^2$, $\mathscr{L}^{-1}\left[\dfrac{1}{s}\,e^{-p\sqrt{s}}\right] = erfc\,(x^*)$, $x^* = x/(2\sqrt{t\,k}\,)$, $p = x/\sqrt{k}$.

Using integration theorem $u(x, t) = \displaystyle\int_0^t erfc\left(\dfrac{x}{2\sqrt{k\tau}}\right)d\tau$.

10. $U(x, s) = \dfrac{g}{s^2} - \dfrac{g}{s^2}\,e^{-p\sqrt{s}}$, $u(x, t) = gt - g\displaystyle\int_0^t erfc\left(\dfrac{x}{2\sqrt{k\tau}}\right)d\tau$, $p = x/\sqrt{k}$.

11. $U(x, s) = \dfrac{u_0}{s} + \dfrac{(u_1 - u_0)}{s}\left[\dfrac{e^{qx} + e^{-qx}}{e^q + e^{-q}}\right] = \dfrac{u_0}{s} + \dfrac{(u_1 - u_0)}{s}[e^{q(x-1)} + e^{-q(x+1)}][1 + e^{-2q}]^{-1}$

$\qquad = \dfrac{u_0}{s} + \dfrac{(u_1 - u_0)}{s}[e^{q(x-1)} + e^{-q(x+1)}]\displaystyle\sum_{n=0}^{\infty}(-1)^n\,e^{-2nq}$

$\qquad = \dfrac{u_0}{s} + (u_1 - u_0)\displaystyle\sum_{n=0}^{\infty}\dfrac{(-1)^n}{s}[e^{-q(1-x+2n)} + e^{-q(1+x+2n)}]$, $q = \sqrt{s/k}$

$\qquad u(x, t) = u_0 + (u_1 - u_0)\displaystyle\sum_{n=0}^{\infty}(-1)^n\left[erfc\left(\dfrac{1 - x + 2n}{2\sqrt{kt}}\right) + erfc\left(\dfrac{1 + x + 2n}{2\sqrt{kt}}\right)\right]$

12. $U(x, s) = \dfrac{1}{s}[e^{-qx} - e^{-q(2-x)}][1 - e^{-2q}]^{-1}$ (as in problem 11)

$\qquad = \dfrac{1}{s}\left[\displaystyle\sum_{n=0}^{\infty}e^{-q(x+2n)} - e^{-q(2-x+2n)}\right]$, $q = \sqrt{s/k}$

$\qquad u(x, t) = \displaystyle\sum_{n=0}^{\infty}\left[erfc\left(\dfrac{x + 2n}{2\sqrt{kt}}\right) - erfc\left(\dfrac{2 - x + 2n}{2\sqrt{kt}}\right)\right]$.

13. $U(x, s) = \dfrac{e^{-x}}{s^2 - 1} + \left(\dfrac{1}{s^2 + 1} - \dfrac{1}{s^2 - 1}\right)e^{-sx}$,

$\qquad u(x, t) = e^{-x}\sinh t + [\sin(t - x) - \sinh(t - x)]\,u_x(t)$.

14. $U(x, s) = \dfrac{\sin 3\pi x}{s^2 + 9\pi^2}$, $u(x, t) = \dfrac{1}{3\pi}\sin 3\pi x \sin 3\pi t$.

15. $U(x, s) = \left(\dfrac{1 + e^{-s\pi}}{s^2 + 1}\right)e^{-sx}$, $u(x, t) = \sin(t - x)u_x(t) - \sin(t - x)u_{\pi+x}(t)$.

16.   For $0 \le x \le b$,

$\qquad I_1 = \dfrac{h}{2b}\displaystyle\int_0^x x^*(\cosh\theta_1 - \cosh\theta_2)\,dx^*$

$\qquad I_2 = \dfrac{h}{2b}\displaystyle\int_x^b x^*(\cosh\theta_1^* - \cosh\theta_2)\,dx^* + \dfrac{h}{2(l - b)}\displaystyle\int_b^l (l - x^*)(\cosh\theta_1^* - \cosh\theta_2)\,dx^*$

Integrating, we obtain

$\qquad I_1 + I_2 = \dfrac{hx}{bp}\sinh pl + \dfrac{hl}{b(l - b)p^2}\sinh px \sinh p(b - l)$

Therefore, $U(x, s) = \dfrac{hx}{bs} - \dfrac{hlc}{b(l-b)s^2}\left[\dfrac{\sinh px \sinh p(l-b)}{\sinh pl}\right]$

For $b \le x \le l$,

$$I_1 = \frac{h}{2b}\int_0^b x^*(\cosh \theta_1 - \cosh \theta_2)\,dx^* + \frac{h}{2(l-b)}\int_b^x (l-x^*)(\cosh \theta_1 - \cosh \theta_2)\,dx^*$$

$$I_2 = \frac{h}{2(l-b)}\int_x^l (l-x^*)(\cosh \theta_1^* - \cosh \theta_2)\,dx^*$$

Integrating, we obtain

$$I_1 + I_2 = \frac{h(l-x)}{(l-b)p}\sinh pl + \frac{hl}{b(l-b)p^2}\sinh pb \sinh p(x-l)$$

Therefore, $U(x, s) = \dfrac{h(l-x)}{s(l-b)} - \dfrac{hlc}{b(l-b)s^2}\left[\dfrac{\sinh pb \sinh p(l-x)}{\sinh pl}\right]$

17. $U(x, s) = \dfrac{s}{s^2-1}(e^{-x} - e^{-sx}),\ u(x, t) = e^{-x}\cosh t - \cosh(t-x)u_x(t).$

18. Substitute $b = l/2$ in Example 8.52.

19. $U(x, s) = \dfrac{g}{s^3}(e^{-sx/c} - 1),\ u(x, t) = -\dfrac{g}{2}t^2 + \dfrac{g}{2}\left(t - \dfrac{x}{c}\right)^2 u_{x/c}(t).$

20. $U(x, s) = \dfrac{g}{s^3}\left[1 - \dfrac{\cosh p(l-x)}{\cosh pl}\right],\ p = s/c.$ Use inversion theorem.

$$u(x, t) = \frac{gt^2}{2} + \frac{gx(2l-x)}{2c^2} - \frac{16gl^2}{\pi^2 c^2}\sum_{n=0}^{\infty}\frac{(-1)^n}{(2n+1)^3}\cos\left[\frac{(2n+1)\pi(l-x)}{2l}\right]\cos\left[\frac{(2n+1)\pi ct}{2l}\right].$$

Chapter 9

Fourier Series, Fourier Integrals and Fourier Transforms

9.1 Introduction

In Chapter 7 (section 7.2.4), we have defined the orthogonal and orthonormal functions. For example, the functions $\cos mx$ and $\sin mx$ are orthogonal over the interval $[-\pi, \pi]$. The orthonormal set of functions corresponding to $\cos mx$ are given by (Example 7.7)

$$\frac{1}{\sqrt{2\pi}}, \ \frac{\cos x}{\sqrt{\pi}}, \ \frac{\cos 2x}{\sqrt{\pi}}, \ \ldots, \ \frac{\cos mx}{\sqrt{\pi}}, \ldots$$

An important application discussed in section 7.2.4 was the series expansion of (suitable) functions in terms of a complete set of orthogonal or orthonormal functions. The expansion of a continuous function $f(x)$, having continuous derivatives over the interval $[-1, 1]$ in terms of Legendre polynomials was discussed in section 7.2.6. This series was called Fourier-Legendre series. The Fourier-Bessel series was discussed in section 7.4.2. A series expansion in terms of the trigonometric functions $\cos mx$ and $\sin mx$ is called a Fourier series. Many functions including some discontinuous periodic functions can be expanded in a Fourier series. Therefore, Fourier series solution method is a powerful tool in solving some ordinary and partial differential equations.

9.2 Fourier Series

Let $\{\phi_0(x), \ \phi_1(x), \ \phi_2(x), \ \ldots\}$ be an orthogonal set of functions, orthogonal with respect to a weight function $W(x) > 0$, on an interval $[a, b]$. Let $f(x)$ be a continuous function defined on the same interval $[a, b]$. Then, $f(x)$ can be expanded in an infinite series of the form (see section 7.2.4)

$$f(x) = c_0\phi_0(x) + c_1\phi_1(x) + c_2\phi_2(x) + \ldots. \tag{9.1}$$

The coefficients c_i, $i = 0, 1, 2, \ldots$ are given by

$$c_i = \left[\int_a^b f(x) \, W(x)\phi_i(x)dx\right] \Big/ \| \phi_i(x)) \|^2, \quad i = 0, 1, 2, \ldots \tag{9.2}$$

where
$$\| \phi_i (x) \|^2 = \int_a^b W(x)\phi_i^2 (x)dx.$$

Consider now, the set of orthogonal functions

$$\left\{1, \cos\left(\frac{\pi x}{l}\right), \cos\left(\frac{2\pi x}{l}\right), \ldots, \sin\left(\frac{\pi x}{l}\right), \sin\left(\frac{2\pi x}{l}\right), \ldots\right\} \qquad (9.3)$$

which are orthogonal on the interval $[-l, l]$ with respect to the weight function $W(x) = 1$. These functions have the following properties

$$\int_{-l}^l \cos\left(\frac{m\pi x}{l}\right)dx = \int_{-l}^l \sin\left(\frac{m\pi x}{l}\right)dx = 0 \qquad (9.4)$$

$$\int_{-l}^l \cos\left(\frac{m\pi x}{l}\right) \cos\left(\frac{n\pi x}{l}\right) dx = \int_{-l}^l \sin\left(\frac{m\pi x}{l}\right) \sin\left(\frac{n\pi x}{l}\right) dx = 0, \; m \neq n, \qquad (9.5)$$

$$\int_{-l}^l \cos\left(\frac{m\pi x}{l}\right) \sin\left(\frac{n\pi x}{l}\right) dx = 0, \text{ for all } m \text{ and } n, \qquad (9.6)$$

$$\int_{-l}^l \cos^2\left(\frac{m\pi x}{l}\right) dx = \int_{-l}^l \sin^2\left(\frac{m\pi x}{l}\right) dx = l, \qquad (9.7)$$

where m and n are integers.

Now, let $f(x)$ be a periodic function of period $2l$ defined on $[-l, l]$, that is $f(x + 2l) = f(x)$ and assume that it can be expanded in an orthogonal series in terms of the trigonometric functions. We shall discuss later in this section, the conditions under which such an expansion is possible. Let the series be written as

$$f(x) = \frac{a_0}{2} + \left[a_1 \cos\left(\frac{\pi x}{l}\right) + a_2 \cos\left(\frac{2\pi x}{l}\right) + \ldots\right] + \left[b_1 \sin\left(\frac{\pi x}{l}\right) + b_2 \sin\left(\frac{2\pi x}{l}\right) + \ldots\right]$$

$$= \frac{a_0}{2} + \sum_{n=1}^{\infty} \left[a_n \cos\left(\frac{n\pi x}{l}\right) + b_n \sin\left(\frac{n\pi x}{l}\right)\right]. \qquad (9.8)$$

The coefficients $a_0, a_1, a_2, \ldots, b_1, b_2 \ldots$ can be determined by using the orthogonal properties of the trigonometric functions given in Eqs. (9.4) to (9.7). Integrating Eq. (9.8) term by term on the interval $[-l, l]$, we obtain

$$\int_{-l}^l f(x)dx = \frac{a_0}{2} \int_{-l}^l dx + \sum_{n=1}^{\infty} \left[a_n \int_{-l}^l \cos\left(\frac{n\pi x}{l}\right)dx + b_n \int_{-l}^l \sin\left(\frac{n\pi x}{l}\right)dx\right] = l a_0$$

since $\cos(n\pi x/l)$ and $\sin(n\pi x/l)$ are orthogonal with respect to $W(x) = 1$, on $[-l, l]$. Therefore,

$$a_0 = \frac{1}{l} \int_{-l}^l f(x)dx.$$

Now, multiply both sides of Eq. (9.8) by $\cos(m\pi x/l)$ and integrate term by term on the interval $[-l, l]$. We obtain

$$\int_{-l}^{l} f(x) \cos\left(\frac{m\pi x}{l}\right) dx = \frac{a_0}{2} \int_{-l}^{l} \cos\left(\frac{m\pi x}{l}\right) dx + \sum_{n=1}^{\infty} \left[a_n \int_{-l}^{l} \cos\left(\frac{n\pi x}{l}\right) \cos\left(\frac{m\pi x}{l}\right) dx \right.$$

$$\left. + b_n \int_{-l}^{l} \cos\left(\frac{m\pi x}{l}\right) \sin\left(\frac{n\pi x}{l}\right) dx \right].$$

Using the orthogonal properties given in Eqs. (9.4) to (9.7), we get

$$\int_{-l}^{l} f(x) \cos\left(\frac{m\pi x}{l}\right) dx = l\, a_m.$$

Therefore,
$$a_m = \frac{1}{l} \int_{-l}^{l} f(x) \cos\left(\frac{m\pi x}{l}\right) dx.$$

Multiplying both sides of Eq. (9.8) by $\sin(m\pi x/l)$ and integrating term by term on the interval $[-l, l]$, we obtain

$$\int_{-l}^{l} f(x) \sin\left(\frac{m\pi x}{l}\right) dx = \frac{a_0}{2} \int_{-l}^{l} \sin\left(\frac{m\pi x}{l}\right) dx + \sum_{n=1}^{\infty} \left[a_n \int_{-l}^{l} \sin\left(\frac{m\pi x}{l}\right) \cos\left(\frac{n\pi x}{l}\right) dx \right.$$

$$\left. + b_n \int_{-l}^{l} \sin\left(\frac{m\pi x}{l}\right) \sin\left(\frac{n\pi x}{l}\right) dx \right].$$

Using the orthogonal properties given in Eqs. (9.4) to (9.7), we get

$$\int_{-l}^{l} f(x) \sin\left(\frac{m\pi x}{l}\right) dx = l b_m.$$

Therefore,
$$b_m = \frac{1}{l} \int_{-l}^{l} f(x) \sin\left(\frac{m\pi x}{l}\right) dx.$$

It can be observed that the expressions for a_0 and a_m can be combined as a single expression. It is to obtain this simplicity of notation that $a_0/2$ is used in Eq. (9.8). This does not mean that the value of a_0 can be obtained after evaluating a_n and setting $n = 0$ in this expression.

The orthogonal series for $f(x)$ given in Eq. (9.8) is called the *Fourier series*. The coefficients a_0, a_n, b_n are called the *Fourier coefficients* on $[-l, l]$. The expressions for the coefficients

$$a_0 = \frac{1}{l} \int_{-l}^{l} f(x)\, dx, \tag{9.9}$$

$$a_n = \frac{1}{l} \int_{-l}^{l} f(x) \cos\left(\frac{n\pi x}{l}\right) dx, \tag{9.10}$$

$$b_n = \frac{1}{l} \int_{-l}^{l} f(x) \sin\left(\frac{n\pi x}{l}\right) dx \tag{9.11}$$

are called the *Euler formulas*.

If the period of the function is 2π, that is $f(x)$ is defined on $[-\pi, \pi]$, then the Euler formulas are simplified as

$$a_0 = \frac{1}{\pi} \int_{-\pi}^{\pi} f(x)\,dx. \tag{9.12}$$

$$a_n = \frac{1}{\pi} \int_{-\pi}^{\pi} f(x)\cos nx\,dx. \tag{9.13}$$

$$b_n = \frac{1}{\pi} \int_{-\pi}^{\pi} f(x)\sin nx\,dx. \tag{9.14}$$

From the definition of definite integrals, we have that *if f (x) is continuous or piecewise continuous (continuous except for a finite number of finite jumps) then the integrals given in Eqs. (9.9) to (9.14) exist and f (x) can be expanded as a Fourier Series.*

Example 9.1 Find the Fourier series expansion of the periodic function

$$f(x) = x, \quad -\pi \le x \le \pi, \quad f(x + 2\pi) = f(x).$$

Solution The Fourier coefficients are obtained as follows.

$$a_0 = \frac{1}{\pi} \int_{-\pi}^{\pi} x\,dx = 0, \qquad\qquad (x \text{ is an odd function on } [-\pi, \pi])$$

$$a_n = \frac{1}{\pi} \int_{-\pi}^{\pi} x\cos nx\,dx = 0, \qquad (x\cos nx \text{ is an odd function on } [-\pi, \pi])$$

$$b_n = \frac{1}{\pi} \int_{-\pi}^{\pi} x\sin nx\,dx = \frac{2}{\pi} \int_{0}^{\pi} x\sin nx\,dx, \quad (x\sin nx \text{ is an even function on } [-\pi, \pi])$$

$$= \frac{2}{\pi}\left[-x\left(\frac{\cos nx}{n}\right) + \left(\frac{\sin nx}{n^2}\right) \right]_0^{\pi} = \frac{2}{\pi}\left[\frac{-\pi \cos n\pi}{n} \right] = \frac{2}{n}(-1)^{n+1}.$$

Therefore, the Fourier expansion of the given function on $[-\pi, \pi]$ is given by

$$x = 2\left[\sin x - \frac{\sin 2x}{2} + \frac{\sin 3x}{3} - \frac{\sin 4x}{4} + \cdots \right].$$

Example 9.2 Find the Fourier series expansion of the following periodic function with period 2π

$$f(x) = \begin{cases} \pi + x, & -\pi < x < 0 \\ 0, & 0 \le x < \pi, \quad f(x + 2\pi) = f(x). \end{cases}$$

Solution The graph of $f(x)$ is given in Fig. 9.1. The Fourier coefficients are obtained as follows.

$$a_0 = \frac{1}{\pi} \int_{-\pi}^{\pi} f(x)\,dx = \frac{1}{\pi} \int_{-\pi}^{0} (\pi + x)\,dx = \frac{1}{\pi}\left[\pi x + \frac{x^2}{2} \right]_{-\pi}^{0} = \frac{\pi}{2}.$$

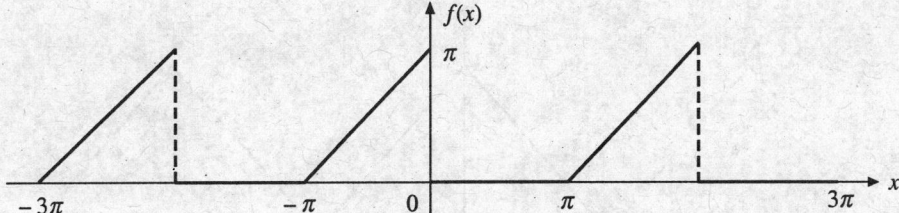

Fig. 9.1. Example 9.2.

$$a_n = \frac{1}{\pi} \int_{-\pi}^{0} (\pi + x) \cos nx \, dx = \frac{1}{\pi} \left[\int_{-\pi}^{0} \pi \cos nx \, dx + \int_{-\pi}^{0} x \cos nx \, dx \right]$$

$$= \frac{1}{\pi} \left[\pi \left(\frac{\sin nx}{n} \right) + \left\{ x \left(\frac{\sin nx}{n} \right) + \frac{\cos nx}{n^2} \right\} \right]_{-\pi}^{0}$$

$$= \frac{1}{\pi} \left[\frac{1}{n^2} (1 - \cos n\pi) \right] = \frac{1}{\pi n^2} [1 - (-1)^n] = \begin{cases} 0, & \text{for } n \text{ even} \\ 2/(\pi n^2), & \text{for } n \text{ odd.} \end{cases}$$

$$b_n = \frac{1}{\pi} \int_{-\pi}^{0} (\pi + x) \sin nx \, dx = \frac{1}{\pi} \left[(\pi + x) \left(-\frac{\cos nx}{n} \right) + \frac{\sin nx}{n^2} \right]_{-\pi}^{0}$$

$$= \frac{1}{\pi} \left[-\frac{\pi}{n} \right] = -\frac{1}{n}.$$

Therefore, the Fourier series expansion is given by

$$f(x) = \frac{\pi}{4} + \sum_{n=1}^{\infty} \left[\frac{1}{\pi n^2} \{1 - (-1)^n\} \cos nx - \frac{1}{n} \sin nx \right]$$

$$= \frac{\pi}{4} + \frac{2}{\pi} \left[\frac{\cos x}{1^2} + \frac{\cos 3x}{3^2} + \dots \right] - \left[\frac{\sin x}{1} + \frac{\sin 2x}{2} + \dots \right].$$

Example 9.3 Find the Fourier series expansion of the following periodic function of period 4

$$f(x) = \begin{cases} 2 + x, & -2 \le x \le 0, \\ 2 - x, & 0 < x \le 2, \end{cases} \quad f(x + 4) = f(x).$$

Solution The graph of $f(x)$ is given in Fig. 9.2. The function is defined on $[-2, 2]$. Using Eqs. (9.9) to (9.11), we obtain the Fourier coefficients as follows.

$$a_0 = \frac{1}{2} \int_{-2}^{2} f(x) dx = \frac{1}{2} \left[\int_{-2}^{0} (2 + x) dx + \int_{0}^{2} (2 - x) dx \right]$$

$$= \frac{1}{2} \left[\left\{ 2x + \frac{x^2}{2} \right\}_{-2}^{0} + \left\{ 2x - \frac{x^2}{2} \right\}_{0}^{2} \right]$$

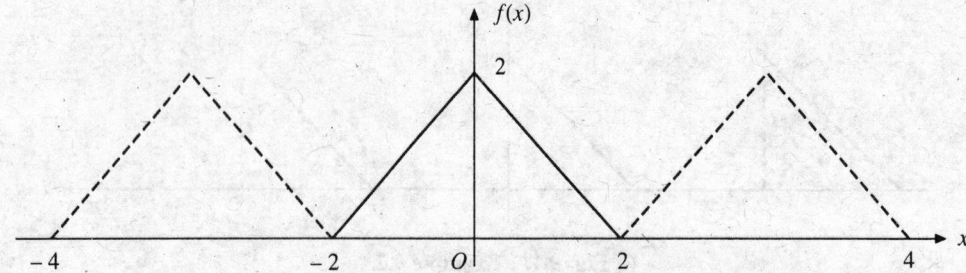

Fig. 9.2.   Example 9.3.

$$= \frac{1}{2}\left[0 - (-4+2) + (4-2) - 0\right] = 2.$$

$$a_n = \frac{1}{2}\left[\int_{-2}^{0}(2+x)\cos\left(\frac{n\pi x}{2}\right)dx + \int_{0}^{2}(2-x)\cos\left(\frac{n\pi x}{2}\right)dx\right]$$

$$= \frac{1}{2}\left[\left\{(2+x)\frac{\sin\left(n\pi x/2\right)}{(n\pi/2)} + \frac{\cos\left(n\pi x/2\right)}{(n\pi/2)^2}\right\}_{-2}^{0}\right.$$

$$\left. + \left\{(2-x)\frac{\sin\left(n\pi x/2\right)}{(n\pi/2)} - \frac{\cos\left(n\pi x/2\right)}{(n\pi/2)^2}\right\}_{0}^{2}\right]$$

$$= \frac{1}{2}\left[\left\{\frac{1}{(n\pi/2)^2} - \frac{\cos\left(n\pi\right)}{(n\pi/2)^2}\right\} + \left\{\frac{1}{(n\pi/2)^2} - \frac{\cos\left(n\pi\right)}{(n\pi/2)^2}\right\}\right]$$

$$= \frac{4}{n^2\pi^2}\left[1 - (-1)^n\right] = \begin{cases} 0, & \text{for } n \text{ even} \\ 8/(n^2\pi^2), & \text{for } n \text{ odd.} \end{cases}$$

$$b_n = \frac{1}{2}\left[\int_{-2}^{0}(2+x)\sin\left(\frac{n\pi x}{2}\right)dx + \int_{0}^{2}(2-x)\sin\left(\frac{n\pi x}{2}\right)dx\right]$$

$$= \frac{1}{2}\left[\left\{-(2+x)\frac{\cos\left(n\pi x/2\right)}{(n\pi/2)} + \frac{\sin\left(n\pi x/2\right)}{(n\pi/2)^2}\right\}_{-2}^{0}\right.$$

$$\left. + \left\{-(2-x)\frac{\cos\left(n\pi x/2\right)}{(n\pi/2)} - \frac{\sin\left(n\pi x/2\right)}{(n\pi/2)^2}\right\}_{0}^{2}\right]$$

$$= \frac{1}{2}\left[-\frac{2}{(n\pi/2)} + \frac{2}{(n\pi/2)}\right] = 0.$$

Therefore, the Fourier series expansion is given by

$$f(x) = 1 + \frac{8}{\pi^2} \sum_{n=1}^{\infty} \frac{1}{(2n-1)^2} \cos\left[(2n-1)\frac{\pi x}{2}\right].$$

9.2.1 Fourier Series Expansions of Even and Odd Functions

Let $f(x)$ be a function defined on $[-l, l]$. Then, $f(x)$ is an even function on $[-l, l]$ if

$$f(-x) = f(x), \quad -l \leq x \leq l. \tag{9.15}$$

The function $f(x)$ is odd if

$$f(-x) = -f(x), \quad -l \leq x \leq l. \tag{9.16}$$

For example, x^{2n}, $\cos(n\pi x/l)$ are even functions on $[-l, l]$, since

$$f(-x) = (-x)^{2n} = [(-x)^2]^n = x^{2n} = f(x),$$

and $\qquad\qquad f(-x) = \cos(-n\pi x/l) = \cos(n\pi x/l) = f(x).$

Similarly, x^{2n+1}, $\sin(n\pi x/l)$ are odd functions on $[-l, l]$, since

$$f(-x) = (-x)^{2n+1} = (-1)^{2n+1} x^{2n+1} = -x^{2n+1} = -f(x)$$

and $\qquad\qquad f(-x) = \sin(-n\pi x/l) = -\sin(n\pi x/l) = -f(x).$

Graphs of even functions $|x|$, x^2 and a typical cosine like function are given in Figs. 9.3. a, b, c.

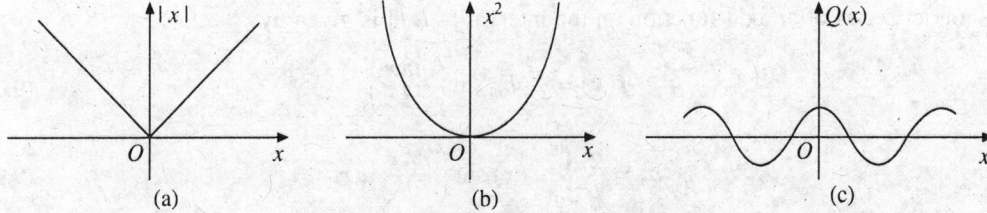

(a)   (b)   (c)

Fig. 9.3.   Graphs of even functions.

Graphs of odd functions x, x^3 and a typical sine like function are given in Figs. 9.4. a, b, c.

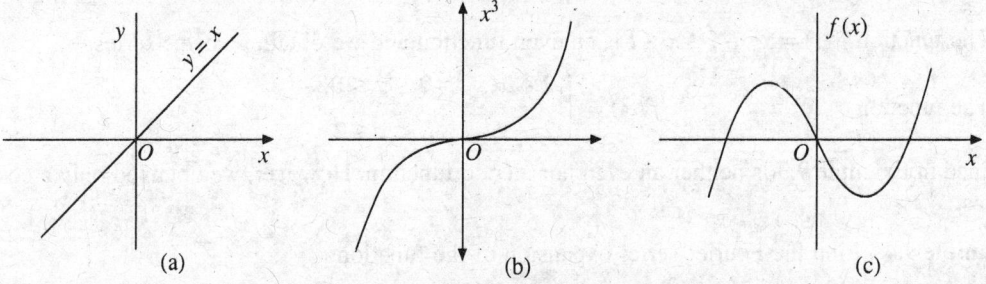

(a)   (b)   (c)

Fig. 9.4.   Graphs of odd functions.

If $f(x)$ is an even function on $[-l, l]$, then we have

$$\int_{-l}^{l} f(x)\,dx = 2 \int_{0}^{l} f(x)\,dx. \tag{9.17}$$

If $f(x)$ is an odd function on $[-l, l]$, then we have

$$\int_{-l}^{l} f(x)dx = 0.$$ (9.18)

The following results can be easily proved from the definition.

(even function) (even function) = even function

(even function) (odd function) = odd function

(odd function) (odd function) = even function.

Therefore, from the definition, we have that

if $f(x)$ is even then $f(x) \cos(n\pi x/l)$ is even and $f(x) \sin(n\pi x/l)$ is odd,

if $f(x)$ is odd then $f(x) \cos(n\pi x/l)$ is odd and $f(x) \sin(n\pi x/l)$ is even.

Hence, if $f(x)$ is an even function on $[-l, l]$, then we have the following Fourier series

$$f(x) = \frac{a_0}{2} + \sum_{n=1}^{\infty} a_n \cos\left(\frac{n\pi x}{l}\right)$$ (9.19)

where $$a_0 = \frac{2}{l} \int_0^l f(x)dx \text{ and } a_n = \frac{2}{l} \int_0^l f(x) \cos\left(\frac{n\pi x}{l}\right)dx.$$ (9.20)

The Fourier series of an odd function on the interval $[-l, l]$ is given by

$$f(x) = \sum_{n=1}^{\infty} b_n \sin\left(\frac{n\pi x}{l}\right)$$ (9.21)

where $$b_n = \frac{2}{l} \int_0^l f(x) \sin\left(\frac{n\pi x}{l}\right)dx.$$ (9.22)

The series given in Eq. (9.19) is called the *Fourier cosine series* and the series given in Eq. (9.21) is called the *Fourier sine series*.

Consider Example 9.1. The function $f(x) = x, -\pi \le x \le \pi$ is an odd function and we obtain a sine series.

The function $f(x) = x^2, -l \le x \le l$ is an even function and we obtain a cosine series.

The function $$f(x) = \begin{cases} 2 + x, & -2 \le x \le 0, \\ 2 - x, & 0 < x \le 2 \end{cases}$$

defined in Example 9.3 is neither an even nor an odd function. However, we obtained only a cosine series.

Example 9.4 Find the Fourier series expansion of the function

$$f(x) = x^2, \quad -2 \le x \le 2.$$

Solution The given function $f(x) = x^2$ is an even function. Therefore,

$$a_0 = \frac{2}{l} \int_0^l f(x)dx = \int_0^2 x^2 dx = \frac{8}{3}.$$

$$a_n = \frac{2}{l} \int_0^l f(x) \cos\left(\frac{n\pi x}{l}\right) dx = \int_0^2 x^2 \cos\left(\frac{n\pi x}{2}\right) dx$$

$$= \left[x^2 \frac{\sin(n\pi x/2)}{(n\pi/2)} \right]_0^2 - 2 \int_0^2 x \frac{\sin(n\pi x/2)}{(n\pi/2)} dx$$

$$= -\frac{4}{n\pi} \left[-x \frac{\cos(n\pi x/2)}{(n\pi/2)} + \frac{\sin(n\pi x/2)}{(n\pi/2)^2} \right]_0^2 = \frac{16}{n^2\pi^2} \cos n\pi = \frac{16(-1)^n}{n^2\pi^2}.$$

Therefore, the Fourier series is given by

$$f(x) = \frac{4}{3} + \frac{16}{\pi^2} \sum_{n=1}^{\infty} \frac{(-1)^n}{n^2} \cos\left(\frac{n\pi x}{2}\right).$$

9.2.2 Convergence of Fourier Series

Many functions including some discontinuous periodic functions can be expanded in a Fourier series. In this section, we shall discuss the conditions under which Fourier series expansion is possible and also find the function to which the series converges.

Let $f(x)$ be piecewise continuous on $[-l, l]$, that is,

(i) $f(x)$ is defined and continuous for all x in $(-l, l)$ except, may be, at a finite number of points in $(-l, l)$.

(ii) At any point $x_0 \in (-l, l)$, where $f(x)$ is not continuous, both the one-sided limits $\lim_{x \to x_0^-} f(x)$ and $\lim_{x \to x_0^+} f(x)$ exist and are finite, that is, the discontinuities are jump discontinuities (Fig. 9.5).

(iii) The one-sided limits $\lim_{x \to -l^+} f(x)$ and $\lim_{x \to l^-}$ exist and are finite.

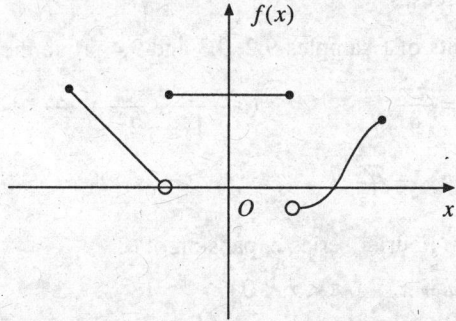

Fig. 9.5. Typical piecewise continuous function.

If both $f(x)$ and $f'(x)$ are piecewise continuous then the function $f(x)$ is also called *piecewise smooth*.

We now state the convergence theorem.

Theorem 9.1 Let $f(x)$ and $f'(x)$ be piecewise continuous on the interval $[-l, l]$. Then, the Fourier series of $f(x)$ on this interval converges to $f(x)$ at a point of continuity. At a point of discontinuity, the Fourier series converges to

$$\frac{1}{2}[f(x+) + f(x-)]$$

where $f(x+)$ and $f(x-)$ are the right and left hand limits respectively.

The proof of the theorem is omitted (see Problem 37, Exercise 9.1, for the proof of Theorem 9.1 for the particular case when $f(x)$, $f'(x)$ and $f''(x)$ are continuous).

Remark 1

At both the end points of the interval $[-l, l]$, the Fourier series converges to

$$\frac{1}{2}[f(-l+) + f(l-)]$$

The series converges to the same number at l and $-l$, since the Fourier series

$$\frac{a_0}{2} + \sum_{n=1}^{\infty} \left[a_n \cos\left(\frac{n\pi x}{l}\right) + b_n \sin\left(\frac{n\pi x}{l}\right) \right]$$

has the same value

$$\frac{a_0}{2} + \sum_{n=1}^{\infty} a_n \cos(n\pi) = \frac{a_0}{2} + \sum_{n=1}^{\infty} (-1)^n a_n$$

at both the end points, $x = l$ and $-l$.

Remark 2

Let the sum upto j terms of the Fourier series be denoted by

$$S_j = \frac{a_0}{2} + \sum_{n=1}^{j} \left[a_n \cos\left(\frac{n\pi x}{l}\right) + b_n \sin\left(\frac{n\pi x}{l}\right) \right], \quad j = 1, 2, 3, \ldots.$$

Then, partial sums S_j give successive approximations to $f(x)$, that is, the approximations represent $f(x)$ closer and closer as j increases.

Example 9.5 Using the results of Examples 9.2, 9.3 and 9.4 prove the following

(i) $\dfrac{1}{1^2} + \dfrac{1}{3^2} + \dfrac{1}{5^2} + \ldots = \dfrac{\pi^2}{8}$,   (ii) $\dfrac{1}{1^2} + \dfrac{1}{2^2} + \dfrac{1}{3^2} \ldots = \dfrac{\pi^2}{6}$,

(iii) $1 - \dfrac{1}{2^2} + \dfrac{1}{3^2} - \dfrac{1}{4^2} + \ldots = \dfrac{\pi^2}{12}$.

Solution In Example 9.2, the Fourier series expansion of

$$f(x) = \begin{cases} \pi + x, & -\pi < x < 0 \\ 0, & 0 \le x < \pi \end{cases}$$

was obtained as

$$f(x) = \frac{\pi}{4} + \frac{2}{\pi}\left[\frac{\cos x}{1^2} + \frac{\cos 3x}{3^2} + \ldots \right] - \left[\frac{\sin x}{1} + \frac{\sin 2x}{2} + \ldots \right].$$

At the point $x = 0$, $f(x)$ is discontinuous. Therefore, the series converges to

$$\frac{1}{2}\left[f(0-)+f(0+)\right]=\frac{1}{2}\left[\pi+0\right]=\frac{\pi}{2}.$$

Setting $x = 0$, we obtain

$$\frac{\pi}{2}=\frac{\pi}{4}+\frac{2}{\pi}\left[\frac{1}{1^2}+\frac{1}{3^2}+\frac{1}{5^2}+\ldots\right] \quad \text{or} \quad \frac{1}{1^2}+\frac{1}{3^2}+\frac{1}{5^2}+\ldots=\frac{\pi^2}{8}.$$

In Example 9.3, the Fourier series expansion of

$$f(x) = \begin{cases} 2+x, & -2 \le x \le 0 \\ 2-x, & 0 < x \le 2 \end{cases}$$

was obtained as

$$f(x)=1+\frac{8}{\pi^2}\left[\frac{1}{1^2}\cos\left(\frac{\pi x}{2}\right)+\frac{1}{3^2}\cos\left(\frac{3\pi x}{2}\right)+\frac{1}{5^2}\cos\left(\frac{5\pi x}{2}\right)+\ldots\right].$$

At the point $x = 0$, the given function is continuous. Therefore, the series converges to $f(0) = 2$. Hence,

$$2=1+\frac{8}{\pi^2}\left[\frac{1}{1^2}+\frac{1}{3^2}+\frac{1}{5^2}+\ldots\right] \quad \text{or} \quad \frac{1}{1^2}+\frac{1}{3^2}+\frac{1}{5^2}+\ldots=\frac{\pi^2}{8}.$$

In Example 9.4, the Fourier series expansion of $f(x) = x^2$, $-2 \le x \le 2$ was obtained as

$$f(x)=\frac{4}{3}+\frac{16}{\pi^2}\sum_{n=1}^{\infty}\frac{(-1)^n}{n^2}\cos\left(\frac{n\pi x}{2}\right).$$

At the end points of the interval $[-2, 2]$, the series converges to

$$\frac{1}{2}\left[f(-2+)+f(2-)\right]=\frac{1}{2}\left[4+4\right]=4.$$

Substituting $x = 2$, we obtain

$$4=\frac{4}{3}+\frac{16}{\pi^2}\sum_{n=1}^{\infty}\frac{(-1)^n}{n^2}\cos(n\pi)=\frac{4}{3}+\frac{16}{\pi^2}\sum_{n=1}^{\infty}\frac{1}{n^2}$$

Hence,
$$\frac{1}{1^2}+\frac{1}{2^2}+\frac{1}{3^2}+\ldots=\frac{\pi^2}{16}\left(4-\frac{4}{3}\right)=\frac{\pi^2}{6}.$$

which is the result given in (ii).

At $x = 0$, the given function is continuous. The series converges to $f(0) = 0$. Therefore,

$$0=\frac{4}{3}+\frac{16}{\pi^2}\sum_{n=1}^{\infty}\frac{(-1)^n}{n^2}=\frac{4}{3}-\frac{16}{\pi^2}\left[\frac{1}{1^2}-\frac{1}{2^2}+\frac{1}{3^2}-\frac{1}{4^2}+\ldots\right]$$

or
$$\frac{1}{1^2}-\frac{1}{2^2}+\frac{1}{3^2}-\frac{1}{4^2}+\ldots=\frac{\pi^2}{12}.$$

which is the result given in (iii).

Theorem 9.2 (Term by term differentiation of Fourier series) Let $f(x)$ be continuous and $f'(x)$ be piecewise continuous on $[-l, l]$. Let $f(-l) = f(l)$. Then, at every $x \in (-l, l)$ where $f''(x)$ exists

$$f'(x)=\frac{\pi}{l}\sum_{n=1}^{\infty}n\left[b_n\cos\left(\frac{n\pi x}{l}\right)-a_n\sin\left(\frac{n\pi x}{l}\right)\right]. \tag{9.23}$$

Note that the right hand side of Eq. (9.23) is the series obtained by differentiating the Fourier series given in Eq. (9.8), term by term.

Theorem 9.3   (Term by term integration of Fourier series) Let $f(x)$ be piecewise continuous on $(-l, l)$ and its Fourier series be given by

$$f(x) = \frac{a_0}{2} + \sum_{n=1}^{\infty} \left[a_n \cos\left(\frac{n\pi x}{l}\right) + b_n \sin\left(\frac{n\pi x}{l}\right) \right].$$

Then, for any $x \in (-l, l)$ we have

$$\int_{-l}^{x} f(x^*) dx^* = \frac{a_0}{2} (x + l) + \frac{l}{\pi} \sum_{n=1}^{\infty} \frac{1}{n} \left[a_n \sin\left(\frac{n\pi x}{l}\right) + b_n \left\{ \cos n\pi - \cos\left(\frac{n\pi x}{l}\right) \right\} \right]. \quad (9.24)$$

Note that the right hand side of Eq. (9.24) is the series obtained by integrating the Fourier series term by term.

Remark 3

The result of Theorem 9.3 holds even if the Fourier series does not converge to the function.

Gibbs phenomenon   To discuss the Gibbs phenomenon, let us consider the Fourier series expansion of the function

$$f(x) = \begin{cases} 1, & -\pi < x < 0, \\ -1, & 0 \le x < \pi. \end{cases}$$

The graph of $f(x)$ is given in Fig. 9.6.

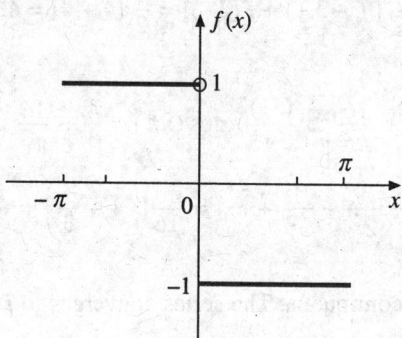

Fig. 9.6. Graph of $f(x)$.

The function is odd. Therefore, we have a sine series. We have

$$b_n = \frac{2}{\pi} \int_{0}^{\pi} - \sin nx\, dx = \frac{2}{n\pi} [\cos n\pi - 1] = \frac{2}{n\pi} [(-1)^n - 1] = \begin{cases} 0, & n \text{ even} \\ -\frac{4}{n\pi}, & n \text{ odd}. \end{cases}$$

Therefore,

$$f(x) = -\frac{4}{\pi} \left[\sin x + \frac{\sin 3x}{3} + \frac{\sin 5x}{5} + \ldots \right].$$

Denote the partial sums of the series as

$$S_1 = -\frac{4}{\pi} \sin x, \quad S_2 = -\frac{4}{\pi} \left(\sin x + \frac{\sin 3x}{3} \right), \quad S_3 = -\frac{4}{\pi} \left(\sin x + \frac{\sin 3x}{3} + \frac{\sin 5x}{5} \right), \dots$$

The typical plots of S_1, S_2, S_3, S_{14} are given in Figs. 9.7 to 9.10.

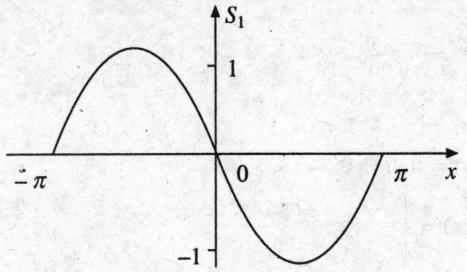

Fig. 9.7. Typical plot of $S_1(x)$.

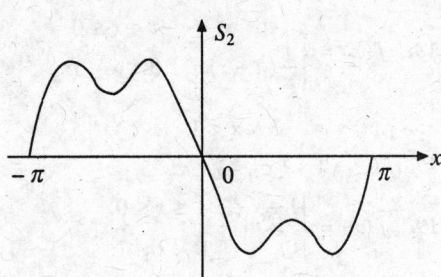

Fig. 9.8. Typical plot of $S_2(x)$.

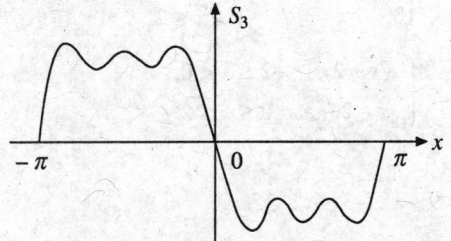

Fig. 9.9. Typical plot of $S_3(x)$.

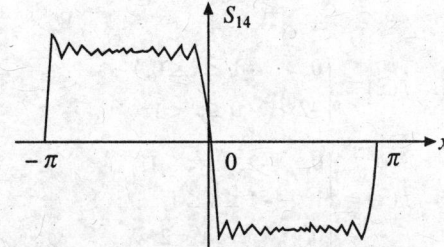

Fig. 9.10. Typical plot of $S_{14}(x)$.

It can be observed that the graph of $S_{14}(x)$ displays spikes near the discontinuities at $x = -\pi$, 0 and π. This oscillatory behaviour of the partial sums S_n for large n, about the true value near a point of discontinuity does not smoothen out even for very large n. This behaviour of the Fourier series near a point of discontinuity is called the *Gibbs phenomenon*.

Exercise 9.1

In the following problems, find the Fourier series of the given function on the given interval.

1. $f(x) = \begin{cases} k, & -\pi < x < 0, \\ 0, & 0 \le x < \pi. \end{cases}$

2. $f(x) = \begin{cases} \pi, & -\pi < x < 0, \\ \pi - x, & 0 \le x < \pi. \end{cases}$

3. $f(x) = \begin{cases} 0, & -\pi < x < -\pi/2, \\ k, & -\pi/2 \le x \le \pi/2, \\ 0, & \pi/2 < x < \pi. \end{cases}$

4. $f(x) = \begin{cases} -k, & -\pi < x < 0, \\ k, & 0 \le x < \pi. \end{cases}$

5. $f(x) = \begin{cases} 2, & -\pi < x < 0, \\ 4, & 0 \le x < \pi. \end{cases}$

6. $f(x) = \begin{cases} -(\pi + x), & -\pi < x < 0, \\ -(\pi - x), & 0 \le x < \pi. \end{cases}$

7. $f(x) = 1 - |x|, -\pi < x < \pi.$

8. $f(x) = x^2, -\pi < x < \pi.$

9. $f(x) = x^3, -\pi < x < \pi.$

10. $f(x) = \begin{cases} x^2, & -\pi < x < 0, \\ -x^2, & 0 \le x < \pi. \end{cases}$

11. $f(x) = \cos x - \sin(x/2), -\pi < x < \pi.$

12. $f(x) = \begin{cases} 0, & -\pi \le x \le 0, \\ \sin x, & 0 \le x \le \pi. \end{cases}$

13. $f(x) = \begin{cases} \cos x, & -\pi < x \le 0, \\ \dfrac{1}{\pi}(\pi - x), & 0 < x < \pi, \end{cases}$

14. $f(x) = x, -2 < x < 2.$

15. $f(x) = x^2, -3 < x < 3.$

16. $f(x) = |x|, -1 < x < 1.$

17. $f(x) = 1 + |x|, -3 < x < 3.$

18. $f(x) = \sinh(\pi x), -1 < x < 1.$

19. $f(x) = \begin{cases} 1+x, & -1 \le x \le 0, \\ 0, & 0 < x \le 1. \end{cases}$

20. $f(x) = \begin{cases} -x, & -2 < x < 0, \\ x^2, & 0 \le x < 2. \end{cases}$

21. $f(x) = \begin{cases} -1, & -1 < x < 0, \\ x, & 0 \le x < 1. \end{cases}$

22. $f(x) = \begin{cases} 0, & -2 \le x \le -1, \\ \cos(\pi x/2), & -1 \le x \le 1, \\ 0, & 1 \le x \le 2. \end{cases}$

23. $f(x) = \begin{cases} 0, & -1 < x \le 0, \\ -2x, & 0 \le x < 1. \end{cases}$

24. $f(x) = \begin{cases} -(x+2), & -2 \le x < 0, \\ x-2, & 0 \le x \le 2. \end{cases}$

25. $f(x) = \begin{cases} 0, & -2 < x < -1, \\ 1, & -1 \le x \le 1, \\ 0, & 1 < x < 2. \end{cases}$

26. $f(x) = 9 - x^2, -3 \le x \le 3.$

27. Let $f(x)$ and $g(x)$ be two integrable functions on the interval $[-l, l]$. It is given that $f(x) = g(x)$ for all x in this interval except at one point $x = c$. Are the Fourier coefficients of these functions different?

28. Find the Fourier series expansion of the function

$$f(x) = \pi + x, -\pi < x < \pi.$$

Hence, show that

$$\frac{\pi}{4} = 1 - \frac{1}{3} + \frac{1}{5} - \frac{1}{7} + \ldots.$$

29. Find the Fourier series expansion of the following periodic function of period 2π

$$f(x) = \begin{cases} 0, & -\pi < x < 0, \\ x^2, & 0 \le x < \pi. \end{cases}$$

Hence, show that

$$\frac{\pi^2}{6} = 1 + \frac{1}{2^2} + \frac{1}{3^2} + \frac{1}{4^2} + \ldots, \quad \text{and} \quad \frac{\pi^2}{12} = 1 - \frac{1}{2^2} + \frac{1}{3^2} - \frac{1}{4^2} + \ldots.$$

30. Manipulate the two series in problem 29 to show that

$$\frac{\pi^2}{8} = \frac{1}{1^2} + \frac{1}{3^2} + \frac{1}{5^2} + \ldots.$$

31. Use the result of Problem 12 to show that

$$\frac{\pi}{4} = \frac{1}{2} + \frac{1}{1 \cdot 3} - \frac{1}{3 \cdot 5} + \frac{1}{5 \cdot 7} - \frac{1}{7 \cdot 9} + \ldots.$$

32. A periodic function of period 2π is defined as

$$f(x) = \begin{cases} 1, & -\pi/2 < x < \pi/2 \\ 0, & \pi/2 \le x < 3\pi/2. \end{cases}$$

Obtain the Fourier series expansion of $f(x)$ and hence, show that

$$\frac{\pi}{4} = 1 - \frac{1}{3} + \frac{1}{5} - \frac{1}{7} + \ldots.$$

33. A periodic function of period 2 is defined as

$$f(x) = 1 + x, -1 < x < 1.$$

Obtain the Fourier series expansion of $f(x)$ and hence, show that

$$\frac{\pi}{4} = 1 - \frac{1}{3} + \frac{1}{5} - \frac{1}{7} + \ldots.$$

34. Obtain the Fourier series expansion of the periodic function

$$f(x) = e^x, -\pi < x < \pi, f(x + 2\pi) = f(x).$$

Hence, find the sum of the series

$$\frac{1}{1 + 2^2} - \frac{1}{1 + 3^2} + \frac{1}{1 + 4^2} - \ldots + \frac{(-1)^n}{1 + n^2} + \ldots.$$

35. Obtain the Fourier series expansion of the following periodic function of period 4

$$f(x) = 4 - x^2, -2 \le x \le 2.$$

Hence, show that $\dfrac{\pi^2}{12} = 1 - \dfrac{1}{2^2} + \dfrac{1}{3^2} - \dfrac{1}{4^2} + \ldots.$

36. Show that $f(x) = |x|, -2 \le x \le 2$ satisfies the hypothesis of Theorem 9.2. Obtain the Fourier series expansion of $f(x)$ and hence find the series expansion of $f'(x)$. Verify that this series is same as the

Fourier series expansion of $f'(x) = \begin{cases} -1, & \text{for} \quad -2 < x < 0 \\ 1, & \text{for} \quad 0 < x < 2. \end{cases}$

37. Prove the convergence (uniform convergence) of the Fourier series of a periodic function $f(x)$ defined over $[-\pi, \pi]$, when $f(x), f'(x)$ and $f''(x)$ are continuous over $[-\pi, \pi]$.

38. Prove that

$$\frac{1}{l} \int_{-l}^{l} f^2(x)dx = \frac{a_0^2}{2} + \sum_{n=1}^{\infty} (a_n^2 + b_n^2).$$

39. Express the Fourier series given in Eq. (9.8) in the form

$$f(x) = \frac{a_0}{2} + \sum_{n=1}^{\infty} A_n \cos\left(\frac{n\pi x}{l} - \alpha_n\right)$$

where α_n is the phase angle defined by $\alpha_n = \tan^{-1}(b_n/a_n)$ and $A_n = \sqrt{a_n^2 + b_n^2}$.

40. Express the Fourier series given in Eq. (9.8) in the form

$$f(x) = \frac{a_0}{2} + \sum_{n=1}^{\infty} A_n \sin\left(\frac{n\pi x}{l} + \beta_n\right)$$

where β_n is the phase angle $\beta_n = \tan^{-1}(a_n/b_n)$. Compare this result with the result given in Problem 39.

9.3   Fourier Half-Range Series

Suppose that a function $f(x)$ is defined on some finite interval. It may also be the case that a periodic function $f(x)$ of period $2l$ is defined only on a half-interval $[0, l]$. It is possible to extend the definition of $f(x)$ to the other half $[-l, 0]$ of the interval $[-l, l]$ so that $f(x)$ is either an even or an odd function. In the first case, we call it an even periodic extension of $f(x)$ and in the second case, we call it an odd periodic extension of $f(x)$. If $f(x)$ is given and an *even periodic extension* is done then $f(x)$ is an even function in $[-l, l]$. Hence, $f(x)$ has a Fourier cosine series. If $f(x)$ is given and an *odd periodic extension* is done then $f(x)$ is an odd function in $[-l, l]$. Hence, $f(x)$ has now a Fourier sine series. Therefore, if a function $f(x)$ is defined only on a half interval $[0, l]$, then it is possible to obtain a Fourier cosine or a Fourier sine series expansion depending on the requirements of a particular problem, by suitable periodic extensions. We have the following results.

Theorem 9.4   (Fourier cosine series) Let $f(x)$ be piecewise continuous on $[0, l]$. Then, the Fourier cosine series expansion of $f(x)$ on the half-range interval $[0, l]$ is given by

$$f(x) = \frac{a_0}{2} + \sum_{n=1}^{\infty} a_n \cos\left(\frac{n\pi x}{l}\right) \tag{9.25}$$

where $\qquad a_0 = \dfrac{2}{l} \displaystyle\int_0^l f(x)\,dx \quad \text{and} \quad a_n = \dfrac{2}{l} \displaystyle\int_0^l f(x) \cos\left(\dfrac{n\pi x}{l}\right) dx.$

The convergence Theorem 9.1 can be extended as follows.

If $x \in [0, l]$ and $f(x)$ has left and right hand derivatives at x, then at x, the Fourier cosine series converges to $[f(x+) + f(x-)]/2$. At a point of continuity, the Fourier cosine series converges to $f(x)$.

If $\lim_{x \to 0^+} f'(x)$ exists, then at $x = 0$, the series converges to $f(0+)$.

If $\lim_{x \to l^-} f'(x)$ exists, then at $x = l$, the series converges to $f(l-)$.

We now, define the Fourier sine series.

Theroem 9.5 (Fourier sine series) Let $f(x)$ be piecewise continuous on $[0, l]$. Then, the Fourier sine series expansion of $f(x)$ on $[0, l]$ is given by

$$f(x) = \sum_{n=1}^{\infty} b_n \sin\left(\frac{n\pi x}{l}\right) \tag{9.26}$$

where $\qquad b_n = \dfrac{2}{l} \displaystyle\int_0^l f(x) \sin\left(\dfrac{n\pi x}{l}\right) dx.$

If $x \in [0, l]$ and $f(x)$ has left and right hand derivatives at x, then at x, the Fourier sine series converges to $[f(x+) + f(x-)]/2$. At both the end points $x = 0$ and l, the series converges to 0.

Example 9.6   Find the Fourier cosine and sine series of the function $f(x) = 1$, $0 \le x \le 2$.

Solution   The Fourier coefficients for the cosine series are obtained as follows,

$$a_0 = \frac{2}{2} \int_0^2 1 \cdot dx = 2$$

$$a_n = \frac{2}{2} \int_0^2 \cos\left(\frac{n\pi x}{2}\right) dx = \left[\frac{\sin(n\pi x/2)}{(n\pi/2)}\right]_0^2 = 0.$$

Therefore, the Fourier cosine series is $f(x) = 1$ itself.
For the Fourier sine series, we have

$$b_n = \frac{2}{2} \int_0^2 \sin\left(\frac{n\pi x}{2}\right) dx = \left[-\frac{\cos(n\pi x/2)}{(n\pi/2)}\right]_0^2 = \frac{2}{n\pi}[1 - \cos n\pi]$$

$$= \frac{2}{n\pi}[1 - (-1)^n] = \begin{cases} 0, & n \text{ even} \\ \dfrac{4}{n\pi}, & n \text{ odd.} \end{cases}$$

Therefore, $\qquad f(x) = \dfrac{4}{\pi}\left[\sin\left(\dfrac{\pi x}{2}\right) + \dfrac{1}{3}\sin\left(\dfrac{3\pi x}{2}\right) + \dfrac{1}{5}\sin\left(\dfrac{5\pi x}{2}\right) + \ldots\right].$

Example 9.7 Find the Fourier cosine series of the function

$$f(x) = \begin{cases} x^2, & 0 \le x \le 2 \\ 4, & 2 \le x \le 4. \end{cases}$$

Solution Note that $f(x)$ is to be extended as an even function. We have

$$a_0 = \frac{2}{4} \int_0^4 f(x)dx = \frac{1}{2}\left[\int_0^2 x^2 dx + \int_2^4 4\,dx\right]$$

$$= \frac{1}{2}\left[\left(\frac{x^3}{3}\right)_0^2 + \{4x\}_2^4\right] = \frac{1}{2}\left[\frac{8}{3} + 8\right] = \frac{16}{3}.$$

$$a_n = \frac{1}{2}\left[\int_0^2 x^2 \cos\left(\frac{n\pi x}{4}\right)dx + \int_2^4 4\cos\left(\frac{n\pi x}{4}\right)dx\right]$$

$$= \frac{1}{2}\left[\left\{\frac{x^2\sin(n\pi x/4)}{(n\pi/4)} + \frac{2x\cos(n\pi x/4)}{(n\pi/4)^2} - \frac{2\sin(n\pi x/4)}{(n\pi/4)^3}\right\}_0^2\right.$$

$$\left. + 4\left\{\frac{\sin(n\pi x/4)}{(n\pi/4)}\right\}_2^4\right]$$

$$= \frac{1}{2}\left[\frac{4\sin(n\pi/2)}{(n\pi/4)} + \frac{4\cos(n\pi/2)}{(n\pi/4)^2} - \frac{2\sin(n\pi/2)}{(n\pi/4)^3} - \frac{4\sin(n\pi/2)}{(n\pi/4)}\right]$$

$$= \frac{32}{n^2\pi^2}\left[\cos\left(\frac{n\pi}{2}\right) - \left(\frac{2}{n\pi}\right)\sin\left(\frac{n\pi}{2}\right)\right].$$

Therefore,

$$f(x) = \frac{8}{3} + \frac{32}{\pi^2}\sum_{n=1}^{\infty}\frac{1}{n^2}\left[\cos\left(\frac{n\pi}{2}\right) - \left(\frac{2}{n\pi}\right)\sin\left(\frac{n\pi}{2}\right)\right]\cos\left(\frac{n\pi x}{4}\right).$$

9.3.1 Complex Form of Fourier Series

The Fourier series (see Eq. 9.8) can be expressed in the complex form. Using the Euler's formulas $e^{ix} = \cos x + i \sin x$ and $e^{-ix} = \cos x - i \sin x$, we obtain

$$f(x) = \frac{a_0}{2} + \sum_{n=1}^{\infty} \left[\frac{1}{2} a_n (e^{in\pi x/l} + e^{-in\pi x/l}) + \frac{1}{2i} b_n (e^{in\pi x/l} - e^{-in\pi x/l}) \right]$$

$$= \frac{a_0}{2} + \sum_{n=1}^{\infty} \left[\frac{1}{2} (a_n - ib_n) e^{in\pi x/l} + \frac{1}{2} (a_n + ib_n) e^{-in\pi x/l} \right].$$

Defining $c_0 = \dfrac{a_0}{2}$, $c_n = \dfrac{1}{2}(a_n - ib_n)$, and $c_{-n} = \dfrac{1}{2}(a_n + ib_n)$, we write $f(x)$ as

$$f(x) = c_0 + \sum_{n=1}^{\infty} [c_n e^{in\pi x/l} + c_{-n} e^{-in\pi x/l}] = \sum_{n=-\infty}^{\infty} c_n e^{in\pi x/l}.$$

Since $f(x)$ is real, c_n and c_{-n} are complex conjugates. From the definition, we have

$$c_0 = \frac{a_0}{2} = \frac{1}{2l} \int_{-l}^{l} f(x) dx. \tag{9.27}$$

$$c_n = \frac{1}{2}(a_n - ib_n) = \frac{1}{2} \left[\frac{1}{l} \int_{-l}^{l} f(x) \cos\left(\frac{n\pi x}{l}\right) dx - \frac{i}{l} \int_{-l}^{l} f(x) \sin\left(\frac{n\pi x}{l}\right) dx \right]$$

$$= \frac{1}{2l} \int_{-l}^{l} f(x) \left[\cos\left(\frac{n\pi x}{l}\right) - i \sin\left(\frac{n\pi x}{l}\right) \right] dx = \frac{1}{2l} \int_{-l}^{l} f(x) e^{-in\pi x/l} dx. \tag{9.28}$$

$$c_{-n} = \frac{1}{2}(a_n + ib_n) = \frac{1}{2l} \int_{-l}^{l} f(x) \left[\cos\left(\frac{n\pi x}{l}\right) + i \sin\left(\frac{n\pi x}{l}\right) \right] dx$$

$$= \frac{1}{2l} \int_{-l}^{l} f(x) e^{in\pi x/l} dx. \tag{9.29}$$

All the coefficients can be expressed in terms of a single expression as

$$c_n = \frac{1}{2l} \int_{-l}^{l} f(x) e^{-in\pi x/l} dx, \ n = 0, \pm 1, \pm 2, \ldots. \tag{9.30}$$

We have the following convergence result.

Theorem 9.6 If $f(x)$ satisfies the hypothesis of Theorem 9.1, then the complex form of the Fourier series converges to $f(x)$ at a point of continuity and to $[f(x+) + f(x-)]/2$ at a point of discontinuity.

Example 9.8 Find the complex Fourier series of the function

$$f(x) = e^{-x}, -\pi < x < \pi.$$

Solution We have $l = \pi$. Hence, the Fourier coefficients are given by

$$c_n = \frac{1}{2\pi} \int_{-\pi}^{\pi} e^{-x} e^{-inx} dx = \frac{1}{2\pi} \int_{-\pi}^{\pi} e^{-(1+in)x} dx$$

$$= \frac{-1}{2\pi(1+in)}\left[e^{-(1+in)x}\right]_{-\pi}^{\pi} = \frac{-1}{2\pi(1+in)}\left[e^{-(1+in)\pi} - e^{(1+in)\pi}\right]$$

$$= \frac{-(1-in)}{2\pi(1+n^2)}\left[e^{-\pi}(\cos n\pi - i\sin n\pi) - e^{\pi}(\cos n\pi + i\sin n\pi)\right]$$

$$= \frac{-(1-in)}{2\pi(1+n^2)}\left[-\cos n\pi(e^{\pi} - e^{-\pi})\right] = \frac{\sinh \pi}{\pi(1+n^2)}(1-in)\cos n\pi.$$

Therefore, the complex form of the Fourier series is

$$f(x) = \frac{\sinh \pi}{\pi} \sum_{n=-\infty}^{\infty} (-1)^n \left(\frac{1-in}{1+n^2}\right) e^{inx}.$$

It is easy to convert the complex exponential form of the Fourier series to a real trigonometric form. We have

$$a_0 = 2c_0, \quad a_n = c_n + c_{-n}, \quad b_n = i\,(c_n - c_{-n}).$$

In many areas of engineering, the complex form of Fourier series is used. Since $f(x)$ is defined on $(-l, l)$ and $f(x + 2l) = f(x)$, the *fundamental period* T of $f(x)$ is $T = 2l$. We define $\omega = 2\pi/T = \pi/l$ as the *fundamental angular frequency*. In terms of ω, we can write the Fourier series as

$$f(x) = \frac{a_0}{2} + \sum_{n=1}^{\infty} [a_n \cos(n\omega x) + b_n \sin(n\omega x)] \tag{9.31}$$

or
$$f(x) = \sum_{-\infty}^{\infty} c_n e^{in\omega x}. \tag{9.32}$$

The plot of the points $(n\omega, |c_n|)$ where ω is the fundamental angular frequency and c_n are the Fourier coefficients defined in Eq. (9.30), is called the *frequency spectrum* of $f(x)$.

Example 9.9 Find the frequency spectrum of the periodic pulse defined by

$$f(x) = \begin{cases} -1, & -1 \le x < 0 \\ 1, & 0 \le x < 1, \ f(x+2) = f(x). \end{cases}$$

Solution We have the period $T = 2l = 2$. The graph of $f(x)$ is given in Fig. 9.11. The fundamental angular frequency is $\omega = \pi$. We have

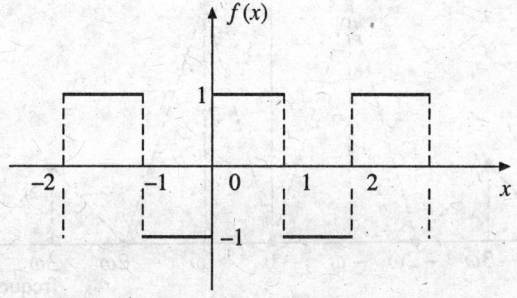

Fig. 9.11.   Graph of $f(x)$ Example 9.9.

$$c_n = \frac{1}{2} \int_{-1}^{1} f(x) e^{-in\pi x} dx = \frac{1}{2} \left[- \int_{-1}^{0} e^{-in\pi x} dx + \int_{0}^{1} e^{-in\pi x} dx \right]$$

$$= \frac{1}{2} \left[\left\{ \frac{e^{-in\pi x}}{in\pi} \right\}_{-1}^{0} - \left\{ \frac{e^{-in\pi x}}{in\pi} \right\}_{0}^{1} \right] = \frac{1}{2} \left[-\frac{i}{n\pi} \{ 1 - e^{in\pi} - e^{-in\pi} + 1 \} \right]$$

$$= -\frac{i}{n\pi} [1 - \cos n\pi].$$

The expression on the right hand side is not valid for $n = 0$. Hence, we compute this term directly to get

$$c_0 = \frac{1}{2} \int_{-1}^{1} f(x) dx = \frac{1}{2} \left[- \int_{-1}^{0} dx + \int_{0}^{1} dx \right] = 0.$$

We have
$$|c_n| = \frac{1}{|n|\pi} [1 - (-1)^n] = \begin{cases} 0, & \text{for } n \text{ even} \\ \dfrac{2}{|n|\pi}, & \text{for } n \text{ odd.} \end{cases}$$

The values of n, $|c_n|$ are tabulated below.

n	-4	-3	-2	-1	0	1	2	3	4
$|c_n|$	0	$\frac{2}{3\pi}$	0	$\frac{2}{\pi}$	0	$\frac{2}{\pi}$	0	$\frac{2}{3\pi}$	0

The frequency spectrum is plotted in Fig. 9.12. On the x-axis, the unit $n\omega = n\pi$ is plotted.

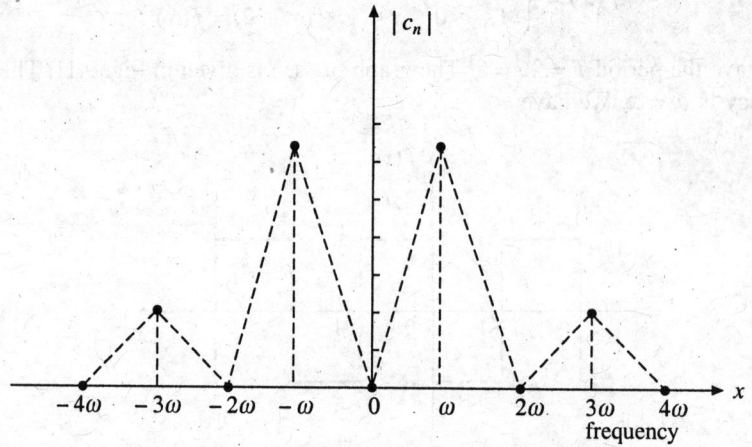

Fig. 9.12. Frequency spectrum. Example 9.9.

Exercise 9.2

In each of the following problems, write the Fourier cosine series and Fourier sine series for the function in the given interval (wherever they are possible).

1. $f(x) = k, 0 \leq x \leq 5$.

2. $f(x) = \cos x, 0 < x \leq \pi/2$.

3. $f(x) = \begin{cases} -1, & 0 \leq x \leq 1, \\ 1, & 1 < x \leq 2. \end{cases}$

4. $f(x) = \begin{cases} x, & 0 < x < 2, \\ 2, & 2 \leq x < 4. \end{cases}$

5. $f(x) = x + x^2, 0 < x < 1$.

6. $f(x) = \begin{cases} x, & 0 < x < \pi/2, \\ \pi - x, & \pi/2 \leq x < \pi. \end{cases}$

7. $f(x) = \begin{cases} \pi - x, & 0 < x < \pi, \\ 0, & \pi \leq x < 2\pi. \end{cases}$

8. $f(x) = \begin{cases} 2, & 0 < x < 2, \\ 4 - x, & 2 \leq x < 4. \end{cases}$

9. $f(x) = \begin{cases} x^2, & 0 \leq x < 1, \\ 1, & 1 \leq x < 2. \end{cases}$

10. $f(x) = \begin{cases} 1, & 0 \leq x \leq 1, \\ -1, & 1 < x < 2, \\ 0, & 2 \leq x < 3. \end{cases}$

11. $f(x) = \begin{cases} x, & 0 \leq x < 1, \\ 1, & 1 \leq x < 2, \\ 3 - x, & 2 \leq x \leq 3. \end{cases}$

12. $f(x) = e^{-x}, 0 \leq x \leq 2$.

13. $f(x) = \cos 3x, 0 \leq x \leq \pi$.

14. $f(x) = 1 + x, 0 \leq x \leq 1$.

15. $f(x) = \sin 3x, 0 \leq x \leq \pi$.

Find the complex form of Fourier series of $f(x)$ on the given interval.

16. $f(x) = \begin{cases} 1, & -2 \leq x < 1, \\ 0, & 1 \leq x < 2. \end{cases}$

17. $f(x) = \begin{cases} x, & -\pi < x \leq 0, \\ 0, & 0 < x < \pi. \end{cases}$

18. $f(x) = e^{-|x|}, -2 < x < 2$.

Find the frequency spectrum of $f(x)$ in the following problems.

19. $f(x) = \begin{cases} 0, & -\pi/2 \leq x < 0 \\ \sin x, & 0 \leq x \leq \pi/2. \end{cases}$

$f(x + \pi) = f(x)$.

20. $f(x) = \begin{cases} -2 \cos x, & -\pi/2 < x < 0 \\ 2 \cos x, & 0 \leq x < \pi/2. \end{cases}$

$f(x + \pi) = f(x)$.

9.4   Fourier Integrals

In the previous sections, we have seen that if $f(x)$ is piecewise smooth on any interval $[-l, l]$ or $[0, l]$ then it can be represented by a Fourier series. If $f(x)$ is a periodic function (so that it is defined on the entire real line) then also $f(x)$ can be represented by a Fourier series. If $f(x)$ is not a periodic function then it cannot be represented by a Fourier series over the entire real line. However, we may be able to represent $f(x)$ in an integral form.

Let $f(x)$ have the following properties.

(P1) $f(x)$ is piecewise continuous on every interval $[-l, l]$.

(P2) $f(x)$ is absolutely integrable on the x-axis, that is $\int_{-\infty}^{\infty} |f(x)|\,dx$ converges.

(P3) At every x on the real line, $f(x)$ has left and right hand derivatives.

Consider the Fourier series representation of $f(x)$ on the interval $[-l, l]$. We have

$$f(x) = \frac{a_0}{2} + \sum_{n=1}^{\infty} \left[a_n \cos\left(\frac{n\pi x}{l}\right) + b_n \sin\left(\frac{n\pi x}{l}\right) \right] \tag{9.33}$$

where

$$a_0 = \frac{1}{l} \int_{-l}^{l} f(t)\,dt, \quad a_n = \frac{1}{l} \int_{-l}^{l} f(t) \cos\left(\frac{n\pi t}{l}\right) dt$$

and

$$b_n = \frac{1}{l} \int_{-l}^{l} f(t) \sin\left(\frac{n\pi t}{l}\right) dt.$$

Set

$$\omega_n = \frac{n\pi}{l} \quad \text{and} \quad \Delta\omega = \omega_n - \omega_{n-1} = \frac{\pi}{l}(n - n + 1) = \frac{\pi}{l}.$$

Then, we can write

$$f(x) = \frac{\Delta\omega}{2\pi} \int_{-l}^{l} f(t)\,dt + \frac{1}{\pi} \sum_{n=1}^{\infty} \left[\left\{ \int_{-l}^{l} f(t) \cos(\omega_n t)\,dt \right\} \cos(\omega_n x) \right.$$

$$\left. + \left\{ \int_{-l}^{l} f(t) \sin(\omega_n t)\,dt \right\} \sin(\omega_n x) \right] \Delta\omega. \tag{9.34}$$

Let $l \to \infty$ so that $(-l, l) \to (-\infty, \infty)$. Then $\Delta\omega \to 0$, and $\dfrac{\Delta\omega}{\pi} \int_{-l}^{l} f(t)\,dt \to 0$, since $\int_{-\infty}^{\infty} |f(x)|\,dx$

converges. As $l \to \infty$, the summation in Eq. (9.34) resembles a Riemann sum of a definite integral. As $\Delta\omega \to 0$, we have (as $f(x)$ satisfies the properties **P1** to **P3**)

$$f(x) = \frac{1}{\pi} \int_{0}^{\infty} \left[\left\{ \int_{-\infty}^{\infty} f(t) \cos(\omega t)\,dt \right\} \cos(\omega x) + \left\{ \int_{-\infty}^{\infty} f(t) \sin(\omega t)\,dt \right\} \sin(\omega x) \right] d\omega. \tag{9.35}$$

This equation is analogous to the Fourier series, wherein the sums are replaced by the integrals.

Denote

$$A(\omega) = \int_{-\infty}^{\infty} f(t) \cos(\omega t)\,dt, \quad B(\omega) = \int_{-\infty}^{\infty} f(t) \sin(\omega t)\,dt. \tag{9.36}$$

Then, from Eq. (9.35), we write

$$f(x) = \frac{1}{\pi} \int_{0}^{\infty} [A(\omega) \cos(\omega x) + B(\omega) \sin(\omega x)]\,d\omega. \tag{9.37}$$

This equation is called the *Fourier integral* of $f(x)$ or Fourier integral representation of $f(x)$.

We have the following convergence result.

Theorem 9.7 If $f(x)$ satisfies the properties **P1** to **P3**, then the Fourier integral representation of $f(x)$ converges to $f(x)$ at a point of continuity and to $[f(x+) + f(x-)]/2$ at a point of discontinuity.

Example 9.10 Find the Fourier integral representation of the function

$$f(x) = \begin{cases} 0, & x < 0 \\ 1, & 0 \le x \le 1 \\ 0, & x > 1. \end{cases}$$

Hence, show that

$$\int_0^\infty \frac{\sin(x/2)}{x} \, dx = \frac{\pi}{2}.$$

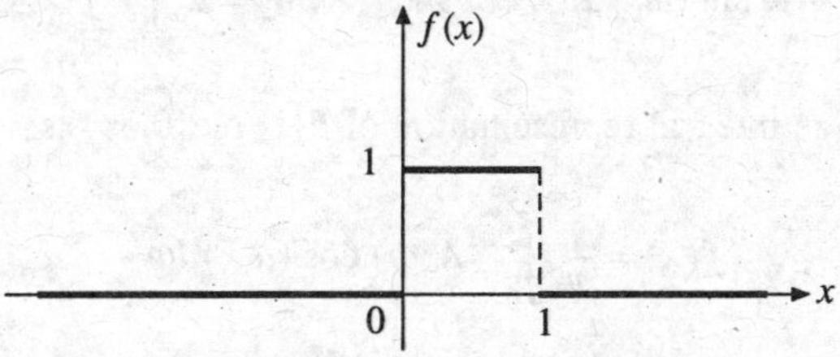

Fig. 9.13.   Example 9.10.

Solution The graph of $f(x)$ is given in Fig. 9.13. Now, $f(x)$ satisfies the hypothesis of Theorem 9.7. We have

$$A(\omega) = \int_{-\infty}^{\infty} f(t) \cos(\omega t) \, dt = \int_0^1 \cos(\omega t) \, dt = \left[\frac{\sin(\omega t)}{\omega} \right]_0^1 = \frac{\sin \omega}{\omega}.$$

$$B(\omega) = \int_{-\infty}^{\infty} f(t) \sin(\omega t) \, dt = \int_0^1 \sin(\omega t) \, dt = \left[\frac{-\cos(\omega t)}{\omega} \right]_0^1 = \frac{1}{\omega}(1 - \cos \omega).$$

Therefore,

$$f(x) = \frac{1}{\pi} \int_0^\infty \frac{1}{\omega} [\sin \omega \cos(\omega x) + (1 - \cos \omega) \sin(\omega x)] d\omega$$

$$= \frac{1}{\pi} \int_0^\infty \frac{2}{\omega} \sin\left(\frac{\omega}{2}\right) \left[\cos\left(\frac{\omega}{2}\right) \cos(\omega x) + \sin\left(\frac{\omega}{2}\right) \sin(\omega x) \right] d\omega$$

$$= \frac{2}{\pi} \int_0^\infty \frac{1}{\omega} \sin\left(\frac{\omega}{2}\right) \cos \omega \left(x - \frac{1}{2} \right) d\omega.$$

This is the Fourier integral representation of the given function.

Let $x = 1/2$. Then $f(1/2) = 1$. Hence

$$1 = \frac{2}{\pi} \int_0^\infty \frac{1}{\omega} \sin\left(\frac{\omega}{2}\right) d\omega \quad \text{or} \quad \int_0^\infty \frac{1}{\omega} \sin\left(\frac{\omega}{2}\right) d\omega = \frac{\pi}{2} \quad \text{or} \quad \int_0^\infty \frac{\sin(x/2)}{x} \, dx = \frac{\pi}{2}.$$

Fourier cosine and sine integrals

Similar to Fourier cosine and sine series defined on half-intervals $[0, l]$, we can define Fourier cosine and sine integral representations of functions defined on the real half-line $[0, \infty)$. We assume that $f(x)$ is defined for $x \geq 0$ and that $\int_0^\infty |f(x)| \, dx$ converges.

To obtain the Fourier cosine integral representation of $f(x)$, we use an even extension of $f(x)$ to the whole real line, that is, we define a function $g(x)$ as

$$g(x) = \begin{cases} f(x), & \text{for } x \geq 0, \\ f(-x), & \text{for } x < 0. \end{cases} \tag{9.38}$$

Since $g(x)$ is an even function, we obtain

$$B(\omega) = \int_{-\infty}^{\infty} g(t) \sin(\omega t) \, dt = 0 \quad \text{and} \quad A(\omega) = 2 \int_0^\infty f(t) \cos(\omega t) dt. \tag{9.39}$$

Therefore, the Fourier cosine integral representation of $f(x)$ on $[0, \infty)$ is

$$f(x) = \frac{1}{\pi} \int_0^\infty A(\omega) \cos(\omega x) d\omega \tag{9.40}$$

where $A(\omega)$ is as defined in Eq. (9.39).

Similarly, to obtain the Fourier sine integral representation of $f(x)$, we use an odd extension of $f(x)$ to the whole real line, that is, we define

$$g(x) = \begin{cases} f(x), & \text{for } x \geq 0 \\ -f(-x), & \text{for } x < 0. \end{cases}$$

Then, we have

$$A(\omega) = \int_{-\infty}^{\infty} g(t) \cos(\omega t) \, dt = 0 \quad \text{and} \quad B(\omega) = 2 \int_0^\infty f(t) \sin(\omega t) \, dt. \tag{9.41}$$

Therefore, the Fourier sine integral representation of $f(x)$ on $[0, \infty)$ is

$$f(x) = \frac{1}{\pi} \int_0^\infty B(\omega) \sin(\omega x) d\omega \tag{9.42}$$

where $B(\omega)$ is as defined in Eq. (9.41).

Theorem 9.8 Let $f(x)$ satisfy the following properties

(i) it is piecewise continuous on each interval $[0, l]$,

(ii) it is absolutely integrable on the real x-axis, and

(iii) at every $x \in (0, \infty)$, $f(x)$ has left and right hand derivatives.

Then, the Fourier cosine and sine integral representations converge to $f(x)$ at a point of continuity and to $[f(x+) + f(x-)]/2$ at a point of discontinuity.

Example 9.11 Find the Fourier cosine and sine integral representations of $f(x) = e^{-kx}$, $x \geq 0$, where k is a positive constant.

Solution We have

$$\int_0^\infty |f(x)|\, dx = \int_0^\infty e^{-kx}\, dx = \lim_{b \to \infty} \left[-\frac{e^{-kx}}{k} \right]_0^b = \lim_{b \to \infty} \left[\frac{1}{k}(1 - e^{-kb}) \right] = \frac{1}{k}.$$

Therefore, the improper integral $\int_0^\infty |f(x)|\, dx$ exists and $f(x)$ satisfies the hypothesis of Theorem 9.8.

To obtain the Fourier cosine integral representation, we compute $A(\omega)$ as

$$A(\omega) = 2 \int_0^\infty f(t) \cos(\omega t)\, dt = 2 \int_0^\infty e^{-kt} \cos(\omega t)\, dt.$$

Now,

$$I = \int_0^\infty e^{-kt} \cos(\omega t)\, dt = \left[\frac{e^{-kt} \sin(\omega t)}{\omega} \right]_0^\infty + \frac{k}{\omega} \int_0^\infty e^{-kt} \sin(\omega t)\, dt$$

$$= \frac{k}{\omega} \left[\left\{ -\frac{e^{-kt} \cos(\omega t)}{\omega} \right\}_0^\infty - \frac{k}{\omega} \int_0^\infty e^{-kt} \cos(\omega t)\, dt \right] = \frac{k}{\omega^2} - \frac{k^2}{\omega^2} I$$

or

$$I = \frac{k}{k^2 + \omega^2}. \qquad (9.42a)$$

Therefore, $A(\omega) = \dfrac{2k}{k^2 + \omega^2}$.

Hence, for all $x \geq 0$, (since $f(x)$ is continuous for $x > 0$ and has a right derivative at 0)

$$e^{-kx} = \frac{2k}{\pi} \int_0^\infty \frac{\cos(\omega x)}{k^2 + \omega^2}\, d\omega.$$

To obtain the Fourier sine integral representation, we compute $B(\omega)$ as

$$B(\omega) = 2 \int_0^\infty e^{-kt} \sin(\omega t)\, dt = \frac{2\omega}{k^2 + \omega^2}.$$

Therefore, for $x > 0$

$$e^{-kx} = \frac{2}{\pi} \int_0^\infty \frac{\omega}{(k^2 + \omega^2)} \sin(\omega x)\, d\omega.$$

Note that the sine integral does not equal the function at $x = 0$. The above two integrals are called the *Laplace integrals*.

Example 9.12 Show that the Fourier integral of $f(x)$ can be written as

$$f(x) = \frac{1}{\pi} \int_0^\infty \left\{ \int_{-\infty}^\infty \cos[\omega(x - t)] f(t)\, dt \right\} d\omega.$$

Solution We have from the definition of Fourier integral (Eq. (9.35))

$$f(x) = \frac{1}{\pi} \int_0^\infty \left[\int_{-\infty}^\infty f(t) \{\cos(\omega t) \cos(\omega x) + \sin(\omega t) \sin(\omega x)\} dt \right] d\omega$$

$$= \frac{1}{\pi} \int_0^\infty \left\{ \int_{-\infty}^\infty \cos[\omega(x-t)] f(t) dt \right\} d\omega.$$

Example 9.13 Find the solution of the integral equation

$$\int_0^\infty f(x) \cos ax \, dx = e^{-a}, \text{ where } a \text{ is a constant.}$$

Solution The given integral is a Fourier cosine integral representation.

Let $$g(x) = \frac{1}{\pi} \int_0^\infty A(\omega) \cos(\omega x) d\omega, \quad \text{where} \quad A(\omega) = 2 \int_0^\infty g(t) \cos(\omega t) dt.$$

Comparing this equation with the given equation, we get $x = a$, $\pi g(a) = e^{-a}$, $f(x) = A(x)$.

Therefore, $$A(\omega) = 2 \int_0^\infty \frac{1}{\pi} e^{-t} \cos(\omega t) dt = \frac{2}{\pi} \left(\frac{1}{1+\omega^2} \right)$$

using Eq. (9.42a) with $k = 1$.

Hence, $$f(x) = A(x) = \frac{2}{\pi(1+x^2)}, \ x > 0.$$

Complex form of Fourier integral representation

The fourier integral (Eq. 9.37) can also be written in equivalent complex form analogus to the complex form of Fourier series. Substituting the expressions for $A(w)$ and $B(\omega)$ from Eq. (9.36) into Eq. (9.37), we obtain

$$f(x) = \frac{1}{\pi} \int_0^\infty \left[\int_{-\infty}^\infty f(t) \{\cos(\omega t) \cos(\omega x) + \sin(\omega t) \sin(\omega x)\} dt \right] d\omega$$

$$= \frac{1}{\pi} \int_0^\infty \left[\int_{-\infty}^\infty f(t) \cos \omega(t-x) dt \right] d\omega = \frac{1}{2\pi} \int_{-\infty}^\infty \int_{-\infty}^\infty f(t) \cos \omega(t-x) dt d\omega \qquad (9.43)$$

since $\cos \omega(t-x)$ is an even function with respect to the variable ω. We add the term

$$\frac{i}{2\pi} \int_{-\infty}^\infty \int_{-\infty}^\infty f(t) \sin \omega(t-x) \, dt \, d\omega$$

whose value is equal to zero, to the right hand side of Eq. (9.43). Then, we have

$$f(x) = \frac{1}{2\pi} \int_{-\infty}^{\infty} \int_{-\infty}^{\infty} f(t) \left[\cos \omega(t-x) + i \sin \omega(t-x) \right] dt \, d\omega$$

$$= \frac{1}{2\pi} \int_{-\infty}^{\infty} \int_{-\infty}^{\infty} f(t) e^{i\omega(t-x)} \, dt \, d\omega = \frac{1}{2\pi} \int_{-\infty}^{\infty} \left[\int_{-\infty}^{\infty} f(t) e^{i\omega t} \, dt \right] e^{-i\omega x} \, d\omega$$

$$= \frac{1}{2\pi} \int_{-\infty}^{\infty} c(\omega) e^{-i\omega x} \, d\omega \tag{9.44}$$

where $c(\omega) = \displaystyle\int_{-\infty}^{\infty} f(t) \, e^{i\omega t} \, dt.$ $\qquad\qquad$ (9.45)

These equations represent the complex form of Fourier integral representation.

Example 9.14 Find the complex form of Fourier integral representation for the function

$$f(x) = \begin{cases} |x|, & -\pi < x < \pi, \\ 0, & \text{elsewhere.} \end{cases}$$

Verify that it is same as that obtained by using the Fourier cosine integral representation.

Solution We have

$$c(\omega) = \int_{-\infty}^{\infty} f(t) e^{i\omega t} \, dt = \int_{-\pi}^{\pi} |t| \, e^{i\omega t} \, dt = \int_{0}^{\pi} t e^{i\omega t} \, dt - \int_{-\pi}^{0} t e^{i\omega t} \, dt$$

$$= \left[\left\{ \frac{t}{i\omega} - \frac{1}{(i\omega)^2} \right\} e^{i\omega t} \right]_{0}^{\pi} - \left[\left\{ \frac{t}{i\omega} - \frac{1}{(i\omega)^2} \right\} e^{i\omega t} \right]_{-\pi}^{0}$$

$$= \left(-\frac{i\pi}{\omega} + \frac{1}{\omega^2} \right) e^{i\omega\pi} - \frac{2}{\omega^2} + \left(\frac{i\pi}{\omega} + \frac{1}{\omega^2} \right) e^{-i\omega\pi}$$

$$= \frac{2}{\omega^2} \cos(\pi\omega) + \frac{2\pi}{\omega} \sin(\pi\omega) - \frac{2}{\omega^2}. \tag{9.46}$$

Therefore,

$$f(x) = \frac{1}{\pi} \int_{-\infty}^{\infty} \frac{1}{\omega^2} \left[\cos(\pi\omega) + \pi\omega \sin(\pi\omega) - 1 \right] e^{-i\omega x} \, d\omega.$$

Directly, we obtain from Eq. (9.39)

$$A(\omega) = 2 \int_{0}^{\pi} t \cos(\omega t) \, dt = 2 \left[\frac{t}{\omega} \sin \omega t + \frac{1}{\omega^2} \cos \omega t \right]_{0}^{\pi}$$

$$= 2 \left[\frac{\pi}{\omega} \sin(\pi\omega) + \frac{1}{\omega^2} \{ \cos(\pi\omega) - 1 \} \right]$$

which is same as Eq. (9.46)

Exercise 9.3

Find the Fourier integral representations of the following functions.

1. $f(x) = \begin{cases} -1, & -2 < x < 0, \\ 1, & 0 < x < 2, \\ 0, & \text{elsewhere.} \end{cases}$

2. $f(x) = \begin{cases} 0, & x < 0, \\ x, & 0 \le x < 2, \\ 0, & x > 2. \end{cases}$

3. $f(x) = \begin{cases} 0, & x < 0, \\ e^{-x}, & x > 0. \end{cases}$

4. $f(x) = \begin{cases} x^2, & -3 \le x \le 3, \\ 0, & |x| > 3. \end{cases}$

5. $f(x) = \begin{cases} e^x, & |x| < 2, \\ 0, & |x| > 2. \end{cases}$

6. $f(x) = \begin{cases} |x|, & -\pi \le x \le \pi, \\ 0, & |x| > \pi. \end{cases}$

7. $f(x) = \begin{cases} 2, & -2 \le x < 1. \\ 1, & 1 \le x < 3. \\ 0, & \text{elsewhere.} \end{cases}$

8. $f(x) = \begin{cases} e^{-|x|}, & |x| < 1 \\ 0, & \text{elsewhere.} \end{cases}$

9. $f(x) = \begin{cases} -(1+x), & -1 < x < 0 \\ 1-x, & 0 \le x \le 1 \\ 0, & |x| > 1. \end{cases}$

10. $f(x) = \begin{cases} \sin x, & -2 \le x \le 0, \\ \cos x, & 0 < x \le 2, \\ 0, & |x| > 2. \end{cases}$

In the following problems, find the appropriate integral representation of the given function.

11. $f(x) = e^{-2x} + e^{-3x}, x > 0.$

    cosine representation.

12. $f(x) = \begin{cases} x^2, & 0 \le x \le 5 \\ 0, & x > 5, \end{cases}$

    cosine representation.

13. $f(x) = \begin{cases} \sin x, & 0 \le x \le \pi \\ 0, & x > \pi, \end{cases}$

    cosine representation.

14. $f(x) = \begin{cases} 0, & 0 < x < 1 \\ 1, & 1 < x < 2 \\ 0, & x > 2, \end{cases}$

    sine representation.

15. $f(x) = \begin{cases} \sinh x, & 0 \le x \le 3 \\ 0, & x > 3, \end{cases}$

    sine representation.

In the following problems, find the complex form of the Fourier integral representation.

16. $f(x) = \begin{cases} 0, & x < 0, \\ e^{-kx}, & x > 0, k > 0. \end{cases}$

17. $f(x) = \begin{cases} 0, & -\infty < x < -1, \\ x, & -1 < x < 0, \\ 1, & 0 < x < 1, \\ 0, & x > 1. \end{cases}$

18. $f(x) = \begin{cases} \sinh x, & |x| \le a, \\ 0, & |x| > a. \end{cases}$
  19. $f(x) = \begin{cases} \cosh 2x, & |x| \le a, \\ 0, & |x| > a. \end{cases}$

20. $f(x) = \begin{cases} 1 + x, & |x| \le 1, \\ 0, & |x| > 1. \end{cases}$
  21. $f(x) = \begin{cases} \sin(\pi x), & -2 \le x \le 2, \\ 0, & |x| > 2. \end{cases}$

22. Define a suitable function $f(x)$ and use the Fourier integral representation to show that

$$\int_0^\infty \frac{\sin ax}{x}\, dx = \frac{\pi}{2}, \quad a > 0.$$

23. Show that

$$x^2 f(x) = \frac{1}{\pi} \int_0^\infty A^*(\omega) \cos(\omega x)\, d\omega, \quad \text{where} \quad A^*(\omega) = -A''(\omega).$$

Assume that $g(x) = x^2 f(x)$ has the Fourier cosine integral representation given by Eqs. (9.39) and (9.40).

24. Show that

$$x f(x) = \frac{1}{\pi} \int_0^\infty B^*(\omega) \sin(\omega x)\, d\omega, \quad \text{where} \quad B^*(\omega) = -A'(\omega).$$

Assume that $g(x) = xf(x)$ has the Fourier sine integral representation given by Eqs. (9.41) and (9.42).

25. If

$$\int_0^\infty f(x) \sin ax\, dx = \begin{cases} 1, & 0 < a < 1 \\ 0, & a > 1. \end{cases}$$

then find $f(x)$.

9.5   Application of Fourier Series: Separation of Variables Solution of Linear Partial Differential Equations

9.5.1   Classification of Linear Second Order Equations

One of the important applications of the Fourier series is to obtain the solution of a class of linear, homogeneous partial differential equations together with the corresponding initial and boundary conditions. This method is also known as the *Fourier method* or the method of *separation of variables*.

Consider the second order, linear homogeneous partial differential equation (*pde*)

$$A \frac{\partial^2 u}{\partial x^2} + B \frac{\partial^2 u}{\partial x \partial y} + C \frac{\partial^2 u}{\partial y^2} + D \frac{\partial u}{\partial x} + E \frac{\partial u}{\partial y} + Fu = 0 \tag{9.47}$$

where A, B, C, D, E, F are functions of x, y or are real constants. The *pde* is said to be a

$$\begin{aligned} parabolic\ equation &\quad \text{if} \quad B^2 - 4AC = 0, \\ hyperbolic\ equation &\quad \text{if} \quad B^2 - 4AC > 0, \\ \text{and} \qquad elliptic\ equation &\quad \text{if} \quad B^2 - 4AC < 0. \end{aligned}$$

The importance of this classification is that it governs the number and type of conditions to be prescribed in order that the problem is well posed and has a unique solution. Note that the lower order (first order) terms do not play any role in the classification of the equation.

Some simple examples of the above equations are the following.

(i) *parabolic equation:* $\dfrac{\partial u}{\partial t} = c^2 \dfrac{\partial^2 u}{\partial x^2}$ (One dimensional heat equation).

$$\dfrac{\partial u}{\partial t} = c^2 \left(\dfrac{\partial^2 u}{\partial x^2} + \dfrac{\partial^2 u}{\partial y^2} \right)$$ (Two dimensional heat equation).

(ii) *hyperbolic equation:* $\dfrac{\partial^2 u}{\partial t^2} = c^2 \dfrac{\partial^2 u}{\partial x^2}$ (One dimensional wave equation).

$$\dfrac{\partial^2 u}{\partial t^2} = c^2 \left(\dfrac{\partial^2 u}{\partial x^2} + \dfrac{\partial^2 u}{\partial y^2} \right)$$ (Two dimensional wave equation).

(iii) *elliptic equation:* $\dfrac{\partial^2 u}{\partial x^2} + \dfrac{\partial^2 u}{\partial y^2} = 0$ (Two dimensional Laplace equation).

Example 9.15   Classify the following partial differential equations.

(a) $\dfrac{\partial^2 u}{\partial x^2} = 5 \dfrac{\partial u}{\partial x} + \dfrac{\partial u}{\partial y}$,

(b) $\dfrac{\partial^2 u}{\partial x^2} + \dfrac{\partial^2 u}{\partial y^2} = 0$.

(c) $\dfrac{\partial^2 u}{\partial x^2} + 3 \dfrac{\partial^2 u}{\partial x \partial y} + \dfrac{\partial^2 u}{\partial y^2} = 0$.

Solution

(a) Write the given equation as $\dfrac{\partial^2 u}{\partial x^2} - 5 \dfrac{\partial u}{\partial x} - \dfrac{\partial u}{\partial y} = 0$. We have $A = 1, B = 0, C = 0, D = -5,$ $E = -1$, and $B^2 - 4AC = 0$. Hence, the given *pde* is a parabolic equation.

(b) We have $A = 1, B = 0, C = 1$ and $B^2 - 4AC = -4 < 0$. Hence, the given *pde* is an elliptic equation.

(c) We have $A = 1, B = 3, C = 1$, and $B^2 - 4AC = 5 > 0$. Hence, the given *pde* is an hyperbolic equation.

One of the important properties of the linear, homogenous *pdes* is that the *superposition principle* for the solutions holds. We now state this principle.

Superposition Principle If $u_1, u_2, \ldots, u_n$ are solutions of a linear, homogeneous partial differential equation, then their linear combination

$$u = c_1 u_1 + c_2 u_2 + \ldots + c_n u_n \tag{9.48}$$

where $c_i, i = 1, 2, \ldots, n$ are arbitrary constants, is also a solution.

If the *pde* has infinite number of solutions $u_1, u_2, \ldots,$ then their linear combination

$$u = c_1 u_1 + c_2 u_2 + \ldots = \sum_{n=1}^{\infty} c_n u_n \tag{9.49}$$

is also a solution.

We now, present the method of separation of variables.

9.5.2   Separation of Variables Method (Fourier Method)

Let x, y be the independent variables and u be the dependent variable of the *pde*. In the method of separation of variables, we find a particular solution in the form of a product of a function of x and a function of y. That is, we write

$$u(x, y) = X(x) \, Y(y). \tag{9.50}$$

Then, we have

$$\frac{\partial u}{\partial x} = \frac{\partial}{\partial x}(XY) = X'Y, \frac{\partial u}{\partial y} = \frac{\partial}{\partial y}(XY) = XY$$

$$\frac{\partial^2 u}{\partial x^2} = X''Y, \ \frac{\partial^2 u}{\partial y^2} = XY'' \tag{9.51}$$

where $\qquad X' = \dfrac{dX}{dx}, Y' = \dfrac{dY}{dy}, X'' = \dfrac{d^2 X}{dx^2}$ and $Y'' = \dfrac{d^2 Y}{dy^2}.$

These expressions are substituted in the given *pde*, the variables X, Y are separated and the solutions for X and Y are determined. We illustrate the method through the following examples.

Example 9.16   Using the method of separation of variables, solve the parabolic partial differential equation

$$\frac{\partial^2 u}{\partial x^2} = 16 \, \frac{\partial u}{\partial y}.$$

Solution   Using $u(x, y) = X(x) \, Y(y)$, we obtain

$$X''Y = 16XY' \quad \text{or} \quad \frac{X''}{16X} = \frac{Y'}{Y}.$$

The variables X, Y are now separated. The left hand side is a function of x alone and the right hand side is a function of y alone. Therefore, both the left and right hand sides must be equal to a constant, say k^2 or $-k^2$. We have the following three cases.

Case 1. $k^2 = 0$. Then, we obtain

$$X'' = 0 \quad \text{and} \quad Y' = 0.$$

Integrating, we obtain $X = A_1 x + B_1$ and $Y = C_1$.

Hence, $\qquad\qquad u(x, y) = (A_1 x + B_1) \, C_1 = Ax + B.$

where $A = A_1 C_1$ and $B = B_1 C_1$ are arbitrary constants.

Case 2.   When the constant is k^2, we obtain

$$X'' - 16k^2 X = 0, \quad Y' - k^2 Y = 0.$$

The solutions of these equations are

$$X(x) = A_1 e^{4kx} + B_1 e^{-4kx} = A_2 \cosh(4kx) + B_2 \sinh(4kx) \quad \text{and} \quad Y(y) = C_1 e^{k^2 y}.$$

Hence, $\qquad\qquad u(x, y) = [A_2 \cosh(4kx) + B_2 \sinh(4kx)]C_1 e^{k^2 y}$

$$= [A \cosh(4kx) + B \sinh(4kx)] \, e^{k^2 y}$$

where $A = A_2 C_1$ and $B = B_2 C_1$ are arbitrary constants.

Case 3. When the constant is $-k^2$, we obtain

$$X'' + 16k^2 X = 0, \quad Y' + k^2 Y = 0.$$

The solutions of these equations are

$$X(x) = A_1 \cos(4\,kx) + B_1 \sin(4kx) \quad \text{and} \quad Y(y) = C_1 e^{-k^2 y}.$$

Hence,

$$u(x, y) = [A_1 \cos(4kx) + B_1 \sin(4kx)]\, C_1 e^{-k^2 y}$$

$$= [A \cos(4kx) + B \sin(4kx)] e^{-k^2 y}$$

where A and B are arbitrary constants.

In the remaining part of this section, we shall be considering the solution of the heat equation, wave equation and Laplace equation by the Fourier method.

9.5.3 Fourier Series Solution of the Heat Equation

One dimensional heat equation

Consider a thin, homogeneous bar or wire of length l. Let the bar coincide with the x-axis on the interval $[0, l]$. Let $u(x, t)$ denote the temperature distribution or heat flow within the bar. We shall study the flow of heat which is produced under various situations.

Ends of the bar kept at zero temperature

We assume that the initial temperature in the bar is $f(x)$ and the ends of the bar are maintained at zero temperature for all time (Fig. 9.14). The boundary value problem (*bvp*) modelling this temperature distribution $u(x, t)$ is given by

$$\frac{\partial u}{\partial t} = c^2 \frac{\partial^2 u}{\partial x^2}, \quad 0 < x < l, t > 0$$

with the initial condition: $\quad u(x, 0) = f(x), 0 < x < l$ and

the boundary conditions: $\quad u(0, t) = u(l, t) = 0, t > 0$

where c^2 is the thermal diffusivity (sometimes, this problem is also called an initial-boundary value problem).

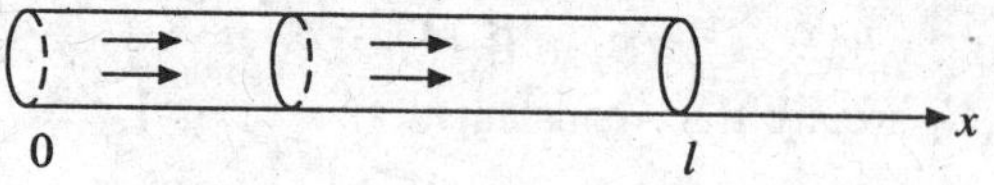

Fig. 9.14. Flow of heat in a bar.

To apply the Fourier method, let

$$u(x, t) = X(x)\, T(t). \tag{9.52}$$

Substituting in the differential equation, we obtain

$$XT' = c^2 X'' T \quad \text{or} \quad \frac{X''}{X} = \frac{T'}{c^2 T}$$

where prime sign(s) denote differentiation with respect to the corresponding variables. Since the left hand side is a function of x and the right side a function of t (where x and t are independent variables), each side must be equal to a constant λ, that is

$$\frac{X''}{X} = \frac{T'}{c^2 T} = \lambda.$$

Therefore, $\qquad X'' - \lambda x = 0 \quad \text{and} \quad T' - c^2 \lambda T = 0.$ $\qquad\qquad$ (9.53)

Using the boundary conditions, we obtain

$$u(0, t) = X(0) \ T(t) = 0, \text{ for all } t$$

and $\qquad\qquad u(l, t) = X(l) \ T(t) = 0, \text{ for all } t.$

Hence, we obtain $X(0) = 0$ and $X(l) = 0$. $\qquad\qquad\qquad\qquad\qquad\qquad\qquad$ (9.54)

Depending on the value of λ, the following three cases arise.

Case 1 $\lambda = 0$. Then, we have $X'' = 0$. The solution is $X(x) = Ax + B$. Substituting in the conditions given in Eq. (9.54), we obtain

$$X(0) = 0 = B \quad \text{and} \quad X(l) = 0 = Al + B.$$

The solution is $A = 0, B = 0$, that is $X(x) = 0$ and $u(x, t) = 0$.

But, $u(x, t) = 0$ is not a solution of the *bvp*. Therefore, the case $\lambda = 0$ cannot be considered.

Case 2 $\lambda = k^2$. Then, we have $X'' - k^2 X = 0$. The solution of this equation is

$$X(x) = Ae^{kx} + Be^{-kx}.$$

Applying the boundary conditions, we obtain

$$X(0) = 0 = A + B \quad \text{and} \quad X(l) = Ae^{kl} + Be^{-kl}.$$

Again, the solution is $A = 0, B = 0$ which gives the trivial solution $u(x, t) = 0$. Hence, λ is not a positive constant.

Case 3 $\lambda = -k^2$. Then, we have $X'' + k^2 X = 0$. The solution of this equation is

$$X(x) = A \cos (kx) + B \sin (kx).$$

Applying the boundary conditions, we obtain

$$X(0) = 0 = A \quad \text{and} \quad X(l) = B \sin (kl) = 0.$$

To obtain non-trivial solutions, we require $B \neq 0$. Hence,

$$\sin (kl) = 0 = \sin (n\pi), \ n = 1, 2, \ldots, \quad \text{or} \quad k = n\pi/l, \ n = 1, 2, \ldots.$$

We obtain $\qquad\qquad X_n(x) = B_n \sin \left(\frac{n\pi x}{l} \right), \ n = 1, 2, \ldots.$

The solution of the equation $T' + c^2 k^2 T = 0$ is

$$T(t) = De^{-c^2 k^2 t} \quad \text{or} \quad T_n(t) = D_n e^{-n^2 \pi^2 c^2 t / l^2}.$$

For each positive integer n, we have the solution as

$$u_n(x, t) = B_n D_n \sin \left(\frac{n\pi x}{l} \right) e^{-n^2 \pi^2 c^2 t / l^2} = d_n \sin \left(\frac{n\pi x}{l} \right) e^{-n^2 \pi^2 c^2 t / l^2}$$

where $d_n = B_n D_n$. By the superposition principle, we have the solution as

$$u(x, t) = \sum_{n=1}^{\infty} d_n \sin\left(\frac{n\pi x}{l}\right) e^{-n^2 \pi^2 c^2 t / l^2}, \ 0 \le x \le l. \tag{9.55}$$

Applying the initial condition, we obtain

$$u(x, 0) = f(x) = \sum_{n=1}^{\infty} d_n \sin\left(\frac{n\pi x}{l}\right), \ 0 \le x \le l.$$

This is a Fourier half-range sine series in $[0, l]$. Hence, d_n are the Fourier coefficients given by (see Eq. (9.26))

$$d_n = \frac{2}{l} \int_0^l f(x) \sin\left(\frac{n\pi x}{l}\right) dx. \tag{9.56}$$

Therefore, the solution of the *bvp* is given by Eq. (9.55) where the coefficients d_n are given by Eq. (9.56).

Temperature distribution in a thin bar or wire with a radiating end

We assume the following

(i) left end of the bar, $x = 0$, is kept at zero temperature.

(ii) right end of the bar $x = l$ is poorly insulated and radiates energy into the surrounding medium. This end radiates at a rate proportional to the temperature at that end of the bar.

(iii) initial temperature of the rod is $f(x)$, $0 \le x \le l$.

The temperature distribution $u(x, t)$ in the bar can be modelled by the following *initial bvp*

$$\frac{\partial u}{\partial t} = c^2 \frac{\partial^2 u}{\partial x^2}, 0 < x < l, t > 0$$

with the boundary conditions: $u(0, t) = 0, \dfrac{\partial u}{\partial x}(l, t) = -\alpha u(l, t), t > 0, \alpha$ positive constant

and the initial condition: $\quad u(x, 0) = f(x), 0 \le x \le l$.

Let $u(x, t) = X(x) T(t)$ be the solution. Substituting in the differential equation, we obtain

$$XT' = c^2 X'' T \ \text{ or } \ \frac{X''}{X} = \frac{T'}{c^2 T} = \lambda.$$

Therefore, we obtain the equations

$$X'' - \lambda X = 0 \ \text{ and } \ T' - \lambda c^2 T = 0. \tag{9.57}$$

Using the boundary conditions, we get

$$u(0, t) = 0 = X(0) T(t) \ \text{ for all } t, \ \text{ or } \ X(0) = 0$$

$$\frac{\partial u}{\partial x}(l, t) = -\alpha u(l, t) \ \text{ gives } X'(l) T(t) + \alpha X(l) T(t) = 0, \text{ for all } t.$$

Hence, $\qquad\qquad\qquad\qquad X'(l) + \alpha X(l) = 0.$

We consider the following three cases.

Case 1 $\lambda = 0$. From Eq. (9.57), we get $X'' = 0$, whose solution is $X(x) = Ax + B$. Using the boundary conditions, we obtain

$$X(0) = 0 = B, \ X'(l) + \alpha X(l) = A + \alpha Al = A(1 + \alpha l) = 0.$$

Hence, we have $A = 0$, $B = 0$, that is, $X = 0$ and $u(x, t) = 0$. Therefore, $\lambda = 0$ is not a valid value.

Case 2 Let $\lambda = k^2$. From Eq. (9.57), we get $X'' - k^2 X = 0$ whose solution is

$$X(x) = A \cosh(kx) + B \sinh(kx).$$

Using the boundary condition $X(0) = 0$, we get $A = 0$. The second boundary condition gives

$$X'(l) + \alpha X(l) = 0 = B[k \cosh(kl) + \alpha \sinh(kl)] = 0.$$

Since $kl > 0$, $\cosh(kl) > 0$, $\sinh(kl) > 0$, we obtain $B = 0$. Again, we have a trivial solution. Hence, $\lambda = k^2$ is not possible.

Case 3 Let $\lambda = -k^2$. From Eq. (9.57), we get $X'' + k^2 X = 0$, whose solution is

$$X(x) = A \cos(kx) + B \sin(kx).$$

Now, $X(0) = 0$ gives $A = 0$. The second condition gives

$$X'(l) + \alpha X(l) = B[k \cos(kl) + \alpha \sin(kl)] = 0.$$

Therefore, non-trivial solutions are obtained when k satisfies the equation

$$k \cos(kl) + \alpha \sin(kl) = 0 \quad \text{or} \quad \tan(kl) = -(k/\alpha). \tag{9.58}$$

Since l and α are given, the roots of this equation can be determined. There are infinite number of roots for this equation which can be written as

$$\tan v = -v/(\alpha l), \ v = kl.$$

Let the roots of this equation be $v_1, v_2, v_3, \ldots$, that is

$$k_n l = v_n \quad \text{or} \quad k_n = \frac{v_n}{l}, n = 1, 2, \ldots.$$

Therefore

$$\lambda_n = -k_n^2 = -\frac{v_n^2}{l^2}, n = 1, 2, \ldots$$

and

$$X_n(x) = B_n \sin\left(\frac{v_n x}{l}\right), n = 1, 2, \ldots.$$

The equation for $T(t)$ is $T' + k^2 c^2 T = 0$, whose solution is

$$T(t) = d e^{-k^2 c^2 t} \quad \text{or} \quad T_n(t) = d_n e^{-v_n^2 c^2 t/l^2}.$$

The solution of the heat equation is therefore

$$u_n(x, t) = X_n(x) T_n(t) = D_n \sin\left(\frac{v_n x}{l}\right) e^{-v_n^2 c^2 t/l^2}, n = 1, 2, \ldots$$

where $D_n = B_n d_n$.

Using the superposition principle, we obtain the solution of the given *initial bvp* as

$$u(x, t) = \sum_{n=1}^{\infty} D_n \sin\left(\frac{v_n x}{l}\right) e^{-v_n^2 c^2 t/l^2}. \qquad (9.59)$$

Applying the initial condition, we obtain

$$u(x, 0) = f(x) = \sum_{n=1}^{\infty} D_n \sin\left(\frac{v_n x}{l}\right). \qquad (9.60)$$

This series is not the standard Fourier half-range series, because v_n's are solutions of a transcendental equation. We shall use the Sturm-Liouville theorem (Theorem 7.1) for the *bvp*

$$X'' + k^2 X = 0, \; X(0) = 0, \; X'(l) + \alpha X(l) = 0.$$

Using this theorem, we have that if $y_m(x) = \sin(v_m x/l)$ and $y_n(x) = \sin(v_n x/l)$ are the eigenfunctions corresponding to distinct eigenvalues v_m, v_n then $y_m(x)$, $y_n(x)$ are orthogonal to each other, on the interval $[0, l]$ with respect to the weight function $p(x) = 1$, that is

$$\int_0^l \sin\left(\frac{v_n x}{l}\right) \sin\left(\frac{v_m x}{l}\right) dx = 0, \; m \neq n. \qquad (9.61)$$

Therefore, multiplying both sides of Eq. (9.60) by $\sin(v_m x/l)$ and integrating on the interval $[0, l]$, we obtain

$$\int_0^l f(x) \sin\left(\frac{v_m x}{l}\right) dx = \sum_{n=1}^{\infty} D_n \int_0^l \sin\left(\frac{v_n x}{l}\right) \sin\left(\frac{v_m x}{l}\right) dx$$

$$= D_m \int_0^l \sin^2\left(\frac{v_m x}{l}\right) dx.$$

Hence, the coefficients D_n in the series solution (Eq. (9.59)) are given by

$$D_n = \left[\int_0^l f(x) \sin\left(\frac{v_n x}{l}\right) dx \right] \Big/ \left[\int_0^l \sin^2\left(\frac{v_n x}{l}\right) dx \right]. \qquad (9.62)$$

Therefore, the solution of the *initial bvp* is given by Eq. (9.59) where the coefficients D_n are given by Eq. (9.62). The roots of the equation $\tan v = -(v/\alpha l)$ can be obtained by a numerical method.

Temperature distribution in a thin, infinite bar

We consider the solution of the heat equation

$$\frac{\partial u}{\partial t} = c^2 \frac{\partial^2 u}{\partial x^2}, \; -\infty < x < \infty, t > 0$$

along with the initial condition

$$u(x, 0) = f(x), \; -\infty < x < \infty,$$

where $f(x)$ is a given function.

Let $u(x, t) = X(x) T(t)$ be the solution. Substituting in the differential equation, we obtain

$$XT' = c^2 X'' T \quad \text{or} \quad \frac{X''}{X} = \frac{T'}{c^2 T} = \lambda$$

since X''/X and $T'/(c^2 T)$ are functions of the independent variables x and t respectively. Primes denote differentiation with respect to the corresponding variables. Therefore, we have the equations

$$X'' - \lambda X = 0 \quad \text{and} \quad T' - \lambda c^2 T = 0. \tag{9.63}$$

If $\lambda = 0$, we obtain $X(x) = Ax + B$. Since $X(x)$ is finite as $x \to \infty$, we get $A = 0$. Now, $X(x) = B$ cannot satisfy the initial condition. Therefore, $\lambda \neq 0$.

If $\lambda = k^2$, we obtain $X(x) = Ae^{kx} + Be^{-kx}$. Again, since $X(x)$ is finite as $x \to \infty$ and $x \to -\infty$, we have $A = 0$ and $B = 0$. Therefore, $\lambda \neq k^2$.

Taking $\lambda = -k^2$, we obtain the solutions of Eq. (9.63) as

$$X(x) = A_1 \cos(kx) + B_1 \sin(kx) \quad \text{and} \quad T(t) = Ce^{-k^2 c^2 t}.$$

Therefore,

$$u(x, t; k) = [A \cos(kx) + B \sin(kx)] e^{-k^2 c^2 t} \tag{9.64}$$

where $A = A_1 C$ and $B = B_1 C$. We note that the region of definition is $-\infty < x < \infty$ and $f(x)$ is not defined as a periodic function on $(-\infty, \infty)$. Therefore, the solution of the problem should be expressible in terms of Fourier integrals. Since A and B are arbitrary, we express them as functions of k, that is $A = A(k)$ and $B = B(k)$, $k > 0$. We write the solution as

$$u(x, t) = \int_0^\infty u(x, t; k)\, dk = \int_0^\infty [A(k) \cos(kx) + B(k) \sin(kx)] e^{-k^2 c^2 t}\, dk. \tag{9.65}$$

We assume that this integral exists and can be differentiated two times with respect to x and once with respect to t.

Using the initial condition, we obtain

$$u(x, 0) = f(x) = \int_0^\infty [A(k) \cos(kx) + B(k) \sin(kx)]\, dk. \tag{9.66}$$

This equation is the Fourier integral defined in Eq. (9.37). Comparing with Eq. (9.35), we obtain

$$A(k) = \frac{1}{\pi} \int_{-\infty}^\infty f(s) \cos(ks)\, ds \quad \text{and} \quad B(k) = \frac{1}{\pi} \int_{-\infty}^\infty f(s) \sin(ks)\, ds$$

where s is the dummy variable of integration.
Therefore, we can write $f(x)$ as

$$f(x) = \frac{1}{\pi} \int_0^\infty \left\{ \int_{-\infty}^\infty f(s) \cos[k(x - s)]ds \right\} dk. \qquad \text{(see Example 9.12)}$$

Hence, we have the solution as

$$u(x, t) = \frac{1}{\pi} \int_0^\infty \left\{ \int_{-\infty}^\infty f(s) \cos[k(x - s)]ds \right\} e^{-k^2 c^2 t}\, dk$$

$$= \frac{1}{\pi} \int_{-\infty}^\infty \left\{ \int_0^\infty \cos[k(x - s)] e^{-k^2 c^2 t}\, dk \right\} f(s)\, ds \tag{9.67}$$

assuming that the order of integration can be interchanged. The inner integral can be evaluated using the complex variable theory.

Denote

$$I = \int_0^\infty \cos\left[k\,(x-s)\right] e^{-k^2 c^2 t}\,dk.$$

Let $z = kc\sqrt{t}$. Then, $dk = dz/(c\sqrt{t})$ and

$$I = \frac{1}{c\sqrt{t}} \int_0^\infty \cos\left[\frac{(x-s)z}{c\sqrt{t}}\right] e^{-z^2}\,dz = \frac{1}{c\sqrt{t}} \int_0^\infty e^{-z^2} \cos(2az)\,dz \tag{9.68}$$

where $2a = (x-s)/(c\sqrt{t})$. From complex variable theory, we have

$$\int_0^\infty e^{-z^2} \cos(2az)\,dz = \frac{\sqrt{\pi}}{2} e^{-a^2} \tag{9.69}$$

(see Example 13.40).

Substituting in Eq. (9.68), we obtain

$$I = \frac{\sqrt{\pi}}{2c\sqrt{t}} e^{-(x-s)^2/(4c^2 t)}. \tag{9.70}$$

Substituting this result in Eq. (9.67), we obtain

$$u\,(x,t) = \frac{1}{2c\sqrt{\pi t}} \int_{-\infty}^\infty f(s)\, e^{-(x-s)^2/(4c^2 t)}\,ds. \tag{9.71}$$

If we set $v = (s-x)/(2c\sqrt{t})$, we have $ds = 2c\sqrt{t}\,dv$. Then, this equation can also be written as

$$u(x,t) = \frac{1}{\sqrt{\pi}} \int_{-\infty}^\infty f(x + 2cv\sqrt{t})e^{-v^2}\,dv. \tag{9.72}$$

Eq. (9.71) or Eq. (9.72) is taken as the solution of the required problem.

9.5.4   Fourier Series Solution of the Wave Equation

One dimensional wave equation

One of the boundary value problems which models the wave phenomena is the problem of a vibrating string.

Consider an elastic string of length l which is fastened at its ends on the x-axis at $x = 0$ and at $x = l$. The string is displaced and then released to vibrate in the x-t plane, where $u(x, t)$ denotes the vertical displacement of the vibrating string. The initial-boundary value problem modelling the motion of the string is given by

$$\frac{\partial^2 u}{\partial t^2} = c^2\,\frac{\partial^2 u}{\partial x^2},\ 0 < x < l, t > 0 \tag{9.73}$$

with the boundary conditions: $u(0, t) = 0,\ u\,(l, t) = 0,\ t > 0$ \hfill (9.74)

and the initial conditions: $u\,(x, 0) = f(x),\ \dfrac{\partial u}{\partial t}(x, 0) = g(x).$ \hfill (9.75)

Here, $f(x)$ denotes initial displacement at time $t = 0$ and $g(x)$ denotes the initial velocity of the string. Let $u(x, t) = X(x) T(t)$ be the solution. Substituting in the differential equation, we obtain

$$XT'' = c^2 X'' T, \quad \text{or} \quad \frac{X''}{X} = \frac{T''}{c^2 T} = \lambda \tag{9.76}$$

since X''/X and $T''/(c^2 T)$ are functions of the independent variables x and t respectively. Primes signs denote differentiation with respect to the corresponding variable. The boundary conditions in Eq. (9.74) give

$$u(0, t) = 0 = X(0) T(t) \quad \text{and} \quad u(l, t) = 0 = X(l) T(t), \text{ for all } t.$$

Therefore, $$X(0) = 0, X(l) = 0.$$

As in the previous sections, it can be shown that $\lambda = -k^2$. Hence, we have the equations

$$X'' + k^2 X = 0 \quad \text{and} \quad T'' + c^2 k^2 T = 0$$

whose solutions are

$$X(x) = A \cos (kx) + B \sin (kx), \, T(t) = D \cos (ckt) + E \sin (ckt). \tag{9.77}$$

Now, $X(0) = 0$ gives $A = 0$. Using the second condition, we obtain

$$X(l) = 0 = B \sin (kl) \quad \text{or} \quad \sin (kl) = \sin (n\pi), n = 1, 2, 3, \ldots .$$

We have, $$k = n\pi/l, n = 1, 2, \ldots \text{ and } X_n(x) = B_n \sin \left(\frac{n\pi x}{l} \right).$$

Therefore, the solution can be written as

$$u_n(x, t) = [D_n \cos (ckt) + E_n \sin (ckt)] \sin \left(\frac{n\pi x}{l} \right)$$

$$= \left[D_n \cos \left(\frac{n\pi ct}{l} \right) + E_n \sin \left(\frac{n\pi ct}{l} \right) \right] \sin \left(\frac{n\pi x}{l} \right), n = 1, 2, \ldots$$

where the arbitrary constant B_n is absorbed in the arbitrary constants D_n and E_n.

Using the superposition principle, we obtain

$$u(x, t) = \sum_{n=1}^{\infty} \left[D_n \cos \left(\frac{n\pi ct}{l} \right) + E_n \sin \left(\frac{n\pi ct}{l} \right) \right] \sin \left(\frac{n\pi x}{l} \right). \tag{9.78}$$

Substituting Eq. (9.78) in the initial condition $u(x, 0) = f(x)$, we get

$$u(x, 0) = f(x) = \sum_{n=1}^{\infty} D_n \sin \left(\frac{n\pi x}{l} \right). \tag{9.79}$$

This series is a Fourier half-range sine series. Hence, we have

$$D_n = \frac{2}{l} \int_0^l f(x) \sin \left(\frac{n\pi x}{l} \right) dx. \tag{9.80}$$

Substituting Eq. (9.78) in the initial condition $\left(\dfrac{\partial u}{\partial t} \right)_{t=0} = g(x)$, we get

$$\frac{\partial u}{\partial t}(x, 0) = g(x) = \sum_{n=1}^{\infty} \left(E_n \frac{n\pi c}{l} \right) \sin\left(\frac{n\pi x}{l} \right).$$

This series is also a Fourier half-range sine series. Hence, we get

$$\frac{n\pi c}{l} E_n = \frac{2}{l} \int_0^l g(x) \sin\left(\frac{n\pi x}{l} \right) dx$$

or
$$E_n = \frac{2}{n\pi c} \int_0^l g(x) \sin\left(\frac{n\pi x}{l} \right) dx. \qquad (9.81)$$

The solution of the *initial bvp* is given by Eq. (9.78), where D_n and E_n are defined by Eqs. (9.80) and (9.81) respectively.

If the string is released from rest, then $g(x) = 0$ for all x in $[0, l]$. Hence, $E_n = 0$ and

$$u(x, t) = \sum_{n=1}^{\infty} D_n \cos\left(\frac{n\pi ct}{l} \right) \sin\left(\frac{n\pi x}{l} \right). \qquad (9.82)$$

Example 9.17 An elastic string of length l which is fastened at its ends $x = 0$ and $x = l$ is picked up at its centre point $x = l/2$ to a height of $l/2$ and released from rest. Find the displacement of the string at any instant of time.

Solution The initial-boundary value problem modelling the motion of the string is given by Eqs. (9.73) to (9.75) where

$$g(x) = 0 \quad \text{and} \quad f(x) = \begin{cases} x, & 0 \le x \le l/2, \\ l - x, & l/2 \le x \le l. \end{cases}$$

The initial position of the string is given in Fig. 9.15. The solution is given by Eq. (9.82) as

$$u(x, t) = \sum_{n=1}^{\infty} D_n \cos\left(\frac{n\pi ct}{l} \right) \sin\left(\frac{n\pi x}{l} \right)$$

where
$$D_n = \frac{2}{l} \int_0^l f(x) \sin\left(\frac{n\pi x}{l} \right) dx. \qquad \text{(see Eq. 9.80)}$$

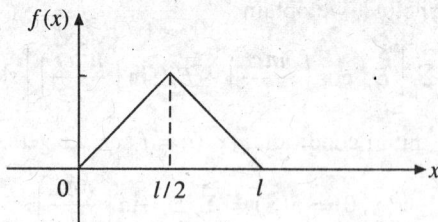

Fig. 9.15.   Function $f(x)$ in Example 9.17.

Substituting for $f(x)$, we obtain

$$D_n = \frac{2}{l} \left[\int_0^{l/2} x \sin\left(\frac{n\pi x}{l} \right) dx + \int_{l/2}^l (l - x) \sin\left(\frac{n\pi x}{l} \right) dx \right]$$

$$= \frac{2}{l} \left[\left\{ -\frac{x \cos (n\pi x/l)}{(n\pi/l)} + \frac{\sin (n\pi x/l)}{(n\pi/l)^2} \right\}_0^{l/2} + \left\{ -(l-x) \frac{\cos (n\pi x/l)}{(n\pi/l)} - \frac{\sin (n\pi x/l)}{(n\pi/l)^2} \right\}_{l/2}^{l} \right]$$

$$= \frac{2}{l} \left[-\frac{(l/2) \cos (n\pi/2)}{(n\pi/l)} + \frac{\sin (n\pi/2)}{(n\pi/l)^2} + \frac{(l/2) \cos (n\pi/2)}{(n\pi/l)} + \frac{\sin (n\pi/2)}{(n\pi/l)^2} \right]$$

$$= \frac{4l}{n^2 \pi^2} \sin \left(\frac{n\pi}{2} \right).$$

Hence, $D_n = 0$ for n even and for n odd we have

$$D_n = D_{2k-1} = \frac{4l \, (-1)^{k+1}}{(2k-1)^2 \, \pi^2}, \ k = 1, 2, 3, \ldots .$$

Therefore,

$$u(x,t) = \frac{4l}{\pi^2} \sum_{k=1}^{\infty} \frac{(-1)^{k+1}}{(2k-1)^2} \cos \left[\frac{(2k-1) \, \pi c t}{l} \right] \sin \left[\frac{(2k-1) \, \pi x}{l} \right].$$

Example 9.18 An elastic string of length l which is fastened at its ends $x = 0$ and $x = l$ is released from its horizontal position (zero initial displacement) with initial velocity $g(x)$ given as

$$g(x) = \begin{cases} x, & 0 \le x \le l/3, \\ 0, & l/3 < x < l. \end{cases}$$

Find the displacement of the string at any instant of time.

Solution The boundary value problem modelling the motion of the string is given by Eqs. (9.73) to (9.75) where $f(x) = 0$ and $g(x)$ is as given. Since $f(x) = 0$, we obtain $D_n = 0$ and the solution is given by

$$u(x,t) = \sum_{n=1}^{\infty} E_n \sin \left(\frac{n\pi c t}{l} \right) \sin \left(\frac{n\pi x}{l} \right)$$

where

$$E_n = \frac{2}{n\pi c} \int_0^l g(x) \sin \left(\frac{n\pi x}{l} \right) dx. \qquad \text{(see Eq. 9.81)}$$

Substituting for $g(x)$, we obtain

$$E_n = \frac{2}{n\pi c} \int_0^{l/3} x \sin \left(\frac{n\pi x}{l} \right) dx = \frac{2}{n\pi c} \left[-\frac{x \cos (n\pi x/l)}{(n\pi/l)} + \frac{\sin (n\pi x/l)}{(n\pi/l)^2} \right]_0^{l/3}$$

$$= \frac{2}{n\pi c} \left[\frac{\sin (n\pi/3)}{(n\pi/l)^2} - \frac{(l/3) \cos (n\pi/3)}{(n\pi/l)} \right].$$

Therefore, the solution is given by

$$u(x,t) = \frac{2l^2}{\pi^2 c} \sum_{n=1}^{\infty} \left[\frac{1}{\pi n^3} \sin\left(\frac{n\pi}{3}\right) - \frac{1}{3n^2} \cos\left(\frac{n\pi}{3}\right) \right] \sin\left(\frac{n\pi ct}{l}\right) \sin\left(\frac{n\pi x}{l}\right).$$

Normal modes of vibration

In the one-dimensional wave equation given in Eq. (9.73), the constant c appearing in it is given by $c = \sqrt{T/\rho}$ where T is the tension in the string and ρ is the mass per unit length of the string. When T is large, then c is large and the vibrating string produces a musical sound. This sound is due to the standing waves produced by the vibrating string. We have shown that the solution of the problem is the superposition of the product solutions called *standing waves* or *normal modes* given by

$$u_n(x,t) = \left[D_n \cos\left(\frac{n\pi ct}{l}\right) + E_n \sin\left(\frac{n\pi ct}{l}\right) \right] \sin\left(\frac{n\pi x}{l}\right).$$

Write $\qquad\qquad\qquad D_n = A_n \sin\phi_n \quad$ and $\quad E_n = A_n \cos\phi_n.$

We have $\qquad\qquad\qquad A_n = (D_n^2 + E_n^2)^{1/2} \quad$ and $\quad \phi_n = \tan^{-1}(D_n/E_n).$

Then, $u_n(x,t)$ can be written as

$$u_n(x,t) = A_n \left[\sin\phi_n \cos\left(\frac{n\pi ct}{l}\right) + \cos\phi_n \sin\left(\frac{n\pi ct}{l}\right) \right] \sin\left(\frac{n\pi x}{l}\right)$$

$$= A_n \sin\left(\frac{n\pi ct}{l} + \phi_n\right) \sin\left(\frac{n\pi x}{l}\right). \qquad (9.83)$$

Therefore, the normal modes are graphs of $\sin(n\pi x/l)$ with a time-varying amplitude given by

$$A_n \sin\left(\frac{n\pi ct}{l} + \phi_n\right). \qquad (9.84)$$

The *frequency* of the vibration is given by $f_n = nc/(2l)$. Therefore, each point on a particular standing wave vibrates with different amplitude but with the same frequency. For $n = 1$, we have

$$u_1(x,t) = A_1 \sin\left(\frac{\pi ct}{l} + \phi_1\right) \sin\left(\frac{\pi x}{l}\right) \qquad (9.85)$$

which is called the *first normal mode*, the *fundamental mode of vibration* or the *first standing wave*. The first three normal modes of vibration, that is $u_1(x,t)$ given in Eq. (9.85) and

$$u_2(x,t) = A_2 \sin\left(\frac{2\pi ct}{l} + \phi_2\right) \sin\left(\frac{2\pi x}{l}\right), \quad u_3(x,t) = A_3 \sin\left(\frac{3\pi ct}{l} + \phi_3\right) \sin\left(\frac{3\pi x}{l}\right)$$

are plotted in Figs. 9.16 *a, b, c.*

The points in the interval $(0, l)$ for which $\sin(n\pi x/l) = 0$, are the points on a standing wave where there is no motion. These points are called the *nodes* of the standing wave. For example, $u_2(x, t) = 0$ for $x = l/2$. Hence, the second standing wave has a node at $x = l/2$. Similarly, $u_3(x, t) = 0$ for $x = l/3$ and $x = 2l/3$. Hence, the third standing wave has two nodes at $x = l/3$ and $x = 2l/3$. Generalising, we have that the nth normal mode of vibration has $(n - 1)$ nodes. The frequency of the fundamental mode of vibration

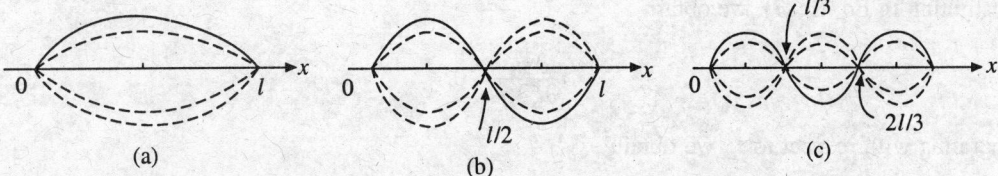

Fig. 9.16. **(a) First normal mode, (b) Second normal mode with one node at** *l/2*. **(c) Third normal mode with two nodes at** *l/3, 2l/3*.

$$f_1 = \frac{c}{2l} = \frac{1}{2l} \left(\frac{T}{\rho} \right)^{1/2} \tag{9.86}$$

is called the *fundamental frequency* or *first harmonic* of the wave. If we consider a stringed musical instrument, the pitch of the sound is directly related to the first harmonic. It can be observed that if the tension in the string is large then the pitch of the sound is high. The frequencies of the remaining normal modes are called *overtones*. The second harmonic is the first overtone, the third harmonic is the second overtone etc.

Vibrations in an infinitely long elastic string (D'Alembert solution)

The vertical displacement $u(x, t)$ of an infinitely long elastic string is governed by the initial value problem

$$\frac{\partial^2 u}{\partial t^2} = c^2 \frac{\partial^2 u}{\partial x^2}, \quad -\infty < x < \infty, t > 0 \tag{9.87}$$

$$u(x, 0) = f(x), \frac{\partial u}{\partial t}(x, 0) = g(x). \tag{9.88}$$

This problem can be solved directly by change of variables. Consider the transformation of variables as

$$\xi = x + ct \quad \text{and} \quad \eta = x - ct. \tag{9.89}$$

Then, by chain rule, we have

$$\frac{\partial u}{\partial x} = \frac{\partial u}{\partial \xi} \frac{\partial \xi}{\partial x} + \frac{\partial u}{\partial \eta} \frac{\partial \eta}{\partial x} = \frac{\partial u}{\partial \xi} + \frac{\partial u}{\partial \eta}$$

$$\frac{\partial^2 u}{\partial x^2} = \frac{\partial}{\partial \xi} \left(\frac{\partial u}{\partial \xi} + \frac{\partial u}{\partial \eta} \right) \frac{\partial \xi}{\partial x} + \frac{\partial}{\partial \eta} \left(\frac{\partial u}{\partial \xi} + \frac{\partial u}{\partial \eta} \right) \frac{\partial \eta}{\partial x} = \frac{\partial^2 u}{\partial \xi^2} + 2 \frac{\partial^2 u}{\partial \xi \partial \eta} + \frac{\partial^2 u}{\partial \eta^2}$$

$$\frac{\partial u}{\partial t} = \frac{\partial u}{\partial \xi} \frac{\partial \xi}{\partial t} + \frac{\partial u}{\partial \eta} \frac{\partial \eta}{\partial t} = c \left(\frac{\partial u}{\partial \xi} - \frac{\partial u}{\partial \eta} \right)$$

$$\frac{\partial^2 u}{\partial t^2} = c \frac{\partial}{\partial \xi} \left(\frac{\partial u}{\partial \xi} - \frac{\partial u}{\partial \eta} \right) \frac{\partial \xi}{\partial t} + c \frac{\partial}{\partial \eta} \left(\frac{\partial u}{\partial \xi} - \frac{\partial u}{\partial \eta} \right) \frac{\partial \eta}{\partial t} = c^2 \left(\frac{\partial^2 u}{\partial \xi^2} - 2 \frac{\partial^2 u}{\partial \xi \partial \eta} + \frac{\partial^2 u}{\partial \eta^2} \right)$$

Substituting in Eq. (9.87), we obtain

$$\frac{\partial^2 u}{\partial \xi \partial \eta} = 0. \tag{9.90}$$

Integrating with respect to η, we obtain

$$\frac{\partial u}{\partial \xi} = h(\xi) \tag{9.91}$$

where $h(\xi)$ is an arbitrary function of ξ. Integrating Eq. (9.91) with respect to ξ, we obtain

$$u(\xi, \eta) = \int h(\xi)\, d\xi + \psi(\eta) \tag{9.92}$$

where $\psi(\eta)$ is an arbitrary function of η. Denoting $\int h(\xi)\, d\xi = \phi(\xi)$, we obtain

$$u(\xi, \eta) = \phi(\xi) + \psi(\eta) \quad \text{or} \quad u(x, t) = \phi(x + ct) + \psi(x - ct). \tag{9.93}$$

Substituting this solution in the initial conditions, we obtain

$$u(x, 0) = \phi(x) + \psi(x) = f(x)$$

$$\frac{\partial u}{\partial t}(x, 0) = c\,[\phi'(x) - \psi'(x)] = g(x).$$

Solving the two equations

$$\phi'(x) + \psi'(x) = f'(x) \quad \text{and} \quad \phi'(x) - \psi'(x) = \frac{1}{c}\, g(x)$$

we obtain

$$\phi'(x) = \frac{1}{2} f'(x) + \frac{1}{2c}\, g(x) \quad \text{and} \quad \psi'(x) = \frac{1}{2} f'(x) - \frac{1}{2c}\, g(x).$$

Integrating, we obtain

$$\phi(x) = \frac{1}{2} f(x) + \frac{1}{2c} \int_{x_0}^{x} g(s)\, ds + k$$

and

$$\psi(x) = \frac{1}{2} f(x) - \frac{1}{2c} \int_{x_0}^{x} g(s)\, ds - k \tag{9.94}$$

[since $\phi(x) + \psi(x) = f(x)$], where x_0 is arbitrary and k is the constant of integration. Substituting these solutions in Eq. (9.93), we obtain

$$u(x, t) = \frac{1}{2}\,[f(x + ct) + f(x - ct)] + \frac{1}{2c} \int_{x_0}^{x+ct} g(s)\, ds - \frac{1}{2c} \int_{x_0}^{x-ct} g(s)\, ds$$

$$= \frac{1}{2}\,[f(x + ct) + f(x - ct)] + \frac{1}{2c} \int_{x-ct}^{x+ct} g(s)\, ds. \tag{9.95}$$

This form of the solution of the wave equation (initial value problem defined by Eqs. (9.87) and (9.88)) is called the *D' Alembert solution*. If the initial velocity $g(x) = 0$, then we obtain

$$u(x, t) = \frac{1}{2}[f(x + ct) + f(x - ct)].$$ (9.96)

This solution can be interpreted as a superposition of two *travelling waves*, one moving to right (that is $f(x - ct)/2$) and the other moving to left (that is $f(x + ct)/2$). Both the waves travel with the same speed c and have the same initial displacement $f(x)$.

Example 9.19 Use D'Alembert solution to find the solution of the initial value problem defining the vibrations of an infinitely long elastic string (Eqs. (9.87) and (9.88)) when $f(x) = \sin x$ and $g(x) = a$, where a is a constant.

Solution We have from the D'Alembert solution (Eq. (9.95))

$$u(x, t) = \frac{1}{2}[\sin(x + ct) + \sin(x - ct)] + \frac{1}{2c}\int_{x-ct}^{x+ct} a\, ds$$

$$= \sin x \cos(ct) + at.$$

Two dimensional wave equation

We consider the motion of a two-dimensional stretched membrane. We assume that the membrane is stretched and fixed along its entire boundary in the x-y plane. The mass of the membrane per unit area is constant. The initial boundary value problem modelling the vibrations of the membrane is given by

$$\frac{\partial^2 u}{\partial t^2} = c^2\left(\frac{\partial^2 u}{\partial x^2} + \frac{\partial^2 u}{\partial y^2}\right)$$ (9.97)

$$u(x, y, t) = 0 \text{ along the boundaries for all } t > 0,$$

$$u(x, y, 0) = f(x, y) \text{ and } \frac{\partial u}{\partial t}(x, y, 0) = g(x, y).$$

Here, $f(x, y)$ and $g(x, y)$ denote the initial displacement and initial velocity respectively.

Let the membrane be a rectangular membrane bounded by the lines $x = 0$, $x = a$, $y = 0$ and $y = b$. We first separate the space and time variables and write

$$u(x, y, t) = F(x, y)\,T(t).$$ (9.98)

Substituting in Eq. (9.97), we obtain

$$T''F = c^2\left(\frac{\partial^2 F}{\partial x^2} + \frac{\partial^2 F}{\partial y^2}\right)T$$

or

$$\frac{1}{F}\left(\frac{\partial^2 F}{\partial x^2} + \frac{\partial^2 F}{\partial y^2}\right) = \frac{T''}{c^2 T} = \lambda$$

since the left hand side is a function of x, y and the right hand side is a function of t. It can be shown that the values $\lambda = 0$ and $\lambda = k^2$ are not possible. Writing $\lambda = -k^2$, we obtain

$$\frac{\partial^2 F}{\partial x^2} + \frac{\partial^2 F}{\partial y^2} + k^2 F = 0, \quad \text{and} \quad T'' + c^2 k^2 T = 0. \tag{9.99}$$

The solution of the second equation is

$$T(t) = P \cos(ckt) + Q \sin(ckt). \tag{9.100}$$

We shall use the method of separation of variables again to solve the first equation in Eq. (9.99). We write

$$F(x, y) = X(x) Y(y).$$

Substituting in the equation, we get

$$X''Y + XY'' + k^2 XY = 0 \tag{9.101}$$

where primes denote differentiation with respect to the corresponding variables. Dividing Eq. (9.101) by XY and separating the variables, we obtain

$$\frac{X''}{X} = -\frac{1}{Y}(Y'' + k^2 Y) = -v^2.$$

Therefore, we have the equations

$$X'' + v^2 X = 0, \quad \text{and} \quad Y'' + (k^2 - v^2) Y = 0. \tag{9.102}$$

The solutions of these equations are

$$X(x) = A \cos(vx) + B \sin(vx)$$

and

$$Y(y) = C \cos(sy) + D \sin(sy), \text{ where } s^2 = k^2 - v^2.$$

From the condition $u(x, y, t) = 0$ on the boundary for all $t > 0$, we obtain that $F(x, y) = 0$ on the boundary. Therefore, we get the conditions

$$X(0) = 0, X(a) = 0, Y(0) = 0 \text{ and } Y(b) = 0.$$

Now, $X(0) = 0$ gives $A = 0$.

$$X(a) = 0 \text{ gives } B \sin(va) = 0 = \sin m\pi, m = 1, 2, \ldots.$$

Hence, $v = m\pi/a, m = 1, 2, \ldots.$

Again, $Y(0) = 0$ gives $C = 0$.

$$Y(b) = 0 \text{ gives } D \sin(sb) = 0 = \sin n\pi, n = 1, 2, \ldots.$$

Hence, $s = n\pi/b, n = 1, 2, \ldots.$

Therefore, $\qquad F_{mn}(x, y) = X_m(x) Y_n(y) = B_{mn} \sin\left(\frac{m\pi x}{a}\right) \sin\left(\frac{n\pi y}{b}\right). \tag{9.103}$

Since $s^2 = k^2 - v^2$, we obtain

$$k_{mn}^2 = k^2 = s_n^2 + v_m^2 = \pi^2 \left(\frac{m^2}{a^2} + \frac{n^2}{b^2}\right). \tag{9.104}$$

Substituting Eq. (9.104) in Eq. (9.100), we get

$$T_{mn} = P_{mn} \cos (c k_{mn} t) + Q_{mn} \sin (c k_{mn} t).$$

Therefore, the solution is given by

$$u_{mn} (x, y, t) = [P_{mn} \cos (ck_{mn}t) + Q_{mn} \sin (c k_{mn}t)] \sin \left(\frac{m\pi x}{a} \right) \sin \left(\frac{n\pi y}{b} \right).$$

where B_{mn} are absorbed in P_{mn} and Q_{mn}.

Using the superposition principle, we obtain the solution of the Eq. (9.97) as

$$u (x, y, t) = \sum_{m=1}^{\infty} \sum_{n=1}^{\infty} [P_{mn} \cos (ck_{mn} t) + Q_{mn} \sin (ck_{mn} t)] \sin \left(\frac{m\pi x}{a} \right) \sin \left(\frac{n\pi y}{b} \right). \quad (9.105)$$

Note that $k_{mn} \neq k_{nm}$.

If the shape of the membrane is a square, then $a = b$ and $k_{mn} = k_{nm}$.

Using the initial condition $u (x, y, 0) = f (x, y)$, we obtain

$$f(x, y) = \sum_{m=1}^{\infty} \sum_{n=1}^{\infty} P_{mn} \sin \left(\frac{m\pi x}{a} \right) \sin \left(\frac{n\pi y}{b} \right). \quad (9.106)$$

This series is a double Fourier series in the variables x and y. If we denote

$$I_m (y) = \sum_{n=1}^{\infty} P_{mn} \sin \left(\frac{n\pi y}{b} \right) \quad (9.107)$$

we obtain

$$f(x, y) = \sum_{m=1}^{\infty} I_m (y) \sin \left(\frac{m\pi x}{a} \right) \quad (9.108)$$

which is a Fourier half-range sine series in terms of the variable x. Therefore,

$$I_m (y) = \frac{2}{a} \int_0^a f(x, y) \sin \left(\frac{m\pi x}{a} \right) dx. \quad (9.109)$$

Now, Eq. (9.107) is a Fourier half-range sine series in terms of the variable y. Therefore,

$$P_{mn} = \frac{2}{b} \int_0^b I_m (y) \sin \left(\frac{n\pi y}{b} \right) dy. \quad (9.110)$$

Substituting Eq. (9.109) in Eq. (9.110), we obtain

$$P_{mn} = \frac{4}{ab} \int_0^b \int_0^a f(x, y) \sin \left(\frac{m\pi x}{a} \right) \sin \left(\frac{n\pi y}{b} \right) dx \, dy. \quad (9.111)$$

Using the second initial condition $\frac{\partial u}{\partial t} (x, y, 0) = g(x, y)$, we get

$$g(x, y) = \sum_{m=1}^{\infty} \sum_{n=1}^{\infty} c Q_{mn} k_{mn} \sin \left(\frac{m\pi x}{a} \right) \sin \left(\frac{n\pi y}{b} \right). \quad (9.112)$$

This series is again a double Fourier half-range series. Following the same procedure, we obtain

$$cQ_{mn}k_{mn} = \frac{4}{ab} \int_0^b \int_0^a g(x,y) \sin\left(\frac{m\pi x}{a}\right) \sin\left(\frac{n\pi y}{b}\right) dx\, dy$$

or

$$Q_{mn} = \frac{4}{abck_{mn}} \int_0^b \int_0^a g(x,y) \sin\left(\frac{m\pi x}{a}\right) \sin\left(\frac{n\pi y}{b}\right) dx\, dy. \qquad (9.113)$$

The solution of the initial boundary value problem is given by $u(x, y, t)$ given in Eq. (9.105) where P_{mn} and Q_{mn} are as defined in Eqs. (9.111) and (9.113).

Example 9.20 Find the deflection $u(x, y, t)$ of a square membrane of side one unit, if the initial velocity is zero and the initial deflection is given by $f(x, y) = \sin(3\pi x)\sin(4\pi y)$. Assume $c = 1$ in the differential equation.

Solution Since the initial velocity $g(x, y) = 0$, we obtain $Q_{mn} = 0$ (see Eq. (9.113)).

We have $\qquad\qquad k_{mn}^2 = \pi^2 (m^2 + n^2)$,

and $P_{mn} = 4 \int_0^1 \int_0^1 \sin(3\pi x)\sin(4\pi y)\sin(m\pi x)\sin(n\pi y)\, dx\, dy$

$$= \int_0^1 \int_0^1 \{\cos[\pi x(m-3)] - \cos[\pi x(m+3)]\}\{\cos[\pi y(n-4)] - \cos[\pi y(n+4)]\}\, dx\, dy$$

$$= 0, \quad \text{for } m \neq 3,\, n \neq 4.$$

For $m = 3$, $n = 4$, we get

$$P_{34} = 4 \int_0^1 \int_0^1 \sin^2(3\pi x)\sin^2(4\pi y)\, dx\, dy$$

$$= \int_0^1 \int_0^1 [1 - \cos(6\pi x)][1 - \cos(8\pi y)]\, dx\, dy = 1.$$

Also $\qquad\qquad k_{34}^2 = \pi^2(9 + 16) = 25\,\pi^2 \quad \text{or} \quad k_{34} = 5\pi.$

Therefore, we have the solution as

$$u(x, y, t) = P_{34}\cos(k_{34}t)\sin(3\pi x)\sin(4\pi y) = \cos(5\pi t)\sin(3\pi x)\sin(4\pi y).$$

9.5.5 Fourier Series Solution of Laplace's Equation

We want to study the steady-state temperature distribution in a thin, flat, rectangular plate. Without any loss of generality let the boundaries of the plate be $x = 0$, $x = a$, $y = 0$ and $y = b$. The differential equation modelling the steady-state temperature distribution is given by the Laplace equation

$$\frac{\partial^2 u}{\partial x^2} + \frac{\partial^2 u}{\partial y^2} = 0,\ 0 < x < a,\, 0 < y < b. \qquad (9.114)$$

Various types of boundary conditions can be imposed to solve this equation.

For example, let the boundary conditions be

$$u(0, y) = 0, u(a, y) = 0, 0 < y < b$$

and

$$u(x, 0) = f(x), u(x, b) = g(x), 0 < x < a.$$

Let the solution be written as $u(x, y) = X(x) Y(y)$. Substituting in the differential equation, we get

$$X''Y + XY'' = 0 \quad \text{or} \quad \frac{X''}{X} = -\frac{Y''}{Y} = \lambda = -k^2$$

since X''/X and Y''/Y are functions of the independent variables x and y respectively. Hence, we have the equations

$$X'' + k^2 X = 0 \quad \text{and} \quad Y'' - k^2 Y = 0$$

whose solutions are

$$X(x) = A \cos(kx) + B \sin(kx) \quad \text{and} \quad Y(y) = C \cosh(ky) + D \sinh(ky).$$

Other values of λ do not give a solution. Using the given boundary conditions, we obtain the conditions $X(0) = 0$ and $X(a) = 0$. Therefore, we obtain

$$X(0) = 0 = A \quad \text{and} \quad X(a) = 0 = \sin(ka) = \sin(n\pi), n = 1, 2, \ldots$$

Hence, $k = n\pi/a, n = 1, 2, \ldots$ The solution is given by

$$u_n(x, y) = X_n(x) Y_n(y) = [C_n \cosh(k_n y) + D_n \sinh(k_n y)] \sin\left(\frac{n\pi x}{a}\right).$$

Using the superposition principle, we have the solution as

$$u(x, y) = \sum_{n=1}^{\infty} [C_n \cosh(k_n y) + D_n \sinh(k_n y)] \sin\left(\frac{n\pi x}{a}\right). \tag{9.115}$$

Applying the boundary condition $u(x, 0) = f(x)$, we get

$$u(x, 0) = f(x) = \sum_{n=1}^{\infty} C_n \sin\left(\frac{n\pi x}{a}\right)$$

which is a Fourier half-range sine series. Therefore,

$$C_n = \frac{2}{a} \int_0^a f(x) \sin\left(\frac{n\pi x}{a}\right) dx. \tag{9.116}$$

Applying the boundary condition $u(x, b) = g(x)$, we get

$$u(x, b) = g(x) = \sum_{n=1}^{\infty} [C_n \cosh(k_n b) + D_n \sinh(k_n b)] \sin\left(\frac{n\pi x}{a}\right)$$

which is again a Fourier half-range sine series. Therefore

$$C_n \cosh(k_n b) + D_n \sinh(k_n b) = \frac{2}{a} \int_0^a g(x) \sin\left(\frac{n\pi x}{a}\right) dx$$

or
$$D_n = \frac{1}{\sinh(k_n b)}\left[\frac{2}{a}\int_0^a g(x)\sin\left(\frac{n\pi x}{a}\right)dx - C_n \cosh(k_n b)\right].$$

Therefore, the steady-state temperature distribution in the rectangular plate is given by

$$u(x,y) = \sum_{n=1}^{\infty}\left[C_n \cosh\left(\frac{n\pi y}{a}\right) + D_n \sinh\left(\frac{n\pi y}{a}\right)\right]\sin\left(\frac{n\pi x}{a}\right)$$

where C_n is given in Eq. (9.116) and

$$D_n = \frac{1}{\sinh(n\pi b/a)}\left[\frac{2}{a}\int_0^a g(x)\sin\left(\frac{n\pi x}{a}\right)dx - C_n \cosh\left(\frac{n\pi b}{a}\right)\right].$$

Exercise 9.4

Classify the following partial differential equations.

1. $\dfrac{\partial^2 u}{\partial x^2} + 3\dfrac{\partial^2 u}{\partial y^2} = \dfrac{\partial u}{\partial x}$.

2. $\dfrac{\partial^2 u}{\partial x \partial y} = 3\dfrac{\partial u}{\partial y}$.

3. $\dfrac{\partial^2 u}{\partial x^2} + \dfrac{\partial^2 u}{\partial y^2} + \dfrac{\partial^2 u}{\partial x \partial y} = 0$.

4. $\dfrac{\partial^2 u}{\partial x^2} + 2\dfrac{\partial^2 u}{\partial x \partial y} + \dfrac{\partial^2 u}{\partial y^2} = 0$.

5. $\dfrac{\partial^2 u}{\partial t^2} = c^2\dfrac{\partial^2 u}{\partial x^2}$.

6. $\dfrac{\partial u}{\partial t} = c^2\dfrac{\partial^2 u}{\partial x^2} + \dfrac{\partial u}{\partial x}$.

7. $(1-y)\dfrac{\partial^2 u}{\partial x^2} + 2x\dfrac{\partial^2 u}{\partial x \partial y} + (1+y)\dfrac{\partial^2 u}{\partial y^2} = 0$.

8. $y\dfrac{\partial^2 u}{\partial x^2} + 2x\dfrac{\partial^2 u}{\partial x \partial y} + y\dfrac{\partial^2 u}{\partial y^2} = 0$.

In the problems **9** to **16**, find the separation of variables solution.

9. $\dfrac{\partial u}{\partial x} = \dfrac{\partial u}{\partial y}$.

10. $4\dfrac{\partial u}{\partial x} + \dfrac{\partial u}{\partial y} = 0$.

11. $y\dfrac{\partial u}{\partial x} + x\dfrac{\partial u}{\partial y} = 0$.

12. $\dfrac{\partial^2 u}{\partial x^2} + \dfrac{\partial^2 u}{\partial y^2} = 0$.

13. $\dfrac{\partial^2 u}{\partial t^2} = c^2\dfrac{\partial^2 u}{\partial x^2}$.

14. $\dfrac{\partial u}{\partial t} = c^2\dfrac{\partial^2 u}{\partial x^2}$.

15. $x^2\dfrac{\partial^2 u}{\partial x^2} + \dfrac{\partial^2 u}{\partial y^2} = 0,\ x \neq 0$.

16. $\dfrac{\partial^2 u}{\partial x^2} + y^2\dfrac{\partial^2 u}{\partial y^2} = 0,\ y \neq 0$.

17. Verify that $u(r,\theta) = (A\cos k\theta + B\sin k\theta)(Cr^k + Dr^{-k})$ is a solution of the differential equation

$$\frac{\partial^2 u}{\partial r^2} + \frac{1}{r}\frac{\partial u}{\partial r} + \frac{1}{r^2}\frac{\partial^2 u}{\partial \theta^2} = 0.$$

18. Verify that $u(r,t) = [A J_0(\lambda r) + B Y_0(\lambda r)]e^{-\lambda^2 t}$. is a solution of the differential equation

$$\frac{\partial u}{\partial t} = \frac{\partial^2 u}{\partial r^2} + \frac{1}{r}\frac{\partial u}{\partial r}.$$

Find the solution of the heat equation

$$\frac{\partial u}{\partial t} = c^2 \frac{\partial^2 u}{\partial x^2}, \ 0 < x < l, t > 0$$

under the boundary conditions $u(0, t) = 0 = u(l, t)$ and the following initial conditions.

19. $u(x, 0) =$ constant temperature T, $0 < x < l$.

20. $u(x, 0) = \begin{cases} x, & 0 \le x < l/2, \\ l - x, & l/2 \le x \le l. \end{cases}$

21. $u(x, 0) = x(l - x), 0 \le x \le l.$

Find the solution of the heat equation

$$\frac{\partial u}{\partial t} = c^2 \frac{\partial^2 u}{\partial x^2}, 0 < x < \pi, t > 0$$

under the boundary conditions $u(0, t) = 0 = u(\pi, t)$ and the following initial conditions.

22. $u(x, 0) =$ constant T, $0 < x < \pi$.

23. $u(x, 0) = \sin x, 0 < x < \pi$.

24. $u(x, 0) = \begin{cases} x, & 0 \le x < \pi/2, \\ \pi - x, & \pi/2 \le x \le \pi. \end{cases}$

25. Find the temperature $u(x, t)$ in a thin rod of length l, if the initial temperature throughout the rod is $f(x)$ and if the ends, $x = 0$ and $x = l$, of the rod are insulated.

26. Solve the problem 25, if $f(x) = \begin{cases} T, & \text{(a constant)}, & 0 < x < l/2, \\ 0, & l/2 < x < l. \end{cases}$

27. Solve the problem 25, if $f(x) = \begin{cases} x, & 0 < x < 2, \\ 0, & 2 < x < 4. \end{cases}$

28. A thin rod of length l has initial temperature equal to a constant T. The right end of the rod, $x = l$, is insulated while the left end of the rod is kept at zero temperature. Find the temperature distribution in the bar.

29. A thin homogeneous rod is of length 12 cm and has thermal diffusivity $c^2 = 6$. The left end of the rod is maintained at zero temperature. The right end of the rod is insulated but is radiating into the surrounding medium such that $(\partial u/\partial x)(12, t) = - u(12, t), t \ge 0$. The initial temperature in the rod is given by

$$f(x) = \begin{cases} x, & 0 \le x < 6 \\ 12 - x, & 6 \le x \le 12. \end{cases}$$

Find the solution of the *bvp* using the Fourier method. Approximate the temperature distribution $u(x, t)$ by finding its first and second partial sums.

30. Find the temperature distribution in a thin, infinite bar if the initial temperature is given by

$$u(x, 0) = f(x) = \begin{cases} T_0, & \text{constant}, |x| < 2 \\ 0, & |x| > 2. \end{cases}$$

31. The temperature distribution in a thin, semi-infinite bar is governed by the boundary value problem

$$\frac{\partial u}{\partial t} = c^2 \frac{\partial^2 u}{\partial x^2}, \ 0 < x < \infty, t > 0$$

$$u(0, t) = g(t), u(x, 0) = f(x), 0 < x < \infty.$$

If $g(t) = 0$, show that the solution can be written as

$$u(x, t) = \frac{1}{\sqrt{\pi}} \left[\int_{-x/\tau}^{\infty} f(x + \tau z) e^{-z^2} dz - \int_{x/\tau}^{\infty} f(\tau z - x) e^{-z^2} dz \right]$$

where $\tau = 2c\sqrt{t}$

32. Find the solution in Problem 31, if $f(x) = 1$. Express the solution in terms of the error function

$$erf(x) = \frac{2}{\sqrt{\pi}} \int_0^x e^{-z^2} dz.$$

Solve the following initial boundary value problems corresponding to the one-dimensional wave equation.

33. $\dfrac{\partial^2 u}{\partial t^2} = c^2 \dfrac{\partial^2 u}{\partial x^2}, 0 < x < 2, t > 0.$ $u(0, t) = u(2, t) = 0.$

$$u(x, 0) = \begin{cases} x, & 0 \le x \le 1, \\ 2 - x, & 1 \le x \le 2, \end{cases} \qquad \frac{\partial u}{\partial t}(x, 0) = 0.$$

34. $\dfrac{\partial^2 u}{\partial t^2} = c^2 \dfrac{\partial^2 u}{\partial x^2}, 0 < x < 4, t > 0.$ $u(0, t) = u(4, t) = 0.$

$$u(x, 0) = 0, \frac{\partial u}{\partial t}(x, 0) = \begin{cases} x, & 0 < x < 2, \\ 0, & 2 < x < 4. \end{cases}$$

35. $\dfrac{\partial^2 u}{\partial t^2} = c^2 \dfrac{\partial^2 u}{\partial x^2}, 0 < x < l, t > 0.$ $u(0, t) = u(l, t) = 0.$

$$u(x, 0) = x(l - x), \frac{\partial u}{\partial t}(x, 0) = 0.$$

36. $\dfrac{\partial^2 u}{\partial t^2} = c^2 \dfrac{\partial^2 u}{\partial x^2}, 0 < x < \pi, t > 0.$ $u(0, t) = u(\pi, t) = 0.$

$$u(x, 0) = 0, \frac{\partial u}{\partial t}(x, 0) = \sin x.$$

37. $\dfrac{\partial^2 u}{\partial t^2} = c^2 \dfrac{\partial^2 u}{\partial x^2}, 0 < x < 1, t > 0.$ $u(0, t) = u(1, t) = 0.$

$$u(x, 0) = \sin 2\pi x, \frac{\partial u}{\partial t}(x, 0) = 0.$$

38. $\dfrac{\partial^2 u}{\partial t^2} = 16 \dfrac{\partial^2 u}{\partial x^2}, 0 < x < 2, t > 0.$ $u(0, t) = u(2, t) = 0.$

$$u(x, 0) = x(2 - x), 0 < x < 2; \frac{\partial u}{\partial t}(x, 0) = 1.$$

39. Reduce the initial boundary value problem

$$\frac{\partial^2 u}{\partial t^2} = \frac{\partial^2 u}{\partial x^2} + 6x, 0 < x < 2, t > 0; u(0, t) = u(2, t) = 0, t > 0.$$

$$u(x, 0) = 0, \frac{\partial u}{\partial t}(x, 0) = 0, 0 < x < 2$$

to the separable form by using the substitution $u(x, t) = U(x, t) + g(x)$. Formulate the new boundary value problem and solve by the method of separation of variables.

40. Reduce the initial boundary value problem

$$\frac{\partial^2 u}{\partial t^2} = \frac{\partial^2 u}{\partial x^2} - \sin x, 0 < x < \pi/2, t > 0; u(0, t) = u(\pi/2, t) = 0, t > 0.$$

$$u(x, 0) = 0, \frac{\partial u}{\partial t}(x, 0) = 0, 0 < x < \pi/2,$$

to the separable form as in Problem 39. Hence, obtain the solution by separation of variables.

41. Solve the vibrating string problem when there is a resistence in the medium which is proportional to the velocity. If the initial displacement is $f(x)$ and the string starts from rest, then the boundary value problem modelling the vibrating string is given by

$$\frac{\partial^2 u}{\partial t^2} = \frac{\partial^2 u}{\partial x^2} - a\frac{\partial u}{\partial t}, 0 < x < \pi, t > 0, 0 < a < 1.$$

$$u(0, t) = u(\pi, t) = 0, t > 0; u(x, 0) = f(x), \frac{\partial u}{\partial t}(x, 0) = 0, 0 < x < \pi.$$

42. The longitudinal displacements in a vibrating elastic bar can be described by the boundary value problem

$$\frac{\partial^2 u}{\partial t^2} = c^2\frac{\partial^2 u}{\partial x^2}, 0 < x < l, t > 0; \frac{\partial u}{\partial x}(0, t) = 0, \frac{\partial u}{\partial x}(l, t) = 0, t > 0.$$

$$u(x, 0) = f(x), \frac{\partial u}{\partial t}(x, 0) = 0, 0 < x < l.$$

(In the case of the elastic bar, $c^2 = E/\rho$ where E is the modulus of elasticity and ρ is the mass per unit volume. The boundary conditions at $x = 0$ and $x = l$ are called the *free-end conditions*). Find the solution of the *bvp* by separation of variables

43. Find the solution in the Problem 42, if $f(x) = x$.

44. Solve the boundary value problem

$$\frac{\partial^2 u}{\partial t^2} + \frac{\partial u}{\partial t} + u = c^2\frac{\partial^2 u}{\partial x^2}, 0 < x < l, t > 0.$$

$$u(0, t) = u(l, t) = 0; u(x, 0) = f(x), \frac{\partial u}{\partial t}(x, 0) = 0, 0 < x < l.$$

(This differential equation is called the *telegraph equation*) using the separation of variables method.

45. Use D'Alembert solution to find the solution of the initial value problem defining the vibrations of an infinitely long elastic string when (i) $f(x) = 0$, $g(x) = \sin 3x$, (ii) $f(x) = \sin 2x$, $g(x) = \cos 2x$.

46. Find the deflection $u(x, y, t)$ of a square membrane of side one unit, if the initial velocity is zero and the initial deflection is given by $f(x, y) = \sin(2\pi x)\sin(3\pi y)$. Assume $c = 2$ in the differential equation.

47. Find the deflection $u(x, y, t)$ of a square membrane of side one unit, if the initial velocity is zero and the initial deflection is given by $f(x, y) = xy$. Assume $c = 1$ in the differential equation.

48. Find the deflection $u(x, y, t)$ of a rectangular membrane with sides $a = 1$, $b = 2$, if the initial velocity is zero and the initial deflection is given by $f(x, y) = \sin(\pi x)\sin(4\pi y)$. Assume $c = 1$ in the differential equation.

49. The boundary value problem governing the steady-state temperature distribution in a flat, thin, rectangular plate is given by

$$\frac{\partial^2 u}{\partial x^2} + \frac{\partial^2 u}{\partial y^2} = 0, 0 < x < a, 0 < y < b.$$

$$u(0, y) = f(y), u(a, y) = g(y), 0 < y < b,$$

$$u(x, 0) = 0, u(x, b) = 0, 0 < x < a.$$

Find the temperature distribution.

50. The boundaries of a thin rectangular plate are $x = 0$, $x = a$, $y = 0$ and $y = b$. The steady state temperature distribution in the plate is to be determined when the vertical edges are insulated. A boundary value problem modelling the temperature distribution is

$$\frac{\partial^2 u}{\partial x^2} + \frac{\partial^2 u}{\partial y^2} = 0, 0 < x < a, 0 < y < b,$$

$$\frac{\partial u}{\partial x}(0, y) = 0, \frac{\partial u}{\partial x}(a, y) = 0, 0 < y < b,$$

$$u(x, 0) = 0, u(x, b) = f(x), 0 < x < a.$$

Find the temperature distribution, if $f(x)$ is given by

$$f(x) = \begin{cases} x, & 0 < x < a/2, \\ a - x, & a/2 < x < a. \end{cases}$$

Solve the Laplace equation $(\partial^2 u/\partial x^2) + (\partial^2 u/\partial y^2) = 0$ for a rectangular or a square plate, subject to the following conditions.

51. $u(0, y) = 0, u(a, y) = 0; u(x, 0) = f(x), u(x, b) = 0.$

52. $u(0, y) = 0, u(a, y) = 0; u(x, 0) = 0, u(x, b) = g(x).$

53. $u(0, y) = 0, u(a, y) = 0; \frac{\partial u}{\partial y}(x, 0) = 0, u(x, b) = g(x).$

54. $u(0, y) = 0, u(a, y) = a - y; \frac{\partial u}{\partial y}(x, 0) = 0, \frac{\partial u}{\partial y}(x, a) = 0.$

55. $u(0, y) = y, \frac{\partial u}{\partial x}(1, y) = -5; u(x, 0) = 0, u(x, 1) = 0.$

9.6 Fourier Transforms

In this section, we define the *Fourier transform* which is an integral transform similar to Laplace transform. Fourier transforms are used in many areas of science, engineering, medicine etc.

Let $f(t)$ be piecewise continuous on $(-\infty, \infty)$. Assume that $f(t)$ is absolutely convergent, that is $\int_{-\infty}^{\infty} |f(t)| \, dt$ converges. Then, the Fourier transform of $f(t)$ denoted by $\mathscr{F}[f(t)]$ is defined as

$$\mathscr{F}[f(t)] = \int_{-\infty}^{\infty} f(t) e^{-i\omega t} \, dt = F(\omega). \tag{9.117}$$

In Laplace transforms, we have defined $\mathscr{L}[f(t)] = F(s)$. Similar notation is also used in Fourier transforms.

Assume now that $\int_{-\infty}^{\infty} |F(\omega)| \, d\omega$ converges. Then, we define the *inverse Fourier transform* of $F(\omega)$ as

$$\mathcal{F}^{-1}[F(\omega)] = \frac{1}{2\pi} \int_{-\infty}^{\infty} F(\omega)\, e^{i\omega t}\, d\omega = f(t). \tag{9.118}$$

The Fourier transform and its inverse are called a *transform pair*. Therefore, the Fourier transform defined in Eq. (9.117) and its inverse defined in Eq. (9.118) form a transform pair.

Example 9.21 Find the Fourier transform of the following functions defined on $(-\infty, \infty)$.

(i) $f(t) = \begin{cases} a, & -l < t < 0, \\ 0, & \text{otherwise},\ a > 0. \end{cases}$ (ii) $f(t) = \begin{cases} a, & -l < t < l, \\ 0, & \text{otherwise},\ a > 0. \end{cases}$

(iii) $f(t) = \begin{cases} a, & -l < t < 0, \\ b, & 0 < t < l, \\ 0, & \text{otherwise},\ a > 0,\ b > 0. \end{cases}$

Solution We use the definition to find the Fourier transform.

(i) $\mathcal{F}[f(t)] = \displaystyle\int_{-\infty}^{\infty} f(t)\, e^{-i\omega t}\, dt = a \int_{-l}^{0} e^{-i\omega t}\, dt = -\frac{a}{i\omega}\, [e^{-i\omega t}]_{-l}^{0}$

$$= -\frac{a}{i\omega}\, [1 - e^{i\omega l}] = \frac{ia}{\omega}\, [1 - e^{i\omega l}].$$

(ii) $\mathcal{F}[f(t)] = a \displaystyle\int_{-l}^{l} e^{-i\omega t}\, dt = -\frac{a}{i\omega}\, [e^{-i\omega t}]_{-l}^{l} = -\frac{a}{i\omega}\, [e^{-i\omega l} - e^{i\omega l}]$

$$= \frac{a}{i\omega}\, [e^{i\omega l} - e^{-i\omega l}] = \frac{2a}{\omega}\, \sin(\omega l).$$

(iii) $\mathcal{F}[f(t)] = a \displaystyle\int_{-l}^{0} e^{-i\omega t}\, dt + b \int_{0}^{l} e^{-i\omega t}\, dt$

$$= -\frac{a}{i\omega}\, [1 - e^{i\omega l}] - \frac{b}{i\omega}\, [e^{-i\omega l} - 1]$$

$$= \frac{1}{i\omega}\, [(b - a) + a e^{i\omega l} - b e^{-i\omega l}].$$

Example 9.22 Find the Fourier transform of the function

$$f(t) = \begin{cases} 0, & t < 0 \\ e^{-\alpha t}, & t \geq 0, \qquad \alpha > 0. \end{cases}$$

Solution The function $f(t)$ has a jump discontinuity at $t = 0$ and is of magnitude 1. Also

$$\int_{-\infty}^{\infty} |f(t)|\, dt = \int_{0}^{\infty} e^{-\alpha t}\, dt = \frac{1}{\alpha}.$$

Therefore, Fourier transform of $f(t)$ exists. We have

$$\mathscr{F}[f(t)] = F(\omega) = \int_{-\infty}^{\infty} f(t)\, e^{-i\omega t}\, dt = \lim_{s \to \infty} \int_0^s e^{-\alpha t}\, e^{-i\omega t}\, dt$$

$$= \lim_{s \to \infty} \left[-\frac{e^{-(\alpha + i\omega)t}}{\alpha + i\omega} \right]_0^s = \frac{1}{\alpha + i\omega}$$

since $\qquad\qquad \lim_{s \to \infty} \left[e^{-(\alpha + i\omega)s} \right] = \lim_{s \to \infty} e^{-\alpha s} \left[\cos(\omega s) - i \sin(\omega s) \right] = 0.$

Hence, $1/(\alpha + i\omega)$ and $f(t) = e^{-\alpha t}\, u_0(t)$ where $u_0(t)$ is the unit step function form a transform pair.

Therefore, $\qquad\qquad\qquad \mathscr{F}[e^{-\alpha t} u_0(t)] = \dfrac{1}{\alpha + i\omega}.$ $\qquad\qquad\qquad$ (9.119)

Example 9.23 Find the Fourier transform of the function $f(t) = e^{-a|t|}, -\infty < t < \infty, a > 0$. Write the inverse transform.

Solution We have $\qquad\qquad\qquad f(t) = \begin{cases} e^{at}, & t < 0, \\ e^{-at}, & t > 0. \end{cases}$

Therefore, $\quad \mathscr{F}[f(t)] = \displaystyle\int_{-\infty}^{0} e^{at}\, e^{-i\omega t}\, dt + \int_{0}^{\infty} e^{-at}\, e^{-i\omega t}\, dt$

$$= \left[\frac{e^{(a - i\omega)t}}{a - i\omega} \right]_{-\infty}^{0} + \left[\frac{e^{-(a + i\omega)t}}{-(a + i\omega)} \right]_{0}^{\infty} = \frac{1}{a - i\omega} + \frac{1}{a + i\omega} = \frac{2a}{a^2 + \omega^2}.$$

The inverse transform is given by

$$\mathscr{F}^{-1} \left[\frac{2a}{a^2 + \omega^2} \right] = e^{-a|t|}.$$

Example 9.24 Find the Fourier transform of $e^{-at^2}, a > 0$.

Solution From the definition, we obtain

$$\mathscr{F}[e^{-at^2}] = \int_{-\infty}^{\infty} e^{-at^2}\, e^{-i\omega t}\, dt = \int_{-\infty}^{\infty} e^{-a[t^2 + (i\omega t/a)]}\, dt$$

$$= \int_{-\infty}^{\infty} e^{-a[\{t + i\omega/(2a)\}^2 + \omega^2/(4a^2)]}\, dt = e^{-\omega^2/(4a)} \int_{-\infty}^{\infty} e^{-a[t + i\omega/(2a)]^2}\, dt$$

$$= e^{-\omega^2/(4a)} \int_{-\infty}^{\infty} e^{-\tau^2}\, \frac{d\tau}{\sqrt{a}} \qquad\qquad \left[\text{setting } \sqrt{a}\left(t + \frac{i\omega}{2a} \right) = \tau \right]$$

$$= \frac{\sqrt{\pi}}{\sqrt{a}}\, e^{-\omega^2/(4a)}$$

since $\displaystyle\int_{-\infty}^{\infty} e^{-\tau^2}\, d\tau = 2 \int_{0}^{\infty} e^{-\tau^2}\, d\tau = \sqrt{\pi}.$

In some applications, the transform pair is defined as

$$\mathscr{F}[f(t)] = \frac{1}{\sqrt{2\pi}} \int_{-\infty}^{\infty} f(t) \, e^{-i\omega t} \, dt = F(\omega) \tag{9.120}$$

and

$$\mathscr{F}^{-1}[F(\omega)] = \frac{1}{\sqrt{2\pi}} \int_{-\infty}^{\infty} F(\omega) \, e^{i\omega t} \, d\omega = f(t). \tag{9.121}$$

Amplitude spectrum   The graph of $(\omega, |F(\omega)|)$ is called the amplitude spectrum of $f(t)$. ω is called the frequency of the transform.

Example 9.25   Find the amplitude spectrum of the function

$$f(t) = \begin{cases} 5, & -2 \le t \le 2 \\ 0, & \text{otherwise.} \end{cases}$$

Solution   We have

$$F(\omega) = \mathscr{F}[f(t)] = \int_{-\infty}^{\infty} f(t) \, e^{-i\omega t} \, dt = \int_{-2}^{2} 5e^{-i\omega t} \, dt$$

$$= -5\left[\frac{e^{-i\omega t}}{i\omega}\right]_{-2}^{2} = -\frac{5}{i\omega}[e^{-2i\omega} - e^{2i\omega}] = \frac{10}{\omega} \sin 2\omega.$$

The graph of ω versus $|F(\omega)|$ is the amplitude spectrum of $f(t)$.

Linearity of Fourier transform

$$\mathscr{F}[af(t) + bg(t)] = a \, \mathscr{F}[f(t)] + b \, \mathscr{F}[g(t)]$$

provided the Fourier transforms of $f(t)$ and $g(t)$ exist.

We now present the shift theorems analogous to the shift theorems of Laplace transforms.

Theorem 9.9   **(Shifting on t-axis)** If $\mathscr{F}[f(t)] = F(\omega)$ and t_0 is any real number then

$$\mathscr{F}[f(t - t_0)] = F(\omega) \, e^{-i\omega t_0}. \tag{9.122}$$

Proof   From the definition, we get

$$\mathscr{F}[f(t - t_0)] = \int_{-\infty}^{\infty} f(t - t_0) e^{-i\omega t} \, dt = e^{-i\omega t_0} \int_{-\infty}^{\infty} f(t - t_0) \, e^{-i\omega(t - t_0)} \, dt.$$

Let $t - t_0 = \tau$. Then,

$$\mathscr{F}[f(t - t_0)] = e^{-i\omega t_0} \int_{-\infty}^{\infty} f(\tau) \, e^{-i\omega \tau} \, d\tau = e^{-i\omega t_0} F(\omega).$$

Remark 4

$$\mathscr{F}^{-1}[e^{-i\omega t_0} F(\omega)] = f(t - t_0). \tag{9.123}$$

Example 9.26 Find $\mathscr{F}^{-1}\left[\dfrac{e^{4i\omega}}{3+i\omega}\right]$.

Solution From Eq. (9.119), we have $\mathscr{F}[e^{-3t}\,u_0(t)] = \dfrac{1}{3+i\omega}$

or $$\mathscr{F}^{-1}\left[\frac{1}{3+i\omega}\right] = e^{-3t}\,u_0(t) = f(t).$$

Using the shift theorem, we get

$$\mathscr{F}^{-1}\left[\frac{e^{-(-4i\omega)}}{3+i\omega}\right] = f(t-(-4)) = e^{-3(t+4)}\,u_{-4}(t) = \begin{cases} 0, & t < -4 \\ e^{-3(t+4)}, & t \geq -4. \end{cases}$$

Theorem 9.10 (Frequency shifting) If $\mathscr{F}[f(t)] = F(\omega)$ and ω_0 is any real number, then

$$\mathscr{F}[e^{i\omega_0 t}\,f(t)] = F(\omega - \omega_0). \tag{9.124}$$

Proof From the definition, we get

$$\mathscr{F}[e^{i\omega_0 t}\,f(t)] = \int_{-\infty}^{\infty} e^{i\omega_0 t}\,e^{-i\omega t}\,f(t)\,dt = \int_{-\infty}^{\infty} e^{-i(\omega-\omega_0)t}\,f(t)\,dt = F(\omega-\omega_0).$$

Remark 5

$$\mathscr{F}^{-1}[F(\omega-\omega_0)] = e^{i\omega_0 t}\,f(t). \tag{9.125}$$

Theorem 9.11 (Modulation theorem) If $\mathscr{F}[f(t)] = F(\omega)$ and ω_0 is any real number, then

$$\mathscr{F}[f(t)\cos(\omega_0 t)] = \frac{1}{2}[F(\omega+\omega_0) + F(\omega-\omega_0)] \tag{9.126}$$

and $$\mathscr{F}[f(t)\sin(\omega_0 t)] = \frac{i}{2}[F(\omega+\omega_0) - F(\omega-\omega_0)]. \tag{9.127}$$

These results can be proved by using Eq. (9.124)

Fourier transforms of derivatives

Let $f(t)$ be continuous and $f^{(k)}(t)$, $k = 1, 2, \ldots, n$ be piecewise continuous on every interval $[-l, l]$

and $\displaystyle\int_{-\infty}^{\infty} |f^{(k-1)}(t)|\,dt$, $k = 1, 2, \ldots, n$ converge. Let $f^{(k)}(t) \to 0$ as $t \to \infty$ for $k = 0, 1, \ldots, n-1$.

If $\mathscr{F}[f(t)] = F(\omega)$, then

$$\mathscr{F}[f^{(n)}(t)] = (i\omega)^n\,F(\omega) \tag{9.128}$$

where $f(x)$ and all its derivatives vanish at infinity.

For example,

$$\mathscr{F}[f'(t)] = \int_{-\infty}^{\infty} f'(t)\,e^{-i\omega t}\,dt$$

$$= [f(t)\,e^{-i\omega t}]_{-\infty}^{\infty} + \int_{-\infty}^{\infty} i\omega f(t)e^{-i\omega t}\,dt = i\omega\,F(\omega).$$

$$\mathscr{F}[f''(t)] = -\omega^2\,F(\omega).$$

Example 9.27 Find the solution of the differential equation

$$y' - 2y = H(t) e^{-2t}, = -\infty < t < \infty.$$

using Fourier transforms, where $H(t) = u_0(t)$ is the unit step function.

Solution Applying the Fourier transform to the differential equation, we get

$$\mathscr{F}[y'] - 2\mathscr{F}[y] = \mathscr{F}[H(t) e^{-2t}]$$

or
$$i\omega Y(\omega) - 2Y(\omega) = \frac{1}{2 + i\omega} \qquad \text{(using Eq. (9.119))}$$

or
$$Y(\omega) = -\frac{1}{(2 + i\omega)(2 - i\omega)} = -\frac{1}{4 + \omega^2},$$

where $\mathscr{F}[y(t)] = Y(\omega)$.

Therefore,
$$y(t) = \mathscr{F}^{-1}\left[-\frac{1}{4 + \omega^2}\right] = -\frac{1}{4} e^{-2|t|}. \qquad \text{(using Example 9.23)}$$

The solution can also be written as

$$y(t) = \begin{cases} -\dfrac{1}{4}e^{2t}, & t < 0, \\[2mm] -\dfrac{1}{4}e^{-2t}, & t > 0, \end{cases}$$

Symmetry property of Fourier transforms

Let $\mathscr{F}[f(t)] = F(\omega)$. Then

$$\mathscr{F}[F(t)] = 2\pi f(-\omega). \qquad (9.129)$$

The result can be proved from the definition. From Eq. (9.118), we have

$$f(t) = \mathscr{F}^{-1}[F(\omega)] = \frac{1}{2\pi}\int_{-\infty}^{\infty} F(\omega)e^{i\omega t}\, d\omega = \frac{1}{2\pi}\int_{-\infty}^{\infty} F(s) e^{ist}\, ds$$

since ω is a dummy variable of integration. Hence, setting $t = -\omega$, we get

$$2\pi f(-\omega) = \int_{-\infty}^{\infty} F(s) e^{-i\omega s}\, ds = \int_{-\infty}^{\infty} F(t) e^{-i\omega t}\, dt = \mathscr{F}[F(t)].$$

Example 9.28 Find the Fourier transform of $f(t) = 1/(5 + it)$.

Solution We shall use the symmetry property to find the Fourier transform. We know that $1/(5 + i\omega)$ is the Fourier transform $H(t) e^{-5t}$ (see Example 9.22). Now, let $b(t) = H(t)e^{-5t}$. Then

$$\mathscr{F}[b(t)] = \mathscr{F}[H(t)e^{-5t}] = \frac{1}{5 + i\omega} = B(\omega).$$

Using the symmetry result, we obtain

$$\mathscr{F}[B(t)] = 2\pi b(-\omega), \text{ or } \mathscr{F}[B(t)] = \mathscr{F}\left[\frac{1}{5 + it}\right] = 2\pi\, b(-\omega) = 2\pi H(-\omega) e^{5\omega}.$$

Therefore,

$$\mathcal{F}[f(t)] = F(\omega) = \begin{cases} 2\pi e^{5\omega}, & \omega \leq 0 \\ 0, & \omega > 0. \end{cases}$$

Differentiation with respect to frequency ω

Theorem 9.12 Let $f(t)$ be piecewise continuous on every interval $[-l, l]$. Let $\displaystyle\int_{-\infty}^{\infty} |t^n f(t)| \, dt$ converge. Then

$$\mathcal{F}[t^n f(t)] = i^n F^{(n)}(\omega). \tag{9.130}$$

Proof From the definition, we have (assuming that integration and differentiation can be interchanged)

$$F'(\omega) = \frac{d}{d\omega} \int_{-\infty}^{\infty} f(t) e^{-i\omega t} \, dt = \int_{-\infty}^{\infty} \frac{d}{d\omega} [f(t) e^{-i\omega t}] \, dt$$

$$= -i \int_{-\infty}^{\infty} [t f(t)] e^{-i\omega t} \, dt = -i \, \mathcal{F}[t f(t)].$$

$$F''(\omega) = -i \frac{d}{d\omega} \int_{-\infty}^{\infty} t f(t) e^{-i\omega t} \, dt = -i \int_{-\infty}^{\infty} \frac{d}{d\omega} [t f(t) e^{-i\omega t}] \, dt$$

$$= (-i)^2 \int_{-\infty}^{\infty} t^2 f(t) e^{-i\omega t} \, dt = (-i)^2 \, \mathcal{F}[t^2 f(t)].$$

By induction we have the result.

In particular $\quad \mathcal{F}[t f(t)] = i F'(\omega) \quad$ and $\quad \mathcal{F}[t^2 f(t)] = -F''(\omega).$

Remark 6
From Eq. (9.130), we have

$$\mathcal{F}^{-1}[F^{(n)}(\omega)] = (-i)^n \, t^n f(t). \tag{9.131}$$

Fourier transform of an integral

Theorem 9.13 Let $f(t)$ be piecewise continuous an every interval $[-l, l]$ and $\displaystyle\int_{-\infty}^{\infty} |f(t)| \, dt$ converge. Let $\mathcal{F}[f(t)] = F(\omega)$ and $F(\omega)$ satisfies $F(0) = 0$. Then

$$\mathcal{F}\left[\int_{-\infty}^{t} f(\tau) \, d\tau\right] = \frac{1}{i\omega} F(\omega). \tag{9.132}$$

Example 9.29 Find the inverse Fourier transform of $(\sqrt{\pi} \, \omega e^{-\omega^2/8})/(4\sqrt{2} i)$.

Solution We have $\dfrac{\sqrt{\pi} \, \omega}{4\sqrt{2} i} e^{-\omega^2/8} = -\dfrac{\sqrt{\pi}}{\sqrt{2} i} \dfrac{d}{d\omega} [e^{-\omega^2/8}].$

Therefore, $\quad \mathscr{F}^{-1}\left[\dfrac{\sqrt{\pi}\,\omega}{4\sqrt{2}\,i}\,e^{-\omega^2/8}\right] = -\dfrac{\sqrt{\pi}}{i\sqrt{2}}\,\mathscr{F}^{-1}\left[\{e^{-\omega^2/8}\}'\right] = -\dfrac{\sqrt{\pi}}{i\sqrt{2}}\,\mathscr{F}^{-1}[F'(\omega)]$

where $F(\omega) = e^{-\omega^2/8}$. Hence, using Eq. (9.130), we obtain

$$\mathscr{F}^{-1}\left[\dfrac{\sqrt{\pi}\,\omega}{4\sqrt{2}\,i}\,e^{-\omega^2/8}\right] = -\dfrac{\sqrt{\pi}}{i\sqrt{2}}\,[-it\,f(t)]$$

where $\quad f(t) = \mathscr{F}^{-1}[F(\omega)] = \mathscr{F}^{-1}[e^{-\omega^2/8}] = \dfrac{\sqrt{2}}{\sqrt{\pi}}\,e^{-2t^2}.$

(see Example 9.24). Therefore,

$$\mathscr{F}^{-1}\left[\dfrac{\sqrt{\pi}\,\omega}{4\sqrt{2}\,i}\,e^{-\omega^2/8}\right] = t\,e^{-2t^2}.$$

Convolution

Theorem 9.14 Let $f(t)$, $g(t)$ be piecewise continuous on every interval $[-l,\, l]$ and let

$$\int_{-\infty}^{\infty} |f(t)|\,dt, \int_{-\infty}^{\infty} |g(t)|\,dt$$

converge. Denote $\mathscr{F}[f(t)] = F(\omega)$ and $\mathscr{F}[g(t)] = G(\omega)$. Then

$$\mathscr{F}[\{f * g\}(t)] = F(\omega)\,G(\omega) \qquad \text{(convolution with respect to time)} \qquad (9.133)$$

and $\qquad \mathscr{F}[f(t)\,g(t)] = \dfrac{1}{2\pi}\,[F * G](\omega) \quad \text{(convolution with respect to frequency)} \qquad (9.134)$

where the convolution $(f * g)(t)$ is defined as

$$(f * g)(t) = \int_{-\infty}^{\infty} f(\tau)\,g(t - \tau)\,d\tau = (g * f)(t).$$

The inverse transforms are given by

$$\mathscr{F}^{-1}[F(\omega)\,G(\omega)] = (f * g)(t) \qquad\qquad (9.135)$$

and $\qquad\qquad \mathscr{F}^{-1}[(F * G)(\omega)] = 2\pi f(t)\,g(t). \qquad\qquad (9.136)$

Example 9.30 Using convolution find $\mathscr{F}^{-1}[1/(12 + 7i\omega - \omega^2)]$.

Solution We have $12 + 7i\omega - \omega^2 = (4 + i\omega)(3 + i\omega)$. Therefore,

$$\mathscr{F}^{-1}\left[\dfrac{1}{(4 + i\omega)(3 + i\omega)}\right] = \mathscr{F}^{-1}\left[\dfrac{1}{4 + i\omega} \cdot \dfrac{1}{3 + i\omega}\right].$$

But $\mathscr{F}^{-1}\left[\dfrac{1}{4 + i\omega}\right] = e^{-4t}\,H(t)$ and $\mathscr{F}^{-1}\left[\dfrac{1}{3 + i\omega}\right] = e^{-3t}\,H(t)$ (see Example 9.22), where $H(t)$

is the unit step function. Using convolution, we obtain

$$\mathscr{F}^{-1}\left[\frac{1}{(4+i\omega)(3+i\omega)}\right] = [e^{-4t}\,H(t)] * [e^{-3t}\,H(t)]$$

$$= \int_{-\infty}^{\infty} e^{-4\tau} H(\tau)\, e^{-3(t-\tau)}\, H(t-\tau)\, d\tau = e^{-3t} \int_{-\infty}^{\infty} e^{-\tau} H(\tau)\, H(t-\tau)\, d\tau$$

But
$$H(\tau)\,H(t-\tau) = \begin{cases} 0, & \text{for } \tau < 0 \text{ and } \tau > t. \\ 1, & \text{for } 0 < \tau < t. \end{cases}$$

Therefore,
$$\mathscr{F}^{-1}\left[\frac{1}{(4+i\omega)(3+i\omega)}\right] = e^{-3t}\int_0^t e^{-\tau}\, d\tau = e^{-3t}\,[1-e^{-t}],\, t \geq 0.$$

Fourier transform of the Dirac-delta function

In section 8.5, we have shown that the Laplace transform of the Dirac-delta function is $\mathscr{L}[\delta(t)] = 1$. We have also proved the filtering property of Dirac-delta function as

$$\int_0^\infty f(t)\,\delta(t-a)\,dt = f(a). \tag{9.137}$$

Similar results can be proved for the Fourier transform of the Dirac-delta function. We have defined the delta function as (see Eq. (8.29))

$$\delta(t) = \lim_{k\to 0} \frac{1}{k}\,[H(t) - H(t-k)]$$

where $H(t)$ is the unit step function.

We have
$$H(t) - H(t-k) = \begin{cases} 0, & t < 0 \\ 1, & 0 \leq t < k \\ 0, & t \geq k. \end{cases}$$

Hence,
$$\mathscr{F}[\delta(t)] = \lim_{k\to 0}\left\{\frac{1}{k}\,\mathscr{F}[H(t) - H(t-k)]\right\} = \lim_{k\to 0}\left\{\frac{1}{k}\int_0^k e^{-i\omega t}\,dt\right\}$$

$$= \lim_{k\to 0}\frac{1}{k}\left[\frac{1-e^{-i\omega k}}{i\omega}\right] = 1.$$

Therefore, Laplace transform and Fourier transform of the Dirac-delta function are both equal to 1. Hence, $\mathscr{F}^{-1}[1] = \delta(t)$.

The filtering property given in Eq. (9.137) gets modified as

$$\int_{-\infty}^\infty f(t)\,\delta(t-a)\,dt = f(a). \tag{9.138}$$

This result can be proved using the definition of the delta function.

Using the time-shift theorem (see Eq. 9.122), we obtain

$$\mathscr{F}[\delta(t-a)] = e^{-ia\omega}F(\omega) = e^{-ia\omega} \tag{9.139}$$

Using the symmetry result (see Eq. 9.129), we obtain

$$\mathscr{F}[1] = 2\pi\delta(-\omega) = 2\pi\delta(\omega). \tag{9.140}$$

Fourier cosine transform

The Fourier cosine transform of $f(t)$ is defined as

$$\mathscr{F}_c[f(t)] = \int_0^\infty f(t)\cos(\omega t)\,dt = F_c(\omega). \tag{9.141}$$

We can compare Fourier cosine transform with the Fourier cosine integral representation of $f(t)$ on $[0, \infty)$ which is given by (see Eq. (9.40))

$$f(t) = \frac{1}{\pi}\int_0^\infty A(\omega)\cos(\omega t)\,d\omega \tag{9.142}$$

where

$$A(\omega) = 2\int_0^\infty f(t)\cos(\omega t)\,dt. \tag{9.143}$$

Comparing Eqs. (9.141) and (9.143), we obtain $A(\omega) = 2F_c(\omega)$. Substituting in Eq. (9.142), we get

$$f(t) = \frac{2}{\pi}\int_0^\infty F_c(\omega)\cos(\omega t)\,d\omega. \tag{9.144}$$

This result can be interpreted as the inverse Fourier cosine transform.

Example 9.31 Find the Fourier cosine transform of $f(t)$, where

(i) $f(t) = \begin{cases} 1, & 0 \le t \le l, \\ 0, & t > l. \end{cases}$
(ii) $f(t) = \begin{cases} t, & 0 \le t \le l \\ 0, & t > l. \end{cases}$

Solution

(i)
$$\mathscr{F}_c[f(t)] = \int_0^\infty f(t)\cos(\omega t)\,dt = \int_0^l \cos(\omega t)\,dt = \frac{\sin(\omega l)}{\omega}.$$

(ii)
$$\mathscr{F}_c[f(t)] = \int_0^\infty f(t)\cos(\omega t)\,dt = \int_0^l t\cos(\omega t)\,dt$$

$$= \left[\frac{t\sin(\omega t)}{\omega} + \frac{\cos(\omega t)}{\omega^2}\right]_0^l = \frac{l}{\omega}\sin(\omega l) + \frac{1}{\omega^2}[\cos(\omega l) - 1].$$

Fourier sine transform

The Fourier sine transform of $f(t)$ is defined as

$$\mathscr{F}_s[f(t)] = \int_0^\infty f(t)\sin(\omega t)\,dt = F_s(\omega). \tag{9.145}$$

Comparing with the Fourier sine integral representation of $f(t)$ (see Eq. (9.42))

$$f(t) = \frac{1}{\pi} \int_0^\infty B(\omega) \sin(\omega t) d\omega, \quad \text{where} \quad B(\omega) = 2 \int_0^\infty f(t) \sin(\omega t) dt$$

we obtain $B(\omega) = 2F_s(\omega)$. Therefore,

$$f(t) = \frac{2}{\pi} \int_0^\infty F_s(\omega) \sin(\omega t) d\omega. \tag{9.146}$$

This result can be interpreted as the inverse Fourier sine transform.

Example 9.32 Find the Fourier sine transform of $f(t)$, where

(i) $f(t) = \begin{cases} 1, & 0 \le t \le l, \\ 0, & t > l. \end{cases}$
(ii) $f(t) = \begin{cases} t, & 0 \le t \le l, \\ 0, & t > l. \end{cases}$

Solution

(i)
$$\mathscr{F}_s[f(t)] = \int_0^\infty f(t) \sin(\omega t) dt = \int_0^l \sin(\omega t) dt$$

$$= \left[-\frac{\cos(\omega t)}{\omega} \right]_0^l = \frac{1}{\omega} [1 - \cos(\omega l)].$$

(ii)
$$\mathscr{F}_s[f(t)] = \int_0^\infty f(t) \sin(\omega t) dt = \int_0^l t \sin(\omega t) dt$$

$$= \left[-\frac{t \cos(\omega t)}{\omega} + \frac{\sin(\omega t)}{\omega^2} \right]_0^l = \frac{\sin(\omega l)}{\omega^2} - \frac{l \cos(\omega l)}{\omega}.$$

Fourier cosine and sine transforms of derivatives

Let $f(t)$ and $f'(t)$ be continuous on the interval $[0, \infty)$. Let $f(t) \to 0, f'(t) \to 0$ as $t \to \infty$ and $f''(t)$ be piecewise continuous on every subinterval $[0, l]$. Then

$$\mathscr{F}_c[f''(t)] = -\omega^2 \mathscr{F}_c[f(t)] - f'(0) \quad \text{and} \quad \mathscr{F}_s[f''(t)] = -\omega^2 \mathscr{F}_s[f(t)] + \omega f(0).$$

From the definition, we have

$$\mathscr{F}_c[f''(t)] = \int_0^\infty f''(t) \cos(\omega t) dt$$

$$= \left[f'(t) \cos(\omega t) + \omega f(t) \sin(\omega t) \right]_0^\infty - \omega^2 \int_0^\infty f(t) \cos(\omega t) dt$$

$$= -\omega^2 \mathscr{F}_c[f(t)] - f'(0) = -\omega^2 F_c(\omega) - f'(0). \tag{9.147}$$

$$\mathscr{F}_s[f''(t)] = \int_0^\infty f''(t) \sin(\omega t)\, dt$$

$$= \left[f'(t) \sin(\omega t) - \omega f(t) \cos(\omega t) \right]_0^\infty - \omega^2 \int_0^\infty f(t) \sin(\omega t)\, dt$$

$$= -\omega^2 F_s[f(t)] + \omega f(0) = -\omega^2 F_s(\omega) + \omega f(0). \tag{9.148}$$

Example 9.33 Prove the following

$$\mathscr{F}_c[f'(t)] = \omega F_s(\omega) - f(0) \quad \text{and} \quad \mathscr{F}_s[f'(t)] = -\omega F_c(\omega)$$

assuming that $f(t) \to 0$ as $t \to \infty$.

Solution From the definition, we have

$$\mathscr{F}_c[f'(t)] = \int_0^\infty f'(t) \cos(\omega t)\, dt = \left[f(t) \cos(\omega t) \right]_0^\infty + \omega \int_0^\infty f(t) \sin(\omega t)\, dt$$

$$= \omega F_s(\omega) - f(0).$$

$$\mathscr{F}_s[f'(t)] = \int_0^\infty f'(t) \sin(\omega t)\, dt = \left[f(t) \sin(\omega t) \right]_0^\infty - \omega \int_0^\infty f(t) \cos(\omega t)\, dt$$

$$= -\omega F_c(\omega).$$

Example 9.34 Find the Fourier cosine and sine transforms of

$$f(t) = e^{-\alpha t}, \; t \geq 0, \; \alpha > 0.$$

Solution We shall use the formulas for derivatives to find the Fourier cosine and sine transforms of $f(t)$. Since $f(t) = e^{-\alpha t}$, we have $f'(t) = -\alpha e^{-\alpha t}$ and $f''(t) = \alpha^2 e^{-\alpha t}$. Using Eq. (9.147), we have

$$\mathscr{F}_c[f''(t)] = \mathscr{F}_c[\alpha^2 e^{-\alpha t}] = \alpha^2 \mathscr{F}_c[e^{-\alpha t}] = \alpha^2 F_c(\omega).$$

Also,

$$\mathscr{F}_c[f''(t)] = -\omega^2 F_c(\omega) - f'(0) = -\omega^2 F_c(\omega) + \alpha.$$

Therefore,

$$\alpha^2 F_c(\omega) = -\omega^2 F_c(\omega) + \alpha, \text{ or } F_c(\omega) = \frac{\alpha}{\alpha^2 + \omega^2}.$$

Now,

$$\mathscr{F}_s[f''(t)] = \alpha^2 \mathscr{F}_s[e^{-\alpha t}] = \alpha^2 F_s(\omega).$$

Also,

$$\mathscr{F}_s[f''(t)] = -\omega^2 F_s(\omega) + \omega f(0) = -\omega^2 F_s(\omega) + \omega.$$

Therefore,

$$\alpha^2 F_s(\omega) = -\omega^2 F_s(\omega) + \omega, \text{ or } F_s(\omega) = \frac{\omega}{\alpha^2 + \omega^2}.$$

Finite Fourier cosine transform

The Fourier cosine and sine transforms are defined on $[0, \infty)$ and were obtained from Fourier cosine and sine integral representations of functions. However, in many applications, we are to deal with problems defined on finite intervals. In this case, we define the finite Fourier cosine and sine transforms and are obtained from the Fourier cosine and sine series.

Let the given function $f(t)$ be piecewise continuous on $[0, \pi]$. Then, the *finite Fourier cosine transform* of $f(t)$ is defined by

$$F_c(n) = \int_0^\pi f(t) \cos(nt)\,dt \tag{9.149}$$

where n is an non-negative integer, $n = 0, 1, 2, \ldots$ This transform is also denoted by $C_n[f(t)]$.

The Fourier cosine half-range series is defined by (see Eq. (9.25))

$$f(t) = \frac{1}{\pi} \int_0^\pi f(t^*)\,dt^* + \frac{2}{\pi} \sum_{n=1}^\infty \left[\int_0^\pi f(t^*) \cos(nt^*)\,dt^* \right] \cos(nt).$$

In terms of the finite Fourier cosine transform, we can write this equation as

$$f(t) = \frac{1}{\pi} \left[F_c(0) + 2 \sum_{n=1}^\infty F_c(n) \cos(nt) \right]. \tag{9.150}$$

This result can be interpreted as the inverse finite Fourier cosine transform.

Formula for the second derivative

We assume that $f(t)$ and $f'(t)$ are continuous and $f''(t)$ is piecewise continuous on $[0, \pi]$. Then

$$C_n[f''(t)] = -n^2 F_c(n) - f'(0) + (-1)^n f'(\pi), \; n = 1, 2, \ldots.$$

This result can be proved using the definition of $C_n[f(t)]$. We have

$$C_n[f''(t)] = \int_0^\pi f''(t) \cos(nt)\,dt$$

$$= \left[f'(t) \cos(nt) + nf(t) \sin(nt) \right]_0^\pi - n^2 \int_0^\pi f(t) \cos(nt)\,dt$$

$$= -n^2 F_c(n) - f'(0) + (-1)^n f'(\pi).$$

Finite Fourier sine transform

Let the function $f(t)$ be piecewise continuous on $[0, \pi]$. Then, the *finite Fourier sine transform* of $f(t)$ is defined as

$$F_s(n) = \int_0^\pi f(t) \sin(nt)\,dt \tag{9.151}$$

where n is an integer and $n = 1, 2, \ldots.$ This transform is also denoted by $S_n[f(t)]$.

The Fourier sine half-range series is defined by (see Eq. (9.26))

$$f(t) = \frac{2}{\pi} \sum_{n=1}^\infty \left[\int_0^\pi f(t^*) \sin(nt^*) \right] \sin(nt).$$

In terms of the finite Fourier sine transform, we can write this equation as

$$f(t) = \frac{2}{\pi} \sum_{n=1}^\infty F_s(n) \sin(nt). \tag{9.152}$$

This result can be interpreted as the inverse finite Fourier sine transform.

Formula for the second derivative

We assume that $f(t)$ and $f'(t)$ are continuous and $f''(t)$ is piecewise continuous on $[0, \pi]$. Then

$$S_n[f''(t)] = -n^2 F_s(n) + nf(0) - n(-1)^n f(\pi), \quad n = 1, 2, \ldots \qquad (9.153)$$

Using the definition, we have

$$S_n[f''(t)] = \int_0^\pi f''(t) \sin(nt) dt$$

$$= \left[f'(t) \sin(nt) - nf(t) \cos(nt) \right]_0^\pi - n^2 \int_0^\pi f(t) \sin(nt) dt$$

$$= -n^2 F_s(n) + nf(0) - n(-1)^n f(\pi).$$

Example 9.35 Find the finite Fourier sine transform of

$$f(t) = \begin{cases} 0, & 0 \le x < \pi/2 \\ 1, & \pi/2 \le x \le \pi. \end{cases}$$

Solution We have

$$S_n[f(t)] = \int_0^\pi f(t) \sin(nt) \, dt = \int_{\pi/2}^\pi \sin(nt) \, dt = -\frac{1}{n} \left[\cos(nt) \right]_{\pi/2}^\pi$$

$$= -\frac{1}{n} [\cos n\pi - \cos(n\pi/2)] = -\frac{1}{n} [(-1)^n - \cos(n\pi/2)].$$

9.6.1 Fourier Transform Solution of Some Partial Differential Equations

In section 8.7, we have studied Laplace transform methods for the solution of some partial differential equations and in section 9.5 we have studied the Fourier series and Fourier integral solution of heat equation, wave equation and Laplace equation. In this section, we shall discuss the Fourier transform solution of some of these partial differential equations.

If the Fourier transform is applied with respect to one of the variables in the differential equation, then we obtain an ordinary differential equation in terms of the other variable. We solve this differential equation. The solution of the given *bvp* is then obtained by taking the inverse Fourier transform.

Example 9.36 The temperature distribution $u(x, t)$ in a thin, homogeneous, infinite bar can be modelled by the initial boundary value problem

$$\frac{\partial u}{\partial t} = c^2 \frac{\partial^2 u}{\partial x^2}, \quad -\infty < x < \infty, t > 0$$

$$u(x, 0) = f(x), \quad u(x, t) \text{ is finite as } x \to \pm\infty.$$

Find $u(x, t), t > 0$.

Solution Since the domain of the bar is $-\infty < x < \infty$, we use the Fourier transform with respect to x. Denote $\mathscr{F}[u(x, t)] = F(\omega, t)$. Taking Fourier transform of the differential equation, we obtain

$$\mathscr{F}\left[\frac{\partial u}{\partial t}\right] = \mathscr{F}\left[c^2 \frac{\partial^2 u}{\partial x^2}\right] = c^2 \mathscr{F}\left[\frac{\partial^2 u}{\partial x^2}\right]. \tag{9.154}$$

Now,

$$\mathscr{F}\left[\frac{\partial u}{\partial t}\right] = \int_{-\infty}^{\infty} \frac{\partial u}{\partial t} e^{-i\omega x}\, dx = \frac{\partial}{\partial t} \int_{-\infty}^{\infty} u(x,t) e^{-i\omega x}\, dx = \frac{\partial F(\omega,t)}{\partial t} = \frac{dF(\omega,t)}{dt}$$

assuming ω as a parameter, and

$$\mathscr{F}\left[\frac{\partial^2 u}{\partial x^2}\right] = -\omega^2 F(\omega,t). \qquad \text{(from Eq. (9.128))}$$

Substituting in Eq. (9.154), we obtain

$$\frac{dF(\omega,t)}{dt} = -c^2 \omega^2 F(\omega,t).$$

The solution of this linear, first order differential equation is $F(\omega,t) = k e^{-c^2 \omega^2 t}$ where k is a parameter (function of ω) to be determined. Writing the transform of the initial condition, we obtain

$$\mathscr{F}[u(x,0)] = \int_{-\infty}^{\infty} f(x) e^{-i\omega x}\, dx = G(\omega).$$

Therefore, $\qquad F(\omega,0) = k = G(\omega) \quad \text{and} \quad F(\omega,t) = G(\omega) e^{-c^2 \omega^2 t}.$

The solution $u(x,t)$ of the *bvp* is the inverse transform of $F(\omega,t)$. The inverse can be computed in a number of ways. We can obtain the inverse directly, using the definition (see Eq. (9.118)). We have

$$u(x,t) = \frac{1}{2\pi} \int_{-\infty}^{\infty} F(\omega,t) e^{i\omega x}\, d\omega = \frac{1}{2\pi} \int_{-\infty}^{\infty} G(\omega) e^{-c^2 \omega^2 t} e^{i\omega x}\, d\omega$$

$$= \frac{1}{2\pi} \int_{-\infty}^{\infty} \left[\int_{-\infty}^{\infty} f(\xi) e^{-i\omega \xi}\, d\xi\right] e^{-c^2 \omega^2 t} e^{i\omega x}\, d\omega$$

$$= \frac{1}{2\pi} \int_{-\infty}^{\infty} \int_{-\infty}^{\infty} f(\xi) e^{-i\omega(\xi - x)} e^{-c^2 \omega^2 t}\, d\xi\, d\omega$$

$$= \frac{1}{2\pi} \left[\int_{-\infty}^{\infty} \int_{-\infty}^{\infty} f(\xi) \cos \omega(\xi - x) e^{-c^2 \omega^2 t}\, d\xi\, d\omega\right.$$

$$\left. - i \int_{-\infty}^{\infty} \int_{-\infty}^{\infty} f(\xi) \sin \omega(\xi - x) e^{-c^2 \omega^2 t}\, d\xi\, d\omega\right] = P - iQ$$

where P and Q are the integrals on the left hand side. The integrand of Q is an odd function of ω. Therefore,

$$u(x, t) = \frac{1}{2\pi} \int_{-\infty}^{\infty} \int_{-\infty}^{\infty} f(\xi) \cos \omega (\xi - x) \, e^{-c^2 \omega^2 t} \, d\xi \, d\omega.$$

Example 9.37 Find the solution in Example 9.36, if

$$f(x) = \begin{cases} v, & -l < x < l \\ 0, & \text{otherwise, } v \text{ a constant.} \end{cases}$$

Solution We have

$$G(\omega) = \int_{-\infty}^{\infty} f(x) e^{-i\omega x} \, dx = v \int_{-l}^{l} e^{-i\omega x} \, dx$$

$$= \frac{v}{i\omega} [e^{i\omega l} - e^{-i\omega l}] = \frac{2v}{\omega} \sin(\omega l).$$

Therefore,

$$u(x, t) = \frac{1}{2\pi} \int_{-\infty}^{\infty} \frac{2v}{\omega} \sin(\omega l) \, e^{-c^2 \omega^2 t} \, e^{i\omega x} \, d\omega$$

$$= \frac{v}{\pi} \int_{-\infty}^{\infty} \frac{1}{\omega} \sin(\omega l) \, [\cos \omega x + i \sin \omega x] \, e^{-c^2 \omega^2 t} \, d\omega$$

$$= \frac{v}{\pi} \int_{-\infty}^{\infty} \frac{1}{\omega} \sin(\omega l) \cos \omega x \, e^{-c^2 \omega^2 t} \, d\omega \tag{9.155}$$

since $[\sin(\omega l) \sin(\omega x)]/\omega$ is an odd function of ω. If the solution of the *bvp* was obtained by the Laplace transform method, then the solution can be expressed in terms of error function. Therefore, Eq. (9.155) gives another form of the solution.

Example 9.38 The temperature distribution $u(x, t)$ in a thin, homogeneous semi-infinite bar can be modelled by the initial boundary value problem

$$\frac{\partial u}{\partial t} = c^2 \frac{\partial^2 u}{\partial x^2}, 0 < x < \infty, t > 0$$

$$u(x, 0) = f(x), x > 0; u(0, t) = 0, t > 0.$$

Find the temperature distribution $u(x, t)$, $t > 0$, $0 < x < \infty$.

Solution The domain of definition of x is $0 < x < \infty$. Hence, Fourier transform cannot be used. We can explore the possibility of using Fourier cosine or sine transforms. Application of Fourier cosine transform requires the information of a derivative (see Eq. (9.147)). Therefore, we can attempt finding solution by Fourier sine transform. Applying the sine transform to the differential equation, we obtain

$$\mathscr{F}_s \left[\frac{\partial u}{\partial t} \right] = c^2 \, \mathscr{F}_s \left[\frac{\partial^2 u}{\partial x^2} \right].$$

Using Eq. (9.148), we obtain

$$\frac{d}{dt} F_s(\omega, t) = c^2 [-\omega^2 F_s(\omega, t) + \omega u(0, t)] = -c^2 \omega^2 F_s(\omega, t).$$

The solution of this linear, first order equation is

$$F_s(\omega, t) = k(\omega) e^{-c^2 \omega^2 t} \tag{9.156}$$

where $k(\omega)$ is any arbitrary function of ω.

Taking sine transform of the boundary condition, we get

$$\mathcal{F}_s[u(x, 0)] = F_s(\omega, 0) = \mathcal{F}_s[f(x)] = G(\omega).$$

Substituting in Eq. (9.156), we get $k(\omega) = G(\omega)$. Therefore,

$$F_s(\omega, t) = G(\omega) e^{-c^2 \omega^2 t}.$$

The inverse transform is given by (see Eq. (9.146))

$$u(x, t) = \frac{2}{\pi} \int_0^\infty F_s(\omega, t) \sin(\omega x) d\omega = \frac{2}{\pi} \int_0^\infty G(\omega) e^{-c^2 \omega^2 t} \sin(\omega x) d\omega$$

$$= \frac{2}{\pi} \int_0^\infty \left[\int_0^\infty f(\xi) \sin(\omega \xi) d\xi \right] e^{-c^2 \omega^2 t} \sin(\omega x) d\omega.$$

Example 9.39 The steady state temperature distribution $u(x, y)$ in a thin, homogeneous semi-infinite plate is governed by the *bvp*

$$\frac{\partial^2 u}{\partial x^2} + \frac{\partial^2 u}{\partial y^2} = 0, 0 < x < l, 0 < y < \infty$$

$$u(0, y) = e^{-2y}, u(l, y) = 0, y > 0; \left(\frac{\partial u}{\partial y} \right)(x, 0) = 0, 0 < x < l.$$

Find the temperature distribution $u(x, y)$, $0 < x < l$, $y > 0$.

Solution   Since the domain of x is finite and the domain of y is $0 < y < \infty$, we can attempt to use Fourier cosine or sine transform (with respect to the variable y). Since the derivative $(\partial u / \partial y)(x, 0)$ is prescribed, we can use the cosine transform. Applying the cosine transform to the differential equation, we obtain

$$\mathcal{F}_c\left[\frac{\partial^2 u}{\partial x^2} \right] + \mathcal{F}_c\left[\frac{\partial^2 u}{\partial y^2} \right] = 0$$

where the transform is taken with respect to the variable y, that is

$$\mathcal{F}_c[u(x, y)] = \int_0^\infty u(x, y) \cos(\omega y) dy = F(x, \omega).$$

Hence, we obtain (using Eq. (9.147))

$$\frac{d^2}{dx^2} F(x, \omega) - \omega^2 F(x, \omega) - \frac{\partial u}{\partial y}(x, 0) = 0 \quad \text{or} \quad \frac{d^2 F}{dx^2} - \omega^2 F = 0.$$

The solution of this ordinary differential equation is

$$F(x, \omega) = A \cosh(\omega x) + B \sinh(\omega x). \qquad (9.157)$$

Taking cosine transform of the other boundary conditions, we obtain

$$\mathscr{F}_c[u(0, y)] = F(0, \omega) = \mathscr{F}_c[e^{-2y}] = \frac{2}{4 + \omega^2} \quad \text{(see Example 9.34)}$$

and
$$\mathscr{F}_c[u(l, y)] = \mathscr{F}_c[0] = 0 = F(l, \omega).$$

Using these conditions in Eq. (9.157), we get

$$F(0, \omega) = A = \frac{2}{4 + \omega^2} \quad \text{and} \quad F(l, \omega) = A \cosh(\omega l) + B \sinh(\omega l) = 0 \cdot$$

We have
$$B = -\frac{A \cosh(\omega l)}{\sinh(\omega l)}$$

and
$$F(x, \omega) = A \cosh(\omega x) + B \sinh(\omega x) = A \left[\cosh(\omega x) - \frac{\cosh(\omega l) \sinh(\omega x)}{\sinh(\omega l)} \right]$$

$$= \frac{A}{\sinh(\omega l)} [\cosh(\omega x) \sinh(\omega l) - \cosh(\omega l) \sinh(\omega x)]$$

$$= \frac{2 \sinh[\omega(l - x)]}{(4 + \omega^2) \sinh(\omega l)} \cdot$$

The inverse transform gives

$$u(x, y) = \frac{2}{\pi} \int_0^\infty F(x, \omega) \cos(\omega y) \, d\omega = \frac{4}{\pi} \int_0^\infty \frac{\sinh[\omega(l - x)] \cos(\omega y)}{(4 + \omega^2) \sinh(\omega l)} \, d\omega.$$

Exercise 9.5

Find the Fourier transform of the following functions.

1. $f(t) = \begin{cases} 1, & |t| \le 1 \\ 0, & |t| > 1. \end{cases}$

2. $f(t) = \begin{cases} -(1 + t), & -1 \le t \le 0, \\ t - 1, & 0 < t \le 1, \\ 0, & |t| > 1. \end{cases}$

3. $f(t) = \begin{cases} \cos t, & -l \le t \le l, \\ 0, & |t| > l. \end{cases}$

4. $f(t) = \begin{cases} e^{\alpha t}, & t < 0 \\ 0, & t > 0, \; \alpha > 0. \end{cases}$

5. $f(t) = \begin{cases} -e^{\alpha t}, & t < 0 \\ e^{-\alpha t}, & t > 0, \; \alpha > 0. \end{cases}$

6. $f(t) = H(t - 3)e^{-4t}.$

7. $f(t) = e^{-a|t+1|}, \; a > 0.$

8. $f(t) = 1/(1 + t^2).$

9. $f(t) = 2e^{-3|t|} \sin(4t)$.

Find the inverse Fourier transform of the following functions.

10. $\dfrac{e^{-i\omega}}{2(1 + i\omega)}$.

11. $\dfrac{2i\omega}{(3 + i\omega)}$.

12. $\dfrac{e^{-(1-i\omega)}}{3 + i\omega}$.

13. $\dfrac{e^{-2i\omega}}{2 + 3i\omega}$.

14. $\dfrac{1}{6 + 5i\omega - \omega^2}$.

15. $\dfrac{i\omega}{(i\omega + 2)(i\omega + 3)}$.

16. $\dfrac{e^{-2i\omega}}{4 + \omega^2}$.

17. $\dfrac{1}{a^4 + \omega^4}$, $a > 0$.

Using frequency convolution show that

18. $\displaystyle\int_{-\infty}^{\infty} \dfrac{d\tau}{(2 - i\tau + i\omega)(2 + i\tau)} = \dfrac{2\pi}{4 + i\omega}$.

19. $\displaystyle\int_{-\infty}^{\infty} \dfrac{d\tau}{(4 - i\tau)(4 - i\tau + i\omega)} = 0$.

Use the time convolution to find the inverse of the following functions.

20. $\dfrac{1}{(i\omega + k)^2}$, $k > 0$.

21. $\dfrac{1}{(i\omega + k)^3}$, $k > 0$.

22. Let $F(\omega)$, $G(\omega)$ be the Fourier transforms of $f(x)$ and $g(x)$ respectively. Then, show that

(i) $\displaystyle\int_{-\infty}^{\infty} F(\omega)\,\overline{G}(\omega)\,d\omega = 2\pi \int_{-\infty}^{\infty} f(x)\,\overline{g}(x)\,dx$,   (ii) $\displaystyle\int_{-\infty}^{\infty} |F(\omega)|^2\,d\omega = 2\pi \int_{-\infty}^{\infty} |f(x)|^2\,dx$

(Parseval's identities in Fourier transforms)

Using Fourier transforms, find the solution of the following differential equations.

23. $y' - 4y = H(t)\,e^{-4t}$, $-\infty < t < \infty$.

24. $y' + 3y = H(t)\,e^{-2t}$, $-\infty < t < \infty$.

25. $y'' + 5y' + 4y = \delta(t - 2)$.

26. $y'' + 3y' + 2y = \delta(t - 3)$.

Find the Fourier cosine and sine transforms of the following functions.

27. $f(t) = \begin{cases} \sin t, & 0 \le t \le l, \\ 0, & t > l. \end{cases}$

28. $f(t) = \begin{cases} \cos t, & 0 \le t \le l \\ 0, & t > l. \end{cases}$

29. $f(t) = \begin{cases} 1 + t, & 0 \le t \le l, \\ 0, & t > l. \end{cases}$

30. $f(t) = \begin{cases} e^{2t} - e^{-2t}, & 1 \le t < 2, \\ 0, & \text{otherwise}. \end{cases}$

The following functions are defined on $[0, \pi]$. Find the finite Fourier cosine and sine transform of the following functions.

31. $f(t) = t$.

32. $f(t) = \sin(at)$, $a > 0$.

33. $f(t) = e^{-t}$.

34. $f(t) = \sinh(at)$, $a > 0$.

Using the Fourier integral transforms, solve the following initial boundary value problems.

35. $\dfrac{\partial u}{\partial t} = c^2 \dfrac{\partial^2 u}{\partial x^2}$, $-\infty < x < \infty$, $t > 0$,

$u(x, 0) = e^{-2|x|}$, $-\infty < x < \infty$.

36. $\dfrac{\partial u}{\partial t} = c^2 \dfrac{\partial^2 u}{\partial x^2}$, $-\infty < x < \infty$, $t > 0$,

$u(x, 0) = \begin{cases} 1, & -1 < x < 0 \\ -1, & 0 < x < 1 \\ 0, & \text{otherwise}. \end{cases}$

37. $\dfrac{\partial u}{\partial t} = c^2 \dfrac{\partial^2 u}{\partial x^2},\ -\infty < x < \infty,\ t > 0.$

$u(x, 0) = e^{-4x^2},\ -\infty < x < \infty.$

It is given that $\mathscr{F}[e^{-at^2}] = \sqrt{\dfrac{\pi}{a}}\, e^{-\omega^2/(4a)}.$

38. $\dfrac{\partial u}{\partial t} = \dfrac{\partial^2 u}{\partial x^2},\ 0 < x < \infty,\ t > 0,$

$u(x, 0) = \begin{cases} 1, & 0 < x \le l, \\ 0, & x > l. \end{cases}$

$u(0, t) = 0,\ t > 0.$

39. $\dfrac{\partial^2 u}{\partial x^2} + \dfrac{\partial^2 u}{\partial y^2} = 0,\ 0 < x < \pi,\ y > 0.$

$u(0, y) = 0,\ u(\pi, y) = 0,\ y > 0.$

$u(x, 0) = \sin x,\ 0 < x < \pi.$

40. $\dfrac{\partial^2 u}{\partial x^2} + \dfrac{\partial^2 u}{\partial y^2} = 0,\ -\infty < x < \infty,\ 0 < y < \pi,$

$u(x, 0) = e^{-2x} H(x),$

$u(x, \pi) = 0,\ -\infty < x < \infty.$

41. $\dfrac{\partial^2 u}{\partial x^2} + \dfrac{\partial^2 u}{\partial y^2} = 0,\ -\infty < x < \infty,\ 0 < y < 1,$

$\dfrac{\partial u}{\partial y}(x, 0) = 0,\ u(x, 1) = e^{-2|x|},\ -\infty < x < \infty.$

9.7 Answers and Hints

Exercise 9.1

In the following problems denote $p_n = 1 - \cos n\pi = 1 - (-1)^n$. The summations are all from $n = 1$ to ∞, except where it is specifically mentioned.

1. $\dfrac{k}{2} - \dfrac{k}{\pi} \Sigma \left[\dfrac{1}{n} p_n \sin(nx)\right].$

2. $\dfrac{3\pi}{4} + \Sigma \left[\dfrac{1}{\pi n^2} p_n \cos(nx) + \dfrac{1}{n} \cos(n\pi) \sin(nx)\right].$

3. $\dfrac{k}{2} + \dfrac{2k}{\pi} \Sigma \left[\dfrac{1}{n} \sin\left(\dfrac{n\pi}{2}\right) \cos(nx)\right].$

4. $\dfrac{2k}{\pi} \Sigma \left[\dfrac{1}{n} p_n \sin(nx)\right].$

5. $3 + \dfrac{2}{\pi} \Sigma \left[\dfrac{1}{n} p_n \sin(nx)\right].$

6. $-\dfrac{\pi}{2} - \dfrac{2}{\pi} \Sigma \left[\dfrac{1}{n^2} p_n \cos(nx)\right].$

7. $\dfrac{1}{2}(2 - \pi) + \dfrac{2}{\pi} \Sigma \left[\dfrac{1}{n^2} p_n \cos(nx)\right].$

8. $\dfrac{\pi^2}{3} + 4 \Sigma \left[\dfrac{1}{n^2} \cos(n\pi) \cos(nx)\right].$

9. $\dfrac{2}{\pi} \Sigma \left[\left(\dfrac{6\pi}{n^3} - \dfrac{\pi^3}{n}\right) \cos(n\pi) \sin(nx)\right].$

10. $\dfrac{2}{\pi} \Sigma \left[\dfrac{\pi^2}{n} \cos n\pi + \dfrac{2}{n^3} p_n\right] \sin(nx).$

11. $\cos x + \dfrac{8}{\pi} \sum\limits_{n=2}^{\infty} \left[\dfrac{n \cos(n\pi)}{(4n^2 - 1)} \sin(nx)\right].$

12. $\dfrac{1}{\pi} + \dfrac{1}{2} \sin x + \dfrac{1}{\pi} \sum\limits_{n=2}^{\infty} \left[\dfrac{(-1)^{n-1} - 1}{(n^2 - 1)} \cos nx\right].$

13. $\dfrac{1}{4} + \dfrac{1}{2\pi^2}(4 + \pi^2) \cos x + \dfrac{1}{\pi} \sin x$

$+ \dfrac{1}{\pi^2} \sum\limits_{n=2}^{\infty} \left[\dfrac{1}{n^2} p_n \cos(nx) + \left\{\dfrac{n\pi}{(n^2 - 1)}((-1)^{n-1} - 1) + \dfrac{\pi}{n}\right\} \sin(nx)\right]$

14. $\dfrac{4}{\pi} \Sigma \left[\dfrac{(-1)^{n+1}}{n} \sin\left(\dfrac{n\pi x}{2} \right) \right].$

15. $3 + \dfrac{36}{\pi^2} \Sigma \left[\dfrac{1}{n^2} \cos(n\pi) \cos\left(\dfrac{n\pi x}{3} \right) \right].$

16. $\dfrac{1}{2} - \dfrac{2}{\pi^2} \Sigma \left[\dfrac{1}{n^2} P_n \cos(n\pi x) \right].$

17. $\dfrac{5}{2} - \dfrac{6}{\pi^2} \Sigma \left[\dfrac{1}{n^2} P_n \cos\left(\dfrac{n\pi x}{3} \right) \right].$

18. $\dfrac{2}{\pi} \sinh \pi \; \Sigma \left[\dfrac{n\,(-1)^{n+1}}{1+n^2} \sin(n\pi x) \right].$

19. $\dfrac{1}{4} + \Sigma \left[\dfrac{1}{n^2 \pi^2} P_n \cos(n\pi x) - \dfrac{1}{n\pi} \sin(n\pi x) \right].$

20. $\dfrac{7}{6} + \Sigma \left[\dfrac{2}{n^2 \pi^2} (5\cos(n\pi) - 1) \cos\left(\dfrac{n\pi x}{2} \right) - \left\{ \dfrac{2}{n\pi} \cos(n\pi) + \dfrac{8 p_n}{(n\pi)^3} \right\} \sin\left(\dfrac{n\pi x}{2} \right) \right].$

21. $-\dfrac{1}{4} + \Sigma \left[-\dfrac{1}{(n\pi)^2} P_n \cos(n\pi x) + \dfrac{1}{n\pi} (1 - 2\cos(n\pi)) \sin(n\pi x) \right].$

22. $\dfrac{1}{\pi} + \dfrac{1}{2} \cos x - \dfrac{2}{\pi} \displaystyle\sum_{n=2}^{\infty} \left[\dfrac{1}{(n^2 - 1)} \cos\left(\dfrac{n\pi}{2} \right) \cos\left(\dfrac{n\pi x}{2} \right) \right].$

23. $-\dfrac{1}{2} + 2 \Sigma \left[\dfrac{1}{(n\pi)^2} P_n \cos(n\pi x) + \dfrac{1}{n\pi} \cos(n\pi) \sin(n\pi x) \right].$

24. $-1 - \dfrac{4}{\pi^2} \Sigma \dfrac{1}{n^2} P_n \cos\left(\dfrac{n\pi x}{2} \right).$

25. $\dfrac{1}{2} + \dfrac{2}{\pi} \Sigma \left[\dfrac{1}{n} \sin\left(\dfrac{n\pi}{2} \right) \cos\left(\dfrac{n\pi x}{2} \right) \right].$

26. $6 - \dfrac{36}{\pi^2} \Sigma \left[\dfrac{1}{n^2} \cos(n\pi) \cos\left(\dfrac{n\pi x}{3} \right) \right].$

27. They are same.

28. $\pi + 2 \Sigma \left[\dfrac{(-1)^{n+1}}{n} \sin(nx) \right].$ Set $x = \dfrac{\pi}{2}$ (point of continuity).

29. $\dfrac{\pi^2}{6} + \Sigma \left[\dfrac{2}{n^2} (-1)^n \cos(nx) + \left\{ \dfrac{\pi}{n} (-1)^{n+1} - \dfrac{2}{\pi n^3} P_n \right\} \sin(nx) \right].$

Set $x = 0$, (point of continuity). Set $x = \pi$ (point of discontinuity).

30. Add the two series.

31. Set $x = \pi/2$, (point of continuity) and re-arrange the terms.

32. $\dfrac{1}{2} + \dfrac{2}{\pi} \Sigma \left[\dfrac{1}{n} \sin\left(\dfrac{n\pi}{2} \right) \cos(nx) \right].$ Set $x = 0.$

33. $1 - \dfrac{2}{\pi} \Sigma \left[\dfrac{1}{n} \cos(n\pi) \sin(n\pi x) \right].$ Set $x = 1/2.$

34. $\dfrac{1}{\pi} \sinh \pi + \dfrac{2 \sinh \pi}{\pi} \Sigma \left[\dfrac{(-1)^n}{n^2 + 1} \{ \cos(nx) - n \sin(nx) \} \right].$ Set $x = 0.$ $\pi/[2 \sinh \pi].$

35. $\dfrac{8}{3} - \dfrac{16}{\pi^2} \Sigma \left[\dfrac{(-1)^n}{n^2} \cos\left(\dfrac{n\pi x}{2} \right) \right].$ Set $x = 0.$

36. $1 - \dfrac{4}{\pi^2} \Sigma \left[\dfrac{1}{n^2} P_n \cos\left(\dfrac{n\pi x}{2} \right) \right], \; \dfrac{2}{\pi} \Sigma \left[\dfrac{1}{n} P_n \sin\left(\dfrac{n\pi x}{2} \right) \right].$

37. Integrate a_n, b_n (Eqs. (9.13), (9.14)), by parts. Using the periodicity and continuity of $f'(x)$, we obtain

$$a_n = -\frac{1}{\pi n^2} \int_{-\pi}^{\pi} f''(x) \cos nx \, dx, \, b_n = -\frac{1}{\pi n^2} \int_{-\pi}^{\pi} f''(x) \sin nx \, dx.$$

Since $f''(x)$ is continuous, $|f''(x)| < L$ for some real constant L., We get

$$|a_n| < \frac{1}{\pi n^2} \int_{-\pi}^{\pi} L \, dx = \frac{2L}{n^2} \quad \text{and} \quad |b_n| < \frac{2L}{n^2}, \text{ for all } n.$$

Hence, the absolute value of each term of the Fourier series is less than or equal to the corresponding term of the series

$$|a_0| + 2L \left(1 + 1 + \frac{1}{2^2} + \frac{1}{2^2} + \dots \right) = |a_0| + 4L \left(1 + \frac{1}{2^2} + \frac{1}{3^2} + \dots \right)$$

Hence, the Fourier series is convergent and by Weirstrass test it is also uniformly convergent.

39. $f(x) = \dfrac{a_0}{2} + \Sigma \left[A_n \left\{ \dfrac{a_n}{A_n} \cos \left(\dfrac{n\pi x}{l} \right) + \dfrac{b_n}{A_n} \sin \left(\dfrac{n\pi x}{l} \right) \right\} \right], A_n = \sqrt{a_n^2 + b_n^2}$.

Set $\cos \alpha_n = a_n/A_n$ and $\sin \alpha_n = b_n/A_n$.

40. In problem 39, set $\cos \beta_n = b_n/A_n$ and $\sin \beta_n = a_n/A_n$. $\beta_n = (\pi/2) - \alpha_n$.

Exercise 9.2

In the following problems denote $p_n = 1 - \cos n\pi = 1 - (-1)^n$. The summations are all from $n = 1$ to ∞, except where it is specifically mentioned.

1. $\dfrac{2k}{\pi} \Sigma \left[\dfrac{1}{n} p_n \sin (n\pi x/5) \right]$.

2. $\dfrac{2}{\pi} - \dfrac{4}{\pi} \Sigma \left[\dfrac{(-1)^n}{4n^2 - 1} \cos (2nx) \right]$, No sine series is possible.

3. $-\dfrac{4}{\pi} \Sigma \left[\dfrac{1}{n} \sin \left(\dfrac{n\pi}{2} \right) \cos \left(\dfrac{n\pi x}{2} \right) \right], \dfrac{2}{\pi} \Sigma \left[\dfrac{1}{n} \left\{ 2 \cos \left(\dfrac{n\pi}{2} \right) - \cos (n\pi) - 1 \right\} \right] \sin \left(\dfrac{n\pi x}{2} \right)$.

4. $\dfrac{3}{2} + \dfrac{8}{\pi^2} \Sigma \left[\dfrac{1}{n^2} \left\{ \cos \left(\dfrac{n\pi}{2} \right) - 1 \right\} \cos \left(\dfrac{n\pi x}{4} \right) \right], \dfrac{4}{\pi} \Sigma \left[\dfrac{2}{\pi n^2} \sin \left(\dfrac{n\pi}{2} \right) - \dfrac{1}{n} \cos (n\pi) \right] \sin \left(\dfrac{n\pi x}{4} \right)$.

5. $\dfrac{5}{6} + \dfrac{2}{\pi^2} \Sigma \left[\dfrac{1}{n^2} \{ 3 \cos (n\pi) - 1 \} \cos (n\pi x) \right], -4 \Sigma \left[\dfrac{1}{n\pi} \cos (n\pi) + \dfrac{1}{(n\pi)^3} p_n \right] \sin (n\pi x)$.

6. $\dfrac{\pi}{4} + \dfrac{2}{\pi} \Sigma \left[\dfrac{1}{n^2} \left\{ 2 \cos \left(\dfrac{n\pi}{2} \right) - \cos (n\pi) - 1 \right\} \cos (nx) \right], \dfrac{4}{\pi} \Sigma \left[\dfrac{1}{n^2} \sin \left(\dfrac{n\pi}{2} \right) \sin (nx) \right]$.

7. $\dfrac{\pi}{4} + \dfrac{4}{\pi} \Sigma \left[\dfrac{1}{n^2} \left\{ 1 - \cos \left(\dfrac{n\pi}{2} \right) \right\} \cos \left(\dfrac{nx}{2} \right) \right], \dfrac{1}{\pi} \Sigma \left[\left\{ \dfrac{2\pi}{n} - \dfrac{4}{n^2} \sin \left(\dfrac{n\pi}{2} \right) \right\} \sin \left(\dfrac{nx}{2} \right) \right]$.

8. $\dfrac{3}{2} + \dfrac{8}{\pi^2} \Sigma \left[\dfrac{1}{n^2} \left\{ \cos \left(\dfrac{n\pi}{2} \right) - \cos (n\pi) \right\} \cos \left(\dfrac{n\pi x}{4} \right) \right], 4 \Sigma \left[\left\{ \dfrac{1}{n\pi} + \dfrac{2}{n^2 \pi^2} \sin \left(\dfrac{n\pi}{2} \right) \right\} \sin \left(\dfrac{n\pi x}{4} \right) \right]$.

9. $\dfrac{2}{3} + \dfrac{8}{\pi^2} \Sigma \left[\dfrac{1}{n^2} \left\{ \cos\left(\dfrac{n\pi}{2}\right) - \dfrac{2}{n\pi} \sin\left(\dfrac{n\pi}{2}\right) \right\} \cos\left(\dfrac{n\pi x}{2}\right) \right]$,

$\Sigma \left[\dfrac{8}{n^2\pi^2} \sin\left(\dfrac{n\pi}{2}\right) + \dfrac{16}{(n\pi)^3} \left\{ \cos\left(\dfrac{n\pi}{2}\right) - 1 \right\} - \dfrac{2}{n\pi} \cos(n\pi) \right] \sin\left(\dfrac{n\pi x}{2}\right)$.

10. $\dfrac{2}{\pi} \Sigma \left[\dfrac{1}{n} \left\{ 2\sin\left(\dfrac{n\pi}{3}\right) - \sin\left(\dfrac{2n\pi}{3}\right) \right\} \cos\left(\dfrac{n\pi x}{3}\right) \right]$,

$\dfrac{2}{\pi} \Sigma \left[\dfrac{1}{n} \left\{ 1 - 2\cos\left(\dfrac{n\pi}{3}\right) + \cos\left(\dfrac{2n\pi}{3}\right) \right\} \sin\left(\dfrac{n\pi x}{3}\right) \right]$.

11. $\dfrac{2}{3} + \dfrac{6}{\pi^2} \Sigma \left[\dfrac{1}{n^2} \left\{ \cos\left(\dfrac{n\pi}{3}\right) - \cos(n\pi) + \cos\left(\dfrac{2n\pi}{3}\right) - 1 \right\} \cos\left(\dfrac{n\pi x}{3}\right) \right]$,

$\dfrac{6}{\pi^2} \Sigma \left[\dfrac{1}{n^2} \left\{ \sin\left(\dfrac{n\pi}{3}\right) + \sin\left(\dfrac{2n\pi}{3}\right) \right\} \sin\left(\dfrac{n\pi x}{3}\right) \right]$.

12. $\dfrac{1}{2}(1 - e^{-2}) + 4\Sigma \dfrac{1}{4 + n^2\pi^2} \left[\{1 - e^{-2}\cos(n\pi)\} \cos\left(\dfrac{n\pi x}{2}\right) \right]$,

$2\pi \Sigma \dfrac{n}{4 + n^2\pi^2} \left[\{1 - e^{-2}\cos(n\pi)\} \sin\left(\dfrac{n\pi x}{2}\right) \right]$.

13. The given function itself is the Fourier cosine series, $\dfrac{2}{\pi} \Sigma \left[\dfrac{n}{(n^2 - 9)} \{(-1)^n + 1\} \sin(nx) \right]$.

14. $\dfrac{3}{2} + \dfrac{2}{\pi^2} \Sigma \left[\dfrac{1}{n^2} \{\cos(n\pi) - 1\} \cos(n\pi x) \right]$, $\dfrac{2}{\pi} \Sigma \left[\dfrac{1}{n} \{1 - 2\cos(n\pi)\} \sin(n\pi x) \right]$.

15. $\dfrac{2}{3\pi} - \dfrac{6}{\pi} \Sigma \left[\dfrac{1}{(n^2 - 9)} \{(-1)^n + 1\} \cos(nx) \right]$.

16. $\dfrac{3}{4} + \dfrac{1}{2\pi i} \displaystyle\sum_{\substack{n=-\infty \\ n\neq 0}}^{\infty} \left[\dfrac{1}{n} (e^{in\pi} - e^{-in\pi/2}) e^{in\pi x/2} \right]$. **17.** $-\dfrac{\pi}{4} + \dfrac{1}{2\pi} \displaystyle\sum_{\substack{n=-\infty \\ n\neq 0}}^{\infty} \left[\dfrac{1}{n^2}(1 - e^{in\pi}) + \dfrac{i\pi}{n} e^{in\pi} \right] e^{inx}$.

18. $\displaystyle\sum_{n=-\infty}^{\infty} \dfrac{2}{4 + n^2\pi^2} [\{1 - (-1)^n e^{-2}\} e^{in\pi x/2}]$. **19.** $\left[n\omega; \dfrac{\sqrt{1 + 4n^2}}{\pi(4n^2 - 1)} \right]$, $\omega = 2, c_0 = \dfrac{1}{\pi}$.

20. $\left[n\omega, \dfrac{8n}{\pi(4n^2 - 1)} \right]$, $\omega = 2$.

Exercise 9.3

1. $A(\omega) = 0$, $B(\omega) = \dfrac{2}{\omega}[1 - \cos(2\omega)]$.

2. $A(\omega) = [2\omega\sin(2\omega) + \cos(2\omega) - 1]/\omega^2$, $B(\omega) = [\sin(2\omega) - 2\omega\cos(2\omega)]/\omega^2$.

3. $A(\omega) = 1/(1 + \omega^2)$, $B(\omega) = \omega/(1 + \omega^2)$.

4. $A(\omega) = 2[(9\omega^2 - 2)\sin(3\omega) + 6\omega\cos(3\omega)]/\omega^3$, $B(\omega) = 0$.

5. $A(\omega) = 2[\omega \sin (2\omega) \cosh 2 + \cos (2\omega) \sinh 2]/(1 + \omega^2)$.

   $B(\omega) = 2[\sin (2\omega) \cosh 2 - \omega \cos (2\omega) \sinh 2]/(1 + \omega^2)$.

6. $A(\omega) = 2[\pi \omega \sin (\pi\omega) + \cos (\pi\omega) - 1]/\omega^2$, $B (\omega) = 0$.

7. $A(\omega) = [\sin \omega + 2 \sin (2\omega) + \sin (3\omega)]/\omega$, $B(\omega) = [2 \cos (2\omega) - \cos \omega - \cos (3\omega)]/\omega$.

8. $A(\omega) = 2[1 + (\omega \sin \omega - \cos \omega) e^{-1}]/(1 + \omega^2)$, $B((\omega) = 0$.

9. $A (\omega) = 0$, $B (\omega) = 2 [\omega - \sin \omega]/\omega^2$.

10. $A(\omega) = \dfrac{1}{2} \left[\dfrac{2}{\omega^2 - 1} + \dfrac{1}{\omega + 1} \{\cos 2 (\omega + 1) + \sin 2 (\omega + 1)\} + \dfrac{1}{\omega - 1} \{\sin 2 (\omega - 1) - \cos 2 (\omega - 1)\} \right]$,

    $B(\omega) = \dfrac{1}{2} \left[\dfrac{2\omega}{\omega^2 - 1} - \dfrac{1}{\omega + 1} \{\sin 2 (\omega + 1) + \cos 2 (\omega + 1)\} + \dfrac{1}{\omega - 1} \{\sin 2 (\omega - 1) - \cos 2 (\omega - 1)\} \right]$

11. $A(\omega) = \dfrac{4}{4 + \omega^2} + \dfrac{6}{9 + \omega^2}$.
    12. $A(\omega) = 2[(25\omega^2 - 2) \sin (5\omega) + 10 \omega \cos (5 \omega)]/\omega^3$.

13. $A(\omega) = -2[1 + \cos (\omega\pi)]/(\omega^2 - 1)$ $\omega \neq 1$, $A(1) = 0$.

14. $B(\omega) = 2[\cos \omega - \cos(2\omega)]/\omega$.

15. $B(\omega) = 2[\sin (3\omega) \cosh 3 - \omega \cos (3\omega) \sinh 3]/(1 + \omega^2)$.

16. $c(\omega) = (k + i\omega)/(k^2 + \omega^2)$.
    17. $c(\omega) = \dfrac{i}{\omega} (1 - 2 \cos \omega) + \dfrac{1}{\omega^2} (1 - e^{-i\omega})$.

18. $2i[\cosh a \sin (\omega a) - \omega \sinh a \cos (\omega a)]/(1 + \omega^2)$.

19. $c(\omega) = \dfrac{1}{2 + i\omega} \sinh \{a (2 + i\omega) + \dfrac{1}{2 - i\omega} \sinh \{a (2 - i\omega)\}$.

20. $c(\omega) = 2i [\sin \omega - \omega e^{i\omega}]/\omega^2$.
    21. $c(\omega) = - 2\pi i \sin (2\omega)/(\pi^2 - \omega^2)$.

22. Let $f (x) = 0$ for $x < 0$ and $x > a$ and $f (x) = 1$ for $0 < x < a$. Substitute $\omega = 2\omega^*$ in the resulting representation.

23. Define $g(t) = t^2 f (t)$ and consider Fourier cosine integral representation of $g(x)$ on $[0, \infty)$.

24. Define $g(t) = t f (t)$ and consider Fourier sine integral representation of $g(x)$ on $[0, \infty)$.

25. Compare with Fourier sine integral representation. Set $x = a$; $\pi g(a) = 1$ for $0 < a < 1$ and 0 for $a > 1$; $B (x) = f (x)$; $f (x) = 2(1 - \cos x)/(\pi x)$, $x > 0$, $0 < a < 1$.

Exercise 9.4

1. Elliptic.
   2. Hyperbolic.
   3. Elliptic.

4. Parabolic.
   5. Hyperbolic.
   6. Parabolic.

7. $x^2 + y^2 > 1$: hyperbolic; $x^2 + y^2 = 1$: parabolic; $x^2 + y^2 < 1$: elliptic.

8. $x^2 > y^2$: hyperbolic; $x^2 = y^2$: parabolic; $x^2 < y^2$: elliptic.

9. $Ae^{\lambda(x+y)}$, A constant.
   10. $Ae^{\lambda(x-4y)}$, A constant.

11. $Ae^{[\lambda(x^2 - y^2)/2]}$, A constant.

12. $\lambda = k^2$: $[A \cosh (kx) + B \sinh (kx)][C \cos (ky) + D \sin (ky)]$; $\lambda = 0$: $(A x + B) (Cy + D)$;

    $\lambda = - k^2$: $[A \cos (kx) + B \sin (kx)] [C \cosh (ky) + D \sinh (ky)]$.

13. $\lambda = 0$: $(Ax + B) (C t + D)$; $\lambda = k^2$: $[A \cosh (kx) + B \sinh (kx)] [C \cosh (kct) + D \sinh (kct)]$;

    $\lambda = - k^2$: $[A \cos (kx) + B \sin (kx)] [C \cos (kct) + D \sin (kct)]$.

14. $\lambda = 0$: $(Ax + B)$; $\lambda = k^2$: $\{A \cosh(kx) + B \sinh(kx)\} \; e^{k^2 c^2 t}$

$\lambda = -k^2 : [A \cos(kx) + B \sin(kx)] e^{-k^2 c^2 t}$.

15. $\lambda = 0$: $(Ax + B)(Cy + D)$; $\lambda = k^2 : [Ax^{m_1} + Bx^{m_2}][C \cos(ky) + D \sin(ky)]$,

$m_1 = [1 + \sqrt{1 + 4k^2}]/2, \; m_2 = [1 - \sqrt{1 + 4k^2}]/2 < 0$.

16. $\lambda = 0$: $(Ax + B)(Cy + D)$; $\lambda = -k^2 : [A \cos(kx) + B \sin(kx)][Cy^{m_1} + Dy^{m_2}]$,

m_1, m_2 are same as in Problem 15.

In the following problems, Σ denotes summation from $n = 1$ to ∞. Also, denote

$p_n = 1 - (-1)^n, \; S_n = \sin(n\pi x/l), \; E_n = e^{-n^2 \pi^2 c^2 t / l^2}$

19. $\dfrac{2T}{\pi} \Sigma \left[\dfrac{1}{n} p_n S_n E_n \right]$.

20. $\dfrac{4l}{\pi^2} \Sigma \left[\dfrac{1}{n^2} \sin\left(\dfrac{n\pi}{2} \right) S_n E_n \right]$.

21. $\dfrac{4l^2}{\pi^3} \Sigma \left[\dfrac{1}{n^3} p_n S_n E_n \right]$.

22. $\dfrac{2T}{\pi} \Sigma \left[\dfrac{1}{n} p_n \sin(nx) e^{-n^2 c^2 t} \right]$

23. $\sin(x) e^{-c^2 t}$.

24. $\dfrac{4}{\pi} \Sigma \left[\dfrac{1}{n^2} \sin\left(\dfrac{n\pi}{2} \right) \sin(nx) e^{-n^2 c^2 t} \right]$

25. $\dfrac{d_0}{2} + \Sigma \left[d_n \cos\left(\dfrac{n\pi x}{l} \right) E_n \right]$, $d_0 = \dfrac{2}{l} \displaystyle\int_0^l f(x) dx, \; d_n = \dfrac{2}{l} \displaystyle\int_0^l f(x) \cos\left(\dfrac{n\pi x}{l} \right) dx$

26. $\dfrac{T}{2} + \dfrac{2T}{\pi} \Sigma \left[\dfrac{1}{n} \sin\left(\dfrac{n\pi}{2} \right) \cos\left(\dfrac{n\pi x}{l} \right) E_n \right]$.

27. $\dfrac{1}{2} + \Sigma d_n \cos\left(\dfrac{n\pi x}{4} \right) e^{-n^2 \pi^2 c^2 t / 16}$, $d_n = \dfrac{4}{n\pi} \left[\sin\left(\dfrac{n\pi}{2} \right) + \dfrac{2}{n\pi} \left\{ \cos\left(\dfrac{n\pi}{2} \right) - 1 \right\} \right]$

28. $\Sigma d_n \sin\left[\dfrac{(2n-1)\pi x}{2l} \right] E_n, \; d_n = \dfrac{4T}{(2n-1)\pi} \left[1 - \cos(2n-1) \dfrac{\pi}{2} \right], \; E_n = e^{-(2n-1)^2 \pi^2 c^2 t / (4l^2)}$.

29. $\Sigma d_n \sin\left(\dfrac{v_n x}{12} \right) e^{-v_n^2 t / 24}$, v_n are positive solutions of $12 \tan v_n + v_n = 0$.

$d_n = \dfrac{144}{v_n} \left[\dfrac{2 \sin(v_n / 2) - \sin(v_n)}{6 v_n - 3 \sin(2 v_n)} \right]$. The values of $v_0, v_1, \ldots$ are determined using the Newton's method.

$u_1 = 4.61823 \sin(0.24201x) e^{-0.35142t}$, $u_2 = 0.58560 \sin(0.48591x) e^{-1.41663t}$; $u_1, u_1 + u_2$ are the first two partial sums.

30. $T_0 [erf(v_2) + erf(v_1)]/2, \; v_1 = (2 + x)/(2c\sqrt{t}), \; v_2 = (2 - x)/(2c\sqrt{t}), \; erf(v) = \dfrac{2}{\sqrt{\pi}} \displaystyle\int_0^v e^{-u^2} du$. (use the result given in Eq. (9.71)).

31. $u(x, t) = \displaystyle\int_0^\infty B(k) \sin(kx) e^{-k^2 c^2 t} \, dk, \; B(k) = \dfrac{2}{\pi} \displaystyle\int_0^\infty f(s) \sin(ks) \, ds$.

Substitute for $B(k)$, set $z = kc\sqrt{t}$ and simplify. Let $\tau = 2c\sqrt{t}$. (use the result given in Eq. (9.64)).

32. $erf(x/(2c\sqrt{t}))$.

33. $\dfrac{8}{\pi^2} \Sigma \left[\dfrac{1}{n^2} \sin\left(\dfrac{n\pi}{2}\right) \cos\left(\dfrac{n\pi ct}{2}\right) \sin\left(\dfrac{n\pi x}{2}\right) \right]$.

34. $\dfrac{8}{\pi^2 c} \Sigma \left[\dfrac{1}{n^2} \left\{ \left(\dfrac{4}{n\pi}\right) \sin\left(\dfrac{n\pi}{2}\right) - 2\cos\left(\dfrac{n\pi}{2}\right) \right\} \sin\left(\dfrac{n\pi ct}{4}\right) \sin\left(\dfrac{n\pi x}{4}\right) \right]$.

35. $\dfrac{4l^2}{\pi^3} \Sigma \left[\dfrac{1}{n^3} P_n \cos\left(\dfrac{n\pi ct}{4}\right) \sin\left(\dfrac{n\pi x}{4}\right) \right]$.

36. $\sin(ct)\sin x/c$.

37. $\cos(2\pi ct)\sin(2\pi x)$.

38. $\Sigma \left[\dfrac{16}{(n\pi)^3} P_n \cos\left(\dfrac{n\pi ct}{2}\right) + \dfrac{4}{(n\pi)^2 c} P_n \sin\left(\dfrac{n\pi ct}{2}\right) \right] \sin\left(\dfrac{n\pi x}{2}\right)$, $c = 4$.

39. Choose $g(0) = 0 = g(2)$ and $g''(x) + 6x = 0$. $(4x - x^3) + \dfrac{96}{\pi^3} \Sigma \left[\dfrac{1}{n^3} (-1)^n \cos\left(\dfrac{n\pi t}{2}\right) \sin\left(\dfrac{n\pi x}{2}\right) \right]$

40. Choose $g(0) = 0 = g(\pi/2)$ and $g''(x) = \sin x$. $\dfrac{2}{\pi} x - \sin x - \dfrac{2}{\pi} \Sigma \left[\dfrac{(-1)^n}{n(4n^2 - 1)} \cos(2nt)\sin(2nx) \right]$

41. $e^{-at/2} \Sigma \left\{ d_n [\cos(s_n t) + \dfrac{a}{2s_n} \sin(s_n t)] \sin(nx) \right\}$, $s_n^2 = n^2 - \dfrac{a^2}{4}$, $d_n = \dfrac{2}{\pi} \displaystyle\int_0^\pi f(x)\sin(nx)\,dx$.

42. $\Sigma \left[C_n \cos\left(\dfrac{n\pi ct}{l}\right) \cos\left(\dfrac{n\pi x}{l}\right) \right]$, $C_n = \dfrac{2}{l} \displaystyle\int_0^l f(x)\cos\left(\dfrac{n\pi x}{l}\right) dx$.

43. $-\dfrac{2l}{\pi^2} \Sigma \left[\dfrac{1}{n^2} P_n \cos\left(\dfrac{n\pi ct}{l}\right) \cos\left(\dfrac{n\pi x}{l}\right) \right]$.

44. $\Sigma \left[C_n \left\{ \cos(s_n t) + \dfrac{1}{2s_n} \sin(s_n t) \right\} e^{-t/2} \sin\left(\dfrac{n\pi x}{l}\right) \right]$, $s_n^2 = \dfrac{1}{4} \left[3 + \dfrac{4n^2 \pi^2 c^2}{l^2} \right]$

$C_n = \dfrac{2}{l} \displaystyle\int_0^l f(x)\sin\left(\dfrac{n\pi x}{l}\right) dx$.

45. (i) $\sin(3x)\sin(3ct)/(3c)$, (ii) $\sin(2x)\cos(2ct) + [\cos(2x)\sin(2ct)]/(2c)$.

46. $\cos(2\sqrt{13}\,\pi t)\sin(2\pi x)\sin(3\pi y)$.

47. $\Sigma\Sigma \left[\dfrac{4}{mn\pi^2} (-1)^{m+n} \cos(k_{mn}t)\sin(m\pi x)\sin(n\pi y) \right]$, $k_{mn} = \pi\sqrt{m^2 + n^2}$.

48. $\cos(\sqrt{17}\,\pi t)\sin(\pi x)\sin(4\pi y)$.

49. $\Sigma \left[A_n \cosh\left(\dfrac{n\pi x}{b}\right) + B_n \sinh\left(\dfrac{n\pi x}{b}\right) \right] \sin\left(\dfrac{n\pi y}{b}\right)$,

$A_n = \dfrac{2}{b} \displaystyle\int_0^b f(y)\sin\left(\dfrac{n\pi y}{b}\right) dy$, $D_n = \dfrac{2}{b} \displaystyle\int_0^b g(y)\sin\left(\dfrac{n\pi y}{b}\right) dy$,

$B_n = \dfrac{1}{\sinh(n\pi a/b)} \left[D_n - A_n \cosh\left(\dfrac{n\pi a}{b}\right) \right]$.

50. $\frac{1}{2} A_0 y + \Sigma \left[A_n \cos \left(\frac{n\pi x}{a} \right) \sinh \left(\frac{n\pi y}{a} \right) \right], A_0 = \frac{2}{ab} \int_0^a f(x) dx,$

$A_n = \frac{2}{a \sinh (n\pi b/a)} \int_0^a f(x) \cos \left(\frac{n\pi x}{a} \right) dx.$

$A_0 = \frac{a}{2b}, A_n = T \left(\frac{a}{n\pi} \right)^2 \left[2 \cos \left(\frac{n\pi}{2} \right) - 1 - \cos n\pi \right], T = \frac{2}{a \sinh (n\pi b/a)}.$

51. $\Sigma A_n \left[\cosh \left(\frac{n\pi y}{a} \right) - \coth \left(\frac{n\pi b}{a} \right) \sinh \left(\frac{n\pi y}{a} \right) \right] \sin \left(\frac{n\pi x}{a} \right), A_n = \frac{2}{a} \int_0^a f(x) \sin \left(\frac{n\pi x}{a} \right) dx.$

52. $\Sigma A_n \sinh \left(\frac{n\pi y}{a} \right) \sin \left(\frac{n\pi x}{a} \right), A_n = \frac{2}{a \sinh (n\pi b/a)} \int_0^a g(x) \sin \left(\frac{n\pi x}{a} \right) dx.$

53. $\Sigma A_n \cosh \left(\frac{n\pi y}{a} \right) \sin \left(\frac{n\pi x}{a} \right), A_n = \frac{2}{a \cosh (n\pi b/a)} \int_0^a g(x) \sin \left(\frac{n\pi x}{a} \right) dx,$

54. $\frac{x}{2} + \frac{2a}{\pi^2 \sinh (n\pi)} \Sigma \frac{1}{n^2} P_n \sinh \left(\frac{n\pi x}{a} \right) \cos \left(\frac{n\pi y}{a} \right), P_n = 1 - (-1)^n.$

55. $\Sigma [A_n \cosh (n\pi x) + B_n \sinh (n\pi x)] \sin (n\pi y)$

$A_n = -\frac{2}{n\pi} \cos (n\pi), B_n = \frac{-1}{n\pi \cosh (n\pi)} \left[\frac{10}{(n\pi)^2} P_n + A_n \sinh (n\pi) \right], P_n = 1 - (-1)^n,$

Exercise 9.5

1. $2 \sin (\omega)/\omega.$

2. $2 [\cos \omega - 1]/\omega^2.$

3. $2 [\omega \cos l \sin (\omega l) - \sin l \cos (\omega l)]/(\omega^2 - 1).$

4. $1/(\alpha - i\omega).$

5. $-2i\omega/(a^2 + \omega^2).$

6. $e^{-3(4+i\omega)}/(4 + i\omega).$

7. $2ae^{i\omega}/(a^2 + \omega^2).$

8. Write $\frac{1}{1 + t^2} = \frac{1}{2} \left[\frac{1}{1 + it} + \frac{1}{1 - it} \right]$. From Example 9.28, $\mathscr{F} \left[\frac{1}{1 + it} \right] = 2\pi e^\omega H(-\omega).$

Define $b(t) = H(-t) e^t$, so that $\mathscr{F}[b(t)] = 1/(1 - i\omega) = B(\omega)$. Use the symmetry property to show

$\mathscr{F} \left[\frac{1}{1 - it} \right] = 2\pi e^{-\omega} H(\omega).$ Hence, $\mathscr{F}[1/(1 + t^2)] = \pi e^{-|\omega|}.$

9. Use frequency modulation theorem, $(6i) \left[\frac{1}{9 + (\omega + 4)^2} - \frac{1}{9 + (\omega - 4)^2} \right]$

10. Use shift theorem, $e^{-(t-1)} H(t - 1)/2.$

11. $F(\omega) = 2 - \frac{6}{3 + i\omega}, 2\delta(t) - 6e^{-3t} H(t).$ **12.** $e^{-(3t+4)} H(t + 1).$

13. $e^{-2(t-2)/3} H(t - 2)/3.$ **14.** $e^{-2t} (1 - e^{-t}) H(t).$

15. $e^{-2t} (3e^{-t} - 2) H(t).$ **16.** $e^{-2|t-2|}/4.$

17. Roots of $\omega^4 + a^4 = 0$ are $a_1 (\pm 1 \pm i), a_1 = a/\sqrt{2}$. Combine two roots each and write

$$F(\omega) = \frac{1}{2\sqrt{2}a^3}\left[\frac{\sqrt{2}\,a - \omega}{(\omega - a_1)^2 + a_1^2} + \frac{\sqrt{2}a + \omega}{(\omega + a_1)^2 + a_1^2}\right] = \frac{1}{2\sqrt{2}a^3}\left[\frac{a_1 - (\omega - a_1)}{(\omega - a_1)^2 + a_1^2} + \frac{a_1 + (\omega + a_1)}{(\omega + a_1)^2 + a_1^2}\right]$$

$$= \frac{1}{2\sqrt{2}a^3}\left[\left\{\frac{a_1}{(\omega - a_1)^2 + a_1^2} + \frac{a_1}{(\omega + a_1)^2 + a_1^2}\right\} + \frac{1}{2i}\left\{\frac{2i(\omega + a_1)}{(\omega + a_1)^2 + a_1^2} - \frac{2i(\omega - a_1)}{(\omega - a_1)^2 + a_1^2}\right\}\right]$$

$$\mathscr{F}^{-1}[F(\omega)] = \frac{1}{2\sqrt{2}a^3}\left[e^{-a_1|t|}\cos(a_1 t) + g(t)\sin(a_1 t)\right] \text{ (see Problem 5 and use modulation}$$

Theorem) where $g(t) = -e^{a_1 t}$ for $t < 0$, and $e^{-a_1 t}$ for $t > 0$. Therefore,

$$\mathscr{F}^{-1}[F(\omega)] = \frac{1}{2\sqrt{2}\,a^3}\,e^{a_1 t}(\cos(a_1 t) - \sin(a_1 t)), \text{ for } t < 0$$

$$= \frac{1}{2\sqrt{2}\,a^3}\,e^{-a_1 t}(\cos(a_1 t) + \sin(a_1 t)), \text{ for } t > 0.$$

18. $F(\tau) = 1/(2 + i\tau)$, $G(\omega - \tau) = 1/[2 + i(\omega - \tau)]$. That is $G(t) = 1/(2 + it)$.

$I = 2\pi\,\mathscr{F}[f(t)\,g(t)] = 2\pi\,\mathscr{F}[e^{-4t}\,H(t)] = 2\pi/(4 + i\omega)$.

19. $I = 2\pi\,\mathscr{F}[\{H(-t)\,e^{4t}\}\{e^{-4t}\,H(t)\}] = 0$.

20. $(f * g)(t) = e^{-kt}\displaystyle\int_{-\infty}^{\infty} H(\tau)\,H(t - \tau)\,d\tau = te^{-kt}\,H(t)$, since $H(\tau)\,H(t - \tau) = 0$ for $\tau < 0$ and $\tau > t$, and 1 for $0 < \tau < t$.

21. Use the result of Problem 20, $t^2 e^{-kt} H(t)/2$.

22. (i) $g(x) = \frac{1}{2\pi}\displaystyle\int_{-\infty}^{\infty} G(w)e^{iwx}\,dw;\quad \overline{g}(x) = \frac{1}{2\pi}\displaystyle\int_{-\infty}^{\infty} \overline{G}(w)e^{-iwx}\,dw$

$$\int_{-\infty}^{\infty} f(x)\overline{g}(x)\,dx = \frac{1}{2\pi}\int_{-\infty}^{\infty} f(x)\left[\int_{-\infty}^{\infty} \overline{G}(w)\,e^{-iwx}\,dw\right]dx$$

$$= \int_{-\infty}^{\infty} \overline{G}(w)\left[\frac{1}{2\pi}\int_{-\infty}^{\infty} f(x)e^{-iwx}\,dx\right]dw = \frac{1}{2\pi}\int_{-\infty}^{\infty} F(w)\,\overline{G}(w)\,dw$$

Hence, $\displaystyle\int_{-\infty}^{\infty} F(w)\,\overline{G}(w)\,dw = 2\pi\int_{-\infty}^{\infty} f(x)\,\overline{g}(x)\,dx$

(ii) set $f(x) = g(x)$, $F(w) = G(w)$

$$\int_{-\infty}^{\infty} |F(w)|^2\,dw = 2\pi\int_{-\infty}^{\infty} |f(x)|^2\,dx$$

23. $F(\omega) = -\dfrac{1}{(4 + i\omega)(4 - i\omega)}$, $y(t) = -\dfrac{1}{8}[e^{-4t}H(t) + e^{4t}H(-t)]\begin{cases} = -\dfrac{1}{8}e^{4t}, t < 0 \\ = -\dfrac{1}{8}e^{-4t}, t > 0. \end{cases}$

24. $F(\omega) = 1/[(2 + i\omega)(3 + i\omega)]$, $y(t) = e^{-2t}(1 - e^{-t})\,H(t)$.

25. $F(\omega) = e^{-2i\omega}/[(4 + i\omega)(1 + i\omega)]$, $y(t) = \dfrac{1}{3}[e^{-(t-2)} - e^{-4(t-2)}]\,H(t - 2)$.

26. $F(\omega) = e^{-3i\omega}/[(2 + i\omega)(1 + i\omega)]$, $y(t) = [e^{-(t-3)} - e^{-2(t-3)}]\,H(t - 3)$.

27. $\dfrac{1}{1 - \omega^2} + \dfrac{\cos(\omega - 1)l}{2(\omega - 1)} - \dfrac{\cos(\omega + 1)l}{2(\omega + 1)}$; for $\omega = 1$, $\dfrac{1}{4}(1 - \cos(2l))$,

$\dfrac{1}{2}\left[\dfrac{\sin{(\omega-1)}\,l}{\omega-1}-\dfrac{\sin{(\omega+1)}l}{\omega+1}\right]$; for $\omega=1$, $\dfrac{1}{4}(2l-\sin{(2l)})$.

28. $\dfrac{1}{2}\left[\dfrac{\sin{(\omega+1)}l}{\omega+1}+\dfrac{\sin{(\omega-1)}l}{\omega-1}\right]$; for $\omega=1$, $\dfrac{1}{4}(2l+\sin{(2l)})$.

$\dfrac{1}{2}\left[\dfrac{2\omega}{\omega^2-1}-\dfrac{\cos{(\omega+1)}\,l}{\omega+1}-\dfrac{\cos{(\omega-1)}l}{\omega-1}\right]$; for $\omega=1$, $\dfrac{1}{4}(1-\cos{(2l)})$.

29. $\left[\dfrac{1}{\omega}(1+l)\sin{(\omega l)}+\dfrac{1}{\omega^2}(\cos{\omega l}-1)\right]$, $\left[\dfrac{1}{\omega}+\dfrac{\sin{(\omega l)}}{\omega^2}-\dfrac{1}{\omega}(1+l)\cos{(\omega l)}\right]$.

30. $A\,[2\cos{(2\omega)}\cosh{(4)}+\omega\sin{(2\,\omega)}\,\sinh{(4)}-2\cos{\omega}\cosh{(2)}-\omega\sin{\omega}\sinh{(2)}]$,

$A[2\sin{(2\omega)}\cosh{(4)}-\omega\cos{2\omega}\sinh{(4)}-2\sin{\omega}\sinh{(2)}+\omega\cos{\omega}\sinh{(2)}]$, $A=2/(4+\omega^2)$.

31. $C_0=\pi^2/2$, $C_n=(\cos{(n\,\pi)}-1)/n^2$, $S_n=-\pi\cos{n\pi}/n$.

32. $C_0=(1-\cos{a\pi})/a$, $C_n=\dfrac{1}{2(n-a)}[\cos{\{(n-a)\pi\}}-1]-\dfrac{1}{2(n+a)}[\cos{\{(n+a)\pi\}}-1]$,

$C_n=0$ if $n=a$ integer. $S_n=\left[\dfrac{\sin{(n-a)\pi}}{2(n-a)}-\dfrac{\sin{(n+a)}\,\pi}{2(n+a)}\right]$, $S_n=\dfrac{\pi}{2}$ if $n=a$ integer.

33. $C_0=1-e^{-\pi}$, $C_n=[1-e^{-\pi}\cos{(n\,\pi)}]/(1+n^2)$. $S_n=n[1-e^{-\pi}\cos{(n\pi)}]/(1+n^2)$.

34. $C_0=[\cosh{(a\pi)}-1]/a$, $C_n=a[\cos{(n\pi)}\cosh{(a\pi)}-1]/(a^2+n^2)$.

$S_n=-n\cos{(n\,\pi)}\sinh{(a\pi)}/(a^2+n^2)$.

35. $u(x,t)=\dfrac{2}{\pi}\displaystyle\int_{-\infty}^{\infty}\dfrac{1}{(4+\omega^2)}e^{i\omega x-c^2\omega^2 t}\,d\omega=\dfrac{2}{\pi}\displaystyle\int_{-\infty}^{\infty}\dfrac{1}{(4+\omega^2)}\cos{(\omega x)}e^{-c^2\omega^2 t}\,d\omega$

(imaginary part is an odd function of ω).

36. $u(x,t)=\dfrac{1}{\pi}\displaystyle\int_{-\infty}^{\infty}\dfrac{1}{\omega}(\cos{\omega}-1)\sin{(\omega x)}\,e^{-c^2\omega^2 t}\,d\omega$, (imaginary part is an odd function of ω).

37. $F(\omega,0)=\dfrac{\sqrt{\pi}}{2}e^{-\omega^2/16}$

$u(x,t)=\dfrac{1}{4\sqrt{\pi}}\displaystyle\int_{-\infty}^{\infty}e^{-\omega^2[(1/16)+c^2 t]}\cos{\omega x}\,d\omega$, (imaginary part is an odd function of ω).

38. Use Fourier sine transform with respect to x.

$F_s(\omega,t)=\dfrac{1}{\omega}(1-\cos{(\omega l)})\,e^{-\omega^2 t}$, $u(x,t)=\dfrac{2}{\pi}\displaystyle\int_{0}^{\infty}\dfrac{1}{\omega}(1-\cos{(\omega l)})\sin{(\omega x)}\,e^{-\omega^2 t}\,d\omega$.

39. Use finite Fourier sine transform with respect to x. $F_s(n,y)=\pi e^{-y}/2$, $u(x,y)=e^{-y}\sin{x}$.

40. Use Fourier transform with respect to x. $F(\omega,y)=\dfrac{\sinh{[\omega(\pi-y)]}}{(2+i\omega)\sinh{(\omega\pi)}}$,

$u(x,y)=\dfrac{1}{2\pi}\displaystyle\int_{-\infty}^{\infty}\dfrac{\sinh{[\omega(\pi-y)]}}{(4+\omega^2)\sinh{(\omega\pi)}}[2\cos{(\omega x)}+\omega\sin{(\omega x)}]\,d\omega$

(imaginary part is an odd function of ω).

41. Use Fourier transform with respect to x.

$F(\omega,y)=\dfrac{4\cosh{(\omega y)}}{(4+\omega^2)\cosh{\omega}}$, $u(x,y)=\dfrac{1}{2\pi}\displaystyle\int_{-\infty}^{\infty}\dfrac{4\cosh{(\omega y)}}{(4+\omega^2)\cosh{\omega}}\cos{(\omega x)}\,d\omega$

(imaginary part is an odd function of ω).

Functions of a Complex Variable: Analytic Functions

10.1   Introduction

While studying the real number system IR, we have seen that there does not exist any real number whose square is a negative real number. Thus, the square root of a negative real number is not a valid arithmetic operation in IR and hence an equation of the form $x^2 + 1 = 0$ has no roots in IR. If these roots are to be determined, we need another number system called the *complex number system* defined on a complex plane. In a first course in Mathematics, you have studied the algebra of complex numbers, polar form of a complex number (defined in the Argand diagram), DeMoivre's theorem, powers and roots of complex numbers. In this chapter, we shall introduce the complex variable and functions of a complex variable. Further, we shall study the concept of limits, continuity and differentiability of the functions of a complex variable. In the process, we are led to the notion of an analytic function which is very important in any application of the complex variable theory.

10.2   Sets of Points in the Complex Plane

Let S be a non-empty set of complex numbers. Let $z_0 \in S$ be any complex number and δ be a positive real number. Then, we define the following:

Circle   The set of points which satisfies the equation

$$|z - z_0| = \delta \quad \text{or} \quad (x - x_0)^2 + (y - y_0)^2 = \delta^2 \qquad (10.1)$$

defines a *circle C* of radius δ with centre at $z_0 = (x_0, y_0)$. This set consists of all points which lie on the boundary of the circle C (Fig. 10.1a). Any point z on this circle has the polar form $z = z_0 + \delta e^{i\theta}$. As θ varies form 0 to 2π, z traverses once over this circle in the counter clockwise direction. If $z_0 = 0$, then the equation $|z| = \delta$ defines a circle of radius δ about the origin.

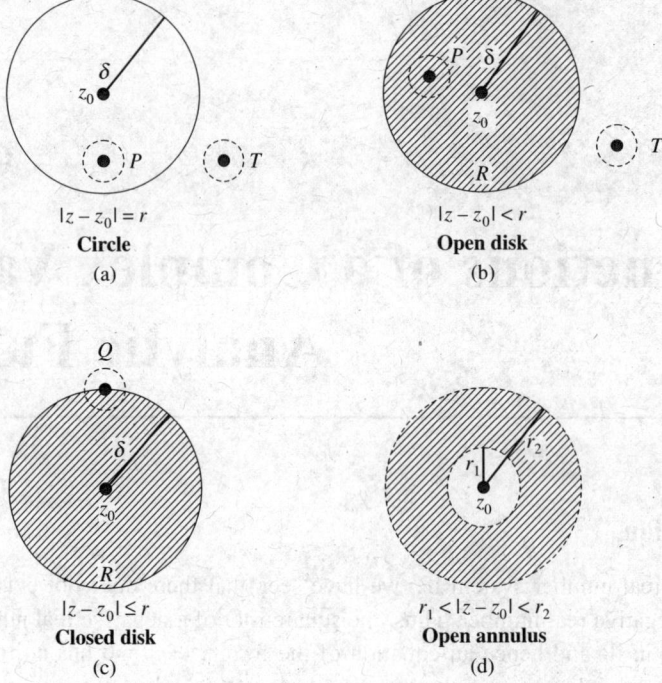

Fig. 10.1.   Circle, disk and annulus in the complex plane.

Open disk   The set of points which satisfies the equation

$$|z - z_0| < \delta \tag{10.2}$$

defines an *open disk* of radius δ with centre at $z_0 = (x_0, y_0)$. This set consists of all points which lie inside C (Fig. 10.1b).

Closed disk   The set of points which satisfies the equation

$$|z - z_0| \leq \delta \tag{10.3}$$

defines a *closed disk* of radius δ with centre at $z_0 = (x_0, y_0)$. This set consists of all points which lie inside and on the boundary of C (Fig. 10.1c).

Annulus   The set of points which lie between two concentric circles C_1: $|z - z_0| = r_1$ and C_2: $|z - z_0| = r_2$ defines an *open annulus* or an *open circular ring*, that is the set of points which satisfies the inequality

$$r_1 < |z - z_0| < r_2. \tag{10.4}$$

(Fig. 10.1d). The set of points which satisfies the inequality

$$r_1 \leq |z - z_0| \leq r_2 \qquad (10.5)$$

defines a *closed annulus*.

Note that the annulus $r_1 \leq |z - z_0| < r_2$ is neither open nor closed.

Neighborhood of a point A δ - *neighborhood* of a point $P(z_0)$ in the complex plane is the set of all points z which lie in the open disk $|z - z_0| < \delta$. Usually, the δ - neighborhood about the point $P(z_0)$ is denoted by $N(P, \delta)$ or $N_\delta(P)$. If we exclude the point z_0 from the open disk $|z - z_0| < \delta$, then it is called the *deleted* neighborhood of the point z_0 and is written as $0 < |z - z_0| < \delta$.

Interior point A point z is an *interior point* of S, if all the points in some δ - neighborhood of z are is S. In Figs. 10.1a and 10.1b, P is an interior point.

Exterior point A point z is an *exterior point* of S, if all the points in some δ - neighborhood of z are outside S. In Fig. 10.1a and 10.1b, T is an exterior point.

Boundary point A point z is a *boundary point* of S, if every δ - neighborhood of z contains at least one point of S and at least one point not in S. For example, for the set of points defined by Im $(z) \geq 1$, the points on the line $y = 1$ are the boundary points. The points on the circle $|z - z_0| = r$ are the boundary points for the disk $|z - z_0| \leq r$. The point Q in Fig. 10.1c is a boundary point. The totality of all the boundary points define the *boundary* of S.

Open set A set S is *open*, if every point of S is an interior point. For example, the sets

$$S = \{z: |z - z_0| < r\}; \quad S = \{z: \text{Re}(z) < 0\}; \quad S = \{z: 1 < |z| < 2\}$$

are open sets.

Closed set A set S is *closed*, if every boundary point of S belongs to S. For example, the sets

$$S = \{z: |z - z_0| \leq r\}; \quad S = \{z: r_1 \leq |z - z_0| \leq r_2\}$$

are closed sets.

Bounded set An open set S is bounded, if there exists a positive real number M, such that $|z| \leq M$ for all $z \in S$. Otherwise, the set S is said to be unbounded. For example, the set

$$S = \{z: |z - z_0| < r\} \text{ is a bounded set}$$

and

$$S = \{z: |z - z_0| > r\} \text{ is an unbounded set.}$$

Connected set An open set S is *connected*, if any two points z_1 and z_2 belonging to S can be joined by a polygonal line (a path which consists of finitely many straight line segments) which is totally contained in S (Fig. 10.2).

Domain An open connected set is called a domain. Usually, a domain is denoted by D.

Region A region is a domain together with all or some or none of its boundary points. Thus, a domain is always a region but a region may or may not be a domain. For example, an open disk is both a domain and a region but a closed disk is a region and not a domain. Usually, a region is denoted by R.

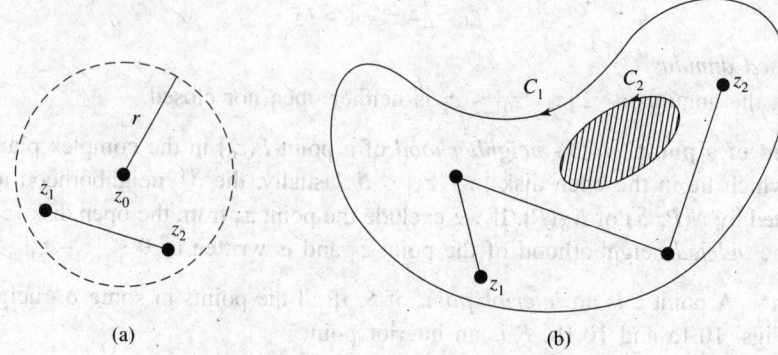

Fig. 10.2.   Connected sets.

Extended complex plane   The complex plane to which the point at $z = \infty$ has been added is called the *extended complex plane*. The complex plane without the point at $z = \infty$ is called the *finite complex plane*.

Example 10.1   Describe the regions

(a) $\{z: \operatorname{Re}(z) \neq 0\}$,

(b) $\{z: \operatorname{Im}(z) > 1\}$,

(c) $\{z: \operatorname{Im}(z) < |z - 1|^2\}$,

(d) $\{z: 0 \leq \operatorname{Arg}(z) \leq \pi/4\}$.

State whether the region is a domain or not.

Solution

(a) The region is the entire complex plane excluding the y-axis. It is open, it is not connected and it is unbounded. It is not a domain (Fig. 10.3a)

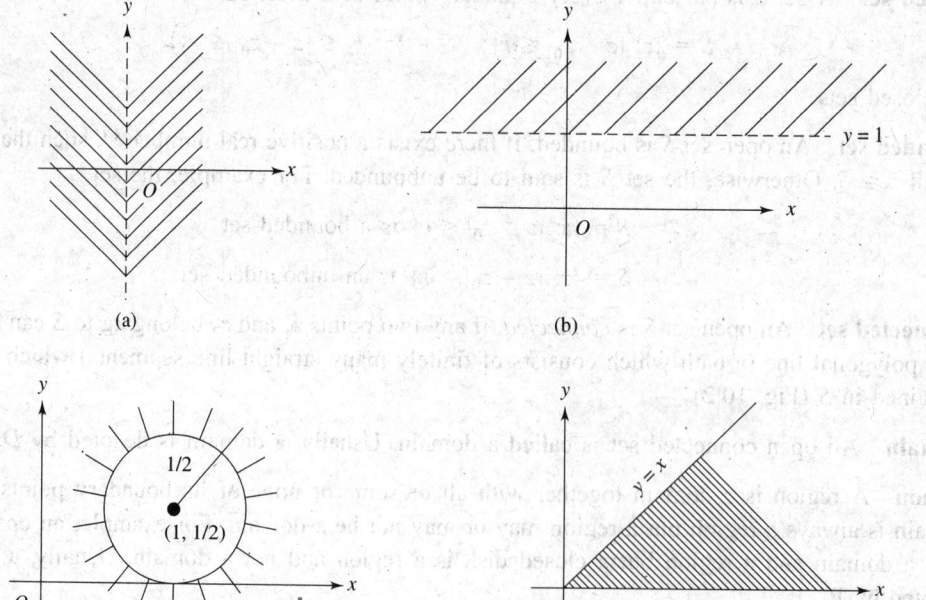

Fig. 10.3.   Example 10.1.

(b) The region is the upper half-plane $y > 1$. It is open, connected and unbounded. It is a domain. (Fig. 10.3b)

(c) The region is given by

$$y < (x-1)^2 + y^2, \quad \text{or} \quad (x-1)^2 + \left(y - \frac{1}{2}\right)^2 > \frac{1}{4}, \quad \text{or} \quad \left|z - \left(1 + \frac{i}{2}\right)\right| > \frac{1}{2}$$

which is the exterior of the circle with centre at $(1, 1/2)$ and radius $1/2$. It is open, connected and unbounded. It is a domain. (Fig. 10.3c)

(d) The region is given by

$$0 \le \tan^{-1}(y/x) \le \pi/4, \quad \text{or} \quad 0 \le y/x \le 1, \quad \text{or} \quad 0 \le y \le x.$$

It is not open, connected and unbounded. It is not a domain. (Fig. 10.3d).

Exercise 10.1

Describe the following regions. Are they open/closed, connected and bounded/unbounded? Determine the regions which are domains.

1. $\text{Re}(z) > 0$.
2. $\text{Re}(z) \le 0$.
3. $\text{Im}(z) \ge 0$.
4. $\text{Im}(z) < 0$.
5. $\text{Re}(z) \ge 0$ and $\text{Im}(z) > 0$.
6. $-2 < \text{Re}(z) < 6$.
7. $-3 < \text{Im}(z) < 4$.
8. $\text{Im}(z - 4i) < 6$.
9. $\text{Re}(z^2) < 4$.
10. $\text{Re}(z) \ge |z - i|^2$.
11. $\text{Im}(z) > [\text{Re}(z)]^2$
12. $\text{Im}(z - i) > \text{Re}(z + 2 + 3i)$.
13. $\text{Im}(3/z) < 1/3$.
14. $0 \le \arg(z) \le \pi/3$.
15. $(\pi/4) \le \arg(z) \le (\pi/2)$.
16. $-(\pi/4) < \arg(z) < (\pi/4)$.
17. $\text{Re}(z) \le 1$ and $0 < \arg(z) < \pi/4$.
18. $|z - 1| \le |z + 1|$.
19. $|2z + 1| < |z + 4|$.
20. $|z - 2i| \le |z + 2i|$.
21. $2 \le |z - 1 - i| \le 3$.
22. $|(1/z)| > 2$.
23. $|z|^2 + 2\,\text{Re}\,|z^2| < 3$.
24. $|z - i| > |z|$ and $|z| < 1$.
25. $|z + i| < |z + 2 - 3i|$ and $|z| > 4$.

10.3   Functions of a Complex Variable

We define the function of a complex variable in a similar way as the function of a real variable. Let S and S^* be two non-empty sets of complex numbers. If there is a rule f, which assigns a complex number w in S^* for each z in S, then f is said to be a complex valued function of a complex variable z and is written as

$$w = f(z). \tag{10.6}$$

The set S is called the *domain of definition* of f. When the set S, on which f is defined is not specified, we consider the so called *natural domain of definition* of the function f to be the set S, which can be the whole of the complex plane or some restricted part of the complex plane, for which $f(z)$ is defined and is in $S^* \subset \mathbb{C}$. For example, the *polynomial function*

$$f(z) = P_n(z) = a_0 z^n + a_1 z^{n-1} + \ldots + a_{n-1} z + a_n, \ a_0 \neq 0 \tag{10.7}$$

where $a_0, a_1, \ldots, a_n$ are real or complex constants, can be computed for any z. Therefore, the natural domain of definition of $P_n(z)$ is the whole of the finite complex plane. However, the *rational function*

$$f(z) = \frac{P(z)}{Q(z)} \tag{10.8}$$

where $P(z)$ and $Q(z)$ are some polynomials in z, is defined for all z except for those values of z for which $Q(z) = 0$. Therefore, the natural domain of definition of the rational function defined in Eq. (10.8) is $\{z : z \in \mathbb{C} \text{ and } Q(z) \neq 0\}$.

The value of w for a given $z = z_0$, that is $f(z_0)$ is called the *image* of the point $z = z_0$ under the rule f. The *inverse image* or the *pre-image* of point w in S^* is a point z in S which has w as its image.

Let $z = x + iy$. The image w of the complex number z is also some complex number, say $u + iv$. Therefore, we can write

$$w = f(z) = u(x, y) + iv(x, y) \tag{10.9}$$

where $u(x, y)$ and $v(x, y)$ are real valued functions of two real variables x and y. Therefore, $w = u + iv$ can be represented as a point (u, v) in the u-v plane, also called the w-plane or the *function plane* (Fig. 10.4).

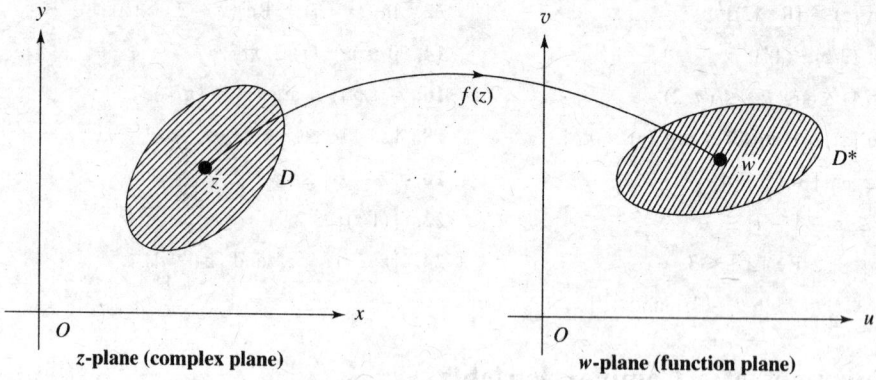

z-plane (complex plane)   **w-plane (function plane)**

Fig. 10.4.   A function $w = f(z)$.

The set of all images in S^* is called the *range* of the function. Thus, the function $w = f(z)$ may be considered as a transformation or a mapping which takes the points from a set S in the z-plane into the points in the set S^* in the w-plane. Therefore, the function $f(z)$ maps a region R in the z-plane into a region R^* in the w-plane.

If for every $z \in D$, there exists an unique image in the w-plane, then the function $f(z)$ is called a *single-valued* function. However, in the theory of complex variables, we come across functions which take more than one value for every z in D. For example, the function $w^n = z$ or $w = z^{1/n}$, n a positive integer takes the n values

$$w_k = |z|^{1/n} \left[\cos\left(\frac{\theta + 2k\pi}{n}\right) + i\sin\left(\frac{\theta + 2k\pi}{n}\right) \right], \ k = 0, 1, \ldots, (n-1)$$

for a given z. Such a function is called a *multiple-valued function*. In such a case, we restrict the discussion to those parts of the domain in which the multiple-valued function behaves like a single-valued function. Each one of these single-valued functions is called a *branch* of the multiple-valued function. Thus, the term function implies a single-valued function unless stated otherwise.

Remark 1

(a) Given any two real valued functions u and v in two real variables x and y, we can write $w = u(x, y) + iv(x, y)$, which defines a complex valued function. However, it may not always be possible to write w as a function of z. For example, if $u(x, y) = 2x + y$ and $v(x, y) = 6xy$, we can write $w = (2x + y) + i(6xy)$, but it is not possible to write w as a function of z alone and obtain $f(z)$.

(b) If $z = r(\cos\theta + i\sin\theta) = re^{i\theta} = (r, \theta)$ is taken in polar form, then the real and the imaginary parts of $f(z)$ can be expressed as real valued functions of the real variables r and θ. Thus, we can also write

$$w = f(z) = f(re^{i\theta}) = u(r, \theta) + iv(r, \theta). \tag{10.10}$$

Example 10.2   Find the image of the point $z = 2 + 3i$ under the transformation $w = f(z) = z(z - 2i)$.

Solution   Substituting $z = 2 + 3i$ in $w = f(z)$, we obtain the corresponding image in the w-plane as

$$w = f(2 + 3i) = (2 + 3i)(2 + 3i - 2i)$$

$$= (2 + 3i)(2 + i) = 1 + 8i.$$

Example 10.3   Find all the pre-images in the z-plane under the transformation $w = z(z - 2i) = -1$.

Solution   Solving the equation $z(z - 2i) = -1$ or $z^2 - 2iz + 1 = 0$ for z, we get

$$z = \frac{2i \pm \sqrt{-8}}{2} = (1 \pm \sqrt{2})i.$$

Hence, the pre-images of $w = -1$ are $z = (1 \pm \sqrt{2})i$.

Example 10.4   For each of the following functions, express $f(z)$ in the form $u(x, y) + iv(x, y)$ by separating into real and imaginary parts. State the natural domain of definition of the function in each case.

(i) $f(z) = |z|^2$,

(ii) $f(z) = \dfrac{1}{z}$,

(iii) $f(z) = \dfrac{1}{1 - |z|^2}$,

(iv) $f(z) = \dfrac{1}{\text{Re}(z)} - i\,[\text{Im}(z)]^2$,

(v) $f(z) = \dfrac{z-1}{1+2z}$.

Solution   Let $z = x + iy$ and $w = f(z) = u(x, y) + iv(x, y)$,   or $w = u + iv$.

(i) $$f(z) = |z|^2 = x^2 + y^2.$$

Therefore, $u = x^2 + y^2$ and $v = 0$.

Since $f(z)$ is defined for all $z \in \mathbb{C}$, the natural domain of definition of $f(z)$ is the whole complex plane.

(ii) $$f(z) = \frac{1}{z} = \frac{1}{x+iy} = \frac{x-iy}{x^2+y^2} = \frac{x}{x^2+y^2} - i\frac{y}{x^2+y^2}, \quad z \neq 0.$$

Therefore, $u = x/(x^2 + y^2)$ and $v = -y/(x^2 + y^2)$ for all $z \neq 0$.

Since $f(z)$ is defined for all $z \in \mathbb{C}$, $z \neq 0$, the natural domain of definition of $f(z)$ is the whole of the complex plane excluding the point $z = 0$.

(iii) $$f(z) = \frac{1}{1-|z|^2} = \frac{1}{1-(x^2+y^2)}, \quad |z| \neq 1.$$

Therefore, $u = 1/[1 - (x^2 + y^2)]$,   and $v = 0$, for all z, $|z| \neq 1$.

Since $f(z)$ is defined for all z, except for those points which lie on the unite circle $|z| = 1$, the domain of definition of $f(z)$ is the whole complex plane excluding the points z for which $|z| = 1$.

(iv) $$f(z) = \frac{1}{\mathrm{Re}(z)} - i\,[\mathrm{Im}\,(z)]^2 = \frac{1}{x} - iy^2, \quad x \neq 0.$$

Therefore, $u = 1/x$ and $v = -y^2$, $x \neq 0$.

Since the function $f(z)$ is defined for all z except when $x = 0$, the domain of definiton of $f(z)$ is the whole complex plane exculding the points lying on the imaginary axis, that is, $x = 0$.

(v) $$f(z) = \frac{z-1}{1+2z} = \frac{(x-1)+iy}{(1+2x)+2iy} = \left[\frac{(x-1)+iy}{(1+2x)+2iy}\right]\left[\frac{(1+2x)-2iy}{(1+2x)-2iy}\right]$$

$$= \frac{[(x-1)(1+2x)+2y^2]+3iy}{(1+2x)^2+4y^2}, \quad z \neq -\frac{1}{2}.$$

Therefore, $$u = \frac{(x-1)(1+2x)+2y^2}{(1+2x)^2+4y^2} \quad \text{and} \quad v = \frac{3y}{(1+2x)^2+4y^2}.$$

Since the function $f(z)$ is defined for all z except at $z = -1/2$, the domain of definition of $f(z)$ is the whole complex plane except the point $z = -1/2$, that is $(-1/2, 0)$.

Example 10.5   Write the function $w = z^2$ in the form $u(x, y) + iv(x, y)$ and $u(r, \theta) + iv(r, \theta)$. Give its graphical representation.

Solution   We have $w = f(z) = z^2$. Substituting $z = x + iy$, we get

$$w = (x + iy)^2 = x^2 - y^2 + 2ixy = u(x, y) + iv(x, y).$$

Therefore, $u(x, y) = x^2 - y^2$ and $v(x, y) = 2xy$.

Substituting $z = r(\cos \theta + i \sin \theta) = re^{i\theta}$, we get

$$w = r^2 e^{2i\theta} = r^2 (\cos 2\theta + i \sin 2\theta) = u(r, \theta) + iv(r, \theta).$$

Therefore, $u(r, \theta) = r^2 \cos 2\theta$ and $v(r, \theta) = r^2 \sin 2\theta$.

Writing $w = R(\cos \phi + i \sin \phi)$, we get $R = r^2$ and $\phi = 2\theta$. The graphical representation of the function $w = z^2$ is given in Fig. 10.5.

A circle $r = a$ in the z-plane is mapped onto the circle $R = a^2$ in the w-plane. The sector $r \leq a$, $0 \leq \theta \leq \pi/2$ in the z-plane is mapped onto the sector $R \leq a^2$, $0 \leq \phi \leq \pi$ in the w-plane.

The image of the hyperbola $x^2 - y^2 = c_1$ is the line $u = c_1$ and the image of the rectangular hyperbola $2xy = c_2$ is the line $v = c_2$ where c_1 and c_2 are real constants.

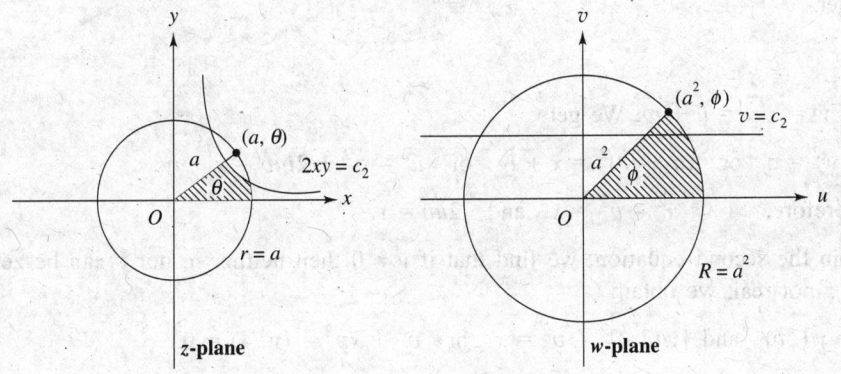

Fig. 10.5.   The function $w = z^2$.

Example 10.6   Find the image of the line $\text{Im}(z) = 1$ under the mapping $w = z^2$.

Solution   For the function $w = z^2 = (x + iy)^2 = u(x, y) + iv(x, y)$, we have $u(x, y) = x^2 - y^2$ and $v(x, y) = 2xy$. Substituting $\text{Im}(z) = y = 1$, we get

$$u(x, y) = x^2 - 1 \quad \text{and} \quad v(x, y) = 2x. \tag{10.11}$$

Eliminating x from Eqs. (10.11), we obtain

$$u = \frac{v^2}{4} - 1, \text{ or } v^2 = 4 (u + 1).$$

Therefore, the image of the line $y = 1$ is the parabola with vertex at $(-1, 0)$ and opening to the right. It intersects the v-axis at $(0, \pm 2)$ (Fig. 10.6).

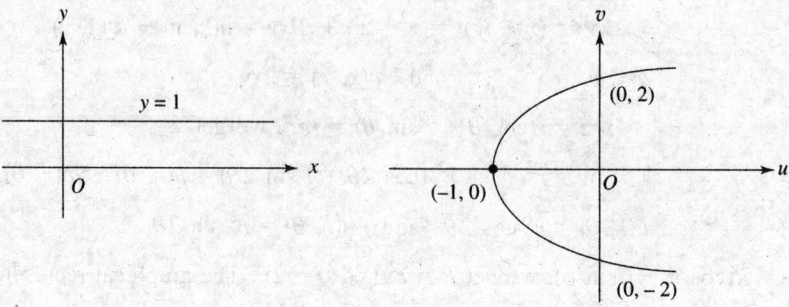

Fig. 10.6.   Example 10.6.

Example 10.7   (a) Show that the equation $w = \sqrt{z}$ has more than one solution and hence defines a multiple-valued function. Write its branches and show that each branch is a single-valued function. (b) Discuss, whether $w = z^{\alpha}$ is a multiple-valued function, when (i) α is an integer, (ii) α is a non-integer.

Solution

(a) Let $w = \sqrt{z} = u + iv$. We get

$$w^2 = z \quad \text{or} \quad (u + iv)^2 = x + iy \quad \text{or} \quad u^2 - v^2 + 2iuv = x + iy.$$

Therefore, $u^2 - v^2 = x$ and $2uv = y$.

From the second equation, we find that if $y \neq 0$ then neither u, nor v can be zero. Hence, for z not real, we obtain

$u = y/(2v)$ and $[y/(2v)]^2 - v^2 = x$ or $v^4 + xv^2 - (y^2/4) = 0$.

Solving for v^2, we get

$$v^2 = \frac{1}{2}\left[-x \pm \sqrt{x^2 + y^2}\right].$$

If we take negative sign, then v^2 becomes negative, which is not possible as v is real. Therefore, we have

$$v^2 = \frac{1}{2}\left[-x + \sqrt{x^2 + y^2}\right] \quad \text{and} \quad u = \frac{y}{2v}.$$

or $v = \pm p$ and $u = y/(2v)$,   where $p = [-x + \sqrt{x^2 + y^2}]^{1/2}\big/\sqrt{2}$.

Thus, for every complex z, w takes two values

$$w_1 = \frac{y}{2p} + ip \quad \text{and} \quad w_2 = -\frac{y}{2p} - ip.$$

Therefore, the function $w = \sqrt{z}$ is a multiple-valued function. It has two branches $w = w_1$ and $w = w_2$, each of which is a single-valued function.

Geometrically, we can interpret this result as follows: Let $z = re^{i\theta}$. Consider a circle of radius r about the origin (Fig. 10.7)

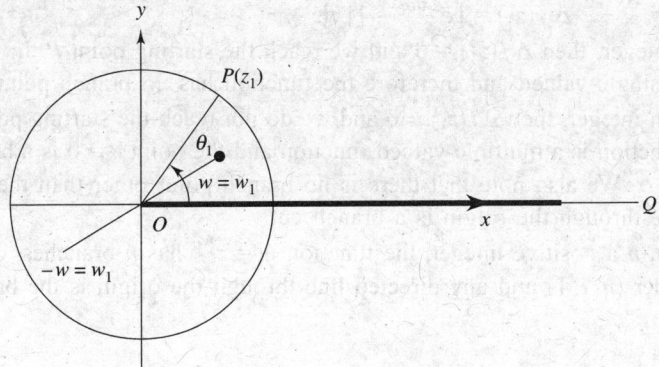

Fig. 10.7.   The function $w = \sqrt{z}$.

Now start at some point $P(z = z_1 = r\, e^{i\theta_1})$ on this circle. Then $w = f(z_1) = \sqrt{r}\, e^{i\theta_1/2}$, where $\theta = \theta_1 = \arg(z_1)$. After one complete revolution $\theta = \theta_1 + 2\pi$, we get

$$w = \sqrt{r}\, e^{i(\theta_1 + 2\pi)/2} = \sqrt{r}\, e^{i\theta_1/2} \cdot e^{\pi i} = -\sqrt{r}\, e^{i\theta_1/2}$$

which is not the same point P. Hence, the graph has two branches. However, after completing one more revolution $\theta = \theta_1 + 4\pi$, we get

$$w = \sqrt{r}\; e^{i(\theta_1 + 4\pi)/2} = \sqrt{r}\; e^{i\theta_1/2} \cdot e^{2\pi i} = re^{i\theta_1/2}$$

which is the same point P.

Therefore, if $0 \le \theta < 2\pi$, we are on one branch of the multiple-valued function $w = \sqrt{z}$ and if $2\pi \le q < 4\pi$, we are on the other branch of this function. Obviously, each one of the branches represents a single-valued function. The first interval $0 \le \theta < 2\pi$ is usually called the *principal range* of θ and the corresponding branch is called the *principal branch* of the multiple-valued function. In order to consider a single-valued function representing $w = \sqrt{z}$, we delete the line $x > 0$, $y = 0$, that is the line OQ in Fig. 10.7, where the point Q is at infinity. This line OQ is called the *branch line* or the *branch cut* and the point O is called the *branch point*. It may be noted that we could have taken any other directed line through O as the branch line by suitably defining the principal range of $\dot{q}$. Since a complete revolution about any point other than $z = 0$ does not produce different values of w, the point $z = 0$ is the only finite branch point for this function.

(b) Let $z = re^{i\theta}$. The point $z = 0$ is the only possible branch point. Now, consider a circle of radius r about $z = 0$ (Fig. 10.7). Take a point $P(z = z_1 = re^{i\theta_1})$ on this circle. Then

$$w = f(z_1) = r^\alpha \, e^{i\theta_1 \alpha}, \quad \text{where} \quad \theta = \theta_1 = \arg(z_1).$$

After one complete revolution $\theta = \theta_1 + 2\pi$, we get

$$w = r^\alpha e^{i(\theta_1 + 2\pi)\alpha} = (r^\alpha \, e^{i\theta_1 \alpha})e^{2\pi i \alpha} = e^{2\pi i \alpha} \, f(z_1).$$

Thus, the difference between the value of f after one complete revolution and its original value $f(z_1)$ is given by

$$\Delta f(z_1) = [e^{2\pi i \alpha} - 1] \, f(z_1).$$

(i) If α is an integer, then $\Delta f(z_1) = 0$ and we reach the starting point P. In this case, the given function is single-valued and therefore the function has no branch points or branch cuts.

(ii) If α is a non-integer, then $\Delta f(z_1) \neq 0$ and we do not reach the starting point P. Thus, in this case, the function is a multiple-valued function and the point $z = 0$ is a branch point for any non-integer α. We also note that there is no branch point other than the origin $z = 0$. Any directed line through the origin is a branch cut.

For $\alpha = 1/n$, n a positive integer, the function $w = z^{1/n}$ has n branches, origin is the branch point of order $(n - 1)$ and any directed line through the origin is the branch cut.

Exercise 10.2

Find the domain of definition for the following functions.

1. $\dfrac{z-i}{z+i}$.

2. $\dfrac{z}{z^2 + 1}$.

3. $\dfrac{1}{z^3 + 1}$.

4. $\dfrac{z+1}{(z^4 + 16)^2}$.

5. $\dfrac{2z + 3i}{z^2 + (1-i)z - i}$.

Find the image of the following functions at the indicated point.

6. $w = z\bar{z}$ at $z = -1 - 2i$.

7. $w = 2z + i\bar{z} + \text{Im}(z)$ at $z = 2 + 7i$.

8. $w = 2z^2 + \bar{z} + 1$ at $z = 3 + 4i$.

9. $w = \dfrac{z-i}{z+i}$ at $z = 2 + 3i$.

10. $w = \dfrac{z}{z^2 + 1}$ at $z = 1 - i$.

11. $w = \dfrac{iz - 2}{z + i}$ at $z = 1 + i$.

Write the following functions in the form $u(x, y) + iv(x, y)$.

12. $\bar{z}^2$.

13. z^3.

14. z^4.

15. $\left| z - 4i \right|^2 + 3\bar{z}$.

16. $\dfrac{\mathrm{Re}\, z}{1-|z|^2}$, $|z| \neq 1$.

17. $\dfrac{z}{z+\bar{z}}$, $\mathrm{Re}(z) \neq 0$.

18. $2\bar{z} + \mathrm{Re}(\bar{z} + 2z) + 5\,\mathrm{Im}(z^2)$.

19. $\mathrm{Re}(z^3) + \mathrm{Im}(z + 1) - 3iz$.

20. $\dfrac{z^2 + i}{|z|}$, $z \neq 0$.

21. $z + \dfrac{1}{z}$, $z \neq 0$.

22. $\dfrac{z-1}{\bar{z}}$, $z \neq 0$.

23. $\dfrac{z}{z+i}$, $z \neq -i$.

24. $(z + i)/(2z - 1)$, $z \neq 1/2$.

25. $z/(z^2 + 1)$, $z \neq \pm i$.

10.4   Elementary Functions

In this section, we introduce a few elementary functions of a complex variable and analyse them. These functions, $f(z)$, reduce to their counterparts $f(x)$, the corresponding functions of real variable when $z = x$ is real.

We now discuss some complex elementary functions which are often used.

10.4.1   Exponential Function

If a and b are real numbers, then we have

$$e^a = 1 + \frac{a}{1!} + \frac{a^2}{2!} + \frac{a^3}{3!} + \frac{a^4}{4!} + \frac{a^5}{5!} + \frac{a^6}{6!} + \ \cdots \qquad (10.12)$$

and

$$e^{a+b} = e^a\, e^b.$$

We now formally write

$$w = e^z = e^{x+iy} = e^x\, e^{iy}. \qquad (10.13)$$

Substituting $a = iy$ in Eq. (10.12), we get

$$e^{iy} = 1 + \frac{iy}{1!} - \frac{y^2}{2!} - \frac{iy^3}{3!} + \frac{y^4}{4!} + \frac{iy^5}{5!} - \frac{y^6}{6!} + \cdots$$

$$= \left(1 - \frac{y^2}{2!} + \frac{y^4}{4!} - \cdots\right) + i\left(y - \frac{y^3}{3!} + \frac{y^5}{5!} - \cdots\right).$$

Using the Maclaurin's series expansion of $\sin y$ and $\cos y$, we obtain the *Euler's formula*

$$e^{iy} = \cos y + i \sin y. \qquad (10.14)$$

Hence, we obtain from Eq. (10.13)

$$w = e^z = e^x\,[\cos y + i \sin y] = u + iv. \qquad (10.15)$$

Therefore, the image of the point $z = (x, y)$ in the z-plane is the point $w = (u, v)$ in the w-plane, where $u = e^x \cos y$ and $v = e^x \sin y$.

When $y = 0$, that is, $z = x$ is real, it reduces to the exponential function of the real variable x. Substituting $x = z$ in the Maclaurin's series for the real exponential function

$$e^x = 1 + x + \frac{x^2}{2!} + \ldots = \sum_{n=0}^{\infty} \frac{1}{n!} x^n$$

we obtain the Maclaurin's series expansion of e^z as

$$e^z = \sum_{n=0}^{\infty} \frac{1}{n!} z^n. \tag{10.16}$$

Taking the limit of the magnitude of the ratio of successive terms in the complex series (10.16), we get

$$\lim_{n \to \infty} \left| \frac{n! z^{n+1}}{(n+1)! z^n} \right| = \lim_{n \to \infty} \left| \frac{z}{n+1} \right| = 0 < 1.$$

Hence, the series expansion for the complex exponential function e^z converges for all z.

The exponential function satisfies the following properties, which can be easily verified.

  (i) e^z is defined for all z. Therefore, its domain of definition is whole of the complex plane.

  (ii) e^z is a single-valued function.   **(iii)** $|e^z| = |e^{x+iy}| = |e^x| \, |e^{iy}| = e^x$ and $\text{Arg}(e^z) = y$.

  (iv) $e^z \neq 0$ for any z. Therefore, its range is the whole of the complex plane except the origin.

  (v) $e^z = 1 = e^{2n\pi i}$ if and only if $z = 2n\pi i$, n any integer.

  (vi) $e^{z_1} = e^{z_2}$ if and only if $z_1 = z_2 + 2n\pi i$, n any integer.

  (vii) $e^{z_1} e^{z_2} = e^{z_1 + z_2}$.

  (viii) $e^{z_1} / e^{z_2} = e^{z_1 - z_2}$.

  (ix) $e^{-z} = 1/e^z$.

  (x) $(e^z)^n = e^{nz}$ and $(e^z)^{-n} = e^{-nz}$, n any positive integer.

Periodicity

Since $e^{2n\pi i} = \cos 2n\pi + i \sin 2n\pi = 1$, we can write $e^z = e^{z + 2n\pi i}$, n any integer.

Therefore, the complex exponential function e^z is *periodic* with *period* $2\pi i$. Since the period is a complex quantity $2\pi i$, all possible values of e^z are assumed in any infinite horizontal strip of width 2π, that is $(2n - 1)\pi < y \leq (2n + 1)\pi$, where n is any positive or negative integer. The strip $-\pi < y \leq \pi$ (Fig. 10.8) is called the *fundamental strip* or the *fundamental region* of e^z.

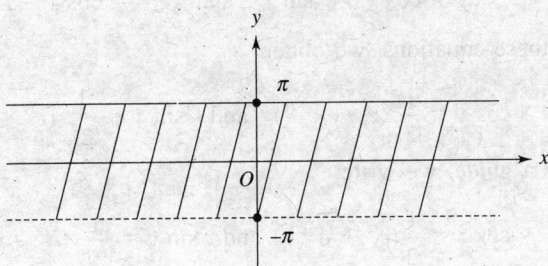

Fig. 10.8.   Fundamental region of e^z.

Example 10.8   Find the values of z for which

  (i) e^z is real,   (ii) e^z is pure imaginary.

Solution   We have

$$e^z = e^{x+iy} = e^x[\cos y + i \sin y] \text{ and } e^x \neq 0 \text{ for any } x.$$

  (i) If e^z is real, then $\sin y = 0$ or $y = \text{Im}(z) = n\pi,$ $n = 0, \pm 1, \pm 2, \ldots$

  (ii) If e^z in pure imaginary, then $\cos y = 0$ or

$$y = \text{Im}(z) = (2n+1)\frac{\pi}{2}, \, n = 0, \pm 1, \pm 2, \ldots$$

Example 10.9   Find all values of z which satisfy $e^z = 1 + i$.

Solution   We have

$$e^z = e^{x+iy} = e^x[\cos y + i \sin y] = 1 + i.$$

Therefore, $e^x \cos y = 1$ and $e^x \sin y = 1.$

Squaring and adding, we get

$$(e^x \cos y)^2 + (e^x \sin y)^2 = 2 \quad \text{or} \quad e^{2x} = 2 \quad \text{or} \quad x = \frac{1}{2} \ln 2$$

and $\tan y = 1$ or $y = \frac{\pi}{4} + n\pi,$ $n = 0, \pm 1, \pm 2, \ldots$

Hence, we obtain

$$z = x + iy = \frac{1}{2} \ln 2 + i\left(\frac{\pi}{4} + n\pi\right), \quad n = 0, \pm 1, \pm 2, \ldots$$

10.4.2   Trigonometric and Hyperbolic Functions

Trigonometric functions

For a real variable y, we get from the Euler's formula (Eq. 10.14)

$$e^{iy} = \cos y + i \sin y \quad \text{and} \quad e^{-iy} = \cos y - i \sin y.$$

Adding and subtracting these equations, we obtain

$$\cos y = \frac{1}{2}(e^{iy} + e^{-iy}) \quad \text{and} \quad \sin y = \frac{1}{2i}(e^{iy} - e^{-iy}).$$

Using these equations as a guide, we write

$$\cos z = \frac{1}{2}(e^{iz} + e^{-iz}) \quad \text{and} \quad \sin z = \frac{1}{2i}(e^{iz} - e^{-iz})$$

Substituting $z = x + iy$, we obtain

$$\cos z = \frac{1}{2}[e^{i(x+iy)} + e^{-i(x+iy)}] = \frac{1}{2}[e^{-y} e^{ix} + e^{y} e^{-ix}]$$

$$= \frac{1}{2}[e^{-y}(\cos x + i \sin x) + e^{y}(\cos x - i \sin x)]$$

$$= \frac{1}{2}[\cos x(e^{-y} + e^{y}) + i \sin x (e^{-y} - e^{y})]$$

$$= \left[\cos x\left(\frac{e^{y} + e^{-y}}{2}\right) - i \sin x\left(\frac{e^{y} - e^{-y}}{2}\right)\right] = \cos x \cosh y - i \sin x \sinh y. \quad (10.17)$$

Similarly, we obtain

$$\sin z = \sin x \cosh y + i \cos x \sinh y. \quad (10.18)$$

The other trigonometric functions are defined as

$$\operatorname{cosec} z = \frac{1}{\sin z}, \quad \sec z = \frac{1}{\cos z}, \quad \tan z = \frac{\sin z}{\cos z}, \quad \cot z = \frac{\cos z}{\sin z},$$

whenever the denominator is not zero.

Complex trigonometric functions satisfy the same identities as real trigonometric functions. For example, it can be easily verified that

(i) $\sin(-z) = -\sin z \quad \text{and} \quad \cos(-z) = \cos z.$

(ii) $\sin^2 z + \cos^2 z = 1.$

(iii) $\sin(z_1 \pm z_2) = \sin z_1 \cos z_2 \pm \cos z_1 \sin z_2.$

(iv) $\cos(z_1 \pm z_2) = \cos z_1 \cos z_2 \mp \sin z_1 \sin z_2.$

(v) $\sin 2z = 2 \sin z \cos z \quad \text{and} \quad \cos 2z = \cos^2 z - \sin^2 z.$

(vi) $\sin \bar{z} = \overline{\sin z}.$

(vii) $\sin(z + 2n\pi) = \sin z, \quad n \text{ any integer.}$

(viii) $\cos(z + 2n\pi) = \cos z, \quad n \text{ any integer.}$

The functions $\sin z$ and $\cos z$ are periodic with period 2π and the function $\tan z$ is periodic with period π.

We further have

$$|\sin z|^2 = |\sin x \cosh y + i \cos x \sinh y|^2$$
$$= \sin^2 x \cosh^2 y + \cos^2 x \sinh^2 y$$
$$= \sin^2 x (1 + \sinh^2 y) + (1 - \sin^2 x) \sinh^2 y = \sin^2 x + \sinh^2 y.$$

Since $\sinh y = (e^y - e^{-y})/2$ is not bounded, $|\sin z|$ and hence $\sin z$ is not a bounded function (unlike its real counterpart $\sin x$ for which $|\sin x| \leq 1$ for a real variable x). Similarly, we find that

$$|\cos z|^2 = \cos^2 x + \sinh^2 y.$$

Therefore, $|\cos z|$ and hence $\cos z$ in not a bounded function.

Now,
$$\sin z = 0 = \Rightarrow |\sin z| = 0 = \sin^2 x + \sinh^2 y, \text{ for all } x, y.$$

Therefore, $\sin x = 0$, and $\sinh y = 0$, or $x = n\pi$ and $y = 0$, n any integer.

Hence, $\sin z = 0$, only when z is real and $z = n\pi$, n any integer. Similarly, we find that $\cos z = 0$, only when z is real and $z = [(2n + 1)\pi]/2$, n any integer.

Hyperbolic functions

Similar to complex trigonometric functions, we define the complex hyperbolic functions as

$$\sinh z = \frac{e^z - e^{-z}}{2}, \quad \cosh z = \frac{e^z + e^{-z}}{2}, \quad \text{cosech } z = \frac{1}{\sinh z}, \quad \text{sech } z = \frac{1}{\cosh z},$$

$$\tanh z = \frac{\sinh z}{\cosh z}, \quad \coth z = \frac{\cosh z}{\sinh z},$$

whenever the denominator is not zero.

Complex hyperbolic functions also satisfy the same identities as real hyperbolic functions. For example,

(i) $\sinh (- z) = - \sinh z$ and $\cosh (- z) = \cosh z$.

(ii) $\cosh^2 z - \sinh^2 z = 1$.

(iii) $\cosh (z_1 + z_2) = \cosh z_1 \cosh z_2 + \sinh z_1 \sinh z_2$.

(iv) $\cosh 2z = \cosh^2 z + \sinh^2 z$.

(v) $\sinh 2z = 2 \sinh z \cosh z$.

We further have the following results:

(a) $\sin iz = \dfrac{1}{2i}[e^{i(iz)} - e^{-i(iz)}] = \dfrac{1}{2i}[e^{-z} - e^z] = \dfrac{i}{2}[e^z - e^{-z}] = i \sinh z,$

$\cos iz = \dfrac{1}{2}[e^{i(iz)} + e^{-i(iz)}] = \dfrac{1}{2}[e^{-z} + e^z] = \cosh z.$

(b) $\sinh z = -i \sin iz = -i \sin (i(x + iy)) = -i \sin (-y + ix) = i \sin (y - ix)$

$$= i[\sin y \cosh x - i \cos y \sinh x] = \sinh x \cos y + i \cosh x \sin y.$$

Similarly, we obtain

$$\cosh z = \cosh x \cos y + i \sinh x \sin y.$$

We also have

$$|\sinh z|^2 = \sinh^2 x \cos^2 y + \cosh^2 x \sin^2 y = \sinh^2 x + \sin^2 y.$$

$$|\cosh z|^2 = \cosh^2 x \cos^2 y + \sinh^2 x \sin^2 y = \sinh^2 x + \cos^2 y.$$

Now, $\sinh z = 0 \Rightarrow |\sinh z| = 0$, that is $\sinh^2 x + \sin^2 y = 0$ for all x, y.

Therefore, $\sinh x = 0$ and $\sin y = 0$ or $x = 0$ and $y = n\pi$.

Hence, $\sinh z = 0$, only when z is pure imaginary, and $z = n\pi i$, where n is any integer. Similarly, we find that $\cosh z = 0$, when z is pure imaginary and $z = [(2n + 1) \pi i]/2$, where n is any integer. We further note that $\sinh z$ and $\cosh z$ are not bounded.

Example 10.10   Express $w = \tan z$ in the form $w = u + iv$ and show that

$$\tan z = \frac{\sin 2x + i \sinh 2y}{\cos 2x + \cosh 2y}.$$

Solution We have

$$\tan z = \frac{\sin z}{\cos z} = \frac{\sin x \cosh y + i \cos x \sinh y}{\cos x \cosh y - i \sin x \sinh y}$$

$$= \frac{[\sin x \cosh y + i \cos x \sinh y][\cos x \cosh y + i \sin x \sinh y]}{[\cos x \cosh y - i \sin x \sinh y][\cos x \cosh y + i \sin x \sinh y]} = \frac{N}{D}$$

where,

$$N = \sin x \cos x(\cosh^2 y - \sinh^2 y) + i \sinh y \cosh y (\cos^2 x + \sin^2 x)$$

$$= \sin x \cos x + i \sinh y \cosh y = \frac{1}{2} (\sin 2x + i \sinh 2y)$$

$$D = \cos^2 x \cosh^2 y + \sin^2 x \sinh^2 y$$

$$= \cos^2 x (1 + \sinh^2 y) + (1 - \cos^2 x) \sinh^2 y$$

$$= \cos^2 x + \sinh^2 y = \frac{1}{2} (1 + \cos 2x) + \frac{1}{2} (\cosh 2y - 1) = \frac{1}{2} (\cos 2x + \cosh 2y).$$

Dividing, we obtain the required result.

Example 10.11   Find all values of z such that $\sin z = 2$.

Solution   Let $z = x + iy$. We have

$$\sin z = \sin x \cosh y + i \cos x \sinh y = 2.$$

Comparing the real and imaginary parts, we get

  (i) $\sin x \cosh y = 2$ and   (ii) $\cos x \sinh y = 0$.

From (ii), we find that either $\cos x = 0$ or $\sinh y = 0$. That is, either $x = [(2n + 1)\pi]/2$ or $y = 0$, n any integer. For $y = 0$, we obtain from (i) $\sin x = 2$, which is not possible.

  For $x = [(2n + 1)\pi]/2$, we obtain form (i)

$$(-1)^n \cosh y = 2.$$

Since $\cosh y > 0$ for every y, n must be even. Thus, $y = \cosh^{-1} 2$.
  Hence, we obtain

$$x = [(4k + 1)\pi]/2, \quad y = \cosh^{-1} 2.$$

and $$z = (4k + 1)\,\frac{\pi}{2} + i \cosh^{-1} 2, \quad k = 0, \pm 1, \pm 2, \ldots$$

Example 10.12   Find all values of z such that $\sinh z = e^{\pi i/3}$.

Solution Since, $e^{\pi i/3} = \cos\dfrac{\pi}{3} + i \sin\dfrac{\pi}{3} = \dfrac{1}{2} + i\dfrac{\sqrt{3}}{2}$, we have for $z = x + iy$

$$\sinh z = \sinh x \cos y + i \cosh x \sin y = \frac{1}{2} + i\frac{\sqrt{3}}{2}.$$

Comparing the real and imaginary parts, we obtain

  (i) $\sinh x \cos y = \dfrac{1}{2}$ and   (ii) $\cosh x \sin y = \dfrac{\sqrt{3}}{2}$,

or $\sinh x = \dfrac{1}{2 \cos y}$ and $\cosh x = \dfrac{\sqrt{3}}{2 \sin y}$.

Substituting in $\cosh^2 x - \sinh^2 x = 1$, we get

$$\frac{3}{4 \sin^2 y} - \frac{1}{4 \cos^2 y} = 1$$

or $3 \cos^2 y - \sin^2 y = 4 \sin^2 y \cos^2 y$,   or $3(1 - \sin^2 y) - \sin^2 y = 4 \sin^2 y\,(1 - \sin^2 y)$

or $4 \sin^4 y - 8 \sin^2 y + 3 = 0$.

Solving for $\sin^2 y$, we get $\sin^2 y = 3/2$ and $\sin^2 y = 1/2$.

Since, y is real $\sin^2 y = 3/2$ is not possible. Thus, we have $\sin^2 y = 1/2$ or $\sin y = \pm\, 1/\sqrt{2}$.

For $\sin y = -\, 1/\sqrt{2}$, we find from (ii) $\cosh x = -\, \sqrt{3/2}$, which is not possible.

Hence, $\sin y = \dfrac{1}{\sqrt{2}} = \sin\!\left(\dfrac{\pi}{4}\right)$ or $y = n\pi + (-1)^n\,\dfrac{\pi}{4}$, n any integer.

When n is even, we get $\sin y = \cos y = 1/\sqrt{2}$ and $\cosh x = \sqrt{3/2}$ and $\sinh x = 1/\sqrt{2}$.

Hence,
$$e^x = \cosh x + \sinh x = \frac{\sqrt{3}+1}{\sqrt{2}} \quad \text{or} \quad x = \ln\left(\frac{\sqrt{3}+1}{\sqrt{2}}\right).$$

When n is odd, we get $\sin y = 1/\sqrt{2}$, $\cos y = -1/\sqrt{2}$ and $\cosh x = \sqrt{3/2}$ and $\sinh x = -1/\sqrt{2}$.

Hence,
$$e^x = \cosh x + \sinh x = \frac{\sqrt{3}-1}{\sqrt{2}} \quad \text{or} \quad x = \ln\left(\frac{\sqrt{3}-1}{\sqrt{2}}\right).$$

Therefore, the solutions are $z = x + iy$, where

$$x = \ln\left(\frac{\sqrt{3}+1}{\sqrt{2}}\right), \quad y = n\pi + \frac{\pi}{4}, \quad n \text{ even integer}$$

and

$$x = \ln\left(\frac{\sqrt{3}-1}{\sqrt{2}}\right), \quad y = n\pi - \frac{\pi}{4}, \quad n \text{ odd integer}.$$

Example 10.13   Find all values of z, such that

$$\sqrt{2}\, \sin z = \cosh \beta + i \sinh \beta, \ \beta \text{ real}.$$

Solution We have

$$\cosh \beta + i \sinh \beta = \cos i\beta + \sin i\beta = \sin\left(\frac{\pi}{2} + i\beta\right) + \sin i\beta$$

$$= 2 \sin\left(\frac{\pi}{4} + i\beta\right) \cos\frac{\pi}{4} = \sqrt{2}\cdot \sin\left(\frac{\pi}{4} + i\beta\right).$$

Therefore, the given equation becomes

$$\sin z = \sin\left(\frac{\pi}{4} + i\beta\right).$$

We obtain
$$z = n\pi + (-1)^n \left(\frac{\pi}{4} + i\beta\right), \ n = 0, \pm 1, \pm 2, \ldots.$$

10.4.3   Logarithm Function

The real valued natural logarithm function of real variable x is defined as

$$y = \ln x, \quad \text{if} \ \ x = e^y, \ \ x > 0.$$

Similarly, we define the logarithm of a complex variable $z = x + iy$ as

$$w = \ln z, \quad \text{if} \ \ z = e^w, \ \ z \ne 0. \tag{10.19}$$

Writing $z = re^{i\theta}$ and $w = u + iv$ in $z = e^w$, we get $re^{i\theta} = e^{u+iv}$.

Since $\quad e^{i\theta} = e^{i(\theta + 2n\pi)}$, n any integer, we obtain $\quad re^{i(\theta + 2n\pi)} = e^{u+iv}$.

Therefore, $\qquad\qquad\qquad |e^w| = e^u = r \quad$ or $\quad u = \ln r$

which is the logarithm of a real variable $r = |z|$ and

$$\arg\,(e^w) = v = \theta + 2n\pi, \ n \text{ any integer.}$$

Hence, for any complex $z \neq 0$, the solutions of the equation $e^w = z$ are given by

$$w = \ln z = \ln r + i(\theta + 2n\pi), \ n \text{ any integer}$$

$$= \ln r + i \arg\,(z), \ z \neq 0. \tag{10.20}$$

Since $\arg(z)$ in Eq.(10.20) can take infinite number of values, $\ln z$ is a multiple-valued function. For a given value of z in its domain of definition, the values of $\ln z$ have the same real part, but their imaginary parts differ by integral multiples of 2π. For each n in Eq. (10.20), we obtain a different branch of the multiple-valued function $\ln z$. If we restrict $\arg\,(z)$ to the interval $-\pi < \arg\,(z) \le \pi$, which is called the *principal argument* of z, then the corresponding *principal branch*, Ln z of $\ln z$, is written as

$$\text{Ln } z = \ln |z| + i \text{ Arg }(z)$$

or $\qquad\qquad\qquad \text{Ln } z = \ln |z| + i\theta, \ -\pi < \theta \le \pi \tag{10.21a}$

or $\qquad\qquad\qquad \text{Ln } z = \ln \sqrt{x^2 + y^2} + i \tan^{-1}\,(y/x). \tag{10.21b}$

The function Ln z is single-valued and is defined for all $z \neq 0$.

Therefore, we have

$$\ln z = \text{Ln } z + 2n\pi i.$$

If $z = x$ is real and positive, then $|z| = x$ and $\text{Arg}(z) = 0$ and we obtain

$$\text{Ln } z = \ln |z| = \ln x, \ x > 0$$

which is the natural logarithm of real positive number x.

If $z = x$ is real and negative, then $|z| = |x|$ and Arg $(z) = \pi$, and we obtain

$$\text{Ln } z = \ln |z| + \pi i = \ln |x| + \pi i, \ x < 0.$$

With this definition, we have

$$\text{Ln } (1) = 0, \qquad\qquad\qquad \ln (1) = 2n\pi i,$$

$$\text{Ln } (- 1 - i) = \ln\sqrt{2} + \left(-\frac{3\pi}{4}\right)i, \qquad \ln (- 1 - i) = \ln \sqrt{2} + \left(2n - \frac{3}{4}\right)\pi i.$$

The following properties of $\ln z$ can be easily verified

   (i) $e^{\ln z} = z$,

   (ii) $\ln (e^z) = z + 2n\pi i$, n any integer,

(iii) $\ln (z_1 z_2) = \ln z_1 + \ln z_2 + 2n\pi i$, n any integer,

(iv) $\ln (z_1/z_2) = \ln z_1 - \ln z_2 + 2n\pi i$, n any integer.

Often, $\ln z$ is also written as $\log z$.

Example 10.14   Find the general and the principal values of

(i) $\log (1 + \sqrt{3}\, i)$,   (ii) $\log (1 - \sqrt{3}\, i)$,   (iii) $\log (-1)$.

Solution

(i) $z = 1 + \sqrt{3}\, i$. We get $|z| = 2$, Arg $(z) = \pi/3$.

Therefore, $\log (1 + \sqrt{3}\, i) = \log 2 + i \left(\dfrac{\pi}{3} + 2n\pi \right)$, n any integer

and $\text{Log } (1 + \sqrt{3}\, i) = \log 2 + i\pi/3$.

(ii) $z = 1 - \sqrt{3}\, i$. We get $|z| = 2$, Arg $(z) = -\pi/3$.

Therefore, $\log (1 - \sqrt{3}\, i) = \log 2 + i \left(-\dfrac{\pi}{3} + 2n\pi \right)$, n any integer

and $\text{Log } (1 - \sqrt{3}\, i) = \log 2 - i\pi/3$.

(iii) $z = -1 = -1 + 0i$. We get $|z| = 1$, Arg $(z) = \pi$.

Therefore, $\log (-1) = \log 1 + i\, (\pi + 2n\pi) = (2n + 1)\, \pi i$, n any integer

and $\text{Log } (-1) = \pi i$.

Example 10.15   Find the real and imaginary parts of Log $[(1 + i)$ Log $i]$.

Solution   We have Log $i = \log 1 + (\pi i/2) = \pi i/2$.

Now, $(1 + i) \text{ Log } i = \dfrac{1}{2}\, (1 + i)\, \pi i = (-1 + i)\, \dfrac{\pi}{2}$.

Therefore, $\text{Log}[(1 + i) \text{ Log } i] = \text{Log} \left(-\dfrac{\pi}{2} + i\dfrac{\pi}{2} \right)$

$$= \log \left[\left(\dfrac{-\pi}{2} \right)^2 + \left(\dfrac{\pi}{2} \right)^2 \right]^{1/2} + i \tan^{-1} (-1) = \log \left(\dfrac{\pi}{\sqrt{2}} \right) + \dfrac{3\pi i}{4}.$$

10.4.4   General Powers of a Complex Number

Let c be a complex constant. Using the definitions of exponential and logarithm functions of the complex variable z, we define

$$w = z^c = e^{c \log z}, \quad z \neq 0. \tag{10.22}$$

Since $\log z$ has infinite number of values for a given z, w as defined by Eq. (10.22), also has infinite number of values. When we use the principal branch Log z of log (z), we obtain from Eq. (10.22), the principal value of z^c.

Remark 2

If $c = n$ is a positive integer, then $w = z^c$ has a unique value. When $c = 1/n$, $w = z^c$ has exactly n values.

Example 10.16   Find the general and principal values of

    (i) $(i)^i$,   (ii) $(-i)^i$,   (iii) $(1 + \sqrt{3}\, i)^{1+i}$.

Solution

(i) $w = (i)^i = e^{i\log i} = e^{i\left[\log 1 + i\left(\frac{\pi}{2} + 2n\pi\right)\right]} = e^{-\left(\frac{\pi}{2} + 2n\pi\right)}, \quad n = 0, \pm 1, \pm 2, \ldots.$

    The principal value of w (for $n = 0$) is $e^{-\pi/2}$.

(ii) $w = (-i)^i = e^{i\log(-i)} = e^{i\left[\log 1 + i\left(-\frac{\pi}{2} + 2n\pi\right)\right]} = e^{\frac{\pi}{2} - 2n\pi}, \quad n = 0, \pm 1, \pm 2, \ldots.$

    The principal value of w (for $n = 0$) is $e^{\pi/2}$.

(iii) $w = (1 + i\sqrt{3})^{1+i} = e^{(1+i)\log(1+i\sqrt{3})} = e^{(1+i)\left[\log 2 + i\left(\frac{\pi}{3} + 2n\pi\right)\right]}$

$$= e^{\left[\log 2 - \left(\frac{\pi}{3} + 2n\pi\right)\right] + i\left[\log 2 + \left(\frac{\pi}{3} + 2n\pi\right)\right]}, \quad n = 0, \pm 1, \pm 2, \ldots$$

The principal value of w (for $n = 0$) is given by

$$w = e^{\left[\log 2 - \frac{\pi}{3}\right] + i\left[\log 2 + \frac{\pi}{3}\right]} = e^{\log 2}\, e^{-\pi/3}\, e^{i\log 2}\, e^{\pi i/3}$$

$$= 2e^{-\pi/3}\,[\cos(\log 2) + i\sin(\log 2)]\,[(1 + \sqrt{3}\, i)/2]$$

$$= (1 + \sqrt{3}\, i)e^{-\pi/3}\,[\cos(\log 2) + i\sin(\log 2)].$$

Example 10.17   Using the principal branch of log z, show that

$$[z^2]^{1/2} = \begin{cases} z, & \text{for } \operatorname{Re} z > 0 \ \ \text{or} \ \ \operatorname{Re} z = 0 \ \text{and} \ \ \operatorname{Im} z > 0, \\ -z, & \text{elsewhere.} \end{cases}$$

Solution We have

$$[z^2]^{1/2} = e^{[\log z^2]/2} = e^{[\log|z|^2 + i\,\operatorname{Arg}(z^2)]/2}$$

$$= e^{[2\log|z| + i\operatorname{Arg}(z^2)]/2} = e^{\log|z|}\, e^{i\operatorname{Arg}(z^2)/2} = |z|\, e^{i\,\operatorname{Arg}(z^2)/2}.$$

Using the range of principal arguments for z^2

$$\text{Arg}(z^2) = \begin{vmatrix} 2\,\text{Arg}\,z, & \text{for Re}\,z > 0 \;\; \text{or} \;\; \text{Re}\,z = 0 \;\; \text{and} \;\; \text{Im}\,z > 0, \\ 2\,\text{Arg}\,z - 2\pi, & \text{for Re}\,z < 0 \;\; \text{and} \;\; \text{Im}\,z \geq 0, \\ 2\,\text{Arg}\,z + 2\pi, & \text{for Re}\,z \leq 0 \;\; \text{and} \;\; \text{Im}\,z < 0, \end{vmatrix}$$

we obtain the required result.

10.4.5   Inverse Trigonometric and Hyperbolic Functions

For a real variable x, we define the inverse sine function as

$$y = \sin^{-1}x, \quad \text{when} \quad x = \sin y.$$

Similarly, we define the inverse sine function for a complex variable z as

$$w = \sin^{-1}z, \quad \text{when} \quad z = \sin w.$$

Writing $\sin w = (e^{iw} - e^{-iw})/(2i)$, we get

$$e^{iw} - e^{-iw} = 2iz \quad \text{or} \quad e^{2iw} - 2ize^{iw} - 1 = 0.$$

Solving for e^{iw}, we obtain

$$e^{iw} = \frac{1}{2}\left[2iz \pm \sqrt{-4z^2 + 4}\right] = iz \pm \sqrt{1 - z^2}.$$

Since $\pm\sqrt{1 - z^2}$ is covered by the double-valued function $\sqrt{1 - z^2}$, we can write

$$e^{iw} = iz + \sqrt{1 - z^2} \quad \text{or} \quad iw = \log\left[iz + \sqrt{1 - z^2}\right].$$

Hence,

$$w = \sin^{-1}z = -i\log\left[iz + \sqrt{1 - z^2}\right]. \tag{10.23}$$

Now, $\sin^{-1}z$ is defined for all z except when

$$iz + \sqrt{1 - z^2} = 0 \quad \text{or} \quad iz = -\sqrt{1 - z^2} \quad \text{or} \quad -z^2 = 1 - z^2$$

which is not possible. Thus, $\sin^{-1}z$ is defined for all z.

Other inverse complex trigonometric functions are defined by the following:

$$\cos^{-1}z = -i\log\left(z + \sqrt{z^2 - 1}\right).$$

$$\tan^{-1}z = -\frac{i}{2}\log\frac{1 + iz}{1 - iz} = \frac{i}{2}\log\left[\frac{i + z}{i - z}\right], \quad z \neq \pm i.$$

$$\operatorname{cosec}^{-1} z = \sin^{-1}\left(\frac{1}{z}\right) = -i \log\left[\frac{i + \sqrt{z^2 - 1}}{z}\right], \quad z \neq 0.$$

$$\sec^{-1} z = \cos^{-1}\left(\frac{1}{z}\right) = -i \log\left[\frac{1 + \sqrt{1 - z^2}}{z}\right], \quad z \neq 0.$$

$$\cot^{-1} z = \tan^{-1}\left(\frac{1}{z}\right) = -\frac{i}{2} \log\left[\frac{z + i}{z - i}\right], \quad z \neq \pm i.$$

Similarly, we define the inverse complex hyperbolic sine function as

$$w = \sinh^{-1} z, \quad \text{when} \quad z = \sinh w.$$

Writing $\sinh w = (e^w - e^{-w})/2$, we get

$$e^w - e^{-w} = 2z \quad \text{or} \quad e^{2w} - 2ze^w - 1 = 0.$$

Solving for e^w, we obtain

$$e^w = \frac{1}{2}\left[2z \pm \sqrt{4z^2 + 4}\right] = z \pm \sqrt{z^2 + 1}.$$

Since $\pm\sqrt{z^2 + 1}$ is covered by the double-valued function $\sqrt{z^2 + 1}$, we can write

$$e^w = z + \sqrt{z^2 + 1} \quad \text{or} \quad w = \log\left[z + \sqrt{z^2 + 1}\right].$$

Hence,
$$w = \sinh^{-1} z = \log\left(z + \sqrt{z^2 + 1}\right).$$

Other inverse complex hyperbolic functions are defined by the following:

$$\cosh^{-1} z = \log\left(z + \sqrt{z^2 - 1}\right).$$

$$\tanh^{-1} z = \frac{1}{2} \log\left(\frac{1 + z}{1 - z}\right), \quad z \neq \pm 1.$$

$$\operatorname{cosech}^{-1} z = \sinh^{-1}\left(\frac{1}{z}\right) = \log\left(\frac{1 + \sqrt{1 + z^2}}{z}\right), \quad z \neq 0.$$

$$\operatorname{sech}^{-1} z = \cosh^{-1}\left(\frac{1}{z}\right) = \log\left(\frac{1 + \sqrt{1 - z^2}}{z}\right), \quad z \neq 0.$$

$$\coth^{-1}z = \tanh^{-1}\left(\frac{1}{z}\right) = \frac{1}{2}\log\left(\frac{z+1}{z-1}\right), \quad z \neq \pm 1.$$

Since $\log z$ is a multiple-valued function, all the inverse trigonometric and hyperbolic functions are also multiple-valued. Using the principal values of both the square root and the logarithm functions, we obtain the principal values of these functions.

Example 10.18 Write $\tan^{-1}z$ in the form $u + iv$.

Solution Let $z = x + iy$ and $\tan^{-1}z = u + iv$. We obtain

$$u + iv = \tan^{-1}(x + iy) \tag{10.24}$$

and

$$u - iv = \tan^{-1}(x - iy). \tag{10.25}$$

Adding Eqs. (10.24) and (10.25), we get

$$2u = \tan^{-1}(x + iy) + \tan^{-1}(x - iy) = \tan^{-1}\left(\frac{2x}{1 - x^2 - y^2}\right).$$

Subtracting Eq. (10.25) from (10.24), we get

$$2iv = \tan^{-1}(x + iy) - \tan^{-1}(x - iy) = \tan^{-1}\left(\frac{2iy}{1 + x^2 + y^2}\right) = i\tanh^{-1}\left(\frac{2y}{1 + x^2 + y^2}\right).$$

Hence, we obtain

$$u = \frac{1}{2}\tan^{-1}\left(\frac{2x}{1 - x^2 - y^2}\right) \quad \text{and} \quad v = \frac{1}{2}\tanh^{-1}\left(\frac{2y}{1 + x^2 + y^2}\right).$$

Example 10.19 Show that

(i) $\tan^{-1}z = -\dfrac{i}{2}\log\left(\dfrac{1 + iz}{1 - iz}\right) = \dfrac{i}{2}\log\left(\dfrac{i + z}{i - z}\right), \quad z \neq \pm i.$

(ii) $\coth^{-1}z = \dfrac{1}{2}\log\left(\dfrac{z+1}{z-1}\right), \quad z \neq \pm 1.$

Solution

(i) Let $w = \tan^{-1}z$. We have $z = \tan w = \sin w/\cos w$

or

$$z = \frac{e^{iw} - e^{-iw}}{i(e^{iw} + e^{-iw})} = \frac{e^{2iw} - 1}{i(e^{2iw} + 1)} \quad \text{or} \quad e^{2iw} = \frac{1 + zi}{1 - zi}.$$

Therefore,

$$w = \frac{1}{2i}\log\left(\frac{1 + zi}{1 - zi}\right) = -\frac{i}{2}\log\left(\frac{1 + zi}{1 - zi}\right).$$

We can also write

$$w = -\frac{i}{2}\log\left[\frac{i(z-i)}{i(-i-z)}\right] = -\frac{i}{2}\log\left(\frac{i-z}{i+z}\right) = \frac{i}{2}\log\left(\frac{i+z}{i-z}\right)$$

This function is not defined, when

$$i + z = 0 \quad \text{or} \quad i - z = 0, \text{ that is, when } z = \pm i.$$

(ii) Let $w = \coth^{-1}z$. We have

$$z = \coth w = \frac{\cosh w}{\sinh w} = \frac{e^w + e^{-w}}{e^w - e^{-w}} = \frac{e^{2w}+1}{e^{2w}-1}.$$

Therefore, we get

$$e^{2w} = \frac{z+1}{z-1} \quad \text{or} \quad w = \frac{1}{2}\log\left(\frac{z+1}{z-1}\right).$$

This function is not defined when $z + 1 = 0$ or $z - 1 = 0$, that is, when $z = \pm 1$.

Example 10.20 Find all values of $\sin^{-1} 2$, treating 2 as complex number. (See Example 10.11).

Solution We have

$$\sin^{-1} z = -i \log [iz + \sqrt{1-z^2}].$$

For $z = 2$, we get $\sqrt{1-z^2} = \sqrt{3}\,i$ and $\sin^{-1} 2 = -i \log [(2+\sqrt{3})i]$

Since $2 + \sqrt{3}$ is positive, we get $\sin^{-1} 2 = -i\left[\log(2+\sqrt{3}) + i\left(\frac{\pi}{2}+2n\pi\right)\right]$, $n = 0, \pm 1, \pm 2, \ldots$

We obtain

$$\sin^{-1} 2 = -i \log (2+\sqrt{3}) + \left(\frac{\pi}{2}+2n\pi\right)$$

$$= -i \cosh^{-1} 2 + \left(\frac{\pi}{2}+2n\pi\right), \quad n = 0, \pm 1, \pm 2, \ldots.$$

since $\cosh^{-1} 2 = \log (2+\sqrt{3}).$

Exercise 10.3

Write the following functions in the form $u + iv$.

1. $e^{\bar{z}}$.

2. e^{z^2}.

3. e^{e^z}.

4. Log (cos z).

5. $\cos^{-1}(e^{i\theta})$, $0 \le \theta \le \pi$.

Find all the values of z which satisfy the following identities.

6. $e^z = -i$.

7. $e^{2z} + e^z + 1 = 0$.

8. $e^{1/z} = 1 - i$.

9. $e^{e^z} = i$.

10. $\cos z = i$.

11. $\sin(i/z) = i$.

12. $\tan z = 2$.

13. $\tanh z = -2$.

14. $e^{z^2} = (1 + i)/\sqrt{2}$

15. $\cosh z + \sinh z = \alpha$, α complex constant.

Find the general values of the following functions.

16. $\log(e^{5\pi i/2})$.

17. $\log(\text{Log } i)$.

18. $\sin\left[1 - \log\left(\dfrac{1+i}{\sqrt{2}} e\right)\right]$.

19. $\tan\left[i \log \dfrac{1 - \sqrt{3}\,i}{1 + \sqrt{3}\,i}\right]$.

20. $(-2)^{\text{Log } i}$.

21. $(-i)^{-i}$.

22. $(1 + i)^{i+1}$.

23. i^{e^i}.

24. $i^{\log(i+1)}$.

Prove the following identities.

25. $\sin^{-1} z = \cos^{-1}\sqrt{1 - z^2} = \tan^{-1} \dfrac{z}{\sqrt{1 - z^2}}$ **26.** $\cos^{-1} z = \sin^{-1}\sqrt{1 - z^2} = \tan^{-1}\dfrac{\sqrt{1 - z^2}}{z}$.

27. $\tan^{-1} z = \sin^{-1}\dfrac{z}{\sqrt{1 + z^2}} = \cos^{-1}\dfrac{1}{\sqrt{1 + z^2}}$.

28. $\sin(\cos^{-1} z) = -i\sqrt{z^2 - 1}$ and $\cos(\sin^{-1} z) = \sqrt{1 - z^2}$.

29. When is the function e^{z^2} (i) real, (ii) pure imaginary?

30. Show that $2\sqrt{2}\,e^{-\pi i/12} = \left(1 + \sqrt{3}\right) + \left(1 - \sqrt{3}\right)i$.

31. If $\cot(\theta + i\phi) = e^{i\alpha}$, then show that

$$\theta = (2n + 1)\frac{\pi}{4} \quad \text{and} \quad \phi = \frac{1}{2}\log\left[\tan\left(\frac{\pi}{4} - \frac{\alpha}{2}\right)\right], \text{ where } n \text{ is an integer.}$$

32. If $\tanh(x + iy) = p + iq$, then show that $\tanh 2x = 2p/(1 + p^2 + q^2)$ and $\tan 2y = 2q/(1 - p^2 - q^2)$.

33. If $\sin^{-1}(u + iv) = \alpha + i\beta$, then show that $\sin^2 \alpha$ and $\cosh^2 \beta$ are the roots of the equation

$$x^2 - (1 + u^2 + v^2)\,x + u^2 = 0.$$

34. If $\sinh^{-1}(x + iy) = u + iv$, then show that

(i) $\dfrac{x^2}{\sinh^2 u} + \dfrac{y^2}{\cosh^2 u} = 1,$ (ii) $\dfrac{x^2}{\cos^2 v} - \dfrac{y^2}{\sin^2 v} = -1.$

35. If $(a + ib)^p = m^{x+iy}$, then show that one of the values of y/x is $2 \tan^{-1} (b/a)/\log(a^2 + b^2)$.

10.5   Limit and Continuity

10.5.1   Limit of a Function

Let $f(z)$ be a single-valued function of z defined on S which includes the δ-neighborhood of a point $z = z_0$. The function $f(z)$ may or may not be defined at $z = z_0$. Then, the function $f(z)$ is said to have a limit $L \in \mathbb{C}$ as $z \to z_0$, if given any arbitrary small real number $\varepsilon > 0$, there exists a real number $\delta > 0$, such that

$$|f(z) - L| < \varepsilon, \quad \text{whenever} \quad 0 < |z - z_0| < \delta. \tag{10.26}$$

where L is finite. If such a number L can be obtained, then it is called the limit of the function $f(z)$ as $z \to z_0$, and we write

$$\lim_{z \to z_0} f(z) = L.$$

If no such number $L \in \mathbb{C}$ exists, then we say that $f(z)$ has no limit as $z \to z_0$. When the limit exists, then for every z in the δ-neighborhood of z_0, the value of $f(z)$ lies in the ε-neighborhood of L (Fig. 10.9).

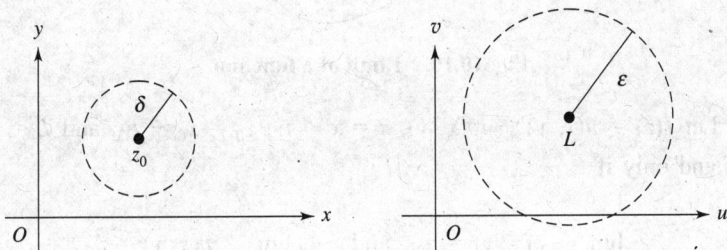

Fig. 10.9.   Limit of a function.

Theorem 10.1   If $\lim\limits_{z \to z_0} f(z)$ exists, then it is unique.

Proof   Let the function $f(z)$ have two different limits L_1 and L_2 as $z \to z_0$, that is

$$\lim_{z \to z_0} f(z) = L_1 \quad \text{and} \quad \lim_{z \to z_0} f(z) = L_2, \quad L_1 \neq L_2.$$

Hence, given an arbitrary real number $\varepsilon > 0$, there exist real numbers $\delta_1 > 0$ and $\delta_2 > 0$, such that

$$|f(z) - L_1| < \varepsilon/2, \quad \text{whenever} \quad 0 < |z - z_0| < \delta_1$$

and $|f(z) - L_2| < \varepsilon/2, \quad \text{whenever} \quad 0 < |z - z_0| < \delta_2.$

If　　$\delta = \min [\delta_1, \delta_2]$, then for $0 < |z - z_0| < \delta$, we have

$$|L_1 - L_2| = |(f(z) - L_2) - (f(z) - L_1)|$$

$$\leq |f(z) - L_2| + |f(z) - L_1| < \frac{\varepsilon}{2} + \frac{\varepsilon}{2} = \varepsilon.$$

This implies $|L_1 - L_2| = 0$　or　$L_1 - L_2 = 0$　or　$L_1 = L_2$.

Remark 3

Since $z \to z_0$ from any direction along any straight line or a curved path (Fig. 10.10), the limit as defined by Eq. (10.26) must be same along all these paths. If we obtain two different limits as $z \to z_0$ along two different paths, we conclude that the limit does not exist.

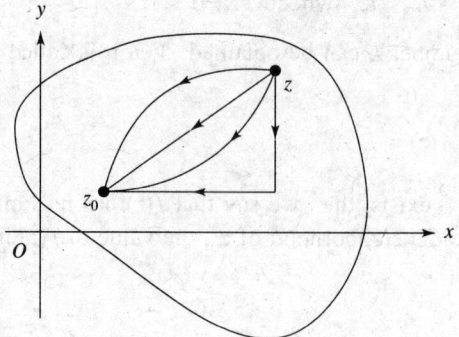

Fig. 10.10.　Limit of a function.

Theorem 10.2　Let $f(z) = u(x, y) + iv(x, y)$, $z = x + iy$, $z_0 = x_0 + iy_0$ and $L = u_0 + iv_0$. Then, $\lim\limits_{z \to z_0} f(z) = L$, if and only if

$$\lim_{x \to x_0, y \to y_0} u(x, y) = u_0 \quad \text{and} \quad \lim_{x \to x_0, y \to y_0} v(x, y) = v_0.$$

Proof　Let $\lim\limits_{z \to z_0} f(z) = L$. Then, $|f(z) - L| < \varepsilon$, whenever

$$0 < |z - z_0| < \delta \quad \text{or} \quad 0 < |(x - x_0) + i(y - y_0)| < \delta.$$

Hence, for $0 < |(x - x_0) + i(y - y_0)| < \delta$, we have

$$|u(x, y) + iv(x, y) - (u_0 + iv_0)| < \varepsilon$$

or　　　　　　$|(u(x, y) - u_0) + i(v(x, y) - v_0)| < \varepsilon.$　　　　　　(10.27)

Since　　　$|u(x, y) - u_0| \leq |(u(x, y) - u_0) + i(v(x, y) - v_0)|$

and　　　　$|v(x, y) - v_0| \leq |(u(x, y) - u_0) + i(v(x, y) - v_0)|$

we can write from Eq. (10.27), that

$$|u(x, y) - u_0| < \varepsilon/2 \quad \text{and} \quad |v(x, y) - v_0| < \varepsilon/2$$

or
$$\lim_{x \to x_0, y \to y_0} u(x, y) = u_0 \quad \text{and} \quad \lim_{x \to x_0, y \to y_0} v(x, y) = v_0. \tag{10.28}$$

Let us now assume that Eqs. (10.28) are satisfied, that is

$$|u(x, y) - u_0| < \varepsilon/2, \quad \text{whenever} \quad 0 < |(x - x_0) + i(y - y_0)| < \delta_1$$

and
$$|v(x, y) - v_0| < \varepsilon/2, \quad \text{whenever} \quad 0 < |(x - x_0) + i(y - y_0)| < \delta_2.$$

Let $\delta = \min(\delta_1, \delta_2)$. Then we have

$$|(u(x, y) - u_0) + i(v(x, y) - v_0)| \leq |u(x, y) - u_0| + |v(x, y) - v_0|$$

$$\leq (\varepsilon/2) + (\varepsilon/2) = \varepsilon,$$

whenever $0 < |(x - x_0) + i(y - y_0)| < \delta.$

Hence, $\qquad\qquad |f(z) - L| < \varepsilon, \quad \text{whenever} \quad 0 < |z - z_0| < \delta.$

Theorem 10.3   If $f(z)$ has a finite limit at z_0, then $f(z)$ is a bounded function in some neighborhood of z_0.

Proof   Since $\lim\limits_{z \to z_0} f(z)$ exists, we have

$$|f(z) - L| < \varepsilon, \quad \text{whenever} \quad 0 < |z - z_0| < \delta.$$

Hence, for any z in the neighborhood of z_0, we find that

$$|f(z)| = |f(z) - L + L| \leq |f(z) - L| + |L| < \varepsilon + |L| = M$$

where $M = \varepsilon + |L|$ is some finite number. Therefore, $f(z)$ is bounded in some neighborhood of z_0.

Remark 4

If (i) $|f(z)| \leq M$, whenever $0 < |z - z_0| < \delta$ and (ii) $\lim\limits_{z \to z_0} g(z) = 0$, then $\lim\limits_{z \to z_0} f(z) \, g(z) = 0$.

For example,

(a) $\lim\limits_{z \to z_0} z e^z = 0$, since $|e^z| < e$ for $|z| \leq \delta$.

(b) $\lim\limits_{z \to 0} z \sin(1/z)$ does not exist as $\sin(1/z)$ is not a bounded function in the neighborhood of $z = 0$.

Theorem 10.4   Let $f(z)$ and $g(z)$ be two functions such that

$$\lim_{z \to z_0} f(z) = L_1 \quad \text{and} \quad \lim_{z \to z_0} g(z) = L_2.$$

Then

(a) $\lim\limits_{z \to z_0} [\alpha f(z)] = \alpha L_1$, where α is a real or a complex constant.

(b) $\lim\limits_{z \to z_0} [f(z) \pm g(z)] = L_1 \pm L_2.$

(c) $\lim\limits_{z \to z_0} [f(z)\, g(z)] = L_1 L_2.$

(d) $\lim\limits_{z \to z_0} [1/g(z)] = 1/L_2, \quad L_2 \neq 0.$

(e) $\lim\limits_{z \to z_0} [f(z)/g(z)] = L_1/L_2, \quad L_2 \neq 0.$

Limit of a function at $z = \infty$

The function $f(z)$ has a limit L as $z \to \infty$, if for any arbitrary small real number $\varepsilon > 0$, there exists a real number $\delta > 0$ such that

$$|f(z) - L| < \varepsilon, \quad \text{whenever} \quad |z| > 1/\delta.$$

Alternately, we substitute $z = 1/\xi$.   Since $\xi \to 0$ as $z \to \infty$, we obtain

$$\lim_{z \to \infty} f(z) = \lim_{\xi \to 0} f(1/\xi).$$

Example 10.21   Using the definition of limits, show that

(i) $\lim\limits_{z \to i} z^2 = -1,$

(ii) $\lim\limits_{z \to -i} (z^2 - z) = i - 1,$

(iii) $\lim\limits_{z \to 2i} [3x + iy^2] = 4i.$

Solution   We want to determine a real number $\delta > 0$ such that for a given real number $\varepsilon > 0$

$$|f(z) - L| < \varepsilon, \quad \text{whenever} \quad 0 < |z - z_0| < \delta.$$

(i) $|z^2 + 1| < \varepsilon$, whenever $0 < |z - i| < \delta.$

Now, $\qquad |z^2 + 1| = |z - i|\,|z + i| = |z - i|\,|z - i + 2i|$

$$\leq |z - i|\,[|z - i| + |2i|] < \delta\,(\delta + 2).$$

For a given ε, we need to determine δ such that

$$\delta(\delta + 2) < \varepsilon, \quad \text{or} \quad (\delta + 1)^2 < \varepsilon + 1, \quad \text{or} \quad \delta < \sqrt{\varepsilon + 1} - 1.$$

With this choice of δ, we get

$$|z^2 + 1| < \varepsilon, \quad \text{whenever} \quad 0 < |z - i| < \delta.$$

Therefore, $\lim\limits_{z \to i} z^2 = -1.$

(ii) $|z^2 - z - (i - 1)| < \varepsilon$, whenever $0 < |z + i| < \delta.$

Now, $|z^2 - z - (i - 1)| = |-(z + i) + (z^2 + 1)| = |-(z + i) + (z + i)(z - i)|$

$$= |(z + i)(z - i - 1)| = |(z + i)(z + i - 2i - 1)|$$

$$\leq |z + i|\,[|z + i| + |2i + 1|] \leq \delta\,[\delta + \sqrt{5}\,].$$

For a given $\varepsilon > 0$, we need to determine $\delta > 0$, such that

$$\delta(\delta + \sqrt{5}) < \varepsilon \quad \text{or} \quad (\delta + \sqrt{5}/2)^2 < \varepsilon + 5/4 \quad \text{or} \quad \delta < \sqrt{\varepsilon + 5/4} - \sqrt{5}/2.$$

With this choice of δ, we get

$$|z^2 - z - (i - 1)| < \varepsilon, \quad \text{whenever} \quad 0 < |z + i| < \delta.$$

Therefore, $\lim\limits_{z \to -i} [z^2 - z] = i - 1.$

(iii) $|3x + iy^2 - 4i| < \varepsilon$, whenever $0 < |z - 2i| < \delta$, or $0 < |x + (y - 2)i| < \delta.$

We may choose $|x| < \delta, \quad |y - 2| < \delta.$

Now, $|3x + iy^2 - 4i| = |3x + i(y - 2)(y + 2)| = |3x + i(y - 2)(y - 2 + 4)|$

$$\leq 3|x| + |y - 2| [|y - 2| + 4] \leq 3\delta + \delta(\delta + 4) = \delta^2 + 7\delta < \varepsilon.$$

From $\delta^2 + 7\delta < \varepsilon$, we obtain

$$\left(\delta + \frac{7}{2}\right)^2 < \varepsilon + \frac{49}{4} \quad \text{or} \quad \delta < \sqrt{\varepsilon + \frac{49}{4}} - \frac{7}{2}.$$

With this choice of δ, we get the result.

Example 10.22 Using the definition of limits, show that $\lim\limits_{z \to \infty} \left[\dfrac{1}{z^2}\right] = 0.$

Solution For a given real number $\varepsilon > 0$, we want to determine a real number $\delta > 0$ such that

$$|f(z) - L| < \varepsilon, \quad \text{whenever} \quad |z| > \frac{1}{\delta}.$$

Therefore, we have

$$\left|\frac{1}{z^2}\right| < \varepsilon, \quad \text{whenever} \quad |z| > \frac{1}{\delta}.$$

Now, $|1/z^2| < \varepsilon$ implies $|z| > 1/\sqrt{\varepsilon}$. Thus, we can take $\delta < \sqrt{\varepsilon}$. With this choice of δ, we find that

$$\left|\frac{1}{z^2}\right| < \varepsilon, \quad \text{whenever} \quad |z| > \frac{1}{\delta} \quad \text{and} \quad \lim\limits_{z \to \infty} \left[\frac{1}{z^2}\right] = 0.$$

Alternative Substituting $z = 1/\xi$, we get $f(1/\xi) = \xi^2$. Since $\xi \to 0$ as $z \to \infty$, we need to determine a real number $\delta > 0$ such that for a given real number $\varepsilon > 0$, we have

$$|f(1/\xi) - 0| < \varepsilon, \quad \text{whenever} \quad 0 < |\xi| < \delta.$$

Now, $\quad |f(1/\xi) - 0| = |\xi^2| < \varepsilon, \quad \text{or} \quad |\xi| < \sqrt{\varepsilon}.$

Thus, we can take $\delta < \sqrt{\varepsilon}$. With this choice of δ, we find that

$$|\xi^2| < \varepsilon, \quad \text{whenever} \quad 0 < |\xi| < \delta.$$

Therefore, $\lim\limits_{\xi \to 0} \xi^2 = 0$ or $\lim\limits_{z \to \infty} [1/z^2] = 0$.

Example 10.23 Find the following limits

(i) $\lim\limits_{z \to 1} \dfrac{z^2 - 1}{z - 1}$,

(ii) $\lim\limits_{z \to 2i} [3x + iy^2]$,

(iii) $\lim\limits_{z \to \infty} \dfrac{z}{2 - iz}$

(iv) $\lim\limits_{z \to \infty} \left[\sqrt{z - 2i} - \sqrt{z - i} \right]$.

Solution

(i) $\lim\limits_{z \to 1} \dfrac{z^2 - 1}{z - 1} = \lim\limits_{z \to 1} \dfrac{(z - 1)(z + 1)}{z - 1} = \lim\limits_{z \to 1} (z + 1) = 2$.

(ii) $z = x + iy = 2i$. Therefore, $x = 0$, $y = 2$.

$$\lim\limits_{z \to 2i} [3x + iy^2] = \lim\limits_{x \to 0, y \to 2} [3x + iy^2] = 4i.$$

(iii) $\lim\limits_{z \to \infty} \left[\dfrac{z}{2 - iz} \right] = \lim\limits_{\xi \to 0} \left[\dfrac{1/\xi}{2 - i/\xi} \right] = \lim\limits_{\xi \to 0} \left[\dfrac{1}{2\xi - i} \right] = -\dfrac{1}{i} = i$.

(iv) $\lim\limits_{z \to \infty} \left[\sqrt{z - 2i} - \sqrt{z - i} \right] = \lim\limits_{z \to \infty} \dfrac{\left[\sqrt{z - 2i} - \sqrt{z - i} \right] \left[\sqrt{z - 2i} + \sqrt{z - i} \right]}{\sqrt{z - 2i} + \sqrt{z - i}}$

$$= \lim\limits_{z \to \infty} \left[\dfrac{-i}{\sqrt{z - 2i} + \sqrt{z - i}} \right] = \lim\limits_{\xi \to 0} \left[\dfrac{-i\sqrt{\xi}}{\sqrt{1 - 2i\xi} + \sqrt{1 - i\xi}} \right] = 0.$$

Example 10.24 Show that the following limits do not exist.

(i) $\lim\limits_{z \to 0} \dfrac{z}{|z|}$

(ii) $\lim\limits_{z \to \infty} \dfrac{\left[\operatorname{Re} z - \operatorname{Im} z \right]^2}{|z|^2}$,

(iii) $\lim\limits_{z \to 0} \left[\dfrac{1}{1 - e^{1/x}} + iy^2 \right]$.

Solution Here, we shall show that as $z \to 0$ along two different paths, we obtain two different limits and therefore the limit does not exist.

(i) When we choose the path $y \to 0$ followed by $x \to 0$, we get

$$\lim_{z \to 0} \frac{z}{|z|} = \lim_{x \to 0} \left[\lim_{y \to 0} \frac{x+iy}{\sqrt{x^2+y^2}} \right] = \lim_{x \to 0} \frac{x}{\sqrt{x^2}} = \lim_{x \to 0} \frac{x}{|x|} = \pm 1.$$

When we choose the path $x \to 0$ followed by $y \to 0$, we get

$$\lim_{z \to 0} \frac{z}{|z|} = \lim_{y \to 0} \left[\lim_{x \to 0} \frac{x+iy}{\sqrt{x^2+y^2}} \right] = \lim_{y \to 0} \frac{iy}{|y|} = \pm i.$$

Hence, the limit does not exist.

(ii) Consider any straight line path $y = mx$ (for different values of m we get different paths). Hence,

$$\lim_{z \to 0} \frac{[\text{Re}\, z - \text{Im}\, z]^2}{|z|^2} = \lim_{z \to 0} \frac{(x-y)^2}{x^2+y^2} = \lim_{x \to 0} \frac{(1-m)^2 x^2}{(1+m^2) x^2} = \frac{(1-m)^2}{1+m^2}.$$

Since the limit depends on m and is not unique, the limit does not exist.

(iii) Consider the path $y \to 0$ followed by $x \to 0$. We get

$$\lim_{z \to 0} \left[\frac{1}{1-e^{1/x}} + iy^2 \right] = \lim_{x \to 0} \left[\frac{1}{1-e^{1/x}} \right] = \begin{cases} 0, & x > 0, \\ 1, & x < 0. \end{cases}$$

Since the limit is not unique, the limit does not exist.

10.5.2 Continuity of a Function

Let $f(z)$ be a single-valued function of z defined in some neighborhood of the point z_0 including the point z_0. Then, $f(z)$ is said to be *continuous* at the point z_0 if for a given real number $\varepsilon > 0$, we can find a real number $\delta > 0$, such that

$$|f(z) - f(z_0)| < \varepsilon, \quad \text{wherever} \quad |z - z_0| < \delta. \tag{10.29}$$

Thus, a function $f(z)$ is continuous at z_0 if the following three conditions are satisfied:

(i) $f(z_0)$ exists,   (ii) $\lim_{z \to z_0} f(z)$ exists,

and   (iii) $\lim_{z \to z_0} f(z) = f(z_0)$.

A function $f(z)$ which is not continuous at z_0 is said to be *discontinuous* at z_0.

A function $f(z)$ is said to be continuous in a domain D if it is continuous at every point in D.

We note the following:

1. If both $f(z_0)$ and $\lim\limits_{z \to z_0} f(z) = L$ exist, but $f(z_0) \neq L$, then the point z_0 is called a point of *removable discontinuity*. In this case, we can redefine the function $f(z)$ at z_0 such that $f(z_0) = L$ and the function can be made continuous at z_0.

2. If the function $f(z) = u(x, y) + iv(x, y)$ is continuous at $z_0 = x_0 + iy_0$, then the real functions $u(x, y)$ and $v(x, y)$ are also continuous at the point (x_0, y_0). Therefore, we can discuss the continuity of a complex valued function by studying the continuity of its real and imaginary parts, that is the real functions $u(x, y)$ and $v(x, y)$.

3. If $f(z)$ and $g(z)$ are continuous at a point z_0, then the functions $f(z) \pm g(z)$, $f(z)\, g(z)$ and $f(z)/g(z)$ where $g(z_0) \neq 0$ are also continuous at z_0.

4. If $f(z)$ is continuous in a closed region S, then it in bounded is S, that is $|f(z)| \leq M$ for all z in S.

5. The function $f(z)$ is continuous at $z = \infty$, if the function $f(1/\xi)$ is continuous at $\xi = 0$.

Composite functions

Let a function $g(z)$ be defined in the neighborhood of a point z_0 and let the image of $g(z)$ in this neighborhood be contained in a region in which $f(z)$ is defined. Then, the composite function $f[g(z)]$ is defined for all z in the neighborhood of the point z_0. We have the following result.

Theorem 10.5   If the function $g(z)$ is continuous at $z = z_0$ and the function $f(z)$ is continuous at $g(z_0)$, then the composite function $f[g(z)]$ is continuous at z_0.

Proof   Let $w = g(z)$ and $w_0 = g(z_0)$. Since $g(z)$ is continuous at z_0, then given a real number $\varepsilon_1 > 0$, we can find a real number $\delta_1 > 0$, such that

$$|g(z) - g(z_0)| < \varepsilon_1, \quad \text{whenever} \quad |z - z_0| < \delta_1$$

or $\qquad |w - w_0| < \varepsilon_1, \qquad \text{whenever} \quad |z - z_0| < \delta_1.$

Since $f(z)$ is continuous at $g(z_0)$, then given a real number $\varepsilon_2 > 0$, we can find a real number $\delta_2 > 0$, such that

$$|f(g(z)) - f(g(z_0))| < \varepsilon_2, \quad \text{whenever} \quad |g(z) - g(z_0)| < \delta_2$$

or $\qquad |f(w) - f(w_0)| < \varepsilon_2, \qquad \text{whenever} \quad |w - w_0| < \delta_2$

that is, $\quad |f(w) - f(w_0)| < \varepsilon_2, \qquad \text{whenever} \quad |z - z_0| < \delta_3$

for some real number $\delta_3 > 0$. Hence, $f(g(z))$ is continuous at $z = z_0$.

Example 10.25   If the function $w = f(z)$ is continuous at $z = z_0$, then show that the function $\overline{f(z)}$ is continuous at $z = z_0$.

Solution   Since $\overline{w - w_0} = \overline{w} - \overline{w_0}$, we have $|w - w_0| = |\overline{w - w_0}| = |\overline{w} - \overline{w_0}|$.

Therefore, $\qquad |\overline{f(z)} - \overline{f(z_0)}| = |f(z) - f(z_0)|$

or $\lim\limits_{z \to z_0} f(z) = f(z_0)$ implies that $\lim\limits_{z \to z_0} \overline{f(z)} = \overline{f(z_0)}$.

Hence, the result.

Example 10.26 Show that the functions

    (i) e^z (ii) $\sin z$,   (iii) $\cos z$.

are continuous for all z.

Solution We can write for $z = x + iy$,

$$e^z = e^x \cos y + ie^x \sin y = u + iv.$$

$$\sin z = \sin x \cosh y + i \cos x \sinh y = u + iv.$$

$$\cos z = \cos x \cosh y - i \sin x \sinh y = u + iv.$$

Since the real valued functions u and v of two real variables x and y are continuous for all x and y in each case, the given functions of a complex variable z are continuous for all z.

Example 10.27 Show that the function $f(z)$ is not continuous at $z = 0$, where

    (i) $f(z) = \begin{vmatrix} \dfrac{\text{Im}\,(z)}{|z|}, & z \neq 0 \\ 0 & z = 0. \end{vmatrix}$ (ii) $f(z) = \begin{vmatrix} \dfrac{\text{Re}\,(z^2)}{|z|^2}, & z \neq 0 \\ 0, & z = 0. \end{vmatrix}$

Solution

    (i) $\lim\limits_{z \to 0} \dfrac{\text{Im}\,(z)}{|z|} = \lim\limits_{x \to 0, y \to 0} \dfrac{y}{\sqrt{x^2 + y^2}}.$

Consider the path $y = mx$. Then

$$\lim_{x \to 0, y \to 0} \frac{y}{\sqrt{x^2 + y^2}} = \lim_{x \to 0} \frac{mx}{\left(\sqrt{1 + m^2}\right)x} = \frac{m}{\sqrt{1 + m^2}}$$

which depends on m. Hence, the limit does not exist at $z = 0$. Therefore, the function is not continuous at $z = 0$.

    (ii) $\lim\limits_{z \to 0} \dfrac{\text{Re}\,(z^2)}{|z|^2} = \lim\limits_{x \to 0, y \to 0} \dfrac{x^2 - y^2}{x^2 + y^2}.$

Consider the path $y = mx$. Then

$$\lim_{x \to 0, y \to 0} \frac{x^2 - y^2}{x^2 + y^2} = \frac{1 - m^2}{1 + m^2}$$

which depends on m. Hence, the limit does not exist at $z = 0$. Therefore, the function is not continuous at $z = 0$.

Alternative Set $z = r\,e^{i\theta}$. Now, $z \to 0$ implies $r \to 0$. We have

$$\lim_{z \to \infty} \frac{\text{Re}(z^2)}{|z|^2} = \lim_{r \to 0} \frac{r^2 \cos 2\theta}{r^2} = \cos 2\theta$$

which depends on θ, that is on the location of the point on the circle $|z| = r$. Since the limit does not exist at $z = 0$, the function is not continuous at $z = 0$.

Example 10.28   From the first principles show, that the function $f(z) = z^k$ is continuous at all points in the finite complex plane for any positive integer k.

Solution   We have

$$|z^k - z_0^k| = |(z - z_0 + z_0)^k - z_0^k|$$

$$= \left| (z - z_0)^k + \binom{k}{1}(z - z_0)^{k-1} z_0 + \binom{k}{2}(z - z_0)^{k-2} z_0^2 + \ldots + \binom{k}{k-1}(z - z_0)\, z_0^{k-1} \right|$$

Setting $z - z_0 = r$, we get

$$|z^k - z_0^k| = \left| r^k + \binom{k}{1} r^{k-1} z_0 + \binom{k}{2} r^{k-2} z_0^2 + \ldots + \binom{k}{k-1} r\, z_0^{k-1} \right|$$

$$\le |r|^k + \binom{k}{1}|r|^{k-1}|z_0| + \binom{k}{2}|r|^{k-2}|z_0|^2 + \cdots + \binom{k}{k-1}|r||z_0|^{k-1}$$

$$\le [|z_0| + |r|]^k - |z_0|^k < \varepsilon.$$

Therefore, $[|z_0| + |r|]^k < \varepsilon + |z_0|^k,$

or $|z_0| + |r| < [\varepsilon + |z_0|^k]^{1/k}$ or $|r| < [\varepsilon + |z_0|^k]^{1/k} - |z_0|.$

If we choose $\delta < [\varepsilon + |z_0|^k]^{1/k} - |z_0|$, we obtain

$$|z^k - z_0^k| < \varepsilon, \quad \text{whenever} \quad |z - z_0| < \delta.$$

Hence, the function $f(z) = z^k$ is continuous at z_0, for any integer k. Since z_0 is an arbitrary point, $f(z)$ is continuous at all points in the finite complex plane.

Example 10.29   Show that the function Ln z is not continuous on the negative real axis including the point $x = 0$.

Solution   Writing $z = x + iy$ and Ln $z = u + iv$, we get

$$u + iv = \text{Ln } z = \text{Ln } (x + iy) = \ln |z| + i\,\text{Arg}(z).$$

Therefore, $u = \ln |z| = \dfrac{1}{2} \ln (x^2 + y^2)$ and $v = \text{Arg } z = \tan^{-1} (y/x)$. Consider now the points on the negative real axis, $z = x < 0$. If we approach the negative real axis from the positive side, that is $y \to 0^+$, we have Arg $z = \pi$. If we approach the negative real axis form the negative side, that is

$y \to 0^-$, we have Arg $(z) = -\pi$. Hence, the imaginary part of Ln z has different values as we approach the negative real axis from two different directions. In fact, Arg (z) has a jump of 2π as it crosses the negative real axis. Also, when $z = 0$, Ln z is not defined. Hence the result.

Remark 5

(a) The function Ln $[f(z)]$ is not continuous at those points where $f(z) = 0$ or where $f(z)$ is real and negative.

(b) The principal value of $[f(z)]^c = e^{c[\text{Ln } f(z)]}$, where c is a complex constant, is not continuous at those points where Im $(f(z)) = 0$ and Re $(f(z)) \le 0$.

(c) Arg $(f(z))$ is not continuous at those points where $f(z) = 0$ or where $f(z)$ is real and negative.

10.5.3 Uniform Continuity

When the continuity of a function $f(z)$ is considered in a region S, then the choice of δ in Eq. (10.29) depends not only on ε, but also on the point $z_0 \in S$. If, for a given S, we can determine a δ independent of the point z_0 and satisfying the requirements given in Eq. (10.29), then the function $f(z)$ is said to be *uniformly continuous* in the region S. Therefore, a function $f(z)$ is said to be uniformly continuous in a region S, if for a given real number $\varepsilon > 0$, there exists a real number $\delta > 0$ depending only on ε such that

$$|f(z_1) - f(z_2)| < \varepsilon, \quad \text{whenever} \quad 0 < |z_1 - z_2| < \delta \tag{10.30}$$

where z_1 and z_2 are any two points in the region S. We note that (i) we do not discuss uniform continuity at a point, and (ii) if $f(z)$ is uniformly continuous in a region S, then $f(z)$ is continuous in S. However, a function may be continuous in a region S but may not be uniformly continuous in S.

Example 10.30 Show that the function $f(z) = z^2$ is uniformly continuous in the region $|z| < 1$.

Solution We need to show that for any given real number $\varepsilon > 0$, we can determine a real number $\delta > 0$ such that

$$|z_2^2 - z_1^2| < \varepsilon, \quad \text{whenever} \quad 0 < |z_2 - z_1| < \delta$$

where z_1 and z_2 are any two points in the region $|z| < 1$. We have

$$|z_2^2 - z_1^2| = |z_2 - z_1| \, |z_2 + z_1| \le [|z_2| + |z_1|] \, |z_2 - z_1| < 2 \, |z_2 - z_1|.$$

Thus, when $|z_2^2 - z_1^2| < \varepsilon$, it follows that $|z_2 - z_1| < \varepsilon/2$. Thus, $\delta < \varepsilon/2$ depends only on ε and not on the choice of the points z_1, z_2 in the given region. Hence, $f(z) = z^2$ is uniformly continuous in the region $|z| < 1$.

Example 10.31 Show that the function $f(z) = 1/z^2$ is

(i) not uniformly continuous in the region $|z| \le 1$,

(ii) uniformly continuous in the region $(1/2) \le |z| \le 1$.

Solution   Let z_1 and z_2 be any two points in the given region. We have

$$|f(z_2) - f(z_1)| = \left| \frac{1}{z_2^2} - \frac{1}{z_1^2} \right| = \left| \frac{z_1^2 - z_2^2}{z_1^2 z_2^2} \right| = \frac{|(z_2 - z_1)||z_2 + z_1|}{|z_1|^2 |z_2|^2}$$

$$\leq \frac{[|z_2| + |z_1|]|z_2 - z_1|}{|z_1|^2 |z_2|^2} \leq \frac{2|z_2 - z_1|}{|z_1|^2 |z_2|^2}$$

(i) In the region $|z| \leq 1$, by choosing either of the two points z_1 or z_2 close to zero, we can make $|f(z_2) - f(z_1)|$ larger than any positive real number. Hence, the function $f(z) = 1/z^2$ is not uniformly continuous in any region which includes the point $z = 0$.

(ii) In the region $(1/2) \leq |z| \leq 1$, we can write

$$|f(z_2) - f(z_1)| \leq \frac{2|z_2 - z_1|}{(1/4)(1/4)} \leq 32 |z_2 - z_1|.$$

Thus, $|f(z_2) - f(z_1)| < \varepsilon$, whenever $|z_2 - z_1| < \varepsilon/32$. Therefore, we can choose $\delta < \varepsilon/32$ for any choice of z_1 and z_2. Hence, $f(z) = 1/z^2$ is uniformly continuous in the region $(1/2) \leq |z| \leq 1$.

Exercise 10.4

Compute the following limits.

1. $\lim\limits_{z \to 1} \dfrac{z^2 + z - 2}{z - 1}$.

2. $\lim\limits_{z \to i} \dfrac{z^3 - iz^2 + z - i}{z - i}$.

3. $\lim\limits_{z \to i} \dfrac{iz^3 - 1}{z - i}$.

4. $\lim\limits_{z \to 0} \dfrac{z^2}{|z|}$.

5. $\lim\limits_{z \to 1-i} [z^2 - \bar{z}^2]$.

6. $\lim\limits_{z \to i} \left[x + \dfrac{i}{1 - x} \right] e^{xy}$.

7. $\lim\limits_{z \to 1+2i} \left[\dfrac{z^2 - 2z + 5}{z^2 + 3 - 4i} \right]$.

8. $\lim\limits_{z \to -i} \dfrac{z^4 - 1}{z + i}$.

9. $\lim\limits_{z \to \infty} \dfrac{z}{2 - iz}$.

10. $\lim\limits_{z \to \infty} \dfrac{iz^3 + iz - 1}{(2z + 3i)(z - i)^2}$.

11. $\lim\limits_{z \to \infty} \sqrt{z} \left[\sqrt{z - 2i} - \sqrt{z - i} \right]$

Show that the following limits do not exist.

12. $\lim\limits_{z \to 0} \dfrac{z^2}{|z|^2}$.

13. $\lim\limits_{z \to 0} \dfrac{\bar{z}}{z}$.

14. $\lim\limits_{z \to 0} \dfrac{\text{Re}(z)}{|z|}$.

15. $\lim\limits_{z \to 0} \dfrac{z}{\text{Re } z}$.

16. $\lim\limits_{z \to 0} \dfrac{\text{Re}(z^2)}{|z|^2}$.

17. $\lim\limits_{z \to 0} \dfrac{\text{Im}(z^2)}{|z|^2}$.

18. $\lim\limits_{z \to 1} \dfrac{z^2 + 1}{z^2 - 3z + 2}$.

19. Let $P(z) = a_0 z^m + a_1 z^{m-1} + \ldots + a_{m-1} z + a_m$, $a_0 \neq 0$ and $Q(z) = b_0 z^n + b_1 z^{n-1} + \ldots + b_{n-1} z + b_n$, $b_0 \neq 0$ be polynomials of degrees m and n respectively. Find

(i) $\lim\limits_{z \to 0} \dfrac{p(z)}{Q(z)}$,

(ii) $\lim\limits_{z \to \infty} \dfrac{P(z)}{Q(z)}$.

Show that the following functions are continuous for all z.

20. Re z.

21. Im z.

22. $|z|$.

23. $\bar{z}$.

24. z^2.

Examine the continuity of the following functions.

25. $f(z) = \begin{vmatrix} \text{Re}(z^3)/|z|^2, & z \neq 0, \\ 0, & z = 0, \end{vmatrix}$ at $z = 0$.

26. $f(z) = \begin{vmatrix} (z^2 + 1)/(z + i), & z \neq -i, \\ 0, & z = -i, \end{vmatrix}$ at $z = -i$.

27. $f(z) = \begin{vmatrix} x^3 y^5 (x + iy)/(x^4 + y^4), & z \neq 0, \\ 0, & z = 0 \end{vmatrix}$ at $z = 0$.

28. $f(z) = \begin{vmatrix} e^{-1/z^2}, & z \neq 0, \\ 0, & z = 0, \end{vmatrix}$ at $z = 0$.

Find the value of $f(z_0)$ so that the function

29. $f(z) = \dfrac{z^2 - 2z + 2}{z^2 + 2i}$ is continuous at $z_0 = 1 - i$.

30. $f(z) = \dfrac{3z^4 - 2z^3 + 8z^2 - 2z + 5}{z - i}$ is continuous at $z = i$.

Where are the following functions not continuous?

31. $\dfrac{z - 1}{z^2 - 4z + 5}$.

32. Arg $(z^2 + 1)$.

33. Arg $\dfrac{z - i}{z + i}$.

34. Ln $\dfrac{z - 1}{z - 2}$.

35. Ln $(1 + z^2)$.

36. Show that the principal branch of the function $\sin^{-1}z$ is continuous at all points in the z-plane except along the cut as shown in Fig. 10.11

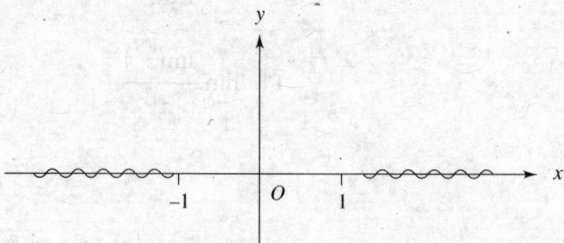

Fig.10.11. Problem 36.

37. Show that the function $f(z) = 3z - 2$ is uniformly continuous in the region $|z| \leq 1$.
38. Show that the function $f(z) = 1/(z - 1)$ is not uniformly continuous in the open disk $|z| < 1$.
39. Show that the function $f(z) = 1/(z^2 + 1)$ is not uniformly continuous in the region $|z| \leq 1$.
40. Show that the function $f(z) = 1/z$ is (i) not uniformly continuous in the region $|z| \leq 1$, (ii) uniformly continuous in the region $p \leq |z| \leq 1$ for any positive real number $p < 1$.

10.6 Differentiability and Analyticity

Let $f(z)$ be a single-valued function defined in a domain D. The function $f(z)$ is said to be differentiable at a point z_0, if

$$\lim_{z \to z_0} \frac{f(z) - f(z_0)}{z - z_0}$$

exists. This limit is called the *derivative* of $f(z)$ at the point $z = z_0$ and is denoted by $f'(z_0)$. Therefore, we have

$$f'(z_0) = \lim_{z \to z_0} \frac{f(z) - f(z_0)}{z - z_0}. \tag{10.31}$$

Substituting $z - z_0 = \Delta z$ in Eq. (10.31), we can write

$$f'(z_0) = \lim_{\Delta z \to 0} \frac{f(z_0 + \Delta z) - f(z_0)}{\Delta z}. \tag{10.32}$$

From the definition of limits, it follows that if $f'(z)$ is the derivative of $f(z)$ at $z \in D$, then for a given real number $\varepsilon > 0$, there exists a real number $\delta > 0$, such that

$$\left| \frac{f(z + \Delta z) - f(z)}{\Delta z} - f'(z) \right| < \varepsilon, \quad \text{whenever} \quad |\Delta z| < \delta. \tag{10.33}$$

If we write $w = f(z)$ and $w + \Delta w = f(z + \Delta z)$, then

$$\frac{\Delta w}{\Delta z} = \frac{f(z + \Delta z) - f(z)}{\Delta z}$$

and we can write

$$f'(z) = \lim_{\Delta z \to 0} \frac{\Delta w}{\Delta z} = \frac{dw}{dz}.$$

From the identity

$$f(z) = \left[\frac{f(z) - f(z_0)}{z - z_0} \right] (z - z_0) + f(z_0), \quad z \neq z_0$$

we obtain

$$\lim_{z \to z_0} f(z) = \lim_{z \to z_0} \left[\left\{ \frac{f(z) - f(z_0)}{z - z_0} \right\} (z - z_0) + f(z_0) \right]$$

$$= \lim_{z \to z_0} \left[\frac{f(z) - f(z_0)}{z - z_0} \right] \lim_{z \to z_0} (z - z_0) + \lim_{z \to z_0} f(z_0).$$

If $f'(z_0)$ exists, then using Eq. (10.31), we get

$$\lim_{z \to z_0} f(z) = f'(z_0) \times 0 + f(z_0) = f(z_0)$$

which shows that the function $f(z)$ is continuous at z_0.

Hence, if $f(z)$ is differentiable at $z = z_0$ then it must be continuous at $z = z_0$.

Remark 6

(a) A function which is not continuous at a point $z = z_0$, is not differentiable at $z = z_0$.

(b) A function which is continuous at a point $z = z_0$, may or may not be differentiable at $z = z_0$.

(c) The rules of differentiation of a function of a real variable x hold also for a function of a complex variable z.

We have the following results for two differentiable functions.

Theorem 10.6 If $f(z)$ and $g(z)$ are two differentiable functions at a point z, then the functions $[f(z) \pm g(z)]$, $[f(z) f(z)]$ and $[f(z)/g(z)]$ are also differentiable at z and

(i) $\dfrac{d}{dz} [f(z) \pm g(z)] = f'(z) \pm g'(z).$

(ii) $\dfrac{d}{dz} [f(z) g(z)] = f(z) g'(z) + g(z) f'(z).$

(iii) $\dfrac{d}{dz} \left[\dfrac{f(z)}{g(z)} \right] = \dfrac{1}{[g(z)]^2} [g(z) f'(z) - f(z) g'(z)], \ g(z) \neq 0.$

Theorem 10.7   If the function $g(z)$ is differentiable at z and the function $f(z)$ is differentiable at $g(z)$, then the composite function $w = f(g(z))$ is differentiable at z and

$$\frac{dw}{dz} = f'[g(z)]\ g'(z) \tag{10.34}$$

where $\qquad\qquad f'[g(z)] = \dfrac{df}{d\xi}$ and $\xi = g(z)$.

Proof   Let $\xi = g(z)$. Then, $w = f(\xi)$. We have

$$f'(\xi) = \lim_{\Delta\xi \to 0} \frac{f(\xi + \Delta\xi) - f(\xi)}{\Delta\xi}, \quad g'(z) = \lim_{\Delta z \to 0} \frac{g(z + \Delta z) - g(z)}{\Delta z}$$

or $\qquad\qquad f'(\xi) = \lim\limits_{\Delta\xi \to 0} \dfrac{\Delta w}{\Delta\xi}, \quad g'(z) = \lim\limits_{\Delta z \to 0} \dfrac{\Delta\xi}{\Delta z}$.

Therefore, we can write

$$\Delta w = f'(\xi)\ \Delta\xi + \varepsilon_1\ \Delta\xi \quad \text{and} \quad \Delta\xi = g'(z)\ \Delta z + \varepsilon_2\ \Delta z$$

where $\qquad\qquad \varepsilon_1$ and $\varepsilon_2 \to 0$ as $\Delta\xi \to 0$ and $\Delta z \to 0$ respectively.

From the above two equations, we obtain

$$\Delta w = f'(\xi)\ [g'(z)\ \Delta z + \varepsilon_2\ \Delta z] + \varepsilon_1\ [g'(z)\ \Delta z + \varepsilon_2\ \Delta z]$$

or $\qquad\qquad \dfrac{\Delta w}{\Delta z} = f'(\xi)\ g'(z) + \varepsilon_3, \quad \text{where} \quad \varepsilon_3 = \varepsilon_2\ f'(\xi) + \varepsilon_1\ g'(z) + \varepsilon_1\varepsilon_2$.

Since $\varepsilon_3 \to 0$ as $\Delta\xi \to 0$ and $\Delta z \to 0$, we obtain

$$\lim_{\Delta z \to 0} \frac{\Delta w}{\Delta z} = \frac{dw}{dz} = f'(\xi)\ g'(z) = f'[g(z)]\ g'(z).$$

Example 10.32   Show that the function $f(z) = \bar{z}$ is continuous at the point $z = 0$ but not differentiable at $z = 0$.

Solution   Let $z = x + iy$. Then, $\bar{z} = x - iy$ and

$$f(z) = \bar{z} = x - iy, \quad \Delta z = \Delta x + i\Delta y, \quad \overline{\Delta z} = \Delta x - i\Delta y.$$

Now, $\qquad\qquad \lim\limits_{z \to 0} f(z) = \lim\limits_{x \to 0, y \to 0} (x - iy) = 0 = f(0).$

Therefore, $f(z) = \bar{z}$ is continuous at $z = 0$.

Now, at $z = 0$ $\quad \lim\limits_{\Delta z \to 0} \dfrac{f(\Delta z) - f(0)}{\Delta z} = \lim\limits_{\Delta z \to 0} \dfrac{\overline{\Delta z}}{\Delta z} = \lim\limits_{\Delta x \to 0, \Delta y \to 0} \left[\dfrac{\Delta x - i\Delta y}{\Delta x + i\Delta y} \right]$

Consider now the path $y = mx$. We have $\Delta y = m\Delta x$.

Hence, $\Delta x \to 0$ implies $\Delta y \to 0$. Therefore

$$\lim_{\Delta z \to 0} \frac{\overline{\Delta z}}{\Delta z} = \lim_{\Delta x \to 0} \frac{\Delta x - im\Delta x}{\Delta x + im\Delta x} = \frac{1 - im}{1 + im}$$

which depends on m. Thus, the limit does not exist. Therefore, the function $f(z) = \bar{z}$ in not differentiable at $z = 0$.

Remark 7

(a) The function $f(z) = \bar{z}$ is not differentiable at any point in the complex plane.

(b) Since $\bar{z}$ is not differentiable at any point, any function $f(\bar{z})$ is also not differentiable at any point.

Example 10.33   Show that the function $f(z) = |z|^2$ is differentiable only at $z = 0$ and no where else.

Solution   We have $f(z) = |z|^2 = z\bar{z}$. Therefore,

$$\lim_{\Delta z \to 0} \frac{f(z + \Delta z) - f(z)}{\Delta z} = \lim_{\Delta z \to 0} \frac{(z + \Delta z)(\bar{z} + \overline{\Delta z}) - z\bar{z}}{\Delta z} = \lim_{\Delta z \to 0} \left[z \frac{\overline{\Delta z}}{\Delta z} + \bar{z} + \overline{\Delta z} \right] = 0 \text{ for } z = 0.$$

Hence, the function $f(z) = |z|^2$ is differentiable at $z = 0$ and $f'(0) = 0$.

For $z \neq 0$, $\lim\limits_{\Delta z \to 0} [\bar{z} + \overline{\Delta z}] = \bar{z}$, but $\lim\limits_{\Delta z \to 0} \frac{\overline{\Delta z}}{\Delta z}$ does not exist (see Example 10.32). Hence, the function $f(z) = |z|^2$ is not differentiable at $z \neq 0$.

Example 10.34   Show that the function $f(z) = z^n$, where n is a positive integer is differentiable at every point in the finite complex plane.

Solution   Let z be any point in the finite complex plane.

Then, $\dfrac{f(z + \Delta z) - f(z)}{\Delta z} = \dfrac{(z + \Delta z)^n - z^n}{\Delta z} = \dfrac{1}{\Delta z}\left[\binom{n}{1} z^{n-1}\Delta z + \binom{n}{2} z^{n-2}(\Delta z)^2 + \dots + \binom{n}{n}(\Delta z)^n \right]$

$$= \binom{n}{1} z^{n-1} + \binom{n}{2} z^{n-2} \Delta z + \dots + \binom{n}{n} (\Delta z)^{n-1}.$$

Taking limits on both sides as $\Delta z \to 0$, we obtain

$$\lim_{\Delta z \to 0} \frac{f(z + \Delta z) - f(z)}{\Delta z} = nz^{n-1}.$$

Therefore, $\dfrac{d}{dz}(z^n) = nz^{n-1}$, where z is any point in the finite complex plane.

Remark 8

Since $f(z) = z^n$, n positive integer, is differentiable at every point in the finite complex plane, any polynomial $P_n(z)$ in z is also differentiable at every point in the finite complex plane.

From the above examples, we find that a complex valued function may have derivative at every point (see Example 10.34), only at a fixed point (see Example 10.33) or at no point (see Remark 7).

We now define the important concept of analyticity of a function $f(z)$ at a point or in a domain D.

Analytic function

A function $f(z)$ of a complex variable z is said to be analytic at a point z_0, if it is differentiable at the point z_0 and also at each point in some neighborhood of the point z_0, Thus, analyticity at a point z_0 means differentiability in some open disk about z_0. A function $f(z)$ is said to be analytic in a domain D, if it is analytic at every point in D.

Note that analyticity implies differentiability but not vice versa. For example, the function

(i) $f(z) = |z|^2$ is differentiable only at $z = 0$ and no where else. Therefore, the function is differentiable at $z = 0$ but not analytic anywhere.

(ii) $f(z) = z^n$, n positive integer, is differentiable at all points and therefore is analytic at every point of the finite complex plane.

An analytic function is also called an *holomorphic* function. A function $f(z)$ which is analytic at every point of the finite complex plane is called an *entire* function. Since the derivative of a polynomial exists at every point, a polynomial of any degree is an entire function.

A function $f(z)$ is said to be analytic at $z = \infty$, if the function $f(1/z)$ is analytic at $z = 0$.

If the functions $f(z)$ and $g(z)$ are analytic in a domain D, then the functions $f(z) \pm g(z)$, $f(z)g(z)$ are also analytic in D. Further, the function $f(z)/g(z)$ is analytic at all points $z \in D$ for which $g(z) \neq 0$. The composition of two analytic functions is also analytic.

We now derive the necessary and sufficient conditions for a function to be analytic.

10.6.1   Cauchy-Riemann Equations

Necessary conditions for a function to be analytic

Theorem 10.8   Suppose that the function $f(z) = u(x, y) + iv(x, y)$ is continuous in some neighborhood of the point $z = x + iy$ and is differentiable at z. Then, the first order partial derivatives of $u(x, y)$ and $v(x, y)$ exist and satisfy the equations

$$u_x = v_y \quad \text{and} \quad u_y = -v_x \tag{10.35}$$

at the point z.

Proof Since $f(z)$ is differentiable at z, we have

$$f'(z) = \lim_{\Delta z \to 0} \frac{f(z+\Delta z) - f(z)}{\Delta z}$$

$$= \lim_{\Delta x \to 0, \Delta y \to 0} \frac{\{u(x+\Delta x, y+\Delta y) + iv(x+\Delta x, y+\Delta y)\} - \{u(x,y) + iv(x,y)\}}{\Delta x + i\Delta y}. \quad (10.36)$$

Since the limit exists, it must have the same value independent of the path along which $\Delta z \to 0$. We consider the following two paths:

(i) Let $\Delta y \to 0$ first and then $\Delta x \to 0$ (Fig. 10.12a). The limit in Eq. (10.36) becomes

$$f'(z) = \lim_{\Delta x \to 0} \left[\frac{u(x+\Delta x, y) - u(x,y)}{\Delta x} + i\frac{v(x+\Delta x, y) - v(x,y)}{\Delta x} \right]$$

$$= \frac{\partial u}{\partial x} + i\frac{\partial v}{\partial x} \quad (10.37)$$

(ii) Let $\Delta x \to 0$ first and then $\Delta y \to 0$ (Fig. 10.12b). The limit in Eq. (10.36) becomes (since $\Delta z = i\Delta y$)

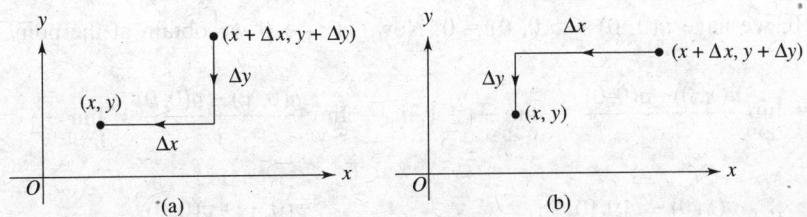

Fig. 10.12. (a) **Path** $\Delta y \to 0$, **then** $\Delta x \to 0$. (b) **Path** $\Delta x \to 0$, **then** $\Delta y \to 0$.

$$f'(z) = \lim_{\Delta y \to 0} \left[\frac{u(x, y+\Delta y) - u(x,y)}{i\,\Delta y} + i\frac{v(x, y+\Delta y) - v(x,y)}{i\,\Delta y} \right]$$

$$= \frac{1}{i}\frac{\partial u}{\partial y} + \frac{\partial v}{\partial y} = \frac{\partial v}{\partial y} - i\frac{\partial u}{\partial y}. \quad (10.38)$$

Since $f(z)$ is differentiable at z, the two limits given in Eqs. (10.37) and (10.38) must be equal.

Therefore,

$$\frac{\partial u}{\partial x} + i\frac{\partial v}{\partial x} = \frac{\partial v}{\partial y} - i\frac{\partial u}{\partial y}.$$

Comparing the real and imaginary parts, we get

$$\frac{\partial u}{\partial x} = \frac{\partial v}{\partial y} \quad \text{and} \quad \frac{\partial u}{\partial y} = -\frac{\partial v}{\partial x}$$

or in short notation $u_x = v_y$ and $u_y = -v_x$.

The equations given in Eq. (10.35) are called the *Cauchy-Riemann equations* and are the necessary conditions for differentiability and analyticity of the function $f(z)$ at a given point. Thus, if the function $f(z)$ does not satisfy the Cauchy-Riemann equations at a point, it is not differentiable and hence not analytic at that point. Further, the derivative $f'(z)$ may not exist, even when it satisfies the Cauchy-Riemann equations.

Example 10.35   Show that the function

$$f(z) = \begin{vmatrix} \dfrac{x^3(1+i) - y^3(1-i)}{x^2 + y^2}, & z \neq 0, \\ 0, & z = 0 \end{vmatrix}$$

satisfies the Cauchy-Riemann equations at $z = 0$ but $f'(0)$ does not exist.

Solution   Writing $f(z) = u(x, y) + iv(x, y)$, we get

$$u(x, y) = \frac{x^3 - y^3}{x^2 + y^2} \quad \text{and} \quad v(x, y) = \frac{x^3 + y^3}{x^2 + y^2}, \quad (x, y) \neq (0, 0).$$

Since $f(0) = 0$, we have $u(0, 0) = v(0, 0) = 0$. Now, as $z \to 0$ we obtain at the point $z = 0$

$$u_x = \lim_{x \to 0} \frac{u(x, 0) - u(0, 0)}{x} = \lim_{x \to 0} \frac{x}{x} = 1, \; u_y = \lim_{y \to 0} \frac{u(0, y) - u(0, 0)}{y} = \lim_{y \to 0} \frac{-y}{y} = -1.$$

$$v_x = \lim_{x \to 0} \frac{v(x, 0) - v(0, 0)}{x} = \lim_{x \to 0} \frac{x}{x} = 1, \; v_y = \lim_{y \to 0} \frac{v(0, y) - v(0, 0)}{y} = \lim_{y \to 0} \frac{y}{y} = 1.$$

Therefore, at $z = 0$, $u_x = v_y$ and $u_y = -v_x$. Thus, the Cauchy-Riemann equations are satisfied at $z = 0$. We now have

$$\lim_{z \to 0} \frac{f(z) - f(0)}{z} = \lim_{z \to 0} \frac{x^3 - y^3 + i(x^3 + y^3)}{(x^2 + y^2)(x + iy)} = \lim_{z \to 0} \frac{(1+i)(x^3 + iy^3)}{(x^2 + y^2)(x + iy)}$$

$$= \lim_{z \to 0} \frac{(1+i)(x^3 + iy^3)(x - iy)}{(x^2 + y^2)(x + iy)(x - iy)} = \lim_{z \to 0} \frac{(1+i)(x^3 + iy^3)(x - iy)}{(x^2 + y^2)^2}.$$

Choosing the path $y = mx$, we get

$$\lim_{x \to 0} \frac{(1+i)(1 + im^3)(1 - im)x^4}{(1 + m^2)^2 x^4} = \frac{(1+i)(1 + im^3)(1 - im)}{(1 + m^2)^2}$$

which depends on m. Therefore, the limit does not exist. Hence, $f'(0)$ does not exist.

We now present the sufficient conditions for a function to be analytic.

Sufficient conditions for a function to be analytic

Theorem 10.9 Suppose that the real and imaginary parts $u(x, y)$ and $v(x, y)$ of the function $f(z) = u(x, y) + iv(x, y)$ are continuous and have continuous first order partial derivatives in a domain D. If u and v satisfy the Cauchy-Riemann equations at all points in D, then the function $f(z)$ is analytic in D and

$$f'(z) = u_x + iv_x = v_y - iu_y. \tag{10.39}$$

Proof Consider a δ-neighborhood of z. Since the partial derivatives of $u(x, y)$ and $v(x, y)$ are continuous, we can write

$$\Delta u = u(x + \Delta x, y + \Delta y) - u(x, y) = u_x \, \Delta x + u_y \, \Delta y + \varepsilon_1 \Delta x + \varepsilon_2 \Delta y$$

$$\Delta v = v(x + \Delta x, y + \Delta y) - v(x, y) = v_x \, \Delta x + v_y \, \Delta y + \varepsilon_3 \Delta x + \varepsilon_4 \Delta y$$

where $\varepsilon_1, \varepsilon_2, \varepsilon_3, \varepsilon_4 \to 0$ as Δx and $\Delta y \to 0$.

Now, $\qquad\qquad \Delta w = f(z + \Delta z) - f(z) = \Delta u + i\Delta v$

$$= (u_x + iv_x) \, \Delta x + (u_y + iv_y) \, \Delta y + (\varepsilon_1 + i\varepsilon_3) \, \Delta x + (\varepsilon_2 + i\varepsilon_4) \, \Delta y.$$

Using the Cauchy-Riemann equations, we obtain

$$\Delta w = (u_x + iv_x) \, \Delta x + (-v_x + iu_x) \, \Delta y + (\varepsilon_1 + i\varepsilon_3) \, \Delta x + (\varepsilon_2 + i\varepsilon_4) \, \Delta y$$

$$= (u_x + iv_x)(\Delta x + i\Delta y) + (\varepsilon_1 + i\varepsilon_3) \, \Delta x + (\varepsilon_2 + i\varepsilon_4) \, \Delta y.$$

Hence, we get

$$\left| \frac{f(z + \Delta z) - f(z)}{\Delta z} - (u_x + iv_x) \right| \le \left| (\varepsilon_1 + i\varepsilon_3) \right| \left| \frac{\Delta x}{\Delta z} \right| + \left| (\varepsilon_2 + i\varepsilon_4) \right| \left| \frac{\Delta y}{\Delta z} \right|.$$

Since $\qquad\qquad \left| \dfrac{\Delta x}{\Delta z} \right| \le 1$ and $\left| \dfrac{\Delta y}{\Delta z} \right| \le 1$, we obtain

$$\lim_{\Delta z \to 0} \frac{f(z + \Delta z) - f(z)}{\Delta z} = f'(z) = u_x + iv_x = v_y - iu_y.$$

Therefore, $f(z)$ is differentiable at an arbitrary point z in D, that is $f(z)$ is analytic in D.

Note that the first order partial derivatives are not continuous in Example (10.35).

Example 10.36 Using the Cauchy-Riemann equations, show that

   (i) $f(z) = |z|^2$ is not analytic at any point.

   (ii) $f(z) = \bar{z}$ is not analytic at any point.

   (iii) $f(z) = 1/z$, $z \ne 0$, is analytic at all points except at the point $z = 0$.

Solution   Let $f(z) = u(x, y) + iv(x, y)$.

(i) $f(z) = |z|^2 = x^2 + y^2$. Therefore, $u(x, y) = x^2 + y^2$ and $v(x, y) = 0$.
We obtain

$$u_x = 2x, \quad u_y = 2y, \quad v_x = 0 \quad \text{and} \quad v_y = 0.$$

The partial derivatives are continuous. Cauchy-Riemann equations are satisfied when

$$u_x = v_y \quad \text{or} \quad 2x = 0 \quad \text{and} \quad u_y = -v_x \quad \text{or} \quad 2y = 0.$$

that is, when $x = 0$ and $y = 0$. Thus, the function is differentiable only at the origin. Therefore, the function is not analytic anywhere.

(ii) $f(z) = \bar{z} = x - iy$. Therefore, $u(x, y) = x$ and $v(x, y) = -y$.

We obtain $u_x = 1, \quad u_y = 0, \quad v_x = 0 \quad \text{and} \quad v_y = -1$.

The Cauchy-Riemann equation $u_x = v_y$ gives $1 = -1$, which is not true for any value of x and y. Therefore, the function $f(z) = \bar{z}$ is not differentiable at any point and hence is not analytic at any point.

(iii) $f(z) = \dfrac{1}{z} = \dfrac{\bar{z}}{z\bar{z}} = \dfrac{x - iy}{x^2 + y^2}, \quad z \neq 0$.

Therefore, $u(x, y) = \dfrac{x}{x^2 + y^2}$ and $v(x, y) = -\dfrac{y}{x^2 + y^2}$.

We obtain

$$\frac{\partial u}{\partial x} = \frac{y^2 - x^2}{(x^2 + y^2)^2} = \frac{\partial v}{\partial y} \quad \text{and} \quad \frac{\partial u}{\partial y} = -\frac{2xy}{(x^2 + y^2)^2} = -\frac{\partial v}{\partial x}.$$

Therefore, the Cauchy-Riemann equations are satisfied at all points except at $z = 0$ and the partial derivatives are also continuous at all points except at $z = 0$. Since, the function $f(z) = 1/z$ is not defined and hence not continuous at $z = 0$, it cannot be differentiable at $z = 0$. Hence, the function is analytic at all points except at $z = 0$.

Example 10.37   Show that the functions
(i) $\sin z$,   (ii) $\cos z$,   (iii) e^z
are analytic in the finite z-plane. Hence, obtain their derivatives.

Solution   Let $z = x + iy$.

(i) $\sin z = \sin x \cosh y + i \cos x \sinh y = u(x, y) + iv(x, y)$.

Therefore, $u(x, y) = \sin x \cosh y$ and $v(x, y) = \cos x \sinh y$.

We get $u_x = \cos x \cosh y = v_y$ and $u_y = \sin x \sinh y = -v_x$.

Since, the Cauchy-Riemann equations are satisfied for all (x, y) and the first order partial derivatives of $u(x, y)$, $v(x, y)$ are continuous everywhere, the given function is analytic for all z in the finite z-plane. We obtain

$$\frac{d}{dz}(\sin z) = u_x + iv_x = \cos x \cosh y - i \sin x \sinh y = \cos z.$$

(ii) $\qquad \cos z = \cos x \cosh y - i \sin x \sinh y = u(x, y) + iv\,(x, y).$

Therefore, $\quad u(x, y) = \cos x \cosh y \quad$ and $\quad v(x, y) = -\sin x \sinh y.$

We get $\qquad u_x = -\sin x \cosh y = v_y \quad$ and $\quad u_y = \cos x \sinh y = -v_x.$

Since the Cauchy-Riemann equations are satisfied for all (x, y) and the first order partial derivatives of $u(x, y)$, $v(x, y)$ are continuous everywhere, the given function is analytic for all z in the finite z-plane. We obtain

$$\frac{d}{dz}(\cos z) = u_x + iv_x = -\sin x \cosh y - i \cos x \sinh y = -\sin z.$$

(iii) $e^z = e^{x+iy} = e^x e^{iy} = e^x(\cos y + i \sin y) = u(x, y) + iv(x, y).$

Therefore, $\qquad u(x, y) = e^x \cos y \quad$ and $\quad v(x, y) = e^x \sin y.$

We get $\qquad u_x = e^x \cos y = v_y \quad$ and $\quad u_y = -e^x \sin y = -v_x.$

Since the Cauchy-Riemann equations are satisfied for all (x, y) and the first order partial derivatives of $u(x, y)$, $v(x, y)$ are continuous everywhere, the given function is analytic for all z in the finite z-plane.

We obtain $\qquad \dfrac{d}{dz}(e^z) = u_x + iv_x = e^x \cos y + ie^x \sin y = e^z.$

Remark 9

The functions $\tan z$ and $\sec z$ are analytic everywhere except at the points $z = (2n + 1)\,\pi/2$, whereas the functions $\cot z$ and $\operatorname{cosec} z$ are analytic every where except at the points $z = n\pi$ where n is any positive or negative integer or zero.

Example 10.38 Show that the function $\operatorname{Ln} z$ is analytic for all z except when $\operatorname{Re} z \le 0$.

Solution $\operatorname{Ln} z$ is not continuous when $x \le 0$ (see Example 10.29). Therefore, the function $\operatorname{Ln} z$ is not differentiable and hence not analytic when $\operatorname{Re} z \le 0$. We now write

$$\operatorname{Ln} z = \frac{1}{2} \ln (x^2 + y^2) + i \tan^{-1}\left(\frac{y}{x}\right) = u(x, y) + iv(x, y).$$

Therefore, $\quad u(x, y) = \dfrac{1}{2} \ln(x^2 + y^2) \quad$ and $\quad v(x, y) = \tan^{-1}\dfrac{y}{x}.$

We get $\qquad u_x = \dfrac{x}{x^2 + y^2} = v_y \quad$ and $\quad u_y = \dfrac{y}{x^2 + y^2} = -v_x.$

Since, the Cauchy-Riemann equations are satisfied for all (x, y) except at $(0, 0)$ and the first order partial derivatives of $u(x, y)$, $v(x, y)$ are continuous everywhere except at $(0, 0)$, the given function is analytic everywhere except when Re $z \le 0$.

Example 10.39   Show that the function Ln $(z - i)$ is analytic everywhere except on the half line $y = 1$, $x \le 0$.

Solution   The function Ln $(z - i) = $ Ln $[x + (y - 1)i]$ is not continuous, when

$$\text{Im } [x + (y - 1)i] = 0 \quad \text{and} \quad \text{Re } [x + (y - 1) \, i] \le 0.$$

We get $y = 1$ and $x \le 0$. Therefore, the function is single valued and continuous for all $z = i + re^{i\theta}$, $r > 0$, $-\pi < \theta < \pi$. The point $z = i$ is the branch point and the line $y = 1$ is the branch cut.

We have

$$\text{Ln } (z - i) = \frac{1}{2} \ln [x^2 + (y - 1)^2] + i \tan^{-1}\left(\frac{y - 1}{x}\right) = u + iv.$$

Therefore,

$$u = \frac{1}{2} \ln [x^2 + (y - 1)^2] \quad \text{and} \quad v = \tan^{-1}\left(\frac{y - 1}{x}\right).$$

We find that

$$\frac{\partial u}{\partial x} = \frac{x}{x^2 + (y - 1)^2} = \frac{\partial v}{\partial y} \quad \text{and} \quad \frac{\partial u}{\partial y} = \frac{y - 1}{x^2 + (y - 1)^2} = -\frac{\partial v}{\partial x}.$$

The Cauchy-Riemann equations are satisfied. Since, the partial derivatives u_x, u_y, v_x, v_y are also continuous, the given function is analytic everywhere except on the half line $y = 1$, $x \le 0$.

Example 10.40   Find the constants, a, b, c such that the function $f(z)$, where

    (i) $f(z) = x - 2ay + i(bx - cy)$,   (ii) $f(z) = -x^2 + xy + y^2 + i(ax^2 + bxy + cy^2)$

is analytic. Express $f(z)$ in terms of z.

Solution   Let $f(z) = u(x, y) + iv(x, y)$. Since, the first order partial derivatives of $u(x, y)$ and $v(x, y)$ in the given problems are continuous everywhere, the functions will be analytic if the Cauchy-Riemann equations are satisfied.

    (i) $u(x, y) = x - 2ay$ and $v(x, y) = bx - cy$.

        Therefore, $u_x = 1$, $u_y = -2a$, $v_x = b$ and $v_y = -c$. Using the Cauchy-Riemann equations $u_x = u_y$ and $u_y = -v_x$ for all x, y, we get $c = -1$ and $b = 2a$, where a is arbitrary. The required analytic function is

$$f(z) = x - 2ay + i(bx - cy) = x - 2ay + i(2ax + y)$$

$$= (x + iy) + 2ai(x + iy) = (1 + 2ai)z.$$

    (ii) $u(x, y) = -x^2 + xy + y^2$ and $v(x, y) = ax^2 + bxy + cy^2$.

        Therefore, $u_x = -2x + y$, $u_y = x + 2y$, $v_x = 2ax + by$,   and $v_y = bx + 2cy$. Using the Cauchy-Riemann equations $u_x = v_y$ and $u_y = -v_x$, we obtain

$$-2x + y = bx + 2cy \quad \text{and} \quad x + 2y = -2ax - by.$$

Solving these equations, we get $b = -2$, $c = 1/2$ and $a = -1/2$.

The required analytic function becomes

$$f(z) = -x^2 + xy + y^2 + \frac{i}{2}(-x^2 - 4xy + y^2)$$

$$= -\frac{1}{2}[2(x^2 - xy - y^2) + i(x^2 + 4xy - y^2)]$$

$$= -\frac{1}{2}[2(x^2 + 2ixy - y^2) + i(x^2 + 2ixy - y^2)]$$

$$= -\frac{1}{2}[2z^2 + iz^2] = -\frac{1}{2}(2 + i)z^2.$$

Example 10.41 Show that if $f(z)$ is analytic and (i) Re $f(z)$ = constant, or (ii) Im $f(z)$ = constant, then $f(z)$ is a constant.

Solution Since the function $f(z) = u(x, y) + iv(x, y)$ is analytic, it satisfies the Cauchy-Riemann equations $u_x = u_y$ and $u_y = -v_x$.

(i) Re $f(z)$ = constant. Therefore, $u(x, y)$ = real constant = c_1. We get $u_x = 0$ and $u_y = 0$. Using the Cauchy-Riemann equations, we obtain $v_x = 0$ and $v_y = 0$. Hence, $v(x, y)$ = real constant = c_2. Therefore, $f(z) = u(x, y) + iv(x, y) = c_1 + ic_2$ is a complex constant.

(ii) Im $f(z)$ = constant. Therefore, $v(x, y)$ = real constant = α_2. We get $v_x = 0$ and $v_y = 0$. Using the Cauchy-Riemann equations, we obtain $u_x = 0$ and $u_y = 0$ Hence, $u(x, y)$ = real constant = α_1. Therefore, $f(z) = u(x, y) + iv(x, y) = \alpha_1 + i\alpha_2$ is a complex constant.

Example 10.42 If $f(z)$ is analytic in a domain D and $|f(z)|$ is a non-zero constant in D, then show that $f(z)$ is constant in D.

Solution Let $f(z) = u(x, y) + iv(x, y)$. Then, $|f(z)|$ = constant gives $u^2 + v^2 = c$, where c is a real constant. Differentiating partially with respect to x and y, we get

$$2u\, u_x + 2v\, v_x = 0 \quad \text{and} \quad 2u\, u_y + 2v\, v_y = 0.$$

Using the Cauchy-Riemann equations, we obtain

$$2uu_x - 2v\, u_y = 0 \tag{10.40}$$

and

$$2uu_y + 2v\, u_x = 0. \tag{10.41}$$

Multiplying Eq. (10.40) by u and Eq. (10.41) by v and adding, we get

$$2u^2 u_x + 2v^2 u_x = 0 \quad \text{or} \quad 2(u^2 + v^2)u_x = 2c\, u_x = 0 \quad \text{or} \quad u_x = 0, \text{ since } c \neq 0.$$

Hence, $u(x, y)$ is independent of x.

Multiplying Eq. (10.40) by v and Eq. (10.41) by u and subtracting, we get

$$2(u^2 + v^2)u_y = 2cu_y = 0 \quad \text{or} \quad u_y = 0, \text{ since } c \neq 0.$$

Hence, $u(x, y)$ is independent of y also. Therefore, $u(x, y) = \text{real constant} = k_1$.

Similarly, writing the equations in terms of v_x, v_y and using the Cauchy-Riemann equations, we get

$$v(x, y) = \text{real constant} = k_2.$$

Hence, we obtain $f(z) = k_1 + ik_2 = $ a complex constant.

Example 10.43   Show that the function

 (i) $f(z) = z/(z + 1)$ is analytic at $z = \infty$.

 (ii) $f(z) = z$ is not analytic at $z = \infty$.

Solution   The function $f(z)$ is analytic at $z = \infty$, if the function $g(z) = f(1/z)$ is analytic at $z = 0$.

 (i) For $f(z) = \dfrac{z}{z+1}$, $g(z) = f\left(\dfrac{1}{z}\right) = \dfrac{1/z}{1 + 1/z} = \dfrac{1}{1+z}$.

   Now, $g(z)$ is differentiable at $z = 0$ and at all points in its neighborhood. Therefore, the function $g(z)$ is analytic at $z = 0$. Hence, the given function $f(z)$ is analytic at $z = \infty$.

 (ii) For $f(z) = z$, $g(z) = f(1/z) = 1/z$.

   Since the function $g(z)$ is not defined at $z = 0$, it is not continuous at $z = 0$. Therefore, $g(z)$ is not differentiable and hence not analytic at $z = 0$. The given function is not analytic at $z = \infty$.

Example 10.44   If $f(z) = u + iv$ is an analytic function of $z = x + iy$ and

$$u + v = (x + y)(2 - 4xy + x^2 + y^2),$$

then, find u, v and the analytic function $f(z)$.

Solution   We have

$$u + v = (x + y)(2 - 4xy + x^2 + y^2) = 2x + 2y - 3x^2y - 3xy^2 + x^3 + y^3.$$

Therefore,

$$u_x + v_x = 2 - 6xy - 3y^2 + 3x^2 \tag{10.42}$$

and

$$u_y + v_y = 2 - 6xy - 3x^2 + 3y^2. \tag{10.43}$$

Using the Cauchy-Riemann equations $u_x = v_y$ and $u_y = -v_x$, we obtain from Eq. (10.43)

$$-v_x + u_x = 2 - 6xy - 3x^2 + 3y^2. \tag{10.44}$$

From Eqs. (10.42) and (10.44), we get

$$u_x = 2 - 6xy = v_y \quad \text{and} \quad v_x = 3x^2 - 3y^2 = -u_y. \tag{10.45}$$

Integrating u_x and v_x partially with respect to x, we get respectively

$$u(x, y) = 2x - 3x^2y + \phi_1(y) \quad \text{and} \quad v(x, y) = x^3 - 3xy^2 + \phi_2(y) \quad (10.46)$$

where $\phi_1(y)$ and $\phi_2(y)$ are arbitrary functions of y. Substituting in Eq. (10.45), we get

$$u_y = -3x^2 + \phi_1'(y) = -v_x = -3x^2 + 3y^2, \quad \text{or} \quad \phi_1'(y) = 3y^2.$$

and

$$v_y = -6xy + \phi_2'(y) = u_x = 2 - 6xy, \quad \text{or} \quad \phi_2'(y) = 2.$$

Integrating, we obtain

$$\phi_1(y) = y^3 + c_1 \quad \text{and} \quad \phi_2(y) = 2y + c_2$$

where c_1 and c_2 are arbitrary constants. Hence, we get

$$u(x, y) = 2x - 3x^2y + y^3 + c_1, \quad v(x, y) = x^3 - 3xy^2 + 2y + c_2$$

and

$$f(z) = u(x, y) + iv(x, y) = 2x - 3x^2y + y^3 + c_1 + i(2y - 3xy^2 + x^3 + c_2)$$

$$= 2(x + iy) + i[x^3 - 3xy^2 - i(y^3 - 3x^2y)] + c_1 + ic_2$$

$$= 2(x + iy) + i(x + iy)^3 + c_1 + ic_2 = 2z + iz^3 + c$$

where $c = c_1 + ic_2$ is a complex constant.

Example 10.45   If $f(z) = u + iv$ is an analytic function of $z = x + iy$ and

$$u - v = e^{-x}[(x - y)\sin y - (x + y)\cos y]$$

then, find u, v and the analytic function $f(z)$.

Solution   From $u - v = e^{-x}[(x - y)\sin y - (x + y)\cos y]$, we get

$$u_x - v_x = -e^{-x}[(x - y)\sin y - (x + y)\cos y] + e^{-x}(\sin y - \cos y)$$

$$= e^{-x}[(1 - x + y)\sin y + (x + y - 1)\cos y] \quad (10.47)$$

$$u_y - v_y = e^{-x}[x\cos y - (\sin y + y\cos y) + x\sin y - (\cos y - y\sin y)]$$

$$= e^{-x}[(x + y - 1)\sin y + (x - y - 1)\cos y]. \quad (10.48)$$

Using the Cauchy-Riemann equations, $u_x = v_y$ and $u_y = -v_x$, we obtain from Eq. (10.48)

$$u_x + v_x = e^{-x}[(1 - x - y)\sin y + (1 - x + y)\cos y]. \quad (10.49)$$

From Eqs. (10.47) and (10.49), we obtain

$$u_x = e^{-x}[(1 - x)\sin y + y\cos y] = v_y \quad (10.50)$$

and

$$v_x = e^{-x}[(1 - x)\cos y - y\sin y] = -u_y. \quad (10.51)$$

Integrating u_x and v_x partially with respect to x, we get respectively

$$u(x, y) = e^{-x}[x\sin y - y\cos y] + \phi_1(y)$$

and

$$v(x, y) = e^{-x}[y\sin y + x\cos y] + \phi_2(y)$$

where $\phi_1(y)$ and $\phi_2(y)$ are arbitrary functions of y. Substituting in Eqs. (10.50) and (10.51), we obtain respectively

$$e^{-x}[(1 - x) \sin y + y \cos y] = e^{-x}[\sin y + y \cos y - x \sin y] + \phi_2'(y)$$

and $e^{-x}[(1 - x) \cos y - y \sin y] = - e^{-x}[x \cos y - \cos y + y \sin y] + \phi_1'(y).$

Simplifying, we obtain

$$\phi_2'(y) = 0 \quad \text{and} \quad \phi_1'(y) = 0$$

which give $\phi_1(y) = c_1$ and $\phi_2(y) = c_2$, where c_1 and c_2 are arbitrary constants.

Hence, we have

$$u(x, y) = e^{-x}(x \sin y - y \cos y) + c_1, \; v(x, y) = e^{-x}(y \sin y + x \cos y) + c_2$$

and
$$f(z) = u(x, y) + iv(x, y)$$

$$= e^{-x}(x \sin y - y \cos y) + c_1 + i[(e^{-x}(y \sin y + x \cos y) + c_2]$$

$$= e^{-x}[(x + iy) \sin y + i(x + iy) \cos y] + c_1 + ic_2$$

$$= ie^{-x}(x + iy) (\cos y - i \sin y) + c_1 + ic_2$$

$$= i(x + iy)e^{-x} e^{-iy} + c_1 + ic_2$$

$$= i(x + iy) e^{-(x+iy)} + c_1 + ic_2 = ize^{-z} + c.$$

where $c = c_1 + ic_2$ is a complex constant.

Polar form of the Cauchy-Riemann equations

Let
$$f(z) = u(r, \theta) + iv(r, \theta), \; z = re^{i\theta}.$$

We have
$$x = r \cos \theta, \; y = r \sin \theta, \; r = \sqrt{x^2 + y^2}, \; \theta = \tan^{-1}(y/x).$$

Using the chain rule of differentiation, we get

$$\frac{\partial u}{\partial r} = \frac{\partial u}{\partial x} \frac{\partial x}{\partial r} + \frac{\partial u}{\partial y} \frac{\partial y}{\partial r} = (\cos \theta) \frac{\partial u}{\partial x} + (\sin \theta) \frac{\partial u}{\partial y} \qquad (10.52)$$

$$\frac{\partial u}{\partial \theta} = \frac{\partial u}{\partial x} \frac{\partial x}{\partial \theta} + \frac{\partial u}{\partial y} \frac{\partial y}{\partial \theta} = (-r \sin \theta) \frac{\partial u}{\partial x} + (r \cos \theta) \frac{\partial u}{\partial y} \qquad (10.53)$$

$$\frac{\partial v}{\partial r} = \frac{\partial v}{\partial x} \frac{\partial x}{\partial r} + \frac{\partial v}{\partial y} \frac{\partial y}{\partial r} = (\cos \theta) \frac{\partial v}{\partial x} + (\sin \theta) \frac{\partial v}{\partial y} \qquad (10.54)$$

$$\frac{\partial v}{\partial \theta} = \frac{\partial v}{\partial x} \frac{\partial x}{\partial \theta} + \frac{\partial v}{\partial y} \frac{\partial y}{\partial \theta} = (-r \sin \theta) \frac{\partial v}{\partial x} + (r \cos \theta) \frac{\partial v}{\partial y}. \qquad (10.55)$$

Using the Cauchy-Riemann equations in cartesian coordinates, $u_x = v_y$ and $u_y = - v_x$, we can write Eqs. (10.54) and (10.55) as

$$\frac{\partial v}{\partial r} = - (\cos \theta) \frac{\partial u}{\partial y} + (\sin \theta) \frac{\partial u}{\partial x}$$

$$= -\frac{1}{r}\left[(-r\sin\theta)\frac{\partial u}{\partial x} + (r\cos\theta)\frac{\partial u}{\partial y}\right] = -\frac{1}{r}\frac{\partial u}{\partial \theta}$$

and
$$\frac{\partial v}{\partial \theta} = (-r\sin\theta)\left(-\frac{\partial u}{\partial y}\right) + (r\cos\theta)\frac{\partial u}{\partial x}$$

$$= r\left[(\cos\theta)\frac{\partial u}{\partial x} + (\sin\theta)\frac{\partial u}{\partial y}\right] = r\frac{\partial u}{\partial r}.$$

Therefore, the Cauchy-Riemann equations in polar coordinates are

$$\frac{\partial v}{\partial r} = -\frac{1}{r}\frac{\partial u}{\partial \theta} \quad \text{and} \quad \frac{\partial v}{\partial \theta} = r\frac{\partial u}{\partial r} \quad \text{or} \quad \frac{1}{r}\frac{\partial v}{\partial \theta} = \frac{\partial u}{\partial r}. \tag{10.56}$$

We also have

$$\frac{\partial u}{\partial x} = \frac{\partial u}{\partial r}\frac{\partial r}{\partial x} + \frac{\partial u}{\partial \theta}\frac{\partial \theta}{\partial x} = \left(\frac{x}{\sqrt{x^2+y^2}}\right)\frac{\partial u}{\partial r} + \left(\frac{-y}{x^2+y^2}\right)\frac{\partial u}{\partial \theta}$$

$$= (\cos\theta)\frac{\partial u}{\partial r} + \frac{1}{r}(-\sin\theta)\frac{\partial u}{\partial \theta}$$

$$\frac{\partial v}{\partial x} = \frac{\partial v}{\partial r}\frac{\partial r}{\partial x} + \frac{\partial v}{\partial \theta}\frac{\partial \theta}{\partial x} = \left(\frac{x}{\sqrt{x^2+y^2}}\right)\frac{\partial v}{\partial r} + \left(\frac{-y}{x^2+y^2}\right)\frac{\partial v}{\partial \theta}$$

$$= (\cos\theta)\frac{\partial v}{\partial r} + \frac{1}{r}(-\sin\theta)\frac{\partial v}{\partial \theta}.$$

Now,
$$f'(z) = \frac{\partial u}{\partial x} + i\frac{\partial v}{\partial x} = \cos\theta\left(\frac{\partial u}{\partial r} + i\frac{\partial v}{\partial r}\right) - \frac{\sin\theta}{r}\left(\frac{\partial u}{\partial \theta} + i\frac{\partial v}{\partial \theta}\right).$$

Using the Cauchy-Riemann equations in polar coordinates, we get.

$$f'(z) = \cos\theta\left(\frac{\partial u}{\partial r} + i\frac{\partial v}{\partial r}\right) - \frac{\sin\theta}{r}\left(-r\frac{\partial v}{\partial r} + ir\frac{\partial u}{\partial r}\right)$$

$$= (\cos\theta - i\sin\theta)\frac{\partial u}{\partial r} + (i\cos\theta + \sin\theta)\frac{\partial v}{\partial r}$$

$$= (\cos\theta - i\sin\theta)\left(\frac{\partial u}{\partial r} + i\frac{\partial v}{\partial r}\right) = e^{-i\theta}\left(\frac{\partial u}{\partial r} + i\frac{\partial v}{\partial r}\right) \tag{10.57}$$

We can also write $f'(z) = \frac{1}{r}e^{-i\theta}\left(\frac{\partial v}{\partial \theta} - i\frac{\partial u}{\partial \theta}\right).$ \hfill (10.58)

Example 10.46 Write $f(z) = z^n$, n any positive integer, in polar form and verify and that the Cauchy-Riemann equations are satisfied. Hence, show that the function $f(z)$ is differentiable and $f'(z) = nz^{n-1}$.

Solution Let $z = re^{i\theta}$, $r \neq 0$. We get

$$z^n = r^n \, e^{in\theta} = r^n(\cos n\theta + i \sin n\theta) = u(r, \theta) + iv(r, \theta).$$

Therefore, we have

$$u(r, \theta) = r^n \cos n\theta \quad \text{and} \quad v(r, \theta) = r^n \sin n\theta.$$

We have
$$u_r = nr^{n-1} \cos n\theta = \frac{1}{r} \, v_\theta \quad \text{and} \quad u_\theta = -nr^n \sin n\theta = -rv_r.$$

Hence, the Cauchy-Riemann equations as given in Eq. (10.56) are satisfied. We also note that the first order partial derivatives are continuous for all r, θ. Therefore, the function $f(z)$ is differentiable for all z. We obtain

$$f'(z) = (u_r + iv_r) \, (\cos \theta - i \sin \theta)$$

$$= (nr^{n-1} \cos n\theta + inr^{n-1} \sin n\theta) \, (\cos \theta - i \sin \theta)$$

$$= nr^{n-1} \, (\cos n\theta + i \sin n\theta) \, (\cos \theta - i \sin \theta)$$

$$= nr^{n-1} \, e^{in\theta} \, e^{-i\theta} = nr^{n-1} \, e^{(n-1)i\theta} = nz^{n-1}$$

where n is any positive integer.

Exercise 10.5

Using the definition of derivatives, obtain $f'(z)$, if it exists, for the following functions $f(z)$.

1. Re z.

2. $z^2 + 1$.

3. iz^3.

4. $\bar{z}^2$.

5. $\dfrac{1}{z}$, $z \neq 0$.

6. $\dfrac{1+z}{1-z}$, $z \neq 1$.

Using the rules of differentiation, obtain $f'(z)$ at the indicated point, for the following functions $f(z)$.

7. $(z^2 - i)^2$ at $z = 1$.

8. ze^z at $z = \pi i$.

9. $(z + 1)^2 \, (z^3 + 2)$ at $z = 0$.

10. $\dfrac{1}{z^3}$ at $z = i$.

11. $\dfrac{z-1}{z+1}$ at $z = -1 - i$.

12. $\dfrac{z^2 - 2z}{z^4 + 1}$ at $z = -i$.

Determine all points (if any) at which the Cauchy-Riemann equations are satisfied and determine all points at which the following functions are differentiable.

13. $z(\text{Im } z)$.

14. $z(\text{Re } z)$.

15. arg (z), $z \neq 0$.

16. z (arg z), $z \neq 0$.

17. z^3.

18. $i \, |z|^2$.

19. $\dfrac{z}{\text{Re} z}$, $x \neq 0$.

20. $\dfrac{\bar{z}}{|z|^2}$, $z \neq 0$.

21. $\dfrac{z + 2i}{1 - iz}$, $z \neq -i$.

22. $(x + i/x)\, e^{xy}$, $x \neq 0$.   23. $e^x(\sin y - i \cos y)$.

24. $\dfrac{1}{2}\, \ln (x^2 + y^2) + i \cot^{-1} (y/x)$, $z \neq 0$.   25. $e^{x^2 - y^2} [\cos 2xy + i \sin 2xy]$.

26. Show that the function $f(z) = |z|^4$ satisfies the Cauchy-Riemann equations only at the origin.

27. Show that for the function

$$f(z) = \begin{vmatrix} (\bar{z})^2/z, & z \neq 0 \\ 0, & z = 0 \end{vmatrix}$$

the Cauchy-Riemann equations are satisfed at the origin. Does $f'(0)$ exist?

28. Show that for the function $f(z) = |xy|^{1/2}$, the Cauchy-Riemann equations are satisfied at the origin. Does $f'(0)$ exist?

29. Show that if $f(z)$ is differentiable at a point z, then

$$|f'(z)|^2 = \begin{vmatrix} u_x & u_y \\ v_x & v_y \end{vmatrix}.$$

30. Show that for the function $w = f(z) = u + iv$, if $\displaystyle\lim_{\Delta z \to 0} \left[\mathrm{Re} \dfrac{\Delta w}{\Delta z} \right]$ exists, then the partial derivatives u_x and v_y exist and $u_x = v_y$.

31. Show that for the function $w = f(z) = u + iv$, if $\displaystyle\lim_{\Delta z \to 0} \left[\mathrm{Im} \dfrac{\Delta w}{\Delta z} \right]$ exists, then the partial derivatives u_y, v_x exist and $u_y = -v_x$.

32. If $f(z) = u(x, y) + iv(x, y)$, where $x = (z + \bar{z})/2$, $y = (z - \bar{z})/(2i)$, is continuous as a function of two variables z and $\bar{z}$, then show that

    (i) $\dfrac{\partial^2 u}{\partial x^2} + \dfrac{\partial^2 u}{\partial y^2} = 4 \dfrac{\partial^2 u}{\partial z \partial \bar{z}}$,

    (ii) $\dfrac{\partial f}{\partial \bar{z}} = 0$ is equivalent to the Cauchy-Riemann equations.

33. If $f'(z) = f(z)$ for all z, then show that $f(z) = ke^z$, where k is an arbitrary constant.

34. Find $f'(z)$ when $f(z) = (r^2 \cos 2\theta + r \cos \theta) + i(r^2 \sin 2\theta + r \sin \theta)$.

35. Find $f'(z)$ when $f(z) = \left(r + \dfrac{1}{r} \right) \cos \theta + i \left(r - \dfrac{1}{r} \right) \sin \theta$, $z \neq 0$.

36. Using polar coordinates, show that the function $f(z) = z^{-2}$ is differentiable at any point $z \neq 0$ and $f'(z) = -2/z^3$.

Where (if anywhere) are the following functions analytic?

37. $(x - iy)^2$.

38. $(x^3 + xy) + i(y^3 - 3x)$.

39. $iz - |z|^2$.

40. $(z + 1)/(z + i)$.

41. e^{z^2}.

42. $\bar{z} e^{z^2}$.

43. $e^{1/z}$.

44. $e^{\bar{z}}$.

45. $z/(e^{\bar{z}} - 1)$.

Find the values of the constants a, b, c, d such that the following functions are analytic.

46. $f(z) = x^2 + axy + by^2 + i(cx^2 + dxy + y^2)$.

47. $f(z) = \cos x \, (\cosh y + a \sinh y) + i \sin x \, (\cosh y + b \sinh y)$.

48. Show that an analytic function $f(z)$ whose derivative is identically zero is constant.

In problems 49 and 50, determine the analytic function $f(z) = u + iv$, where

49. $u + v = e^x \, (\cos y + \sin y)$.

50. $u - v = e^{-2xy} \sin (x^2 - y^2) + e^{2xy} \cos (x^2 - y^2)$.

51. Find an analytic function $f(z)$ such that $\text{Re}[\, f'(z)] = 3x^2 - 4y - 3y^2$ and $f(1 + i) = 0$.

52. Let $f(z) = u + iv$ and $g(z) = v + iu$ be analytic functions for all z. Let $f(0) = 1$ and $g(0) = i$. Obtain the value of $h(z)$ at $z = 1 + i$, where $h(z) = f'(z) + g'(z) + 2f(z) \, g(z)$.

53. If the function $w = f(z) = u(r, \theta) + iv(r, \theta)$ is analytic, then show that $\dfrac{dw}{dz} = e^{-i\theta} \dfrac{\partial w}{\partial r}$.

54. It is given that a function $f(z)$ and its conjugate $\overline{f(z)}$ are both analytic. Determine the function $f(z)$.

55. It $f(z) = u + iv$ is an analytic function of z and ϕ is a function of u and v, then show that

$$\left(\frac{\partial \phi}{\partial x}\right)^2 + \left(\frac{\partial \phi}{\partial y}\right)^2 = \left[\left(\frac{\partial \phi}{\partial u}\right)^2 + \left(\frac{\partial \phi}{\partial v}\right)^2\right] |f'(z)|^2.$$

10.7 Harmonic Functions

A real valued function $\phi(x, y)$ of two variables x and y that has continuous second order partial derivatives in a domain D and satisfies the *Laplace* equation

$$\frac{\partial^2 \phi}{\partial x^2} + \frac{\partial^2 \phi}{\partial y^2} = 0$$

is said to be *harmonic* in D.

We now prove the harmonic property of the real and imaginary parts of an analytic function.

Theorem 10.10   If $f(z) = u(x, y) + iv(x, y)$ is analytic in a domain D, then the real valued functions $u(x, y)$ and $v(x, y)$ satisfy the Laplace equation

$$u_{xx} + u_{yy} = 0 \quad \text{and} \quad v_{xx} + v_{yy} = 0$$

respectively in D, that is $u(x, y)$ and $v(x, y)$ are harmonic in D.

Proof Assume that $u(x, y)$ and $v(x, y)$ have continuous second order partial derivatives so that $u_{xy} = u_{yx}$ and $v_{xy} = v_{yx}$. Since the function $f(z)$ is analytic, it satisfies the Cauchy-Riemann equations $u_x = v_y$ and $u_y = -v_x$. Differentiating $u_x = v_y$ with respect to x, and $u_y = -v_x$ with respect to y, we get

$$u_{xx} = v_{yx} \quad \text{and} \quad u_{yy} = -v_{xy}.$$

Adding these two equations, we get $u_{xx} + u_{yy} = 0$. Similarly, differentiating $v_y = u_x$ with respect to y and $v_x = -u_y$ with respect to x, we get

$$v_{yy} = u_{xy} \quad \text{and} \quad v_{xx} = -u_{yx}.$$

Adding these two equations, we get $v_{xx} + v_{yy} = 0$.

Thus, if $f(z) = u(x, y) + iv(x, y)$ is analytic in a domain D, then its real and imaginary parts $u(x, y)$ and $v(x, y)$ are harmonic functions. The function $v(x, y)$ is called the *conjugate harmonic function* of $u(x, y)$ in D. The conjugate harmonic function $v(x, y)$ for a given $u(x, y)$ can be obtained by using the Cauchy-Riemann equations. The conjugate harmonic function is unique except for an additive constant.

Remark 10

The converse of Theorem 10.10 may not hold. That is, if $u(x, y)$ and $v(x, y)$ are any two harmonic functions in D, then $u + iv$ need not be analytic in D. However, $(u_y - v_x) + i(u_x + v_y)$ is analytic in D. For example, consider

$$u(x, y) = \text{Re } z^2 = x^2 - y^2 \quad \text{and} \quad v(x, y) = \text{Im } z^3 = 3x^2y - y^3.$$

It can be verified that $u(x, y)$ and $v(x, y)$ are harmonic functions, but $u + iv$ is not analytic as it does not satisfy the Cauchy-Riemann equations except at the origin.

Now, let $\qquad\qquad U = u_y - v_x \quad \text{and} \quad V = u_x + v_y,$

Since $\qquad\qquad\qquad U_x = u_{xy} - v_{xx}, \quad U_y = u_{yy} - v_{xy},$

and $\qquad\qquad\qquad V_x = u_{xx} + v_{xy}, \quad V_y = u_{xy} + v_{yy},$

we get $\qquad\qquad\qquad U_x = V_y \quad \text{as} \quad v_{xx} + v_{yy} = 0,$

and $\qquad\qquad\qquad U_y = -V_x \quad \text{as} \quad u_{xx} + u_{yy} = 0.$

Therefore, $U + iV$ is analytic.

Example 10.47 Show that the function $u(x, y) = 2x + y^3 - 3x^2y$ is harmonic. Find its conjugate harmonic function $v(x, y)$ and the corresponding analytic function $f(z)$.

Solution We have $u(x, y) = 2x + y^3 - 3x^2y$.

Therefore, $\qquad u_x = 2 - 6xy, \quad u_{xx} = -6y, \quad u_y = 3y^2 - 3x^2, \quad u_{yy} = 6y.$

Since $u_{xx} + u_{yy} = -6y + 6y = 0$, the function $u(x, y)$ is harmonic. From the Cauchy-Riemann equation $v_y = u_x$, we get

$$v_y = u_x = 2 - 6xy.$$

Integrating partially with respect to y, we obtain

$$v = 2y - 3xy^2 + \phi(x)$$

where $\phi(x)$ is an arbitrary function of x. Using the Cauchy-Riemann equation $v_x = -u_y$, we get

$$-3y^2 + \phi'(x) = -(3y^2 - 3x^2) \quad \text{or} \quad \phi'(x) = 3x^2.$$

Integrating with respect to x, we obtain

$$\phi(x) = x^3 + c, \quad \text{where } c \text{ is a constant.}$$

Hence, we obtain

$$v(x, y) = 2y - 3xy^2 + x^3 + c$$

and

$$f(z) = u(x, y) + iv(x, y)$$

$$= 2x + y^3 - 3x^2y + i(2y - 3xy^2 + x^3 + c)$$

$$= 2(x + iy) + y^3 - 3x^2y + i(x^3 - 3xy^2) + ic$$

$$= 2(x + iy) + i[x^3 - 3xy^2 - i(y^3 - 3x^2y)] + ic$$

$$= 2(x + iy) + i(x + iy)^3 + ic = 2z + iz^3 + ic.$$

Example 10.48   Show that the function $v(x, y) = e^x \sin y$ is harmonic. Find its conjugate harmonic function $u(x, y)$ and the corresponding analytic function $f(z)$.

Solution   We have $v(x, y) = e^x \sin y$.

Therefore,

$$v_x = e^x \sin y, \quad v_{xx} = e^x \sin y, \quad v_y = e^x \cos y, \quad v_{yy} = -e^x \sin y.$$

Since $v_{xx} + v_{yy} = 0$, the given function $v(x, y)$ is harmonic. From the Cauchy-Riemann equation $u_x = v_y$, we get

$$u_x = v_y = e^x \cos y.$$

Integrating partially with respect to x, get $u(x, y) = e^x \cos y + \phi(y)$ where $\phi(y)$ is an arbitrary function of y. Using the Cauchy-Riemann equation $u_y = -v_x$, we get

$$-e^x \sin y + \phi'(y) = -e^x \sin y, \quad \text{or} \quad \phi'(y) = 0.$$

Integrating with respect to y, we obtain $\phi(y) = c$, c any arbitrary constant.

Hence, we obtain $u(x, y) = e^x \cos y + c$

and

$$f(z) = u(x, y) + iv(x, y) = e^x \cos y + ie^x \sin y + c$$

$$= e^x(\cos y + i \sin y) + c = e^x e^{iy} + c = e^{x+iy} + c = e^z + c.$$

Example 10.49 Show that the real and imaginary parts of an analytic function, $f(z) = u(r, \theta) + iv(r, \theta)$, satisfy the Laplace equation in polar form

$$\frac{\partial^2 u}{\partial r^2} + \frac{1}{r}\frac{\partial u}{\partial r} + \frac{1}{r^2}\frac{\partial^2 u}{\partial \theta^2} = 0 \quad \text{and} \quad \frac{\partial^2 v}{\partial r^2} + \frac{1}{r}\frac{\partial v}{\partial r} + \frac{1}{r^2}\frac{\partial^2 v}{\partial \theta^2} = 0 \tag{10.59}$$

respectively.

Solution   Assume that $u(r, \theta)$ and $v(r, \theta)$ have continuous second order partial derivatives so that $u_{r\theta} = u_{\theta r}$ and $v_{r\theta} = v_{\theta r}$. Since the function $f(z)$ is analytic, it satisfies the Cauchy-Riemann equations

$$u_r = \frac{1}{r}v_\theta \quad \text{and} \quad u_\theta = -rv_r.$$

From $u_r = \frac{1}{r}v_\theta$ or $v_\theta = ru_r$, we get, $v_{\theta r} = u_r + ru_{rr}$.

From $u_\theta = -r\,v_r$ or $v_r = -\frac{1}{r}u_\theta$, we get, $v_{r\theta} = -\frac{1}{r}u_{\theta\theta}$.

Since $v_{\theta r} = v_{r\theta}$, we obtain

$$u_r + ru_{rr} = -\frac{1}{r}\,u_{\theta\theta} \quad \text{or} \quad u_{rr} + \frac{1}{r}\,u_r + \frac{1}{r^2}\,u_{\theta\theta} = 0.$$

Similarly, differentiating u_r with respect to θ, u_θ with respect to r and equating the resulting equations, we obtain the second result.

Example 10.50   Show that the function $u(r, \theta) = r^2 \cos 2\theta$ is harmonic. Find its conjugate harmonic function and the corresponding analytic function $f(z)$.

Solution   We have $z = re^{i\theta}$ and $u(r, \theta) = r^2 \cos 2\theta$.

Therefore, $u_r = 2r \cos 2\theta$, $u_{rr} = 2 \cos 2\theta$, $u_\theta = -2r^2 \sin 2\theta$, $u_{\theta\theta} = -4r^2 \cos 2\theta$.

Since $u_{rr} + \frac{1}{r}\,u_r + \frac{1}{r^2}\,u_{\theta\theta} = 2 \cos 2\theta + 2 \cos 2\theta - 4 \cos 2\theta = 0,$

the function $u(r, \theta)$ satisfies the Laplace equation in polar form and therefore is harmonic. From the Cauchy-Riemann equation $v_\theta = ru_r$, we get

$$v_\theta = ru_r = 2r^2 \cos 2\theta.$$

Integrating partially with respect to θ, we obtain

$$v(r, \theta) = r^2 \sin 2\theta + \phi(r)$$

where $\phi(r)$ is an arbitrary function of r. Using the Cauchy-Riemann equation $v_r = -\frac{1}{r}u_\theta$, we get

$$2r \sin 2\theta + \phi'(r) = 2r \sin 2\theta \quad \text{or} \quad \phi'(r) = 0.$$

Integrating with respect to r, we get $\phi(r) = c$, where c is an arbitrary real constant. Hence, we obtain

$$v(r, \theta) = r^2 \sin 2\theta + c.$$

Therefore,
$$f(z) = u(r, \theta) + iv\ (r, \theta) = r^2 \cos 2\theta + i[r^2 \sin 2\theta + c]$$
$$= r^2 [\cos 2\theta + i \sin 2\theta] + ic$$
$$= r^2 e^{2i\theta} + ic = z^2 + ic.$$

Example 10.51 If u is a harmonic function, then show that $w = u^2$ is not a harmonic function, unless u is a constant.

Solution Since u is an harmonic function, $u_{xx} + u_{yy} = 0$. Now, $w = u^2$ will be a harmonic function if $w_{xx} + w_{yy} = 0$. We have

$$w_x = 2uu_x, \quad w_y = 2uu_y \quad \text{and} \quad w_{xx} = 2\ u_x^2 + 2uu_{xx}, \quad w_{yy} = 2u_y^2 + 2uu_{yy}.$$

Therefore,
$$w_{xx} + w_{yy} = 2(u_x^2 + u_y^2) + 2u(u_{xx} + u_{yy}) = 2(u_x^2 + u_y^2).$$

Now, $w_{xx} + w_{yy} = 0$, only when $u_x = 0$ and $u_y = 0$ that is u is a constant function. Hence, $w = u^2$ is not a harmonic function unless u is a constant.

Example 10.52 If $f(z)$ is an analytic function of z, then show that

$$\nabla^2[\text{Re}(f(z))]^p = p(p - 1)\ [\text{Re}\ f(z)]^{p-2}\ |f'(z)|^2, \quad p > 1$$

where
$$\nabla^2 = \frac{\partial^2}{\partial x^2} + \frac{\partial^2}{\partial y^2} \text{ is the Laplacian operator.}$$

Solution Since $f(z) = u + iv$ is analytic, we have $u_x = v_y$, $u_y = -v_x$, $u_{xx} + u_{yy} = 0$, $v_{xx} + v_{yy} = 0$ and $f'(z) = u_x + iv_x$. Now, $\text{Re}[f(z)] = u$.

Therefore,
$$\nabla^2[\text{Re}\ (f(z))]^p = \nabla^2[u^p] = \frac{\partial^2}{\partial x^2}\ (u^p) + \frac{\partial^2}{\partial y^2}\ (u^p)$$

$$= \frac{\partial}{\partial x}\ (pu^{p-1}u_x) + \frac{\partial}{\partial y}\ (pu^{p-1}u_y)$$

$$= p[(p - 1)\ u^{p-2}\ (u_x)^2 + u^{p-1}\ u_{xx}] + p[(p - 1)u^{p-2}\ (u_y)^2 + u^{p-1}\ u_{yy}]$$

$$= p(p - 1)u^{p-2}\ \left(u_x^2 + u_y^2\right) + pu^{p-1}\ (u_{xx} + u_{yy})$$

$$= p(p - 1)u^{p-2}\ \left(u_x^2 + u_y^2\right) = p(p - 1)\ [\text{Re}\ f(z)]^{p-2}\ |f'(z)|^2, p > 1.$$

Level curves

Let $w = f(z) = u(x, y) + iv\ (x, y)$ be an analytic function. Then, $u(x, y)$ and $v(x, y)$ are harmonic functions. The curves $u(x, y) = p$, $v(x, y) = q$, where p and q are real constants, are called *level curves* generated by the analytic function $f(z)$. For example, if $f(z) = z$, then $u(x, y) = x$ and $v(x, y) = y$. The level curves are given by $x = p$ and $y = q$ which are straight lines parallel to the y- and the *x*-axis, respectively. The interesting property of these level curves $u(x, y) = p$ and $v(x, y) = q$ is that they form an orthogonal family of curves.

From $u(x, y) = p$, we get $u_x + \dfrac{dy}{dx}\, u_y = 0$.

Therefore, m_1 = slope of the curve $u(x, y) = p$ is $- u_x/u_y$.

From $v(x, y) = q$, we get $v_x + \dfrac{dy}{dx}\, v_y = 0$.

Therefore, m_2 = slope of the curve $v(x, y) = q$ is $- v_x/v_y$.
Using the Cauchy-Riemann equations, we have

$$m_1 m_2 = \frac{u_x v_x}{u_y v_y} = - 1.$$

Hence, the level curves are orthogonal to each other.

Exercise 10.6

1. If $u(x, y) = \mathrm{Re}(z^2) = x^2 - y^2 =$ and $v(x, y) = \mathrm{Im}\,(z^3) = 3x^2 y - y^3$ are harmonic functions in a domain D, show that the function $(u_y - v_x) + i(u_x + v_y)$ is analytic in D.

2. If $f(z)$ is an analytic function, show that $|\,f(z)|$ is not a harmonic function.

3. If $f(z)$ is an analytic function, show that $\log |\,f(z)|$ is a harmonic function.

4. If $f(z)$ is an analytic, function, then show that $w = \arg [\,f(z)]$ is a harmonic function.

5. Show that $u(x^2 - y^2, 2xy)$ is harmonic, if and only if $u(x, y)$ is a harmonic function.

6. If $f(z)$ is an analytic function of z, then show that

$$\nabla^2 \left[|f(z)|^p\right] = p^2 \,|\,f(z)|^{p-2}\,|f'(z)|^2$$

where $\qquad \nabla^2 = \dfrac{\partial^2}{\partial x^2} + \dfrac{\partial^2}{\partial y^2}.$

Show that the given function $u(x, y)$ is harmonic. Find the corresponding conjugate harmonic function $v(x, y)$ and construct the analytic function $f(z) = u + iv$.

7. $y^3 - 3x^2 y.$

8. $\ln (x^2 + y^2)$, $\mathrm{Re}\, z > 0$.

9. $x/(x^2 + y^2)$, $z \neq 0$.

10. $e^x \cos y.$

11. $\sinh x \cos y.$

12. $e^{-x}[x \cos y + y \sin y].$

13. $e^x[(x^2 - y^2)\cos y - 2xy \sin y].$

Show that the given function $v(x, y)$ is harmonic. Find the corresponding conjugate harmonic function $u(x, y)$ and construct the analytic function $f(z) = u + iv$.

14. $y + 3x^2 y - y^3.$

15. $\ln (x^2 + y^2) + x + y$, $z \neq 0$.

16. $(x - y)/(x^2 + y^2)$, $z \neq 0$.

17. $\arg (z)$, $z \neq 0$.

18. $- \sin x \sinh y.$

19. $e^{-x}(y \cos y - x \sin y).$

Show that the given function is harmonic. Find the corresponding conjugate harmonic function and construct the analytic function $f(z) = u + iv$.

20. $u = \ln r$.

21. $u = (r + (1/r)) \cos \theta, r \neq 0$.

22. $u = r^n \cos n\theta$.

23. $v = 3r^2 \sin 2\theta - 2r \sin \theta$.

24. Consider the analytic function $f(z) = z^2$. Find its level curves. Show that these curves are mutually orthogonal.

25. Consider the analytic function $f(z) = e^z$, $z \neq 0$. Find its level curves. Show that these curves are mutually orthogonal.

10.8 Answers and Hints

Exercise 10.1

1. Right-half plane $x > 0$ (excluding the y-axis); it is open, connected and unbounded; defines a domain.

2. Left-half plane $x \leq 0$ (including the y-axis); it is not open, connected and unbounded; not a domain.

3. Upper-half plane $y \geq 0$ (including the x-axis); it is not open, connected and unbounded; not a domain.

4. Lower-half plane $y < 0$ (excluding the x-axis); it is open, connected and unbounded; defines a domain.

5. First quadrant $(x \geq 0, y > 0)$, (including the y-axis and excluding the x-axis); it is connected, unbounded and not open; not a domain.

6. $-2 < x < 6$; it is open, connected and unbounded; defines a domain.

7. $-3 < y < 4$; it is open, connected and unbounded; defines a domain.

8. $y < 10$; it is open, connected and unbounded; defines a domain.

9. $x^2 - y^2 < 4$; it is open, connected and unbounded; defines a domain.

10. $(x - 1/2)^2 + (y - 1)^2 \leq 1/4$, or $|z - (1/2 + i)| \leq 1/2$; it is connected, bounded and not open; not a domain.

11. $y > x^2$; it is open, connected and unbounded; defines a domain.

12. $x - y + 3 < 0$; it is open, connected and unbounded; defines a domain.

13. $x^2 + (y + 9/2)^2 > 81/4$; it is open, connected and unbounded; defines a domain.

14. $0 \leq \theta \leq \pi/3$ or $0 \leq y \leq \sqrt{3}x$; it is connected, unbounded and not open; not a domain.

15. $\pi/4 \leq \theta \leq \pi/2$ or $x \leq y < \infty$; it is connected, unbounded and not open; not a domain.

16. $-\pi/4 < \theta < \pi/4$ or $-x < y < x$; it is open, connected and unbounded; defines a domain.

17. $x \leq 1, 0 < \theta < \pi/4$ or $x \leq 1, 0 < y < 1$; it is connected, unbounded and not open; not a domain.

18. $x \geq 0$; it is connected, unbounded and not open; not a domain.

19. $[x - (2/3)]^2 + y^2 < (7/3)^2$; it is open, connected and bounded; defines a domain.

20. $y \geq 0$; it is connected, unbounded and not open; not a domain.

21. Annulus; it is connected, bounded and not open; not a domain.

22. $|z| < 1/2$; it is open, connected and bounded; defines a domain.

23. $x^2 - (1/3) y^2 < 1$; it is open, connected and unbounded; defines a domain.

24. $y < 1/2$ and $x^2 + y^2 < 1$; it is open, connected and bounded; defines a domain.

25. $x - 2y + 3 > 0$ and $x^2 + y^2 > 16$; it is open, connected and unbounded; defines a domain.

Exercise 10.2

1. Whole of the complex plane except $z = -i$.   **2.** Whole of the complex plane except $z = \pm i$.

3. Whole of the complex plane except $z = -1$, $e^{\pm i\pi/3}$.

4. Whole of the complex plane except $z_k = 2\left[\cos\dfrac{(2k+1)\pi}{4} + i\sin\left(\dfrac{(2k+1)\pi}{4}\right)\right]$, $k = 0, 1, 2, 3$.

5. Whole of the complex plane except $z = i, -1$.

6. 5.   **7.** $18 + 16\,i$.   **8.** $-10 + 44i$.

9. $(3 - i)/5$.   **10.** $(3 + i)/5$.   **11.** $(-1 + 7i)/5$.

12. $x^2 - y^2 - 2ixy$.   **13.** $x^3 - 3xy^2 + i(3x^2y - y^3)$.

14. $x^4 + y^4 - 6x^2y^2 + 4ixy(x^2 - y^2)$.   **15.** $x^2 + (y - 4)^2 + 3x - 3iy$.

16. $x/(1 - x^2 - y^2) + 0i$.   **17.** $(1/2) + iy/(2x)$.

18. $5x(1 + 2y) - 2iy$.   **19.** $x^3 - 3xy^2 + 4y - 3ix$.

20. $[x^2 - y^2 + (2xy + 1)i]/\sqrt{x^2 + y^2}$.   **21.** $[x(x^2 + y^2 + 1) + iy(x^2 + y^2 - 1)]/(x^2 + y^2)$.

22. $[(x^2 - y^2 - x) + iy(2x - 1)]/(x^2 + y^2)$.   **23.** $[x^2 + y^2 + y - ix]/[x^2 + (y + 1)^2]$.

24. $[2x^2 + 2y^2 - x + 2y + i(2x - y - 1)]/[(2x - 1)^2 + 4y^2]$.

25. $[x^3 + xy^2 + x + i(y - x^2y - y^3)]/[(x^2 - y^2 + 1)^2 + 4x^2y^2]$.

Exercise 10.3

1. $e^x (\cos y - i \sin y)$.   **2.** $e^{x^2 - y^2} [\cos (2xy) + i \sin (2xy)]$.

3. $e^{e^x \cos y} [\cos (e^x \sin y) + i \sin (e^x \sin y)]$.

4. $\text{Log} (\cos z) = \text{Log} (\cos x \cosh y - i \sin x \sinh y) = \log r + i\theta$.
   where $r^2 = \cos^2 x \cosh^2 y + \sin^2 x \sinh^2 y = \cos^2 x + \sinh^2 y$ and $\theta = -\tan^{-1} (\tan x \tanh y)$.

5. Let $\cos^{-1} (e^{i\theta}) = u + iv$ or $e^{i\theta} = \cos (u + iv)$. Therefore,
   (i) $\cos u \cosh v = \cos \theta$ and   (ii) $\sin u \sinh v = -\sin \theta$.
   Eliminating θ, we get $\sinh^2 v = \sin^2 u$. From (ii), we obtain $\sin^2 u \sinh^2 v = \sin^2 \theta$ or

   $\sinh^4 v = \sin^2 \theta$. Therefore, $\sinh v = \sqrt{\sin \theta}$. Using $\sinh v = (e^v - e^{-v})/2$, we get

   $v = \log [\sqrt{\sin \theta} + \sqrt{1 + \sin \theta}\,]$. From (ii), we get $\sin^2 u = \sin \theta$ or $u = \sin^{-1} (\sqrt{\sin \theta})$.

6. $(4n - 1)\,\pi i/2$, n any integer.

7. $e^z = (-1 \pm \sqrt{3}\,i)/2$; $z = (2\pi i/3)(3n \pm 1)$, n any integer,

8. $1/z = \log (1 - i) = \log \sqrt{2} + i(8n - 1)\pi/4$. Therefore,

   $z = [\log \sqrt{2} - i(8n - 1)\pi/4]/[(\log \sqrt{2}\,)^2 + ((8n - 1)\,\pi/4)^2]$, n any integer.

9. $e^z = \log i = ia$, where $a = (\pi/2) + 2n\pi$, n any integer.

Therefore, $\quad z = \log (ia) = \begin{vmatrix} \log|a| + i((\pi/2) + 2k\pi), & a > 0 \\ \log|a| + i((-\pi/2) + 2k\pi), & a < 0 \end{vmatrix}$

where n and k are any integers.

10. Using $\cos z = (e^{iz} + e^{-iz})/2$, we get $e^{iz} = (1 \pm \sqrt{2})i$. Therefore,

$$z = -i \log [(1 \pm \sqrt{2})\, i] = -i \,[\log |1 \pm \sqrt{2}| + i((\pm \pi/2) + 2n\pi)].$$

11. Let $i/z = t_1 + it_2$. We get $\sin (t_1 + it_2) = i$. Therefore, $\sin t_1 \cosh t_2 = 0$ and $\cos t_1 \sinh t_2 = 1$ or $t_1 = n\pi$ and $t_2 = (-1)^n \sinh^{-1}1$ and $z = (t_2 + it_1)/\left(t_1^2 + t_2^2\right)$, n any integer.

12. $\tan z = (e^{iz} - e^{-iz})/[i(e^{iz} + e^{-iz})] = 2$. We get $e^{2iz} = (1 + 2i)/(1 - 2i) = (-3 + 4i)/5$, $z = [2n\pi - \tan^{-1}(4/3)]/2$, n any integer.

13. $\tanh z = (e^z - e^{-z})/(e^z + e^{-z}) = -2$. We get $e^{2z} = -1/3$. $z = [\log (1/3) + i(2n + 1)\pi]/2$, n any integer.

14. $z^2 = \log ((1 + i)/\sqrt{2}) = ia$, where $a = (\pi/4) + 2n\pi$, n any integer. We get $x^2 - y^2 = 0$ and $2xy = a$.

Therefore, $y = \pm x$. When $a > 0$, x and y are of the same sign. We get $x = y = \pm \sqrt{a/2}$ and $z = \pm (1 + i) \sqrt{(a/2)}$. When $a < 0$, x and y are of opposite signs. We get $x = \pm \sqrt{|a|/2}$, $y = \mp \sqrt{|a|/2}$ and $z = \pm \sqrt{|a|/2}\ (1 - i)$.

15. $\cosh z + \sinh z = e^z = \alpha$; $z = \log \alpha = \log |\alpha| + i[2n\pi + \text{Arg}(\alpha)]$.

16. $e^{5\pi i/2} = i$, $\log i = (4n + 1)\pi i/2$, n any integer.

17. $\text{Log } i = \pi i/2$, $\log (\pi i/2) = \log (\pi/2) + (4n + 1)\, \pi i/2$, n any integer.

18. $\log ((1 + i)e/\sqrt{2}) = \log e + \log ((1 + i)/\sqrt{2}) = 1 + (8n + 1)\pi i/4$.

Therefore, given expression has the value $\sin (-(8n + 1)\,\pi i/4) = -i \sinh ((8n + 1)\pi/4)$, n any integer.

19. $\log [(1 - \sqrt{3}i)/(1 + \sqrt{3}i)] = \log (-(1 + \sqrt{3}i)/2) = [2n\pi - (2\pi/3)]i$, n any integer. Therefore, the given expression has the value $\tan [(1 - 3n)\,(2\pi/3)]$.

20. $(-2)^{\text{Log}\, i} = (-2)^{(\pi i/2)} = e^{\pi i [\log(-2)]/2}$

$= e^{\pi i [\log 2 + i(2n+1)\pi]/2} = e^{-\pi^2 (2n+1)/2}\,[\cos p + i \sin p]$, where $p = (\pi \log 2)/2$, n any integer.

21. $(-i)^{-i} = e^{-i\log(-i)} = e^{-i[(4n - 1)\pi i/2]} = e^{(4n-1)\pi/2}$.

22. $(1 + i)^{1+i} = e^{(1+i)\log(1+i)} = e^{\alpha}[\cos \beta + i \sin \beta]$, where $\alpha = \ln \sqrt{2} - (8n + 1)\,\pi/4$, $\beta = \ln \sqrt{2} + (8n + 1)\,\pi/4$, n any integer.

23. $i^{e^i} = e^{e^i \log i} = e^{ai(\cos 1 + i \sin 1)} = e^{-a \sin 1}[\cos (a \cos 1) + i \sin (a \cos 1)]$

where $a = (4n + 1)\pi/2$, n any integer.

24. $i^{\log(i + 1)} = e^{\log(i + 1)\log i} = e^{ib[\log \sqrt{2} + ia]} = e^{-ab}[\cos (b \log \sqrt{2}) + i \sin (b \log \sqrt{2})]$, where $a = (8n + 1)\pi/4$ and $b = (4k + 1)\,\pi/2$, n, k any integers.

25. $\cos^{-1} \sqrt{1 - z^2} = -i \log \left[\sqrt{1 - z^2} + \sqrt{1 - z^2 - 1}\right] = -i \log [iz + \sqrt{1 - z^2}] = \sin^{-1} z$,

$\tan^{-1} \dfrac{z}{\sqrt{1 - z^2}} = -\dfrac{i}{2} \log \dfrac{\sqrt{1 - z^2} + iz}{\sqrt{1 - z^2} - iz} = -i \log [iz + \sqrt{1 - z^2}] = \sin^{-1} z.$

28. $\sin (\cos^{-1} z) = \dfrac{1}{2i}\left[e^{i\cos^{-1}z} - e^{-i\cos^{-1}z}\right] = \dfrac{1}{2i}\left[e^{\log(z+\sqrt{z^2-1})} - e^{-\log(z+\sqrt{z^2-1})}\right]$

$$= \dfrac{1}{2i}\left[\left(z+\sqrt{z^2-1}\right) - 1/\left(z+\sqrt{z^2-1}\right)\right] = \dfrac{1}{2i}\left[\left(z+\sqrt{z^2-1}\right) - \left(z-\sqrt{z^2-1}\right)\right] = -i\sqrt{z^2-1}.$$

Similarly prove the other part.

29. $e^{z^2} = e^{x^2-y^2+2ixy} = e^{x^2-y^2}\,[\cos (2xy) + i\sin (2xy)].$

    (i) real, when $\sin (2xy) = 0$ or $xy = n\pi/2,$

    (ii) pure imaginary, when $\cos (2xy) = 0$ or $xy = (2n + 1)\ \pi/4$; where n is any integer.

30. Write $(\pi/12) = (\pi/3) - (\pi/4).$

31. We have $\tan (\theta + i\phi) = e^{-i\alpha}$ and $\tan (\theta - i\phi) = e^{i\alpha}$. We get

$\tan 2\theta = \tan [(\theta + i\phi) + (\theta - i\phi)] = \infty = \tan \pi/2.$

Therefore, $2\theta = (\pi/2) + n\pi$ or $\theta = (2n + 1)\ \pi/4,$ n any integer

and $\tan 2i\phi = \tan [(\theta + i\phi) - (\theta - i\phi)] = -i\sin \alpha,$ or $\tanh 2\phi = -\sin \alpha.$

Now, $\tanh 2\phi = \dfrac{\sinh 2\phi}{\cosh 2\phi} = \dfrac{e^{2\phi} - e^{-2\phi}}{e^{2\phi} + e^{-2\phi}} = -\sin \alpha$ or $e^{4\phi} = \dfrac{1 - \sin\alpha}{1 + \sin\alpha}$

or $e^{4\phi} = \tan^2\left(\dfrac{\pi}{4} - \dfrac{\alpha}{2}\right).$ Hence, $\phi = \dfrac{1}{2}\ \log\left[\tan\left(\dfrac{\pi}{4} - \dfrac{\alpha}{2}\right)\right].$

32. Write $\tanh z = -i\tan iz = -i\tan (-y + ix) = p + iq.$ Therefore, we have

$\tan (-y + ix) = pi - q$ and $\tan (-y - ix) = -pi - q.$

Now, $\tan 2ix = i\tanh 2x = \tan [(-y + ix) - (-y - ix)] = 2pi/(1 + p^2 + q^2)$

or $\tanh 2x = 2p/(1 + p^2 + q^2)$

and $\tan (-2y) = \tan [(-y + ix) + (-y - ix)] = -2q/(1 - p^2 - q^2)$

or $\tan 2y = 2q/(1 - p^2 - q^2).$

33. We have $u + iv = \sin (\alpha + i\beta).$ Therefore, $u = \sin \alpha \cosh \beta$ and $v = \cos \alpha \sinh \beta.$ We get

$$u^2 + v^2 = \sin^2 \alpha \cosh^2 \beta + \cos^2 \alpha \sinh^2 \beta = \sin^2 \alpha + \cosh^2\beta - 1.$$

From the given equation we have

sum of roots $= \sin^2 \alpha + \cosh^2 \beta = 1 + u^2 + v^2$ and product of roots $= \sin^2 \alpha \cosh^2 \beta = u^2.$

34. We have $x + iy = \sinh (u + iv) = -i\sin (iu - v) = -i[i\sinh u \cos v - \cosh u \sin v]$

Therefore, $x = \sinh u \cos v$ and $y = \cosh u \sin v.$

    (i) $\dfrac{x}{\sinh u} = \cos v$ and $\dfrac{y}{\cosh u} = \sin v,$ square and add.

    (ii) $\dfrac{x}{\cos v} = \sinh u$ and $\dfrac{y}{\sin v} = \cosh u.$ Using $\cosh^2 u - \sinh^2 u = 1,$ we get the result.

35. We have $p \log (a + ib) = (x + iy) \log m$

or $\quad p[(\log(a^2 + b^2))/2 + i \tan^{-1} (b/a)] = (x + iy) \log m$

Therefore, $x = p[\log (a^2 + b^2)]/(2 \log m)$, $\quad y = p[\tan^{-1} (b/a)]/ \log m$.

or $\quad y/x = 2 \tan^{-1} (b/a)/\log (a^2 + b^2)$.

Exercise 10.4

1. 3.
           2. 0.
           3. $f(z) = (iz^3 - i^4)/(z - i)$, $-3i$.

4. $\lim\limits_{z \to 0} |f(z)| = \lim\limits_{z \to 0} |z| = 0$, or set $z = re^{i\theta}$.

5. $-4i$.
           6. i.

7. $f(z) = \dfrac{(z-1)^2 - (2i)^2}{z^2 - (1+2i)^2} = \dfrac{(z-1+2i)(z-1-2i)}{(z+1+2i)(z-1-2i)}$, $\quad \lim\limits_{z \to 1+2i} f(z) = \dfrac{4i}{2+4i} = \dfrac{2}{5} (2 + i)$.

8. $4i$.
           9. Set $z = 1/\xi$, i.
           10. Set $z = 1/\xi$, $i/2$.

11. Multiply and divide by $\left[\sqrt{z-2i} + \sqrt{z-i}\right]$, set $z = 1/\xi$, $-i/2$.

12. Choose two different paths and show that limit is not unique. Alternately, take $y = mx$ and show that limit depends on m.

18. $\lim\limits_{z \to 1} f(z) \to \infty$, therefore, limit does not exist.

19. (i) a_m/b_n,
        (ii) Set $z = 1/\xi$, 0 if $n > m$; does not exist if $n < m$; a_0/b_0 if $m = n$.

20. Show that $\lim\limits_{z \to z_0} f(z) = f(z_0)$, where z_0 is any arbitrary point.

25. Set $z = re^{i\theta}$; $r \to 0$ as $z \to 0$, $\lim\limits_{z \to 0} f(z) = f(0)$; continuous.

26. $\lim\limits_{z \to -i} f(z) = -2i \neq f(-i)$; not continuous.

27. $\lim\limits_{z \to 0} f(z) = f(0)$; continuous.

28. Set $z = re^{i\theta}$, $r \to 0$ as $z \to 0$, $\lim\limits_{z \to 0} f(z) = f(0)$; continuous.

29. $f(z) = \dfrac{(z-1)^2 - i^2}{z^2 - (1-i)^2}$; $f(1 - i) = (1 - i)/2$.

30. $f(z) = \dfrac{(z-i)(3z^3 - (2-3i)z^2 + (5-2i)z + 5i)}{z-1}$, $f(i) = 4 + 4i$.

31. $z = 2 \pm i$.

32. Not continuous at points where $\text{Im}(z^2 + 1) = 0$ and $\text{Re}(z^2 + 1) \leq 0$; $x = 0$ and $|y| \geq 1$.

33. Not continuous at points where $\text{Im}\left(\dfrac{z-i}{z+i}\right) = 0$ and $\text{Re}\left(\dfrac{z-i}{z+i}\right) \leq 0$; $x = 0$ and $-1 \leq y \leq 1$.

34. Ln $[\, f(z)\,]$ is not continuous at points where $f(z) = 0$ or when $f(z)$ is real and negative; $y = 0$ and $1 \leq x < 2$.

35. $x = 0$, $|y| \geq 1$.

36. Write $\sin^{-1} z = -i\, \text{Ln}\left(iz + \sqrt{1-z^2}\right)$. The principal value of $\sqrt{1-z^2}$ is not continuous at points where $\text{Im}\,(1-z^2) = 0$ and $\text{Re}\,(1-z^2) < 0$ which give $y = 0$ and $|x| > 1$, or the cuts as shown in Fig. 10.11. Now, $\sin^{-1} z$ is continuous for all z except at the points where

$$\text{Im}\left(iz + \sqrt{1-z^2}\right) = 0 \quad \text{and} \quad \text{Re}\left(iz + \sqrt{1-z^2}\right) \leq 0 \tag{10.59}$$

Since $\lambda = iz + \sqrt{1-z^2}$ is not zero for any z, we get

$$(\lambda - iz)^2 = 1 - z^2, \quad \text{or} \quad z = (\lambda^2 - 1)/(2\lambda i). \tag{10.60}$$

The first condition in Eq (10.59) gives $\text{Im}(\lambda) = 0$, that is, λ is real. Therefore, from Eq. (10.60), we find that z is pure imaginary. Taking $z = iy$, we obtain $\text{Re}\left(iz + \sqrt{1-z^2}\right) = -y + \sqrt{1+y^2}$ which is positive for any real y. Thus, the second condition in Eq. (10.59) is never satisfied. Hence, $\sin^{-1} z$ is continuous at all points except at points along the cut as shown in Fig. 10.11.

Exercise 10.5

Let $L = \lim\limits_{\Delta z \to 0}\left[\dfrac{f(z + \Delta z) - f(z)}{\Delta z}\right]$ in Problems 1 to 6.

1. $L = \lim\limits_{\Delta z \to 0} \dfrac{\Delta x}{\Delta x + i\Delta y}$ limit does not exist, not differentiable.

2. $L = \lim\limits_{\Delta z \to 0} [2z + \Delta z] = 2z$, $f'(z) = 2z$.

3. $L = \lim\limits_{\Delta z \to 0} [i(3z^2 + 3z\Delta z + (\Delta z)^2] = 3iz^2$, $f'(z) = 3iz^2$.

4. $L = \lim\limits_{\Delta z \to 0}\left[\dfrac{2\bar{z}\,\overline{\Delta z} + (\overline{\Delta z})^2}{\Delta z}\right]$, limit does not exist, not differentiable.

5. $L = \lim\limits_{\Delta z \to 0}\left[\dfrac{-1}{z(z + \Delta z)}\right] = -\dfrac{1}{z^2}$, $f'(z) = -\dfrac{1}{z^2}$, $z \neq 0$.

6. $L = \lim\limits_{\Delta z \to 0}\left[\dfrac{2}{(1-z)(1-z-\Delta z)}\right] = \dfrac{2}{(1-z)^2}$, $f'(z) = \dfrac{2}{(1-z)^2}$.

7. $4(1 - i)$.   **8.** $-(1 + \pi i)$.   **9.** 4.

10. -3.   **11.** -2.   **12.** 1.

13. Differentiable only at origin, $f'(0) = 0$.

14. Differentiable only at origin, $f'(0) = 0$.

15. Cauchy-Riemann equations are not satisfied for any point; no where differentiable.

16. Not differentiable any where.

17. Differentiable everywhere, $f'(z) = 3z^2$.

18. Differentiable only at origin, $f'(0) = 0$.   **19.** Not differentiable.

20. $f(z) = \bar{z}/(z\bar{z}) = 1/z$, $z \neq 0$, differentiable everywhere, $f'(z) = -1/z^2$, $z \neq 0$

21. $u = -\dfrac{x}{x^2 + (y+1)^2}$, $v = \dfrac{x^2 + y^2 + 3y + 2}{x^2 + (y+1)^2}$, differentiable everywhere except at $z = -i$,
    $f'(z) = -1/(1 - iz)^2$.

22. Not differentiable.

23. $f(z) = -ie^z$. Differentiable for all z, $f'(z) = -ie^z$.

24. Not differentiable.

25. $f(z) = e^{z^2}$. Differentiable for all z, $f'(z) = 2ze^{z^2}$.

26. $f(z) = (x^2 + y^2)^2 = u + iv$; $u = (x^2 + y^2)^2$, $v = 0$; Cauchy-Riemann equations are satisfied only at the origin.

27. $f(z) = u + iv$; $u = \dfrac{x^3 - 3xy^2}{x^2 + y^2}$, $v = \dfrac{y^3 - 3x^2y}{x^2 + y^2}$, $(x, y) \neq (0, 0)$; $u(0, 0) = v(0, 0) = 0$;

$$u_x(0, 0) = \lim_{\Delta x \to 0} \frac{u(\Delta x, 0) - u(0, 0)}{\Delta x} = 1, \ u_y(0, 0) = \lim_{\Delta y \to 0} \frac{u(0, \Delta y) - u(0, 0)}{\Delta y} = 0;$$

$$v_x(0, 0) = \lim_{\Delta x \to 0} \frac{v(\Delta x, 0) - v(0, 0)}{\Delta x} = 0, \ v_y(0, 0) = \lim_{\Delta y \to 0} \frac{v(0, \Delta y) - v(0, 0)}{\Delta y} = 1;$$

The Cauchy-Riemann equations are satisfied. Now

$$\lim_{z \to 0} \frac{f(z) - f(0)}{z} = \lim_{\Delta z \to 0} \frac{[(\overline{\Delta z})^2/\Delta z] - 0}{\Delta z} = \lim_{\Delta z \to 0} \left[\frac{\overline{\Delta z}}{\Delta z}\right]^2.$$

Since the limit does not exist, $f'(0)$ does not exist.

28. $f(z) = u + iv$, $u = \sqrt{|xy|}$, $v = 0$; $u(0, 0) = v(0, 0) = 0$; $u_x(0, 0) = \lim_{\Delta x \to 0} \dfrac{u(\Delta x, 0) - u(0, 0)}{\Delta x} = 0$.

Similarly, we obtain $u_y(0, 0) = 0$, $v_x(0, 0) = 0$ and $v_y(0, 0) = 0$. The Cauchy-Riemann equations are satisfied at the origin. Show directly that $f'(0)$ does not exist.

29. From $f'(z) = u_x + iv_x$, we get $|f'(z)|^2 = u_x^2 + v_x^2 = u_x u_x + v_x v_x = u_x v_y - v_x u_y$ which is the right hand side.

30. Let $\Delta w = \Delta u + i\Delta v$ and $\Delta z = \Delta x + i\Delta y$. Now

$$\lim_{\Delta z \to 0}\left[\text{Re}\,\frac{\Delta w}{\Delta z}\right] = \lim_{\Delta z \to 0} T, \text{ where } T = \frac{\Delta u \Delta x + \Delta v \Delta y}{(\Delta x)^2 + (\Delta y)^2}.\text{ Since the limit exists, it is same irrespective of}$$

the path along which $\Delta z \to 0$. We find that

$$\lim_{\Delta z \to 0} T = \lim_{\Delta x \to 0}\left[\lim_{\Delta y \to 0} T\right] = u_x \text{ and } \lim_{\Delta z \to 0} T = \lim_{\Delta y \to 0}\left[\lim_{\Delta x \to 0} T\right] = v_y. \text{ Therefore, } u_x \text{ and } v_y \text{ exist and } u_x = v_y.$$

31. $\lim\limits_{\Delta z \to 0}\left[\text{Im}\dfrac{\Delta w}{\Delta z}\right] = \lim\limits_{\Delta z \to 0} T$, where $T = \dfrac{\Delta v \Delta x - \Delta u \Delta y}{(\Delta x)^2 + (\Delta y)^2}$. Since the limit exists, it is same irrespective of

the path along which $\Delta z \to 0$. We find that

$$\lim_{\Delta z \to 0} T = \lim_{\Delta x \to 0}\left[\lim_{\Delta y \to 0} T\right] = v_x \text{ and } \lim_{\Delta z \to 0} T = \lim_{\Delta y \to 0}\left[\lim_{\Delta x \to 0} T\right] = -u_y.$$

Therefore, v_x and u_y exist and $v_x = -u_y$.

32. (i) $\dfrac{\partial u}{\partial x} = \dfrac{\partial u}{\partial z}\dfrac{\partial z}{\partial x} + \dfrac{\partial u}{\partial \bar{z}}\dfrac{\partial \bar{z}}{\partial x} = \dfrac{\partial u}{\partial z} + \dfrac{\partial u}{\partial \bar{z}}$　or　$\dfrac{\partial}{\partial x} = \dfrac{\partial}{\partial z} + \dfrac{\partial}{\partial \bar{z}};$

$$\dfrac{\partial u}{\partial y} = \dfrac{\partial u}{\partial z}\dfrac{\partial z}{\partial y} + \dfrac{\partial u}{\partial \bar{z}}\dfrac{\partial \bar{z}}{\partial y} = i\left(\dfrac{\partial u}{\partial z} - \dfrac{\partial u}{\partial \bar{z}}\right)\quad\text{or}\quad \dfrac{\partial}{\partial y} = i\left(\dfrac{\partial}{\partial z} - \dfrac{\partial}{\partial \bar{z}}\right)$$

$$\dfrac{\partial^2 u}{\partial x^2} = \left(\dfrac{\partial}{\partial z} + \dfrac{\partial}{\partial \bar{z}}\right)^2 u = \dfrac{\partial^2 u}{\partial z^2} + 2\dfrac{\partial^2 u}{\partial z \partial \bar{z}} + \dfrac{\partial^2 u}{\partial \bar{z}^2}\quad\text{and}$$

$$\dfrac{\partial^2 u}{\partial y^2} = \left[i\left(\dfrac{\partial}{\partial z} - \dfrac{\partial}{\partial \bar{z}}\right)\right]^2 u = -\dfrac{\partial^2 u}{\partial z^2} + 2\dfrac{\partial^2 u}{\partial z \partial \bar{z}} - \dfrac{\partial^2 u}{\partial \bar{z}^2}.$$

Adding the last two equations, we obtain the result.

(ii) $\dfrac{\partial f}{\partial \bar{z}} = \left[\dfrac{\partial u}{\partial x}\dfrac{\partial x}{\partial \bar{z}} + \dfrac{\partial u}{\partial y}\dfrac{\partial y}{\partial \bar{z}}\right] + i\left[\dfrac{\partial v}{\partial x}\dfrac{\partial x}{\partial \bar{z}} + \dfrac{\partial v}{\partial y}\dfrac{\partial y}{\partial \bar{z}}\right]$

$$= \dfrac{1}{2}\left[\dfrac{\partial u}{\partial x} - \dfrac{1}{i}\dfrac{\partial u}{\partial y}\right] + \dfrac{1}{2}i\left[\dfrac{\partial v}{\partial x} - \dfrac{1}{i}\dfrac{\partial v}{\partial y}\right] = \dfrac{1}{2}\left(\dfrac{\partial u}{\partial x} - \dfrac{\partial v}{\partial y}\right) + \dfrac{i}{2}\left(\dfrac{\partial v}{\partial x} + \dfrac{\partial u}{\partial y}\right)$$

Now, $\dfrac{\partial f}{\partial \bar{z}} = 0$ gives $\dfrac{\partial u}{\partial x} = \dfrac{\partial v}{\partial y}$　and　$\dfrac{\partial u}{\partial y} = -\dfrac{\partial v}{\partial x}$.

33. Let $f(z) = u + iv$. Then $f'(z) = u_x + iv_x$. From $f'(z) = f(z)$, we get $u_x = u$ and $v_x = v$. Integrating with respect to x, we obtain $u = e^x g(y)$ and $v = e^x h(y)$, where $g(y)$ and $h(y)$ are arbitrary functions. Since $f'(z)$ exists, $f(z)$ satisfies the Cauchy-Riemann equations $u_x = v_y$ and $u_y = -v_x$. From $u_x = v_y$, we get $e^x g(y) = e^x h'(y)$ or $h'(y) = g(y)$ and from $u_y = -v_x$, we get $e^x g'(y) = -e^x h(y)$ or $g'(y) = -h(y)$.

Therefore, we obtain $\dfrac{d^2 h}{dy^2} = -h(y)$. Solving this differential equation, we get

$h(y) = c_1 \cos y + c_2 \sin y$　and　$g(y) = -c_1 \sin y + c_2 \cos y,$

where c_1, c_2 are arbitrary constants. Hence, $f(z) = u + iv = (c_1 + ic_2) e^x(\cos y + i \sin y) = ke^z$

where $k = c_1 + ic_2$ is an arbitrary constant.

34. $f'(z) = 2z + 1.$ **35.** $f'(z) = 1 - 1/z^2,\ z \neq 0.$

36. $f(z) = u + iv = \dfrac{1}{r^2}\cos 2\theta - \dfrac{i}{r^2}\sin 2\theta,\ f'(z) = -\dfrac{2}{z^3},\ z \neq 0.$

37. No where.   **38.** No where.   **39.** No where.

40. Everywhere except at $z = -i$.   **41.** Everywhere.   **42.** No where.

43. Everywhere except at $z = 0$.   **44.** No where.   **45.** Everywhere except when $e^z = 1$.

46. $a = 2,\ b = -1,\ c = -1,\ d = 2$.   **47.** $a = -1,\ b = -1$.

48. $f'(z) = u_x + iv_x = v_y - iu_y \equiv 0$. Thus, $u_x = u_y = 0$ and $v_x = v_y = 0$. Therefore, u and v are constants and hence $f(z) = u + iv$ is a constant.

49. $e^z + c$, c is a complex constant.

50. $-i\,(\sin z^2 + \cos z^2) + c$, c is a complex constant.

51. We have $f'(z) = u_x + iv_x$. Therefore, Re $[f'(z)] = u_x = 3x^2 - 4y - 3y^2$. We get $u = x^3 - 4xy - 3xy^2 + h(y)$, where $h(y)$ is an arbitrary function. Since $u_x = v_y$, we get $v_y = 3x^2 - 4y - 3y^2$ and $v = 3x^2y - 2y^2 - y^3 + k(x)$, where $k(x)$ is an arbitrary function. Now, $u_y = -v_x$ gives $h'(y) = 4x - k'(x)$. Since $h(y)$ is a function of y alone, we get $k'(x) - 4x = 0$ and $h'(y) = 0$, that is $h(y) = c_1$ and $k(x) = 2x^2 + c_2$ where c_1 and c_2 are constants. $f(z) = u + iv = z^3 + 2iz^2 + c_1 + ic_2$. Now, $f(1 + i) = 0$ gives $c_1 = 6$ and $c_2 = -2$.

52. Since $f(z)$ and $g(z)$ are analytic functions, we have $u_x = v_y$, $u_y = -v_x$ and $v_x = u_y$, $v_y = -u_x$. Therefore, $u_x = u_y = v_x = v_y = 0$ for all x and y. Thus, u and v are constants. Hence, $f(z) = c_1$ and $g(z) = c_2$ where c_1 and c_2 are constants. Using $f(0) = 1$ and $g(0) = i$, we get $c_1 = 1$ and $c_2 = i$. We also get $f'(z) = g'(z) = 0$. Hence, $h(1 + i) = 2i$.

53. Use Eqs. (10.56), (10.57).

54. We have $f(z) = u + iv$, $\overline{f(z)} = u - iv$. Since both $f(z)$ and $\overline{f(z)}$ satisfy the Cauchy-Riemann equations, we get $u_x = u_y = 0$ and $v_x = v_y = 0$. Therefore, u and v are both constants. Hence, $f(z)$ is a constant function.

55. We have $\phi_x = \phi_u u_x + \phi_v v_x$ and $\phi_y = \phi_u u_y + \phi_v v_y = -\phi_u v_x + \phi_v u_x$.

Squaring and adding, we get $\phi_x^2 + \phi_y^2 = (\phi_u^2 + \phi_v^2)(u_x^2 + v_x^2) = (\phi_u^2 + \phi_v^2)\,|f'(z)|^2$.

Exercise 10.6

1. Let $S = u_y - v_x$ and $T = u_x + v_y$. S and T are continuous. Show that S and T satisfy the Cauchy-Riemann equations.

2. Let $f(z) = u + iv$. Then, $w = |f(z)| = \sqrt{u^2 + v^2}$. Show that $w_{xx} + w_{yy} \neq 0$.

3. Let $f(z) = u + iv$. Then, $w = \ln|f(z)| = [\ln(u^2 + v^2)]/2$. Show that $w_{xx} + w_{yy} = 0$.

4. Let $f(z) = u + iv$. Then, $w = \arg[f(z)] = \tan^{-1}(v/u)$. Differentiate and show that $w_{xx} + w_{yy} = 0$.

5. Let $X = x^2 - y^2$ and $Y = 2xy$. We get

$$u_x = u_X X_x + u_Y Y_x = 2(xu_X + yu_Y),\quad u_y = u_X X_y + u_Y Y_y = 2(-yu_X + xu_Y)$$

$$u_{xx} = 2u_X + 4(x^2 u_{XX} + y^2 u_{YY}) + 8xyu_{XY},\quad u_{yy} = -2u_X + 4(y^2 u_{XX} + x^2 u_{YY}) - 8xy\,u_{XY}.$$

Therefore, $u_{xx} + u_{yy} = 4(x^2 + y^2)(u_{XX} + u_{YY}) = 0$. Hence, the result.

6. $\nabla^2[|f(z)|^p] = \nabla^2(u^2 + v^2)^{p/2} = \dfrac{\partial^2}{\partial x^2}(u^2 + v^2)^{p/2} + \dfrac{\partial^2}{\partial y^2}(u^2 + v^2)^{p/2}$. Differentiate two times and use

   the results $u_x = v_y$, $u_y = -v_x$, $u_{xx} + u_{yy} = 0$, $v_{xx} + v_{yy} = 0$ and $|f'(z)|^2 = u_x^2 + v_x^2$.

7. $v = x^3 - 3xy^2 + c$; $f(z) = iz^3 + ic$, c a real constant.

8. $v = 2\tan^{-1}(y/x) + c$; $f(z) = 2\ln z + ic$, c a real constant.

9. $v = -y/(x^2 + y^2) + c$; $f(z) = (1/z) + ic$, c a real constant.

10. $v = e^x \sin y + c$; $f(z) = e^z + ic$, c a real constant.

11. $v = \sin y \cosh x + c$; $f(z) = \sinh z + ic$, c a real constant.

12. $v = e^{-x}[y \cos y - x \sin y] + c$; $f(z) = ze^{-z} + ic$, c a real constant.

13. $v = e^x[(x^2 - y^2)\sin y + 2xy \cos y]$; $f(z) = z^2 e^z + ic$, c a real constant.

14. $u = x^3 - 3xy^2 + x + c$; $f(z) = z^3 + z + c$, c a real constant.

15. $u = x - y - 2\tan^{-1}(y/x) + c$; $f(z) = 2i \ln z + (1 + i)z + c$, c a real constant.

16. $u = (x + y)/(x^2 + y^2) + c$; $f(z) = (1 + i)/z + c$, c a real constant.

17. $u = [\ln(x^2 + y^2)]/2 + c$; $f(z) = \ln z + c$, c a real constant.

18. $u = \cos x \cosh y + c$; $f(z) = \cos z + c$, c a real constant.

19. $u = e^{-x}(x \cos y + y \sin y) + c$; $f(z) = ze^{-z} + c$, c a real constant.

20. $v = \theta + c$; $f(z) = \ln|z| + i \arg z + ic$, c a real constant.

21. $v = (r - (1/r))\sin \theta + c$; $f(z) = z + (1/z) + ic$, c a real constant.

22. $v = r^n \sin n\theta + c$; $f(z) = z^n + ic$, c a real constant.

23. $u = 3r^2 \cos 2\theta - 2r \cos \theta + c$; $f(z) = 3z^2 - 2z + c$, c a real constant.

24. Level curves are $u = x^2 - y^2 = p$ and $v = xy = q$ where p and q are constants. From the first curve, we get $2x - 2y (dy/dx) = 0$. Therefore, slope of the first curve is $m_1 = x/y$. From the second curve, we get $y + x (dy/dx) = 0$. Therefore, slope of the second curve is $m_2 = -y/x$. Since, $m_1 m_2 = -1$, the curves are mutually orthogonal.

25. The level curves are $u = e^x \cos y = p$ and $v = e^x \sin y = q$ where p and q are constants. Slope of the first curve is $m_1 = \cot y$ and slope of the second curve is $m_2 = -\tan y$. Since, $m_1 m_2 = -1$, the curves are mutually orthogonal.

or $\qquad e^{4\phi} = \tan^2\left(\dfrac{\pi}{4} - \dfrac{\alpha}{2}\right)$. Hence, $\phi = \dfrac{1}{2}\log\left[\tan\left(\dfrac{\pi}{4} - \dfrac{\alpha}{2}\right)\right]$.

32. Write $\tanh z = -i\tan iz = -i\tan(-y + ix) = p + iq$. Therefore, we have

$$\tan(-y + ix) = pi - q \quad \text{and} \quad \tan(-y - ix) = -pi - q.$$

Now, $\qquad \tan 2ix = i\tanh 2x = \tan[(-y + ix) - (-y - ix)] = 2pi/(1 + p^2 + q^2)$

or $\qquad \tanh 2x = 2p/(1 + p^2 + q^2)$

and $\qquad \tan(-2y) = \tan[(-y + ix) + (-y - ix)] = -2q/(1 - p^2 - q^2)$

or $\qquad \tan 2y = 2q/(1 - p^2 - q^2)$.

33. We have $u + iv = \sin(\alpha + i\beta)$. Therefore, $u = \sin\alpha\cosh\beta$ and $v = \cos\alpha\sinh\beta$. We get

$$u^2 + v^2 = \sin^2\alpha\cosh^2\beta + \cos^2\alpha\sinh^2\beta = \sin^2\alpha + \cosh^2\beta - 1.$$

From the given equation we have

sum of roots $= \sin^2\alpha + \cosh^2\beta = 1 + u^2 + v^2$ and product of roots $= \sin^2\alpha\cosh^2\beta = u^2$

34. We have $x + iy = \sinh(u + iv) = -i\sin(iu - v) = -i[i\sinh u\cos v - \cosh u\sin v]$

Therefore, $\qquad x = \sinh u\cos v \quad \text{and} \quad y = \cosh u\sin v$

(i) $\dfrac{x}{\sinh u} = \cos v$ and $\dfrac{y}{\cosh u} = \sin v$, square and add,

(ii) $\dfrac{x}{\cos v} = \sinh u$ and $\dfrac{y}{\sin v} = \cosh u$. Using $\cosh^2 u - \sinh^2 u = 1$, we get the result.

35. We have $p\log(a + ib) = (x + iy)\log m$

or $\qquad p[(\log(a^2 + b^2))/2 + i\tan^{-1}(b/a)] = (x + iy)\log m$

Therefore, $\qquad x = p[\log(a^2 + b^2)]/(2\log m), \; y = p[\tan^{-1}(b/a)]/\log m$

or $\qquad y/x = 2\tan^{-1}(b/a)/\log(a^2 + b^2)$.

Exercise 10.5

1. 3.   **2.** 0.   **3.** $f(z) = (iz^3 - i^4)/(z - i), -3i$.

4. $\displaystyle\lim_{z\to 0}|f(z)| = \lim_{z\to 0}|z| = 0$, or set $z = re^{i\theta}$.

5. $-4i$.   **6.** i.

7. $f(z) = \dfrac{(z-1)^2 - (2i)^2}{z^2 - (1 + 2i)^2} = \dfrac{(z - 1 + 2i)(z - 1 - 2i)}{(z + 1 + 2i)(z - 1 - 2i)}, \; \displaystyle\lim_{z\to 1+2i} f(z) = \dfrac{4i}{2 + 4i} = \dfrac{2}{5}(2 + i)$.

8. $4i$.   **9.** Set $z = 1/\xi, i$.   **10.** Set $z = 1/\xi, i/2$.

11. Multiply and divide by $[\sqrt{z - 2i} + \sqrt{z - i}]$, set $z = 1/\xi, -i/2$.

12. Choose two different paths and show that limit is not unique. Alternately, take $y = mx$ and show that limit depends on m.

18. $\displaystyle\lim_{z\to 1} f(z) \to \infty$, therefore, limit does not exist.

19. (i) a_m/b_n,   (ii) Set $z = 1/\xi$, 0 if $n > m$; does not exist if $n < m$; a_0/b_0 if $m = n$.

20. Show that $\displaystyle\lim_{z\to z_0} f(z) = f(z_0)$, where z_0 is any arbitrary point.

25. Set $z = re^{i\theta}$; $r \to 0$ as $z \to 0$, $\displaystyle\lim_{z\to 0} f(z) = f(0)$; continuous .

26. $\lim\limits_{z \to -i} f(z) = -2i \neq f(-i)$; not continuous.

27. $\lim\limits_{z \to 0} f(z) = f(0)$; continuous.

28. Set $z = re^{i\theta}$, $r \to 0$ as $z \to 0$, $\lim\limits_{z \to 0} f(z) = f(0)$; continuous.

29. $f(z) = \dfrac{(z-1)^2 - i^2}{z^2 - (1-i)^2}$; $f(1-i) = (1-i)/2$.

30. $f(z) = \dfrac{(z-i)(3z^3 - (2-3i)z^2 + (5-2i)z + 5i)}{z-i}$, $f(i) = 4 + 4i$.

31. $z = 2 \pm i$.

32. Not continuous at points where $\text{Im}(z^2 + 1) = 0$ and $\text{Re }(z^2 + 1) \leq 0$; $x = 0$ and $|y| \geq 1$.

33. Not continuous at points where $\text{Im}\left(\dfrac{z-i}{z+i}\right) = 0$ and $\text{Re}\left(\dfrac{z-i}{z+i}\right) \leq 0$; $x = 0$ and $-1 \leq y \leq 1$.

34. Ln $[f(z)]$ is not continuous at points where $f(z) = 0$ or when $f(z)$ is real and negative; $y = 0$ and $1 \leq x < 2$.

35. $x = 0, |y| \geq 1$.

36. Write $\sin^{-1} z = -i \text{ Ln }(iz + \sqrt{1 - z^2})$. The principal value of $\sqrt{1 - z^2}$ is not continuous at points where $\text{Im }(1 - z^2) = 0$ and $\text{Re }(1 - z^2) < 0$ which give $y = 0$ and $|x| > 1$, or the cuts as shown in Fig. 10.15. Now $\sin^{-1}z$ is continuous for all z except at the points where

$$\text{Im}(iz + \sqrt{1 - z^2}) = 0 \text{ and } \text{Re}(iz + \sqrt{1 - z^2}) \leq 0 \qquad (10.71)$$

Since $\lambda = iz + \sqrt{1 - z^2}$ is not zero for any z, we get

$$(\lambda - iz)^2 = 1 - z^2, \text{ or } z = (\lambda^2 - 1)/(2\lambda i). \qquad (10.72)$$

The first condition in Eq. (10.71) gives $\text{Im}(\lambda) = 0$, that is, λ is real. Therefore, from Eq. (10.72), we find that z is pure imaginary. Taking $z = iy$, we obtain $\text{Re }(iz + \sqrt{1 - z^2}) = -y + \sqrt{1 + y^2}$ which is positive for any real y. Thus, the second condition in Eq. (10.71) is never satisfied. Hence, $\sin^{-1} z$ is continuous at all points except at points along the cut as shown in Fig. 10.15.

Exercise 10.6

Let $L = \lim\limits_{\Delta z \to 0} \left[\dfrac{f(z + \Delta z) - f(z)}{\Delta z} \right]$ in problems **1** to **6**.

1. $L = \lim\limits_{\Delta z \to 0} \dfrac{\Delta x}{\Delta x + i\Delta y}$, limit does not exist, not differentiable.

2. $L = \lim\limits_{\Delta z \to 0} [2z + \Delta z] = 2z, f'(z) = 2z$.

3. $L = \lim\limits_{\Delta z \to 0} \left[i(3z^2 + 3z\Delta z + (\Delta z)^2 \right] = 3iz^2, f'(z) = 3iz^2$.

4. $L = \lim\limits_{\Delta z \to 0} \left[\dfrac{2\bar{z}\overline{\Delta z} + (\overline{\Delta z})^2}{\Delta z} \right]$, limit does not exist, not differentiable.

5. $L = \lim\limits_{\Delta z \to 0} \left[\dfrac{-1}{z(z + \Delta z)} \right] = -\dfrac{1}{z^2}, f'(z) = -\dfrac{1}{z^2}, z \neq 0$.

6. $L = \lim\limits_{\Delta z \to 0} \left[\dfrac{2}{(1-z)(1-z-\Delta z)} \right] = \dfrac{2}{(1-z)^2}, f'(z) = \dfrac{2}{(1-z)^2}.$

7. $4(1-i).$ **8.** $-(1+\pi i).$ **9.** $4.$

10. $-3.$ **11.** $-2.$ **12.** $1.$

13. Differentiable only at origin, $f'(0) = 0.$ **14.** Differentiable only at origin, $f'(0) = 0.$

15. Cauchy-Riemann equations are not satisfied for any point; no where differentiable.

16. Not differentiable any where.   **17.** Differentiable everywhere, $f'(z) = 3z^2.$

18. Differentiable only at origin, $f'(0) = 0.$ **19.** Not differentiable.

20. $f(z) = \bar{z}/(z\bar{z}) = 1/z, z \neq 0,$ differentiable everywhere, $f'(z) = -1/z^2.$

21. $u = -\dfrac{x}{x^2 + (y+1)^2}, v = \dfrac{x^2 + y^2 + 3y + 2}{x^2 + (y+1)^2},$ differentiable everywhere except at $z = -i, f'(z) = -1/(1-iz)^2.$

22. Not differentiable.

23. $f(z) = -ie^z.$ Differentiable for all $z, f'(z) = -ie^z.$

24. Not differentiable.

25. $f(z) = e^{z^2}.$ Differentiable for all $z, f'(z) = 2ze^{z^2}.$

26. $f(z) = (x^2 + y^2)^2 = u + iv, u = (x^2 + y^2)^2, v = 0;$ Cauchy-Riemann equations are satisfied only at the origin.

27. $f(z) = u + iv; u = \dfrac{x^3 - 3xy^2}{x^2 + y^2}, v = \dfrac{y^3 - 3x^2y}{x^2 + y^2}, (x, y) \neq (0, 0); u(0, 0) = v(0, 0) = 0;$

$u_x(0, 0) = \lim\limits_{\Delta x \to 0} \dfrac{u(\Delta x, 0) - u(0, 0)}{\Delta x} = 1, u_y(0, 0) = \lim\limits_{\Delta y \to 0} \dfrac{u(0, \Delta y) - u(0, 0)}{\Delta y} = 0,$

$v_x(0, 0) = \lim\limits_{\Delta x \to 0} \dfrac{v(\Delta x, 0) - v(0, 0)}{\Delta x} = 0, v_y(0, 0) = \lim\limits_{\Delta y \to 0} \dfrac{v(0, \Delta y) - v(0, 0)}{\Delta y} = 1;$

The Cauchy-Riemann equations are satisfied. Now

$\lim\limits_{z \to 0} \dfrac{f(z) - f(0)}{z} = \lim\limits_{\Delta z \to 0} \dfrac{[(\overline{\Delta z})^2 / \Delta z] - 0}{\Delta z} = \lim\limits_{\Delta z \to 0} \left[\dfrac{\overline{\Delta z}}{\Delta z} \right]^2$

Since the limit does not exist, $f'(0)$ does not exist.

28. $f(z) = u + iv, u = \sqrt{|xy|}, v = 0; u(0, 0) = v(0, 0) = 0; u_x(0, 0) = \lim\limits_{\Delta x \to 0} \dfrac{u(\Delta x, 0) - u(0, 0)}{\Delta x} = 0.$

Similarly, we obtain $u_y(0, 0) = 0, v_x(0, 0) = 0$ and $v_y(0, 0) = 0.$ The Cauchy-Riemann equations are satisfied at the origin. Show directly that $f'(0)$ does not exist.

29. From $f'(z) = u_x + iv_x,$ we get $|f'(z)|^2 = u_x^2 + v_x^2 = u_x u_x + v_x v_x = u_x v_y - v_x u_y$ which is the right hand side.

30. Let $\Delta w = \Delta u + i\Delta v$ and $\Delta z = \Delta x + i\Delta y.$ Now

$\lim\limits_{\Delta z \to 0} \left[\text{Re} \, \dfrac{\Delta w}{\Delta z} \right] = \lim\limits_{\Delta z \to 0} T, \quad \text{where } T = \dfrac{\Delta u \Delta x + \Delta v \Delta y}{(\Delta x)^2 + (\Delta y)^2}.$ Since the limit exists, it is same irrespective of the path along which $\Delta z \to 0.$ We find that

$\lim\limits_{\Delta z \to 0} T = \lim\limits_{\Delta x \to 0} \left[\lim\limits_{\Delta y \to 0} T \right] = u_x$ and $\lim\limits_{\Delta z \to 0} T = \lim\limits_{\Delta y \to 0} \left[\lim\limits_{\Delta x \to 0} T \right] = v_y.$ Therefore, u_x and v_y exist and $u_x = v_y.$

31. $\lim_{\Delta z \to 0}\left[\operatorname{Im}\dfrac{\Delta w}{\Delta z}\right] = \lim_{\Delta z \to 0} T$, where $T = \dfrac{\Delta v \Delta x - \Delta u \Delta y}{(\Delta x)^2 + (\Delta y)^2}$. Since the limit exists, it is same irrespective of the path along which $\Delta z \to 0$. We find that

$$\lim_{\Delta z \to 0} T = \lim_{\Delta x \to 0}\left[\lim_{\Delta y \to 0} T\right] = v_x \quad \text{and} \quad \lim_{\Delta z \to 0} T = \lim_{\Delta y \to 0}\left[\lim_{\Delta x \to 0} T\right] = -u_y$$

Therefore, v_x and u_y exist and $v_x = -u_y$.

32. (i) $\dfrac{\partial u}{\partial x} = \dfrac{\partial u}{\partial z}\dfrac{\partial z}{\partial x} + \dfrac{\partial u}{\partial \bar{z}}\dfrac{\partial \bar{z}}{\partial x} = \dfrac{\partial u}{\partial z} + \dfrac{\partial u}{\partial \bar{z}}$ or $\dfrac{\partial}{\partial x} = \dfrac{\partial}{\partial z} + \dfrac{\partial}{\partial \bar{z}}$;

$\dfrac{\partial u}{\partial y} = \dfrac{\partial u}{\partial z}\dfrac{\partial z}{\partial y} + \dfrac{\partial u}{\partial \bar{z}}\dfrac{\partial \bar{z}}{\partial y} = i\left(\dfrac{\partial u}{\partial z} - \dfrac{\partial u}{\partial \bar{z}}\right)$ or $\dfrac{\partial}{\partial y} = i\left(\dfrac{\partial}{\partial z} - \dfrac{\partial}{\partial \bar{z}}\right)$

$\dfrac{\partial^2 u}{\partial x^2} = \left(\dfrac{\partial}{\partial z} + \dfrac{\partial}{\partial \bar{z}}\right)^2 u = \dfrac{\partial^2 u}{\partial z^2} + 2\dfrac{\partial^2 u}{\partial z \partial \bar{z}} + \dfrac{\partial^2 u}{\partial \bar{z}^2}$ and $\dfrac{\partial^2 u}{\partial y^2} = \left[i\left(\dfrac{\partial}{\partial z} - \dfrac{\partial}{\partial \bar{z}}\right)\right]^2 u = -\dfrac{\partial^2 u}{\partial z^2} + 2\dfrac{\partial^2 u}{\partial z \partial \bar{z}} - \dfrac{\partial^2 u}{\partial \bar{z}^2}$

Adding the last two equations, we obtain the result.

(ii) $\dfrac{\partial f}{\partial \bar{z}} = \left[\dfrac{\partial u}{\partial x}\dfrac{\partial x}{\partial \bar{z}} + \dfrac{\partial u}{\partial y}\dfrac{\partial y}{\partial \bar{z}}\right] + i\left[\dfrac{\partial v}{\partial x}\dfrac{\partial x}{\partial \bar{z}} + \dfrac{\partial v}{\partial y}\dfrac{\partial y}{\partial \bar{z}}\right]$

$= \dfrac{1}{2}\left[\dfrac{\partial u}{\partial x} - \dfrac{1}{i}\dfrac{\partial u}{\partial y}\right] + \dfrac{1}{2}i\left[\dfrac{\partial v}{\partial x} - \dfrac{1}{i}\dfrac{\partial v}{\partial y}\right] = \dfrac{1}{2}\left(\dfrac{\partial u}{\partial x} - \dfrac{\partial v}{\partial y}\right) + \dfrac{i}{2}\left(\dfrac{\partial v}{\partial x} + \dfrac{\partial u}{\partial y}\right)$

Now $\dfrac{\partial f}{\partial \bar{z}} = 0$ gives $\dfrac{\partial u}{\partial x} = \dfrac{\partial v}{\partial y}$ and $\dfrac{\partial u}{\partial y} = -\dfrac{\partial v}{\partial x}$.

33. Let $f(z) = u + iv$. Then $f'(z) = u_x + iv_x$. From $f'(z) = f(z)$, we get $u_x = u$ and $v_x = v$. Integrating with respect to x, we obtain $u = e^x g(y)$ and $v = e^x h(y)$, where $g(y)$ and $h(y)$ are arbitrary functions. Since $f'(z)$ exists, $f(z)$ satisfies the Cauchy-Riemann equations $u_x = v_y$ and $u_y = -v_x$. From $u_x = v_y$, we get $e^x g(y) = e^x h'(y)$ or $h'(y) = g(y)$ and from $u_y = -v_x$, we get $e^x g'(y) = -e^x h(y)$ or $g'(y) = -h(y)$. Therefore, we obtain

$$\dfrac{d^2 h}{dy^2} = -h(y).$$

Solving this differential equation, we get

$$h(y) = c_1 \cos y + c_2 \sin y \text{ and } g(y) = -c_1 \sin y + c_2 \cos y,$$

where c_1, c_2 are arbitrary constants. Hence,

$$f(z) = u + iv = (c_1 + ic_2) e^x (\cos y + i \sin y) = ke^z$$

where $k = c_1 + ic_2$ is an arbitrary constant.

34. $f'(z) = 2z + 1$.   **35.** $f'(z) = 1 - 1/z^2, z \neq 0$.

36. $f(z) = u + iv = \dfrac{1}{r^2}\cos 2\theta - \dfrac{i}{r^2}\sin 2\theta, f'(z) = -\dfrac{2}{z^3}, z \neq 0$.

37. No where.   **38.** No where.   **39.** No where.

40. Every where except at $z = -i$.   **41.** Everywhere.   **42.** No where.

43. Everywhere except at $z = 0$.   **44.** No where.   **45.** Everywhere except when $e^z = 1$.

46. $a = 2, b = -1, c = -1, d = 2$.   **47.** $a = -1, b = -1$.

48. $f'(z) = u_x + iv_x = v_y - iu_y \equiv 0$. Thus $u_x = u_y = 0$ and $v_x = v_y = 0$. Therefore u and v are constants and hence $f(z) = u + iv$ is a constant.

49. $e^z + c$, c is a complex constant

50. $-i(\sin z^2 + \cos z^2) + c$, c is a complex constant.

51. We have $f'(z) = u_x + iv_x$. Therefore, $\text{Re}[f'(z)] = u_x = 3x^2 - 4y - 3y^2$. We get $u = x^3 - 4xy - 3xy^2 + h(y)$, where $h(y)$ is an arbitrary function. Since $u_x = v_y$, we get $v_y = 3x^2 - 4y - 3y^2$ and $v = 3x^2y - 2y^2 - y^3 + k(x)$, where $k(x)$ is an arbitrary function. Now $u_y = -v_x$ gives $h'(y) = 4x - k'(x)$. Since $h(y)$ is a function of y alone, we get $k'(x) - 4x = 0$ and $h'(y) = 0$, that is $h(y) = c_1$ and $k(x) = 2x^2 + c_2$ where c_1 and c_2 are constants. $f(z) = u + iv = z^3 + 2iz^2 + c_1 + ic_2$. Now, $f(1 + i) = 0$ gives $c_1 = 6$ and $c_2 = -2$.

52. Since $f(z)$ and $g(z)$ are analytic functions, we have $u_x = v_y$, $u_y = -v_x$ and $v_x = u_y$, $v_y = -u_x$. Therefore, $u_x = u_y = v_x = v_y = 0$ for all x and y. Thus, u and v are constants. Hence $f(z) = c_1$ and $g(z) = c_2$ where c_1 and c_2 are constants. Using $f(0) = 1$ and $g(0) = i$, we get $c_1 = 1$ and $c_2 = i$. We also get $f'(z) = g'(z) = 0$. Hence $h(1 + i) = 2i$.

53. Use Eqs. (10.68), (10.69).

54. We have $f(z) = u + iv$, $\overline{f(z)} = u - iv$. Since both $f(z)$ and $\overline{f(z)}$ satisfy the Cauchy-Riemann equations, we get $u_x = u_y = 0$ and $v_x = v_y = 0$. Therefore, u and v are both constants. Hence, $f(z)$ is a constant function.

55. We have $\phi_x = \phi_u u_x + \phi_v v_x$ and $\phi_y = \phi_u u_y + \phi_v v_y = -\phi_u v_x + \phi_v u_x$.

Squaring and adding we get $\phi_x^2 + \phi_y^2 = (\phi_u^2 + \phi_v^2)(u_x^2 + v_x^2) = (\phi_u^2 + \phi_v^2)\,|f'(z)|^2$.

Exercise 10.7

1. Let $S = u_y - v_x$ and $T = u_x + v_y$. S and T are continuous. Show that S and T satisfy the Cauchy-Riemann equations.

2. Let $f(z) = u + iv$. Then, $w = |f(z)| = \sqrt{u^2 + v^2}$. Show that $w_{xx} + w_{yy} \neq 0$.

3. Let $f(z) = u + iv$. Then, $w = \ln|f(z)| = [\ln(u^2 + v^2)]/2$. Show that $w_{xx} + w_{yy} = 0$.

4. Let $f(z) = u + iv$. Then, $w = \arg[f(z)] = \tan^{-1}(v/u)$. Differentiate and show that $w_{xx} + w_{yy} = 0$.

5. Let $X = x^2 - y^2$ and $Y = 2xy$. We get

$$u_x = u_X X_x + u_Y Y_x = 2(xu_X + yu_Y),\ u_y = u_X X_y + u_Y Y_y = 2(-yu_X + xu_Y)$$

$$u_{xx} = 2u_X + 4(x^2 u_{XX} + y^2 u_{YY}) + 8xy u_{XY},\ u_{yy} = -2u_X + 4(y^2 u_{XX} + x^2 u_{YY}) - 8xy\,u_{XY}.$$

Therefore, $u_{xx} + u_{yy} = 4(x^2 + y^2)(u_{XX} + u_{YY}) = 0$. Hence, the result.

6. $\nabla^2[|f(z)|^p] = \nabla^2(u^2 + v^2)^{p/2} = \dfrac{\partial^2}{\partial x^2}(u^2 + v^2)^{p/2} + \dfrac{\partial^2}{\partial y^2}(u^2 + v^2)^{p/2}$. Differentiate two times

and use the results $u_x = v_y$, $u_y = -v_x$, $u_{xx} + u_{yy} = 0$, $v_{xx} + v_{yy} = 0$ and $|f'(z)|^2 = u_x^2 + v_x^2$.

7. $v = x^3 - 3xy^2 + c$; $f(z) = iz^3 + ic$, c a real constant.

8. $v = 2\tan^{-1}(y/x) + c$; $f(z) = 2\ln z + ic$, c a real constant.

9. $v = -y/(x^2 + y^2) + c$; $f(z) = (1/z) + ic$, c a real constant.

10. $v = e^x \sin y + c$; $f(z) = e^z + ic$, c a real constant.

11. $v = \sin y \cosh x + c$; $f(z) = \sinh z + ic$, c a real constant.

12. $v = e^{-x}[y \cos y - x \sin y] + c$; $f(z) = ze^{-z} + ic$, c a real constant.

13. $v = e^x[(x^2 - y^2)\sin y + 2xy \cos y]$; $f(z) = z^2 e^z + ic$, c a real constant.

14. $u = x^3 - 3xy^2 + x + c$, $f(z) = z^3 + z + c$, c a real constant.

15. $u = x - y - 2\tan^{-1}(y/x) + c$; $f(z) = 2i \ln z + (1 + i)z + c$, c a real constant.

16. $u = (x + y)/(x^2 + y^2) + c; f(z) = (1 + i)/z + c$, c a real constant.

17. $u = [\ln (x^2 + y^2)]/2 + c; f(z) = \ln z + c$, c a real constant.

18. $u = \cos x \cosh y + c; f(z) = \cos z + c$, c a real constant.

19. $u = e^{-x} (x \cos y + y \sin y) + c; f(z) = z e^{-z} + c$, c a real constant.

20. $v = \theta + c; f(z) = \ln |z| + i \arg z + ic$, c a real constant.

21. $v = (r - (1/r)) \sin \theta + c; f(z) = z + (1/z) + ic$, c a real constant.

22. $v = r^n \sin n\theta + c; f(z) = z^n + ic$, c a real constant.

23. $u = 3r^2 \cos 2\theta - 2r \cos \theta + c, f(z) = 3z^2 - 2z + c$, c a real constant.

24. Level curves are $u = x^2 - y^2 = p$ and $v = xy = q$ where p and q are constnats. From the first curve, we get $2x - 2y (dy/dx) = 0$. Therefore, slope of the first curve is $m_1 = x/y$. From the second curve, we get $y + x (dy/dx) = 0$. Therefore, slope of the second curve is $m_2 = - y/x$. Since, $m_1 m_2 = - 1$, the curves are mutually orthogonal.

25. The level curves are $u = e^x \cos y = p$ and $v = e^x \sin y = q$ where p and q are constants. Slope of the first curve is $m_1 = \cot y$ and slope of the second curve is $m_2 = - \tan y$. Since, $m_1 m_2 = -1$, the curves are mutually orthogonal.

Chapter 11

Integration of Complex Functions

11.1 Introduction

Integration of functions of a complex variable plays a very important role in many areas of science and engineering. Using integration, we shall prove a very important result in the theory of analytic functions, which is: if a function $f(z)$ is analytic in a domain D, then it possesses derivatives of all orders in D, that is, $f'(z), f''(z), \ldots$ are all analytic functions in D. Such a result does not exist in the real variable theory. Also, complex integration approach can be used to evaluate many improper integrals of a real variable, which cannot be evaluated using real integral calculus. The concept of definite integrals for functions of a real variable does not directly extend to the case of complex variables. In the case of a real variable, the path of integration in the definite integral $\displaystyle\int_a^b f(x)\, dx$ is along the x-axis from $x = a$ to $x = b$. That is, the path of integration is along a straight line. In complex integration, the path could be along any curve from $z = a$ to $z = b$.

11.2 Definite Integrals

A function $\phi(t)$ of the form

$$\phi(t) = \phi_1(t) + i\phi_2(t) \tag{11.1}$$

where $\phi_1(t)$ and $\phi_2(t)$ are real functions of t is called a complex function of the real variable t. The function $\phi(t)$ is integrable over an interval $[a, b]$, if the real functions $\phi_1(t)$ and $\phi_2(t)$ are integrable over this interval. We write

$$\int_a^b \phi(t)\, dt = \int_a^b [\phi_1(t) + i\phi_2(t)]\, dt = \int_a^b \phi_1(t)\, dt + i\int_a^b \phi_2(t)\, dt. \tag{11.2}$$

The real integrals $\displaystyle\int_a^b \phi_1(t)\, dt$ and $\displaystyle\int_a^b \phi_2(t)\, dt$ can be evaluated by using the familiar rules of integration of real functions.

Remark 1

If either a or b or both a and b are infinite or if $\phi_1(t)$ or $\phi_2(t)$ or both $\phi_1(t)$ and $\phi_2(t)$ have infinite discontinuity at a or at b (a, b finite) or at some point in this interval, then the integral (11.2) is called an *improper integral*.

Example 11.1   Evaluate $\displaystyle\int_a^b \phi(t)\,dt$, where

(i) $\phi(t) = t + it^2$, $a = 0$, $b = 1$,

(ii) $\phi(t) = te^{-it}$, $a = 0$, $b = \pi$,

(iii) $\phi(t) = (t - i)^{-1}$, $t \neq i$, $a = 0$, $b = 1$,

(iv) $\phi(t) = te^{-t^2} + 2i/\sqrt{t}$, $a = 0$, $b = 1$.

Solution   When $\phi(t) = \phi_1(t) + i\,\phi_2(t)$, we have

$$I = \int_a^b \phi(t)\,dt = \int_a^b \phi_1(t)\,dt + i\int_a^b \phi_2(t)\,dt.$$

(i) $\phi_1(t) = t$, $\phi_2(t) = t^2$, $a = 0$, $b = 1$.

Therefore,
$$I = \int_0^1 t\,dt + i\int_0^1 t^2\,dt = \frac{1}{2} + \frac{1}{3}\,i.$$

(ii) $\phi(t) = te^{-it} = t(\cos t - i\sin t)$. Hence, $\phi_1(t) = t\cos t$, $\phi_2(t) = -t\sin t$, $a = 0$, $b = \pi$.

Therefore,
$$I = \int_0^\pi t\cos t\,dt - i\int_0^\pi t\sin t\,dt$$

$$= \left[t\sin t + \cos t\right]_0^\pi - i\left[-t\cos t + \sin t\right]_0^\pi = -2 - i\pi.$$

(iii) $\phi(t) = \dfrac{1}{t - i} = \dfrac{t + i}{(t - i)(t + i)} = \dfrac{t + i}{t^2 + 1}$.

Hence,
$$\phi_1(t) = \frac{t}{t^2 + 1},\quad \phi_2(t) = \frac{1}{t^2 + 1},\quad a = 0, b = 1.$$

Therefore,
$$I = \int_0^1 \frac{t}{t^2 + 1}\,dt + i\int_0^1 \frac{1}{t^2 + 1}\,dt$$

$$= \frac{1}{2}\left[\ln(t^2 + 1)\right]_0^1 + i\left[\tan^{-1} t\right]_0^1 = \frac{1}{2}\ln 2 + \frac{\pi i}{4}.$$

(iv) $\phi_1(t) = te^{-t^2}$, $\phi_2(t) = 2/\sqrt{t}$.

Therefore,
$$I = \int_0^1 te^{-t^2}\,dt + 2i\int_0^1 \frac{dt}{\sqrt{t}}.$$

Since $\phi_2(t)$ becomes infinite at $a = 0$, we write

$$I = \int_0^1 te^{-t^2}\,dt + 2i\left[\lim_{\varepsilon \to 0}\int_\varepsilon^1 \frac{dt}{\sqrt{t}}\right] = -\frac{1}{2}\left[e^{-t^2}\right]_0^1 + 2i\left\{\lim_{\varepsilon \to 0}\left[(2\sqrt{t})\right]_\varepsilon^1\right\}$$

$$= -\frac{1}{2}(e^{-1} - 1) + 2i\left[\lim_{\varepsilon \to 0}(2 - 2\sqrt{\varepsilon})\right] = -\frac{1}{2}(e^{-1} - 1) + 4i.$$

11.2.1   Curves in the Complex Plane

Let $x(t)$ and $y(t)$ be two continuous functions of a real variable t, $a \leq t \leq b$. If we write

$$z = z(t) = x(t) + i\, y(t), \quad a \leq t \leq b \tag{11.3}$$

then, as the parameter t varies from $t = a$ to $t = b$, the point $z = z(t)$ traces a curve C in the complex plane starting at the point $z(a) = x(a) + i\, y(a)$ and terminating at the point $z(b) = x(b) + i\, y(b)$. The direction of the curve C, as t increases is taken as the positive direction.

The curve C is said to be closed if $z(a) = z(b)$. If the curve is not closed, it is often called an *arc*. A curve C is said to be *simple* (Fig. 11.1a) if it does not intersect itself, that is $z(t_1) \neq z(t_2)$, whenever $t_1 \neq t_2$, with the only exception that $z(a) = z(b)$. Thus, except when $z(a) = z(b)$, there is one-to-one correspondence between the points on the simple curve and the value of t in the interval $a \leq t \leq b$. A curve C which intersects itself is not simple (Fig. 11.1b).

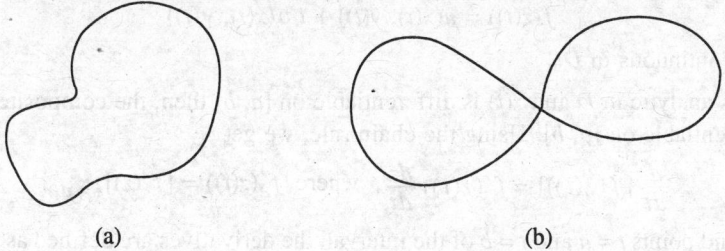

(a)   (b)

Fig. 11.1.   (a) Simple curve (b) not simple curve.

Let $z(t)$ as defined by Eq. (11.3) be a simple curve. Then, it has the following properties.

(i) $z(t)$ has a limit at $t = t^*$, if both $x(t)$ and $y(t)$ have limits at $t = t^*$ and

$$\lim_{t \to t^*} z(t) = \lim_{t \to t^*} x(t) + i \lim_{t \to t^*} y(t).$$

At the end values $t = a$ and $t = b$, the limits are to be taken as one sided limits.

(ii) $z(t)$ is continuous in the interval $[a, b]$, if and only if both $x(t)$ and $y(t)$ are continuous in this interval.

(iii) $z(t)$ is piecewise continuous, if it consists of a finite number of continuous arcs, that is there are a finite number of subintervals $t_{k-1} < t < t_k$, $k = 1, 2, ..., N$, $t_0 = a$, $t_N = b$, on each of which $z(t)$ is continuous and has a limit as $t \to t_{k-1}$ and $t \to t_k$ through values in this interval. Thus, a piecewise continuous function is continuous on $[a, b]$ except for, at most, a finite number of finite jumps in the interval.

(iv) $z(t)$ is differentiable in the interval $[a, b]$, if both $x(t)$ and $y(t)$ are differentiable in this interval. We write

$$\frac{d}{dt}[z(t)] = z'(t) = \frac{d}{dt}[x(t)] + i\frac{d}{dt}[y(t)] = x'(t) + i\, y'(t).$$

If $z'(t)$ exists and is continuous in the interval $[a, b]$, then $z(t)$ is said to be continuously differentiable in $[a, b]$.

(v) The curve defined by $z = z(t)$ is said to be *smooth*, if $z(t)$ is continuously differentiable and $z'(t) \neq 0$ for all t, $a \leq t \leq b$. Geometrically, this means that the curve C has a unique tangent at each of the points whose direction varies continuously as we traverse C.

(vi) The curve defined by $z = z(t)$ is called a *contour*, if it is smooth or piecewise smooth, that is, it consists of a finite number of smooth arcs.

(vii) If C is the curve $z = z(t)$, $a \leq t \leq b$, then

$$-C : z = z(-t), -b \leq t \leq -a.$$

Thus, the curves C and $-C$ have the same trace but opposite directions. The counter clockwise (anti-clockwise) direction of the curve is taken to be the positive direction.

(viii) Let $f(z) = u(x, y) + i\, v\,(x, y)$ be a complex valued function defined in a domain D which contains the curve $z = z(t)$, $a \leq t \leq b$, where $z(t)$ is continuous on $[a, b]$. Then

(a) if $f(z)$ is continuous in the domain D, then the composite function

$$f(z(t)) = u(x(t), y(t)) + i\, v(x(t), y(t))$$

is also continuous in D.

(b) If $f(z)$ is analytic in D and $z(t)$ is differentiable on $[a, b]$ then, the composite function $f(z(t))$ is differentiable on $[a, b]$. Using the chain rule, we get

$$\frac{d}{dt}[f(z(t))] = f'(z(t))\frac{dz}{dt}, \quad \text{where } f'(z(t)) = [f'(z)]_{z = z(t)}.$$

At the end points $t = a$ and $t = b$ of the interval, the derivatives are defined as one sided limits.

Remark 2

Let C be a simple closed contour. Then, C separates the complex plane into two distinct regions, the inside and outside of C (with C as the common boundary), one side of which is bounded and the other side is unbounded. This result is called the *Jordan curve lemma*.

Example 11.2 Give a parametric representation of the simple, closed, smooth curve, whose trace is the ellipse $9x^2 + y^2 = 9$ traversed in the (i) anti-clockwise direction, (ii) clockwise direction.

Solution From the given equation of the ellipse, we have

$$\frac{x^2}{1} + \frac{y^2}{9} = 1.$$

We take $x = \cos t$ and $y = 3 \sin t$, $0 \leq t \leq 2\pi$ as the parametric form of the given ellipse. The real functions $\sin t$ and $\cos t$ are continuous and differentiable and

$$z'(t) = x'(t) + iy'(t) = -\sin t + 3i \cos t \neq 0, \, 0 \leq t \leq 2\pi.$$

Therefore, the curve

$$C : z(t) = \cos t + 3i \sin t, \, 0 \leq t \leq 2\pi$$

is a smooth curve. Since $z(0) = z(2\pi) = 1$, the curve is also closed. The curve C is traversed in the anti-clockwise direction. In the clockwise direction, we have

$$-C : z(-t) = \cos(-t) + 3i \sin(-t) = \cos t - 3i \sin t, \, -2\pi \leq t \leq 0.$$

Example 11.3 Find a parametric representation for the simple, smooth curve, whose trace is the line $y = 3x$ with the initial point at $1 + 3i$ and the terminating point at $2 + 6i$.

Solution Taking t as a parameter, we can write the parametric representation of the curve as

$$C: z(t) = x(t) + iy(t) = t + 3it, \ 1 \le t \le 2.$$

For $t = 1$, we have $z(1) = 1 + 3i$ and for $t = 2$, we have $z(2) = 2 + 6i$, which are the end points of the straight line path.

Example 11.4 Trace the curve $C: z(t) = (1 + e^{-t}) + i(1 + 2e^{-t})$ for (i) $0 \le t < \infty$, (ii) $-\infty < t < \infty$.

Solution Since $1 + 2e^{-t} = 2(1 + e^{-t}) - 1$, we can write the curve as part of the line segment $y = 2x - 1$, where $x = 1 + e^{-t}$.

(i) The initial point corresponding to $t = 0$ is $2 + 3i$ or $(x, y) = (2, 3)$. As $t \to \infty$, we obtain $(x, y) \to (1, 1)$ which does not lie on the curve. The graph of the curve is given in Fig. 11.2 (i). The curve is simple, smooth but not closed.

(ii) As $t \to -\infty$, $x(t) \to \infty$ and $y(t) \to \infty$, therefore, $z(t) \to \infty$. As $t \to \infty$, $x(t) \to 1$ and $y(t) \to 1$, therefore, $z(t) \to 1 + i$. Both the points do not lie on the curve. The graph of the curve is given in Fig. 11.2 (ii). The curve is simple, smooth but not closed.

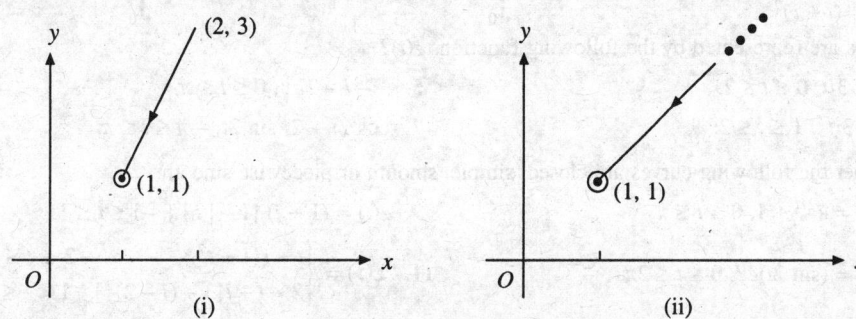

Fig. 11.2. Example 11.4.

Example 11.5 State whether the following curves

(a) $z(t) = \begin{cases} t, & -1 \le t < 1, \\ e^{i(t-1)}, & 1 \le t \le \pi + 1, \end{cases}$

(b) $z(t) = (1 - \cos t)\, e^{it}, \ 0 \le t \le 2\pi$

are closed, simple, smooth or piecewise smooth.

Solution

(a) $z(-1) = -1$, $z(\pi + 1) = -1$. Since $z(-1) = z(\pi + 1)$, the curve is closed.

Also $$\lim_{t \to 1^-} z(t) = \lim_{t \to 1^+} z(t) = 1.$$

Therefore, the curve is continuous in the interval $-1 \le t \le (\pi + 1)$. Since, the curve is one-to-one in the whole interval, it is simple. The derivative of $z(t)$ is given by

$$z'(t) = \begin{cases} 1, & -1 < t < 1 \\ ie^{(t-1)i}, & 1 < t < (\pi + 1). \end{cases}$$

We have

$$\lim_{t \to 1^-} z'(t) = 1, \quad \lim_{t \to 1^+} z'(t) = i, \quad \text{and} \quad \lim_{t \to -1^+} z'(t) = 1, \quad \lim_{t \to (\pi+1)^-} z'(t) = -i.$$

Thus, the curve is not differentiable at $t = 1$ and also at the other point where the curves meet. Also, $z(t)$ is continuously differentiable in the intervals $-1 < t < 1$ and $1 < t < (\pi + 1)$ and $z'(t) \neq 0$ for any t in these intervals. Hence, the given curve is closed, simple and piecewise smooth.

(b) $z(0) = z(2\pi) = 0$. Therefore, the curve is closed. The curve is one-to-one and hence simple. The derivative of $z(t)$ is given by

$$z'(t) = [\sin t + i (1 - \cos t)]e^{it},\ 0 \le t \le 2\pi.$$

Now, $z'(t)$ is continuous for all t in the given interval and $z'(0) = 0$ and $z'(2\pi) = 0$. Hence, the curve is simple, closed but not smooth.

Exercise 11.1

Evaluate the following integrals.

1. $\displaystyle\int_0^1 \dfrac{dt}{(t+i)^2}$
2. $\displaystyle\int_0^{2\pi} e^{int}\, dt.$
3. $\displaystyle\int_0^{\infty} e^{-(1-i)t}\, dt.$

What curves are represented by the following functions $z(t)$?

4. $2 + 3it,\ 0 \le t \le 2.$

5. $1 - i - 2e^{it},\ 0 \le t \le \pi.$

6. $t + 3it^2,\ 1 \le t \le 2.$

7. $\cos 2t + 2i \sin 2t,\ -\pi \le t \le \pi.$

State whether the following curves are closed, simple, smooth or piecewise smooth.

8. $z(t) = e^{2it} + 1,\ 0 \le t \le \pi.$

9. $z(t) = (1 + i)\,[1 - |t|\,],\ -1 \le t \le 1.$

10. $z(t) = (\sin 2t)e^{it},\ 0 \le t \le 2\pi.$

11. $z(t) = \begin{cases} t + i(1 + t^2), & -2 \le t \le 1 \\ 2 - t + i[3 - (t - 2)^2], & 1 < t \le 4. \end{cases}$

12. $z(t) = \begin{cases} (1 + i)t, & 0 \le t \le 2 \\ 4i + (1 - i)t, & 2 < t \le 3 \\ 6 + i - t, & 3 < t \le 6. \end{cases}$

Represent the following curves in the form $z = z(t)$.

13. $|z - 3 + 4i| = 4.$

14. $y = 1/x,\ 1 \le x \le 3.$

Find a simple, piecewise smooth curve in the form $z = z(t)$, whose trace and direction is as given below.

15. The circle $|z - i| = 2$, traversed clockwise.

16. The upper half of the ellipse $\dfrac{(x - x_0)^2}{a^2} + \dfrac{(y - y_0)^2}{b^2} = 1$, traversed counter-clockwise

17. The branch of the hyperbola $xy = 1$, lying in the first quadrant and traversed clockwise.

18. The boundary of the disk $|z| < 1$, Im $z > 0$ traversed counter-clockwise.

19. The triangle with vertices at $z = 0$, $z = 1$ and $z = 1 + i$, traversed counter-clockwise.

20. The square with vertices at $z = 1 \pm i$, $z = -1 \pm i$ traversed counter-clockwise.

11.2.2   Contour Integrals (Line Integrals in the Complex Plane)

Let $z(t) = x(t) + i\,y(t)$, $a \le t \le b$ represent a simple and smooth curve C. Let $f(z)$ be a continuous function defined at each point of C. Partition the interval $a \le t \le b$ as $a = t_0 < t_1 < t_2 \ldots < t_n = b$. Corresponding to these values of t, we obtain the points $z_0, z_1, \ldots, z_n$ on the curve, where $z_k = z(t_k)$, $k = 0, 1, \ldots, n$ (Fig. 11.3).

Let ξ_k be any arbitrary point in the interval (z_{k-1}, z_k), that is $z_{k-1} < \xi_k < z_k$. Now, form the sum

$$S_n = \sum_{k=1}^{n} f(\xi_k)\, \Delta z_k \qquad (11.4)$$

where $\Delta z_k = z_k - z_{k-1}$ and $|\Delta z_k| = |z_k - z_{k-1}|$ is the length of the chord PQ (Fig. 11.3). Let n, the number of subdivisions increase such that the largest of $|\Delta z_k|$ decreases, that is, as $n \to \infty$, $\max |\Delta z_k| \to 0$. If the sum S_n in Eq. (11.4) approaches a limit as $n \to \infty$, regardless of the choice of points z_k and ξ_k, then $f(z)$ is said to be integrable along the curve C and the limit of S_n denoted by

$$\lim_{\max |\Delta z_k| \to 0} S_n = \int_C f(z)\, dz \qquad (11.5)$$

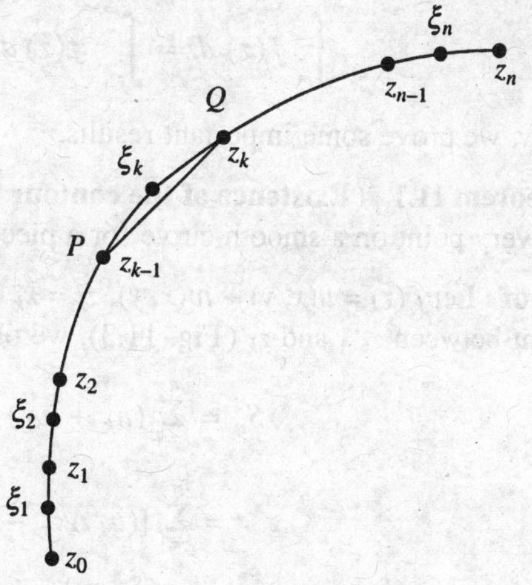

Fig. 11.3.   Contour integral.

is called the *contour integral* or the *line integral* of $f(z)$ along the curve C. When C is a closed curve, we often write $\int_C f(z)\, dz$ as $\oint_C f(z)\, dz$. The curve C is called the *path of integration*.

Remark 3

This result also holds when C is a piecewise smooth curve and $f(z)$ is a piecewise continuous function defined at each point of C.

Properties of contour integrals

Let $f(z)$ and $g(z)$ be two continuous functions defined on a piecewise smooth curve C and α, β be arbitrary complex constants. Using the definition of the contour integral given in Eq. (11.5), the following results can be easily proved.

P1
$$\int_{C^*} f(z)\, dz = -\int_C f(z)\, dz$$

where C^* denotes the curve traversed in the opposite direction of C.

P2
$$\int_C [\alpha f(z) + \beta g(z)]\, dz = \alpha \int_C f(z)\, dz + \beta \int_C f(z)\, dz$$

P3 If C consists of two piecewise smooth arcs C_1 and C_2 joined end-to-end (Fig. 11.4), then

$$\int_C f(z)\, dz = \int_{C_1} f(z)\, dz + \int_{C_2} f(z)\, dz$$

$$(11.6)$$

Similarly, if C consists of n piecewise smooth arcs $C_1, C_2 \ldots, C_n$ joined end-to-end, that is, the terminal point of C_k coincides with the initial point of C_{k+1} for $k = 1, 2, \ldots, n-1$, then

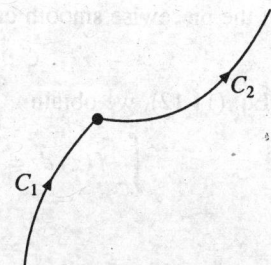

Fig. 11.4.   Integration along piecewise smooth arc.

$$\int_C f(z)\, dz = \int_{C_1} f(z)\, dz + \int_{C_2} f(z)\, dz + \ldots + \int_{C_n} f(z)\, dz. \tag{11.7}$$

Now, we prove some important results.

Theorem 11.1 **(Existence of the contour integral)** If $f(z)$ is continuous (or piecewise continuous) at every point on a smooth curve (or a piecewise smooth curve) C, then $f(z)$ is integrable.

Proof Let $f(z) = u(x, y) + iv(x, y)$, $z_k = x_k + iy_k$ and $\Delta z_k = \Delta x_k + i\Delta y_k$. If $\xi_k = \delta_k + i\eta_k$ is an arbitrary point between z_{k-1} and z_k (Fig. 11.3), we obtain from Eq. (11.4)

$$S_n = \sum_{k=1}^{n} (u_k + iv_k)(\Delta x_k + i\Delta y_k)$$

$$= \sum_{k=1}^{n} [(u_k \Delta x_k - v_k \Delta y_k) + i(v_k \Delta x_k + u_k \Delta y_k)] \tag{11.8}$$

where $u_k = u(\delta_k, \eta_k)$ and $v_k = v(\delta_k, \eta_k)$.

Now, as $n \to \infty$ such that max $\Delta x_k \to 0$ and max $\Delta y_k \to 0$, the first part in Eq. (11.8) approaches the line integral

$$\int_C (u\,dx - v\,dy) \tag{11.9}$$

and the second part in Eq. (11.8) approaches the line integral

$$\int_C (v\,dx + u\,dy). \tag{11.10}$$

Since $f(z)$ is continuous, the real functions $u(x, y)$ and $v(x, y)$ are continuous and the real integrals defined in Eqs. (11.9) and (11.10) exist. Hence, we obtain from Eq. (11.8)

$$\lim_{n\to\infty} S_n = \int_C f(z)\, dz = \int_C (u\,dx - v\,dy) + i \int_C (v\,dx + u\,dy) \tag{11.11}$$

or

$$\int_C f(z)\, dz = \int_C (u + iv)(dx + i\,dy). \tag{11.12}$$

Remark 4

From Eq. (11.11), we observe that a complex line integral can be expressed in terms of two real line integrals, which can be evaluated by using the rules of integration for real functions.

Remark 5

Suppose that the piecewise smooth curve C is represented in the parametric form as

$$C : z(t) = x(t) + i\, y(t),\ a \le t \le b.$$

Then, from Eq. (11.12), we obtain

$$\int_C f(z)\, dz = \int_C [u(t) + iv(t)] \left[\frac{dx(t)}{dt} + i\, \frac{dy(t)}{dt} \right] dt$$

$$= \int_C [u(t) + iv(t)] \frac{dz}{dt}\, dt$$

or

$$\int_C f(z)\, dz = \int_a^b f(z(t))\, z'(t)\, dt. \tag{11.13}$$

Theorem 11.2 Let C be a piecewise smooth curve

$$C : z = z(t) = x(t) + iy(t), a \le t \le b.$$

Then,

$$\left| \int_C f(z)\, dz \right| \le ML$$

where L is the length of the curve C and $|f(z)| \le M$ everywhere on C.

Proof From the definition of a contour integral (see Eq. (11.4)), we get

$$|S_n| = \left| \sum_{k=1}^{n} f(\xi_k) \Delta z_k \right| \le \sum_{k=1}^{n} |f(\xi_k)| \, |\Delta z_k|$$

$$\le M \sum_{k=1}^{n} |z_k - z_{k-1}| \tag{11.14}$$

where $|f(z)| \le M$ on C. Now, $\sum_{k=1}^{n} |z_k - z_{k-1}| = L^*$, is the sum of the lengths of the chords whose end points are $z_0, z_1, \ldots, z_n$ (Fig. 11.3). Since a straight line path is the shortest distance between any two points, $|z_k - z_{k-1}|$ does not exceed the length of the arc joining the points z_{k-1} and z_k. Therefore, as $n \to \infty$ such that $\max |\Delta z_k| \to 0$, we have $L^* \to L$, the length of the curve C. Hence, from Eq. (11.14), we obtain

$$\left| \int_C f(z)\, dz \right| = \lim_{n \to \infty} |S_n| \le ML. \tag{11.15}$$

This inequality is called the *ML-inequality*.

Remark 6

For a smooth curve C

$$C : z = z(t) = x(t) + iy(t),\ a \le t \le b$$

the length of the curve C is given by

$$L = \int_a^b |z'(t)|\, dt = \int_a^b \sqrt{[x'(t)]^2 + [y'(t)]^2}\, dt. \tag{11.16}$$

Example 11.6 Using the definition of contour integral, show that for any smooth curve between z_0 and z_m

(i) $\displaystyle \int_C dz = z_m - z_0,$ (ii) $\displaystyle \int_C z\, dz = \frac{1}{2}(z_m^2 - z_0^2).$

Solution

(i) We have $f(z) = 1$. Using Eqs. (11.4) and (11.5), we get

$$I = \lim_{\max |\Delta z_k| \to 0} [\Delta z_1 + \Delta z_2 + \ldots + \Delta z_m]$$

$$= [(z_1 - z_0) + (z_2 - z_1) + \ldots + (z_m - z_{m-1})] = z_m - z_0.$$

(ii) We have $f(z) = z$. Using Eqs. (11.4) and (11.5), we get

$$I = \lim_{\max |\Delta z_k| \to 0} [\xi_1 \Delta z_1 + \xi_2 \Delta z_2 + \ldots + \xi_m \Delta z_m].$$

Since, the integral exists for any choice of ξ_k, we first choose in the limit, $\xi_k = z_{k-1}$. We get

$$I = \lim [z_0 \, \Delta z_1 + z_1 \, \Delta z_2 + \ldots + z_{m-1} \, \Delta z_m]. \tag{11.17}$$

Now, choose in the limit, $\xi_k = z_k$. We get

$$I = \lim [z_1 \Delta z_1 + z_2 \, \Delta z_2 + \ldots + z_m \, \Delta z_m]. \tag{11.18}$$

Adding Eqs. (11.17) and (11.18) and using $\Delta z_k = z_k - z_{k-1}$, we obtain

$$2I = \lim [(z_1 + z_0)(z_1 - z_0) + (z_2 + z_1)(z_2 - z_1) + \ldots + (z_m + z_{m-1})(z_m - z_{m-1})]$$

$$= (z_1^2 - z_0^2) + (z_2^2 - z_1^2) + \ldots + (z_m^2 - z_{m-1}^2) = z_m^2 - z_0^2.$$

Therefore, $I = \dfrac{1}{2} (z_m^2 - z_0^2)$.

Example 11.7 Evaluate the integral $I = \displaystyle\int_C z^n dz, n = 0, \pm 1, \pm 2, \ldots$ where $C : |z| = r$ is traversed in the counter clockwise direction.

Solution Represent the circle $C : |z| = r$ in the parametric form as

$$z(t) = r(\cos t + i \sin t), \ 0 \le t \le 2\pi.$$

Then

$$z'(t) = r(- \sin t + i \cos t)$$

and

$$f[z(t)] = [z(t)]^n = r^n (\cos t + i \sin t)^n = r^n (\cos nt + i \sin nt).$$

Using Eq. (11.13), we obtain

$$I = \int_C f(z) \, dz = \int_0^{2\pi} f[z(t)] \, z'(t) \, dt$$

$$= r^{n+1} \int_0^{2\pi} [\cos nt + i \sin nt][- \sin t + i \cos t] \, dt$$

$$= r^{n+1} \int_0^{2\pi} [- \sin (n + 1)t + i \cos (n + 1)t] \, dt \tag{11.19}$$

$$= \frac{r^{n+1}}{n + 1} [\cos (n + 1)t + i \sin (n + 1)t]_0^{2\pi}, n \ne - 1 \tag{11.20}$$

$$= 0.$$

For $n = - 1$, we obtain from Eq. (11.19)

$$I = \int_0^{2\pi} i \, dt = 2\pi i.$$

Hence,

$$\int_C z^n \, dz = \begin{cases} 2\pi i, & n = - 1, \\ 0, & n \ne - 1. \end{cases}$$

Thus, we have the interesting result that

$$\int_C \frac{dz}{z} = 2\pi i,$$

where C is any circle $|z| = r$ traversed in the positive direction.

Example 11.8 Evaluate the integral $I = \int_C (z - z_0)^n \, dz, n = 0, \pm 1, \pm 2, \ldots$, where $C: |z - z_0| = r$ is traversed in the counter-clockwise direction.

Solution   Set $z - z_0 = w$. The given problem reduces to $I = \int_C w^n \, dw$, where C is the circle $|w| = r$, which is same as Example 11.7. Hence, we have

$$\int_C (z - z_0)^n \, dz = \begin{cases} 2\pi i, & n = -1, \\ 0, & n \neq -1. \end{cases}$$

Thus, we have the interesting result that $I_k = \int_C \frac{dz}{z - z_0} = 2\pi i$, where C is any circle $|z - z_0| = r$ traversed in the positive direction.

Example 11.9   Evaluate the integral $I_k = \int_C f_k(z) \, dz, k = 1, 2, 3$, where

(i) $f_1(z) = z$,   (ii) $f_2(z) = \bar{z}$,   (iii) $f_3(z) = (z - i)^n, n = 0, \pm 1, \pm 2, \ldots$

and C is the closed curve

$$C : z = z(t) = i + e^{it}, 0 \leq t \leq 2\pi.$$

Solution   On C, we have $z'(t) = ie^{it}$. Since $z'(t) \neq 0$ for any t, C is a smooth curve.

(i) $f_1(z) = z = i + e^{it}$. Hence,

$$I_1 = \int_0^{2\pi} f_1(z(t)) \, z'(t) \, dt = \int_0^{2\pi} (i + e^{it}) i e^{it} \, dt$$

$$= \int_0^{2\pi} [-e^{it} + ie^{2it}] \, dt = \left[-\frac{1}{i} e^{it} + \frac{1}{2} e^{2it} \right]_0^{2\pi} = 0.$$

(ii) $f_2(z) = \bar{z} = -i + e^{-it}$. Hence,

$$I_2 = \int_0^{2\pi} f_2(z(t)) \, z'(t) \, dt = \int_0^{2\pi} (-i + e^{-it}) i e^{it} \, dt$$

$$= \int_0^{2\pi} (e^{it} + i) \, dt = \left[\frac{1}{i} e^{it} + it \right]_0^{2\pi} = 2\pi i.$$

(iii) $f_3(z) = (z - i)^n = e^{int}$. Hence,

$$I_3 = \int_0^{2\pi} f_3(z(t)) \, z'(t) \, dt = \int_0^{2\pi} e^{int} \cdot ie^{it} \, dt = i \int_0^{2\pi} e^{i(n+1)t} \, dt.$$

For $n = -1$, we obtain, $I_3 = i \int_0^{2\pi} dt = 2\pi i$.

For $n \neq -1$, we obtain, $I_3 = \frac{i}{(n+1)i} \left[e^{i(n+1)t} \right]_0^{2\pi} = 0$.

Example 11.10 Evaluate the following integrals

(a) $\int_C z^2 dz$, where C is the arc of the circle $|z| = 2$ from $\theta = 0$ to $\theta = \pi/3$.

(b) $\int_C \frac{2z+3}{z} dz$, where C is

   (i)   upper half of the circle $|z| = 2$ in the clockwise direction,

   (ii) lower half of the circle $|z| = 2$ in the anti-clockwise direction,

   (iii) is the circle $|z| = 2$ in the anti-clockwise direction.

(c) $\int_C |z| \, dz$, where C is the left half of the unit circle in the clockwise direction.

Solution

(a) Let $z = 2e^{i\theta}$. We have $dz = 2ie^{i\theta} d\theta$. Hence,

$$I = \int_C z^2 dz = \int_0^{\pi/3} [2e^{i\theta}]^2 \, (2ie^{i\theta}) \, d\theta = 8i \int_0^{\pi/3} e^{3i\theta} d\theta$$

$$= \frac{8i}{3i} \left[e^{3i\theta} \right]_0^{\pi/3} = \frac{8}{3} (e^{\pi i} - 1) = -\frac{16}{3}.$$

(b) Let $z = 2e^{i\theta}$. We have $dz = 2ie^{i\theta} d\theta$. Hence,

$$I = \int_C \left[\frac{2(2e^{i\theta}) + 3}{2e^{i\theta}} \right] (2ie^{i\theta}) \, d\theta = i \int_C (4e^{i\theta} + 3) \, d\theta = 4e^{i\theta} + 3i\theta.$$

Integrating along the required paths (Fig. 11.5), we obtain

(i) $I = [(4e^{i\theta} + 3i\theta)]_\pi^0 = 8 - 3\pi i$.

(ii) $I = [(4e^{i\theta} + 3i\theta)]_{-\pi}^0 = 8 + 3\pi i$.

(iii) $I = [(4e^{i\theta} + 3i\theta)]_0^{2\pi} = 6\pi i$.

(c) Let $z = e^{i\theta}$. We have $dz = ie^{i\theta} d\theta$ and $|z| = 1$.

Hence, $I = \int_{-\pi/2}^{\pi/2} ie^{i\theta} d\theta = [e^{i\theta}]_{-\pi/2}^{\pi/2} = 2i$.

Example 11.11 Evaluate the integral

$$I = \int_C \text{Re}(z^2) \, dz$$

from 0 to $2 + 4i$ along the

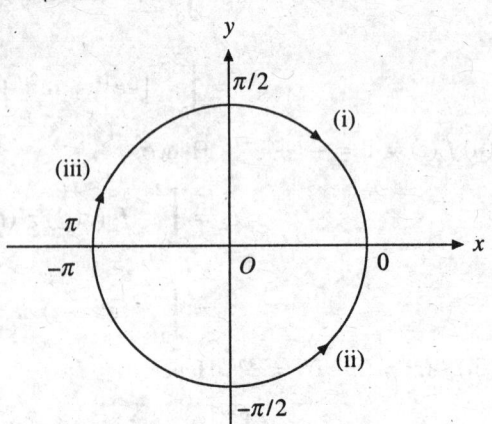

Fig. 11.5. Example 11.10.

(a) line segment joining the points (0, 0) and (2, 4),

(b) x-axis from 0 to 2, and then vertically to 2 + 4i,

(c) parabola $y = x^2$.

Solution   The curve C is sketched in Fig. 11.6.

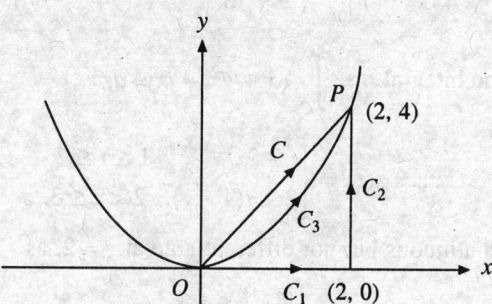

Fig. 11.6.   Example 11.11.

(a) The curve C is the line segment joining the points $O(0, 0)$ and $P(2, 4)$. The equation of this line OP in given by

$$z = z(t) = t + 2it = (1 + 2i)t, \; 0 \le t \le 2.$$

Therefore, $z'(t) = 1 + 2i$ and

$$f(z(t)) = \text{Re} \, (z^2(t)) = \text{Re} \, [t^2(1 + 2i)^2] = \text{Re} \, [(-3 + 4i)t^2] = -3t^2.$$

Hence, we obtain

$$I = \int_C f(z(t)) \, z'(t) \, dt = \int_0^2 (-3t^2) \, (1 + 2i) \, dt = -8(1 + 2i).$$

(b) The integration is first carried along C_1 and then along C_2 (Fig. 11.6). Thus, the path of integration consists of two curves C_1 and C_2. The parametric form of C_1 is given by $z = z(t) = t$, $0 \le t \le 2$. The parametric form of C_2 is given by $z = z(t) = 2 + 2it, 0 \le t \le 2$. We have

$$\text{for } C_1 : z'(t) = 1, f(z(t)) = \text{Re}(z^2) = t^2$$

$$\text{for } C_2 : z'(t) = 2i, f(z(t)) = \text{Re} \, [(2 + 2it)^2] = 4 - 4t^2.$$

Hence, we obtain

$$I = \int_{C_1} f(z(t)) \, z'(t) \, dt + \int_{C_2} f(z(t)) \, z'(t) \, dt$$

$$= \int_0^2 t^2 dt + \int_0^2 (4 - 4t^2) 2i \, dt = \left[\frac{t^3}{3} + 2i \left(4t - \frac{4}{3} t^3 \right) \right]_0^2$$

$$= \frac{8}{3} + 2i \left(8 - \frac{32}{3} \right) = \frac{8}{3} (1 - 2i).$$

(c) The parameteric form of the curve $y = x^2$ can be written as

$$z = z(t) = t + it^2, \; 0 \le t \le 2.$$

We have $z'(t) = 1 + 2i \, t$, and $f(z(t)) = \text{Re}[z^2(t)] = \text{Re} \, (t + it^2)^2 = t^2 - t^4.$

Hence, we obtain

$$I = \int_C f(z(t)) \, z'(t) \, dt = \int_0^2 (t^2 - t^4)(1 + 2it) \, dt$$

$$= \left[\frac{1}{3} t^3 - \frac{1}{5} t^5 + 2i \left(\frac{t^4}{4} - \frac{t^6}{6} \right) \right]_0^2 = \left(\frac{8}{3} - \frac{32}{5} \right) + 2i \left(\frac{16}{4} - \frac{64}{6} \right) = -\frac{56}{15} - \frac{40}{3} i.$$

Example 11.12 Evaluate the integral $I = \int_C (x + y^2 - ixy) \, dz$,

where
$$C : z = z(t) = \begin{cases} t - 2i, & 1 \le t \le 2, \\ 2 - i(4 - t), & 2 < t \le 3. \end{cases}$$

Solution The curve C is continuous but not differentiable at $z = 2$, as

$$\frac{dz}{dt} = \begin{cases} 1, & 1 < t \le 2, \\ i, & 2 < t < 3. \end{cases}$$

Also, $dz/dt \ne 0$ for any t. Therefore, the curve C is piecewise smooth.

On the interval $[1, 2]$, we have $z = t - 2i$, that is $x = t, \, y = -2, \, z'(t) = 1$ and

$$f(z) = x + y^2 - ixy = t + 4 + 2it.$$

On the interval $[2, 3]$, we have $z = 2 - i(4 - t)$, that is $x = 2, \, y = t - 4, \, z'(t) = i$ and

$$f(z) = x + y^2 - ixy = 2 + (t - 4)^2 - 2i(t - 4).$$

Hence,
$$I = \int_C (x + y^2 - ixy) \, dz = \int_1^2 f(z) \, z'(t) \, dt + \int_2^3 f(z) \, z'(t) \, dt$$

$$= \int_1^2 [4 + t + 2it] \, dt + \int_2^3 [2 + (t - 4)^2 - 2i(t - 4)]i \, dt$$

$$= \int_1^2 [4 + t + 2it] \, dt + \int_2^3 [t^2 - (8 + 2i)t + 18 + 8i]i \, dt$$

$$= \left[4t + \frac{1}{2} (1 + 2i)t^2 \right]_1^2 + i \left[\frac{1}{3} t^3 - (4 + i)t^2 + (18 + 8i)t \right]_2^3$$

$$= 4 + \frac{3}{2} (1 + 2i) + \left[\frac{19}{3} - 5(4 + i) + (18 + 8i) \right] i$$

$$= \left(4 + \frac{3}{2} - 3 \right) + i \left(3 + \frac{19}{3} - 2 \right) = \frac{5}{2} + \frac{22}{3} i.$$

Example 11.13 Evaluate the integral $I = \int_C \sqrt{z} \, dz$, where $C : \, z = z(t) = e^{it}, \, 0 \le t \le 2\pi$.

Solution We note that the integrand is a multiple valued function. To obtain the correct result, we take the principal value of $\sqrt{z(t)}$ for every t in the interval $[0, 2\pi]$. Hence, we must represent $z(t) = e^{it}$ such that its argument lies in the principal range

$$-\pi < \arg(e^{it}) \leq \pi \tag{11.21}$$

for each t in the interval $0 \leq t \leq 2\pi$.

Since

$$\arg(e^{it}) = t + 2n\pi, n = 0, \pm 1, \pm 2, \ldots,$$

we must choose n, possibly different for different values of t such that the Eq. (11.21) is satisfied. We find that

$$\arg(e^{it}) = \begin{cases} t, & 0 \leq t \leq \pi, \\ t - 2\pi, & \pi < t < 2\pi. \end{cases}$$

Therefore, we get

$$\text{(principal branch) } \sqrt{z(t)} = \begin{cases} e^{it/2}, & 0 \leq t \leq \pi, \\ e^{i(t-2\pi)/2}, & \pi < t \leq 2\pi. \end{cases}$$

We also get

$$\frac{dz}{dt} = \begin{cases} ie^{it}, & 0 \leq t \leq \pi, \\ ie^{i(t-2\pi)}, & \pi < t \leq 2\pi. \end{cases}$$

Hence, we obtain

$$I = \int_0^\pi e^{it/2} (ie^{it})\, dt + \int_\pi^{2\pi} e^{i(t-2\pi)/2}(ie^{i(t-2\pi)})\, dt$$

$$= i \int_0^\pi e^{3it/2}\, dt + i \int_\pi^{2\pi} e^{3i(t-2\pi)/2}\, dt$$

$$= \frac{2}{3}(e^{3\pi i/2} - 1) + \frac{2}{3}(1 - e^{-3\pi i/2}) = \frac{2}{3}(-i - 1) + \frac{2}{3}(1 - i) = -\frac{4i}{3}.$$

Example 11.14 Find the length of the curve $C : z = (1 - i)t^2, -1 \leq t \leq 1$.

Solution   From the given curve $C : z(t) = (1 - i) t^2$, we get

$$z'(t) = 2(1 - i)t \text{ and } |z'(t)| = |2(1 - i)\, t| = 2\sqrt{2}\, |t|.$$

Hence, the length of the curve C is given by

$$L = \int_C |z'(t)|\, dt = \int_{-1}^1 2\sqrt{2}\, |t|\, dt = 4\sqrt{2} \int_0^1 t = 2\sqrt{2}.$$

Alternative   If we write $z(t) = x(t) + iy(t) = (1 - i)t^2$, we get $x(t) = t^2$ and $y(t) = -t^2$. Therefore, $x'(t) = 2t$ and $y'(t) = -2t$.

Hence,

$$L = \int_{-1}^1 \sqrt{[x'(t)]^2 + [y'(t)]^2}\, dt = \int_{-1}^1 \sqrt{4t^2 + 4t^2}\, dt$$

$$= \int_{-1}^1 2\sqrt{2}\, |t|\, dt = 2\sqrt{2}.$$

Example 11.15 Evaluate the integral $I = \displaystyle\int_C (z/\bar{z})\, dz$, where C is the boundary of the half annulus as given in Fig. 11.7.

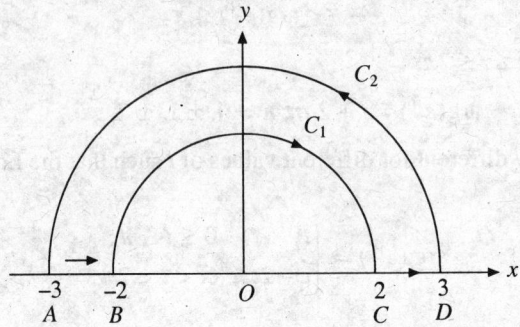

Fig. 11.7. Example 11.15.

Solution Let $z = x + iy$ and $f(z) = z/\bar{z}$. We have

along $AB : y = 0, z = x, f(z) = 1, -3 \le x \le -2, dz = dx$

along $C_1 : z = 2e^{i\theta}, f(z) = e^{2i\theta}, \pi \le \theta \le 0, dz = 2ie^{i\theta}\, d\theta$

along $CD : y = 0, z = x, f(z) = 1, 2 \le x \le 3, dz = dx$

along $C_2 : z = 3e^{i\theta}, f(z) = e^{2i\theta}, 0 \le \theta \le \pi, dz = 3ie^{i\theta}\, d\theta$.

Hence, we obtain

$$I = \int_{-3}^{-2} dx + 2i \int_{\pi}^{0} e^{3i\theta}\, d\theta + \int_{2}^{3} dx + 3i \int_{0}^{\pi} e^{3i\theta}\, d\theta$$

$$= 1 + \frac{4}{3} + 1 - 2 = \frac{4}{3}.$$

Example 11.16 Find an upper bound for the absolute value of the integral $I = \displaystyle\int_C e^z\, dz$, where C is the line segment joining the points $(0, 0)$ and $(1, 2\sqrt{2})$.

Solution The integrand is $e^z = e^{x+iy}$, $0 \le x \le 1$ and $0 \le y \le 2\sqrt{2}$. Now,

$$|f(z)| = |e^x e^{iy}| = |e^x| \le e, \text{ for } 0 \le x \le 1.$$

Therefore, $M = \max |f(z)| = e$.

L = length of C = distance between the points $(0, 0)$ and $(1, 2\sqrt{2})$

$$= \sqrt{(1)^2 + (2\sqrt{2})^2} = \sqrt{9} = 3.$$

Using the *ML*-inequality given by Eq. (11.15), we get

$$|I| \le ML = 3\,e.$$

Example 11.17 Find an upper bound for the absolute value of the integral $I = \displaystyle\int_C e^{(\bar{z})^2}\, dz$, $C : |z| = 1$, where C is traversed in the anti-clockwise direction.

Solution The integrand is $f(z) = e^{(\bar{z})^2} = e^{x^2 - y^2} e^{-2ixy}$. On $|z| = 1$, we have $x = \cos\theta$, $y = \sin\theta$, $0 \le \theta \le 2\pi$.

Therefore, $$|f(z)| = e^{x^2 - y^2} = e^{\cos^2\theta - \sin^2\theta} = e^{\cos 2\theta} \leq e$$

since $|\cos 2\theta| \leq 1$. Therefore, $M = \max |f(z)| = e$. Also, length of the curve is $L = 2\pi$. Using the ML-inequality given by Eq. (11.15), we get $|I| \leq ML = 2\pi e$.

Example 11.18   Find an upper bound for the absolute value of the integral $I = \displaystyle\int_C (e^{2z} - z^2)\, dz$, where C is the contour given in Fig. 11.8.

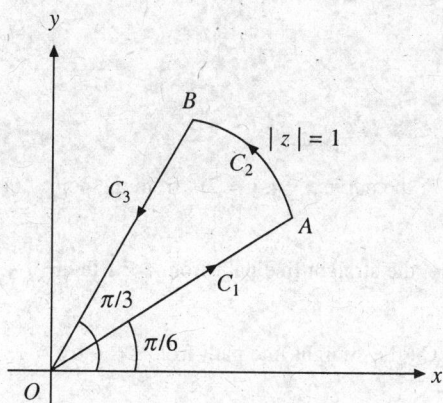

Fig. 11.8.   Contour for Example 11.18.

Solution   We can write

$$I = \int_C f(z)\, dz = \int_{C_1} f(z)\, dz + \int_{C_2} f(z)\, dz + \int_{C_3} f(z)\, dz = I_1 + I_2 + I_3$$

where $f(z) = e^{2z} - z^2$. We have

$$|f(z)| = \left| e^{2(x + iy)} - (x + iy)^2 \right| \leq \left| e^{2(x + iy)} \right| + |x + iy|^2 \leq e^{2x} + 1.$$

Along C_1: Equation of the line joining $O(0, 0)$ and $A(\sqrt{3}/2, 1/2)$ is $y = x/\sqrt{3}, 0 \leq x \leq \sqrt{3}/2$ and $OA = 1$. Therefore,

$$M_1 = \max |f(z)| = e^{\sqrt{3}} + 1, L_1 = \text{length of } OA = 1$$

and $$|I_1| \leq M_1 L_1 = e^{\sqrt{3}} + 1.$$

Along C_2: $x = \cos\theta, y = \sin\theta, \dfrac{\pi}{6} \leq \theta \leq \dfrac{\pi}{3}$. Therefore,

$$M_2 = \max |f(z)| = e^{2\cos(\pi/6)} + 1 = e^{\sqrt{3}} + 1,$$

$$L_2 = \text{length of the arc } AB = \int_{\pi/6}^{\pi/3} \sqrt{\left(\frac{dx}{d\theta}\right)^2 + \left(\frac{dy}{d\theta}\right)^2}\, d\theta = \int_{\pi/6}^{\pi/3} d\theta = \frac{\pi}{3} - \frac{\pi}{6} = \frac{\pi}{6},$$

and
$$| I_2 | \leq M_2 L_2 = \frac{\pi}{6} (e^{\sqrt{3}} + 1).$$

Along C_3 : Equation of the line joining B $(1/2, \sqrt{3}/2)$ and $O(0, 0)$ is $y = \sqrt{3}x$, $1/2 \leq x \leq 0$ and $OB = 1$. Therefore,

$M_3 = \max | f(z) | = e + 1$, $L_3 = $ length of $OB = 1$ and $| I_3 | \leq M_3 L_3 = e + 1$.

Hence,
$$| I | \leq | I_1 | + | I_2 | + | I_3 | = (e + 1) + \left(1 + \frac{\pi}{6} \right) (e^{\sqrt{3}} + 1).$$

Exercise 11.2

Evaluate the following integrals.

1. $\displaystyle\int_C (x^2 - 2ixy)\, dz$, $\quad C$: the curve $x = t$, $y = 2t^2$ from the point $A(1, 2)$ to $B(2, 8)$.

2. $\displaystyle\int_C (x^2 + iy^3)\, dz$, $\quad C$: the straight line path from $z = 1$ to $z = 1 + 2i$.

3. $\displaystyle\int_C (z^2 + z + 2)\, dz$, $\quad C$: the straight line path from $z = i$ to $z = 1$.

4. $\displaystyle\int_C z\, dz$, $\quad C$: left half of the ellipse $\dfrac{x^2}{9} + \dfrac{y^2}{1} = 1$ from $z = i$ to $z = -i$.

5. $\displaystyle\int_C z^i\, dz$, $\quad C : z = e^{it}$, $-\dfrac{\pi}{2} \leq t \leq \dfrac{\pi}{2}$. $\qquad$ 6. $\displaystyle\int_C \dfrac{z}{1 + z^4}\, dz$, $\quad C : z = it$, $0 \leq t \leq 1$.

7. $\displaystyle\int_C (\operatorname{Re} z)\, dz$, $\quad C$: the semicircle $| z | = 1$, $0 \leq \arg z \leq \pi$.

8. $\displaystyle\int_C | z |\, dz$, $\quad C : | z | = 1$, $0 \leq \arg z \leq \pi$. $\qquad$ 9. $\displaystyle\int_C (\bar{z} + i)^{-1}\, dz$, $\quad C : z = i + e^{it}$, $-\dfrac{\pi}{2} \leq t \leq \dfrac{\pi}{2}$.

10. $\displaystyle\int_C z | z |^2\, dz$, $\quad C : z = t + it^3$, $-1 \leq t \leq 1$.

11. $\displaystyle\int_C z\, (\operatorname{Im} z)\, dz$ $\quad C : z = t$, $0 \leq t \leq 1$ and $z = 1 + i(t - 1)$, $1 \leq t \leq 2$.

12. $\displaystyle\int_C f(z)\, dz$, $\quad f(z) = \begin{cases} z, & \operatorname{Re} z > 0, \\ z^2, & \operatorname{Re} z < 0, \end{cases}$ $C : z = e^{it}$, $0 \leq t \leq 2\pi$.

13. $\displaystyle\int_C z^2\, dz$, $\quad C$: (i) $| z | = 1$, $\quad$ (ii) $| z - 2 | = 1$.

14. $\displaystyle\int_C e^{-z}\, dz$, $\quad C$: the straight line path from $z = \pi i$ to $z = 1$.

15. $\displaystyle\int_C | z |\, dz$, $\quad C$: (i) the straight line path from $z = -i$ to $z = i$, (ii) the circle $| z - 1 | = 1$.

16. $\int_C \bar{z}\, dz$, C : (i) upper half of the circle $|z| = 1$ from $z = 1$ to $z = -1$,

        (ii) lower half of the circle $|z| = 1$ from $z = 1$ to $z = -1$.

17. $\int_C \dfrac{z+2}{3z}\, dz$, C : the sector of the circle $|z| = 3$, (i) $0 \le \theta \le \pi/3$, (ii) $\pi \le \theta \le 2\pi$ (iii) $-\pi \le \theta \le \pi$.

18. $\int_C \bar{z}^2\, dz$, C : (i) the straight line path from $z = 0$ to $z = 3 + 2i$

        (ii) real axis from $(0, 0)$ to $(3, 0)$ and then along vertical line from $(3, 0)$ to $(3, 2)$.

19. $\int_C (z^2 + 2iz)\, dz$, C : (i) the straight line path from $z = 2 + 3i$ to $z = 3 - 2i$

        (ii) the straight line path from $z = 2 + 3i$ to $z = 3 + 3i$ and then along the straight line path from $z = 3 + 3i$ to $z = 3 - 2i$.

20. $\int_C z^2\, dz$, C : (i) the line $x = 2y$, (ii) the parabola $x = y^2$ from the point $(0, 0)$ to the point $(4, 2)$.

21. $\int_C |z|^2\, dz$, C : the boundary of the square with vertices at $(0, 0)$, $(2, 0)$, $(2, 2)$ and $(0, 2)$ in that order.

22. $\int_C (\operatorname{Re} z^2)\, dz$, C : the boundary of the triangle with vertices at $(0, 0)$, $(1, 0)$, and $(0, 1)$ in that order.

23. $\int_C |z|\,\bar{z}\, dz$, C : a closed contour consisting of the upper semicircle $|z| = 1$ and the line segment $-1 \le x \le 1$.

24. $\int_C \dfrac{dz}{\sqrt{z}}$, C : (i) the semicircle $|z| = 1$, $y \ge 0$ in the anti-clockwise direction,

        (ii) the semicircle $|z| = 1$, $y \le 0$ in the clockwise direction.

25. $\int_C z^n (\operatorname{Ln} z)\, dz$, C : the circle $|z| = 1$, in the anti-clockwise direction and n is an even integer.

26. $\int_C (2\bar{z} - 3z)\, dz$, C : the hypocycloid $x^{2/3} + y^{2/3} = a^{2/3}$, in the anti-clockwise direction.

Compute the length of the following curves.

27. $z(t) = (1 - i)e^{-it}$, $0 \le t \le \pi/2$.   28. $z(t) = (t + \sin t) + i(1 + \cos t)$, $0 \le t \le 2\pi$.

Obtain an upper bound for the absolute value of the following integrals.

29. $\int_C \dfrac{z}{z+1}\, dz$, C : upper half of the circle $|z| = 2$.

30. $\int_C e^{z^2}\, dz$, C : the broken lines from $z = 0$ to $z = 1$ and then from $z = 1$ to $z = 1 + i$.

31. $\int_C \dfrac{e^{iz}}{1 + z^2}\, dz$, C : the semicircle $z = Re^{it}$, $0 \le t \le \pi$, $R > 1$.

32. $\displaystyle\int_C \frac{dz}{z^4}$, C : the line segment from $z = 1 + i$ to $z = 2$.

33. $\displaystyle\int_C \frac{\text{Ln}\, z}{z^2}\, dz$, C : the circle $|z| = R$, $R > 1$.

34. $\displaystyle\int_C e^z\, dz$, C : the triangle with vertices at $(0, 0)$, $(1, 0)$ and $(0, 1)$ in that order.

35. $\displaystyle\int_C \sin(z^2)\, dz$, C : the circle $|z| = 2$.

11.3 Cauchy Integral Theorem

We first define the following:

Simply connected domain A connected domain is *simply connected* if every simple closed curve inside D encloses only points of D, or any simple closed curve which lies in D can be shrunk to a point inside D without leaving D.

For example, a plain sheet of paper with no 'holes' in it is a simply connected domain (Fig. 11.9a). If any closed curve C is shrunk, then it shrinks to a point lying inside D.

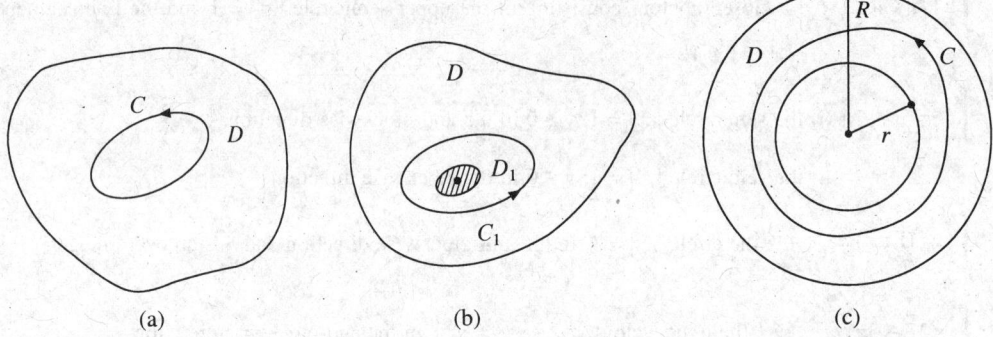

Fig. 11.9. **(a) Simply connected domain, (b) and (c) Multiply connected domains.**

Multiply connected domain A domain which is not simply connected is called a *multiply connected* domain. For example, a plain sheet of paper from which some interior parts are removed is a multiply connected domain (Fig. 11.9b). In this figure, let D_1 be removed from D. If a closed curve C_1 is shurnk, then it cannot be shrunk to a point as D_1 is not a part of D. Another simple example of a multiply connected domain is the annulus $r < |z| < R$ (Fig. 11.9c) because every closed curve C between the circles $|z| = R$ and $|z| = r$ cannot be shrunk to a point. Thus, we may say that a multiply connected domain has 'holes' in it. If it has one hole it is called *doubly connected*, if it has two holes it is called *triply connected* and so on (Fig. 11.10).

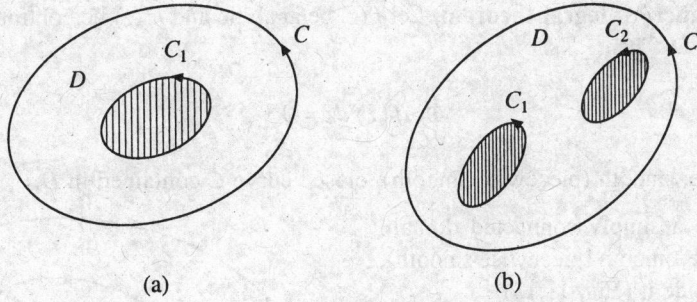

Fig. 11.10. (a) Doubly connected domain and (b) Triply connected domain.

Any multiply connected domain can be converted into a simply connected domain by introducing sufficient number of cuts in the domain. Consider for example, a doubly connected domain as shown in Fig. 11.11a. Introduce a cut in the domain along PQ (Fig. 11.11b).

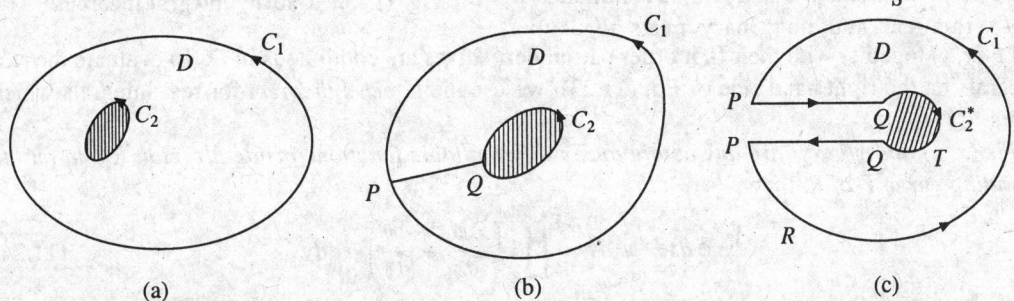

Fig. 11.11. Doubly connected domain and its conversion to a simply connected domain.

The new domain can be viewed as in Fig. 11.11c. The new domain is simply connected with the bounding curve as *PRSPQTQP*. Note that the connected sets in both the domains are same.
If the domain is triply connected, then we introduce two cuts to convert it into a simply connected domain (Fig. 11.12).

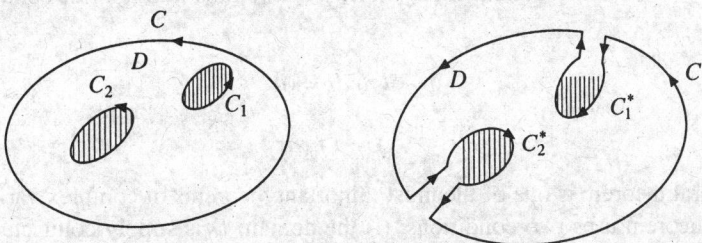

Fig. 11.12. Triply connected domain and its conversion to a simply connected domain.

In general, by introducing the required number of cuts, a multiply connected domain can be converted into a simply connected domain. This concept is very useful in evaluating integrals along a curve C which is bounding a multiply connected domain.

Theorem 11.3 (Cauchy integral theorem) Let $f(z)$ be analytic and $f'(z)$ be continuous in a simply connected domain D. Then,

$$\oint_C f(z)\, dz = 0 \tag{11.22}$$

along every simple, smooth (piecewise smooth), closed curve C contained in D.

Proof   Let D be a simply connected domain containing a simple, smooth (piecewise smooth), closed curve C inside it (Fig. 11.13).

Let $f(z) = u(x, y) + iv(x, y)$. Then,

$$\oint_C f(z)\, dz = \oint_C (u\, dx - v\, dy) + i \oint_C (v\, dx + u\, dy) \tag{11.23}$$

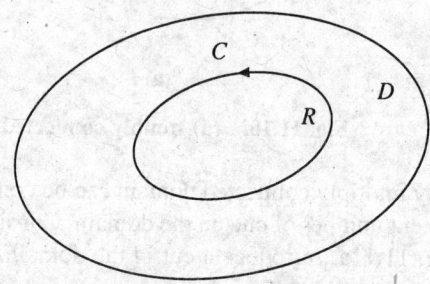

Fig. 11.13.   **Cauchy integral theorem.**

Since $f(z)$ is analytic in D and $f'(z)$ is continuous in D, the real and imaginary parts $u(x, y)$ and $v(x, y)$ together with their first order partial derivatives are continuous in D. To evaluate the real integrals on the right hand side of Eq. (11.23), we use the *Green's theorem* for real integrals which states:

If $\phi(x, y)$ and $\psi(x, y)$ are any continuously differentiable functions inside a region R and on the bounding curve Γ of R, then

$$\oint_\Gamma \phi\, dx + \psi\, dy = \iint_R \left(\frac{\partial \psi}{\partial x} - \frac{\partial \phi}{\partial y} \right) dx\, dy. \tag{11.24}$$

(This theorem will be proved in Chapter 15.)

The real functions $u(x, y)$ and $v(x, y)$ satisfy these conditions. Using the Green's theorem, we obtain from Eq. (11.23).

$$\oint_C f(z)\, dz = \iint_R \left(-\frac{\partial v}{\partial x} - \frac{\partial u}{\partial y} \right) dx\, dy + i \iint_R \left(\frac{\partial u}{\partial x} - \frac{\partial v}{\partial y} \right) dx\, dy. \tag{11.25}$$

Since the function $f(z)$ in analytic in D, which includes R where R is the region bounded by C, it satisfies the Cauchy-Riemann equations $u_x = v_y$ and $u_y = -v_x$. Substituting these conditions, we obtain from Eq. (11.25)

$$\oint_C f(z)\, dz = 0\cdot$$

Remark 7

The Cauchy integral theorem is one of the most important theorems of complex variable theory. The Cauchy integral theorem uses two conditions: (i) the domain D is simply connected and (ii) $f(z)$ is analytic and $f'(z)$ is continuous in D. The condition that D is simply connected is a necessary condition. The theorem cannot be applied for multiply connected domains. However, the conditions that $f(z)$ is analytic and $f'(z)$ is continuous in D are sufficient. For example

$$\oint_C \frac{dz}{(z - z_0)^n} = 0, \; n \neq 1, \; \text{where } C : |z - z_0| = r$$

(see Example 11.8). Here, $f(z) = 1/(z - z_0)^n$, $n \neq 1$, is not differentiable at $z = z_0$ and hence not analytic at z_0 in D.

The requirement of the condition that $f'(z)$ is continuous in D, in the Cauchy integral theorem, can be relaxed. This result is the consequence of the *Cauchy-Goursat theorem* which is stated as follows:

Theorem 11.4 (Cauchy-Goursat theorem) Let $f(z)$ be analytic in a simply connected domain D.

Then, $\displaystyle\oint_C f(z)\,dz = 0$ along every simple, smooth, closed curve C in D.

Example 11.19 Can the Cauchy integral theorem be applied for evaluating the following integrals? Hence, evaluate these integrals.

   (i) $\displaystyle\oint_C e^{\sin z^2}\,dz;\ \ C: |z| = 1,$ (ii) $\displaystyle\oint_C \tan z\,dz, \ C: |z| = 1,$

   (iii) $\displaystyle\oint_C \frac{e^z}{z^2 + 9}\,dz,\ \ C: |z| = 2,$ (iv) $\displaystyle\oint_C \frac{dz}{z^3 - 1}, \ C:$ a triangle with vertices at $0, \pm\frac{1}{4} + \frac{i}{2}$.

Solution Denote each of the integrals by I.

   (i) The integrand $f(z) = e^{\sin z^2}$ is analytic for all z and $f'(z)$ is continuous inside C. Hence, Cauchy integral theorem can be applied. Therefore, $I = 0$.

   (ii) The integrand $f(z) = \tan z = \sin z / \cos z$ is analytic for all z except at the points $z = \pm \pi/2$, $\pm 3\pi/2, \ldots$. All these points lie outside C. Also $f'(z)$ is continuous inside C. Hence, Cauchy integral theorem is applicable and $I = 0$.

   (iii) The integrand $f(z) = e^z/(z^2 + 9)$ is analytic every where except at the points $z = \pm 3i$. These points lie outside C and $f'(z)$ in continuous inside C. Hence, Cauchy integral theorem is applicable and $I = 0$.

   (iv) The integrand $f(z) = 1/(z^3 - 1)$ is analytic for all z except when $z^3 = 1$ or at the points $z = 1$, ω and ω^2, where $\omega = (-1 + i\sqrt{3})/2$. Since all these points lie outside C and $f'(z)$ is continuous inside C, Cauchy integral theorem is applicable and $I = 0$.

Example 11.20 Evaluate the integral $\displaystyle I = \oint_C \frac{3z + 5}{z^2 + 2z}\,dz, \ C: |z| = 1$.

Solution We have $\displaystyle f(z) = \frac{3z + 5}{z^2 + 2z} = \frac{3z + 5}{z(z + 2)} = \frac{1}{2}\left[\frac{5}{z} + \frac{1}{z + 2}\right].$

Therefore, $\displaystyle I = \oint_C \frac{3z + 5}{z^2 + 2z}\,dz = \frac{5}{2}\oint_C \frac{dz}{z} + \frac{1}{2}\oint_C \frac{1}{z + 2}\,dz = I_1 + I_2.$

In the first integral I_1, the integrand is not analytic at $z = 0$ which lies inside C. Therefore, Cauchy integral theorem cannot be applied. However, using the result of Example 11.7, we get

$$I_1 = \frac{5}{2}(2\pi i) = 5\pi i.$$

In the second integral I_2, the integrand is analytic everywhere except at $z = -2$, which lies outside the circle $|z| = 1$. Therefore, by the Cauchy integral theorem, we get $I_2 = 0$.

Hence, $$I = I_1 + I_2 = 5\pi i.$$

Independence of path

Let $f(z)$ be analytic in a simply connected domain D and C be any path joining two points z_1 and z_2 in D. The path C lies entirely in D. Then, $\int_C f(z)\,dz$ is independent of the path C and depends only on the end points z_1 and z_2.

Let z_1 and z_2 be any two points in D (Fig. 11.14(a)).

Consider two paths C_1 and C_2 in D from z_1 to z_2 such that apart from the initial point z_1 and the final point z_2, the curves C_1 and C_2 do not intersect each other (or do not have common points).

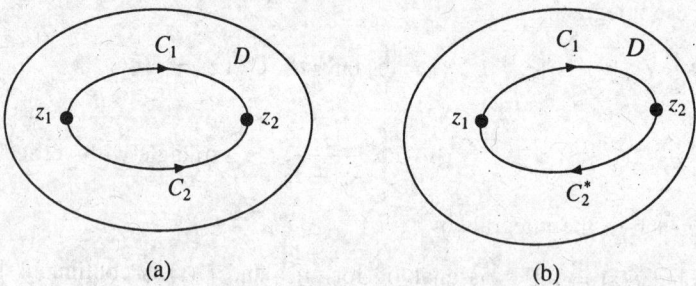

(a)   (b)

Fig. 11.14. Independence of path.

Let C_2^* denote the path C_2 with its orientation reversed (Fig. 11.14(b)). Consider the path along C_1 from z_1 to z_2 and along C_2 from z_2 to z_1, that is, along C_2^*, which forms a simple closed path. Since, $f(z)$ is analytic in D, the Cauchy integral theorem can be applied for this path. We get

$$\int_{C_1} f(z)\,dz + \int_{C_2^*} f(z)\,dz = 0$$

or

$$\int_{C_1} f(z)\,dz = -\int_{C_2^*} f(z)\,dz = \int_{C_2} f(z)\,dz$$

which shows that the integrals of $f(z)$ along C_1 and C_2 are equal. But, the paths C_1 and C_2 joining the points z_1 and z_2 are arbitrary. Hence, the integral of $f(z)$ is independent of the path of integration in D and depends only on the end points z_1 and z_2.

Remark 8

When the integral is independent of the path of integration, we can write the integral in terms of its end points, that is

$$\int_C f(z)\,dz = \int_{z_1}^{z_2} f(z)\,dz.$$

where C is any path joining the points z_1 and z_2.

Example 11.21 Verify that the value of the integral $I = \int_C z^2\,dz$ in each of the following cases is same and is equal to $-(11 + 2i)/3$.

(i) C is the straight line path joining the points $A(0, 0)$ and $B(1, 2)$,

(ii) C is the straight line path from $A(0, 0)$ to $P(1, 0)$ followed by the straight line path from $P(1, 0)$ to $B(1, 2)$,

(iii) C is the parabolic path $y = 2x^2$.

Solution Since the integrand z^2 is analytic in the entire finite complex plane, the value of the integral is independent of the path of integration. We have $z = x + iy$ and $dz = dx + idy$.

(i) Equation of the line AB is $y = 2x$, $0 \leq x \leq 1$. We get $z = x(1 + 2i)$. Therefore,

$$I = \int_C z^2 dz = \int_0^1 (-3 + 4i)(1 + 2i)x^2 dx = -\frac{1}{3}(11 + 2i).$$

(ii) Along AP : $y = 0$, $0 \leq x \leq 1$. We get $z = x$ and $dz = dx$. Along PB : $x = 1$, $0 \leq y \leq 2$. We get $z = 1 + iy$ and $dz = idy$.

Therefore, $\quad I = \int_0^1 x^2 dx + \int_0^2 (1 - y^2 + 2iy)\, idy = -\frac{1}{3}(11 + 2i)$.

(iii) Along the parabola $y = 2x^2$, $0 \leq x \leq 1$, we get $dy = 4x\, dx$, $dz = (1 + 4xi)dx$, $z^2 = x^2 - 4x^4 + 4ix^3$.

Therefore, $\quad I = \int_0^1 (x^2 - 4x^4 + 4ix^3)(1 + 4xi)\, dx$

$$= \int_0^1 [(x^2 - 20x^4) + i(8x^3 - 16x^5)]dx = -\frac{1}{3}(11 + 2i).$$

In all the three cases, the value of the integral is same.

Deformation of path

Let z_1, z_2 be any two points in D. Consider any two paths C_1 and C_2 in D, joining the points z_1 and z_2. When $\int_C f(z)\, dz$ is independent of the path of integration, we can imagine that the path C_2 was obtained from C_1 by a continuous deformation (Fig. 11.15).

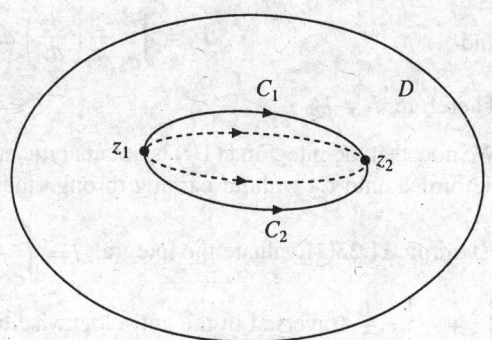

This deformation of the path of integration, keeping the end points fixed, can be continued so long as the deformed path does not contain any point where $f(z)$ is not analytic. The value of the integral remains the same under the deformation. Thus, contour integral of an analytic function $f(z)$ along any path is equal to contour integral of the same function along any other path into which

Fig. 11.15. Deformation of path.

the first path can be continuously deformed without passing through a point where $f(z)$ is not analytic.

Example 11.22 Evaluate the integrals

$$I_1 = \int_{C_1} \frac{dz}{z} \quad \text{and} \quad I_2 = \int_{C_2} \frac{dz}{z}$$

where C_1 and C_2 are the upper and the lower semi-circles of the unit circle $|z| = 1$, respectively and are traversed in opposite directions. Show that $I_1 \neq I_2$. Explain the reason.

Solution   The curves C_1 and C_2 are given in Fig. 11.16.

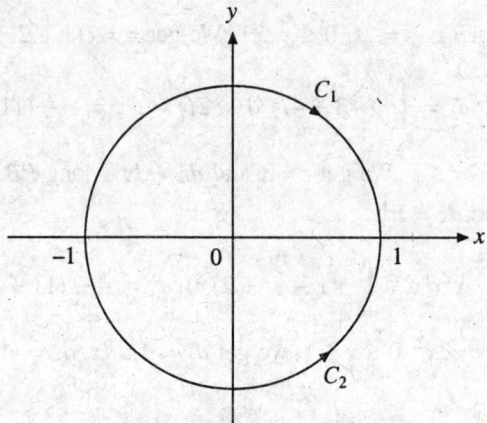

Fig. 11.16.   Example 11.22.

We have

$C_1 : z = e^{it}$, t varies from π to 0, and $C_2 : z = e^{it}$, t varies from $-\pi$ to 0.

We obtain

$$I_1 = \int_{C_1} \frac{1}{z} \left(\frac{dz}{dt} \right) dt = \int_{\pi}^{0} \frac{1}{e^{it}} (i\, e^{it}) dt = -\pi i$$

and

$$I_2 = \int_{C_2} \frac{1}{z} \left(\frac{dz}{dt} \right) dt = \int_{-\pi}^{0} \frac{1}{e^{it}} (i\, e^{it}) dt = \pi i.$$

Therefore, $I_1 \neq I_2$.

We note that the integrand $(1/z)$ is not analytic at $z = 0$. Therefore, the curve C_1 cannot be continuously deformed onto C_2 without passing through the point $z = 0$ at which $f(z) = 1/z$ is not analytic.

Example 11.23   Evaluate the integral $I = \int_C \frac{z\, dz}{1 + z^2}$ where C is the upper semi-circle of the circle $\left| z + \frac{1}{2} \right| = \frac{1}{2}$ traversed in the anti-clockwise direction. Use the principle of deformation of path.

Solution   The given path C is along the upper portion of the circle $\left| z + \frac{1}{2} \right| = \frac{1}{2}$ from $(0, 0)$ to $(-1, 0)$ as given in Fig. 11.17.

The integrand $f(z) = z/(1 + z^2)$ is analytic in any domain which does not contain the points $z = \pm i$. Let C_1 be any other path joining the points $z = 0$ and $z = -1$ which is obtained by continuously deforming C without passing through the points $z = \pm i$. Therefore, by the principle of deformation of path, we have

$$I = \int_C \frac{z}{1 + z^2}\, dz = \int_{C_1} \frac{z}{1 + z^2}\, dz$$

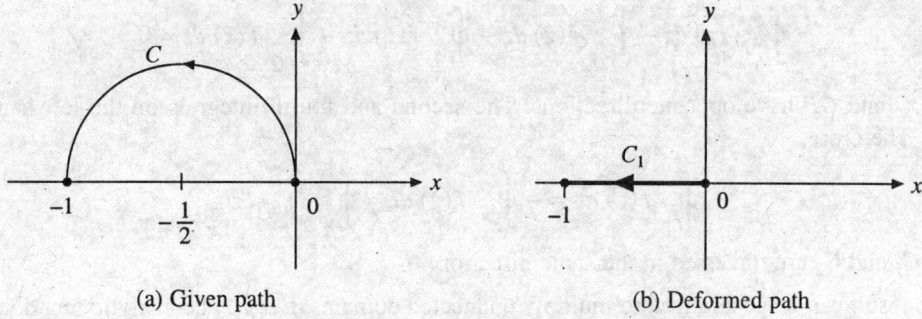

(a) Given path   (b) Deformed path

Fig. 11.17.   Example 11.23.

Choose C_1 as the straight line path along the negative real axis from $(0, 0)$ to $(-1, 0)$. Hence, we obtain

$$I = \int_0^{-1} \frac{x \, dx}{1 + x^2} = \frac{1}{2} \ln 2.$$

11.3.1   Extension of Cauchy Integral Theorem for Multiply Connected Domains

Consider a doubly connected domain D, whose outer and inner boundary curves are C_1 and C_2, respectively (Fig. 11.18).

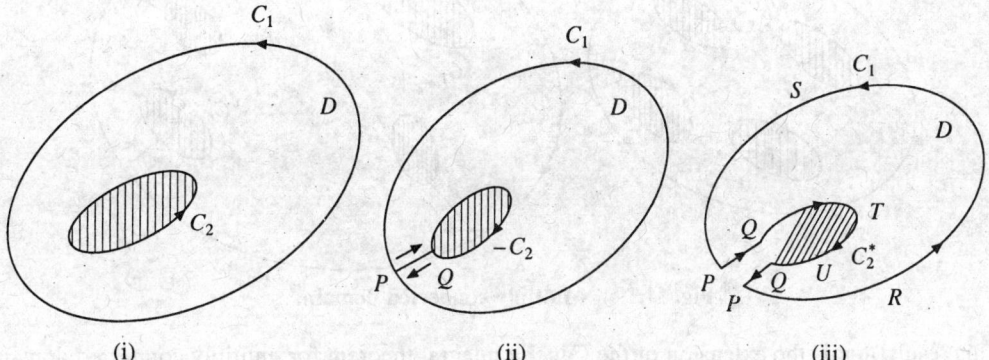

(i)   (ii)   (iii)

Fig. 11.18.   Doubly connected domain.

Let $f(z)$ be analytic in the doubly connected domain D and on C_1 and C_2. Then, we can prove that

$$\oint_{C_1} f(z) dz = \oint_{C_2} f(z) \, dz \tag{11.26}$$

where both the curves C_1 and C_2 are traversed in the same direction. In section 11.3, we have discussed the procedure of converting a doubly connected domain into a simply connected domain by introducing a cut in the domain. The simply connected domain (Fig. 11.18iii) is bounded by a simple closed curve *PRSPQTUQP*. Since $f(z)$ is analytic in D, we can apply the Cauchy integral theorem to obtain

$$\int_{PRSPQTUQP} f(z) \, dz = 0 \quad \text{or} \quad \left[\int_{PRSP} + \int_{PQ} + \int_{QTUQ} + \int_{QP} \right] f(z) dz = 0$$

or

$$\oint_{C_1} f(z)\,dz + \int_{PQ} f(z)\,dz + \oint_{C_2^*} f(z)\,dz - \int_{PQ} f(z)\,dz = 0$$

since PQ and QP have opposite directions. The second and fourth integrals on the left hand side cancel. Therefore,

$$\oint_{C_1} f(z)\,dz = -\oint_{C_2^*} f(z)\,dz = \oint_{C_2} f(z)\,dz$$

where C_1 and C_2 are traversed in the same direction.

This result can be generalized to multiply connected domains. Let $f(z)$ be analytic in a domain D bounded by non-intersecting simple closed curves $C, C_1, C_2, \ldots, C_n$ where $C_1, C_2, \ldots, C_n$ lie inside C. Let $f(z)$ be analytic on C and C_i, $i = 1, 2, \ldots, n$. We introduce n cuts and convert D to a simply connected domain (Fig. 11.19). Applying the Cauchy integral theorem, we get

$$\oint_{C} f(z)\,dz = \oint_{C_1} f(z)\,dz + \oint_{C_2} f(z)\,dz + \ldots + \oint_{C_n} f(z)\,dz \qquad (11.27)$$

where all the curves are traversed in the positive direction (anti-clockwise direction).

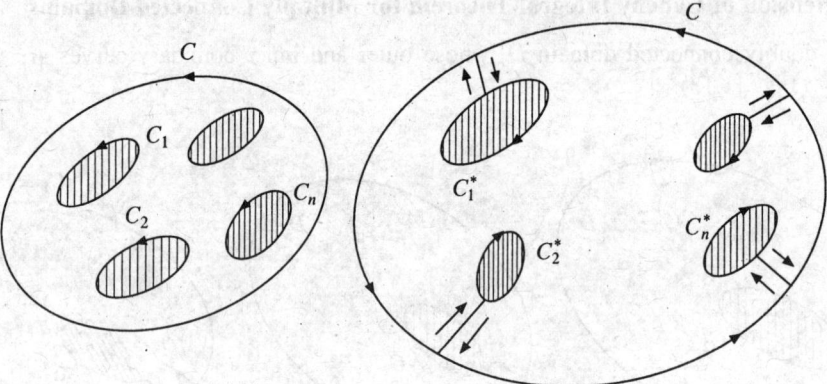

Fig. 11.19.　Multiply connected domain.

This result, that is the extension of the Cauchy integral theorem for multiply connected domains, has important applications. For example, consider the evaluation of the integral $I = \oint_{C} \dfrac{dz}{z}$, where C is the curve as given in Fig. 11.20. Let C_1 be any circle $|z| = r$ lying completely inside C. Let D be the domain bounded by the closed curves C and C_1. Now, $f(z) = 1/z$ is analytic inside D and on C and C_1. Hence, using Eq. (11.26), we get

$$\oint_{C} \frac{dz}{z} = \oint_{C_1} \frac{dz}{z} = 2\pi i. \qquad \text{(see Example 11.7).}$$

Example 11.24　Using the Cauchy integral theorem and its extension evaluate $\oint_{C} \dfrac{dz}{z(z+2)}$ where C is any rectangle containing the points $z = 0$ and $z = -2$ inside it.

Solution　The integrand $f(z) = 1/[z(z + 2)]$ is not analytic at $z = 0$ and $z = -2$. Both these points lie

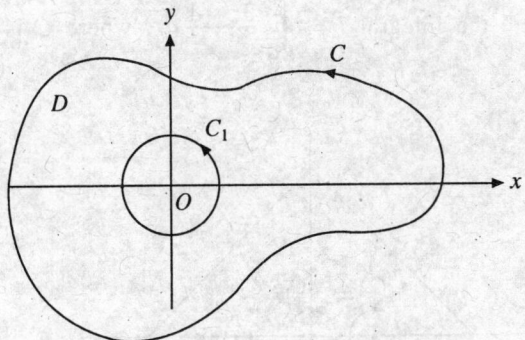

Fig. 11.20.   Doubly connected domain.

inside C. Enclose the points $z = 0$ and $z = -2$ by circles C_1 and C_2 of radii r_1 and r_2 respectively, such that both these circles lie inside C and do not intersect with each other (Fig. 11.21).

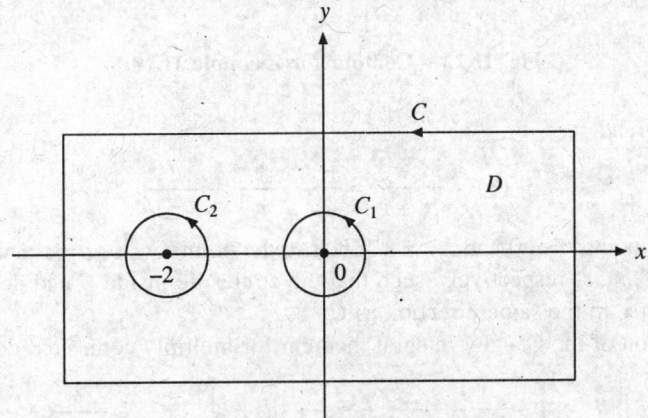

Fig. 11.21.   Example 11.24.

Therefore, the domain D bounded by C, C_1 and C_2 is a multiply connected domain. Hence, by the extension of the Cauchy integral theorem for multiply connected domains, we get

$$I = \oint_C \frac{dz}{z(z+2)} = \oint_{C_1} \frac{dz}{z(z+2)} + \oint_{C_2} \frac{dz}{z(z+2)}$$

$$= \frac{1}{2} \oint_{C_1} \left[\frac{1}{z} - \frac{1}{z+2} \right] dz + \frac{1}{2} \oint_{C_2} \left[\frac{1}{z} - \frac{1}{z+2} \right] dz.$$

By Cauchy integral theorem, we obtain $\oint_{C_1} \frac{dz}{z+2} = 0$ and $\oint_{C_2} \frac{dz}{z} = 0$.

Also $\oint_{C_1} \frac{dz}{z} = 2\pi i$ and $\oint_{C_2} \frac{dz}{z+2} = 2\pi i$.   (see Examples 11.7 and 11.8)

Therefore, $I = \frac{1}{2}(2\pi i) + \frac{1}{2}(-2\pi i) = 0.$

Example 11.25 Evaluate the integral $I = \oint_C \dfrac{3z-1}{z^3-z}\, dz$, where C is the contour as shown in Fig. 11.22.

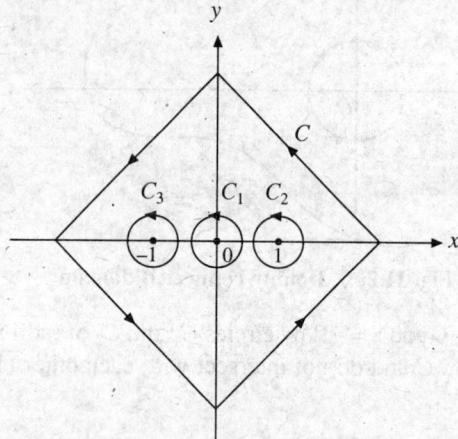

y

C_3 C_1 C_2

C

-1 0 1 x

Fig. 11.22. Contour for Example 11.25.

Solution The integrand

$$f(z) = \frac{3z-1}{z^3-z} = \frac{1}{z} - \frac{2}{z+1} + \frac{1}{z-1}$$

is not analytic at the points $z = 0$ and $z = \pm 1$. Enclose the points $z = 0$, $z = 1$ and $z = -1$ by circles C_1, C_2, C_3 of radii r_1, r_2, r_3 respectively, such that the circles lie inside C and do not intersect with each other and traverse in the same direction as C.

Using the extension of the Cauchy integral theorem for multiply connected domains, we get

$$I = \oint_C f(z)\, dz = \oint_{C_1} \left(\frac{1}{z} - \frac{2}{z+1} + \frac{1}{z-1} \right) dz + \oint_{C_2} \left(\frac{1}{z} - \frac{2}{z+1} + \frac{1}{z-1} \right) dz$$

$$+ \oint_{C_3} \left(\frac{1}{z} - \frac{2}{z+1} + \frac{1}{z-1} \right) dz$$

$$= (2\pi i - 0 + 0) + (0 - 0 + 2\pi i) + (0 - 2(2\pi i) + 0) = 0$$

(using Examples 11.7 and 11.8).

Example 11.26 If $0 < r < R$, evaluate the integral $I = \oint_C \dfrac{R+z}{z(R-z)}\, dz$, where $C : |z| = r$ and deduce that

(i) $\displaystyle\int_0^{2\pi} \frac{d\theta}{R^2 - 2rR\cos\theta + r^2} = \frac{2\pi}{R^2 - r^2},$ (ii) $\displaystyle\int_0^{2\pi} \frac{\sin\theta\, d\theta}{R^2 - 2rR\cos\theta + r^2} = 0.$

Solution We have

$$\frac{R+z}{z(R-z)} = \frac{1}{z} + \frac{2}{R-z}.$$

Therefore,

$$I = \oint_C \frac{R+z}{z(R-z)}\, dz = \oint_C \frac{dz}{z} + 2\oint_C \frac{dz}{R-z}.$$

Now, for $C : |z| = r, 0 < r < R, \oint_C \frac{dz}{R-z} = 0$ and $\oint_C \frac{dz}{z} = 2\pi i$.

Hence,

$$2\pi i = \oint_C \frac{R+z}{z(R-z)} \, dz.$$

Substituting $z = re^{i\theta}$ on the right hand side, we get

$$2\pi i = \int_0^{2\pi} \frac{R + re^{i\theta}}{re^{i\theta}(R - re^{i\theta})} (rie^{i\theta}) \, d\theta$$

$$= i \int_0^{2\pi} \frac{(R + re^{i\theta})(R - re^{-i\theta})}{(R - re^{i\theta})(R - re^{-i\theta})} \, d\theta = i \int_0^{2\pi} \frac{R^2 - r^2 + 2irR \sin\theta}{R^2 - 2rR \cos\theta + r^2} \, d\theta.$$

Comparing the real and imaginary parts on both sides, we obtain

$$\int_0^{2\pi} \frac{2rR \sin\theta \, d\theta}{R^2 - 2rR \cos\theta + r^2} = 0 \quad \text{or} \quad \int_0^{2\pi} \frac{\sin\theta}{R^2 - 2rR \cos\theta + r^2} \, d\theta = 0$$

and $\quad \displaystyle\int_0^{2\pi} \frac{(R^2 - r^2) \, d\theta}{R^2 - 2rR \cos\theta + r^2} = 2\pi \quad \text{or} \quad \int_0^{2\pi} \frac{d\theta}{R^2 - 2rR \cos\theta + r^2} = \frac{2\pi}{R^2 - r^2}.$

Example 11.27 By integrating $f(z) = 1/(R - z)$ over $C : |z| = r, 0 < r < R$ and using the result of Example 11.26, show that

$$\int_0^{2\pi} \frac{R \cos\theta}{R^2 - 2rR \cos\theta + R^2} \, d\theta = \frac{2\pi r}{R^2 - r^2}.$$

Solution Using the Cauchy integral theorem, we get

$$\oint_C \frac{dz}{R - z} = 0, \text{ where } C : |z| = r \text{ and } R > r.$$

Substituting $z = re^{i\theta}$ in the above integral, we obtain

$$0 = \int_0^{2\pi} \frac{rie^{i\theta}}{R - re^{i\theta}} d\theta = \int_0^{2\pi} \frac{ire^{i\theta}(R - re^{-i\theta})}{(R - re^{i\theta})(R - re^{-i\theta})} d\theta$$

$$= \int_0^{2\pi} \frac{ir(Re^{i\theta} - r)}{R^2 - 2rR \cos\theta + r^2} d\theta = \int_0^{2\pi} \frac{irR(\cos\theta + i \sin\theta) - ir^2}{R^2 - 2rR \cos\theta + r^2} d\theta.$$

Comparing the imaginary parts on both sides, we get

$$\int_0^{2\pi} \frac{(rR \cos\theta - r^2) \, d\theta}{R^2 - 2rR \cos\theta + r^2} = 0, r \neq 0$$

or $\quad \displaystyle\int_0^{2\pi} \frac{R \cos\theta \, d\theta}{R^2 - 2rR \cos\theta + r^2} = r \int_0^{2\pi} \frac{d\theta}{R^2 - 2rR \cos\theta + r^2} = \frac{2\pi r}{R^2 - r^2}$

(using the result of Example 11.26).

11.3.2   Use of Indefinite Integral in the Evaluation of Line Integrals

Let $f(z)$ be an analytic function in a simply connected domain D, z_0 be a fixed point inside D and z be any other point inside D. We define

$$F(z) = \int_{z_0}^{z} f(z^*)\, dz^*. \tag{11.28}$$

Therefore, the line integral of $f(z)$ from any point z_0 in D to any other point z in D is independent of the path of integration in D. Hence, the integral becomes a function of z as given in Eq. (11.28).

We now show that (i) $F(z)$ is analytic (ii) $F'(z) = f(z)$, and (iii) $\int_{z_0}^{z_1} f(z)\, dz = F(z_1) - F(z_0)$.

Let z be fixed. Take a point $z + \Delta z$ in D so that the whole line segment with end points z and $z + \Delta z$ is in D (Fig. 11.23).

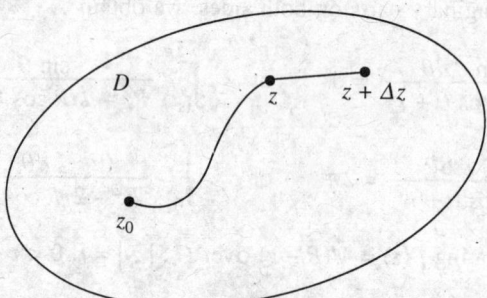

Fig. 11.23.   Indifinite integration.

We use this line segment as the path of integration. Using Eq. (11.28), we can write

$$\frac{F(z + \Delta z) - F(z)}{\Delta z} = \frac{1}{\Delta z}\left[\int_{z_0}^{z+\Delta z} f(z^*)\, dz^* - \int_{z_0}^{z} f(z^*)\, dz^*\right] = \frac{1}{\Delta z}\int_{z}^{z+\Delta z} f(z^*)\, dz^*. \tag{11.29a}$$

We can also write

$$\int_{z}^{z+\Delta z} f(z)\, dz^* = f(z)\int_{z}^{z+\Delta z} dz^* = \Delta z f(z) \quad \text{or} \quad f(z) = \frac{1}{\Delta z}\int_{z}^{z+\Delta z} f(z)\, dz^* \tag{11.29b}$$

since z is fixed, we treat $f(z)$ as a constant.

Subtracting Eq. (11.29b) from Eq(11.29a), we get

$$\frac{F(z + \Delta z) - F(z)}{\Delta z} - f(z) = \frac{1}{\Delta z}\int_{z}^{z+\Delta z} [f(z^*) - f(z)]\, dz^*.$$

Since $f(z)$ is analytic, it is continuous. Therefore, for a given real number $\varepsilon > 0$, there exists a real number $\delta > 0$, such that

$$|f(z^*) - f(z)| < \varepsilon, \text{ whenever } |z^* - z| < \delta.$$

Letting $|\Delta z| < \delta$ and using the *ML*-inequality, we get

$$\left|\frac{F(z + \Delta z) - F(z)}{\Delta z} - f(z)\right| = \frac{1}{|\Delta z|}\left|\int_{z}^{z+\Delta z} [f(z^*) - f(z)]\, dz^*\right| \leq \frac{1}{|\Delta z|}\,\varepsilon\,|\Delta z| = \varepsilon.$$

By the definition of limit and derivative, we obtain

$$F'(z) = \lim_{\Delta z \to 0} \frac{F(z + \Delta z) - F(z)}{\Delta z} = f(z).$$

Since z is any arbitrary point in D, this implies that $F(z)$ is analytic in D and is an indefinite integral of $f(z)$ in D, written as

$$F(z) = \int f(z) \, dz.$$

Let $H(z)$ be another indefinite integral of $f(z)$. Therefore, $H'(z) = f(z)$. Then,

$$H'(z) - F'(z) = f(z) - f(z) = 0.$$

Hence,

$$F(z) = H(z) + c = \int_{z_0}^{z} f(z^*) \, dz^*$$

where c is an arbitrary constant. Letting $z = z_0$, we find that

$$H(z_0) + c = 0 \quad \text{or} \quad c = - H(z_0).$$

Now, letting $z = z_1$, we obtain

$$\int_{z_0}^{z_1} f(z) \, dz = H(z_1) + c = H(z_1) - H(z_0) = [H(z)]_{z_0}^{z_1}$$

$$= F(z_1) - F(z_0) = [F(z)]_{z_0}^{z_1}.$$

Alternative Proof

Let $f(z) = u + iv$ and $F(z) = U + iV$. Then, from Eq. (11.28), we get

$$U + iV = \int_{z_0}^{z} (u + iv)(dx + idy)$$

$$= \int_{(x_0, y_0)}^{(x, y)} (u dx - v dy) + i \int_{(x_0, y_0)}^{(x, y)} (u dy + v dx).$$

Comparing the real and imaginary parts, we obtain

$$U = \int_{(x_0, y_0)}^{(x, y)} (u dx - v dy) \quad \text{and} \quad V = \int_{(x_0, y_0)}^{(x, y)} (u dy + v \, dx).$$

Differentiating partially with respect to x and y, we get

$$\frac{\partial U}{\partial x} = u, \quad \frac{\partial U}{\partial y} = -v \quad \text{and} \quad \frac{\partial V}{\partial x} = v, \quad \frac{\partial V}{\partial y} = u.$$

We have

$$\frac{\partial U}{\partial x} = \frac{\partial V}{\partial y} = u \quad \text{and} \quad \frac{\partial U}{\partial y} = -\frac{\partial V}{\partial x} = -v.$$

Hence, U and V satisfy the Cauchy-Riemann equations. Since $f(z)$ is analytic, u, v are continuous and therefore, the partial derivatives $\partial U/\partial x, \partial U/\partial y, \partial V/\partial x, \partial V/\partial y$ are also continuous. Hence, $F(z)$ is an analytic function. We also have

$$F'(z) = U_x + iV_x = u + iv = f(z).$$

Example 11.28   Evaluate the following integrals.

(i) $\displaystyle\int_1^{2-i} z\,dz,$ (ii) $\displaystyle\int_{-\pi i}^{\pi i} e^{2z}\,dz,$ (iii) $\displaystyle\int_0^i \sinh \pi z\, dz,$

(iv) $\displaystyle\int_0^{1+2i} z\,e^z dz,$ (v) $\displaystyle\int_0^1 z^2 e^{z^3}\,dz.$

Solution   If $f(z)$ is analytic in a simply connected domain D which contains the curve joining the points z_0 and z_1 in D, then

$$I = \int_{z_0}^{z_1} f(z)\,dz = F(z_1) - F(z_0), \quad \text{where}\quad F(z) = \int f(z)\,dz.$$

In all the given problems (i) – (v), $f(z)$ is analytic in the entire complex plane. Therefore, the integrals can be evaluated by indefinite integration.

(i) $F(z) = \displaystyle\int z\,dz = \frac{z^2}{2}$. Therefore, $I = F(2-i) - F(1) = \frac{1}{2}(2-i)^2 - \frac{1}{2} = 1 - 2i.$

(ii) $F(z) = \displaystyle\int e^{2z}\,dz = \frac{1}{2}\,e^{2z}$. Therefore, $I = F(\pi i) - F(-\pi i) = \frac{1}{2}[e^{2\pi i} - e^{-2\pi i}] = 0.$

(iii) $F(z) = \displaystyle\int \sinh \pi z\, dz = \frac{1}{\pi}\cosh \pi z = \frac{1}{2\pi}(e^{\pi z} + e^{-\pi z})$. Therefore,

$$I = F(i) - F(0) = \frac{1}{2\pi}[e^{\pi i} + e^{-\pi i}) - (1+1)] = -\frac{2}{\pi}.$$

(iv) $F(z) = \displaystyle\int z e^z dz$. Integrating by parts, we obtain $F(z) = (z-1)e^z$. Therefore,

$$I = F(1+2i) - F(0) = 2i\,e^{1+2i} + 1.$$

(v) $F(z) = \displaystyle\int z^2 e^{z^3}\,dz$. Substituting $z^3 = T$, we get

$$F(z) = \frac{1}{3}\int e^T\,dT = \frac{1}{3}\,e^T = \frac{1}{3}\,e^{z^3}. \text{Therefore,}\quad I = F(1) - F(0) = \frac{1}{3}(e-1).$$

Exercise 11.3

In each of the following problems, verify the Cauchy integral theorem. All curves are traversed in the positive direction.

1. $\displaystyle\oint_C (2z^2 - iz + 3)\,dz,\ C:|z| = 1.$

2. $\oint_C z\,dz$, C : boundary of the triangle with vertices at $1, 1 + i, i$.

3. $\oint_C z^2\,dz$, C : boundary of the square with vertices at $-1 - i, 1 - i, 1 + i, -1 + i$.

4. $\oint_C z^3\,dz$, C : boundary of the rhombus with vertices at $-1, -i, 1, i$.

5. $\oint_C \cos 2z\,dz$ C : boundary of the square with vertices at $1 \pm i, -1 \pm i$.

6. $\oint_C \sinh z\,dz$, C : boundary of the rectangle with vertices at $2 \pm i, -2 \pm i$.

In each of the following problems, indicate whether Cauchy integral theorem can be applied and evaluate the integral.

7. $\oint_C \sec z\,dz$, $C : |z| = 1$.

8. $\oint_C \dfrac{z^2 + 1}{\cosh z}\,dz$, $C : |z| = 1$.

9. $\oint_C z^2 \cos z\,dz$, $C : |z - i| = 1$.

10. $\oint_C z e^z dz$, $C : |z + 2i| = 4$.

11. $\oint_C z e^{1/(z+2)}\,dz$, $C : |z| = 1$.

12. $\oint_C \dfrac{e^z}{1 + e^z}\,dz$, $C : |z| = 1$.

13. $\oint_C \dfrac{dz}{e^z - e^2}$, $C : |z| = 1$.

14. $\oint_C \dfrac{e^z}{z^2 - 5iz - 6}\,dz$, $C : |z| = 1$.

15. $\oint_C \dfrac{e^z}{z^3 + 8}\,dz$, $C : |z| = 1$.

16. $\oint_C \dfrac{\mathrm{Ln}\,z}{z + 1}\,dz$, $C : \left|z - \dfrac{1}{2}\right| = \dfrac{1}{4}$.

17. $\oint_C \dfrac{\cosh^2 2z}{(z + 3i)(z^2 + 16)}\,dz$, $C : |z| = 2$.

18. $\oint_C \dfrac{z^3 + 5z - 7}{\cos z}\,dz$, C : rectangle with vertices at $\pm \pi i/8, \pi(1 \pm i)/8$.

19. $\oint_C \left[z + \dfrac{3}{z^2}\right]\,dz$, $C : |z| = 1$.

20. $\oint_C [\mathrm{Re}\,z + z]\,dz$, $C : |z| = 2$.

21. $\oint_C \left[\dfrac{e^z}{(z + 3)(z + 2)} + 3\bar{z}\right]\,dz$, $C : |z| = 1$.

Using the extension of the Cauchy-integral theorem for multiply connected domains, evaluate the following integrals.

22. $\oint_C \dfrac{7z + 12}{z^2 + z - 2}\,dz$, $C : |z - 2| = 2$.

23. $\oint_C \dfrac{(z - 1)\,dz}{z(z + i)(z + 3i)}$, $C : |z + i| = \dfrac{1}{2}$.

24. $\oint_C \dfrac{dz}{z^3 + 3iz^2}$, $C : |z| = 1$.

25. $\oint_C \dfrac{2z}{z^2 + \pi^2}\,dz$, $C : |z| = 4$.

26. $\oint_C \dfrac{2z - 3}{z^2 - 3z - 18}\,dz$, $C : |z| = 8$.

27. $\oint_C \dfrac{2z - 1}{z^2 - z}\,dz$, $C : |z| = 2$.

28. $\oint_C \dfrac{dz}{(z-1)(z-2)(z+4)}$, $C : |z| = 3$.　　**29.** $\oint_C \dfrac{dz}{(z^2+1)(z-1)}$, $C : |z| = 2$.

30. $\oint_C \dfrac{dz}{(z-1)(z-2)(z-3)}$, $C : |z| = 4$.

31. Show that the integral $\displaystyle\int_C (z^2 - z)\, dz$, where C is the path joining the points $z = 0$ and $z = 4 + 2i$, is independent of the path of integration. Evaluate the integral by taking a suitable path.

32. Show that the integral $\displaystyle\int_C e^{-2z}\, dz$, where C is the path joining the points $z = 1 + 2\pi i$ and $z = 3 + 4\pi i$, is independent of the path of integration. Evaluate the integral by taking a suitable path.

33. Evaluate the integral $\displaystyle\int_C (z^2 + 2)^2\, dz$, where C is the arc of the cycloid $x = a(\theta - \sin\theta)$, $y = a(1 - \cos\theta)$ joining the points $(0, 0)$ and $(2\pi a, 0)$.

34. By evaluating $\displaystyle\oint_C e^z\, dz$, $C : |z| = 1$, show that

(i) $\displaystyle\int_0^{2\pi} e^{\cos\theta} \cos(\theta + \sin\theta)\, d\theta = 0$,　　(ii) $\displaystyle\int_0^{2\pi} e^{\cos\theta} \sin(\theta + \sin\theta)\, d\theta = 0$.

35. Let n be a positive integer. Then show that

(i) $\displaystyle\int_0^{2\pi} e^{\sin n\theta} \cos(\theta - \cos n\theta)\, d\theta = 0$,　　(ii) $\displaystyle\int_0^{2\pi} e^{\sin n\theta} \sin(\theta - \cos n\theta)\, d\theta = 0$.

Evaluate the following integrals.

36. $\displaystyle\int_1^{2-i} z^3\, dz$.　　**37.** $\displaystyle\int_{i/2}^{i} e^{-\pi z}\, dz$.　　**38.** $\displaystyle\int_0^{\pi} \cos\left(\dfrac{z}{2}\right) dz$.

39. $\displaystyle\int_2^{2+i} (z - 2)^3\, dz$.　　**40.** $\displaystyle\int_0^{1+2i} z \sin(z^2)\, dz$.　　**41.** $\displaystyle\int_0^{\pi/2} z \sin z\, dz$.

42. $\displaystyle\int_0^{1+\pi i} (z^2 + \cosh 2z)\, dz$.　　**43.** $\displaystyle\int_i^{1} \dfrac{dz}{(1+z)^2}$.　　**44.** $\displaystyle\int_{-i}^{i} z^4 e^{\pi z^5}\, dz$.

45. $\displaystyle\int_0^{i} (z^2 + 1)^3\, dz$.　　**46.** $\displaystyle\int_0^{\pi i} \cos^2 z\, dz$.　　**47.** $\displaystyle\int_0^{1} \dfrac{\tan^{-1} z}{1+z^2}\, dz$.

48. $\displaystyle\int_0^{\pi} z \cos 3z\, dz$.　　**49.** $\displaystyle\int_{\alpha}^{\beta} \dfrac{dz}{z^2}$, $\alpha \neq 0, \beta \neq 0$.　　**50.** $\displaystyle\int_1^{2} z^2\, \text{Ln}\, z\, dz$.

11.4 Cauchy Integral Formula

Cauchy integral formula is a representation of an analytic function $f(z)$ at any interior point z_0 of a simply connected domain D as a contour integral, that is as an integral evaluated along the boundary of a simple closed curve C which lies inside D and encloses the point z_0.

Theorem 11.5 Let $f(z)$ be analytic in a simply connected domain D. Let z_0 be any point in D and C be any simple closed curve in D enclosing the point $z = z_0$. Then,

$$f(z_0) = \frac{1}{2\pi i} \oint_C \frac{f(z)}{z - z_0} \, dz \tag{11.30}$$

where C is traversed in the anti-clockwise direction.

Proof The given function $f(z)$ is analytic in a simply connected domain D and C is any simple closed curve in D enclosing the point $z = z_0$. Let $C_1 : |z - z_0| = \rho$ be a circle with infinitesimal radius ρ, such that it lies entirely inside C. Consider now, the function $f(z)/(z - z_0)$. This function is analytic everywhere in D except at the point $z = z_0$. However, this function is analytic in the domain bounded by the simple closed curve C and the circle C_1, that is, in a doubly connected domain (Fig. 11.24). Hence, by the extension of the Cauchy integral theorem for multiply connected domains we have

$$\oint_C \frac{f(z)}{z - z_0} \, dz = \oint_{C_1} \frac{f(z)}{z - z_0} \, dz. \tag{11.31}$$

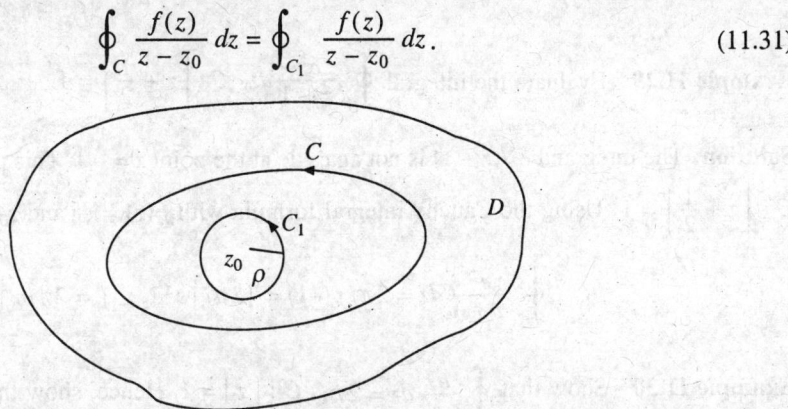

Fig. 11.24.   Cauchy integral formula.

Now, writing $f(z) = [f(z) - f(z_0)] + f(z_0)$ in Eq. (11.31), we obtain

$$\oint_C \frac{f(z)}{z - z_0} \, dz = \oint_{C_1} \frac{f(z_0) + [f(z) - f(z_0)]}{z - z_0} \, dz$$

$$= f(z_0) \oint_{C_1} \frac{dz}{z - z_0} + \oint_{C_1} \frac{f(z) - f(z_0)}{z - z_0} \, dz$$

$$= 2\pi i \, f(z_0) + I_1 \tag{11.32}$$

where $\qquad I_1 = \oint_{C_1} \frac{f(z) - f(z_0)}{z - z_0} \, dz.$

We have $\qquad |I_1| = \left| \oint_{C_1} \frac{f(z) - f(z_0)}{z - z_0} \, dz \right| \leq \oint_{C_1} \left| \frac{f(z) - f(z_0)}{z - z_0} \right| |dz|$

$$= \oint_{C_1} \frac{|f(z) - f(z_0)|}{|z - z_0|} |dz|. \tag{11.33}$$

Now, $f(z)$ is analytic and hence continuous in D. Therefore, for any given real number $\varepsilon > 0$, there exists a real number $\delta > 0$ such that

$$|f(z) - f(z_0)| < \varepsilon, \text{ whenever } |z - z_0| < \delta.$$

Choosing the radius of the circle C_1 such that $\rho < \delta$, we obtain from Eq. (11.33)

$$|I_1| \le \oint_{C_1} \frac{|f(z) - f(z_0)|}{|z - z_0|} |dz| < \oint_{C_1} \frac{\varepsilon}{\rho} |dz| = \frac{\varepsilon}{\rho} (2\pi\rho) = 2\pi\varepsilon.$$

Thus, the value of $|I_1|$ can be made arbitrarily small. Therefore, the value of the integral I_1 is zero. Hence, we obtain from Eq. (11.32)

$$\oint_C \frac{f(z)}{z - z_0} \, dz = 2\pi i f(z_0), \text{ or } f(z_0) = \frac{1}{2\pi i} \oint_C \frac{f(z)}{z - z_0} \, dz.$$

Example 11.29   Evaluate the integral $\oint_C \dfrac{e^z}{z+1} \, dz$, $C : \left| z + \dfrac{1}{2} \right| = 1$.

Solution   The integrand $e^z/(z+1)$ is not analytic at the point $z = -1$. This point lies inside the circle $C : \left| z + \dfrac{1}{2} \right| = 1$. Using the Cauchy integral formula with $f(z) = e^z$ and $z_0 = -1$, we obtain

$$\oint_C \frac{e^z}{z+1} \, dz = 2\pi i f(-1) = \{2\pi i \, [e^z]_{z=-1}\} = 2\pi i e^{-1}.$$

Example 11.30   Show that $\oint_C \dfrac{e^z}{z} \, dz = 2\pi i$, $C : |z| = 1$. Hence, show that

$$\int_0^{2\pi} e^{\cos\theta} \cos(\sin\theta) \, d\theta = 2\pi, \quad \text{and} \quad \int_0^{2\pi} e^{\cos\theta} \sin(\sin\theta) \, d\theta = 0.$$

Solution   The integrand e^z/z in not analytic at the point $z = 0$ which lies inside C. Using the Cauchy integral formula with $f(z) = e^z$ and $z_0 = 0$, we get

$$\oint_C \frac{e^z}{z} \, dz = 2\pi i f(0) = 2\pi i \{[e^z]_{z=0}\} = 2\pi i.$$

Now, let $z = e^{i\theta}$. We have

$$\oint_C \frac{e^z}{z} \, dz = \int_0^{2\pi} \frac{e^{e^{i\theta}} (i e^{i\theta})}{e^{i\theta}} \, d\theta = i \int_0^{2\pi} e^{\cos\theta + i \sin\theta} \, d\theta$$

or

$$i \int_0^{2\pi} e^{\cos\theta} [\cos(\sin\theta) + i \sin(\sin\theta)] \, d\theta = 2\pi i.$$

Comparing the real and imaginary parts, we obtain

$$\int_0^{2\pi} e^{\cos\theta} \sin(\sin\theta) \, d\theta = 0 \quad \text{and} \quad \int_0^{2\pi} e^{\cos\theta} \cos(\sin\theta) \, d\theta = 2\pi.$$

Example 11.31 Evaluate the integral $\displaystyle\oint_C \frac{dz}{2-\bar{z}}$, $C:|z|=1$.

Solution We write $2-\bar{z} = 2 - \dfrac{z\bar{z}}{z} = 2 - \dfrac{|z|^2}{z} = 2 - \dfrac{1}{z} = \dfrac{2z-1}{z}$, since $|z|^2 = 1$ on C.

Therefore,
$$I = \oint_C \frac{dz}{2-\bar{z}} = \oint_C \frac{z}{2z-1}\,dz = \frac{1}{2}\oint_C \frac{z}{z-(1/2)}\,dz.$$

The integrand $z/[z-(1/2)]$ is not analytic at the point $z=1/2$ which lies within C. Using the Cauchy integral formula with $f(z)=z$ and $z_0=1/2$, we obtain

$$\frac{1}{2}\oint_C \frac{z\,dz}{z-(1/2)} = \frac{1}{2}\left[2\pi i\,f\left(\frac{1}{2}\right)\right] = \frac{\pi i}{2}.$$

Example 11.32 Evaluate the integral $\displaystyle\oint_C \frac{dz}{z(z^2+4)}$, $C:|z|=1$.

Solution The integrand $1/[z(z^2+4)]$ is not analytic at the points $z=0$ and $z=\pm 2i$. The points $z=\pm 2i$ lie outside C. Using the Cauchy integral formula with $f(z)=1/(z^2+4)$ and $z_0=0$, we obtain

$$\oint_C \frac{dz}{z(z^2+4)}\,dz = \oint_C \frac{[1/(z^2+4)]}{z}\,dz = 2\pi i\,f(0) = \frac{2\pi i}{4} = \frac{\pi i}{2}.$$

Example 11.33 Evaluate the integral $\displaystyle\oint_C \frac{z^2+1}{z(2z-1)}\,dz$, $C:|z|=1$.

Solution Denote the given integral by I. The integrand of I is not analytic at the points $z=0$ and $z=1/2$ both of which lie inside C. We write

$$\frac{z^2+1}{z(2z-1)} = (z^2+1)\left[\frac{1}{z-(1/2)} - \frac{1}{z}\right]$$

Therefore,
$$\oint_C \frac{z^2+1}{z(2z-1)}\,dz = \oint_C \frac{z^2+1}{z-(1/2)}\,dz - \oint_C \frac{z^2+1}{z}\,dz = I_1 - I_2.$$

Now, the integrands of I_1 and I_2 are not analytic at a point which lies inside C. Using the Cauchy integral formula to I_1 and I_2, we get

$$I_1 = \oint_C \frac{z^2+1}{z-(1/2)}\,dz = \left\{2\pi i\,[z^2+1]_{z=1/2}\right\} = \frac{5\pi i}{2}.$$

$$I_2 = \oint_C \frac{z^2+1}{z}\,dz = \left\{2\pi i\,[z^2+1]_{z=0}\right\} = 2\pi i.$$

Hence,
$$I = I_1 - I_2 = \frac{5\pi i}{2} - 2\pi i = \frac{\pi i}{2}.$$

Example 11.34 Evaluate the integral $\displaystyle\oint_C \frac{dz}{z(z+2)}$, where C is a rectangle enclosing the points $z=0$ and $z=-2$.

Solution Let the given integral be denoted by I. The integrand of I is not analytic at the points $z = 0$ and $z = -2$ both of which lie inside C. We can evaluate I using the procedure given in Example 11.33 by using the partial fractions. However, we can evaluate I alternatively as follows.

Enclose the points $z = 0, -2$ by circles $C_1 : |z| = \delta_1$ and $C_2 : |z + 2| = \delta_2$ respectively such that C_1 and C_2 do not intersect each other and lie inside C. Using the extension of the Cauchy integral theorem for multiply connected domains, we get

$$I = \oint_C \frac{dz}{z(z+2)} = \oint_{C_1} \frac{[1/(z+2)]}{z} \, dz + \oint_{C_2} \frac{(1/z)}{z+2} \, dz.$$

Using the Cauchy integral formula, we obtain

$$I = 2\pi i \left[\frac{1}{z+2}\right]_{z=0} + 2\pi i \left[\frac{1}{z}\right]_{z=-2} = \pi i - \pi i = 0.$$

Example 11.35 Evaluate the integral $\oint_C \frac{dz}{z^2 + 4}$ where

(i) $C : |z - 2i| = 1$, (ii) $C : |z + 2i| = 1$, (iii) $C : |z| = 4$,

(iv) C is a piecewise smooth curve with initial point $z = 0$ and the terminal point $z = 2$ and the curve does not pass through the points $z = \pm 2i$.

Solution Let the given integral be denoted by I. The integrand is not analytic at the points $z = \pm 2i$.

(i) The point $z = -2i$ lies outside C. Taking $f(z) = 1/(z + 2i)$, $z_0 = 2i$ and using the Cauchy integral formula, we obtain

$$I = \oint_C \frac{[1/(z+2i)]}{z - 2i} \, dz = 2\pi i \left[\frac{1}{z+2i}\right]_{z=2i} = 2\pi i \left(\frac{1}{4i}\right) = \frac{\pi}{2}.$$

(ii) The point $z = 2i$ lies outside C. Taking $f(z) = 1/(z - 2i)$, $z_0 = -2i$ and using the Cauchy integral formula, we obtain

$$I = \oint_C \frac{[1/(z-2i)]}{z + 2i} \, dz = 2\pi i \left(\frac{1}{z - 2i}\right)_{z=-2i} = 2\pi i \left(-\frac{1}{4i}\right) = -\frac{\pi}{2}.$$

(iii) Both the points $z = \pm 2i$ lie inside C. Enclose the points $z = -2i$ and $z = 2i$ by circles $C_1 : |z + 2i| = \delta_1$ and $C_2 : |z - 2i| = \delta_2$ respectively such that C_1 and C_2 do not intersect each other and lie inside C. Using the extension of the Cauchy integral theorem for multiply connected domains, we get

$$\oint_C \frac{dz}{z^2 + 4} = \oint_{C_1} \frac{dz}{z^2 + 4} + \oint_{C_2} \frac{dz}{z^2 + 4} = \frac{\pi}{2} - \frac{\pi}{2} = 0$$

(using the results in parts (i) and (ii)).

(iv) Since C is not a closed curve, Cauchy integral theorem or Cauchy integral formula cannot be used. The integrand is analytic for all z except when $z = \pm 2i$. The curve C does not pass through the points $z = \pm 2i$. Therefore, the integral is independent of the path of integration. Choose the path of integration along x-axis from $x = 0$ to $x = 2$. Hence, we have $z = x$ and $dz = dx$. Therefore,

$$I = \oint_C \frac{dz}{z^2 + 4} = \int_0^2 \frac{dx}{x^2 + 4} = \frac{1}{2} \left[\tan^{-1}\left(\frac{x}{2}\right)\right]_0^2 = \frac{1}{2} [\tan^{-1} 1 - \tan^{-1} 0] = \frac{\pi}{8}.$$

Example 11.36 Consider the function $g(z)$ defined by $g(z) = \oint_{|\xi|=1} \dfrac{e^{\xi}+1}{\xi-z}\,d\xi, \; |z| \neq 1$. Show that

(i) $g(z) = 0$ when $|z| > 1$,   (ii) $g(z) = 2\pi i\,(e^z + 1)$ when $|z| < 1$.

Solution

(i) When $|z| > 1$, the integrand is analytic inside and on the unit circle $|\xi| = 1$. Hence, by Cauchy integral theorem $g(z) = 0$.

(ii) When $|z| < 1$, the integrand is not analytic at the point $\xi = z$ which lies inside $C : |\xi| = 1$. Using the Cauchy integral formula with $f(\xi) = e^{\xi} + 1$ and $\xi_0 = z$, we get

$$g(z) = \{2\pi i\,[f(\xi)]_{\xi=z}\} = 2\pi i\,(e^z + 1).$$

Example 11.37 Let $f(z)$ be analytic within and on a circle $C : |z| = R$. Let $z = re^{i\theta}$ be any point inside C. Choosing a point z^* outside the circle C at a distance of R^2/r from the origin on the same ray as z, prove the *Poisson integral formula* for the circle

$$u(r, \theta) = \frac{1}{2\pi}\int_0^{2\pi}\frac{(R^2 - r^2)\,u(R,\phi)\,d\phi}{R^2 - 2rR\cos(\phi - \theta) + r^2}, \; r < R.$$

Solution $z = re^{i\theta}$ is a point inside C. By Cauchy integral formula, we get

$$f(z) = \frac{1}{2\pi i}\oint_C \frac{f(\xi)}{\xi - z}\,d\xi. \tag{11.34}$$

Since ξ is a point on C, let $\xi = Re^{i\phi}$. Therefore, $|z| = r$ and $|\xi| = R$. Now, z^* is a point outside the circle C at a distance R^2/r from the origin on the same ray as z. We have

$$z^* = \frac{R^2}{r}\,e^{i\theta} = \frac{R^2}{re^{-i\theta}} = \frac{\xi\bar{\xi}}{\bar{z}}.$$

Since z^* is a point outside C, we have by Cauchy integral theorem

$$\oint_C \frac{f(\xi)}{\xi - z^*}\,d\xi = 0 \;\text{ or }\; \frac{1}{2\pi i}\oint_C \frac{f(\xi)}{\xi - z^*}\,d\xi = 0. \tag{11.35}$$

Now $$\xi - z^* = \xi - \frac{\xi\bar{\xi}}{\bar{z}} = \frac{\xi(\bar{z} - \bar{\xi})}{\bar{z}}.$$

Subtracting Eq. (11.35) from (11.34), we obtain

$$f(z) = \frac{1}{2\pi i}\oint_C\left[\frac{1}{\xi - z} - \frac{1}{\xi - z^*}\right]f(\xi)\,d\xi$$

$$= \frac{1}{2\pi i}\oint_C\left[\frac{1}{\xi - z} - \frac{\bar{z}}{\xi(\bar{z} - \bar{\xi})}\right]f(\xi)\,d\xi = \frac{1}{2\pi i}\oint_C\left[\frac{z\bar{z} - \xi\bar{\xi}}{\xi(\xi - z)(\bar{z} - \bar{\xi})}\right]f(\xi)\,d\xi.$$

Substituting $z = re^{i\theta}$, $\xi = Re^{i\phi}$, we get

$$f(re^{i\theta}) = \frac{1}{2\pi i} \int_0^{2\pi} \frac{(r^2 - R^2)f(Re^{i\phi})\,Rie^{i\phi}}{(Re^{i\phi} - re^{i\theta})(re^{-i\theta} - Re^{-i\phi})\,Re^{i\phi}}\,d\phi$$

$$= \frac{1}{2\pi} \int_0^{2\pi} \frac{(R^2 - r^2)f(Re^{i\phi})}{(Re^{i\phi} - re^{i\theta})(Re^{-i\phi} - re^{-i\theta})}\,d\phi$$

$$= \frac{1}{2\pi} \int_0^{2\pi} \frac{(R^2 - r^2)f(Re^{i\phi})}{R^2 - 2rR\cos(\phi - \theta) + r^2}\,d\phi. \tag{11.36}$$

Comparing the real parts in Eq. (11.36), we obtain

$$u(r, \theta) = \frac{1}{2\pi} \int_0^{2\pi} \frac{(R^2 - r^2)\,u(R, \phi)}{R^2 - 2rR\cos(\phi - \theta) + r^2}\,d\phi.$$

Exercise 11.4

Evaluate the following integrals.

1. $\oint_C \dfrac{e^z + 1}{z}\,dz$, $C: |z| = 1$.

2. $\oint_C \dfrac{\tan z + \sec z}{z}\,dz$, $C: |z| = 1$.

3. $\oint_C \dfrac{e^{2z}}{z + \pi i}\,dz$, $C:$ (i) $|z - 1| = 4$, (ii) $|z - 2| + |z + 2| = 6$.

4. $\oint_C \dfrac{z^2 - 3z - i}{z - 2i}\,dz$, $C: |z| = 3$.

5. $\oint_C \dfrac{2z^3 + 3z + 5}{z - \xi}\,dz$, $C: |z| = 3$ and (i) $\xi = 2$, (ii) $\xi = 4$.

6. $\oint_C \dfrac{\sin z}{4z - \pi}\,dz$, $C: |z| = 1$.

7. $\oint_C \dfrac{e^z}{z(1 + z)^3}\,dz$, $C: |z| = \dfrac{1}{2}$.

8. $\oint_C \dfrac{e^z + \sin \pi z}{(z - 1)(z - 3)^2(z + 4)}\,dz$, $C: |z| = 2$.

9. $\oint_C \dfrac{\tan^{-1}(zi)}{z + i}\,dz$, $C:$ boundary of the triangle with vertices at -1, 1 and $-2i$.

10. $\oint_C (z + \bar{z})\,dz$, $C: |z| = r$.

11. $\oint_C \dfrac{z + 1}{z^2 - 9}\,dz$, $C:$ (i) $|z - 3| = 1$, (ii) $|z + 3| = 1$.

12. $\oint_C \dfrac{z^2 + 1}{z^2 + 9}\,dz$, $C:$ (i) $|z - 2i| = 2$, (ii) $|z - 2i| = 6$.

13. $\oint_C \dfrac{e^{\pi i z}}{z^2 - 3z + 2}\,dz$, $C: |z - 3| = 3$.

14. $\oint_C \dfrac{dz}{2z^2 + 5z - 3}$, $C:$ the ellipse $x^2 + 4y^2 = 16$.

15. $\oint_C \dfrac{\cos 2z}{z^2 - 4z + 5}\,dz$, $C: |z - 1 - 2i| = 2$.

16. $\oint_C \dfrac{3z + 5}{z^2 + z}\,dz$, $C:$ (i) $|z| = \dfrac{1}{2}$, (ii) $|z| = 2$.

17. $\displaystyle\oint_C \frac{\sin(iz)}{z^2+1}\,dz$, C : boundary of the triangle with vertices at $1-2i, -1-2i, 2i$.

18. $\displaystyle\oint_C \frac{\bar{z}_0}{(z-z_0)(z\bar{z}_0-1)}\,dz$, $C : |z|=1$ and (i) $|z_0|<1$, (ii) $|z_0|>1$.

19. $\displaystyle\oint_C \frac{z-3}{z^3+z}\,dz$, $C : |z|=2$.   **20.** $\displaystyle\oint_C \frac{dz}{z(z^2-1)}$, $C : |z|=2$.

11.5 Cauchy Integral Formula for Derivatives

Using the Cauchy integral formula, we now establish that if $f(z)$ is analytic in a domain D, then its derivatives of all orders exist and are also analytic in D. This is a very fundamental result in the case of functions of a complex variable. Such a result does not exist for functions of real variables.

Theorem 11.6 Let $f(z)$ be analytic in a simply connected domain D. Let z_0 be any point in D and C be any simple closed curve in D enclosing the point $z=z_0$. Then, $f(z)$ has derivatives of all orders in D which are also analytic in D. Further

$$f^{(n)}(z_0) = \frac{n!}{2\pi i}\oint_C \frac{f(z)}{(z-z_0)^{n+1}}\,dz, \quad n = 1, 2, \ldots \tag{11.37}$$

where C is traversed in the anti-clockwise direction.

Proof The result can be proved by the mathematical induction. Consider the case $n=1$. We are to prove that

$$f'(z_0) = \frac{1}{2\pi i}\oint_C \frac{f(z)}{(z-z_0)^2}\,dz. \tag{11.38}$$

By the definition of a derivative, we have

$$f'(z_0) = \lim_{\Delta z_0 \to 0}\frac{f(z_0+\Delta z_0)-f(z_0)}{\Delta z_0}.$$

Now, since the interior of C is an open set, if z_0 is inside C, then $z_0+\Delta z_0$ is also inside C for $|\Delta z_0|$ sufficiently small. Using the Cauchy integral formula, we get

$$f(z_0) = \frac{1}{2\pi i}\oint_C \frac{f(z)}{z-z_0}\,dz \quad \text{and} \quad f(z_0+\Delta z_0) = \frac{1}{2\pi i}\oint_C \frac{f(z)}{z-(z_0+\Delta z_0)}\,dz.$$

Hence,

$$\frac{f(z_0+\Delta z_0)-f(z_0)}{\Delta z_0} = \frac{1}{2\pi i\,\Delta z_0}\oint_C \left[\frac{1}{z-(z_0+\Delta z_0)}-\frac{1}{z-z_0}\right]f(z)\,dz$$

$$= \frac{1}{2\pi i}\oint_C \frac{f(z)\,dz}{(z-z_0)(z-z_0-\Delta z_0)}.$$

We now write

$$\frac{1}{(z-z_0)(z-z_0-\Delta z_0)} = \frac{1}{(z-z_0)^2}\left[\frac{z-z_0-\Delta z_0+\Delta z_0}{z-z_0-\Delta z_0}\right] = \frac{1}{(z-z_0)^2}\left[1+\frac{\Delta z_0}{z-z_0-\Delta z_0}\right]$$

$$= \frac{1}{(z-z_0)^2}+\frac{\Delta z_0}{(z-z_0)^2(z-z_0-\Delta z_0)}$$

Therefore,

$$f'(z_0) = \frac{1}{2\pi i} \lim_{\Delta z_0 \to 0} \left[\oint_C \frac{f(z)}{(z-z_0)^2} \, dz + \oint_C \frac{\Delta z_0 \, f(z)}{(z-z_0)^2 (z-z_0-\Delta z_0)} \, dz \right]$$

$$= \frac{1}{2\pi i} \oint_C \frac{f(z)}{(z-z_0)^2} \, dz + \frac{1}{2\pi i} \lim_{\Delta z_0 \to 0} \oint_C \frac{\Delta z_0 \, f(z)}{(z-z_0)^2 (z-z_0-\Delta z_0)} \, dz. \quad (11.39)$$

Since $f(z)$ is analytic and hence continuous on C, it is bounded on C, that is $|f(z)| \le M$ for all z on C. Let d be the smallest distance from z_0 to any point z on C (Fig. 11.25). Therefore, for any z on C, we have the following:

(i) $|z - z_0| \ge d$ or $|z - z_0|^2 \ge d^2$ or $\dfrac{1}{|z - z_0|^2} \le \dfrac{1}{d^2}$

(ii) $|z - z_0 - \Delta z_0| \ge ||z - z_0| - |\Delta z_0|| \ge d - |\Delta z_0|$ or $\left| \dfrac{1}{z - z_0 - \Delta z_0} \right| \le \dfrac{1}{d - |\Delta z_0|}$

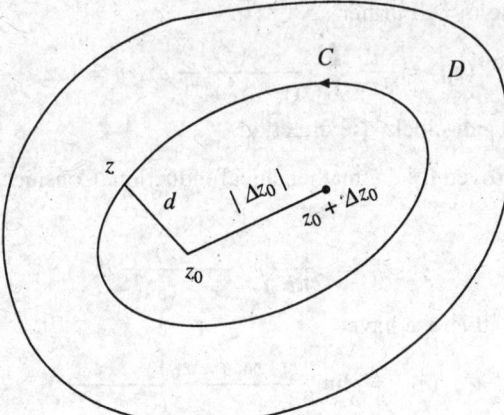

Fig. 11.25.   Theorem 11.6.

Let L be the length of the curve C. Using the ML-inequality, we get

$$\left| \frac{1}{2\pi i} \oint_C \frac{\Delta z_0 \, f(z)}{(z-z_0)^2 (z-z_0-\Delta z_0)} \, dz \right| \le \frac{ML|\Delta z_0|}{2\pi d^2 (d - |\Delta z_0|)} \to 0 \text{ as } |\Delta z_0| \to 0.$$

Hence, $f'(z_0)$ exists and from Eq. (11.39), we obtain

$$f'(z_0) = \frac{1}{2\pi i} \oint_C \frac{f(z)}{(z-z_0)^2} \, dz.$$

Since z_0 is any point in D, $f'(z)$ is analytic in D. Hence, we have the result that if $f(z)$ is analytic, then $f'(z)$ is also analytic and $f'(z_0)$ is given by Eq. (11.38). Repeating the above procedure with $f(z)$ replaced $f'(z)$, we establish the existence and analyticity of $f''(z)$. Continuing this procedure, we prove that derivatives of all orders of an analytic function $f(z)$ exist and these derivatives are all analytic. We obtain

$$f''(z_0) = \frac{2!}{2\pi i} \oint_C \frac{f(z)}{(z - z_0)^3} \, dz$$

$$f'''(z_0) = \frac{3!}{2\pi i} \oint_C \frac{f(z)}{(z - z_0)^4} \, dz$$

$$\cdots \quad \cdots \quad \cdots \quad \cdots \quad \cdots \quad \cdots$$

$$f^{(n)}(z_0) = \frac{n!}{2\pi i} \oint_C \frac{f(z)}{(z - z_0)^{n+1}} \, dz, \quad n = 1, 2, \ldots.$$

Though it is not rigorous, these results can be obtained by repeated differentiation with respect to z_0 of the Cauchy integral formula.

Remark 9

If $f(z) = u(x, y) + iv(x, y)$ is an analytic function in D, then all the partial derivatives of the real functions $u(x, y)$ and $v(x, y)$ are continuous at every point in D.

As a consequence of the above Theorem 11.6, we prove the converse of the Cauchy integral theorem which is known as Morera's theorem.

Theorem 11.7 (Morera's theorem). If $f(z)$ is continuous in a simply connected domain D and if

$$\oint_C f(z) \, dz = 0.$$

for every simple closed curve C in D, then $f(z)$ is analytic in D.

Proof Let z be an arbitrary point in D and consider a fixed point z_0 in D. While defining the indefinite integration in section 11.3.2, it was shown that if $f(z)$ is continuous and its integral around any closed path in D is zero then $F(z)$ defined by

$$F(z) = \int_{z_0}^{z} f(\xi) \, d\xi$$

is analytic in D and $F'(z) = f(z)$. Since $F(z)$ is analytic, $F'(z) = f(z)$ is also analytic in D.

Using the Cauchy integral formula for derivatives, it is possible to find a bound for $f^{(n)}(z_0)$. This result is known as the *Cauchy inequality*.

Cauchy inequality

Let $f(z)$ be analytic within and on a circle $C : |z - z_0| = r$. Then,

$$|f^{(n)}(z_0)| \le \frac{M(n!)}{r^n}, \text{ where } |f(z)| \le M \text{ on } C. \tag{11.40}$$

Using Eq. (11.37) and the ML-inequality, we obtain

$$|f^{(n)}(z_0)| = \left| \frac{n!}{2\pi i} \oint_C \frac{f(z)}{(z - z_0)^{n+1}} \, dz \right| \le \frac{n!}{2\pi} \frac{M}{r^{n+1}} (2\pi r) = \frac{n! M}{r^n}.$$

For $n = 0$, Cauchy inequality gives $|f(z_0)| \le M$. Note that z_0 is the centre of the circle C and M is the bound of $f(z)$ on C. As a consequence, we have the following result:

Theorem 11.8 (Maximum modulus theorem) If $f(z)$ is non-constant and analytic within and on a simple closed curve C, then $|f(z)|$ assumes its maximum value on C.

Proof Suppose that the maximum of $|f(z)|$ occurs at a point ξ contained in D whose boundary is C. Enclose the point ξ by some circle C_1 with centre at ξ and radius ρ (Fig. 11.26).

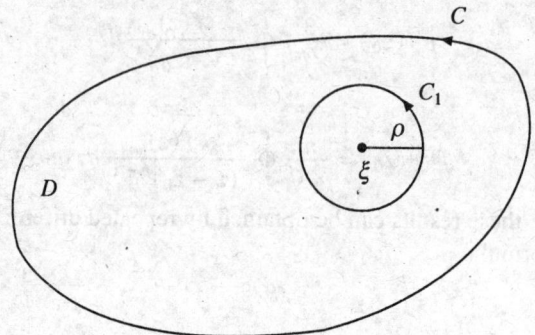

Fig. 11.26.   Theorem 11.8.

By Cauchy inequality, we have

$$|f(\xi)| \le M, \text{ where } M = \max |f(z)| \text{ on } C_1$$

(note that the equality holds only for a constant function $f(z)$). That is, the value of $|f(z)|$ at the centre $z = \xi$ is smaller than M. But, ξ is an arbitrary point in D and we have assumed that max $|f(z)|$ occurs at ξ, which is a contradiction. Therefore, ξ must be a point on the bounding curve C.

Similar result is true for the minimum value of $|f(z)|$ of an analytic function. We have the following result.

Theorem 11.9 (Minimum Modulus theorem) If $f(z)$ is non-constant and analytic within and on a simple closed curve C, then $|f(z)|$ takes its minimum value on C.

As a consequence of the above theorems, we conclude that if $f(z)$ is analytic inside and on a simple closed curve C, then (if $f(z) \ne 0$ and not a constant) the maximum and minimum values of $|f(z)|$ occur on the boundary of C.

Let $f(z)$ be analytic in the entire complex plane. Then, we have the following result.

Theorem 11.10 (Liouville's theorem) If an entire function $f(z)$ is bounded for all values of z in the finite complex plane, then $f(z)$ must be a constant.

Proof Since $f(z)$ is bounded for all finite z, we have $|f(z)| \le M$. Let C be the circle $|z - z_0| = r$. Obviously, the function $f(z)$ is analytic within and on C. Using the Cauchy integral formula for derivatives, we have

$$f'(z) = \frac{1}{2\pi i} \oint_C \frac{f(\xi)}{(\xi - z)^2} \, d\xi$$

and the Cauchy inequality (see Eq. 11.40) gives

$$|f'(z)| \le M/r.$$

Since $f(z)$ is analytic in the finite complex plane, we can take r as large as possible. Hence, $|f'(z)| \to 0$, that is $f'(z) = 0$ for all z. Therefore, $f(z)$ is a constant.

Example 11.38 If $f(z)$ is an analytic function within and on a simple closed contour C and z_0 is any point inside C, then show that

$$\oint_C \frac{f(z)}{(z-z_0)^2}\,dz = \oint_C \frac{f'(z)}{z'-z_0}\,dz.$$

Solution Since $f(z)$ is analytic, $f'(z)$ is also analytic. Using the Cauchy integral formula, we get

$$\oint_C \frac{f'(z)}{z-z_0}\,dz = 2\pi i\,f'(z_0).$$

Using the Cauchy integral formula for derivatives $(n = 1)$, we get

$$\oint_C \frac{f(z)}{(z-z_0)^2} = 2\pi i\,f'(z_0)$$

and hence the result.

Example 11.39 Show that $\displaystyle\oint_C \frac{dz}{(z^2+4)^2} = \frac{\pi}{16},\ C:|z-i|=2.$

Solution The integrand is not analytic at the points $z = \pm 2i$. The point $z = 2i$ lies inside C. We have

$$I = \oint_C \frac{dz}{(z^2+4)^2} = \oint_C \frac{[1/(z+2i)^2]}{(z-2i)^2}\,dz.$$

Taking $f(z) = 1/[z+2i]^2$ and using the Cauchy integral formula for derivatives $(n = 1)$, we obtain

$$I = \{2\pi i\,[f'(z)]_{z=2i}\} = 2\pi i\left[\frac{-2}{(z+2i)^3}\right]_{z=2i} = 2\pi i\left[\frac{-2}{(4i)^3}\right] = \frac{\pi}{16}.$$

Example 11.40 Evaluate the integral $\displaystyle I = \oint_C \frac{3z^4+5z^2+2}{(z+1)^4}\,dz$ where C is any simple closed contour containing the point $z = -1$ inside it.

Solution The integrand is not analytic at the point $z = -1$ which lies inside C. Using the Cauchy integral formula for derivatives $(n = 3)$ with $f(z) = 3z^4 + 5z^2 + 2$, we obtain

$$I = \frac{2\pi i}{3!}\left[\frac{d^3}{dz^3}(3z^4+5z^2+2)\right]_{z=-1} = \left\{\frac{2\pi i}{6}\,[72z]_{z=-1}\right\} = -24\pi i.$$

Example 11.41 Evaluate the integral $\displaystyle I = \oint_C \frac{e^z}{z^2(z+1)^3}\,dz,\ C:|z|=2.$

Solution The integrand is not analytic at the points $z = 0$ and $z = -1$ both of which lie inside C. Enclose points $z = 0$ and $z = -1$ by circles $C_1:|z| = \delta_1$ and $C_2:|z+1| = \delta_2$ respectively, such that they do not intersect and lie inside C. Using the extension of the Cauchy integral theorem for multiply connected domains, we obtain

$$I = \oint_C \frac{e^z}{z^2(z+1)^3}\,dz = \oint_{C_1} \frac{[e^z/(z+1)^3]}{z^2}\,dz + \oint_{C_2} \frac{[e^z/z^2]}{(z+1)^3}\,dz = I_1 + I_2.$$

For evaluating I_1, we have $f(z) = e^z/(z+1)^3$ and for evaluating I_2, we have $f(z) = e^z/z^2$. Using the Cauchy integral formula for derivatives, we obtain

$$I = 2\pi i \left[\frac{d}{dz}\left(\frac{e^z}{(z+1)^3} \right) \right]_{z=0} + \frac{2\pi i}{2!}\left[\frac{d^2}{dz^2}\left(\frac{e^z}{z^2} \right) \right]_{z=-1}$$

$$= 2\pi i \left[\frac{\{(z+1)^3 - 3(z+1)^2\}e^z}{(z+1)^6} \right]_{z=0} + \pi i \left[\frac{d}{dz}\left\{ \frac{(z-2)e^z}{z^3} \right\} \right]_{z=-1}$$

$$= 2\pi i\,(-2) + \pi i \left[\frac{\{z^3(z-1) - 3z^2(z-2)\}e^z}{z^6} \right]_{z=-1}$$

$$= -4\pi i + 11\,\pi i e^{-1} = (11\,e^{-1} - 4)\,\pi i.$$

Example 11.42　Verify that the maximum and minimum modulus theorems hold for the functions (i) $f(z) = e^z$ and (ii) $f(z) = z^2 + 1$ where C is the circle $|z| = 1$ and D is the domain inside C.

Solution

(i) For $f(z) = e^z$, we have $|f(z)| = e^x$. Now x takes the values in the interval $-1 \le x \le 1$. Also, e^x is maximum when $x = 1$ and minimum when $x = -1$. The maximum value is e and the minimum value is e^{-1}. The points $x = 1$ and $x = -1$ correspond to the points $z = 1$ and $z = -1$ respectively, which lie on the boundary of C.

(ii) For $f(z) = z^2 + 1$, we have $g(x, y) = |f(z)|^2 = (x^2 - y^2 + 1)^2 + 4x^2 y^2$. The points on C satisfy $x^2 + y^2 = 1$ and the points in D satisfy $x^2 + y^2 < 1$. The function $g(x, y)$ attains local extremum

when $g_x = 4x(x^2 + y^2 + 1) = 0$　and　$g_y = 4y(x^2 + y^2 - 1) = 0$.

The only point, in D satisfying these equations is $x = 0$, $y = 0$, that is $z = 0$. In this case, we have $|f(0)| = 1$.

To determine the extreme values on the boundary of C, we substitute $y^2 = 1 - x^2$ in $g(x, y)$. We get

$$h(x) = 4x^4 + 4x^2(1 - x^2) = 4x^2, \quad -1 \le x \le 1.$$

Now $h(x)$ has maximum at $x = \pm 1$. From $y^2 = 1 - x^2$, we get $y = 0$. Therefore, maximum occurs at the points $z = \pm 1$, which lie on the boundary of the circle C. We obtain

$$\text{maximum value} = |f(\pm 1)| = 2.$$

Again, $h(x)$ has minimum at $x = 0$. From $y^2 = 1 - x^2$, we get $y = \pm 1$. Therefore, minimum occurs at the points $z = \pm i$ which lie on the boundary of the circle C. We obtain

$$\text{minimum value} = |f(\pm i)| = 0.$$

Thus, maximum and minimum values of $|f(z)|$ occur on the boundary.

Example 11.43　The function $f(z) = e^z$ is analytic within and on $C : |z| = 1$. Using the Cauchy inequality find a bound on $f^{(n)}(0)$.

Solution　The centre of the circle is $z = 0$. From Example 11.42 (i), we have $|f(z)| \le e$ on C. Hence, by Cauchy inequality, we obtain

$$|f^{(n)}(0)| \le \frac{n!\,M}{r} = \frac{e(n!)}{1} = e(n!).$$

It can be observed that the Cauchy inequality provides a very rough upper bound as the actual value is $f^{(n)}(0) = 1$.

Example 11.44 Using the Liouville's theorem prove the *fundamental theorem of algebra* which states that if $f(z) = P(z)$ is a non-constant polynomial in z with complex coefficients, then there exists at least one complex number ξ such that $P(\xi) = 0$.

Solution If possible, let $P(z) \neq 0$ for all z. Then $1/P(z)$ is differentiable for all z. Now, $\mid P(z) \mid \to \infty$ as $\mid z \mid \to \infty$. Hence, there exists an r such that $\mid P(z) \mid > 1$ for $\mid z \mid > r$. Therefore, $1/P(z)$ is bounded in $\mid z \mid > r$.

Since $1/P(z)$ is differentiable for all z, $1/P(z)$ is continuous in $\mid z \mid \leq r$. Hence, $1/P(z)$ is bounded for all z, that is in the entire complex plane. By Liouville's theorem, we obtain that

$$\frac{1}{P(z)} = \text{constant. or} \quad P(z) = \text{constant,}$$

which is a contradiction. Hence, these exists atleast one complex number ξ such that $P(\xi) = 0$.

Exercise 11.5

Evaluate the following integrals.

1. $\displaystyle\oint_C \frac{z^2 + 1}{z^2 (z - 2)} \, dz, \quad C : \mid z \mid = 1.$

2. $\displaystyle\oint_C \frac{\cos^2 z}{(z + \pi/2)^2} \, dz, \quad C : \mid z \mid = 2.$

3. $\displaystyle\oint_C \frac{\text{Ln}\,(z + i)}{z(z - 1)^2} \, dz, \quad C : \left| z - \frac{5}{2} \right| = 2.$

4. $\displaystyle\oint_C \frac{e^z}{(z - 1)^2 (z^2 + 9)} \, dz, \quad C : \mid z \mid = 2.$

5. $\displaystyle\oint_C \frac{e^{zi} \cos z}{(z - \pi)^2} \, dz, \quad C : \mid z \mid = 4.$

6. $\displaystyle\oint_C \frac{\cos (z^4)}{(z + 2)^2} \, dz, \quad C : \mid z \mid = 3.$

7. $\displaystyle\oint_C \frac{dz}{(z - 2i)^2 (z + 2i)}, \, C : \text{the ellipse } x^2 + 4(y - 2)^2 = 4.$

8. $\displaystyle\oint_C \frac{z}{(z^2 - 6z + 25)^2} \, dz, \quad C : \mid z - (3 + 4i) \mid = 4.$

9. $\displaystyle\oint_C \frac{e^z - 3z}{z^4 + 8iz^2} \, dz, \, C : \text{the rectangle with vertices at } \pm 1 \pm i.$

10. $\displaystyle\oint_C \frac{e^z}{(z - 1 - i)^3} \, dz, \quad C : \mid z - 1 \mid = 3.$

11. $\displaystyle\oint_C \frac{\cos z - \sin z}{(z + i)^3} \, dz, \quad C : \mid z \mid = 2.$

12. $\displaystyle\oint_C \frac{z^3 - \sin (3z)}{(z - \pi/2)^3} \, dz, \quad C : \mid z \mid = 2.$

13. $\displaystyle\oint_C \frac{dz}{e^z (z - i)^3}, \quad C : \mid z \mid = 2.$

14. $\displaystyle\oint_C \frac{\text{Ln}\, z}{(z - 1)^3} \, dz, \quad C : \mid z - 1 \mid = \frac{1}{2}.$

15. $\displaystyle\oint_C \frac{z e^z}{(z + a)^3} \, dz, \, C : \text{any simple closed curve enclosing the point } z = -a.$

16. $\displaystyle\oint_C \frac{\sin z}{z^m} \, dz, \quad C : \mid z \mid = 1, m = 2, 3, 4, \ldots.$

17. $\displaystyle\oint_C \left[\frac{e^{iz}}{z^3} + \frac{z^4}{(z + i)^2} \right] dz, \quad C : \mid z \mid = 2.$

18. $\displaystyle\oint_C \frac{z^2 - z - 1}{z(z - i)^2} \, dz, \quad C : \left| z - \frac{i}{2} \right| = 1.$

19. $\displaystyle\oint_C \frac{z - 2}{(z - 1)^3 (z + 2)^3} \, dz, \quad C : \mid z \mid = 3.$

20. $\oint_C \dfrac{dz}{(z^2+4)^3}$, $C:|z-1|=4$.

21. $\oint_C \dfrac{e^z\,dz}{z(1+z)^3}$, C : the square with vertices at $\pm 2 \pm 2i$.

22. $\oint_C \dfrac{e^{z^2}\,dz}{z^2(z-i)^2}$, C : the square with vertices at $\pm 3 \pm 3i$.

23. $\oint_C \dfrac{dz}{(z^2+3+4i)^3}\,dz$, $C:|z|=4$.

24. If the function $g(z)$ is analytic on and inside a simple closed contour C and z_0 is a point inside C, then show that

(i) $\oint_C \dfrac{g(z)}{(z-z_0)^{n+1}}\,dz = \dfrac{1}{n!}\oint_C \dfrac{g^{(n)}(z)}{z-z_0}\,dz$,   (ii) $g^{(n)}(z_0) = \dfrac{k!}{2\pi i}\oint_C \dfrac{g^{(n-k)}(z)}{(z-z_0)^{k+1}}\,dz, k \le n$.

25. If the function $f(z)$ is analytic inside and on a simple closed curve C containing the point $z = a$ inside it, then show that

(i) $f(a) = \dfrac{1}{2\pi}\displaystyle\int_0^{2\pi} f(a+e^{i\theta})\,d\theta$,   (ii) $f'(a) = \dfrac{1}{2\pi}\displaystyle\int_0^{2\pi} e^{-i\theta} f(a+e^{i\theta})\,d\theta$,

(iii) $f^{(n)}(a) = \dfrac{n!}{2\pi}\displaystyle\int_0^{2\pi} e^{-in\theta} f(a+e^{i\theta})\,d\theta$, $n = 2, 3, \ldots$.

The closed curve C contains the circle $|z-a|=1$ within it.

In the following problems, verify the maximum and minimum modulus theorems.

26. $f(z) = z+5$, $C:|z|=4$.   **27.** $f(z) = 3z^2+2$, $C:|z|=\dfrac{1}{2}$.

28. $f(z) = e^{2z}$, $C:|z|=1$.

Let $f(z)$ be analytic within and on a circle $C:|z-z_0|=r$. Using the Cauchy inequality, find a bound on $f^{(n)}(0)$, where

29. $f(z) = z^2+1$, $C:|z|=1$.   **30.** $f(z) = e^{3z}$, $C:|z|=1$.

11.6   Answers and Hints

Exercise 11.1

1. $-(1+i)/2$.   **2.** 2π, if $n = 0$, and 0 otherwise.

3. $(1+i)/2$.

4. Straight line path from $(2, 0)$ to $(2, 6)$; $x = 2, y = 3t$.

5. Lower semi-circle, radius 2, centre at $1-i$, $x = 1-2\cos t$, $y = -1-2\sin t$.

6. The parabola $y = 3x^2$ from $(1, 3)$ to $(2, 12)$.

7. The ellipse $4x^2+y^2 = 4$.   **8.** Simple, closed, smooth.

9. Simple, closed, piecewise smooth, (not differentiable at $t = 0$).

10. Simple, closed, smooth.

11. Simple, not closed, piecewise smooth (not differentiable at $t = 1$).

12. Not simple ($z = 1 + i$ for $t = 1$ and $t = 5$), not closed, piecewise smooth (not differentiable at $t = 2$ and $t = 3$).

13. $z(t) = 3 - 4i + 4e^{it}$, $0 \le t \le 2\pi$.   14. $z(t) = t + i/t$, $1 \le t \le 3$.

15. $z = i + 2e^{it}$, $0 \le t \le 2\pi$ in the counter clockwise direction. Therefore, $z = i + 2e^{-it}$, $-2\pi \le t \le 0$ in the clockwise direction.

16. $z = z_0 + a \cos t + ib \sin t$, $0 \le t \le \pi$, $z_0 = (x_0, y_0)$.

17. $z = t + i/t$, $0 < t < \infty$ in the counter clockwise direction. Therefore, $z = -t - i/t$, $-\infty < t < 0$ in the clockwise direction.

18. $z = e^{it}$, $0 < t < \pi$.   19. $z = \begin{cases} t; & 0 \le t \le 1, \\ 1 + i(t - 1), & 1 \le t \le 2, \\ (1 + i)(3 - t), & 2 \le t \le 3. \end{cases}$

20. $z = \begin{cases} 1 + it, & -1 \le t \le 1, \\ 2 + i - t, & 1 \le t \le 3, \\ -1 + i(4 - t), & 3 \le t \le 5, \\ t - 6 - i, & 5 \le t \le 7. \end{cases}$

Exercise 11.2

1. $z = t + 2it^2$, $I = 1523/15$.   2. $I = \int_0^2 (1 + iy^3)i\, dy = 2i - 4$.

3. Parametric form of $C : z = t + (1 - t)i$, $0 \le t \le 1$, $I = 5(2 - i)/3$.

4. Parametric form of $C : z = (3 \cos \theta + i \sin \theta)$, $\pi/2 \le \theta \le -\pi/2$, $I = 0$.

5. $z = e^{it}$, $I = (1 + i) \cosh (\pi/2)$.   6. $z = it$, $I = -\pi/8$.

7. $z = e^{i\theta}$, $I = \pi i/2$.   8. $z = e^{i\theta}$, $I = -2$.

9. $z = i + e^{it}$, $I = 0$.   10. $z = t + it^3$, $I = 0$.

11. $I = (3i - 2)/6$.

12. $z = \cos t + i \sin t$, Re $z = \cos t > 0$, when $0 < t < \pi/2$ and $3\pi/2 < t < 2\pi$, and Re $z = \cos t < 0$ when $\pi/2 < t < 3\pi/2$. Hence

$$I = \int_0^{\pi/2} e^{it}(ie^{it})dt + \int_{\pi/2}^{3\pi/2} e^{2it}(ie^{it})dt + \int_{3\pi/2}^{2\pi} e^{it}(ie^{it})dt = 2i/3.$$

13. (i) $z = e^{i\theta}$, $I = 0$, (ii) $z = 2 + e^{i\theta}$, $I = 0$.

14. Parametric form of $C : z = x + \pi i(1 - x)$, $0 \le x \le 1$; $I = -(1 + e^{-1})$

15. (i) Parametric form of $C : z = iy$, $-1 \le y \le 1$; $I = i$. (ii) $z = 1 + e^{i\theta}$, $|z| = 2 \left| \cos \dfrac{\theta}{2} \right|$. Therefore,

$|z| = 2 \cos \dfrac{\theta}{2}$, $0 < \theta < \pi$ and $|z| = -2 \cos \dfrac{\theta}{2}$, $\pi < \theta < 2\pi$; $I = 8i/3$.

16. $z = e^{i\theta}$; (i) $I = \pi i$, (ii) $I = -\pi i$.

17. $z = 3e^{i\theta}$; (i) $I = [-9 + (4\pi + 9\sqrt{3})i]/18$, (ii) $I = 2(\pi i + 3)/3$, (iii) $I = 4\pi i/3$.

18. (i) Parametric form of $C : z = [1 + (2i/3)]x$, $0 \le x \le 3$, $I = (39 - 26i)/3$, (ii) $I = (63 + 46i)/3$

19. (i) Parametric form of $C : z = x + (13 - 5x)i$, (i) $I = (109 - 25i)/3$, (ii) Path 1: $y = 3, 2 \leq x \leq 3, z = x + 3i$, $I_1 = (-26/3) + 20i$, Path 2 : $x = 3, 3 \leq y \leq -2, z = 3 + iy, I_2 = 45 - (85/3)i; I = I_1 + I_2 = (109 - 25i)/3$.

20. (i) $z = (2 + i)y; I = (16 + 88i)/3$, (ii) $z = y^2 + iy; I = (16 + 88i)/3$.

21. Let the vertices of the square be $O(0,0)$, $B(2, 0)$, $C(2, 2)$ and $D(0, 2)$. Along the line $OB : y = 0$, $0 \leq x \leq 2$; along the line $BC : x = 2; 0 \leq y \leq 2$; along the line $CD : y = 2 , 2 \leq x \leq 0$; along the line $DO : x = 0, 2 \leq y \leq 0; I = 8(-1 + i)$.

22. Let the vertices of the triangle be $O(0, 0)$, $B(1, 0)$ and $C(0, 1)$. Along the line $OB: y = 0, 0 \leq x \leq 1$; along the line $BC : y = 1 - x, 1 \leq x \leq 0$; along the line $CO : x = 0, 1 \leq y \leq 0; I = (1 + i)/3$.

23. Along the circle $z = e^{i\theta}, 0 \leq \theta \leq \pi$; along the line $y = 0, -1 \leq x \leq 1; I = \pi i$.

24. (i) $z = e^{i\theta}, 0 \leq \theta \leq \pi ; I = 2(i - 1)$.   (ii) $z = e^{i\theta}, -\pi \leq \theta \leq 0; I = -2(i + 1)$.

25. $z = e^{i\theta}$, Ln $z = $ Ln $|z| + i\theta = i\theta; I = 2\pi i/(n + 1)$.

26. $x = a \cos^3\theta, y = a \sin^3\theta, z = a (\cos^3\theta + i \sin^3 \theta) = 3ai(e^{i\theta} - e^{-3i\theta})/4, I = 3\pi a^2i/2$.

27. $L = \displaystyle\int_C |z'(t)| \, dt = \pi/\sqrt{2}$.

28. $L = 2\displaystyle\int_0^{2\pi} |\cos (t/2)| \, dt = 2\int_0^{\pi} \cos (t/2) \, dt - 2\int_{\pi}^{2\pi} \cos (t/2)dt = 8$.

29. $|f(z)| \leq \dfrac{|z|}{|z|-1} = 2 = M; |I| \leq ML = 4\pi$.

30. Path 1: $y = 0, z = x, 0 \leq x \leq 1, |f (z)| \leq e$; Path 2 : $x = 1, z = 1 + iy, 0 \leq y \leq 1, |f (z)| \leq e; M = e, L = 2; |I| \leq 2e$.

31. $|f(z)| \leq 1/(R^2 - 1), L = \pi R; |I| \leq \pi R/(R^2 - 1)$.

32. Equation of the line is $y = -x + 2, |z^4| = (x^2 + y^2)^2 = [2(x - 1)^2 + 2]^2 \geq 4, |f (z)| \leq 1/4, M = 1/4$, $L = \sqrt{2}; |I| \leq 1/2\sqrt{2}$.

33. $z = Re^{i\theta}, |f (z)| \leq (\pi + \text{Ln } R)/R^2 = M, L = 2\pi R; |I| \leq 2\pi (\pi + \text{Ln } R)/R$.

34. Let the vertices of the triangle be $O(0, 0)$, $A(1, 0)$ and $B(0, 1)$.

Along $OA : z = x, 0 \leq x \leq 1, |f (z)| \leq e$; Along AB : $z = x + (1 - x)i, 1 \leq x \leq 0, |f (z)| \leq e$.

Along $BO : z = iy, 1 \leq y \leq 0, |f (z)| = 1, M = e, L = 2 + \sqrt{2}; |I| \leq e(2 + \sqrt{2})$.

35. $|f(z)| = |\sin z^2| \leq \dfrac{1}{2} \left[|e^{iz^2}| + |e^{-iz^2}| \right] = \dfrac{1}{2}[e^{-2xy} + e^{2xy}] = \dfrac{1}{2}[e^{-4\sin 2\theta} + e^{4\sin 2\theta}] \leq \dfrac{1}{2}(1 + e^4) = M$, $L = 4\pi; |I| \leq 2\pi (1 + e^4)$.

Exercise 11.3

In problems **1** to **18**, Cauchy integral theorem is applicable, $I = 0$. In problems **19–21**, write the given integral as $I = I_1 + I_2$.

19. By Cauchy integral theorem $I_1 = 0$. Using example 11.7, $I_2 = 0; I = 0$.

20. $I_1 = \displaystyle\oint_C x \, dz = 4\pi i$, (set $z = 2e^{i\theta}$), $I_2 = 0; I = 4\pi i$.

21. $I_1 = 0$ (by Cauchy integral theorem), $I_2 = 6\pi i; I = 6\pi i$.

22. $f (z) = \dfrac{19}{3(z - 1)} + \dfrac{2}{3(z + 2)}$ is not analytic at $z = 1$ and $z = -2$. Only $z = 1$ lies inside $C, I = 38\pi i/3$.

23. $f(z) = \dfrac{1}{3z} - \dfrac{i+1}{2(z+i)} + \dfrac{3i+1}{6(z+3i)}$ is not analytic at $z = 0, -i$ and $-3i$, $z = -i$ lies inside C; $I = \pi(1-i)$.

24. $f(z) = \dfrac{1}{9}\left[\dfrac{1}{z} - \dfrac{3i}{z^2} - \dfrac{1}{z+3i}\right]$ is not analytic at $z = 0$ and $z = -3i$, $z = 0$ lies inside C; $I = 2\pi i/9$.

In problems **25–30**, use partial fractions to simplify the integrand $f(z)$. Enclose the points $z_0, z_1, \ldots$ inside C, at which $f(z)$ is not analytic by non-intersecting circles $|z - z_0| = r_1$, $|z - z_1| = r_2, \ldots$ which lie inside C. Use the extension of the Cauchy integral theorem and the Cauchy integral theorem.

25. $f(z) = \dfrac{1}{z+i\pi} + \dfrac{1}{z-i\pi}$; the points at which $f(z)$ is not analytic inside C are $z = \pm \pi i$. $I = 4\pi i$.

26. $f(z) = \dfrac{1}{z+3} + \dfrac{1}{z-6}$; the points at which $f(z)$ is not analytic inside C are $z = -3$ and $z = 6$. $I = 4\pi i$.

27. $f(z) = \dfrac{1}{z} + \dfrac{1}{z-1}$; the points at which $f(z)$ is not analytic inside C are $z = 0$ and $z = 1$. $I = 4\pi i$.

28. $f(z) = \dfrac{1}{30(z+4)} + \dfrac{1}{6(z-2)} - \dfrac{1}{5(z-1)}$; the points at which $f(z)$ is not analytic inside C are $z = 1$ and $z = 2$. $I = -\pi i/15$.

29. $f(z) = \dfrac{1}{2(z-1)} - \dfrac{1+i}{4(z+i)} - \dfrac{1-i}{4(z-i)}$; the points at which $f(z)$ is not analytic inside C are $z = 1, \pm i$. $I = 0$.

30. $f(z) = \dfrac{1}{2(z-1)} - \dfrac{1}{z-2} + \dfrac{1}{2(z-3)}$; the points at which $f(z)$ is not analytic inside C are $z = 1$, $z = 2$ and $z = 3$. $I = 0$.

In problems **31–34**, the integrand is analytic inside any domain which contains the given points. Choose the straight line path and integrate.

31. Choose the straight line path joining $(0, 0)$, $(4, 2)$. $I = (-2 + 64i)/3$.

32. Choose the straight line path joining $(1, 2\pi)$ and $(3, 4\pi)$. $I = (e^{-2} - e^{-6})/2$.

33. Choose the path $y = 0$, $0 \le x \le 2\pi a$. $I = 8\pi a\,(12\pi^4 a^4 + 20\pi^2 a^2 + 15)/15$.

34. By Cauchy integral theorem $I = 0$. Let $z = e^{i\theta}$. Then

$$\oint_C e^z dz = i\int_0^{2\pi} e^{\cos\theta}[\cos(\theta + \sin\theta) + i\sin(\theta + \sin\theta)]\,d\theta = 0$$

Compare the real and imaginary parts.

35. Consider $I = \oint_C e^{-iz^n} dz$. By Cauchy integral theorem $I = 0$. Let $z = e^{i\theta}$, and compare the real and imaginary parts.

36. $-(2 + 6i)$.

37. $(1 - i)/\pi$.

38. 2.

39. 1/4.

40. $[1 - \cos(-3 + 4i)]/2$.

41. 1.

42. $\dfrac{1}{3}(\pi i + 1)^3 + \dfrac{1}{2}\sinh 2$.

43. $-\dfrac{i}{2}$.

44. 0.

45. $16\,i/35$.

46. $i[2\pi + \sinh 2\pi]/4$.

47. $\pi^2/32$.

48. $-2/9$.

49. $(\beta - \alpha)/(\alpha\beta)$.

50. $[24(\text{Ln }2) - 7]/9$.

Exercise 11.4

1. $4\pi i$.

2. $2\pi i$.

3. (i) $z = -\pi i$ lies inside C, $2\pi i$, (ii) C is the ellipse $5x^2 + 9y^2 = 45$, $z = -\pi i$ lies outside C, 0.

4. $2\pi (7 - 4i)$.

5. (i) $54\,\pi i$, (ii) 0.

6. $\pi i/(2\sqrt{2})$.

7. $z = 0$ lies inside C, $2\pi i$.

8. $z = 1$ lies inside C, $\pi i e/10$.

9. $z = -i$ lies inside C, $\pi^2 i/2$.

10. $f(z) = z + \dfrac{z\bar{z}}{z} = \dfrac{z^2 + r^2}{z}$, $z = 0$ lies inside C, $2\pi i\, r^2$.

11. (i) $z = 3$ lies inside C, $4\pi i/3$; (ii) $z = -3$ lies inside C, $2\pi i/3$.

12. (i) $z = 3i$ lies inside C, $-8\pi/3$; (ii) both $z = \pm 3i$ lie inside C, 0.

13. Both the points $z = 1$ and $z = 2$ lie inside C, $4\pi i$.

14. Both the points $z = 1/2$ and $z = -3$ lie inside C, 0.

15. C is the circle $(x - 1)^2 + (y - 2)^2 = 4$, $z = 2 + i$ lies inside C, $\pi \cos (4 + 2i)$.

16. (i) $z = 0$ lies inside C, $10\pi i$; (ii) both $z = 0$ and $z = -1$ lie inside C, $6\pi i$.

17. Both $z = \pm i$ lie inside C, $-2\pi \sin 1$.

18. $f(z) = \dfrac{|z_0|^2}{(z - z_0)(z\,|z_0|^2 - z_0)}$ is not analytic at $z = z_0$ and $z = z_0/|z_0|^2$,

(i) when $|z_0| < 1$, $z = z_0$ lies inside C, $\dfrac{2\pi i\,|z_0|^2}{z_0\,[|z_0|^2 - 1]}$

(ii) when $|z_0| > 1$, $z = z_0\,|z_0|^2$ lies inside C, $\dfrac{2\pi i\,|z_0|^2}{z_0\,[1 - |z_0|^2]}$.

19. The points $z = 0, \pm i$ lie inside C, 0.

20. The points $z = 0, \pm 1$ lie inside C, 0.

Exercise 11.5

1. $z = 0$ lies inside C, $n = 1$, $-\pi i/2$.

2. $z = \pi/2$ lies inside C, $n = 1$, 0.

3. $z = 1$ lies inside C, $n = 1$, $\pi[(2 + \pi) + 2i(1 - \text{Ln } 2)]/2$.

4. $z = 1$ lies inside C, $n = 1$, $4\pi i e/25$.

5. $z = \pi$ lies inside C, $n = 1$, -2π.

6. $z = -2$ lies inside C, $n = 1$, $64\pi i \sin (16)$.

7. $z = 2i$ lies inside C, $n = 1$, $\pi i/8$.

8. $z = 3 + 4i$ lies inside C, $n = 1$, $3\pi/128$.

9. $z = 0$ lies inside C, $n = 1$, $-\pi/2$.

10. $z = 1 + i$ lies inside C, $n = 2$, $\pi i\, e^{1+i}$.

11. $z = -i$ lies inside C, $n = 2$, $-\pi i\,(\cosh 1 + i \sinh 1)$.

12. $z = \pi/2$ lies inside C, $n = 2$, $\pi i(3\pi - 9)$.   **13.** $z = i$ lies inside C, $n = 2$, $\pi i\, e^{-i}$.

14. $z = 1$ lies inside C, $n = 2$, $-\pi i$.

15. $z = -a$ lies inside C, $n = 2$, $\pi i(2 - a)\,e^{-a}$.

16. $z = 0$ lies inside C, $n = m - 1$, $\dfrac{2\pi i}{(m - 1)!} \sin\left[(m - 1)\dfrac{\pi}{2}\right]$.

17. $I = I_1 + I_2;\ I_1 = \displaystyle\oint_C \dfrac{e^{iz}}{z^3}\, dz,\ I_2 = \oint_C \dfrac{z^4}{(z + i)^2}\, dz,\ I = -\pi (8 + i)$.

18. Both $z = 0$ and $z = i$ lie inside C, $2\pi i$.   **19.** Both $z = 1$ and $z = -2$ lie inside C, 0.

20. Both $z = 2i$ and $z = -2i$ lie inside C, 0.

21. Both $z = 0$ and $z = -1$ lie inside C, $\pi i(2 - 5e^{-1})$.

22. Both $z = 0$ and $z = i$ lie inside C, $4\pi(2e^{-1} - 1)$.

23. Both $z = 1 - 2i$ and $-1 + 2i$ lie inside C, 0.

24. (i) L.H.S. $= \dfrac{2\pi i}{n!} g^{(n)}(z_0) = $ R.H.S.

   (ii) Apply Cauchy integral formula for derivatives to the right hand side.

25. Substitute $z = a + e^{i\theta}$ in the Cauchy integral formula for derivatives.

26. $|f(z)|^2 = g(x, y) = (x + 5)^2 + y^2$. For local extremum $g_x = 2(x + 5) = 0$ and $g_y = 2y = 0$. We get the point $(-5, 0)$. This point is not inside C. Hence, there is no extreme value inside C. On $|z| = 4$, $x^2 + y^2 = 16$. Substituting $y^2 = 16 - x^2$, we obtain $|f(z)|^2 = 10x + 41$, $-4 \le x \le 4$.

Maximum $|f(z)| = 9$ at $x = 4$ or $z = 4$

Minimum $|f(z)| = 1$ at $x = -4$ or $z = -4$.

Thus, the maximum and minimum values occur on the boundary.

27. $|f(z)|^2 = g(x, y) = [3(x^2 - y^2) + 2]^2 + 36x^2y^2$, $C : x^2 + y^2 = 1/4$. From $g_x = 0$ and $g_y = 0$, we get the point $(0, 0)$ inside C and $f(0) = 2$. On $C : y^2 = (1/4) - x^2$, $|f(z)|^2 = 24x^2 + (25/16)$, $-1/2 \le x \le 1/2$.

Maximum $|f(z)|$ occurs at $x = \pm 1/2$, $y = 0$ and max $|f(z)| = 11/4$.

Minimum $|f(z)|$ occurs at $x = 0$, $y = \pm 1/2$ and min $|f(z)| = 5/4$. Thus, maximum and minimum values occur on the boundary.

28. $|f(z)| = e^{2x}$, $-1 \le x \le 1$. No extreme value inside C. Maximum $|f(z)| = e^2$ at $x = 1$ and minimum $|f(z)| = e^{-2}$ at $x = -1$, both of them occur on the boundary.

29. $M = 2$, $r = 1$, $|f^{(n)}(0)| \le 2(n!)$.

30. $M = e^3$, $r = 1$, $|f^{(n)}(0)| \le e^3 (n!)$.

Chapter 12

Power Series, Taylor and Laurent Series

12.1 Introduction

An infinite series of the form $\sum\limits_{n=0}^{\infty} a_n (z - z_0)^n$ in which the coefficients a_i's are constants is called a *power series*. In this chapter, we shall prove one of the important results of the complex variable theory, that every complex function $f(z)$ which is analytic in a domain D can be represented by a power series valid in some circular region R about a point z_0. Both the circular region R and the point z_0 lie inside D. Conversely, every power series represents an analytic function in some domain. We shall show that a power series is a *Taylor's series*. If $f(z)$ is not analytic at a point z_0, we can still expand $f(z)$ in an infinite series having both positive and negative powers of $(z - z_0)$. This series is called the *Laurent's series*. In order to obtain and analyse these series, we need some knowledge of sequences and series of real or complex numbers and functions. We first briefly review a few basic concepts of sequences and series.

12.2 Infinite Sequences

A sequence is a special class of a function f whose domain is the set of positive consecutive integers $\mathbb{N}$ and the range is the ordered list of the corresponding values of the function. A sequence is denoted by $\{a_n\}$ or $\{a_1, a_2, \ldots, a_n\}$, where $a_n = f(n)$, $n \in \mathbb{N}$, is called the nth term (element) or the general term of the sequence. A sequence which has infinite number of terms is called an *infinite sequence*. If the elements of a sequence are real numbers or real functions, it is called a *real sequence* and if the elements are complex numbers or complex functions, it is called a *complex sequence*.

12.2.1 Real Sequences

Let $\{a_n\}$, $a_n \in \mathbb{R}$ be an infinite real sequence. We define the following concepts of real sequences.

Bounded sequence A sequence $\{a_n\}$ is said to be *bounded above* if $a_n \leq M$ for all n and *bounded below* if $a_n \geq m$ for all n, where M and m are, respectively, called the *upper bound* and the *lower bound* of the sequence. Thus, a sequence $\{a_n\}$ is said to be *bounded* if there exist two real numbers m and M such that

$$m \leq a_n \leq M \quad \text{for all } n. \tag{12.1}$$

For example, the sequence

(i) $\{1, -1, 1, -1, \dots\}$ is both bounded above and bounded below

(ii) $\{1, 2, 3, \dots\}$ is not bounded above.

Limit point of a sequence A point a is called a *limit point* or a *cluster point* of the sequence $\{a_n\}$, if for a given positive real number $\varepsilon > 0$, $|a_n - a| < \varepsilon$ for infinitely many n.

For example, the sequence

(i) $\{(-1)^n\} = \{1, -1, 1, -1, \dots\}$ has limit points 1 and -1.

(ii) $\{i^n\} = \{i, -1, -i, 1, \dots\}$ has limit points ± 1 and $\pm i$.

(iii) $\left\{\dfrac{1}{n}\right\} = \left\{1, \dfrac{1}{2}, \dfrac{1}{3}, \dots\right\}$ has only one limit point 0.

(iv) $\{(-1)^n n\} = \{-1, 2, -3, \dots\}$ has no limit point.

Convergence of a sequence

A sequence $\{a_n\}$ is said to be *convergent* or *converges* to l or has a *limit* l, if for a given positive real number $\varepsilon > 0$ (no matter how small), there exists an $N \in \mathbb{N}$, such that

$$|a_n - l| < \varepsilon \quad \text{for all} \quad n \geq N \tag{12.2}$$

or symbolically

$$\lim_{n \to \infty} a_n = l. \tag{12.3}$$

The value of N will in general depend on ε. If no such limit l exists, then the sequence is said to be *divergent*. Thus, for a convergent sequence, we can find an open interval $(l - \varepsilon, l + \varepsilon)$ of length 2ε which contains all a_n, $n \geq N$ (Fig. 12.1).

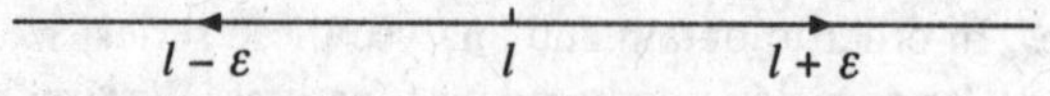

Fig. 12.1. Convergence of a real sequence to l.

For example, the sequence

(i) $\left\{\dfrac{1}{n}\right\}$ converges to 0. In this case $\left|\dfrac{1}{n} - 0\right| < \varepsilon$ is satisfied when $n > 1/\varepsilon$. Thus, if we choose $N > 1/\varepsilon$, condition (12.2) holds.

(ii) $\{a_n\}$, $a_n = 1$ for all n, has limit 1 and hence is convergent.

(iii) $\{n\}$ diverges as the $\lim_{n \to \infty} n$ does not exist.

Theorem 12.1 A convergent sequence has a unique limit.

Proof Let the sequence $\{a_n\}$ converge to two limits l_1 and l_2, $l_1 \neq l_2$. By the definition of convergence, for a given positive real number $\varepsilon > 0$, there exist N_1 and $N_2 \in \mathbb{N}$ such that

$$|a_n - l_1| < \varepsilon/2, n \geq N_1 \quad \text{and} \quad |a_n - l_2| < \varepsilon/2, n \geq N_2.$$

Let $N = \max(N_1, N_2)$. Then,

$$|l_1 - l_2| = |(a_n - l_2) - (a_n - l_1)| \leq |a_n - l_2| + |a_n - l_1| < \frac{\varepsilon}{2} + \frac{\varepsilon}{2} = \varepsilon.$$

Since ε is arbitrarily small, $|l_1 - l_2| \to 0$ or $l_1 = l_2$.

Hence, the sequence has a unique limit.

Remark 1

From the definitions of convergence and the limit point of a sequence, it follows that if a sequence $\{a_n\}$ converges to l, then l is also a limit point of the sequence $\{a_n\}$, but the converse may not be true. For example, the sequence $\{(-1)^n\} = \{1, -1, \ldots\}$ has two limit points 1 and -1, none of which is the limit of the sequence. In fact the sequence does not converge. Let if possible, the sequence converge to l. Let $\varepsilon = 1$. Then,

$$|a_n - l| < 1, n > N.$$

For $n = 2m$ and $2m + 1$, we obtain $|(-1)^{2m} - l| < 1$ and $|(-1)^{2m+1} - l| < 1$

or $$|1 - l| < 1 \quad \text{and} \quad |1 + l| < 1.$$

Therefore, $2 = |(1 - l) + (1 + l)| \le |1 - l| + |1 + l| < 1 + 1 = 2$

which is not possible. Hence, the series does not converge.

Theorem 12.2 Every convergent sequence is bounded.

Proof Let the sequence $\{a_n\}$ converge to l. Then we have

$$|a_n - l| < \varepsilon, n \ge N \quad \text{or} \quad l - \varepsilon < a_n \le l + \varepsilon, n \ge N.$$

Let $m = \min \{|a_1|, |a_2|, \ldots, |a_{N-1}|, |l| - \varepsilon\}$ and

$$M = \max\{|a_1|, |a_2|, \ldots, |a_{N-1}|, |l| + \varepsilon\}.$$

Therefore, $m \le a_n \le M$ for all n.

Hence, the sequence is bounded.

The converse of this theorem is not true, that is, the sequence may be bounded but not convergent. For example, the sequence $\{(-1)^n\}$ is bounded but not convergent.

Remark 2

A sequence which is not bounded cannot be convergent. For example, the sequence $\{\ln n\}$ is not bounded (for any $M > 0$, $\ln n > M$ if $n > e^M$) and hence not convergent. Combining the theorems 12.1 and 12.2, we have the following result.

Theorem 12.3 Every convergent sequence is bounded and has a unique limit.

Range set: The range set of a sequence $\{a_n\}$ is the non-empty set $S : \{a_n : n \in \text{IN}\}$ consisting of all distinct elements of the sequence without repetition and without regard to the position of the term.

Remark 3

Since S is a set, no two elements of S are equal. Note that the terms of a sequence may be equal.

Theorem 12.4 (Bolzano-Weierstrass theorem) Every bounded sequence has a limit point.

Proof Let S be the range set of the sequence $\{a_n\}$. If S is a finite set, then there exists an element of S which repeats in the sequence and hence it is the limit point. Now, suppose that S is an infinite set. Since the sequence is bounded, by the Bolzano-Weierstrass theorem of sets, S has a limit point, say α. Then, for every given positive real number $\varepsilon > 0$, every neighborhood $(\alpha - \varepsilon, \alpha + \varepsilon)$ of α contains an infinite number of elements of S, that is $a_n \in (\alpha - \varepsilon, \alpha + \varepsilon)$ for an infinite number of values of n. Hence, α is a limit point of the sequence.

As a consequence of this theorem, we have the following results.

Theorem 12.5 Every bounded sequence with a unique limit is convergent.

Theorem 12.6 A necessary and sufficient condition for the convergence of a sequence is that it is bounded and has a unique limit.

Example 12.1 Show that the sequence $\{a_n\}$, $a_n = r^n$, $n = 1, 2, \ldots$ and r^n is real, is convergent for $-1 < r \leq 1$ and divergent otherwise.

Solution We discuss the following cases:

(i) Let $r > 1$. Then for any $h > 0$ and $r = 1 + h$, we have

$$a_n = r^n = (1 + h)^n \geq 1 + nh, \quad n = 1, 2, \ldots.$$

Let $M > 1$ be an arbitrary positive real number. Then

$$1 + nh > M \quad \text{when} \quad n > (M - 1)/h.$$

Let N be the nearest integer greater than $(M - 1)/h$. Thus, for every $M > 1$ there exists a positive integer $N \in \mathbb{N}$ such that $a_N > M$. Therefore, the sequence is not bounded and hence not convergent. (Note that in this case $\lim\limits_{n \to \infty} a_n = \infty$.)

(ii) Let $r = 1$. In this case, $a_n = 1$ for all n. Therefore, the sequence is bounded with a unique limit 1. Hence, the sequence is convergent.

(iii) Let $-1 < r < 1$ or $|r| < 1$. Let $r \neq 0$. Then, for some $h > 0$,

$$|r| = \frac{1}{1 + h} \quad \text{and} \quad |a_n| = |r^n| = |r|^n = \frac{1}{(1 + h)^n} \leq \frac{1}{1 + nh}, n = 1, 2, \ldots$$

Thus, for a given positive real number $\varepsilon > 0$,

$$\frac{1}{1 + nh} < \varepsilon \quad \text{when} \quad n > \left(\frac{1}{\varepsilon} - 1\right)\Big/h.$$

Let N be an integer greater than $[(1/\varepsilon) - 1]/h$. Therefore, for a given $\varepsilon > 0$, there exists a positive integer $N \in \mathbb{N}$, such that $|r^n| < \varepsilon$, for all $n > N$. Therefore, the sequence is convergent. (Note that in this case $\lim\limits_{n \to \infty} a_n = 0$.)
When $r = 0$, the sequence converges to $a_0 = 0$.

(iv) Let $r = -1$. In this case, the sequence $\{a_n\}$ is such that

$$a_n = \begin{cases} 1, & n = 2, 4, 6, \ldots \\ -1, & n = 1, 3, 5, \ldots. \end{cases}$$

Thus, the sequence has two limit points. Therefore, the sequence is not convergent.

(v) Let $r < -1$. Then, $t = -r > 1$. Hence, $a_n = (-1)^n t^n$. As in case (i), for arbitrary positive real $M > 1$, we can find $N \in \mathbb{N}$ such that

$$|a_n| = |t|^n > M \quad \text{for all} \quad n > N.$$

Therefore, the sequence is not convergent.

Hence, the sequence $\{r^n\}$ is convergent only when $-1 < r \leq 1$.

Cauchy sequence A real sequence $\{a_n\}$ is said to be a *Cauchy sequence* if for a positive real number $\varepsilon > 0$, there exists an $N \in \text{IN}$ such that

$$|a_n - a_m| < \varepsilon \quad \text{for all} \quad n, m \geq N. \tag{12.4a}$$

Thus, all terms of a Cauchy sequence can be made close to one another after a certain number of terms of the sequence. We can also write Eq. (12.4a) as

$$|a_{n+p} - a_n| < \varepsilon \quad \text{for all} \quad n \geq N \text{ and } p = 1, 2, \ldots . \tag{12.4b}$$

Example 12.2 Show that the sequence $\{a_n\}$, where $a_n = 1/n$, $n = 1, 2, \ldots$ is a Cauchy sequence.

Solution We have

$$|a_n - a_m| = \left| \frac{1}{n} - \frac{1}{m} \right| \leq \left| \frac{1}{n} \right| + \left| \frac{1}{m} \right| < \frac{2}{N}, \; n, m \geq N.$$

Therefore, $|a_n - a_m| < \varepsilon$ for $n, m \geq N$ and $N > 2/\varepsilon$. Hence, the given sequence is a Cauchy sequence.

Example 12.3 Show that the sequence $\{a_n\}$, where $a_n = 1 + (-1)^n$, $n = 1, 2, \ldots$ is not a cauchy sequence.

Solution For a given N, we find that

$$|a_{N+1} - a_N| = 2.$$

Thus, for a given positive real number $\varepsilon > 0$, there exists an $N \in \text{IN}$ such that for $m, n \geq N$, $|a_n - a_m| \geq \varepsilon$. Hence, the given sequence is not a Cauchy sequence.

Theorem 12.7 (Cauchy convergence criterion) A necessary and sufficient condition for the convergence of a sequence $\{a_n\}$ is that it is a Cauchy sequence.

Proof Let the sequence $\{a_n\}$ converge to l. Then, for every given positive real number $\varepsilon > 0$, there exists an $N \in \text{IN}$ such that $|a_n - l| < \varepsilon/2$ and $|a_m - l| < \varepsilon/2$ for all $n, m \geq N$. Therefore,

$$|a_n - a_m| = |(a_n - l) - (a_m - l)| \leq |a_n - l| + |a_m - l| \leq \frac{\varepsilon}{2} + \frac{\varepsilon}{2} = \varepsilon.$$

Hence, the sequence $\{a_n\}$ is a Cauchy sequence.

Now, let the sequence $\{a_n\}$ be a Cauchy sequence. Then, for a given positive real number $\varepsilon > 0$, there exists an $N \in \text{IN}$ such that

$$|a_n - a_m| < \varepsilon \quad \text{for all} \quad n, m \geq N. \tag{12.4c}$$

We now show that the sequence $\{a_n\}$ is bounded and converges to a limit. From Eq. (12.4c), we get for $m = N$

$$|a_n - a_N| < \varepsilon \quad \text{or} \quad a_N - \varepsilon < a_n < a_N + \varepsilon.$$

Choosing $\qquad m = \min(|a_1|, |a_2|, \ldots, |a_{N-1}|, |a_N| - \varepsilon)$

and $\qquad M = \max(|a_1|, |a_2|, \ldots |a_{N-1}|, |a_N| + \varepsilon)$

we find that $m \leq a_n \leq M$ and therefore the sequence $\{a_n\}$ is bounded. From the Bolzano-Weierstrass theorem, the sequence $\{a_n\}$ has at least one limit point l. We now show that the sequence converges to l. From Eq. (12.4c), we can write

$$|a_n - a_N| < \varepsilon/3, n \geq N \quad \text{and} \quad |a_{n_1} - a_N| < \varepsilon/3, n_1 \geq N.$$

As l is the limit point of the sequence $\{a_n\}$, we get $|a_{n_1} - l| < \varepsilon/3, n_1 \geq N$.

Hence, $|a_n - l| = |a_n - a_N + a_N - a_{n_1} + a_{n_1} - l|$

$$\leq |a_n - a_N| + |a_{n_1} - a_N| + |a_{n_1} - l| < \frac{\varepsilon}{3} + \frac{\varepsilon}{3} + \frac{\varepsilon}{3} = \varepsilon$$

for all $n \geq N$. Hence, the sequence $\{a_n\}$ converges to l.

Example 12.4 Discuss the convergence of the sequence $\{a_n\}$, where $a_n = 1 + \frac{1}{2} + \ldots + \frac{1}{n}$.

Solution We have

$$|a_{2n} - a_n| = \left| \frac{1}{n+1} + \frac{1}{n+2} + \ldots + \frac{1}{2n} \right|, \ n \geq N.$$

Since $n + k \leq n + n$, we get $\frac{1}{n+k} \geq \frac{1}{2n}$ for $k = 1, 2, \ldots, n$.

Hence, $|a_{2n} - a_n| > n\left(\frac{1}{2n} \right) = \frac{1}{2}$ for $n \geq N$.

Therefore, for any $\varepsilon \leq 1/2$, the given sequence is not a Cauchy sequence. Hence, the sequence $\{a_n\}$ is not convergent.

12.2.2 Complex Sequences

Let $\{z_n\}$, where $z_n \in \mathbb{C}$, be a complex sequence. A complex sequence $\{z_n\}$ is said to be *bounded* if there exists a finite real positive number M such that $|z_n| \leq M$ for all n. Thus, a complex sequence is bounded if there exists some disk with centre at $(0, 0)$ and radius M which contains all z_n's.

If $\lim_{n \to \infty} z_n = L$, where L is a complex number, then the sequence $\{z_n\}$ is said to be *convergent*. If no such L exists, the sequence is said to be *divergent*. Using the definition of limit, we say that a sequence $\{z_n\}$ converges to a number L if for a given real positive number $\varepsilon > 0$, there exists an $N \in \mathbb{N}$ such that

$$|z_n - L| < \varepsilon, \quad \text{for all} \quad n \geq N. \quad (12.5)$$

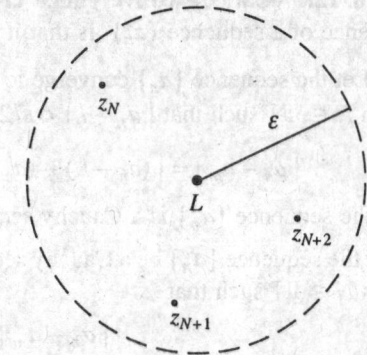

Thus, there exists an open disc of radius ε and centre at $z = L$ such that all z_n, $n \geq N$ lie in this disk and only a finite number of elements z_1, $z_2, \ldots, z_{N-1}$ do not lie inside the disk (Fig. 12.2).

Fig. 12.2. Convergence of a complex sequence to L.

Theorem 12.8 A complex sequence $\{z_n\}$, $z_n = x_n + iy_n$ converges to a complex number $L = a + ib$ if and only if the real sequences $\{x_n\}$ and $\{y_n\}$ of the real and imaginary parts of the sequence $\{z_n\}$ converge to $a = \text{Re}(L)$ and $b = \text{Im}(L)$, respectively.

Proof Let the sequence $\{z_n\}$ converge to L, that is, $\lim_{n \to \infty} z_n = L = a + ib$. Then, for a given real positive number $\varepsilon > 0$, there exists an $N \in \mathbb{N}$ such that

$$|z_n - L| < \varepsilon \quad \text{for all} \quad n \geq N$$

or $\qquad |(x_n + iy_n) - (a + ib)| < \varepsilon \quad \text{or} \quad |(x_n - a) + i(y_n - b)| < \varepsilon, \ n \geq N.$

Since $|x_n - a| \le |(x_n - a) + i(y_n - b)| < \varepsilon$ and $|y_n - b| \le |(x_n - a) + i(y_n - b)| < \varepsilon$

we obtain $|x_n - a| < \varepsilon$ and $|y_n - b| < \varepsilon$, for all $n \ge N$.

Hence, $\lim\limits_{n \to \infty} x_n = a$ and $\lim\limits_{n \to \infty} y_n = b$. (12.6)

Now, let Eq. (12.6) hold. Therefore, for a given real positive number $\varepsilon > 0$, there exist integers N_1 and $N_2 \in$ IN such that

$$|x_n - a| < \varepsilon/2, n \ge N_1 \quad \text{and} \quad |y_n - b| < \varepsilon/2, n \ge N_2.$$

Let $N = \max (N_1, N_2)$. Then,

$$|x_n - a| < \varepsilon/2 \quad \text{and} \quad |y_n - b| < \varepsilon/2 \quad \text{for all} \quad n \ge N.$$

Since $|(x_n + iy_n) - (a + ib)| = |(x_n - a) + i(y_n - b)| \le |x_n - a| + |y_n - b| < \dfrac{\varepsilon}{2} + \dfrac{\varepsilon}{2} = \varepsilon$

we obtain

$$|z_n - L| < \varepsilon, \text{ for all } n \ge N.$$

Hence, the sequence $\{z_n\}$ converges to L.

Remark 4
As a consequence of the above theorem, the convergence of a complex sequence $\{z_n\}$, $z_n = x_n + iy_n$ can be discussed by considering the convergence of the real sequences $\{x_n\}$ and $\{y_n\}$.

Theorem 12.9 If the sequence $\{z_n\}$ converges to L, then L is unique.

Theorem 12.10 Every convergent sequence $\{z_n\}$ is bounded.

Example 12.5 Show that the sequence. $\{z_n\}$, where $z_n = \left(3 + \dfrac{4}{n}\right) + i\left(1 - \dfrac{3}{n}\right)$, $n = 1, 2, \ldots$

is convergent.

Solution we have $z_n = x_n + iy_n = \left(3 + \dfrac{4}{n}\right) + i\left(1 - \dfrac{3}{n}\right)$.

Therefore, $x_n = 3 + (4/n)$ and $y_n = 1 - (3/n)$.

Since $\lim\limits_{n \to \infty} x_n = 3$ and $\lim\limits_{n \to \infty} y_n = 1$, the sequence $\{z_n\}$ converges to $3 + i$.

Example 12.6 Show that the sequence $\{z_n\}$, where $z_n = \sinh \dfrac{n\pi}{4} + i \cosh \dfrac{n\pi}{4}$, $n = 1, 2, \ldots$
is divergent.

Solution We have

$$z_n = x_n + iy_n = \sinh \frac{n\pi}{4} + i \cosh \frac{n\pi}{4}.$$

Therefore, $x_n = \sinh \dfrac{n\pi}{4}$ and $y_n = \cosh \dfrac{n\pi}{4}$.

Now, the functions $\sinh (n\pi/4)$ and $\cosh (n\pi/4)$ are not bounded. Therefore, the real sequences $\{x_n\}$ and $\{y_n\}$ are not bounded and hence the sequence $\{z_n\}$ is not bounded. We conclude that the given sequence is not convergent.

Cauchy sequence A complex sequence $\{z_n\}$ is called a *Cauchy sequence*, if for a given real positive number $\varepsilon > 0$, there exists an $N \in \mathbb{N}$ such that

$$|z_n - z_m| \le \varepsilon \quad \text{for all} \quad n, m \ge N.$$

When the complex sequence $\{z_n\}$ is a Cauchy sequence, the real sequences of its real parts and imaginary parts, that is the sequences $\{x_n\}$ and $\{y_n\}$ are also Cauchy sequences.

Theorem 12.11 (Cauchy convergent criterion) A complex sequence $\{z_n\}$ is convergent if and only if it is a Cauchy sequence.

Proof Suppose that the sequence is convergent. Therefore, for a given real positive number $\varepsilon > 0$, there exists an $N \in \mathbb{N}$ such that

$$|z_n - L| < \varepsilon/2 \quad \text{for all} \quad n \ge N.$$

Therefore, for $m > N$ and $n > N$

$$|z_n - z_m| = |(z_n - L) - (z_m - L)| \le |z_n - L| + |z_m - L| < \frac{\varepsilon}{2} + \frac{\varepsilon}{2} = \varepsilon.$$

Hence, the sequence $\{z_n\}$ is a Cauchy sequence.

Now, let the sequence $\{z_n\}$ be a Cauchy sequence. Let $z_n = x_n + iy_n$.

Now, $|x_n - x_m| \le |z_n - z_m| < \varepsilon,$ for all $n, m \ge N$

and $|y_n - y_m| \le |z_n - z_m| < \varepsilon,$ for all $n, m \ge N.$

Since the real sequences $\{x_n\}$ and $\{y_n\}$ are Cauchy sequences and hence convergent, we conclude that the complex sequence $\{z_n\}$ is convergent.

Example 12.7   Suppose that the complex sequence $\{z_n\}$ is convergent. Prove that the sequence $\{|z_n|\}$ is also convergent.

Solution   Since the sequence $\{z_n\}$ is convergent, it is a Cauchy sequence, that is

$$|z_n - z_m| < \varepsilon, n, m \ge N.$$

Now $$||z_n| - |z_m|| \le |z_n - z_m| < \varepsilon, \; n, m \ge N.$$

Therefore, the sequence $\{|z_n|\}$ is a Cauchy sequence. Hence, it is convergent.

Example 12.8 Without evaluating the integral, show that the sequence $\{z_n\}$, where $z_n = \displaystyle\int_C \frac{z^n}{n+z}\, dz$ and C is a straight line path joining the points $z = 0$ and $z = (1 + i)/\sqrt{2}$, is convergent.

Solution   The length of the line joining the points $z = 0$ and $z = (1 + i)/\sqrt{2}$, that is, the points $(0, 0)$ and $(1/\sqrt{2}, 1/\sqrt{2})$ is $L = 1$. Now, on C

$$|z| \le 1 \text{ and } |n + z| \ge |n| - |z| \ge n - 1.$$

Therefore, for any z on C, we have

$$\left| \frac{z^n}{n+z} \right| = \frac{|z|^n}{|n+z|} \le \frac{1}{n-1}.$$

Using the *ML*-inequality of integrals with $M = 1/(n-1)$ and $L = 1$, we get

$$\lim_{n\to\infty} |z_n| = \lim_{n\to\infty} \left| \int_C \frac{z^n}{n+z}\, dz \right| \leq \lim_{n\to\infty} \frac{1}{n-1} = 0.$$

Hence, the given sequence converges to 0.

Example 12.9 Let z_1 and z_2 be any two complex numbers and $z_n = (z_{n-1} + z_{n-2})/2$, $n \geq 3$. Show that the sequence $\{z_n\}$ is a Cauchy sequence and hence convergent.

Solution Let $z_1 = a_1 + ib_1$ and $z_2 = a_2 + ib_2$ where a_1, a_2, b_1 and b_2 are finite real numbers. We have

$$|z_2 - z_1| = |(a_2 - a_1) + i(b_2 - b_1)| = \sqrt{(a_2 - a_1)^2 + (b_2 - b_1)^2} = c$$

$$|z_3 - z_2| = \left| \frac{1}{2}(z_1 + z_2) - z_2 \right| = \frac{1}{2}|z_2 - z_1| = \frac{c}{2}$$

$$|z_4 - z_3| = \left| \frac{1}{2}(z_2 + z_3) - \frac{1}{2}(z_1 + z_2) \right| = \left| \frac{1}{4}(z_1 + 3z_2) - \frac{1}{2}(z_1 + z_2) \right|$$

$$= \frac{1}{4}|z_2 - z_1| = \frac{c}{4} = \frac{c}{2^2}$$

$$\cdots\cdots\cdots\cdots\cdots\cdots\cdots\cdots\cdots$$

$$|z_n - z_{n-1}| = \frac{c}{2^{n-2}}.$$

Now, for $n > m$, we get

$$|z_n - z_m| = |(z_n - z_{n-1}) + (z_{n-1} - z_{n-2}) + \ldots + (z_{m+1} - z_m)|$$

$$\leq |(z_n - z_{n-1})| + |(z_{n-1} - z_{n-2})| + \ldots + |(z_{m+1} - z_m)|$$

$$= c\left[\frac{1}{2^{n-2}} + \frac{1}{2^{n-3}} + \ldots + \frac{1}{2^{m-1}} \right]$$

$$= \frac{2c}{2^{m-1}}\left[1 - \left(\frac{1}{2} \right)^{n-m} \right] \to 0 \quad \text{as} \quad m \to \infty.$$

Hence, we can find an $N \in \mathbb{N}$ such that

$$|z_n - z_m| < \varepsilon, \quad n, m \geq N.$$

Therefore, the sequence $\{z_n\}$ is a Cauchy sequence and hence convergent.

12.2.3 Sequence of Functions

Let S be a set of real or complex numbers and $\{f_n(z)\}$, $z \in S$, be a sequence of single-valued, real or complex functions. A sequence of functions $\{f_n(z)\}$ is called a Cauchy sequence, if for a given real positive number $\varepsilon > 0$, there exists an $N \in \mathbb{N}$ such that

$$|f_n(z) - f_m(z)| < \varepsilon \quad \text{for all} \quad n, m \geq N. \tag{12.7a}$$

For a fixed z in S, $f_n(z)$ becomes a number and we obtain a sequence of numbers (real or complex). A given sequence of functions $\{f_n(z)\}$ is said to converge to a function $f(z)$ over S if for every real positive number $\varepsilon > 0$, there exists an $N \in \mathbb{N}$ such that

$$|f_n(z) - f(z)| < \varepsilon \quad \text{for all} \quad n \geq N. \tag{12.7b}$$

Here, the choice of N may depend on ε and also on each value of $z \in S$. Hence, the convergence as defined above is called the *point-wise convergence*. That is, we are studying the convergence of the sequence $\{f_n(z)\}$ for a fixed $z \in S$.

Example 12.10 Find the limit of the sequence $\left\{\dfrac{\sin nx}{n}\right\}$ over $\mathbb{R}$.

Solution We have $f_n(x) = (\sin nx)/n$, $|\sin nx| \leq 1$ for all $x \in \mathbb{R}$ and $1/n \to 0$ as $n \to \infty$.

Therefore,
$$|f_n(x) - 0| = \left|\frac{\sin nx}{n}\right| \leq \frac{1}{n} < \varepsilon$$

for all $n > 1/\varepsilon$ and for all x. Hence, $\lim\limits_{n \to \infty} \dfrac{\sin nx}{n} = 0$ for all x.

Example 12.11 Find the limit of the sequence $\{f_n(z)\}$, where $f_n(z) = \dfrac{n^2 z}{1 + n^2 z^2}$ and $z \in \mathbb{C}$ is a complex variable.

Solution We have $\lim\limits_{n \to \infty} f_n(z) = \lim\limits_{n \to \infty} \dfrac{n^2 z}{1 + n^2 z^2} = \lim\limits_{n \to \infty} \dfrac{z}{(1/n^2) + z^2} = \dfrac{1}{z}$, $z \neq 0$.

Therefore, we may expect the limit of the sequence $\{f_n(z)\}$ to be $1/z$. Now,

$$\left|f_n(z) - \frac{1}{z}\right| = \left|\frac{n^2 z}{1 + n^2 z^2} - \frac{1}{z}\right| = \left|\frac{1}{z(1 + n^2 z^2)}\right| \leq \frac{1}{|z|[|n^2 z^2| - 1]} < \varepsilon.$$

For $z \neq 0$, we choose n such that

$$|z|[|n^2 z^2| - 1] > \frac{1}{\varepsilon} \quad \text{or} \quad n > \left[\frac{1}{\varepsilon|z|} + 1\right]^{1/2} \frac{1}{|z|} = N.$$

Such a value of N ensures convergence.

When $z = 0$, $f_n(0) = 0$ for all n and the sequence $\{f_n(0)\}$ converges to 0.

Therefore, the limit of the given sequence is $f(z)$, where

$$f(z) = \begin{cases} 0, & z = 0, \\ 1/z, & z \neq 0. \end{cases}$$

The choice of N depends both on ε and z.

Example 12.12 Show that the sequence $\{f_n(x)\}$, where $f_n(x) = x/(n + x)$ is Cauchy convergent on $[0, \infty)$.

Solution We have

$$|f_n(x) - f_m(x)| = \left|\frac{x}{n + x} - \frac{x}{m + x}\right| = |x|\left|\frac{m - n}{(n + x)(m + x)}\right|$$

$$\leq |x|\left|\frac{m + n}{mn}\right| < \frac{2}{N}|x| < \varepsilon, \quad m, n > N.$$

Now, for a given $\varepsilon > 0$, we can find an $N > 2|x|/\varepsilon$ for which $|f_n(x) - f_m(x)| < \varepsilon$. Hence, the given sequence is a Cauchy sequence and therefore Cauchy convergent.

Remark 5

The limit function $f(x)$ of the sequence $\{f_n(x)\}$ may not possess the same properties like continuity and differentiability at a point, as the members $f_n(x)$ of the sequence.

Example 12.13   Discuss the continuity of the limit function $f(x)$ of the sequence $\{f_n(x)\}$ on $[0, 1]$, where $f_n(x) = \tan^{-1}(nx)$.

Solution   The limit function is obtained as

$$f(x) = \begin{cases} 0, & x = 0, \\ \pi/2, & x \neq 0. \end{cases}$$

The limit function is not continuous at $x = 0$, even though each member function $f_n(x)$ is continuous at $x = 0$.

Example 12.14   Discuss the differentiability of the limit function $f(x)$ of the sequence $\{f_n(x)\}$ over $\mathbb{R}$, where $f_n(x) = \dfrac{2x}{\pi} \tan^{-1}(nx)$.

Solution   The limit function is $f(x) = \begin{cases} 0, & x = 0, \\ x, & x \neq 0. \end{cases}$

Therefore, $f'(x) = \begin{cases} 0, & x = 0, \\ 1, & x \neq 0. \end{cases}$

For the member functions $f_n(x)$, we have $\lim\limits_{n \to \infty} f_n'(x) = \begin{cases} 0, & x = 0, \\ 1, & x \neq 0. \end{cases}$

Both the functions $f_n(x)$ and the limit function $f(x)$ have discontinuous derivative at $x = 0$.

12.2.4   Uniform Convergence

Let $\{f_n(z)\}$ be a sequence of the complex valued functions defined in a domain D. The sequence $\{f_n(z)\}$ is said to be *uniformly convergent* in D and converges to a function $f(z)$, if for a given real positive number $\varepsilon > 0$, there exists an $N \in \mathbb{N}$ such that

$$|f_n(z) - f(z)| < \varepsilon \quad \text{for all} \quad n \geq N \tag{12.8}$$

where N depends on ε but not on z.

We observe that uniform convergence is different from the point-wise convergence in the following respects:

(i)   Point-wise convergence is examined over a set of points, whereas uniform convergence is examined over a domain (an interval for a sequence of real functions).

(ii)   To prove uniform convergence, an integer $N \in \mathbb{N}$ is to be determined which depends only on ε and is independent of z in D, where as in point-wise convergence N depends, in general, on both ε and z.

(iii)   Uniform convergence implies point-wise convergence, but the converse in general is not true.

Theorem 12.12 (Cauchy criterion for uniform convergence)   A sequence $\{f_n(z)\}$ is said to be uniformly Cauchy convergent, if for every given real positive number $\varepsilon > 0$ and for all z in D, there exists an $N \in \mathbb{N}$ independent of z such that

$$|f_n(z) - f_m(z)| < \varepsilon \quad \text{for all} \quad n, m \geq N. \tag{12.9}$$

The same definition holds when $\{f_n(x)\}$ is a sequence of real functions defined over a closed interval.

Example 12.15 Show that the sequence $\{f_n(z)\}$, where $f_n(z) = 1/(nz)$ is (i) not uniformly convergent in the region $0 < |z| < 1$ (ii) uniformly convergent in the region $\varepsilon_0 < |z| < 1$, $\varepsilon_0 > 0$.

Solution Since $\lim_{n \to \infty} f_n(z) = \lim_{n \to \infty} [1/(nz)] = 0$ for every $z \neq 0$, the sequence $\{f_n(z)\}$ has limit zero. Therefore, $f(z) = 0$, $z \neq 0$.

(i) Now,
$$|f_n(z) - f(z)| = \left|\frac{1}{nz}\right| < \varepsilon, \text{ if } n > \frac{1}{\varepsilon|z|} = N.$$

Since $N = 1/(\varepsilon|z|)$ depends on both ε and z, the sequence is only point-wise convergent and not uniformly convergent.

(ii) In the region $\varepsilon_0 < |z| < 1$, $\varepsilon_0 > 0$ and fixed, we find that

$$|f_n(z) - f(z)| = \left|\frac{1}{nz}\right| < \frac{1}{n\varepsilon_0} < \varepsilon \quad \text{if} \quad n > \frac{1}{\varepsilon\varepsilon_0}.$$

Thus, $N > 1/(\varepsilon\varepsilon_0)$ depends only on ε and is independent of z. Therefore, the given sequence is uniformly convergent.

Example 12.16 Discuss whether the sequence $\{f_n(x)\}$, where $f_n(x) = \int_0^x \dfrac{e^{-t}}{t^2 + n}\, dt$ is uniformly convergent over $x \geq 0$.

Solution We have

$$|f_n(x) - f_m(x)| = \left| \int_0^x e^{-t} \left(\frac{1}{t^2 + n} - \frac{1}{t^2 + m} \right) dt \right|$$

$$\leq \int_0^x \frac{e^{-t}|m - n|}{(t^2 + n)(t^2 + m)}\, dt < \frac{(m + n)}{mn} \int_0^x e^{-t}\, dt$$

$$< \frac{(m + n)}{mn}(1 - e^{-x}) < \frac{(m + n)}{mn}, \quad m, n \geq N.$$

Therefore, $|f_n(x) - f_m(x)| < \dfrac{(m + n)}{mn} < \dfrac{2}{N} < \varepsilon$ for all $N \geq \dfrac{2}{\varepsilon}$ and $x > 0$.

For $x = 0$, $f_n(x) = 0$.
Hence, the given sequence is uniformly convergent for all $x \geq 0$.

Exercise 12.1

Discuss the convergence of the real sequence $\{a_n\}$, where a_n is as given in the following problems:

1. $\dfrac{n}{n + \sqrt{n}}$.

2. $\dfrac{n}{1 + n + 3n^2}$.

3. $\dfrac{(-1)^n}{n} + 1$.

4. $\dfrac{1}{\ln(n + 1)}$.

5. $\dfrac{n}{\ln(n + 1)}$.

6. $\dfrac{e^n}{n}$.

7. $\sqrt{n + 1} - \sqrt{n}$.

8. $\sqrt[n]{n}$,

9. $\sqrt[n]{1/n}$.

10. $\dfrac{5^n}{n!}$.

11. $\dfrac{1}{(n + 1)^2} + \dfrac{1}{(n + 2)^2} + \cdots \dfrac{1}{(n + n)^2}$.

12. $\left(1 + \dfrac{1}{n}\right)^n$.

Discuss the convergence of the complex sequence $\{z_n\}$, where z_n is as given in the following problems:

13. $\dfrac{(-1)^n}{2n + 3i}$.

14. $(-1)^n + \dfrac{i}{n + 1}$.

15. $\dfrac{\cos in}{n}$.

16. $e^{-n\alpha}$, Re $\alpha > 0$.

17. $\dfrac{2}{n} + \left(\dfrac{n^2 + 1}{n}\right)i$.

18. $\dfrac{n - i}{n + i}$.

19. $\dfrac{n^2 - 2n - 1}{in^2 + 2}$.

20. $(i)^n \cos n\pi$.

21. $(i)^{2/n}$.

22. $e^{1 - n^2 - 2in}$.

23. $\left(\dfrac{n + 1}{n^2 - n + 1}\right) + \left(\dfrac{n - 1}{n + 2}\right)i$.

24. $\dfrac{i}{2} z_{n-1} + 1$, $z_0 = 0$.

Find whether the sequence $\{a_n\}$ is a Cauchy sequence or not, where a_n is as defined in the following problems.

25. $(-1)^n$.

26. $\dfrac{n + 1}{n}$.

27. $\sqrt{n}$.

28. $\ln n$.

29. $1 + \dfrac{1}{3} + \ldots + \dfrac{1}{2n - 1}$.

30. $1 + \dfrac{1}{4} + \dfrac{1}{7} + \ldots + \dfrac{1}{3n - 2}$.

31. Show that the sequence $\{f_n(x)\}$, where $f_n(x) = x^n$, is uniformly convergent on $[0, k]$, $k < 1$ and only pointwise convergent on $[0, 1]$.

32. Show that the sequence $\{f_n(x)\}$, where $f_n(x) = \tan^{-1} nx$, $x \geq 0$ is uniformly convergent in any interval $[a, b]$, $a > 0$ and is only pointwise convergent in $[0, b]$.

33. Show that the sequence $\{f_n(x)\}$, where $f_n(x) = nxe^{-nx^2}$, $x \geq 0$ is not uniformly convergent on $[0, a]$, $a > 0$.

34. Show that the sequence $\{f_n(z)\}$, where $f_n(z) = 1/(1 + nz)$ is uniformly convergent for all z such that $|z| \geq 2$.

35. Find the region in which the sequence $\{f_n(z)\}$, where $f_n(z) = n^{-z}$ is uniformly convergent.

12.3 Infinite Series

Let a_k, $k = 1, 2, \ldots$ be a set of real or complex numbers. Then

$$\Sigma a_k = \sum_{k=1}^{\infty} a_k = a_1 + a_2 + \ldots \tag{12.10}$$

is called an infinite series of numbers and a_k is called its kth term. We define the partial sum S_n of the series as

$$S_n = \sum_{k=1}^{n} a_k = a_1 + a_2 + \ldots + a_n.$$

The remainder of the series (after the nth term) is defined as

$$r_n = \sum_{k=n+1}^{\infty} a_k = a_{n+1} + a_{n+2} + \ldots.$$

Consider now, the sequence $\{S_n\}$ of the partial sums. The series given in Eq. (12.10) is said to be convergent if the sequence $\{S_n\}$ is convergent. The limit S of the sequence $\{S_n\}$ is called the sum of the series. Then, we have $S = S_n + r_n$. The infinite series given in Eq. (12.10) is convergent, if for a given real positive number $\varepsilon > 0$, there exists an $N \in \mathbb{IN}$ such that

$$|S_n - S| < \varepsilon \quad \text{for all} \quad n \geq N, \text{ that is } \lim_{n \to \infty} S_n = S$$

or

$$|r_n| < \varepsilon \quad \text{for all} \quad n \geq N, \text{ that is } \lim_{n \to \infty} r_n = 0.$$

Since a series can be considered in terms of a sequence of partial sums, various results regarding series follow from those of the sequences. We have the following results.

Theorem 12.13   A necessary condition for a series $\sum a_k$ to be convergent is $\lim_{n \to \infty} a_n = 0$.

Proof   Suppose that the series $\sum a_k$ is convergent. Then

$$\lim_{n \to \infty} S_n = S \quad \text{and} \quad \lim_{n \to \infty} S_{n-1} = S.$$

Since $a_n = S_n - S_{n-1}$, we get

$$\lim_{n \to \infty} a_n = \lim_{n \to \infty} S_n - \lim_{n \to \infty} S_{n-1} = S - S = 0.$$

Theorem 12.14 (Cauchy criterion for convergence)   The series $\sum a_k$ is convergent if and only if for any given real positive number $\varepsilon > 0$, there exists an $N \in \text{IN}$ such that

$$| S_n - S_m | < \varepsilon \quad \text{for all} \quad n, m \geq N. \tag{12.11}$$

(or $| S_{n+p} - S_n | < \varepsilon$ for all $n \geq N$, $p = 1, 2, \ldots$)

Theorem 12.15   The series $\sum a_k$, where $a_k = x_k + i y_k$, of complex numbers converges to $S = A + iB$ if and only if the series of the real parts $\sum x_k$ converges to A and the series of imaginary parts $\sum y_k$ converges to B.

Example 12.17   Discuss the convergence of the *geometric series* $\sum_{n=0}^{\infty} r^n$, where r is any real number.

Solution   We have

$$S_n = 1 + r + r^2 + \ldots + r^{n-1} = \frac{1 - r^n}{1 - r}$$

or

$$S_n - \frac{1}{1 - r} = - \frac{r^n}{1 - r}.$$

Therefore,

$$\left| S_n - \frac{1}{1 - r} \right| = \left| - \frac{r^n}{1 - r} \right| = \frac{| r |^n}{| 1 - r |}.$$

Since $| r |^n \to 0$ when $| r | < 1$, the geometric series converges to $1/(1 - r)$ when $| r | < 1$.

When $| r | > 1$, $| r |^n \to \infty$ as $n \to \infty$. The geometric series diverges in this case.

When $r = 1$, each term of the series is 1. Hence, $S_n = n \to \infty$ as $n \to \infty$ and the series is divergent.

When $r = - 1$, the terms in the series alternate in sign. Since the sequence $\{S_n\}$ has two limits 0 and 1, the sequence $\{S_n\}$ does not converge.

Hence, the geometric series is convergent when $| r | < 1$.

Remark 6

The series $\sum_{n=0}^{\infty} z^n$ converges to $1/(1 - z)$ if $| z | < 1$.

Example 12.18   Show that $\sum_{n=1}^{\infty} z^n = \dfrac{z}{1 - z}$ when $|z| < 1$. Writing $z = re^{i\theta}$, $0 < r < 1$, show that

(i) $\sum_{n=1}^{\infty} r^n \cos n\theta = \dfrac{r \cos \theta - r^2}{1 - 2r \cos \theta + r^2}$ and   (ii) $\sum_{n=1}^{\infty} r^n \sin n\theta = \dfrac{r \sin \theta}{1 - 2r \cos \theta + r^2}$.

Solution   We have

$$\sum_{n=1}^{\infty} z^n = z + z^2 + \ldots$$

which is a geometric series and converges to $z/(1 - z)$, when $|z| < 1$.

Substituting $z = re^{i\theta}, 0 < r < 1$, on both sides of

$$\sum_{n=1}^{\infty} z^n = \frac{z}{1 - z}$$

we obtain

$$\sum_{n=1}^{\infty} r^n (\cos n\theta + i \sin n\theta) = \frac{re^{i\theta}}{1 - re^{i\theta}} = \left[\frac{re^{i\theta}}{1 - re^{i\theta}}\right]\left[\frac{1 - re^{-i\theta}}{1 - re^{-i\theta}}\right]$$

$$= \frac{r(\cos \theta + i \sin \theta) - r^2}{1 - 2r \cos \theta + r^2}.$$

Comparing the real and imaginary parts on both sides, we obtain

$$\sum_{n=1}^{\infty} r^n \cos n\theta = \frac{r \cos \theta - r^2}{1 - 2r \cos \theta + r^2}$$

and

$$\sum_{n=1}^{\infty} r^n \sin n\theta = \frac{r \sin \theta}{1 - 2r \cos \theta + r^2}.$$

Example 12.19 Discuss the convergence of the *harmonic series* $\sum_{n=1}^{\infty} \frac{1}{n}$.

Solution We have

$$S_n = 1 + \frac{1}{2} + \ldots + \frac{1}{n}$$

and $|S_{n+p} - S_n| = \left| \frac{1}{n+1} + \frac{1}{n+2} + \ldots + \frac{1}{n+p} \right| > \frac{1}{n+p} + \frac{1}{n+p} + \ldots + \frac{1}{n+p} = \frac{p}{n+p}$

Now, $\frac{p}{n+p} = \frac{1}{2}$ when $p = n$. That is $|S_{2n} - S_n| > \frac{1}{2}$.

Thus, for $\varepsilon < 1/2$, we cannot have $|S_{n+p} - S_n| < \varepsilon$, for all $n \geq N$. Therefore, the series is not Cauchy convergent and hence not convergent.

Remark 7

Note that the necessary condition for convergence, $\lim_{n \to \infty} a_n = \lim_{n \to \infty} \frac{1}{n} = 0$ is satisfied by the harmonic series.

12.3.1 Tests for Convergence

The following results are frequently used to test the convergence of an infinite series.

1. Comparison test Let $\sum_{n=0}^{\infty} a_n$ and $\sum_{n=0}^{\infty} b_n$ be two real series with positive terms and $a_n \leq k b_n$ for any real positive k and $n = 1, 2, \ldots$. Then,

(i) convergence of the series $\sum_{n=0}^{\infty} b_n$ implies convergence of the series $\sum_{n=0}^{\infty} a_n$,

(ii) divergence of the series $\sum_{n=0}^{\infty} a_n$ implies divergence of the series $\sum_{n=0}^{\infty} b_n$.

Remark 8

If the series $\sum_{n=0}^{\infty} a_n$ converges, no conclusion can be drawn about the series $\sum_{n=0}^{\infty} b_n$ and if the series $\sum_{n=0}^{\infty} b_n$ diverges, no conclusion can be drawn about the series $\sum_{n=0}^{\infty} a_n$.

2. Limit comparison test   Let $\sum_{n=0}^{\infty} a_n$ and $\sum_{n=0}^{\infty} b_n$ be two real series with positive terms and

$$\lim_{n \to \infty} \frac{a_n}{b_n} = l, 0 < l < \infty. \tag{12.12}$$

Then, both the series $\sum_{n=0}^{\infty} a_n$ and $\sum_{n=0}^{\infty} b_n$ converge or diverge together.

Example 12.20   Show that the *p-series* $\sum_{n=1}^{\infty} 1/n^p$, $p > 0$ is (i) convergent if $p > 1$ and (ii) divergent if $p \le 1$.

Solution

(i) Since the convergence or divergence of a series of positive terms is not altered by a rearrangement of terms, we can write

$$S_n = \frac{1}{1^p} + \frac{1}{2^p} + \ldots + \frac{1}{n^p}$$

$$= \frac{1}{1^p} + \left(\frac{1}{2^p} + \frac{1}{3^p} \right) + \left(\frac{1}{4^p} + \frac{1}{5^p} + \frac{1}{6^p} + \frac{1}{7^p} \right) + \ldots$$

$$< \frac{1}{1^p} + \frac{2}{2^p} + \frac{4}{4^p} + \ldots$$

$$= \frac{1}{1^p} + \left(\frac{1}{2^{p-1}} \right) + \left(\frac{1}{2^{p-1}} \right)^2 + \ldots$$

which is a geometric series with common ratio $r = 1/2^{p-1}$. Therefore, the series is convergent if

$$r = \frac{1}{2^{p-1}} < 1 \quad \text{or} \quad p > 1.$$

(ii) When $0 < p \le 1$, $\dfrac{1}{n} \le \dfrac{1}{n^p}$.

Since the harmonic series is divergent, the *p*-series is also divergent for $0 < p \le 1$, by the comparison test.

Example 12.21   Discuss the convergence of the series $\sum_{n=1}^{\infty} \dfrac{1}{n(n+1)}$.

Solution   We have

$$S_n = \frac{1}{1.2} + \frac{1}{2.3} + \ldots + \frac{1}{n(n+1)} = \left(1 - \frac{1}{2} \right) + \left(\frac{1}{2} - \frac{1}{3} \right) + \ldots + \left(\frac{1}{n} - \frac{1}{n+1} \right) = 1 - \frac{1}{n+1}.$$

Since $\lim_{n \to \infty} S_n = 1$, the given series is convergent and the sum of the series is 1.

Alternative Consider the two series $\Sigma a_n = \Sigma \dfrac{1}{n(n+1)}$ and $\Sigma b_n = \Sigma \dfrac{1}{n^2}$. We have

$$\lim_{n\to\infty} \frac{a_n}{b_n} = \lim_{n\to\infty} \frac{n^2}{n(n+1)} = 1.$$

The series $\Sigma (1/n^2)$ is convergent. Therefore, by limit comparison test, the given series is also convergent.

3. D'Alembert's test (Ratio test) Let Σa_n be a real series of positive terms or a complex series. Let

$$\lim_{n\to\infty} \left| \frac{a_{n+1}}{a_n} \right| = l. \tag{12.13}$$

Then, the series Σa_n is (i) convergent if $l < 1$ and (ii) divergent if $l > 1$. The ratio test fails when $l = 1$ (see Appendix 2 for proof).

For example, in the series

(i) $\Sigma \dfrac{1}{n}$, $\lim\limits_{n\to\infty} \dfrac{a_{n+1}}{a_n} = 1$ and the series is divergent.

(ii) $\Sigma \dfrac{1}{n^2}$, $\lim\limits_{n\to\infty} \dfrac{a_{n+1}}{a_n} = 1$ and the series is convergent.

4. Raabe's test Let Σa_n be a real series of positive terms or a complex series. Let

$$\lim_{n\to\infty} n\left[1 - \left| \frac{a_{n+1}}{a_n} \right| \right] = l \tag{12.14}$$

Then, the series Σa_n is (i) convergent (absolutely convergent) if $l > 1$ and (ii) divergent if $l < 1$. The Raabe's test fails when $l = 1$. (see Appendix 2 for proof)

Raabe's test may often give conclusive result, when the ratio test fails.

Example 12.22 Discuss the convergence of the series

(i) $\Sigma \dfrac{1}{n^2}$,

(ii) $\Sigma \dfrac{1 \cdot 4 \cdot 7 \ldots (3n-2)}{2 \cdot 5 \cdot 8 \ldots (3n-1)}$,

(iii) $\Sigma \dfrac{z^n}{n(n+2)}$,

(iv) $\Sigma \dfrac{z^{n-1}}{(n-1)!}$,

(v) $\Sigma n! \, z^n$.

using the ratio test and/or Raabe's test.

Solution

(i)
$$\left| \frac{a_{n+1}}{a_n} \right| = \left(\frac{n}{n+1} \right)^2 \quad \text{and} \quad \lim_{n\to\infty} \left| \frac{a_{n+1}}{a_n} \right| = 1.$$

Therefore, the ratio test fails.

Now, we use the Raabe's test. We get

$$\lim_{n \to \infty} n \left[1 - \left| \frac{a_{n+1}}{a_n} \right| \right] = \lim_{n \to \infty} n \left[1 - \frac{n^2}{(n+1)^2} \right]$$

$$= \lim_{n \to \infty} n \left[\frac{2n+1}{(n+1)^2} \right] = \lim_{n \to \infty} \frac{2 + (1/n)}{(1 + (1/n))^2} = 2 > 1.$$

Hence, the series is convergent.

(ii)
$$\left| \frac{a_{n+1}}{a_n} \right| = \frac{3n+1}{3n+2} \quad \text{and} \quad \lim_{n \to \infty} \left| \frac{a_{n+1}}{a_n} \right| = 1.$$

Therefore, the ratio test fails.

Now, we use the Raabe's test. We get

$$\lim_{n \to \infty} n \left[1 - \left| \frac{a_{n+1}}{a_n} \right| \right] = \lim_{n \to \infty} n \left[1 - \frac{3n+1}{3n+2} \right]$$

$$= \lim_{n \to \infty} \left[\frac{n}{3n+2} \right] = \lim_{n \to \infty} \left[\frac{1}{3 + (2/n)} \right] = \frac{1}{3} < 1.$$

Hence, the series is divergent.

(iii)
$$\left| \frac{a_{n+1}}{a_n} \right| = \frac{n(n+2)}{(n+1)(n+3)} |z| \quad \text{and} \quad \lim_{n \to \infty} \left| \frac{a_{n+1}}{a_n} \right| = |z|.$$

Therefore, by ratio test, the series is convergent when $|z| < 1$ and divergent when $|z| > 1$. The ratio test fails when $|z| = 1$.

Using the Raabe's test when $|z| = 1$, we get

$$\lim_{n \to \infty} n \left[1 - \left| \frac{a_{n+1}}{a_n} \right| \right] = \lim_{n \to \infty} n \left[1 - \frac{n(n+2)}{(n+1)(n+3)} \right]$$

$$= \lim_{n \to \infty} \left[\frac{n(2n+3)}{(n+1)(n+3)} \right] = \lim_{n \to \infty} \left[\frac{2 + (3/n)}{(1 + (1/n))(1 + (3/n))} \right] = 2 > 1.$$

Therefore, the series is convergent when $|z| = 1$. Hence, the given series is convergent when $|z| \le 1$ and divergent when $|z| > 1$.

(iv) $\left| \dfrac{a_{n+1}}{a_n} \right| = \dfrac{1}{n} |z|$ and $\lim_{n \to \infty} \left| \dfrac{a_{n+1}}{a_n} \right| = 0 < 1$, for all z. Hence, by ratio test, the series is convergent for all z.

(v) $\left| \dfrac{a_{n+1}}{a_n} \right| = (n+1) |z|$ and $\lim_{n \to \infty} \left| \dfrac{a_{n+1}}{a_n} \right| = \infty > 1$, for all z. Hence, by ratio test, the series is divergent for all z.

5. Cauchy root test Let $\Sigma \, a_n$ be a real series of positive terms or a complex series. Let

$$\lim_{n\to\infty} (|a_n|)^{1/n} = l. \tag{12.15}$$

Then, the series is convergent if $l < 1$ and divergent if $l > 1$. The test fails when $l = 1$. (see Appendix 2 for proof)

Example 12.23 Discuss the convergence of the series $\Sigma \, a_n$, where $a_n = \left(1 + \dfrac{1}{n^p}\right)^{-n^{p+1}}$, $p > 0$ using the Cauchy root test.

Solution We have

$$(a_n)^{1/n} = \left(1 + \frac{1}{n^p}\right)^{-n^p} = \frac{1}{(1 + 1/n^p)^{n^p}}$$

Now,

$$\lim_{n\to\infty} (a_n)^{1/n} = \lim_{n\to\infty} \frac{1}{[1 + (1/n^p)]^{n^p}} = \frac{1}{e} < 1,$$

Hence, the series is convergent.

6. Cauchy integral test Let f be a non-negative decreasing function on $[1, \infty)$. Then, the series $\sum\limits_{n=1}^{\infty} f(n)$ and the improper integral $\displaystyle\int_1^{\infty} f(x)\,dx$ converge or diverge together. As an example, let us discuss the convergence of the series

$$\sum_{n=2}^{\infty} \frac{1}{n[\log n]^p}$$

Now, $f(x) = 1/[x(\log x)^p]$ is a non-negative decreasing function for $x \in [2, \infty)$. Hence, by Cauchy integral test, the given series and the integral $\displaystyle\int_2^{\infty} f(x)\,dx$ converge or diverge together. Now,

$$\int_2^t f(x)\,dx = \int_2^t \frac{dx}{x(\log x)^p} = \int_{\log 2}^{\log t} \frac{du}{u^p} \quad (\text{set } u = \log x)$$

$$= \left[\frac{u^{-p+1}}{1 - p}\right]_{\log 2}^{\log t} = \frac{1}{1 - p}[(\log t)^{-p+1} - (\log 2)^{-p+1}] = F(t)$$

If $p < 1$, $F(t) \to \infty$ as $t \to \infty$. The integral diverges. Therefore, the given series is divergent.

If $\quad p = 1$, $F(t) = \displaystyle\int_2^t \frac{dx}{x \log x} = [\log (\log x)]_2^t = [\log (\log t) - \log (\log 2)] \to \infty$ as $t \to \infty$.

The integral diverges. Therefore, the given series is divergent.

If $\quad p > 1$, $F(t) = -\dfrac{1}{p - 1}\left[\dfrac{1}{(\log t)^{p-1}} - \dfrac{1}{(\log 2)^{p-1}}\right] \to \dfrac{1}{(p - 1)(\log 2)^{p-1}}$ as $t \to \infty$

The integral converges. Therefore, the given series is convergent.

7. Logarithmic test Let $\Sigma \, a_n$ be a series of positive terms and

$$\lim_{n \to \infty} \left[n \log \left(\frac{a_n}{a_{n+1}} \right) \right] = l.$$

Then, if $l > 1$, the series converges. If $l < 1$, the series diverges. If $l = 1$, the test fails.
As an example, let us discuss the convergence of the series

$$1 + \frac{x}{2} + \frac{2!}{3^2} x^2 + \frac{3!}{4^3} x^3 + \cdots$$

By ratio test, we get

$$\lim_{n \to \infty} \frac{a_{n+1}}{a_n} = \lim_{n \to \infty} \left[\frac{(n+1)! x^{n+1}}{(n+2)^{n+1}} \right] \left[\frac{(n+1)^n}{n! x^n} \right]$$

$$= \lim_{n \to \infty} \frac{[1 + (1/n)]^n \, [1 + (1/n)]}{[1 + (2/n)]^n [1 + (2/n)]} x = \frac{x(e)}{e^2} = \frac{x}{2}.$$

If $x < e$, the series converges and if $x > e$, the series diverges. If $x = e$, the test fails and we use the logarithmic test. We have

$$\frac{a_n}{a_{n+1}} = \frac{(n+2)^{n+1}}{e(n+1)^{n+1}} = \frac{1}{e} \left[1 + \frac{1}{n+1} \right]^{n+1}$$

$$\log \left(\frac{a_n}{a_{n+1}} \right) = -1 + (n+1) \left[\frac{1}{n+1} - \frac{1}{2(n+1)^2} + \frac{1}{3(n+1)^3} - \cdots \right]$$

$$= -\frac{1}{2(n+1)} + \frac{1}{3(n+1)^2} - \cdots$$

$$\lim_{n \to \infty} \left[n \log \left(\frac{a_n}{a_{n+1}} \right) \right] = \lim_{n \to \infty} \left[-\frac{n}{2(n+1)} + \frac{n}{3(n+1)^2} - \cdots \right] = -\frac{1}{2} < 1.$$

Therefore, the series diverges for $x = e$. The given series is convergent for $x < e$ and divergent for $x \ge e$.

Alternating series

A real series in which the terms are alternately positive and negative is called an *alternating series* and is of

the form $\sum\limits_{n=1}^{\infty} (-1)^{n-1} a_n$, $a_n > 0$. The following theorem gives a sufficient condition for the convergence of an alternating series.

Theorem 12.6 (Leibnitz theorem) Let $\sum\limits_{n=1}^{\infty} (-1)^{n-1} a_n, a_n > 0$ be an alternating series such that

 (i) the sequence $\{a_n\}$ is non-increasing, that is $a_{n+1} \le a_n$ for all n, and

 (ii) $\lim\limits_{n \to \infty} a_n = 0$.

Then, the series $\sum (-1)^{n-1} a_n$ is convergent.

(see Appendix 2 for proof.)

Absolutely convergent series

Let $\sum a_n$ be an arbitrary series of real or complex numbers. If the series of positive terms $\sum |a_n|$ is convergent, then we say that the series $\sum a_n$ is *absolutely convergent*. If the series $\sum a_n$ is convergent but $\sum |a_n|$ is divergent, then the series is called *conditionally convergent*.

For example, consider the series

$$\sum (-1)^{n+1} a_n = \frac{1}{1} - \frac{1}{2} + \frac{1}{3} - \cdots$$

where $a_n = 1/n$ and $a_{n+1} = 1/(n+1)$.

Since $a_{n+1} \le a_n$ for all n and $\lim\limits_{n \to \infty} a_n = 0$, this series is convergent by Leibnitz theorem.

Now, the corresponding series of positive terms

$$\Sigma \, |(-1^{n+1})a_n| = \frac{1}{1} + \frac{1}{2} + \frac{1}{3} - \cdots$$

is divergent (harmonic series). Hence the series

$$\frac{1}{1} - \frac{1}{2} + \frac{1}{3} - \cdots$$

is conditionally convergent.

Theorem 12.17 An absolutely convergent series is convergent.

Proof Suppose that the series $\Sigma \, a_n$ is absolutely convergent, that is, the series $\Sigma \, |a_n|$ is convergent. Therefore, the series $\Sigma \, |a_n|$ is Cauchy convergent. Hence, given a real positive number $\varepsilon > 0$, there exists an $N \in$ IN such that

$$|S_{n+p} - S_n| = ||\,a_{n+1}| + |a_{n+2}| + \ldots + |a_{n+p}\,|| < \varepsilon, \quad n > N, \quad p \in \text{IN}$$

Now, $\qquad |a_{n+1} + a_{n+2} + \ldots + a_{n+p}| < |a_{n+1}| + |a_{n+2}| + \ldots + |a_{n+p}| < \varepsilon.$

Therefore, the series $\Sigma \, a_n$ is Cauchy convergent and hence convergent.

Remark 9 Divergence of the series $\Sigma \, |a_n|$ does not imply the divergence of the series $\Sigma \, a_n$.

Example 12.24 Discuss the convergence of the series $\overset{\infty}{\underset{n=1}{\Sigma}} a_n$, where

(i) $a_n = \dfrac{\cos(2n - 3i)}{n^p}$
$\qquad\qquad$
(ii) $a_n = \left(\dfrac{3n - 2}{np + 1}\right)^n (3 - 4i)^n,$

with $p > 0$

Solution

(i) $\qquad a_n = \dfrac{\cos(2n - 3i)}{n^p} = \dfrac{\cos(2n)\cosh 3 - i \sin(2n)\sinh 3}{n^p}$

$$|a_n| = \frac{[\cos^2(2n)\cosh^2 3 + \sin^2(2n)\sinh^2 3]^{1/2}}{n^p} \le \frac{\cosh^2 3 + \sinh^2 3}{n^p} = \frac{A}{n^p}$$

where $A = \cosh^2 3 + \sinh^2 3$ is a finite constant.

The series $\Sigma \, b_n = A \overset{\infty}{\underset{n=1}{\Sigma}} (1/n^p)$ is the p-series, which converges when $p > 1$ and diverges when $0 < p \le 1$. Hence, the given series converges absolutely if $p > 1$.

(ii) $|a_n| = \left(\dfrac{3n - 2}{np + 1}\right)^n |(3 - 4i)^n| = \left(\dfrac{3n - 2}{np + 1}\right)^n (5^n) = \left(\dfrac{15n - 10}{np + 1}\right)^n.$

Using the Cauchy root test, we find that

$$\lim_{n \to \infty} |a_n|^{1/n} = \lim_{n \to \infty} \left(\frac{15n - 10}{np + 1}\right) = \frac{15}{p}.$$

Therefore, the given series converges absolutely if $p > 15$.

12.3.2 Uniform Convergence of Series of Functions

Let $f_1(z) + f_2(z) + \ldots$ be a series of single-valued complex functions defined in a domain D (or a series of real functions defined on a closed interval). Let

$$S_n(z) = f_1(z) + f_2(z) + \ldots + f_n(z)$$

be the nth partial sum. Now, consider the sequence $\{S_n(z)\}$ of partial sums. If at a point, $z = z_0$ in D, the sequence $\{S_n(z)\}$ of partial sums converges to $f(z_0)$, then we say that the series $\Sigma f_k(z_0)$ converges to $f(z_0)$. This convergence is called *pointwise convergence*.

If for each z in D, the sequence $\{S_n(z)\}$ of partial sums converges to $f(z)$, then we say that the series $\Sigma f_k(z)$ converges uniformly to $f(z)$. Thus, for a given real positive number $\varepsilon > 0$, there exists an $N \in$ IN independent of z, but dependent on ε such that

$$|S_n(z) - f(z)| < \varepsilon \quad \text{for all} \quad n > N.$$

Thus, a series which is uniformly convergent is also pointwise convergent. In the following theorem, sufficient conditions are given in order that a series is uniformly convergent.

Theorem 12.18 (*Weierstrass's M-test*) Let $\Sigma f_n(z)$ be an infinite series defined in some domain D of the complex plane and let $\{M_n\}$ be a sequence of positive terms, where $|f_n(z)| \le M_n$ for all n and for all z in D. If the series ΣM_n is convergent, then the series $\Sigma f_n(z)$ is uniformly and absolutely convergent.

Proof Let $S_n = M_1 + M_2 + \dots M_n$. Since the series of positive terms ΣM_n is convergernt, it is Cauchy convergent. Therefore,

$$|S_{n+p} - S| = M_{n+1} + M_{n+2} + \dots + M_{n+p} < \varepsilon, n > N, p \in \text{IN}.$$

Now, $\qquad |f_{n+1}(z) + f_{n+2}(z) + \dots + f_{n+p}(z)| \le |f_{n+1}(z)| + |f_{n+2}(z)| + \dots + |f_{n+p}(z)|$

$$\le M_{n+1} + M_{n+2} + \dots + M_{n+p} < \varepsilon.$$

for all $n > N$ and $p \in \text{IN}$ and for all z in D. Therefore, the given series is uniformly and absolutely convergent in D.

Example 12.25 Show that the series $\Sigma \dfrac{z^n - 1}{n^2 + |z|^2}$ converges uniformly in disk $|z| < 1$.

Solution We have $\quad |f_n(z)| = \left| \dfrac{z^n - 1}{n^2 + |z|^2} \right| \le \dfrac{|z|^n + 1}{n^2 + |z|^2} < \dfrac{2}{n^2}$ for all z in $|z| < 1$.

Since, the series $\Sigma 1/n^2$ is convergent, the given series is uniformly convergent.

Example 12.26 Test for uniform convergence the series $\displaystyle\sum_{n=0}^{\infty} e^{inz}$, $\text{Im}(z) > 0$.

Solution We have

$$|e^{inz}| = |e^{in(x+iy)}| = |e^{inx - ny}| = e^{-ny}.$$

Now, consider the series of positive terms $\displaystyle\sum_{n=0}^{\infty} e^{-ny}$.

Using the Cauchy root test, we find that

$$\lim_{n \to \infty} (a_n)^{1/n} = \lim_{n \to \infty} (e^{-ny})^{1/n} = e^{-y} < 1, \text{ when } y > 0.$$

Therefore, the series $\displaystyle\sum_{n=0}^{\infty} e^{-ny}$, $y > 0$ is convergent. Using the Weierstrass M-test, we conclude that the given series is uniformly convergent.

Example 12.27 Show that the geometric series $1 + z + z^2 + \dots$ is

(i) uniformly convergent in any closed disk $|z| \le r < 1$.

(ii) not uniformly convergent in the open disk $|z| < 1$.

Solution We have

$$S_n(z) = 1 + z + \dots z^{n-1} \text{ and } f(z) = S(z) = \lim_{n \to \infty} S_n(z) = \dfrac{1}{1-z}, |z| < 1.$$

(i) In the closed disk $|z| \le r < 1$, we have

$$|1 - z| \ge 1 - |z| \ge 1 - r \quad \text{or} \quad \dfrac{1}{1 - |z|} \le \dfrac{1}{1 - r}.$$

Hence, $\qquad |S_n(z) - S(z)| = |z^n + z^{n+1} + \dots| = \left| \dfrac{z^n}{1-z} \right| \le \dfrac{r^n}{1-r}.$

Since $r < 1$, the right hand side can be made as small as necessary by choosing n large enough.

Therefore, $\quad |S_n(z) - f(z)| < \varepsilon, \quad n > N \in \text{IN}$ and for all z.

Hence, the given series is uniformly convergent.

(ii) In the open disk $|z| < 1$. we can find a z for a given n and a real k (no matter how large) such that

$$\left| \frac{z^n}{1-z} \right| = \frac{|z|^n}{1-|z|} > k$$

by taking $|z|$ sufficiently close to 1. Thus, no single $N \in \mathbb{IN}$ will make $|S_n(z) - S(z)| < \varepsilon$ for every z in the open disk $|z| < 1$. In this case, N depends both on z and ε. Hence, the convergence is not uniform.

Uniformly convergent series have the following properties:

1. If the functions $f_n(z)$ are continuous in a domain D and the series $\Sigma f_n(z)$ converges uniformly to $f(z)$ in this domain, then $f(z)$ is continuous in D.

2. $\int_C f(z)\, dz = \sum\limits_{k=1}^{\infty} \int_C f_k(z)\, dz$, where C is any piecewise smooth curve in D.

3. $f'(z) = \sum\limits_{k=1}^{\infty} f_k'(z)$ for all z in D.

Exercise 12.2

Test the convergence of the following series.

1. $\Sigma \dfrac{n^2+1}{n^2}$.

2. $\Sigma \left(1 + \dfrac{1}{n}\right)^{-n}$

3. $\Sigma \tan^{-1} nx$.

4. $\Sigma \sin \left(\dfrac{1}{n^2}\right)$.

5. $\Sigma \dfrac{\sin^n |z\pi|}{n(n+1)}$.

6. $\Sigma \dfrac{1}{n^n}$.

7. $\Sigma \left[\dfrac{\sqrt{n+1}-\sqrt{n}}{n^p}\right], p > 0$.

8. $\Sigma [\sqrt{n^4+1}-\sqrt{n^4-1}]$.

9. $\Sigma \dfrac{1}{n!}$.

10. $\Sigma \dfrac{\sqrt{n^4+1}-n^2}{\sqrt[4]{n^4+1}+n}$.

11. $\Sigma \dfrac{x^{n+1}}{(n+1)\sqrt{n}}$.

12. $\Sigma \dfrac{nx^n}{(n+1)^n}$.

13. $\Sigma \dfrac{1}{[\log n]^n}$.

14. $\Sigma \dfrac{(n+1)^n}{n^{n+1}} x^n, (x > 0)$.

15. $\sum\limits_{n=2}^{\infty} \dfrac{1}{n \log n}$.

16. $\Sigma \dfrac{n^n x^n}{n!}$.

17. $\Sigma \dfrac{(-1)^{n-1}}{(2n-2)!}$.

18. $\Sigma \dfrac{\cos (n\pi)}{n^2+1}$.

19. $\Sigma \dfrac{(-1)^{n-1} n}{5n+1}$.

20. $\Sigma n!\, z^n$.

21. $\Sigma \dfrac{(2i)^n n!}{n^n}$.

22. $\Sigma \left(\dfrac{3}{4} i\right)^n n^3$.

23. $\Sigma \dfrac{i^n}{n!}$.

24. $\dfrac{n^2-1}{n^2+1} x^n$.

25. $\Sigma \dfrac{n^3}{2^n}$.

26. $\Sigma \dfrac{i^n}{n(2^n)}$.

27. $\Sigma e^{(1-in)^2}$.

Find the regions in which the following series converge.

28. $\Sigma \left(\dfrac{z-i}{z+i}\right)^n$.

29. $\Sigma \left(\dfrac{iz-1}{2+i}\right)^n$.

30. $\Sigma n\, z^n$.

31. Σe^{inz}.

32. $\Sigma \dfrac{(2n)!}{2^n (n!)^2} \left(\dfrac{1-z}{z}\right)^n$.

Test whether the following series converge absolutely or not.

33. $\Sigma \dfrac{(-1)^n}{n^3}$.

34. $\Sigma \dfrac{\sin nx}{n^2}$.

35. $\Sigma \dfrac{e^{in}}{n^2}$.

36. $\sum \dfrac{(n+2)^2}{(2i)^n}$.

37. $\sum \left(\dfrac{1-i}{3}\right)^n$.

38. $\sum \dfrac{(-1)^n(1+i)}{2^n}$.

39. Show that the series $\sum 1/(x^2 + n^2)$ is uniformly convergent.

40. Show that the series $\sum f_n(x)$ whose sum to n terms is given by $S_n(x) = nx/(1 + n^2x^2)$ is not uniformly convergent on any interval containing zero.

41. Show that the series $\sum 1/(z^2 + n^2)$ is uniformly convergent in $|z| < R$.

42. Show that the series $\sum (\sin nz)/n^2$ is uniformly convergent in $|z| \leq 1$.

12.4 Power Series

An infinite series of the form

$$\sum_{n=0}^{\infty} a_n(z - z_0)^n = a_0 + a_1(z - z_0) + a_2(z - z_0)^2 + \ldots + a_n(z - z_0)^n + \ldots \qquad (12.16)$$

is called a *power series* about $z = z_0$ where $a_0, a_1, \ldots$ are real or complex constants. It is also written as $\sum a_n(z - z_0)^n$. The point z_0 is called the centre of the power series. The power series always converges at the point $z = z_0$, since for $z = z_0$, the series reduces to a constant a_0.

When $z_0 = 0$, we obtain a power series in powers of z, in the form

$$\sum a_n z^n = a_0 + a_1 z + \ldots + a_n z^n + \ldots.$$

The convergence result of a power series is given in the following theorem.

Theorem 12.19 (Abel's theorem) If the power series given in Eq. (12.16) converges at a point $z = z^* \neq z_0$, then it converges absolutely for all z in the disk $|z - z_0| < |z^* - z_0|$.

Proof At the fixed point $z = z^*$, the power series becomes an infinite series of complex numbers. Since the series $\sum a_n(z^* - z_0)^n$ converges, we get

$$\lim_{n \to \infty} a_n(z^* - z_0)^n = 0.$$

Therefore, the terms of the series $\sum a_n(z^* - z_0)^n$ are bounded, that is

$$|a_n||z^* - z_0|^n \leq M, \ n = 0, 1, \ldots$$

for a finite positive constant M. Now,

$$|a_n(z - z_0)^n| = |a_n||z - z_0|^n \leq M\left|\frac{z - z_0}{z^* - z_0}\right|^n = Mk^n \quad \text{where} \quad k = \left|\frac{z - z_0}{z^* - z_0}\right|.$$

Consider the series of positive terms $\sum Mk^n$, which is a geometric series with common ratio k. This series converges when $k < 1$. Therefore, the series $\sum a_n(z - z_0)^n$ converges absolutely if

$$k = \left|\frac{z - z_0}{z^* - z_0}\right| < 1 \quad \text{or} \quad |z - z_0| < |z^* - z_0|.$$

Hence, the result.

Remark 10

If the power series given in Eq. (12.16) diverges at a point $z = z^*$, then it diverges for all z, where $|z - z_0| > |z^* - z_0|$.

Now, consider the set of all points in the complex plane at which the series is convergent. Let R be the radius of the circle with centre at z_0 that contains all the points at which the series is convergent. Then, the series is convergent for all z for which $|z - z_0| < R$ and divergent for all z for which $|z - z_0| > R$. The real number R is called the *radius of convergence* and the circle $|z - z_0| = R$ is called the *circle of convergence* of the power series. If $R = 0$, the series is convergent only at the point $z = z_0$. If $R = \infty$, the series is convergent for all z.

Theorem 12.20 Let $\sum a_n (z - z_0)^n$ be a power series and let $L^* = \lim\limits_{n \to \infty} \left| \dfrac{a_{n+1}}{a_n} \right|$. Then, the radius of convergence of the power series is $R = 1/L^*$.

Proof Let z_n denote the nth term of the power series. We obtain

$$\left| \frac{z_{n+1}}{z_n} \right| = \left| \frac{a_{n+1}}{a_n} (z - z_0) \right| = \left| \frac{a_{n+1}}{a_n} \right| |z - z_0|.$$

Therefore, $$\lim_{n \to \infty} \left| \frac{z_{n+1}}{z_n} \right| = \lim_{n \to \infty} \left| \frac{a_{n+1}}{a_n} \right| |z - z_0| = L^* |z - z_0|.$$

Using the ratio test, we find that the series converges if

$$L^* |z - z_0| < 1 \quad \text{or} \quad |z - z_0| < \frac{1}{L^*} = R.$$

The series diverges if $|z - z_0| > \dfrac{1}{L^*} = R$. Therefore, the radius of convergence of the power series is

$$R = \frac{1}{L^*}, \quad \text{where} \quad L^* = \lim_{n \to \infty} \left| \frac{a_{n+1}}{a_n} \right|, \quad \text{or} \quad R = \lim_{n \to \infty} \left| \frac{a_n}{a_{n+1}} \right|.$$

Remark 11

(a) We can also use the Cauchy's root test and write $L^* = \lim\limits_{n \to \infty} [\, |a_n| \,]^{1/n}$.

(b) For the real power series $\sum a_n(x - x_0)^n$, a_n real constants, the circle of convergence reduces to the interval of convergence $x_0 - R < x < x_0 + R$ on the real axis.

Theorem 12.21   The power series $\sum a_n(z - z_0)^n$ with a non-zero radius of convergence R is uniformly convergent in the circle $|z - z_0| \le r < R$.

Proof Let $S_n = \sum\limits_{k=1}^{n} a_k (z - z_0)^k$ be the nth partial sum of the power series. Then, for $|z - z_0| \le r$

and any positive integers n and m, we have

$$|S_{n+m} - S_n| = | a_{n+1} (z - z_0)^{n+1} + a_{n+2}(z - z_0)^{n+2} + \ldots + a_{n+m}(z - z_0)^{n+m} |$$

$$\le | a_{n+1} | |z - z_0|^{n+1} + | a_{n+2} | |z - z_0|^{n+2} + \ldots + | a_{n+m} | |z - z_0|^{n+m}$$

$$\le | a_{n+1} | r^{n+1} + | a_{n+2} | r^{n+2} + \ldots + | a_{n+m} | r^{n+m} < \varepsilon. \tag{12.17}$$

Since the power series converges absolutely in the region $|z - z_0| \le r < R$, we have, by using the Cauchy convergence criterion that for any given real positive number $\varepsilon > 0$, there exists an $N \in$ IN such that for $m, n \ge N$ and for all z in this region, equation (12.17) holds. Hence, the power series is uniformly convergent within its circle of convergence.

Let the power series $\sum a_n(z - z_0)^n$ converge to the function $f(z)$. Then we have the following results.

Theorem 12.22 A power series $\sum a_n(z - z_0)^n$ represents an analytic function within its circle of convergence.

For example, the series $\sum z^n$ represents the function $1/(1 - z)$ which is analytic within its circle of convergence $|z| = 1$.

Theorem 12.23 A power series $f(z) = \sum a_n(z - z_0)^n$ can be integrated term by term within its circle of convergence, that is

$$\int_C f(z) \, dz = \sum_{n=0}^{\infty} \int_C a_n(z - z_0)^n \, dz$$

for every contour C lying entirely within the circle of convergence.

The radius of convergence of the integrated series is same as the radius of convergence of the original power series.

Theorem 12.24 A power series $f(z) = \sum a_n(z - z_0)^n$ can be differentiated term by term within its circle of convergence. The radius of convergence of the differentiated series

$$f'(z) = \sum n\, a_n(z - z_0)^{n-1} = a_1 + 2\, a_2(z - z_0) + \ldots + n\, a_n(z - z_0)^{n-1} + \ldots$$

is same as that of the original power series.

Theorem 12.25 Suppose that the two power series $\sum a_n(z - z_0)^n$ and $\sum b_n(z - z_0)^n$ have the same radius of convergence R and converge to the same sum within the circle of convergence $|z - z_0| = R$. Then these two series are identical.

Remark 12

(a) If $f(z) = \sum a_n(z - z_0)^n$ and $g(z) = \sum b_n(z - z_0)^n$ are two power series with radii of convergence R_1 and R_2 respectively, then the radius of convergence of the series

$$f(z) \pm g(z) = \sum a_n(z - z_0)^n \pm \sum b_n(z - z_0)^n$$

is $R = \min(R_1, R_2)$.

(b) Let $f(z) = \sum a_n(z - z_0)^n$ and $g(z) = \sum b_n(z - z_0)^n$ be two power series with radii of convergence R_1 and R_2 respectively. The *Cauchy product* of these series is defined as

$$f(z)\, g(z) = \sum_{n=0}^{\infty} \left(\sum_{r=0}^{n} a_r\, b_{n-r} \right) (z - z_0)^n = \sum_{n=0}^{\infty} c_n (z - z_0)^n \tag{12.18}$$

where $c_n = \sum_{r=0}^{n} a_r\, b_{n-r}$. The radius of convergence of the product series is $R = \min(R_1, R_2)$.

Example 12.28 Find the radius of convergence and the circle of convergence of the following power series

(i) $\sum \dfrac{(n!)^2\, z^n}{(2n)!}$,

(ii) $\sum \dfrac{n!}{2^n} (z + 1 - i)^n$,

(iii) $\sum \left(1 + \dfrac{2}{n} \right)^{n^2} z^n$,

(iv) $\sum n^{\ln n}\, z^n$.

Solution

(i) $\quad L^* = \lim\limits_{n\to\infty} \left| \dfrac{a_{n+1}}{a_n} \right| = \lim\limits_{n\to\infty} \dfrac{[(n+1)!]^2}{(2n+2)!} \dfrac{(2n)!}{(n!)^2} = \lim\limits_{n\to\infty} \dfrac{(n+1)^2}{(2n+2)(2n+1)} = \dfrac{1}{4}.$

Therefore, radius of convergence $= R = 1/L^* = 4$. Since the centre of the power series is $z_0 = 0$, the circle of convergence is $|z| = 4$.

(ii) $\quad L^* = \lim\limits_{n\to\infty} \left| \dfrac{a_{n+1}}{a_n} \right| = \lim\limits_{n\to\infty} \dfrac{(n+1)!}{2^{n+1}} \cdot \dfrac{2^n}{n!} = \lim\limits_{n\to\infty} \dfrac{n+1}{2} = \infty.$

Therefore, the radius of convergence is $R = 1/L^* = 0$. The power series is convergent only at its centre, that is, at $z_0 = -1 + i$.

(iii) $\quad L^* = \lim\limits_{n\to\infty} [\, |a_n|^{1/n}] = \lim\limits_{n\to\infty} \left[\left(1 + \dfrac{2}{n}\right)^{n^2} \right]^{1/n} = \lim\limits_{n\to\infty} \left(1 + \dfrac{2}{n}\right)^n = \lim\limits_{n\to\infty} \left[\left(1 + \dfrac{2}{n}\right)^{n/2} \right]^2 = e^2.$

Therefore, the radius of convergence is $R = 1/L^* = 1/e^2 = e^{-2}$. Since the centre of the power series is $z_0 = 0$, the circle of convergence is $|z| = e^{-2}$.

(iv) $\quad L^* = \lim\limits_{n\to\infty} |a_n|^{1/n} = \lim\limits_{n\to\infty} [n^{\ln n}]^{1/n}.$

Taking logarithms on both sides, we get

$$\ln L^* = \lim\limits_{n\to\infty} \dfrac{1}{n} \ln [n^{\ln n}] = \lim\limits_{n\to\infty} \dfrac{[\ln n]^2}{n}.$$

Using L' Hospitals rule, we obtain

$$\ln L^* = \lim\limits_{n\to\infty} \left[\dfrac{2 \ln n}{n} \right] = \lim\limits_{n\to\infty} \dfrac{2}{n} = 0.$$

Therefore, $L^* = e^0 = 1$ and radius of convergence is $R = 1/L^* = 1$. Since the centre of the power series is $z_0 = 0$, the circle of convergence is $|z| = 1$.

Example 12.29 Show that the following power series is convergent in the given region.

(i) $\sum \dfrac{z^{2n}}{4^n n^\alpha}$, $\alpha > 0$, $|z| < 2$,

(ii) $\sum \dfrac{n! z^n}{n^p}$, p a positive integer, $z = 0$,

(iii) $\sum \dfrac{z^{n^2}}{2^n}$, $|z| < 1$.

Solution In all the problems $z_0 = 0$. Let z_n denote the nth term of the power series.

(i) $\quad \left| \dfrac{z_{n+1}}{z_n} \right| = \left| \dfrac{4^n n^\alpha}{4^{n+1}(n+1)^\alpha} \cdot \dfrac{z^{2n+2}}{z^{2n}} \right| = \left| \left(\dfrac{n}{n+1} \right)^\alpha \cdot \dfrac{z^2}{4} \right|.$

Therefore, $\lim\limits_{n\to\infty} \left| \dfrac{z_{n+1}}{z_n} \right| = \dfrac{z^2}{4}.$

For convergence $|[z^2/4]| < 1$ or $|z| < 2$.

(ii) $\left| \dfrac{z_{n+1}}{z_n} \right| = \left| \dfrac{(n+1)! \, n^p}{(n+1)^p \, n!} \cdot \dfrac{z^{n+1}}{z^n} \right| = \left| (n+1) \left(\dfrac{n}{n+1} \right)^p z \right|.$

Therefore, $\displaystyle\lim_{n \to \infty} \left| \dfrac{z_{n+1}}{z_n} \right| = \infty$ and $R = 0$.

The series is convergent only at $z = 0$.

(iii) $\left| \dfrac{z_{n+1}}{z_n} \right| = \left| \dfrac{2^n \, z^{(n+1)^2}}{2^{n+1} \, z^{n^2}} \right| = \left| \dfrac{z^{2n+1}}{2} \right| = \dfrac{|z|^{2n+1}}{2}.$

Therefore, $\displaystyle\lim_{n \to \infty} \left| \dfrac{z_{n+1}}{z_n} \right| = 0$ when $|z| < 1$.

Hence, the series is convergent for $|z| < 1$.

Example 12.30 Find the radius of convergence of the series $\displaystyle\sum_{n=1}^{\infty} a_n (z - z_0)^n$, where

(i) $a_n = \dfrac{(-1)^{n+1} n}{5n^2 + 3}$, (ii) $a_n = \left(\dfrac{3n+4}{5n+6} \right)^n$.

Solution

(i) $R = \displaystyle\lim_{n \to \infty} \left| \dfrac{a_n}{a_{n+1}} \right| = \lim_{n \to \infty} \left| \dfrac{(-1)^{n+1} n}{5n^2 + 3} \cdot \dfrac{5(n+1)^2 + 3}{(-1)^{n+2} (n+1)} \right| = 1.$

Therefore, the series converges when $|z - z_0| < 1$ and diverges when $|z - z_0| > 1$.

(ii) $R = \dfrac{1}{\displaystyle\lim_{n \to \infty} \{ |a_n|^{1/n} \}} = \lim_{n \to \infty} \left(\dfrac{3n+4}{5n+6} \right)^{-1} = \dfrac{5}{3}.$

Therefore, the series converges, when $|z - z_0| < 5/3$ and diverges when $|z - z_0| > 5/3$.

Example 12.31 Find the radius of convergence of the series $\displaystyle\sum_{n=1}^{\infty} (n + a^n) z^n$, where a is a real or a complex constant.

Solution We can consider the series as the sum of two series $\sum n z^n$ and $\sum a^n z^n$. We find that the radius of convergence of the first series is $R_1 = 1$ and that of the second series is $R_2 = 1/|a|$. Hence, the radius of convergence of the given series is

$$R = \min(R_1, R_2) = \min(1, 1/|a|)$$

If $|a| \le 1$, $R = 1$ and if $|a| > 1$, $R = 1/|a|$.

Example 12.32 Find the radius of convergence of the power series $\displaystyle\sum_{n=0}^{\infty} \dfrac{1}{n!} \left(\dfrac{iz - 1}{2 + i} \right)^n$, and determine the analytic function $f(z)$ that the series represents inside its circle of convergence.

Solution We have

$$\sum \dfrac{1}{n!} \left(\dfrac{iz - 1}{2 + i} \right)^n = \sum \dfrac{1}{n!} \dfrac{i^n}{(2 + i)^n} \left(z - \dfrac{1}{i} \right)^n = \sum \dfrac{1}{n!} \left(\dfrac{i}{2 + i} \right)^n (z + i)^n$$

$$= e^{i(z + i)/(2 + i)}.$$

Therefore, the series represents the function $f(z) = e^{i(z+i)/(2+i)}$. Using the ratio test, we obtain

$$L^* = \lim_{n \to \infty} \left| \frac{a_{n+1}}{a_n} \right| = \lim_{n \to \infty} \left| \frac{i}{2+i} \cdot \frac{1}{n+1} \right| = 0.$$

Hence, radius of convergence $R = 1/L^* = \infty$. The series is convergent in the whole of complex plane.

Exercises 12.3

Find the radius of convergence of the following power series.

1. $\sum n^n z^n$.

2. $\sum \dfrac{3^n}{4^n + 5^n} z^n$.

3. $\sum (n + 1 + 2^n) z^n$.

4. $\sum \dfrac{e^{-n}}{n!} (iz - 1)^n$.

5. $\sum \dfrac{4^n n(n+1)}{9^n} z^{2n-1}$.

6. $\sum \dfrac{(-1)^n}{\sqrt{n}} (2z + 1)^n$.

7. $\sum \left(1 - \dfrac{\pi}{n} \right)^{n^2} z^n$.

8. $\sum \left[\tan^{-1} \left(\dfrac{n+1}{n} \right) \right]^n z^n$.

9. $\sum 2^{\sqrt{n}} z^{2n}$.

10. $\sum (\log n)^n z^n$.

11. $\sum \dfrac{(1+i)^n}{n+2} (z + 3i)^n$.

12. $\sum \dfrac{i^n}{3^{n+1}} (z - i)^n$.

13. $\sum (\cos in) z^n$.

14. $\sum z^{n!}$.

15. $\sum z^{2^n}$.

If the radius of convergence of the power series $\sum c_n z^n$ is R, $0 < R < \infty$, find the radius of convergence of the following power series, where k is any natural number.

16. $\sum \dfrac{c_n}{n!} z^n$.

17. $\sum c_n^k z^n$.

18. $\sum c_n z^{kn}$.

19. $\sum n^k c_n z^n$.

20. $\sum n^n c_n z^n$.

Using the results on term by term differentiation/integration of a power series, reduce the following power series to series with simpler coefficients. Hence, or otherwise obtain the radius of convergence of the given series.

21. $\displaystyle\sum_{n=2}^{\infty} \dfrac{n(n-1)}{2^n} (z - 1)^{n-2}$.

22. $\displaystyle\sum_{n=0}^{\infty} \dfrac{2^n}{n(n+1)} z^{n+2}$.

23. $\displaystyle\sum_{n=0}^{\infty} \dfrac{(-3)^n (z - i)^{n+3}}{(n+1)(n+2)(n+3)}$.

24. Show that $1/(1 - z)^2 = \sum (n + 1)z^n$ and find its radius of convergence.

25. Using Cauchy product of two power series show that if

$$f(z) = \sum \frac{z^n}{n!} \quad \text{for all} \quad z, \text{ then} \quad [f(z)]^2 = f(2z).$$

12.5   Taylor Series

In the previous section we have stated the result that every power series with non-zero radius of convergence defines an analytic function. The converse of this result is also true, that is, every function which is analytic inside a certain circle $|z - z_0| < R$ can be expressed as a power series which converges to this function inside this circle. This series is called the *Taylor's series*. We shall now prove this result.

Theorem 12.26 (Taylor's theorem)   A function $f(z)$ that is analytic inside a circle $|z - z_0| = R$ may be represented inside this circle as a convergent power series

$$f(z) = \sum a_n (z - z_0)^n \text{ where } a_n = \frac{1}{n!} f^{(n)}(z_0).$$

Also, this series is unique.

Proof   Choose an arbitrary point z inside the circle $|z - z_0| = R$ and construct a circle C_r of radius $r < R$ centred at the point z_0 and containing the point z (Fig. 12.3). Since z is an interior point of the domain $|z - z_0| = r$ in which the function $f(z)$ is analytic, we obtain from the Cauchy integral formula

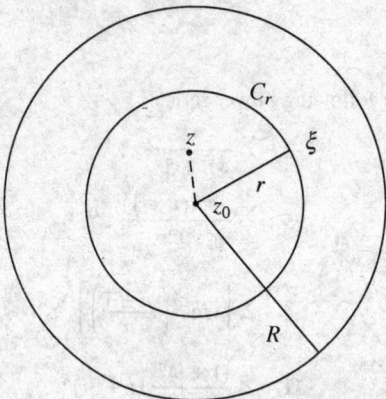

Fig. 12.3.   Taylor's theorem.

$$f(z) = \frac{1}{2\pi i} \oint_{C_r} \frac{f(\xi)}{\xi - z} \, d\xi \tag{12.19}$$

where ξ is a variable point on C_r. Since $\xi \neq z$, we can write

$$\frac{1}{\xi - z} = \frac{1}{(\xi - z_0) - (z - z_0)} = \frac{1}{\xi - z_0} \left[1 - \frac{z - z_0}{\xi - z_0} \right]^{-1}. \tag{12.20}$$

Let $u = (z - z_0)/(\xi - z_0)$. Since $|u| < 1$, we can expand the right hand side of Eq. (12.20) in binomial series as

$$\frac{1}{\xi - z} = \frac{1}{(\xi - z_0)} [1 + u + u^2 + \ldots + u^n] + T_n(z) \tag{12.21}$$

where

$$T_n(z) = \frac{1}{(\xi - z_0)} [u^{n+1} + u^{n+2} + \ldots]$$

$$= \frac{1}{(\xi - z_0)} \left[\frac{u^{n+1}}{1 - u} \right] = \frac{1}{\xi - z_0} \left[\frac{u^{n+1}}{1 - (z - z_0)/(\xi - z_0)} \right]$$

$$= \frac{u^{n+1}}{\xi - z_0} \left[\frac{\xi - z_0}{\xi - z} \right] = \frac{1}{\xi - z} \left[\frac{z - z_0}{\xi - z_0} \right]^{n+1}.$$

Substituting the expression for $1/(\xi - z)$ given in Eq. (12.21) into Eq. (12.19), we get

$$f(z) = \frac{1}{2\pi i} \oint_{C_r} \left[\frac{1}{\xi - z_0} \left\{ 1 + \left(\frac{z - z_0}{\xi - z_0} \right) + \ldots + \left(\frac{z - z_0}{\xi - z_0} \right)^n \right\} + T_n(z) \right] f(\xi) \, d\xi$$

$$= \frac{1}{2\pi i} \oint_{C_r} \frac{f(\xi)}{\xi - z_0} \, d\xi + \frac{z - z_0}{2\pi i} \oint_{C_r} \frac{f(\xi)}{(\xi - z_0)^2} \, d\xi + \cdots$$

$$+ \frac{(z - z_0)^n}{2\pi i} \oint_{C_r} \frac{f(\xi)}{(\xi - z_0)^{n+1}} \, d\xi + R_n(z) \tag{12.22}$$

where $\qquad R_n(z) = \dfrac{1}{2\pi i} \oint_{C_r} T_n(z) f(\xi) \, d\xi = \dfrac{(z - z_0)^{n+1}}{2\pi i} \oint_{C_r} \dfrac{f(\xi)}{(\xi - z_0)^{n+1}(\xi - z)} \, d\xi. \tag{12.23}$

Using the Cauchy integral formula for derivatives, we obtain from Eq. (12.22)

$$f(z) = f(z_0) + (z - z_0) f'(z_0) + \cdots + \frac{1}{n!}(z - z_0)^n f^{(n)}(z_0) + R_n(z). \tag{12.24}$$

This representation of $f(z)$ is called the *Taylor's formula* and $R_n(z)$ is called the remainder term. Since $f(z)$ is an analytic function, it has derivatives of all orders and we can take n as large as desired. If we let $n \to \infty$, we obtain from Eq. (12.24)

$$f(z) = f(z_0) + (z - z_0) f'(z_0) + \cdots + \frac{1}{n!}(z - z_0)^n f^{(n)}(z_0) + \cdots$$

or $\qquad f(z) = \displaystyle\sum_{n=0}^{\infty} a_n (z - z_0)^n, \quad \text{where} \quad a_n = \dfrac{1}{n!} f^{(n)}(z_0) \tag{12.25}$

provided $\lim\limits_{n \to \infty} R_n(z) = 0$.

The series given by Eq. (12.25) is called the Taylor's series expansion of $f(z)$ about the point z_0. Now we prove that $\lim\limits_{n \to \infty} R_n(z) = 0$.

Since, $\xi - z \neq 0$ and $f(z)$ is analytic within and on C_r, it follows that $f(\xi)/(\xi - z)$ is bounded. Therefore, there exists a positive real number M such that

$$\left| \frac{f(\xi)}{\xi - z} \right| \leq M \quad \text{for all} \quad \xi \text{ on } C_r.$$

We also have $|\xi - z_0| = r$ and the length of C_r is $2\pi r$. We have

$$|R_n(z)| = \frac{|z - z_0|^{n+1}}{|2\pi i|} \left| \oint_{C_r} \frac{f(\xi)}{(\xi - z_0)^{n+1}(\xi - z)} \, d\xi \right|$$

$$\leq \frac{|z - z_0|^{n+1}}{2\pi} \oint_{C_r} \frac{|f(\xi)|}{|\xi - z_0|^{n+1} |\xi - z|} \, |d\xi|$$

Using the *ML*-inequality, we get

$$|R_n(z)| \leq \frac{|z - z_0|^{n+1}}{2\pi} \cdot \frac{M}{r^{n+1}} (2\pi r) = Mr \left| \frac{z - z_0}{r} \right|^{n+1}.$$

Now, z is any point inside C_r. Therefore, $|(z - z_0)/r| < 1$. Hence,

$$|R_n(z)| \to 0 \text{ as } n \to \infty.$$

Since z is any point inside C_r, the series (12.25) converges for every point in the given domain. Hence, the convergence is uniform.

To prove the uniqueness of the expression given in Eq. (12.25), suppose that

$$f(z) = \Sigma \, b_n \, (z - z_0)^n \qquad (12.26)$$

where at least one coefficient $b_n \neq a_n$. The power series (12.26) converges in the circle $|z - z_0| < R$. Using the formula given in Eq. (12.25), we obtain

$$b_n = \frac{1}{n!} f^{(n)}(z_0)$$

which is same as a_n for all n. Hence, the Taylor series expansion of $f(z)$ is unique.

Remark 13

(a) If a function $f(z)$ is not analytic at a point $z = z_0$, it cannot be expanded in Taylor's series about the point $z = z_0$.

(b) Let a function $f(z)$ be analytic in a domain D except at the points $z_1, z_2, \ldots, z_n$. Suppose that $f(z)$ is to be expanded in Taylor's series about a point z_0 in D. The radius of convergence of the Taylor's series is the distance between z_0 and the nearest point at which $f(z)$ is not analytic.

(c) If $f(z)$ is analytic in a domain D and the point z_0 lies inside D, then the radius of convergence of the Taylor's series (Eq. (12.25)) is not less than the distance from the point z_0 to the boundary of the domain D.

(d) If M^* is the maximum value of $|f(z)|$ on the circle $C_r : |z - z_0| = r$, then we can write

$$|a_n| = \left| \frac{1}{2\pi i} \oint_{C_r} \frac{f(z)}{(z - z_0)^{n+1}} \, dz \right| \leq \frac{1}{2\pi} \oint_{C_r} \frac{|f(z)|}{|z - z_0|^{n+1}} \, |dz|.$$

Using the *ML*-inequality, we obtain

$$|a_n| \leq \frac{M^*}{2\pi} \cdot \frac{1}{r^{n+1}} (2\pi r) = \frac{M^*}{r^n}.$$

(e) If $z_0 = 0$ in Eq. (12.25), then

$$f(z) = \sum_{n=0}^{\infty} a_n z^n, \quad \text{where} \quad a_n = \frac{f^{(n)}(0)}{n!}.$$

This series is called the *Maclaurin's series*.

(f) Since the Taylor's series representation of an analytic function is unique, we can also obtain this expansion directly by using the binomial series expansions, whenever it is possible.

(g) Suppose f and g are analytic at $z = z_0$ and that $f(z_0) = g(z_0) = 0$, while $g'(z_0) \neq 0$. Then

$$\lim_{z \to z_0} \frac{f(z)}{g(z)} = \frac{f'(z_0)}{g'(z_0)}.$$

We have

$$\frac{f(z)}{g(z)} = \frac{f(z_0) + (z - z_0)f'(z_0) + \frac{1}{2!}(z - z_0)^2 \, f''(z_0) + \ldots}{g(z_0) + (z - z_0)\, g'(z_0) + \frac{1}{2!}(z - z_0)^2 \, g''(z_0) + \ldots}.$$

Therefore, $\displaystyle\lim_{z \to z_0} \frac{f(z)}{g(z)} = \lim_{z \to z_0} \frac{f'(z_0) + \frac{1}{2}(z - z_0)f''(z_0) + \cdots}{g'(z_0) + \frac{1}{2}(z - z_0)g''(z_0) + \cdots} = \frac{f'(z_0)}{g'(z_0)}.$

The following Maclaurin's series expansions for some standard functions can be easily verified

$$e^z = 1 + z + \frac{z^2}{2!} + \cdots + \frac{z^n}{n!} + \cdots; \quad R = \infty$$

$$\sin z = z - \frac{z^3}{3!} + \frac{z^5}{5!} - \cdots + (-1)^{n-1}\frac{z^{2n-1}}{(2n-1)!} + \cdots, \quad R = \infty$$

$$\cos z = 1 - \frac{z^2}{2!} + \frac{z^4}{4!} - \cdots + (-1)^{n-1}\frac{z^{2n-2}}{(2n-2)!} + \cdots, \quad R = \infty$$

$$\ln(1+z) = z - \frac{z^2}{2} + \frac{z^3}{3} - \cdots + (-1)^{n-1}\frac{z^n}{n} + \cdots, \quad R = 1$$

$$(1+z)^p = 1 + pz + \frac{p(p-1)}{2!}z^2 + \cdots + \frac{p(p-1)\cdots(p-n+1)}{n!}z^n + \cdots, \quad R = 1$$

for any real p.

Example 12.33   Expand the function $f(z) = 1/z$ about $z = 2$ in Taylor's series. Obtain its radius of convergence.

Solution   We have $z_0 = 2$. From Eq. (12.25), we get

$$f(z) = \frac{1}{z} = \sum_{n=0}^{\infty} a_n(z-2)^n, \text{ where } a_n = \frac{f^{(n)}(2)}{n!}.$$

From $f(z) = 1/z$, we get

$$f'(z) = -\frac{1}{z^2}, \; f''(z) = \frac{2}{z^3}, \ldots, \; f^{(n)}(z) = \frac{(-1)^n n!}{z^{n+1}}.$$

Therefore, we get $f^{(n)}(2) = \dfrac{(-1)^n n!}{2^{n+1}}$ and $a_n = \dfrac{(-1)^n}{2^{n+1}}.$

Hence, we obtain the series

$$\frac{1}{z} = \frac{1}{2} - \frac{1}{2^2}(z-2) + \frac{1}{2^3}(z-2)^2 - \cdots + \frac{(-1)^n}{2^{n+1}}(z-2)^n + \cdots.$$

Now, we have

$$L^* = \lim_{n \to \infty}\left|\frac{a_{n+1}}{a_n}\right| = \lim_{n \to \infty}\left|\frac{(-1)^{n+1}}{2^{n+2}} \cdot \frac{2^{n+1}}{(-1)^n}\right| = \frac{1}{2}.$$

Therefore, radius of convergence is $R = 1/L^* = 2$.

Alternative   We can obtain the series expansion as

$$\frac{1}{z} = \frac{1}{2+z-2} = \frac{1}{2}\left[1 + \frac{z-2}{2}\right]^{-1} = \frac{1}{2} - \frac{z-2}{2^2} + \frac{(z-2)^2}{2^3} - \cdots$$

which converges when $\left|\dfrac{z-2}{2}\right| < 1$, or $|z-2| < 2$.

Example 12.34 Obtain the Taylor series expansion of $f(z) = e^z$, (i) about the point $z = 0$, (ii) about the point $z = 2$. Show that the radius of convergence in each case is $R = \infty$.

Solution We have $f(z) = e^z$ and $f^{(n)}(z) = e^z$, $n = 1, 2, \ldots$. From Eq. (12.25), we get

$$f(z) = \sum_{n=0}^{\infty} a_n (z - z_0)^n, \quad a_n = \frac{f^{(n)}(z_0)}{n!} = \frac{e^{z_0}}{n!}.$$

(i) Here, $z_0 = 0$ and $a_n = \dfrac{f^{(n)}(0)}{n!} = \dfrac{1}{n!}$.

Therefore, $e^z = 1 + z + \dfrac{z^2}{2!} + \ldots + \dfrac{z^n}{n!} + \ldots$.

Now, $\quad L^* = \lim\limits_{n \to \infty} \left| \dfrac{a_{n+1}}{a_n} \right| = \lim\limits_{n \to \infty} \left| \dfrac{n!}{(n+1)!} \right| = \lim\limits_{n \to \infty} \left| \dfrac{1}{n+1} \right| = 0$. Hence, $R = \infty$.

(ii) Here, $z_0 = 2$ and $a_n = \dfrac{e^2}{n!}$.

Therefore, $e^z = e^2 \left[1 + (z - 2) + \dfrac{(z-2)^2}{2!} + \ldots + \dfrac{(z-2)^n}{n!} + \ldots \right]$.

Now, $L^* = \lim\limits_{n \to \infty} \left| \dfrac{a_{n+1}}{a_n} \right| = \lim\limits_{n \to \infty} \left| \dfrac{n!}{(n+1)!} \right| = 0$. Hence, $R = \infty$.

Remark 14

Since e^z is an entire function, it possesses Taylor's series expansion about any point and converges for all z.

Example 12.35 Obtain the Taylor series expansion of

$$f(z) = \frac{1}{z^2 + (1 + 2i)z + 2i} \quad \text{about} \quad z = 0.$$

Also find its radius of convergence.

Solution We have

$$f(z) = \frac{1}{z^2 + (1 + 2i)z + 2i} = \frac{1}{(z + 2i)(z + 1)} = \frac{1}{(1 - 2i)} \left[\frac{1}{z + 2i} - \frac{1}{z + 1} \right].$$

The function is analytic at $z = 0$. We have

$$\frac{1}{z + 2i} = \frac{1}{2i} \left[1 + \frac{z}{2i} \right]^{-1} = \frac{1}{2i} \sum_{n=0}^{\infty} (-1)^n \left(\frac{z}{2i} \right)^n, \quad \left| \frac{z}{2i} \right| < 1, \text{ or } |z| < 2, R_1 = 2$$

and $\qquad \dfrac{1}{z + 1} = [1 + z]^{-1} = \sum (-1)^n z^n, \quad |z| < 1, R_2 = 1.$

Hence, $$f(z) = \frac{1}{(1-2i)} \left[\frac{1}{2i} \sum_{n=0}^{\infty} (-1)^n \left(\frac{z}{2i} \right)^n - \sum_{n=0}^{\infty} (-1)^n z^n \right]$$

$$= \frac{1}{(1-2i)} \left[\sum_{n=0}^{\infty} \left\{ \left(\frac{1}{2i} \right)^{n+1} - 1 \right\} (-1)^n z^n \right].$$

The radius of convergence $R = \min (R_1, R_2) = 1$, which is the distance between the point $z = 0$ and the nearest point $z = -1$ at which the function $f(z)$ is not analytic.

Example 12.36   Obtain the terms upto z^3 in the Taylor series expansion of

$$f(z) = (z^2 + \sin^2 z)/(1 - \cos z)$$

about the point $z = 0$. Find its radius of convergence.

Solution   The function $f(z)$ is not analytic when $1 - \cos z = 0$, that is when $z = 2n\pi$, n any integer. At $z = 0$, $f(z)$ has a limiting value 4. We have

$$\sin z = z - \frac{z^3}{3!} + \frac{z^5}{5!} - \ldots, \quad \sin^2 z = z^2 - \frac{z^4}{3} + \frac{2}{45} z^6 - \ldots$$

$$\cos z = 1 - \frac{z^2}{2!} + \frac{z^4}{4!} - \ldots.$$

Therefore,

$$f(z) = \frac{z^2 + \sin^2 z}{1 - \cos z} = \frac{z^2 + [z^2 - (z^4/3) + (2z^6/45) - \ldots]}{1 - [1 - (z^2/2!) + (z^4/4!) - \ldots]}$$

$$= \frac{z^2 [2 - (z^2/3) + (2z^4/45) - \ldots]}{(z^2/2) [1 - (z^2/12) + (z^4/360) - \ldots]}$$

$$= 2 \left[2 - \frac{z^2}{3} + \frac{2z^4}{45} - \ldots \right] \left[1 - \left(\frac{z^2}{12} - \frac{z^4}{360} + \ldots \right) \right]^{-1}$$

$$= 2 \left[2 - \frac{z^2}{3} + \frac{2z^4}{45} - \ldots \right] \left[1 + \frac{z^2}{12} + \frac{3}{720} z^4 - \ldots \right]^{-1}$$

$$= 2 \left[2 - \frac{z^2}{6} + O(z^4) \right] = 4 - \frac{z^2}{3} + O(z^4).$$

The distance between $z = 0$ and the nearest points $z = \pm 2\pi$ at which the function $f(z)$ is not analytic is 2π. Therefore, $R = 2\pi$.

Example 12.37   Obtain the Taylor series expansion of the principal logarithm function $f(z) = \text{Ln } z$ about the point $z = (-1 + i)$. Obtain its radius of convergence.

Solution   We write

$$f(z) = \text{Ln } z = \text{Ln } [(-1 + i) + \{z - (-1 + i)\}] = \text{Ln } (-1 + i) + \text{Ln } \left[1 + \frac{z - (-1 + i)}{-1 + i} \right].$$

We can expand the logarithmic function and obtain the Taylor's series expansion.

Alternately, we can expand $f'(z) = 1/z$ in Taylor's series about the point $z = -1 + i$ and then integrate it term by term. We have

$$F(z) = f'(z) = \frac{1}{z} = \frac{1}{(-1 + i) + z - (-1 + i)} = \frac{1}{(i - 1)} \left[1 + \frac{z + (1 - i)}{i - 1}\right]^{-1}$$

$$= \frac{1}{(i - 1)} \sum_{n=0}^{\infty} \frac{(-1)^n [z + (1 - i)]^n}{(i - 1)^n}, \quad \left|\frac{z + (1 - i)}{i - 1}\right| < 1, \text{ or } |z + (1 - i)| < \sqrt{2}.$$

Integrating term by term, we obtain

$$f(z) = \text{Ln } z = \frac{1}{(i - 1)} \sum_{n=0}^{\infty} \frac{(-1)^n [z + (1 - i)]^{n+1}}{(n + 1)(i - 1)^n} + k$$

where k is the constant of integration. Substituting $z = i - 1$ on both sides, we get

$$k = \text{Ln } (i - 1) = \frac{1}{2} \ln 2 + i \tan^{-1}(-1) = \frac{1}{2} \ln 2 + \frac{3\pi i}{4}.$$

Hence, we obtain

$$\text{Ln } z = \left(\frac{1}{2} \ln 2 + \frac{3\pi i}{4}\right) + \frac{1}{(i - 1)} \sum_{n=0}^{\infty} \frac{(-1)^n [z + (1 - i)]^{n+1}}{(n + 1)(i - 1)^n}.$$

The point nearest to $i - 1$ at which $f(z) = \text{Ln } z$ is not analytic is $z = -1$. Therefore, the radius of convergence is the distance between the points $z = i - 1$ and $z = -1$. We obtain $R = 1$. Hence, the Taylor series expansion is valid in the open disk $|z + (1 - i)| < 1$.

Exercise 12.38   For the series

$$\cos z = 1 - \frac{z^2}{2!} + \frac{z^4}{4!} - \ldots + (-1)^n \frac{z^{2n}}{(2n)!} + \ldots$$

(i) show that the series converges uniformly in $|z| < r$ for any r,

(ii) find the bound on the remainder, if the series is terminated at the term $z^{2m}/(2m!)$.

Solution   Since $\cos z$ is analytic for all z, the radius of convergence is $R = \infty$. Hence, the series is uniformly convergent in $|z| < r$ for any r.

Let $z = x + iy$. We get

$$|\cos z| = \left|\frac{e^{iz} + e^{-iz}}{2}\right| \leq \frac{1}{2} |e^{iz}| + \frac{1}{2} |e^{-iz}| \leq \frac{1}{2} (e^y + e^{-y})$$

$$\leq \frac{1}{2} (e^{|y|} + 1) \leq \frac{1}{2} (e^r + 1).$$

The remainder in the series is given by (using Eq. (12.23))

$$|R_m(z)| = \left|\cos z - \sum_{k=0}^{m} \frac{(-1)^k z^{2k}}{(2k)!}\right| = \frac{|z^{2m+1}|}{2\pi} \left|\int_{C_r} \frac{\cos \xi}{\xi^{2m+1}(\xi - z)} d\xi\right|$$

where C_r is the circle of convergence and z is a point inside C_r.

Using the *ML*-inequality for $\xi = r_1 > r$, we obtain

$$|R_m(z)| \le \frac{1}{2\pi} r^{2m+1} \left[\frac{1}{2} (1 + e^{r_1}) \right] \frac{2\pi r_1}{(r_1 - r) r_1^{2m+1}}$$

$$= \frac{1}{2} (1 + e^{r_1}) \left(\frac{r}{r_1} \right)^{2m+1} \left(\frac{r_1}{r_1 - r} \right).$$

For example, for $r_1 = 2r$, we obtain

$$|R_m(z)| \le \frac{1 + e^{2r}}{2^{2m+1}}.$$

Example 12.39 Find the Maclaurin's series expansion of $[(1 - z)(1 + z^2)^2]^{-1}$.

Solution Let $f(z) = \dfrac{1}{1 - z}$ and $g(z) = \dfrac{1}{(1 + z^2)^2}$. We have

$$f(z) = [1 - z]^{-1} = 1 + z + z^2 + \ldots = \sum_{n=0}^{\infty} z^n; \; |z| < 1.$$

$$g(z) = [1 + z^2]^{-2} = 1 - 2z^2 + 3z^4 - \ldots = \sum_{n=0}^{\infty} (-1)^n (n + 1) z^{2n}; \; |z| < 1.$$

Using the Cauchy product formula for the two series, we get

$$f(z) g(z) = 1 + z - z^2 - z^3 + 2z^4 + 2z^5 - \ldots$$

which converges when $|z| < 1$.

Exercise 12.4

Obtain the Taylor's series expansions of the following functions about the point z_0. Find also the radius of convergence of each series.

1. $\sqrt{z + i}$, $z_0 = 0$.

2. $\left(\dfrac{z}{z + 1} \right)^2$, $z_0 = 1$.

3. $\dfrac{1}{2z + z^2}$, $z_0 = 1$.

4. $\dfrac{z}{z^2 - 4z + 13}$, $z_0 = 0$.

5. $\dfrac{1}{(z^2 - 1)(z^2 - 2)}$, $z_0 = 0$.

6. $\cos z$, $z_0 = \pi/2$.

7. $\sin z$, $z_0 = \pi$.

8. $\text{Ln} \, (2 + iz)$, $z_0 = i$.

9. $\text{Ln} \left(\dfrac{1 + z}{1 - z} \right)$, $z_0 = 0$.

10. $\sin (2z + z^2)$, $z_0 = -1$.

11. $\sinh z$, $z_0 = \pi i/2$.

12. $\cosh z$, $z_0 = \pi i$.

13. $(a + z)^b$, a, b constants, $b > 0$, $z_0 = 0$.

Obtain the first three non-zero terms in the Taylor's series expansion for the following functions about the point $z_0 = 0$.

14. $\tan z$.

15. $1/(2 + e^z)$.

16. $\sqrt{1 + 4z + 5z^2}$.

17. $z \cot z$.

18. $\text{Ln} \, (1 + \sqrt{1 - z})$.

19. $e^{z \cos z}$.

20. $\text{Ln} \, (1 + e^z)$.

21. $e^z/(1 + e^z)$.

22. $e^{z/(1-z)}$.

23. $\sin^{-1}z$.

24. $\sinh^{-1}z$.

25. $\sqrt{\cos z}$.

26. $(1 + z)^z$.

27. e^{e^z}.

Using the Taylor's series expansion of the integrand, evaluate the following integrals.

28. $\int_0^z e^{-\xi^2}\, d\xi$.

29. $\int_0^z e^{-2\xi}\, d\xi$.

30. $\int_0^z \dfrac{\sin \xi}{\xi}\, d\xi$.

31. $\int_0^z \dfrac{d\xi}{\sqrt{1 + \xi^2}}$.

32. $\int_0^z \sin \xi^2\, d\xi$.

33. Let $f(z)$ be analytic at $z = 0$ with $f(0) = 1$. Assume that $f(z)$ satisfies $[f(z)]^2 = f(2z)$ in some neighborhood of $z = 0$ and $f(z) \neq 1$. Write the Taylor's series $f(z) = \sum a_n z^n$ and show that

$$a_0 = 1,\ a_2 = \frac{1}{2!} a_1^2,\ a_3 = \frac{1}{3!} a_1^3, \ldots,\ a_n = \frac{1}{n!} a_1^n,\ \text{where } a_1 \text{ is arbitrary.}$$

34. If $|w| \leq 1/2$, show that (a) $|\text{Ln}(1 - w)| \leq 2|w|$ and   (b) $|\text{Ln}(1 - w) + w| \leq |w|^2$.

35. Show that an even function can have only even powers and an odd function can have only odd powers in its Taylor's series expansion about $z = 0$.

36. Compute the limit $\lim_{z \to 0}\left[(1 - e^z)^{-1} \int_C \dfrac{d\xi}{1 + \xi^2}\right]$, where C is the straight line path from $\xi = 0$ to $\xi = z$ and z is a point in the neighborhood of 0.

37. If $P(z)$ is an arbitrary polynomial of degree ≤ 3, find a_n such that

$$\frac{P(z)}{(z^2 + 1)(z - 1)(z - 2)} = \sum_{n=0}^{\infty} a_n z^n,\ |z| < 1.$$

38. Prove that the coefficients a_n in the expansion $\dfrac{1}{1 - z - z^2} = \sum_{n=0}^{\infty} a_n z^n$ satisfy the relation $a_n = a_{n-1} + a_{n-2}, n \geq 2$. Assuming the solution of this recurrence relation as $a_n = A\xi^n$, find a_n and the radius of convergence of the series.
(The numbers a_n are called the *Fibonacci numbers*.)

39. If the series expansion of $z/(e^z - 1)$ in powers of z is written as $\dfrac{z}{e^z - 1} = \sum_{n=0}^{\infty} \dfrac{1}{n!} B_n z^n$, then show that B_n satisfy the relation

$$B_0 = 1 \quad \text{and} \quad {}^{n+1}C_0 B_0 + {}^{n+1}C_1 B_1 + \ldots + {}^{n+1}C_n B_n = 0.$$

(The numbers B_n are called the *Bernoulli numbers*.)

40. If the expansion of the function $\sec z$ is written in the form $\sec z = \sum_{n=0}^{\infty} (-1)^n \dfrac{E_{2n}}{(2n)!} z^{2n}$, then show that the numbers E_{2n} satisfy the relation

$$E_0 = 1 \quad \text{and} \quad {}^{2n}C_0 E_0 + {}^{2n}C_2 E_2 + \ldots + {}^{2n}C_{2n} E_{2n} = 0.$$

(The numbers E_n are called the *Euler's numbers*.)

12.6   Laurent Series

It was shown in the previous section that if a function $f(z)$ is analytic in some domain D, then it can be expanded in Taylor's series about any point $z_0 \in D$. Suppose now, that the function $f(z)$ is not analytic at the points $z_0, z_1, \ldots, z_k \in D$. Then, we can expand the function $f(z)$ about any of these points, say z_0, in a series which contains both the positive and negative powers of $z - z_0$. This series is called the *Laurent's series*.

Consider a series of the form

$$\sum_{n=-\infty}^{\infty} c_n (z - z_0)^n = \sum_{n=0}^{\infty} c_n (z - z_0)^n + \sum_{n=1}^{\infty} \frac{c_{-n}}{(z - z_0)^n} \qquad (12.27)$$

where z_0 is a fixed point in the complex plane and the coefficients c_n are real or complex numbers. The region of convergence of the series given in Eq. (12.27) is the intersection of the regions of convergence of the two series

$$S_1 = \sum_{n=0}^{\infty} c_n (z - z_0)^n \text{ and } S_2 = \sum_{n=1}^{\infty} \frac{c_{-n}}{(z - z_0)^n}. \qquad (12.28)$$

The region of convergence of the series S_1 may be the entire complex plane or a circle of radius R with centre at z_0, that is $|z - z_0| = R$. Inside this circle of convergence, this series converges to some analytic function $f_1(z)$. Therefore, we have

$$f_1(z) = \sum_{n=0}^{\infty} c_n (z - z_0)^n, \ |z - z_0| < R. \qquad (12.29)$$

To determine the region of convergence of the series S_2, let $\xi = 1/(z - z_0)$. Then, we obtain another

series $S_2^* = \sum_{n=1}^{\infty} c_{-n} \xi^n$, which is a power series in ξ and converges to some analytic function $\phi(\xi)$ within its circle of convergence. Let its radius of convergence be $1/r$. Therefore, we have

$$\phi(\xi) = \sum_{n=1}^{\infty} c_{-n} \xi^n, \ |\xi| < \frac{1}{r} \qquad (12.30)$$

that is,
$$f_2(z) = \sum_{n=1}^{\infty} \frac{c_{-n}}{(z - z_0)^n}, \ |z - z_0| > r. \qquad (12.31)$$

Thus, the region of convergence of the series S_2 is the region exterior to the circle $|z - z_0| > r$. Let $r < R$. Then, the intersection of the two regions of convergence is $r < |z - z_0| < R$. Therefore, the series of positive and negative powers given by equation (12.27) converges in the annulus $r < |z - z_0| < R$, and converges to the function $f_1(z) + f_2(z)$ which is analytic in the annulus, that is

$$f(z) = f_1(z) + f_2(z) = \sum_{n=-\infty}^{\infty} c_n (z - z_0)^n, \ r < |z - z_0| < R. \qquad (12.32)$$

The series given in Eq. (12.32) is called the *Laurent series*. We state and prove the result in the following theorem.

Theorem 12.27 (Laurent series) Let a function $f(z)$ be analytic inside the annulus $r < |z - z_0| < R$ and on the bounding circles $C_1 : |z - z_0| = r$ and $C_2 : |z - z_0| = R$. Then, $f(z)$ can be represented by the Laurent series as

$$f(z) = c_0 + c_1(z - z_0) + c_2(z - z_0)^2 + \ldots + \frac{c_{-1}}{z - z_0} + \frac{c_{-2}}{(z - z_0)^2} + \ldots$$

$$= \sum_{n=0}^{\infty} c_n (z - z_0)^n + \sum_{n=1}^{\infty} \frac{c_{-n}}{(z - z_0)^n} = \sum_{n=-\infty}^{\infty} c_n (z - z_0)^n$$

where $\qquad c_n = \dfrac{1}{2\pi i} \oint_C \dfrac{f(\xi)}{(\xi - z_0)^{n+1}} \, d\xi$ and $c_{-n} = \dfrac{1}{2\pi i} \oint_C (\xi - z_0)^{n-1} f(\xi) \, d\xi$

or $\qquad c_n = \dfrac{1}{2\pi i} \oint_C \dfrac{f(\xi)}{(\xi - z_0)^{n+1}} \, d\xi, \; n = 0, \pm 1, \pm 2, \dots$ $\qquad$ (12.33)

and the circles C_1, C_2 and $C : |z - z_0| = \rho$, $r < \rho < R$ are traveresed in the anti-clockwise direction.

Proof Let C be the circle $|z - z_0| = \rho$, $r < \rho < R$ inside the annulus. Let z be an arbitrary fixed point inside the annulus $r < |z - z_0| < R$ and inside C (Fig. 12.4).

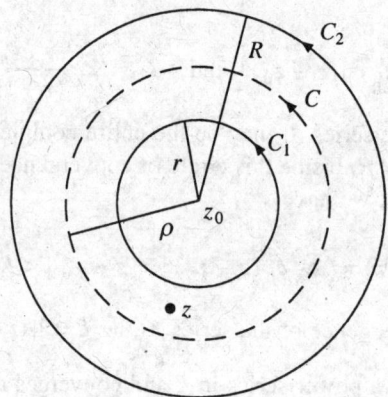

Fig. 12.4. The Laurent series.

Using the extension of the Cauchy integral formula for multiply connected domains, we obtain

$$f(z) = \frac{1}{2\pi i} \oint_{C_2} \frac{f(\xi)}{\xi - z} \, d\xi - \frac{1}{2\pi i} \oint_{C_1} \frac{f(\xi)}{\xi - z} \, d\xi = I_1 + I_2 \qquad (12.34)$$

where both C_1 and C_2 are traversed in the anti-clockwise direction. We also have

$$\left| \frac{z - z_0}{\xi - z_0} \right| < 1 \text{ for } \xi \text{ on } C_2 \text{ and } \left| \frac{\xi - z_0}{z - z_0} \right| < 1 \text{ for } \xi \text{ on } C_1.$$

Consider the integral along the path C_2. We have

$$\frac{1}{\xi - z} = \frac{1}{(\xi - z_0) - (z - z_0)} = \frac{1}{(\xi - z_0)} \left[1 - \frac{z - z_0}{\xi - z_0} \right]^{-1} = \frac{1}{(\xi - z_0)} [1 - u]^{-1}$$

where $|u| = |(z - z_0)/(\xi - z_0)| < 1$. Therefore,

$$\frac{1}{\xi - z} = \frac{1}{(\xi - z_0)} \left[1 + u + u^2 + \dots + u^n + \sum_{k=n+1}^{\infty} u^k \right] = \frac{1}{(\xi - z_0)} \left[\sum_{k=0}^{n} u^k \right] + T_n(z) \quad (12.35)$$

where $\qquad T_n(z) = \dfrac{1}{(\xi - z_0)} [u^{n+1} + u^{n+2} + u^{n+3} + \dots] = \dfrac{u^{n+1}}{(\xi - z_0)} [1 + u + u^2 + \dots]$

$$= \frac{u^{n+1}}{(\xi - z_0)} \left(\frac{1}{1 - u} \right) = \frac{u^{n+1}}{(\xi - z_0)} \left[\frac{1}{1 - \{(z - z_0)/(\xi - z_0)\}} \right] = \frac{u^{n+1}}{\xi - z}.$$

Integrating Eq. (12.35) term by term, we get

$$I_1 = \frac{1}{2\pi i} \oint_{C_2} \frac{f(\xi)}{\xi - z} d\xi$$

$$= \frac{1}{2\pi i} \oint_{C_2} \frac{f(\xi)}{\xi - z_0} d\xi + \frac{(z - z_0)}{2\pi i} \oint_{C_2} \frac{f(\xi)}{(\xi - z_0)^2} d\xi + \ldots$$

$$+ \frac{(z - z_0)^n}{2\pi i} \oint_{C_2} \frac{f(\xi)}{(\xi - z_0)^{n+1}} d\xi + R_n(z) \tag{12.36}$$

where

$$R_n(z) = \frac{(z - z_0)^{n+1}}{2\pi i} \oint_{C_2} \frac{f(\xi)}{(\xi - z_0)^{n+1}(\xi - z)} d\xi. \tag{12.37}$$

Since $\xi - z \neq 0$ and $f(z)$ is analytic in $r < |z - z_0| < R$, it follows that $f(\xi)/(\xi - z)$ is continuous on C_2 and hence bounded. That is, there exists a real positive number M^*, such that

$$\left| \frac{f(\xi)}{\xi - z} \right| \leq M^* \quad \text{for all} \quad \xi \text{ on } C_2.$$

Now, $|R_n(z)| = \left| \dfrac{(z - z_0)^{n+1}}{2\pi i} \oint_{C_2} \dfrac{f(\xi)}{(\xi - z_0)^{n+1}(\xi - z)} d\xi \right|$

$$\leq \frac{|(z - z_0)^{n+1}|}{2\pi} \frac{M^*}{R^{n+1}} (2\pi R) = \left| \frac{z - z_0}{R} \right|^{n+1} (M^* R) \to 0 \text{ as } n \to \infty$$

since $|(z - z_0)/R| < 1$.

Since the integrands in Eq. (12.36) are analytic in the annulus $r < |z - z_0| < R$, the value of the integrand does not change under an arbitrary deformation of path in the domain of analyticity of the integrand function. Therefore, we can integrate over C instead of C_2.

Hence, we obtain from Eq. (12.36)

$$I_1 = \frac{1}{2\pi i} \oint_C \frac{f(\xi)}{\xi - z_0} d\xi + \frac{z - z_0}{2\pi i} \oint_C \frac{f(\xi)}{(\xi - z_0)^2} d\xi + \ldots + \frac{(z - z_0)^n}{2\pi i} \oint_C \frac{f(\xi)}{(\xi - z_0)^{n+1}} d\xi + \ldots$$

$$= \sum_{n=0}^{\infty} c_n (z - z_0)^n \tag{12.38}$$

where $c_n = \dfrac{1}{2\pi i} \oint_C \dfrac{f(\xi)}{(\xi - z_0)^{n+1}} d\xi$.

Consider now, integration along the path C_1.

Since ξ is on C_1 and z is inside the annulus, we have $|(\xi - z_0)/(z - z_0)| < 1$.

Let $v = (\xi - z_0)/(z - z_0)$. Therefore, we can write

$$\frac{1}{\xi - z} = \frac{1}{(\xi - z_0) - (z - z_0)} = -\frac{1}{(z - z_0)} \left[1 - \frac{\xi - z_0}{z - z_0} \right]^{-1} = -\frac{1}{(z - z_0)} [1 - v]^{-1}$$

$$= -\frac{1}{(z - z_0)} [1 + v + v^2 + \ldots + v^n] + T_n^*(z) \tag{12.39}$$

where $T_n^*(z) = -\dfrac{1}{(z - z_0)}[v^{n+1} + v^{n+2} + \ldots] = -\dfrac{v^{n+1}}{(z - z_0)}[1 + v + v^2 + \ldots]$

$$= -\frac{v^{n+1}}{(z - z_0)}\left(\frac{1}{1 - v}\right) = -\frac{v^{n+1}}{(z - z_0)}\left[\frac{1}{1 - \{(\xi - z_0)/(z - z_0)\}}\right] = -\frac{v^{n+1}}{z - \xi}.$$

Integrating Eq. (12.39) term by term, we get

$$I_2 = -\frac{1}{2\pi i}\oint_{C_1}\frac{f(\xi)}{\xi - z}\,d\xi$$

$$= \frac{1}{2\pi i}\oint_{C_1}\frac{f(\xi)}{z - z_0}\,d\xi + \frac{1}{2\pi i}\oint_{C_1}\frac{(\xi - z_0)f(\xi)}{(z - z_0)^2}\,d\xi + \ldots$$

$$+ \frac{1}{2\pi i}\oint_{C_1}\frac{(\xi - z_0)^n f(\xi)}{(z - z_0)^{n+1}}\,d\xi + R_n^*(z) \tag{12.40}$$

where $$R_n^*(z) = \frac{1}{2\pi i}\oint_{C_1}\left(\frac{\xi - z_0}{z - z_0}\right)^{n+1}\frac{f(\xi)}{z - \xi}\,d\xi.$$

Since $f(z)$ is analytic in the annulus $r < |z - z_0| < R$ and $z - \xi \neq 0$, it follows that $f(\xi)/(z - \xi)$ is continuous and hence bounded on C_1. Therefore, there exists a real positive number N, such that

$$\left|\frac{f(\xi)}{z - \xi}\right| \le N \quad \text{for all} \quad \xi \quad \text{on} \quad C_1.$$

Hence, $$|R_n^*(z)| \le \frac{N}{2\pi}\left|\left(\frac{r}{z - z_0}\right)^{n+1}\right|(2\pi r) \to 0 \text{ as } n \to \infty,$$

since $|r/(z - z_0)| < 1$.

Since the integrands in Eq. (12.40) are analytic in the annulus $r < |z - z_0| < R$, by the principle of deformation of path, we can integrate along any contour C in the domain of analyticity of the integrand function instead of C_1. Therefore, we obtain from Eq. (12.40)

$$I_2 = \frac{1}{2\pi i}\oint_C\frac{f(\xi)}{z - z_0}\,d\xi + \frac{1}{2\pi i}\oint_C\frac{(\xi - z_0)}{(z - z_0)^2}f(\xi)\,d\xi + \ldots + \frac{1}{2\pi i}\oint_C\frac{(\xi - z_0)^{n-1}}{(z - z_0)^n}f(\xi)\,d\xi + \ldots$$

$$= \sum_{n=1}^{\infty}\frac{c_{-n}}{(z - z_0)^n} \tag{12.41}$$

where $$c_{-n} = \frac{1}{2\pi i}\oint_C(\xi - z_0)^{n-1}f(\xi)\,d\xi.$$

Combining equations (12.38) and (12.41), we obtain

$$f(z) = \sum_{n=0}^{\infty}c_n(z - z_0)^n + \sum_{n=1}^{\infty}\frac{c_{-n}}{(z - z_0)^n} = \sum_{n=-\infty}^{\infty}c_n(z - z_0)^n. \tag{12.42}$$

Remark 15

(a) The coefficients of the positive powers of $z - z_0$ in the Laurent series cannot be replaced by the derivative expressions $f^{(n)}(z_0)/n!$, although they are identical in form as in Taylor's series, since $f(z)$ is not analytic throughout the region inside C_2 and the Cauchy integral formula for derivatives cannot be used.

(b) In the Laurent series, let $r \to 0$. Then, $f(z)$ is analytic in $|z - z_0| < R$ except at the point $z = z_0$. Therefore, the region of convergence of the Laurent series in this case is $0 < |z - z_0| < R$. If $f(z)$ is analytic at z_0 also, then the Laurent series is same as the Taylor's series.

(c) In the Laurent series, let $R \to \infty$. Then, the region of convergence of the Laurent series is $|z - z_0| > r$.

(d) Laurent series expansion of a function $f(z)$ in the annulus $r < |z - z_0| < R$ is unique. However, $f(z)$ may have different Laurent series valid in different annuli about the same centre. The number of Laurent series, that a function $f(z)$ can have, depends on the number of points at which the function is not analytic. For example, suppose $f(z)$ is not analytic at the points z_0, $z_1, \ldots, z_n$ with $|z_0| < |z_1| < \ldots < |z_n|$ and we seek Laurent series expansions about $z = z_0$. Let $|z_1 - z_0| = R_1, |z_2 - z_0| = R_2, \ldots, |z_n - z_0| = R_n$ and R_i's are distinct. Then, $n + 1$ different Laurent series can be obtained in the regions, $0 < |z - z_0| < R_1$, $R_1 < |z - z_0| < R_2, \ldots,$ $R_{n-1} < |z - z_0| < R_n$ and $|z - z_0| > R_n$.

Let $z = \xi$ be any point at which $f(z)$ is analytic. Of the points $z_0, z_1, \ldots, z_n$ at which $f(z)$ is not analytic, let z_0 be the nearest point to ξ. Let $|z_0 - \xi| = R_0, |z_1 - \xi| = R_1, \ldots, |z_n - \xi| = R_n$ and R_i's be distinct. Then, we have a Taylor's series expansion valid in $|z - \xi| < R_0$ and $n + 1$ Laurent series expansions valid in the annuli $R_0 < |z - \xi| < R_1, \ldots, R_{n-1} < |z - \xi| < R_n$ and $|z - \xi| > R_n$. Therefore, the number of Laurent series expansions that a function $f(z)$ can have about any point is at most equal to the number of points at which $f(z)$ is not analytic in the finite complex plane. The number of Laurent series expansions would be less if some of the points z_i have the same magnitude, that is they lie on the same circle.

For example, if $f(z)$ is not analytic at three points z_0, z_1 and z_2, where $|z_0| < |z_1| < |z_2|$, then we obtain three Laurent series expansions about $z = z_0$ as shown in Fig. 12.5.

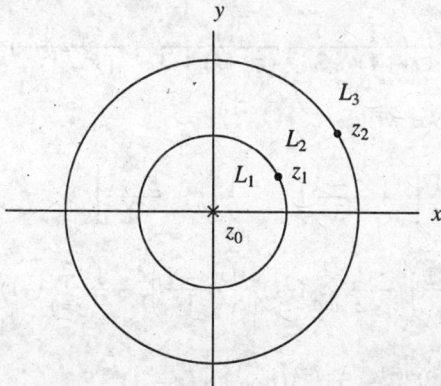

Fig. 12.5. Laurent series expansions.

(e) If $|f(z)| < M$ for all z in the annulus $r < |z - z_0| < R$, then

$$|c_n| \le \frac{M}{2\pi \rho^{n+1}} (2\pi\rho) = \frac{M}{\rho^n}, \quad n = 0, \pm 1, \pm 2, \ldots \quad (12.43)$$

valid for all ρ, $r < \rho < R$.

(f) It is difficult to obtain the Laurent series expansions using Eq. (12.33). Since the Laurent series is unique, it can be obtained by various other methods like binomial expansions etc.

Example 12.40   Find all possible Taylor's and Laurent series expansions for the function $f(z) = 1/(1 - z)$ about $z = 0$.

Solution   The function $f(z) = 1/(1-z)$ is not analytic at $z = 1$. Therefore, we consider the two regions (i) $|z| < 1$ and (ii) $|z| > 1$. The given function $f(z) = 1/(1-z)$ is analytic inside $|z| < 1$. Hence, we obtain a Taylor's series expansion in this region. We have the Taylor's series as

$$\frac{1}{1-z} = (1-z)^{-1} = 1 + z + z^2 + \ldots + z^n + \ldots, \quad |z| < 1$$

In the region $|z| > 1$, we obtain a Laurent series expansion. We write

$$\frac{1}{1-z} = -\frac{1}{z}\left[1 - \frac{1}{z}\right]^{-1} = -\frac{1}{z}\left[1 + \frac{1}{z} + \frac{1}{z^2} + \ldots\right]$$

$$= -\left[\frac{1}{z} + \frac{1}{z^2} + \ldots + \frac{1}{z^n} + \ldots\right], \quad |z| > 1.$$

Example 12.41   Find all possible Taylor's and Laurent series expansions of the function $f(z) = 1/[(z + 1)(z + 2)^2]$ about the point $z = 1$.

Solution   The given function is not analytic at the points $z = -1$ and $z = -2$. The distances between the point $z = 1$ and the points $z = -1$, $z = -2$ are 2 and 3 respectively. Therefore, we consider the regions (i) $|z - 1| < 2$, (ii) $2 < |z - 1| < 3$ and (iii) $|z - 1| > 3$.

In the region, $|z - 1| < 2$, the function $f(z)$ is analytic. Therefore, we obtain a Taylor's series expansion in this region. In the other regions, we obtain Laurent series expansions. We write

$$f(z) = \frac{1}{(z+1)(z+2)^2} = \frac{1}{z+1} - \frac{1}{z+2} - \frac{1}{(z+2)^2}$$

$$= \frac{1}{(z-1)+2} - \frac{1}{(z-1)+3} - \frac{1}{[(z-1)+3]^2}.$$

(i) In the region $|z - 1| < 2$, we write

$$f(z) = \frac{1}{2}\left[1 + \frac{z-1}{2}\right]^{-1} - \frac{1}{3}\left[1 + \frac{z-1}{3}\right]^{-1} - \frac{1}{9}\left[1 + \frac{z-1}{3}\right]^{-2}$$

$$= \frac{1}{2}\sum_{n=0}^{\infty}(-1)^n \left(\frac{z-1}{2}\right)^n - \frac{1}{3}\sum_{n=0}^{\infty}(-1)^n \left(\frac{z-1}{3}\right)^n - \frac{1}{9}\sum_{n=0}^{\infty}(-1)^n(n+1)\left(\frac{z-1}{3}\right)^n$$

$$= \sum_{n=0}^{\infty}(-1)^n \left[2^{-n-1} - 3^{-n-1} - 3^{-n-2}(n+1)\right](z-1)^n$$

$$= \sum_{n=0}^{\infty}(-1)^n \left[2^{-n-1} - (n+4)3^{-n-2}\right](z-1)^n.$$

The first series is valid in $|z-1| < 2$ and the second and third series are valid in $|z-1| < 3$. Hence, the sum is valid in $|z-1| < 2$.

(ii) In the region $2 < |z-1| < 3$, we write

$$f(z) = \frac{1}{(z-1)} \left[1 + \frac{2}{z-1} \right]^{-1} - \frac{1}{3} \left[1 + \frac{z-1}{3} \right]^{-1} - \frac{1}{9} \left[1 + \frac{z-1}{3} \right]^{-2}$$

$$= \frac{1}{(z-1)} \left[\sum_{n=0}^{\infty} (-1)^n \left(\frac{2}{z-1} \right)^n \right] - \frac{1}{3} \sum_{n=0}^{\infty} (-1)^n \left(\frac{z-1}{3} \right)^n - \frac{1}{9} \sum_{n=0}^{\infty} \left[(-1)^n (n+1) \left(\frac{z-1}{3} \right)^n \right]$$

$$= \sum_{n=0}^{\infty} (-1)^n \frac{2^n}{(z-1)^{n+1}} - \frac{1}{9} \sum_{n=0}^{\infty} (-1)^n (n+4) \left(\frac{z-1}{3} \right)^n.$$

The first series is valid in $|z-1| > 2$ and the second series is valid in $|z-1| < 3$. Hence, the sum is valid in the annulus $2 < |z-1| < 3$.

(iii) In the region $|z-1| > 3$, we write

$$f(z) = \frac{1}{(z-1)} \left[1 + \frac{2}{z-1} \right]^{-1} - \frac{1}{(z-1)} \left[1 + \frac{3}{z-1} \right]^{-1} - \frac{1}{(z-1)^2} \left[1 + \frac{3}{z-1} \right]^{-2}$$

$$= \frac{1}{(z-1)} \sum_{n=0}^{\infty} (-1)^n \left(\frac{2}{z-1} \right)^n - \frac{1}{(z-1)} \sum_{n=0}^{\infty} (-1)^n \left(\frac{3}{z-1} \right)^n - \frac{1}{(z-1)^2} \sum_{n=0}^{\infty} (-1)^n (n+1) \left(\frac{3}{z-1} \right)^n$$

$$= \sum_{n=0}^{\infty} (-1)^n \left[\frac{(2^n - 3^n)}{(z-1)^{n+1}} - \frac{3^n (n+1)}{(z-1)^{n+2}} \right].$$

Example 12.42 Obtain the first three terms of the Laurent series expansion of the function $f(z) = 1/(e^z - 1)$ about the point $z = 0$ valid in the region $0 < |z| < 2\pi$.

Solution The given function is not analytic when $e^z = 1$, that is, at $z = 0$ and $z = 2n\pi i$, $n = \pm 1$, $\pm 2, \ldots$. The required Laurent series expansion is about the point $z = 0$. Its region of convergence is $0 < |z| < 2\pi$. We have

$$\frac{1}{e^z - 1} = \frac{1}{z + (z^2/2!) + \ldots + (z^n/n!) + \ldots} = \frac{1}{z} \left[1 + \left(\frac{z}{2!} + \frac{z^2}{3!} + \ldots \right) \right]^{-1}$$

$$= \frac{1}{z} \left[1 - \left(\frac{z}{2} + \frac{z^2}{6} + \frac{z^3}{24} + \ldots \right) + \left(\frac{z}{2} + \frac{z^2}{6} + \frac{z^3}{24} + \ldots \right)^2 \right.$$

$$\left. - \left(\frac{z}{2} + \frac{z^2}{6} + \frac{z^3}{24} + \ldots \right)^3 + \ldots \right]$$

$$= \frac{1}{z} \left[1 - \left(\frac{z}{2} + \frac{z^2}{6} + \frac{z^3}{24} + \ldots \right) + \left(\frac{z^2}{4} + \frac{z^3}{6} + \ldots \right) - \left(\frac{z^3}{8} + \ldots \right) \right]$$

$$= \frac{1}{z} \left[1 - \frac{z}{2} + \left(-\frac{1}{6} + \frac{1}{4} \right) z^2 + \left(-\frac{1}{24} + \frac{1}{6} - \frac{1}{8} \right) z^3 + \ldots \right] \approx \frac{1}{z} - \frac{1}{2} + \frac{z}{12}.$$

Example 12.43 Show that the function $f(z) = \text{Ln}\,[z/(z-1)]$ is analytic in the region $|z| > 1$. Obtain its Laurent series expansion about $z = 0$ valid in this region.

Solution The function $f(z) = \text{Ln}\,[z/(z-1)]$ is not analytic when

$$\text{Im}\left(\frac{z}{z-1}\right) = -\frac{y}{(x-1)^2 + y^2} = 0 \quad \text{and} \quad \text{Re}\left(\frac{z}{z-1}\right) = \frac{x(x-1) + y^2}{(x-1)^2 + y^2} \le 0.$$

These conditions are satisfied when $y = 0$ and $0 \le x \le 1$. Thus, the given function is analytic in the region $|z| > 1$. In the region $|z| > 1$, consider the function

$$f(z) = \frac{1}{z} - \frac{1}{z-1}$$

Then,

$$\int_C f(z)\, dz = \int_C \frac{dz}{z} - \int_C \frac{dz}{z-1} = \text{Ln}\left(\frac{z}{z-1}\right), \quad C : |z| = r > 1.$$

We write

$$f(z) = \frac{1}{z} - \frac{1}{z}\left[1 - \frac{1}{z}\right]^{-1} = \frac{1}{z} - \frac{1}{z}\left[1 + \frac{1}{z} + \dots + \frac{1}{z^n} + \dots\right]$$

$$= -\frac{1}{z^2} - \frac{1}{z^3} - \dots - \frac{1}{z^{n+1}} - \dots.$$

Integrating term by term, we obtain the Laurent series expansion as

$$\text{Ln}\left(\frac{z}{z-1}\right) = \frac{1}{z} + \frac{1}{2z^2} + \dots + \frac{1}{nz^n} + \dots + k$$

where k is the constant of integration. Letting $z \to \infty$, we obtain $k = 0$. Therefore, we get

$$\text{Ln}\left(\frac{z}{z-1}\right) = \frac{1}{z} + \frac{1}{2z^2} + \dots + \frac{1}{nz^n} + \dots.$$

which is valid in the region $1 < |z| < \infty$ or $|z| > 1$.

Alternately, we can write

$$\text{Ln}\left(\frac{z}{z-1}\right) = -\ln\left(1 - \frac{1}{z}\right) = \frac{1}{z} + \frac{1}{2z^2} + \dots + \frac{1}{nz^n} + \dots.$$

Example 12.44 Find the Laurent series expansion of the function $f(z) = e^{1/z}$ in the region $|z| > 0$ and deduce that

$$\frac{1}{\pi}\int_0^\pi e^{\cos\theta} \cos(\sin\theta - n\theta)\, d\theta = \frac{1}{n!}, \quad n = 0, 1, 2, \dots.$$

Solution The function $f(z) = e^{1/z}$ is not analytic at $z = 0$. In the region $|z| > 0$, we obtain the Laurent series expansion of $e^{1/z}$ as

$$e^{1/z} = 1 + \frac{1}{z} + \frac{1}{2! \, z^2} + \frac{1}{3! \, z^3} + \ldots + \frac{1}{n! \, z^n} + \ldots = 1 + \sum_{n=1}^{\infty} c_{-n} z^{-n}$$

where $c_{-n} = 1/n!$.

We also have from Eq. (12.33)

$$c_{-n} = \frac{1}{2\pi i} \int_C e^{1/\xi} \, \xi^{n-1} \, d\xi, \quad C : |\xi| = \rho, \rho > 0.$$

Let $\rho = 1$. Substituting $\xi = e^{i\theta}$, we obtain

$$c_{-n} = \frac{1}{2\pi} \int_{-\pi}^{\pi} e^{\cos \theta} e^{i(n\theta - \sin \theta)} d\theta = \frac{1}{n!}.$$

Comparing the real parts, we get

$$\frac{1}{2\pi} \int_{-\pi}^{\pi} e^{\cos \theta} \cos(n\theta - \sin \theta) d\theta = \frac{1}{n!}.$$

Since the integrand is an even function, we obtain the required result.

Exercise 12.5

Find all the Taylor's and Laurent series expansions of the following functions about the given point.

1. $1/(z - 3)$, $z_0 = 0$.
2. $z^2 e^{1/z}$, $z_0 = 0$.
3. $1/[z(z - 1)]$, $z_0 = 0$.
4. $1/(z^2 - 1)$, $z_0 = 1$.
5. $1/(z^2 + 1)$, $z_0 = i$.
6. $\tan z / z^2$, $z_0 = 0$.
7. $e^{1/z}/z^2$, $z_0 = 0$.
8. $(\sin z^2)/z^4$, $z_0 = 0$.
9. $\sin(1/z)$, $z_0 = 0$.
10. $e^z/(z + 1)$, $z_0 = -1$.
11. $(e^{z^2} - 1)/z^4$, $z_0 = 0$.
12. $(2z - 3)/(z^2 - 3z + 2)$, $z_0 = 0$.
13. $1/(z^2 + 1)^2$, $z_0 = i$.
14. $1/[(a - z)(b - z)]$, $|a| < |b|$, $z_0 = 0$.
15. $1/[z^2 + (3i - 1)z - 3i]$, $z_0 = 1$.
16. $1/[(z + 1)^2 (z + 3)]$, $z_0 = -1$.
17. $z(z + i)/(z^2 + 1)^2$, $z_0 = 0$.
18. $1/[(z + 2)(z^2 + 1)]$, $z_0 = -2$.
19. $1/[z(z^2 - 3z + 2)]$, $z_0 = 0$.
20. $(z^2 + 1)/[z(z^2 + 3z - 10)]$, $z_0 = 0$.
21. $(z + 3) \sin[1/(z - 2)]$, $z_0 = 2$.
22. $\sinh(1/z)/z^2$, $z_0 = 0$.

In the following problems, obtain the Laurent series expansions about the given point. Find also the region of convergence of the series.

23. $\operatorname{cosec} z$, $z_0 = n\pi$, n integer.
24. $e^{-1/(z-2)}$, $z_0 = 2$.
25. $\cos[z/(1 - z)]$, $z_0 = 1$.
26. $\sin[(z^2 - 6z)/(z - 3)^2]$, $z_0 = 3$.
27. $z^2 \cos[1/(z - 1)]$, $z_0 = 1$.
28. $e^{z/(z-2)}$, $z_0 = 2$.
29. $[1 + e^z]/(\sin z + z \cos z)$, $z_0 = 0$.
30. The series expansions of the functions $1/(1 - z)$ and $1/(z - 1)$ are

$$\frac{1}{1-z} = 1 + z + z^2 + \ldots; \quad \frac{1}{z-1} = \frac{1}{z}\left(1 + \frac{1}{z} + \frac{1}{z^2} + \ldots\right)$$

Adding we get $(1 + z + z^2 + \ldots) + \frac{1}{z}\left(1 + \frac{1}{z} + \frac{1}{z^2} + \ldots\right) = 0$.

Is this result true? If not give the reason.

31. Find the Laurent series expansion of $f(z) = 1/[2z^2 - iz]$ about $z_0 = 0$ which converges at $z = i$.

32. Find the Laurent series expansion of $f(z) = 1/(z^2 + 2iz + 3)$ about $z_0 = 0$ which converges at $z = 4i$.

33. Show that $\cosh\left(z + \frac{1}{z}\right) = a_0 + \sum\limits_{n=1}^{\infty} a_n\left(z^n + \frac{1}{z^n}\right)$, where $a_n = \frac{1}{2\pi}\int_0^{2\pi} \cos n\theta \cosh(2\cos\theta)\,d\theta$.

34. Show that the Laurent series expansion of the function $f(z) = e^{u(z-1/z)/2}, |z| > 0$ is $\sum\limits_{n=-\infty}^{\infty} c_n z^n$, where

$$c_n = \frac{1}{2\pi}\int_0^{2\pi} \cos(n\theta - u\sin\theta)\,d\theta$$

(Note that $c_n = J_n(u)$ is the Bessel function of the first kind.)

35. Show that $\sin\left(z + \frac{1}{z}\right) = a_0 + \sum\limits_{n=1}^{\infty}\left(z^n + \frac{1}{z^n}\right)a_n$, where $a_n = \frac{1}{2\pi}\int_0^{2\pi} \cos n\theta \sin(2\cos\theta)\,d\theta$.

12.7 Answers and Hints

Exercise 12.1

1. Converges to 1.
2. Converges to 0.
3. Converges to 1.
4. Converges to 0.
5. Diverges.

6. The sequence is unbounded and hence divergent.

7. $a_n = 1/[\sqrt{n+1} + \sqrt{n}]$, converges to 0.

8. Let $a_n = 1 + h_n$ where $h_n > 0$. From $a_n = \sqrt[n]{n}$, we obtain $n = a_n^n = (1 + h_n)^n = 1 + nh_n + \frac{1}{2}n(n-1)h_n^2 + \ldots + h_n^n > \frac{1}{2}n(n-1)h_n^2$ or $0 < h_n < \sqrt{2/(n-1)} < \varepsilon$, that is $n > 1 + 2/\varepsilon^2$. Hence, for $N > 1 + 2/\varepsilon^2$, $0 < h_n < \varepsilon$ for all $n \geq N$. Therefore, $h_n \to 0$ and $\lim \sqrt[n]{n} = 1$.

9. $a_n = 1/(n^{1/n})$, converges to 1.
10. $a_n \leq \frac{5^5}{5!}\left(\frac{5}{6}\right)^{n-5}$ if $n \geq 5$, converges to 0.

11. Since $\dfrac{1}{(n+n)^2} + \dfrac{1}{(n+n)^2} + \ldots + \dfrac{1}{(n+n)^2} < \dfrac{1}{(n+1)^2} + \ldots + \dfrac{1}{(n+n)^2} < \dfrac{1}{n^2} + \dfrac{1}{n^2} + \ldots + \dfrac{1}{n^2}$,

we get $\dfrac{n}{(2n)^2} < a_n < \dfrac{n}{n^2}$ or $\dfrac{1}{4n} < a_n < \dfrac{1}{n}$, converges to 0.

12. $a_n = \left(1 + \dfrac{1}{n}\right)^n = 1 + n \cdot \dfrac{1}{n} + \dfrac{n(n-1)}{2!}\dfrac{1}{n^2} + \ldots + \dfrac{n(n-1)(n-2)\ldots 1}{n!} \cdot \dfrac{1}{n^n}$

$= 1 + 1 + \dfrac{1}{2!}\left(1 - \dfrac{1}{n}\right) + \ldots + \dfrac{1}{n!}\left(1 - \dfrac{1}{n}\right)\left(1 - \dfrac{2}{n}\right)\ldots\left(1 - \dfrac{n-1}{n}\right)$

Therefore, $2 < a_n < 1 + 1 + \dfrac{1}{2!} + \dfrac{1}{3!} + \ldots + \dfrac{1}{n!} < 1 + 1 + \dfrac{1}{2} + \dfrac{1}{2^2} + \ldots + \dfrac{1}{2^{n-1}} < 3$.

Since $\{a_n\}$ is bounded and has unique limit, it converges. The unique limit lies between 2 and 3. This limit is denoted by e.

13. Converges to 0.   **14.** Diverges, has two limits ± 1.   **15.** Diverges.

16. Converges to 0.   **17.** Diverges.   **18.** Converges to 1.

19. Converges to $-i$.   **20.** Diverges, has four limits $\pm 1, \pm i$.

21. Converges to 1.   **22.** Converges to 0.   **23.** Converges to i.

24. $z_n = \dfrac{i}{2} z_{n-1} + 1 = \left(\dfrac{i}{2}\right)^n z_0 + 1 + \left(\dfrac{i}{2}\right) + \left(\dfrac{i}{2}\right)^2 + \ldots + \left(\dfrac{i}{2}\right)^{n-1}$, converges to $\dfrac{2}{2-i}$ or $\dfrac{2}{5}(2+i)$.

25. Not Cauchy convergent as $|a_n - a_m| = 0$ or 2.

26. Cauchy convergent as $|a_n - a_m| \le \dfrac{2}{N} < \varepsilon, N > \dfrac{2}{\varepsilon}$.

27. Not Cauchy convergent as $|a_{2n} - a_n| < (\sqrt{2} - 1)\sqrt{n} \not< \varepsilon$ for all n.

28. Not Cauchy convergent as $|a_{2n} - a_n| = \ln 2 \not< \varepsilon$.   **29.** Not Cauchy convergent as

$$|a_{2n} - a_n| = \frac{1}{2n+1} + \frac{1}{2n+3} + \ldots + \frac{1}{4n-1} > \frac{1}{4n-1} + \ldots + \frac{1}{4n-1} = \frac{n}{4n-1} > \frac{1}{4}.$$

30. Not Cauchy convergent as

$$|a_{2n} - a_n| = \frac{1}{3n+1} + \frac{1}{3n+4} + \ldots + \frac{1}{6n-2} > \frac{1}{6n-2} + \ldots + \frac{1}{6n-2} = \frac{n}{6n-2} > \frac{1}{6}.$$

31. $\lim\limits_{n\to\infty} f_n(x) = f(x)$, where $f(x) = 0$, $0 \le x < 1$ and $f(x) = 1$, $x = 1$. Therefore, the sequence is pointwise convergent. Let $\varepsilon > 0$ and $0 < x \le k < 1$. $|f_n(x) - f(x)| = x^n < \varepsilon$ or $n > [\ln(1/\varepsilon)]/[\ln(1/x)]$ which increases with x. Its maximum value is $m = [\ln(1/\varepsilon)]/[\ln(1/k)]$. For $N > m$, $|f_n(x) - f(x)| < \varepsilon$ for all $n > N$ and $0 < x \le k < 1$. Hence, the sequence is uniformly convergent on $[0, k]$, $k < 1$. As $x \to 1$, $m \to \infty$ and therefore, the sequence is not uniformly convergent on $[0, 1]$.

32. $\lim\limits_{n\to\infty} f_n(x) = f(x)$, where $f(x) = \pi/2$, $x > 0$ and $f(x) = 0$, $x = 0$. We have $|f_n(x) - f(x)| = |\tan^{-1} nx - (\pi/2)| < \varepsilon$ or $\cot^{-1} nx < \varepsilon$ or $n > (\cot \varepsilon)/x$. Now, $[\cot \varepsilon]/x$ decreases as x increases and its maximum value is $(\cot \varepsilon)/a$. For any integer $N > (\cot \varepsilon)/a$, $|f_n(x) - f(x)| < \varepsilon$, $x > 0$, $n > N$. Hence, $\{f_n(x)\}$ converges uniformly on $[a, b]$. Since $(\cot \varepsilon)/x \to \infty$ as $x \to 0$, the sequence $\{f_n(x)\}$ is not uniformly convergent on $[0, b]$.

33. $\lim\limits_{n\to\infty} f_n(x) = f(x) = 0$ for all $x \ge 0$. We have $|f_n(x) - f(x)| = nx e^{-nx^2}$. It attains its maximum at $x = 1/\sqrt{2n}$ and the maximum value is $\sqrt{n/(2e)}$ which tends to ∞ as $n \to \infty$. Hence the sequence $\{f_n\}$ is not uniformly convergent on $[0, a]$.

34. $\lim\limits_{n\to\infty} f_n(z) = f(z) = 0$ for all z. Now $|f_n(z) - f(z)| = 1/|1 + nz| < \varepsilon$ when $n > (1 - \varepsilon)/[\varepsilon|z|]$. Therefore, the sequence converges pointwise to 0 for $|z| > 2$. The maximum value of n occurs at $z = 2$. For $N \ge (1 - \varepsilon)/(2\varepsilon)$, $|f_n(z) - f(z)| < \varepsilon$, $n \ge N$. Hence, the sequence is uniformly convergent.

35. $f_n(z) = n^{-z} = e^{-z \log n}$ and $|f_n(z)| = e^{-x \log n} = n^{-x}$. Now, $\lim\limits_{n\to\infty} n^{-x} = 0$ when $x > 0$ or Re $z > 0$. For any real number $a > 0$, $n^{-a} < \varepsilon$ when $n > \varepsilon^{-1/a}$. For $N > \varepsilon^{-1/a}$, $|f_n(z) - f(z)| < \varepsilon$, $n \ge N$. Hence, the sequence is uniformly convergent in the region Re $z \ge a > 0$.

Exercise 12.2

1. $\lim a_n = 1 \neq 0$, divergent.

2. $\lim a_n = e^{-1} \neq 0$, divergent.

3. $\lim a_n = \pi/2 \neq 0$, divergent.

In problems **4** to **10**, use comparison test.

4. $|a_n| < 1$; $a_n < 1/n^2$, converges.

5. $|a_n| < \dfrac{1}{n(n+1)} < \dfrac{1}{n^2}$, convergent.

6. $n^n > 2^n$ for all $n > 2$; $\dfrac{1}{n^n} < \dfrac{1}{2^n}$, convergent.

7. Rationalise; $1/[n^p(\sqrt{n+1} + \sqrt{n})] < 1/[2n^{p+1/2}]$, convergent for $p > 1/2$.

8. Rationalize, $a_n = 2/[\sqrt{n^4+1} + \sqrt{n^4-1}] < 2/n^2$, convergent.

9. $1 + \dfrac{1}{2!} + \dfrac{1}{3!} + \ldots < 1 + \dfrac{1}{2} + \dfrac{1}{2^2} + \ldots$, convergent.

10. $a_n = \dfrac{1}{[\sqrt[4]{n^4+1} + n][\sqrt{n^4+1} + n^2]} < \dfrac{1}{4n^3}$, convergent.

In the following problems, let $\left|\dfrac{a_{n+1}}{a_n}\right| = l$ and $\lim |a_n|^{1/n} = l_1$.

11. $l = x$, $|x| < 1$: convergent; $|x| > 1$: divergent; $x = 1$: $a_n = \dfrac{1}{(n+1)\sqrt{n}} < \dfrac{1}{n^{3/2}}$, convergent;

$x = -1$: alternating series, absolutely convergent.

12. Converges for all x ($l = 0$).

13. Convergent.

14. $0 < x < 1$, convergent; $x \geq 1$, divergent, (use $\lim\limits_{n \to \infty} n^{1/n} = 1$).

15. Divergent.

16. $x < 1/e$, convergent; $x \geq 1/e$, divergent.

17. Convergent.

18. Convergent.

19. Divergent.

20. Diverges for all z ($l = \infty$).

21. Convergent ($l = 2/e < 1$).

22. Convergent ($l = 3/4 < 1$).

23. Convergent ($l = 0$).

24. $l = x$; convergent for $|x| < 1$, divergent for $|x| > 1$; for $x = \pm 1$, $a_n = \pm\dfrac{n^2-1}{n^2+1}$, $\lim a_n = \pm 1 \neq 0$, divergent.

25. Convergent ($l_1 = 1/2$).

26. Convergent ($l_1 = 1/2$).

27. Convergent ($l_1 = 0$).

28. $l = \left|\dfrac{z-i}{z+i}\right| < 1$ gives $\operatorname{Im} z > 0$.

29. $l = \left|\dfrac{iz-1}{2+i}\right| < 1$ gives $|z+i| < \sqrt{5}$.

30. $l = |z| < 1$.

31. $l = e^{-y} < 1$ gives $y > 0$.

32. $l = 2\left|\dfrac{1-z}{z}\right| < 1$ gives $\left|z - \dfrac{4}{3}\right| < \dfrac{2}{3}$.

33. Absolutely convergent.   **34.** Absolutely convergent.   **35.** Absolutely convergent.
36. Absolutely convergent.   **37.** Absolutely convergent.   **38.** Absolutely convergent.

39. Take $M_n = 1/n^2$ in Weierstrass M-test, uniformly convergent.

40. $S(x) = \lim\limits_{n \to \infty} S_n(x) = 0$ for all x. Now, $S_n(x)$ attains maximum value $1/2$ at $x = 1/n$ and $1/n \to 0$ as $n \to \infty$. Consider the interval $[a, b]$ containing $x = 0$. Let $M_n = \max |S_n(x) - S(x)| = \max\left|\dfrac{nx}{1 + n^2 x^2}\right| = \dfrac{1}{2}$ which does not tend to zero as $n \to \infty$. The given series is not uniformly convergent.

41. $\left|\dfrac{1}{n^2 + z^2}\right| \leq \dfrac{1}{n^2 - |z|^2} = \dfrac{1}{n^2 - R^2} < \varepsilon$ or $n > \sqrt{R^2 + 1/\varepsilon}$. Hence, series is uniformly convergent in $|z| < R$.

42. $|\sin z| = [\sinh^2 y + \sin^2 x]^{1/2}$. Therefore, $|f_n(z)| = \dfrac{[\sin^2 x + \sinh^2 y]^{1/2}}{n^2} \le \dfrac{[1 + \sinh^2 y]^{1/2}}{n^2} \le \dfrac{2}{n^2}$,

since $|z| \le 1$ gives $x^2 + y^2 \le 1$, that is $|y| \le 1$, $|x| \le 1$. Hence, the series is uniformly convergent by Weierstrass M-test as $\sum M_n = 2 \sum (1/n^2)$ is convergent.

Exercise 12.3

1. $R = 0$.   **2.** $R = 5/3$.

3. Write the given power series as the sum of the two power series $\sum (n + 1)z^n$ and $\sum 2^n z^n$. The radius of convergence of the first series is $R_1 = 1$ and that of the second series is $R_2 = 1/2$. The radius of convergence of the given series is $R = \min (R_1, R_2) = 1/2$.

4. $R = \infty$.   **5.** $R = 3/2$.   **6.** $|z + 1/2| < 1/2$, $R = 1/2$.

Use root test in problems **7** to **10**.

7. $R = e^{\pi}$.   **8.** $R = 4/\pi$.   **9.** $R = 1$.   **10.** $R = 0$

Use ratio test in problems **11** to **25**

11. $|z + 3i| < 1/\sqrt{2}$, $R = 1/\sqrt{2}$.   **12.** $|z - i| < 3$, $R = 3$.   **13.** $R = 1/e$.

14. $|z| < 1$. $R = 1$.   **15.** $|z| < 1$, $R = 1$.   **16.** ∞.

17. R^k.   **18.** $R^{1/k}$.   **19.** R.   **20.** 0.

21. Consider the power series $\sum (z - 1)^n/2^n$. Differentiating two times term by term, we obtain the given power series. $R = 2$.

22. Consider the power series $\sum 2^n z^n$. Integrating, term by term, two times, we obtain the given power series. $R = 1/2$.

23. Consider the power series $\sum (-3)^n (z - i)^n$. Integrating term by term three times, we obtain the given power series, $R = 1/3$.

24. Take $\dfrac{1}{1 - z} = (1 - z)^{-1} = 1 + z + z^2 + \dots$. Differentiating term by term, we obtain the given power series. $R = 1$.

25. $\sum [f(z)]^2 = \sum c_k z^k$, where

$$c_k = \frac{1}{k!} + \frac{1}{1!\,(k - 1)!} + \frac{1}{2!\,(k - 2)!} + \dots + \frac{1}{k!\,1!} = \frac{1}{k!}\,[^k C_0 + {}^k C_1 + \dots + {}^k C_k] = \frac{2^k}{k!}$$

Exercise 12.4

1. $f(z) = \dfrac{(1 + i)}{\sqrt{2}} \left[\displaystyle\sum_{r=0}^{\infty} {}^{(1/2)} C_r \left(\dfrac{z}{i} \right)^r \right]$, $R = 1$.

2. $f(z) = \dfrac{1}{4} [1 + 2(z - 1) + (z - 1)^2] \left[1 + \left(\dfrac{z - 1}{2} \right) \right]^{-2} = \dfrac{1}{4} \left[1 + \displaystyle\sum_{n=1}^{\infty} \dfrac{(-1)^n (n - 3)(z - 1)^n}{2^n} \right]$, $R = 2$.

3. $f(z) = \dfrac{1}{2} \left[\dfrac{1}{z} - \dfrac{1}{z + 2} \right] = \dfrac{1}{2} \displaystyle\sum_{n=0}^{\infty} (-1)^n [1 - (1/3)^{n+1}] (z - 1)^n$, $R = 1$.

4. $f(z) = \dfrac{i}{6} \left[\dfrac{B}{z - B} - \dfrac{A}{z - A} \right]$, $A = 2 + 3i$, $B = 2 - 3i$, $f(z) = \dfrac{i}{6} \displaystyle\sum_{n=0}^{\infty} \left[\dfrac{B^n - A^n}{13^n} \right] z^{n+1}$, $R = \sqrt{13}$,

5. $f(z) = \dfrac{1}{1-z^2} - \dfrac{1}{2-z^2} = \sum\limits_{n=0}^{\infty} \left[1 - \dfrac{1}{2^{n+1}}\right] z^{2n}, \; R = 1.$

6. $f(z) = -\sin\left(z - \dfrac{\pi}{2}\right) = \sum\limits_{n=1}^{\infty} \dfrac{(-1)^n [z - (\pi/2)]^{2n-1}}{(2n-1)!}, \; R = \infty.$

7. $f(z) = -\sin(z - \pi) = \sum\limits_{n=1}^{\infty} \dfrac{(-1)^n [z - \pi]^{2n-1}}{(2n-1)!}, \; R = \infty.$

8. $f(z) = \sum\limits_{n=1}^{\infty} \dfrac{(-1)^{n-1} i^n (z-i)^n}{n}$, Ln $(2 + iz)$ is analytic except when $x = 0$, $y \geq 2$. The distance between $z = i$ and the nearest point $z = 2i$ at which the function $f(z)$ is not analytic is 1, $R = 1$.

9. $f(z) = \text{Ln}(1 + z) - \text{Ln}(1 - z) = 2 \sum\limits_{n=1}^{\infty} \dfrac{z^{2n-1}}{2n-1}, \; R = 1.$

10. $f(z) = \sin[(z + 1)^2 - 1] = \sin T \cos 1 - \cos T \sin 1, \; T = (z + 1)^2$

$\qquad f(z) = \sum\limits_{n=0}^{\infty} \dfrac{\sin[(n\pi/2) - 1]}{n!} (z + 1)^{2n}, \; R = \infty.$

11. $f(z) = \sinh z = \dfrac{(e^z - e^{-z})}{2} = \dfrac{i}{2} [e^{(z - \pi i/2)} + e^{-(z - \pi i/2)}] = i \sum\limits_{n=0}^{\infty} \dfrac{[z - (\pi i/2)]^{2n}}{(2n)!}, \; R = \infty.$

12. $f(z) = \cosh z = \dfrac{e^z + e^{-z}}{2} = -\dfrac{1}{2} [e^{(z - \pi i)} + e^{-(z - \pi i)}] = -\sum\limits_{n=0}^{\infty} \dfrac{(z - \pi i)^{2n}}{(2n)!}, \; R = \infty.$

13. $f(z) = a^b \left(1 + \dfrac{z}{a}\right)^b = a^b \sum\limits_{n=0}^{\infty} {}^b C_n \left(\dfrac{z}{a}\right)^n, R = |a|.$ **14.** $z + \dfrac{z^3}{3} + \dfrac{2}{15} z^5.$

15. $1 - \dfrac{z}{3} - \dfrac{z^2}{18}.$ **16.** $1 + 2z + \dfrac{1}{2} z^2.$

17. $1 - \dfrac{1}{3} z^2 - \dfrac{1}{45} z^4.$ **18.** $\ln 2 - \dfrac{1}{4} z - \dfrac{3}{32} z^2.$

19. $1 + z + \dfrac{z^2}{2}.$ **20.** $\ln 2 + \dfrac{1}{2} z + \dfrac{1}{8} z^2.$

21. $\dfrac{1}{2} + \dfrac{1}{4} z - \dfrac{1}{48} z^3.$ **22.** $1 + z + \dfrac{3}{2} z^2.$

23. $z + \dfrac{z^3}{6} + \dfrac{3}{40} z^5.$ **24.** $\sinh^{-1} z = \ln(z + \sqrt{z^2 + 1}) \approx z - \dfrac{z^3}{6} + \dfrac{3}{40} z^5.$

25. $1 - \dfrac{1}{4} z^2 - \dfrac{1}{96} z^4.$ **26.** $(1 + z)^z = e^{z \ln(1+z)} \approx 1 + z^2 - \dfrac{1}{2} z^3.$

27. $e(1 + z + z^2).$ **28.** $\sum\limits_{n=0}^{\infty} \dfrac{(-1)^n z^{2n+1}}{n!(2n+1)}.$

29. $\sum\limits_{n=0}^{\infty} \dfrac{(-1)^n 2^n z^{n+1}}{(n+1)!}.$ **30.** $\sum\limits_{n=0}^{\infty} \dfrac{(-1)^n z^{2n+1}}{(2n+1)!(2n+1)}.$

31. $z + \sum\limits_{n=1}^{\infty} (-1)^n \dfrac{1 \cdot 3 \cdot 5 \dots (2n-1)}{2^n (n!)} \left(\dfrac{z^{2n+1}}{2n+1}\right).$ **32.** $\sum\limits_{n=0}^{\infty} \dfrac{(-1)^n}{(2n+1)!} \left(\dfrac{z^{4n+3}}{4n+3}\right).$

34. (a) $|\text{Ln}(1-w)| = \left| -w - \dfrac{w^2}{2} - \dfrac{w^3}{3} - \ldots \right| = |w| \left| 1 + \dfrac{w}{2} + \dfrac{w^2}{3} + \ldots \right|$

For $|w| \le 1/2$ and $n \ge 2$, $1 + \dfrac{w}{2} + \dfrac{w^2}{3} + \ldots \le 1 + \dfrac{1}{2}\left(\dfrac{1}{2}\right) + \dfrac{1}{3}\left(\dfrac{1}{2}\right)^2 + \ldots$

$$\le 1 + \dfrac{1}{2} + \dfrac{1}{2^2} + \ldots = \dfrac{1}{1-(1/2)} = 2.$$

(b) $|\text{Ln}(1-w) + w| = \left| -\dfrac{1}{2}w^2 - \dfrac{1}{3}w^3 - \ldots \right| \le \dfrac{1}{2}|w|^2 \left[1 + \dfrac{2}{3}|w| + \dfrac{2}{4}|w|^2 + \ldots \right]$

$$\le \dfrac{1}{2}|w|^2 \left[1 + \dfrac{2}{3}\left(\dfrac{1}{2}\right) + \dfrac{2}{4}\left(\dfrac{1}{2^2}\right) + \ldots \right]$$

$$\le \dfrac{1}{2}|w|^2 \left[1 + \dfrac{1}{2} + \dfrac{1}{2^2} + \ldots \right] = |w|^2.$$

35. Let $f(z) = a_0 + a_1 z + a_2 z^2 + \ldots$;

$f(z) = f(-z)$ gives $a_1 z + a_3 z^3 + \ldots = 0$ and $f(z) = -f(-z)$ gives $a_0 + a_2 z^2 + a_4 z^4 + \ldots = 0$.

36. $\displaystyle\int_0^z \dfrac{d\xi}{1+\xi^2} = z - \dfrac{z^3}{3} + \ldots$ and $1 - e^z = -z - \dfrac{1}{2!}z^2 - \ldots$, limit $= -1$.

37. $\dfrac{P(z)}{(z^2+1)(z-1)(z-2)} = \dfrac{A}{z+i} + \dfrac{B}{z-i} + \dfrac{C}{z-1} + \dfrac{D}{z-2} = \Sigma a_n z^n$,

$A = P(-i)/[2(3-i)]$, $B = P(i)/[2(3+i)]$, $C = -P(1)/2$, $D = P(2)/5$ and

$$a_n = -A(i)^{n+1} + B(-1)^n(i)^{n+1} - C - 2^{-(n+1)}D.$$

38. Cross multiplying and comparing, we get $a_n = a_{n-1} + a_{n-2}$, $n \ge 2$ with $a_0 = a_1 = 1$.

Let $a_n = A\xi^n$. Therefore, $\xi^2 - \xi - 1 = 0$ or $\xi = (1 \pm \sqrt{5})/2$.

and $a_n = c_1[(1+\sqrt{5})/2]^n + c_2[(1-\sqrt{5})/2]^n$. Using $a_0 = a_1 = 1$, we get

$$a_n = \dfrac{1}{\sqrt{5}}\left[\left(\dfrac{\sqrt{5}+1}{2}\right)^{n+1} + (-1)^n\left(\dfrac{\sqrt{5}-1}{2}\right)^{n+1} \right] \text{ and } R = \dfrac{\sqrt{5}-1}{2}.$$

39. Cross multiply and compare the coefficients of z and z^{n+1}.

40. Write $\sec z = 1/\cos z$. Write the Taylor's series expansion of $\cos z$. Cross multiply and compare the constant term and the coefficient of z^{2n}.

Exercise 12.5

Denote T.S : Taylor series, L.S: Laurent series.

1. T.S: $-\dfrac{1}{3}\displaystyle\sum_{n=0}^{\infty}\left(\dfrac{z}{3}\right)^n$, $|z| < 3$, L.S: $\displaystyle\sum_{n=0}^{\infty}\dfrac{3^n}{z^{n+1}}$, $|z| > 3$.

2. L.S: $\displaystyle\sum_{n=0}^{\infty}\dfrac{1}{n!\, z^{n-2}}$, $|z| > 0$.

3. L.S : $-\left[\dfrac{1}{z} + \displaystyle\sum_{n=0}^{\infty} z^n \right]$, $0 < |z| < 1$, L.S: $\displaystyle\sum_{n=0}^{\infty}\dfrac{1}{z^{n+2}}$, $|z| > 1$.

4. L.S: $\dfrac{1}{2(z-1)}\sum\limits_{n=0}^{\infty}(-1)^n\left(\dfrac{z-1}{2}\right)^n, 0<|z-1|<2;$ L.S: $\dfrac{1}{(z-1)^2}\sum\limits_{n=0}^{\infty}(-1)^n\left(\dfrac{2}{z-1}\right)^n, |z-1|>2.$

5. L.S: $-\dfrac{i}{2(z-i)}\sum\limits_{n=0}^{\infty}(-1)^n\left(\dfrac{z-i}{2i}\right)^n, 0<|z-i|<2;$ L.S: $\dfrac{1}{(z-i)^2}\sum\limits_{n=0}^{\infty}(-1)^n\left(\dfrac{2i}{z-i}\right)^n, |z-i|>2.$

6. L.S: $\dfrac{1}{z}+\dfrac{z}{3}+\dfrac{2z^3}{15}+\dfrac{17z^5}{315}+\dots, 0<|z|<\pi/2.$ **7.** L.S: $\sum\limits_{n=0}^{\infty}\dfrac{1}{n!\,z^{n+2}}, |z|>0.$

8. L.S: $\sum\limits_{n=1}^{\infty}\dfrac{(-1)^{n-1}z^{4n-6}}{(2n-1)!}, |z|>0.$ **9.** L.S: $\sum\limits_{n=0}^{\infty}(-1)^n\dfrac{1}{(2n+1)!}\left(\dfrac{1}{z}\right)^{2n+1}, |z|>0.$

10. L.S: $e^{-1}\sum\limits_{n=0}^{\infty}\dfrac{(z+1)^{n-1}}{n!}, |z+1|>0.$ **11.** L.S: $\sum\limits_{n=1}^{\infty}\dfrac{z^{2n-4}}{n!}, |z|>0.$

12. T.S: $-\sum\limits_{n=0}^{\infty}\left(1+\dfrac{1}{2^{n+1}}\right)z^n, |z|<1;$ L.S: $\sum\limits_{n=0}^{\infty}\left[\dfrac{1}{z^{n+1}}-\dfrac{z^n}{2^{n+1}}\right], 1<|z|<2$ L.S: $\sum\limits_{n=0}^{\infty}\dfrac{2^n+1}{z^{n+1}}, |z|>2.$

13. L.S: $-\dfrac{1}{4(z-i)^2}\sum\limits_{n=0}^{\infty}(-1)^n(n+1)\left(\dfrac{z-i}{2i}\right)^n, 0<|z-i|<2;$

L.S: $\sum\limits_{n=0}^{\infty}\dfrac{(-1)^n(n+1)(2i)^n}{(z-i)^{n+4}}, |z-i|>2.$

14. T.S: $\dfrac{1}{(a-b)}\sum\limits_{n=0}^{\infty}\left[\dfrac{a^{n+1}-b^{n+1}}{a^{n+1}b^{n+1}}\right]z^n, |z|<a;$ L.S: $\dfrac{1}{(a-b)}\sum\limits_{n=0}^{\infty}\left[\dfrac{z^n}{b^{n+1}}+\dfrac{a^n}{z^{n+1}}\right], |a|<|z|<|b|;$

L.S: $\dfrac{1}{(a-b)}\sum\limits_{n=0}^{\infty}\dfrac{(a^n-b^n)}{z^{n+1}}, |z|>|b|.$

15. L.S: $\dfrac{(1-3i)}{10}\left[\dfrac{1}{z-1}-\sum\limits_{n=0}^{\infty}\dfrac{(-1)^n(z-1)^n}{(1+3i)^{n+1}}\right], 0<|z-1|<\sqrt{10};$

L.S: $\dfrac{1}{(z-1)^2}\sum\limits_{n=0}^{\infty}\dfrac{(-1)^n(1+3i)^n}{(z-1)^n}, |z-1|>\sqrt{10}.$

16. L.S: $\sum\limits_{n=0}^{\infty}\dfrac{(-1)^n(z+1)^{n-2}}{2^{n+1}}, 0<|z+1|<2;$ L.S: $\sum\limits_{n=0}^{\infty}\dfrac{(-1)^n2^n}{(z+1)^{n+3}}, |z+1|>2.$

17. T.S: $\dfrac{1}{4}\sum\limits_{n=0}^{\infty}[(-1)^n+1-2(n+1)]\left(\dfrac{z}{i}\right)^n, |z|<1;$

L.S: $\sum\limits_{n=0}^{\infty}\left\{\dfrac{1}{4}[(-1)^n-1]\left(\dfrac{i}{z}\right)^{n+1}+\left(\dfrac{n+1}{2}\right)\dfrac{i^n}{z^{n+2}}\right\}, |z|>1.$

18. L.S: $\dfrac{1}{5(z+2)}-\dfrac{1}{(6i-8)}\sum\limits_{n=0}^{\infty}\left(\dfrac{z+2}{2+i}\right)^n+\dfrac{1}{(6i+8)}\sum\limits_{n=0}^{\infty}\left(\dfrac{z+2}{2-i}\right)^n, 0<|z+2|<\sqrt{5};$

L.S: $\dfrac{1}{5(z+2)}+\sum\limits_{n=0}^{\infty}\left[\dfrac{(2+i)^n}{4i-2}-\dfrac{(2-i)^n}{4i+2}\right]\left[\dfrac{1}{z+2}\right]^{n+1}, |z+2|>\sqrt{5}.$

19. L.S: $\dfrac{1}{2z} + \sum\limits_{n=0}^{\infty} [1 - 2^{-n-2}] z^n$, $0 < |z| < 1$;

L.S: $\dfrac{1}{2z} - \sum\limits_{n=0}^{\infty} z^{-n-1} - \dfrac{1}{4} \sum\limits_{n=2}^{\infty} 2^{-n} z^n$, $1 < |z| < 2$; L.S: $\sum\limits_{n=2}^{\infty} (2^{n-1} - 1) z^{-n-1}$, $|z| > 2$.

20. L.S: $-\dfrac{1}{10z} + \sum\limits_{n=0}^{\infty} \left[-\dfrac{5}{28} (2^{-n}) + \dfrac{26}{175} (-1)^n 5^{-n} \right] z^n$, $0 < |z| < 2$;

L.S: $-\dfrac{1}{10z} + \dfrac{5}{14} \sum\limits_{n=0}^{\infty} \dfrac{2^n}{z^{n+1}} + \dfrac{26}{175} \sum\limits_{n=0}^{\infty} (-1)^n \left(\dfrac{z}{5} \right)^n$, $2 < |z| < 5$;

L.S: $-\dfrac{1}{10z} + \sum\limits_{n=0}^{\infty} \left[\dfrac{26}{35} (-5)^n + \dfrac{5}{14} (2)^n \right] \dfrac{1}{z^{n+1}}$, $|z| > 5$.

21. L.S: $\sum\limits_{n=0}^{\infty} \dfrac{(-1)^n}{(2n+1)!} \left[\dfrac{1}{(z-2)^{2n}} + \dfrac{5}{(z-2)^{2n+1}} \right]$, $|z-2| > 0$.

22. L.S: $\sum\limits_{n=1}^{\infty} \dfrac{1}{(2n-1)! \, z^{2n+1}}$, $|z| > 0$.

23. Write $\sin z = \sin [n\pi + (z - n\pi)] = (-1)^n \sin (z - n\pi)$,

$\operatorname{cosec} z = (-1)^n \left[\dfrac{1}{z - n\pi} + \dfrac{z - n\pi}{6} + \dfrac{7}{360} (z - n\pi)^3 + \ldots \right]$, $0 < |z - n\pi| < \pi$.

24. $\sum\limits_{n=0}^{\infty} \dfrac{(-1)^n}{n! \, (z-2)^n}$, $|z-2| > 0$.

25. Write $f(z) = \cos \left(1 - \dfrac{1}{1-z} \right) = (\cos 1) \cos \left(\dfrac{1}{1-z} \right) + (\sin 1) \sin \left(\dfrac{1}{1-z} \right)$

$= (\cos 1) \sum\limits_{n=0}^{\infty} \dfrac{(-1)^n}{(2n)! \, (z-1)^{2n}} - (\sin 1) \sum\limits_{n=0}^{\infty} \dfrac{(-1)^n}{(2n+1)! \, (z-1)^{2n+1}}$, $|z-1| > 0$.

26. Write $f(z) = \sin \left[1 - \dfrac{9}{(z-3)^2} \right] = (\sin 1) \sum\limits_{n=0}^{\infty} \dfrac{(-1)^n}{(2n)!} \left(\dfrac{3}{z-3} \right)^{4n} - (\cos 1) \sum\limits_{n=1}^{\infty} \dfrac{(-1)^{n+1}}{(2n-1)!} \left(\dfrac{3}{z-3} \right)^{4n-2}$

$|z-3| > 0$.

27. $\sum\limits_{n=0}^{\infty} \dfrac{(-1)^n}{(2n)!} \left[\dfrac{(z-1)^2 + 2(z-1) + 1}{(z-1)^{2n}} \right]$, $|z-1| > 0$.

28. Write $f(z) = e^{1 + 2/(z-2)} = e \sum\limits_{n=0}^{\infty} \dfrac{1}{n!} \left(\dfrac{2}{z-2} \right)^n$, $|z-2| > 0$.

29. Write $f(z) = \dfrac{1 + (1 + z + z^2/2! + \ldots)}{(z - z^3/3! + \ldots) + z(1 - z^2/2! + \ldots)} = \dfrac{1}{2z} \left[2 + z + \dfrac{7}{6} z^2 + \ldots \right]$, $|z| > 0$.

30. No. The first series is valid for $|z| < 1$ and the second series is valid for $|z| > 1$. There is no common point where both the series are valid.

31. Function is not analytic at $z = 0$ and $z = i/2$. For convergence at $z = i$, we obtain the Laurent series expansion in the region $|z| > 1/2$. $f(z) = \sum\limits_{n=0}^{\infty} \dfrac{(i)^n}{2^{n+1} \, z^{n+2}}$.

32. Function is not analytic at $z = i$ and $z = -3i$. For convergence at $z = 4i$, we obtain the Laurent series expansion in the region $|z| > 3$. $f(z) = \frac{1}{4} \sum_{n=0}^{\infty} (i)^{n-1} \left[\frac{1 + (-1)^{n-1} 3^n}{z^{n+1}} \right]$.

33. Function is analytic in the region $|z| > 0$. We have

$$\cosh\left(z + \frac{1}{z}\right) = 1 + \frac{1}{2!}\left(z + \frac{1}{z}\right)^2 + \frac{1}{4!}\left(z + \frac{1}{z}\right)^4 + \ldots, \text{ which can be written as}$$

$$\cosh\left(z + \frac{1}{z}\right) = a_0 + \sum_{n=1}^{\infty} a_n\left(z^n + \frac{1}{z^n}\right).$$

From the Laurent series expansion, we have

$$\cosh\left(z + \frac{1}{z}\right) = \sum_{n=-\infty}^{\infty} a_n z^n \text{ where } a_n = \frac{1}{2\pi i} \oint_C \frac{\cosh(z + z^{-1})dz}{z^{n+1}}, C : |z| = r.$$

Taking $r = 1$ and $z = e^{i\theta}$, we obtain

$$a_n = \frac{1}{2\pi} \int_0^{2\pi} \cosh(2\cos\theta)\cos n\theta d\theta - \frac{i}{2\pi} \int_0^{2\pi} \cosh(2\cos\theta)\sin n\theta d\theta$$

The value of the second integral is zero. Hence, the result.

34. $c_n = \frac{1}{2\pi i} \oint_C \frac{e^{u(z - 1/z)/2}}{z^{n+1}} dz$. Let $C : |z| = 1$ and $z = e^{i\theta}$. We obtain

$$c_n = \frac{1}{2\pi} \int_0^{2\pi} \cos(u\sin\theta - n\theta)d\theta + \frac{i}{2\pi} \int_0^{2\pi} \sin(u\sin\theta - n\theta)d\theta.$$

The value of the second integral is zero. Hence, the result.

35. We have $a_n = \frac{1}{2\pi i} \oint_C \frac{\sin(z + z^{-1})dz}{z^{n+1}}$, $C : |z| = r$. Taking $r = 1$ and $z = e^{i\theta}$, we obtain

$$a_n = \frac{1}{2\pi} \int_0^{2\pi} \sin(2\cos\theta)\cos n\theta\, d\theta - \frac{i}{2\pi} \int_0^{2\pi} \sin(2\cos\theta)\sin n\theta\, d\theta.$$

The value of the second integral is zero. Hence, the result.

Chapter 13

Zeros, Singularities and Residues

13.1 Introduction

In this chapter, we shall define zeros and singularities of complex functions. There are different types of singularities which can be classified using the Laurent series. Residue of a function $f(z)$ at a singularity is also defined using the Laurent series. The concept of residues has extensive applications. One application of the residue theorem is the evaluation of complex integrals and also evaluation of certain real integrals.

13.2 Zeros and Singularities of Complex Functions

Let $f(z)$ be a complex function. A point $z = z_0$ is called a *zero* of $f(z)$ if $f(z)$ is analytic at z_0 and $f(z_0) = 0$. If $f(z_0) = 0, f'(z_0) = 0, \ldots, f^{(m-1)}(z_0) = 0$ and $f^{(m)}(z_0) \neq 0$, then $z = z_0$ is called a zero of order m. When $m = 1$, $f(z)$ is said to have a *simple zero* at $z = z_0$. Thus, the order of the first non-vanishing derivative of an analytic function at $z = z_0$ is the order of the zero. For example, the function

(i) $\sin z$ has simple zeros at $z = n\pi, n = 0, \pm 1, \pm 2, \ldots$, since at each of these points $\sin z = 0$ but its derivative $\cos z \neq 0$.

(ii) $(1 - e^z)^2$ has zeros of order 2 (double zeros) at $z = 2n\pi i, n = 0, \pm 1, \pm 2, \ldots$, since at each of these points $f(z) = (1 - e^z)^2 = 0$ and $f'(z) = -2e^z(1 - e^z) = 0$ but $f''(z) = 2e^z(-1 + 2e^z) \neq 0$.

If the function $f(z)$ is analytic at $z = z_0$, we can write the Taylor series expansion of $f(z)$ about the point z_0 as

$$f(z) = f(z_0) + (z - z_0) f'(z_0) + \ldots + \frac{(z - z_0)^m}{m!} f^{(m)}(z_0) + \frac{(z - z_0)^{m+1}}{(m + 1)!} f^{(m+1)}(z_0) + \ldots.$$

When $z = z_0$ is a zero of order m, we obtain

$$f(z) = \frac{(z - z_0)^m}{m!} f^{(m)}(z_0) + \frac{(z - z_0)^{m+1}}{(m + 1)!} f^{(m+1)}(z_0) + \ldots$$

$$= (z - z_0)^m \left[\frac{1}{m!} f^{(m)}(z_0) + \frac{(z - z_0)}{(m + 1)!} f^{(m+1)}(z_0) + \ldots \right].$$

$$= (z - z_0)^m g(z). \tag{13.1}$$

where $g(z)$ is analytic at z_0 and $g(z_0) \neq 0$. This is an alternative way of defining a zero of order m. When a function $f(z)$ has a zero at $z = z_0$, but has no other zero in some neighborhood $N(z_0, \delta)$ of the point z_0, then the zero $z = z_0$ is called an *isolated zero*. Thus, the zero $z = z_0$ is an isolated zero of $f(z)$, if we can find a small real number $\delta > 0$ such that there is no other zero of $f(z)$ in the domain $|z - z_0| < \delta$. If no such δ exists, then the point $z = z_0$ is called a *non-isolated zero*.

Singular points of a function

A point $z = z_0$, at which the function $f(z)$ is not defined or the function is not analytic, is called a *singular point* of $f(z)$. A rational function $f(z) = P(z) / Q(z)$ has a singular point at $z = z_0$, whenever $Q(z_0) = 0$ and $P(z_0) \neq 0$.

For example, the function $f(z) = \dfrac{(z + 1)(z - 2)}{(z - 3)(z + 2)}$

   (i) has simple zeros at $z = -1$ and $z = 2$,

   (ii) has singular points at $z = -2$ and $z = 3$.

The singular point $z = z_0$ is called an *isolated singular point* of $f(z)$, if we can find a small real number $\delta > 0$ such that there is no other singular point of $f(z)$ in the neighborhood of z_0, that is in the domain $|z - z_0| < \delta$. If no such δ exists, then the singular point $z = z_0$ is called a *non-isolated singular point*.

Consider the following examples:

   (i) $f(z) = 1/(e^z - 1)$ has singular points when

$$e^z - 1 = 0, \text{ that is, at } z = 2n\pi i, \ n = 0 \pm 1, \pm 2, \ldots.$$

Each one of these points is an isolated singular point, since the neighborhoods of these points, $|z - z_0| < \delta$, do not contain other singular points of $f(z)$.

   (ii) $f(z) = \tan\left(\dfrac{1}{z}\right) = \dfrac{\sin(1/z)}{\cos(1/z)}$ has singular points when

$$\cos\left(\frac{1}{z}\right) = 0, \quad \text{that is } \frac{1}{z} = (2n + 1)\frac{\pi}{2} \quad \text{or} \quad z = \frac{2}{\pi(2n + 1)}, n = 0, \pm 1, \pm 2, \ldots.$$

Each one of these points, is an isolated singular point. Since the function is not defined at $z = 0$, the point $z = 0$ is also a singular point of $f(z)$. We also note that

$$\lim_{n \to \infty} \frac{2}{\pi(2n + 1)} = 0.$$

Therefore, the neighborhood of the point $z = 0$, that is $|z| < \delta$ contains many singular points of $f(z)$ other than $z = 0$. Thus, $z = 0$ is a non-isolated singular point of $f(z)$.

Laurent series expansion of the function $f(z)$ can be used to classify the isolated singular points. Let $z = z_0$ be an isolated singular point of $f(z)$. Then, there exists a deleted neighborhood of the point z_0, $0 < |z - z_0| < R$ inside which $f(z)$ is analytic. Hence, in this region we can expand the function $f(z)$ as a Laurent series. Thus, we have

$$f(z) = \sum_{n=0}^{\infty} a_n(z - z_0)^n + \sum_{n=1}^{\infty} \frac{b_n}{(z - z_0)^n} \tag{13.2}$$

The sum

$$\sum_{n=0}^{\infty} a_n(z - z_0)^n = a_0 + a_1(z - z_0) + \ldots + a_n(z - z_0)^n + \ldots \tag{13.3}$$

is called the *analytic part* of the Laurent series, where as the sum

$$\sum_{n=1}^{\infty} \frac{b_n}{(z - z_0)^n} = \frac{b_1}{z - z_0} + \frac{b_2}{(z - z_0)^2} + \ldots + \frac{b_n}{(z - z_0)^n} + \ldots \qquad (13.4)$$

is called the *principal part* of the Laurent series.
We define the following:

Removable singularity   If the function $f(z)$ is not defined at $z = z_0$, but $\lim\limits_{z \to z_0} f(z)$ exists, then the

point $z = z_0$ is called a removable singular point. Note that in this case, the principal part of the Laurent series is zero and $\lim\limits_{z \to z_0} f(z) = a_0$. Thus, if we define $f(z_0) = a_0$, the function $f(z)$ becomes an

analytic function at $z = z_0$.

For example, the function $f(z) = \sin z / z$ has a removable singularity at $z = 0$, since $\lim\limits_{z \to 0} (\sin z / z) = 1$ exists.

Pole   If the principal part of the Laurent series expansion of the function $f(z)$ has only a finite number of terms, that is, it is of the form

$$\frac{b_1}{z - z_0} + \frac{b_2}{(z - z_0)} + \ldots + \frac{b_m}{(z - z_0)^m}$$

where m is a finite integer and $b_{m+1}, b_{m+2}, \ldots$ are all zero, then $z = z_0$ is called a *pole* of order m.
If $m = 1$, then $z = z_0$ is called a *simple pole*. Thus, a point $z = z_0$ is a pole if

$$\lim_{z \to z_0} f(z) \to \infty \text{ and } \lim_{z \to z_0} [(z - z_0)^k f(z)] \text{ exists for } k \geq 1.$$

The smallest value of k for which this limit exists defines the order of the pole.

For example, the function $f(z) = \dfrac{4}{(z - 2)^5} + \dfrac{3}{(z + 2)^4}$ has a pole of order 5 at $z = 2$ and a pole of

order 4 at $z = -2$.

Essential singular point   If the principal part of the Laurent series expansion of the function $f(z)$ has infinite number of terms, then the point $z = z_0$ is called an *essential singular point*.

For example, $z = 0$ is an essential singular point of the function $e^{1/z}$, since the principal part of its Laurent series expansion about $z = 0$

$$e^{1/z} = 1 + \frac{1}{z} + \frac{1}{2! z^2} + \ldots + \frac{1}{n! z^n} + \ldots$$

has infinite number of terms.

To determine whether the point $z = \infty$ is a singular point of $f(z)$, we substitute $z = 1/w$ in the given function and discuss the singularity of the resulting function $f(1/w)$ at $w = 0$.

For example, the function e^z has an isolated essential singularity at $z = \infty$, since

$$e^{1/w} = 1 + \frac{1}{w} + \frac{1}{2! w^2} + \ldots$$

has an isolated essential singularity at $w = 0$.

Remark 1

(a) If $f(z)$ has a zero of nth order at $z = z_0$, then $1/f(z)$ has a pole of order n at $z = z_0$.

(b) Let $f(z)$ have a zero of order n at $z = z_0$, then $f^2(z)$ has a zero of order $2n$ at $z = z_0$ and the derivative $f'(z)$ has a zero of order $(n-1)$ at $z = z_0$. In general, $[f(z)]^m$ (mth power of $f(z)$) has a zero of order mn at $z = z_0$ and $f^{(m)}(z)$ (mth order derivative of $f(z)$) has a zero of order $(n-m)$, $m < n$, where $m > 0$ is an integer.

(c) A function which is analytic everywhere in the finite complex plane except at a finite number of poles is called a *meromorphic function*. For example, the function

$$f(z) = e^z/[(z-1)^2(z+2)^3]$$

which is analytic everywhere except at the poles $z = 1$ (order 2) and $z = -2$ (order 3) is a meromorphic function.

Example 13.1 Show that the function $f(z) = \displaystyle\int_0^z e^\xi \sin^3 \xi \, d\xi$ has a zero of order 4 at $z = 0$.

Solution The integrand is an analytic function for all z. We have

$$f(z) = \int_0^z e^\xi \sin^3 \xi \, d\xi, \quad f(0) = 0,$$

$$f'(z) = e^z \sin^3 z, \quad f'(0) = 0,$$

$$f''(z) = e^z(\sin^3 z + 3 \sin^2 z \cos z), \quad f''(0) = 0,$$

$$f'''(z) = e^z(-2\sin^3 z + 6\sin^2 z \cos z + 6 \sin z \cos^2 z), \quad f'''(0) = 0,$$

$$f^{iv}(z) = e^z(6\cos^3 z - 8\sin^3 z - 12\sin^2 z \cos z + 18 \sin z \cos^2 z), \quad f^{iv}(0) \neq 0.$$

Hence, $z = 0$ is a zero of order 4 of $f(z)$.

Example 13.2 Classify the singular points of the function $f(z) = \dfrac{z^2 + iz + 2}{(z^2 + 1)^2 \, (z + 3)}$.

Solution The singular points of $f(z) = P(z)/Q(z)$ are the points where

$$Q(z) = (z^2 + 1)^2 (z + 3) = 0.$$

Therefore, $z = \pm i$ and $z = -3$ are the singular points of $f(z)$. To determine the type of singularity, we separate out the analytic and the principal parts of the Laurent series expansion of $f(z)$. For the singular point $z = -3$, we write $f(z)$ as

$$f(z) = \frac{g_1(z)}{z + 3}, \text{ where } g_1(z) = \frac{z^2 + iz + 2}{(z^2 + 1)^2}.$$

Since $g_1(z)$ is analytic at $z = -3$, we can expand $g_1(z)$ in Taylor series about the point $z = -3$. We obtain

$$g_1(z) = g_1(-3) + (z + 3) \ g_1'(-3) + \text{higher powers of } (z + 3)$$

where $g_1(-3) = (11 - 3i)/100 \neq 0$.

Thus, in some deleted neighborhood of the point $z = -3$, we get

$$f(z) = \frac{g_1(-3)}{z+3} + \text{(constant and positive powers of } (z+3)).$$

We conclude that $z = -3$ is a simple pole as the principal part of the Laurent series is the single term $g_1(-3)/(z+3)$.

For the singular point $z = i$, we write $f(z)$ as

$$f(z) = \frac{g_2(z)}{(z-i)^2}, \quad \text{where} \quad g_2(z) = \frac{z^2 + iz + 2}{(z+i)^2(z+3)}.$$

Now, $g_2(z)$ is analytic at $z = i$. Since $g_2(i) = 0$ and $g_2'(i) = -3i/[4(3+i)] \neq 0$, the Taylor series expansion of $g_2(z)$ about $z = i$ is of the form

$$g_2(z) = (z-i)\ g_2'(i) + \text{higher powers of } (z-i).$$

Therefore, we can write $f(z) = \dfrac{g_2'(i)}{z-i} + \text{(constant and positive powers of } (z-i)).$

Hence, $z = i$ is a simple pole of $f(z)$ as the principal part of the Laurent series is the single term $g_2'(i)/(z-i)$.

For the singular point $z = -i$, we write $f(z)$ as

$$f(z) = \frac{g_3(z)}{(z+i)^2}, \quad \text{where} \quad g_3(z) = \frac{z^2 + iz + 2}{(z-i)^2(z+3)}.$$

Now, $g_3(z)$ is analytic at $z = -i$. Since $g_3(-i) \neq 0$, we obtain the Taylor series expansion of $g_3(z)$ about $z = -i$ as

$$g_3(z) = g_3(-i) + (z+i)\ g_3'(-i) + \text{higher powers of } (z+i)$$

Therefore, we can write $f(z) = \dfrac{g_3(-i)}{(z+i)^2} + \dfrac{g_3'(-i)}{z+i} + \text{(constant and positive powers of } (z+i)).$

Hence, $z = -i$ is a pole of order 2 as the principal part of the Laurent series has two terms

$$\frac{g_3(-i)}{(z+i)^2} + \frac{g_3'(-i)}{z+i}.$$

Example 13.3 Classify the singular point $z = 0$ of the functions

(i) $\dfrac{e^z}{z + \sin z}$,   (ii) $\dfrac{e^z}{z - \sin z}$

and obtain the principal part of the Laurent series expansion of $f(z)$ in each case.

Solution

(i) Write the denominator of $f(z)$ as

$$z + \sin z = z + \left(z - \frac{z^3}{3!} + \frac{z^5}{5!} - \dots \right) = 2z - \frac{z^3}{3!} + \dots = z\, g(z)$$

where $g(z) = 2 - \dfrac{z^2}{3!} + \dots$ and $g(0) \neq 0.$

Therefore, we can write

$$f(z) = \frac{e^z}{z\,g(z)}, \quad g(0) \neq 0.$$

Hence, $z = 0$ is a simple pole. To obtain the principal part of the Laurent series expansion of $f(z)$ about $z = 0$, we write

$$f(z) = \frac{e^z}{z + \sin z} = \frac{1 + z + (z^2/2!) + \ldots}{2z\,[1 - (z^2/12) + \ldots]}$$

$$= \frac{1}{2z}\left[1 + z + \frac{z^2}{2} + \ldots\right]\left[1 - \left(\frac{z^2}{12} + \ldots\right)\right]^{-1}$$

$$= \frac{1}{2z}\left[1 + z + \frac{z^2}{2} + \ldots\right]\left[1 + \frac{z^2}{12} + \ldots\right]$$

$$= \frac{1}{2z}\left[1 + z + \frac{7z^2}{12} + \ldots\right] = \frac{1}{2z} + \frac{1}{2} + \frac{7z}{24} + \ldots.$$

Therefore, the principal part of the Laurent series is $1/(2z)$.

(ii) Write the denominator of $f(z)$ as

$$z - \sin z = z - \left(z - \frac{z^3}{3!} + \frac{z^5}{5!} - \ldots\right) = \frac{z^3}{3!} - \frac{z^5}{5!} + \ldots = z^3 g(z)$$

where

$$g(z) = \frac{1}{6}\left(1 - \frac{z^2}{20} + \frac{z^4}{840} - \ldots\right) \quad \text{and} \quad g(0) \neq 0.$$

Therefore, we can write

$$f(z) = \frac{e^z}{z^3 g(z)}, \quad g(0) \neq 0.$$

Hence, $z = 0$ is a pole of order 3. To determine the principal part of the Laurent series expansion of $f(z)$ about $z = 0$, we write

$$f(z) = \frac{e^z}{z - \sin z} = \frac{6[1 + z + (z^2/2!) + (z^3/3!) + \ldots]}{z^3\left(1 - \frac{z^2}{20} + \frac{z^4}{840} - \ldots\right)}$$

$$= \frac{6}{z^3}\left[1 + z + \frac{z^2}{2} + \frac{z^3}{6} + \ldots\right]\left[1 - \left(\frac{z^2}{20} - \frac{z^4}{840} + \ldots\right)\right]^{-1}$$

$$= \frac{6}{z^3}\left[1 + z + \frac{z^2}{2} + \frac{z^3}{6} + \ldots\right]\left[1 + \frac{z^2}{20} + \frac{11z^4}{8400} + \ldots\right]$$

$$= \frac{6}{z^3}\left[1 + z + \frac{11}{20}z^2 + \frac{13}{60}z^3 + \ldots\right].$$

Therefore, the principal part of the Laurent series is $\dfrac{6}{z^3} + \dfrac{6}{z^2} + \dfrac{33}{10z}$.

Example 13.4 Show that the function

(i) cosec z has a simple pole at $z = 0$,

(ii) $\dfrac{1}{z^2 - 1}$ has simple poles at $z = 1$ and $z = -1$.

Solution We first obtain the Laurent series expansion of the function $f(z)$ to classify the singular points.

(i) $f(z) = 1/\sin z$. We find that $z = 0$ is a singular point of $f(z)$. Now,

$$f(z) = \frac{1}{\sin z} = \frac{1}{z - (z^3/3!) + (z^5/5!) - \ldots} = \frac{1}{z}\left[1 - \left(\frac{z^2}{3!} - \frac{z^4}{5!} + \ldots\right)\right]^{-1}$$

$$= \frac{1}{z} + \frac{z}{3!} + \text{ higher powers of } z.$$

The principal part of the Laurent series is the single term $1/z$. Hence, $z = 0$ is a simple pole.

(ii) $f(z) = 1/(z^2 - 1)$. We find that $z = \pm 1$ are the singular points of $f(z)$.

We write for $z = 1$

$$f(z) = \frac{1}{1 - z^2} = \frac{1}{2}\left[\frac{1}{1 - z} + \frac{1}{1 + z}\right] = -\frac{1}{2}\left[\frac{1}{z - 1}\right] + \frac{1}{2}\left[\frac{1}{2 + (z - 1)}\right]$$

$$= -\frac{1}{2(z - 1)} + \frac{1}{4}\left[1 + \frac{z - 1}{2}\right]^{-1}$$

$$= -\frac{1}{2(z - 1)} + \frac{1}{4}\left[1 - \frac{z - 1}{2} + \text{ higher powers of } (z - 1)\right]$$

valid in $|z - 1| < 2$. Therefore, the principal part of the Laurent series is $-1/[2(z - 1)]$. Hence $z = 1$ is a simple pole.

Now, we write for $z = -1$

$$f(z) = \frac{1}{2(z + 1)} + \frac{1}{2[2 - (z + 1)]}$$

$$= \frac{1}{2(z + 1)} + \frac{1}{4}\left[1 - \frac{z + 1}{2}\right]^{-1}$$

$$= \frac{1}{2(z + 1)} + \frac{1}{4}\left[1 + \frac{z + 1}{2} + \text{ higher powers of } (z + 1)\right]$$

valid in $|z + 1| < 2$. Therefore, the principal part of the Laurent series is $1/[2(z + 1)]$. Hence, $z = -1$ is a simple pole.

Example 13.5 Show that the function $f(z) = \dfrac{z^4 + 2z + 1}{z^2 + 5z + 2}$ has a pole of order 2 at $z = \infty$.

Solution Writing $z = 1/w$, we get

$$f\left(\frac{1}{w}\right) = \frac{1}{w^2}\left[\frac{1 + 2w^3 + w^4}{1 + 5w + 2w^2}\right] = \frac{1}{w^2}g(w), \; g(0) \neq 0.$$

Therefore, the function $f(1/w)$ has a pole of order 2 at $w = 0$. Hence, the function $f(z)$ has a pole of order 2 at $z = \infty$.

Example 13.6   Show that the function $\sin z/z^r$, $r \geq 2$ is a positive integer, has a pole of order $(r-1)$ at $z = 0$.

Solution   We have

$$f(z) = \frac{\sin z}{z^r} = \frac{1}{z^r}\left[z - \frac{z^3}{3!} + \ldots\right] = \frac{1}{z^{r-1}}\,g(z),$$

where

$$g(z) = 1 - \frac{z^2}{3!} + \ldots, \quad \text{and} \quad g(0) \neq 0.$$

Hence, $f(z)$ has a pole of order $(r-1)$ at $z = 0$.

Example 13.7   Discuss the singularities of the function $f(z) = \dfrac{(z^2 - 1)(z-2)^3}{(\sin \pi z)^3}$ in the extended complex plane.

Solution   The function $f(z)$ is analytic everywhere in the finite complex plane except at the points where $\sin \pi z = 0$, that is at $z = 0, \pm 1, \pm 2, \ldots$. The zeros of $\sin \pi z$ at these points are of order 1, since $\dfrac{d}{dz}(\sin \pi z) = \pi \cos \pi z \neq 0$ at these points. Therefore, the zeros of $(\sin \pi z)^3$ are of order 3 at each of these points. Hence, $f(z)$ has a pole of order 3 at all positive and negative integral values of z except at $z = -1$, 1 and 2 where the numerator vanishes.

Now, $g(z) = z^2 - 1 = (z-1)(z+1)$ has zeros of order 1 at $z = 1$ and $z = -1$. Therefore, $f(z)$ has poles of order 2 at $z = 1$ and $z = -1$.

At the point $z = 2$, we get (using L' Hospital's rule)

$$\lim_{z \to 2} \frac{z-2}{\sin \pi z} = \lim_{z \to 2} \frac{1}{\pi \cos \pi z} = \frac{1}{\pi}.$$

Therefore,

$$\lim_{z \to 2} f(z) = \lim_{z \to 2}(z^2 - 1)\left[\frac{z-2}{\sin \pi z}\right]^3 = \frac{3}{\pi^3}.$$

Hence, $z = 2$ is a removable singular point of $f(z)$.

To test the singularity at $z = \infty$, we substitute $z = 1/w$ and obtain

$$f\left(\frac{1}{w}\right) = \frac{(1 - w^2)(1 - 2w)^3}{w^5[\sin(\pi/w)]^3}$$

Now, $f(1/w)$ has singular points at $w = 0$ and $w = 1/n$, where $n \geq 2$ is an integer. (Note that $w = 1$ is not a pole as $f(1)$ is of the form 0/0 and $\lim\limits_{w \to 1} f(1/w)$ exists.) Since $\lim\limits_{n \to \infty}(1/n) = 0$, the singularity of $f(1/w)$ at $w = 0$ is not isolated. Hence, the singularity of $f(z)$ at $z = \infty$ is not isolated.

Exercise 13.1

Classify the singular points of the following functions in the finite complex plane.

1. $\dfrac{\sin z}{e^z - 1}$.

2. $z^2 \cos \dfrac{1}{z}$.

3. $\dfrac{\cot z}{z}$

4. $\dfrac{(1 - e^z)\cos(1/(z-8))}{z^2(z-8)(z^2 - 16)}$.

5. $\dfrac{\sin z^2}{z(e^z - 1)}$.

6. $\dfrac{1 - \cos z}{\sin^3 z}$.

7. $\dfrac{1}{\sin z - \cos z}$.

8. $\dfrac{1}{(2 \sin z + 1)^2}$.

9. $\dfrac{z^2}{e^{1/z} - 1}$.

10. $\dfrac{e^{z-1} \sin (1/(z-1))}{z^2 (z^2 - 4)}$.

11. $\dfrac{e^{(z-1)^{-2}} \sin z \cos^3 z}{z(z - \pi/2)}$.

12. $(z - \pi) e^{1/z} \cot z$.

13. $e^{z/(1-z)}$.

14. $\dfrac{z}{e^{z^2} - 1}$.

15. $\dfrac{\text{Ln} (1 + z^m)}{z^n}$, m, n positive integers.

16. $\dfrac{e^{z^2} - 1}{z^3}$.

17. $e^{\tan (1/z)}$.

18. $\sin (1/[\sin (1/z)])$.

19. $\displaystyle\sum_{n=0}^{\infty} n(-1)^n z^n$.

20. $\displaystyle\sum_{n=0}^{\infty} (\sin z)^n$.

Classify the singular points of (a) $f(z)$ and (b) $1/f(z)$ in the extended complex plane, where $f(z)$ is defined as the following.

21. e^z.

22. $\dfrac{z^5}{z + z^3}$.

23. $\dfrac{\cos z}{z}$.

24. $e^{\cosh z}$.

25. $\dfrac{z(z - \pi)^2}{[\sin (z)]^2}$.

Find the principal part in the Laurent series expansion of the following functions at each isolated singular point.

26. $\dfrac{z}{e^z - 1}$.

27. $\dfrac{z^2 + z + 1}{(z - i)(z + 1)^2}$.

28. $\dfrac{\sin z}{z(z - 1)}$.

29. $\dfrac{\text{Ln } z}{(z - 1)^2 (z^2 + 4)}$.

30. $\dfrac{1 - \cos z}{z^4}$.

31. Show that an analytic function is bounded in some deleted neighborhood of a removable singular point.

32. Show that an analytic function cannot be bounded in the neighborhood of an isolated singular point.

33. If $g(z)$ is analytic at 0 and 0 is a removable sigularity of $g(z)/z^n$, what can be said about $g(z)$ at $z = 0$?

34. If $f(z)$ has a pole of order N at $z = z_0$, what is the nature of singularity for the function $(z - z_0)^k f(z)$ at z_0, where k can take values $0, \pm 1, \pm 2, \ldots$?

35. The function $f(z)$ has a pole of order N at $z = z_0$. What is the nature of singularity for the function $f'(z)/f(z)$ at $z = z_0$? Find the principal part in its Laurent series expansion.

36. Let $f(z)$ be analytic at z_0.

   (i) If $f(z)$ has a zero of order $N \geq 1$, then show that $1/f(z)$ has a pole of order N at z_0.

   (ii) If $f(z)$ has a pole of order $N \geq 1$, then show that $1/f(z)$ has a zero of order N at z_0.

   (iii) If $f(z)$ is non-vanishing in a deleted neighborhood of z_0 and has an essential singularity at z_0, then show that $1/f(z)$ also has an essential singularity at z_0.

37. The function $f(z)$ has a pole of order m at z_0. Show that the nth derivative $f^{(n)}(z)$ has a pole of order $(m + n)$ at z_0.

38. Let $f(z)$ be an entire function and not a constant. Then, show that $f(z)$ has a pole, or has an essential singularity at $z = \infty$.

39. Show that a polynomial of order N has a pole of order N at $z = \infty$.

40. Show that a rational function is either analytic at $z = \infty$ or has a pole at $z = \infty$.

13.3   Residues

If a function $f(z)$ is analytic inside and on a simple closed contour C, then by Cauchy integral theorem, we have

$$\oint_C f(z)\,dz = 0.$$

However, when the integrand $f(z)$ has one or more isolated singular points inside C, then the Cauchy integral theorem cannot be used and the value of the complex integral $\oint_C f(z)\,dz$ may not be zero.

Each of these isolated singular points inside C contributes to the value of the complex integral. These contributions are called *residues*. The theory of residues provides a simpler way of evaluating complex integrals. The theory of residues can also be used to evaluate certain types of real integrals.

Residue of an analytic function at an isolated singular point

Let $f(z)$ be analytic for all z except at the point $z = z_0$. Let z_0 be an isolated singular point of $f(z)$. Then, $f(z)$ can be expanded in Laurent series about the point z_0 in the form

$$f(z) = a_0 + a_1(z - z_0) + a_2(z - z_0)^2 + \ldots + \frac{b_1}{z - z_0} + \frac{b_2}{(z - z_0)^2} + \ldots \qquad (13.5)$$

which converges in some region $R : 0 < |z - z_0| < r$. The coefficients b_n are defined by

$$b_n = \frac{1}{2\pi i} \oint_C \frac{f(z)}{(z - z_0)^{-n+1}}\,dz, \quad n = 1, 2, \ldots$$

where C is a simple closed curve in the annulus $0 < |z - z_0| < r$. In particular, when $n = 1$, we get

$$b_1 = \frac{1}{2\pi i} \oint_C f(z)\,dz \quad \text{or} \quad \oint_C f(z)\,dz = 2\pi i b_1. \qquad (13.6)$$

The coefficient of $(z - z_0)^{-1}$, that is b_1 in the Laurent series expansion of $f(z)$, given in Eq. (13.5) is called the *residue* of $f(z)$ at z_0 and is written as

$$b_1 = \operatorname*{Res}_{z = z_0} f(z). \qquad (13.7)$$

Therefore, if the residue b_1 at $z = z_0$ can be determined by some method, then the value of the contour integral $\oint_C f(z)\,dz$ is given by $2\pi i b_1$.

Now, we discuss methods to find the residue of a function $f(z)$ at an isolated singular point.

Residue at a removable singular point

If $z = z_0$ is a removable singular point of $f(z)$, then there is no term in the principal part of the Laurent series expansion of $f(z)$ about z_0. Therefore, in this case $b_1 = 0$ and $\oint_C f(z)\,dz = 0$.

Residue at a simple pole

Let z_0 be a simple pole of $f(z)$. Then, in the neighborhood of the point z_0, we can write $f(z)$ as

$$f(z) = \sum_{n=0}^{\infty} a_n (z - z_0)^n + \frac{b_1}{z - z_0}.$$

(13.8)

Multiplying both sides of Eq. (13.8) by $(z - z_0)$ and taking limits as $z \to z_0$, we obtain

$$b_1 = \lim_{z \to z_0} (z - z_0) f(z).$$

(13.9)

Alternative

Since $z = z_0$ is a simple pole, $f(z)$ is of the form

$$f(z) = \frac{\phi(z)}{\psi(z)}.$$

where $\phi(z)$ and $\psi(z)$ have no common terms, $\phi(z_0) \neq 0$, $\psi(z_0) = 0$ and $\psi'(z_0) \neq 0$. Since $\psi(z)$ is analytic, we can write its Taylor series expansion as

$$\psi(z) = \psi(z_0) + (z - z_0) \, \psi'(z_0) + \frac{1}{2!} (z - z_0)^2 \, \psi''(z_0) + \ldots$$

$$= (z - z_0) \, \psi'(z_0) + \text{higher powers of } (z - z_0).$$

Therefore,

$$b_1 = \lim_{z \to z_0} [(z - z_0) f(z)] = \lim_{z \to z_0} \left[(z - z_0) \frac{\phi(z)}{\psi(z)} \right] = \frac{\phi(z_0)}{\psi'(z_0)}, \quad \psi'(z_0) \neq 0.$$

(13.10)

Residue at a pole of order m

Let $z = z_0$ be a pole of order m of $f(z)$. Then, in the neighborhood of the point $z = z_0$, we can write

$$f(z) = \sum_{n=0}^{\infty} a_n (z - z_0)^n + \frac{b_1}{z - z_0} + \frac{b_2}{(z - z_0)^2} + \ldots + \frac{b_m}{(z - z_0)^m}.$$

Multiplying both sides by $(z - z_0)^m$, we get

$$(z - z_0)^m f(z) = \sum_{n=0}^{\infty} a_n (z - z_0)^{m+n} + b_1 (z - z_0)^{m-1} + b_2 (z - z_0)^{m-2} + \ldots + b_m.$$

Differentiating both sides $(m - 1)$ times and then taking limits as $z \to z_0$, we obtain

$$b_1 = \frac{1}{(m - 1)!} \lim_{z \to z_0} \frac{d^{m-1}}{dz^{m-1}} \left[(z - z_0)^m f(z) \right].$$

(13.11)

Residue at an isolated essential singular point

When $z = z_0$ is an isolated essential singular point, we need to expand $f(z)$ in Laurent series about the point $z = z_0$ and obtain the residue (coefficient of $(z - z_0)^{-1}$) directly. In this case, this is the only way of computing residue at z_0.

Residue at $z = \infty$

Let $z = \infty$ be an isolated singular point of $f(z)$. Let C be the circle $|z| = r$ traversed in the clockwise direction such that all the finite singularities of $f(z)$ lie inside C. The isolated singularity at $z = \infty$ is the only singularity in the domain $|z| > r$. Note that C is negatively oriented to show that the domain being considered is $|z| > r$.

By definition, the residue at $z = \infty$ is given by

$$\text{(residue at } z = \infty) = \frac{1}{2\pi i} \oint_C f(z)\, dz.$$

Let $z = 1/w$. Then, $dz = -(1/w^2)\, dw$ and

$$\frac{1}{2\pi i} \oint_C f(z)\, dz = \frac{1}{2\pi i} \oint_{C^*} f\left(\frac{1}{w}\right)\left(-\frac{1}{w^2}\right) dw = -\frac{1}{2\pi i} \oint_{C^*} g(w)\, dw$$

where C^* is a circle described about the origin in the anticlockwise direction (since the transformation $z = 1/w$ changes the orientation of C) and

$$g(w) = \frac{1}{w^2} f\left(\frac{1}{w}\right). \tag{13.12a}$$

Now, $z = \infty$ gives $w = 0$, and $w = 0$ is the only isolated singularity of $g(w)$ inside C^*. Hence, we obtain

$$\underset{z=\infty}{\text{Res}}\, f(z) = -\underset{w=0}{\text{Res}}\, g(w)$$

$$= -\text{ (coefficient of } (1/w) \text{ in the Laurent series expansion of } g(w) \text{ about } w = 0) \tag{13.12b}$$

Example 13.8 Compute the residues at the singular points of $f(z)$, where $f(z)$ is given by

(i) $\dfrac{z}{(z+1)(z-2)}$,

(ii) $\dfrac{z^2}{z^2 - 2z + 2}$.

Solution

(i) The function $f(z) = \dfrac{z}{(z+1)(z-2)}$ has simple poles at $z = -1$ and $z = 2$.

Hence,

$$\underset{z=-1}{\text{Res}}\, f(z) = \lim_{z \to -1}[(z+1)f(z)] = \lim_{z \to -1}\left[\frac{z}{z-2}\right] = \frac{1}{3}.$$

$$\underset{z=2}{\text{Res}}\, f(z) = \lim_{z \to 2}[(z-2)f(z)] = \lim_{z \to 2}\left[\frac{z}{z+1}\right] = \frac{2}{3}.$$

Alternative

$z = -1$ and $z = 2$ are simple poles of $f(z)$ and $f(z)$ is a rational function of the form $\phi(z)/\psi(z)$, with $\phi(z) = z$ and $\psi(z) = (z+1)(z-2) = z^2 - z - 2$ and $\psi'(z) = 2z - 1$. Using the formula given in Eq. (13.10), we obtain

$$\underset{z=-1}{\text{Res}}\, f(z) = \frac{\phi(-1)}{\psi'(-1)} = \frac{1}{3} \quad \text{and} \quad \underset{z=2}{\text{Res}}\, f(z) = \frac{\phi(2)}{\psi'(2)} = \frac{2}{3}.$$

(ii) The singular points of $f(z)$ are the roots of the equation $z^2 - 2z + 2 = 0$. We have $z_1 = 1 + i$ and $z_2 = 1 - i$, which are simple poles. We write

$$f(z) = \frac{z^2}{z^2 - 2z + 2} = \frac{z}{(z - z_1)(z - z_2)}$$

Hence,

$$\operatorname*{Res}_{z=z_1} f(z) = \lim_{z \to z_1} [(z - z_1)f(z)] = \frac{z_1^2}{z_1 - z_2} = \frac{(1 + i)^2}{2i} = 1.$$

$$\operatorname*{Res}_{z=z_2} f(z) = \lim_{z \to z_2} [(z - z_2)f(z)] = \frac{z_2^2}{z_2 - z_1} = \frac{(1 - i)^2}{-2i} = 1.$$

Example 13.9 Compute the residues at the singular points $z = 1, -2$ of

$$f(z) = \frac{1 + z + z^2}{(z - 1)^2 (z + 2)}.$$

Solution The given function has a simple pole at $z = -2$ and a pole of order 2 at $z = 1$. We obtain

$$\operatorname*{Res}_{z=-2} f(z) = \lim_{z \to -2} [(z + 2)f(z)] = \lim_{z \to -2} \left[\frac{1 + z + z^2}{(z - 1)^2} \right] = \frac{1 - 2 + 4}{9} = \frac{1}{3}.$$

$$\operatorname*{Res}_{z=1} f(z) = \lim_{z \to 1} \frac{d}{dz} [(z - 1)^2 f(z)] = \lim_{z \to 1} \frac{d}{dz} \left[\frac{1 + z + z^2}{z + 2} \right]$$

$$= \lim_{z \to 1} \frac{d}{dz} \left[z - 1 + \frac{3}{z + 2} \right] = \lim_{z \to 1} \left[1 - \frac{3}{(z + 2)^2} \right] = \frac{2}{3}.$$

Example 13.10 Compute the residues at all the singular points of $f(z)$, where $f(z)$ is given by

(i) cot z, (ii) sec z.

Solution The function

(i) $f(z) = \cot z = \dfrac{\cos z}{\sin z} = \dfrac{\phi(z)}{\psi(z)}$ has singular points where

$$\sin z = 0, \text{ that is, at } z = z_k = k\pi, \ \ k = 0, \pm 1, \pm 2, \ldots.$$

Since $\psi'(z) = \cos z \neq 0$ at $z = z_k$, $f(z)$ has simple poles at z_k. Using the formula given in Eq. (13.10), we obtain

$$\operatorname*{Res}_{z=z_k} f(z) = \frac{\phi(z_k)}{\psi'(z_k)} = \frac{\cos z_k}{\cos z_k} = 1 \text{ for all } k.$$

(ii) $f(z) = \dfrac{1}{\cos z} = \dfrac{\phi(z)}{\psi(z)}$ has singular points where

$$\cos z = 0 \quad \text{or} \quad \text{at } z = z_k = (2k + 1)\frac{\pi}{2}, \ \ k = 0, \pm 1, \pm 2, \ldots.$$

Since $\psi'(z) = -\sin z \neq 0$ at z_k, $f(z)$ has simple poles at z_k. Using the formula given in Eq. (13.10), we obtain

$$\operatorname*{Res}_{z=z_k} f(z) = \frac{\phi(z_k)}{\psi'(z_k)} = \frac{1}{-\sin z_k} = -\frac{1}{\sin[(2k + 1)\pi/2]} = (-1)^{k+1}.$$

Example 13.11 Find the residues at all the singular points of

(i) $f(z) = \dfrac{1}{z^3 + z^5}$, (ii) $f(z) = \dfrac{z^2}{(z^2 + 1)^2}$.

Solution The function

(i) $f(z) = \dfrac{1}{z^3 + z^5} = \dfrac{1}{z^3(1 + z^2)} = \dfrac{1}{z^3(z + i)(z - i)}$

has simple poles at $z = \pm i$ and a pole of order 3 at $z = 0$.

Hence, $\underset{z=i}{\text{Res}} f(z) = \lim_{z \to i} [(z - i) f(z)] = \lim_{z \to i} \left[\dfrac{1}{z^3(z + i)} \right] = \dfrac{1}{i^3(2i)} = \dfrac{1}{2}$.

$\underset{z=-i}{\text{Res}} f(z) = \lim_{z \to -i} [(z + i) f(z)] = \lim_{z \to -i} \left[\dfrac{1}{z^3(z - i)} \right] = \dfrac{1}{(-i)^3(-2i)} = \dfrac{1}{2}$.

$\underset{z=0}{\text{Res}} f(z) = \dfrac{1}{2!} \lim_{z \to 0} \dfrac{d^2}{dz^2} [z^3 f(z)] = \dfrac{1}{2} \lim_{z \to 0} \dfrac{d^2}{dz^2} \left[\dfrac{1}{1 + z^2} \right]$

$= \dfrac{1}{2} \lim_{z \to 0} \dfrac{d^2}{dz^2} [1 - z^2 + z^4 - \ldots]$

$= \dfrac{1}{2} \lim_{z \to 0} [-2 + 12z^2 - \ldots] = -1$.

(ii) $f(z) = \dfrac{z^2}{(z^2 + 1)^2} = \dfrac{z^2}{(z + i)^2(z - i)^2}$

has poles of order 2 at $z = \pm i$. Hence,

$\underset{z=i}{\text{Res}} f(z) = \lim_{z \to i} \dfrac{d}{dz} [(z - i)^2 f(z)] = \lim_{z \to i} \dfrac{d}{dz} \left[\dfrac{z^2}{(z + i)^2} \right]$

$= \lim_{z \to i} \left[\dfrac{(z + i)^2(2z) - z^2 \{2(z + i)\}}{(z + i)^4} \right] = -\dfrac{i}{4}$.

$\underset{z=-i}{\text{Res}} f(z) = \lim_{z \to -i} \dfrac{d}{dz} [(z + i)^2 f(z)] = \lim_{z \to -i} \dfrac{d}{dz} \left[\dfrac{z^2}{(z - i)^2} \right]$

$= \lim_{z \to -i} \left[\dfrac{(z - i)^2(2z) - z^2 \{2(z - i)\}}{(z - i)^4} \right] = \dfrac{i}{4}$.

Example 13.12 Compute the residues at all the singular points of $f(z) = z^2/(z^n - 1)$, n any positive integer.

Solution The function $f(z) = \dfrac{z^2}{z^n - 1} = \dfrac{\phi(z)}{\psi(z)}$ has simple poles at

$z = z_k = (1)^{1/n} = e^{(2\pi ki)/n}, \quad k = 0, 1, \ldots n - 1$.

Using the formula given in Eq. (13.10), we obtain

$$\operatorname*{Res}_{z \to z_k} f(z) = \frac{\phi(z_k)}{\psi'(z_k)} = \frac{z_k^2}{n z_k^{n-1}} = \frac{z_k^3}{n z_k^n} = \frac{1}{n} e^{(6\pi ik)/n}, \ k = 0, 1, \ldots, n-1.$$

Example 13.13 Compute the residues at all the singular points of

(i) $f(z) = z \sin\left(\dfrac{1}{z}\right),$ (ii) $f(z) = z \cos\left(\dfrac{1}{z}\right).$

Solution The point $z = 0$ is an isolated essential singular point of $f(z)$. Write the Laurent series expansion of $f(z)$ about $z = 0$ to compute the residue.

(i) $f(z) = z \sin\left(\dfrac{1}{z}\right) = z\left[\dfrac{1}{z} - \dfrac{1}{3! \, z^3} + \ldots\right] = 1 - \dfrac{1}{3! \, z^3} + \ldots.$

Therefore, $\operatorname*{Res}_{z=0} f(z) = $ coefficient of $\left(\dfrac{1}{z}\right)$ in the Laurent series $= 0$.

(ii) $f(z) = z \cos\left(\dfrac{1}{z}\right) = z\left[1 - \dfrac{1}{2! \, z^2} + \dfrac{1}{4! \, z^4} - \ldots\right] = z - \dfrac{1}{2! \, z} + \dfrac{1}{4! \, z^3} - \ldots.$

Therefore, $\operatorname*{Res}_{z=0} f(z) = $ coefficient of $\left(\dfrac{1}{z}\right)$ in the Laurent series $= -\dfrac{1}{2}$.

Example 13.14 Obtain the residue at $z = \infty$ for the function $f(z)$, where $f(z)$ is given by

(i) $\dfrac{z^4 + z^2}{z^3},$ (ii) $\dfrac{z}{e^{-z^2} + 1},$ (iii) $z^3 \cos\left(\dfrac{1}{z}\right).$

Solution Let $z = 1/w$. Then, $g(w) = (1/w^2)\, f(1/w)$. We have

$$\operatorname*{Res}_{z=\infty} f(z) = - \text{ (residue of } g(w) \text{ at } w = 0).$$

(i) $g(w) = \dfrac{1}{w^2} f\left(\dfrac{1}{w}\right) = \dfrac{(1/w^4) + (1/w^2)}{(1/w^3)\, w^2} = \dfrac{1}{w} + \dfrac{1}{w^3}.$

Therefore, $\operatorname*{Res}_{w=0} g(w) = 1$ and $\operatorname*{Res}_{z=\infty} f(z) = -1$.

(ii) $g(w) = \dfrac{1}{w^2} f\left(\dfrac{1}{w}\right) = \dfrac{1/w}{w^2\,(e^{-1/w^2} + 1)} = \dfrac{1}{w^3\left[2 - \dfrac{1}{w^2} + \dfrac{1}{2! \, w^4} - \ldots\right]}$

$= \dfrac{1}{2w^3}\left[1 - \left(\dfrac{1}{2w^2} - \ldots\right)\right]^{-1} = \dfrac{1}{2w^3}\left[1 + \dfrac{1}{2w^2} - \ldots\right].$

Therefore, $\operatorname*{Res}_{w=0} g(w) = 0$ and $\operatorname*{Res}_{z=\infty} f(z) = 0$.

(iii) $g(w) = \dfrac{1}{w^2} f\left(\dfrac{1}{w}\right) = \dfrac{1}{w^5} \cos w = \dfrac{1}{w^5}\left[1 - \dfrac{w^2}{2!} + \dfrac{w^4}{4!} - \cdots\right] = \dfrac{1}{w^5} - \dfrac{1}{2w^3} + \dfrac{1}{24w} - \dfrac{w}{6!} + \cdots$

Therefore, $\underset{w=0}{\text{Res}}\, g(w) = \dfrac{1}{24}$ and $\underset{z=\infty}{\text{Res}}\, f(z) = -\dfrac{1}{24}.$

Example 13.15   Find the residue of the function $f(z) = \cos\left(\dfrac{z^2 + 6z + 5}{z + 2}\right)$ at $z = -2.$

Solution   The point $z = -2$ is an isolated essential singular point of $f(z)$. The residue at $z = -2$ is the coefficient of $1/(z + 2)$ in the Laurent series expansion of $f(z)$ about $z = -2$. We write

$$f(z) = \cos\left(\dfrac{z^2 + 6z + 5}{z + 2}\right) = \cos\left(z + 4 - \dfrac{3}{z + 2}\right)$$

$$= \cos(z + 4)\cos\left(\dfrac{3}{z + 2}\right) + \sin(z + 4)\sin\left(\dfrac{3}{z + 2}\right)$$

$$= \cos(z + 2 + 2)\cos\left(\dfrac{3}{z + 2}\right) + \sin(z + 2 + 2)\sin\left(\dfrac{3}{z + 2}\right)$$

$$= [\cos(z + 2)\cos 2 - \sin(z + 2)\sin 2]\cos\left(\dfrac{3}{z + 2}\right)$$

$$+ [\sin(z + 2)\cos 2 + \cos(z + 2)\sin 2]\sin\left(\dfrac{3}{z + 2}\right)$$

$$= \cos 2 \cos(z + 2)\cos\left(\dfrac{3}{z + 2}\right) - \sin 2 \sin(z + 2)\cos\left(\dfrac{3}{z + 2}\right)$$

$$+ \cos 2 \sin(z + 2)\sin\left(\dfrac{3}{z + 2}\right) + \sin 2 \cos(z + 2)\sin\left(\dfrac{3}{z + 2}\right)$$

$$= \cos 2\left[1 - \dfrac{(z + 2)^2}{2!} + \dfrac{(z + 2)^4}{4!} - \cdots\right]\left[1 - \dfrac{1}{2!}\left(\dfrac{3}{z + 2}\right)^2 + \dfrac{1}{4!}\left(\dfrac{3}{z + 2}\right)^4 - \cdots\right]$$

$$- \sin 2\left[(z + 2) - \dfrac{1}{3!}(z - 2)^3 \cdots\right]\left[1 - \dfrac{1}{2!}\left(\dfrac{3}{z + 2}\right)^2 + \dfrac{1}{4!}\left(\dfrac{3}{z + 2}\right)^4 - \cdots\right]$$

$$+ \cos 2\left[(z + 2) - \dfrac{1}{3!}(z - 2)^3 + \cdots\right]\left[\left(\dfrac{3}{z + 2}\right) - \dfrac{1}{3!}\left(\dfrac{3}{z + 2}\right)^3 + \cdots\right]$$

$$+ \sin 2\left[1 - \dfrac{1}{2!}(z + 2)^2 + \dfrac{1}{4!}(z + 2)^4 - \cdots\right]\left[\left(\dfrac{3}{z + 2}\right) - \dfrac{1}{3!}\left(\dfrac{3}{z + 2}\right)^3 + \cdots\right].$$

Note that the first and the third products do not contain $(z + 2)^{-1}$ term. From the second and the fourth products, collecting the coefficients of $(z + 2)^{-1}$, we obtain

$$\underset{z=-2}{Res}\, f(z) = -\sin 2 \left[-\frac{3^2}{1!\,2!} - \frac{3^4}{3!\,4!} - \frac{3^6}{5!\,6!} - \dots \right] + \sin 2 \left[\frac{3}{1!\,1!} + \frac{3^3}{2!\,3!} + \frac{3^5}{4!\,5!} - \dots \right]$$

$$= \sin 2 \left[\sum_{n=1}^{\infty} \frac{3^{2n}}{(2n-1)!\,(2n)!} + \sum_{n=0}^{\infty} \frac{3^{2n+1}}{(2n)!\,(2n+1)!} \right].$$

Exercise 13.2

Compute the residues at all the isolated singular points in the finite complex plane of the function $f(z)$ defined in the following problems.

1. $\dfrac{1}{z(z-1)}$.

2. $\dfrac{z}{z^2+1}$.

3. $\dfrac{z^4+3}{z^2-z}$.

4. $\dfrac{\sin z}{z^2+1}$.

5. $\dfrac{\sin z}{z^2(z-\pi)}$.

6. $\tan z$.

7. $\coth z$.

8. $\dfrac{\sqrt{z}}{\sin \sqrt{z}}$.

9. $\dfrac{1}{\sin(1/z)}$.

10. $\dfrac{e^{iz}+\cos z}{(z-\pi)^4}$.

11. $z^3 \cos(1/z)$.

12. $\sin z \cos(2/z)$.

13. $e^{z-1/z}$.

14. $\dfrac{\sin(1/z)}{z-1}$.

15. $z^n \sin(1/z)$, n integer.

16. $\sin\left(\dfrac{z}{z+1}\right)$.

17. $\dfrac{ze^{iz}}{(z-\pi)^2}$.

18. $\dfrac{\cos 2z}{(z+1)^2}$.

19. $\dfrac{z-3}{z^2 \sin(z-1)}$.

20. $\dfrac{\sin z}{z^3(z-1)}$.

21. $\dfrac{z^3+5}{(z^4-1)(z+1)}$.

22. $\dfrac{e^{imz}}{z(z^2+a^2)^2}$.

23. $\dfrac{\sin 2z}{(z+1)^3}$.

24. $\dfrac{z^2}{(z^2+1)^3}$.

25. $\dfrac{z^{3n}}{(z+1)^n}$.

26. $\dfrac{1}{\text{Log}^2 z}$.

27. $\cot^2 z$.

28. $\dfrac{\cot z}{z}$.

29. $\dfrac{z^2}{\sin^3 z}$.

30. $\dfrac{1}{(1+z^2)^{n+1}}$.

Find the residue at $z = \infty$ for the following functions.

31. $\dfrac{z^2}{z^3+2}$.

32. $\dfrac{1}{z^3+z^5}$.

33. $\dfrac{e^z}{z^2(z^2+1)}$.

34. $\dfrac{\sin 2z}{(z+1)^3}$.

35. $\sin z \sin(1/z)$.

36. $z^3 \sin(1/(z-2))$.

37. $\sin(e^{1/z})$.

38. Give a simple example of a function which has a non-removable singularity at $z = 0$ and whose residue at that point is zero.

39. Find the residue of the function $h(z) = f(z)\, g(z)$ at $z = a$, if $g(z)$ is analytic at $z = a$ and $f(z)$

(i) has a simple pole with residue A at $z = a$.

(ii) has a pole of order m with the principal part of the Laurent series expansion as

$$\frac{b_1}{z-a} + \frac{b_2}{(z-a)^2} + \dots + \frac{b_m}{(z-a)^m}.$$

40. Find the residue of the function $h(z) = f'(z)/f(z)$ at $z = a$ if

(i) $z = a$ is a zero of order n of $f(z)$, (ii) $z = a$ is a pole of order n of $f(z)$.

13.4 Evaluation of Contour Integrals Using Residues

In this section, we shall discuss the application of residues in evaluating contour integrals of the form

$$I = \oint_C f(z)\, dz$$

where C is a simple, piecewise smooth closed curve. We use the result given in the following theorem for evaluating the contour integrals.

Theorem 13.1 (Residue theorem) Let C be a simple, piecewise smooth closed curve and let $f(z)$ be analytic inside and on C except at a finite number of isolated singularities $z_1, z_2, \ldots, z_n$ lying inside C. Then

$$\oint_C f(z)\, dz = 2\pi i \ [\text{sum of residues of } f(z) \text{ at isolated singularities}]$$

$$= 2\pi i \sum_{k=1}^{n} \left[\operatorname*{Res}_{z=z_k} f(z) \right]. \tag{13.13}$$

Proof Since $z_1, z_2, \ldots, z_n$ are isolated singularities, there is a set of non-intersecting circles C_k with centres at z_k, $k = 1, 2, \ldots, n$ such that each circle C_k encloses only one singular point z_k and lies inside C (Fig. 13.1). By the extension of the Cauchy integral theorem for multiply connected domains, we can write

$$\oint_C f(z)\, dz = \oint_{C_1} f(z)\, dz + \oint_{C_2} f(z)\, dz + \ldots + \oint_{C_n} f(z)\, dz \tag{13.14}$$

where all the curves $C, C_1, C_2, \ldots, C_n$ are traversed in the anti-clockwise direction. Using Eqs. (13.6) and (13.7), we obtain

$$\oint_{C_k} f(z)\, dz = 2\pi i \left[\operatorname*{Res}_{z=z_k} f(z) \right].$$

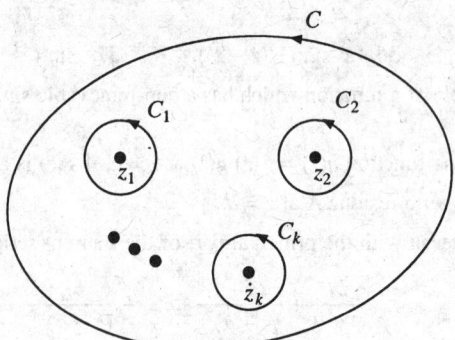

Fig. 13.1. Residue theorem.

Substituting this result in Eq. (13.14), we get

$$\oint_C f(z)\,dz = 2\pi i \sum_{k=1}^{n} \left[\underset{z=z_k}{\mathrm{Res}}\, f(z) \right].$$

Remark 2

Let $z_1, z_2, \ldots z_n$ be the isolated singularities of $f(z)$ in the finite complex plane with the corresponding residues $r_1, r_2, \ldots, r_n$ respectively. Let r_∞ be the residue at $z = \infty$. Since

$$r_\infty = \underset{z=\infty}{\mathrm{Res}}\, f(z) = -\frac{1}{2\pi i} \oint_C f(z)\,dz$$

where C encloses all the isolated singularities in the finite complex plane, we have

$$\sum_{k=1}^{n} r_k + r_\infty = 0.$$

Example 13.16 Evaluate the contour integral $I = \oint_C \dfrac{dz}{e^z - 1}$, $C : |z| = 1$.

Solution The integrand $f(z) = 1/(e^z - 1)$ has singularities when $e^z - 1 = 0$ or $e^z = 1 = e^{2n\pi i}$, or $z = 2n\pi i$, $n = 0 \pm 1, \pm 2, \ldots$ which are simple poles. Only the simple pole $z = 0$ lies inside C. Hence, by residue theorem, we get

$$I = 2\pi i \left[\underset{z=0}{\mathrm{Res}}\, f(z) \right] = 2\pi i \lim_{z \to 0} \left[\frac{z}{e^z - 1} \right] = 2\pi i.$$

Example 13.17 Use the residue theorem to evaluate the integral

$$I = \oint_C \frac{e^z - 1}{z(z - 1)(z - i)^2}\,dz$$

where (i) $C : |z| = 1/2$,   (ii) $C : |z| = 2$

Solution The integrand $f(z) = (e^z - 1)/[z(z - 1)(z - i)^2]$ has a removable singularity at $z = 0$, simple pole at $z = 1$ and a pole of order 2 at $z = i$.

(i) Since $z = 0$ is the only singular point which lies inside C and $z = 0$ is a removable singular point, we get

$$I = 2\pi i \left[\underset{z=0}{\mathrm{Res}}\, f(z) \right] = 2\pi i(0) = 0.$$

(ii) All the singular points $z = 0$, $z = 1$ and $z = i$ lie inside C. We get

$$\underset{z=0}{\mathrm{Res}}\, f(z) = 0. \quad \underset{z=1}{\mathrm{Res}}\, f(z) = \lim_{z \to 1} [(z - 1) f(z)] = \lim_{z \to 1} \left[\frac{e^z - 1}{z(z - i)^2} \right] = \frac{e - 1}{(1 - i)^2} = \frac{i(e - 1)}{2}.$$

$$\underset{z=i}{\mathrm{Res}}\, f(z) = \lim_{z \to i} \frac{d}{dz} [(z - i)^2\, f(z)] = \lim_{z \to i} \frac{d}{dz} \left[\frac{e^z - 1}{z(z - 1)} \right]$$

$$= \lim_{z \to i} \left[\frac{z(z - 1)e^z - (e^z - 1)(2z - 1)}{z^2(z - 1)^2} \right]$$

$$= \frac{i(i - 1)e^i - (e^i - 1)(2i - 1)}{i^2(i - 1)^2} = -\frac{1}{2}(3e^i - 2 - i).$$

Hence,
$$I = 2\pi i \left[\operatorname*{Res}_{z=0} f(z) + \operatorname*{Res}_{z=1} f(z) + \operatorname*{Res}_{z=i} f(z) \right]$$

$$= 2\pi i \left[0 + \frac{i(e-1)}{2} - \frac{1}{2}(3e^i - 2 - i) \right] = \pi i (ie - 3e^i + 2)$$

Example 13.18   Evaluate the integral $I = \oint_C \dfrac{e^z}{(z+1)^n}\, dz,\ C : |z| = 2.$

Solution   The integrand $f(z) = e^z/[z+1]^n$ has a pole of order n at $z = -1$, which lies inside C. we get

$$\operatorname*{Res}_{z=-1} f(z) = \frac{1}{(n-1)!} \lim_{z \to -1} \frac{d^{n-1}}{dz^{n-1}} [(z+1)^n f(z)] = \frac{1}{(n-1)!} \lim_{z \to -1} \frac{d^{n-1}}{dz^{n-1}} [e^z]$$

$$= \frac{1}{(n-1)!} \lim_{z \to -1} [e^z] = \frac{e^{-1}}{(n-1)!}$$

Hence,
$$I = 2\pi i \left[\operatorname*{Res}_{z=-1} f(z) \right] = \frac{2\pi i}{e(n-1)!}$$

Example 13.19   Evaluate the integrals

(i) $I = \oint_C z e^{1/z} dz,\ C : |z| = 1,$ (ii) $I = \oint_C \sin\left(\dfrac{1}{z}\right) dz,\ C : |z| = 1.$

Solution

(i) The integrand $f(z) = z e^{1/z}$ has an isolated essential singularity at $z = 0$ which lies inside C. The Laurent series expansion of $f(z)$ about $z = 0$ is given by

$$f(z) = z e^{1/z} = z + 1 + \frac{1}{2z} + \frac{1}{6z^2} + \dots .$$

Therefore,
$$\operatorname*{Res}_{z=0} f(z) = \frac{1}{2}.$$

Using the residue theorem, we get $I = 2\pi i \left[\operatorname*{Res}_{z=0} f(z) \right] = \pi i.$

(ii) The integrand $f(z) = \sin(1/z)$ has an isolated essential singularity at $z = 0$ which lies inside C. The Laurent series expansion of $f(z)$ about $z = 0$ is obtained as

$$f(z) = \sin\left(\frac{1}{z}\right) = \frac{1}{z} - \frac{1}{3! z^3} + \dots .$$

Therefore,
$$\operatorname*{Res}_{z=0} f(z) = 1 \quad \text{and} \quad I = 2\pi i.$$

Example 13.20   Evaluate the integral $I = \oint_C \dfrac{dz}{z^4 + 1},\ C : |z - 1| = 1.$

Solution   The integrand $f(z) = 1/(1 + z^4)$ has simple poles at

$$z_1 = \frac{1+i}{\sqrt{2}}, \ z_2 = \frac{-1-i}{2}, \ z_3 = \frac{1-i}{\sqrt{2}} \ \text{ and } \ z_4 = \frac{-1+i}{\sqrt{2}}.$$

Only z_1 and z_3 lie inside C (Fig. 13.2). We get

$$\operatorname*{Res}_{z=z_1} f(z) = \frac{1}{4z_1^3} = -\frac{z_1}{4}, \ \text{ since } z_1^4 = -1,$$

$$\operatorname*{Res}_{z=z_3} f(z) = \frac{1}{4z_3^3} = -\frac{z_3}{4}, \ \text{ since } z_3^4 = -1.$$

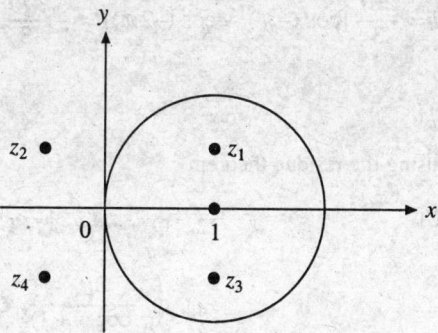

Fig. 13.2.   Example 13.20.

Hence,
$$I = 2\pi i \left[\operatorname*{Res}_{z=z_1} f(z) + \operatorname*{Res}_{z=z_3} f(z) \right] = -\frac{2\pi i}{4}[z_1 + z_3] = -\frac{\pi i}{\sqrt{2}}.$$

Example 13.21   Evaluate the integral $I = \oint_C (z+1) \cot\left(\frac{z}{2}\right) dz, \ C: |z| = 1.$

Solution   The integrand $f(z) = (z+1) \cot (z/2)$ has a simple pole at $z = 0$ which lies inside C. We get

$$\operatorname*{Res}_{z=0} f(z) = \lim_{z \to 0} \left[\frac{z(z+1)\cos(z/2)}{\sin(z/2)} \right] = 2 \lim_{z \to 0} \left[\frac{(z/2)}{\sin(z/2)} \right] \lim_{z \to 0} [(z+1)\cos(z/2)] = 2.$$

Hence,
$$I = 2\pi i \left[\operatorname*{Res}_{z=0} f(z) \right] = 4\pi i.$$

Example 13.22   Evaluate the integral $I = \oint_C \frac{z \cosh \pi z}{z^4 + 5z^2 + 4} dz, \ C : |z| = 4.$

Solution   The integrand has singular points where $z^4 + 5z^2 + 4 = 0$. We get

$$z^2 = -1, -4 \ \text{ or } z = \pm i, \pm 2i.$$

These singular points are simple poles and lie inside C. Let $z = a$ be a pole. Then

$$\operatorname*{Res}_{z=a} f(z) = \operatorname*{Res}_{z=a} \left[\frac{\phi(z)}{\psi(z)} \right] = \frac{\phi(a)}{\psi'(a)} = \frac{a \cosh \pi a}{4a^3 + 10a} = \frac{\cosh \pi a}{4a^2 + 10}$$

Hence,

$$I = 2\pi i \left[\operatorname*{Res}_{z=i} f(z) + \operatorname*{Res}_{z=-i} f(z) + \operatorname*{Res}_{z=2i} f(z) + \operatorname*{Res}_{z=-2i} f(z) \right]$$

$$= 2\pi i \left[\frac{\cosh \pi i}{6} + \frac{\cosh \pi i}{6} - \frac{\cosh 2\pi i}{6} - \frac{\cosh 2\pi i}{6} \right]$$

$$= \frac{2\pi i}{3} \left[\cosh \pi i - \cosh 2\pi i \right].$$

Since, $\cosh z = \cos iz$, we obtain

$$I = \frac{2\pi i}{3} \left[\cos (-\pi) - \cos (-2\pi) \right] = -\frac{4\pi i}{3}.$$

Exercise 13.3

Evaluate the following integrals using the residue theorem.

1. $\oint_C \dfrac{2z}{e^z - 1} \, dz,\ C : |z| = 1.$

2. $\oint_C \dfrac{1 - e^{z^2}}{z^3} \, dz,\ C : |z - 2| = 1.$

3. $\oint_C \dfrac{z^2}{1 + e^{zi}} \, dz,\ C : |z| = 4.$

4. $\oint_C \dfrac{dz}{\operatorname{Log}(z + 1)},\ C : |z| = \dfrac{1}{2}.$

5. $\oint_C \dfrac{2dz}{z^2 + 4iz - 1},\ C : |z| = 1.$

6. $\oint_C \dfrac{e^{-z^2}}{\sin 2z} \, dz,\ C : |z| = 1.$

7. $\oint_C \dfrac{e^{2z}}{\cos \pi z} \, dz,\ C : |z| = 1.$

8. $\oint_C \tanh z \, dz,\ C : |z| = 4.$

9. $\oint_C \dfrac{\sin z}{\cos 3z} \, dz,\ C : |z| = 1.$

10. $\oint_C \dfrac{\sinh 2z}{z^2 + 9} \, dz,\ C : |z| = 5.$

11. $\oint_C \dfrac{\sin z \, e^{1/(z-1)^2}}{z^2(z^2 - 1)} \, dz,\ C : \left| z + \dfrac{1}{2} \right| = 1.$

12. $\oint_C \dfrac{z}{\sin z} \, dz,\ C : |z| = 5.$

13. $\oint_C \dfrac{z^3}{z^4 - 1} \, dz,\ C : |z| = 2.$

14. $\oint_C \tan \pi z \, dz,\ C : |z| = 2.$

15. $\oint_C \dfrac{z^3}{2z^4 + 1} \, dz,\ C : |z| = 1.$

16. $\oint_C \dfrac{z \, dz}{\sin z (1 - \cos z)},\ C : |z| = 5.$

17. $\oint_C \dfrac{3 + 2z}{1 + \cos z} \, dz,\ C : |z| = 4.$

18. $\oint_C \dfrac{e^z (z^2 + 4) dz}{(z - i)^3},\ C : |z| = 2.$

19. $\oint_C \dfrac{2 - 3z}{1 - \sin z} \, dz,\ C : |z - 4| = 3.$

20. $\oint_C \dfrac{dz}{(z - 1)(z - 2)^2},\ C : |z - 2| = \dfrac{1}{2}.$

21. $\oint_C \dfrac{e^z dz}{(z^2 + z - 3/4)^2},\ C : |z| = 2.$

22. $\oint_C \dfrac{\cos z \, dz}{1 + z^2},\ C : |z - i| = r$ (i) $r < 2,$ (ii) $r > 2.$

23. $\oint_C \dfrac{1 - z}{z^3(1 + z)} \, dz,\ C : |z| = \dfrac{1}{2}.$

24. $\oint_C \dfrac{dz}{\cos^2 z},\ C : |z - \pi| = \pi.$

25. $\oint_C (2z - z^2)e^{1/z}\, dz$, $C : |z| = 1$.

26. $\oint_C (z-1)^2\, e^{1/z^2}\, dz$, $C : |z| = 2$.

27. $\oint_C e^{1/z} \sin(1/z)\, dz$, $C : |z| = 1$.

28. $\oint_C e^{1/z^n}\, dz$, $C : |z| = 1$, n integer.

29. $\oint_C z^n e^{1/z}\, dz$, $C : |z| = 1$.

30. $\oint_C \tan(1/z)\, dz$, $C : |z| = 1$.

31. $\oint_C \sin^2(1/z)\, dz$, $C : |z| = 1$.

32. $\oint_C (1 + z + z^2)\left[e^{2/z} + e^{1/(z-1)} + e^{1/(z-2)}\right] dz$, $C : |z| = \dfrac{1}{2}$.

33. Evaluate the integral $I = \oint_C \dfrac{f(z)}{z g(z)}\, dz$, where C is a simple, closed contour bounding a domain which contains the point $z = 0$. The functions $f(z)$ and $g(z)$ are analytic in the closed domain D. The function $g(z)$ does not vanish on C and has only simple zeros $z_1, z_2, \ldots, z_n$ in D, none of which coincides with $z = 0$.

34. Let a function $f(z)$ be analytic throughout a simply connected domain D and let z_0 be the only zero of $f(z)$ inside D. If C is a positively oriented simple closed contour in D that encloses z_0, then show that

$$\frac{1}{2\pi i} \oint_C \frac{f'(z)}{f(z)}\, dz = m$$

where the positive integer m is the order of the zero of $f(z)$.

35. Let $f(z)$ be analytic inside and on a simple, smooth closed curve C except for poles inside C at b_1, $b_2, \ldots, b_n$. Suppose that $f(z)$ has zeros at $a_1, a_2, \ldots, a_m$ inside C but none on C. Then, show that

$$\frac{1}{2\pi i} \oint_C \frac{f'(z)}{f(z)}\, dz = \sum_{k=1}^{n} \alpha_k - \sum_{k=1}^{n} \beta_k$$

where α_k is the order of a_k and β_k is the order of b_k.

13.5   Evaluation of Real Integrals Using Residues

One of the important applications of the residue theorem is its use in the evaluation of certain real integrals. These integrals are first transformed to associated contour integrals. The contour integrals are then evaluated by using the residue theorem. The values of the real integrals are then obtained from these results.

13.5.1   Real Definite Integrals Involving Trigonometric Functions

Consider an integral of the form

$$I = \int_0^{2\pi} F(\sin\theta,\, \cos\theta)\, d\theta \tag{13.15}$$

where F is a real rational function of $\sin\theta$ and $\cos\theta$.
Setting $z = e^{i\theta} = \cos\theta + i\sin\theta$, we get

$$\cos \theta = \frac{1}{2}(e^{i\theta} + e^{-i\theta}) = \frac{1}{2}\left(z + \frac{1}{z}\right) = \frac{z^2 + 1}{2z}$$

$$\sin \theta = \frac{1}{2i}(e^{i\theta} - e^{-i\theta}) = \frac{1}{2i}\left(z - \frac{1}{z}\right) = \frac{z^2 - 1}{2zi}$$

and

$$dz = ie^{i\theta}d\theta, \text{ or } d\theta = \frac{dz}{ie^{i\theta}} = \frac{dz}{iz}. \tag{13.16}$$

As θ varies from 0 to 2π, the variable z traverses once around the unit circle $|z| = 1$ in the anti-clockwise direction. Then, the integral given in Eq. (13.15) can be written as

$$I = \oint_C f(z)\frac{dz}{iz}, \ C : |z| = 1 \tag{13.17}$$

where $f(z) = F\left(\frac{z^2 - 1}{2iz}, \frac{z^2 + 1}{2z}\right)$ is a rational function of z. The contour integral given in Eq. (13.17) can be evaluated by using the residue theorem.

Example 13.23 Evaluate the integral $I = \int_0^{2\pi} \frac{d\theta}{2 + \sin \theta}$.

Solution Substitute $z = e^{i\theta}$. We have $d\theta = dz/(iz)$, $\sin \theta = (z^2 - 1)/(2zi)$ and

$$2 + \sin \theta = 2 + \frac{z^2 - 1}{2zi} = \frac{z^2 + 4zi - 1}{2zi}.$$

Therefore,

$$I = \int_0^{2\pi} \frac{d\theta}{2 + \sin \theta} = \oint_C \frac{2zi}{z^2 + 4zi - 1}\left(\frac{dz}{iz}\right) = 2\oint_C \frac{dz}{z^2 + 4zi - 1} = \oint_C f(z)dz$$

where $f(z) = 2/(z^2 + 4iz - 1)$ and C is $|z| = 1$. Setting $z^2 + 4zi - 1 = 0$, we find that the integrand has simple poles at $z_1 = (-2 + \sqrt{3})i$ and $z_2 = -(2 + \sqrt{3})i$. Only the pole z_1 lies inside the contour $C : |z| = 1$.

Hence, $\operatorname*{Res}_{z = z_1} f(z) = 2 \lim_{z \to z_1}\left[\frac{z - z_1}{(z - z_1)(z - z_2)}\right] = \frac{2}{z_1 - z_2} = \frac{1}{\sqrt{3}i}$

and $I = 2\pi i\left[\operatorname*{Res}_{z = z_1} f(z)\right] = 2\pi/\sqrt{3}.$

Example 13.24 Evaluate the integral $I = \int_0^{2\pi} \frac{d\theta}{a + \cos \theta}, a > 1.$

Solution Substitute $z = e^{i\theta}$. We have $d\theta = dz/(iz)$, $\cos \theta = (z^2 + 1)/(2z)$ and

$$a + \cos \theta = a + \frac{z^2 + 1}{2z} = \frac{z^2 + 2az + 1}{2z}.$$

Therefore, $I = \int_0^{2\pi} \frac{d\theta}{a + \cos \theta} = \oint_C \frac{2z}{z^2 + 2az + 1}\left(\frac{dz}{iz}\right) = \oint_C f(z)dz$

where $f(z) = 2/[(z^2 + 2az + 1)i]$ and C is $|z| = 1$. The integrand has singular points when $z^2 + 2az + 1 = 0$, that is, when $z_1 = -a + \sqrt{a^2 - 1}$ and $z_2 = -(a + \sqrt{a^2 - 1})$. Since $a^2 > 1$, only the singular point z_1, which is a simple pole, lies inside C. Hence,

$$\operatorname*{Res}_{z=z_1} f(z) = \frac{2}{i} \lim_{z \to z_1} \left[\frac{(z - z_1)}{(z - z_1)(z - z_2)} \right] = \frac{2}{i(z_1 - z_2)} = \frac{1}{i\sqrt{a^2 - 1}}$$

and
$$I = 2\pi i \left[\operatorname*{Res}_{z=z_1} f(z) \right] = \frac{2\pi}{\sqrt{a^2 - 1}}.$$

Example 13.25 Evaluate the integral $I = \displaystyle\int_0^\pi \sin^4\theta d\theta$.

Solution We have

$$\int_0^{2\pi} \sin^4\theta\, d\theta = \int_0^\pi \sin^4\theta\, d\theta + \int_\pi^{2\pi} \sin^4\theta\, d\theta.$$

Substituting $\theta = 2\pi - \phi$, in the second integral and simplifying, we obtain

$$\int_0^{2\pi} \sin^4\theta\, d\theta = \int_0^\pi \sin^4\theta\, d\theta + \int_0^\pi \sin^4\phi\, d\phi = 2\int_0^\pi \sin^4\theta\, d\theta.$$

Therefore,

$$I = \frac{1}{2} \int_0^{2\pi} \sin^4\theta d\theta.$$

Substituting $z = e^{i\theta}$, we have $d\theta = dz/(iz)$, $\sin \theta = (z^2 - 1)/(2zi)$, and

$$I = \frac{1}{2} \oint_C \left(\frac{z^2 - 1}{2zi} \right)^4 \left(\frac{dz}{iz} \right) = \oint_C f(z)dz$$

where $f(z) = (z^2 - 1)^4/(32z^5 i)$ and C is $|z| = 1$. The integrand has a pole of order 5 at $z = 0$. The residue at $z = 0$ is given by

$$\operatorname*{Res}_{z=0} f(z) = \left(\frac{1}{4!} \right) \lim_{z \to 0} \frac{d^4}{dz^4} \left[z^5 f(z) \right]$$

$$= \frac{1}{768i} \lim_{z \to 0} \frac{d^4}{dz^4} [z^8 - 4z^6 + 6z^4 - 4z^2 + 1]$$

$$= \frac{144}{768i} = \frac{3}{16i}.$$

Therefore, $I = 2\pi i \left[\operatorname*{Res}_{z=0} f(z) \right] = \dfrac{3\pi}{8}$.

Example 13.26 Evaluate the integral $I = \displaystyle\int_0^{2\pi} \frac{d\theta}{1 - 2a \cos \theta + a^2}$, where a is a complex constant and (i) $|a| < 1$, (ii) $|a| > 1$.

Solution Substitute $z = e^{i\theta}$. We have $d\theta = dz/(iz)$, $\cos\theta = (z^2 + 1)/(2z)$ and

$$1 - 2a\cos\theta + a^2 = 1 + a^2 - \frac{2a(z^2 + 1)}{2z} = -\frac{a}{z}\left[(z^2 + 1) - \left(\frac{a^2 + 1}{a}\right)z\right]$$

Therefore,

$$I = -\frac{1}{a}\oint_C \frac{z}{z^2 - \left(\dfrac{a^2 + 1}{a}\right)z + 1}\left(\frac{dz}{iz}\right) = \oint_C f(z)dz$$

where

$$f(z) = -\frac{1}{ai}\left[\frac{1}{z^2 - \left(\dfrac{a^2 + 1}{a}\right)z + 1}\right] \quad \text{and} \quad C \text{ is } |z| = 1.$$

Setting $z^2 - \left(\dfrac{a^2 + 1}{a}\right)z + 1 = 0$, we find that the singular points of the integrand are $z_1 = a$ and $z_2 = 1/a$.

(i) When $|a| < 1$, only the singular point $z_1 = a$, which is a simple pole, lies inside C. Hence,

$$\operatorname*{Res}_{z=a} f(z) = -\frac{1}{ia}\lim_{z \to a}\left[\frac{(z - a)}{(z - a)(z - (1/a))}\right] = \frac{-1}{ia[a - (1/a)]} = \frac{1}{i(1 - a^2)}.$$

Therefore, $I = 2\pi i\left[\operatorname*{Res}_{z=a} f(z)\right] = \dfrac{2\pi}{1 - a^2}$.

(ii) When $|a| > 1$, only the singular point $z_2 = 1/a$, which is a simple pole, lies inside C. Hence,

$$\operatorname*{Res}_{z=(1/a)} f(z) = -\frac{1}{ia}\lim_{z \to (1/a)}\left[\frac{(z - (1/a))}{(z - a)(z - (1/a))}\right] = -\frac{1}{ia[(1/a) - a]} = \frac{1}{i(a^2 - 1)}.$$

Therefore, $I = 2\pi i\left[\operatorname*{Res}_{z=(1/a)} f(z)\right] = \dfrac{2\pi}{a^2 - 1}$.

Example 13.27 Evaluate the integral $I = \displaystyle\int_0^{2\pi} e^{\cos\theta}\cos(\sin\theta)d\theta.$

Solution The integrand can be written as

$$e^{\cos\theta}\cos(\sin\theta) = \frac{1}{2}e^{\cos\theta}[e^{i\sin\theta} + e^{-i\sin\theta}] = \frac{1}{2}[e^{\cos\theta + i\sin\theta} + e^{\cos\theta - i\sin\theta}]$$

$$= \frac{1}{2}[e^{e^{i\theta}} + e^{e^{-i\theta}}] = \frac{1}{2}[e^z + e^{1/z}]$$

where $z = e^{i\theta}$. Therefore, setting $z = e^{i\theta}$ we have $d\theta = dz/iz$ and

$$I = \int_0^{2\pi} e^{\cos\theta}\cos(\sin\theta)d\theta = \frac{1}{2}\oint_C (e^z + e^{1/z})\frac{dz}{iz} = \oint_C f(z)dz$$

where $f(z) = (e^z + e^{1/z})/(2iz)$ and C is $|z| = 1$.

Now, $z = 0$ is an essential singularity of the integrand. The Laurent series expansion of $f(z)$ is given by

$$f(z) = \frac{1}{2iz}\left[e^z + e^{1/z}\right] = \frac{1}{2iz}\left[\left(1 + z + \frac{z^2}{2!} + \ldots\right) + \left(1 + \frac{1}{z} + \frac{1}{2!z^2} + \ldots\right)\right]$$

$$= \frac{1}{2i}\left[\left(1 + \frac{z}{2!} + \ldots\right) + \left(\frac{2}{z} + \frac{1}{z^2} + \ldots\right)\right]$$

Hence, $\operatorname*{Res}\limits_{z=0} f(z) = \frac{1}{i}$.

Therefore, $I = 2\pi i\left[\operatorname*{Res}\limits_{z=0} f(z)\right] = 2\pi$.

13.5.2   Improper Real Integrals of the Form $\displaystyle\int_{-\infty}^{\infty} f(x)dx$

A real integral $\displaystyle\int_{a}^{b} f(x)dx$ is called an improper integral if either (i) one or both of the limits of integration are not finite, or (ii) the integrand has infinite discontinuity at a or at b (a, b finite) or at some point c, $a < c < b$.

Now,
$$I = \int_{-\infty}^{\infty} f(x)dx \tag{13.18}$$

is defined as

$$I = \lim_{R\to-\infty}\int_{R}^{0} f(x)\,dx + \lim_{S\to\infty}\int_{0}^{S} f(x)\,dx \tag{13.19}$$

where $(R, S) \to (-\infty, \infty)$ arbitrarily. If the limits in Eq. (13.19) do not exist then the integral I does not exist. However, if $(R, S) \to (-\infty, \infty)$ along $R = S$, then we say that the *Cauchy principal value* (*CPV*) of the integral I exists, that is

$$CPV(I) = \lim_{R\to\infty}\int_{-R}^{R} f(x)dx. \tag{13.20}$$

We note the following:

   (i) If $CPV(I)$ exists, it does not mean that I converges.

   (ii) If I exists, then $CPV(I)$ exists and in this case

$$CPV(I) = I. \tag{13.21}$$

For example, if $f(x) = x$, then

$$CPV(I) = \lim_{R\to\infty}\left[\frac{x^2}{2}\right]_{-R}^{R} = 0$$

but $\displaystyle\lim_{R\to-\infty}\int_{R}^{0} xdx = \lim_{R\to-\infty}\left[\frac{x^2}{2}\right]_{R}^{0} = -\infty$ does not exist.

In fact, both the integrals on the right hand side of Eq. (13.19) do not exist, even though $CPV(I)$ exists.

Consider the case when $f(x) = p(x)/q(x)$ is a rational function, where $p(x)$ and $q(x)$ have no common factors. To evaluate the improper integral given in Eq. (13.18), we consider the associated contour integral

$$\int_C f(z)dz \qquad (13.22)$$

where $f(z) = p(z)/q(z), p(z), q(z)$ have real coefficients and have no common factors and C is the path $C_R \cup C_1$ in the upper half plane and traversed in the anti-clockwise direction. Therefore, C is the path from A to B along the semicircle C_R and then from B to A along the real axis (Fig. 13.3).

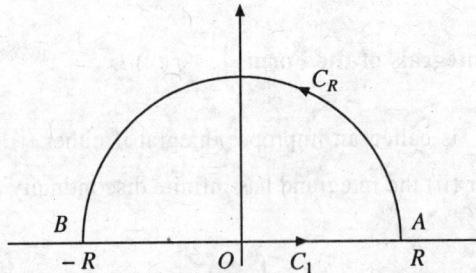

Fig. 13.3.   Contour for the integral in Eq. (13.22) in z-plane.

We assume that $f(z)$ satisfies the following conditions:

(i) $f(z)$ is analytic in the upper half of the z-plane except at a finite number of poles $z_1, z_2, \ldots, z_n$ in this half plane, none of which lies on the real axis. It implies that the contour C encloses all the poles in the upper half plane.

(ii) $zf(z)$ converges uniformly to zero as $R \to \infty$ through values for which $0 \le \theta \le \pi$. As a corollary, we obtain the condition that the degree of the denominator $q(z)$ in $f(z)$ is at least two units higher than the degree of the numerator $p(z)$.

Now, I can be written as

$$I = \int_C f(z)dz = \int_{C_R} f(z)dz + \int_{-R}^R f(x)dx \qquad (13.23)$$

Using the residue theorem for evaluating a contour integral, we obtain

$$\int_C f(z)dz = \int_{C_R} f(z)dz + \int_{-R}^R f(x)dx = 2\pi i \left[\sum_{k=1}^n \operatorname*{Res}_{z=z_k} f(z) \right] \qquad (13.24)$$

where z_k's are the poles of $f(z)$ lying in the upper half plane.

We now determine the contribution of the integral $\int_{C_R} f(z)dz$.

Let $z = Re^{i\theta}$. Then, $dz = Rie^{i\theta}d\theta = izd\theta$

and
$$\left| \int_{C_R} f(z)dz \right| = \left| \int_0^\pi f(z) \, izd\theta \right| \le \int_0^\pi |zf(z)| \, |d\theta|. \qquad (13.25)$$

Since, by assumption, $z f(z)$ converges uniformly to zero as $R \to \infty$, $0 \le \theta \le \pi$, we have that given a real number $\varepsilon > 0$, there exists a radius R^*, depending on ε but not on θ, such that

$$| z f(z) | < \varepsilon, \quad \text{whenever} \quad R > R^* \quad \text{for all } \theta \text{ in } 0 \le \theta \le \pi.$$

Hence, from Eq. (13.25), we obtain

$$\left| \int_{c_R} f(z)dz \right| \le \varepsilon \int_0^\pi | d\theta | = \varepsilon \pi.$$

Since, ε is arbitrarily small, we have $\displaystyle \lim_{R \to \infty} \int_{C_R} f(z)dz = 0$.

Therefore, $\displaystyle \lim_{R \to \infty} \int_C f(z)dz = \lim_{R \to \infty} \int_{C_R} f(z)dz + \lim_{R \to \infty} \int_{-R}^R f(x)dx = \int_{-\infty}^\infty f(x)dx$

Using Eq. (13.24), we obtain

$$\int_{-\infty}^\infty f(x)dx = 2\pi i \left[\sum_{k=1}^n \operatorname*{Res}_{z=z_k} f(z) \right]. \tag{13.26}$$

Consider now, the second part of the condition (ii) on $f(z) = p(z)/q(z)$, that is the degree of $q(z)$ is at least two units higher than the degree of the numerator $p(z)$. Let

$$f(z) = \frac{a_n z^n + a_{n-1} z^{n-1} + \ldots + a_1 z + a_0}{b_m z^m + b_{m-1} z^{m-1} + \ldots + b_1 z + b_0}, \quad a_n \ne 0, b_m \ne 0 \tag{13.27}$$

where $m - n = d \ge 2$ and d is an integer. Rewrite $f(z)$ as

$$f(z) = \frac{a_n + a_{n-1} z^{-1} + \ldots + a_0 z^{-n}}{b_m + b_{m-1} z^{-1} + \ldots + b_0 z^{-m}} \left(\frac{z^n}{z^m} \right).$$

For sufficiently large R^*, such that $| z | = R \ge R^*$, we get

$$|f(z)| < \frac{M}{| z |^d} = \frac{M}{R^d}$$

where M is any number greater than $| a_n / b_m |$. Hence, using the *ML*-inequality of integrals, we obtain

$$\left| \int_{c_R} f(z)dz \right| \le \frac{M}{R^d} (\pi R) = \frac{\pi M}{R^{d-1}} \to 0 \text{ as } R \to \infty, \ d \ge 2.$$

Example 13.28 Evaluate the integral $\displaystyle \int_0^\infty \frac{x^2 + 2}{(x^2 + 1)(x^2 + 4)} dx$.

Solution Since the integrand is an even function, we have

$$\int_0^\infty \frac{x^2 + 2}{(x^2 + 1)(x^2 + 4)} dx = \frac{1}{2} \int_{-\infty}^\infty \frac{x^2 + 2}{(x^2 + 1)(x^2 + 4)} dx.$$

Consider the contour integral

$$I = \int_C f(z)\, dz, \quad \text{where} \quad f(z) = \frac{z^2 + 2}{2(z^2 + 1)(z^2 + 4)}$$

and C is the path $C_R \cup C_1$ as given in Fig. 13.3. The integrand has simple poles at $z = \pm i$ and $z = \pm 2i$. Therefore, $f(z)$ is analytic in the upper half plane except for the simple poles at $z_1 = i$ and $z_2 = 2i$. We find that

$$\operatorname*{Res}_{z=i} f(z) = \lim_{z \to i} \left[\frac{z^2 + 2}{2(z^2 + 4)(z + i)} \right] = \frac{1}{12i}$$

and

$$\operatorname*{Res}_{z=2i} f(z) = \lim_{z \to 2i} \left[\frac{z^2 + 2}{2(z^2 + 1)(z + 2i)} \right] = \frac{1}{12i}.$$

We now write

$$\int_C f(z)dz = \int_{-R}^{R} f(x)dx + \int_{C_R} f(z)dz. \tag{13.28}$$

Along $C_R : |z| = R$ in the upper half plane, encircling the poles $z = i, 2i$, we have

$$|f(z)| = \frac{|2 + z^2|}{2|z^2 + 1||z^2 + 4|} \le \frac{R^2 + 2}{2(R^2 - 1)(R^2 - 4)}$$

since $|z^2 + 1| \ge |z|^2 - 1 = R^2 - 1$ and $|z^2 + 4| \ge |z|^2 - 4 = R^2 - 4$.

Hence,

$$\left| \int_{C_R} f(z)dz \right| \le \frac{(R^2 + 2)\pi R}{2(R^2 - 1)(R^2 - 4)} \to 0 \text{ as } R \to \infty.$$

Letting $R \to \infty$ in Eq. (13.28), we obtain

$$I = \frac{1}{2} \int_{-\infty}^{\infty} \frac{(x^2 + 2)dx}{(x^2 + 1)(x^2 + 4)} = \int_{0}^{\infty} \frac{(x^2 + 2)dx}{(x^2 + 1)(x^2 + 4)} = 2\pi i \left[\operatorname*{Res}_{z=i} f(z) + \operatorname*{Res}_{z=2i} f(z) \right]$$

$$= 2\pi i \left(\frac{1}{12i} + \frac{1}{12i} \right) = \frac{\pi}{3}.$$

Therefore, $\displaystyle \int_{0}^{\infty} \frac{x^2 + 2}{(x^2 + 1)(x^2 + 4)} dx = \frac{\pi}{3}.$

Example 13.29   Evaluate the integral $\displaystyle \int_{-\infty}^{\infty} \frac{dx}{1 + x^4}$.

Solution   Consider the contour integral

$$I = \int_C f(z)dz, \quad \text{where} \quad f(z) = \frac{1}{1 + z^4}$$

and C is the path $C_R \cup C_1$ as given in Fig. 13.3. Setting $1 + z^4 = 0$, we obtain $z^2 = \pm i$. The singular points of $f(z)$ are given by $z = \pm (1 + i)/\sqrt{2}, z = \pm (1 - i)/\sqrt{2}$, which are simple poles. The poles $z_1 = (1 + i)/\sqrt{2} = e^{\pi i/4}$ and $z_2 = (-1 + i)/\sqrt{2} = e^{3\pi i/4}$ lie inside the upper half plane.

We obtain

$$\operatorname*{Res}_{z=z_1} f(z) = \lim_{z \to z_1} \left[\frac{1}{4z^3} \right] = \frac{1}{4z_1^3} = \frac{z_1}{4z_1^4} = -\frac{1+i}{4\sqrt{2}}$$

and

$$\operatorname*{Res}_{z=z_2} f(z) = \lim_{z \to z_2} \left[\frac{1}{4z^3} \right] = \frac{1}{4z_2^3} = \frac{z_2}{4z_2^4} = -\frac{-1+i}{4\sqrt{2}}.$$

We now write

$$\int_C f(z)dz = \int_{-R}^R f(x)dx + \int_{C_R} f(z)dz. \tag{13.29}$$

Let $R > 1$, so that the singular points z_1 and z_2 lie inside $C : |z| = R$. Then

$$|f(z)| = \frac{1}{|1 + z^4|} \le \frac{1}{R^4 - 1}.$$

Using the *ML*-inequality, we obtain

$$\left| \int_{C_R} f(z)dz \right| \le \frac{1}{R^4 - 1}(\pi R) \to 0 \text{ as } R \to \infty.$$

Letting $R \to \infty$ in Eq. (13.29), we get

$$\int_{-\infty}^{\infty} f(x)dx = \int_{-\infty}^{\infty} \frac{dx}{1 + x^4} = 2\pi i \left[\operatorname*{Res}_{z=z_1} f(z) + \operatorname*{Res}_{z=z_2} f(z) \right]$$

$$= -\frac{2\pi i}{4\sqrt{2}}[(1 + i) + (-1 + i)] = \frac{\pi}{\sqrt{2}}.$$

Example 13.30   Evaluate the integral $\displaystyle\int_0^{\infty} \frac{x^2 dx}{(x^2 + a^2)^2}$, $a > 0$.

Solution   Since the integral is an even function, we can write

$$\int_0^{\infty} \frac{x^2 dx}{(x^2 + a^2)^2} = \frac{1}{2} \int_{-\infty}^{\infty} \frac{x^2}{(x^2 + a^2)^2} dx.$$

Consider the contour integral

$$I = \int_C f(z)dz, \text{ where } f(z) = \frac{z^2}{2(z^2 + a^2)^2}$$

and C is the path $C_R \cup C_1$ as given in Fig. 13.3. The integrand $f(z)$ is analytic in the upper half plane except for the pole of order 2 at $z = ai$. We find that

$$\operatorname*{Res}_{z=ai} f(z) = \lim_{z \to ai} \frac{d}{dz} [(z - ai)^2 f(z)] = \lim_{z \to ai} \frac{d}{dz} \left[\frac{z^2}{2(z + ai)^2} \right] = -\frac{i}{8a}.$$

We now write

$$I = \int_{-R}^R f(x)dx + \int_{C_R} f(z)dz. \tag{13.30}$$

Let $R > a$. Then

$$|f(z)| = \frac{|z|^2}{2|z^2 + a^2|^2} \le \frac{R^2}{2(R^2 - a^2)^2}, \quad \text{since} \quad |z^2 + a^2| \ge R^2 - a^2.$$

Hence,

$$\left| \int_{C_R} f(z)dz \right| \le \frac{R^2}{2(R^2 - a^2)^2} (\pi R) \to 0 \text{ as } R \to \infty.$$

Letting $R \to \infty$ in Eq. (13.30), we obtain

$$I = \frac{1}{2} \int_{-\infty}^{\infty} \frac{x^2}{(x^2 + a^2)^2} \, dx = \int_{0}^{\infty} \frac{x^2}{(x^2 + a^2)^2} \, dx = 2\pi i \left[\operatorname*{Res}_{z=ai} f(z) \right] = \frac{\pi}{4a}.$$

Example 13.31 Evaluate the integral $\displaystyle\int_{-\infty}^{\infty} \frac{dx}{(x^2 + a^2)^3}, a > 0$.

Solution   Consider the contour integral

$$I = \int_{C} f(z)dz, \quad \text{where} \quad f(z) = \frac{1}{(z^2 + a^2)^3}$$

and C is the path $C_R \cup C_1$ as given in Fig. 13.3.

The integrand $f(z)$ is analytic in the upper half plane except for the pole of order 3 at $z = ai$. We find that

$$\operatorname*{Res}_{z=ai} f(z) = \frac{1}{2!} \lim_{z \to ai} \frac{d^2}{dz^2} [(z - ai)^3 f(z)] = \frac{1}{2} \lim_{z \to ai} \frac{d^2}{dz^2} \left[\frac{1}{(z + ai)^3} \right]$$

$$= \frac{1}{2} \lim_{z \to ai} \left[\frac{12}{(z + ai)^5} \right] = \frac{3}{16a^5 i}.$$

We now write

$$I = \int_{C} f(z)dz = \int_{-R}^{R} f(x)dx + \int_{C_R} f(z)dz. \tag{13.31}$$

Let $R > a$. Then $|f(z)| = \dfrac{1}{(z^2 + a^2)^3} \le \dfrac{1}{(R^2 - a^2)^3}$, since $|z^2 + a^2| \ge R^2 - a^2$.

Using the *ML*-inequality, we obtain

$$\left| \int_{C_R} f(z)dz \right| \le \frac{\pi R}{(R^2 - a^2)^3} \to 0 \text{ as } R \to \infty.$$

Letting $R \to \infty$ in Eq. (13.31), we obtain

$$I = \int_{-\infty}^{\infty} f(x) \, dx = \int_{-\infty}^{\infty} \frac{dx}{(x^2 + a^2)^3} = 2\pi i \left[\operatorname*{Res}_{z=ai} f(z) \right] = \frac{3\pi}{8a^5}.$$

13.5.3 Improper Real Integrals of the form $\int_{-\infty}^{\infty} \cos ax f(x)dx$ and $\int_{-\infty}^{\infty} \sin ax f(x)dx$.

We now consider the evaluation of improper real integrals of the forms

$$\int_{-\infty}^{\infty} \cos (ax) f(x)dx \text{ and } \int_{-\infty}^{\infty} \sin (ax) f(x)dx, \quad a > 0. \tag{13.32}$$

These integrals are called *Fourier integrals*. Since

$$\int_{-\infty}^{\infty} \cos (ax) f(x)dx = \int_{-\infty}^{\infty} \text{Re}(e^{aix}) f(x)dx = \text{Re} \int_{-\infty}^{\infty} e^{aix} f(x)dx$$

and

$$\int_{-\infty}^{\infty} \sin (ax) f(x)dx = \int_{-\infty}^{\infty} \text{Im}(e^{aix}) f(x)dx = \text{Im} \int_{-\infty}^{\infty} e^{aix} f(x)dx$$

it is sufficient to consider the evaluation of the integral

$$I = \int_{-\infty}^{\infty} e^{aix} f(x)dx. \tag{13.33}$$

The real and imaginary parts of I give the values of the integrals in Eq. (13.32).

To evaluate the improper real integral given in Eq. (13.33), we consider the associated contour integral

$$I = \int_C e^{aiz} f(z)dz \tag{13.34}$$

where $f(z) = p(z)/q(z)$ and C is the path $C_R \cup C_1$ as defined in Fig. 13.3. We assume that $f(z)$ satisfies the same conditions as discussed in the previous section 13.5.2. Now, write I as

$$I = \int_C e^{aiz} f(z)dz = \int_{C_R} e^{aiz} f(z)dz + \int_{-R}^{R} e^{aix} f(x)dx. \tag{13.35}$$

Using the residue theorem, we obtain

$$\int_C e^{aiz} f(z)dz = \int_{C_R} e^{aiz} f(z)dz + \int_{-R}^{R} e^{aix} f(x)dx = 2\pi i \left[\sum_{k=1}^{n} \operatorname*{Res}_{z=z_k} (e^{aiz} f(z)) \right] \tag{13.36}$$

where z_k's are the poles of $f(z)$ lying in the upper half plane. Now, consider the contribution of the integral $\int_{C_R} e^{aiz} f(z)dz$. We have

$$\left| \int_{C_R} e^{aiz} f(z)dz \right| \leq \int_{C_R} |e^{aiz} f(z)| |dz|.$$

Now,

$$|e^{aiz}| = |e^{ai(x+iy)}| = |e^{aix} e^{-ay}| = e^{-ay} \leq 1 \tag{13.37}$$

since $a > 0$ and $y \geq 0$. Hence,

$$\left| \int_{C_R} e^{aiz} f(z)dz \right| \leq \int_{C_R} |e^{-ay} f(z)| |dz| \leq \int_{C_R} |f(z)| |dz| \tag{13.38}$$

which is the same integral as in Eq. (13.25). Hence, as $R \to \infty$ the contribution of the integral $\int_{C_R} e^{aiz} f(z)\, dz$ is zero. Therefore, as $R \to \infty$, we obtain from Eq. (13.35)

$$\lim_{R \to \infty} \int_C e^{aiz} f(z)dz = \lim_{R \to \infty} \int_{-R}^{R} e^{aix} f(x)dx.$$

Hence,

$$\int_{-\infty}^{\infty} e^{aix} f(x)dx = 2\pi i \left[\sum_{k=1}^{n} \operatorname*{Res}_{z=z_k} (e^{aiz} f(z)) \right]. \tag{13.39}$$

Comparing the real and imaginary parts on both sides, we get

$$\int_{-\infty}^{\infty} \cos (ax) f(x)dx = \operatorname{Re} \left[2\pi i \sum_{k=1}^{n} \operatorname*{Res}_{z=z_k} (e^{aiz} f(z)) \right] \tag{13.40}$$

and

$$\int_{-\infty}^{\infty} \sin (ax) f(x)dx = \operatorname{Im} \left[2\pi i \sum_{k=1}^{n} \operatorname*{Res}_{z=z_k} (e^{aiz} f(z)) \right]. \tag{13.41}$$

Remark 3

The justification of considering the contour $C = C_R \cup C_1$ in the upper half plane can be seen from Eq. (13.37). If, for example, the contour is taken as the lower half plane, then $e^{-ay}, y < 0$ is unbounded, that is, $| e^{aiz} |$ is unbounded and the result given in Eq. (13.39) is not valid.

Example 13.32   Evaluate the integrals

(i) $I_1 = \displaystyle\int_{-\infty}^{\infty} \frac{\cos ax}{b^2 + x^2}\, dx,$

(ii) $I_2 = \displaystyle\int_{-\infty}^{\infty} \frac{\sin ax}{b^2 + x^2}\, dx, a > 0, b > 0.$

Solution   Consider the contour integral

$$I = \int_C \frac{e^{aiz}}{b^2 + z^2}\, dz = \int_C e^{aiz} f(z)dz$$

where $f(z) = 1/(b^2 + z^2)$ and C is the path $C_R \cup C_1$ as given in Fig. 13.3. The function $f(z)$ is analytic in the upper half plane except for the simple pole at $z_1 = bi$. We now write

$$I = \int_C \frac{e^{aiz}}{b^2 + z^2} dx = \int_{C_R} \frac{e^{aiz}}{b^2 + z^2} dz + \int_{-R}^{R} \frac{e^{aix}}{b^2 + x^2}\, dx = 2\pi i \left[\operatorname*{Res}_{z=bi} e^{aiz} f(z) \right].$$

We find that

$$\operatorname*{Res}_{z=bi} [e^{aiz} f(z)] = \lim_{z \to bi} \left[\frac{(z - bi)e^{aiz}}{z^2 + b^2} \right] = \frac{e^{-ab}}{2bi}.$$

Now,

$$\left| \int_{C_R} e^{aiz} f(z)dz \right| \leq \int_{C_R} | e^{aiz} | | f(z) | | dz | \leq \int_{C_R} |f(z)| | dz |$$

since $| e^{aiz} | = e^{-ay} \leq 1$. Also, $| z^2 + b^2 | \geq | z |^2 - b^2 = R^2 - b^2$. Therefore,

$$\int_{C_R} |f(z)| | dz | = \int_{C_R} \frac{| dz |}{| z^2 + b^2 |} \leq \int_{C_R} \frac{| dz |}{R^2 - b^2} = \frac{\pi R}{R^2 - b^2} \to 0 \text{ as } R \to \infty.$$

Hence, as $R \to \infty$, we obtain

$$\int_{-\infty}^{\infty} \frac{e^{aix}}{b^2 + x^2} dx = 2\pi i \left[\operatorname*{Res}_{z=bi} f(z) \right] = 2\pi i \left(\frac{e^{-ab}}{2bi} \right) = \frac{\pi}{b} e^{-ab}.$$

Comparing the real and imaginary parts on both sides, we get

$$I_1 = \int_{-\infty}^{\infty} \frac{\cos ax}{b^2 + x^2} dx = \frac{\pi e^{-ab}}{b} \quad \text{and} \quad I_2 = \int_{-\infty}^{\infty} \frac{\sin ax}{b^2 + x^2} dx = 0.$$

Remark 4

$$\int_0^{\infty} \frac{\cos ax}{b^2 + x^2} dx = \frac{1}{2} \int_{-\infty}^{\infty} \frac{\cos ax}{b^2 + x^2} dx = \frac{\pi}{2b} e^{-ab}.$$

$$\int_0^{\infty} \frac{\sin ax}{b^2 + x^2} dx = \frac{1}{2} \int_{-\infty}^{\infty} \frac{\sin ax}{b^2 + x^2} dx = 0.$$

Example 13.33 Evaluate the integral $\displaystyle\int_{-\infty}^{\infty} \frac{\sin^2 2x}{1 + x^2} dx$.

Solution   We have $\sin^2 2x = \dfrac{1}{2}(1 - \cos 4x) = -\operatorname{Re}\left[\dfrac{1}{2}(e^{4ix} - 1)\right]$

Therefore,

$$\int_{-\infty}^{\infty} \frac{\sin^2 2x}{1 + x^2} dx = -\frac{1}{2} \operatorname{Re} \int_{-\infty}^{\infty} \frac{(e^{4ix} - 1)}{1 + x^2} dx.$$

Consider the corresponding contour integral

$$I = \int_C \frac{e^{4iz} - 1}{1 + z^2} dz = \int_C (e^{4iz} - 1) f(z) dz.$$

where $f(z) = 1/(1 + z^2)$ and C is the path $C_R \cup C_1$ as given in Fig. 13.3. The function $f(z) = 1/(1 + z^2)$ is analytic in the upper half plane except for the simple pole at $z = i$. We find that

$$\operatorname*{Res}_{z=i} (e^{4iz} - 1) f(z) = \lim_{z \to i} \left[\frac{(z - i)(e^{4iz} - 1)}{1 + z^2} \right] = \lim_{z \to i} \left[\frac{e^{4iz} - 1}{z + i} \right] = \frac{e^{-4} - 1}{2i}.$$

We now write

$$I = \int_{C_R} \frac{e^{4iz} - 1}{1 + z^2} dz + \int_{-R}^{R} \frac{e^{4ix} - 1}{1 + x^2} dx = 2\pi i \left[\operatorname*{Res}_{z=i} \left(\frac{e^{4iz} - 1}{1 + z^2} \right) \right] = \pi(e^{-4} - 1).$$

Now,

$$\left| \int_{C_R} \frac{e^{4iz} - 1}{1 + z^2} dz \right| \le \int_{C_R} \frac{|e^{4iz} - 1|}{|z^2 + 1|} |dz| \le \frac{2(\pi R)}{R^2 - 1} \to 0 \text{ as } R \to \infty$$

since $|e^{4iz} - 1| \le |e^{4iz}| + 1 = e^{-4y} + 1 \le 2$ and $|z^2 + 1| \ge |z|^2 - 1 = R^2 - 1.$

Hence, as $R \to \infty$, we obtain

$$\int_{-\infty}^{\infty} \frac{e^{4ix} - 1}{1 + x^2} \, dx = \pi(e^{-4} - 1).$$

Therefore,

$$\int_{-\infty}^{\infty} \frac{\sin^2 2x}{1 + x^2} \, dx = -\frac{1}{2} \operatorname{Re} \int_{-\infty}^{\infty} \frac{e^{4ix} - 1}{1 + x^2} \, dx = \frac{\pi}{2} (1 - e^{-4}).$$

Example 13.34 Evaluate the integral $\displaystyle\int_0^\infty \frac{x \sin x}{(x^2 + a^2)^2} \, dx$.

Solution Since the integrand is an even function, we write

$$\int_0^\infty \frac{x \sin x}{(x^2 + a^2)} \, dx = \frac{1}{2} \int_{-\infty}^\infty \frac{x \sin x}{(x^2 + a^2)^2} \, dx = \frac{1}{2} \operatorname{Im} \int_{-\infty}^\infty \frac{x e^{ix}}{(x^2 + a^2)^2} \, dx.$$

Now, consider the contour integral

$$I = \int_C f(z) \, e^{iz} dz, \quad \text{where } f(z) = z/(z^2 + a^2)^2.$$

and C is the path $C_R \cup C_1$ as given in Fig. 13.3. The function $f(z)$ is analytic in the upper half plane except for the pole of order 2 at $z = ai$. We find that

$$\operatorname*{Res}_{z=ai} [f(z)e^{iz}] = \lim_{z \to ai} \frac{d}{dz} \left[\frac{(z - ai)^2 \, z e^{iz}}{(z^2 + a^2)^2} \right] = \lim_{z \to ai} \frac{d}{dz} \left[\frac{z e^{iz}}{(z + ai)^2} \right] = \frac{e^{-a}}{4a}.$$

We now write

$$I = \int_{C_R} f(z) \, e^{iz} dz + \int_{-R}^R f(x) e^{ix} \, dx = 2\pi i \left[\operatorname*{Res}_{z=ai} (f(z)e^{iz}) \right] = \frac{\pi i e^{-a}}{2a}.$$

Now,

$$\left| \int_{C_R} \frac{z e^{iz}}{(z^2 + a^2)^2} \, dz \right| \leq \frac{R\,(\pi R)}{(R^2 - a^2)^2} \to 0 \text{ as } R \to \infty$$

since $|e^{iz}| = e^{-y} \leq 1$ and $|z^2 + a^2| \geq R^2 - a^2$.

Hence, as $R \to \infty$, we obtain

$$I = \int_{-\infty}^\infty f(x) e^{ix} dx = \int_{-\infty}^\infty \frac{x e^{ix}}{(x^2 + a^2)^2} \, dx = \frac{\pi i e^{-a}}{2a}$$

Therefore,

$$\int_0^\infty \frac{x \sin x}{(x^2 + a^2)^2} \, dx = \frac{1}{2} \operatorname{Im} \int_{-\infty}^\infty \frac{x e^{ix}}{(x^2 + a^2)^2} \, dx = \frac{\pi e^{-a}}{4a}.$$

Example 13.35 Evaluate the integral $\displaystyle\int_0^\infty \frac{\sin ax \sin bx}{x^2 + \alpha^2} \, dx, 0 < a < b, \alpha > 0$.

Solution Since $\sin ax \sin bx = \frac{1}{2} [\cos (b - a)x - \cos(b + a)x]$, the given integral can be written as

$$I = \frac{1}{2} \int_0^\infty \frac{\cos{(b-a)x}}{x^2 + \alpha^2} \, dx - \frac{1}{2} \int_0^\infty \frac{\cos{(b+a)}}{x^2 + \alpha^2} \, dx.$$

Since both the integrands are even functions, we can write

$$I = \frac{1}{4} \int_{-\infty}^\infty \frac{\cos{(b-a)x}}{x^2 + \alpha^2} \, dx - \frac{1}{4} \int_{-\infty}^\infty \frac{\cos{(b+a)x}}{x^2 + \alpha^2} \, dx = \frac{1}{4} \left[\text{Re}\,(I_1) - \text{Re}\,(I_2) \right]$$

where $$I_1 = \int_{-\infty}^\infty \frac{e^{(b-a)ix}}{x^2 + \alpha^2} \, dx \quad \text{and} \quad I_2 = \int_{-\infty}^\infty \frac{e^{(b+a)ix}}{x^2 + \alpha^2} \, dx.$$

Now, consider the corresponding contour integrals

$$I_1^* = \int_C \frac{e^{(b-a)iz}}{z^2 + \alpha^2} \, dz = \int_C e^{(b-a)iz} f(z) dz$$

and $$I_2^* = \int_C \frac{e^{(b+a)iz}}{z^2 + \alpha^2} dz = \int_C e^{(b+a)iz} f(z) dz$$

where C is the path $C_R \cup C_1$ as given in Fig. 13.3. The function $f(z) = 1/(z^2 + \alpha^2)$ in I_1^* and I_2^* has a simple pole at $z = \alpha i$ in the upper half plane. The residues are given by

$$\operatorname*{Res}_{z=\alpha i} \left[f(z)\, e^{(b-a)iz} \right] = \frac{e^{-(b-a)\alpha}}{2\alpha i} \quad \text{and} \quad \operatorname*{Res}_{z=\alpha i} \left[f(z) e^{(b+a)iz} \right] = \frac{e^{-(b+a)\,\alpha}}{2\alpha i}.$$

Now, $$\left| \int_{C_R} \frac{e^{(b-a)iz}}{z^2 + \alpha^2} \, dz \right| \leq \frac{\pi R}{R^2 - \alpha^2} \quad \text{and} \quad \left| \int_{C_R} \frac{e^{(b+a)iz}}{z^2 + \alpha^2} \, dz \right| \leq \frac{\pi R}{R^2 - \alpha^2}$$

since $\left| e^{(b-a)iz} \right| = \left| e^{-(b-a)y} \, e^{(b-a)ix} \right| = e^{-(b-a)y} \leq 1, \left| e^{(b+a)iz} \right| = e^{-(b+a)y} \leq 1$

and $\left| z^2 + \alpha^2 \right| \geq R^2 - \alpha^2$.

Therefore, as $R \to \infty$, $\pi R/(R^2 - \alpha^2) \to 0$.

Hence, as $R \to \infty$, we obtain

$$I_1 = 2\pi i \left[\frac{e^{-(b-a)\alpha}}{2\alpha i} \right] = \frac{\pi}{\alpha} e^{-(b-a)\,\alpha}, \; I_2 = 2\pi i \left[\frac{e^{-(b+a)\alpha}}{2\alpha i} \right] = \frac{\pi}{\alpha} e^{-(b+a)\,\alpha}$$

and

$$I = \frac{1}{4} \left[\text{Re}\,(I_1) - \text{Re}\,(I_2) \right] = \frac{\pi}{4\alpha} \left[e^{-(b-a)\alpha} - e^{-(b+a)\alpha} \right] = \frac{\pi}{2\alpha} e^{-b\alpha} \sinh{(a\alpha)}.$$

13.5.4 Improper Integrals with Singular Points on the Real Axis

We consider the integral

$$I = \int_a^b f(x) dx. \tag{13.42a}$$

In this case, there exists a point $c, a < c < b$ at which the integrand $f(x)$ has no finite limit (c is a point of infinite discontinuity), that is, $\lim_{x \to c} |f(x)| = \infty$. Then, we define the *Cauchy principal value*

(*CPV*) of I as

$$CPV(I) = \lim_{\varepsilon \to 0} \left[\int_a^{c-\varepsilon} f(x)dx + \int_{c+\varepsilon}^b f(x)dx \right].$$ (13.42b)

If the limit on the right hand side exists, then $CPV(I)$ exists. Often $CPV(I)$ may exist, even though I itself may not exist. In this section, we assume that the limit on the right hand side of Eq. (13.42b) exists.

Consider evaluation of improper integrals of the form given in Eq. (13.42a), in which either a or b or both are infinite, where $f(x)$ satisfies the following properties

(i) $f(x)$ is a rational function $f(x) = p(x)/q(x)$, where $p(x)$ and $q(x)$ have no common factors.

(ii) The corresponding function of the complex variable $f(z)$ with real coefficients is analytic in the upper half plane except at a finite number of poles in this half plane.

(iii) The function $f(z)$ has a finite number of simple poles $z_1, z_2, \ldots, z_m$ on the real axis.

(iv) The degree of the denominator $q(z)$ is at least two units higher than the degree of the numerator $p(z)$.

We consider the corresponding contour integral

$$I = \int_C f(z)dz = \int_C \frac{p(z)}{q(z)} \, dz.$$ (13.43)

Let the integrand have simple poles $z_1, z_2, \ldots, z_m$ on the real axis. Enclose these poles with semi-circles C_j with centres at z_j, and infinitesimal radius ε such that these semi-circles are non-intersecting. We, then consider the *indented contour* which is the union of the semi-circles $C_R, C_1, C_2, \ldots, C_m$ and line segments $l_1, l_2, \ldots, l_{m+1}$ (Fig. 13.4), that is

$$C = C_R \cup l_1 \cup C_1 \cup l_2 \ldots \cup C_m \cup l_{m+1}.$$ (13.44)

(Note the orientation of C_R and $C_j, j = 1, 2, \ldots, m$.)

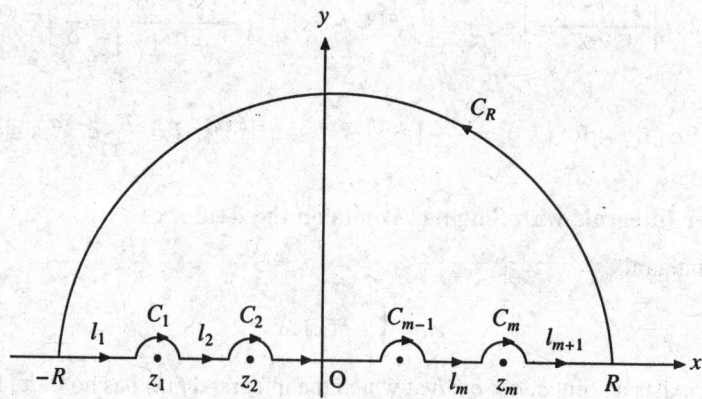

Fig. 13.4.   **Indented contour.**

The radius R is large enough such that C contains all the poles in the upper half plane. Therefore, we have

$$\int_C f(z)dz = \int_{C_R} f(z)dz + \sum_{j=1}^{m} \int_{C_j} f(z)dz + \sum_{k=1}^{m+1} \int_{l_k} f(x)dx \qquad (13.45)$$

$$= 2\pi i \sum_{k=1}^{n} \left[\operatorname*{Res}_{z=z_k^*} f(z) \right] \qquad (13.46)$$

where z_k^*, $k = 1, 2, \ldots, n$ are the poles in the upper half plane. Since the degree of the denominator $q(z)$ in $f(z) = p(z)/q(z)$ is at least two units higher than the degree of the numerator $p(z)$, we have from the previous sections 13.5.2 and 13.5.3 that

$$\left| \int_{C_R} f(z)\, dz \right| = \left| \int_{C_R} \frac{p(z)}{q(z)}\, dz \right| \to 0 \text{ as } R \to \infty. \qquad (13.47)$$

Let now, $\varepsilon \to 0$. When the radius ε of the semicircles C_j tend to zero, the semicircles contract to a point and the line segments l_j expand to cover the entire line $-R$ to R on the real axis.

Since $f(z)$ has simple poles at $z_1, z_2, \ldots, z_m$ on the real line, its Laurent series expansion about a pole z_j can be written as

$$f(z) = a_0 + a_1 (z - z_j) + \ldots + \frac{b_1}{z - z_j} = \frac{b_1}{z - z_j} + g(z). \qquad (13.48)$$

Now, any point on C_j can be written as $z = z_j + \varepsilon e^{i\theta}$, $\pi \le \theta \le 0$, taking into consideration the orientation of C_j. Then, we have from Eq. (13.48)

$$\int_{C_j} f(z)dz = b_1 \int_{\pi}^{0} \frac{i \varepsilon e^{i\theta}}{\varepsilon e^{i\theta}} d\theta + \int_{\pi}^{0} g(z_j + \varepsilon e^{i\theta}) i \varepsilon e^{i\theta} d\theta$$

$$= -\pi i b_1 + i\varepsilon \int_{\pi}^{0} g(z_j + \varepsilon e^{i\theta}) e^{i\theta} d\theta$$

$$= -\pi i \left[\operatorname*{Res}_{z=z_j} f(z) \right] + I_1. \qquad (13.49)$$

We now have

$$|I_1| = \left| i\varepsilon \int_{\pi}^{0} g(z_j + \varepsilon e^{i\theta}) e^{i\theta} d\theta \right| \le \varepsilon \int_{0}^{\pi} |g(z_j + \varepsilon e^{i\theta})| |d\theta| \le M\pi\varepsilon \qquad (13.50)$$

where $|g(z_j + \varepsilon e^{i\theta})| \le M$ on C_j.

It follows that $|I_1| \to 0$ as $\varepsilon \to 0$ and therefore, $\lim_{\varepsilon \to 0} I_1 = 0$.

Hence, we have as $\varepsilon \to 0$

$$\int_{C_j} f(z)dz = -\pi i \left[\operatorname*{Res}_{z=z_j} f(z) \right]. \qquad (13.51)$$

Therefore, as $R \to \infty$ and $\varepsilon \to 0$, we obtain from Eqs. (13.45) and (13.46)

$$\int_C f(z)dz = (CPV)\int_{-\infty}^{\infty} f(x)dx - \pi i \sum_{j=1}^{m}\left[\operatorname*{Res}_{z=z_j} f(z)\right] = 2\pi i \sum_{k=1}^{n}\left[\operatorname*{Res}_{z=z_k^*} f(z)\right]$$

or

$$(CPV)\int_{-\infty}^{\infty} f(x)dx = 2\pi i \sum_{k=1}^{n}\left[\operatorname*{Res}_{z=z_k^*} f(z)\right] + \pi i \sum_{j=1}^{m}\left[\operatorname*{Res}_{z=z_j} f(z)\right]. \qquad (13.52)$$

Example 13.36 Evaluate the integral $\int_0^{\infty}\dfrac{\sin x}{x}\,dx$.

Solution Since $\sin x = \operatorname{Im} e^{ix}$, we consider the contour integral

$$I = \int_C \frac{e^{iz}}{z}dz.$$

The integrand has a simple pole at $z = 0$ and has no singular point in the upper half plane. Enclose the point $z = 0$ by a semicircle in the upper half plane with centre at $z = 0$ and radius ε. The contour C is given by $C = C_R \cup l_1 \cup C_1 \cup l_2$, where l_1 is the line segment from $-R$ to $-\varepsilon$ and l_2 is the line segment from ε to R (Fig. 13.5). Note that in this example $f(z) = p(z)/q(z)$ does not satisfy the condition that the degree of $q(z)$ is two units higher than the degree of $p(z)$.

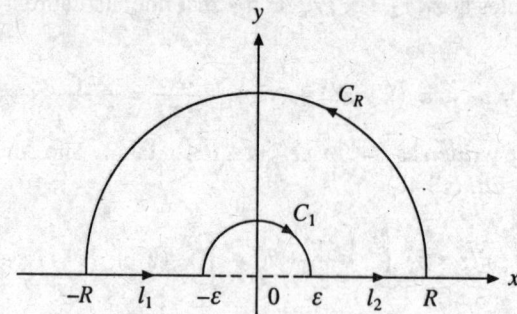

Fig. 13.5. Example 13.36.

Since, there is no singular point inside C, by Cauchy integral theorem, we have

$$\int_C \frac{e^{iz}}{z}\,dz = 0.$$

Hence,

$$\int_C \frac{e^{iz}}{z}dz = \int_{C_R}\frac{e^{iz}}{z}\,dz + \int_{-R}^{-\varepsilon}\frac{e^{ix}}{x}dx + \int_{C_1}\frac{e^{iz}}{z}dz + \int_{\varepsilon}^{R}\frac{e^{ix}}{x}\,dx$$

$$= I_1 + I_2 + I_3 + I_4 = 0.$$

Now, $|I_1| = \left|\int_{C_R}\dfrac{e^{iz}}{z}\,dz\right| \le \int_{C_R}\dfrac{|e^{iz}|}{|z|}\,|dz| \le \dfrac{1}{R}\int_0^{\pi} e^{-R\sin\theta}\,(R\,d\theta) \le \dfrac{\pi}{R} \to 0$ as $R \to \infty$

since $\displaystyle\int_0^{\pi} e^{-R\sin\theta}\,d\theta = 2\int_0^{\pi/2} e^{-R\sin\theta}\,d\theta$ and use the Jordan's lemma (see Problem 16, Ex. 13.4).

$$I_3 = \int_{C_1}\frac{e^{iz}}{z}\,dz = -\pi i\left[\operatorname*{Res}_{z=0}\left(\frac{e^{iz}}{z}\right)\right] = -\pi i.$$

Therefore, as $R \to \infty$ and $\varepsilon \to 0$, we obtain

$$\int_{-\infty}^{0} \frac{e^{ix}}{x} dx - \pi i + \int_{0}^{\infty} \frac{e^{ix}}{x} dx + 0 = 0$$

or $(CPV) \int_{-\infty}^{\infty} \frac{e^{ix}}{x} dx = \pi i.$

Comparing the imaginary part on both sides, we have

$$(CPV) \int_{-\infty}^{\infty} \frac{\sin x}{x} dx = \pi.$$

Note that $(\sin x)/x$ has removable singularity at $x = 0$ and the integral $\int_{-\infty}^{\infty} \frac{\sin x}{x} dx$ converges. Hence,
$CPV(I) = I$.

Therefore, $\int_{-\infty}^{\infty} \frac{\sin x}{x} dx = \pi.$

Since $(\sin x/x)$ is an even function, we obtain

$$\int_{0}^{\infty} \frac{\sin x}{x} dx = \frac{\pi}{2}.$$

Example 13.37   Evaluate the integral $I = \int_{0}^{\infty} \frac{\sin ax}{x(x^2 + b^2)} dx,\ a > 0, b > 0.$

Solution   We note that the integrand of I has removable singularity at $x = 0$. The integral converges
and hence $CPV(I) = I$.

Since $\sin ax = \text{Im}(e^{iax})$, consider the corresponding contour integral

$$I = \int_{C} f(z) dz,\ \text{where}\ f(z) = \frac{e^{iaz}}{z(z^2 + b^2)}$$

and C is as given in Fig. 13.6.

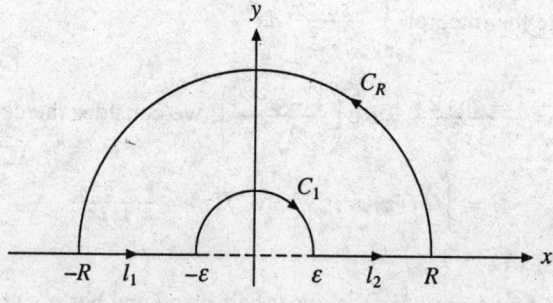

Fig. 13.6.   Example 13.37.

The integrand has a simple pole at $z = bi$ in the upper half plane and a simple pole at $z = 0$ on the real axis. We have

$$\operatorname*{Res}_{z=bi} f(z) = \lim_{z \to bi} \left[\frac{(z - bi) e^{iaz}}{(z^2 + b^2)z} \right] = -\frac{e^{-ab}}{2b^2}.$$

By residue theorem, we get

$$\int_C f(z)dz = \int_{C_R} f(z)dz + \int_{-R}^{-\varepsilon} f(x)dx + \int_{C_1} f(z)dz + \int_{\varepsilon}^{R} f(z)dz = 2\pi i \left(\frac{-e^{-ab}}{2b^2} \right)$$

or

$$I_1 + I_2 + I_3 + I_4 = -\frac{\pi i e^{-ab}}{b^2}$$

where I_1, I_2, I_3 and I_4 are the integrals on the left hand side.

Now

$$\left| \int_{C_R} f(z)dz \right| = \left| \int_{C_R} \frac{e^{iaz}}{z(z^2 + b^2)} dz \right| \le \frac{(\pi R)}{R(R^2 - b^2)} \to 0 \text{ as } R \to \infty.$$

Using the result given in Eq. (13.51), we obtain

$$I_3 = \int_{C_1} f(z)dz = -\pi i \left[\operatorname*{Res}_{z=0} \frac{e^{iaz}}{z(z^2 + b^2)} \right] = -\frac{\pi i}{b^2}.$$

Therefore, as $R \to \infty$ and $\varepsilon \to 0$, we get

$$\int_{-\infty}^{\infty} \frac{e^{iax}}{x(x^2 + b^2)}dx - \frac{\pi i}{b^2} = -\frac{\pi i e^{-ab}}{b^2}$$

or

$$\int_{-\infty}^{\infty} \frac{e^{iax}}{x(x^2 + b^2)}dx = \frac{\pi i}{b^2}(1 - e^{-ab}).$$

Comparing the imaginary part on both sides, we obtain

$$\int_{-\infty}^{\infty} \frac{\sin ax}{x(x^2 + b^2)} dx = \frac{\pi}{b^2}(1 - e^{-ab}) \quad \text{or} \quad \int_{0}^{\infty} \frac{\sin ax}{x(x^2 + b^2)} dx = \frac{\pi}{2b^2}(1 - e^{-ab})$$

since the integrand is an even function.

Example 13.38 Evaluate the integral $\displaystyle\int_{0}^{\infty} \frac{\sin^2 x}{x^2} dx$.

Solution Since $\dfrac{\sin^2 x}{x^2} = \dfrac{1 - \cos 2x}{2x^2} = \operatorname{Re} \left(\dfrac{1 - e^{2ix}}{2x^2} \right)$, we consider the contour integral as

$$I = \int_C f(z) dz, \quad \text{where} \quad f(z) = \frac{1 - e^{2iz}}{2z^2}$$

and C is as given in Fig. 13.7.

The integrand has no singular point in the upper half plane and has a simple pole at $z = 0$ on the real axis. We have by the Cauchy integral theorem

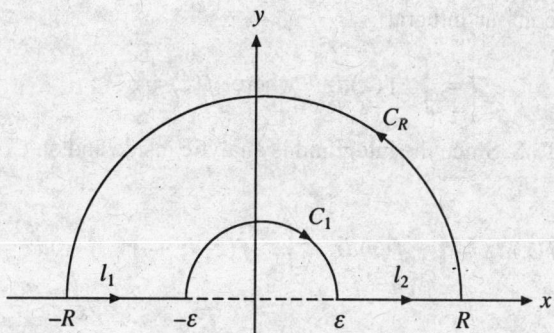

Fig. 13.7.   Example 13.38.

$$\int_C f(z)dz = \int_{C_R} f(z)dz + \int_{-R}^{-\varepsilon} f(x)dx + \int_{C_1} f(z)dz + \int_{\varepsilon}^{R} f(x)dx = 0.$$

Now
$$\left| \int_{C_R} f(z)dz \right| \leq \int_{C_R} \left| \frac{1 - e^{2iz}}{2z^2} \right| |dz| \leq \frac{2}{2R^2}(\pi R) \rightarrow 0 \text{ as } R \rightarrow \infty.$$

since $|1 - e^{2iz}| \leq 1 + e^{-2y} \leq 2$.

Using the result given in Eq. (13.51), we obtain

$$\int_{C_1} f(z)dz = -\pi i \left[\operatorname*{Res}_{z=0} f(z) \right] = -\pi i \lim_{z \to 0} \left[\frac{1 - e^{2iz}}{2z} \right] = -\pi.$$

Hence, when $R \rightarrow \infty$ and $\varepsilon \rightarrow 0$, we obtain

$$(CPV) \int_{-\infty}^{\infty} \frac{1 - e^{2ix}}{2x^2} dx \doteq \pi.$$

Comparing the real part on both sides, we get

$$(CPV) \int_{-\infty}^{\infty} \frac{\sin^2 x}{x^2} dx = \pi.$$

Since $(\sin x/x)$ has removable singularity at $x = 0$ and the integral converges, $CPV(I) = I$. Hence,

$$\int_{-\infty}^{\infty} \frac{\sin^2 x}{x^2} dx = \pi \quad \text{or} \quad \int_0^{\infty} \frac{\sin^2 x}{x^2} dx = \frac{\pi}{2}.$$

Some special improper real integrals

Example 13.39 Integrating $\int_C e^{-z^2} dz$, where C consists of three parts C_1, $C_2 : |z| = R$, $0 \leq \arg z \leq \pi/4$ and C_3, as shown in Fig. 13.8, and letting $R \rightarrow \infty$, evaluate the integrals (*Fresnel's integrals*)

$$\int_0^{\infty} \sin x^2 \, dx \quad \text{and} \quad \int_0^{\infty} \cos x^2 \, dx.$$

It is given that $\int_0^{\infty} e^{-x^2} dx = \sqrt{\pi}/2$.

Solution Consider the contour integral

$$I = \int_C f(z)dz, \quad \text{where} \quad f(z) = e^{-z^2}$$

and C is as given in Fig. 13.8. Since the integrand is analytic inside and on C, we have by the Cauchy integral theorem

$$\int_C f(z)dz = \int_{C_1} f(z)dz + \int_{C_2} f(z)dz + \int_{C_3} f(z)dz = 0. \tag{13.53}$$

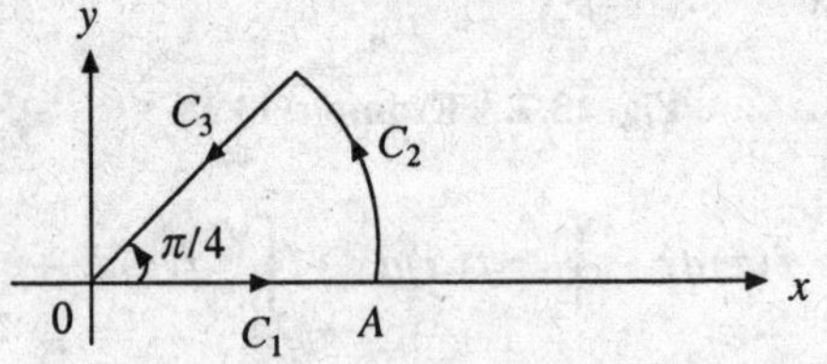

Fig. 13.8. Example 13.39.

For evaluating along C_1, we have $z = x$ and $0 \le x \le R$. Therefore,

$$\int_{C_1} f(z)dz = \int_0^R f(x)dx = \int_0^R e^{-x^2} dx.$$

Letting $R \to \infty$, we obtain $\displaystyle\int_{C_1} f(z)dz = \int_0^\infty e^{-x^2} dx = \frac{\sqrt{\pi}}{2}$.

For evaluating along C_2, let $z = Re^{i\theta}$, $0 \le \theta \le \pi/4$. Therefore,

$$\int_{C_2} f(z)dz = \int_0^{\pi/4} e^{-R^2(\cos 2\theta + i\sin 2\theta)} (iRe^{i\theta})d\theta$$

$$= iR \int_0^{\pi/4} e^{-R^2\cos 2\theta} e^{-iR^2\sin 2\theta} e^{i\theta} d\theta.$$

We have

$$\left| \int_{C_2} f(z)dz \right| \le R \int_0^{\pi/4} e^{-R^2\cos 2\theta} d\theta = R \int_0^{\pi/4} e^{-R^2\sin 2\theta} d\theta = \frac{R}{2} \int_0^{\pi/2} e^{-R^2\sin\theta} d\theta.$$

Using the *Jordan's Lemma* (see Problem 16 in Exercise 13.4), we get

$$\left| \int_{C_2} f(z)dz \right| \le \frac{R}{2} \int_0^{\pi/2} e^{-R^2\sin\theta} d\theta < \frac{R}{2} \int_0^{\pi/2} e^{-2aR^2\theta/\pi} d\theta$$

$$= \frac{R\pi}{(-4aR^2)}(e^{-aR^2} - 1) = \frac{\pi}{4aR}(1 - e^{-aR^2}) \to 0 \text{ as } R \to \infty$$

since $\dfrac{2\theta}{\pi} \le \sin\theta \le 1$ for $0 \le \theta \le \dfrac{\pi}{2}$.

For evaluating along C_3, let $z = Re^{i\pi/4}$ and z varies from R to 0. We get

$$e^{-z^2} = e^{-R^2 e^{i\pi/2}} = e^{-iR^2} \text{ and } dz = e^{i\pi/4}\,dR.$$

Therefore,

$$\int_{C_3} f(z)dz = \int_R^0 e^{-iR^2} e^{i\pi/4}\,dR = \frac{(1+i)}{\sqrt{2}} \int_R^0 e^{-iR^2}\,dR.$$

Letting $R \to \infty$, we obtain

$$\int_{C_3} f(z)dz = -\left(\frac{1+i}{\sqrt{2}}\right) \int_0^\infty e^{-iR^2}\,dR = -\left(\frac{1+i}{\sqrt{2}}\right) \int_0^\infty e^{-ix^2}\,dx$$

since R is a dummy variable of integration. Therefore, we get from Eq. (13.53)

$$\frac{\sqrt{\pi}}{2} - \frac{1+i}{\sqrt{2}} \int_0^\infty e^{-ix^2}\,dx = 0$$

or

$$\int_0^\infty e^{-ix^2}\,dx = \frac{\sqrt{\pi}}{2}\left(\frac{\sqrt{2}}{1+i}\right) = \frac{\sqrt{\pi}(1-i)}{2\sqrt{2}}.$$

Comparing the real and imaginary parts on both sides, we obtain

$$\int_0^\infty \cos x^2\,dx = \frac{\sqrt{\pi}}{2\sqrt{2}} \text{ and } \int_0^\infty \sin x^2\,dx = \frac{\sqrt{\pi}}{2\sqrt{2}}.$$

Example 13.40 Integrating $\displaystyle\int_C e^{-z^2}\,dz$, where C is the boundary of the rectangle with vertices at $-a, a, a+ib, -a+ib$, and letting $a \to \infty$, evaluate the integral $\displaystyle\int_0^\infty e^{-x^2} \cos 2bx\,dx$. It is given that $\displaystyle\int_0^\infty e^{-x^2}\,dx = \frac{\sqrt{\pi}}{2}$.

Solution Let $I = \displaystyle\int_C f(z)dz$, where $f(z) = e^{-z^2}$ and C is as given in Fig. 13.9.

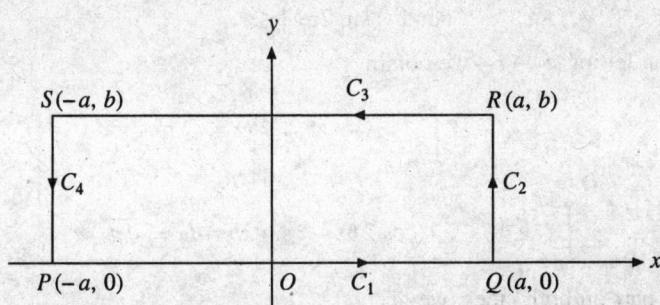

Fig. 13.9. Example 13.40.

The integrand is analytic inside and on C. Therefore, by the Cauchy integral theorem

$$\int_C f(z)dz = \int_{C_1} f(z)dz + \int_{C_2} f(z)dz + \int_{C_3} f(z)dz + \int_{C_4} f(z)dz = 0. \tag{13.54}$$

For evaluating along C_1, we have $z = x, -a \le x \le a$. Therefore,

$$\int_{C_1} f(z)dz = \int_{-a}^{a} e^{-x^2} dx = 2 \int_{0}^{a} e^{-x^2} dx.$$

Letting $a \to \infty$, we get

$$\int_{C_1} f(z)dz = 2 \int_{0}^{\infty} e^{-x^2} dx = \sqrt{\pi}. \qquad (13.55)$$

For evaluating along C_2, we have $z = a + iy, 0 \le y \le b$. Therefore,

$$\int_{C_2} f(z)dz = \int_{0}^{b} e^{-(a+iy)^2} idy = i \int_{0}^{b} e^{-(a^2-y^2)} e^{-2aiy} dy. \qquad (13.56)$$

For evaluating along C_3, we have $z = x + ib, a \le x \le -a$. Therefore,

$$\int_{C_3} f(z)dz = -\int_{-a}^{a} e^{-(x+ib)^2} dx = -\int_{-a}^{a} e^{-(x^2-b^2)} e^{-2ibx} dx. \qquad (13.57)$$

For evaluating along C_4, we have $z = -a + iy, b \le y \le 0$. Therefore,

$$\int_{C_4} f(z)dz = -\int_{0}^{b} e^{-(-a+iy)^2} idy = -i \int_{0}^{b} e^{-(a^2-y^2)} e^{2iay} dy. \qquad (13.58)$$

Combining Eqs. (13.56) and (13.58), we obtain

$$I^* = \int_{C_2} f(z)dz + \int_{C_4} f(z)dz = -i \int_{0}^{b} e^{-(a^2-y^2)} [e^{2aiy} - e^{-2aiy}] dy$$

$$= 2 \int_{0}^{b} e^{-(a^2-y^2)} \sin 2ay \, dy.$$

Now, $|I^*| \le 2 \int_{0}^{b} e^{-(a^2-y^2)} |\sin 2ay| \, |dy| \le 2be^{-a^2+b^2} \to 0$ as $a \to \infty$

since, $e^{-(a^2-y^2)} = e^{-a^2+y^2} \le e^{-a^2+b^2}$ and $|\sin 2ay| \le 1$.

From Eq. (13.54), on letting $a \to \infty$, we obtain

$$\sqrt{\pi} - \int_{-\infty}^{\infty} e^{-(x^2-b^2)} e^{-2ibx} dx = 0$$

or

$$\int_{-\infty}^{\infty} e^{-(x^2-b^2)} (\cos 2bx - i \sin 2bx) dx = \sqrt{\pi}.$$

Comparing the real part on both sides, we get

$$\int_{-\infty}^{\infty} e^{-(x^2-b^2)} \cos 2bx \, dx = \sqrt{\pi}$$

or

$$\int_{0}^{\infty} e^{-x^2} \cos 2bx \, dx = \frac{\sqrt{\pi}}{2} e^{-b^2}.$$

Exercise 13.4

Evaluate the following real definite integrals using contour integration.

1. $\displaystyle\int_0^{2\pi} \frac{d\theta}{3 + 2\sin\theta}.$

2. $\displaystyle\int_0^{2\pi} \frac{\sin\theta}{3 + \cos\theta}\, d\theta.$

3. $\displaystyle\int_0^{2\pi} \frac{\cos\theta}{5 + 4\cos 2\theta}\, d\theta.$

4. $\displaystyle\int_0^{2\pi} \frac{d\theta}{13 + 4\cos\theta + 3\sin\theta}.$

5. $\displaystyle\int_0^{\pi} \frac{\cos 2\theta}{1 - 2a\cos\theta + a^2}\, d\theta,\ |a| < 1.$

6. $\displaystyle\int_0^{2\pi} \frac{\cos 3\theta}{5 - 3\cos\theta}\, d\theta.$

7. $\displaystyle\int_0^{2\pi} \frac{\cos 2\theta \cos\theta}{1 - 2a\cos\theta + a^2}\, d\theta,\ 0 < a < 1.$

8. $\displaystyle\int_0^{2\pi} \frac{\sin^2\theta}{5 + 4\cos\theta}\, d\theta.$

9. $\displaystyle\int_0^{2\pi} \frac{\cos^2 2\theta}{5 + 4\cos 2\theta}\, d\theta.$

10. $\displaystyle\int_0^{2\pi} \frac{d\theta}{\cos^2\theta + 4\sin^2\theta}.$

11. $\displaystyle\int_0^{2\pi} \frac{d\theta}{(a + b\cos\theta)^2},\ a, b\ \text{real},\ a > b > 0.$

12. $\displaystyle\int_{-\pi}^{\pi} \frac{\cos n\theta}{1 - 2a\cos\theta + a^2}\, d\theta,\ |a| < 1.$

13. $\displaystyle\int_0^{2\pi} \cos^{2n}\theta\, d\theta.$

14. $\displaystyle\int_0^{2\pi} \cos^{2n+1}\theta\, d\theta.$

15. $\displaystyle\int_0^{2\pi} e^{\cos\theta} \cos(n\theta - \sin\theta)\, d\theta.$

16. Prove the *Jordan's lemma*. $\displaystyle\int_0^{\pi/2} e^{-R\sin\theta}\, d\theta < \frac{\pi}{2R}$

where R is any positive real number.

Obtain the Cauchy principal value of the following improper real integrals using contour integration.

17. $\displaystyle\int_{-\infty}^{\infty} \frac{dx}{x^2 + 9}.$

18. $\displaystyle\int_{-\infty}^{\infty} \frac{dx}{2x^2 + 2x + 5}.$

19. $\displaystyle\int_{-\infty}^{\infty} \frac{dx}{(x^2 + a^2)(x^2 + b^2)},\ a > 0,\ b > 0,\ a \neq b.$

20. $\displaystyle\int_{-\infty}^{\infty} \frac{x + 2}{(x^2 + 1)(x^2 + 4)}\, dx.$

21. $\displaystyle\int_{-\infty}^{\infty} \frac{dx}{(x^2 + a^2)^2},\ a > 0.$

22. $\displaystyle\int_{-\infty}^{\infty} \frac{dx}{(x^2 + 4x + 13)^2}.$

23. $\displaystyle\int_{-\infty}^{\infty} \frac{dx}{(x^4 + 5x^2 + 4)^2}.$

24. $\displaystyle\int_{-\infty}^{\infty} \frac{dx}{(x^2 + 9)(x^2 + 4)^2}.$

25. $\displaystyle\int_{-\infty}^{\infty} \frac{x^2 + a^2}{x^4 + a^4}\, dx,\ a > 0.$

26. $\displaystyle\int_{-\infty}^{\infty} \frac{x^2 + 1}{5x^4 + 26x^2 + 5}\, dx.$

27. $\displaystyle\int_{-\infty}^{\infty} \frac{x^2}{x^6 + a^6}\, dx,\ a > 0.$

28. $\int_{-\infty}^{\infty} \frac{x^2}{(x^2+1)^3} dx.$

29. $\int_{-\infty}^{\infty} \frac{dx}{(x^2+1)(2x^4+5x^2+2)}$

30. $\int_{-\infty}^{\infty} \frac{dx}{(1+x^2)^{n+1}}.$

31. $\int_{-\infty}^{\infty} \frac{x^{2m}}{1+x^{2n}} dx, \ m \neq n, \ m, n \text{ positive integers}, n \geq 2.$

32. $\int_{-\infty}^{\infty} \frac{\cos 2x}{x^2+4} dx.$

33. $\int_{-\infty}^{\infty} \frac{x \sin x}{x^2+4x+5} dx.$

34. $\int_{-\infty}^{\infty} \frac{x \cos x}{x^2-2x+2} dx.$

35. $\int_{-\infty}^{\infty} \frac{(a+bx)\cos \alpha x}{x^2+2Bx+A} dx, \ A > B^2, \alpha > 0.$

36. $\int_{-\infty}^{\infty} \frac{x \cos ax \sin bx}{x^2+\alpha^2} dx, \ 0 < a < b, \alpha > 0.$

37. $\int_{-\infty}^{\infty} \frac{\cos 2x}{(x^2+a^2)(x^2+b^2)} dx, \ a > 0, b > 0.$

38. $\int_{-\infty}^{\infty} \frac{\cos ax}{(x^2+b^2)^2} dx, \ b > 0, a > 0.$

39. $\int_{-\infty}^{\infty} \frac{\cos^2 x}{(1+x^2)^2} dx.$

40. $\int_{-\infty}^{\infty} \frac{x^3 \sin x}{(x^2+a^2)^2} dx.$

41. $\int_{-\infty}^{\infty} \frac{x^3 \sin mx}{(x^2+a^2)(x^2+b^2)} dx, \ m \text{ positive integer}.$

42. $\int_{-\infty}^{\infty} \frac{x^3 \sin ax}{x^4+2(\cos 2\alpha)b^2 x^2+b^4} dx, \ a > 0, b > 0, 0 < \alpha < \frac{\pi}{2}.$

43. $\int_{-\infty}^{\infty} \frac{\sin mx}{x} dx, \ m \text{ positive integer}.$

44. $\int_{-\infty}^{\infty} \frac{\cos 4x - \cos 2x}{x^2} dx.$

45. $\int_{0}^{\infty} \frac{\sin^2 mx}{x^2} dx, \ m \text{ positive integer}.$

46. $\int_{-\infty}^{\infty} \frac{\sin x}{x(x^2+9)} dx.$

47. $\int_{-\infty}^{\infty} \frac{dx}{(x+1)(x^2+1)}.$

48. $\int_{-\infty}^{\infty} \frac{dx}{x^2+ix}.$

49. $\int_{-\infty}^{\infty} \frac{(x+1)dx}{x(x-2)(x^2+16)}.$

50. $\int_{-\infty}^{\infty} \frac{(3x+5)dx}{x(x+2)(x^2+1)}.$

13.6 Answers and Hints

Exercise 13.1

1. $z = 0$: removable singularity, $z = 2\pi in, n = \pm 1, \pm 2, \ldots$: simple poles.

2. $z = 0$: essential singularity.

3. $z = 0$: pole of order 2, $z = n\pi, n = \pm 1, \pm 2, \ldots$: simple poles.

4. $z = 0$: simple pole; $z = \pm 4$: simple poles, $z = 8$: essential singularity.

5. $z = 0$: removable singularity, $z = 2n\pi i$, $n = \pm 1, \pm 2, \ldots$: simple poles.

6. $z = 0$ and $z = \pm 2n\pi$: simple poles, $z = \pm (2n - 1)\pi$: poles of order 3, where $n = 1, 2, 3, \ldots$.

7. $z = (\pi/4) + n\pi$, $n = 0, \pm 1, \pm 2, \ldots$: simple poles.

8. $z = 2n\pi - (\pi/6)$, $z = (2n + 1)\pi + (\pi/6)$, $n = 0, \pm 1, \pm 2, \ldots$: poles of order 2.

9. $z = 0$: essential singularity, $z = 1/(2n\pi i)$, $n = \pm 1, \pm 2, \ldots$: simple poles.

10. $z = 0$: pole of order 2, $z = \pm 2$: simple poles, $z = 1$: essential singularity.

11. $z = 0$, $z = (\pi/2)$: removable singularity, $z = 1$: essential singularity.

12. $z = 0$: essential singularity, $z = \pi$: removable singularity, $z = n\pi$, n integer and $n \neq 0, 1$: simple poles.

13. $z = 1$: essential singularity.

14. $z = \pm \sqrt{\pi n} \, (1 + i)$, $n > 0$ and integer, $z = \pm \sqrt{\pi |n|} \, (1 - i)$, $n < 0$ and integer : simple poles; $z = 0$: simple pole.

15. $z = 0$: removable singularity if $m \geq n$, $z = 0$: pole of order $n - m$ if $m < n$.

16. $z = 0$: simple pole.

17. $z = 2/[(2n + 1)\pi]$, $n = 0, \pm 1, \pm 2, \ldots$: essential singularity. $z = 0$ is a non-isolated essential singularity.

18. $z = 1/(n\pi)$, $n = \pm 1, \pm 2, \ldots$: essential singularity ($z = 0$ is a non-isolated essential singularity).

19. $f(z) = -z/(1 + z)^2$, $z = -1$: pole of order 2.

20. $f(z) = 1/(1 - \sin z)$, $z = (4n + 1)\pi/2$, $n = 0, \pm 1, \pm 2, \ldots$: simple poles.

21. (a) Essential singularity at $z = \infty$, (b) essential singularity at $z = \infty$.

22. (a) $z = 0$: removable singularity, $z = \pm i$: simple poles, pole of order 2 at $z = \infty$.

    (b) Pole of order 4 at $z = 0$, removable singularity at $z = \infty$.

23. (a) Simple pole at $z = 0$, essential singularity at $z = \infty$.

    (b) Simple poles at $z = (2n + 1)\pi/2$, $n = 0, \pm 1, \pm 2, \ldots$, essential singularity at $z = \infty$.

24. (a) Essential singularity at $z = \infty$, (b) essential singularity at $z = \infty$.

25. (a) Removable singularity at $z = \pi$, simple pole at $z = 0$, poles of order 2 at $z = n\pi$, n any integer and $n \neq 0$ or $n \neq 1$, essential singularity at $z = \infty$.

    (b) Removable singularity at $z = 0$ and $z = \pi$, essential singularity at $z = \infty$.

26. $2\pi i n /(z - 2\pi i n)$ at $z = 2\pi i n$, $n = \pm 1, \pm 2, \ldots$.

27. $\dfrac{1}{2(z - i)}$ at $z = i$, $\dfrac{1}{2(z + 1)} - \dfrac{1 - i}{2(z + 1)^2}$ at $z = -1$.

28. $z = 0$ is a removable singularity; $\sin 1/(z - 1)$ at $z = 1$.

29. $\dfrac{1}{5(z - 1)}$ at $z = 1$, $\dfrac{\text{Log } 2 + (\pi i)/2}{4(4 - 3i)(z - 2i)}$ at $z = 2i$, $\dfrac{\text{Log } 2 - (\pi i)/2}{4(4 + 3i)(z + 2i)}$ at $z = -2i$.

30. $1/(2z^2)$ at $z = 0$.

31. Since z_0 is a removable singularity of $f(z)$, $\lim\limits_{z \to z_0} f(z)$ exists and is finite. Hence, $f(z)$ is bounded in some deleted neighborhood of z_0.

32. Let z_0 be an isolated singularity of $f(z)$. Let C be $|z - z_0| = r$. Then, $f(z)$ is analytic inside and on a circle C of radius r except at $z = z_0$. Therefore, $f(z)$ has a Laurent series expansion

$$f(z) = \sum_{n=-\infty}^{\infty} c_n (z - z_0)^n, \, 0 < |z - z_0| < r$$

where

$$c_{-n} = \frac{1}{2\pi i} \oint_C \frac{f(z)}{(z - z_0)^{-n+1}} \, dz, \, n = 1, 2, \ldots.$$

Let $f(z)$ be bounded. Then

$$|c_{-n}| \leq \frac{1}{2\pi} \oint_C |z - z_0|^{n-1} |f(z)| |dz| \leq \frac{r^{n-1}}{2\pi} (2\pi r) M = Mr^n$$

where $|f(z)| < M$. Since r can be made arbitrarily small, we obtain $c_{-n} = 0$ for all n. Therefore, Laurent series reduces to Taylor series and $f(z)$ is analytic at $z = z_0$ which is a contradiction.

33. $d^k g/dz^k = 0$ at $z = 0$, $k = 0, 1, \ldots, n - 1$.

34. Pole of order $N - k$ if $k \leq N - 1$, removable singularity if $k \geq N$,

35. Simple pole, $- N/(z - z_0)$.

36. (i) Write $f(z) = (z - z_0)^N g(z)$, $g(z_0) \neq 0$ and consider

$$h(z) = \frac{1}{f(z)} = \frac{[1/g(z)]}{(z - z_0)^N} = \frac{\psi(z)}{(z - z_0)^N}, \, \psi(z_0) \neq 0.$$

Hence, $h(z)$ has a pole of order N at z_0.

(ii) Write $f(z) = \frac{\phi(z)}{(z - z_0)^N}$, $\phi(z_0) \neq 0$ and consider

$$h(z) = \frac{1}{f(z)} = \frac{(z - z_0)^N}{\phi(z)} = (z - z_0)^N \psi(z), \, \psi(z_0) \neq 0$$

$\psi(z)$ is analytic at z_0. Hence, the function $h(z)$ has a zero of order N at z_0.

(iii) If z_0 is not an essential singular point of $\phi(z) = 1/f(z)$, then z_0 is either a pole or a removable singularity of $\phi(z)$. If z_0 is a pole of $\phi(z)$, then z_0 is a zero of $f(z)$, which is a contradiction. If z_0 is a removable singularity, then

$$\text{either} \quad \lim_{z \to z_0} \phi(z) = 0 \quad \text{or} \quad \lim_{z \to z_0} \phi(z) \neq 0.$$

If $\lim_{z \to z_0} \phi(z) = 0$, then $f(z)$ has a pole at z_0, which is a contradiction.

If $\lim_{z \to z_0} \phi(z) \neq 0$, then $f(z)$ has a removable singular point, which is a contradiction.

Hence, z_0 is an essential singular point of $\phi(z)$.

37. Write $f(z) = g(z)/(z - z_0)^m$, $g(z)$ is analytic at z_0 and $g(z_0) \neq 0$. Write Taylor series expansion of $g(z)$ about $z = z_0$ and differentiate $f(z)$, n times.

38. Since $f(z)$ is an entire function and not a constant, then it has a Taylor series expansion of the form

$$f(z) = \sum_{n=0}^{\infty} a_n z^n, \, 0 < |z| < \infty.$$

Let $z = 1/w$. Then,

$$f(z) = f(1/w) = g(w) = \sum_{n=0}^{\infty} a_n (1/w)^n, \, 0 < |w| < \infty.$$

Therefore, Laurent series expansion of $g(w)$ contains only non-positive powers of w. Thus, if $f(z)$ is not constant, $w = 0$ is either a pole or an essential singular point of $g(w)$, with the same result for $f(z)$ at $z = \infty$.

39. Let $P(z) = \sum\limits_{n=0}^{N} a_n z^n$, $a_N \neq 0$. Let $z = 1/w$. Then $P(z) = P(1/w) = Q(w) = \sum\limits_{n=0}^{N} a_n (1/w)^N$ has a pole of

order N at $w = 0$, with the same result for $P(z)$ at $z = \infty$.

40. Let $f(z) = \dfrac{a_N z^N + a_{N-1} z^{N-1} + \ldots + a_1 z + a_0}{b_M z^M + b_{M-1} z^{M-1} + \ldots + b_1 z + b_0}$, $a_N \neq 0, b_M \neq 0$.

Let $z = 1/w$. Then

$$f(z) = f\left(\frac{1}{w}\right) = g(w) = w^{M-N}\left[\frac{a_N + a_{N-1}w + \ldots + a_0 w^N}{b_M + b_{M-1}w + \ldots + b_0 w^M}\right].$$

If $M - N \geq 0$, then $g(w)$ is analytic at $w = 0$ or $f(z)$ is analytic at $z = \infty$. If $M - N < 0$, then $g(w)$ has a pole of order $N - M$ at $w = 0$ or $f(z)$ has a pole of order $N - M$ at $z = \infty$.

Exercise 13.2

1. Simple poles at $z = 0, 1$; Res $(z = 0) = -1$, Res $(z = 1) = 1$.

2. Simple poles at $z = \pm i$, Res $(z = i) = 1/2$; Res $(z = -i) = 1/2$.

3. Simple poles at $z = 0, 1$, Res $(z = 0) = -3$; Res $(z = 1) = 4$.

4. Simple poles at $z = \pm i$, Res $(z = i) = (\sinh 1)/2$; Res $(z = -i) = (\sinh 1)/2$.

5. Simple pole at $z = 0$, Res $(z = 0) = -1/\pi$; $z = \pi$ is a removable singularity, Res $(z = \pi) = 0$.

6. Simple poles at $z_k = [(2k + 1)\pi/2]$, Res $(z = z_k) = -1$.

7. Simple poles at $z_k = k\pi i$, Res $(z = z_k) = 1$.

8. Removable singularity at $z = 0$, Res $(z = 0) = 0$.

9. Simple poles at $z_k = 1/k\pi$, Res $(z = 1/(k\pi)) = (-1)^{k+1}/(k^2\pi^2)$; (non-isolated singularity at $z = 0$).

10. Pole of order 4 at $z = \pi$, Res $(z = \pi) = i/6$.

11. Essential singularity at $z = 0$, Res $(z = 0) = 1/24$.

12. Essential singularity at $z = 0$, Res $(z = 0) = -\sum\limits_{n=1}^{\infty} \dfrac{2^{2n}}{(2n-1)!(2n)!}$.

13. Essential singularity at $z = 0$, Res $(z = 0) = \sum\limits_{n=0}^{\infty} \dfrac{(-1)^{n+1}}{n!(n+1)!}$.

14. Simple pole at $z = 1$, Res $(z = 1) = \sin 1$; essential singularity at $z = 0$, Res $(z = 0) = \sum\limits_{n=0}^{\infty} \dfrac{(-1)^{n+1}}{(2n+1)!}$.

15. Essential singularity at $z = 0$; if $n < 0$: Res $(z = 0) = 0$; if $n > 0$ and odd : Res $(z = 0) = 0$; if $n = 0$ or

even: Res $(z = 0) = \dfrac{(-1)^{n/2}}{(n+1)!}$.

16. Essential singularity at $z = -1$. Write $f(z) = \sin\left(1 - \dfrac{1}{z+1}\right)$, Res $(z = -1) = -\cos 1$.

17. Pole of order 2 at $z = \pi$, Res $(z = \pi) = -(1 + \pi i)$.

18. Pole of order 2 at $z = -1$, Res $(z = -1) = 2 \sin 2$.

19. Pole of order 2 at $z = 0$, Res $(z = 0) = (3 \cos 1 - \sin 1)/\sin^2 1$; simple poles at $z_k = 1 + k\pi$, Res $(z = z_k) = (-1)^k (k\pi - 2)/[(1 + k\pi)^2]$.

20. Simple pole at $z = 1$, Res $(z = 1) = \sin 1$; pole of order 2 at $z = 0$, Res $(z = 0) = -1$.

21. Simple pole at $z = 1$, Res $(z = 1) = 3/4$; simple pole at $z = i$, Res $(z = i) = (3 + 2i)/4$; simple pole at $z = -i$, Res $(z = -i) = (3 - 2i)/4$; pole of order 2 at $z = -1$, Res $(z = -1) = -9/4$.

22. Simple pole at $z = 0$, Res $(z = 0) = 1/a^4$; pole of order 2 at $z = \pm ai$, Res $(z = ai) = -(2 + ma) \, e^{-ma}/(4a^4)$; Res $(z = -ai) = -(2 - ma) \, e^{ma}/(4a^4)$.

23. Pole of order 3 at $z = -1$, Res $(z = -1) = 2 \sin 2$.

24. Pole of order 3 at $z = \pm i$, Res $(z = i) = -i/16$; Res $(z = -i) = i/16$.

25. Pole of order n at $z = -1$, Res $(z = -1) = -(3n)!/[(n-1)!\,(2n)!]$

26. Pole of order 2 at $z = 1$, write Log $(z) =$ Log $(1 + (z - 1))$, Res $(z = 1) = 1$.

27. Pole of order 2 at $z = z_k = k\pi$, write

$$\sin z = \sum_{m=0}^{\infty} \frac{1}{m!} (z - z_k)^m \sin^{(m)}(z_k) = (-1)^k \left[\frac{(z - z_k)}{1!} - \frac{(z - z_k)^3}{3!} + \dots \right]$$

$$\operatorname*{Res}_{z = z_k} f(z) = \lim_{z \to z_k} \frac{d}{dz} \left(\left\{ 1 - \left[\frac{(z - z_k)^2}{3!} - \frac{(z - z_k)^4}{5!} + \dots \right] \right\}^{-2} \cos^2 z \right) = 0.$$

28. Pole of order 2 at $z = 0$, Res $(z = 0) = 0$; simple poles at $z = k\pi$, Res $(z = k\pi) = 1/(k\pi)$.

29. Simple pole at $z = 0$, Res $(z = 0) = 1$.

30. Poles of order $n + 1$ at $z = \pm i$, Res $(z = i) = -\dfrac{(2n)!\,i}{(n!)^2 \, 2^{2n+1}}$; Res $(z = -i) = \dfrac{i\,(2n)!}{(n!)^2 \, 2^{2n+1}}$.

31. Res $(z = \infty) = -1$.   **32.** Res $(z = \infty) = 0$.

33. Res $(z = \infty) = -\left[\dfrac{1}{3!} - \dfrac{1}{5!} + \dfrac{1}{7!} - \dots \right] = \sin 1 - 1$.

34. $g(w) = \dfrac{1}{w^2} f\left(\dfrac{1}{w}\right) = w \sin\left(\dfrac{2}{w}\right)(1 + w)^{-3}$

$$= w \left[\frac{2}{w} - \frac{1}{3!}\left(\frac{2}{w}\right)^3 + \frac{1}{5!}\left(\frac{2}{w}\right)^5 - \dots \right][1 - 3w + 6w^2 - 10w^3 + 15w^4 - \dots].$$

Coefficient of $(1/w)$ in $g(w) = (3)\dfrac{2^3}{3!} - \dfrac{2^5}{5!}(10) + \dfrac{2^7}{7!}(21) - \dots$

$$= \frac{2^3}{2!} - \frac{2^6}{4!} + \frac{3 \cdot 2^7}{6!} - \dots = 2\left[2 - \frac{2^3}{3!} + \frac{2^5}{5!} - \dots\right] = 2 \sin 2.$$

Hence, Res $(z = \infty) = -2 \sin 2$.

35. Res $(z = \infty) = 0$.

36. $g(w) = \dfrac{1}{w^2} f\left(\dfrac{1}{w}\right) = \dfrac{1}{w^5} \sin\left(\dfrac{w}{1 - 2w}\right) = \dfrac{1}{w^5}\left[\left(\dfrac{w}{1 - 2w}\right) - \dfrac{1}{3!}\left(\dfrac{w}{1 - 2w}\right)^3 + \dfrac{1}{5!}\left(\dfrac{w}{1 - 2w}\right)^5 - \dots \right]$

$$= \frac{1}{w^5}\left[w(1 - 2w)^{-1} - \frac{w^3}{3!}(1 - 2w)^{-3} + \frac{w^5}{5!}(1 - 2w)^{-5} - \dots \right]$$

Coefficient of $(1/w)$ in $g(w) = 7$. Hence, Res $(z = \infty) = -7$.

37. $g(w) = \dfrac{1}{w^2} f\left(\dfrac{1}{w}\right) = \sin(e^w)$, Res $(z = \infty) = -\cos 1$.

38. $f(z) = 1/z^2$.

39. (i) $f(z) = \dfrac{A}{z-a} + A_1 + A_2(z-a) + \dots$; Res $(z=a) = A\,g(a)$.

(ii) $f(z) = a_0 + a_1(z-a) + \dots + \dfrac{b_1}{z-a} + \dfrac{b_2}{(z-a)^2} + \dots + \dfrac{b_m}{(z-a)^m}$

$g(z) = g(a) + (z-a)\,g'(a) + \dots + \dfrac{(z-a)^m}{m!}\,g^{(m)}(a) + \dots$

Res $(z=a)$ = coefficient of $(z-a)^{-1}$ in $f(z)\,g(z) = b_1 g(a) + \dfrac{b_2}{1!}\,g'(a) + \dots \dfrac{b_m}{(m-1)!}\,g^{(m-1)}(a)$.

40. (i) $f(z) = (z-a)^n\,g(z)$, $g(a) \neq 0$; $f'(z)/f(z)$ has a simple pole at $z = a$; Res $(z=a) = n$.

(ii) If $f(z)$ has a pole of order n, then $f'(z)$ has a pole of order $n+1$. Therefore, $h(z)$ has a simple pole at $z = a$; Res $(z=a) = -n$.

Exercise 13.3

1. $z = 0$, removable singular point inside C, $I = 0$.

2. Integrand has no singular points inside C, $I = 0$.

3. $z = \pm\,\pi$, simple poles inside C, $I = -4\pi^3$.   **4.** $z = 0$, simple pole inside C, $I = 2\pi i$.

5. $z = (-2 + \sqrt{3})i$, simple pole inside C, $I = 2\pi/\sqrt{3}$.

6. $z = 0$, simple pole inside C, $I = \pi i$.

7. $z = \pm\,1/2$, simple poles inside C, $I = -2i(e^1 - e^{-1}) = -4i \sinh 1$.

8. $z = \pm\,\pi i/2$, simple poles inside C, $I = 4\pi i$.   **9.** $z = \pm\,\pi/6$, simple poles inside C, $I = -2\pi i/3$.

10. $z = \pm\,3i$, simple poles inside C, $I = (2\pi i \sin 6)/3$.

11. $z = 0, -1$, simple poles inside C, $I = \pi i[e^{1/4} \sin 1 - 2e]$.

12. $z = 0$, removable singular point, $z = \pm\,\pi$, simple poles, $I = 0$.

13. $z = \pm\,1, \pm\,i$, simple poles inside C, $I = 2\pi i$.

14. $z = \pm\,1/2, \pm\,3/2$, simple poles inside C, residue at each pole is $-1/\pi$, $I = -8i$.

15. $z = \pm\,(1 + i)/2^{3/4}, \pm\,(1 - i)/2^{3/4}$, simple poles inside C, residue at each pont is $1/8$, $I = \pi i$.

16. $z = \pm\,\pi$, simple poles; $z = 0$ pole of order 2 inside C, $I = 0$.

17. $z = \pm\,\pi$, poles of order 2 inside C, $I = 16\pi i$.

18. $z = i$, pole of order 3 inside C, $I = \pi(-4 + 5i)\,e^i$.

19. $z = \pi/2$, pole of order 2 inside C, $I = -12\pi i$.

20. $z = 2$, pole of order 2 inside C, $I = -2\pi i$.   **21.** $z = 1/2, -3/2$, poles of order 2 inside C, $I = \pi i e^{-3/2}$.

22. If $r < 2$, $z = i$, pole of order 2 inside C. $I = \pi(\cosh 1 - \sinh 1)/2$,

If $r > 2$, $z = \pm\,i$ poles of order 2 inside C, $I = 0$.

23. $z = 0$, pole of order 3 inside C, $I = 4\pi i$.   **24.** $z = \pi/2, 3\pi/2$, poles of order 2 inside C, $I = 0$.

25. $z = 0$, essential singular point inside C, $I = 5\pi i/3$.

26. $z = 0$, essential singular point inside C, $I = -4\pi i$.

27. $z = 0$, essential singular point inside C, $I = 2\pi i$.

28. If $n \leq 0$, there is no singular point inside C. In other cases, $z = 0$ is an essential singular point inside C; $\operatorname*{Res}_{z=0} f(z) = 0$ if $n \leq 0$ or $n \geq 2$, $\operatorname*{Res}_{z=0} f(z) = 1$ if $n = 1$, $I = 2\pi i$, if $n = 1$, otherwise $I = 0$.

29. $z = 0$, essential singular point inside C,

$$\operatorname*{Res}_{z=0} f(z) = \frac{1}{(n+1)!} \text{ if } n \geq -1 \text{ and } 0 \text{ if } n < -1, \quad I = \frac{2\pi i}{(n+1)!} \text{ if } n \geq -1 \text{ and } 0 \text{ if } n < -1.$$

30. $z = 0$, essential singular point inside C, $I = 2\pi i$.

31. $z = 0$, essential singular point inside C, $I = 0$.

32. $z = 0$, essential singular point inside C, $I = 32\pi i/3$.

33. Integrand has simple poles at $0, z_1, z_2, \ldots, z_k$. Therefore $f(z_k) \neq 0$.

$$I = 2\pi i \left[\frac{f(0)}{g(0)} + \sum_{k=1}^{n} \frac{f(z_k)}{z_k g'(z_k)} \right].$$

34. Let $f(z) = (z - z_0)^m g(z)$, $g(z_0) \neq 0$ then

$$F(z) = \frac{f'(z)}{f(z)} = \frac{m}{z - z_0} + \frac{g'(z)}{g(z)}, \quad \operatorname*{Res}_{z=z_0} f(z) = m.$$

35. Let $f(z) = (z - a_1)^{\alpha_1} (z - a_2)^{\alpha_2} \ldots (z - a_m)^{\alpha_m} g(z)$, where $g(z)$ is analytic everywhere except at poles $b_1, b_2, \ldots, b_n$ and $g(a_k) \neq 0$, $k = 1, 2, \ldots, m$. We have

$$\ln f(z) = \alpha_1 \ln (z - a_1) + \alpha_2 \ln (z - a_2) + \ldots + \alpha_m \ln (z - a_m) + \ln g(z)$$

and

$$\frac{f'(z)}{f(z)} = \sum_{k=1}^{m} \frac{\alpha_k}{(z - a_k)} + \frac{g'(z)}{g(z)}$$

Therefore,

$$\sum_{k=1}^{m} \operatorname*{Res}_{z \to a_k} \left[\frac{f'(z)}{f(z)} \right] = \sum_{k=1}^{m} \alpha_k$$

Now, let

$$f(z) = h(z)/[(z - b_1)^{\beta_1} (z - b_2)^{\beta_2} \ldots (z - b_n)^{\beta_n}],$$

where $h(z)$ is analytic everywhere and $h(b_k) \neq 0$, $k = 1, 2, \ldots, n$. We have

$$\ln f(z) = \ln h(z) - [\beta_1 \ln (z - b_1) + \ldots + \beta_n \ln (z - b_n)]$$

and

$$\frac{f'(z)}{f(z)} = \frac{h'(z)}{h(z)} - \sum_{k=1}^{n} \frac{\beta_k}{(z - b_k)}$$

Therefore,

$$\sum_{k=1}^{n} \operatorname*{Res}_{z \to b_k} \left[\frac{f'(z)}{f(z)} \right] = - \sum_{k=1}^{n} \beta_k$$

Hence, the result.

Exercise 13.4

1. Integrand has a simple pole at $z = (-3 + \sqrt{5})i/2$ which lies inside $|z| = 1$; $I = 2\pi/\sqrt{5}$.

2. Integrand has simple poles at $z = 0$ and $(-3 + 2\sqrt{2})$ which lie inside $|z| = 1$; $I = 0$.

3. Integrand has simple poles at $z = \pm i/\sqrt{2}$ which lie inside $|z| = 1$; $I = 0$.

4. Integrand has a simple pole at $z = -i(3 - 4i)/25$ which lies inside $|z| = 1$; $I = \pi/6$.

5. Let $f(\theta) = \cos 2\theta/(1 - 2a \cos \theta + a^2)$.

$$I = \int_0^{2\pi} f(\theta)d\theta = 2 \int_0^{\pi} f(\theta)d\theta, \quad \text{or} \quad \int_0^{\pi} f(\theta)d\theta = I/2. \text{ Substituting } z = e^{i\theta}, \text{ we obtain}$$

$$I = \int_C f(z)dz, \quad \text{where } f(z) = -\frac{1}{4ai} \left[\frac{(z^4 + 1)}{z^2[z^2 - \{(a^2 + 1)/a\}z + 1]} \right]$$

The integrand has a pole of order 2 at $z = 0$ and a simple pole at $z = a$ which lie inside $|z| = 1$; $I = \pi a^2/(1 - a^2)$.

6. Write $\cos 3\theta = (z^3 + z^{-3})/2$ and $I = I_1 + I_2$, where I_1 corresponds to z^3 and I_2 corresponds to z^{-3}. Integrand of I_1 has a simple pole at $z = 1/3$ inside $|z| = 1$, and the integrand of I_2 has a simple pole at $z = -1/3$, and a pole of order 3 at $z = 0$ inside $|z| = 1$; $I = \pi/54$.

7. Integrand has a simple pole at $z = a$ and a pole of order 3 at $z = 0$ which lie inside $|z| = 1$; $I = \pi a(a^2 + 1)/(1 - a^2)$.

8. Integrand has a simple pole at $z = -1/2$ and a pole of order 2 at $z = 0$ which lie inside $|z| = 1$; $I = \pi/4$.

9. Integrand has a pole of order 3 at $z = 0$ and simple poles at $z = \pm i/\sqrt{2}$ which lie inside $|z| = 1$; $I = 5\pi/12$.

10. Integrand has simple poles at $z = \pm 1/\sqrt{3}$ which lie inside $|z| = 1$; $I = \pi$.

11. Integrand has a pole of order 2 at $z = (-a + \sqrt{a^2 - b^2})/b$ which lies inside $|z| = 1$; $I = 2\pi a/(a^2 - b^2)^{3/2}$.

12. Write $\cos n\theta = (z^n + z^{-n})/2$ and $I = I_1 + I_2$, where I_1 corresponds to z^n and I_2 corresponds to z^{-n}. Integrand of I_1 has a simple pole at $z = a$ inside $|z| = 1$ and the integrand of I_2 has a simple pole at $z = a$ and a pole of order n at $z = 0$ inside $|z| = 1$; $I = 2\pi a^n/(1 - a^2)$.

13. Integrand has a pole of order $(2n + 1)$ at $z = 0$ inside $|z| = 1$. $I = \left[2\pi \binom{2n}{n} \right] / 2^{2n}$.

14. Integrand has a pole of order $(2n + 2)$ at $z = 0$ inside $|z| = 1$, $I = 0$.

15. We have $I = \dfrac{1}{2i} \displaystyle\int_C z^{n-1} e^{1/z} dz + \dfrac{1}{2i} \displaystyle\int_C \dfrac{e^z}{z^{n+1}} dz$, $C : |z| = 1$. First integral has an essential singularity at $z = 0$ inside $|z| = 1$ and the second integral has a pole of order $(n + 1)$ at $z = 0$ inside $|z| = 1$; $I = 2\pi/n!$.

16. For $0 \leq \theta \leq \dfrac{\pi}{2}$, we have $1 \geq \dfrac{\sin \theta}{\theta} \geq \dfrac{2}{\pi}$. Therefore, $e^{-R \sin \theta} \leq e^{-(2R\theta)/\pi}$, $0 \leq \theta \leq \dfrac{\pi}{2}$.

Hence, $\displaystyle\int_0^{\pi/2} e^{-R \sin \theta} d\theta \leq \int_0^{\pi/2} e^{-(2R\theta)/\pi} d\theta = \left[\dfrac{\pi}{-2R} e^{-(2R\theta)/\pi} \right]_0^{\pi/2}$

$= -\dfrac{\pi}{2R} \left[e^{-R} - 1 \right] = \dfrac{\pi}{2R} (1 - e^{-R}) < \dfrac{\pi}{2R}.$

In problems **17** to **31**, we consider the contour integral $\displaystyle\int_C f(z)dz$, where C is the upper half plane.

17. Integrand has a simple pole at $z = 3i$ inside C, $I = \pi/3$.

18. Integrand has a simple pole at $z = (-1 + 3i)/2$ inside C, $I = \pi/3$.

19. integrand has simple poles at $z = ai$ and $z = bi$ inside C, $I = \pi/[ab(a + b)]$.

20. Integrand has simple poles at $z = i$ and $z = 2i$ inside C, $I = \pi/3$.

21. Integrand has a pole of order 2 at $z = ai$ inside C, $I = \pi/(2a^3)$.

22. Integrand has a pole of order 2 at $z = -2 + \sqrt{3}i$ inside C, $I = \pi/54$.

23. Integrand has poles of order 2 at $z = i$, $z = 2i$ inside C, $I = 11\pi/432$.

24. Integrand has a simple pole at $z = 3i$ and a pole of order 2 at $z = 2i$; $I = 7\pi/1200$.

25. Integrand has simple poles at $z = a(1 + i)/\sqrt{2}$ and $z = a(-1 + i)/\sqrt{2}$ inside C, $I = \pi\sqrt{2}/a$.

26. Integrand has simple poles at $z = \sqrt{5}i$ and $z = i/\sqrt{5}$ inside C, $I = \pi\sqrt{5}/15$.

27. Integrand has simple poles at $z_k = a\, e^{(2k+1)\pi i/6}$, $k = 0, 1, \ldots 5$, the poles z_0, z_1 and z_2 lie inside C, $I = \pi/(3a^3)$.

28. Integrand has a pole of order 3 at $z = i$ inside C, $I = \pi/8$.

29. Integrand has simple poles at $z = i/\sqrt{2}, z = \sqrt{2}i$ and $z = i$ inside C, $I = \pi(5\sqrt{2} - 6)/6$.

30. Integrand has a pole of order $n + 1$ at $z = i$ inside C, $I = [\pi(2n)!]/[(n!)^2\, 2^{2n}]$.

31. Integrand $f(z)$ has simple poles at $z_k = e^{[(2k+1)\pi i/(2n)]}$, $k = 0, 1, \ldots, (2n - 1)$. Only the poles z_k, $k = 0, 1, \ldots, (n - 1)$ lie inside C. Let $\alpha = e^{(2m+1)\pi i/(2n)}$. Then, $\alpha^{2n} = -1$.

$$\operatorname*{Res}_{z = z_k} f(z) = \frac{z_k^{2m}}{(2n)z_k^{2n-1}} = \frac{z_k^{2m+1}}{2n\, z_k^{2n}} = -\frac{1}{2n}\alpha^{(2k+1)}, \quad k = 0, 1, \ldots, (n - 1).$$

Therefore, $I = \dfrac{-2\pi i}{2n}[\alpha + \alpha^3 + \ldots + \alpha^{2n-1}] = -\dfrac{\pi i\alpha}{n}\left[\dfrac{1 - \alpha^{2n}}{1 - \alpha^2}\right]$

$$= -\frac{2\pi i\alpha}{n(1 - \alpha^2)} = \frac{2\pi i}{n(\alpha - \alpha^{-1})} = \frac{\pi}{[n \sin\{(2m + 1)\pi/(2n)\}]}.$$

32. $\pi e^{-4}/2$.

33. $\pi e^{-1}(\cos 2 + 2 \sin 2)$.

34. $\pi e^{-1}(\cos 1 - \sin 1)$.

35. $\pi e^{-q\alpha}[(a - bB)\cos \alpha B + bq \sin \alpha B]/q$, $q = \sqrt{A - B^2}$.

36. $\pi e^{-b\alpha} \cosh(a\alpha)$.

37. $\pi[ae^{-2b} - be^{-2a}]/[ab(a^2 - b^2)]$.

38. $\pi(ab + 1)e^{-ab}/(2b^3)$.

39. $\pi(1 + 3e^{-2})/4$.

40. $(\pi/2)(2 - a)e^{-a}$.

41. $\pi[a^2 e^{-ma} - b^2 e^{-mb}]/(a^2 - b^2)$.

42. $\pi e^{-ab\cos\alpha} \sin(2\alpha - ab \sin \alpha)/(\sin 2\alpha)$.

43. $I^* = \displaystyle\int_C \frac{e^{imz}}{z}dz$, $I = \operatorname{Im}(I^*) = \pi$.

44. $I^* = \displaystyle\int_C \frac{e^{4iz} - e^{2iz}}{z^2}\,dz$, $I = \operatorname{Re}(I^*) = -2\pi$.

45. $f(x) = \operatorname{Re}[(1 - e^{2imx})/x^2]$. $I^* = \displaystyle\int_C \frac{1 - e^{2imz}}{z^2}\,dz$.

$z = 0$ is a simple pole on the real axis. Also, $(CPV)(I) = I$. We get $I^* = m\pi$ and $I = \operatorname{Re}(I^*) = m\pi$.

46. $\pi(1 - e^{-3})/9$. **47.** $\pi/2$. **48.** π.

49. $I = \displaystyle\int_C \frac{(z + 1)dz}{z(z - 2)(z^2 + 16)}$; integrand has a simple pole at $z = 4i$ in the upper half plane and simple poles at $z = 0$ and $z = 2$ on the real axis; $I = -3\pi/80$.

50. $I = \displaystyle\int_C \frac{(3z + 5)dz}{z(z + 2)(z^2 + 1)}$; integrand has a simple pole at $z = i$ in the upper half plane and simple poles at $z = 0$ and $z = -2$ on the real axis; $I = \pi/5$.

Chapter 14

Bilinear Transformations and Conformal Mapping

14.1 Introduction

A function $w = f(z) = u(x, y) + iv(x, y)$ defined in a certain domain D of the z-plane defines a *mapping* or a *transformation* which maps the domain D of the z-plane into a domain D^* of the w-plane. Thus, given a point $P(z_0)$, $z_0 \in D$, in the z-plane, there corresponds a point $Q(w_0) = f(z_0)$, $Q \in D^*$, in the w-plane. If to each point $P \in D$ there corresponds exactly one point $Q \in D^*$ and Q is the image of only one point P, then the mapping is said to be *one-to-one*. Now, consider the set of values of z on a segment of any curve C lying inside D and passing through a point z_0 in the z-plane. Corresponding to this set of values of z, there exists an image set of values in the w-plane, on a segment of the image curve C^* lying inside D^* and passing through the point $w_0 = f(z_0)$. The angle of inclination of the curve C in the z-plane (or C^* in the w-plane) is defined as the angle of inclination of the directed line tangent at the point z_0 on C (or at the point w_0 on C^*) with the positive x-axis (or u-axis). Let C_1, C_2 be two curves contained in D and passing through a point $z_0 \in D$ and let C_1^*, C_2^* be the corresponding image curves passing through the point $w_0 \in D^*$ in the w-plane. Then, *the mapping $w = f(z)$ is said to be conformal at $z = z_0$ if it is one-to-one and preserves the angle between the curves C_1 and C_2 both in magnitude and direction* (Fig. 14.0).

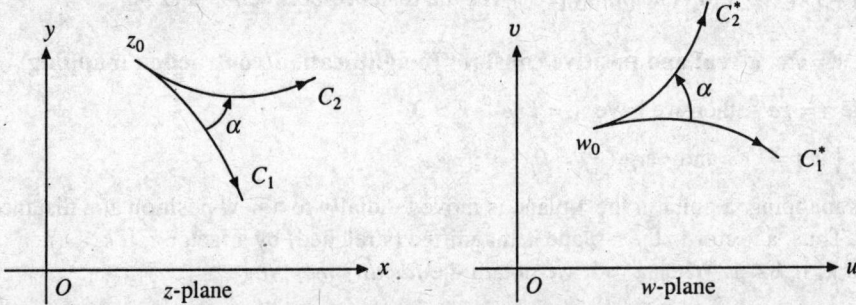

Fig. 14.0. Conformal mapping.

Before we derive conditions for testing whether a mapping $w = f(z)$ is conformal in D, we define a few important transformations.

14.2   Linear and Inverse Transformations

14.2.1   Linear Transformation

A mapping or a transformation of the form

$$w = az + b, \tag{14.1}$$

where a, b are real or complex constants, is called a *linear function* or a *linear mapping* or a *linear transformation*. The linear mapping defined by Eq. (14.1), when written in the form

$$w = a\left[z + \frac{b}{a}\right] = |a|\, e^{i\phi}\left[z + \frac{b}{a}\right]$$

where $a = |a|\, e^{i\phi}$, $\phi = \arg(a)$, can be considered as a composition of the following three mappings:

(i) $w_1 = z + (b/a)$ or $w_1 = z + c$, where $c = b/a$ is a real or a complex constant.

(ii) $w_2 = k w_1$, where $k = |a|$ is a real positive constant.

(iii) $w = e^{i\phi} w_2$.

Alternatively, we can write $w = |a|\, e^{i\phi} z + b$ and consider it as the composition of the following mappings:

(i) $w_1 = |a|\, z$,   (ii) $w_2 = e^{i\phi} w_1$ and   (iii) $w = w_2 + b$.

We discuss each of these mappings separately.

Mapping $w = z + c$ (translation mapping)

Let $z = x + iy$, $c = c_1 + ic_2$ and $w = u + iv$. Then, we get

$$u + iv = (x + iy) + (c_1 + ic_2) = (x + c_1) + i(y + c_2)$$

or $u = x + c_1$ and $v = y + c_2$.

Therefore, the point (x, y) in the z-plane is mapped as the point $(u, v) = (x + c_1, y + c_2)$ in the w-plane. Thus, the image in the w-plane of any region of the z-plane is the translation of that region in the direction of the vector $c_1 + ic_2$.

For any two points z_1 and z_2 in the z-plane, we get $w_1 = z_1 + c$ and $w_2 = z_2 + c$. Since, $|w_1 - w_2| = |z_1 - z_2|$, this mapping preserves the distance between z_1 and z_2.

Mapping $w = kz$, k real and positive constant (magnification/contraction mapping)

If we write $z = re^{i\theta}$, then we have $w = kre^{i\theta}$, $k > 0$.

Therefore, $|w| = kr$ and $\arg(w) = \theta$.

Under this mapping, a point in the z-plane is moved radially to a new position at a distance kr from the origin. Thus, a vector in the z-plane is magnified (stretched) by a factor k, if $k > 1$ and contracted by a factor k, if $k < 1$. When $k = 1$, we obtain the *identity mapping* $w = z$.

Mapping $w = e^{i\phi} z$ (rotation mapping)

If we write $z = re^{i\theta}$, we get

$$w = re^{i(\theta + \phi)}, \quad |w| = r \quad \text{and} \quad \arg(w) = \theta + \phi.$$

Under this mapping, a point $z(r, \theta)$ in the z-plane is mapped as a point $w(r, \theta + \phi)$ in the w-plane. Therefore, this mapping rotates a region in the z-plane through an angle ϕ in the w-plane, in the anti-clockwise direction, if $\phi > 0$ and in the clockwise direction if $\phi < 0$.

Example 14.1   Let a rectangular region $OABC$ with vertices at $O(0, 0)$, $A(1, 0)$, $B(1, 2)$, $C(0, 2)$ be defined in the z-plane. Find the image of this region in the w-plane, under the mapping $w = z + 2 + i$.

Solution   Writing $z = x + iy$ and $w = u + iv = z + 2 + i = (x + 2) + i(y + 1)$, we get

$$u = x + 2 \quad \text{and} \quad v = y + 1.$$

The point (x, y) in the z-plane is mapped as the point $(u, v) = (x + 2, y + 1)$ in the w-plane. Therefore, the point

$O(0, 0)$ is mapped as $O'(2, 1)$, $A(1, 0)$ is mapped as $A'(3, 1)$,

$B(1, 2)$ is mapped as $B'(3, 3)$, $C(0, 2)$ is mapped as $C'(2, 3)$.

Thus, the image of the rectangular region $OABC$ in the z-plane is the rectangular region $O'A'B'C'$ in the w-plane (Fig. 14.1).

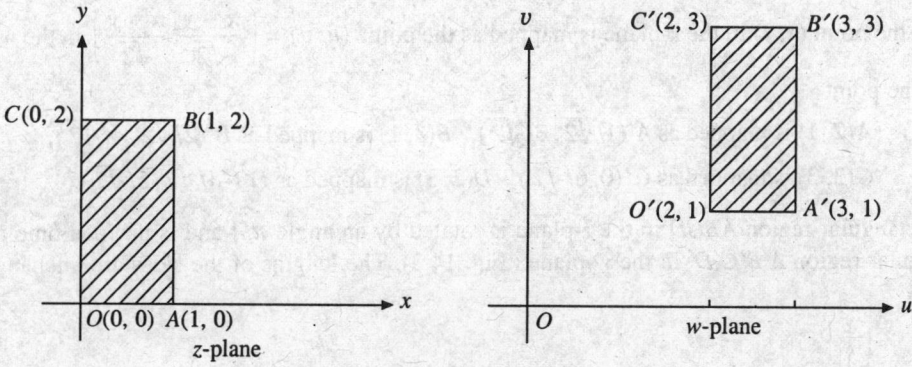

Fig. 14.1.   Example 14.1.

Example 14.2   Let a rectangular region $ABCD$ with vertices at $A(2, 1)$, $B(3, 1)$, $C(3, 3)$, $D(2, 3)$ be defined in the z-plane. Find the image of this region in the w-plane under the mapping $w = 2z$.

Solution   Writing $z = x + iy$ and $w = u + iv$, we get $u = 2x$ and $v = 2y$. The point (x, y) in the z-plane is mapped as the point $(u, v) = (2x, 2y)$ in the w-plane. Therefore, the point

$A(2, 1)$ is mapped as $A'(4, 2)$, $B(3, 1)$ is mapped as $B'(6, 2)$,

$C(3, 3)$ is mapped as $C'(6, 6)$, $D(2, 3)$ is mapped as $D'(4, 6)$.

Thus, the image of the rectangular region $ABCD$ in the z-plane is another rectangular region $A'B'C'D'$ in the w-plane (Fig. 14.2). The region is magnified in both u and v directions.

Example 14.3   Let a rectangular region $ABCD$ with vertices at $A(2, 1)$, $B(3, 1)$, $C(3, 3)$, $D(2, 3)$ be defined in the z-plane. Find the image of this region in the w-plane under the mapping $w = e^{i\pi/4}z$.

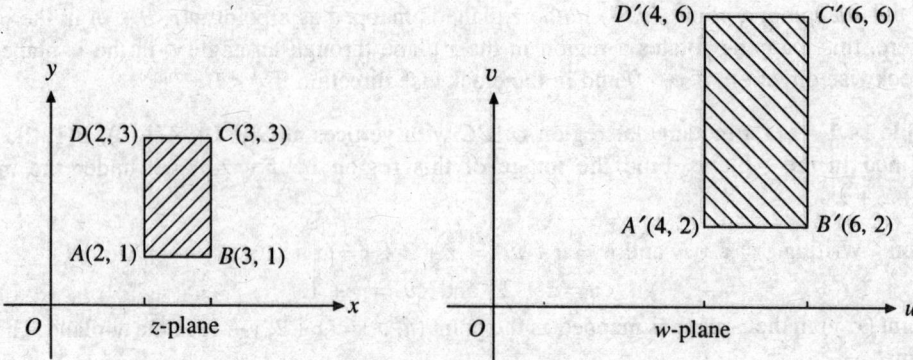

Fig. 14.2.　Example 14.2.

Solution　Writing $z = re^{i\theta}$, we obtain $w = re^{i(\theta + \pi/4)}$. Therefore, any point in the z-plane is rotated through an angle $\pi/4$ in the w-plane. Writing $z = x + iy$, $w = u + iv$, we get

$$w = u + iv = \left(\cos \frac{\pi}{4} + i \sin \frac{\pi}{4} \right)(x + iy) = \frac{1}{\sqrt{2}}(x - y) + \frac{1}{\sqrt{2}}i(x + y).$$

Hence, the point (x, y) in the z-plane is mapped as the point $(u, v) = \left(\dfrac{x - y}{\sqrt{2}}, \dfrac{x + y}{\sqrt{2}} \right)$ in the w-plane.

Thus, the point

$A(2, 1)$ is mapped as $A'(1/\sqrt{2}, 3/\sqrt{2})$,　$B(3, 1)$ is mapped as $B'(2/\sqrt{2}, 4/\sqrt{2})$,

$C(3, 3)$ is mapped as $C'(0, 6/\sqrt{2})$,　$D(2, 3)$ is mapped as $D'(-1/\sqrt{2}, 5/\sqrt{2})$.

The rectangular region $ABCD$ in the z-plane is rotated by an angle $\pi/4$ and is mapped onto another rectangular region $A'B'C'D'$ in the w-plane (Fig. 14.3). The lengths of the sides are unchanged.

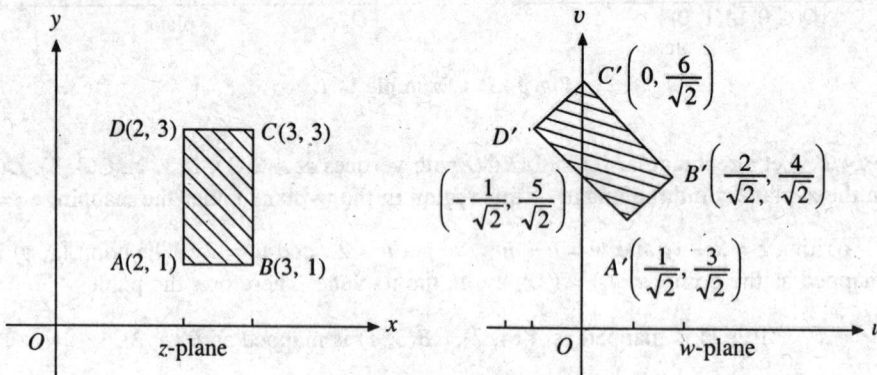

Fig. 14.3.　Example 14.3.

Example 14.4　Let a rectangular region with vertices at $O(0, 0)$, $A(1, 0)$, $B(1, 2)$, $C(0, 2)$ be defined in the z-plane. Find the image of this region in the w-plane under the mapping $w = (1 - i)z - 2i$.

Solution　We write the given mapping as

$$w = (1 - i)\left[z - \frac{2i}{1 - i}\right] = \sqrt{2}\,e^{-\pi i/4}[z + (1 - i)].$$

This mapping may be considered as the composition of the following three mappings

(i) $w_1 = z + (1 - i)$, (translation)

(ii) $w_2 = \sqrt{2}\,w_1$, (magnification)

(iii) $w = e^{-\pi i/4}w_2$, (rotation).

The first mapping translates the rectangular region $OABC$ in the z-plane, under the mapping $w_1 = z + (1 - i)$ onto the rectangle $O'A'B'C'$ in the w_1-plane.
Thus the point

$O(0, 0)$ in mapped as $O'(1, -1)$, $A(1, 0)$ is mapped as $A'(2, -1)$,

$B(1, 2)$ is mapped as $B'(2, 1)$, $C(0, 2)$ is mapped as $C'(1, 1)$ (Fig. 14.4(ii)).

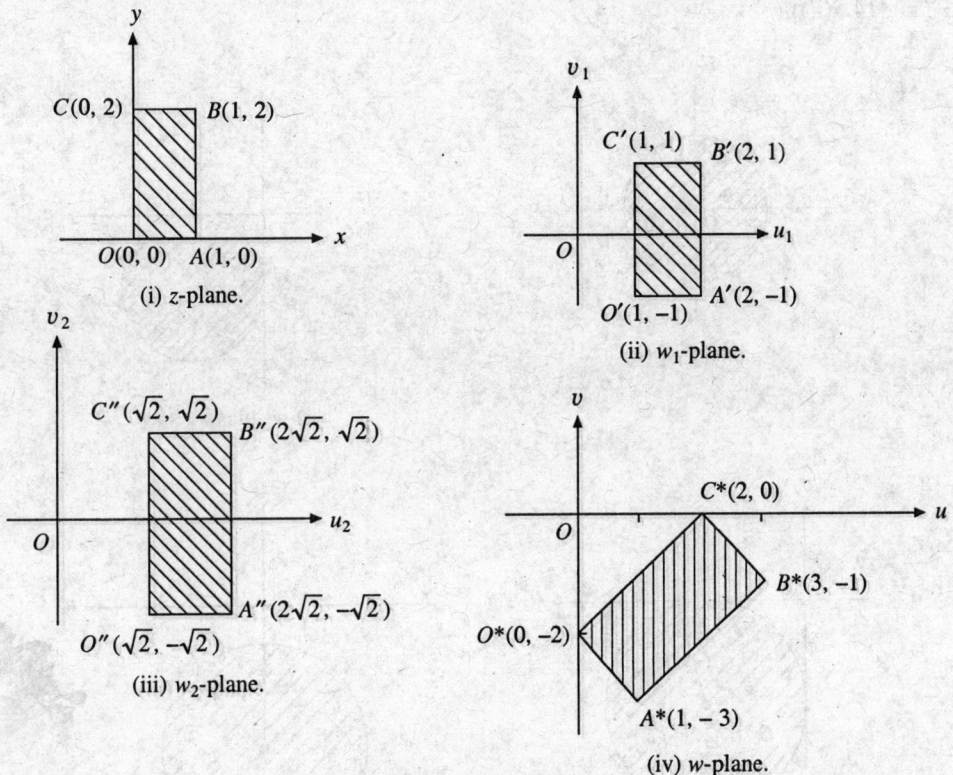

Fig. 14.4.   Example 14.4.

The second mapping magnifies the rectangular region $O'A'B'C'$ in the w_1-plane, under the mapping $w_2 = \sqrt{2}\,w_1$ onto the rectangle $O''A''B''C''$ in the w_2-plane (Fig. 14.4(iii)).

The third mapping rotates the rectangular region $O''A''B''C''$ in the w_2-plane by an angle $-\pi/4$, under the mapping $w = e^{-\pi i/4} w_2$ onto the rectangle $O^*A^*B^*C^*$ in the w-plane (that is rotation through an angle $\pi/4$ in the clockwise direction).

Thus the point

$$O''(\sqrt{2}, -\sqrt{2}) \text{ is mapped as } O^*(0, -2), \qquad A''(2\sqrt{2}, -\sqrt{2}) \text{ is mapped as } A^*(1, -3),$$

$$B''(2\sqrt{2}, \sqrt{2}) \text{ is mapped as } B^*(3, -1), \ C''(\sqrt{2}, \sqrt{2}) \text{ is mapped as } C^*(2, 0) \text{ (Fig. 14.4(iv))}.$$

Example 14.5 Find the image of the closed unit disk $|z| \le 1$ under the mapping $w = (1 - i)z - 2i$.

Solution We write the given mapping as

$$w = (1 - i)\left[z - \frac{2i}{1 - i} \right] = \sqrt{2} e^{-\pi i/4} [z + (1 - i)].$$

Therefore, the given mapping is a composition of the following three mappings:

(i) $w_1 = z + (1 - i)$, (translation).

$|z| \le 1$ gives $|w_1 - (1 - i)| \le 1$, which is a closed unit disk with centre at $(1, -1)$ (Fig 14.5(ii)).

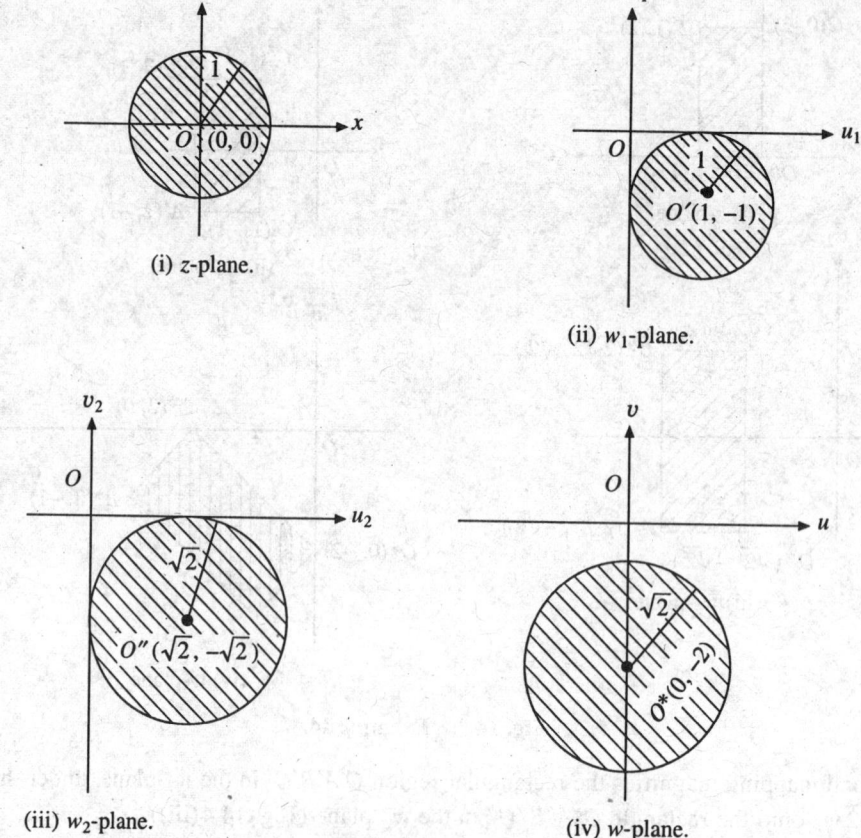

(i) z-plane.

(ii) w_1-plane.

(iii) w_2-plane.

(iv) w-plane.

Fig. 14.5. Example 14.5.

(ii) $w_2 = \sqrt{2}\, w_1$, (magnification).

The required region is $|w_2 - \sqrt{2}(1 - i)| \le \sqrt{2}$, which is the region within and on the circle with centre at $(\sqrt{2}, -\sqrt{2})$ and radius $\sqrt{2}$ (Fig. 14.5(iii)).

(iii) $w = e^{-\pi i/4} w_2$, (rotation).

The region defined in part (ii) is rotated through an angle of $-\pi/4$. The centre $(\sqrt{2}, -\sqrt{2})$ in the w_2-plane is mapped onto $(0, -2)$ in the w-plane. The required region is $|w + 2i| \le \sqrt{2}$ (Fig. 14.5(iv)).

Alternatively, from the given mapping, we have

$$z = \frac{w + 2i}{1 - i}.$$

Hence, $|z| = \dfrac{|w + 2i|}{|1 - i|} = \dfrac{1}{\sqrt{2}} |w + 2i|$.

Therefore, the unit disk $|z| \le 1$ is mapped as $|w + 2i| \le \sqrt{2}$ or $u^2 + (v + 2)^2 \le 2$.

14.2.2   Inverse Transformation

Consider the mapping (transformation)

$$w = \frac{1}{z}. \tag{14.2}$$

The inverse mapping is given by $z = 1/w$. The mapping $w = 1/z$ is one-to-one between the non-zero points of the z-plane and the non-zero points of the w-plane. Write the mapping given by Eq. (14.2) as

$$w = \frac{\bar{z}}{z\bar{z}} = \frac{\bar{z}}{|z|^2},$$

This mapping can be considered as the composition of the two mappings

(i) $w_1 = \dfrac{z}{|z|^2}$,   and   (ii) $w = \bar{w}_1$.

In the first mapping, we have

$$|w_1| = \frac{1}{|z|} \quad \text{and} \quad \arg(w_1) = \arg(z).$$

Consider now, the mapping with respect to the unit circle $|z| = 1$. The points interior to the unit circle $|z| = 1$ are mapped onto the points exterior to the unit circle $|w_1| = 1$ and the points exterior to the unit circle $|z| = 1$ are mapped onto the non-zero points interior to the unit circle $|w_1| = 1$. The points on the circle $|z| = 1$ are mapped onto the points on the unit circle $|w_1| = 1$. Similar discussions hold if we consider the mapping with respect to the circle $|z| = r$.

The point $z = 0$ is mapped as $w_1 = \infty$ in the extended w_1-plane and the point $z = \infty$ in the z-plane is mapped onto $w_1 = 0$. Thus, the image in the w-plane, of a region R in the z-plane under this mapping will be a bounded set if $z = 0$ is not inside or on the boundary of the given region R in the z-plane.

The second mapping, $w = \bar{w}_1$, is reflection with respect to the real axis in the w-plane.

Example 14.6 Find the image in the w-plane, of the disk $|z - 1| \le 1$ under the mapping $w = 1/z$.

Solution We have $w = 1/z$ or $z = 1/w$. The given region $|z - 1| \le 1$ is mapped as

$$\left|\frac{1}{w}-1\right| \leq 1 \quad \text{or} \quad |1-w| \leq |w|.$$

Let $w = u + iv$. Then, we obtain

$$|(1-u)-iv| \leq |u+iv| \quad \text{or} \quad (1-u)^2 + v^2 \leq u^2 + v^2$$

or $\qquad\qquad\qquad\qquad 1 - 2u \leq 0 \quad \text{or} \quad u \geq 1/2.$

Thus, the image of the region $|z - 1| \leq 1$, under the mapping $w = 1/z$ is the half plane $u \geq 1/2$ in the w-plane. The bounding circle $|z - 1| = 1$ is mapped as the line $u = 1/2$.

Example 14.7 Find the image of the region bounded by the lines $x - y < 2$ and $x + y > 2$ under the mapping $w = 1/z$.

Solution The given region is shown in Fig. 14.6a.

Substituting $z = x + iy$ and $w = u + iv$ in $w = 1/z$ and comparing the real and imaginary parts, we get

$$x = \frac{u}{u^2 + v^2} \quad \text{and} \quad y = -\frac{v}{u^2 + v^2}.$$

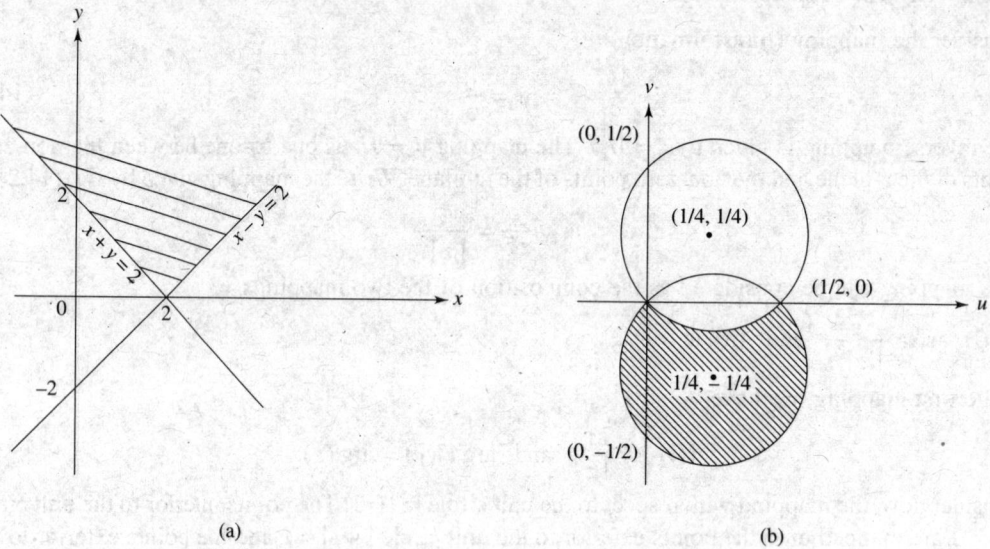

(a) (b)

Fig. 14.6. (a) Region in Exaple 14.7. (b) Image in Example 14.7.

The region $x - y < 2$ is transformed as

$$\frac{u+v}{u^2+v^2} < 2, \quad \text{or} \quad u^2 + v^2 - \frac{1}{2}(u+v) > 0, \quad \text{or} \quad \left(u-\frac{1}{4}\right)^2 + \left(v-\frac{1}{4}\right)^2 > \frac{1}{8}.$$

The boundary of this region is the circle

$$\left|w - \frac{1}{4}(1+i)\right| = \frac{1}{2\sqrt{2}} \qquad\qquad (14.3)$$

with centre at (1/4, 1/4) and radius $1/(2\sqrt{2})$.

$$\frac{u-v}{u^2+v^2}>2 \quad \text{or} \quad u^2+v^2-\frac{1}{2}(u-v)<0 \quad \text{or} \quad \left(u-\frac{1}{4}\right)^2+\left(v+\frac{1}{4}\right)^2<\frac{1}{8}.$$

The boundary of this region is the circle

$$\left| w-\frac{1}{4}(1-i)\right| = \frac{1}{2\sqrt{2}} \tag{14.4}$$

with centre at $(1/4, -1/4)$ and radius $1/(2\sqrt{2})$.

The required image is the exterior of the circle given in Eq. (14.3) and the interior of the circle given in Eq. (14.4) as shown in Fig. 14.6b.

Example 14.8 Show that under the mapping $w = 1/z$, all circles and straight lines in the z-plane are transformed to circles and straight lines in the w-plane.

Solution The equation

$$a(x^2+y^2)+bx+cy+d=0 \tag{14.5}$$

represents a circle if $a \neq 0$ and a straight line if $a = 0$, in the z-plane. Substituting $z = x + iy$ and $w = u + iv$ in $w = 1/z$ and comparing the real and imaginary parts, we get

$$x = \frac{u}{u^2+v^2} \quad \text{and} \quad y = -\frac{v}{u^2+v^2}.$$

Substituting in Eq. (14.5), we obtain

$$\frac{a}{u^2+v^2}+\frac{bu}{u^2+v^2}-\frac{cv}{u^2+v^2}+d=0$$

or

$$a+bu-cv+d(u^2+v^2)=0. \tag{14.6}$$

If $d \neq 0$, Eq. (14.6) can be written as

$$\left(u+\frac{b}{2d}\right)^2+\left(v-\frac{c}{2d}\right)^2=\frac{b^2+c^2-4ad}{4d^2}$$

which is the equation of a circle.

If $d = 0$, we obtain from Eq. (14.6), $a + bu - cv = 0$. We observe the following.

(i) A circle $(a \neq 0)$ not passing through the origin $(d \neq 0)$ in the z-plane, is transformed into a circle not passing through the origin in the w-plane.

(ii) A circle $(a \neq 0)$ passing through the origin $(d = 0)$ in the z-plane is transformed into a straight line not passing through the origin in the w-plane.

(iii) A straight line $(a = 0)$ not passing through the origin $(d \neq 0)$ in the z-plane, is transformed into a circle passing through the origin in the w-plane.

(iv) A straight line $(a = 0)$ passing through the origin $(d = 0)$ in the z-plane, is transformed into a straight line passing through the origin in the w-plane.

Remark 1

(a) When the circle given in Eq. (14.5) with centre at $(-b/(2a), -c/(2a))$ and radius $\sqrt{b^2+c^2-4ad}/(2a)$ is transformed into the circle given in Eq. (14.6) with centre at $(-b/(2d), c/(2d))$ and radius $\sqrt{b^2 + c^2 - 4ad}/(2d)$, the centre of the original circle is not mapped onto the centre of the image circle and their radii are unequal.

(b) The equation of the circle given in Eq. (14.5) can also be written as

$$az\bar{z} + \bar{s}z + s\bar{z} + d = 0, \ a \neq 0$$

where $s = (b + ic)/2$, a and d are real constants and $s\bar{s} - ad > 0$. This equation represents a straight line if $a = 0$ and $s \neq 0$. Under the mapping $w = 1/z$, the image of a disk D in the z-plane, is the interior of a disk or the exterior of a disk or a half plane depending upon whether the point $z = 0$ lies outside the disk D or inside the disk D or on the boundary of the disk D respectively.

(c) Consider the mapping of the open disk $D : |z - z_0| < r$, bounded by the circle $C : |z - z_0| = r$ under the transformation $w = c/(z + d)$. Then, the image of D in the w-plane is the interior of an open disk or the exterior of an open disk or a half plane depending on whether the point $z = -d$ respectively lies outside, inside or on the boundary C in the z-plane.

Example 14.9 Under the mapping $w = i/(z - i)$, find the images of the following regions

(i) Im $z < 0$, (ii) $|z| > 1$.

Solution From $w = \dfrac{i}{z - i}$, we get $z = i\left(\dfrac{1 + w}{w}\right)$.

(i) Writing $w = u + iv$, we obtain

$$z = i\left[\frac{1 + u + iv}{u + iv}\right] = \left[\frac{i(1 + u + iv)(u - iv)}{(u + iv)(u - iv)}\right] = \frac{i}{u^2 + v^2}[u + u^2 + v^2 - iv].$$

The mapping of the region Im $z < 0$ is given by

$$u + u^2 + v^2 < 0 \ \text{ or } \ \left(u + \frac{1}{2}\right)^2 + v^2 < \frac{1}{4} \ \text{ or } \ \left|w + \frac{1}{2}\right| < \frac{1}{2},$$

which is a circle with centre at $(-1/2, 0)$ and radius $1/2$.

(ii) We have

$$|z| = \left|\frac{i(1 + w)}{w}\right| = \left|\frac{1 + w}{w}\right|$$

Hence, $|z| > 1$ gives $|1 + w| > |w|$.

Substituting $w = u + iv$, we obtain

$$(1 + u)^2 + v^2 > u^2 + v^2 \ \text{ or } \ 2u + 1 > 0 \ \text{ or } \ u > -1/2.$$

which is the half plane $u > -1/2$.

Example 14.10 Find the image of the region $|z - i| < 2$ under the mapping $w = (1 + i)/(z + i)$.

Solution From $\quad w = \dfrac{1 + i}{z + i}$, or $w(z + i) = 1 + i$, we have $z = \dfrac{1 + i(1 - w)}{w}$.

Hence, $|z - i| = \left|\dfrac{1 + i(1 - w)}{w} - i\right| = \left|\dfrac{1 + i(1 - 2w)}{w}\right| < 2$

gives $| 1 + i(1 - 2w) | < 2 | w |$ or $(1 + 2v)^2 + (1 - 2u)^2 < 4(u^2 + v^2)$

or $2v - 2u + 1 < 0$

where $w = u + iv$. Therefore, interior of the disk $| z - i | < 2$ is mapped as the half plane $2v - 2u + 1 < 0$ (Fig. 14.7). The bounding circle C is mapped as the line C^*.

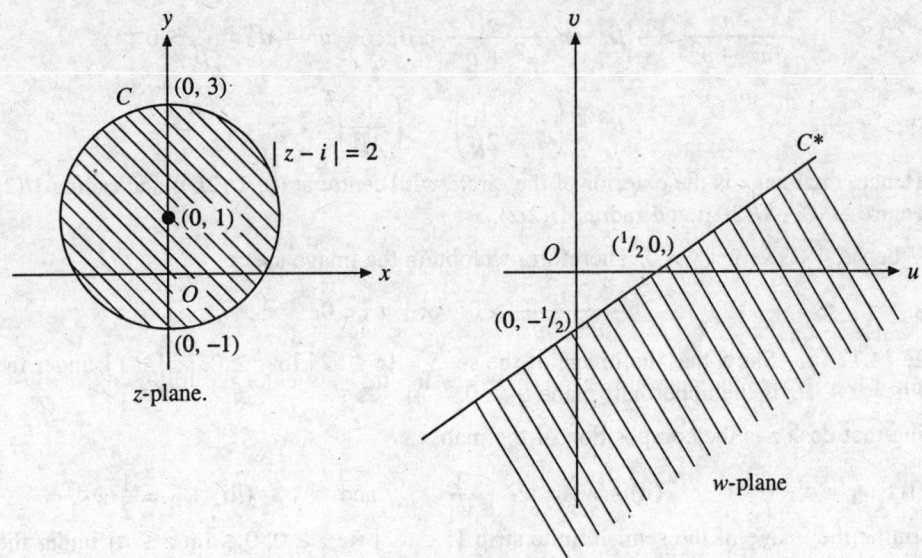

Fig. 14.7.   Example 14.10.

Alternative

Substituting $z = x + iy$ and $w = u + iv$, we get

$$z = x + iy = \frac{1 + i(1 - w)}{w} = \frac{[(1 + v) + i(1 - u)][u - iv]}{(u + iv)(u - iv)}$$

$$= \frac{(u + v) + i(u - v - u^2 - v^2)}{u^2 + v^2}$$

Comparing the real and imaginary parts, we obtain

$$x = \frac{u + v}{u^2 + v^2} \quad \text{and} \quad y = \frac{u - v - u^2 - v^2}{u^2 + v^2}.$$

Substituting the values of x and y in $| z - i | < 2$, that is $x^2 + (y - 1)^2 < 4$, we obtain the image as $2v - 2u + 1 < 0$.

Example 14.11   Find the image of the half plane $y > \alpha$ under the mapping $w = 1/z$, when

(i) $\alpha > 0$,   (ii) $\alpha < 0$,   (iii) $\alpha = 0$.

Solution   We have $w = 1/z$ or $z = 1/w$.

Writing $z = x + iy$ and $w = u + iv$, we get $x + iy = \dfrac{u - iv}{u^2 + v^2}$.

Therefore, $x = \dfrac{u}{u^2 + v^2}$ and $y = \dfrac{-v}{u^2 + v^2}$.

(i) $y > \alpha$, $\alpha > 0$ gives $\dfrac{-v}{u^2 + v^2} > \alpha$ or $u^2 + \left(v + \dfrac{1}{2\alpha}\right)^2 < \left(\dfrac{1}{2\alpha}\right)^2$.

Therefore, the image of any point in the half plane lies inside the circle having radius $1/(2\alpha)$ and centre at $(0, -1/(2\alpha))$.

(ii) $y > \alpha$, $\alpha < 0$. Let $\alpha = -\beta$, $\beta > 0$. Then, we have

$$-\frac{v}{u^2 + v^2} > -\beta, \quad \text{or} \quad \frac{v}{u^2 + v^2} < \beta, \quad \text{or} \quad u^2 + v^2 - \frac{v}{\beta} > 0$$

or
$$u^2 + \left(v - \frac{1}{2\beta}\right)^2 > \left(\frac{1}{2\beta}\right)^2.$$

Hence, the image is the exterior of the circle with centre at $(0, 1/(2\beta))$ and radius $1/(2\beta)$, or centre at $(0, -1/(2\alpha))$ and radius $1/(2\alpha)$.

(iii) When $\alpha = 0$, we get $y > 0$. Therefore, we obtain the image as

$$-\frac{v}{u^2 + v^2} > 0 \quad \text{or} \quad v < 0.$$

Example 14.12 (a)   Show that the image of the set $S = \{z \in \mathbb{C} \mid \text{Im } z \geq 0, \mid z \mid \geq 1\}$ under the map $w = u + iv = z + 1/z$ is the upper half plane $v \geq 0$.

(b) Noting that $\cosh z$ is the composition of the maps

(i) $w_1 = e^z$,   (ii) $w_2 = w_1 + \dfrac{1}{w_1}$ and   (iii) $w = \dfrac{1}{2} w_2$

determine the image of the semi-infinite strip $\{z \in \mathbb{C} \mid \text{Re } z \geq 0, 0 \leq \text{Im } z \leq \pi\}$ under the map $w = \cosh z$.

Solution

(a) The set S is given in Fig. 14.8(i).

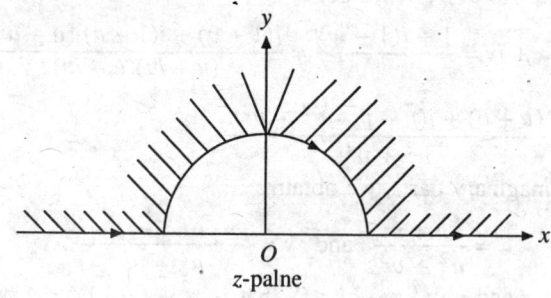

z-palne

Fig. 14.8(i).   Region in Example 14 .12 (a).

Let $z = re^{i\theta}$. Then $w = re^{i\theta} + \dfrac{1}{r}e^{-i\theta}$.

For $z = x$, $x \geq 1$, $w = x + (1/x)$ is real and positive. Since $(dw/dx) > 0$ for all x, w is increasing for $x \in [1, \infty)$. Therefore, as z moves to right along x-axis, w moves to right along u-axis starting at $u = 2$ (Fig. 14.8(ii)).

For $z = -x$, $x \geq 1$, $w = -\left[x + \dfrac{1}{x}\right]$ is real and negative. Since $(dw/dx) < 0$ for all x, w is decreasing for $x \in (-\infty, -1)$. Therefore, as z moves to left along x-axis, w moves to left along u-axis starting at $u = -2$.

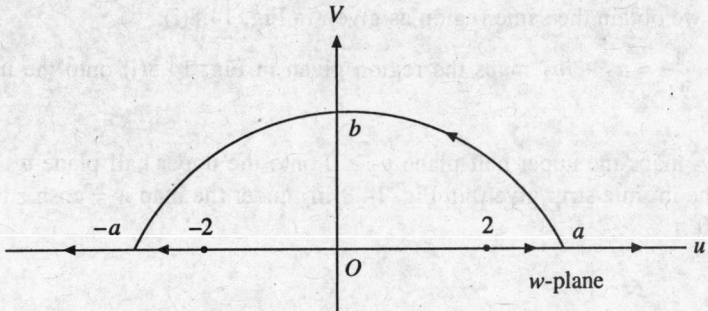

Fig. 14.8(ii). **Image in Example 14.12(a).**

For points on the boundary of the upper half of the circle $|z| = 1$, we get $z = e^{i\theta}$ and $w = e^{i\theta} + e^{-i\theta} = 2 \cos \theta$, $\pi \leq \theta \leq 0$. These points are mapped onto $-2 \leq u \leq 2$.

For points above the boundary of the upper half of the circle $|z| = 1$, that is,

$$z = r_1 e^{i\theta}, \quad r_1 > 1, 0 \leq \theta \leq \pi,$$

we get $\qquad w = \left(r_1 + \dfrac{1}{r_1} \right) \cos \theta + i \left(r_1 - \dfrac{1}{r_1} \right) \sin \theta = a \cos \theta + ib \sin \theta = u + iv$

where $a = r_1 + \dfrac{1}{r_1} > 0$, $b = r_1 - \dfrac{1}{r_1} > 0$ and $a > b$.

Therefore, we obtain

$$u = a \cos \theta \quad \text{and} \quad v = b \sin \theta \quad \text{or} \quad \frac{u^2}{a^2} + \frac{v^2}{b^2} = 1, v > 0$$

which is an ellipse with foci at $\pm \sqrt{a^2 - b^2} = \pm 2$.

Therefore, for a fixed $r > 1$, the points on this semicircle ($0 \leq \theta \leq \pi$) are mapped onto the ellipse $(u^2/a^2) + (v^2/b^2) = 1$, $(0 \leq \theta \leq \pi)$ (Fig. 14.8(ii)). As r increases, the ellipse gets larger and larger with the same foci at ± 2. Hence as r increases, the image of the given set is the whole of the upper half plane $v \geq 0$.

(b) The semi infinite strip $\{z \in \mathbb{C}, \text{Re } z \geq 0, 0 \leq \text{Im } z \leq \pi\}$ is given in Fig. 14.8(iii).

(i) Let $z = x + iy$ and $w_1 = re^{i\theta}$. We have

$$w_1 = re^{i\theta} = e^{x+iy} = e^x e^{iy}.$$

Therefore, $r = e^x > 1$, $\theta = y$, $0 \leq \theta \leq \pi$.

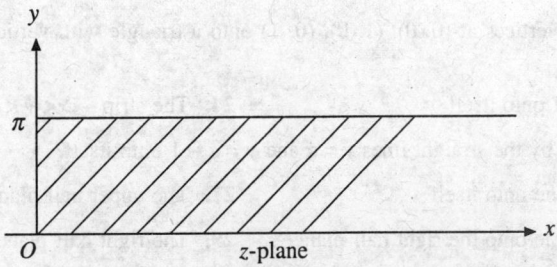

Fig. 14.8(iii). **Region in Example 14.12(b).**

Therefore, we obtain the same region as given in Fig. 14.8(i).

(ii) $w_2 = w_1 + \dfrac{1}{w_1} = u_2 + iv_2$ maps the region given in Fig. 14.8(i) onto the upper half plane $v_2 \geq 0$.

(iii) $w = (1/2)w_2$ maps the upper half plane $v_2 \geq 0$ onto the upper half plane $v \geq 0$. Hence, the image of the infinite strip given in Fig. 14.8(iii) under the map $w = \cosh z$ is the upper half plane $v \geq 0$.

Exercise 14.1

Find the images of the following regions or curves in the z-plane onto the w-plane under the given mappings.

1. The circle $|z - 1| = 1$; $w = iz + 4$. 2. The quarter plane $x > 0$, $y > 0$; $w = i(z + 1)$.

3. The quarter plane $x > 0$, $y > 0$; $w = (1 + i)z + 1$.

4. The semi-circular region $|z| < 1$, Im $z > 0$; $w = z + b$, b complex constant.

5. The rectangle with vertices at $(1, 1)$, $(3, 1)$, $(3, 3)$, $(1, 3)$;

   (i) $w = z + (1 + 2i)$, (ii) $w = (\sqrt{2}\, e^{-\pi i/4})z + (1 + 2i)$.

6. The triangle with vertices at $(0, 1)$, $(1, -1)$, $(1, 1)$;

   (i) $w = z + (1 - 2i)$, (ii) $w = iz + 4$.

7. The half plane Im $z > 0$; $w = (1 + i)z + 2$. 8. The half plane Re $z > 0$; $w = (1 - i)z + 2$.

9. The unit disk $|z| < 1$; $w = (1 + i)z + 2i$. (discuss the mapping as a composite transformation also).

10. The infinite strip $0 < y < 1/(2c)$, $c > 0$; $w = 1/z$. 11. The region $x > 1$, $y > 0$; $w = 1/z$.

12. The region $x + y > 1$; $w = 1/z$. 13. The line Re $z = a$, $a \neq 0$; $w = 1/z$.

14. The line Im $z = b$, $b \neq 0$; $w = 1/z$. 15. The wedge Re $z > 1$ and Im $z > 1$; $w = 1/z$.

16. The infinite strip $x \geq 0$, $0 \leq y \leq 1$; $w = i/z$. 17. The disk $|z - 2i| < 2$; $w = 1/z$.

18. The disk $|z - 1| < 1$; $w = i/z$. 19. The region $|z| > 1$, Im $z < 0$; $w = i/(z - i)$.

20. The circle $|z| = a$, $a > 1$; $w = z + (1/z)$.

Find a linear mapping of the form $w = az + b$ which maps the given domain/region in the z-plane into a given domain/region in the w-plane.

21. The disk $|z - i| < 1$ onto the disk $|w - 1| < 2$.

22. The triangle with vertices at $(0, 0)$, $(1, 0)$, $(0, 1)$ onto a triangle with vertices at $(1, 1)$, $(0, 0)$, $(2, 0)$ respectively.

23. The strip $0 < x < 1$ onto itself. 24. The strip $-2 < y < 1$ onto itself.

25. The strip bounded by the straight lines $y = x$ and $y = x - 1$ onto itself.

26. The upper half plane onto itself. 27. The upper half plane onto lower half plane.

28. The upper half plane onto the right half plane. 29. The right half plane onto itself.

30. Find an inverse mapping which maps the disk $|z - 2| < 3$ onto $|w - i| > 2$.

14.3   Bilinear Transformations

The transformation

$$w = \frac{az + b}{cz + d}, \quad ad - bc \neq 0, \tag{14.7}$$

where a, b, c, d are real or complex constants, is called a *bilinear* or a *linear fractional transformation*. This is also called a *Möbius transformation*. We note the following:

(i) for each z for which $cz + d \neq 0$, we obtain a unique value of w.

(ii) if both $c = 0$ and $d = 0$, then w is not defined for any z.

(iii) if $a = c = 0$, or $b = d = 0$, then $w = $ constant for all z, that is, it maps the whole z-plane onto a point.

(iv) if $a = b = 0$, we get $w = 0$ for all z.

(v) if $c = 0$ and $d \neq 0$, then we get $w = (az + b)/d$, which is a linear mapping.

(vi) if $a = 0$, $b \neq 0$ and $c \neq 0$, then we get

$$w = \frac{b}{c} \left[\frac{1}{z + (d/c)} \right]$$

which can be written as a composition of the following mappings

(a) translation: $w_1 = z + (d/c)$

(b) inversion: $w_2 = 1/w_1$

(c) rotation and magnification/contraction: $w = (b/c)w_2$.

(vii) if $a \neq 0$, $c \neq 0$, then we can write Eq. (14.7) as

$$w = \frac{a}{c} + \frac{bc - ad}{c} \left[\frac{1}{cz + d} \right] = \frac{a}{c} + p \left[\frac{1}{z + (d/c)} \right] \tag{14.8}$$

where $p = (bc - ad)/c^2$. This mapping can be considered as the composition of the following mappings.

(a) translation: $w_1 = z + (d/c)$

(b) inversion: $w_2 = 1/w_1$

(c) rotation and magnification/contraction; $w_3 = pw_2$

(d) translation: $w = w_3 + (a/c)$.

Therefore, the general bilinear transformation can be considered as a composition of linear and inversion mappings.

Now, under linear mapping, the images of the regions in the z-plane retain the similar shape in the w-plane. Under inversion, the totality of the circles and straight lines in the z-plane are mapped as the totality of circles and straight lines in the w-plane. Therefore, we have the following result.

Theorem 14.1   Every bilinear transformation of the form given in Eq. (14.7) maps the circles and straight lines in the z-plane onto circles and straight lines in the w-plane.

The proof is left as an exercise.

The necessity of the condition $ad - bc \neq 0$, given in Eq. (14.7) shall be discussed in the next section on conformal mapping.

Solving for z, we obtain the inverse mapping of the bilinear transformation given in Eq. (14.7) as

$$w(cz + d) = az + b \quad \text{or} \quad z = \frac{b - wd}{cw - a}. \tag{14.9}$$

Mapping in the extended complex plane   If $c \neq 0$, then under the mapping given in Eq. (14.7), the image of the point $z = -d/c$ becomes $w = \infty$. Similarly, when $c = 0$ the image of the point $z = \infty$ is $w = \infty$. In these cases, we consider the extended z-plane and the extended w-plane. From Eq. (14.9), we obtain that each point w, for which $cw - a \neq 0$, is the image of exactly one point in the z-plane. If $c \neq 0$, then $w = a/c$ is the image of the point $z = \infty$. Thus, the bilinear mapping is an one-to-one mapping from the extended z-plane onto the extended w-plane.

Remark 2

The interior and the exterior of a circle are called *complimentary domains* of each other, because together with the circle, they cover the entire extended plane. In the case of a straight line, the complimentary domains are the two half planes, one on each side of the line.

Fixed points of a mapping

Fixed points of a mapping $w = f(z)$ are the points which are mapped onto themselves. These points are *fixed* or *invariant* under the given transformation. Fixed points are obtained by solving the equation $w = f(z) = z$.

The fixed points of some mappings are given below.

 (i) The identity mapping $w = z$ has every point as its fixed point.

 (ii) The mapping $w = \bar{z}$ has infinite number of fixed points, since $z = \bar{z}$ gives $y = 0$. Thus, $z = x$ for any x is a fixed point.

 (iii) The translation mapping $w = z + c$ has no fixed points in the finite z-plane.

 (iv) The rotation and magnification/contraction mapping $w = az$, a any complex constant has $z = 0$ as the fixed point.

 (v) The linear mapping $w = az + c$ has one fixed point in the finite plane at $z = c/(1 - a)$, $a \neq 1$.

 (vi) The inversion mapping $w = 1/z$ has two fixed points at $z = \pm 1$.

The bilinear transformation $w = (az + b)/(cz + d)$ has a fixed point, if

$$z = \frac{az + b}{cz + d}, \quad \text{or} \quad cz^2 + (d - a)z - b = 0 \tag{14.10}$$

which is a quadratic in z. Hence, the bilinear transformation has at most two fixed points.

If the roots of Eq. (14.10) are equal, then $(d - a)^2 + 4bc = 0$. In this case, the mapping has a single fixed point at $z = (a - d)/(2c)$.

If $c = 0$, the bilinear transformation (Eq. (14.7)) reduces to a linear transformation $w = (az + b)/d$ and has one fixed point at $z = b/(d - a)$, $d \neq a$ in the finite plane.

We summarize these results in the following theorem.

Theorem 14.2   A bilinear transformation $w = (az + b)/(cz + d)$, which is not an identity mapping, has at most two fixed points. If it has more than two fixed points, then it must be the identity mapping $w = z$.

The bilinear transformation given in Eq. (14.7) has four constants a, b, c and d. Since, we can divide the numerator and denominator by one of these non-zero constants (without changing the mapping), the bilinear transformation has only three constants. Therefore, we can obtain a unique mapping if three conditions are imposed on the mapping. For example, we may be given that three distinct points in the z-plane map onto three distinct points in the w-plane oriented in the same direction. In the following theorem, we describe a method for finding such a bilinear transformation.

Theorem 14.3 The transformation $w = f(z)$ which maps three distinct points z_1, z_2, z_3 in the z-plane onto three given distinct points w_1, w_2, w_3 in the w-plane is unique and is given by

$$\frac{w - w_1}{w - w_2} \cdot \frac{w_3 - w_2}{w_3 - w_1} = \frac{z - z_1}{z - z_2} \cdot \frac{z_3 - z_2}{z_3 - z_1}. \tag{14.11}$$

If one of these points is the point ∞, then the ratio of the two differences containing this point is replaced by 1.

Proof Let the required transformation be

$$w = f(z) = \frac{az + b}{cz + d}, \ ad - bc \neq 0.$$

We are given that

$$w_j = \frac{az_j + b}{cz_j + d}, \ j = 1, 2, 3, \ z_i \neq z_j, \ i \neq j.$$

The differences in Eq. (14.11) become

$$w - w_1 = \frac{az + b}{cz + d} - \frac{az_1 + b}{cz_1 + d} = \frac{(ad - bc)(z - z_1)}{(cz + d)(cz_1 + d)},$$

$$w - w_2 = \frac{az + b}{cz + d} - \frac{az_2 + b}{cz_2 + d} = \frac{(ad - bc)(z - z_2)}{(cz + d)(cz_2 + d)},$$

$$w_3 - w_1 = \frac{az_3 + b}{cz_3 + d} - \frac{az_1 + b}{cz_1 + d} = \frac{(ad - bc)(z_3 - z_1)}{(cz_3 + d)(cz_1 + d)},$$

$$w_3 - w_2 = \frac{az_3 + b}{cz_3 + d} - \frac{az_2 + b}{cz_2 + d} = \frac{(ad - bc)(z_3 - z_2)}{(cz_3 + d)(cz_2 + d)}.$$

Hence, we have

$$\frac{w - w_1}{w - w_2} = \frac{(z - z_1)(cz_2 + d)}{(z - z_2)(cz_1 + d)}, \text{ and } \frac{w_3 - w_1}{w_3 - w_2} = \frac{(z_3 - z_1)(cz_2 + d)}{(z_3 - z_2)(cz_1 + d)}.$$

Dividing these two equations, we obtain

$$\frac{w - w_1}{w - w_2} \cdot \frac{w_3 - w_2}{w_3 - w_1} = \frac{z - z_1}{z - z_2} \cdot \frac{z_3 - z_2}{z_3 - z_1}$$

or

$$\frac{w - w_1}{w - w_2} : \frac{w_3 - w_1}{w_3 - w_2} = \frac{z - z_1}{z - z_2} : \frac{z_3 - z_1}{z_3 - z_2}.$$

If one of the points is ∞, say $z_1 = \infty$, then

$$\frac{z - z_1}{z_3 - z_1} = \lim_{z_1 \to \infty} \left[\frac{(z/z_1) - 1}{(z_3/z_1) - 1} \right] = 1.$$

If $w_1 = \infty$, we obtain

$$\frac{w - w_1}{w_3 - w_1} = \lim_{w_1 \to \infty} \left[\frac{(w/w_1) - 1}{(w_3/w_1) - 1} \right] = 1.$$

Remark 3

(a) Eq. (14.11) can also be written as

$$\frac{w - w_1}{w - w_3} \cdot \frac{w_2 - w_3}{w_2 - w_1} = \frac{z - z_1}{z - z_3} \cdot \frac{z_2 - z_3}{z_2 - z_1}. \tag{14.12}$$

(b) Two points z_1 and z_2 are said to be inverse points with respect to the circle $|z| = 1$ if and only if $\bar{z}_1 z_2 = 1$. If follows that the inverse of any point z_1, with respect to the circle $|z| = 1$ is $1/\bar{z}_1$. If $z_1 = 0$, then its inverse point is $z = \infty$.

Example 14.13 Find the images of the regions (i) $x \geq 2$, (ii) $y \geq 1$, in the z-plane under the mapping $w = (4z + 1)/(z - 2 - i)$.

Solution From $w = \dfrac{4z + 1}{z - 2 - i}$, we get $z = \dfrac{(2 + i)w + 1}{w - 4}$.

Let $w = u + iv$ and $z = x + iy$. We obtain

$$x + iy = \frac{(2 + i)(u + iv) + 1}{(u - 4) + iv} = \frac{[(2 + i)(u + iv) + 1][(u - 4) - iv]}{(u - 4)^2 + v^2}.$$

Comparing the real and imaginary parts, we get

$$x = \frac{2(u^2 + v^2) - 7u + 4v - 4}{(u - 4)^2 + v^2}, \quad \text{and} \quad y = \frac{u^2 + v^2 - 4u - 9v}{(u - 4)^2 + v^2}.$$

(i) The image of the region $x \geq 2$ is obtained as

$$2(u^2 + v^2) - 7u + 4v - 4 \geq 2[(u - 4)^2 + v^2], \quad \text{or} \quad 9u + 4v - 36 \geq 0.$$

(ii) The image of the region $y \geq 1$ is obtained as

$$u^2 + v^2 - 4u - 9v \geq (u - 4)^2 + v^2, \quad \text{or} \quad 4u - 9v - 16 \geq 0.$$

Exercise 14.14 Show that the bilinear transformation $w = (iz + 2)/(4z + i)$ transforms the real axis in the z-plane onto a circle in the w-plane. Find the centre and the radius of the circle in the w-plane. Find the point in the z-plane which is mapped onto the centre of the circle in the w-plane.

Solution From $w = \dfrac{iz + 2}{4z + i}$, we get $z = \dfrac{2 - iw}{4w - i}$.

Substituting $z = x + iy$ and $w = u + iv$, we obtain

$$x + iy = \frac{2 - i(u + iv)}{4(u + iv) - i} = \frac{[(2 + v) - iu][4u - i(4v - 1)]}{[4u + i(4v - 1)][4u - i(4v - 1)]}$$

$$= \frac{[4u(2 + v) - u(4v - 1)] - i[4u^2 + (2 + v)(4v - 1)]}{16u^2 + (4v - 1)^2}.$$

Comparing the real and imaginary parts, we get

$$x = \frac{9u}{16u^2 + (4v - 1)^2}, \text{ and } y = -\frac{4u^2 + 4v^2 + 7v - 2}{16u^2 + (4v - 1)^2}.$$

Therefore, the image of the real axis, that is, $y = 0$ is obtained as

$$4u^2 + 4v^2 + 7v - 2 = 0 \text{ or } u^2 + \left(v + \frac{7}{8}\right)^2 = \left(\frac{9}{8}\right)^2$$

which is a circle with centre at $(0, -7/8)$ and radius $9/8$.

For $u = 0$ and $v = -7/8$, we get $x = 0$ and $y = 1/4$. Thus, the centre of the circle in the w-plane is the image of the point $(0, 1/4)$ in the z-plane.

Example 14.15 Find the image of the closed half disk $|z| \le 1$, Im $(z) \ge 0$ under the bilinear transformation $w = z/(z + 1)$.

Solution The given half disk is the intersection of the unit disk $|z| \le 1$ and the upper half plane Im $(z) \ge 0$ (Fig. 14.9). Hence, its image is also the intersection of the images of these two regions.

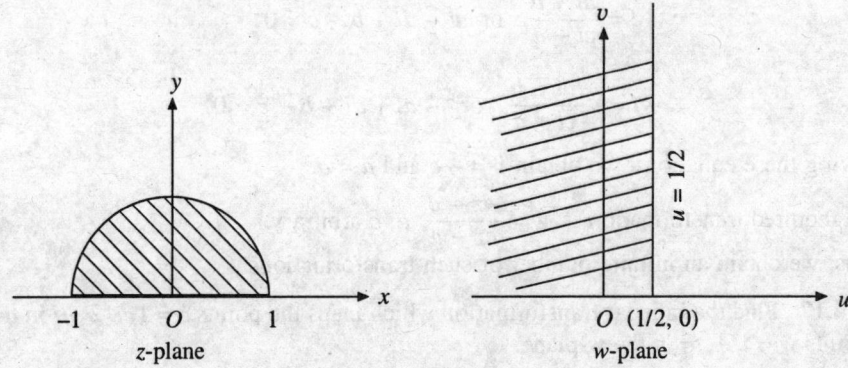

Fig. 14.9. Example 14.15.

From the given mapping, we obtain $z = w/(1 - w)$. Hence, $|z| = 1$ gives

$$|z| = \frac{|w|}{|1 - w|} = 1, \text{ or } |w| = |1 - w| \text{ or } u^2 + v^2 = (1 - u)^2 + v^2 \text{ or } u = \frac{1}{2}.$$

Since $z = 0$ is mapped as $w = 0$, the closed unit disk $|z| \le 1$, is mapped as $u \le 1/2$.

Substituting $z = x + iy$ and $w = u + iv$ in the mapping $z = w/(1 - w)$, we get

$$x + iy = \frac{u + iv}{1 - (u + iv)} = \frac{(u + iv)[(1 - u) + iv]}{[(1 - u) - iv][(1 - u) + iv]}$$

$$= \frac{(u - u^2 - v^2) + iv}{(1 - u)^2 + v^2}.$$

Comparing the real and imaginary parts, we obtain

$$x = \frac{u - u^2 - v^2}{(1 - u)^2 + v^2}, \text{ and } y = \frac{v}{(1 - u)^2 + v^2}.$$

Therefore, the image of the region Im $(z) \geq 0$, that is, $y \geq 0$ is obtained as $v \geq 0$. Hence, the required image is the region $u \leq 1/2$ and $v \geq 0$ (Fig. 14.9).

Example 14.16 Find all the bilinear transformations whose fixed points are (i) -1 and 1, (ii) i and $-i$.

Solution Let the bilinear transformation be $w = (az + b)/(cz + d)$.

(i) Since 1 and -1 are the fixed points, we have

$$1 = \frac{a + b}{c + d}, \quad \text{or} \quad a - c + b - d = 0$$

and

$$-1 = \frac{-a + b}{-c + d}, \quad \text{or} \quad a + c - b - d = 0.$$

Solving these equations, we obtain $a = d$ and $b = c$.

The required transformation is $w = \dfrac{az + b}{bz + a}$, a, b arbitrary.

(ii) Since i and $-i$ are the fixed points, we have

$$i = \frac{ai + b}{ci + d}, \quad \text{or} \quad ai - di + b + c = 0$$

and

$$-i = \frac{-ai + b}{-ci + d}, \quad \text{or} \quad -ai + di + b + c = 0.$$

Solving these equations, we obtain $b = -c$ and $a = d$.

The required transformation is $w = \dfrac{az - c}{cz + a}$, a, c arbitrary.

Thus, we obtain an infinite number of such transformations.

Example 14.17 Find the bilinear transformation which maps the points $z = 1, i, 2 + i$ in the z-plane onto the points $w = i, 1, \infty$ in the w-plane.

Solution Let the bilinear transformation be given by $w = \dfrac{az + b}{cz + d}$. Since the points $z = 1, i, 2 + i$ are mapped as $w = i, 1, \infty$, respectively, we obtain

$$i = \frac{a + b}{c + d}, \quad 1 = \frac{ia + b}{ic + d} \quad \text{and} \quad (2 + i)c + d = 0.$$

Solving these equations, we obtain

$$a = (2 + i)c, \quad b = -(2i + 1)c \quad \text{and} \quad d = -(2 + i)c$$

where c is arbitrary.

Therefore, the required transformation becomes

$$w = \frac{(2 + i)z - (2i + 1)}{z - (2 + i)}.$$

Alternative

The bilinear transformation which maps the points z_1, z_2, z_3 in the z-plane onto the points w_1, w_2, w_3 in the w-plane is given by

$$\frac{w - w_1}{w - w_2} \cdot \frac{w_3 - w_2}{w_3 - w_1} = \frac{z - z_1}{z - z_2} \cdot \frac{z_3 - z_2}{z_3 - z_1}.$$

We have the transformation as

$$\frac{w - i}{w - 1} \cdot \frac{\infty - 1}{\infty - i} = \frac{z - 1}{z - i} \cdot \frac{2 + i - i}{2 + i - 1}.$$

Since $w_3 = \infty$, the ratio of the two differences containing this point is replaced by 1 (see Theorem 14.3). Therefore, we obtain

$$\frac{w - i}{w - 1} = \frac{z - 1}{z - i} \cdot \frac{2}{1 + i} = \frac{(1 - i)(z - 1)}{z - i}$$

or

$$w - i = (w - 1) \left[\frac{(1 - i)(z - 1)}{z - i} \right]$$

or

$$w \left[1 - \frac{(1 - i)(z - 1)}{z - i} \right] = i - \frac{(z - 1)(1 - i)}{z - i}$$

or

$$w \left[\frac{iz + 1 - 2i}{z - i} \right] = \frac{2iz - z + 2 - i}{z - i}$$

or

$$w = \frac{i(2z + iz - 2i - 1)}{i(z - i - 2)} = \frac{(2 + i)z - (2i + 1)}{z - (2 + i)}.$$

Example 14.18 Find a bilinear map which maps the upper half of the z-plane onto the right half of the w-plane.

Solution Let the bilinear transformation be

$$w_1 = \frac{az + b}{cz + d}.$$

This bilinear transformation must map the boundaries of the half planes onto each other. Thus, the mapping should map the real z-axis onto the imaginary w-axis. To find a, b, c, d, we choose any three distinct values of z on the x-axis, that is real values for z, and three corresponding distinct images on the v-axis. For example, we may choose $z = 0, 1, \infty$ and the corresponding images as $w_1 = 0, i, 2i$ respectively.

Since, $z = 0$ maps onto $w_1 = 0$,   we get $b = 0$.

$z = 1$ maps onto $w_1 = i$,   we get $a = i(c + d)$.

$z = \infty$ maps onto $w_1 = 2i$,   we get $a = 2ci$.

Solving these equations, we obtain $a = 2ci, d = c, b = 0$ and c is arbitrary. Therefore, the transformation becomes

$$w_1 = \frac{2iz}{z + 1}. \tag{14.13}$$

This transformation maps the whole real axis in the z-plane onto the whole imaginary axis in the w_1-plane. Hence, the image of the upper half of the z-plane under this mapping is either the right or the left half of the w_1-plane. To check onto which half plane it is mapped, choose any point, say

$z = i$ in the upper half of the z-plane and find its corresponding image in the w_1-plane under the mapping given in Eq. (14.13). The point $z = i$ maps as $w_1 = -1 + i$, which lies in the left half w_1-plane. Hence, the mapping given in Eq. (14.13) maps the upper half of z-plane onto left half of w_1-plane. Now, rotate this mapping, in the anti-clockwise direction by an angle of π radians to obtain the right half plane. Hence, the required mapping is obtained as

$$w = e^{\pi i} w_1 = -w_1 = -\frac{2iz}{z+1}.$$

Note that this mapping is not unique.

Example 14.19 Find the image of the half plane $x + y > 0$ under the bilinear transformation $w = (z - 1)/(z + i)$.

Solution Since $z = -i$ lies outside the half plane $x + y > 0$ (Fig. 14.10), the image of the half plane $x + y > 0$ must be an open disk whose boundary is the image of the line $x + y = 0$ (see also Remark 1c). Now choose three distinct points $0, 1 - i, \infty$ on the line $x + y = 0$. The corresponding images under the mapping $w = (z - 1)/(z + i)$ are $w = i, -i, 1$ respectively. These images lie on the unit disk $|w| = 1$. Therefore, the line $x + y = 0$ is mapped as $|w| = 1$. Now, consider any point in the given half plane, say, $z = 1$. Its image is $w = 0$, which lies inside $|w| = 1$. Hence, the image of the region $x + y > 0$ is the interior of the unit disk (Fig. 14.10).

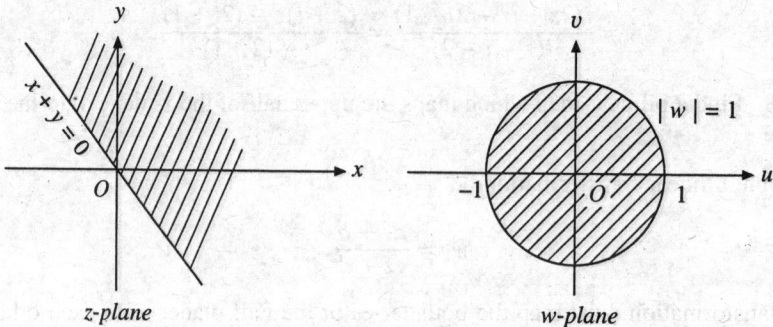

Fig. 14.10. Example 14.19.

Alternative

The mapping can also be obtained directly. From the mapping

$$w = \frac{z-1}{z+i}, \quad \text{we get} \quad z = \frac{1+iw}{1-w}. \tag{14.14}$$

Substituting $z = x + iy$ and $w = u + iv$ in Eq. (14.14), we obtain

$$x + iy = \frac{1 + i(u + iv)}{1 - (u + iv)} = \frac{[(1 - v) + iu][(1 - u) + iv]}{[(1 - u) - iv][(1 - u) + iv]}.$$

Comparing the real and imaginary parts, we obtain

$$x = \frac{1 - u - v}{(1 - u)^2 + v^2}, \quad \text{and} \quad y = \frac{v + u - u^2 - v^2}{(1 - u)^2 + v^2}.$$

Therefore, $x + y > 0$ gives

$$1 - u^2 - v^2 > 0 \quad \text{or} \quad u^2 + v^2 < 1, \quad \text{or} \quad |w| < 1.$$

Example 14.20   Find a transformation $w = f(z)$ which maps the upper half plane Im $(z) \geq 0$ onto the unit disk $|w| \leq 1$.

Solution   It is sufficient to consider the images of three distinct points on the real axis, $y = 0$. Let the three points on the x-axis be p, 0, $-p$ and their images respectively be 1, $-i$, -1. Let the required bilinear transformation be $w = (az + b)/(cz + d)$. Substituting the above points, we obtain

$$1 = \frac{ap + b}{cp + d}, \quad -i = \frac{b}{d}, \quad -1 = \frac{-ap + b}{-cp + d}.$$

The solution of these equations is $a = d/p$, $b = -id$, $c = -id/p$. The required transformation is

$$w = \frac{(d/p)z - id}{(-id/p)z + d} = \frac{z - ip}{-iz + p}. \tag{14.15}$$

Without loss of generality, we can assume $p = 1$. The transformation becomes

$$w = \frac{z - i}{-iz + 1} \tag{14.16}$$

Writing $w = u + iv$ and $z = x + iy$, we obtain

$$(u + iv)[-ix + y + 1] = x + i(y - 1).$$

Comparing the real and imaginary parts, we get

$$x = u(y + 1) + vx, \quad \text{or} \quad x = \frac{u(y + 1)}{1 - v}$$

$$y - 1 = v(y + 1) - ux = v(y + 1) - \frac{u^2(y + 1)}{1 - v}$$

or $\quad \left[1 - \frac{v - v^2 - u^2}{1 - v}\right] y = 1 + \frac{v - v^2 - u^2}{1 - v}, \quad \text{or} \quad y = -\frac{u^2 + v^2 - 1}{u^2 + (v - 1)^2}$

Now, Im $(z) = y = 0$ gives $u^2 + v^2 - 1 = 0$, or $|w| = 1$. Therefore, the real axis, that is the boundary of the upper half plane, is mapped as the unit circle $|w| = 1$. Further, $y > 0$ gives $|w| < 1$. Hence, the region in the upper half plane is mapped onto the interior of the unit circle $|w| = 1$ (Fig. 14.11).

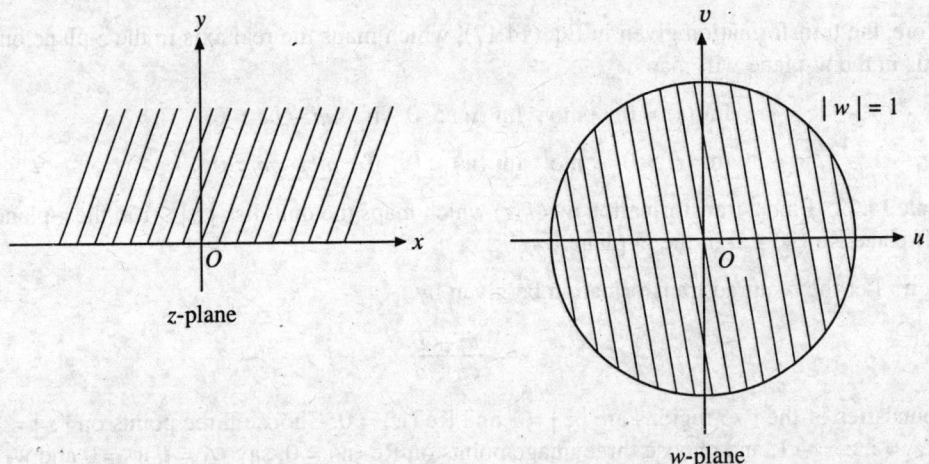

Fig. 14.11.   Example 14.20.

Example 14.21   Find a transformation $w = f(z)$ which maps the real axis in the z-plane onto the real axis in the w-plane.

Solution   Let x_1, x_2, x_3 be three distinct points on the real axis in the z-plane. Let the images of these three points on the real axis of the w-plane be $u_1 = 0$, $u_2 = 1$ and $u_3 = \infty$ respectively. Substituting in the bilinear transformation $w = (az + b)/(cz + d)$, we obtain

$$ax_1 + b = 0, \quad \frac{ax_2 + b}{cx_2 + d} = 1, \quad cx_3 + d = 0.$$

The solution of this system is

$$a = -\frac{b}{x_1}, \quad c = \frac{b(x_1 - x_2)}{x_1(x_2 - x_3)}, \quad d = -\frac{bx_3(x_1 - x_2)}{x_1(x_2 - x_3)}.$$

The bilinear transformation becomes

$$w = \frac{az + b}{cz + d} = \frac{(x_1 - z)(x_2 - x_3)}{(x_1 - x_2)(z - x_3)}. \tag{14.17}$$

We find that $ad - bc = \dfrac{b^2(x_1 - x_2)(x_3 - x_1)}{x_1^2(x_2 - x_3)} \neq 0$.

Further, a, b, c, d can be taken as real when x_1, x_2, x_3 are real. We also have

$$\overline{w} = \frac{a\overline{z} + b}{c\overline{z} + d}$$

and

$$w - \overline{w} = \frac{az + b}{cz + d} - \frac{a\overline{z} + b}{c\overline{z} + d} = \frac{(ad - bc)(z - \overline{z})}{(cz + d)(c\overline{z} + d)}$$

or

$$\text{Im}(w) = \frac{ad - bc}{|cz + d|^2}\, \text{Im}(z).$$

Therefore, the transformation given in Eq. (14.17), which maps the real axis in the z-plane onto the real axis in the w-plane will map

$$\text{Im}(z) > 0 \quad \text{onto} \quad \text{Im}(w) > 0 \quad \text{if} \quad ad - bc > 0,$$

$$\text{Im}(z) > 0 \quad \text{onto} \quad \text{Im}(w) < 0 \quad \text{if} \quad ad - bc < 0.$$

Example 14.22   Find a transformation $w = f(z)$ which maps the unit disk $|z| \leq 1$ in the z-plane onto the half plane $\text{Re}(w) \geq 0$ in the w-plane.

Solution   Let the required transformation be given by

$$w = \frac{az + b}{cz + d}$$

The boundaries of the two regions are $|z| = 1$ and $\text{Re}(w) = 0$. Choose three points on $|z| = 1$, say $z_1 = 1$, $z_2 = i$, $z_3 = -1$; and choose three image points on $\text{Re}(w) = 0$, say $w_1 = i$, $w_2 = 0$ and $w_3 = -i$, respectively.

Since $z = 1$ maps onto $w = i$,   we get $a + b = i(c + d)$
$\qquad$ $z = i$ maps onto $w = 0$,   we get $ai + b = 0$
$\qquad$ $z = -1$ maps onto $w = -i$, we get $-a + b = -i(-c + d)$.

Solving these equations, we obtain $a = id$, $b = d$ and $c = -id$. Therefore, the transformation which maps $|z| = 1$ onto Re $(w) = 0$ becomes

$$w = \frac{1 + iz}{1 - iz}. \tag{14.18}$$

Substituting $w = u + iv$ and $z = x + iy$, we obtain

$$u + iv = \frac{(1 + ix) - y}{(1 - ix) + y} = \frac{[(1 - y) + ix][(1 + y) + ix]}{(1 + y)^2 + x^2}.$$

Comparing the real parts, we get

$$u = \frac{1 - x^2 - y^2}{x^2 + (1 + y)^2}.$$

For $|z| = x^2 + y^2 \leq 1$, we obtain Re $(w) = u \geq 0$. Therefore, the mapping given in Eq. (14.18) maps the region $|z| \leq 1$ onto the region Re $(w) \geq 0$.

Example 14.23 Find a transformation $w = f(z)$ which maps the unit disk $|z| \leq 1$ in the z-plane onto the unit disk $|w| \leq 1$ in the w-plane.

Solution Let the transformation be given by

$$w = \frac{az + b}{cz + d}. \tag{14.19}$$

Now, $w = 0$ and $w = \infty$ are the inverse points for $|w| = 1$ and they are the images of the points $z = -b/a$ and $z = -d/c$ respectively. Therefore, $z = -b/a$ and $z = -d/c$ are the inverse points for $|z| = 1$. Thus, if we write (see Remark 3)

$$\alpha = -\frac{b}{a}, \quad \text{then} \quad \frac{1}{\bar{\alpha}} = -\frac{d}{c} \quad \text{or} \quad \bar{\alpha} = -\frac{c}{d}.$$

Substituting in Eq (14.19), we obtain

$$w = \frac{a(z + b/a)}{c(z + d/c)} = \frac{a(z - \alpha)}{c(z - (1/\bar{\alpha}))} = \frac{a\bar{\alpha}}{c}\left(\frac{z - \alpha}{\bar{\alpha}z - 1}\right).$$

We can write

$$w = k\left(\frac{z - \alpha}{\bar{\alpha}z - 1}\right), \quad \text{where} \quad k = \frac{a\bar{\alpha}}{c}. \tag{14.20}$$

We have from Eq. (14.20)

$$|w| = |k|\left|\frac{z - \alpha}{\bar{\alpha}z - 1}\right| = |k|\left|\frac{z - \alpha}{\bar{\alpha}z - z\bar{z}}\right| = \frac{|k|}{|z|}\left|\frac{z - \alpha}{\bar{z} - \bar{\alpha}}\right| = |k|$$

since $z\bar{z} = |z|^2 = 1$.

Hence, the transformation given by Eq. (14.20) maps $|z| = 1$ onto $|w| = 1$ when $|k| = 1$. In this case, we further have

$$\overline{w} = \overline{k} \left(\frac{\overline{z} - \overline{\alpha}}{\alpha \overline{z} - 1} \right)$$

and
$$w\overline{w} - 1 = k\overline{k} \left(\frac{z - \alpha}{\overline{\alpha}z - 1} \right)\left(\frac{\overline{z} - \overline{\alpha}}{\alpha \overline{z} - 1} \right) - 1 = \frac{(1 - \alpha\overline{\alpha})(z\overline{z} - 1)}{\mid \alpha\overline{z} - 1 \mid^2},$$

since $k\overline{k} = \mid k \mid^2 = 1$.

This shows that the bilinear transformation given in Eq. (14.20) with $\mid k \mid = 1$, (which maps $\mid z \mid = 1$ onto $\mid w \mid = 1$)

$$\text{maps } \mid z \mid < 1 \quad \text{onto} \quad \mid w \mid < 1, \text{ if } \mid \alpha \mid < 1, \text{ and}$$

$$\text{maps } \mid z \mid < 1 \quad \text{onto} \quad \mid w \mid > 1, \text{ if } \mid \alpha \mid > 1,$$

Without any loss of generality, we can assume $k = 1$. Then, the required bilinear transformation is given by

$$w = \frac{z - \alpha}{\overline{\alpha}z - 1}. \tag{14.21}$$

This transformation maps $\mid z \mid < 1$ onto $\mid w \mid < 1$ if $\mid \alpha \mid < 1$ and maps $\mid z \mid < 1$ onto $\mid w \mid > 1$, if $\mid \alpha \mid > 1$. Further, the point $z = \alpha$ maps onto the centre $w = 0$.

Example 14.24 Find the image of the annulus $1 < \mid z \mid < 2$ under the mapping $w = z/(z - 1)$.

Solution We have $w = \dfrac{z}{z - 1}$ or $z = \dfrac{w}{w - 1} = \dfrac{u + iv}{(u - 1) + iv}$

The region $\mid z \mid > 1$ is mapped onto the region

$$\left| \frac{u + iv}{u - 1 + iv} \right| > 1 \quad \text{or} \quad u^2 + v^2 > (u - 1)^2 + v^2 \quad \text{or} \quad u > \frac{1}{2}.$$

The region $\mid z \mid < 2$ is mapped onto the region

$$\left| \frac{u + iv}{u - 1 + iv} \right| < 4 \quad \text{or} \quad u^2 + v^2 < 4[(u - 1)^2 + v^2]$$

or
$$(u - 4/3)^2 + v^2 > (4/3)^2.$$

Thus, the required region is on the right of the plane $u = 1/2$ and outside the circle $(u - 4/3)^2 + v^2 = 9/16$. The points on the line $u = 1/2$ and on the boundary of the circle are not included.

Exercise 14.2

Find the images of the following regions under the given mappings,

1. Re $z = 2$; $w = (2z - i)/(z + i)$.
2. The first quadrant $x > 0$, $y > 0$; $w = (z - i)/(z + i)$.
3. The strip $0 < x < 1$; $w = (z - 1)/z$.
4. The half plane $x + 2y > 1$; $w = (z + 2i)/(z + 1)$.
5. The wedge $y < 1$, $x + y > 2$; $w = z/(z - 1 - i)$.
6. The disk $\mid z - 2i \mid \leq 2$; $w = 2z/[2(1 + i) - z]$.
7. The disk $\mid z - 1 \mid \leq 2$; $w = (z - 1)/(z - 2)$.
8. The circle $\mid z + i \mid = 1$; $w = [(1 - 2i)z + 2]/z$.

9. The line $\left(\dfrac{z+\bar{z}}{2}\right) - \left(\dfrac{z-\bar{z}}{2i}\right) = 1;\ w = (1 - iz)/z$.

10. The line $(z + \bar{z}) - 2i(z - \bar{z}) = -4;\ w = (z - 5)/(z + i)$.

Find the fixed points of the following bilinear transformations.

11. $w = \dfrac{-2 + (2 - i)z}{z - i}$.

12. $w = \dfrac{1 - 3iz}{z - i}$.

13. $w = \dfrac{2z - 1 + 3i}{z + 3i}$.

14. $w = \dfrac{2(z + i)}{z + 2i}$.

15. $w = \dfrac{z(i + 1) - 3}{z + i - 3}$.

16. $w = \dfrac{az + b}{a - bz},\ b \neq 0$

Find a bilinear transformation which maps respectively the given points.

17. $-1, 0, 1$ in the z-plane onto the points $1, i, -1$ in the w-plane.

18. $i, -1, 0$ in the z-plane onto the points $1, i, -1$ in the w-plane.

19. $1, i, -i$ in the z-plane onto the points $\infty, 0, 1$ in the w-plane.

20. $0, 1, \infty$ in the z-plane onto the points $-i, 1, i$ in the w-plane.

21. $0, 1 + i, 2i$ in the z-plane onto the points $1, i, \infty$ in the w-plane.

22. $0, 1, \infty$ in the z-plane onto the points $-i, \infty, 1$ in the w-plane.

23. Show that the composition of two bilinear transformations is also a bilinear transformation.

24. Find the bilinear transformation such that the points $1, i$ are its fixed points and the point 0 is mapped as -1.

25. Find the bilinear transformation such that the point $z = i$ is the only fixed point and the point 1 is mapped as ∞.

26. Show that a bilinear transformation which possesses two distinct fixed points z_1 and z_2 can be represented as

$$\frac{w - z_1}{w - z_2} = \alpha\left(\frac{z - z_1}{z - z_2}\right),\ \alpha \text{ complex constant.}$$

Deduce that when $z_1 \to z_2$, the transformation becomes.

$$\frac{1}{w - z_2} = \frac{1}{z - z_2} + k.$$

where k is a complex constant.

27. Show that the bilinear transformation $w = (z - a)/(az - 1)$, $a < 1$ and real, maps the region $|z| < 1$ onto $|w| < 1$. Find which part of the w-plane corresponds to the region $|z| < 1$ and Im $(z) > 0$.

28. Given the mapping $w = i(2z + 3i)/(z + 2i)$, find the image in the w-plane of a region within a circle with centre at $z = -2i$ and radius 3 in the z-plane. Find also the image of the region in the w-plane corresponding to the region Re $(z) \geq 0$ and Im $(z) \geq -2$.

29. Find a bilinear transformation which maps the right half plane Re $(z) \geq 0$ onto the unit disk $|w| \leq 1$.

30. Find a bilinear trnasformation which maps the disk $|z - 1| > 2$ in the z-plane onto the disk $|w| < 1$ in the w-plane.

31. Let $r \neq 1$ be a given positive number and let z_0 be any complex number with $|z_0| < r$. Show that the bilinear transformation

$$w = A\left(\frac{z - z_0}{z\bar{z}_0 - r^2}\right),\ |A| = r^2,\ |z_0| < r$$

maps the disk $|z| < r$ onto the disk $|w| < r$ in the w-plane.

32. Find a bilinear transformation which maps the unit disk $|z| < 1$ in the z-plane onto the region $u + v \geq 0$ in the w-plane.

33. Find a bilinear transformation that maps the disk $|z| < 1$ in the z-plane onto the upper half plane in the w-plane.

34. Find the bilinear transformation which maps the circle $|z| = 1$ onto the circle $|w + 1| = 1$ and maps the points $z = 0$ and $z = 1$ as $w = -3/2$ and $w = -2$ respectively.

35. Find the bilinear transformation which maps the real axis, Im $(z) = 0$, in the z-plane onto the unit circle $|w| = 1$ in the w-plane.

14.4   Conformal Mapping

Let $w = f(z)$ be an analytic function and define a mapping from a region D in the z-plane onto a region D^* in the w-plane. *A mapping $w = f(z)$ is said to be conformal if it preserves the angle between two curves both in magnitude and direction.*

Let C be a continuous curve in the z-plane passing through a point $z_0 \in D$. Let

$$z(t) = x(t) + iy(t), \, a \leq t \leq b \tag{14.22}$$

be the parametric representation of the curve C. The increasing values of t is taken as the positive direction of the curve C. We assume that $z(t)$ defined in Eq. (14.22) is differentiable and the derivative $\dot{z}(t) \neq 0$ and is continuous on C. Therefore, C is a smooth curve. Let C^* be the image curve of C in the w-plane. Then

$$w = f[z(t)], \, a \leq t \leq b \tag{14.23}$$

is the parametric representation of the image curve C^* in the w-plane under the mapping of an analytic function $w = f(z)$.

Let $z_1 = z(t_1)$ be a point in the neighborhood of the point $z_0 = z(t_0)$ on C. Let $\Delta t = t_1 - t_0$. Now, $\dot{z}(t)$ is the tangent vector to the curve C at the point t. Then, arg $(\dot{z}(t_0))$ is defined by

$$\arg(\dot{z}(t_0)) = \arg\left[\lim_{\Delta t \to 0} \frac{z_1 - z_0}{\Delta t}\right].$$

Denote $\theta_0 = \arg(\dot{z}(t_0))$, that is, θ_0 is the angle of inclination of the directed tangent to C at the point z_0 (Fig. 14.12). Using Eq. (14.23), we obtain

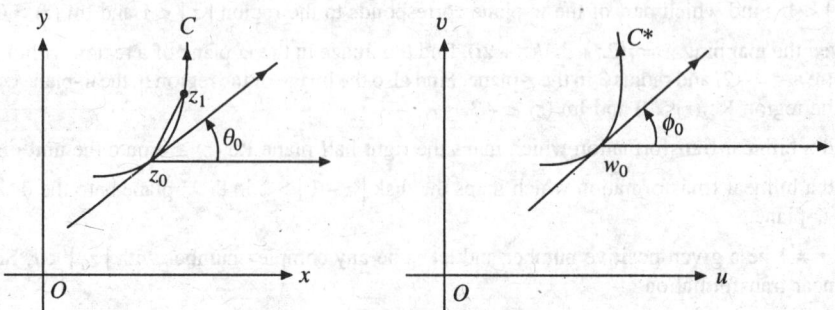

Fig. 14.12. Angle made by a curve with real x-axis.

$$\dot{w}(t) = f'[z(t)]\dot{z}(t). \qquad (14.24)$$

Hence,
$$\dot{w}(t_0) = f'[z_0]\dot{z}(t_0) \qquad (14.25)$$

where $z_0 = z(t_0)$. Writing $\dot{w}(t_0) = \rho_0 e^{i\phi_0}$, $f'(z_0) = Re^{i\psi_0}$ and $\dot{z}(t_0) = r_0 e^{i\theta_0}$, we get

$$\rho_0 e^{i\phi_0} = Rr_0 e^{i(\psi_0 + \theta_0)}, \quad \text{or} \quad \phi_0 = \psi_0 + \theta_0 = \theta_0 + \arg(f'(z_0))$$

where ϕ_0 is the angle of inclination of the directed tangent to C^* at the point $w_0 = f(z_0)$. Now, if $f'(z_0) \neq 0$, then from Eq. (14.25) $\dot{w}(t_0) \neq 0$. Hence, the curve C^* has a unique tangent at the point $w_0 = f(z_0)$. We have

$$\psi_0 = \phi_0 - \theta_0 \quad \text{or} \quad \arg(f'(z_0)) = \arg(\dot{w}(t_0)) - \arg(\dot{z}(t_0)). \qquad (14.26)$$

Therefore, under the mapping $w = f(z)$, the tangent line to a smooth curve C in the z-plane is rotated by the angle $\psi_0 = \arg(f'(z_0))$ in the w-plane. The angle ψ_0 is called the *angle of rotation* and $|f'(z_0)|$ is called the *scale factor* at the point z_0. The angle of rotation ψ_0 is independent of C as $\arg(f'(z_0))$ depends only on the point z_0. Therefore, the mapping $w = f(z)$ rotates all the curves passing through the point z_0 by the same angle $\arg(f'(z_0))$.

Now, let C_1 and C_2 be two smooth curves passing through the point z_0 and θ_1 and θ_2 be the angles of inclination of directed line tangents to C_1 and C_2 respectively at the point z_0. Under the mapping $w = f(z)$, let C_1^* and C_2^* be the images of C_1 and C_2 respectively. The image curves pass through the point $w_0 = f(z_0)$. Let ϕ_1, ϕ_2 be the angles of inclination of the directed line tangent to C_1^* and C_2^* respectively, at w_0, which is the image of the point z_0. Using Eq. (14.26), we obtain that under the mapping $w = f(z)$, the curves C_1 and C_2 are rotated in the same direction by the same angle $\arg(f'(z_0))$, (Fig. 14.13).

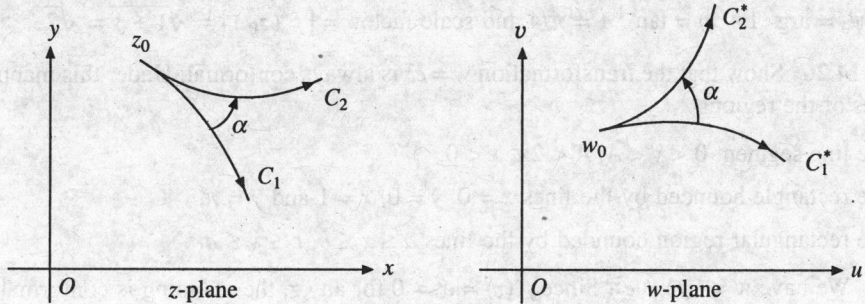

Fig. 14.13. Conformal mapping.

Therefore,
$$\arg[f'(z_0)] = \psi_0 = \phi_1 - \theta_1 \quad \text{and} \quad \arg[f'(z_0)] = \psi_0 = \phi_2 - \theta_2.$$

Subtracting, we obtain
$$\phi_2 - \phi_1 = \theta_2 - \theta_1.$$

We state this result in the following theorem.

Theorem 14.4 At each point z where $f(z)$ is analytic and $f'(z) \neq 0$, the mapping $w = f(z)$ is conformal.

The points at which $f'(z) = 0$, that is the points at which the mapping is not conformal are called the *critical points* of the mapping.

A mapping that preserves the magnitude of the angle but not necessarily the direction is called an *isogonal mapping*. The following are some examples.

1. The mapping given by the analytic function $w = f(z) = z^2$ is conformal everywhere except when $f'(z) = 2z = 0$. The point $z = 0$ is the critical point and any curve passing through the point $z = 0$ is not mapped conformally.

2. The linear mapping $w = f(z) = az + b$, $a \neq 0$ is always conformal, since $f'(z) = a \neq 0$ for all z.

3. The inversion mapping $w = f(z) = 1/z$ is conformal in the finite complex plane except at the point $z = 0$, where $f(z)$ is not analytic.

4. The bilinear mapping $w = (az + b)/(cz + d)$ is conformal if $ad - dc \neq 0$, since $(dw/dz) = (ad - bc)/(cz + d)^2$.

Remark 4

Since conformal maps are one-to-one, the relative order of points is preserved by a conformal mapping. That is, if A, B, C are points on a curve C in that order in the z-plane, then A', B', C', the image points of A, B, C respectively, will appear on the image curve C^* in the same order in the w-plane.

Example 14.25 Determine the angle of rotation at the point $z = (1 + i)/2$ under the mapping $w = z^2$. Find its scale factor also.

Solution The angle of rotation is given by $\psi_0 = \arg[f'(z_0)]$. We have $f(z) = z^2$ and $z_0 = (1 + i)2$.

Therefore,
$$f'(z) = 2z \quad \text{and} \quad f'((1 + i)/2)) = 1 + i.$$

Hence, $\psi_0 = \arg(1 + i) = \tan^{-1} 1 = \pi/4$ and scale factor $= |f'(z_0)| = \sqrt{1 + 1} = \sqrt{2}$.

Example 14.26 Show that the transformation $w = e^z$ is always conformal. Under this mapping, find the images of the regions

(i) the line segment $0 < y < A$, $A < 2\pi$, $x < 0$,

(ii) the rectangle bounded by the lines $x = 0$, $y = 0$, $x = 1$ and $y = \pi$,

(iii) the rectangular region bounded by the lines $a \leq x \leq b$, $c \leq y \leq d$.

Solution We have $w = f(z) = e^z$. Since $f'(z) = e^z \neq 0$ for any z, the mapping is conformal at every point z. Setting $z = x + iy$, we have $w = e^{x+iy}$ and $|w| = e^x$.

If we write $w = Re^{i\phi}$, then we have $R = e^x$ and $\phi = y$.

(i) For $-\infty < x < 0$, $0 < y < A$, $A < 2\pi$, we get $0 < |w| < 1$ and $0 < \arg(w) < A$. The image curves are given in Fig. 14.14(i).

(ii) We consider the images of the four sides of the rectangle separately (Fig. 14.14(ii)). The line OA is given by $z = t$, $0 \leq t \leq 1$. Its images curve in the w-plane is $w = e^t$, $0 \leq t \leq 1$, w is real and takes values $1 < w < e$ on the positive u-axis.

The line AB is given by $z = 1 + it$, $0 \leq t \leq \pi$. Its image curve in the w-plane is $w = e^{1+it}$, $0 \leq t \leq \pi$. We find that $|w| = e$. As t varies in the interval $0 \leq t \leq \pi$, w moves on the upper semicircle $|w| = e$ in anti-clockwise direction.

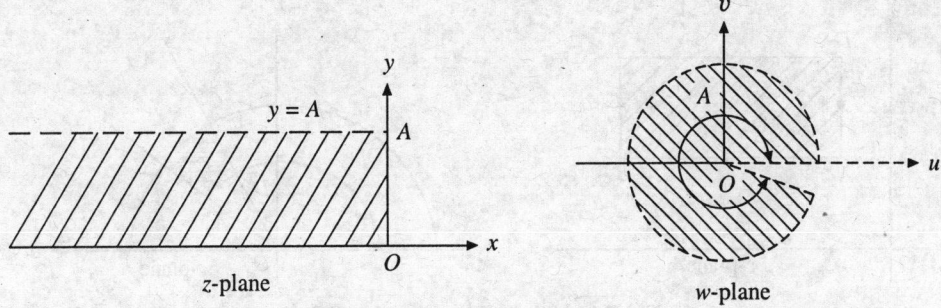

Fig. 14.14(i).   Example 14.26(i).

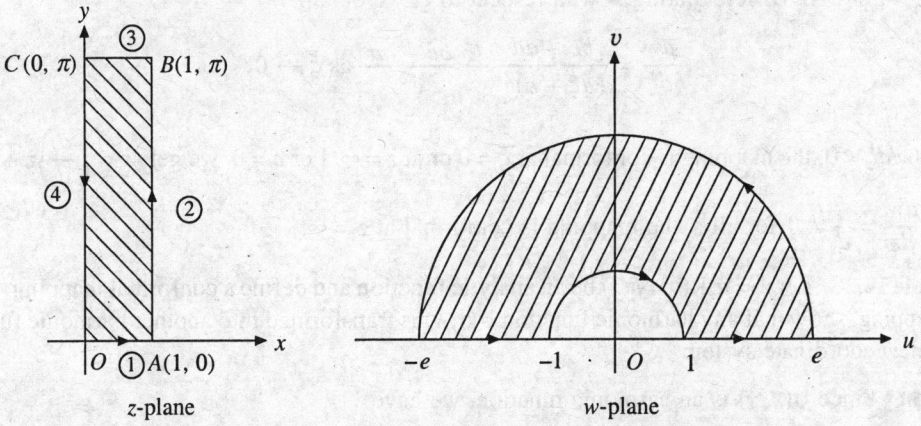

Fig. 14.14(ii).   Example 14.26(ii).

The line BC is given by $z = -t + \pi i$, $-1 \le t \le 0$. Its image curve in the w-plane is $w = e^{-t+\pi i} = -e^{-t}$, $-1 \le t \le 0$. We find that w is real and takes values $-e < w < -1$ on the negative u-axis.

The line CO is given by $z = -it$, $-\pi \le t \le 0$. Its image curve in the w-plane is given by $w = e^{-it}$, $-\pi \le t \le 0$. We find that $|w| = 1$. As t varies in the interval $-\pi \le t \le 0$, w moves on the semicircle $|w| = 1$ is the clockwise direction.

The required region is given in Fig. 14.14(ii). The angles at the vertices O, A, B, C of the rectangle are preserved both in magnitude and direction, under the given transformation.

(iii) The vertical line AD, that is $x = a$, $c \le y \le d$ is mapped onto the arc $A'D'$: $R = e^a$, $c \le \phi \le d$. The vertical line BC, that is $x = b$, $c \le y \le d$ is mapped onto the arc $B'C'$: $R = e^b$, $c \le \phi \le d$.

The horizontal line AB, that is, $y = c$, $a \le x \le b$ is mapped onto the ray $A'B'$: $\phi = c$ and the horizontal line DC, that is, $y = d$, $a \le x \le b$ is mapped onto the ray $D'C'$: $\phi = d$, $a \le x \le b$. The required region is given in Fig. 14.14(iii). Note that the mapping is one-to-one if $d - c < 2\pi$.

Example 14.27 Show that the bilinear transformation $w = \dfrac{az + b}{cz + d}$, $ad - bc \ne 0$ is conformal at $z = \infty$.

Solution   For $c \ne 0$, let $z = 1/\xi$. We get

$$w = \frac{b\xi + a}{d\xi + c}.$$

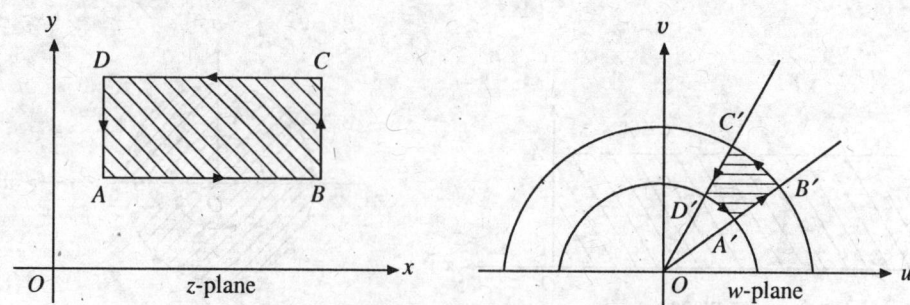

Fig. 14.14(iii).   Example 14.26(iii).

As $z \to \infty$, $\xi \to 0$. Differentiating w with respect to ξ, we obtain

$$\frac{dw}{d\xi} = \frac{bc - ad}{(d\xi + c)^2} \to \frac{bc - ad}{c^2} \text{ as } \xi \to 0.$$

Since $dw/d\xi \neq 0$, the mapping is conformal at $\xi = 0$ or at $z = \infty$. For $c = 0$, we get $w = \left(\frac{a}{d}\right) z + \left(\frac{b}{d}\right)$.

Since $\dfrac{dw}{dz} = \dfrac{a}{d} \neq 0$ for all z, the mapping is conformal at $z = \infty$.

Example 14.28   Let $w = u + iv = f(z)$ be an analytic function and define a conformal mapping. Under this mapping, show that any harmonic function $\phi(x, y)$ is transformed into another harmonic function in the new coordinate system.

Solution   Since $\phi(x, y)$ is an harmonic function, we have

$$\frac{\partial^2 \phi}{\partial x^2} + \frac{\partial^2 \phi}{\partial y^2} = 0.$$

Also, the conformal mapping $w = u + iv$ is analytic. Therefore, it satisfies the Cauchy-Riemann equations

$$\frac{\partial u}{\partial x} = \frac{\partial v}{\partial y}, \ \frac{\partial u}{\partial y} = -\frac{\partial v}{\partial x} \quad \text{and} \quad \frac{\partial^2 u}{\partial x^2} + \frac{\partial^2 u}{\partial y^2} = 0, \ \frac{\partial^2 v}{\partial x^2} + \frac{\partial^2 v}{\partial y^2} = 0.$$

Now,

$$\frac{\partial \phi}{\partial x} = \frac{\partial \phi}{\partial u} \frac{\partial u}{\partial x} + \frac{\partial \phi}{\partial v} \frac{\partial v}{\partial x} = \frac{\partial \phi}{\partial u} \frac{\partial u}{\partial x} - \frac{\partial \phi}{\partial v} \frac{\partial u}{\partial y}$$

$$\frac{\partial \phi}{\partial y} = \frac{\partial \phi}{\partial u} \frac{\partial u}{\partial y} + \frac{\partial \phi}{\partial v} \frac{\partial v}{\partial y} = \frac{\partial \phi}{\partial u} \frac{\partial u}{\partial y} + \frac{\partial \phi}{\partial v} \frac{\partial u}{\partial x}$$

$$\frac{\partial^2 \phi}{\partial x^2} = \frac{\partial \phi}{\partial u} \frac{\partial^2 u}{\partial x^2} + \frac{\partial u}{\partial x} \left[\frac{\partial^2 \phi}{\partial u^2} \frac{\partial u}{\partial x} + \frac{\partial^2 \phi}{\partial u \partial v} \frac{\partial v}{\partial x} \right] - \frac{\partial \phi}{\partial v} \frac{\partial^2 u}{\partial x \partial y} - \frac{\partial u}{\partial y} \left[\frac{\partial^2 \phi}{\partial u \partial v} \frac{\partial u}{\partial x} + \frac{\partial^2 \phi}{\partial v^2} \frac{\partial v}{\partial x} \right]$$

$$= \frac{\partial \phi}{\partial u} \frac{\partial^2 u}{\partial x^2} + \frac{\partial^2 \phi}{\partial u^2} \left(\frac{\partial u}{\partial x} \right)^2 - 2 \frac{\partial^2 \phi}{\partial u \partial v} \frac{\partial u}{\partial x} \frac{\partial u}{\partial y} - \frac{\partial \phi}{\partial v} \frac{\partial^2 u}{\partial x \partial y} + \frac{\partial^2 \phi}{\partial v^2} \left(\frac{\partial u}{\partial y} \right)^2.$$

Similarly, we obtain

$$\frac{\partial^2 \phi}{\partial y^2} = \frac{\partial \phi}{\partial u}\frac{\partial^2 u}{\partial y^2} + \frac{\partial^2 \phi}{\partial u^2}\left(\frac{\partial u}{\partial y}\right)^2 + 2\frac{\partial^2 \phi}{\partial u \partial v}\frac{\partial u}{\partial x}\frac{\partial u}{\partial y} - \frac{\partial \phi}{\partial v}\frac{\partial^2 u}{\partial x \partial y} + \frac{\partial^2 \phi}{\partial v^2}\left(\frac{\partial u}{\partial x}\right)^2.$$

Adding these two equations, we obtain

$$\frac{\partial^2 \phi}{\partial x^2} + \frac{\partial^2 \phi}{\partial y^2} = \frac{\partial \phi}{\partial u}\left(\frac{\partial^2 u}{\partial x^2} + \frac{\partial^2 u}{\partial y^2}\right) + \left[\frac{\partial^2 \phi}{\partial u^2} + \frac{\partial^2 \phi}{\partial v^2}\right]\left[\left(\frac{\partial u}{\partial x}\right)^2 + \left(\frac{\partial u}{\partial y}\right)^2\right].$$

Since $u(x, y)$ and $\phi(x, y)$ are harmonic functions, we obtain

$$\left[\frac{\partial^2 \phi}{\partial u^2} + \frac{\partial^2 \phi}{\partial v^2}\right]\left[\left(\frac{\partial u}{\partial x}\right)^2 + \left(\frac{\partial u}{\partial y}\right)^2\right] = 0.$$

Now, $f'(z) = \dfrac{\partial u}{\partial x} - i\dfrac{\partial u}{\partial y}$ and $|f'(z)|^2 = \left(\dfrac{\partial u}{\partial x}\right)^2 + \left(\dfrac{\partial u}{\partial y}\right)^2 \neq 0$, since the mapping is conformal.

Hence, $\dfrac{\partial^2 \phi}{\partial u^2} + \dfrac{\partial^2 \phi}{\partial v^2} = 0$, that is $\phi(u, v)$ is also an harmonic function.

Exercise 14.3

1. Find the angle of rotation produced by the following transformations.

   (i) $w = 2z - 1 + 2i$ at its fixed point,

   (ii) $w = iz + 1$ at its fixed point.

   Determine the scale factor in each case.

2. Find the angle of rotation produced by the transformation $w = 1/z$ at the points (i) $z = 1$, (ii) $z = i$. Obtain the scale factor in each case.

3. Find the angle of rotation, at a non-zero point $z_0 = r_0 e^{i\theta_0}$, produced by the transformation $w = z^n$, n positive integer. Obtain its scale factor also.

4. Using the Cauchy-Riemann equations, express the scale factor of the mapping $w = w(z) = u + iv$ as a 2×2 determinant.

5. Determine the points where the following mappings $w = f(z)$ are not conformal.

   (i) $\cos z$ (ii) $\sin \pi z$ (iii) $\cosh z$ (iv) $e^{z^5 + 80z}$.

6. Is the mapping $w = \bar{z}$ conformal?

7. Find a conformal mapping which maps $\text{Im}(z) < 2$ onto $\text{Im}(w) > -3$.

8. Find a conformal mapping which takes the upper half plane $\text{Im}(z) > 0$ into the wedge $0 < \arg(w) < \pi/4$.

9. Show that for any given θ and z_0 with $\text{Im}(z_0) > 0$, the mapping $w = e^{i\theta}(z - z_0)/(z - \bar{z}_0)$ is a conformal mapping of $\text{Im}(z) > 0$ onto $|w| < 1$.

10. The interior of a square with vertices at $0, 1, 1 + i$ and i in the z-plane is mapped onto a region R^* in the w-plane under the mapping $w = f(z) = z + 2 - 3i$. Verify that the mapping is conformal and the interior angles of the mapped region R^* are at right angles.

11. Obtain the images of the circle $(x - 3)^2 + (y - 2)^2 = 2$ and the line $x + 2y = 8$ under the transformation $w = 1/z$. Show that the images of the circle and the line intersect at the same angle in the w-plane as in the z-plane.

12. Under the mapping $w = f(z) = z^2$, find the image of the region bounded by the lines $x = 1$, $y = $ and $x + y = 1$. Is the mapping conformal?

13. Under the mapping $w = f(z) = e^{-z}$, find the image in the w-plane of the rectangle $R : 0 < a \le x \le b$, $0 < c \le y \le d < 2\pi$ in the z-plane. Is the mapping conformal?

14. Under the mapping $w = f(z) = \sin z$, find the image in the w-plane of the strip $(-\pi/2) < \text{Re}(z) < (\pi/2)$, $\text{Im}(z) > 0$ in the z-plane. Is the mapping conformal?

15. Under the mapping $w = f(z) = \text{Log } z$, find the image in the w-plane of the region $0 < |z| < R_0$, $0 < \arg(z) < \theta_0$ in the z-plane. Is the mapping conformal?

14.5   Answers and Hints

Exercise 14.1

1. $|z - 1| = |(w - 4 - i)/i| = |w - 4 - i| = 1$, circle with centre at $(4, 1)$ and radius 1.

2. $u < 0$, $v > 1$.

3. $x = (u + v - 1)/2$, $y = (v - u + 1)/2$; $u + v > 1$, $v - u > -1$.

4. Let $b = b_1 + ib_2$, $|z| = |w - b| < 1$; circle with centre at (b_1, b_2), radius 1 and $v > b_2$.

5. (i) Maps as a rectangle with vertices at $(2, 3)$, $(4, 3)$, $(4, 5)$, $(2, 5)$,

    (ii) Maps as a rectangle with vertices at $(3, 2)$, $(5, 0)$, $(7, 2)$, $(5, 4)$.

6. (i) Maps as a triangle with vertices at $(1, -1)$, $(2, -3)$, $(2, -1)$,

    (ii) Maps as a triangle with vertices at $(3, 0)$, $(5, 1)$, $(3, 1)$.

7. $v > u - 2$.   **8.** $u > v + 2$.

9. $|z| = |(w - 2i)/(1 + i)| < 1$ gives $|w - 2i| < \sqrt{2}$; circle with centre at $(0, 2)$ and radius $\sqrt{2}$. Write $w = (1 + i)[z + (1 + i)] = \sqrt{2}e^{\pi i/4}[z + (1 + i)]$. Define $w_1 = z + (1 + i)$, (translation); $w_2 = \sqrt{2}w_1$, (magnification); $w = e^{\pi i/4}w_2$, (rotation by angle $\pi/4$).

10. $|w + ic| > c$, $v < 0$.   **11.** $|w - (1/2)| < 1/2$, $v < 0$.

12. $|w - (1 - i)/2| < 1/\sqrt{2}$.   **13.** $|w - 1/(2a)| = 1/(2a)$.

14. $|w + i/(2b)| = 1/(2b)$.

15. Image is the intersection of the regions $|w - (1/2)| < 1/2$ and $|w + (i/2)| < 1/2$.

16. $|w - (1/2)| \ge 1/2$, $u \ge 0$, $v \ge 0$.

17. $v < -1/4$ (note that $z = 0$ lies on the boundary of the given region).

18. $v > 1/2$ (note that $z = 0$ lies on the boundary of the given region).

19. $|w + (1/2)| < 1/2$, $\text{Re}(w) > (-1/2)$.

20. Using $|z|^2 = x^2 + y^2 = a^2$ and $(1/z) = (x - iy)/a^2$, we obtain the image as $(u^2/A^2) + (v^2/B^2) = 1$, where $A = (a^2 + 1)/a$ and $B = (a^2 - 1)/a$. The image is an ellipse.

21. $|z - i| = |(w - b - ai)/a| < 1$. Comparing with $|(w - 1)/2| < 1$, we obtain one of the solutions as $w = 2(z - i) + 1$.

22. $w = (1 + i)(1 - z)$.   **23.** $w = z + bi$, b real.

24. $w = z + b$, b real.   **25.** $w = z + b(1 + i)$, b real.

26. $w = az + b$, a, b real, $a > 0$.   **27.** $w = -az + b$, a, b real, $a > 0$.

28. $w = -i(az + b)$ or $e^{-\pi i/2}(az + b)$, a, b real, $a > 0$.

29. $w = az + bi$, a, b, real, $a > 0$.

30. Write $w_1 = az + b$; $|(z - 2)/3| = |(w_1 - 2a - b)/(3a)| < 1$. Comparing with $|(w_1 - i)/2| < 1$, we get $2a + b = i$ and $3a = 2$. The solution is $w_1 = (2z - 4 + 3i)/3$. The required mapping is $w = 1/w_1$.

Exercise 14.2

1. $|w - (2 - (3/4)i)| = 3/2$.

2. $|w| < 1$ and Im $(w) < 0$.

3. $|w - (1/2)| > 1/2$ and Re $(w) < 1$.

4. $|w - [(7/4) + i]| < 5/4$.

5. $u - v < 1$ and $u > 1$.

6. $u + v \geq 0$.

7. $|w - (4/3)| \geq (2/3)$.

8. $v = -1$.

9. $|w - (1 - i)/2| = 1/\sqrt{2}$.

10. $7u - 9v - 7 = 0$.

11. $1 - i$, $1 + i$.

12. $-i$.

13. 1, $1 - 3i$.

14. $1 - i$.

15. 1, 3.

16. i, $-i$.

17. $w = (z - i)/(iz - 1)$.

18. $w = [(3i - 1)z + (i + 1)]/[(i + 1)(z - 1)]$.

19. $w = [(1 - i)(z - i)]/\{2(z - 1)\}$.

20. $w = [1 + iz]/(z + i)$.

21. $w = [(1 + 3i)z + 2(1 - i)]/[(1 + i)(z - 2i)]$. **22.** $w = (z + i)/(z - 1)$.

23. Let $w = (az + b)/(cz + d)$ and $z = (\alpha s + \beta)(\gamma s + \delta)$. Substituting for z, we obtain another bilinear transformation.

24. $w = [(1 + 2i)z - i]/(z + i)$.

25. $w = [(2i - 1)z + 1]/(z - 1)$.

26. Suppose that the bilinear transformation $w = (az + b)/(cz + d)$ having two fixed points z_1, z_2 maps any point $z = \gamma$ onto $w = \delta$. Therefore, the points z_1, z_2, γ are mapped onto z_1, z_2, δ. The transformation is given by

$$\frac{w - z_1}{w - z_2} = \alpha\left(\frac{z - z_1}{z - z_2}\right), \text{ where } \alpha = \left(\frac{\delta - z_1}{\delta - z_2}\right)\left(\frac{\gamma - z_2}{\gamma - z_1}\right) \tag{14.27}$$

Let $z_1 = z_2 + \varepsilon$. Substituting in Eq. (14.27), we obtain

$$1 - \frac{\varepsilon}{w - z_2} = \alpha\left[1 - \frac{\varepsilon}{z - z_2}\right] \text{ and } \alpha = 1 + \frac{(\delta - \gamma)\varepsilon}{(\gamma - z_2)(\delta - z_2)} + 0(\varepsilon^2)$$

Simplifying and taking limits as $\varepsilon \to 0$, we get

$$\frac{1}{w - z_2} = \frac{1}{z - z_2} - k$$

where $k = (\delta - \gamma)/[(\gamma - z_2)(\delta - z_2)]$.

27. Solving for z, we obtain $z = (w - a)/(wa - 1)$. Now, $|z| < 1$ gives $|w - a| < |wa - 1|$. Simplifying, we obtain $u^2 + v^2 < 1$ or $|w| < 1$. Further $y > 0$ gives $v < 0$.

28. $|w - 2i| > 1/3$. Also, $x = u/[u^2 + (v - 2)^2]$, $y + 2 = (2 - v)/[u^2 + (v - 2)^2]$;

$x \geq 0$ gives $u \geq 0$ and $y \geq -2$ gives $v \leq 2$.

29. Choose the points on the boundary line $x = 0$ as $-i$, 0, i and the corresponding images in the w-plane as -1, i, 1 respectively. We obtain the mapping $w = i(1 - z)/(1 + z)$. The point $z = 1$ maps as $w = 0$, which is inside the circle $|w| = 1$. Hence, w is the required mapping.

30. The bilinear transformation $w = (z - \alpha)/(\bar{\alpha}z - 1)$, $|\alpha| < 1$ maps $|z| < 1$ onto $|w| < 1$. Replacing z by $1/z$, we find that the transformation $w = (1 - \alpha z)/(\bar{\alpha} - z)$ maps $|z| > 1$ onto $|w| < 1$. Now, replacing z by $(z - 1)/2$, we obtain the required transformation as

$$w = [(2 + \alpha) - \alpha z]/[(2\bar{\alpha} + 1) - z].$$

31. The bilinear transformation $w = k(z - \alpha)/[\overline{\alpha}z - 1]$, $|k| = 1$, $|\alpha| < 1$ maps $|z| < 1$ onto $|w| < 1$. Replacing z by z/r, w by w/r and α by α/r, we obtain the transformation

$$w = A(z - \alpha)/[\overline{\alpha}z - r^2], \text{ where } |A| = |kr^2| = r^2 \text{ and } \alpha = z_0.$$

32. In Example 14.22, we obtained the transformation $w = (1 + iz)/(1 - iz)$ which maps $|z| \le 1$ onto Re $(w) \ge 0$. Now, the region $u + v \ge 0$ is the right half-plane rotated counter clockwise by an angle $\pi/4$. The required mapping is

$$w = e^{\pi i/4}(1 + iz)/(1 - iz).$$

33. Choose the points $1, i, -1$ on the boundary of the region $|z| = 1$ and the corresponding images $0, 1, \infty$ respectively on the real axis Im $(w) = 0$. The required transformation is $w = i(1 - z)/(1 + z)$. The point $z = 0$ (centre of the circle) is mapped as $w = i$.

34. The transformation $w = k(z - \alpha)/(\overline{\alpha}z - 1)$, $|k| = 1$, $|\alpha| < 1$ maps $|z| = 1$ onto $|w| = 1$. Replacing $w = w + 1$, we find that the transformation $w + 1 = k(z - \alpha)/(\overline{\alpha}z - 1)$ maps $|z| = 1$ onto $|w + 1| = 1$. Now, $z = 0$ maps as $w = -3/2$ and $z = 1$ maps as $w = -2$. We obtain $k = 1$ and $\alpha = -1/2$. The transformation becomes $w = -3(z + 1)/(z + 2)$.

35. Let the transformation be $w = (az + b)/(cz + d)$. Now, the points $w = 0$ and $w = \infty$ are the inverse points for $|w| = 1$ and are the images of the points $z = -b/a$ and $z = -d/c$ respectively. Thus, if we write $-b/a = \alpha$, then $-d/c = \overline{\alpha}$. The transformation becomes $w = A(z - \alpha)/(z - \overline{\alpha})$, where $A = a/c$ and $|A| = 1$. Alternately, Eq. (14.16) gives the transformation where the points $1, 0, -1$ on the real axis are mapped as $1, -i, -1$ in the w-plane. This is a particular case of the above transformation.

Exercise 14.3

1. (i) Fixed point: $1 - 2i$, $\theta = 0$, scale factor $= 2$. (ii) Fixed point: $(1 + i)/2$, $\theta = \pi/2$, scale factor $= 1$.

2. (i) $\theta = \pi$, scale factor $= 1$; (ii) $\theta = 0$, scale factor $= 1$.

3. $\theta = (n - 1)\theta_0$, scale factor $= nr_0^{n-1}$.

4. Scale factor $= |f'(z)|^2 = \left(\dfrac{\partial u}{\partial x}\right)^2 + \left(\dfrac{\partial v}{\partial x}\right)^2 = \dfrac{\partial u}{\partial x}\dfrac{\partial v}{\partial y} - \dfrac{\partial u}{\partial y}\dfrac{\partial v}{\partial x} = \begin{vmatrix} \partial u/\partial x & \partial u/\partial y \\ \partial v/\partial x & \partial v/\partial y \end{vmatrix}$

5. (i) $z = k\pi$, $k = 0, \pm 1, \pm 2, \ldots$; (ii) $z = (2k + 1)/2$, $k = 0, \pm 1, \pm 2, \ldots$;

   (iii) $z = k\pi i$, $k = 0, \pm 1, \pm 2, \ldots$; (iv) $z = \pm\sqrt{2}(1 + i)$, $z = \pm\sqrt{2}(1 - i)$.

6. No. The mapping preserves the magnitude of the angle but not its direction.

7. One such mapping is $w = -z - i$.   8. $w = z^{1/4}$.

9. $|w|^2 = |z - z_0|^2/|z - \overline{z}_0|^2 = [(x - x_0)^2 + (y - y_0)^2]/[(x - x_0)^2 + (y + y_0)^2] < 1$ gives $y > 0$. Since $f'(z) = e^{i\theta}(2iy_0)/(z - \overline{z}_0)^2 \ne 0$ for any z, the mapping is conformal.

10. Since $f'(z) = 1 \ne 0$, the mapping is conformal. The vertices $A(0, 0)$, $B(1, 0)$, $C(1, 1)$, $D(0, 1)$ of the given square are mapped as the vertices $A'(2, -3)$, $B'(3, -3)$, $C'(3, -2)$, $D'(2, -2)$ of a square in that order. The interior of the square $ABCD$ is mapped as the interior of the square $A'B'C'D'$ in the w-plane. The direction of the mapping of the points is preserved. The angles between the sides are preserved and are equal to $\pi/2$.

11. The circle and the line in the z-plane intersect at $(2, 3)$ and $(22/5, 9/5)$. The angle between the circle and the line in the z-plane at the point $(2, 3)$ is $\tan^{-1}(3)$. The given circle and the line in the z-plane are mapped as the circles $11(u^2 + v^2) - 6u + 4v + 1 = 0$ and $8(u^2 + v^2) - u + 2v = 0$ respectively in the w-plane. The point $(2, 3)$ is mapped as $(2/13, -3/13)$ which is the point of intersection of the circles in

the w-plane. The angle between these circles at the point (2/13, – 3/13) is $\tan^{-1}(3)$. Similar result holds at the other point.

12. The regions in the z-plane and the w-plane are given in Fig. 14.15.

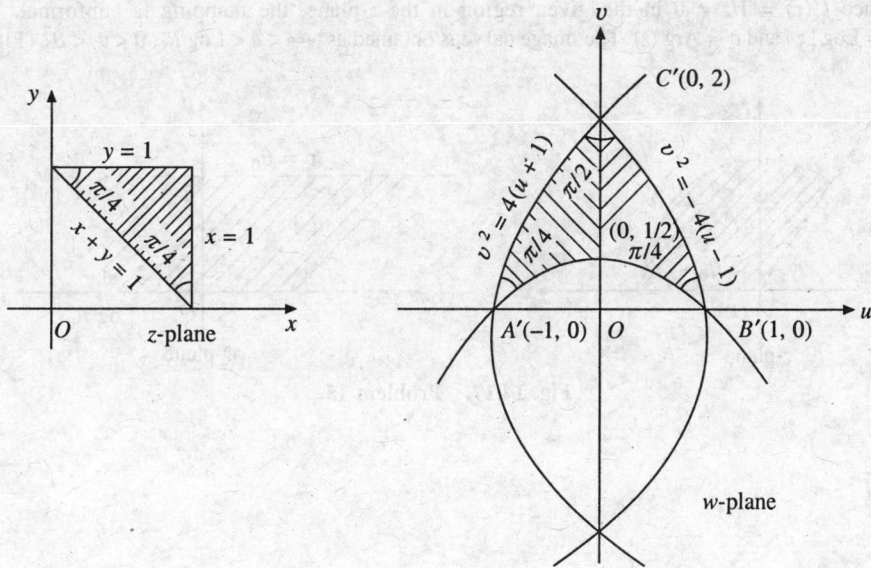

Fig. 14.15. Problem 12.

The line $x = 1$ is mapped as the parabola $v^2 = -4(u-1)$ with vertex at (1, 0), and opening to left. The line $y = 1$ is mapped as the parabola $v^2 = 4(u+1)$ with vertex at (–1, 0) and opening to right. The line $x + y = 1$ is mapped as the parabola $u^2 = -2[v - (1/2)]$ with vertex at (0, 1/2) and opening downward. Since the point $z = 0$ at which $f'(z) = 0$ is not inside the region considered, the mapping is conformal. The angle between the intersecting curves are preserved both in magnitude and direction.

13. Since $f'(z) \neq 0$ for any z, the mapping is conformal. The regions in the z-plane and the w-plane are given in Fig. 14.16. The line $x = a$ is mapped onto $|w| = e^{-a}$ and the line $x = b$ is mapped onto $|w| = e^{-b}$. The wedge $0 < c \leq y \leq d < 2\pi$ is mapped as $-d \leq \theta \leq -c$. The image is as given in Fig. 14.16.

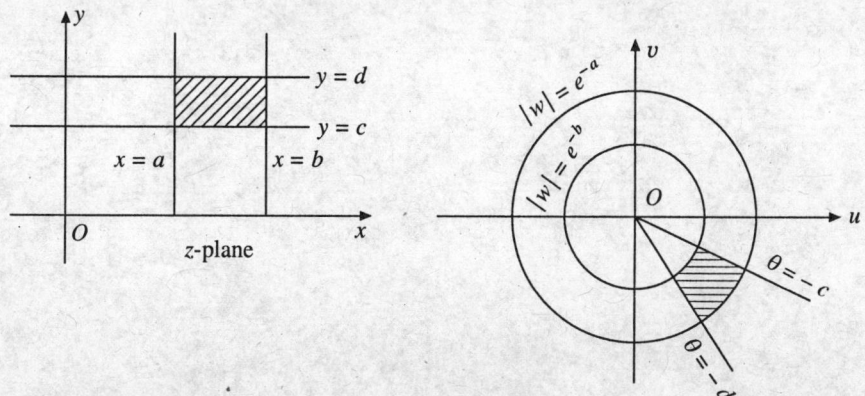

Fig. 14.16. Problem 13.

14. The mapping is conformal, since $z = \pm \pi/2$ are not included in the domain. The boundary line $x = -\pi/2$, $y > 0$ or $z = -(\pi/2) + it$, $0 \le t < \infty$ is mapped as $w = -\cosh t$, $0 \le t < \infty$. The line $x = \pi/2$, $y > 0$ or $z = (\pi/2) + it$, $0 \le t < \infty$ is mapped as $w = \cosh t$, $0 \le t < \infty$. The line $y = 0$, $-\pi/2 < x < \pi/2$ is mapped as $-1 < u < 1$, $v = 0$. The image is the region in the upper half w-plane.

15. Since $f'(z) = 1/z \ne 0$ in the given region in the z-plane, the mapping is conformal. We have $u = \text{Log} \, |z|$ and $v = \text{Arg} \, (z)$. The image curve is obtained as $-\infty < u < \text{Log} \, R_0$, $0 < v < \theta_0$. (Fig. 14.17).

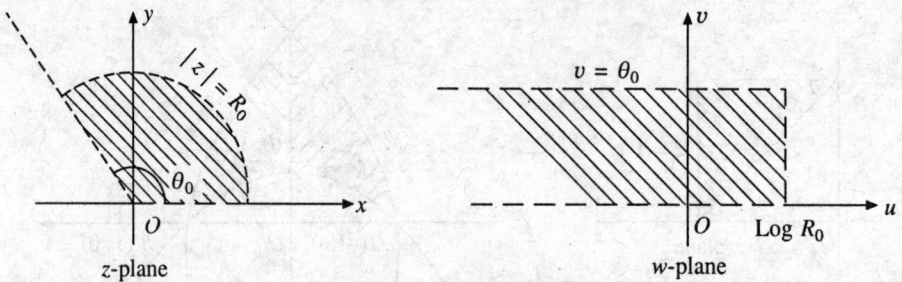

Fig. 14.17.   Problem 15.

Chapter 15

Vector Differential and Integral Calculus

15.1   Introduction

In this chapter, we shall study the vector differential and integral calculus. We often call this study as vector analysis or vector field theory. We first introduce few concepts.

Scalar function   A scalar function $f(x, y, z)$ is a function defined at each point in a certain domain D in space. Its value is real and depends only on the point $P(x, y, z)$ in space, but not on any particular coordinate system being used. For every point $(x, y, z) \in D, f$ has a real value. We say that a *scalar field f* is defined in D. For example, The distance function in 3-D space which defines the distance between the points $P(x, y, z)$ and $P_0(x_0, y_0, z_0)$

$$f(P) = f(x, y, z) = \sqrt{(x - x_0)^2 + (y - y_0)^2 + (z - z_0)^2} \qquad (15.1)$$

defines a scalar field. In this case, the domain D is the whole of the 3-D space.

Vector function   A function $\mathbf{v} = \mathbf{v}(P) = v_1 \mathbf{i} + v_2 \mathbf{j} + v_3 \mathbf{k}$ defined at each point $P \in D$ is called a vector function. We say that a *vector field* is defined in D. In cartesian coordinates, we can write

$$\mathbf{v} = v_1(x, y, z)\,\mathbf{i} + v_2(x, y, z)\,\mathbf{j} + v_3(x, y, z)\,\mathbf{k}. \qquad (15.2)$$

For example, the velocity field $\mathbf{v}(P)$ defined at any point P on a rotating body defines a vector field.

If the scalar and vector fields depend on time also, then we denote them as $f(P, t)$ and $\mathbf{v}(P, t)$, respectively. Both the fields are independent of the choice of the coordinate systems.

Level surfaces   Let $f(x, y, z)$ be a single valued continuous scalar function defined at every point $P \in D$. Then $f(x, y, z) = c$, a constant, defines the equation of a surface and is called a level surface of the function. For different values of c, we obtain different surfaces, no two of which intersect. For example, if $f(x, y, z)$ represents temperature in a medium, then $f(x, y, z) = c$ represents a surface on which the temperature is a constant c. Such surfaces are called *isothermal* surfaces.

Example 15.1   Find the level surfaces of the scalar fields in space, defined by the following functions

(i) $f(x, y, z) = (x^2 + y^2 + z^2)$,   (ii) $f(x, y, z) = z - \sqrt{x^2 + y^2}$.

Solution

(i) We find that $f(x, y, z) = c$ gives $x^2 + y^2 + z^2 = c$. Therefore, the level surfaces are spheres of radius $\sqrt{c}$.

(ii) We find that $f(x, y, z) = c$ gives $z - \sqrt{x^2 + y^2} = c$ or $x^2 + y^2 = (z - c)^2$. The level surfaces are cones.

15.2 Parametric Representations, Continuity and Differentiability of Vector Functions

Parametric representation of curves

We recall that a curve C in the two dimensional x-y plane can be parametrised by $x = x(t)$, $y = y(t)$, $a \le t \le b$. Then, the position vector of a point P on the curve C can be written as

$$\mathbf{r}(t) = x(t)\,\mathbf{i} + y(t)\,\mathbf{j}. \tag{15.3}$$

Therefore, the position vector of a point on a curve defines a vector function (Fig. 15.1a). Similarly, a three-dimensional curve or a space curve C can be parametrised as

$$\mathbf{r}(t) = x(t)\,\mathbf{i} + y(t)\,\mathbf{j} + z(t)\,\mathbf{k}, \; a \le t \le b \tag{15.4}$$

(Fig. 15.1b).

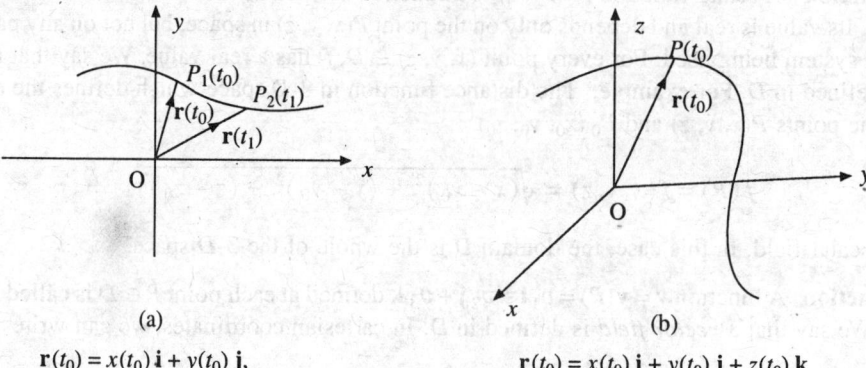

(a)

$$\mathbf{r}(t_0) = x(t_0)\,\mathbf{i} + y(t_0)\,\mathbf{j},$$

(b)

$$\mathbf{r}(t_0) = x(t_0)\,\mathbf{i} + y(t_0)\,\mathbf{j} + z(t_0)\,\mathbf{k}$$

Fig. 15.1. Position vector of a point.

The following are the parametric forms of some of the curves

Straight line $\mathbf{r}(t) = \mathbf{a} + t\,\mathbf{b} = (a_1 + tb_1)\mathbf{i} + (a_2 + tb_2)\mathbf{j} + (a_3 + tb_3)\mathbf{k}$ $\tag{15.5}$

This represents the position vector of a point on the line L which passes through the point A with position vector $\mathbf{a}$ and has the direction of the vector $\mathbf{b}$. There are also alternate ways of writing the parametric form of a straight line. For example, consider the line $L: x + y = 1$ in the first quadrant (Fig. 15.2). We can write the parametric form as

$$\mathbf{r}(t) = t\,\mathbf{i} + (1 - t)\,\mathbf{j}, \; 0 \le t \le 1.$$

Circle   The parametric form of the circle $x^2 + y^2 = a^2$, can be written as

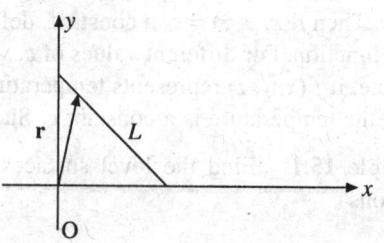

Fig. 15.2. Line $x + y = 1$.

$$\mathbf{r}(t) = a \cos t\, \mathbf{i} + a \sin t\, \mathbf{j}, \, 0 \le t \le 2\pi. \tag{15.6}$$

Ellipse The parametric form of the ellipse $\dfrac{x^2}{a^2} + \dfrac{y^2}{b^2} = 1$ can be written as

$$\mathbf{r}(t) = a \cos t\, \mathbf{i} + b \sin t\, \mathbf{j}, \, 0 \le t \le 2\pi. \tag{15.7}$$

Parabola Consider the parabola as $y^2 = 4ax$. Then, we can take $y = t$ as a parameter and write the parametric form of the parabola as

$$\mathbf{r}(t) = (t^2/4a)\, \mathbf{i} + t\, \mathbf{j}, \, -\infty < t < \infty. \tag{15.8}$$

Circle in a plane in 3-dimensions Consider the circle as $x^2 + y^2 = a^2$, $z = d$ which lies in the plane $z = d$. Then, we can write the parametric form as

$$\mathbf{r}(t) = a \cos t\, \mathbf{i} + a \sin t\, \mathbf{j} + d\, \mathbf{k}. \tag{15.9}$$

Helix The curve traced by the vector function

$$\mathbf{r}(t) = a \cos t\, \mathbf{i} + a \sin t\, \mathbf{j} + ct\, \mathbf{k}, \, a > 0 \tag{15.10}$$

is called a *circular helix*. The curve lies on the cylinder $x^2 + y^2 = a^2$. If $c > 0$, it is shaped in the right-handed direction and if $c < 0$, it is shaped in the left-handed direction (Fig. 15.3).

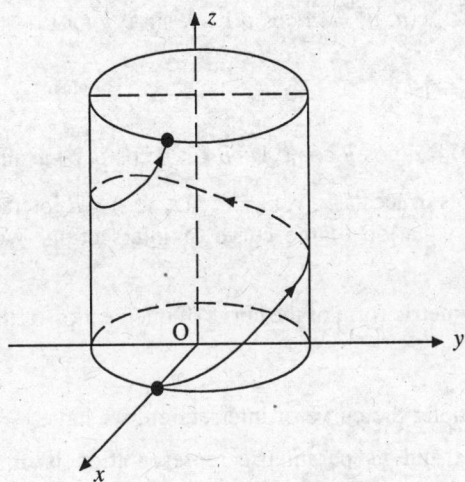

Fig. 15.3. Circular Helix.

The vector function

$$\mathbf{r}(t) = a \cos t\, \mathbf{i} + b \sin t\, \mathbf{j} + ct\, \mathbf{k}, \, a > 0, b > 0, c > 0 \tag{15.11}$$

describes an elliptical helix.

Parametric representation of surfaces

Parametric representation of surfaces can be done using two parameters. Let $f(x, y, z) = c$ or $g(x, y, z) = 0$ be the equation of a surface. Let an explicit representation of the surface be written as

$$z = h(x, y). \tag{15.12}$$

Then, if we set $x = u$, $y = v$, then the parametric form of the surface can be written as

$$\mathbf{r}(u, v) = u\,\mathbf{i} + v\,\mathbf{j} + h(u, v)\,\mathbf{k}. \tag{15.13}$$

Alternately, we can choose u, v as two independent parameters and write $x = x(u, v)$, $y = y(u, v)$, $z = z(u, v)$. Then the parametric representation of the surface can be written as

$$\mathbf{r}(u, v) = x(u, v)\,\mathbf{i} + y(u, v)\,\mathbf{j} + z(u, v)\,\mathbf{k}; \; (u, v) \in D. \tag{15.14}$$

The following are the parametric representations of some surfaces.

Cylinder $\quad x^2 + y^2 = a^2$.

$$\mathbf{r}(u, v) = a \cos u\,\mathbf{i} + a \sin u\,\mathbf{j} + v\mathbf{k}. \tag{15.15}$$

Sphere $\quad x^2 + y^2 + z^2 = a^2$

$$\mathbf{r}(u, v) = a \cos u \cos v\,\mathbf{i} + a \sin u \cos v\,\mathbf{j} + a \sin v\,\mathbf{k},\, 0 \le u \le 2\pi, \; -\pi/2 \le v \le \pi/2. \tag{15.16}$$

Paraboloid of revolution $\quad z = x^2 + y^2$.

$$\mathbf{r}(u, v) = u \cos v\,\mathbf{i} + u \sin v\mathbf{j} + u^2\mathbf{k}. \tag{15.17}$$

Cone of revolution $\quad z^2 = x^2 + y^2$.

$$\mathbf{r}(u, v) = u \cos v\,\mathbf{i} + u \sin v\,\mathbf{j} + u\mathbf{k}. \tag{15.18}$$

Ellipsoid $\quad \dfrac{x^2}{a^2} + \dfrac{y^2}{b^2} + \dfrac{z^2}{c^2} = 1$.

$$\mathbf{r}(u, v) = a \cos u \cos v\,\mathbf{i} + b \sin u \cos v\,\mathbf{j} + c \sin v\,\mathbf{k}. \tag{15.19}$$

Curve of intersection Two surfaces $f(x, y, z) = c$, $g(x, y, z) = d$ intersect along a curve. It is often possible to parametrise the equation of the curve of intersection. We illustrate the same in the following example.

Example 15.2 Find the parametric form of the curve of intersection of the plane $y = x$ and the surface $z = \sqrt{16 - x^2 - y^2}$.

Solution Let $x = t$. Then, along the curve of intersection, we have $y = t$ and $z = \sqrt{16 - 2t^2}$. The curve is in the plane $y = x$ and its parametric representation is $\mathbf{r}(t) = t\mathbf{i} + t\mathbf{j} + \sqrt{16 - 2t^2}\,\mathbf{k}$, $0 \le t \le 2\sqrt{2}$.

Limit, continuity and differentiability of vector functions

The concepts of limit, continuity and differentiability of calculus can easily be used for vector functions. Let the vector function be written in its parametric form, that is $\mathbf{v} = \mathbf{v}(t)$. In the cartesian system, we can write $\mathbf{v}(t) = v_1(t)\,\mathbf{i} + v_2(t)\mathbf{j} + v_3(t)\,\mathbf{k}$. Then, we define the following.

Limit The vector function $\mathbf{v}(t)$ has the limit l as $t \to a$, if $\mathbf{v}(t)$ is defined in some neighborhood of a, except possibly at $t = a$, and

$$\lim_{t \to a} |\,\mathbf{v}(t) - l\,| = 0. \tag{15.20}$$

We write $\lim\limits_{t\to a} \mathbf{v}(t) = l$. In the cartesian system, this implies that limits of the component functions $v_1(t)$, $v_2(t)$ and $v_3(t)$ exist as $t \to a$ and

$$\lim_{t\to a} v_1(t) = l_1, \ \lim_{t\to a} v_2(t) = l_2, \ \lim_{t\to a} v_3(t) = l_3$$

where $\quad l = l_1\mathbf{i} + l_2\mathbf{j} + l_3\mathbf{k}.$

Continuity A vector function $\mathbf{v}(t)$ is said to be continuous at $t = a$, if

(i) $\mathbf{v}(t)$ is defined in some neighborhood of a, (ii) $\lim\limits_{t\to a} \mathbf{v}(t)$ exists, and (iii) $\lim\limits_{t\to a} \mathbf{v}(t) = \mathbf{v}(a)$.

In cartesian system, this implies that $\mathbf{v}(t)$ is continuous at $t = a$, if and only if the component functions $v_1(t)$, $v_2(t)$ and $v_3(t)$ are continuous at $t = a$.

Differentiability A vector function $\mathbf{v}(t)$ is said to be differentiable at a point , if the limit

$$\lim_{\Delta t \to 0} \frac{\mathbf{v}(t + \Delta t) - \mathbf{v}(t)}{\Delta t}$$

exists. If the limit exists, then we write it as $\mathbf{v}'(t)$ or as $d\mathbf{v}/dt$. In the cartesian system, this implies that the component functions $v_1(t)$, $v_2(t)$ and $v_3(t)$ are differentiable at a point t, that is the limits

$$\lim_{\Delta t \to 0} \frac{v_i(t + \Delta t) - v_i(t)}{\Delta t}, \ i = 1, 2, 3 \ \text{exist.}$$

Therefore, $\qquad\qquad \mathbf{v}'(t) = v_1'(t)\,\mathbf{i} + v_2'(t)\,\mathbf{j} + v_3'(t)\,\mathbf{k}.$ \hfill (15.21)

Let $\mathbf{v}(t) = \mathbf{r}(t) = x(t)\mathbf{i} + y(t)\,\mathbf{j} + z(t)\,\mathbf{k}$ be the parametric representation of a curve C. Then

$$\frac{d\mathbf{r}}{dt} = \mathbf{r}'(t) = \frac{dx(t)}{dt}\mathbf{i} + \frac{dy(t)}{dt}\mathbf{j} + \frac{dz(t)}{dt}\mathbf{k}. \tag{15.22}$$

Geometric representation of $\mathbf{r}'(t)$ Let $\mathbf{r}'(t) \neq \mathbf{0}$. The vectors

$\Delta\mathbf{r} = \mathbf{r}(t + \Delta t) - \mathbf{r}(t)$ and $\dfrac{\Delta\mathbf{r}}{\Delta t} = \dfrac{\mathbf{r}(t + \Delta t) - \mathbf{r}(t)}{\Delta t}$

are parallel. Now, let $\Delta t \to 0$. Then $\mathbf{r}(t + \Delta t) \to \mathbf{r}(t)$. If limit $\Delta\mathbf{r}/\Delta t$ exists as $\Delta t \to 0$, then the limiting position of this vector, that is $\lim\limits_{\Delta t \to 0} (\Delta\mathbf{r}/\Delta t) = d\mathbf{r}/dt$ is the tangent line to the curve at the point P (Fig. 15.4). Therefore, $\mathbf{r}'(t)$ represents the tangent vector to the curve C.

The unit vector in the direction of the tangent is given by $\mathbf{r}'(t)/ |\mathbf{r}'(t)|$.

Smooth curve Let $\mathbf{r}(t)$ denote the position vector of a point P on the curve C. Let $\mathbf{r}(t)$ have continuous first derivative, that is the component functions $x(t)$, $y(t)$ and $z(t)$ have continuous first derivatives. Let $\mathbf{r}'(t) \neq \mathbf{0}$, for

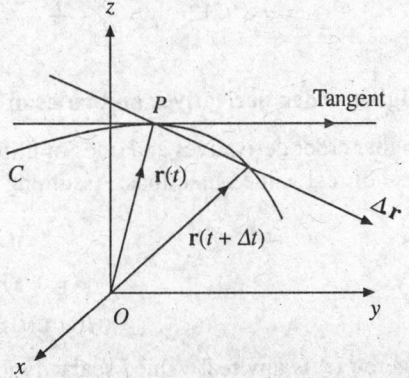

Fig. 15.4. Tangent vector.

all $t \in (a, b)$. Then, $\mathbf{r}(t)$ defines a smooth function on (a, b). The curve C traced by $\mathbf{r}(t)$ is called a *smooth curve*.

Example 15.3 Represent the parabola $y = 1 - 2x^2$, $-1 \le x \le 1$ in parametric form. Hence, find $\mathbf{r}'(0)$ and $\mathbf{r}'(1/\sqrt{2})$.

Solution Let $x = \sin t$. Then $y = 1 - 2\sin^2 t = \cos 2t$, and $-\pi/2 \le t \le \pi/2$. Hence,

$$\mathbf{r}(t) = \sin t\, \mathbf{i} + \cos 2t\, \mathbf{j}, \quad -\pi/2 \le t \le \pi/2.$$

For $x = 1/\sqrt{2}$, we get $t = \pi/4$. Therefore, $\mathbf{r}'(t) = \cos t\, \mathbf{i} - 2\sin 2t\, \mathbf{j}$, and $\mathbf{r}'(0) = \mathbf{i}$, $\mathbf{r}'(\pi/4) = (\mathbf{i}/\sqrt{2}) - 2\,\mathbf{j}$. The tangent line at $t = 0$ is parallel to the x-axis. (Note that $t = 0$ gives $x = 0, y = 1$ which is the vertex of the parabola.)

Example 15.4 Find the tangent vector to the curve whose parametric representation is $x = \cos t$, $y = \sin t$, $z = t$, $-\pi \le t \le \pi$. Hence, find the unit tangent vector.

Solution The position vector of a point on the curve is $\mathbf{r}(t) = \cos t\, \mathbf{i} + \sin t\, \mathbf{j} + t\, \mathbf{k}$. Therefore, the tangent vector is $\mathbf{r}'(t) = -\sin t\, \mathbf{i} + \cos t\, \mathbf{j} + \mathbf{k}$. We have $|\mathbf{r}'(t)| = \sqrt{\sin^2 t + \cos^2 t + 1} = \sqrt{2}$. Hence, the unit tangent vector is $\hat{\mathbf{r}}'(t) = (-\sin t\, \mathbf{i} + \cos t\, \mathbf{j} + \mathbf{k})/\sqrt{2}$.

Example 15.5 Find the tangent vector to the curve whose parametric representation is $x = t^3$, $y = (t + 1)/t$, $z = t^2 + 1$, at $t = 2$. Hence, find the parametric representation of the tangent vector.

Solution The position vector of a point on the given curve is

$$\mathbf{r}(t) = t^3 \mathbf{i} + \left(1 + \frac{1}{t}\right)\mathbf{j} + (t^2 + 1)\,\mathbf{k}, \quad t \ne 0.$$

Therefore, the tangent vector is

$$\mathbf{r}'(t) = (3t^2)\,\mathbf{i} - \frac{1}{t^2}\,\mathbf{j} + 2t\,\mathbf{k} \quad \text{and} \quad \mathbf{r}'(2) = 12\,\mathbf{i} - \frac{1}{4}\,\mathbf{j} + 4\,\mathbf{k}.$$

The position vector of the point at which $\mathbf{r}'(2)$ is the tangent is $\mathbf{r}(2) = 8\mathbf{i} + (3/2)\,\mathbf{j} + 5\,\mathbf{k}$. Therefore, we require the position vector of a point on the line passing through the point whose position vector is $\mathbf{r}(2)$ and has the direction of $\mathbf{r}'(2)$. Hence, parametric form of the line is given by (see Eq. (15.5))

$$x = 8 + 12t, \quad y = \frac{3}{2} - \frac{t}{4}, \quad z = 5 + 4t, \quad \text{or} \quad \mathbf{r}'(t) = \left(8, \frac{3}{2}, 5\right) + t\left(12, -\frac{1}{4}, 4\right).$$

Higher order derivatives and rules of differentiation

Higher order derivatives and rules of differentiation for vector functions have the same form as in the case of real valued functions. Assuming that the derivatives exist, we have the following results.

$$\mathbf{v}''(t) = v_1''(t)\,\mathbf{i} + v_2''(t)\,\mathbf{j} + v_3''(t)\,\mathbf{k} \tag{15.23}$$

$$(\mathbf{u} + \mathbf{v})' = \mathbf{u}' + \mathbf{v}' \tag{15.24}$$

$$(f(t)\,\mathbf{u}(t))' = f'(t)\,\mathbf{u}(t) + f(t)\,\mathbf{u}'(t), \tag{15.25}$$

where $f(t)$ is any real valued scalar function.

$$(\mathbf{u}(t) \cdot \mathbf{v}(t))' = \mathbf{u}(t) \cdot \mathbf{v}'(t) + \mathbf{u}'(t) \cdot \mathbf{v}(t) \tag{15.26}$$

$$(\mathbf{u}(t) \times \mathbf{v}(t))' = \mathbf{u}(t) \times \mathbf{v}'(t) + \mathbf{u}'(t) \times \mathbf{v}(t) \tag{15.27}$$

where • and × represent the dot and cross products, respectively. Note that the cross product of two vectors is not commutative.

Example 15.6 Find $\mathbf{v}'(t)$ in each of the following cases.

(i) $\mathbf{v}(t) = (\cos t + t^2)(t\mathbf{i} + \mathbf{j} + 2\mathbf{k})$, (ii) $v(t) = (3t\,\mathbf{i} + 5t^2\mathbf{j} + 6\mathbf{k}) \cdot (t^2\mathbf{i} - 2t\mathbf{j} + t\mathbf{k})$

(iii) $\mathbf{v}(t) = (t\mathbf{i} + e^t\mathbf{j} - t^2\mathbf{k}) \times (t^2\mathbf{i} + \mathbf{j} + t^3\mathbf{k})$

Solution

(i) Using Eq. (15.25), we obtain

$$\mathbf{v}'(t) = (\cos t + t^2)'(t\mathbf{i} + \mathbf{j} + 2\mathbf{k}) + (\cos t + t^2)(t\mathbf{i} + \mathbf{j} + 2\mathbf{k})'$$

$$= (-\sin t + 2t)(t\mathbf{i} + \mathbf{j} + 2\mathbf{k}) + (\cos t + t^2)(\mathbf{i})$$

$$= (3\,t^2 - t\sin t + \cos t)\,\mathbf{i} + (2t - \sin t)(\mathbf{j} + 2\mathbf{k})$$

(ii) Using Eq. (15.26) we obtain

$$v'(t) = (3t\,\mathbf{i} + 5t^2\mathbf{j} + 6\mathbf{k})' \cdot (t^2\mathbf{i} - 2t\mathbf{j} + t\mathbf{k}) + (3t\,\mathbf{i} + 5t^2\mathbf{j} + 6\mathbf{k}) \cdot (t^2\mathbf{i} - 2t\mathbf{j} + t\,\mathbf{k})'$$

$$= (3\mathbf{i} + 10t\mathbf{j}) \cdot (t^2\mathbf{i} - 2t\mathbf{j} + t\,\mathbf{k}) + (3t\,\mathbf{i} + 5t^2\mathbf{j} + 6\mathbf{k}) \cdot (2t\mathbf{i} - 2\mathbf{j} + \mathbf{k})$$

$$= 3t^2 - 20t^2 + 6t^2 - 10t^2 + 6 = 6 - 21t^2.$$

(iii) $\mathbf{v}'(t) = (t\,\mathbf{i} + e^t\mathbf{j} - t^2\mathbf{k}) \times (t^2\mathbf{i} + \mathbf{j} + t^3\mathbf{k})' + (t\mathbf{i} + e^t\mathbf{j} - t^2\mathbf{k})' \times (t^2\mathbf{i} + \mathbf{j} + t^3\mathbf{k})$

$$= (t\,\mathbf{i} + e^t\mathbf{j} - t^2\mathbf{k}) \times (2t\mathbf{i} + 3t^2\mathbf{k}) + (\mathbf{i} + e^t\,\mathbf{j} - 2t\mathbf{k}) \times (t^2\mathbf{i} + \mathbf{j} + t^3\mathbf{k})$$

$$= [3t^2e^t\mathbf{i} - 5t^3\mathbf{j} - 2t\,e^t\mathbf{k}] + [(t^3e^t + 2t)\mathbf{i} - 3t^3\mathbf{j} + (1 - t^2e^t)\mathbf{k}]$$

$$= [t^2e^t(3 + t) + 2t]\,\mathbf{i} - 8t^3\mathbf{j} + [1 - te^t(2 + t)]\,\mathbf{k}$$

Example 15.7 Prove that $[\mathbf{v}(t) \times \mathbf{v}'(t)]' = \mathbf{v}(t) \times \mathbf{v}''(t)$.

Solution Using Eq. (15.27), we obtain

$$[\mathbf{v}(t) \times \mathbf{v}'(t)]' = \mathbf{v}(t) \times [\mathbf{v}'(t)]' + \mathbf{v}'(t) \times \mathbf{v}'(t) = \mathbf{v}(t) \times \mathbf{v}''(t)$$

since the second term on the right hand side is a null vector.

Length of a space curve Let the curve C be represented in the parametric form as $\mathbf{r} = \mathbf{r}(t)$, $a \le t \le b$. In cartesian system, we have $\mathbf{r}(t) = x(t)\,\mathbf{i} + y(t)\,\mathbf{j} + z(t)\,\mathbf{k}$. Then, the length of the curve is given by

$$\text{length} = l = \int_a^b [(x'(t))^2 + (y'(t))^2 + (z'(t))^2]^{1/2}\,dt = \int_a^b [\mathbf{r}'(t) \cdot \mathbf{r}'(t)]^{1/2}\,dt. \quad (15.28)$$

Note that the integrand is the norm of $\mathbf{r}'(t)$, that is

$$\|\mathbf{r}'(t)\| = [(x'(t))^2 + (y'(t))^2 + (z'(t))^2]^{1/2}$$

Then, we can write

$$l = \int_a^b \| \mathbf{r}'(t) \| \, dt \tag{15.29}$$

The notation $| \mathbf{r}'(t) |$ may also be used instead of $\| \mathbf{r}'(t) \|$.

Now, define the real valued function $s(t)$ as

$$s(t) = \int_a^t [(x'(\xi))^2 + (y'(\xi))^2 + (z'(\xi))^2]^{1/2} \, d\xi = \int_a^t \| \mathbf{r}'(\xi) \| \, d\xi \tag{15.30}$$

Then, $s(t)$ is the arc length of the curve from its initial point $(x(a), y(a), z(a))$ to an arbitrary point $(x(t), y(t), z(t))$ on the curve C. Therefore, $s(t)$ is the length function. Using Eq. (15.30), it is possible to solve for t as a function of s, that is $t = t(s)$. Then, the curve C can be parametrised in terms of the arc length s as

$$\mathbf{r}(s) = \mathbf{r}(t(s)) = x(t(s)) \, \mathbf{i} + y(t(s)) \, \mathbf{j} + z(t(s)) \, \mathbf{k} \tag{15.31}$$

Remark 1

If the position vector $\mathbf{r}(t)$ of a point on a curve C is expressed in terms of the arc length s, that is $\mathbf{r} = \mathbf{r}(s)$, then $\mathbf{r}'(s)$ is a unit vector.

Example 15.8 Find the length of the Helix traced by

$$\mathbf{r}(t) = a \cos t \, \mathbf{i} + a \sin t \, \mathbf{j} + ct \, \mathbf{k}, \ \ a > 0, \ \ 0 \le t \le 2\pi.$$

Solution We have $x(t) = a \cos t$, $y(t) = a \sin t$, $z(t) = ct$ and $x'(t) = -a \sin t$, $y'(t) = a \cos t$, $z'(t) = c$. Therefore,

$$s = \text{arc length} = \int_0^{2\pi} [a^2 \sin^2 t + a^2 \cos^2 t + c^2]^{1/2} \, dt = (2\pi)(a^2 + c^2)^{1/2}.$$

Example 15.9 In Example 15.8, express the position vector $\mathbf{r}(t)$ in terms of the arc length s.

Solution We have

$$s = \int_0^t (a^2 + c^2)^{1/2} \, dt = t(a^2 + c^2)^{1/2} \ \text{ or } \ t = s/(a^2 + c^2)^{1/2}.$$

Therefore,

$$\mathbf{r}(s) = a \cos(s^*) \, \mathbf{i} + a \sin(s^*) \, \mathbf{j} + c \, s^* \, \mathbf{k}, \ \ s^* = s/(a^2 + c^2)^{1/2}.$$

It can be verified that $| \mathbf{r}'(s) | = 1$.

15.2.1 Motion of a Body or Particle on a Curve

Suppose that a body or a particle moves along a curve C. Then, the position vector of the particle or the body at time t is given by $\mathbf{r}(t) = x(t) \, \mathbf{i} + y(t) \, \mathbf{j} + z(t) \, \mathbf{k}$. Assume that $x(t)$, $y(t)$ and $z(t)$ are twice differentiable. Let t vary over the interval $[a, b]$. Then, at time t the particle has travelled a distance $s(t)$ along the curve from the initial point $(x(a), y(a), z(a))$ to the point $(x(t), y(t), z(t))$. This distance is given by (see Eq. (15.30))

$$s(t) = \int_a^t [(x'(\xi))^2 + (y'(\xi))^2 + (z'(\xi))^2]^{1/2} \, d\xi. \tag{15.32}$$

Then, the vectors

$$v(t) = r'(t) = x'(t)\,i + y'(t)\,j + z'(t)\,k \tag{15.33}$$

$$a(t) = r''(t) = x''(t)\,i + y''(t)\,j + z''(t)\,k \tag{15.34}$$

are called the *velocity* and *acceleration* of the particle respectively. The scalar quantity $|\,v(t)\,|$ is called the *speed* of the particle, that is

$$|\,v(t)\,| = [(x'(t))^2 + (y'(t))^2 + (z'(t))^2]^{1/2} \tag{15.35}$$

gives the speed of the particle. If $v(t) \neq 0$, then the velocity vector is tangential to the curve of motion of the particle. Therefore, we can interpret that at any given instant of time, the particle is moving in the direction of the tangent to the curve. Comparing Eqs. (15.32) and (15.35), we find that speed is related to the arc length where $s'(t) = |\,v(t)\,|$. Therefore, we may also write

$$s(t) = \int_a^t |\,v\,(\xi)\,|\,d\xi. \tag{15.36}$$

Example 15.10 The position vector of a moving particle is given by $r(t) = t^3\,i + t\,j + t^2\,k$. Determine the velocity, speed and acceleration of the particle in the direction of the motion.

Solution We have $r'(t) = 3t^2\,i + j + 2t\,k$ and $r''(t) = 6t\,i + 2k$. Hence,

velocity $= v(t) = r'(t) = 3t^2\,i + j + 2t\,k$, speed $= |\,v(t)\,| = (9\,t^4 + 4t^2 + 1)^{1/2}$,

acceleration $= a(t) = r''(t) = 6t\,i + 2\,k$.

Exercise 15.1

Find the level surfaces of the scalar fields defined by the following functions.

1. $f = x + y + z$.   2. $f = x^2 + y^2 + z^2$.
3. $f = x^2 + y^2 - z$.   4. $f = x^2 + 9y^2 + 16z^2$.

Find the parametric representation of the straight line through the point P and has the direction b in the following problems.

5. $P(1, 2, 3)$, $b = i + 2j + 2k$.   6. $P(1, -1, 1)$, $b = i - j$.

Find the parametric representations of the following straight lines/curves. Use the indicated representation whereever given.

7. $x = y$, $y = z$.   8. $x + y + z = 3$, $y - z = 0$.
9. $x - y + z = 5$, $x - 2y + 3z = 3$.   10. $2x + y + 2z = 2$, $y + z = 0$.
11. $x = y^2 + z^2$, $y = z$.   12. $y^2 + z^2 = 9$, $x = 9 - y^2$, $y = 3\sin t$.
13. $y = x^2 + z^2$, $y = 4$, $x = 2\sin t$.
14. $(y - a)^2 + (z - a)^2 = 2a^2$, $x = 0$, $y - a = \sqrt{2}\,a\sin t$.

In the following problems, find the indicated derivative using the differentiation rules. Assume that all the given vector functions are differentiable.

15. $u(t) = 5t^2\,i + t\,j + t^3\,k$, $f(t) = \sin t$, $[f(t)\,u(t)]'$.
16. $u(t) = [\sin(2t)\,i - \cos(2t)\,j + t\,k]$, $v(t) = [\cos(2t)\,i - \sin(2t)\,j + t^2\,k]$, $[u(t) \cdot v(t)]'$.

17. $\mathbf{u}(t) = 6t^2\mathbf{i} - t\mathbf{j} + 3t^2\mathbf{k}$, $\mathbf{v}(t) = t\,\mathbf{i} + t^2\mathbf{j} + 2t\,\mathbf{k}$, $[\mathbf{u}(t) \times \mathbf{v}(t)]'$.

18. $\mathbf{u}(t) = (1 - t)\,\mathbf{i} + t^2\mathbf{j} + e^t\,\mathbf{k}$, $\mathbf{v}(t) = (1 + t)\,\mathbf{i} + e^t\,\mathbf{j} + t\,\mathbf{k}$, $[\mathbf{u}(t) \times \mathbf{v}(t)]'$.

19. $[t^2\mathbf{u}(t^2)]'$.   20. $[\mathbf{u}(at) + \mathbf{v}(a/t)]'$.

21. $[\mathbf{u}(t) \times \mathbf{u}''(t)]'$.   22. $[\mathbf{u}(t) \cdot (\mathbf{u}'(t) \times \mathbf{u}''(t))]'$.

In the following problems, find the parametric equation of the tangent line to the given curve at the indicated point.

23. $x = t$, $y = 2t^2$, $z = 3t^3$; $t = 2$.   24. $x = \sin t$, $y = \cos t$, $z = t$, $t = \pi/4$.

25. $x = t^2 - 1$, $y = t + 1$, $z = t/(t + 1)$, $t = 2$.   26. $x = t$, $y = e^t$, $z = 1$, $t = 1$.

Find the lengths of the following curves. In problems **29** to **31**, express $\mathbf{r}(t)$ as a function of the arc length.

27. $\mathbf{r}(t) = a \cos^3 t\,\mathbf{i} + a \sin^3 t\,\mathbf{j}$, $0 \le t \le \pi/2$ (one cusp of hypocycloid).

28. $\mathbf{r}(t) = t \sin t\,\mathbf{i} + t \cos t\,\mathbf{j} + \sqrt{3}\,t\,\mathbf{k}$, $0 \le t \le \pi$.

29. $\mathbf{r}(t) = a \cos t\,\mathbf{i} + a \sin t\,\mathbf{j}$, $0 \le t \le 2\pi$.

30. $\mathbf{r}(t) = \cos t\,\mathbf{i} + \sin t\,\mathbf{j} + 3t\,\mathbf{k}$, $-2\pi \le t \le 2\pi$.

31. $\mathbf{r}(t) = (t^2/2)\,\mathbf{i} + (t^3/3)\,\mathbf{k}$, $0 \le t \le 2$.   32. $\mathbf{r}(t) = t\,\mathbf{i} + (t^2/2)\,\mathbf{j}$, $0 \le t \le 1$.

33. Let $\mathbf{T}$ denote the unit tangent to the curve $x = 2t$, $y = 3t + 4$, $z = 3t$. Show that $d\mathbf{T}/dt = \mathbf{0}$. Interpret the result.

34. The position vector of a moving particle is $\mathbf{r}(t) = (\cos t + \sin t)\,\mathbf{i} + (\sin t - \cos t)\,\mathbf{j} + t\,\mathbf{k}$. Determine the velocity, speed and acceleration of the particle in the direction of the motion.

35. If a particle moves with constant speed c, then show that its acceleration vector $\mathbf{a}(t)$ is perpendicular to the velocity vector $\mathbf{v}(t)$.

15.3   Gradient of a Scalar Field and Directional Derivative

Let $f(x, y, z)$ be a real valued function defining a scalar field. To define the gradient of a scalar field, we first introduce a vector operator called *del* operator denoted by ∇. We define the vector differential operator in two and three dimensions as

$$\nabla = \mathbf{i}\frac{\partial}{\partial x} + \mathbf{j}\frac{\partial}{\partial y} \quad \text{and} \quad \nabla = \mathbf{i}\frac{\partial}{\partial x} + \mathbf{j}\frac{\partial}{\partial y} + \mathbf{k}\frac{\partial}{\partial z}.$$

The *gradient* of a scalar field $f(x, y, z)$, denoted by ∇f or grad (f) is defined as

$$\nabla f = \mathbf{i}\frac{\partial f}{\partial x} + \mathbf{j}\frac{\partial f}{\partial y} + \mathbf{k}\frac{\partial f}{\partial z}. \tag{15.37}$$

Note that the *del* operator ∇ operates on a scalar field and produces a vector field.

Example 15.11   Find the gradient of the following scalar fields

(i) $f(x, y) = y^2 - 4xy$ at $(1, 2)$,   (ii) $x^2y^2 + xy^2 - z^2$ at $(3, 1, 1)$.

Solution

(i) We have $\nabla f(x, y) = \left(\mathbf{i}\dfrac{\partial}{\partial x} + \mathbf{j}\dfrac{\partial}{\partial y} \right)(y^2 - 4xy)$

$$= \mathbf{i}\frac{\partial}{\partial x}(y^2 - 4xy) + \mathbf{j}\frac{\partial}{\partial y}(y^2 - 4xy) = -4y\,\mathbf{i} + (2y - 4x)\,\mathbf{j}$$

At (1, 2), we obtain $\nabla f(x, y) = -8\,\mathbf{i}$.

(ii) We have $\nabla f(x, y, z) = \left(\mathbf{i}\dfrac{\partial}{\partial x} + \mathbf{j}\dfrac{\partial}{\partial y} + \mathbf{k}\dfrac{\partial}{\partial z} \right)(x^2 y^2 + xy^2 - z^2)$

$$= (2xy^2 + y^2)\mathbf{i} + (2x^2 y + 2xy)\mathbf{j} - 2z\mathbf{k}.$$

At (3, 1, 1) we obtain $\nabla f(x, y) = 7\mathbf{i} + 24\mathbf{j} - 2\mathbf{k}$.

Example 15.12   If $\mathbf{r} = x\mathbf{i} + y\mathbf{j} + z\mathbf{k}$, $|\mathbf{r}| = r$ and $\hat{\mathbf{r}} = \mathbf{r}/r$, then show that $\mathrm{grad}\,(1/r) = -\hat{\mathbf{r}}/r^2$.

Solution We have $r^2 = x^2 + y^2 + z^2$. Therefore,

$$\mathrm{grad}\left(\frac{1}{r}\right) = \left(\mathbf{i}\frac{\partial}{\partial x} + \mathbf{j}\frac{\partial}{\partial y} + \mathbf{k}\frac{\partial}{\partial z} \right)\left(\frac{1}{r}\right) = \mathbf{i}\left(-\frac{1}{r^2}\frac{\partial r}{\partial x} \right) + \mathbf{j}\left(-\frac{1}{r^2}\frac{\partial r}{\partial y} \right) + \mathbf{k}\left(-\frac{1}{r^2}\frac{\partial r}{\partial z} \right)$$

$$= -\frac{1}{r^2}\left(\frac{x}{r}\mathbf{i} + \frac{y}{r}\mathbf{j} + \frac{z}{r}\mathbf{k} \right) = -\frac{1}{r^2}\left(\frac{\mathbf{r}}{r} \right) = -\frac{\hat{\mathbf{r}}}{r^2}$$

where $\hat{\mathbf{r}} = (x\mathbf{i} + y\mathbf{j} + z\mathbf{k})/r$ is the unit vector.

Geometrical representation of the gradient   Let $f(P) = f(x, y, z)$ be a differentiable scalar field. Let $f(x, y, z) = k$ be a level surface and $P_0(x_0, y_0, z_0)$ be a point on it. There are infinite number of smooth curves on the surface passing through the point P_0. Each of these curves has a tangent at P_0. The totality of all these tangent lines form a tangent plane to the surface at the point P_0. A vector normal to this plane at P_0 is called the *normal vector* to the surface at this point.

Consider now a smooth curve C on the surface passing through a point P on the surface. Let $x = x(t)$, $y = y(t)$, $z = z(t)$ be the parametric representation of the curve C. Any point P on C has the position vector $\mathbf{r} = x(t)\mathbf{i} + y(t)\mathbf{j} + z(t)\mathbf{k}$. Since the curve lies on the surface, we have

$$f(x(t), y(t), z(t)) = k.$$

Then, $\quad \dfrac{d}{dt}f(x(t), y(t), z(t)) = 0$

By chain rule, we have $\quad \dfrac{\partial f}{\partial x}\dfrac{dx}{dt} + \dfrac{\partial f}{\partial y}\dfrac{dy}{dt} + \dfrac{\partial f}{\partial z}\dfrac{dz}{dt} = 0$

or $\quad \left(\mathbf{i}\dfrac{\partial f}{\partial x} + \mathbf{j}\dfrac{\partial f}{\partial y} + \mathbf{k}\dfrac{\partial f}{\partial z} \right) \cdot \left(\mathbf{i}\dfrac{dx}{dt} + \mathbf{j}\dfrac{dy}{dt} + \mathbf{k}\dfrac{dz}{dt} \right) = 0$

or $\quad \nabla f \cdot \mathbf{r}'(t) = 0.$ \hfill (15.38)

Let $\nabla f(P) \neq \mathbf{0}$ and $\mathbf{r}'(t) \neq \mathbf{0}$. Now, $\mathbf{r}'(t)$ is a tangent vector to C at the point P and lies in the tangent plane to the surface at P. Hence, $\nabla f(P)$ is orthogonal to every tangent vector at P. Therefore, $\nabla f(P)$ is the vector normal to the surface $f(x, y, z) = k$ at the point P (Figs. 15.5 and 15.6).

Remark 2

In two variables, let $f(x, y) = k$ be a level curve C of a differentiable scalar field $f(x, y)$. Let $P_0(x_0, y_0)$ be a point on it. If this level curve is parametrised as $x = x(t)$, $y = y(t)$ then $x_0 = x(t_0)$ and $y_0 = y(t_0)$. Then, $\nabla f(x_0, y_0)$ is the normal vector to the curve C at the point P_0 (Fig. 15.7).

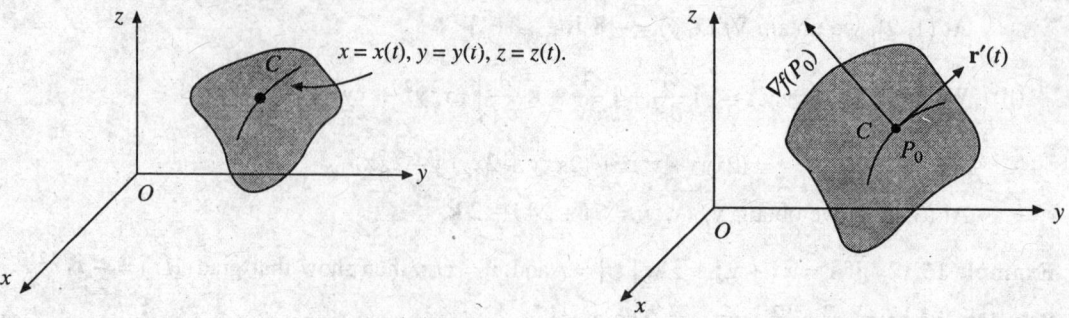

Fig. 15.5 **A smooth curve on the surface.**   **Fig. 15.6** **Normal vector to a surface.**

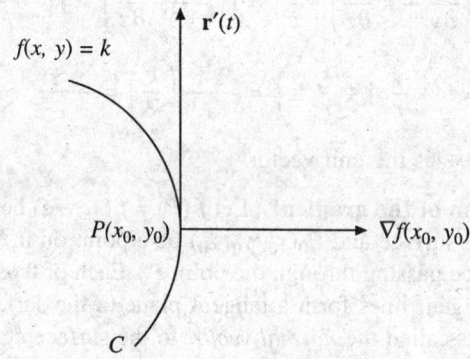

Fig. 15.7 **Normal vector to a curve C**

Remark 3 The unit normal vector is $\hat{\mathbf{n}} = \text{grad } f/|\text{grad } f|$.

Remark 4

(a) The gradient vector (normal vector) at a point $P_0 (x_0, y_0, z_0)$ on the surface $f(x, y, z) = k$ can be used to derive the equation of the tangent plane at the point P_0, to the surface. Let (x, y, z) be any point on the tangent plane. Then, $(x - x_0)\mathbf{i} + (y - y_0)\mathbf{j} + (z - z_0)\mathbf{k}$ is a vector in the tangent plane. Hence, this vector is orthogonal to the gradient vector at P_0. Therefore,

$$\nabla f(P_0) \cdot [(x - x_0)\mathbf{i} + (y - y_0)\mathbf{j} + (z - z_0)\mathbf{k}] = 0$$

or $\qquad (x - x_0)\, \dfrac{\partial f}{\partial x}\, (P_0) + (y - y_0)\, \dfrac{\partial f}{\partial y}\, (P_0) + (z - z_0)\, \dfrac{\partial f}{\partial z}\, (P_0) = 0 \qquad (15.39)$

is the equation of the tangent plane at P_0.

(b) (i) Angle between two curves is the angle between their tangents at the common point.

(ii) Angle between two surfaces is the angle between their normals at the common point.

Example 15.13 Find a unit normal vector to the surface $xy^2 + 2yz = 8$ at the point $(3, -2, 1)$.

Solution Let $f(x, y, z) = xy^2 + 2yz = 8$. Then,

$$\frac{\partial f}{\partial x} = y^2, \quad \frac{\partial f}{\partial y} = 2xy + 2z \quad \text{and} \quad \frac{\partial f}{\partial z} = 2y.$$

Therefore,
$$\nabla f = \mathbf{i}\frac{\partial f}{\partial x} + \mathbf{j}\frac{\partial f}{\partial y} + \mathbf{k}\frac{\partial f}{\partial z} = y^2\mathbf{i} + 2(xy + z)\mathbf{j} + 2y\mathbf{k}.$$

At $(3, -2, 1)$, we obtain the normal vector as $\nabla f(3, -2, 1) = 4\mathbf{i} - 10\mathbf{j} - 4\mathbf{k}$. The unit normal vector at $(3, -2, 1)$ is given by

$$\frac{(4\mathbf{i} - 10\mathbf{j} - 4\mathbf{k})}{\sqrt{16 + 100 + 16}} = \frac{2\mathbf{i} - 5\mathbf{j} - 2\mathbf{k}}{\sqrt{33}}.$$

Example 15.14   Find the normal vector and the equation of the tangent plane to the surface $z = \sqrt{x^2 + y^2}$ at the point $(3, 4, 5)$.

Solution   Let $f(x, y, z) = z - \sqrt{x^2 + y^2} = 0$ be the surface. Then, the normal vector is given by

$$\nabla f = \left(\mathbf{i}\frac{\partial}{\partial x} + \mathbf{j}\frac{\partial}{\partial y} + \mathbf{k}\frac{\partial}{\partial z}\right)(z - \sqrt{x^2 + y^2})$$

$$= -\frac{x}{\sqrt{x^2 + y^2}}\mathbf{i} - \frac{y}{\sqrt{x^2 + y^2}}\mathbf{j} + \mathbf{k} = -\frac{x}{z}\mathbf{i} - \frac{y}{z}\mathbf{j} + \mathbf{k}, \quad (z \neq 0).$$

At $(3, 4, 5)$, the normal vector is given by

$$\nabla f(3, 4, 5) = -\frac{3}{5}\mathbf{i} - \frac{4}{5}\mathbf{j} + \mathbf{k}$$

The tangent plane at the point $(3, 4, 5)$ is given by

$$-\frac{3}{5}(x - 3) - \frac{4}{5}(y - 4) + (z - 5) = 0 \quad \text{or} \quad 3x + 4y - 5z = 0.$$

Example 15.15   Find the angle between the surfaces $x \log z = y^2 - 1$ and $x^2 y = 2 - z$ at the point $(1, 1, 1)$.

Solution   The angle between two surfaces at a common point is the angle between their normals at that point. We have

$$f_1(x, y, z) = x \log z - y^2 + 1 = 0, \quad \nabla f_1(x, y, z) = (\log z)\mathbf{i} - 2y\mathbf{j} + (x/z)\mathbf{k}$$

$$\nabla f_1(1, 1, 1) = -2\mathbf{j} + \mathbf{k} = \mathbf{n}_1$$

$$f_2(x, y, z) = x^2 y + z - 2 = 0, \quad \nabla f_2(x, y, z) = 2xy\mathbf{i} + x^2\mathbf{j} + \mathbf{k}$$

$$\nabla f_2(1, 1, 1) = 2\mathbf{i} + \mathbf{j} + \mathbf{k} = \mathbf{n}_2$$

Therefore,
$$\cos\theta = \left|\frac{\mathbf{n}_1 \cdot \mathbf{n}_2}{|\mathbf{n}_1||\mathbf{n}_2|}\right| = \frac{1}{\sqrt{5}\sqrt{6}} = \frac{1}{\sqrt{30}}, \quad \text{or} \quad \theta = \cos^{-1}\left(\frac{1}{\sqrt{30}}\right).$$

Properties of Gradient   Let f and g be any two differentiable scalar fields. Then,

$$\nabla(f + g) = \nabla f + \nabla g \tag{15.40}$$

$$\nabla(c_1 f + c_2 g) = c_1\nabla f + c_2\nabla g, \quad c_1, c_2 \text{ arbitrary constants} \tag{15.41}$$

$$\nabla(f g) = f\nabla g + g\nabla f \tag{15.42}$$

$$\nabla\left(\frac{f}{g}\right) = \frac{g\,\nabla f - f\,\nabla g}{g^2}, \, g \neq 0 \tag{15.43}$$

Directional derivative

Let $f(P) = f(x, y, z)$ be a differentiable scalar field. Then, $\partial f/\partial x$, $\partial f/\partial y$, $\partial f/\partial z$ denote the rates of change of f in the directions of x, y and z axis, respectively. If $f(x, y, z) = k$ is a level surface and P_0 is any point on it, then $\partial f/\partial x$, $\partial f/\partial y$, $\partial f/\partial z$ at $P_0 (x_0, y_0, z_0)$ denote the slopes of the tangent lines in the directions of $\mathbf{i}$, $\mathbf{j}$ and $\mathbf{k}$ respectively. It is natural to give the definition of derivative in any direction which we shall call as the directional derivative.

Let $\hat{\mathbf{b}} = b_1\mathbf{i} + b_2\mathbf{j} + b_3\mathbf{k}$ be any unit vector. Let P_0 be any point P_0: $\mathbf{a} = a_1\mathbf{i} + a_2\mathbf{j} + a_3\mathbf{k}$.

Then, the position vector of any point Q on the line passing through P_0 and in the direction of $\hat{\mathbf{b}}$ is given by

$$\mathbf{r} = \mathbf{a} + t\hat{\mathbf{b}} = (a_1 + b_1 t)\mathbf{i} + (a_2 + b_2 t)\mathbf{j} + (a_3 + b_3 t)\mathbf{k} = x(t)\mathbf{i} + y(t)\mathbf{j} + z(t)\mathbf{k}.$$

That is, the point $Q\,(a_1 + b_1 t, a_2 + b_2 t, a_3 + b_3 t)$ is on this line. Now, the vector from the point P_0 to Q is given by $t\hat{\mathbf{b}}$. Since $|\hat{\mathbf{b}}| = 1$, the distance from P_0 to Q is t. Then,

$$\frac{\partial f}{\partial t} = \lim_{t \to 0} \frac{f(Q) - f(P_0)}{t}$$

if it exists, is called the *directional derivative* of f at the point P_0 in the direction of $\hat{\mathbf{b}}$ (Fig. 15.8).

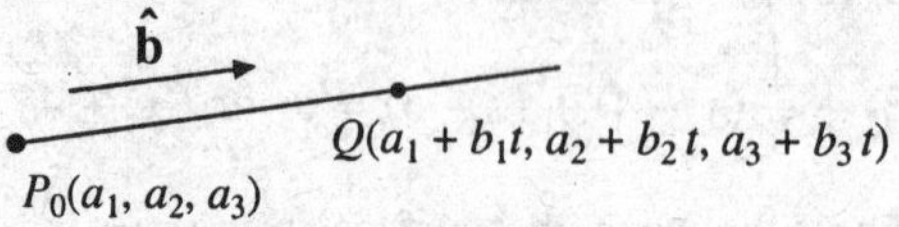

$$\hat{\mathbf{b}} \rightarrow$$

$$Q(a_1 + b_1 t, a_2 + b_2 t, a_3 + b_3 t)$$

$$P_0(a_1, a_2, a_3)$$

Fig. 15.8.   Directional derivative.

Therefore, $\dfrac{\partial}{\partial t} f(x(t), y(t), z(t))$ is the rate of change of f with respect to the distance t.

We have (by chain rule)

$$\frac{\partial f}{\partial t} = \frac{\partial f}{\partial x}\frac{dx}{dt} + \frac{\partial f}{\partial y}\frac{dy}{dt} + \frac{\partial f}{\partial z}\frac{dz}{dt} \tag{15.44}$$

where dx/dt, dy/dt and dz/dt are evaluated at $t = 0$, that is, at the point P_0. We write Eq. (15.44) as

$$\frac{\partial f}{\partial t} = \left(\mathbf{i}\frac{\partial f}{\partial x} + \mathbf{j}\frac{\partial f}{\partial y} + \mathbf{k}\frac{\partial f}{\partial z}\right) \cdot \left(\mathbf{i}\frac{dx}{dt} + \mathbf{j}\frac{dy}{dt} + \mathbf{k}\frac{dz}{dt}\right) = \nabla f \cdot \frac{d\mathbf{r}}{dt}.$$

But $d\mathbf{r}/dt = \hat{\mathbf{b}}$ (a unit vector). Therefore, the directional derivative of f in the direction of $\hat{\mathbf{b}}$ is given by

$$\textit{directional derivative} = \nabla f \cdot \hat{\mathbf{b}} = \text{grad}\,(f) \cdot \hat{\mathbf{b}} \tag{15.45}$$

which is denoted by $D_{\mathbf{b}}(f)$. Note that $\hat{\mathbf{b}}$ is a unit vector. If the direction is specified by a vector $\mathbf{u}$, then we have $\hat{\mathbf{b}} = \mathbf{u}/|\mathbf{u}|$. Some authors use s in place of t to denote length in the above derivation.

Remark 5

The length and direction of grad f is independent of the choice of the coordinate system.

Example 15.16   Find the directional derivative of $f(x, y, z) = xy^2 + 4xyz + z^2$ at the point $(1, 2, 3)$ in the direction of $3\mathbf{i} + 4\mathbf{j} - 5\mathbf{k}$.

Solution   We have

$$\nabla f = (y^2 + 4yz)\,\mathbf{i} + (2xy + 4xz)\,\mathbf{j} + (4xy + 2z)\,\mathbf{k}.$$

At the point $(1, 2, 3)$, we have $\nabla f = 28\,\mathbf{i} + 16\,\mathbf{j} + 14\,\mathbf{k}$. The unit vector in the given direction is

$$\hat{\mathbf{b}} = (3\,\mathbf{i} + 4\,\mathbf{j} - 5\,\mathbf{k})/5\sqrt{2}.$$

Therefore, $D_{\mathbf{b}}(1, 2, 3) = \nabla f \cdot \mathbf{b} = \dfrac{1}{5\sqrt{2}}(28\,\mathbf{i} + 16\,\mathbf{j} + 14\,\mathbf{k}) \cdot (3\,\mathbf{i} + 4\,\mathbf{j} - 5\,\mathbf{k}) = \dfrac{78}{5\sqrt{2}}.$

Example 15.17   Find the directional derivative of $f(x, y) = x^2y^3 + xy$ at $(2, 1)$, in the direction of a unit vector which makes an angle of $\pi/3$ with x-axis.

Solution   We have

$$\nabla f = (2xy^3 + y)\,\mathbf{i} + (3x^2y^2 + x)\,\mathbf{j}, \quad \text{and} \quad \text{at } (2, 1), \quad \nabla f = 5\mathbf{i} + 14\mathbf{j}.$$

The unit vector is given by $\hat{\mathbf{b}} = \cos\theta\,\mathbf{i} + \sin\theta\,\mathbf{j} = \dfrac{1}{2}\,\mathbf{i} + \dfrac{\sqrt{3}}{2}\,\mathbf{j}$, since $\theta = \pi/3$. Therefore,

$$\text{directional derivative} = (5\,\mathbf{i} + 14\,\mathbf{j}) \cdot \left(\dfrac{1}{2}\,\mathbf{i} + \dfrac{\sqrt{3}}{2}\,\mathbf{j}\right) = \dfrac{5 + 14\sqrt{3}}{2}.$$

Maximum rate of change of a scalar field

Using the definition of scalar product, we have from Eq. (15.45)

$$D_{\mathbf{b}}f = \nabla f \cdot \hat{\mathbf{b}} = |\nabla f|\,|\hat{\mathbf{b}}|\cos\theta = |\nabla f|\cos\theta$$

since $|\hat{\mathbf{b}}| = 1$ and θ is the angle between the vectors ∇f and $\hat{\mathbf{b}}$. Since $-1 \leq \cos\theta \leq 1$, we have $-|\nabla f| \leq D_{\mathbf{b}}f \leq |\nabla f|$. Therefore, *the maximum value of the directional derivative is* $|\nabla f|$ *and it occurs when* $\theta = 0$, *that is,* $\hat{\mathbf{b}}$ *has the direction of* ∇f. *This direction is the direction of the normal vector. The minimum value of the directional derivative is* $-|\nabla f|$ *and it occurs when* $\theta = \pi$, *that is,* $\hat{\mathbf{b}}$ *and* ∇f *have opposite directions.* We may also say that the gradient vector ∇f points in the direction in which f increases most rapidly and $-\nabla f$ points in the direction in which f decreases most rapidly. Sometimes, the maximum rate of change is denoted by $\partial f/\partial n$, where $\mathbf{n}$ is the unit normal.

Conservative vector field

A vector field $\mathbf{v}$ is said to be conservative if the vector function can be written as the gradient of a scalar function f, that is, $\mathbf{v} = \nabla f$. In such a vector field, the work done in moving a particle from a point P to a point Q depends only on the points P and Q, and is independent of path along which the particle is displaced from P to Q. It may be noted that not every vector field is conservative (we shall have more discussion on conservative vector fields in section 15.5).

Example 15.18 Show that the vector field defined by the vector function $\mathbf{v} = xyz(yz\mathbf{i} + xz\mathbf{j} + xy\mathbf{k})$ is conservative.

Solution If the given vector field is conservative, then it can be expressed as the gradient of a scalar function $f(x, y, z)$. Therefore,

$$\nabla f = \left(\mathbf{i} \frac{\partial f}{\partial x} + \mathbf{j} \frac{\partial f}{\partial y} + \mathbf{k} \frac{\partial f}{\partial z} \right) = \mathbf{v} = xyz \, (yz\mathbf{i} + xz\mathbf{j} + xy\mathbf{k}).$$

Comparing, we obtain

$$\frac{\partial f}{\partial x} = xy^2z^2, \quad \frac{\partial f}{\partial y} = x^2yz^2, \quad \frac{\partial f}{\partial z} = x^2y^2z.$$

Integrating the first equation, we obtain $f(x, y, z) = \frac{1}{2} x^2y^2z^2 + g(y, z)$. Substituting in the second equation, we get

$$\frac{\partial f}{\partial y} = x^2yz^2 = x^2yz^2 + \frac{\partial g}{\partial y}, \quad \text{or} \quad \frac{\partial g}{\partial y} = 0 \quad \text{or} \quad g = g\,(z).$$

Substituting in the third equation, we get

$$\frac{\partial f}{\partial z} = x^2y^2z = x^2y^2z + \frac{dg}{dz}, \quad \text{or} \quad \frac{dg}{dz} = 0 \quad \text{or} \quad g = k, \text{constant}.$$

Hence, $f(x, y, z) = \frac{1}{2}x^2y^2z^2 + k$. Therefore, there exists a scalar function $f(x, y, z)$ such that $\nabla f = \mathbf{v}$ and the vector field $\mathbf{v}$ is conservative.

Exercise 15.2

In problems **1** to **8**, compute the gradient of the scalar function and evaluate it at the given point.

1. $x^3 - 3x^2y^2 + y^3$, (1, 2).
2. $x \sin(yz) + y \sin(xz) + z \sin(xy)$, $(0, \pi/4, 1)$.
3. $\sin(xyz)$, $(1, -1, \pi)$.
4. $\ln(x^2 + y^2 + z^2)$, $(3, -4, 5)$.
5. $(x^2 + y^2 + z^2)^{1/2}$, (1, 1, 1).
6. $e^{xy}(x + y + z)$, (2, 1, 1).
7. $\ln(x + y + z)$, $(1, 2, -1)$.
8. $x^3 + y^3 \sin 4y + z^2$, $(1, \pi/3, 1)$.

In problems **9** to **14**, find the normal vector and the unit normal vector to the given curve/surface at the indicated point.

9. $y^2 = 16x$, (4, 8).
10. $x^2 + y^2 = 25$, (3, 4)
11. $x^2 - y^2 = 12$, (4, 2).
12. $x^2 + 2y^2 + z^2 = 4$, (1, 1, 1).
13. $z = xy$, $(-1, -2, 2)$.
14. $z^2 = x^2 - y^2$, $(2, 1, \sqrt{3})$.

Prove the following properties of gradient (f and g are scalar functions).

15. $\nabla(fg) = f\nabla g + g\nabla f$.
16. $\nabla(f/g) = (g\nabla f - f\nabla g)/g^2$, $g \neq 0$.
17. $\nabla^2(fg) = f\nabla^2 g + 2\nabla f \cdot \nabla g + g\nabla^2 f$.
18. $\nabla(g^m) = mg^{m-1}\nabla g$.
19. If $\mathbf{r} = x\,\mathbf{i} + y\,\mathbf{j} + z\,\mathbf{k}$, show that $(\mathbf{u} \cdot \nabla)\,\mathbf{r} = \mathbf{u}$.
20. If $u = u(x, y, z, t)$, $x = x(t)$, $y = y(t)$, $z = z(t)$, show that

In problems **21** to **26**, find the directional derivative of the given scalar function at the given point in the indicated direction.

21. xyz, $(1, 4, 3)$, in the direction of the line from $(1, 2, 3)$ to $(1, -1, -3)$.

22. $\sqrt{xy^2 + 2x^2z}$, $(2, -2, 1)$, in the direction of negative z-axis

23. $x^2y - y^2z - xyz$, $(1, -1, 0)$, in the direction $(\mathbf{i} - \mathbf{j} + 2\,\mathbf{k})$.

24. $(x^2 + y^2 + z^2)^{3/2}$, $(-1, 1, 2)$, in the direction $\mathbf{i} - 2\,\mathbf{j} + \mathbf{k}$.

25. $x^2 + y^2 + 2z^2$, $(1, 1, 2)$, in the direction of grad f.

26. $2x^2 + y^2 + z^2$, $(1, 2, 3)$, in the direction of the line $x/3 = y/4 = z/5$.

In problems **27** to **29**, find a vector that gives the direction of maximum rate of increase. Find the maximum rate.

27. $e^{2y} \cos x$, at $(\pi/4, 0)$.

28. $3x^2 + y^2 + 2z^2$, at $(0, 1, 2)$.

29. $6xyz$, at $(-1, 2, 1)$.

30. $x^2y^2z^2 + xz^2 + x^2y$, at $(1, 2, -1)$.

In problems **31** to **34**, find a vector that gives the direction of minimum rate of increase. Find the minium rate.

31. $x^3 - xy^2 + y^3$, $(-2, 1)$.

32. $\tan(x^2 + y^2)$, $(\sqrt{\pi/3}, \sqrt{\pi/3})$.

33. $x^2 - y^2 + z^2$, $(1, 2, 1)$.

34. $\sqrt{xy}\, e^z$, $(4, 4, 1)$.

35. If $f(x, y) = x^2 - xy - y + y^2$, find all points where the directional derivative in the direction $\mathbf{b} = (\mathbf{i} + \sqrt{3}\,\mathbf{j})/2$ is zero.

36. It is given that $\nabla f(P) = 3\mathbf{i} + 4\mathbf{j}$. Find a unit vector $\hat{\mathbf{b}}$ such that (i) $D_\mathbf{b} f(P)$ is maximum and (ii) $D_\mathbf{b} f(P)$ is minimum.

37. Suppose $D_\mathbf{b} f(P) = 1$, $D_\mathbf{u} f(P) = 3$, $\mathbf{b} = (3\mathbf{i} + 4\mathbf{j})/5$, $\mathbf{u} = (4\mathbf{i} - 3\mathbf{j})/5$. Find $\nabla f(P)$.

38. Find the values of the constants a, b and c such that the maximum value of the directional derivative of $f(x, y, z) = axy^2 + byz + cx^2z^2$ at $(1, -1, 1)$ is in the direction parallel to the axis of y and has magnitude 6.

39. The temperature at a point (x, y, z) in space is given by $T(x, y, z) = x^2 + y^2 - z$. A mosquito located at $(4, 4, 2)$ desires to fly in such a direction that it gets cooled faster. Find the direction in which it should fly.

In problems **40** to **45**, find the equation of the tangent plane to the graph of the equation at the given point.

40. $x^2 - 3y^2 - z^2 = 2$, $(3, 1, 2)$.

41. $z = 16 - x^2 - y^2$, $(1, 3, 6)$.

42. $xy + yz + zx = -1$, $(1, -1, 2)$.

43. $z = 2e^{-x} \sin(2y)$, $(0, \pi/12, 1)$.

44. $(x^2/a^2) + (y^2/b^2) + (z^2/c^2) = 1$, (x_0, y_0, z_0).

45. $(x^2/a^2) - (y^2/b^2) + (z^2/c^2) = 1$, (x_0, y_0, z_0).

In problems **46** to **49**, find a scalar function f such that $\mathbf{v} = \nabla f$.

46. $\mathbf{v} = xy(2yz\,\mathbf{i} + 2xz\,\mathbf{j} + xy\,\mathbf{k})$.

47. $\mathbf{v} = (x\mathbf{i} + y\mathbf{j} + z\mathbf{k})/\sqrt{x^2 + y^2 + z^2}$.

48. $\mathbf{v} = 12x\,\mathbf{i} - 15y^2\,\mathbf{j} + \mathbf{k}$.

49. $\mathbf{v} = e^{xyz}(yz\,\mathbf{i} + xz\,\mathbf{j} + xy\,\mathbf{k})$.

50. Does there exist a function $f(x, y, z) \neq 0$ such that $\nabla f = \mathbf{0}$ for all (x, y, z)? If it exists, does the level surface $f(x, y, z) = k$ have a normal vector?

In problems **51** and **53**, find the angle between the two surfaces at the indicated point of intersection.

51. $z = x^2 + y^2$, $z = 2x^2 - 3y^2$, $(2, 1, 5)$.

52. $x^2 + y^2 = 4$, $x^2 + y^2 + z^2 = 12$, $(2, 2, -2)$.

53. $x^2 + y^2 + z^2 = 9$, $z + 3 = x^2 + y^2$, $(-2, 1, 2)$.

In problems **54** and **55**, find the parametric equations for the normal line at the given point.

54. $z = 3x^2 - 2y^2$, $(2, 1, 10)$.

55. $x^2 + 2y^2 + 4z^2 = 10$, $(2, 1, -1)$.

15.4 Divergence and Curl of a Vector Field

In the previous section, we have defined the gradient operator which, when operated on a scalar field, produces a vector field. We shall now discuss two other important vector operations.

Let $\mathbf{v} = v_1(x, y, z)\,\mathbf{i} + v_2(x, y, z)\,\mathbf{j} + v_3(x, y, z)\,\mathbf{k}$ define a vector field.

Divergence of vector field v

Divergence of $\mathbf{v}$, denoted by div $\mathbf{v}$, is defined as the scalar field

$$\text{div } \mathbf{v} = \frac{\partial v_1}{\partial x} + \frac{\partial v_2}{\partial y} + \frac{\partial v_3}{\partial z}. \tag{15.46}$$

We observe that div $\mathbf{v}$ can also be written in terms of the gradient operator as

$$\text{div } \mathbf{v} = \nabla \cdot \mathbf{v} = \left(\mathbf{i}\frac{\partial}{\partial x} + \mathbf{j}\frac{\partial}{\partial y} + \mathbf{k}\frac{\partial}{\partial z} \right) \cdot (v_1\,\mathbf{i} + v_2\,\mathbf{j} + v_3\,\mathbf{k})$$

$$= \frac{\partial v_1}{\partial x} + \frac{\partial v_2}{\partial y} + \frac{\partial v_3}{\partial z}.$$

Note that $\nabla \cdot \mathbf{v}$ is just a notation for div $\mathbf{v}$ and it is not a scalar product in the usual sense, since $\nabla \cdot \mathbf{v} \neq \mathbf{v} \cdot \nabla$. In fact

$$\mathbf{v} \cdot \nabla = v_1\frac{\partial}{\partial x} + v_2\frac{\partial}{\partial y} + v_3\frac{\partial}{\partial z}$$

is a scalar operator.

Example 15.19 Find the divergence of the vector field

$$\mathbf{v} = (x^2 y^2 - z^3)\,\mathbf{i} + 2xyz\,\mathbf{j} + e^{xyz}\,\mathbf{k}$$

Solution We have

$$\text{div } \mathbf{v} = \frac{\partial}{\partial x}(x^2 y^2 - z^3) + \frac{\partial}{\partial y}(2xyz) + \frac{\partial}{\partial z}(e^{xyz})$$

$$= 2xy^2 + 2xz + xye^{xyz}.$$

Curl of vector field v

Curl of a vector field $\mathbf{v}$, denoted by curl $\mathbf{v}$, is defined as the vector field

$$\text{curl } \mathbf{v} = \left(\frac{\partial v_3}{\partial y} - \frac{\partial v_2}{\partial z} \right)\mathbf{i} + \left(\frac{\partial v_1}{\partial z} - \frac{\partial v_3}{\partial x} \right)\mathbf{j} + \left(\frac{\partial v_2}{\partial x} - \frac{\partial v_1}{\partial y} \right)\mathbf{k}. \tag{15.47}$$

We observe that curl $\mathbf{v}$ can also be written in terms of the gradient operator as

$$\text{curl } \mathbf{v} = \nabla \times \mathbf{v} = \begin{vmatrix} \mathbf{i} & \mathbf{j} & \mathbf{k} \\ \partial/\partial x & \partial/\partial y & \partial/\partial z \\ v_1 & v_2 & v_3 \end{vmatrix}. \tag{15.48}$$

Note that $\nabla \times \mathbf{v}$ is just a notation for curl $\mathbf{v}$ and it is not a vector product in the usual sense, since $\nabla \times \mathbf{v} \neq -\mathbf{v} \times \nabla$.

Sometimes, curl **v** is also written as

$$\text{curl } \mathbf{v} = \Sigma \left(\frac{\partial v_3}{\partial y} - \frac{\partial v_2}{\partial z} \right) \mathbf{i}$$

where Σ denotes summation obtained by the cyclic rotation of the unit vectors **i, j, k,** the components v_1, v_2, v_3 and the independent variables x, y, z, respectively.

Example 15.20 Find the curl of the vector field

$$\mathbf{v} = (x^2 y^2 - z^3)\,\mathbf{i} + 2xyz\,\mathbf{j} + e^{xyz}\,\mathbf{k}.$$

Solution   We have

$$\text{curl } \mathbf{v} = \begin{vmatrix} \mathbf{i} & \mathbf{j} & \mathbf{k} \\ \partial/\partial x & \partial/\partial y & \partial/\partial z \\ x^2 y^2 - z^3 & 2xyz & e^{xyz} \end{vmatrix}$$

$$= \mathbf{i}(xze^{xyz} - 2xy) - \mathbf{j}(yze^{xyz} + 3z^2) + \mathbf{k}\,(2yz - 2x^2 y).$$

There are two fundamental relations between the gradient, divergence and curl vectors. We prove these.

Curl of gradient   Let f be a differentiable scalar field. Then

$$\text{curl (grad } f) = 0 \quad \text{or} \quad \nabla \times (\nabla f) = 0 \tag{15.49}$$

Proof.   From the definition, we have

$$\nabla \times (\nabla f) = \begin{vmatrix} \mathbf{i} & \mathbf{j} & \mathbf{k} \\ \partial/\partial x & \partial/\partial y & \partial/\partial z \\ \partial f/\partial x & \partial f/\partial y & \partial f/\partial z \end{vmatrix}$$

$$= \mathbf{i}\left(\frac{\partial^2 f}{\partial y\,\partial z} - \frac{\partial^2 f}{\partial y\,\partial z} \right) + \mathbf{j}\left(\frac{\partial^2 f}{\partial x\,\partial z} - \frac{\partial^2 f}{\partial x\,\partial z} \right) + \mathbf{k}\left(\frac{\partial^2 f}{\partial x\,\partial y} - \frac{\partial^2 f}{\partial x\,\partial y} \right) = 0.$$

Divergence of curl   Let **v** be a differentiable vector field. Then

$$\text{div (curl } \mathbf{v}) = 0 \quad \text{or} \quad \nabla \cdot (\nabla \times \mathbf{v}) = 0. \tag{15.50}$$

Proof   From the definition, we have for $\mathbf{v} = v_1 \mathbf{i} + v_2 \mathbf{j} + v_3 \mathbf{k}$

$$\nabla \cdot (\nabla \times \mathbf{v}) = \left(\mathbf{i}\frac{\partial}{\partial x} + \mathbf{j}\frac{\partial}{\partial y} + \mathbf{k}\frac{\partial}{\partial z} \right) \cdot \left[\mathbf{i}\left(\frac{\partial v_3}{\partial y} - \frac{\partial v_2}{\partial z} \right) + \mathbf{j}\left(\frac{\partial v_1}{\partial z} - \frac{\partial v_3}{\partial x} \right) + \mathbf{k}\left(\frac{\partial v_2}{\partial x} - \frac{\partial v_1}{\partial y} \right) \right]$$

$$= \frac{\partial}{\partial x}\left(\frac{\partial v_3}{\partial y} - \frac{\partial v_2}{\partial z} \right) + \frac{\partial}{\partial y}\left(\frac{\partial v_1}{\partial z} - \frac{\partial v_3}{\partial x} \right) + \frac{\partial}{\partial z}\left(\frac{\partial v_2}{\partial x} - \frac{\partial v_1}{\partial y} \right) = 0.$$

Example 15.21   Prove that div $(f\mathbf{v}) = f\,(\text{div } \mathbf{v}) + (\text{grad } f) \cdot \mathbf{v}$, where f is a scalar function.

Solution   We have

$$\nabla \cdot (f\mathbf{v}) = \left(\mathbf{i}\frac{\partial}{\partial x} + \mathbf{j}\frac{\partial}{\partial y} + \mathbf{k}\frac{\partial}{\partial z} \right) \cdot (f v_1 \mathbf{i} + f v_2 \mathbf{j} + f v_3 \mathbf{k})$$

$$= \frac{\partial}{\partial x}(f v_1) + \frac{\partial}{\partial y}(f v_2) + \frac{\partial}{\partial z}(f v_3)$$

$$= \left(v_1 \frac{\partial f}{\partial x} + f \frac{\partial v_1}{\partial x} \right) + \left(v_2 \frac{\partial f}{\partial y} + f \frac{\partial v_2}{\partial y} \right) + \left(v_3 \frac{\partial f}{\partial z} + f \frac{\partial v_3}{\partial z} \right)$$

$$= f \left(\frac{\partial v_1}{\partial x} + \frac{\partial v_2}{\partial y} + \frac{\partial v_3}{\partial z} \right) + (v_1 \mathbf{i} + v_2 \mathbf{j} + v_3 \mathbf{k}) \cdot \left(\mathbf{i} \frac{\partial f}{\partial x} + \mathbf{j} \frac{\partial f}{\partial y} + \mathbf{k} \frac{\partial f}{\partial z} \right)$$

$$= f (\nabla \cdot \mathbf{v}) + \mathbf{v} \cdot (\nabla f) = f (\nabla \cdot \mathbf{v}) + \nabla f \cdot \mathbf{v}.$$

Example 15.22   If $\mathbf{r} = x\mathbf{i} + y\mathbf{j} + z\,\mathbf{k}$ and $r = |\,\mathbf{r}\,|$, show that div $(\mathbf{r}/r^3) = 0$.

Solution   We have

$$\nabla \cdot \left(\frac{\mathbf{r}}{r^3} \right) = \left(\mathbf{i} \frac{\partial}{\partial x} + \mathbf{j} \frac{\partial}{\partial y} + \mathbf{k} \frac{\partial}{\partial z} \right) \cdot \left(\frac{x}{r^3} \mathbf{i} + \frac{y}{r^3} \mathbf{j} + \frac{z}{r^3} \mathbf{k} \right)$$

$$= \frac{\partial}{\partial x} \left(\frac{x}{r^3} \right) + \frac{\partial}{\partial y} \left(\frac{y}{r^3} \right) + \frac{\partial}{\partial z} \left(\frac{z}{r^3} \right) = \frac{3}{r^3} - \frac{3}{r^4} \left(x \frac{\partial r}{\partial x} + y \frac{\partial r}{\partial y} + z \frac{\partial r}{\partial z} \right).$$

Since $r^2 = x^2 + y^2 + z^2$, we obtain $\dfrac{\partial r}{\partial x} = \dfrac{x}{r}$, $\dfrac{\partial r}{\partial y} = \dfrac{y}{r}$ and $\dfrac{\partial r}{\partial z} = \dfrac{z}{r}$.

Therefore,

$$\nabla \cdot \left(\frac{\mathbf{r}}{r^3} \right) = \frac{3}{r^3} - \frac{3}{r^5} (x^2 + y^2 + z^2) = \frac{3}{r^3} - \frac{3}{r^3} = 0.$$

Example 15.23   If $\mathbf{a}$ is a constant vector and $\mathbf{r} = x\,\mathbf{i} + y\,\mathbf{j} + z\,\mathbf{k}$, show that curl $(\mathbf{a} \times \mathbf{r}) = 2\,\mathbf{a}$.

Solution   Let $\mathbf{a} = a_1\mathbf{i} + a_2\mathbf{j} + a_3\mathbf{k}$ where a_1, a_2 and a_3 are constants. We have

$$\nabla \times (\mathbf{a} \times \mathbf{r}) = \left(\mathbf{i} \frac{\partial}{\partial x} + \mathbf{j} \frac{\partial}{\partial y} + \mathbf{k} \frac{\partial}{\partial z} \right) \times (\mathbf{a} \times \mathbf{r})$$

$$= \mathbf{i} \times \frac{\partial}{\partial x} (\mathbf{a} \times \mathbf{r}) + \mathbf{j} \times \frac{\partial}{\partial y} (\mathbf{a} \times \mathbf{r}) + \mathbf{k} \times \frac{\partial}{\partial z} (\mathbf{a} \times \mathbf{r})$$

$$= \mathbf{i} \times \left(\mathbf{a} \times \frac{\partial r}{\partial x} \right) + \mathbf{j} \times \left(\mathbf{a} \times \frac{\partial r}{\partial y} \right) + \mathbf{k} \times \left(\mathbf{a} \times \frac{\partial r}{\partial z} \right) \text{ (since } \mathbf{a} \text{ is a constant vector)}$$

$$= \mathbf{i} \times (\mathbf{a} \times \mathbf{i}) + \mathbf{j} \times (\mathbf{a} \times \mathbf{j}) + \mathbf{k} \times (\mathbf{a} \times \mathbf{k})$$

$$= (\mathbf{i} \cdot \mathbf{i}) \, \mathbf{a} - (\mathbf{i} \cdot \mathbf{a}) \, \mathbf{i} + (\mathbf{j} \cdot \mathbf{j}) \, \mathbf{a} - (\mathbf{j} \cdot \mathbf{a}) \, \mathbf{j} + (\mathbf{k} \cdot \mathbf{k}) \, \mathbf{a} - (\mathbf{k} \cdot \mathbf{a}) \, \mathbf{k}$$

$$= 3\mathbf{a} - (a_1\mathbf{i} + a_2\mathbf{j} + a_3\mathbf{k}) = 2\mathbf{a}.$$

Remark 6

Let $\mathbf{v}$ denote the velocity of a fluid in a medium. If div $(\mathbf{v}) = 0$, then the fluid is said to be *incompressible*. In electromagnetic theory, if div $(\mathbf{v}) = 0$, then the vector field $\mathbf{v}$ is said to be *solenoidal*.

Remark 7

Some books use the word *rotation* in place of curl, that is, curl $(\mathbf{v})$ is written as rot $(\mathbf{v})$. In fluid mechanics, curl $(\mathbf{v})$ where $\mathbf{v}$ is the velocity, measures the vorticity of the fluid. If curl $(\mathbf{v}) = \mathbf{0}$, then $\mathbf{v}$ is said to be an *irrotational field*.

Physical interpretation of divergence   We shall present an interpretation in fluid mechanics. Consider the flow of a compressible fluid of density $\rho(x, y, z, t)$, (density is mass per unit volume) and velocity

$$\mathbf{v}(x, y, z, t) = v_1(x, y, z, t)\,\mathbf{i} + v_2(x, y, z, t)\mathbf{j} + v_3(x, y, z, t)\,\mathbf{k}.$$

Therefore, the density and velocity vary from point to point and also with respect to time. Consider an infinitesimal volume element (parallelopiped of sides Δx, Δy, Δz) placed in the fluid as given in Fig. 15.9. The fluid enters the elemental volume through the faces and goes out from the other faces.

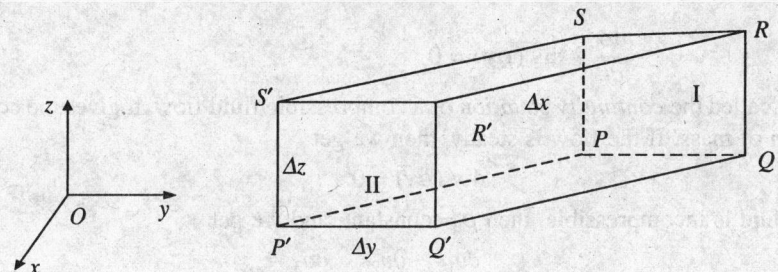

Fig. 15.9.   Flow through a parallelopiped.

The face $PQRS$ is denoted as I and the face $P'\,Q'\,R'\,S'$ is denoted as II. Let us now compute the loss of fluid as it flows through the element in time Δt. We assume the following

(Volume of the fluid flowing through an element of surface area Δs in time Δt) $\approx$ (component of fluid velocity normal to the surface $\times$ area of surface $\Delta S \times \Delta t$).

The area of face I is $\Delta y\,\Delta z$ and the direction of the normal is $-\mathbf{i}$. Therefore, the mass of the fluid entering through the face I in time Δt, is approximately equal to

$$- (\rho v_1)(x, y, z, t)\,\Delta y\,\Delta z\,\Delta t.$$

The area of the face II is $\Delta y\,\Delta z$ and the direction of the normal is $\mathbf{i}$. Hence, the mass of the fluid leaving this face in time Δt is approximately equal to

$$(\rho v_1)\,(x + \Delta x, y, z, t)\,\Delta y\,\Delta z\,\Delta t.$$

Therefore, the approximate loss of mass as the fluid flows through the faces, perpendicular to the y-z plane, is

$$[(\rho v_1)\,(x + \Delta x, y, z, t) - (\rho v_1)\,(x, y, z, t)]\,\Delta y\,\Delta z\,\Delta t. \tag{15.51a}$$

Similarly, the approximate losses of mass through other faces of the elemental volume $\Delta V (= \Delta x\,\Delta y\,\Delta z)$, are

$$[(\rho v_2)\,(x, y + \Delta y, z, t) - (\rho v_2)\,(x, y, z, t)]\,\Delta x\,\Delta z\,\Delta t \tag{15.51b}$$

and

$$[(\rho v_3)\,(x, y, z + \Delta z, t) - (\rho v_3)\,(x, y, z, t)]\,\Delta x\,\Delta y\,\Delta t. \tag{15.51c}$$

Therefore, adding Eqs. (15.51a) to (15.51c), we obtain the total loss of mass of the fluid during the time Δt as

$$\left[\frac{1}{\Delta x}\,\{(\rho v_1)\,(x + \Delta x, y, z, t) - (\rho v_1)\,(x, y, z, t)\} + \frac{1}{\Delta y}\,\{(\rho v_2)\,(x, y + \Delta y, z, t) - (\rho v_2)\,(x, y, z, t)\} \right.$$

$$\left. + \frac{1}{\Delta z}\,\{(\rho v_3)\,(x, y, z + \Delta z, t) - (\rho v_3)\,(x, y, z, t)\}\right]\,\Delta V\,\Delta t \tag{15.52a}$$

This loss of mass is due to the rate of change of density with respect to time and hence is equal to

$$-\frac{\partial \rho}{\partial t} \, \Delta V \, \Delta t. \tag{15.52b}$$

Equate the expressions in Eqs. (15.52a) and (15.52b), let $\Delta x \to 0$, $\Delta y \to 0$, $\Delta z \to 0$, $\Delta t \to 0$, and divide the resulting equation by $\Delta V \, \Delta t$. In the limit, we get

$$\frac{\partial}{\partial x}(\rho v_1) + \frac{\partial}{\partial y}(\rho v_2) + \frac{\partial}{\partial z}(\rho v_3) = -\frac{\partial \rho}{\partial t}$$

or

$$\frac{\partial \rho}{\partial t} + \text{div}\,(\rho \mathbf{v}) = 0. \tag{15.53}$$

This equation is called the *continuity equation* of a compressible fluid flow. It gives the condition for the conservation of mass. If the flow is steady, then we get

$$\text{div}\,(\rho \mathbf{v}) = 0.$$

Further, if the fluid is incompressible, then $\rho = $ constant, and we get

$$\text{div}\,(\mathbf{v}) = \frac{\partial v_1}{\partial x} + \frac{\partial v_2}{\partial y} + \frac{\partial v_3}{\partial z} = 0.$$

Therefore, when the fluid is incompressible, the divergence of the velocity vector vanishes. It states that the amount of fluid that enters and leaves a given volume is same, that is, there is no loss in the mass of the fluid.

Physical interpretation of curl   Let a rigid body rotate with the uniform angular velocity $\Omega = a\mathbf{i} + b\mathbf{j} + c\mathbf{k}$, about an axis l through the origin O (Fig. 15.10). Let the position vector of any point $P(x, y, z)$ on the rotating body be $\mathbf{r} = x\mathbf{i} + y\mathbf{j} + z\mathbf{k}$. The tangential (linear) velocity $\mathbf{v}$ of the point $P(x, y, z)$ is given by

$$\mathbf{v} = \Omega \times \mathbf{r} = (a\mathbf{i} + b\mathbf{j} + c\mathbf{k}) \times (x\mathbf{i} + y\mathbf{j} + z\mathbf{k})$$

$$= (bz - cy)\,\mathbf{i} + (cx - az)\,\mathbf{j} + (ay - bx)\,\mathbf{k}.$$

Now, $$\text{curl } \mathbf{v} = \nabla \times \mathbf{v} = \begin{vmatrix} \mathbf{i} & \mathbf{j} & \mathbf{k} \\ \partial/\partial x & \partial/\partial y & \partial/\partial z \\ bz - cy & cx - az & ay - bx \end{vmatrix} = 2(a\mathbf{i} + b\mathbf{j} + c\mathbf{k}) = 2\Omega.$$

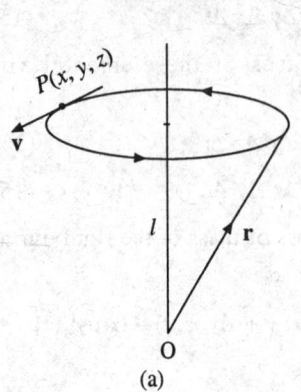

(a)

(b)

Fig. 15.10.   Angular velocity.

Therefore, the angular velocity of the point $P(x, y, z)$ is given by $\Omega = (\text{curl } \mathbf{v}/2)$. Hence, the angular velocity of a uniformly rotating body is equal to one-half of the curl of the linear velocity. Because of this interpretation, the notation *rotation* or *rot* for curl is also used.

Remark 8

A force field $\mathbf{F}$ is said to be *conservative* if it is derivable from a potential function f, that is $\mathbf{F} = \text{grad } f$. Then, curl $(\mathbf{F}) = \text{curl } (\text{grad } f) = \mathbf{0}$. Therefore, if $\mathbf{F}$ is conservative then curl $(\mathbf{F}) = \mathbf{0}$ and there exists a scalar potential function f such that $\mathbf{F} = \text{grad } f$.

*See Appendix 3 for expressions of grad f, div(**v**), curl(**v**) etc. in polar coordinate systems.*

Example 15.24   Prove the following identities

(i) curl $(f \mathbf{v}) = (\text{grad } f) \times \mathbf{v} + f \text{ curl } \mathbf{v}$

(ii) div $(\text{grad } f) = \nabla^2 f$, where $\nabla^2 = \dfrac{\partial^2}{\partial x^2} + \dfrac{\partial^2}{\partial y} + \dfrac{\partial^2}{\partial z^2}$ is the Laplacian operator

(iii) curl $(\text{curl } \mathbf{v}) = \nabla(\nabla \cdot \mathbf{v}) - \nabla^2 \mathbf{v}$, or grad $(\text{div } \mathbf{v}) = \nabla \times (\nabla \times \mathbf{v}) + \nabla^2 \mathbf{v}$.

where f is a scalar function.

Solution   We have

(i) curl $(f \mathbf{v}) = \nabla \times (f \mathbf{v}) = \nabla \times (f v_1 \mathbf{i} + f v_2 \mathbf{j} + f v_3 \mathbf{k})$

$$= \Sigma \left[\frac{\partial}{\partial y}(f v_3) - \frac{\partial}{\partial z}(f v_2) \right] \mathbf{i} = \Sigma \left[f \left(\frac{\partial v_3}{\partial y} - \frac{\partial v_2}{\partial z} \right) + v_3 \frac{\partial f}{\partial y} - v_2 \frac{\partial f}{\partial z} \right] \mathbf{i}$$

$$= f \left[\left(\frac{\partial v_3}{\partial y} - \frac{\partial v_2}{\partial z} \right) \mathbf{i} + \left(\frac{\partial v_1}{\partial z} - \frac{\partial v_3}{\partial x} \right) \mathbf{j} + \left(\frac{\partial v_2}{\partial x} - \frac{\partial v_1}{\partial y} \right) \mathbf{k} \right]$$

$$+ \left[\left(v_3 \frac{\partial f}{\partial y} - v_2 \frac{\partial f}{\partial z} \right) \mathbf{i} + \left(v_1 \frac{\partial f}{\partial z} - v_3 \frac{\partial f}{\partial x} \right) \mathbf{j} + \left(v_2 \frac{\partial f}{\partial x} - v_1 \frac{\partial f}{\partial y} \right) \mathbf{k} \right]$$

$$= f(\text{curl } \mathbf{v}) + \left(\mathbf{i} \frac{\partial f}{\partial x} + \mathbf{j} \frac{\partial f}{\partial y} + \mathbf{k} \frac{\partial f}{\partial z} \right) \times (\mathbf{i} v_1 + \mathbf{j} v_2 + \mathbf{k} v_3)$$

$$= f(\text{curl } \mathbf{v}) + (\text{grad } f) \times \mathbf{v}.$$

(ii)

$$\text{div}(\text{grad } f) = \left(\mathbf{i} \frac{\partial}{\partial x} + \mathbf{j} \frac{\partial}{\partial y} + \mathbf{k} \frac{\partial}{\partial z} \right) \cdot \left(\mathbf{i} \frac{\partial f}{\partial x} + \mathbf{j} \frac{\partial f}{\partial y} + \mathbf{k} \frac{\partial f}{\partial z} \right)$$

$$= \frac{\partial^2 f}{\partial x^2} + \frac{\partial^2 f}{\partial y^2} + \frac{\partial^2 f}{\partial z^2} = \nabla^2 f.$$

(iii)

$$\text{curl}(\text{curl } \mathbf{v}) = \nabla \times (\nabla \times \mathbf{v}) = \left(\Sigma \mathbf{i} \frac{\partial}{\partial x} \right) \times \left[\Sigma \mathbf{i} \left(\frac{\partial v_3}{\partial y} - \frac{\partial v_2}{\partial z} \right) \right]$$

$$= \Sigma \mathbf{i} \left[\frac{\partial}{\partial y} \left(\frac{\partial v_2}{\partial x} - \frac{\partial v_1}{\partial y} \right) - \frac{\partial}{\partial z} \left(\frac{\partial v_1}{\partial z} - \frac{\partial v_3}{\partial x} \right) \right]$$

$$= \Sigma \mathbf{i} \left[\frac{\partial^2 v_2}{\partial y \partial x} + \frac{\partial^2 v_3}{\partial z \partial x} - \left(\frac{\partial^2 v_1}{\partial y^2} + \frac{\partial^2 v_1}{\partial z^2} \right) \right]$$

$$= \Sigma \, \mathbf{i} \left[\frac{\partial}{\partial x} \left(\frac{\partial v_2}{\partial y} + \frac{\partial v_3}{\partial z} \right) - \left(\frac{\partial^2 v_1}{\partial y^2} + \frac{\partial^2 v_1}{\partial z^2} \right) \right]$$

$$= \Sigma \, \mathbf{i} \left[\frac{\partial}{\partial x} \left(\frac{\partial v_1}{\partial x} + \frac{\partial v_2}{\partial y} + \frac{\partial v_3}{\partial z} \right) - \left(\frac{\partial^2 v_1}{\partial x^2} + \frac{\partial^2 v_1}{\partial y^2} + \frac{\partial^2 v_1}{\partial z 2} \right) \right]$$

$$= \left(\Sigma \, \mathbf{i} \frac{\partial}{\partial x} \right) (\nabla \cdot \mathbf{v}) - \left(\frac{\partial^2}{\partial x^2} + \frac{\partial^2}{\partial y^2} + \frac{\partial^2}{\partial z^2} \right) (\Sigma \, \mathbf{i} \, v_1)$$

$$= \nabla (\nabla \cdot \mathbf{v}) - \nabla^2 \mathbf{v}.$$

Hence, grad (div $\mathbf{v}$) = curl (curl $\mathbf{v}$) + $\nabla^2 \mathbf{v}$.

Exercise 15.3

In problems **1** to **8**, compute div ($\mathbf{v}$), curl ($\mathbf{v}$) and verify that div (curl $\mathbf{v}$) = 0.

1. $\mathbf{v} = x \mathbf{i} + 2y \mathbf{j} + z \mathbf{k}$.
2. $\mathbf{v} = xy \mathbf{i} + yz \mathbf{j} + zx \mathbf{k}$.

3. $\mathbf{v} = (x^2 + y^2 + z^2)^{3/2} (x \mathbf{i} + y \mathbf{j} + z \mathbf{k})$.
4. $\mathbf{v} = x^2 \mathbf{i} + y^2 \mathbf{j} + z^2 \mathbf{k}$.

5. $\mathbf{v} = xe^{-y} \mathbf{i} + 2ze^{-y} \mathbf{j} + xy^2 \mathbf{k}$.
6. $\mathbf{v} = xyz \mathbf{i} + 2x^2 y \mathbf{j} + (xz^2 - y^2 z) \mathbf{k}$.

7. $\mathbf{v} = (x^2 - y^2) \mathbf{i} + 4xy \mathbf{j} + (x^2 - xy) \mathbf{k}$.
8. $\mathbf{v} = (x^2 + yz) \mathbf{i} + (y^2 + zx) \mathbf{j} + (z^2 + xy) \mathbf{k}$.

In problems **9** to **12**, compute grad f and verify that curl (grad f) = **0**.

9. $f(x, y, z) = x + y - 2z^2$.
10. $f(x, y, z) = x \sin (x + y + z)$.

11. $f(x, y, z) = e^{x+y+z}$.
12. $f(x, y, z) = 16x \, y^3 z^2$.

Show that the vectors in problems **13**, **14** are solenoidal.

13. $(2x + 3y) \mathbf{i} + (x - y) \mathbf{j} - (x + y + z) \mathbf{k}$.
14. $e^{x+y-2z} (\mathbf{i} + \mathbf{j} + \mathbf{k})$.

15. If $\mathbf{v} = - (x + y + 2) \mathbf{i} - 2 \mathbf{j} + (x + y) \mathbf{k}$, show that $\mathbf{v} \cdot$ curl $\mathbf{v}$ = 0.

16. If $f = x^2 + y^2 + z^2$ and $\mathbf{r} = x \mathbf{i} + y \mathbf{j} + z \mathbf{k}$, show that div ($f \mathbf{r}$) = 5$f$.

In problems **17** to **25**, **a** is a constant vector and $\mathbf{r} = x \mathbf{i} + y \mathbf{j} + z \mathbf{k}$. Then prove the given identities.

17. grad ($\mathbf{a} \cdot \mathbf{r}$) = $\mathbf{a}$.
18. div ($\mathbf{a} \times \mathbf{r}$) = 0.

19. curl ($\mathbf{a} \times \mathbf{r}$) = 2$\mathbf{a}$.
20. div [($\mathbf{a} \cdot \mathbf{r}$) $\mathbf{r}$] = 4 ($\mathbf{a} \cdot \mathbf{r}$).

21. $\nabla \cdot [(\mathbf{r} \cdot \mathbf{r}) \, \mathbf{a}] = 2(\mathbf{r} \cdot \mathbf{a})$.
22. $\mathbf{a} \times$ (curl $\mathbf{r}$) = **0**.

23. $\nabla \times \left[\dfrac{1}{r^3} (\mathbf{a} \times \mathbf{r}) \right] = \dfrac{3}{r^5} (\mathbf{a} \cdot \mathbf{r}) \mathbf{r} - \dfrac{\mathbf{a}}{r^3}$, $r = | \mathbf{r} |$.

24. $\nabla \times (\mathbf{a} \times \mathbf{v}) = \mathbf{a} (\nabla \cdot \mathbf{v}) - (\mathbf{a} \cdot \nabla) \, \mathbf{v}$, $\mathbf{v}$ is any vector.

25. $\nabla \cdot (\mathbf{a} \times \mathbf{v}) = - \mathbf{a} \cdot (\nabla \times \mathbf{v})$, $\mathbf{v}$ is any vector.

In problems **26** to **29**, show that the vector field $\mathbf{v}$ is irrotational and find a scalar function $f(x, y, z)$ such that $\mathbf{v}$ = grad f.

26. $(y^2 - x^2 + y) \mathbf{i} + x (2y + 1) \mathbf{j}$.
27. $3x^2 y^2 z^4 \mathbf{i} + 2x^3 y z^4 \mathbf{j} + 4x^3 y^2 z^3 \mathbf{k}$.

28. $e^{xy} (y \mathbf{i} + x \mathbf{j}) + 2e^z \mathbf{k}$.
29. $\cos (x^2 + y^2 + z^2) (x \mathbf{i} + y \mathbf{j} + z \mathbf{k})$.

30. Find the constants a, b and c such that $\mathbf{v} = (3x + ay + z) \mathbf{i} + (2x - y + bz) \mathbf{j} + (x + cy + z) \mathbf{k}$ is irrotational.

Assuming continuity of partial derivatives verify the identities given in problems **31** to **36**.

31. $\nabla \cdot (f \, \nabla g) = f \, \nabla^2 g + \nabla f \cdot \nabla g$.
32. $\nabla \cdot (f \nabla g) - \nabla \cdot (g \nabla f) = f \, \nabla^2 g - g \nabla^2 f$.

33. $\nabla \cdot (\mathbf{u} \times \mathbf{v}) = \mathbf{v} \cdot (\nabla \times \mathbf{u}) - \mathbf{u} \cdot (\nabla \times \mathbf{v})$.
34. $\nabla \cdot [(f \nabla g) \times (g \nabla f)] = 0$.

35. $\nabla \cdot (\nabla f \times \nabla g) = 0$.

36. $\nabla \times (\mathbf{u} \times \mathbf{v}) = (\mathbf{v} \cdot \nabla) \, \mathbf{u} - (\mathbf{u} \cdot \nabla) \, \mathbf{v} + (\nabla \cdot \mathbf{v}) \, \mathbf{u} - (\nabla \cdot \mathbf{u}) \, \mathbf{v}$.

37. If **E** and **H** are irrotational vector fields, show that **E** × **H** is a solenoidal vector field.

38. If $\nabla \cdot \mathbf{E} = 0$, $\nabla \cdot \mathbf{H} = 0$, $\nabla \times \mathbf{E} = -\dfrac{1}{c}\dfrac{\partial \mathbf{H}}{\partial t}$, $\nabla \times \mathbf{H} = \dfrac{1}{c}\dfrac{\partial \mathbf{E}}{\partial t}$ show that **E** and **H** satisfy the wave equation

$$\frac{\partial^2 f}{\partial t^2} = c^2 \, \nabla^2 f \cdot$$ (**E** and **H** are the electric and magnetic fields in empty space).

39. Let $f(x, y, z)$ be a solution of the Laplace equation $\nabla^2 f = 0$. Then, show that ∇f is a vector which is both irrotational and solenoidal.

40. Let $f(x, y, z)$ be a solution of the Poisson equation $\nabla^2 f = c$, a constant. If $\mathbf{v} = \nabla f$, show that curl $\mathbf{v} = \mathbf{0}$, but div $\mathbf{v} \neq 0$.

15.5   Line Integrals and Green's Theorem

In chapter 11 (section 11.2.2), we have introduced line integral along a given curve, in the context of complex variables. The definition and method of evaluation of line integrals discussed there are valid in the present context also. We briefly present the method of evaluation of line integrals.

Let C be a simple curve. Let the parametric representation of C be written as

$$x = x(t), \ \ y = y(t), \ \ z = z(t), \ \ a \le t \le b. \tag{15.54}$$

Therefore, the position vector of a point on the curve C can be written as

$$\mathbf{r}(t) = x(t)\,\mathbf{i} + y(t)\,\mathbf{j} + z(t)\,\mathbf{k}, \ a \le t \le b. \tag{15.55}$$

Line integral with respect to arc length

Let C be a simple smooth curve whose parametric representation is as given in Eqs. (15.54) and (15.55). Let $f(x, y, z)$ be continuous on C. Then, we define the line integral of f over C with respect to the arc length s by

$$\int_C f(x, y, z)\, ds = \int_a^b f[x(t), y(t), z(t)] \sqrt{x'(t)^2 + y'(t)^2 + z'(t)^2} \ dt \tag{15.56}$$

since $ds = \dfrac{ds}{dt}\, dt = \sqrt{\left(\dfrac{dx}{dt}\right)^2 + \left(\dfrac{dy}{dt}\right)^2 + \left(\dfrac{dz}{dt}\right)^2}\ dt.$

This result is true since the arc length along the curve from an initial point $(x(a), y(a), z(a))$ to any point $(x(t), y(t), z(t))$ is given by

$$s(t) = \int_a^t \sqrt{\left(\frac{dx}{d\eta}\right)^2 + \left(\frac{dy}{d\eta}\right)^2 + \left(\frac{dz}{d\eta}\right)^2} \ d\eta$$

and ds is as given above.

The initial point of C is given by $(x(a), y(a), z(a))$ and the terminal point of C is given by $(x(b), y(b), z(b))$.

Example 15.25   Evaluate $\displaystyle\int_C x^2 y\, ds$, where C is the curve defined by $x = 3 \cos t$, $y = 3 \sin t$, $0 \le t \le \pi/2$.

Solution   We have $ds = \dfrac{ds}{dt}\, dt = \sqrt{\left(\dfrac{dx}{dt}\right)^2 + \left(\dfrac{dy}{dt}\right)^2}\; dt = \sqrt{9}\, dt = 3\, dt$.

Therefore, $\displaystyle\int_C x^2 y\, ds = \int_0^{\pi/2} 27 \cos^2 t \sin t\, (3dt) = -27 \left[\cos^3 t\right]_0^{\pi/2} = 27.$

Example 15.26   Evaluate $\displaystyle\int_C (x^2 + yz)\, ds$, where C is the curve defined by $x = 4y$, $z = 3$ from $(2, 1/2, 3)$ to $(4, 1, 3)$.

Solution   Let $x = t$. Then, $y = t/4$ and $z = 3$. Therefore, the curve C is represented by $x = t$, $y = t/4$, $z = 3$, $2 \le t \le 4$. We have $ds = \sqrt{17}/4$.

Hence, $\displaystyle\int_C (x^2 + yz)\, ds = \frac{\sqrt{17}}{4} \int_2^4 \left(t^2 + \frac{3}{4}t\right) dt = \frac{\sqrt{17}}{4}\left[\frac{1}{3}t^3 + \frac{3}{8}t^2\right]_2^4 = \frac{139\sqrt{17}}{24},$

Line integral of vector fields

Let C be a smooth curve whose parametric representation is as given in Eqs. (15.54) and (15.55). Let

$$\mathbf{v}(x, y, z) = v_1(x, y, z)\mathbf{i} + v_2(x, y, z)\mathbf{j} + v_3(x, y, z)\mathbf{k}$$

be a vector field that is continuous on C. Then, the line integral of $\mathbf{v}$ over C is defined by

$$\int_C \mathbf{v} \cdot d\mathbf{r} = \int_C v_1\, dx + v_2\, dy + v_3\, dz$$

$$= \int_a^b \mathbf{v}\left[x(t), y(t), z(t)\right] \cdot \frac{d\mathbf{r}}{dt}\, dt. \tag{15.57}$$

If $\mathbf{v} = v_1(x, y, z)\,\mathbf{i}$, then Eq. (15.57) reduces to

$$\int_C \mathbf{v} \cdot d\mathbf{r} = \int_C v_1\, dx = \int_a^b v_1\left[x(t), y(t), z(t)\right] \frac{dx}{dt}\, dt \tag{15.58}$$

Similarly, if $\mathbf{v} = v_2(x, y, z)\,\mathbf{j}$ or $\mathbf{v} = v_3(x, y, z)\,\mathbf{k}$, we respectively obtain

$$\int_C \mathbf{v} \cdot d\mathbf{r} = \int_C v_2\, dy = \int_a^b v_2\left[x(t), y(t), z(t)\right] \frac{dy}{dt}\, dt \tag{15.59}$$

and $$\int_C \mathbf{v} \cdot d\mathbf{r} = \int_C v_3\, dz = \int_a^b v_3\left[x(t), y(t), z(t)\right] \frac{dz}{dt}\, dt. \tag{15.60}$$

If the curve C is piecewise smooth containing the arcs $C_1, C_2, \ldots, C_n$, then we write

$$\int_C \mathbf{v} \cdot d\mathbf{r} = \int_{C_1} \mathbf{v} \cdot d\mathbf{r} + \int_{C_2} \mathbf{v} \cdot d\mathbf{r} + \ldots + \int_{C_n} \mathbf{v} \cdot d\mathbf{r}. \tag{15.61}$$

Example 15.27 Evaluate the line integral of $v = xy\,i + y^2 j + e^z k$ over the curve C whose parametric representation is given by $x = t^2$, $y = 2t$, $z = t$, $0 \le t \le 1$.

Solution The position vector of any point on C is given by $r = t^2\,i + 2t\,j + t\,k$. We have

$$\int_C v \cdot dr = \int_C v \cdot \frac{dr}{dt}\,dt = \int_0^1 (2t^3 i + 4t^2 j + e^t k) \cdot (2t i + 2j + k)\,dt$$

$$= \int_0^1 (4t^4 + 8t^2 + e^t)\,dt = \left[\frac{4}{5}t^5 + \frac{8}{3}t^3 + e^t\right]_0^1 = \frac{37}{15} + e.$$

Example 15.28 Evaluate the line integral of $v = x^2\,i - 2y\,j + z^2\,k$ over the straight line path from $(-1, 2, 3)$ to $(2, 3, 5)$.

Solution The parametric representation of the straight line is given by

$$r(t) = (-i + 2j + 3k) + t(3i + j + 2k) = (-1 + 3t)\,i + (2 + t)j + (3 + 2t)\,k, \quad 0 \le t \le 1$$

[If a and b are the two points, then the parametric representation of the line joining them is $r = a + t(b - a)$]. Therefore, $\dfrac{dr}{dt} = 3i + j + 2k$ and

$$\int_C v \cdot dr = \int_C v \cdot \frac{dr}{dt}\,dt = \int_0^1 [3(-1 + 3t)^2 - 2(2 + t) + 2(3 + 2t)^2]\,dt$$

$$= \int_0^1 (17 + 4t + 35t^2)\,dt = \left[17t + 2t^2 + \frac{35}{3}t^3\right]_0^1 = \frac{92}{3}.$$

Example 15.29 Evaluate the integral $\displaystyle\int_C (x^2 + yz)dz$, where C is given by $x = t$, $y = t^2$, $z = 3t$, $1 \le t \le 2$.

Solution We have (using Eq. (15.60))

$$\int_C (x^2 + yz)\,dz = \int_1^2 (t^2 + 3t^3)\,3\,dt = 3\left[\frac{1}{3}t^3 + \frac{3}{4}t^4\right]_1^2$$

$$= 3\left[\frac{7}{3} + \frac{45}{4}\right] = \frac{163}{4}.$$

Line integral of scalar fields

Let C be a smooth curve whose parametric representation is as given in Eqs. (15.54) and (15.55). Let $f(x, y, z)$, $g(x, y, z)$ and $h(x, y, z)$ be scalar fields which are continuous at points over C. Then, we define a line integral as

$$\int_C f(x, y, z)\,dx + g(x, y, z)\,dy + h(x, y, z)\,dz$$

$$= \int_a^b \left[f(x(t), y(t), z(t))\frac{dx}{dt} + g(x(t), y(t), z(t))\frac{dy}{dt} + h(x(t), y(t), z(t))\frac{dz}{dt}\right]dt. \quad (15.62)$$

This line integral does not contain any vector field, but involves three scalar fields. However, if we define $\mathbf{v} = f(x, y, z)\,\mathbf{i} + g(x, y, z)\,\mathbf{j} + h(x, y, z)\,\mathbf{k}$, and $d\mathbf{r} = \mathbf{i}\,dx + \mathbf{j}\,dy + \mathbf{k}\,dz$, then the line integral in Eq. (15.62) is same as the line integral in Eq. (15.57).

If C is a closed curve, then we usually write

$$\int_C \mathbf{v} \cdot d\mathbf{r} = \oint_C \mathbf{v} \cdot d\mathbf{r}. \tag{15.63}$$

Example 15.30   Evaluate

$$\int_C (x + y)dx - x^2 dy + (y + z)\,dz$$

where C is $x^2 = 4y$, $z = x$, $0 \le x \le 2$.

Solution   We parametrise C as $x = t$, $y = t^2/4$, $z = t$, $0 \le t \le 2$. Therefore,

$$\int_C (x + y)\,dx - x^2 dy + (y + z)dz = \int_0^2 \left[\left(t + \frac{t^2}{4} \right) - t^2 \left(\frac{t}{2} \right) + \left(\frac{t^2}{4} + t \right) \right] dt$$

$$= \int_0^2 \left(2t + \frac{t^2}{2} - \frac{t^3}{2} \right) dt = \left(t^2 + \frac{t^3}{6} - \frac{t^4}{8} \right)_0^2 = \frac{10}{3}.$$

Application of line integrals

Work done by a force   Let $\mathbf{v}(x, y, z) = v_1(x, y, z)\,\mathbf{i} + v_2(x, y, z)\,\mathbf{j} + v_3(x, y, z)\,\mathbf{k}$ be a vector function defined and continuous at every point on C. Then, the integral of the tangential component of $\mathbf{v}$ along the curve C from a point P to the point Q is given by

$$\int_P^Q \mathbf{v} \cdot d\mathbf{r} = \int_{C*} \mathbf{v} \cdot d\mathbf{r} = \int_{C*} v_1\,dx + v_2\,dy + v_3\,dz.$$

where $C*$ is the part of C, whose initial and terminal points are P and Q.

Let now $\mathbf{v} = \mathbf{F}$, a variable force acting on a particle which moves along a curve C. Then, the work W done by the force $\mathbf{F}$ in displacing the particle from the point P to the point Q along the curve C is given by

$$W = \int_P^Q \mathbf{F} \cdot d\mathbf{r} = \int_{C*} \mathbf{F} \cdot d\mathbf{r} \tag{15.64}$$

where $C*$ is the part of C, whose initial and terminal points are P and Q.

Suppose that $\mathbf{F}$ is a conservative vector field. Then, $\mathbf{F}$ can be written as $\mathbf{F} = \text{grad}\,f$, where f is a scalar potential (field). Then, the work done

$$W = \int_{C*} \mathbf{F} \cdot dr = \int_{C*} (\text{grad}\,f) \cdot dr$$

$$= \int_{C*} \left(\frac{\partial f}{\partial x}\,dx + \frac{\partial f}{\partial y}\,dy + \frac{\partial f}{\partial z}\,dz \right) = \int_P^Q df = \left[f(x, y, z) \right]_P^Q \tag{15.65}$$

Therefore, work done depends only on the initial and terminal points of the curve $C*$, that is the work done is independent of the path of integration. The units of work depend on the units of $|\mathbf{F}|$ and on the units of distance.

Example 15.31 Find the work done by the force $\mathbf{F} = -xy\,\mathbf{i} + y^2\,\mathbf{j} + z\,\mathbf{k}$ in moving a particle over the circular path $x^2 + y^2 = 4$, $z = 0$ from $(2, 0, 0)$ to $(0, 2, 0)$.

Solution The parametric representation of the given curve is $x = 2\cos t$, $y = 2\sin t$, $z = 0$, $0 \le t \le \pi/2$. Therefore, work done W is given by

$$W = \int_C \mathbf{F} \cdot d\mathbf{r} = \int_C -xy\,dx + y^2\,dy + z\,dz$$

$$= \int_0^{\pi/2} [-4\sin t \cos t\,(-2\sin t) + 4\sin^2 t\,(2\cos t)]\,dt = 16\int_0^{\pi/2} \sin^2 t \cos t\,dt$$

$$= 16\left[\frac{1}{3}\sin^3 t\right]_0^{\pi/2} = \frac{16}{3}.$$

Example 15.32 Show that the vector field $\mathbf{F} = 2x\,(y^2 + z^3)\,\mathbf{i} + 2x^2y\,\mathbf{j} + 3x^2z^2\,\mathbf{k}$ is conservative. Find its scalar potential and the work done in moving a particle from $(-1, 2, 1)$ to $(2, 3, 4)$.

Solution We have

$$\text{curl }(\mathbf{F}) = \begin{vmatrix} \mathbf{i} & \mathbf{j} & \mathbf{k} \\ \partial/\partial x & \partial/\partial y & \partial/\partial z \\ 2x\,(y^2 + z^3) & 2x^2y & 3x^2z^2 \end{vmatrix} = \mathbf{i}\,(0) - \mathbf{j}\,(6xz^2 - 6xz^2) + \mathbf{k}\,(4xy - 4xy) = \mathbf{0}.$$

Therefore, the vector field $\mathbf{F}$ is conservative. We have $\mathbf{F} = \text{grad }f$. Hence,

$$\frac{\partial f}{\partial x} = 2x\,(y^2 + z^3),\ \frac{\partial f}{\partial y} = 2x^2y,\ \frac{\partial f}{\partial z} = 3x^2z^2.$$

Integrating the first equation, we get

$$f(x, y, z) = x^2(y^2 + z^3) + g(y, z).$$

Substituting in the second equation, we get

$$\frac{\partial f}{\partial y} = 2x^2y = 2x^2y + \frac{\partial g}{\partial y},\ \text{or}\ \frac{\partial g}{\partial y} = 0,\ \text{or}\ g = h(z).$$

Substituting in the third equation, we get

$$\frac{\partial f}{\partial z} = 3x^2z^2 = 3x^2z^2 + \frac{dh}{dz},\ \text{or}\ \frac{dh}{dz} = 0,\ \text{or}\ h(z) = c, \text{constant}.$$

Therefore, the scalar potential is given by $f(x, y, z) = x^2(y^2 + z^3) + c$.

Hence, the work done by $\mathbf{F}$ in moving a particle from $P\,(-1, 2, 1)$ to $Q(2, 3, 4)$ is (see Eq. (15.65))

$$W = \int_P^Q \mathbf{F} \cdot d\mathbf{r} = \left[f(x, y, z)\right]_P^Q = \left[x^2\,(y^2 + z^3)\right]_{(-1,2,1)}^{(2,3,4)} = 287.$$

Circulation A line integral of a vector field **v** around a simple closed curve C is defined as the circulation of **v** around C.

$$\text{circulation} = \oint_C \mathbf{v} \cdot d\mathbf{r} = \oint_C \mathbf{v} \cdot \frac{d\mathbf{r}}{ds}\, ds = \oint_C \mathbf{v} \cdot \mathbf{T}\, ds \tag{15.66}$$

where **T** is the tangent vector to C. For example, in fluid mechanics, let **v** represent the velocity field of a fluid and C be a closed curve in its domain. Then, circulation gives the amount by which the fluid tends to turn the curve by rotating or circulating around C. If $\oint_C \mathbf{v} \cdot \mathbf{T}\, ds > 0$, then the fluid tends to rotate C in the anti-clockwise direction, while if $\oint_C \mathbf{v} \cdot \mathbf{T}\, ds < 0$, then the fluid tends to rotate C in the clockwise direction. If **v** acts in the direction perpendicular to **T** at every point on C, then $\oint_C \mathbf{v} \cdot \mathbf{T}\, ds = 0$, that is the curve does not move at all.

15.5.1 Line Integrals Independent of the Path

We have seen that the value of $\displaystyle\int_C \mathbf{F} \cdot d\mathbf{r}$ or $\displaystyle\int_C f\, dx + g\, dy + h\, dz$ depends not only on the end points P and Q of the curve C but also on the path of C. We shall now derive the conditions under which the line integral is independent of the path of integration, that is, it depends only on the end points P and Q of the curve C.

Let $\phi(x, y, z)$ be a differentiable scalar function. The differential of $\phi(x, y, z)$ is defined by

$$d\phi = \frac{\partial \phi}{\partial x}\, dx + \frac{\partial \phi}{\partial y}\, dy + \frac{\partial \phi}{\partial z}\, dz = (\text{grad } \phi) \cdot d\mathbf{r}. \tag{15.67}$$

Therefore, a differential expression $f(x, y, z)\, dx + g(x, y, z)\, dy + h(x, y, z)\, dz$ is an exact differential, if there exists a scalar function $\phi(x, y, z)$ such that

$$d\phi = f(x, y, z)\, dx + g(x, y, z)\, dy + h(x, y, z)\, dz. \tag{15.68}$$

We now present the result on the independence of the path of a line integral.

Theorem 15.1 Let C be a curve in a simply connected domain D in space. Let f, g and h be continuous functions having continuous first partial derivatives in D. Then, $\displaystyle\int_C f\, dx + g\, dy + h\, dz$ is independent of path C if and only if the integrand is an exact differential in D.

Proof (a) Suppose that $f\, dx + g\, dy + h\, dz$ is an exact differential in D. Then, there exists a scalar function ϕ such that

$$f = \frac{\partial \phi}{\partial x},\; g = \frac{\partial \phi}{\partial y},\; \text{and}\; h = \frac{\partial \phi}{\partial z}. \tag{15.69}$$

Let the parametric form of C be $\mathbf{r}(t) = x(t)\, \mathbf{i} + y(t)\mathbf{j} + z(t)\mathbf{k}$, $a \le t \le b$. The initial and terminal points P and Q of C correspond to $t = a$ and $t = b$ respectively. Then

$$\int_C f\, dx + g\, dy + h\, dz = \int_P^Q \left(\frac{\partial \phi}{\partial x}\, dx + \frac{\partial \phi}{\partial y}\, dy + \frac{\partial \phi}{\partial z}\, dz \right)$$

$$= \int_a^b \left(\frac{\partial \phi}{\partial x} \frac{dx}{dt} + \frac{\partial \phi}{\partial y} \frac{dy}{dt} + \frac{\partial \phi}{\partial z} \frac{dz}{dt} \right) dt$$

$$= \int_a^b \frac{d\phi}{dt} \, dt = \left[\phi(x(t), y(t), z(t)) \right]_a^b = \phi(Q) - \phi(P).$$

Therefore, the line integral is independent of the path C.

(b) Suppose that the line integral is independent of the path C. Choose a fixed point $P(x_0, y_0, z_0)$ and an arbitrary point Q in D (Fig. 15.11). Define a function $\phi(x, y, z)$ as

$$\phi(x, y, z) = \phi(x_0, y_0, z_0) + \int_{C^*} (f \, dx^* + g \, dy^* + h \, dz^*) \tag{15.70}$$

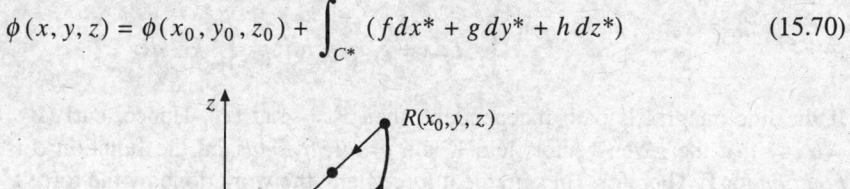

Fig. 15.11. Path in theorem 15.1.

where C^* is any arbitrary path from P to Q. Since P is a fixed point, the right hand side is a function of Q, that is, it depends on x, y, z. We need to show that the relations given in Eq. (15.69) hold. We prove the first relation. Since the line integral is independent of the path, let us choose the path from $P(x_0, y_0, z_0)$ to $R(x_0, y, z)$, then from $R(x_0, y, z)$ to $Q(x, y, z)$. Note that RQ is in a direction parallel to the x-axis.

Then, $$\phi(x, y, z) = \phi(x_0, y_0, z_0) + \int_P^R (f \, dx^* + g \, dy^* + h \, dz^*) + \int_R^Q (f \, dx^* + g \, dy^* + h \, dz^*). \tag{15.71}$$

We note that the coordinates of P and R do not depend on x. Now, take the partial derivative of Eq. (15.71) with respect to x. The derivatives of the first two terms on the right hand side are zero. Now, since on the line segment RQ, y and z are constant we have $dy = 0$, $dz = 0$ and the second integral on the right hand side can be written as

$$\int_{x'}^x f(x^*, y, z) \, dx^*.$$

Hence, the partial derivative of this integral with respect to x is $f(x, y, z)$. Therefore, we have $\partial \phi / \partial x = f(x, y, z)$. Similarly, choosing suitable paths of integration we can show that $\partial \phi / \partial y = g(x, y, z)$ and $\partial \phi / \partial z = h(x, y, z)$. Therefore, integrand of the given integral is an exact differential.

We now state the conditions for testing the path independence.

Theorem 15.2   Let C be a curve in a simply connected domain D in space. Let f, g and h be continuous functions having continuous first partial derivatives in D. Then, $\int_C f\,dx + g\,dy + h\,dz$ is independent of path C if and only if

$$\frac{\partial f}{\partial y} = \frac{\partial g}{\partial x}, \frac{\partial f}{\partial z} = \frac{\partial h}{\partial x}, \text{ and } \frac{\partial g}{\partial z} = \frac{\partial h}{\partial y}. \tag{15.72}$$

Remark 9

If we define $\mathbf{F} = f\,\mathbf{i} + g\,\mathbf{j} + h\,\mathbf{k}$ then we can write

$$\int_C f\,dx + g\,dy + h\,dz = \int_C \mathbf{F} \cdot d\mathbf{r}.$$

If the line integral is path independent, then $\mathbf{F} = \text{grad}\,(\phi)$. Hence, curl $(\mathbf{F}) = $ curl (grad ϕ) $= \mathbf{0}$. We say that the given vector field $\mathbf{F}$ is a *gradient field* and the function ϕ is called the *potential function* for $\mathbf{F}$. Therefore, in a gradient force field, the work done by the force $\mathbf{F}$ in moving a particle from a position P to a position Q is independent of the path of integration, that is, it is same for all paths. Such a force field is also called a *conservative field*, that is

total energy = kinetic energy + potential energy = constant.

Remark 10

If $\mathbf{F}$ is a conservative force field, the work done along any simple closed path is zero.

Remark 11

In two dimensions, the conditions for testing the path independence of $\int_C f\,dx + g\,dy$ reduce to $\partial f/\partial y = \partial g/\partial x$.

Example 15.33   Show that $\int_C \dfrac{x\,dx + y\,dy}{\sqrt{x^2 + y^2}}$ is independent of any path of integration which does not pass through the origin. Find the value of the integral from the point $P\,(-1, 2)$ to the point $Q\,(2, 3)$.

Solution   We have $f(x, y) = \dfrac{x}{\sqrt{x^2 + y^2}}$ and $g\,(x, y) = \dfrac{y}{\sqrt{x^2 + y^2}}$.

Now, $\dfrac{\partial f}{\partial y} = -\dfrac{xy}{(x^2 + y^2)^{3/2}}$ and $\dfrac{\partial g}{\partial x} = -\dfrac{xy}{(x^2 + y^2)^{3/2}}$.

Since $\partial f/\partial y = \partial g/\partial x$, the integral is independent of any path of integration which does not pass through the origin. Also, the integrand is an exact differential. Therefore, there exists a function $\phi\,(x, y)$ such that

$$\frac{\partial \phi}{\partial x} = f(x, y) = \frac{x}{\sqrt{x^2 + y^2}} \quad \text{and} \quad \frac{\partial \phi}{\partial y} = g\,(x, y) = \frac{y}{\sqrt{x^2 + y^2}}.$$

Integrating the first equation with respect to x, we get $\phi\,(x, y) = \sqrt{x^2 + y^2} + h\,(y)$. Substituting in the second equation, we obtain

$$\frac{\partial \phi}{\partial y} = \frac{y}{\sqrt{x^2 + y^2}} = \frac{y}{\sqrt{x^2 + y^2}} + \frac{dh}{dy}, \quad \text{or} \quad \frac{dh}{dy} = 0, \quad \text{or} \quad h(y) = k, \text{constant}.$$

Hence, $\phi(x, y) = \sqrt{x^2 + y^2} + k.$

Therefore, $\displaystyle \int_C \frac{x\,dx + y\,dy}{\sqrt{x^2 + y^2}} = \int_{(-1,2)}^{(2,3)} d(\sqrt{x^2 + y^2}) = \left[\sqrt{x^2 + y^2}\right]_{(-1,2)}^{(2,3)} = \sqrt{13} - \sqrt{5},$

Example 15.34 Show that $\displaystyle \int_C (yz - 1)dx + (z + xz + z^2)\,dy + (y + xy + 2yz)dz$ is independent of the path of integration from $(1, 2, 2)$ to $(2, 3, 4)$. Evaluate the integral.

Solution We have $f(x, y, z) = yz - 1$, $g(x, y, z) = z + xz + z^2$, and $h(x, y, z) = y + xy + 2yz$.

Now, $\displaystyle \frac{\partial f}{\partial y} = z = \frac{\partial g}{\partial x}, \quad \frac{\partial f}{\partial z} = y = \frac{\partial h}{\partial x}, \quad \text{and} \quad \frac{\partial g}{\partial z} = 1 + x + 2z = \frac{\partial h}{\partial y}.$

The integral is independent of path of integration. Also, the integrand is an exact differential. Therefore, there exists a function $\phi(x, y, z)$ such that

$$\frac{\partial \phi}{\partial x} = yz - 1, \quad \frac{\partial \phi}{\partial y} = z + xz + z^2, \quad \text{and} \quad \frac{\partial \phi}{\partial z} = y + xy + 2yz.$$

Integrating the first equation with respect to x, we get

$$\phi(x, y, z) = xyz - x + h(y, z).$$

Substituting in the second equation, we get

$$\frac{\partial \phi}{\partial y} = z + xz + z^2 = xz + \frac{\partial h}{\partial y}(y, z), \quad \text{or} \quad \frac{\partial h}{\partial y} = z + z^2.$$

Integrating, we get

$$h(y, z) = yz + yz^2 + s(z), \quad \text{and} \quad \phi(x, y, z) = xyz - x + yz + yz^2 + s(z).$$

Substituting in the third equation, we get

$$\frac{\partial \phi}{\partial z} = xy + y + 2yz + \frac{ds}{dz} = y + xy + 2yz, \quad \text{or} \quad \frac{ds}{dz} = 0, \quad \text{or} \quad s = k, \text{constant}.$$

Therefore, $\phi(x, y, z) = xyz - x + yz + yz^2 + k.$

The value of the integral is

$$\int_C (yz - 1)dx + (z + xz + z^2)\,dy + (y + xy + 2yz)\,dz = \int_{(1,2,2)}^{(2,3,4)} d(xyz - x + yz + yz^2)$$

$$= \left[xyz - x + yz + yz^2\right]_{(1,2,2)}^{(2,3,4)} = 82 - 15 = 67.$$

15.5.2 Green's Theorem

In chapter 2, we have defined the definite integral and discussed methods for evaluating it. The vector field theory provides three important theorems. One of the theorems is the Green's theorem, which provides a relationship between a double integral over a region R and the line integral over the closed curve C bounding R. Green's theorem is also called the *first fundamental theorem* of integral vector calculus.

Theorem 15.3 (Green's theorem)

Let C be a piecewise smooth simple closed curve bounding a region R. If f, g, $\partial f / \partial y$ and $\partial g / \partial x$ are continuous on R, then

$$\oint_C f(x, y)\, dx + g(x, y)\, dy = \iint_R \left(\frac{\partial g}{\partial x} - \frac{\partial f}{\partial y} \right) dx\, dy \tag{15.73}$$

the integration being carried in the positive direction (counter clockwise direction) of C.

Proof We shall prove Green's theorem for a particular case of the region R.

Let the region R be simultaneously expressed in the following forms (Figs. 15.12 and 15.13)

$$R : u_1(x) \le y \le u_2(x), \qquad a \le x \le b \qquad \text{(Fig. 15.12)}, \tag{15.74}$$

$$R : v_1(y) \le x \le v_2(y), \qquad c \le y \le d \qquad \text{(Fig. 15.13)}. \tag{15.75}$$

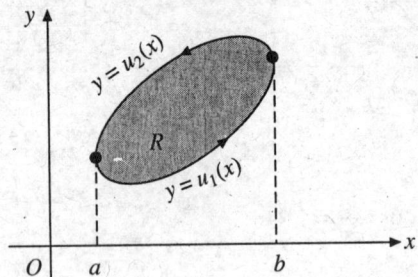

Fig. 15.12. Region R (Green's theorem).

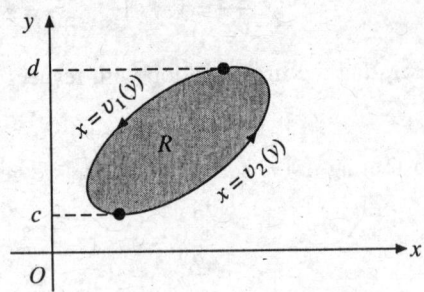

Fig. 15.13. Region R (Green's theorem).

Using Fig. 15.13, we obtain

$$\iint_R \frac{\partial g}{\partial x}\, dx\, dy = \int_c^d \left[\int_{v_1(y)}^{v_2(y)} \frac{\partial g}{\partial x}\, dx \right] dy = \int_c^d \left[g(v_2(y), y) - g(v_1(y), y) \right] dy$$

$$= \int_c^d g(v_2(y), y)\, dy + \int_d^c g(v_1(y), y)\, dy = \oint_C g(x, y)\, dy$$

the integration being carried in the counter clockwise direction.

Using Fig. 15.12, we obtain

$$\iint_R \frac{\partial f}{\partial y}\, dx\, dy = \int_a^b \left[\int_{u_1(x)}^{u_2(x)} \frac{\partial f}{\partial y}\, dy \right] dx = \int_a^b \left[f(x, u_2(x)) - f(x, u_1(x)) \right] dx$$

$$= \int_a^b f(x, u_2(x))\, dx + \int_b^a f(x, u_1(x))\, dx = - \oint_C f(x, y)\, dx$$

the integration being carried in the counter clockwise direction.
Therefore,

$$\iint_R \left(\frac{\partial g}{\partial x} - \frac{\partial f}{\partial y} \right) dx\, dy = \oint_C f(x, y)\, dx + g(x, y)\, dy.$$

The proof can be extended to more general regions R. The region R is decomposed into a finite number of subregions $R_1, R_2, \ldots, R_n$ such that each region can be expressed in both the forms given in Eqs. (15.74) and (15.75) (Figs. 15.14a, b). We illustrate the theorem through the following examples.

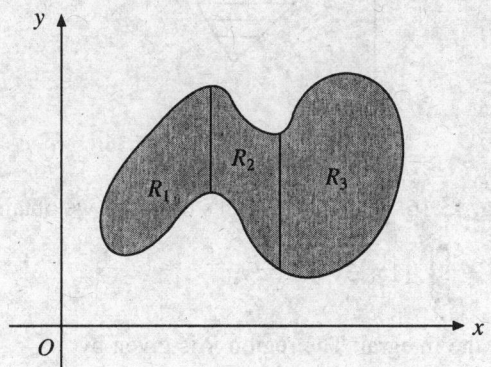

| Fig. 15.14a. Decomposition of R. | Fig. 15.14b. Decomposition of R. |

Example 15.35 Evaluate $\oint_C (x^2 + y^2)\, dx + (y + 2x)\, dy$, where C is the boundary of the region in the first quadrant that is bounded by the curves $y^2 = x$ and $x^2 = y$.

Solution. The curves intersect at $(0, 0)$ and $(1, 1)$. The bounding curve C is along $OQPSO$ (Fig. 15.15). We have $f(x, y) = x^2 + y^2$ and $g(x, y) = y + 2x$.

Using the Green's theorem, we obtain

$$\oint_C (x^2 + y^2)\, dx + (y + 2x)\, dy = \iint_R (2 - 2y)\, dx\, dy$$

$$= \int_0^1 \left[\int_{x^2}^{\sqrt{x}} (2 - 2y)\, dy \right] dx = \int_0^1 \left[2y - y^2 \right]_{x^2}^{\sqrt{x}} dx$$

$$= \int_0^1 (2\sqrt{x} - x - 2x^2 + x^4)\, dx = \frac{4}{3} - \frac{1}{2} - \frac{2}{3} + \frac{1}{5} = \frac{11}{30}.$$

Example 15.36 Find the work done by the force $\mathbf{F} = (x^2 - y^3)\, \mathbf{i} + (x + y)\, \mathbf{j}$ in moving a particle along the closed path C containing the curves $x + y = 0$, $x^2 + y^2 = 16$ and $y = x$ in the first and fourth quadrants.

Solution The work done by the force is given by

$$W = \oint_C \mathbf{F} \cdot d\mathbf{r} = \oint_C (x^2 - y^3)\, dx + (x + y)\, dy.$$

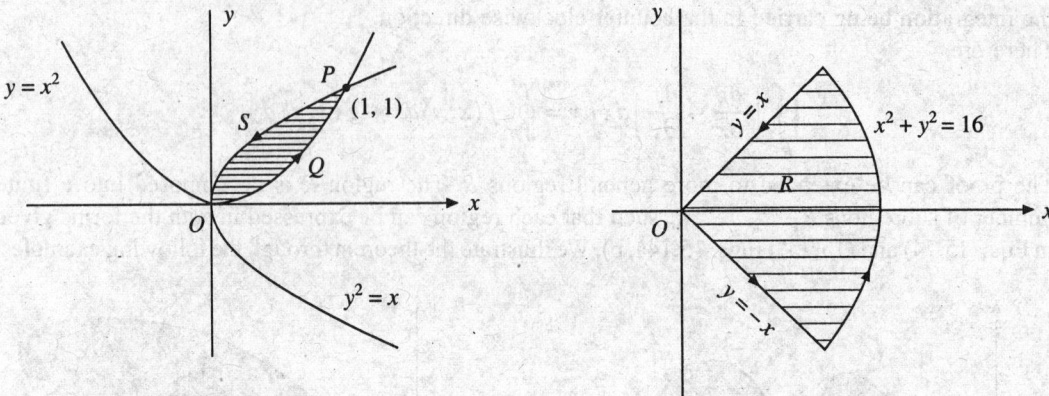

Fig. 15.15.   Example 15.35.   **Fig. 15.16.   Example 15.36.**

The closed path C bounds the region R as given in Fig. 15.16. Using the Green's theorem, we obtain

$$\oint_C (x^2 - y^3)\, dx + (x + y)\, dy = \iint_R (1 + 3y^2)\, dx\, dy.$$

It is convenient to use polar coordinates to evaluate the integral. The region R is given by

$$R : x = r \cos\theta,\ y = r \sin\theta,\ 0 \le r \le 4,\ -\pi/4 \le \theta \le \pi/4.$$

Therefore,

$$\iint_R (1 + 3y^2)\, dx\, dy = \int_{-\pi/4}^{\pi/4} \left[\int_0^4 (1 + 3r^2 \sin^2\theta)\, r\, dr \right] d\theta = \int_{-\pi/4}^{\pi/4} \left[\frac{r^2}{2} + \frac{3}{4} r^4 \sin^2\theta \right]_0^4 d\theta$$

$$= \int_{-\pi/4}^{\pi/4} (8 + 192 \sin^2\theta)\, d\theta = 2 \int_0^{\pi/4} [8 + 96\, (1 - \cos 2\theta)]\, d\theta$$

$$= 2 \left[104\, \theta - 48 \sin 2\theta \right]_0^{\pi/4} = 52\pi - 96.$$

Example 15.37   Verify the Green's theorem for $f(x, y) = e^{-x} \sin y$, $g(x, y) = e^{-x} \cos y$ and C is the square with vertices at $(0, 0)$, $(\pi/2, 0)$, $(\pi/2, \pi/2)$, $(0, \pi/2)$.

Solution   We can write the line integral as

$$\oint_C f\, dx + g\, dy = \left[\int_{C_1} + \int_{C_2} + \int_{C_3} + \int_{C_4} \right] (f\, dx + g\, dy)$$

where C_1, C_2, C_3 and C_4 are the boundary lines as given in Fig. 15.17. We have

along C_1: $y = 0$, $0 \le x \le \pi/2$ and

$$\int_{C_1} e^{-x} (\sin y\, dx + \cos y\, dy) = 0,$$

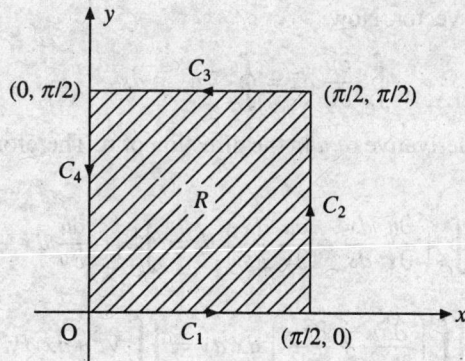

Fig. 15.17. Example 15.37.

along C_2: $x = \pi/2$, $0 \le y \le \pi/2$ and

$$\int_{C_2} e^{-x} (\sin y \, dx + \cos y \, dy) = \int_0^{\pi/2} e^{-\pi/2} \cos y \, dy = e^{-\pi/2},$$

along C_3: $y = \pi/2$, $\pi/2 \le x \le 0$ and

$$\int_{C_3} e^{-x} (\sin y \, dx + \cos y \, dy) = \int_{\pi/2}^0 e^{-x} dx = e^{-\pi/2} - 1,$$

along C_4: $x = 0$, $\pi/2 \le y \le 0$ and

$$\int_{C_4} e^{-x} (\sin y \, dx + \cos y \, dy) = \int_{\pi/2}^0 \cos y \, dy = -1.$$

Therefore, $$\oint_C f \, dx + g \, dy = e^{-\pi/2} + e^{-\pi/2} - 1 - 1 = 2(e^{-\pi/2} - 1).$$

Using Green's theorem, we obtain

$$\oint_C f \, dx + g \, dy = \iint_R (-2e^{-x} \cos y) \, dx \, dy = \int_0^{\pi/2} \int_0^{\pi/2} -2e^{-x} \cos y \, dx \, dy = 2(e^{-\pi/2} - 1).$$

Example 15.38 Using the Green's theorem, show that

$$\oint_C \frac{\partial u}{\partial n} \, ds = \iint_R \nabla^2 u \, dx \, dy$$

where ∇^2 is the Laplace operator $\partial^2/\partial x^2 + \partial^2/\partial y^2$ and $\mathbf{n}$ is the unit outward normal to C.

Solution Let the position vector of a point on C, in terms of the arc length s be $\mathbf{r}(s) = x(s)\mathbf{i} + y(s)\mathbf{j}$. Then, the tangent vector to C is given by

$$\mathbf{T} = \frac{d\mathbf{r}}{ds} = \frac{dx}{ds}\mathbf{i} + \frac{dy}{ds}\mathbf{j}$$

and the normal vector $\mathbf{n}$ is given by (since $\mathbf{n} \cdot \mathbf{T} = 0$)

$$\mathbf{n} = \frac{dy}{ds}\mathbf{i} - \frac{dx}{ds}\mathbf{j}.$$

Note that **n** is the unit normal vector. Now,

$$\oint_C \frac{\partial u}{\partial n}\, ds = \oint_C \nabla u \cdot \mathbf{n}\, ds$$

since $\partial u/\partial n$ is the directional derivative of u in the direction of **n**. Therefore, using Green's theorem, we obtain

$$\oint_C \frac{\partial u}{\partial n}\, ds = \oint_C \left(\frac{\partial u}{\partial x} \frac{dy}{ds} - \frac{\partial u}{\partial y} \frac{dx}{ds} \right) ds = \oint_C \left(-\frac{\partial u}{\partial y}\, dx + \frac{\partial u}{\partial x}\, dy \right)$$

$$= \iint_R \left(\frac{\partial^2 u}{\partial x^2} + \frac{\partial^2 u}{\partial y^2} \right) dx\, dy = \iint_R \nabla^2 u\, dx\, dy.$$

Exercise 15.4

In problems **1** to **8**, evaluate the line integral $\displaystyle\int_C f(x, y)\, ds$ or $\displaystyle\int_C f(x, y, z)\, ds$.

1. $f(x, y, z) = 2x + 3y$, C is given by $x = t$, $y = 2t$, $z = 3t$, $0 \le t \le 3$.

2. $f(x, y, z) = xy^2 z$, C is the line segment from $(1, 2, 2)$ to $(2, 3, 5)$.

3. $f(x, y) = x^2 - y^2$, C is the closed curve $x = 3 \cos t$, $y = 3 \sin t$, $0 \le t \le 2\pi$.

4. $f(x, y, z) = 3x + 2z$, C is the parabola $y = z^2$, $x = 1$ for $0 \le z \le 4$.

5. $f(x, y, z) = x + 2y + z$, C is given by $x = 5 \cos t$, $y = 3$, $z = 5 \sin t$, $0 \le t \le \pi$.

6. $f(x, y, z) = y + z$, C is given by $x = t^2$, $y = t$, $z = 2$, $0 \le t \le 3$.

7. $f(x, y, z) = x + y^2 + yz$, C is the curve $y = 2x$, $z = 2$ from $(1, 2, 2)$ to $(3, 6, 2)$.

8. $f(x, y, z) = x^2 + y + z^2$, C is the curve $2x = y = z$ for $2 \le x \le 4$.

In problems **9** to **15**, evaluate the line integral $\displaystyle\int_C \mathbf{v} \cdot d\mathbf{r}$.

9. $\mathbf{v} = x\mathbf{i} + y\mathbf{j} + z\mathbf{k}$, C is the line segment from $(1, 2, 2)$ to $(3, 6, 6)$.

10. $\mathbf{v} = x\mathbf{i} + y\mathbf{j} - z\mathbf{k}$, C is the circle $x^2 + y^2 = 9$, $z = 0$ going around once in the anti-clockwise direction.

11. $\mathbf{v} = x\mathbf{i} + (\sin y)\mathbf{j} + \mathbf{k}$, C is given by $x = t^2$, $y = t$, $z = 2t$, $0 \le t \le 1$.

12. $\mathbf{v} = x^2 y\, \mathbf{i} - xy^2\, \mathbf{j}$, $\mathbf{r}(t) = t\mathbf{i} + t^2\mathbf{j}$, $0 \le t \le 3$.

13. $\mathbf{v} = e^x \mathbf{i} + x e^{xy} \mathbf{j} + \mathbf{k}$, $\mathbf{r}(t) = t\mathbf{i} + t^2\mathbf{j} + t^3\mathbf{k}$, $0 \le t \le 2$.

14. $\mathbf{v} = z\mathbf{i} + x\mathbf{j} + y\mathbf{k}$, $\mathbf{r}(t) = (\cos t)\, \mathbf{i} + (\sin t)\, \mathbf{j} + t\mathbf{k}$, $0 \le t \le 2\pi$.

15. $\mathbf{v} = y\mathbf{i} + x\mathbf{j} + xyz^2 \mathbf{k}$, C is the circle $x^2 + y^2 - 2y = 2$, $z = 1$ going around once in the anti-clockwise direction.

In problems **16** to **18**, evaluate the line integrals $\displaystyle\int_C f(x, y)\, dx$ and $\displaystyle\int_C f(x, y)\, dy$

16. $f(x, y) = xy$, $x = 3 \cos t$, $y = 3 \sin t$, $0 \le t \le \pi/4$.

17. $f(x, y) = x^2 + 2x^2 y + 3y^2$, $x = t$, $y = 2t^2$, $0 \le t \le 2$

18. $f(x, y) = x + y$, $x = 2 \cos t$, $y = 3 \sin t$, $0 \le t \le \pi/2$.

In problems **19** to **21**, evaluate the line integrals $\int_C f(x, y, z)\,dx, \int_C f(x, y, z)\,dy$ and $\int_C f(x, y, z)\,dz$.

19. $f(x, y, z) = xyz, x = 2\cos t, y = 2\sin t, z = t, 0 \le t \le \pi/2$.

20. $f(x, y, z) = x + y + z, x = t, y = t, z = t^2, 1 \le t \le 2$.

21. $f(x, y, z) = x^2 - y - z, x = t^2, y = t, z = 2t, 0 \le t \le 1$.

In problems **22** to **27**, evaluate the line integrals $\int_C f(x, y)\,dx + g(x, y)\,dy$.

22. $f(x, y) = y, g(x, y) = x, C$ is $y = 2x^2$ from $(0, 0)$ to $(2, 8)$.

23. $f(x, y) = xy, g(x, y) = x + 2y, C$ is (i) $y = x - 1$, (ii) $x = y^2 + 1$ from $(1, 0)$ to $(2, 1)$.

24. $f(x, y) = y, g(x, y) = x^2, C$ is the line segment from $(1, 1)$ to $(3, 5)$.

25. $f(x, y) = 2xy^2, g(x, y) = 2x + 3y, C$ is the parabola $x = y^2$ from $(1, -1)$ to $(4, 2)$.

26. $f(x, y) = y, g(x, y) = -2x, C$ is given by $x = 3\cos t, y = 2\sin t, 0 \le t \le \pi$.

27. $f(x, y) = (x + y)/(x^2 + y^2), g(x, y) = -(x - y)/(x^2 + y^2), C$ is the circle $x^2 + y^2 = 9$ going around once in the anti-clockwise direction.

In problems **28** to **30**, evaluate the line integrals $\int_C f\,dx + g\,dy + h\,dz$.

28. $f = 2x + y, g = y^2, h = (x + z), C$ is $y^2 = 2x, z = x, 0 \le x \le 2$.

29. $f = z, g = x, h = y, C$ consists of line segments from $(0, 0, 0)$ to $(1, 2, 3)$ and then from $(1, 2, 3)$ to $(3, 4, 5)$.

30. $f = x/(yz), g = y/(xz), h = z/(xy), C$ is given by $x = \cos t, y = \sin t, z = \cos t, \pi/6 \le t \le \pi/3$.

31. Let C be a smooth directed plane curve. Let $\mathbf{T}$ and $\mathbf{n}$ denote the unit tangent and unit normal fields respectively to C, satisfying $\mathbf{n} = \mathbf{T} \times \mathbf{k}$. If $\mathbf{u} = f\,\mathbf{i} + g\,\mathbf{j}$ and $\mathbf{v} = -g\,\mathbf{i} + f\,\mathbf{j}$, show that $\int_C \mathbf{u} \cdot \mathbf{n}\,ds = \int_C \mathbf{v} \cdot \mathbf{T}\,ds$.

In problems **32** to **36**, find the work done by the force $\mathbf{F}$ in moving a particle from a point P to the point Q.

32. $\mathbf{F} = x^2\,\mathbf{i} + yz\,\mathbf{j} + z\,\mathbf{k}, C$ is the line segment from $(1, 2, 2)$ to $(3, 4, 2)$.

33. $\mathbf{F} = yz\,\mathbf{i} + zx\,\mathbf{j} + xy\,\mathbf{k}, C$ is the curve with parametric representation $\mathbf{r}(t) = t\,\mathbf{i} + t^2\,\mathbf{j} + t^3\,\mathbf{k}, 1 \le t \le 2$.

34. $\mathbf{F} = (2x + y)\,\mathbf{i} + (4y - x)\,\mathbf{j}, C$ is taken once round the triangle with vertices at $(2, 2), (4, 2)$ and $(4, 4)$ taken counter clockwise.

35. $\mathbf{F} = 16y\,\mathbf{i} + (3x^2 + 2)\,\mathbf{j}, C$ is the right half of the ellipse $x^2 + a^2y^2 = a^2$ from $(0, 1)$ to $(0, -1)$.

36. $\mathbf{F} = (2x - y - z)\,\mathbf{i} + (x + y - z^2)\,\mathbf{j} + (3x - 2y + 4z)\,\mathbf{k}, C$ is taken once round the circle $x^2 + y^2 = 16$ in the x-y plane, in the anti-clockwise direction.

37. Find the circulation of $\mathbf{F} = e^x[(\sin y)\,\mathbf{i} + (\cos y)\,\mathbf{j}]$ around the curve C, where C is the rectangle with vertices $(0, 0), (2, 0), (2, \pi/2)$ and $(0, \pi/2)$, taken in that order.

Show that in the problems **38** to **45**, the line integrals are independent of path of integration Evaluate the integrals.

38. $\int_P^Q 2xy^2\,dx + (2x^2y + 1)\,dy, \ P:(-1, 2), \ Q:(2, 3)$.

39. $\int_P^Q (1 - \sin x \sin y)\,dx + (1 + \cos x \cos y)\,dy, \ P:(\pi/4, \pi/4), \ Q:(\pi/2, 0)$.

40. $\int_{P}^{Q} \dfrac{-x\,dy+y\,dx}{x^2}$, $P:(1,1)$, $Q:(3,4)$ on any path not crossing the y-axis.

41. $\int_{P}^{Q} (y^3 + 2xy^2)\,dx + (3xy^2 + 2x^2y + 1)\,dy$, $P:(-2,1)$, $Q:(1,2)$,

42. $\int_{P}^{Q} e^{yz}[\sin y\,dx + (xz\sin y + x\cos y)dy + xy\sin y\,dz]$, $P:(1,\pi/4,2)$, $Q:(2,\pi/2,4)$.

43. $\int_{P}^{Q} (3x^2 + 2xyz)\,dx + (1 + x^2z)\,dy + x^2y\,dz$, $P:(1,1,1)$, $Q:(-2,-3,-4)$.

44. $\int_{P}^{Q} (2xz + y)\,dx + (x + z)\,dy + (x^2 + y)dz$, $P:(-1,2,3)$, $Q:(2,2,4)$.

45. $\int_{P}^{Q} yz\cos(x+y-z)(dx+dy-dz) + \sin(x+y-z)(z\,dy + y\,dz)$, $P:(\pi/2,1,1)$, $Q:(\pi,2,1)$.

In problems **46** to **50**, find whether the given vector field **F** is conservative (gradient field). If it is, find the potential function.

46. $\mathbf{F} = \cosh(x+y)\,(\mathbf{i}+\mathbf{j})$. **47.** $\mathbf{F} = (2x + ye^{xy})\mathbf{i} + (2y + xe^{xy})\mathbf{j}$.

48. $\mathbf{F} = yz\,\mathbf{i} + xz\,\mathbf{j} + xy\,\mathbf{k}$. **49.** $\mathbf{F} = 2xy\,\mathbf{i} + (x^2 + 2yz)\,\mathbf{j} + y^2\,\mathbf{k}$.

50. $\mathbf{F} = (y^3 - 3x^2z)\mathbf{i} + 3xy^2\mathbf{j} - x^3\mathbf{k}$.

51. Let C be a positively oriented simple closed path enclosing a simply connected region R. Use Green's theorem to show that

$$\text{area of the region} = R = \oint_{C} x\,dy = -\oint_{C} y\,dx = \frac{1}{2}\oint_{C} x\,dy - y\,dx.$$

In problems **52** to **56**, verify Green's theorem by evaluating both integrals.

52. $\oint_{C} (x+y)dx + x^2\,dy$, C is the triangle with vertices at $(0,0)$, $(2,0)$ and $(2,4)$, taken in that order.

53. $\oint_{C} (xy^2 - 2xy)dx + (x^2y + 3)\,dy$, C is the rectangle with vertices at $(-1,0)$, $(1,0)$, $(1,1)$ and $(-1,1)$, taken in that order.

54. $\oint_{C} x^3\,dy - y^3\,dx$, C is the circle $x = 2\cos\theta$, $y = 2\sin\theta$, $0 \le \theta \le 2\pi$.

55. $\oint_{C} (xy^2 + 2xy)dx + x^2y\,dy$, C is the boundary of the region enclosing $y^2 = 4x$, $x = 3$.

56. $\oint_{C} (x-y)\,dx + 3xy\,dy$, C is the boundary of the region enclosing $x^2 = 4y$ and $y^2 = 4x$.

In problems **57** to **65**, evaluate the line integral using the Green's theorem. The orientation of C is anti-clockwise unless otherwise stated.

57. $\oint_{C} 3x^2y\,dx - 2xy^2\,dy$, C is the boundary of the region $x^2 + y^2 \le 16$, $x \ge 0$, $y \ge 0$.

58. $\oint_C y^2\,dx + x^2\,dy$, C is the circle $(x + 1)^2 + (y - 2)^2 = 16$.

59. $\oint_C (x^2 + y^2)\,dx + (5x^2 - 3y)\,dy$, C is the boundary of the region enclosing $x^2 = 4y$, $y = 4$.

60. $\oint_C xy^2\,dx + 5x^3\,dy$, C is the rectangle with vertices at $(-1, 0)$, $(2, 0)$, $(2, 2)$ and $(-1, 2)$.

61. $\oint_C (x^2 - y^3)\,dx + (x^3 + y^2)\,dy$, C is the ellipse $x^2 + 4y^2 = 64$.

62. $\oint_C (y^3 - xy)\,dx + (xy + 3xy^2)\,dy$, C is the boundary of the region in the first quadrant enclosed by $x = 0$, $y = 1 - x^2$, and $y = x^2$.

63. $\oint_C e^x (\sin y\,dx + \cos y\,dy)$, C is the ellipse $4(x + 1)^2 + 9(y - 3)^2 = 36$.

64. $\oint_C xy\,dx + x^3\,dy$, C is the boundary of the region enclosed by $y = 0$, $x^2 + y^2 = 4$, $y \geq 0$.

65. $\oint_C x^2 y\,dx + y^2\,dy$, C is the boundary of the region enclosed by $y^2 = x$ and $y = x$.

66. Evaluate $\int_C x\,dy - y\,dx$, where C is the line segment joining the points (a_1, b_1) and (a_2, b_2). Using this result and the result in problem **51**, show that the area of a polygon with vertices at (a_1, b_1), (a_2, b_2), $\ldots$, (a_n, b_n) taken in the anti-clockwise direction is

Area $= [(a_1 b_2 - a_2 b_1) + (a_2 b_3 - a_3 b_2) + \ldots + (a_{n-1} b_n - a_n b_{n-1}) + (a_n b_1 - a_1 b_n)]/2$.

Use the result of problem **51** to find the area bounded by the given closed curves in problems **67** to **69**.

67. Ellipse : $x = a\cos\theta$, $y = b\sin\theta$, $0 \leq \theta \leq 2\pi$, a and b are positive constants.

68. Circle : $x = a\cos\theta$, $y = a\sin\theta$, $0 \leq \theta \leq 2\pi$, a is a positive constant.

69. Hypocycloid : $x = a\cos^3\theta$, $y = a\sin^3\theta$, $a > 0$, $0 \leq \theta \leq 2\pi$.

In problems **70** to **72** evaluate $\int_C (\partial u/\partial n)\,ds$ using the result given in example 15.38.

70. $u = x^2 + y^2$, C is the boundary of the region $x^2 + y^2 \leq 9$, $x \geq 0$, $y \geq 0$.

71. $u = x^3 - xy^2$, C is the rectangle with vertices at $(1, 1)$, $(3, 1)$, $(3, 4)$ and $(1, 4)$.

72. $u = xy^3 + x^2 y^2$, C is the circle $x^2 + y^2 = 4$.

In problems **73** and **74**, use Green's Theorem to evaluate the double integral in terms of a line integral (choose appropriate functions f or g or f and g).

73. $\iint_R y^2\,dx\,dy$, R is the region enclosed by the ellipse $x^2 + 4y^2 = 16$.

74. $\iint_R (y - x)\,dx\,dy$, R is the region enclosed by the circle $x^2 + y^2 = 4$.

75. Find the work done by the force $\mathbf{F} = (x^2 - y^2)\,\mathbf{i} + (x^2 + y^2)\,\mathbf{j}$ in moving a particle along a closed path C bounding the regions $x^2 + y^2 \leq 16,\ x^2 + y^2 \geq 4,\ x \geq 0$.

76. If f and g are differentiable scalar fields, show that

$$\oint_C \left(f\frac{\partial g}{\partial n} - g\frac{\partial f}{\partial n} \right) ds = \iint_R (f\nabla^2 g - g\nabla^2 f)\,dx\,dy.$$

where $\mathbf{n}$ is the outward unit normal to C.

77. Let f satisfy the Laplace equation in a region R. Then, show that

$$\oint_C f\frac{\partial f}{\partial n}\,ds = \iint_R \left[\left(\frac{\partial f}{\partial x}\right)^2 + \left(\frac{\partial f}{\partial y}\right)^2 \right] dx\,dy.$$

78. If f and g are differentiable scalar fields, then show that

$$\oint_C f\,dg = \iint_R \left(\frac{\partial f}{\partial x}\frac{\partial g}{\partial y} - \frac{\partial f}{\partial y}\frac{\partial g}{\partial x} \right) dx\,dy.$$

79. If $\mathbf{F} = g\,\mathbf{i} - f\,\mathbf{j}$, $\mathbf{n}$ is the outer unit normal to the curve C, then using the Green's theorem show that

$$\oint_C \mathbf{F}\cdot\mathbf{n}\,ds = \iint_R (\nabla\cdot\mathbf{F})\,dx\,dy.$$

80. If $\mathbf{F} = g\,\mathbf{i} - f\,\mathbf{j}$, $\mathbf{T}$ is the unit tangent vector to C, then using the Green's theorem show that

$$\oint_C \mathbf{F}\cdot\mathbf{T}\,ds = \iint_R (\nabla\times\mathbf{F})\cdot\mathbf{k}\,dx\,dy.$$

15.6   Surface Area and Surface Integrals

15.6.1   Surface Area

Consider a surface S whose equation is $f(x, y, z) = c$. We may also write the equation as $f(x, y, z) = 0$. We have shown in section 15.3 that

$$\mathbf{N} = \operatorname{grad}(f) \quad \text{and} \quad \mathbf{n} = \frac{\operatorname{grad}(f)}{|\operatorname{grad}(f)|} \tag{15.76}$$

are the normal and unit normal vectors respectively to the surface S. We assume that $f(x, y, z)$ has continuous first order partial derivatives at each point (x, y, z) in its domain and at least one of them is not equal to zero. Then, a unique normal exists at each point of the surface S. We then say that S is a *smooth surface*. A *piecewise smooth surface* consists of a number of surfaces each of which is a smooth surface. For example, the surface of a sphere is a smooth surface while the surfaces of a closed cylinder or a cube are piecewise smooth surfaces.

In section 15.1, we have discussed the parametric representation of a surface. If u, v are two independent parameters taking values in a region R in the u-v plane and if we write $x = x(u, v)$, $y = y(u, v)$, $z = z(u, v)$, then the parametric representation of the surface S can be written as

$$\mathbf{r} = \mathbf{r}(u, v) = x(u, v)\,\mathbf{i} + y(u, v)\,\mathbf{j} + z(u, v)\,\mathbf{k}. \tag{15.77}$$

If the equation of the surface is in the form $z = f(x, y)$, then the parametric representation of the surface S can be written as

$$\mathbf{r}(u, v) = u\,\mathbf{i} + v\,\mathbf{j} + f(u, v)\,\mathbf{k} \tag{15.78}$$

where $u = x$ and $v = y$, or simply as

$$\mathbf{r}(x, y) = x\,\mathbf{i} + y\,\mathbf{j} + f(x, y)\,\mathbf{k}. \tag{15.79}$$

Consider now a surface S whose parametric representation is $\mathbf{r} = \mathbf{r}(u, v)$. Let C be a curve on S. Then, the parametric representation of the curve C can be written as

$$u = f(t),\ v = g(t),\ t \text{ a real parameter} \tag{15.80}$$

that is ,

$$\mathbf{r} = \mathbf{r}(t) = \mathbf{r}[u(t), v(t)] = \mathbf{r}[f(t), g(t)] \tag{15.81}$$

We assume that $f(t)$ and $g(t)$ have continuous first order derivatives with respect to t. Then, the tangent vector to C for any value of the parameter t is given by

$$\frac{d\mathbf{r}}{dt} = \frac{\partial \mathbf{r}}{\partial u}\frac{du}{dt} + \frac{\partial \mathbf{r}}{\partial v}\frac{dv}{dt} = \mathbf{r}_u\frac{du}{dt} + \mathbf{r}_v\frac{dv}{dt}. \tag{15.82}$$

Now, since u, v are independent parameters, $\partial \mathbf{r}/\partial u$ and $\partial \mathbf{r}/\partial v$ are independent vectors and hence they determine a plane. Consider the point $P(\mathbf{r})$ on the surface S. At this point, we have two independent vectors $\mathbf{r}_u$ and $\mathbf{r}_v$. Let $\mathbf{r}^*$ be the position vector of any point in the tangent plane (Fig. 15.18). Then, the vector $\mathbf{r}^* - \mathbf{r}$ can be written as a linear combination of the vectors $\mathbf{r}_u$ and $\mathbf{r}_v$. Alternately, the vectors $\mathbf{r}^* - \mathbf{r}$, $\mathbf{r}_u$ and $\mathbf{r}_v$ are coplanar. Therefore, the equation of the *tangent plane at P* is given by

$$(\mathbf{r}^* - \mathbf{r}) \cdot (\mathbf{r}_u \times \mathbf{r}_v) = [\mathbf{r}^* - \mathbf{r}\ \ \mathbf{r}_u\ \ \mathbf{r}_v] = 0 \quad (15.83)$$

where $[\ldots]$ is the scalar triple product.

 (Refer to Eq. (15.39) which is another form of the equation of the tangent plane at a point P on the surface S). Now,

$$\frac{\partial \mathbf{r}}{\partial u} \times \frac{\partial \mathbf{r}}{\partial v} = \mathbf{r}_u \times \mathbf{r}_v$$

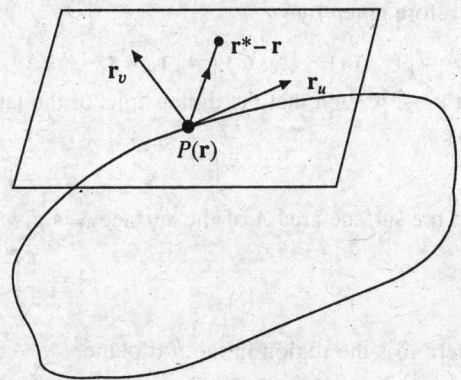

Fig. 15.18.   Tangent plane.

is a vector perpendicular to both the vectors $\partial \mathbf{r}/\partial u$ and $\partial \mathbf{r}/\partial v$ and hence it is the normal vector to the tangent plane, that is normal vector to the surface S at a point P. This gives the representation of the normal vector in terms of the parametric representation while we have earlier shown that grad (f) is the normal vector to the surface S. Since S is a smooth surface, a unique normal exists. This implies that $\mathbf{r}_u \times \mathbf{r}_v \neq \mathbf{0}$. The unit normal vector at a point P on the surface S is given by

$$\mathbf{n} = \frac{\mathbf{r}_u \times \mathbf{r}_v}{|\mathbf{r}_u \times \mathbf{r}_v|}. \tag{15.84}$$

In two dimensions, the element ds of the curve C is given by

$$ds^2 = dx^2 + dy^2 = d\mathbf{r} \cdot d\mathbf{r}$$

where $\mathbf{r} = x\,\mathbf{i} + y\,\mathbf{j}$. This definition of the linear element can be extended to three dimensions, that is to the element ds of the curve C on the surface S. From Eq. (15.82), we have

$$d\mathbf{r} = \frac{\partial \mathbf{r}}{\partial u}\,du + \frac{\partial \mathbf{r}}{\partial v}\,dv$$

Therefore, $ds^2 = d\mathbf{r} \cdot d\mathbf{r} = (\mathbf{r}_u \, du + \mathbf{r}_v \, dv) \cdot (\mathbf{r}_u \, du + \mathbf{r}_v \, dv)$

$$= \mathbf{r}_u^2 \, du^2 + 2\mathbf{r}_u \cdot \mathbf{r}_v \, du \, dv + \mathbf{r}_v^2 \, dv^2 \qquad (15.85)$$

where $\mathbf{r}_u^2 = \mathbf{r}_u \cdot \mathbf{r}_u$ and $\mathbf{r}_v^2 = \mathbf{r}_v \cdot \mathbf{r}_v$.

This differential form of ds is called the *first fundamental form* of S.

Surface Area   Consider now, a surface S in its parameteric form $\mathbf{r} = \mathbf{r}(u, v)$. Let the surface S be
divided into finite number of parts $S_1, S_2, \ldots, S_n$.
Consider one typical part S_k. Let $P(u, v)$ be any
point on S_k. Consider the tangent plane at $P(u, v)$
to the element surface S_k (Fig. 15.19). In the figure,
$PQRS$ is an element of the surface and $PQ'R'S'$ is
an element (a parallelogram) in the tangent plane
at P. Now, we approximate the area of element of
the surface by the area of the element of the tangent
plane, that is by the area of the parallelogram
with sides $\mathbf{r}_u \, \Delta u$ and $\mathbf{r}_v \, \Delta v$. The approximation is
therefore given by

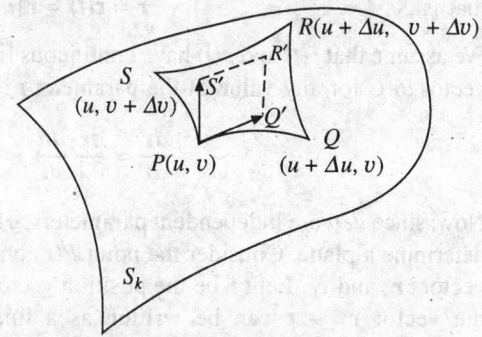

$$\Delta A = |\,(\mathbf{r}_u \Delta u) \times (\mathbf{r}_v \Delta v)\,| = |\,\mathbf{r}_u \times \mathbf{r}_v\,| \, \Delta u \Delta v.$$

Fig. 15.19.   Surface and tangent elements.

Let $n \to \infty$, such that the dimensions of the largest part of the surface tends to zero. In the limit, we
have

$$dA = |\,\mathbf{r}_u \times \mathbf{r}_v\,| \, du \, dv \qquad (15.86)$$

and the surface area A of the surface S is given by

$$A = \iint_R |\,\mathbf{r}_u \times \mathbf{r}_v\,| \, du \, dv \qquad (15.87)$$

where R is the region in the u-v plane.

Since $|\,\mathbf{a} \times \mathbf{b}\,|^2 = a^2 b^2 - (\mathbf{a} \cdot \mathbf{b})^2$, we obtain

$$|\,\mathbf{r}_u \times \mathbf{r}_v\,|^2 = \mathbf{r}_u^2 \, \mathbf{r}_v^2 - (\mathbf{r}_u \cdot \mathbf{r}_v)^2. \qquad (15.88)$$

Hence, the surface area of S is given by

$$A = \iint_R \sqrt{\mathbf{r}_u^2 \, \mathbf{r}_v^2 - (\mathbf{r}_u \cdot \mathbf{r}_v)^2} \, du \, dv \qquad (15.89)$$

This representation can be simplified if the equation of the surface is given by $z = f(x, y)$. In this case,
the parametric representation is given by

$$\mathbf{r}(u, v) = u\,\mathbf{i} + v\,\mathbf{j} + f(u, v)\mathbf{k} \qquad (15.90)$$

where $u = x$ and $v = y$. Therefore,

$$\mathbf{r}_u = \mathbf{i} + f_u \mathbf{k}, \ \mathbf{r}_v = \mathbf{j} + f_v \mathbf{k}, \ \mathbf{r}_u \times \mathbf{r}_v = -f_u \mathbf{i} - f_v \mathbf{j} + \mathbf{k}$$

and

$$|\,\mathbf{r}_u \times \mathbf{r}_v\,| = \sqrt{1 + f_u^2 + f_v^2}. \qquad (15.91)$$

Therefore, the surface area is given by

$$A = \iint_R \sqrt{1 + f_u^2 + f_v^2} \, du \, dv. \qquad (15.92)$$

Since $u = x$, $v = y$, we obtain A as

$$A = \iint\limits_{R^*} \sqrt{1 + f_x^2 + f_y^2} \, dx \, dy \tag{15.93}$$

and
$$dA = \sqrt{1 + f_x^2 + f_y^2} \, dx \, dy. \tag{15.94}$$

Note that R^* is the orthogonal projection of $S(z = f(x, y))$ on the x-y plane.

We now derive an alternate form for the element surface area ΔA using projections, which is of use in evaluating surface integrals.

(a) Let $z = f(x, y)$ be the equation of the surface. Let the element area ΔA be projected on the x-y plane (Fig. 15.20). Let **N** be the normal vector to the element area. Now,

$$\mathbf{N} \cdot \mathbf{k} = (\mathbf{r}_u \times \mathbf{r}_v) \cdot \mathbf{k} = (-f_u \mathbf{i} - f_v \mathbf{j} + \mathbf{k}) \cdot \mathbf{k} = 1. \tag{15.95}$$

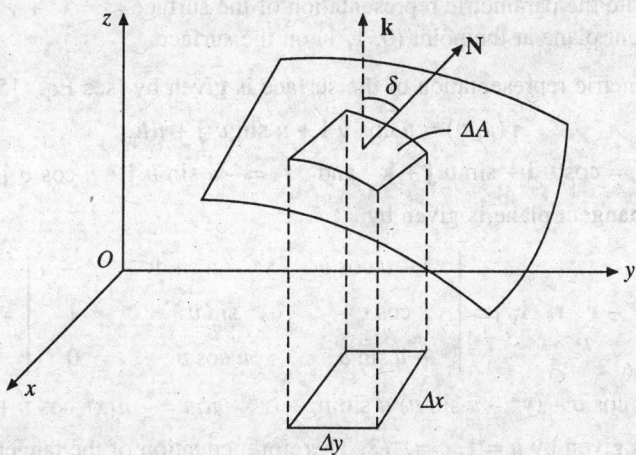

Fig. 15.20.   Projection of element surface area.

Let $\delta \, (< \pi/2)$ be the angle which the normal vector **N** makes with the positive direction of z-axis. Then, using Eq. (15.95), we obtain

$$\mathbf{N} \cdot \mathbf{k} = |\mathbf{N}| \cos \delta = 1.$$

Therefore, $\sec \delta = |\mathbf{N}| = |\mathbf{r}_u \times \mathbf{r}_v| = \sqrt{1 + f_x^2 + f_y^2}$, $dA = (\sec \delta) \, dx \, dy$ \hfill (15.96)

and
$$A = \iint\limits_{R^*} \sqrt{1 + f_x^2 + f_y^2} \, dx \, dy. \tag{15.97}$$

Similarly, we have the following cases.

(b) Equation of the surface: $x = g(y, z)$. We project S on the y-z plane. Then
$$dA = (\sec \alpha) \, dy \, dz, \quad (\alpha < \pi/2) \tag{15.98}$$

where α is the angle which the normal vector **N** makes with the positive direction of x-axis. We have

$$A = \iint\limits_{R^*} \sqrt{1 + g_y^2 + g_z^2} \, dy \, dz \tag{15.99}$$

where R^* is the projection of S on the y-z plane.

(c) Equation of the surface: $y = h(x, z)$. We project S on the x-z plane. Then

$$dA = (\sec \beta) \, dx \, dz, \quad (\beta < \pi/2) \tag{15.100}$$

where β is the angle which the normal vector $\mathbf{N}$ makes with the positive direction of y-axis. We have

$$A = \iint\limits_{R^*} \sqrt{1 + h_x^2 + h_z^2} \, dx \, dz, \tag{15.101}$$

where R^* is the projection of S on the x-z plane.

For example, consider the lateral surface S of an open cylinder $x^2 + y^2 = a^2, 0 < z < k$. We can project the surface on the y-z or on the x-z planes. The projections are rectangular regions on the corresponding planes. We cannot project the surface S on the x-y plane.

Example 15.39 Write the parametric representation of the surface $z^2 = x^2 + y^2$ and hence find the equation of the tangent plane at the point $(0, 1, 1)$ on the surface.

Solution The parametric representation of the surface is given by (see Eq. (15.18))

$$\mathbf{r}(u, v) = u \cos v \, \mathbf{i} + u \sin v \, \mathbf{j} + u\mathbf{k}.$$

We have $\quad \mathbf{r}_u = \cos v \, \mathbf{i} + \sin v \, \mathbf{j} + \mathbf{k} \quad$ and $\quad \mathbf{r}_v = - u \sin v \, \mathbf{i} + u \cos v \, \mathbf{j}.$

The equation of the tangent plane is given by

$$[\mathbf{r}^* - \mathbf{r} \quad \mathbf{r}_u \quad \mathbf{r}_v] = \begin{vmatrix} x^* - u \cos v & y^* - u \sin v & z^* - u \\ \cos v & \sin v & 1 \\ - u \sin v & u \cos v & 0 \end{vmatrix} = 0$$

or $- (x^* - u \cos v) u \cos v - (y^* - u \sin v) u \sin v + (z^* - u)u = - u[x^* \cos v + y^* \sin v - z^*] = 0$.

The point $(0, 1, 1)$ is given by $u = 1, v = \pi/2$. Therefore, equation of the tangent plane is given by $z^* = y^*$, that is, $z = y$.

Example 15.40 Obtain the first fundamental form (differential form) of the sphere

$$x^2 + y^2 + z^2 = a^2.$$

Solution The parametric representation of the sphere is given by (see Eq. (15.16))

$$\mathbf{r} (u, v) = a \cos u \cos v \, \mathbf{i} + a \sin u \cos v \, \mathbf{j} + a \sin v \, \mathbf{k}$$

We have $\quad \mathbf{r}_u = - a \sin u \cos v \, \mathbf{i} + a \cos u \cos v \, \mathbf{j}$

$$\mathbf{r}_v = - a \cos u \sin v \, \mathbf{i} - a \sin u \sin v \, \mathbf{j} + a \cos v \, \mathbf{k}$$

The first fundamental form of the surface is given by

$$ds^2 = \mathbf{r}_u^2 \, du^2 + 2 \mathbf{r}_u \cdot \mathbf{r}_v \, du \, dv + \mathbf{r}_v^2 \, dv^2$$

$$= (a^2 \sin^2 u \cos^2 v + a^2 \cos^2 u \cos^2 v) \, du^2 + 2(a^2 \sin u \cos u \sin v \cos v$$

$$- a^2 \sin u \cos u \sin v \cos v) du \, dv + (a^2 \cos^2 u \sin^2 v + a^2 \sin^2 u \sin^2 v + a^2 \cos^2 v) \, dv^2$$

$$= (a^2 \cos^2 v) \, du^2 + a^2 \, dv^2.$$

Example 15.41 The cylinder $y^2 + z^2 = 9$ intersects the sphere $x^2 + y^2 + z^2 = 25$. Find the surface area of the portion of the sphere cut by the cylinder above the y-z plane and within the cylinder.

Solution   We define the sphere as

$$x = g(y, z) = \sqrt{25 - y^2 - z^2}.$$

We project the sphere on the y-z plane.

We have

$$g_y = \frac{-y}{\sqrt{25 - y^2 - z^2}}, \quad g_z = \frac{-z}{\sqrt{25 - y^2 - z^2}}$$

and

$$1 + g_y^2 + g_z^2 = 1 + \frac{y^2 + z^2}{25 - y^2 - z^2} = \frac{25}{25 - y^2 - z^2}.$$

Therefore, the surface area within the cylinder is given by

$$A = \iint_R \sqrt{1 + g_y^2 + g_z^2}\, dy\, dz = \iint_R \frac{5}{\sqrt{25 - y^2 - z^2}}\, dy\, dz$$

where R is the projection of the cut portion of the surface of the sphere on the y-z plane which is given by $y^2 + z^2 \leq 9$, $x = 0$. In polar coordinates, the region R is given by $y = r\cos\theta$, $z = r\sin\theta$, $0 \leq r \leq 3$, $0 \leq \theta \leq 2\pi$, Therefore,

$$A = \int_0^3 \int_0^{2\pi} \frac{5}{\sqrt{25 - r^2}}\, r\, dr\, d\theta = 5 \int_0^{2\pi} \left[-\sqrt{25 - r^2} \right]_0^3 d\theta = 10\,\pi.$$

15.6.2   Surface Integral

We have shown in chapter 2, that double and triple integrals are generalisations of the definite integral $\int_a^b f(x)\, dx$ to two and three dimensions respectively. The surface area integral (Eq. (15.93)) is a generalisation of the arc length integral $\int_a^b \sqrt{1 + (y')^2}\, dx$. We shall now present a generalisation of the line integral $\int_C f(x, y)\, ds$ to three dimensions. This generalisation is called the surface integral.

Let $g(x, y, z)$ be a given function defined in the three dimensional space and let S be a surface which is the graph of a function $z = f(x, y)$, or $y = h_1(x, z)$, or $x = h(y, z)$. We assume that (i) $g(x, y, z)$ is continuous at all points on S, (ii) S is smooth and bounded and (iii) the projection R of the surface S on x-y plane, or x-z plane, or y-z plane respectively can be expressed in the forms as assumed in the proof of the Green's theorem. For example, the projection R on the x-y plane can be expressed in the forms (see Eqs. (15.74), (15.75))

$$R : u_1(x) \leq y \leq u_2(x); \quad a \leq x \leq b, \tag{15.102}$$

or

$$R : v_1(y) \leq x \leq v_2(y), \quad c \leq y \leq d. \tag{15.103}$$

The surface integral can be defined in a similar way as the double integral is defined. Subdivide S into n parts $S_1, S_2, \ldots, S_n$ of areas $\Delta A_1, \Delta A_2, \ldots, \Delta A_n$. The projection R of S is therefore partitioned into n rectangles $R_1, R_2, \ldots, R_n$. We choose an arbitrary point $P_k(x_k, y_k, z_k)$ on each element of the surface area S_k and form the sum

$$I_n = \sum_{k=1}^{n} g(x_k, y_k, z_k) \Delta A_k. \tag{15.104}$$

Let $n \to \infty$, such that the largest element of the surface area shrinks to a point. This implies that as $n \to \infty$, the length of the longest diagonal of the projected rectangles tends to zero. In the limit as $n \to \infty$, the sequence $\{I_n\}$ has a limiting value which is independent of the way S is subdivided and the choice of P_k on S_k. This limiting value is called the *surface integral* of $g(x, y, z)$ over S. That is, we define the surface integral as

$$\iint\limits_{S} g(x, y, z) \, dA = \lim_{|d| \to 0} \sum_{k=1}^{n} g(x_k, y_k, z_k) \Delta A_k \tag{15.105}$$

where $|d|$ is the length of the longest diagonal of the projected rectangles.

The surface integral can be evaluated in any one of the following ways.

(i) Let S be represented in parametric form as $\mathbf{r} = \mathbf{r}(u, v)$. Then, using Eqs.(15.86) and (15.88), we write

$$\iint\limits_{S} g(x, y, z) \, dA = \iint\limits_{R^*} g[x(u, v), y(u, v), z(u, v)] \, |\, \mathbf{r}_u \times \mathbf{r}_v \,| \, du \, dv \tag{15.106}$$

$$= \iint\limits_{R^*} g[x(u, v), y(u, v), z(u, v)] \, [\mathbf{r}_u^2 \, \mathbf{r}_v^2 - (\mathbf{r}_u \cdot \mathbf{r}_v)^2]^{1/2} \, du \, dv \tag{15.107}$$

where R^* is the region corresponding to S in the u-v plane.

(ii) Let S be represented in the form $z = f(x, y)$. Then, using Eq. (15.94) We write

$$\iint\limits_{S} g(x, y, z) \, dA = \iint\limits_{R} g[x, y, f(x, y)] \, [1 + f_x^2 + f_y^2]^{1/2} \, dx \, dy \tag{15.108}$$

where R is the orthogonal projection of S on the x-y plane.

(iii) Let S be represented in the form $y = h_1(x, z)$. Then, we write

$$\iint\limits_{S} g(x, y, z) \, dA = \iint\limits_{R} g[x, h_1(x, z), z] \, [1 + (h_1)_x^2 + (h_1)_z^2]^{1/2} \, dx \, dz \tag{15.109}$$

where R is the orthogonal projection of S on the x-z plane.

(iv) Let S be represented in the form $x = h(y, z)$. Then we write

$$\iint\limits_{S} g(x, y, z) \, dA = \iint\limits_{R} g[h(y, z), y, z] \, [1 + h_y^2 + h_z^2]^{1/2} \, dy \, dz \tag{15.110}$$

where R is the orthogonal projection of S on the y-z plane.

If S is piecewise smooth and consists of the surfaces $S_1, S_2, \ldots, S_k$, then

$$\iint\limits_{S} g(x, y, z) \, dA = \iint\limits_{S_1} g(x, y, z) \, dA + \iint\limits_{S_2} g(x, y, z) \, dA + \ldots + \iint\limits_{S_k} g(x, y, z) \, dA.$$

We now present some of the important applications of the surface integrals.

Mass of a surface Let $\rho(x, y, z)$ denote the density of a surface S at any point or mass per unit surface area. Then, the *mass m* of the surface is given by

$$m = \iint_S \rho(x, y, z)\, dA.$$ (15.111)

Moment of inertia Let $\rho(x, y, z)$ denote the density of a surface S at a point. Then, the *moment of inertia I* of the mass m with respect to a given axis l is defined by the surface integral

$$I = \iint_S \rho(x, y, z)\, d^2\, dA$$

where d is the distance of the point (x, y, z) from the reference axis l. If the surface is homogeneous, then $\rho(x, y, z) = $ constant and from Eq. (15.111), $\rho(x, y, z) = m/A$ where A is the surface area of S. Then,

$$I = \frac{m}{A} \iint_S d^2\, dA$$ (15.112)

Example 15.42 Find the mass of the surface of the cone $z = 2 + \sqrt{x^2 + y^2}$, $2 \le z \le 7$, in the first octant, if the density $\rho(x, y, z)$ at any point of the surface is proportional to its distance from the x-y plane.

Solution The density is given by $\rho(x, y, z) = cz$, c constant. We have

$$z = f(x, y) = 2 + \sqrt{x^2 + y^2}, \, f_x = \frac{x}{\sqrt{x^2 + y^2}}, \, f_y = \frac{y}{\sqrt{x^2 + y^2}}$$

$$dA = \sqrt{1 + f_x^2 + f_y^2}\, dx\, dy = \sqrt{1 + \frac{x^2}{x^2 + y^2} + \frac{y^2}{x^2 + y^2}}\, dx\, dy = \sqrt{2}\, dx\, dy.$$

The projection of S on the x-y plane is given by (Fig. 15.21) $R : x^2 + y^2 = 25$, in the first quadrant.

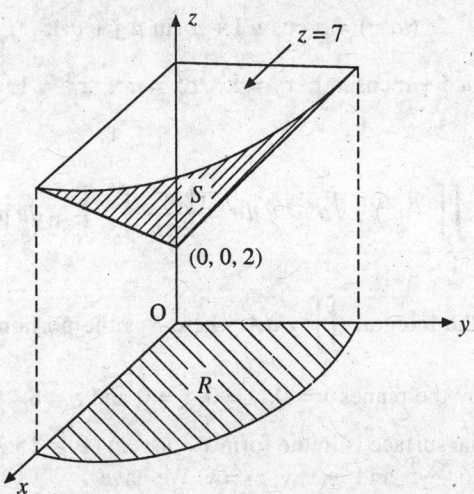

Fig. 15.21. Example 15.42.

Therefore, mass of the surface is given by

$$m = \iint\limits_{S} c\, z\, dA = \iint\limits_{R} c\, [2 + \sqrt{x^2 + y^2}\,]\, \sqrt{2}\, dx\, dy$$

$$= c\sqrt{2} \iint\limits_{R} [2 + \sqrt{x^2 + y^2}\,]\, dx\, dy.$$

Substituting $x = r \cos\theta$, $y = r \sin\theta$, $0 \le r \le 5$, $0 \le \theta \le \pi/2$, we obtain

$$m = c\sqrt{2} \int_0^5 \int_0^{\pi/2} (2 + r)\, r\, dr\, d\theta = c\sqrt{2} \int_0^{\pi/2} \left(r^2 + \frac{r^3}{3} \right)_0^5 d\theta$$

$$= c\sqrt{2} \left(25 + \frac{125}{3} \right) \frac{\pi}{2} = \frac{100\sqrt{2}}{3}\, \pi c.$$

Example 15.43 Find the moment of inertia of the homogeneous cylindrical lamina $x^2 + y^2 = a^2$, $0 \le z \le b$ of mass m about the z-axis.

Solution Since the surface is homogeneous, the density $\rho(x, y, z) = $ constant and the moment of inertia is given by (see Eq. (15.112))

$$I = \frac{m}{A} \iint\limits_{S} d^2\, dA$$

where $A = $ surface area of cylinder $= 2\pi a b$,

   $d = $ distance of any point on S from z-axis $= \sqrt{x^2 + y^2}$.

Now, the parametric form of the surface S is

$$\mathbf{r}(u, v) = a \cos u\, \mathbf{i} + a \sin u\, \mathbf{j} + v\, \mathbf{k}$$

We have $\mathbf{r}_u = -a \sin u\, \mathbf{i} + a \cos u\, \mathbf{j}$, $\mathbf{r}_v = \mathbf{k}$, $\mathbf{r}_u^2 = a^2$, $\mathbf{r}_v^2 = 1$, $\mathbf{r}_u \cdot \mathbf{r}_v = 0$.

Therefore,

$$I = \frac{m}{2\pi a b} \iint\limits_{R} (a^2) \cdot \sqrt{a^2}\, du\, dv = \frac{ma^2}{2\pi b} \int_0^b \int_0^{2\pi} du\, dv = ma^2.$$

Example 15.44 Evaluate the integral $\iint\limits_{S} y\, dA$ where S is the portion of the cylinder $x = 6 - y^2$

in the first octant bounded by the planes $x = 0$, $y = 0$, $z = 0$ and $z = 8$.

Solution The equation of the surface is in the form $x = h(y, z)$ (Fig. 15.22). Therefore, we shall use Eq. (15.110) with $h(y, z) = 6 - y^2$ and $g(x, y, z) = y$. We have

$$h_y = -2y, \quad h_z = 0, \quad (1 + h_y^2 + h_z^2)^{1/2} = (1 + 4y^2)^{1/2}$$

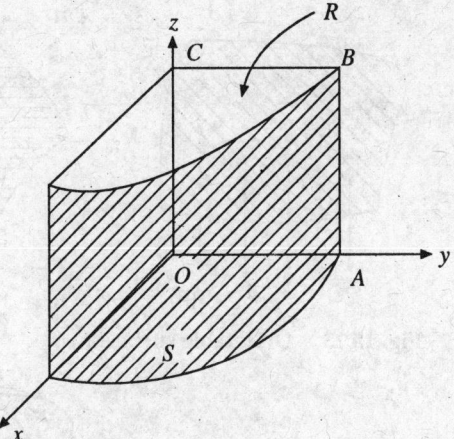

Fig. 15.22.   Example 15.44.

The projection of S on the y-z plane is the rectangle $OABC$ with sides $y = 0$, $y = \sqrt{6}$, $z = 0$ and $z = 8$. Therefore,

$$\iint_S y\, dA = \iint_R y\, (1 + 4y^2)^{1/2}\, dy\, dz = \int_0^{\sqrt{6}} \int_0^8 y\, (1 + 4y^2)^{1/2}\, dy\, dz$$

$$= 8 \left[\frac{(1 + 4y^2)^{3/2}}{8(3/2)} \right]_0^{\sqrt{6}} = \frac{2}{3} \left[(25)^{3/2} - 1 \right] = \frac{248}{3}.$$

Orientable surfaces   Before we try to evaluate a surface integral of a vector field we need the concept of an orientable surface. Without mathematical rigour, we may say that S is an orientable surface if it has two sides (which may be painted in two different colours). A surface which has only one side is not an orientable surface. A famous example of a non-orientable surface is the Möbius strip (Fig. 15.24b). To construct a Möbius strip, we can cut out a long strip of paper (Fig. 15.24a), give one of the ends a half twist and then attach the ends.

A smooth surface S is said to be orientable if there exists a continuous unit normal vector field $\mathbf{n}$ defined at each point (x, y, z) on the surface (Figs. 15.23a, b). We then say that the vector field $\mathbf{n}(x, y, z)$ is the orientation of S. An orientable surface has two orientations since a unit normal to a surface S at (x, y, z) can be $\mathbf{n}(x, y, z)$ or $-\mathbf{n}(x, y, z)$ (Fig. 15.23c). We usually call them, outward and inward normals. They are also sometimes called upward and downward orientations. The Möbius strip is not an orientable surface, since if a unit normal vector $\mathbf{n}$ starts at a point P_0 on the surface and moves once round the curve C, then the resulting normal vector points in the opposite direction. Most of the surfaces that we encounter are orientable surfaces.

Let S be an orientable surface. We choose a unit normal $\mathbf{n}$ and orient the surface S. Then, $\mathbf{n}$ can be written as

$$\mathbf{n} = \cos \alpha\, \mathbf{i} + \cos \beta\, \mathbf{j} + \cos \gamma\, \mathbf{k} \tag{15.113}$$

where α, β, γ are the angles which the unit normal $\mathbf{n}$ makes with the positive directions of x, y and z axis, respectively. From Eq. (15.113), we have

$$\mathbf{n} \cdot \mathbf{i} = \cos \alpha, \quad \mathbf{n} \cdot \mathbf{j} = \cos \beta \text{ and } \mathbf{n} \cdot \mathbf{k} = \cos \gamma. \tag{15.114}$$

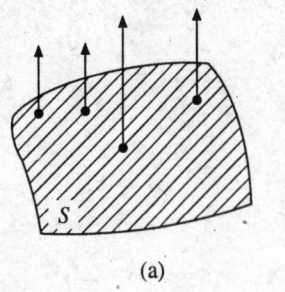
(a)

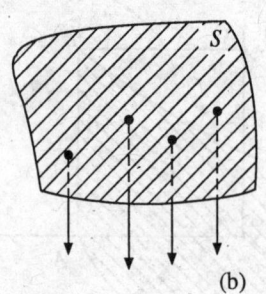
(b)

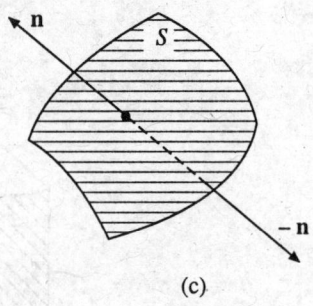
(c)

Fig. 15.23.   Orientable surfaces.

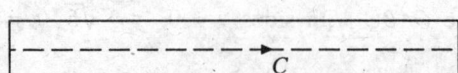

Fig. 15.24a.   Strip of paper.

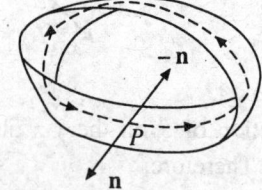

Fig. 15.24b.   Möbius strip.

Let $\mathbf{v}(x, y, z) = v_1(x, y, z)\,\mathbf{i} + v_2(x, y, z)\,\mathbf{j} + v_3(x, y, z)\,\mathbf{k}$ be a vector field. Then,

$$\iint_S \mathbf{v} \cdot \mathbf{n}\, dA = \iint_S (v_1 \cos \alpha + v_2 \cos \beta + v_3 \cos \gamma)\, dA$$

$$= \iint_S v_1 (\cos \alpha)\, dA + v_2 (\cos \beta)\, dA + v_3 (\cos \gamma)\, dA$$

$$= \iint_S v_1\, dy\, dz + v_2\, dz\, dx + v_3\, dx\, dy \tag{15.115}$$

where
$$(\cos \alpha)\, dA = dy\, dz, \quad \text{or} \quad (\mathbf{n} \cdot \mathbf{i})\, dA = dy\, dz, \quad \text{or} \quad dA = \frac{dy\, dz}{\mathbf{n} \cdot \mathbf{i}} \tag{15.116}$$

$$(\cos \beta)\, dA = dz\, dx, \quad \text{or} \quad (\mathbf{n} \cdot \mathbf{j})\, dA = dz\, dx, \quad \text{or} \quad dA = \frac{dz\, dx}{\mathbf{n} \cdot \mathbf{j}} \tag{15.117}$$

and
$$(\cos \gamma)\, dA = dx\, dy, \quad \text{or} \quad (\mathbf{n} \cdot \mathbf{k})\, dA = dx\, dy, \quad \text{or} \quad dA = \frac{dx\, dy}{\mathbf{n} \cdot \mathbf{k}} \tag{15.118}$$

Eqs. (15.116) to (15.118) give expressions for the element surface area in terms of its projections on the cordinate planes (see Eqs. (15.96), (15.98) and (15.100)). If the surface S is represented in the parametric form

$$\mathbf{r}(u, v) = x(u, v)\,\mathbf{i} + y(u, v)\,\mathbf{j} + z(u, v)\,\mathbf{k}$$

then the unit normal vector is given by (see Eq. 15.84))

$$\mathbf{n} = \frac{\mathbf{r}_u \times \mathbf{r}_v}{|\,\mathbf{r}_u \times \mathbf{r}_v\,|}.$$

Flux of a vector field v through a surface *S*

Let $\mathbf{v}(x, y, z) = v_1(x, y, z)\,\mathbf{i} + v_2(x, y, z)\,\mathbf{j} + v_3(x, y, z)\,\mathbf{k}$ be a vector field representing the velocity field of a fluid. The flux of the velocity vector field $\mathbf{v}$ through the area ΔA is approximated by $(\mathbf{v} \cdot \mathbf{n})\,\Delta A$, where $\mathbf{n}$ is a unit vector normal to the surface (see physical interpretation of divergence). The total volume of the fluid flowing through S per unit time is called the *flux of* $\mathbf{v}$ *through S*. It is given by

$$\text{flux} = \iint_S (\mathbf{v} \cdot \mathbf{n})\,dA.$$

(15.119)

Example 15.45   Evaluate the surface integral $\displaystyle\iint_S \mathbf{F} \cdot \mathbf{n}\,dA$ where $\mathbf{F} = 6z\,\mathbf{i} + 6\,\mathbf{j} + 3y\,\mathbf{k}$ and S is the portion of the plane $2x + 3y + 4z = 12$, which is in the first octant.

Solution   Let $f(x, y, z) = 2x + 3y + 4z - 12 = 0$ be the surface. Then

$$\text{grad}\,f = 2\mathbf{i} + 3\mathbf{j} + 4\mathbf{k}, \ \mathbf{n} = \frac{\text{grad}\,f}{|\,\text{grad}\,f\,|} = \frac{1}{\sqrt{29}}\,(2\mathbf{i} + 3\mathbf{j} + 4\mathbf{k}).$$

Consider the projection of S on the x-y plane. The projection of the portion of the plane ABC in the first octant is the triangle bounded by $x = 0$, $y = 0$ and $2x + 3y = 12$ (Fig. 15.25). We have

$$dA = \frac{dx\,dy}{\mathbf{n} \cdot \mathbf{k}} = \frac{dx\,dy}{4/\sqrt{29}}.$$

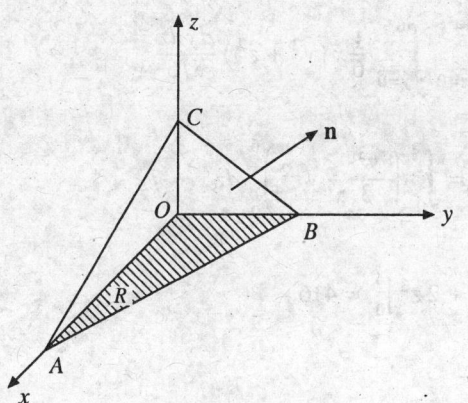

Fig. 15.25.   Example 15.45.

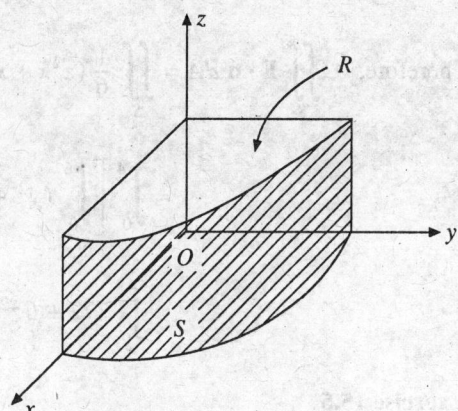

Fig. 15.26.   Example 15.46.

Therefore, $\displaystyle\iint_S \mathbf{F} \cdot \mathbf{n}\,dA = \iint_S \frac{1}{\sqrt{29}}\,(12z + 18 + 12y)\,dA.$

From the equation of the surface, we get $4z = 12 - 2x - 3y$. Hence,

$$\iint_S \mathbf{F} \cdot \mathbf{n}\,dA = \iint_S \frac{1}{\sqrt{29}}\,(54 - 6x + 3y)\,dA = \frac{1}{4}\iint_R (54 - 6x + 3y)\,dx\,dy$$

$$= \frac{1}{4} \int_{x=0}^{6} \left[\int_{y=0}^{(12-2x)/3} (54 - 6x + 3y)\, dy \right] dx = \frac{1}{4} \int_{0}^{6} \left[(54 - 6x)y + \frac{3}{2} y^2 \right]_{0}^{(12-2x)/3} dx$$

$$= \frac{1}{4} \int_{0}^{6} \left[(54 - 6x)\frac{1}{3}(12 - 2x) + \frac{1}{6}(12 - 2x)^2 \right] dx = \frac{1}{6} \int_{0}^{6} (360 - 102x + 7x^2)\, dx$$

$$= \frac{1}{6} \left[360x - 51x^2 + \frac{7}{3} x^3 \right]_{0}^{6} = 138.$$

Example 15.46 Evaluate the surface integral $\displaystyle\iint_S \mathbf{F} \cdot \mathbf{n}\, dA$ where $\mathbf{F} = z^2 \mathbf{i} + xy\mathbf{j} - y^2 \mathbf{k}$ and S is the

portion of the surface of the cylinder $x^2 + y^2 = 36$, $0 \le z \le 4$ included in the first octant.

Solution   Let $f(x, y, z) = x^2 + y^2 - 36 = 0$ be the surface. Then

$$\operatorname{grad} f = 2x\mathbf{i} + 2y\mathbf{j}, \quad \mathbf{n} = \frac{\operatorname{grad} f}{|\operatorname{grad} f|} = \frac{2(x\mathbf{i} + y\mathbf{j})}{\sqrt{4(x^2 + y^2)}} = \frac{1}{6}(x\mathbf{i} + y\mathbf{j}).$$

The projection of S on x-y plane cannot be considered (Fig. 15.26). Project S on the y-z plane. The projection is a rectangle with sides of lengths 6 and 4. We have

$$dA = \frac{dy\, dz}{\mathbf{n} \cdot \mathbf{i}} = \frac{dy\, dz}{x/6}.$$

Therefore, $\displaystyle\iint_S \mathbf{F} \cdot \mathbf{n}\, dA = \iint_S \frac{1}{6}(z^2 x + xy^2)\, dA = \int_{z=0}^{4} \int_{y=0}^{6} \frac{1}{6} x(y^2 + z^2) \frac{dy\, dz}{x/6}$

$$= \int_{0}^{4} \left[\int_{0}^{6} (y^2 + z^2)\, dy \right] dz = \int_{0}^{4} \left(\frac{y^3}{3} + yz^2 \right)_{0}^{6} dz$$

$$= \int_{0}^{4} (72 + 6z^2)\, dz = \left[72z + 2z^3 \right]_{0}^{4} = 416.$$

Exercise 15.5

In problems **1** to **6**, find the equation of the tangent plane at the indicated point using the parametric representation of the surface.

1. $x^2 + z^2 = 25$, $P(3, -2, 4)$.   2. $x^2 + y^2 + z^2 = 16$, $P(-2, -3, \sqrt{3})$.

3. $z = x^2 + y^2$, $P(1, 3, 10)$.   4. $z^2 = x^2 + y^2$, $P(1/\sqrt{2}, 1/\sqrt{2}, 1)$.

5. $36x^2 + 16y^2 + 9z^2 = 144$, $P(0, 3\sqrt{2}/2, 2\sqrt{2})$.

6. $x = z^2$, $P(4, 5, 2)$.

In problems **7** to **10**, obtain the first fundamental form of the surface.

7. Cylinder $x^2 + y^2 = a^2$.

8. Paraboloid of revolution $z = x^2 + y^2$.

9. Cone of revolution $z^2 = x^2 + y^2$.

10. $z = \cos(xy)$.

In problems **11** to **20**, find the surface area of the given surfaces.

11. $z^2 = x^2 + y^2$, $0 \le z \le 4$.

12. $z = x^2 + y^2$, $0 \le z \le 9$.

13. $x^2 + y^2 + z^2 = a^2$.

14. $x^2 + y^2 = 16$, $0 \le z \le 2$.

15. Portion of the plane $3x + 4y + 2z = 24$ that is bounded by the coordinate planes in the first octant.

16. Portion of the cylinder $x^2 + z^2 = 25$ above the first octant and bounded by the planes $x = 0$, $x = 3$, $y = 0$ and $y = 6$.

17. $z = 16 - x^2 - y^2$, above the x-y plane.

18. Portion of the sphere $x^2 + y^2 + z^2 = 8$, above the x-y plane and within the cone $z^2 = x^2 + y^2$.

19. Portion of the sphere $x^2 + y^2 + z^2 = 16$, above the x-y plane and within the cylinder $x^2 + y^2 = 4y$.

20. Portion of the cylinder $y^2 + z^2 = a^2$ that is within the cylinder $x^2 + y^2 = a^2$.

In problems **21** to **23**, find the mass of the surface with the given density.

21. $z = 2 + x^2 + y^2$, $2 \le z \le 6$ in the first octant. Density $\rho(x, y, z)$ at any point P is directly proportional to its distance from the x-y plane.

22. $2x + 3y + 4z = 12$, in the first octant. Density at $\rho(x, y, z)$ is directly proportional to the square of its distance from the x-z plane.

23. $z = \sqrt{x^2 + y^2}$, $0 \le z \le 3$, $\rho(x, y, z) = x^2 + y^2$.

In problems **24** and **25**, find the moment of inertia of a lamina S of given density about the given axis.

24. $S : y^2 + z^2 = a^2$, $0 \le x \le b$, $\rho(x, y, z) = 1$, about the line $x = b/2$ in the x-y plane.

25. $S: y^2 = x^2 + z^2$, $0 \le y \le b$, $\rho(x, y, z) = 1$, about the y-axis.

In problems **26** to **31**, evaluate the surface integrals $\displaystyle\iint_S g(x, y, z)\, dA$.

26. $g(x, y, z) = x^2 \sqrt{z}$. S is the portion of the cylinder $z = y^2$ in the first octant bounded by $z = 0$, $z = 6$, $x = 0$, $x = 4$.

27. $g(x, y, z) = z$. S is the portion of the paraboloid $z = 1 + x^2 + y^2$ in the first octant bounded by $x = 0$, $y = x$, $z = 5$.

28. $g(x, y, z) = 6xyz$. S is the portion of the plane $x + y + z = 1$ in the first octant.

29. $g(x, y, z) = z^2$. S is the portion of the sphere $x^2 + y^2 + z^2 = 16$ in the first octant.

30. $g(x, y, z) = 2x + 3y + 5z$. S is the cone $z = \sqrt{x^2 + y^2}$ between $z = 2$ and $z = 4$.

31. $g(x, y, z) = x^2 z$. S is the cone $z = \sqrt{x^2 + y^2}$ cut by the cylinder $x^2 + y^2 = 16$, in the first octant.

In problems **32** to **35**, find the flux of the vector field **v** through the given surface. Assume that the surface S is oriented in the upward direction, unless stated otherwise.

32. $\mathbf{v}(x, y, z) = x\mathbf{i} + y\mathbf{j} + z\mathbf{k}$. S is the portion of the sphere $x^2 + y^2 + z^2 = 36$ lying between the planes $z = \sqrt{11}$ and $z = \sqrt{20}$.

33. $\mathbf{v}(x, y, z) = xy\mathbf{i} + z\mathbf{j} + 3yz\mathbf{k}$. S is the portion of the plane $x + y + z = 1$ included in the first octant.

34. $\mathbf{v}(x, y, z) = x^2\mathbf{i} + 4z\mathbf{j} + 3y\mathbf{k}$. S is the portion of the cylinder $y^2 + z^2 = 25$ included in the first octant and bounded by the planes $x = 0$, $x = 5$, $y = 0$, $z = 0$.

35. $\mathbf{v}(x, y, z) = 2xz\mathbf{i} - y^2\mathbf{j} + 4yz\mathbf{k}$. S is the closed unit cube bounded by the planes $x = 0$, $x = 1$, $y = 0$, $y = 1$, $z = 0$ and $z = 1$. (Take the outward normals as the orientation of the surfaces).

15.7 Divergence Theorem of Gauss and Stokes's Theorem

15.7.1 Divergence Theorem of Gauss

The Green's theorem derived in section 15.5 can be written in a vector form (see Problem 79, Exercise 15.4). Let C be a curve in two dimensions which is written in the parametric form $\mathbf{r} = \mathbf{r}(s)$. Then, the unit tangent and unit normal vectors to C are given by

$$\mathbf{T} = \frac{dx}{ds}\mathbf{i} + \frac{dy}{ds}\mathbf{j}, \quad \mathbf{n} = \frac{dy}{ds}\mathbf{i} - \frac{dx}{ds}\mathbf{j}.$$

Then,

$$f\,dx + g\,dy = \left(f\frac{dx}{ds} + g\frac{dy}{ds} \right)ds = (g\,\mathbf{i} - f\mathbf{j}) \cdot \left(\frac{dy}{ds}\mathbf{i} - \frac{dx}{ds}\mathbf{j} \right)ds = (\mathbf{v} \cdot \mathbf{n})\,ds$$

where $\mathbf{v} = g\,\mathbf{i} - f\,\mathbf{j}$. Also

$$\frac{\partial g}{\partial x} - \frac{\partial f}{\partial y} = \left(\mathbf{i}\frac{\partial}{\partial x} + \mathbf{j}\frac{\partial}{\partial y} \right) \cdot (g\,\mathbf{i} - f\,\mathbf{j}) = \nabla \cdot \mathbf{v}$$

Hence, Green's theorem can be written in a vector form as

$$\oint_C (\mathbf{v} \cdot \mathbf{n})\,ds = \iint_R (\nabla \cdot \mathbf{v})\,dx\,dy \tag{15.120}$$

This result is a particular case of the Gauss's *divergence theorem*. Extension of the Green's theorem to three dimensions can be done under the following generalisations.

(i) A region R in the plane $\rightarrow$ a three dimensional solid D.

(ii) The closed curve C enclosing R in the plane $\rightarrow$ the closed surface S enclosing the solid D.

(iii) The unit outer normal $\mathbf{n}$ to C $\rightarrow$ the unit outer normal $\mathbf{n}$ to S.

(iv) A vector field $\mathbf{v}$ in the plane $\rightarrow$ a vector field $\mathbf{v}$ in the three dimensional space.

(v) The line integral $\oint_C (\mathbf{v} \cdot \mathbf{n})\,ds$ $\rightarrow$ a surface integral $\iint_S (\mathbf{v} \cdot \mathbf{n})\,dA$.

(vi) The double integral $\iint_R \nabla \cdot \mathbf{v}\,dx\,dy$ $\rightarrow$ a triple (volume) integral $\iiint_D \nabla \cdot \mathbf{v}\,dV$.

The above generalisations give the following divergence theorem.

Theorem 15.4 (**Divergence theorem of Gauss**) Let D be a closed and bounded region in the three dimensional space whose boundary is a piecewise smooth surface S that is oriented outward. Let $\mathbf{v}(x, y, z) = v_1(x, y, z)\mathbf{i} + v_2(x, y, z)\mathbf{j} + v_3(x, y, z)\mathbf{k}$ be a vector field for which v_1, v_2 and v_3 are continuous and have continuous first order partial derivatives in some domain containing D. Then,

$$\iint\limits_{S} (\mathbf{v} \cdot \mathbf{n}) dA = \iiint\limits_{D} \nabla \cdot \mathbf{v} \, dV = \iiint\limits_{D} \text{div} (\mathbf{v}) dV \qquad (15.121)$$

where $\mathbf{n}$ is the outer unit normal vector to S.

Proof In terms of the components of $\mathbf{v}$, the left and right hand sides of Eq. (15.121) can be written as

$$\iint\limits_{S} (\mathbf{v} \cdot \mathbf{n}) dA = \iint\limits_{S} v_1 (\mathbf{i} \cdot \mathbf{n}) dA + \iint\limits_{S} v_2 (\mathbf{j} \cdot \mathbf{n}) dA + \iint\limits_{S} v_3 (\mathbf{k} \cdot \mathbf{n}) dA$$

$$\iiint\limits_{D} \nabla \cdot \mathbf{v} \, dV = \iiint\limits_{D} \frac{\partial v_1}{\partial x} dV + \iiint\limits_{D} \frac{\partial v_2}{\partial y} dV + \iiint\limits_{D} \frac{\partial v_3}{\partial z} dV$$

where $dV = dx \, dy \, dz$. To prove the divergence theorem it is sufficient to show that

$$\iint\limits_{S} v_1 (\mathbf{i} \cdot \mathbf{n}) dA = \iiint\limits_{D} \frac{\partial v_1}{\partial x} dV, \qquad (15.122)$$

$$\iint\limits_{S} v_2 (\mathbf{j} \cdot \mathbf{n}) dA = \iiint\limits_{D} \frac{\partial v_2}{\partial y} dV, \qquad (15.123)$$

and

$$\iint\limits_{S} v_3 (\mathbf{k} \cdot \mathbf{n}) dA = \iiint\limits_{D} \frac{\partial v_3}{\partial z} dV. \qquad (15.124)$$

We shall prove Eq. (15.124). The other results are proved in a similar manner.

We shall prove the theorem for the special case of the region D whose bounding surface can be written as follows (Fig. 15.27).

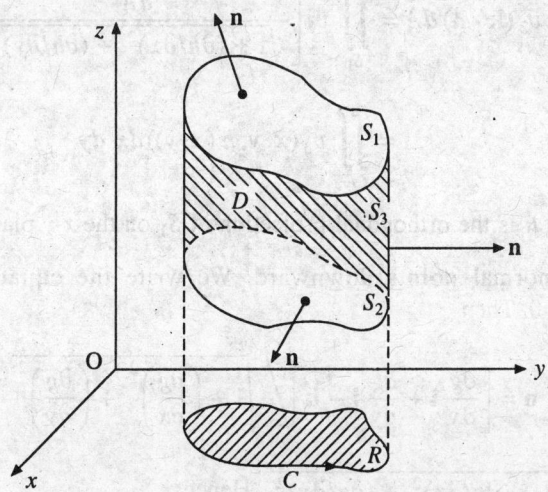

Fig. 15.27. Surface S in divergence theorem.

Top surface $S_1 : z = h(x, y)$, (x, y) in R

Bottom surface $S_2 : z = g(x, y)$, (x, y) in R

Side (vertical) surface $S_3 : g(x, y) \le z \le h(x, y)$, (x, y) in R

where R is the orthogonal projection of S in the x-y plane.

Now,

$$\iiint_D \frac{\partial v_3}{\partial z} dV = \iint_R \left[\int_{g(x,y)}^{h(x,y)} \frac{\partial v_3}{\partial z} dz \right] dx\, dy$$

$$= \iint_R [v_3(x, y, h(x, y)) - v_3(x, y, g(x, y))] dx\, dy \qquad (15.125)$$

We write

$$\iint_S v_3 (\mathbf{k} \cdot \mathbf{n}) dA = \iint_{S_1} v_3 (\mathbf{k} \cdot \mathbf{n}) dA + \iint_{S_2} v_3 (\mathbf{k} \cdot \mathbf{n}) dA + \iint_{S_3} v_3 (\mathbf{k} \cdot \mathbf{n}) dA.$$

We evaluate the surface integrals on the right hand side separately.

On S_1: The outward normal points upward. We write the equation of the surface as $f(x, y, z) = z - h(x, y) = 0$. Then

$$\mathbf{n} = \left[-\frac{\partial h}{\partial x} \mathbf{i} - \frac{\partial h}{\partial y} \mathbf{j} + \mathbf{k} \right] \Big/ \sqrt{1 + \left(\frac{\partial h}{\partial x}\right)^2 + \left(\frac{\partial h}{\partial y}\right)^2}$$

so that $\mathbf{k} \cdot \mathbf{n} = 1 \big/ \sqrt{1 + (\partial h/\partial x)^2 + (\partial h/\partial y)^2}$. Hence,

$$\iint_{S_1} v_3 (\mathbf{k} \cdot \mathbf{n}) dA = \iint_{S_1} v_3 \left[\frac{dA}{\sqrt{1 + (\partial h/\partial x)^2 + (\partial h/\partial y)^2}} \right]$$

$$= \iint_R v_3(x, y, h(x, y)) dx\, dy \qquad (15.126)$$

(see Eq. (15.94)), where R is the orthogonal projection of S_1 on the x-y plane.

On S_2: The outward normal points downward. We write the equation of the surface as $f(x, y, z) = g(x, y) - z = 0$. Then

$$\mathbf{n} = \left[\frac{\partial g}{\partial x} \mathbf{i} + \frac{\partial g}{\partial y} \mathbf{j} - \mathbf{k} \right] \Big/ \sqrt{1 + \left(\frac{\partial g}{\partial x}\right)^2 + \left(\frac{\partial g}{\partial y}\right)^2}$$

so that $\mathbf{k} \cdot \mathbf{n} = -1 \big/ \sqrt{1 + (\partial g/\partial x)^2 + (\partial g/\partial y)^2}$. Hence,

$$\iint_{S_2} v_3(\mathbf{k} \cdot \mathbf{n})dA = \iint_{S_2} -v_3 \left[\frac{dA}{1 + (\partial g/\partial x)^2 + (\partial g/\partial y)^2} \right]$$

$$= -\iint_{R} v_3(x, y, g(x, y))\, dx\, dy \tag{15.127}$$

where, again, R is the orthogonal projection of S_2 on the x-y plane.

On S_3: Since the surface is vertical, the outward normal $\mathbf{n}$ is perpendicular to $\mathbf{k}$, that is, $\mathbf{k} \cdot \mathbf{n} = 0$.

Therefore, $\displaystyle\iint_{S_3} v_3(\mathbf{k} \cdot \mathbf{n})dA = 0.$ $\tag{15.128}$

Adding Eqs. (15.126) to (15.128), we obtain

$$\iint_{S} v_3(\mathbf{k} \cdot \mathbf{n})dA = \iint_{R} [v_3(x, y, h(x, y)) - v_3(x, y, g(x, y))]\, dx\, dy. \tag{15.129}$$

From Eqs. (15.125) and (15.129), we obtain

$$\iint_{S} v_3(\mathbf{k} \cdot \mathbf{n})\, dA = \iiint_{D} \frac{\partial v_3}{\partial z}\, dV.$$

Expressing the bounding surface of the region D in a suitable manner, similar to the particular case given in Fig. 15.27, Eqs. (15.122) and (15.123) can be proved.

Remark 12

The given domain D can be subdivided into finitely many special regions such that each such region can be described in the required manner. In the proof of the divergence theorem, the special region D has a vertical surface. This type of region is not required in the proof. The region may have a vertical surface on a part of the region, the other part may be simply a curve. Also, the region may not have any vertical surface. For example, the region bounded by a sphere or an ellipsoid has no vertical surface. The divergence theorem holds in all these cases. The divergence theorem also holds for the region D bounded by two closed surfaces.

Remark 13

In terms of the components of $\mathbf{v}$, divergence theorem can be written as

$$\iint_{S} v_1\, dy\, dz + v_2\, dz\, dx + v_3\, dx\, dy = \iiint_{D} \left(\frac{\partial v_1}{\partial x} + \frac{\partial v_2}{\partial y} + \frac{\partial v_3}{\partial z} \right) dx\, dy\, dz \tag{15.130}$$

(see Eqs. (15.116) to (15.118)), or as

$$\iint_{S} (v_1 \cos\alpha + v_2 \cos\beta + v_3 \cos\gamma)\, dA = \iiint_{D} \left(\frac{\partial v_1}{\partial x} + \frac{\partial v_2}{\partial y} + \frac{\partial v_3}{\partial z} \right) dx\, dy\, dz. \tag{15.131}$$

Remark 14

(Physical interpretation of divergence) We have shown in section 15.4, that if $\mathbf{v}$ is the velocity field of a fluid, then div($\mathbf{v}$) gives the flux per unit volume. The same interpretation is also obtained through the divergence theorem. Consider a small sphere S_r of radius r, centered at a fixed point $P_0(x_0, y_0, z_0)$. Let D_r be the spherical volume bounded by S_r. Then by divergence theorem, we obtain

$$\iint\limits_{S_r} (\mathbf{v} \cdot \mathbf{n}) \, dA = \iiint\limits_{D_r} (\nabla \cdot \mathbf{v}) \, dV$$

Now, approximate $(\nabla \cdot \mathbf{v}) \, (P(x, y, z)) \approx (\nabla \cdot \mathbf{v}) \, (P_0(x_0, y_0, z_0))$, where (x, y, z) is any arbitrary point in D_r. We obtain

$$\iint\limits_{S_r} (\mathbf{v} \cdot \mathbf{n}) \, dA = (\nabla \cdot \mathbf{v})(P_0) \iiint\limits_{D_r} dV = V_r (\nabla \cdot \mathbf{v})(P_0)$$

where V_r is the volume of the small sphere. Let $r \to 0$, so that the sphere D_r tends to the point P_0. Then,

$$(\nabla \cdot \mathbf{v})(P_0) = \lim_{r \to 0} \frac{1}{V_r} \iint\limits_{S_r} (\mathbf{v} \cdot \mathbf{n}) \, dA \qquad (15.132)$$

We conclude that div $(\mathbf{v})$ is flux per unit volume.

Example 15.47   Let D be the region bounded by the closed cylinder $x^2 + y^2 = 16$, $z = 0$ and $z = 4$. Verify the divergence theorem if $\mathbf{v} = 3x^2 \mathbf{i} + 6y^2 \mathbf{j} + z \mathbf{k}$.

Solution   We have $\nabla \cdot \mathbf{v} = 6x + 12y + 1$. Therefore,

$$\iiint\limits_{D} (\nabla \cdot \mathbf{v}) \, dV = \int_{z=0}^{4} \int_{x=-4}^{4} \int_{y=-\sqrt{16-x^2}}^{\sqrt{16-x^2}} (6x + 12y + 1) \, dy \, dx \, dz.$$

Since x, y are odd functions, we obtain

$$\iiint\limits_{D} (\nabla \cdot \mathbf{v}) \, dV = (4)(2)(2) \int_{x=0}^{4} \int_{y=0}^{\sqrt{16-x^2}} dy \, dx = 16 \int_{0}^{4} \sqrt{16 - x^2} \, dx$$

$$= 16 \left[\frac{1}{2} x \sqrt{16 - x^2} + \frac{16}{2} \sin^{-1}\left(\frac{x}{4}\right) \right]_{0}^{4} = 64 \pi.$$

The surface consists of three parts, S_1 (top), S_2 (bottom) and S_3 (vertical), (Fig. 15.28).
On S_1 : $z = 4$, $\mathbf{n} = \mathbf{k}$.

$$\iint\limits_{S_1} (\mathbf{v} \cdot \mathbf{n}) \, dA = \iint\limits_{S_1} z \, dA = 4 \iint\limits_{S_1} dA = 4 \text{ (area of circular region with radius 4)} = 64 \pi.$$

On S_2 : $z = 0$, $\mathbf{n} = -\mathbf{k}$.

$$\iint\limits_{S_2} (\mathbf{v} \cdot \mathbf{n}) \, dA = \iint\limits_{S_2} -z \, dA = 0.$$

On S_3 : $x^2 + y^2 = 16$, $\mathbf{n} = \dfrac{2x \mathbf{i} + 2y \mathbf{j}}{2\sqrt{x^2 + y^2}} = \dfrac{1}{4}(x \mathbf{i} + y \mathbf{j})$

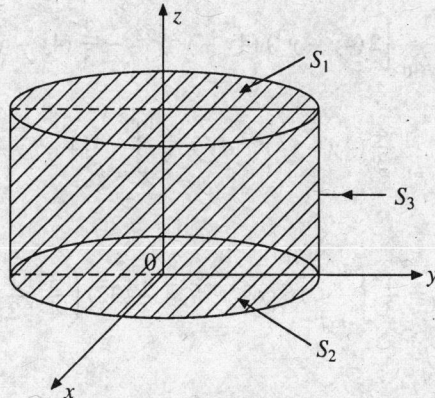

Fig. 15.28.   Example 15.47.

$$\iint_{S_3} (\mathbf{v} \cdot \mathbf{n})\, dA = \frac{1}{4} \iint_{S_3} (3x^3 + 6y^3)\, dA.$$

Using the cylindrical coordinates, we write $x = 4 \cos \theta$, $y = 4 \sin \theta$, $dA = 4\, d\theta\, dz$.
Therefore,

$$\iint_{S_3} (\mathbf{v} \cdot \mathbf{n})\, dA = \frac{1}{4} \int_{z=0}^{4} \int_{\theta=0}^{2\pi} [192 \cos^3\theta + 384 \sin^3\theta]\, 4\, d\theta\, dz$$

$$= 192 \int_{0}^{2\pi} [(\cos 3\theta + 3 \cos \theta) + 2(3 \sin\theta - \sin 3\theta)]\, d\theta = 0.$$

Hence, $\displaystyle \iint_{S} (\mathbf{v} \cdot \mathbf{n})\, dA = \iiint_{D} (\nabla \cdot \mathbf{v})\, dV.$

Example 15.48   Use the divergence theorem to evaluate $\displaystyle \iint_{S} (\mathbf{v} \cdot \mathbf{n})\, dA$, where

$\mathbf{v} = x^2 z\, \mathbf{i} + y\, \mathbf{j} - xz^2\, \mathbf{k}$ and S is the boundary of the region bounded by the paraboloid $z = x^2 + y^2$ and
the plane $z = 4y$.

Solution   We have

$$\iint_{S} (\mathbf{v} \cdot \mathbf{n})\, dA = \iiint_{D} \nabla \cdot \mathbf{v}\, dV = \iiint_{D} (2xz + 1 - 2xz)\, dV = \iiint_{D} dV$$

$$= \int_{y=0}^{4} \int_{x=-\sqrt{4y-y^2}}^{\sqrt{4y-y^2}} \int_{z=x^2+y^2}^{4y} dz\, dx\, dy$$

since the projection of S on the x-y plane is $x^2 + y^2 = 4y$. Therefore,

$$\iint_{S} (\mathbf{v} \cdot \mathbf{n})\, dA = \int_{y=0}^{4} \int_{x=-\sqrt{4y-y^2}}^{\sqrt{4y-y^2}} (4y - x^2 - y^2)\, dx\, dy$$

$$= \int_{y=0}^{4} \left[2(4y - y^2)(4y - y^2)^{1/2} - \frac{2}{3}(4y - y^2)^{3/2} \right] dy$$

$$= \int_{y=0}^{4} \frac{4}{3}(4y - y^2)^{3/2} dy = \frac{4}{3} \int_{y=0}^{4} [4 - (y - 2)^2]^{3/2} dy$$

Set $y - 2 = 2 \sin t$. We obtain

$$\iint_{S} (\mathbf{v} \cdot \mathbf{n}) dA = \frac{4}{3} \int_{-\pi/2}^{\pi/2} 16 \cos^4 t \, dt = \frac{4}{3}(32) \left(\frac{3}{4} \cdot \frac{1}{2} \cdot \frac{\pi}{2} \right) = 8\pi.$$

Green's Identities (formulas)

Divergence theorem can be used to prove some important identities, called *Green's identities* which are of use in solving partial differential equations. Let f and g be scalar functions which are continuous and have continuous first and second order partial derivatives in some region of the three dimensional space. Let S be a piecewise smooth surface bounding a domain D in this region. Let the functions f and g be such that $\mathbf{v} = f$ grad g. Then, we have

$$\nabla \cdot (f \nabla g) = f \nabla^2 g + \nabla f \cdot \nabla g$$

By divergence theorem, we obtain

$$\iint_{S} (\mathbf{v} \cdot \mathbf{n}) \, dA = \iint_{S} f(\nabla g \cdot \mathbf{n}) dA = \iiint_{D} \nabla \cdot (f \nabla g) dV$$

$$= \iiint_{D} (f \nabla^2 g + \nabla f \cdot \nabla g) \, dV.$$

Now, $\nabla g \cdot \mathbf{n}$ is the directional derivative of g in the direction of the unit normal vector $\mathbf{n}$. Therefore, it can be denoted by $\partial g/\partial n$. We have the *Green's first identity* as

$$\iint_{S} f(\nabla g \cdot \mathbf{n}) dA = \iint_{S} f \frac{\partial g}{\partial n} \, dA = \iiint_{D} (f \nabla^2 g + \nabla f \cdot \nabla g) dV. \qquad (15.133)$$

Interchanging f and g, we obtain

$$\iint_{S} g(\nabla f \cdot \mathbf{n}) dA = \iint_{S} g \frac{\partial f}{\partial n} \, dA = \iiint_{D} (g \nabla^2 f + \nabla g \cdot \nabla f) dV.$$

Subtracting the two results, we obtain the *Green's second identity* as

$$\iint_{S} (f \nabla g - g \nabla f) \cdot \mathbf{n} \, dA = \iint_{S} \left(f \frac{\partial g}{\partial n} - g \frac{\partial f}{\partial n} \right) dA = \iiint_{D} (f \nabla^2 g - g \nabla^2 f) dV. \qquad (15.134)$$

Let $f = 1$ in Eq. (15.133). Then, we obtain

$$\iint\limits_{S} \nabla g \cdot \mathbf{n}\, dA = \iint\limits_{S} \frac{\partial g}{\partial n}\, dA = \iiint\limits_{D} \nabla^2 g\, dV. \tag{15.135}$$

If g is a harmonic function, then $\nabla^2 g = 0$ and Eq. (15.135) gives

$$\iint\limits_{S} \nabla g \cdot \mathbf{n}\, dA = \iint\limits_{S} \frac{\partial g}{\partial n}\, dA = 0.$$

This equation gives a very important property of the solutions of Laplace equation, that is of harmonic functions. It states that if $g(x, y, z)$ is a harmonic function, that is, it is a solution of the equation

$$\frac{\partial^2 g}{\partial x^2} + \frac{\partial^2 g}{\partial y^2} + \frac{\partial^2 g}{\partial z^2} = 0$$

then, the integral of the normal derivative of g over any piecewise smooth closed orientable surface is zero.

15.7.2   Stokes's Theorem

The Green's theorem derived in section 15.5 can be written in a vector form different from Eq. (15.120), (see Problem 80, Exercise 15.4). Let C be a curve in two dimensions which is written in the parametric form $\mathbf{r} = \mathbf{r}(s)$. Then, the unit tangent vector to C is given by

$$\mathbf{T} = \frac{dx}{ds}\mathbf{i} + \frac{dy}{ds}\mathbf{j}$$

Let $\mathbf{v}$ be written in the form $\mathbf{v} = g\,\mathbf{i} - f\,\mathbf{j}$. Then,

$$\mathbf{v} \cdot \mathbf{T} = (g\mathbf{i} - f\mathbf{j}) \cdot \left(\frac{dx}{ds}\mathbf{i} + \frac{dy}{ds}\mathbf{j} \right) = g\,\frac{dx}{ds} - f\,\frac{dy}{ds}.$$

By Green's theorem, we have

$$\oint_C \mathbf{v} \cdot d\mathbf{r} = \oint_C \mathbf{v} \cdot \mathbf{T}\, ds = \oint_C g\, dx - f\, dy = \iint\limits_{R} -\left(\frac{\partial f}{\partial x} + \frac{\partial g}{\partial y} \right) dx\, dy = \iint\limits_{R} (\nabla \times \mathbf{v}) \cdot \mathbf{k}\, dx\, dy.$$

$$\tag{15.136}$$

This result can be considered as a particular case of the Stokes's theorem. Extension of the Green's theorem to three dimensions can be done under the following generalisations.

(i) The closed curve C enclosing R in the plane $\rightarrow$ the closed curve C bounding an open smooth orientable surface S (open two sided surface).

(ii) The unit normal $\mathbf{n}$ to $C \rightarrow$ the unit outward or inward normal $\mathbf{n}$ to S.

(iii) Counter clockwise direction of $C \rightarrow$ the direction of C is governed by the direction of the normal $\mathbf{n}$ to S. If $\mathbf{n}$ is taken as outward normal, then C is oriented as right handed screw and if $\mathbf{n}$ is taken as inward normal, then C is oriented as left handed screw (Fig. 15.29. a,b).

We now prove the Stokes's theorem.

Theorem 15.5 (Stokes's theorem)   Let S be a piecewise smooth orientable surface bounded by a piecewise smooth simple closed curve C. Let $\mathbf{v}(x, y, z) = v_1(x, y, z)\,\mathbf{i} + v_2(x, y, z)\,\mathbf{j} + v_3(x, y, z)\,\mathbf{k}$ be a vector function which is continuous and has continuous first order partial derivatives in a domain which contains S. If C is traversed in the positive direction, then

$$\oint_C \mathbf{v} \cdot d\mathbf{r} = \oint_C (\mathbf{v} \cdot \mathbf{T})\,ds = \iint_S (\nabla \times \mathbf{v}) \cdot \mathbf{n}\,dA \tag{15.137}$$

where $\mathbf{n}$ is the unit normal vector to S in the direction of orientation of C.
In terms of components of $\mathbf{v}$ we have

$$\oint_C [v_1(x, y, z)\,dx + v_2(x, y, z)\,dy + v_3(x, y, z)\,dz] = \iint_S (\nabla \times \mathbf{v}) \cdot \mathbf{n}\,dA. \tag{15.138}$$

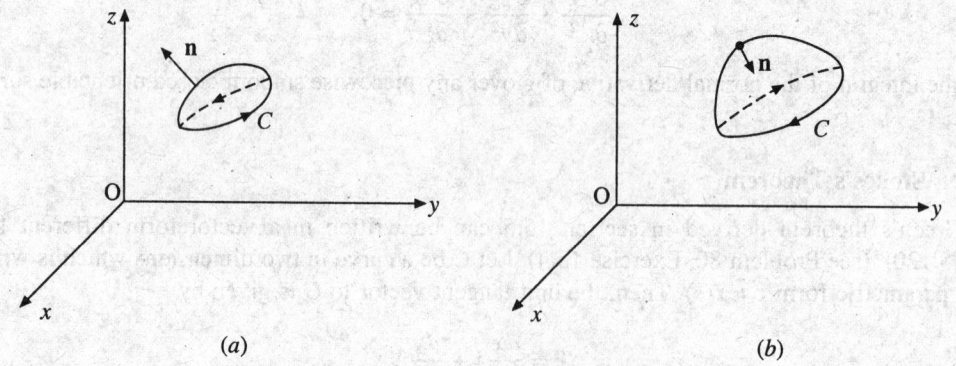

(a)   (b)

Fig. 15.29.   Direction of C in Stokes's theorem.

Proof   We shall prove the theorem for the special case when the equation of the surface can be written simultaneously in the forms

$$z = f(x, y), \quad y = g(x, z) \quad \text{and} \quad x = h(y, z)$$

where f, g, h are continuous functions and have continuous first order partial derivatives.

Let the surface S be oriented upward. Consider the case when the equation of the surface is written as $z = f(x, y)$. If we write $g(x, y, z) = z - f(x, y) = 0$ then the unit normal is given by

$$\mathbf{n} = \frac{-(\partial f / \partial x)\,\mathbf{i} - (\partial f / \partial y)\,\mathbf{j} + \mathbf{k}}{\sqrt{1 + (\partial f / \partial x)^2 + (\partial f / \partial y)^2}}. \tag{15.139}$$

Also $\mathbf{n} = \cos\alpha\,\mathbf{i} + \cos\beta\,\mathbf{j} + \cos\gamma\,\mathbf{k}$, where α, β, γ are the angles which the unit normal makes with the positive directions of x, y and z axis respectively. Comparing, we have

$$\frac{\cos\alpha}{-(\partial f / \partial x)} = \frac{\cos\beta}{-(\partial f / \partial y)} = \frac{\cos\gamma}{1}. \tag{15.140}$$

We need to show that

$$\oint_C [v_1(x, y, z)\,dx + v_2(x, y, z)\,dy + v_3(x, y, z)\,dz] = \iint_S \left[\left(\frac{\partial v_3}{\partial y} - \frac{\partial v_2}{\partial z} \right) \cos\alpha \right.$$

$$\left. + \left(\frac{\partial v_1}{\partial z} - \frac{\partial v_3}{\partial x} \right) \cos\beta + \left(\frac{\partial v_2}{\partial x} - \frac{\partial v_1}{\partial y} \right) \cos\gamma \right] dA. \tag{15.141}$$

Using the equation of the surface as $z = f(x, y)$, we shall prove that

$$\oint_C v_1(x, y, z)\, dx = \iint_S \left(\frac{\partial v_1}{\partial z} \cos \beta - \frac{\partial v_1}{\partial y} \cos \gamma \right) dA. \tag{15.142}$$

Let R be the projection of S and C^* be the projection of the bounding curve C on the x-y plane. Then,

$$\oint_C v_1(x, y, z)\, dx = \oint_{C^*} v_1[x, y, f(x, y)]\, dx$$

$$= \iint_R -\frac{\partial}{\partial y} v_1[x, y, f(x, y)]\, dx\, dy \qquad \text{(by Green's theorem)}$$

$$= -\iint_R \left(\frac{\partial v_1}{\partial y} + \frac{\partial v_1}{\partial z} \frac{\partial f}{\partial y} \right) dx\, dy \qquad \text{(by chain rule)}$$

$$= -\iint_R \left[\frac{\partial v_1}{\partial y} - \frac{\partial v_1}{\partial z} \frac{\cos \beta}{\cos \gamma} \right] dx\, dy \qquad \text{(using Eq. (15.140))}$$

$$= -\iint_S \left[\frac{\partial v_1}{\partial y} - \frac{\partial v_1}{\partial z} \frac{\cos \beta}{\cos \gamma} \right] \cos \gamma\, dA$$

since $(\cos \gamma) dA = dx\, dy$ (see Eq. (15.118)). Hence,

$$\oint_C v_1(x, y, z)\, dx = -\iint_S \left(\frac{\partial v_1}{\partial y} \cos \gamma - \frac{\partial v_1}{\partial z} \cos \beta \right) dA = \iint_S \left(\frac{\partial v_1}{\partial z} \cos \beta - \frac{\partial v_1}{\partial y} \cos \gamma \right) dA$$

and the result in Eq. (15.142) is proved.

Similarly, assuming the equation of the surface as $y = g(x, z)$ and $x = h(y, z)$ we can prove the equality of the terms corresponding to the components $v_2(x, y, z)$ and $v_3(x, y, z)$.

Remark 15

As in divergence theorem, the theorem holds if the given surface S can be subdivided into finitely many special surfaces such that each of these surfaces can be described in the required manner.

Remark 16

To prove the Stokes's theorem, it is not necessary that the equation of the surface should be simultaneously written in the forms $z = f(x, y)$, $y = g(x, z)$ and $x = h(y, z)$. For example, if we take the equation of the surface as $z = f(x, y)$ and assume that $f(x, y)$ has continuous second order partial derivatives then the theorem can be easily proved (see Problem 35, Exercise 15.6).

Remark 17

(Physical interpretation of curl) In section 15.4, we have shown that in rigid body rotation, if $\mathbf{v}$ denotes the tangential (linear) velocity of a point on it, then curl $\mathbf{v}$ represents the angular velocity of the uniformly rotating body. In section 15.5, we have shown that a line integral of a vector field $\mathbf{v}$ around a simple closed curve C defines the circulation of $\mathbf{v}$ around C. For example, if $\mathbf{v}$ denotes the velocity of a fluid, then circulation gives the amount by which the fluid tends to turn the curve by rotating or circulating around C. Therefore, circulation (line integral) is closely related to curl of the vector field. To see this, let C_r be a small circle with centre at $P^*(x^*, y^*, z^*)$. Then, by Stokes's theorem, we have

$$\oint_{C_r} \mathbf{v} \cdot d\mathbf{r} = \iint_{S_r} \text{curl } \mathbf{v} \cdot \mathbf{n} \, dA$$

where S_r is a small surface whose bounding curve is C_r. Let $P(x, y, z)$ be any arbitrary point on C_r. We approximate curl $\mathbf{v}(P) \approx$ curl $\mathbf{v}(P^*)$. Then, we have

$$\oint_{C_r} \mathbf{v} \cdot d\mathbf{r} = \iint_{S_r} [\text{curl } \mathbf{v}(P^*)] \cdot \mathbf{n} \, (P^*) \, dA = [\text{curl } \mathbf{v}(P^*) \cdot \mathbf{n}(P^*)] \iint_{S_r} dA$$

$$= [\text{curl } \mathbf{v}(P^*) \cdot \mathbf{n}(P^*)] \, A_r$$

where A_r is the surface area of S_r. Let the radius r of C_r tend to zero. Then, the approximation curl $\mathbf{v}(P) \approx$ curl $\mathbf{v}(P^*)$ becomes more accurate and in the limit as $r \to 0$, we get

$$\text{curl } \mathbf{v}(P^*) \cdot \mathbf{n}(P^*) = \lim_{r \to 0} \frac{1}{A_r} \oint_{C_r} \mathbf{v} \cdot d\mathbf{r}. \qquad (15.143)$$

The left hand side of Eq. (15.143) is the normal component of curl $\mathbf{v}$. The right hand side of Eq. (15.143) is circulation of $\mathbf{v}$ per unit area. The left hand side is maximum when the circle C_r is positioned such that the normal to surface, $\mathbf{n}(P^*)$ points in the same direction as curl $\mathbf{v}(P^*)$.

Remark 18

Stokes's theorem states that the value of the surface integral is same for any surface as long as the bounding curve, bounding the projection R on any coordinate plane, is the same curve C. Hence, in the degenerate case, when S coincides with R, we can take $\mathbf{n} = \mathbf{k}$ or $\mathbf{j}$ or $\mathbf{i}$ depending on whether the projection is taken on the x-y plane or x-z plane or y-z plane (Fig. 15.30).

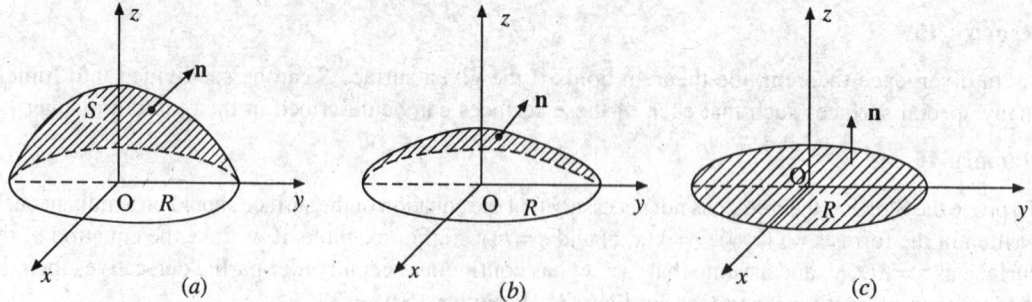

Fig. 15.30.   Application of Stoke's theorem.

Example 15.49   Verify Stokes's theorem for the vector field $\mathbf{v} = (3x - y)\mathbf{i} - 2yz^2\mathbf{j} - 2y^2z\,\mathbf{k}$, where S is the surface of the sphere $x^2 + y^2 + z^2 = 16, z > 0$.

Solution   Consider projection of S on the x-y plane. The projection is the circular region $x^2 + y^2 \leq 16, z = 0$ and the bounding curve C is the circle $z = 0, x^2 + y^2 = 16$. We have

$$\oint_C \mathbf{v} \cdot d\mathbf{r} = \oint_C (3x - y)dx - 2yz^2 dy - 2y^2 z\, dz = \oint_C (3x - y)dx$$

since $z = 0$. Setting $x = 4 \cos \theta$, $y = 4 \sin \theta$, we obtain

$$\oint_C (3x - y)dx = \int_0^{2\pi} 4(3 \cos \theta - \sin \theta)(-4 \sin \theta)d\theta = -16 \int_0^{2\pi} \left[\frac{3}{2} \sin 2\theta - \frac{1}{2}(1 - \cos 2\theta)\right]d\theta$$

$$= 16\left(\frac{1}{2}\right)2\pi = 16\pi.$$

Now, $\quad \nabla \times \mathbf{v} = \begin{vmatrix} \mathbf{i} & \mathbf{j} & \mathbf{k} \\ \partial/\partial x & \partial/\partial y & \partial/\partial z \\ 3x - y & -2yz^2 & -2y^2 z \end{vmatrix} = \mathbf{i}(-4yz + 4yz) - \mathbf{j}(0) + \mathbf{k}(1) = \mathbf{k}$

$$\mathbf{n} = \frac{2(x\mathbf{i} + y\mathbf{j} + z\mathbf{k})}{2\sqrt{x^2 + y^2 + z^2}} = \frac{1}{4}(x\mathbf{i} + y\mathbf{j} + z\mathbf{k}), \quad (\nabla \times \mathbf{v}) \cdot \mathbf{n} = \frac{z}{4}.$$

Therefore, $\quad \displaystyle\iint_S (\nabla \times \mathbf{v}) \cdot \mathbf{n}\, dA = \iint_S \frac{z}{4} dA = \iint_R \frac{z}{4} \frac{dx\, dy}{\mathbf{n} \cdot \mathbf{k}} = \iint_R \frac{z}{4} \frac{dx\, dy}{(z/4)}$

$$= \iint_R dx\, dy = 16\pi$$

which is the area of the circular region in the x-y plane. Hence, Stokes's theorem is proved.

Example 15.50 Evaluate $\displaystyle\oint_C 2y^3 dx + x^3 dy + z\, dz$ where C is the trace of the cone $z = \sqrt{x^2 + y^2}$ intersected by the plane $z = 4$ and S is the surface of the cone below $z = 4$.

Solution We have $\mathbf{v} = 2y^3\, \mathbf{i} + x^3\, \mathbf{j} + z\, \mathbf{k}$ and

$$\text{curl } \mathbf{v} = \begin{vmatrix} \mathbf{i} & \mathbf{j} & \mathbf{k} \\ \partial/\partial x & \partial/\partial y & \partial/\partial z \\ 2y^3 & x^3 & z \end{vmatrix} = \mathbf{i}(0) - \mathbf{j}(0) + \mathbf{k}(3x^2 - 6y^2).$$

If the outward normal to S is taken, then it points downwards. Then, the orientation of C is taken as given in Fig. 15.31. Alternately, if the inward normal to S is taken, then C is oriented in the counter clockwise direction.

Let $f(x, y, z) = \sqrt{x^2 + y^2} - z = 0$ be taken as the equation of the surface. Then, the normal and unit normal are given by

$$\mathbf{N} = \frac{x\mathbf{i} + y\mathbf{j}}{\sqrt{x^2 + y^2}} - \mathbf{k} = \frac{x\mathbf{i} + y\mathbf{j} - z\mathbf{k}}{z} \quad \text{and} \quad \mathbf{n} = \frac{(x\mathbf{i} + y\mathbf{j} - z\mathbf{k})/z}{\sqrt{(x^2 + y^2 + z^2)/z^2}} = \frac{x\mathbf{i} + y\mathbf{j} - z\mathbf{k}}{\sqrt{2}\,z}$$

except at the origin.

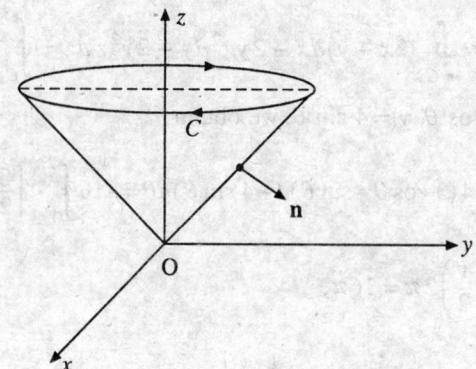

Fig. 15.31.   Example 15.50.

We have $\displaystyle\iint_S (\nabla \times \mathbf{v}) \cdot \mathbf{n}\, dA = \iint_S -\frac{(3x^2 - 6y^2)}{\sqrt{2}}\, dA = -\iint_R \frac{(3x^2 - 6y^2)}{\sqrt{2}}\, \frac{dx\, dy}{(-1/\sqrt{2})}$

since $dx\, dy = (\mathbf{n} \cdot \mathbf{k})\, dA$. Therefore, substituting $x = r \cos\theta$, $y = r \sin\theta$, we obtain

$$\iint_S (\nabla \times \mathbf{v}) \cdot \mathbf{n}\, dA = \iint_R (3x^2 - 6y^2)\, dx\, dy = \int_{r=0}^4 \int_{2\pi}^0 (3\cos^2\theta - 6\sin^2\theta)\, r^3\, dr\, d\theta$$

$$= \frac{3}{2} \int_0^4 \int_{2\pi}^0 [(1 + \cos 2\theta) - 2(1 - \cos 2\theta)]\, r^3\, dr\, d\theta = \frac{3}{2} \int_0^4 \int_{2\pi}^0 (3\cos 2\theta - 1)\, r^3\, dr\, d\theta$$

$$= \frac{3}{2} \left[\frac{r^4}{4} \right]_0^4 \left[\frac{3 \sin 2\theta}{2} - \theta \right]_{2\pi}^0 = 192\pi.$$

The bounding curve C is given by $x^2 + y^2 = 16$, $z = 4$. Now, setting $x = 4 \cos\theta$, $y = 4 \sin\theta$, we obtain

$$\oint_C 2y^3\, dx + x^3\, dy + z\, dz = \oint_C 2y^3\, dx + x^3\, dy$$

$$= \int_{2\pi}^0 64\, [2 \sin^3\theta\, (-4 \sin\theta) + \cos^3\theta\, (4 \cos\theta)]\, d\theta$$

$$= -256 \int_0^{2\pi} [\cos^4\theta - 2\sin^4\theta]\, d\theta = -1024 \int_0^{\pi/2} (\cos^4\theta - 2\sin^4\theta)\, d\theta$$

$$= -1024 \left[\frac{3}{4} \cdot \frac{1}{2} \cdot \frac{\pi}{2} - 2\left(\frac{3}{4} \cdot \frac{1}{2} \cdot \frac{\pi}{2} \right) \right] = 192\,\pi.$$

Hence, Stokes's theorem is verified.

Exercise 15.6

In problems **1** to **5**, verify the divergence theorem.

1. $\mathbf{v} = (x^2 - 2yz)\,\mathbf{i} + (y^2 - 3zx)\,\mathbf{j} + (z^2 - xy)\,\mathbf{k}$. D is the region bounded by $0 \le x \le a$, $0 \le y \le b$, $0 \le z \le c$.

2. $\mathbf{v} = 6x^2y\,\mathbf{i} - y^2\mathbf{j} + 8xz^2\mathbf{k}$. D is the region in the first octant bounded by the cylinder $y^2 + z^2 = 16$ and $x = 3$.

3. $\mathbf{v} = x\mathbf{i} + y\mathbf{j} + z\,\mathbf{k}$. D is the region bounded by the sphere $x^2 + y^2 + z^2 = 16$.

4. $\mathbf{v} = 2xy\,\mathbf{i} + 6yz\mathbf{j} + 3zx\,\mathbf{k}$. D is the region bounded by the coordinate planes and the plane $x + y + z = 2$.

5. $\mathbf{v} = x^2\mathbf{i} + 2y^2\mathbf{j} + 3z^2\mathbf{k}$. D is the region bounded by the cylinder $x^2 + y^2 = 9$, $z = 0$ and $z = 3$.

In problems **6** to **12**, use the divergence theorem to evaluate the surface integral $\displaystyle\iint\limits_{S} \mathbf{v} \cdot \mathbf{n}\, dA$.

6. $\mathbf{v} = 2x^3\mathbf{i} + 3y^3\mathbf{j} + z^3\mathbf{k}$. D is the region bounded by $x^2 + y^2 + z^2 = 9$.

7. $\mathbf{v} = z^2\mathbf{i} + (y - 1)^2\mathbf{j} + xy^2\mathbf{k}$. D is the region bounded by the cylinder $y^2 + z^2 = 16$ and the planes $x = 1$, $x = 5$.

8. $\mathbf{v} = x^2y\,\mathbf{i} + y^3\mathbf{j} + xz^2\mathbf{k}$. D is the region bounded by the parallelopiped $0 \le x \le 2$, $0 \le y \le 3$, $0 \le z \le 4$.

9. $\mathbf{v} = 2xy^2\mathbf{i} + x^2y\,\mathbf{j} + x^3\mathbf{k}$. D is the region bounded by the cone $z = \sqrt{x^2 + y^2}$ and the plane $z = 4$.

10. $\mathbf{v} = 5xy\,\mathbf{i} + 3y\mathbf{j} + x^2\mathbf{k}$. D is the region bounded by $x + y = 4$, $x = 0$, $y = 0$, $z = 0$ and $z = 4$.

11. $\mathbf{v} = x^3\mathbf{i} + y^3\mathbf{j} + z^2\mathbf{k}$. D is the region bounded within by $z = \sqrt{16 - x^2 - y^2}$, $x^2 + y^2 = 4$, $z = 0$.

12. $\mathbf{v} = x^2\mathbf{i} + y^2\mathbf{j} + z^2\mathbf{k}$. D is the region bounded by the hemisphere $x^2 + y^2 + z^2 = a^2$, $z \ge 0$.

In problems **13** to **16**, evaluate the surface integrals by divergence theorem.

13. $\displaystyle\iint\limits_{S} (yz\,dy\,dz + zx\,dz\,dx + xy\,dx\,dy)$. S is surface of the cube $0 \le x \le 1$, $0 \le y \le 1$, $0 \le z \le 1$.

14. $\displaystyle\iint\limits_{S} (xy\,dy\,dz + yz\,dz\,dx + zx\,dx\,dy)$. S is the surface of the parallelopiped $0 \le x \le 4$, $0 \le y \le 3$, $0 \le z \le 4$.

15. $\displaystyle\iint\limits_{S} (x^3\,dy\,dz + y^3\,dz\,dx + z^3\,dx\,dy)$. S is the surface of the sphere $x^2 + y^2 + z^2 = a^2$.

16. $\displaystyle\iint\limits_{S} (x\,dy\,dz + y\,dz\,dx + z\,dx\,dy)$. S is the surface of the sphere $(x - 2)^2 + (y - 2)^2 + (z - 2)^2 = 4$.

In problems **17** to **21**, S is the boundary of a closed and bounded region D and S is a orientable surface.

17. Show that $\displaystyle\iint\limits_{S} \mathbf{r} \cdot \mathbf{n}\, dA = 3V$, where V is the volume of the bounded region and $\mathbf{r} = x\mathbf{i} + y\mathbf{j} + z\mathbf{k}$.

18. Show that $\displaystyle\iint\limits_{S} (\mathbf{a} \cdot \mathbf{n})\, dA = 0$, where $\mathbf{a}$ is a constant vector.

19. Show that $\iint\limits_S (\text{curl } \mathbf{v} \cdot \mathbf{n})\, dA = 0$.

20. Show that $\iint\limits_S r^n\, (\mathbf{r} \cdot \mathbf{n})\, dA = (n+3) \iiint\limits_D r^n\, dV,\ n \neq -3,$ and $r^2 = x^2 + y^2 + z^2$.

21. Show that $\iint\limits_S (\nabla r^2) \cdot \mathbf{n}\, dA = 6V$, where V is the volume of the bounded region and $r^2 = x^2 + y^2 + z^2$.

22. Show that $\iint\limits_S f \dfrac{\partial f}{\partial n}\, dA = \iiint\limits_D |\text{grad } f|^2\, dV$, where $f(x, y, z)$ is a harmonic function.

23. Show that $\iint\limits_S \left(f \dfrac{\partial g}{\partial n} - g \dfrac{\partial f}{\partial n} \right) dA = 0$, where $f(x, y, z)$ and $g(x, y, z)$ are harmonic functions.

In problems **24** to **26**, verify the Stokes's theorem. Assume that the surface S is oriented upward.

24. $\mathbf{v} = x^3\, \mathbf{i} + x^2 y\, \mathbf{j}$. C is the boundary of the rectangle whose sides are $x = 0$, $x = 3$, $y = 0$, $y = 4$ in the plane $z = 0$.

25. $\mathbf{v} = z\, \mathbf{i} + x\, \mathbf{j} + z\, \mathbf{k}$. S is the portion of the sphere $x^2 + y^2 + z^2 = 9$ above the x-y plane.

26. $\mathbf{v} = z\, \mathbf{i} + (2x + z)\, \mathbf{j} + x\, \mathbf{k}$. C is the boundary of the triangle with vertices at $(1, 0, 0)$, $(0, 2, 0)$ and $(0, 0, 3)$.

27. Let S be the surface of the sphere $x^2 + y^2 + z^2 = a^2$. Show that $\iint\limits_S (\nabla \times \mathbf{v}) \cdot \mathbf{n}\, dA = 0$ where $\mathbf{v}$ is any differentiable vector field.

In problems **28** to **30**, evaluate the integral $\iint\limits_S (\nabla \times \mathbf{v}) \cdot \mathbf{n}\, dA$ by Stokes's theorem.

28. $\mathbf{v} = (x^2 - y^2)\, \mathbf{i} + (y^2 - x^2)\, \mathbf{j} + z\, \mathbf{k}$. S is the portion of the surface $x^2 + y^2 - 2by + bz = 0$, b constant, whose boundary lies in the x-y plane.

29. $\mathbf{v} = (x + y)\, \mathbf{i} + (y + z)\, \mathbf{j} + (z + x)\, \mathbf{k}$. S is the portion of the cone $z = \sqrt{x^2 + y^2}$ for $x^2 + y^2 \le 4$.

30. $\mathbf{v} = 2yz\, \mathbf{i} + 3z\, x\, \mathbf{j} + xy\, \mathbf{k}$. S is the paraboloid $z = x^2 + y^2$ for $x^2 + y^2 \le 4$.

In problems **31** to **34**, evaluate $\oint\limits_C \mathbf{v} \cdot d\mathbf{r}$ using the Stokes's theorem. Assume C is oriented in the counter clockwise direction as viewed from above, in problem **31** to **33**.

31. $\mathbf{v} = 3y\, \mathbf{i} + 4z\, \mathbf{j} + 2x\, \mathbf{k}$. C is the intersection of the surface of the sphere $x^2 + y^2 + z^2 = 16$, $x \ge 0$ and the cylinder $y^2 + z^2 = 4$.

32. $\mathbf{v} = (3x + 2z)\, \mathbf{i} + (x + 3y)\, \mathbf{j} + (2y - 3z)\, \mathbf{k}$. C is the curve of intersection of the plane $6x + 3y + 4z = 12$ with the coordinate planes.

33. $\mathbf{v} = x\, \mathbf{i} + z\, \mathbf{j} + y\, \mathbf{k}$. C is the boundary of the ellipsoid $y = (\sqrt{144 - 36x^2 - 9z^2})/4$ in the plane $y = 0$.

34. $\mathbf{v} = y\, \mathbf{i} + z\, \mathbf{j} + x\, \mathbf{k}$. C is the curve of intersection of the sphere $x^2 + y^2 + z^2 - 2x - 2y = 0$ and the plane $x + y = 2$. The direction of the curve is taken such that the curve C begins at $(2, 0, 0)$, goes below the x-y plane and then comes to $(2, 0, 0)$.

35. Prove the Stokes's theorem by taking the equation of the surface as $z = f(x, y)$ and assuming that $f(x, y)$ has continuous second order partial derivatives.

15.8   Answers and Hints

Exercise 15.1

1. Parallel planes.
2. Concentric Spheres.
3. Paraboloids of revolution.
4. Ellipsoids.
5. $\mathbf{r}(t) = (1 + t)\mathbf{i} + 2(1 + t)\mathbf{j} + (3 + 2t)\mathbf{k}$.
6. $\mathbf{r}(t) = (1 + t)(\mathbf{i} - \mathbf{j}) + \mathbf{k}$.
7. $\mathbf{r}(t) = t(\mathbf{i} + \mathbf{j} + \mathbf{k})$.
8. $\mathbf{r}(t) = (3 - 2t)\mathbf{i} + t\mathbf{j} + t\mathbf{k}$.
9. $(7 + t)\mathbf{i} + 2(1 + t)\mathbf{j} + t\mathbf{k}$.
10. $(1 - t/2)\mathbf{i} - t\mathbf{j} + t\mathbf{k}$.
11. $2t^2\mathbf{i} + t\mathbf{j} + t\mathbf{k}$.
12. $9\cos^2 t\,\mathbf{i} + 3\sin t\,\mathbf{j} \pm 3\cos t\,\mathbf{k}$.
13. $2\sin t\,\mathbf{i} + 4\mathbf{j} \pm 2\cos t\,\mathbf{k}$
14. $a(1 + \sqrt{2}\sin t)\mathbf{j} + a(1 \pm \sqrt{2}\cos t)\mathbf{k}$.
15. $(5t^2\cos t + 10t\sin t)\mathbf{i} + (t\cos t + \sin t)\mathbf{j} + (t^3\cos t + 3t^2\sin t)\,\mathbf{k}$.
16. $4\cos 4t + 3t^2$.
17. $-4t(1 + 3t^2)\mathbf{i} - 27t^2\mathbf{j} + 2t(1 + 12\,t^2)\mathbf{k}$.
18. $(3t^2 - 2e^{2t})\mathbf{i} - [(1 - 2t) - (2 + t)e^t]\mathbf{j} - [te^t + t(2 + 3t)]\mathbf{k}$.
19. $2t\,\mathbf{u}(t^2) + 2t^3\mathbf{u}'(t^2)$.
20. $a\mathbf{u}'(at) - (a/t^2)\mathbf{v}'(a/t)$.
21. $\mathbf{u}(t) \times \mathbf{u}'''(t) + \mathbf{u}'(t) \times \mathbf{u}''(t)$.
22. $\mathbf{u}(t) \cdot \mathbf{u}'(t) \times \mathbf{u}'''(t)$.
23. $x(t) = 2 + t$, $y(t) = 8(1 + t)$, $z(t) = 12(2 + 3t)$.
24. $x(t) = (1 + t)/\sqrt{2}$, $y(t) = (1 - t)/\sqrt{2}$, $z(t) = (\pi/4) + t$.
25. $x(t) = 3 + 4t$, $y(t) = 3 + t$, $z(t) = (6 + t)/9$.
26. $x(t) = 1 + t$, $y(t) = (1 + t)e$, $z(t) = 1$.
27. $3a/2$.
28. $(p\pi/2) + 2\log(\pi + p) - 2\log 2$, $p = \sqrt{4 + \pi^2}$.
29. $2\pi a$, $\mathbf{r}(s) = a\cos(s/a)\mathbf{i} + a\sin(s/a)\mathbf{j}$, $0 \le s \le 2\pi a$.
30. $4\pi\sqrt{10}$, $\mathbf{r}(s) = \cos(s/\sqrt{10})\mathbf{i} + \sin(s/\sqrt{10})\mathbf{j} + (3s/\sqrt{10})\mathbf{k}$, $-2\pi\sqrt{10} \le s \le 2\pi\sqrt{10}$.
31. $(5\sqrt{5} - 1)/3$, $\mathbf{r}(s) = [(s^*)^2/2]\mathbf{i} + [(s^*)^3/3]\mathbf{k}$, $s^* = [(3s + 1)^{2/3} - 1]^{1/2}$, $0 \le s \le (5\sqrt{5} - 1)/3$.
32. $[\sqrt{2} + \log(1 + \sqrt{2})]/2$.
33. Unit tangent vector $= (2\mathbf{i} + 3\mathbf{j} + 3\mathbf{k})/\sqrt{22}$. The given curve is a straight line. $\mathbf{T}$ is independent of t and $d\mathbf{T}/dt = \mathbf{0}$.
34. $\mathbf{v}(t) = (\cos t - \sin t)\mathbf{i} + (\cos t + \sin t)\mathbf{j} + \mathbf{k}$, speed $= \sqrt{3}$, $\mathbf{a}(t) = -[(\sin t + \cos t)\mathbf{i} + (\sin t - \cos t)\mathbf{j}]$
35. $|\mathbf{v}(t)|^2 = c^2$ or $\mathbf{v}(t) \cdot \mathbf{v}(t) = c^2$. Then, $d[\mathbf{v}(t) \cdot \mathbf{v}(t)]/dt = 0$ or $\mathbf{a}(t) \cdot \mathbf{v}(t) = 0$ for all t.

Exercise 15.2

1. $-21\,\mathbf{i}$.
2. $\mathbf{i}(\pi + \sqrt{2})/2$.
3. $\pi(\mathbf{i} - \mathbf{j}) + \mathbf{k}$.
4. $(3\mathbf{i} - 4\mathbf{j} + 5\mathbf{k})/25$.
5. $(\mathbf{i} + \mathbf{j} + \mathbf{k})/\sqrt{3}$.
6. $e^2(5\mathbf{i} + 9\mathbf{j} + \mathbf{k})$.
7. $(\mathbf{i} + \mathbf{j} + \mathbf{k})/2$.
8. $3\mathbf{i} - (\pi^2/54)(9\sqrt{3} + 4\pi)\mathbf{j} + 2\,\mathbf{k}$.

9. $16(i - j)$, $(i - j)/\sqrt{2}$.

10. $2(3i + 4j)$, $(3i + 4j)/5$.

11. $4(2i - j)$, $(2i - j)/\sqrt{5}$.

12. $2(i + 2j + k)$, $(i + 2j + k)/\sqrt{6}$.

13. $-(2i + j + k)$, $-(2i + j + k)/\sqrt{6}$.

14. $2(2i - j - \sqrt{3}k)$, $(2i - j - \sqrt{3}k)/\sqrt{8}$.

21. $-11/\sqrt{5}$.

22. -1.

23. $-3i/\sqrt{6}$.

24. -3.

25. $2\sqrt{18}$.

26. $58/5\sqrt{2}$.

27. $(-i + 2j)/\sqrt{2}$, $\sqrt{5/2}$.

28. $2(j + 4k)$, $2\sqrt{17}$.

29. $6(2i - j - 2k)$, 18.

30. $13i + 5j - 10k$, $\sqrt{294}$.

31. $-(11i + 7j)$, $-\sqrt{170}$.

32. $-8(\sqrt{\pi/3})(i + j)$, $-8\sqrt{2\pi/3}$.

33. $-2(i - 2j + k)$, $-2\sqrt{6}$.

34. $-(e/2)(i + j + 8k)$, $-\sqrt{66}(e/2)$.

35. All points on the line $(2 - \sqrt{3})x + (2\sqrt{3} - 1)y = \sqrt{3}$.

36. $(3i + 4j)/5$, $-(3i + 4j)/5$.

37. $3i - j$.

38. $a = -2$, $b = 2$, $c = 1$.

39. In the direction of maximum rate of decrease, $-(8i + 8j - k)$.

40. $3x - 3y - 2z = 2$.

41. $2x + 6y + z = 26$.

42. $x + 3y + 2 = 0$.

43. $x - 2\sqrt{3}y + z = (6 - \pi\sqrt{3})/6$.

44. $(xx_0/a^2) + (yy_0/b^2) + (zz_0/c^2) = 1$.

45. $(xx_0/a^2) - (yy_0/b^2) + (zz_0/c^2) = 1$.

46. $x^2y^2z + c$.

47. $\sqrt{x^2 + y^2 + z^2} + c$.

48. $6x^2 - 5y^3 + z + c$.

49. $e^{xyz} + c$.

50. $f = k$, constant, No.

51. $\cos^{-1}(\sqrt{21/101})$.

52. $\cos^{-1}(\sqrt{2/3})$.

53. $\cos^{-1}(8/(3\sqrt{21}))$.

54. $x(t) = 2(1 + 6t)$, $y(t) = 1 - 4t$, $z(t) = 10 - t$.

55. $x(t) = 2(1 + 2t)$, $y(t) = 1 + 4t$, $z(t) = -(1 + 8t)$.

Exercise 15.3

1. $4, 0$.

2. $x + y + z$, $-(iy + jz + kx)$.

3. $6(x^2 + y^2 + z^2)^{3/2}$, 0.

4. $2(x + y + z)$, 0.

5. $(1 - 2z)e^{-y}$, $[2i(xy - e^{-y}) - jy^2 + kxe^{-y}]$.

6. $yz + 2x^2 + 2xz - y^2$, $-[2yzi + (z^2 - xy)j + (xz - 4xy)k]$.

7. $6x$, $-[ix + j(2x - y) - 6yk]$.

8. $2(x + y + z)$, 0.

9. $i + j - 4zk$.

10. $(i + j + k)x\cos(x + y + z) + i\sin(x + y + z)$.

11. $(\mathbf{i} + \mathbf{j} + \mathbf{k})\, e^{x+y+z}$.

12. $16(y^3z^2\mathbf{i} + 3xy^2z^2\mathbf{j} + 2xy^3z\,\mathbf{k})$.

16. $f\,(\text{div } \mathbf{r}) + \nabla f \cdot \mathbf{r} = 5f$.

23. $\displaystyle \sum \mathbf{i}\left[-\frac{3y}{r^5}(a_1 y - a_2 x) - \frac{3z}{r^5}(a_1 z - a_3 x) + \frac{2a_1}{r^3}\right]$

$$= \sum \mathbf{i}\left[-\frac{3}{r^5}\{a_1(x^2 + y^2 + z^2) - x\,(a_1 x + a_2 y + a_3 z)\} + \frac{2a_1}{r^3}\right]$$

$$= \sum \mathbf{i}\left[-\frac{3}{r^3}a_1 + \frac{3x}{r^5}(a_1 x + a_2 y + a_3 z) + \frac{2a_1}{r^3}\right] = \frac{3}{r^5}(\mathbf{a}\cdot\mathbf{r})\mathbf{r} - \frac{1}{r^3}\mathbf{a}$$

24. $\displaystyle \sum \mathbf{i}\left[a_1\frac{\partial v_2}{\partial y} - a_2\frac{\partial v_1}{\partial y} - a_3\frac{\partial v_1}{\partial z} + a_1\frac{\partial v_3}{\partial z}\right]$

$$= \sum \mathbf{i}\left[a_1\left(\frac{\partial v_1}{\partial x} + \frac{\partial v_2}{\partial y} + \frac{\partial v_3}{\partial z}\right) - \left(a_1\frac{\partial v_1}{\partial x} + a_2\frac{\partial v_1}{\partial y} + a_3\frac{\partial v_1}{\partial z}\right)\right] = (\nabla \cdot \mathbf{v})\,\mathbf{a} - (\mathbf{a}\cdot\nabla)\,\mathbf{v}$$

26. $x(y^2 + y) - (x^3/3) + c$.

27. $x^3 y^2 z^4 + c$.

28. $e^{xy} + 2e^z + c$.

29. $(1/2)\sin(x^2 + y^2 + z^2) + c$.

30. $a = 2,\ b = c,\ c$ arbitrary.

38. $\displaystyle \nabla \times (\nabla \times \mathbf{E}) = \nabla \times \left(-\frac{1}{c}\frac{\partial \mathbf{H}}{\partial t}\right) = -\frac{1}{c}\frac{\partial}{\partial t}(\nabla \times \mathbf{H}) = -\frac{1}{c^2}\frac{\partial^2 \mathbf{E}}{\partial t^2}$

$\nabla \times (\nabla \times \mathbf{E}) = \nabla(\nabla \cdot \mathbf{E}) - \nabla^2 \mathbf{E} = -\nabla^2 \mathbf{E}$. Therefore, $\dfrac{\partial^2 \mathbf{E}}{\partial t^2} = c^2\nabla^2\mathbf{E}$. Similarly for $\mathbf{H}$.

Exercise 15.4

1. $36\sqrt{14}$.

2. $2291\sqrt{11}/60$.

3. 0.

4. $\dfrac{3}{4}\left[8\sqrt{65} + \log(8 + \sqrt{65})\right] + \dfrac{1}{6}[(65)^{3/2} - 1]$.

5. $10(3\pi + 5)$.

6. $[3\sqrt{37} + (1/2)\log(6 + \sqrt{37})] + [(37)^{3/2} - 1]/12$.

7. $164\sqrt{5}/3$.

8. 316.

9. 36.

10. 0.

11. $(7/2) - \cos 1$.

12. $-20169/35$.

13. $(2e^8 + 3e^2 + 19)/3$.

14. 3π.

15. 0.

16. $-9\sqrt{2}/4,\ 9(4 - \sqrt{2})/4$.

17. $1576/15,\ 2096/3$.

18. $-(8 + 3\pi)/4,\ (9 + 3\pi)/2$.

19. $4(4 - 3\pi)/9,\ 16/9,\ \pi/2$.

20. $16/3,\ 16/3,\ 101/6$.

21. $-5/3,\ -13/10,\ -13/5$.

22. 16.

23. $10/3,\ 17/5$.

24. $70/3$.

25. $105/2$.

26. -9π.

27. -2π.

28. $40/3$.

29. $47/2$.

30. $-[\pi - 3\log\sqrt{3}]/3$.

31. $\mathbf{T} = \dfrac{d\mathbf{r}}{ds} = \dfrac{dx}{ds}\mathbf{i} + \dfrac{dy}{ds}\mathbf{j}, \ \mathbf{n} = \dfrac{dy}{ds}\mathbf{i} - \dfrac{dx}{ds}\mathbf{j}.$

32. 62/3.

33. 63.

34. -4.

35. $4(2\pi a - a^2 - 1)$.

36. $32\,\pi$

37. 0

38. 33.

39. $-1/2$

40. $-1/3$.

41. 11.

42. $(2\sqrt{2}\,e^{2\pi} - e^{\pi/2})/\sqrt{2}$.

43. 34.

44. 21.

45. $-(1 + 2\sin 1)$.

46. $\sinh(x + y)$.

47. $x^2 + y^2 + e^{xy}$.

48. xyz.

49. $x^2y + y^2z$.

50. $xy^3 - x^3z$.

51. In Green's theorem set

$f = 0, \ g = x; \ g = 0, \ f = -y; \ f = -y/2, \ g = x/2.$

52. 20/3.

53. 0.

54. 48π

55. $-144\sqrt{3}/5$.

56. 512/15.

57. -80π

58. -96π.

59. $-512/5$.

60. 84.

61. 1920π.

62. $(3 + 4\sqrt{2})/24$.

63. 0.

64. 6π.

65. $-1/28$.

67. $\pi\,ab$.

68. πa^2.

69. $3\pi a^2/8$.

70. 9π.

71. 48.

72. 16π.

73. $f = 0, \ g = xy^2, \ 8\pi$.

74. $f = g = xy, \ 0$.

75. 224/3.

76. $\displaystyle\oint_C f\,\frac{\partial g}{\partial n}\,ds = \oint_C f\,\nabla g \cdot \mathbf{n}\,ds = \oint_C f\,\frac{\partial g}{\partial x}\,dy - f\,\frac{\partial g}{\partial y}\,dx$ (see example 15.38)

$$= \iint_R (f\,\nabla^2 g + \nabla f \cdot \nabla g)\,dx\,dy$$

Interchange f and g, and substract the two results.

77. Use $g = f$ in the first result of problem **76**.

78. Write $dg = \dfrac{\partial g}{\partial x}\,dx + \dfrac{\partial g}{\partial y}\,dy$ and use Green's theorem.

79. $\mathbf{n} = \dfrac{dy}{ds}\mathbf{i} - \dfrac{dx}{ds}\mathbf{j}.$

80. $\mathbf{T} = \dfrac{dx}{ds}\mathbf{i} + \dfrac{dy}{ds}\mathbf{j}.$

Exercise 15.5

1. $\mathbf{r} = u\,\mathbf{i} + v\,\mathbf{j} + \sqrt{25 - u^2}\;\mathbf{k};\, 3x + 4z = 25.$

2. $\mathbf{r} = (4\cos u \cos v)\mathbf{i} + (4\sin u \cos v)\mathbf{j} + (4\sin v)\,\mathbf{k};\, u = \tan^{-1}(3/2),$
   $2x + 3y - \sqrt{3}z + 16 = 0.$

3. $\mathbf{r} = u\cos v\,\mathbf{i} + u\sin v\,\mathbf{j} + u^2\mathbf{k};\, u = \sqrt{10},\, \tan v = 3;\, 2\,x + 6y - z = 10 .$

4. $\mathbf{r} = u\cos v\,\mathbf{i} + u\sin v\,\mathbf{j} + u\mathbf{k};\, u = 1,\, v = \pi/4;\, x + y - \sqrt{2}\,z = 0.$

5. $\mathbf{r} = 2\cos u\cos v\,\mathbf{i} + 3\sin u\cos v\,\mathbf{j} + 4\sin v\,\mathbf{k};\, u = \pi/2,\, v = \pi/4;\, 4y + 3z - 12\,\sqrt{2} = 0.$

6. $\mathbf{r} = u^2\mathbf{i} + v\,\mathbf{j} + u\,\mathbf{k};\, u = 2,\, v = 5;\, x - 4z + 4 = 0.$

7. $ds^2 = a^2 du^2 + dv^2.$ 8. $ds^2 = (1 + 4u^2)\,du^2 + u^2 dv^2.$

9. $ds^2 = 2du^2 + u^2 dv^2.$

10. $ds^2 = [1 + v^2\sin^2(uv)]\,du^2 + 2uv\sin^2(uv)\,du\,dv + [1 + u^2\sin^2(uv)]dv^2$

11. Project S on x-y plane; Projection: Region inside $x^2 + y^2 = 16$; $16\pi\sqrt{2}$.

12. Project S on x-y plane; Projection; Region inside $x^2 + y^2 = 9$; $\pi[37\sqrt{37} - 1]/6$.

13. Project both portions (above and below) on x-y plane; Projection: Region inside $x^2 + y^2 = a^2$; $4\pi a^2$.

14. Project both portions on y-z plane, Projection: Region inside the rectangle with sides of length 2 and 8; 16π.

15. Project on x-y plane; Projection : Region bounded by $x = 0$, $y = 0$ and $3x + 4y = 24$; $12\sqrt{29}$.

16. Project on x-y plane; Projection: rectangle with sides of length 3 and 6; $30\sin^{-1}(3/5)$.

17. Project on x-y plane; Projection : Region inside $x^2 + y^2 = 16$; $\pi[65\sqrt{65} - 1]/6$.

18. Project the intersected surface on x-y plane; Projection: Region inside $x^2 + y^2 = 4$; $4\sqrt{8}\,\pi(\sqrt{2} - 1)$.

19. Area = 2 (Area in the first octant). Project on x-y plane; Projection : Region inside $x^2 + (y - 2)^2 = 4$; $16(\pi - 2)$ (Fig. 15.32).

20. Area = 8 (Area in the first octant). Project $z = \sqrt{a^2 - y^2}$ on x-y plane. Projection: Region inside $x^2 + y^2 = a^2$; $8a^2$. (Fig. 15.33).

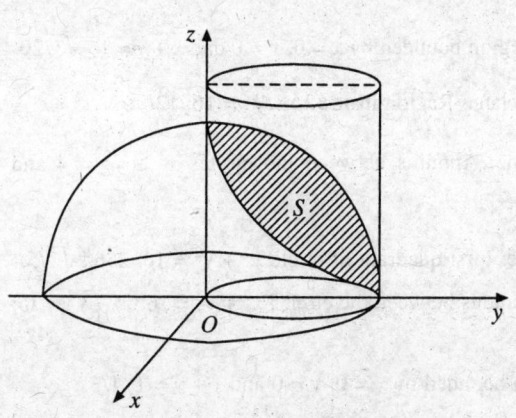

Fig. 15.32. Problem 19.

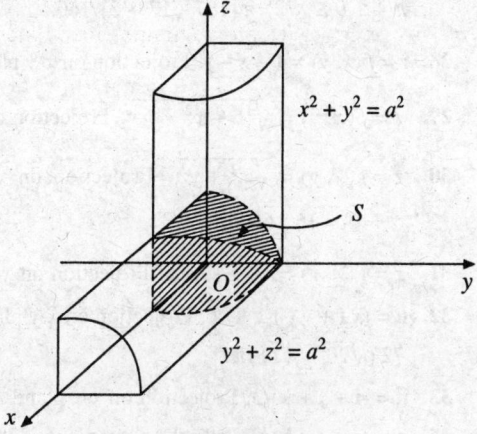

Fig. 15.33. Problem 20.

21. $\rho(x, y, z) = cz$, c constant. Projection on x-y plane : Region inside $x^2 + y^2 = 4$ in the first quadrant; $(731\sqrt{17} - 19)\pi c/240$.

22. $\rho(x, y, z) = cy^2$, c constant. Projection on x-y plane: Region bounded by $x = 0$, $y = 0$ and $2x + 3y = 12$; $8\sqrt{29}\,c$.

23. $81\pi\sqrt{2}/2$.

24. Point on the given line is $(b/2, y, 0)$, $d^2 = (x - (b/2))^2 + z^2$.

Parametric form: $\mathbf{r}(u, v) = v\mathbf{i} + a\cos u\mathbf{j} + a\sin u\,\mathbf{k}$, $0 \le u \le 2\pi$, $0 \le v \le b$; $\pi a\,(6a^2b + b^3)/6$.

25. $d^2 = x^2 + z^2$. $\mathbf{r}(u, v) = u\cos v\mathbf{i} + u\mathbf{j} + u\sin v\,\mathbf{k}$; $\pi b^4\sqrt{2}/2$.

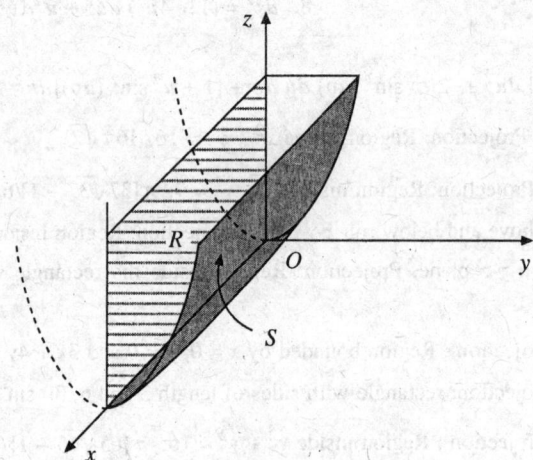

Fig. 15.34.   Problem 26.

26. $y = h(x, z) = \sqrt{z}$; 1984/9 (Fig. 15.34).

27. $z = f(x, y) = 1 + x^2 + y^2$. Projection on x-y plane: Region inside the circular portion $x^2 + y^2 = 4$, $\pi/4 \le \theta \le \pi/2$; $\pi[187\sqrt{17} - 3]/160$.

28. $z = f(x, y) = 1 - x - y$. Projection on x-y plane: Region bounded by $x = 0$, $y = 0$ and $x + y = 1$; $\sqrt{3}/20$.

29. $z = f(x, y) = \sqrt{16 - x^2 - y^2}$. Projection on x-y plane: Region inside $x^2 + y^2 = 16$; $128\pi/3$.

30. $z = f(x, y) = \sqrt{x^2 + y^2}$. Projection on x-y plane: Annulus between the circles $x^2 + y^2 = 4$ and $x^2 + y^2 = 16$; $560\ \pi\sqrt{2}/3$.

31. $z = f(x, y) = \sqrt{x^2 + y^2}$. Projection on x-y plane: First quadrant of circle $x^2 + y^2 = 16$. $256\pi\sqrt{2}/5$.

32. $\mathbf{n} = (x\,\mathbf{i} + y\,\mathbf{j} + z\,\mathbf{k})/6$. Projection on x-y plane: Annulus between the circles $x^2 + y^2 = 16$, $x^2 + y^2 = 25$; $72\,(\sqrt{20} - \sqrt{11})\pi$.

33. $\mathbf{n} = (\mathbf{i} + \mathbf{j} + \mathbf{k})/3$. Projection on x-y plane: Region bounded by $x = 0$, $y = 0$ and $x + y = 1$; 1/3.

34. $\mathbf{n} = (y\,\mathbf{j} + z\,\mathbf{k})/5$. Project on x-y plane; 875/2.

35. $x = 0, \mathbf{n} = -\mathbf{i}; x = 1, \mathbf{n} = \mathbf{i}; y = 0, \mathbf{n} = -\mathbf{j}; y = 1, \mathbf{n} = \mathbf{j}; z = 0, \mathbf{n} = -\mathbf{k}; z = 1, \mathbf{n} = \mathbf{k}; 2.$

Exercise 15.6

1. $abc(a + b + c).$

2. 2560. (Fig. 15.35)

3. $256\pi.$

4. 22/3.

5. $243\pi.$

6. $5832\pi/5.$

7. $-128\pi.$

8. 384.

9. $3072\pi/5.$

10. 928/3.

11. $2\pi(2188 - 1056\sqrt{3})/5.$

12. $\pi a^4/2.$

13. 0.

14. 264.

15. $12\pi a^5/5.$

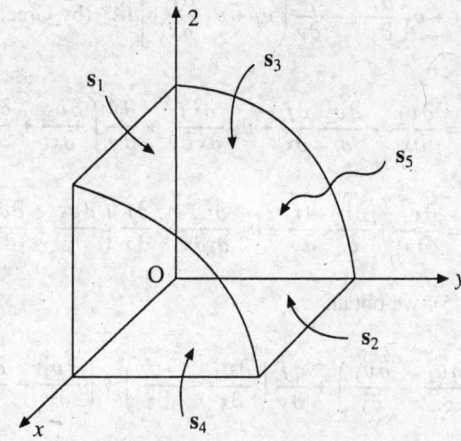

Fig. 15.35. Problem 2.

16. $32\pi.$

22. Set $f = g$ in Eq. (15.133).

23. Use Eq. (15.134)

24. 72.

25. $9\pi.$

26. $-1.$

27. Consider that S is made into portions (say above and below x-y plane). If the outward unit normal $\mathbf{n}$ to S is taken as the direction of the normal, then $\mathbf{n}$ points upward for upper part of the surface and points downwards for lower part of the surface. Therefore, orientations of C are in opposite directions.

28. $2\pi b^3$

29. $-4\pi.$

30. $16\pi.$

31. Consider projection on y-z plane, $-16\pi.$

32. 22.

33. 0.

34. $-2\pi\sqrt{2}.$

35. Write $g(x, y, z) = z - f(x, y) = 0.$ Then

$$\mathbf{n} = \left(-\frac{\partial f}{\partial x}\mathbf{i} - \frac{\partial f}{\partial y}\mathbf{j} + \mathbf{k}\right) \bigg/ \sqrt{1 + \left(\frac{\partial f}{\partial x}\right)^2 + \left(\frac{\partial f}{\partial y}\right)^2}, \text{ and}$$

$$\iint_S \operatorname{curl} \mathbf{v} \cdot \mathbf{n} \, dA = \iint_R \left[-\left(\frac{\partial v_3}{\partial y} - \frac{\partial v_2}{\partial z}\right)\frac{\partial f}{\partial x} - \left(\frac{\partial v_1}{\partial z} - \frac{\partial v_3}{\partial x}\right)\frac{\partial f}{\partial y} + \left(\frac{\partial v_2}{\partial x} - \frac{\partial v_1}{\partial y}\right)\right] dA^* \quad (15.144)$$

Let the parametric representation of C^*, the projection of C, be $x = x(t), y = y(t), z = f(x(t), y(t)),$ $a \le t \le b.$ Then,

$$\oint_C \mathbf{v} \cdot d\mathbf{r} = \oint_C v_1\,dx + v_2\,dy + v_3\,dz$$

$$= \int_a^b \left[v_1 \frac{dx}{dt} + v_2 \frac{dy}{dt} + v_3 \left(\frac{\partial f}{\partial x} \frac{dx}{dt} + \frac{\partial f}{\partial y} \frac{dy}{dt} \right) \right] dt$$

$$= \int_{C^*} \left(v_1 + v_3 \frac{\partial f}{\partial x} \right) dx + \left(v_2 + v_3 \frac{\partial f}{\partial y} \right) dy$$

$$= \iint_R \left[\frac{\partial}{\partial x} \left(v_2 + v_3 \frac{\partial f}{\partial y} \right) - \frac{\partial}{\partial y} \left(v_1 + v_3 \frac{\partial f}{\partial x} \right) \right] dA^* \text{ (by Green's theorem)} \qquad (15.145)$$

Now, $\dfrac{\partial}{\partial x} \left(v_2 + v_3 \dfrac{\partial f}{\partial y} \right) = \dfrac{\partial v_2}{\partial x} + \dfrac{\partial v_2}{\partial z} \dfrac{\partial f}{\partial x} + v_3 \dfrac{\partial^2 f}{\partial x \partial y} + \dfrac{\partial f}{\partial y} \left(\dfrac{\partial v_3}{\partial x} + \dfrac{\partial v_3}{\partial z} \dfrac{\partial f}{\partial x} \right)$

and $\dfrac{\partial}{\partial y} \left(v_1 + v_3 \dfrac{\partial f}{\partial x} \right) = \dfrac{\partial v_1}{\partial y} + \dfrac{\partial v_1}{\partial z} \dfrac{\partial f}{\partial y} + v_3 \dfrac{\partial^2 f}{\partial y \partial x} + \dfrac{\partial f}{\partial x} \left(\dfrac{\partial v_3}{\partial y} + \dfrac{\partial v_3}{\partial z} \dfrac{\partial f}{\partial y} \right).$

Substituting in Eq. (15.145), we obtain

$$\oint_C \mathbf{v} \cdot d\mathbf{r} = \iint_R \left[\frac{\partial f}{\partial x} \left(\frac{\partial v_2}{\partial z} - \frac{\partial v_3}{\partial y} \right) + \frac{\partial f}{\partial y} \left(\frac{\partial v_3}{\partial x} - \frac{\partial v_1}{\partial z} \right) + \left(\frac{\partial v_2}{\partial x} - \frac{\partial v_1}{\partial y} \right) \right] dA^*$$

which is same as Eq. (15.144).

Chapter 16

Partial Differential Equations

16.1 Introduction

Let z be a dependent variable and x, y be independent variables, that is, $z = z(x, y)$. Then, we can define the first order partial derivatives $p = \partial z/\partial x = z_x$ and $q = \partial z/\partial y = z_y$, the second order partial derivatives $r = \partial^2 z/\partial x^2 = z_{xx}$, $s = \partial^2 z/\partial x\partial y = z_{xy}$, $t = \partial^2 z/\partial y^2 = z_{yy}$ etc., if they exist.

An equation containing x, y, z, p, q defines a first order partial differential equation, that is

$$f(x, y, z, p, q) = 0. \tag{16.1}$$

This equation is linear, if it is linear in p, q. An equation containing x, y, z, p, q, r, s, t defines a second order partial differential equation, that is

$$g(x, y, z, p, q, r, s, t) = 0. \tag{16.2}$$

This equation is linear, if it is linear in p, q, r, s, t. Most of the mathematical models describing the physical processes contain a partial differential equation or a system of partial differential equations. We have the following examples (sec section 9.5)

One-dimensional heat conduction equation: $u_t = u_{xx}$.

Laplace equation: $u_{xx} + u_{yy} = 0$.

One-dimensional wave equation: $u_{tt} = u_{xx}$.

The solutions of partial differential equations are of the form $f(x, y, z) = c$, which represents a surface in three dimensions, where c is a constant. Therefore, we shall first discuss a few properties relating to surfaces and curves in space.

Let C be a curve in space. Then, its parametric form can be written as

$$C : x = x(t), \ y = y(t), \ z = z(t), \ a \le t \le b. \tag{16.3}$$

and

$$\left(\frac{dx}{dt}\right)_{t_0} = x'(t_0), \left(\frac{dy}{dt}\right)_{t_0} = y'(t_0), \left(\frac{dz}{dt}\right)_{t_0} = z'(t_0), a \le t \le b$$

give the direction ratios of the tangent to the curve C at a point $P(t_0)$.

If the parametric form of the curve C is written in terms of the arc length s, measured from a fixed point, that is

$$C: x = x(s), y = y(s), z = z(s), 0 \le s \le s_0 \tag{16.4}$$

then, (dx/ds), (dy/ds), (dz/ds) are the direction cosines of the tangent to the curve C at any point s.

Let

$$S : f(x, y, z) = c, c \text{ constant} \tag{16.5}$$

be the equation of a surface. Then, the normal to the surface is given by (see Eq. (15.38))

$$\mathbf{N} = \operatorname{grad} f = \mathbf{i}\frac{\partial f}{\partial x} + \mathbf{j}\frac{\partial f}{\partial y} + \mathbf{k}\frac{\partial f}{\partial z}.$$

Therefore, the direction ratios of the normal at a point P on the surface are $\partial f/\partial x$, $\partial f/\partial y$ and $\partial f/\partial z$. If the equation of the surface is given in the form

$$S: z = f(x, y), \quad \text{or} \quad F(x, y, z) = f(x, y) - z = 0$$

then, the normal to the surface S is given by

$$\mathbf{N} = \mathbf{i}\frac{\partial f}{\partial x} + \mathbf{j}\frac{\partial f}{\partial y} - \mathbf{k} = \mathbf{i}\frac{\partial z}{\partial x} + \mathbf{j}\frac{\partial z}{\partial y} - \mathbf{k} = \mathbf{i}p + \mathbf{j}q - \mathbf{k}$$

where $p = \partial z/\partial x$ and $q = \partial z/\partial y$. Therefore, the direction ratios of the normal at a point P on the surface S are p, q, -1.

Let $S: f(x, y, z) = c$ be a surface on which a curve $C : x = x(s)$, $y = y(s)$, $z = z(s)$ lies. That is

$$f[x(s), y(s), z(s)] = c.$$

Then, we have

$$\frac{df}{ds} = \frac{\partial f}{\partial x}\frac{dx}{ds} + \frac{\partial f}{\partial y}\frac{dy}{ds} + \frac{\partial f}{\partial z}\frac{dz}{ds} = 0$$

or

$$\left(\mathbf{i}\frac{\partial f}{\partial x} + \mathbf{j}\frac{\partial f}{\partial y} + \mathbf{k}\frac{\partial f}{\partial z}\right) \cdot \left(\mathbf{i}\frac{dx}{ds} + \mathbf{j}\frac{dy}{ds} + \mathbf{k}\frac{dz}{ds}\right) = 0$$

showing that the normal to S is perpendicular to the tangent line.

Tangent to a curve in space

Let C be the curve of intersection of the surfaces

$$S_1: f(x, y, z) = c_1 \quad \text{and} \quad S_2: g(x, y, z) = c_2. \tag{16.6}$$

Let $P_0 = P(x_0, y_0, z_0)$ be any point on C.

Now, equation of the tangent plane to the surface S_1 is (see Eq. (15.39))

$$(x - x_0)\frac{\partial f}{\partial x}(P_0) + (y - y_0)\frac{\partial f}{\partial y}(P_0) + (z - z_0)\frac{\partial f}{\partial z}(P_0) = 0. \tag{16.7}$$

Equation of the tangent plane at P_0 to the surface S_2 is

$$(x - x_0)\frac{\partial g}{\partial x}(P_0) + (y - y_0)\frac{\partial g}{\partial y}(P_0) + (z - z_0)\frac{\partial g}{\partial z}(P_0) = 0. \tag{16.8}$$

These two planes intersect along a line, which is the tangent to the curve C at P_0.

Solving Eqs. (16.7) and (16.8) simultaneously, we get

$$\frac{x - x_0}{(\partial f/\partial y)(\partial g/\partial z) - (\partial f/\partial z)(\partial g/\partial y)} = \frac{y - y_0}{(\partial f/\partial z)(\partial g/\partial x) - (\partial f/\partial x)(\partial g/\partial z)}$$

$$= \frac{z - z_0}{(\partial f/\partial x)(\partial g/\partial y) - (\partial f/\partial y)(\partial g/\partial x)} \qquad (16.9\text{i})$$

We define the Jacobians

$$\frac{\partial(f, g)}{\partial(y, z)} = \begin{vmatrix} \partial f/\partial y & \partial f/\partial z \\ \partial g/\partial y & \partial g/\partial z \end{vmatrix} = \frac{\partial f}{\partial y}\frac{\partial g}{\partial z} - \frac{\partial f}{\partial z}\frac{\partial g}{\partial y},$$

$$\frac{\partial(f, g)}{\partial(z, x)} = \begin{vmatrix} \partial f/\partial z & \partial f/\partial x \\ \partial g/\partial z & \partial g/\partial x \end{vmatrix} = \frac{\partial f}{\partial z}\frac{\partial g}{\partial x} - \frac{\partial f}{\partial x}\frac{\partial g}{\partial z},$$

$$\frac{\partial(f, g)}{\partial(x, y)} = \begin{vmatrix} \partial f/\partial x & \partial f/\partial y \\ \partial g/\partial x & \partial g/\partial y \end{vmatrix} = \frac{\partial f}{\partial x}\frac{\partial g}{\partial y} - \frac{\partial f}{\partial y}\frac{\partial g}{\partial x}. \qquad (16.9\text{ii})$$

We find from Eq. (16.9i), that the direction ratios of the tangent line to the curve C at a point P are given by

$$\frac{\partial(f, g)}{\partial(y, z)}, \quad \frac{\partial(f, g)}{\partial(z, x)}, \quad \frac{\partial(f, g)}{\partial(x, y)}.$$

16.2   Formation of First and Second Order Equations

1. *Elimination of arbitrary constants* In chapter 4, we have discussed how to eliminate an arbitrary constant c from $f(x, y, c) = 0$ and obtain a first order ordinary differential equation $F(x, y, dy/dx) = 0$. This differential equation governs the family of curves described by $f(x, y, c) = 0$.

Now, Consider an equation

$$f(x, y, z, a, b) = 0 \qquad (16.10)$$

where a and b are arbitrary constants. This equation represents a two parameter family of surfaces. To eliminate the arbitrary constants a, b, in Eq. (16.10), we need three equations. One equation is given by Eq. (16.10) itself and the other two equations can be obtained by differentiating Eq. (16.10) partially with respect to x and y. Differentiating Eq. (16.10) partially, we obtain

$$\frac{\partial f}{\partial x} + \frac{\partial f}{\partial z}\frac{\partial z}{\partial x} = 0, \quad \text{or} \quad \frac{\partial f}{\partial x} + p\frac{\partial f}{\partial z} = 0 \qquad (16.11\text{i})$$

$$\frac{\partial f}{\partial y} + \frac{\partial f}{\partial z}\frac{\partial z}{\partial y} = 0, \quad \text{or} \quad \frac{\partial f}{\partial y} + q\frac{\partial f}{\partial z} = 0. \qquad (16.11\text{ii})$$

On eliminating a, b from the three equations (Eqs. (16.10), (1611i) and (16.11ii)), we obtain an equation containing $x, y, z, p = \partial z/\partial x$ and $q = \partial z/\partial y$, that is, an equation of the form

$$F(x, y, z, p, q) = 0 \qquad (16.12)$$

which is a first order partial differential equation.

Remark 1

It may not always be possible to eliminate the arbitrary constants a, b using the equations (16.10) and (16.11). Then, we find the second order partial derivatives and eliminate the arbitrary constants. However, this higher order partial differential equation may not be unique (see Example 16.3).

Remark 2

If we eliminate $n + 1$ constants $a_1, a_2, \ldots, a_{n+1}$ from the equation

$$f(x, y, z, a_1, a_2, \ldots, a_{n+1}) = 0$$

it may be possible to obtain an nth order partial differential equation (see Example 16.4).

Example 16.1   Obtain a partial differential equation that governs the family of surfaces

$$z = (x - \alpha)^2 + (y - \beta)^2. \tag{16.13}$$

Solution   Differentiating Eq. (16.13) partially with respect to x and y, we obtain

$$p = \frac{\partial z}{\partial x} = 2(x - \alpha), \quad q = \frac{\partial z}{\partial y} = 2(y - \beta).$$

Substituting for $(x - \alpha)$ and $(y - \beta)$ in Eq. (16.13), we obtain

$$4z = p^2 + q^2$$

which is the required partial differential equation.

Example 16.2   Eliminate the arbitrary constants a and b from $z = ax + by + a^2 b^2$, to obtain the partial differential equation governing it.

Solution   Differentiating $z = ax + by + a^2 b^2$ partially with respect to x and y, we get

$$p = \frac{\partial z}{\partial x} = a, \quad q = \frac{\partial z}{\partial y} = b.$$

Hence, the required partial differential equation is obtained as

$$z = px + qy + p^2 q^2.$$

Example 16.3   Obtain a partial differential equation by eliminating the arbitrary constants c and ω from $z = ce^{\omega t} \cos(\omega x)$.

Solution   Differentiating $z = ce^{\omega t} \cos(\omega x)$ partially with respect to x and t, we get

$$z_x = -c\omega e^{\omega t} \sin(\omega x), \quad z_t = c\omega e^{\omega t} \cos(\omega x). \tag{16.14}$$

It is not easy to eliminate c and ω from Eqs. (16.14) and the given equation. Differentiating Eqs. (16.14) again partially with respect to x and t, we obtain

$$z_{xx} = -c\omega^2 e^{\omega t} \cos(\omega x), \quad z_{tt} = c\omega^2 e^{\omega t} \cos(\omega x)$$

From the last two equations, we get the required partial differential eqation as

$$z_{xx} + z_{tt} = 0. \tag{16.15}$$

We note that the Eq. (16.15) is not unique.
From Eqs. (16.14), we get

$$z_{xt} = -c\omega^2 e^{\omega t}\sin(\omega x) = \omega z_x$$

and
$$z_{xx} = -c\omega^2 e^{\omega t}\cos(\omega x) = -\omega z_t.$$

Eliminating ω from the above two equations, we obtain another partial differential equation

$$z_t z_{xt} + z_x z_{xx} = 0.$$

Example 16.4 Obtain a partial differential equation by eliminating the arbitrary constants c_1, c_2 and c_3 from $z = c_1 x^2 + 2c_2 xy + c_3 y^2$.

Solution Differentiating $z = c_1 x^2 + 2c_2 xy + c_3 y^2$, partially with respect to x and y, we get

$$z_x = 2c_1 x + 2c_2 y, \quad z_y = 2c_2 x + 2c_3 y,$$

$$z_{xx} = 2c_1, \quad z_{xy} = 2c_2, \quad z_{yy} = 2c_3.$$

Substituting the values of c_1, c_2, c_3 in the given equation, we obtain

$$z = \frac{1}{2}(x^2 z_{xx} + 2xy\, z_{xy} + y^2 z_{yy})$$

which is the required partial differential equation.

2. *Elimination of arbitrary functions* Partial differential equations can also be formed by eliminating arbitrary functions from a given equation. A first order partial differential equation can be derived from an equation containing one arbitrary function. A second order partial differential equation can be derived from an equation containing two arbitrary functions.

However, it may not always be possible to derive an nth order partial differential equation from an equation containing n arbitrary functions. We may require higher order partial derivatives than n to eliminate the n arbitrary functions and thus obtain an higher order partial differential equation than n. But, such a higher order equation may not be unique.

Example 16.5 Eliminate the arbitrary function from $z = f(x^2 + y^2)$ to obtain a first order partial differential equation.

Solution we have $z = f(u)$, where $u = x^2 + y^2$. Differentiating partially with respect to x and y, we get

$$\frac{\partial z}{\partial x} = \frac{dz}{du}\frac{\partial u}{\partial x} = 2x\, z'(u), \quad \frac{\partial z}{\partial y} = \frac{dz}{du}\frac{\partial u}{\partial y} = 2y z'(u).$$

Eliminating $z'(u)$, we get

$$y\frac{\partial z}{\partial x} - x\frac{\partial z}{\partial y} = 0.$$

Example 16.6 Obtain a second order partial differential equation by eliminating the arbitrary functions from

$$u = f(x + ct) + g(x - ct).$$

Solution Differentiating the given equation partially with respect to x and t, we obtain

$$\frac{\partial u}{\partial x} = f'(x + ct) + g'(x - ct), \quad \frac{\partial u}{\partial t} = cf'(x + ct) - cg'(x - ct),$$

$$\frac{\partial^2 u}{\partial x^2} = f''(x + ct) + g''(x - ct), \quad \frac{\partial^2 u}{\partial t^2} = c^2 f''(x + ct) + c^2 g''(x - ct).$$

Hence, we obtain the equation

$$\frac{\partial^2 u}{\partial t^2} = c^2 \frac{\partial^2 u}{\partial x^2}.$$

Consider now, the case when two functions $u = u(x, y, z)$, $v = v(x, y, z)$ are related through the relation $\phi(u, v) = 0$, where ϕ is some arbitrary function of u and v. Differentiating $\phi(u, v) = 0$ partially with respect to x and y, we get

$$\frac{\partial \phi}{\partial u}\left[\frac{\partial u}{\partial x} + \frac{\partial u}{\partial z}p\right] + \frac{\partial \phi}{\partial v}\left[\frac{\partial v}{\partial x} + \frac{\partial v}{\partial z}p\right] = 0$$

and

$$\frac{\partial \phi}{\partial u}\left[\frac{\partial u}{\partial y} + \frac{\partial u}{\partial z}q\right] + \frac{\partial \phi}{\partial v}\left[\frac{\partial v}{\partial y} + \frac{\partial v}{\partial z}q\right] = 0.$$

Since both $\partial\phi/\partial u$ and $\partial\phi/\partial v$ cannot be zero, non-trivial solution exists for the above equations. Hence, we have

$$\begin{vmatrix} \dfrac{\partial u}{\partial x} + p\dfrac{\partial u}{\partial z} & \dfrac{\partial v}{\partial x} + p\dfrac{\partial v}{\partial z} \\[2mm] \dfrac{\partial u}{\partial y} + q\dfrac{\partial u}{\partial z} & \dfrac{\partial v}{\partial y} + q\dfrac{\partial v}{\partial z} \end{vmatrix} = 0$$

or

$$\left[\frac{\partial u}{\partial x} + p\frac{\partial u}{\partial z}\right]\left[\frac{\partial v}{\partial y} + q\frac{\partial v}{\partial z}\right] - \left[\frac{\partial v}{\partial x} + p\frac{\partial v}{\partial z}\right]\left[\frac{\partial u}{\partial y} + q\frac{\partial u}{\partial z}\right] = 0$$

or

$$\left[\frac{\partial u}{\partial x}\frac{\partial v}{\partial y} - \frac{\partial u}{\partial y}\frac{\partial v}{\partial x}\right] + p\left[\frac{\partial u}{\partial z}\frac{\partial v}{\partial y} - \frac{\partial u}{\partial y}\frac{\partial v}{\partial z}\right] + q\left[\frac{\partial u}{\partial x}\frac{\partial v}{\partial z} - \frac{\partial u}{\partial z}\frac{\partial v}{\partial x}\right] = 0. \tag{16.16}$$

Then, Eq. (16.16) can be written as a first order partial differential equation given by

$$p\frac{\partial(u, v)}{\partial(y, z)} + q\frac{\partial(u, v)}{\partial(z, x)} = \frac{\partial(u, v)}{\partial(x, y)} \tag{16.17i}$$

or as

$$P(x, y, z)\,p + Q(x, y, z)\,q = R(x, y, z) \tag{16.17ii}$$

where

$$P(x, y, z) = \frac{\partial(u, v)}{\partial(y, z)}, \quad Q(x, y, z) = \frac{\partial(u, v)}{\partial(z, x)}, \quad R(x, y, z) = \frac{\partial(u, v)}{\partial(x, y)}$$

are the Jacobians as defined in Eq. (16.9ii).

Example 16.7 Obtain the partial differential equation governing the equations

$$\phi(u, v) = 0, \quad u = xyz, \quad v = x + y + z.$$

Solution Differentiating $\phi(u, v) = 0$ partially with respect to x and y, we get

$$\frac{\partial \phi}{\partial u}\left[\frac{\partial u}{\partial x} + \frac{\partial u}{\partial z}p\right] + \frac{\partial \phi}{\partial v}\left[\frac{\partial v}{\partial x} + \frac{\partial v}{\partial z}p\right] = 0$$

and

$$\frac{\partial \phi}{\partial u}\left[\frac{\partial u}{\partial y} + \frac{\partial u}{\partial z}q\right] + \frac{\partial \phi}{\partial v}\left[\frac{\partial v}{\partial y} + \frac{\partial v}{\partial z}q\right] = 0$$

or

$$\frac{\partial \phi}{\partial u}[yz + xyp] + \frac{\partial \phi}{\partial v}[1 + p] = 0$$

and

$$\frac{\partial \phi}{\partial u}[xz + xyq] + \frac{\partial \phi}{\partial v}[1 + q] = 0.$$

Since both $\partial\phi/\partial u$ and $\partial\phi/\partial v$ are not zero, we get

$$(1 + q)(yz + xy\,p) - (1 + p)(xz + xy\,q) = 0$$

or

$$px(y - z) + qy(z - x) = z(x - y)$$

which is the required partial differential equation.

We can also use Eq. (16.17) directly to obtain the partial differential equation.

Exercise 16.1

In the following problems, eliminate the arbitrary constants to obtain a partial differential equation.

1. $z = (x + c)(y + d)$.
2. $z = ax + by + a^4 + b^4$.
3. $(x - a)^2 + y^2 + (z - b)^2 = 16$.
4. $z = ax + by^2 - axy$.
5. $z = ce^{-\omega t}\sin(\omega x)$.
6. $z = (x + ay)^2 + by$.

In the following problems, eliminate the arbitrary functions to obtain a partial differential equation.

7. $z = f(ax + y)$, a constant.
8. $z = xy + f(x^2 - y^2)$.
9. $z = f(x/y)$.
10. $z = f(x + iy) + g(x - iy)$, $i^2 = -1$.
11. $f(x + y + z, x^2 + y^2 + z^2) = 0$.
12. $x^2 + y^2 + z^2 = g(2x - y)$.
13. $xyz = f(x + y + z)$.
14. $x^2 + y^2 + z^2 = f(xy)$.
15. $f(x + yz, x^2 + y^2 - z^2) = 0$.

16.3   Solution of First Order Equations

We have shown in the previous section that by eliminating the arbitrary constants a and b from a relation of the form $f(x, y, z, a, b) = 0$, we can obtain a first order partial differential equation given by

$$F(x, y, z, p, q) = 0 \qquad\qquad (16.18)$$

where $p = \partial z/\partial x$ and $q = \partial z/\partial y$. We have also shown that by eliminating the arbitrary function ϕ from the relation $\phi(u, v) = 0$, where $u = u(x, y, z)$, $v = v(x, y, z)$, we can obtain a first order partial differential equation as given in Eq. (16.18). We now define the following:

Complete integral or complete solution   Any relation of the form $f(x, y, z, a, b) = 0$ which contains two arbitrary constants and satisfies Eq. (16.18) is called a *complete integral* or a *complete solution* of Eq. (16.18). Such a solution is also called an *integral surface* of Eq. (16.18).

General integral or general solution   A relation of the form $\phi(u, v) = 0$, where ϕ is an arbitrary function of $u = u(x, y, z), v = v(x, y, z)$ and satisfies Eq. (16.18) is called a *general integral* or a *general solution* of Eq. (16.18)

Particular integral or particular solution   The solution obtained by determining the arbitrary constants in the complete integral or the arbitrary function in the general integral by using some specified condition is called a *particular integral* or a *particular solution*. For example, it may be specified that the integral surface passes through the circle $x^2 + y^2 = c^2$, $z = 0$.

Singular integral or singular solution   The equation of the envelope of a two parameter family of surfaces representing the complete integral of a given partial differential equation is called a *singular integral* or a *singular solution*. If $f(x, y, z, a, b) = 0$ is the complete integral, then the equation of the surface obtained by eliminating a and b from the equations $f = 0$, $\partial f/\partial a = 0$, $\partial f/\partial b = 0$ is the envelope and hence the singular solution. Note that, the singular solution cannot be obtained from the complete integral by assigning some values to a and b. Also, a singular solution may not exist for every partial differential equation.

16.3.1   Lagrange's Equation

The linear first order partial differential equation of the form

$$P(x, y, z)\, p + Q(x, y, z)q = R(x, y, z) \tag{16.19}$$

is called the Lagrange's equation in two independent variables x, y.

Theorem 16.1   The general solution of the equation $Pp + Qq = R$ is given by $\phi(u, v) = 0$ where ϕ is an arbitrary function and $u(x, y, z) = c_1$, $v(x, y, z) = c_2$ are two linearly independent solutions of the equations

$$\frac{dx}{P} = \frac{dy}{Q} = \frac{dz}{R} \tag{16.20}$$

(These equations are called the *auxiliary* or *subsidiary equations*).

Proof   Since $u(x, y, z) = c_1$ is a solution of Eq. (16.20), we have

$$du = 0, \text{ or } \quad \frac{\partial u}{\partial x}dx + \frac{\partial u}{\partial y}dy + \frac{\partial u}{\partial z}\,dz = 0. \tag{16.21}$$

From Eq. (16.20), we can write

$$dx = kP, \, dy = kQ \text{ and } dz = kR, \, k \text{ any constant.}$$

Substituting in Eq. (16.21) and cancelling k, we get

$$P\frac{\partial u}{\partial x} + Q\frac{\partial u}{\partial y} + R\frac{\partial u}{\partial z} = 0. \tag{16.22}$$

Similarly, since $v(x, y, z) = c_2$ is a solution of Eq. (16.20), we obtain

$$P\frac{\partial v}{\partial x} + Q\frac{\partial v}{\partial y} + R\frac{\partial v}{\partial z} = 0. \tag{16.23}$$

Solving Eqs. (16.22) and (16.23) for P, Q, R, we get

$$\frac{P}{\partial(u, v)/\partial(y, z)} = \frac{Q}{\partial(u, v)/\partial(z, x)} = \frac{R}{\partial(u, v)/\partial(x, y)} \tag{16.24}$$

where

$$\frac{\partial(u, v)}{\partial(y, z)} = \frac{\partial u}{\partial y}\frac{\partial v}{\partial z} - \frac{\partial u}{\partial z}\frac{\partial v}{\partial y}, \quad \frac{\partial(u, v)}{\partial(z, x)} = \frac{\partial u}{\partial z}\frac{\partial v}{\partial x} - \frac{\partial u}{\partial x}\frac{\partial v}{\partial z},$$

and

$$\frac{\partial(u, v)}{\partial(x, y)} = \frac{\partial u}{\partial x}\frac{\partial v}{\partial y} - \frac{\partial u}{\partial y}\frac{\partial v}{\partial x}.$$

But, we have shown in the previous section (see Eqs. (16.17i), (16.17ii)) that the relation $\phi(u, v) = 0$ leads to the equation

$$p\frac{\partial(u, v)}{\partial(y, z)} + q\frac{\partial(u, v)}{\partial(z, x)} = \frac{\partial(u, v)}{\partial(x, y)}. \tag{16.25}$$

Setting each term in Eq. (16.24) equal to $1/c$, where c is a constant, we obtain

$$\frac{\partial(u, v)}{\partial(y, z)} = cP, \quad \frac{\partial(u, v)}{\partial(z, x)} = cQ, \quad \frac{\partial(u, v)}{\partial(x, y)} = cR. \tag{16.26}$$

Substituting in Eq. (16.25) and cancelling c, we get

$$Pp + Qq = R.$$

Therefore, $\phi(u, v) = 0$, where $u(x, y, z) = c_1$, $v(x, y, z) = c_2$ are two linearly independent solutions of Eq. (16.20), is the general integral of $Pp + Qq = R$.

Geometrical Interpretation

First consider the equations (16.20), that is

$$\frac{dx}{P} = \frac{dy}{Q} = \frac{dz}{R}. \tag{16.27}$$

The solution of these equations can be obtained by considering them in pairs or by any other method. The solutions are called the *integral curves*. From Eqs. (16.27), we conclude that the tangent to an integral curve, at any point (x, y, z) on it has direction ratios P, Q and R, that is, the direction $\mathbf{T} = (P, Q, R)$ is tangential to the integral curve. If P, Q, R are constants, then we obtain a two parameter family of straight lines. If P, Q, R are functions of x, y, z, then we obtain a system of curves, each of which is generated by a moving point (x, y, z) whose direction changes continuously.

Now, consider the Lagrange's equation

$$P(x, y, z)\, p + Q(x, y, z)\, q = R(x, y, z). \tag{16.28}$$

In vector notation, we can write this equation as

$$(\mathbf{i}\, P + \mathbf{j}\, Q + \mathbf{k}\, R) \cdot (p\mathbf{i} + q\mathbf{j} - \mathbf{k}) = 0. \tag{16.29}$$

Suppose that S: $z = f(x, y)$ or $F(x, y, z) = f(x, y) - z = 0$ is an integral surface of Eq. (16.28). Then, the normal $\mathbf{N}$ to this surface is given by

$$N = \text{grad } (F) = p\mathbf{i} + q\mathbf{j} - \mathbf{k}$$

Therefore, the direction ratios of the normal are p, q, -1. From Eq. (16.29), we find that the normal $\mathbf{N}$ to the integral surface is perpendicular to the direction of the tangent line whose direction ratios are P, Q and R.

Remark 3

The method used for solving the Lagrange's equation can be extended to any number of variables. If the given differential equation is

$$Q_1 p_1 + Q_2 p_2 + \ldots + Q_m p_m = R \qquad (16.30i)$$

where $Q_i = Q_i(x_1, x_2, \ldots, x_m)$ and $p_i = \partial z / \partial x_i$, $i = 1, 2, \ldots, m$, then the auxiliary equations are given by

$$\frac{dx_1}{Q_1} = \frac{dx_2}{Q_2} = \ldots = \frac{dx_m}{Q_m} = \frac{dz}{R}. \qquad (16.30 \text{ ii})$$

Solution of the auxiliary equations

The solutions of the auxiliary equations (16.27) are of the form $u(x, y, z) = c_1$ and $v(x, y, z) = c_2$, where c_1 and c_2 are arbitrary constants. We now discuss some methods for determining these solutions.

1. Using the methods of solution of ordinary differential equations, we solve taking any two pairs. This method is generally used when one of the variables is absent from one term of Eq. (16.27).
2. We solve one pair and if necessary use this solution to get the second solution.
3. We use multipliers and write each term in Eq. (16.27) equal to

$$\frac{a\,dx + b\,dy + c\,dz}{aP + bQ + cR}.$$

If possible, we can find some simple functions a, b, c such that $aP + bQ + cR = 0$. Then, the numerator gives $adx + bdy + c\,dz = 0$ which can be integrated if the left hand side is an exact differential. Alternately, we may have the case when $aP + bQ + cR \neq 0$ and the numerator is the differential of $aP + bQ + cR$, that is, $a\,dx + b\,dy + c\,dz = d(aP + bQ + cR)$.

The solution of the Lagrange's equation $Pp + Qq = R$ is then given by $\phi(c_1, c_2) = 0$ or $\phi(u, v) = 0$. The solution can also be written as $c_2 = \phi(c_1)$ or $c_1 = \phi(c_2)$.

We now illustrate through some examples.

Example 16.8   Find the general solution of the Lagrange's equation

$$2yzp + zx\,q = 3xy.$$

Solution   The subsidiary equations are given by

$$\frac{dx}{2yz} = \frac{dy}{zx} = \frac{dz}{3xy}.$$

Taking the first pair of terms, we get

$$x\,dx = 2y\,dy, \text{ whose solution is } (x^2/2) = y^2 + c, \text{ or } x^2 = 2y^2 + c_1$$

where $c_1 = 2c$. Taking the second and third terms, we get

$3y \, dy = z \, dz$, whose solution is $(3y^2/2) = (z^2/2) + k$, or $3y^2 = z^2 + c_2$ where $c_2 = 2k$. The general solution is given by

$$\phi(c_1, c_2) = 0, \quad \text{or} \quad \phi(x^2 - 2y^2, \, 3y^2 - z^2) = 0$$

or as

$$3y^2 - z^2 = f(x^2 - 2y^2).$$

Example 16.9   Find the general solution of the partial differential equation

$$xy^2 p + y^3 q = (zxy^2 - 4x^3).$$

Solution   The subsidiary equations are

$$\frac{dx}{xy^2} = \frac{dy}{y^3} = \frac{dz}{x(zy^2 - 4x^2)}.$$

Taking the first pair of terms, we get

$$\frac{dx}{xy^2} = \frac{dy}{y^3}, \quad \text{or} \quad \frac{dx}{x} = \frac{dy}{y} \quad \text{which on integration gives } \ln y = \ln x + \ln a, \text{ or } y = ax.$$

Using this relation, we get

$$\frac{dx}{a^2 x^3} = \frac{dz}{x(a^2 x^2 z - 4x^2)}, \quad \text{or} \quad dx = \frac{dz}{[z - (4/a^2)]}.$$

Integrating, we get

$$x = \ln\left[z - \left(\frac{4}{a^2}\right)\right] + b, \quad \text{or} \quad x = \ln\left[z - \left(\frac{4x^2}{y^2}\right)\right] + b.$$

The general solution is given by

$$\phi(a, b) = 0, \text{ or } \quad \phi\left[\frac{y}{x}, \; x - \ln\left(z - \frac{4x^2}{y^2}\right)\right] = 0.$$

Example 16.10   Find the general solution of the partial differential equation

$$(x + y^2)p + yq = z + x^2.$$

Solution   The subsidiary equations are

$$\frac{dx}{x + y^2} = \frac{dy}{y} = \frac{dz}{z + x^2}.$$

Taking the first pair of terms, we get

$$\frac{dx}{x + y^2} = \frac{dy}{y}, \quad \text{or} \quad \frac{dx}{dy} = \frac{x + y^2}{y}, \quad \text{or} \quad \frac{dx}{dy} - \frac{x}{y} = y.$$

The integrating factor is $1/y$. Integrating the equation

$$\frac{1}{y}\frac{dx}{dy} - \frac{x}{y^2} = 1, \text{ we get } \frac{x}{y} = y + a, \text{ or } x = y(y + a).$$

Using this relation, we get

$$\frac{dy}{y} = \frac{dz}{z + y^2(y + a)^2},$$

or

$$\frac{dz}{dy} = \frac{z + y^2(y + a)^2}{y} = \frac{z}{y} + y(y + a)^2$$

or

$$\frac{dz}{dy} - \frac{z}{y} = y(y + a)^2.$$

The integrating factor is $1/y$. Integrating the equation

$$\frac{1}{y}\frac{dz}{dy} - \frac{z}{y^2} = (y + a)^2, \text{ we get } \frac{z}{y} = \frac{1}{3}(y + a)^3 + b$$

or

$$\frac{z}{y} = \frac{1}{3}\left(\frac{x^3}{y^3}\right) + b.$$

The general solution is given by

$$\phi(a, b) = 0, \text{ or } \phi\left(\frac{x - y^2}{y}, \frac{3zy^2 - x^3}{3y^3}\right) = 0.$$

Example 16.11   Find the general solution of the partial differential equation

$$(y + z)\, p + (z + x)\, q = x + y.$$

Solution   The subsidiary equations are

$$\frac{dx}{y + z} = \frac{dy}{z + x} = \frac{dz}{x + y}.$$

Now,   each term $= \dfrac{dx - dy}{-(x - y)} = \dfrac{dx + dy + dz}{2(x + y + z)} = \dfrac{dx - dz}{-(x - z)}.$

Integrating the first pair of terms, we get $-2 \ln (x - y) = \ln[(x + y + z)\, a]$

or $(x - y)^2 (x + y + z)\, a = 1, \text{ or } (x - y)^2 (x + y + z) = c$

where $c = 1/a$.

Using the first and the third terms, we get

$$\ln(x - y) = \ln [(x - z)\, b], \text{ or } x - y = b(x - z).$$

The general solution is given by

$$\phi(b, c) = 0, \text{ or } \phi\left(\frac{x - y}{x - z}, (x - y)^2 (x + y + z)\right) = 0.$$

Example 16.12   Find the general solution of the partial differential equation

$$2xzp + 2yzq = z^2 - x^2 - y^2$$

Solution   The subsidiary equations are

$$\frac{dx}{2xz} = \frac{dy}{2yz} = \frac{dz}{z^2 - x^2 - y^2}.$$

From the first pair of terms, we get

$$\ln y = \ln ax, \quad \text{or} \quad y = ax.$$

Now, each term is equal to

$$\frac{x\,dx + y\,dy + z\,dz}{2x^2 z + 2y^2 z + z^3 - x^2 z - y^2 z} = \frac{x\,dx + y\,dy + z\,dz}{z(x^2 + y^2 + z^2)}.$$

From the first term and this term, we get

$$\frac{dx}{2xz} = \frac{x\,dx + y\,dy + z\,dz}{z(x^2 + y^2 + z^2)}, \quad \text{or} \quad \frac{dx}{x} = \frac{2(x\,dx + y\,dy + z\,dz)}{x^2 + y^2 + z^2}.$$

Integrating, we get

$$\ln (x^2 + y^2 + z^2) = \ln x + \ln b, \quad \text{or} \quad x^2 + y^2 + z^2 = bx.$$

The general solution is given by

$$\phi(a, b) = 0, \quad \text{or} \quad \phi\left(\frac{y}{x}, \frac{x^2 + y^2 + z^2}{x} \right) = 0.$$

We may also write the solution as

$$\frac{x^2 + y^2 + z^2}{x} = \psi\left(\frac{y}{x} \right), \quad \text{or} \quad x^2 + y^2 + z^2 = x\psi\left(\frac{y}{x} \right).$$

Particular solution passing through a given curve

The solution of the Lagrange's equation $Pp + Qq = R$ is given by $\phi(u, v) = 0$ where $u(x, y, z) = c_1$ and $v(x, y, z) = c_2$ are the linearly independent solutions of the auxiliary equation (16.20).
Let the integral surfaces pass through a curve C whose parametric form can be written as

$$x = x(t), \, y = y(t), \, z = z(t).$$

Since the curve lies on the integral surface, we have

$$v(x(t), y(t), z(t)) = c_1 \quad \text{and} \quad v(x(t), y(t), z(t)) = c_2.$$

If we eliminate t from these two equations, we obtain a relation between c_1 and c_2. Substituting $c_1 = u(x, y, z)$, $c_2 = v(x, y, z)$ in this relation, we obtain the particular solution.

Example 16.13 Find the general solution of the partial differential equation

$$(3 - 2yz)\, p + x(2z - 1)\, q = 2\, x\, (y - 3).$$

Hence, obtain the particular solution which passes through the curve $z = 0$, $x^2 + y^2 = 4$.

Solution The subsidiary equations are

$$\frac{dx}{3 - 2yz} = \frac{dy}{x(2z - 1)} = \frac{dz}{2x(y - 3)}.$$

From the second and third terms, we get

$$(2y - 6)\, dy = (2z - 1)dz.$$

Integrating, we get $y^2 - 6y = z^2 - z + a.$

Now, each term is equal to

$$\frac{x\,dx + z\,dz}{3x - 2xyz + 2xyz - 6xz} = \frac{x\,dx + z\,dz}{-3x(2z - 1)}.$$

From the second term and this term, we get

$$\frac{dy}{x(2z - 1)} = \frac{x\,dx + z\,dz}{-3x(2z - 1)}, \text{ or } -3dy = x\,dx + z\,dz.$$

Integrating, we get $x^2 + z^2 + 6y = b$.

The general solution is

$$\phi(a, b) = 0, \text{ or } \phi(y^2 - 6y - z^2 + z, \; x^2 + z^2 + 6y) = 0.$$

The integral surface passes through the curve $z = 0$, $x^2 + y^2 = 4$. Hence, we have

$$y^2 - 6y = a \quad \text{and} \quad x^2 + 6y = b.$$

Adding, we get $x^2 + y^2 = a + b$. Hence $a + b = 4$. Substituting the values of a and b, we obtain

$$(y^2 - 6y - z^2 + z) + (x^2 + z^2 + 6y) = 4, \text{ or } x^2 + y^2 + z = 4.$$

The particular solution is $x^2 + y^2 + z = 4$.

Surfaces orthogonal to a given family of surfaces

We shall use the property that if two surfaces cut each other orthogonally, then their normals at every point on the curve of intersection are perpendicular to each other.

Let S_1: $f(x, y, z) = c$ be the given family of surfaces. Then the normal at any point on the surface S_1 is given by

$$\mathbf{N}_1 = \operatorname{grad} f = \mathbf{i}\,\frac{\partial f}{\partial x} + \mathbf{j}\,\frac{\partial f}{\partial y} + \mathbf{k}\,\frac{\partial f}{\partial z}. \tag{16.31}$$

Let S_2: $z = g(x, y)$ be the surface which cuts the family of surfaces S_1 orthogonally. Then, the normal at any point on the surface S_2 is given by

$$\mathbf{N}_2 = \mathbf{i}\,\frac{\partial g}{\partial x} + \mathbf{j}\,\frac{\partial g}{\partial y} - \mathbf{k}$$

$$= \mathbf{i}\,\frac{\partial z}{\partial x} + \mathbf{j}\,\frac{\partial z}{\partial y} - \mathbf{k} = \mathbf{i}p + \mathbf{j}q - \mathbf{k}. \tag{16.32}$$

At all points on the curve of intersection, we have that $\mathbf{N}_1$ is perpendicular to $\mathbf{N}_2$. Therefore,

$$\mathbf{N}_1 \cdot \mathbf{N}_2 = \left(\mathbf{i}\,\frac{\partial f}{\partial x} + \mathbf{j}\,\frac{\partial f}{\partial y} + \mathbf{k}\,\frac{\partial f}{\partial z}\right) \cdot (\mathbf{i}p + \mathbf{j}q - \mathbf{k})$$

$$= p\,\frac{\partial f}{\partial x} + q\,\frac{\partial f}{\partial y} - \frac{\partial f}{\partial z} = 0$$

or
$$p\left(\frac{\partial f}{\partial x}\right) + q\left(\frac{\partial f}{\partial y}\right) = \left(\frac{\partial f}{\partial z}\right). \tag{16.33}$$

Hence, surfaces orthogonal to $f(x, y, z) = c$ are the surfaces generated by the solution curves of the equations

$$\frac{dx}{\partial f/\partial x} = \frac{dy}{\partial f/\partial y} = \frac{dz}{\partial f/\partial z}. \tag{16.34}$$

Example 16.14   Find the equation of the surface which cuts orthogonally the system of surfaces

$$2xz + 3yz = a(z + 2), \text{ where } a \text{ is an arbitrary constant}$$

and passes through the circle $z = 0$, $x^2 + y^2 = 9$.

Solution   We have $f(x, y, z) = \dfrac{2xz + 3yz}{z + 2} = a$.

The equations governing the orthogonal family are

$$\frac{dx}{2z/(z + 2)} = \frac{dy}{3z/(z + 2)} = \frac{dz}{2(2x + 3y)/(z + 2)^2}.$$

or

$$\frac{dx}{2z(z + 2)} = \frac{dy}{3z(z + 2)} = \frac{dz}{2(2x + 3y)}.$$

The first two terms give the solution $3x - 2y = c_1$.

Now, each term is equal to

$$\frac{x\,dx + y\,dy - [(2z + z^2)/2]dz}{(2xz^2 + 4xz + 3yz^2 + 6yz) - (2x + 3y)(2z + z^2)} = \frac{x\,dx + y\,dy - [(2z + z^2)/2]dz}{0}$$

Integrating, we get $\dfrac{x^2}{2} + \dfrac{y^2}{2} - \dfrac{z^2}{2} - \dfrac{z^3}{6} = c$

or

$$3(x^2 + y^2 - z^2) - z^3 = c_2, \text{ where } c_2 = 6c.$$

The orthogonal surfaces are given by $c_2 = \phi(c_1)$, or

$$3(x^2 + y^2 - z^2) - z^3 = \phi(3x - 2y).$$

Since, the required surface passes through $z = 0$, $x^2 + y^2 = 9$, we have $\phi = 27$. Therefore, the orthogonal surface is obtained as $3(x^2 + y^2 - z^2) - z^3 = 27$.

Exercise 16.2

In Problems **1** to **11**, obtain the general solution of the Lagrange's equations.

1. $2p + q = 3z + \sin(x - 2y)$.
2. $x(y - z)p + y(z - x)q = z(x - y)$.
3. $(y + z)p + (x + z)q = x + y$.
4. $2xzp + 2yzq + x^2 + y^2 - z^2 = 0$.
5. $(2y^2 + z)p + (y + 2x)q = 4xy - z$.
6. $x(y^2 + z^2)p + y(x^2 - z^2)q + z(x^2 + y^2) = 0$.
7. $y(x - z)p + (z^2 - xz - x^2)q = y(2x - z)$.
8. $(x + 2y^2 + z)p + y(2x - 1)q + 2(x^2 + y^2 + xz) = 0$.

9. $y^2(x + y)p + x^2(x + y) q = (x^2 + y^2) z.$

10. $x(z - 3y^3)p + y (3x^3 - z)q = 3(y^3 - x^3)z.$

11. $(y^2 - x)p + yq = z + xy.$

12. In Problem 5, find the particular solution, which passes through,
(a) the curve $z = 1$, $[x + (1/2)]^2 - [y + (1/2)]^2 = 1$, (b) the straight line $z = 1$, $y = x$.

13. In Problem 7, find the particular solution, which passes through the ellipse $z = 0$, $2x^2 + 4y^2 = 1$.

14. Find the equation of the surface which cuts orthogonally the family of spheres

$$x^2 + y^2 + z^2 = cy, \ c \neq 0 \text{ and arbitrary}$$

and passes through the circle $z = 1$, $x^2 + y^2 = 4$.

15. Find the equation of the system of surfaces which cut orthogonally the family of cones $z^2 = c(x^2 + y^2)$. Obtain the particular surface which passes through the circle $z = 3$, $x^2 + y^2 = 9$.

16.3.2   Non-linear First Order Equations

Consider the first order non-linear partial differential equation

$$F (x, y, z, p, q) = 0. \tag{16.35}$$

The solution of this equation can be categorized into the following three types.

(a) We have seen earlier in section 16.2, that from the equation $f (x, y, z, a, b) = 0$, where a, b, are arbitrary constants, we can derive a first order partial differential equation of the form as given in Eq. (16.35) by eliminating a, b, from the equations $f(x, y, z, a, b) = 0$, $(\partial f/\partial x) = 0$ and $(\partial f/\partial y) = 0$. Therefore, the solutions of Eq. (16.35) are the two parameter family of surfaces

$$f (x, y, z, a, b) = 0. \tag{16.36}$$

This solution is called the *complete integral* or *complete solution*.

(b) Using the complete integral obtained from Eq. (16.36), we deduce a one parameter family of surfaces (under a given condition on the solution) by choosing a particular case $a = g_1(b)$ or as $b = g(a)$. The one parameter family is given by

$$f (x, y, z, a, g(a)) = 0. \tag{16.37}$$

Now, we find the envelope of the Eq. (16.37) by eliminating the arbitrary constant a from equation (16.37) and the equation $(\partial f/\partial a) = 0$. The eliminant or the envelope is a surface $\phi(x, y, z) = 0$. This envelope is also a solution of Eq. (16.35). When $g(a)$ is arbitrary, this solution is called the *general solution* or *general integral* of Eq. (16.35). If a particular value is used for $g(a)$, then the solution is called a *particular integral*.

(c) We try to obtain the envelope of the two parameter family of surfaces defined by Eq. (16.36), by eliminating the arbitrary constants a, b from equation (16.36) and the equations $(\partial f/\partial a) = 0$ and $(\partial f/\partial b) = 0$. If this envelope exists, then it is called the *singular solution* or *singular integral* of Eq. (16.35). Obviously, not all first order differential equations may have singular solutions.

We now derive a general method for finding the solution of Eq. (16.35).

16.3.3   Charpit's Method

Consider the first order partial differential equation in two variables x, y as

$$F (x, y, z, p, q) = 0. \tag{16.38}$$

We try to find another equation

$$g(x, y, z, p, q) = 0 \tag{16.39}$$

such that the solution of Eq. (16.38) is also a solution of Eq. (16.39). If such an equation exists, then the equations (16.38) and (16.39) are said to be compatible. We solve Eqs. (16.38), (16.39) for p and q as $p = P(x, y, z)$, $q = Q(x, y, z)$. Since $z = z(x, y)$, we have

$$dz = \frac{\partial z}{\partial x} dx + \frac{\partial z}{\partial y} dy = p\, dx + q\, dy.$$

Hence, we integrate the equation

$$dz = P\, dx + Q\, dy. \tag{16.40}$$

The integral of this equation satisfies Eqs. (16.38), (16.39) as p and q are obtained from them. This implies that z, p, q taken as functions of x and y satisfy the Eqs. (16.38) and (16.39).

Setting the derivatives of $F(x, y, z, p, q)$ and $g(x, y, z, p, q)$ with respect to x and y to zero, we obtain

$$F_x + F_z p + F_p p_x + F_q q_x = 0, \; F_y + F_z q + F_p p_y + F_q q_y = 0 \tag{16.41}$$

and

$$g_x + g_z p + g_p p_x + g_q q_x = 0, \; g_y + g_z q + g_p p_y + g_q q_y = 0 \tag{16.42}$$

We note that since the solution of p, q (obtained from Eqs. (16.38), (16.39)) exists, we have

$$F_p g_q - F_q g_p \neq 0. \tag{16.43}$$

We also have
$$q_x = \frac{\partial}{\partial x}\left(\frac{\partial z}{\partial y}\right) = \frac{\partial}{\partial y}\left(\frac{\partial z}{\partial x}\right) = p_y.$$

Substituting in Eqs. (16.41) and (16.42), we get

$$F_p p_x + F_q p_y + (F_x + F_z p) = 0, \qquad F_p p_y + F_q q_y + (F_y + F_z q) = 0,$$

$$g_p p_x + g_q p_y + (g_x + g_z p) = 0, \qquad g_p p_y + g_q q_y + (g_y + g_z q) = 0.$$

Eliminating p_x, p_y and q_y from these equations we get

$$\begin{vmatrix} F_p & F_q & 0 & F_x + F_z p \\ 0 & F_p & F_q & F_y + F_z q \\ g_p & g_q & 0 & g_x + g_z p \\ 0 & g_p & g_q & g_y + g_z q \end{vmatrix} = 0.$$

Expanding the determinant in terms of the elements of the last column, we get

$$- (F_x + F_z p)\, (-g_p)\, (F_p g_q - F_q g_p) + (F_y + F_z q)\, (g_q)\, (F_p g_q - F_q g_p)$$

$$- (g_x + g_z p)\, (F_p)\, (F_p g_q - F_q g_p) + (g_y + g_z q)\, (-F_q)\, (F_p g_q - F_q g_p) = 0.$$

Cancelling the factor $F_p g_q - F_q g_p$ (see Eq. (16.43)) and simplifying, we get

$$F_p g_x + F_q g_y + (pF_p + qF_q)g_z - (F_x + p F_z)\, g_p - (F_y + qF_z)g_q = 0. \tag{16.44}$$

This is a linear, first order partial differential equation for g with x, y, z, p and q as independent variables. Hence, by Lagranges's method the auxiliary equations are given by (see Eqs. (16.30))

$$\frac{dx}{F_p} = \frac{dy}{F_q} = \frac{dz}{pF_p + qF_q} = \frac{dp}{-(F_x + pF_z)} = \frac{dq}{-(F_y + qF_z)}. \tag{16.45}$$

These equations are called the *Charpit's equations*. Note that we need not solve all the equations in Eq. (16.45). But, the solution must contain p or q or both p and q. This gives the required compatible equation $g(x, y, z, p, q) = 0$. We solve for p, q from Eq. (16.38) and $g(x, y, z, p, q) = 0$, and then integrate the equation (16.40).

When the partial differential equation given in Eq. (16.38) is of some special form, then the Charpit's equations are simplified and the required solution for g is obtained very easily. We discuss the following cases.

Case 1   *Equations which are independent of x, y, z*
We consider the given differential equation as $F(p, q) = 0$. Then, the Charpit's equations (16.45) become

$$\frac{dx}{F_p} = \frac{dy}{F_q} = \frac{dz}{pF_p + qF_q} = \frac{dp}{0} = \frac{dq}{0}. \tag{16.46}$$

We obtain a solution as $p = a$, constant or as $q = a$, constant. The choice of $p = a$ or $q = a$ depends on the given problem. Then, we solve for q from $F(a, q) = 0$, which gives $q = \phi(a)$.
Now, we integrate the equation (see Eq. (16.40))

$$dz = pdx + qdy \quad \text{or} \quad dz = a\, dx + \phi(a)\, dy$$

which gives
$$z = ax + \phi(a) y + b, \tag{16.47}$$

where a, b are arbitrary constants.

Example 16.15   Find the complete integral of the partial differential equation $p^2 - 3q^2 = 5$.

Solution   Using the last term in the Charpit's equations (16.46), we get $q = a$, where a is an arbitrary constant. Hence, from the given equation, we get $p^2 = 5 + 3a^2$, or $p = \sqrt{5 + 3a^2}$. We have

$$dz = pdx + qdy = \sqrt{5 + 3a^2}\, dx + ady.$$

Integrating, we get $z = \sqrt{5 + 3a^2}\, x + ay + b$, where a and b are arbitrary constants.

Case 2   *Equations which are independent of x, y*
We consider the given differential equation as $F(z, p, q) = 0$. Then, the Charpit's equations (16.45) become

$$\frac{dx}{F_p} = \frac{dy}{F_q} = \frac{dz}{pF_p + qF_q} = \frac{dp}{-pF_z} = \frac{dq}{-qF_z}. \tag{16.48}$$

From the last two terms, we get

$$\ln p = \ln q + \ln a, \text{ or } p = aq \tag{16.49}$$

where a is an arbitrary constant. Now, we solve for p and q from the equations $p = aq$ and $F(z, p, q) = 0$. The complete integral is obtained from $dz = pdx + qdy$.

Example 16.16   Find the complete integral of the partial differential equation $p(3 + q) = 2qz$.

Solution   The given equation is independent of x, y. Hence, from the Charpit's equations (16.48), we have $p = aq$, where a is an arbitrary constant. Substituting in the given equation, we get

$$aq\,(3 + q) = 2qz, \quad \text{or} \quad a(3 + q) = 2z, \quad \text{or} \quad q = (2z - 3a)/a, \quad \text{and} \quad p = aq = 2z - 3a.$$

Hence, we have

$$dz = p\,dx + q\,dy, \quad \text{or} \quad dz = (2z - 3a)dx + \frac{1}{a}(2z - 3a)dy$$

or

$$\frac{dz}{2z - 3a} = dx + \frac{1}{a}dy.$$

Integrating, we get

$$\frac{1}{2}\ln(2z - 3a) = x + \frac{1}{a}y + b^*, \quad \text{or} \quad \ln(2z - 3a) = 2x + \frac{2}{a}y + b$$

or

$$2z - 3a = ce^{2x + (2/a)y}$$

where $b = 2b^*$ and $c = e^b$ are arbitrary constants.

Case 3   *Separable equation of the form $f(x, p) = g(y, q)$*

Note that the differential equation is independent of z. We write the given equation as

$$F = f(x, p) - g(y, q) = 0. \tag{16.50}$$

Then, the Charpit's equations (16.45) become

$$\frac{dx}{f_p} = \frac{dy}{-g_q} = \frac{dz}{pf_p - qg_q} = \frac{dp}{-f_x} = \frac{dq}{g_y}. \tag{16.51}$$

Consider the equation

$$\frac{dx}{f_p} = \frac{dp}{-f_x}, \quad \text{or} \quad f_x dx + f_p dp = 0, \quad \text{or} \quad d[f(x, p)] = 0.$$

Integrating, we obtain $f(x, p) = a$, where a is an arbitrary constant. From the given equation, we also obtain $g(y, q) = a$. We solve for p, q from the equations $f(x, p) = a$ and $g(y, q) = a$. The complete integral is obtained from $dz = p\,dx + q\,dy$.

Example 16.17   Find the complete integral of the partial differential equation

$$p^2 q^2 = 9p^2 y^2 (x^2 + y^2) - 9x^2 y^2.$$

Solution   We write the given equation as

$$p^2 q^2 + 9x^2 y^2 = 9p^2 y^2 (x^2 + y^2).$$

Dividing by $p^2 y^2$, we obtain

$$\frac{q^2}{y^2} + \frac{9x^2}{p^2} = 9(x^2 + y^2), \quad \text{or} \quad 9\left(x^2 - \frac{x^2}{p^2}\right) = \frac{q^2}{y^2} - 9y^2$$

or

$$\frac{9x^2}{p^2}(p^2 - 1) = \frac{1}{y^2}(q^2 - 9y^4).$$

This is a separable equation. Setting each side equal to a constant a^2, we get

$$9x^2(p^2 - 1) = a^2p^2, \quad \text{or} \quad p^2(9x^2 - a^2) = 9x^2, \quad \text{or} \quad p = \frac{3x}{(9x^2 - a^2)^{1/2}}.$$

and $\qquad q^2 - 9y^4 = a^2y^2, \quad \text{or} \quad q^2 = a^2y^2 + 9y^4, \quad \text{or} \quad q = y(9y^2 + a^2)^{1/2}.$

Hence, we have

$$dz = pdx + q\,dy, \quad \text{or} \quad dz = \frac{3x}{(9x^2 - a^2)^{1/2}}dx + y(9y^2 + a^2)^{1/2}\,dy.$$

Integrating, we obtain

$$z = \int \frac{3x}{(9x^2 - a^2)^{1/2}}dx + \int y(9y^2 + a^2)^{1/2}\,dy + b$$

$$= \frac{1}{3}(9x^2 - a^2)^{1/2} + \frac{1}{27}(9y^2 + a^2)^{3/2} + b$$

where a and b are arbitrary constants.

Example 16.18   Find the complete integral of the partial differential equation

$$2\sqrt{p} + 3\sqrt{q} = 6x + 2y.$$

Solution   We write the given equation as $2\sqrt{p} - 6x = 2y - 3\sqrt{q}$, which is a separable equation. Equating both sides to a constant a, we obtain

$$2\sqrt{p} - 6x = a, \quad \text{or} \quad 4p = (a + 6x)^2$$

and $\qquad\qquad 2y - 3\sqrt{q} = a, \quad \text{or} \quad 9q = (2y - a)^2.$

Hence, we have

$$dz = pdx + qdy, \quad \text{or} \quad dz = \frac{1}{4}(6x + a)^2\,dx + \frac{1}{9}(2y - a)^2\,dy.$$

Integrating, we obtain

$$z = \int \frac{1}{4}(6x + a)^2\,dx + \int \frac{1}{9}(2y - a)^2\,dy + b$$

$$= \frac{1}{72}(6x + a)^3 + \frac{1}{54}(2y - a)^3 + b$$

where a and b are arbitrary constants.

Case 4   *Clairaut's equation*
An equation of the form $z = px + qy + f(p, q)$, which is linear in x and y is called a Clairaut's equation. Let

$$F(x, y, z, p, q) = px + qy + f(p, q) - z = 0. \tag{16.52}$$

Then, the Charpit's equations (16.45) become

$$\frac{dx}{x + f_p} = \frac{dy}{y + f_q} = \frac{dz}{p(x + f_p) + q(y + f_q)} = \frac{dp}{0} = \frac{dq}{0}.$$

We have a solution as $p = a$, where a is an arbitrary constant. We can also take another solution as $q = b$, since

$$z = ax + by + f(a, b) \tag{16.53}$$

satisfies the given differential equation (16.52), where a and b are arbitrary constants.

Example 16.19 Find the complete integral of the partial differential equation

$$p^2 q^2 (px + qy - z) = 2.$$

Solution We write the given equation as

$$z = px + qy - \frac{2}{p^2 q^2}$$

which is a Clairaut's equation. Hence, the solution is

$$z = ax + by - \frac{2}{a^2 b^2}$$

where a and b are arbitrary constants.

Example 16.20 Using a transformation, reduce the partial differential equation

$$y + 2zq = q(4xp + yq)$$

to Clairaut's form and hence find its complete integral.

Solution We write the given equation as

$$z = \frac{1}{2q}[q(4xp + yq) - y] = 2xp + \frac{y}{2}q - \frac{y}{2q}. \tag{16.54}$$

If this equation is to be reduced to Clairaut's form, then it should be of the form

$$z = Pu + Qv + f(P, Q) \tag{16.55}$$

where u, v are the new independent variables and $P = \partial z / \partial u$ and $Q = \partial z / \partial v$.

Comparaing the first two terms on the right hand side of the equations (16.54) and (16.55), we get

$$Pu = 2xp, \text{ and } Qv = \frac{y}{2}q$$

or
$$\frac{\partial z}{\partial u}u = 2x\frac{\partial z}{\partial x}, \text{ and } \frac{\partial z}{\partial v}v = \frac{y}{2}\frac{\partial z}{\partial y}$$

or
$$\frac{dx}{2x} = \frac{du}{u}, \text{ and } \frac{dv}{v} = \frac{2dy}{y}.$$

The solution of these equations are $u = \sqrt{x}$ and $v = y^2$. Hence, these are the required transformations. The third term on the right hand side of Eq. (16.54) becomes

$$-\frac{y}{2q} = -\frac{y}{2(2Qv/y)} = -\frac{y^2}{4Qv} = -\frac{1}{4Q}.$$

The required Clairaut's form becomes

$$z = Pu + Qv - \frac{1}{4Q}.$$

Hence, the solution is given by

$$z = au + bv - \frac{1}{4b} = a\sqrt{x} + by^2 - \frac{1}{4b}$$

where a and b are arbitrary constants.

Now, we give some examples which do not fall into the above four cases. We solve these problems using the Charpit's equations (16.45).

Example 16.21   Find the complete integrals of the following partial differential equations.

(i) $(p^2 + q^2)x = pz$,   (ii) $px + qy + z = xq^2$.

Solution

(i) Let $F(x, y, z, p, \dot{q}) = (p^2 + q^2)x - pz = 0$.

The Charpit's equations (16.45) become

$$\frac{dx}{2px - z} = \frac{dy}{2qx} = \frac{dz}{p(2px - z) + 2q^2x} = \frac{dp}{-(p^2 + q^2 - p^2)} = \frac{dq}{-(-pq)}.$$

Using the last two terms, we get

$$\frac{dp}{-q^2} = \frac{dq}{pq}, \quad \text{or} \quad p\,dp + q\,dq = 0.$$

Integrating, we get $p^2 + q^2 = a^2$, where a is an arbitrary constant.
From the given equation, we get $a^2x - pz = 0$, or $p = a^2x/z$.

Hence, $q^2 = a^2 - p^2 = a^2 - \dfrac{a^4x^2}{z^2} = \dfrac{a^2}{z^2}(z^2 - a^2x^2)$.

Therefore, we have

$$dz = p\,dx + q\,dy = \frac{a^2x}{z}dx + \frac{a}{z}(z^2 - a^2x^2)^{1/2}\,dy.$$

or $\dfrac{z\,dz - a^2x\,dx}{(z^2 - a^2x^2)^{1/2}} = a\,dy.$

Integrating, we get $(z^2 - a^2x^2)^{1/2} = ay + b$, or $z^2 - a^2x^2 = (ay + b)^2$

where b is an arbitrary constant.

(ii) Let $F(x, y, z, p, q) = px + qy + z - xq^2 = 0$.

The Charpit's equations (16.45) become

$$\frac{dx}{x} = \frac{dy}{y - 2xq} = \frac{dz}{px + q(y - 2xq)} = \frac{dp}{-(p - q^2 + p)} = \frac{dq}{-(q + q)}.$$

Using the first and last terms, we get $\dfrac{dx}{x} = -\dfrac{dq}{2q}.$

Integrating, we get $qx^2 = a$, where a is an arbitrary constant. Substituting $q = a/x^2$, in the given differential equation, we get

$$px + \frac{ay}{x^2} + z - x\left(\frac{a^2}{x^4}\right) = 0, \quad \text{or} \quad p = \frac{1}{x^4}(a^2 - axy - x^3z).$$

Therefore, we have

$$dz = pdx + qdy = \frac{1}{x^4}(a^2 - axy - x^3z)dx + \frac{ady}{x^2}$$

$$= -\frac{z}{x}dx + \frac{a^2}{x^4}dx + \frac{a}{x^3}(xdy - ydx)$$

or $\qquad xdz + zdx = \frac{a^2}{x^3}dx + \frac{a}{x^2}(xdy - ydx)$

Integrating, we get $xz = -\frac{a^2}{2x^2} + a\left(\frac{y}{x}\right) + b$, where b is an arbitrary constant.

Example 16.22 Find the singular solutions of the following differential equations, if they exist.

(i) $6yz - 6pxy - 3qy^2 + pq = 0$, $\qquad\qquad$ (ii) $px + qy + z = xq^2$.

Solution

(i) Let $F = 6yz - 6pxy - 3qy^2 + pq = 0$.

The Charpit's equations (16.45) become

$$\frac{dx}{-(6xy - q)} = \frac{dy}{-(3y^2 - p)} = \frac{dz}{-[p(6xy - q) + q(3y^2 - p)]}$$

$$= \frac{dp}{-(-6py + 6py)} = \frac{dq}{-(6z - 6px - 6qy + 6qy)}.$$

Using the fourth term and any other term, we get $p = a$, where a is an arbitrary constant. From the given differential equation, we get

$$q(3y^2 - a) = 6y(z - ax), \quad \text{or} \quad q = \frac{by(z - ax)}{3y^2 - a}.$$

Therefore, we have

$$dz = pdx + qdy = adx + \frac{6y(z - ax)}{3y^2 - a}dy$$

or $\qquad (3y^2 - a)(dz - adx) - 6y(z - ax)dy = 0$

or $\qquad d\left[\frac{z - ax}{3y^2 - a}\right] = 0.$

Integrating, we get $\qquad \frac{z - ax}{3y^2 - a} = b \quad \text{or} \quad (z - ax) = b(3y^2 - a)$ $\qquad\qquad$ (16.56)

where b is an arbitrary constant.

Now, we find the envelope, if it exists, of the two parameter family of surfaces given by Eq. (16.56).

Let $\qquad\qquad \phi(a, b) = z - ax - b(3y^2 - a) = 0.$ $\qquad\qquad$ (16.57)

Differentiating Eq. (16.57) partially with respect to a and b and setting the expressions obtained to zero, we get

$$\frac{\partial \phi}{\partial a} = -x + b = 0, \quad \frac{\partial \phi}{\partial b} = -(3y^2 - a) = 0.$$

We obtain $x = b$, $y^2 = a/3$. Substituting in Eq. (16.57), we get $z - 3xy^2 = 0$. Therefore, $z = 3xy^2$ is the singular solution of the given partial differential equation.

(ii) The complete integral of the given partial differential equation is (see Example 16.21 ii)

$$xz = -\frac{a^2}{2x^2} + \frac{ay}{x} + b, \quad \text{or} \quad \phi(a, b) = 2bx^2 + 2axy - a^2 - 2x^3z = 0. \qquad (16.58)$$

Differentiating partially with respect to a and b, we get

$$\frac{\partial \phi}{\partial a} = 2xy - 2a, \quad \frac{\partial \phi}{\partial b} = 2x^2. \qquad (16.59)$$

Setting $(\partial \phi / \partial a) = 0$, $(\partial \phi / \partial b) = 0$, we get $a = xy$ and $x = 0$. This implies $a = 0$. The equation (16.58) is satisfied always and the eliminant of a, b between the equations (16.58), (16.59) cannot be obtained. Hence, there is no singular solution for the given partial differential equation.

Particular Integrals from General Integrals

We have defined the general integral of the partial differential equation (16.38) as the envelope of the one parameter family of surfaces $f(x, y, z, a, g(a)) = 0$ which is derived from the complete integral

$$f(x, y, z, a, b) = 0 \qquad (16.60)$$

by choosing $b = g(a)$. All members of the one parameter family of surfaces touch the envelope. We wish to find that particular integral which passes through a given curve C whose parametric form is given by

$$C: x = x(t), \ y = y(t), \ z = z(t). \qquad (16.61)$$

Therefore, the required surface S is the envelope of that one parameter family, that is, of that subsystem of the complete integral given in Eq. (16.60) whose members touch the given curve. Now, the points of intersection of the surface given in Eq. (16.60) and the curve C are given by

$$h(t) = f[x(t), y(t), z(t), a, b] = 0. \qquad (16.62)$$

If the surface is to touch C, then this equation (16.62) must have two equal roots. When Eq. (16.62) is a quadratic, then we require the condition that the discriminant of Eq. (16.62) is zero. Alternately, $h(t) = 0$ and $(dh/dt) = 0$ must have a common root. The eliminant of t from these two equations gives a relation $\phi(a, b) = 0$. We solve for b and choose one of these values for b (we may also solve for a and choose one of these values). Each value for b gives a different subsystem of Eq. (16.60). The envelope of each of these subsystems is a solution of the given problem, that is, a particular solution passing through C.

Example 16.23   Find a complete integral of the partial differential equation $ypq + xp^2 = 1$. Hence, find a particular solution which passes through the curve $x = 0$, $y - z = 0$.

Solution   Let $F = ypq + xp^2 - 1 = 0$. The Charpit's equations (16.45) are obtained as

$$\frac{dx}{yq + 2px} = \frac{dy}{yp} = \frac{dz}{p(yq + 2px) + ypq} = \frac{dp}{-(p^2)} = \frac{dq}{-(pq)}.$$

From the last two terms, we obtain $p = a^2q$, where a is an arbitrary constant. From the given equation, we get

$$a^2 yq^2 + a^4 xq^2 - 1 = 0, \quad \text{or} \quad q^2 = \frac{1}{a^2(y + a^2 x)}, \quad \text{or} \quad q = \frac{1}{a(y + a^2 x)^{1/2}}.$$

Therefore, we have

$$dz = pdx + qdy = \frac{a\,dx}{(y + a^2 x)^{1/2}} + \frac{dy}{a(y + a^2 x)^{1/2}}$$

or

$$a\,dz = \frac{a^2 dx + dy}{(a^2 x + y)^{1/2}}.$$

Integrating, we obtain the complete integral $az + b = 2(a^2 x + y)^{1/2}$, or $(az + b)^2 = 4(a^2 x + y)$, where b is an arbitrary constant. The parametric form of the given curve is $x = 0$, $y = z = t$. Hence, we have

$$(at + b)^2 = 4t, \quad \text{or} \quad a^2 t^2 + 2(ab - 2)\,t + b^2 = 0.$$

This equation has equal roots when

$$4(ab - 2)^2 - 4a^2 b^2 = 0, \quad \text{or} \quad -4ab + 4 = 0, \quad \text{or} \quad ab = 1.$$

Substituting $b = 1/a$ in the complete integral, we get

$$4(a^2 x + y) = \left(az + \frac{1}{a} \right)^2 = \frac{1}{a^2}(a^2 z + 1)^2$$

or

$$4a^2(a^2 x + y) - (a^2 z + 1)^2 = 0. \tag{16.63}$$

Differentiating partially with respect to a, we get

$$16a^3 x + 8ay - 2(a^2 z + 1)(2az) = 0$$

or

$$4a^2 x + 2y - a^2 z^2 - z = 0, \quad \text{or} \quad a^2 = \frac{z - 2y}{4x - z^2}.$$

Simplifying Eq. (16.63), we get

$$4a^4 x + 4a^2 y - (a^4 z^2 + 2a^2 z + 1) = 0, \quad \text{or} \quad a^4(4x - z^2) + 2a^2(2y - z) - 1 = 0.$$

Substituting the value of a^2, we get

$$\frac{(z - 2y)^2}{4x - z^2} - \frac{2(z - 2y)^2}{4x - z^2} - 1 = 0 \quad \text{or} \quad \frac{(z - 2y)^2}{4x - z^2} + 1 = 0$$

or

$$(z - 2y)^2 + (4x - z^2) = 0, \quad \text{or} \quad x + y^2 - yz = 0$$

which is the required equation of the particular integral.

Exercise 16.3

Find the complete integrals of the following first order partial differential equations.

1. $2p + 3q + 4pq = 0$.

2. $\sqrt{p} + 3\sqrt{q} = 1$.

3. $p^2 + q^2 = 5z$.

4. $q(1 + p) = pz$.

5. $p^2 + q^2 z^2 = 1$.

6. $p^2 x(1 + y^2) = qy$.

7. $\sqrt{p} + \sqrt{q} = 2x^2 + y$.

8. $p^3 + q^3 = 216z$.

9. $4x^2 p^2 + 9y^2 q^2 = z^2$.

10. $z = px + qy + [p/(p + q)]$.

11. Using the transformations $z^2 = u$, reduce the partial differential equation $z^2(2p^2 + 3q^2) = 8x^2 + 27y^2$ to separable form and hence solve the equation.

12. Using the transformation $z^{3/2} = u$, reduce the partial differential equation $z(p^2 + q^2) = x + y$ to separable form and hence solve the equation.

13. Using the transformations $x^2 = u$, $y^2 = v$ reduce the partial differential equation

    $2xyz = px^2 y + qxy^2 + 4pq$ to Clairaut's form and hence solve the equation.

14. Find a suitable transformation to reduce the partial differential equation

    $4(y)^{-1/2} + qz = q(3xp + 2yq)$ to Clairaut's form. Hence, find the solution of the given equation.

Solve the following partial differential equations.

15. $2xz - px^2 - 2xy\,q + pq = 0$.

16. $yq + 3xp = 2(z - y^2 p^2)$.

17. $qxy + p(q + x) = xz$.

18. $(p^2 + q^2)x = pz$.

In Problems **19** to **21**, find the singular solutions of the given partial differential equations.

19. $2xz - px^2 - 2xyq + pq = 0$.

20. $z = px + qy + p^2 + q^2$.

21. $4xyz - 6pq - 2px^2 y - 2qxy^2 = 0$ (use the substitutions $x^2 = u$, $y^2 = v$ and reduce the equation to Clairaut's form).

22. Find a complete integral of the partial differential equation $px + q^2 y = z$. Hence, find a particular solution which passes through the curve $x = 1$, $y + z = 0$.

16.3.4   Higher Order Linear Equations with Constant Coefficients

In Chapter 5, we have discussed methods for solving the higher order linear ordinary differential equations with constant coefficients. These methods can be extended to solve the higher order linear partial differential equations with constant coefficients. Consider the partial differential equation

$$A_0 \frac{\partial^n z}{\partial x^n} + A_1 \frac{\partial^{n-1} z}{\partial x^{n-1} \partial y} + A_2 \frac{\partial^{n-2} z}{\partial x^{n-2} \partial y^2} + \ldots + A_{n-1} \frac{\partial^n z}{\partial x \partial y^{n-1}} + A_n \frac{\partial^n z}{\partial y^n} = f(x, y)$$

where A_i, $i = 0, 1, \ldots, n$ are constants.   (16.64)

Denote $p = \dfrac{\partial z}{\partial x} = Dz$, $q = \dfrac{\partial z}{\partial y} = D'z$, $D^2 z = \dfrac{\partial^2 z}{\partial x^2}$, $DD'z = \dfrac{\partial^2 z}{\partial x \partial y}$ etc.

Then, Eq. (16.64) can be written as

$$F(D, D')z = \sum_{r=0}^{n} A_r D^{n-r} (D')^r z = f(x, y).$$ (16.65)

As in the case of ordinary differential equations the general solution of Eq. (16.65) can be written as

$$z = (\text{complementary function}) + (\text{particular integral})$$

where the complementary function is the solution of the homogeneous equation $F(D, D')z = 0$ and the particular integral is a solution satisfying the non-homogeneous equation (16.65) and does not contain any arbitrary constants.

Complementary function

Consider the linear, homogeneous equation

$$F(D, D')\, z = 0. \tag{16.66}$$

If $F(D, D')$ can be factorised into linear factors of the form $(a_i D + b_i D' + c_i)$, then $F(D, D')$ is said to be *reducible*. Otherwise, it is said to be *irreducible*. $F(D, D')$ may contain some reducible and some irreducible factors. We have the following examples.

$$(D^2 - D'^2)z = (D - D')(D + D')\, z = 0 \tag{16.67i}$$

$$(D^3 - 6D^2D' + 11DD'^2 - 6D'^3)z = (D - D')(D - 2D')(D - 3D')\, z = 0 \tag{16.67ii}$$

$$(2D^2 - D')\, z = 0 \tag{16.68i}$$

$$(D^3 + D^2D' + D'^3)\, z = 0 \tag{16.68ii}$$

$$(D^3 - D^2D' + DD' - D'^2)z = (D - D')(D^2 + D')z = 0. \tag{16.69}$$

Eqs. (16.67) are the reducible forms, Eqs. (16.68) are irreducible forms and Eq. (16.69) has one reducible and one irreducible factor.

We will discuss only the equations which are in reducible form. We consider the following cases.

Distinct factors

Consider the solution of the partial differential equation with one linear factor, that is, the solution of

$$(a_i D + b_i D' + c_i)\, z = 0, \quad \text{or} \quad a_i p + b_i q = -c_i z. \tag{16.70}$$

Since, this is a Lagrange's equation, the auxiliary equations are given by

$$\frac{dx}{a_i} = \frac{dy}{b_i} = \frac{dz}{-c_i z}.$$

The first two terms give the solution as

$$b_i x - a_i y = k_1, \ k_1 \text{ arbitrary constant.} \tag{16.71}$$

The first and third terms give the solution as

$$a_i \ln z = -c_i x + k_3, \quad \text{or} \quad z = k_2 e^{-(c_i x)/a_i} \tag{16.72}$$

where k_2, k_3 are arbitrary constants and $k_2 = e^{k_3/a_i}$. Hence, the solution of Eq. (16.70) is given by

$$k_2 = \phi_i(k_1), \quad \text{or} \quad ze^{(c_i x)/a_i} = \phi_i(b_i x - a_i y)$$

or

$$z = e^{-(c_i x)/a_i}\, \phi_i\, (b_i x - a_i y). \tag{16.73}$$

If the linear factor is of the form $(a_i D + b_i D')$, then Eq. (16.73) simplifies to

$$z = \phi_i(b_i x - a_i y). \tag{16.74}$$

If the linear factor is of the form $(b_i D' + c_i)$, then we have the differential equation as $(b_i D' + c_i)z = 0$ or $b_i q_i + c_i z = 0$. The solutions of the auxiliary equations are

$$b_i x = k_1 \quad \text{and} \quad z = k_2 e^{-(c_i y)/b_i}$$

Hence, the solution of Eq. (16.70), when $a_i = 0$, is given by

$$k_2 = \phi_i(k_1) \quad \text{or} \quad z e^{(c_i y)/b_i} = \phi_i(b_i x)$$

or

$$z = e^{-(c_i y)/b_i} \phi_i(b_i x). \tag{16.75}$$

Similarly, if the linear factor is if the form $(a_i D + c_i)$, then the solution of Eq. (16.70), when $b_i = 0$ is given by

$$z = e^{-(c_i x)/a_i} \phi_i(a_i y). \tag{16.76}$$

Corresponding to each linear factor of $F(D, D')$ of the above forms, we have the solution in either of the forms given in Eqs. (16.73), (16.74), (16.75) or (16.76). Therefore, if $F(D, D')$ can be written as

$$F(D, D') = \prod_{i=1}^{n} (a_i D + b_i D' + c_i)$$

then, the complementary function is given by

$$z = e^{-(c_1 x)/a_1} \phi_1(b_1 x - a_1 y) + e^{-(c_2 x)/a_2} \phi_2(b_2 x - a_2 y) + \ldots + e^{-(c_n x)/a_n} \phi_n(b_n x - a_n y).$$

$$= \sum_{i=1}^{n} e^{-(c_i x)/a_i} \phi_i(b_i x - a_i y). \tag{16.77}$$

Similarly, the complementary function in the other cases can be written.

Remark 4

A method of factorising $F(D, D')$ is to set symbolically, if possible, $m = (D/D')$ and find the factors of the equation in m. This is possible, when the total degree of each term in $F(D, D')$ is same.

Example 16.24 Find the solutions of the following partial differential equations

(i) $(D^2 - D'^2)z = 0$,

(ii) $(D^3 - 6D^2D' + 11DD'^2 - 6D'^3)z = 0$,

(iii) $3r + 7s + 2t + 7p + 4q + 2z = 0$,

(iv) $2r - s - t - p + q = 0$.

Solution

(i) We have $(D^2 - D'^2)z = (D + D')(D - D')z = 0$.

For the factor $D - D'$, we have $a_1 = 1, b_1 = -1$.
For the factor $D + D'$, we have $a_2 = 1, b_2 = 1$.
Therefore, using Eq. (16.74), we obtain the solution as

$$z = \phi_1^*(-x - y) + \phi_2(x - y) = \phi_1(x + y) + \phi_2(x - y).$$

(ii) We have

$$(D^3 - 6D^2D' + 11\, DD'^2 - 6D'^3)z = (D - D')\,(D - 2D')\,(D - 3D')\,z = 0.$$

For the factor $D - D'$, we have $a_1 = 1$, $b_1 = -1$.
For the factor $D - 2D'$, we have $a_2 = 1$, $b_2 = -2$.
For the factor $D - 3D'$, we have $a_3 = 1$, $b_3 = -3$.
If we set $(D/D') = m$, we get the equation as $m^3 - 6m^2 + 11m - 6 = 0$, whose roots are $m = 1$, 2, 3. Hence, the factors are $D - D'$, $D - 2D'$, $D - 3D'$.
Using Eq. (16.74), the solution can be written as

$$z = \phi_1^*(-x - y) + \phi_2^*(-2x - y) + \phi_3^*(-3x - y)$$

$$= \phi_1(x + y) + \phi_2(2x + y) + \phi_3(3x + y).$$

(iii) We have

$$(3D^2 + 7DD' + 2D'^2 + 7D + 4D' + 2)z = (3D + D' + 1)(D + 2D' + 2)\,z = 0.$$

For the factor $3D + D' + 1$, we have $a_1 = 3$, $b_1 = 1$, $c_1 = 1$.
For the factor $D + 2D' + 2$, we have $a_2 = 1$, $b_2 = 2$, $c_2 = 2$.
Using Eq. (16.73), the solution can be written as

$$z = e^{-x/3}\phi_1(x - 3y) + e^{-2x}\phi_2(2x - y).$$

(iv) We have $(2D^2 - DD' - D'^2 - D + D')\,z = (D - D')(2D + D' - 1)\,z = 0.$

For the factor $D - D'$, we have $a_1 = 1$, $b_1 = -1$, $c_1 = 0$.
For the factor $2D + D' - 1$, we have $a_2 = 2$, $b_2 = 1$, $c_2 = -1$.
Using Eqs. (16.74) and (16.73), the solution can be written as

$$z = \phi_1^*(-x - y) + e^{x/2}\phi_2(x - 2y)$$

$$= \phi_1(x + y) + e^{x/2}\phi_2(x - 2y).$$

Multiple factors

Let a factor be of multiplicity 2, that is, we have either of the factors

$$(a_i D + b_i D' + c_i)^2, \quad \text{or} \quad (a_i D + c_i)^2, \quad \text{or} \quad (b_i D' + c_i)^2.$$

First, consider the equation

$$(a_i D + b_i D' + c_i)^2 z = 0. \tag{16.78}$$

Let $(a_i D + b_i D' + c_i)\,z = u$. Then, Eq. (16.78) simplifies to

$$(a_i D + b_i D' + c_i)u = 0.$$

Using Eq. (16.73), the solution of this equation is obtained as

$$u = e^{-(c_i x)/a_i}\,\phi_i(b_i x - a_i y).$$

Hence, we have

$$(a_i D + b_i D' + c_i)z = u = e^{-(c_i x)/a_i}\,\phi_i(b_i x - a_i y)$$

or
$$a_i p + b_i q = -c_i z + e^{-(c_i x)/a_i} \phi_i(b_i x - a_i y).$$

This is a Lagrange's equation. The auxiliary equations are given by

$$\frac{dx}{a_i} = \frac{dy}{b_i} = \frac{dz}{-c_i z + e^{-X} \phi_i(b_i x - a_i y)}$$

where $X = c_i x/a_i$. Using the first two terms, we obtain $b_i x - a_i y = k_1$. Using this solution, we have

$$\frac{dx}{a_i} = \frac{dz}{-c_i z + e^{-X} \phi_i(k_1)}, \quad \text{or} \quad \frac{dz}{dx} = \frac{1}{a_i}(-c_i z + e^{-X} \phi_i(k_1))$$

or
$$\frac{dz}{dx} + \left(\frac{c_i}{a_i}\right) z = \frac{1}{a_i} e^{-X} \phi_i(k_1). \tag{16.79}$$

The integrating factor of this linear differential equation is given by

$$I.F. = e^{\int (c_i/a_i) dx} = e^{(c_i x)/a_i} = e^X.$$

Multiplying Eq. (16.79) by the *I.F.* and integrating, we get

$$e^X z = \frac{1}{a_i}[x \phi_i^*(k_1) + k_2^*] = x \phi_i(k_1) + k_2$$

where a_i is absorbed in the arbitrary function and $k_2 = k_2^*/a_i$. Hence, we have the solution as
$k_2 = \psi(k_1)$,

or
$$e^X z - x \phi_i(k_1) = \psi(k_1)$$

or
$$e^{(c_i x/a_i)} z = x \phi_i(b_i x - a_i y) + \psi(b_i x - a_i y).$$

Therefore, the general solution is given by

$$z = e^{-(c_i x)/a_i} [x \phi_i(b_i x - a_i y) + \psi_i(b_i x - a_i y)]. \tag{16.80}$$

Now, consider the equation

$$(a_i D + c_i)^2 z = 0. \tag{16.81}$$

Let $(a_i D + c_i) z = u$. Then, Eq. (16.81) simplifies to $(a_i D + c_i)u = 0$, whose solution is given as

$$u = e^{-X} \phi_i(a_i y), \quad X = c_i x/a_i. \tag{16.82}$$

Hence, we have

$$(a_i D + c_i) z = e^{-X} \phi_i(a_i y), \quad \text{or} \quad a_i p = -c_i z + e^{-X} \phi_i(a_i y).$$

This is again a Lagrange's equation. The auxiliary equations are

$$\frac{dx}{a_i} = \frac{dy}{0} = \frac{dz}{-c_i z + e^{-X} \phi_i(a_i y)}$$

From the first two terms, we get $a_i y = k_1$.
    Using this solution, we have

$$\frac{dx}{a_i} = \frac{dz}{-c_i z + e^{-X} \phi_i(k_1)}, \quad \text{or} \quad \frac{dz}{dx} = \frac{1}{a_i}[-c_i z + e^{-X} \phi_i(k_1)]$$

or
$$\frac{dz}{dx} + \left(\frac{c_i}{a_i}\right) z = \frac{1}{a_i} e^{-X} \phi_i(k_1). \tag{16.83}$$

The integrating factor of this equation is $I.F. = e^{(c_i x)/a_i} = e^X$. Multiplying Eq. (16.83) by the $I.F.$ and integrating, we get

$$e^X z = \frac{1}{a_i}[x\phi_i^*(k_1) + k_2^*] = x\phi_i(k_1) + k_2.$$

where a_i is absorbed in the arbitrary function and $k_2 = k_2^*/a_i$ is an arbitrary function of $a_i y$. Therefore, the general solution is given by

$$z = e^{-(c_i x)/a_i}[x\phi_i(a_i y) + \psi_i(a_i y)]. \tag{16.84}$$

Similarly, in the case of the quadratic factor $(b_i D' + c_i)^2$, we obtain the solution as

$$z = e^{-(c_i y)/b_i}[y\phi_i(b_i x) + \psi_i(b_i x)]. \tag{16.85}$$

If the factor is of multiplicity m, say $(a_1 D + b_1 D' + c_1)^m$, then we obtain the general solution as

$$z = e^{-(c_1 x)/a_1}[\psi_1(b_1 x - a_1 y) + x\psi_2(b_1 x - a_1 y) + \ldots + x^{m-1}\psi_m(b_1 x - a_1 y)]. \tag{16.86}$$

Similarly, we can write the general solution in the other cases.

Example 16.25 Find the solutions of the following partial differential equations

(i) $(D^4 - 2D^2 D'^2 + D'^4) z = 0$, (biharmonic equation),

(ii) $(4D^3 - 3DD'^2 + D'^3) z = 0$.

Solution

(i) We have

$$(D^4 - 2D^2 D'^2 + D'^4) z = (D^2 - D'^2)^2 z = (D - D')^2 (D + D')^2 z = 0.$$

For the factor $(D - D')^2$, we have $a_1 = 1$, $b_1 = -1$, $c_1 = 0$.
For the factor $(D + D')^2$, we have $a_2 = 1$, $b_2 = 1$, $c_2 = 0$.
If we set $m = D/D'$, we get $m^4 - 2m^2 + 1 = 0$. The roots are $m^2 = 1$, or $m = \pm 1$ which are double roots. Hence, the factors are $(D - D')^2$ and $(D + D')^2$.
Therefore, using Eq. (16.80), we obtain the general solution as

$$z = [x\phi_1^*(-x - y) + \phi_2^*(-x - y)] + [x\psi_1(x - y) + \psi_2(x - y)]$$

or $\quad z = x\phi_1(x + y) + \phi_2(x + y) + x\psi_1(x - y) + \psi_2(x - y).$

(ii) We have

$$(4D^3 - 3DD'^2 + D'^3)z = (D + D')(4D^2 - 4DD' + D'^2)z$$
$$= (D + D')(2D - D')^2 z = 0.$$

For the factor $D + D'$, we have $a_1 = 1$, $b_1 = 1$, $c_1 = 0$.
For the factor $(2D - D')^2$, we have $a_2 = 2$, $b_2 = -1$, $c_2 = 0$.
Using Eqs. (16.74) and (16.80), we obtain the general solution as

$$z = \phi_1(x - y) + x\psi_1^*(-x - 2y) + \psi_2^*(-x - 2y)$$

or $$z = \phi_1(x - y) + x\psi_1(x + 2y) + \psi_2(x + 2y).$$

Particular integral

Consider the non-homogeneous equation (16.65)

$$F(D, D')z = \sum_{r=0}^{n} A_r D^{n-r} (D')^r z = f(x, y). \tag{16.87}$$

The particular integral is now written as

$$z = [F(D, D')]^{-1} f(x, y). \tag{16.88}$$

Symbolically, we shall perform all of the operations as we do on an algebraic function like $(ax + by + c)^{-1}$ including expanding as an infinite series.

For simple forms of $f(x, y)$, we shall develop rules for finding a particular integral.

Case 1 $f(x, y) = e^{ax + by}$

Since, $$(D')^r e^{ax+by} = b^r e^{ax+by}, \quad D^{n-r} e^{ax+by} = a^{n-r} e^{ax+by},$$

$$D^{n-r} (D')^r e^{ax+by} = a^{n-r} b^r e^{ax+by},$$

we get $$F(D, D') e^{ax+by} = \sum_{r=0}^{n} A_r a^{n-r} b^r e^{ax+by} = F(a, b)e^{ax+by}.$$

Hence, we write

$$e^{ax+by} = [F(D, D')]^{-1} F(a, b)e^{ax+by}, \quad \text{or} \quad [F(D, D')]^{-1} e^{ax+by} = \frac{1}{F(a, b)} e^{ax+by}.$$

Therefore, the particular integral in this case is

$$z = [F(D, D')]^{-1} e^{ax+by} = \frac{1}{F(a, b)} e^{ax+by}, \text{ if } F(a, b) \neq 0. \tag{16.89}$$

If $F(a, b) = 0$, then we write the particular integral as

$$z = \phi(x, y)e^{ax+by}. \tag{16.90}$$

Applying the Leibnitz theorem, we get

$$D^{n-r}[\phi e^{ax+by}] = a^{n-r} e^{ax+by} \phi + {}^nC_1 a^{n-r-1} e^{ax+by} D\phi + \ldots + e^{ax+by} D^{n-r}\phi$$

$$= e^{ax+by}[a^{n-r} + {}^nC_1 a^{n-r-1}D + {}^nC_2 a^{n-r-2}D^2 + \ldots + D^{n-r}]\phi$$

$$= e^{ax+by}(D + a)^{n-r}\phi$$

$$(D')^r[\phi e^{ax+by}] = b^r e^{ax+by}\phi + {}^nC_1 b^{r-1} e^{ax+by} D'\phi + \ldots + e^{ax+by}(D')^r \phi$$

$$= e^{ax+by}[b^r + {}^nC_1 b^{r-1}D' + {}^nC_2 b^{r-2}(D')^2 + \ldots + (D')^r]\phi$$

$$= e^{ax+by}(D' + b)^r\phi$$

and
$$D^{n-r}(D')^r[\phi e^{ax+by}] = e^{ax+by}(D + a)^{n-r}(D' + b)^r \phi.$$

Hence,
$$F(D, D')[\phi e^{ax+by}] = \sum_{r=0}^{n} A_n e^{ax+by} (D + a)^{n-r} (D' + b)^r \phi$$

$$= e^{ax+by} F(D + a, D' + b) \phi.$$

Now, from Eq. (16.87), we get

$$F(D, D')z = F(D, D') [\phi e^{ax+by}]$$

$$= e^{ax+by} F(D + a, D' + b) \phi = e^{ax+by}.$$

Cancelling e^{ax+by}, we get

$$F(D + a, D' + b) \phi = 1 \qquad (16.91\text{i})$$

and
$$\phi(x, y) = [F(D + a, D' + b)]^{-1} (1). \qquad (16.91\text{ii})$$

We expand the operator $[F(D + a, D' + b)]^{-1}$ in an infinite series symbolically, and determine $\phi(x, y)$. Then, the particular integral is given by Eq. (16.90). Note that D^{-1} and $(D')^{-1}$ mean integral with respect to x and y respectively, keeping the other variable as constant.

Example 16.26 Find the general solution of the partial differential equation

$$[2D^2 - DD' - (D')^2 + D - D']z = e^{2x+3y}.$$

Solution We write

$$F(D, D')z = [2D^2 - DD' - (D')^2 + D - D']z$$

$$= [(2D + D' + 1) (D - D')]z = e^{2x+3y}.$$

The homogeneous equation is

$$(2D + D' + 1) (D - D') z = 0$$

For the factor $D - D'$, we have $a_1 = 1$, $b_1 = -1$, $c_1 = 0$.
For the factor, $2D + D' + 1$, we have $a_2 = 2$, $b_2 = 1$, $c_2 = 1$.
Therefore, using Eqs. (16.73) and (16.74), we get the complementary function as

$$z = e^{-x/2}\phi_1(x - 2y) + \phi_2 (x + y).$$

Since $F(2, 3) = (4 + 3 + 1) (2 - 3) = -8 \neq 0$, we obtain the particular integral as

$$z = \frac{1}{F(2, 3)} e^{2x+3y} = -\frac{1}{8}e^{2x+3y}.$$

Therefore, the general solution of the given differential equation is given by

$$z = e^{-x/2}\phi_1(x - 2y) + \phi_2(x + y) - \frac{1}{8}e^{2x+3y}.$$

Example 16.27 Find a particular integral of the differential equation

$$(4D^2 + 3 DD' - D'^2 - D - D')z = 3e^{(x+2y)/2}.$$

Solution   We have

$$F(D, D') = 4D^2 + 3DD' - D'^2 - D - D'$$

and

$$F\left(\frac{1}{2}, 1\right) = 4\left(\frac{1}{4}\right) + 3\left(\frac{1}{2}\right)(1) - 1 - \frac{1}{2} - 1 = 0.$$

We write the particular integral as $z = \phi(x, y)\, e^{(x+2y)/2}$.
We get

$$F(D, D')[\phi\, e^{(x+2y)/2}] = e^{(x+2y)/2} F\left(D + \frac{1}{2}, D' + 1\right)\phi = 3e^{(x+2y)/2}.$$

Now,

$$F\left(D + \frac{1}{2}, D' + 1\right) = 4\left(D + \frac{1}{2}\right)^2 + 3\left(D + \frac{1}{2}\right)(D' + 1) - (D' + 1)^2 - \left(D + \frac{1}{2}\right) - (D' + 1)$$

$$= 6D - \frac{3}{2}D' + 3DD' + 4D^2 - (D')^2 \qquad (16.92)$$

$$= 6D\left[1 - \frac{1}{4}D^{-1}D' + \frac{1}{6}\left\{3D' + 4D - D^{-1}(D')^2\right\}\right]$$

Therefore,

$$F\left(D + \frac{1}{2}, D' + 1\right)\phi = 3$$

or

$$\phi = \left[F\left(D + \frac{1}{2}, D' + 1\right)\right]^{-1} (3)$$

$$= \frac{1}{2}D^{-1}\left[1 - \frac{1}{4}D^{-1}D' + \frac{1}{6}\left\{3D' + 4D - D^{-1}(D')^2\right\}\right]^{-1} (1)$$

$$= \frac{1}{2}D^{-1}\left[1 + \frac{1}{4}D^{-1}D' - \frac{1}{6}\left\{3D' + 4D - D^{-1}(D')^2\right\} + \text{other terms}\right](1)$$

$$= \frac{1}{2}D^{-1}(1) = \frac{x}{2}.$$

Hence, the particular integral is given by $z = \frac{x}{2}\, e^{(x+2y)/2}$
If we write Eq. (16.92) as

$$F\left(D + \frac{1}{2}, D' + 1\right) = -\frac{3}{2}D'\left[1 - \left\{4(D')^{-1}D + 2D + \frac{8}{3}(D')^{-1}D^2 - \frac{2}{3}D'\right\}\right]$$

then, we obtain

$$\phi = \left[F\left(D + \frac{1}{2}, D' + 1\right)\right]^{-1} (3)$$

$$= -\frac{2}{3}(D')^{-1}\left[1 - \left\{4(D')^{-1}D + 2D + \frac{8}{3}(D')^{-1}D^2 - \frac{2}{3}D'\right\}\right]^{-1} (3)$$

$$= -2(D')^{-1}(1) = -2y.$$

Hence, another particular integral can be written as

$$z = -2y \, e^{(x+2y)/2}.$$

Case 2 $f(x, y) = \sin(ax + by)$, or $f(x, y) = \cos(ax + by)$

Since, $D^2 \sin(ax + by) = -a^2 \sin(ax + by)$, $(D')^2 \sin(ax + by) = -b^2 \sin(ax + by)$,

$DD' \sin(ax + by) = -ab \sin(ax + by)$

we get $F(D^2, DD', (D')^2) \sin(ax + by) = F(-a^2, -ab, -b^2) \sin(ax + by)$.

Hence, $\sin(ax + by) = [F(D^2, DD', (D')^2)]^{-1} F(-a^2, -ab, -b^2) \sin(ax + by)$

or $[F(D^2, DD', (D')^2)]^{-1} \sin(ax + by) = \dfrac{1}{F(-a^2, -ab, -b^2)} \sin(ax + by)$.

Therefore, the particular integral in this case is

$$z = \frac{\sin(ax + by)}{F(-a^2, -ab, -b^2)}, \quad \text{if } F(-a^2, -ab, -b^2) \neq 0. \tag{16.93}$$

Similarly, if $f(x, y) = \cos(ax + by)$, we get the particular integral as

$$z = \frac{\cos(ax + by)}{F(-a^2, -ab, -b^2)}, \quad \text{if } F(-a^2, -ab, -b^2) \neq 0. \tag{16.94}$$

If $F(-a^2, -ab, -b^2) = 0$, then we shall follow the procedure as given in *Case 4*.

Example 16.28 Find the particular integral of the differential equation

$$2\frac{\partial^2 z}{\partial x^2} - 3\frac{\partial^2 z}{\partial x \partial y} + \frac{\partial^2 z}{\partial y^2} = \sin(x - 2y).$$

Solution We have

$$F(D^2, DD', (D')^2)z = [2D^2 - 3DD' + (D')^2]z = \sin(x - 2y).$$

We have the right hand side as $\sin(ax + by)$, where $a = 1$, $b = -2$.

Hence, $F(D^2, DD', (D')^2) \sin(x - 2y) = [2(-1) - 3(2) + (-4)] \sin(x - 2y) = -12 \sin(x - 2y)$.

The particular integral is given by

$$z = \frac{\sin(x - 2y)}{F(-a^2, -ab, -b^2)} = -\frac{1}{12} \sin(x - 2y).$$

Case 3 $F(x, y) = x^m y^n$, or a polynomial in x, y. We write the particular integral as

$$z = [F(D, D')]^{-1} x^m y^n \tag{16.95}$$

We expand $[F(D, D')]^{-1}$ as an infinite series and operate on $x^m y^n$. If $F(D, D')$ does not contain the constant term, that is, a term independent of D and D', then we expand $[F(D, D')]^{-1}$ in powers of $(D')^{-1}D$ if $m < n$ and in powers of $D^{-1}D'$ if $m > n$.

Example 16.29   Find the particular integrals of the following partial differential equations

(a) $\dfrac{\partial^2 z}{\partial x^2} - \dfrac{\partial^2 z}{\partial y^2} = x^2 + y^2,$

(b) $\dfrac{\partial^2 z}{\partial x^2} + 2\dfrac{\partial^2 z}{\partial x \partial y} + \dfrac{\partial^2 z}{\partial y^2} = 3x + 2y,$

(c) $4\dfrac{\partial^3 z}{\partial x^3} - 3\dfrac{\partial^3 z}{\partial x \partial y^2} + \dfrac{\partial^3 z}{\partial y^3} = 6x^2 y^2,$

(d) $2\dfrac{\partial^2 z}{\partial x^2} + 3\dfrac{\partial^2 z}{\partial x \partial y} + \dfrac{\partial^2 z}{\partial y^2} + \dfrac{\partial z}{\partial x} + \dfrac{\partial z}{\partial y} = x - y.$

(e) $\dfrac{\partial^2 z}{\partial x^2} - \dfrac{2\partial^2 z}{\partial x \partial y} + \dfrac{\partial^2 z}{\partial y^2} = 2x\cos y.$

Solution

(a) We have $[D^2 - (D')^2]\, z = x^2 + y^2.$

We write $[D^2 - (D')^2] = D^2[1 - (D^{-1})^2 (D')^2]$

and $[D^2 - (D')^2]^{-1} = D^{-2}[1 - (D^{-1})^2\, (D')^2]^{-1}$

$= D^{-2}[1 + (D^{-1})^2(D')^2 + (D^{-1})^4\, (D')^4 + \ldots].$

Therefore, the particular integral is given by

$z = [D^2 - (D')^2]^{-1}\, (x^2 + y^2)$

$= D^{-2}[1 + (D^{-1})^2(D')^2 + (D^{-1})^4\, (D')^4 + \ldots]\, (x^2 + y^2)$

$= D^{-2}[(x^2 + y^2) + (D^{-1})^2\, (2)] = D^{-2}[x^2 + y^2 + x^2].$

$= D^{-2}[2x^2 + y^2] = \dfrac{x^4}{6} + \dfrac{x^2 y^2}{2}.$

We may also write

$[D^2 - (D')^2] = -(D')^2\, [1 - (D')^{-2}D^2].$

In this case, we obtain a different form of the particular integral as

$z = -\dfrac{y^2}{6}[3x^2 + y^2].$

(b) We have $[D^2 + 2DD' + (D')^2]z = 3x + 2y,$ or $[D + D']^2 z = 3x + 2y.$

We write $[D + D']^2 = [D(1 + D^{-1}D')]^2$

and $[D + D']^{-2} = D^{-2}(1 + D^{-1}D')^{-2}$

The particular integral is given by

$z = D^{-2}(1 + D^{-1}D')^{-2}\, (3x + 2y) = D^{-2}\, [1 - 2(D^{-1}D') + 3(D^{-1}D')^2 - \ldots]\, (3x + 2y)$

$= D^{-2}[3x + 2y - 2D^{-1}(2)] = D^{-2}[3x + 2y - 4x] = D^{-2}[-x + 2y]$

$= -\dfrac{x^3}{6} + x^2 y.$

We may also write

$[D + D']^2 = [D'\, (1 + (D')^{-1}D)]^2$

The particular integral is given by

$$z = (D')^{-2} [1 + (D')^{-1}D]^{-2} (3x + 2y)$$

$$= (D')^{-2}[1 - 2(D')^{-1}D + 3((D')^{-1}D)^2 - \ldots] (3x + 2y)$$

$$= (D')^{-2}[3x + 2y - 2(D')^{-1}(3)] = (D')^{-2} [3x + 2y - 6y]$$

$$= (D')^{-2} [3x - 4y] = \frac{3xy^2}{2} - \frac{2y^3}{3}.$$

(c) We have $\qquad [4D^3 - 3D(D')^2 + (D')^3]z = 6x^2y^2.$

We write

$$4D^3 - 3D(D')^2 + (D')^3 = 4D^3\left[1 - \frac{3}{4}D^{-2}(D')^2 + \frac{1}{4}D^{-3}(D')^3\right]$$

The particular integral is given by

$$z = \frac{1}{4}D^{-3}\left[1 - \left\{\frac{3}{4}D^{-2}(D')^2 - \frac{1}{4}D^{-3}(D')^3\right\}\right]^{-1} (6x^2y^2)$$

$$= \frac{3}{2}D^{-3}\left[1 + \frac{3}{4}D^{-2}(D')^2 - \frac{1}{4}D^{-3}(D')^3 + \ldots\right](x^2y^2)$$

$$= \frac{3}{2}D^{-3}\left[x^2y^2 + \frac{3}{2}D^{-2}(x^2)\right]$$

$$= \frac{3}{2}D^{-3}\left[x^2y^2 + \frac{1}{8}x^4\right] = \frac{1}{80}x^5y^2 + \frac{1}{1120}x^7.$$

(d) We have $\quad [2D^2 + 3DD' + (D')^2 + D + D'] z = x - y.$

We write $\quad [2D^2 + 3DD' + (D')^2 + D + D'] = D[1 + D^{-1}D' + 3D' + 2D + D^{-1}(D')^2]$

The particular integral is given by

$$z = D^{-1}[1 + \{D^{-1}D' + 3D' + 2D + D^{-1}(D')^2\}]^{-1}(x - y)$$

$$= D^{-1}[1 - \{D^{-1}D' + 3D' + 2D + D^{-1}(D')^2\} + \ldots] (x - y)$$

$$= D^{-1}[x - y - \{D^{-1}(-1) + 3(-1) + 2(1)\}]$$

$$= D^{-1}[x - y + x + 1] = x^2 + (1 - y) x.$$

(e) We have $[D^2 - 2DD' + (D')^2]z = 2x \cos y.$

We write $\qquad [D^2 - 2DD' + (D'^2)] = (D - D')^2 = (D')^2 [1 - (D')^{-1}D]^2$

The particular integral is given by

$$z = (D')^{-2}[1 - (D')^{-1}D]^{-2}(2x \cos y)$$

$$= (D')^{-2}[1 + 2(D')^{-1}D + 3(D')^{-2}D^2 + \ldots] (2x \cos y)$$

$$= (D')^{-2}[2x \cos y + 2(D')^{-1}(2 \cos y)] = (D')^{-2} [2x \cos y + 4 \sin y]$$

$$= -2x \cos y - 4 \sin y = -2 (x \cos y + 2 \sin y).$$

Case 4 Let $f(x, y)$ be not of any one of the forms as given in the above cases or the case of failure in *case 2*.

Assume that $F(D, D')$ is reducible, that is, it can be factorised. Then, we can use the procedure adopted in solving Eq. (16.70) and the case of multiple factors. Now, consider one of the factors as

$$(a_1 D + b_1 D') z = a_1 p + b_1 q = f(x, y). \tag{16.96}$$

Since, this is a Lagrange's equation, the auxiliary equations are given by

$$\frac{dx}{a_1} = \frac{dy}{b_1} = \frac{dz}{f(x, y)}.$$

The first two terms give the solution as

$$b_1 x - a_1 y = c, \ c \text{ arbitrary constant.} \tag{16.97}$$

Consider now, the first and third terms

$$\frac{dx}{a_1} = \frac{dz}{f(x, y)} = \frac{dz}{f[x, (b_1 x - c)/a_1]}$$

or $\qquad a_1 \, dz = f[x, (b_1 x - c)/a_1] \, dx.$

Integrating, we get

$$a_1 z = \int f\left[x, \frac{b_1 x - c}{a_1}\right] dx + c_1 = F(x, c) + c_1$$

where c_1 is an arbitrary constant. After integration, we replace c by $b_1 x - a_1 y$ as given in Eq. (16.97). Since a particular integral is required, we set $c_1 = 0$. Therefore,

$$z = \frac{1}{a_1} F(x, b_1 x - a_1 y). \tag{16.98}$$

Hence, $\qquad z = [a_1 D + b_1 D']^{-1} f(x, y)$

$$= \frac{1}{a_1} \int f\left[x, \frac{b_1 x - c}{a_1}\right] dx = \frac{1}{a_1} F(x, b_1 x - a_1 y). \tag{16.99}$$

We repeat the procedure for each factor to obtain the required particular integral.

Example 16.30   Find the solution of the following differential equations

(a) $[2D^2 + 5DD' + 3(D')^2]z = ye^x$,   (b) $[D^2 + DD' - 2(D')^2]z = 8 \ln(x + 5y)$.

Solution

(a) We write

$$[2D^2 + 5DD' + 3(D')^2]z = (2D + 3D')(D + D')z = ye^x.$$

Using Eq. (16.74), we obtain the complementary function as

$$z = \phi_1(3x - 2y) + \phi_2(x - y).$$

The particular integral is given by

$$z = (2D + 3D')^{-1}(D + D')^{-1}(ye^x).$$

We first obtain $(D + D')^{-1} (ye^x)$ as in *case* 4. For the sake of completeness, we repeat the procedure used in this case. Denote

$$u = (D + D')^{-1} (ye^x) \text{ or } (D + D') u = ye^x.$$

The auxiliary equations are

$$\frac{dx}{1} = \frac{dy}{1} = \frac{du}{ye^x}.$$

The first two terms give $y = x + c$. Using the first and third terms, we get

$$\frac{dx}{1} = \frac{du}{(x + c)e^x}$$

and $\qquad u = \displaystyle\int (x + c)e^x \, dx = (x + c - 1)e^x = (y - 1)e^x.$

Now, denote $\qquad z = (2D + 3D')^{-1}u = (2D + 3D')^{-1} (y - 1) e^x$

or $\qquad (2D + 3D') z = (y - 1)e^x.$

The auxiliary equations are

$$\frac{dx}{2} = \frac{dy}{3} = \frac{dz}{(y - 1)e^x}.$$

The first two terms give $2y = 3x + c_1$. The first and third terms give

$$\frac{dx}{2} = \frac{dz}{[\{(3x + c_1)/2\} - 1]e^x}$$

and $\qquad z = \dfrac{1}{4} \displaystyle\int (3x + c_1 - 2)e^x \, dx = \dfrac{1}{4}(3x + c_1 - 5)e^x = \dfrac{1}{4}(2y - 5)e^x$

which is the required particular integral. The general solution of the differential equation is

$$z = \phi_1(3x - 2y) + \phi_2(x - y) + \frac{1}{4}(2y - 5)e^x.$$

(b) We write $[D^2 + DD' - 2(D')^2] z = (D + 2D') (D - D') z = 8 \ln (x + 5y).$
Using Eq. (16.74), we obtain the complementary function as

$$z = \phi_1(2x - y) + \phi_2(x + y).$$

The particular integral is given by

$$z = (D + 2 D')^{-1} (D - D')^{-1} (8 \ln (x + 5y))$$

Denote $\qquad u = (D - D')^{-1} (8 \ln (x + 5y)), \text{ or } (D - D')u = 8 \ln (x + 5y).$

The auxiliary equations are

$$\frac{dx}{1} = \frac{dy}{-1} = \frac{du}{8 \ln(x + 5y)}.$$

The first and second terms give $x + y = c$. The first and third terms give

$$\frac{dx}{1} = \frac{du}{8\ln(5c - 4x)}$$

and

$$u = 8\int \ln(5c - 4x)dx = 8\left[x\ln(5c - 4x) + \int \frac{4x}{5c - 4x}dx\right]$$

$$= 8\left[x\ln(5c - 4x) - \int\left(1 + \frac{5c}{4x - 5c}\right)dx\right]$$

$$= 8\left[x\ln(5c - 4x) - x + \int \frac{5c}{5c - 4x}dx\right]$$

$$= 8\left[\left(x - \frac{5c}{4}\right)\ln(5c - 4x) - x\right]$$

$$= -2(x + 5y)\ln(x + 5y) - 8x.$$

Now, denote $z = (D + 2D')^{-1}u$

or

$$(D + 2D')z = u = -[8x + 2(x + 5y)\ln(x + 5y)].$$

The auxiliary equations are

$$\frac{dx}{1} = \frac{dy}{2} = \frac{dz}{-[8x + 2(x + 5y)\ln(x + 5y)]}.$$

The first two terms give $2x - y = c_1$. The first and third terms give

$$\frac{dx}{1} = \frac{dz}{-[8x + 2(11x - 5c_1)\ln(11x - 5c_1)]}$$

and

$$z = -\int\left[8x + 2(11x - 5c_1)\ln(11x - 5c_1)\right]dx$$

$$= -\left[4x^2 + \frac{2}{11}\left\{\frac{1}{2}(11x - 5c_1)^2\ln(11x - 5c_1) - \frac{1}{4}(11x - 5c_1)^2\right\}\right]$$

$$= -\left[4x^2 + \frac{1}{22}(11x - 5c_1)^2\left\{2\ln(11x - 5c_1) - 1\right\}\right]$$

$$= -\left[4x^2 + \frac{1}{22}(x + 5y)^2\left\{2\ln(x + 5y) - 1\right\}\right]$$

which is the required particular integral.
The general solution of the differential equation is

$$z = \phi_1(2x - y) + \phi_2(x + y) - \left[4x^2 + \frac{1}{22}(x + 5y)^2\left\{2\ln(x + 5y) - 1\right\}\right]$$

Exercise 16.4

Find the solutions of the following homogeneous partial differential equations.

1. $2r - 5s + 2t = 0$.

2. $[D^3 - D^2D' - 4D(D')^2 + (4D')^3] z = 0$.

3. $[2D^2 + 5DD' + 3(D')^2 + D + D']z = 0$.

4. $2r + 7s + 3t = 0$.

5. $2r - s - 6t = 0$.

6. $r - t + p - q = 0$.

7. $[D^2 - DD' - 2(D')^2 + 2D - 4D'] z = 0$.

8. $[D^2 - (D')^2 - 2D' - 1]z = 0$.

9. $[2D^2 + 3DD' + 2D - 2(D')^2 - D'] z = 0$.

10. $(4D^2 + 4D + 1) z = 0$.

11. $[4 (D')^2 + 12D' + 9] z = 0$.

12. $[D^3 - 3D^2D' + 3D (D')^2 + (D')^3] z = 0$.

13. $[D^3 + D^2D' - D (D')^2 - (D')^3]z = 0$.

14. $[4D^3 - 8D^2D' + 4D^2 - 8DD' + D - 2D'] z = 0$.

15. $[4D^4 - 4D^3D' - 3D^2(D')^2 + 2D (D')^3 + (D')^4] z = 0$.

Find the general solutions of the following partial differential equations.

16. $[D^2 + DD' - 2(D')^2]z = 5e^{x+2y}$.

17. $[3D^2 + 10DD' + 3(D')^2]z = e^{x-y}$.

18. $[D^2 + DD' - 2(D')^2 - D - 2D']z = 16 \, e^{2x-3y}$.

19. $[D^2 + 3DD' + 2D'^2]z = e^{x-y}$

20. $[D^2 + 5DD' + 6(D')^2 + D + 2D']z = 4e^{2x-y}$.

21. $[2D^2 - 5DD' + 3(D')^2 + D - D']z = 12e^{x+y}$.

22. $[2D^2 + 5DD' - 3(D')^2]z = \sin (2x - y)$.

23. $[D^2 + 3DD' + 2(D')^2]z = 84 \cos (x + 3y)$.

24. $[3D^2 + 7DD' + 2(D')^2]z = 3x^2 + 2y^2$.

25. $[2D^2 - 7DD' + 6(D')^2] z = xy$.

26. $[4D^2 + 4DD' + (D')^2]z = 4y \cos 2x$.

27. $[2D^2 + 7DD' + 6(D')^2]z = 144 \, x \sin 2y$.

28. $[D^2 - DD' - 2(D')^2]z = 16x \, e^{2y}$.

29. $[6D^2 + 5DD' - 6(D')^2]z = 132 \log (x + 3y)$.

30. $[D^2 - (D')^2]z = \cos (x + y)$.

16.3.5 Quasi-linear Second Order Equations: Monge's Method

Most of the nonlinear second order equations of the form

$$F (x, y, z, p, q, r, s, t) = 0 \qquad (16.100)$$

cannot be integrated exactly. Monge's method gives the solution of a quasi-linear second order equation of the form

$$R\frac{\partial^2 z}{\partial x^2} + S\frac{\partial^2 z}{\partial x \partial y} + T\frac{\partial^2 z}{\partial y^2} = V, \quad \text{or} \quad Rr + Ss + Tt = V \qquad (16.101)$$

where R, S, T, V are functions of x, y, z, p and q.

The method reduces the given equation (16.101) into an equivalent system of two equations from which we determine p or q or both p, q. If both p and q are determined, then we integrate $dz = pdx + qdy$ to find the solution. If p or q is determined, then, the solution is obtained following the procedure of solving a Lagrange's equation.

We have

$$dp = \frac{\partial p}{\partial x}dx + \frac{\partial p}{\partial y}dy = rdx + sdy, \quad dq = \frac{\partial q}{\partial x}dx + \frac{\partial q}{\partial y}dy = sdx + tdy. \qquad (16.102)$$

Multiply Eq. (16.101) by $dx\, dy$. We get

$$Rr\, dx\, dy + Ss\, dx\, dy + Tt\, dx\, dy = V\, dx\, dy.$$

Eliminate $r\, dx$ and $t\, dy$ using the equations (16.102). We have

$$r\, dx = dp - sdy, \quad tdy = dq - sdx.$$

Hence, $\qquad R\, dy\,(dp - s\, dy) + Ss\, dx\, dy + T\, dx\,(dq - sdx) = V\, dxdy$

or $\qquad (R\, dp\, dy + T\, dq\, dx - Vdx\, dy) + s\,(Sdx\, dy - Rdy^2 - Tdx^2) = 0.$

Since this equation holds for arbitrary values of s, we set the expressions in the first and second brackets to zero, that is

$$R\, dp\, dy + \;T\, dq\, dx - Vdx\, dy = 0 \qquad (16.103\text{i})$$

$$R\, dy^2 - S\, dy\, dx + T\, dx^2 = 0. \qquad (16.103\text{ii})$$

These equations are called *Monge's subsidiary equations*. Using these equations, we try to find one or two relations between x, y, z, p and q, each relation containing an arbitrary function. These relations are called the *intermediate integrals*. We solve these relations for p and q. Then, we integrate

$$dz = p\, dx + q\, dy$$

to obtain the solution of the given partial differential equation.

Now, consider the equation (16.103ii). We can write is as

$$R\left(\frac{dy}{dx}\right)^2 - S\left(\frac{dy}{dx}\right) + T = 0$$

Solving, we get $\dfrac{dy}{dx} = \dfrac{1}{2R}[S \pm \sqrt{S^2 - 4RT}].$

We may get two distinct values for (dy/dx), in which case (16.103ii) can be factorised into distinct factors. We may use both the values of dy/dx to obtain the solution. We may also use only one of these values and then use the methods of solving the first order partial differential equations like Lagrange's equation, to obtain the solution.

If $S^2 - 4RT = 0$, we get only one value for dy/dx. In this case, (16.103ii) becomes a perfect square. We use the methods of solving the first order partial differential equations, to obtain the solution. If $R = 0$ or $T = 0$, then also we get only one value for dy/dx.

We illustrate the above procedure through the following examples.

Examples 16.31   Solve the partial differential equation

$$r - t\sin^2 x - p\cot x = 0.$$

Solution   We have $R = 1$, $T = -\sin^2 x$, $S = 0$ and $V = p\cot x$.

Monge's subsidiary equations become

$$R\, dy^2 - S\, dy\, dx + T\, dx^2 = 0, \quad \text{or} \quad dy^2 - \sin^2 x\, dx^2 = 0. \qquad (16.104\text{i})$$

$$R \, dp \, dy + T \, dq \, dx - V dx \, dy = 0, \quad \text{or} \quad dp \, dy - \sin^2 x \, dq \, dx - p \cot x \, dx dy = 0 \qquad \text{(16.104 ii)}$$

The solutions of the equations (16.104) are

$$dy = \sin x \, dx, \quad \text{or} \quad y = -\cos x + c_1 \qquad \text{(16.105i)}$$

and
$$dy = -\sin x \, dx, \text{or} \quad y = \cos x + c_2. \qquad \text{(16.105 ii)}$$

We shall illustrate both the procedures, that is, using both the relations given in Eq. (16.105) and using one of the relations given in Eq. (16.105), to obtain the solution.

Substituting Eq. (16.105i) in Eq. (16.104ii), we get

$$\sin x \, dp \, dx - \sin^2 x \, dq \, dx - p \cot x \sin x \, (dx)^2 = 0$$

or
$$\sin x \, [dp - \sin x \, dq - p \cot x \, dx] \, dx = 0$$

or
$$dp - \sin x \, dq - p \cot x \, dx = 0.$$

or
$$\text{cosec } x \, dp - dq - p \cot x \, \text{cosec } x \, dx = 0$$

or
$$d \, [p \, \text{cosec } x] - dq = 0.$$

Integrating, we get $p \, \text{cosec } x - q = c_3.$ \hfill (16.106)

From, Eqs. (16.106) and (16.105i), we get

$$p \, \text{cosec } x - q = f \, (y + \cos x). \qquad \text{(16.107)}$$

Now, we use Eq. (16.105 ii) in Eq. (16.104 ii). We get

$$-\sin x \, dp \, dx - \sin^2 x \, dq \, dx + p \cot x \sin x \, (dx)^2 = 0$$

or
$$-\sin x \, [dp + \sin x \, dq - p \cot x \, dx] \, dx = 0$$

or
$$dp + \sin x \, dq - p \cot x \, dx = 0$$

or
$$\text{cosec } x \, dp + dq - p \cot x \, \text{cosec } x \, dx = 0$$

or
$$d \, [p \, \text{cosec } x] + dq = 0.$$

Integrating, we get $p \, \text{cosec } x + q = c_4.$ \hfill (16.108)

From, Eqs. (16.108) and (16.105 ii), we get

$$p \, \text{cosec } x + q = f \, (y - \cos x). \qquad \text{(16.109)}$$

Solving Eqs. (16.107) and (16.109), we get

$$p = \frac{1}{2} \sin x \, [f(y + \cos x) + g(y - \cos x)]$$

and
$$q = \frac{1}{2} [g(y - \cos x) - f(y + \cos x)].$$

Now, $dz = pdx + q \, dy$ gives

$$dz = \frac{1}{2} \sin x \, [f(y + \cos x) + g(y - \cos x)] dx + \frac{1}{2} [g(y - \cos x) - f(y + \cos x)] dy$$

$$= \frac{1}{2} f(y + \cos x)[\sin x \, dx - dy] + \frac{1}{2} g(y - \cos x)[\sin x \, dx + dy]$$

$$= -\frac{1}{2} f(y + \cos x) \, d[y + \cos x] + \frac{1}{2} g(y - \cos x) \, d[y - \cos x].$$

Integrating, we get

$$z = \phi (y + \cos x) + \psi (y - \cos x).$$

Suppose now, that we use only one of the relations given in Eq. (16.105). Consider equation (16.105i). We can solve the equation, $p \csc x - q = f(y + \cos x)$, using the Lagrange's method.

We have

$$\frac{dx}{\csc x} = \frac{dy}{-1} = \frac{dz}{f(y + \cos x)}.$$

The first two terms give

$$\sin x \, dx = -dy, \quad \text{or} \quad y = \cos x + d_1. \tag{16.110i}$$

The second and third terms give

$$\frac{dy}{-1} = \frac{dz}{f(2y - d_1)}, \quad \text{(using (16.110i))}.$$

We have $\qquad\qquad f(2y - d_1) \, dy + dz = 0.$

Integrating, we get

$$\frac{1}{2} F(2y - d_1) + z = d_2^* \quad \text{or} \quad F(2y - d_1) + 2z = d_2.$$

or $\qquad\qquad\qquad F(y + \cos x) + 2z = d_2 \tag{16.110 ii}$

where $F = \int f \, dy$.

Using Eqs. (16.110 ii) and (16.109), we get the solution as

$$F(y + \cos x) + 2z = \phi (y - \cos x), \quad \text{or} \quad z = \psi_1 (y + \cos x) + \psi_2 (y - \cos x).$$

Example 16.32   Solve the partial differential equation

$$x^2 r + 2xys + y^2 t = 0.$$

Solution   We have $R = x^2$, $S = 2xy$, $T = y^2$ and $V = 0$.

The first subsidiary equation is given by

$$R \, dy^2 - S \, dy \, dx + T \, dx^2 = 0, \quad \text{or} \quad x^2 \, dy^2 - 2 \, xy \, dy \, dx + y^2 dx^2 = 0$$

or $\qquad\qquad\qquad (x \, dy - y \, dx)^2 = 0.$

We have one equation $x \, dy - y \, dx = 0$, whose solution is $y = ax$.

The second subsidiary equation is given by

$$R \, dp \, dy + T \, dq \, dx - V \, dy \, dx = 0, \quad \text{or} \quad x^2 \, dp \, dy + y^2 dq \, dx = 0.$$

Substituting $y = ax$, and $x \, dy = y \, dx$, we get

$$(x \, dp) \, (y \, dx) + a^2x^2 \, dq \, dx = 0, \quad \text{or} \quad ax^2 \, [dp + a \, dq] \, dx = 0$$

or

$$dp + a \, dq = 0.$$

Integrating, we get

$$p + aq = c, \quad \text{or} \quad p + \left(\frac{y}{x}\right)q = c.$$

Hence, from $c = f(a)$, we obtain $p + \left(\frac{y}{x}\right)q = f\left(\frac{y}{x}\right)$ or $xp + yq = xf\left(\frac{y}{x}\right)$.

The auxiliary equations are

$$\frac{dx}{x} = \frac{dy}{y} = \frac{dz}{xf(y/x)}.$$

The first two terms give $y = d_1x$.
The first and third terms give

$$\frac{dx}{x} = \frac{dz}{xf(d_1)}, \quad \text{or} \quad f(d_1) \, dx = dz.$$

Integrating, we get

$$z = xf(d_1) + d_2 = xf(y/x) + d_2.$$

Hence, we obtain the solution as $d_2 = g(d_1)$

$$z - xf(y/x) = g(y/x), \quad \text{or} \quad z = g(y/x) + xf(y/x).$$

Example 16.33 Solve the partial differential equation

$$rq^2 - 2psq + tp^2 = qr - ps.$$

Solution The given equation is

$$rq(q - 1) + ps(1 - 2q) + tp^2 = 0.$$

We have $R = q(q - 1)$, $S = p(1 - 2q)$, $T = p^2$ and $V = 0$.
The subsidiary equations become

$$R \, dy^2 - S \, dy \, dx + T \, dx^2 = 0, \quad \text{or} \quad q(q - 1) \, dy^2 - p(1 - 2q) \, dy \, dx + p^2dx^2 = 0$$

and

$$R \, dp \, dy + T \, dq \, dx - V \, dx \, dy = 0, \quad \text{or} \quad q(q - 1) \, dp \, dy + p^2 \, dq \, dx = 0. \tag{16.111}$$

We have

$$S^2 - 4RT = p^2(1 - 2q)^2 - 4p^2q(q - 1) = p^2.$$

and

$$\frac{dy}{dx} = \frac{p(1 - 2q) \pm p}{2q(q - 1)} = -\frac{p}{q}, \, -\frac{p}{q - 1}.$$

Therefore, the first equation can be factorised as

$$(p \, dx + q \, dy) \, [p \, dx + (q - 1) \, dy] = 0.$$

Hence, we have

$$p \, dx + q \, dy = 0 \tag{16.112}$$

and

$$p \, dx + (q - 1) \, dy = 0. \tag{16.113}$$

Using Eq. (16.112) and $dz = p \, dx + q \, dy = 0$, we get $z = a$. \tag{16.114}

Substituting Eq. (16.112) in Eq. (16.111), we get

$$q(q-1)dp\,dy - qp\,dq\,dy = 0 \quad \text{or} \quad \frac{dp}{p} - \frac{dq}{q-1} = 0.$$

Integrating, we get
$$\frac{p}{q-1} = c. \tag{16.115}$$

Now, Eqs. (16.114) and (16.115) give $\dfrac{p}{q-1} = f(z)$, or $p - qf(z) = -f(z)$.

This is a Lagrange's equation. The auxiliary equations are given by

$$\frac{dx}{1} = \frac{dy}{-f(z)} = \frac{dz}{-f(z)}.$$

The second and third terms give $y = z + c_1$.   (16.116)
The first and third terms give

$$x = -\int \frac{dz}{f(z)} + c_2 = -F(z) + c_2. \tag{16.117}$$

Hence, from Eqs. (16.116) and (16.117), the required solution is given by
$$x + F(z) = \phi(y-z).$$

Example 16.34   Find the solution of the partial differential equation
$$qr - ps = p^3.$$

Solution   We have $R = q$, $S = -p$, $T = 0$ and $V = p^3$.
The subsidiary equations become

$$R\,dy^2 - S\,dy\,dx + T\,dx^2 = 0, \quad \text{or} \quad q\,dy^2 + p\,dy\,dx = 0, \quad \text{or} \quad pdx + q\,dy = 0 \tag{16.118}$$

and $R\,dp\,dy + T\,dq\,dx - V\,dx\,dy = 0$, or $q\,dp\,dy - p^3\,dx\,dy = 0$, or $q\,dp - p^3 dx = 0$. (16.119)

From Eq. (16.118), we get $dz = pdx + q\,dy = 0$, or $z = c_1$.   (16.120)

Substituting $p\,dx = -q\,dy$ in Eq. (16.119), we get

$$qdp + p^2 qdy = 0, \quad \text{or} \quad dp + p^2 dy = 0, \quad \text{or} \quad (1/p^2)\,dp + dy = 0.$$

Integrating, we get $(-1/p) + y = c_2$.   (16.121)
    From Eqs. (16.120) and (16.121), we get

$$y - \frac{1}{p} = f(z), \quad \text{or} \quad p[y - f(z)] = 1.$$

This is a Langrange's equation. The auxiliary equations are given by

$$\frac{dx}{y - f(z)} = \frac{dy}{0} = \frac{dz}{1}.$$

The second and third terms give $y = d_1$.   (16.122)
The first and third terms give

$$\frac{dx}{d_1 - f(z)} = \frac{dz}{1}, \quad \text{or} \quad dx = [d_1 - f(z)]dz.$$

Integrating, we get $x = d_1 z - F(z) + d_2$, or $x = yz - F(z) + d_2$ \hfill (16.123)

where $\int f(z)\, dz = F(z)$.

From Eq. (16.122) and (16.123), we obtain the solution as

$$x - yz + F(z) = G(y), \quad \text{or} \quad x = yz - F(z) + G(y).$$

Remarks 5

1. Reduction of the variable coefficient second order partial differential equation to its canonical form is given Appendix 3.
2. Classification of the second order variable coefficient partial differential equations is given in section 9.5.1.
3. Fourier series solutions (method of separation of variables) of the heat conduction equation, wave equation and Laplace equation are given in sections 9.5.2 to 9.5.5.
4. Fourier transform solution of some partial differential equations is given in section 9.6.1.
5. Laplace transform method for the solution of some partial differential equations is given in section 8.7.

Exercise 16.5

Find the solution of the following partial differential equations.

1. $r = 25t$.
2. $r - (\sec^4 x)\, t = 2p \tan x$.
3. $(\cos^3 y)\, r - (\cos y)\, t = q \sin y$.
4. $q^2 r - 2pq\, s + p^2 t = p^2 q$.
5. $r - 2ys + y^2 t + p = 0$.
6. $y^2 r - 2ys + t = p$.
7. $q(q+1)r + p^2 t - p(1+2q)s = 0$.
8. $yqs - ypt = pq$.
9. $z(ps - qr) = p^2 q$.
10. $(q+1)r = (p+1)s$.

16.4   Answers and Hints

Exercise 16.1

1. $z = pq$.
2. $z = px + qy + p^4 + q^4$.
3. $(p^2 + q^2 + 1)\, y^2 = 16q^2$.
4. $2z(1-y) = px\,(2-y) + qy\,(1-y)$.
5. $z_{xx} + z_{tt} = 0$.
6. $4z = 4px + 4qy - p^2$.
7. $p = aq$.
8. $py + qx = x^2 + y^2$.
9. $px + qy = 0$.
10. $z_{xx} + z_{yy} = 0$.
11. $p(y-z) + q(z-x) = x - y$.
12. $z(p+2q) + x + 2y = 0$.
13. $px(y-z) + qy(z-x) = z(x-y)$.
14. $z(px-qy) = y^2 - x^2$.
15. $p(y^2 + z^2) - q(z+xy) = xz - y$.

Exercise 16.2

In the solution of all the problems the multipliers (a, b, c) means $[a\, dx + b\, dy + c\, dz]/[aP + bQ + cR]$.

1. $\phi\,[(x-2y),\ \{3z + \sin(x-2y)\}e^{-3x/2}] = 0$.

2. multipliers: $(1, 1, 1)$ and $(1/x, 1/y, 1/z)$; $\phi\,[x + y + z, xyz] = 0$.

3. multipliers; $(1, 1, 1)$, $(1, -1, 0)$ and $(0, 1, -1)$;

$\phi\,[(x - y)^2\,(x + y + z),\,(y - z)/(x - y)] = 0$.

4. multiplier: (x, y, z);

$$\phi\left[\frac{1}{x}\,(x^2 + y^2 + z^2),\,\frac{x}{y}\right] = 0,\quad \text{or } x^2 + y^2 + z^2 = xf\,(x/y).$$

5. multipliers: $(-1, 2y, -1)$ and $(-2x, z, y)$; $\phi\,[y^2 - x - z,\,yz - x^2] = 0$.

6. multipliers: $(x, -y, z)$ and $(1/x, 1/y, 1/z)$; $\phi\,[x^2 - y^2 + z^2,\,xyz] = 0$.

7. multipliers: (x, y, z), $(x, -y, 0)$ and $(z, 0, x)$; $\phi\,[x^2 + y^2 + z^2,\,x^2 - y^2 - 2xz] = 0$.

8. multipliers: $(x, -y, 1/2)$ and $(y, x + z, y)$; $\phi\,[x^2 - y^2 + z,\,xy + yz] = 0$.

9. multipliers: $(1, 1, 0)$ and $(x^2, -y^2, 0)$; $\phi\left[x^3 - y^3,\,\dfrac{x + y}{z}\right] = 0$.

10. multipliers: $(x^2, y^2, 1/3)$ and $(1/x, 1/y, 1/z)$; $\phi\,[x^3 + y^3 + z,\,xyz] = 0$.

11. First two terms give $\dfrac{dx}{dy} + \dfrac{x}{y} = y$. Solving it, we get $3xy = y^3 + c_1$.

Last two terms give $\dfrac{dz}{dy} - \dfrac{z}{y} = \dfrac{1}{3}\left(y^2 + \dfrac{c_1}{y}\right)$. Solving it, we get $z = \dfrac{1}{6}y^3 - \dfrac{1}{3}c_1 + c_2 y$, or

$c_2 = \dfrac{1}{2y}\,(2z - y^3 + 2xy)$; $\phi\left[3xy - y^3,\,\dfrac{1}{y}\,(2z - y^3 + 2xy)\right] = 0$.

12. (a) $y^2 - x^2 + yz - x - z + 2 = 0$,   (b) $y^2 - x^2 + yz - x - z + 1 = 0$.

13. $2x^2 + 4y^2 + 3z^2 + 2xz = 1$.

14. $\phi\left[\dfrac{z}{x},\,\dfrac{x^2 + y^2 + z^2}{z}\right] = 0$; $x^2 + y^2 + z^2 = 5z$.

15. $\phi\left[\dfrac{y}{x},\,x^2 + y^2 + z^2\right] = 0$; $x^2 + y^2 + z^2 = 18$.

Exercise 16.3

1. $z = ax - [2a/(3 + 4a)]y + b$.

2. $z = a^2 x - [(1 - a)/3]^2 y + b$.

3. $4cz = (ax + y + b)^2$, $c = (1 + a^2)/5$.

4. $(az - 1)b = e^{ax+y}$.

5. $z\sqrt{a^2 + z^2} + a^2 \ln\left|z + \sqrt{z^2 + a^2}\,\right| = 2(ax + y + b)$.

6. $2z = 4a\sqrt{x} + a^2(y^2 + 2\ln|y|) + b$.

7. $15z = 15a^2 x + 20ax^3 + 12x^5 + 15a^2 y - 15ay^2 + 5y^3 + b$.

8. $(1 + a^3)z^2 = 64(ax + y + b)^3$.

9. Use the transformation $x = e^u$, $y = e^v$, $z = e^w$, $z = bx^{a/2}y^{(\sqrt{1 - a^2})/3}$.

10. $z = ax + by + a/(a + b)$.

11. $z^2 = 2x\sqrt{x^2 + A^2} + 2A^2 \ln |x + \sqrt{x^2 + A^2}| + 3y\sqrt{y^2 - B^2} - 3B^2 \ln |y + \sqrt{y^2 - B^2}| + b$

where $A^2 = a^2/32$, $B^2 = a^2/108$.

12. $27z^{3/2} = (9x + 4a^2)^{3/2} + (9y - 4a^2)^{3/2} + b$.   13. $z = ax^2 + by^2 + 8ab$.

14. Write $Pu + Qv = f(P, Q)$; $x = u^3$, $y = v^2$; $z = ax^{1/3} + by^{1/2} - 8/b$.

15. $z = ay + b(x^2 - a)$.   16. $z = a^2 + by^2 + (ax/y)$.

17. $z = ay + be^x(a + x)^{-a}$.   18. $z^2 = a^2x^2 + (ay + b)^2$.

19. $z = yx^2$.   20. $4z = -(x^2 + y^2)$.

21. $z = au + bv + 6ab = ax^2 + by^2 + 6ab$; $6z = -x^2y^2$.

22. $xy = z(x - 2)$.

Exercise 16.4

1. $z = \phi(2x + y) + \psi(x + 2y)$.   2. $z = \phi_1(x + y) + \phi_2(2x - y) + \phi_3(2x + y)$.

3. $z = \phi_1(x - y) + e^{-x/2}\phi_2(3x - 2y)$.   4. $z = \phi_1(x - 2y) + \phi_2(3x - y)$.

5. $z = \phi_1(2x + y) + \phi_2(3x - 2y)$.   6. $z = \phi_1(x + y) + e^{-x}\phi_2(x - y)$.

7. $z = \phi_1(2x + y) + e^{-2x}\phi_2(x - y)$.   8. $z = e^x\phi_1(x + y) + e^{-x}\phi_2(x - y)$.

9. $z = \phi_1(x + 2y) + e^{-x}\phi_2(2x - y)$.   10. $z = e^{-x/2}[\phi_1(2y) + x\phi_2(2y)]$.

11. $z = e^{-3y/2}[\phi_1(2x) + y\phi_2(2x)]$.   12. $z = \phi_1(x + y) + x\phi_2(x + y) + x^2\phi_3(x + y)$.

13. $z = \phi_1(x + y) + \phi_2(x - y) + x\phi_3(x - y)$.   14. $z = \phi_1(2x + y) + e^{-x/2}[\psi_1(2y) + x\psi_2(2y)]$.

15. $z = \phi_1(x + y) + x\phi_2(x + y) + \psi_1(x - 2y) + x\psi_2(x - 2y)$.

16. $z = \phi_1(x + y) + \phi_2(2x - y) - e^{x+2y}$.   17. $z = \phi_1(x - 3y) + \phi_2(3x - y) - \dfrac{1}{4}e^{x-y}$.

18. $z = \phi_1(2x - y) + e^x\phi_2(x + y) - e^{2x-3y}$.

19. C.F: $z = \phi_1(x - y) + \phi_2(2x - y)$.   P.I: $-xe^{x-y}$, or $-ye^{x-y}$.

20. C.F: $z = \phi_1(2x - y) + e^{-x}\phi_2(3x - y)$,   P.I: $2x^2e^{2x-y}$, or $(y^2/3)e^{2x-y}$.

21. C.F: $z = \phi_1(x + y) + e^{-x/2}\phi_2(3x + 2y)$.   P.I: $3x^2 e^{x+y}$, or $2y^2e^{x+y}$.

22. $z = \phi_1(3x - y) + \phi_2(x + 2y) + \dfrac{1}{5}\sin(2x - y)$.

23. $z = \phi_1(x - y) + \phi_2(2x - y) - 3\cos(x + 3y)$.

24. $z = \phi_1(2x - y) + \phi_2(x - 3y) + \dfrac{x^2}{324}[113x^2 + 108y^2 - 168xy]$.

25. $z = \phi_1(2x + y) + \phi_2(3x + 2y) + \dfrac{1}{96}x^3(8y + 7x)$.

26. $z = \phi_1(x - 2y) + x\phi_2(x - 2y) + \dfrac{1}{8}[\sin(2x) - 2y\cos(2x)]$.

27. $z = \phi_1(2x - y) + \phi_2(3x - 2y) - \dfrac{1}{2}[12x\sin(2y) + 7\cos(2y)]$.

28. $z = \phi_1(x - y) + \phi_2(2x + y) + \dfrac{1}{2}(1 - 4x)e^{2y}$.

29. $z = \phi_1(2x + 3y) + \phi_2(3x - 2y) - [11x^2 + (x + 3y)^2 \{2 \ln (x + 3y) - 1\}].$

30. $z = \phi_1 (x + y) + \phi_2 (x - y) + \dfrac{1}{4} [2x \sin (x + y) + \cos (x + y)].$

Exercise 16.5

1. $z = F (y - 5x) + G (y + 5x).$

2. $z = F (y - \tan x) + G(y + \tan x).$

3. $z = F (x + \sin y) + G(x - \sin y).$

4. $x - f (z) = e^y g(z),$ or as $e^y = x\phi (z) + \psi(z).$

5. $z + f (e^x y) e^{-x} = g(e^x y),$ or $z = e^{-x} f (x + \ln y) + g (x + \ln y),$

or $z = yf (x + \ln y) + g (x + \ln y).$

6. $z = - yf (y^2 + 2x) + g (y^2 + 2x).$

7. $x = g (z) + \phi (y + z).$

8. $\ln y = G (x) + F (z).$

9. $x = z \, \phi (y) + \psi (z).$

10. $z = \phi (y) + \psi (x + y + z).$

Chapter 17

Z Transformation

17.1 Introduction

In the area of digital signal processing in Electrical engineering, we encounter sequences of discrete signals. If we represent a sequence in signal processing as $\{f_n\}$, then $f_n = (nT)$ where $n = 0, 1, 2, \ldots$ is a non-negative integer and T is called the sampling period. Therefore, the elements of the sequence $\{f_n\}$ are functions of discrete non-negative integer valued arguments. Such sequences are also encountered in many application areas of digital filters. The operation of such discrete systems is governed by difference equations. One of the important tools for solving the difference equations is Z transforms. The Z transform of a sequence $\{f_n\}$ can be defined as a functional transformation of sequences. We usually define the Z transform of a sequence $\{f_n\}$, $n \geq 0$, where f_n is a real or complex number.

Z transforms have properties similar to Laplace transforms. For almost every result in Laplace transform, there is a corresponding result in Z transform.

17.2 Basic Theory of Z Transforms

Let IN be the set of non-negative integers and $\{f_n\} = \{f_0, f_1, \ldots, f_n, \ldots\}$ be an infinite sequence, where $f_n = f(n)$, $n \in$ IN is the general term of the sequence. We assume that $f_n = 0$ for $n < 0$. The elements of the sequence may be real or complex numbers. Therefore, f_n is a function of the discrete integer valued (non-negative) arguments. The Z transform of the sequence $\{f_n\}$ is defined as

$$Z\{f_n\} = \sum_{n=0}^{\infty} f_n z^{-n} = F(z), \tag{17.1}$$

whenever the series is convergent. Note that z may be real or complex. The Z transform is a series in powers of $(1/z)$. We may also call it a power series in terms of the variable $(1/z)$, that is a Laurent series. The region of convergence of the series given in Eq. (17.1) is the set of values of z for which the series is convergent. This region of convergence can be determined by using the tests of convergence as discussed in section 12.4. For a given sequence $\{f_n\}$, there exists a number R such that the series

$\sum_{n=0}^{\infty} f_n z^{-n}$ converges for all z such that $|1/z| < R$, that is, $|z| > 1/R$ and diverges for all $|z| < 1/R$.

The Z transform $F(z)$, valid in its region of convergence, is unique.

The *inverse Z transform* is defined as

$$f_n = Z^{-1}[F(z)].$$ (17.2)

Given a function $F(z)$, we write it as a sum of factors. The inverse Z transform is then obtained by finding the inverse Z transform of each of these factors,

Remark 1

It is possible to consider Z transforms of the sequences $\{f_n\}$ defined on integer valued arguments, that is, the arguments can take values $n = \ldots, -2, -1, 0, 1, 2, \ldots$. In this case, we call $\{f_n\}$ as a two sided sequence and the corresponding transform as two sided Z transform.

Z transforms of some standard sequences

Using the definition, we now obtain the Z transforms of some standard sequences defined on the non-negative set of values $n = 0, 1, 2, \ldots$

1. Let $f_0 = 1, f_n = 0, n = 1, 2, \ldots$ (Fig. 17.1). A sequence of this type plays a role similar to Dirac-delta function $\delta(t)$ of continuous argument t. Then,

$$Z\{f_n\} = 1$$ (17.3i)

and $Z^{-1}[1] = f_n$, such that $f_0 = 1, f_n = 0, n = 1, 2, \ldots$ (17.3ii)

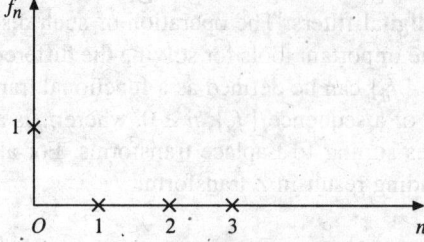

Fig. 17.1.   The sequence $\{f_n\}$.

2. Let $f_n = 1, n = 0, 1, 2, \ldots$ (Fig. 17.2). This sequence is also called the unit sequence or the discrete unit step sequence. Then,

$$Z\{1\} = \sum_{n=0}^{\infty} z^{-n} = 1 + \frac{1}{z} + \frac{1}{z^2} + \ldots = \frac{1}{1 - (1/z)} = \frac{z}{z-1}.$$ (17.4i)

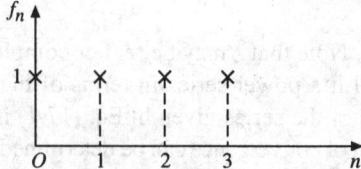

Fig. 17.2 The sequence $\{f_n\}$.

The region of convergence is $|1/z| < 1$, or $|z| > 1$ and

$$Z^{-1}\left[\frac{z}{z-1}\right] = 1.$$ (17.4 ii)

3. Let $f_n = a^n$, where a is any real or complex number.

Then,
$$Z\{a^n\} = \sum_{n=0}^{\infty} a^n z^{-n} = \sum_{n=0}^{\infty} \left(\frac{a}{z}\right)^n = \frac{1}{1-(a/z)} = \frac{z}{z-a}. \qquad (17.5\text{i})$$

The region of convergence is $|a/z| < 1$, or $|z| > a$ and

$$Z^{-1}\left[\frac{z}{z-a}\right] = a^n. \qquad (17.5\text{ii})$$

When a is a negative real number $a = -p$, $p > 0$, we have

$$Z\{(-p)^n\} = \frac{z}{z+p} \qquad (17.6\text{i})$$

and
$$Z^{-1}\left[\frac{z}{z+p}\right] = (-p)^n. \qquad (17.6\text{ii})$$

4. Let $f_n = a^{-n}$, where a is any real or complex number. Then,

$$Z\{a^{-n}\} = \sum a^{-n} z^{-n} = \sum_{n=0}^{\infty} \left(\frac{1}{az}\right)^n = \frac{1}{1-1/(az)} = \frac{az}{az-1}. \qquad (17.7\text{i})$$

The region of convergence is $|1/(az)| < 1$ or $|z| > 1/|a|$ and

$$Z^{-1}\left[\frac{az}{az-1}\right] = a^{-n}. \qquad (17.7\text{ii})$$

We could have also obtained this result by replacing a by $1/a$ in Eq. (17.5i)

5. Let $f_n = e^{an}$, where a is any real or complex number. Then,

$$Z\{e^{an}\} = \sum_{n=0}^{\infty} e^{an} z^{-n} = \sum_{n=0}^{\infty} \left(\frac{e^a}{z}\right)^n = \frac{1}{1-(e^a/z)} = \frac{z}{z-e^a}. \qquad (17.8\text{i})$$

The region of convergence is $|e^a/z| < 1$ or $|z| > |e^a|$ and

$$Z^{-1}\left[\frac{z}{z-e^a}\right] = e^{an}. \qquad (17.8\text{ii})$$

We could have also obtained this result by replacing a by e^a in Eq. (17.5i).

6. Let $f_n = \dfrac{a^n}{n!}$, where a is any real or complex number. Then

$$Z\left\{\frac{a^n}{n!}\right\} = \sum_{n=0}^{\infty} \frac{a^n z^{-n}}{n!} = 1 + \frac{1}{1!}\left(\frac{a}{z}\right) + \frac{1}{2!}\left(\frac{a}{z}\right)^2 + \ldots = e^{a/z}. \qquad (17.9\text{i})$$

The region of convergence is $|z| > 0$ and

$$Z^{-1}[e^{a/z}] = \frac{a^n}{n!}. \qquad (17.9\text{ii})$$

For $a = 1$, we have

$$Z\left\{\frac{1}{n!}\right\} = e^{1/z} \quad \text{and} \quad Z^{-1}[e^{1/z}] = \frac{1}{n!}.$$

7. Let $f_n = \dfrac{e^{an}}{n!}$. Then

$$Z\left\{\frac{e^{an}}{n!}\right\} = \sum_{n=0}^{\infty} \frac{e^{an} z^{-n}}{n!} = \sum_{n=0}^{\infty} \frac{1}{n!}\left(\frac{e^a}{z}\right)^n = e^{e^a/z}. \tag{17.10i}$$

The region of convergence is $|z| > 0$ and

$$Z^{-1}[e^{e^a/z}] = \frac{e^{an}}{n!}. \tag{17.10ii}$$

We could have also obtained this result by replacing a by e^a in Eq. (17.9i)

Example 17.1   Determine Z transform, $Z\{f_n\}$, where f_n is defined as follows

(i) 2^n,   (ii) $(-3)^n$,   (iii) e^{-3n},   (iv) $\dfrac{e^{-n}}{n!}$.

Solution

(i) Substituting $a = 2$ in Eq. (17.5i), we get $Z\{2^n\} = \dfrac{z}{z-2}$.

(ii) Substituting $a = -3$ in Eq. (17.5i) or $p = 3$ in (17.6i), we get $Z\{(-3)^n\} = \dfrac{z}{z+3}$.

(iii) Substituting $a = -3$ in Eq. (17.8i), we get $Z\{e^{-3n}\} = \dfrac{ze^3}{ze^3 - 1}$.

(iv) Substituting $a = -1$ in Eq. (17.10i), we get $Z\left\{\dfrac{e^{-n}}{n!}\right\} = e^{1/(ez)}$.

Example 17.2   Let the sequence $\{f_n\}$ be defined as

(i) $f_n = 1$, $n \geq 1$, $f_0 = 0$

(ii) $f_n = \begin{vmatrix} 2, n = 0, 2, 4, \ldots, 2k, \ldots \\ 1, n = 1, 3, 5, \ldots, 2k+1, \ldots \end{vmatrix}$

Find $Z\{f_n\}$.

Solution

(i)
$$Z\{f_n\} = \sum_{n=0}^{\infty} f_n z^{-n} = f_0 + f_1 z^{-1} + f_2 z^{-2} + \ldots$$

$$= \frac{1}{z} + \frac{1}{z^2} + \ldots + \frac{1}{z^n} + \ldots = \frac{1}{z}\left[1 + \frac{1}{z} + \frac{1}{z^2} + \ldots\right]$$

$$= \frac{(1/z)}{1 - (1/z)} = \frac{1}{z-1}.$$

(ii)
$$Z\{f_n\} = \sum_{n=0}^{\infty} f_n z^{-n} = f_0 + f_1 z^{-1} + f_2 z^{-2} + \ldots + f_{2k} z^{-2k} + f_{2k+1} z^{-(2k+1)} + \ldots$$

$$= 2\left[1 + \frac{1}{z^2} + \frac{1}{z^4} + \ldots\right] + \left[\frac{1}{z} + \frac{1}{z^3} + \ldots\right]$$

$$= 2\left[\frac{1}{1 - (1/z^2)}\right] + \frac{(1/z)}{1 - (1/z^2)} = \frac{2z^2}{z^2 - 1} + \frac{z}{z^2 - 1}$$

$$= \frac{2z^2 + z}{z^2 - 1}.$$

Example 17.3 Find the inverse Z transform of $F(z)$, where $F(z)$ is given by

(i) $\dfrac{z}{z + 2}$, (ii) $\dfrac{3z}{3z - 1}$, (iii) $e^{-2/z}$, (iv) $e^{1/(2z)}$.

Solution

(i) $F(z) = z/(z + 2)$. Using Eq. (17.6ii) with $p = 2$, we get $Z^{-1}[z/(z + 2)] = (-2)^n$.

(ii) $F(z) = 3z/(3z - 1)$. Using Eq. (17.7ii) with $a = 3$, we get $Z^{-1}\{3z/(3z - 1)\} = 3^{-n}$.

(iii) $F(z) = e^{-2/z}$. Using Eq. (17.9ii) with $a = -2$, we get $Z^{-1}[e^{-2/z}] = \dfrac{(-2)^n}{n!}$.

(iv) $F(z) = e^{1/(2z)}$. Using Eq. (17.9ii) with $a = 1/2$, we get $Z^{-1}[e^{1/(2z)}] = \dfrac{2^{-n}}{n!}$.

Existence of Z transform

Let $\{f_n\}$ be a sequence of exponential type, that is, there exist numbers $N > 0$, $t_0 \geq 0$ and $n_0 \geq 0$, such that

$$|f_n| < N e^{nt_0} \quad \text{for all } n \geq n_0. \tag{17.11}$$

Now, we state the existence theorem.

Theorem 17.1 (Existence theorem) Every sequence $\{f_n\}$ of exponential type has the Z transform.
   This is a necessary and sufficient condition for the existence of Z transform.
We now prove some important results of Z transforms.

Theorem 17.2 (Linearity property) Let $\{f_n\}$ and $\{g_n\}$ be two sequences such that $Z\{f_n\} = F(z)$ and $Z\{g_n\} = G(z)$ with regions of convergence $|z| > (1/R_1)$ and $|z| > (1/R_2)$ respectively. Then, for any constants α, β

$$Z\{\alpha f_n + \beta g_n\} = \alpha Z\{f_n\} + \beta Z\{g_n\} = \alpha F(z) + \beta G(z)$$

for all $|z| > (1/R)$, where $(1/R) = \max[(1/R_1), (1/R_2)]$.

Proof Using the definition, we have

$$Z\{\alpha f_n + \beta g_n\} = \sum_{n=0}^{\infty} [\alpha f_n + \beta g_n] z^{-n} = \alpha \sum_{n=0}^{\infty} f_n z^{-n} + \beta \sum_{n=0}^{\infty} g_n z^{-n}$$

$$= \alpha F(z) + \beta G(z).$$

Remark 2

$$Z^{-1}[\alpha\, F\,(z) + \beta G\,(z)] = \alpha Z^{-1}[F\,(z)] + \beta Z^{-1}\,[g\,(z)] = \alpha f_n + \beta g_n.$$

Example 17.4 Obtain the Z transform of the sequence $\{f_n\}$, where f_n is given by

(i) $\cos(n\theta)$, \qquad\qquad (ii) $\sin(n\theta)$.

Hence, write the inverse Z transform.

Solution

We write $\cos(n\theta) = \mathrm{Re}\,(e^{in\theta})$ and $\sin(n\theta) = \mathrm{Im}\,(e^{in\theta})$, where Re and Im are the real and imaginary parts respectively.

Substituting $a = i\theta$ in Eq. (17.8i), we get

$$Z\{e^{in\theta}\} = \frac{z}{z - e^{i\theta}} = \frac{z}{(z - \cos\theta) - i\sin\theta} = \frac{z[(z - \cos\theta) + i\sin\theta]}{(z - \cos\theta)^2 + \sin^2\theta}$$

Hence, \qquad $Z\{\cos(n\theta) + i\sin(n\theta)\} = Z\{\cos(n\theta)\} + i\,Z\{\sin(n\theta)\}$

$$= \frac{z[(z - \cos\theta) + i\sin\theta]}{(z - \cos\theta)^2 + \sin^2\theta} = \frac{z[(z - \cos\theta) + i\sin\theta]}{z^2 - 2(\cos\theta)z + 1}$$

Comparing the real and imaginary parts on both sides, we obtain

(i)
$$Z\{\cos(n\theta)\} = \frac{z(z - \cos\theta)}{z^2 - 2(\cos\theta)z + 1}. \tag{17.12i}$$

(ii)
$$Z\{\sin(n\theta)\} = \frac{(\sin\theta)z}{z^2 - 2(\cos\theta)z + 1}. \tag{17.12ii}$$

The region of convergence is $|z| > |e^{i\theta}| = 1$ and

$$Z^{-1}\left[\frac{z(z - \cos\theta)}{(z - \cos\theta)^2 + \sin^2\theta}\right] = \cos(n\theta). \tag{17.12iii}$$

$$Z^{-1}\left[\frac{(\sin\theta)z}{(z - \cos\theta)^2 + \sin^2\theta}\right] = \sin(n\theta). \tag{17.12iv}$$

Example 17.5 Obtain the Z transform of the sequence $\{f_n\}$, where f_n is given by

(i) $\cosh(n\theta)$, \qquad\qquad (ii) $\sinh(n\theta)$.

Solution

(i) We write $\cosh(n\theta) = (e^{n\theta} + e^{-n\theta})/2$. Therefore, we get

$$Z\{\cosh(n\theta)\} = \frac{1}{2}[Z\{e^{n\theta}\} + Z\{e^{-n\theta}\}].$$

Setting $a = \theta$ and $a = -\theta$ in Eq. (17.8i), we obtain

$$Z\{\cosh(n\theta)\} = \frac{1}{2}\left[\frac{z}{z - e^{\theta}} + \frac{z}{z - e^{-\theta}}\right] = \frac{1}{2}\left[\frac{2z^2 - (e^{\theta} + e^{-\theta})z}{z^2 - (e^{\theta} + e^{-\theta})z + 1}\right]$$

$$= \frac{1}{2}\left[\frac{2z^2 - 2(\cosh\theta)z}{z^2 - 2(\cosh\theta)z + 1}\right] = \frac{z(z - \cosh\theta)}{z^2 - 2(\cosh\theta)z + 1}. \tag{17.13i}$$

(ii) We write $\sinh(n\theta) = (e^{n\theta} - e^{-n\theta})/2$. Therefore, we get

$$Z\{\sinh(n\theta)\} = \frac{1}{2}[Z\{e^{n\theta}\} - Z\{e^{-n\theta}\}].$$

Setting $a = \theta$ and $a = -\theta$ in Eq. (17.8i), we obtain

$$Z\{\sinh(n\theta)\} = \frac{1}{2}\left[\frac{z}{z - e^{\theta}} - \frac{z}{z - e^{-\theta}}\right] = \frac{1}{2}\left[\frac{(e^{\theta} - e^{-\theta})z}{z^2 - (e^{\theta} + e^{-\theta})z + 1}\right]$$

$$= \frac{(\sinh\theta)z}{z^2 - 2(\cosh\theta)z + 1}. \tag{17.13ii}$$

Example 17.6 Find the inverse Z transform of $F(z)$, where $F(z)$ is given by

(i) $\dfrac{z}{(z + 2)(z + 3)}$, (ii) $\dfrac{7z - 11z^2}{(z - 1)(z - 2)(z + 3)}$.

Solution We factorise $F(z)/z$, using partial fractions.

(i)

$$\frac{F(z)}{z} = \frac{1}{(z + 2)(z + 3)} = \frac{1}{z + 2} - \frac{1}{z + 3}$$

or

$$F(z) = \frac{z}{z + 2} - \frac{z}{z + 3}.$$

Therefore, using Eq. (17.6ii), we obtain

$$Z^{-1}[F(z)] = Z^{-1}\left[\frac{z}{z + 2}\right] - Z^{-1}\left[\frac{z}{z + 3}\right] = (-2)^n - (-3)^n.$$

(ii)

$$\frac{F(z)}{z} = \frac{7 - 11z}{(z - 1)(z - 2)(z + 3)} = \frac{A}{z - 1} + \frac{B}{z - 2} + \frac{C}{z + 3}.$$

Comparing both sides, we get

$$7 - 11z = A(z - 2)(z + 3) + B(z - 1)(z + 3) + C(z - 1)(z - 2).$$

Substituting $z = 1, 2$ and -3 respectively, we obtain $A = 1$, $B = -3$, $C = 2$. Hence, we have

$$\frac{F(z)}{z} = \frac{1}{z - 1} - \frac{3}{z - 2} + \frac{2}{z + 3} \quad \text{or} \quad F(z) = \frac{z}{z - 1} - \frac{3z}{z - 2} + \frac{2z}{z + 3}.$$

Therefore,

$$Z^{-1}[F(z)] = Z^{-1}\left[\frac{z}{z - 1}\right] - 3Z^{-1}\left[\frac{z}{z - 2}\right] + 2Z^{-1}\left[\frac{z}{z + 3}\right]$$

$$= 1 - 3(2)^n + 2(-3)^n.$$

Example 17.7   Find the inverse Z transform $Z^{-1}\left[\dfrac{z^2}{z^2+1}\right]$.

Solution   Let $F(z) = z^2/(z^2+1)$. Then,

$$\frac{F(z)}{z} = \frac{z}{z^2+1} = \frac{1}{2}\left[\frac{1}{z+i} + \frac{1}{z-i}\right] = \frac{1}{2}\left[\frac{1}{z-e^{-\pi i/2}} + \frac{1}{z-e^{\pi i/2}}\right].$$

Therefore,

$$Z^{-1}[F(z)] = \frac{1}{2}\left\{Z^{-1}\left[\frac{z}{z-e^{-\pi i/2}}\right] + Z^{-1}\left[\frac{z}{z-e^{\pi i/2}}\right]\right\}.$$

Using Eq. (17.8i), we obtain

$$Z^{-1}[F(z)] = \frac{1}{2}\left[e^{-n\pi i/2} + e^{n\pi i/2}\right] = \cos\left(\frac{n\pi}{2}\right).$$

Theorem 17.3   (**Translation theorem/shifting theorem**) Let $F(z)$ be the Z transform of the sequence $\{f_n\}$ valid in the region $|z| > (1/R)$ and k be a positive integer. Then,

(i) $\qquad\qquad Z\{f_{n-k}\} = z^{-k}F(z),\, n \geq k \qquad\qquad$ (*shifting to the right*) $\qquad$ (17.14i),

(ii) $\qquad\qquad Z\{f_{n+k}\} = z^k\left[F(z) - \sum_{n=0}^{k-1} f_n z^{-n}\right] \qquad$ (*shifting to the left*) $\qquad$ (17.14ii)

   valid in the region $|z| > (1/R)$.

Proof   We use the defintion of Z transform to prove the results.

(i) $\qquad\qquad Z\{f_{n-k}\} = \sum_{n=0}^{\infty} f_{n-k}z^{-n} = \sum_{n=k}^{\infty} f_{n-k}z^{-n},\,$ (since $f_m = 0, m < 0$).

   Writing $n - k = m$, we get

$$Z\{f_{n-k}\} = \sum_{m=0}^{\infty} f_m z^{-k-m} = z^{-k}\sum_{m=0}^{\infty} f_m z^{-m} = z^{-k}F(z).$$

(ii) $\qquad\qquad Z\{f_{n+k}\} = \sum_{n=0}^{\infty} f_{n+k}z^{-n} = z^k\sum_{n=0}^{\infty} f_{n+k}z^{-(n+k)}$

$$= z^k\left[\sum_{n=0}^{k-1} f_n z^{-n} + \sum_{m=n}^{\infty} f_m z^{-m} - \sum_{n=0}^{k-1} f_n z^{-n}\right]$$

$$= z^k\left[\sum_{n=0}^{\infty} f_n z^{-n} - \sum_{n=0}^{k-1} f_n z^{-n}\right] = z^k\left[F(z) - \sum_{n=0}^{k-1} f_n z^{-n}\right]$$

since $\sum_{n=0}^{k-1} f_n z^{-n} + \sum_{n=0}^{\infty} f_{n+k} z^{-(n+k)} = \left(f_0 + \dfrac{f_1}{z} + \ldots + \dfrac{f_{k-1}}{z^{k-1}}\right) + \left(\dfrac{f_k}{z^k} + \dfrac{f_{k+1}}{z^{k+1}} + \ldots\right) = \sum_{n=0}^{\infty} f_n z^{-n}.$

Example 17.8   Determine the Z transform $Z\left\{\dfrac{1}{(n+p)!}\right\},\, p = 1, 2, 3.$

Solution   Let $f_n = 1/(n!)$. Then, we have

$$Z\{f_n\} = Z\left\{\frac{1}{n!}\right\} = e^{1/z}. \quad \text{(set } a = 1 \text{ in Eq. (17.9i))}.$$

Using Theorem 17.3 for $k = p$, we get

$$Z\{f_{n+p}\} = Z\left\{\frac{1}{(n+p)!}\right\} = z^p\left[Z\left\{\frac{1}{n!}\right\} - \sum_{n=0}^{p-1} f_n z^{-n}\right] = z^p\left[e^{1/z} - \sum_{n=0}^{p-1} f_n z^{-n}\right].$$

Setting $p = 1, 2, 3$, we obtain

$$p = 1: \quad Z\left\{\frac{1}{(n+1)!}\right\} = z\left[e^{1/z} - f_0\right] = z\left[e^{1/z} - 1\right].$$

$$p = 2: \quad Z\left\{\frac{1}{(n+2)!}\right\} = z^2\left[e^{1/z} - f_0 - \frac{f_1}{z}\right] = z^2\left[e^{1/z} - 1 - \frac{1}{z}\right].$$

$$p = 3: \quad Z\left\{\frac{1}{(n+3)!}\right\} = z^3\left[e^{1/z} - f_0 - \frac{f_1}{z} - \frac{f_2}{z^2}\right] = z^3\left[e^{1/z} - 1 - \frac{1}{z} - \frac{1}{2z^2}\right].$$

Example 17.9   Determine the Z transforms

(i) $Z\{\cos (n+1)\theta\}$,   (ii) $Z\{\sin (n+1)\,\theta\}$.

Solution

(i) Let $f_n = \cos (n\theta)$. Then,

$$Z\{f_n\} = Z\{\cos (n\theta)\} = \frac{z(z - \cos \theta)}{z^2 - 2(\cos \theta)z + 1}.$$

Using Theorem 17.3 for $k = 1$, we obtain

$$Z\{f_{n+1}\} = Z\{\cos (n+1)\theta\} = z[Z\{\cos (n\theta)\} - f_0]$$

$$= z\left[\frac{z(z - \cos \theta)}{z^2 - 2(\cos \theta)z + 1} - 1\right] = z\left[\frac{(\cos \theta)z - 1}{z^2 - 2(\cos \theta)z + 1}\right].$$

since $f_0 = 1$.

(ii) Let $f_n = \sin (n\theta)$. Then,

$$Z\{f_n\} = Z\{\sin (n\theta)\} = \frac{(\sin \theta)z}{z^2 - 2(\cos \theta)z + 1}.$$

Using Theorem 17.3 for $k = 1$, we obtain

$$Z\{f_{n+1}\} = z[Z\{\sin (n\theta)\} - f_0] = \frac{(\sin \theta)z^2}{z^2 - 2(\cos \theta)z + 1}.$$

since $f_0 = 0$.

Example 17.10　Determine the Z transforms

(i) $Z\{e^{n+2}\}$,　　　　　　　　　　(ii) $Z\{e^{n-2}\}$.

Solution Let $f_n = e^n$. We have

$$Z\{f_n\} = Z\{e^n\} = \frac{z}{z-e}.$$

Using Theorem 17.3 for $k = 2$, we obtain

(i)　$Z\{e^{n+2}\} = z^2\left[Z\{e^n\} - f_0 - \frac{f_1}{z}\right] = z^2\left[\frac{z}{z-e} - 1 - \frac{e}{z}\right] = \frac{ze^2}{z-e}$.

(ii)　$Z\{e^{n-2}\} = z^{-2}[Z\{e^n\}] = z^{-2}\left[\frac{z}{z-e}\right] = \frac{1}{z(z-e)}$.

Theorem 17.4　(**Scaling property of Z transforms**) Let the Z transform $Z\{f_n\} = F(z)$ exist in the region $|z| > (1/R)$. If a is any real or complex constant, then

(i)　　　　　　　　　　　　　$Z\{a^{-n}f_n\} = F(az)$　　　　　　　　　　(17.15i)

(ii)　　　　　　　　　　　　$Z\{a^n f_n\} = F(z/a)$.　　　　　　　　　(17.15ii)

Proof　Using the definition of Z transform, we get

(i)　　　　　$Z\{a^{-n}f_n\} = \sum_{n=0}^{\infty} a^{-n} f_n z^{-n} = \sum_{n=0}^{\infty} f_n (az)^{-n} = F(az)$.

(ii)　　　　　$Z\{a^n f_n\} = \sum_{n=0}^{\infty} a^n f_n z^{-n} = \sum_{n=0}^{\infty} f_n \left(\frac{z}{a}\right)^{-n} = F\left(\frac{z}{a}\right)$.

Example 17.11　Determine the Z transform

(i) $Z\{e^{\alpha n} \cos(\beta n)\}$,　　　　　　(ii) $Z\{e^{-\alpha n} \sin(\beta n)\}$.

Solution

(i)　Let $f_n = \cos(\beta n)$. We have

$$Z\{\cos(\beta n)\} = \frac{z(z - \cos \beta)}{z^2 - 2(\cos \beta)z + 1} = F(z).$$

Using Theorem 17.4 for $a = e^\alpha$, we get

$$Z\{e^{\alpha n} \cos(\beta n)\} = F\left(\frac{z}{e^\alpha}\right) = \frac{(z/e^\alpha)[(z/e^\alpha) - \cos \beta]}{(z/e^\alpha)^2 - 2(\cos \beta)(z/e^\alpha) + 1}$$

$$= \frac{z(z - e^\alpha \cos \beta)}{z^2 - 2(\cos \beta)e^\alpha z + e^{2\alpha}}.$$

(ii)　Let $f_n = \sin(\beta n)$. We have

$$Z\{\sin(\beta n)\} = \frac{(\sin \beta)z}{z^2 - 2(\cos \beta)z + 1} = F(z).$$

Using Theorem 17.4 for $a = e^{\alpha}$, we get

$$Z\{e^{-\alpha n} \sin(\beta n)\} = F(e^{\alpha} z) = \frac{(\sin \beta) z e^{\alpha}}{e^{2\alpha} z^2 - 2(\cos \beta) z e^{\alpha} + 1}.$$

Theorem 17.5 (Initial value theorem) Let the Z transform $Z\{f_n\} = F(z)$ exist in the region $|z| > (1/R)$. Then,

$$f_p = \lim_{z \to \infty} \left[z^p \left\{ F(z) - f_0 - \frac{f_1}{z} - \frac{f_2}{z^2} - \ldots - \frac{f_{p-1}}{z^{p-1}} \right\} \right], p = 0, 1, 2, \ldots \qquad (17.16)$$

Proof Using the definition of Z transform, we have

$$F(z) = \sum_{n=0}^{\infty} f_n z^{-n} = f_0 + \frac{f_1}{z} + \frac{f_2}{z^2} + \ldots + \frac{f_p}{z^p} + \frac{f_{p+1}}{z^{p+1}} + \ldots$$

or

$$z^p F(z) = z^p \left[f_0 + \frac{f_1}{z} + \ldots + \frac{f_{p-1}}{z^{p-1}} + \frac{f_p}{z^p} + \frac{f_{p+1}}{z^{p+1}} + \ldots \right]. \qquad (17.17)$$

Taking limits on both sides as $z \to \infty$, we obtain the result.

We have the following initial values.
Set $p = 0$ in Eq. (17.17). We get

$$F(z) = f_0 + \frac{f_1}{z} + \frac{f_2}{z^2} + \ldots$$

Taking limits as $z \to \infty$, we obtain

$$f_0 = \lim_{z \to \infty} F(z). \qquad (17.18\text{i})$$

Set $p = 1$ in Eq. (17.17). We get

$$z F(z) = z f_0 + f_1 + \frac{f_2}{z} + \ldots$$

Taking limits as $z \to \infty$, we obtain

$$f_1 = \lim_{z \to \infty} [z F(z) - z f_0] = \lim_{z \to \infty} [z \{ F(z) - f_0 \}]. \qquad (17.18\text{ii})$$

Set $p = 2$ in Eq. (17.17). We get

$$z^2 F(z) = z^2 f_0 + z f_1 + f_2 + \frac{f_3}{z} + \ldots$$

Taking limits as $z \to \infty$, we obtain

$$f_2 = \lim_{z \to \infty} [z^2 F(z) - z^2 f_0 - z f_1] = \lim_{z \to \infty} \left[z^2 \{ F(z) - f_0 - \frac{f_1}{z} \} \right]. \qquad (17.18\text{iii})$$

Theorem 17.6 (Final value theorem) Let the Z transform $Z\{f_n\} = F(z)$ exist in the region $|z| > (1/R)$. Then,

$$\lim_{n \to \infty} f_n = \lim_{z \to 1} (z - 1) F(z). \qquad (17.19)$$

Proof   Using the definition of Z transform, we have

$$Z\{f_{n+1} - f_n\} = Z\{f_{n+1}\} - Z\{f_n\} = Z\{f_{n+1}\} - F(z).$$

Using the shifting theorem, we obtain

$$Z\{f_{n+1}\} = z\,[Z\,\{f_n\} - f_0] = z[F\,(z) - f_0].$$

Therefore, we obtain

$$Z\{f_{n+1} - f_n\} = \sum_{n=0}^{\infty} (f_{n+1} - f_n)z^{-n} = z[F(z) - f_0] - F(z) = (z - 1)\,F(z) - zf_0$$

or $(z - 1)\,F(z) - zf_0 = (f_1 - f_0) + (f_2 - f_1)z^{-1} + \cdots + (f_{n+1} - f_n)z^{-n} + \cdots$

Taking limits on both sides as $z \to 1$, we get

$$\lim_{z \to 1}[(z - 1)\,F(z) - zf_0] = \lim_{z \to 1}[(z - 1)F(z)] - f_0 = \left[\lim_{n \to \infty} f_n\right] - f_0.$$

Hence, $$\lim_{n \to \infty} f_n = \lim_{z \to 1} (z - 1)\,F(z).$$

Example 17.12   Let $Z\{f_n\} = F(z) = [3z^2 - 4z + 7]/(z - 1)^3$. Find f_0, f_1, f_2 and f_3.

Solution   We have from Theorem 17.5

$$f_0 = \lim_{z \to \infty} F(z) = \lim_{z \to \infty}\left[\frac{3z^2 - 4z + 7}{(z - 1)^3}\right] = \lim_{z \to \infty}\left[\frac{(3/z) - (4/z^2) + (7/z^3)}{[1 - (1/z)]^3}\right] = 0.$$

$$f_1 = \lim_{z \to \infty} z[F(z) - f_0] = \lim_{z \to \infty}\left[\frac{3z^3 - 4z^2 + 7z}{(z - 1)^3}\right] = \lim_{z \to \infty}\left[\frac{3 - (4/z) + (7/z^2)}{[1 - (1/z)]^3}\right] = 3.$$

$$f_2 = \lim_{z \to \infty} z^2\left[F(z) - f_0 - \frac{f_1}{z}\right] = \lim_{z \to \infty}\left[\frac{3z^4 - 4z^3 + 7z^2}{(z - 1)^3} - 3z\right]$$

$$= \lim_{z \to \infty}\left[\frac{5z^3 - 2z^2 + 3z}{(z - 1)^3}\right] = \lim_{z \to \infty}\left[\frac{5 - (2/z) + (3/z^2)}{[1 - (1/z)]^3}\right] = 5.$$

$$f_3 = \lim_{z \to \infty} z^3\left[F(z) - f_0 - \frac{f_1}{z} - \frac{f_2}{z^2}\right] = \lim_{z \to \infty}\left[\frac{3z^5 - 4z^4 + 7z^3}{(z - 1)^3} - 3z^2 - 5z\right]$$

$$= \lim_{z \to \infty}\left[\frac{13z^3 - 12z^2 + 5z}{(z - 1)^3}\right] = \lim_{z \to \infty}\left[\frac{13 - (12/z) + (5/z^2)}{[1 - (1/z)]^3}\right] = 13.$$

Example 17.13   Let $Z\{f_n\} = F(z) = \left[\dfrac{3z^3 + 5z^2 - 7z + 1}{(z + 2)^2(z - 1)}\right]$. Find $\lim\limits_{n \to \infty} f_n$.

Solution   Using Theorem 17.6, we obtain

$$\lim_{n \to \infty} f_n = \lim_{z \to 1} (z-1) F(z) = \lim_{z \to 1} \left[\frac{(z-1)(3z^3 + 5z^2 - 7z + 1)}{(z+2)^2 (z-1)} \right]$$

$$= \lim_{z \to 1} \left[\frac{3z^3 + 5z^2 - 7z + 1}{(z+2)^2} \right] = \frac{2}{9}.$$

Theorem 17.7 (Differentiation of Z transform) Let Z transform $Z\{f_n\} = F(z)$ exist in the region $|z| > (1/R)$. Then,

$$Z\{n f_n\} = -z \frac{d}{dz} [F(z)] \qquad (17.20)$$

which is also convergent in the region $|z| > (1/R)$.

Proof   We have

$$Z\{f_n\} = F(z) = \sum_{n=0}^{\infty} f_n z^{-n} = f_0 + f_1 z^{-1} + f_2 z^{-2} + \dots$$

Differentiating both sides with respect to z, (assuming that the term by term differentiation of the infinite series is allowed), we get

$$\frac{d}{dz} [F(z)] = -f_1 z^{-2} - 2 f_2 z^{-3} - 3 f_3 z^{-4} - \dots$$

$$= -z^{-1} \sum_{n=0}^{\infty} n f_n z^{-n} = -z^{-1} Z\{n f_n\}.$$

Hence, we obtain

$$Z\{n f_n\} = -z \frac{d}{dz} [F(z)] = -z \frac{d}{dz} [Z\{f_n\}].$$

Remark 3

$$Z\{n^p f_n\} = -z \frac{d}{dz} [Z\{n^{p-1} f_n\}], \ p = 1, 2, \dots.$$

Example 17.14   Determine (i) $Z\{n\}$, (ii) $Z\{n^2\}$, (iii) $Z\{n^3\}$ and (iv) $Z\{n^4\}$.

Solution   Let $f_n = 1$. Then $Z\{f_n\} = z/(z-1)$. From Theorem 17.7, we get

(i)
$$Z\{n\} = -z \frac{d}{dz} [Z\{1\}] = -z \frac{d}{dz} \left[\frac{z}{z-1} \right] = \frac{z}{(z-1)^2}.$$

(ii)
$$Z\{n^2\} = -z \frac{d}{dz} [Z\{n\}] = -z \frac{d}{dz} \left[\frac{z}{(z-1)^2} \right] = \frac{z^2 + z}{(z-1)^3}.$$

(iii)
$$Z\{n^3\} = -z \frac{d}{dz} [Z\{n^2\}] = -z \frac{d}{dz} \left[\frac{z^2 + z}{(z-1)^3} \right] = \frac{z^3 + 4z^2 + z}{(z-1)^4}.$$

(iv)
$$Z\{n^4\} = -z \frac{d}{dz} [Z\{n^3\}] = -z \frac{d}{dz} \left[\frac{z^3 + 4z^2 + z}{(z-1)^4} \right] = \frac{z(z^3 + 11z^2 + 11z + 1)}{(z-1)^5}.$$

Example 17.15 Let $f_n = a^n$. Determine (i) $Z\{nf_n\}$ and (ii) $Z\{n^2 f_n\}$.

Solution We have $Z\{a^n\} = z/(z - a)$. Using Theorem 17.7, we get

(i)
$$Z\{na^n\} = -z\frac{d}{dz}[Z\{a^n\}] = -z\frac{d}{dz}\left[\frac{z}{z - a}\right] = \frac{az}{(z - a)^2}.$$

(ii)
$$Z\{n^2 a^n\} = -z\frac{d}{dz}[Z\{na^n\}] = -z\frac{d}{dz}\left[\frac{az}{(z - a)^2}\right] = \frac{az(z + a)}{(z - a)^3}.$$

Example 17.16 Let $f_n = \dfrac{(n + 1)a^n}{n!}$. Find $Z\{f_n\}$.

Solution We have $f_n = \dfrac{a^n}{n!} + n\left(\dfrac{a^n}{n!}\right)$ and from Eq. (17.9i), we have

$$Z\left\{\frac{a^n}{n!}\right\} = e^{a/z}.$$

Using Theorem 17.7, we obtain

$$Z\left\{n\left(\frac{a^n}{n!}\right)\right\} = -z\frac{d}{dz}\left[Z\left\{\frac{a^n}{n!}\right\}\right] = -z\frac{d}{dz}[e^{a/z}] = \frac{a}{z}e^{a/z}$$

Hence,
$$Z\{f_n\} = Z\left\{\frac{a^n}{n!}\right\} + Z\left\{n\left(\frac{a^n}{n!}\right)\right\} = \left(1 + \frac{a}{z}\right)e^{a/z}.$$

Example 17.17 Find the inverse Z transform $Z^{-1}[F(z)]$, where $F(z)$ is given by

(i) $\dfrac{z^3 + 2z^2 + 29z}{(z - 1)(z + 3)^2}$,

(ii) $\dfrac{z^3 + 5z^2 + 6z}{(z - 2)(z - 3)^3}$.

Solution

(i) We decompose $F(z)/z$ into partial fractions as

$$\frac{F(z)}{z} = \frac{z^2 + 2z + 29}{(z - 1)(z + 3)^2} = \frac{A}{z - 1} + \frac{B}{z + 3} + \frac{C}{(z + 3)^2}.$$

Therefore, $z^2 + 2z + 29 = A(z + 3)^2 + B(z - 1)(z + 3) + C(z - 1)$.
When $z = 1$, we get $A = 2$ and when $z = -3$, we get $C = -8$. Comparing the coefficients of z^2 on both sides, we get $A + B = 1$ or $B = -1$. Hence, we have

$$\frac{F(z)}{z} = \frac{2}{z - 1} - \frac{1}{z + 3} - \frac{8}{(z + 3)^2}$$

or
$$F(z) = \frac{2z}{z - 1} - \frac{z}{z + 3} - \frac{8z}{(z + 3)^2}.$$

We note that
$$\frac{d}{dz}\left[\frac{z}{z + 3}\right] = \frac{3}{(z + 3)^2}.$$

Now $Z^{-1}\left[\dfrac{z}{z+3}\right] = (-3)^n$ and using Eq. (17.20), we get $Z^{-1}\left[\dfrac{3z}{(z+3)^2}\right] = -n(-3)^n$. Taking inverse Z transform on both sides, we obtain

$$f_n = 2 - (-3)^n + \frac{8n}{3}(-3)^n = 2 + \frac{1}{3}(8n - 3)(-3)^n.$$

(ii) We have $Z\{2^n\} = \dfrac{z}{z-2}$, $Z\{3^n\} = \dfrac{z}{z-3}$, $Z\{n3^n\} = \dfrac{3z}{(z-3)^2}$ and $Z\{n^2 3^n\} = \dfrac{3z(z+3)}{(z-3)^3}$.

We decompose $F(z)$ as a linear combination of the right hand sides of the above factors as

$$F(z) = \frac{z^3 + 5z^2 + 6z}{(z-2)(z-3)^3} = \frac{Az}{z-2} + \frac{Bz}{z-3} + \frac{3Cz}{(z-3)^2} + \frac{3Dz(z+3)}{(z-3)^3}$$

or $\quad z^3 + 5z^2 + 6z = Az(z-3)^3 + Bz(z-2)(z-3)^2 + 3Cz(z-2)(z-3)$

$$+ 3Dz(z+3)(z-2).$$

When $z = 2$, we get $A = -20$ and when $z = 3$, we get $D = 5/3$. Comparing the coefficients of z^4 and z, we get $A + B = 0$, or $B = -A = 20$ and $6 = -27A - 18B + 18C - 18D$, or $C = -8$. Hence, we have

$$F(z) = -20\left[\frac{z}{z-2}\right] + 20\left[\frac{z}{z-3}\right] - 8\left[\frac{3z}{(z-3)^2}\right] + \frac{5}{3}\left[\frac{3z(z+3)}{(z-3)^3}\right].$$

Taking inverse Z transforms on both sides, we obtain

$$f_n = -20(2^n) + 20(3^n) - 8(n3^n) + \frac{5}{3}(n^2 3^n)$$

$$= -20(2^n) + \frac{1}{3}(60 - 24n + 5n^2)(3^n).$$

Convolution of sequences

Let $\{f_n\}$ and $\{g_n\}$ be two sequences. Then the *convolution* of these sequences is defined by

$$\{f_n\} * \{g_n\} = \{f_n * g_n\} = \sum_{k=0}^{n} f_k\, g_{n-k}$$

$$= f_0\, g_n + f_1\, g_{n-1} + \ldots + f_n\, g_0. \tag{17.21}$$

The following results on convolution of sequences can be easily verified.

(i) $\{f_n * g_n\} = \{g_n * f_n\}$

(ii) $\{(u_n + v_n) * g_n\} = \{u_n * g_n\} + \{v_n * g_n\}$.

(iii) $\{(u_n * v_n)\} * \{g_n\} = \{u_n\} * \{v_n * g_n\}$

for any sequences $\{f_n\}$, $\{g_n\}$, $\{u_n\}$ and $\{v_n\}$.

Theorem 17.8 (**Convolution theorem**) Let $\{f_n\}$, $\{g_n\}$ be any two sequences. Let the Z transform of $\{f_n\}$, $Z\{f_n\} = F(z)$ exist in the region $|z| > (1/R_1)$ and the Z transform of $\{g_n\}$, $Z\{g_n\} = G(z)$ exist in the region $|z| > (1/R_2)$. Then,

$$Z\{f_n * g_n\} = F(z)\, G(z) \tag{17.22}$$

valid in the region $|z| > (1/R)$, where $(1/R) = \max\,[(1/R_1),\,(1/R_2)]$.

Proof   We have

$$F(z)\,G(z) = Z\{f_n\}Z\{g_n\} = \left[\sum_{n=0}^{\infty} f_n z^{-n}\right]\left[\sum_{n=0}^{\infty} g_n z^{-n}\right]$$

$$= [f_0 + f_1 z^{-1} + \ldots + f_n z^{-n} + \ldots][g_0 + g_1 z^{-1} + \ldots + g_n z^{-n} + \ldots]$$

$$= f_0\, g_0 + (f_0\, g_1 + f_1\, g_0)\, z^{-1} + (f_0\, g_2 + f_1\, g_1 + f_2\, g_0)\, z^{-2}$$

$$+ \ldots + (f_0\, g_n + f_1\, g_{n-1} + \ldots + f_n\, g_0)\, z^{-n} + \ldots$$

$$= \sum_{n=0}^{\infty}\left[\sum_{k=0}^{n} f_k g_{n-k}\right] z^{-n} = \sum_{n=0}^{\infty} \{f_n * g_n\} z^{-n} = Z\{f_n * g_n\}.$$

Remark 4

$$Z^{-1}[F(z)\,G(z)] = \{f_n * g_n\}. \tag{17.23}$$

Example 17.18   Using the convolution of sequences, show that

$$\left\{\frac{1}{n!}\right\} * \left\{\frac{1}{n!}\right\} = \left\{\frac{2^n}{n!}\right\}.$$

Solution   We have

$$\left\{\frac{1}{n!}\right\} * \left\{\frac{1}{n!}\right\} = \sum_{k=0}^{n}\left(\frac{1}{k!}\right)\left(\frac{1}{(n-k)!}\right)$$

$$= \frac{1}{n!} + \frac{1}{1!(n-1)!} + \frac{1}{2!(n-2)!} + \ldots + \frac{1}{n!}$$

$$= \frac{1}{n!}\left[1 + n + \frac{n(n-1)}{2!} + \ldots + 1\right]$$

$$= \frac{1}{n!}\,[{}^nC_0 + {}^nC_1 + {}^nC_2 + \ldots + {}^nC_n] = \frac{2^n}{n!}.$$

Example 17.19   Find the Z transform $Z\{f_n * g_n\}$, where

(i) $f_n = p^n$, $g_n = q^n$;   (ii) $f_n = p^n$, $g_n = p^n$.

Verify the convolution theorem in each case.

Solution

(i) $$\{f_n * g_n\} = \sum_{k=0}^{n} p^k q^{n-k} = q^n + pq^{n-1} + \ldots + p^n = \frac{q^{n+1} - p^{n+1}}{q - p}.$$

We have $$Z\{p^n\} = \frac{z}{z-p} \quad \text{and} \quad Z\{q^n\} = \frac{z}{z-q}.$$

Therefore, $$Z\{f_n * g_n\} = Z\{p^n\}Z\{q^n\} = \frac{z^2}{(z-p)(z-q)}.$$

Now,
$$\frac{z^2}{(z-p)(z-q)} = \frac{1}{q-p}\left[\frac{qz}{z-q} - \frac{pz}{z-p}\right].$$

Hence,
$$Z^{-1}\left[\frac{z^2}{(z-p)(z-q)}\right] = \frac{q}{q-p}Z^{-1}\left[\frac{z}{z-q}\right] - \frac{p}{q-p}Z^{-1}\left[\frac{z}{z-p}\right]$$

$$= \frac{q}{q-p}(q^n) - \frac{p}{q-p}(p^n) = \frac{q^{n+1} - p^{n+1}}{q-p} = \{f_n * g_n\}.$$

(ii)
$$\{f_n * g_n\} = \sum_{k=0}^{n} p^k p^{n-k} = \sum_{k=0}^{n} p^n = p^n \sum_{k=0}^{n} 1 = (n+1)p^n.$$

We have
$$Z\{p^n\} = \frac{z}{z-p}.$$

Therefore,
$$Z\{f_n * g_n\} = Z\{p^n\}Z\{p^n\} = \frac{z^2}{(z-p)^2}.$$

Now,
$$\frac{z^2}{(z-p)^2} = \frac{z}{z-p} + \frac{pz}{(z-p)^2}.$$

Hence,

$$Z^{-1}\left[\frac{z^2}{(z-p)^2}\right] = Z^{-1}\left[\frac{z}{z-p}\right] + Z^{-1}\left[\frac{pz}{(z-p)^2}\right]$$

$$= p^n + np^n = (n+1)p^n.$$

Example 17.20   Using convolution theorem, find $Z^{-1}\left[\dfrac{z^2}{(z-1)(z-3)}\right]$.

Solution   Let $F(z) = z/(z-1)$ and $G(z) = z/(z-3)$. We have
$$f_n = Z^{-1}[F(z)] = 1^n \text{ and } g_n = Z^{-1}[G(z)] = 3^n.$$
Using Eq. (17.23), we get

$$Z^{-1}\left[\frac{z^2}{(z-1)(z-3)}\right] = Z^{-1}[F(z)\,G(z)] = \{f_n * g_n\} = \{(1)^n * (3)^n\}$$

$$= \sum_{k=0}^{n} (1)^k\, 3^{n-k} = 3^n + 3^{n-1} + \ldots + 1 = \frac{1}{2}(3^{n+1} - 1).$$

Example 17.21   Verify convolution theorem when $\{f_n\} = \{n\}$ and $\{g_n\} = \{n^2\}$.

Solution   We have $f_n = n$ and $g_n = n^2$. Therefore,

$$\{f_n * g_n\} = \sum_{k=0}^{n} k^2(n-k) = n\sum_{k=0}^{n} k^2 - \sum_{k=0}^{n} k^3$$

$$= n\left[\frac{n(n+1)(2n+1)}{6}\right] - \frac{n^2(n+1)^2}{4}$$

$$= n^2(n+1)\left[\frac{2n+1}{6} - \frac{n+1}{4}\right] = \frac{n^2(n+1)}{12}(4n+2-3n-3)$$

$$= \frac{n^2}{12}(n+1)(n-1) = \frac{1}{12}(n^4 - n^2).$$

From Example 17.14, we have

$$Z\{n^2\} = \frac{z^2+z}{(z-1)^3} \quad\text{and}\quad Z\{n^4\} = \frac{z(z^3+11z^2+11z+1)}{(z-1)^5}.$$

Hence, $Z\{f_n * g_n\} = \frac{1}{12}[Z\{n^4\} - Z\{n^2\}] = \frac{1}{12}\left[\frac{z^4+11z^3+11z^2+z}{(z-1)^5} - \frac{z^2+z}{(z-1)^3}\right]$

$$= \frac{1}{12(z-1)^5}[(z^4+11z^3+11z^2+z)-(z^2+z)(z-1)^2] = \frac{z^3+z^2}{(z-1)^5}.$$

Now, $F(z) = Z\{f_n\} = Z\{n\} = \frac{z}{(z-1)^2}$, and $G(z) = Z\{g_n\} = Z\{n^2\} = \frac{z^2+z}{(z-1)^3}.$

Therefore, $$Z\{f_n * g_n\} = F(z)G(z) = \frac{z^3+z^2}{(z-1)^5}.$$

Exercise 17.1

Obtain Z transform of the sequence $\{f_n\}$, where f_n is given as follows.

1. a^{2n}
2. $e^{ni\theta}$.
3. $\sin(n\pi/2)$.
4. $\cos(n\pi/3)$.

5. $\sin(n\pi/6)$.
6. $Aa^n + Bb^n$.
7. $n+1$.
8. $(n+1)^2$.

9. $1 - a^{2n}$.
10. a^{n+1}.
11. ne^n.
12. n^2e^n.

13. $a^{-n}\cos(\beta n)$.
14. $a^n\cos(\beta n)$.
15. $a^n e^{\beta n}$.
16. $n\sin(\beta n)$.

17. $n\cos(\beta n)$.
18. $\dfrac{\cos(\beta n)}{n!}$.
19. $\dfrac{2^n e^{-n}}{n!}$.
20. $\dfrac{a^n\sin(\beta n)}{n!}$.

In the following problems determine f_0, f_1, f_2 in the sequence $\{f_n\}$, when $Z\{f_n\} = F(z)$ is as given below.

21. $\dfrac{1}{z-2}$.
22. $\dfrac{1}{(z-1)(z-3)}$.
23. $\dfrac{z^2}{z^2+1}$.
24. $\dfrac{(z-1)^2(z+2)}{(z+3)(z+5)^2}$.

25. $\dfrac{z^3+5z^2+3z-1}{(z-1)^3(z+2)}$

26. Let $Z\{f_n\} = \dfrac{z^2-3z+5}{(z-1)(z+2)}$. Find $\lim\limits_{n\to\infty} f_n$.

27. Let $Z\{f_n\} = \dfrac{z^3-3z^2+2z+1}{(z+3)^2(z-1)}$. Find $\lim\limits_{n\to\infty} f_n$.

Find the inverse Z transform of $F(z)$, where $F(z)$ is given as below.

28. $\dfrac{z}{(z-1)(z-2)}$.
29. $\dfrac{z^2}{(z-1)(z-2)}$.
30. $\dfrac{z^2-z}{(z+1)(z+2)(z+3)}$.

31. $\dfrac{z^2}{(z+2)(z-1)^2}$.

32. $\dfrac{z^3+z}{(z-2)^2(z+3)^2}$.

33. $\dfrac{z^2+5z}{(z-1)^3}$.

34. $\dfrac{3z^2-2z}{(z-1)^3(z+2)}$.

35. $\dfrac{z}{(z-1)^4}$.

36. $\dfrac{6z^2}{(z-1)^4(z+5)}$.

Verify convolution theorem in the following problems.

37. $f_n = 1$, $g_n = n$.

38. $f_n = 2^n$, $g_n = 3^n$,

39. $f_n = e^n$, $g_n = e^n$.

40. $f_n = 1$, $g_n = n(3^n)$.

17.3 Solution of Difference Equations using Z Transforms

Let n, $n+1$, ..., $n+k$ be a set of $k+1$ positive integers and y_n, y_{n+1}, ..., y_{n+k} be the corresponding values of a function $y(x)$ at these points, that is $y_{n+i} = y(n+i)$, $i = 0, 1, \ldots, k$.
A relation of the form

$$y_{n+k} + a_1 y_{n+k-1} + \ldots + a_k y_n = f(n) \tag{17.24}$$

where $a_1, a_2, \ldots, a_k$ are constants, is called a linear, constant coefficient difference equation of order k. Note that the order of a difference equation is the difference between the largest and the smallest arguments in the difference equation. The difference equation is called homogeneous if $f(n) = 0$, and nonhomogeneous if $f(n) \neq 0$. A linear difference equation with constant coefficients can be solved using Z transforms. We follow the following steps:

1. Let $F(z) = Z\{y_n\}$.
2. Apply Z transform to both sides in Eq. (17.24) using the linearity principle and Theorem 17.3.
3. Simplify and obtain $F(z)$.
4. Take the inverse Z transform of $F(z)$ and determine y_n, which is the required solution.

The solution y_n of the difference equation depends on k values $y_0, y_1, \ldots, y_{k-1}$.

Remark 5

A difference equation of the form

$$a_0 y_{n+k} + a_1 y_{n+k-1} + \ldots + a_k y_n = f(n), \, a_0 \neq 0$$

can be brought to the form as given in Eq. (17.25) by dividing both sides by a_0.

Example 17.22 Solve the following difference equations using Z transforms.

(i) $y_{n+1} - 5 y_n = 0$.

(ii) $y_{n+2} - 3y_{n+1} + 2y_n = 0$, $y_0 = -1$, $y_1 = 2$.

(iii) $y_{n+2} + 5y_{n+1} + 4 y_n = 2^n$, $y_0 = 1$, $y_1 = -4$.

(iv) $y_{n+2} - 2y_{n+1} + y_n = n$, $y_0 = 1$, $y_1 = 1$.

Solution Let $Z\{y_n\} = F(z)$. Apply Z transforms on both sides and use Theorem 17.3.

(i) We have $Z\{y_{n+1}\} - 5Z\{y_n\} = 0$.

Therefore, $z[F(z) - y_0] - 5F(z) = 0$, or $F(z) = \left(\dfrac{z}{z-5}\right) y_0$.

Taking inverse Z transform, we obtain

$$y_n = Z^{-1}[F(z)] = y_0\, Z^{-1}\left[\frac{z}{z-5}\right] = (5^n)\, y_0$$

and y_0 is arbitrary.

(ii) We have $Z\{y_{n+2}\} - 3Z\{y_{n+1}\} + 2Z\{y_n\} = 0$.

Therefore, $\qquad z^2\left[F(z) - y_0 - \dfrac{y_1}{z}\right] - 3z[F(z) - y_0] + 2F(z) = 0$

or $\qquad\qquad (z^2 - 3z + 2)\, F(z) - (z^2 - 3z)\, y_0 - zy_1 = 0$

or $\qquad\qquad (z^2 - 3z + 2)\, F(z) - (z^2 - 3z)\, (-1) - z(2) = 0$

or $\qquad\qquad F(z) = \dfrac{-z^2 + 5z}{(z-1)(z-2)} = \dfrac{3z}{z-2} - \dfrac{4z}{z-1}.$

Taking inverse Z transform, we obtain

$$y_n = Z^{-1}[F(z)] = Z^{-1}\left[\frac{3z}{z-2}\right] - Z^{-1}\left[\frac{4z}{z-1}\right]$$

$$= 3(2^n) - 4(1)^n = 3(2^n) - 4.$$

(iii) We have $Z\{y_{n+2}\} + 5Z\{y_{n+1}\} + 4Z\{y_n\} = Z\{2^n\}$.

Hence, we obtain

$$z^2\left[F(z) - y_0 - \frac{y_1}{z}\right] + 5z[F(z) - y_0] + 4F(z) = \frac{z}{z-2}$$

or $\qquad\qquad (z^2 + 5z + 4)\,F(z) - (z^2 + 5z)y_0 - zy_1 = \dfrac{z}{z-2}$

or $\qquad (z^2 + 5z + 4)\,F(z) = \dfrac{z}{z-2} + (z^2 + 5z) - 4z = \dfrac{z}{z-2} + z^2 + z$

or $\qquad\qquad F(z) = \dfrac{z[1 + (z+1)(z-2)]}{(z^2 + 5z + 4)(z-2)} = \dfrac{z(z^2 - z - 1)}{(z+4)(z+1)(z-2)}.$

We write $\qquad \dfrac{F(z)}{z} = \dfrac{z^2 - z - 1}{(z+4)(z+1)(z-2)} = \dfrac{A}{z+4} + \dfrac{B}{z+1} + \dfrac{C}{z-2}.$

Hence, $A\,(z+1)\,(z-2) + B\,(z+4)\,(z-2) + C\,(z+1)\,(z+4) = z^2 - z - 1.$

Setting $z = 2$, $z = -1$ and $z = -4$ respectively, we get $C = 1/18$, $B = -1/9$ and $A = 19/18$.

Therefore, we have $F(z) = \dfrac{19}{18}\left(\dfrac{z}{z+4}\right) - \dfrac{1}{9}\left(\dfrac{z}{z+1}\right) + \dfrac{1}{18}\left(\dfrac{z}{z-2}\right).$

Taking inverse Z transform, we obtain

$$y_n = Z^{-1}[F(z)] = \frac{19}{18}\,Z^{-1}\left[\frac{z}{z+4}\right] - \frac{1}{9}\,Z^{-1}\left[\frac{z}{z+1}\right] + \frac{1}{18}\,Z^{-1}\left[\frac{z}{z-2}\right]$$

$$= \frac{19}{18}(-4)^n - \frac{1}{9}(-1)^n + \frac{1}{18}(2^n).$$

(iv) We have $Z\{y_{n+2}\} - 2 Z\{y_{n+1}\} + Z\{y_n\} = Z\{n\}$.

Hence, we obtain

$$z^2 \left[F(z) - y_0 - \frac{y_1}{z} \right] - 2z[F(z) - y_0] + F(z) = \frac{z}{(z-1)^2}$$

or

$$(z^2 - 2z + 1) F(z) - (z^2 - 2z) - z = \frac{z}{(z-1)^2}$$

or

$$F(z) = \frac{z}{(z-1)^4} + \frac{z(z-1)}{(z-1)^2} = \frac{z[1 + (z-1)^3]}{(z-1)^4} = \frac{z(z^3 - 3z^2 + 3z)}{(z-1)^4}.$$

Since $Z\{1\} = \frac{z}{z-1}$, $Z\{n\} = \frac{z}{(z-1)^2}$, $Z\{n^2\} = \frac{z^2+z}{(z-1)^3}$, $Z\{n^3\} = \frac{z^3 + 4z^2 + z}{(z-1)^4}$,

we write $F(z)$ as a linear combination of the right hand side factors, that is

$$F(z) = \frac{z^4 - 3z^3 + 3z^2}{(z-1)^4} = \frac{A(z^3 + 4z^2 + z)}{(z-1)^4} + \frac{B(z^2 + z)}{(z-1)^3} + \frac{Cz}{(z-1)^2} + \frac{Dz}{z-1}.$$

Hence, $z^4 - 3z^3 + 3z^2 = A(z^3 + 4z^2 + z) + B(z^2 + z)(z-1) + Cz(z-1)^2 + Dz(z-1)^3$

$$= Dz^4 + (A + B + C - 3D)z^3 + (4A - 2C + 4D)z^2 + (A - B + C - D)z.$$

Setting $z = 1$, we get $A = 1/6$. Comparing the coefficients of z^4, z^3 and z^2, we obtain

$$D = 1, \quad A + B + C - 3D = -3, \quad 4A - 2C + 3D = 3.$$

We get $\qquad C = 1/3 \quad$ and $\quad B = -1/2$.

Hence, $\qquad F(z) = \frac{1}{6}\left[\frac{z^3 + 4z^2 + z}{(z-1)^4} \right] - \frac{1}{2}\left[\frac{z^2 + z}{(z-1)^3} \right] + \frac{1}{3}\left[\frac{z}{(z-1)^2} \right] + \frac{z}{z-1}.$

Taking the inverse Z transform, we get

$$y_n = \frac{1}{6}n^3 - \frac{1}{2}n^2 + \frac{1}{3}n + 1 = \frac{1}{6}(6 + 2n - 3n^2 + n^3).$$

Example 17.23 Solve the difference equation

$$y_{n+3} - 6y_{n+2} + 12y_{n+1} - 8y_n = 1, \ y_0 = 1, \ y_1 = 1, \ y_2 = 2$$

using Z transforms.

Solution Let $Z\{y_n\} = F(z)$. Taking Z transforms on both sides and using Theorem 17.3, we get

$$z^3 \left[F(z) - y_0 - \frac{y_1}{z} - \frac{y_2}{z^2} \right] - 6z^2 \left[F(z) - y_0 - \frac{y_1}{z} \right] + 12z[F(z) - y_0] - 8 F(z) = \frac{z}{z-1}$$

or $\qquad (z^3 - 6z^2 + 12z - 8) F(z) - (z^3 - 6z^2 + 12z) y_0 - (z^2 - 6z) y_1 - z y_2 = \frac{z}{z-1}.$

or $(z^3 - 6z^2 + 12z - 8) F(z) - (z^3 - 6z^2 + 12z) - (z^2 - 6z) - 2z = \dfrac{z}{z-1}$

or $(z-2)^3 F(z) = \dfrac{z}{z-1} + (z^3 - 5z^2 + 8z) = \dfrac{z^4 - 6z^3 + 13z^2 - 7z}{z-1}$

or $F(z) = \dfrac{z^4 - 6z^3 + 13z^2 - 7z}{(z-1)(z-2)^3}.$

Since $Z\{1\} = \dfrac{z}{z-1}, \ Z\{2^n\} = \dfrac{z}{z-2}, \ Z\{n2^n\} = \dfrac{2z}{(z-2)^2}$ and $Z\{n^2 2^n\} = \dfrac{2z(z+2)}{(z-2)^3}$

(see Example 17.15), we write $F(z)$ as a linear combination of the right hand side factors, that is

$$F(z) = \frac{z^4 - 6z^3 + 13z^2 - 7z}{(z-1)(z-2)^3} = \frac{Az}{z-1} + \frac{Bz}{z-2} + \frac{2Cz}{(z-2)^2} + \frac{2Dz(z+2)}{(z-2)^3}.$$

Hence, $Az(z-2)^3 + Bz(z-1)(z-2)^2 + 2Cz(z-1)(z-2) + 2Dz(z+2)(z-1)$

$$= z^4 - 6z^3 + 13z^2 - 7z.$$

Setting $z = 2$, we get $16D = 6$, or $D = 3/8$. Setting $z = 1$, we get $A = -1$. Comparing the coefficients of z^4 and z on both sides, we get

$$A + B = 1 \text{ and } - 8A - 4B + 4C - 4D = -7.$$

Solving, we obtain $B = 2$ and $C = -11/8$.
Therefore, we have

$$F(z) = -\left[\frac{z}{z-1}\right] + 2\left[\frac{z}{z-2}\right] - \frac{11}{8}\left[\frac{2z}{(z-2)^2}\right] + \frac{3}{8}\left[\frac{2z(z+2)}{(z-2)^3}\right].$$

Taking the inverse Z transforms, we get

$$y_n = -1 + 2(2^n) - \frac{11}{8}(n2^n) + \frac{3}{8}(n^2 2^n) = \frac{2^n}{8}(16 - 11n + 3n^3) - 1$$

$$= 2^{n-3}(16 - 11n + 3n^2) - 1.$$

Exercise 17.2

Solve the following difference equations using Z transforms.

1. $y_{n+1} - 3y_n = 0, \ y_0 = 1.$

2. $y_{n+2} + 4y_n = n + 1, \ y_0 = -1.$

3. $y_{n+2} - y_n = 1, \ y_0 = 1, \ y_1 = 2.$

4. $y_{n+2} + 5y_{n+1} + 6y_n = 0, \ y_0 = 1, \ y_1 = 1.$

5. $y_{n+2} - 6y_{n+1} + 9y_n = 3^n, \ y_0 = 0, \ y_1 = 1.$

6. $y_{n+2} + 10y_{n+1} + 25y_n = n, \ y_0 = 1, \ y_1 = -5.$

7. $y_{n+3} - 6y_{n+2} + 11y_{n+1} - 6y_n = 0, \ y_0 = 0, \ y_1 = 1, \ y_2 = 2.$

8. $y_{n+3} + 2y_{n+2} - y_{n+1} - 2y_n = 1, \ y_0 = 1, \ y_1 = -1, \ y_2 = 3.$

9. $y_{n+3} - 3y_{n+1} - 2y_n = 3^n, \ y_0 = 2, \ y_1 = 1, \ y_2 = 6.$

10. $y_{n+3} - 9y_{n+2} + 27y_{n+1} - 27y_n = 2^n, \ y_0 = -1, \ y_1 = -3, \ y_2 = 2.$

17.4   Application of *Z*-Transforms to find the Sum of Series

We consider the application of *Z*-transforms to find the sum of some suitable finite or infinite series.

Sum of a finite series

Let the given $n + 1$ term series be

$$h_n = f_0 + f_1 + f_2 + \cdots + f_n. \tag{17.25}$$

We can interpret that h_k represents the partial sum of $k + 1$ terms. If possible, we derive a two term difference equation connecting h_{n+1}, h_n or a three term difference equation connecting h_{n+2}, h_{n+1}, h_n. Alternately, we derive a two term difference equation connecting h_{n-1}, h_n or a three term difference equation connecting h_{n-2}, h_{n-1}, h_n. Then, we use *Z*-transform to solve this difference equation and obtain the sum of the given series.

Sum of an infinite series

Let the given series be

$$h = f_0 + f_1 + f_2 + \cdots \tag{17.26}$$

We have the following theorem to find the sum of the series.

Theorem 17.9   Let the series $\displaystyle\sum_{n=0}^{\infty} f_n$ converge. Let $Z\{f_n\} = F(z)$. Then,

$$\sum_{n=0}^{\infty} f_n = \lim_{z \to 1^+} F(z), \quad z \text{ real.} \tag{17.27}$$

Proof   From the definition, we have

$$Z\{f_n\} = f_0 + \frac{f_1}{z} + \frac{f_2}{z^2} + \frac{f_3}{z^3} + \cdots = F(z).$$

Taking the limit as $z \to 1^+$, we get

$$\lim_{z \to 1^+} F(z) = f_0 + f_1 + f_2 + \cdots, \quad z \text{ real.}$$

We illustrate the above method through some examples.

Example 17.24   Using *Z*-transforms, find the sum of the series

$$1 - \frac{1}{2} + \frac{1}{3} - \frac{1}{4} + \cdots$$

Solution   We know that the series is convergent but not absolutely convergent. Denote

$$f_n = \frac{(-1)^n}{n+1}, \quad n = 0, 1, 2, \cdots$$

so that the given series is $\quad f_0 + f_1 + f_2 + \cdots$

From the definition, we obtain

$$Z\{f_n\} = Z\left\{\frac{(-1)^n}{n+1}\right\} = 1 - \frac{1}{2z} + \frac{1}{3z^2} - \frac{1}{4z^3} + \cdots$$

$$= z\left[\frac{1}{z} - \frac{1}{2z^2} + \frac{1}{3z^3} - \frac{1}{4z^4} + ...\right] = z\ln\left(1 + \frac{1}{z}\right) = F(z).$$

Hence, $\displaystyle\sum_{n=0}^{\infty} \frac{(-1)^n}{n+1} = \lim_{z \to 1^+} F(z) = \lim_{z \to 1^+}\left[z\ln\left(1 + \frac{1}{z}\right)\right] = \ln 2.$

Example 17.25 Using Z-transforms, find the sum of the series

$$2x + \frac{4x^2}{2} + \frac{8x^3}{3} + \frac{16x^4}{4} + \cdots, \quad -\frac{1}{2} < x < \frac{1}{2}.$$

Solution Denote

$$f_n = \frac{(2x)^{n+1}}{n+1}$$

so that the given series is $f_0 + f_1 + f_2 + \cdots$

From the definition, we obtain

$$Z\{f_n\} = Z\left\{\frac{(2x)^{n+1}}{n+1}\right\} = 2x + \frac{(2x)^2}{2z} + \frac{(2x)^3}{3z^2} + \frac{(2x)^4}{4z^3} + \cdots.$$

$$= z\left[\frac{(2x/z)}{1} + \frac{(2x/z)^2}{2} + \frac{(2x/z)^3}{3} + ...\right] = -z\ln\left(1 - \frac{2x}{z}\right) = F(z).$$

Hence, $\displaystyle\sum_{n=0}^{\infty} \frac{(2x)^{n+1}}{n+1} = \lim_{z \to 1^+} F(z) = \lim_{z \to 1^+}\left[-z\ln\left(1 - \frac{2x}{z}\right)\right] = -\ln(1 - 2x).$

Example 17.26 Using Z-transforms, find the sum of the series

(i) $1 + \cos\alpha + \cos 2\alpha + \cdots + \cos n\alpha$.

(ii) $\sin\alpha + \sin 2\alpha + \cdots + \sin n\alpha$.

Solution

(i) Let $h_n = 1 + \cos\alpha + \cos 2\alpha + \cdots + \cos n\alpha$, with $h_0 = 1$.

The given series satisfies the difference equation $h_{n+1} - h_n = \cos[(n + 1)\alpha]$.

Taking Z-transforms on both sides, we get

$$Z\{h_{n+1}\} - Z\{h_n\} = Z\{\cos[(n + 1)\alpha]\} = \frac{z[z\cos\alpha - 1]}{z^2 - 2z\cos\alpha + 1}.$$

Let $Z\{h_n\} = F(z)$. We obtain

$$z[F(z) - h_0] - F(z) = \frac{z[z\cos\alpha - 1]}{z^2 - 2z\cos\alpha + 1}$$

$$(z - 1)F(z) = z + \frac{z[z\cos\alpha - 1]}{z^2 - 2z\cos\alpha + 1}$$

$$F(z) = \frac{z}{z-1} + \frac{z[z\cos\alpha - 1]}{(z-1)(z^2 - 2z\cos\alpha + 1)}$$

$$= \frac{z}{z-1} + \frac{1}{2}\left[\frac{(z^2 + z)}{z^2 - 2z\cos\alpha + 1} - \frac{z}{z-1}\right]$$

$$= \frac{z}{2(z-1)} + \frac{1}{2}\left[\frac{z(z-\cos\alpha) + z(1+\cos\alpha)}{z^2 - 2z\cos\alpha + 1}\right]$$

$$= \frac{z}{2(z-1)} + \frac{1}{2}\left[\frac{z(z-\cos\alpha)}{z^2 - 2z\cos\alpha + 1} + \frac{(1+\cos\alpha)}{\sin\alpha}\cdot\frac{z\sin\alpha}{z^2 - 2z\cos\alpha + 1}\right].$$

Taking the inverse *Z*-transform, we get

$$h_n = \frac{1}{2} + \frac{1}{2}\left[\cos n\alpha + \left(\frac{1+\cos\alpha}{\sin\alpha}\right)\sin n\alpha\right].$$

Simplifying, we may also write it as

$$h_n = \frac{1}{2}\left[1 + \frac{\sin\{n + (1/2)\}\alpha}{\sin(\alpha/2)}\right].$$

(ii) Let $h_n = \sin\alpha + \sin 2\alpha + \cdots + \sin n\alpha$, with $h_0 = 0$.

The given series satisfies the difference equation $h_{n+1} - h_n = \sin[(n+1)\alpha]$.

Taking *Z*-transforms on both sides, we get

$$Z[h_{n+1}] - Z\{h_n\} = Z\{\sin[(n+1)\alpha]\} = \frac{z^2\sin\alpha}{z^2 - 2z\cos\alpha + 1}.$$

Let $Z\{h_n\} = F(z)$. We obtain

$$z[F(z) - h_0] - F(z) = \frac{z^2\sin\alpha}{z^2 - 2z\cos\alpha + 1},$$

$$F(z) = \frac{z^2\sin\alpha}{(z-1)(z^2 - 2z\cos\alpha + 1)}.$$

$$= \frac{\sin\alpha}{2(1-\cos\alpha)}\left[\frac{z}{z-1} + \frac{z - z^2}{z^2 - 2z\cos\alpha + 1}\right]$$

$$= \frac{\sin\alpha}{2(1-\cos\alpha)}\left[\frac{z}{(z-1)} + \frac{z}{z^2 - 2z\cos\alpha + 1} - \frac{z(z-\cos\alpha) + z\cos\alpha}{z^2 - 2z\cos\alpha + 1}\right].$$

$$= \frac{1}{2(1-\cos\alpha)}\left[\frac{z}{(z-1)}\sin\alpha + \frac{z\sin\alpha}{z^2 - 2z\cos\alpha + 1} - \frac{z(z-\cos\alpha)\sin\alpha}{z^2 - 2z\cos\alpha + 1} - \frac{z\sin\alpha\cos\alpha}{z^2 - 2z\cos\alpha + 1}\right].$$

Taking the inverse *Z*-transform, we get

$$h_n = \frac{1}{2(1-\cos\alpha)}\ [\sin\alpha + \sin n\alpha - \sin\alpha\cos n\alpha - \cos\alpha\sin n\alpha]$$

$$= \frac{1}{2(1-\cos\alpha)}\ [\sin\alpha + \sin n\alpha - \sin\{(n+1)\alpha\}].$$

We may also write it as

$$h_n = \frac{1}{2}\left[\sin n\alpha + \left(\frac{\sin\alpha}{1-\cos\alpha}\right)(1-\cos n\alpha)\right].$$

Example 17.27   Using Z-transforms, find the sum of the series

$$\sin^2\theta + \sin^2 2\theta + \sin^2 3\theta + \cdots + \sin^2 n\theta.$$

Solution   Let $h_n = \sin^2\theta + \sin^2 2\theta + \sin^2 3\theta + \cdots + \sin^2 n\theta$, with $h_0 = 0$.
The given series satisfies the difference equation

$$h_{n+1} - h_n = \sum_{k=1}^{n+1}\sin^2(k\theta) - \sum_{k=1}^{n}\sin^2(k\theta) = \sin^2[(n+1)\theta].$$

Taking Z-transforms on both sides, we get

$$Z\{h_{n+1}\} - Z\{h_n\} = Z\{\sin^2[(n+1)\theta]\}.$$

Let $Z\{h_n\} = F(z)$. We obtain

$$Z\{\sin^2[(n+1)\theta]\} = \frac{1}{2}Z\{1 - \cos[2(n+1)\theta]\}$$

$$= \frac{1}{2}\left[\frac{z}{z-1} - \frac{z\{z\cos(2\theta)-1\}}{z^2 - 2z\cos(2\theta)+1}\right].$$

Hence,

$$z[F(z) - h_0] - F(z) = \frac{1}{2}\left[\frac{z}{z-1} - \frac{z\{z\cos(2\theta)-1\}}{z^2 - 2z\cos(2\theta)+1}\right]$$

$$F(z) = \frac{1}{2}\cdot\frac{z}{(z-1)^2} - \frac{1}{2}\cdot\frac{z\{z\cos(2\theta)-1\}}{(z-1)\{z^2 - 2z\cos(2\theta)+1\}}$$

$$= \frac{1}{2}\cdot\frac{z}{(z-1)^2} - \frac{1}{4}\left[\frac{z^2+z}{\{z^2 - 2z\cos(2\theta)+1\}} - \frac{z}{z-1}\right]$$

$$= \frac{1}{4}\cdot\frac{z}{z-1} + \frac{1}{2}\cdot\frac{z}{(z-1)^2} - \frac{1}{4}\left[\frac{z\{z-\cos(2\theta)\}+z\{1+\cos(2\theta)\}}{\{z^2 - 2z\cos(2\theta)+1\}}\right].$$

$$= \frac{1}{4}\cdot\frac{z}{z-1} + \frac{1}{2}\cdot\frac{z}{(z-1)^2}$$

$$- \frac{1}{4}\left[\frac{z\{z-\cos(2\theta)\}}{\{z^2 - 2z\cos(2\theta)+1\}} + \frac{\{1+\cos(2\theta)\}}{\sin(2\theta)}\cdot\frac{z\sin(2\theta)}{\{z^2 - 2z\cos(2\theta)+1\}}\right].$$

Taking the inverse Z-transform, we get

$$h_n = \frac{1}{4} + \frac{n}{2} - \frac{1}{4}\left[\cos(2n\theta) + \left(\frac{\{1+\cos(2\theta)\}}{\sin(2\theta)}\right)\sin(2n\theta)\right]$$

$$= \frac{1}{4} + \frac{n}{2} - \frac{1}{4\sin(2\theta)}[\cos(2n\theta)\sin(2\theta) + \sin(2n\theta)\cos(2\theta) + \sin(2n\theta)]$$

$$= \frac{1}{4} + \frac{n}{2} - \frac{1}{4\sin(2\theta)}[\sin\{2(n+1)\theta\} + \sin(2n\theta)]$$

$$= \frac{1}{4} + \frac{n}{2} - \frac{1}{4\sin(2\theta)}[2\sin\{(2n+1)\theta\}\cos\theta]$$

$$= \frac{1}{4} + \frac{n}{2} - \frac{\sin\{(2n+1)\theta\}}{4\sin\theta}.$$

Alternate solution

Write
$$h_n = \sin^2\theta + \sin^2 2\theta - \sin^2 3\theta + \cdots + \sin^2 n\theta$$

$$= \frac{1}{2}[(1-\cos 2\theta) + (1-\cos 4\theta) + (1-\cos 6\theta) + \cdots + (1-\cos 2n\theta)]$$

$$= \frac{n}{2} - \frac{1}{2}[\cos 2\theta + \cos 4\theta + \cos 6\theta + \cdots + \cos 2n\theta]$$

$$= \frac{n}{2} - \frac{1}{2}[\{1 + \cos\phi + \cos 2\phi + \cos 3\phi + \cdots + \cos n\phi\} - 1], \quad \phi = 2\theta.$$

Using the result of Example 17.26(i), we obtain

$$h_n = \frac{n}{2} + \frac{1}{2} - \frac{1}{4}\left[1 + \frac{\sin\{n+(1/2)\}\phi}{\sin(\phi/2)}\right] = \frac{n}{2} + \frac{1}{4} - \frac{\sin(2n+1)\theta}{4\sin\theta}.$$

Example 17.28 Using Z-transforms, find the sum of the series

$$1 + \cos^2\alpha + \cos^2 2\alpha + \cos^2 3\alpha + \cdots + \cos^2 n\alpha.$$

Solution Let $h_n = 1 + \cos^2\alpha + \cos^2 2\alpha + \cos^2 3\alpha + \cdots + \cos^2 n\alpha$, with $h_0 = 1$.

The given series satisfies the difference equation

$$h_{n+1} - h_n = \sum_{k=0}^{n+1}\cos^2(k\alpha) - \sum_{k=0}^{n}\cos^2(k\alpha) = \cos^2[(n+1)\alpha].$$

Taking Z-transforms on both sides, we get

$$Z\{h_{n+1}\} - Z\{h_n\} = Z\{\cos^2[(n+1)\alpha]\}.$$

Let $Z\{h_n\} = F(z)$. Now,

$$Z\{\cos^2[(n+1)\alpha]\} = \frac{1}{2}Z\{1 + \cos[2(n+1)\alpha]\}$$

$$= \frac{1}{2}\left[\frac{z}{z-1} + \frac{z\{z\cos(2\alpha)-1\}}{z^2-2z\cos(2\alpha)+1}\right].$$

Hence,

$$z[F(z)-h_0]-F(z) = (z-1)\,F(z)-z = \frac{1}{2}\left[\frac{z}{z-1} + \frac{z\{z\cos(2\alpha)-1\}}{z^2-2z\cos(2\alpha)+1}\right].$$

$$F(z) = \frac{z}{z-1} + \frac{z}{2(z-1)^2} + \frac{z\{z\cos(2\alpha)-1\}}{2\{z^2-2z\cos(2\alpha)+1\}(z-1)}$$

$$= \frac{z}{z-1} + \frac{z}{2(z-1)^2} + \frac{1}{4}\left[\frac{z^2+z}{\{z^2-2z\cos(2\alpha)+1\}} - \frac{z}{(z-1)}\right]$$

$$= \frac{3}{4}\cdot\frac{z}{z-1} + \frac{z}{2(z-1)^2} + \frac{1}{4}\left[\frac{z\{z-\cos(2\alpha)\}+z\{1+\cos(2\alpha)\}}{\{z^2-2z\cos(2\alpha)+1\}}\right]$$

$$= \frac{3}{4}\cdot\frac{z}{z-1} + \frac{z}{2(z-1)^2}$$

$$+ \frac{1}{4}\left[\frac{z\{z-\cos(2\alpha)\}}{\{z^2-2z\cos(2\alpha)+1\}} + \frac{\{1+\cos(2\alpha)\}}{\sin(2\alpha)}\cdot\frac{z\sin(2\alpha)}{\{z^2-2z\cos(2\alpha)+1\}}\right]$$

Taking the inverse Z-transform, we get

$$h_n = \sum_{k=0}^{\infty}\cos^2(k\alpha) = \frac{3}{4} + \frac{n}{2} + \frac{1}{4}\left[\cos(2n\alpha) + \frac{\{1+\cos(2\alpha)\}}{\sin(2\alpha)}\sin(2n\alpha)\right]$$

or

$$h_n = \frac{3}{4} + \frac{n}{2} + \frac{\sin\{(2n+1)\alpha\}}{4\sin\alpha}.$$

Alternate solution

Write

$$h_n = 1 + \cos^2\alpha + \cos^2 2\alpha + \cos^2 3\alpha + \cdots + \cos^2 n\alpha$$

$$= 1 + \frac{1}{2}\left[(1+\cos 2\alpha)+(1+\cos 4\alpha)+(1+\cos 6\alpha)+\cdots+(1+\cos 2n\alpha)\right]$$

$$= 1 + \frac{n}{2} + \frac{1}{2}\left[\cos 2\alpha + \cos 4\alpha + \cos 6\alpha + \cdots + \cos 2n\alpha\right]$$

$$= \frac{1}{2} + \frac{n}{2} + \frac{1}{2}\left[1 + \cos\phi + \cos 2\phi + \cos 3\phi + \cdots + \cos n\phi\right], \quad \phi = 2\alpha.$$

Using the result of Example 17.26(i), we obtain

$$h_n = \frac{1}{2} + \frac{n}{2} + \frac{1}{4}\left[1 + \frac{\sin\{n+(1/2)\}\phi}{\sin(\phi/2)}\right] = \frac{n}{2} + \frac{3}{4} + \frac{\sin\{(2n+1)\alpha\}}{4\sin\alpha}.$$

17.5 Answers and Hints

Exercise 17.1

1. $z/(z - a^2)$.

2. $z/(z - e^{i\theta})$.

3. $z/(z^2 + 1)$.

4. $\dfrac{z(z - (1/2))}{z^2 - z + 1}$.

5. $\dfrac{z}{2[z^2 - \sqrt{3}\,z + 1]}$.

6. $\dfrac{Az}{z - a} + \dfrac{Bz}{z - b}$.

7. $z^2/(z - 1)^2$.

8. $z^2(z + 1)/(z - 1)^3$.

9. $z(1 - a^2)/[(z - 1)\,(z - a^2)]$.

10. $az/(z - a)$.

11. $ez/(z - e)^2$.

12. $ez(e + z)/(z - e)^3$.

13. $\dfrac{az(az - \cos\beta)}{a^2 z^2 - 2a(\cos\beta)z + 1}$.

14. $\dfrac{z(z - a\cos\beta)}{z^2 - 2a(\cos\beta)z + a^2}$.

15. $\dfrac{z}{z - ae^{\beta}}$.

16. $\dfrac{z(z^2 - 1)\sin\beta}{[z^2 - 2(\cos\beta)z + 1]^2}$.

17. $\dfrac{z[(z^2 + 1)\cos\beta - 2z]}{[z^2 - 2(\cos\beta)z + 1]^2}$.

18. $e^{(\cos\beta)/z}\cos\left(\dfrac{\sin\beta}{z}\right)$.

19. $e^{2/(ez)}$.

20. $e^{(a\cos\beta)/z}\sin\left(\dfrac{a\sin\beta}{z}\right)$.

21. $f_0 = 0, f_1 = 1, f_2 = 2$.

22. $f_0 = 0, f_1 = 1, f_2 = 4$.

23. $f_0 = 1, f_1 = 0, f_2 = -1$.

24. $f_0 = 1, f_1 = -13, f_2 = 111$.

25. $f_0 = 0, f_1 = 1, f_2 = 6$.

26. 1.

27. 1/16.

28. $F(z) = \dfrac{z}{z - 2} - \dfrac{z}{z - 1}$, $f_n = 2^n - 1$.

29. $F(z) = \dfrac{2z}{z - 2} - \dfrac{z}{z - 1}$, $f_n = 2^{n+1} - 1$.

30. $F(z) = \dfrac{3z}{z + 2} - \dfrac{z}{z + 1} - \dfrac{2z}{z + 3}$, $f_n = 3(-2)^n + (-1)^{n+1} - 2(-3)^n$.

31. $F(z) = \dfrac{1}{9}\left[-\dfrac{2z}{z + 2} + \dfrac{2z}{z - 1} + \dfrac{3z}{(z - 1)^2}\right]$, $f_n = \dfrac{1}{9}[2 + 3n - 2(-2)^n]$.

32. $F(z) = \dfrac{1}{25}\left[\dfrac{2z}{z - 2} + \dfrac{5z}{(z - 2)^2} - \dfrac{2z}{z + 3} + \dfrac{10z}{(z + 3)^2}\right]$, $f_n = \dfrac{1}{150}[(12 + 15n)2^n - (12 + 20n)(-3)^n]$.

33. $F(z) = -\dfrac{2z}{(z - 1)^2} + \dfrac{3(z^2 + z)}{(z - 1)^3}$; $f_n = -2n + 3n^2$.

34. $F(z) = \dfrac{1}{54}\left[\dfrac{16z}{z + 2} - \dfrac{16z}{z - 1} + \dfrac{39z}{(z - 1)^2} + \dfrac{9(z^2 + z)}{(z - 1)^3}\right]$; $f_n = \dfrac{1}{54}[16(-2)^n - 16 + 39n + 9n^2]$.

35. $F(z) = \dfrac{z}{3(z - 1)^2} - \dfrac{(z^2 + z)}{2(z - 1)^3} + \dfrac{(z^3 + 4z^2 + z)}{6(z - 1)^4}$, $f_n = \dfrac{1}{6}(2n - 3n^2 + n^3)$.

36. $F(z) = \dfrac{1}{216}\left[-\dfrac{5z}{z + 5} + \dfrac{5z}{z - 1} - \dfrac{48z}{(z - 1)^2} - \dfrac{18(z^2 + z)}{(z - 1)^3} + \dfrac{36(z^3 + 4z^2 + z)}{(z - 1)^4}\right]$,

$f_n = \dfrac{1}{216}[-5(-5)^n + 5 - 48n - 18n^2 + 36n^3]$.

Exercise 17.2

1. $F(z) = \dfrac{z}{z-3}$, $y_n = 3^n$.

2. $F(z) = -\dfrac{29}{25}\left[\dfrac{z}{z+4}\right] + \dfrac{4}{25}\left[\dfrac{z}{z-1}\right] + \dfrac{1}{5}\left[\dfrac{z}{(z-1)^2}\right]$, $y_n = \dfrac{1}{25}[-29(-4)^n + 4 + 5n]$.

3. $F(z) = \dfrac{5}{4}\left(\dfrac{z}{z-1}\right) - \dfrac{1}{4}\left(\dfrac{z}{z+1}\right) + \dfrac{1}{2}\left(\dfrac{z}{(z-1)^2}\right)$, $y_n = \dfrac{1}{4}[5 - (-1)^n + 2n]$.

4. $F(z) = 4\left(\dfrac{z}{z+2}\right) - 3\left(\dfrac{z}{z+3}\right)$, $y_n = (-1)^n[2^{n+2} - 3^{n+1}]$.

5. $F(z) = \dfrac{5}{18}\left[\dfrac{3z}{(z-3)^2}\right] + \dfrac{1}{18}\left[\dfrac{3z(z+3)}{(z-3)^3}\right]$, $y_n = \dfrac{1}{18}(5n + n^2)3^n$.

6. $F(z) = -\dfrac{1}{108}\left[\dfrac{z}{z-1}\right] + \dfrac{109}{108}\left[\dfrac{z}{z+5}\right] + \dfrac{1}{36}\left[\dfrac{z}{(z-1)^2}\right] + \dfrac{1}{180}\left[\dfrac{5z}{(z+5)^2}\right]$,

   $y_n = \dfrac{1}{540}[(545 - 3n)(-5)^n + 5(3n - 1)]$.

7. $F(z) = -\dfrac{3}{2}\left[\dfrac{z}{z-1}\right] + 2\left[\dfrac{z}{z-2}\right] - \dfrac{1}{2}\left[\dfrac{z}{z-3}\right]$, $y_n = \dfrac{1}{2}[2^{n+2} - 3^n - 3]$.

8. $F(z) = \dfrac{1}{4}\left[\dfrac{z}{z+1}\right] + \dfrac{5}{9}\left[\dfrac{z}{z+2}\right] + \dfrac{1}{6}\left[\dfrac{z}{(z-1)^2}\right] + \dfrac{7}{36}\left[\dfrac{z}{z-1}\right]$,

   $y_n = \dfrac{1}{36}[7 + 9(-1)^n + 20(-2)^n + 6n]$.

9. $F(z) = \left[\dfrac{z}{z-2}\right] + \dfrac{1}{16}\left[\dfrac{z}{z-3}\right] + \dfrac{15}{16}\left[\dfrac{z}{z+1}\right] - \dfrac{1}{4}\left[\dfrac{z}{(z+1)^2}\right]$,

   $y_n = \dfrac{1}{16}[16(2^n) + 3^n + (15 + 4n)(-1)^n]$.

10. $F(z) = -\left[\dfrac{z}{z-2}\right] - 3\left[\dfrac{z}{(z-3)^2}\right] + 2\left[\dfrac{(z^2 + 3z)}{(z-3)^3}\right]$, $y_n = \dfrac{1}{3}[(2n^2 - 3n)3^n - 3(2^n)]$.

Chapter 18

Numerical Methods

18.1   Introduction

In the previous chapters, we have discussed methods which can be used to obtain the exact or analytical solutions of various types of problems. However, there are many problems in science and engineering, which cannot be solved exactly. For example, for the simple looking problems

$$\int_0^1 \frac{dx}{x + \sin x} \qquad \text{(definite integral)}$$

$$y' = x + \cos y \quad \text{(differential equation)}$$

$$\tan x = x \qquad \text{(root of a non-linear equation)}$$

there are no methods which can produce the exact solution. Even when a method exists to solve a problem, it may be too complicated to use the method or in most cases, we may have to add an infinite number of terms. In such situations, we use numerical methods.

Further, the mathematical models of many physical problems are either nonlinear or too complex. A way of solving such a problem is through numerical methods. Application of these methods produce a large system of linear or nonlinear algebraic equations for solution. The solution of this large system, may be thousands of equations, is obtained by some suitable numerical method.

In this chapter, we shall discuss a few numerical methods to solve certain types of problems and also determine the maximum possible error in the numerical solutions.

Errors in Computation

Since the numerical methods give approximate solutions, these solutions contain a number of errors. The following four types of errors arise in computation.

1. Machine Error   Computers have a finite word length–usually a 32 bit or a 64 bit word length. Since, most of the numbers cannot be represented exactly in, say 32 bits, errors are automatically introduced when these numbers are used in computation. The accuracy of a computer is called *machine epsilon*. The cumulative effect of these errors plays an important role in the total error in computation.

2. Inherent Errors   These are errors, which are present in the problem itself. Such errors arise due to wrong formulation of the problem or unsuitable solution procedure or inaccuracies in the data or due to simplified (invalid) assumptions in the mathematical formulation of the problem. In such cases, the problem or the solution procedure is modified suitably.

3. Round-off Errors   Real numbers like 2/3, $\sqrt{\pi}$, $\sqrt{3}$ etc. contain an infinite number of digits. In scientific and engineering computation, we express a real number x in the *floating point from* as

$$x = \pm 0 \cdot d_1 d_2 \ldots d_n \ldots \times 10^m \tag{18.1}$$

where $d_1, d_2, \ldots, d_n, \ldots$ are natural numbers between 0 and 9, m is a positive or a negative integer and is called the *exponent*. Each digit $d_1, d_2, \ldots d_n \ldots$ except the leading zeros, is called significant digit. Since, we cannot retain infinite number of digits in a number, we round-off the number to, say, n significant digits. To round-off a number to n significant digits, we discard all digits to the right of the nth digit, if the $(n + 1)$th digit is less than 5. If the $(n + 1)$th digit is more than 5, then we increase the nth digit by 1. If the $(n + 1)$th digit is equal to 5, then we increase the nth digit by 1 if it is odd and leave the nth digit unchanged if it is even. For example,

$1.6583 = 0.16583 \times 10^1$

$\qquad = 0.1658 \times 10^1$ (rounded to four significant digits)

$\qquad = 0.166 \times 10^1$ (rounded to three significant digits)

$\qquad = 0.17 \times 10^1$ (rounded to two significant digits)

$\qquad = 0.2 \times 10^1$ (rounded to one significant digit).

Note that the number $0.00230543 = 0.230543 \times 10^{-2}$, has six significant digits.

Let x be an exact number and x^* be its approximation.

If $\qquad\qquad\qquad |x - x^*| \le 0.5 \times 10^{-k}$ or $|x - x^*| \le 5 \times 10^{-(k+1)}$

then, x^* represents x accurate to k significant digits.

We define, $\qquad$ *absolute error* $= |x - x^*|$.

$$\text{relative error} = \frac{|x - x^*|}{|x|} \quad \text{or} \quad \frac{|x - x^*|}{|x^*|}.$$

$$\text{percentage error} = \frac{|x - x^*|}{|x|} \times 100.$$

In numerical computations, it is very important that the cumulative effect of round-off errors must tend to zero or atleast remains bounded. Otherwise, the round-off errors may dominate the solution and the solution may become meaningless or even overflow may occur.

Example 18.1   The number $\pi = 3.14159265 \ldots$, is approximated by 22/7. Find upto how many digits is this approximation accurate. Obtain the absolute and the relative errors.

Solution   We have $x = 3.14159265 \ldots$ $x^* = \dfrac{22}{7} = 3.14285714 \ldots,$

$$| x - x^* | = \left| \pi - \frac{22}{7} \right| = 0.00126 < 5 \times 10^{-3}.$$

Hence, the approximation is accurate to 2 decimal digits. We obtain

$$\text{absolute error} = | x - x^* | = 0.00126.$$

$$\text{relative error} = \frac{| x - x^* |}{| x |} \approx 0.0004.$$

4. Truncation Error Truncation error arises due to the use of approximate formulas which are generally obtained by truncating an infinite series, that is taking only a finite number of terms in an infinite series. We consider the Taylor series with a remainder to study the truncation error. Let

$$f(x) = f(x_0) + (x - x_0) f'(x_0) + \ldots + \frac{(x - x_0)^{m-1}}{(m - 1)!} f^{(m-1)}(x_0)$$

$$+ \frac{(x - x_0)^m}{(m)!} f^{(m)}(x_0) + \ldots \tag{18.2}$$

denote the Taylor series expansion of $f(x)$ about $x = x_0$, $x_0 \in [a, b]$. If we retain the first m terms, we get the approximation

$$f(x) \approx P_{m-1}(x) = f(x_0) + (x - x_0) f'(x_0) + \ldots + \frac{(x - x_0)^{m-1}}{(m - 1)!} f^{(m-1)}(x_0).$$

The neglected part of the series

$$\frac{(x - x_0)^m}{m!} f^{(m)}(x_0) + \frac{(x - x_0)^{m+1}}{(m + 1)!} f^{(m+1)}(x_0) + \ldots$$

also forms an infinite series. The first term in this series is called the principal part of the truncation error or simply the *truncation error* (T.E). We write the truncation error as

$$T.E = \frac{(x - x_0)^m}{m!} f^{(m)}(\xi), \quad x_0 < \xi < x. \tag{18.3}$$

Since ξ is an unknown function of x, we obtain the bound on T.E as

$$| T.E | \leq \frac{1}{m!} \max_{[a, b]} | x - x_0 |^m M_m \tag{18.4}$$

where

$$M_m = \max_{[a, b]} | f^{(m)}(x) |.$$

Assume that the value of M_m or its estimate is available. Then, we can use (18.4) to determine the maximum absolute error or an upper bound on the error. Suppose that we require $|T.E| \leq \varepsilon$. Then, we can determine

(i) the number of terms m for a given x an ε,

(ii) $|x - x_0|$ for a given m and ε. This gives an interval about x_0 in which the Taylor polynomial approximation $P_{m-1}(x)$ is valid to the prescribed accuracy.

Example 18.2 Obtain a polynomial approximation $P(x)$ to e^{-x}, using Taylor series expansion about $x_0 = 0$. Determine

(i) maximum error for $x \in [-1, 1]$, when the first four terms are used in the approximation. Also find the maximum error when $x = 0.3$.

(ii) The least number of terms required in the approximation such that error $\leq 5 \times 10^{-4}$ for $x \in [-1, 1]$.

(iii) x, when the approximation obtained from the first four terms is accurate to 5×10^{-4}.

Solution We have $f(x) = e^{-x}, f^{(r)}(x) = (-1)^r e^{-x}, r = 1, 2, \ldots$
Hence, Taylor series expansion gives

$$f(x) = e^{-x} = 1 - x + \frac{x^2}{2!} - \frac{x^3}{3!} + \ldots$$

(i) When the first four terms in the expansion are used, we get

$$P(x) = 1 - x + \frac{x^2}{2!} - \frac{x^3}{3!}$$

$$T.E = \frac{x^4}{4!} f^{(4)}(\xi) = \frac{x^4}{24} e^{-\xi}$$

and

$$|T.E| \leq \frac{1}{24} \max_{-1 \leq x \leq 1} |x|^4 \cdot \max_{-1 \leq x \leq 1} |e^{-x}|.$$

Now, $\max\limits_{-1 \leq x \leq 1} |x|^4 = 1$ and $\max\limits_{-1 \leq x \leq 1} e^{-x} = e.$

Therefore, $|T.E| \leq \dfrac{e}{24} = 0.1133$, or maximum error $= 0.1133$.

For $x = 0.3$, we get

$$|T.E| \leq \frac{(0.3)^4}{24} \cdot e = 0.00092.$$

(ii) If we use the first m terms in the approximation,

$$f(x) = e^{-x} \approx 1 - x + \frac{x^2}{2!} - \ldots + \frac{(-1)^{m-1}}{(m-1)!} x^{m-1}$$

then, $|T.E| = \left| \dfrac{x^m}{m!} f^{(m)}(\xi) \right| \leq \dfrac{1}{m!} \max\limits_{-1 \leq x \leq 1} |x|^m \max\limits_{-1 \leq x \leq 1} |f^{(m)}(x)| = \dfrac{e}{m!}.$

We choose m such that

$$\frac{e}{m!} \leq 5 \times 10^{-4}, \quad \text{or} \quad m! \geq 2000e.$$

We find that the least m for which the inequality is satisfied is $m = 8$. Hence, we need the first 8 terms to achieve the given accuracy.

(iii) From (i), we get

$$|T.E| = \frac{|x|^4}{24} |e^{-\xi}| \leq \frac{|x|^4}{24} \max_{-1 \leq x \leq 1} |e^{-x}| = \frac{|x|^4 e}{24}.$$

We choose $|x|$ such that

$$\frac{|x|^4 e}{24} \leq 5 \times 10^{-4}, \quad \text{or} \quad |x|^4 \leq \frac{120}{e} 10^{-4} = 0.00441.$$

Hence, $|x| \leq 0.2577$. Therefore, the approximation with the given accuracy is valid in the interval $[-0.2577, 0.2577]$.

Exercise 18.1

1. Write the following numbers in floating point form : 48.61416, 2.3748, 0.0436, 1.03092. Round these numbers to (i) 4 significant digits, (ii) 2 significant digits.

2. The approximations to the number 2/3 are given as 0.6, 0.66 and 0.67. Which of these is the best approximation?

3. Find to how many digits is the value 355/113, an accurate approximation to $\pi = 3.14159265...$

4. An approximate value of $\sqrt{2} = 1.414214...$ is given by 1.414. Find the absolute error and the relative error in the approximation.

5. If $x = 0.178693 \times 10^1$, $y = 0.178439 \times 10^1$ are accurate to six significant digits, then find the number of significant digits in $z = x - y$.

6. Obtain a second degree polynomial approximation to $f(x) = \sqrt{1+x}$, using Taylor series expansion about $x = 0$. Find a bound on the truncation error when (i) $x = 0.2$, (ii) $x \in [0, 1]$.

7. Obtain an nth degree polynomial approximation to $f(x) = 1/(1 + x)$, using Taylor series expansion about $x = 0$. Find the number of terms required in the expansion so that $|T.E| \leq 5 \times 10^{-3}$ for $x \in [0, 1/2]$.

8. Using Taylor series expansion for $f(x) = e^x$ about $x = 0$, find the value of e correct to four decimal places.

9. Find the first three non-vanishing terms in the Taylor series expansion of $f(x) = \cos(2x)$ about $x = \pi/4$. Use this polynomial to approximate $\cos(\pi/3)$. Obtain a bound on the truncation error in using this approximation.

10. Calculate the value of $\sin 20°$ with an error less than 1×10^{-5} in magnitude using Taylor series expansion of $f(x) = \sin x$ about $x = 0$.

18.2   Solution of Algebraic and Transcendental Equations

In many engineering applications, we need to find a root or all the roots of a polynomial equation

$$f(x) = P_n(x) = a_0 x^n + a_1 x^{n-1} + \ldots + a_{n-1} x + a_n = 0$$

or a root of a transcendental equation, which is a combination of polynomials, exponential functions, trigonometric functions etc.

For example,

$$\tan x = x, \quad x\, e^x - 1 = 0, \quad \cos x - x e^x = 0$$

are transcendental equations.

Zero or root   A number ξ, for which $f(\xi) \equiv 0$ is called a root or a zero of $f(x) = 0$. Geometrically, the root of an equation $f(x) = 0$ is the value of x at which the graph of the function $y = f(x)$ intersects the x-axis (see Fig. 18.1). We assume that $f(x)$ is a continuous function in some interval which contains the root.

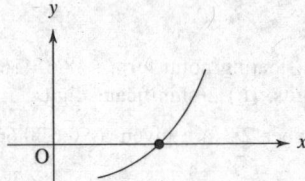

Fig. 18.1.   Root of $f(x) = 0$.

Simple root   A number ξ is a simple root of $f(x) = 0$, if $f(\xi) \equiv 0$ and $f'(\xi) \neq 0$. We may also write $f(x)$ as

$$f(x) = (x - \xi)\, g(x), \quad g(\xi) \neq 0.$$

Multiple root   A number ξ is a multiple root of multiplicity m if

$$f(\xi) \equiv 0 \equiv f'(\xi) \equiv \ldots \equiv f^{(m-1)}(\xi) \text{ and } f^{(m)}(\xi) \neq 0. \tag{8.5a}$$

We may also write $f(x)$ as

$$f(x) = (x - \xi)^m\, g(x), \quad g(\xi) \neq 0. \tag{8.5b}$$

A polynomial equation of degree n has exactly n roots, real or complex, simple or multiple, where as a transcendental equation may have one root, infinite number of roots or no root.

The methods for finding a root are classified as (i) direct methods, and (ii) iterative methods.

Direct methods   These methods give the solution in a finite number of steps. Therefore, it is possible to give the total operation count (additions, substractions, divisions, multiplications) for any method.

For example, a method for finding one root or both the roots of $a_0 x^2 + a_1 x + a_2 = 0$, is

$$x = \frac{1}{2a_0}\left[-a_1 \pm \sqrt{a_1^2 - 4a_0 a_2}\,\right], \ a_0 \neq 0$$

for which we can find the total number of operations. However, there are no direct methods available for finding a root of a transcendental equation.

Iterative methods   These methods are based on successive approximations. Starting with one or more initial approximations, we obtain a sequence of approximations $x_0, x_1, ..., x_k, ...$ which in the limit as $k \to \infty$, converge to the exact root. An iterative method for finding a root of the equation $f(x) = 0$ is of the form

$$x_{k+1} = \phi(x_k), \; k = 0, 1, 2,... \tag{18.6}$$

where x_0 is called the *initial approximation* and $\phi(x)$ is called the *iteration function*. Starting with some initial approximation x_0 to the root, we generate a sequence of iterates as $x_1 = \phi(x_0)$, $x_2 = \phi(x_1)$, ... This sequence is said to converge to the exact root ξ, if

$$\lim_{k \to \infty} x_k = \xi, \quad \text{or} \quad \lim_{k \to \infty} |x_k - \xi| = 0. \tag{18.7}$$

If we define the error of approximation as $x_k - \xi$, then (18.7) gives

$$\lim_{k \to \infty} |\text{error of approximation}| = 0.$$

A method of the form (18.6) is also called a *general one-point iteration method*. A method of the form

$$x_{k+1} = \phi(x_{k-1}, x_k), \; k = 1, 2, ... \tag{18.8}$$

is called a *general two-point iteration method*, since it requires two initial approximations x_0 and x_1 to start the method.

When to stop iteration   Since, we cannot perform infinite number of iterations, we stop the iteration when

$$|f(x_k)| \le \varepsilon, \quad \text{or} \quad |x_{k+1} - x_k| \le \varepsilon \tag{18.9}$$

where ε is the prescribed error and x_k, x_{k+1} are two consecutive iterates. Hence, to use an iterative method, we need one or more initial approximations and the iteration function $\phi(x)$. We, now describe methods for finding suitable initial approximations.

18.2.1   Initial Approximations

We, generally use the following method to obtain one or more initial approximations to the root.

Graphic method   We sketch a rough graph of the function $y = f(x)$ and find its point of intersection with the x-axis. Any point in its neighbourhood may be taken as an initial approximation. If we write the equation $f(x) = 0$ in an equivalent form, $\phi_1(x) = \phi_2(x)$, where $\phi_1(x)$ and $\phi_2(x)$ are simpler functions compared to $f(x)$, then the point of intersection of the graphs of the functions $y = \phi_1(x)$ and $y = \phi_2(x)$ gives the root of $f(x) = 0$. Any point in its neighbourhood may be taken as an initial approximation to the root. However, the method is not a practical one.

The most suitable method to find an approximation to the root is the intermediate value theorem, which we state below.

Intermediate value theorem   *If f(x) is continuous in some interval [a, b] and f(a) f(b) < 0, then there is atleast one real root or an odd number of real roots of f(x) = 0 in the interval (a, b).*

We prepare a table of values of the function $y = f(x)$ for various values of x. From the changes of signs of $f(x)$ in this table of values, we obtain an interval (a, b) which contains the root. We may take either of the end points or a suitable point in the interval (a, b) as an initial approximation. However, the values of $f(x)$ at the end points a, b, usually suggest a suitable choice of the approximation. For example, if $f(a) = -1$ and $f(b) = 4$, the root is closer to a than to b.

Example 18.3   Find an initial approximation to a root of the equation $2x^3 - x^2 + 2x - 1 = 0$, using (i) graphic method, (ii) intermediate value theorem.

Solution   We prepare a table of values of the function $y = 2x^3 - x^2 + 2x - 1$.

x	-2	-1	0	1	2
$f(x)$	-25	-6	-1	2	15

(i) We plot these points and obtain the graph of $y = f(x)$ as given in Fig. 18.2. The graph intersects the x-axis between $x = 0$ and $x = 1$. Any point in the interval $(0, 1)$ can be taken as an initial approximation.

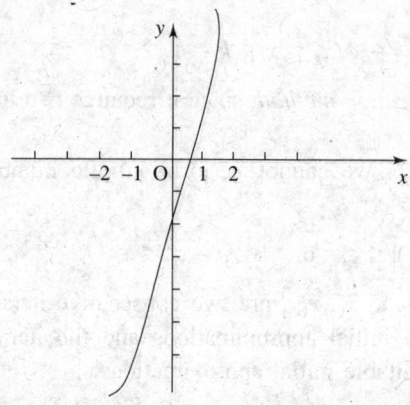

Fig. 18.2.   Graph of $y = f(x)$.

(ii) From the above table of values, we find that $f(0) = -1$ and $f(1) = 2$. Since, $f(0) f(1) < 0$, we conclude that a root exists in the interval $(0,1)$.

Example 18.4   Write the equation $xe^x - 1 = 0$ in the form $x = e^{-x}$ and find an interval which contains the root of the equation.

Solution   We plot the graphs of the functions $y = x$ and $y = e^{-x}$ as given in Fig. 18.3. We find that the two graphs intersect at a point in the interval $(0, 1)$. Hence, a root of the given equation lies in the interval $(0, 1)$.

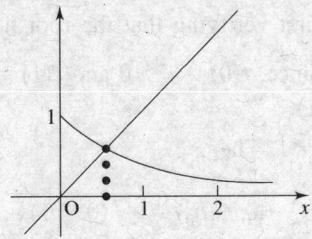

Fig. 18.3.   Graphs of $y = x$ and $y = e^{-x}$.

18.2.2   Bisection Method

This method is based on the repeated application of the intermediate value theorem to obtain an approximation to the root. Suppose that a root of $f(x) = 0$ lies in the interval $I_0 = (a_0, b_0)$, that is, $f(a_0) f(b_0) < 0$. We bisect this interval and obtain $c_1 = (a_0 + b_0)/2$. Then, the root lies in the interval

$$I_1 = (a_0, c_1), \text{ if } f(a_0) f(c_1) < 0.$$

Otherwise, it lies in the interval $I_1 = (c_1, b_0)$. Thus, the length of the interval I_1, is one half of that of I_0. Continuing this procedure, we obtain a nested set of subitervals $I_0 \supset I_1 \supset I_2 \ldots$ such that each of these subintervals contains the root. After repeating the bisection procedure n times, we obtain an interval I_n of length $(b_0 - a_0)/2^n$, which contains the root. The mid-point of the last interval is taken as the required approximation to the root.

Note that the method does not use the value of $f(x)$, but only its sign. Hence, if an accuracy for the root is prescribed, we can determine in advance for all equations, the number of iterations. After the nth iteration, we have

$$\frac{b_0 - a_0}{2^n} \leq \varepsilon.$$

Taking logarithms, we get

$$n \geq \text{int} \left[\frac{\log(b_0 - a_0) - \log \varepsilon}{\log 2} \right] \tag{18.10}$$

where int [...] stands for the next integer. For example, if $b_0 - a_0 = 1$ and $\varepsilon = 10^{-2}$, we get $n \geq 7$, that is seven iterations are required.

We observe the following:

   (i) The cost of the method is one function evaluation per iteration.
  (ii) The method never fails as the root lies in the interval being considered.
 (iii) The value of $f(x_k)$ is not used, but only its sign.
 (iv) At each iteration, the length of the interval is reduced by a factor of 2.
  (v) The method has linear convergence (see 18.2.6).
 (vi) The method is very slow, if high accuracy is required.

Example 18.5 Perform three iterations of the bisection method to obtain the smallest positive root of the equation $x^3 - 5x + 1 = 0$, first verifying that the root lies in $(0, 1)$.

Solution Let $f(x) = x^3 - 5x + 1$. Since, $f(0) = 1 > 0$ and $f(1) = -3 < 0$, the smallest positive root of $f(x) = 0$ lies in the interval $(0, 1)$.

First iteration Let $a_0 = 0$ and $b_0 = 1$. Then.

$$c_1 = \frac{1}{2}\,(a_0 + b_0) = \frac{1}{2}\,(0 + 1) = 0.5.$$

Since, $f(c_1) = -1.375 < 0$ and $f(a_0)\,f(c_1) < 0$, the root lies in the interval $(0, 0.5)$.

Second iteration Let $a_1 = 0$ and $b_1 = 0.5$. Then,

$$c_2 = \frac{1}{2}\,(a_1 + b_1) = 0.25.$$

Since, $f(c_2) = -0.2344 < 0$ and $f(a_1)\,f(c_2) < 0$, the root lies in the interval $(0, 0.25)$.

Third iteration Let $a_2 = 0$ and $b_2 = 0.25$. Then,

$$c_3 = \frac{1}{2}\,(a_2 + b_2) = 0.125.$$

Since, $f(c_3) = 0.3770 > 0$ and $f(c_3)\,f(b_2) < 0$, the root lies in the interval $(0.125, 0.25)$.

The approximation to the root is taken as $x \approx \dfrac{1}{2}\,(0.125 + 0.25) = 0.1875$.

18.2.3 Secant and Regula-falsi Methods

Secant method Let x_{k-1}, x_k be two approximations to the root $f(x) = 0$. Then, $P(x_{k-1}, f(x_{k-1}))$, $Q(x_k, f(x_k))$ are points on the curve $y = f(x)$, (see Fig. 18.4). Join the points P and Q. We approximate the curve $y = f(x)$ by the secant (chord) PQ and take the point of intersection of the secant PQ with the x-axis as the next approximation to the root. Denote $f(x_{k-1}) = f_{k-1}$ and $f(x_k) = f_k$. The equation of the straight line PQ is given by

$$\frac{y - f_k}{f_{k-1} - f_k} = \frac{x - x_k}{x_{k-1} - x_k}.$$

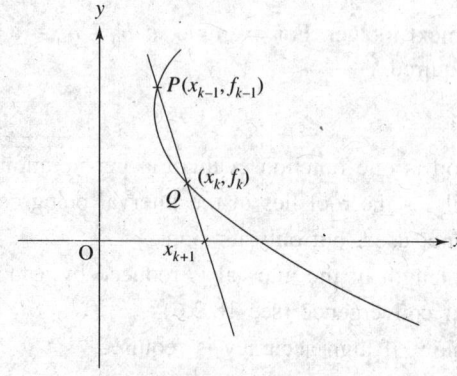

Fig. 18.4. Secant method.

Substituting $y = 0$ and solving for x, we get

$$x = x_k - \left(\frac{x_k - x_{k-1}}{f_k - f_{k-1}}\right) f_k.$$

The next approximation, x_{k+1}, to the root is written as

$$x_{k+1} = x_k - \left(\frac{x_k - x_{k-1}}{f_k - f_{k-1}}\right) f_k, \quad \text{or} \quad x_{k+1} = \frac{x_{k-1} f_k - x_k f_{k-1}}{f_k - f_{k-1}} \qquad (18.11)$$

$k = 1, 2, 3,\ldots$ This method is a two point iteration method and is called the *secant method* or the *chord method*. At any stage of the iterations, we do not test whether the root lies in the interval (x_{k-1}, x_k). That is, we use the last two consecutive iterates to obtain the next approximation to the root.

We observe the following:

  (i) The method requires two initial approximations to the root.

 (ii) The method may sometimes fail if x_0, x_1 are far away from the root.

(iii) The cost of the method is one evaluation of $f(x)$ per iteration.

(iv) The method has super linear convergence. The order of convergence is 1.62 (see 18.2.6).

Regula-falsi method   In this method, we require that the initial approximations x_0 and x_1 to the root are so chosen that $f(x_0) f(x_1) < 0$ and for each $k, f(x_{k-1}) f(x_k) < 0$. That is, at every stage of iteration the root lies in the interval (x_{k-1}, x_k). The method (18.11), with this condition, is called the *regula-falsi method* or the *method of false position*, (see, Figs. 18.5(a), (b)).

We observe the following:

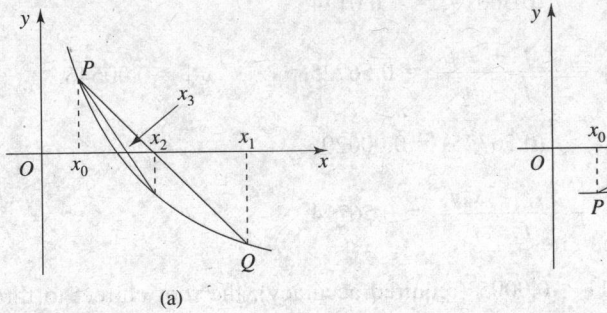

(a)   (b)

Fig. 18.5(a), (b). Regula-falsi method.

  (i) The method requires two initial approximations to the root.

 (ii) The method always converges to the root.

(iii) The cost of the method is one evaluation of $f(x)$ per iteration.

(iv) If the root lies initially in (x_0, x_1), then one of the end points is fixed for all iterations. In Fig. 18.5(a), the end point x_0 is fixed, while in Fig. 18.5(b), the end point x_1 is fixed. Then the method is of the from

$$x_{k+1} = \frac{x_0 f_k - x_k f_0}{f_k - f_0}, \quad k = 1, 2, \ldots \qquad (18.11\text{a})$$

This is the disadvantage of the method. However, it can be speeded up by inserting a bisection iteration after few iterations of the method of false position.

(v) The method has linear convergence (see 18.2.6).

(vi) If secant method converges, it converges faster than the regula-falsi method.

Example 18.6   A root of the equation $xe^x - 1 = 0$ lies in the interval (0.5, 0.1). Determine this root correct to three decimal places using (i) secant method, (ii) regula-falsi method.

Solution   Since the accuracy of three decimal places is required, we iterate until

$$| x_{k+1} - x_k| \leq 5 \times 10^{-4} = 0.0005.$$

We need to retain at least 5 decimal digits in our computation.

(i) *Secant method*   We have $x_0 = 0.5$, $x_1 = 1.0$, $f_0 = -0.17564$, $f_1 = 1.71828$. Using the method (18.11), we get the following results.

$k = 1$:
$$x_2 = \frac{x_0 f_1 - x_1 f_0}{f_1 - f_0} = 0.54637, \quad |x_2 - x_1| = 0.45363,$$

$$f_2 = f(0.54637) = -0.05643.$$

$k = 2$:
$$x_3 = \frac{x_1 f_2 - x_2 f_1}{f_2 - f_1} = 0.56079, \quad |x_3 - x_2| = 0.01442,$$

$$f_3 = f(0.56079) = -0.01746.$$

$k = 3$:
$$x_4 = \frac{x_2 f_3 - x_3 f_2}{f_3 - f_2} = 0.56725, \quad |x_4 - x_3| = 0.00646,$$

$$f_4 = f(0.56725) = 0.00029.$$

$k = 4$:
$$x_5 = \frac{x_3 f_4 - x_4 f_3}{f_4 - f_3} = 0.56714.$$

Since $| x_5 - x_4| = 0.00011 < 0.0005$ (required accuracy), the root correct to three decimal is 0.567.

(ii) *Regula-falsi method*   The root lies in (0.5, 1.0). We have $x_0 = 0.5$, $x_1 = 1.0$, $f_0 = -0.17564$, $f_1 = 1.71828$.

$k = 1$:
$$x_2 = \frac{x_0 f_1 - x_1 f_0}{f_1 - f_0} = 0.54637, \quad |x_2 - x_1| = 0.45363,$$

$$f_2 = f(0.54637) = -0.05643.$$

Since $f_1 f_2 < 0$, the root lies in the interval (x_2, x_1).

$k = 2$: $\qquad\qquad x_3 = \dfrac{x_1 f_2 - x_2 f_1}{f_2 - f_1} = 0.56079, \quad |x_3 - x_2| = 0.01442,$

$$f_3 = f(0.56079) = -0.01746.$$

Since $f_1 f_3 < 0$, the root lies in the interval (x_3, x_1).

$k = 3$: $\qquad\qquad x_4 = \dfrac{x_1 f_3 - x_3 f_1}{f_3 - f_1} = 0.56521, \quad |x_4 - x_3| = 0.00442,$

$$f_4 = f(0.56521) = -0.00533.$$

Since $f_1 f_4 < 0$, the root lies in the interval (x_4, x_1).

$k = 4$: $\qquad\qquad x_5 = \dfrac{x_1 f_4 - x_4 f_1}{f_4 - f_1} = 0.56654, \quad |x_5 - x_4| = 0.00133,$

$$f_5 = f(0.56654) = -0.00167.$$

Since $f_1 f_5 < 0$, the root lies in the interval (x_5, x_1).

$k = 5$: $\qquad\qquad x_6 = \dfrac{x_1 f_5 - x_5 f_1}{f_5 - f_1} = 0.56696, \quad |x_6 - x_5| = 0.00042 < 0.0005.$

Hence, the root correct to three decimal places is 0.567. Note that the right end point x_1, of the initial interval, is fixed in all iterations.

18.2.4 Newton-Raphson Method

Let x_0 be an initial approximation to the root of the equation $f(x) = 0$. Usually, if root $\in (a, b)$, we take x_0 as any point in (a, b). Then, $P(x_0, f_0)$, where $f_0 = f(x_0)$ is a point on the curve $y = f(x)$, (see Fig. 18.6). Draw the tangent to the curve at P. We approximate the curve $y = f(x)$ by the tangent to the curve at the point $P(x_0, f_0)$ and take the point of intersection of this tangent with the x-axis as the next approximation to the root. The equation of the tangent to the curve $y = f(x)$ at the point $P(x_0, f_0)$ is given by

$$y - f_0 = (x - x_0) f_0', \quad f_0' = f'(x_0).$$

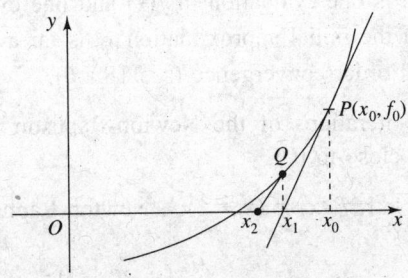

Fig. 18.6. Newton-Raphson method.

Setting $y = 0$ and solving for x, we get

$$x = x_0 - \frac{f_0}{f_0'}, \quad f_0' \neq 0.$$

Hence, we obtain the next approximation to the root as

$$x_1 = x_0 - \frac{f_0}{f_0'}.$$

The iteration method is written as

$$x_{k+1} = x_k - \frac{f_k}{f_k'}, \quad k = 0, 1, 2, \ldots \qquad (18.12)$$

This method is called the *Newton-Raphson method or the tangent method*.

Alternative

Let x_k be an approximation to the root of the equation $f(x) = 0$. Let h be an increment in x_k so that $x_k + h$ is the exact root, that is $f(x_k + h) \equiv 0$. Writing the Taylor series expansion, we get

$$f(x_k) + h f'(x_k) + \frac{h^2}{2!} f''(x_k) + \ldots \equiv 0. \qquad (18.13)$$

Since h is small, we neglect the terms containing h^2 and higher powers of h. We obtain from (18.13), the approximation

$$f_k + h f_k' = 0, \quad \text{or} \quad h = -f_k/f_k'.$$

Hence, we obtain the next approximation to the root as

$$x_{k+1} = x_k + h = x_k - \frac{f_k}{f_k'}, \quad k = 0, 1, 2, \ldots \qquad (18.14)$$

which is same as (18.12).

We observe the following

   (i) The method requires one initial approximation.
  (ii) The cost of the method is one evaluation of $f(x)$ and one evaluation of $f'(x)$ per iteration.
 (iii) The method may fail if the initial approximation x_0 is far away from the root.
 (iv) The method has second order convergence (see 18.2.6).

Example 18.7 Perform three iterations of the Newton-Raphson method to find a root of the equation $xe^x - 1 = 0$, which is close to 0.5.

Solution We have $f(x) = xe^x - 1$, $f'(x) = (x + 1)e^x$. Newton-Raphson method gives the iteration

$$x_{k+1} = x_k - \frac{f_k}{f_k'} = x_k - \frac{(x_k e^{x_k} - 1)}{(x_k + 1)e^{x_k}}.$$

We obtain the following results.

$k = 0$: $x_0 = 0.5,\ f_0 = -0.1756,\ f_0' = 2.4731.$

$$x_1 = 0.5 - \frac{-0.1756}{2.4731} = 0.5710.$$

$k = 1$: $x_1 = 0.5710,\ f_1 = 0.0107,\ f_1' = 2.7807.$

$$x_2 = 0.5710 - \frac{0.0107}{2.7807} = 0.5672.$$

$k = 2$: $x_2 = 0.5672,\ f_2 = 0.0002,\ f_2' = 2.7635.$

$$x_3 = 0.5672 - \frac{0.0002}{2.7635} = 0.5671.$$

Example 18.8 Using Newton-Raphson method, derive formulas to find (i) $1/N$, (ii) $N^{1/q}$, $N > 0$, q integer. Hence, find $1/18$, $(18)^{1/3}$ to four decimals. Use suitable initial approximations.

Solution

(i) Let $x = \dfrac{1}{N}$, $f(x) = \dfrac{1}{x} - N = 0$. We have $f'(x) = -\dfrac{1}{x^2}$. Newton-Raphson method gives

$$x_{k+1} = x_k - \frac{f_k}{f_k'} = x_k - \frac{[(1/x_k) - N]}{[-1/x_k^2]} = 2x_k - Nx_k^2.$$

We have $N = 18$, $x = 1/18$. Let $x_0 = 0.05$. The method gives

$$x_{k+1} = 2x_k - 18x_k^2, \quad k = 0, 1, 2,\ldots$$

We obtain

$$x_1 = 2x_0 - 18x_0^2 = 0.055, \quad x_2 = 2x_1 - 18x_1^2 = 0.05555,$$

$$x_3 = 2x_2 - 18x_2^2 = 0.055555.$$

Since $|x_3 - x_2| = 0.000005$, we get $(1/18) \approx 0.0556$.

(ii) Let $x = N^{1/q}$, $f(x) = x^q - N = 0$. We have $f'(x) = qx^{q-1}$.

The method gives

$$x_{k+1} = x_k - \frac{x_k^q - N}{qx_k^{q-1}} = \frac{(q-1)x_k^q + N}{qx_k^{q-1}}.$$

We have $N = 18$, $q = 3$. Let $x_0 = 2.5$. We obtain the method as

$$x_{k+1} = \frac{2x_k^3 + 18}{3\,x_k^2}, \quad k = 0, 1, 2,\ldots$$

We obtain, $\quad x_1 = \dfrac{2x_0^3 + 18}{3x_0^2} = 2.62667, \quad x_2 = \dfrac{2x_1^3 + 18}{3x_1^2} = 2.62075,$

$$x_3 = \frac{2x_2^3 + 18}{3x_2^2} = 2.62074.$$

Since $|x_3 - x_2| = 0.00001$, we get the approximation as $\sqrt[3]{18} \approx 2.6207$.

18.2.5 General Iteration Method

The method is also called *successive approximation method* or *fixed point iteration method*. We write the equation $f(x) = 0$ in an equivalent form as

$$x = \phi(x) \tag{18.15}$$

and write an iteration method as

$$x_{k+1} = \phi(x_k), \; k = 0, 1, 2, \ldots \tag{18.16}$$

The function $\phi(x)$ is called the *iteration function*.

Note that we can write the equation $f(x) = 0$ in the form $x = \phi(x)$ in several ways and can have several iteration methods of the form (18.16).

For example, consider the equation $x^3 - 5x + 1 = 0$, which has a root in the interval (0, 1). We write it in the form $x = \phi(x)$ and the corresponding iteration method in the following ways:

(i) $\qquad x = \dfrac{1}{5}\,(x^3 + 1)$, and $x_{k+1} = \dfrac{1}{5}(x_k^3 + 1), \quad k = 0, 1, 2, \ldots$ $\qquad$ (18.17)

(ii) $\qquad x = (5x - 1)^{1/3}$, and $x_{k+1} = (5x_k - 1)^{1/3}, \quad k = 0, 1, 2, \ldots$ $\qquad$ (18.18)

(iii) $\qquad x = x^3 - 4x + 1$, and $x_{k+1} = x_k^3 - 4x_k + 1, \quad k = 0, 1, 2, \ldots$ $\qquad$ (18.19)

If we take $x_0 = 1.0$, we obtain from

Method (18.17): $x_1 = 0.4$, $x_2 = 0.2128$, $x_3 = 0.2019$, …, which is converging to the root in (0, 1).

Method (18.18): $x_1 = 1.5874$, $x_2 = 1.9072$, $x_3 = 2.0437$, …, which is not converging to the root in (0, 1).

Method (18.19): $x_1 = -2$, $x_2 = 1$, $x_3 = -2$, …, which is not converging to any root and the iteration values oscillate.

Hence, the convergence of the method of the form (18.16) depends on the suitable choice of the iteration function $\phi(x)$ and the initial approximation x_0.

Condition for convergence

We write $f(x) = 0$ as $x = \phi(x)$, where $\phi(x)$ is also continuous in the interval in which the root lies. We write the iteration method as

$$x_{k+1} = \phi(x_k), \quad k = 0, 1, 2, \ldots \tag{18.20a}$$

Let ξ be the exact root. That is

$$\xi = \phi(\xi). \tag{18.20b}$$

Define, error of approximation as

$$\varepsilon_k = \xi - x_k, \quad k = 0, 1, 2, \ldots \tag{18.20c}$$

Subtracting (18.20a) from (18.20b), we get

$$\xi - x_{k+1} = \phi(\xi) - \phi(x_k)$$

$$= (\xi - x_k)\, \phi'(\alpha_k), \quad \text{(using mean value theorem)}$$

or

$$\varepsilon_{k+1} = \phi'(\alpha_k)\, \varepsilon_k, \quad x_k < \alpha_k < \xi.$$

Using this equation recursively, we get

$$\varepsilon_{k+1} = \phi'(\alpha_k)\, \phi'(\alpha_{k-1})\varepsilon_{k-1}$$

$$\vdots$$

$$= \phi'(\alpha_k)\, \phi'(\alpha_{k-1}) \ldots \phi'(\alpha_0)\, \varepsilon_0.$$

The initial error ε_0 is known and $|\varepsilon_0|$ is a finite quantity. We have

$$|\varepsilon_{k+1}| = |\phi'(\alpha_k)|\, |\phi'(\alpha_{k-1})| \ldots |\phi'|(\alpha_0)|\, |\varepsilon_0|.$$

Let

$$|\phi'(\alpha_m)| \le c, \quad m = 0, 1, \ldots, k.$$

Then,

$$|\varepsilon_{k+1}| \le c^{k+1}\, |\varepsilon_0|.$$

In the limit, as $k \to \infty$, the right hand side $\to 0$, if and only if $c < 1$. Hence, the general iteration method (18.16) converges to the root if

$$|\phi'(x_k)| \le c < 1, \quad k = 0, 1, 2, \ldots \tag{18.21}$$

Let us now study the three methods (18.17) to (18.19).

Method (18.17):

$$\phi(x) = \frac{1}{5}\,(x^3 + 1), \quad \phi'(x) = \frac{3x^2}{5}, \quad |\phi'(x)| < 1, \quad 0 < x < 1.$$

Therefore, the method converges to the root in (0, 1).

Method (18.18):

$$\phi(x) = (5x - 1)^{1/3}, \quad \phi'(x) = \frac{5}{3(5x - 1)^{2/3}}, \quad \max |\phi'(x)| = \frac{5}{3} > 1, \quad 0 < x < 1.$$

The iterations do not converge to the root in (0, 1).

Method (18.19):

$$\phi(x) = x^3 - 4x + 1, \quad \phi'(x) = 3x^2 - 4, \quad |\phi'(x)| > 1, \quad 0 < x < 1.$$

The iterations do not converge to the root in (0, 1).

How to find φ(x)   Given an equation $f(x) = 0$, it is not always possible to find a suitable iteration function $\phi(x)$ such that the iterations converge. We use the following procedure in such cases. Write $f(x) = 0$ as $x = x + \alpha\, f(x) = \phi(x)$, where $\phi(x) = x + \alpha f(x)$, and α is a parameter to be determined. Now, for convergence, we require

$$|\phi'(x)| < 1, \quad \text{or} \quad |1 + \alpha f'(x)| < 1$$

for all iterates including the initial approximation x_0. Substituting $x = x_0$, we get the condition as $|1 + \alpha f'(x_0)| < 1$. We determine the interval I for α, in which this inequality holds. Then, any choice of α in this interval I gives convergence. A suitable choice can give faster convergence. We illustrate the method through the following example.

Example 18.9   Evaluate $\sqrt{5}$ using the equation $x^2 - 5 = 0$ by applying the fixed point iteration method.

Solution   Let $f(x) = x^2 - 5$. We write $f(x) = 0$ as $x = \phi(x)$, where $\phi(x) = x + \alpha(x^2 - 5)$. The root of $f(x) = x^2 - 5 = 0$ lies in (2, 3), since $f(2) f(3) < 0$. Let $x_0 = 2.2$. Then

$$|\phi'(x_0)| = |1 + 2\alpha\, x_0| = |1 + 4.4\alpha| < 1$$

gives $-1 < 1 + 4.4\alpha < 1$. The right hand inequality gives $\alpha < 0$ and the left hand inequality gives $\alpha > (-1/2.2)$. Therefore, α lies in the interval $(-0.45, 0)$. Choose $\alpha = -0.25$. Then, the iteration method is given by

$$x_{k+1} = \phi(x_k) = x_k - 0.25(x_k^2 - 5), \quad k = 0,1,2, \ldots$$

With $x_0 = 2.2$, we get $x_1 = 2.24$, $x_2 = 2.2356$, $x_3 = 2.2361$, $x_4 = 2.2361$. Hence, $\sqrt{5} \approx 2.2361$.

18.2.6   Rate of Convergence

Now, we study the rate at which an iteration method converges if the initial approximation is close to the exact root. Let x_k be the kth iterate and ξ be the exact root. Then, $\varepsilon_k = x_k - \xi$, is the error in the kth iterate.

Order of the method   An iterative method is said to be of order p or has the rate of convergence p, if p is the largest positive real number for which there exists a positive finite constant $C \neq 0$, such that

$$|\varepsilon_{k+1}| \leq C\, |\varepsilon_k|^p. \tag{18.22}$$

The constant C is called the *asymptotic error constant* and depends on the derivatives of $f(x)$ at $x = \xi$. Equation (18.22) is also called the *error equation*.

Bisection method

Let the root of $f(x) = 0$ lie in the interval $I_0 = (a_0, b_0)$. We obtain a sequence of intervals $I_0 = (a_0, b_0)$, $I_1 = (a_1, b_1)$, $I_2 = (a_2, b_2)$, ... such that the length of the interval I_k is one half of the length of the interval I_{k-1}. For any k, we take the the mid-point of the interval I_k, that is $x_k = (a_k + b_k)/2$, as an approximation to the root. Since ξ lies in the the interval (a_k, b_k) for all k, we have

$$|\xi - x_{k+1}| \le \frac{1}{2} |\xi - x_k|, \quad \text{or} \quad |\varepsilon_{k+1}| \le \frac{1}{2} |\varepsilon_k|.$$

Comparing with (18.22), we find that $p = 1$ and $C = 1/2$. Hence, we may conclude that the rate of convergence is 1, that is the method has linear convergence.

Secant Method

Substituting $x_m = \xi + \varepsilon_m$, $m = k - 1, k, k + 1$, in the secant method

$$x_{k+1} = x_k - \left[\frac{x_k - x_{k-1}}{f_k - f_{k-1}} \right] f_k$$

we get

$$\xi + \varepsilon_{k+1} = \xi + \varepsilon_k - \left[\frac{\xi + \varepsilon_k - \xi - \varepsilon_{k-1}}{f(\xi + \varepsilon_k) - f(\xi + \varepsilon_{k-1})} \right] f(\xi + \varepsilon_k)$$

or

$$\varepsilon_{k+1} = \varepsilon_k - \left[\frac{\varepsilon_k - \varepsilon_{k-1}}{f(\xi + \varepsilon_k) - f(\xi + \varepsilon_{k-1})} \right] f(\xi + \varepsilon_k). \tag{18.23}$$

Since ξ is a simple root of $f(x) = 0$, we have $f(\xi) = 0$ and $f'(\xi) \ne 0$. Expanding in Taylor series and setting $f(\xi) = 0$. we obtain

$$f(\xi + \varepsilon_{k-1}) = \varepsilon_{k-1} f'(\xi) + \frac{1}{2} \varepsilon_{k-1}^2 f''(\xi) + \dots$$

$$f(\xi + \varepsilon_k) = \varepsilon_k f'(\xi) + \frac{1}{2} \varepsilon_k^2 f''(\xi) + \dots$$

$$f(\xi + \varepsilon_k) - f(\xi + \varepsilon_{k-1}) = (\varepsilon_k - \varepsilon_{k-1}) f'(\xi) + \frac{1}{2} (\varepsilon_k^2 - \varepsilon_{k-1}^2) f''(\xi) \dots$$

$$= (\varepsilon_k - \varepsilon_{k-1}) f'(\xi) [1 + A(\varepsilon_k + \varepsilon_{k-1}) + \dots]$$

where $A = f''(\xi)/[2 f'(\xi)]$. Substituting in (18.23), we get (after cancelling the factor $(\varepsilon_k - \varepsilon_{k-1}) f'(\xi)$)

$$\varepsilon_{k+1} = \varepsilon_k - [1 + A (\varepsilon_{k-1} + \varepsilon_k) + \dots]^{-1} [\varepsilon_k + A \varepsilon_k^2 + \dots]$$

$$= \varepsilon_k - [1 - A (\varepsilon_{k-1} + \varepsilon_k) + \ldots] [\varepsilon_k + A \varepsilon_k^2 + \ldots]$$

$$= \varepsilon_k - [\varepsilon_k - A \varepsilon_{k-1} \varepsilon_k + \ldots] = A \varepsilon_k \varepsilon_{k-1} + \ldots$$

Neglecting higher powers of ε_{k-1} and ε_k, we obtain the error equation

$$\varepsilon_{k+1} = A \varepsilon_k \varepsilon_{k-1}, \text{ and } |\varepsilon_{k+1}| = A_1 |\varepsilon_k| |\varepsilon_{k-1}| \qquad (18,24)$$

where $A_1 = |A|$. From (18.22), we obtain

$$|\varepsilon_{k+1}| = C |\varepsilon_k|^p, \quad C > 0$$

and

$$|\varepsilon_k| = C |\varepsilon_{k-1}|^p, \quad \text{or} \quad |\varepsilon_{k-1}| = C^{-1/p} |\varepsilon_k|^{1/p}.$$

Substituting in (18.24), we get

$$C |\varepsilon_k|^p = A_1 C^{-1/p} |\varepsilon_k|^{1+(1/p)}. \qquad (18.25)$$

Comparing the powers of $|\varepsilon_k|$ on both sides in (18.25), we get

$$p = 1 + \frac{1}{p} \quad \text{or} \quad p^2 - p - 1 = 0.$$

Solving, we get $p = (1 \pm \sqrt{5})/2$. Since $p > 0$, we get $p = [(1 + \sqrt{5})/2] = 1.62$. Comparing the constant terms in (18.25), we get

$$C^{1+(1/p)} = A_1, \quad \text{or} \quad C = A_1^{p/(p+1)}.$$

Hence, the secant method has super linear rate of convergence 1.62.

Regula-falsi method

We have noted earlier that in regula-falsi method, one of the end points x_0 or x_1 of the initial interval (x_0, x_1) is always fixed and the other end point varies with k. If the point x_0 is fixed, then the method is given by

$$x_{k+1} = \frac{x_0 f_k - x_k f_0}{f_k - f_0} = x_k - \frac{(x_k - x_0)}{(f_k - f_0)} f_k.$$

Following the above procedure, we obtain,

$$\varepsilon_{k+1} = A \varepsilon_0 \varepsilon_k, \quad \text{and} \quad |\varepsilon_{k+1}| = A^* |\varepsilon_k|$$

where $A^* = |A| |\varepsilon_0|$. Note that ε_0 is fixed.

Hence, regula-falsi method has linear rate of convergence.

Newton-Raphson method

Substituting $x_k = \xi + \varepsilon_k$ in the Newton-Raphson method

$$x_{k+1} = x_k - \frac{f(x_k)}{f'(x_k)}$$

we get, after cancelling ξ on both sides,

$$\varepsilon_{k+1} = \varepsilon_k - \frac{f(\xi + \varepsilon_k)}{f'(\xi + \varepsilon_k)} = \varepsilon_k - \frac{\varepsilon_k f'(\xi) + \frac{1}{2}\varepsilon_k^2 f''(\xi) + \ldots}{f'(\xi) + \varepsilon_k f''(\xi) + \ldots}$$

$$= \varepsilon_k - [\varepsilon_k + C^*\varepsilon_k^2 + \ldots][1 + 2C^*\varepsilon_k + \ldots]^{-1}$$

where $C^* = f''(\xi)/[2f'(\xi)]$, and $f'(\xi)$ is cancelled. Simplifying, we get

$$\varepsilon_{k+1} = \varepsilon_k - [\varepsilon_k + C^*\varepsilon_k^2 + \ldots][1 - 2C^*\varepsilon_k + \ldots] = \varepsilon_k - [\varepsilon_k - C^*\varepsilon_k^2 + \ldots] = C^*\varepsilon_k^2 + \ldots$$

Neglecting the higher order terms, we get

$$\varepsilon_{k+1} = C^*\varepsilon_k^2, \quad \text{or} \quad |\varepsilon_{k+1}| \le C\,|\varepsilon_k|^2, \quad \text{where} \quad C = |C^*|.$$

Comparing with (18.22), we get $p = 2$. Hence, the Newton-Raphson method is of order 2 or has quadratic convergence. The error constant is given by $C = |f''(\xi)/\{2f'(\xi)\}|$.

General iteration method (fixed point iteration method)

Write $f(x) = 0$ as $x = \phi(x)$. The iteration method is defined as $x_{k+1} = \phi(x_k)$, $k = 0, 1, 2, \ldots$ We have earlier shown that for the fixed point iteration method

$$|\varepsilon_{k+1}| \le c^{k+1}|\varepsilon_k| = A|\varepsilon_k|, \quad \text{if} \quad \phi'(x_k) \ne 0, \quad \text{that is } \phi'(\xi) \ne 0.$$

The iterations converge if $|\phi'(\xi)| \le c < 1$. In this case, the method has linear rate of convergence.

Now, let $\phi'(\xi) = 0$ and $\phi''(\xi) \ne 0$.

Then, since $\xi = \phi(\xi)$, we get

$$x_{k+1} - \xi = \phi(x_k) - \xi = \phi[\xi + (x_k - \xi)] - \xi$$

$$= [\phi(\xi) + (x_k - \xi)\,\phi'(\xi) + \frac{1}{2}\,(x_k - \xi)^2\,\varphi''(\xi) + \ldots] - \xi$$

$$= \frac{1}{2}\,(x_k - \xi)^2\,\varphi''(\xi) + \ldots$$

Neglecting the higher order terms, we get

$$|\varepsilon_{k+1}| \le \frac{1}{2}\,|\phi''(\xi)|\,|\varepsilon_k|^2.$$

Hence, the method is of order 2 or has quadratic convergence.

Now, let $\xi = \varphi(\xi)$, $\phi'(\xi) = 0 = \ldots = \phi^{(p-1)}(\xi)$, $\varphi^{(p)}(\xi) \ne 0$.

Then, from the above equation, we get

$$|\varepsilon_{k+1}| \le \frac{1}{p!}\,|\phi^{(p)}(\xi)|\,|\varepsilon_k|^p.$$

Hence, the method is of order p.

18.2.7   System of Nonlinear Equations

We can extend the Newton-Raphson method derived for finding a root of the equation $f(x) = 0$ to a system of nonlinear equations. First, consider the system of two nonlinear equations

$$f(x, y) = 0, \quad g(x, y) = 0. \tag{18.26}$$

Let $x = \xi$ and $y = \eta$ be the exact solution. That is $f(\xi, \eta) \equiv 0$ and $g(\xi, \eta) \equiv 0$. Let (x_k, y_k) be a suitable approximation to the root (ξ, η). Let $\Delta x, \Delta y$ be increments in x_k and y_k respectively, so that $(x_k + \Delta x, y_k + \Delta y)$ becomes the exact solution, that is $f(x_k + \Delta x, y_k + \Delta y) \equiv 0$ and $g(x_k + \Delta x, y_k + \Delta y) \equiv 0$. Expanding in Taylor series about the point (x_k, y_k), we get

$$f(x_k, y_k) + \left[\Delta x \frac{\partial}{\partial x} + \Delta y \frac{\partial}{\partial y}\right] f(x_k, y_k) + \frac{1}{2!}\left[\Delta x \frac{\partial}{\partial x} + \Delta y \frac{\partial}{\partial y}\right]^2 f(x_k, y_k) + \ldots = 0.$$

$$g(x_k, y_k) + \left[\Delta x \frac{\partial}{\partial x} + \Delta y \frac{\partial}{\partial y}\right] g(x_k, y_k) + \frac{1}{2!}\left[\Delta x \frac{\partial}{\partial x} + \Delta y \frac{\partial}{\partial y}\right]^2 g(x_k, y_k) + \ldots = 0.$$

Neglecting the terms containing the second and higher powers of Δx and Δy, we obtain

$$f(x_k, y_k) + \Delta x f_x(x_k, y_k) + \Delta y f_y(x_k, y_k) = 0,$$

$$g(x_k, y_k) + \Delta x g_x(x_k, y_k) + \Delta y g_y(x_k, y_k) = 0, \tag{18.27}$$

where f_x, f_y, g_x and g_y denote partial derivatives. Solving the equations (18.27) for Δx and Δy, we get

$$\Delta x = -\frac{1}{D_k}[fg_y - gf_y]_{(x_k, y_k)}, \quad \Delta y = -\frac{1}{D_k}[gf_x - fg_x]_{(x_k, y_k)} \tag{18.28}$$

where $\quad D_k = [f_x g_y - g_x f_y]_{(x_k, y_k)}.$

Hence, we obtain the next approximation as

$$x_{k+1} = x_k + \Delta x, \quad y_{k+1} = y_k + \Delta y, \quad k = 0, 1, 2, \ldots$$

Alternately, writing equations (18.27) in matrix form, we get

$$\begin{bmatrix} f_k & f_y \\ g_x & g_y \end{bmatrix}_{(x_k, y_k)} \begin{bmatrix} \Delta x \\ \Delta y \end{bmatrix} = -\begin{bmatrix} f \\ g \end{bmatrix}_{(x_k, y_k)}, \quad \text{or} \quad \mathbf{J}_k \Delta \mathbf{x} = -\mathbf{F}_k \tag{18.29}$$

where $\quad \mathbf{J}_k = \begin{bmatrix} f_x & f_y \\ g_x & g_y \end{bmatrix}_{(x_k, y_k)}, \quad \mathbf{F}_k = \begin{bmatrix} f \\ g \end{bmatrix}_{(x_k, y_k)}, \quad \text{and} \quad \Delta \mathbf{x} = \begin{bmatrix} \Delta x \\ \Delta y \end{bmatrix}.$

The matrix $\mathbf{J}$ is called the *Jocobian matrix*. The solution of the system (18.29) is given by

$$\Delta \mathbf{x} = -\mathbf{J}_k^{-1} \mathbf{F}(x_k, y_k)$$

where
$$\mathbf{J}_k^{-1} = \frac{1}{D_k} \begin{bmatrix} g_y & -f_y \\ -g_x & f_x \end{bmatrix}_{(x_k, y_k)}.$$

Therefore,
$$\begin{bmatrix} \Delta x \\ \Delta y \end{bmatrix} = - \mathbf{J}_k^{-1} \begin{bmatrix} f(x_k, y_k) \\ g(x_k, y_k) \end{bmatrix} \tag{18.30}$$

which is same as (18.28). The method has quadratic convergence.

The method can be easily extended to a system of n equations in n unknowns
$$f_1(x_1, x_2, ..., x_n) = 0,$$
$$f_2(x_1, x_2, ..., x_n) = 0,$$
$$\cdots\cdots\cdots\cdots\cdots$$
$$f_n(x_1, x_2, ..., x_n) = 0.$$

If $\mathbf{x}^{(0)} = [x_1^{(0)}, x_2^{(0)}, ..., x_n^{(0)}]$ is an initial approximation to the solution vector, then we write the method as
$$\mathbf{x}^{(k+1)} = \mathbf{x}^{(k)} - \mathbf{J}_k^{-1} F(\mathbf{x}^{(k)}), \quad k = 0, 1, 2, ... \tag{18.31}$$

where
$$\mathbf{J}_k = \begin{bmatrix} \partial f_1/\partial x_1 & \partial f_1/\partial x_2 & \cdots & \partial f_1/\partial x_n \\ \partial f_2/\partial x_1 & \partial f_2/\partial x_2 & \cdots & \partial f_2/\partial x_n \\ \cdot & \cdot & \cdot & \cdot \\ \partial f_n/\partial x_1 & \partial f_n/\partial x_2 & \cdots & \partial f_n/\partial x_n \end{bmatrix}_{(\mathbf{x}^{(k)})}$$

and
$$\mathbf{F}(\mathbf{x}^{(k)}) = [f_1, f_2, ..., f_n]^T_{(\mathbf{x}^{(k)})}.$$

We have observed in the one-dimensional case that the convergence of the Newton-Raphson method depends on the initial approximation. In the case of higher dimensions, this requirement becomes more important. A necessary and sufficient condition for convergence, is
$$\rho(\mathbf{J}_k^{-1}) < 1 \tag{18.32}$$

where ρ denotes the spectral radius (largest eigenvalue in magnitude) of the matrix $\mathbf{J}_k^{-1}$. We stop the iterations, when
$$|x_i^{(k+1)} - x_i^{(k)}| \le \varepsilon, \quad \text{for all } i$$

where ε is the prescribed error tolerance.

Remark 1

The solution of the system as given in (18.31) may be computationally expensive and prone to round off errors. We use the form of the system as given in (18.29), which can be solved by any direct or iterative method for the increment. The next iterated value is given by $\mathbf{x}^{(k+1)} = \mathbf{x}^{(k)} + \Delta \mathbf{x}$.

Example 18.10 Set up the Newton-Raphson iteration scheme in matrix form to solve the system of equations

$$3x^2 + y^2 = 4, \quad x^2 + xy + y^2 = 3.$$

Perform two iterations, starting with $(x_0, y_0) = (0.8, 0.8)$.

Solution Let $f(x, y) = 3x^2 + y^2 - 4 = 0$ and $g(x, y) = x^2 + xy + y^2 - 3 = 0$.

We have

$$\mathbf{J}_k = \begin{bmatrix} f_x & f_y \\ g_x & g_y \end{bmatrix}_{(x_k, y_k)} = \begin{bmatrix} 6x_k & 2y_k \\ 2x_k + y_k & x_k + 2y_k \end{bmatrix}.$$

$$\mathbf{J}_k^{-1} = \frac{1}{D_k} \begin{bmatrix} x_k + 2y_k & -2y_k \\ -(2x_k + y_k) & 6x_k \end{bmatrix}, \quad D_k = 6x_k^2 + 8x_k\,y_k - 2y_k^2.$$

From (18.31), we obtain the iteration scheme

$$\begin{bmatrix} x_{k+1} \\ y_{k+1} \end{bmatrix} = \begin{bmatrix} x_k \\ y_k \end{bmatrix} - \mathbf{J}_k^{-1} \begin{bmatrix} 3x_k^2 + y_k^2 - 4 \\ x_k^2 + x_k\,y_k + y_k^2 - 3 \end{bmatrix}.$$

Starting with $x_0 = 0.8$, $y_0 = 0.8$, we obtain

$$\begin{bmatrix} x_1 \\ y_1 \end{bmatrix} = \begin{bmatrix} 0.8 \\ 0.8 \end{bmatrix} - \frac{1}{7.68} \begin{bmatrix} 2.4 & -1.6 \\ -2.4 & 4.8 \end{bmatrix} \begin{bmatrix} -1.44 \\ -1.08 \end{bmatrix} = \begin{bmatrix} 1.025 \\ 1.025 \end{bmatrix}.$$

$$\begin{bmatrix} x_2 \\ y_2 \end{bmatrix} = \begin{bmatrix} 1.025 \\ 1.025 \end{bmatrix} - \frac{1}{12.6075} \begin{bmatrix} 3.075 & -2.05 \\ -3.075 & 6.15 \end{bmatrix} \begin{bmatrix} 0.2025 \\ 0.1519 \end{bmatrix} = \begin{bmatrix} 1.0003 \\ 1.0003 \end{bmatrix}.$$

18.2.8 Roots of Polynomial Equations

In many applications, we require all the roots, real and complex, of a polynomial equation of degree n

$$P_n(x) = a_0 x^n + a_1 x^{n-1} + \ldots + a_{n-1} x + a_n = 0. \tag{18.33}$$

where $a_0, a_1, \ldots, a_n$ are real constants.

Methods like Newton-Raphson method can be used to extract a root. If complex arithmetic is done, we can find a complex root. However, complex roots occur in pairs as the coefficients in the polynomial equation are real. Therefore, complex pairs can be obtained as the roots of a quadratic equation. We now present two efficient methods to find one or all roots of a polynomial equation.

18.2.8.1 Birge-Vieta method

Newton-Raphson method is an efficient method for finding a root of an algebraic or a polynomial equation. Birge-Vieta method is same as Newton-Raphson method as applied to polynomial equations, where $f(x_n)$ and $f'(x_n)$ are evaluated by using *synthetic division procedure* (recurrence relations).

Let $x = \alpha$ be an approximate root of the equation (18.33). Then, we can write

$$P_n(x) = (x - \alpha) \, Q_{n-1}(x) + g(\alpha), \tag{18.34}$$

where $Q_{n-1}(x) = b_0 x^{n-1} + b_1 x^{n-2} + \ldots + b_{n-2} x + b_{n-1}$ is a polynomial of degree $n - 1$ and $g(\alpha)$ is the remainder. Since α is not the exact root, $g(\alpha) \neq 0$. The problem is to find the value of α, say α^*, such that $P_n(\alpha^*) = g(\alpha^*) = 0$. Starting with an intial approximation α_0, we write the Newton-Raphson method to find the root as

$$\alpha_{n+1} = \alpha_n - \frac{P_n(\alpha_n)}{P_n'(\alpha_n)}, \quad n = 0, 1, 2, \ldots$$

From (18.34), we get

$$P_n(x) = a_0 x^n + a_1 x^{n-1} + \ldots + a_{n-1} x + a_n$$
$$= (x - \alpha)(b_0 x^{n-1} + b_1 x^{n-2} + \ldots + b_{n-2} x + b_{n-1}) + g(\alpha).$$

Multiplying the terms on the right hand side and comparing the powers of x, we get

$$b_0 = a_0, \; b_1 = a_1 + \alpha b_0, \; b_2 = a_2 + \alpha b_1, \; \ldots, \; b_{n-1} = a_{n-1} + \alpha b_{n-2}, \; g = a_n + \alpha b_{n-1}.$$

Hence, we have the recurrence relation,

$$b_k = a_k + \alpha b_{k-1}, \quad k = 1, 2, \ldots, n$$

where $b_0 = a_0$, and $P_n(\alpha) = g(\alpha) = b_n$. $\tag{18.35}$

Differentiating the recurrence relation with respect to α, we get

$$\frac{db_k}{d\alpha} = b_{k-1} + \alpha \frac{db_{k-1}}{d\alpha}. \tag{18.36}$$

Denote $c_{k-1} = (db_k/d\alpha)$. Then, we can write (18.36) as

$$c_{k-1} = b_{k-1} + \alpha c_{k-2}, \quad \text{or} \quad c_k = b_k + \alpha c_{k-1}, \quad k = 1, 2, \ldots, n-1$$

and $c_0 = \dfrac{db_1}{d\alpha} = b_0$. Differentiating (18.35) with respect to α, we get

$$P_n'(\alpha) = \frac{dg(\alpha)}{d\alpha} = \frac{db_n}{d\alpha} = b_{n-1} + \alpha \frac{db_{n-1}}{d\alpha} = b_{n-1} + \alpha \, c_{n-2} = c_{n-1}.$$

We obtain the Birge-Vieta method as

$$\alpha_{n+1} = \alpha_n - \frac{b_n}{c_{n-1}}, \quad n = 0, 1, 2, \ldots \tag{18.37}$$

The computation of the coefficients is given below

α	a_0	a_1	a_2	$\ldots$	a_{n-1}	a_n
		$+ \alpha b_0$	$+ \alpha b_1$	$\ldots$	$+ \alpha b_{n-2}$	$+ \alpha b_{n-1}$
	b_0	b_1	b_2	$\ldots$	b_{n-1}	$b_n = g$
		$+ \alpha c_0$	$+ \alpha c_1$	$\ldots$	$+ \alpha c_{n-2}$	
	c_0	c_1	c_2	$\ldots$	$c_{n-1} = \dfrac{dg}{d\alpha}$	

Eq.(18.37) is iterated until the prescribed accuracy is obtained. If the deflated polynomial $Q_{n-1}(x)$ is required, we obtain the coefficients $b_0, b_1, \ldots b_{n-1}$, from the first step of the synthetic division procedure. If another root is required, we repeat the above procedure on $Q_{n-1}(x)$.

Example 18.11 Using the Birge-Vieta method, find the smallest positive root of the polynomial equation $3x^3 - 8x^2 - 31x + 60 = 0$. Find the deflated polynomial.

Solution Since, $f(1) f(2) < 0$, the root lies in $(1, 2)$. Let $x_0 = 1.5$. We have the following table of values.

1.5	3	− 8	− 31	60
		4.5	− 5.25	− 54.375
	3	− 3.5	− 36.25	$b_3 = 5.625$
		4.5	1.5	
	3	1.0	− 34.75 = c_2	

First iteration gives $x_1 = x_0 - \dfrac{b_3}{c_2} = 1.5 - \dfrac{5.625}{(-34.75)} = 1.66187.$

1.66187	3	− 8	− 31	60
		4.98561	− 5.00952	− 59.84314
	3	− 3.01439	− 36.00952	$b_3 = 0.15686$
		4.98561	3.27591	
	3	1.97122	− 32.73361 = c_2	

Second iteration gives $x_2 = x_1 - \dfrac{b_3}{c_2} = 1.66187 - \dfrac{0.15686}{(-32.73361)} = 1.66666.$

Since $f(x_2) = f(1.66666) = 0.0002$, the root may be taken as 1.66666. The exact root is 5/3. We perform one step of the above procedure to get the deflated polynomial.

1.66666	3	− 8	− 31	60
		4.99998	− 5.00001	− 59.99978
	3	− 3.00002	− 36.00001	…

The deflated polynomial is

$$Q_2(x) = b_0 x^2 + b_1 x + b_2 = 3x^2 - 3.00002x - 36.00001 \approx 3(x^2 - x - 12).$$

18.2.8.2   Bairstow's method

Bairstow's method uses the following procedure:

  (i) Extract a quadratic factor from the given polynomial. The roots of the quadratic may be real or complex.

 (ii) Deflate the polynomial $P_n(x)$ using the extracted quadratic to obtain a polynomial $P_{n-2}(x)$ of degree $n-2$.

(iii) Repeat the extraction of a quadratic factor from $P_{n-2}(x)$, and again deflate it. Continue until a quadratic or a linear polynomial is obtained. The roots of the quadratics give pairs of real roots or complex pairs.

It is suggested to start with an even degree polynomial (when n is large). If the polynomial is of odd degree, it can be multiplied by x (a root is zero).

Let $x^2 + \alpha x + \beta$ be the quadratic factor to be extracted. Then

$$P_n(x) = (x^2 + \alpha x + \beta) P_{n-2}(x) + h x + g,$$

where $P_{n-2}(x)$ is a polynomial of degee $n-2$ and $hx + g = h(\alpha, \beta) x + g(\alpha, \beta)$ is the remainder. The problem is to determine α, β such that $h(\alpha, \beta) = 0$ and $g(\alpha, \beta) = 0$. The solution of these equations are obtained by an iterative procedure. If (α_0, β_0) is an initial approximation and $(\alpha_0 + \Delta\alpha, \beta_0 + \Delta\beta)$ is the true solution, then we obtain (see Eq. (18.28)) the iteration procedure

$$\Delta\alpha \approx -\frac{1}{D_k}[hg_\beta - gh_\beta]_{(a_k, \beta_k)}, \quad \Delta\beta \approx -\frac{1}{D_k}[gh_\alpha - hg_\alpha]_{(\alpha_k, \beta_k)}, \tag{18.38}$$

where $D_k = [h_\alpha g_\beta - h_\beta g_\alpha]_{(a_k, \beta_k)}$. The next approximation is given by $\alpha_{k+1} = \alpha_k + \Delta\alpha$, $\beta_{k+1} = \beta_k + \Delta\beta$. The iteration is continued until the required accuracy is attained, that is $|\alpha_{k+1} - \alpha_k| < \varepsilon, |\beta_{k+1} - \beta_k| < \varepsilon$.

The values of $h(\alpha_k, \beta_k)$, $g(\alpha_k, \beta_k)$ and their partial derivatives at (α_k, β_k) can be obtained using synthetic division procedure. We write

$$P_n(x) = (x^2 + \alpha x + \beta) P_{n-2}(x) + hx + g$$

$$= (x^2 + \alpha x + \beta)(b_0 x^{n-2} + b_1 x^{n-3} + \ldots + b_{n-3}x + b_{n-2}) + hx + g.$$

Compare the powers of x. The first step of synthetic division procedure is obtained using the recurrence formula

$$b_k = a_k - \alpha b_{k-1} - \beta b_{k-2}, \quad k = 1, 2, \ldots, n; \quad b_0 = a_0, \quad b_{-1} = 0, \tag{18.39}$$

and $h(\alpha_k, \beta_k) = a_{n-1} - \alpha b_{n-2} - \beta b_{n-3} = b_{n-1}, \quad g(\alpha_k, \beta_k) = a_n - \beta b_{n-2} = b_n + \alpha b_{n-1}.$

Differentiating (18.39) partially with respect to α and β, we obtain the second step of synthetic division procedure as

$$c_k = b_k - \alpha c_{k-1} - \beta c_{k-2}, \quad k = 1, 2, \ldots, n-1; \quad c_0 = b_0, \quad c_{-1} = 0, \tag{18.40}$$

where the notations $c_{k-1} = -(\partial b_k/\partial\alpha)$, and $c_{k-2} = -(\partial b_k/\partial\beta)$ are used . At (α_k, β_k), the partial derivatives are defined by

$$h_\alpha = -c_{n-2}, \quad h_\beta = -c_{n-3}, \quad g_\alpha = b_{n-1} - c_{n-1} - \alpha\, c_{n-2}, \quad g_\beta = -(c_{n-2} + \alpha\, c_{n-3}).$$

Substituting in (18.38), we obtain the approximations for $\Delta\alpha$, $\Delta\beta$, and $\alpha_{k+1} = \alpha_k + \Delta\alpha$, $\beta_{k+1} = \beta_k + \Delta\beta$. We repeat the procedure until the required accuracy is obtained. The pair of real or complex roots are the roots of $x^2 + \alpha x + \beta = 0$. The synthetic division procedure is given below.

$-\alpha$	a_0	a_1	a_2	$\cdots$	a_{n-1}	a_n
$-\beta$		$-\alpha b_0$	$-\alpha b_1$	$\cdots$	$-\alpha b_{n-2}$	$-\alpha b_{n-1}$
			$-\beta b_0$	$\cdots$	$-\beta b_{n-3}$	$-\beta b_{n-2}$
	b_0	b_1	b_2	$\cdots$	b_{n-1}	b_n
		$-\alpha c_0$	$-\alpha c_1$	$\cdots$	$-\alpha c_{n-2}$	
			$-\beta c_0$	$\cdots$	$-\beta c_{n-3}$	
	c_0	c_1	c_2	$\cdots$	c_{n-1}	

If all the roots are required, we repeat the steps (ii) and (iii). The first step of the synthetic division procedure gives the deflated polynomial.

Example 18.12   Obtain the complex pair of roots of the equation $x^3 - 1 = 0$, using the Bairstow method. Assume the initial approximation as $(\alpha_0, \beta_0) = (0.9, 0.9)$.

Solution   The given equation has one real root and a complex pair of roots. With the given initial approximation, we obtain the synthetic division procedure as follows:

-0.9	1.0	0.0	0.0	-1.0
-0.9		-0.9	0.81	0.081
			-0.9	0.81
	$1.0 = b_0$	$-0.9 = b_1$	$-0.09 = b_2$	$-0.109 = b_3$
		-0.9	1.62	
			-0.9	
	$1.0 = c_0$	$-1.8 = c_1$	$0.63 = c_2$	

We obtain $h(\alpha_0, \beta_0) = b_2 = -0.09$, $\quad g(\alpha_0, \beta_0) = b_3 + \alpha_0 b_2 = -0.109 + 0.9(-0.09) = -0.19$.

$$h_\alpha(\alpha_0, \beta_0) = -c_1 = 1.8, \quad h_\beta(\alpha_0, \beta_0) = -c_0 = -1.0.$$

$$g_\alpha(\alpha_0, \beta_0) = b_2 - c_2 - \alpha_0 c_1 = -0.09 - 0.63 - 0.9\,(-1.8) = 0.9,$$

$$g_\beta(\alpha_0, \beta_0) = -(c_1 + \alpha_0 c_0) = -(-1.8 + 0.9) = 0.9.$$

$$D_k = [h_\alpha\, g_\beta - h_\beta\, g_\alpha]_{(\alpha_0, \beta_0)} = 1.8\,(0.9) - 0.9\,(-1) = 2.52.$$

$$\Delta\alpha = -\frac{1}{D_k}\,[hg_\beta - gh_\beta] = -\frac{1}{2.52}\,[(-0.09)(0.9) + 0.19(-1.0)] = 0.10754.$$

$$\Delta\beta = -\frac{1}{D_k}\,[gh_\alpha - hg_\alpha] = -\frac{1}{2.52}\,[(-0.19)(1.8) + (0.09)(0.9)] = 0.10357.$$

$$\alpha_1 = \alpha_0 + \Delta\alpha = 0.9 + 0.10754 = 1.00754, \quad \beta_1 = \beta_0 + \Delta\beta = 0.9 + 0.10357 = 1.00357.$$

For the next iteration, we have

-1.00754	1.0	0.0	0.0	-1.0
-1.00357		-1.00754	1.01514	-0.011657
			-1.00357	1.011137
	$1.0 = b_0$	$-1.00754 = b_1$	$0.01157 = b_2$	$-0.00052 = b_3$
		-1.00754	2.03027	
			-1.00357	
	$1.0 = c_0$	$-2.01508 = c_1$	$1.03827 = c_2$	

We obtain $h(\alpha_1,\,\beta_1) = b_2 = 0.01157, \quad g(\alpha_1,\,\beta_1) = b_3 + \alpha_0 b_2 = 0.01114.$

$$h_\alpha(\alpha_1,\,\beta_1) = -c_1 = 2.01508, \quad h_\beta(\alpha_1,\,\beta_1) = -c_0 = -1.0.$$

$$g_\alpha(\alpha_1,\,\beta_1) = b_2 - c_2 - \alpha_1 c_1 = 1.00357, \quad g_\beta(\alpha_1,\,\beta_1) = -(c_1 + \alpha_1 c_0) = 1.00754.$$

$$D_k = [h_\alpha g_\beta - h_\beta g_\alpha]_{(a_1,\,\beta_1)} = 3.03384.$$

$$\Delta\alpha = -\frac{1}{D_k}\,[hg_\beta - gh_\beta] = -0.000751. \quad \Delta\beta = -\frac{1}{D_k}\,[gh_\alpha - hg_\alpha] = -0.00357.$$

$$\alpha_2 = \alpha_1 + \Delta\alpha = 1.00003, \quad \beta_2 = \beta_1 + \Delta\beta = 1.0.$$

The quadratic factor is $x^2 + x + 1$ and the roots of $x^2 + x + 1 = 0$ are $(-1 \pm i\sqrt{3})/2.$

Exercise 18.2

1. Find an interval of unit length which contains the smallest positive root of the equation $x^3 - 5x - 1 = 0$. Hence, determine the number of iterations required by the bisection method so that $|\text{error}| < 10^{-3}$.

2. Find an interval of unit length which contains the smallest negative root in magnitude of the equation $x^3 + x + 12 = 0$. Hence, determine the number of iterations required by the bisection method so that we obtain the root correct to 5×10^{-3} (two decimal places).

3. Find an interval of unit length which contains the smallest positive root of the equation $x^3 - 3x - 1 = 0$. Take the end points of this interval as initial approximations and perform three iterations of the secant method.

4. Find an interval of unit length which contains the smallest negative root in magnitude of the equation $2x^3 + 3x^2 + 2x + 5 = 0$. Using the end points of this interval as initial approximations, perform four iterations of the regula-falsi method.

5. Find an interval of unit length which contains the smallest positive root of the equation $e^x - 2x^2 = 0$. Taking the mid-point of this interval as initial approximation, perform two iterations of the Newton-Raphson method.

6. Using three iterations of the Newton-Raphson method, obtain approximate values of $\sqrt{31}$, $(101)^{1/3}$ and $1/17$. Use suitable initial approximations.

7. The equation $f(x) = \ln x - x + 3 = 0$ has a root in the interval $(4, 5)$. Obtain the root correct to three decimal places using (i) secant method with $x_0 = 4$, $x_1 = 5$, (ii) regula-falsi method with $x_0 = 4$, $x_1 = 5$, (iii) Newton-Raphson method with $x_0 = 4.5$.

8. The equation $x^3 - 5x - 1 = 0$ has a root in the interval $(-1, 0)$. Write this equation in an equivalent form $x = \phi(x)$ so that the general iteration method $x_{k+1} = \phi(x_k)$ is convergent. Hence, perform four iterations of this method starting with $x_0 = -0.5$.

9. The equation $f(x) = 3x^3 + 4x^2 + 4x + 1 = 0$ has a root in the interval $(-1, 0)$. Write this equation in an equivalent from as $x = x + \alpha f(x) = \phi(x)$ and write the iteration method as $x_{k+1} = \phi(x_k)$. Determine the values of α such that the method converges, when $x_0 = -0.5$. Use a suitable value of α and perform four iterations starting with $x_0 = -0.5$.

10. The method $x_n^* = x_n - \dfrac{f(x_n)}{f'(x_n)}$, $x_{n+1} = x_n^* - \dfrac{f(x_n^*)}{f'(x_n)}$, $n = 0, 1, 2, \ldots$ is used to find a simple root

   of the equation $f(x) = 0$. Perform two iterations of this method to find a root of the equation $x^3 + x^2 + 3x + 4 = 0$, starting with $x_0 = -1.5$.

11. The equation $x^2 + ax + b = 0$ has two real roots α and β. Show that the method

   (i) $x_{k+1} = -\dfrac{1}{x_k}(ax_k + b)$ converges to α, if $|\alpha| > |\beta|$.

   (ii) $x_{k+1} = -\dfrac{b}{x_k + a}$ converges to α, if $|\alpha| < |\beta|$.

   (iii) $x_{k+1} = -\dfrac{1}{a}(x_k^2 + b)$ converges to α, if $2|\alpha| < |\alpha + \beta|$.

12. The method $x_{k+1} = x_k - \dfrac{x_k - x_0}{f_k - f_0} f_k$, $k = 0, 1, \ldots$ is used to find a simple root of $f(x) = 0$. Determine

   the rate of convergence of the method.

13. An iteration method is defined by

   $$x_{n+1} = \frac{x_n}{2a}(3a - x_n^2), \quad a > 0, \quad n = 0, 1, \ldots$$

   Find the quantity to which the method converges. Hence, determine the rate of convergence of the method. Also, obtain the asymptotic error constant.

14. Show that the method

   $$x_{n+1} = \frac{1}{8} x_n \left[6 + \frac{3a}{x_n^2} - \frac{x_n^2}{a}\right],$$

   converges to $\sqrt{a}$. Find the rate of convergence of the method.

15. The equation $f(x) = 0$ has a simple root in the interval $(1, 2)$. The function $f(x)$ is such that $|f'(x)| \geq 4$ and $|f''(x)| \leq 3$, for all x in $(1, 2)$. Assuming that the Newton-Raphson method converges for all initial approximations in $(1, 2)$, determine the maximum number of iterations required to obtain root correct to 6 decimal places.

16. The method $x_{n+1} = px_n + \dfrac{qa}{x_n^2}$, $n = 0, 1, \ldots$ is used to find an approximate value of $a^{1/3}$, $a > 0$.

Determine p and q, so that the order of the method is as high as possible. Find the method.

17. Find the order of convergence of the *Steffenson* method

$$x_{k+1} = x_k = \frac{f_k}{g_k}, \quad \text{where} \quad g_k = \frac{f(x_k + f_k) - f_k}{f_k}$$

where $f_k = f(x_k)$. Perform two iterations of this method to obtain the root of $x - 1 + e^{-2x} = 0$, $x_0 = 0.7$.

In problems **18** to **21**, set up the Newton-Raphson iteration scheme in matrix from and hence solve the given system of equations.

18. $x^3 + 2y^3 = 10$, $3x^2 + 4y^2 = 16$, starting with $x_0 = 1.8$, $y_0 = 0.8$, iterate three times. The exact solution is $x = 2$, $y = 1$.

19. $x^2y + y^3 = 10$, $xy^2 - x^2 = 3$, starting with $x_0 = 0.8$, $y_0 = 2.2$, iterate three times. The exact solution is $x = 1$, $y = 2$.

20. $x^2 + y^2 = 1.12$, $xy = 0.23$, staring with $x_0 = 1.0$, $y_0 = 0.2$, iterate two times.

21. $y\cos(xy) + 1 = 0$, $\sin(xy) + x - y = 0$, starting with $x_0 = 1$, $y_0 = 2$, iterate two times.

22. Using the Birge-Vieta method, find a real root of the equation $3x^3 + 8x^2 + 8x + 5 = 0$.

23. Using the Birge-Vieta method, find the smallest positive root of the equation $3x^4 + 2x^3 - 13x^2 - 8x + 4 = 0$.

24. Using the Birge-Vieta method, find a negative root of the equation $7x^3 + 37x^2 + 81x + 115 = 0$.

In Problems **25** to **28**, find the roots using Bairstow's method. Assume the intial approximation to (α, β) as given.

25. $x^4 + x^3 - 2x^2 + 2x + 4 = 0$. $(\alpha_0, \beta_0) = (2.5, 1.9)$.

26. $x^3 - 7x^2 + 25x - 39 = 0$. $(\alpha_0, \beta_0) = (-3.9, 12.8)$.

27. $x^3 + 2x^2 - 5x - 6 = 0$. $(\alpha_0, \beta_0) = (0.9, -5.8)$.

28. $x^3 - 6x^2 + 11x - 6 = 0$. $(\alpha_0, \beta_0) = (-2.8, 1.9)$.

18.3   Solution of Linear System of Equations

In Chapter 3 (section 3.4), we have defined the elementary row transformations and used them to find the rank of a matrix. We have also discussed a few direct methods for solving the linear system of equations

$$a_{11}x_1 + a_{12}x_2 + \ldots + a_{1n}x_n = b_1,$$

$$a_{21}x_1 + a_{22}x_2 + \ldots + a_{2n}x_n = b_2,$$

$$\ldots \quad \ldots \quad \ldots \quad \ldots \quad \ldots$$

$$a_{n1}x_1 + a_{n2}x_2 + \ldots + a_{nn}x_n = b_n$$

or, in matrix form

$$\mathbf{Ax} = \mathbf{b}. \tag{18.41}$$

where, the coefficient matrix $\mathbf{A}$, right hand side vector $\mathbf{b}$ and solution vector $\mathbf{x}$ are respectively given by

$$\mathbf{A} = \begin{bmatrix} a_{11} & a_{12} & \cdots & a_{1n} \\ a_{21} & a_{22} & \cdots & a_{2n} \\ \cdot\cdot & \cdot\cdot & \cdot\cdot & \cdot\cdot \\ a_{n1} & a_{n2} & \cdots & a_{nn} \end{bmatrix}, \quad \mathbf{b} = \begin{bmatrix} b_1 \\ b_2 \\ \vdots \\ b_n \end{bmatrix}, \quad \mathbf{x} = \begin{bmatrix} x_1 \\ x_2 \\ \vdots \\ x_n \end{bmatrix}.$$

The matrix $[\mathbf{A} \mid \mathbf{b}] = \begin{bmatrix} a_{11} & a_{12} & \cdots & a_{1n} & b_1 \\ a_{21} & a_{22} & \cdots & a_{2n} & b_2 \\ \cdots & \cdots & \cdots & \cdots & \cdots \\ a_{n1} & a_{n2} & \cdots & a_{nn} & b_n \end{bmatrix}$

is called the *augmented matrix* of the system.

The system of equations (18.41) is said to be

(a) *consistent* (it has atleast one solution) if rank $(\mathbf{A})$ = rank $[\mathbf{A} \mid \mathbf{b}] = r$.
   (i) If $r = n$, then the system has a unique solution.
   (ii) If $r < n$, then the system has $n - r$ parameter family of solutions, that is, infinite number of solutions.

(b) *inconsistent* (it has no solution) if rank $(\mathbf{A}) \neq$ rank $[\mathbf{A} \mid \mathbf{b}]$.

We assume that the system of equations (18.41) has a unique solution, that is rank $(\mathbf{A}) = n$, or $|\mathbf{A}| \neq 0$ or $\mathbf{A}$ is non-singular. We now discuss some of the direct and iterative methods for the solution of the system of equations (18.41).

Direct Method

We shall derive and illustrate the method for a 3×3 system of equations.

First, consider the system of equations

$$\begin{aligned} a_{11}x_1 &= b_1 \\ a_{22}x_2 &= b_2 \\ a_{33}x_3 &= b_3 \end{aligned} \qquad (18.42a)$$

or $\mathbf{Dx} = \mathbf{b}$, where $\mathbf{D}$ is a diagonal matrix. The solution of the system is $x_i = b_i/a_{ii}$, $i = 1, 2, 3$. Next, consider the system of equations

$$\begin{aligned} a_{11}x_1 &= b_1 \\ a_{21}x_1 + a_{22}x_2 &= b_2 \\ a_{31}x_1 + a_{32}x_2 + a_{33}x_3 &= b_3 \end{aligned} \qquad (18.42b)$$

or $\mathbf{Lx} = \mathbf{b}$, where $\mathbf{L}$ is a lower triangular matrix. The solution of the system is obtained by *forward substitution* as

$$x_1 = \frac{b_1}{a_{11}}, \quad x_2 = \frac{1}{a_{22}}(b_2 - a_{21}x_1), \quad x_3 = \frac{1}{a_{33}}(b_3 - a_{31}x_1 - a_{32}x_2).$$

Now, consider the system of equations

$$a_{11}x_1 + a_{12}x_2 + a_{13}x_3 = b_1$$

$$a_{22}x_2 + a_{23}x_3 = b_2$$

$$a_{33}x_3 = b_3 \qquad\qquad (18.42c)$$

or $\mathbf{Ux} = \mathbf{b}$, where $\mathbf{U}$ is an upper triangular matrix. The solution of the system is obtained by *backward substitution* as

$$x_3 = \frac{b_3}{a_{33}}, \quad x_2 = \frac{1}{a_{22}}(b_2 - a_{23}\,x_3), \quad x_1 = \frac{1}{a_{11}}(b_1 - a_{12}x_2 - a_{13}x_3).$$

Hence, if the given system is in either of the three forms (18.42a) to (18.42c), then the solution is directly obtained. All the direct methods reduce the given system of equations to one of the above three forms and the solution is obtained.

In sections 3.4.4 and 3.4.5, we have discussed the Gauss elimination method and Gauss-Jordan method for the solution of a system of equations. Gauss elimination method reduces the augmented matrix $[\mathbf{A}|\mathbf{b}]$ to $[\mathbf{U}|\mathbf{c}]$, where $\mathbf{U}$ is upper triangular matrix (see 18.42c). Gauss-Jordan method reduces the augmented matrix $[\mathbf{A}|\mathbf{b}]$ to $[\mathbf{I}|\mathbf{c}]$, where $\mathbf{I}$ is an identity matrix (see 18.42a). We now derive some more methods for the solution of a system of equations.

18.3.1 LU Decomposition Method

This method is also called the *factorization method* or *triangularization method*. The method can be used to solve a system of equations or to find the inverse of a matrix. In this method, the matrix $\mathbf{A}$ is decomposed or factorized as the product of a lower triangular matrix $\mathbf{L}$ and an upper triangular matrix $\mathbf{U}$. We write the matrix $\mathbf{A}$ as

$$\mathbf{A} = \mathbf{LU} \qquad\qquad (18.43)$$

where $\mathbf{L} = \begin{bmatrix} l_{11} & 0 & 0 & \cdots & 0 \\ l_{21} & l_{22} & 0 & \cdots & 0 \\ l_{31} & l_{32} & l_{33} & 0 & 0 \\ \cdot & \cdot\cdot & \cdot\cdot & \cdot\cdot & \cdot \\ l_{n1} & l_{n2} & l_{n3} & \cdots & l_{nn} \end{bmatrix}$, and $\mathbf{U} = \begin{bmatrix} u_{11} & u_{12} & u_{13} & \cdots & u_{1n} \\ 0 & u_{22} & u_{23} & \cdots & u_{2n} \\ 0 & 0 & u_{33} & \cdots & u_{3n} \\ \cdot\cdot & \cdot\cdot & \cdot\cdot & \cdots & \cdot\cdot \\ 0 & 0 & 0 & \cdots & u_{nn} \end{bmatrix}$.

Multiplying the matrices $\mathbf{L}$ and $\mathbf{U}$ and comparing the elements of the product matrix with the corresponding elements of $\mathbf{A}$, we get

$$l_{i1}u_{1j} + l_{i2}u_{2j} + \ldots + l_{in}u_{nj} = a_{ij}, \quad i, j = 1, 2, \ldots, n \qquad\qquad (18.44)$$

where $l_{ij} = 0, j > i$ and $u_{ij} = 0, i > j$.

There are $n(n + 1)/2$ unknowns in each of the matrices $\mathbf{L}$ and $\mathbf{U}$, that is, a total of $n(n + 1)$ unknowns. Comparing the elements of $\mathbf{A} = \mathbf{LU}$, we obtain n^2 equations from (18.44). Hence, there are n arbitrary unknowns. To obtain a unique solution, we can choose the values for n elements in either $\mathbf{L}$ or $\mathbf{U}$ arbitrarily. The simplest choices are

(i) $l_{ii} = 1$, $i = 1, 2, \ldots, n$ (the method is called the *Doolittle method*),

(ii) $u_{ii} = 1$, $i = 1, 2, \ldots, n$ (the method is called the *Crout's method*).

We find l_{ij}, u_{ij} from (18.44) in the order first column of $\mathbf{L}$ followed by the first row of $\mathbf{U}$, second column of $\mathbf{L}$ followed by the second row of $\mathbf{U}$, $\ldots,(n - 1)$th column of $\mathbf{L}$ followed by $(n - 1)$th row of $\mathbf{U}$, nth column of $\mathbf{L}$. We write the given system of equations as

$$\mathbf{Ax} = \mathbf{LUx} = \mathbf{b}. \tag{18.45}$$

Set
$$\mathbf{Ux} = \mathbf{z}. \tag{18.46}$$

Then, from (18.45) we get, $\mathbf{Lz} = \mathbf{b}.$ $\tag{18.47}$

We first solve the system (18.47) using forward substitution method to determine $\mathbf{z}$ and then solve the system (18.46) using the backward substitution to determine $\mathbf{x}$.

The inverse of the matrix $\mathbf{A}$ is obtained as

$$\mathbf{A}^{-1} = (\mathbf{LU})^{-1} = \mathbf{U}^{-1}\mathbf{L}^{-1}. \tag{18.48}$$

Remark 2

The method fails when any of the diagonal elements l_{ii} in $\mathbf{L}$ or u_{ii} in $\mathbf{U}$ becomes zero. A sufficient condition which guarantees the $\mathbf{LU}$ decomposition of the matrix $\mathbf{A}$ is that the matrix $\mathbf{A}$ is positive definite.

Remark 3

If u_{ii} are chosen as $u_{ii} = 1$, $i = 1, 2, \ldots, n$, then the first column of $\mathbf{L}$ is $l_{i1} = a_{i1}$, $i = 1, 2, \ldots, n$ and the first row of $\mathbf{U}$ is $u_{1j} = a_{1j}/a_{11}$, $j = 2, 3, \ldots, n$. If l_{ii} are chosen as $l_{ii} = 1$, $i = 1, 2, \ldots, n$, then the first row of $\mathbf{U}$ is $u_{1j} = a_{1j}$, $j = 1, 2, \ldots, n$ and the first column of $\mathbf{L}$ is $l_{i1} = a_{i1}/a_{11}$, $i = 2, 3, \ldots, n$.

Remark 4

$\mathbf{L}^{-1}$ can be computed from $\mathbf{LL}^{-1} = \mathbf{I}$, using forward substitution and $\mathbf{U}^{-1}$ can be computed from $\mathbf{UU}^{-1} = \mathbf{I}$, using backward substitution. Note that the inverse of a lower triangular matrix is a lower triangular matrix and inverse of an upper triangular matrix is an upper triangular matrix.

Example 18.13 Solve the system of equations $\mathbf{Ax} = \mathbf{b}$, where

$$\mathbf{A} = \begin{bmatrix} 4 & 1 & 1 \\ 1 & 4 & -2 \\ 3 & 2 & -4 \end{bmatrix}, \mathbf{b} = \begin{bmatrix} 4 \\ 4 \\ 6 \end{bmatrix}$$

by factorization method. Take $u_{ii} = 1$, $i = 1, 2, 3$. Also find $\mathbf{A}^{-1}$.

Solution We write

$$\begin{bmatrix} 4 & 1 & 1 \\ 1 & 4 & -2 \\ 3 & 2 & -4 \end{bmatrix} = \begin{bmatrix} l_{11} & 0 & 0 \\ l_{21} & l_{22} & 0 \\ l_{31} & l_{32} & l_{33} \end{bmatrix} \begin{bmatrix} 1 & u_{12} & u_{13} \\ 0 & 1 & u_{23} \\ 0 & 0 & 1 \end{bmatrix}$$

$$= \begin{bmatrix} l_{11} & l_{11}u_{12} & l_{11}u_{13} \\ l_{21} & l_{21}u_{12} + l_{22} & l_{21}u_{13} + l_{22}u_{23} \\ l_{31} & l_{31}u_{12} + l_{32} & l_{31}u_{13} + l_{32}u_{23} + l_{33} \end{bmatrix}.$$

Comparing the corresponding elements on both sides, we get

first column of **L:** $l_{11} = 4; \ l_{21} = 1; \ l_{31} = 3;$

first row of **U:** $l_{11}u_{12} = 1, \ u_{12} = 1/4; \ l_{11}u_{13} = 1, \ u_{13} = 1/4;$

second column of **L:** $l_{21}u_{12} + l_{22} = 4, \ l_{22} = 15/4; \ l_{31}u_{12} + l_{32} = 2, \ l_{32} = 5/4;$

second row of **U:** $l_{21}u_{13} + l_{22}u_{23} = -2, \ u_{23} = \dfrac{4}{15}\left(-2 - \dfrac{1}{4}\right) = -\dfrac{3}{5};$

third column of **L:** $l_{31}u_{13} + l_{32}u_{23} + l_{33} = -4, \ l_{33} = -4 - \dfrac{3}{4} + \dfrac{3}{4} = -4.$

Therefore, we obtain

$$\mathbf{L} = \begin{bmatrix} 4 & 0 & 0 \\ 1 & 15/4 & 0 \\ 3 & 5/4 & -4 \end{bmatrix}, \quad \mathbf{U} = \begin{bmatrix} 1 & 1/4 & 1/4 \\ 0 & 1 & -3/5 \\ 0 & 0 & 1 \end{bmatrix}.$$

From the system **Lz = b**

$$\begin{bmatrix} 4 & 0 & 0 \\ 1 & 15/4 & 0 \\ 3 & 5/4 & -4 \end{bmatrix}\begin{bmatrix} z_1 \\ z_2 \\ z_3 \end{bmatrix} = \begin{bmatrix} 4 \\ 4 \\ 6 \end{bmatrix}$$

we obtain by forward substitution

$$z_1 = 1, \quad z_2 = \frac{4}{15}(4 - 1) = \frac{4}{5}, \quad z_3 = -\frac{1}{4}\left(6 - 3z_1 - \frac{5}{4}z_2\right) = \frac{1}{4}(6 - 3 - 1) = -\frac{1}{2}.$$

From the system **Ux = z**

$$\begin{bmatrix} 1 & 1/4 & 1/4 \\ 0 & 1 & -3/5 \\ 0 & 0 & 1 \end{bmatrix}\begin{bmatrix} x_1 \\ x_2 \\ x_3 \end{bmatrix} = \begin{bmatrix} 1 \\ 4/5 \\ -1/2 \end{bmatrix}$$

we obtain by backward substitution

$$x_3 = -\frac{1}{2}, \quad x_2 = \frac{4}{3} + \frac{3}{5}x_3 = \frac{4}{5} - \frac{3}{10} = \frac{1}{2}, \quad x_1 = 1 - \frac{1}{4}x_2 - \frac{1}{4}x_3 = 1 - \frac{1}{8} + \frac{1}{8} = 1.$$

Now, from $\mathbf{LL}^{-1} = \mathbf{I}$

$$\begin{bmatrix} 4 & 0 & 0 \\ 1 & 15/4 & 0 \\ 3 & 5/4 & -4 \end{bmatrix} \begin{bmatrix} l'_{11} & 0 & 0 \\ l'_{21} & l'_{22} & 0 \\ l'_{31} & l'_{32} & l'_{33} \end{bmatrix} = \begin{bmatrix} 1 & 0 & 0 \\ 0 & 1 & 0 \\ 0 & 0 & 1 \end{bmatrix}$$

we get by forward substitution

$$\mathbf{L}^{-1} = -\frac{1}{60} \begin{bmatrix} -15 & 0 & 0 \\ 4 & -16 & 0 \\ -10 & -5 & 15 \end{bmatrix}.$$

From $\mathbf{UU}^{-1} = \mathbf{I}$

$$\begin{bmatrix} 1 & 1/4 & 1/4 \\ 0 & 1 & -3/5 \\ 0 & 0 & 1 \end{bmatrix} \begin{bmatrix} u'_{11} & u'_{12} & u'_{13} \\ 0 & u'_{22} & u'_{23} \\ 0 & 0 & u'_{33} \end{bmatrix} = \begin{bmatrix} 1 & 0 & 0 \\ 0 & 1 & 0 \\ 0 & 0 & 1 \end{bmatrix}$$

we get by backward substitution

$$\mathbf{U}^{-1} = \begin{bmatrix} 1 & -1/4 & -2/5 \\ 0 & 1 & 3/5 \\ 0 & 0 & 1 \end{bmatrix}.$$

For a 3×3 system, we may directly obtain $\mathbf{L}^{-1}$, $\mathbf{U}^{-1}$ using the adjoint method. Now

$$\mathbf{A}^{-1} = \mathbf{U}^{-1} \mathbf{L}^{-1} = -\frac{1}{60} \begin{bmatrix} 1 & -1/4 & -2/5 \\ 0 & 1 & 3/5 \\ 0 & 0 & 1 \end{bmatrix} \begin{bmatrix} -15 & 0 & 0 \\ 4 & -16 & 0 \\ -10 & -5 & 15 \end{bmatrix} = -\frac{1}{60} \begin{bmatrix} -12 & 6 & -6 \\ -2 & -19 & 9 \\ -10 & -5 & 15 \end{bmatrix}.$$

18.3.2 Choleski Decomposition Method

When the coefficient matrix $\mathbf{A}$ of the system $\mathbf{Ax} = \mathbf{b}$ is symmetric, then this property should reflect in the reduction of the amount of computation (and also memory storage in the computer) being done. If $\mathbf{A}$ is symmetric, then the decomposition can be written as

$$\mathbf{A} = \mathbf{LL}^{T} \tag{18.49}$$

where $\mathbf{L}$ is the Lower triangular matrix

$$\mathbf{L} = \begin{pmatrix} l_{11} & 0 & 0 & \dots & 0 \\ l_{21} & l_{22} & 0 & \dots & 0 \\ l_{31} & l_{32} & l_{33} & 0 & 0 \\ .. & .. & .. & .. & .. \\ l_{n1} & l_{n2} & l_{n3} & \dots & l_{nn} \end{pmatrix}.$$

This method is also called the *square root* method. Let us illustrate the decomposition for a 3×3 system. We write

$$
\begin{bmatrix} a_{11} & a_{12} & a_{13} \\ a_{12} & a_{22} & a_{23} \\ a_{13} & a_{23} & a_{33} \end{bmatrix} = \begin{bmatrix} l_{11} & 0 & 0 \\ l_{21} & l_{22} & 0 \\ l_{31} & l_{32} & l_{33} \end{bmatrix} \begin{bmatrix} l_{11} & l_{21} & l_{31} \\ 0 & l_{22} & l_{32} \\ 0 & 0 & l_{33} \end{bmatrix}
$$

$$
= \begin{bmatrix} l_{11}^2 & l_{11}l_{21} & l_{11}l_{31} \\ l_{21}l_{11} & l_{21}^2 + l_{22}^2 & l_{21}l_{31} + l_{22}l_{32} \\ l_{31}l_{11} & l_{31}l_{21} + l_{32}l_{22} & l_{31}^2 + l_{32}^2 + l_{33}^2 \end{bmatrix}.
$$

Comparing the elements, we obtain

$$
l_{11}^2 = a_{11}, \ l_{11} = \sqrt{a_{11}} \ ; \ l_{11}l_{21} = a_{12}, \ l_{21} = \frac{a_{12}}{l_{11}} ; \ l_{11}l_{31} = a_{13}, \ l_{31} = \frac{a_{13}}{l_{11}} ;
$$

$$
l_{21}^2 + l_{22}^2 = a_{22}, \ l_{22} = \sqrt{a_{22} - \frac{a_{12}^2}{a_{11}}} = \sqrt{\frac{a_{11}a_{22} - a_{12}^2}{a_{11}}} ;
$$

$$
l_{31}l_{21} + l_{32}l_{22} = a_{23}, \ l_{32} = \frac{1}{l_{22}} [a_{23} - l_{31}l_{21}];
$$

$$
l_{31}^2 + l_{32}^2 + l_{33}^2 = a_{33}, \ l_{33} = \sqrt{a_{33} - l_{31}^2 - l_{32}^2} .
$$

The remaining part of the solution procedure is same as in the **LU** decomposition method. We have

$$
\mathbf{L}\mathbf{L}^T \mathbf{x} = \mathbf{b}. \quad \text{Set } \mathbf{L}^T \mathbf{x} = \mathbf{z}. \quad \text{Then, we have } \mathbf{L}\mathbf{z} = \mathbf{b}.
$$

We first solve $\mathbf{L}\mathbf{z} = \mathbf{b}$. Then, solve $\mathbf{L}^T \mathbf{x} = \mathbf{z}$. Again, one forward substitution and one backward substitution is required to obtain $\mathbf{x}$.

To find $\mathbf{A}^{-1}$, we have

$$
\mathbf{A}^{-1} = (\mathbf{L}\mathbf{L}^T)^{-1} = (\mathbf{L}^T)^{-1} \mathbf{L}^{-1} = (\mathbf{L}^{-1})^T \mathbf{L}^{-1} \tag{18.50}
$$

The reduction in the amount of computation is obvious from this step. We require the inverse of the lower triangular matrix $\mathbf{L}$ to find $\mathbf{A}^{-1}$.

Remark 5

A sufficient condition which guarantees the $\mathbf{L}\mathbf{L}^T$ decomposition of the matrix $\mathbf{A}$ is that the matrix is positive definite. This implies that all the leading minors should be positive. If 1×1 minor a_{11} is

positive, then $l_{11} = \sqrt{a_{11}}$ exists. If 2×2 minor $\begin{vmatrix} a_{11} & a_{12} \\ a_{12} & a_{22} \end{vmatrix} = a_{11}a_{22} - a_{12}^2$ is positive then l_{22} exists, etc.

Example 18.14   Solve the system of equations

$$4x_1 - x_2 - x_3 = 3$$

$$-x_1 + 4x_2 - 3x_3 = -0.5$$

$$-x_1 - 3x_2 + 5x_3 = 0.$$

using the Choleski method.

Solution   We decompose the coefficient matrix as

$$\begin{bmatrix} 4 & -1 & -1 \\ -1 & 4 & -3 \\ -1 & -3 & 5 \end{bmatrix} = \begin{bmatrix} l_{11} & 0 & 0 \\ l_{21} & l_{22} & 0 \\ l_{31} & l_{32} & l_{33} \end{bmatrix} \begin{bmatrix} l_{11} & l_{21} & l_{31} \\ 0 & l_{22} & l_{32} \\ 0 & 0 & l_{33} \end{bmatrix}$$

$$= \begin{bmatrix} l_{11}^2 & l_{11}l_{21} & l_{11}l_{31} \\ l_{21}l_{11} & l_{21}^2 + l_{22}^2 & l_{21}l_{31} + l_{22}l_{32} \\ l_{31}l_{11} & l_{31}l_{21} + l_{32}l_{22} & l_{31}^2 + l_{32}^2 + l_{33}^2 \end{bmatrix}.$$

Comparing the elements, we get

$$l_{11}^2 = 4, \quad l_{11} = 2; \quad l_{11}l_{21} = -1, \quad l_{21} = -1/2; \quad l_{11}l_{31} = -1, \quad l_{31} = -1/2;$$

$$l_{21}^2 + l_{22}^2 = 4, \quad l_{22}^2 = 4 - \frac{1}{4} = \frac{15}{4}, \quad l_{22} = \frac{\sqrt{15}}{2};$$

$$l_{31}l_{21} + l_{32}l_{22} = -3, \quad \frac{1}{4} + \frac{\sqrt{15}}{2} l_{32} = -3, \quad l_{32} = \frac{2}{\sqrt{15}}\left(-3 - \frac{1}{4}\right) = -\frac{13}{2\sqrt{15}};$$

$$l_{31}^2 + l_{32}^2 + l_{33}^2 = 5, \quad \frac{1}{4} + \frac{169}{60} + l_{33}^2 = 5, \quad l_{33}^2 = 5 - \frac{1}{4} - \frac{169}{60} = \frac{116}{60}, \quad l_{33} = \sqrt{\frac{29}{15}}.$$

Hence,

$$L = \begin{bmatrix} 2 & 0 & 0 \\ -1/2 & \sqrt{15}/2 & 0 \\ -1/2 & -13/(2\sqrt{15}) & \sqrt{29/15} \end{bmatrix}.$$

Solving $Lz = b$

$$\begin{bmatrix} 2 & 0 & 0 \\ -1/2 & \sqrt{15}/2 & 0 \\ -1/2 & -13/(2\sqrt{15}) & \sqrt{29/15} \end{bmatrix} \begin{bmatrix} z_1 \\ z_2 \\ z_3 \end{bmatrix} = \begin{bmatrix} 3 \\ -1/2 \\ 0 \end{bmatrix}$$

we get

$$z_1 = \frac{3}{2}; \quad -\frac{z_1}{2} + \frac{\sqrt{15}}{2} z_2 = -\frac{1}{2}, \quad z_2 = \frac{2}{\sqrt{15}}\left(-\frac{1}{2} + \frac{3}{4}\right) = \frac{1}{2\sqrt{15}};$$

$$-\frac{z_1}{2} - \frac{13}{2\sqrt{15}} z_2 + \sqrt{\frac{29}{15}} z_3 = 0, \quad z_3 = \sqrt{\frac{15}{29}}\left(\frac{3}{4} + \frac{13}{60}\right) = \frac{58}{60}\sqrt{\frac{15}{29}}.$$

Solving $\mathbf{L}^T\mathbf{x} = \mathbf{z}$

$$\begin{bmatrix} 2 & -1/2 & -1/2 \\ 0 & \sqrt{15}/2 & -13/(2\sqrt{15}) \\ 0 & 0 & \sqrt{29/15} \end{bmatrix}\begin{bmatrix} x_1 \\ x_2 \\ x_3 \end{bmatrix} = \begin{bmatrix} 3/2 \\ 1/(2\sqrt{15}) \\ (58/60)\sqrt{15/29} \end{bmatrix}$$

we get

$$\sqrt{\frac{29}{15}} x_3 = \frac{58}{60}\sqrt{\frac{15}{29}}, \quad x_3 = \frac{15}{29}\cdot\frac{58}{60} = \frac{1}{2},$$

$$\frac{\sqrt{15}}{2} x_2 = \frac{1}{2\sqrt{15}} + \frac{13}{4\sqrt{15}} = \frac{15}{4\sqrt{15}}, \quad x_2 = \frac{1}{2},$$

$$2x_1 - \frac{1}{4} - \frac{1}{4} = \frac{3}{2}, \quad x_1 = \frac{1}{2}\left(\frac{3}{2} + \frac{1}{2}\right) = 1.$$

Iterative methods

In iterative methods, an initial approximation to the solution vector $\mathbf{x}^{(0)}$ is assumed and it is improved by iteration. The iteration method is of the form

$$\mathbf{x}^{(k+1)} = \mathbf{H}\,\mathbf{x}^{(k)} + \mathbf{c}, \quad k = 0, 1, 2, \ldots \tag{18.51}$$

where $\mathbf{H}$ is called the *iteration matrix*. A suitable initial approximation may be taken as $x_i = b_i/a_{ii}$, $i = 1, 2, \ldots, n$. Starting with the initial approximation $\mathbf{x}^{(0)}$, we generate the sequence of iterates $\mathbf{x}^{(1)}, \mathbf{x}^{(2)}, \ldots, \mathbf{x}^{(k)}, \mathbf{x}^{(k+1)}, \ldots$, which in the limit converges to the exact solution. The iteration is stopped when for a given error tolerance ε

$$\left| x_i^{(k+1)} - x_i^{(k)} \right| \le \varepsilon \quad \text{for all } i. \tag{18.52}$$

The convergence properties of the iteration method depends on the iteration matrix $\mathbf{H}$. A necessary and sufficient condition for convergence is $\rho(\mathbf{H}) < 1$, where $\rho(\mathbf{H})$ is the *spectral radius* of $\mathbf{H}$. If the

coefficient matrix $\mathbf{A}$ of the given system is *diagonally dominant*, that is, $|a_{ii}| \ge \sum_{\substack{j=1 \\ j \ne i}}^{n} |a_{ij}|$, for all i, then this condition is automatically satisfied and convergence is guaranteed. If the iteration method converges, then

$$v = - \log_{10}(\rho(\mathbf{H})) \tag{18.53}$$

is called the *rate of convergence* or the *order* of the method.

18.3.3 Gauss-Jacobi Iteration Method

We assume that the diagonal elements a_{ii} of the matrix $\mathbf{A}$, are non-zero. We write the system of equations (18.41) as

$$a_{11}x_1 = b_1 - (a_{12}x_2 + a_{13}x_3 + \ldots + a_{1n}x_n)$$

$$a_{22}x_2 = b_2 - (a_{21}x_1 + a_{23}x_3 + \ldots + a_{2n}x_n)$$

$$\ldots \quad \ldots \quad \ldots \quad \ldots$$

$$a_{nn}x_n = b_n - (a_{n1}x_1 + a_{n2}x_2 + \ldots + a_{n,n-1}x_{n-1}).$$

The *Gauss-Jacobi iteration method* or simply the *Jocobi iteration method* is defined as

$$x_1^{(k+1)} = \frac{1}{a_{11}}[b_1 - (a_{12}\, x_2^{(k)} + a_{13}\, x_3^{(k)} + \ldots + a_{1n}x_n^{(k)})]$$

$$x_2^{(k+1)} = \frac{1}{a_{22}}\, [b_2 - (a_{21}\, x_1^{(k)} + a_{23}\, x_3^{(k)} + \ldots + a_{2n}\, x_n^{(k)})]$$

$$\vdots$$

$$x_n^{(k+1)} = \frac{1}{a_{nn}}\, [b_n - (a_{n1}x_1^{(k)} + a_{n2}\, x_2^{(k)} + \ldots + a_{n,n-1}\, x_{n-1}^{(k)})], \tag{18.54}$$

$k = 0, 1, 2, \ldots$

Starting with the initial approximation $\mathbf{x}^{(0)} = (x_1^{(0)}, x_2^{(0)}, \ldots, x_n^{(0)})^T$, we generate the sequence of iterates $\mathbf{x}^{(1)}, \mathbf{x}^{(2)}, \ldots$, which in the limit converges to the exact solution $\mathbf{x}$. Since we replace all the elements simultaneously, this method is also called the method of *simultaneous displacement*.

We can also write the method (18.54) in matrix from. Let $\mathbf{A} = \mathbf{L} + \mathbf{D} + \mathbf{U}$, where $\mathbf{D}$ is the diagonal part, $\mathbf{L}$ and $\mathbf{U}$ are respectively the strictly lower triangular and upper triangular parts of the matrix $\mathbf{A}$. For example, for a 3×3 system, we have

$$\begin{bmatrix} a_{11} & a_{12} & a_{13} \\ a_{21} & a_{22} & a_{23} \\ a_{31} & a_{32} & a_{33} \end{bmatrix} = \begin{bmatrix} 0 & 0 & 0 \\ a_{21} & 0 & 0 \\ a_{31} & a_{32} & 0 \end{bmatrix} + \begin{bmatrix} a_{11} & 0 & 0 \\ 0 & a_{22} & 0 \\ 0 & 0 & a_{33} \end{bmatrix} + \begin{bmatrix} 0 & a_{12} & a_{13} \\ 0 & 0 & a_{23} \\ 0 & 0 & 0 \end{bmatrix} = \mathbf{L} + \mathbf{D} + \mathbf{U}.$$

We write the system of equations $\mathbf{Ax} = \mathbf{b}$ as

$$(\mathbf{L} + \mathbf{D} + \mathbf{U})\mathbf{x} = \mathbf{b} \quad \text{or} \quad \mathbf{Dx} = - (\mathbf{L} + \mathbf{U})\mathbf{x} + \mathbf{b}$$

and write the iteration method as

$$\mathbf{Dx}^{(k+1)} = -(\mathbf{L} + \mathbf{U})\mathbf{x}^{(k)} + \mathbf{b}, \quad \text{or} \quad \mathbf{x}^{(k+1)} = -\mathbf{D}^{-1}(\mathbf{L} + \mathbf{U})\mathbf{x}^{(k)} + \mathbf{D}^{-1}\mathbf{b}, \tag{18.55}$$

$k = 0, 1, 2, \ldots$

The method (18.55) is of the form (18.51), where the iteration matrix $\mathbf{H}$ and the vector $\mathbf{c}$ are given by

$$\mathbf{H} = -\mathbf{D}^{-1}(\mathbf{L} + \mathbf{U}) \quad \text{and} \quad \mathbf{c} = \mathbf{D}^{-1}\mathbf{b}. \tag{18.56}$$

The iteration is stopped when $\left| x_i^{(k+1)} - x_i^{(k)} \right| \le \varepsilon$, for all i, where ε is a given error tolerance. If any diagonal element is zero, we interchange rows to obtain a non-zero diagonal element. However, convergence properties are affected by such interchanges.

Example 18.15 Perform three iterations of the Gauss-Jacobi iteration method for solving the system of equations

$$\begin{bmatrix} 6 & 1 & 2 \\ 1 & 4 & 3 \\ 2 & 1 & 8 \end{bmatrix} \begin{bmatrix} x_1 \\ x_2 \\ x_3 \end{bmatrix} = \begin{bmatrix} 6 \\ -4 \\ 8 \end{bmatrix}.$$

Take the initial approximation as $\mathbf{x}^{(0)} = [1.3, -1.9, 0.8]^T$. Compare with the exact solution $x_1 = 1$, $x_2 = -2$, $x_3 = 1$.

Solution We write the Jacobi iteration method as

$$x_1^{(k+1)} = \frac{1}{6}[6 - x_2^{(k)} - 2x_3^{(k)}], \quad x_2^{(k+1)} = \frac{1}{4}[-4 - x_1^{(k)} - 3x_3^{(k)}], \quad x_3^{(k+1)} = \frac{1}{8}[8 - 2x_1^{(k)} - x_2^{(k)}],$$

$k = 0, 1, 2, \ldots$

Starting with $x_1^{(0)} = 1.3$, $x_2^{(0)} = -1.9$, $x_3^{(0)} = 0.8$, we get the following results.

$k = 0$: $\quad x_1^{(1)} = \frac{1}{6}[6 - x_2^{(0)} - 2x_3^{(0)}] = 1.0500, \quad x_2^{(1)} = \frac{1}{4}[-4 - x_1^{(0)} - 3x_3^{(0)}] = -1.9250,$

$\quad x_3^{(1)} = \frac{1}{8}[8 - 2x_1^{(0)} - x_2^{(0)}] = 0.9125.$

$k = 1$: $\quad x_1^{(2)} = \frac{1}{6}[6 - x_2^{(1)} - 2x_3^{(1)}] = 1.0167, \quad x_2^{(2)} = \frac{1}{4}[-4 - x_1^{(1)} - 3x_3^{(1)}] = -1.9469,$

$\quad x_3^{(2)} = \frac{1}{8}[8 - 2x_1^{(1)} - x_2^{(1)}] = 0.9781.$

$k = 2$: $\quad x_1^{(3)} = \frac{1}{6}[6 - x_2^{(2)} - 2x_3^{(2)}] = 1.9984, \quad x_2^{(3)} = \frac{1}{4}[4 - x_1^{(2)} - 3x_3^{(2)}] = -1.9878,$

$\quad x_3^{(3)} = \frac{1}{8}[8 - 2x_1^{(2)} - x_2^{(2)}] = 0.9892.$

Comparing with the exact solution, the errors in magnitude are 0.0016, 0.0122 and 0.0108. The maximum absolute error is 0.0122.

18.3.4 Gauss-Seidel Iteration Method

In the Jacobi iteration method, when we evaluate $x_2^{(k+1)}$, both the vaues $x_1^{(k)}$ and $x_1^{(k+1)}$ are available. But, we use only the previous iteration value $x_1^{(k)}$. Similarly, when we evaluate $x_3^{(k+1)}$, though the values $x_1^{(k+1)}$, $x_2^{(k+1)}$ are available, we still use the previous iteration values $x_1^{(k)}$ and $x_2^{(k)}$. This is the disadvantage of the Jacobi method . If we use the latest available values of each variable in (18.54), then the method is called *Gauss-Seidel iteration method*. We write the method as

$$x_1^{(k+1)} = \frac{1}{a_{11}}[b_1 - (a_{12}x_2^{(k)} + a_{13}x_3^{(k)} + \ldots + a_{1n}x_n^{(k)})]$$

$$x_2^{(k+1)} = \frac{1}{a_{22}}[b_2 - (a_{21}x_1^{(k+1)} + a_{23}x_3^{(k)} + \ldots + a_{2n}x_n^{(k)})]$$

$$\vdots$$

$$x_n^{(k+1)} = \frac{1}{a_{nn}}[b_n - (a_{n1}x_1^{(k+1)} + a_{n2}x_2^{(k+1)} + \ldots + a_{n,n-1}\ x_{n-1}^{(k+1)})]. \qquad (18.57)$$

This method is also called the *method of successive displacement*, since the elements are replaced successively.

We can also write this method in matrix form. We write the system $\mathbf{Ax} = \mathbf{b}$ as

$$(\mathbf{L} + \mathbf{D} + \mathbf{U})\ \mathbf{x} = \mathbf{b}, \quad \text{or} \quad (\mathbf{L} + \mathbf{D})\mathbf{x} = \mathbf{b} - \mathbf{Ux},$$

where $\mathbf{L}, \mathbf{D}, \mathbf{U}$ have the same meaning as in Jacobi method. The iteration method is written as

$$(\mathbf{D} + \mathbf{L})\ \mathbf{x}^{(k+1)} = \mathbf{b} - \mathbf{Ux}^{(k)}$$

or

$$\mathbf{x}^{(k+1)} = -(\mathbf{D} + \mathbf{L})^{-1}\ \mathbf{Ux}^{(k)} + (\mathbf{D} + \mathbf{L})^{-1}\ \mathbf{b}, \quad k = 0, 1, 2, \ldots \qquad (18.58)$$

Comparing with the iteration method $\mathbf{x}^{(k+1)} = \mathbf{Hx}^{(k)} + \mathbf{c}$, we obtain

$$\mathbf{H} = -(\mathbf{D} + \mathbf{L})^{-1}\ \mathbf{U}, \quad \text{and} \quad \mathbf{c} = (\mathbf{D} + \mathbf{L})^{-1}\ \mathbf{b}, \qquad (18.58)$$

where $\mathbf{H}$ is the iteration matrix.

We state the following results:

(a) If the matrix $\mathbf{A}$ is strictly diagonally dominant, then both the Gauss-Jacobi and Gauss-Seidel iteration methods for solving the system of equations $\mathbf{Ax} = \mathbf{b}$ converge for any initial approximate vector $\mathbf{x}^{(0)}$. If no better approximation is known, we may take $\mathbf{x}^{(0)} = \mathbf{0}$.

(b) Let the coefficient matrix $\mathbf{A}$ be strictly diagonally dominant and has the '*property A*'. Then, the Gauss-Seidel method converges two times faster than the Gauss-Jacobi method.

Example 18.16   Perform three iterations of the Gauss-Seidel iteration method for solving the system of equations

$$\begin{bmatrix} 4 & 0 & 2 \\ 0 & 5 & 2 \\ 5 & 4 & 10 \end{bmatrix} \begin{bmatrix} x_1 \\ x_2 \\ x_3 \end{bmatrix} = \begin{bmatrix} 6 \\ -3 \\ 11 \end{bmatrix}.$$

Take the components of the approximate initial vector as $x_i^{(0)} = b_i / a_{ii}$, $i = 1, 2, 3$. Compare with the exact solution $\mathbf{x} = [1, -1, 1]^T$.

Solution   We write the Gauss-Seidel method as

$$x_1^{(k+1)} = \frac{1}{4}[6 - 2x_3^{(k)}], \quad x_2^{(k+1)} = \frac{1}{5}[-3 - 2x_3^{(k)}],$$

$$x_3^{(k+1)} = \frac{1}{10}[11 - 5x_1^{(k+1)} - 4x_2^{(k+1)}], \quad k = 0, 1, 2.$$

The initial approximation is $x_1^{(0)} = 1.5$, $x_2^{(0)} = -0.6$, $x_3^{(0)} = 1.1$. For $k = 0, 1, 2$, we have the following results.

$k = 0$:
$$x_1^{(1)} = \frac{1}{4}[6 - 2(1.1)] = 0.95, \quad x_2^{(1)} = \frac{1}{5}[-3 - 2(1.1)] = -1.04,$$

$$x_3^{(1)} = \frac{1}{10}[11 - 5(0.95) - 4(-1.04)] = 1.041.$$

$k = 1$:
$$x_1^{(2)} = \frac{1}{4}[6 - 2(1.041)] = 0.9795, \quad x_2^{(2)} = \frac{1}{5}[-3 - 2(1.041)] = -1.0164,$$

$$x_3^{(2)} = \frac{1}{10}[11 - 5(0.9795) - 4(-1.0164)] = 1.0168.$$

$k = 2$:
$$x_1^{(3)} = \frac{1}{4}[6 - 2(1.0168)] = 0.9916, \quad x_2^{(3)} = \frac{1}{5}[-3 - 2(1.0168)] = -1.0067,$$

$$x_3^{(3)} = \frac{1}{10}[11 - 5(0.9916) - 4(-1.0067)] = 1.0069.$$

The errors in magnitude of the components of the solution vector are 0.0084, 0.0067, 0.0069. The maximum absolute error is 0.0084.

Example 18.17   Set up the Jacobi iteration scheme in matrix form to solve the system of equations

$$\begin{bmatrix} 4 & -1 & 0 \\ -1 & 4 & -1 \\ 0 & -1 & 4 \end{bmatrix} \begin{bmatrix} x \\ y \\ z \end{bmatrix} = \begin{bmatrix} 3 \\ 2 \\ 3 \end{bmatrix}.$$

(i) Show that the iteration scheme is convergent.

(ii) Find the rate of convergence of the method.

Solution   We have the method as $\mathbf{x}^{(k+1)} = \mathbf{H}\mathbf{x}^{(k)} + \mathbf{c}$, where

$$\mathbf{H} = -\mathbf{D}^{-1}(\mathbf{L} + \mathbf{U}) = -\begin{bmatrix} 4 & 0 & 0 \\ 0 & 4 & 0 \\ 0 & 0 & 4 \end{bmatrix}^{-1}\begin{bmatrix} 0 & -1 & 0 \\ -1 & 0 & -1 \\ 0 & -1 & 0 \end{bmatrix}$$

$$= -\begin{bmatrix} 1/4 & 0 & 0 \\ 0 & 1/4 & 0 \\ 0 & 0 & 1/4 \end{bmatrix}\begin{bmatrix} 0 & -1 & 0 \\ -1 & 0 & -1 \\ 0 & -1 & 0 \end{bmatrix} = \begin{bmatrix} 0 & 1/4 & 0 \\ 1/4 & 0 & 1/4 \\ 0 & 1/4 & 0 \end{bmatrix}.$$

$$\mathbf{c} = \mathbf{D}^{-1}\,\mathbf{b} = \begin{bmatrix} 1/4 & 0 & 0 \\ 0 & 1/4 & 0 \\ 0 & 0 & 1/4 \end{bmatrix}\begin{bmatrix} 3 \\ 2 \\ 3 \end{bmatrix} = \begin{bmatrix} 3/4 \\ 1/2 \\ 3/4 \end{bmatrix}.$$

Hence, the Jacobi iteration scheme in matrix form becomes

$$\mathbf{x}^{(k+1)} = \begin{bmatrix} 0 & 1/4 & 0 \\ 1/4 & 0 & 1/4 \\ 0 & 1/4 & 0 \end{bmatrix}\mathbf{x}^{(k)} + \begin{bmatrix} 3/4 \\ 1/2 \\ 3/4 \end{bmatrix}.$$

Now, we find the eigenvalues of $\mathbf{H}$.

$$|\mathbf{H} - \lambda\mathbf{I}| = \begin{bmatrix} -\lambda & 1/4 & 0 \\ 1/4 & -\lambda & 1/4 \\ 0 & 1/4 & -\lambda \end{bmatrix} = -\lambda\left(\lambda^2 - \frac{1}{8}\right) = 0.$$

Eigenvalues of $\mathbf{H}$ are 0, $\pm\,(1/\sqrt{8})$ and $\rho(\mathbf{H}) = (1/\sqrt{8}) < 1$. Hence, the Jacobi iteration method converges.

Rate of convergence $= v = -\log_{10}[\rho(\mathbf{H})] = -\log_{10}\left(\dfrac{1}{\sqrt{8}}\right) = 0.4515$.

Example 18.18   Set up the Gauss-Seidel iteration scheme in matrix form for solving the system of equations

$$\begin{bmatrix} -2 & 1 & 5 \\ 4 & -8 & 1 \\ 4 & -1 & -2 \end{bmatrix}\begin{bmatrix} x \\ y \\ z \end{bmatrix} = \begin{bmatrix} 15 \\ -21 \\ -2 \end{bmatrix}.$$

(i) Show that the method is divergent.

(ii) Exchange the first and third equations and set up the Gauss-Seidel method for solving the new system. Show that the method is convergent.

Solution

(i) We have the Gauss-Seidel method as $\mathbf{x}^{(k+1)} = \mathbf{H}\mathbf{x}^{(k)} + \mathbf{c}$, where

$$\mathbf{H} = -(\mathbf{D} + \mathbf{L})^{-1}\,\mathbf{U} = -\begin{bmatrix} -2 & 0 & 0 \\ 4 & -8 & 0 \\ 4 & -1 & -2 \end{bmatrix}^{-1}\begin{bmatrix} 0 & 1 & 5 \\ 0 & 0 & 1 \\ 0 & 0 & 0 \end{bmatrix}$$

$$= \frac{1}{32}\begin{bmatrix} 16 & 0 & 0 \\ 8 & 4 & 0 \\ 28 & -2 & 16 \end{bmatrix}\begin{bmatrix} 0 & 1 & 5 \\ 0 & 0 & 1 \\ 0 & 0 & 0 \end{bmatrix} = \frac{1}{32}\begin{bmatrix} 0 & 16 & 80 \\ 0 & 8 & 44 \\ 0 & 28 & 138 \end{bmatrix}.$$

Now, we find the eigenvalues of **H**.

$$|\mathbf{H} - \lambda\mathbf{I}| = \begin{vmatrix} -\lambda & 1/2 & 5/2 \\ 0 & (1/4)-\lambda & 11/8 \\ 0 & 7/8 & (69/16)-\lambda \end{vmatrix} = \lambda[16\lambda^2 - 73\lambda - 2] = 0.$$

The eigenvalues are $\lambda = 0, \dfrac{73 \pm \sqrt{5457}}{32}$ or $0, -0.003, 4.59$. Since, $\rho(\mathbf{H}) = 4.59 > 1$, the method is divergent.

(ii) After exchanging the first and third equations, the coefficient matrix becomes

$$\mathbf{A} = \begin{bmatrix} 4 & -1 & -2 \\ 4 & -8 & 1 \\ -2 & 1 & 5 \end{bmatrix}; \text{ and}$$

$$\mathbf{H} = -(\mathbf{D} + \mathbf{L})^{-1}\,\mathbf{U} = -\begin{bmatrix} 4 & 0 & 0 \\ 4 & -8 & 0 \\ -2 & 1 & 0 \end{bmatrix}^{-1}\begin{bmatrix} 0 & -1 & -2 \\ 0 & 0 & 1 \\ 0 & 0 & 0 \end{bmatrix}$$

$$= \frac{1}{160}\begin{bmatrix} -40 & 0 & 0 \\ -20 & 20 & 0 \\ -12 & -4 & -32 \end{bmatrix}\begin{bmatrix} 0 & -1 & -2 \\ 0 & 0 & 1 \\ 0 & 0 & 0 \end{bmatrix} = \frac{1}{160}\begin{bmatrix} 0 & 40 & 80 \\ 0 & 20 & 60 \\ 0 & 12 & 20 \end{bmatrix}.$$

Now, we find the eigenvalues of **H**.

$$|\mathbf{H} - \lambda\mathbf{I}| = \begin{vmatrix} -\lambda & 1/4 & 1/2 \\ 0 & (1/8)-\lambda & 3/8 \\ 0 & 3/40 & (1/8)-\lambda \end{vmatrix} = \lambda\left[\lambda^2 - \frac{1}{4}\lambda - \frac{1}{80}\right] = 0.$$

The eigenvalues are $\lambda = 0$, $\lambda = \dfrac{20 \pm \sqrt{720}}{160}$ or $0, 0.293, -0.043$.

Since $\rho(\mathbf{H}) = 0.293 < 1$, the method is convergent.

Exercise 18.3

1. Using the decomposition method, solve the system of equations (choose $l_{ii} = 1$, $i = 1, 2, 3$).

$$x - y + 5z = 5,$$
$$2x - 3y + z = 0,$$
$$x + 3y + 7z = 11.$$

2. Using the decomposition method, solve the system of equations (choose $u_{ii} = 1$, $i = 1, 2, 3$).

$$\begin{bmatrix} 2 & -4 & 3 \\ 4 & 1 & -6 \\ 5 & 8 & -4 \end{bmatrix} \begin{bmatrix} x_1 \\ x_2 \\ x_3 \end{bmatrix} = \begin{bmatrix} 1 \\ -1 \\ 9 \end{bmatrix}.$$

3. Find the inverse of the following matrices using **LU** decomposition method

   (i) $\begin{bmatrix} 2 & 3 & 1 \\ 1 & 2 & 3 \\ 3 & 1 & 2 \end{bmatrix}$.
   (ii) $\begin{bmatrix} 1 & 1 & 1 \\ 3 & 1 & -3 \\ 1 & -2 & -5 \end{bmatrix}$.

   Chose $l_{11} = l_{22} = l_{33} = 1$, in each case.

4. Find the **LU** decomposition of the matrix (choose $u_{ii} = 1$, $i = 1, 2, 3, 4$).

$$\begin{bmatrix} 1 & 0 & 5 & 2 \\ -1 & 4 & 1 & 0 \\ 3 & 0 & 4 & 1 \\ -2 & 1 & 1 & 3 \end{bmatrix}$$

5. Using the Choleski decomposition method, solve the following system of equations.

   (i) $x_1 - x_2 + x_3 = 0.5$,
   $-x_1 + 4x_2 - 2x_3 = 1.5$,
   $x_1 - 2x_2 + 3x_3 = -1.0$.

   (ii) $4x_1 - 2x_2 + x_3 = 2.5$,
   $-2x_1 + 3x_2 - x_3 = -2.0$,
   $x_1 - x_2 + 4x_3 = -1.0$.

6. Perform four iterations of the (i) Jacobi iteration method, (ii) Gauss-Seidel iteration method, to solve the system of equations

   (a) $\begin{bmatrix} 4 & 1 & 2 \\ 1 & 5 & 1 \\ 2 & 1 & 4 \end{bmatrix} \mathbf{x} = \begin{bmatrix} -1 \\ 5 \\ 3 \end{bmatrix}$.
   (b) $\begin{bmatrix} 4 & -1 & 1 \\ -1 & 4 & -1 \\ 1 & -1 & 4 \end{bmatrix} \mathbf{x} = \begin{bmatrix} 4 \\ 2 \\ 4 \end{bmatrix}$.

   Take the initial approximation $\mathbf{x}^{(0)} = \mathbf{0}$, in each case.

7. Set up the Jacobi iteration scheme in matrix form to solve the following system of equations.

(i) $\begin{bmatrix} 2 & -1 & 0 \\ -1 & 2 & -1 \\ 0 & -1 & 2 \end{bmatrix} \mathbf{x} = \begin{bmatrix} 3 \\ -4 \\ 3 \end{bmatrix}.$ **(ii)** $\begin{bmatrix} 3 & 2 & 0 \\ 2 & 3 & -1 \\ 0 & -1 & 2 \end{bmatrix} \mathbf{x} = \begin{bmatrix} 5 \\ 4 \\ 1 \end{bmatrix}.$

Show that the iteration scheme converges. Hence, find the rate of convergence of the method. Starting with $x_i = b_i/a_{ii}$, $i = 1, 2, 3$, perform three iterations of the method.

8. Set up the Gauss-Jacobi iteration scheme to solve the system of equations.

$$\begin{bmatrix} 3 & -6 & 2 \\ -4 & 1 & -1 \\ 1 & -3 & 7 \end{bmatrix} \mathbf{x} = \begin{bmatrix} 23 \\ -15 \\ 16 \end{bmatrix}.$$

Show that the iteration scheme is divergent.

9. Find the necessary and sufficient condition on k, so that (i) Jacobi iteration method, (ii) Gauss-Seidel iteration method, converges for the solution of the system of equations $\mathbf{Ax} = \mathbf{b}$, where

$$\mathbf{A} = \begin{bmatrix} 1 & 0 & k \\ 2 & 1 & 3 \\ k & 0 & 1 \end{bmatrix}, \quad \mathbf{b} \text{ arbitary}, k \neq 0.$$

If both methods converge, show that the rate of convergence of the Gauss-Seidel method is twice that of the Jocobi method.

10. Set up the Gauss-seidel iteration scheme in matrix form to solve the system of equations

$$\begin{bmatrix} 1 & 2 & 4 \\ 2 & 1 & 2 \\ 4 & 2 & 1 \end{bmatrix} \mathbf{x} = \begin{bmatrix} -1 \\ 5 \\ 3 \end{bmatrix}.$$

Show that the iteration scheme diverges.

11. Set up the Gauss-Seidel iteration scheme in matrix form to solve the system of equations

$$\begin{bmatrix} 4 & 0 & 2 \\ 0 & 5 & 2 \\ 5 & 4 & 10 \end{bmatrix} \mathbf{x} = \begin{bmatrix} 4 \\ -3 \\ 2 \end{bmatrix}.$$

Show that the iteration scheme is convergent. Hence, find the rate of convergence of the method. Starting with the initial approximation $\mathbf{x}^{(0)} = \mathbf{0}$, perform four iterations of the method.

18.4   Interpolation

In this section, we consider the problem of approximating a given function $f(x)$ by polynomials $P(x)$. These approximating polynomials are generally used for two purposes. The first use is the reconstruction of the function $f(x)$, when it is not given explicitly and only the values of $f(x)$ are given at a set of distinct points. The second use is to perform the required operations like finding roots, derivatives, integrals etc. which were intended for $f(x)$, using the approximating polynomial $P(x)$. The deviation of $f(x)$ from $P(x)$, that is, $f(x) - P(x)$, $x \in [a, b]$, is called the *error of approximation*.

Let $f(x)$ be a continuous function in some interval $[a, b]$, and defined at $n + 1$ distinct points $x_0, x_1, \ldots, x_n$ such that $a = x_0 < x_1 < x_2 \ldots < x_n = b$. These points may be equispaced, that is, $x_{i+1} - x_i = h$, $i = 0, 1, \ldots, n - 1$, or non-equispaced. The problem of polynomial interpolation is to find a polynomial $P_n(x)$ of degree $\leq$ n, which fits the given distinct data, exactly. That is

$$P_n(x_i) = f(x_i), \ i = 0, 1, 2, \ldots, n. \tag{18.59}$$

Such a polynomial is called the *interpolating polynomial*. The conditions (18.59) are called *interpolating conditions*. The points $x_0, x_1, \ldots, x_n$ are called *nodal points*, *arguments* or *tabular points*. If $P_n(x)$ is determined, then for non-tabular point $x = x^* \neq x_i$, we predict that $f(x^*) \approx P_n(x^*)$.

Through two distinct points, we can always pass a straight line (a polynomial of degree 1). Through three distinct points we can either pass a parabola (a polynomial degree 2) or a straight line if all the three points lie on the line. In general, through $n + 1$ distinct points a polynomial of degree $\leq n$ can always be fitted.

The polynomial $P_n(x)$ obtained by using conditions (18.59) is unique. That is, for a given data, the interpolating polynomials obtained in two different ways may look different in form but are identical otherwise. For example, the quadratic polynomial $3x^2 + 2x + 5$ can be written in an alternate form about $x = 1$ as $10 + 8(x - 1) + 3(x - 1)^2$.

18.4.1 Lagrange Interpolation

Consider the following data values

x	x_0	x_1	x_2	$\cdots$	x_n
$f(x)$	f_0	f_1	f_2	$\cdots$	f_n

where $f_i = f(x_i)$. We can fit a polynomial of degree $\leq n$ to this data. This polynomial must use all the ordinates $f_0, f_1, \ldots, f_n$. Hence, the polynomial can be written as a linear combination of $f_0, f_1, f_2, \ldots, f_n$ as

$$P_n(x) = l_0(x) f_0 + l_1(x) f_1 + \ldots + l_n(x) f_n \tag{18.60}$$

where $l_i(x)$, $i = 0, 1, \ldots, n$ are polynomials of degree n.

Since $P_n(x)$ fits the data exactly, we get at $x = x_0$

$$f(x_0) \equiv P_n(x_0) = l_0(x_0) f_0 + l_1(x_0) f_1 + \ldots + l_n(x_0) f_n.$$

The only possibility is $l_0(x_0) = 1$ and $l_i(x_0) = 0$, $i \neq 0$.

At a point $x = x_i$, we get

$$f(x_i) \equiv P_n(x_i) = l_0(x_i) f_0 + \ldots + l_i(x_i) f_i + \ldots + l_n(x_i) f_n.$$

Hence, we require $l_i(x_i) = 1$ and $l_j(x_i) = 0, j \neq i$.

Therefore, $l_i(x)$, the polynomials of degree n satisfy the conditions

$$l_i(x_j) = 0, \quad i \neq j$$
$$= 1, \quad i = j. \tag{18.61}$$

The polynomial of degree n having the above property can easily be written. Since $l_i(x) = 0$ at $x = x_0, \ldots, x_{i-1}, x_{i+1}, \ldots, x_n$, we have that $(x - x_0), (x - x_1), \ldots, (x - x_{i-1}), (x - x_{i+1}), \ldots, (x - x_n)$ are factors of $l_i(x)$. Since the product of these factors is a polynomial of degree n, we can write

$$l_i(x) = A(x - x_0) (x - x_1) \ldots (x - x_{i-1}) (x - x_{i+1}) \ldots (x - x_n) \tag{18.62}$$

where A is a constant.

Now, since $l_i(x_i) = 1$, we get

$$A = 1/[(x_i - x_0) (x_i - x_1) \ldots (x_i - x_{i-1}) (x_i - x_{i+1}) \ldots (x_i - x_n)].$$

Hence,

$$l_i(x) = \frac{(x - x_0)(x - x_1)\ldots(x - x_{i-1})(x - x_{i+1})\ldots(x - x_n)}{(x_i - x_0)(x_i - x_1)\ldots(x_i - x_{i-1})(x_i - x_{i+1})\ldots(x_i - x_n)}. \tag{18.63}$$

The polynomial (18.60), where $l_i(x)$ are defined by (18.63) is called the *Lagrange interpolating polynomial*. The polynomials $l_i(x)$ are called *Lagrange fundamental polynomials*.

Linear interpolation   For $n = 1$, we have the data values

x	x_0	x_1
$f(x)$	f_0	f_1

The Lagrange fundamental polynomials are given by

$$l_0(x) = \frac{x - x_1}{x_0 - x_1}, \quad l_1(x) = \frac{x - x_0}{x_1 - x_0}.$$

Hence,

$$P_1(x) = l_0(x) f_0 + l_1(x) f_1$$

$$= \frac{x - x_1}{x_0 - x_1} f_0 + \frac{x - x_0}{x_1 - x_0} f_1 \tag{18.64}$$

is the linear Lagrange interpolating polynomial which fits the given data.

Quadratic interpolation   For $n = 2$, we have the data values

x	x_0	x_1	x_2
$f(x)$	f_0	f_1	f_2

The Lagrange fundamentals polynomials are obtained as

$$l_0(x) = \frac{(x - x_1)(x - x_2)}{(x_0 - x_1)(x_0 - x_2)}, \quad l_1(x) = \frac{(x - x_0)(x - x_2)}{(x_1 - x_0)(x_1 - x_2)}, \quad l_2(x) = \frac{(x - x_0)(x - x_1)}{(x_2 - x_0)(x_2 - x_1)}.$$

Hence, the quadratic Lagrange interpolating polynomial is given by

$$P_2(x) = l_0(x)f_0 + l_1(x)f_1 + l_2(x)f_2. \tag{18.65}$$

The Lagrange interpolating polynomial of any degree for a given data can be obtained. However, it is very laborious to collect and simplify the coefficients of x^i, $i = 0, 1; ..., n$. Further, assume that, we have the interpolating polynomial $P_n(x)$ of degree n based on the $(n + 1)$ distinct points (x_i, f_i), $i = 0, 1, 2, ..., n$. Now, suppose that we add one more data point (x_{n+1}, f_{n+1}). We need to find all l_i's for the new data and the previously determined lower order polynomial $P_n(x)$ is of no use in constructing $P_{n+1}(x)$. This is a disadvantage of the Lagrange interpolation. It is desirable that the polynomial $P_{n+1}(x)$ of degree $(n + 1)$ is obtained by adding one extra term of degree $(n + 1)$ to the previously calculated polynomial $P_n(x)$ of degree n. That is

$$P_{n+1}(x) = P_n(x) + \text{(one term of degree } (n + 1)).$$

Such an interpolating polynomial is said to possess *permanence property*. Lagrange interpolating polynomial does not possess this property. However, the Lagrange polynomial is very useful for theoretical analysis of methods constructed by using $f(x_i) = P_n(x_i)$.

Example 18.19   Find the Lagrange interpolating polynomial that fits the following data values.

(i)

x	2.5	3.5
$f(x)$	6	8

Interpolate at $x = 3$.

(ii)

x	1	2	4
$f(x)$	1	7	61

Determine the approximate value of $f(3)$.

(iii)

x	−1	2	3	4
$f(x)$	−1	11	31	69

Interpolate at $x = 1.5$.

Solution

(i) We have two data values. Hence, we can fit a linear Lagrange interpolating polynomial. We have

$$l_0(x) = \frac{x - 3.5}{2.5 - 3.5} = -(x - 3.5), \quad l_1(x) = \frac{x - 2.5}{3.5 - 2.5} = (x - 2.5).$$

Hence, $P_1(x) = l_0(x)f_0 + l_1(x)f_1$

$$= -(x - 3.5)(6) + (x - 2.5)(8) = 2x + 1.$$

At $x = 3$, we get $f(3) \approx P_1(3) = 7$.

(ii) We have three data values. We can fit an interpolating polynomial of degree ≤ 2. We have

$$P_2(x) = l_0(x)f_0 + l_1(x)f_1 + l_2(x)f_2$$

where $l_0(x) = \dfrac{(x - 2)(x - 4)}{(1 - 2)(1 - 4)} = \dfrac{1}{3}(x^2 - 6x + 8),$

$$l_1(x) = \frac{(x - 1)(x - 4)}{(2 - 1)(2 - 4)} = -\frac{1}{2}(x^2 - 5x + 4),$$

$$l_2(x) = \frac{(x-1)(x-2)}{(4-1)(4-2)} = \frac{1}{6}\,(x^2 - 3x + 2).$$

Therefore, we obtain

$$P_2(x) = \frac{1}{3}\,(x^2 - 6x + 8)\,(1) - \frac{1}{2}\,(x^2 - 5x + 4)\,(7) + \frac{1}{6}\,(x^2 - 3x + 2)\,(61)$$

$$= \left(\frac{1}{3} - \frac{7}{2} + \frac{61}{6}\right) x^2 + \left(-2 + \frac{35}{2} - \frac{61}{2}\right) x + \left(\frac{8}{3} - 14 + \frac{61}{3}\right)$$

$$= 7x^2 - 15x + 9.$$

At $x = 3$, we get $f(3) \approx P_2(3) = 27$.

(iii) We have four nodal points. We can fit a polynomial of degree ≤ 3. We write

$$P_3(x) = l_0(x)f_0 + l_1(x)f_1 + l_2(x)f_2 + l_3(x)f_3$$

where

$$l_0(x) = \frac{(x-2)(x-3)(x-4)}{(-1-2)(-1-3)(-1-4)} = -\frac{1}{60}(x^3 - 9x^2 + 26x - 24)$$

$$l_1(x) = \frac{(x+1)(x-3)(x-4)}{(2+1)(2-3)(2-4)} = \frac{1}{6}(x^3 - 6x^2 + 5x + 12),$$

$$l_2(x) = \frac{(x+1)(x-2)(x-4)}{(3+1)(3-2)(3-4)} = -\frac{1}{4}(x^3 - 5x^2 + 2x + 8),$$

$$l_3(x) = \frac{(x+1)(x-2)(x-3)}{(4+1)(4-2)(4-3)} = \frac{1}{10}\,(x^3 - 4x^2 + x + 6)$$

Therefore, we obtain

$$P_3(x) = -\frac{1}{60}(x^3 - 9x^2 + 26x - 24)\,(-1) + \frac{1}{6}(x^3 - 6x^2 + 5x + 12)\,(11)$$

$$-\frac{1}{4}(x^3 - 5x^2 + 2x + 8)\,(31) + \frac{1}{10}(x^3 - 4x^2 + x + 6)\,(69)$$

$$= x^3 + x + 1.$$

At $x = 1.5$, we get $f(1.5) \approx P_3(1.5) = 5.875$.

If the expression for the polynomial is not required, then we can find $l_0(1.5)$, $l_1(1.5)$, $l_2(1.5)$, $l_3(1.5)$ and then find $P_3(1.5)$.

Error of interpolation

We assume that $f(x)$ has continuous derivatives of orders up to $n + 1$ for all $x \in [a, b]$. Since $f(x)$ is approximated by $P_n(x)$, the computed results contain errors. The difference $f(x) - P_n(x)$ is called the *error of interpolation* or *truncation error*. We write

$$E_n(f, x) = f(x) - P_n(x). \tag{18.66}$$

Obviously, at the tabular point x_i, $E_n(f, x_i) = 0$, $i = 0, 1, \ldots, n$ and $E_n(f, x^*) \neq 0$ for $x^* \neq x_i$.
To derive an expression for the error, we define an auxiliary function

$$\phi(t) = f(t) - P_n(t) - [f(x) - P_n(x)] \frac{(t - x_0)(t - x_1) \ldots (t - x_n)}{(x - x_0)(x - x_1) \ldots (x - x_n)}. \tag{18.67}$$

We observe that $\phi(t) = 0$ at $t = x$ and $t = x_i$, $i = 0, 1, \ldots, n$. Therefore, $\phi(t)$ has $n + 2$ zeros. Applying the Rolle's theorem repeatedly for $\phi(t)$, $\phi'(t)$, $\ldots$, $\phi^{(n)}(t)$, we obtain $\phi^{(n+1)}(\xi) = 0$, where

$$\min(x_0, x_1, \ldots, x_n) < \xi < \max (x_0, x_1, \ldots, x_n).$$

Differentiating (18.67), with respect to t, $(n + 1)$ times, we get

$$\phi^{(n+1)}(t) = f^{(n+1)}(t) - \frac{(n+1)![f(x) - P_n(x)] \cdot}{(x - x_0)(x - x_1) \ldots (x - x_n)} \tag{18.68}$$

since $P_n(t)$ is a polynomial of degree n. Setting $\phi^{(n+1)}(\xi) = 0$, we obtain from (18.68) the error in Lagrange interpolation as

$$E_n(f, x) = f(x) - P_n(x) = \frac{1}{(n + 1)!} \left[(x - x_0)(x - x_1) \ldots (x - x_n) \right] f^{(n+1)}(\xi). \tag{18.69}$$

The bound for the error is given by

$$|E_n(f, x)| \leq \frac{1}{(n + 1)!} \left[\max_{a \leq x \leq b} |(x - x_0)(x - x_1) \ldots (x - x_n)| \right] M_{n+1} \tag{18.70}$$

where $$M_{n+1} = \max_{a \leq x \leq b} |f^{(n+1)}(x)|.$$

Example 18. 20   For the data (x_0, f_0), (x_1, f_1), show that the error in linear interpolation is bounded by $h^2 M_2 / 8$, where $h = x_1 - x_0$ and $M_2 = \max_{x_0 \leq x \leq x_1} |f''(x)|$.

Solution   The error bound for the linear interpolation is given by

$$|E_1| \leq \frac{1}{2!} \max_{x_0 \leq x \leq x_1} |(x - x_0)(x - x_1)| M_2, \quad \text{where} \quad M_2 = \max_{x_0 \leq x \leq x_2} |f''(x)|.$$

Let $g(x) = (x - x_0)(x - x_1)$. Setting $g'(x) = 0 = 2x - (x_0 + x_1)$, we get the stationary point as $x^* = (x_0 + x_1)/2$. The required maximum value is given by

$$|g(x^*)| = |(x^* - x_0)(x^* - x_1)| = \frac{(x_1 - x_0)^2}{4} = \frac{h^2}{4}.$$

Therefore, $$|E_1| \leq \frac{h^2}{8} M_2.$$

Example 18.21 Consider the equispaced data (x_0, f_0), (x_1, f_1), (x_2, f_2). Show that the error in quadratic interpolation is bounded by $h^3 M_3 / (9\sqrt{3})$, where $M_3 = \max |f'''(x)|$, $x \in [x_0, x_2]$.

Solution The bound for the error in quadratic interpolation is given by

$$|\text{Error}| \leq \frac{1}{6} \max |(x - x_0)(x - x_1)(x - x_2)| M_3, \quad \text{where} \quad M_3 = \max_{x_0 \leq x \leq x_2} |f'''(x)|.$$

Since the data is equispaced, we can assume without any loss of generality that $x_0 = -h$, $x_1 = 0$ and $x_2 = h$. Then

$$\max |(x - x_0)(x - x_1)(x - x_2)| = \max |x(x^2 - h^2)|.$$

Let $g(x) = x(x^2 - h^2)$. Setting $g'(x) = 0 = 3x^2 - h^2$, we get the stationary points as $x = \pm h/\sqrt{3}$. The required maximum is

$$\left| g\left(\pm \frac{h}{3} \right) \right| = \left| \frac{h}{\sqrt{3}} \left(\frac{h^3}{3} - h^2 \right) \right| = \frac{2h^3}{3\sqrt{3}}.$$

Therefore $$|\text{Error}| \leq \frac{h^3}{9\sqrt{3}} M_3.$$

Example 18.22 Determine the step size in an equidistant table for $f(x) = (1 + x)^6$ in $[0, 2]$, if the error in magnitude in linear interpolation is $\leq 5 \times 10^{-5}$.

Solution We have the bound for the error in linear interpolation as

$$|\text{Error}| \leq \frac{h^2}{8} \max_{0 \leq x \leq 2} |f''(x)|.$$

Now, $f'(x) = 6(1 + x)^5$, $f''(x) = 30(1 + x)^4$ and $\max |f''(x)| = 2430$, in $[0, 2]$. Hence, we find h such that

$$\frac{h^2}{8}(2430) \leq 5 \times 10^{-5} \quad \text{or} \quad h^2 \leq \frac{4}{243}(10^{-5}) \quad \text{or} \quad h \leq 0.000406.$$

Example 18.23 Determine the step size in an equidistant table for $f(x) = \sin x$ in $[0, \pi/4]$, if the error in magnitude in quadratic interpolation is $\leq 5 \times 10^{-8}$.

Solution We have the bound for the error in quadratic interpolation as

$$|\text{Error}| \leq \frac{h^3}{9\sqrt{3}} \max_{0 \leq x \leq \pi/4} |f'''(x)|.$$

Now, $f'(x) = \cos x$, $f''(x) = -\sin x$, $f'''(x) = -\cos x$, and $|f'''(x)| \leq 1$ for $0 \leq x \leq \pi/4$. Hence, we find h such that

$$\frac{h^3}{9\sqrt{3}} \leq 5 \times 10^{-8}, \quad \text{or} \quad h^3 \leq 45\sqrt{3} \times 10^{-8}, \quad \text{or} \quad h \leq 0.009.$$

18.4.2 Newton's Divided Difference Interpolation

Divided differences

Let the data (x_i, f_i), $i = 0, 1, 2, \ldots, n$ be given. Then, we define the divided differences as follows.

First divided difference The first divided difference of any two consecutive data values is defined as

$$f[x_0, x_1] = \frac{f(x_1) - f(x_0)}{x_1 - x_0}, \ldots,$$

$$f[x_i, x_{i+1}] = \frac{f(x_{i+1}) - f(x_i)}{x_{i+1} - x_i}, \quad i = 0, 1, \ldots, n - 1. \tag{18.71}$$

Second divided difference The second divided difference using three consecutive data values is defined as

$$f[x_0, x_1, x_2] = \frac{f[x_1, x_2] - f[x_0, x_1]}{x_2 - x_0}.$$

Simplifying, we get

$$f[x_0, x_1, x_2] = \frac{1}{(x_2 - x_0)}\left[\frac{f_2 - f_1}{x_2 - x_1} - \frac{f_1 - f_0}{x_1 - x_0}\right]$$

$$= \frac{f_0}{(x_0 - x_1)(x_0 - x_2)} - \frac{f_1}{(x_2 - x_0)}\left[\frac{1}{x_2 - x_1} + \frac{1}{x_1 - x_0}\right] + \frac{f_2}{(x_2 - x_0)(x_2 - x_1)}.$$

$$= \frac{f_0}{(x_0 - x_1)(x_0 - x_2)} + \frac{f_1}{(x_1 - x_0)(x_1 - x_2)} + \frac{f_2}{(x_2 - x_0)(x_2 - x_1)}.$$

In general, we have

$$f[x_i, x_{i+1}, x_{i+2}] = \frac{f[x_{i+1}, x_{i+2}] - f[x_i, x_{i+1}]}{x_{i+2} - x_i}$$

$$= \frac{f_i}{(x_i - x_{i+1})(x_i - x_{i+2})} + \frac{f_{i+1}}{(x_{i+1} - x_i)(x_{i+1} - x_{i+2})} + \frac{f_{i+2}}{(x_{i+2} - x_i)(x_{i+2} - x_{i+1})}. \tag{18.72}$$

The *nth divided difference* using all the data values (x_i, f_i), $i = 0, 1, 2, \ldots, n$, is defined as

$$f[x_0, x_1, \ldots, x_n] = \frac{f[x_1, \ldots, x_n] - f[x_0, \ldots, x_{n-1}]}{x_n - x_0} \tag{18.73}$$

which can also be expressed as a linear combination of $f_0, f_1, \ldots, f_n$.

The divided differences are given in Table 18.1 for the data values (x_i, f_i), $i = 0, 1, 2, 3$. The *divided differences are symmetric with respect to their arguments*, that is, $f[x_0, x_1] = f[x_1, x_0]$, $f[x_0, x_1, x_2] = f[x_2, x_1, x_0]$ etc.

Table 18.1. Divided differences (*d · d*) table.

x	$f(x)$	First d · d	Second d · d	Third d · d
x_0	f_0			
x_1	f_1	$f[x_0, x_1]$		
x_2	f_2	$f[x_1, x_2]$	$f[x_0, x_1, x_2]$	
x_3	f_3	$f[x_2, x_3]$	$f[x_1, x_2, x_3]$	$f[x_0, x_1, x_2, x_3]$

Example 18.24 Find the second divided difference of $f(x) = 1/x$, using the points x_0, x_1, x_2.

Solution We have

$$f[x_0, x_1] = \frac{f_1 - f_0}{x_1 - x_0} = \frac{1}{(x_1 - x_0)}\left[\frac{1}{x_1} - \frac{1}{x_0}\right] = \frac{x_0 - x_1}{x_0 x_1 (x_1 - x_0)} = -\frac{1}{x_0 x_1}.$$

$$f[x_1, x_2] = \frac{f_2 - f_1}{x_2 - x_1} = -\frac{1}{x_1 x_2}.$$

Now,

$$f[x_0, x_1, x_2] = \frac{f[x_1, x_2] - f[x_0, x_1]}{x_2 - x_0}$$

$$= \frac{1}{(x_2 - x_0)}\left[-\frac{1}{x_1 x_2} + \frac{1}{x_0 x_1}\right] = \frac{x_2 - x_0}{x_0 x_1 x_2 (x_2 - x_0)} = \frac{1}{x_0 x_1 x_2}.$$

Divided difference interpolation

We mentioned in the introduction that interpolating polynomials obtained in two different ways may look different in form but are identical otherwise. For example, we can write the linear Lagrange interpolating polynomial in terms of the first divided difference as

$$P_1(x) = \frac{x - x_1}{x_0 - x_1} f_0 + \frac{x - x_0}{x_1 - x_0} f_1 = \frac{(x - x_0) + (x_0 - x_1)}{x_0 - x_1} f_0 + \frac{x - x_0}{x_1 - x_0} f_1$$

$$= f_0 + (x - x_0)\left[\frac{f_1 - f_0}{x_1 - x_0}\right] = f_0 + (x - x_0) f[x_0, x_1]. \tag{18.74}$$

We now write the interpolating polynomial $P_n(x)$ of degree n, based on the $n + 1$ nodal points $x_0, x_1, ..., x_n$ as

$$P_n(x) = a_0 + (x - x_0)\, a_1 + (x - x_0)\, (x - x_1)\, a_2 + ... + (x - x_0)\, (x - x_1) ... (x - x_{n-1})\, a_n. \qquad (18.75)$$

Substituting $x = x_0$, $x = x_1$, ..., $x = x_n$, and using the interpolating conditions $P_n(x_i) = f_i$, we get

$$P_n(x_0) = f_0 = a_0.$$

$$P_n(x_1) = f_1 = a_0 + (x_1 - x_0)a_1$$

or

$$a_1 = \frac{f_1 - f_0}{x_1 - x_0} = f[x_0, x_1].$$

$$P_n(x_2) = f_2 = a_0 + (x_2 - x_0)\, a_1 + (x_2 - x_0)\, (x_2 - x_1)\, a_2$$

or

$$a_2 = \frac{1}{(x_2 - x_0)(x_2 - x_1)}\left[f_2 - f_0 - \frac{(x_2 - x_0)(f_1 - f_0)}{x_1 - x_0} \right]$$

$$= \frac{f_0}{(x_0 - x_1)(x_0 - x_2)} + \frac{f_1}{(x_1 - x_0)(x_1 - x_2)} + \frac{f_2}{(x_2 - x_0)(x_2 - x_1)}$$

$$= f[x_0, x_1, x_2].$$

Using induction, we can prove that $a_n = f[x_0, x_1, ..., x_n]$.

Hence, we obtain

$$P_n(x) = f_0 + (x - x_0)\, f[x_0, x_1] + (x - x_0)\, (x - x_1)\, f[x_0, x_1, x_2] + ...$$

$$+ (x - x_0)\, (x - x_1) ... (x - x_{n-1})\, f[x_0, x_1, ..., x_n]. \qquad (18.76)$$

This form of the polynomial is called *Newton's divided difference interpolating polynomial*. Since the interpolating polynomial based on the same data is unique, the Lagrange interpolating polynomial (18.60) and Newton's divided difference interpolating polynomial (18.76) are identical. Therefore, the truncation error associated with the divided difference interpolating polynomial (18.76) is same as given in (18.69).

We note that the *Newton's divided difference interpolating polynomial possesses the permanence property*.

Remark 6

If a large data (x_i, f_i), $i = 0, 1, ..., m$ is represented by $P_n(x)$, a polynomial of degree n, $(n < m)$, then its nth order divided differences are all equal and divided differences of order $n + 1$ or higher are all zero. Conversely, if all the nth order divided differences obtained from a given data are equal, that is, all $(n + 1)$th order divided differences vanish, then the data represents an nth degree polynomial. For example, for a given data, if all the divided differences of order ≥ 3 are zero, then the data is represented by a quadratic polynomial. This is the advantage of the divided difference interpolating polynomial.

Example 18.25 Using divided differences, show that the data

x	-3	-2	-1	1	2	3
$f(x)$	18	12	8	6	8	12

represents a second degree polynomial. Hence, determine the interpolating polynomial.

Solution We construct the divided difference table for the given data.

x	$f(x)$	First d · d	Second d · d	Third d · d
-3	18			
-2	12	-6		
-1	8	-4	1	
1	6	-1	1	0
2	8	2	1	0
3	12	4	1	0

Since all the third order divided differences are zero, the data represents a second degree polynomial, Using Newton's divided difference interpolating polynomial, we get

$$P_2(x) = f_0 + (x - x_0) f[x_0, x_1] + (x - x_0)(x - x_1) f[x_0, x_1, x_2]$$

$$= 18 + (x + 3)(-6) + (x + 3)(x + 2)(1) = x^2 - x + 6.$$

18.4.3 Finite Difference Operators

Let the tabular points $x_0, x_1, \ldots, x_n$ be equispaced with step length h, that is

$$x_i = x_0 + ih, \quad i = 1, 2, \ldots, n, \quad \text{or} \quad x_{i+1} - x_i = h \text{ for all } i.$$

For equispaced data, we define the following operators.

Shift operator E:

$$E f(x_i) = f(x_i + h), \quad E^r f(x_i) = f(x_i + rh), \quad r \text{ any real number.}$$

Forward difference operator Δ:

$$\Delta f(x_i) = f(x_i + h) - f(x_i) = f_{i+1} - f_i.$$

We define for any positive integer n

$$\Delta^n f(x_i) = \Delta^{n-1}[\Delta f(x_i)] = \Delta^{n-1}[f_{i+1} - f_i]$$

$$= \Delta^{n-1} f_{i+1} - \Delta^{n-1} f_i = \Delta[\Delta^{n-1} f_i].$$

In particular, we have

$$\Delta^2 f(x_i) = \Delta[\Delta f(x_i)] = \Delta[f_{i+1} - f_i]$$

$$= \Delta f_{i+1} - \Delta f_i = f_{i+2} - 2f_{i+1} + f_i$$

$$\Delta^3 f_i = \Delta[\Delta^2 f_i] = \Delta f_{i+2} - 2\Delta f_{i+1} + \Delta f_i$$

$$= f_{i+3} - 3f_{i+2} + 3f_{i+1} - f_i.$$

Since $\Delta f(x_i) = (E - 1) f(x_i)$, we get $\Delta = E - 1$ or $E = 1 + \Delta$.

Hence, we write

$$\Delta^n f(x_i) = (E - 1)^n f(x_i) = \sum_{k=0}^{n} (-1)^k \frac{n!}{k!(n-k)!} f_{i+n-k}. \qquad (18.77)$$

(see Table 18.2).

Backward difference operator ∇:

$$\nabla f(x_i) = f(x_i) - f(x_i - h) = f_i - f_{i-1}.$$

We define for any positive integer n

$$\nabla^n f(x_i) = \nabla^{n-1}[\nabla f(x_i)] = \nabla^{n-1}[f_i - f_{i-1}]$$
$$= \nabla^{n-1} f_i - \nabla^{n-1} f_{i-1} = \nabla[\nabla^{n-1} f_i].$$

In particular, we have

$$\nabla^2 f(x_i) = \nabla[\nabla f(x_i)] = \nabla[f_i - f_{i-1}]$$
$$= \nabla f_i - \nabla f_{i-1} = f_i - 2f_{i-1} + f_{i-2}.$$

$$\nabla^3 f(x_i) = \nabla[\nabla^2 f_i] = \nabla f_i - 2 \nabla f_{i-1} + \nabla f_{i-2}$$
$$= f_i - 3f_{i-1} + 3f_{i-2} - f_{i-3}.$$

Since $\nabla f(x_i) = (1 - E^{-1}) f(x_i)$, we get $\nabla = 1 - E^{-1}$, or $E = (1 - \nabla)^{-1}$.

Hence, we write

$$\nabla^n f(x_i) = (1 - E^{-1})^n f(x_i) = \sum_{k=0}^{n} (-1)^k \frac{n!}{k!(n-k)!} f_{i-k} \qquad (18.78)$$

(see Table 18.2).

We note that $\Delta^n f_i = \nabla^n f_{i+n}$.

Table 18.2. Forward and backward differences.

x	$f(x)$	$\Delta f / \nabla f$	$\Delta^2 f / \nabla^2 f$	$\Delta^3 f / \nabla^3 f$
x_0	f_0	$\Delta f_0 = f_1 - f_0 = \nabla f_1$		
x_1	f_1		$\Delta^2 f_0 = \Delta f_1 - \Delta f_0$ $= \nabla f_2 - \nabla f_1 = \nabla^2 f_2$	
		$\Delta f_1 = f_2 - f_1 = \nabla f_2$		$\Delta^3 f_0 = \Delta^2 f_1 - \Delta^2 f_0$ $= \nabla^2 f_3 - \nabla^2 f_2 = \nabla^3 f_3$
x_2	f_2		$\Delta^2 f_1 = \Delta f_2 - \Delta f_1$ $= \nabla f_3 - \nabla f_2 = \nabla^2 f_3$	
		$\Delta f_2 = f_3 - f_2 = \nabla f_3$		
x_3	f_3			

Central difference operator δ:

$$\delta f(x_i) = f\left(x_i + \frac{h}{2}\right) - f\left(x_i - \frac{h}{2}\right) = f_{i+1/2} - f_{i-1/2}.$$

Note that the ordinates on the right hand side are not data values. However,

$$\delta f_{i+1/2} = \delta f\left(x_i + \frac{h}{2}\right) = f(x_i + h) - f(x_i) = f_{i+1} - f_i$$

uses the data values.

We define for any positive integer n

$$\delta^n f(x_i) = \delta^{n-1}[\delta f(x_i)] = \delta^{n-1}[f_{i+1/2} - f_{i-1/2}]$$

$$= \delta^{n-1} f_{i+1/2} - \delta^{n-1} f_{i-1/2} = \delta[\delta^{n-1} f_i]$$

In particular, we have

$$\delta^2 f(x_i) = \delta(\delta f_i) = \delta(f_{i+1/2} - f_{i-1/2})$$

$$= \delta f_{i+1/2} - \delta f_{i-1/2} = f_{i+1} - 2f_i + f_{i-1}.$$

$$\delta^3 f(x_i) = \delta(\delta^2 f_i) = \delta(f_{i+1} - 2f_i + f_{i-1})$$

$$= f_{i+3/2} - 3f_{i+1/2} + 3f_{i-1/2} - f_{i-3/2}.$$

Since, $\delta f(x_i) = (E^{1/2} - E^{-1/2}) f(x_i)$, we get $\delta = E^{1/2} - E^{-1/2}$.

Hence, we write

$$\delta^n f(x_i) = (E^{1/2} - E^{-1/2})^n f(x_i) = \sum_{k=0}^{n} (-1)^k \frac{n!}{k!(n-k)!} f_{i+(n/2)-k}. \quad (18.79)$$

Note that all even order central differences contain the tabular values (see Table 18.3).

We note that $\Delta^n f_i = \nabla^n f_{i+n} = \delta^n f_{i+(n/2)}$.

Table 18.3.   Central differences.

x	$f(x)$	δf	$\delta^2 f$	$\delta^3 f$
x_0	f_0			
		$\delta f_{1/2} = f_1 - f_0$		
x_1	f_1		$\delta^2 f_1 = \delta f_{3/2} - \delta f_{1/2}$	
		$\delta f_{3/2} = f_2 - f_1$		$\delta^3 f_{3/2} = \delta^2 f_2 - \delta^2 f_1$
x_2	f_2		$\delta^2 f_2 = \delta f_{5/2} - \delta f_{3/2}$	
		$\delta f_{5/2} = f_3 - f_2$		
x_3	f_3			

Relation between differences and derivatives

Using Taylor series expansion, we get

$$\Delta f(x) = f(x + h) - f(x) = hf'(x) + \frac{h^2}{2} f''(x) + \dots$$

Neglecting the higher order terms, we get

$$\Delta f(x) \approx hf'(x), \text{ or } f'(x) \approx \frac{1}{h} \Delta f(x).$$

The approximation is of first order or $O(h)$ approximation.

$$\Delta^2 f(x) = f(x + 2h) - 2f(x + h) + f(x)$$

$$= \left[f(x) + 2hf'(x) + 2h^2 f''(x) + \frac{8h^3}{6} f'''(x) + \dots \right]$$

$$- 2\left[f(x) + hf'(x) + \frac{h^2}{2} f''(x) + \frac{h^3}{6} f'''(x) \dots \right] + f(x)$$

$$= h^2 f''(x) + h^3 f'''(x) + \dots$$

Neglecting the higher order terms, we get

$$\Delta^2 f(x) \approx h^2 f''(x), \text{ or } f''(x) \approx \frac{1}{h^2} \Delta^2 f(x).$$

Again, the approximation is of first order or $O(h)$ approximation.
Similarly, we obtain

$$\nabla f(x) \approx hf'(x), \text{ or } f'(x) \approx \frac{1}{h} \nabla f(x), \qquad [O(h) \text{ approximation}].$$

$$\nabla^2 f(x) \approx h^2 f''(x), \text{ or } f''(x) \approx \frac{1}{h^2} \nabla^2 f(x), \qquad [O(h) \text{ approximation}].$$

We have
$$\delta f(x) = f\left(x + \frac{h}{2}\right) - f\left(x - \frac{h}{2}\right)$$

$$= \left[f(x) + \frac{h}{2} f'(x) + \frac{h^2}{8} f''(x) + \frac{h^3}{48} f'''(x) + \dots \right]$$

$$- \left[f(x) - \frac{h}{2} f'(x) + \frac{h^2}{8} f''(x) - \frac{h^3}{48} f'''(x) + \dots \right]$$

$$= hf'(x) + \frac{h^3}{24} f'''(x) + \dots$$

Hence, $\delta f(x) \approx h f'(x)$, or $f'(x) \approx \dfrac{1}{h} \delta f(x)$, $[O(h^2)$ approximation].

$$\delta^2 f(x) = f(x + h) - 2f(x) + f(x - h) = h^2 f''(x) + \frac{h^4}{12} f^{iv}(x) + \dots$$

Hence, $\qquad\qquad \delta^2 f(x) = h^2 f''(x)$, or $f''(x) \approx \dfrac{1}{h^2} \delta^2 f(x)$, $[O(h^2)$ approximation].

Note that the central differences give approximations of higher order because of symmetry of arguments.

The first divided difference can be written as

$$f[x_i, x_{i+1}] = \frac{f(x_{i+1}) - f(x_i)}{x_{i+1} - x_i} = \frac{1}{h} \Delta f_i = \frac{1}{h} \nabla f_{i+1}.$$

Finally, $Ef(x) = f(x + h) = f(x) + h f'(x) + \dfrac{h^2}{2!} f''(x) + \dots$

$$= \left[1 + hD + \frac{h^2 D^2}{2!} + \dots \right] f(x) = e^{hD} f(x)$$

where D is the derivative operator and $D^r f(x) = d^r f / dx^r$, $r = 1, 2, \dots$ Therefore, symbolically

$$E = e^{hD}, \text{ or } hD = \ln(E).$$

Hence, $\qquad\qquad hD = \ln(1 + \Delta)$

$$= \ln (1 - \nabla)^{-1} = -\ln(1 - \nabla).$$

$$\delta = E^{1/2} - E^{-1/2} = e^{hD/2} - e^{-hD/2} = 2 \sinh(hD/2).$$

Example 18.26 Prove the following relations

(i) $\Delta \left(\dfrac{1}{f_i} \right) = - \dfrac{\Delta f_i}{f_i f_{i+1}}$,

(ii) $\Delta \left(\dfrac{f_i}{g_i} \right) = \dfrac{g_i \Delta f_i - f_i \Delta g_i}{g_i g_{i+1}}$,

(iii) $\Delta (f_i^2) = (f_i + f_{i+1}) \Delta f_i$,

(iv) $\Delta + \nabla = \dfrac{\Delta}{\nabla} - \dfrac{\nabla}{\Delta}$.

Solution

(i) $\Delta \left(\dfrac{1}{f_i} \right) = \dfrac{1}{f_{i+1}} - \dfrac{1}{f_i} = - \dfrac{f_{i+1} - f_i}{f_i f_{i+1}} = - \dfrac{\Delta f_i}{f_i f_{i+1}}$.

(ii) $\Delta \left(\dfrac{f_i}{g_i} \right) = \dfrac{f_{i+1}}{g_{i+1}} - \dfrac{f_i}{g_i} = \dfrac{g_i f_{i+1} - f_i g_{i+1}}{g_i g_{i+1}}$

$$= \frac{g_i(f_{i+1} - f_i) - f_i(g_{i+1} - g_i)}{g_i \, g_{i+1}} = \frac{g_i \Delta f_i - f_i \Delta g_i}{g_i \, g_{i+1}}.$$

(iii) $\Delta(f_i^2) = f_{i+1}^2 - f_i^2 = (f_{i+1} + f_i)(f_{i+1} - f_i) = (f_{i+1} + f_i)\,\Delta f_i.$

(iv) Using $\Delta = E - 1$, $\nabla = 1 - E^{-1}$, we obtain

L.H.S. $= \Delta + \nabla = E - 1 + 1 - E^{-1} = E - E^{-1}.$

R.H.S. $= \dfrac{\Delta}{\nabla} - \dfrac{\nabla}{\Delta} = \dfrac{E-1}{1-E^{-1}} - \dfrac{1-E^{-1}}{E-1}$

$$= \frac{E(1-E^{-1})}{1-E^{-1}} - \frac{E^{-1}(E-1)}{E-1} = E - E^{-1} = \text{L.H.S.}$$

18.4.4 Newton's Forward Difference Interpolation

For equispaced data with spacing h, we obtain

$$f[x_0, x_1] = \frac{f(x_1) - f(x_0)}{x_1 - x_0} = \frac{1}{h}\,\Delta f_0.$$

$$f[x_0, x_1, x_2] = \frac{f[x_1, x_2] - f(x_0, x_1)}{x_2 - x_0} = \frac{1}{2h}\left[\frac{1}{h}\Delta f_1 - \frac{1}{h}\Delta f_0\right] = \frac{1}{2!h^2}\,\Delta^2 f_0.$$

$$\cdots \qquad \cdots \qquad \cdots \qquad \cdots$$

$$f[x_0, x_1, \ldots, x_n] = \frac{1}{n!h^n}\,\Delta^n f_0.$$

Replacing divided differences by forward differences in the divided difference interpolating polynomial (18.76), we get

$$P_n(x) = f_0 + (x - x_0)\,\frac{\Delta f_0}{1!h} + (x - x_0)\,(x - x_1)\,\frac{\Delta^2 f_0}{2!h^2} + \ldots$$

$$+ (x - x_0)\,(x - x_1)\,\ldots\,(x - x_{n-1})\,\frac{\Delta^n f_0}{n!h^n} \tag{18.80}$$

which is called the *Newton's forward difference interpolating polynomial*. Since the polynomial is unique, the error of interpolation is same as given in (18.69).

Let $x - x_0 = hs$. Then, $x - x_i = x - x_0 - ih = sh - ih = (s - i)h$, $i = 1, 2, \ldots, n$. Hence, we can write (18.80) as

$$P_n(x_0 + hs) = f_0 + s\Delta f_0 + \frac{s(s-1)}{2!}\,\Delta^2 f_0 + \ldots + \frac{s(s-1)\ldots(s-n+1)}{n!}\,\Delta^n f_0 \tag{18.81}$$

where $s = (x - x_0)/h$. The coefficients of Δf_0, $\Delta^2 f_0$, ..., $\Delta^n f_0$ are binomial coefficients. The error of interpolation (18.69) becomes

$$E_n(f, s) = \frac{s(s-1)\ldots(s-n)}{(n+1)!} \, h^{n+1} f^{(n+1)}(\xi), \quad 0 < \xi < n. \tag{18.82}$$

Again, the coefficient in the error term is the next binomial coefficient.

18.4.5 Newton's Backward Difference Interpolation

Since the divided differences are symmetric with respect to the arguments, we can take the arguments in the order x_n, x_{n-1}, ..., x_1, x_0 and write the Newton's divided difference interpolating polynomial in the form

$$P_n(x) = f_n + (x - x_n) f[x_n, x_{n-1}] + (x - x_n)(x - x_{n-1}) f[x_n, x_{n-1}, x_{n-2}]$$

$$+ \ldots + (x - x_n)(x - x_{n-1}) \ldots (x - x_1) f[x_n, x_{n-1}, \ldots, x_0]. \tag{18.83}$$

Now, due to symmetry

$$f[x_n, x_{n-1}] = f[x_{n-1}, x_n] = \frac{1}{h} \, \nabla f_n,$$

$$f[x_n, x_{n-1}, x_{n-2}] = f[x_{n-2}, x_{n-1}, x_n] = \frac{1}{2! h^2} \, \nabla^2 f_n, \ldots$$

$$\cdots \qquad \qquad \cdots \qquad \qquad \cdots \qquad \qquad \cdots$$

$$f[x_n, x_{n-1}, \ldots, x_0] = \frac{1}{n! h^n} \, \nabla^n f_n.$$

Replacing the divided differences in terms of backward differences in (18.83), we obtain

$$P_n(x) = f_n + (x - x_n) \frac{\nabla f_n}{1! h} + (x - x_n)(x - x_{n-1}) \frac{\nabla^2 f_n}{2! h^2} + \ldots$$

$$+ (x - x_n)(x - x_{n-1}) \ldots (x - x_1) \frac{\nabla^n f_n}{n! h^n} \tag{18.84}$$

which is called the *Newton's backward difference interpolating polynomial*.

Let $x - x_n = hs$. Then,

$$x - x_i = x - x_n + x_n - x_i = x - x_n + (x_0 + nh) - (x_0 + ih)$$

$$= hs + h(n - i) = h(s + n - i), \quad i = n - 1, n - 2, \ldots, 0.$$

Substituting in (18.84), we obtain

$$P_n(x_n + hs) = f_n + s \nabla f_n + \frac{s(s+1)}{2!} \, \nabla^2 f_n + \ldots + \frac{s(s+1)\ldots(s+n-1)}{n!} \, \nabla^n f_n. \tag{18.85}$$

The error of interpolation becomes

$$E_n(f, s) = \frac{s(s+1)\ldots(s+n)}{(n+1)!} h^{n+1} f^{(n+1)}(\xi), \ 0 < \xi < n. \tag{18.86}$$

Remark 7

We often use the forward difference formula for interpolating at the values near the top of the table (since all the forward differences are available), while we use the backward difference formula for interpolating at the values near the bottom of the table (since all the backward differences are available).

Example 18.27 . For the data

x	-4	-2	0	2	4	6
$f(x)$	-139	-21	1	23	141	451

construct the forward and backward difference tables. Using the corresponding interpolation, show that the interpolating polynomial is same.

Solution The step length is $h = 2$, and $x_0 = -4$, $x_1 = -2$, $x_2 = 0$, $x_3 = 2$, $x_4 = 4$, $x_5 = 6$. We have the following difference Table 18.4.

Table 18.4   Difference table, Example 18.27.

x	$f(x)$	$\Delta f / \nabla f$	$\Delta^2 f / \nabla^2 f$	$\Delta^3 f / \Delta^3 f$	$\Delta^4 f \nabla^4 f$
-4	-139				
		118			
-2	-21		-96		
		22		96	
0	1		0		0
		22		96	
2	23		96		0
		118		96	
4	141		192		
		310			
6	451				

(a) Using the Newton's forward difference interpolating polynomial (18.80), we get

$$P_3(x) = -139 + (x+4)\left[\frac{118}{2}\right] + (x+4)(x+2)\left[\frac{-96}{2 \times 4}\right] + (x+4)(x+2) \, x\left[\frac{96}{6 \times 8}\right].$$

$$= -139 + 59(x+4) - 12(x+4)(x+2) + 2x(x+4)(x+2)$$

$$= 2x^3 + 3x + 1.$$

(b) Using the Newton's backward difference interpolating polynomial (18.84), we get

$$P_3(x) = 451 + (x - 6)\left[\frac{310}{2}\right] + (x - 6)(x - 4)\left[\frac{192}{2 \times 4}\right] + (x - 6)(x - 4)(x - 2)\left[\frac{96}{6 \times 8}\right]$$

$$= 451 + 155(x - 6) + 24(x - 6)(x - 4) + 2(x - 6)(x - 4)(x - 2)$$

$$= 2x^3 + 3x + 1.$$

18.4.6 Spline Interpolation

In the earlier years, smooth curves (for engineering and other models) were drawn using an instrument called a spline. The instrument was generally used to draw curves which are continuous, have continuous first derivatives $f'(x)$ (slopes) and continuos second derivatives $f''(x)$ (curvature). The linear and quadratic spline interpolations are not normally used. One of the useful splines is the cubic spline, which has many applications like plotting of curves, solution of differential equations etc. We discuss below the derivation of cubic splines.

18.4.6.1 Cubic splines

Consider a data given over an interval $[a, b]$, and a subdivision of the interval as $a = x_0 < x_1 < ... < x_n = b$. The abscissas (called nodes or knots) may or may not be uniformly spaced. Denote $h_i = x_i - x_{i-1}$. Let the $n + 1$ data values be (x_k, f_k), $k = 0, 1\ 2, ...\ n$.

The cubic spline is defined as a function $S(x)$, which has the following properties:

  (i) On each subinterval (x_{i-1}, x_i), $1 \le i \le n$, $S(x)$ is a cubic polynomial, denoted by $s_i(x)$.

 (ii) $S(x)$ satisfies the interpolating conditions: $S(x_i) = f_i$, $i = 0, 1, 2, ..., n$.

 (iii) $S(x)$, $S'(x)$, $S''(x)$ are continuous on (a, b).

The third property implies the following:

$$s_i(x_i) = s_{i+1}(x_i), \quad s'_i(x_i) = s'_{i+1}(x_i), \quad s''_i(x_i) = s''_{i+1}(x_i), \quad i = 1, 2, ..., n - 1. \tag{18.87}$$

Example 18.28 Find whether the following function is a cubic spline.

$$f(x) = \begin{cases} 2x^3 + x^2 - 3x, & -1 \le x \le 0, \\ -3x^3 + x^2 - 3x, & 0 < x \le 1. \end{cases}$$

Solution The given function is a cubic in both intervals. We have

$$\lim_{x \to 0^+} f(x) = 0 = \lim_{x \to 0^-} f(x),$$

$$f'(x) = \begin{cases} 6x^2 + 2x - 3, & -1 \le x \le 0, \\ -9x^2 + 2x - 3, & 0 < x \le 1. \end{cases} \qquad \lim_{x \to 0^+} f'(x) = -3 = \lim_{x \to 0} f'(x),$$

$$f''(x) = \begin{cases} 12x + 2, & -1 \le x \le 0, \\ -18x + 2, & 0 < x \le 1. \end{cases} \qquad \lim_{x \to 0^+} f''(x) = 2 = \lim_{x \to 0} f''(x).$$

The given function and its first and second derivatives are continuous at $x = 0$ and at the end points. Hence, $f(x)$ is a cubic spline.

It is possible to derive a cubic spline in three ways.

The first way is a direct method. we assume a cubic polynomial of the form, $a_i x^3 + b_i x^2 + c_i x + d_i$, $i = 1, 2, \ldots, n$ in each of the subintervals (x_{i-1}, x_i). We satisfy the interpolating conditions and continuity conditions of $f(x), f'(x)$, and $f''(x)$. The resulting equations are solved for the parameters. However, it is a laborious method.

The second method is used when the slopes at the knots $(f'_I = m_i)$ are also required. We start with the cubic Hermite interpolating polynomial, and use the continuity requirement of $S''(x)$ at the nodal points. This gives $n - 1$ equations for computing the slopes f'_i, $i = 0, 1, 2, \ldots, n$. We require two more conditions. If f''_0, f''_n are prescribed, then we obtain two additional equations to compute f'_0, f'_n. The solution of the resulting tridiagonal system of equations gives the values of the slopes f'_i, $i = 0, 1, 2, \ldots, n$. Substituting in the Hermite polynomial, we obtain the cubic spline.

We discuss the third method in which the second derivatives are to be computed at the knots. Since $S(x)$ is a cubic polynomial, its first derivative is a quadratic polynomial and the second derivative is a linear polynomial. Denote $f''_i = s''_i = M_i$. Assume the data as (x_i, M_i), $i = 0, 1, 2, \ldots$, where M_i are unknowns.

On $[x_{i-1}, x_i]$, the linear Lagrange interpolating polynomial that fits the data (x_{i-1}, M_{i-1}), (x_i, M_i) is given by

$$s''_i(x) = \frac{x_i - x}{x_i - x_{i-1}} M_{i-1} + \frac{x - x_{i-1}}{x_i - x_{i-1}} M_i = \frac{1}{h_i} [(x_i - x) M_{i-1} + (x - x_{i-1}) M_i]. \tag{18.88}$$

Integrating two times, we get

$$s_i(x) = \frac{1}{6h_i} [(x_i - x)^3 M_{i-1} + (x - x_{i-1})^3 M_i] + p_i (x_i - x) + q_i (x - x_{i-1}),$$

where the contribution of the constants of integration is written in a suitable form. Interpolating conditions give

$$s_i(x_{i-1}) = f_{i-1} = \frac{h_i^2}{6} M_{i-1} + p_i h_i, \quad \text{or} \quad p_i = \frac{1}{h_i} \left[f_{i-1} - \frac{h_i^2}{6} M_{i-1} \right].$$

$$s_i(x_i) = f_i = \frac{h_i^2}{6} M_i + q_i h_i, \quad \text{or} \quad q_i = \frac{1}{h_i} \left[f_i - \frac{h_i^2}{6} M_i \right].$$

Hence, the cubic spline on the interval $[x_{i-1}, x_i]$ is given by

$$s_i(x) = \frac{1}{6h_i} [(x_i - x)^3 M_{i-1} + (x - x_{i-1})^3 M_i]$$

$$+ \frac{1}{h_i} (x_i - x) \left[f_{i-1} - \frac{h_i^2}{6} M_{i-1} \right] + \frac{1}{h_i} (x - x_{i-1}) \left[f_i - \frac{h_i^2}{6} M_i \right]. \tag{18.89}$$

Setting $i = i + 1$ in (18.89), we get the spline valid in the interval $[x_i, x_{i+1}]$ as

$$s_{i+1}(x) = \frac{1}{6h_{i+1}} [(x_{i+1} - x)^3 M_i + (x - x_i)^3 M_{i+1}]$$

$$+ \frac{1}{h_{i+1}} (x_{i+1} - x) \left[f_i - \frac{h_{i+1}^2}{6} M_i \right] + \frac{1}{h_{i+1}} (x - x_i) \left[f_{i+1} - \frac{h_{i+1}^2}{6} M_{i+1} \right]. \qquad (18.90)$$

Since the first derivative is continuous, we require $\lim\limits_{\varepsilon \to 0} s_i'(x_i - \varepsilon) = \lim\limits_{\varepsilon \to 0} s_{i+1}'(x_i + \varepsilon)$.

Using (18.89) and (18.90), we obtain

$$\frac{h_i}{2} M_i - \frac{1}{h_i} f_{i-1} + \frac{h_i}{6} M_{i-1} + \frac{1}{h_i} f_i - \frac{h_i}{6} M_i$$

$$= -\frac{h_{i+1}}{2} M_i - \frac{1}{h_{i+1}} f_i + \frac{h_{i+1}}{6} M_i + \frac{1}{h_{i+1}} f_{i+1} - \frac{h_{i+1}}{6} M_{i+1},$$

or $\quad \dfrac{h_i}{6} M_{i-1} + \dfrac{1}{3} [h_i + h_{i+1}] M_i + \dfrac{h_{i+1}}{6} M_{i+1} = \dfrac{1}{h_{i+1}} [f_{i+1} - f_i] - \dfrac{1}{h_i} [f_i - f_{i-1}], \qquad (18.91)$

$$i = 1, 2, ..., n - 1.$$

When the mesh spacing is uniform, that is $h_i = h_{i+1} = h$, $x_i = x_0 + ih$, Eq. (18.91) simplifies to

$$M_{i-1} + 4M_i + M_{i+1} = \frac{6}{h^2} [f_{i-1} - 2f_i + f_{i+1}], \quad i = 1, 2, ... n - 1. \qquad (18.92)$$

Eq.(18.92) gives $n - 1$ equations for the solution of $n + 1$ unknowns $M_0, M_1, ..., M_n$. We require two more equations to uniquely determine M_i. For example, we may prescribe any of the following conditions:

(i) *Natural spline*: $M_0 = 0 = M_n$. Alternately, we may prescribe the values of M_0, M_n.

(ii) The values of the first derivatives may be prescribed at x_0, x_n.

The solution of the resulting tridiagonal system of equations gives the values of the second derivatives f_i'', $i = 1, 2, ..., n - 1$ or f_i'', $i = 0, 1, 2, ..., n$. Substituting in Eq. (18.89), we obtain the cubic spline. We note the following:

(i) cubic spline is a global spline. If any data value is changed, then the splines in all the intervals change.

(ii) cubic splines provide better approximation (compared to general polynomial approximations) to the behavior of functions which change rapidly.

(iii) cubic spline $S(x)$ gives the smoothest interpolating function (compared to any polynomial approximation) to a given set of data values in the sense that of all functions

$p(x) \in C^2 [a, b]$, $S(x)$ minimizes the integral $\int_a^b [p''(x)]^2 dx$.

Example 18.29   Construct the cubic spline approximation for the function defined by the data, ن the intervals [1.0, 1.5], and [1.5, 2.0]

x	0	0.5	1.0	1.5	2.0
$f(x)$	1.0	2.0625	4.0	9.0625	21.0

Assume $M(0) = 0$, $M(2.0) = 48$. Interpolate at 1.2 and 1.6.

Solution   Let the abscissas be denoted as $x_0 = 0$, $x_1 = 0.5$, $x_2 = 1.0$, $x_3 = 1.5$, $x_4 = 2$, and the second derivatives at the nodal points be denoted by M_0, M_1, M_2, M_3, M_4. We have $M_0 = 0$ and $M_4 = 48$. With $h = 0.5$, we obtain the following equations:

$$M_{i-1} + 4M_i + M_{i+1} = 24 \, [f_{i-1} - 2f_i + f_{i+1}], \quad i = 1, 2, 3.$$

$i = 1$: $4M_1 + M_2 \qquad\qquad = 24 \, (1 - 4.125 + 4) = 21.$

$i = 2$: $M_1 + 4M_2 + M_3 = 24 \, (2.0625 - 8 + 9.0625) = 75.$

$i = 3$: $M_2 + 4M_3 + M_4 = 24 \, (4 - 18.125 + 21) = 165,$ or $M_2 + 4M_3 = 165 - 48 = 117.$

We have the tridiagonal system of equations

$$\begin{bmatrix} 4 & 1 & 0 \\ 1 & 4 & 1 \\ 0 & 1 & 4 \end{bmatrix} \begin{bmatrix} M_1 \\ M_2 \\ M_3 \end{bmatrix} = \begin{bmatrix} 21 \\ 75 \\ 117 \end{bmatrix}.$$

The solution is $M_1 = 33/14$, $M_2 = 81/7$, $M_3 = 369/14$. The cubic spline in the interval $[x_{i-1}, x_i]$ is

$$s_i(x) = \frac{1}{3} \, [(x_i - x)^3 \, M_{i-1} + (x - x_{i-1})^3 \, M_i] + 2(x_i - x) \left[f_{i-1} - \frac{1}{24} M_{i-1} \right] + 2 \, (x - x_{i-1}) \left[f_i - \frac{1}{24} M_i \right].$$

On [1, 1.5], we have $x_2 = 1$, $x_3 = 1.5$, and

$$s_3(x) = \frac{1}{3} \left[(1.5 - x)^3 \left(\frac{81}{7} \right) + (x - 1)^3 \left(\frac{369}{14} \right) \right] + 2 \, (1.5 - x) \left[4 - \left(\frac{1}{24} \right) \left(\frac{81}{7} \right) \right]$$

$$+ \, 2 \, (x - 1) \left[9.0625 - \left(\frac{1}{24} \right) \left(\frac{369}{14} \right) \right]$$

$$= \frac{1}{42} \, [162(1.5 - x)^3 + 369 \, (x - 1)^3] + \frac{197}{28} \, (1.5 - x) + \frac{223}{14} \, (x - 1).$$

We get $s_3 \, (1.2) = 5.4709.$

On [1.5, 2], we have $x_3 = 1.5$, $x_4 = 2$, and

$$s_4(x) = \frac{1}{3} \left[(2 - x)^3 \left(\frac{369}{14} \right) + (x - 1.5)^3 (48) \right] + 2 \, (2 - x) \left[9.0625 - \left(\frac{1}{24} \right) \left(\frac{369}{14} \right) \right]$$

$$+ \, 2 \, (x - 1.5) \left[21 - \left(\frac{1}{24} \right) (48) \right]$$

$$= \frac{1}{42} \left[369 \, (2-x)^3 + 672 \, (x-1.5)^3 \right] + \frac{223}{14} \, (2-x) + 38 \, (x-1.5).$$

We get $s_4(1.6) = 10.7497$.

It may be noted that the data represents the values of the function $f(x) = x^4 + 2x + 1$, and the exact values are $f(1.2) = 5.4736$, $f(1.6) = 10.7536$.

18.4.6.2 B-splines

B-splines is an important tool in computer graphics, solution of differential equations and various other areas of engineering and science. *B*-splines form a basis for some spline spaces. *B*-splines form an important class of functions. Most other spline functions can be written as a linear combination of *B*-splines. The importance of *B*-splines is that they are *local splines*. If any knot t_i (data value) is changed, then we need to change only *B*-splines derived by using that knot.

Let the interval $[a, b]$ be subdivided into n parts such that $a = t_0 < t_1 < t_2 < \ldots < t_n = b$.

Often, for theoretical purposes, *B*-splines are considered over an infinite set of points $\ldots < t_{-2} < t_{-1} < t_0 < t_1 < t_2 < \ldots$. Denote the *B*-spline by $B_i^k(x)$, where k is the degree of the spline and i corresponds to the knot t_i. Another notation for the *B*-spline is $B_{k,\, i}(x)$. *B*-splines can be derived using the knots (nodes) in the forward direction or in the backward direction or symmetrically placed knots. For our discussion, we shall use the representation using knots in the forward direction. The approximating *B*-spline function over the whole interval $[t_0, t_n]$ is written as a linear combination of the basis *B*-spline functions over the subintervals. The coefficients are determined by satisfying the data. However, approximating functions for a given data in term of *B*-splines are available in literature, which may not satisfy the interpolating conditions, that is, curve of the approximating function may not pass through any of the data points.

B-spline of degree zero

Consider the interval $[t_i, t_{i+1}]$. On this interval, *B*-spline of degree zero is defined as a constant as follows

$$B_i^0(x) = 1, \text{ for } t_i \le x < t_{i+1},$$

$$= 0, \text{ elsewhere.} \tag{18.93}$$

(see Fig. 18.7)

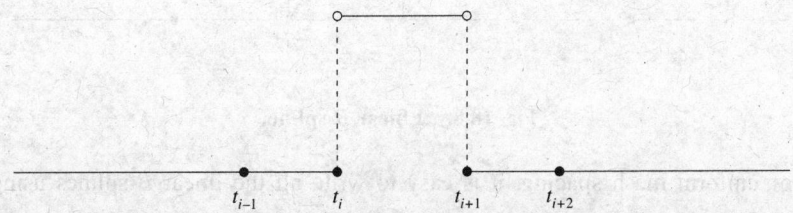

Fig. 18.7. *B*-spline of degree 0.

Note that $B_i^0(x)$ is continuous over the interval $[t_i, t_{i+1}]$. The interval $[t_i, t_{i+1}]$ is called the support of $B_i^0(x)$. A spline approximation of degree zero for a given data can be written in terms of B-splines $B_i^0(x)$, that is, $B_i^0(x)$ form a basis for such a spline approximation.

B-spline of degree one (Linear B-splines)

Consider the knots t_i, t_{i+1}, t_{i+2}. On the interval $[t_i, t_{i+2}]$, B-spline of degree one (linear spline) is defined as follows

$$B_i^1(x) = \frac{x - t_i}{t_{i+1} - t_i}, \qquad t_i \le x < t_{i+1}$$

$$= \frac{t_{i+2} - x}{t_{i+2} - t_{i+1}}, \qquad t_{i+1} \le x < t_{i+2}$$

$$= 0, \qquad \text{elsewhere.} \qquad (18.94)$$

Consider the case of uniform spacing of knots, $t_{i+1} - t_i = h$, $i = 0, 1, 2, ..., n - 1$. Setting $u = [(x - t_i)/h]$, we obtain

$$B_i^1(u), = u, \qquad 0 \le u < 1$$

$$= 2 - u, \qquad 1 \le u < 2$$

$$= 0, \qquad \text{elsewhere.} \qquad (18.95)$$

$B_i^1(x)$ is plotted in Fig. 18.8. The Linear B-splines are same as the linear shape functions used in piecewise linear interpolation.

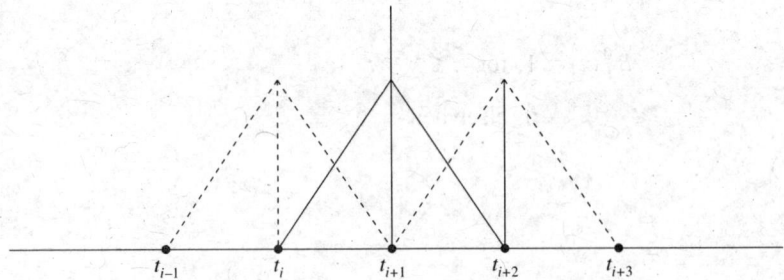

Fig. 18.8.   Linear B-spline.

For the case of uniform mesh spacing, it is easy to write all the linear B-splines using (18.95).

Since $\dfrac{x - t_{i+1}}{h} = \dfrac{x - t_i - h}{h} = u - 1$, we can write the spline $B_{i+1}^1(u)$ as

$$B^1_{i+1}(x) = B^1_{i+1}(u) = u-1, \quad 1 \le u < 2,$$
$$= 3-u, \quad 2 \le u < 3,$$
$$= 0, \quad \text{elsewhere.} \tag{18.96}$$

Since $\dfrac{x-t_{i-1}}{h} = \dfrac{x-t_i+h}{h} = u+1$, we can write the spline $B^1_{i-1}(u)$ as

$$B^1_{i-1}(x) = B^1_{i-1}(u) = u+1, \quad -1 \le u < 0,$$
$$= 1-u, \quad 0 \le u < 1,$$
$$= 0, \quad \text{elsewhere.} \tag{18.97}$$

Therefore, $B^1_{i+1}(u)$ and $B^1_{i-1}(u)$ can be obtained from (18.95) by replacing u in (18.95) by $u-1$ and $u+1$ respectively. The linear B-splines over the whole interval $[t_0, t_n]$ are plotted in Fig 18.9. Note that for any $t \in (t_i, t_{i+1})$, only the two splines B^1_{i-1} and B^1_i contribute.

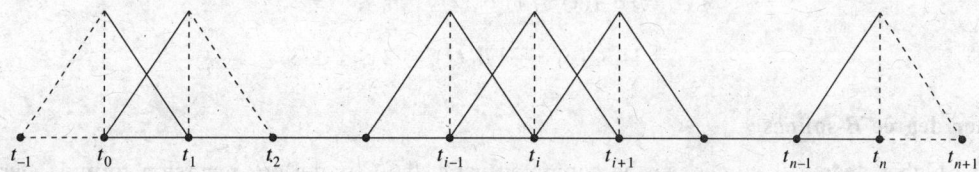

Fig. 18.9. Linear *B*-splines over the whole interval $[t_0, t_n]$.

Linear *B*-spline approximation

The linear B-spline approximation over $[t_0, t_n]$ is written as (see Fig.18.9)

$$S(x) = \alpha_{-1} B^1_{-1}(x) + \alpha_0 B^1_0(x) + \alpha_1 B^1_1(x) + \ldots + \alpha_{n-1} B^1_{n-1}(x) = \sum_{i=-1}^{n-1} \alpha_i B^1_i(x). \tag{18.98}$$

We have $n+1$ unknowns and $n+1$ data values. Setting $S(t_i) = f_i$, we can determine uniquely, the values of the coefficients α_i.

At $t = t_i$, only two basis functions $B^1_{i-1}(t_i)$, $B^1_i(t_i)$ contribute. Therefore,

$$S(t_i) = f_i = \alpha_{i-1} B^1_{i-1}(t_i) + \alpha_i B^1_i(t_i)$$
$$= \alpha_{i-1}(1) + \alpha_i(0) = \alpha_{i-1}, \quad i = 0, 1, 2, \ldots, n \tag{18.99}$$

where (t_i, f_i), $i = 0, 1, 2, \ldots, n$ are the data values. Hence, the coefficients α_i in (18.98) are the given ordinates. Using (18.98), we can write the expressions in each subinterval.

Example 18.30 For the following data, fit an approximating linear *B*-spline approximation.

$$
\begin{array}{cccc}
x & 1 & 2 & 3 \\
y & 3 & 11 & 31
\end{array}
$$

Solution Let $t_0 = 1$, $t_1 = 2$, $t_2 = 3$, $y_1 = 3$, $y_2 = 11$, $y_3 = 31$. Using (18.99), we get $\alpha_{-1} = y_0 = 3$, $\alpha_0 = y_1 = 11$, $\alpha_1 = y_2 = 31$, and the spline approximation is

$$
S(x) = \alpha_{-1} B_{-1}^1(x) + \alpha_0 B_0^1(x) + \alpha_1 B_1^1(x) = 3 B_{-1}^1(x) + 11 B_0^1(x) + 31 B_1^1(x).
$$

Let us now write the contributions of B_i^1 on the given intervals. We have $h = 1$ and $u = x - 1$.

	On (1, 2)	On (2, 3)
$B_{-1}^1(x)$	$2 - x$	0
$B_0^1(x)$	$x - 1$	$3 - x$
$B_1^1(x)$	0	$x - 2$

Therefore, we have

$$
S(x) = 3 B_{-1}^1(x) + 11 B_0^1(x) \qquad 1 \le x \le 2,
$$

$$
= 11 B_0^1(x) + 31 B_1^1(x), \qquad 2 \le x \le 3.
$$

Higher degree *B*-splines

All the higher degree *B*-splines can be computed using the Cox-deBoor recursion formula starting with $B_i^0(x)$. The recursion formula for kth degree spline is given by

$$
B_i^k(x) = \frac{x - t_i}{t_{i+k} - t_i} B_i^{k-1}(x) + \frac{t_{i+k+1} - x}{t_{i+k+1} - t_{i+1}} B_{i+1}^{k-1}(x), \quad k \ge 1. \tag{18.100}
$$

Denote $V_i^k(x) = \dfrac{x - t_i}{t_{i+k} - t_i}$. Then, $1 - V_{i+1}^k(x) = 1 - \dfrac{x - t_{i+1}}{t_{i+k+1} - t_{i+1}} = \dfrac{t_{i+k+1} - x}{t_{i+k+1} - t_{i+1}}$.

Eq. (18.100) can be written as

$$
B_i^k(x) = V_i^k(x) \, B_i^{k-1}(x) + [1 - V_{i+1}^k(x)] \, B_{i+1}^{k-1}(x).
$$

The functions $V_i^k(x)$ and $1 - V_{i+1}^k(x)$ are ratios of lengths. They are similar to the local or natural coordinates used in linear interpolation.

For uniform spacing of knots, $t_{i+1} - t_i = h$, (18.100) simplifies to

$$
B_i^k(x) = B_i^k(u) = \frac{1}{kh} [(x - t_i) \, B_i^{k-1}(x) + (t_{i+k+1} - x) \, B_{i+1}^{k-1}(x)]
$$

$$
= \frac{1}{k} [u B_i^{k-1}(u) + (k + 1 - u) \, B_{i+1}^{k-1}(u)] \tag{18.101}
$$

where $u = [(x - t_i)/h]$. Using $B_i^0(x)$, we can compute $B_i^1(x)$ which is a spline polynomial of degree 1. Using $B_i^1(x)$, we can compute $B_i^2(x)$ which is a spline polynomial of degree 2, etc. For computing the quadratic B-splines, $B_i^2(x)$, we use the knots t_i, t_{i+1}, t_{i+2}, t_{i+3}.

For computing the cubic B-splines, $B_i^3(x)$, we use the knots t_i, t_{i+1}, t_{i+2}, t_{i+3}, t_{i+4}.
Consider the case of uniform mesh spacing.

Quadratic B-splines

Using the expressions for $B_i^1(x)$, and $B_{i+1}^1(x)$ (see Eqs.(18.95),(18.96)), we obtain

$$\begin{aligned}
B_i^2(u) &= [u^2/2], & 0 \le u < 1, \\
&= -u^2 + 3u - (3/2), & 1 \le u < 2, \\
&= [(3 - u)^2/2], & 2 \le u < 3, \\
&= 0, \text{ elsewhere.} & (18.102)
\end{aligned}$$

The quadratic spline $B_i^2(x)$ is a bell shaped curve (see Fig. 18.10).

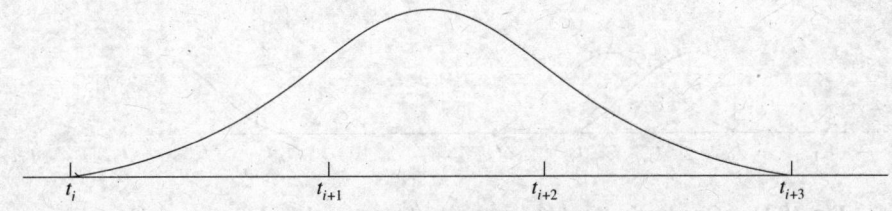

Fig. 18.10.   Quadratic B-spline.

The expressions for $B_{i-2}^2(u)$, $B_{i-1}^2(u)$, $B_{i+1}^2(u)$, $B_{i+2}^2(u)$ can be obtained by replacing u by $u + 2$, $u + 1$, $u - 1$, $u - 2$ in $B_i^2(u)$ respectively.

$$\begin{aligned}
B_{i-2}^2(u) &= [(u + 2)^2/2], & -2 \le u < -1, \\
&= -(u + 2)^2 + 3(u + 2) - (3/2), & -1 \le u < 0, & (18.103) \\
&= [(1 - u)^2/2], & 0 \le u < 1, \\
&= 0, \text{ elsewhere.}
\end{aligned}$$

$$\begin{aligned}
B_{i-1}^2(u) &= [(u + 1)^2/2], & -1 \le u < 0, \\
&= -(u + 1)^2 + 3(u + 1) - (3/2), & 0 \le u < 1, & (18.104) \\
&= [(2 - u)^2/2], & 1 \le u < 2, \\
&= 0, \text{ elsewhere.}
\end{aligned}$$

$$B_{i+1}^2(u) = [(u-1)^2/2], \qquad\qquad 1 \le u < 2,$$
$$= -(u-1)^2 + 3(u-1) - (3/2), \quad 2 \le u < 3, \qquad (18.105)$$
$$= [(4-u)^2/2], \qquad\qquad 3 \le u < 4,$$
$$= 0, \text{ elsewhere.}$$

$$B_{i+2}^2(u) = [(u-2)^2/2], \qquad\qquad 2 \le u < 3,$$
$$= -(u-2)^2 + 3(u-2) - (3/2), \quad 3 \le u < 4, \qquad (18.106)$$
$$= [(5-u)^2/2], \qquad\qquad 4 \le u < 5,$$
$$= 0, \text{ elsewhere.}$$

The quadratic B-splines over the whole interval $[t_0, t_n]$ are plotted in Fig.18.11. Note that for any $t \in (t_i, t_{i+1})$, only the three splines $B_{i-2}^2(x)$, $B_{i-1}^2(x)$, $B_i^2(x)$ contribute.

Quadratic B-spline approximation

Consider constructing a spline approximation using quadratic basis splines. The spline over (t_0, t_n) is written as (see Fig 18.11)

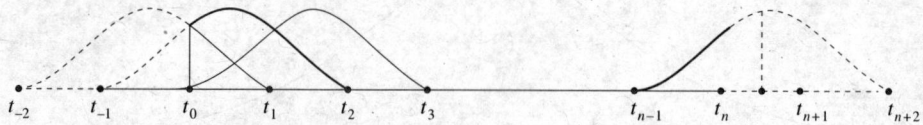

t_{-2} t_{-1} t_0 t_1 t_2 t_3 t_{n-1} t_n t_{n+1} t_{n+2}

Fig. 18.11.   Quadratic B-spline approximation over the whole interval $[t_0, t_n]$.

$$S(x) = \alpha_{-2} B_{-2}^2(x) + \alpha_{-1} B_{-1}^2(x) + \alpha_0 B_0^2(x) + \dots + \alpha_{n-1} B_{n-1}^2(x) = \sum_{i=-2}^{n-1} \alpha_i B_i^2(x) \qquad (18.107)$$

There are $n+2$ parameters, α_i, $i = -2, -1, 0, 1, \dots, n-1$, to be determined. The interpolation conditions $S(t_i) = f_i$, $i = 0, 1, \dots, n$ give $n+1$ equations. An extra condition is required to obtain α_i uniquely. We may prescribe the condition $S'(t_0) = f_0'$, or $S'(t_n) = f_n'$, to obtain an extra equation. Now, at $t = t_i$, the contributions of the three splines B_{i-2}^2, B_{i-1}^2, B_i^2 are $B_{i-2}^2(t_i) = B_{i-2}^2(u=0) = (1/2)$, $B_{i-1}^2(t_i) = B_{i-1}^2(u=0) = (1/2)$, and $B_i^2(t_i) = B_i^2(u=0) = 0$.

Substituting $x = t_i$ in (18.107), the interpolating condition $S(t_i) = f_i$ gives

$$\frac{1}{2}\alpha_{i-2} + \frac{1}{2}\alpha_{i-1} = f_i, \quad i = 0,1, 2, \dots, n. \qquad (18.108)$$

If the condition $S'(t_0) = f_0'$ is used, we get

$$(B_{-2}^2)'(t_0)\alpha_{-2} + (B_{-1}^2)'(t_0)\alpha_{-1} + (B_0^2)'(t_0)\alpha_0 = -\frac{1}{h}\alpha_{-2} + \frac{1}{h}\alpha_{-1} = f_0' \qquad (18.109)$$

since, on $0 \le u < 1$, $(B_0^2)'(u) = (u/h)$, $(B_0^2)'(0) = 0$,

$$(B_{-2}^2)'(u) = -[1 - u)/h], \quad (B_{-2}^2)'(0) = -(1/h),$$

$$(B_{-1}^2)'(u) = [-2(u + 1) + 3]/h, \quad (B_{-1}^2)'(0) = (1/h).$$

In matrix form, we get the system of equations as

$$\begin{bmatrix} -1 & 1 & & & & \\ & 1 & 1 & & & \\ & & 1 & 1 & & \\ & & & 1 & 1 & \\ \cdots & \cdots & \cdots & \cdots & \cdots & \cdots \\ & & & & 1 & 1 \end{bmatrix} \begin{bmatrix} \alpha_{-2} \\ \alpha_{-1} \\ \alpha_0 \\ \alpha_1 \\ \cdots \\ \alpha_{n-1} \end{bmatrix} = \begin{bmatrix} h f_0' \\ 2 f_0 \\ 2 f_1 \\ 2 f_1 \\ \cdots \\ 2 f_n \end{bmatrix} \tag{18.110}$$

The first two equations give α_{-2}, α_{-1}. Substituting in the forward direction we get the values of the remaining parameters. Substituting the values of α_i in (18.107), we obtain the quadratic *B*-spline approximation.

If the condition $S'(t_n) = f_n'$ is used, we get in place of Eq. (18.109), the equation

$$-\frac{1}{h} \alpha_{n-2} + \frac{1}{h} \alpha_{n-1} = f_n'. \tag{18.111}$$

The system of equations is now given by (18.108) and (18.111).

Example 18.31   For the following data values fit a quadratic *B*-spline

x	0	1	2	3
$f(x)$	1	6	31	94

Use the condition $f_0' = 1$.

Solution   We have $h = 1$, $u = x$. The spline approximation in $(t_0, t_3) = (0, 3)$ is given by

$$S(x) = \sum_{i=-2}^{2} \alpha_i B_i^2(x)$$

$$= \alpha_{-2} B_{-2}^2(x) + \alpha_{-1} B_{-1}^2(x) + \alpha_0 B_0^2(x) + \alpha_1 B_1^2(x) + \alpha_2 B_2^2(x).$$

Using (18.101) – (18.105), we obtain the contributions of the *B*-splines as follows.

	On (0, 1)	On (1, 2)	On (2, 3)
$B_{-2}^2(x)$	$0.5(1 - x)^2$	0	0
$B_{-1}^2(x)$	$-(x + 1)^2 + 3(x + 1) - 1.5$	$0.5(2 - x)^2$	0
$B_0^2(x)$	$0.5x^2$	$-x^2 + 3x - 1.5$	$0.5(3 - x)^2$
$B_1^2(x)$	0	$0.5(x - 1)^2$	$-(x - 1)^2 + 3(x - 1) - 1.5$
$B_2^2(x)$	0	0	$0.5(x - 2)^2$

Using the condition $f'_0 = 1$, the system of equations (18.110) becomes

$$\begin{bmatrix} -1 & 1 & 0 & 0 & 0 \\ 1 & 1 & 0 & 0 & 0 \\ 0 & 1 & 1 & 0 & 0 \\ 0 & 0 & 1 & 1 & 0 \\ 0 & 0 & 0 & 1 & 1 \end{bmatrix} \begin{bmatrix} \alpha_{-2} \\ \alpha_{-1} \\ \alpha_0 \\ \alpha_1 \\ \alpha_2 \end{bmatrix} = \begin{bmatrix} 1 \\ 2 \\ 12 \\ 62 \\ 188 \end{bmatrix}$$

The first two equations give $\alpha_{-2} = 0.5$, $\alpha_{-1} = 1.5$. From the remaining equations, we get

$$\alpha_0 = 12 - \alpha_{-1} = 10.5, \quad \alpha_1 = 62 - \alpha_0 = 51.5, \quad \alpha_2 = 188 - \alpha_1 = 136.5.$$

The quadratic *B*-spline approximation is given by

$$S(x) = 0.5 B^2_{-2}(x) + 1.5 B^2_{-1}(x) + 10.5 B^2_0(x) + 51.5 B^2_1(x) + 136.5 B^2_2(x).$$

Therefore, we have

$$S(x) = 0.5 B^2_{-2}(x) + 1.5 B^2_{-1}(x) + 10.5 B^2_0(x), \quad 0 \le x \le 1,$$

$$= 1.5 B^2_{-1}(x) + 10.5 B^2_0(x) + 51.5 B^2_1(x), \quad 1 \le x \le 2,$$

$$= 10.5 B^2_0(x) + 51.5 B^2_1(x) + 136.5 B^2_2(x), \quad 2 \le x \le 3.$$

Cubic *B*-spline

We consider again the uniform mesh case. From (18.101), we obtain

$$B^3_i(x) = B^3_i(u) = \frac{1}{3} [u B^2_i(u) + (4 - u) B^2_{i+1}(u)]. \tag{18.112}$$

Substituting the values of $B^2_i(u)$, and $B^2_{i+1}(u)$, we obtain the cubic *B*-spline as

$$B^3_i(x) = B^3_i(u) = (u^3/6), \quad 0 \le u < 1,$$

$$= [- 3(u - 1)^3 + 3(u - 1)^2 + 3(u - 1) + 1]/6, \quad 1 \le u < 2,$$

$$= [- 3(3 - u)^3 + 3(3 - u)^2 + 3(3 - u) + 1]/6. \quad 2 \le u < 3, \tag{18.113}$$

$$= [(4 - u)^3/6], \quad 3 \le u < 4,$$

$$= 0, \text{ elsewhere.}$$

Note that $B^3_i(t_i) = B^3_i(u = 0) = 0$, $B^3_i(t_{i+1}) = B^3_i(u = 1) = (1/6)$, $B^3_i(t_{i+2}) = B^3_i(u = 2) = (2/3)$, $B^3_i(t_{i+3}) = B^3_i(u = 3) = (1/6)$. The graph of $B^3_i(u)$ is a bell shaped curve (see Fig. 18.12). It is easy to verify that $B^3_i(u)$ is continuous and has continuous first and second derivatives on [0, 4],

Further, $\left[\dfrac{d}{du} B_i^3(u)\right] = 0$, at $u = 2$. The tangent to the graph of $B_i^3(u)$ at the middle node $u = 2$ is parallel to the axis.

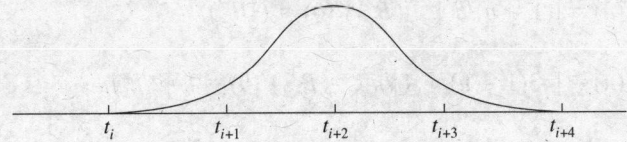

Fig. 18.12.   Cubic B-spline.

The expressions for $B_{i-3}^3(u)$, $B_{i-2}^3(u)$, $B_{i-1}^3(u)$, $B_{i+1}^3(u)$, $B_{i+2}^3(u)$ can be obtained by raplacing u by

$u + 3$, $u + 2$, $u + 1$, $u - 1$, $u - 2$ in $B_i^3(u)$ respectively.

Note that for any $t \in (t_i, t_{i+1})$, only the four splines $B_{i-3}^3(x)$, $B_{i-2}^3(x)$, $B_{i-1}^3(x)$, $B_i^3(x)$ contribute.

Cubic B-spline approximation

The cubic B-spline over (t_0, t_n) is written as

$$S(x) = \alpha_{-3} B_{-3}^3(x) + \alpha_{-2} B_{-2}^3(x) + \alpha_{-1} B_{-1}^3(x) + \ldots + \alpha_{n-1} B_{n-1}^3(x) = \sum_{i=-3}^{n-1} \alpha_i B_i^3(x). \qquad (18.114)$$

There are $n + 3$ parameters, α_i, $i = -3, -2, -1, 0, 1, \ldots n - 1$, to be determined. The interpolating conditions $S(t_i) = f_i$, $i = 0, 1, \ldots, n$ give $n + 1$ equations. In order that unique solution exists, we require two additional conditions at the end points of the interval (t_0, t_n). For example, we may assume natural cubic spline conditions $S''(t_0) = 0 = S''(t_n)$. In the following, we derive the equations when the values of $S''(t_0)$, $S''(t_n)$ are prescribed. Now, at $t = t_i$, only four basis spline functions contribute:

$$B_{i-3}^3(t_i) = B_{i-3}^3(u = 0) = 1/6, \quad B_{i-2}^3(t_i) = B_{i-2}^3(u = 0) = 2/3,$$

$$B_{i-1}^3(t_i) = B_{i-1}^3(u = 0) = 1/6, \quad B_i^3(t_i) = B_i^3(u = 0) = 0.$$

The interpolating condition $S(t_i) = f_i$ gives

$$\frac{1}{6} \alpha_{i-3} + \frac{2}{3} \alpha_{i-2} + \frac{1}{6} \alpha_{i-1} = f_i, \quad i = 0, 1, 2, \ldots, n.$$

The condition $S''(t_0) = f_0''$ gives the equation

$$(B_{-3}^3)''(t_0)\alpha_{-3} + (B_{-2}^3)''(t_0)\alpha_{-2} + (B_{-1}^3)''(t_0)\alpha_{-1} = f_0''.$$

On $0 \leq u < 1$, we get

$$B^3_{i-3}(u) = [(1 - u)^3/6], \quad B^3_{i-2}(u) = [-3(1 - u)^3 + 3(1 - u)^2 + 3(1 - u) + 1]/6,$$

$$B^3_{i-1}(u) = [-3u^3 + 3u^2 + 3u + 1]/6, \quad B^3_i(u) = u^3/6.$$

$$(B^3_{-3})''(u) = [(1 - u)/h^2], \quad (B^3_{-3})''(0) = (1/h^2),$$

$$(B^3_{-2})''(u) = [-3(1 - u) + 1]/h^2, \quad (B^3_{-2})''(0) = (-2/h^2),$$

$$(B^3_{-1})''(u) = [(-3u + 1)/h^2], \quad (B^3_{-1})''(0) = (1/h^2),$$

$$(B^3_0)''(u) = [u/h^2], \quad (B^3_0)''(0) = 0,$$

Therefore, the equation valid at $t = t_0$, becomes

$$\alpha_{-3} - 2\alpha_{-2} + \alpha_{-1} = h^2 f_0''$$

Similarly, the condition $S''(t_n) = f_n''$ gives the equation

$$\alpha_{n-3} - 2\alpha_{n-2} + \alpha_{n-1} = h^2 f_n''.$$

In matrix form, we get the system of equations as

$$
\begin{bmatrix}
1 & -2 & 1 & & & \\
1 & 4 & 1 & & & \\
 & 1 & 4 & 1 & & \\
 & & 1 & 4 & 1 & \\
\cdots & \cdots & \cdots & \cdots & \cdots & \cdots \\
 & & & 1 & -2 & 1
\end{bmatrix}
\begin{bmatrix}
\alpha_{-3} \\
\alpha_{-2} \\
\alpha_{-1} \\
\alpha_0 \\
\cdots \\
\alpha_{n-1}
\end{bmatrix}
=
\begin{bmatrix}
h^2 f_0'' \\
6 f_0 \\
6 f_1 \\
6 f_2 \\
\cdots \\
h^2 f_n''
\end{bmatrix}
\tag{18.115}
$$

Subtracting the first two equations, we obtain α_{-2}. Subtracting the last two equations, we obtain α_{n-2}. Solve this system and substitute the values of α_i in (18.114) to obtain the cubic *B*-spline approximation.

In the general case, when different conditions on $S(x)$ are prescribed at $t = t_0$, and $t = t_n$, the system can be reduced to a tridiagonal system of equations. The interior rows have 1/6, 4/6 and 1/6 (or (1, 4, 1) if multiplied by 6) on the sub-diagonal, diagonal and super-diagonal elements.

Remark 8

Using the abscissas of the given data, the *B*-splines of required order can be constructed. The ordinates play the part when we determine the *B*-spline approximation to the data.

Remark 9

It is useful to consider a cubic *B*-spline at t_i, using the symmetrically placed knots at $t_{i-2}, t_{i-1}, t_i,$ $t_{i+1}, t_{i+2}.$

Example 18.32 For the following data values fit a cubic *B*-spline

$$
\begin{array}{cccccc}
x & 0 & 1 & 2 & 3 & 4 \\
f(x) & 1 & 1 & 17 & 82 & 257
\end{array}
$$

Use the conditions $f''(0) = 0$, $f''(4) = 192$.

Solution We have $h = 1$, and $u = x$. The cubic *B*-spline approximation on $(t_0, t_n) = (0, 4)$ is given by

$$S(x) = \alpha_{-3} B_{-3}^3(x) + \alpha_{-2} B_{-2}^3(x) + \alpha_{-1} B_{-1}^3(x) + \alpha_0 B_0^3(x) + \alpha_1 B_1^3(x) + \alpha_2 B_2^3(x) + \alpha_3 B_3^3(x).$$

The contribution of splines at $t = t_0 = 0$ is given in Fig. 18.13.

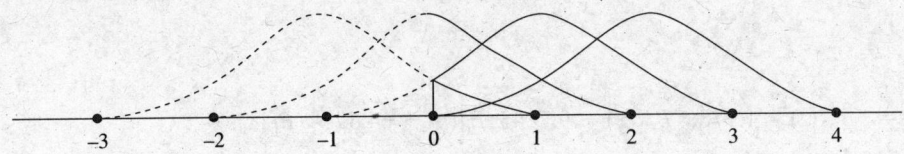

Fig. 18.13. Example 18.32. Splines at $t = 0$.

Using the conditions $f''(0) = 0$, $f''(4) = 192$, the system of equations (18.115) becomes

$$
\begin{bmatrix}
1 & -2 & 1 & 0 & 0 & 0 & 0 \\
1 & 4 & 1 & 0 & 0 & 0 & 0 \\
0 & 1 & 4 & 1 & 0 & 0 & 0 \\
0 & 0 & 1 & 4 & 1 & 0 & 0 \\
0 & 0 & 0 & 1 & 4 & 1 & 0 \\
0 & 0 & 0 & 0 & 1 & 4 & 1 \\
0 & 0 & 0 & 0 & 1 & -2 & 1
\end{bmatrix}
\begin{bmatrix}
\alpha_{-3} \\
\alpha_{-2} \\
\alpha_{-1} \\
\alpha_0 \\
\alpha_1 \\
\alpha_2 \\
\alpha_3
\end{bmatrix}
=
\begin{bmatrix}
0 \\
6 \\
6 \\
102 \\
492 \\
1542 \\
192
\end{bmatrix}
$$

Subtracting the first and second equations, we get $\alpha_{-2} = 1$. Subtracting the sixth and seventh equations, we get $\alpha_2 = 225$. The solution of the remaining equations give

$$\alpha_{-3} = 3.1786, \quad \alpha_{-1} = -1.1786, \quad \alpha_0 = 9.7143, \quad \alpha_1 = 64.3214, \quad \alpha_3 = 577.6786.$$

The cubic *B*-spline approximation is given by

$$S(x) = 3.1786\, B_{-3}^3(x) + B_{-2}^3(x) - 1.1786\, B_{-1}^3(x) + 9.7143\, B_0^3(x).$$

$$+ 64.3214\, B_1^3 + 225\, B_2^3(x) + 577.6786\, B_3^3(x).$$

The contribution of the cubic *B*-splines on each interval can be written using (18.112). We have $h = 1$, $t_0 = 0$, and $u = x$.

On [0, 1], we have the following contributions:

$B^3_{-3}(x) = (1 - x)^3/6; \ B^3_{-2}(x) = [-3(1 - x)^3 + 3(1 - x)^2 + 3(1 - x) + 1]/6;$

$B^3_{-1}(x) = (-3x^3 + 3x^2 + 3x + 1)/6; \ B^3_0(x) = (x^3/6); \ B^3_1(x) = 0; \ B^3_2(x) = 0; \ B^3_3(x) = 0.$

$S(x) = 3.1786 \ B^3_{-3}(x) + B^3_{-2}(x) - 1.1786 \ B^3_{-1}(x) + 9.7143 \ B^3_0(x).$

On [1, 2], we have the following contributions:

$B^3_{-3}(x) = 0; \ B^3_{-2}(x) = (2 - x)^3/6; \ B^3_{-1}(x) = [-3(2 - x)^3 + 3(2 - x)^2 + 3(2 - x) + 1]/6;$

$B^3_0(x) = [-3(x - 1)^3 + 3(x - 1)^2 + 3(x - 1) + 1]/6; \ B^3_1(x) = (x - 1)^3/6; \ B^3_2(x) = 0;$

$B^3_3(x) = 0.$

$S(x) = B^3_{-2}(x) - 1.1786 \ B^3_{-1}(x) + 9.7143 \ B^3_0(x) + 64.3214 \ B^3_1(x).$

On [2, 3], we have the following contributions:

$B^3_{-3}(x) = 0, \ B^3_{-2}(x) = 0; \ B^3_{-1}(x) = (3 - x)^3/6;$

$B^3_0(x) = [-3(3 - x)^3 + 3(3 - x)^2 + 3(3 - x) + 1]/6;$

$B^3_1(x) = [-3(x - 2)^3 + 3(x - 2)^2 + 3(x - 2) + 1]/6;$

$B^3_2(x) = [(x - 2)^3/6]; \ B^3_3(x) = 0.$

$S(x) = -1.1786 \ B^3_{-1}(x) + 9.7143 \ B^3_0(x) + 64.3214 \ B^3_1(x) + 225 \ B^3_2(x).$

On [3, 4], we have the following contributions:

$B^3_{-3}(x) = 0; \ B^3_{-2}(x) = 0; \ B^3_{-1}(x) = 0; \ B^3_0(x) = (4 - x)^3/6;$

$B^3_1(x) = [-3(4 - x)^3 \ 3(4 - x)^2 + 3(4 - x) + 1]/6;$

$B^3_2(x) = [-3(x - 3)^3 + 3(x - 3)^2 + 3(x - 3) + 1]/6; \ B^3_3(x) = [(x - 3)^3/6].$

$S(x) = 9.7143 B^3_0(x) + 64.3214 \ B^3_1(x) + 225 \ B^3_2(x) + 577.6786 \ B^3_3(x).$

Exercise 18.4

1. Find the Lagrange interpolating polynomial that fits the following data

(i)

x	0	1	2
$f(x)$	2	1	12

(ii)

x	-2	1	0	2
$f(x)$	3	-3	1	-1

2. Let $f(x) = \ln(1 + x)$, $x_0 = 1$ and $x_1 = 1.2$. Use Lagrange linear interpolating polynomial to obtain an approximate value of $\ln(1.1)$. Find the maximum absolute error at this point.

3. Let $f(x) = e^{-x}$, $x_0 = 1$, $x_1 = 1.2$ and $x_2 = 1.5$. Use Lagrange quadratic interpolating polynomial to obtain an approximate value of $f(1.3)$. Find the maximum absolute error at this point.

4. Using the Lagrange interpolating polynomial with remainder, show that

$$\sum_{k=0}^{n} l_k(0)\, x^{n+1}_k = (-1)^n\, x_0\, x_1\, \ldots\, x_n$$

where $l_k(x)$ are Lagrange fundamental polynomials.

5. Prove that $\displaystyle\sum_{k=0}^{n} l_k(x)\, x^{j}_k = x^j$, $j = 0, 1, \ldots, n$, where $l_k(x)$ are Lagrange fundamental polynomials.

6. Construct the divided difference table for the following data

   (i)

x	-3	-1	0	2	3
$f(x)$	-9	5	3	11	33

   (ii)

x	0	1	2	4	5	6
$f(x)$	1	14	15	5	6	19

   Hence, obtain the interpolating polynomial that fits the given data.

7. Show by induction that the nth divided difference of $f(x) = 1/x$, based on the points $x_0, x_1, \ldots, x_n$ is $(-1)^n/(x_0\, x_1\, \ldots\, x_n)$.

8. If $f(x) = u(x)\, v(x)$, show that

   $$f[x_0, x_1] = u(x_1)\, v[x_0, x_1] + v(x_0)\, u[x_0, x_1].$$

9. Using divided differences, show that the following data represents a second degree polynomial

x	-3	-2	-1	1	2	3	5
$f(x)$	18	12	8	6	8	12	26

   Determine this polynomial and obtain the approximate value of $f(0)$.

10. Prove the following relations

    (i) $\nabla - \Delta = -\Delta\nabla$.

    (ii) $\displaystyle\sum_{k=0}^{n} \Delta^2 f_k = \Delta f_{n+1} - \Delta f_0$

    (iii) $\Delta(f_i\, g_i) = f_i\, \Delta g_i + g_{i+1}\, \Delta f_i$.

    (iv) $\Delta[f_i/g_i] = (g_i\, \Delta f_i - f_i\, \Delta g_i)/(g_i\, g_{i+1})$.

11. If $f(x) = e^{ax}$, then show by induction that $\Delta^n e^{ax} = (e^{ah} - 1)^n e^{ax}$.

12. Construct the forward difference table for the data

    (i)

x	-4	-2	0	2	4	6
$f(x)$	-67	-9	1	11	69	223

    (ii)

x	-3	-2	-1	0	1	2
$f(x)$	-2	-4	-4	-2	2	8

    Determine the corresponding interpolating polynomial.

13. Construct the backward difference table for the data

    (i)

x	-3	-2	-1	0	1	2	3
$f(x)$	13	7	3	1	1	3	7

    (ii)

x	-4	-2	0	2	4	6
$f(x)$	261	19	1	15	253	1291

    Determine the corresponding interpolating polynomial.

14. A function $f(x)$ is tabulated on $[0, 1]$ and the derivatives satisfy $|f^{(m)}(x)| \le m!$, $m = 0, 1, 2, \ldots$ for all $x \in [0, 1]$. Let $P_n(x)$ be the interpolating polynomial for $f(x)$ at $1, q, q^2 \ldots, q^n$, $0 < q < 1$. Show that
    $$\lim_{n \to \infty} P_n(0) = f(0).$$

15. A table of values for the following functions is to be constructed at equispaced points. Find the largest step size h that can be used so that error in magnitude is less than 5×10^{-4} using (a) linear interpolation, (b) quadratic interpolation.

    (i) $1/[1+x]$, $1 \le x \le 2$, (ii) $1/[1+x^2]$, $0 \le x \le 1$,

    (iii) $\sqrt{x}$, $4 \le x \le 6$, (iv) $x^2 e^x$, $-1 \le x \le 1$.

16. Find whether the following functions are cubic splines.

    (i) $f(x) = \begin{cases} x^3 - 2x + 1, & 1 \le x \le 2, \\ x^3 - 3, & 2 < x \le 3. \end{cases}$ **(ii)** $f(x) = \begin{cases} 2x^3 - 3x^2, & -1 \le x \le 0, \\ x^3 - 3x^2, & 0 < x \le 1. \end{cases}$

17. For the data given below, construct cubic splines on all or on the indicated intervals.

    (i)

x	1	2	3
$f(x)$	4	33	162

    Assume $M(1) = 24$, $M(3) = 216$. Interpolate at 1.1 and 2.5.

    (ii)

x	1	2	3	4
$f(x)$	3	40	189	576

    Find the cubic spline valid on $[3, 4]$. Assume $M(1) = 30$, $M(4) = 408$.

    (iii)

x	0	1	2	3	4
$f(x)$	0	-81	-192	-81	0

    Find the cubic splines valid on the intervals $[2, 3]$ and $[3, 4]$. Assume $M(0) = 0$, $M(4) = 0$. Interpolate at 2.5 and 3.8.

18. Find the natural cubic spline for $f(x) = \cos x$, $-(\pi/2) \le x \le (\pi/2)$ over two sub-intervals. Interpolate at $x = \pi/4$. Compare with the exact solution.

For the following data, fit the linear *B*-spline approximation.

19.
x	0	1	2	3
y	1	5	21	55

20.
x	0	1	2	3
y	6	11	18	32

For the following data, fit the quadratic *B*-spline approximation.

21.

x	0	1	2	3
y	1	5	21	55

Use the condition $y_0' = 1$.

22.

x	1	2	3	4
y	5	25	91	269

Use the condition $y_0' = 0$.

23.

x	0.1	0.2	0.3	0.4
y	1.21	1.56	2.41	4.36

Use the condition $y_n' = 2.4$.

For the following data, fit the cubic *B*-spline approximation.

24.

x	0.1	0.2	0.3	0.4
y	1.21	1.56	2.41	4.36

Use the conditions $y''(0.1) = 12$, $y''(0.4) = 192$.

25.

x	1	1.5	2	2.5	3
y	4	10.5625	24	49.5625	94

Use the conditions $y''(1) = 12$, $y''(3) = 108$.

18.5 Least Squares Approximation

It is also called the *method of least squares*. Let (x_i, f_i), $i = 1, \ldots, N$ be the given set of data. Here, we do not fit data exactly. That is the graph of the approximating function $P(x)$ does not pass through any of the data points. This implies that we need an approximation to the function $f(x)$ which represents the data.

18.5.1 Approximation by Polynomials

Let
$$P(x) = a_0 x^n + a_1 x^{n-1} + \ldots + a_n \tag{18.116}$$

be the polynomial approximation to $f(x)$ which represents the data.

Then, at $x = x_k$, the error is given by
$$E(x_k) = f(x_k) - P(x_k).$$

The method of least squares requires that the sum of squares of the errors over the entire data is a minimum. That is

$$J = \sum_{k=1}^{N} [E(x_k)]^2 = \sum_{k=1}^{N} [f(x_k) - P(x_k)]^2$$

$$= \sum_{k=1}^{N} [f(x_k) - \{a_0 x_k^n + a_1 x_k^{n-1} + \ldots + a_n\}]^2 = \text{minimum}.$$

The stationary points are given by

$$\frac{\partial J}{\partial a_i} = 0, \; i = 0, 1, 2, \ldots, n.$$

This gives a system of $n + 1$ equations in $n + 1$ unknowns $a_0, a_1, ..., a_n$. These equations are called the *normal equations* of the system. We solve for a_i and substitute in (18.116) to obtain the least squares polynomial approximation. Substituting the values of a_i in J, we obtain the least squares error.

Least squares approximation can also be obtained for a continuous function defined on $[\alpha, \beta]$. If $P(x)$ in (18.116) is the approximation, then the error is defined as

$$E(x) = f(x) - P(x).$$

The method of least squares requires that

$$I = \int_\alpha^\beta [E(x)]^2 \, dx = \int_\alpha^\beta [f(x) - P(x)]^2 \, dx = \text{minimum}.$$

Extremum exists when

$$\frac{\partial I}{\partial a_i} = 0, \ i = 0, 1, 2, ..., n. \tag{18.117}$$

This gives a system of $n + 1$ equations for finding the values of $a_0, a_1, a_2, ..., a_n$. These equations are also called the *normal equations*. We evaluate the integrals in (18.117) and determine a_i. Substituting a_i in (18.116), we obtain the least squares approximation. Substituting $P(x)$ in I and evaluating the integral, we obtain the value of the least squares error.

Let us now derive the least squares linear and quadratic polynomials for a given data.

Least squares linear polynomial approximation

This approximation is also called the *linear regression* of y on x. Let, $P(x) = a + bx$ be the linear polynomial approximation to the given data. Then, error of approximation at a tabular point $x = x_k$ is given by

$$E(x_k) = f(x_k) - (a + bx_k).$$

The method of least squares requires that

$$J = \sum_{k=1}^N [E(x_k)]^2 = \sum_{k=1}^N [f(x_k) - (a + bx_k)]^2 = \text{minimum}.$$

The stationary points are given by

$$\frac{\partial J}{\partial a} = 0 = \sum_{k=1}^N 2[f(x_k) - (a + bx_k)](-1),$$

$$\frac{\partial J}{\partial b} = 0 = \sum_{k=1}^N 2\,[f(x_k) - (a + bx_k)]\,(-x_k).$$

Hence, we obtain the normal equations

$$a\sum 1 + b \sum x_k = \sum f_k, \quad \text{or} \quad Na + b \sum x_k = \sum f_k$$

and
$$a \sum x_k + b \sum x_k^2 = \sum x_k f_k.$$

Solving, we obtain the values of a and b. Substituting in $P(x) = a + bx$, we get the least squares linear polynomial approximation. Substituting the values of a and b in J, we obtain the least squares error. We find that

$$\frac{\partial^2 J}{\partial a^2} = 1 > 0 \quad \text{and} \quad \frac{\partial^2 J}{\partial b^2} = x_k^2 > 0$$

which shows that J has a minimum value.

If a continuous function $f(x)$, $\alpha \le x \le \beta$, is given, then we require

$$I = \int_\alpha^\beta [f(x) - (a + bx)]^2 \, dx = \text{minimum}.$$

The stationary points are given by

$$\frac{\partial I}{\partial a} = 2 \int_\alpha^\beta [f(x) - (a + bx)] \, (-1) dx = 0$$

or
$$a \int_\alpha^\beta dx + b \int_\alpha^\beta x dx = \int_\alpha^\beta f(x) dx,$$

or
$$(\beta - \alpha)a + \frac{1}{2} (\beta^2 - \alpha^2)b = \int_\alpha^\beta f(x) \, dx, \tag{18.118}$$

and
$$\frac{\partial I}{\partial b} = 2 \int_\alpha^\beta [f(x) - (a + bx)] \, (- x) \, dx = 0$$

or
$$a \int_\alpha^\beta x dx + b \int_\alpha^\beta x^2 dx = \int_\alpha^\beta x f(x) \, dx$$

or
$$\frac{1}{2} (\beta^2 - \alpha^2) \, a + \frac{1}{3} (\beta^3 - \alpha^3)b = \int_\alpha^\beta x f(x) \, dx. \tag{18.119}$$

Equations (18.118), (18.119) are the normal equations. Solving for a, b and substituting in $P(x) = a + bx$, we get the least squares approximation. Substituting the values of a and b in I, we obtain the least squares error.

Least squares quadratic polynomial approximation

Let $P(x) = a + bx + cx^2$ be the quadratic approximation to the given data. Then, the error of approximation at a tabular Point $x = x_k$ is given by

$$E(x_k) = f(x_k) - (a + bx_k + cx_k^2).$$

The method of least squares requires that

$$J = \sum_{k=1}^{N} [E(x_k)]^2 = \sum_{k=1}^{N} [f(x_k) - (a + bx_k + cx_k^2)]^2 = \text{minimum.}$$

The stationary points are given by

$$\frac{\partial J}{\partial a} = 0 = \sum_{k=1}^{N} 2[f(x_k) - (a + bx_k + cx_k^2)](-1),$$

$$\frac{\partial J}{\partial b} = 0 = \sum_{k=1}^{N} 2[f(x_k) - (a + bx_k + cx_k^2)](-x_k),$$

$$\frac{\partial J}{\partial c} = 0 = \sum_{k=1}^{N} 2[f(x_k) - (a + bx_k + cx_k^2)](-x_k^2).$$

Hence, the normal equations are given by

$$Na + b \sum x_k + c \sum x_k^2 = \sum f_k$$

$$a \sum x_k + b \sum x_k^2 + c \sum x_k^3 = \sum x_k f_k$$

$$a \sum x_k^2 + b \sum x_k^3 + c \sum x_k^4 = \sum x_k^2 f_k.$$

We can solve this system for a, b, c using the Gauss elimination method. Substitution of a, b, c in $P(x) = a + bx + cx^2$ gives the required least squares approximation.

For a continuous function $f(x)$, $\alpha \leq x \leq \beta$, we require

$$I = \int_{\alpha}^{\beta} [f(x) - (a + bx + cx^2)]^2 = \text{minimum.}$$

The stationary points are given by

$$\frac{\partial I}{\partial a} = \frac{\partial I}{\partial b} = \frac{\partial I}{\partial c} = 0.$$

We obtain the normal equations

$$\int_{\alpha}^{\beta} [f(x) - (a + bx + cx^2)] \, dx = 0$$

$$\int_{\alpha}^{\beta} [f(x) - (a + bx + cx^2)] \, x dx = 0$$

$$\int_{\alpha}^{\beta} [f(x) - (a + bx + cx^2)] x^2 dx = 0.$$

Solving these equations for a, b, c and substituting in $P(x) = a + bx + cx^2$, we get the least squares approximation. Substituting $P(x)$ in I, we obtain the least squares error.

Example 18.33 Fit a least squares linear approximation (regression of $y = f(x)$ on x) and quadratic approximation for the data

x	-1	0	1	2	3
$f(x)$	0	1	2	9	28

Solution We have $N = 5$ and the following data values.

x	-1	0	1	2	3	:	$\Sigma x_k = 5$
$f(x)$	0	1	2	9	28	:	$\Sigma f_k = 40$
x^2	1	0	1	4	9	:	$\Sigma x_k^2 = 15$
x^3	-1	0	1	8	27	:	$\Sigma x_k^3 = 35$
x^4	1	0	1	16	81	:	$\Sigma x_k^4 = 99$
$xf(x)$	0	0	2	18	84	:	$\Sigma x_k f_k = 104$
$x^2 f(x)$	0	0	2	36	252	:	$\Sigma x_k^2 f_k = 290.$

Linear fit: We have the normal equations as

$$5a + 5b = 40, \quad 5a + 15b = 104$$

whose solution is $a = 1.6$, $b = 6.4$. The least squares linear fit is $P(x) = 1.6 + 6.4x$.

Quadratic fit: We have the normal equations as

$$5a + 5b + 15c = 40$$

$$5a + 15b + 35c = 104$$

$$15a + 35b + 99c = 290$$

whose solution is $a = -1.4$, $b = 0.4$, $c = 3$. The least squares quadratic fit is $P(x) = 3x^2 + 0.4x - 1.4$.

Example 18.34 Obtain the least squares polynomial approximations of degree one and two for $f(x) = \sqrt{x}$ on $[0, 1]$.

Solution For $n = 1$, we have

$$I = \int_0^1 [\sqrt{x} - a - bx]^2 \, dx = \text{minimum}.$$

The normal equations are obtained from

$$\frac{\partial I}{\partial a} = 0, \quad \frac{\partial I}{\partial b} = 0.$$

We get

$$\int_0^1 (x^{1/2} - a - bx) \, dx = 0, \quad \text{or} \quad a + \frac{b}{2} = \frac{2}{3},$$

$$\int_0^1 (x^{1/2} - a - bx)x \, dx = 0, \quad \text{or} \quad \frac{a}{2} + \frac{b}{3} = \frac{2}{5}.$$

Solving these equations, we obtain $a = 4/15$ and $b = 4/5$. Thus, the first degree least squares polynomial approximation is

$$P(x) = \frac{4}{15}(1 + 3x).$$

For $n = 2$, we have

$$I = \int_0^1 [\sqrt{x} - a - bx - cx^2]^2 \, dx = \text{minimum}.$$

The normal equations are obtained from

$$\frac{\partial I}{\partial a} = 0, \quad \frac{\partial I}{\partial b} = 0, \quad \frac{\partial I}{\partial c} = 0.$$

We get

$$\int_0^1 [x^{1/2} - a - bx - cx^2] \, dx = 0, \quad \text{or} \quad a + \frac{b}{2} + \frac{c}{3} = \frac{2}{3},$$

$$\int_0^1 [x^{1/2} - a - bx - cx^2] \, x \, dx = 0, \quad \text{or} \quad \frac{a}{2} + \frac{b}{3} + \frac{c}{4} = \frac{2}{5},$$

$$\int_0^1 [x^{1/2} - a - bx - cx^2] \, x^2 \, dx = 0, \quad \text{or} \quad \frac{a}{3} + \frac{b}{4} + \frac{c}{5} = \frac{2}{7}.$$

Solving these equations, we get

$$a = \frac{6}{35}, \quad b = \frac{48}{35}, \quad c = -\frac{20}{35}.$$

The least squares second degree polynomial approximation is

$$P(x) = \frac{1}{35} \left(6 + 48x - 20x^2 \right).$$

It is not necessary to fit only polynomials to the given data. Depending on the source of the data, we can obtain approximations using exponential functions, power function, trigonometric functions, logarithmic function etc.

18.5.2 Approximation by Exponential Function

Consider an approximation of the form

$$f(x) = ae^{bx} \tag{18.120}$$

We can directly obtain the normal equations. However, taking logarithm on both sides of (18.120), we obtain

$$\ln f(x) = \ln (ae^{bx}) = \ln a + bx$$

which can be written as $y = A + bx$, where $y = \ln f(x)$ and $A = \ln a$. The data is now (x_i, y_i), $i = 1, 2, \ldots N$. The problem reduces to linear least squares approximation.

Remark 10

We can also take logarithm to the base 10 in (18.120). The approximation becomes

$$y = A + Bx$$

where $y = \log_{10} f(x)$, $A = \log_{10} a$ and $B = b \log_{10} e$.

Example 18.35 Obtain the least squares approximation of the form $f(x) = ae^{bx}$ to the data

x	0.5	1.0	2.0	2.5	3.0
$f(x)$	0.57	1.46	5.10	7.65	9.20

Solution Taking logarithm of $f(x) = ae^{bx}$, we obtain

$$\ln f(x) = \ln a + bx \text{ or } y = A + bx$$

where $\qquad y = \ln f(x)$ and $A = \ln a$.

Now, we need to obtain linear least squares approximation to the data (x_i, y_i). We get

x	0.5	1.0	2.0	2.5	3.0
y	-0.5621	0.3784	1.6292	2.0347	2.2192

We have, $N = 5$, $\sum x_k = 9.0$, $\sum x_k^2 = 20.5$, $\sum y_k = 5.6994$, $\sum x_k y_k = 15.1001$.
The normal equations are

$$NA + b\sum x_k = \sum y_k, \quad A\sum x_k + b\sum x_k^2 = \sum x_k y_k$$

or $\qquad 5A + 9b = 5.6994, \quad 9A + 20.5b = 15.1001$

whose solution is $A = -0.8867$ and $b = 1.1259$. Hence, $a = e^A = 0.4120$. The required approximation is $f(x) = ae^{bx} = 0.412\ e^{1.1259x}$.

18.5.3 Approximation by Power Function

Consider an approximation of the form

$$f(x) = ax^b. \tag{18.121}$$

Taking logarithm on both sides, we obtain

$$\ln f(x) = \ln a + b \ln x \quad \text{or} \quad y = A + bt$$

where $y = \ln f(x)$, $A = \ln a$ and $t = \ln x$. The data is now (t_i, y_i), $i = 1, 2, \ldots, N$. The problem reduces to linear least squares approximation.

Remark 11

We can also take logarithm to the base 10 in (18.121). The approximation becomes $y = A + bt$, where $y = \log_{10} f(x)$, $A = \log_{10} a$ and $t = \log_{10} x$.

Example 18.36 Obtain the least squares approximation of the form $f(x) = ax^b$ to the data

x	0.5	0.6	0.7	0.8	1.0
$f(x)$	0.3136	0.4515	0.6146	0.8027	1.2542

Solution   Taking logarithm on both sides, we obtain

$$\ln f(x) = \ln a + b \ln x, \quad \text{or} \quad y = A + bt$$

where $y = \ln f(x)$, $A = \ln a$ and $t = \ln x$. Now, we form the data (t_i, y_i).

x	0.5	0.6	0.7	0.8	1.0
$t = \ln x$	-0.6931	-0.5108	-0.3567	-0.2231	0.0000
$f(x)$	0.3136	0.4515	0.6146	0.8027	1.2542
$y = \ln f(x)$	-1.1596	-0.7952	-0.4868	-0.2198	0.2265

We have $N = 5$, $\sum t_k = -1.7837$, $\sum t_k^2 = 0.9183$, $\sum y_k = -2.4349$, $\sum t_k y_k = 1.4326$.

The normal equations are

$$5A - 1.7837b = -2.4349, \quad -1.7837\,A + 0.9183\,b = 1.4326$$

whose solution is $A = 0.2265$ and $b = 2.0$. Hence, $a = e^A = 1.2542$. The required approximation is $f(x) = 1.2542\, x^2$.

Exercise 18.5

In the following problems, construct the linear least squares approximations. Find the least squares errors.

1.

x	-2	-1	0	1
$f(x)$	6	3	2	2

2.

x	1	2	3	4	5
$f(x)$	0.25	0.42	0.56	0.72	1

3.

x	0.5	0.6	0.7	0.8	0.9
$f(x)$	0.5	0.92	1.42	2.0	2.8

In the following problems, construct the least squares quadratic approximations.

4.

x	1	2	3	4	5	6
$f(x)$	1000	800	500	300	150	90

5.

x	1.5	2.5	3.5	4.5	5.5
$f(x)$	3.2	6.2	10	12	14

6.

x	1	2	3	4	5
$f(x)$	2.5	4.5	3.7	5.0	4.2

In the following problems, obtain linear least squares polynomial approximation to the given function on the given intervals.

7. $f(x) = x^2$, $x \in [1, 2]$, **8.** $f(x) = x^3$, $x \in [0, 1]$.

9. $f(x) = \sqrt{x}$, $x \in [0, 4]$.

In the following problems, obtain the quadratic least squares polynomial approximations on the given intervals.

10. $f(x) = x^3 + 2$, $x \in [1, 2]$. **11.** $f(x) = x^{3/2}$, $x \in [0, 1]$.

Obtain an approximation of the form $f(x) = ae^{bx}$ for the following data.

12.

x	0.5	0.6	0.7	0.8	0.9
$f(x)$	1.1809	1.0164	0.8748	0.7530	0.6481

13.

x	0.6	0.7	0.8	0.9	1.0
$f(x)$	1.0465	0.9856	0.9282	0.8741	0.8232

Obtain an approximation of the from $f(x) = ax^b$ for the following data.

14.

x	1	2	3	4
$f(x)$	3.0	8.4853	15.5885	24.000

15.

x	1	2	3	4
$f(x)$	1.75	7.00	15.75	28.00

16. A physicist studying a decaying process decides to fit an approximation of the form $f(t) = ae^{-t} + be^{-2t}$ to the data

t	0.5	0.6	0.7	0.8	0.9
$f(x)$	1.2409	1.0943	0.9668	0.8557	0.7586

Determine a, b using the method of least squares.

17. A chemist wants to approximate the following data

x	0	0.5	1.0	2.0
$f(x)$	2.95	3.9445	5.5043	11.7869

using the function $f(x) = ae^x + c$. Determine a, c using the method of least squares.

18. A mathematical model of a periodic process in an experiment is taken as $f(t) = a + b \cos t$, and a data of N points (x_i, f_i), $i = 1, 2, ..., N$ is given. Using the method of least squares, find the normal equations to determine a and b. Use these equations to find a, b for the following data

t (radians)	0.5	1.0	1.5	2.0	2.5
$f(t)$	0.9882	0.5552	0.3338	-0.0275	-0.3509

19. In an experiment, a scientist finds that the process starts decaying and stabilizes to a certain value after a long time. The scientist obtains the following data

t	0.5	0.7	0.9	1.1	1.3
$f(t)$	3.6525	2.9962	2.0945	1.6532	1.3654

Obtain an approximation of the form $f(t) = a + be^{-t}$, using the method of least squares.

20. Using the method of least squares, obtain a function of the from $y = a + (b/x)$ for the following data

x	1	2	3	4	5
y	6	4.5	3.5	2.7	2.0

18.6　Numerical Differentiation and Integration

Let a function $y = f(x)$ be defined at a set of $n + 1$ distinct points $x_0, x_1, ..., x_n$ lying in some interval $[a, b]$, $a \le x_0 < x_1 ... < x_n \le b$. From the given tabular data, we require approximation to the derivative $f^{(r)}(x^*)$, $r \ge 1$, where x^* is a tabular or a non-tabular point. In most practical applications, we have $r = 1$ or 2.

In many science and engineering applications, we require the value of the definite integral $\int_a^b f(x)dx$, where $f(x)$ is either given explicitly or defined by a tabular data (x_1, f_i) $i = 0, 1, ..., n$.

For our discussion, we assume that the nodal points are equispaced and

$$a = x_0 < x_1 < x_2 < ... < x_n = b, \text{ where } x_i = x_0 + ih, i = 1, 2, ..., n.$$

18.6.1　Numerical Differentiation Methods

Interpolation approach

We approximate the function $f(x)$ by Newton's forward or backward difference interpolating polynomial $P_n(x)$ and differentiate it the required number of times to get an approximation to the derivative. The error of approximation in the rth order derivative at the point $x = x^*$, is defined as

$$E_r(f, x^*) = f^{(r)}(x^*) - P_n^{(r)}(x^*). \tag{18.122}$$

Order A numerical differentiation method is said to be of order p, if

$$|E_r(f, x^*)| = |f^{(r)}(x^*) - P_n^{(r)}(x^*)| \le ch^p \tag{18.123}$$

where c is a constant independent of h.

Let us now construct a few differentiation methods using the Newton's forward difference interpolating polynomial

$$f(x) \approx P_n(x) = f_0 + \frac{(x - x_0)}{h} \Delta f_0 + \frac{(x - x_0)(x - x_1)}{2!h^2} \Delta^2 f_0$$

$$+ \frac{(x - x_0)(x - x_1)(x - x_2)}{3!h^3} \Delta^3 f_0 + ... \tag{18.124}$$

Differentiating with respect to x, we get

$$f'(x) \approx P_n'(x) = \frac{1}{h} \Delta f_0 + \frac{1}{2!h^2} [(x - x_0) + (x - x_1)] \Delta^2 f_0 +$$

$$\frac{1}{3!h^3} \left[(x - x_1)(x - x_2) + (x - x_0)(x - x_2) + (x - x_0)(x - x_1)\right] \Delta^3 f_0 + \dots$$

and $f''(x) \approx P''_n(x) = \dfrac{1}{h^2} \Delta^2 f_0 + \dfrac{2}{3!h^3} \left[(x - x_0) + (x - x_1) + (x - x_2)\right] \Delta^3 f_0 + \dots$

Using these relations, we can derive the expressions for the first and second order derivatives at any nodal or non-nodal point $x = x^*$. For example, at $x = x_0$, we get

$$f'(x_0) \approx \frac{1}{h} \left[\Delta f_0 - \frac{1}{2} \Delta^2 f_0 + \frac{1}{3} \Delta^3 f_0 - \frac{1}{4} \Delta^4 f_0 + \cdots\right]. \tag{18.125}$$

$$f''(x_0) \approx \frac{1}{h^2} \left[\Delta^2 f_0 - \Delta^3 f_0 + \frac{11}{12} \Delta^4 f_0 - \cdots\right]. \tag{18.126}$$

Therefore, approximations for $f'(x_0)$ can be written as

(1) $f'(x_0) = \dfrac{1}{h} \Delta f_0 = \dfrac{1}{h} [f(x_1) - f(x_0)].$ $\hspace{2cm}$ (18.127)

(2) $f'(x_0) = \dfrac{1}{h} \left[\Delta f_0 - \dfrac{1}{2} \Delta^2 f_0\right] = \dfrac{1}{h} \left[\{f(x_1) - f(x_0)\} - \dfrac{1}{2} \{f(x_2) - 2f(x_1) + f(x_0)\}\right]$

$$= \frac{1}{2h} [-3f(x_0) + 4f(x_1) - f(x_2)]. \tag{18.128}$$

We can replace x_0 by x_k and write the formulas for any tabular point $x = x_k$.

Error in the formula (18.127) at $x = x_k$.

We have $E_1(f, x_k) = f'(x_k) - \dfrac{1}{h} [f(x_k + h) - f(x_k)]$

$$= f'(x_k) - \frac{1}{h} \left[hf'(x_k) + \frac{h^2}{2} f''(x_k) + \cdots\right] = \frac{h}{2} f''(x_k) - \cdots$$

Therefore, the method (18.127) is of order 1 or $O(h)$ formula.

Error in the formula (18.128) at $x = x_k$

We have

$$E_1(f, x_k) = f'(x_k) - \frac{1}{2h} [-3f(x_k) + 4f(x_k + h) - f(x_k + 2h)]$$

$$= f'(x_k) - \frac{1}{2h} \left[-3f(x_k) + 4\left\{f(x_k) + hf'(x_k) + \frac{h^2}{2} f''(x_k) + \frac{h^3}{6} f'''(x_k) + \cdots\right\}\right.$$

$$- \left\{ f(x_k) + 2h\,f'(x_k) + \frac{4h^2}{2}\,f''(x_k) + \frac{8h^3}{6}\,f'''(x_k) + \cdots \right\} \right]$$

$$= f'(x_k) - \frac{1}{2h} \left[2h\,f'(x_k) - \frac{2}{3}h^3\,f'''(x_k) + \cdots \right] = \frac{h^2}{3}\,f'''(x_k) + \ldots$$

Therefore, the method (18.128) is of order 2, or $O(h^2)$ formula.

Now, an approximation to $f''(x_k)$ can be written from (18.126) as

(3)
$$f''(x_k) = \frac{1}{h^2}\,\Delta^2 f_k = \frac{1}{h^2}\,[f(x_k) - 2f(x_{k+1}) + f(x_{k+2})]. \tag{18.129}$$

Error of approximation is given by

$$E_2(f,\,x_k) = f''(x_k) - \frac{1}{h^2}\,[f(x_k) - 2f(x_k + h) + f(x_k + 2h)]$$

$$= - h f'''(x_k) + \ldots$$

Therefore, the method (18.129) is of order 1.

Similarly, we can use the Newton's backward difference interpolating polynomial

$$f(x) \approx P_n(x) = f_n + \frac{(x - x_n)}{h}\,\nabla f_n + \frac{(x - x_n)(x - x_{n-1})}{2!h^2}\,\nabla^2 f_n$$

$$+ \frac{(x - x_n)(x - x_{n-1})(x - x_{n-2})}{3!h^3}\,\Delta^3 f_n + \ldots$$

Differentiating with respect to x, we get

$$f'(x) \approx P'_n(x) = \frac{1}{h}\,\nabla f_n + \frac{1}{2!h^2}\,[(x - x_n) + (x - x_{n-1})]\,\nabla^2 f_n$$

$$+ \frac{1}{3!h^3}\,[(x - x_{n-1})(x - x_{n-2}) + (x - x_n)(x - x_{n-1}) + (x - x_n)(x - x_{n-2})]\,\nabla^3 f_n + \ldots$$

$$f''(x) \approx P''(x) = \frac{1}{h^2}\,\nabla^2 f_n + \frac{2}{3!h^3}\,[(x - x_n) + (x - x_{n-1}) + (x - x_{n-2})]\,\nabla^3 f_n + \ldots$$

Using these relations, we can derive the expressions for the first and second derivatives at any nodal or non-nodal point $x = x^*$. For example, at $x = x_n$, we get

$$f'(x_n) = \frac{1}{h} \left[\nabla f_n + \frac{1}{2}\nabla^2 f_n + \frac{1}{3}\nabla^3 f_n + \ldots \right]. \tag{18.130}$$

$$f''(x_n) = \frac{1}{h^2}\,[\nabla^2 f_n + \nabla^3 f_n + \ldots]. \tag{18.131}$$

We obtain the following approximations.

(4) $$f'(x_n) = \frac{1}{h} \nabla f_n = \frac{1}{h} [f(x_n) - f(x_{n-1})].$$ (18.132).

The error of approximation is given by

$$E_1 (f, x_n) = f'(x_n) - \frac{1}{h} [f(x_n) - f(x_n - h)]$$

$$= \frac{h}{2} f''(\xi), \quad x_{n-1} < \xi < x_n.$$

Hence, the method (18.132) is of order 1.

(5) $$f'(x_n) = \frac{1}{h} \left[\nabla f_n + \frac{1}{2} \nabla^2 f_n \right]$$

$$= \frac{1}{h} [(f_n - f_{n-1}) + \frac{1}{2} (f_n - 2f_{n-1} + f_{n-2})]$$

$$= \frac{1}{2h} [3f(x_n) - 4 f(x_{n-1}) + f(x_{n-2})].$$ (18.133)

The error of approximation is given by

$$E_1(f, x_n) = f'(x_n) - \frac{1}{2h} [3 f(x_n) - 4 f(x_n - h) + f(x_n - 2h)]$$

$$= \frac{h^2}{3} f'''(\xi), \quad x_{n-2} < \xi < x_n.$$

Hence the method (18.133) is of order 2.

(6) $$f''(x_n) = \frac{1}{h^2} \nabla^2 f_n = \frac{1}{h^2} [f(x_n) - 2 f(x_{n-1}) + f(x_{n-2})].$$ (18.134)

The error of appoximation is given by

$$E_2(f, x_n) = f''(x_n) - \frac{1}{h^2} [f(x_n) - 2 f(x_n - h) + f(x_n - 2h)].$$

$$= h f'''(\xi), \quad x_{n-2} < \xi < x_n.$$

Hence, the method (18.134) is of order 1.

Example 18.37 The following data is given.

x	-3	-2.5	-2	-1	1
$f(x)$	-25	-14.125	-7	-1	-1

Using the method $f'(x_0) = [-3f_0 + 4f_1 - f_2]/(2h)$, obtain an approximate value of $f'(-3)$ with (i) $h = 2$, (ii) $h = 1$ and (iii) $h = 1/2$. The exact value of $f'(-3)$ is 26.

Solution   We have $x_0 = -3$.

(i) For $h = 2$, we get $x_0 = -3$, $x_1 = -1$, $x_2 = 1$.

$$f'(-3) = \frac{1}{4} [-3f(-3) + 4f(-1) - f(1)]$$

$$= \frac{1}{4} [-3(-25) + 4(-1) - (-1)] = 18.$$

(ii) For $h = 1$, we get $x_0 = -3$, $x_1 = -2$, $x_2 = -1$.

$$f'(-3) = \frac{1}{2} [-3f(-3) + 4f(-2) - f(-1)]$$

$$= \frac{1}{2} [-3(-25) + 4(-7) - (-1)] = 24.$$

(iii) For $h = 1/2$, we get $x_0 = -3$, $x_1 = -2.5$, $x_2 = -2$.

$$f'(-3) = [-3f(-3) + 4f(-2.5) - f(-2)]$$

$$= [-3(-25) + 4(-14.125) - (-7)] = 25.5.$$

Example 18.38   The following data represents e^{-x}.

x	-1	-0.5	0	1
$f(x)$	2.7183	1.6487	1	0.3679

Obtain an approximate value of $f''(-1)$ using the method $f''(x_0) = [f_0 - 2f_1 + f_2]/h^2$, with (i) $h = 1$, (ii) $h = 1/2$. Find the maximum error in each case.

Solution   We have $x_0 = -1$.

(i) For $h = 1$, we get $x_0 = -1$, $x_1 = 0$, $x_2 = 1$.

$f''(-1) = [f(-1) - 2f(0) + f(1)] = [2.7183 - 2 + 0.3679] = 1.0862.$

(ii) For $h = 1/2$, we get $x_0 = -1$, $x_1 = -1/2$, $x_2 = 0$.

$$f''(-1) = 4\left[f(-1) - 2f\left(-\frac{1}{2}\right) + f(0) \right] = 4[2.7183 - 2(1.6487) + 1] = 1.6836.$$

The magnitude of the error of approximation is given by (see (18.129))

$$|E_2(f, x_0)| = |-hf'''(\xi)| \le h \max_{x_0 \le x \le x_2} |f'''(x)|.$$

We have $f(x) = e^{-x}$.

For $h = 1$: $\displaystyle\max_{-1 \le x \le 1} |f'''(x)| = e$, and $|E_2(f, -1)| \le e$.

For $h = 1/2$: $\displaystyle\max_{-1 \le x \le 0} |f'''(x)| = e$, and $|E_2(f, -1)| \le \dfrac{e}{2}$.

Example 18.39 A function $f(x)$ representing the following data has a minimum in the interval $(0.5, 0.8)$. Find this point of minimum and the minimum value.

x	0.5	0.6	0.7	0.8
$f(x)$	1.3254	1.1532	0.9432	1.0514

Solution We construct the forward difference interpolating polynomial and differentiate it to find the point of minimum

x	$f(x)$	Δf	$\Delta^2 f$	$\Delta^3 f$
0.5	1.3254			
		-0.1722		
0.6	1.1532		-0.0378	
		-0.2100		0.3560
0.7	0.9432		0.3182	
		0.1082		
0.8	1.0514			

We have the Newton's forward difference interpolating polynomial as

$$f(x) = 1.3254 + (x - 0.5)\left(-\frac{0.1722}{0.1}\right) + (x - 0.5)(x - 0.6)\left(-\frac{0.0378}{0.02}\right)$$

$$+ (x - 0.5)\,(x - 0.6)\,(x - 0.7)\left(-\frac{0.3560}{0.006}\right)$$

Differentiating and setting it to zero, we get

$$f'(x) = 0 = -1.722 + (2x - 1.1)(-1.89) + (3x^2 - 3.6x + 1.07)(178/3)$$

$$= 178\,x^2 - 217.38\,x + 63.8437.$$

The roots of this equation are $x = 0.7297,\ 0.4915$ of which the first root lies in the interval $(0.5, 0.8)$. The point of minimum is $x \approx 0.7297$ and the minimum value is

$$f(0.7297) \approx 0.9260.$$

(Note that $f''(0.7297) > 0$).

Method of undetermined parameters

We express the required derivative $f^{(r)}(x_k)$, $r \ge 1$ as a linear combination of the values of $f(x)$ at the prescribed tabular points. The terms are expanded using Taylor series about x_k and various order derivatives of $f(x)$ at $x = x_k$ are compared. This gives the required number of equations for the parameters to be determined. Solution of this system of equations gives the required method. The leading non-vanishing term gives the error term. We illustrate this method through the following example. Consider the method

$$f'(x_k) = \frac{1}{h} \left[af(x_k + h) + bf(x_k) + cf(x_k - h) \right].$$

Writing the Taylor series expansion about x_k, we get

$$f'(x_k) = \frac{1}{h} \left[a \left\{ f(x_k) + hf'(x_k) + \frac{h^2}{2} f''(x_k) + \frac{h^3}{6} f'''(x_k) + \ldots \right\} \right.$$

$$\left. + bf(x_k) + c \left\{ f(x_k) - hf'(x_k) + \frac{h^2}{2} f''(x_k) - \frac{h^3}{6} f'''(x_k) + \ldots \right\} \right]$$

$$= \frac{1}{h} \left[(a+b+c) f(x_k) + h(a-c) f'(x_k) + \frac{h^2}{2} (a+c) f''(x_k) \right.$$

$$\left. + \frac{h^3}{6} (a-c) f'''(x_k) + \ldots \right].$$

We have three parameters a, b, c to be determined. Comparing the coefficients of $f(x_k)$, $f'(x_k)$ and $f''(x_k)$, we get

$$a + b + c = 0, \quad a - c = 1, \quad a + c = 0,$$

whose solution is $a = 1/2$, $b = 0$, $c = -1/2$.

The method is

$$f'(x_k) = \frac{1}{2h} \left[f(x_k + h) - f(x_k - h) \right].$$

The error term is given by

$$E(f, x_k) = -\frac{1}{h} \left[\frac{h^3}{6} (a-c) f'''(\xi) \right] = -\frac{h^2}{6} f'''(\xi), \quad x_{k-1} < \xi < x_{k+1}.$$

The order of the method is 2.

18.6.2   Numerical Integration Methods

The problem of numerical integration is to evaluate the integral

$$I = \int_a^b f(x)dx \tag{18.135}$$

where $f(x)$ is either given explicitly or defined by a data of $n + 1$ values $(x_i, f(x_i))$, $i = 0, 1, 2, \ldots, n$. We assume that the tabular points are equispaced with spacing $h = (b - a)/n$. The points are given by $x_0 = a$, $x_i = x_0 + ih$, $i = 1, 2, ..., n$ and $x_n = b$. It is possible to construct numerical intergration methods when the limits are finite or one of the limits is infinite or both the limits are infinite. The integral I is approximated by a linear combination of the values of $f(x)$ at the tabular points as

$$I = \int_a^b f(x)dx \doteq \lambda_0 f(x_0) + \lambda_1 f(x_1) + ... + \lambda_n f(x_n).$$ (18.136)

The tabular points x_k's are called *nodes* or *abscissas* and coefficients λ_k's are called *weights* of the integration method. The method (18.136) is also called a *quadrature rule* or an *integration rule*.

We define the error of approximation as

$$R_n(f, x) = \int_a^b f(x)dx - \sum_{k=0}^n \lambda_k f(x_k).$$ (18.137)

Order of a method

An integration rule is said to be of order p, if it produces exact results for all polynomials of degree $\leq p$. That is, it produces exact results for $f(x) = 1, x, ..., x^p$. Therefore,

$$R_n(f, x^m) = \int_a^b x^m dx - \sum_{k=0}^n \lambda_k x_k^m = 0, \quad \text{for} \quad m = 0, 1, 2, ..., p.$$

The error term is obtained for $f(x) = x^{p+1}$. We define

$$c = \int_a^b f(x)dx - \sum_{k=0}^n \lambda_k x_k^{p+1}$$ (18.138)

where c is called the *error constant*. Then, the error term is given by

$$R_n(f, x) = \frac{c}{(p+1)!} f^{(p+1)}_{(\xi)}, \quad a < \xi < b.$$ (18.139)

18.6.2.1 Newton-Cotes quadrature rules

Consider the case when the abscissas are prescribed and are equi-spaced. That is, the interval (a, b) is divided into n equal parts. The mesh spacing is given by $h = (b - a)/n$, and $x_0 = a$, $x_i = x_0 + ih$, $i = 1, 2, 3, ..., n$ and $x_n = b$. Then the integration rule (18.136) has $n + 1$ unknowns $\lambda_0, \lambda_1, ..., \lambda_n$. Replace the integrand of the rule (18.136) by the Lagrange interpolating polynomial interpolating at the $n + 1$ point (x_k, f_k), $k = 0, 1, 2, 3, ..., n$. We obtain

$$I = \int_a^b f(x)\, dx = \int_a^b \sum_{k=0}^n l_k(x) f_k \, dx = \sum_{k=0}^n f_k \int_a^b l_k(x)dx = \sum_{k=0}^n \lambda_k f_k,$$ (18.140)

where

$$l_k(x) = \frac{(x - x_0)(x - x_1)...(x - x_{k-1})(x - x_{k+1})...(x - x_n)}{(x_k - x_0)(x_k - x_1)...(x_k - x_{k-1})(x_k - x_{k+1})...(x_k - x_n)}$$

are the Lagrange fundamental polynomials, and

$$\lambda_k = \int_a^b l_k(x)\, dx. \tag{18.141}$$

Let $x = x_0 + sh$. Then,

$$l_k(x) = \frac{(sh)[(s-1)h]\dots[\{s-(k-1)\}h][\{s-(k+1)\}h]\dots[\{s-n\}h]}{(kh)[(k-1)h]\dots(h)(-h)(-2h)\dots[(n-k)h]}$$

$$= \frac{(-1)^{n-k}}{k!(n-k)!}\, s\,(s-1)\,(s-2)\,\dots\,(s-k+1)\,(s-k-1)\,\dots\,(s-n). \tag{18.142}$$

The error term can be obtained by using (18.138) and (18.139)

Trapezoidal rule

Consider the case $n = 1$, that is $h = b - a$. The data values are $(a, f(a))$, $(b, f(b))$, and

$$\int_a^b f(x)\, dx = \lambda_0 f(a) + \lambda_1 f(b).$$

We obtain from (18.141) and (18.142)

$$\lambda_0 = -h \int_0^1 (s-1)\, ds = \frac{h}{2}, \quad \lambda_1 = h \int_0^1 s\, ds = \frac{h}{2}.$$

The rule is given by

$$\int_a^b f(x)\, dx = \frac{h}{2}\, [f(a) + f(b)] = \frac{(b-a)}{2}\, [f(a) + f(b)]. \tag{18.143}$$

This rule is called the trapezoidal rule.

Geometrical interpretation We note that, $\int_a^b f(x)\, dx$ defines the area under the curve $y = f(x)$, above the x-axis, between the lines $x = a$ and $x = b$. The end points on the curve are $(a, f(a))$, $(b, f(b))$, (see Fig. 18.14). The right hand side of (18.143) is the area of the trapezoid $ABQP$. Therefore, trapezoidal rule approximates the area under the curve by the area of the trapezoid.

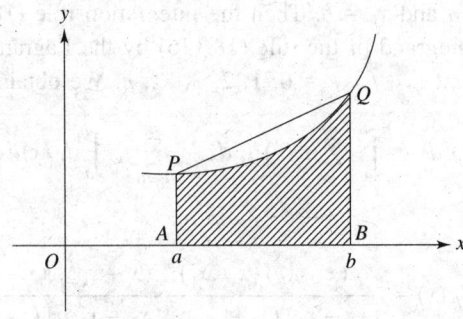

Fig. 18.14.   Trapezoidal rule.

Remark 12

The rule can also be obtained by replacing the integrand $f(x)$ by the Newton's forward difference interpolating polynomial up to the Δf_0 term.

Error term We first show that the trapezoidal rule integrates exactly polynomials of degree ≤ 1, that is, $R_1(f, x) = 0$ for $f(x) = 1$ and x. Now, substituting $f(x) = 1$, x in (18.143), we get

$$f(x) = 1: \quad b - a = \frac{(b-a)}{2} \ (2), \quad \text{which is true.}$$

$$f(x) = x: \quad \frac{1}{2} \ (b^2 - a^2) = \frac{(b-a)}{2} \ (a + b) = \frac{1}{2} \ (b^2 - a^2), \quad \text{which is true.}$$

Let $f(x) = x^2$. Then, from (18.138), we get

$$c = \int_a^b x^2 \, dx - \frac{(b-a)}{2} \ (a^2 + b^2) = \frac{1}{3} \ (b^3 - a^3) - \frac{1}{2} \ (b^3 + a^2b - ab^2 - a^3)$$

$$= \frac{1}{2} \ (a^3 - 3a^2b + 3ab^2 - b^3) = - \frac{1}{6} \ (b - a)^3.$$

The expression for the error is given by

$$R_1 \ (f, x) = \frac{c}{2!} \ f''(\xi) = - \frac{(b-a)^3}{12} \ f''(\xi) = - \frac{h^3}{12} \ f''_\xi \qquad (18.144)$$

and

$$|R_1(f, x)| \leq \frac{(b-a)^3}{12} \ M_2, \quad \text{where} \quad M_2 = \max_{a \leq x \leq b} \ |f''(x)|. \qquad (18.145)$$

Hence, the method is of order 1.

Composite formula If the length of the interval $[a, b]$ is large, then $b - a$ is large and the error estimate (18.144) becomes meaningless. Therefore, to have meaningful results, we subdivide $[a, b]$ into a number of sub-intervals of equal length and apply the trapezoidal rule to evaluate each integral. Let $[a, b]$ be sub-divided into N parts of equal length h.

Then, $h = (b - a)/N$. Let the points be

$$a = x_0, \quad x_1 = x_0 + h, \quad x_2 = x_0 + 2h, \ ..., \quad x_N = x_0 + Nh = b.$$

We write

$$\int_a^b f(x) \, dx = \int_{x_0}^{x_N} f(x) \, dx = \int_{x_0}^{x_1} f(x) \, dx + \int_{x_1}^{x_2} f(x) \, dx + \ ... \ + \int_{x_{N-1}}^{x_N} f(x) \, dx.$$

Now, using the trapezoidal rule to evaluate each integral, we obtain

$$\int_a^b f(x) \, dx = \frac{h}{2} \ [\{f(x_0) + f(x_1)\} + \{f(x_1) + f(x_2)\} + \ ... \ + \{f(x_{N-1}) + f(x_N)\}]$$

$$= \frac{h}{2} \left[f(x_0) + 2 \left\{ f(x_1) + f(x_2) + \ldots + f(x_{N-1}) \right\} + f(x_N) \right]. \tag{18.146}$$

This formula is called the *composite trapezoidal rule* and is of order 1.

The error expression (18.144) becomes

$$R_1(f, x) = -\frac{h^3}{12} \left[f''(\xi_1) + f''(\xi_2) + \ldots + f''(\xi_N) \right], \quad x_{N-1} < \xi_n < x_N.$$

Let $M_2 = \max_{[x_0, x_n]} |f''(x)|$. Then

$$|R_1(f, x)| \le \frac{h^3}{12} \left[|f''(\xi_1)| + |f''(\xi_2)| + \ldots + |f''(\xi_N)| \right]$$

$$\le \frac{Nh^3}{12} M_2 = \frac{(b-a)h^2}{12} M_2 \frac{(b-a)^3}{12N^2} M_2 \tag{18.147}$$

since $Nh = b - a$. This expression is a realistic representation of the error in the trapezoidal rule. If an error tolerance ε is prescribed, it is possible to find N, the number of sub-intervals required to achieve that accuracy. We get from (18.147)

$$\frac{(b-a)^3}{12N^2} M_2 \le \varepsilon, \quad \text{or} \quad N^2 \ge \frac{(b-a)^3 M_2}{12\varepsilon}. \tag{18.148}$$

Example18.40   Evaluate the following integrals using trapezoidal rule with $N = 2, 4$. Compare with the exact solution.

(i) $\displaystyle\int_0^1 \frac{dx}{3 + 2x}$.   (ii) $\displaystyle\int_0^2 \frac{dx}{x^2 + 2x + 10}$.

Find the bound on the error. Also, find the number of sub-intervals required if the error is to be less than 5×10^{-4}.

Solution

(i) We have $a = 0$, $b = 1$ and $f(x) = 1/(3 + 2x)$. For $N = 2$, we get $h = 1/2$. The abscissas are $x_0 = 0$, $x_1 = 1/2$ and $x_2 = 1.0$. Hence,

$$I = \frac{h}{2} \left[f(0) + 2f\left(\frac{1}{2}\right) + f(1) \right] = \frac{1}{4} \left[\frac{1}{3} + 2\left(\frac{1}{4}\right) + \frac{1}{5} \right] = 0.25833.$$

For $N = 4$, we get $h = 1/4$. The abscissas are $x_0 = 0$, $x_1 = 1/4$, $x_2 = 1/2$, $x_3 = 3/4$ and $x_4 = 1.0$. Hence,

$$I = \frac{h}{2} \left[f(0) + 2 \left\{ f\left(\frac{1}{4}\right) + f\left(\frac{1}{2}\right) + f\left(\frac{3}{4}\right) \right\} + f(1) \right]$$

$$= \frac{1}{8}\left[\frac{1}{3} + 2\left\{\frac{2}{7} + \frac{1}{4} + \frac{2}{8}\right\} + \frac{1}{5}\right] = 0.25615.$$

The exact solution is 0.5 ln (5/3) = 0.25541. The actual errors in magnitude are respectively 0.00292 and 0.00074.

We have $f(x) = \dfrac{1}{3+2x}$, $f'(x) = -\dfrac{2}{(3+2x)^2}$, $f''(x) = \dfrac{8}{(3+2x)^3}$ and

$$M_2 = \max_{[0,1]}\left[\frac{8}{(3+2x)^3}\right] = \frac{8}{27}.$$

Hence, $|\text{Error}| \le \dfrac{(b-a)h^2}{12}M_2 = \dfrac{2h^2}{81}$.

For $h = 1/2$, $|\text{Error}| \le 0.00617$, and for $h = 1/4$, $|\text{Error}| \le 0.00154$.

Note that the actual errors are smaller than the maximum magnitude errors.

If $\varepsilon = 5 \times 10^{-4}$, then the number of sub-intervals required is given by

$$N^2 \ge \frac{(b-a)^3 M_2}{12\,\varepsilon} = \frac{10^4}{60}\left(\frac{8}{27}\right) = 49.38 \quad \text{or} \quad N \ge 7.03.$$

Since N is an integer, we require $N = 8$.

(ii) We have $a = 0$, $b = 2$ and $f(x) = 1/(x^2 + 2x + 10)$. For $N = 2$, we get $h = 1$. The abscissas are $x_0 = 0$, $x_1 = 1$ and $x_2 = 2$. Hence,

$$I = \frac{h}{2}\,[f(0) + 2f(1) + f(2)] = \frac{1}{2}\left[\frac{1}{10} + \frac{2}{13} + \frac{1}{18}\right] = 0.15470.$$

For $N = 4$, we get $h = 1/2$. The abscissas are $x_0 = 0$, $x_1 = 1/2$, $x_2 = 1$, $x_3 = 3/2$ and $x_4 = 2$. Hence,

$$I = \frac{h}{2}\left[f(0) + 2\left\{f\left(\frac{1}{2}\right) + f(1) + f\left(\frac{3}{2}\right)\right\} + f(2)\right]$$

$$= \frac{1}{4}\,[0.1 + 2(0.08889 + 0.07692 + 0.06557) + 0.05556] = 0.15458.$$

The exact solution is $\dfrac{1}{3}\left[\dfrac{\pi}{4} - \tan^{-1}\left(\dfrac{1}{3}\right)\right] = 0.15455.$

The actual errors in magnitude are respectively 0.00015 and 0.00003. We have

$$f(x) = \frac{1}{x^2 + 2x + 10}, \quad f'(x) = -\frac{2(x+1)}{(x^2 + 2x + 10)^2}, \quad f''(x) = \frac{6(x^2 + 2x - 2)}{(x^2 + 2x + 10)^3}$$

$$\max_{[0,2]} |x^2 + 2x - 2| = 6 \quad \text{and} \quad \min_{[0,2]} (x^2 + 2x + 10)^3 = 1000.$$

Hence,
$$M_2 \leq \frac{36}{1000}.$$

Therefore, $|\text{Error}| \leq \dfrac{(b-a)h^2}{12} M_2 = \dfrac{2}{12}\left(\dfrac{36}{1000}\right)h^2 = \dfrac{3}{500} h^2.$

For $h = 1$, $|\text{Error}| \leq 0.006$ and for $h = 1/2$, $|\text{Error}| \leq 0.0015.$

If $\varepsilon = 5 \times 10^{-4}$, then the number of sub-intervals required is given by

$$N^2 \geq \frac{(b-a)^3 M_2}{12\varepsilon} = \frac{8}{60}\left(\frac{36}{1000}\right)10^4 = 48, \quad \text{or} \quad N \approx 7.$$

Simpson's rule (Simpson's 1/3rd rule)

We have shown that the trapezoidal rule of integration is of order 1, that is it integrates exactly polynomials of degree ≤ 1. However, in many applications, we require methods that give higher accuracy. One such method is the *Simpson's rule* of integration.

Consider the case $n = 2$, that is $h = (b - a)/2$. The abscissas are $x_0 = a$, $x_1 = x_0 + h = (a + b)/2$, $x_2 = b$. Integrand $f(x)$ is approximated by a quadratic polynomial. From (18.140), the rule is written as

$$\int_a^b f(x)\,dx = \lambda_0 f(a) + \lambda_1 f\left(\frac{a+b}{2}\right) + \lambda_2 f(b).$$

.We obtain from (18.141) and (18.142)

$$\lambda_0 = \frac{h}{2}\int_0^2 (s-1)(s-2)\,ds = \frac{h}{3}, \quad \lambda_1 = -h\int_0^2 s(s-2)\,ds = \frac{4h}{3},$$

$$\lambda_2 = \frac{h}{2}\int_0^2 s(s-1)\,ds = \frac{h}{3}.$$

The rule is given by

$$\int_a^b f(x)\,dx = \frac{h}{3}\,[f(x_0) + 4f(x_1) + f(x_2)]$$

$$= \frac{(b-a)}{6}\left[f(a) + 4f\left(\frac{a+b}{2}\right) + f(b)\right]. \tag{18.149}$$

This rule is called the Simpson's 1/3rd rule or simply Simpson's rule.

Geometrical interpretation The abscissas are $x_0 = a$, $x_1 = (a + b)/2$, $x_2 = b$. The points on the curve are $P(a, f(a))$, $Q((a + b)/2, f\{(a + b)/2\})$, and $R(b, f(b))$. The area under the curve $y = f(x)$, above the x–axis, and between the lines $x = a$ and $x = b$ is approximated by the area under the parabola passing through the points P, Q, R.

Remark 13

The rule can also be obtained by replacing the integrand $f(x)$ by the Newton's forward difference interpolating polynomial up to the $\Delta^2 f_0$ term.

Error term We first show that the Simpson's 1/3rd rule integrates exactly polynomials of degree ≤ 3, that is, $R_2 (f, x) = 0$ for $f(x) = 1, x, x^2$ and x^3. Substituting $f(x) = 1, x, x^2$ and x^3 in (18.149), we get

$$f(x) = 1: \qquad b - a = \frac{(b-a)}{6} \text{ (6), which is true.}$$

$$f(x) = x: \qquad \frac{1}{2} (b^2 - a^2) = \frac{(b-a)}{6}\left[a + 4\left(\frac{a+b}{2}\right) + b\right] = \frac{1}{2} (b^2 - a^2), \text{ which is true.}$$

$$f(x) = x^2: \qquad \frac{1}{3} (b^3 - a^3) = \frac{(b-a)}{6}\left[a^2 + 4\left(\frac{a+b}{2}\right)^2 + b^2\right]$$

$$= \frac{(b-a)}{3} [a^2 + ab + b^2] = \frac{1}{3} (b^3 - a^3), \text{ which is true.}$$

$$f(x) = x^3: \qquad \frac{1}{4} (b^4 - a^4) = \frac{(b-a)}{6}\left[a^3 + 4\left(\frac{a+b}{2}\right)^3 + b^3\right]$$

$$= \frac{(b-a)}{4} [a^3 + a^2 b + ab^2 + b^3] = \frac{1}{4} (b^4 - a^4), \text{ which is true.}$$

Let $f(x) = x^4$. Then, from (18.138), we get

$$c = \int_a^b x^4 dx - \frac{(b-a)}{6}\left[a^4 + 4\left(\frac{a+b}{2}\right)^4 + b^4\right]$$

$$= \frac{1}{5} (b^5 - a^5) - \frac{(b-a)}{6}\left[a^4 + \frac{1}{4}(a^4 + 4a^3 b + 6a^2 b^2 + 4ab^3 + b^4) + b^4\right]$$

$$= \frac{1}{120} [24(b^5 - a^5) - 5(b - a)\{5a^4 + 4a^3 b + 6a^2 b^2 + 4ab^3 + 5b^4\}]$$

$$= -\frac{(b-a)}{120} [b^4 - 4ab^3 + 6a^2 b^2 - 4a^3 b + a^4] = -\frac{(b-a)^5}{120}.$$

The expression for the error is given by

$$R_2(f, x) = \frac{c}{4!} f^{(4)}(\xi) = -\frac{(b-a)^5}{2880} f^{(4)}(\xi) = -\frac{h^5}{90} f^{(4)}(\xi) \qquad (18.150)$$

since $h = (b - a)/2$ and $x_0 < \xi < x_2$.

Therefore, Simpson's rule is of order 3, which is two orders higher than that of the trapezoidal rule.

Composite rule   As in trapezoidal rule, if the length of the interval $[a, b]$ is large, then $(b - a)$ is large and the error estimate (18.150) becomes meaningless. We note that the application of Simpson's rule requires three points. Therefore, we subdivide $[a, b]$ into an even number of sub-intervals. Let the number of sub-intervals be $2N$ so that $h = (b - a)/(2N)$, and we have odd number of points. Let the abscissas be

$$a = x_0, \quad x_1 = x_0 + h, \quad x_2 = x_0 + 2h, \ldots, \quad x_{2N} = x_0 + 2Nh = b.$$

We write

$$\int_a^b f(x)dx = \int_{x_0}^{x_{2N}} f(x)dx = \int_{x_0}^{x_2} f(x)dx + \int_{x_2}^{x_4} f(x)dx + \ldots + \int_{x_{2N-2}}^{x_{2N}} f(x)dx.$$

We note that there are N integrals. Now, using the Simpson's rule to evaluate each integral, we obtain

$$\int_a^b f(x)dx = \frac{h}{3} [\{f(x_0) + 4 f(x_1) + f(x_2)\} + \{f(x_2) + 4 f(x_3) + f(x_4)\}$$

$$+ \ldots + \{f(x_{2N-2}) + 4 f(x_{2N-1}) + f(x_{2N})\}]$$

$$= \frac{h}{3} [f(x_0) + 4\{f(x_1) + f(x_3) + \ldots + f(x_{2N-1})\}$$

$$+ 2 \{f(x_2) + f(x_4) + \ldots + f(x_{2N-2})\} + f(x_{2N})]. \qquad (18.151)$$

This formula in called *composite Simpson's rule*.

The error expression (18.150) becomes

$$R_2(f, x) = -\frac{h^5}{90} [f^{(4)}(\xi_1) + f^{(4)}(\xi_2) + \ldots + f^{(4)}(\xi_N)]$$

where $x_0 < \xi_1 < x_2$, $x_2 < \xi_2 < x_4$ etc. Let

$$M_4 = \max_{[x_0, x_{2N}]} |f^{(4)}(x)|.$$

Then,

$$|R_2(f, x)| \le \frac{h^5}{90} \left[\left| f^{(4)}(\xi_1) \right| + \left| f^{(4)}(\xi_2) \right| + \ldots + \left| f^{(4)}(\xi_N) \right| \right]$$

$$\le \frac{Nh^5}{90} M_4 = \frac{(b-a)}{180} h^4 M_4 = \frac{(b-a)^5}{2880 N^4} M_4 \qquad (18.152)$$

since $Nh = (b - a)/2$. If the error tolerance ε is prescribed, it is possible to find the number of sub-intervals required to achieve that accuracy. We get from (18.152)

$$\frac{(b-a)^5 M_4}{2880 N^4} \leq \varepsilon, \quad \text{or} \quad N^4 \geq \frac{(b-a)^5 M_4}{2880 \varepsilon} \tag{18.153}$$

where N is an integer.

Example 18.41 Evaluate the following integrals using Simpson's 1/3rd rule with two and four sub-intervals. Compare with the exact solution.

(i) $\displaystyle \int_0^1 \frac{dx}{3 + 2x}$.

(ii) $\displaystyle \int_0^2 \frac{dx}{x^2 + 2x + 10}$.

Solution

(i) We have $a = 0$, $b = 1$ and $f(x) = 1/(3 + 2x)$. For two sub-intervals, we get $h = 1/2$. The abscissas are $x_0 = 0$, $x_1 = 1/2$ and $x_2 = 1.0$. Hence,

$$I = \frac{h}{3}\left[f(0) + 4f\left(\frac{1}{2}\right) + f(1) \right] = \frac{1}{6}\left[\frac{1}{3} + 4\left(\frac{1}{4}\right) + \frac{1}{5} \right] = 0.25556.$$

For four sub-intervals, we get $h = 1/4$. The abscissas are $x_0 = 0$, $x_1 = 1/4$, $x_2 = 1/2$ $x_3 = 3/4$ and $x_4 = 1$. Hence,

$$I = \frac{h}{3}\left[f(0) + 4\left\{ f\left(\frac{1}{4}\right) + f\left(\frac{3}{4}\right) \right\} + 2f\left(\frac{1}{2}\right) + f(1) \right]$$

$$= \frac{1}{12}\left[\frac{1}{3} + 4\left(\frac{2}{7} + \frac{2}{9}\right) + 2\left(\frac{1}{4}\right) + \frac{1}{5} \right] = 0.25542.$$

The exact solution is $0.5 \ln (5/3) = 0.25541$. The errors in magnitude are respectively 0.00015 and 0.00001.

(ii) We have $a = 0$, $b = 2$ and $f(x) = 1/(x^2 + 2x + 10)$. For two sub-intervals, we get $h = 1$. The abscissas are $x_0 = 0$, $x_1 = 1$ and $x_2 = 2$. Hence,

$$I = \frac{h}{3}\left[f(0) + 4f(1) + f(2) \right] = \frac{1}{3}\left[\frac{1}{10} + 4\left(\frac{1}{13}\right) + \frac{1}{18} \right] = 0.15442.$$

For four sub-intervals, we get $h = 1/2$. The abscissas are $x_0 = 0$, $x_1 = 1/2$, $x_2 = 1$, $x_3 = 3/2$ and $x_4 = 2$. Hence,

$$I = \frac{h}{3}\left[f(0) + 4\left\{ f\left(\frac{1}{2}\right) + f\left(\frac{3}{2}\right) \right\} + 2f(1) + f(2) \right]$$

$$= \frac{1}{\cdot 6} \left[\frac{1}{10} + 4(0.08889 + 0.06557) + 2(0.07692) + 0.05556 \right] = 0.15454.$$

The exact solution is 0.15455. The errors in magnitude respectively are 0.00013, 0.00001.

Simpson's 3/8th rule

We can derive a formula by taking $n = 3$, that is, approximating the integrand $f(x)$ by a cubic polynomial. We have $h = (b - a)/3$. The abscissas are $x_0 = a$, $x_1 = x_0 + h$, $x_2 = x_0 + 2h$, $x_3 = x_0 + 3h = b$. From (18.140), the rule is given by

$$\int_a^b f(x)\, dx = \lambda_0 f(x_0) + \lambda_1 f(x_1) + \lambda_2 f(x_2) + \lambda_3 f(x_3).$$

We obtain from (18.141) and (18.142)

$$\lambda_0 = -\frac{h}{6} \int_0^3 (s-1)(s-2)(s-3)\, ds = \frac{3h}{8}, \quad \lambda_1 = \frac{h}{2} \int_0^3 s(s-2)(s-3)\, ds = \frac{9h}{8},$$

$$\lambda_2 = -\frac{h}{2} \int_0^3 s(s-1)(s-3)\, ds = \frac{9h}{8}, \quad \lambda_3 = \frac{h}{6} \int_0^3 s(s-1)(s-2)\, ds = \frac{3h}{8}.$$

The rule is given by

$$\int_a^b f(x)\, dx = \frac{3h}{8} \left[f(x_0) + 3f(x_1) + 3f(x_2) + f(x_3) \right] \tag{18.154}$$

This rule is called the *Simpson's 3/8*th *rule*.

The error term is given by

$$R_3(f, x) = -\frac{3}{80} h^5 f^{(4)}(\xi), \quad x_0 < \xi < x_3. \tag{18.155}$$

Composite rule   As in the case of Simpson's 1/3rd rule, if the length of the interval $[a, b]$ is large, then the error expression given in (18.155) becomes meaningless. In this case, we subdivide $[a, b]$ into a number of sub-intervals of equal length such that the number of sub-intervals is divisible by 3. Let the number of sub-intervals be $3N$. Therefore, the number of sub-intervals can be 6, 9, 12 etc. Then, the number of nodal points are 7, 10, 13 etc. We evaluate each of the integrals by the Simpson's 3/8th rule and simplify.

Let $h = (b - a)/(3N)$. The abscissas are given by

$$a = x_0,\ x_1 = x_0 + h,\ x_2 = x_0 + 2h,\ x_3 = x_0 + 3h,\ \ldots,\ x_{3N} = x_0 + 3Nh = b.$$

We write,

$$\int_a^b f(x)\, dx = \int_{x_0}^{x_{3N}} f(x)\, dx = \int_{x_0}^{x_3} f(x)\, dx + \int_{x_3}^{x_6} f(x)\, dx + \int_{x_6}^{x_9} f(x)\, dx + \ldots + \int_{x_{3N-3}}^{x_{3N}} f(x)\, dx$$

$$= \frac{3h}{8} \left[\{f(x_0) + 3f(x_1) + 3f(x_2) + f(x_3)\} + \{f(x_3) + 3f(x_4) + 3f(x_5) + f(x_6)\} \right.$$

$$\left. + \ldots + \{f(x_{3N-3}) + 3f(x_{3N-2}) + 3f(x_{3N-1}) + f(x_{3N})\} \right]$$

$$= \frac{3h}{8} \left[f(x_0) + 3f(x_1) + 3f(x_2) + 2f(x_3) + 3f(x_4) + 3f(x_5) + 2f(x_6) \right.$$

$$\left. + \ldots + 2f(x_{3N-3}) + 3f(x_{3N-2}) + 3f(x_{3N-1}) + f(x_{3N}) \right] \tag{18.156}$$

The error expression becomes

$$R_3(f, x) = -\frac{3}{80} h^5 [f^{(4)}(\xi_1) + f^{(4)}(\xi_2) + \ldots + f^{(4)}(\xi_N)], \quad x_0 < \xi_1 < x_3, \text{ etc.} \tag{18.157}$$

Then, $$|R_3(f, x)| \le \frac{3Nh^5}{80} M_4, \quad \text{where} \quad M_4 = \max_{[x_0, x_{3N}]} |f^{(4)}(x)|.$$

Remark 14

Simpson's 3/8th rule has a disadvantage. From the error expression (18.157), we conclude that the rule integrates exactly polynomials of degree ≤ 3, which is same as for the Simpson's 1/3rd rule. Simpson's 1/3rd rule requires computation of three function evaluations, while Simpson's 3/8th rule requires computation of four function evaluations. Further, the error coefficients in the Simpson's 1/3rd and 3/8th rules are −1/90 and −3/80 respectively. Therefore, computationally Simpson's 1/3rd rule is superior to the Simpson's 3/8th rule.

Example 18.42 Using Simpson's 3/8th rule, evaluate the following integrals with 3 and 6 sub-intervals. Compare with the exact solution.

(i) $$\int_0^1 \frac{dx}{4 + 3x},$$ (ii) $$\int_0^3 \frac{dx}{1 + x^2}.$$

Solution

(i) Three sub-intervals: $h = 1/3$. The abscissas are 0, 1/3, 2/3, 1. We have the following data.

x	0	1/3	2/3	1
$f(x)$	0.250000	0.200000	0.166667	0.142857

$$I_1 = \frac{3h}{8} [f(0) + 3f(1/3) + 3f(2/3) + f(1)]$$

$$= \frac{3}{24} [0.250000 + 3(0.200000) + 3(0.166667) + 0.142857] = 0.186607.$$

Six sub-intervals: $h = 1/6$. The abscissas are 0, 1/6, 2/6, 3/6, 4/6, 5/6, 1.
We use the previous data values and the following data values.

x	1/6	1/2	5/6
$f(x)$	0.222222	0.181818	0.153846

Hence,

$$I_2 = \frac{3h}{8} \left[f(0) + 3f(1/6) + 3f(1/3) + 2f(1/2) + 3f(2/3) + 3f(5/6) + f(1) \right]$$

$$= \frac{3}{48} \left[0.250000 + 3(0.222222) + 3(0.200000) + 2(0.181818) \right.$$

$$\left. + 3(0.166667) + 3(0.153846) + 0.142867 \right] = 0.186544.$$

The exact solution is

$$I = \left[\frac{1}{3} \log(4 + 3x) \right]_0^1 = \frac{1}{3} \left[\log 7 - \log 4 \right] = 0.186539.$$

The magnitudes of errors in the solution are 0.000068 and 0.000005 respectively.

(ii) Three subintervals: $h = 1$. The points are 0, 1, 2, 3. We have the following data.

x	0	1	2	3
$f(x)$	1.0	0.5	0.2	0.1

Hence,

$$I_1 = \frac{3h}{8} \left[f(0) + 3f(1) + 3f(2) + f(3) \right]$$

$$= \frac{3}{8} \left[1.0 + 3(0.5) + 3(0.2) + 0.1 \right] = 1.2.$$

Six subintervals: $h = 1/2$. The points are 0, 1/2, 1, 3/2, 2, 5/2, 3.
We use the previous data values and the following data values.

x	1/2	3/2	5/2
$f(x)$	0.8	0.307692	0.137931

Hence,

$$I_2 = \frac{3h}{8} \left[f(0) + 3f(1/2) + 3f(1) + 2f(3/2) + 3f(2) + 3f(5/2) + f(3) \right]$$

$$= \frac{3}{16} \left[1.0 + 3(0.8) + 3(0.5) + 2(0.307692) + 3(0.2) + 3(0.137931) + 0.1 \right] = 1.242971.$$

The exact solution is

$$I = \left[\tan^{-1} x \right]_0^3 = \tan^{-1} 3 = 1.249046.$$

The magnitudes of errors in the solution are 0.049046 and 0.006075 respectively.

Boole's rule

We can derive a formula by taking $n = 4$. We have $h = (b - a)/4$. The abscissas are

$$x_0 = a, \ x_1 = x_0 + h, \ x_2 = x_0 + 2h, \ x_3 = x_0 + 3h, \text{ and } x_4 = x_0 + 4h = b.$$

From (18.140), the rule is given by

$$\int_a^b f(x)\, dx = \lambda_0 f(x_0) + \lambda_1 f(x_1) + \lambda_2 f(x_2) + \lambda_3(x_3) + \lambda_4 f(x_4).$$

We obtain from (18.142) and (18.142)

$$\lambda_0 = \frac{h}{24} \int_0^4 (s-1)(s-2)(s-3)(s-4)\, ds = \frac{14h}{45}, \quad \lambda_1 = -\frac{h}{6} \int_0^4 s(s-2)(s-3)(s-4)\, ds = \frac{64h}{45},$$

$$\lambda_2 = \frac{h}{4} \int_0^4 s(s-1)(s-3)(s-4)\, ds = \frac{8h}{15}, \quad \lambda_3 = -\frac{h}{6} \int_0^4 s(s-1)(s-2)(s-4)\, ds = \frac{64h}{45},$$

$$\lambda_4 = \frac{h}{24} \int_0^4 s(s-1)(s-2)(s-3)\, ds = \frac{14h}{45}.$$

The rule is given by

$$\int_a^b f(x)\, dx = \frac{2h}{45}\left[7f(x_0) + 32f(x_1) + 12f(x_2) + 32f(x_3) + 7f(x_4)\right]. \tag{18.158}$$

This rule is called the *Boole's rule*.

The error term is given by

$$R_4(f, x) = -\frac{8}{945} h^6 f^{(6)}(\xi), \quad x_0 < \xi < x_4. \tag{18.159}$$

The formula integrates exactly polynomials of degree ≤ 5. (Order = 5).

18.6.2.2 Romberg integration

Romberg integration is a powerful tool which uses the method of extrapolation. Generally, to obtain accurate results, we compute the integral by a Newton-Cotes formula for a number of values of step lengths, each time reducing the step length. We stop the computation, when the specified accuracy is obtained, that is, when the magnitude of the difference of successive values of the integral obtained by reducing values of the step lengths is less than a given accuracy. Convergence may be obtained after computing the values of the integral using a number of step lengths. In many cases, convergence may be slow. It may be noted that while computing the value of the integral with a particular step length, the values of the integral obtained earlier by using larger step lengths were not used. Romberg integration method is derived by studying the error of the method that is being used. The method uses the values of the integrals obtained with various step lengths, to refine the solution such that the new values are of higher order, that is as if the results are obtained by using a higher order method than the order of the method used. We illustrate the Romberg method for the trapezoidal rule and the Simpson's 1/3rd rule.

Romberg method for the trapezoidal rule

Let the integral (18.135) be computed by the composite trapezoidal rule. Let I denote the exact value of the integral and I_T denote the value obtained by the composite trapezoidal rule. The error, $I - I_T$, in the composite trapezoidal rule in computing the integral is given by

$$I - I_T = c_1 h^2 + c_2 h^4 + c_3 h^6 + \ldots, \quad \text{or} \quad I = I_T + c_1 h^2 + c_2 h^4 + \ldots \quad (18.160)$$

where $c_1, c_2, \ldots$ are independent of h.

Let I be evaluated using two step lengths h and qh, $0 < q < 1$. Let these values be denoted by $I_T(h)$ and $I_T(qh)$. The error equations become

$$I - I_T(h) = c_1 h^2 + c_2 h^4 + \ldots \quad (18.161)$$

$$I - I_T(qh) = c_1 q^2 h^2 + c_2 q^4 h^4 + \ldots \quad (18.162)$$

Eliminating c_1 from these two equations (neglecting the higher order terms), we obtain

$$(1 - q^2)\, I - I_T(qh) + q^2 I_T(h) = c_2 q^2 h^4 (q^2 - 1).$$

Solving for I, we obtain

$$I = \frac{I_T(qh) - q^2 I_T(h)}{(1 - q^2)} - c_2 q^2 h^4.$$

Note that the error on the right hand side is now of order $O(h^4)$.

Neglecting the $O(h^4)$ error term, we obtain the new approximation to the value of the integral as

$$I \approx I_T^{(1)}(h) = \frac{I_T(qh) - q^2 I_T(h)}{(1 - q^2)}. \quad (18.163)$$

This computed result is of order, $O(h^4)$, which is higher than the order of the trapezoial rule, which is of order $O(h^2)$. For $q = 1/2$, (computations are done with step lengths h and $h/2$), the formula (18.163) simplifies to

$$I_T^{(1)}(h) \approx \frac{I_T(h/2) - (1/4) I_T(h)}{1 - (1/4)}$$

$$= \frac{4 I_T(h/2) - I_T(h)}{4 - 1}. \quad (18.164)$$

For ease in computations, we normally use the sequence of step lengths as h, $h/2$, $h/2^2$, $h/2^3$, ... Formula (18.164) gives the first column of extrapolated values, which are of order $O(h^4)$. Repeating the extrapolation procedure, we obtain the Romberg method as (when the step lengths are reduced by the factor 2)

$$I_T^m(h) \approx \frac{4^m I_T^{(m-1)}(h/2) - I_T^{(m-1)}(h)}{4^m - 1}, \quad m = 1, 2, \ldots \quad (18.165)$$

where $I_T^{(0)} = I_T(h)$. The computed result is of order $O(h^{2m+2})$.

The extrapolations using three step lengths h, $h/2$, $h/2^2$, are given in Table 18.5.

Table 18.5. Romberg method for trapezoidal rule.

Step length	Value of I: $O(h^2)$	Value of I: $O(h^4)$	Value of I: $O(h^6)$
h	$I(h)$		
		$I^{(1)}(h) = \dfrac{4I(h/2) - I(h)}{3}$	
$h/2$	$I(h/2)$		$I^{(2)}(h) = \dfrac{16I^{(1)}(h/2) - I^{(1)}(h)}{15}$
		$I^{(1)}(h/2) = \dfrac{4I(h/4) - I(h/2)}{3}$	
$h/4$	$I(h/4)$		

Note that the most accurate value is the values at the end of each column.

Romberg method for the Simpson's 1/3rd rule

We can apply the same procedure as in the trapezoidal rule to obtain the Romberg's integration procedure for the Simpson's 1/3rd rule. The error, $I - I_S$, in the composite Simpson's 1/3rd rule in computing the integral is given by

$$I - I_S = c_1 h^4 + c_2 h^6 + c_3 h^8 + \dots, \quad \text{or} \quad I = I_S + c_1 h^4 + c_2 h^6 + c_3 h^8 + \dots \tag{18.166}$$

Following the extrapolation procedure described above, we obtain the Romberg integration procedure for the composite Simpson's 1/3rd rule as

$$I_S^{(m)}(h) \approx \frac{4^{m+1} I_S^{(m-1)}(h/2) - I_S^{(m-1)}(h)}{4^{m+1} - 1}, \quad m = 1, 2, \dots \tag{18.167}$$

where $I_S^{(0)}(h) = I_S(h)$. The computed result is of order $O(h^{2m+4})$.

The extrapolations using three step lengths h, $h/2$, $h/2^2$, are given in Table 18.6. Note that the most accurate value is the value at the end of each column.

Table 18.6. Romberg method for Simpson's 1/3rd rule.

Step length	Value of I: $O(h^4)$	Value of I: $O(h^6)$	Value of I: $O(h^8)$
h	$I(h)$		
		$I^{(1)}(h) = \dfrac{16I(h/2) - I(h)}{15}$	
$h/2$	$I(h/2)$		$I^{(2)}(h) = \dfrac{64I^{(1)}(h/2) - I^{(1)}(h)}{63}$
		$I^{(1)}(h/2) = \dfrac{16I(h/4) - I(h/2)}{15}$	
$h/4$	$I(h/4)$		

Example 18.43   Compute the extrapolated values of the integral given in Example 18.40(i) by Romberg method and the trapezoidal rule. Compare with the exact solution.

Solution   We obtain the following extrapolated values as given in Table 18.7,

Table 18.7.   Extrapolated values using the trapezoidal rule.

h	$O(h^2)$ values	$O(h^4)$ values
0.5	$I(h) = 0.25833$	
		$I^{(1)}(h) = \dfrac{4I(h/2) - I(h)}{3} = 0.25542$
0.25	$I(h/2) = 0.25615$	

The magnitudes of the errors in the computed and extrapolated values are 0.00292, 0.00074 and 0.00001 respectively.

18.6.2.3   Gauss quadrature rules

The formulas that we have derived in the previous section (Newton-Cotes rules) are formulas based on uniform mesh spacing. Gaussian integration rules are formulas based on non-uniform mesh spacing. Consider the general intergration rule as

$$I = \int_a^b w(x) f(x) \, dx = \lambda_0 f(x_0) + \lambda_1 f(x_1) + \ldots + \lambda_n f(x_n). \tag{18.168}$$

Here, $w(x)$ is called the weight function and x_i's are called abscissas of the formula. When the abscissas are not prescribed in advance and they are also to be determined, then the formulas using lesser number of abscissas can produce higher order methods compared to the Newton-Cotes formulas. Gaussian integration rules can be derived when the limits are finite or one of the limits is infinite or both the limits are infinite. Gaussian integration rules depend on the limits of integration and the expression for the weight function $w(x)$. Since the formula (18.168) has $2n + 2$ parameters ($n + 1$ weights and $n + 1$ abscissas), the formula can be made exact for polynomials of degree $\leq 2n + 1$.

The derivation of the integration rule can be done using the following theorem.

Theorem 18.1   If the abscissas x_i's are chosen as the zeros of an orthogonal polynomial, orthogonal with respect to the weight function $w(x)$ over $[a, b]$, then the integration rule (18.168) has order $2n + 1$. The weights are given by

$$\lambda_i = \int_a^b w(x) \, l_i(x) \, dx,$$

where $l_i(x)$ are Lagrange fundamental polynomials. Further, $\lambda_i > 0$.

1. Gauss-Legendre integration rules

Weight function $= w(x) = 1$. Limits of integration $= [-1, 1]$.

Abscissas = Zeros of the corresponding Legendre polynomial.

2. Gauss-Chebyshev integration rules

Weight function $= w(x) = 1/\sqrt{1-x^2}$. Limits of integration $= [-1, 1]$.

Abscissas = Zeros of the corresponding Chebyshev polynomial.

3. Gauss-Laguerre integration rules

Weight function $= w(x) = e^{-x}$, Limits of intergration $= [0, \infty)$.

Abscissas = Zeros of the corresponding Laguerre polynomial.

4. Gauss-Hermite integration rules

Weight function $= w(x) = e^{-x^2}$. Limits of integration $(-\infty, \infty)$.

Abscissas = Zeros of the corresponding Hermite polynomial.

Note that we can also derive the above methods using the method of undetermined parameters.

Order of a method

A method is said to be of order p if it integrates exactly polynomials of degree $\leq p$. That is, it integrates exactly for $f(x) = 1, x, x^2, ..., x^p$.

For the Gauss-Legendre and Gauss-Chebyshev formulas, define

$$c = \int_{-1}^{1} w(x)\, x^{p+1}\, dx - [\lambda_0(x_0^{p+1}) + \lambda_1(x_1^{p+1}) + ... + \lambda_n(x_n^{p+1})].$$

Then, error in the formula $= \dfrac{c}{(n+1)!}\, f^{(n+1)}(\xi),\quad -1 < \xi < 1$.

Similarly, we define errors in Gauss-Laguerre and Gauss-Hermite formulas.

1. Gauss-Legendre quadrature rules

To derive the rules, we reduce the interval $[a, b]$ to $[-1, 1]$ by using a linear transformation. Write the transformation as $x = pt + q$. We get, $a = -p + q$, $b = p + q$, whose solution is $p = (b-a)/2$, $q = (b+a)/2$. The linear transformation is given by $x = [(b-a)t + (b+a)]/2$. Hence,

$$\int_{a}^{b} f(x)\, dx = \left(\frac{b-a}{2}\right) \int_{-1}^{1} f\left[\frac{1}{2}\{(b-a)t + (b+a)\}\right] dt = \int_{-1}^{1} g(t)\, dt.$$

Without loss of generality, we may write this integral as $\int_{-1}^{1} f(x)\, dx$.

Gauss-Legendre one point rule

We use the method of undetermined parameters. The one point rule is given by

$$\int_{-1}^{1} f(x)\, dx = \lambda_0 f(x_0),\quad \lambda_0 \neq 0.$$

The method has two parameters λ_0, x_0. Making the formula exact for $f(x) = 1, x$, we get

$$f(x) = 1: \quad \int_{-1}^{1} dx = 2 = \lambda_0, \quad f(x) = x: \quad \int_{-1}^{1} x\,dx = 0 = \lambda_0\,x_0.$$

Since, $\lambda_0 \neq 0$, we get $x_0 = 0$. Therefore, the Gauss-Legendre one point rule is given by

$$\int_{-1}^{1} f(x)\,dx = 2\,f(0). \tag{18.169}$$

The error term is given by, error $= (1/3)\,f''(\xi)$, $-1 < \xi < 1$. Since the error term contains $f''(\xi)$, Gauss-Lagendre one point rule integrates exactly polynomials of degree less than or equal to 1. Therefore, the results obtained from this rule can be compared with the results obtained from the trapezoidal rule. However, we require two function evaluations in the trapezoidal rule whereas we need only one function evaluation in the Gauss-Legendre one point rule. The corresponding Legendre polynomial is $P_1(x) = x$, whose zero is $x = 0$.

Gauss-Legendre two point rule

The two point rule is given by

$$\int_{-1}^{1} f(x)\,dx = \lambda_0 f(x_0) + \lambda_1 f(x_1), \quad \lambda_0 \neq 0, \lambda_1 \neq 0, \text{ and } x_0 \neq x_1.$$

The method has four parameters $\lambda_0, \lambda_1, x_0, x_1$. Making the formula exact for $f(x) = 1, x, x^2, x^3$, we get

$$f(x) = 1: \quad \int_{-1}^{1} dx = 2 = \lambda_0 + \lambda_1, \qquad f(x) = x: \quad \int_{-1}^{1} x\,dx = 0 = \lambda_0 x_0 + \lambda_1 x_1.$$

$$f(x) = x^2: \quad \int_{-1}^{1} x^2 dx = \frac{2}{3} = \lambda_0 x_0^2 + \lambda_1 x_1^2, \quad f(x) = x^3: \quad \int_{-1}^{1} x^3 dx = 0 = \lambda_0 x_0^3 + \lambda_1 x_1^3.$$

The solution of these equations is $\lambda_0 = \lambda_1 = 1$, and $x_0 = \pm\left(1/\sqrt{3}\right) = -x_1$.

Therefore, the Gauss-Legendre two point rule is given by

$$\int_{-1}^{1} f(x)\,dx = f\left(\frac{-1}{\sqrt{3}}\right) + f\left(\frac{1}{\sqrt{3}}\right). \tag{18.170}$$

The error term is given by, error $= (1/135)\,f^{(4)}(\xi)$, $-1 < \xi < 1$. Since the error term contains $f^{(4)}(\xi)$, Gauss-Legendre two point rule integrates exactly polynomials of degree less than or equal to 3. Therefore, the results obtained from this rule can be compared with the results obtained from the Simpson's 1/3rd rule. However, we require three function evaluations in the Simpson's 1/3rd rule whereas we need only two function evaluations in the Gauss-Legendre two point rule. The corresponding Legendre polynomial is $P_2(x) = (3x^2 - 1)/2$, whose zeros are $x = \pm 1/\sqrt{3}$.

Gauss-Legendre three point rule

The three point rule is given by

$$\int_{-1}^{1} f(x)\,dx = \lambda_0 f(x_0) + \lambda_1 f(x_1) + \lambda_2 f(x_2), \quad \lambda_0 \neq 0, \lambda_1 \neq 0, \lambda_2 \neq 0, \text{ and } x_0 \neq x_1 \neq x_2.$$

The method has six unknowns λ_0, λ_1, λ_2, x_0, x_1, x_2. Making the formula exact for $f(x) = 1, x, x^2$, x^3, x^4, x^5, we get the Gauss-Legendre three point rule as

$$\int_{-1}^{1} f(x)\, dx = \frac{1}{9}\left[5f\left(-\sqrt{3/5}\right) + 8f(0) + 5f\left(\sqrt{3/5}\right)\right]. \tag{18.171}$$

The error term is given by, error $= (1/15750) f^{(6)}(\xi)$, $-1 < \xi < 1$. Since the error term contains $f^{(6)}(\xi)$, Gauss-Legendre three point rule integrates exactly polynomials of degree less than or equal to 5. Further, the error coefficient is very small $(1/15750)$. Therefore, the results obtained from this rule are very accurate. The corresponding Legendre polynomial is $P_3(x) = x(5x^2 - 3)/2$, whose zeros are $x = 0, \pm\sqrt{3/5}$.

Example 18.44 Evaluate the following integrals using Gauss-Legendre one point, two point and three point rules. Compare with the exact solution.

(i) $\displaystyle\int_{1}^{2} \frac{dx}{1+x^2}.$ (ii) $\displaystyle\int_{0}^{2} \frac{dx}{x^2 + 2x + 10}$

Solution (i) The transformation from $[1, 2]$ to $[-1, 1]$ is given by

$x = \{[(b-a)t + (b+a)]/2\} = (t+3)/2$. We have $dx = dt/2$.

Hence, $\displaystyle\int_{1}^{2} \frac{dx}{1+x^2} = \frac{1}{2}\int_{-1}^{1} \frac{4\,dt}{4+(t+3)^2} = \int_{-1}^{1}\frac{2\,dt}{4+(t+3)^2} = \int_{-1}^{1} f(t)\, dt.$

One point rule: $\quad I_1 = 2f(0) = 2\left(\dfrac{2}{13}\right) = 0.307692.$

Two point rule: $\quad I_2 = f\left(-\dfrac{1}{\sqrt{3}}\right) + f\left(\dfrac{1}{\sqrt{3}}\right) = 0.202650 + 0.119066 = 0.321716.$

Three point rule: $\quad I_3 = \dfrac{1}{9}\left[5f\left(-\sqrt{3/5}\right) + 8f(0) + 5f\left(\sqrt{3/5}\right)\right]$

$$= \frac{1}{9}\,[5(0.223403) + 8(0.153846) + 5(0.109604)] = 0.321756.$$

Exact solution: $\quad I = \tan^{-1}(2) - (\pi/4) = 0.321751.$

(ii) The transformation from $[0, 2]$ to $[-1, 1]$ is given by

$x = \{[(b-a)t + (b+a)]/2\} = t + 1$. We have $dx = dt$.

Hence, $\displaystyle\int_{0}^{2} \frac{dx}{x^2 + 2x + 10} = \int_{-1}^{1} \frac{dt}{(t+1)^2 + 2(t+1) + 10} = \int_{-1}^{1} f(t)\,dt.$

One point rule: $\quad I_1 = 2f(0) = (2/13) = 0.153846.$

Two point rule: $\quad I_2 = f(-1/\sqrt{3}) + f(1/\sqrt{3}) = 0.090712 + 0.063927 = 0.154639.$

Three point rule:
$$I_3 = \frac{1}{9} \left[5f\left(-\sqrt{3/5}\right) + 8f(0) + 5f\left(\sqrt{3/5}\right) \right]$$

$$= \frac{1}{9} \left[5(0.095223) + 8(0.076923) + 5(0.059886) \right] = 0.154548.$$

Exact solution:
$$I = \left[\{(\pi/4) + \tan^{-1}(1/3)\}/3 \right] = 0.154549.$$

The magnitudes of errors in the computed values are 0.000703, 0.00009, and 0.000001 respectively.

2. Gauss-Chebyshev integration rules

The weight function is given by $w(x) = 1/\sqrt{1-x^2}$. The integration rule is given by

$$I = \int_{-1}^{1} \frac{f(x)}{\sqrt{1-x^2}} \, dx = \lambda_0 f(x_0) + \lambda_1 f(x_1) + \ldots + \lambda_n f(x_n).$$

Gauss-Chebyshev one point rule

We use the method of undetermined parameters. The one point rule is given by

$$I = \int_{-1}^{1} \frac{f(x)}{\sqrt{1-x^2}} \, dx = \lambda_0 f(x_0), \quad \lambda_0 \neq 0.$$

The method has two parameters. λ_0, x_0. Making the formula exact for $f(x) = 1, x$, we get

$$f(x) = 1: \qquad \int_{-1}^{1} \frac{dx}{\sqrt{1-x^2}} = \left[\sin^{-1} x \right]_{-1}^{1} = \pi = \lambda_0$$

$$f(x) = x: \qquad \int_{-1}^{1} \frac{x}{\sqrt{1-x^2}} \, dx = 0 = \lambda_0 x_0.$$

We get $x_0 = 0$. Therefore, the Gauss-Chebyshev one point rule is given by

$$\int_{-1}^{1} \frac{f(x)}{\sqrt{1-x^2}} \, dx = \pi f(0). \tag{18.172}$$

The error term is given by, error $= (\pi/4) f''(\xi), -1 < \xi < 1$. Since the error term contains $f''(\xi)$ Gauss-Chebyshev one point rule integrates exactly polynomials of degree less than or equal to 1. The corresponding Chebyshev polynomial is $T_1(x) = x$, whose zero is $x = 0$.

Gauss-Chebyshev two point rule

The two point rule is given by

$$\int_{-1}^{1} \frac{f(x)}{\sqrt{1-x^2}} \, dx = \lambda_0 f(x_0) + \lambda_1 f(x_1), \quad \lambda_0 \neq 0, \lambda_1 \neq 0, \text{ and } x_0 \neq x_1.$$

The method has four parameters $\lambda_0, \lambda_1, x_0, x_1$. Making the formula exact for $f(x) = 1, x, x^2, x^3$, we get

$$f(x) = 1: \int_{-1}^{1} \frac{dx}{\sqrt{1-x^2}} = \pi = \lambda_0 + \lambda_1, \qquad f(x) = x: \int_{-1}^{1} \frac{x\,dx}{\sqrt{1-x^2}} = 0 = \lambda_0 x_0 + \lambda_1 x_1,$$

$$f(x) = x^2: \int_{-1}^{1} \frac{x^2\,dx}{\sqrt{1-x^2}} = \frac{\pi}{2} = \lambda_0 x_0^2 + \lambda_1 x_1^2, \quad f(x) = x^3: \int_{-1}^{1} \frac{x^3}{\sqrt{1-x^2}}\,dx = 0 = \lambda_0 x_0^3 + \lambda_1 x_1^3.$$

The solution of these equations is $\lambda_0 = \lambda_1 = \pi/2$, and $x_0 = \pm (1/\sqrt{2}) = -x_1$.

Therefore, the Gauss-Chebyshev two point rule is given by

$$\int_{-1}^{1} \frac{f(x)}{\sqrt{1-x^2}}\,dx = \frac{\pi}{2}\left[f\left(-\frac{1}{\sqrt{2}}\right) + f\left(\frac{1}{\sqrt{2}}\right)\right]. \tag{18.173}$$

The error term is given by, error $= (\pi/192) f^{(4)}(\xi)$, $-1 < \xi < 1$. Since the error term contains $f^{(4)}(\xi)$, Gauss-Chebyshev two point rule integrates exactly polynomials of degree less than or equal to 3. The corresponding Chebyshev polynomial is $T_2(x) = 2x^2 - 1$, whose zeros are $x = \pm 1/\sqrt{2}$.

Gauss-Chebyshev three point rule

Following the above procedure, the three point rule is given by

$$\int_{-1}^{1} \frac{f(x)}{\sqrt{1-x^2}}\,dx = \frac{\pi}{3}\left[f\left(-\frac{\sqrt{3}}{2}\right) + f(0) + f\left(\frac{\sqrt{3}}{2}\right)\right]. \tag{18.174}$$

The error term is given by, error $= (\pi/23040) f^{(6)}(\xi)$, $-1 < \xi < 1$. Since the error term contains $f^{(6)}(\xi)$, Gauss-Chebyshev three point rule integrates exactly polynomials of degree less than or equal to 5. The corresponding Chebyshev polynomial is $T_3(x) = x(4x^2 - 3)$, whose zeros are $x = 0$, $\pm \sqrt{3}/2$. Note that the error constant is small.

We give below the one-point, two-point and three point Gauss-Laguerre and Gauss-Hermite integration rules.

3. Gauss-Laguerre integration rules

Gauss-Laguerre one point rule:

$$\int_{0}^{\infty} e^{-x} f(x)\,dx = f(1). \tag{18.175}$$

Error $= (1/2) f''(\xi), \quad 0 < \xi < \infty.$ Order $= 1$.

The corresponding Laguerre polynomial is $L_1(x) = x - 1$, whose zero is $x = 1$.

Gauss-Laguerre two point rule:

$$\int_{0}^{\infty} e^{-x} f(x)\,dx = \frac{1}{4}\left[(2 + \sqrt{2}) f(2 - \sqrt{2}) + (2 - \sqrt{2}) f(2 + \sqrt{2})\right]. \tag{18.176}$$

$$\text{Error} = (1/6)\, f^{(4)}(\xi), \quad 0 < \xi < \infty. \quad \text{Order} = 3.$$

The corresponding Laguerre polynomial is $L_2(x) = x^2 - 4x + 2$, whose zeros are $x = 2 \pm \sqrt{2}$.

Gauss-Laguerre three point rule:

$$\int_0^\infty e^{-x} f(x)\, dx = 0.711093\, f(0.415775) + 0.278518\, f(2.294280)$$

$$+ 0.010389\, f(6.289945) \tag{18.177}$$

Order = 5. The corresponding Laguerre polynomial is $L_3(x) = x^3 - 9x^2 + 18x - 6$.

4. Gauss-Hermite integration rules

Gauss-Hermiteone ponit rule:

$$\int_{-\infty}^\infty e^{-x^2} f(x)\, dx = \sqrt{\pi}\, f(0). \tag{18.178}$$

$$\text{Error} = (\sqrt{\pi}/4)\, f''(\xi), \quad -\infty < \xi < \infty. \quad \text{Order} = 1.$$

The corresponding Hermite polynomial is $H_1(x) = 2x$, whose zero is $x = 0$.

Gauss-Hermite two point rule:

$$\int_{-\infty}^\infty e^{-x^2} f(x)\, dx = \frac{\sqrt{\pi}}{2}\left[f\left(-\frac{1}{\sqrt{2}}\right) + f\left(\frac{1}{\sqrt{2}}\right) \right] \tag{18.179}$$

$$\text{Error} = (\sqrt{\pi}/48)\, f^{(4)}(\xi), \quad -\infty < \xi < \infty. \quad \text{Order} = 3.$$

The corresponding Hermite polynomial is $H_2(x) = 2(2x^2 - 1)$, whose zeros are $x = \pm 1/\sqrt{2}$.

Gauss-Hermite three point rule:

$$\int_{-\infty}^\infty e^{-x^2} f(x)\, dx = \frac{\sqrt{\pi}}{6}\left[f\left(-\sqrt{\frac{3}{2}}\right) + 4f(0) + f\left(\sqrt{\frac{3}{2}}\right) \right] \tag{18.180}$$

$$\text{Error} = (\sqrt{\pi}/960)\, f^{(6)}(\xi), \quad -\infty < \xi < \infty. \quad \text{Order} = 5.$$

The corresponding Hermite polynomial is $H_3(x) = 4(2x^3 - 3x)$, whose zeros are $x = 0$, $x = \pm \sqrt{3/2}$.

The abscissas and weights of the Gauss formulas, accurate to about 15 decimal places, are available in Tables of functions.

Example 18.45 Evaluate the integral using $\int_0^1 \dfrac{\cos x \, dx}{\sqrt{1-x^2}}$ the Gauss-Chebyshev two point and three point rules.

Solution We have $I = \dfrac{1}{2} \int_{-1}^1 \dfrac{\cos x \, dx}{\sqrt{1-x^2}}$, and $f(x) = (\cos x)/2.$

Two point formula: $\qquad I = \dfrac{\pi}{4} \, (2) \cos (1/\sqrt{2}) = 1.19419.$

Three point formula: $\qquad I = \dfrac{\pi}{6} \, [2 \cos (\sqrt{3}/2) + 1] = 1.20203.$

Exercise 18.6

1. The following table of values for $\ln(1 + x)$ is given.

x	1.0	1.2	1.4
$f(x)$	0.6931	0.7885	0.8755

   Find an approximate value of $f'(1.0)$, using the following formulas.

   (a) $f'(x_k) = (f_{k+1} - f_k)/h$, with $h = 0.4$ and 0.2.

   (b) $f'(x_k) = (-3f_k + 4f_{k+1} - f_{k+2})/(2h)$, with $h = 0.2$.

   Compare with the exact value 0.5.

2. Using the data

x	1	2	2.5	3	3.5	4	5
$y(x)$	1	32	97.656	243	525.219	1024	3125

   and the methods

   (a) $y'(x_0) = [y(x_0 + h) - y(x_0 - h)]/(2h)$,

   (b) $y''(x_0) = [y(x_0 + h) - 2y(x_0) + y(x_0 - h)]/h^2$,

   determine approximate values of $y'(3)$ and $y''(3)$ with $h = 2$, $h = 1$ and $h = 1/2$.

3. Find the error term and the order of the following methods.

   (i) $y'(x_0) = [y(x_0 + h) - y(x_0 - h)]/(2h)$,

   (ii) $y''(x_0) = [y(x_0 - h) - 2y(x_0) + y(x_0 + h)]/h^2$.

4. Prove the operator relation $hD = 2 \sin h^{-1} (\delta/2)$, where D is the derivative operator and δ is the central difference operator.

5. Find the values of $\alpha_0, \alpha_1, \alpha_2$ in the differentiation method

   $$f'(x_0) = \alpha_0 \, f(x_0) + \alpha_1 f(x_0 - h) + \alpha_2 f(x_0 - 2h)$$

   so that the method is of highest possible order. Find the error term and the order of the method.

6. Evaluate $\int_1^2 \cos x\,dx$ using the trapezoidal rule with (i) $h = 1$, (ii) $h = 1/2$. Compare with the exact solution. Find the maximum error in each case.

7. Find the minimum number of intervals required to evaluate $\int_0^1 \ln(1 + x)\,dx$, using trapezoidal rule with an accuracy of 10^{-3}.

8. Using the following ·data

x	1	2	3	4
$f(x)$	0.3679	0.1353	0.0498	0.0183

and the trapezoidal rule with $n = 1$ and $n = 3$, determine an approximate value of $\int_1^4 f(x)dx$.

9. Evaluate the integral $\int_0^{\pi/2} e^{-x} \cos x\,dx$, using trapezoidal rule with $h = \pi/2$ and $h = \pi/4$.

10. Obtain the approximate value of $\int_{-1}^1 \sqrt{1 - x^2} \cos x\,dx$, using the trapezoidal rule with $h = 1$, $1/2$ and $1/4$.

11. Evaluate $\int_1^2 \dfrac{x^2}{1+x^3}\,dx$ using the Simpson's 1/3rd rule with two and four sub-intervals. Compare with the exact solution.

12. Evaluate $\int_1^2 \sqrt{1+4x^2}\,\sin x\,dx$ using the Simpson's 1/3rd rule with $h = 1/2$ and $1/4$.

13. Find the minimum number of intervals required to evaluate $\int_0^1 \ln(1 + x)dx$ using the Simpson's 1/3rd rule with an accuracy of 10^{-6}.

14. Determine the maximum error in evaluating $\int_0^{\pi/2} e^{-x} \cos x\,dx$, using the Simpson's 1/3rd rule with $h = \pi/8$.

15. Using the Simpson's 3/8th rule, evaluate the following integrals with 4 and 7 nodal points. Compare with the exact solution.

   (i) $\int_1^2 \dfrac{dx}{5 + 3x}$. (ii) $\int_0^1 \dfrac{dx}{1 + x^2}$. (iii) $\int_0^2 \dfrac{(x + 1)\,dx}{x^2 + 2x + 2}$.

16. Using the Gauss-Legendre one point, two point and three point rules, evalnate the following integrals. Compare with the exact solution wherever possible.

   (i) $\int_0^2 \dfrac{(x + 1)^2}{1 + (x + 1)^4}\,dx$. (ii) $\int_{-1}^1 \dfrac{dx}{25 + x^2}$.

   (iii) $\int_0^2 \dfrac{dx}{5 + 4x}$. (iv) $\int_0^2 \dfrac{dx}{x^2 + 2x + 10}$.

17. Using the Gauss-Chebyshev two point and three point rules, evalnate the following integrals. Compare with the exact solution wherever possible.

    (i) $\int_0^1 \sqrt{1-x^2} \cos x \, dx.$ (ii) $\int_0^1 (1-x^2)^{3/2} \, dx.$

18. Using the Gauss-Laguerre two and three point rules, evalnate the following integrals. Compare with the exact solution wherever possible.

    (i) $\int_0^\infty e^{-3x} \, dx.$ (ii) $\int_0^\infty \dfrac{e^{-x}}{3+5x} \, dx.$

19. Using the Gauss-Hermite two and three point rules, evalnate the following integrals. Compare with the exact solution wherever possible.

    (i) $\int_{-\infty}^\infty e^{-x^2} \, dx.$ (ii) $\int_{-\infty}^\infty \dfrac{e^{-x^2}}{x^2+2x+10} \, dx.$

18.7 Numerical Solution of Ordinary Differential Equations

There are many ordinary differential equations (variable coefficients and nonlinear), which cannot be solved exactly using analytical techniques. In this section, we consider the numerical solution of initial value problems and two point boundary value problems.

18.7.1 Initial Value Problems

A higher order ordinary differential equation (along with the initial conditions) can be reduced to a system of first order ordinary differential equations, which in vector notation can be written as $\mathbf{y}' = \mathbf{f}(x, \mathbf{y})$, $\mathbf{y}(x_0) = \mathbf{y}_0$. Therefore, it is sufficient to construct numerical methods for the solution of the initial value problem (IVP)

$$y' = f(x, y), \quad y(x_0) = y_0, \tag{18.181}$$

and write it in the vector form to solve the original problem.

Starting from the initial point x_0, we need the solution at different points in some interval $[x_0, b]$. We partition the interval $[x_0, b]$ into a finite number of sub-intervals using the points $x_0, x_1, \ldots, x_N$ such that $x_0 < x_1 < x_2 < \ldots < x_N = b$. These points are called the *grid points* or the *mesh points*. We assume that the grid points are equispaced with spacing h, which is called the *step size* or the *step length*. Hence, the grid points are defined by

$$x_i = x_0 + ih, \quad i = 0, 1, 2, \ldots, N.$$

Order of the method   We denote the numerical solution at $x = x_n$ by y_n and the exact solution by $y(x_n)$. If y_{n+1} is the numerical solution at the nodal point x_{n+1}, obtained by using a certain numerical method, then we define the error or *truncation error* (T. E) at $x = x_{n+1}$ as

$$T.E = y(x_{n+1}) - y_{n+1}.$$

If this error term can be written in the form

$$T.~E = ch^{p+1} + O(h^{p+2})$$

or as
$$\frac{1}{h}(T.E) = ch^p + O(h^{p+1})$$
(18.182)

where c is a constant independent of h, then the method is said to be of *order p*.

Single step method   If we use only the values y_n, $f(x_n, y_n)$ to obtain the solution value y_{n+1} at the next nodal point $x = x_{n+1}$, then the method is called an *explicit single step method*. The method is of the form

$$y_{n+1} = y_n + \phi(x_n, y_n, h).$$
(18.183)

If the current values y_{n+1} and $f(x_{n+1}, y_{n+1})$ are also used, then the method is called an *implicit single step method*. The method is of the form

$$y_{n+1} = y_n + \phi(x_n, x_{n+1}, y_n, y_{n+1}, h).$$
(18.184)

Multi step method   If we use k previously calculated values, y_n, $f(x_n, y_n)$; y_{n-1}, $f(x_{n-1}, y_{n-1})$; ...; y_{n-k+1}, $f(x_{n-k+1}, y_{n-k+1})$ to obtain the solution value y_{n+1} at the next nodal point $x = x_{n+1}$, then the method is called an *explicit multi step method* or a *k*-step multi step method. The method is of the form

$$y_{n+1} = y_n + \phi(x_{n-k+1}, ..., x_n, \ y_{n-k+1}, ..., y_n, h).$$
(18.185)

If the current values y_{n+1} and $f(x_{n+1}, y_{n+1})$ are also used, then the method is called an *implicit multi step method*. The method is of the form

$$y_{n+1} = y_n + \phi(x_{n-k+1}, ..., x_{n+1}, \ y_{n-k+1}, ..., y_{n+1}, h).$$
(18.186)

We now derive a few single step and multi step methods to solve the initial value problem (18.181). We assume that $y(x)$ has continuous derivatives of the required order for all $x \in [x_0, b]$.

18.7.1.1   Single step methods

Taylor series method

We write the Taylor series expansion of $y(x)$ about the point x_n as

$$y(x) = y(x_n) + (x - x_n)\, y'(x_n) + ... + \frac{(x - x_n)^p}{p!}\, y^{(p)}(x_n) + ...$$

Substituting $x = x_{n+1}$ and noting that $x_{n+1} - x_n = h$, we get

$$y(x_{n+1}) = y(x_n) + hy'(x_n) + ... + \frac{h^p}{p!}\, y^{(p)}(x_n) + ...$$
(18.187)

If we retain terms upto pth power of h in (18.187), we get an approximation y_{n+1} to $y(x_{n+1})$ as

$$y_{n+1} = y_n + hy'_n + ... + \frac{h^p}{p!} y^{(p)}_n.$$
(18.188)

This method is called the *Taylor series method*. The error of approximation is given by

$$\text{Error} = y\,(x_{n+1}) - y_{n+1}$$

$$= \frac{h^{p+1}}{(p+1)!}\, y^{(p+1)}(x_n) + \frac{h^{p+2}}{(p+2)!}\, y^{(p+2)}(x_n) + \dots$$

which is also an infinite series.

The principal part of the error term, called the truncation error (*T.E*) is defined as

$$T.E = \frac{h^{p+1}}{(p+1)!}\, y^{(p+1)}(\xi),\quad x_n < \xi < x_{n+1}$$

or

$$\frac{1}{h}\,(T.E) = \frac{h^p}{(p+1)!}\, y^{(p+1)}(\xi). \tag{18.189}$$

Therefore, the Taylor series method (18.188) is of order p. The higher order derivatives in (18.188) are obtained by differentiating the differential equation (18.181) repeatedly.

Euler method

For $p = 1$, we obtain from (18.188)

$$y_{n+1} = y_n + hy_n', \quad \text{or} \quad y_{n+1} = y_n + hf(x_n, y_n), \quad n = 0, 1, \dots \tag{18.190}$$

which is called the *Euler* method. It is an explicit single step method. The truncation error of the Euler method is given by

$$T.E = \frac{h^2}{2}\, y''(\xi),\quad x_n < \xi < x_{n+1} \tag{18.191}$$

and the bound of *T.E* is given by

$$|T.E| \le \frac{h^2}{2} M_2, \quad \text{where} \quad M_2 = \max_{[x_n, x_{n+1}]} |y''(x)|. \tag{18.192}$$

Therefore, Euler method is of order 1.

Example 18.46   Obtain the approximate value of $y(1.3)$ for the initial value problem

$$y' = -2xy^2, \quad y(1) = 1$$

using (i) Euler method, (ii) Taylor series second order method with step size $h = 0.1$. Compare with the exact solution $y = 1/x^2$.

Solution   We have $f(x, y) = -2xy^2$, and $h = 0.1$.

(i) Euler method gives

$$y_{n+1} = y_n + hf(x_n, y_n) = y_n - 2hx_n y_n^2.$$

We have the following results.

$n = 0$: $x_0 = 1, y_0 = 1.$

$$y(1.1) \approx y_1 = y_0 - 2hx_0\, y_0^2 = 1 - 2(0.1)(1)(1) = 0.8.$$

$n = 1$: $x_1 = x_0 + h = 1.1, y_1 = 0.8.$

$$y(1.2) \approx y_2 = y_1 - 2hx_1\, y_1^2$$
$$= 0.8 - 2(0.1)\,(1.1)\,(0.8)^2 = 0.6592.$$

$n = 2$: $x_2 = x_0 + 2h = 1.2, \quad y_2 = 0.6592.$

$$y(1.3) \approx y_3 = y_2 - 2hx_2\, y_2^2$$
$$= 0.6592 - 2(0.1)\,(1.2)\,(0.6592)^2 = 0.5449.$$

The exact solution is $y(1.3) = 0.5917$.

(ii) Taylor series second order method is given by

$$y_{n+1} = y_n + hy_n' + \frac{h^2}{2}\, y_n''.$$

From the given differential equation, we get

$$y' = -2xy^2, \quad y'' = -2y^2 - 4xyy'.$$

We have the following results.

$n = 0$: $x_0 = 1, y_0 = 1, y_0' = -2, y_0'' = 6, x_1 = x_0 + h = 1.1.$

$$y(1.1) \approx y_1 = y_0 + hy_0' + \frac{h^2}{2}\, y_0''$$

$$= 1 + (0.1)(-2) + \frac{(0.1)^2}{2}(6) = 0.83.$$

$n = 1$: $x_1 = 1.1, y_1 = 0.83, y_1' = -1.5156, y_1'' = 4.1572, x_2 = x_0 + 2h = 1.2$

$$y(1.2) \approx y_2 = y_1 + hy_1' + \frac{h^2}{2}y_1''$$

$$= 0.83 + (0.1)\,(-1.5156) + \frac{(0.1)^2}{2}(4.1572) = 0.6992.$$

$n = 2$: $x_2 = 1.2, y_2 = 0.6992, y_2' = -1.1733, y_2'' = 2.9600, x_3 = x_0 + 3h = 1.3.$

$$y(1.3) \approx y_3 = y_2 + hy_2' + \frac{h^2}{2}\, y_2''$$

$$= 0.6992 + (0.1)\,(-1.1733) + \frac{(0.1)^2}{2}(2.9600) = 0.5967.$$

Exact solution is $y(1.3) = 0.5917$.

18.7.1.2   Runge-Kutta methods

Taylor series method of arbitrary order for the solution of the initial value problem (18.181) can be written. However, from application point of view, Taylor series method has a major disadvantage. The method requires evaluation of derivatives of higher order for the function $f(x, y)$ of two variables x and y. These higher order derivatives have to be obtained manually for each problem. Therefore, we need to develop methods which do not require the evaluation of higher order derivatives. One such class of methods is the Runge-Kutta methods. To illustrate the idea behind the Runge-Kutta methods, we integrate $y' = f(x, y)$ in the interval $[x_n, x_{n+1}]$ to obtain

$$\int_{x_n}^{x_{n+1}} \left(\frac{dy}{dx}\right) dx = \int_{x_n}^{x_{n+1}} f(x, y)dx.$$

Integrating, we obtain

$$y(x_{n+1}) = y(x_n) + \int_{x_n}^{x_{n+1}} f(x, y)dx. \tag{18.193}$$

We note that $f(x, y)$ is the slope of the solution curve and it changes continuously in (x_n, x_{n+1}). Let the continuously varying slope $f(x, y)$ in this interval be approximated by the fixed slope at $x = x_n$, that is $f(x, y) \approx f(x_n, y_n)$ in $[x_n, x_{n+1}]$. Then, from (18.193), we obtain

$$y_{n+1} = y_n + f(x_n, y_n) \int_{x_n}^{x_{n+1}} dx = y_n + hf(x_n, y_n) \tag{18.194}$$

which is the *Euler method* (18.190).

Now, approximate $f(x, y)$ in $[x_n, x_{n+1}]$ by the slope at the point $x = x_{n+1}$, that is, $f(x, y) \approx f(x_{n+1}, y_{n+1})$. Then, we obtain

$$y_{n+1} = y_n + f(x_{n+1}, y_{n+1}) \int_{x_n}^{x_{n+1}} dx = y_n + hf(x_{n+1}, y_{n+1}) \tag{18.195}$$

which is an implicit method. This method is called the *backward Euler method*. The nonlinear equation for y_{n+1} can be solved using the Newton-Raphson method.

Now, approximate $f(x, y)$ in $[x_n, x_{n+1}]$ by the mean of the slopes at $x = x_n$ and $x = x_{n+1}$, that is, $f(x, y) \approx [f(x_n, y_n) + f(x_{n+1}, y_{n+1}]/2$. Then, we obtain the implicit method

$$y_{n+1} = y_n + \frac{h}{2}[f(x_n, y_n) + f(x_{n+1}, y_{n+1})] \tag{18.196}$$

Note that we can interpret the second term on the right hand side as evaluation of the intergral in (18.193) by the trapezoidal rule.

If we approximate y_{n+1} on the right hand side of (18.196) by the Euler method (18.190), then we obtain the explicit method

$$y''' = \left(\frac{\partial^2 f}{\partial x^2} + \frac{\partial^2 f}{\partial x \partial y}\frac{dy}{dx}\right) + f\left(\frac{\partial^2 f}{\partial x \partial y} + \frac{\partial^2 f}{\partial y^2}\frac{dy}{dx}\right) + \frac{\partial f}{\partial y}\left(\frac{\partial f}{\partial x} + \frac{\partial f}{\partial y}\frac{dy}{dx}\right)$$

$$= \left(\frac{\partial^2 f}{\partial x^2} + 2f\frac{\partial^2 f}{\partial x \partial y} + f^2\frac{\partial^2 f}{\partial y^2}\right) + \frac{\partial f}{\partial y}\left(\frac{\partial f}{\partial x} + f\frac{\partial f}{\partial y}\right).$$

The exact solution at $x = x_{n+1}$, gives

$$y(x_{n+1}) = y(x_n) + hy'(x_n) + \frac{h^2}{2!}y''(x_n) + \frac{h^3}{3!}y'''(x_n) + \ldots$$

$$= y_n + hf_n + \frac{h^2}{2}\left[\frac{\partial f}{\partial x} + f\frac{\partial f}{\partial y}\right]_n + \frac{h^3}{6}\left[\left(\frac{\partial^2 f}{\partial x^2} + 2f\frac{\partial^2 f}{\partial x \partial y} + f^2\frac{\partial^2 f}{\partial y^2}\right) + \frac{\partial f}{\partial y}\left(\frac{\partial f}{\partial x} + f\frac{\partial f}{\partial y}\right)\right]_n + \ldots \quad (18.200)$$

We also have from (18.199)

$$y_{n+1} = y_n + w_1 k_1 + w_2 k_2$$

$$= y_n + h(w_1 + w_2)f_n + h^2 w_2\left(\alpha\frac{\partial f}{\partial x} + \beta f_n\frac{\partial f}{\partial y}\right)$$

$$+ \frac{h^3 w_2}{2}\left[\alpha^2\frac{\partial^2 f}{\partial x^2} + 2\alpha\beta f_n\frac{\partial^2 f}{\partial x \partial y} + \beta^2 f_n^2\frac{\partial^2 f}{\partial y^2}\right] + \ldots \quad (18.201)$$

Comparing the coefficients of h and h^2 in (18.200) and (18.201), we get

$$w_1 + w_2 = 1, \quad w_2\,\alpha = 1/2, \quad w_2\,\beta = 1/2$$

whose solution is $\beta = \alpha$, $w_2 = \dfrac{1}{2\alpha}$, $w_1 = 1 - \dfrac{1}{2\alpha}$, $\alpha \neq 0$ is arbitrary. It is not possible to compare the coefficient of h^3, as there are five terms in (18.200) and three terms in (18.201). The two stage Runge-Kutta method is given by

$$y_{n+1} = y_n + \left(1 - \frac{1}{2\alpha}\right)k_1 + \frac{1}{2\alpha}k_2$$

$$k_1 = hf(x_n, y_n), \quad k_2 = hf(x_n + \alpha h, y_n + \alpha k_1). \quad (18.202)$$

The truncation error is given by

$$T.E = y(x_{n+1}) - y_{n+1}$$

$$= h^3\left[\left(\frac{1}{6} - \frac{\alpha}{4}\right)\left(\frac{\partial^2 f}{\partial x^2} + 2f\frac{\partial^2 f}{\partial x \partial y} + f^2\frac{\partial^2 f}{\partial y^2}\right) + \frac{1}{6}\frac{\partial f}{\partial y}\left(\frac{\partial f}{\partial x} + f\frac{\partial f}{\partial y}\right)\right]_n + O(h^4) \quad (18.203)$$

where $\alpha = \beta$, $w_2 \alpha = 1/2$, $w_2 \beta = 1/2$ are used in simplification. Therefore, the two-stage Runge-Kutta method (18.202) is of second order for all α. For different values of α, we obtain different methods. For $\alpha = 1/2$, we get $w_1 = 0$, $w_2 = 1$ and the method becomes

$$y_{n+1} = y_n + k_2, \ n = 0, 1, 2, \ldots$$

where

$$k_1 = hf(x_n, y_n), \quad k_2 = hf\left(x_n + \frac{h}{2}, y_n + \frac{1}{2}k_1\right). \tag{18.204}$$

which is called the *modified Euler-Cauchy method.*

For $\alpha = 1$, we get $w_1 = w_2 = 1/2$ and the method becomes

$$y_{n+1} = y_n + \frac{1}{2}(k_1 + k_2), \ n = 0, 1, 2, \ldots$$

where

$$k_1 = hf(x_n, y_n), \quad k_2 = hf(x_n + h, y_n + k_1), \tag{18.205}$$

which is *Euler-Cauchy method or Heun's method* (18.197)

Four stage explicit Runge-Kutta method

Without going into detailed derivation, we list the simplest four stage Runge-Kutta method as

$$y_{n+1} = y_n + \frac{1}{6}(k_1 + 2k_2 + 2k_3 + k_4), \ n = 0, 1, 2, \ldots \tag{18.206}$$

where

$$k_1 = hf(x_n, y_n), \quad k_2 = hf\left(x_n + \frac{h}{2}, y_n + \frac{1}{2}k_1\right)$$

$$k_3 = hf\left(x_n + \frac{h}{2}, y_n + \frac{1}{2}k_2\right), \quad k_4 = hf(x_n + h, y_n + k_3),$$

which is called the *classical Runge-Kutta method* and is of fourth order.

We note that we require two function evaluations per step for a second order explicit Runge-Kutta method and four function evaluations per step for a fourth order explicit Runge-Kutta method.

Example 18.47 Find the approximate value of $y(1.4)$ for the initial value problem

$$y' = x^2 + y^2, \ y(1) = 2$$

with $h = 0.2$, using the (i) Heun's method, and (ii) modified Euler-Cauchy method.

Solution We have $f(x, y) = x^2 + y^2$, $h = 0.2$.

(i) Heun's method is given by

$$y_{n+1} = y_n + \frac{1}{2}(k_1 + k_2)$$

$$k_1 = hf(x_n, y_n), \quad k_2 = hf(x_n + h, y_n + k_1).$$

We have the following results.

$n = 0$: $x_0 = 1,\quad y_0 = 2.$

$k_1 = hf(x_0, y_0) = 0.2\, f(1, 2) = 1,$

$k_2 = hf(x_0 + h,\, y_0 + k_1) = 0.2\, f(1.2, 3) = 2.088,$

$y(1.2) \approx y_1 = y_0 + \dfrac{1}{2}(k_1 + k_2) = 2 + \dfrac{1}{2}(1 + 2.088) = 3.544.$

$n = 1$: $x_1 = 1.2,\quad y_1 = 3.544.$

$k_1 = hf(x_1, y_1) = 0.2\, f(1.2, 3.544) = 2.8000,$

$k_2 = hf(x_1 + h,\, y_1 + k_1) = 0.2\, f(1.4, 6.344) = 8.4413,$

$y(1.4) \approx y_2 + \dfrac{1}{2}(k_1 + k_2) = 3.544 + \dfrac{1}{2}(2.8000 + 8.4413) = 9.1647.$

(ii) Modified Euler-Cauchy method is given by

$$y_{n+1} = y_n + k_2$$

$$k_1 = hf(x_n, y_n),\quad k_2 = hf\left(x_n + \frac{h}{2},\, y_n + \frac{1}{2}k_1\right).$$

We have the following results.

$n = 0$: $x_0 = 1,\quad y_0 = 2,\quad h = 0.2.$

$k_1 = hf(x_0, y_0) = 0.2\, f(1, 2) = 1,$

$k_2 = hf\left(x_0 + \dfrac{h}{2},\, y_0 + \dfrac{1}{2}k_1\right) = 0.2\, f(1.1, 2.5) = 1.492,$

$y(1.2) \approx y_1 = y_0 + k_2 = 3.492.$

$n = 1$: $x_1 = 1.2,\quad y_1 = 3.492$

$k_1 = hf(x_1, y_1) = 0.2\, f(1.2, 3.492) = 2.7268,$

$k_2 = hf\left(x_1 + \dfrac{h}{2},\, y_1 + \dfrac{1}{2}k_1\right) = 0.2\, f(1.3, 4.8554) = 5.0530,$

$y(1.4) \approx y_2 = y_1 + k_2 = 3.492 + 5.0530 = 8.545.$

Example 18.48   Solve the initial value problem

$$y' = x(y - x),\ y(2) = 3$$

in the interval [2, 2.4] using the classical Runge-Kutta fourth order method with the step size $h = 0.2$.

Solution We have $f(x, y) = x (y - x)$, $h = 0.2$.

$n = 0$: $\qquad\qquad\qquad\qquad x_0 = 2, \quad y_0 = 3.$

$$k_1 = h f(x_0, y_0) = 0.2 \, f(2, 3) = 0.4,$$

$$k_2 = hf\left(x_0 + \frac{h}{2}, y_0 + \frac{1}{2} k_1\right) = 0.2 \, f(2.1, 3.2) = 0.4620,$$

$$k_3 = hf\left(x_0 + \frac{h}{2}, y_0 + \frac{1}{2} k_2\right) = 0.2 \, f(2.1, 3.231) = 0.4750,$$

$$k_4 = h f(x_0 + h, y_0 + k_3) = 0.2 \, f(2.2, 3.4750) = 0.5610,$$

$$y(2.2) \approx y_1 = y_0 + \frac{1}{6}(k_1 + 2k_2 + 2k_3 + k_4)$$

$$= 3 + \frac{1}{6}[0.4 + 2(0.4620 + 0.4750) + 0.5610] = 3.4725.$$

$n = 1$: $\qquad\qquad\qquad\qquad x_1 = 2.2, \quad y_1 = 3.4725.$

$$k_1 = h f(x_1, y_1) = 0.2 \, f(2.2, 3.4725) = 0.5599,$$

$$k_2 = hf\left(x_1 + \frac{h}{2}, y_1 + \frac{1}{2} k_1\right) = 0.2 \, f(2.3, 3.7525) = 0.6682,$$

$$k_3 = hf\left(x_1 + \frac{h}{2}, y_1 + \frac{1}{2} k_2\right) = 0.2 \, f(2.3, 3.8066) = 0.6930,$$

$$k_4 = h f(x_1 + h, y_1 + k_3) = 0.2 \, f(2.4, 4.1655) = 0.8474,$$

$$y(2.4) \approx y_2 = y_1 + \frac{1}{6}(k_1 + 2k_2 + 2k_3 + k_4)$$

$$= 3.4725 + \frac{1}{6}[0.5599 + 2(0.6682 + 0.6930) + 0.8474] = 4.1608.$$

18.7.1.3 Multi step methods

For the solution of the initial value problem (18.181), we have earlier defined an explicit k-step multi step method as (see (18.185))

$$y_{n+1} = y_n + \phi(x_{n-k+1}, \ldots, x_n, y_{n-k+1}, \ldots, y_n, h)$$

and an implicit k-step multi step method as (see (18.186))

$$y_{n+1} = y_n + \phi(x_{n-k+1}, \ldots, x_{n+1}, y_{n-k+1}, \ldots, y_{n+1}, h).$$

Both the methods are not self-starting since they require k previous values $y_n, y_{n-1}, \ldots, y_{n-k+1}$, while the Runge-Kutta methods are self-starting. The k values that are required for starting the application of the multi step method are obtained using some single step method like Euler, Taylor series or Runge-Kutta methods. Hence, we assume that these k values are known.

For constructing the multistep methods, we start with the equation (18.193).

$$y(x_{n+1}) = y(x_n) + \int_{x_n}^{x_{n+1}} f(x, y)dx. \tag{18.207}$$

To drive the methods, we replace $f(x, y)$ in the integral by some suitable interpolating polynomial.

Adams-Bashforth methods

These are explicit methods. Through the k data values (x_n, f_n), (x_{n-1}, f_{n-1}), ..., (x_{n-k+1}, f_{n-k+1}), we fit the Newton's backward difference interpolating polynomial of degree $k - 1$ as

$$P_{k-1}(x) = f_n + \frac{(x - x_n)}{h} \nabla f_n + \frac{(x - x_n)(x - x_{n-1})}{2!h^2} \nabla^2 f_n + \ldots + \frac{(x - x_n)\ldots(x - x_{n-k+2})}{(k - 1)!h^{k-1}} \nabla^{k-1} f_n. \tag{18.208}$$

The error of interpolation is given by

$$T. E = \frac{(x - x_n)\ldots(x - x_{n-k+1})}{k!} f^{(k)}(\xi) \tag{18.209}$$

where ξ lies in some interval containing the points x_n, x_{n-1}, ..., x_{n-k+1} and x. Replacing $f(x, y)$ by $P_{k-1}(x)$ in (18.207), we get

$$y_{n+1} = y_n + \int_{x_n}^{x_{n+1}} \left[f_n + \frac{(x - x_n)}{h} \nabla f_n + \frac{(x - x_n)(x - x_{n-1})}{2!h^2} \nabla^2 f_n + \ldots \right]dx.$$

Set $x - x_n = hs$. Then, $(x - x_i) = x - [x_n - (n - i)h] = (s + n - i) h$, $dx = hds$. Hence,

$$y_{n+1} = y_n + h \int_0^1 [f_n + s\nabla f_n + \frac{1}{2} s(s + 1) \nabla^2 f_n + \ldots]ds$$

$$= y_n + h\left[f_n + \frac{1}{2} \nabla f_n + \frac{5}{12} \nabla^2 f_n + \ldots \right]. \tag{18.210}$$

The error term can be obtained by integrating (18.209) or can be obtained directly. By choosing different values for k, we get different methods.

$k = 1$: We get a single step method. Retaining one term inside the bracket in (18.210), we get the method

$$y_{n+1} = y_n + h f_n \tag{18.211}$$

which is the *Euler method*. The method is of order 1.

$k = 2$: Retaining two terms inside the bracket in (18.210), we get the method

$$y_{n+1} = y_n + h\left[f_n + \frac{1}{2} \nabla f_n \right] = y_n + h\left[f_n + \frac{1}{2} (f_n - f_{n-1}) \right]$$

$$= y_n + \frac{h}{2}[3f_n - f_{n-1}] \qquad (18.212)$$

The error term is given by

$$T.E = y(x_{n+1}) - \left[y_n + \frac{h}{2}(3f_n - f_{n-1}) \right]$$

$$= \left[y_n + hy'_n + \frac{h^2}{2}y''_n + \frac{h^3}{6}y'''_n ... \right] - \left[y_n + \frac{3h}{2}y'_n - \frac{h}{2}\left(y'_n - hy''_n + \frac{h^2}{2}y'''_n - ... \right) \right]$$

$$= \frac{5}{12}h^3 y'''(\xi), \quad x_{n-1} < \xi < x_{n+1}.$$

The method is of order 2.

$k = 3$: Retaining three terms inside the bracket in (18.210), we get the method

$$y_{n+1} = y_n + \frac{h}{12}[23f_n - 16f_{n-1} + 5f_{n-2}]. \qquad (18.213)$$

The error term is given by

$$T.E = y(x_{n+1}) - \left[y_n + \frac{h}{12}(23y'_n - 16y'_{n-1} + 5y'_{n-2}) \right]$$

$$= \frac{3}{8}h^4 y^{(4)}(\xi), \quad x_{n-2} < \xi < x_{n+1}.$$

The method is of order 3.

In general, a k-step explicit method of the form (18.210) gives a method of order k. We require one function evaluation per step for the application of the multi step method.

Adams-Moulton methods

These are implicit methods. Through the $k + 1$ data values (x_{n+1}, y_{n+1}), (x_n, f_n), (x_{n-1}, f_{n-1}), ..., (x_{n-k+1}, f_{n-k+1}), we fit the Newton's backward difference interpolating polynomial of degree k as

$$P_k(x) = f_{n+1} + \frac{(x - x_{n+1})}{h} \nabla f_{n+1} + \frac{(x - x_{n+1})(x - x_n)}{2!h^2} \nabla^2 f_{n+1} + ...$$

$$+ \frac{(x - x_{n+1})(x - x_n) ... (x - x_{n-k+2})}{k!h^k} \nabla^k f_{n+1}. \qquad (18.214)$$

Note that we are interpolating using the current data point (x_{n+1}, f_{n+1}) also.

The error of interpolation is given by

$$T.E = \frac{1}{(k+1)!}(x - x_{n+1})(x - x_n) ... (x - x_{n-k+1}) f^{(k+1)}(\xi) \qquad (18.215)$$

where ξ lies in some interval containing the points $x_{n+1}, x_n, \ldots, x_{n-k+1}$ and x. Replacing $f(x, y)$ by $P_k(x)$ in (18.207) , we get

$$y_{n+1} = y_n + \int_{x_n}^{x_{n+1}} \left[f_{n+1} + \frac{(x - x_{n+1})}{h} \nabla f_{n+1} + \frac{(x - x_{n+1})(x - x_n)}{2! h^2} \nabla^2 f_{n+1} + \ldots \right] dx.$$

Set $x - x_n = hs$. Then $(x - x_i) = (s + n - i)h$, $dx = hds$. Hence,

$$y_{n+1} = y_n + h \int_0^1 \left[f_{n+1} + (s - 1) \nabla f_{n+1} + \frac{(s-1)s}{2!} \nabla^2 f_{n+1} + \ldots \right] ds$$

$$= y_n + h \left[f_{n+1} - \frac{1}{2} \nabla f_{n+1} - \frac{1}{12} \nabla^2 f_{n+1} - \ldots \right]. \tag{18.216}$$

The error term can be obtained by integrating (18.215) or can be obtained directly. By choosing different values for k, we get different methods.

$k = 0$: We get a single step method. Retaining one term inside the bracket in (18.216), we get the method

$$y_{n+1} = y_n + hf_{n+1} \tag{18.217}$$

which is the *backward Euler method*. The method (18.217) is of order 1.

$k = 1$: Retaining two terms inside the bracket in (18.216), we get the method

$$y_{n+1} = y_n + h \left[f_{n+1} - \frac{1}{2} \nabla f_{n+1} \right] = y_n + h \left[f_{n+1} - \frac{1}{2} (f_{n+1} - f_n) \right]$$

$$= y_n + \frac{h}{2} [f_{n+1} + f_n] \tag{18.218}$$

which is also called the *trapezoidal method*. The error is obtained as

$$T.E = y(x_{n+1}) - y_{n+1} = y(x_n + h) - \left[y_n + \frac{h}{2} (y'_{n+1} + y'_n) \right]$$

$$= -\frac{h^3}{12} y'''(\xi), \quad x_n < \xi < x_{n+1}.$$

This method (18.218) is of order 2.

$k = 2$: Retaining three terms inside the bracket in (18.216), we get the method

$$y_{n+1} = y_n + \frac{h}{12} [5f_{n+1} + 8f_n - f_{n-1}]. \tag{18.219}$$

The error term is obtained as

$$T.E = y(x_{n+1}) - y_{n+1} = y(x_n + h) - \left[y_n + \frac{h}{12} (5y'_{n+1} + 8y'_n - y'_{n-1}) \right]$$

$$= \frac{h^4}{24} y^{(4)}(\xi), \quad x_{n-1} < \xi < x_{n+1}$$

The method (18.219) is of order 3.

In general, a k-step multi step method of the form (18.216) is of the order (k + 1). Since, the methods are implicit, we obtain a nonlinear equation for the solution of y_{n+1}. The solution can be obtained using Newton-Raphson method. For linear problems, we obtain the solution directly.

Milne-Simpson methods

These are mostly implicit methods. If we integrate $y' = f(x, y)$ in $[x_{n-1}, x_{n+1}]$, we get

$$y(x_{n+1}) = y(x_{n-1}) + \int_{x_{n-1}}^{x_{n+1}} f(x, y)dx. \tag{18.220}$$

Now, we replace $f(x, y)$ in (18.220) by the interpolating polynomial (18.214). Using the transformation $x - x_n = hs$, we obtain

$$y_{n+1} = y_{n-1} + h \int_{-1}^{1} \left[f_{n+1} + (s-1)\nabla f_{n+1} + \frac{(s-1)s}{2} \nabla^2 f_{n+1} + \ldots \right] ds$$

$$= y_{n-1} + \left[2f_{n+1} - 2\nabla f_{n+1} + \frac{1}{3} \nabla^2 f_{n+1} + 0 \cdot \nabla^3 f_{n+1} + \ldots \right]. \tag{18.221}$$

By choosing different values for k, we get different methods.

$k = 0$: Retaining one term inside the bracket in (18.221), we obtain the method

$$y_{n+1} = y_{n-1} + 2h f_{n+1}. \tag{18.222}$$

The method is of order 1.

$k = 1$: Retaining two terms inside the bracket in (18.221), we get the method

$$y_{n+1} = y_{n-1} + h \left[2f_{n+1} - 2(f_{n+1} - f_n) \right] = y_{n-1} + 2hf_n. \tag{18.223}$$

This is an explicit method. The method is of order 2.

$k = 2$: Retaining three terms inside the bracket in (18.221), we obtain the method

$$y_{n+1} = y_{n-1} + h \left[2f_{n+1} - 2(f_{n+1} - f_n) + \frac{1}{3}(f_{n+1} - 2f_n + f_{n-1}) \right]$$

$$= y_{n-1} + \frac{h}{3} [f_{n+1} + 4f_n + f_{n-1}]. \tag{18.224}$$

This method is called the *Milne-Simpson method or simply Milne's method*. Note that the second term on the right hand side of (18.224) is the Simpson's rule for evaluating the integral on the right hand side of (18.220). The error term is given by

$$T.E = y(x_{n+1}) - y_{n+1} = y(x_n + h) - \left[y_{n-1} + \frac{h}{3}(y'_{n+1} + 4y'_n + y'_{n-1}) \right]$$

$$= -\frac{h^5}{90} y^{(5)}(\xi), \quad x_{n-1} < \xi < x_{n+1}. \tag{18.224}$$

The method is of order 4.

18.7.1.4 Predictor-corrector methods

In the previous sections, we have derived explicit single step methods, explicit and implicit multi step methods for the solution of the initial value problem $y' = f(x, y), y(x_0) = y_0$. If we perform the analysis for stability (numerical stability), then we find that all the explicit methods converge only if h is sufficiently small. Hence, if integration of the initial value problem is to be performed over a large interval, then one may require to use the method for thousands or even millions of steps, which may not be economical. On the other hand, most implicit methods have strong stability properties, that is one can choose sufficiently large values of h for integration. However, we need to solve a nonlinear equation by iteration at each step, which is also expensive. Hence, we combine the explicit and implicit methods to define a new class of methods called *predictor-corrector methods*. The explicit methods are called the predictors, which predict or give an initial approximation $y^{(p)}_{n+1}$. The implicit methods are called the correctors which use $y^{(p)}_{n+1}$ as an initial approximation to compute y_{n+1} and is iterated to obtain the solution $y^{(c)}_{n+1}$. Therefore, we define predictor-corrector method ($P - C$ sets) as follows.

P: Predict an approximation $y^{(p)}_{n+1}$ to compute y_{n+1} using explicit methods.

C: Correct using an implicit method to obtain $y^{(c)}_{n+1}$. The corrector is used 1 or 2 or 3 times depending on the combination of the predictor and corrector methods.

The order of the predictor method is less than or equal to the order of the corrector method. If both are of the same order, then we may need only 1 or 2 iterations of corrector. However, if the predictor method is of lower order, then we need to use more iterations of the corrector. For example, if the predictor is of first order and the corrector is of second order (as in Example 2 given below), then the result obtained by the application of the predictor and one iteration of the corrector is of first order. If the corrector is iterated the second time, then the order of the result goes up by one, that is, we obtain a second order result. If the corrector is iterated a third time, then the combined truncation error may reduce. Further iterations do not improve the result.

The following are some examples of the $P - C$ sets.

Example 1

P: $\qquad y^{(p)}_{n+1} = y_n + hf(x_n, y_n),$ $\qquad$ (Euler method)

C: $\qquad y^{(c)}_{n+1} = y_n + hf(x_{n+1}, y^{(p)}_{n+1}).$ $\qquad$ (Backward Euler method) $\qquad$ (18.225)

We have $\qquad y^{(0)}_{n+1} = y_n + hf(x_n, y_n)$

$\qquad\qquad y^{(1)}_{n+1} = y_n + hf(x_{n+1}, y^{(0)}_{n+1}),$

$$y_{n+1}^{(2)} = y_n + hf(x_{n+1},\ y_{n+1}^{(1)}),\ \text{etc.}$$

Both predictor and corrector methods are of first order.

Example 2

P: $y_{n+1}^{(p)} = y_n + hf(x_n,\ y_n),$ (Euler method)

C: $y_{n+1}^{(c)} = y_n + \dfrac{h}{2}[\ f(x_n,\ y_n) + f(x_{n+1},\ y_{n+1}^{(p)})].$ (Trapezoidal method)   (18.226)

We have $y_{n+1}^{(0)} = y_n + hf(x_n,\ y_n),$

$$y_{n+1}^{(1)} = y_n + \frac{h}{2}[f(x_n,\ y_n) + f(x_{n+1},\ y_{n+1}^{(0)})],$$

$$y_{n+1}^{(2)} = y_n + \frac{h}{2}[f(x_n,\ y_n) + f(x_{n+1},\ y_{n+1}^{(1)})],\ \text{etc.}$$

The predictor is of first order and the corrector is of second order.

Example 3 Milne's method

P: $y_{n+1}^{(p)} = y_{n-3} + \dfrac{4h}{3}\ (2f_n - f_{n-1} + 2f_{n-2}),$

C: $y_{n+1}^{(c)} = y_{n-1} + \dfrac{h}{3}\ [\ f_{n-1} + 4f_n + f(x_{n+1},\ y_{n+1}^{(p)})].$ (18.227)

Both methods are of fourth order. To apply this method, we need starting values y_1, y_2, y_3, which can be obtained by Taylor series method of order 4 or any other single step method. y_0 is the given initial value. Hence, the Milne's method gives

P: $y_4^{(0)} = y_0 + \dfrac{4h}{3}\ (2f_3 - f_2 + 2f_1),$

C: $y_4^{(1)} = y_2 + \dfrac{h}{3}[f_2 + 4f_3 + f(x_4,\ y_4^{(0)})],$

$$y_4^{(2)} = y_2 + \frac{h}{3}[f_2 + 4f_3 + f(x_4,\ y_4^{(1)})],\ \text{etc. (18.228)}$$

Example 4   Adam's fourth order $P - C$ method

The $P - C$ method is the combination of Adams-Bashforth method of fourth order and Adams-Moulton method of fourth order.

P: Adams-Bashforth method of fourth order:

$$y_{n+1}^{(p)} = y_n + \frac{h}{24}\ [55f_n - 59\ f_{n-1} + 37f_{n-2} - 9f_{n-3}].$$

C: Adams-Moulton method of fourth order:

$$y_{n+1}^{(c)} = y_n + \frac{h}{24} [f_{n-2} - 5f_{n-1} + 19f_n + 9f(x_{n+1}, y_{n+1}^{(p)})]. \qquad (18.229)$$

To apply the method, we need the starting values y_n, y_{n-1}, y_{n-2}, y_{n-3}. That is, to compute approximations to y_4, we require the starting values y_3, y_2, y_1, and y_0. The starting values are obtained using the Taylor series method of fourth order or Runge-Kutta method of fourth order or any other single step method. Hence, we start the method as follows:

P:
$$y_4^{(0)} = y_3 + \frac{h}{24} [55f_3 - 59f_2 + 37f_1 - 9f_0],$$

C:
$$y_4^{(1)} = y_3 + \frac{h}{24} [f_2 - 5f_3 + 19f_3 + 9f(x_4, y_4^{(0)})],$$

$$y_4^{(2)} = y_3 + \frac{h}{24} [f_2 - 5f_3 + 19f_3 + 9f(x_4, y_4^{(1)})], \text{ etc.} \qquad (18.230)$$

Note that both in Milne's method and Adam's method, computations can be economized by storing the values of the terms that do not vary (only the value of one term changes in each iteration).

Example 18.49 Obtain the approximate value of $y(0.3)$ for the initial value problem

$$y' = x^2 + y^2, \quad y(0) = 1$$

using the methods

$$y_{n+1} = y_n + hf(x_n, y_n), \quad \text{as predictor}$$

and

$$y_{n+1} = y_n + \frac{h}{2} [f(x_n, y_n) + f(x_{n+1}, y_{n+1})], \quad \text{as corrector},$$

with $h = 0.1$. Perform two corrector iterations per step.

Solution Since both the predictor and corrector methods are single step methods, we do not require any starting value. We have $f(x, y) = x^2 + y^2$, and $h = 0.1$.

$n = 0$:
$$x_0 = 0, \ x_1 = 0.1, \ y_0 = 1, \ f(x_0, y_0) = x_0^2 + y_0^2 = 1.$$

P:
$$y_1^{(0)} = y_0 + hf(x_0, y_0) = 1 + 0.1(1) = 1.1.$$

C:
$$y_1^{(k+1)} = y_0 + \frac{h}{2} [f(x_0, y_0) + x_1^2 + (y_1^{(k)})^2].$$

$$k = 0: \quad y_1^{(1)} = y_0 + \frac{0.1}{2} [f(x_0, y_0) + x_1^2 + (y_1^{(0)})^2]$$

$$= 1 + 0.05[1 + 0.01 + 1.21] = 1.111.$$

$$k = 1: \quad y_1^{(2)} = y_0 + \frac{0.1}{2}[f(x_0, y_0) + x_1^2 + (y_1^{(1)})^2]$$

$$= 1 + 0.05[1 + 0.01 + (1.111)^2] = 1.1122.$$

Therefore, $\quad y(0.1) \approx y_1^{(2)} = 1.1122.$

$n = 1: \quad x_1 = 0.1, x_2 = 0.2, y_1 = 1.1122, f(x_1, y_1) = x_1^2 + y_1^2 = 1.2470.$

$P: \quad y_2^{(0)} = y_1 + hf(x_1, y_1) = 1.1122 + 0.1(1.2470) = 1.2369.$

$C: \quad y_2^{(k+1)} = y_1 + \frac{h}{2}[f(x_1, y_1) + x_2^2 + (y_2^{(k)})^2]$

$$k = 0: \quad y_2^{(1)} = y_1 + \frac{h}{2}[f(x_1, y_1) + x_2^2 + (y_2^{(0)})^2]$$

$$= 1.1122 + 0.05[1.2470 + (0.2)^2 + (1.2369)^2] = 1.2530.$$

$$k = 1: \quad y_2^{(2)} \qquad\qquad\qquad = y_1 + \frac{h}{2}[f(x_1, y_1) + x_2^2 + (y_2^{(1)})^2]$$

$$= 1.1122 + 0.05[1.2470 + (0.2)^2 + (1.2530)^2] = 1.2550.$$

Therefore, $\quad y(0.2) \approx y_2^{(2)} = 1.2550.$

$n = 2: \quad x_2 = 0.2, x_3 = 0.3, y_2 = 1.2550, f(x_2, y_2) = x_2^2 + y_2^2 = 1.6150.$

$P: \quad y_3^{(0)} = y_2 + hf(x_2, y_2) = 1.2550 + 0.1(1.6150) = 1.4165.$

$C: \quad y_3^{(k+1)} = y_2 + \frac{h}{2}[f(x_2, y_2) + x_3^2 + (y_3^{(k)})^2]$

$$k = 0: \quad y_3^{(1)} = y_2 + \frac{h}{2}[f(x_2, y_2) + x_3^2 + (y_3^{(0)})^2]$$

$$= 1.2550 + 0.05[1.6150 + (0.3)^2 + (1.4165)^2] = 1.4406.$$

$$k = 1: \quad y_3^{(2)} = y_2 + \frac{h}{2}[f(x_2, y_2) + x_3^2 + (y_3^{(1)})^2]$$

$$= 1.2550 + 0.05[1.6150 + (0.3)^2 + (1.4406)^2] = 1.4440$$

Therefore, $\quad y(0.3) \approx y_3^{(2)} = 1.4440.$

Example 18.50 Using Milne's predictor-corrector method find $y(0.8)$ and $y(1.0)$ for the initial value problem

$$y' = x^2 + y^2, \ y(0) = 1, \text{ with } h = 0.2.$$

Calculate the required initial values using the Euler method. Perform two iterations of the corrector.

Solution Milne's predictor – corrector method is given by (18.228). We need starting values for $y(0.2)$, $y(0.4)$ and $y(0.6)$. These values are to be obtained by the Euler method

$$y_{n+1} = y_n + hf(x_n, y_n) = y_n + h(x_n^2 + y_n^2).$$

We have the following results.

$n = 0$: $\qquad\qquad x_0 = 0, \ y_0 = 1.$

$\qquad\qquad\qquad y(0.2) \approx y_1 = y_0 + h(x_0^2 + y_0^2) = 1 + 0.2(1) = 1.2.$

$n = 1$: $\qquad\qquad x_1 = 0.2, \ y_1 = 1.2.$

$\qquad\qquad\qquad y(0.4) \approx y_2 = y_1 + h(x_1^2 + y_1^2) = 1.2 + 0.2(0.04 + 1.44) = 1.496.$

$n = 2$: $\qquad\qquad x_2 = 0.4, \ y_2 = 1.496.$

$\qquad\qquad\qquad y(0.6) \approx y_3 = y_2 + h(x_2^2 + y_2^2) = 1.496 + 0.2[0.16 + (1.496)^2] = 1.9756.$

We have $x_4 = 0.8$, $f_1 = x_1^2 + y_1^2 = 1.48$, $f_2 = x_2^2 + y_2^2 = 2.3980$, $f_3 = x_3^2 + y_3^2 = 4.2630$.

We have the following results from the predictor-corrector method.

P: $\qquad\qquad y_4^{(0)} = y_0 + \dfrac{4h}{3}[2f_3 - f_2 + 2f_1]$

$\qquad\qquad\qquad = 1 + \dfrac{0.8}{3}[2(4.2630) - 2.3980 + 2(1.48)] = 3.4235.$

C: $\qquad\qquad y_4^{(1)} = y_2 + \dfrac{h}{3}[f_2 + 4f_3 + f(0.8, 3.4235)]$

$\qquad\qquad\qquad = 1.496 + \dfrac{0.2}{3}[2.3980 + 4(4.2630) + 0.64 + (3.4235)^2] = 3.6167.$

$\qquad\qquad\quad y_4^{(2)} = y_2 + \dfrac{h}{3}[f_2 + 4f_3 + f(0.8, 3.6167)]$

$\qquad\qquad\qquad = 1.496 + \dfrac{0.2}{3}[2.3980 + 4(4.2630) + 0.64 + (3.6167)^2] = 3.7074.$

Therefore, $\qquad\qquad y(0.8) \approx y_4^{(2)} = 3.7074.$

Now, $\qquad\qquad\qquad x_5 = 1.0, \ f_4 = x_4^2 + y_4^2 = 0.64 + (3.7074)^2 = 14.3848.$

P: $\qquad\qquad y_5^{(0)} = y_1 + \dfrac{4h}{3}[2f_4 - f_3 + 2f_2]$

$$= 1.2 + \frac{0.8}{3}[2(14.3848) - 4.2630 + 2(2.3980)] = 9.0140.$$

C: $$y_5^{(1)} = y_3 + \frac{h}{3}[f_3 + 4f_4 + f(1.0, 9.0140)]$$

$$= 1.9756 + \frac{0.2}{3}[4.2630 + 4(14.3848) + 1.0 + (9.0140)^2] = 11.5792.$$

$$y_5^{(2)} = y_3 + \frac{h}{3}[f_3 + 4f_4 + f(1.0, 11.5792)]$$

$$= 1.9756 + \frac{0.2}{3}[4.2630 + 4(14.3848) + 1.0 + (11.5792)^2] = 15.1009.$$

Therefore, $\qquad y(1.0) \approx y_5^{(2)} = 15.1009.$

Example 18.51   Using Adam's fourth order *P-C* method, find $y(1.4)$ for the initial value problem

$$y' = x^2(1 + y^2), \ y(1) = 1, \text{ with } h = 0.1.$$

Calculate the starting values using Euler method.

Solution   We need the starting values $y(1.1)$, $y(1.2)$ and $y(1.3)$. Euler method gives

$$y_{n+1} = y_n + h \, f(x_n, y_n) = y_n + h x_n^2 (1 + y_n^2).$$

We have the following results:

$n = 0$: $\qquad x_0 = 1, y_0 = 1.$

$\qquad\qquad y_1 = y_0 + h f(x_0, y_0) = 1 + 0.1(1 + 1) = 1.2.$

$n = 1$: $\qquad x_1 = 1.1, y_1 = 1.2.$

$\qquad\qquad y_2 = y_1 + h f(x_1, y_1) = 1.2 + 0.1(1.1)^2 (1 + (1.2)^2) = 1.49524.$

$n = 2$: $\qquad x_2 = 1.2, y_2 = 1.49524.$

$\qquad\qquad y_3 = y_2 + h f(x_2, y_2) = 1.49524 + 0.1(1.2)^2 (1 + (1.49524)^2) = 1.961187.$

We have $\qquad x_3 = 1.3, f_3 = 8.190170, f_2 = 4.659469, f_1 = 2.9524, f_0 = 2.$

P: $$y_4^{(0)} = y_3 + \frac{h}{24} [55f_3 - 59f_2 + 37f_1 - 9f_0]$$

$$= 1.961187 + \frac{0.1}{24} [55(8.19017) - 59(4.659469) + 37(2.9524) - 18] = 3.07281.$$

C: $$y_4^{(1)} = y_3 + \frac{h}{24} [f_1 - 5f_2 + 19f_3 + 9f(1.4, y_4^{(0)})]$$

$$= 1.961187 + \frac{0.1}{24} [2.9524 - 5(4.659469) + 19(8.19017) + 9f (1.4, y_4^{(0)})]$$

$$= 2.524804 + \frac{1.764}{24} [1 + (y_4^{(0)})^2].$$

With $y_4^{(0)} = 3.07281$, we get the following approximations: 3.292303, 3.39499, 3.445462, 3.4708378, 3.483738, 3.490332, 3.493712. Two decimal place accuracy is obtained at this stage. It may be noted that we have used the first order Euler method to compute the required initial values.

18.7.2 Two Point Boundary Value Problems

Consider the solution of the boundary value problem in $[a, b]$

$$y'' = f (x, y, y'),$$ (18.231)

under the following boundary conditions

(i) $y(a) = A$, $y(b) = B$, or (18.232)

(ii) $\alpha y(a) + \beta y'(a) = A$, $\gamma y(a) + \delta y'(a) = B$. (18.233)

If $\alpha = 0$, and $\gamma = 0$ in (18.233), then the slopes of the solution curve are prescribed at the end points. Assume that the solution of the problem exists.

Subdivide $[a, b]$ into n subintervals of equal mesh size h, where $h = (b - a)/n$. The nodal points are $x_0 = a$, $x_1 = x_0 + h$, ..., $x_n = x_0 + nh = b$. Replace the derivatives at the nodal points by suitable finite difference approximations (see section 18.4.3).

$$y_i'' = \frac{1}{h^2} \delta^2 y_i, \qquad [O(h^2) \text{ approximation}],$$ (18.234)

$$y_i' = \frac{1}{2h} (y_{i+1} - y_{i-1}), \qquad [O(h^2) \text{ approximation}],$$ (18.235a)

$$y_i' = \frac{1}{h} \Delta y_i = \frac{1}{h} (y_{i+1} - y_i), \quad [O(h) \text{ approximation}],$$ (18.235b)

$$y_i' = \frac{1}{h} \nabla y_i = \frac{1}{h} (y_i - y_{i-1}), \quad [O(h) \text{ approximation}].$$ (18.235c)

Consider the boundary conditions (18.232). Substituting (18.234) and (18.235a) in (18.231), we get a second order finite difference approximation to the boundary value problem as

$$y_{i-1} - 2 y_i + y_{i+1} = h^2 f [x_i, y_i, (y_{i+1} - y_{i-1})/h], \quad i = 1, 2, 3, ..., n - 1$$ (18.236)

$y_0 = A$, $y_n = B$. The solution of the system of $(n - 1) \times (n - 1)$ nonlinear difference equations can be obtained by an iterative method like Newton's method. The linearized equations give rise to a tridiagonal system. Using (18.235b) or (18.235c), and (18.234) in (18.231), we get a first order approximation. However, in some applications in Fluid dynamics, the forward difference approximation (18.235b) is used. Consider now the solution of the linear boundary value problem

$$a(x)\, y'' + b(x)y' + c(x)y = r(x), \quad a \le x \le b. \tag{18.237}$$

The boundary conditions are same as given in (18.232) or (18.233). When the boundary conditions (18.233) are prescribed, then y_0 and y_n are also to be determined. If $O(h)$ approximations are to be used, then we use the approximation (18.235b) at the left boundary point and (18.235c) at the right boundary point. If we use (18.235a) at the boundary points, then we get a second order approximation. In this case, we use both the approximations to the differential equation and the boundary condition at the boundary point so that the external point can be eliminated using these two equations. The difference approximation to the differential equation in this case is given by

$$\frac{a(x_i)}{h^2}(y_{i-1} - 2y_i + y_{i+1}) + \frac{b(x_i)}{2h}(y_{i+1} - y_{i+1}) + c(x_i)\, y_i = r(x_i),$$

or $\quad a(x_i)(y_{i-1} - 2y_i + y_{i+1}) + \dfrac{h}{2} b(x_i)(y_{i+1} - y_{i-1}) + c(x_i)h^2\, y_i = h^2\, r(x_i),$

or $\quad \left[a(x_i) - \dfrac{h}{2} b(x_i)\right] y_{i-1} - [2a(x_i) - c(x_i)h^2]\, y_i + [a(x_i) + \dfrac{h}{2} b(x_i)]y_{i+1} = h^2 r(x_i). \tag{18.238}$

If the boundary condition $\alpha y(a) + \beta y'(a) = A$, is prescribed, we write the second order approximation at x_0 as

$$\alpha y_0 + \frac{\beta}{2h}(y_1 - y_{-1}) = A, \quad \text{or} \quad y_{-1} = y_1 - \frac{2h}{\beta}(A - \alpha y_0). \tag{18.239}$$

At x_0, we get from (18.238)

$$\left[a(x_0) - \frac{h}{2} b(x_0)\right] y_{-1} - [2a(x_0) - c(x_0)h^2]\, y_0 + [a(x_0) + \frac{h}{2} b(x_0)]\, y_1 = h^2 r(x_0). \tag{18.240}$$

Eliminating y_{-1} from (18.239) and (18.240), we get the equation valid at x_0. Similarly, the boundary condition at x_n is treated to get the difference equation valid at x_n. The system of equations in all cases is tridiagonal.

In the numerical solution of the initial value problems, we are concerned with the study of the numerical stability/ instability of the solutions. In the case of boundary value problems, we are concerned with the convergence of the system of difference equations $\mathbf{Ax} = \mathbf{b}$. Convergence depends on the properties of the matrix $\mathbf{A}$. A sufficient condition for convergence is that the system is irreducibly diagonal dominant.

Example 18.52 Find the numerical solution of the boundary value problem

$$2x^2 y'' + 3xy' - y = x, \quad y(1) = 1, \ y(4) = 41/16.$$

Use second order approximations with $h = 0.5$.

Compare with the exact solution $y(x) = 0\ 0.25\ [25\sqrt{x} + (1/x)] + 0.5x$.

Solution The mesh points are $x_0 = 1$, $x_1 = 1.5$, $x_2 = 2$, $x_3 = 2.5$, $x_4 = 3$, $x_5 = 3.5$, $x_6 = 4$. From the given boundary conditions, we have $y_0 = 1$, $y_6 = 41/16$. With $h = 0.5$, we obtain

$$y_{n+1} = y_n + \frac{h}{2}[f(x_n, y_n) + f(x_{n+1}, y_n + hf_n)] \tag{18.197}$$

where $f_n = f(x_n, y_n)$. This method is called the *Heun's method or Euler-Cauchy method*. We can rewrite this method using the following notation

$$k_1 = hf(x_n, y_n),$$

$$k_2 = hf(x_{n+1}, y_n + k_1), \tag{18.198}$$

$$y_{n+1} = y_n + \frac{1}{2}(k_1 + k_2).$$

The philosophy behind the Runge-Kutta methods is to consider a weighted average of slopes or approximate slopes at a number of points in $[x_n, x_{n+1}]$. If we use v slopes, then the method is written as follows

$$y_{n+1} = y_n + w_1 k_1 + w_2 k_2 + \ldots + w_v k_v.$$

where $k_i = h$ (slope or approximate slope at a point in $[x_n, x_{n+1}]$). The method is also called a *v*-stage method.

Two stage explicit method

We write the method as

$$y_{n+1} = y_n + w_1 k_1 + w_2 k_2, \quad n = 0, 1, 2, \ldots \tag{18.199}$$

where

$$k_1 = hf(x_n, y_n)$$

$$k_2 = hf(x_n + \alpha h, y_n + \beta k_1)$$

and α, β, w_1 and w_2 are parameters to be determined such that the method (18.199) compares with the Taylor series method upto certain powers of h. Hence, to determine these parameters, we expand k_1, k_2 and y_{n+1} in powers of h such that both sides agree upto certain powers of h. We obtain from (18.199)

$$k_1 = hf(x_n, y_n) = hf_n,$$

$$k_2 = hf(x_n + \alpha h, y_n + \beta h f_n)$$

$$= h\left[f_n + \left(\alpha h \frac{\partial f}{\partial x} + \beta h f_n \frac{\partial f}{\partial y} \right) + \frac{1}{2}\left(\alpha^2 h^2 \frac{\partial^2 f}{\partial x^2} + 2\alpha\beta h^2 f_n \frac{\partial^2 f}{\partial x \partial y} + \beta^2 h^2 f_n^2 \frac{\partial^2 f}{\partial y^2} \right) + \ldots \right]$$

$$= h\left[f_n + h\left(\alpha \frac{\partial f}{\partial x} + \beta f_n \frac{\partial f}{\partial y} \right) + \frac{h^2}{2}\left(\alpha^2 \frac{\partial^2 f}{\partial x^2} + 2\alpha\beta f_n \frac{\partial^2 f}{\partial x \partial y} + \beta^2 f_n^2 \frac{\partial^2 f}{\partial y^2} \right) + \ldots \right]$$

and partial derivatives are evaluated at the point (x_n, y_n).

From the given differential equation $y' = f(x, y)$, we get

$$y'' = \frac{\partial f}{\partial x} + \frac{\partial f}{\partial y} \frac{dy}{dx} = \frac{\partial f}{\partial x} + f \frac{\partial f}{\partial y}.$$

$$8x_i^2(y_{i-1} - 2y_i + y_{i+1}) + 3x_i(y_{i+1} - y_{i-1}) - y_i = x_i, \quad \text{or}$$

$$(8x_i^2 - 3x_i) y_{i-1} - (16x_i^2 + 1)y_i + (8x_i^2 + 3x_i) y_{i+1} = x_i.$$

At $x_1 = 1.5$, we get $\quad -37 \ y_1 + 22.5 \ y_2 = -12$.

At $x_2 = 2.0$, we get $\quad 26 \ y_1 - 65 \ y_2 + 38 \ y_3 = 2$.

At $x_3 = 2.5$, we get $42.5 \ y_2 - 101 \ y_3 + 57.5 \ y_4 = 2.5$.

At $x_4 = 3.0$, we get $\quad 63 \ y_3 - 145 \ y_4 + 81 \ y_5 = 3$.

At $x_5 = 3.5$, we get $87.5 \ y_4 - 197 \ y_5 = -274.53125$.

We have the tridiagonal system of equations

$$\begin{bmatrix} -37 & 22.5 & 0 & 0 & 0 \\ 26 & -65 & 38 & 0 & 0 \\ 0 & 42.5 & -101 & 57.5 & 0 \\ 0 & 0 & 63 & -145 & 81 \\ 0 & 0 & 0 & 87.5 & -197 \end{bmatrix} \begin{bmatrix} y_1 \\ y_2 \\ y_3 \\ y_4 \\ y_5 \end{bmatrix} = \begin{bmatrix} -12 \\ 2 \\ 2.5 \\ 3 \\ -274.53125 \end{bmatrix}$$

The solution of the system is

$y_1 = 1.224115$, $y_2 = 1.479657$, $y_3 = 1.746072$, $y_4 = 2.016833$, $y_5 = 2.289361$.

Exact solution is

$y_1 = 1.222853$, $y_2 = 1.478553$, $y_3 = 1.745284$, $y_4 = 2.016346$, $y_5 = 2.289136$.

Exercise 18.7

In the following initial value problems, find the approximate values of $y(x)$ at the given points using the Euler method.

1. $y' = xy + x^2y^2 + 1$, $y(1) = 2$, $h = 0.1$, $x \in [1, 1.3]$.

2. $y' = 3x^2 + \sqrt{y}$, $y(1) = 1$, $h = 0.2$, $x \in [1, 2]$.

In the following initial value problems, find the approximate values of $y(x)$ at the given points using the Taylor series method of given order.

3. $y' = 1 + y^2$, $y(0) = 1$, $h = 0.1$, order 2, $x \in [0, 0.3]$.

4. $y' = x^2 + \sqrt{y}$, $y(1) = 1$, $h = 0.1$, order 3, $x \in [1, 1.2]$.

5. $y' = x + \sin y$, $y(1) = 0$, $h = 0.2$, order 4, at $x = 1.4$.

In the following initial value problems, find the approximate values of $y(x)$ at the given points using Heun's method.

6. $y' = x^2 + 2y^2$, $y(1) = 1$, $h = 0.2$, $x \in [1.0, 1.4]$.

7. $y' = \dfrac{y-x}{y+x}$, $y(1) = 2$, $h = 0.1$, $x \in [1.0, 1.2.]$.

In the following initial value problems, find the approximate values of $y(x)$ at the given points using classical fourth order Runge-Kutta method.

8. $y' = x^2 + y^2$, $y(1) = 2$, $h = 0.1$, $x \in [1.0, 1.2]$.

9. $y' = \sqrt{x+y}$, $y(0.4) = 0.41$, $h = 0.2$, $x \in [0.4, 0.8]$.

10. $y' = 2 + \sqrt{xy}$, $y(2) = 1$, $h = 0.1$, $x \in [2.0, 2.2]$.

11. Find an approximate value of $y(0.8)$ for the initial value problem

$$y' = -2xy^2, \ y(0) = 1$$

using the Adams-Bashforth method of second order

$$y_{n+1} = y_n + \frac{h}{2}\ (3f_n - f_{n-1})$$

with $h = 0.2$. Compute the starting value using Heun's method with the same step size.

12. Find an approximate value of $y(0.6)$ for the initial value problem

$$y' = x + 2y, \ y(0) = 1$$

using Adams-Moulton third order method

$$y_{n+1} = y_n + \frac{h}{12}\ [5f_{n+1} + 8f_n - f_{n-1}]$$

with $h = 0.2$. Compute the starting value using Taylor series third order method with the same step size.

13. Find an approximate value of $y(1.0)$ for the initial value problem

$$y' = x - 2y, \ y(0) = 1$$

using Milne-Simpson fourth order method

$$y_{n+1} = y_{n-1} + \frac{h}{3}\ [f_{n+1} + 4f_n + f_{n-1}]$$

with $h = 0.2$. Compute the staring value using classical Runge-Kutta fourth order method with the same step length.

14. Solve the initial value problem

$$y' = 1 + y^2, \ y(0) = 1$$

with $h = 0.2$ on the interval $[0, 0.6]$ using the predictor-corrector method

P: $$y_{n+1} = y_n + \frac{h}{2}[3\ f_n - f_{n-1}],$$

C: $$y_{n+1} = y_n + \frac{h}{2}[\ f_{n+1} + f_n].$$

Compute the starting value using second order Taylor series method with the same step length. Perform two iterations of the corrector.

15. Solve the initial value problem

$$y' = (1 + x^2) (y - 2), \ y(1) = 0, \ x \in [1.0, 1.5]$$

with $h = 0.1$ using the Milne's predictor-corrector method. Perform two iterations of the corrector. Compute the starting values using Euler method with the same step size.

16. Find an approximation to $y(1.4)$ using the Adams fourth order P-C method for the initial value problem

$$y' = x^2 (1 + y), \ y(1) = 1, \ \text{with} \ h = 0.1.$$

It is given that $y(1.1) = 1.2333$, $y(1.2) = 1.5993$, $y(1.3) = 1.9807$.

17. Find an approximation to $y(0.8)$ using the Adams fourth order P-C method for the initial value problem

$$y' = 1 + y^2, \ y(0) = 0, \ \text{with} \ h = 0.2.$$

Compute the required initial values by Euler method.

18. Find the solution of the boundary value problem

$$y'' + 4y' + 3y = 6e^{-x}, \ y(0) = 1, \ y(1) = 1/e.$$

Use second order approximations with $h = 1/3$.

19. Find the solution of boundary value problem

$$4x^2 y'' + y = \ln x, \ x > 0, \ y(1) = 0, \ y(e) = 5.$$

Use second order approximations with $h = (e - 1)/3$. Compare with the exact solution

$$y(x) = 4(\ln x - 1) \sqrt{x} + \ln x + 4.$$

20. Find the solution of the boundary value problem

$$(x^2 - 1) y'' - xy' - 3y = - (6x + 3), \ y(0) = 1, \ y(1) = 2.$$

Use second order approximations with $h = 0.25$. Compare with the exact solution

$$y(x) = x^3 + 1.$$

18.8 Answers and Hints

Exercise 18.1

1. 0.4861416×10^2, 0.23748×10^1, 0.436×10^{-1}, 0.103092×10^1.

   (i) 0.4861×10^2, 0.2375×10^1, 0.4360×10^{-1}, 0.1031×10^1.

   (ii) 0.49×10^2, 0.24×10^1, 0.44×10^1, 0.10×10^1.

2. 0.67.   3. Six significant digits

4. 0.000214, 0.000151.   5. Three.

6. $P_2(x) = 1 + \dfrac{x}{2} - \dfrac{x^2}{8}$; $|T.E| \leq \dfrac{|x|^3}{16} (1 + \xi)^{-5/2}$, 0.0005, 0.0625.

7. $P_n(x) = 1 - x + x^2 + \ldots + (-1)^n x^n$; $T.E = \dfrac{(-1)^{n+1} x^{n+1}}{(1 + \xi)^{n+1}}$, $n \geq 7$. Eight terms.

8. For $0 \le x \le 1$, $|T.E| \le [3e/(n+1)!] \le 5 \times 10^{-5}$ gives $n \ge 8$. With $n = 8$, we get $e \approx 2.7183$.

9. $\cos(2x) = -2\left(x - \dfrac{\pi}{4}\right) + \dfrac{4}{3}\left(x - \dfrac{\pi}{4}\right)^3 - \dfrac{4}{15}\left(x - \dfrac{\pi}{4}\right)^5$; 0.5000,

$$|T.E| \le \frac{64}{6!}\left(x - \frac{\pi}{4}\right)^6 \left[\max_{(\pi/6) \le x \le (\pi/4)} |\cos 2x|\right] \text{ gives at } x = \frac{\pi}{6}. \ |T.E| \le 0.000014.$$

10. $|T.E| \le \dfrac{|x|^{2n}}{(2n)!}$. For $x = \dfrac{\pi}{9}$, we get $n \ge 3$. For $n = 3$, we get $\sin(\pi/9) \approx 0.342020$.

Exercise 18.2

1. $(2, 3)$, $n \ge 10$. **2.** $(-3, -2)$, $n \ge 8$.

3. $(1, 2)$, 1.75, 1.86777, 1.88060.

4. $(-2, -1)$, -1.5714, -1.7146, -1.7413, -1.7459.

5. $(1, 2)$, 1.4879, 1.4880.

6. $f(x) = x^2 - 31 = 0$, $x_0 = 5$, 5.6, 5.5679, 5.5678; $f(x) = x^3 - 101 = 0$, $x_0 = 4$, 4.7708, 4.6597, 4.6570;

   $f(x) = (1/x) - 17 = 0$, $x_0 = 0.05$, 0.0575, 0.0588, 0.0588.

7. (i) 4.505, (ii) 4.505, (iii) 4.505.

8. $x = (x^3 - 1)/5$, -0.5, -0.225, -0.2023, -0.2016, -0.2016.

9. $|1 + \alpha(9x^2 + 8x + 4)| < 1$. With $x_0 = -0.5$, we get $-(8/9) < \alpha < 0$. $\alpha = -0.5$, -0.3125, -0.3370,

   -0.3327, -0.3334.

10. -1.2313, -1.2225.

11. $\alpha + \beta = -a$, $\alpha\beta = b$ (i) $|\phi'(x)| = \left|\dfrac{b}{x^2}\right|$. For $x = \alpha$, $\left|\dfrac{\alpha\beta}{\alpha^2}\right| < 1$ gives $|\alpha| > |\beta|$.

(ii) $|\phi'(x)| = \left|\dfrac{b}{(a+x)^2}\right|$. For $x = \alpha$, $\left|\dfrac{\alpha\beta}{[-(\alpha+\beta)+\alpha]^2}\right| < 1$ gives $|\alpha| < |\beta|$.

(iii) $|\phi'(x)| = \left|\dfrac{2x}{a}\right|$. For $x = \alpha$, $\left|\dfrac{2\alpha}{-(\alpha+\beta)}\right| < 1$ gives $2|\alpha| < |\alpha + \beta|$.

12. $|\varepsilon_{k+1}| \le C |\varepsilon_0| |\varepsilon_k|$, order: 1, $C = |f''(\xi)/[2f'(\xi)]|$.

13. $\sqrt{a}$, order: 2, $C = 3/(2|\xi|)$. **14.** Order: 3, $C = 1/(2\xi^2)$

15. $C = 3/8$, $|\varepsilon_0| \le 1$, $\varepsilon_n = C^{2^n - 1} \varepsilon_0^{2^n}$, $n \ge 4$.

16. $p = 2/3$, $q = 1/3$, order: 2, $C = 1/|\xi|$.

17. Order: 2, 0.8114, 0.7971.

18. $[2.0815, 0.9062]^T$, $[2.0117, 0.9874]^T$, $[2.0003, 0.9997]^T$.

19. $[0.9013, 2.0409]^T$, $[0.9772, 2.0080]^T$, $[0.9983, 2.0006]^T$.

20. $[1.0354, 0.2229]^T$, $[1.0347, 0.2223]^T$.

21. $[1.0797, 1.9454]^T$, $[1.0863, 1.9437]^T$.

22. -1.66667.   **23.** $1/3$.

24. -3.28571 **25.** $-1, -2, 1 \pm i$.

26. $3, 2 \pm 3i$.   **27.** $-1, 2, 3$.

28. $1, 2, 3$.

Exercise 18.3

1. $\mathbf{Lz} = \mathbf{b}$ gives $\mathbf{z} = [5, -10, -34]^T$. $\mathbf{Ux} = \mathbf{z}$ gives $\mathbf{x} = [1, 1, 1]^T$.

2. $\mathbf{Lz} = \mathbf{b}$ gives $\mathbf{z} = [1/2, -1/3, 1]^T$. $\mathbf{Ux} = \mathbf{z}$ gives $\mathbf{x} = [1, 1, 1]^T$.

3. (i) $\mathbf{L}^{-1} = \begin{bmatrix} 1 & 0 & 0 \\ -1/2 & 1 & 0 \\ -5 & 7 & 1 \end{bmatrix}$, $\mathbf{U}^{-1} = \dfrac{1}{18}\begin{bmatrix} 9 & -54 & 7 \\ 0 & 36 & -5 \\ 0 & 0 & 1 \end{bmatrix}$, $\mathbf{A}^{-1} = \mathbf{U}^{-1}\mathbf{L}^{-1} = \dfrac{1}{18}\begin{bmatrix} 1 & -5 & 7 \\ 7 & 1 & -5 \\ -5 & 7 & 1 \end{bmatrix}$.

(ii) $\mathbf{L}^{-1} = \begin{bmatrix} 1 & 0 & 0 \\ -3 & 1 & 0 \\ 7/2 & -3/2 & 1 \end{bmatrix}$, $\mathbf{U}^{-1} = \dfrac{1}{6}\begin{bmatrix} 6 & 3 & 4 \\ 0 & -3 & -6 \\ 0 & 0 & 2 \end{bmatrix}$, $\mathbf{A}^{-1} = \mathbf{U}^{-1}\mathbf{L}^{-1} = \dfrac{1}{6}\begin{bmatrix} 11 & -3 & 4 \\ -12 & 6 & -6 \\ 7 & -3 & 2 \end{bmatrix}$.

4. $\mathbf{L} = \begin{bmatrix} 1 & 0 & 0 & 0 \\ -1 & 4 & 0 & 0 \\ 3 & 0 & -11 & 0 \\ -2 & 1 & 19/2 & 24/11 \end{bmatrix}$, $\mathbf{U} = \begin{bmatrix} 1 & 0 & 5 & 2 \\ 0 & 1 & 3/2 & 1/2 \\ 0 & 0 & 1 & 5/11 \\ 0 & 0 & 0 & 1 \end{bmatrix}$.

5. (i) $[3/2, 1/2, -1/2]^T$.   **(ii)** $[1/2, -1/2, -1/2]^T$

6. (i) (a) $-0.25, 1.0, 0.75$; $-0.875, 0.9, 0.625$; $-0.7875, 1.05, 0.9625$; $-0.9938, 0.965, 0.8813$.

       (b) $1, 0.5, 1$; $0.875, 1.0, 0.875$; $1.0313, 0.9375, 1.0313$; $0.9766, 1.0157, 0.9766$.

7. (i) $\mathbf{H}_J = -\mathbf{D}^{-1}(\mathbf{L} + \mathbf{U})$, $\rho(\mathbf{H}_J) = 1/\sqrt{2} < 1$, coverges; $v = 0.1505$; $0.5, -0.5, 0.5$; $1.25, -1.5, 1.25$; $0.75, -0.75, 0.75$.

(ii) $\rho(\mathbf{H}_J) = \sqrt{11/18} < 1$, converges; $v = 0.1069$; $0.7778, 0.3889, 1.1667$; $1.4074, 1.2037, 0.6944$; $0.8642, 0.6265, 1.1018$.

8. $\rho(\mathbf{H}_J) = 2.83 > 1$, diverges.

9. (i) $\rho(\mathbf{H}_J) = |k|$. For convergences $|k| < 1$, $v_J = -\log|k|$.

　　(ii) $\rho(\mathbf{H}_{GS}) = k^2$. For convergence $|k| < 1$, $v_{GS} = -2\log|k| = 2v_J$.

10. $\rho(\mathbf{H}_{GS}) = 4 > 1$, diverges.

11. $\rho(\mathbf{H}_{GS}) = 0.41 < 1$, converges, $v = 0.3872$; 1, −0.6, −0.06; 1.03, −0.576, −0.0846; 1.0423, −0.5662, −0.0947; 1.0474, −0.5621, − 0.0988.

Excersice 18.4

1. (i) $6x^2 - 7x + 2$, 　　　　　　(ii) $x^3 - 5x + 1$.

2. $0.477x + 0.2161$, 0.7408, $M_2 = 0.25$, 0.00125.

3. $0.1463x^2 - 0.6554x + 0.877$, 0.2722, $M_3 = 1/e$, $0.001/e$.

4. $f(x) = \sum l_k(x) f_k + \dfrac{1}{(n+1)!}[(x-x_0)(x-x_1)\ldots(x-x_n)] f^{(n+1)}(\xi)$. Take $f(x) = x^{n+1}$, substitute and then set $x = 0$.

5. In problem 4, set $f(x) = x^j$. Since $f^{(n+1)}(x^j) = 0$ for $j = 0, 1, \ldots, n$, we get the result.

6. (i) $x^3 + x^2 - 2x + 3$, 　　　　　(ii) $x^3 - 9x^2 + 21x + 1$.

8. $f[x_0, x_1] = \dfrac{1}{(x_1 - x_0)}[u(x_1)\,v(x_1) - u(x_0)\,v(x_0)] = \dfrac{1}{(x_1 - x_0)}[u(x_1)\{v(x_1) - v(x_0)\} + v(x_0)\{u(x_1) - u(x_0)\}]$

　　　　　　$= u(x_1)\,v[x_0, x_1] + v(x_0)\,u[x_0, x_1]$.

9. $x^2 - x + 6$, 6.

12. (i) $x^3 + x + 1$, 　　　　　　　(ii) $x^2 + 3x - 2$.

13. (i) $x^2 - x + 1$, 　　　　　　　(ii) $x^4 - x + 1$.

14. $|E_n(f, x)| \le \left| \dfrac{(x-1)(x-q)\ldots(x-q^n)(n+1)!}{(n+1)!} \right|$

　　$|E_n(f, 0)| \le (1)(q)(q^2)\ldots(q^n) = q^{[n(n+1)/2]}$. Since $q < 1$, we have the result.

15. (a) (i) $M_2 = 0.25$, 0.1265, 　　　(ii) $M_2 = 4$, 0.0316,

　　　　(iii) $M_2 = 1/32$, 0.3578, 　　　(iv) $M_2 = 7e$, 0.0145.

　　(b) (i) $M_3 = 3/8$, 0.02749, 　　　(ii) $M_3 = 16/\sqrt{3}$, 0.0945.

　　　　(iii) $M_3 = 3/256$, 0.8729, 　　(iv) $M_3 = 13e$, 0.0604.

16. (i) No. 　　　　　　　　　　(ii) Yes.

17. (i) $s_1(x) = 4(2-x)^3 + 15(x-1)^3 + 18(x-1)$; $s_2(x) = 15(3-x)^3 + 36(x-2)^3 + 18(3-x) + 126(x-2)$. $f(1.1) = 4.731$, $f(2.5) = 78.375$.

(ii) $s_3(x) = 38.2(4 - x)^3 + 68(x - 3)^3 - 150.8(4 - x) + 508\ (x - 3)$.

(iii) $s_3(x) = [474\ (3 - x)^3 - 171(x - 2)^3 - 1818(3 - x) - 396(x - 2)]/7$.

$\qquad s_4(x) = -\ [171(4 - x)^3 + 396(4 - x)]/7.\ f\ (2.5) = -152.732,\ f\ (3.8) = -11.5097$.

18. $s_2(x) = -\ (4/\pi^3)[(\pi/2) - x]^3 + (3/\pi)\ [(\pi/2) - x)].\ f\ (\pi/4) = 0.6875$.

19. $u = x.\ \alpha_{-1} = 1,\ \alpha_0 = 5,\ \alpha_1 = 21,\ \alpha_2 = 55$.

20. $u = x.\ \alpha_{-1} = 6,\ \alpha_0 = 11,\ \alpha_1 = 18,\ \alpha_2 = 32$.

21. $u = x.\ \alpha_{-2} = 0.5,\ \alpha_{-1} = 1.5,\ \alpha_0 = 8.5,\ \alpha_1 = 33.5,\ \alpha_2 = 76.5$.

22. $u = x - 1.\ \alpha_{-2} = 5,\ \alpha_{-1} = 5,\ \alpha_0 = 45,\ \alpha_1 = 137,\ \alpha_2 = 401$.

23. $u = 10x - 1.\ \alpha_{-2} = -\ 0.12,\ \alpha_{-1} = 2.54,\ \alpha_0 = 0.58,\ \alpha_1 = 4.24,\ \alpha_2 = 4.48$.

24. $u = 10x - 1.\ \alpha_{-3} = 1.016,\ \alpha_{-2} = 1.19,\ \alpha_{-1} = 1.484,\ \alpha_0 = 2.234,\ \alpha_1 = 4.04,\ \alpha_2 = 7.766$.

25. $u = 2(x - 1).\ \alpha_{-3} = 15/28,\ \alpha_{-2} = 3.5,\ \alpha_{-1} = 265/28,\ \alpha_0 = 1233/56,\ \alpha_1 = 1301/28,$
$\alpha_2 = 89.5,\ \alpha_3 = 4467/28$.

Exercise 18.5

1. $2.6 - 1.3x,\ 2.3$.

2. $0.05 + 0.18x,\ 0.0064$.

3. $-2.448 + 5.68x,\ 0.05264$.

4. $1360.0002 - 353.9304x + 23.2146x^2$.

5. $-3.2926 + 4.64x - 0.2714x^2$

6. $1.06 + 1.89x - 0.25x^2$

7. $-(13/6) + 3x$.

8. $-0.2 + 0.9x$.

9. $2(8 + x)/15$.

10. $(103 - 132x + 90x^2)/20$.

11. $(-2 + 48x + 60x^2)/105$.

12. $2.5\ e^{-1.5x}$

13. $1.5e^{-0.6x}$.

14. $3x^{1.5}$

15. $1.75x^2$.

16. $a = 1.4998,\ b = 0.9004$.

17. $a = 1.37381,\ c = 1.66534$.

18. $a = 0.25902,\ b = 0.75081$.

19. $a = -\ 0.67418,\ b = 7.1546$.

20. $a = 1.62207,\ b = 4.63782$.

Exercise 18.6

1. (a) 0.456, 0.477;

(b) 0.4980.

2. (a) 781, 496, 427.563;

(b) 660, 570, 547.5.

3. (i) $-(h^2/6)\ y'''(\xi),\ 2$;

(ii) $-(h^2/12)\ y^{(4)}\ (\xi),\ 2$.

4. $\delta = E^{1/2} - E^{-1/2} = e^{hD/2} - e^{-hD/2} = 2\ \sinh(hD/2)$.

5. $\alpha_0 = 3/(2h),\ \alpha_1 = -2/h,\ \alpha_2 = 1/(2h),\ (h^2/2)\ f'''(\xi);\ 2$.

6. 0.0621, 0.0664; Actual errors: 0.0057, 0.0014; 0.0450, 0.0113.

7. $n = 10$.

8. 0.5793, 0.3782

9. 0.7854, 0.6459

10. 1.0, 1.26, 1.3411

11. 0.8148, 0.7317; exact: 0.7324.

12. 3.0414, 3.0406.

13. $n = 2N = 14$.

14. $M_4 = 4$, 0.00083.

15. (i) 0.10616, 0.10615. Exact: 0.10615. (ii) 0.784616, 0.785396, Exact: 0.785398.

(iii) 0.802432, 0.804505. Exact: 0.804712.

16. (i) 0.470588, 0.544341, 0.536422. (ii) 0.08, 0.078947, 0.078958. Exact: 0.078958.

(iii) 0.222222. 0.237885, 0.238822. Exact:0.238878.

(iv) 0.153846, 0.154639, 0.154548. Exact: 0.154549.

17. (i) 0.59709, 0.69321. (ii) 0.39270, 0.58905. Exact: 0.58905.

18. (i) 0.26466, 0.312423. Exact: 1/3. (ii) 0.15126, 0.15956.

19. (i) $\sqrt{\pi}$, $\sqrt{\pi}$. Exact: $\sqrt{\pi}$. (ii) 0.17192, 0.17198.

Exercise 18.7

1. 2.7, 3.9791, 6.8366.

2. 1.8, 2.9323, 4.4508, 6.4087, 8.8590.

3. 1.22, 1.4992, 1.8726.

4. 1.2128, 1.4575.

5. 0.2427, 0.5809.

6. 1.956, 5.9104.

7. 2.0316, 2.0597.

8. 2.6398, 3.7711.

9. 0.6104, 0.8491

10. 1.3551, 1.7372.

11. 0.96, 0.8494, 0.7131, 0.5878.

12. Since the given *IVP* is linear, we obtain a linear equation for y_{n+1}. We get

$y_{n+2} = 1.52\, y_{n+1} - 0.04\, y_n + 0.02(5x_{n+2} + 8x_{n+1} - x_n)$. 1.5133, 2.3322, 3.6044.

13. As in Problem **12**, we get

$y_{n+2} = [-\, 8y_{n+1} + 13y_n + (x_{n+2} + 4x_{n+1} + x_n)]/17$.

0.688, 0.5115, 0.4266, 0.4022, 0.4193.

14. 1.48; 2.2371, 2.3995, 2.4748; 4.2931, 5.1303, 5.9193.

15. −0.4, −0.8066, −1.1359; −1.2056, −1.2737, −1.2498; −1.4931, −1.4383, −1.5015.

16. 2.575578 (four iterations)

17. 0.970433 (four iterations).

18. 0.09488, 0.18149.

19. 1.709984, 3.378286. Exact: 1.708026, 3.377031.

20. 1.02273, 1.13636, 1.43182. Exact: 1.01563, 1.125, 1.42188.

Chapter 19

Theory of Probability

19.1   Introduction

The word "*probable*" means "likely", or "most likely to be true", or "likely, though not certain to occur". In other words, we have an uncertain situation, where we cannot predict in advance what is going to happen.

Suppose, we conduct a certain experiment like tossing a coin/coins, throwing a die etc. There could be several possible outcomes for the experiment. We may also have a certain statement, like getting "two heads in the tossing of 3 coins simultaneously" etc., that is associated with a particular outcome of the experiment. For each possible outcome, the statement is either "true" or "false". We want to assign to the statement a numerical value that in some sense, suggests how much chance the statement has of "turning out to be true", that is, the experiment produces the outcome for which the statement has the value "true". This measure of "chance" or "likelihood" for the statement to be true is called the *probability* of the statement. The probability theory provides a mathematical model to study the uncertain situations. Consider the following simple examples.

Suppose, we toss a coin on the floor. Then, either a head or a tail may show up. We cannot predict in advance whether a head or a tail will show up.

Consider now a dice. A, dice is a cube which has six faces. Let these faces be numbered as 1, 2, 3, 4, 5, and 6. If the dice is thrown, then any one of the six faces can turn up. Again, we cannot predict in advance which number is going to turn up.

Consider a pack of 52 playing cards. Assume that we have shuffled the pack of cards and a card is drawn. Again, we are not sure which card it will be.

19.2   Random Experiments and Events

Random means "haphazard". Hence, any experiment happening under uncertain situations is called a *random trial* or a *random experiment*. It may also be called an *experiment of chance*. The three examples which we have described in the previous section – result of tossing a coin, result of throwing a dice, result of drawing a card from a pack of 52 playing cards – are random events or random experiments. In all these experiments, there are more than one possible outcomes, but we are not sure which one of these outcomes actually occurs. Thus, the theory of probability may be defined as that branch of mathematics where we investigate and discuss rules for random experiments.

Event

Any question that we ask with regard to a random experiment defines an event.

Elementary events

Suppose that we have conducted a random experiment. It is completely defined when we know all the possible outcomes. Each outcome of this experiment is called an elementary event. Suppose that a dice is thrown. The appearance of any number i ($i = 1, 2, 3, 4, 5, 6$) is an elementary event and there are six elementary events in this experiment. If a coin is tossed, then either a head (H) or a tail (T) may turn up. Therefore, head or tail are the two elementary events of the experiment of tossing a coin.

Sample space

Sample space of a random experiment is the set of all the possible outcomes, that is, the set of all elementary events of that experiment. We assume that this set is finite. Let E denote the random experiment and $e_1, e_2, ..., e_m$ denote all the possible out-comes. Then, the sample space S of the experiment is the set $S = \{e_1, e_2, ..., e_m\}$. Obviously, each element of the set is a possible outcome and each outcome of a trial corresponds to only one element of the set S.

Consider the following random experiments:

1. A coin is tossed. There are two possible outcomes, either a head (H) or a tail (T) may turn up. Hence, the sample space of the experiment contains two elementary events H, T and therefore, $S = \{H, T\}$.
2. Consider the random experiment of throwing a dice and noting the resulting number. The experiment can result in turning up of any one of the six numbers 1, 2, 3, 4, 5, 6, which are the elementary events of the experiment. Hence, the sample space of the experiment is $S = \{1, 2, 3, 4, 5, 6\}$.
3. Let two coins be thrown simultaneously. Then, the first coin may show up either H or T and the second coin may also show up either H or T. Denote by HT, the case of a head turning up on the first coin and a tail turning up on the second coin. Similarly, we define HH, TH and TT. The possible outcomes of the experiment are HH, HT, TH and TT, which are the elementary events of the experiment. Hence, the sample space is $S = \{HH, HT, TH, TT\}$.
4. When two random experiments, having m outcomes $e_1, e_2, ..., e_m$ and n outcomes $p_1, p_2, ..., p_n$ respectively, are conducted simultaneously, the sample space consists of mn elementary events and hence is the set

$$S = \{(e_1, p_1), (e_1, p_2), ..., (e_1, p_n), ..., (e_m, p_1), (e_m, p_2), ..., (e_m, p_n)\}$$

or the set

$$S = \{(e_i, p_j), i = 1, 2, ..., m; j = 1, 2, ..., n\}.$$

The ordered pair (e_i, p_j) means that e_i is the outcome of the first experiment and p_j is the outcome of the second experiment.

Example 19.1 If three coins are tossed simultaneously, then describe the sample space.

Solution Let H and T denote head and tail respectively. On each of the coins, either H or T can turn up. Hence, the sample space is

$$S = \{HHH, HHT, HTH, HTT, THH, THT, TTH, TTT\}.$$

Example 19.2 Two balls are to be drawn simultaneously from a set of 3 red and 2 white balls. Find the sample space.

Solution Let the red and white balls be denoted by R_1, R_2, R_3 and W_1, W_2 respectively. Since, any two balls can be drawn, the sample space is given by

$$S = \{R_1R_2, R_1R_3, R_2R_3, R_1W_1, R_1W_2, R_2W_1, R_2W_2, R_3W_1, R_3W_2, W_1W_2\}.$$

There are 10 elementary events. Alternately, since, any two balls may be drawn out of 5 balls, the number of ways they can be drawn is $C(5, 2) = 5C_2 = 10$. Hence, the experiment has 10 elementary events.

Sure event (Universal event)

The sample space S of an experiment is the set of all possible outcomes. Since, a set is also a subset of itself, S can also be considered as representing an event associated with the experiment. Now, since every outcome belongs to S, the event represented by the set S always occurs. Therefore, the event represented by S is called a sure event.

Impossible event

An empty set ϕ is always a subset of a set S. Hence, the empty set ϕ can always be considered as representing an event of an experiment. But, there is no outcome of the experiment which can belong to ϕ. Hence, the event represented by ϕ is called an impossible event.

Consider again, the example of throwing two dice simultaneously. Let an event B be defined as "the sum of the numbers on the faces is greater than or equal to 2". Since, the sum of the smallest numbers of the two faces is 2, the sum of the two numbers appearing on the two faces is always greater than or equal to 2. Hence, the set of outcomes is same as the sample space S. Therefore, the event B is a sure event. Define another event A as "the sum of the numbers is greater than 12". Since, the sum of the largest numbers of the two faces is 12, the sum of the two numbers appearing on the two faces can never be greater than 12. Hence, the set of outcomes of the event A is ϕ. Therefore, A is an impossible event.

Equally likely events

Let S be the sample space of a random experiment. If all the elementary events of S have the same chance of occuring, then the events are said to be equally likely events.

Mutually exclusive events

Consider the example of throwing two dice simultaneously, and the sum of the two numbers is noted. Define the events A and B as

A: the sum of the two numbers appearing on the dice is less than or equal to 4,

B: the sum of the two numbers appearing on the dice is greater than 9.

The set E_1 of outcomes of the event A is $E_1 = \{2, 3, 4\}$. The set E_2 of outcomes of the event B is $E_2 = \{10, 11, 12\}$. The sample space of the experiment is

$$S = \{2, 3, 4, 5, 6, 7, 8, 9, 10\ 11, 12\}.$$

We find that $E_1 \cap E_2 = \phi$ and hence, the sets E_1, E_2 are disjoint. We then say that the events A and B are mutually exclusive. That is, when one event has occured, the other event cannot occur or the two events cannot occur together. If the events associated with an experiment are mutually exclusive, then the subsets of the sample space representing the events are disjoint. The converse is also true. That is, if the subsets representing the events of an experiment are disjoint, then the events are said to be mutually exclusive.

Mutually exhaustive events

Let S be the sample space of a random experiment and A_1, A_2, ..., A_m be the events defined on the sample space. If $A_1 \cup A_2 \cup ... \cup A_m = S$, then the events are said to be exhaustive. If further $A_i \cap A_j = \phi$, $i \neq j$, then the events are said to be mutually exclusive and exhaustive.

Combination of events

We now consider combination of events defined in an experiment. This can be done by using the operations "*or*", "*and*", "*not*". Let us first consider an example. Consider the experiment of throwing two dice simultaneously and noting the total of the numbers that have turned up. Define the events

A : the sum of the two numbers is less than or equal to 5,

B : the sum of the numbers satisfy, $4 \leq \text{sum} \leq 8$,

C : the sum of the numbers satisfy, $5 < \text{sum} < 8$.

Let E_1, E_2 and E_3 denote the sets of outcomes of the events A, B and C respectively. Now, $E_1 = \{2, 3, 4, 5\}$, $E_2 = \{4, 5, 6, 7, 8\}$, $E_3 = \{6, 7\}$, while the sample space is $S = \{2, 3, ..., 12\}$.

We define the event *A or B* as the event which occurs when either A or B or both occur. In set notation, we denote this operation as $A \cup B$. The set representing $A \cup B$ in the above example is $E = E_1 \cup E_2 = \{2, 3, 4, 5, 6, 7, 8\}$.

We define a new event *A and B*, as the event which occurs when A and B both occur. In set notation, we denote this operation by $A \cap B$. In the above example, the set representing $A \cap B$ is $E = E_1 \cap E_2 = \{4, 5\}$.

We define the event *not A*, as the event which occurs when A does not occur. If E_1 is the set representing the event A, then the set representing *not A* contains all elements of the sample space S which do not belong to E_1. Thus, *not A* is represented by the *complement* in S of the set E_1, which can be written as E_1^c or $\overline{E}_1$. Often, *not A* is also called the *complementary event* of A, or the *negation of A*.

Suppose that E_2, E_3 are the subsets of a sample space and we have $E_3 \subset E_2$. The subsets E_2 and E_3 represent the outcomes of the events B and C respectively. We define the event $C \subset B$ as "the event C implies the event B" or if the event C occurs, then the event B must occur.

We also define an event $A - B$ as "the event A but *not B*".

We summarize the combination of events as follows:

(i) $A \cup B$: either A or B or both,   (ii) $A \cap B$: both A and B,   (iii) A^c or $\overline{A}$: not A, (iv) $A - B$: A but *not B*,   (v) $A \cap B = \phi$: mutually exclusive events A and B.

Example 19.3 A coin is tossed thrice. If event A denotes the "number of heads is odd" and event B denotes the "number of tails is odd", then find the cases favourable to $A \cap B$.

Solution The sample space is

$$S = \{HHH, HHT, HTH, HTT, THH, THT, TTH, TTT\}.$$

The events A and B are

$$A = \{HHH, HTT, THT, TTH\},$$

$$B = \{HHT, HTH, THH, TTT\}.$$

Hence, $A \cap B = \{0\} = \phi$.

Remark 1

Since $A \cap \overline{A} = \phi$, the event A and its complement $\overline{A}$ are mutually exclusive.

19.3 Probability of an Event

To every event in a random experiment we attach a numerical value which is called its probability. We denote the probabilities of the events A and B by $P(A)$ and $P(B)$ respectively. We say that $P(A) > P(B)$, if the event A is more likely to occur, than the event B. Obviously, every event is more likely to occur than the impossible event and is less likely to occur than the sure event. Hence, the impossible event must have the smallest probability and the sure event must have the largest probability. By convention, we assign the value 0 to the probability of the impossible event and the value 1 to the probability of the sure event. Hence, the probability of an event A satisfies the inequality

$$0 \leq P(A) \leq 1. \tag{19.1}$$

This result is an *axiom* of calculus of probability.

We now need a rule to compute the value of the probability of an event.

Let the sample space S contain n elementary events e_i, that is

$$S = \{e_1, e_2, e_3, ..., e_n\}. \tag{19.2}$$

By definition, $P(e_i) \geq 0$, $i = 1, 2, ..., n$. Assume now, that all the elementary events are equally likely to occur when the experiment is performed.

Let the set of outcomes E represent an event A. If the experiment produces an outcome which belongs to E, then the event A is said to have occured. Let n and m, $(m \leq n)$ be the number of elementary events in the sample space S and E respectively. Since, all the elementary events are equally likely to occur, we define the probability of the event A as

$$P(A) = p = \frac{\text{number of elementary events favourable to E}}{\text{total number of equally likely elementary events}} = \frac{m}{n}. \tag{19.3}$$

This result is called the statistical definition of probability. We further have

$$P(\phi) = 0 \text{ and } P(S) = 1. \tag{19.4}$$

This result is also an *axiom* of calculus of probability.

Example 19.4 Three coins are tossed simultaneously. What is the probability that atleast two tails occur?

Solution The sample space contains 8 elementary events,

$$S = (HHH, HHT, HTH, THH, HTT, THT, TTH, TTT\}.$$

Define the event A as

A : atleast two tails occur.

The favourable cases are *HTT, THT, TTH* and *TTT*.

Hence, $$P(A) = \frac{4}{8} = \frac{1}{2}.$$

Example 19.5 What is the probability that a number selected from the numbers 1, 2, ..., 20 is an even number, when each of the given numbers is equally likely to be selected?

Solution From the numbers 1, 2, ..., 20, the even numbers are 2, 4, 6, 8, 10, 12, 14, 16, 18 and 20, or a total of 10 even numbers. Since each number is equally likely to be selected, the required probability is

$$p = \frac{10}{20} = \frac{1}{2}.$$

We now present another axiom of probability.

Theorem 19.1 If A and B are two mutually exclusive events of a random experiment, then

$$P(A \cup B) = P(A) + P(B). \tag{19.5}$$

Proof Let n be the total number of elementary events in the sample space S. Let m_1 and m_2 be the number of elementary events in the sets E_1 and E_2 associated with the events A and B respectively. Since, A and B are mutually exclusive, the sets E_1 and E_2 are disjoint. Hence, the number of elementary events in $E_1 \cup E_2$ is $m_1 + m_2$. Therefore,

$$P(A \cup B) = \frac{m_1 + m_2}{n} = \frac{m_1}{n} + \frac{m_2}{n} = P(A) + P(B).$$

This result is called the *addition rule for probabilities of mutually exclusive events*.

We summarise the axioms of probability.

Axioms of probability

Let S be the sample space and A and B be two mutually exclusive events. Then

    (i) $0 \leq P(A) \leq 1, 0 \leq P(B) \leq 1$.
    (ii) $P(S) = 1$.
    (iii) $P(A \cup B) = P(A) + P(B)$.

Remark 2

    1. If $A_1, A_2, ..., A_k$ are k mutually exclusive events, then

$$P(A_1 \cup A_2 \cup ... \cup A_k) = P(A_1) + P(A_2) + ... + P(A_k). \tag{19.6}$$

    2. If $A_1, A_2, ..., A_k$ are k mutually exclusive and exhaustive events, that is,

$$A_1 \cup A_2 \cup ... \cup A_k = S, \quad \text{then} \quad P(A_1 \cup A_2 \cup ... \cup A_k) = P(S) = 1$$

or $$P(A_1) + P(A_2) + ... + P(A_k) = 1.$$

Theorem 19.2 (*Law of addition of probabilities*) If A and B are any two events associated with a random experiment, then

$$P(A \cup B) = P(A) + P(B) - P(A \cap B). \tag{19.7}$$

Proof Consider the Venn diagram (see Fig. 19.1) of the subsets A and B of the sample space S. We have

$$A \cup B = I + II + III \tag{19.8}$$

Since I, II, III are mutually exclusive, using Theorem 19.1, we get

$$P(A \cup B) = P(I) + P(II) + P(III). \tag{19.9}$$

Now, $P(I) = P(A) - P(A \cap B) \tag{19.10i}$

$P(II) = P(A \cap B), \tag{19.10ii}$

and $P(III) = P(B) - P(A \cap B). \tag{19.10iii}$

Substituting in (19.9), we obtain

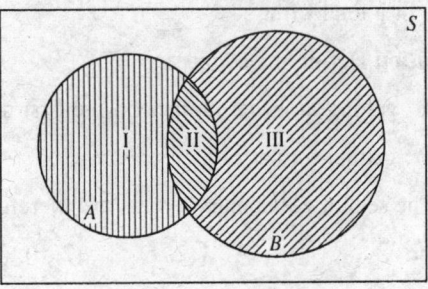

Fig. 19.1 Venn diagram.

$$P(A \cup B) = P(A) + P(B) - P(A \cap B).$$

Remark 3

1. If A and B are mutually exclusive events, then $A \cap B = \phi$. Hence, $P(A \cap B) = P(\phi) = 0$. Then, (19.7) reduces to

$$P(A \cup B) = P(A) + P(B)$$

which is same as (19.5).

2. If A, B, C are any three events, then

$$P(A \cup B \cup C) = P(A) + P(B) + P(C) - P(A \cap B) - P(B \cap C)$$
$$- P(C \cap A) + P(A \cap B \cap C). \tag{19.11}$$

Theorem 19.3 If A is any event associated with an experiment, then

$$P(not\ A) = P(\overline{A}) = 1 - P(A). \tag{19.12}$$

Proof The events A and $\overline{A}$ are mutually exclusive and $A \cup \overline{A} = S$, where S is the sample space. Hence

$$P(A \cup \overline{A}) = P(A) + P(\overline{A}) = P(S) = 1.$$

Therefore, $P(\overline{A}) = 1 - P(A).$

Theorem 19.4 If $A \subset B$, that is, the event A implies the event B, then

$$P(A) \le P(B). \tag{19.13}$$

Proof We are given that $A \subset B$ (see the Venn diagram, Fig. 19.2). We write $B = A + (B - A)$, where $B - A$ means "B but *not* A", (shaded portion in Fig. 19.2).

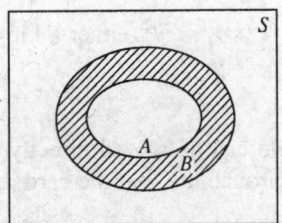

Fig. 19.2 Venn diagram.

The events A and $B - A$ are mutually exclusive. Hence,

$$P(B) = P[A + (B - A)] = P(A) + P(B - A).$$

Since, $P(B - A) \geq 0$, we obtain $P(B) \geq P(A)$.

Example 19.6 Two dice are tossed once. Find the probability of getting an even number on the first dice or a total of 8.

Solution Define the events

A : getting an even number on the first dice,

B : getting a total of 8.

The sets of elementary events are therefore

$$A : \{(2, 1), ..., (2, 6), (4, 1), ..., (4, 6), (6, 1), ..., (6, 6)\},$$

$$B : \{(2, 6), (3, 5), (4, 4), (5, 3) (6, 2)\}$$

and

$$A \cap B : \{(2, 6), (4, 4), (6, 2)\}.$$

Hence,

$$P(A) = \frac{18}{36} = \frac{1}{2}, \quad P(B) = \frac{5}{36}, \quad \text{and} \quad P(A \cap B) = \frac{3}{36} = \frac{1}{12}.$$

Therefore,

P (even number on the first dice or a total of 8) $= P(A \cup B)$

$$= P(A) + P(B) - P(A \cap B) = \frac{1}{2} + \frac{5}{36} - \frac{1}{12} = \frac{5}{9}.$$

Example 19.7 From a pack of well shuffled cards, one card is drawn. Find the probability that this card is either a king or an ace.

Solution Define the events as

A : card drawn is a king,

B : card drawn is an ace.

The events A and B are mutually exclusive. There are 52 cards in a pack, 4 kings and 4 aces. Hence,

$$P(A) = \frac{4}{52} = \frac{1}{13}, \quad P(B) = \frac{4}{52} = \frac{1}{13}.$$

Since, A and B are mutually exclusive

$$P \text{ (either a king or an ace)} = P(A \cup B) = P(A) + P(B)$$

$$= \frac{1}{13} + \frac{1}{13} = \frac{2}{13}.$$

This example can be solved directly also. The number of kings and aces in a pack of cards is 8. Hence, the probability that the card drawn is either a king or an ace is 8/52 = 2/13.

Example 19.8 From a pack of well shuffled cards, four cards are drawn. Find the probability that there are two hearts and two diamonds.

Solution Total number of possibilities = $52C_4$.

Two hearts can be chosen in $13C_2$ ways. Two diamonds can be chosen in $13C_2$ ways. Hence,

$$\text{required probability} = \frac{(13C_2)\,(13C_2)}{(52C_4)}.$$

Example 19.9 A bag contains 4 red and 3 black balls. A second bag contains 2 red and 4 black balls. One bag is selected at random. From the selected bag one ball is drawn. find the probability that the ball drawn is red.

Solution We define the following events:

   A : The ball is drawn from the first bag,

   B : The ball is drawn from the second bag.

Since, both bags are equally likely to be selected, we have

$$P(A) = P(B) = 1/2.$$

Now, probability of drawing a red ball from the first bag is $p_1 = 4/7$ and the probability of drawing a red ball from the second bag is $p_2 = 2/6 = 1/3$. Hence, the required probability is

$$P(A)\,p_1 + P(B)\,p_2 = \frac{1}{2}\cdot\frac{4}{7} + \frac{1}{2}\cdot\frac{1}{3} = \frac{19}{42}.$$

Example 19.10 If A, B, C are mutually exclusive and exhaustive events associated with a random experiment and $P(B) = 0.6\ P(A)$ and $P(C) = 0.2\ P(A)$, then find $P(A)$.

Solution Since, A, B, C are mutually exclusive and exhaustive events, we have

$$P(A) + P(B) + P(C) = 1$$

or $\qquad\qquad\qquad P(A) + 0.6\ P(A) + 0.2\ P(A) = 1$

or $\qquad\qquad\qquad\qquad\qquad 1.8\ P(A) = 1$

or $\qquad\qquad\qquad\qquad\qquad P(A) = \dfrac{1}{1.8} = \dfrac{5}{9}.$

Example 19.11 The probability that atleast one of the events A and B occurs is 0.8 and the probability that both the events occur simultaneously is 0.25. Find the probability $P(\overline{A}) + P(\overline{B})$.

Solution We are given that $P(A \cup B) = 0.8$ and $P(A \cap B) = 0.25$.

Since, $\qquad\qquad\qquad P(A \cup B) = P(A) + P(B) - P(A \cap B)$

we have $\qquad P(A) + P(B) = P(A \cup B) + P(A \cap B) = 0.8 + 0.25 = 1.05.$

Also, $\qquad P(A) + P(B) = 1 - P(\overline{A}) + 1 - P(\overline{B}) = 2 - P(\overline{A}) - P(\overline{B}).$

We obtain $\qquad P(\overline{A}) + P(\overline{B}) = 2 - [P(A) + P(B)] = 2 - 1.05 = 0.95.$

Example 19.12 A bag contains 6 red and 5 blue balls and another bag contains 5 red and 8 blue balls. A ball is drawn from the first bag and without seeing its colour is put in the second bag. A ball is then drawn from the second bag. Find the probability that the ball drawn is blue in colour.

Solution Two cases arise. The ball drawn from the first bag is either red or blue.

Case 1: Let the ball drawn from the first bag be red. The probability of drawing a red ball from this bag is $p_1 = 6/11$.

The ball is now put in the second bag. The second bag has now 6 red and 8 blue balls. The probability of drawing a ball of blue colour is $p_2 = 8/14$.

Probability of these two events occuring simultaneouely is

$$p_3 = \frac{6}{11} \cdot \frac{8}{14} = \frac{24}{77}.$$

Case 2 : Let the ball drawn from the first bag be blue. The probability of drawing a blue ball from this bag is $p_4 = 5/11$.

The ball is now put in the second bag. The second bag has now 5 red and 9 blue balls. The probability of drawing a ball of blue colour is $p_5 = 9/14$.

Probability of these two events occuring simultaneously is

$$p_6 = \frac{5}{11} \cdot \frac{9}{14} = \frac{45}{154}.$$

The required probability is

$$p = p_3 + p_6 = \frac{24}{77} + \frac{45}{154} = \frac{93}{154}.$$

Example 19.13 Prove the Boole's inequality $P\left[\bigcup_{i=1}^{n} A_i\right] \leq \sum_{i=1}^{n} P(A_i)$ where $A_1, A_2, ..., A_n$ are n events.

Solution We shall prove the result by induction.

For $n = 1$: $\qquad\qquad\qquad\qquad P(A_1) = P(A_1),$

For $n = 2$: $\qquad\qquad\qquad P(A_1 \cup A_2) = P(A_1) + P(A_2) - P(A_1 \cap A_2)$

$$\leq P(A_1) + P(A_2).$$

Let the result be true for $n = r$.

$$P\left[\bigcup_{i=1}^{r} A_i\right] \leq \sum_{i=1}^{r} P(A_i).$$

Then, for $n = r + 1$, we obtain

$$P\left[\bigcup_{i=1}^{r+1} A_i\right] = P\left[\bigcup_{i=1}^{r} A_i \cup A_{r+1}\right] \leq P\left[\bigcup_{i=1}^{r} A_i\right] + P(A_{r+1})$$

$$\leq \left[\sum_{i=1}^{r} P(A_i)\right] + P(A_{r+1}) = \left[\sum_{i=1}^{r+1} P(A_i)\right].$$

Example 19.14 The word ASSASIN is given. It is rearranged so that the three S's come consecutively. Find the probability of this event.

Solution There are 7 letters in the given word. The total permutations of the 7 letters, in which there are 3 of first kind (S), 2 of second kind (A), one each of third and fourth kinds (I, N) is

$$\frac{7!}{3!\,2!\,1!\,1!} = 420.$$

There are 5 possible combinations of occuring of 3 S's consecutively.

$$
\begin{array}{l}
S\ S\ S - - - - \\
- S\ S\ S - - - \\
- - S\ S\ S - - \\
- - - S\ S\ S - \\
- - - - S\ S\ S
\end{array}
$$

In each case, the total permutations of the 4 letters (2 of one kind), 1 each of third and fourth kinds is

$$\frac{4!}{2!\,1!\,1!} = 12.$$

$$\text{Required probability} = \frac{5(12)}{420} = \frac{1}{7}.$$

Exercise 19.1

1. Three coins are tossed simultaneously. Derive the sample space.
2. Two dice are thrown simultaneously and the sum is noted. Obtain the sample space.
3. (i) Three males and two females appear in an interview. Two candidates are to be selected. Find the set A representing the event "atleast one female is selected."
   (ii) Three dice are thrown simultaneously and the total of the numbers appearing on the faces is noted. An event A is defined as "the sum is greater than 16". Find the set representing this event.
   (iii) A coin is tossed three times. An event A is defined as "atmost one tail appear". Find the set representing this event.
4. Two dice are thrown simultaneously. Find the probability of
   (i) getting a sum as 8, (ii) getting a sum of atleast 9,
   (iii) getting the two numbers as equal (doublets),
   (iv) getting a sum of atmost 5, (v) getting a total of 7 or 11.
   (vi) getting an even number as sum, (vii) getting a total of 9 or 11.
5. A and B are two mutually exclusive events of a random experiment. If $P(A \cup B) = 0.75$ and $P(\overline{A}) = 0.6$, then find $P(B)$.
6. A committee of 5 principals is to be constituted from a group of 6 male principals and 8 lady principals. If the selection is made randomly, find the probability that there are 3 lady principals and 2 male principals.
7. One hundred tokens are numbered 1, 2, 3, ..., 100. The tokens are put in a box, shuffled well and one token is drawn. Find the probability that the drawn number is divisible by 3 or 7.
8. Bag A contains 5 white and 7 black balls. Bag B contains 6 white and 4 black balls. One ball is drawn from bag A and two balls are drawn from bag B. Find the probability that out of the 3 balls drawn, two are white and one is black.
9. There are three bags. The first bag contains 3 green, 2 red and 5 black balls. The second bag contains 3 green, 4 red and 3 black balls. The third bag contains 4 green, 3 red and 3 black balls. A bag is selected at random and one ball is drawn from this bag. Find the probability that the ball drawn is red.

10. A dice is tossed two times. Find the probability of getting 4, 5 or 6 in the first toss and 3, 4 or 5 in the second toss.

11. A bag contains 4 white and 5 black balls and another bag contains 3 white and 4 black balls. A ball is taken out from the first bag and without seeing its colour is put in the second bag. A ball is taken out from the latter. Find the probability that the ball drawn is white.

12. A box contains 8 white and 12 black balls. A second box contains 12 white and 13 black balls. One box is selected at random. From the selected box, one ball is drawn. Find the probability that the drawn ball is white.

13. A bag contains 2 blue, 3 white and 5 black balls. If 3 balls are drawn one by one with replacement, then what is the probability that atleast one is black?

14. A bag contains 5 white, 7 red and 8 black balls. If four balls are drawn one by one with replacement, what is the probability that atleast one is white?

15. A bag contains 8 red and 12 white balls. Two balls are drawn simultaneously. What is the probability that one is red and the other is white?

16. A bag contains 8 blue, 11 red and 6 green balls. If five balls are drawn one by one with replacement, what is the probability that none is red?

17. Two dice are thrown once. Find the probability of getting "an odd number on the first dice" or "a total of 7".

18. A dice is thrown 3 times and the sum of the three numbers is found to be 16. Find the probability that 5 appears on the third throw.

19. Four cards are drawn at a time from a pack of 52 playing cards. Find the probability of getting
    (i) all the 4 cards of the same suit.
    (ii) all the 4 cards of the same number (assume figures like king, queen etc.as numbers).

20. A card is drawn at random from a well shuffled pack of 52 cards. Find the probability that it is not an ace, king or queen.

21. A coin is tossed 3 times. Event A denotes "the number of tails is even" and event B denotes "the number of heads is even". Find the cases that are favourable to $A \cap B$.

22. A and B are two mutually exclusive events for which $P(A) = a$, $P(B) = 0.4$ and $P(A \cup B) = 0.7$. Find the value of a.

23. A coin is tossed 5 times. What is the probability that head apears even number of times (take zero as an even number)?

24. A coin is tossed 5 times. What is the probability that tail appears an odd number of times?

25. An unbiased dice with faces marked 1, 2, 3, 4, 5 and 6 is rolled four times. Out of the four face values obtained, find the probability that the minimum face value is not less than 2 and the maximum face value is not greater than 5.

26. A box contains 2 one rupee coins, 5 fifty paise coins and 8 ten paise coins. Five coins are taken out of the box at random. Find the probability that the total value of these 5 coins is less than 3 rupees.

27. The probability that a player A wins a game against B is 0.7. The players may play either a "best of three games" or a "best of five games". If A has been asked to choose, then which option should he choose, so that he wins the match?

28. Two cards are drawn from a pack of 52 playing cards, one by one without replacement. Find the probability that one of these is a king and the other is a jack of opposite colour.

29. A total of $(2n + 1)$ tickets are numbered consecutively. Three tickets are drawn at random. Find the probability that these numbers are in A.P.

30. Three faces of a fair dice are yellow, two faces red and one face blue. The dice is tossed three times. Find the probability that the colours yellow, red and blue appear in the first, second and the third tosses respectively.

19.4   Conditional Probability

We assume that the outcomes of the random experiment are equally likely. To understand the concept of conditional probability, consider the following two examples.

Suppose that a bag contains 10 blue, 15 yellow and 20 green balls. We assume that except for colour, all the balls are identical. Hence, we can assume that elementary events of the experiment are equally likely to occur. Let us define an event A as

A : the ball is blue.

Then, $P(A) = 10/45 = 2/9$. Now suppose we are told after the ball has been drawn that the ball drawn is not green. Because of this extra information (condition) given to us, we need to change the value of $P(A)$. Now, since the ball is not green, the total number of balls can be considered as 25 only. Hence, $P(A) = 10/25 = 2/5$.

Consider the experiment of throwing two dice simultaneously and noting the sum of the numbers. Define the event A as

A : the sum is even.

The sample space of the experiment is $S = \{2, 3, 4, \ldots, 12\}$. The subset of elementary events favouring A is $E = \{2, 4, 6, 8, 10, 12\}$. Therefore $P(A) = 6/11$. Suppose that, after the dice have been thrown, we are told that the sum of the numbers that appeared is greater than 5. We define a new event B as

B : the sum is greater than 5.

When B has occured, the elementary events of the new sample space are 6, 7, 8, 9, 10, 11 and 12. Of these, 6, 8, 10 and 12 are even sums. Hence, given that B has occured, $P(A) = 4/7$.

The extra information given in the above two examples can be considered as another event. This means, that if after the experiment has been conducted, we are told that a particular event has occured, then we need to revise the values of the probabilities of the previous events. In otherwords, we wish to find the probability of the random event A under the condition that the random event B has occured. We call this changed probability as the *conditional probability* of A when B has occured. We denote this conditional probability by $P(A/B)$.

Let S be the sample space of an experiment having n elementary events. Let m_1 denote the number of elementary events favourable to B, which are also favourable to A and m_2 denote the number of elementary events in the sample space which are favourable to B. Obviously, $m_2 \geq m_1$.

Hence,
$$P(A/B) = \frac{m_1}{m_2}. \tag{19.14}$$

However, m_1 represents the event A *and* B, that is, m_1 denotes the number of elementary events favourable to both the events A and B. Dividing the numerator and denominator in (19.14) by the total number of elementary events n, we have

$$P(A/B) = \frac{(m_1/n)}{(m_2/n)} = \frac{P(A \text{ and } B)}{P(B)} = \frac{P(A \cap B)}{P(B)} \tag{19.15}$$

where $P(B) \neq 0$. Similarly, the conditional probability of B when A has occured is given by

$$P(B/A) = \frac{P(B \text{ and } A)}{P(A)} = \frac{P(B \cap A)}{P(A)} = \frac{P(A \cap B)}{P(A)} \tag{19.16}$$

where $P(A) \neq 0$. From (19.15) and (19.16), we have the following result.

Theorem 19.5 (*Multiplication law of probabilities*) Let $P(A/B)$ and $P(B/A)$ denote the conditional probabilities of A when B has occured, and of B when A has occured respectively. Then,

$$P(A \cap B) = P(B)\, P(A/B) = P(A)\, P(B/A). \tag{19.17}$$

Example 19.15 For two events A and B, $P(A) = 0.5$, $P(B) = 0.6$ and $P(A \cup B) = 0.8$. Find the conditional probability $P(A/B)$ and $P(B/A)$.

Solution We have

$$P(A \cup B) = P(A) + P(B) - P(A \cap B).$$

Hence, $$P(A \cap B) = P(A) + P(B) - P(A \cup B)$$

$$= 0.5 + 0.6 - 0.8 = 0.3.$$

We have

$$P(A/B) = \frac{P(A \cap B)}{P(B)} = \frac{0.3}{0.6} = \frac{1}{2}$$

and

$$P(B/A) = \frac{P(B \cap A)}{P(A)} = \frac{0.3}{0.5} = 0.6.$$

Example 19.16 Let A and B be two events such that $P(A) = 0.6$, $P(B) = 0.4$ and $P(A \cap B) = 0.25$. Find $P(A/B)$ and $P(B/A)$.

Solution
$$P(A/B) = \frac{P(A \cap B)}{P(B)} = \frac{0.25}{0.4} = \frac{5}{8}.$$

$$P(B/A) = \frac{P(B \cap A)}{P(A)} = \frac{0.25}{0.6} = \frac{5}{12}.$$

Example 19.17   A dice is thrown twice and the sum of the numbers appearing is noted to be 8. What is the conditional probability that the number 5 has appeared atleast once?

Solution Define the events

 A : the number 5 appears atleast once.

 B : the sum of the numbers appearing is 8.

The sets of elementary events are therefore

$A = \{(5, 1), (5, 2), (5, 3), (5, 4), (5, 5), (5, 6), (1, 5),(2, 5), (3, 5), (4, 5), (6, 5)\}$

$B = \{(2, 6), (3, 5),(4, 4), (5, 3), (6, 2)\}$

and $A \cap B = \{(5, 3), (3, 5)\}$.

 The total outcomes of throwing two die is $6 \times 6 = 36$. Hence,

$$P(A) = \frac{11}{36}, \quad P(B) = \frac{5}{36}, \quad P(A \cap B) = \frac{2}{36},$$

$$P(A/B) = \frac{P(A \cap B)}{P(B)} = \frac{2/36}{5/36} = \frac{2}{5}.$$

Example 19.18 If $P(B) \neq 1$, then show that $P(\overline{A}/\overline{B}) = \dfrac{1 - P(A \cup B)}{P(\overline{B})}$.

Solution We have

$$P(\overline{A}/\overline{B}) = \frac{P(\overline{A} \cap \overline{B})}{P(\overline{B})} \quad \text{and} \quad \overline{A \cup B} = \overline{A} \cap \overline{B}.$$

Hence, $\quad P(\overline{A}/\overline{B}) = \frac{P(\overline{A} \cap \overline{B})}{P(\overline{B})} = \frac{P(\overline{A \cup B})}{P(\overline{B})} = \frac{1 - P(A \cup B)}{P(\overline{B})}.$

Example 19.19 Two cards are drawn one after the other from a well-shuffled deck of 52 cards. Find the probability that both are spade cards, if the first card is (i) replaced, (ii) not replaced.

Solution Let

$$A_1 = \text{spade card on first draw,}$$

$$A_2 = \text{spade card on second draw.}$$

We need to find $P(A_1 \cap A_2)$.

(i) Let the first card be replaced. Then, we have

$$P(A_1 \cap A_2) = P(A_1)\, P(A_2/A_1) = \left(\frac{13}{52}\right)\left(\frac{13}{52}\right) = \frac{1}{16}.$$

since there are 13 spade cards.

(ii) Consider the case when the drawn first card is not replaced.

$$P(A_1 \cap A_2) = P(A_1)\, P(A_2/A_1) = \left(\frac{13}{52}\right)\left(\frac{12}{51}\right) = \frac{1}{17}.$$

19.5 Independent Events and Independent Experiments

Two events A and B are said to be independent, if the information that one of them has occured does not change the probability of occurence of the other event. That is, the probability of event A does not depend on occurence or non-occurence of the event B. Hence, in this case the conditional probability $P(A/B)$ is same as $P(A)$. That is,

$$P(A/B) = P(A), \quad \text{and} \quad P(B/A) = P(B). \tag{19.18}$$

Substituting in (19.17) we get

$$P(A \cap B) = P(A/B)\, P(B) = P(A)\, P(B). \tag{19.19}$$

Therefore, we have that the events A and B are independent, if and only if $P(A \cap B) = P(A)\, P(B)$. It should be noted that the physical description of the events of an experiment, may not always give information about the events being independent or not. Only when the formula (19.19) is verified, we can decide that the events A and B are independent.

Remark 4

1. If the events A and B are independent, then $\overline{A}$ and $\overline{B}$ are also independent. We have

$$P(A \cap B) = P(A)\, P(B).$$

$$P(\overline{A} \cap \overline{B}) = P(\overline{A \cup B}) = 1 - P(A \cup B)$$

$$= 1 - [P(A) + P(B) - P(A \cap B)] = 1 - P(A) - P(B) + P(A \cap B)$$

$$= 1 - P(A) - P(B) + P(A) P(B) = 1 - P(A) - P(B) [1 - P(A)]$$

$$= [1 - P(A)] [1 - P(B)] = P(\overline{A}) P(\overline{B}).$$

Hence, the events $\overline{A}$ and $\overline{B}$ are also independent.

2. For the independence of three events A_1, A_2, A_3, we must have

$$P(A_i \cap A_j) = P(A_i) P(A_j), i \neq j, i, j = 1, 2, 3$$

and $\qquad P(A_1 \cap A_2 \cap A_3) = P(A_1) P(A_2) P(A_3).$

Pairwise independence of events A_1, A_2, A_3 alone does not imply independence of events. That is, if A_1, A_2, A_3 are pairwise independent, that is

$$P(A_1 \cap A_2) = P(A_1) P (A_2), P(A_2 \cap A_3) = P(A_2) P(A_3),$$

and $\qquad P(A_1 \cap A_3) = P (A_1) P(A_3),$

then, it need not always imply that $P(A_1 \cap A_2 \cap A_3) = P(A_1) P(A_2) P(A_3)$.

Again, if $P(A_1 \cap A_2 \cap A_3) = P(A_1) P(A_2)P(A_3)$, then it need not imply that they are pairwise independent.

Example 19.20 Events A and B are independent. Examine if the events

(i) $\overline{A}$ and B, (ii) A and $\overline{B}$.

are independent?

Solution Since the events A and B are independent, we have

$$P(A \cap B) = P(A) P(B).$$

(i) The events $A \cap B$ and $\overline{A} \cap B$ are mutually exclusive and

$$(A \cap B) \cup (\overline{A} \cap B) = B.$$

We obtain from addition theorem

$$P(B) = P(A \cap B) + P(\overline{A} \cap B)$$

or $\qquad P(\overline{A} \cap B) = P(B) - P(A \cap B) = P(B) - P(A) P(B)$

$$= P(B) [1 - P(A)] = P(B) P(\overline{A}).$$

Therefore, the events $\overline{A}$ and B are independent.

(ii) The events $A \cap B$ and $A \cap \overline{B}$ are mutually exclusive and

$$(A \cap B) \cup (A \cap \overline{B}) = A.$$

We obtain from addition theorem

$$P(A) = P(A \cap B) + P(A \cap \overline{B})$$

or
$$P(A \cap \overline{B}) = P(A) - P(A \cap B) = P(A) - P(A)\,P(B)$$
$$= P(A)\,[1 - P(B)] = P(A)\,P(\overline{B}).$$

Therefore, events A and $\overline{B}$ are independent.

Example 19.21 Events E and F are given to be independent. Find $P(F)$ if it is given that $P(E) = 0.4$ and $P(E \cup F) = 0.55$.

Solution Since, E and F are independent, we have $P(E \cap F) = P(E)\,P(F)$. Hence,
$$P(E \cup F) = P(E) + P(F) - P(E)\,P(F) = 0.55$$

or
$$[1 - P(E)]\,P(F) = 0.55 - P(E)$$

or
$$[1 - 0.4]P(F) = 0.55 - 0.4 = 0.15$$

or
$$P(F) = \frac{0.15}{0.60} = \frac{1}{4}.$$

Example 19.22 A and B are two events such that $P(A) = 0.4$ and $P(A \cup \overline{B}) = 0.7$. If A and B are independent events, then find $P(B)$.

Solution Since, A and B are independent, A and $\overline{B}$ are also independent. We get
$$P(A \cup \overline{B}) = P(A) + P(\overline{B}) - P(A \cap \overline{B})$$
$$= P(A) + P(\overline{B}) - P(A)\,P(\overline{B}).$$

Hence, we have
$$0.7 = 0.4 + P(\overline{B}) - 0.4\,P(\overline{B})$$

or
$$0.6\,P(\overline{B}) = 0.3; \quad \text{or} \quad P(\overline{B}) = 0.5.$$

Hence,
$$P(B) = 1 - P(\overline{B}) = 0.5.$$

Example 19.23 Given that $P(A \cap \overline{B}) = 1/4$, and $P(A \cup B) = 3/4$, find (i) $P(B)$, (ii) $P(A)$.

Solution We have

(i)
$$P(A \cap \overline{B}) = P(A) - P(A \cap B).$$
$$P(A \cup B) = P(A) + P(B) - P(A \cap B)$$
$$= [P(A) - P(A \cap B)] + P(B) = P(A \cap \overline{B}) + P(B).$$

Therefore,
$$P(B) = P(A \cup B) - P(A \cap \overline{B}) = \frac{3}{4} - \frac{1}{4} = \frac{1}{2}.$$

(ii)
$$P(\overline{A} \cap B) = P(B) - P(A \cap B).$$
$$P(A \cup B) = P(A) + P(B) - P(A \cap B) = P(A) + P(\overline{A} \cap B).$$

Therefore,

$$P(A) = P(A \cup B) - P(\overline{A} \cap B) = \frac{3}{4} - \frac{1}{3} = \frac{5}{12}.$$

Example 19.24 The probability that the student P fails in an examination is 0.25 and the probability that the student Q fails in the same examination is 0.3. Find the probability that either P or Q fails in the examination.

Solution Define the following events

A : student P fails in the examination,

B : student Q fails in the examination.

The events A and B are independent. Since $P(A) = 0.25$ and $P(B) = 0.3$, we have

$$P(A \cap B) = P(A)\, P(B) = (0.25)\,(0.3) = 0.075.$$

Therefore,

$$P(A \text{ or } B) = P(A \cup B) = P(A) + P(B) - P(A \cap B)$$

$$= 0.25 + 0.3 - 0.075 = 0.475.$$

Example 19.25 A can solve 90 percent of the problems given in a book and B can solve 75 percent. What is the probability that atleast one of them will solve a problem selected at random from the book?

Solution The probability that A can solve the problem is $P(A) = 0.9$, and the probability that B can solve the problem is $P(B) = 0.75$. Since the events are independent, the probability that atleast one of them will solve the problem is given by

$$P(A \cup B) = P(A) + P(B) - P(A \cap B) = P(A) + P(B) - P(A)\,P(B)$$

$$= 0.9 + 0.75 - (0.9)\,(0.75) = 0.975.$$

Example 19.26 A problem in a question paper is given to 3 students in a class to be solved. The probabilities of their solving the problem are 0.5, 0.7 and 0.8 respectively. Find the probability that the problem will be solved.

Solution We have

$$P(\text{problem will be solved}) = 1 - P(\text{problem will not be solved})$$

$$= 1 - P(\overline{A})\,P(\overline{B})\,P(\overline{C})$$

$$= 1 - [1 - P(A)]\,[1 - P(B)][1 - P(C)]$$

$$= 1 - [1 - 0.5]\,[1 - 0.7][1 - 0.8]$$

$$= 1 - (0.5)\,(0.3)\,(0.2) = 1 - 0.03 = 0.97.$$

Example 19.27 A is a set containing n elements. A subset C of A is chosen at random. The set A is reconstructed by replacing the elements of C. A subset D of A is again chosen at random. Find the probability that C and D have no common elements.

Solution Since the elements are replaced, the sets C and D are independent. Let x be an arbitrary element of A. Then the following cases arise:

(i) $x \in C$, $x \in D$, that is $x \in C \cap D$.

(ii) $x \in C$, $x \notin D$

(iii) $x \notin C$, $x \in D$, and

(iv) $x \notin C$, $x \notin D$.

These four cases are equally likely, mutually exclusive and exhaustive.

Hence, $P(C \cap D) = 1/4 = q$ and $p = 1 - q = 3/4$.

Since there are n elements,

$$P \ (C \text{ and } D \text{ have no common elements}) = p^n = (3/4)^n.$$

Exercise 19.2

1. A die is thrown twice and the sum of numbers appearing is observed to be 7. What is the conditional probability that the number (*i*) 3 has appeared atleast once, (*ii*) 5 has appeared atleast once?

2. A die is thrown twice and the sum of numbers appearing is noted to be 8. What is the conditional probability that the number 2 has appeared atleast once?

3. If $P(A) = 0.4$, $P(B) = 0.6$ and $P(A/B) = 0.5$, then find $P(B/A)$ and $P(A \cup B)$.

4. If $P \ (not \ A) = 0.7$, $P(B) = 0.7$ and $P(B/A) = 0.5$, then find $P \ (A/B)$ and $P(A \cup B)$.

5. The probability that the candidates A and B fail in an examination is 0.3 and 0.4 respectively. Find the probability that (i) both A and B fail, (ii) either A or B fails.

6. The faces of a dice have been coloured. Four faces are coloured white, one face is coloured black and the other face is coloured red. The dice is thrown four times. Find the probability that the colours
   (i)   white, black, red and white appear in the first, second, third and fourth throws respectively.
   (ii) red, white, black and black appear in the first, second, third and fourth throws respectively.

7. Player A plays two chess matches each with the players B and C. A win, a draw and a loss are given 2, 1, 0 points respectively. The probability that A wins, draws or loses in any match is 0.6, 0.3, and 0.1 respectively. Outcomes of the matches are assumed to be independent. Find the probability that A gets atleast 6 points.

8. A student A can solve 80% of the problems in a question paper and student B can solve 60% of the problems from the same question paper. If a problem from the question paper is selected at random, what is the probability that atleast one of them will solve the problem?

9. If A and B are two independent events such that $P(\overline{B}) = 0.6$ and $P(A \cup B) = 0.8$, then find $P(A)$.

10. If $P \ (not \ B) = 0.65$, $P(A \cup B) = 0.85$, and A and B are independent events, then find $P(A)$.

11. A and B are two independent events such that $P(A) = 0.6$, $P(A \cup B) = 0.8$. Find $P(B)$.

12. A coin is tossed three times. Events A and B are defined as
    A : the second throw results in tail
    B : the third throw results in head.
    Prove that the events A and B are independent.

13. A factory manufactured a certain item in large numbers of which 10% are found to be defective. Two items are chosen at random one at a time with replacement. Events A, B, C are defined as follows

    A : both items are defective or both items are non-defective,

    B : the first item is defective,

    C : second item is non-defective.

    Determine whether the events A, B, C are (i) pairwise independent, (ii) independent.

14. A lot contains 50 defective and 50 non-defective bulbs. Two bulbs are drawn at random, one at a time, with replacement. The events A, B, C are defined as

    A = the first bulb is defective,

    B = the second bulb is non-defective,

    C = the two bulbs are both defective or both non-defective.

Determine whether

(i) A, B, C are pairwise independent,

(ii) A, B, C are independent.

15. A pair of dice is thrown together till a sum of 4 or 8 is obtained. Determine the probability that the sum 4 appears before 8.

19.6   Theorem of Total Probability and Baye's Theorem

Theorem 19.6 Let B_1, B_2, ..., B_n be a set of exhaustive and mutually exclusive events of the sample space S with $P(B_k) \neq 0$, $k = 1, 2, ..., n$ and let A be any event of S. Then

$$P(A) = \sum_{i=1}^{n} P(B_i \cap A) = \sum_{i=1}^{n} P(B_i)\, P(A/B_i).$$ (19.20)

Proof Since B_1, B_2, ..., B_n are exhaustive and mutually exclusive events of S, we have

$$S = \bigcup_{i=1}^{n} B_i,$$

and $B_i \cap B_j = \phi$, $i \neq j$, (see Fig. 19.3). Therefore, $A \cap B_1$, $A \cap B_2$, ..., $A \cap B_n$ are also mutually exclusive and

$$A = (A \cap B_1) \cup (A \cap B_2) \cup ... \cup (A \cap B_n).$$

Hence, by additive rule for mutually exclusive events (see Eq. 19.6), we obtain

$$P(A) = P[(A \cap B_1) \cup (A \cap B_2) \cup ... \cup (A \cap B_n)]$$

$$= P(A \cap B_1) + P(A \cap B_2) + ... + P(A \cap B_n).$$

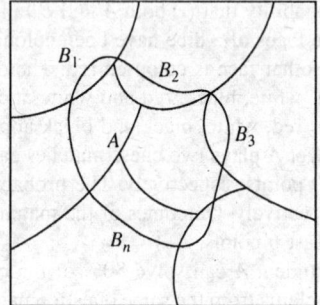

Fig. 19.3 Total Probability.

Using the multiplicative rule (see Eq. 19.16), we get

$$P(A) = \sum_{i=1}^{n} P(B_i \cap A) = \sum_{i=1}^{n} P(B_i)\, P(A/B_i).$$

Example 19.28 Two players A and B participate in a game of throwing two dice. The first player who gets a sum of 7 is awarded the prize. If A starts the game, find the probabilities of their winning.

Solution A sum of 7 is obtained if the numbers appearing on the dice are (1, 6), (2, 5), (3, 4), (4, 3), (5, 2), (6, 1).

Hence, $$P(\text{sum is } 7) = \frac{6}{36} = \frac{1}{6} = p.$$

$$P(\text{not getting sum as } 7) = 1 - \frac{1}{6} = \frac{5}{6} = q.$$

Player A wins, if he gets a sum of 7 in the first or third or fifth, ..., throw. Player B wins, if he gets a sum of 7 in the second or fourth or sixth, ..., throw.

Player A starts the game.

$$P(\text{sum of 7 in first throw}) = p = \frac{1}{6}$$

$$P(\text{sum of 7 in third throw}) = q^2 p = \left(\frac{5}{6}\right)^2 \left(\frac{1}{6}\right).$$

$$P(\text{sum of 7 in fifth throw}) = q^4 p = \left(\frac{5}{6}\right)^4 \left(\frac{1}{6}\right), \ldots$$

$$P(A \text{ wins}) = p + q^2 p + q^4 p + \ldots$$

$$= p(1 + q^2 + q^4 + \ldots) = \frac{p}{1 - q^2} = \frac{1}{6}\left(\frac{36}{11}\right) = \frac{6}{11}.$$

$$P(B \text{ wins}) = 1 - P(A \text{ wins}) = \frac{5}{11}.$$

Example 19.29 A factory has four independent units A, B, C and D which produce 40%, 30%, 20%, and 10% of identical items, respectively. The percentages of defective items produced by these units are 2%, 1%, 0.5% and 0.25% respectively. If an item is selected at random, find the probability that the item is defective.

Solution Let the defective item be denoted by M. By the theorem of total probability, we obtain

$$P(\text{defective item}) = P(A)\,P(M/A) + P(B)\,P(M/B) + P(C)\,P(M/C) + P(D)\,P(M/D)$$

$$= 0.4(0.02) + 0.3(0.01) + 0.2(0.005) + 0.1(0.0025) = 0.01225.$$

Example 19.30 Box A contains 6 red and 4 blue balls, box B contains 3 red and 7 blue balls. Two balls are drawn at random from box A and placed in box B. Now, a ball is drawn at random from the box B. Find the probability that it is a blue ball.

Solution Two balls are drawn at random from box A. The possibilities are 2 red, 1 red and 1 blue, 2 blue balls. Denote these possibilities as events B_1, B_2 and B_3 respectively. Note that B_1, B_2, B_3 are exhaustive and mutually exclusive events. We have

$$P(B_1) = \frac{6C_2}{10C_2} = \frac{6 \times 5}{10 \times 9} = \frac{1}{3}, \quad P(B_2) = \frac{6 \times 4}{10C_2} = \frac{24}{45} = \frac{8}{15}, \quad P(B_3) = \frac{4C_2}{10C_2} = \frac{6}{45} = \frac{2}{15}.$$

Denote, E = event of drawing a blue ball from box B after transfer. Now,

$$P(E/B_1) = P(\text{drawing a blue ball/box } B \text{ contains 5 red and 7 blue balls}) = \frac{7}{12}.$$

$$P(E/B_2) = P(\text{drawing a blue ball/box } B \text{ contains 4 red and 8 blue balls}) = \frac{8}{12} = \frac{2}{3}$$

$$P(E/B_3) = P(\text{drawing a blue ball/box } B \text{ contains 3 red and 9 blue balls}) = \frac{9}{12} = \frac{3}{4}.$$

Therefore,

$$P(E) = P(B_1)\, P(E/B_1) + P(B_2)\, P(E/B_2) + P(B_3)\, P(E/B_3)$$

$$= \frac{1}{3} \cdot \frac{7}{12} + \frac{8}{15} \cdot \frac{2}{3} + \frac{2}{15} \cdot \frac{3}{4} = \frac{117}{180} = 0.65.$$

Theorem 19.7 (Baye's Theorem) Let $B_1, B_2, ..., B_n$ be a set of exhaustive and mutually exclusive events of the sample space S with $P(B_k) \neq 0$, $k = 1, 2, ..., n$ and A be any event of S with $P(A) \neq 0$. Then,

$$P(B_k / A) = \frac{P(B_k)\, P(A/B_k)}{\sum_{k=1}^{n} P(B_k) P(A/B_k)}, \quad k = 1, 2, ..., n \qquad (19.21)$$

Proof Using the definition of conditional probability, we have

$$P(B_k / A) = \frac{P(B_k \cap A)}{P(A)}. \qquad (19.22)$$

The theorem of total probability gives

$$P(A) = \sum_{k=1}^{n} P(B_k) P(A/B_k). \qquad (19.23)$$

Also, the multiplication law (see Eq. 19.17) gives

$$P(B_k \cap A) = P(B_k)\, P(A/B_k), \qquad (19.24)$$

Substituting (19.23), (19.24) in (19.22), we obtain

$$P(B_k / A) = \frac{P(B_k) P(A/B_k)}{\sum_{k=1}^{n} P(B_k) P(A/B_k)}.$$

Remark 5

1. Note that the denominator in Eq. (19.21) gives the total probability.
2. $P(B_k)$ is known before the experiment and is known as *priori* probability.
3. $P(B_k/A)$ is called *posteriori* probability. It gives the probability of occurance of the event B_k with respect to the occurance of the event A.
4. Baye's theorem is also called the theorem of *probability of causes*.

Example 19.31 *A* bag *A* contains 8 white and 4 black balls. *A* second bag *B* contains 5 white and 6 black balls. One ball is drawn at random from bag *A* and is placed in bag *B*. Now, a ball is drawn at random from bag *B*. It is found that this ball is white. Find the probability that a black ball has been transferred from bag *A*.

Solution Denote the events as

　　　B_1 : transfer of white ball to bag *B*;　B_2 : transfer of black ball to bag *B*.

We have, $$P(B_1) = \frac{8}{12} = \frac{2}{3}, \quad P(B_2) = \frac{4}{12} = \frac{1}{3}.$$

Let, E = event of drawing a white ball from bag B after transfer.

Then, $P(E/B_1) = P$(drawing of a white ball/B contains 6 white and 6 black balls) $= \frac{6}{12} = \frac{1}{2}$.

$P(E/B_2) = P$(drawing of a white ball/B contains 5 white and 7 black balls) $= \frac{5}{12}$.

Hence, $$P(E) = P(B_1)P(E/B_1) + P(B_2)\,P(E/B_2)$$

$$= \frac{2}{3} \cdot \frac{1}{2} + \frac{1}{3} \cdot \frac{5}{12} = \frac{17}{36}.$$

We require probability that a black ball is transferred.

$$P(B_2/E) = \frac{P(B_2)P(E/B_2)}{P(E)} = \frac{(5/36)}{(17/36)} = \frac{5}{17}.$$

Example 19.32 Four boxes A, B, C, D contain fuses. The boxes contain 5000, 3000, 2000 and 1000 fuses respectively. The percentages of fuses in the boxes which are defective are 3%, 2%, 1% and 0.5% respectively. One fuse is selected at random arbitrarily from one of the boxes. It is found to be a defective fuse. Find the probability that it has come from box D.

Solution Selection of a box is done at random. Therefore,

$$P(A) = \frac{1}{4} = P(B) = P(C) = P(D).$$

Denote E as the event that a defective fuse is found.

We have, $P(E/A) = 0.03, P(E/B) = 0.02, P(E/C) = 0.01, P(E/D) = 0.005.$

Hence, $$P(E) = P(A)\,P(E/A) + P(B)\,P(E/B) + P(C)\,P(E/C) + P(D)\,P(E/D)$$

$$= \frac{1}{4}\,(0.03 + 0.02 + 0.01 + 0.005) = 0.01625.$$

Now, we require the probability that the defective fuse has come from box D.

$$P(D/E) = \frac{P(D)\,P(E/D)}{P(E)} = \frac{(1/4)(0.005)}{0.01625} = \frac{5}{65} = \frac{1}{13}.$$

Exercise 19.3

1. An experiment with the tossing of a coin is performed. If the coin shows head, 1 die is thrown and the outcome (number) is noted. But, if the coin shows tail, then 2 die are thrown and the outcome (sum of numbers) is noted. Find the probability that the recorded number is 5.

2. A company has 4 machines A, B, C and D manufacturing bulbs. The machines A, B, C, D produce 50%,

25%, 15% and 10% bulbs respectively. The percentages of defective bulbs produced by the machines A, B, C and D are 2%, 1%, 1%, and 0.5% respectively. Out of the output, one bulb is chosen at random. What is the probability that it is defective?

3. There are two bags containing balls. In the first bag, there are 3 white and 2 red balls. In the second bag, there are 1 white and 4 red balls. A bag is arbitrarily chosen and a ball is drawn. Find the probability that the drawn ball is white.

4. There are 4 candidates for the post of Incharge of constructing a flyover. The probabilities that they will be selected are 0.3, 0.25, 0.35 and 0.1. The probabilities that the above selected candidates can get the job done on time are 0.4, 0.7, 0.3 and 0.2 respectively. What is probability that the job is done on time?

5. A bag contains 6 red and 2 black balls and a second bag contains 4 red and 6 black balls. One ball is drawn at random from the first bag and transferred to the second bag. Now, a ball is drawn from the second bag. Find the probability that it is a red ball.

6. A bag A contains 3 red and 7 white balls. A second bag B contains 5 red and 4 white balls. One ball is drawn at random from the first bag and transferred to the second bag. Now, a ball is drawn from the second bag. It is found that the drawn ball is white. Find the probability that a red ball was transferred to bag B.

7. Three identical bags contain the following coloured balls. First bag : 3 black, 2 white; Second bag: 2 black, 3 white; Third bag: 3 black, 3 white balls. One of the bags is randomly selected and one ball is drawn. It turns out to be a white ball. Find the probability of drawing a white ball again, if the first white ball is not returned to the bag.

8. There are n boxes each containing s white balls and r red balls. One ball is transferred from the first box to the second box, then one ball is transferred from the second box to the third box, and so on. If p_k is the probability of drawing a white ball from the kth box, show that

$$p_k = \frac{1}{s+r+1}[s + p_{k-1}].$$

9. Box 1 contains 5000 bulbs of which 5% are defective. Box 2 contains 4000 bulbs of which 3% are defective. Two bulbs are drawn without replacement from a randomly selected box. (i) Find the probability that both bulbs are defective. (ii) Assuming that both the bulbs are defective, find the probability that they came from box 2.

10. For a certain binary communication channel, the probability that a transmitted '0' is received as '0' is 0.99 and the probability that a transmitted '1' is received as '1' is 0.95. If the probability that a '0' is transmitted is 0.6, find the probability that (i) a '0' is received and (ii) a '0' was transmitted given that a '0' was received.

19.7   Random Variables

Let S be the sample space corresponding to a random experiment E. A *random variable (RV)* on a sample space S defines a function that assigns a real number $X(s)$, $s \in S$. That is, a random variable defines a function $X : S \to R$ where S is the domain and range is $R = (-\infty, \infty)$. The set of all possible values of X, which is a subset of R is called the *range space* R_X. We shall be considering the case when the RV X takes a value x and its corresponding probability $P(X = x)$ is defined.

For example, if the event A consists of two tosses of a coin, then we may consider the random variable as the number of heads, 0, 1, 2. The corresponding probabilities are 1/4, 2/4 and 1/4.

19.7.1   One Dimensional Random Variables

Discrete random variable

If a random variable X takes a finite number or countably infinite number of values, then X is called a discrete random variable. We denote the possible values taken by X as $x_1, x_2, \ldots, x_n, \ldots$, which terminate in the finite case. Therefore, a real valued function X defined on a discrete sample space S is called a discrete random variable. In the above experiment of tossing of two coins, the possible values of the random variable (number of heads) are 0, 1, 2.

Probability function of a discrete random variable

Let X be a discrete RV which takes values $x_1, x_2, x_3, \ldots$, and let $P\,(X = x_i) = p_i$. Then, p_i is called the probability function if it satisfies the following conditions

(i) $p_i \geq 0$ for all i, and

(ii) $\sum_i p_i = 1$.

The collection of pairs (x_i, p_i), $i = 1, 2, 3, \ldots$

X	x_1	x_2	$\cdots$	x_n	$\cdots$
$P\,(X = x_i)$	p_1	p_2	$\cdots$	p_n	$\cdots$

is called the *probability distribution* or the *discrete probability distribution* of the discrete random variable X.

The probability function is also called *probability mass function* or *point probability function*.

Example 19.33 From a lot of 12 items containing 3 defective items, a sample of 4 items are drawn at random without replacement. Let a random variable X denote the number of defective items in the sample. Find the probability distribution of X.

Solution The lot contains 9 non-defective and 3 defective items. Since X denotes the number of defective items, x can take the values 0, 1, 2, 3. Four items are drawn without replacement.

For $\qquad x = 0, \qquad p(x) = \dfrac{9C_4}{12C_4} = \dfrac{14}{55}.$

For $\qquad x = 1, \qquad p(x) = \dfrac{(9C_3)(3C_1)}{12C_4} = \dfrac{28}{55}.$

For $\qquad x = 2, \qquad p(x) = \dfrac{(9C_2)(3C_2)}{12C_4} = \dfrac{12}{55}.$

For $\qquad x = 3, \qquad p(x) = \dfrac{(9C_1)(3C_3)}{12C_4} = \dfrac{1}{55}.$

We have the following probability distribution.

x	0	1	2	3
$p\,(x)$	$\dfrac{14}{55}$	$\dfrac{28}{55}$	$\dfrac{12}{55}$	$\dfrac{1}{55}$

Distribution function

Let X be a random variable. Then, the function $F(X)$ defined by

$$F(X) = P(X \le x)$$

is called the *distribution function* of X. It has the following properties.

   (i) $0 \le F(X) \le 1$,   (ii) If $x_1 < x_2$, then $F(x_1) \le F(x_2)$,
  (iii) $P(a \le X \le b) = F(b) - F(a)$.

In the case of a discrete RV, the discrete distribution function is defined by

$$F(X) = \sum_{x_i \le x} p(x_i).$$

From this definition, we have

$$p(x_i) = P(X = x_i) = F(x_i) - F(x_{i-1}).$$

$F(x)$ is also called the *cumulative distribution function of X*.

Mean and variance of a discrete random variable

Mean and variance are also called the expected values, which are denoted by $E[g(x)]$ where $g(x)$ is a random variable. If X is a discrete random variable, the mean μ_x and variance σ_x^2 are defined as

$$\mu_x = E(X) = \sum_i x_i p_i \qquad (19.25\text{i})$$

$$\sigma_x^2 = E[(x - \mu_x)^2] = \sum_i (x_i - \mu_x)^2 p_i = E(X^2) - [E(X)]^2. \qquad (19.25\text{ii})$$

Example 19.34 A random variable X has the following probability distribution

x	0	1	2	3	4
$p(x)$	c	$2c$	$2c$	c^2	$5c^2$

Find the value of c. Evaluate $P(X < 3)$, $P(0 < X < 4)$. Determine the distribution function of X. Find the mean and variance of X.

Solution Since $\sum_{x=0}^{4} p(x) = 1$, we get

$$c + 2c + 2c + c^2 + 5c^2 = 1, \quad \text{or} \quad 6c^2 + 5c - 1 = 0,$$

or $(6c - 1)(c + 1) = 0$,   or $c = 1/6, -1$.
Since $p(x) \ge 0$, the possible value is $c = 1/6$. Now,

$$P(X < 3) = P(X = 0) + P(x = 1) + P(x = 2)$$

$$= c + 2c + 2c = 5c = 5/6.$$

$$P(0 < X < 4) = P(X = 1) + P(X = 2) + P(X = 3)$$

$$= 2c + 2c + c^2 = \frac{25}{36}.$$

We have the following results for the probability distribution and distribution function.

x	0	1	2	3	4
$p(x)$	$\frac{1}{6}$	$\frac{2}{6}$	$\frac{2}{6}$	$\frac{1}{36}$	$\frac{5}{36}$
$F(x)$	$\frac{1}{6}$	$\frac{3}{6}$	$\frac{5}{6}$	$\frac{31}{36}$	1

$$\text{mean} = \mu_X = \Sigma\, x_i\, p_i = 0 + \frac{2}{6} + \frac{4}{6} + \frac{3}{36} + \frac{20}{36} = \frac{59}{36}.$$

Variance can be obtained by either of the formulas in (19.25ii). We have

$$\text{Variance} = \sigma_X^2 = E(x^2) - [E(x)]^2$$

$$= \left[0\left(\frac{1}{6}\right) + 1\left(\frac{2}{6}\right) + 4\left(\frac{2}{6}\right) + 9\left(\frac{1}{36}\right) + 16\left(\frac{5}{36}\right) \right] - \left(\frac{59}{36}\right)^2 = 1.4529.$$

Continuous random variables

A random variable X is said to be a continuous random variable if it takes all possible values in a given interval I_X. For example, age, height etc. are continuous random variables.

Probability density function

Consider a small interval $\left(x - \frac{dx}{2}, x + \frac{dx}{2} \right)$ of length dx about the point x. Let $f(x)$ be a continuous function so that $f(x)\,dx$ represents probability that X falls in this interval. That is

$$P\left(x - \frac{dx}{2} \le X \le x + \frac{dx}{2} \right) = f(x)dx. \tag{19.26}$$

Now, $f(x)\,dx$ represents the area under the curve $y = f(x)$, x-axis and the ordinates $x - (dx/2)$ and $x + (dx/2)$ (see Fig. 19.4). The function $f(x)$ is called the *probability density function (pdf)* of X and the curve $y = f(x)$ is called the *probability curve* or *probability density curve*. The expression $f(x)dx = dF(x)$ is called the probability differential.

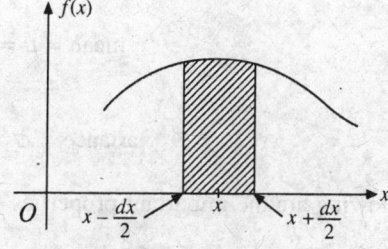

Fig. 19.4. Probability curve.

The *pdf* $f(x)$ satisfies the following conditions

   (i) $f(x) \geq 0$ for all $x \in I_X,$ (19.27i)

   (ii) $\int_{I_X} f(x)dx = 1.$ (19.27ii)

From the definition, we get

$$P(a \leq X \leq b) = \int_a^b f(x)dx.$$ (19.27iii)

Now,
$$P(X = a) = P(a \leq X \leq a) = \int_a^a f(x)dx = 0.$$

This implies that

$$P(a \leq x \leq b) = P(a < x \leq b) = P(a \leq x < b) = P(a < x < b).$$

Therefore, it does not matter if we include or do not include the end points of the interval I_X. Further, unlike in the case of a discrete random variable, we cannot represent a continuous random variable by a table.

Mean and variance

If X is a continuous random variable and $f(x)$ is the *pdf* of X, then we define

$$\text{mean} = \mu_X = \int_{I_X} x f(x)dx,$$

$$\text{variance} = \sigma_X^2 = \int_{I_X} (x - \mu_x)^2 f(x)dx.$$

Cumulative distribution function (*cdf*)

If X is a continuous random variable, then $F(x) = P(X \leq x)$ is called the *cdf* as in the case of a discrete random variables. That is,

$$F(x) = P(-\infty < X \leq x) = \int_{-\infty}^x f(x)dx.$$ (19.28)

The probability distribution of a continuous random variable is known if either its *pdf* or *cdf* is given. The mean and variance of the distribution defined on $I_X = [a, b]$, are defined as

$$\text{mean} = \mu = \int_a^b x f(x)dx.$$ (19.29)

$$\text{variance} = \sigma^2 = \int_a^b x^2 f(x)dx - (\mu)^2.$$ (19.30)

The following are the important properties of the *cdf* $F(x)$.

   (i) $0 \leq F(x) \leq 1, -\infty < x < \infty.$
   (ii) $F(x)$ is a non-decreasing function, that is , if $x_1 < x_2$, then $F(x_1) \leq F(x_2),$

(iii) $F(\infty) = 1$ and $F(-\infty) = 0$.

(iv) $F(x)$ is a continuous function of x on the right and may have countable number of discontinuities.

(v) $f(x) = F'(x)$ at all points where $F(x)$ is differentiable.

Example 19.35 If the density function of a continuous random variable X is given by

$$f(x) = 0, \quad x < 0$$

$$= ax, \quad 0 \le x \le 2,$$

$$= (4 - x)a, \quad 2 \le x \le 4$$

$$= 0, \quad x > 4$$

(i) Find the value of a.　　　(ii) Find the *cdf* of X.　　　(iii) Find $P(X > 2.5)$.

Solution

(i) Since $f(x)$ is a *pdf*, we have

$$\int_{Ix} f(x)dx = 1, \quad \text{that is,} \quad \int_0^4 f(x)dx = 1.$$

Note that $f(x)$ is a continuous function. We have

$$\int_0^2 ax\,dx + \int_2^4 a(4-x)dx = 1$$

or　　　　　　　　　　　$2a + 2a = 1, \quad \text{or} \quad a = \dfrac{1}{4}.$

(ii)　　　　　　　$F(x) = P(-\infty < x \le x) = \int_{-\infty}^x f(x)dx.$

We have

$x < 0:$　　　$F(x) = 0.$

$0 \le x \le 2:$　　$F(x) = \displaystyle\int_0^x \frac{x}{4}\,dx = \frac{x^2}{8}.$

$2 \le x \le 4:$　　$F(x) = \displaystyle\int_0^2 \frac{x}{4}\,dx + \int_2^x \frac{1}{4}(4-x)\,dx$

$$= \frac{1}{2} + \frac{1}{8}(8x - x^2 - 12) = \frac{1}{8}(8x - x^2 - 8).$$

$x > 4:$　　　$F(x) = 1.$

(iii)　　　　　$P(X > 2.5) = \displaystyle\int_{2.5}^4 \frac{1}{4}(4-x)\,dx$

$$= \frac{1}{4}\left[4x - \frac{x^2}{2}\right]_{2.5}^4 = \frac{9}{32}.$$

Example 19.36 If the probability density function of a continuous random variable is given by $f(x) = e^{-x}$, $0 \le x < \infty$, find the mean and variance.

Solution We have

$$\text{mean} = \mu = \int_0^\infty x e^{-x} dx = -[(x+1)e^{-x}]_0^\infty = 1.$$

$$\text{variance} = \int_0^\infty x^2 e^{-x} dx - (\mu)^2$$

$$= \left[(-x^2 e^{-x})_0^\infty + 2\int_0^\infty x e^{-x} dx \right] - 1 = 2 - 1 = 1.$$

Example 19.37 A continuous random variable X has *pdf*

$$f(x) = \frac{3}{4}(x^2 + 1), \quad 0 \le x \le 1.$$

Find a such that $P(X \le a) = P(X > a)$.

Solution We verify that the total probability is 1.

$$P(0 \le X \le 1) = \int_0^1 \frac{3}{4}(x^2 + 1)dx = \frac{3}{4}\left(\frac{1}{3} + 1\right) = 1.$$

Since $P(X \le a) = P(X > a)$, each must be equal to 1/2.

$$P(X \le a) = \int_0^a \frac{3}{4}(x^2 + 1)dx = \frac{1}{2}$$

or

$$\frac{3}{4}\left(\frac{a^3}{3} + a\right) = \frac{1}{2} \quad \text{or} \quad a^3 + 3a - 2 = 0.$$

A value of a in $(0, 1)$ is 0.5961.

Example 19.38 A continuous random variable X has the probability distribution

$$dF = a\, e^{-2|x|}\, dx, \quad -\infty < x < \infty.$$

Find the value of a. Find the standard deviation of the distribution.

Solution Since $\int_{-\infty}^\infty f(x)dx = 1$, we have

$$a \int_{-\infty}^\infty e^{-2|x|} dx = 2a \int_0^\infty e^{-2x} dx = 2a\left(\frac{1}{2}\right) = a = 1.$$

Therefore, $a = 1$.

$$\text{Mean} = \int_{-\infty}^{\infty} x e^{-2|x|}\, dx = 0 \quad \text{(integrand is an odd function).}$$

Hence,

$$\text{variance} = \int_{-\infty}^{\infty} x^2 e^{-2|x|}\, dx - 0$$

$$= 2 \int_{0}^{\infty} x^2 e^{-2x}\, dx = 2\left[-\frac{x^2 e^{-2x}}{2} - \frac{x e^{-2x}}{2} - \frac{e^{-2x}}{4} \right]_{0}^{\infty} = \frac{1}{2}.$$

Example 19.39 Verify that the following is a distribution function

$$
\begin{aligned}
F(x) &= 0, &&\text{for } x < 0, \\
&= x/3, &&\text{for } 0 \le x < 1, \\
&= 1/3, &&\text{for } 1 \le x < 2, \\
&= x/6, &&\text{for } 2 \le x < 6, \\
&= 1, &&\text{for } x \ge 6.
\end{aligned}
$$

Solution We find that $F(x)$ is continuous at all points. If $F(x)$ is a distribution function, then

$$
\begin{aligned}
f(x) = \frac{d}{dx} F(x) &= 0, &&\text{for } x < 0, \\
&= 1/3, &&\text{for } 0 \le x < 1, \\
&= 0, &&\text{for } 1 \le x < 2, \\
&= 1/6, &&\text{for } 2 \le x < 6, \\
&= 0, &&\text{for } x \ge 6,
\end{aligned}
$$

must be a *pdf*. $F(x)$ has four discontinuities, $F(-\infty) = 0$ and $F(\infty) = 1$. We need to show that

$$\int_{-\infty}^{\infty} f(x)\, dx = 1.$$

We have

$$\int_{0}^{1} \frac{1}{3}\, dx + \int_{2}^{6} \frac{1}{6}\, dx = \frac{1}{3} + \frac{1}{6}(4) = 1.$$

Hence, $F(x)$ is a distribution function as it satisfies all the required properties.

19.7.2 Two Dimensional Random Variables

In the earlier part of this section, we have considered a random variable X on a sample space S, that is, the outcome of a random experiment has only one characteristic and assumes a single real value. But, in many practical situations, the outcome of an experiment may have 2 or more characteristics. We shall consider now the case of 2 random variables or a bivariate experiment.

Let X and Y be two random variables on a sample space S associated with a random experiment. That is, $X = X(s)$ and $Y = Y(s)$, $s \in S$ are two functions each assigning a real number. We denote the two-dimensional random variable as (X, Y).

If the values assigned for (X, Y) are finite or countably infinite, then (X, Y) is called a *two-dimensional discrete random variable*. In this case, the possible values can be represented by (x_i, y_j), $i = 1, 2, \ldots m$, and $j = 1, 2, \ldots, n$. If the random variable (X, Y) can assume all possible values in a region R in the xy-plane, then (X, Y) is called a *two-dimensional continuous random variable*.

Joint probability function or mass function of (X, Y)

Let (X, Y) be a two-dimensional discrete random variable and let

$$P(X = x_i, Y = y_j) = P(X = x_i \cap Y = y_j) = p(x_i, y_j) = p_{ij}.$$

Then, p_{ij} is called the joint probability function of (X, Y), if the following conditions are satisfied.

(i) $p_{ij} \geq 0$, for all i, j

(ii) $\sum_i \sum_j p_{ij} = 1$.

The set of triplets $\{x_i, y_j, p_{ij}\}$, $i = 1, 2, \ldots, m$; $j = 1, 2, \ldots, n$, is called the joint probability distribution. The joint probability function can be represented in the form of a table (Table 19.1).

Table 19.1. Joint probability function.

Y \ X	y_1	y_2	.	.	.	y_j	.	.	.	y_n	Total
x_1	p_{11}	p_{12}	.		.	p_{1j}	.	.	.	p_{1n}	p_{1*}
x_2	p_{21}	p_{22}			.	p_{2j}				p_{2n}	p_{2*}
:	:	:	:	:	:	:	:	:	:	:	:
x_i	p_{i1}	p_{i2}	.	.	.	p_{ij}	.	.	.	p_{in}	p_{i*}
:	:	:	:	:	:	:	:	:	:	.	.
x_m	p_{m1}	p_{m2}	.		.	p_{mj}	.		.	p_{mn}	p_{m*}
Total	p_{*1}	p_{*2}	.		.	p_{*j}	.	.	.	p_{*n}	1

Marginal probability function

We assume that the joint distribution of a two-dimensional discrete random variable (X, Y) is given. Then, the probability distribution of X or the *marginal probability function* of X is defined as

$$P(X = x_i) = P[(X = x_i \cap Y = y_1) \text{ or } (X = x_i \cap Y = y_2) \text{ or } \ldots \text{ or } (X = x_i \cap Y = y_n)]$$

$$= p_{i1} + p_{i2} + \ldots + p_{ij} + \ldots + p_{in}$$

$$= \sum_{j=1}^{n} p_{ij} = \sum_{j=1}^{n} p(x_i, y_j) = p_{i*}. \tag{19.31}$$

Note that

$$\sum_{i=1}^{m} p_{i*} = p_{1*} + p_{2*} + \ldots + p_{m*} = \sum_{i} \sum_{j} p_{ij} = 1$$

(see the last column of Table 19.1).

Similarly, the probability distribution of Y or the *marginal probability function* of Y is defined as

$$P\,(Y = y_j) = P\,[(X = x_1 \cap Y = y_j) \text{ or } (X = x_2 \cap Y = y_j) \text{ or } \ldots \text{ or } (X = x_m \cap Y = y_j)]$$

$$= p_{1j} + p_{2j} + \ldots + p_{ij} + \ldots + p_{mj}$$

$$= \sum_{i=1}^{m} p_{ij} = \sum_{i=1}^{m} p(x_i, y_j) = p_{*j}. \tag{19.32}$$

Note that, $\sum_{j=1}^{n} p_{*j} = p_{*1} + p_{*2} + \ldots + p_{*n} = \sum_{i} \sum_{j} p_{ij} = 1$, (see the last row of Table 19.1).

The *conditional probability function* of X given $Y = y_j$, is given by

$$P(X = x_i \mid Y = y_j) = \frac{P[X = x_i \cap Y = y_j]}{P(Y = y_j)} = \frac{p_{ij}}{p_{*j}}. \tag{19.33}$$

The set $\{x_i, (p_{ij}/p_{*j})\}$ is called the conditional probability distribution of X, given $Y = y_j$.

Similarly, the *conditional probability function* of Y given $X = x_i$ is given by

$$P(Y = y_j \mid X = x_i) = \frac{P[Y = y_j \cap X = x_i]}{P(X = x_i)} = \frac{p_{ij}}{p_{i*}}. \tag{19.34}$$

The set $\{(p_{ij}/p_{i*}), y_j\}$ is called the *conditional probability distribution* of Y, given $X = x_i$.

The two random variables X and Y are said to be *independent,* if

$$P\,(X = x_i, Y = y_j) = P\,(X = x_i)\,P(Y = y_j). \tag{19.35}$$

Cumulative distribution function (cdf)

Let (X, Y) be a two-dimensional discrete random variable.

Then, the *cdf* of (X, Y) is defined by

$$F\,(x, y) = P\,(X \le x \text{ and } Y \le y) = \sum_{i} \sum_{j} p_{ij}, \; y_j \le y, x_i \le x.$$

The properties of a *cdf* are the following

  (i) $F\,(-\infty, y) = 0$, $F\,(x, -\infty) = 0$, $F\,(-\infty, \infty) = 1$.   (19.36i)

  (ii) $F\,(x, y)$ is a monotonic non-decreasing function.   (19.36 ii)

  (iii) $P\,(a < X < b, Y \le y) = F\,(b, y) - F\,(a, y)$.   (19.36iii)

The ratio $[f(x, y)/f_Y(y)]$ is called the *conditional density* of X given Y and is denoted by $f(x \mid y)$. Similarly, the ratio $[f(x, y)/f_X(x)]$ is called the *conditional density* of Y given X and is denoted by $f(y \mid x)$.

Independent random variables

Two continuous random variables are *independent* if $f(x, y) = f_X(x) f_Y(y)$, where $f_X(x)$ and $f_Y(y)$ are marginal *pdf* of X and Y respectively. Therefore, the random variables are independent if and only if $f(x, y)$ can be written as a product of a non-negative function of x alone and a non-negative function of y alone.

Expected values

Let (X, Y) be a discrete random variable with probability mass function p_{ij}. Then, we define

$$E\{g(X, Y)\} = \sum_i \sum_j g(x_i, y_j) p_{ij}.$$

Let (X, Y) be a continuous random variable with joint *pdf* $f(x, y)$. Then,

$$E\{g(X, Y)\} = \int_{-\infty}^{\infty} \int_{-\infty}^{\infty} g(x, y) f(x, y) dx dy.$$

Example 19.40 For the following bivariate probability distribution of X and Y

X \ Y	0	1	2	3
0	$\frac{1}{32}$	$\frac{2}{32}$	$\frac{3}{32}$	$\frac{2}{32}$
1	$\frac{2}{32}$	$\frac{1}{8}$	$\frac{1}{4}$	$\frac{1}{32}$
2	$\frac{3}{32}$	$\frac{2}{16}$	0	$\frac{1}{16}$

find (i) $P(X \le 1, Y = 3)$, (ii) $P(Y \le 2)$, (iii) $P(X \le 1 \mid Y \le 2)$.

Solution The marginal distributions are given below (Table 19.2).

Table 19.2. Marginal distributions.

X \ Y	0	1	2	3	$P_X(x) = p_{i*}$
0	$\frac{1}{32}$	$\frac{2}{32}$	$\frac{3}{32}$	$\frac{2}{32}$	$\frac{8}{32}$
1	$\frac{2}{32}$	$\frac{1}{8}$	$\frac{1}{4}$	$\frac{1}{32}$	$\frac{15}{32}$
2	$\frac{3}{32}$	$\frac{2}{16}$	0	$\frac{1}{16}$	$\frac{9}{32}$
$P_Y(y) = p_{*j}$	$\frac{6}{32}$	$\frac{10}{32}$	$\frac{11}{32}$	$\frac{5}{32}$	1

(i)
$$P(X \leq 1, Y = 3) = P(X = 0, Y = 3) + P(X = 1, Y = 3)$$

$$= \frac{2}{32} + \frac{1}{32} = \frac{3}{32}.$$

(ii)
$$P(Y \leq 2) = P(Y = 0) + P(Y = 1) + P(Y = 2)$$

$$= \frac{6}{32} + \frac{10}{32} + \frac{11}{32} = \frac{27}{32}.$$

(iii)
$$P(X \leq 1 \mid Y \leq 2) = \frac{P(X \leq 1, Y \leq 2)}{P(Y \leq 2)}.$$

Now, $P(X \leq 1, Y \leq 2) = P(X \leq 1, Y = 0) + P(X \leq 1, Y = 1) + P(X \leq 1, Y = 2)$

$$= \frac{3}{32} + \frac{6}{32} + \frac{11}{32} = \frac{20}{32}.$$

Hence,
$$P(X \leq 1 \mid Y \leq 2) = \frac{(20/32)}{(27/32)} = \frac{20}{27}.$$

Example 19.41 The joint probability mass function of (X, Y) is given by $p(x, y) = K (3x + 5y)$, $x = 1, 2, 3; y = 0, 1, 2$. Find the marginal distributions and conditional probability distribution of X, $P(X = x_i \mid Y = 2)$, $P(X \leq 2 \mid Y \leq 1)$.

Solution The joint probability mass function of (X, Y) and marginal distributions are given in Table 19.3.

Table 19.3. Marginal distributions

X \ Y	0	1	2	p_{i*}
1	$3K$	$8K$	$13K$	$24K$
2	$6K$	$11K$	$16K$	$33K$
3	$9K$	$14K$	$19K$	$42K$
p_{*j}	$18K$	$33K$	$48K$	$99K$

Since $99K = 1$, we obtain $K = 1/99$.
The marginal distributions are

$$p_{1*} = 24K = \frac{24}{99}, \quad p_{2*} = 33K = \frac{33}{99}, \quad p_{3*} = 42K = \frac{42}{99},$$

$$p_{*1} = 18K = \frac{18}{99}, \quad p_{*2} = 33K = \frac{33}{99}, \quad p_{*3} = 48K = \frac{48}{99}.$$

The conditional probabilities are

$$P(X = x_1 \mid Y = 2) = \frac{p_{12}}{p_{*2}} = \frac{13K}{48K} = \frac{13}{48},$$

19.7.3   Functions of Random Variables

Function of one random variable

Let X be a random variable defined on the sample space S and let g be a function such that $Y = g(X)$ is also a random variable defined on S. Further, let $f_X(x)$ be the *pdf* of X. We assume that $y = g(x)$ is a strictly monotonic (increasing or decreasing) function of x, and $g(x)$ is differentiable for all x. Then, the *pdf* of the random variable Y is given by

$$f_Y(y) = \left| \frac{dx}{dy} \right| f_X(x). \tag{19.45}$$

where x is expressed in terms of y, that is, $x = g^{-1}(y)$ and x is single valued. If x takes the values x_1, $x_2, \ldots, x_m$, then the above formula is extended as

$$f_Y(y) = f_X(x_1) \left| \frac{dx_1}{dy} \right| + f_X(x_2) \left| \frac{dx_2}{dy} \right| + \ldots + f_X(x_m) \left| \frac{dx_m}{dy} \right|. \tag{19.46}$$

Example 19.44 Let $F(x)$ and $f(x)$ be the distribution and the density function of X, respectively. Then, find the distribution and density function of $Y = aX + b$, $a \neq 0$, b real.

Solution Let $G(y)$ and $g_Y(y)$ denote the distribution and density functions of Y, respectively.

When $a > 0 : G(y) = P(Y \leq y) = P(aX + b \leq y)$

$$= P\left(X \leq \frac{y - b}{a} \right) = F\left(\frac{y - b}{a} \right).$$

When $a < 0 : G(y) = P(Y \leq y) = P(aX + b \leq y)$

$$= P\left(X \geq \frac{y - b}{a} \right) = 1 - P\left(X \leq \frac{y - b}{a} \right)$$

$$= 1 - F\left(\frac{y - b}{a} \right).$$

Now,

$$g_Y(y) = \frac{dG(y)}{dy}.$$

Hence,

when $a > 0$:

$$g_Y(y) = \frac{1}{a} f\left(\frac{y - b}{a} \right),$$

when $a < 0$:

$$g_Y(y) = -\frac{1}{a} f\left(\frac{y - b}{a} \right).$$

Combining the two results, we get $g_Y(y) = \dfrac{1}{|a|} f\left(\dfrac{y - b}{a} \right)$.

Example 19.45 The continuous random variable X has *pdf*

$$f(x) = x/2, \qquad 0 < x < 2$$
$$= 0, \qquad \text{elsewhere.}$$

Find the *pdf* of $Y = 3X + 2$.

Solution From $Y = 3X + 2$, we get $x = (y - 2)/3$, and x is single valued. For $0 < x < 2$, we get $2 < y < 8$. Therefore,

$$g(y) = \left|\frac{dx}{dy}\right| f_X(x) = \frac{1}{3}\left(\frac{x}{2}\right) = \frac{1}{18}(y - 2), \quad 2 < y < 8.$$

Example 19.46 The continuous random variable X has *pdf*

$$f_X(x) = \frac{2}{25}(x + 2), \quad \text{for } -2 < x < 3$$

$$= 0, \quad \text{elsewhere.}$$

Find the *pdf* of $Y = X^2$.

Solution We note that $y = x^2 \geq 0$ and this transformation is not monotonic in $(-2, 3)$. Hence, we divide this interval into two parts, a symmetric part $(-2, 2)$ and $(2, 3)$. Let $G(y)$ and $g_Y(y)$ be the distribution and density functions of Y, respectively.

When $-2 < x < 2$, that is, for $0 < y < 4$, we get

$$G(y) = P(Y < y) = P(X^2 < y)$$

$$= P(-\sqrt{y} < X < \sqrt{y}) = F(\sqrt{y}) - F(-\sqrt{y}).$$

Differentiating with respect to y, we get

$$g_Y(y) = \frac{1}{2\sqrt{y}}[f(\sqrt{y}) + f(-\sqrt{y})] = \frac{1}{2\sqrt{y}} \cdot \left(\frac{2}{25}\right)[(2 + \sqrt{y}) + (2 - \sqrt{y})] = \frac{4}{25\sqrt{y}}.$$

When $2 < x < 3$, $y = x^2$ is strictly increasing, and

$$g_Y(y) = \left|\frac{dx}{dy}\right| f_X(x) = \left(\frac{1}{2\sqrt{y}}\right)\left(\frac{2}{25}\right)(2 + \sqrt{y}) = \frac{1}{25\sqrt{y}}(2 + \sqrt{y}).$$

19.7.4 Transformations of Two Dimensional Random Variables

Let the random variables U and V be defined by the transformations $u = u(x, y)$ and $v = v(x, y)$, where u and v are continuous and differentiable functions. We assume that the *Jacobian* of the transformation

$$J = \frac{\partial(x, y)}{\partial(u, v)} = \begin{vmatrix} \partial x/\partial u & \partial x/\partial v \\ \partial y/\partial u & \partial y/\partial v \end{vmatrix} \tag{19.47}$$

is either > 0 or < 0 through the entire x-y plane, so that the unique inverse transformation $x = x(u, v)$, $y = y(u, v)$ exists.

We now present a result on the joint *pdf* of (U, V).

Theorem 19.8 The joint *pdf* $g_{UV}(u, v)$ of the transformed variables U and V is given by

$$g_{UV}(u, v) = |J| f_{XY}(x, y) \qquad (19.48)$$

where $|J|$ is the magnitude of the Jacobian J, and is expressed in terms of u, v.

Proof We have

$$P\left(x - \frac{dx}{2} < X \le x + \frac{dx}{2}, \ y - \frac{dy}{2} < Y \le y + \frac{dy}{2}\right)$$

$$= P\left(u - \frac{du}{2} < U \le u + \frac{du}{2}, \ v - \frac{dv}{2} < V \le v + \frac{dv}{2}\right)$$

Then, $\qquad\qquad f_{XY}(x, y)\, dx\, dy = g_{UV}(u, v)\, du\, dv.$

But, under the given transformation $x = x(u, v)$, $y = y(u, v)$

$$dx\, dy = \left|\frac{\partial(x, y)}{\partial(u, v)}\right| du\, dv = |J|\, du\, dv.$$

Therefore, $\qquad\qquad g_{UV}(u, v)\, du\, dv = f_{XY}(x, y)\, |J|\, du\, dv$

or $\qquad\qquad\qquad g_{UV}(u, v) = |J| f_{XY}(x, y).$

As an application of the transformation of variables, we prove the following result on independent random variables.

Theorem 19.9 If X and Y are independent, continuous random variables, then the *pdf* of $U = X + Y$ is given by the convolution of f_X and f_Y, that is,

$$f_U(u) = \int_{-\infty}^{\infty} f_X(v) f_Y(u - v)\, dv. \qquad (19.49)$$

Proof Let $f_{XY}(x, y)$ denote the joint *pdf* of the independent, continuous random variables X and Y. Consider the transformation of the variables as

$$u = x + y, \ v = x, \text{ that is, } x = v, \ y = u - v.$$

Now, the Jacobian is given by

$$J = \frac{\partial(x, y)}{\partial(u, v)} = \begin{vmatrix} \partial x/\partial u & \partial x/\partial v \\ \partial y/\partial u & \partial y/\partial v \end{vmatrix} = \begin{vmatrix} 0 & 1 \\ 1 & -1 \end{vmatrix} = -1.$$

Therefore, the joint *pdf* of the random variables U and V is given by (since X and Y are independent)

$$g_{UV}(u, v) = |J| f_{XY}(x, y)$$

$$= f_X(x) f_Y(y) = f_X(v) f_Y(u - v).$$

The marginal density of U is given by

$$f_U(u) = \int_{-\infty}^{\infty} g_{UV}(u, v)dv = \int_{-\infty}^{\infty} f_X(v)f_Y(u - v)dv.$$

Example 19.47 The *pdf* of the random variables (X, Y) is given by

$$f(x, y) = \frac{1}{4}e^{-(x+y)/2}, \quad x > 0, \quad y > 0,$$

$$= 0, \quad \text{elsewhere.}$$

Find the distribution of $(X - Y)/4$.

Solution We use the transformations

$$u = \frac{1}{4}(x - y), \quad v = y; \quad \text{or} \quad x = 4u + v, \quad y = v.$$

For $y > 0$, we get $v > 0$. For $x > 0$, we get $4u + v > 0$, or $v > -4u$ if $u < 0$ and $v > 0$, if $u \geq 0$. The Jacobian of transformation is

$$J = \begin{vmatrix} \partial x/\partial u & \partial x/\partial v \\ \partial y/\partial u & \partial y/\partial v \end{vmatrix} = \begin{vmatrix} 4 & 1 \\ 0 & 1 \end{vmatrix} = 4.$$

The joint *pdf* of the random variables (U, V) is given by

$$g_{UV}(u, v) = \frac{4}{4}e^{-(4u+2v)/2} = e^{-(2u+v)}, \quad -\infty < u < \infty$$

and under the above conditions $v > -4u$ if $u < 0$ and $v > 0$, if $u \geq 0$. The marginal *pdf* of U is given by

for $u < 0$:
$$f_U(u) = \int_{-4u}^{\infty} e^{-(2u+v)}dv = e^{-2u}(e^{4u}) = e^{2u},$$

for $u > 0$:
$$f_U(u) = \int_0^{\infty} e^{-(2u+v)}dv = e^{-2u}.$$

Combining the two results, we get $f_U(u) = e^{-2|u|}$, $-\infty < u < \infty$.

Example 19.48 Let X and Y be independent random variables having the density functions

$$\left.\begin{array}{l} f_X(x) = e^{-x}, \quad x \geq 0 \\ \quad\quad = 0 \quad, \quad x < 0 \end{array}\right\} \quad \text{and} \quad \left.\begin{array}{l} f_Y(y) = 2e^{-2y}, \quad y \geq 0, \\ \quad\quad = 0 \quad\quad, \quad y < 0. \end{array}\right\}$$

Find the density function of their sum $U = X + Y$.

Solution We use the transformation

$$u = x + y, \quad v = x, \quad \text{or} \quad x = v, \quad y = u - v.$$

Now, $x \geq 0$ gives $v \geq 0$ and $y \geq 0$ gives $u - v \geq 0$ or $v \leq u$. From the two results, we get $0 \leq v \leq u$. Hence, the density function of U is given by

$$f_U(u) = \int_0^u e^{-v}[2e^{-2(u-v)}]dv = 2e^{-2u}\int_0^u e^v dv$$

$$= 2e^{-2u}(e^u - 1), \quad u > 0.$$

Example 19.49 The joint *pdf* of the continuous random variables X and Y is given by

$$f(x, y) = 2x\, e^{-y}, \quad 0 < x < 1, y > 0$$

$$= 0 \quad \text{, elsewhere}$$

Find the distribution of $X + Y$.

Solution We use the transformations

$$u = x + y, \quad v = y, \quad \text{or} \quad x = u - v, \quad y = v.$$

The Jacobian of the transformation is $J = 1$.
Now, $y > 0$ gives $v > 0$ and $0 < x < 1$ gives $0 < u - v < 1$. The region is given in Fig. 19.9. We have the joint *pdf* as

$$g_{UV}(u, v) = 2(u - v)e^{-v}, \quad 0 < v < u, u > 0.$$

We divide the region into two parts R_1 and R_2.

For $R_1 : 0 < u \le 1, v > 0.$

$$f_U(u) = \int_0^u g_{UV}(u, v)dv = 2\int_0^u (u - v)e^{-v}dv$$

$$= 2\left[e^{-v}(-u + v + 1)\right]_0^u = 2[e^{-u} + u - 1].$$

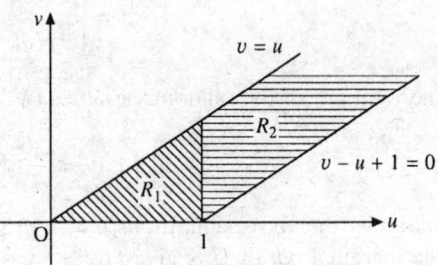

Fig. 19.9. Region in Example 19.49.

For $R_2 : 1 < u < \infty, u - 1 < v < u.$

$$f_U(u) = 2\int_{u-1}^u (u - v)e^{-v}dv$$

$$= 2[e^{-v}(-u + v + 1)]_{u-1}^u = 2e^{-u}.$$

Hence,
$$f_U(u) = 2(e^{-u} + u - 1), \quad 0 < u \le 1$$

$$= 2e^{-u}, \quad 1 < u < \infty$$

$$= 0, \text{ elsewhere.}$$

Exercise 19.4

1. A lot of 10 fuses contains 2 defectives fuses. A sample of 3 fuses are drawn at random without replacement. A random variable X denotes the number of defective fuses in the sample. Find the probability distribution of X.

2. A random variable X has the following probability function

x	1	2	3	4	5
$p(x)$	c	c	$3c$	$c^2 + c$	$6c^2$

Find the value of c. Evaluate $P(X < 3)$ and $P(1 < X < 4)$. Determine the distribution function of X.

3. Let X be a random variable such that $P(X = -1) = 2P(X = -2)$, $P(X = 0) = P(X = 1) = P(X = 2)$ and $P(X > 0) = P(X < 0)$. Obtain probability mass function of X and its distribution function. Find also the mean of X.

4. Let $p(x)$ be the probability function of a discrete random variable X, which assumes the values x_1, x_2, x_3, x_4 such that $p(x_1) = 3p(x_2) = 2p(x_3) = 4p(x_4)$. Find the probability distribution and cumulative probability distribution of X. Find the mean of X.

5. The probability function of an infinite discrete random variable is given by $P(X = K) = 1/2^K$, $K = 1, 2, 3, \ldots$. Show that it represents the probability of a discrete random variable. Find its mean and variance.

6. Can a function defined as

$$f(x) = |x|, \quad -1 \leq x \leq 1,$$
$$= 0, \text{ elsewhere}$$

be the *pdf* of a continuous random variable?

7. A continuous random variable X has the *pdf*

$$f(x) = ax^3, \quad 0 \leq x \leq 1$$
$$= 0 \quad , \quad \text{elsewhere.}$$

Determine a and find $P(X < 1/4)$ and $P(X > 1/2)$.

8. A continuous random variable X has the *pdf*

$$f(x) = a + bx, \qquad 0 \leq x \leq 1,$$
$$= 0 \qquad \text{elsewhere.}$$

If the mean of the distribution is $1/3$, find the values of a, b,

9. It is given that $f(x)$ and $g(x)$ are two *pdfs*. Find the conditions on a, b such that $a f(x) + b g(x)$, $a > 0$, $b > 0$ is also a *pdf*.

10. The mileage in thousands of kilometers which car owners get with a particular brand of tyre is a random variable having *pdf*

$$f(x) = \frac{1}{50} e^{-x/50}, \qquad \text{for } x > 0,$$
$$= 0 \qquad , \qquad \text{for } x \leq 0.$$

Find the probability that a tyre lasts (i) atmost 10,000 kms (ii) between 15,000 kms to 30,000 kms.

11. A *pdf* is defined by $f(x) = ae^{-bx}$, $0 \leq x < \infty$, $b > 0$, where b is prescribed and a is a constant. Find the value of the constant. Find the mean and standard deviation of the distribution.

12. A continuous random variable X is distributed over the interval $[0, 2]$ with *pdf* $f(x) = ax^2 + bx$, where a, b are constants. If the mean of the distribution is $1/2$, find the values of the constants a, b.

13. A random variable has the *pdf*

$$f(x) = 3x^2, 0 < x < 1,$$
$$= 0, \text{ elsewhere.}$$

Find (i) $P\left(\frac{1}{4} < X < \frac{1}{2}\right)$, (ii) $P\left(X < \frac{3}{4} \mid X > \frac{1}{4}\right)$.

14. The life in hours of a particular brand of electric bulbs has the *pdf*

$$f(x) = \frac{1000}{x^2}, \qquad \text{for } x \geq 1000,$$
$$= 0 \qquad , \qquad \text{for } x < 1000.$$

Find the distribution function. What is the probability that bulbs in a particular place is not to be replaced in the first 1500 hours of lighting? If two bulbs are chosen at random, find the probability that both the original bulbs have to be replaced during the first 1500 hours.

15. A continuous random variable X has the *pdf*

$$f(x) = \frac{K}{x^3} \quad , \qquad 15 < x < 25,$$

$$= 0 \quad , \qquad \text{elsewhere,}$$

where K is a constant. Determine the constant K and the distribution function $F(x)$. Find the probability of an event defined by $15 \le X \le 20$.

16. A continuous random variable has the distribution function

$$F(x) = 0 \quad , \qquad \text{for } x \le 2,$$

$$= a(x-2)^3 \quad , \qquad \text{for } 2 \le x \le 4,$$

$$= 1 \quad , \qquad \text{for } x > 4.$$

Find a and the *pdf*.

17. A bombing plane carrying two bombs flies directly above the target. If the bomb falls within 20 feet of the target, the target is destroyed. The points of impact of a bomb within the target site has the *pdf*

$$f(x) = (40 + x)/1600 \quad , \qquad \text{for } -40 \le x < 0,$$

$$= (40 - x)/1600 \quad , \qquad \text{for } 0 \le x \le 40,$$

$$= 0 \quad , \qquad \text{elsewhere,}$$

where x is the vertical deviation in feet above the target. If both the bombs are used, what is the probability that the target is destroyed?

18. A continuous random variable X has *pdf* $f(x) = 2x/9$, $0 \le x \le 3$. If two independent determinations of X are made, find the probability that both the determinations are greater than 2. If 3 independent determinations are made, what is the probability that exactly 2 of these are greater than 2?

19. Given

$$f(x) = 0 \quad , \qquad \text{for } x < 0,$$

$$= Ke^{-x/4} \quad , \qquad \text{for } x \ge 0,$$

determine the value of K that makes $f(x)$ a *pdf*. Hence, find $P(X < 5)$ and $P(X > 6)$.

20. The distribution function of a continuous random variable X is given by $F(x) = 1 - (2x + 1)e^{-2x}$, $x \ge 0$. Find the density function, mean and variance of the distribution.

21. X and Y are two random variables having the joint density function, $f(x, y) = (3x + 2y)/45$, where x and y assume integer values 0, 1 and 2. Find the conditional distribution of Y for $X = x$.

22. The joint *pdf* of the random variables (X, Y) is given by

$$f(x, y) = Ke^{-(x+y)}, \quad x > 0, \quad y > 0.$$

Find the value of K. Prove that X, Y are independent. Find $P(X > 2)$ and $P(2 < X + Y < 3)$.

23. The joint density function of a continuous random variable (X, Y) is given by $f(x, y) = Kxye^{-(x^2+y^2)}$, $x > 0$, $y > 0$. Find the value of K. Are the random variables independent?

24. The joint probability distribution function of X and Y is given by

$$F(x, y) = 1 - e^{-y} - e^{-x} + e^{-(x+y)}, \quad x > 0, \quad y > 0,$$

$$= 0, \quad \text{elsewhere.}$$

Are X and Y independent? Find the marginal distributions. Also find $P(X \le 2 \cap Y \le 2)$ and $P(X + Y \le 2)$.

25. Let $F(x, y)$ be the distribution function of X and Y. Prove that for all x, y

$$F_X(x) + F_Y(y) - 1 \le F_{XY}(x, y) \le \sqrt{F_X(x) F_Y(y)}.$$

26. Find the value of c such that $f(x, y) = cxy$, $1 \leq x \leq y \leq 2$ is a probability density function.

27. Find the value of c such that

$$f(x, y) = cy(y - x), \quad \text{for } 0 < x < 2, \quad -x < y < x$$

$$= 0, \quad \text{elsewhere},$$

is a probability density function. Find the marginal distributions $f_X(x)$, $f_Y(y)$.

28. If the joint distribution function of X and Y is

$$F(x, y) = (1 - e^{-x})(1 - e^{-y}), \quad \text{for } x > 0, \quad y > 0$$

$$= 0, \quad \text{elsewhere},$$

find $P(1 < X < 4)$ and $P(1 < X < 3, 1 < Y < 3)$.

29. The joint density function of X and Y is given by

$$f(x, y) = K, \quad \text{for } x^2 + y^2 \leq R^2$$

$$= 0, \quad \text{elsewhere}.$$

Find K and $f_X(x)$.

30. The random variable X has the *pdf*

$$f(x) = e^{-x}, \quad 0 \leq x < \infty$$

$$= 0, \quad \text{elsewhere}.$$

Find the density function of the variable $Y = 4X + 3$.

31. Let X be a given random variable. The second random variable Y is defined as $Y = (X + |X|)/4$. Determine the density and the distribution function of Y in terms of the density and distribution function of X.

32. Let the *pdf* of a random variable X be

$$f(x) = x/2, \quad 0 < x < 2$$

$$= 0, \quad \text{elsewhere}.$$

Find the *pdf* of $Y = 64X^3$.

33. Let X and Y be independent continuous random variables having the *pdf*

$$f_X(x) = e^{-x}, \quad x > 0 \qquad \qquad f_Y(y) = e^{-y}, \quad y > 0$$
$$= 0, \quad \text{elsewhere} \qquad \text{and} \qquad = 0, \quad \text{elsewhere}.$$

Find the *pdf* of $U = X - Y$.

34. Let the continuous random variable (X, Y) have the joint density function

$$f(x, y) = 4xye^{-(x^2 + y^2)}, \quad x \geq 0, \quad y \geq 0,$$

$$= 0, \quad \text{elsewhere}.$$

Show that the density function of the random variable $U = \sqrt{X^2 + Y^2}$ is

$$h(u) = 2u^3 e^{-u^2}, \quad 0 \leq u < \infty,$$

$$= 0, \quad \text{elsewhere}.$$

35. The joint density function of the random variables X, Y is given by $f(x, y) = e^{-(x+y)}$, $x > 0$, $y > 0$. Find the *pdf* of $U = (X + Y)/2$.

36. Let X and Y be two independent random variables with *pdf*

$$f(x) = e^{-x}, \quad x \geq 0 \quad \text{and} \quad g(y) = 2e^{-2y}, \ y \geq 0.$$

Find the probability distribution of $Z = X/Y$.

37. Let X and Y be two independent random variables with *pdf*

$$f_X(x) = 2e^{-2x}, \quad x \geq 0 \ ; \quad f_Y(y) = e^{-y}, \quad y \geq 0.$$

Find the density function of the random variables $U = X/(X+Y)$ and $V = X+Y$. Are the random variables U, V independent?

19.8   Some Probability Distributions

In this section, we shall discuss the following discrete and continuous probability distributions. (i) Binomial distribution, (ii) Poisson distribution, (iii) Uniform distribution and (iv) Normal distribution.

19.8.1   Binomial Distribution

Binomial distribution is a discrete distribution. Consider a random experiment E. Let A be an event or a trial associated with the experiment E. If event A happens, then we call it a success, otherwise it is a failure. We associate with the happening of the event A, a probability of success $P(A) = p$, and the probability of failure defined as $P(\overline{A}) = 1 - p = q$. We assume the following.

(i)   Each trial results in two disjoint outcomes, a success or a failure.
(ii)   The number of trials made 'n' is finite.
(iii)   The trials are independent.
(iv) $P(\text{success}) = p$ is a constant for each trial.

For example, throwing of dice, tossing of coins, drawing a card or cards from a pack of cards give rise to binomial distribution.

We associate a random variable X with the number of times that the event A has occured out of n trials. Therefore, the possible values of X are $0, 1, 2, \ldots, n$. Then, X is called a *binomial random variable* with parameters n and p. We denote the binomial distribution by $B(n, p)$.

A random variable X is said to follow the binomial distribution if the probability mass function is given by

$$P(X = r) = p(r) = nC_r \, p^r q^{n-r}, \quad r = 0, 1, 2, \ldots, n$$

where $q = 1 - p$.

Remark 7

(i)   Assignment of probability in binomial distribution can be explained from the fact

$$\sum_{r=0}^{n} P(X = r) = \sum_{r=0}^{n} p(r) = \sum_{r=0}^{n} nC_r \, p^r q^{n-r} = (p + q)^n = 1.$$

(ii)   Assume that n trials constitute an experiment E. If the experiment is repeated N times, then the *frequency function* of the binomial distribution is defined by

$$f(r) = Np(r) = (N) \, nC_r \, p^r \, q^{n-r}, \quad r = 0, 1, 2, \ldots, n.$$

Mean and variance of binomial distribution

We have

$$E(X) = \sum_{r=0}^{n} x_r p_r = \sum_{r=0}^{n} r p_r = \sum_{r=0}^{n} (r) nC_r p^r q^{n-r}$$

$$= \sum_{r=0}^{n} r \left(\frac{n!}{r!(n-r)!} \right) p^r q^{n-r}$$

$$= np \sum_{r=1}^{n} \frac{(n-1)!}{(r-1)![(n-1)-(r-1)]!} p^{r-1} q^{(n-1)-(r-1)}$$

$$= np \sum_{r=1}^{n} (n-1)C_{r-1} p^{r-1} q^{(n-1)-(r-1)}$$

$$= np (q+p)^{n-1} = np. \tag{19.50}$$

$$E(X^2) = \sum_{r=0}^{n} x_r^2 p_r = \sum_{r=0}^{n} r^2 p_r = \sum_{r=0}^{n} [r(r-1)+r] p_r$$

$$= \sum_{r=0}^{n} r(r-1) p_r + \sum_{r=0}^{n} r p_r$$

$$= \sum_{r=0}^{n} r(r-1) \left(\frac{n!}{r!(n-r)!} \right) p^r q^{n-r} + np$$

$$= n(n-1) p^2 \sum_{r=2}^{n} (n-2)C_{r-2} p^{r-2} q^{(n-2)-(r-2)} + np$$

$$= n(n-1) p^2 (q+p)^{n-2} + np$$

$$= n(n-1) p^2 + np. \tag{19.51}$$

$$\text{Var}(X) = E(X^2) - [E(X)]^2$$

$$= n(n-1) p^2 + np - n^2 p^2 = np - np^2$$

$$= np(1-p) = npq. \tag{19.52}$$

Recurrence formula for the moments of bionomial distribution

By the definition, the kth central moment μ_k is defined by

$$\mu_k = E\{X - E(X)\}^k. \tag{19.53}$$

For the bionomial distribution $B(n, p)$, we have $X = r$, $E(X) = np$, $q = 1 - p$ and

$$\mu_k = \sum_{r=0}^{n} (r - np)^k nC_r p^r q^{n-r}. \tag{19.54}$$

Differentiating (19.54) with respect to p, where $q = 1 - p$, we get

$$\frac{d\mu_k}{dp} = \sum_{r=0}^{n} nC_r [(-nk)(r - np)^{k-1} p^r q^{n-r}$$

$$+ (r - np)^k \{rp^{r-1}q^{n-r} - p^r(n - r)q^{n-r-1}\}]$$

$$= -nk\mu_{k-1} + \sum_{r=0}^{n} nC_r p^{r-1} q^{n-r-1} (r - np)^k [rq - (n - r)p].$$

Now, $rq - (n - r)p = r(1 - p) - (n - r)p = r - np.$

Hence, $$\frac{d\mu_k}{dp} = -nk\mu_{k-1} + \sum_{r=0}^{n} nC_r p^{r-1} q^{n-r-1} (r - np)^{k+1}$$

$$= -nk\mu_{k-1} + \frac{1}{pq} \sum_{r=0}^{n} nC_r p^r q^{n-r} (r - np)^{k+1}$$

$$= -nk\mu_{k-1} + \frac{1}{pq} \mu_{k+1}.$$

Therefore, $$\mu_{k+1} = pq\left[\frac{d\mu_k}{dp} + nk\,\mu_{k-1}\right]. \tag{19.55}$$

From the moments of lower order, we can compute moments of higher order using this recurrence formula.

We compute the following moments. We have $\mu_0 = 1$, $\mu_1 = 0$. Then,

$k = 1$: $$\mu_2 = pq\left[\frac{d\mu_1}{dp} + n\mu_0\right] = npq \tag{19.56}$$

which is the variance of the binomial distribution.

$k = 2$: $$\mu_3 = pq\left[\frac{d\mu_2}{dp} + 2n\mu_1\right] = pq\left[\frac{d}{dp}\{np(1 - p)\}\right]$$

$$= npq[1 - 2p] = npq(q - p), \tag{19.57}$$

which can be taken as a measure of skewness.

$k = 3$: $$\mu_4 = pq\left[\frac{d\mu_3}{dp} + 3n\mu_2\right] = npq[1 + 3pq(n - 2)] \tag{19.58}$$

which can be taken as a measure of kurtosis.

Moment generating function of binomial distribution

Let X be a binomial variate. Then, the moment generating function (MGF) of X is defined by

$$M_X(t) = E(e^{tX}) = \sum_{r=0}^{n} e^{tr} nC_r p^r q^{n-r}$$

$$= \sum_{r=0}^{n} nC_r (e^t p)^r q^{n-r} = (q + pe^t)^n.$$

Remark 8

In general, the sum of two independent binomial variates is not a binomial variate.

Let $X \sim B(n_1, p_1)$ and $Y \sim B(n_2, p_2)$ be independent random variables. Then

$$M_X(t) = (q_1 + p_1 e^t)^{n_1}, \quad \text{and} \quad M_Y(t) = (q_2 + p_2 e^t)^{n_2}.$$

Now, $\qquad\qquad M_{X+Y}(t) = M_X(t)\, M_Y(t) \quad \text{(since } X, Y \text{ are independent)}$

$$= (q_1 + p_1 e^t)^{n_1} (q_2 + p_2 e^t)^{n_2}.$$

This expression cannot be written in the form $(q + pe^t)^n$. Hence, $M_{X+Y}(t)$ is not a binomial variate.

Example 19.50 Two players A and B play tennis games. Their chances of winning a game are in the ratio 3 : 2 respectively. Find A's chance of winning at least two games out of four games played.

Solution Let

X = random variable denoting the number of successes.

p = probability that A wins a game.

Then, $p = 3/5$ and $q = 2/5$.

$$P(X = r) = p(r) = 4C_r \left(\frac{3}{5}\right)^r \left(\frac{2}{5}\right)^{4-r}, r = 0, 1, 2, 3, 4.$$

Hence, $\qquad\qquad P(X \geq 2) = 1 - P(X < 2) = 1 - [P(X = 0) + P(X = 1)]$

$$= 1 - \frac{1}{5^4}[16 + 4(3)8] = \frac{513}{625}.$$

Example 19.51 A target is to be destroyed in a bombing exercise. There is 75% chance that any one bomb will strike the target. Assume that two direct hits are required to destroy the target completely. How many bombs must be dropped in order that the chance of destroying the target is $\geq 99\%$?

Solution Let

X = random variable representing the number of bombs to be used.

p = probability that the bombs hits the target = $3/4$, $\ q = 1/4$

n = number of bombs required to destroy the target.

Here, $X \sim B(n, 3/4)$. Hence,

$$p(r) = nC_r \left(\frac{3}{4}\right)^r \left(\frac{1}{4}\right)^{n-r}, \quad r = 0, 1, 2, \ldots, n.$$

We require $\qquad\qquad P(X \geq 2) \geq 0.99, \quad \text{or} \ \ 1 - P(X \leq 1) \geq 0.99,$

or $\qquad\qquad [1 - \{P(X = 0) + P(X = 1)\}] \geq 0.99.$

or
$$\left(\frac{1}{4}\right) + n\left(\frac{3}{4}\right)\left(\frac{1}{4}\right)^{n-1} \leq 0.01, \quad \text{or} \quad 1 + 3n \leq (0.01)4^n.$$

We obtain $n = 6$.

Example 19.52 An irregular six faced dice is thrown 12 times. The expectation that it will give six even numbers is twice the expectation that it will give 5 even numbers. If 1000 sets, each of exactly 12 trials are made, how many sets are expected not to give any even number?

Solution Let

X = number of even numbers obtained in 12 throws.

p = probability of getting an even number.

It is given that

$$P(X = 6) = 2P(X = 5), \quad \text{or} \quad 12C_6\, p^6 q^6 = 2[12C_5 p^5 q^7]$$

or
$$\frac{7}{6}p = 2q, \quad \text{or} \quad 7p = 12(1-p), \quad \text{or} \quad p = \frac{12}{19}.$$

Hence, $q = 7/19$.

Now, P (getting no even number) $= P(X = 0) = q^{12}$.

Number of sets having no even numbers out of $N = 1000$ sets

$$= N \cdot P(X = 0) = 1000\, q^{12} = 1000\left(\frac{7}{19}\right)^{12} \approx 0.0063.$$

19.8.2 Poisson Distribution

Poisson distribution is a discrete distribution. Poisson distribution is a limiting case of the binomial distribution. Poisson distribution can be used under the following conditions.

(i) Each trial results in two mutually exclusive outcomes, termed as success and a failure.
(ii) n, the number of trials is very large, that is $n \to \infty$,
(iii) The constant probability of success, p, is very small, that is, $p \to 0$,
(iv) $np = \lambda$, where λ is a positive real number, is finite.

Therefore, $p = \lambda/n$ and $q = 1 - (\lambda/n)$.

Consider the following examples:

(i) Number of defective items out of lots produced in a large manufacturing factory.
(ii) The number of vehicles passing, per minute, during the peak hours in a day through a particular traffic junction.
(iii) Number of printing mistakes on each page of a book. etc.

Let X be a discrete random variable taking values $0, 1, 2, \ldots$. If the probability mass function is given by

$$P(X = r) = \frac{e^{-\lambda} \lambda^r}{r!}, \quad r = 0, 1, 2, \ldots, \quad \lambda > 0 \tag{19.59}$$

then, X is said to follow *Poisson distribution* with parameter λ. We denote the Poisson distribution by $P(\lambda)$, and X is called a Poisson variate.

Remark 9

(i) Assignment of probability in Poisson distribution can be explained from the fact

$$\sum_{r=0}^{\infty} \frac{e^{-\lambda} \lambda^r}{r!} = e^{-\lambda} \sum_{r=0}^{\infty} \frac{\lambda^r}{r!} = e^{-\lambda} (e^{\lambda}) = 1.$$

(ii) The distribution function is given by

$$F(x) = P(X \le x) = \sum_{r=0}^{x} p(r) = e^{-\lambda} \sum_{r=0}^{x} \frac{\lambda^r}{r!}, \quad x = 0, 1, 2, \ldots \tag{19.60}$$

(iii) Unlike in binomial distribution, Poisson distribution occurs when there are events which do not occur as outcomes of a definite number of trials.

(iv) The binomial probabilities may be computed approximately by computing the corresponding Poisson probabilities, whenever n is large and p is small.

Poisson distribution as a limiting case of binomial distribution

Let the random variable X be binomially distributed with parameters n and p.
Then, for $r = 0, 1, 2, \ldots, n$ and $p = \lambda/n$ we get,

$$p(X = r) = nC_r \, p^r q^{n-r} = \frac{n(n-1)(n-2) \ldots (n-r+1)}{r!} \left(\frac{\lambda}{n}\right)^r \left[1 - \frac{\lambda}{n}\right]^{n-r}$$

$$= \frac{\lambda^r}{r!} \left[1\left(1 - \frac{1}{n}\right)\left(1 - \frac{2}{n}\right) \ldots \left(1 - \frac{r-1}{n}\right)\right]\left(1 - \frac{\lambda}{n}\right)^n \left(1 - \frac{\lambda}{n}\right)^{-r}. \tag{19.61}$$

Now, $\lim_{n \to \infty} \left(1 - \frac{\lambda}{n}\right)^n = e^{-\lambda}$, $\lim_{n \to \infty} \left(1 - \frac{\lambda}{n}\right)^{-r} = 1$, $\lim_{n \to \infty} \left(1 - \frac{a}{n}\right) = 1$, when a is finite.

Therefore, taking limits as $n \to \infty$ in (19.61), we obtain

$$\lim_{\substack{n \to \infty \\ np = \lambda = \text{finite}}} P(X = r) = \frac{\lambda^r}{r!} \, (1) \, e^{-\lambda} = \frac{e^{-\lambda} \lambda^r}{r!}$$

which is the probability mass function of the Poisson variate X. It is possible to derive the Poisson distribution independently, without considering it as the limiting case of the binomial distribution.

Mean and variance of the Poisson distribution

From the definition, we have

$$E(X) = \sum_{r=0}^{\infty} x_r p_r = \sum_{r=0}^{\infty} \frac{r e^{-\lambda} \lambda^r}{r!} = \lambda e^{-\lambda} \sum_{r=1}^{\infty} \frac{\lambda^{r-1}}{(r-1)!} = \lambda e^{-\lambda} e^{\lambda} = \lambda. \qquad (19.62)$$

$$E(X^2) = \sum_{r=0}^{\infty} x_r^2 p_r = \sum_{r=0}^{\infty} \frac{r^2 e^{-\lambda} \lambda^r}{r!} = \sum_{r=0}^{\infty} [r(r-1) + r] \frac{e^{-\lambda} \lambda^r}{r!}$$

$$= \lambda^2 e^{-\lambda} \sum_{r=2}^{\infty} \frac{\lambda^{r-2}}{(r-2)!} + \lambda = \lambda^2 e^{-\lambda} e^{\lambda} + \lambda = \lambda^2 + \lambda.$$

Hence, variance $(X) = E(X^2) - [E(X)]^2$

$$= \lambda^2 + \lambda - \lambda^2 = \lambda. \qquad (19.63)$$

Therefore, for Poisson distribution, mean = variance = λ.

Recurrence formula for the moments of Poisson distribution

By the definition, the kth central moment μ_k is defined by

$$\mu_k = E\{X - E(X)\}^k.$$

For Poisson distribution $P(\lambda)$, we have $X = r$, $E(X) = \lambda$, and

$$\mu_k = \sum_{r=0}^{\infty} (r - \lambda)^k \frac{e^{-\lambda} \lambda^r}{r!}. \qquad (19.64)$$

Differentiating (19.64) with respect to λ, we get

$$\frac{d\mu_k}{d\lambda} = \sum_{r=0}^{\infty} \frac{1}{r!} [e^{-\lambda} \{-k(r-\lambda)^{k-1} \lambda^r + (r-\lambda)^k (r\lambda^{r-1} - \lambda^r)\}]$$

$$= -k\mu_{k-1} + \sum_{r=0}^{\infty} \frac{(r-\lambda)^{k+1} e^{-\lambda} \lambda^{r-1}}{r!} = -k\mu_{k-1} + \frac{1}{\lambda} \mu_{k+1}$$

or $\qquad\qquad \mu_{k+1} = \lambda \left[k\mu_{k-1} + \frac{d\mu_k}{d\lambda} \right]. \qquad (19.65)$

From the moments of lower order, we can compute moments of higher order using this recurrence formula. With $\mu_0 = 1$, $\mu_1 = 0$, we obtain the following moments

$$\mu_2 = \lambda \left[\mu_0 + \frac{d\mu_1}{d\lambda} \right] = \lambda.$$

$$\mu_3 = \lambda \left[2\mu_1 + \frac{d\mu_2}{d\lambda} \right] = \lambda.$$

$$\mu_4 = \lambda \left[3\mu_2 + \frac{d\mu_3}{d\lambda} \right] = \lambda[3\lambda + 1].$$

Recurrence formula for the probabilities

For Poisson distribution with parameter λ, we have

$$p(r) = \frac{e^{-\lambda} \lambda^r}{r!}, \quad r = 0, 1, 2, \dots$$

Setting $r = r + 1$, we get

$$p(r + 1) = \frac{e^{-\lambda} \lambda^{r+1}}{r + 1!} = \frac{\lambda}{r + 1} \cdot p(r). \tag{19.66}$$

This formula is also called "fitting of Poisson distribution". If we know $p(0) = e^{-\lambda}$, then we can calculate all the other probabilities, $p(1), p(2)$ etc.

Moment generating function (MGF) of Poisson distribution

Let X be a Poisson variate. Then, the MGF of X is defined by

$$M_X(t) = E(e^{tX}) = \sum_{r=0}^{\infty} e^{tr} \left(\frac{e^{-\lambda} \lambda^r}{r!} \right)$$

$$= e^{-\lambda} \left[1 + \lambda e^t + \frac{(\lambda e^t)^2}{2!} + \dots \right] = e^{-\lambda} e^{\lambda e^t} = e^{\lambda(e^t - 1)}.$$

Remark 10

If $X_1, X_2, \dots, X_n$ are independent Poisson variates with parameters $\lambda_1, \lambda_2, \dots \lambda_n$, then $\sum_{i=1}^{n} X_i$ is also a Poisson variate with parameter $\lambda_1 + \lambda_2 + \dots + \lambda_n$. The converse of this result is also true. That is, if $X_1, X_2, \dots, X_n$ are independent and $\sum_{i=1}^{n} X_i$ has a Poisson distribution, then each of the random variables has a Poisson distribution.

Let $X_1 \sim P(\lambda_1), X_2 \sim P(\lambda_2), \dots, X_n \sim P(\lambda_n)$.

Then,
$$M_{X_i}(t) = e^{\lambda_i(e^t - 1)}, \quad i = 1, 2, \dots, n.$$

Now,

$$M_{X_1 + X_2 + \dots + X_n}(t) = M_{X_1}(t) M_{X_2}(t) \dots M_{X_n}(t) = e^{\lambda_1(e^t - 1)} \cdot e^{\lambda_2(e^t - 1)} \dots e^{\lambda_n(e^t - 1)}$$

$$= e^{(\lambda_1 + \lambda_2 + \dots + \lambda_n)(e^t - 1)} = e^{\lambda(e^t - 1)}$$

since X_i are independent and $\lambda = \lambda_1 + \lambda_2 + \dots + \lambda_n$. Hence, $\sum_{i=1}^{n} X_i$ is also a Poisson variate with parameter $\lambda = \lambda_1 + \lambda_2 + \dots + \lambda_n$.

However, the difference of two independent Poisson variates is not a Poisson variate.

Example 19.53 The number of emergency admissions each day to a hospital is found to have Poisson distribution with mean 4. Find the probability that on a particular day there will be no emergency admissions.

Solution We have mean $\lambda = 4$. Hence,

$$P(X = 0) = e^{-\lambda} = e^{-4}.$$

Example 19.54 In a book of 600 pages, there are 60 typographical errors. Assuming Poisson law for the number of errors per page, find the probability that a randomly chosen 4 pages will contain no errors.

Solution Average number of errors per page $= \lambda = \dfrac{60}{600} = \dfrac{1}{10}$. Probability of r errors per page is

$$P(X = r) = \frac{e^{-\lambda} \lambda^r}{r!} = \frac{e^{-1/10} (1/10)^r}{r!}$$

and

$$P(X = 0) = e^{-1/10} = e^{-0.1}.$$

Probability that a random sample of 4 pages have no errors $= [P(x = 0)]^4 = e^{-0.4} \approx 0.6703$.

Example 19.55 If X and Y are independent variates with means λ_1 and λ_2 respectively, then find the probability that (i) $X + Y = K$, and (ii) $X = Y$.

Solution Since X and Y are independent variates with means, λ_1 and λ_2, we get

$$P(X = r) = \frac{e^{-\lambda_1} \lambda_1^r}{r!}, \quad r = 0, 1, 2, \ldots$$

$$P(Y = s) = \frac{e^{-\lambda_2} \lambda_2^s}{s!}, \quad s = 0, 1, 2, \ldots$$

(i) By the addition property of Poisson distribution, we have that $X + Y$ has Poisson distribution $P(\lambda_1 + \lambda_2)$.

Hence, $\qquad P(X + Y = K) = \dfrac{1}{K!} e^{-(\lambda_1 + \lambda_2)} (\lambda_1 + \lambda_2)^K, \quad K = 0, 1, 2, \ldots$

This result can also be obtained directly. Since X, Y are independent, we get

$$P(X + Y = K) = \sum_{r=0}^{K} P(X = r) P(Y = K - r)$$

$$= \sum_{r=0}^{K} \frac{e^{-\lambda_1} \lambda_1^r}{r!} \cdot \frac{e^{-\lambda_2} \lambda_2^{K-r}}{(K-r)!} = \frac{1}{K!} e^{-(\lambda_1 + \lambda_2)} \sum_{r=0}^{K} \frac{K!}{r!(K-r)!} \lambda_1^r \lambda_2^{K-r}$$

$$= \frac{1}{K!} e^{-(\lambda_1 + \lambda_2)} \sum_{r=0}^{K} KC_r \lambda_1^r \lambda_2^{K-r} = \frac{1}{K!} e^{-(\lambda_1 + \lambda_2)} (\lambda_1 + \lambda_2)^K.$$

(ii)
$$P(X = Y) = \sum_{r=0}^{\infty} P(X = r \cap Y = r) = \sum_{r=0}^{\infty} P(X = r)P(Y = r)$$

$$= e^{-(\lambda_1 + \lambda_2)} \sum_{r=0}^{\infty} \frac{(\lambda_1 \lambda_2)^r}{(r!)^2}$$

since X and Y are independent.

Now, we shall discuss the following distributions of continuous random variables:

(i) Uniform (rectangular) distribution,
(ii) Normal (Gaussian) distribution.

19.8.3 Uniform (Rectangular) Distribution

Let X be a continuous random variable. Then, X is said to have a continuous uniform distribution in any finite interval (a, b), if its probability density function is constant over the entire interval. Let

$$f(x) = K, \quad a < x < b,$$

$$= 0, \quad \text{elesewhere}$$

Since,
$$\int_a^b f(x)dx = 1, \quad \text{we get } K \int_a^b dx = 1, \quad \text{or } K(b - a) = 1.$$

We obtain
$$K = 1/(b - a),$$

and
$$f(x) = 1/(b - a), \quad a < x < b,$$

$$= 0, \quad \text{elesewhere.} \tag{19.67}$$

When X has uniform distribution in (a, b), we usually denote it as $U(a, b)$. We can consider a and b as two parameters of the uniform distribution.

The distribution function $F(x)$ is obtained as

$$F(x) = 0, \qquad \text{for} \quad -\infty < x < a,$$

$$= \frac{x - a}{b - a}, \qquad \text{for} \quad a \le x \le b,$$

$$= 1. \qquad \text{for} \quad b < x < \infty. \tag{19.68}$$

Note that $f(x)$ is not continuous at $x = a$ and $x = b$ and $F(x)$ is not differentiable at these points. The *pdf* of the distribution is given by $f(x)$.

Moments

The moments μ_r' of the uniform distribution $U(a, b)$ about the origin are given by

$$\mu_r' = \int_a^b x^r f(x)dx = \frac{1}{b - a} \int_a^b x^r dx = \frac{1}{(b - a)} \left[\frac{b^{r+1} - a^{r+1}}{r + 1} \right] \tag{19.69}$$

Hence,

$$\text{Mean} = E(X) = \mu_1' = \frac{1}{2}(b + a). \tag{19.70}$$

Now,
$$J = \begin{vmatrix} \dfrac{1}{2} & \dfrac{1}{2} \\ \dfrac{1}{2} & -\dfrac{1}{2} \end{vmatrix} = -\dfrac{1}{2}.$$

We obtain $g(u, v) = |J| f(x_1, x_2) = \dfrac{1}{2}$.

For the distribution of u, we subdivide the region R^* into two parts R_1 and R_2.
In R_1, we have

$$g_1(u) = \int_{-u}^{u} \dfrac{1}{2} dv = \dfrac{1}{2}[v]_{-u}^{u} = u.$$

In R_2, we have

$$g_2(u) = \int_{u-2}^{2-u} \dfrac{1}{2} dv = 2 - u.$$

Hence, the distribution is given by

$$g(u) = u, \quad 0 < u < 1,$$
$$= 2 - u, \quad 1 < u < 2.$$

19.8.4   Normal Distribution

Normal distribution is also called *Gaussian distribution*. It is the most important probability model in statistical analysis. Originally, normal distribution was derived as a limiting case of the binomial distribution, when n is very large and p and q are not very small.

A continuous random variable X is said to have a normal or Gaussian distribution with parameters μ (mean) and σ^2 (variance), if its probability density function is given by

$$f(x) = \dfrac{1}{\sigma\sqrt{2\pi}} e^{-(x-\mu)^2/(2\sigma^2)}, \quad -\infty < x < \infty, \ -\infty < \mu < \infty, \ \sigma > 0. \tag{19.72}$$

Usually, we say that X follows $N(\mu, \sigma^2)$ or sometimes as $N(\mu, \sigma)$. We can verify that $f(x)$ is a probability density function. We have

$$\int_{-\infty}^{\infty} f(x) dx = \dfrac{1}{\sigma\sqrt{2\pi}} \int_{-\infty}^{\infty} e^{-(x-\mu)^2/(2\sigma^2)} dx$$

Let $\dfrac{x - \mu}{\sigma\sqrt{2}} = t$, Then $dx = \sigma\sqrt{2}\, dt$ and

$$\int_{-\infty}^{\infty} f(x) dx = \dfrac{1}{\sigma\sqrt{2\pi}} \int_{-\infty}^{\infty} e^{-t^2} \sigma\sqrt{2}\, dt = \dfrac{1}{\sqrt{\pi}} \int_{-\infty}^{\infty} e^{-t^2} dt = \dfrac{1}{\sqrt{\pi}}(\sqrt{\pi}) = 1,$$

since
$$\int_{-\infty}^{\infty} e^{-t^2} dt = \sqrt{\pi}.$$

Standard normal distribution

Let $Z = (X - \mu)/\sigma$ in (19.72). Then,

$$E(Z) = 0 \quad \text{and} \quad \text{variance } (Z) = 1.$$

We call Z as the *standard normal variate* and we write $Z \sim N(0, 1)$. The probability density function is given by

$$\phi(z) = \frac{1}{\sqrt{2\pi}} e^{-z^2/2}, \quad -\infty < z < \infty. \tag{19.73}$$

The corresponding distribution function, $\Phi(z)$ is given by

$$\Phi(z) = P(Z \leq z) = \int_{-\infty}^{z} \phi(u)\,du = \frac{1}{\sqrt{2\pi}} \int_{-\infty}^{z} e^{-u^2/2}\,du. \tag{19.74}$$

Properties of the distribution function $\Phi(z)$

(i) We have
$$\Phi(-z) = 1 - \Phi(z). \tag{19.75}$$

$$\Phi(-z) = P(Z \leq -z) = P(Z \geq z) \quad \text{(by symmetry)}$$

$$= 1 - P(Z \leq z) = 1 - \Phi(z).$$

(ii)
$$P(a \leq X \leq b) = \Phi\left(\frac{b - \mu}{\sigma}\right) - \Phi\left(\frac{a - \mu}{\sigma}\right) \tag{19.76}$$

where $X \sim N(\mu, \sigma^2)$. Since, $Z = (X - \mu)/\sigma$, we get

$$P(a \leq X \leq b) = P\left(\frac{a - \mu}{\sigma} \leq Z \leq \frac{b - \mu}{\sigma}\right)$$

$$= P\left(Z \leq \frac{b - \mu}{\sigma}\right) - P\left(Z \leq \frac{a - \mu}{\sigma}\right)$$

$$= \Phi\left(\frac{b - \mu}{\sigma}\right) - \Phi\left(\frac{a - \mu}{\sigma}\right).$$

Properties of normal distribution $N(\mu, \sigma^2)$

Graph of $y = f(x)$

The graph of $y = f(x)$ is a bell shaped curve, called the normal probability curve. The curve is symmetric about the mean, that is, about $x = \mu$. The maximum occurs at $x = \mu$ and the maximum value is $1/(\sigma\sqrt{2\pi})$. The ordinate $f(x)$ decreases rapidly as $x \to \pm\infty$, and x-axis is an asymptote of the curve. The graph is concave downward at $x = \mu$ and concave upward for large x. The concavity changes at the points of inflection given by the solutions of $y''(x) = 0$. These points are given by $x = \mu \pm \sigma$. Hence, if σ is large, the curve tends to be flat and if σ is small, the curve tends to be more peaked at $x = \mu$. (see Fig. 19.12).

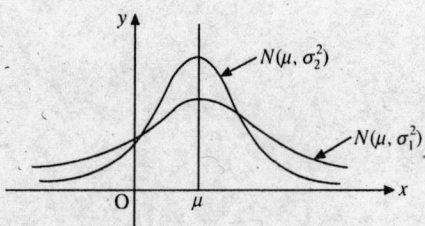

Fig. 19.12. Normal probability curve. $\sigma_1 > \sigma_2$.

$$M_X(t) = \frac{1}{\sqrt{2\pi}} \int_{-\infty}^{\infty} e^{t(\mu + \sigma z)} e^{-z^2/2} \, dz = \frac{e^{\mu t}}{\sqrt{2\pi}} \int_{-\infty}^{\infty} e^{-(z^2 - 2\sigma t z)/2} \, dz$$

$$= \frac{e^{\mu t}}{\sqrt{2\pi}} \int_{-\infty}^{\infty} e^{-[(z - \sigma t)^2 - \sigma^2 t^2]/2} \, dz$$

$$= \frac{e^{[\mu t + (\sigma^2 t^2/2)]}}{\sqrt{2\pi}} \int_{-\infty}^{\infty} e^{-(z - \sigma t)^2/2} \, dz$$

$$= \frac{e^{[\mu t + (\sigma^2 t^2/2)]}}{\sqrt{\pi}} \int_{-\infty}^{\infty} e^{-u^2} \, du, \quad \left[\text{set } \frac{z - \sigma t}{\sqrt{2}} = u \right]$$

$$= e^{[\mu t + (\sigma^2 t^2/2)]} .$$

Hence, $M_X(t) = e^{[\mu t + (\sigma^2 t^2/2)]} .$ (19.78)

Expanding the right hand side, we get

$$M_X(t) = 1 + \frac{t}{1!}\left[\mu + \frac{\sigma^2 t}{2}\right] + \frac{t^2}{2!}\left[\mu + \frac{\sigma^2 t}{2}\right]^2 + \frac{t^3}{3!}\left[\mu + \frac{\sigma^2 t}{2}\right]^3 + \dots$$

$$= 1 + \mu t + \frac{t^2}{2!}(\mu^2 + \sigma^2) + \frac{t^3}{3!}(\mu^3 + 3\mu\sigma^2) + \dots$$

Hence,

$$E(X) = \text{coefficient of } \frac{t}{1!} = \mu.$$

$$E(X^2) = \text{coefficient of } \frac{t^2}{2!} = \sigma^2 + \mu^2.$$

$$E(X^3) = \text{coefficient of } \frac{t^3}{3!} = \mu^3 + 3\mu\sigma^2, \text{ etc.}$$

Since the kth order central moment is defined by

$$\mu_k = E\{(X - \mu)^k\} = \frac{1}{\sigma\sqrt{2\pi}} \int_{-\infty}^{\infty} (x - \mu)^k e^{-(x-\mu)^2/(2\sigma^2)} \, dx$$

we obtain $\mu_1 = 0$, $\mu_2 = \sigma^2$, $\mu_3 = 0$, $\mu_4 = 3\sigma^4$, etc.

Let us now deduce the MGF of the standard normal distribution $N(0, 1)$. The standard normal variate is defined by $Z = (X - \mu)/\sigma$ and

$$M_Z(t) = \frac{1}{\sigma\sqrt{2\pi}} \int_{-\infty}^{\infty} e^{[(x-\mu)t/\sigma]} e^{-(x-\mu)^2/(2\sigma^2)} \, dx$$

$$= e^{-\mu t/\sigma} \left[\frac{1}{\sigma\sqrt{2\pi}} \int_{-\infty}^{\infty} e^{xt/\sigma} e^{-(x-\mu)^2/(2\sigma^2)} dx \right]$$

$$= e^{-\mu t/\sigma} M_X\left(\frac{t}{\sigma}\right) = e^{-\mu t/\sigma} e^{\left[\frac{\mu t}{\sigma} + \frac{\sigma^2}{2}\left(\frac{t^2}{\sigma^2}\right)\right]} = e^{t^2/2}. \tag{19.79}$$

Now, we prove an important property of the normal distribution.

Let X_i, $i = 1, 2, \ldots, n$ be n independent normal variables with mean μ_i and variance σ_i^2 respectively, that is, $X_i \sim N(\mu_i, \sigma_i^2)$. Then, we show that

$$\sum_{i=1}^{n} a_i X_i \sim N\left[\sum_{i=1}^{n} a_i \mu_i, \sum_{i=1}^{n} a_i^2 \sigma_i^2\right] \tag{19.80}$$

which is called the *additive property* of the normal distribution. We have

$$M_{X_i}(t) = e^{\left[\mu_i t + (\sigma_i^2 t^2/2)\right]}$$

The MGF of $a_1 X_1 + a_2 X_2 + \ldots + a_n X_n$, where a_i are constants, is given by (since X_i are independent)

$$M_{a_1 X_1 + \ldots + a_n X_n}(t) = M_{a_1 X_1}(t) \cdot M_{a_2 X_2}(t) \ldots M_{a_n X_n}(t)$$

$$= M_{X_1}(a_1 t) M_{X_2}(a_2 t) \ldots M_{X_n}(a_n t)$$

$$= e^{\left[a_1 \mu_1 t + (\sigma_1^2 a_1^2 t^2/2)\right]} e^{\left[a_2 \mu_2 t + (\sigma_2^2 a_2^2 t^2/2)\right]} \ldots e^{\left[a_n \mu_n t + (\sigma_n^2 a_n^2 t^2/2)\right]}$$

$$= e^{\left[(a_1 \mu_1 + a_2 \mu_2 + \ldots + a_n \mu_n)t + (a_1^2 \sigma_1^2 + a_2^2 \sigma_2^2 + \ldots + a_n^2 \sigma_n^2)(t^2/2)\right]}$$

which is the MGF of a normal variate with mean $\sum_{i=1}^{n} a_i \mu_i$ and variance $\sum_{i=1}^{n} a_i^2 \sigma_i^2$. Hence the result in Eq. (19.80) is proved.

Particular cases

(i) Let $\qquad\qquad a_1 = 1 = a_2$ and $a_3 = a_4 = \ldots = a_n = 0$. Then,

$$X_1 + X_2 \sim N(\mu_1 + \mu_2, \sigma_1^2 + \sigma_2^2). \tag{19.81}$$

(ii) Let $\qquad\qquad a_1 = 1, a_2 = -1$ and $a_3 = a_4 = \ldots = a_n = 0$. Then,

$$X_1 - X_2 \sim N(\mu_1 - \mu_2, \sigma_1^2 + \sigma_2^2). \tag{19.82}$$

Therefore, the sum and the difference of two independent normal variates is also a normal variate. (In the case of Poisson distribution the difference $X_1 - X_2$ is not a Poisson variate).

(iii) Let $\qquad\qquad a_1 = a_2 = \ldots a_n = 1/n$. Then,

$$\frac{1}{n}\sum_{i=1}^{n} X_i \sim N\left[\frac{1}{n}\sum_{i=1}^{n} \mu_i, \frac{1}{n^2}\sum_{i=1}^{n} \sigma_i^2\right]. \tag{19.83}$$

But, the left hand side is the mean $\overline{X}$. Therefore,

$$\overline{X} \sim N\left(\mu, \frac{\sigma^2}{n}\right). \tag{19.84}$$

Therefore, if X_i are independent and are identically distributed normal variates with mean μ and variance σ^2, then, their mean $\overline{X}$ is also a normal variate and $\overline{X} \sim N\left(\mu, \frac{\sigma^2}{n}\right)$.

Mean deviation from mean

The mean deviation (MD) from mean is defined as

$$\text{MD about mean} = E\left[\,|\,x - E\,(x)\,|\,\right]. \tag{19.85}$$

For the normal distribution $N\,(\mu, \sigma^2)$, we obtain

$$\text{MD about the mean} = \int_{-\infty}^{\infty} |x - \mu| f(x) dx$$

$$= \frac{1}{\sigma\sqrt{2\pi}} \int_{-\infty}^{\infty} |x - \mu| e^{-(x-\mu)^2/(2\sigma^2)} dx.$$

Let $z = (x - \mu)/\sigma$. Then, $dx = \sigma dz$ and

$$\text{MD about the mean} = \frac{1}{\sqrt{2\pi}} \int_{-\infty}^{\infty} \sigma |z|\, e^{-z^2/2} dz$$

$$= \frac{2\sigma}{\sqrt{2\pi}} \int_0^{\infty} z e^{-z^2/2}\, dz = \sigma\sqrt{\frac{2}{\pi}}\left[-e^{-z^2/2}\right]_0^{\infty} = \sqrt{\frac{2}{\pi}}\,\sigma \tag{19.86}$$

which is approximately equal to (0.7979σ) or $4\sigma/5$.

Quartile deviation of $N\,(\mu, \sigma^2)$

The first quartile Q_1 and the third quartile Q_3 are defined by

$$\int_{-\infty}^{Q_1} f(x) dx = \frac{1}{4}, \quad \text{and} \quad \int_{-\infty}^{Q_3} f(x) dx = \frac{3}{4}.$$

Since normal probability curve is symmetric about $x = \mu$, these equations are equivalent to the equations

$$\int_{Q_1}^{\mu} f(x) dx = \frac{1}{4}, \text{ and } \int_{\mu}^{Q_3} f(x) dx = \frac{1}{4}.$$

Hence, the quartile deviation (QD) is given by

$$QD = \frac{1}{2}(Q_3 - Q_1).$$

Now, for the normal distribution $N(\mu, \sigma^2)$, we have $\dfrac{1}{\sigma\sqrt{2\pi}} \displaystyle\int_{Q_1}^{\mu} e^{-(x-\mu)^2/(2\sigma^2)} = \dfrac{1}{4}.$

Setting $z = (x - \mu)/\sigma$, we get

$$\frac{1}{\sqrt{2\pi}} \int_{(Q_1 - \mu)/\sigma}^{0} e^{-z^2/2} dz = \frac{1}{4}, \quad \text{or} \quad \int_{(Q_1 - \mu)/\sigma}^{0} \phi(z) dz = \frac{1}{4}.$$

Since $(Q_1 - \mu) < 0$, we get

$$\int_{0}^{(\mu - Q_1)/\sigma} \phi(z) dz = 0.25.$$

From the table of standard normal areas, we get

$$\frac{\mu - Q_1}{\sigma} = 0.674, \quad \text{or} \quad Q_1 = \mu - 0.674\sigma.$$

By symmetry, we get $Q_3 = \mu + 0.674\sigma$.

Hence, $$QD = \frac{1}{2}(Q_3 - Q_1) = 0.674\sigma \approx \frac{2}{3}\sigma.$$

Normal probability integral

Normal probability integral gives the area property of the normal distribution. Let $X \sim N(\mu, \sigma^2)$. Then, the probability that X will lie between μ and x_1 is given by

$$P(\mu < X < x_1) = \int_{\mu}^{x_1} f(x) dx = \frac{1}{\sigma\sqrt{2\pi}} \int_{\mu}^{x_1} e^{-(x-\mu)^2/(2\sigma^2)} dx.$$

Let $z = (x - \mu)/\sigma$. Then $dx = \sigma dz$ and the limits $(\mu, x_1) \rightarrow (0, z_1)$ where $z_1 = (x_1 - \mu)/\sigma$. Therefore,

$$P(\mu < X < x_1) = P(0 < Z < z_1) = \frac{1}{\sqrt{2\pi}} \int_{0}^{z_1} e^{-z^2/2} dz = \int_{0}^{z_1} \phi(z) dz \qquad (19.87)$$

where $\phi(z) = e^{-z^2/2}/\sqrt{2\pi}$ is the probability density function of the standard normal variate (see Eq. (19.73)). This definite integral $\int_{0}^{z_1} \phi(z) dz$ is called the *normal probability integral*. It gives the area under the standard normal curve $y = \phi(z)$, between the ordinates at $Z = 0$ and $Z = z_1$ (see Fig. 19.13). These areas are tabulated for different values of z_1, usually at intervals of 0.01 (see Table 19.5).

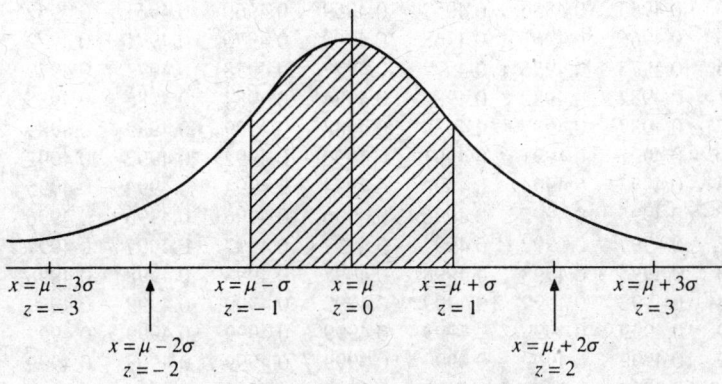

Fig. 19.13. Normal probability integral.

Table 19.5

Areas under the Standard Normal Curve from 0 to z

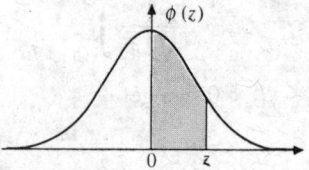

z	0	1	2	3	4	5	6	7	8	9
0.0	0.0000	0.0040	0.0080	0.0120	0.0160	0.0199	0.0239	0.0279	0.0319	0.0359
0.1	0.0398	0.0438	0.0478	0.0517	0.0557	0.0596	0.0636	0.0675	0.0714	0.0754
0.2	0.0793	0.0832	0.0871	0.0910	0.0948	0.0987	0.1026	0.1064	0.1103	0.1141
0.3	0.1179	0.1217	0.1255	0.1293	0.1331	0.1368	0.1406	0.1443	0.1480	0.1517
0.4	0.1554	0.1591	0.1628	0.1664	0.1700	0.1736	0.1772	0.1808	0.1844	0.1879
0.5	0.1915	0.1950	0.1985	0.2019	0.2054	0.2088	0.2123	0.2157	0.2190	0.2224
0.6	0.2258	0.2291	0.2324	0.2357	0.2389	0.2422	0.2454	0.2486	0.2518	0.2549
0.7	0.2580	0.2612	0.2642	0.2673	0.2704	0.2734	0.2764	0.2794	0.2823	0.2852
0.8	0.2881	0.2910	0.2939	0.2967	0.2996	0.3023	0.3051	0.3078	0.3106	0.3133
0.9	0.3159	0.3186	0.3212	0.3238	0.3264	0.3289	0.3315	0.3340	0.3365	0.3389
1.0	0.3413	0.3438	0.3461	0.3485	0.3508	0.3531	0.3554	0.3577	0.3599	0.3621
1.1	0.3643	0.3665	0.3686	0.3708	0.3729	0.3749	0.3770	0.3790	0.3810	0.3830
1.2	0.3849	0.3869	0.3888	0.3907	0.3925	0.3944	0.3962	0.3980	0.3997	0.4015
1.3	0.4032	0.4049	0.4066	0.4082	0.4099	0.4115	0.4131	0.4147	0.4162	0.4177
1.4	0.4192	0.4207	0.4222	0.4236	0.4251	0.4265	0.4279	0.4292	0.4306	0.4319
1.5	0.4332	0.4345	0.4357	0.4370	0.4382	0.4394	0.4406	0.4418	0.4429	0.4441
1.6	0.4452	0.4463	0.4474	0.4484	0.4495	0.4505	0.4515	0.4525	0.4535	0.4545
1.7	0.4554	0.4564	0.4573	0.4582	0.4591	0.4599	0.4608	0.4616	0.4625	0.4633
1.8	0.4641	0.4649	0.4656	0.4664	0.4671	0.4678	0.4686	0.4693	0.4699	0.4706
1.9	0.4713	0.4719	0.4726	0.4732	0.4738	0.4744	0.4750	0.4756	0.4761	0.4767
2.0	0.4772	0.4778	0.4783	0.4788	0.4793	0.4798	0.4803	0.4808	0.4812	0.4817
2.1	0.4821	0.4826	0.4830	0.4834	0.4838	0.4842	0.4846	0.4850	0.4854	0.4857
2.2	0.4861	0.4864	0.4868	0.4871	0.4875	0.4878	0.4881	0.4884	0.4887	0.4890
2.3	0.4893	0.4896	0.4898	0.4901	0.4904	0.4906	0.4909	0.4911	0.4913	0.4916
2.4	0.4918	0.4920	0.4922	0.4925	0.4927	0.4929	0.4931	0.4932	0.4934	0.4936
2.5	0.4938	0.4940	0.4941	0.4943	0.4945	0.4946	0.4648	0.4949	0.4951	0.4952
2.6	0.4953	0.4955	0.4956	0.4957	0.4959	0.4960	0.4961	0.4962	0.4963	0.4964
2.7	0.4965	0.4966	0.4967	0.4968	0.4969	0.4970	0.4971	0.4972	0.4973	0.4974
2.8	0.4974	0.4975	0.4976	0.4977	0.4977	0.4978	0.4979	0.4979	0.4980	0.4981
2.9	0.4981	0.4982	0.4982	0.4983	0.4984	0.4984	0.4985	0.4985	0.4986	0.4986
3.0	0.4987	0.4987	0.4987	0.4988	0.4988	0.4989	0.4989	0.4989	0.4990	0.4990
3.1	0.4990	0.4991	0.4991	0.4991	0.4992	0.4992	0.4992	0.4992	0.4993	0.4993
3.2	0.4993	0.4993	0.4994	0.4994	0.4994	0.4994	0.4994	0.4995	0.4995	0.4995
3.3	0.4995	0.4995	0.4995	0.4996	0.4996	0.4996	0.4996	0.4996	0.4996	0.4997
3.4	0.4997	0.4997	0.4997	0.4997	0.4997	0.4997	0.4997	0.4997	0.4997	0.4998
3.5	0.4998	0.4998	0.4998	0.4998	0.4998	0.4998	0.4998	0.4998	0.4998	0.4998
3.6	0.4998	0.4998	0.4999	0.4999	0.4999	0.4999	0.4999	0.4999	0.4999	0.4999
3.7	0.4999	0.4999	0.4999	0.4999	0.4999	0.4999	0.4999	0.4999	0.4999	0.4999
3.8	0.4999	0.4999	0.4999	0.4999	0.4999	0.4999	0.4999	0.4999	0.4999	0.4999
3.9	0.5000	0.5000	0.5000	0.5000	0.5000	0.5000	0.5000	0.5000	0.5000	0.5000

The probability that a random value of X lies in the interval $(\mu - \sigma, \mu + \sigma)$ is given by

$$P(\mu - \sigma < X < \mu + \sigma) = \int_{\mu-\sigma}^{\mu+\sigma} f(x)dx$$

or

$$P(-1 < Z < 1) = \int_{-1}^{1} \phi(z)dz \quad \left(\text{where } z = \frac{x - \mu}{\sigma}\right)$$

$$= 2\int_{0}^{1} \phi(z)dz \quad \text{(because of symmetry)}$$

$$= 2(0.3413) = 0.6826 \quad \text{(from Table 19.4)}.$$

Thus, approximately there is a probability of 68%, that a normal variate lies in the interval $(\mu - \sigma, \mu + \sigma)$.

Now,

$$P(\mu - 2\sigma < X < \mu + 2\sigma) = \int_{\mu-2\sigma}^{\mu+2\sigma} f(x)dx$$

or

$$P(-2 < Z < 2) = \int_{-2}^{2} \phi(z)dz = 2\int_{0}^{2} \phi(z)dz = 2(0.4772) = 0.9544.$$

Thus, approximately, there is a probability of 95%, that a normal variate lies in the interval $(\mu - 2\sigma, \mu + 2\sigma)$.

Now,

$$P(\mu - 3\sigma < X < \mu + 3\sigma) = \int_{\mu-3\sigma}^{\mu+3\sigma} f(X)dX$$

or

$$p(-3 < Z < 3) = \int_{-3}^{3} \phi(z)dz = 2\int_{0}^{3} \phi(z)dz = 2(0.49865) = 0.9973.$$

We many conclude that in all probability (99.7%), a normal variate lies in the interval $(\mu - 3\sigma, \mu + 3\sigma)$.

We note that in the normal probability tables, the areas under the standard normal curve are given. Therefore, in the numerical problems we convert the required area (given with respect to the variate X) to the form $P(0 < Z < z)$, since these areas are tabulated.

Normal distribution as a limiting form of binomial distribution

Standard normal distribution can be considered as a limiting form of the standard binomial distribution under the following conditions

(i) n is very large, $n \to \infty$
(ii) neither p nor q is very small.

Let $X \sim B(n, p)$. Then, the standard binomial variate is defined as

$$Z = \frac{X - np}{\sqrt{npq}}, \quad X = 0, 1, 2, \ldots, n.$$

Now, when $X = 0$, $$Z = -\frac{np}{\sqrt{npq}} = -\sqrt{\frac{np}{q}}, \text{ and}$$

when $X = n$, $$Z = \frac{n - np}{\sqrt{npq}} = \frac{nq}{npq} = \sqrt{\frac{nq}{p}}.$$

Therefore, as $n \rightarrow \infty$, Z varies from $-\infty$ to ∞ with increment of $1/\sqrt{npq}$.

Now, MGF of $X \sim B(n, p)$ is given by $M_X(t) = (q + pe^t)^n$.

For the standard binomial variable $Z = (X - np)/\sqrt{npq}$, we have

$$M_Z(t) = M_{[(X-np)/\sqrt{npq}]}(t) = e^{-npt/\sqrt{npq}} [q + pe^{t/\sqrt{npq}}]^n.$$

Now, we expand the right hand side as an infinite series.

We have $$\log M_Z(t) = \frac{-npt}{\sqrt{npq}} + n \log [q + pe^{t/\sqrt{npq}}]$$

Now, $$e^{t/\sqrt{npq}} = 1 + \frac{t}{\sqrt{npq}} + \frac{t^2}{2!(\sqrt{npq})^2} + \frac{t^3}{3!(\sqrt{npq})^3} + \dots$$

$$q + pe^{t/\sqrt{npq}} = 1 - p + p\left[1 + \frac{t}{\sqrt{npq}} + \dots\right]$$

$$= 1 + \left\{\frac{pt}{\sqrt{npq}} + \frac{pt^2}{2(npq)} + \frac{pt^3}{6(npq)^{3/2}} + \dots\right\}$$

$$\log [q + pe^{t/\sqrt{npq}}] = \left\{\frac{pt}{\sqrt{npq}} + \frac{pt^2}{2(npq)} + \dots\right\} - \frac{1}{2}\left\{\frac{pt}{\sqrt{npq}} + \frac{pt^2}{2(npq)} + \dots\right\}^2 + \dots$$

$$= \frac{pt}{\sqrt{npq}} + \frac{p(1 - p)t^2}{2npq} + \dots = \frac{pt}{\sqrt{npq}} + \frac{t^2}{2n} + \dots$$

Therefore, $$\log M_Z(t) = -\frac{npt}{\sqrt{npq}} + n\left[\frac{pt}{\sqrt{npq}} + \frac{t^2}{n} + \dots\right]$$

$$= \frac{t^2}{2} + \text{ terms containing } \frac{1}{\sqrt{n}} \text{ and lower order in } n$$

and $$\lim_{n \to \infty} \log M_Z(t) = \frac{t^2}{2}.$$

Hence, $$M_Z(t) = e^{t^2/2}$$

which is the MGF of the standard normal distribution. We conclude that in the limit as $n \to \infty$, the distribution of Z is a continuous distribution ranging from $-\infty$ to ∞ and having mean 0 and variance 1.

Importance of normal distribution

Normal distribution plays a very significant role in the statistical theory. The following are some examples.

(i) Many distributions like binomial, Poisson etc. can be approximated by normal distribution.

(ii) Many sampling distributions like Student's t distribution, Chi - square distribution etc. tend to normal distribution when the sample size is large. Tests of significance for small samples are based on the assumption that samples have been drawn from normal distributions.

(iii) Statistical quality control in industry is based on the theory of normal distribution.

Example 19.59 The random variable X is normally distributed with mean equal to 9 and standard deviation 3.

(a) Find the probabilities, (i) $X \geq 15$, (ii) $X \leq 15$, (iii) $0 \leq X \leq 9$.

(b) Find x^*, when $P(X > x^*) = 0.16$.

Solution We have $\mu = 9$, $\sigma = 3$ and $X \sim N(9, 9)$. The standard normal variate is $Z = (X - 9)/3$.

(a) (i) $X = 15$, $Z = \{(15 - 9)/3] = 2$.

$P(X \geq 15) = P(Z \geq 2) = 0.5 - P(0 \leq Z \leq 2) = 0.5 - 0.4772 = 0.0228$.

(ii) $P(X \leq 15) = 1 - P(X \geq 15) = 1 - 0.0228 = 0.9772$.

(iii) $P(0 \leq X \leq 9) = P(-3 \leq Z \leq 0) = P(0 \leq Z \leq 3) = 0.4987$.

(b) $P(X > x^*) = P(Z > z^*) = 0.16$.

Hence, $P(0 < Z < z^*) = 0.5 - 0.16 = 0.34$.

From tables, we obtain $z^* \approx 1.0$. Now

$$\frac{x^* - \mu}{\sigma} = 1.0, \quad \text{or} \quad x^* = 9 + 3 = 12.$$

Example 19.60 There are 500 students taking a Mathematics course in an engineering college. The probability for any student to need a particular book from the college library on any day is 0.07. How many copies of the book should be kept in the library so that the probability may be greater than 0.95 that none of the students needing a copy from the library has to go back disappointed? Assume normal distribution.

Solution We have $n = 500$, $p = 0.07$, $\mu = np = 35$, $\sigma^2 = npq = 32.55$, $\sigma \approx 5.7$.

We need to find x^* such that

$$P(X < x^*) > 0.95.$$

Set $z_1 = (x^* - 35)/5.7$. Hence, we require $P(0 < Z < z_1) > 0.45$.

From Table 19.4, we get $z_1 > 1.65$, or $x^* > 35 + 5.7 (1.65) > 44.4 \approx 45$. Therefore, the library should keep atleast 45 copies.

Example 19.61 In a distribution which is exactly normal, 12% of the items are under 30 and 85% are under 60. Find the mean and standard deviation of the distribution.

Solution Let $X \sim N(\mu, \sigma^2)$. Then,

$$P(X < 60) = 0.85, \text{ or } P(X > 60) = 0.15; \text{ and } P(X < 30) = 0.12.$$

Obviously, $X = 30$ is to the left of $X = \mu$ and $X = 60$ is to the right of $X = \mu$ (see Fig. 19.14).

When $\qquad X = 30, \ Z = \dfrac{30 - \mu}{\sigma} = -z_1.$

When $\qquad X = 60, \ Z = \dfrac{60 - \mu}{\sigma} = z_2.$

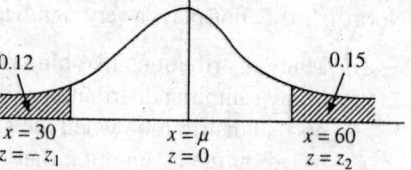

Therefore, $P(0 < Z < z_1) = 0.38$ and $P(0 < Z < z_2) = 0.35.$

Fig. 19.14. Example 19.64.

From the Table 19.4, we obtain $z_1 \approx 1.175$, and $z_2 \approx 1.0365$.

Hence, $\qquad\qquad\qquad\qquad 30 - \mu = -\sigma z_1 = -1.175\sigma$

and $\qquad\qquad\qquad\qquad 60 - \mu = \sigma z_2 = 1.0365\,\sigma.$

The solution of these equations is $\sigma = 13.5655$ and $\mu = 45.9395$.

Example 19.62 In an examination, the candidates are awarded the following grades depending on the marks scored by them:

> distinction: $\geq 80\%$, first class : $60\% \leq$ marks $< 80\%$, second class: $45\% \leq$ marks $< 60\%$,
>
> third class: $30\% \leq$ marks $< 45\%$, fail: $< 30\%$.

It was found that 8% of the students failed and 8% have scored distinction. Find the average marks obtained by the candidates. Deduce the percentage of students placed in the second class. Assume normal distribution of marks.

Solution Let X denote the marks obtained by the candidates in the examination. Let $X \sim N(\mu, \sigma^2)$. We have $P(X < 30) = 0.08$ and $P(X \geq 80) = 0.08$ (see Fig. 19.15).

The standard normal variate is $\qquad\qquad Z = \dfrac{X - \mu}{\sigma}.$

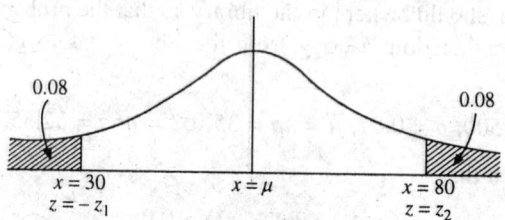

Fig. 19.15. Example 19.65.

We note that $X = 30$ is to the left of $X = \mu$ and $X = 80$ is to the right of $X = \mu$.

For $X = 30$, $\quad Z = \dfrac{30 - \mu}{\sigma} = -z_1$.

For $X = 80$, $\quad Z = \dfrac{80 - \mu}{\sigma} = z_2$.

Since $P(X < 30) = 0.08$, we have $P(Z < -z_1) = 0.08$, or $P(0 < Z < z_1) = 0.50 - 0.08 = 0.42$.
Since $P(X \geq 80) = 0.08$, we have $P(0 < Z < z_2) = 0.42$.
From Table 19.4, we get $z_1 = 1.405 = z_2$.

Hence, $\qquad \dfrac{30 - \mu}{\sigma} = -1.405$ and $\dfrac{80 - \mu}{\sigma} = 1.405$.

Adding, we get $110 - 2\mu = 0$ or $\mu = 55$, and $\sigma = \dfrac{80 - \mu}{1.405} = \dfrac{25}{1.405} \approx 17.79$.

Let p be the probability that a candidate is placed in second class.

Then, $\qquad\qquad p = P(45 \leq X < 60) = P(-0.56 \leq Z < 0.28)$

$$= P(-0.56 \leq Z < 0) + P(0 < Z < 0.28)$$

$$= P(0 < Z \leq 0.56) + P(0 < Z < 0.28)$$

$$= 0.2123 + 0.1103 = 0.3226.$$

Therfore, 32% candidates (approximately) obtained second class.

Example 19.63 It is given that X and Y are independent normal variates with means 2, 5 and variances 4, 9 respectively. Find the value of K such that

$$P(X + Y \leq K) = P(3X - Y \geq 3K).$$

Solution We have $X \sim N(2, 4)$, $Y \sim N(5, 9)$ and X, Y are independent. Define $U = X + Y$, and $V = 3X - Y$. Then, U and V are also normal variates and

$$U = X + Y \sim N(2 + 5, 4 + 9) = N(7, 13).$$

$$V = 3X - Y \sim N(3(2) - 5, 9(4) + 9) = N(1, 45).$$

Define, $Z = (U - 7)/\sqrt{13}$. Now, $Z \sim N(0, 1)$ and for $U = K$, $Z = [(K - 7)/\sqrt{13}] = z_1$.
Define, $Z = (V - 1)/\sqrt{45}$. Again, $Z \sim N(0, 1)$ and for $V = 3K$, $Z = (3K - 1)/\sqrt{45} = z_2$.
We require

$$P(X + Y \leq K) = P(3X - Y \geq 3K)$$

or $\qquad\qquad\qquad P(Z \leq z_1) = P(Z \geq z_2)$.

Since the normal probability curve is symmetric about $Z = 0$, this implies that $z_1 = -z_2$. Hence,

$$\frac{K - 7}{\sqrt{13}} = -\left[\frac{3K - 1}{\sqrt{45}}\right]; \quad \text{or} \quad \sqrt{45}(K - 7) = \sqrt{13}(1 - 3K)$$

or $\qquad\qquad\qquad K = \dfrac{7\sqrt{45} + \sqrt{13}}{\sqrt{45} + 3\sqrt{13}} \approx 2.885.$

Exercise 19.5

1. A factory has 20 machines out of which 6 are old and 14 are new. The probability that any old machine needs repair on any particular day is 1/10 and the probability that any new machine needs repair is 1/50. Assume that no machine is to be repaired two or more times in any day. Find the probability that just 2 old machines and no new machines need repair on any particular day.

2. Out of 1600 families with four children each, how many families would be expected to have (i) atleast one boy, (ii) exactly 2 boys and 2 girls.

3. The mean and variance of a binomial distribution are 4.5 and 1.125 respectively. Find $P(X \geq 1)$.

4. If X is binomially distributed with parameters n and p, what is the distribution of $Y = n - X$?

5. Two dice are thrown n times. Let X denote the number of throws in which the number on the first dice exceeds the number on the second dice. Find the distribution of X.

6. Let X denote the number of girls in a family of four children. If p denotes the probability that a girl is there in a family, find the least value of p such that $P(X = 0) > P(X = 1)$.

7. Let $X \sim B(n, p)$. Assuming that n is fixed, find the value of p such that the variance (X) is maximised. What is the maximum value of variance?

8. A fair coin is tossed $2n$ times. Show that the probability of obtaining more heads than tails is

$$\frac{1}{2}[1 - (2n)C_n p^{2n}], \quad p = \frac{1}{2}.$$

9. A aircraft can fly if atleast one half of its engines function properly. Each engine functions independently with probability p. Find the values of p when a 4 engine plane is to be preferred in camparison with a 2 engine plane.

10. A poisson variate satisfies $P(X = 1) = 0.5 \, P(X = 2)$. Find $P(X = 4)$.

11. If X is a Poisson variate such that

$$P(X = 2) = 3P(X = 4) + 45P(X = 6)$$

find the mean and variance of X.

12. If X is a Poisson variate with mean λ, find the expectation of e^{-KX}.

13. In a book of 800 pages, 300 typographical errors are noticed. Assuming Poisson law for the number of errors per page, find the probability that a randomly chosen 4 pages will contain no errors.

14. The number of telephone calls, X_1, coming into a telephone exchange between 9.00 A.M and 10.00 A.M., is a random variable with Poisson distribution with parameter 3. The number of telephone calls, X_2, coming into the same telephone exchange between 10.00 A.M and 11.00 A.M. is also a random variable with Poisson distribution with parameter 5. If X_1, X_2 are independent, what is the probability that more than 4 calls come in between 9.00 and 11.00 A.M.?

15. The first print of a book containing 1000 pages, was corrected by a proof reader. After correcting 80 pages of the proof, he found that there are, on an average, 3 errors per 6 pages. Assuming Poisson law for the number of errors per page, find the number of pages that are expected to contain 0, 1, 2, 3 errors, in the whole book.

16. In a certain factory manufacturing electric bulbs, there is a chance of 1/500 for any bulbs to be defective. The bulbs are packed in packets of 50. Use the Poisson distribution to calculate the approximate number of packets containing no defective, one, two, three defective bulbs in a consignment of 10,000 packets.

17. Let X and Y be independent Poisson variates. Show that the conditional distribution of X, given the value of $X + Y$, is a binomial distribution.

18. An insurance company has discovered that only about 0.05% of the population is involved in a certain type of accident each year. The company has 20,000 policy holders covered for this type of accident. Using Poisson distribution find the probability that not more than 4 of its policy holders are involved in such an accident in the next year.

19. Red blood cell deficiency in human population may be detected by examining a specimen blood under a microscope. Assume that for normal persons, the average red blood cell count, for a fixed volume of

blood, should be 20. Using Poisson distribution, find the probability that a specimen blood of a person will contain less than 12 red blood cells.

20. The continuous random variable X is uniformly distributed with mean 1.5 and variance 27/4. Find $P(X > 0)$.

21. The continuous random variable X is uniformly distributed on [0, 1]. Find the *pdf* of $(-3 \log X)$.

22. A thread of length 10 cm is given. It is randomly cut into 2 pieces. Assuming uniform distribution find the probability that the length of the longer piece is three times the length of the shorter piece.

23. Let X be uniformly distributed over [0, 1]. Find the distribution of $Y = 1/X$. Find $E(1/X)$ if it exists.

24. Suppose X is uniformly distributed over $(-a, a)$, $a > 0$. Determine a such that (i) $P(X > 1) = 1/4$, (ii) $P(X < 1/2) = 0.6$.

25. Show that if X is a random variable with a continuous distribution function F, then $F(X)$ has a uniform distribution on [0, 1].

26. Let X_1 and X_2 be two independent uniformly distributed variables on [0, 1]. Then, find the distributions of (i) $X_1 X_2$, (ii) $X_1 - X_2$.

27. A continuous random variable X is normally distributed with mean 16 and standard deviation 5. Find the probability that (i) $X \geq 25$, (ii) $X \leq 25$, (iii) $0 \leq X \leq 16$. Also, find a x^* such that $P(X > x^*) = 0.2$.

28. A continuous random variable X is normally distributed with mean 25 and standard deviation 8. Find the probability that (i) $20 \leq X \leq 40$, (ii) $|X - 25| \leq 5$.

29. Prove that for the normal distribution $N(\mu, \sigma^2)$, the quartile deviation, the mean deviation and standard deviation are approximately in the ratios $10 : 12 : 15$.

30. A large group of men attend a selection camp for army cadets. It was found that the heights of 10% of them are below 165cm and the heights of 30% of them are between 165cm and 180cm. Assuming a normal distribution, find the mean height and standard deviation of the group of men.

31. Let X be normally distributed such that $\sigma^2 = \mu^2$, $\mu > 0$. Find $P(X < -\mu \mid X < \mu)$ in terms of the cumulative distribution function of $N(0, 1)$.

32. It is given that X and Y are independent normal variates and $X \sim N(1, 4)$, $Y \sim N(3, 16)$. Find the value of K such that

$$P(2X + Y \leq K) = P(4X - Y \geq 2K).$$

33. The marks obtained in Mathematics by the students in a class are approximately normally distributed with mean 62 and variance 36. If 3 students are selected at random, find the probability that at least 1 of them would score more than 80 marks.

34. In a normal population with mean 12 and standard deviation 4, it is known that 750 observations exceed 15. Find the total number of observations in the population.

35. A minimum height is to be prescribed for eligibility to enter the military services, such that 40% of the applicants will have a fair chance of qualifying the height eligibility. The heights are normally distributed with mean 165 cm and standard deviation 8 cm. Determine the minimum height specification.

19.9 Answers and Hints

Exercise 19.1

1. $S = \{HHH, HHT, HTH, HTT, THH, THT, TTH, TTT\}$.

2. $S = \{2, 3, 4, 5, 6, 7, 8, 9, 10, 11, 12\}$.

3. (i) $A = \{F_1 M_1, F_1 M_2, F_1 M_3, F_2 M_1, F_2 M_2, F_2 M_3, F_1 F_2\}$.
   (ii) $A = \{(6, 6, 6), (6, 6, 5), (6, 5, 6), (5, 6, 6)\}$
   (iii) $A = \{HHH, HHT, HTH, THH\}$.

4. (i) Favourable cases: (6, 2), (5, 3), (4, 4), (3, 5), (2, 6). $P(A) = 5/36$,
   (ii) 5/18,   (iii) 1/6,   (iv) 5/18,

(v)   Events are mutually exclusive.
P(total of 7 or 11) = P(total is 7) + P(total is 11) = 8/36.

(vi)   P (sum is even) = P(odd number on the first dice and an odd number on the second dice) + P(even number on the first dice and an even number on the second dice) = (9/36) + (9/36) = 1/2.

(vii)   1/6.

5. $P(A \cup B) = P(A) + P(B) = 1 - P(\overline{A}) + P(B), P(B) = 0.35.$ **6.**   60/143.

7. $S = \{1, 2, 3, \ldots, 100\}$. A : number is divisible by 3, B : number is divisible by 7. A and B are mutually exclusive
$P(A \text{ or } B) = P(A \cup B) = P(A) + P(B) = (33/100) + (14/100) = (47/100).$

8.   P(2 white and 1 black) = $P(A_w \, B_w \, B_b \text{ or } A_b \, B_w \, B_w) = \dfrac{5}{12} \cdot \dfrac{6 \times 4}{10C_2} + \dfrac{7}{12} \cdot \dfrac{6C_2}{10C_2} = \dfrac{15}{36}.$

9. A : first bag, B : second bag, C : third bag. $P(A) = P(B) = P(C) = 1/3$.
P (red ball is chosen from first bag) = 2/10, P (red ball is chosen from second bag) = 4/10,
P (red ball is chosen from third bag) = 3/10.

P (ball drawn is red) = $\dfrac{1}{3} \cdot \dfrac{2}{10} + \dfrac{1}{3} \cdot \dfrac{4}{10} + \dfrac{1}{3} \cdot \dfrac{3}{10} = \dfrac{9}{30} = \dfrac{3}{10}.$

10. A : getting 4, 5, 6 on first toss. $P(A) = 1/2$. B : getting 3, 4 or 5 on second toss, $P(B) = 1/2$.
$A \cap B = \{4, 5\}$; $P(A \cap B) = 1/3,$

$P(A \cup B) = P(A) + P(B) - P(A \cap B) = (1/2) + (1/2) - (1/3) = (2/3).$

11.   *Case* 1 : White ball is transferred. $p_1 = 2/9$.
Case 2 : Black ball is transferred. $p_2 = 5/24$. Probability = $p_1 + p_2 = 31/72$.

12. A : box 1, B : box 2. P(ball is white) = P(box A *and* white ball) + P (box B *and* white ball) =
$\dfrac{1}{2} \cdot \dfrac{8}{20} + \dfrac{1}{2} \cdot \dfrac{12}{25} = \dfrac{11}{25}.$

13.   P(atleast one is black) = 1 − P(no black ball in 3 draws) = $1 - (1/2)^3 = 7/8$.

14.   P(atleast one is white) = 1 − P(no white ball in 4 draws) = $1 - (3/4)^4 = 175/256$.

15.   P(1 red and 1 white ball) = $\dfrac{8C_1 \times 12C_1}{20C_2} = \dfrac{48}{95}.$ **16.** $\left(\dfrac{14}{25}\right)^5.$

17. A : odd number on the first dice = $\{1, 3, 5\}$, $P(A) = 1/2$.
B : total of 7 = $\{(6, 1), (5, 2), (4, 3), (3, 4), (2, 5), (1, 6)\}$, $P(B) = 1/6$.
$A \cap B = \{(5, 2), (3, 4), (1, 6)\}$, $P(A \cap B) = 1/12$, $P(A \cup B) = P(A) + P(B) - P(A \cap B) = 7/12$.

18.   1/3.

19.   **(i)**   P(all cards of same suit) = $\dfrac{4(13C_4)}{52C_4} = \dfrac{44}{4165}.$ **(ii)** $\dfrac{13}{52C_4}$

20.   P(not ace, king or queen) = 1 − P(ace, king or queen) = 1 − (12/52) = 10/13.

21. ϕ.   **22.** $a = 0.3$

23.   P(0, 2 or 4 heads) = 1/2.   **24.**   P(1, 3 or 5 tails) = 1/2.

25.   Face value on each dice is not less than 2 and not greater than 5. The probability for each dice is 2/3. Total probability is $(2/3)^4 = 16/81$.

26.   Total number of coins = 15, number of ways of selecting 5 coins = $15C_5$.

A : Coins of worth < 3 rupees. That is, $\overline{A}$: coins of worth ≥ 3 rupees. Coins of worth ≥ 3 rupees can be obtained only in the following cases:
1 rupee coin + 4 fifty paise coins + no ten paise coin,
2 rupee coins + 3 fifty paise coins + no ten paise coin,

2 rupee coins + 2 fifty paise coins + 1 ten paise coin.

$$P(\bar{A}) = \frac{2C_1 \times 5C_4 + 2C_2 \times 5C_3 + 2C_3 \times 5C_2 \times 10C_1}{15C_5} = \frac{40}{1001}. \quad P(A) = 1 - P(\bar{A}) = 961/1000.$$

27. P(A wins) = 0.7, P(A loses) = 0.3. W : win, L : loss.
Best of three games : WW, WLW, LWW. $P = (0.7)^2 + 2(0.7)^2 (0.3) = 0.784$.

Best of five games : 3 straight wins, four games where last one is a win and a loss in the earlier 3 games, five games where last one is a win and two losses in the earlier 4 games

$$P = (0.7)^3 + 3C_1(0.7)^3 (0.3) + 4C_2(0.7)^3 (0.3)^2 = 0.83692.$$

A should choose to play best of five games.

28. $\left(\dfrac{4C_1}{52C_1}\right)\left(\dfrac{2C_1}{51C_1}\right) + \left(\dfrac{4C_1}{52C_1}\right)\left(\dfrac{2C_1}{51C_1}\right) = \dfrac{16}{2652}.$

29. Possibilities = $(2n + 1)C_3$. Drawn numbers are to be in A.P with common difference: 1, 2, 3, . . . , $n - 1, n$. Total possibilities = $(2n - 1) + (2n - 3) + ... + 3 + 1 = \dfrac{n}{2}[1 + (2n - 1)] = n^2$. Required Probability $= 3n/(4n^2 - 1)$.

30. 1/36.

Exercise 19.2

1. (a) A : 3 appeared atleast once, $P(A) = 11/36$, B : sum of the numbers is 7, $P(B) = 6/36$,

     $P(A/B) = 1/3$.

   (b) A : 5 has appeared atleast once. $P(A/B) = 1/3$.

2. A : 2 has appeared atleast once, B : sum of the numbers is 8. $P(A/B) = 2/5$.

3. $P(B/A) = 0.75$, $P(A \cup B) = 0.7$.

4. $P(A/B) = 3/14$, $P(A \cap B) = 0.15$, $P(A \cup B) = 0.85$.

5. The events A fails and B fails are independent.

   (i)   P(A fails and B fails) = $(0.3)(0.4) = 0.12$.

   (ii)   P (either A or B fails) = P(A fails and B passes) + P(A passes and B fails)

     $= P(A)P(\bar{B}) + P(\bar{A})P(B) = 0.46.$

6. **(i)** $P(W, B, R, W) = 1/81$,   (ii) $P(R, W, B, B) = 1/324$.

7. Total matches played by A is 4. Denote W, D, L for win, draw and loss respectively.

     P(points ≥ 6) =   P(6 points) + P(7 points) + P(8 points)

              =   P(2W, 2D) + P(3W, 1L) + P(3W, 1D) + P(4W)

              = $(0.6)^2(0.3)^2 + (0.6)^3 (0.1) + (0.6)^3 (0.3) + (0.6)^4 = 0.2484.$

8. 0.92.

9. $P(B) = 0.4$, $P(A \cap B) = 0.4\, P(A)$, $P(A \cup B) = P(A) + P(B) - P(A \cap B)$, $P(A) = 2/3$.

10. P $(B) = 0.35$, P $(A \cap B) = 0.35$ P (A), P $(A) = 10/13$.   **11.** 1/2.

12. P $(A) = 1/2$, P$(B) = 1/2$, P$(A \cap B) = 1/4 = P(A)$ P(B). A, B are independent.

13. $S = \{DD, DN, ND, NN\}$ where D, N denote defective and non-defective item respectively.

   $A = \{DD, NN\}$, $B = \{DD, DN\}$, $C = \{DN, NN\}$. $A \cap B = \{DD\}$, $A \cap C = \{NN\}$, $B \cap C = \{DN\}$, $A \cap B \cap C = \phi$.

   P$(A) = 1/2$, P$(B) = 1/2$, P$(C) = 1/2$. P$(A \cap B) = 1/4 = P(A)$ P(B), P$(A \cap C) = 1/4 = P(A)$ P(C), P$(B \cap C) = 1/4 = P(B)$ P (C).

(i) A, B, C are pairwise independent.

(ii) $P(A \cap B \cap C) = 0 \neq P(A)\,P(B)\,P(C)$. A, B, C are not independent.

14. $S = \{DD, DN, ND, NN\}$ where D, N denote defective and non-defective bulbs respectively

$A = \{DD, DN\}$, $B = \{DN, NN\}$, $C = \{DD, NN\}$. Also, $P(A) = 1/2 = P(B) = P(C)$. $P(A \cap B) = 1/4 = P(A)\,P(B)$; $P(A \cap C) = 1/4 = P(A)\,P(C)$; $P(B \cap C) = 1/4 = P(B)\,P(C)$. We also have $P(A \cap B \cap C) = 0 \neq P(A)\,P(B)\,P(C)$. Therefore, A, B, C are pairwise independent. A, B, C are not independent.

15. $p_1 = P(\text{sum is 4}) = 1/12$, $P(\text{sum is 8}) = 5/36$.
$P(\text{sum is 4 or 8}) = 2/9$, $P(\text{sum is not 4 or 8}) = (7/9) = q$

$P(\text{getting a sum 4 before 8}) = p_1 + qp_1 + q^2 p_1 + \ldots = p_1(1 + q + q^2 + \ldots) = \dfrac{p_1}{1 - q} = \dfrac{3}{8}$.

Exercise 19.3

1. $P(5) = P(H)\,P(5/H) + P(T)\,P(5/T) = \dfrac{1}{2}\left[\dfrac{1}{6} + \dfrac{4}{36}\right] = \dfrac{5}{36}$.

2. Denote E as the event of getting a defective bulb.

$P(E) = P(A)\,P(E/A) + P(B)\,P(E/B) + P(C)\,P(E/C) + P(D)\,P(E/D)$

$\qquad = 0.5\,(0.02) + 0.25(0.01) + 0.15(0.01) + 0.01\,(0.05) = 0.019$.

3. $P(W) = P(A)\,P(W/A) + P(B)\,P(W/B) = \dfrac{1}{2}\left[\dfrac{3}{5} + \dfrac{1}{5}\right] = 0.4$.

4. $P(\text{Job is done}) = \sum P(A_i)\,P(\text{Job is done}/A_i)$

$\qquad = 0.3(0.4) + 0.25(0.7) + 0.35\,(0.3) + 0.1(0.2) = 0.42$.

5. Two possibilities are there. (i) B_1 : Transfer of red ball to the second bag, and (ii) B_2 : Transfer of black ball to the second bag. Let A be the event of drawing a red ball from the second bag after transfer.

$$P(B_1) = \dfrac{3}{4},\quad P(B_2) = \dfrac{1}{4},\quad P(A/B_1) = \dfrac{5}{11},\quad P(A/B_2) = \dfrac{4}{11}.$$

$$P(A) = P(B_1)P(A/B_1) + P(B_2)P(A/B_2) = \dfrac{3}{4}\cdot\dfrac{5}{11} + \dfrac{1}{4}\cdot\dfrac{4}{11} = \dfrac{19}{44}.$$

6. B_1 : Red ball is transferred, B_2 : white ball is transferred. A : event of drawing a white ball after transfer.

$$P(B_1) = \dfrac{3}{10},\quad P(B_2) = \dfrac{7}{10},\quad P(A/B_1) = \dfrac{2}{5},\quad P(A/B_2) = \dfrac{1}{2}.$$

$$P(A) = \dfrac{3}{10}\cdot\dfrac{2}{5} + \dfrac{7}{10}\cdot\dfrac{1}{2} = \dfrac{47}{100}.\quad P(B_1/A) = \dfrac{(3/10)(2/5)}{(47/100)} = \dfrac{12}{47}.$$

7. $B_i = $ Event of selecting ith bag, $i = 1, 2, 3$. $A = $ Event of drawing a white ball.

$$P(B_1) = P(B_2) = P(B_3) = 1/3,\quad P(A/B_1) = 2/5,\quad P(A/B_2) = 3/5,\quad P(A/B_3) = 1/2.$$

$$P(A) = \sum P(B_i)\,P(A/B_i) = 1/2.$$

$E = $ Event of drawing a white ball again (knowing that the first draw is a white ball)

$$P(E/B_1 \cap A) = 1/4,\quad P(E \mid B_2 \cap A) = 1/2,\quad P(E \mid B_3 \cap A) = 2/5.$$

$$P(E/A) = \dfrac{\sum[P(B_i)\,P(A/B_i)\,P(E/B_i \cap A)]}{\sum[P(B_i)\,P(A/B_i)]} = \dfrac{2}{5}.$$

8. A : Event of drawing a white ball from kth box. Let p_{k-1} be the probability of drawing a white ball from the $(k-1)$ th box. Drawing a white ball from kth box can happen in two ways (i) A white ball is transferred from the $(k-1)$th box and then a white ball is drawn from the kth box, (ii) A red ball is transferred from the $(k-1)$th box and then a white ball is drawn from the kth box.

(i) P(drawing a white ball from kth box) $= \dfrac{s+1}{s+r+1}$. Hence, Prob(case (i)) $= \left(\dfrac{s+1}{s+r+1}\right)p_{k-1}$.

(ii) P(drawing a red ball from $(k-1)$th box) $= 1 - p_{k-1}$.

P(drawing a white ball from kth box) $= \dfrac{s}{s+r+1}$.

Hence, Prob(case (ii)) $= \left(\dfrac{s}{s+r+1}\right)(1 - p_{k-1})$.

Since, case (i) and case (ii) are mutually exclusive

$$p_k = \left(\frac{s+1}{(s+r+1)}\right)p_{k-1} + \left(\frac{s}{s+r+1}\right)(1 - p_{k-1}) = \frac{1}{s+r+1}(p_{k-1} + s).$$

9. B_1 : Event that box 1 is selected, B_2 : Event that box 2 is selected. A : Event that 2 defective bulbs are drawn. No. of defective bulbs in box 1 = 250; in box 2 = 120.

$$P(B_1) = \frac{1}{2} = P(B_2), P(A/B_1) = \frac{250C_2}{5000C_2} = \frac{249}{99980}. \; P(A/B_2) = \frac{120C_2}{4000C_2} = \frac{357}{399900}.$$

(i) $P(A) = \Sigma P\,(B_i)\,P(A/B_i) = 0.001694$.

(ii) P(defective bulbs$/B_2$) $= \dfrac{P(B_2)P(A/B_2)}{P(A)} = 0.26495$.

10. A = Event of transmitting '0', $\overline{A}$ = Event of transmitting '1'.
    B = Event of receiving '0', $\overline{B}$ = Event of receving '1'.

$P(A) = 0.6, P(\overline{A}) = 0.4, P(B/A) = 0.99, P(\overline{B}/A) = 0.95, P(B/\overline{A}) = 0.05$.

(i) $P(B) = P(A)P(B/A) + P(\overline{A})P(B/\overline{A}) = 0.6(0.99) + 0.4(0.05) = 0.614$.

(ii) $P(A/B) = \dfrac{P(A)P(B/A)}{P(B)} = \dfrac{0.6\,(0.99)}{0.614} = \dfrac{297}{307}$.

Exercise 19.4

1.

x	0	1	2
$p(x)$	7/15	7/15	1/15

2. $c = 1/7, P(X < 3) = 2/7, P(1 < X < 4) = 4/7$.

x	1	2	3	4	5
$F(X)$	1/7	2/7	5/7	43/49	1

3. Let $P\,(x = -2) = C, P(x = 0) = K$. We have $3C + 3K = 1$ and $3C = 2K$. we get $C = 2/15$ and $K = 1/5$.

x	-2	-1	0	1	2
$p(x)$	2/15	4/15	1/5	1/5	1/5
$F(X)$	2/15	6/15	9/15	12/15	1; $\mu_x = 1/15$.

4. Let $p(x_2) = C$. We get $C = 4/25$.

x	x_1	x_2	x_3	x_4
$p(x)$	12/25	4/25	6/25	3/25
$F(X)$	12/25	16/25	22/25	1

5. Since $\sum\limits_{K=1}^{\infty} P(X = x_K) = \sum\limits_{K=1}^{\infty} p_K = 1$, it represents a probability function.

$$\mu_x = \sum\limits_{K=1}^{\infty} x_K p_K = \frac{1}{2}[1 + 2t + 3t^2 + 4t^3 + \ldots] = \frac{1}{2}(1-t)^{-2} = 2, \text{ since } t = 1/2.$$

$$\sigma_X^2 = E(X^2) - [E(X)]^2 = \left(\sum\limits_{K=1}^{\infty} \frac{K^2}{2^K}\right) - 4 = 6 - 4 = 2,$$

since $\sum \dfrac{K^2}{2^K} = \sum K(K+1)t^{K} - \sum K t^K = 2t[1 + 3t + 6t^2 + \ldots] - t(1 + 2t + 3t^2 + \ldots)$

$$= 2t(1-t)^{-3} - t(1-t)^{-2} = 6.$$

6. Yes. $\displaystyle\int_{-1}^{1} f(x)dx = \int_{-1}^{1} |x|\, dx = 1.$

7. $a = 4$, 1/256, 15/16.   **8.** $a = 2, b = -2$.

9. $0 \le a \le 1$, $0 \le b \le 1$ and $a + b = 1$.

10. X be the continuous random variable denoting the mileage in thousands of km, $1 - e^{-0.2}$, $e^{-0.3} - e^{-0.6}$.

11. $a = b$, $1/b$, $1/b$.   **12.** $a = -15/8, b = 3$.

13. (i) 7/64,   (ii) 26/63.

14. $P(X \le 1500) = 1/3$; $P(1$ bulb is to be replaced$) = 2/3$; P (two bulbs are to be replaced) $= 4/9$.

15. $K = \dfrac{5625}{8}$; $F(x) = \dfrac{5625}{16}\left(\dfrac{1}{225} - \dfrac{1}{x^2}\right)$; $\dfrac{175}{256}$.

16. $f(x) = F'(x)$, $a = \dfrac{1}{8}$, $f(x) = \dfrac{3}{8}(x-2)^2$, $2 \le x \le 4$.

17. $P(|X| < 20) = 3/4$; probability that a bomb does not destroy the target $= 1/4$; probability that both bombs do not destroy the target $= 1/16$; probability that the target is destroyed $= 15/16$.

18. $P(X > 2) = 5/9$; probability that both determinations are geater than 2 $= 25/81$; probability that exactly 2 determinations are greater than 2 $= 100/243$.

19. $K = 1/4$; $P(X < 5) = 1 - e^{-1.25}$; $P(X > 6) = e^{-1.5}$.

20. $f(x) = F'(x) = 4xe^{-2x}$, $x \ge 0$; $E(x) = 1$; $E(X^2) = 3/2$; $V(X) = 1/2$.

21.

X \\ Y	0	1	2
0	0	1/3	2/3
1	1/5	1/3	7/15
2	1/4	1/3	5/12

22. $K = 1$; $P(X > 2) = e^{-2}$; $P(2 < X + Y < 3) = \int_0^3 \int_0^{3-x} e^{-(x+y)} dx dy - \int_0^2 \int_0^{2-x} e^{-(x+y)} dx dy$

$= 3e^{-2} - 4e^{-3}$.

23. Set $x = r \cos \theta$, $y = r \sin \theta$; $K = 4$; independent

24. $f(x, y) = F_{XY}(x, y) = e^{-(x-y)}$, for $x > 0$, $y > 0$ and 0 elsewhere. X and Y are independent.

$P(X \le 2 \cap Y \le 2) = (1 - e^{-2})^2$; $P(X + Y \le 2) = 1 - 3e^{-2}$.

25. Let $A = \{X \le x\}$, $B = \{Y \le y\}$. Then, $P(A) = F_X(x)$, $P(B) = F_y(y)$, $P(A \cap B) = F_{XY}(x, y)$.
Now, $P(A \cap B) \le P(A)$ and $P(A \cap B) \le P(B)$ imply $F_{XY}(x, y) \le F_X(x)$ and $F_{XY}(x, y) \le F_Y(y)$. Therefore,
$F_{XY}(x, y) \le \sqrt{F_X(x) F_Y(y)}$. $P(A \cup B) = P(A) + P(B) - P(A \cap B) \le 1$ gives $F_X(x) + F_y(y) - 1 \le F_{XY}(x, y)$.

26. $c = 8/9$.

27. $c = 3/8$; $f_X(x) = \int_{-x}^x f(x, y) dy = (x^3/4)$, in $0 < x < 2$,

$$f_Y(y) = \int_{-y}^2 f(x, y) dx = \frac{3}{16}[4y^2 - 4y + 3y^3], -2 < y \le 0,$$

$$= \int_y^2 f(x, y) dx = \frac{3}{16}[4y^2 - 4y - y^3], 0 \le y < 2.$$

28. $f(x, y) = \partial^2 F/\partial x \partial y = e^{-x} e^{-y}$; $P(1 < X < 4) = e^{-1} - e^{-4}$; $P(1 < X < 3, 1 < Y < 3) = (1 - e^{-3})^2$.

29. $K = 1/(\pi R^2)$; $f_X(x) = (2/K) \sqrt{R^2 - x^2}$, $-R \le x \le R$, and 0 elsewhere.

30. $[e^{-(y-3)/4}]/4$, $3 \le y < \infty$.

31. $F_Y(y) = [F(2y) - F(0)]/[1 - F_X(0)]$, for $y \ge 0$; and 0 for $y < 0$.

$g_Y(y) = 2f_X(2y)/[1 - F_X(0)]$, for $y \ge 0$; and 0 for $y < 0$.

32. $y = 64x^3$ is a strictly increasing function in $(0, 2)$. $f_Y(y) = y^{-1/3}/96$, $0 < y < 512$.

33. $x = v$, $y = v - u$. For $u > 0$, $f_U(u) = e^{-u}/2$. For $u < 0$, $f_U(u) = e^u/2$. Hence, $f_U(u) = e^{-|u|}/2$.

34. $u = \sqrt{x^2 + y^2}$, $v = x$ or $x = v$, $y^2 = u^2 - v^2$. We get $g_{UV}(u, v) = 4uv \ e^{-u^2}$, $u \ge 0, 0 \le v \le u$. Also $f_U(u)$

$= 2u^3 e^{-u^2}$, $u \ge 0$ and 0 elsewhere.

35. $x = v$, $y = 2u - v$; $u > 0$, $0 < v < 2u$; $g_{UV}(x, y) = 2e^{-2u}$; $f_U(u) = 4ue^{-2u}$, $0 < v < 2u$ and 0 elsewhere.

36. $x = vz$, $y = v$; $v \ge 0$, $z \ge 0$; $g_{ZV}(z, v) = 2ve^{-(2+z)v}$; $f_Z(z) = 2/(2 + z)^2$, $z \ge 0$.

37. $f_{XY}(x, y) = 2e^{-(2x+y)}$; $x = uv$, $y = v(1 - u)$; $0 \le u \le 1$, $v \ge 0$; $f_{UV}(u, v) = 2ve^{-(u+1)v}$, $0 \le u \le 1$, $v \ge 0$ and 0 elsewhere. Since $f_{UV}(u, v) \ne f_U(u) f_V(v)$, U and V are not independent.

Exercise 19.5

1. Let p_1, p_2 denote respectively, the probabilities that old and new machines need repair. Then, $p_1 = 1/10$, $q_1 = 9/10$, $p_2 = 1/50$, $q_2 = 49/50$.

P(just 2 old machines need repair) $= P_1(2) \ P_2(0) = [6C_2 p_1^2 q_1^4][14C_0 p_2^0 q_2^{14}] \approx 0.0742$.

2. (i) $1600(15/16) = 1500$.   (ii) $1600 (6/16) = 600$.

3. $q = 0.25$, $p = 0.75$, $n = 6$, $P(X \geq 1) = 1 - P(X = 0) \approx 0.9998$.

4. $Y = n - X$ represents the number of failures in n independent trials with constant probability q. $Y \sim B(n, q)$.

5. $B(n, 5/12)$.

6. $q^4 > 4C_1 pq^3$; $p < 1/5$.

7. Maximise $V = npq = np(1 - p)$. We get $p = 1/2$. Max $(V) = n/4$.

8. Number of heads $= X$, number of tails $= Y$; $p = q = 1/2$. Also, $P(X > Y) = P(X < Y)$, From, $P(X > Y) + P(X < Y) + P(X = Y) = 1$, we get $P(X > Y) = 1/2 [1 - (2n)C_n p^{2n}]$, $p = 1/2$.

9. Four engine plane: $P(X = 2 \text{ or } X = 3 \text{ or } X = 4) = 1 - P(X = 0 \text{ or } X = 1) = 1 - (q^4 + 4pq^3)$. Two engine plane: $P(X = 1 \text{ or } X = 2) = 1 - P(X = 0) = 1 - q^2$. We require $1 - q^4 - 4pq^3 \geq 1 - q^2$. We obtain $(2/3) < p < 1$.

10. $\lambda = 4$; $32e^{-4}/3$.   11. $45\lambda^4 + 90\lambda^2 - 360 = 0$, mean $=$ variance $= \lambda = \sqrt{2}$.   12. $e^{-\lambda(1 - e^{-K})}$.

13. $\lambda = 3/8$; $P(X = 0) = e^{-3/8}$; $P(\text{no errors on 4 pages}) = e^{-1.5}$.

14. $X = X_1 + X_2$; $\lambda = \lambda_1 + \lambda_2 = 8$; $P(X > 4) = 1 - P(X \leq 4) \approx 0.9004$.

15. $\lambda = 0.5$. For $X = 0, 1, 2, 3$ the expected number of pages respectively are 606, 303, 76 and 13 (approx.).

16. $\lambda = np = 1/10$; $N(X = 0) \approx 9048$; $N(X = 1) = 904$; $N(X = 2) = 45$; $N(X = 3) = 2$.

18. $\lambda = np = 10$; $P(\text{Not more than 4 meet an accident}) \approx 0.0293$.

19. $\lambda = 20$; $P = \sum\limits_{r=0}^{11} [e^{-20} (20)^r / r!]$.

20. $p(x) = 1/9$, $-3 < x < 6$; $P(X > 0) = 2/3$.

21. $Y = -3 \log X$; $X \sim [0, 1]$, $Y \sim [0, \infty)$; $G_Y(y) = 1 - e^{-y/3}$; $g_Y(y) = [e^{-y/3}/3]$, $0 < y < \infty$.

22. $f(x) = 1/10$, $0 < x < 10$; required probability $= 1/4$.

23. $f(x) = 1$, $0 \leq x \leq 1$; $G_Y(y) = 1 - (1/y)$; $g_Y(y) = (1/y^2)$, $1 \leq y < \infty$; $E(1/X)$ does not exist

24. $f(x) = 1/(2a)$, $-a < x < a$; (i) $a = 2$; (ii) $a = 5/2$.

25. $G_Y(y) = P(Y \leq y) = P[F(x) \leq y] = P[X \leq F^{-1}(y)]$. The inverse exists as $F(X)$ is non-decreasing and continuous. Since $P(X \leq x) = F(x)$, we get $G_Y(y) = F[F^{-1}(y)] = y$; $g_Y(y) = 1$; Y satisfies $0 \leq y \leq 1$. Therefore, Y is a continuous uniform variate on $[0, 1]$.

26. (i)   Let $u = x_1 x_2$, $v = x_1$ or $x_1 = v$, $x_2 = u/v$; $J = -1/v$; $g(u, v) = 1/v$;

$$g(u) = \int_u^1 (1/v) \, dv = -\log u, \quad 0 < u < 1.$$

   (ii) $g_1(v) = \int_v^{2-v} \frac{du}{2} = 1 - v$, $0 < v < 1$, $g_2(v) = \int_{-v}^{2+v} \frac{du}{2} = 1 + v$, $-1 < v < 0$.

27. $X \sim N(16, 25)$; $P(X \geq 25) = P(Z \geq 1.8) = 0.0359$; $P(X \leq 25) = 0.9641$; $P(0 \leq X \leq 16) = 0.4993$; $x^* = 20.2$.

28. $X \sim N(25, 64)$; $P(20 \leq X \leq 40) = P(-0.625 \leq Z \leq 1.875) = 0.7036$;
$P(|X - 25| \leq 5) = P(20 \leq X \leq 30) = 0.4680$.

29. $QD \approx 2\sigma/3$; $MD = 4\sigma/5$.

30. Both the points $X = 165$ and $X = 180$ are to the left of $X = \mu$. Hence, Z is negative; $\mu = 183.6946$, $\sigma = 14.5858$.

31. $P(X < -\mu \mid X < \mu) = \dfrac{P(Z < -2)}{P(Z < 0)} = 2P(Z > 2) = 2[1 - \Phi(2)]$.

32. $U = 2X + Y \sim N(5, 32)$; $V = 4X - Y \sim N(1, 80)$; $P(Z \le z_1) = P(Z \ge z_2)$, where $z_1 = (K - 5)/\sqrt{32}$, $z_2 = (2K - 1)/\sqrt{80}$ gives $z_1 = -z_2$; $K = (5\sqrt{5} + \sqrt{2})/(\sqrt{5} + 2\sqrt{2})$.

33. $X \sim N(62, 36)$; $P(X > 80) = 0.0013 = p$; $q = 0.9987$; $n = 3$; P (atleast one scores > 80 marks) $= 1 - q^3 \approx 0.0039$.

34. $X \sim N(12, 16)$. To find M such that $M[P(X > 15)] = 750$; $0.2266M = 750$, $M \approx 3310$.

35. To find x^* such that $P(X > x^*) = 0.4$; $P(Z > z_1) = 0.4$, where $z_1 = (x^* - 165)/8$ is positive; $z_1 \approx 0.254$; $x^* = 167.03$cm.

Chapter 20

Mathematical Statistics - Tests of Significance

20.1 Introduction

In this chapter, we shall first study random sampling and large sample tests. Then, we shall study the exact sampling distributions and small sample tests.

20.2 Sampling and Standard Error

First, we define a few concepts.

Population or Universe In statistical investigations, we normally study or investigate certain characteristics of a group of individuals rather than of the individuals. Such a group of individuals is called a *population* or a *universe.* The individuals may be objects – animate or inanimate. The population may be finite or infinite. For example, an industry may be manufacturing electric fuses. The totality of the fuses manufactured is called a population. If the population is infinite or very large, then it is not possible to study a characteristic or some characteristics using the entire population. Therefore, we select a finite subset or subsets of the population, called *samples,* for studying the characteristics. The number of individuals in the sample is called the *sample size.* The process of selecting these samples is called sampling. The main objective in sampling is to draw inference about the population after studying the characteristics of the samples or using the information obtained from the samples.

For example, we may arrive at a decision of accepting or rejecting particular items by examining some samples of the items. However, the acceptance or rejection based on the characteristics of the sample or samples gives rise to an error called the *sampling error*, which is unavoidable and is inherent in such studies. In our day-to-day life, we use many sampling tests. Sampling has its advantages, particularly saving of costs and time.

Random Sampling In random sampling, the samples are selected at random, which excludes the possibility of any biasedness. Thus, each item of the population has equal chance of being included in the samples.

Parameters and Statistics As mentioned earlier, the interest in statistical analysis is to study one or more characteristics of the population. The statistical measures or constants of the population, like mean μ and variance σ^2, are called the *parameters* of the population. The statistical measures computed from the samples, like mean $\bar{x}$ and variance s^2, are called the *statistics*. Most often, parameter values are not known and the statistics obtained from the samples are used for the analysis of the population. However, the statistics based on different samples can vary from one sample to another sample.

One of the fundamental problems of the sampling theory is to find out whether these variations in the statistic of the samples (which may be due to fluctuations in the sampling processes) are significant or insignificant.

Consider a sample of size n. Let the characterisation of the sample to be studied be denoted by the variable X. Then, X_i, $i = 1, 2, \ldots, n$ are called the *sample observations* and the sample is denoted or referred to as $(X_1, X_2, \ldots, X_n)$. Therefore, we shall be considering a statistic $t = t (X_1, X_2, \ldots, X_n)$.

Sampling distribution of a statistic

Consider a population of size N and let r samples be drawn (using random sampling) each of size n. Now, we compute some statistic t, say the mean $\bar{x}$ or variance s^2, for each of the r samples. The values of the statistic can be given in the form of a frequency table. We call this data of values of the statistic as a *sampling distribution*. The standard deviation of this sampling distribution of a statistic is called the *standard error* (SE) of the statistic.

Consider now, the case of large samples. Generally, if the size of the sample $n \geq 30$, then the sample is considered as a large sample. If the sample is large, then the following assumptions hold.

(i) The sampling distribution of a statistic is approximately normal. Note that the distribution of the population may or may not be normal.
(ii) Sample statistics approximate the corresponding population parameters. Hence, sample statistics may be used to calculate the standard error of the sampling distribution.

Now, consider a sample of size n drawn from a population whose mean is μ and the variance is σ^2. Let the sample observations be denoted by $X_1, X_2, \ldots, X_n$, which are independent and identically distributed. Then, each $X_i \sim N(\mu, \sigma^2)$. We have the following standard errors (SE) in different cases.

Case 1. Standard error of the difference between sample mean and population mean

Let the sample mean be $\bar{X}$. Since, X_i are independent normal variates with mean μ_i and variance σ_i^2, we have that $\Sigma c_i X_i$ is also a normal variate with mean $\mu = \Sigma c_i \mu_i$ and variance $\sigma^2 = \Sigma c_i^2 \sigma_i^2$.

Now, let $c_i = 1/n$, $\mu_i = \mu$ and $\sigma_i = \sigma$. Then,

$$\Sigma c_i X_i = \frac{1}{n} \Sigma X_i = \bar{X} = \Sigma c_i \mu_i = \frac{1}{n}(\mu_1 + \mu_2 + \ldots + \mu_n) = \frac{1}{n}(n\mu) = \mu$$

and

$$\Sigma c_i^2 \sigma_i^2 = \frac{1}{n^2}(\sigma_1^2 + \sigma_2^2 + \ldots + \sigma_n^2) = \frac{1}{n^2}(n\sigma^2) = \frac{\sigma^2}{n}.$$

Therefore, the mean $\bar{X}$ of the n independent normal variates (with same mean μ and same variance σ^2) follows the normal distribution $\bar{X} \sim N(\mu, \sigma^2/n)$. Hence, the standard error is $\sigma / \sqrt{n}$.

Remark 1

When n is large, the above result holds good even if the population is not normal. This observation is due to the following *central limit theorem.*

"If X_i, $i = 1, 2, \ldots, n$ are independent random variables such that $E(X_i) = \mu_i$ and $V(X_i) = \sigma_i^2$, then under certain conditions, the random variable $X = X_1 + X_2 + \ldots + X_n$ is approximately normal with mean $\mu = \mu_1 + \mu_2 + \ldots + \mu_n$ and variance $\sigma^2 = \sigma_1^2 + \sigma_2^2 + \ldots + \sigma_n^2$."

Case 2. Standard error of the difference between the means of two samples.

Let n_1 and n_2 be the sizes of two independent random samples drawn from a given population or from two different populations. Let $\overline{X}_1$ and $\overline{X}_2$ be respectively the means of the samples and let the populations have the same mean μ and variances σ_1^2 and σ_2^2 respectively. Then

$$\overline{X}_1 \sim N\left(\mu, \frac{\sigma_1^2}{n_1}\right), \quad \text{and} \quad \overline{X}_2 \sim N\left(\mu, \frac{\sigma_2^2}{n_2}\right).$$

Therefore, $\overline{X}_1 - \overline{X}_2$ is also normal and

$$E(\overline{X}_1 - \overline{X}_2) = E(\overline{X}_1) - E(\overline{X}_2) = \mu - \mu = 0.$$

$$V(\overline{X}_1 - \overline{X}_2) = V(\overline{X}_1) + V(\overline{X}_2) = \frac{\sigma_1^2}{n_1} + \frac{\sigma_2^2}{n_2}.$$

Hence,
$$\overline{X}_1 - \overline{X}_2 \sim N\left(0, \frac{\sigma_1^2}{n_1} + \frac{\sigma_2^2}{n_2}\right) \tag{20.1}$$

and standard error is $\sqrt{\dfrac{\sigma_1^2}{n_1} + \dfrac{\sigma_2^2}{n_2}}.$ $\tag{20.2}$

Case 3. Standard error of the difference between sample standard deviation and population standard deviation.

Let s denote the standard deviation of a sample of size n drawn from a population with the standard deviation σ. Then, $s \sim N(\sigma, \sigma^2/(2n))$. Hence, the standard error is $\sigma/\sqrt{2n}$.

Case 4. Standard error of difference between standard deviations of two samples.

Let s_1, s_2 denote the standard deviations of two samples of sizes n_1 and n_2 drawn from two normal populations with standard deviations σ_1 and σ_2 respectively. Then,

$$s_1 \sim N\left(\sigma_1, \frac{\sigma_1^2}{2n_1}\right), \quad \text{and} \quad s_2 \sim N\left(\sigma_2, \frac{\sigma_2^2}{2n_2}\right).$$

Therefore, $\qquad s_1 - s_2 \sim N\left(\sigma_1 - \sigma_2, \dfrac{\sigma_1^2}{2n_1} + \dfrac{\sigma_2^2}{2n_2}\right).$ $\tag{20.3}$

Hence, the standard error is $\sqrt{\dfrac{\sigma_1^2}{2n_1} + \dfrac{\sigma_2^2}{2n_2}}.$ $\tag{20.4}$

Case 5. Standard error of the difference between sample proportion and population proportion.

Let P and p denote the population proportion and sample proportion respectively. Let the variable X denote the number of successes in n independent Bernoulli trials and $P(\text{successes}) = P$. Then, $X \sim B(n, P)$ with mean $E(X) = nP$ and variance $V(X) = nPQ$, where $Q = 1 - P$. When n is large, $X \sim N(nP, nPQ)$. Therefore, X/n follows the normal distribution

$$N\left(\frac{nP}{n}, \frac{nPQ}{n^2}\right) = N\left(P, \frac{PQ}{n}\right). \tag{20.5}$$

But, $p = X/n$ denotes the proportion of successes in the sample of n trials. We conclude from (20.5), that the standard error is $\sqrt{PQ/n}$ and $p \sim N(P, PQ/n)$.

Case 6. Standard error of the difference between two sample proportions.

Let p_1, p_2 denote the proportions of successes in two samples of sizes n_1 and n_2 respectively, that is, $p_1 = X_1/n$ and $p_2 = X_2/n$. Let the samples be drawn from two populations with population proportions P_1 and P_2 respectively. We define $Q_1 = 1 - P_1$ and $Q_2 = 1 - P_2$. Then,

$$p_1 \sim N\left(P_1, \frac{P_1 Q_1}{n_1}\right) \quad \text{and} \quad p_2 \sim N\left(P_2, \frac{P_2 Q_2}{n_2}\right).$$

Hence, $p_1 - p_2$ also defines a normal distribution and

$$E(p_1 - p_2) = E(p_1) - E(p_2) = P_1 - P_2.$$

$$V(p_1 - p_2) = V(p_1) + V(p_2) = \frac{P_1 Q_1}{n_1} + \frac{P_2 Q_2}{n_2}.$$

Therefore,
$$p_1 - p_2 \sim N\left(P_1 - P_2, \frac{P_1 Q_1}{n_1} + \frac{P_2 Q_2}{n_2}\right). \tag{20.6}$$

The standard error is given by $\sqrt{\dfrac{P_1 Q_1}{n_1} + \dfrac{P_2 Q_2}{n_2}}$. \hfill (20.7)

We discuss below the tests of significance for large and small samples. The corresponding tests are called the *parametric* and *non-parametric tests*.

20.3 Tests of Significance for Large Samples (Parametric Tests)

In sampling theory, we shall mainly, be concerned with the following two types of problems.

1. Estimation Some characteristic of the population, in which we are interested, is not known. We choose a random sample and obtain information about this characteristic, which is taken as an estimate of the characteristic of the population. This is called the problem of *estimation*.

2. Testing of hypothesis Some information about a characteristic of the population is known. We wish to know whether this information can be accepted. We choose a random sample and obtain information about this characteristic. Based on this information, we conclude whether the available information of the characteristic of the population can be accepted or rejected. We also wish to know that if it can be accepted then, to what degree of confidence it can be accepted. This is called the problem of *testing of hypothesis*.

Tests of significance

Let θ be a parameter of the population and θ_0 be the corresponding sample statistic. Since θ_0 is obtained based on a random sample, there will be some deviation (difference) between θ and θ_0. This may be due to the reason that the selection of the sample is not completely random. If this difference is large, then, we say that the difference is significant. If θ_1 is the statistic obtained from a second random sample, we wish to know whether the difference of θ_0 and θ_1 is significant. The methods that are used to decide whether the difference is significant or not, are called the *tests of significance*.

Null Hypothesis and alternate hypothesis

A population is given to us and we wish to have information about a characteristic of the population. We start with the assumption that there is no significant difference between the sample statistic and the corresponding population parameter or between two sample statistics. This assumption that there is no significant difference is called a *null hypothesis* and is denoted by H_0. A hypothesis that is different from the null hypothesis is called an *alternate hypothesis* and is denoted by H_1.

The methods that are used to decide whether to accept or reject a null hypothesis or an alternate hypothesis are called *tests of hypothesis*.

Consider the following example.

Let the null hypothesis be defined as
H_0 : The population has an assumed value of mean μ_0, that is, $\mu = \mu_0$.
The alternate hypothesis can be defined as any of the following.

(i) $H_1 : \mu \neq \mu_0$, that is, $\mu > \mu_0$ or $\mu < \mu_0$.
(ii) $H_1 : \mu > \mu_0$.
(iii) $H_1 : \mu < \mu_0$.

The alternate hypothesis (i) is called a *two tailed alternative*, (ii) is called the *right tailed alternative*, and (iii) is called the *left tailed alternative*.

The alternatives (ii) and (iii) are also called *single tailed tests*, where as (i) is called a *two tailed test*. This example can be generalised.

Let θ be a population parameter and θ_0 be the corresponding statistic obtained from a random sample. Then, we define the null hypothesis as $H_0 : \theta = \theta_0$, and the alternative hypothesis as

(i) $H_1 : \theta \neq \theta_0$, (two tailed alternative),   (20.8)
or (ii) $H_1 : \theta > \theta_0$, (right tailed alternative),   (20.9)
or (iii) $H_1 : \theta < \theta_0$, (left tailed alternative).   (20.10)

We define the tests given in (20.9) and (20.10) as single tailed tests and the test given in (20.8) as two tailed test. As mentioned earlier, for large n, most of the distributions (Binomial, Poisson, Hypergeometric, t, F, Chi-square distributions) can be approximated by a normal distribution.

Test statistic

1. In the large sample theory, *SE* forms the basis of the testing of hypothesis. For a large sample, if t is any statistic, then it follows a normal distribution. The corresponding population parameter is mean $= E(t)$ and standard deviation equals $SE(t)$. That is, for large samples,

$$Z = \frac{t - E(t)}{\sqrt{V(t)}} \sim N(0, 1).$$

Approximating $\sqrt{V(t)} \approx SE(t)$, we obtain

$$Z = \frac{t - E(t)}{SE(t)} \sim N(0, 1), \tag{20.11}$$

which is called a *test statistic*.

The test statistics in the various cases discussed earlier are given in Table 20.1.

Since $\sqrt{V(t)} \approx SE(t)$, we can determine the approximate limits within which the popultion parameter may lie.

Remark 2

(a) Test for difference between sample mean and population mean:

If the population standard deviation is not known, then we approximate $\sigma \approx s$, where s is the standard deviation of the sample. The test statistic is given by

$$Z = \frac{\overline{X} - \mu}{\sigma/\sqrt{n}}.$$

(b) Test for difference between means:

(i) If $\sigma_1^2 = \sigma_2^2 = \sigma$, that is, the samples are drawn from populations with same standard deviation, then the test statistic is

$$Z = \frac{\overline{X}_1 - \overline{X}_2}{\sigma\sqrt{(1/n_1) + (1/n_2)}} \sim N(0, 1). \tag{20.11a}$$

(ii) In (20.11a), If σ is not known and the samples are large, then we approximate σ^2 by

$$\hat{\sigma}^2 = \frac{n_1 s_1^2 + n_2 s_2^2}{n_1 + n_2}$$

where s_1^2, s_2^2 are the variances of the samples.

(iii) If $\sigma_1^2 \neq \sigma_2^2$ and σ_1^2, σ_2^2 are not known, then we approximate

$$Z = \frac{\overline{X}_1 - \overline{X}_2}{\sqrt{(s_1^2/n_1) + (s_2^2/n_2)}}$$

where s_1^2, s_2^2 are the variances of the samples.

(c) Test for difference between standard deviations:

If for two populations, their variances σ_1^2, σ_2^2 are unknown, then we approximate by their corresponding variances, $\sigma_1^2 \approx s_1^2$ and $\sigma_2^2 \approx s_2^2$. The test statistic is given by

$$Z = \frac{s - \sigma}{\sigma/\sqrt{2n}},$$

Critical region, critical values and level of significance

Let the sample statistic t lie in a certain region R in the sample space. If we decide that the difference between the parameter of the population and the sample statistic is significant (that is, the null hypothesis is rejected), then, this region R is called the *critical region* or *region of rejection*. The complementary region $\overline{R}$ is called the *region of acceptance*.

Table 20.1. Test statistics.

Case	Description (test of significance of difference between)	$SE = \sigma^*$	Statistic Z
1	sample mean ($\overline{X}$) and population mean (μ)	$\sigma/\sqrt{n}$	$(\overline{X} - \mu)/\sigma^*$
2	means of two samples, $\overline{X}_1, \overline{X}_2$	$\sqrt{\dfrac{\sigma_1^2}{n_1} + \dfrac{\sigma_2^2}{n_2}}$	$(\overline{X}_1 - \overline{X}_2)/\sigma^*$
3	sample standard deviation(s) and population standard deviation(s)	$\sigma/\sqrt{2n}$	$(s - \sigma)/\sigma^*$
4	samples standard deviations s_1, s_2	$\sqrt{\dfrac{\sigma_1^2}{2n_1} + \dfrac{\sigma_2^2}{2n_2}}$	$(s_1 - s_2)/\sigma^*$
5	sample proportion p and population proportion P	$\sqrt{PQ/n}$	$(p - P)/\sigma^*$
6	two sample proportions p_1, p_2	$\sqrt{\dfrac{P_1 Q_1}{n_1} + \dfrac{P_2 Q_2}{n_2}}$	$(p_1 - p_2)/\sigma^*$

Let $t = t(X_1, X_2, \ldots, X_n)$ be the value of the statistic obtained using a random sample of size n. Let R be the critical region (region of rejection) and $\overline{R}$ be the region of acceptance. Define

$$P(t \in R \mid H_0) = \alpha, \quad P(t \in \overline{R} \mid H_1) = \beta. \tag{20.12}$$

Then, α gives the probability that a random value of the statistic lies in the region R. We call α as the *level of significance*. We note that level of significance is always fixed in advance before the characterstic of the random sample is studied. The level of significance is usually expressed as a percentage. That is, the total area of the critical region is written as $\alpha\%$ level of significance.

The value of the test statistic Z, which separates the rejected region (critical region) and the accepted region is called the *critical value* or the *significant value* of Z. We denote this value by z_α, where α is the level of significance. The critical value depends on the following factors.

(i) The prescribed level of significance.

(ii) The test being used for checking the alternative hypothesis, that is, whether two tailed or single tailed test is being used.

As given in (20.11), for large samples, the test statistic follows a normal distribution

$$Z = \frac{t - E(t)}{SE(t)} \sim N(0, 1). \tag{20.13}$$

In the two tailed or single tailed tests, the critical region is given by the portion of the area under the probability curve of the sampling distribution of the test statistic Z. In the case of two tailed test, the critical region is given by the portion of the area lying at both the ends (symmetrically) of the probability curve. In the case of the left tailed test, the critical region is in the left tail under the probability curve while in the right tailed test, the critical region is in the right tail under the probability curve.

Let a level of significance α be prescribed. We denote by z_α, the critical value of the test statistic Z at the given level of significance α. For the three tests, we have the following results.

   (i) Two tailed test: The critical value z_α is obtained from the equation (using the tables of area under the standard normal curve)

$$P(|Z| > z_\alpha) = \alpha \qquad (20.14)$$

(see Fig. 20.1). The total area of the critical region under the probability curve is α. Because of symmetry of the probability curve, we get

$$P(Z > z_\alpha) = P(Z < -z_\alpha).$$

Therefore, $P(|Z| > Z_\alpha)$

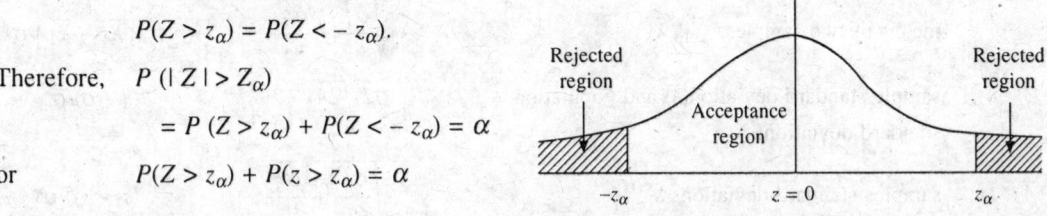

$$= P(Z > z_\alpha) + P(Z < -z_\alpha) = \alpha$$

or $P(Z > z_\alpha) + P(z > z_\alpha) = \alpha$

or $P(Z > z_\alpha) = \dfrac{\alpha}{2}.$

Fig. 20.1 Two tailed test with level of significance α.

This implies that the area under each tail is $\alpha/2$. We call this value $Z = z_\alpha$ as the upper critical value and $Z = -z_\alpha$ as the lower critical value. The acceptance region is given by $(-z_\alpha, z_\alpha)$. Some books use the notation $z_{\alpha/2}, -z_{\alpha/2}$ and the acceptance region is taken as $(-z_{\alpha/2}, z_{\alpha/2})$.

   (ii) Right tailed test: The critical value z_α is obtained from the equation

$$P(Z > z_\alpha) = \alpha \qquad (20.15)$$

(see Fig. 20.2). The total area of the critical region, α, is the area of the right tail under the probability curve.

   (iii) Left tailed test: The critical value z_α is obtained from the equation

$$P(Z < -z_\alpha) = \alpha \qquad (20.16)$$

(see Fig. 20.3). The total area of the critical region, α, is the area of the left tail under the probability curve.

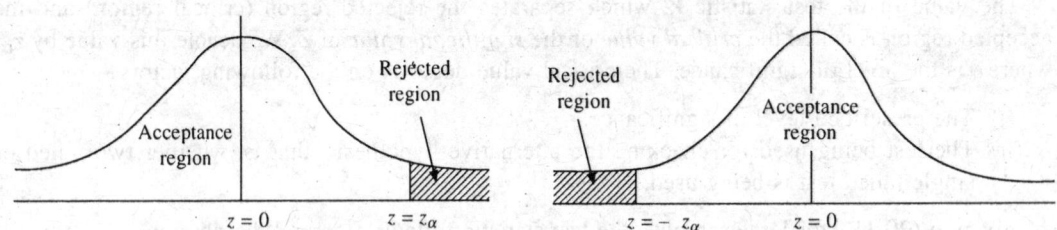

Fig. 20.2 Right tailed test.   **Fig. 20.3 Left tailed test.**

Now, let z_α be the critical value of Z corresponding to the level of significance α in the right tailed test, that is, $P(Z > z_\alpha) = \alpha$. Due to symmetry, we have

$$P(|Z| > z_\alpha) = P(Z > z_\alpha) + P(Z < -z_\alpha) = P(Z > z_\alpha) + P(Z > z_\alpha) = 2\alpha.$$

Therefore, the critical value of Z for a single tailed test at level of significance α is same as the critical value of Z for a two tailed test at the level of significance 2α.

   Consider the following example. From the normal distribution, we know that the area under the standard normal curve between $t = -2.58$ and 2.58 is 0.99. Denote, $E_1 = E(t) - (2.58)\ SE(t)$ and $E_2 = E(t) + (2.58)\ SE(t)$. Now, the area under the standard normal curve of t between E_1 and E_2 is 0.99, that is, 99% of the values of the variate t lie in the interval $[E_1, E_2]$. Therefore, only 1% of the values of t lie outside this interval. The regions $(-\infty, E_1)$ and (E_2, ∞) constitute the critical regions if we take that the difference between t and $E(t)$ is significant. From the normal distribution, we get

$$P\left[E(t) - (2.58)\ SE(t) < t < E(t) + (2.58)\ SE(t)\right] = 0.99$$

or
$$P\left[\left|\frac{t - E(t)}{SE(t)}\right| < 2.58\right] = 0.99$$

or
$$P[|\,Z\,| > 2.58] = 0.01 \quad \text{or} \quad 1\%. \tag{20.17}$$

We conclude that in this case, the level of significance is 0.01 or 1% and the critical values are ± 2.58.

The critical values for the commonly used values of level of significance : 1%, 2%, 5% and 10% are given in Table 20.2.

Table 20.2 Critical value (CV) z_a of Z.

Test		Level of significance			
	CV	1%	2%	5%	10%
Two tailed	$|\,z_\alpha\,|$	2.58	2.33	1.96	1.645
Right tailed	z_α	2.33	2.055	1.645	1.28
Left tailed	z_α	–2.33	–2.055	–1.645	–1.28

Confidence interval

We would like to determine an interval in which the population parameter is supposed to lie. The procedure to determine this interval is called *interval estimation* and the interval is called the *confidence interval* for that population parameter. The end points of this interval are called the *confidence limits*. Consider the case discussed above in (20.17), in which the level of significance was taken as 0.01. We have

$$P[\,|\,Z\,| \le 2.58] = 0.99$$

or
$$P\left[\left|\frac{t - E(t)}{SE(t)}\right| \le 2.58\right] = 0.99$$

or
$$P\left[t - (2.58)\ SE(t) \le E(t) \le t + (2.58)\ SE(t)\right] = 0.99. \tag{20.18}$$

We conclude that with 99% confidence, the population parameter $E(t)$ will lie in the interval $[t - (2.58)\ SE(t),\ t + (2.58)\ SE(t)]$. This interval is called the confidence interval and the end points of the interval are called confidence limits.

Similarly, the 98% confidence interval is given by

$$[t - (2.33)\ SE(t),\ t + (2.33)\ SE(t)], \tag{20.19}$$

and the 95% confidence interval is given by

$$[t - (1.96)\ SE(t),\ t + (1.96)\ SE(t)]. \tag{20.20}$$

Errors in testing of hypothesis

As mentioned earlier, in sampling theory, we draw conclusions about the population parameters on the basis of investigations of random samples. Also, the level of significance is fixed in advance, which may make the region of rejection larger or smaller. Because of these reasons, two types of errors can arise in the testing of hypothesis.

Type I error We reject the null hypothesis H_0, when it is true. That is, we reject a consignment of items, when the items are good.

$$P[\text{Reject a consignment of items when they are good}] = P[\text{Reject } H_0 \mid H_0] = K_1 \quad (20.21)$$

This type of error is called the *producer's risk*.

Type II error We accept the null hypothesis, when it is not true. That is, we accept a consignment of items, when the items are not good.

$$P[\text{Accept a consignment of items when they are not good}] = P[\text{Accept } H_0 \mid H_1] = K_2. \quad (20.22)$$

This type of error is called the *consumer's risk.*

We now discuss the procedure for testing of hypothesis.

1. Define the null hypothesis H_0.
2. Define the alternative hypothesis H_1. Based on this alternative hypothesis, decide the test to be used, two tailed test or single tailed (right or left) test.
3. Fix the suitable level of significance α, which depends on the particular problem. This level of significance is prescribed in advance before the random sample is drawn. From the standard normal tables, we obtain the value of z_α for this level of significance α.
4. Compute the test statistic $Z = \dfrac{t - E(t)}{SE(t)}$.
5. Compare the values of $|Z|$ and z_α. If $|Z| < z_\alpha$, we accept H_0, that is H_1 is rejected for the given level of significance α. This implies that the difference $|t - E(t)|$ is due to some fluctuations in sampling and is not significant. If $|Z| > z_\alpha$, we reject the null hypothesis H_0 and accept the alternate hypothesis H_1 for the given level of significance α. This implies that the difference $|t - E(t)|$ is significant for the given level of significance α.

We note the following.
From the standard normal probability tables, we have

$$P(-3 \leq Z \leq 3) = P(|Z| \leq 3) = 0.9973$$

or $$P(|Z| > 3) = 0.0027.$$

Therefore, we should almost expect a standard normal variate to lie between ± 3. This implies that if the test statistic Z satisfies $|Z| > 3$, then the null hypothesis H_0 is always rejected.

Consider now, the case of test for a sample proportion (*case 5*). From the above result, the probable limits for a normal variate X are $E(X) \pm 3\sqrt{V(X)}$. Hence, the probable limits for the observed proportion (p) of successes are $E(p) \pm 3\, SE(p)$, or $P \pm 3\sqrt{PQ/n}$. If P is not known, then we take $P \approx p$ and the approximate limits for the proportion of population are $p \pm 3\sqrt{pq/n}$.

Example 20.1 In a large lot of electric bulbs, the mean life and standard deviation of the bulbs are 360 hours and 90 hours respectively. A sample of 625 bulbs is chosen. It is found that the mean life and standard deviation of the bulbs in the sample are 355 hours and 90 hours respectively. Can we conclude that the sample is drawn from the given population? Test at 5% level of significance. If we assume that the population is normal and its mean is unknown, find the 98% confidence limits of the mean.

Solution Define

Null hypothesis H_0 :

The sample has been drawn from the population with mean $\mu = 360$ hours and $S.D = \sigma = 90$ hours.

Alternate hypothesis $H_1 : \mu \neq 360$ hours.

We shall use the two tailed test. Define the test statistic as (see Table 20.1)

$$Z = \frac{\bar{x} - \mu}{(\sigma/\sqrt{n})} \sim N(0, 1).$$

From the given data, we have $\bar{x} = 355$ hours, $\mu = 360$ hours, $n = 625$ and $\sigma = 90$ hours. We get

$$Z = \frac{(355 - 360)25}{90} \approx -1.389,$$

Since, $|Z| = 1.389 < 1.96$, we accept the null hypothesis, that the sample is drawn from the population (at 5% level of significance), (see Table 20.2).

The 98% confidence interval for μ is given by (see 20.19)

$$\left(\bar{x} - 2.33 \left(\frac{\sigma}{\sqrt{n}} \right), \bar{x} + 2.33 \left(\frac{\sigma}{\sqrt{n}} \right) \right) \text{ or } \left(355 - \frac{2.33(90)}{25}, 355 + \frac{2.33(90)}{25} \right)$$

or   (346.612, 363.388).
The 98% confidence limits are 346.612 and 363·388.

Example 20.2 Let E be the permissible error in the means $\bar{x}$ and μ of a sample and population respectively. Let the standard deviation of the population be σ. Find the minimum sample size n if (i) $P(|\bar{x} - \mu| < E) > 0.95$, (ii) $P(|\bar{x} - \mu| < E) > 0.98$, and (iii) $P(|\bar{x} - \mu| < E) > 0.99$. Hence, find the minimum sample size to estimate the mean within 5 units of the true mean, if $\sigma = 30$ and with 98% confidence.

Solution We know by central limit theorem that $\bar{X} \sim N(\mu, \sigma^2/n)$ for large n. That is

$$Z = \frac{\bar{x} - \mu}{\sigma/\sqrt{n}} \sim N(0, 1).$$

(i) We know that $P(|Z| \leq 1.96) = 0.95$, or $P\left(\left| \frac{\bar{x} - \mu}{\sigma/\sqrt{n}} \right| \leq 1.96 \right) = 0.95$

or $P\left(|\bar{x} - \mu| \leq \frac{1.96\sigma}{\sqrt{n}} \right) = 0.95.$

Since E is the permissible error, that is $|\bar{x} - \mu| \le E$ and $P(|\bar{x} - \mu| \le E) > 0.95$, we obtain

$$E > \frac{1.96\sigma}{\sqrt{n}}, \quad \text{or} \quad n > \left(\frac{1.96\sigma}{E}\right)^2 \approx \frac{3.8416\sigma^2}{E^2}.$$

Hence, the minimum sample size is $3.8416\sigma^2/E^2$.

(ii) We know that $P(|Z| \le 2.33) = 0.98$. Following the above procedure, we get

$$n > \left(\frac{2.33\,\sigma}{E}\right)^2 \approx \frac{5.4289\,\sigma^2}{E^2}.$$

(iii) We know that $P(|Z| \le 2.58) = 0.99$. We obtain

$$n > \left(\frac{2.58\sigma}{E}\right)^2 \approx \frac{6.6564\sigma^2}{E^2}.$$

We are given $\sigma = 30$, $E = 5$ and 98% confidence. Hence,

$$n \approx \frac{5.4289\,(30)^2}{(5)^2} = 195.4 \approx 196.$$

Example 20.3 The sizes and means of two independent random samples are 400, 225; 3.5 and 3.0 respectively. Can we conclude that the samples are drawn from the same popultion with standard deviation 1.5?

Solution We have $n_1 = 400$, $\bar{x}_1 = 3.5$, $n_2 = 225$, $\bar{x}_2 = 3.0$, $\sigma = 1.5$. Define,
Null hypothesis H_0: $\mu_1 \doteq \mu_2$ and $\sigma = 1.5$ (samples are drawn from the same population).
Alternate hypothesis H_1: $\mu_1 \ne \mu_2$, (two tailed test).

We have
$$Z = \frac{\bar{x}_1 - \bar{x}_2}{\sqrt{\sigma^2\left(\dfrac{1}{n_1} + \dfrac{1}{n_2}\right)}} = \frac{3.5 - 3.0}{(1.5)\sqrt{\dfrac{1}{400} + \dfrac{1}{225}}} = 4.0.$$

Since, $|Z| = 4 > 3$, H_0 is rejected. We conclude that in all probability, the samples are not drawn from the same population.

Example 20.4 The number of students in a class is 100. The average marks scored by 64 boys is 66 with standard deviation of 10 while the average marks scored by 36 girls is 70 with standard deviation of 8. Test at 1% level of significance whether the girls performed better than boys.

Solution We have $n_1 = 36$, $\bar{x}_1 = 70$, $\sigma_1 = 8$; $n_2 = 64$, $\bar{x}_2 = 66$, $\sigma_2 = 10$ and level of significance = 1%. Define

Null hypothesis H_0: $\mu_1 = \mu_2$, (boys and girls performed equally well).

Alternate hypothesis H_1: $\mu_1 > \mu_2$, (right tailed test: girls performed better than boys).

We have the test statistic as

$$Z = \frac{\bar{x}_1 - \bar{x}_2}{\sqrt{\dfrac{s_1^2}{n_1} + \dfrac{s_2^2}{n_2}}} = \frac{70 - 66}{\sqrt{\dfrac{64}{36} + \dfrac{100}{64}}} \approx 2.189.$$

Since, $Z = 2.189 < 2.33$, we conclude that at 1% level of significance, the difference between μ_1 and μ_2 is not significant. The null hypothesis H_0 is accepted and H_1 is rejected. We cannot conclude that girls performed better than boys.

Example 20.5 The mean height of 80 boys, who participated in the athletic competetion in a college was 167 cm with a standard deviation of 9 cm. The mean height of the remaining 160 boys who did not participate in the atheletic competetion was 163 cm with a standard deviation of 10 cm. Test the hypothesis at 5% level of significance, whether the students who participated in atheletics are taller than the other students.

Solution We have $n_1 = 80$, $\bar{x}_1 = 167$ cm, $s_1 = 9$ cm; $n_2 = 160$, $\bar{x}_2 = 163$ cm and $s_2 = 10$ cm. Define

Null hypotheis $H_0 : \mu_1 = \mu_2$ (no significant difference between means of participating and non-participating students).

Alternate hypothesis $H_1 : \mu_1 > \mu_2$ (right tailed test : students participating in athletics are taller). The test statistic is given by

$$Z = \frac{\bar{x}_1 - \bar{x}_2}{\sqrt{\dfrac{s_1^2}{n_1} + \dfrac{s_2^2}{n_2}}} = \frac{167 - 163}{\sqrt{\dfrac{81}{80} + \dfrac{100}{160}}} = 3.126.$$

We find $Z = 3.126 > 1.645$. Hence, we conclude that at 5% level of significance for right tailed test, the difference of means is significant and H_0 is rejected. Therefore, in all probability, the students who participated in athletics are taller than the non-participating students.

Example 20.6 From a large lot of mangoes, a random sample of 600 mangoes was drawn and 60 were found to be bad. Find the standard error of the proportion of bad mangoes in this sample. Hence, find the 3σ limits for the percentage of bad mangoes in this lot.

Solution We have $n = 600$, $X =$ number of bad mangoes in the sample $= 60$,

$$p = \text{proportion of bad samples} = \frac{60}{600} = 0.1, \ q = 1 - p = 0.9.$$

Since the proportion P of bad mangoes in the lot is not known, we assume $\hat{P} = p = 0.1$, $\hat{Q} = q = 0.9$. Therefore,

$$SE = \sqrt{\frac{\hat{P}\hat{Q}}{n}} = \sqrt{\frac{(0.1)(0.9)}{600}} = 0.0122.$$

Therefore, the limits for the proportion of bad mangoes in the lot are

$$\hat{P} \pm 3\sqrt{\frac{\hat{P}\hat{Q}}{n}} = 0.1 \pm 3(0.0122) = 0.0634, 0.1366.$$

The percentage of bad mangoes is in the range $(0.0634, 0.1366)$.

Example 20.7 In a random sample of 200 people in a city, 108 like to purchase imported watches and the remaining like to purchase local watches. Can we conclude that both the imported and local watches are popular in the city? (Use 2% level of significance).

Solution We have $n = 200$,

p = sample proportion of people who would like to purchase imported watches

$$= \frac{108}{200} = 0.54.$$

Define

Null hypothesis H_0 : both imported and local watches are popular.

Alternate hypothesis H_1 : $P \neq 0.5$ (two tailed test).

Therefore, P = population proportion of people purchasing imported watches = 0.5

$$Q = 1 - P = 0.5.$$

The test statistic is

$$Z = \frac{p - P}{\sqrt{PQ/n}} = \frac{0.54 - 0.5}{\sqrt{(0.5)^2/200}} = 1.13.$$

We find $|Z| = 1.13 < 2.33$. At 2% level of significance, the difference between population and sample proportions is not significant and the null hypothesis is accepted. We conclude that both the imported and local watches are popular.

Example 20.8 Before a big increase in the price of petrol, 400 persons out of a sample of 1000 persons were found to purchase big sized cars. After the increase in the price of petrol, 280 persons out of a sample of 800 persons were found to purchase big sized cars. Find whether there is a significant decrease in the purchase of big cars. Test at 5% and 2% levels of significance.

Solution We have $n_1 = 1000$, $p_1 = \frac{400}{1000} = 0.4$; $n_2 = 800$, $p_2 = \frac{280}{800} = 0.35$. Define

Null Hypothesis H_0 : $P_1 = P_2$ (no significant decrease in the purchase of big cars).

Alternate hypothesis H_1 : $P_1 > P_2$ (significant decrease in the purchase of big cars : right tailed test). Since the population proportion is not given, we approximate

$$P \approx \hat{P} = \frac{n_1 p_1 + n_2 p_2}{n_1 + n_2} = \frac{400 + 280}{1800} = 0.3778.$$

$$\hat{Q} = 1 - \hat{P} = 1 - 0.3778 = 0.6222.$$

The test statistic is given by

$$Z = \frac{p_1 - p_2}{\sqrt{\hat{P}\hat{Q}\left(\frac{1}{n_1} + \frac{1}{n_2}\right)}} = \frac{0.4 - 0.35}{\sqrt{(0.3778)(0.6222)\left(\frac{1}{1000} + \frac{1}{800}\right)}} = 2.17.$$

Since, $Z = 2.17 > 1.645$ (at 5% level of significance) and $Z = 2.17 > 2.055$ (at 2% level of significance), we reject the null hypothesis H_0 at both levels of significance. We conclude that there is a significant decrease in the purchase of big cars.

Example 20.9 A manufacturer of an electronic item finds the standard deviation of the life of the items to be 60 hours. The manufacturer wants to adopt a new process for producing the same items, which improves the life time of the items. A random sample of 200 items produced by the new process is selected and it was found that the standard deviation is 52 hours. Should the manufacturer adopt the new process? (Test at 2% level of significance).

Solution We have $\sigma = 60$ hours, $n = 200$, $s = 52$ hours. Define

Null hypothesis $H_0 : s = \sigma$ (no significant variation in standard deviation, that is, the new process has not significantly improved the life time of items).

Alternate hypothesis $H_1 : s < \sigma$ (left tailed test: the new process can be adopted).

The test statistic is given by

$$Z = \frac{s - \sigma}{\sigma/\sqrt{2n}} = \frac{52 - 60}{60/\sqrt{400}} = -2.667.$$

Since $|Z| = 2.667 > 2.055$, the null hypothesis H_0 is rejected at 2% level of significance. We accept the alternative hypothesis H_1. The manufacturer is justified in adopting the new process.

Example 20.10 Two independent random samples of sizes 1200 and 800 are drawn. The standard deviations of the samples are found to be 36 and 37 respectively. Find the possibility that the two samples are drawn from populations with the same standard deviation. Test at 5% level of significance.

Solution We have $n_1 = 1200$, $s_1 = 36$, $n_2 = 800$, $s_2 = 37$. Define

Null hypothesis H_0: $\sigma_1 = \sigma_2$ (the two samples are drawn from populations with the same standard deviation).

Alternate hypothesis $H_1 : \sigma_1 \neq \sigma_2$ (two tailed test).

Since σ is not known, we approximate it by

$$\sigma \approx \sigma^* = \sqrt{\frac{n_1 s_1^2 + n_2 s_2^2}{n_1 + n_2}} = 36.403.$$

Then, the test statistic is given by

$$Z = \frac{s_1 - s_2}{\sigma\left(\sqrt{\frac{1}{2n_1} + \frac{1}{2n_2}}\right)} = -0.851.$$

Since $|Z| = 0.851 < 1.96$, we accept the null hypothesis H_0 at 5% level of significance. We can conclude that the samples might have come from populations with the same standard deviation.

Exercise 20.1

1. A sample of 400 electric fuses is taken from a big lot of electric fuses. The mean life of the fuses in this sample is found to be 265 days. Can we assume that this sample has come from a population of fuses with mean life 280 days and variance 900 days? Assume 5% level of significance.
2. The mean life of an electronic equipment is 600 days with standard deviation of 120 days. The manufacturer instals new machines to improve the quality of production and claims that the new process has increased the mean life of the equipment. To test his claim, a sample of 100 items is chosen and it is found that the mean life has increased to 625 days. Can we accept the claim at 5% level of significance?
3. The mean value of a random sample of 144 items is 75 with standard deviation 15. Find the 95%

confidence limits for the population mean. Assume normal approximation to the sample. Also find the minimum sample size to estimate the mean within 4 units of the true mean at 95% confidence limits.

4. A normal population has mean 0.2 and standard deviation 2. Find the probability that mean of a sample of size 625 will be negative.

5. A normal population has mean 0.05 and standard deviation 1.5. Find the probability that mean of a sample of size 625 drawn from this population will be negative.

6. The sizes of two independent random samples are 1600 and 900 and their means are 35 and 33 respectively. Can we conclude that the samples are drawn from the same population with standard deviation 20? Test at 1% level of significance.

7. Two industries A and B manufacture textile machines. In a sample of 300 workers from the industry A, it was found that average weekly salary is Rs. 1500 with standard deviation Rs. 500. From a sample of 325 workers from the industry B, it was found that average weekly salary is Rs. 1550 with standard deviation Rs. 510. Are the average weekly wages in industry B higher than the average weekly wages in industry A? Test at 5% level of significance.

8. Two normal populations have the same standard deviation. A random sample is drawn from each of the populations. The samples are of sizes 400 and 900 respectively. It is found that the sample means are 15 and 16 respectively and the corresponding standard deviations are 1.4 and 1.5. Test the significance of the difference of means at 5% level of significance.

9. Two populations have the same mean, and the standard deviations are σ and 3σ respectively. Random samples each of size 1000 are drawn from the populations. Show that the difference of the means of the samples will not exceed 0.3σ in all possibility.

10. In a survey, an estimate of the mean income μ of population in a large city is to be obtained. It was decided to approximate μ by the mean of a random sample. The error in the estimation is to be less than or equal to Rs. 100 with probability 0.95. If the standard deviation is known to be Rs. 1200, how large a sample should be taken?

11. The following data gives the number of harvested mangoes stored in open and closed godowns.

	Sample size	S.D	mean
Inside	1000	45	210
Outside	800	42	200

Assume that the two random samples are drawn from a normal population. Examine whether the mean value of the harvested mangoes is affected by weather conditions. Test at 5% level of significance.

12. From a big consignment of electrical bulbs, a random sample of 1000 bulbs were drawn and tested. A total of 40 bulbs were found to be defective. Find the standard error of the proportion of defective bulbs in this sample. Hence, find the limits for the percentage of defective bulbs in this consignment.

13. The fatality rate of sun strokes in a certain rural population is 0.03%. A rural hospital treats some of these patients every year. In a particular year, 500 patients suffering from sun stroke were admitted in this rural hospital and only 2 died. Can we accept the claim of the hospital that they are efficient? Test at 1% level of significance.

14. A random sample of size 400 is drawn from a population with replacement. Find the bound on the standard error of the sample proportion.

15. In a large city, the number of students in two schools who said that their best subject is mathematics is noted. In the first school, 30% of a random sample of 600 students said that their best subject is mathematics. In the second school, 28% of a random sample of 400 students said that their best subject is mathematics. Is the difference between the proportions significant? Test at 5% level of significance.

16. In an entrance examination, 5000 candidates appeared. Above a certain cut off point of total marks scored, there are 6% candidates, who are put in List 1. The remaining candidates are put in List 2. Analysis of the results had shown that the Question numbered 10 was correctly answered by 70% of the candidates in List 1 and 60% of the candidates in List 2. On the basis of this analysis, can we conclude that Question number 10 is not good at separating the ability of the candidates? Test at 5% level of significance.

17. A soft drink manufacturer has two brands of soft drinks A and B. The company claims that brand A outsells brand B by 10%. Two independent random samples of people who regularly consume soft drinks are chosen. In the first sample of size 300, 80 people preferred brand A and in the second sample of size 200, 50 people preferred brand A. Can we accept the manufacturer's claim of 10% difference in sales of A over B? Test at 2% level of significance.

18. In two cities, candidates were going through a procedure for recruitment as cadets to the armed forces. One random sample each was drawn from the two cities recruitment centres. The data relates to the heights of the candidates appearing at the selection.

	Standard deviation (in cm.)	Number in samples
City A	4	500
City B	3.75	600

Is the difference between standard deviations significant? Test at 5% level of significance.

19. The variability of yield of a cereal in two independent sets of plots is given below

Number of plots	S.D
50	40 kg
80	42 kg

Examine whether the difference in the variability is significant at 5% level of significance.

20. The standard deviation of a random sample of size 600 is 3.6 and that of another independent sample of size 800 is 3.7. Can we assume that the two samples are drawn from a population with standard deviation 3.5?

20.4   Tests of Significance for Small Samples

In the previous sections, we have discussed tests of significance for large random samples, that is, when $n \geq 30$. When n is small (that is , $n < 30$), in many cases the statistics are not normal even though the parent population is normal. Further, the approximations of population parameters by the corresponding sample statistics are not valid. Therefore, we shall discuss different tests of significance which are applicable to small samples. However, these tests of significance for small samples can also be applied to the cases of large samples. We shall assume that the parent population is normally distributed.

It is interesting to note that these tests were first proposed by Gosset (1908) and later, Fisher (1926) modified and developed these tests which are in use today. We shall discuss the following tests: (i) t-test and (ii) F-test.

In t-test, we shall be using the Beta function, which is defined as follows. The improper integral

$$\beta(m, n) = \int_0^1 x^{m-1}(1-x)^{n-1}dx, \ m > 0, n > 0 \tag{20.23}$$

is called the Beta function. It is related to the Gamma function

$$\Gamma(\alpha) = \int_0^\infty x^{\alpha-1}e^{-x}dx, \ \alpha > 0 \tag{20.24}$$

as
$$\beta(m, n) = \frac{\Gamma(m)\Gamma(n)}{\Gamma(m+n)} = \beta(n, m). \tag{20.25}$$

We note the following values of the Gamma function.

$$\Gamma(\alpha + 1) = \alpha\Gamma(\alpha), \quad \Gamma(m + 1) = m! \quad \text{for any positive integer } m, \quad \Gamma(1/2) = \sqrt{\pi}.$$

20.4.1 Student's *t*–Distribution

The *t*-distribution was first proposed by Gosset whose penname was Student.

Let x_i, $i = 1, 2, \ldots, n$ be a random sample of size n drawn from a normal population with mean μ and variance σ^2.

A random variable X follows the *t*-distribution, if its probability density function (*pdf*) is given by

$$f(t) = \frac{1}{\sqrt{v}\,\beta\left(\frac{1}{2}, \frac{v}{2}\right)}\left(1 + \frac{t^2}{v}\right)^{-(v+1)/2}, \quad -\infty < t < \infty \tag{20.26}$$

where t is a suitably defined statistic and v denotes the number of degrees of freedom of the *t*-distribution. Note that the number of degrees of freedom is also denoted by n in place of v. Usually, we denote $t \sim t_v$, if the *t*-distribution has v degrees of freedom. For $v = 1$, we get

$$f(t) = \frac{(1 + t^2)^{-1}}{\beta\left(\frac{1}{2}, \frac{1}{2}\right)} = \frac{1}{\pi(1 + t^2)}, \quad -\infty < t < \infty.$$

Graph and properties of *t*-distribution

1. Since $f(-t) = f(t)$, the probability curve is symmetric about the line $t = 0$. As $t \to \infty$, $f(t) \to 0$ rapidly and *t*-axis is an asymptote of the curve. The curve resembles the standard normal probability curve and is bell shaped (see Fig. 20.4). As the number of degrees of freedom increases, the *t*-distribution curve moves closer to the standard normal probability curve. We have

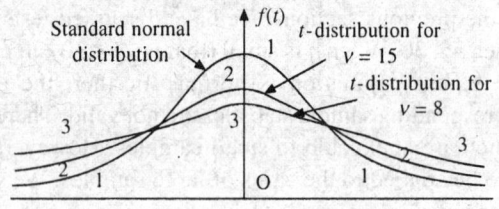

Fig. 20.4. *t*-distribution.

$$\lim_{v \to \infty} f(t) = \lim_{v \to \infty} \frac{1}{\sqrt{v}\,\beta\left(\frac{1}{2}, \frac{v}{2}\right)} \cdot \lim_{v \to \infty}\left[\left(1 + \frac{t^2}{v}\right)^v\right]^{-1/2} \cdot \lim_{v \to \infty}\left[1 + \frac{t^2}{v}\right]^{-1/2}$$

$$= \frac{1}{\sqrt{2\pi}}e^{-t^2/2}, \quad -\infty < t < \infty$$

since,

$$\lim_{v \to \infty} \frac{1}{\sqrt{v}\,\beta\left(\frac{1}{2}, \frac{v}{2}\right)} = \lim_{v \to \infty} \frac{1}{\sqrt{v}} \cdot \frac{\Gamma((v + 1)/2)}{\Gamma(v/2)\Gamma(1/2)}$$

$$= \lim_{v \to \infty} \frac{1}{\sqrt{v}}\left[\frac{1}{\sqrt{\pi}}\left(\frac{v}{2}\right)^{1/2}\right] = \frac{1}{\sqrt{2\pi}}.$$

Note that $\qquad \lim\limits_{n \to \infty} \dfrac{\Gamma(n + \alpha)}{\Gamma(n)} = n^{\alpha}$.

2. Since $f(t)$ is symmetric about the line $t = 0$, all the moments of odd order about the origin vanish. Hence, mean of the t-distribution is zero. Therefore, central moments coincide with moments about origin. The variance is obtained as

$$\text{variance} = \frac{n}{n - 2} > 1, \quad \text{for} \quad n > 2$$

which tends to 1 as $n \to \infty$.

Applications of t-distribution

The following are some of the uses of the t-distribution

1. To test the significance of the difference between the mean of a small sample and the mean of the population.
2. To test the significance of the difference between the means of two small random samples.
3. To test the significance of the coefficient of correlation in a small sample. Note that the coefficient of correlation in the population is assumed as zero.

Remark 3

The number of degrees of freedom of a statistic is the number of independent variates used to compute this statistic. For example, if n is the number of observations in the small sample and m is the number of constraints on them (or m values are already available), then the number of degrees of freedom is given by $v = n - m$. For calculating the mean $\bar{x}$, we use all the observed values $x_1, x_2, ..., x_n$. Therefore, the mean $\bar{x}$ has n degrees of freedom. Since the standard deviation of the sample depends on the mean, the standard deviation has $n - 1$ degrees of freedom.

Critical (Significant) values of t

As in the case of normal distribution and large samples, we define the critical or significant values of t at level of significance α with v degrees of freedom as

$$P[\,|t| > t_v(\alpha)] = \alpha \qquad\qquad (20.27)$$

for two tailed test (see Fig. 20.5),

$$P[t > t_v(\alpha)] = \alpha \qquad\qquad (20.28)$$

for the right tailed test, and

$$P[t < -t_v(\alpha)] = \alpha \qquad\qquad (20.29)$$

for the left tailed test.

Since t-distribution is symmetric about $t = 0$, we get from (20.27)

$$P[\,|t| > t_v(\alpha)]$$

$$= P[t > t_v(\alpha)] + P[t < -t_v(\alpha)] = \alpha$$

or $\qquad 2P[t > t_v(\alpha)] = \alpha$, or $P[t > t_v(\alpha)] = \alpha/2$.

This implies that $P[t > t_v(2\alpha)] = \alpha$. We conclude

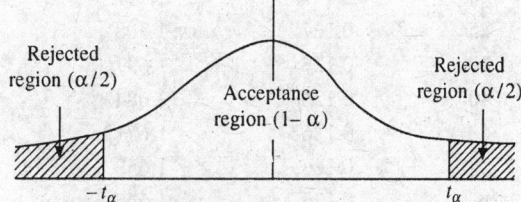

Fig. 20.5. Two tailed test, t-distribution.

that the critical values for a single tailed test at level of significance α is the same as for a two tailed test at level of significance 2α (corresponding to the same degrees of freedom). Some books use the notation $t_{\alpha/2}$, $-t_{\alpha/2}$ and the acceptance region is taken as $(-(t_{\alpha/2}, t_{\alpha/2})$

The values of $t_v(\alpha)$ (two tailed test) have been tabulated and are known as Fisher and Yate's tables. A table of values of $t_v(\alpha)$, at a few levels of significance is given in Table 20.3.

Table 20.3

t-**Table**

v	Probability				
	0.9	0.1	0.05	0.02	0.01
1	0.158	6.314	12.706	31.821	63.657
2	0.142	2.920	4.303	6.965	9.925
3	0.137	2.353	3.182	4.541	5.841
4	0.134	2.132	2.776	3.747	4.604
5	0.132	2.015	2.571	3.365	4.032
6	0.131	1.943	2.447	3.143	3.707
7	0.130	1.895	2.365	2.998	3.496
8	0.130	1.860	2.306	2.896	3.355
9	0.129	1.833	2.262	2.821	3.250
10	0.129	1.812	2.228	2.764	3.169
11	0.129	1.796	2.201	2.718	3.106
12	0.128	1.782	2.179	2.681	3.055
13	0.128	1.771	2.160	2.650	3.012
14	0.128	1.761	2.145	2.624	2.977
15	0.128	1.753	2.131	2.602	2.947
16	0.128	1.746	2.120	2.583	2.921
17	0.128	1.740	2.110	2.567	2.898
18	0.127	1.734	2.101	2.552	2.878
19	0.127	1.729	2.093	2.539	2.861
20	0.127	1.725	2.086	2.528	2.845
21	0.127	1.721	2.080	2.518	2.831
22	0.127	1.717	2.074	2.508	2.819
23	0.127	1.714	2.069	2.500	2.807
24	0.127	1.711	2.064	2.492	2.797
25	0.127	1.708	2.060	2.485	2.787
30	0.127	1.697	2.042	2.457	2.750
40	0.126	1.684	2.021	2.423	2.704
60	0.126	1.671	2.000	2.390	2.660
120	0.126	1.658	1.980	2.358	2.617
∞	0.126	1.645	1.960	2.326	2.576

For example, we have the following probabilities:

$$t_5(0.10) \text{ for two tailed test} = t_5(0.05) \text{ for a single tail test} = 2.015,$$

$$t_8(0.02) \text{ for two tailed test} = t_8(0.01) \text{ for a single tail test} = 2.896.$$

We discuss the following applications of *t-test*.

t-test

Case 1 *Test of significance for t-distribution of the difference between mean of a small sample and mean of population.*

We assume the following:

   (a) The random sample is drawn from a normal population.
   (b) The standard deviation σ of the parent population is not known.

We know that if $\bar{x}$ is the mean of a sample of size n drawn from a normal population $N(\mu, \sigma^2)$, then the test statistic is (see Table 20.1, case 1)

$$Z = \frac{\bar{x} - \mu}{\sigma/\sqrt{n}} \sim N(0, 1).$$

If the standard deviation σ of the population is not known, then we have to determine its estimate by using the sample standard deviation. Using the sample variance s^2, we obtain

$$s^2 = \frac{1}{n} \sum_i (x_i - \bar{x})^2, \text{ or } ns^2 = (n - 1)\sigma^2$$

or
$$\sigma^2 = \left(\frac{n}{n - 1}\right)s^2.$$

Therefore, an unbiased estimate of σ with $(n - 1)$ degrees of freedom is given by

$$\sigma^2 = \left(\frac{n}{n - 1}\right)s^2 \tag{20.30}$$

where s is the sample standard deviation. Now, for large n, $[n/(n - 1)] \to 1$ and hence $\sigma \to s$, and we are justified in approximating the test statistic as $Z = (\bar{x} - \mu)/(s/\sqrt{n})$, which follows $N(0, 1)$. For n small, we write the test statistic as

$$t = \frac{\bar{x} - \mu}{s/\sqrt{n - 1}}. \tag{20.31}$$

Note that t does not follow $N(0, 1)$, but follows a *t*-distribution with $v = n - 1$ degrees of freedom. In the notation used earlier $t \sim t_{n-1}$, with $n - 1$ degrees of freedom, that is, we obtain the value of $t_v(\alpha)$ for a given level of significance α and $v = n - 1$ degrees of freedom from the table. If the computed value of t satisfies $|t| < t_v(\alpha)$, then the null hypothesis H_0 (there is no significant difference between sample mean and population mean) is accepted at α level of significance, otherwise, H_0 is rejected.

   Sometimes, the statistic (20.31) is defined as

$$t = \frac{\bar{x} - \mu}{S/\sqrt{n}}$$

where $\qquad \bar{x} = \dfrac{1}{n} \sum_i x_i$ and $S^2 = \dfrac{1}{n-1} \sum_i (x_i - \bar{x})^2$.

Confidence intervals

Let us compute the 95% confidence interval of μ. We have

$$P\left[\left| \frac{\bar{x} - \mu}{s/\sqrt{n-1}} \right| \leq t_{n-1}(0.05)\right] = 0.95$$

and

$$\left| \frac{\bar{x} - \mu}{s/\sqrt{n-1}} \right| \leq t_{n-1}(0.05).$$

This gives $\qquad \bar{x} - \left(\dfrac{s}{\sqrt{n-1}}\right) t_{n-1}(0.05) \leq \mu \leq \bar{x} + \left(\dfrac{s}{\sqrt{n-1}}\right) t_{n-1}(0.05) \qquad (20.32)$

where $t_{n-1}(0.05)$ is the 5% critical value of t for $n-1$ degrees of freedom for a two tailed test. Similar results hold for other values of confidence intervals.

Case 2 Test of significance of the difference between the means of two small random samples drawn from the same normal population.

We assume the following:

   (i) Parent population from which random samples are drawn is normally distributed.
  (ii) The population variances are equal and unknown.

We know that if $\bar{x}_1, \bar{x}_2$ are the means of two random samples of sizes n_1 and n_2 respectively, drawn from a normal population $N(\mu, \sigma^2)$, then the test statistic (see Table 20.1, *case 2*) is given by

$$Z = \frac{\bar{x}_1 - \bar{x}_2}{\sigma\sqrt{\dfrac{1}{n_1} + \dfrac{1}{n_2}}} \sim N(0, 1).$$

If σ was not known, then we approximated σ by

$$\sigma^2 = \frac{n_1 s_1^2 + n_2 s_2^2}{n_1 + n_2}$$

where s_1 and s_2 are the standard deviations of the samples. However, this approximation is not valid for small samples. When n_1 and n_2 are small, σ is approximated by

$$\sigma^2 \approx \frac{n_1 s_1^2 + n_2 s_2^2}{n_1 + n_2 - 2} \qquad (20.33)$$

with $n_1 + n_2 - 2$ degrees of freedom. The test statistic becomes

$$t = \frac{\bar{x}_1 - \bar{x}_2}{\sqrt{\left(\dfrac{n_1 s_1^2 + n_2 s_2^2}{n_1 + n_2 - 2}\right)\left(\dfrac{1}{n_1} + \dfrac{1}{n_2}\right)}} \qquad (20.34)$$

Note that t does not follow $N(0, 1)$ but follows a t-distribution with $v = n_1 + n_2 - 2$ degrees of freedom. We have the following particular cases.

1. If $n_1 = n_2 = n$, that is sample sizes are same, then the test statistic simplifies to

$$t = \frac{\bar{x}_1 - \bar{x}_2}{\sqrt{(s_1^2 + s_2^2)/(n - 1)}} \qquad (20.35)$$

with $v = 2n - 2$ degrees of freedom.

2. Let $n_1 = n_2 = n$ and the two samples are not independent, that is, they are related in some way. This implies that the pairs of observation (x_i, y_i) belong to same sample unit. Then, the statistic t defined in (20.35) cannot be used. In this case, we consider the increments $d_i = x_i - y_i$, $\bar{d} = \bar{x} - \bar{y}$ and define the null hypothesis as $H_0 : \bar{d} = 0$ and alternate hypothesis as $H_1 : \bar{d} \neq 0$. The test equation is defined as

$$t = \frac{\bar{d}}{s/\sqrt{n - 1}} \qquad (20.36)$$

with $n - 1$ degrees of freedom and variance of d is

$$s^2 = \frac{1}{n} \sum_i (d_i - \bar{d})^2.$$

Example 20.11 The annual rainfall at a certain place is normally distributed with mean 45 cm. The rainfall during the last five years are 48 cm, 42 cm, 40 cm, 44 cm and 43 cm. Can we conclude that the average rainfall during the last five years is less than the normal rainfall? Test at 5% level of significance.

Solution We have the mean and standard deviation of the small sample as

$$\bar{x} = \frac{1}{n} \sum_i x_i = \frac{1}{5}(48 + 42 + 40 + 44 + 43) = 43.4.$$

$$s^2 = \left(\frac{1}{n} \sum_i x_i^2\right) - \bar{x}^2 = \frac{1}{5}(9453) - (43.4)^2 = 7.04.$$

Define

Null hypothesis $H_0 : \bar{x} = \mu$ (there is no significant difference in the rain fall).

Alternate hypothesis $H_1 : \bar{x} < \mu$.

We use the left tailed test with 5% level of significance. Now, $t(0.05)$ for one tailed test $= t(0.1)$ for two tailed test with $n = 5$. The value of t for $P = 0.05$ and $v = 4$ is 2.132. The test statistic is given by

$$t = \frac{\bar{x} - \mu}{s/\sqrt{n - 1}} = \frac{43.4 - 45}{1.3266} = -1.206.$$

Since, $|t| = 1.206 < 2.132$, we accept the null hypothesis. There is no significant difference in the rainfall.

Example 20.12 The heights of 8 males participating in an atheletic championship are found to be 175 cm, 168 cm, 165 cm, 170 cm, 167 cm, 160 cm, 173 cm and 168 cm. Can we conclude that the average height is greater than 165 cm? Test at 5% level of significance.

Solution We define

Null hypothesis $H_0 : \mu = 165$ cm.

Alternate hypothesis $H_1 : \mu > 165$ cm.

We use the right tailed test with 5% level of significance. We have $n = 8$.
Since, $t(0.05)$ for one tailed test $= t(0.1)$ for two tailed test, we have for 7 degrees of freedom and $P = 0.05$, $t = 1.895$. We compute the sample mean and standard deviation. We have

$$\bar{x} = \frac{1}{8} (175 + 168 + 165 + 170 + 167 + 160 + 173 + 168) = 168.25.$$

$$s^2 = \frac{1}{8}(\sum_i x_i^2) - \bar{x}^2 = \frac{1}{8}(226616) - (168.25)^2 = 18.9375.$$

The test statistic is given by

$$t = \frac{\bar{x} - \mu}{s/\sqrt{n-1}} = \frac{168.25 - 165}{4.3517/2.6458} = 1.976.$$

Since, $t = 1.976 > 1.895$, we reject the null hypothesis and accept the alternative hypothesis. The average height is greater than 165 cm.

Example 20.13 A random sample of 17 values from a normal population has a mean of 105 cm and the sum of squares of deviations from this mean is 1225 sq.cm. Is the assumption of a mean of 110 cm for the normal population reasonable? Test under 5% and 1% levels of significance. Also obtain the 95% and 99% confidence limits.

Solution We have $n = 17$ and degrees of freedom $= 16$. Then, for $P = 0.05$, $t = 2.12$ and for $P = 0.01$, $t = 2.921$.

We have, $\qquad \bar{x} = 105$ cm, $\quad s^2 = \frac{1}{n} \sum_i (x_i - \bar{x})^2 = \frac{1225}{17} = 72.0588$, $s = 8.4887$.

Define

Null hypothesis $H_0 : \mu = 110$ cm (assumption that the mean of the population is 110 cm is valid).

Alternative hypothesis $H_1 : \mu \neq 110$ cm.
We have the test statistic as

$$t = \frac{\bar{x} - \mu}{s/\sqrt{n-1}} = \frac{105 - 110}{(8.4887/4)} = -2.3561.$$

Now, $|t| = 2.3561 > 2.12$. Therefore, H_0 is rejected at 5% level of significance. Data is not consistent with the assumption. Now, $|t| = 2.3561 < 2.921$. Therefore, H_0 is accepted at 1% level of significance. At this level of significance, the data is consistent with the assumption that the population has mean 110 cm.

The confidence limits at 5% level of significance are

$$\bar{x} \pm \left(\frac{s}{\sqrt{n-1}} \right) 2.12, \quad \text{or} \quad 105 \pm \left(\frac{8.4887}{4} \right) 2.12, \quad \text{or} \quad (100.501, 109.499).$$

The confidence limits at 1% level of significance are

$$\bar{x} \pm \left(\frac{s}{\sqrt{n-1}} \right) 2.921 \quad \text{or} \quad 105 \pm \left(\frac{8.4887}{4} \right) 2.921, \quad \text{or} \quad (98.801, 111.199).$$

Example 20.14 The following two independent random samples of sizes 10 and 8 are given.

Sample 1	165	162	164	169	170	166	165	163	168	171
Sample 2	160	164	163	168	171	162	166	163		

Is the difference between the sample means significant at 5% level of significance?

Solution We have $n_1 = 10$, $n_2 = 8$. We first compute the means and variances of the samples.

	Sample 1			Sample 2		
x_1	$u = x_1 - 165$	u^2	x_2	$v = x_2 - 165$	v^2	
165	0	0	160	-5	25	
162	-3	9	164	-1	1	
164	-1	1	163	-2	4	
169	4	16	168	3	9	
170	5	25	171	6	36	
166	1	1	162	-3	9	
165	0	0	166	1	1	
163	-2	4	163	-3	9	
168	3	9				
171	6	36				
Total	13	101		-4	94	

Sample 1 $\quad \bar{x}_1 = 165 + \bar{u} = 165 + \dfrac{1}{10}(13) = 166.3$

$$s_1^2 = \frac{1}{n_1} \sum_i u_i^2 - \left(\frac{1}{n_i} \sum u_i \right)^2 = \frac{101}{10} - \left(\frac{13}{10} \right)^2 = 8.41.$$

Sample 2 $\quad \bar{x}_2 = 165 + \bar{v} = 165 + \dfrac{1}{8}(-4) = 164.5.$

$$s_2^2 = \frac{1}{n_2} \sum_i v_i^2 - \left(\frac{1}{n_2} \sum v_i \right)^2 = \frac{94}{8} - (0.5)^2 = 11.5.$$

Define, $H_0 : \bar{x}_1 = \bar{x}_2$ and $H_1 : \bar{x}_1 \neq \bar{x}_2$, (two tailed test).

We have, degrees of freedom $= n_1 + n_2 - 2 = 16$. From the Table 20.3, $t_{16}(0.05) = 2.12$.

The variance of the population is approximated by

$$\sigma^2 \approx \frac{n_1 s_1^2 + n_2 s_2^2}{n_1 + n_2 - 2} = \frac{10(8.41) + 8(11.5)}{16} = 11.00625.$$

The test statistic is given by (see (20.34))

$$t = \frac{\bar{x}_1 - \bar{x}_2}{\sqrt{11.00625\left(\frac{1}{10} + \frac{1}{8}\right)}} = \frac{166.3 - 164.5}{1.5737} = 1.144.$$

Since, $|t| = 1.144 < 2.12$, we accept the null hypothesis that there is no significant difference between the means of the two samples.

Example 20.15 The scores of 10 candidates obtained in tests before and after attending some coaching classses are given below

Before :	54	76	92	65	75	78	66	82	80	78
After:	60	80	86	72	80	72	66	88	82	73

Is the coaching for the test effective? Test at 5% level of significance.

Solution The data relates to the marks obtained by the same set of students. Hence, we can regard that the marks are correlated. The test statistic, in this case, is given, by (20.36).
If x_i, y_i denote the marks obtained in the two tests, we obtain the values of $d_i = x_i - y_i$ as
$-6, -4, 6, -7, -5, 6, 0, -6, -2, 5$.

We find
$$\bar{d} = \frac{1}{n} \Sigma d_i = -\frac{13}{10} = -1.3,$$

$$s_d^2 = \frac{1}{n} \Sigma d_i^2 - \bar{d}^2 = \frac{1}{10}(263) - 1.69 = 24.61.$$

We define
Null hypothesis $H_0 : \bar{d} = 0$ (the students have not benefitted from coaching).
Alternate hypothesis $H_1 : \bar{d} < 0$ (the students have benefitted from coaching).
We shall use the one tailed test. Now, $t (0.05)$ for one tailed test $= t(0.1)$ for two tailed test, with the degrees of freedom $= n - 1 = 9$. The value of t for $P = 0.05$ and $v = 9$ is 1.833. The test statistic is given by (see (20.36))

$$t = \frac{\bar{d}}{s_d / \sqrt{n - 1}} = \frac{-1.3}{4.9608/3} = -0.786.$$

We find $|t| = 0.786 < 1.833$. Hence, we accept the null hypothesis that the students have not benefitted from coaching.

20.4.2 Snedecor's F-Distribution

Snedecor's F-distribution is often called F-distribution. A random variable F is said to follow the F-distribution with (v_1, v_2) degrees of freedom, if its probability density fuction is given by

$$f(F) = \frac{(v_1/v_2)^{v_1/2} \, F^{(v_1/2)-1}}{\beta\left(\dfrac{v_1}{2}, \dfrac{v_2}{2}\right)[1 + (v_1/v_2)F]^{(v_1+v_2)/2}}, \quad F > 0. \tag{20.37}$$

We usually denote $F \sim F(v_1, v_2)$. Note that the sampling distribution does not contain any population parameters and it depends only on the degrees of freedom v_1 and v_2.

Graph and properties of F-distribution

1. The approximate graph of F-distribution is given in Fig. 20.6. For large values of degrees of freedom, v_1 and v_2, F tends to the normal variate $N\left[1, 2\left(\dfrac{1}{v_1} + \dfrac{1}{v_2}\right)\right]$.

   The probability $P(F)$ increases steadily until it reaches its peak and then decreases slowly. The F-axis is the asymptote of the probability curve.

2. The mean of the F-distribution is $v_2/(v_2 - 2)$, $v_2 > 2$, which is independent of v_1.

3. The variance of the F-distribution is

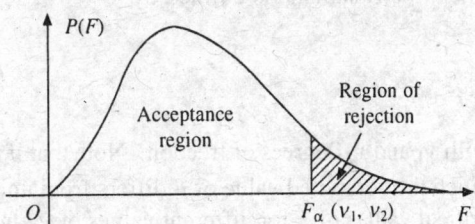

Fig. 20.6 Probability curve of F-distribution.

$$\frac{2v_2^2(v_1 + v_2 - 2)}{v_1(v_2 - 2)^2(v_2 - 4)}, \quad v_2 > 4. \tag{20.38}$$

4. The Students's t-distribution and F-distribution are related. If a statistic t follows t-distribution with v degrees of freedom, then t^2 follows Snedecor's F-distribution with $(1, v)$ degrees of freedom.

5. If $F(v_1, v_2)$ is an F-variate with v_1 and v_2 degrees of freedom, then $F(v_2, v_1)$ is distributed as $1/F(v_1, v_2)$ variate. Also

$$P[F(v_1, v_2) \geq K] = P\left[F(v_2, v_1) \leq \frac{1}{K}\right]. \tag{20.39}$$

6. If $v_1 = v_2$, the median of the distribution is at $F = 1$.

Application of F-distribution

F-distribution is used to test the equality of the variances of the populations from which two small random samples are drawn.

Remark 4

To test if two small random samples have been drawn from the same normal population, we must apply both the F-test and t-test. Using F-test, we test the equality of the population variances. If equality of variances is shown, then we apply t-test for the significance of the differences of the two sample means.

F-test

We use F-test to find whether the difference between population variances is significant or not. To test this significance, we first need estimates of the population variances $\hat{\sigma}_1^2$ and $\hat{\sigma}_2^2$, based on the small sample variances s_1^2 and s_2^2, From (20.30), we have

$$\hat{\sigma}_1^2 = \left(\frac{n_1}{n_1 - 1}\right)s_1^2, \text{ with } v_1 = n_1 - 1 \text{ degrees of freedom,} \tag{20.40}$$

$$\hat{\sigma}_2^2 = \left(\frac{n_2}{n_2 - 1}\right)s_2^2, \text{ with } v_2 = n_2 - 1 \text{ degrees of freedom.} \tag{20.41}$$

Then, the *F*-statistic is defined by

$$F = \frac{\hat{\sigma}_1^2}{\hat{\sigma}_2^2} \tag{20.42}$$

with v_1 and v_2 degrees of freedom. Note that if $\hat{\sigma}_1^2 = \hat{\sigma}_2^2$, then $F = 1$. Therefore, the *F*-test finds how much the computed value of *F* differs from unity. The *F*-tables, (Tables 20.4a and 20.4b) provide the critical values of *F* for different values of v_1 and v_2 at 1% and 5% levels of significance respectively. If *F* is the statistic and $F_\alpha(v_1, v_2)$ is the tabulated critical value of *F* at a given level of significance α, then

$$P\left[F > F_\alpha(v_1, v_2)\right] = \alpha \tag{20.43}$$

(see Fig. 20.6). This implies that if $F > F_\alpha(v_1, v_2)$, then the difference between *F* and 1 is significant at the given level of significance α. Then, we conclude that the samples are not drawn from the same population or from populations with the same variance. If $F < F_\alpha(v_1, v_2)$, then the difference is not significant at the given level of significance and the null hypothesis is accepted.

Remark 5

We shall always choose v_1, v_2 such that $F > 1$. That is, we take larger of the estimates of σ^2 of the samples as $\hat{\sigma}_1^2$ and the corresponding degree of freedom as v_1,

Example 20.16 Two random samples of sizes 9 and 7 gave the sum of squares of deviations from their respective means as 175 and 95 respectively. Can they be regarded as drawn from normal populations with the same variance?

Solution We have

$$n_1 = 9, \ \Sigma(x_i - \bar{x})^2 = n_1 s_1^2 = 175, \ \hat{\sigma}_1^2 = \frac{n_1 s_1^2}{n_1 - 1} = \frac{175}{8} = 21.875.$$

$$n_2 = 7, \ \Sigma(y_i - \bar{y})^2 = n_2 s_2^2 = 95, \ \hat{\sigma}_2^2 = \frac{n_2 s_2^2}{n_2 - 1} = \frac{95}{6} = 15.8333.$$

Now, $\hat{\sigma}_1^2 > \hat{\sigma}_2^2$. Hence, we take $v_1 = n_1 - 1 = 8$, and $v_2 = n_2 - 1 = 6$. We define

Null hypothesis $H_0 : \hat{\sigma}_1^2 = \hat{\sigma}_2^2$.

Alternate hypothesis $H_1 : \hat{\sigma}_1^2 \neq \hat{\sigma}_2^2$.

At 5% level of significance, we have from Table 20.4b, $F_{0.05}(8, 6) = 4.15$.

Now, the F-statistic is given by $F = \dfrac{\hat{\sigma}_1^2}{\hat{\sigma}_2^2} = \dfrac{21.875}{15.8333} = 1.381 < 4.15$.

Therefore, we accept the null hypothesis H_0. The two random samples might have come from two normal populations with the same variance.

Table 20.4a

F-Table : 1% level of significance.

v_2	\multicolumn{10}{c}{v_1}
	1	2	3	4	5	6	8	12	24	∞
1	4052	4999	5403	5625	5764	5859	5981	6106	6234	6366
2	98.49	99.01	99.17	99.25	99.30	99.33	99.36	99.42	99.46	99.50
3	34.12	30.81	29.46	28.71	28.24	27.91	27.49	27.05	26.60	26.12
4	21.20	18.00	16.69	15.98	15.52	15.21	14.80	14.37	13.93	13.46
5	16.26	13.27	12.06	11.39	10.97	10.67	10.27	9.89	9.47	9.02
6	13.74	10.92	9.78	9.15	8.75	8.47	8.10	7.72	7.31	6.88
7	12.25	9.55	8.45	7.85	7.46	7.19	6.84	6.47	6.07	5.65
8	11.26	8.65	7.59	7.01	6.63	6.37	6.03	5.67	5.28	4.86
9	10.56	8.02	6.99	6.42	6.06	5.80	5.47	5.11	4.73	4.31
10	10.04	7.56	6.55	5.99	5.64	5.39	5.06	4.71	4.33	3.91
11	9.65	7.20	6.22	5.67	5.32	5.07	4.74	4.40	4.02	3.60
12	9.33	6.93	5.95	5.41	5.06	4.82	4.50	4.16	3.78	3.36
13	9.07	6.70	5.74	5.20	4.86	4.62	4.30	3.96	3.59	3.16
14	8.86	6.51	5.56	5.03	4.69	4.46	4.14	3.80	3.43	3.00
15	8.68	6.36	5.42	4.89	4.56	4.32	4.00	3.67	3.29	2.87
16	8.53	6.23	5.29	4.77	4.44	4.20	3.89	3.55	3.18	2.75
17	8.40	6.11	5.18	4.67	4.34	4.10	3.79	3.45	3.08	2.65
18	8.28	6.01	5.09	4.58	4.25	4.01	3.71	3.37	3.00	2.57
19	8.18	5.93	5.01	4.50	4.17	3.94	3.63	3.30	2.92	2.49
20	8.10	5.85	4.94	4.43	4.10	3.87	3.56	3.23	2.86	2.42
21	8.02	5.78	4.87	4.37	4.04	3.81	3.51	3.17	2.80	2.36
22	7.94	5.72	4.82	4.31	3.99	3.76	3.45	3.12	2.75	2.31
23	7.88	5.66	4.76	4.26	3.94	3.71	3.41	3.07	2.70	2.26
24	7.82	5.61	4.72	4.22	3.90	3.67	3.36	3.03	2.66	2.21
25	7.77	5.57	4.68	4.18	3.86	3.63	3.32	2.99	2.62	2.17
26	7.72	5.53	4.64	4.14	3.82	3.59	3.29	2.96	2.58	2.13
27	7.68	5.49	4.60	4.11	3.79	3.56	3.26	2.93	2.55	2.10
28	7.64	5.45	4.57	4.07	3.76	3.53	3.23	2.90	2.52	2.06
29	7.60	5.42	4.54	4.04	3.73	3.50	3.20	2.87	2.49	2.03
30	7.56	5.39	4.51	4.02	3.70	3.47	3.17	2.84	2.47	2.01
40	7.31	5.18	4.31	3.83	3.51	3.29	2.99	2.66	2.29	1.81
60	7.08	4.98	4.13	3.65	3.34	3.12	2.82	2.50	2.12	1.60
120	6.85	4.79	3.95	3.48	3.17	2.96	2.66	2.34	1.95	1.38
∞	6.64	4.60	3.78	3.32	3.02	2.80	2.51	2.18	1.79	1.00

Table 20.4b

F-Table : 5% level of significance.

v_2	v_1									
	1	2	3	4	5	6	8	12	24	∞
1	161.4	199.5	215.7	224.6	230.2	234.0	238.9	243.9	249.0	253.4
2	18.51	19.00	19.16	19.25	19.30	19.33	19.37	19.41	19.45	19.50
3	10.13	9.55	9.28	9.12	9.01	8.94	8.84	8.74	8.64	8.53
4	7.71	6.94	6.59	6.39	6.26	6.16	6.04	5.91	5.77	5.63
5	6.61	5.79	5.41	5.19	5.05	4.95	4.82	4.68	4.53	4.36
6	5.99	5.14	4.76	4.53	4.39	4.28	4.15	4.00	3.84	3.67
7	5.59	4.74	4.35	4.12	3.97	3.87	3.73	3.57	3.41	3.23
8	5.32	4.46	4.07	3.84	3.69	3.58	3.44	3.28	3.12	2.93
9	5.12	4.26	3.86	3.63	3.48	3.37	3.23	3.07	2.90	2.71
10	4.96	4.10	3.71	3.48	3.33	3.22	3.07	2.91	2.74	2.54
11	4.84	3.98	3.59	3.36	3.20	3.09	2.95	2.79	2.61	2.40
12	4.75	3.88	3.49	3.26	3.11	3.00	2.85	2.69	2.50	2.30
13	4.67	3.80	3.41	3.18	3.02	2.92	2.77	2.60	2.42	2.21
14	4.60	3.74	3.34	3.11	2.96	2.85	2.70	2.53	2.35	2.13
15	4.54	3.68	3.29	3.06	2.90	2.79	2.64	2.48	2.29	2.07
16	4.49	3.63	3.24	3.01	2.85	2.74	2.59	2.42	2.24	2.01
17	4.45	3.59	3.20	2.96	2.81	2.70	2.55	2.38	2.19	1.96
18	4.41	3.55	3.16	2.93	2.77	2.66	2.51	2.34	2.15	1.92
19	4.38	3.52	3.13	2.90	2.74	2.63	2.48	2.31	2.11	1.88
20	4.35	3.49	3.10	2.87	2.71	2.60	2.45	2.28	2.08	1.84
21	4.32	3.47	3.07	2.84	2.68	2.57	2.42	2.25	2.05	1.81
22	4.30	3.44	3.05	2.82	2.66	2.55	2.40	2.23	2.03	1.78
23	4.28	3.42	3.03	2.80	2.64	2.53	2.38	2.20	2.00	1.76
24	4.26	3.40	3.10	2.78	2.62	2.51	2.36	2.18	1.98	1.73
25	4.24	3.38	2.99	2.76	2.60	2.49	2.34	2.16	1.96	1.71
26	4.22	3.37	2.98	2.74	2.59	2.47	2.32	2.15	1.95	1.69
27	4.21	3.35	2.96	2.73	2.57	2.46	2.30	2.13	1.93	1.67
28	4.20	3.34	2.95	2.71	2.56	2.44	2.29	2.12	1.91	1.65
29	4.18	3.33	2.93	2.70	2.54	2.43	2.28	2.10	1.90	1.64
30	4.17	3.32	2.92	2.69	2.53	2.42	2.27	2.09	1.89	1.62
40	4.08	3.23	2.84	2.61	2.45	2.34	2.18	2.00	1.79	1.51
60	4.00	3.15	2.76	2.52	2.37	2.25	2.10	1.92	1.70	1.39
120	3.92	3.07	2.68	2.45	2.29	2.17	2.02	1.83	1.61	1.25
∞	3.84	2.99	2.60	2.37	2.21	2.09	1.94	1.75	1.52	1.00

Example 20.17 Two random samples of sizes 9 and 6 gave the following values of the variable.

Sample 1	15	22	28	26	18	17	29	21	24
Sample 2	8	12	9	16	15	10			

Test the difference of the estimates of the population variances at 5% level of significance.

Solution We have

$$n_1 = 9, \quad \bar{x} = 22.2222, \quad s_1^2 = 21.7294, \quad \hat{\sigma}_1^2 = \frac{n_1 s_1^2}{n_1 - 1} = 24.4456.$$

$$n_1 = 6, \quad \bar{y} = 11.6667, \quad s_2^2 = 8.8881, \quad \hat{\sigma}_2^2 = \frac{n_2 s_2^2}{n_2 - 1} = 10.6657.$$

Now, $\hat{\sigma}_1^2 > \hat{\sigma}_2^2$. We take $v_1 = n_1 - 1 = 8, \ v_2 = n_2 - 1 = 5$.

We define

Null hypothesis $H_0 : \hat{\sigma}_1^2 = \hat{\sigma}_2^2$.

Alternate hypothesis $H_1 : \hat{\sigma}_1^2 \neq \sigma_2^2$.

At 5% level of significance, we have from Table 20.4b, $F_{0.05}(8, 5) = 4.82$.

The F-statistic is given by

$$F = \frac{\hat{\sigma}_1^2}{\hat{\sigma}_2^2} = \frac{24.4456}{10.6657} = 2.29 < 4.82.$$

Therefore, we accept the null hypothesis. There is no significant difference between the population variances.

Example 20.18 The values in two random samples are given below.

Sample 1	15	25	16	20	22	24	21	17	19	23		
Sample 2	35	31	25	38	26	29	32	34	33	27	29	31

Can we conclude that the two samples are drawn from the same population? Test at 5% level of significance.

Solution We shall use both the F-test and t-test to draw a conclusion. We have

$$n_1 = 10, \quad \bar{x}_1 = 20.2, \quad s_1^2 = 10.56, \quad \hat{\sigma}_1^2 = \frac{n_1 s_1^2}{n_1 - 1} = 11.7333.$$

$$n_2 = 12, \quad \bar{x}_2 = 30.8333, \quad s_2^2 = 13.6409, \quad \hat{\sigma}_2^2 = \frac{n_2 s_2^2}{n_2 - 1} = 14.8810.$$

Note that $\hat{\sigma}_2^2 > \hat{\sigma}_1^2$. We define

Null hypothesis $H_0 : \hat{\sigma}_1^2 = \hat{\sigma}_2^2$.

Alternate hypothesis $H_1 : \hat{\sigma}_1^2 \neq \hat{\sigma}_2^2$.

At 5% level of significance, $F_{0.05}(11, 9) = 3.11$. The F-statistic is given by

$$F = \frac{\hat{\sigma}_2^2}{\hat{\sigma}_1^2} = \frac{14.8810}{11.7333} = 1.268 < 3.11.$$

We accept the null hypothesis. We conclude that the difference between the estimates of population variances is not significant.

Now, we use the t-test. Define, $H_0 : \mu_1 = \mu_2$ and $H_1 : \mu_1 \neq \mu_2$.

The estimate of the population variance is given by

$$\hat{\sigma}^2 = \frac{n_1 s_1^2 + n_2 s_2^2}{n_1 + n_2 - 2} = \frac{10(10.56) + 12(13.6409)}{20} = 13.4645.$$

The t-statistic is given by

$$t = \frac{\bar{x}_1 - \bar{x}_2}{\hat{\sigma} \sqrt{\frac{1}{n_1} + \frac{1}{n_2}}} = -6.76$$

with $v = n_1 + n_2 - 2 = 20$ degrees of freedom. At 5% level of significance, we get from Table 20.3 the value of t as 2.086. Since, $|t| = 6.76 > 2.086$, we reject the null hypothesis. The means of the two samples and hence the means of the two populations differ significantly. We conclude that the samples are not from the same population.

Exercise 20.2

1. The I.Q of the students in an elementary school were tested. A random sample of 7 students had the following I.Q's : 85, 96, 105, 102, 82, 89, 90. Does the data support the claim of a population mean of I.Q 100? Test at 5% level of significance. Find a range in which most of the mean I.Q values lie at the same level of significance.

2. A number of candidates appeared at a selection trial for recruitment in the army. A random sample of heights of 10 candidates are 162 cm, 170 cm, 168 cm, 169 cm, 173 cm, 171 cm, 165 cm, 166 cm, 161 cm and 160 cm. Can we conclude that the average height of the candidates is greater than 165 cm? Test at 5% level of significance.

3. A random sample of 13 values from a normal population has mean 65.5 cm. The sum of squares of deviations from this mean is 625.5. Can we assume that the random sample is drawn from a normal population with mean 66.5 cm? Test at 5% and 1% level of significance. Find the confidence limits at these levels of significance.

4. A certain injection administered to each of the five patients resulted in the following changes in the blood pressure: 4, 3, –1, 0, 1. Can we conclude that the injection is accompanied by an increase in the blood pressure? Test at 5% level of significance.

5. A company manufactures electric fuses. The mean lifetime of a sample of 17 fuses is found to be 665 days with standard deviation of 65 days. The company claims that the mean life of the electric fuses is 700 days. Can their claim be accepted at 5% level of significance?

6. Two independent samples of sizes 7 and 8 have the following values

| Sample 1 | 55 | 49 | 65 | 60 | 56 | 59 | 54 | |
| Sample 2 | 45 | 69 | 70 | 49 | 54 | 57 | 59 | 48 |

Find whether the difference between the sample means is significant. Test at 5% level of significance.

7. Measurements of the fat contents of two brands of ice cream A and B yield the following sample data

| Brand A | 13.1 | 13.6 | 12.8 | 14.0 | 14.1 | 13.9 |
| Brand B | 12.7 | 13.3 | 14.0 | 12.9 | 13.5 | 13.0 |

Test the null hypothesis $H_0 : \mu_1 = \mu_2$ (where μ_1, μ_2 are the true average fat contents of the ice creams A and B respectively) against the alternative hypothesis $H_1 : \mu_1 \neq \mu_2$. Test at 5% level of significance.

8. Two companies A and B manufacture batteries for cell phones. Random samples of sizes 9 and 8 were chosen from the two companies respectively. When tested for length of life of the batteries, the following data was obtained.

	mean	S.D
Battery from A	780 days	50 days
Battery from B	750 days	45 days.

Company A claims that their batteries are superior. Can their claim be accepted at 5% level of significance?

9. Samples of 12 supervisors of workshop in factory A and 10 supervisors of workshop in factory B were chosen at random and the following data was obtained. The average monthly salary of the supervisors were Rs. 8500 and Rs. 8000 respectively in factories A and B with corresponding standard deviation of salaries as Rs. 180 and Rs. 170. Can we conclude that the average salary of supervisors in factory A is more than in factory B? Test at 5% level of significance.

10. Two independent groups of 8 children in an elementary school were given a memory test. The children were asked to repeat the digits of a number from memory after hearing the number. The following are the results of such a test.

Group A	5	6	4	7	6	4	8	5
Group B	4	7	7	5	3	4	6	4

Find whether the difference between mean scores of the two groups is significant. Test at 5% level of significance.

11. The following data represents the marks obtained (out of 100 marks) by 10 students before and after attending some special coaching sessions.

Before	36	60	80	70	62	61	66	68	56	50
After	45	70	76	80	66	58	70	68	64	58

Is the coaching effective? Test at 5% level of significance.

12. Eight soldiers visit a shooting range on two consecutive days and fire 100 shots in the range. On the first day their scores were 57, 64, 49, 72, 66, 68, 75 and 60. On the second day, their corresponding scores were 60, 70, 54, 76, 60, 65, 80 and 60. Examine whether their performance has improved on the second day after practice. Test at 5% level of significance.

13. In one sample of 15 items, the sum of squares of deviations of the sample values from sample mean is 92.5 and in another sample of 13 items, the sum of squares of deviations of the sample values from sample mean is 100. Test whether this difference of variances is significant at 5% level of significance.

14. A random sample of size 9 gave an estimated population variance as 15.6 and another random sample of size 11 gave an estimated population variance as 18.2. Can we conclude that the samples have come from populations with the same variance? Test at 5% level of significance.

15. Two random samples of sizes 7 and 6 have the following values of the variable

Sample 1	12	16	18	22	19	28	21
Sample 2	10	15	14	19	24	22	

At 5% level of significance, do the estimates of population variances differ significantly?

16. Two random samples gave the following data

Sample	Size	Sample mean	Sample variance
1	11	15.6	3.2
2	8	10.5	1.5

Test at 5% level of significance whether the two samples could have come from the same population.

17. Two random samples gave the following data

Sample	Size	Sample mean	Sample Variance
1	9	23.5	5.6
2	8	21.5	4.0

Test at 5% level of significance whether the two samples could have come from the same populaton.

18. Two random samples have the following values

Sample 1	31	22	28	29	27	35

Sample 2	27	21	25	23	26

Test at 5% level of significance whether the two samples could have come from the same population.

20.5   Chi-Square Distribution (Non–Parametric Test)

The chi-square distribution is denoted as χ^2-distribution. We know that if a random variable $X \sim N(\mu, \sigma^2)$, then the standard normal variate is $Z = [(X - \mu)/\sigma] \sim N(0, 1)$. The square of the standard normal variate Z

$$Z^2 = \left(\frac{X - \mu}{\sigma}\right)^2$$

is called a chi-square variate with 1 degree of freedom. Generalising, we have that if $X_i, i = 1, 2, \ldots, n$ are n independent normal variables, $N(\mu_i, \sigma_i^2)$, then

$$\chi^2 = \sum_{i=1}^{n}\left[\frac{X_i - \mu_i}{\sigma_i}\right]^2 \tag{20.44}$$

is a chi-square variate with n degrees of freedom. The distribution is called a chi-square distribution and is denoted by $\chi^2(n)$. The probability density function of $X \sim \chi^2(n)$ with n degrees of freedom is given by

$$f(x) = \frac{1}{2^{n/2}\Gamma(n/2)}e^{-x/2}\, x^{(n/2)-1}, \, 0 \le x < \infty. \tag{20.45}$$

or

$$f(\chi^2) = \frac{1}{2^{n/2}\Gamma(n/2)}\, e^{-\chi^2/2}\, (\chi^2)^{(n/2)-1}, \, \, 0 \le \chi^2 < \infty. \tag{20.46}$$

Graph and properties of χ^2 - distribution

1. From Eq. (20.45), we get

$$\log f(x) = -\frac{x}{2} + \left(\frac{n}{2} - 1\right)\log x - \log[2^{n/2}\Gamma(n/2)]$$

Differentiating, we get

$$\frac{f'(x)}{f(x)} = -\frac{1}{2} + \left(\frac{n}{2} - 1\right)\left(\frac{1}{x}\right) = \frac{1}{2x}(n - 2 - x)$$

or

$$f'(x) = \frac{1}{2x}(n - 2 - x)f(x) \tag{20.47}$$

Setting $f'(x) = 0$, we get $x = n - 2$ (since $f(x) \ne 0$). Since $x > 0$, and $f(x) \ge 0$, we obtain

$$f'(x) < 0 \text{ for } n - 2 \leq 0 \text{ and for all } x.$$

This implies that for $n = 1, 2$ degrees of freedom, the χ^2-probability curve is monotonically decreasing. For $n > 2$, we get

$$f'(x) = \begin{cases} > 0, & \text{for} \quad x < (n-2) \\ = 0, & \text{for} \quad x = n - 2 \\ < 0, & \text{for} \quad x > (n-2). \end{cases}$$

Hence, for $n > 2$, $f(x)$ is monotonically increasing for $0 < x < n - 2$, and monotonically decreasing for $n - 2 < x < \infty$. At $x = n - 2$, it attains the maximum value. For $n \geq 1$, $f(x)$ decreases rapidly as x increases and $\to 0$ as $x \to \infty$. The x-axis is an asymptote to the curve. For $n = 2$, the curve meets the y-axis at $(0, 0.5)$. As n becomes smaller, the curve is skewed more to the right. The rough sketch of the curve is given in Fig. 20.7.

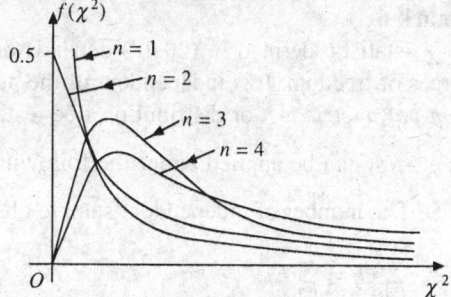

Fig. 20.7. χ^2-probability curve.

2. The mean and variance of the χ^2-distribution are n and $n/2$ respectively.
3. As the number of degrees of freedom $n \to \infty$, the χ^2-distribution tends to the normal distribution.
4. If $X_1, X_2, \ldots, X_m$ are m independent χ^2-variates with degrees of freedom $n_1, n_2, \ldots, n_m$ respectively, then $\sum_{i=1}^{m} X_i$ is also a χ^2-variate with $n_1 + n_2 + \ldots + n_m$ degrees of freedom. The converse is also true. If, X_i, $i = 1, 2, \ldots, m$ are χ^2-variates with degrees of freedom $n_1, n_2, \ldots, n_m$ respectively and $\sum X_i$ is a χ^2-variate with degree of freedom $n_1 + n_2 + \ldots + n_m$, then X_i are independent.

Applications of χ^2-distribution

1. χ^2-distribution is used to test the *goodness of fit*. For example, suppose that we have fitted a binomial or a Poisson distribution to a given data of a sample. We use the χ^2-distribution to test whether this fitting of the binomial or Poisson distribution to the data is acceptable.
2. χ^2-distribution is used to test the independence of the attributes of a population. For example, suppose that a population has two *characteristics* or *attributes* or *traits*. χ^2-distribution can be used to test whether the two attributes are dependent (related) or independent, based on a random sample drawn from a population.

χ^2 - test for goodness of fit

Let a distribution be given. Suppose that O_i and E_i, $i = 1, 2, \ldots, n$ are the observed and expected frequencies of the ith class with $\sum_{i=1}^{n} O_i = \sum_{i=1}^{n} E_i$. The expected frequencies are computed using the hypothesis assumed about the population. Then, it is known that

$$\chi^2 = \sum_{i=1}^{n} \left[\frac{(O_i - E_i)^2}{E_i} \right], \quad \text{with} \quad \sum_{i=1}^{n} O_i = \sum_{i=1}^{n} E_i \qquad (20.48)$$

follows a χ^2 - distribution with $n-1$ degrees of freedom, which is equal to the number of independent frequencies. Simplifying, we get

$$\chi^2 = \sum_{i=1}^{n} \frac{1}{E_i}[O_i^2 + E_i^2 - 2O_iE_i] = \sum_{i=1}^{n}\left[\frac{O_i^2}{E_i} + E_i - 2O_i\right]$$

$$= \sum_{i=1}^{n}\left(\frac{O_i^2}{E_i}\right) + \sum_{i=1}^{n} E_i - 2\sum_{i=1}^{n} O_i = \sum_{i=1}^{n}\left(\frac{O_i^2}{E_i}\right) - N \qquad (20.49)$$

since $\sum_{i=1}^{n} O_i = \sum_{i=1}^{n} E_i = N$ is the total frequency. We can use either of the expressions in (20.48) or (20.49).

Remark 6

The χ^2-statistic defined in (20.48) depends only on the observed and expected freqencies and the degrees of freedom. It is independent of the population parameters. Hence, the test is also known as a *non-parametric test* or distribution free test.

The χ^2-test can be applied under the following conditions.

1. The number of independent sample observations, N, should be reasonably large, say, $N \geq 50$.

2. $\sum_{i=1}^{n} O_i = \sum_{i=1}^{n} E_i$.

3. Individual theoretical frequencies of the classes should not be small. Ideally, $O_i \geq 10$. If a theoretical frequency is less than 10, then it is combined with the preceeding or the succeeding frequencies such that the combined frequency is ≥ 10. This procedure is also called the *method of pooling*. The degrees of freedom is adjusted accordingly. Some authors (books) prescribe this limit as 5. If a cell frequency is less than 5, then the distribution cannot maintain its character of continuity (X^2 is a cantinuous distribution).

4. The number of classes, n, should neither be too small or too large, Generally $4 \leq n \leq 16$.

Critical values

Let $\chi_n^2(\alpha)$ denote the value of the χ^2-distribution with n degrees of freedom at α level of significance. Then, the critical value of χ^2 is given by

$$P[\chi^2 > \chi_n^2(\alpha)] = \alpha \qquad (20.50)$$

that is, the area to the right of this point is α. The critical values of χ^2- distribution are available for a range of values of the degree of freedom n and levels of significance α. A few of these values are given in Table 20.5. If the computed $\chi^2 < \chi_n^2(\alpha)$, then the null hypothesis H_0 is accepted, that is, the difference between the observed and expected (values) frequencies is not significant at α level of significance. We conclude that the given sample is drawn from the hypothetical population. If the computed $\chi^2 > \chi_n^2(\alpha)$, then the null hypothesis is rejected and we conclude that the random sample is not from the hypothetical population. The critical value of χ^2 increases as the number of free parameters increases and the level of significance α decreases.

χ^2–test for independence of attributes

Consider the case when two attributes P, Q of the population are to be investigated for independence. The attributes P and Q are divided into m and k classes respectively. Therefore, the attributes are

Table 20.5

χ^2–Table – goodness of fit.

n	0.99	0.95	0.10	0.05	0.02	0.01
			Probability			
1	0.000157	0.00393	2.706	3.841	5.412	6.635
2	0.0201	0.103	4.605	5.991	7.824	9.210
3	0.115	0.352	6.251	7.815	9.837	11.345
4	0.297	0.711	7.779	9.488	11.668	13.277
5	0.554	1.145	9.236	11.070	13.388	15.086
6	0.872	1.635	10.645	12.592	15.033	16.812
7	1.238	2.167	12.017	14.067	16.622	18.475
8	1.646	2.733	13.362	15.507	18.168	20.090
9	2.088	3.325	14.684	16.919	19.670	21.666
10	2.558	3.940	15.987	18.307	21.161	23.209
11	3.053	4.575	17.275	19.675	22.618	24.725
12	3.571	5.226	18.549	21.026	24.054	26.217
13	4.107	5.892	19.812	22.362	25.472	27.688
14	4.660	6.571	21.064	23.685	26.873	29.141
15	5.229	7.261	22.307	24.996	28.259	30.578
16	5.812	7.962	23.542	26.296	29.633	32.000
17	6.408	8.672	24.768	27.587	30.995	33.409
18	7.015	9.390	25.989	28.869	32.346	34.805
19	7.633	10.117	27.204	30.114	33.687	36.191
20	8.260	10.851	28.412	31.410	35.020	37.566
21	8.897	11.581	29.615	32.671	36.343	38.932
22	9.542	12.338	30.813	33.924	37.659	40.289
23	10.196	13.091	32.007	35.172	38.968	41.638
24	10.856	13.848	33.196	36.415	40.270	42.980
25	11.524	14.611	34.382	37.652	41.566	44.314
26	12.198	15.379	35.563	38.885	42.856	45.642
27	12.879	16.151	36.741	40.113	44.140	46.963
28	13.565	16.928	37.916	41.337	45.419	48.278
29	14.256	17.708	39.087	42.557	46.693	49.558
30	14.953	18.493	40.256	43.773	47.962	50.892

divided into mk classes $P_1Q_1, P_1 Q_2, \ldots, P_1 Q_k, \ldots, P_m Q_1, P_m Q_2, \ldots, P_m Q_k$. The tabular form of this data, which look like the elements in a matrix, is called an $m \times k$ *contingency table*. The observed frequencies are denoted by O_{ij}, $i = 1, 2, \ldots, m; j = 1, 2, \ldots, k$, (see Table 20.6). We define the null hypothesis as H_0 : the two attributes P and Q are independent. The expected frequency for each cell is computed as

$$E_{ij} = \frac{\text{(total observed frequencies in the } i\text{th row)} \times \text{(total observed frequencies in the } j\text{th column)}}{\text{(Total frequencies} = N)}$$

$$= \frac{(O_{i*})(O_{*j})}{N}, \quad i = 1, 2, \ldots, m; j = 1, 2, \ldots, k. \tag{20.51}$$

Table 20.6 Contingency table.

Q \ P	Q_1	Q_2	...	Q_j	...	Q_k	row total
P_1	O_{11}	O_{12}	...	O_{1j}	...	O_{1k}	O_{1*}
P_2	O_{21}	O_{22}	...	O_{2j}	...	O_{2k}	O_{2*}
. .	. .		. .	. .		. .	
P_i	O_{i1}	O_{i2}	...	O_{ij}	...	O_{ik}	O_{i*}
. .	. .	. .		. .		. .	
P_m	O_{m1}	O_{m2}	...	O_{mj}	...	O_{mk}	O_{m*}
Column total	O_{*1}	O_{*2}	...	O_{*j}	...	O_{*k}	N

Then, we sum E_{ij} over all the cells to obtain

$$\chi^2 = \sum_{i=1}^{m} \sum_{j=1}^{k} \left[\frac{(O_{ij} - E_{ij})^2}{E_{ij}} \right].$$ (20.52)

We may also use the expression given in (20.49). We have from the contingency table

$$\text{number of degrees of freedom} = n = (m - 1)(k - 1)$$ (20.53)

Conclusion If $\chi^2 < \chi_n^2(\alpha)$, then H_0 is accepted at the level of significance α. We conclude that the attributes are independent. If $\chi^2 > \chi_n^2(\alpha)$, then H_1 is accepted at the level of significance α. We conclude that the attributes are dependent in some way.

Remark 7
Consider the t-statistic defined in (20.31). Then, $t^2/(n-1)$ can be written as the ratio of two independent χ^2-variates with 1 and $(n-1)$ degrees of freedom respectively.

Remark 8
Let X and Y be two independent chi-square variates with n_1 and n_2 degrees of freedom respectively. Then, the F-statistic can be written as

$$F = \frac{(X/n_1)}{(Y/n_2)}.$$

χ^2-test for population variance

We assume that the population from which the random sample is drawn, is normal. We want to test whether a random sample that has been drawn is from a normal population with a specified value of variance σ_0^2. We define the null hypothesis as H_0: Population variance is $\sigma^2 = \sigma_0^2$. The chi-square statistic is defined as

$$\chi^2 = \sum_{i=1}^{n} \left[\frac{(x_i - \bar{x})^2}{\sigma_0^2} \right] = \frac{1}{\sigma_0^2} \sum_{i=1}^{n} [x_i^2 - 2x_i\bar{x} + \bar{x}^2]$$

$$= \frac{1}{\sigma_0^2} \left[\sum_{i=1}^n x_i^2 - 2\bar{x} \sum_{i=1}^n x_i + n\bar{x}^2 \right]$$

$$= \frac{1}{\sigma_0^2} \left[\sum_{i=1}^n x_i^2 - 2n\bar{x}^2 + n\bar{x}^2 \right] = \frac{ns^2}{\sigma_0^2}, \tag{20.54}$$

with $n - 1$ degrees of freedom. If $\chi^2 < \chi_n^2(\alpha)$, then H_0 is accepted at α level of significance (the random sample is drawn from a normal population with variance σ_0^2). If $\chi^2 > \chi_n^2(\alpha)$, then H_0 is rejected.

Remark 9

If the sample size is large, say, $n > 30$, then we can use the Fisher's approximation

$$\sqrt{2\chi^2} \sim N(\sqrt{2n-1}, 1), \quad \text{or} \quad Z = \sqrt{2\chi^2} - \sqrt{2n-1} \sim N(0, 1). \tag{20.55}$$

We can then apply the normal distribution test.

Example 20.19 A sample survey of 500 families with 4 children has been made regarding the number of boys and girls in the families. The following data was obtained.

No. of families	35	100	200	125	40
No. of boys	4	3	2	1	0
No. of girls	0	1	2	3	4

Is the data consistent with the hypothesis that the male and female births are equally possible? Test at 5% level of significance.

Solution We define the null hypothesis as H_0 : male and female births are equally possible.

Therefore, $\qquad P(\text{male birth}) = p = \frac{1}{2},$ and $P(\text{female birth}) = q = \frac{1}{2}.$

Now, using the binomial law, we get

$$\text{probability that a family of 4 children has } r \text{ male children} = 4C_r \left(\frac{1}{2} \right)^4.$$

Therefore, expected number of families having r male children $= \frac{1}{16}(500)(4C_r) = (31.25)4C_r.$ We have the following data

E_i	31	125	188	125	31
O_i	35	100	200	125	40
$O_i - E_i$	4	−25	12	0	9

where E_i are converted into whole numbers such that $\Sigma E_i = 500$. Now,

$$\chi^2 = \sum_i \left[\frac{(O_i - E_i)^2}{E_i} \right] = \frac{16}{31} + \frac{625}{125} + \frac{144}{188} + 0 + \frac{81}{31} = 8.89.$$

We note that the values of p and q are not obtained from the given data. The given data has been used to obtain E_i. Hence, degrees of freedom $= n - 1 = 5 - 1 = 4$, and $\chi_4^2(0.05) = 9.488$.

Since, $\chi^2 = 8.89 \lesssim 9.488$, we accept the null hypothesis H_0. We conclude that male and female births are equally possible.

Example 20.20 Fit a Poisson distribution for the following data and test the goodness of fit at 5% level of significance.

x	0	1	2	3	4	5
$f(x)$	110	170	130	60	23	7

Solution The Poisson distribution is given by

$$P(X = r) = \frac{e^{-\lambda} \lambda^r}{r!}, \quad r = 0, 1, 2, \ldots$$

The mean of the Poisson distribution is given by λ. Now,

$$\bar{x} = \frac{\Sigma x f}{\Sigma f} = \frac{737}{500} = 1.474 = \lambda.$$

The expected frequencies are given by

$$E_i = \frac{500 \, e^{-\lambda} \lambda^i}{i!}$$

For $i = 6, 7, \ldots$, E_i are small and are neglected. The values of E_i are adjusted to make them whole numbers such that $\Sigma E_i = 500$. We have the frequencies

E_i	115	169	124	61	22	9
O_i	110	170	130	60	23	7

Grouping the classes so that each class frequency is ≥ 10, we obtain

E_i	116	169	124	61	31
O_i	110	170	130	60	30
$O_i - E_i$	-5	1	6	-1	-1

The chi-square statistic is given by

$$\chi^2 = \sum_i \left[\frac{(O_i - E_i)^2}{E_i} \right] = \frac{25}{116} + \frac{1}{169} + \frac{36}{124} + \frac{1}{61} + \frac{1}{31} = 0.560.$$

Now, the given data has been used to find ΣE_i and λ. Hence, degrees of freedom $= n = 5 - 2 = 3$.

Now,
$$\chi_3^2(0.05) = 7.815.$$

Since, $\chi^2 = 0.560 < 7.815$, we accept the null hypothesis. We conclude that the fitting of Poisson distribution to the given data is satisfactory.

Example 20.21 Fit a normal curve to the data

Class	30–35	35–40	40–45	45–50	50–55	55–60	60–65	65–70
Frequency	3	32	80	120	114	72	39	5

Compute the expected normal frequencies and test the goodness of fit at 5% level of significance.

Solution First, we compute the mean and standard deviation of the distribution. We obtain $\Sigma f_i = 465$, $\mu = 50.1021$, $\sigma^2 = 50.6482$ or $\sigma = 7.1168$. The normal curve is given by

$$f(x) = \frac{1}{7.1168\sqrt{2\pi}} e^{-\left[\left(\frac{x-50.1021}{7.1168}\right)^2 / 2\right]}$$

We construct the following table of values to compute expected normal frequencies.

Class	Lower class boundary	$Z = (X - \mu)/\sigma$	$\phi(z)$	$\phi_{z+1} - \phi_z$	Expected frequency
Below 30	$-\infty$	$-\infty$	0	0.0024	1.1 ⎱
30–35	30	-2.8246	0.0024	0.0145	6.7 ⎰ = 8
35–40	35	-2.1220	0.0169	0.0610	28.4 = 28
40–45	40	-1.4195	0.0779	0.1588	73.8 = 74
45–50	45	-0.7169	0.2367	0.2576	119.8 = 120
50–55	50	-0.0143	0.4943	0.2600	120.9 = 121
55–60	55	0.6882	0.7543	0.1635	76.03 = 76
60–65	60	1.3908	0.9178	0.0640	29.76 = 30
65–70	65	2.0933	0.9818	0.0156	7.25 = 8
Over 70	70	2.7959	0.9974		

Grouping the classes so that each class frequency is ≥ 10, we obtain

E_i	36	74	120	121	76	38
O_i	35	80	120	114	72	44
$O_i - E_i$	-1	6	0	-7	-4	6

The χ^2-statistic is given by

$$\chi^2 = \sum_i \left[\frac{(O_i - E_i)^2}{E_i}\right] = \frac{1}{36} + \frac{36}{74} + \frac{49}{121} + \frac{16}{76} + \frac{36}{38} = 2.077,$$

The given data has been used to find μ, σ and ΣE_i. Hence, degrees of freedom $= n = 6 - 3 = 3$. Now, $\chi_3^2(0.05) = 7.815$. Since $\chi^2 = 2.077 < 7.815$, we accept the null hypothesis. We conclude that the fitting of normal distribution to the given data is safisfactory and can be accepted.

Example 20.22 Let the entries in the 2×2 contingency table be $\begin{array}{|c|c|} \hline a & b \\ \hline c & d \\ \hline \end{array}$. Prove that the chi-square value is given by

$$\chi^2 = \frac{N(ad - bc)^2}{(a+b)(c+d)(a+c)(b+d)}, \quad N = a + b + c + d.$$

Solution The contingency table is given by

a	b	a + b
c	d	c + d
a + c	b + d	a + b + c + d = N

The expected values are given by

$$E(a) = \frac{(a+b)(a+c)}{N}, \quad E(b) = \frac{(b+d)(a+b)}{N}$$

$$E(c) = \frac{(a+c)(c+d)}{N}, \quad \text{and } E(d) = \frac{(b+d)(c+d)}{N}.$$

Now,

$$a - E(a) = a - \frac{(a+b)(a+c)}{N} = \frac{a(a+b+c+d) - (a+b)(a+c)}{N}$$

$$= \frac{ad - bc}{N}.$$

$$b - E(b) = b - \frac{(b+d)(a+b)}{N} = \frac{b(a+b+c+d) - (b+d)(a+b)}{N}$$

$$= -\frac{ad - bc}{N}$$

$$c - E(c) = c - \frac{(a+c)(c+d)}{N} = \frac{c(a+b+c+d) - (a+c)(c+d)}{N}$$

$$= -\frac{ad - bc}{N}$$

$$d - E(d) = d - \frac{(b+d)(c+d)}{N} = \frac{d(a+b+c+d) - (b+d)(c+d)}{N}$$

$$= \frac{ad - bc}{N}.$$

Therefore,

$$\chi^2 = \frac{(ad-bc)^2}{N^2} \left[\frac{1}{E(a)} + \frac{1}{E(b)} + \frac{1}{E(c)} + \frac{1}{E(d)} \right]$$

$$= \frac{(ad-bc)^2}{N} \left[\frac{1}{(a+b)(a+c)} + \frac{1}{(a+b)(b+d)} + \frac{1}{(a+c)(c+d)} + \frac{1}{(b+d)(c+d)} \right]$$

$$= \frac{(ad-bc)^2}{N} \left[\frac{(b+d)(c+d) + (a+c)(c+d) + (a+b)(b+d) + (a+b)(a+c)}{(a+b)(a+c)(b+d)(c+d)} \right]$$

$$= \frac{(ad - bc)^2}{N} \left[\frac{(b + d)(c + d + a + b) + (a + c)(c + d + a + b)}{(a + b)(a + c)(b + d)(c + d)} \right]$$

$$= \frac{N(ad - bc)^2}{(a + b)(a + c)(b + d)(c + d)}.$$

Yate's correction

In a 2×2 contingency table4 (see Example 20.22), the number of degrees of freedom is $(n - 1)(n - 1) = (2 - 1)(2 - 1) = 1$. If any of the theoretical frequencies is lass than 5, then we usually apply the pooling method. But, if we pool the frequencies in a 2×2 table, then the number of degrees of freedom becomes 0, which is meaningless. In such cases, we apply the *yate's correction for continuity*. In this correction procedure, we add (subtract) 1/2 to the cell frequency which is lass than 5 and adjust the remaining frequencies accordingly, so that the row ad column sums do not change. On the resulting frequencies, the X^2-test for goodness of fit is applied (without pooling).

Yate's correction to a 2×2 contingency table gives the following

		Sum
$a \mp \frac{1}{2}$	$b \pm \frac{1}{2}$	$a + b$
$c \pm \frac{1}{2}$	$d \mp \frac{1}{2}$	$c + d$
Sum $\quad a + c$	$b + d$	$N = a + b + c + d$

The corrected value of X^2 is given by

$$\chi^2 = \frac{N[\{a \mp (1/2)\}\{d \mp (1/2)\} - \{b \pm (1/2)\}\{c \pm (1/2)\}]^2}{(a + b)(a + c)(b + d)(c + d)}$$

$$= \frac{N[(ad - bc) \mp (1/2)(a + b + c + d)]^2}{(a + b)(a + c)(b + d)(c + d)} = \frac{N[\,|ad - bc\,| - (N/2)]^2}{(a + b)(a + c)(b + d)(c + d)}$$

Remark 10

1. If N is large, then the above estimate of X^2 may not deviate much from the original estimate. However, if N is smale, then the variation in the estimates may be significant and the probability may be over stated.

2. Yate's correction can be applied to every 2×2 contingency table, even if none of the frequencies is lass than 5.

Example 20.23 Test the hypothesis that $\sigma = 20$, given that $s = 30$, for a random sample of size 80 from a normal population.

Solution We have $n = 80$, $s = 30$. We define the null hypothesis as $H_0 : \sigma = 20$. Now, from (20.53),

$$\chi^2 = \frac{ns^2}{\sigma_0^2} = \frac{(80)(900)}{400} = 180.$$

Using (20.54), (since $n > 30$), we get

$$Z = \sqrt{2\chi^2} - \sqrt{2n-1} \sim N(0, 1)$$

or
$$Z = \sqrt{360} - \sqrt{159} = 6.36.$$

Since, $|Z| = 6.36 > 3$, it is significant at all levels of significance. We conclude that $\sigma \ne 20$.

Exercise 20.3

1. A sample survey of 360 families with 5 children has been made regarding the number of boys and girls in the families. The following data was obtained

No. of families	10	30	100	80	60	80
No. of boys	5	4	3	2	1	0
No. of girls	0	1	2	3	4	5

Can we conclude that the male and female births are equally possible? Test at 5% level of significance.

2. A sample survey of 200 families with twins has been made regarding the gender of the twins. The following data was obtained.

No. of families	40	100	60
No. of boys	2	1	0
No. of girls	0	1	2

Can we conclude that among twins, male and female births are equally possible? Test at 2% level of significance.

3. Fit a binomial distribution for the data and test the goodness of fit at 5% level of significance.

x	0	1	2	3	4	5
$f(x)$	15	26	35	48	32	24

4. Fit a Poisson distribution for the following data and test the goodness of fit at 5% level of significance.

x	0	1	2	3	4	5	6	7
$f(x)$	62	133	142	90	50	15	6	2

5. Fit a Poisson distribution for the following data and test the goodness of fit at 5% level of significance.

x	0	1	2	3	4	5	6	7
$f(x)$	30	100	110	85	60	20	11	4

6. A random sample of 400 people is drawn to check the smoking habits of people in a town. The following data was obtained.

	Literates	Illiterates
Smokers	70	120
Non-smokers	110	100

Based on this data, can we conclude that there is no relation between literacy and smoking habit. Test at 5% level of significance.

7. Two candidates A and B contested for an election for the post of college student's union president. The college has large number of students from the urban and rural areas. A sample of 500 students was drawn to find the preferences of the students. The following data was obtained

Area	Votes for	
	A	B
Urban	120	80
Rural	125	175

Determine whether the nature of the area is related to the voting preferences of the students.

8. A dice is thrown 102 times and the following distribution of faces is obtained

Face	1	2	3	4	5	6
Frequency	15	25	16	20	12	14

Can we conclude that all faces are equally likely to occur? Test at 5% level of significance.

9. A sample of 22 values shows that the standard deviation is 5.5. Is this compatible with the hypothesis that the sample is drawn from a normal population with standard deviation 4. Test at 5% level of significance.

10. Test the hypothesis that $\sigma = 8$ given that $s = 9$ for a random sample of size 200 drawn from a normal population.

11. Fit a normal curve to the data

class	30–35	35–40	40–45	45–50	50–55	55–60	60–65
frequency	3	32	125	230	120	35	5

Compute the expected normal frequencies and test the goodness of fit at 5% level of significance.

20.6 Answers and Hints

Exercise 20.1

1. No, $|Z| = 10$, (two tailed test).

2. $H_0 : \bar{x} = \mu; H_1 : \bar{x} > \mu$ (right tailed test); $Z = 2.083$; H_0 is rejected and H_1 is accepted (claim of improved quality).

3. $\bar{x} \pm (1.96\sigma)/\sqrt{n}$; 72.55, 77.45; $n > [3.8416\sigma^2/E^2] \approx 55$.

4. $Z = (\bar{x} - \mu)/(\sigma/\sqrt{n})$; $\bar{x} = 0.2 + 0.08Z$; $P(\bar{x} < 0) = P(Z < -2.5) = P(Z \geq 2.5) = 0.0062$.

5. $P(\bar{x} < 0) = P(Z < -0.8333) = P(Z > 0.8333) = 0.2024$.

6. Two tailed test; $|Z| = 2.4 < 2.58$; yes, the samples may be from the same population.

7. $Z = -1.237$, (left tailed test). There is no significant difference between wages in industries A and B.

8. $Z = -11.62$; (two tailed test); Null hypothesis $H_0 : \mu_1 = \mu_2$ is rejected.

9. $E(\bar{X}_1 - \bar{X}_2) = 0$, $V(\bar{X}_1 - \bar{X}_2) = \sigma^2/100$, $\bar{X}_1 - \bar{X}_2 \sim N(0, \sigma^2/100)$. $P(|Z| \leq 3) = 0.9974$, where $Z = (\bar{X}_1 - \bar{X}_2)/(\sigma/10) = 0.3\sigma/(\sigma/10) = 3$.

10. $n > (z_\alpha \sigma/E)^2 \approx 554$.

11. $|Z| = 4.86 > 1.96$, (two tailed test). The null hypotheis $H_0 : \mu_1 = \mu_2$ is rejected. The mean value is affcted by the weather.

12. $\hat{P} \approx P = 0.04, \hat{Q} = 0.96$; $S.E = \sqrt{\hat{P}\hat{Q}/n} = 0.0062$; limits are $\hat{P} \pm 3\sqrt{\hat{P}\hat{Q}/n}$, or (0.0214, 0.0586).

13. $Z = (p - P)/\sqrt{PQ/n} = -3.408$, (left tailed test). Since $|Z| > 2.33$, $H_0 : p = P$ is rejected and accept the claim of the hospital.

14. P is a constant. $p \sim N(P, PQ/n)$; SE of $p = \sqrt{PQ}/20$; $(\sqrt{P} - \sqrt{Q})^2 \geq 0$ gives $\sqrt{PQ} < 1/2$. Hence, SE of $p \leq 0.025$.

15. Two tailed test. $P \approx \hat{P} = (n_1 p_1 + n_2 p_2)/(n_1 + n_2) = 0.292$; $Z = \left[(p_1 - p_2) \middle/ \sqrt{\hat{P}\hat{Q}\left(\dfrac{1}{n_1} + \dfrac{1}{n_2}\right)}\right] = 0.681$

    < 1.96. The null hypothesis $H_0 : p_1 = p_2$ (no significant difference) is accepted.

16. $H_0 : P_1 = P_2$ (question 10 is not good in separating the ability of the students); two tailed test; $|Z| = 3.44 > 1.96$, we reject the null hypothesis. Question 10 is good in discriminating the students.

17. $p_1 = 80/300$, $p_2 = 50/200$; $P \approx \hat{P} = [(n_1 p_1 + n_2 p_2)/(n_1 + n_2)] = 0.26$; $H_0 : P_1 - P_2 = 0.1$; two tailed test;

    $Z = \left[\{(p_1 - p_2) - (P_1 - P_2)\} \middle/ \sqrt{\hat{P}\hat{Q}\left(\dfrac{1}{n_1} + \dfrac{1}{n_2}\right)}\right] = -2.08.$ Since, $|Z| = 2.08 < 2.33$, we accept the

    null hypothesis. The claim of 10% difference in sales is valid (two tailed test).

18. $H_0 : \sigma_1 = \sigma_2$; two tailed test. $Z = \left[(s_1 - s_2) \middle/ \sqrt{\dfrac{s_1^2}{2n_1} + \dfrac{s_2^2}{2n_2}}\right] = 1.5.$ Since $|Z| = 1.5 < 1.98$, we accept

    H_0. The difference between standard deviations is not signifiant.

19. $H_0 : \sigma_1 = \sigma_2$; two tailed test; $Z = [(s_1 - s_2)/\sqrt{s_1^2/(2n_1) + s_2^2/(2n_2)}] = -0.38$. Since, $|Z| = 0.38 < 1.96$, we accept H_0. The variation is not significant.

20. Yes. $|Z| = 0.748$.

Exercise 20.2

1. $\bar{x} = 92.71$, $s^2 = 64.14$, $H_0 : \bar{x} = \mu$; $|t| = 2.23 < 2.447$. H_0 is accepted (the data is from a normal population with mean 100). (84.71, 100.71).

2. $\bar{x} = 166.5$, $s = 4.2249$, $H_0 : \bar{x} = \mu$; $H_1 : \bar{x} > \mu$; $|t| = 1.065 < 1.833$. H_0 is accepted.

3. $H_0 : \mu = 66.5$, $H_1 : \mu \neq 66.5$. $|t| = 0.4994$ is less than 2.179 and 3.055. H_0 is accepted at both levels of significance. (61.137, 69.86), (59.383, 71.617).

4. $d_i = x_i - y_i$ are given, $\bar{d} = 1.4$, $s^2 = 3.44$. $H_0 : d = 0$; $H_1 : d < 0$. $|t| = 1.5097 < 2.776$. H_0 is accepted (injection will not result in increase of blood pressure).

5. $H_0 : \bar{x} = \mu$ and $H_1 : \bar{x} < \mu$. $|t| = 2.154 > 1.746$. H_0 is rejected and H_1 is accepted. Company's claim cannot be accepted.

6. $|t| = 0.1213 < 2.16$. No significant difference between means.

7. $|t| = 1.2095 < 2.228$. No significant difference between average fat content.

8. $|t| = 1.215 < 1.753$. $H_0 : \bar{x}_1 = \bar{x}_2$ is accepted (claim of company A is not accepted).

9. $|t| = 6.343 > 1.725$. $H_0 : \mu_1 = \mu_2$ is rejected (average salary of supervisors in factory A is more than in factory B).

10. $|t| = 0.86 < 2.145$. H_0 is accepted (no significant difference in means).

11. $d_i = x_i - y_i$, $\bar{d} = -4.6$, $s^2 = 25.44$. $|t| = 2.736 > 1.833$. $H_0 : \bar{d} = 0$ is rejected and $H_1 : \bar{d} < 0$ is accepted (coaching was effective).

12. $d_i = x_i - y_i$, $\bar{d} = -1.75$, $s^2 = 16.4375$. $|t| = 1.142 < 1.895$. $H_0 : \bar{d} = 0$ is accepted (no significant improvement).

13. $F = 1.26 < 2.53$; not significant.

14. $F = 1.167 < 3.36$; not significant.

15. $F = 1.10 < 4.39$; not significant.

16. $F = 2.05 < 3.65$; difference of variances is not significant. $t = 6.587 > 2.11$; difference of means is significant. The samples are not from the same population.

17. $F = 1.378 < 3.73$; difference of variances is not significant. $t = 1.756 < 2.131$; difference of means is not significant. The random samples are from the same population.

18. $F = 3.218 < 6.26$; difference of variances is not significant. $t = 1.958 < 2.262$; difference of means is not significant. The random samples are from the same population.

Exercise 20.3

1. H_0 : male and female births are equally possible. $E_i = (11, 56, 113, 113, 56, 11)$.
   $\chi^2 = 456.4 > \chi_5^2 (0.05)$. H_0 is rejected.

2. H_0 : male and female births are equally possible; $p = q = 1/2$. $E_i = (50, 100, 50)$.
   $\chi^2 = 4 < 7.82 = \chi_2^2 (0.02)$. H_0 is accepted.

3. $\bar{x} = 2.71 = 5p$; $E_i = 180 (0.458 + 0.542)^5$. Adjusted $E_i = (25, 51, 60, 44)$, adjusted $O_i = (41, 35, 48, 56)$.
   $\chi^2 = 20.932 > 5.99 = \chi_2^2 (0.05)$. H_0 is rejected. Binomial distribution is not satisfactory.

4. $E_i = 500 \, e^{-2.024} (2.024)^r / r!$. Adjusted $E_i = (66, 134, 135, 91, 46, 28)$, $O_i = (62, 133, 142, 90, 50, 23)$.
   $\chi^2 = 1.86 < 9.488 = \chi_4^2 (0.05)$. H_0 is accepted. Poisson distribution is satisfactory.

5. $E_i = 420 \, e^{-2.4} (2.4)^r / r!$. Adjusted $E_i = (38, 91, 110, 88, 53, 25, 15)$, $O_i = (30, 100, 110, 85, 60, 20, 15)$.
   $\chi^2 = 4.6 < 11.07 = \chi_5^2 (0.05)$. H_0 is accepted. Poisson distribution is satisfactory.

6. $E_i = (86, 105, 94, 115)$. $\chi^2 = 9.8 > 3.84 = \chi_1(0.05)$. H_0 is rejected. There is a relationship between literacy and smoking habits.

7. $E_i = (98, 102, 147, 153)$. $\chi^2 = 16.14 > 3.84 = \chi_1^2 (0.05)$. H_0 is rejected. Voting pattern is related to the nature of the area.

8. $\chi^2 = 6.59 < 11.07 = \chi_5^2(0.05)$. H_0 is accepted. All faces are equally likely to occur.

9. $n = 22 < 30$. $\chi^2 = 41.6 > 32.6 = \chi_{21}^2 (0.05)$. Population standard deviation is not 4.

10. $n = 200 > 30$. $Z = \sqrt{2\chi^2} - \sqrt{2n - 1} = 2.525$. Significant. Population standard deviation is not 8.

11. $\mu = 47.5636$, $s = 5.1562$, $\chi^2 = 5.22$, d.f. $= 2$, $\chi^2 < \chi_2^2 (0.05) = 5.991$. Fitting of normal distribution is accepted.

Tests for Convergence

D'Alembert's test (ratio test)

Let $\sum a_n$ be a real series of positive terms or a complex series. Let

$$\lim_{n \to \infty} \left| \frac{a_{n+1}}{a_n} \right| = l.$$

Then, the series $\sum a_n$ is convergent if $l < 1$ and divergent if $l > 1$. The test fails when $l = 1$.

Proof

Case 1 Let

$$\lim_{n \to \infty} \left| \frac{a_{n+1}}{a_n} \right| = l < 1.$$

By the definition of a limit, we can find a positive integer N, such that

$$\left| \frac{a_{n+1}}{a_n} \right| \leq r \quad \text{for all } n \geq N$$

where $l < r < 1$. Hence, leaving out the first N terms, we have

$$|a_{N+1}| \leq r |a_N|$$

$$|a_{N+2}| \leq r |a_{N+1}| \leq r^2 |a_N|$$

$$|a_{N+3}| \leq r |a_{N+2}| \leq r^3 |a_N|, \ldots$$

Adding the above inequalities, we obtain

$$|a_{N+1}| + |a_{N+2}| + |a_{N+3}| + \ldots \leq (r + r^2 + r^3 + \ldots) |a_N|$$

$$= \frac{r}{1 - r} |a_N|$$

which is a finite quantity, since $0 < r < 1$. Hence, by comparison test, the series $\sum |a_N|$ is convergent. Therefore, the given series $\sum a_n$ is convergent.

Case 2 Let

$$\lim_{n \to \infty} \left| \frac{a_{n+1}}{a_n} \right| = l > 1.$$

By the definition of a limit, there exists a positive integer N, such that

$$\left| \frac{a_{n+1}}{a_n} \right| > 1 \quad \text{for all } n \geq N.$$

Hence, leaving out the first N terms, we have

$$|a_{N+1}| > |a_N|$$

$$|a_{N+2}| > |a_{N+1}| > |a_N|$$

$$|a_{N+3}| > |a_{N+2}| > |a_N|, \ldots$$

Therefore,

$$|a_{N+1}| + |a_{N+2}| + \ldots + |a_{N+m}| > m|a_N|.$$

As $m \to \infty$, the right hand side $\to \infty$. Hence, the given series is divergent when $l > 1$.

Cauchy root test

Let $\sum a_n$ be a real series of positive terms or a complex series. Let

$$\lim_{n \to \infty} (|a_n|)^{1/n} = l.$$

Then, the series $\sum a_n$ is convergent if $l < 1$ and divergent if $l > 1$. The test fails when $l = 1$.

Proof

Case 1 Let

$$\lim_{n \to \infty} (|a_n|)^{1/n} = l < 1.$$

By the definition of a limit, we can find a positive integer N, such that

$$(|a_n|)^{1/n} < r \quad \text{for all } n \geq N$$

where $l < r < 1$. Hence, we have

$$|a_n| < r^n \quad \text{for all } n \geq N.$$

and

$$|a_1| + |a_2| + \ldots + |a_m| < r + r^2 + \ldots + r^m.$$

Since the geometric series $\sum r^n$ converges when $r < 1$, the series $\sum a_n$ is absolutely convergent and hence convergent.

Case 2 Let

$$\lim_{n \to \infty} (|a_n|)^{1/n} = l > 1.$$

By the definition of a limit, there exists a positive integer N, such that

$$(|a_n|)^{1/n} > 1, \quad \text{or} \quad |a_n| > 1 \quad \text{for all } n \geq N.$$

Therefore, leaving out the first N terms, we have

$$|a_N| > 1, |a_{N+1}| > 1, \ldots, |a_{N+m}| > 1$$

and

$$|a_N| + |a_{N+1}| + \ldots + |a_{N+m}| > m.$$

As $m \to \infty$, the right hand side $\to \infty$.

Hence, the given series $\sum a_n$ is divergent.

Raabe's test

Let $\sum a_n$ be a real series of positive terms or a complex series. Let

$$\lim_{n \to \infty} n \left[1 - \left| \frac{a_{n+1}}{a_n} \right| \right] = l.$$

Then, the series $\sum a_n$ is convergent if $l > 1$ and divergent if $l < 1$, the test fails when $l = 1$.

Proof

Case 1 Let $l > 1$. Choose a number p such that $l > p > 1$. Now, compare the series $\sum |a_n|$ with the p-series $\sum v_n = 1/n^p$. The p-series is convergent when $p > 1$.

Let, from and after some term $n = N$

$$\left| \frac{a_{n+1}}{a_n} \right| < \frac{v_{n+1}}{v_n} = \frac{n^p}{(1+n)^p}.$$

Then,

$$\left| \frac{a_{n+1}}{a_n} \right| < \left[1 + \frac{1}{n} \right]^{-p} = 1 - \frac{p}{n} + \frac{p(p+1)}{(2!)n^2} - \cdots$$

or

$$n \left[1 - \left| \frac{a_{n+1}}{a_n} \right| \right] > p - \frac{p(p+1)}{2n} + \cdots$$

Taking the limit, we obtain

$$\lim_{n \to \infty} n \left[1 - \left| \frac{a_{n+1}}{a_n} \right| \right] > p > 1.$$

Therefore, $\left| \dfrac{a_{n+1}}{a_n} \right| < \dfrac{v_{n+1}}{v_n}$ if $p > 1$, which is true. But the p-series is convergent for $p > 1$.

Hence, by comparison test (of series), the series $\sum a_n$ is absolutely convergent and hence convergent.

Case 2 Let $l < 1$. Choose a number p such that $l < p < 1$. Let, from and after some term $n = N$

$$\left| \frac{a_{n+1}}{a_n} \right| > \frac{v_{n+1}}{v_n} = \frac{n^p}{(1+n)^p}$$

$$= \left[1 + \frac{1}{n} \right]^{-p} = 1 - \frac{p}{n} + \frac{p(p+1)}{(2!)n^2} - \cdots$$

Therefore,

$$n \left[1 - \left| \frac{a_{n+1}}{a_n} \right| \right] < p - \frac{p(p+1)}{2n} + \cdots$$

Taking the limit, we obtain

$$\lim_{n \to \infty} n \left[1 - \left| \frac{a_{n+1}}{a_n} \right| \right] < p < 1.$$

Therefore, $\left| \dfrac{a_{n+1}}{a_n} \right| > \dfrac{v_{n+1}}{v_n}$ if $p < 1$, which is true. But the p-series is divergent for $p < 1$.

Hence by comparison test (of series), the given series $\sum a_n$ is divergent.

Leibnitz theorem

Let $\sum_{n=1}^{\infty} (-1)^{n-1} a_n$, $a_n > 0$ be an alternating series such that

  (i) the sequence $\{a_n\}$ is non-increasing, that is $a_{n+1} \leq a_n$ for all n, and

  (ii) $\lim_{n \to \infty} a_n = 0$

Then, the series $\sum (-1)^{n-1} a_n$ is convergent

Proof   Consider the sequence of partial sums

$$S_1 = a_1, \; S_2 = a_1 - a_2, \ldots \quad \text{and} \quad S_n = a_1 - a_2 + \ldots + (-1)^{n-1} a_n.$$

Write
$$S_{2m+2} = (a_1 - a_2) + (a_3 - a_4) + \ldots + (a_{2m-1} - a_{2m}) + (a_{2m+1} - a_{2m+2})$$
$$S_{2m+1} = a_1 - (a_2 - a_3) - (a_4 - a_5) \ldots - (a_{2m} - a_{2m+1})$$
and
$$S_{2m} = (a_1 - a_2) + (a_3 - a_4) + \ldots + (a_{2m-1} - a_{2m})$$
$$S_{2m} = a_1 - (a_2 - a_3) - (a_4 - a_5) - \ldots - (a_{2m-2} - a_{2m-1}) - a_{2m}.$$

Since $a_{n+1} \leq a_n$, we find that $S_{2m+2} - S_{2m} = a_{2m+1} - a_{2m+2} \geq 0$, or $S_{2m+2} \geq S_m$,

$$S_{2m} > 0 \quad \text{and} \quad S_{2m} < a_1 \quad \text{for all } m.$$

Thus, the monotonic increasing sub-sequence $\{S_{2m}\}$ of $\{S_n\}$, a sequence of positive terms, is bounded above, has a limit and is therefore convergent.

Let $\lim_{m \to \infty} S_{2m} = S$. We will now show that the sequence $\{S_{2m+1}\}$ also converges to S. We have

$$S_{2m+1} = S_{2m} + a_{2m+1}.$$

Taking the limits, we obtain

$$\lim_{m \to \infty} S_{2m+1} = \lim_{m \to \infty} S_{2m} + \lim_{m \to \infty} a_{2m+1} = S + 0 = S.$$

Hence, the sequence $\{S_{2m+1}\}$ converges to S.

Since, both the sequences $\{S_{2m}\}$ and $\{S_{2m+1}\}$ converge to S, for a given real positive number $\varepsilon > 0$, there exist N_1 and $N_2 \in \mathbb{IN}$, such that

$$|S_{2m} - S| < \varepsilon, \; 2m \geq N_1 \quad \text{and} \quad |S_{2m+1} - S| < \varepsilon, \; 2m + 1 \geq N_2$$

or
$$|S_n - S| < \varepsilon \text{ for all } n \geq N, \quad \text{where} \quad n = \max(2m, 2m + 1)$$

and $N = \max(N_1, N_2)$. Hence, the sequence $\{S_n\}$ converges to S.

Canonical form of Variable Coefficient Second Order Partial Differential Equations

Using the notation of sections 9.5.1 and 16.1, we write the variable coefficient second order, partial differential equation as

$$A\frac{\partial^2 u}{\partial x^2} + B\frac{\partial^2 u}{\partial x \partial y} + C\frac{\partial^2 u}{\partial y^2} + f\left(x, y, u, \frac{\partial u}{\partial x}, \frac{\partial u}{\partial y}\right) = 0 \tag{A 2.1}$$

or
$$A\,r + B\,s + C\,t + f(x,\ y,\ u,\ p,\ q) = 0$$

where A, B, C are functions of x, y.

Consider a transformation of variables as

$$\xi = \xi(x,\ y),\ \eta = \eta(x,\ y). \tag{A 2.2}$$

Then, $u\ (x,\ y) = u(\xi,\ \eta)$. We have

$$u_x = u_\xi\,\xi_x + u_\eta\eta_x,\ u_y = u_\xi\,\xi_y + u_\eta\eta_y,$$

$$u_{xx} = u_\xi\,\xi_{xx} + [u_{\xi\xi}\,\xi_x + u_{\xi\eta}\eta_x]\,\xi_x + u_\eta\eta_{xx} + [u_{\xi\eta}\,\xi_x + u_{\eta\eta}\eta_x]\,\eta_x$$

$$u_{xy} = u_\xi\,\xi_{xy} + [u_{\xi\xi}\,\xi_y + u_{\xi\eta}\eta_y]\,\xi_x + u_\eta\eta_{xy} + [u_{\xi\eta}\,\xi_y + u_{\eta\eta}\eta_y]\,\eta_x$$

$$u_{yy} = u_\xi\,\xi_{yy} + [u_{\xi\xi}\,\xi_y + u_{\xi\eta}\eta_y]\xi_y + u_\eta\,\eta_{yy} + [u_{\xi\eta}\,\xi_y + u_{\eta\eta}\eta_y]\,\eta_y.$$

Substituting these expressions in Eq. (A 2.1) and collecting the terms, we get

$$R\,u_{\xi\xi} + S\,u_{\xi\eta} + T\,u_{\eta\eta} = F(\xi,\ \eta,\ u,\ u_\xi,\ u_\eta) \tag{A 2.3}$$

where
$$R = A\xi_x^2 + B\xi_x\xi_y + C\xi_y^2,\ T = A\eta_x^2 + B\eta_x\eta_y + C\eta_y^2,$$

and
$$S = 2A\,\xi_x\eta_x + B\,(\xi_x\,\eta_y + \xi_y\eta_x) + 2C\,\xi_y\,\eta_y. \tag{A 2.4}$$

The remaining terms get absorbed in $F(\xi,\ \eta,\ u,\ u_\xi,\ u_\eta)$. We need to find the solutions $\xi(x,\ y) = a.$, $\eta(x,\ y) = b$, where a and b are arbitrary constants, such that Eq. (A 2.3) takes a simpler form. The values of R and T depend on the value of the discriminant $B^2 - 4AC$.

We set $R = 0$, $T = 0$, that is,

$$A\xi_x^2 + B\xi_x\xi_y + C\xi_y^2 = 0,\ \text{and}\ A\eta_x^2 + B\eta_x\eta_y + C\eta_y^2 = 0. \tag{A 2.5}$$

From $\xi(x, y) = a$, $\eta(x, y) = b$, we have

$$d\xi = \xi_x\, dx + \xi_y\, dy = 0, \quad \text{or} \quad \frac{dy}{dx} = -\frac{\xi_x}{\xi_y}$$

and $\qquad\qquad d\eta = \eta_x\, dx + \eta_y\, dy = 0, \quad \text{or} \quad \frac{dy}{dx} = -\frac{\eta_x}{\eta_y}.$ \hfill (A 2.6)

From Eq. (A 2.5), we have

$$A(\xi_x/\xi_y)^2 + B(\xi_x/\xi_y) + C = 0, \quad A(\eta_x/\eta_y)^2 + B(\eta_x/\eta_y) + C = 0. \tag{A 2.7}$$

Using Eq. (A 2.6), we find that both the equations (A 2.7) reduce to

$$A\left(\frac{dy}{dx}\right)^2 - B\left(\frac{dy}{dx}\right) + C = 0 \tag{A 2.8}$$

whose solution is $\qquad\qquad \dfrac{dy}{dx} = \dfrac{1}{2A}[B \pm \sqrt{B^2 - 4AC}\,].$ \hfill (A 2.9)

We consider the following cases:

Case 1 $B^2 - 4AC > 0$.

Since $B^2 - 4AC > 0$, we have two real and distinct values for dy/dx. Solving these two equations, we obtain

$$\xi(x, y) = y - f_1(x) = a, \quad \eta(x, y) = y - f_2(x) = b,$$

which are called the *characteristics* of the given equation. The characteristics form the new coordinate system. Eq. (A 2.3) reduces to

$$S\, u_{\xi\eta} = F(\xi, \eta, u, u_\xi, u_\eta), \quad \text{or} \quad u_{\xi\eta} = G(\xi, \eta, u, u_\xi, u_\eta). \tag{A 2.10}$$

This equation is called the *canonical form* of the given partial differential equation. The equation has two real characteristics and is called an *hyperbolic partial differential equation*. The domain of dependence of the solution is the open domain between the characteristics. The differential equation, along with the prescribed conditions, is called an *initial value problem*. The prescribed conditions are called the initial conditions.

The simplest example is the wave equation $u_{tt} = c^2 u_{xx}$.

Case 2 $B^2 - 4AC = 0$.

In this case, Eq. (A 2.8) is a perfect square and we have one equation $(dy/dx) = B/(2A)$. Hence, the equation has one characteristic. Therefore, we cannot set both the equations in (A 2.7) to zero. Let the first equation be satisfied. Then, we have $B^2 - 4AC = 0$ or $C = B^2/(4A)$ and

$$\frac{dy}{dx} = -\frac{\xi_x}{\xi_y} = \frac{B}{2A}, \quad \text{or} \quad A\left(\frac{\xi_x}{\xi_y}\right) = -\frac{B}{2}.$$

Substituting in Eq. (A 2.4), we get

$$S = \xi_y \left[2A \left(\frac{\xi_x}{\xi_y} \right) \eta_x + B \left\{ \left(\frac{\xi_x}{\xi_y} \right) \eta_y + \eta_x \right\} + 2C \eta_y \right]$$

$$= \xi_y \left[2 \left(-\frac{B}{2} \right) \eta_x + B \left\{ -\frac{B}{2A} \eta_y + \eta_x \right\} + 2 \left(\frac{B^2}{4A} \right) \eta_y \right] = 0.$$

The equation (A 2.3) reduces to

$$Tu_{\eta\eta} = F(\xi, \eta, u, u_\xi, u_\eta), \quad \text{or} \quad u_{\eta\eta} = G(\xi, \eta, u, u_\xi, u_\eta) \tag{A 2.11}$$

which is the required canonical form.

If the second equation in Eq. (A 2.7) is satisfied, then the canonical form is obtained as

$$u_{\xi\xi} = G_1(\xi, \eta, u, u_\xi, u_\eta). \tag{A 2.12}$$

In this case, the equation is called a *parabolic partial differential equation*. The domain of dependence of the solution is the open domain on one side of the characteristic. The differential equation along with the prescribed (initial) conditions is called an *initial value problem*.

The simplest example is the *heat conduction equation* $u_t = c^2 u_{xx}$.

Case 3 $B^2 - 4AC < 0$.

In this case, Eq. (A 2.8) has no real solutions as the right hand side of Eq. (A 2.9) is a complex pair. Therefore, the given equation has no real characteristics. To get a real canonical form, we make one more transformation as

$$\alpha = \frac{1}{2} (\xi + \eta), \quad \beta = \frac{i}{2} (\eta - \xi). \tag{A 2.13}$$

Then, $u(\xi, \eta) = u(\alpha, \beta)$ and

$$u_\xi = u_\alpha \alpha_\xi + u_\beta \beta_\xi = \frac{1}{2} u_\alpha = \frac{i}{2} u_\beta,$$

$$u_{\xi\eta} = \frac{1}{2} \left[\frac{1}{2} u_{\alpha\alpha} + \frac{i}{2} u_{\alpha\beta} \right] - \frac{i}{2} \left[\frac{1}{2} u_{\alpha\beta} + \frac{i}{2} u_{\beta\beta} \right] = \frac{1}{4} (u_{\alpha\alpha} + u_{\beta\beta}).$$

Hence, the canonical form reduces to

$$u_{\alpha\alpha} + u_{\beta\beta} = G(\alpha, \beta, u, u_\alpha, u_\beta) \tag{A 2.14}$$

In this case, the equation is called an *elliptic partial differential equation*. The domain of dependence of the solution is, usually, a closed domain. The differential equation along with the prescribed boundary conditions is called an *elliptic boundary value problem*.

The simplest example is the *Laplace equation* $u_{xx} + u_{yy} = 0$ or a *Poisson equation* $u_{xx} + u_{yy} = g(x, y)$.

Orthogonal Curvilinear Coordinate Systems

Consider the transformation of coordinate system, from orthogonal cartesian coordinate system to an orthogonal curvilinear coordinate system, $x = x(u, v, w)$, $y(u, v, w)$ and $z = z(u, v, w)$. Then, the position vector, $\mathbf{r}$, of any point can be written as

$$\mathbf{r} = x(u, v, w)\, du + y(u, v, w)\, dv + z(u, v, w)\, dw. \tag{A 3.1}$$

and

$$d\mathbf{r} = \frac{\partial \mathbf{r}}{\partial u} du + \frac{\partial \mathbf{r}}{\partial v} dv + \frac{\partial \mathbf{r}}{\partial w} dw. \tag{A 3.2}$$

Let $\hat{\mathbf{a}}$, $\hat{\mathbf{b}}$, $\hat{\mathbf{c}}$ be the unit vectors in the directions of u, v and w respectively. But $\partial\mathbf{r}/\partial u$, $\partial\mathbf{r}/\partial v$, $\partial\mathbf{r}/\partial\omega$ are vectors in the directions of u, v and w respectively. Hence,

$$\hat{\mathbf{a}} = \frac{\partial\mathbf{r}/\partial u}{|\,\partial\mathbf{r}/\partial u\,|}, \quad \hat{\mathbf{b}} = \frac{\partial\mathbf{r}/\partial v}{|\,\partial\mathbf{r}/\partial v\,|}, \quad \hat{\mathbf{c}} = \frac{\partial\mathbf{r}/\partial w}{|\,\partial\mathbf{r}/\partial w\,|}. \tag{A 3.3}$$

Note that $\hat{\mathbf{a}}$, $\hat{\mathbf{b}}$, $\hat{\mathbf{c}}$ are mutually orthogonal. Denote

$$h_1 = |\partial\mathbf{r}/\partial u|, \quad h_2 = |\partial\mathbf{r}/\partial v| \quad \text{and} \quad h_3 = |\partial\mathbf{r}/\partial w|. \text{ Then,}$$

$$\frac{\partial\mathbf{r}}{\partial u} = \hat{\mathbf{a}}\, h_1, \quad \frac{\partial\mathbf{r}}{\partial v} = \hat{\mathbf{b}}\, h_2, \quad \frac{\partial\mathbf{r}}{\partial w} = \hat{\mathbf{c}}\, h_3 \tag{A 3.4}$$

and

$$d\mathbf{r} = h_1\hat{\mathbf{a}}\, du + h_2\hat{\mathbf{b}}\, dv + h_3\hat{\mathbf{c}}\, dw. \tag{A 3.5}$$

Element of arc ds in u, v, w system

$$ds^2 = dx^2 + dy^2 + dz^2 = d\mathbf{r}\cdot d\mathbf{r} = h_1^2\, du^2 + h_2^2\, dv^2 + h_3^2\, dw^2. \tag{A 3.6}$$

We have the following expressions

(i) $\text{grad } f = \nabla f = \displaystyle\sum \left(\frac{\hat{\mathbf{a}}}{h_1} \frac{\partial f}{\partial u} \right).$ \hfill (A 3.7)

(ii) $\text{div } (v) = \nabla \cdot v = \dfrac{1}{h_1 h_2 h_3} \displaystyle\sum \frac{\partial}{\partial u} (h_2 h_3 v_1).$ \hfill (A 3.8)

(iii) $\nabla^2 f = \nabla \cdot (\nabla f) = \dfrac{1}{h_1 h_2 h_3} \displaystyle\sum \frac{\partial}{\partial u} \left(\frac{h_2 h_3}{h_1} \frac{\partial f}{\partial u} \right).$ \hfill (A 3.9)

(iv) Curl $(v) = \nabla \times v = \dfrac{1}{h_1 h_2 h_3} \begin{vmatrix} \hat{\mathbf{a}}\, h_1 & \hat{\mathbf{b}}\, h_2 & \hat{\mathbf{c}}\, h_3 \\ \partial/\partial u & \partial/\partial v & \partial/\partial w \\ h_1 v_1 & h_2 v_2 & h_3 v_3 \end{vmatrix}$ (A 3.10)

Expressions in cylindrical polar coordinate system r, θ, z

We have the transformation as $x = r \cos\theta$, $y = r \sin\theta$, $z = z$. We first show that the cylindrical polar coordinate system is orthogonal. We have

$$\mathbf{r} = x\,\mathbf{i} + y\,\mathbf{j} + z\,\mathbf{k} = r\cos\theta\,\mathbf{i} + r\sin\theta\,\mathbf{j} + z\,\mathbf{k}.$$

$$\hat{\mathbf{a}} = \frac{\partial\mathbf{r}/\partial r}{|\partial\mathbf{r}/\partial r|} = (\cos\theta)\,\mathbf{i} + (\sin\theta)\,\mathbf{j}, \quad \hat{\mathbf{b}} = \frac{\partial\mathbf{r}/\partial\theta}{|\partial\mathbf{r}/\partial\theta|} = -(\sin\theta)\,\mathbf{i} + (\cos\theta)\,\mathbf{j}, \quad \hat{\mathbf{c}} = \frac{\partial\mathbf{r}/\partial z}{|\partial\mathbf{r}/\partial z|} = \mathbf{k}.$$

Now, $\hat{\mathbf{a}} \cdot \hat{\mathbf{b}} = 0 = \hat{\mathbf{b}} \cdot \hat{\mathbf{c}} = \hat{\mathbf{c}} \cdot \hat{\mathbf{a}}$. Hence, $\hat{\mathbf{a}}, \hat{\mathbf{b}}, \hat{\mathbf{c}}$ are mutually orthogonal.

we have
$$ds^2 = dx^2 + dy^2 + dz^2$$
$$= (\cos\theta\, dr - r\sin\theta\, d\theta)^2 + (\sin\theta\, dr + r\cos\theta\, d\theta)^2 + dz^2$$
$$= dr^2 + r^2\, d\theta^2 + dz^2.$$

Hence, $\qquad h_1 = 1, \; h_2 = r, \; h_3 = 1.$

(i) grad $f = \nabla f = \hat{\mathbf{a}}\,\dfrac{\partial f}{\partial r} + \dfrac{\hat{\mathbf{b}}}{r}\,\dfrac{\partial f}{\partial\theta} + \hat{\mathbf{c}}\,\dfrac{\partial f}{\partial z}.$ (A 3.11)

(ii) div $(v) = \nabla \cdot v = \dfrac{1}{r}\left[\dfrac{\partial}{\partial r}(r\,v_1) + \dfrac{\partial}{\partial\theta}(v_2) + \dfrac{\partial}{\partial z}(r\,v_3)\right]$

$\qquad\qquad\qquad = \dfrac{1}{r}\dfrac{\partial}{\partial r}(r\,v_1) + \dfrac{1}{r}\dfrac{\partial v_2}{\partial\theta} + \dfrac{\partial v_3}{\partial z}.$ (A 3.12)

(iii) $\nabla^2 f = \dfrac{1}{r}\left[\dfrac{\partial}{\partial r}\left(r\,\dfrac{\partial f}{\partial r}\right) + \dfrac{\partial}{\partial\theta}\left(\dfrac{1}{r}\dfrac{\partial f}{\partial\theta}\right) + \dfrac{\partial}{\partial z}\left(r\,\dfrac{\partial f}{\partial z}\right)\right]$ (A 3.13)

$\qquad\qquad = \dfrac{\partial^2 f}{\partial r^2} + \dfrac{1}{r}\dfrac{\partial f}{\partial r} + \dfrac{1}{r^2}\dfrac{\partial^2 f}{\partial\theta^2} + \dfrac{\partial^2 f}{\partial z^2}.$

(iv) Curl $(v) = \dfrac{1}{r}\begin{vmatrix} \hat{\mathbf{a}} & \hat{\mathbf{b}}\,r & \hat{\mathbf{c}} \\ \partial/\partial r & \partial/\partial\theta & \partial/\partial z \\ v_1 & r\,v_2 & v_3 \end{vmatrix}$ (A 3.14)

Expressions in spherical polar coordinate system r, φ, θ

We have the transformation as

$$x = r \sin \phi \cos \theta, \quad y = r \sin \phi \sin \theta, \quad z = r \cos \phi,$$

$$0 \le r \le a, \quad 0 \le \theta \le 2\pi, \quad 0 \le \phi \le \pi,$$

where ϕ is the vertical angle measured from z-axis and θ is the angle in the x-y plane measured from the x-axis. We first show that the spherical polar coordinate system is orthogonal. We have

$$\mathbf{r} = r \sin \phi \cos \theta \, \mathbf{i} + r \sin \phi \sin \theta \, \mathbf{j} + r \cos \phi \, \mathbf{k},$$

$$\hat{\mathbf{a}} = \frac{\partial \mathbf{r}/\partial r}{|\partial \mathbf{r}/\partial r|} = \sin \phi \cos \theta \, \mathbf{i} + \sin \phi \sin \theta \, \mathbf{j} + \cos \phi \, \mathbf{k},$$

$$\hat{\mathbf{b}} = \frac{\partial \mathbf{r}/\partial \phi}{|\partial \mathbf{r}/\partial \phi|} = \cos \phi \cos \theta \, \mathbf{i} + \cos \phi \sin \theta \, \mathbf{j} - \sin \phi \, \mathbf{k}$$

$$\hat{\mathbf{c}} = \frac{\partial \mathbf{r}/\partial \theta}{|\partial \mathbf{r}/\partial \theta|} = \frac{1}{r \sin \phi} [- r \sin \phi \sin \theta \, \mathbf{i} + r \sin \phi \cos \theta \, \mathbf{j}]$$

$$= - \sin \theta \, \mathbf{i} + \cos \theta \, \mathbf{j}.$$

Now, $\hat{\mathbf{a}} \cdot \hat{\mathbf{b}} = 0 = \hat{\mathbf{b}} \cdot \hat{\mathbf{c}} = \hat{\mathbf{c}} \cdot \hat{\mathbf{a}}$. Hence, $\hat{\mathbf{a}}, \hat{\mathbf{b}}, \hat{\mathbf{c}}$ are mutually orhogonal.

We have

$$ds^2 = dx^2 + dy^2 + dz^2 = dr^2 + r^2 \, d\phi^2 + r^2 \sin^2 \phi \, d\theta^2.$$

Hence,

$$h_1 = 1, \quad h_2 = r \text{ and } h_3 = r \sin \phi.$$

(i) $\text{grad} f = \hat{\mathbf{a}} \dfrac{\partial f}{\partial r} + \dfrac{\hat{\mathbf{b}}}{r} \dfrac{\partial f}{\partial \phi} + \dfrac{\hat{\mathbf{c}}}{r \sin \phi} \dfrac{\partial f}{\partial \theta}.$ (A 3.15)

(ii) $\text{div} (\boldsymbol{v}) = \dfrac{1}{r^2 \sin \phi} \left[\dfrac{\partial}{\partial r} (r^2 \sin \phi \, v_1) + \dfrac{\partial}{\partial \phi} (r \sin \phi \, v_2) + \dfrac{\partial}{\partial \theta} (r v_3) \right]$

$$= \frac{1}{r^2} \frac{\partial}{\partial r} (r^2 v_1) + \frac{1}{r \sin \phi} \frac{\partial}{\partial \phi} (\sin \phi \, v_2) + \frac{1}{r \sin \phi} \frac{\partial v_3}{\partial \theta}.$$ (A 3.16)

(iii) $\nabla^2 f = \dfrac{1}{r^2 \sin \phi} \left[\dfrac{\partial}{\partial r} \left(r^2 \sin \phi \dfrac{\partial f}{\partial r} \right) + \dfrac{\partial}{\partial \phi} \left(\dfrac{r \sin \phi}{r} \dfrac{\partial f}{\partial \phi} \right) + \dfrac{\partial}{\partial \theta} \left(\dfrac{r}{r \sin \phi} \dfrac{\partial f}{\partial \theta} \right) \right]$

$$= \frac{1}{r^2} \left(r^2 \frac{\partial^2 f}{\partial r^2} + 2r \frac{\partial f}{\partial r} \right) + \frac{1}{r^2 \sin \phi} \left(\sin \phi \frac{\partial^2 f}{\partial \phi^2} + \cos \phi \frac{\partial f}{\partial \phi} \right) + \frac{1}{r^2 \sin^2 \phi} \frac{\partial^2 f}{\partial \theta^2}$$

$$= \frac{\partial^2 f}{\partial r^2} + \frac{2}{r} \frac{\partial f}{\partial r} + \frac{1}{r^2} \frac{\partial^2 f}{\partial \phi^2} + \frac{\cot \phi}{r^2} \frac{\partial f}{\partial \phi} + \frac{1}{r^2 \sin^2 \varphi} \frac{\partial^2 f}{\partial \theta^2}.$$ (A 3.17)

(iv) $\text{Curl} (\boldsymbol{v}) = \dfrac{1}{r^2 \sin \phi} \begin{vmatrix} \hat{\mathbf{a}} & \hat{\mathbf{b}} r & \hat{\mathbf{c}} r \sin \phi \\ \partial/\partial r & \partial/\partial \phi & \partial/\partial \theta \\ v_1 & r v_2 & r \sin \phi \, v_3 \end{vmatrix}$ (A 3.18)

Some Reference Textbooks

The following is a brief list of textbooks related to the topics covered in the present textbook. There are many other textbooks which are not reported here.

1. Ahlfors, L.V. *Complex Analysis*, Third Edition, McGraw-Hill, 1979.
2. Athanasios Papoulis, *Probability, Random Variables and Stochastic Processes*, Third Edition, McGraw-Hill, 1991.
3. Bolton, W. *Fourier Series*, Longman Scientific and Technical Publications, 1995.
4. Bracewell, R. *Fourier Transforms and Its Applications*, McGraw-Hill, 1965.
5. Bronson, R. *Matrix Methods. An Introduction*, Academic Press, 1969.
6. Carslaw, H.S and Jaeger, J.C. *Operational Methods in Applied Mathematics*, Dover Publications, 1963.
7. Champeney, D.C. *Fourier Transforms and their Physical Applications*, Academic Press, 1973.
8. Churchill, R.V. *Fourier Series and Boundary Value Problems*, McGraw-Hill, 1963.
9. Churchill, R.V., Brown, J.W and Verhey, R.F. *Complex Variables and Applications*, Third Edition, McGraw-Hill, 1974.
10. Flanders, H., Korfhage, R.R and Price, J.J. *Calculus,* Academic Press, 1970.
11. Greenberg, M.D *Application of Green's Functions in Science and Engineering*, Prentice-Hall, 1971.
12. Gupta, S.C. and Kapoor, V.K. *Fundamentals of Mathematical Statistics*, Sultan chand & Sons, 1994.
13. Hildebrand, F.B. *Advanced Calculus for Applications*, Prentice-Hall, 1962.
14. Hoffman, K. and Kunze, R. *Linear Algebra*, Second Edition, Prentice-Hall of India, 1991.
15. Jain, M.K., Iyengar S.R.K. and Jain R.K., *Numerical Methods for Scientific and Engineering Computation*, Fourth-Edition, New Age Internation (P) Limited, 2003.
16. Johnson, R.E. and Kiokemeister, F.L. *Calculus with Analytic Geometry*, Fourth Edition, Allyn & Bacon, 1969.
17. Kreyszig, E. *Advanced Engineering Mathematics*, Eighth Edition, John Wiley, 1999.
18. Malik, S.C. and Savita Arora, *Mathematical Analysis*, Second Edition, Wiley Eastern Limited, 1991.
19. O'Neil, P.V. *Advanced Engineering Mathematics*, Fourth Edition, Brooks/Cole Publishing, 1995.
20. Peyton Z. Peebles, Jr., *Probability, Random Variables and Random Signal Principles*, Third Edition, McGraw-Hill, 1980.
21. Piaggio, H.T.H. *An Elementary Treatise on Differential Equations and their Applications*, G. Bell and Sons, 1956.

22. Pinkus Allan and Samy Zafrany, *Fourier Series and Integral Transforms*, Cambridge University Press, 1997.
23. Pipes, L.A and Harvill, L.R. *Applied Mathematics for Engineers and Physicists*, McGraw-Hill, 1970.
24. Piskunov, N. *Differential and Integral Calculus, Vol 1*, Mir Publishers, 1974.
25. Piskunov, N. *Differential and Integral Calculus, Vol 2*, Mir Publishers, 1974.
26. Rabenstein, A.L. *Elementary Differential Equations with Linear Algebra*. Academic Press, 1970.
27. Ray Wylie, C. *Advanced Engineering Mathematics*, Sixth Edition, McGraw-Hill, 1995.
28. Richard A. Johnson, *Probability and Statistics for Engineers*, Prentice Hall of India, 2002.
29. Ronald E. Walpole, Raymond H. Myers, Sharon L. Myers, Keying Lee, *Probability and Statistics for Engineers and Scientists*, Seventh Edition, Pearson Education, 2002.
30. Rudin, W. *Real and Complex Analysis*. Third Edition, McGraw-Hill, 1987.
31. Sneddon, Ian. *Elements of Partial Differential Equations*, McGraw-Hill, 1985.
32. Staib, J.H. *An Introduction to Matrices and Linear Transformations*, Addison-Wesley, 1969.
33. Strum, R.D. and Ward, J.R. *Laplace Transform Solution of Differential Equations*, A Programmed Text, Prentice-Hall, 1968.
34. Thomas Jr. G.B, and Finney, R.L. *Calculus and Analytic Geometry*, Sixth Edition, Narosa Publishing House, 1985.
35. Veerarajan, T. *Probability, Statistics and Random Processes*, Tata Mc-Graw-Hill, 2002.
36. Vich, Robert. *Z Transform Theory and Applications*, D. Reidel Publishing Company, 1987.
37. Watson, E.J. *Laplace Transforms and Applications*, Van Nostrand Reinhold, 1981.
38. William H. Hines, Douglas C. Montgomery, David M. Goldsman, Connie M. Borror, *Probability and Statistics in Engineering,* John Wiley & Sons, 2003.
39. Zelinsky, D. *A First Course in Linear Algebra*, Academic Press, 1968.
40. Zill, D.G and Cullen, M.R. *Advanced Engineering Mathematics*, Second Edition, CBS Publishers, Indian Edition, 2000.

Index

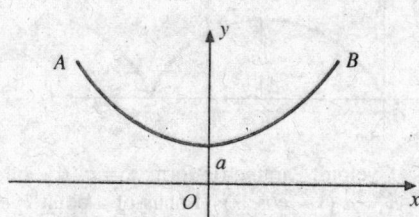

Fig. 1 Catenary, $y = a \cosh (x/a)$.

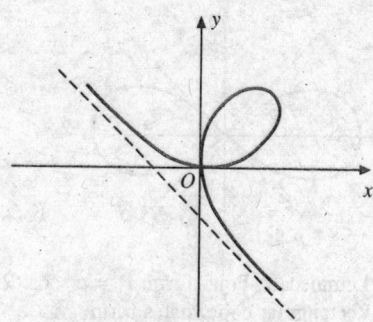

Fig. 2. Folium of Descartes, $x^3 + y^3 = 3axy$. Parametric equations: $x = 3at/(1 + t^3)$, $y = 3at^2/(1 + t^3)$.

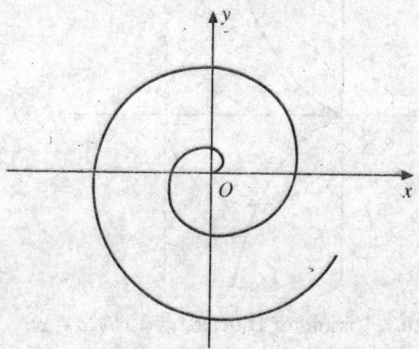

Fig. 3. Spiral of Archimedes, $r = a\theta$.

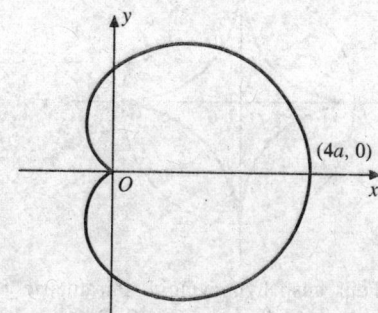

Fig. 4. Cardioid, $r = 2a (1 + \cos \theta)$. (Locus of a point on a circle of radius a rolling on the outside of a fixed circle of radius a. The fixed circle has centre at $(a, 0)$ and touches y-axis at the origin).

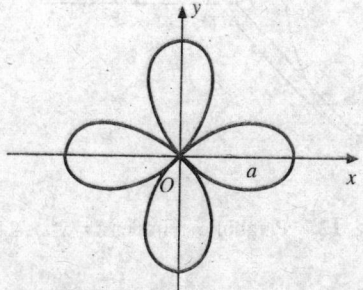

Fig. 5. Four leaved rose, $r = a \cos (2\theta)$.

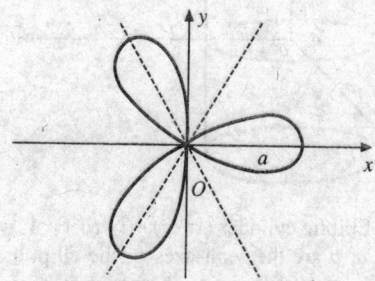

Fig. 6. Three leaved rose, $r = a \cos (3\theta)$.

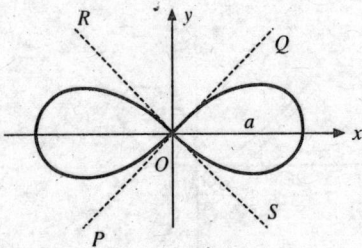

Fig. 7. Lemniscate. Polar form: $r^2 = a^2 \cos(2\theta)$. Rectangular coordinates form: $(x^2 + y^2)^2 = a^2(x^2 - y^2)$.

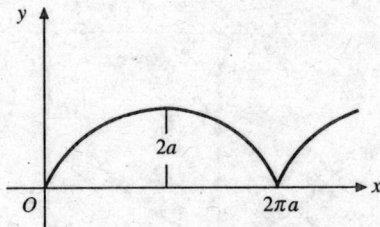

Fig. 8. Cycloid. Parametric form: $x = a(\theta - \sin\theta)$, $y = a(1 - \cos\theta)$. (Locus of a point P on a circle of radius a rolling along x-axis. Initially, the circle has centre at $(0, a)$ and touches x-axis at the origin).

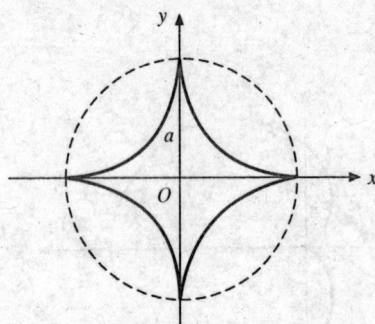

Fig. 9. Four cusp hypocycloid. Parametric form: $x = a\cos^3\theta$, $y = a\sin^3\theta$. Rectangular coordinates form: $x^{2/3} + y^{2/3} = a^{2/3}$. (Locus of a point P on a circle of radius $a/4$, rolling on the inside of a circle of radius a).

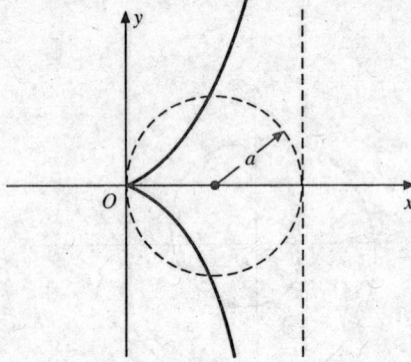

Fig. 10. Cissoid of Diocles, $y^2 = x^3/(2a - x)$.

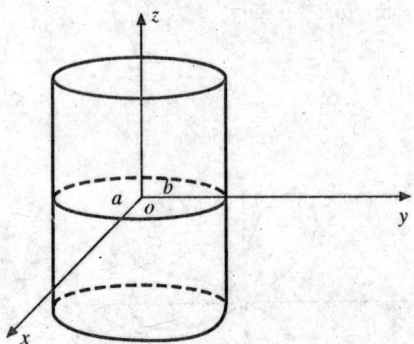

Fig. 11. Elliptic cylinder $(x^2/a^2) + (y^2/b^2) = 1$, where a, b are the semi-axes of the elliptic cross section. When $a = b$, we get the circular cylinder $x^2 + y^2 = a^2$.

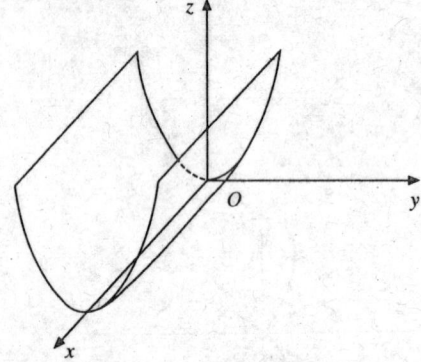

Fig. 12. Parabolic cylinder $z = y^2$.

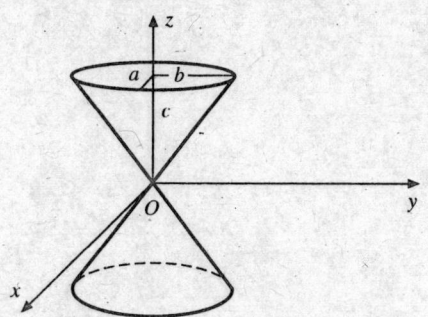

Fig. 13. Elliptic cone (with z-axis as axis).
$(x^2/a^2) + (y^2/b^2) = (z^2/c^2)$.
When $a = b$, we get a right circular cone.

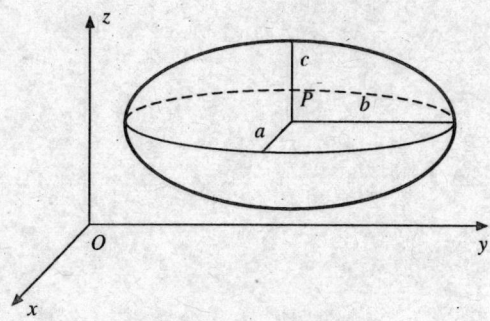

Fig. 14. Ellipsoid with centre at $P(x_0, y_0, z_0)$ and semi-axes a, b, c.

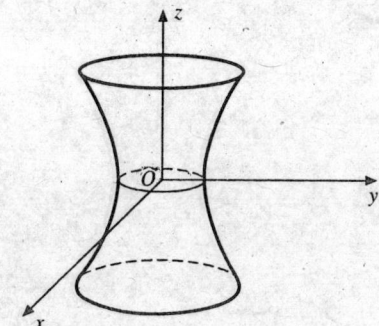

Fig. 15. Hyperboloid of one sheet,
$(x^2/a^2) + (y^2/b^2) - (z^2/c^2) = 1$.

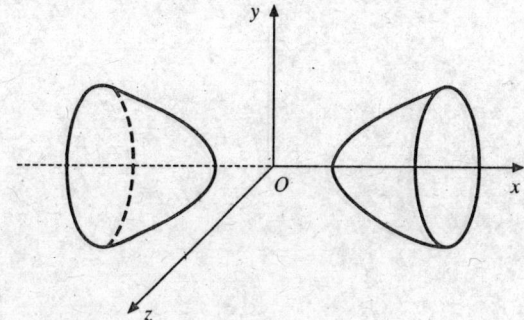

Fig. 16. Hyperboloid of two sheets,
$(x^2/a^2) - (y^2/b^2) - (z^2/c^2) = 1$.

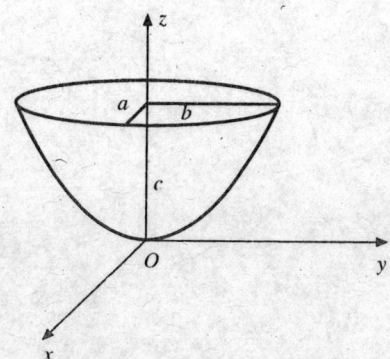

Fig. 17. Elliptic paraboloid.
$(x^2/a^2) + (y^2/b^2) = (z/c)$.

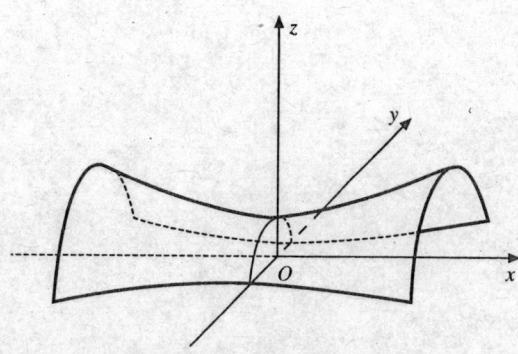

Fig. 18. Hyperbolic paraboloid.
$(x^2/a^2) - (y^2/b^2) = (z/c)$.